Applied Probability and Statistics (Continued)

CHAKRAVARTI, LAHA and ROY · Handbook of Methods of Applied Statistics, Vol. II
CHERNOFF and MOSES · Elementary Decision Theory
CHIANG · Introduction to Stochastic Processes in Biostatistics
CLELLAND, deCANI, BROWN, BURSK, and MURRAY · Basic Statistics with Business Applications
COCHRAN · Sampling Techniques, *Second Edition*
COCHRAN and COX · Experimental Designs, *Second Edition*
COX · Planning of Experiments
COX and MILLER · The Theory of Stochastic Processes
DAVID · Order Statistics
DEMING · Sample Design in Business Research
DODGE and ROMIG · Sampling Inspection Tables, *Second Edition*
DRAPER and SMITH · Applied Regression Analysis
ELANDT-JOHNSON · Probability Models and Statistical Methods in Genetics
GOLDBERGER · Econometric Theory
GUTTMAN, WILKS and HUNTER · Introductory Engineering Statistics, *Second Edition*
HAHN and SHAPIRO · Statistical Models in Engineering
HALD · Statistical Tables and Formulas
HALD · Statistical Theory with Engineering Applications
HOEL · Elementary Statistics, *Third Edition*
HUANG · Regression and Econometric Methods
JOHNSON and LEONE · Statistics and Experimental Design: In Engineering and the Physical Sciences, Volumes I and II
LANCASTER · The Chi Squared Distribution
MILTON · Rank Order Probabilities: Two-Sample Normal Shift Alternatives
PRABHU · Queues and Inventories: A Study of Their Basic Stochastic Processes
RAO and MITRA · Generalized Inverse of Matrices and Its Applications
SARD and WEINTRAUB · A Book of Splines
SARHAN and GREENBERG · Contributions to Order Statistics
SEAL · Stochastic Theory of a Risk Business
SEARLE · Linear Models
THOMAS · An Introduction to Applied Probability and Random Processes
WHITTLE · Optimization under Constraints
WILLIAMS · Regression Analysis
WOLD and JUREEN · Demand Analysis
WONNACOTT and WONNACOTT · Introduction to Econometric Methods
YOUDEN · Statistical Methods for Chemists
ZELLNER · An Introduction to Bayesian Inference in Econometrics

Tracts on Probability and Statistics

BILLINGSLEY · Ergodic Theory and Information
BILLINGSLEY · Convergence of Probability Measures
CRAMÉR and LEADBETTER · Stationary and Related Stochastic Processes
JARDINE and SIBSON · Mathematical Taxonomy
RIORDAN · Combinatorial Identities
TAKÁCS · Combinatorial Methods in the Theory of Stochastic Processes

A WILEY PUBLICATION IN APPLIED STATISTICS

A Book of Splines

ARTHUR SARD

and

SOL WEINTRAUB

Queens College
The City University of New York

John Wiley & Sons, Inc.,
New York · London · Sydney · Toronto

Library of Congress Catalog Card Number: 71–148509

ISBN 0 471 75415 3

Printed in the United States of America

10 9 8 7 6 5 4 3 2 1

To our wives Marguerite and Rita

Preface

Spline approximation is new, having been discovered in four independent ways in the years 1946 to 1966. Four different criteria of optimality lead to the same approximation. This fact is strong evidence that splines are important and useful.

Our Table 1 gives spline approximations explicitly. It provides the reader with optimal approximations of a function, or of derivatives of a function, or of integrals when values of the input function at regularly spaced arguments are known. Table 2 provides optimal appraisals of the error of approximation, in terms of the norm of the nth derivative of the input function, where $n =$ 1, 2, 3, 4, 5, or 8. Table 3 lists Lagrangian polynomials.

Chapter 1 describes Tables 1 and 3 with a minimum of symbolism. Chapter 2 describes Tables 1, 2, and 3 and is independent of Chapter 1. Both chapters contain worked examples. Either chapter may be read without the other.

The theory of spline approximation for a wide class of splines—much wider than those tabulated—is set forth in Chapter 3. Although the proofs use powerful tools of functional analysis, the definitions, theorems, and lemmas—that is, the statements without the proofs—may be of interest to all readers.

The appendix offers a FORTRAN program for the use of Table 1.

We hope that *A Book of Splines* will be a contribution to both the theory and the practice of approximation.

La Jolla, California
Flushing, New York
October, 1970

ARTHUR SARD
SOL WEINTRAUB

The writing of this book was supported in part by the United States Atomic Energy Commission under Contract AT(30-1)-3964.

Contents

A Book of Splines

CHAPTER 1†

Use of the Tables: First Introduction

1.1. INTERPOLATION

Given a set of data ($m + 1$ points),

$$x_0 \quad x_1 \quad x_2 \cdots x_i \cdots x_m,$$
$$y_0 \quad y_1 \quad y_2 \cdots y_i \cdots y_m,$$

obtained from experimental observations or from a table of values, the interpolator wishes to find a y-value for a value of x not given in the set.

We assume in this chapter that the x-values are equally spaced; that is, the differences of any pair of successive x-values are equal. In this case we can restate the problem in the equivalent form: given a set of data ($m + 1$ points),

$$0 \quad 1 \quad 2 \cdots i \quad \cdots m,$$
$$y_0 \quad y_1 \quad y_2 \cdots y_i \cdots y_m,$$

the interpolator wishes to find a value of y for a value other than $0, 1, \ldots, m$.

Let u be this other value. It is then desired to find $y(u)$. The solution of this problem is

$$y(u) \approx \beta^0(u)y_0 + \beta^1(u)y_1 + \cdots + \beta^i(u)y_i + \cdots + \beta^m(u)y_m,$$

where the β's are the splines found in Table 1 and $\approx$ indicates approximate equality.

1.2. EXAMPLES

Example 1. Given the following set,

$$x\text{-values:} \quad 0 \quad 10 \quad 20,$$
$$y\text{-values:} \quad 10 \quad 0 \quad 10,$$

what is the y-value when $x = 5$?

† Chapters 1 and 2 are independent introductions. Either one may be read alone. Chapter 1 describes some of the uses of Tables 1 and 3 in elementary terms. Chapter 2 describes the uses of all the tables.

The equivalent problem is

$$u\text{-values:} \quad 0 \quad 1 \quad 2,$$
$$y\text{-values:} \quad 10 \quad 0 \quad 10.$$

What is the y-value when $u = 0.5$?

The first page of Table 1 is reproduced below and the pertinent entries are underlined. The answer is

$$y(0.5) \approx 0.40625(10) + 0.6875(0) + (-0.09375)(10)$$
$$= 3.125.$$

We have used the portion of Table 1 headed $n = 2$, $m = 2$. The reason for $m = 2$ is this: we have $3 = m + 1$ points; hence $m = 2$. The reason for

```
              N = 2          M = 2                        P = 1,0        Q = 3

I = 0      H = -1,0
0             1,000000000000D+00    2,500000000000D-01    -1,00   2,250000000000D+00
1            -1,250000000000D+00   -5,000000000000D-01     -,75   1,937500000000D+00
                                                           -,50   1,625000000000D+00
0    -1,0     2,500000000000D-01    1,250000000000D-01     -,25   1,312500000000D+00
1     0,0    -5,000000000000D-01   -2,500000000000D-01      ,25   6,914062500000D-01
2     1,0     2,500000000000D-01    1,250000000000D-01      ,50   4,062500000000D-01
                                                            ,75   1,679687500000D-01
        0    -2,500000000000D-01   -2,500000000000D-01     1,25  -8,203125000000D-02
        1    -1,250000000000D+00   -5,000000000000D-01     1,50  -9,375000000000D-02

I = 1      H = 0,0
0             0,                    1,500000000000D+00    -1,00  -1,500000000000D+00
1             1,500000000000D+00    0,                     -,75  -1,125000000000D+00
                                                           -,50  -7,500000000000D-01
0    -1,0    -5,000000000000D-01   -2,500000000000D-01     -,25  -3,750000000000D-01
1     0,0     1,000000000000D+00    5,000000000000D-01      ,25   3,671875000000D-01
2     1,0    -5,000000000000D-01   -2,500000000000D-01      ,50   6,875000000000D-01
                                                            ,75   9,140625000000D-01
        0     1,500000000000D+00    1,500000000000D+00     1,25   9,140625000000D-01
        1     1,500000000000D+00    0,                     1,50   6,875000000000D-01

I = 2      H = 1,0
0             0,                   -7,500000000000D-01    -1,00   2,500000000000D-01
1            -2,500000000000D-01    5,000000000000D-01     -,75   1,875000000000D-01
                                                           -,50   1,250000000000D-01
0    -1,0     2,500000000000D-01    1,250000000000D-01     -,25   6,250000000000D-02
1     0,0    -5,000000000000D-01   -2,500000000000D-01      ,25  -5,859375000000D-02
2     1,0     2,500000000000D-01    1,250000000000D-01      ,50  -9,375000000000D-02
                                                            ,75  -8,203125000000D-02
        0    -2,500000000000D-01   -2,500000000000D-01     1,25   1,679687500000D-01
        1    -2,500000000000D-01    5,000000000000D-01     1,50   4,062500000000D-01
```

Figure 1. First page of Table 1.

$n = 2$ is discussed in Section 2.2. It suffices to note here that $n = 2$ yields a cubic spline, that is, a particular broken polynomial of degree† 3. The spline is represented by a different polynomial in each of the intervals $-\infty < u \leq 0$, $0 \leq u \leq 1$, $1 \leq u \leq 2$, ..., $m - 1 \leq u \leq m$, and $m \leq u < \infty$. The spline is continuous along with its first $2n - 2 = 2$ derivatives at the transitions. The spline reduces to degree $n - 1 = 1$ on the two infinite segments.

For general n the spline is a broken polynomial of degree $2n - 1$. It reduces to degree $n - 1$ on the two infinite segments and is continuous along with its first $2n - 2$ derivatives on the infinite line. In Table 1, $m \geq n$. The case $m = n-1$ corresponds to Lagrangian interpolation and is discussed in Section 1.7. In most of the illustrations in this chapter we consider only the case $n = 2$, as the procedure for $n = 2$ exemplifies that for $n > 2$.

Example 1 (continued). What is the value of y when $u = 1.75$?

The value 1.75 is not listed in the table (the u-values are tabulated only to 1.50 in this portion of the table—see the next to last column in Figure 1). However, because of symmetry (see Section 2.3), we use the value $m - u$ and take the coefficients in reverse order:

$$y(u) \approx \beta^m(m - u)y_0 + \beta^{m-1}(m - u)y_1 + \cdots + \beta^0(m - u)y_m.$$

Thus $m - u = 2 - 1.75 = 0.25$ and $y(1.75) \approx -0.05859375(10) + 0.3671875(0) + 0.6914025(10) = 6.328125$. Here we have used the values at $u = 0.25$ in Figure 1 in reverse order.

Example 2

$$u: \; 0 \quad 1 \quad 2,$$

$$y: \; 5 \quad 0 \quad 10.$$

Find the value of y when $u = 1.25$ and when $u = 1.75$.

Solution

$$y(1.25) \approx -0.08203125(5) + 0.9140625(0) + 0.16796875(10) = 1.26953125,$$

$$y(1.75) \approx -0.05859375(5) + 0.3671875(0) + 0.69140625(10) = 6.62109375.$$

Example 3

$$u: \; 0 \quad 1 \quad 2 \quad 3$$

$$y: \; 5 \quad 0 \quad 10 \quad 100.$$

† We say that a function is a polynomial of degree r if it is a polynomial of strict degree less than or equal to r. Thus the coefficient of the term of degree r may be zero.

Find the value of y when $u = 1.50$ and when $u = 2.50$.

Solution. Use Table 1 at $n = 2$, $m = 3$.

$$\begin{aligned} y(1.50) &\approx -0.075(5) + 0.575(0) + 0.575(10) + (-0.075)(100) \\ &= -2.125, \\ y(2.50) &\approx (0.025)(5) + (-0.15)(0) + 0.725(10) + 0.4(100) \\ &= 47.375. \end{aligned}$$

1.3. SPLINE FUNCTIONS

Naturally only a few values of u can be tabulated. It is often desired to find the value of y for values of u other than those listed or obtainable by the methods in the preceding section.

Table 1 may be used to write the entire spline, that is, the explicit function that best approximates the given data. The splines are piecewise polynomials in u, and one can use the explicit formulas to find y for any value of u.

The solution, as before, is

$$y(u) \approx \beta^0(u)y_0 + \beta^1(u)y_1 + \cdots + \beta^i(u)y_i + \cdots + \beta^m(u)y_m;$$

now, however, the β's are functions of u rather than just numbers. The β's again may be found in the table, as we now explain.

Example 4

$$\begin{array}{llll} u: & 0 & 1 & 2, \\ y: & 10 & 0 & 10. \end{array}$$

Find the spline function that approximates this data.

Solution. We again reproduce Table 1 for $n = 2$ and $m = 2$; the appropriate values are boxed.

From the table we read

$$\begin{aligned} \beta^0(u) &= 0.25 - 0.5u + 0.125\,|u|^3 - 0.25\,|u-1|^3 + 0.125\,|u-2|^3, \\ \beta^1(u) &= 1.5 - 0.25\,|u|^3 + 0.5\,|u-1|^3 - 0.25\,|u-2|^3, \\ \beta^2(u) &= -0.75 + 0.5u + 0.125\,|u|^3 - 0.25\,|u-1|^3 + 0.125\,|u-2|^3; \end{aligned}$$

and the answer to the problem is

$$y(u) \approx \beta^0(u)(10) + \beta^1(u)(0) + \beta^2(u)(10).$$

The reader may check by substitution that when $u = 0.5$, $y = 3.125$, as in Example 1 of the preceding section.

Before explaining the general pattern we present another example.

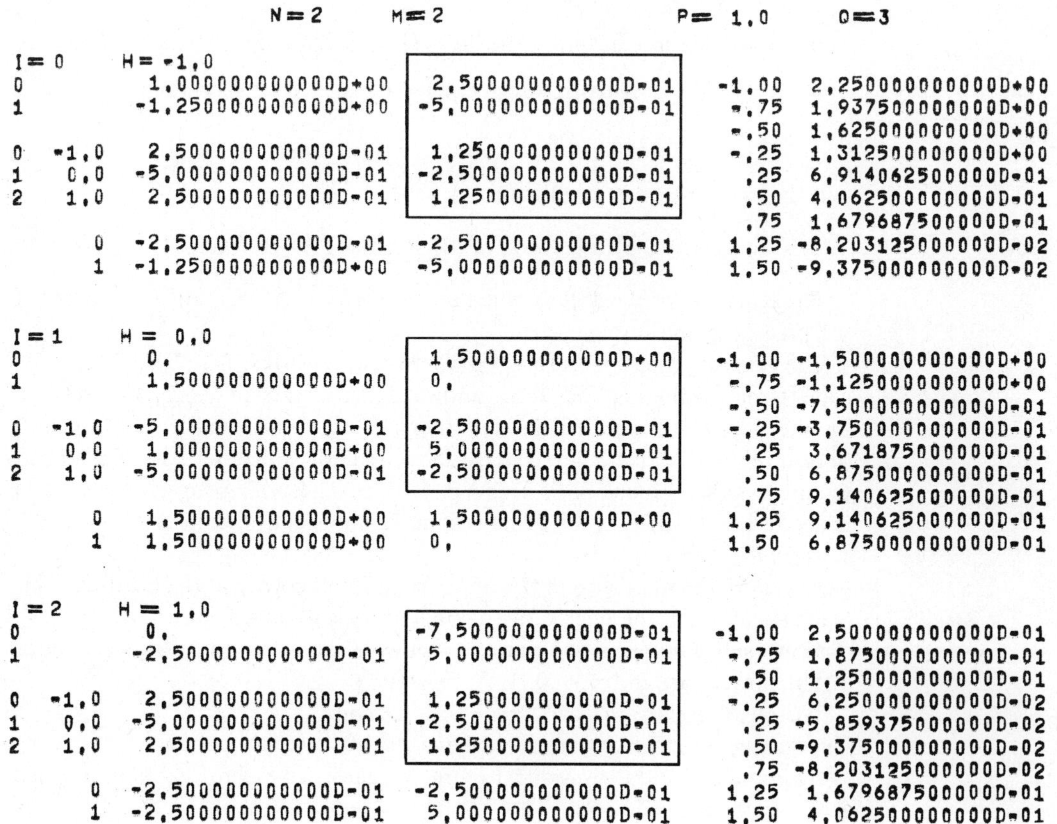

		N=2	M=2	P= 1,0	Q=3
I= 0		H= -1,0			
0		1,000000000000D+00	2,500000000000D-01	-1,00	2,250000000000D+00
1		-1,250000000000D+00	-5,000000000000D-01	-,75	1,937500000000D+00
				-,50	1,625000000000D+00
0	-1,0	2,500000000000D-01	1,250000000000D-01	-,25	1,312500000000D+00
1	0,0	-5,000000000000D-01	-2,500000000000D-01	,25	6,914062500000D-01
2	1,0	2,500000000000D-01	1,250000000000D-01	,50	4,062500000000D-01
				,75	1,679687500000D-01
	0	-2,500000000000D-01	-2,500000000000D-01	1,25	-8,203125000000D-02
	1	-1,250000000000D+00	-5,000000000000D-01	1,50	-9,375000000000D-02
I= 1		H= 0,0			
0		0,	1,500000000000D+00	-1,00	-1,500000000000D+00
1		1,500000000000D+00	0,	-,75	-1,125000000000D+00
				-,50	-7,500000000000D-01
0	-1,0	-5,000000000000D-01	-2,500000000000D-01	-,25	-3,750000000000D-01
1	0,0	1,000000000000D+00	5,000000000000D-01	,25	3,671875000000D-01
2	1,0	-5,000000000000D-01	-2,500000000000D-01	,50	6,875000000000D-01
				,75	9,140625000000D-01
	0	1,500000000000D+00	1,500000000000D+00	1,25	9,140625000000D-01
	1	1,500000000000D+00	0,	1,50	6,875000000000D-01
I= 2		H= 1,0			
0		0,	-7,500000000000D-01	-1,00	2,500000000000D-01
1		-2,500000000000D-01	5,000000000000D-01	-,75	1,875000000000D-01
				-,50	1,250000000000D-01
0	-1,0	2,500000000000D-01	1,250000000000D-01	-,25	6,250000000000D-02
1	0,0	-5,000000000000D-01	-2,500000000000D-01	,25	-5,859375000000D-02
2	1,0	2,500000000000D-01	1,250000000000D-01	,50	-9,375000000000D-02
				,75	-8,203125000000D-02
	0	-2,500000000000D-01	-2,500000000000D-01	1,25	1,679687500000D-01
	1	-2,500000000000D-01	5,000000000000D-01	1,50	4,062500000000D-01

Figure 2. Table 1 reproduced for $n = 2$, $m = 2$.

Example 5

$$u\colon\quad 0 \quad 1 \quad 2 \quad 3,$$

$$y\colon\quad 5 \quad 0 \quad 10 \quad 100.$$

Find the spline function that approximates this data.

Here $m = 3$ and we reproduce the portion of Table 1 for $n = 2$, $m = 3$ in Figure 3. The appropriate values are again boxed.

From Figure 3 we read

$$\beta^0(u) = 0.6 - 0.667u + 0.1333\,|u|^3 - 0.3\,|u-1|^3 + 0.2\,|u-2|^3 - 0.033\,|u-3|^3,$$

$$\beta^1(u) = -0.6 + u - 0.3\,|u|^3 + 0.8\,|u-1|^3 - 0.7\,|u-2|^3 + 0.2\,|u-3|^3,$$

$$\beta^2(u) = 2.4 - u + 0.2\,|u|^3 - 0.7\,|u-1|^3 + 0.8\,|u-2|^3 - 0.3\,|u-3|^3,$$

$$\beta^3(u) = -1.4 + 0.667u - 0.0333\,|u|^3 + 0.2\,|u-1|^3 - 0.3\,|u-2|^3$$
$$+ 0.1333\,|u-3|^3;$$

and the answer to our problem is

$$y(u) \approx \beta^0(u)(5) + \beta^1(u)(0) + \beta^2(u)(10) + \beta^3(u)(100).$$

The reader may check that when $u = 1.5$, $y = -2.125$, as in Example 3 of the preceding section.

The general pattern is as follows. There are $m + 1$ β's: $\beta^0, \beta^1, \ldots \beta^i, \ldots \beta^m$, called† the cardinal splines for n and m. Each section in Table 1 is headed by the i-value (upper left-hand corner). Thus, for example, in Figure 3 ($n = 2$, $m = 3$) we read $i = 0, 1, 2, 3$. Directly under the letter i are two strings of integers (see Figure 3) 0, 1 and 0, 1, 2, 3. The first string is called λ; in general, $\lambda = 0, 1, \ldots, n-1$. The second string is called v; in general, $v = 0, 1, \ldots, m$.

Also note the number q in the upper right-hand corner of the heading. This $q = 2n - 1$ and is the degree of the polynomials involved.

Now each β involves two parts, a "power" part and an "absolute" part.

The powers are u^λ, $\lambda = 0, 1, \ldots, n-1$.

The absolutes are $|u - v|^q$, $v = 0, 1, \ldots, m$.

The numbers in Table 1 are the *coefficients* of these two parts.

As an illustration consider Figure 3 again. The third section, headed $i = 2$, yields the following.

The "power" part is

u^0 with coefficient 2.4,

u^1 with coefficient -1.0

The "absolute" part is

$|u - 0|^3$ with coefficient 0.2,

$|u - 1|^3$ with coefficient -0.7,

$|u - 2|^3$ with coefficient 0.8,

$|u - 3|^3$ with coefficient -0.3.

† The superscripts here are not exponents but indices which distinguish different functions. Capital letters in the tables correspond to lower-case letters in the text.

N=2 M=3 P= 1,5 Q=3

I=0	H=-1,5				
0		1,000000000000D+00	6,000000000000D-01	-1,00	2,266666666667D+00
1		-1,266666666667D+00	-6,666666666667D-01	-,75	1,950000000000D+00
				-,50	1,633333333333D+00
0	-1,5	2,666666666667D-01	1,333333333333D-01	-,25	1,316666666667D+00
1	-,5	-6,000000000000D-01	-3,000000000000D-01	,25	6,875000000000D-01
2	,5	4,000000000000D-01	2,000000000000D-01	,50	4,000000000000D-01
3	1,5	-6,666666666667D-02	-3,333333333333D-02	,75	1,625000000000D-01
				1,25	-7,187500000000D-02
	0	-9,000000000000D-01	-4,000000000000D-01	1,50	-7,500000000000D-02
	1	-1,266666666667D+00	-6,666666666667D-01	1,75	-4,062500000000D-02

I=1	H=-,5				
0		0,	-6,000000000000D-01	-1,00	-1,600000000000D+00
1		1,600000000000D+00	1,000000000000D+00	-,75	-1,200000000000D+00
				-,50	-8,000000000000D-01
0	-1,5	-6,000000000000D-01	-3,000000000000D-01	-,25	-4,000000000000D-01
1	-,5	1,600000000000D+00	8,000000000000D-01	,25	3,906250000000D-01
2	,5	-1,400000000000D+00	-7,000000000000D-01	,50	7,250000000000D-01
3	1,5	4,000000000000D-01	2,000000000000D-01	,75	9,468750000000D-01
				1,25	8,531250000000D-01
	0	2,400000000000D+00	9,000000000000D-01	1,50	5,750000000000D-01
	1	1,600000000000D+00	1,000000000000D+00	1,75	2,593750000000D-01

I=2	H= ,5				
0		0,	2,400000000000D+00	-1,00	4,000000000000D-01
1		-4,000000000000D-01	-1,000000000000D+00	-,75	3,000000000000D-01
				-,50	2,000000000000D-01
0	-1,5	4,000000000000D-01	2,000000000000D-01	-,25	1,000000000000D-01
1	-,5	-1,400000000000D+00	-7,000000000000D-01	,25	-9,375000000000D-02
2	,5	1,600000000000D+00	8,000000000000D-01	,50	-1,500000000000D-01
3	1,5	-6,000000000000D-01	-3,000000000000D-01	,75	-1,312500000000D-01
				1,25	2,593750000000D-01
	0	-6,000000000000D-01	9,000000000000D-01	1,50	5,750000000000D-01
	1	-4,000000000000D-01	-1,000000000000D+00	1,75	8,531250000000D-01

I=3	H=1,5				
0		0,	-1,400000000000D+00	-1,00	-6,666666666667D-02
1		6,666666666667D-02	6,666666666667D-01	-,75	-5,000000000000D-02
				-,50	-3,333333333333D-02
0	-1,5	-6,666666666667D-02	-3,333333333333D-02	-,25	-1,666666666667D-02
1	-,5	4,000000000000D-01	2,000000000000D-01	,25	1,562500000000D-02
2	,5	-6,000000000000D-01	-3,000000000000D-01	,50	2,500000000000D-02
3	1,5	2,666666666667D-01	1,333333333333D-01	,75	2,187500000000D-02
				1,25	-4,062500000000D-02
	0	1,000000000000D-01	-4,000000000000D-01	1,50	-7,500000000000D-02
	1	6,666666666667D-02	6,666666666667D-01	1,75	-7,187500000000D-02

Figure 3. Table 1 reproduced for $n = 2$, $m = 3$.

Thus we have the function $\beta^2(u)$ given in Example 5. For convenience we call the coefficients of the power part b's and the coefficients of the absolute part c's. We can then write a formula for each β^i:

$$\beta^i(u) = \sum_{\lambda=0}^{n-1} b_\lambda^i u^\lambda + \sum_{\nu=0}^{m} c_\nu^i |u - \nu|^q;$$

u is any real number, $i = 0, \ldots, m$; the superscripts i are indices. Table 1 gives the coefficients b_λ^i, c_ν^i.

1.4. OTHER FORMS OF THE SPLINES

In addition to the formulas described in the preceding section, Table 1 gives three other forms of the same spline functions.

Powers and Plusses

We first define the "plus symbol":

$$t_+ = t \quad \text{if } t \geq 0,$$
$$t_+ = 0 \quad \text{if } t < 0;$$

for example,

$$(u-3)_+ = u - 3 \quad \text{if } u \geq 3,$$
$$(u-3)_+ = 0 \quad \text{if } u < 3.$$
$$(5)_+^3 = 125,$$
$$(-5)_+^3 = 0.$$

Example 6

$$u: \quad 0 \quad 1 \quad 2,$$
$$y: \quad 10 \quad 0 \quad 10.$$

Find the spline function in terms of powers and plusses that approximates this data.

The table for $n = 2$, $m = 2$ is again reproduced in Figure 4. The appropriate values are between [brackets]. The solution is $y(u) \approx \beta^0(u)y_0 + \beta^1(u)y_1 + \beta^2(u)y_2$.

From Figure 4 we read

$$\beta^0(u) = 1.0 - 1.25u + 0.25(u)_+^3 - 0.5(u-1)_+^3 + 0.25(u-2)_+^3,$$
$$\beta^1(u) = 1.5u - 0.5(u)_+^3 + (u-1)_+^3 - 0.5(u-2)_+^3,$$
$$\beta^2(u) = -0.25u + 0.25(u)_+^3 - 0.5(u-1)_+^3 + 0.25(u-2)_+^3;$$

```
              N = 2          M = 2                          P = 1.0        Q = 3

I = 0       H = -1.0
0             1.000000000000D+00    2.500000000000D-01     -1.00  2.250000000000D+00
1            -1.250000000000D+00   -5.000000000000D-01      -.75  1.937500000000D+00
                                                            -.50  1.625000000000D+00
0   -1.0      2.500000000000D-01    1.250000000000D-01      -.25  1.312500000000D+00
1    0.0     -5.000000000000D-01   -2.500000000000D-01       .25  6.914062500000D-01
2    1.0      2.500000000000D-01    1.250000000000D-01       .50  4.062500000000D-01
                                                             .75  1.679687500000D-01
        0    -2.500000000000D-01   -2.500000000000D-01      1.25 -8.203125000000D-02
        1    -1.250000000000D+00   -5.000000000000D-01      1.50 -9.375000000000D-02

I = 1       H = 0.0
0             0.                    1.500000000000D+00     -1.00 -1.500000000000D+00
1             1.500000000000D+00    0.                      -.75 -1.125000000000D+00
                                                            -.50 -7.500000000000D-01
0   -1.0     -5.000000000000D-01   -2.500000000000D-01      -.25 -3.750000000000D-01
1    0.0      1.000000000000D+00    5.000000000000D-01       .25  3.671875000000D-01
2    1.0     -5.000000000000D-01   -2.500000000000D-01       .50  6.875000000000D-01
                                                             .75  9.140625000000D-01
        0     1.500000000000D+00    1.500000000000D+00      1.25  9.140625000000D-01
        1     1.500000000000D+00    0.                      1.50  6.875000000000D-01

I = 2       H = 1.0
0             0.                   -7.500000000000D-01     -1.00  2.500000000000D-01
1            -2.500000000000D-01    5.000000000000D-01      -.75  1.875000000000D-01
                                                            -.50  1.250000000000D-01
0   -1.0      2.500000000000D-01    1.250000000000D-01      -.25  6.250000000000D-02
1    0.0     -5.000000000000D-01   -2.500000000000D-01       .25 -5.859375000000D-02
2    1.0      2.500000000000D-01    1.250000000000D-01       .50 -9.375000000000D-02
                                                             .75 -8.203125000000D-02
        0    -2.500000000000D-01   -2.500000000000D-01      1.25  1.679687500000D-01
        1    -2.500000000000D-01    5.000000000000D-01     -1.50  4.062500000000D-01
```

Figure 4. Table 1 reproduced for $n = 2$, $m = 2$.

and the answer is

$$y(u) \approx \beta^0(u)(10) + \beta^1(u)(0) + \beta^2(u)(10).$$

Example 7

$$u:\quad 0 \quad 1 \quad 2 \quad 3,$$

$$y:\quad 5 \quad 0 \quad 10 \quad 100.$$

Find the spline function in terms of powers and plusses that approximates this data. Table 1 for $n = 2$, $m = 3$ is reproduced in Figure 5 and the appropriate numbers are between [brackets]. From the table we read

$$\beta^0(u) = 1.0 - 1.267u + 0.2667(u)_+^3 - 0.6(u-1)_+^3$$
$$+ 0.4(u-2)_+^3 - 0.0667(u-3)_+^3,$$

$$\beta^1(u) = 0.0 + 1.6u - 0.6(u)^3_+ + 1.6(u-1)^3_+ - 1.4(u-2)^3_+ + 0.4(u-3)^3_+,$$

$$\beta^2(u) = 0.0 - 0.4u + 0.4(u)^3_+ - 1.4(u-1)^3_+ + 1.6(u-2)^3_+ - 0.6(u-3)^3_+,$$

$$\beta^3(u) = 0.0 - 0.0667u - 0.0667(u)^3_+ + 0.4(u-1)^3_+ - 0.6(u-2)^3_+ + 0.2667(u-3)^3_+;$$

and the answer is

$$y(u) \approx \beta^0(u)(5) + \beta^1(u)(0) + \beta^2(u)(10) + \beta^3(u)(100).$$

Using a notation similar to the one in the preceding section, we can write the general β^i as

$$\beta^i(u) = \sum_{\lambda=0}^{n-1} \tilde{b}^i_\lambda u^\lambda + \sum_{\nu=0}^{m} \tilde{c}^i_\nu (u-\nu)^q_+;$$

u is any real number, $i = 0, \ldots, m$, and the coefficients $\tilde{b}^i_\lambda$ and $\tilde{c}^i_\nu$ are found in Table 1.

Symmetric Forms

The splines have symmetric properties which are best exhibited by centering the data about the midpoint rather than about the initial point of the range 0 to m. Table 1 provides two additional forms of the splines with data in the symmetric range $-p$ to p, where $p = m/2$. For example, if the original data are

x:	0	10	20,
y:	5	70	90,

we may consider the equivalent problem

t:	−1	0	1,
y:	5	70	90.

If the original data are

x:	0	10	20	30,
y:	5	0	10	100,

we may consider the equivalent problem

t:	−1.5	−0.5	+0.5	1.5
y:	5	0	10	100.

N=2 M=3 P = 1,5 Q=3

I=0		H=-1,5			
0		1,000000000000D+00	6,000000000000D-01	-1.00	2,266666666667D+00
1		-1,266666666667D+00	-6,666666666667D-01	-,75	1,950000000000D+00
				-,50	1,633333333333D+00
0	-1,5	2,666666666667D-01	1,333333333333D-01	-,25	1,316666666667D+00
1	-,5	-6,000000000000D-01	-3,000000000000D-01	,25	6,875000000000D-01
2	,5	4,000000000000D-01	2,000000000000D-01	,50	4,000000000000D-01
3	1,5	-6,666666666667D-02	-3,333333333333D-02	,75	1,625000000000D-01
				1,25	-7,187500000000D-02
	0	-9,000000000000D-01	-4,000000000000D-01	1,50	-7,500000000000D-02
	1	-1,266666666667D+00	-6,666666666667D-01	1,75	-4,062500000000D-02
I=1		H=-,5			
0		0,	-6,000000000000D-01	-1,00	-1,600000000000D+00
1		1,600000000000D+00	1,000000000000D+00	-,75	-1,200000000000D+00
				-,50	-8,000000000000D-01
0	-1,5	-6,000000000000D-01	-3,000000000000D-01	-,25	-4,000000000000D-01
1	-,5	1,600000000000D+00	8,000000000000D-01	,25	3,906250000000D-01
2	,5	-1,400000000000D+00	-7,000000000000D-01	,50	7,250000000000D-01
3	1,5	4,000000000000D-01	2,000000000000D-01	,75	9,468750000000D-01
				1,25	8,531250000000D-01
	0	2,400000000000D+00	9,000000000000D-01	1,50	5,750000000000D-01
	1	1,600000000000D+00	1,000000000000D+00	1,75	2,593750000000D-01
I=2		H= ,5			
0		0,	2,400000000000D+00	-1,00	4,000000000000D-01
1		-4,000000000000D-01	-1,000000000000D+00	-,75	3,000000000000D-01
				-,50	2,000000000000D-01
0	-1,5	4,000000000000D-01	2,000000000000D-01	-,25	1,000000000000D-01
1	-,5	-1,400000000000D+00	-7,000000000000D-01	,25	-9,375000000000D-02
2	,5	1,600000000000D+00	8,000000000000D-01	,50	-1,500000000000D-01
3	1,5	-6,000000000000D-01	-3,000000000000D-01	,75	-1,312500000000D-01
				1,25	2,593750000000D-01
	0	-6,000000000000D-01	9,000000000000D-01	1,50	5,750000000000D-01
	1	-4,000000000000D-01	-1,000000000000D+00	1,75	8,531250000000D-01
I=3		H=1,5			
0		0,	-1,400000000000D+00	-1,00	-6,666666666667D-02
1		6,666666666667D-02	6,666666666667D-01	-,75	-5,000000000000D-02
				-,50	-3,333333333333D-02
0	-1,5	-6,666666666667D-02	-3,333333333333D-02	-,25	-1,666666666667D-02
1	-,5	4,000000000000D-01	2,000000000000D-01	,25	1,562500000000D-02
2	,5	-6,000000000000D-01	-3,000000000000D-01	,50	2,500000000000D-02
3	1,5	2,666666666667D-01	1,333333333333D-01	,75	2,187500000000D-02
				1,25	-4,062500000000D-02
	0	1,000000000000D-01	-4,000000000000D-01	1,50	-7,500000000000D-02
	1	6,666666666667D-02	6,666666666667D-01	1.75	-7,187500000000D-02

Figure 5. Table 1 reproduced for $n = 2$, $m = 3$.

If the original data are

$$x: \quad 1 \quad 2 \quad 3 \quad 4 \quad 5,$$
$$y: \quad 2 \quad 3 \quad 5 \quad 7 \quad 11,$$

we may consider the equivalent problem

$$t: \quad -2 \quad -1 \quad 0 \quad 1 \quad 2,$$
$$y: \quad 2 \quad 3 \quad 5 \quad 7 \quad 11.$$

As may be noted from these examples, the numbers are relabeled from

$$x_0 \quad x_1 \quad x_2 \cdots x_m,$$
$$y_0 \quad y_1 \quad y_2 \cdots y_m,$$

to

$$-p \quad -p+1 \quad -p+2 \cdots p,$$
$$y_{-p} \quad y_{-p+1} \quad y_{-p+2} \cdots y_p.$$

***Powers and Absolutes* (*symmetric form*)**

Example 8

$$t: \quad -1 \quad 0 \quad 1,$$
$$y: \quad 2 \quad 3 \quad 5.$$

Find the spline function in symmetric form, using powers and absolutes, that approximates the data.

The relevant portion of the table is reproduced in Figure 6 and the appropriate values are $\boxed{\text{boxed}}$.

The solution is

$$y(t) \approx \alpha^{-1}(t)y_{-1} + \alpha^0(t)y_0 + \alpha^1(t)y_1,$$

where

$$\alpha^{-1}(t) = 0.25 \quad - 0.5t + 0.125\,|t-(-1)|^3 - 0.25\,|t-0|^3 + 0.125\,|t-1|^3,$$
$$\alpha^0(t) = 1.5 \quad - 0.25\,|t-(-1)|^3 + 0.5\,|t-0|^3 - 0.25\,|t-1|^3,$$
$$\alpha^1(t) = -0.25 + 0.5t + 0.125\,|t-(-1)|^3 - 0.25\,|t-0|^3 + 0.125\,|t-1|^3;$$

and the answer is

$$y(t) \approx \alpha^{-1}(t)(2) + \alpha^0(t)(3) + \alpha^1(t)(5).$$

```
              N = 2          M = 2                          P = 1.0        Q = 3

I = 0     H = -1.0
0              1.000000000000D+00     2.500000000000D-01     -1.00  2.250000000000D+00
1             -1.250000000000D+00    -5.000000000000D-01      -.75  1.937500000000D+00
                                                              -.50  1.625000000000D+00
0  /-1.0\ μ    2.500000000000D-01  | 1.250000000000D-01 |    -.25  1.312500000000D+00
1  | 0.0 |    -5.000000000000D-01  |-2.500000000000D-01 | d^h_μ .25  6.914062500000D-01
2  \ 1.0/      2.500000000000D-01  | 1.250000000000D-01 |     .50  4.062500000000D-01
                                                               .75  1.679687500000D-01
   (0) λ      -2.500000000000D-01  |-2.500000000000D-01 | a^h_λ 1.25 -8.203125000000D-02
   (1)        -1.250000000000D+00  |-5.000000000000D-01 |     1.50 -9.375000000000D-02

I = 1     H =  0.0
0              0.                     1.500000000000D+00     -1.00 -1.500000000000D+00
1              1.500000000000D+00     0.                      -.75 -1.125000000000D+00
                                                              -.50 -7.500000000000D-01
0  /-1.0\     -5.000000000000D-01  |-2.500000000000D-01 |    -.25 -3.750000000000D-01
1  | 0.0 |     1.000000000000D+00  | 5.000000000000D-01 |     .25  3.671875000000D-01
2  \ 1.0/     -5.000000000000D-01  |-2.500000000000D-01 |     .50  6.875000000000D-01
                                                               .75  9.140625000000D-01
   (0)         1.500000000000D+00  | 1.500000000000D+00 |    1.25  9.140625000000D-01
   (1)         1.500000000000D+00  | 0.                 |    1.50  6.875000000000D-01

I = 2     H =  1.0
0              0.                    -7.500000000000D-01     -1.00  2.500000000000D-01
1             -2.500000000000D-01     5.000000000000D-01      -.75  1.875000000000D-01
                                                              -.50  1.250000000000D-01
0  /-1.0\      2.500000000000D-01  | 1.250000000000D-01 |    -.25  6.250000000000D-02
1  | 0.0 |    -5.000000000000D-01  |-2.500000000000D-01 |     .25 -5.859375000000D-02
2  \ 1.0/      2.500000000000D-01  | 1.250000000000D-01 |     .50 -9.375000000000D-02
                                                               .75 -8.203125000000D-02
   (0)        -2.500000000000D-01  |-2.500000000000D-01 |    1.25  1.679687500000D-01
   (1)        -2.500000000000D-01  | 5.000000000000D-01 |    1.50  4.062500000000D-01
```

Figure 6. Table 1 for $n = 2$, $m = 2$. The boxes contain d^h_μ and a^h_λ. The parentheses contain μ and λ.

This equation, aside from approximating $y(t)$ for any value of t, should, of course, give the exact value of y for $t = -1$, $t = 0$, and $t = 1$. It does so. Thus,

if $t = -1$,

$$\alpha^{-1}(-1) = -0.25 + 0.5 - 0.25 + 1 = 1,$$

$$\alpha^{0}(-1) = 1.5 + 0.5 - 2 = 0,$$

$$\alpha^{1}(-1) = -0.25 - 0.5 - 0.25 + 1 = 0,$$

and

$$y(-1) = 1(2) + 0(3) + 0(3) + 0(5) = 2;$$

if $t = 0$,

$$\alpha^{-1}(0) = -0.25 + 0.125 + 0.125 = 0,$$

$$\alpha^{0}(0) = 1.5 - 0.25 - 0.25 = 1,$$

$$\alpha^{1}(0) = -0.25 + 0.125 + 0.125 = 0,$$

and

$$y(0) = 0(2) + 1(3) + 0(5) = 3;$$

if $t = 1$,

$$\alpha^{-1}(1) = -0.25 - 0.5 + 1 - 0.25 = 0,$$

$$\alpha^{0}(1) = 1.5 - 2 + 0.5 = 0,$$

$$\alpha^{1}(1) = -0.25 + 0.5 + 1 - 0.25 = 1,$$

and

$$y(1) = 0(2) + 0(3) + 1(5) = 5.$$

Our approximation may also be used for extrapolation, for example, if $t = 2$,

$$\alpha^{-1}(2) = -0.25 - 1 + 3.375 - 2 + 0.125 = 0.25,$$

$$\alpha^{0}(2) = 1.5 - 6.75 + 4 - 2.5 = -1.5,$$

$$\alpha^{1}(2) = -0.25 + 1 + 3.375 - 2 + 0.125 = 2.25,$$

and

$$y(2) \approx (0.25)(2) - (1.5)(3) + (2.25)(5) = 7.25.$$

The general formula for α^h is

$$\alpha^h(t) = \sum_{\lambda=0}^{n-1} a_\lambda^h t^\lambda + \sum_{\mu=-p}^{p} d_\mu^h |t - \mu|^q,$$

t is any number, $h = -p, -p+1, \ldots, p$.
In Figure 6 the values of $\mu: -1, 0, 1$ and $\lambda: 0, 1$ are noted in parentheses and the numbers in the boxes are the coefficients d_μ^h and a_λ^h.

Example 9

t:	-1.5	-0.5	$+5$	1.5,
y:	1	1	2	3.

Find the spline in symmetric form, using powers and absolutes, that approximates the data. Using Table 1 for $n = 2$, $m = 3$, the reader may check that

$$\alpha^{-1.5}(t) = -0.4 - 0.667t + 0.1333\,|t + 1.5|^3 - 0.3\,|t + 0.5|^3 + 0.2\,|t - 0.5|^3 - 0.0333\,|t - 1.5|^3,$$

$$\alpha^{-0.5}(t) = 0.9 + t - 0.3\,|t + 1.5|^3 + 0.8\,|t + 0.5|^3 - 0.7\,|t - 0.5|^3 + 0.2\,|t - 1.5|^3,$$

$$\alpha^{0.5}(t) = 0.9 - t + 0.2\,|t + 1.5|^3 - 0.7\,|t + 0.5|^3 + 0.8\,|t - 0.5|^3 - 0.3\,|t - 1.5|^3,$$

$$\alpha^{1.5}(t) = -0.4 + 0.667t - 0.033\,|t + 1.5|^3 + 0.2\,|t + 0.5|^3 - 0.3\,|t - 0.5|^3 + 0.133\,|t - 1.5|^3;$$

and the answer is

$$y(t) \approx \alpha^{-1.5}(t)(1) + \alpha^{-0.5}(t)(1) + \alpha^{0.5}(t)(2) + \alpha^{1.5}(t)(3).$$

Powers and Plusses (symmetric argument). The fourth form of the splines is in terms of symmetric argument and plusses:

$$\alpha^h(t) = \sum_{\lambda=0}^{n-1} \tilde{a}^h_\lambda t^\lambda + \sum_{\mu=-p}^{p} \tilde{d}^h_\mu (t - \mu)^q_+;$$

t is any number, $h = -p, -p + 1, \ldots, p$.

Example 10

t:	-1	0	1,
y:	2	3	5.

Find the spline, using powers and plusses, that approximates the data.

The relevant portion of the table is reproduced in Figure 7 and the appropriate values are between [brackets].

From Figure 7 we have

$$\alpha^{-1}(t) = -0.25 - 1.25t + 0.25(t + 1)^3_+ - 0.5(t)^3_+ + 0.25(t - 1)^3_+,$$

$$\alpha^{0}(t) = 1.5 + 1.5t - 0.5(t + 1)^3_+ + 1.0(t)^3_+ - 0.5(t - 1)^3_+,$$

$$\alpha^{1}(t) = -0.25 - 0.25t + 0.25(t + 1)^3_+ - 0.5(t)^3_+ + 0.25(t - 1)^3_+;$$

and the answer is

$$y(t) \approx \alpha^{-1}(t)(2) + \alpha^{0}(t)(3) + \alpha^{1}(t)(5);$$

```
                N=2        M=2                              P = 1.0        Q = 3

I = 0      H = -1.0
0            1.000000000000D+00     2.500000000000D-01     -1.00   2.250000000000D+00
1           -1.250000000000D+00    -5.000000000000D-01      -.75   1.937500000000D+00
       μ                                                    -.50   1.625000000000D+00
0   (-1.0) [ 2.500000000000D-01] d̃^h_μ 1.250000000000D-01   -.25   1.312500000000D+00
1   ( 0.0) [-5.000000000000D-01]   -2.500000000000D-01       .25   6.914062500000D-01
2   ( 1.0) [ 2.500000000000D-01]    1.250000000000D-01       .50   4.062500000000D-01
                                                             .75   1.679687500000D-01
     λ (0) [-2.500000000000D-01]   -2.500000000000D-01      1.25  -8.203125000000D-02
       (1) [-1.250000000000D+00]   -5.000000000000D-01      1.50  -9.375000000000D-02
                                ã^h_λ

I = 1      H = 0.0
0            0.                     1.500000000000D+00     -1.00  -1.500000000000D+00
1            1.500000000000D+00     0.                      -.75  -1.125000000000D+00
                                                            -.50  -7.500000000000D-01
0   (-1.0) [-5.000000000000D-01]   -2.500000000000D-01      -.25  -3.750000000000D-01
1   ( 0.0) [ 1.000000000000D+00]    5.000000000000D-01       .25   3.671875000000D-01
2   ( 1.0) [-5.000000000000D-01]   -2.500000000000D-01       .50   6.875000000000D-01
                                                             .75   9.140625000000D-01
       (0) [ 1.500000000000D+00]    1.500000000000D+00      1.25   9.140625000000D-01
       (1) [ 1.500000000000D+00]    0.                      1.50   6.875000000000D-01

I = 2      H = 1.0
0            0.                    -7.500000000000D-01     -1.00   2.500000000000D-01
1           -2.500000000000D-01     5.000000000000D-01      -.75   1.875000000000D-01
                                                            -.50   1.250000000000D-01
0   (-1.0) [ 2.500000000000D-01]    1.250000000000D-01      -.25   6.250000000000D-02
1   ( 0.0) [-5.000000000000D-01]   -2.500000000000D-01       .25  -5.859375000000D-02
2   ( 1.0) [ 2.500000000000D-01]    1.250000000000D-01       .50  -9.375000000000D-02
                                                             .75  -8.203125000000D-02
       (0) [-2.500000000000D-01]   -2.500000000000D-01      1.25   1.679687500000D-01
       (1) [-2.500000000000D-01]    5.000000000000D-01      1.50   4.062500000000D-01
```

Figure 7. Table 1 for $n = 2$, $m = 2$. The brackets contain $\tilde{d}^h_\mu$ and $\tilde{a}^h_\lambda$. The parentheses contain μ and λ.

for example, if $t = 2$,

$$\alpha^{-1}(2) = -0.25 - 2.5 + 6.75 - 4.0 + 0.25 = 0.25,$$

$$\alpha^{0}(2) = 1.5 + 3.0 - 13.5 + 8.0 - 0.5 = -1.5,$$

$$\alpha^{1}(2) = -0.25 - 0.5 + 6.75 - 4.0 + 0.25 = 2.25,$$

and

$$y(2) \approx (0.25)(2) - 1.5(3) + 2.25(5) = 7.25.$$

Example 11

$$t: \quad -1.5 \quad -0.5 \quad 0.5 \quad 1.5,$$
$$y: \quad 1 \quad 1 \quad 2 \quad 3.$$

Using powers and plusses, find the spline that approximates this data.

Using Table 1 for $n = 2$, $m = 3$, the reader may check that

$$\alpha^{-1.5}(t) = -0.9 - 1.267t + 0.2667(t+1.5)_+^3 - 0.6(t+0.5)_+^3 + 0.4(t-0.5)_+^3 - 0.0667(t-1.5)_+^3,$$

$$\alpha^{-0.5}(t) = 2.4 + 1.6t - 0.6(t+1.5)_+^3 + 1.6(t+0.5)_+^3 - 1.4(t-0.5)_+^3 + 0.4(t-1.5)_+^3,$$

$$\alpha^{0.5}(t) = -0.6 - 0.4t + 0.4(t+1.5)_+^3 - 1.4(t+0.5)_+^3 + 1.6(t-0.5)_+^3 - 0.6(t-1.5)_+^3,$$

$$\alpha^{1.5}(t) = 0.1 - 0.067t - 0.0667(t+1.5)_+^3 + 1.6(t+0.5)_+^3 - 0.6(t-0.5)_+^3 + 0.0267(t-1.5)_+^3,$$

and

$$y(t) \approx \alpha^{-1.5}(t)(1) + \alpha^{-0.5}(t)(1) + \alpha^{0.5}(t)(2) + \alpha^{1.5}(t)(3).$$

SUMMARY OF THE FOUR FORMS

1. Powers and absolutes unshifted: $y(u) \approx \sum_{i=0}^{m} \beta^i(u) y_i$,

$$\beta^i(u) = \sum_{\lambda=0}^{n-1} b_\lambda^i u^\lambda + \sum_{\nu=0}^{m} c_\nu^i |u - \nu|^q, \qquad i = 0, \ldots, m; \quad q = 2n - 1.$$

2. Powers and plusses unshifted: $y(u) \approx \sum_{i=0}^{m} \beta^i(u) y_i$,

$$\beta^i(u) = \sum_{\lambda=0}^{n-1} \tilde{b}_\lambda^i u^\lambda + \sum_{\nu=0}^{m} \tilde{c}_\nu^i (u - \nu)_+^q, \qquad i = 0, \ldots, m.$$

The functions β^i in (1) and (2) are identical.

3. Powers and absolutes, centered at midpoint: $y(t) \approx \sum_{h=-p}^{p} \alpha^h(t) y_h$,

$$\alpha^h(t) = \sum_{\lambda=0}^{n-1} a_\lambda^h t^\lambda + \sum_{\mu=-p}^{p} d_\mu^h |t - \mu|^q, \qquad h = -p, -p+1, \ldots, p, \quad p = \frac{m}{2}.$$

4. Powers and plusses, centered at midpoint: $y(t) \approx \sum_{h=-p}^{p} \alpha^h(t) y_h$,

$$\alpha^h(t) = \sum_{\lambda=0}^{n-1} \tilde{a}_\lambda^h t^\lambda + \sum_{\mu=-p}^{p} \tilde{d}_\mu^h (t-\mu)_+^q, \qquad h = -p, -p+1, \ldots, p.$$

The functions α^h in (3) and (4) are identical. Furthermore, they are merely a shifted form of the functions β^i in (1) and (2):

$$\alpha^h(t) = \beta^i(u), \qquad h = i - p, \quad t = u - p, \quad i = 0, \ldots, m;$$

u is any real number. Figure 8 indicates the location of all the coefficients in Table 1.

```
                    N=2        M=3                              P = 1,5      Q=3
                                 ~ i
                                 b
                                   λ
I=0       H=-1,5
0]λ         1,00000000000000D+00]↙ 6,00000000000000D-01] i   -1,00   2,26666666666667D+00
1]         -1,26666666666667D+00]  -6,66666666666667D-01] b    -,75   1,95000000000000D+00
                                                           λ   -,50   1,63333333333333D+00
0]   -1,5]   2,66666666666667D-01]~c i  1,33333333333333D-01] c i   -,25   1,31666666666667D+00
1]ν   -,5]  -6,00000000000000D-01]=  ν -3,00000000000000D-01]   ν    ,25   6,87500000000000D-01
2]     ,5]μ  4,00000000000000D-01]~d h  2,00000000000000D-01]=  h    ,50   4,00000000000000D-01
3]    1,5]  -6,66666666666667D-02]   μ -3,33333333333333D-02] d μ    ,75   1,62500000000000D-01
                                                                    1,25  -7,18750000000000D-02
      0]λ -9,00000000000000D-01]   -4,00000000000000D-01]  h       1,50  -7,50000000000000D-02
      1]  -1,26666666666667D+00]↖  -6,66666666666667D-01] a λ      1,75  -4,06250000000000D-02
                                 ~ h
                                 a
                                   λ
```

Figure 8. Part of Table 1 for $n = 2$, $m = 3$.

1.5. DERIVATIVES

The splines may be used to approximate the derivative dy/dx as well as y. Since

$$y(u) \approx \beta^0(u) y_0 + \beta^1(u) y_1 + \cdots + \beta^m(u) y_m,$$

we find

$$\frac{dy}{du} \approx \frac{d\beta^0(u)}{du} y_0 + \frac{d\beta^1(u)}{du} y_1 + \cdots + \frac{d\beta^m(u)}{du} y_m;$$

and, since $dy/du = (dy/dx)(dx/du)$,

$$\frac{dy}{dx} \approx \left(\frac{dx}{du}\right)^{-1} \left[\frac{d\beta^0(u)}{du} y_0 + \frac{d\beta^1(u)}{du} y_1 + \cdots + \frac{d\beta^m(u)}{du} y_m\right].$$

Example 12

$$x{:}\quad 0 \quad 2 \quad 4,$$

$$y{:}\quad 0 \quad 4 \quad 16.$$

The equivalent problem is

$$u: \quad 0 \quad 1 \quad 2,$$
$$y: \quad 0 \quad 4 \quad 16,$$

where $u = \frac{1}{2}x$ and therefore $dx/du = 2$. From Table 1 ($n = 2$, $m = 2$), using the first form of the interpolating polynomial, we find

$$\beta^0(u) = \tfrac{1}{4} - \tfrac{1}{2}u + \tfrac{1}{8}|u-0|^3 - \tfrac{1}{4}|u-1|^3 + \tfrac{1}{8}|u-2|^3,$$
$$\beta^1(u) = \tfrac{3}{2} - \tfrac{1}{4}|u-0|^3 + \tfrac{1}{2}|u-1|^3 - \tfrac{1}{4}|u-2|^3,$$
$$\beta^2(u) = -\tfrac{3}{4} + \tfrac{1}{2}u + \tfrac{1}{8}|u-0|^3 - \tfrac{1}{4}|u-1|^3 + \tfrac{1}{8}|u-2|^3.$$

Now let us use the following notation (the signum function):

$$\begin{aligned} \text{sig } z &= 1 && \text{if } z > 0, \\ \text{sig } z &= 0 && \text{if } z = 0, \\ \text{sig } z &= -1 && \text{if } z < 0; \end{aligned}$$

for example,

$$\begin{aligned} \text{sig}(u-2) &= 1 && \text{if } \quad u > 2, \\ \text{sig}(u-2) &= -1 && \text{if } \quad u < 2. \end{aligned}$$

Differentiating the above β's, we have

$$\frac{d\beta^0(u)}{du} = -\tfrac{1}{2} + \tfrac{3}{8}|u|^2 \text{ sig } u - \tfrac{3}{4}|u-1|^2 \text{ sig}(u-1) + \tfrac{3}{8}|u-2|^2 \text{ sig}(u-2),$$

$$\frac{d\beta^1(u)}{du} = -\tfrac{3}{4}|u|^2 \text{ sig } u + \tfrac{3}{2}|u-1|^2 \text{ sig}(u-1) - \tfrac{3}{4}|u-2|^2 \text{ sig}(u-2),$$

$$\frac{d\beta^2(u)}{du} = \tfrac{1}{2} + \tfrac{3}{8}|u|^2 \text{ sig } u - \tfrac{3}{4}|u-1|^2 \text{ sig}(u-1) + \tfrac{3}{8}|u-2|^2 \text{ sig}(u-2),$$

and

$$\frac{dy}{dx} \approx \tfrac{1}{2}\left[\frac{d\beta^0(u)}{du}(0) + \frac{d\beta^1(u)}{du}(4) + \frac{d\beta^2(u)}{du}(16)\right];$$

for example, when $u = 1$, ($x = 2$), we can use this expression to approximate dy/dx at $x = 2$. Thus

$$\left(\frac{d\beta^0}{du}\right)_{u=1} = -\tfrac{1}{2} + \tfrac{3}{8}(1) + \tfrac{3}{8}(-1) = -\tfrac{1}{2},$$

$$\left(\frac{d\beta^1}{du}\right)_{u=1} = -\tfrac{3}{4}(1) - \tfrac{3}{4}(-1) = 0,$$

$$\left(\frac{d\beta^2}{du}\right)_{u=1} = \tfrac{1}{2} + \tfrac{3}{8}(1) + \tfrac{3}{8}(-1) = \tfrac{1}{2},$$

and

$$\left(\frac{dy}{dx}\right)_{x=2} \approx \tfrac{1}{2}[0 + 0 + \tfrac{1}{2}(16)] = 4.$$

Example 12 (continued). We now illustrate the same procedure by using the second form of the splines (i.e., powers and plusses unshifted) on the given data.

From Table 1 for $n = 2$, $m = 2$ (or see Figure 4).

$$\beta^0(u) = 1 - 1.25u + \tfrac{1}{4}(u)^3_+ - \tfrac{1}{2}(u-1)^3_+ + \tfrac{1}{4}(u-2)^3_+,$$
$$\beta^1(u) = 1.5u - \tfrac{1}{2}(u)^3_+ + 1.0(u-1)^3_+ - \tfrac{1}{2}(u-2)^3_+,$$
$$\beta^2(u) = -\tfrac{1}{4}u + \tfrac{1}{4}(u)^3_+ - \tfrac{1}{2}(u-1)^3_+ + \tfrac{1}{4}(u-2)^3_+.$$

Differentiating,

$$\frac{d\beta^0(u)}{du} = -1.25 + \tfrac{3}{4}(u)^2_+ - \tfrac{3}{2}(u-1)^2_+ + \tfrac{3}{4}(u-2)^2_+,$$

$$\frac{d\beta^1(u)}{du} = 1.5 - \tfrac{3}{2}(u)^2_+ + 3(u-1)^2_+ - \tfrac{3}{2}(u-2)^2_+,$$

$$\frac{d\beta^2(u)}{du} = -\tfrac{1}{4} + \tfrac{3}{4}(u)^2_+ - \tfrac{3}{2}(u-1)^2_+ + \tfrac{3}{4}(u-2)^2_+,$$

and

$$\frac{dy}{dx} \approx \frac{1}{2}\left[\frac{d\beta^0(u)}{du}(0) + \frac{d\beta^1(u)}{du}(4) + \frac{d\beta^2(u)}{du}(16)\right],$$

and, if $u = 1$ $(x = 2)$,

$$\frac{d\beta^0}{du_1} = -\tfrac{5}{4} + \tfrac{3}{4} = -\tfrac{1}{2},$$

$$\frac{d\beta^1}{du} = 1.5 - \tfrac{3}{2} = 0,$$

$$\frac{d\beta^2}{du} = -\tfrac{1}{4} + \tfrac{3}{4} = \tfrac{1}{2},$$

and $(dy/dx)_{x=2} \approx \tfrac{1}{2}[0 + 0 + \tfrac{1}{2}(16)] = 4$.

1.6. INTEGRALS

To approximate the integral $\int y(x)\,dx$ we calculate the integral of the spline approximation of y; that is, we use the formula

$$\int_c^d y(x)\,dx \approx \sum_{i=0}^{m} y(x_i) \int_c^d \beta^i\left(\frac{x - x_0}{h}\right) dx = h \sum_{i=0}^{m} y(x_i) \int_\gamma^\delta \beta^i(u)\,du,$$

where

$$u = \frac{x - x_0}{h}, \qquad \gamma = \frac{c - x_0}{h}, \qquad \delta = \frac{d - x_0}{h}.$$

Example 13

$$x: \quad 0 \quad 2 \quad 4,$$

$$y: \quad 0 \quad 4 \quad 16.$$

Approximate $\int_0^4 y(x)\,dx$. Using $u = (x - 0)/2$, we have

$$u = 0 \quad 1 \quad 2,$$

$$y = 0 \quad 4 \quad 16.$$

Table 1 for $n = 2$, $m = 2$ (absolutes unshifted; see Example 12) yields, to within a constant,

$$\int \beta^0(u)\,du = \tfrac{1}{4}u - \tfrac{1}{4}u^2 + \tfrac{1}{32}|u - 0|^4 \operatorname{sig}(u) - \tfrac{1}{16}|u - 1|^4 \operatorname{sig}(u - 1) + \tfrac{1}{32}|u - 2|^4 \operatorname{sig}(u - 2),$$

$$\int \beta^1(u)\,du = \tfrac{3}{2}u - \tfrac{1}{16}|u - 0|^4 \operatorname{sig}(u) + \tfrac{1}{8}|u - 1|^4 \operatorname{sig}(u - 1) - \tfrac{1}{16}|u - 2|^4 \operatorname{sig}(u - 2),$$

$$\int \beta^2(u)\,du = -\tfrac{3}{4}u + \tfrac{1}{4}u^2 + \tfrac{1}{32}|u - 0|^4 \operatorname{sig}(u) - \tfrac{1}{16}|u - 1|^4 \operatorname{sig}(u - 1) + \tfrac{1}{32}|u - 2|^4 \operatorname{sig}(u - 2).$$

At $u = 0$ we have

$$\left[\int \beta^0(u)\,du\right]_0 = \tfrac{1}{16} - \tfrac{1}{2} = -\tfrac{7}{16},$$

$$\left[\int \beta^1(u)\,du\right]_0 = -\tfrac{1}{8} + 1 = \tfrac{7}{8},$$

$$\left[\int \beta(u)\,du\right]_0 = \tfrac{1}{16} - \tfrac{1}{2} = -\tfrac{7}{16}.$$

At this lower limit therefore we have

$$(0)(-\tfrac{7}{16}) + (4)(\tfrac{7}{8}) + (16)(-\tfrac{7}{16}) = -\tfrac{7}{2}.$$

At $u = 2$ we have

$$\left[\int \beta^0(u)\,du\right]_2 = \tfrac{1}{2} - 1 + \tfrac{1}{2} - \tfrac{1}{16} = -\tfrac{1}{16},$$

$$\left[\int \beta^1(u)\,du\right]_2 = 3 - 1 + \tfrac{1}{8} = \tfrac{17}{8},$$

$$\left[\int \beta^2(u)\,du\right]_2 = -\tfrac{3}{2} + 1 + \tfrac{1}{2} - \tfrac{1}{16} = -\tfrac{1}{16}.$$

At this upper limit we have

$$(0)(-\tfrac{1}{16}) + 4(\tfrac{17}{8}) + (16)(-\tfrac{1}{16}) = \tfrac{15}{2}.$$

Taking the difference of the limits and multiplying by the step size $h = 2$, we obtain

$$\int_0^4 y(x)\,dx \approx 2\,[\tfrac{15}{2} - (-\tfrac{7}{2})] = 22.$$

Alternatively, we might use plusses, as in the second part of Example 12.
In this example the data might have arisen from the function $y = x^2$, in which case the exact integral would have been

$$\int_0^4 y(x)\,dx = \tfrac{64}{3} = 21.33\ldots.$$

1.7. LAGRANGIANS

In this section we take $m = n - 1$. Then the spline approximation that fits the table

$$\begin{array}{llll} x\colon & x_0 & x_1 \cdots x_m, \\ y\colon & y_0 & y_1 \cdots y_m, \end{array}$$

is simply the polynomial of minimal degree that fits. Thus the spline approximation of the value of y at x is

$$y_0 L_0(x) + y_1 L_1(x) + \cdots + y_m L_m(x)$$

for all numbers x, where

$$L_i(x) = \frac{(x - x_0)(x - x_1)\cdots\widehat{(x - x_i)}\cdots(x - x_m)}{(x_i - x_0)(x_i - x_1)\cdots\widehat{(x_i - x_i)}\cdots(x_i - x_m)}, \quad i = 0, \ldots, m.$$

In this expression the roof indicates omission of the factor beneath it. After expansion L_i may be written in the standard form of a polynomial:

$$L_i(x) = b_{i,0} + b_{i,1}x + b_{i,2}x^2 + \cdots + b_{i,m}x^m.$$

Table 3 gives the coefficients $b_{i,0}, \ldots, b_{i,m}$ for $m = 1, \ldots, 9$ in the two cases that are considered in Table 1, namely, regularly spaced arguments with unit step with $x_0 = 0$ and with $x_0 = -p = -m/2$. Of course, any set of regularly spaced arguments may be reduced to either of these cases by a transformation.

The arrangement of Table 3 is like that of Table 1 except that the two cases are printed side by side instead of one above the other. Table 3 is included only as a possible convenience; the reader could readily calculate the Lagrangians for himself.

Example 14

$$\begin{array}{llll} x: & 0 & 1 & 2, \\ y: & -3 & -2 & 5. \end{array}$$

Approximate the function y and, in particular, its value when $x = 0.5$. From Table 3 with $n = 3$ we read that

$$\begin{aligned} L_0(x) &= 1 - 1.5x + 0.5x^2, \\ L_1(x) &= 2x - x^2, \\ L_2(x) &= -0.5x + 0.5x^2. \end{aligned}$$

The polynomial of minimal degree that fits the data is

$$-3(1 - 1.5x + 0.5x^2) - 2(2x - x^2) + 5(-0.5x + 0.5x^2) = -3 - 2x + 3x^2.$$

At $x = 0.5$, y is approximately -3.25.

Example 15

$$\begin{array}{lllll} x: & -1.5 & -0.5 & 0.5 & 1.5, \\ y: & 11 & -4 & -5 & -4. \end{array}$$

Approximate the function y and, in particular, its value at $x = 0$. From Table 3 with $n = 4$ we read that

$$\begin{aligned} L_0(x) &= -0.0625 + 0.04167x + 0.25x^2 - 0.16667x^3, \\ L_1(x) &= 0.5625 - 1.125x - 0.25x^2 + 0.5x^3, \\ L_2(x) &= L_1(-x), \\ L_3(x) &= L_0(-x). \end{aligned}$$

The approximation of the function y is $11L_0 - 4L_1 - 5L_2 - 4L_3$; at $x = 0$ it is -5.5.

CHAPTER 2

Use of the Tables: Second Introduction

2.1. INTRODUCTION

Let x denote a real function defined on a subset of the reals R. Let $s_0, s_1, \ldots, s_m$ be fixed numbers, all in the domain of definition of x. We shall consider, to start with, the approximation of x in terms of the values $x(s_0)$, $x(s_1)$, $\ldots$, $x(s_m)$ of x at the abscissae $s_0, \ldots, s_m$. We call the values $x(s_0), \ldots, x(s_m)$ *observations* of the function x.

The approximation of x will involve a choice by the reader of a positive integer n, which is called the *type* of the approximation. For each $n \leq m + 1$ and each set of observations there is an approximation ξ of x, called the *spline approximation of type n based on* $x(s_0), \ldots, x(s_m)$. The spline ξ is a function defined for all real arguments, whether or not x is so defined. The spline ξ is the optimal approximation of x in a precise and apt sense that will be described. Table 1 permits the reader to write ξ explicitly and easily, in the case in which $s_0, \ldots, s_m$ are regularly spaced, just as a table of Lagrangian polynomials permits one to write the polynomial of least degree that fits the observations.

Other things being equal, one would prefer to use a type n as small as possible to simplify the splines. The choice of n is discussed in Section 2.12.

The name spline is due to Schoenberg [2] and refers to the fact that the splines just described are articulated functions which are smooth, along with a number of their derivatives, but which allow a jump in a higher derivative [the $(2n - 1)$th, in the case described] at $s_0, \ldots, s_m$. In Chapter 3 we use the word spline more generally to describe a function with specific minimal properties.

There is an optimal appraisal of the error $x(t) - \xi(t)$ for each real t in terms of the quantity

$$\text{(1)} \qquad \int_I x_n(s)^2 \, ds \underset{\text{def}}{=} \|x_n\|^2,$$

where x_n denotes the nth derivative of x; I is the smallest interval which contains $s_0, \ldots, s_m$, and t; and x is assumed to have a continuous nth derivative (or somewhat less) in I. Table 2 provides this appraisal.

If G is a functional of a sort to be described, then $G\xi$ approximates Gx. Table 2 provides an optimal appraisal of the error $Gx - G\xi$. In particular Gx may be the jth derivative of x at t, $j = 0, \ldots, n-1$, $t \in \mathsf{R}$; then the number $\xi_j(t)$, which is accessible to us, being the value of the jth derivative of ξ at t, approximates $x_j(t)$. Similarly Gx may be an integral $\int_u^v x(s)\,ds$, u, $v \in \mathsf{R}$; then $\int_u^v \xi(s)\,ds$, which is accessible to us, approximates $\int_u^v x(s)\,ds$. In each case appraisals of the error are accessible to us. Thus our tables may be used for interpolation and extrapolation, approximate differentiation, approximate integration (ordinary or Stieltjes), and in other ways (see Section 2.10).

2.2. PLUSSES AND ABSOLUTES

The formulas for the splines involve either absolute values or the *plus function*, whose values t_+ are

$$t_+ = \begin{cases} t & \text{if } t \geq 0, \\ 0 & \text{if } t < 0. \end{cases} \tag{2}$$

There is a simple relation between plusses and absolutes. Let r denote a non-nonegative integer. Then

$$\left.\begin{aligned} |t|^r &= 2t_+^r - t^r \quad \text{if } r \text{ is odd}, \\ |t|^r &= t^r \qquad\quad\;\; \text{if } r \text{ is even}, \end{aligned}\right\} \quad t \in \mathsf{R}. \tag{3}$$

(In the case $r = 0 = t$ we take $0^0 = 1$.)

The *signum function* sig is useful. Its values are

$$\operatorname{sig} t = \operatorname{signum} t = \begin{cases} 1 & \text{if } t > 0, \\ 0 & \text{if } t = 0, \\ -1 & \text{if } t < 0. \end{cases} \tag{4}$$

Then

$$|t| = t \operatorname{sig} t, \qquad t \in \mathsf{R}. \tag{5}$$

The following formulas for derivatives and integrals will be useful.

$$D_t t_+^r = r t_+^{r-1}, \qquad r = 1, 2, \ldots; t \in \mathsf{R}, \tag{6}$$

except for the isolated subcase $r = 1$ and $t = 0$, where only one-sided derivatives exist, 0 on the left and 1 on the right.

$$\int t_+^r \, dt = \frac{t_+^{r+1}}{r+1} + \text{constant}, \qquad r = 0, 1, \ldots . \tag{7}$$

$$D_t |t|^r = r\,|t|^{r-1} \operatorname{sig} t, \qquad r = 1, 2, \ldots; t \in \mathsf{R}, \tag{8}$$

except for the isolated subcase $r = 1$ and $t = 0$, where only one-sided derivatives exist, -1 on the left and 1 on the right.

$$\int |t|^r \, dt = \frac{|t|^{r+1} \operatorname{sig} t}{r+1} + \text{constant}, \qquad r = 0, 1, \ldots . \tag{9}$$

If r is even, then $|t|^r = t^r$ and the last two formulas are not needed.

2.3. THE SPLINES

Let m, n be integers such that

$$m + 1 \geq n \geq 1. \tag{10}$$

Put

$$q = 2n - 1. \tag{11}$$

For any function x whose domain includes the integers $0, 1, \ldots, m$ there is a spline approximation ξ based on the values $x(0), \ldots, x(m)$ and of type n. This approximation ξ is obtainable from Table 1 in the following way:

$$\xi = \sum_{i=1}^{m} x(i)\beta^{m,i}, \tag{12}$$

where $\beta^{m,i}$ is a function which we call the ith *cardinal spline* for m and n. Although $\beta^{m,i}$ depends on n, we do not need an index n in its symbol. Table 1 gives the functions $\beta^{m,i}$ in two forms (absolutes and plusses):

$$\begin{aligned} \beta^{m,i}(u) &= \sum_{\lambda=0}^{n-1} b_\lambda^{m,i} u^\lambda + \sum_{\nu=0}^{m} c_\nu^{m,i} |u - \nu|^q \\ &= \sum_{\lambda=0}^{n-1} \tilde{b}_\lambda^{m,i} u^\lambda + \sum_{\nu=0}^{m} \tilde{c}_\nu^{m,i} (u - \nu)_+^q, \; u \in \mathsf{R}, \qquad i = 0, \ldots, m; \end{aligned} \tag{13}$$

that is, Table 1 gives the numerical coefficients $b_\lambda^{m,i}$, $c_\nu^{m,i}$, $\tilde{b}_\lambda^{m,i}$, $\tilde{c}_\nu^{m,i}$ which enter in these formulas for

$$n = 2, 3, 4, 5 \text{ and } 8;$$
$$m = n, n + 1, \ldots, 20.$$

The printed pattern for each m, n consists of $m + 1$ blocks of information headed successively $i = 0, 1, \ldots, m$. The block for each i has the following structure:

Pattern of the Part of Table 1 that refers to $\beta^{m,i}$

$i=$		$\cdots$			
0		$\tilde{b}_0^{m,i}$	$b_0^{m,i}$	-1.00	$\beta^{m,i}(-1.00)$
1		$\tilde{b}_1^{m,i}$	$b_1^{m,i}$	-0.75	$\beta^{m,i}(-0.75)$
$\vdots$		$\vdots$	$\vdots$	$\vdots$	$\vdots$
λ		$\tilde{b}_\lambda^{m,i}$	$b_\lambda^{m,i}$		
$\vdots$		$\vdots$	$\vdots$		
$n-1$		$\tilde{b}_{n-1}^{m,i}$	$b_{n-1}^{m,i}$		
0		$\tilde{c}_0^{m,i}$	$c_0^{m,i}$		
1		$\tilde{c}_1^{m,i}$	$c_1^{m,i}$		
$\vdots$		$\vdots$	$\vdots$		
ν	$\vdots$	$\tilde{c}_\nu^{m,i}$	$c_\nu^{m,i}$		
$\vdots$		$\vdots$	$\vdots$		
m		$\tilde{c}_m^{m,i}$	$c_m^{m,i}$		
	$\vdots$	$\vdots$	$\vdots$		

Besides giving the coefficients $b_\lambda^{m,i}$, $c_\nu^{m,i}$, $\tilde{b}_\lambda^{m,i}$, $\tilde{c}_\nu^{m,i}$ that enter in the two forms of the function $\beta^{m,i}$, the block gives the numerical values $\beta^{m,i}(u)$ for the indicated values of u in the last two columns. The printed values $\beta^{m,i}(u)$ are correct to the last figure, since they are based on 30-digit versions of the coefficients of $\beta^{m,i}$ rather than on the printed 13-digit versions. The values $\beta^{m,i}(u)$ may, of course, be calculated by the reader, using the printed coefficients, subject to the effect of the rounding errors in the last figure of the coefficients.

The function $\beta^{m,i}$ has the following properties, among others:

(14) $$\beta^{m,i}(u) = \beta^{m,m-i}(m-u), \qquad u \in \mathsf{R}, \qquad i = 0, \ldots, m,$$

(15) $$\beta^{m,i}(\nu) = \delta^{i,\nu}, \qquad \nu = 0, \ldots, m,$$

where $\delta^{i,\nu}$ is the Kronecker delta; that is,

(16) $$\delta^{i,\nu} = \begin{cases} 0 & \text{if } i \neq \nu, \\ 1 & \text{if } i = \nu. \end{cases}$$

Consequently we have sometimes not carried the tabulation of values $\beta^{m,i}(u)$ in the last columns beyond $u = m/2$, and we have omitted the abscissae $u = 0, 1, \ldots$. It will always be true in Table 1 that

(17) $$\tilde{c}_\nu^{m,i} = 2c_\nu^{m,i}.$$

Illustration 1. $n = 2$, $m = 2$. Here $q = 2n - 1 = 3$, and

$$\begin{aligned}
\beta^{2,0}(u) &= 1 - 1.25u + 0.25u_+^3 - 0.5(u-1)_+^3 + 0.25(u-2)_+^3 \\
&= 0.25 - 0.5u + 0.125\,|u|^3 - 0.25\,|u-1|^3 + 0.125\,|u-2|^3, \; u \in \mathsf{R}, \\
\beta^{2,1}(u) &= 1.5u - 0.5u_+^3 + (u-1)_+^3 - 0.5(u-2)_+^3 \\
&= 1.5 - 0.25\,|u|^3 + 0.5\,|u-1|^3 - 0.25\,|u-2|^3, \\
\beta^{2,2}(u) &= \quad - 0.25u + 0.25u_+^3 - 0.5(u-1)_+^3 + 0.25(u-2)_+^3 \\
&= -0.75 + 0.5u + 0.125\,|u|^3 - 0.25\,|u-1|^3 + 0.125\,|u-2|^3.
\end{aligned}$$

The reader may use (3) to check that the two forms given for each function $\beta^{2,0}$, $\beta^{2,1}$, $\beta^{2,2}$ are equivalent. The reader may check the entries in the last column of Table 1; that is, he may calculate that

$$\beta^{2,0}(-1) = 2.25, \qquad \beta^{2,0}(-0.75) = 1.9375,$$
$$\beta^{2,0}(-0.5) = 1.625, \qquad \beta^{2,0}(-0.25) = 1.312;$$
$$\beta^{2,0}(0.25) = 6.9140625, \ldots, \beta^{2,0}(1.5) = -0.09375;$$

also that

$$\beta^{2,1}(-1) = -1.5, \qquad \beta^{2,1}(-0.75) = -1.125, \ldots,$$

and that

$$\beta^{2,2}(-1) = 0.25, \ldots.$$

The reader may check that

$$\beta^{2,i}(v) = \delta^{i,v}, \qquad i, v = 0, 1, 2.$$

Also, that

$$\beta^{2,0}(u) = \beta^{2,2}(2-u), \qquad \beta^{2,1}(u) = \beta^{2,1}(2-u), \qquad u \in \mathsf{R}.$$

The cardinal splines $\beta^{2,0}$, $\beta^{2,1}$, $\beta^{2,2}$ are used as follows: if the function x takes on the values $x(0)$, $x(1)$, $x(2)$ at the abscissae 0, 1, 2, the spline approximation ξ of x based on these values and of type 2 is

$$\xi = x(0)\beta^{2,0} + x(1)\beta^{2,1} + x(2)\beta^{2,2}.$$

Thus, for all $u \in \mathsf{R}$,

$$\xi(u) = x(0)\beta^{2,0}(u) + x(1)\beta^{2,1}(u) + x(2)\beta^{2,2}(u).$$

This function ξ is the optimal approximation of x in a sense that is described in Section 2.7 and Chapter 3. If $0 < u < 2$, then $\xi(u)$ is an interpolatory approximation of $x(u)$. If $u < 0$ or $u > 2$, then $\xi(u)$ is extrapolatory; for example,

$$\xi(-1) = 2.25x(0) - 1.5x(1) + 0.25x(2),$$
$$\xi(0.25) = 0.69140625x(0) + 0.3671875x(1) - 0.05859375x(2),$$

and, by (14),

$$\xi(1.75) = -0.05859375x(0) + 0.3671875x(1) + 0.69140625x(2).$$

Because of (15),

$$\xi(v) = x(v), \qquad v = 0, 1, 2.$$

Furthermore, the value $x_1(u)$ of the first derivative of x at u may be approximated by its counterpart $\xi_1(u)$, and integrals $\int_a^b x(s)\,ds$ may be approximated by their counterparts $\int_a^b \xi(s)\,ds$. These approximations are optimal. Optimal appraisals of the errors in these approximations in terms of $\int_I x_2(s)^2\,ds$ are given in Sections 2.2 and 2.9, where I is the smallest interval that contains u, 0, 2 or a, b, 0, 2, respectively. Thus, for example,

$$\xi_1(u) = x(0)\beta_1^{2,0}(u) + x(1)\beta_1^{2,1}(u) + x(2)\beta_1^{2,2}(u), \; u \in \mathsf{R},$$

where, by (6) and (8),

$$\begin{aligned}\beta_1^{2,0}(u) &= \frac{d\beta^{2,0}(u)}{du} = -1.25 + 0.75u_+^2 - 1.5(u-1)_+^2 + 0.75(u-2)_+^2 \\ &= -0.5 + 0.375u^2 \;\operatorname{sig} u - 0.75(u-1)^2 \operatorname{sig} u \\ &\quad + 0.375(u-2)^2 \operatorname{sig}(u-2), \\ \beta_1^{2,1}(u) &= 1.5 - 1.5u_+^2 + 3(u-1)_+^2 - 1.5(u-2)_+^2, \\ \beta_2^{2,2}(u) &= -0.25 + 0.75u_+^2 - 1.5(u-1)_+^2 + 0.75(u-2)_+^2.\end{aligned}$$

Also, for example,

$$\int_0^1 \xi(s)\,ds = 0.4375x(0) + 0.625x(1) - 0.0625x(2),$$

since, by (7) and (2),

$$\int_0^1 \beta^{2,0}(s)\,ds = 1 - \frac{1.25}{2} + \frac{0.25}{4} = 0.4375,$$

$$\int_0^1 \beta^{2,1}(s)\,ds = \frac{1.5}{2} - \frac{0.5}{4} = 0.625,$$

$$\int_0^1 \beta^{2,2}(s)\,ds = -\frac{0.25}{2} + \frac{0.25}{4} = -0.0625.$$

Also, for example,

$$\int_0^2 \xi(s)\,ds = \frac{3x(0) + 10x(1) + 3x(2)}{8},$$

recovering the known formula in Sard [1, p. 53], since

$$\int_0^2 \beta^{2,0}(s)\,ds = 0.375 = \int_0^2 \beta^{2,2}(s)\,ds,$$

$$\int_0^2 \beta^{2,1}(s)\,ds = 1.25.$$

As u_+^3 and $|u|^3$ are polynomials for ≥ 0 and polynomials for ≤ 0, we may express the cardinal splines between abscissae as polynomials. Thus, for example,

$$\beta^{2,0}(u) = 1 - 1.25u \quad \text{if } u \leq 0,$$
$$\beta^{2,0}(u) = 1 - 1.25u + 0.25u^3 \quad \text{if } 0 \leq u \leq 1,$$
$$\beta^{2,0}(u) = 1.5 - 2.75u + 1.5u^2 - 0.25u^3 \quad \text{if } 1 \leq u \leq 2,$$
$$\beta^{2,0}(u) = -0.5 + 0.25u \quad \text{if } 2 \leq u.$$

These formulas agree with the coefficients of $x(0)$ in the formula B_2^2 in Sard [1, p. 94].

As u_+^3 and $|u|^3$ are functions of class C_2 on $(-\infty, \infty)$, so are the splines $\beta^{2,0}$, $\beta^{2,1}$, $\beta^{2,2}$. The third derivatives of these functions have jumps at 0, 1, 2.

Illustration 2. $n = 2$, $m = 3$. Here $q = 3$ and

$$\begin{aligned}\beta^{3,0}(u) &= 1 - 1.266666666667u + 0.2666666666667u_+^3 - 0.6(u-1)_+^3 \\ &\quad + 0.4(u-2)_+^3 - 0.06666666666667(u-3)_+^3 \\ &= 0.6 - 0.6666666666667u + 0.1333333333333\,|u|^3 \\ &\quad - 0.3\,|u-1|^3 + 0.2\,|u-2|^3 - 0.03333333333333\,|u-3|^3, \\ &\quad u \in \mathsf{R}, \\ \beta^{3,1}(u) &= 1.6u - 0.6u_+^3 + 1.6(u-1)_+^3 - 1.4(u-2)_+^3 + 0.4(u-3)_+^3, \\ \beta^{3,2}(u) &= \beta^{3,1}(3-u) = -0.4u + 0.4u_+^3 - 1.4(u-1)_+^3 \\ &\quad + 1.6(u-2)_+^3 - 0.6(u-3)_+^3, \\ \beta^{3,3}(u) &= \beta^{3,0}(3-u) = 0.06666666666667u - 0.06666666666667u_+^3 \\ &\quad - 0.4(u-1)_+^3 - 0.6(u-2)_+^3 + 0.2666666666667(u-3)_+^3.\end{aligned}$$

The spline approximation ξ of x based on $x(0)$, $x(1)$, $x(2)$, $x(3)$ of type 2 is

$$\xi = x(0)\beta^{3,0} + x(1)\beta^{3,1} + x(2)\beta^{3,2} + x(3)\beta^{3,3}.$$

Illustration 3. $n = 3$, $m = 3$. Then $q = 5$ and

$$\begin{aligned}\beta^{3,0}(u) &= 1 - 1.689393939394u + 0.7045454545455u^2 - 0.01515151515152u_+^5 \\ &\quad + 0.04545454545455(u-1)_+^5 - 0.04545454545455(u-2)_+^5 \\ &\quad + 0.01515151515152(u-3)_+^5, \qquad u \in \mathsf{R},\end{aligned}$$

$$\beta^{3,1}(u) = 2.568181818182u - 1.613636363636u^2 + 0.04545454545455u_+^5$$
$$- 0.1363636363636(u-1)_+^5 + 0.1363636363636(u-2)_+^5$$
$$- 0.04545454545455(u-3)_+^5,$$
$$\beta^{3,2}(u) = \beta^{3,1}(3-u) = -1.068181818182u + 1.113636363636u^2$$
$$- 0.04545454545455u_+^5 + 0.1363636363636(u-1)_+^5$$
$$- 0.1363636363636(u-2)_+^5 + 0.04545454545455(u-3)_+^5,$$
$$\beta^{3,3}(u) = \beta^{3,0}(3-u) = 0.1893939393939u - 0.2045454545455u^2$$
$$- 0.01515151515152u_+^5 - 0.04545454545455(u-1)_+^5$$
$$+ 0.04545454545455(u-2)_+^5 - 0.01515151515152(u-3)_+^5.$$

2.4. RANGE OF PARAMETERS IN TABLE 1

The range is

$$n = 2, 3, 4, 5, \text{ and } 8,$$
$$m = n, n+1, \ldots, 20.$$

As regards $n = 1$, a table is not needed. The spline approximation of x, based on $x(0), \ldots, x(m)$ and of type 1, is the piecewise linear function which interpolates to these values, which is constant below 0 and constant above m, and which is continuous on $(-\infty, \infty)$.

If $m = n - 1$, the spline approximation of x, based on $x(0), \ldots, x(m)$ and of type n, is precisely the Lagrangian polynomial of degree† $n - 1$ that interpolates to these values. The case $m = n - 1$ is considered separately in Section 2.7.

2.5. THE SPLINES, SYMMETRIC FORM

Put

(18) $$p = \frac{m}{2},$$
$$t = u - p, \qquad u \in \mathrm{R},$$

(19) $$h = i - p,$$
$$\mu = \nu - p.$$

Then, when u, i, or ν takes on the values $0, 1, \ldots, m$, the variable t, h, or μ takes on the regularly spaced, symmetric values

(20) $$-p, -p+1, -p+2, \ldots, p.$$

† By a polynomial of degree r in u we mean an expression of the form $c_0 u^r + c_1 u^{r-1} + \cdots + c_r$, whether $c_0 = 0$ or not.

The values (20) are integers if m is even and odd multiples of 1/2 if m is odd.

The change from u to t in (19) is a shift to a centered origin. The splines shift in a natural way.

Suppose that y is a function whose domain includes the abscissae $-p, \ldots, p$. Then the spline approximation η of y, based on the values $y(-p), \ldots, y(p)$ and of type n, is

$$\eta = \sum_{h=-p}^{p} y(h)\alpha^{p,h}, \tag{21}$$

where the cardinal splines $\alpha^{p,h}$ are the following functions:

$$\begin{aligned} \alpha^{p,h}(t) &= \sum_{\lambda=0}^{n-1} a_\lambda^{p,h} t^\lambda + \sum_{\mu=-p}^{p} d_\mu^{p,h}|t-\mu|^q \\ &= \sum_{\lambda=0}^{n-1} \tilde{a}_\lambda^{p,h} t^\lambda + \sum_{\mu=-p}^{p} \tilde{d}_\lambda^{p,h}(t-\mu)_+^q, \qquad t \in \mathrm{R},\ h = -p, \ldots, p; \end{aligned} \tag{22}$$

and the coefficients $a_\lambda^{p,h}$, $d_\mu^{p,h}$, $\tilde{a}_\lambda^{p,h}$ $\tilde{d}_\mu^{p,h}$ are given in Table 1 for the parameters within its range.

The printed pattern for each m, n consists of $m+1 = 2p+1$ blocks of information, headed successively $i=0$, $h=-p$; $i=1$, $h=-p+1, \ldots$; $i=m$, $h=p$. The block for each i, $h=i-p$ has the following structure:

Pattern of Table 1

$i=$	$h=i-p$				
0		$\tilde{b}_0^{m,i}$	$b_0^{m,i}$	-1.00	$\beta^{m,i}(-1.00)$
1		$\tilde{b}_1^{m,i}$	$b_1^{m,i}$	-0.75	$\beta^{m,i}(-0.75)$
⋮		⋮	⋮	⋮	⋮
λ		$\tilde{b}_\lambda^{m,i}$	$b_\lambda^{m,i}$		
⋮		⋮	⋮		
$n-1$		$\tilde{b}_{n-1}^{m,i}$	$b_{n-1}^{m,i}$		
0	$-p$	$\tilde{c}_0^{m,i} = \tilde{d}_{-p}^{p,h}$	$c_0^{m,i} = d_{-p}^{p,h}$		
1	$-p+1$	$\tilde{c}_1^{m,i} = \tilde{d}_{-p+1}^{p,h}$	$c_1^{m,i} = d_{-p+1}^{p,h}$		
⋮	⋮	⋮	⋮		
ν	$\mu=\nu-p$	$\tilde{c}_\nu^{m,i} = \tilde{d}_\mu^{p,h}$	$c_\nu^{m,i} = d_\mu^{p,h}$		
⋮	⋮	⋮	⋮		
m	p	$\tilde{c}_m^{m,i} = \tilde{d}_p^{p,h}$	$c_m^{m,i} = d_p^{p,h}$		
	0	$\tilde{a}_0^{p,h}$	$a_0^{p,h}$		
	1	$\tilde{a}_1^{p,h}$	$a_1^{p,h}$		
	⋮	⋮	⋮		
	λ	$\tilde{a}_\lambda^{p,h}$	$a_\lambda^{p,h}$		
	⋮	⋮	⋮		
	$n-1$	$\tilde{a}_{n-1}^{p,h}$	$a_{n-1}^{p,h}$		

The following relations hold.

(23) $\beta^{m,i}(u) = \alpha^{p,h}(t)$, all $t, u \in \mathsf{R}$; $i = 0, \ldots, m$; $h = -p, \ldots, p$;

here and elsewhere throughout this section the equalities (18 and 19) are in force. Also

(24) $c_\nu^{m,i} = d_\mu^{p,h}, \qquad \tilde{c}_\nu^{m,i} = \tilde{d}_\mu^{p,h}, \qquad \nu = 0, \ldots, m, \qquad \mu = -p, \ldots, p.$

Thus the coefficients of the plusses in the centered cardinal splines $\alpha^{p,h}$ are the same as those in the uncentered cardinal splines $\beta^{m,i}$. Likewise for the coefficients of the absolutes. Also

(25) $$\alpha^{p,h}(t) = \alpha^{p,-h}(-t), \qquad t \in \mathsf{R}.$$

Also, the $(m+1) \times (m+1)$ matrix $(d_\mu^{p,h})$, $h, \mu = -p, \ldots, p$, is symmetric in both its diagonals; that is,

(26) $$d_\mu^{p,h} = d_{-\mu}^{p,-h} = d_h^{p,\mu}.$$

Likewise for $(\tilde{d}_\mu^{p,h}) = 2(d_{\mu,}^{p,h})$. Also

(27) $$a_\lambda^{p,h} = (-1)^\lambda a_\lambda^{p,-h}, \qquad \lambda = 0, \ldots, n-1.$$

In the particular case

$$m = n = 2p,$$

with which Table 1 starts for each n,

(28) $$d_\mu^{p,h} = (-1)^n d_{-\mu}^{p,h}, \qquad n = 2p.$$

Illustration 4. $n = 2$, $m = 2$, symmetric form. Here $p = 1$, $q = 3$, and

$$\begin{aligned}
\alpha^{1,-1}(t) &= -0.25 - 1.25t + 0.25(t+1)_+^3 - 0.5t_+^3 + 0.25(t-1)_+^3 \\
&= -0.25 - 0.5t + 0.125\,|t+1|^3 - 0.25\,|t|^3 + 0.125\,|t-1|^3, \qquad t \in \mathsf{R}, \\
\alpha^{1,0}(t) &= \alpha^{1,0}(-t) = 1.5 + 1.5t - 0.5(t+1)_+^3 + t_+^3 - 0.5(t-1)_+^3 \\
&= 1.5 - 0.25\,|t+1|^3 + 0.5\,|t|^3 - 0.25\,|t-1|^3, \\
\alpha^{1,1}(t) &= \alpha^{1,-1}(-t) = -0.25 - 0.25t + 0.25(t+1)_+^3 - 0.5t_+^3 + 0.25(t-1)_+^3 \\
&= -0.25 + 0.5t + 0.125\,|t+1|^3 - 0.25\,|t|^3 + 0.125\,|t-1|^3.
\end{aligned}$$

The spline approximation of y, based on $y(-1)$, $y(0)$, $y(1)$, of type 2 is

$$\eta = y(-1)\alpha^{1,-1} + y(0)\alpha^{1,0} + y(1)\alpha^{1,1}.$$

The reader may compare the present illustration with Illustration 1 and verify that

$$\alpha^{1,h}(t) = \beta^{2,h+1}(t+1), \qquad t \in \mathsf{R}, \qquad h = -1, 0, 1.$$

Illustration 9 below includes the symmetric cardinals for $n = 2$, $m = 3$. The choice of formula among the four equivalent forms is discussed in Section 2.11.

2.6. THE SPLINES, GENERAL POSITION

If z is a function whose domain includes the regularly spaced abscissae $a, a + w, a + 2w, \ldots, a + mw$, the spline approximation ζ of z, based on $z(a), \ldots, z(a + mw)$, of type n, may be obtained from Table 1 as follows.

Put

$$u = \frac{s - a}{w}, \qquad s \in \mathsf{R}; \tag{29}$$

and define the function x by the relation

$$x(u) = z(s).$$

Then ζ, the spline approximation of z, is given by the relation

$$\zeta(s) = \xi(u), \tag{30}$$

where ξ is the spline approximation of x, based on $x(0), \ldots, x(m)$, of type n.

Alternatively, we may use the splines with centered origin, starting with the change of variable

$$t = \frac{s - (a + b)/2}{w}, \qquad s \in \mathsf{R}, \tag{31}$$

where

$$b = a + mw. \tag{32}$$

Illustration 5. $n = 2$, $m = 2$. The spline approximation ζ of z, based on $z(10), z(15), z(20)$, of type 2, is given by the relation

$$\zeta(s) = z(10)\beta^{2,0}\left(\frac{s - 10}{5}\right) + z(15)\beta^{2,1}\left(\frac{s - 10}{5}\right) + z(20)\beta^{2,2}\left(\frac{s - 10}{5}\right), \qquad s \in \mathsf{R},$$

where $\beta^{2,0}, \beta^{2,1}, \beta^{2,2}$ are the cardinal splines of type 2, given in Illustration 1. Alternatively,

$$\zeta(s) = z(10)\alpha^{1,-1}\left(\frac{s - 15}{5}\right) + z(15)\alpha^{1,0}\left(\frac{s - 15}{5}\right) + z(20)\alpha^{1,1}\left(\frac{s - 15}{5}\right), \qquad s \in \mathsf{R},$$

where $\alpha^{1,-1}, \alpha^{1,0}, \alpha^{1,1}$ are the centered cardinals of type 2, given in Illustration 4.

2.7. THE CASE $m = n - 1$. LAGRANGIAN POLYNOMIALS

If $m = n - 1$, the cardinal splines $\beta^{m,i}$, $i = 0, \ldots, m$, are the Lagrangian polynomials of degree m such that

$$\beta^{m,i}(j) = \delta^{i,j}, \qquad j = 0, \ldots, m,$$

and the cardinal splines $\alpha^{p,h}$, $h = -p, \ldots, p$, $p = m/2$, are the Lagrangian polynomials of degree m such that

$$\alpha^{p,h}(g) = \delta^{h,g}, \qquad g = -p, \ldots, p.$$

Table 3 gives these polynomials for $n = 2, \ldots, 10$ as a possible convenience. Of course, the polynomials can be computed from the formulas

$$\beta^{m,i}(u) = \frac{u(u-1)\cdots\widehat{(u-i)}\cdots(u-m)}{i(i-1)\cdots\widehat{(i-i)}\cdots(i-m)}, \qquad u \in \mathrm{R}, \quad i = 0, \ldots, m,$$

$$\alpha^{p,h}(t) = \frac{(t+p)(t+p-1)\cdots\widehat{(t-h)}\cdots(t-p)}{(h+p)(h+p-1)\cdots\widehat{(h-h)}\cdots(h-p)}, \qquad t \in \mathrm{R}, \quad h = -p, \ldots, p,$$

where the roof indicates omission of the indicated factor.

The pattern of Table 3 is similar to that of Table 1, except that the coefficients for $\beta^{m,i}$ and $\alpha^{p,h}$ are put side by side instead of one above the other, and the coefficients of absolutes and plusses, being zero, are absent. For each $n = 2, \ldots, 10$ there are n blocks of information, labeled successively $i = 0$, $h = -p$; $i = 1$, $h = -p + 1$; $\ldots$; $i = m$, $h = p$; where $m = n - 1 = 2p$. The block for each i, $h = i - p$ has the following structure:

Pattern of Table 3

	$m = n - 1 = 2p$		
	$i =$	$h = i - p$	
0	$b_0^{m,i}$	$a_0^{p,h}$	0
1	$b_1^{m,i}$	$a_1^{p,h}$	1
$\vdots$	$\vdots$	$\vdots$	$\vdots$
λ	$b_\lambda^{m,i}$	$a_\lambda^{p,h}$	λ
$\vdots$	$\vdots$	$\vdots$	$\vdots$
m	$b_m^{m,i}$	$a_m^{p,h}$	m

Here

$$\beta^{m,i}(u) = \sum_{\lambda=0}^{m} b_\lambda^{m,i} u^\lambda, \qquad u \in \mathrm{R}, \quad i = 0, \ldots, m,$$

$$\alpha^{p,h}(t) = \sum_{\lambda=0}^{m} a_\lambda^{p,h} t^\lambda, \qquad t \in \mathrm{R}, \quad h = -p, \ldots, p.$$

Illustration 6. $n = 3$, $m = n - 1 = 2$, $p = 1$. The cardinals $\beta^{2,i}$, $i = 0, 1, 2$, are

$$\begin{aligned} \beta^{2,0}(u) &= 1 - 1.5u + u^2, \\ \beta^{2,1}(u) &= 2u - u^2, \\ \beta^{2,2}(u) &= -0.5u + 0.5u^2, \qquad u \in \mathsf{R}. \end{aligned}$$

The unique polynomial of degree 2 which takes on the values $x(0)$, $x(1)$, $x(2)$ at 0, 1, 2, respectively, is

$$x(0)\beta^{2,0} + x(1)\beta^{2,1} + x(2)\beta^{2,2}.$$

Similarly, the cardinals $\alpha^{1,h}$, $h = -1, 0, 1$, are

$$\begin{aligned} \alpha^{1,-1}(t) &= -0.5t + 0.5t^2, \\ \alpha^{1,0}(t) &= 1 - t^2, \\ \alpha^{1,1}(t) &= 0.5t + 0.5t^2, \qquad t \in \mathsf{R}. \end{aligned}$$

The unique polynomial of degree 2, which takes on the values $y(-1)$, $y(0)$, $y(1)$ at -1, 0, 1, respectively, is

$$y(-1)\alpha^{1,-1} + y(0)\alpha^{1,0} + y(1)\alpha^{1,1}.$$

Illustration 7. $n = 4$. $m = n - 1 = 3$. $p = 1.5$. The cardinals $\beta^{3,i}$, $i = 0$, 1, 2, 3, are

$$\begin{aligned} \beta^{3,0}(u) &= \frac{6 - 11u + 6u^2 - u^3}{6}, \\ \beta^{3,1}(u) &= 3u - 2.5u^2 + 0.5u^3, \\ \beta^{3,2}(u) &= -1.5u + 2u^2 - 0.5u^3, \\ \beta^{3,3}(u) &= \frac{2u - 3u^2 + u^3}{6}, \qquad u \in \mathsf{R}. \end{aligned}$$

The cardinals $\alpha^{1.5,h}$, $h = -1.5, -0.5, 0.5, 1.5$, are

$$\begin{aligned} \alpha^{1.5,-1.5}(t) &= \frac{-3 + 2t + 12t^2 - 8t^3}{48}, \\ \alpha^{1.5,-1.5}(t) &= 0.5625 - 1.125t - 0.25t^2 + 0.5t^3, \\ \alpha^{1.5,.5}(t) &= \alpha^{1.5,-.5}(-t), \\ \alpha^{1.5,1.5}(t) &= \alpha^{1.5,-1.5}(-t), \qquad t \in \mathsf{R}. \end{aligned}$$

2.8. APPRAISALS

First we discuss the symmetric case with steps of unity. Take m, n with $m+1 \geq n \geq 1$. Put $p = m/2$, $q = 2n-1$, and

$$\bar{m} = m + 1. \tag{33}$$

Let ξ be the spline approximation of a function x, based on the observations $x(-p), x(-p+1), \ldots, x(p)$, of type n. If we approximate $x(t)$, for $t \in \mathsf{R}$, by $\xi(t)$, the error is $x(t) - \xi(t)$. Let I be the smallest interval that contains $-p$, p, and t. Assume that the nth derivative x_n of x is continuous on I (actually a somewhat weaker assumption, described in Section 3.2 is sufficient). Then the error $x(t) - \xi(t)$ satisfies the following appraisal:

$$|x(t) - \xi(t)|^2 \leq J(t)\,[\|x_n\|^2 - \omega^* B_{2,2}\,\omega] \leq J(t)\|x_n\|^2, \qquad t \in \mathsf{R}, \tag{34}$$

where

$$\|x_n\|^2 = \int_I x_n(s)^2\, ds, \tag{35}$$

ω is the $\bar{m} \times 1$ matrix (a single column) of observations

$$\omega = \begin{pmatrix} x(-p) \\ x(-p+1) \\ \vdots \\ x(p) \end{pmatrix}, \tag{36}$$

ω^* is the transpose of ω, and $J(t)$ and $B_{2,2}$ are entities independent of x, which will be described below. By symmetry

$$J(t) = J(-t), \qquad t \in \mathsf{R}. \tag{37}$$

The appraisal (34) is useful if a bound on $\|x_n\|$ is known or assumed. Note that (34) implies that $\xi = x$ if x is a polynomial of degree $n-1$.

The number $\xi(t)$ is the optimal approximation of $x(t)$ in the following sense (among others to be described in Chapter 3). If $\xi(t)$ in (34) is replaced by another number, then (34) becomes false for some function x for which ω and $\|x_n\|$ are unchanged.

Furthermore, the appraisal (34) is optimal in the following sense. If the number $J(t)$ is replaced by a smaller number, both inequalities in (34) become false for some function x.

We now define certain matrices. B *is a symmetric* $(n+\bar{m}) \times (n+\bar{m})$ *matrix given by Table* 2, *which enumerates each row of* B *up to and including the diagonal element*. We define matrices $B_{1,1}$, $B_{1,2}$, $B_{2,1}$, $B_{2,2}$ by partitioning B as follows:

(38)
$$B = \underset{(n+\overline{m})\times(n+\overline{m})}{B} = \begin{pmatrix} \underset{n\times n}{B_{1,1}} & \underset{n\times \overline{m}}{B_{1,2}} \\ \underset{\overline{m}\times n}{B_{2,1}} & \underset{\overline{m}\times \overline{m}}{B_{2,2}} \end{pmatrix}.$$

Here we indicate the size of a matrix by a gloss underneath its symbol; for example, $B_{1,2}$ is a matrix of n rows and $\overline{m}$ columns. Because $B = B^*$,

(39)
$$B_{1,1} = B^*_{1,1}, \qquad B_{2,2} = B^*_{2,2},$$
$$B_{1,2} = B^*_{2,1}, \qquad B_{2,1} = B^*_{1,2}.$$

An asterisk attached to a symbol for a matrix indicates transposition.

The matrix $B_{2,2}$ enters in (34); $B_{2,2}$ may be read from Table 2, since $B_{2,2}$ is the last $\overline{m} \times \overline{m}$ submatrix of B.

Put τ equal to the $(n + \overline{m}) \times 1$ matrix (single column)

(40)
$$\tau = \underset{(n+\overline{m})\times 1}{\tau} = \begin{pmatrix} 1 \\ t \\ t^2 \\ \vdots \\ t^{n-1} \\ |t+p|^q \\ |t+p-1|^q \\ |t+p-2|^q \\ \vdots \\ |t-p|^q \end{pmatrix}, \qquad t \in \mathsf{R}.$$

Then the number $J(t)$, which enters in (34), is

(41)
$$J(t) = \frac{(-1)^{n+1}\tau^* B \tau}{2q!}.$$

The matrices $B_{2,1}$ and $B_{2,2}$ have a simple relation to Table 1: $B_{2,1}$ is the matrix of coefficients of powers of t in the expressions for the centered cardinals in terms of absolutes and $B_{2,2}$ is the matrix of coefficients of absolutes in the same expressions; that is,

(42)
$$B_{2,1} = (a_\lambda^{p,h}), \qquad h = 0, \ldots, m, \quad \lambda = 0, \ldots, n-1;$$

and

(43)
$$B_{2,2} = (d_\mu^{p,h}), \qquad h, \mu = 0, \ldots, m.$$

The range of parameters in Table 2 is $n = 1, 2, 3, 4, 5$, and 8; $m = n - 1, n, \ldots, 20$.

Illustration 8. $n = 2$, $m = 2$. Here $p = 1$, $q = 3$. The cardinal splines $\alpha^{1,-1}$, $\alpha^{1,0}$, $\alpha^{1,1}$ are given in Illustration 4. The spline approximation ξ of x, based on $x(-1)$, $x(0)$, $x(1)$, of type 2, is

$$\xi = x(-1)\alpha^{1,-1} + x(0)\alpha^{1,0} + x(1)\alpha^{1,1}.$$

We now obtain the appraisal (34) of the error $x(t) - \xi(t)$, $t \in \mathsf{R}$. From Table 2

$$B = \begin{pmatrix} 0.5 & 0 & -0.25 & 1.5 & -0.25 \\ 0 & 4 & -0.5 & 0 & 0.5 \\ -0.25 & -0.5 & 0.125 & -0.25 & 0.125 \\ 1.5 & 0 & -0.25 & 0.5 & -0.25 \\ -0.25 & 0.5 & 0.125 & -0.25 & 0.125 \end{pmatrix}.$$

By (40)

$$\tau^* = (1 \quad t \quad |t+1|^3 \quad |t|^3 \quad |t-1|^3).$$

The calculation of $\tau^* B \tau$, which enters in (41), may conveniently use a pattern of bordering:

	1	t	$\|t+1\|^3$	$\|t^3\|$	$\|t-1\|^3$
1	0.5	.	.	.	.
t	0	4	.	.	.
$\|t+1\|^3$	−0.25	−0.5	0.125	.	.
$\|t\|^3$	1.5	0	−0.25	0.5	.
$\|t-1\|^3$	−0.25	0.5	0.125	−0.25	0.125

By (41)

$$\begin{aligned} -2(3!)J(t) = \tau^* B \tau = {} & \\ & 0.5 \\ & + 4t^2 \\ & -0.5\,|t+1|^3 - t\,|t+1|^3 + 0.125\,|t+1|^6 \\ & +3\,|t|^3 \qquad -0.5\,|t|^3\,|t+1|^3 \qquad + 0.5\,|t|^6 \\ & -0.5\,|t-1|^3 + t\,|t-1|^3 + 0.25\,|t-1|^3\,|t+1|^3 - 0.5\,|t-1|^3\,|t|^3 \\ & + 0.125\,|t-1|^6, \qquad t \in \mathsf{R}. \end{aligned}$$

This expression may be transformed. Thus

$$12J(t) = 12J(-t) = 2t^2 - 2t^3 - 2.5t^4 + 3t^5 - 0.5t^6, \qquad 0 \le t \le 1;$$

$$12J(t) = 12J(-t) = -0.5 + 5t - 8.5t^2 + 4t^3, \qquad 1 \le t.$$

These values agree with J_2^2 as given in Sard [1, p. 94], if we put $u = t + 1$ and if we correct the following typographical error in J_2^2. In the range $0 \le u \le 1$ the printed $(-u)^2$ in the expression for J_2^2 should be deleted.

The extremes of (34) give the appraisal

$$|x(t) - \xi(t)|^2 \le J(t)\,\|x_2\|^2, \qquad t \in \mathrm{R},$$

where x_2 is the second derivative of x,

$$\|x_2\|^2 = \int_I x_2(s)^2\, ds,$$

I is the smallest interval containing t, -1, and 1, and $J(t)$ is the number just calculated.

The bound $J(t)$ vanishes at $t = -1, 0, 1$, confirming the fact that ξ interpolates to x at the abscissae.

A better appraisal, at the cost of slightly more calculation, is the first part of (34), obtained as follows. The matrix $B_{2,2} = B_{2,2}$ is the last 3×3 submatrix of B:

$$B_{2,2} = \begin{pmatrix} 0.125 & \cdot & \cdot \\ -0.25 & 0.5 & \cdot \\ 0.125 & -0.25 & 0.125 \end{pmatrix}.$$

The matrix $B_{2,2}$ will always be symmetric in both its diagonals. In the case $m = n$ (as in the present illustration) $B_{2,2}$ is also symmetric or skew-symmetric in its midrow and in its midcolumn, by (28).

The observations (36) are given by ω, where

$$\omega^* = \big(x(-1) \quad x(0) \quad x(1)\big).$$

Hence

$$\begin{aligned}\omega^* B_{2,2}\,\omega = {} & 0.125x(-1)^2 \\ & - 0.5x(-1)x(0) + 0.5x(0)^2 \\ & + 0.25x(-1)x(1) - 0.5x(0)x(1) + 0.125x(1)^2,\end{aligned}$$

and

$$|x(t) - \xi(t)|^2 \le J(t)[\|x_2\|^2 - \omega^* B_{2,2}\,\omega], \qquad t \in \mathrm{R}.$$

Here $\xi(t)$ is the optimal approximation of $x(t)$, and the inequality is the optimal appraisal of the error.

Illustration 9. $n = 2$, $m = 3$. Here $p = 1.5$, $q = 3$. The centered cardinal splines are

$$\begin{aligned}\alpha^{1.5,\,-1.5}(t) = {} & -0.4 - 0.6666666666667t + 0.1333333333333\,|t + 1.5|^3 \\ & - 0.3\,|t + 0.5|^3 + 0.2\,|t - 0.5|^3 - 0.03333333333333\,|t - 1.5|^3,\end{aligned}$$

$$\alpha^{1.5,-.5}(t) = 0.9 + t - 0.3|t+1.5|^3 + 0.8|t+0.5|^3 - 0.7|t-0.5|^3 + 0.2|t-1.5|^3,$$

$$\alpha^{1.5,.5}(t) = \alpha^{1.5,-.5}(-t) = 0.9 - t + 0.2|t+1.5|^3 - 0.7|t+0.5|^3 + 0.8|t-0.5|^3 - 0.3|t-1.5|^3,$$

$$\alpha^{1.5,1.5}(t) = \alpha^{1.5,-1.5}(-t) = -0.4 + 0.6666666666667t - 0.03333333333333 \times |t+1.5|^3 + 0.2|t+0.5|^3 - 0.3|t-0.5|^3 + 0.1333333333333|t-1.5|^3, \qquad t \in \mathsf{R}.$$

The spline approximation ξ of x, based on $x(-1.5)$, $x(-0.5)$, $x(0.5)$, $x(1.5)$, of type 2, is

$$\xi = x(-1.5)\alpha^{1.5,-1.5} + x(-0.5)\alpha^{1.5,-.5} + x(0.5)\alpha^{1.5,.5} + x(1.5)\alpha^{1.5,1.5}.$$

From Table 2,

$$B = B^* = \begin{pmatrix} 2.7 & \cdot & \cdot & \cdot & \cdot & \cdot \\ 0 & 7.33333 & \cdot & \cdot & \cdot & \cdot \\ -0.4 & -0.666667 & 0.133333 & \cdot & \cdot & \cdot \\ 0.9 & 1 & -0.3 & 0.8 & \cdot & \cdot \\ 0.9 & -1 & 0.2 & -0.7 & 0.8 & \cdot \\ -0.4 & 0.666667 & -0.0333333 & 0.2 & -0.3 & 0.133333 \end{pmatrix}.$$

By (40) and (41),

$$\tau^* = (1 \quad t \quad |t+1.5|^3 \quad |t+0.5|^3 \quad |t-0.5|^3 \quad |t-1.5|^3),$$

$$\begin{aligned} -12J(t) = {} & 2.7 + 7.33333t^2 - 0.8|t+1.5|^3 - 1.33333t|t+1.5|^3 \\ & + 0.133333|t+1.5|^6 + 1.8|t+0.5|^3 + 2t|t+0.5|^3 \\ & - 0.6|t+1.5|^3|t+0.5|^3 + 0.8|t+0.5|^6 \\ & + 1.8|t-0.5|^3 - 2t|t-0.5|^3 + 0.4|t+1.5|^3|t-0.5|^3 \\ & - 1.4|t+0.5|^3|t-0.5|^3 + 0.8|t-0.5|^6 - 0.8|t-1.5|^3 \\ & + 1.33333t|t-1.5|^3 - 0.0666667|t+1.5|^3|t-1.5|^3 \\ & + 0.4|t+0.5|^3|t-1.5|^3 - 0.6|t-0.5|^3|t-1.5|^3 \\ & + 0.133333|t-1.5|^6, \qquad t \in \mathsf{R}. \end{aligned}$$

Also $B_{2,2}$ is the last 4×4 submatrix of B, and

$$\omega^* = (x(-1.5) \quad x(-0.5) \quad x(0.5) \quad x(1.5)).$$

Then

$$|x(t) - \xi(t)|^2 \le J(t)[\|x_2\|^2 - \omega^* B_{2,2}\omega] \le J(t)\|x_2\|^2, \qquad t \in \mathsf{R},$$

where

$$\|x_2\|^2 = \int_I x_2(s)^2\, ds,$$

I is the smallest interval containing -1.5, 1.5, and t, providing that the second derivative x_2 is continuous on I.

2.9. APPRAISAL OF THE ERROR, GENERAL POSITION

Let ξ be the spline approximation of a function x, based on $x(a)$, $x(a+w)$, $\ldots$, $x(a+mw)$, of type n, where $m+1 \geq n \geq 1$. Put $p = m/2$, $q = 2n - 1$. Put

$$b = a + mw,$$

$$t = \frac{s-(a+b)/2}{w}, \qquad s \in \mathsf{R},$$

and define the function y by the relation

$$y(t) = x(s).$$

Then, as we have seen in Section 2.6,

$$\xi(s) = \eta(t),$$

where η is the spline approximation of y, based on $y(-p), \ldots, y(p)$, of type n.

The appraisal (34) with x replaced by y implies the following appraisal of $x(s) - \xi(s)$:

$$(44) \qquad |x(s)-\xi(s)|^2 \leq J(s)\left[\|x_n\|^2 - \frac{\omega^* B_{2,2}\,\omega}{w^q}\right] \leq J(s)\,\|x_n\|^2, \qquad s \in \mathsf{R},$$

where x_n is the nth derivative of x, assumed continuous on I, the smallest interval containing a, b, and s;

$$(45) \qquad \|x_n\|^2 = \int_I x_n(t)^2\,dt,$$

$$(46) \qquad \underset{1\times\bar{m}}{\omega^*} = (x(a) \quad x(a+w) \cdots x(b)),$$

$$(47) \qquad J(s) = \frac{(-1)^{n+1} w^q \tau^* B \tau}{2q!},$$

$$(48) \qquad \underset{1\times(n+\bar{m})}{\tau^*} = \Bigg(1 \quad \frac{s-(a+b)/2}{w} \quad \left[\frac{s-(a+b)/2}{w}\right]^2 \cdots \left[\frac{s-(a+b)/2}{w}\right]^{n-1}$$

$$\left|\frac{s-a}{w}\right|^q \quad \left|\frac{s-a-w}{w}\right|^q \cdots \left|\frac{s-b}{w}\right|^q\Bigg),$$

and B and $B_{2,2}$ are the matrices of Section 2.8 [cf. (38)].

Illustration 10. $n = 2$, $m = 2$. Consider the spline approximation ζ of the function z in Illustration 5. Applying (44), we see that

$$(49) \qquad |z(s) - \zeta(s)|^2 \le K(s)\left(\|z_2\|^2 - \frac{\omega^* B_{2,2}\,\omega}{5^3}\right) \le K(s)\,\|z_2\|^2, \qquad s \in \mathsf{R},$$

where

$$\omega^* = \big(z(10)\; z(15)\; z(20)\big),$$

$\underset{5\times 5}{B}$ and $\underset{3\times 3}{B_{2,2}}$ are given in Illustration 8,

$$-12K(s) = 5^3 \tau^* B \tau,$$

$$\tau^* = \left(1 \quad \frac{s-15}{5} \quad \left|\frac{s-10}{5}\right|^3 \quad \left|\frac{s-15}{5}\right|^3 \quad \left|\frac{s-20}{5}\right|^3\right),$$

and

$$\|z_2\|^2 = \int_I z_2(t)^2\, dt;$$

I is the smallest interval containing 10, 20, and s. Thus

$$K(s) = 5^3 J\left(\frac{s-15}{5}\right), \qquad s \in \mathsf{R},$$

where J is the function in Illustration 8. For example, s may equal 12.5. Then the approximation of $z(12.5)$ is

$$\zeta(12.5) = 0.40625z(10) + 0.6875z(15) - 0.09375z(20);$$

the coefficients here are given in Table 1, being $\beta^{2,0}$ (0,5), $\beta^{2,1}$ (0.5), and $\beta^{2,2}(0.5)$, since $(12.5 - 10)/5 = 0.5$, and

$$K(12.5) = 125J(-0.5),$$

where J is the function in Illustration 8. Hence

$$K(12.5) = 125J(0.5) = \frac{125(0.1796875)}{12} = 1.871744792.$$

Also (cf. Illustration 8),

$$(50) \qquad \omega^* B_{2,2}\,\omega = 0.125z(10)^2 - 0.5z(10)z(15) + 0.5z(15)^2 + 0.25z(10)z(20) - 0.5z(15)z(20) + 0.125z(20)^2;$$

$$\|z_2\|^2 = \int_{10}^{20} z_2(t)^2\, dt.$$

Then

$$|z(12.5) - \zeta(12.5)|^2 \le 1.8717\left[\|z_2\|^2 - \frac{\omega^* B_{2,2}\,\omega}{125}\right] \le 1.8717\,\|z_2\|^2,$$

providing that z_2 is continuous on [10, 20].

As another example s may equal 7.5. Then the approximation of $z(7.5)$ is

$$\zeta(7.5) = 1.625z(10) - 0.75z(15) + 0.125z(20).$$

The coefficients here are given in Table 1, being $\beta^{2,0}(-0.5)$, $\beta^{2,1}(-0.5)$, $\beta^{2,2}(-0.5)$, since $(7.5 - 10)/5 = -0.5$, and

$$K(7.5) = 125J(-1.5) = 125J(1.5) = \frac{125 \quad 1.375}{12} = 14.32292;$$

J here is the function in Illustration 8. Also $\omega^* B_{2,2}\,\omega$ is as above and

$$\|z_2\|^2 = \int_{7.5}^{20} z_2(t)^2\,dt.$$

Then

$$|z(7.5) - \zeta(7.5)|^2 \le 14.323\left[\|z_2\|^2 - \frac{\omega^* B_{2,2}\,\omega}{125}\right] \le 14.323\,\|z_2\|^2,$$

providing that z_2 is continuous on [7.5, 20].

It is interesting to compare the appraisals for the interpolation ($s = 12.5$) and the extrapolation ($s = 7.5$). We would expect the interpolation to be better; the appraisals confirm this, since

$$\int_{10}^{20} z_2(t)^2\,dt \le \int_{7.5}^{20} z_2(t)^2\,dt$$

and

$$1.8717 < 14.323.$$

2.10. APPRAISAL OF ERROR, GENERAL FUNCTIONALS

If G is a functional of a sort described at the end of this section, then the optimal approximation of Gx, based on observations $x(a)$, $x(a + w), \ldots, x(a + mw)$, of type n, is simply $G\xi$, where ξ is the corresponding spline approximation of x, providing that x has a continuous nth derivative on I, the smallest interval containing a, $a + mw$, and the support of G. The optimal appraisal of the error $Gx - G\xi$ is

$$|Gx - G\xi|^2 \le J \times \left(\|x_n\|^2 - \frac{\omega^* B_{2,2}\,\omega}{w^q}\right) \le J\,\|x_n\|^2, \tag{51}$$

where

$$\|x_n\|^2 = \int_I x_n(s)^2\, ds, \tag{52}$$

$$\underset{1\times\bar{m}}{\omega^*} = (x(a)\, x(a+w) \cdots x(a+mw)), \tag{53}$$

$$J = \frac{(-1)^n}{2q!}\,[G_t G_s(|t-s|^q) - w^q \tau^* B\tau]. \tag{54}$$

The subscript s in G_s indicates that the functional G is to operate on its argument as a function of s;

$$\underset{1\times(n+\bar{m})}{\tau^*} = \left(G[1]\, G_s\left\{\left[\frac{s-(a+b)/2}{w}\right]\right\} \cdots G_s\left\{\left[\frac{s-(a+b)/2}{w}\right]^{n-1}\right\}\right. \tag{55}$$

$$\left. G_s\left[\left|\frac{s-a}{w}\right|^q\right] G_s\left[\left|\frac{s-a-w}{w}\right|^q\right] \cdots G_s\left[\left|\frac{s-b}{w}\right|^q\right]\right),$$

$$b = a + mw,$$

and $B_{2,2}$ and B are the matrices of Section 2.8.

If we take $Gx = x(u)$, $u \in \mathrm{R}$, then (51) and (54) reduce to (44) and (47), respectively, since $G_t G_s[|t-s|^q] = |u-u|^q = 0$.

The functional G may be an interpolatory or extrapolatory differentiation of order $\leq n-1$:

$$Gx = x_\lambda(t), \qquad t \in \mathrm{R}, \qquad \lambda \leq n-1,$$

in which case I is the smallest interval containing a, $a+mw$, and t; or G may be a Stieltjes integral of a derivative of order $\leq n-1$:

$$Gx = \int_t^u x_\lambda(s)\, d\mu(s), \qquad t, u \in \mathrm{R}, \qquad \lambda \leq n-1,$$

where μ is a function of bounded variation, in which case I is the smallest interval containing a, $a+mw$, and the part of the support of μ in the interval $[t, u]$. Alternatively, G may be a finite linear combination of functionals of the above sort. In the notation of Section 3.2, G may be any element of X^*.

Illustration 11. $n = 2$, $m = 2$. The spline approximation ξ of a function x, based on the observations $x(10)$, $x(15)$, $x(20)$, of type 2, is

$$\xi(s) = x(10)\alpha^{1,-1}\left(\frac{s-15}{5}\right) + x(15)\alpha^{1,0}\left(\frac{s-15}{5}\right) \tag{56}$$

$$+\, x(20)\alpha^{1,1}\left(\frac{s-15}{5}\right), \qquad s \in \mathrm{R},$$

where $\alpha^{1,-1}$, $\alpha^{1,0}$, $\alpha^{1,1}$ are given in Illustration 4. Suppose that G is a functional of the sort just described. Then $G\xi$ approximates Gx, and by (51)

$$|Gx - G\xi|^2 \leq J\left(\|x_2\|^2 - \frac{\omega^* B_{2,2}\,\omega}{5^3}\right) \leq J\,\|x_2\|^2, \tag{57}$$

where $\|x_2\|^2 = \int_I x_2(s)^2\,ds$, I is the smallest interval containing 10, 20, and the support of G, $\omega^* B_{2,2}\,\omega$ is given by (50) with z replaced by x, and J is given by (54) and (55), with $n = 2$, $q = 3$, $m = 2$, $w = 5$, $a = 10$, $b = 20$.

We now consider several instances of G.

Interpolation or Extrapolation

If $Gx = x(s)$, $s \in \mathsf{R}$, then (57) reduces to (49), with changed notation.

Approximate Differentiation

If $Gx = x_1(s)$, $s \in \mathsf{R}$, then $G\xi = \xi_1(s)$ is the approximation of $x_1(s)$. Now, by Illustration 4 and the formula (6) for the derivative of the plus function,

$$\begin{aligned}
\alpha_1^{1,-1}(t) &= -1.25 + 0.75(t+1)_+^2 - 1.5t_+^2 + 0.75(t-1)_+^2\,, \qquad t \in \mathsf{R},\\
\alpha_1^{1,0}(t) &= 1.5 - 1.5(t+1)_+^2 + 3t_+^2 - 1.5(t-1)_+^2\,,\\
\alpha_1^{1,1}(t) &= -0.25 + 0.75(t+1)_+^2 - 1.5t_+^2 + 0.75(t-1)_+^2\,;
\end{aligned}$$

and by (56)

$$\begin{aligned}
\xi_1(s) &= \frac{x(10)}{5}\,\alpha_1^{1,-1}\left(\frac{s-15}{5}\right) + \frac{x(15)}{5}\,\alpha_1^{1,0}\left(\frac{s-15}{5}\right)\\
&\quad + \frac{x(20)}{5}\,\alpha_1^{1,1}\left(\frac{s-15}{5}\right), \qquad s \in \mathsf{R}.
\end{aligned}$$

Thus $\xi_1(s)$ is easily accessible to us in terms of the observations.

By (57)

$$|x_1(s) - \xi_1(s)| \leq J\left(\|x_2\|^2 - \frac{\omega^* B_{2,2}\,\omega}{125}\right) \leq J\,\|x_2\|^2,$$

where $\|x_2\|^2 = \int_I x_2(t)^2\,dt$, I is the smallest interval containing 10, 20, and s, and by (54) and (55)

$$J(s) = \frac{1}{12}(0 - 125\tau^* B\tau),$$

$$\underset{1\times 5}{\tau^*} =$$

$$\left(0 \quad \frac{1}{5} \quad \frac{3(s-10)^2\,\mathrm{sig}(s-10)}{125} \quad \frac{3(s-15)^2\,\mathrm{sig}(s-15)}{125} \quad \frac{3(s-20)^2\,\mathrm{sig}(s-20)}{125}\right),$$

and $\underset{5\times 5}{B}$ is given in Illustration 8.

Approximate Integration

If $Gx = \int_u^v x(s)\,ds$, $u, v \in \mathsf{R}$, then $\int_u^v \xi(s)\,ds$ is the optimal approximation of Gx. By (56)

$$\int_n^v \xi(s)\,ds = 5x(10)\int_{(u-15)/5}^{(v-15)/5} \alpha^{1,-1}(t)\,dt + 5x(15)\int_{(u-15)/5}^{(v-15)/5} \alpha^{1,0}(t)\,dt$$
$$+ 5x(20)\int_{(u-15)/5}^{(v-15)/5} \alpha^{1,1}(t)\,dt.$$

Now the coefficients of $x(10)$, $x(15)$, and $x(20)$ may be calculated by using Illustration 4 and formula (7) or (9); for example,

$$\int_a^b \alpha^{1,-1}(t)\,dt = -0.25(b-a) - 0.625(b^2-a^2)$$
$$+ 0.0625[(b+1)_+^4 - (a+1)_+^4] - 0.125[b_+^4 - a_+^4]$$
$$+ 0.0625[(b-1)_+^4 - (a-1)_+^4], \qquad a, b \in \mathsf{R}.$$

Thus the approximation $\int_u^v \xi(s)\,ds$ is easily accessible to us in terms of the observations.

The appraisal (57) becomes

$$\left|\int_u^v x(s)\,ds - \int_u^v \xi(s)\,ds\right|^2 \le J\left(\|x_2\|^2 - \frac{\omega^* B_{2,2}\,\omega}{125}\right) \le J\|x_2\|^2,$$

where $\|x_2\|^2 = \int_I x_2(s)^2\,ds$, I is the smallest interval containing 10, 20, u, and v, and by (54) and (55)

$$12J = \frac{|v-u|^5}{10} - 125\tau^* B\tau,$$

$$\underset{1\times 5}{\tau^*} = \Bigg[v-u \quad \frac{(v-15)^2-(u-15)^2}{10} \quad \frac{(v-10)^4\,\mathrm{sig}(v-10)-(u-10)^4\,\mathrm{sig}(u-10)}{500}$$
$$\frac{(v-15)^4\,\mathrm{sig}(v-15)-(u-15)^4\,\mathrm{sig}(u-15)}{500}$$
$$\frac{(v-20)^4\,\mathrm{sig}(v-20)-(u-20)^4\,\mathrm{sig}(u-20)}{500}\Bigg],$$

and $\underset{5\times 5}{B}$ is given in Illustration 8.

2.11. CHOICE OF FORMULA

Table 1 gives four formulas for each cardinal spline, formulas which are equivalent except for the effect of rounding errors. For many values of the parameters and arguments the four formulas give effectively identical results.

We now compare the four formulas, taking into account the unavoidable rounding errors in their coefficients.

The tabulated coefficients $b_\lambda^{m,i}$, $c_\nu^{m,i}$, $a_\lambda^{p,h}$, and their curled counterparts of (13) and (22) are correct to the last printed figure except when the true value is zero, in which case the error is in an acceptably remote decimal place. For any $u \in \mathsf{R}$, m, i and corresponding $t = u - p$, $h = i - p$, $p = m/2$, the calculated value of

$$\beta^{m,i}(u) = \alpha^{p,h}(t), \qquad i = 0, \ldots, m,$$

based on the printed coefficients, will contain an error due to the errors in the coefficients. As the factors u^λ, $(u - v)^q$, t^λ in (13) and (22) can be large, there can be magnification of errors.

It is advantageous to use a plus form when it produces many zeros. It is advantageous to have u or t small. It is advantageous to use the $\alpha^{p,h}$ form with absolutes when symmetry is to be emphasized. The case $u \geq p$ may be reduced to the case $u \leq p$ by the use of (14), that is, by reflection in the midpoint p of the interval $[0, m]$. Of course (25) permits us to reduce the case $t \leq 0$ to $t \geq 0$.

To gain insight into the effect of rounding errors we have run what may be called the $\delta^{i,j}$ test: using the tabulated coefficients in Table 1 for the four forms, we have calculated $\beta^{m,i}(j)$ and $\alpha^{p,h}(g)$, $i, j = 0, \ldots, m$; $h = i - p$; $g = j - p$; $p = m/2$; and we have observed the deviations of the calculated results from the true values $\delta^{i,j}$. The following table gives the maximum absolute deviations in our most extreme case $n = 8$, $m = 20$. Here the first entry at $j = 1$, for example, asserts that the indicated maximum is $8 \cdot 10^{-12}$ when $\beta^{20,i}(1)$ is calculated from (13), using plusses and Table 1, and the maximum is $6 \cdot 10^{-2}$ when $\beta^{20,i}(1)$ is calculated from (13), using absolutes and Table 1.

For $0 \leq u \leq 10 = p$ good results are obtained from $\beta^{20,i}(u)$ in terms of plusses, although the worst $\alpha^{10,h}(-2)$ in plusses is better than the worst $\beta^{20,i}(8)$ in plusses; likewise $\alpha^{10,h}(-1)$ in plusses is better than $\beta^{20,i}(9)$ in plusses; and $\alpha^{10,h}(0)$ in absolutes is better than $\beta^{20,i}(10)$ in plusses; $i = 0, \ldots, 20$; $h = i - 10$.

For $p \leq u \leq m$ we may reduce to the case $0 \leq u \leq p$ by reflection in the midpoint of $[0, m]$.

Thus for the extreme case $n = 8$, $m = 20$, the deviation can be kept to the order of magnitude of 10^{-6} or less by using $\beta^{m,i}(u)$ or $\beta^{m,m-i}(m - u)$ in plusses; $i = 0, \ldots, m$; $0 \leq u \leq p$.

2.12. CHOICE OF TYPE

Consider the approximation of Gx in terms of $x(s_0), \ldots, x(s_m)$, where G is a functional of the sort described in Section 2.11 and x is a function with

$\delta^{i,j}$ ***Test***

$n = 8, m = 20, p = 10, q = 15$					
$\max_{i=0,\ldots,20} \lvert\beta^{20,i}(j) - \delta^{i,j}\rvert$ when the calculation is based on Table 1 and			$\max_{h=-10,\ldots,10} \lvert\alpha^{10,h}(g) - \delta^{h,g}\rvert$ when the calculation is based on Table 1 and		
j	Plusses	Absolutes	g	Plusses	Absolutes
0	0	2E-1	−10	7E-6	2E-1
1	8E-12	6 -2	−9	4 -6	6 -2
2	9 -11	3 -2	−8	2 -6	3 -2
3	6 -10	9 -3	−7	1 -6	9 -3
4	2 -9	3 -3	−6	7 -7	3 -3
5	9 -9	9 -4	−5	4 -7	9 -4
6	3 -8	2 -4	−4	2 -7	3 -4
7	6 -8	4 -5	−3	1 -7	7 -5
8	1 -7	2 -5	−2	4 -8	2 -5
9	3 -7	5 -5	−1	2 -7	3 -6
10	2 -6	7 -5	0	2 -6	7 -7
11	1 -5	9 -5	1	1 -5	3 -6
12	5 -5	1 -4	2	5 -5	2 -5
13	2 -4	1 -4	3	2 -4	7 -5
14	8 -4	3 -4	4	8 -4	3 -4
15	2 -3	9 -4	5	2 -3	9 -4
16	7 -3	3 -3	6	7 -3	3 -3
17	2 -2	9 -3	7	2 -2	9 -3
18	4 -2	2 -2	8	4 -2	2 -2
19	1 -1	6 -2	9	1 -1	6 -2
20	2 -1	2 -1	10	2 -1	2 -1

continuous nth derivative x_n on I, the smallest interval that contains $s_0, \ldots, s_m$ and the support of G.

It is appropriate to approximate Gx by $G\xi$, where ξ is the spline approximation of x, of type n, based on $x(s_0), \ldots, x(s_m)$, if $\|x_n\|$ is known or if a bound on $\|x_n\|$ is known or assumed to be known, where

$$\|x_n\|^2 = \int_I x_n(s)^2 \, ds.$$

An appraisal like (51) is then apt.

If we have bounds on $\|x_n\|$ for several values of n, we may consider the splines of the several types and the appraisals of the several errors and choose the n that gives the best appraisal.

In many engineering situations the user will have no firm knowledge about $\|x_n\|$. Then he may proceed tentatively and look for indirect evidence that $G\xi$ is an acceptable or inacceptable approximation of Gx for a value of n which has been chosen tentatively. We may describe n as the order of the derivative of the input function x relative to which $G\xi$ is the optimal approximation of Gx. We would use $n = n_0$ tentatively if we knew or assumed that the n_0th derivative of the input were controlled.

Knowledge of the observations $x(s_0), \ldots, x(s_m)$ alone is inadequate for any firm statement about the error in any approximation of Gx. We may say something about the error only in terms of the observations and some additional information about x. That information may be Ux, where U is a known operator, or some measure of the size of Ux.

In spline approximation of type n we take $Ux = x_n$ and we suppose $\|x_n\|$, as defined above, bounded by a known bound before we can get an appraisal. Other approaches are possible; for example, we can still take $Ux = x_n$, and suppose that a bound on

$$\sup_{s \in I} |x_n(s)| \tag{58}$$

is known or assumed. Much conventional approximation and appraisal uses the norms (58). It is a mathematical fact, however, that the theory for $\int_I x_n(s)^2\, ds$ has many strengths and beauties (see Chapter 3) which the theory for $\sup |x_n(s)|$ does not have [Meinguet 1]. In many situations it is as easy and apt for the user to assume information about $\int_I x_n^2$ as about $\sup |x_n|$.

Operators U other than n-fold differentiation may be used but are not considered in this book (see Section 3.15).

CHAPTER 3

The Theory of Spline Approximation relative to the *n*th Derivative

3.1. INTRODUCTION

A spline approximation, in our usage of the term, is one that has certain minimizing properties, to be described. The spline ξ that approximates an unknown function x is based on observations $F^0x, \ldots, F^mx$ of x; the observations are known scalars. The minimizing properties of the spline involve not only the observations but also other properties of x, such as the existence of its nth derivative x_n and the knowledge or assumed knowledge of a bound on $\int_I x_n^2$, where I is a suitable interval. Our splines are not a priori functions that are piecewise of specified character, with specified types of articulation at points of transition.

More generally, the observations that are the basis of the approximation may be functions instead of scalars; the minimizing properties may refer to Ux, where U is a suitable operator, not necessarily n-fold differentiation; and x may be a function of several variables. In this book, however, we confine our attention to scalar observations of a function x of one variable and to minimizing properties relative to $\int_I x_n^2$.

The splines of our tables are based on the particular observations

$$F^ix \underset{\text{def}}{=} x(c + iw), \qquad i = 0, \ldots, m,$$

where c and w are constants, $w > 0$. Our consideration, in this chapter, of arbitrary scalar observations will permit the reader to calculate splines and to devise programs for their calculation for the widest class of splines relative to $\int_I x_n^2$.

3.2. HYPOTHESES

Let I be a compact interval of the reals R and n be a positive integer. Let $L^2(I)$ denote, as usual, the space of functions y defined on almost all of I, Lebesgue measurable, and such that

$$\int_I y(s)^2\, ds < \infty,$$

two functions being considered as the same element of $L^2(I)$ if equal almost everywhere.

Let X denote the space of functions x on I whose $(n-1)$th derivative x_{n-1} is absolutely continuous on I and whose nth derivative x_n is an element of $L^2(I)$. Thus $x \in X$ iff $x_n \in L^2(I)$ and Taylor's formula

$$x(s) = \sum_{\lambda=0}^{n-1} x_\lambda(a)\frac{(s-a)^\lambda}{\lambda!} + \int_a^s \frac{(s-t)^{n-1}}{(n-1)!} x_n(t)\, dt, \qquad s \in I, \qquad a \in I,$$

holds.

By a *functional* Ψ on X, we mean a map of X into R. For each $x \in X$, Ψx denotes the number that corresponds to x under the map Ψ. *Let X^* denote the set of functionals Ψ of the following form*:

$$\text{(1)} \qquad \Psi x = \sum_{\lambda=0}^{n-1} \int_I x_\lambda(s)\, d\sigma^\lambda(s) + \int_I x_n(s)\psi(s)\, ds, \qquad x \in X,$$

where $\sigma^0, \ldots, \sigma^{n-1}$ are functions of bounded variation on I and $\psi \in L^2(I)$. The fixed functions $\sigma^0, \ldots, \sigma^{n-1}$, and ψ determine Ψ. As an element of $L^2(I)$, ψ need only be defined almost everywhere. We may and shall take ψ so that

$$\psi(s) = \frac{d}{ds}\int^s \psi(t)\, dt;$$

that is, $\psi(s)$ is unambiguously defined for each $s \in I$ for which the integral is differentiable. If $x \in X$, then the derivatives x_λ, $\lambda = 0, \ldots, n-1$, are continuous on I and $x_n \in L^2(I)$. Consequently the scalar Ψx in (1) is well defined. By the *support* of Ψ we mean the smallest closed set containing all points $s \in I$ at which $d\sigma^\lambda(s) \neq 0$ for some $\lambda = 0, \ldots, n-1$ or at which $\psi(s) \neq 0$.

We call X^* the set of *admissible functionals*. In particular, the value $x_j(s_0)$ of the jth derivative of x at $s_0 \in I$, $j \leq n-1$, for fixed s_0 and j, is an admissible functional obtainable by taking $\psi = 0$, $\sigma^\lambda = 0$ for $\lambda \neq j$, and

$$\sigma^j(s) = \begin{cases} 0 & \text{if } s < s_0, \\ 1 & \text{if } s > s_0. \end{cases}$$

Let $F^0, \ldots, F^m$ be $m+1$ admissible functionals, given and fixed throughout the chapter. For $x \in X$, we call $F^0x, \ldots, F^mx$ the *observations* of x. Our approximations will be based on the observations. Put

$$q = 2n - 1$$

and

$$(2) \qquad f^i(t) = F^i_s(|t-s|^q), \qquad t \in \mathsf{R}, \qquad i = 0, \ldots, m.$$

Here the subscript s attached to F^i indicates that F^i is to act on its argument as a function of s, so that s is a bound variable, absorbed in the operation of calculating $f^i(t)$. Thus, another form of the definition of $f^i(t)$ is the following:

$$f^i(t) = F^i y,$$

where y is the function for which

$$y(s) = |t-s|^q, \qquad s \in I.$$

Clearly $y \in X$ and therefore $f^i(t)$ is defined for each $t \in \mathsf{R}$. Thus f^i is a well-defined function on R to R, $i = 0, \ldots, m$. We call $f^0, \ldots, f^m$ the *key functions.*

Let I_0 be the smallest closed interval that contains the supports of $F^0, \ldots, F^m$. Then $I_0 \subset I$, and I_0 may be less than I. Let a be an element of I_0, fixed throughout the chapter.

Define matrices P, Φ, A as follows:

$$(3) \qquad P = \underset{\bar{m}\times n}{P} \underset{\text{def}}{=} (F^i_s[(s-a)^\lambda]), \qquad i = 0, \ldots, m;$$

$$\lambda = 0, \ldots, n-1; \qquad \bar{m} = m+1;$$

here the gloss $\bar{m} \times n$ under P indicates that P is a matrix of $\bar{m}$ rows and n columns. Thus $F^i_s[(s-a)^\lambda]$ is the element in the ith row and λth column of P; the scalar $F^i_s[(s-a)^\lambda]$ is the result of operating on the power functions $(s-a)^\lambda$ by the functional F^i. Also put

$$(4) \qquad \Phi = \underset{\bar{m}\times\bar{m}}{\Phi} \underset{\text{def}}{=} (F^i_t F^j_s[|t-s|^q]) = (F^i f^j), \qquad j = 0, \ldots, m.$$

(It is not immediately evident that Φ is well defined, but we shall show later that it is, since $f^j \in X$.) We indicate the transpose of a matrix by an asterisk. We shall show later (Lemma 3) that

$$\Phi = \Phi^*.$$

Put

$$(5) \qquad A = \underset{(n+\bar{m})\times(n+\bar{m})}{A} \underset{\text{def}}{=} \begin{pmatrix} \underset{n\times n}{0} & \underset{n\times\bar{m}}{P^*} \\ \underset{\bar{m}\times n}{P} & \underset{\bar{m}\times\bar{m}}{\Phi} \end{pmatrix} = A^*.$$

Thus A is the $(n + \bar{m})$ square matrix built up as indicated out of 0, P^*, P, and Φ.

Our final hypothesis is that A is *invertible.*

In using the theory of spline approximation one need not verify the invertibility of A initially, for the calculations that one proceeds to execute involve the inversion of A. If the calculations go through, then A is invertible. If the calculations break down, then A is not invertible and the theory does not apply. In the latter case one would in practice modify the choice of the functionals $F^0, \ldots, F^m$.

Lemmas 5 and 6 below imply that A is invertible iff $F^0, \ldots, F^m$ are independent and P is of rank n.

It will be convenient to introduce column matrices as follows:

$$F = \underset{\bar{m}\times 1}{F} = (F^i), \qquad i = 0, \ldots, m, \tag{6}$$

$$f = \underset{\bar{m}\times 1}{f} = (f^i), \tag{7}$$

$$\omega = \underset{\bar{m}\times 1}{\omega} = Fx = (F^i x), \qquad x \in X. \tag{8}$$

Thus F is the column of functionals which provide the observations, f is the column of key functions, and ω is the column of observations of x, $x \in X$. And (4) may be written

$$\Phi = F(f^*). \tag{9}$$

3.3. CONCLUSIONS

Put

$$B = \underset{(n+\bar{m})\times(n+\bar{m})}{B} = A^{-1} = \begin{pmatrix} \underset{n\times n}{B_{1,1}} & \underset{n\times \bar{m}}{B_{1,2}} \\ \underset{\bar{m}\times n}{B_{2,1}} & \underset{\bar{m}\times \bar{m}}{B_{2,2}} \end{pmatrix} = B^*. \tag{10}$$

Thus $B_{1,1}$, $B_{1,2}$, $B_{2,1}$, $B_{2,2}$ are the matrices of the indicated sizes, obtained by partitioning B. The matrix B exists by our hypothesis that A is invertible, and

$$B^*_{1,1} = B_{1,1}, \qquad B^*_{2,2} = B_{2,2},$$
$$B^*_{1,2} = B_{2,1}, \qquad B^*_{2,1} = B_{1,2}.$$

Put

$$\eta(u) = \underset{n\times 1}{\eta(u)} = \begin{pmatrix} 1 \\ u-a \\ (u-a)^2 \\ \vdots \\ (u-a)^{n-1} \end{pmatrix}, \qquad u \in \mathsf{R}. \tag{11}$$

Thus η is the column of the n functions whose values at $u \in \mathsf{R}$ are 1, $(u - a)$,

$(u-a)^2, \ldots, (u-a)^{n-1}$, respectively. Then (3) may be written

$$\underset{\overline{m}\times n}{P} = F(\eta^*). \tag{12}$$

Put

$$\beta = \underset{\overline{m}\times 1}{\beta} = (B_{2,1} \quad B_{2,2})\begin{pmatrix}\eta \\ f\end{pmatrix} = B_{2,1}\eta + B_{2,2}f = (\beta^i), \qquad i = 0, \ldots, m. \tag{13}$$

Thus β is a column of $\overline{m}$ functions on R. These functions are called the *cardinal splines*. More fully, and more exactly, the restrictions to I of the functions β are called the cardinal splines based on the functionals F and relative to the norm $\int_I x_n^2$. It is advantageous to define the functions β on R, as we have done, because we can then replace I by any compact interval that contains I.

Define the operator Π on X to X as follows:

$$\Pi x = \beta^*\omega = \beta^* F x = \omega^*\beta, \qquad x \in X; \tag{14}$$

here $\beta^*\omega$ is a 1×1 matrix and so equals its own transpose

$$(\beta^*\omega)^* = \beta^*\omega = \omega^*\beta,$$

and (14) is understood as asserting that Πx is the restriction to I of the function $\omega^*\beta$. We call Πx the *spline approximation of* x, based on Fx and relative to $\int_I x_n^2$. Observe that for each $x \in X$, Πx is determined by the observations $\omega = Fx$. Let M be the set of all images Πx, $x \in X$; that is,

$$M = \Pi X = \{y \in X : y = \Pi x \quad \text{and} \quad x \in X\}. \tag{15}$$

We call M the *space of splines*; M is a subspace of X.

In this chapter we shall establish the following conclusions.

I. Description of Π and M

The operator Π is linear and continuous on X onto $M \subset X$ and

$$\Pi^2 = \Pi.$$

The subspace M is the span of the elements of β and is of dimension $\overline{m} = m + 1$. Let N be the intersection of the kernels of the functionals F; that is,

$$N = \left\{x \in X : Fx = \underset{\overline{m}\times 1}{0}\right\} = \{x \in X : F^0 x = \cdots = F^m x = 0\}. \tag{16}$$

Then the splines are precisely those elements x of X for which $\int_I x_n \zeta_n = 0$ whenever $\zeta \in N$; that is,

$$M = \left\{x \in X : \int_I x_n \zeta_n = 0 \quad \text{whenever} \quad \zeta \in N\right\}. \tag{17}$$

II. Geometric Property of the Spline Approximation

For any admissible functional $G \in X^*$ the best approximation of Gx, $x \in X$, is $G\Pi x$ in the following sense. In the first place the approximation is subject to the appraisal

(18)
$$|Gx - G\Pi x|^2 \leq J\left[\int_I x_n^2 - 2(-1)^n q!(Fx)^* B_{2,2}(Fx)\right] \leq J \int_I x_n^2, \qquad \text{all } x \in X,$$

with the first equality attainable for any prescribed Fx and $\int_I x_n^2$ for which the middle bracket is nonnegative, where J is a known scalar independent of x:

$$J = (-1)^n \frac{G_t G_s[|t-s|^q] - \gamma^* B \gamma}{2q!}, \qquad q = 2n-1, \tag{19}$$

and

$$\gamma = \underset{(n+\bar{m})\times 1}{\gamma} = \begin{pmatrix} \underset{n\times 1}{G\eta} \\ \underset{\bar{m}\times 1}{Gf} \end{pmatrix}. \tag{20}$$

In the second place, as regards any other approximation $\alpha \in \mathrm{R}$ of Gx, $x \in X$, the following is true. If $\alpha \neq G\Pi x$, then the difference $|Gy - \alpha|^2$ exceeds the middle number of (18) for some $y \in X$ for which $\int_I y_n^2 = \int_I x_n^2$, $Fy = Fx$, and $\Pi y = \Pi x$.

An equivalent statement is the following: let $d \geq 0$ and $\omega_{\bar{m}\times 1} \in \mathrm{R}^m$ be prescribed; let

$$\Gamma = \left\{x \in \mathrm{X}: Fx = \omega \text{ and } \int_I x_n^2 \leq d^2\right\}.$$

By virtue of (14), Πx is the same, say ξ^0, for all $x \in \Gamma$. Then the set

$$G\Gamma = \{y \in \mathrm{R}: y = Gx \text{ and } x \in \Gamma\}$$

is the closed interval with midpoint $G\xi^0$ and length equal to twice the square root of $J[d^2 - 2(-1)^n q!\,\omega^* B_{2,2}\,\omega]$.

III. Interpolating Property of the Splines

For each $x \in X$ there is one and only one $\xi \in M$ such that $F\xi = Fx$. Furthermore, $\xi = \Pi x$.

IV. Minimal Deviation among Interpolators

For each $x \in X$ the integral $\int_I y_n^2$ is minimal among all $y \in X$ such that $Fy = Fx$ iff $y = \Pi x$.

V. Minimal Quotient

For any $G \in X^*$ define the set $\mathscr{A}$ of *admissible approximations* of G as follows:

$$\text{(21)} \quad \mathscr{A} = \{A \in X^* : Ax = Gx \text{ whenever } x \in X \text{ and } x_n = 0\}$$
$$= \{A \in X^* : Ax = Gx \text{ whenever } x \text{ is a polynomial of degree } n-1\}.$$

Put $R = G - A$. For each $A \in \mathscr{A}$, there exists a linear continuous functional Q on $L^2(I)$ such that

$$\text{(22)} \qquad Rx = Q(x_n), \quad \text{all } x \in X,$$

by Peano's theorem.

Put

$$\text{(23)} \qquad A_0 = G\Pi.$$

Then $\mathscr{A}_0 \in \mathscr{A}$, and the Banach norm $\|Q\|$ of Q is minimal among $A \in \mathscr{A}$ iff $A = A_0$, in which case

$$\text{(24)} \qquad \min\|Q\|^2 = J = \sup_{\substack{\zeta \in N, \\ \int_I \zeta_n{}^2 \neq 0}} \frac{|G\zeta|^2}{\int_I \zeta_n^2}.$$

VI. Minimal Deviation in M

For each $x \in X$ the integral $\int_I (x_n - y_n)^2$ is minimal among $y \in M$ iff $y_n = \xi_n$, where $\xi = \Pi x$; that is, iff $y - \xi$ is a polynomial of degree $n - 1$.

VII. Analytic Description of M

The splines are precisely those functions ξ on R which are such that

(i) ξ is a linear combination of the key functions $f^0, \ldots, f^m$ plus a polynomial of degree $n - 1$, and

(ii) the nth derivative $\xi_n(u)$ vanishes for all u outside I_0. Equivalently,

$$\text{(25)} \quad M = \left\{x : x = c^*f + b^*\eta,\ c = \underset{\bar{m} \times 1}{c} \in \mathrm{R}^{\bar{m}},\ b = \underset{n \times 1}{b} \in \mathrm{R}^n;\ \text{and } P^*c = \underset{n \times 1}{0}\right\},$$

where f, η, and P are the matrices (7), (11), and (12).

VIII. Enlargement of I_0

Conclusions I, III, IV, VI, and VII hold with I replaced by I_0. The splines are independent of G. The interval I is a compact interval which contains I_0 and the support of G. If I is strictly larger than I_0, we may say that the approximation of Gx, $x \in X$, in terms of the observations, is extrapolatory, since the observations are based entirely on the values $x(s)$ with $s \in I_0$. Spline approximation includes extrapolatory as well as interpolatory processes.

We now proceed to establish the eight conclusions stated.

3.4. THE SPACE X

We suppose for the rest of the chapter that the hypotheses of Section 3.2 are in force. In particular, the interval I and the positive integer n are given, the space X is the space of functions with absolutely continuous $(n-1)$th derivative on I and with nth derivative in $L^2(I)$, the admissible functionals $F^0, \ldots, F^m$, and G are given, the matrix A is invertible, and a is a fixed element of I_0, the smallest compact interval containing the supports of $F^0, \ldots, F^m$.

Introduce norms in X as follows:

$$\|x\|_X^2 = \sum_{\lambda=0}^{n-1} |x_\lambda(a)|^2 + \int_I x_n^2, \qquad x \in X, \tag{26}$$

$$|||x|||^2 = \max\left[\int_I x_n^2, \quad \sup_{\substack{s \in I,\\ \lambda=0,\ldots,n-1}} |x_\lambda(s)|^2\right], \tag{27}$$

where, as usual, x_λ denotes the λth derivative of x. Then Taylor's formulas for $x, x_1, \ldots, x_{n-1}$ about $s = a$ in terms of x_n imply that $\|x\|_X$ and $|||x|||$ are equivalent; that is, that constants B, B' exist such that

$$\|x\|_X \le B\,|||x||| \le B'\,\|x\|_X, \qquad x \in X. \tag{28}$$

Taylor's formula implies also that X *is* now *a Banach space*: linear, normed, and complete [Sard 1, pp. 122 and 201]. Since $|||x|||$ is independent of a, the topology of X is in fact independent of a.

We have defined the admissible functionals X^* as those of the form (1). The definition implies that if $\Psi \in X^*$, then Ψ is a linear functional on X, bounded relative to the norm (27) and therefore continuous. Conversely, if Ψ is a linear continuous functional on X, then $\Psi \in X^*$ (in other words X^* *as defined in* Section 3.2 *is precisely the adjoint* (*dual*) *space of* X, justifying our asterisk). This may be proved as follows: think of X as a real Hilbert space with inner product

$$(x, y)_X = \sum_{\lambda=0}^{n-1} x_\lambda(a) y_\lambda(a) + \int_I x_n y_n, \qquad x, y \in X.$$

This is a valid inner product and it induces our norm $\|x\|_X$, since $\|x\|_X^2 = (x, x)_X$. Now if Ψ is a linear continuous functional on X, there exists $\phi \in X$ such that

$$\Psi x = (x, \phi)_X = \sum_{\lambda=0}^{n-1} x_\lambda(a)\phi_\lambda(a) + \int_I x_n \phi_n, \qquad x \in X \tag{29}$$

[Sard 1, p. 451]. This formula shows that Ψ is an admissible functional: we take the functions $\sigma_\lambda(s)$ in (1) as atoms of jump $\phi_\lambda(a)$ at $s = a$.

Putting $k_\lambda = \phi_\lambda(a)$ and $\psi = \phi_n$, we have the following lemma.

Lemma 1 (standard form for admissible functionals). *A necessary and sufficient condition that Ψ be an admissible functional is that constants k_λ, $\lambda = 0, 1, \ldots, n-1$, and a function $\psi \in L^2(I)$ exist such that*

$$\Psi x = \sum_{\lambda=0}^{n-1} k_\lambda x_\lambda(a) + \int_I x_n(s)\psi(s)\, ds, \qquad x \in X. \tag{30}$$

In this case

$$\|\Psi\|^2 = \sum_{\lambda=0}^{n-1} |k_\lambda|^2 + \int_I \psi^2. \tag{31}$$

Proof. All has been proved except the formula (31) for the Banach norm $\|\Psi\|$ of Ψ. But (31) is a consequence of Schwarz's inequality for the Hilbert space X:

$$(x, y)_X^2 \le \|x\|_X^2 \|y\|_X^2, \qquad x, y \in X,$$

with equality iff x and y are collinear.

Corollary. *A necessary and sufficient condition that a functional Ψ be admissible and vanish for degree $n-1$ is that a function $\psi \in L^2(I)$ exist such that*

$$\Psi x = \int_I x_n \psi, \qquad x \in X. \tag{32}$$

Proof. Suppose that (32) holds. If x is a polynomial of degree $n-1$, then $x_n = 0$ and $\Psi x = 0$.

Conversely, if Ψ is admissible, than (30) holds. In (30) put x equal to the function for which

$$x(s) = \frac{s^\lambda}{\lambda!}, \qquad \lambda = 0, \ldots, \text{ or } n-1, \qquad s \in \mathsf{R}.$$

Then

$$\Psi x = k_\lambda = 0,$$

if Ψ vanishes for degree $n-1$.

Lemma 2. *Suppose that $\Psi \in X^*$. Put*

$$y(t) = \Psi_s[|t-s|^q], \qquad t \in \mathsf{R}, \qquad q = 2n-1. \tag{33}$$

Then $y \in X$; *and*

$$\text{(34)}\quad y_\lambda(a) = (-1)^n q! \int_I \psi(s) \frac{(a-s)^{n-1-\lambda}}{(n-1-\lambda)!} \operatorname{sig}(a-s)\, ds, \qquad \lambda = 0, \ldots, n-1;$$

$$\text{(35)}\quad y_n(t) = q! \sum_{\lambda=0}^{n-1} (-1)^\lambda k_\lambda \frac{(t-a)^{n-1-\lambda}}{(n-1-\lambda)!} \operatorname{sig}(t-a) + 2q!(-1)^n \psi(t)$$

for almost all $t \in I$;

where k_λ *and* ψ *enter in* (30).

The conclusion $y \in X$ is an informal statement (what Bourbaki calls an abuse of language but what is rather a use of informal language). The function y is defined on R. The restriction of y to I is an element of X.

Proof. What is involved here is a calculation. Put

$$z(t, s) = \frac{|t-s|^q}{q!}, \qquad t, s \in \mathsf{R}.$$

Then

$$\frac{\partial^\lambda z}{\partial s^\lambda} = (-1)^\lambda \frac{(t-s)^{q-\lambda}}{(q-\lambda)!} \operatorname{sig}(t-s), \qquad \lambda < q;$$

and by (30)

$$\text{(36)}\qquad \frac{y(t)}{q!} = \Psi_s z(t, s) = \sum_{\lambda<n} k_\lambda (-1)^\lambda \frac{(t-a)^{q-\lambda}}{(q-\lambda)!} \operatorname{sig}(t-a)$$

$$+ \int_I \psi(s)(-1)^n \frac{(t-s)^{n-1}}{(n-1)!} \operatorname{sig}(t-s)\, ds, \qquad t \in \mathsf{R}.$$

We want to show that $y \in X$; that is, that y_{n-1} is absolutely continuous and that $y_n \in L^2(I)$. Now $q - \lambda = 2n - 1 - \lambda > n - 1$, so $n - 1$ differentiations of the sigma term can be taken without complication, and the nth derivative thereof exists everywhere except at $t = a$, where right and left nth derivatives exist. Thus the sigma term is an element of X. The integral term, if $t \in I$, equals

$$-\int^t \psi(s) \frac{(s-t)^{n-1}}{(n-1)!}\, ds + \int_t \psi(s) \frac{(s-t)^{n-1}}{(n-1)!}\, ds,$$

where the omitted limits of integration are the constant extremities of I. Hence the $(n-1)$th derivative, as to t, of the integral term is

$$(-1)^n \int^t \psi(s)\, ds - (-1)^n \int_t \psi(s)\, ds,$$

and the nth derivative is $2(-1)^n \psi(t)$ almost everywhere.

Thus differentiations of (36) as to t yield (34) and (35).

Corollary. *If*

$$\Psi x = \int_I x_n \psi, \qquad x \in X,$$

where $\psi \in L^2(I)$, *then*

$$\Psi_t \Psi_s[|t-s|^q] = 2q!(-1)^n \int_I \psi^2, \qquad q = 2n-1. \tag{37}$$

Proof. Our hypothesis is that (30) holds with $k_\lambda = 0$, all λ. Put x equal to the function y of (33) and use (35).

We have defined the key functions $f_{\bar{m}\times 1} = (f^i)$ as

$$f^i(t) = F_s^i[|t-s|^q], \qquad t \in \mathsf{R}, \qquad i = 0, \ldots, m.$$

As the given functionals $F_{\bar{m}\times 1} = (F^i)$ are admissible, Lemma 2 assures that the key functions are elements of X. Accordingly, the matrix

$$\underset{\bar{m}\times\bar{m}}{\Phi} = (F^i f^j), \qquad i, j = 0, \ldots, m;$$

is well defined.

Lemma 3. *The matrix* Φ *is symmetric:*

$$\Phi = (F^i f^j) = (F^j f^i) = (F_t^i F_s^j[|t-s|^q]) = \Phi^*, \qquad i, j = 0, \ldots, m.$$

Proof. It will be sufficient to show that $F_t^i F_s^j[|t-s|^q]$ is symmetric in i, j. By Lemma 1 there are constants k_λ^i and functions $\phi^i \in L^2(I_0)$ such that

$$F^i x = \sum_{\lambda=0}^{n-1} k_\lambda^i x_\lambda(a) + \int_{I_0} x_n \phi^i, \qquad x \in X,\ i = 0, \ldots, m. \tag{38}$$

We have replaced I with I_0 here because I_0 contains the supports of all F^i; in other words, the functions ϕ^i vanish outside I_0. By Lemma 2

$$\begin{aligned}\frac{1}{q!} F_t^i F_s^j[|t-s|^q] = {} & \sum_{\lambda<n} k_\lambda^i \int_{I_0} \phi^j(s)(-1)^n \frac{(a-s)^{n-1-\lambda}}{(n-1-\lambda)!} \operatorname{sig}(a-s)\, ds \\ & + \int_{I_0} \phi^i(t) \sum_{\lambda<n} k_\lambda^j (-1)^\lambda \frac{(t-a)^{n-1-\lambda}}{(n-1-\lambda)!} \operatorname{sig}(t-a)\, dt \\ & + \int_{I_0} 2(-1)^n \phi^i(t)\phi^j(t)\, dt.\end{aligned}$$

Now

$$(-1)^\lambda (t-a)^{n-1-\lambda} \operatorname{sig}(t-a) = (-1)^n (a-t)^{n-1-\lambda} \operatorname{sig}(a-t).$$

Hence the expression is indeed symmetric in i, j.

Lemma 4 (on the completeness condition). *A necessary and sufficient condition that there exist a constant B such that*

$$\|x\|_X^2 \le B^2\left[\sum_{i=0}^{m}|F^i x|^2 + \int_I x_n^2\right], \qquad \textit{all } x \in X, \tag{39}$$

is that the matrix $P_{\bar{m}\times n}$ *be of rank n.*

Proof. Suppose first that P, given by (3), is of rank $< n$. Then scalars $x_\lambda(a)$, not all zero, exist such that

$$\sum_{\lambda=0}^{n-1}\frac{x_\lambda(a)}{\lambda!}F^i[(s-a)^\lambda] = 0, \qquad i = 0, \ldots, m.$$

Let x be the polynomial

$$x(s) = \sum_{\lambda<n}\frac{x_\lambda(a)(s-a)^\lambda}{\lambda!}, \qquad s \in \mathsf{R}.$$

Then $\int_I x_n^2 = 0$ and $F^i x = 0$, all i; whereas

$$\|x\|_X^2 = \sum_{\lambda<n}|x_\lambda(a)|^2 + 0 > 0.$$

Thus (39) is impossible.

Now suppose that P is of rank n. We shall show that (39) holds for some B. For any $x \in X$,

$$x(s) = \sum_{\lambda<n}\frac{x_\lambda(a)}{\lambda!}(s-a)^\lambda + y(s), \qquad s \in I,$$

$$y(s) \underset{\text{def}}{=} \int_a^s \frac{(s-t)^{n-1}}{(n-1)!}x_n(t)\,dt,$$

$$y_\lambda(a) = 0, \qquad \lambda = 0, \ldots, n-1,$$

$$y_n(s) = x_n(s),$$

and

$$F^i x = \sum_{\lambda<n}\frac{x_\lambda(a)}{\lambda!}F^i[(s-a)^\lambda] + F^i y, \qquad i = 0, \ldots, m.$$

We may write the last relation in matrix form:

$$P\cdot\left(\frac{x_\lambda(a)}{\lambda!}\right) = F(x-y),$$

where $(x_\lambda(a)/\lambda!)$ is a column matrix of n elements. Since P is of rank n, P^*P is of rank n and therefore invertible. Hence

$$\left(\frac{x_\lambda(a)}{\lambda!}\right) = (P^*P)^{-1}P^*\cdot F(x-y),$$

and for suitable B

$$|x_\lambda(a)| \le B\left[\sum_{i=0}^{m} |F^i x - F^i y|^2\right]^{1/2}, \qquad \text{all } \lambda.$$

In the following relations each B is a suitable constant, not necessarily the same as a preceding B. By Minkowski's inequality,

$$|x_\lambda(a)| \le B\left[\left(\sum_{i\le m} |F^i x|^2\right)^{1/2} + (\sum |F^i y|^2)^{1/2}\right].$$

By (38)

$$F^i y = \int_{I_0} x_n \phi^i;$$

hence

$$|F^i y|^2 \le B^2 \int_{I_0} x_n^2,$$

$$\sum_{\le m} |F^i y|^2 \le B^2 \int_{I_0} x^2,$$

$$|x_\lambda(a)| \le B\left[\left(\sum_{i\le m} |F^i x|^2\right)^{1/2} + \left(\int_{I_0} x_n^2\right)^{1/2}\right] \le 2B\left(\sum_{i\le m} |F^i x|^2 + \int_{I_0} x_n^2\right)^{1/2},$$

$$|x_\lambda(a)|^2 \le B^2\left(\sum_{i\le m} |F^i x|^2 + \int_{I_0} x_n^2\right),$$

$$\|x\|_X^2 = \sum_{\lambda<n} |x_\lambda(a)|^2 + \int_I x_n^2 \le B^2\left(\sum_i |F^i x|^2 + \int_I x_n^2\right),$$

as was to be shown.

Lemma 5. *If A is invertible, then $\bar{m} = m + 1 \ge n$, P is of rank n, the completeness condition* (39) *holds, and the functionals $F_{\bar{m}\times 1} = (F^i)$ are independent.*

Proof. The matrix A is given by (5). If P were of rank $< n$, then by Laplace's development $\det A = 0$. Hence $P_{\bar{m}\times n}$ is of rank n; hence $\bar{m} \ge n$. The completeness condition holds, by Lemma 4.

Suppose that $\sum_{i=0}^{m} c_i F^i = 0$, $c_i \in \mathsf{R}$. We shall show that $c_i = 0$, all i. Now

$$A = \begin{pmatrix} \underset{n^2}{0} & \underset{n\times\bar{m}}{(F^j[(s-a)^\mu])} \\ \underset{\bar{m}\times n}{(F^i[(s-a)^\lambda])} & \underset{\bar{m}^2}{(F^i F^j[|t-s|^q])} \end{pmatrix} \qquad \begin{matrix} \lambda, \mu = 0, \ldots, n-1, \\ \\ i, j = 0, \ldots, m. \end{matrix}$$

Hence the product of the row

$$\begin{pmatrix} \underset{1\times n}{0} & \underset{1\times \overline{m}}{(c_i)} \end{pmatrix}$$

into A vanishes. As A is invertible, it follows that $c_i = 0$, all i.

Thus our hypothesis in Section 3.2 that A is invertible assures that the conclusions of Lemma 5 hold.

In the next lemma we set aside for the moment the hypothesis that A is invertible.

Lemma 6. *If the admissible functionals $F_{\overline{m}\times 1} = (F^i)$ are independent and P is of rank n, then A is invertible.*

Proof. Assume, as will be absurd, that A is not invertible. Then constants $b_{n\times 1} = (b_\lambda)$ and $c_{\overline{m}\times 1} = (c_i)$ exist, not all zero, such that

$$\underset{(n+\overline{m})\times 1}{0} = A\cdot\begin{pmatrix} b \\ c \end{pmatrix} = \begin{pmatrix} 0 & P^* \\ P & \Phi \end{pmatrix}\begin{pmatrix} b \\ c \end{pmatrix} = \begin{pmatrix} \underset{n\times 1}{0} \\ \underset{\overline{m}\times 1}{0} \end{pmatrix},$$

$$P^*c = 0,$$

$$Pb + \Phi c = 0, \tag{40}$$

$$\sum_{j=0}^{m} c_j F^j[(s-a)^\lambda] = 0, \qquad \lambda = 0, \ldots, n-1,$$

$$\sum_{\lambda=0}^{n-1} b_\lambda F^i[(s-a)^\lambda] + \sum_{j=0}^{m} c_j F_t^i F_s^j[|t-s|^q] = 0, \quad i = 0, \ldots, m.$$

Put

$$E = \sum_{j\le m} c_j F^j.$$

Then

$$E[(s-a)^\lambda] = 0, \qquad \lambda = 0, \ldots, n-1, \tag{41}$$

$$F_t^i E_s[|t-s|^q] = -\sum_{\lambda<n} b_\lambda F^i[(s-a)^\lambda], \qquad i = 0, \ldots, m.$$

Multiply the last relation by c_i and sum:

$$E_t E_s[|t-s|^q] = -\sum_{\lambda<n} b_\lambda E[(s-a)^\lambda] = 0.$$

On the other hand, by Lemma 1, for suitable constants k_λ and a function $e \in L^2(I)$,

$$Ex = \sum_{\lambda<n} k_\lambda x_\lambda(a) + \int_I x_n e = \int_I x_n e, \qquad x \in X,$$

because of (41). Hence by the corollary of Lemma 2

$$E_t E_s[|t-s|^q] = 2q!(-1)^n \int_I e^2.$$

Hence $e(s) = 0$ for almost all $s \in I$, $E = 0$, and $c_j = 0$, all j, since the functionals F^j are independent.

As P is of rank n, $P^*P_{n\times n}$ is invertible. Then (40) implies that

$$\underset{n\times 1}{b} = -(P^*P)^{-1}P^*\Phi c = 0,$$

a contradiction.

3.5. THE SPACE $\mathscr{X}$

We now construct another Hilbert space $\mathscr{X}$, by introducing a different inner product in X. The space $\mathscr{X}$, due to Golomb-Weinberger, is a tool of proof.

Define a new inner product on X as

$$(x, y) = \sum_{i=0}^{m} F^i x F^i y + \int_I x_n y_n, \qquad x, y \in X. \tag{42}$$

Note that (x, y) and $(x, y)_X$ are quite different. The completeness condition (39), in force by virtue of Lemma 4, ensures that $(x, x) = 0$ iff $x = 0$. Let $\mathscr{X}$ denote the space X with the inner product (42). As sets,

$$\mathscr{X} = X.$$

The completeness condition (39) may be written

$$\|x\|_X^2 \le B^2(x, x), \qquad x \in X.$$

Conversely, for suitable B'

$$(x, x) \le B'^2 \|x\|_X^2, \qquad x \in X,$$

because the functionals F are continuous on X. Hence the norms $\|x\|_X$ and $\|x\|_{\mathscr{X}} = (x, x)^{1/2}$ are equivalent. As X is complete relative to $\|x\|_X$, it follows that $\mathscr{X}$ is complete relative to $\|x\|_{\mathscr{X}}$. Hence $\mathscr{X}$ is a Hilbert space.

Since the topologies of X and $\mathscr{X}$ are equivalent, any functional which is linear and continuous on one space is linear and continuous on the other. Thus admissible functionals are linear continuous functionals on $\mathscr{X}$ and vice versa.

3.6. THE SPLINES M

Let N be the set of elements of X for which all observations vanish:

$$N = \left\{x \in X : Fx = \underset{\bar{m}\times 1}{0}\right\} = \bigcap_{i=0}^{m} \ker F^i, \tag{43}$$

and let M be the set of elements of X whose nth derivative is orthogonal in $L^2(I)$ to the nth derivative of all elements of N:

$$M = \left\{x \in X : \int_I x_n \zeta_n = 0 \quad \text{whenever } \zeta \in N\right\}. \tag{44}$$

In particular, any polynomial of degree $n - 1$ is an element of M.

Lemma 7. *The spaces M and N constitute an orthogonal decomposition of $\mathscr{X}$:*

$$M = N^{\perp}, \qquad N = M^{\perp}, \qquad \mathscr{X} = M \oplus N.$$

Furthermore

$$M = \operatorname{span}\{\phi^0, \ldots, \phi^m\},$$

where $\phi^i \in \mathscr{X}$ is the dual of the functional F^i, $i = 0, \ldots, m$. The space M is of dimension $m + 1$, and N is infinite dimensional.

Proof. We show first that $M = N^{\perp}$. Thus

$$\begin{aligned} N^{\perp} &= \{x \in \mathscr{X} : (x, \zeta) = 0 \quad \text{whenever } \zeta \in N\} \\ &= \left\{x \in X : \int_I x_n \zeta_n = 0 \quad \text{whenever } \zeta \in N\right\} = M, \end{aligned}$$

since

$$(x, \zeta) = \sum_{i=0}^{m} F^i x F^i \zeta + \int_I x_n \zeta_n = \int_I x_n \zeta_n$$

whenever $\zeta \in N$.

As N is closed, it follows that $M^{\perp} = N$ and $\mathscr{X} = M \oplus N$.

Let ϕ^i be the dual (sometimes called the representer) of F^i; that is, $\phi^i \in \mathscr{X}$ and

$$F^i x = (x, \phi^i), \qquad x \in \mathscr{X}, \qquad i = 0, \ldots, m.$$

Then $x \perp \operatorname{span}\{\phi^0, \ldots, \phi^m\}$, $x \in \mathscr{X}$, iff $(x, \phi^i) = 0 = F^i x$, all $i = 0, \ldots, m$; that is, iff $x \in N$. Hence

$$N^{\perp} = \operatorname{span}\{\phi^0, \ldots, \phi^m\} = M.$$

Since the functionals $F^0, \ldots, F^m$ are independent, so are their duals $\phi^0, \ldots, \phi^m$. Hence M is of dimension $m + 1$. As $\mathscr{X}$ is infinite dimensional, it follows that N is. This completes the proof.

We call M the *space of splines*, relative to the functionals F and the norm $\int_I x_n^2$. For any $x \in \mathscr{X}$ put ξ equal to the projection of x on M:

$$\xi = \text{proj}_M\, x, \qquad x \in \mathscr{X}.$$

We call ξ the *spline approximation* of x. Since $\xi \in \mathscr{X}$, ξ itself is a function on I. The operator Π of the introduction will turn out to be essentially the operator proj_M; that is, the restriction of Πx to the interval I will equal $\text{proj}_M\, x$, $x \in \mathscr{X}$. If $G \in X^*$, that is, if G is an admissible functional, put

$$A_0 = G\, \text{proj}_M; \qquad A_0 x = G\, \text{proj}_M\, x = G\xi, \qquad x \in \mathscr{X}. \tag{45}$$

We call A_0 the *optimal* or *spline-based approximation of* G. The justification of the name optimal will appear in Theorems 1 to 5. Put

$$R_0 = G - A_0 = G\, \text{proj}_N. \tag{46}$$

Then $R_0 x$, $x \in \mathscr{X}$, is the remainder in the approximation of Gx by $A_0 x$.

Lemma 8. *There exists a linear continuous functional Q_0 on $L^2(I)$ such that*

$$R_0 x = Q_0(x_n), \qquad x \in \mathscr{X}. \tag{47}$$

Equivalently, there exists a kernel function $\kappa \in L^2(I)$ such that

$$R_0 x = \int_I x_n \kappa, \qquad x \in \mathscr{X}. \tag{48}$$

In particular, $R_0 x = 0$ if x is a polynomial of degree $n - 1$.

Proof. We use the quotient theorem [Sard 2, 1, p. 311]. Let $U: \mathscr{X} \to L^2(I)$ be the operator of n-fold differentiation:

$$Ux = x_n, \qquad x \in \mathscr{X}.$$

Then $R_0 x = 0$ whenever $Ux = 0$, $x \in \mathscr{X}$, because $x_n = 0$ implies that $x \in M$ by Lemma 7 and therefore that

$$R_0 x = G\, \text{proj}_N\, x = 0.$$

Furthermore $\mathscr{X}$ and $L^2(I)$ are Banach spaces and U maps $\mathscr{X}$ onto all of $L^2(I)$. The quotient theorem ensures the first conclusion of the lemma. The second conclusion follows from the standard form of a linear continuous functional on $L^2(I)$:

$$Q_0(y) = (y, \kappa)_{L^2(I)} = \int_I y\kappa, \qquad y \in L^2(I).$$

Lemma 9. *For each $x \in \mathscr{X}$, $A_0 x$ is a linear combination of the observations $F^0 x, \ldots, F^m x$.*

Proof. We shall show that constants $c_{\bar{m}\times 1} = (c_i)$ exist such that

$$A_0 x = \sum_{i=0}^{m} c_i F^i x, \qquad x \in \mathscr{X}.$$

The proof uses the quotient theorem.

Let $V : \mathscr{X} \to \mathsf{R}^{\bar{m}}$ be the operator which carries $x \mapsto Fx = (F^0 x, \ldots, F^m x)$. Now if $Vx = 0$, $x \in \mathscr{X}$, then $x \in N$ and

$$A_0 x = G \operatorname{proj}_M x = 0.$$

The range $V\mathscr{X}$ of V is a linear subspace of $\mathsf{R}^{\bar{m}}$ and therefore closed. Hence there exists a linear continuous functional W on $V\mathscr{X}$ such that

$$A_0 = WV$$

by the quotient theorem. Extend the functional W, defined on $V\mathscr{X}$, to the functional W_0, defined on $\mathsf{R}^{\bar{m}}$, as follows:

$$W_0 = W \operatorname{proj}_{V\mathscr{X}}.$$

Then $W_0 y = Wy$ whenever $y \in V\mathscr{X}$. Hence

$$A_0 x = WVx = W_0 Vx, \qquad x \in \mathscr{X},$$

because $Vx \in V\mathscr{X}$.

If $y = (y^0, \ldots, y^m) \in \mathsf{R}^{\bar{m}}$, then

$$W_0(y^0, y^1, \ldots, y^m) = W_0(y^0, 0, \ldots, 0) + W_0(0, y^1, 0, 0, \ldots, 0) + \cdots$$
$$+ W_0(0, \ldots, 0, y^m) = c_0 y^0 + c_1 y^1 + \cdots + c_m y^m$$

for suitable constants c_i, since $W(0, \ldots, 0, y^i, 0, \ldots, 0)$ is the value of a linear continuous functional on $y^i \in \mathsf{R}^1$ and is therefore a suitable constant c_i times y^i. The conclusion of the lemma follows.

Consider a fixed $s_0 \in I$. The functional G_0 defined as

$$G_0 x = x(s_0), \qquad x \in \mathscr{X},$$

is an admissible functional. The optimal approximation A_0 of G_0 is the functional such that

$$A_0 x = G_0 \operatorname{proj}_M x = G_0 \xi = \xi(s_0),$$

where

$$\xi = \operatorname{proj}_M x.$$

Thus *each value* $\xi(s_0)$, $s_0 \in I$, *of the spline* ξ *is the optimal approximation of the value* $x(s_0)$ of $x \in \mathscr{X}$. The spline ξ, which is an element of $\mathscr{X}$, approximates the function x because each value of ξ approximates the corresponding value of x.

Corollary. *The spline* $\xi = \text{proj}\, x$, $x \in \mathscr{X}$, *is a linear combination of the observations* $F^0x, \ldots, F^mx$:

$$\xi = (F^0x)\beta^0 + \cdots + (F^mx)\beta^m,$$

where the functions $\beta^0, \ldots, \beta^m$ *are elements of* M, *independent of* G *and of* x.

Proof. (An alternative proof of this corollary is in Section 3.10, in which explicit formulas for $\beta^0, \ldots, \beta^m$ are given.)

We have seen that, for each $s_0 \in I$,

$$\xi(s_0) = \sum_{i=0}^{m} c_i(s_0)F^ix,$$

where $c_i(s_0) \in \mathsf{R}$. This is just Lemma 9 for the functional $G_0 x = x(s_0)$. It remains to show that the function $c_i\colon s \mapsto c_i(s)$ is an element of M, independent of G. This is indeed the case: Lemma 9 ensures that $c_i(s_0)$ is independent of $x \in \mathscr{X}$. If we take x as a function for which $F^ix = \delta^{i,\,i_0}$, $i = 0, \ldots, m$, as is possible because $\mathscr{X}$ is infinite dimensional, we see that $\xi(s) = c_{i_0}(s)$, $s \in I$. As $\xi \in M$, so is c_{i_0}, $i_0 = 0, \ldots, m$.

Lemma 10. *If* $x \in \mathscr{X}$ *and* $\xi = \text{proj}_M\, x$, *then*

(49)

$$\int_I x_n^2 - \int_I \xi_n^2 = \int_I (x_n - \xi_n)^2 = \|x\|^2 - \|\xi\|^2 = \|x - \xi\|^2 = \|\text{proj}_N\, x\|^2 \geq 0.$$

Proof. The last two equalities and the inequality are immediate, and

$$\|x - \xi\|^2 = \int_I (x_n - \xi_n)^2 + \sum_{i=0}^{m} |F^i(x - \xi)|^2 = \int_I (x_n - \xi_n)^2,$$

since $x - \xi \in N$. Finally

$$\|x\|^2 - \|\xi\|^2 = \int_I x_n^2 + \sum |F^ix|^2 - \int_I \xi_n^2 - \sum |F^i\xi|^2 = \int_I x_n^2 - \int_I \xi_n^2,$$

since $Fx = F\xi$.

In the next theorem we consider $G \in X^*$, $A_0 = G\, \text{proj}_M$, and $R_0 = G - A_0 = G\, \text{proj}_N$.

Theorem 1 (geometric condition).† *The functional* A_0 *is the optimal approximation of* G *in the sense that*

(50) $$|Gx - A_0 x|^2 \leq J\left[\int_I x_n^2 - \int_I \xi_n^2\right], \qquad \textit{all } x \in \mathscr{X},$$

† [Golomb–Weinberger 1; Schoenberg 10, 11; Ahlberg–Nilson 3.]

where

$$\xi = \text{proj}_M x,$$

(51) $$J = \sup_{0 \neq \zeta \in N} \frac{|G\zeta|^2}{\|\zeta\|^2} = \sup \frac{|G\zeta|^2}{\int_I \zeta_n^2};$$

and, furthermore, for any $x \in \mathscr{X}$ *and* $\alpha \in \mathsf{R}$, *the quantity* $|Gy - \alpha|^2$ *attains values strictly greater than the right member of* (50) *for some* $y \in \mathscr{X}$ *for which* $Fy = Fx$, $\int_I y_n^2 = \int_I x_n^2$, *and* $\text{proj}_M y = \xi$, *unless* $\alpha = A_0 x$. *Equality is attainable in* (50) *for any prescribed* Fx *and* $\int_I x_n^2$ *for which the right member is nonnegative.*

An alternative statement is the following.

Let $\omega \in \mathsf{R}^{\bar{m}}$ *and* $d \geq 0$ *be prescribed, and let* Γ *be the set of elements* $x \in \mathscr{X}$ *such that*

$$Fx = \omega, \qquad \int_I x_n^2 \leq d^2.$$

If Γ *is nonempty, then its image* $G\Gamma$ *under* G *is the closed interval of* R *of center* $G\xi$ *and length*

$$2J^{1/2}\left[d^2 - \int_I \xi_n^2\right]^{1/2}, \qquad \xi = \text{proj}_M x, \qquad x \in \Gamma.$$

Here ξ, $G\xi$, *and* $\int_I \xi_n^2$ *are constant for all* $x \in \Gamma$.

In (51) and elsewhere the omitted domain of a supremum is the same as that of the preceding supremum. Note that *J is the square of the Banach norm of the restriction of G to N.*

Proof. The alternative statement implies the first, since, for any $x \in \mathscr{X}$, we may put $d^2 = \int_I x^2$. To prove the alternative statement, consider

$$\Gamma = \left\{x \in \mathscr{X} : \; Fx = \omega \; \text{ and } \; \int_I x_n^2 \leq d^2\right\},$$

where $\omega \in \mathsf{R}^{\bar{m}}$ and $d \geq 0$ are given and fixed. Suppose that Γ is not empty. Put $\xi^0 = \text{proj}_M x^0$ for some $x^0 \in \Gamma$. Then $\xi^0 = \text{proj}_M x$ for all $x \in \Gamma$ by the corollary of Lemma 9, and $A_0 x = G\xi^0$ is constant for all $x \in \Gamma$.

Now Γ is a convex set. For if $x, y \in \Gamma$ and $\alpha + \beta = 1$, $\alpha \geq 0$, $\beta \geq 0$, then

$$F(\alpha x + \beta y) = (\alpha + \beta)\omega = \omega$$

and

$$\left|\int_I (\alpha x_n + \beta y_n)^2\right|^{1/2} \leq \alpha \left|\int_I x_n^2\right|^{1/2} + \beta \left|\int_I y_n^2\right|^{1/2} \leq (\alpha + \beta)\, d = d.$$

Hence $\alpha x + \beta y \in \Gamma$. (As a matter of fact Γ is a hypercircle in $\mathscr{X}$: the intersection of a ball in $\mathscr{X}$ with a translation of the subspace N.)

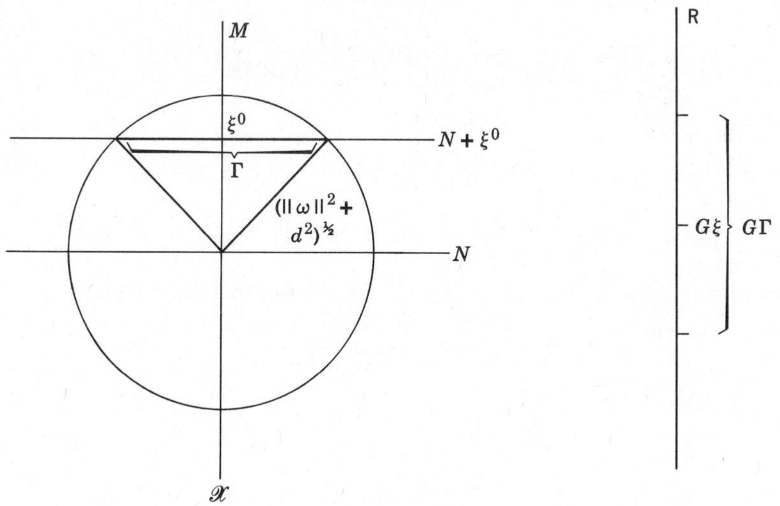

Figure 9. The geometry of Theorem 1.

M is n-dimensional.

N is ∞-dimensional.

$\Gamma = \{x \in \mathscr{X}: Fx = \omega \text{ and } \int_I x_n^2 \le d^2\} = \{x \in \mathscr{X}: Fx = \omega \text{ and } \|x\|^2 \le d^2 + \|\omega\|^2\}$.

$\xi^0 = \text{Proj}_M\, x, \qquad x \in \Gamma$.

It follows that the image $G\Gamma$ of Γ under G is convex and therefore an interval, since $G\Gamma \subset$ R. Now

$$|Gx - A_0 x|^2 = |Gx - G\xi|^2 = |G \,\text{proj}_N x|^2 \le J\|\text{proj}_N x\|^2$$

$$= J\left[\int_I x_n^2 - \int_I \xi_n^2\right], \qquad x \in \mathscr{X},\ \xi = \text{proj}_M x,$$

where $J = \|G \upharpoonright N\|^2$ is the square of the Banach norm of the restriction of G to N; and J is given by (51). (Note that by Lemma 7 N surely contains elements other than the origin.) Thus

$$|Gx - G\xi^0|^2 \le J\left[d^2 - \int_I \xi_n^2\right], \qquad x \in \Gamma.$$

This implies that Gx lies in the interval of center $G\xi^0$ and half-length $J^{1/2}[d^2 - \int_I \xi_n^2]^{1/2}$.

To complete the proof it will be sufficient to show that the endpoints of that interval are elements of $G\Gamma$. Take $\zeta^0 \in N$ such that

$$G\zeta^0 = J^{1/2}, \qquad \|\zeta^0\|^2 = 1 = \int_I \zeta_n^{0\,2},$$

as is possible by Riesz's theorem on linear continuous functionals on the Hilbert space N [Sard 1, p. 451]. Put

$$c = \pm\left[d^2 - \int_I \xi_n^{0\,2}\right]^{1/2},$$

$$y^0 = \xi^0 + c\zeta^0.$$

Recall that $F\xi^0 = \omega$, since $x^0 - \xi^0 \in N$ and $x^0 \in \Gamma$. Hence $Fy^0 = \omega$, and

$$\|y^0\|^2 = \|\xi^0\|^2 + c^2\,\|\zeta^0\|^2 = \|\omega\|^2 + \int_I \xi_n^{0\,2} + c^2 = \|\omega\|^2 + d^2,$$

where $\|\omega\|$ is the Euclidean length of $\omega \in \mathsf{R}^{\overline{m}}$. On the other hand,

$$\|y^0\|^2 = \|\omega\|^2 + \int_I y_n^{0\,2},$$

by (42). Hence

$$\int_I y_n^{0\,2} = d^2$$

and $y^0 \in \Gamma$. Furthermore

$$Gy^0 - G\xi^0 = G(c\zeta^0) = cJ^{1/2} = \pm J^{1/2}\left[d^2 - \int_I \xi_n^{0\,2}\right]^{1/2}.$$

Thus the two choices of sign of c make Gy^0 either endpoint of the stated interval. This completes the proof.

Remark. The relation (50) implies that

$$|Gx - A_0 x|^2 \le J\int_I x_n^2, \qquad x \in \mathscr{X}. \tag{52}$$

Equality in (52) occurs if $\int_I \xi_n^2 = 0$ and if equality occurs in (51). It is a consequence of Lemma 14 below that $\int_I \xi_n^2 = 0$ iff $(Fx)^* B_{2,2} Fx = 0$.

3.7. RELATION TO PEANO'S REPRESENTATION OF THE REMAINDER

Another approach to the splines and optimal approximation is related to Lemmas 8 and 9.

Suppose that $G \in X^*$. We say that a functional A on X is an *admissible approximation of* G if Ax, $x \in X$, is a linear combination of the observations Fx and if A is exact for degree $n - 1$. We denote the set of admissible approximations by $\mathscr{A}$. Thus

$$\mathscr{A} = \{A : Ax = c^* Fx, \qquad \text{all } x \in X, \ \underset{\overline{m}\times 1}{c} \in \mathsf{R}^{\overline{m}};$$

and $Ax = Gx$ whenever x is a polynomial of degree $n - 1\}$.

Put $R = G - A$; then

$$\mathscr{A} = \{A : Ax = c^*Fx, \text{ all } x \in X;\ c \in \mathsf{R}^{\bar{m}};\ \text{and } Rx = 0 \text{ whenever } x_n = 0\}.$$

If $A \in \mathscr{A}$, then $A = c^*F$ for some $c \in \mathsf{R}^{\bar{m}}$ and therefore $A \in X^*$. Hence $R \in X^*$. Since $Rx = 0$ whenever x is a polynomial of degree $n - 1$, the Corollary of Lemma 1 implies that a kernel $\kappa \in L^2(I)$ exists such that

(53) $$Rx = \int_I x_n \kappa \underset{\text{def}}{=} Q(x_n), \qquad \text{all } x \in X.$$

Thus Q, defined by (53), is a linear continuous functional on $L^2(I)$. By Schwartz's inequality

$$|Rx|^2 \le \int_I x_n^2 \int_I \kappa^2, \qquad x \in X,$$

with equality attainable. Hence the square of the Banach norm of Q is

$$\|Q\|^2 = \int_I \kappa^2.$$

Thus for each $A \in \mathscr{A}$ the remainder $R = G - A = QD^n$, where D is differentiation, and

$$Qy = \int_I y\kappa, \qquad y \in L^2(I).$$

We may think of Q as the quotient of R by D^n.

Theorem 2 (minimal quotient).† *For $A \in \mathscr{A}$, the norm $\|Q\|$ of the quotient is minimal iff $A = A_0$, in which case*

(54) $$\|Q\|^2 = J = \|G\restriction N\|^2.$$

Here Q, A_0, J, and N have been defined in (53), (45), (51), and (43).

Proof. For any $A \in \mathscr{A}$

$$\|Q\| = \sup_{\substack{x \in \mathfrak{X},\\ Ux \neq 0}} \frac{|Rx|}{\|Ux\|} \ge \sup_{\substack{\zeta \in N,\\ U\zeta \neq 0}} \frac{|R\zeta|}{\|U\xi\|} = \sup \frac{|G\zeta|}{\|\zeta\|} = \|G\restriction N\| = J^{1/2},$$

where

(55) $$Ux = x_n, \ \|Ux\|^2 = \int_I x_n^2,$$

† [Sard 3, 1, Chapter 2; Schoenberg 10, 11; Secrest 1, 2, 3; de Boor-Lynch 1; Ahlberg–Nilson 3.]

since $N \subset X$ and since

$$R\zeta = G\zeta - A\zeta = G\zeta, \qquad \zeta \in N,$$

because $A\zeta = 0$ by the definition of $\mathscr{A}$. The last equality in the statement about $\|Q\|$ is a consequence of (51). Of course, $\|U\zeta\| = \|\zeta\|$, for $\zeta \in N$.

Now $A_0 \in \mathscr{A}$, by Lemmas 8 and 9. If $A = A_0$, then $Q = Q_0$ and $\|Q_0\|^2 \geq J$ by the above inequality. Furthermore

$$\|Q_0\|^2 = \sup_{\substack{x \in \mathscr{X},\\ Ux \neq 0}} \frac{|R_0 x|^2}{\|Ux\|^2} = \sup_{\substack{x \in \mathscr{X},\\ Ux \neq 0,\\ \mathrm{proj}_N x \neq 0}} \left(\frac{|G\, \mathrm{proj}_N x|^2}{\|\mathrm{proj}_N x\|^2} \frac{\|\mathrm{proj}_N x\|^2}{\|Ux\|^2} \right),$$

since $R_0 = G\, \mathrm{proj}_N$ and the excluded case $\mathrm{proj}_N x = 0$ would give $R_0 x = 0$. Now

$$0 \leq \frac{\|\mathrm{proj}_N x\|^2}{\|Ux\|^2} = 1 - \frac{\|U\xi\|^2}{\|Ux\|^2} \leq 1$$

by Lemma 10. Hence

$$\|Q_0\|^2 \leq \sup_{\substack{x \in \mathscr{X},\\ \mathrm{proj}_N x \neq 0}} \frac{|G\, \mathrm{proj}_N x|^2}{\|\mathrm{proj}_N x\|^2} = \sup_{0 \neq \zeta \in N} \frac{|G\zeta|^2}{\|\zeta\|^2} = \|G \restriction N\|^2 = J.$$

Hence $\|Q_0\|^2 = J$, and A_0 minimizes $\|Q\|$ among admissible approximations.

Suppose that $A_0 \neq A \in \mathscr{A}$. We shall show that

$$\|Q\|^2 > J.$$

To this end let $\xi^0 \in M$ be such that $R\xi^0 \neq 0$. (Such a ξ^0 surely exists. Otherwise $R\xi = 0$ for all $\xi \in M$; and, since $R\zeta = G\zeta$ for all $\zeta \in N$, it would follow that $Rx = R_0 x$ for all $x \in \mathscr{X}$; that is, $A = A_0$.) Then $U\xi^0 \neq 0$, since $R = QU$.

Now

$$J = \sup_{\substack{\zeta \in N,\\ \|\zeta\| = \|U\zeta\| = 1}} |G\zeta|^2.$$

If $J = 0$, then $G\zeta = 0$ for all $\zeta \in N$; and $R_0 x = G(x - \xi) = 0$ for all $x \in \mathscr{X}$, where $\xi = \mathrm{proj}_M x$. Hence $Q_0 = 0$, since the quotient in the quotient theorem is unique. On the other hand, $Q \neq 0$, else $R\xi^0$ would vanish. Therefore $\|Q\| > \|Q_0\| = 0$, as was to be shown, if $J = 0$. Assume that $J > 0$.

There is an element $\zeta^0 \in N$ such that

$$G\zeta^0 = \|G\| = J^{1/2}, \qquad \|\zeta^0\| = \|U\zeta^0\| = 1; \tag{56}$$

in fact, ζ^0 is $g/\|g\|$, where g is the dual of $G\restriction N$. Put

$$t = \frac{R\xi^0}{J^{1/2}\,\|U\xi^0\|^2},$$

$$x^0 = t\xi^0 + \zeta^0 \in \mathscr{X}.$$

Then

$$Rx^0 = tR\xi^0 + R\zeta^0 = tR\xi^0 + G\zeta^0 = tR\xi^0 + J^{1/2},$$

since $A\zeta^0 = 0$, and

$$Ux^0 = tU\xi^0 + U\zeta^0.$$

Now

$$tR\xi^0 = \frac{|R\xi^0|^2}{J^{1/2}\,\|U\xi^0\|^2} > 0;$$

hence

$$|Rx^0| = Rx^0 = \frac{|R\xi^0|^2 + J\,\|U\xi^0\|^2}{J^{1/2}\,\|U\xi^0\|^2}.$$

Furthermore

$$(tU\xi^0, U\zeta^0) = t(U\xi^0, U\zeta^0) = t(\xi^0, \zeta^0) = 0$$

by (42) and the fact that $\xi^0 \in M$, $\zeta^0 \in N$. Hence by (56)

$$\|Ux^0\|^2 = |t|^2\,\|U\xi^0\|^2 + 1 = \frac{|R\xi^0|^2 + J\,\|U\xi^0\|^2}{J\,\|U\xi^0\|^2}$$

and

$$\frac{|Rx^0|^2}{\|Ux^0\|^2} = \frac{|R\xi^0|^2 + J\,\|U\xi^0\|^2}{\|U\xi^0\|^2} > J.$$

Hence

$$\|Q\|^2 = \sup_{\substack{x \in \mathscr{X},\\ Ux \neq 0}} \frac{|Rx|^2}{\|Ux\|^2} > J,$$

as was to be shown.

3.8. INTERPOLATORY AND OTHER PROPERTIES

The next theorems refer to the splines but not to the functional G. The theorems are immediate consequences of the fact that the spline approximation of x is $\mathrm{proj}_M\, x$, $x \in \mathscr{X}$.

If $Fy = Fx$, we may say that y interpolates to x (relative to the functionals F).

Theorem 3 (spline interpolation).† *For any $x \in \mathscr{X}$ there is one and only one $\xi \in M$ such that $F\xi = Fx$. Furthermore $\xi = \operatorname{proj}_M x$.*

Proof. The condition $F\xi = Fx$ may be written $x - \xi \in N$. Now $x = (x - \xi) + \xi$ and $\xi \in M = N^{\perp}$. The decomposition theorem for the Hilbert space $\mathscr{X}$ implies that ξ exists and that $\xi = \operatorname{proj}_M x$.

Theorem 4 (optimal interpolation).‡ *For each $x \in \mathscr{X}$ the integral $\int_I y_n^2$ is minimal among all $y \in \mathscr{X}$ such that $Fy = Fx$ iff*

$$y = \xi = \operatorname{proj}_M x.$$

Thus among all elements of $\mathscr{X}$ which interpolate to x the spline ξ is smoothest in the sense that $\int_I \xi_n^2$ is minimal. Walsh, Ahlberg, and Nilson describe this as the minimal curvature property.

Proof. Put $\xi = \operatorname{proj}_M x$. Then $F\xi = Fx$, by Theorem 3. Now consider $y \in \mathscr{X}$ such that $Fy = Fx$, that is, such that $y - x \in N$. Then $\xi = \operatorname{proj}_M y$ and

$$0 \leq \|y - \xi\|^2 = \int_I y_n^2 - \int_I \xi_n^2$$

by Lemma 10. Hence

$$\int_I y_n^2 \geq \int_I \xi_n^2$$

with equality iff $y = \xi$.

Theorem 5 (approximation of x_n).§ *For each $x \in \mathscr{X}$, the integral $\int_I (x_n - \eta_n)^2$ is minimal among all $\eta \in M$ iff $\eta_n = \xi_n$, where $\xi = \operatorname{proj}_M x$.*

Thus $\xi = \eta$ minimizes $\int_I (x_n - \eta_n)^2$. Other elements of M may do so also, but ξ is the only element that interpolates to x. The condition $\eta_n = \xi_n$ states that η differs from ξ by a polynomial of degree $n - 1$. Walsh, Ahlberg, and Nilson call this theorem the best approximation property.

Proof. Put $\xi = \operatorname{proj}_M x$. If $\eta \in M$, then

$$\eta - x = (\eta - \xi) + (\xi - x), \qquad \eta - \xi \in M, \qquad \xi - x \in N.$$

† [Golomb–Weinberger 1; de Boor 2; Schoenberg 8, 10, 12; Ahlberg–Nilson–Walsh 5.]

‡ [Holladay 1; Golomb–Weinberger 1; Walsh–Ahlberg–Nilson 1; de Boor 2; Schoenberg 8, 10, 12; Ahlberg–Nilson–Walsh 5.]

§ [Walsh–Ahlberg–Nilson 1; de Boor 2; Schoenberg 8, 12; Ahlberg–Nilson–Walsh 5.]

Hence

$$\|\eta - x\|^2 = \|\eta - \xi\|^2 + \|\xi - x\|^2,$$

$$\int_I (\eta_n - x_n)^2 + \sum_{i=0}^{m} |F^i(\eta - x)|^2 = \int_I (\eta_n - \xi_n)^2 + \sum_{i=0}^{m} |F^i(\eta - \xi)|^2$$

$$+ \int_I (\xi_n - x_n)^2 + 0.$$

Now $F(\eta - x) = F(\eta - \xi)$, since $x - \xi \in N$, and

$$\int_I (\eta_n - x_n)^2 = \int_I (\eta_n - \xi_n)^2 + \int_I (\xi_n - x_n)^2 \geq \int_I (\xi_n - x_n)^2,$$

with equality iff $\eta_n - \xi_n = 0$.

3.9. ANALYTIC DESCRIPTION OF THE SPLINES

We have defined the *key functions* in (2) as

(57) $$\underset{\bar{m} \times 1}{f} = (f^i), \quad f^i(u) = F_s^i[|u - s|^q], \quad u \in \mathsf{R}, q = 2n - 1, i = 0, \ldots, m.$$

It is convenient also to define the *curled key functions* as follows:

(58) $$\underset{\bar{m} \times 1}{\tilde{f}} = (\tilde{f}^i), \quad \tilde{f}^i(u) = F_s^i[(u - s)_+^q].$$

Since

$$|u - s|^q = 2(u - s)_+^q - (u - s)^q,$$

it follows that $f^i(u)$ equals $2\tilde{f}^i(u)$ plus a polynomial in u of degree q.

In this section we shall show that each spline is a linear combination of key functions (or, equivalently, curled key functions), subject to a side condition, plus a polynomial of degree $n - 1$.

Let L be the set of functions $\theta : \mathsf{R} \to \mathsf{R}$ such that

(59) $$\theta = c^*f + b^*\eta$$

and

(60) $$P^*c = 0,$$

where $c_{\bar{m} \times 1} \in \mathsf{R}^{\bar{m}}$ is a column of $\bar{m}$ constants, f is the column (57) of key functions, $b_{n \times 1} \in \mathsf{R}^n$ is a column of n constants, η is the column (11) of powers, that is,

(61) $$\eta(u) = \begin{pmatrix} 1 \\ u - a \\ (u - a)^2 \\ \vdots \\ (u - a)^{n-1} \end{pmatrix}, \quad u \in \mathsf{R},$$

$a \in I_0$ is the constant that entered in our earlier Taylor formulas, and $P_{\bar{m} \times n}$ is the matrix (3) = (12). Recall that I_0 is the smallest compact interval that contains the supports of the functionals F. Thus $\theta \in L$ iff

(62) $$\theta(u) = \sum_{i=0}^{m} c_i f^i(u) + \sum_{j=0}^{n-1} b_j(u-a)^j, \qquad u \in , \mathsf{R}$$

and

(63) $$\sum_{i=0}^{m} c_i F_s^i[(s-a)^\lambda] = 0, \qquad \lambda = 0, \ldots, n-1.$$

The set L may be described alternatively as follows:

(64) $$\begin{aligned} L &= \{\theta : \theta = c^*f + b^*\eta \text{ and } \theta_n(u) = 0 \quad \text{for all} \quad u \geq I_0\} \\ &= \{\theta : \theta = c^*f + b^*\eta \text{ and } \theta_n(u) = 0 \quad \text{for all} \quad u \leq I_0\}. \end{aligned}$$

That the nth derivative θ_n vanishes above or below I_0 means that θ is a polynomial of degree $n-1$ there.

Proof of (64). We first show that a function θ of the form (62) satisfies (63) iff $\theta_n(u) = 0$ for all $u \geq I_0$.

Assume that $u \geq I_0$ and $s \in I_0$. Then $|u-s| = u - s$ and

$$f^i(u) = F_s^i[|u-s|^q] = F_s^i[(u-s)^q] = F_s^i[(u-a)-(s-a)]^q, \; i = 0, \ldots, m,$$

because I_0 contains the support of F^i (the only values of the argument $|u-s|^q$ that are material are those for which $s \in I_0$). Thus for $u \geq I_0$,

$$\sum_{i=0}^{m} c_i f^i(u) = \sum_{i=0}^{m} \sum_{h=0}^{q} C_{q,h}(-1)^h(u-a)^{q-h} c_i F_s^i[(s-a)^h].$$

This polynomial in u of degree $q = 2n-1$ reduces to a polynomial of degree $n-1$ iff (63) holds, as was to be shown.

Similarly for $u \leq I_0$. This establishes (64).

Let $\tilde{L}$ be the set of functions $\theta : \mathsf{R} \to \mathsf{R}$ such that

(65) $$\theta = \tilde{c}^*\tilde{f} + \tilde{b}^*\eta$$

and

(66) $$P^*\tilde{c} = 0,$$

where $\tilde{c}_{\bar{m} \times 1} \in \mathsf{R}^{\bar{m}}$, $\tilde{b} \in \mathsf{R}^n$, $\tilde{f}_{\bar{m} \times 1}$ is the column (58) of curled key functions, and $\eta_{n \times 1}$ is the column (61) of powers. Thus $\theta \in \tilde{L}$ iff

(67) $$\theta(u) = \sum_{i=0}^{m} \tilde{c}_i \tilde{f}^i(u) + \sum_{j=0}^{n-1} \tilde{b}_j(u-a)^j, \qquad u \in \mathsf{R},$$

and

$$(68) \qquad \sum_{i=0}^{m} \tilde{c}_i F_s^i[(s-a)^\lambda] = 0, \qquad \lambda = 0, \ldots, n-1.$$

Then

$$(69) \qquad \tilde{L} = \{\theta : \theta = \tilde{c}^* \tilde{f} + \tilde{b}^* \eta \text{ and } \theta_n(u) = 0 \quad \text{for} \quad u \geq I_0\}.$$

The proof of (69) is similar to that of (64) and starts as follows. We show that a function θ of the form (67) satisfies (68) iff $\theta_n(u) = 0$ for all $u \geq I_0$. Thus assume that $u \geq I_0$ and $s \in I_0$. Then

$$(u-s)_+ = u - s$$

and

$$\tilde{f}^i(u) = F_s^i[(u-s)_+^q] = F_s^i[(u-s)^q] = F_s^i[(u-a)-(s-a)]^q, \qquad i = 0, \ldots, m.$$

We continue as before.

The condition $\theta_n(u) = 0$ for all $u \leq I_0$ is satisfied automatically because

$$(70) \qquad (u-s)_+ = 0 \quad \text{if} \quad u \leq I_0, \qquad s \in I_0.$$

The condition is uninformative as regards $\tilde{c}$.

Lemma 11. *The sets L and $\tilde{L}$ are identical*:

$$(71) \qquad L = \tilde{L}.$$

Furthermore

$$(72) \qquad L \ni c^* f + b^* \eta = \tilde{c}^* \tilde{f} + \tilde{b}^* \eta \in \tilde{L}$$

iff

$$(73) \qquad \tilde{c} = 2c,$$

$$(74) \qquad \tilde{b}^* \eta = b^* \eta - c^*[F_s(u-s)^q],$$

$$(75) \qquad b^* \eta = \tilde{b}^* \eta + \frac{\tilde{c}^*[F_s(u-s)^q]}{2}.$$

Thus we may shift back and forth from (62), (63) to (67), (68).

Proof. Suppose that $\theta \in L$ and that (62), (63) hold. Now

$$|u-s|^q = 2(u-s)_+^q - (u-s)^q, \qquad u, s \in \mathrm{R}.$$

Hence by (57) and (58)

$$\theta(u) = \sum_{i=0}^{m} c_i f^i(u) + \sum_{j=0}^{n-1} b_j(u-a)^j$$
$$= \sum_i 2c_i F_s^i[(u-s)_+^q] - \sum_i c_i F_s^i[(u-s)^q]$$
$$+ \sum_j b_j(u-a)^j = \sum_i \tilde{c}_i \tilde{f}^i(u) + \sum_j \tilde{b}_j(u-a)^j,$$

with $\tilde{c}$, $\tilde{b}$ defined by (73), (74). Note that $c^*F_s[(u-s)^q]$ is a polynomial in u of degree $n-1$, by (63), and that (74) is therefore a valid definition of $\tilde{b}$. Hence $\theta \in \tilde{L}$ and $L \subset \tilde{L}$.

Similarly $\tilde{L} \subset L$, completing the proof.

Lemma 12 (interpolation).† *For any* $\omega_{\overline{m}\times 1} \in \mathsf{R}^{\overline{m}}$ *there exists one and only one* $\theta \in L$ *such that*

(76) $$F\theta = \omega.$$

Furthermore,

(77) $$\theta = (\eta^* B_{1,2} + f^* B_{2,2})\omega = \omega^*(B_{2,1}\eta + B_{2,2}f),$$

(78) $$\theta = \eta^* b + f^* c,$$

(79), (80) $$\begin{cases} P^*c = 0, \\ Pb + \Phi c = \omega, \end{cases}$$

(81), (82) $$\begin{cases} b = B_{1,2}\,\omega, \\ c = B_{2,2}\,\omega. \end{cases}$$

Proof. The matrices Φ, P, A, B, $B_{1,1}$, $B_{1,2}$, $B_{2,1}$, $B_{2,2}$ have been defined in (3) to (12).

For any $\theta \in L$ (59) and (60) hold. Since a 1×1 matrix is its own transpose, we may write (78) instead of (59). The relation (79) is precisely (60). Now (78) implies that

$$F\theta = F\eta^* b + Ff^* c = Pb + \Phi c.$$

Hence the interpolatory condition (76) is (80). Now (79) and (80) may be written

$$A \cdot \begin{pmatrix} b \\ c \end{pmatrix} = \begin{pmatrix} 0 \\ \omega \end{pmatrix},$$

where $\begin{pmatrix} b \\ c \end{pmatrix}$ is the $n + \overline{m}$ column matrix obtained by putting c below b and $\begin{pmatrix} 0 \\ \omega \end{pmatrix}$

† [Schoenberg 8, 10, 11, 12.]

is similar. Since $B = A^{-1}$,

$$\begin{pmatrix} b \\ c \end{pmatrix} = B\begin{pmatrix} 0 \\ \omega \end{pmatrix} = \begin{pmatrix} B_{1,1} & B_{1,2} \\ B_{2,1} & B_{2,2} \end{pmatrix}\begin{pmatrix} 0 \\ \omega \end{pmatrix} = \begin{pmatrix} B_{1,2}\,\omega \\ B_{2,2}\,\omega \end{pmatrix},$$

which yields (81), (82). These, with (78), imply the first part of (77), which in turn implies the second part of (77) because $\theta^* = \theta$.

Thus for each $\omega \in \mathsf{R}^{\bar{m}}$ there is one and only one element of L that satisfies (76). The proof is complete.

Corollary. *For $\theta \in L$ the coefficients c, b in* (59) *and $\tilde{c}$, $\tilde{b}$ in* (65) *are unique.*

Proof. Take $\omega = 0_{\bar{m}\times 1}$ in the lemma.

For our next proofs we introduce standard forms for the functionals F. By Lemma 1 there are constants k_λ^i and functions $\phi^i \in L^2(I_0)$ such that

$$F^i x = \sum_{\lambda=0}^{n-1} k_\lambda^i x_\lambda(a) + \int_{I_0} x_n \phi^i, \qquad \text{all } x \in X,\ i = 0, \ldots, m. \tag{83}$$

Then $F_s^i[(s-a)^\lambda] = k_\lambda^i \lambda!$, $\lambda = 0, \ldots, n-1$. Hence condition (63) is equivalent to

$$\sum_{i=0}^{m} c_i k_\lambda^i = 0, \qquad \lambda = 0, \ldots, n-1 \tag{84}$$

and condition (68) is equivalent to

$$\sum_{i=0}^{m} \tilde{c}_i k_\lambda^i = 0, \qquad \lambda = 0, \ldots, n-1. \tag{85}$$

Lemma 13 (on the nth derivative of an element of L). *If*

$$\theta = c^* f + b^* \eta = \tilde{c}^* \tilde{f} + \tilde{b}^* \eta \in L, \tag{86}$$

then $\theta \in X$ and

$$\theta_n(u) = (-1)^n q! \sum_{i=0}^{m} \tilde{c}_i \phi^i(u) = (-1)^n q!\, 2 \sum_{i=0}^{m} c_i \phi^i(u), \qquad \text{almost all } u \in \mathsf{R}, \tag{87}$$

$$\int_I \theta_n x_n = \int_{I_0} \theta_n x_n = (-1)^n q! \sum_{i=0}^{m} \tilde{c}_i F^i x = (-1)^n q!\, 2 \sum_i c_i F^i x, \qquad \text{all } x \in X, \tag{88}$$

and

$$\int_I \theta_n^2 = \int_{I_0} \theta_n^2 = (-1)^n q! \sum_i \tilde{c}_i F^i \theta = (-1)^n q!\, 2 \sum_i c_i F^i \theta. \tag{89}$$

Proof. It will be convenient to use $\tilde{f}$ instead of f. For each $u \in \mathsf{R}$ put

$$y(s) = \frac{(u-s)_+^q}{q!}, \qquad s \in \mathsf{R};$$

then

$$\frac{\tilde{f}(u)}{q!} = F_s\left[\frac{(u-s)_+^q}{q!}\right] = Fy, \qquad u \in \mathsf{R}.$$

Now

$$y_\lambda(s) = (-1)^\lambda \frac{(u-s)_+^{q-\lambda}}{(q-\lambda)!}, \qquad \lambda = 0, \ldots, n.$$

Hence by (83)

$$\frac{\tilde{f}^i(u)}{q!} = \int_{I_0} (-1)^n \frac{(u-s)_+^{n-1}}{(n-1)!} \phi^i(s)\, ds$$

$$+ \sum_{\lambda=0}^{n-1} k_\lambda^i (-1)^\lambda \frac{(u-a)_+^{q-\lambda}}{(q-\lambda)!}, \qquad u \in \mathsf{R},\ i = 0, \ldots, m.$$

Suppose first that $u \in I_0$. Then

$$\tilde{f}^i(u) = (-1)^n q! \int^u \frac{(u-s)^{n-1}}{(n-1)!} \phi^i(s)\, ds + \sum_{\lambda=0}^{n-1} (-1)^\lambda k_\lambda^i \frac{(u-a)_+^{q-\lambda}}{(q-\lambda)!},$$

where the lower limit of the integral is the initial point of I_0. Hence by (85)

$$\theta(u) = \tilde{b}^* \eta(u) + (-1)^n q! \sum_{i=0}^m \tilde{c}_i \int^u \frac{(u-s)^{n-1}}{(n-1)!} \phi^i(s)\, ds + 0, \qquad u \in I_0,$$

and

$$\theta_n(u) = (-1)^n q! \sum_{i=0}^m \tilde{c}_i \phi^i(u)$$

for almost all u in I_0. Furthermore the last formula is valid if $u \notin I_0$, for in that case $\theta_n(u) = 0$ by (69) and (70), and $\phi^i(u) = 0$, since I_0 contains the support of F^i. This establishes (87) and the fact that $\theta \in X$.

If $x \in X$, then

$$\int_I \theta_n x_n = \int_{I_0} \theta_n x_n = (-1)^n q! \sum_{i=0}^m \tilde{c}_i \int_{I_0} x_n \phi^i,$$

since $\theta_n(u) = 0$ for $u \notin I_0$. On the other hand, by (83) and (85)

$$\sum_{i=0}^m \tilde{c}_i F^i x = 0 + \sum_{i=0}^m \tilde{c}_i \int_{I_0} x_n \phi^i.$$

This establishes (88). Finally, for (89) put $x = \theta$ in (88).

Theorem 6. $M = L = \tilde{L}$.

Proof. The space M of splines was defined in (44) as

$$M = \left\{\theta \in X : \int_I \theta_n \zeta_n = 0 \quad \text{whenever } \zeta \in N\right\},$$

where

$$N = \bigcap_i \ker F^i = \{\zeta \in X : F\zeta = 0\}.$$

Now $L \subset M$. For if $\theta \in L$ and $\zeta \in N$ then

$$\int_I \theta_n \zeta_n = (-1)^n q!\,2 \sum_{i=0}^{m} c_i F^i \zeta = 0$$

by (88); hence $\theta \in M$.

Next $M \subset L$. For if $\xi \in M$ take $\theta \in L$ so that $F\theta = F\xi$, as is possible by Lemma 12. Then $\theta \in M$, since $L \subset M$. Hence $\theta - \xi \in M$. By construction, however, $\theta - \xi \in N$. Hence $\theta - \xi = 0$ by Lemma 7 and $\xi = \theta \in L$. Thus $M = L$. Finally, $L = \tilde{L}$ by Lemma 11.

Theorem 6, Lemma 12, and the earlier Theorem 3 permit us to characterize

$$\xi = \text{proj}_M\, x, \qquad x \in \mathscr{X}, \tag{90}$$

uniquely as follows:

$$\xi = \eta^* b + f^* c, \qquad \underset{n \times 1}{b} \in \mathrm{R}^n, \quad \underset{\bar{m} \times 1}{c} \in \mathrm{R}^{\bar{m}}, \tag{91}$$

where

$$\begin{aligned} P^* c &= 0, \\ Pb + \Phi c &= Fx. \end{aligned} \tag{92}$$

Likewise ξ may be characterized uniquely as follows:

$$\xi = \eta^* \tilde{b} + \tilde{f}^* \tilde{c}, \quad \underset{n \times 1}{\tilde{b}} \in \mathrm{R}^n, \quad \underset{\bar{m} \times 1}{\tilde{c}} \in \mathrm{R}^{\bar{m}}, \tag{93}$$

where

$$\begin{aligned} P^* \tilde{c} &= 0, \\ P\tilde{b} + \tilde{\Phi} \tilde{c} &= Fx, \end{aligned} \tag{94}$$

and

$$\underset{\bar{m} \times \bar{m}}{\tilde{\Phi}} \underset{\text{def}}{=} F\tilde{f}^* = (F^i \tilde{f}^j), \qquad i, j = 0, \ldots, m. \tag{95}$$

Thus finding ξ in the form (91) when the observations of x are known is equivalent to solving (92) for b, c, and finding ξ in the form (93) is equivalent to solving (94) for $\tilde{b}$, $\tilde{c}$.

Lemma 14.† *If* $\xi = \text{proj}_M\, x$, $x \in \mathscr{X}$, *then*

$$\xi = \eta^* B_{1,2}(Fx) + f^* B_{2,2}(Fx) = (Fx)^* B_{2,1}\eta + (Fx)^* B_{2,2} f \tag{96}$$

and

$$\int_I \xi_n^2 = (-1)^n q!\, 2(Fx)^* B_{2,2}(Fx) = \int_{I_0} \xi_n^2. \tag{97}$$

Proof. By Theorem 3 $F\xi = Fx$. Hence Lemma 12 implies (96). By Lemma 13

$$\int_I \xi_n^2 = (-1)^n q!\, 2(Fx)^* c = \int_{I_0} \xi_n^2,$$

where $c = B_{2,2}(Fx)$. This establishes (97) and completes the proof.

Corollary. *For any* $x \in X$, *the observations* Fx *and the n-th derivative* x_n *satisfy the inequalities*

$$\int_{I_0} x_n^2 \geq (-1)^n q!\, 2(Fx)^* B_{2,2}(Fx) \geq 0.$$

Equality at the left occurs iff x *equals* $(Fx)^* B_{2,2} f$ *to within a polynomial of degree* $n - 1$.

This Corollary affords a sharp lower bound on the L^2-norm of the n-th derivative of a function in terms of the observations of the function.

Proof. We may take $I = I_0$ in Lemma 10. Then

$$\int_{I_0} x_n^2 \geq \int_{I_0} \xi_n^2 \geq 0,$$

which implies the inequalities. Equality at the left occurs iff

$$x_n = \xi_n = (Fx)^* B_{2,2} f_n.$$

3.10. THE CARDINAL SPLINES

We define the *cardinal* splines

$$\beta = \underset{\bar{m} \times 1}{\beta} = (\beta^i), \qquad i = 0, \ldots, m, \tag{98}$$

as particular splines corresponding to observations that consist of unity once

† [Golomb–Weinberger 1; Secrest 1, 2, 3.]

and zero m times; that is, β^i, $i = 0, \ldots, m$, is the element of M such that

(99) $$F^j \beta^i = \delta^{j,i}, \qquad j = 0, \ldots, m,$$

where

$$\delta^{j,i} = \begin{cases} 0 & \text{if } j \neq i, \\ 1 & \text{if } j = i. \end{cases}$$

In matrix form

(100) $$F\beta^* = \underset{\bar{m} \times \bar{m}}{id} .$$

Lemma 15. *The cardinal splines are*

(101) $$\beta = B_{2,1}\eta + B_{2,2}f.$$

For any $x \in \mathscr{X}$,

(102) $$\text{proj}_M x = (Fx)^*\beta = \beta^* Fx.$$

Proof. For each $i = 0, \ldots, m$, put $Fx = (\delta^{j,i})$ in (96). Because of (99) $\xi = \beta^i$ and β^i is thus seen to be the ith element of the matrix on the right of (101). This establishes (101). For any $x \in \mathscr{X}$ (96) is precisely (102). This completes the proof.

Now (101) is the relation (13), and (102) shows that

(103) $$\Pi = \text{proj}_M,$$

where Π is the operator defined in (14).

A number of the statements of Section 3.3 now follow. Thus (16) and (17) are (43) and (44). The stated properties I of the operator Π are properties of proj_M. The interpolating property III of the splines is Theorem 3. The minimal deviation property IV is Theorem 4. The minimal quotient property V is Theorem 2. The minimal property VI is Theorem 5. The description VII of the splines is a consequence of Theorem 6. Property VIII on the enlargement of I_0 is a consequence of (64) because (64) characterizes $L = M$ in terms of I_0 alone.

It remains to discuss the geometric property II and the calculation of J.

3.11. THE BOUNDS J THAT ENTER IN THE APPRAISALS

We have seen that

(104) $$|Gx - G\xi|^2 \leq J\left[\int_I x_n^2 - \int_I \xi_n^2\right], \qquad x \in \mathscr{X},$$

with equality attainable, where $\xi = \text{proj}_M x$ is the spline approximation of x,

$G \in X^*$ is a given admissible functional, and

$$J = \|G \restriction N\|^2 = \|Q\|^2 = \sup_{0 \neq \zeta \in N} \frac{|G\zeta|^2}{\int_I \zeta_n^2}, \tag{105}$$

$$Gx - G\xi = Q(x_n) \tag{106}$$

[cf. Theorems 1, 2 and (51), (50), (53), (54). The subscript n indicates the nth derivative.]

Theorem 7.† *The bound J is given by the formula*

$$2q!(-1)^n J = G_u G_s[|u - s|^q] - \gamma^* B \gamma, \qquad q = 2n - 1, \tag{107}$$

where

$$\underset{(n+\bar{m}) \times 1}{\gamma} = \begin{pmatrix} G\eta \\ Gf \end{pmatrix}. \tag{108}$$

Recall that $B = A^{-1}$ is the matrix (10) and f, the matrix (7). Our proof is due to [Golomb-Weinberger 1]; cf. [Secrest 3] also.

Proof. Suppose first that G is not a linear combination of the elements $F^0, \ldots, F^m$ of F. Then, for some $\zeta \in N$, $G\zeta \neq 0$. (Otherwise $Gx = G\xi$ for all $x \in \mathscr{X}$, since $x - \xi \in N$; and $Gx = G\xi = A_0 x =$ a linear combination of Fx, by Lemma 9, contrary to hypothesis.) Now (105) implies that

$$J = \sup_{0 \neq \zeta \in N} \frac{|G\zeta|^2}{\int_I \zeta_n^2} = \sup_{\substack{G\zeta = 1, \\ \zeta \in N}} \frac{1}{\int_I \zeta_n^2} = \frac{1}{\inf\limits_{\substack{G\zeta = 1, \\ \zeta \in N}} \int_I \zeta_n^2}, \tag{109}$$

since we may replace ζ by $\zeta/G\zeta$. The condition $G\zeta = 1$, $\zeta \in N$, may be written

$$G\zeta = 1, F^0\zeta = \cdots = F^m\zeta = 0, \qquad \zeta \in \mathscr{X}.$$

Consider new splines relative to the augmented observations $G, F^0, \ldots, F^m$. Theorem 4 implies that

$$\inf_{\substack{G\zeta = 1, \\ \zeta \in N}} \int_I \zeta_n^2 = \int_I \rho_n^2, \tag{110}$$

where ρ is the spline based on the augmented observations

$$G\rho = 1, \qquad F^0\rho = \cdots = F^m\rho = 0. \tag{111}$$

It is permissible to consider splines relative to the augmented observations and to apply our theorems because the new matrix A is indeed invertible. Thus let a superscript $\#$ refer to the augmented observations. Then, putting

† [Golomb-Weinberger 1; Secrest 1, 2, 3; Greville 6, 9.]

(112) $$g(u) = G_s[|u - s|^q], \qquad u \in \mathrm{R},$$

we have

$$\underset{(\bar{m}+1)\times n}{P^{\#}} = \begin{pmatrix} \underset{\bar{m}\times n}{P} \\ \underset{1\times n}{(G\eta)^*} \end{pmatrix},$$

$$\underset{(\bar{m}+1)^2}{\Phi^{\#}} = \begin{pmatrix} \underset{\bar{m}\times\bar{m}}{\Phi} & \underset{\bar{m}\times 1}{Fg} \\ \underset{1\times\bar{m}}{(Gf)^*} & \underset{1\times 1}{Gg} \end{pmatrix},$$

$$A^{\#} = \begin{pmatrix} \underset{n\times n}{0} & \underset{n\times(\bar{m}+1)}{P^{\#*}} \\ \underset{(\bar{m}+1)\times n}{P^{\#}} & \underset{(\bar{m}+1)^2}{\Phi} \end{pmatrix} = \begin{pmatrix} 0 & P^* & G\eta \\ P & \Phi & Fg \\ (G\eta)^* & (Gf)^* & Gg \end{pmatrix}.$$

Note that $\Phi^{\#} = \Phi^{\#*}$ because

(113) $$(Gf)^{**} = Gf = G_u F_s[|u - s|^q] = F_u G_s[|u - s|^q] = Fg,$$

as in the proof of Lemma 3. Now P is of rank n and $\bar{m} \geq n$ by Lemma 5. Hence $P^{\#}$ is of rank n. Hence by Lemma 6 $A^{\#}$ is invertible, as was to be shown, because G, F are independent.

Thus

(114) $$J = \left(\int_I \rho_n^2\right)^{-1},$$

where ρ is the spline based on the augmented observations (111). Now, by (91) and (92)

$$\rho = \eta^* b + f^* c + gd, \qquad b \in \mathrm{R}^n,\ c \in \mathrm{R}^{\bar{m}},\ d \in \mathrm{R}^1,$$

where

(115) $$\begin{aligned} P^*c + (G\eta)\,d &= \underset{n\times 1}{0}, \\ Pb + \quad \Phi c + (Fg)\,d &= \underset{\bar{m}\times 1}{0}, \\ (G\eta)^*b + (Gf)^*c + (Gg)\,d &= \underset{1\times 1}{1}. \end{aligned}$$

We have here $n + \bar{m} + 1$ equations in the unknowns b, c, d, and the matrix of coefficients is $A^{\#}$. By Cramer's rule and the usual expansion of a bordered determinant

(116) $$\frac{1}{d} = \frac{\det A^{\#}}{\det A} = Gg - \gamma^* B \gamma,$$

where $B^{-1} = A = \begin{pmatrix} 0 & P^* \\ P & \Phi \end{pmatrix}$ and

$$\underset{(n+\bar{m})\times 1}{\gamma} = \begin{pmatrix} \underset{n\times 1}{G\eta} \\ \\ \underset{\bar{m}\times 1}{Fg} \end{pmatrix}.$$

Furthermore, by (89)

$$\int_I \rho_n^2 = (-1)^n q!\,2d.$$

This, with (114) and (116), implies (107), since

$$Gg = G_u[g(u)] = G_u G_s[|u-s|^q].$$

It remains to consider the case in which G is dependent on F. Then $Gx = \sum_{i=0}^m c_i F^i x$, $x \in X$, for suitable $c \in \mathsf{R}^{\bar{m}}$. Hence an exact approximation of Gx in terms of the observation exists. The optimal approximation $A_0 x = G \operatorname{proj}_M x$ must therefore be exact and $J = 0$. On the other hand, the determinant of $A^{\#}$ must vanish, since its last row is a linear combination of the preceding $\bar{m}$ rows. Thus the right member of (116) vanishes and (107) yields the correct value $J = 0$. This completes the proof.

Property II of Section 3.3 is a consequence of Lemma 14 and Theorems 1 and 7.

3.12. METHODS OF CALCULATION

We have now established the theory of spline approximation, relative to scalar observations and an L^2-norm of the nth derivative of the input.

In practice, one may wish to calculate

$$\xi = \operatorname{proj}_M x, \qquad x \in \mathscr{X},$$

in terms of the observations Fx. One may wish to calculate $G\xi$. If a bound on $\int_I x_n^2$ is known or assumed, one may wish to calculate J and perhaps

$$\int_I \xi_n^2 = (-1)^n q!\,2(Fx)^* B_{2,2}(Fx),$$

in order to use the appraisal (50) or (52).

For all these calculations it is sufficient to know the matrix

$$B = A^{-1} = \begin{pmatrix} B_{1,1} & B_{1,2} \\ B_{2,1} & B_{2\ 2} \end{pmatrix} = B^*,$$

by virtue of (96), (107), and (97).† Cf. [Secrest 1, 2, 3].

† For the particular functionals

$$F^i x = x(i-p), \qquad i = 0, 1, \ldots, m = 2p,$$

of Chapters 1 and 2, Table 2 gives the elements of B to 6 figures, and Table 1 gives, among other things, $B_{2,1}$ and $B_{2,2}$ to 13 figures [cf. (38), (42), and (43) of Chapter 2].

If one wishes to avoid the direct inversion of the matrix A, there are alternative procedures.

One may use the duals of the functions G, F (cf. Lemma 7). Once the duals are known, the calculation of ξ, $\int_I \xi_n^2$, and J is straightforward. The entire calculation, including that of the duals, is described in [Golomb–Weinberger 1]. (See also [de Boor-Lynch 1].)

One may use Theorem 2. One first calculates Q for any admissible approximation of G by Peano's theorem and then minimizes $\|Q\|^2$. What is involved is the minimization of a quadratic function of $\bar{m}$ variables subject to n linear constraints. Calculations of this sort are described in [Sard 1, Chapter 2].

One may calculate ξ by using Theorems 3 and 6. One solves (92) for the particular observations Fx. Equivalently, one may solve (94)† [Schoenberg 2, 16; Greville 6, 9]. See also [Schumaker 6].

One may calculate ξ by use of Theorem 4: one minimizes an integral subject to linear constraints [Holladay 1, Greville 4, Jerome and Schumaker 1]. One may calculate J from (51), using techniques of the calculus of variations.

3.13. THE MATRIX B

We have seen that the matrix

$$B = \begin{pmatrix} \underset{n \times n}{B_{1,1}} & \underset{n \times \bar{m}}{B_{1,2}} \\ \underset{\bar{m} \times n}{B_{2,1}} & \underset{\bar{m} \times \bar{m}}{B_{2,2}} \end{pmatrix} = B^* = A^{-1}$$

contains all the information that we need for spline approximation of type n based on Fx, $x \in X$.

The matrix B need not be definite. However, *the matrix $B_{2,2}$ is positive or negative semidefinite*, according as n is even or odd, by virtue of Lemmas 14 and 12.

† A number of authors have reported that equations (94) are difficult to solve. It may be that the lack of symmetry of the plus function is a source of difficulty. Equations (92) involve absolute values and are more symmetrical than (94). The direct inversion of the matrix A provides the solution of (92) for all observations. Now A is not particularly easy to invert: its diagonal consists entirely of zeros in the cases of Chapters 1 and 2 and its leading $n \times n$ minor consists entirely of zeros in all cases. Thus A is quite unlike the identity. It is our opinion, however, that the difficulties of inverting A are inherent in the problem of spline approximation and that alternative methods of calculation have comparable difficulties of their own. Among the distinct calculations of our Table 1 the direct inversion of A was best; likewise for Table 2. Cf. Section 3.14.

Since $BA = id$, it follows from (5) that

$$B_{2,2}P = \underset{\bar{m}\times n}{0}, \tag{117}$$

$$B_{2,1}P^* + B_{2,2}\Phi = \underset{\bar{m}\times\bar{m}}{id}, \tag{118}$$

and

$$B_{1,1}P^* + B_{1,2}\Phi = \underset{n\times\bar{m}}{0}. \tag{119}$$

Hence

$$B_{1,1} = B^*_{1,1} = -(P^*P)^{-1}P^*\Phi B_{2,1}, \tag{120}$$

since $B^*_{1,2} = B_{2,1}$. Thus we may calculate $B_{1,1}$ from $B_{2,1}$. Rounding errors accumulate in the matrix multiplications.

The determinant of $B_{2,2}$ *vanishes*; otherwise (117) would imply that $P = 0$.

Let us say that the *problem is symmetric* if $a = 0$,

$$F^{p+h}[s^\lambda] = (-1)^\lambda F^{p-h}[s^\lambda], \qquad h = -p, -p+1, \ldots, p, p = \frac{m}{2},$$
$$\lambda = 0, \ldots, n-1,$$

and

$$F_t^{p+h}F_s^{p+i}[|t-s|^q] = F_t^{p-i}F_s^{p-h}[|t-s|^q], \qquad h, i = -p, \ldots, p.$$

Thus the column of P of index λ is symmetric or skew-symmetric, according as λ is even or odd; and the matrix Φ is symmetric in its second diagonal (as well as its principal diagonal). In Chapters 1 and 2 the problem with centered origin is symmetric.

Lemma 16. *Suppose that the problem is symmetric. Then the matrix* $B_{2,2}$ *is symmetric in both its diagonals, the columns of* $B_{2,1}$ *are alternatively symmetric and skew-symmetric, and the odd elements of* $B_{1,1}$ (i.e., *elements for which the sum of row and column index is odd*) *vanish.*

We omit the proof. The reader may confirm the lemma by inspection of Table 2.

3.14. THE CASE OF CHAPTERS 1 AND 2

If the functionals Fx are simply values of x, that is, if

$$F^ix = x(s_i), \qquad s_i \in I_0, i = 0, \ldots, m,$$

the key functions are

$$f^i(t) = |t - s_i|^q, \qquad q = 2n-1, t \in \mathsf{R}.$$

and

$$\bar{f}^i(t) = (t - s_i)_+^q .$$

The key functions are therefore piecewise polynomials of degree q, with transitions at s_i, and with continuous $(2n - 2)$th derivative at the transitions. Theorem 6 and (64) now imply the following. *The splines M are precisely the functions on* R *which are piecewise polynomials of degree* $2n - 1$, *with transitions at* $s_0, \ldots, s_m$, *and which are of class* C_{2n-2} *on* R, *of degree* $n - 1$ *above* I_0, *and of degree* $n - 1$ *below* I_0 [Schoenberg 8, 10, 12].

Now suppose that the arguments $s_0, \ldots, s_m$ are regularly spaced and, indeed, that

$$s_i = i, \qquad i = 0, \ldots, m,$$

as in Chapters 1 and 2. Theorem 3 and the description of the splines just given lead to the following recursive formulas of the splines [Greville, as communicated to us by Schoenberg]. We use the notation of Section 2.3, in which $\beta^{m,i}$, $i = 0, \ldots, m$, are the cardinal splines; that is, $\beta^{m,i}$ is the spline that takes on the values

$$\beta^{m,i}(j) = \delta^{i,j}, \qquad j = 0, \ldots, m.$$

Then

(121)
$$\beta^{m+1,m+1}(u) = \frac{\beta^{m,m}(u-1) - k_m \beta^{m,m}(m-u)}{1 - k_m^2},$$
$$u \in \mathrm{R},\ m = n, n+1, \ldots,$$

where

$$k_m = \beta^{m,m}(-1),$$

and

(122)
$$\beta^{m+1,j} = \beta^{m,j} - \beta^{m,j}(m+1)\beta^{m+1,m+1}, \qquad j = 0, \ldots, m.$$

If the cardinals are known for $m = n$, these equations generate the cardinals for all higher m. One cannot start with the Lagrangian case $m = n - 1$, else all the splines would be pure polynomials of degree $n - 1$. One may verify that $k_{n-1} = \pm 1$, making (121) invalid in the case $m = n - 1$. We omit the proof of (121), (122).

In the present case of regular intervals the splines of type n may alternatively be calculated in terms of the solutions of a particular algebraic equation of degree n, independent of m [Weinberger 1; Nilson 1, 2, 3]. See also [Greville 10].

Splines of type n were first obtained in [Meyers-Sard 2], which is based on [Sard 3]. The name spline was not used, and the formulas for the splines were called preparatory formulas. In [Sard 1] the spline formulas are called special formulas. Optimal integration formulas were first obtained in [Sard 3]. See also [Holladay 1; Sard 1, Chapter 2].

Table 1 gives the cardinal splines in four forms, which are related in elementary fashion: any one of the forms determines the others (cf. (3), (13), (22), (23) of Chapter 2).

In preparing the present book we calculated each cardinal in three different ways: (1) inversion of the matrix A, (2) use of the duals of the functionals F [Golomb-Weinberger 1], and (3) the recursions (121), (122).

The three methods gave concurrent results to more than the number of significant figures printed in Table 1.

It appears to us that the method (1) of inversion is best and the method (3) of recursion is next best. Our opinion is based in part on the result of the following calculation.

In the case of methods (2) and (3), we determined $B_{1,1}$ from $B_{2,1}$ by (120). The three methods thus gave us three versions of $B = A^{-1}$. We then postmultiplied each approximate B by A and observed how close the product was to the identity. The residuals were least under method (1). These residuals were less than $4\ 10^{-12}$ for the values of n, m included in our tables, except for the cases $n = 8$, $m \geq 12$. In the extreme case $n = 8$, $m = 20$, the residuals in the last 21 rows were less than 10^{-15}. It is these residuals which affect the validity of Table 1. The residuals in the first eight rows were less than or equal to 8×10^{-5}. This indicates that the errors in $B_{1,1}$ are of greater influence than those in $B_{2,1}$ and $B_{2,2}$. The errors in $B_{1,1}$, however, do not affect Table 1.

In our calculations we took advantage of the power of the Control Data Corporation's 6600 computer and used 30-digit versions of the elements of B rather than the 13-digit versions printed in Table 1 or the 6-digit versions in Table 2.

By Lemma 16 the odd elements of $B_{1,1}$ vanish in the symmetric case. In our calculations the values produced for these odd elements were less than or equal to 10^{-12} except in the cases $n = 8$, $m \geq 12$. In the extreme case $n = 8$, $m = 20$, one zero of $B_{1,1}$ appeared as $2\ \ 10^{-3}$. The nonzero elements of $B_{1,1}$ in this case ranged from 4×10^5 to 4×10^{16}, numbers compared to which 2×10^{-3} is small. In Table 2 (which gives B in the symmetric case) we have printed the true values, namely, zero, of the odd elements of $B_{1,1}$. As a result, the matrix B, as printed in Table 2, times A deviates more from the identity than if we had not made the change.

3.15. GENERALIZATIONS

This chapter has considered splines relative to $\int_I x_n^2$, and arbitrary admissible scalar observations Fx, $x \in X$. One may generalize by using $\| Ux \|^2$ instead of $\int_I x_n^2$, where U is a linear homogeneous differential operator of order n

[Greville 5; de Boor-Lynch 1] or where U is a type of linear continuous operator and $\| \ \|$ is a preHilbert norm [Atteia 1 to 5].†

A further generalization allows the observations Fx to be functions or maps rather than scalars [Sard 5]. Such observations constitute more information about x than can be given by a finite number of scalars. One would therefore expect these generalized splines to be more powerful than those based on scalars. Cf. [Gordon 3].

† See also [Schoenberg 13, 17; Ahlberg–Nilson–Walsh 1 to 4; Ahlberg-Nilson 2, 3; Ahlberg 1; Karlin-Ziegler 1, 2; Schultz-Varga 1; de Boor 7; Jerome–Schumaker 2; Mangasarian-Schumaker 1; Laurent 5; Ritter 1].

APPENDIX

A FORTRAN Program for Spline Approximation

1. DESCRIPTION

Given a table,

$$\begin{array}{llll} \text{argument } x: & x_0 & x_1 \cdots x_m, \\ \text{function } y: & y_0 & y_1 \cdots y_m, \end{array}$$

of $m + 1$ values of a function at regularly spaced arguments, the program calculates the values of the spline approximation of y at any specified set

$$x_{\min}, x_{\min} + \Delta x, x_{\min} + 2\Delta x, \ldots, x_{\max},$$

of regularly spaced arguments. The program also calculates the corresponding values of the approximations of the derivative of y and the integral of y from a specified lower limit. The input to the program consists of the cardinal splines obtained from Table 1 and the above data.

2. INPUT

a. First card: N, M, NSETS (FORMAT 313)

N is the type of the spline

M is one less than the number of points

NSETS is the number of different problems to be run with these values of N and M. If only one problem is to be run NSETS is put equal to 1.

b. Next k Cards: The Spline (FORMAT 4E20.12)

These values are obtained from Table 1 by using the plusses unshifted form. The spline N, M consists of $(M + 1)(N + M + 1)$ values. These values are punched on k cards, where

$$k = (M + 1)\left\{\left[\frac{N-1}{4}\right] + \left[\frac{M}{4}\right] + 2\right\}.$$

The brackets denote the greatest integer function. (For $N = 3$, $M = 10$, there are 44 cards.) For each coefficient, first the powers are punched and then (starting a new card) the plusses are punched: for example, the spline for $N = 3$, $M = 10$ is punched as follows:

```
 1,000000000000E+00 -1,727557439443E+00  7,461136136019E-01
-1,855617415868E-02  6,445799755652E-02 -8,585580533766E-02  5,802664319795E-02
-2,589028012677E-02  1,118177345416E-02 -4,804539250392E-03  2,041612633867E-03
-8,181676359465E-04  2,621717554306E-04 -4,523208847693E-05
 0,                  2,781181076170E+00 -1,845639073726E+00
 6,445799755652E-02 -2,424617792433E-01  3,626919645784E-01 -2,874191333196E-01
 1,479559170060E-01 -6,472034285843E-02  2,784389365479E-02 -1,183331520434E-02
 4,742211409832E-03 -1,519585335325E-03  2,621717554306E-04
 0,                 -1,520635292438E+00  1,606491097776E+00
-8,585580533766E-02  3,626919645784E-01 -6,396961721261E-01  6,311583865767E-01
-4,071795741589E-01  1,996239247711E-01 -8,679202843011E-02  3,692424584582E-02
-1,479898549315E-02  4,742211409832E-03 -8,181676359465E-04
 0,                  6,686634381114E-01 -7,266900813094E-01
 5,802664319795E-02 -2,874191333196E-01  6,311583865767E-01 -8,211140599814E-01
 7,120293491467E-01 -4,419340367172E-01  2,141550839074E-01 -9,203477608577E-02
 3,692424584582E-02 -1,183331520434E-02  2,041612633867E-03
 0,                 -2,884859632717E-01  3,143762433985E-01
-2,589028012677E-02  1,479559170060E-01 -4,071795741589E-01  7,120293491467E-01
-8,570125386871E-01  7,271019733103E-01 -4,474072563718E-01  2,141550839074E-01
-8,679202843011E-02  2,784389365479E-02 -4,804539250392E-03
 0,                  1,241846569290E-01 -1,353664303831E-01
 1,118177345416E-02 -6,472034285843E-02  1,996239247711E-01 -4,419340367172E-01
 7,271019733103E-01 -8,625065839197E-01  7,271019733103E-01 -4,419340367172E-01
 1,996239247711E-01 -6,472034285843E-02  1,118177345416E-02
 0,                 -5,334160267810E-02  5,814614192849E-02
-4,804539250392E-03  2,784389365479E-02 -8,679202843011E-02  2,141550839074E-01
-4,474072563718E-01  7,271019733103E-01 -8,570125386871E-01  7,120293491467E-01
-4,071795741589E-01  1,479559170060E-01 -2,589028012677E-02
 0,                  2,266591927197E-02 -2,470753190583E-02
 2,041612633867E-03 -1,183331520434E-02  3,692424584582E-02 -9,203477608577E-02
 2,141550839074E-01 -4,419340367172E-01  7,120293491467E-01 -8,211140599814E-01
 6,311583865767E-01 -2,874191333196E-01  5,802664319795E-02
 0,                 -9,083241198335E-03  9,901408834281E-03
-8,181676359465E-04  4,742211409832E-03 -1,479898549315E-02  3,692424584582E-02
-8,679202843011E-02  1,996239247711E-01 -4,071795741589E-01  6,311583865767E-01
-6,396961721261E-01  3,626919645784E-01 -8,585580533766E-02
 0,                  2,910611908365E-03 -3,172783663796E-03
 2,621717554306E-04 -1,519585335325E-03  4,742211409832E-03 -1,183331520434E-02
 2,784389365479E-02 -6,472034285843E-02  1,479559170060E-01 -2,874191333196E-01
 3,626919645784E-01 -2,424617792433E-01  6,445799755652E-02
 0,                 -5,021633609618E-04  5,473954494387E-04
-4,523208847693E-05  2,621717554306E-04 -8,181676359465E-04  2,041612633867E-03
-4,804539250392E-03  1,118177345416E-02 -2,589028012677E-02  5,802664319795E-02
-8,585580533766E-02  6,445799755652E-02 -1,855617415868E-02
```

c. Next card: ORX,STEP,QUADX,XMIN,DELX,XMAX,IDENF (FORMAT 6F10.5, 20A1)

ORX	is x_0, the first x value of the *given* data.
STEP	is the x increment of the *given* data.
QUADX	is the lower limit of the integral to be approximated.
XMIN	is the first x value of the desired output.
DELX	is the x increment of the desired output.
XMAX	is the last x value of the desired output.
IDENF	is an identification of the problem.

d. Next j cards: Function values (FORMAT 4E20.12)

The given y values are punched in the same format as the spline. There should be exactly $M + 1$ y values punched on $j = [M/4] + 1$ cards (for $M = 10$ there are three cards).

If more than one set of data is to be run for a spline (i.e., if NSETS > 1), successive sections (c) and (d) of the input should be punched for each set, with no repetition of sections (a) and (b). The new sections (c) and (d) should follow immediately with no intervening blank cards.

If a different spline is also to be run, the entire input for it should be punched. This also should follow immediately, with no intervening blank cards.

The last card of the entire input should be blank.

If a user is not interested in the integral approximation, he may put QUADX equal to zero and disregard the last portion of the output.

3. PROGRAM

The FORTRAN program, as compiled on the CDC-6600, follows.

```
      PROGRAM PLUSS(INPUT,OUTPUT)
      DIMENSION YVALS(21),UV(21) , P(21,10), A(21,21), DV(21)
      COMMON YVALS,P,A,QP,STEP,MF,NN ,EM,N
      DIMENSION IDENF(20)
C
C           READ CONTROL CARD....     N,  M  AND NUMBER OF SETS OF DATA
 1    READ 89, N,M,NSETS
      IF (N) 500,500,2
 2    MP = M+1
      EM = M
      HEM = .5*EM
      NN1 = N+N-1
      NN = N+N
      QP = NN
      NN2 = N+N-2
      Q = NN1
      PRINT 919, N, M
 919  FORMAT (1H1, 10X, 2HN=, I3, 10X, 2HM=, I3//)
C
C
C            READ THE SPLINE
C
 31   DO 32  J = 1,MP
      READ 93,(P(J,I),I=1,N)
      PRINT93,(P(J,I),I=1,N)
      READ 93,(A(J,I) ,I=1,MP)
      PRINT93,(A(J,I) ,I=1,MP)
 32   CONTINUE
C
C
      DO 100 NS = 1,NSETS
C
C        READ CONTROL CARD
C        ORX     FIRST X-VALUE OF GIVEN DATA
C        STEP    X-INCREMENT OF GIVEN DATA
C        QUADX   LEFT LIMIT OF INTEGRAL
C        XMIN    FIRST X-VALUE OF DESIRED OUTPUT
C        DELX    X-INCREMENT OF DESIRED OUTPUT
C        XMAX    LAST X-VALUE OF DESIRED OUTPUT
C        IDENF IS PROBLEM TITLE
C
      READ 91,ORX,STEP,QUADX,XMIN,DELX,XMAX, (IDENF(IE),IE=1,20)
      PRINT 90,N,M , (IDENF(IE),IE=1,20)
C
C             READ GIVEN Y-VALUES
      READ 93, (YVALS(K),K = 1, MP)
      ORG = ORX
      DO 5 K = 1, MP
      PRINT 94, ORX, YVALS(K)
 5    ORX= ORX + STEP
      USET =(XMIN-ORG)/STEP
      DELU = DELX/STEP
      UMAX =(XMAX-ORG)/STEP
      PRINT 95
      QUADU = (QUADX-ORG)/STEP
      PRINT 96,QUADX
      U = USET
```

```
      CALL QUA (QUADU,U,SG,1)
      DO 234 JX = 1,999
      USAV = U
 56   INV = 1
      IF (U-HEM) 58,58,57
 57   USAV = U
      U = EM-U
      INV = -1
 58   CONTINUE
C
      DO 232 J = 1,MP
      UV(J) = P(J,1) + P(J,2)*U
      DV(J) = P(J,2)
      IF (N-2) 50,50,40
 40   DO 227 I = 3,N
      G = I
      E = I-1
      P2 = U**(I-2)
      P1 = U*P2
      PWR = P(J,I)
      UV(J) = UV(J) + PWR*P1
      DV(J) = DV(J) + PWR*P2*E
 227  CONTINUE
 50   DO 226 I = 1,MP
      EI = I-1
      AB = A(J,I)
      ARG = U-EI
      IF (ARG) 226,226,25
 25   A2 = ARG**NN2
      A1 = A2*ARG
      UV(J) = UV(J) + AB*A1
      DV(J) = DV(J) +      Q*AB*A2
 226  CONTINUE
      DV(J) = DV(J)/STEP
 232  CONTINUE
C
      SM = 0
      SD = 0
C
      IF (INV-1) 328,228,328
 228  DO 229 L = 1,MP
      SM = SM + UV(L)*YVALS(L)
 229  SD = SD+DV(L)*YVALS(L)
      GO TO 330
C
 328  DO 329 L = 1,MP
      I = MP-L+1
      SM = SM+ UV(I)*YVALS(L)
 329  SD = SD + DV(I)*YVALS(L)
      SD = -SD
 330  U = USAV
C
      CALL QUA (QUADU,U,SG,4)
      X =    U*STEP        + ORG
      PRINT 231,X,SM,SD,SG
 233  U = U + DELU
      IF (    U-UMAX) 234,234,100
 234  CONTINUE
```

```
 100   CONTINUE
C
       GO TO 1
C
  89   FORMAT (3I3)
  90   FORMAT (1H1,10X,2HN=,I2,5X,2HM=,I2,20X, 10HPROBLEM,.,.,20A1,
     1      ///2X, 10HINPUT DATA, 13X, 1HX,19X,1HY/)
  91   FORMAT(6F10.5,20A1)
  93   FORMAT (4E20.12)
  94   FORMAT (12X,2(2X,E20.13))
  95   FORMAT ( ////2X,20HSPLINE APPROXIMATION,//
     1 7X,1HX,13X,                 8HFUNCTION,11X,10HDERIVATIVE,10X
     1 ,8HINTEGRAL)
  96   FORMAT (55X, 4HFROM,E10.3,2X,4HTO X)
  97   FORMAT (55X, E20.13)
  99   FORMAT (I5,5X,7F10.6)
 230   FORMAT (I5,2E20.13)
 231   FORMAT ( E13.5,3(3X,E17.10))
 886   FORMAT (          I5, 3(3X,E17.10))
C
 500   CALL EXIT
       END
```

```
      SUBROUTINE QUA (QUADU,U,SG,KT)
      DIMENSION GV(21), YVALS(21), P(21,10),A(21,21)
      DIMENSION UG(50), GL(50), GR(50)
      COMMON YVALS,P,A,QP,STEP,MP,NN ,EM,N
      HEM = ,5*EM
      UG(1) = QUADU
      UG(2) = EM-QUADU
      UG(3) = HEM
      UG(4) = U
      UG(5) = EM-U
C
      DO 734  IU= KT,5
      U= UG(IU)
      DO 732  J = 1,MP
      GV(J) = P(J,1)*U  + ,5*P(J,2)*U*U
      IF(N-2) 750,750,740
 740  DO 727 I = 3,N
      G=I
      PN = U**I
 727  GV(J) = GV(J) + P(J,I)*PN/G
 750  DO 726 I = 1,MP
      EI = I-1
      ARG = U-EI
      IF(ARG) 726,726,725
 725  AN = ARG**NN
      GV(J) = GV(J) + A(J,I)*AN/QP
 726  CONTINUE
      GV(J) = GV(J)*STEP
 732  CONTINUE
C
      GL(IU)=0
      GR(IU) = 0
      DO 729 L = 1,MP
      I= MP+L+1
      GL(IU) = GL(IU) + GV(L)*YVALS(L)
 729  GR(IU) = GR(IU) + GV(I)*YVALS(L)
 734  CONTINUE
C
      U = UG(4)
      IF (QUADU-HEM) 11,11,21
 11   IF(U-HEM) 12,12,31
 12   SG= GL(4) -GL(1)
      GO TO 60
 21   IF(U-HEM) 41,41,22
 22   SG = GR(2) -GR(5)
      GO TO 60
 31   SG= GL(3) - GL(1)   +GR(3)-GR(5)
      GO TO 60
 41   SG= GR(2)-GR(3)+GL(4)-GL(3)
 60   CONTINUE
 93   FORMAT (5E16,8)
      RETURN
      END
```

4. OUTPUT

The sample output is shown for four problems.

Problem 1. Here the argument (x value) represents the day number in 1970. The function is the true distance from Earth to the planet Venus in astronomical units. (SOURCE: The American Ephemeris and Nautical Almanac, 1970.)

Problem 2. The argument represents the atomic number of the elements. The function represents the atomic weight. (SOURCE: Chemical Rubber Publishing Company Tables 1969.)

Problem 3. The argument represents an integer x. The function represents the number of primes less than x.

Problem 4. The argument is a number x. The function is $y = (x^2 + 1)^{-1}$.

N=3 M=10 PROBLEM,,, VENUS

INPUT DATA

X	Y
2,6400000000000E+02	5,2662700000000E-01
2,7400000000000E+02	4,5325900000000E-01
2,8400000000000E+02	3,8554900000000E-01
2,9400000000000E+02	3,2765300000000E-01
3,0400000000000E+02	2,8607800000000E-01
3,1400000000000E+02	2,6859600000000E-01
3,2400000000000E+02	2,8002300000000E-01
3,3400000000000E+02	3,1705300000000E-01
3,4400000000000E+02	3,7173800000000E-01
3,5400000000000E+02	4,3719200000000E-01
3,6400000000000E+02	5,0860500000000E-01

SPLINE APPROXIMATION

X	FUNCTION	DERIVATIVE	INTEGRAL FROM 0, TO X
2,59000E+02	5,6501742232E-01	-7,7904760471E-03	5,3781524187E+02
2,64000E+02	5,2662700000E-01	-7,5656928798E-03	5,4054388463E+02
2,69000E+02	4,8986177780E-01	-7,3396254273E-03	5,4308338614E+02
2,74000E+02	4,5325900000E-01	-7,0955779826E-03	5,4543943125E+02
2,79000E+02	4,1850910383E-01	-6,7895218144E-03	5,4761821565E+02
2,84000E+02	3,8554900000E-01	-6,3730717586E-03	5,4962749430E+02
2,89000E+02	3,5501226497E-01	-5,8165543752E-03	5,5147773877E+02
2,94000E+02	3,2765300000E-01	-5,0980935632E-03	5,5318290608E+02
2,99000E+02	3,0434491803E-01	-4,1917768422E-03	5,5476101335E+02
3,04000E+02	2,8607800000E-01	-3,0806108530E-03	5,5623475551E+02
3,09000E+02	2,7386966597E-01	-1,7724582275E-03	5,5763189755E+02
3,14000E+02	2,6859600000E-01	-3,2099894560E-04	5,5898503370E+02
3,19000E+02	2,7070825828E-01	1,1592148075E-03	5,6033020602E+02
3,24000E+02	2,8002300000E-01	2,5417932497E-03	5,6170415076E+02
3,29000E+02	2,9581600969E-01	3,7399727680E-03	5,6314125082E+02
3,34000E+02	3,1705300000E-01	4,7182816031E-03	5,6467138600E+02
3,39000E+02	3,4266359346E-01	5,4959715174E-03	5,6631905863E+02
3,44000E+02	3,7173800000E-01	6,1079072683E-03	5,6810378812E+02
3,49000E+02	4,0348867438E-01	6,5676611033E-03	5,7004089758E+02
3,54000E+02	4,3719200000E-01	6,8951996697E-03	5,7214191890E+02
3,59000E+02	4,7281085008E-01	7,1448718815E-03	5,7441515790E+02
3,64000E+02	5,0860500000E-01	7,3718139378E-03	5,7686697541E+02

N= 3   M=10   PROBLEM,,,ATOMIC WEIGHTS

INPUT DATA

X	Y
1,000000000000E+00	1,008000000000E+00
1,000000000000E+01	2,018300000000E+01
1,900000000000E+01	3,909600000000E+01
2,800000000000E+01	5,869000000000E+01
3,700000000000E+01	8,548000000000E+01
4,600000000000E+01	1,067000000000E+02
5,500000000000E+01	1,329100000000E+02
6,400000000000E+01	1,569000000000E+02
7,300000000000E+01	1,808800000000E+02
8,200000000000E+01	2,072100000000E+02
9,100000000000E+01	2,310000000000E+02

SPLINE APPROXIMATION

X	FUNCTION	DERIVATIVE	INTEGRAL FROM 1,000E+00 TO X
1,00000E+00	1,0080000000E+00	2,0095337907E+00	0,
4,00000E+00	7,1695153129E+00	2,0976856933E+00	1,2200044695E+01
7,00000E+00	1,3584051549E+01	2,1751651301E+00	4,3271942366E+01
1,00000E+01	2,0183000000E+01	2,2145286181E+00	9,3892425335E+01
1,30000E+01	2,6790742219E+01	2,1746440088E+00	1,6438284273E+02
1,60000E+01	3,3144396302E+01	2,0513673864E+00	2,5437905200E+02
1,90000E+01	3,9096000000E+01	1,9282199638E+00	3,6283420444E+02
2,20000E+01	4,4878890610E+01	1,9682079564E+00	4,8876798211E+02
2,50000E+01	5,1177370037E+01	2,2731642908E+00	6,3262240584E+02
2,80000E+01	5,8690000000E+01	2,7430903137E+00	7,9706708367E+02
3,10000E+01	6,7506478240E+01	3,0851353845E+00	9,8610191678E+02
3,40000E+01	7,6816403127E+01	3,0530306870E+00	1,2026108142E+03
3,70000E+01	8,5480000001E+01	2,6865230402E+00	1,4463346437E+03
4,00000E+01	9,2922980733E+01	2,3074292100E+00	1,7142277697E+03
4,30000E+01	9,9643349909E+01	2,2367995207E+00	2,0031307061E+03
4,60000E+01	1,0670000000E+02	2,5120814682E+00	2,3124358711E+03
4,90000E+01	1,1481367283E+02	2,8838734991E+00	2,6444236415E+03
5,20000E+01	1,2380296601E+02	3,0630529816E+00	3,0022131937E+03
5,50000E+01	1,3291000000E+02	2,9679041772E+00	3,3873558563E+03
5,80000E+01	1,4146795108E+02	2,7322400914E+00	3,7991021047E+03
6,10000E+01	1,4985829245E+02	2,5476148826E+00	4,2354810828E+03
6,40000E+01	1,5690000000E+02	2,5055362686E+00	4,6948997944E+03
6,70000E+01	1,6451816318E+02	2,5884154392E+00	5,1769639580E+03
7,00000E+01	1,7248657934E+02	2,7278036833E+00	5,6823657672E+03
7,30000E+01	1,8088000000E+02	2,8626290154E+00	6,2123139122E+03
7,60000E+01	1,8960932624E+02	2,9450889150E+00	6,7679856997E+03
7,90000E+01	1,9846931734E+02	2,9472338313E+00	7,3501020375E+03
8,20000E+01	2,0721000000E+02	2,8675506872E+00	7,9586810949E+03
8,50000E+01	2,1561641293E+02	2,7298927095E+00	8,5930243645E+03
8,80000E+01	2,2356267934E+02	2,5651240478E+00	9,2519168234E+03
9,10000E+01	2,3100000000E+02	2,3927643947E+00	9,9338901935E+03
9,40000E+01	2,3791894038E+02	2,2198625279E+00	1,0637398280E+04
9,70000E+01	2,4431917517E+02	2,0469606610E+00	1,1360885130E+04

N=3 M=10 PROBLEM,,,PRIMES

INPUT DATA

X	Y
0,	0,
1,000000000000E+04	1,229000000000E+03
2,000000000000E+04	2,262000000000E+03
3,000000000000E+04	3,245000000000E+03
4,000000000000E+04	4,203000000000E+03
5,000000000000E+04	5,133000000000E+03
6,000000000000E+04	6,057000000000E+03
7,000000000000E+04	6,935000000000E+03
8,000000000000E+04	7,837000000000E+03
9,000000000000E+04	8,713000000000E+03
1,000000000000E+05	9,592000000000E+03

SPLINE APPROXIMATION

X	FUNCTION	DERIVATIVE	INTEGRAL FROM 0, TO X
0,	0,	1,3565968202E-01	0,
5,00000E+03	6,4587756370E+02	1,2273605557E-01	1,6416491811E+06
1,00000E+04	1,2290000000E+03	1,1085571184E-01	6,3536869567E+06
1,50000E+04	1,7603085738E+03	1,0246461362E-01	1,3844496444E+07
2,00000E+04	2,2620000000E+03	9,8956038524E-02	2,3907498806E+07
2,50000E+04	2,7547004321E+03	9,8316804693E-02	3,6450472245E+07
3,00000E+04	3,2450000000E+03	9,7647452707E-02	5,1451082087E+07
3,50000E+04	3,7292569461E+03	9,5883877068E-02	6,8890421788E+07
4,00000E+04	4,2030000000E+03	9,3659497373E-02	8,8725765725E+07
4,50000E+04	4,6680341521E+03	9,2700326947E-02	1,1090537086E+08
5,00000E+04	5,1330000001E+03	9,3408476342E-02	1,3540636624E+08
5,50000E+04	5,6000400452E+03	9,2910327517E-02	1,6223993221E+08
6,00000E+04	6,0570000000E+03	8,9562980020E-02	1,9138965564E+08
6,50000E+04	6,4971675076E+03	8,7093548248E-02	2,2278034719E+08
7,00000E+04	6,9350000000E+03	8,8608697345E-02	2,5635747282E+08
7,50000E+04	7,3843569211E+03	9,0768752679E-02	2,9215121117E+08
8,00000E+04	7,8370000000E+03	8,9737306001E-02	3,3020682872E+08
8,50000E+04	8,2792767238E+03	8,7281903884E-02	3,7050276234E+08
9,00000E+04	8,7130000000E+03	8,6610062340E-02	4,1298484998E+08
9,50000E+04	9,1485718973E+03	8,7808990066E-02	4,5763623025E+08
1,00000E+05	9,5920000000E+03	8,9587071910E-02	5,0448393842E+08
1,05000E+05	1,0044483985E+04	9,1406521904E-02	5,5357135786E+08
1,10000E+05	1,0506065219E+04	9,3225971898E-02	6,0494394035E+08
1,15000E+05	1,0976743704E+04	9,5045421893E-02	6,5864717213E+08
1,20000E+05	1,1456519438E+04	9,6864871887E-02	7,1472653947E+08
1,25000E+05	1,1945392422E+04	9,8684321882E-02	7,7322752860E+08
1,30000E+05	1,2443362657E+04	1,0050377188E-01	8,3419562577E+08
1,35000E+05	1,2950430141E+04	1,0232322187E-01	8,9767631725E+08
1,40000E+05	1,3486594875E+04	1,0414267187E-01	9,6371508927E+08
1,45000E+05	1,3991856860E+04	1,0596212186E-01	1,0323574281E+09
1,50000E+05	1,4526216094E+04	1,0778157185E-01	1,1036488199E+09

N=3 M=10 PROBLEM...1/(1+X**2)

INPUT DATA

X	Y
-5,0000000000000E+00	3,8453845380000E-02
-4,0000000000000E+00	5,8823529410000E-02
-3,0000000000000E+00	1,0000000000000E-01
-2,0000000000000E+00	2,0000000000000E-01
-1,0000000000000E+00	5,0000000000000E-01
0,	1,0000000000000E+00
1,0000000000000E+00	5,0000000000000E-01
2,0000000000000E+00	2,0000000000000E-01
3,0000000000000E+00	1,0000000000000E-01
4,0000000000000E+00	5,8823529410000E-02
5,0000000000000E+00	3,8453845380000E-02

SPLINE APPROXIMATION

X	FUNCTION	DERIVATIVE	INTEGRAL FROM 0, TO X
-6,00000E+00	-1,5217770582E-02	7,1458777337E-02	-1,3956233876E+00
-5,50000E+00	1,6064827743E-02	5,3671615962E-02	-1,3950410575E+00
-5,00000E+00	3,8453845380E-02	3,5884454587E-02	-1,3810408233E+00
-4,50000E+00	5,2020294543E-02	1,8807415343E-02	-1,3580635579E+00
-4,00000E+00	5,8823529410E-02	1,1672085927E-02	-1,3301950726E+00
-3,50000E+00	6,9146925214E-02	3,6358295781E-02	-1,2987173132E+00
-3,00000E+00	1,0000000000E-01	8,7573998389E-02	-1,2575229209E+00
-2,50000E+00	1,5059484100E-01	1,0453939447E-01	-1,1952268954E+00
-2,00000E+00	2,0000000000E-01	1,0166035806E-01	-1,1074404450E+00
-1,50000E+00	2,8160458574E-01	2,6852500679E-01	-9,9050405844E-01
-1,00000E+00	5,0000000000E-01	6,0414241065E-01	-8,0229270015E-01
-5,00000E-01	8,2916936245E-01	6,0768338591E-01	-4,7019491610E-01
0,	1,0000000000E+00	2,2326099303E-11	0,
5,00000E-01	8,2916936245E-01	-6,0768338591E-01	4,7019491610E-01
1,00000E+00	5,0000000000E-01	-6,0414241065E-01	8,0229270015E-01
1,50000E+00	2,8160458574E-01	-2,6852500679E-01	9,9050405844E-01
2,00000E+00	2,0000000000E-01	-1,0166035806E-01	1,1074404450E+00
2,50000E+00	1,5059484100E-01	-1,0453939447E-01	1,1952268954E+00
3,00000E+00	1,0000000000E-01	-8,7573998389E-02	1,2575229209E+00
3,50000E+00	6,9146925214E-02	-3,6358295781E-02	1,2987173132E+00
4,00000E+00	5,8823529410E-02	-1,1672085927E-02	1,3301950726E+00
4,50000E+00	5,2020294543E-02	-1,8807415343E-02	1,3580635579E+00
5,00000E+00	3,8453845380E-02	-3,5884454587E-02	1,3810408233E+00
5,50000E+00	1,6064827743E-02	-5,3671615962E-02	1,3950410575E+00
6,00000E+00	-1,5217770582E-02	-7,1458777337E-02	1,3956233876E+00
6,50000E+00	-5,5393949595E-02	-8,9245938712E-02	1,3783410234E+00
7,00000E+00	-1,0446370929E-01	-1,0703310009E-01	1,3387471746E+00

Bibliography†

Books on splines are [Ahlberg, Nilson and Walsh 1; Greville 1; Schoenberg 1].

Ahlberg, J. H. (*see also* Walsh)

1. Splines in the complex plane, pp. 1–27 of [Schoenberg 1, 1969].

Ahlberg, J. H., and E. N. Nilson (*see also* Walsh)

1. Convergence properties of the spline fit, *SIAM J.*, 11 (1963), 95–104.
2. Orthogonality properties of spline functions, *J. Math. Anal. Appl.*, 11 (1965), 321–337.
3. The approximation of linear functionals, *SIAM J. Numer. Anal.*, 3 (1966), 173–182.
4. Existence of complex polynomial splines on fine uniform meshes, Brown University Report No. NR 044–309/6–26–68 (1968), Providence, R.I.
5. Polynomial splines on the real line, *J. Approximation Theory*, 3 (1970), 398–409.

Ahlberg, J. H., E. N. Nilson, and J. L. Walsh

1. *The theory of splines and their applications*, Academic Press, New York, 1967.
2. Fundamental properties of generalized splines, *Proc. Natl. Acad. Sci., U.S.*, 52 (1964), 1412–1419.
3. Convergence properties of generalized splines, *Proc. Natl. Acad. Sci., U.S.*, 54 (1965), 344–350.
4. Extremal, orthogonality, and convergence properties of multidimensional splines, *J. Math. Anal. Appl.*, 12 (1965), 27–48.
5. Best approximation and convergence properties of higher order spline approximations, *J. Math. Mech.*, 14 (1965), 231–243.
6. Complex cubic splines, *Trans. Amer. Math. Soc.*, 129 (1967), 391–413.
7. Cubic splines on the real line, *J. Approximation Theory*, 1 (1968), 5–10.
8. Properties of analytic splines, *J. Math. Anal. Appl.*, to appear.

Anselone, P. M., and P. J. Laurent

1. A general method for the construction of interpolating or smoothing spline functions, *Numer. Math.*, 12 (1968), 66–82.

† Closing date March 1, 1970.

Asker, B.

1. The spline curve, a smooth interpolating function used in numerical design of ship lines, *Nord. Tidskr. Inform. Behandlung* (BIT), 2 (1962), 76–82.

Atkinson, K. E.

1. On the order of convergence of natural cubic spline interpolation, *SIAM J. Numer. Anal.*, 5 (1968), 89–101.

Atteia, M.

1. Généralisation de la définitions et des propriétés des "spline fonctions," *Compt. Rend. Acad. Sci., Paris*, 260 (1965), 3350–3553.
2. Splines-fonctions généralisées, *Compt. Rend. Acad. Sci., Paris*, 261 (1965), 2149–2152.
3. Existence et détermination des fonctions "spline" a plusiers variables, *Compt. Rend. Acad. Sci., Paris*, Ser. A–B, 262 (1966), 575–578.
4. Étude de certains noyeaux et théorie des fonctions "spline" en analyse numérique, thèse, Université de Grenoble, 1966.
5. Fonctions "spline" définies sur un ensemble convexe, *Numer. Math.*, 12 (1968), 192–210.

Aubin, J. P.

1. Interpolation et approximation optimales et "spline functions," *J. Math. Anal. Appl.*, 24 (1968), 1–24.
2. Best approximation of linear operators in Hilbert spaces, *SIAM J. Numer. Anal.*, 5 (1968), 518–521.

Aumann, G.

1. Theoretical foundations of approximations of functions, section I/I of [Sauer and Szabó 1, 1968].

Barrodale, I., and A. Young

1. A note on numerical procedures for approximation by spline functions, *Comput. J.*, 9 (1966), 318–320.

Birkhoff, G.

1. Local spline approximations by moments, *J. Math. Mech.*, 16 (1967), 987–990.
2. Piecewise bicubic interpolation and approximation in polygons, pp. 185–221 of [Schoenberg 1, 1969].

Birkhoff, G., and C. R. de Boor

1. Error bounds for spline interpolation, *J. Math. Mech.*, 13 (1964), 827–835.
2. Piecewise polynomial interpolation and approximation, pp. 164–190 of [Garabedian 1, 1965].

Birkhoff, G., C. R. de Boor, B. Swartz, and B. Wendroff

1. Rayleigh-Ritz approximation by piecewise cubic polynomials, *SIAM J. Numer. Anal.*, 3 (1966), 188–203.

Birkhoff, G., and H. L. Garabedian

1. Smooth surface interpolation, *J. Math. and Phys.*, 39 (1960), 258–268.

Birkhoff, G., and W. J. Gordon

1. The draftsman's and related equations, *J. Approximation Theory*, 1 (1968), 199–208.

Birkhoff, G., M. H. Schultz, and R. S. Varga

1. Piecewise Hermite interpolation in one and two variables with applications to partial differential equations, *Numer. Math.*, 11 (1968), 232–256.

de Boor, C. R. (*see also* Birkhoff)

1. Bicubic spline interpolation, *J. Math. and Phys.*, 41 (1962), 212–218.
2. Best approximation properties of spline functions of odd degree, *J. Math. Mech.*, 12 (1963), 747–749.
3. The method of projections as applied to the numerical solution of two point boundary problems using cubic splines, thesis, University of Michigan, 1966.
4. On local spline approximation by moments, *J. Math. Mech.*, 17 (1967/68), 729–736.
5. On the convergence of odd-degree spline interpolation, *J. Approximation Theory*, 1 (1968), 452–463.
6. On uniform approximation by splines, *J. Approximation Theory*, 1 (1968), 219–235.
7. On the approximation by γ-polynomials, pp. 157–183 of [Schoenberg 1, 1969].

de Boor, C. R., and R. E. Lynch

1. On splines and their minimum properties, *J. Math. Mech.*, 15 (1966), 953–969.

de Boor, C., and J. R. Rice

1. Least square cubic spline approximation. I: Fixed knots II. Variable knots. Rpts. CSD TR 20 and 21, Computer Science Department, Purdue University, Lafayette, Ind., 1968.

Bulirsch, R., and H. Rutishauser

1. Interpolation and approximate integration, section H in [Sauer and Szabó 1, 1968].

Butzer, P. L., and J. Korevaar

1. (Editors) *On approximation theory. Proceedings of the conference at Oberwolfach*, 1963, Birkhäuser Verlag, Basel, 1964.

Butzer, P. L., and B. Sz.-Nagy

1. (Editors) *Abstract spaces and approximation. Proceeding of the conference at Oberwolfach*, 1968, Birkhäuser Verlag, Basel, 1969.

Carasso, C.

1. Méthodes numériques pour l'obtention de fonctions-spline, thèse, Université de Grenoble, 1966.
2. Méthode générale de construction de fonctions spline, *Rev. Française Informat., Recherche Opérationelle*, 1 (1967), No. 5, 119–127.

Carasso, C., and P. J. Laurent

1. On the numerical construction and the practical use of interpolating spline functions, pp. 86-89 of *Information Processing* 68, *Proceedings IFIP Congress, Edinburgh* 1968, *Mathematics*, *Software*, North Holland, Amsterdam, 1969.

Carlson, R. E., and C. A. Hall

1. Bicubic spline approximation of smooth functions in polygonal regions, Report WAPD-T-2177, 1968.

Cheney, E. W. (*see also* Schurer)

1. *Introduction to approximation theory*, McGraw-Hill, New York, 1966.

Cheney, E. W., and F. Schurer

1. A note on the operators arising in spline approximation, *J. Approximation Theory*, 1 (1968), 94–102.

Cherruault, Y.

1. *Approximation d'opérateurs linéaires et applications*, Dunod, Paris, 1968.

Collatz, L.

1. *The numerical treatment of differential equations*, 3rd ed., Springer Verlag, Berlin-Heidelberg, 1960.
2. *Functional analysis and numerical mathematics*, Academic Press, New York, 1966.

Curry, H. B., and Schoenberg, I. J.

1. On Pólya frequency functions IV: The fundamental spline functions and their limits, *J. Analyse Math.*, 17 (1966), 71–107.

Curtis, A. R.

1. Approximating functions of one variable by cubic splines, in [Hayes 1, 1970].

Curtis, A. R., and M. J. D. Powell

1. Error analysis for equal interval interpolation by cubic splines, Atomic Energy Research Establishment, Harwell, R5600, 1967.
2. Using cubic splines to approximate functions of one variable to prescribed accuracy, Atomic Energy Research Establishment, Harwell, R5602, 1967.

Davis, P. J.

1. *Interpolation and approximation*, Blaisdell, Waltham, Mass., 1963.

Davis, P. J., and P. Rabinowitz

1. *Numerical integration*, Blaisdell, Waltham, Mass., 1967.

Eastman, W. L. (*see* Esch)

Esch, R. E., and W. L. Eastman

1. Computational methods for best approximation, Tech. Report SEG.-TR-67-30, Sperry Rand Research Center, Sudbury, Mass., 1967.
2. Computational methods for best approximation and associated numerical analyses, Tech. Report SRRC-CR-68-18, Sperry Rand Research Center, Sudbury, Mass., 1968.
3. Computational methods for best spline function approximation. *J. Approximation Theory*, 2 (1969), 85–96.

FitzGerald, C. H., and L. Schumaker

1. A differential equation approach to interpolation at extremal points, Tech. Summary Report #731, Mathematics Research Center, Madison, Wis., 1967.

Garabedian, H. L. (*see also* Birkhoff)

1. (Editor) *Approximation of functions*, Elsevier, Amsterdam and New York, 1965.

Golomb, M.

1. Lectures on theory of approximation (mimeographed notes, 1962), Argonne National Laboratory, University of Chicago.
2. Approximation by periodic spline interpolants on uniform meshes, *J. Approximation Theory*, 1 (1968), 26–65.

3. Splines, n-widths, and optimal approximations, Mathematics Research Center, Technical Summary Report No. 784, Madison, Wis., 1968.
4. Spline interpolation near discontinuities, pp. 51–74 of [Schoenberg 1, 1969].

Golomb, M., and H. F. Weinberger
1. Optimal approximation and error bounds, pp. 117–190 of [Langer 1, 1959].

Gordon, W. J. (*see also* Birkhoff)
1. Spline-blended surface interpolation through curve networks, *J. Math. Mech.*, 18 (1969), 931–952.
2. "Blending-function" methods of bivariate and multivariate interpolation and approximation, *SIAM J. Numer. Anal.*, 1969.
3. Distributive lattices and the approximation of multivariate functions, pp. 223–277 of [Schoenberg 1, 1969].
4. Cardinal expansions of multivariate functions, *SIAM Rev.*

Gordon, W. J., and D. H. Thomas
1. Computation of cardinal functions for spline interpolation, General Motors Research Report, Warren, Mich., 1968.

Greville, T. N. E.
1. (Editor) *Theory and applications of spline functions*, Academic Press, New York, 1969.
2. The general theory of osculatory interpolation, *Trans. Actuarial Soc. Amer.*, 45 (1944), 202–265.
3. On smoothing a finite table: a matrix approach, *J. SIAM*, 5 (1957), 137–154.
4. Numerical procedures for interpolation by spline functions, *SIAM J. Numer. Anal.*, 1 (1964), 53–68.
5. Interpolation by generalized spline functions, Tech. Summary Rept. #476, Mathematics Research Center, Madison, 1964.
6. Spline functions, interpolation and numerical quadrature, pp. 156–168, Vol. II of [Ralston and Wilf 1, 1967].
7. On the normalization of the B-splines and the location of the nodes for the case of unequally spaced knots, pp. 286–290 of [Schoenberg 16, 1967].
8. Data fitting by spline functions, Trans. 12th Conf. Army Mathematicians, Report 67–1, U.S. Army Research Office, Durham, N.C., 1967, pp. 65–90 = Tech. Summary Report #893, Mathematics Research Center, Madison, Wis., 1968.
9. Introduction to spline functions, pp. 1–35 of [1 above].
10. Table for third-degree spline interpolation with equally spaced arguments, *Math. Comp.*, 24 (1970), 179–183.

Hall, C. A. (*see also* Carlson)
1. On error bounds for spline interpolation, *J. Approximation Theory*, 1 (1968), 209–218.
2. Bicubic interpolation over triangles, *J. Math. Mech.*, 19 (1969), 1–12.
3. Error bounds for periodic quintic splines, Comm. ACM, 12 (1969), 450–452.

Handscomb, D. C.
1. (Editor), *Methods of numerical approximation*, Pergamon Press, London, 1966.
2. Spline functions, pp. 163–167 of [1 above].

3. Optimal approximation of linear functionals, pp. 169–176 of [1 above].
4. Optimal approximation by means of spline functions, pp. 177–181 of [1 above].

Hayes, J. G.
1. (Editor), *Numerical approximation to functions and data*, Oxford University Press, 1970.

Hedstrom, G. W., and R. S. Varga
1. Applications of Besov spaces to spline approximation, *J. Approximation Theory*, to appear.

Hobby, C. R., and J. R. Rice
1. Approximation from a curve of functions, *Arch. Rat. Mech. Anal.*, 24 (1967), 91–106.

Holladay, J. C.
1. A smoothest curve approximation, *Math. Tables Aids Computation*, 11 (1957), 233–243.

Hulme, B. L.
1. Piecewise bicubic methods for plate bending problems, thesis, Harvard University, 1969.

Innanen, K. A.
1. An example of precise interpolation with a spline function, *J. Computational Phys.*, 1 (1967), 303–304.

Jerome, J. W., and L. L. Schumaker
1. A note on obtaining natural spline functions by the abstract approach of Atteia and Laurent, *SIAM J. Numer. Anal.*, 5 (1968), 657–663.
2. On Lg.-splines, *J. Approximation Theory*, 2 (1969), 29–49.

Jerome, J. W., and R. S. Varga
1. Generalizations of spline functions and applications to nonlinear boundary value and eigenvalue problems, pp. 103–155 of [Greville 1, 1969].

Johnson, R. S.
1. On monosplines of least deviation, *Trans. Amer. Math. Soc.*, 96 (1960), 458–477.

Joly, J. L.
1. Théorèmes de convergence des fonctions-" spline " générales d'interpolation et d'ajustement, *Compt. Rend. Acad. Sci., Paris*, Sér. A-B264 (1967), A126–A128.

Karlin, S.
1. *Total positivity and applications to analysis*, Stanford University Press, Calif., 1968.
2. Best quadrature formulas and interpolation by splines satisfying boundary conditions, pp. 447–466 of [Schoenberg 1, 1969].
3. The fundamental theorem of algebra for monosplines satisfying certain boundary conditions and applications to optimal quadrature formulas, pp. 467–484 of [Schoenberg 1, 1969].

Karlin, S., and J. M. Karon
1. A variation-diminishing generalized spline approximation method, *J. Approximation Theory*, 1 (1968), 255–268.
2. A remark on B-splines, *J. Approximation Theory*, 3 (1970), 455.

Karlin, S., and L. L. Schumaker

1. The fundamental theorem of algebra for Tchebycheffian monosplines, *J. Analyse Math.*, 20 (1967), 233–270.

Karlin, S., and W. J. Studden

1. *Tchebycheff systems: with applications in analysis and statistics*, Wiley-Interscience, New York, 1966.

Karlin, S., and Z. Ziegler

1. Chebyshevian spline functions. *SIAM J. Numer. Anal.*, 3 (1966), 514–543.
2. Chebyshevian spline functions, pp. 137–149 of [Shisha 1, 1967].

Karon, J. M. (*see* Karlin).

Kautsky, J.

1. Optimal quadrature formulae and minimal monosplines in Lq, *J. Austral. Math. Soc.*, 11 (1970), 48-56.

Korevaar, J. (*see* Butzer)

Korovkin, P. P.

1. *Linear operators and approximation theory*, translated from the Russian, Hindustan Publishing Co., Delhi, 1960.

Krylov, V. I.

1. *Approximate calculation of integrals*, Macmillan, New York, 1962.

Langer, R. E.

1. (Editor), *On numerical approximation*, University of Wisconsin Press, Madison, 1959.

Laurent, P. J. (*see also* Anselone, Carasso)

1. Représentation de données expérimentales à l'aide de fonctions-spline d'ajustement et évaluation optimale de fonctionelles linéares continues, *Apl., Mat.* 13 (1968), 154–162.
2. Théorèmes de caracterisation en approximation convexe, *Mathematica*, 10 (33) (1968), 95–111.
3. Propriétés des fonctions-spline et meilleure approximation au sens de Sard, to appear in *Mathematica*, 1969.
4. Construction of spline functions in a convex set, pp. 415–446, of [Schoenberg 1, 1969].

Lorentz, G. G.

1. *Approximation of functions*, Holt, Rinehart and Winston, New York, 1966.

Loscalzo, F. R.

1. Numerical solution of ordinary differential equations by spline functions (SPLINDIF), Tech. Summary Report #842, Mathematics Research Center, Madison, Wis., 1968.
2. On the use of spline functions for the numerical solution of ordinary differential equations, Tech. Summary Report #869, Mathematics Research Center, Madison, Wis., 1968.
3. An introduction to the applications of spline functions to initial value problems, pp. 37–64 of [Greville 1, 1969].

Loscalzo, F. R., and I. J. Schoenberg

1. On the use of spline functions for the approximation of solutions of ordinary

differential equations, Tech. Summary Report #723, Mathematics Research Center, Madison, Wis., 1967.

Loscalzo, F. R., and T. D. Talbot

1. Spline function approximations for solutions of ordinary differential equations, *SIAM J. Numer. Anal.*, 4 (1967), 433–455.
2. Spline function approximations for solutions of ordinary differential equations, *Bull. Amer. Math. Soc.*, 73 (1967), 438–442.

Lynch, R. E. (*see* de Boor)

Mangasarian, O. L., and L. L. Schumaker

1. Splines via optimal control, pp. 119–155 of [Schoenberg 1, 1969].

Marsden, M.

1. An identity for spline functions with applications to variation diminishing spline approximation, *J. Approximation Theory*, 3 (1970), 7–49.

Marsden, M., and I. J. Schoenberg

1. On variation diminishing spline approximation methods, *Mathematica*, 8 (31) 1 (1966), 61–82.

Meinardus, G.

1. *Approximation of functions: theory and numerical methods*, Springer Verlag, New York, 1967.

Meinguet, J.

1. Optimal approximation and error bounds in seminormed spaces, *Numer. Math.*, 10 (1967), 370–388.
2. Optimal approximation and interpolation in normed spaces, in [Hayes 1, 1970].

Meir, A., and A. Sharma (*see also* Sharma)

1. One sided spline approximation, *Studia Sci. Math. Hungar.*, 3 (1968), 211–218.
2. Convergence of a class of interpolatory splines, *J. Approximation Theory*, 1 (1968), 243–250.
3. On uniform approximation by cubic splines, *J. Approximation Theory*, 2 (1969), 270–274.
4. Multipoint expansions of finite differences, pp. 389–404 of [Schoenberg 1, 1969].

Meyers, L. F., and A. Sard

1. Best approximate integration formulas, *J. Math. and Phys.*, 29 (1950), 118–123.
2. Best interpolation formulas, *ibid.*, 198–206.

Morin, M.

1. Méthodes de calcul des fonctions spline dans un convexe, thèse, Grenoble, 1969, to appear.

Nikolskii, S. M.

1. *Quadrature formulas*, Moscow, 1958 (Russian).
2. Concerning estimation for approximate quadrature formulas, *Usp. Mat. Nauk*. 5, 2 (36) (1950), 165–177 (Russian).
3. Quadrature formulas, *Izv. Akad. Nauk SSSR Ser. Mat.*, 16 (1952), 181–196 (Russian).

4. *Formule de cuadratura*, translation of [1] into Romanian, Editura Technica, Bucharest, 1964.

Nilson, E. N. (*see also* Ahlberg, Walsh)

1. A simple algorithm for cubic splines, in mss., 1969.
2. A complete set of polynomial cardinal splines, in mss., 1969.
3. Cubic splines on uniform meshes, *Comm. ACM*, 13 (1970), 255–258.

Nitsche, J.

1. Orthogonal Reihenentwicklung nach linearen Spline-Funktionen, *J. Approximation Theory*, 2 (1969), 66–78.
2. Verfahren von Ritz und Spline-Interpolation bei Sturm-Liouville-Randwertproblemen, *Numer. Math.*, 13 (1969), 260–265.
3. Interpolation in Sobolevschen Funktionenräumen, *Numer. Math.*, 13 (1969), 334–343.
4. Sätze vom Jackson-Bernstein-Typ für die Approximation mit Spline-Funktionen, *Math. Z.*, 109 (1969), 97–106.
5. Eine Bemerkung zur kubischen Spline-Interpolation, pp. 367–372 of [Butzer, Sz.-Nagy 1, 1969].
6. Lineare Spline-Funktionen und die Methoden von Ritz für elliptische Randwertprobleme, *Arch. Rational Mech. Anal.*, 36 (1970), 348–355.
7. Umkehrsätze für Spline-Approximationen, to appear.

Nord, S.

1. Approximation properties of the spline fit, *Nord. Tidskr. Inform. Behandlung* (BIT), 7 (1967), 132–144.

Norlund, N. E.

1. *Vorlesungen über Differenzenrechnung*, Springer Verlag, Berlin, 1924.

Peano, G.

1. Resto nelle formule di quadratura espresso con un integrale definito, *Rend. Accad. Lincei* (5^a), 22_1 (1913), 562–569 = Opere Scelte I, Roma, 1957, pp. 410–418.
2. Residuo in formulas de quadratura, *Mathesis* (4), 4 [34] (1914), 5–10 = Opere Scelte I, pp. 419–425.

Powell, M. J. D. (*see also* Curtis)

1. On best L_2 spline approximations, *Numerische Mathematik, Differential Gleichungen, Approximationstheorie*, Birkhäuser Verlag, Basel (1968), 317–339.
2. Curve fitting by splines in one variable, in [Hayes 1, 1970].
3. The local dependence of least squares cubic splines, *SIAM J. Numer. Anal.*, 6 (1969), 398–413.
4. A comparison of spline approximations with classical interpolation methods, pp. 95–98 of *Information Processing* 68, *Proceedings IFIP Congress, Edinburgh* 1968, *Mathematics, Software*, North-Holland, Amsterdam, 1969.

Rabinowitz, P. (*see* Davis)

Ralston, A., and H. S. Wilf

1. (Editors) *Mathematical methods for digital computers*, Wiley, New York, 1967.

Reinsch, C. H.

1. Smoothing by spline functions, *Numer. Math.*, 10 (1967), 177–183.

Rice, J. R. (*see also* de Boor, Hobby)

1. *The approximation of functions*. I. *Linear theory*, 1964. II. *Nonlinear and multivariate theory*, 1969. Addison-Wesley, Reading, Mass.
2. Characterization of Chebyshev approximations by splines, *SIAM J. Numer. Anal.*, 4 (1967), 557–565.
3. On the degree of convergence of nonlinear spline approximation, pp. 349–365 of [Schoenberg 1, 1969].

Ritter, K.

1. Generalized spline interpolation and nonlinear programming, pp. 75–117 of [Schoenberg 1, 1969].
2. Two dimensional splines and their extremal properties, *Z. Angew. Math. Mech.*, 49 (1969), 597–608.
3. Two dimensional spline functions and best approximation of linear functionals, *J. Approximation Theory*, 3 (1970), 352–368.

Rivlin, T. J.

1. *An introduction to the approximation of functions*, Blaisdell, Waltham, Mass., 1969.

Rutishauser, H. (*see* Bulirsch)

Sard, A. (*see also* Meyers)

1. *Linear approximation*, American Mathematical Society, Providence, Rhode Island, 1963.
2. Integral representations of remainders, *Duke Math. J.*, 15 (1948), 333–345.
3. Best approximate integration formulas; best approximation formulas, *Am. J. Math.*, 71 (1949), 80–91.
4. Uses of Hilbert spaces in approximation, pp. 17–26 of [Garabedian 1, 1965].
5. Optimal approximation, *J. Functional Analysis*, 1 (1967), 222–224, and 2 (1968), 368–369.

Sauer, R., and I. Szabó

1. (Editors), *Mathematisch Hilfsmittel des Ingenieurs* III, Springer Verlag, Berlin-New York, 1968.

Schoenberg, I. J. (*see also* Curry, Loscalzo, Marsden)

1. (Editor), *Approximations with special emphasis on spline functions*, Academic Press, New York, 1969.
2. Contributions to the problem of approximation of equidistant data by analytic functions. Part A—On the problem of smoothing or graduation. A first class of analytic approximation formulae, *Quart. Appl. Math.*, 4 (1946), 45–99. Part B—On the problem of osculatory interpolation. A second class of analytic approximation formulae, *ibid.* 112–141.
3. On Pólya frequency functions I: The totally positive functions and their Laplace transforms, *J. Analyse Math.*, 1 (1951), 331–374.
4. On Pólya frequency functions II: Variation diminishing integral operators of the convolution type, *Acta Sci. Math. Szeged*, 12 (1950), 97–106.
5. On the zeros of generating functions of multiply positive sequences and functions, *Ann. of Math.*, 62 (1955), 447–471.
6. On variation diminishing approximation methods, pp. 249–274 of [Langer 1, 1959].

7. Spline functions, convex curves, and mechanical quadrature, *Bull. Amer. Math. Soc.*, 64 (1958), 352–357.
8. Spline interpolation and the higher derivatives, *Proc. Natl. Acad. Sci. U.S.*, 51 (1964), 24–28.
9. Spline functions and the problem of graduation, *Proc. Natl. Acad. Sci. U.S.*, 52 (1964), 947–950.
10. Spline interpolation and best quadrature formulae, *Bull. Am. Math. Soc.*, 70 (1964), 143–148.
11. On best approximations of linear operators, *Indag. Math.*, 26 (1964), 155–163.
12. On interpolation by spline functions and its minimal properties, pp. 109–129 of [Butzer and Koreevar 1, 1964].
13. On trigonometric spline interpolation, *J. Math. Mech.*, 13 (1964), 795–826.
14. On monosplines of least deviation and best quadrature formulae, I and II, *SIAM J. Numer. Anal.*, 2 (1965), 144–170; 3 (1966), 321–328.
15. On Hermite-Birkhoff interpolation, *J. Math. Anal. Appl.*, 16 (1966), 538–543.
16. On spline functions, pp. 255–291 of [Shisha 1, 1967].
17. On the Ahlberg–Nilson extension of spline interpolation. The g-splines and their optimal properties, *J. Math. Anal. Appl.*, 21 (1968), 207–231.
18. On spline interpolation at all integer points of the real axis, *Mathematica*, 10 (33) (1968), 151–170.
19. Spline interpolation and the higher derivatives, pp. 281–295 of *Abhandlungen aus Zahlentheorie und Analysis, zur Errinnerung an Edmund Landau*, VEB. Deutscher Verlag der Wissenschaften, Berlin, 1968.
20. Cardinal interpolation and spline functions, *J. Approximation Theory*, 2 (1969), 167–206.
21. Monosplines and quadrature formulae, pp. 157–207 of [Greville 1, 1969].
22. A second look at approximate quadrature formulae and spline interpolation, *Advances in Math.*, 4 (1970), 277–300.

Schoenberg, I. J., and Anne Whitney

1. On Pólya frequency functions III: the positivity of translation determinants with an application to the interpolation problem by spline curves, *Trans. Amer. Math. Soc.*, 74 (1953), 246–259.

Schultz, M. H.

1. L^2-approximation theory of even order multivariate splines, *SIAM J. Numer. Anal.*, 6 (1969), 467–475.
2. Multivariate spline functions and elliptic problems, pp. 279–347 of [Schoenberg 1, 1969].
3. L^2-multivariate approximation theory, *SIAM J. Numer. Anal.*, 6 (1969), 184–209.
4. Approximation theory of multivariate spline functions in Sobolev spaces, *SIAM J. Numer. Anal.*, 6 (1969), 570–582.
5. Multivariate L-spline interpolation, *J. Approximation Theory*, 2 (1969), 127–135.
6. Error bounds for polynomial spline interpolation, to appear.

7. Elliptic spline functions and the Rayleigh–Ritz–Galerkin method, *Math. Comp.*, 24 (1970), 65–80.

Schultz, M. H., and R. S. Varga (*see also* Birkhoff)

1. L-splines, *Numer. Math.*, 10 (1967), 345–369.

Schumaker, L. L. (*see also* FitzGerald, Jerome, Karlin, Mangasarian)

1. On some problems involving Tchebycheff systems and spline functions, thesis, Stanford University, 1965.
2. On computing best spline approximations, Tech. Summary Report #833, Mathematics Research Center, Madison, Wis., 1968.
3. On the smoothness of best spline approximations, *J. Approximation Theory*, 2 (1969), 410–418.
4. Uniform approximation by Tchebycheffian spline functions. I: Fixed knots, *J. Math. Mech.*, 18 (1968/69), 369–377; II: Free knots, *SIAM J. Numer. Anal.*, 5 (1968), 647–656.
5. Approximation by splines, pp. 65–85 of [Greville 1, 1969].
6. Some algorithms for the computation of interpolating and approximating spline functions, pp. 87–102 of [Greville 1, 1969].

Schurer, F. (*see also* Cheney)

1. A note on interpolating periodic quintic splines with equally spaced nodes, *J. Approximation Theory*, 1 (1968), 493–500.

Schurer, F. and E. W. Cheney

1. On interpolating cubic splines with equally-spaced nodes, *Indag. Math.*, 30 (1968), 517–524.

Schwartz, L.

1. *Théorie des distributions*, Nouvelle édition, Hermann, Paris, 1966.

Schweikert, D. G.

1. An interpolation curve using a spline in tension, *J. Math. and Phys.*, 45 (1966), 312–317.

Secrest, D.

1. Best approximate integration formulas and best error bounds, *Math. Comp.*, 19 (1965), 79–83.
2. Numerical integration of arbitrarily spaced data and estimation of errors, *SIAM J. Numer Anal.* 2 (1965), 52–68.
3. Error bounds for interpolation and differentiation by the use of spline functions, *SIAM J. Numer. Anal.*, 2 (1965), 440–447.

Sharma, A., and A. Meir (*see also* Meir)

1. Degree of approximation of spline interpolation, *J. Math. Mech.*, 15 (1966), 759–767.
2. Convergence of a class of interpolatory splines, pp. 373– of [Butzer, Sz.-Nagy 1, 1969].

Sonneveld, P.

1. Errors in cubic spline interpolation, *J. Engrg. Math.*, 3 (1969), 107–117.

Späth, H.

1. Exponential spline interpolation, *Computing* (*Arch. Elektron. Rechnen*), 4 (1969), 225–233.

Shisha, O.
1. (Editor) *Inequalities*, Academic Press, New York, 1967.
Starkweather, W. (*see* Theilheimer)
Stern, M. D.
1. Optimal quadrature formulae, *Comput, J.*, 9 (1967), 396–403.
2. Some round-off properties of optimal quadrature formulae, Atomic Energy Establishment, Harwell, T.P., 223, 1966.
3. Best approximate integration formulas and Sard's hypothesis, same establishment, T.P., 225, 1966.
4. Some new optimal quadrature formulae, same establishment, T.P., 227, 1966.
Studden, W. J. (*see* Karlin)
Subbotin, J. N.
1. Piecewise-polynomial interpolation (Russian), *Mat. Zametki*, 1 (1967), 63–70.
Swartz, B. (*see also* Birkhoff)
1. $O(h^{2n+2-l})$ bounds on some spline interpolation errors, *Bull. Amer. Math. Soc.*, 74 (1968), 1072–1078.
Szabó, I. (*see* Sauer)
Sz.-Nagy, B. (*see* Butzer)
Talbot, T. D. (*see* Loscalzo)
Theilheimer, F., and W. Starkweather
1. The fairing of ship lines on a high-speed computer, *Math. of Comp.*, 15 (1961), 338–355.
Thomas, D. H. (*see* Gordon)
Todd, J.
1. (Editor) *Survey of numerical analysis*, McGraw-Hill, New York, 1962.
Varga, R. S. (*see also* Birkhoff, Hedstrom, Jerome, Schultz)
1. Error bounds for spline interpolation, pp. 367–388 of [Schoenberg 1, 1969].
Walsh, J. L., J. H. Ahlberg, and E. N. Nilson (*see also* Ahlberg)
1. Best approximation properties of the spline fit, *J. Math. Mech.*, 11 (1962), 225–234.
Weinberger, H. F. (*see also* Golomb)
1. Optimal approximation for functions prescribed at equally spaced points, *J. Res. Nat. Bur. Standards*, Sect. B., 65 (1961), 99–104.
Wendroff, B. (*see* Birkhoff)
Wilf, H. S. (*see* Ralston)
Whitney, A. (*see* Schoenberg)
Young, A. (*see* Barrodale)
Ziegler, Z. (*see also* Karlin)
1. One-sided L_1-approximation by splines of an arbitrary degree, pp. 405–413 of [Schoenberg 1, 1969].
2. On the best order of L_p-approximation by splines, to appear.

Table 1. The cardinal splines

$n = 2, 3, 4, 5$, and 8.
$m = n, n + 1, n + 2, \ldots, 20$.
$p = m/2$.
$q = 2n - 1$.

CARDINAL SPLINES N=2

```
                 N= 2      M= 2                          P=  1.0      Q= 3

I= 0     H= -1.0
0          1.000000000000D+00   2.500000000000D-01    -1.00  2.250000000000D+00
1         -1.250000000000D+00  -5.000000000000D-01     -.75  1.937500000000D+00
                                                       -.50  1.625000000000D+00
0  -1.0    2.500000000000D-01   1.250000000000D-01     -.25  1.312500000000D+00
1   0.0   -5.000000000000D-01  -2.500000000000D-01      .25  6.914062500000D-01
2   1.0    2.500000000000D-01   1.250000000000D-01      .50  4.062500000000D-01
                                                        .75  1.679687500000D-01
       0  -2.500000000000D-01  -2.500000000000D-01     1.25 -8.203125000000D-02
       1  -1.250000000000D+00  -5.000000000000D-01     1.50 -9.375000000000D-02


I= 1     H=  0.0
0          0.                   1.500000000000D+00    -1.00 -1.500000000000D+00
1          1.500000000000D+00   0.                     -.75 -1.125000000000D+00
                                                       -.50 -7.500000000000D-01
0  -1.0   -5.000000000000D-01  -2.500000000000D-01     -.25 -3.750000000000D-01
1   0.0    1.000000000000D+00   5.000000000000D-01      .25  3.671875000000D-01
2   1.0   -5.000000000000D-01  -2.500000000000D-01      .50  6.875000000000D-01
                                                        .75  9.140625000000D-01
       0   1.500000000000D+00   1.500000000000D+00     1.25  9.140625000000D-01
       1   1.500000000000D+00   0.                     1.50  6.875000000000D-01


I= 2     H=  1.0
0          0.                  -7.500000000000D-01    -1.00  2.500000000000D-01
1         -2.500000000000D-01   5.000000000000D-01     -.75  1.875000000000D-01
                                                       -.50  1.250000000000D-01
0  -1.0    2.500000000000D-01   1.250000000000D-01     -.25  6.250000000000D-02
1   0.0   -5.000000000000D-01  -2.500000000000D-01      .25 -5.859375000000D-02
2   1.0    2.500000000000D-01   1.250000000000D-01      .50 -9.375000000000D-02
                                                        .75 -8.203125000000D-02
       0  -2.500000000000D-01  -2.500000000000D-01     1.25  1.679687500000D-01
       1  -2.500000000000D-01   5.000000000000D-01     1.50  4.062500000000D-01




                 N= 2      M= 3                          P=  1.5      Q= 3

I= 0     H= -1.5
0          1.000000000000D+00   6.000000000000D-01    -1.00  2.266666666667D+00
```

CARDINAL SPLINES N=2

```
1           -1.266666666667D+00  -6.666666666667D-01     -.75  1.950000000000D+00
                                                         -.50  1.633333333333D+00
0   -1.5     2.666666666667D-01   1.333333333333D-01     -.25  1.316666666667D+00
1    -.5    -6.000000000000D-01  -3.000000000000D-01      .25  6.875000000000D-01
2     .5     4.000000000000D-01   2.000000000000D-01      .50  4.000000000000D-01
3    1.5    -6.666666666667D-02  -3.333333333333D-02      .75  1.625000000000D-01
                                                         1.25 -7.187500000000D-02
       0    -9.000000000000D-01  -4.000000000000D-01     1.50 -7.500000000000D-02
       1    -1.266666666667D+00  -6.666666666667D-01     1.75 -4.062500000000D-02


I= 1     H=  -.5
0             0.                 -6.000000000000D-01    -1.00 -1.600000000000D+00
1             1.600000000000D+00  1.000000000000D+00     -.75 -1.200000000000D+00
                                                         -.50 -8.000000000000D-01
0   -1.5    -6.000000000000D-01  -3.000000000000D-01     -.25 -4.000000000000D-01
1    -.5     1.600000000000D+00   8.000000000000D-01      .25  3.906250000000D-01
2     .5    -1.400000000000D+00  -7.000000000000D-01      .50  7.250000000000D-01
3    1.5     4.000000000000D-01   2.000000000000D-01      .75  9.468750000000D-01
                                                         1.25  8.531250000000D-01
       0     2.400000000000D+00   9.000000000000D-01     1.50  5.750000000000D-01
       1     1.600000000000D+00   1.000000000000D+00     1.75  2.593750000000D-01


I= 2     H=   .5
0             0.                  2.400000000000D+00    -1.00  4.000000000000D-01
1            -4.000000000000D-01 -1.000000000000D+00     -.75  3.000000000000D-01
                                                         -.50  2.000000000000D-01
0   -1.5     4.000000000000D-01   2.000000000000D-01     -.25  1.000000000000D-01
1    -.5    -1.400000000000D+00  -7.000000000000D-01      .25 -9.375000000000D-02
2     .5     1.600000000000D+00   8.000000000000D-01      .50 -1.500000000000D-01
3    1.5    -6.000000000000D-01  -3.000000000000D-01      .75 -1.312500000000D-01
                                                         1.25  2.593750000000D-01
       0    -6.000000000000D-01   9.000000000000D-01     1.50  5.750000000000D-01
       1    -4.000000000000D-01  -1.000000000000D+00     1.75  8.531250000000D-01


I= 3     H=  1.5
0             0.                 -1.400000000000D+00    -1.00 -6.666666666667D-02
1             6.666666666667D-02  6.666666666667D-01     -.75 -5.000000000000D-02
                                                         -.50 -3.333333333333D-02
0   -1.5    -6.666666666667D-02  -3.333333333333D-02     -.25 -1.666666666667D-02
1    -.5     4.000000000000D-01   2.000000000000D-01      .25  1.562500000000D-02
2     .5    -6.000000000000D-01  -3.000000000000D-01      .50  2.500000000000D-02
3    1.5     2.666666666667D-01   1.333333333333D-01      .75  2.187500000000D-02
                                                         1.25 -4.062500000000D-02
       0     1.000000000000D-01  -4.000000000000D-01     1.50 -7.500000000000D-02
       1     6.666666666667D-02   6.666666666667D-01     1.75 -7.187500000000D-02
```

CARDINAL SPLINES N=2

```
                N= 2      M= 4                         P=  2.0      Q= 3

I= 0     H= -2.0
0          1.000000000000D+00   4.642857142857D-01   -1.00  2.267857142857D+00
1         -1.267857142857D+00  -6.250000000000D-01    -.75  1.950892857143D+00
                                                      -.50  1.633928571429D+00
0   -2.0   2.678571428571D-01   1.339285714286D-01    -.25  1.316964285714D+00
1   -1.0  -6.071428571429D-01  -3.035714285714D-01     .25  6.872209821429D-01
2    0.0   4.285714285714D-01   2.142857142857D-01     .50  3.995535714286D-01
3    1.0  -1.071428571429D-01  -5.357142857143D-02     .75  1.621093750000D-01
4    2.0   1.785714285714D-02   8.928571428571D-03    1.25 -7.114955357143D-02
                                                      1.50 -7.366071428571D-02
       0  -1.535714285714D+00  -7.857142857143D-01    1.75 -3.934151785714D-02
       1  -1.267857142857D+00  -6.250000000000D-01    2.25  1.925223214286D-02


I= 1     H= -1.0
0          0.                   2.142857142857D-01   -1.00 -1.607142857143D+00
1          1.607142857143D+00   7.500000000000D-01    -.75 -1.205357142857D+00
                                                      -.50 -8.035714285714D-01
0   -2.0  -6.071428571429D-01  -3.035714285714D-01    -.25 -4.017857142857D-01
1   -1.0   1.642857142857D+00   8.214285714286D-01     .25  3.922991071429D-01
2    0.0  -1.571428571429D+00  -7.857142857143D-01     .50  7.276785714286D-01
3    1.0   6.428571428571D-01   3.214285714286D-01     .75  9.492187500000D-01
4    2.0  -1.071428571429D-01  -5.357142857143D-02    1.25  8.487723214286D-01
                                                      1.50  5.669642857143D-01
       0   3.214285714286D+00   1.714285714286D+00    1.75  2.516741071429D-01
       1   1.607142857143D+00   7.500000000000D-01    2.25 -1.155133928571D-01


I= 2     H=  0.0
0          0.                  -8.571428571429D-01   -1.00  4.285714285714D-01
1         -4.285714285714D-01   0.                    -.75  3.214285714286D-01
                                                      -.50  2.142857142857D-01
0   -2.0   4.285714285714D-01   2.142857142857D-01    -.25  1.071428571429D-01
1   -1.0  -1.571428571429D+00  -7.857142857143D-01     .25 -1.004464285714D-01
2    0.0   2.285714285714D+00   1.142857142857D+00     .50 -1.607142857143D-01
3    1.0  -1.571428571429D+00  -7.857142857143D-01     .75 -1.406250000000D-01
4    2.0   4.285714285714D-01   2.142857142857D-01    1.25  2.767857142857D-01
                                                      1.50  6.071428571429D-01
       0  -8.571428571429D-01  -8.571428571429D-01    1.75  8.839285714286D-01
       1  -4.285714285714D-01   0.                    2.25  8.839285714286D-01


I= 3     H=  1.0
0          0.                   3.214285714286D+00   -1.00 -1.071428571429D-01
1          1.071428571429D-01  -7.500000000000D-01    -.75 -8.035714285714D-02
                                                      -.50 -5.357142857143D-02
0   -2.0  -1.071428571429D-01  -5.357142857143D-02    -.25 -2.678571428571D-02
1   -1.0   6.428571428571D-01   3.214285714286D-01     .25  2.511160714286D-02
```

CARDINAL SPLINES N=2

```
2    0.0  -1.571428571429D+00  -7.857142857143D-01     .50   4.017857142857D-02
3    1.0   1.642857142857D+00   8.214285714286D-01     .75   3.515625000000D-02
4    2.0  -6.071428571429D-01  -3.035714285714D-01    1.25  -6.529017857143D-02
                                                      1.50  -1.205357142857D-01
       0   2.142857142857D-01   1.714285714286D+00    1.75  -1.155133928571D-01
       1   1.071428571429D-01  -7.500000000000D-01    2.25   2.516741071429D-01


I= 4     H=  2.0
0          0.                  -2.035714285714D+00   -1.00   1.785714285714D-02
1         -1.785714285714D-02   6.250000000000D-01    -.75   1.339285714286D-02
                                                      -.50   8.928571428571D-03
0   -2.0   1.785714285714D-02   8.928571428571D-03    -.25   4.464285714286D-03
1   -1.0  -1.071428571429D-01  -5.357142857143D-02     .25  -4.185267857143D-03
2    0.0   4.285714285714D-01   2.142857142857D-01     .50  -6.696428571429D-03
3    1.0  -6.071428571429D-01  -3.035714285714D-01     .75  -5.859375000000D-03
4    2.0   2.678571428571D-01   1.339285714286D-01    1.25   1.088169642857D-02
                                                      1.50   2.008928571429D-02
       0  -3.571428571429D-02  -7.857142857143D-01    1.75   1.925223214286D-02
       1  -1.785714285714D-02   6.250000000000D-01    2.25  -3.934151785714D-02




                  N= 2      M= 5                         P=  2.5      Q= 3

I= 0     H= -2.5
0          1.000000000000D+00   5.119617224880D-01   -1.00   2.267942583732D+00
1         -1.267942583732D+00  -6.363636363636D-01    -.75   1.950956937799D+00
                                                      -.50   1.633971291866D+00
0   -2.5   2.679425837321D-01   1.339712918660D-01    -.25   1.316985645933D+00
1   -1.5  -6.076555023923D-01  -3.038277511962D-01     .25   6.872009569378D-01
2    -.5   4.306220095694D-01   2.153110047847D-01     .50   3.995215311005D-01
3     .5  -1.148325358852D-01  -5.741626794258D-02     .75   1.620813397129D-01
4    1.5   2.870813397129D-02   1.435406698565D-02    1.25  -7.109748803828D-02
5    2.5  -4.784688995215D-03  -2.392344497608D-03    1.50  -7.356459330144D-02
                                                      1.75  -3.924940191388D-02
       0  -2.169856459330D+00  -1.078947368421D+00    2.25   1.906399521531D-02
       1  -1.267942583732D+00  -6.363636363636D-01    2.50   1.973684210526D-02


I= 1     H= -1.5
0          0.                  -7.177033492823D-02   -1.00  -1.607655502392D+00
1          1.607655502392D+00   8.181818181818D-01    -.75  -1.205741626794D+00
                                                      -.50  -8.038277511962D-01
0   -2.5  -6.076555023923D-01  -3.038277511962D-01    -.25  -4.019138755981D-01
1   -1.5   1.645933014354D+00   8.229665071770D-01     .25   3.924192583732D-01
2    -.5  -1.583732057416D+00  -7.918660287081D-01     .50   7.278708133971D-01
3     .5   6.889952153110D-01   3.444976076555D-01     .75   9.493869617225D-01
4    1.5  -1.722488038278D-01  -8.612440191388D-02    1.25   8.484599282297D-01
5    2.5   2.870813397129D-02   1.435406698565D-02    1.50   5.663875598086D-01
                                                      1.75   2.511214114833D-01
```

```
          0   4.019138755981D+00    1.973684210526D+00     2.25 -1.143839712919D-01
          1   1.607655502392D+00    8.181818181818D-01     2.50 -1.184210526316D-01


I= 2    H=  -.5
0             0.                    2.870813397129D-01    -1.00  4.306220095694D-01
1            -4.306220095694D-01   -2.727272727273D-01     -.75  3.229665071770D-01
                                                           -.50  2.153110047847D-01
0   -2.5      4.306220095694D-01    2.153110047847D-01     -.25  1.076555023923D-01
1   -1.5     -1.583732057416D+00   -7.918660287081D-01      .25 -1.009270334928D-01
2    -.5      2.334928229665D+00    1.167464114833D+00      .50 -1.614832535885D-01
3     .5     -1.755980861244D+00   -8.779904306220D-01      .75 -1.412978468900D-01
4    1.5      6.889952153110D-01    3.444976076555D-01     1.25  2.780352870813D-01
5    2.5     -1.148325358852D-01   -5.741626794258D-02     1.50  6.094497607656D-01
                                                           1.75  8.861393540670D-01
          0  -1.076555023923D+00   -3.947368421053D-01     2.25  8.794108851675D-01
          1  -4.306220095694D-01   -2.727272727273D-01     2.50  5.986842105263D-01


I= 3    H=   .5
0             0.                   -1.076555023923D+00    -1.00 -1.148325358852D-01
1             1.148325358852D-01    2.727272727273D-01     -.75 -8.612440191388D-02
                                                           -.50 -5.741626794258D-02
0   -2.5     -1.148325358852D-01   -5.741626794258D-02     -.25 -2.870813397129D-02
1   -1.5      6.889952153110D-01    3.444976076555D-01      .25  2.691387559809D-02
2    -.5     -1.755980861244D+00   -8.779904306220D-01      .50  4.306220095694D-02
3     .5      2.334928229665D+00    1.167464114833D+00      .75  3.767942583732D-02
4    1.5     -1.583732057416D+00   -7.918660287081D-01     1.25 -6.997607655502D-02
5    2.5      4.306220095694D-01    2.153110047847D-01     1.50 -1.291866028708D-01
                                                           1.75 -1.238038277512D-01
          0   2.870813397129D-01   -3.947368421053D-01     2.25  2.686154306220D-01
          1   1.148325358852D-01    2.727272727273D-01     2.50  5.986842105263D-01


I= 4    H=  1.5
0             0.                    4.019138755981D+00    -1.00  2.870813397129D-02
1            -2.870813397129D-02   -8.181818181818D-01     -.75  2.153110047847D-02
                                                           -.50  1.435406698565D-02
0   -2.5      2.870813397129D-02    1.435406698565D-02     -.25  7.177033492823D-03
1   -1.5     -1.722488038278D-01   -8.612440191388D-02      .25 -6.728468899522D-03
2    -.5      6.889952153110D-01    3.444976076555D-01      .50 -1.076555023923D-02
3     .5     -1.583732057416D+00   -7.918660287081D-01      .75 -9.419856459330D-03
4    1.5      1.645933014354D+00    8.229665071770D-01     1.25  1.749401913876D-02
5    2.5     -6.076555023923D-01   -3.038277511962D-01     1.50  3.229665071770D-02
                                                           1.75  3.095095693780D-02
          0  -7.177033492823D-02    1.973684210526D+00     2.25 -6.324760765550D-02
          1  -2.870813397129D-02   -8.181818181818D-01     2.50 -1.184210526316D-01


I= 5    H=  2.5
0             0.                   -2.669856459330D+00    -1.00 -4.784688995215D-03
1             4.784688995215D-03    6.363636363636D-01     -.75 -3.588516746411D-03
                                                           -.50 -2.392344497608D-03
0   -2.5     -4.784688995215D-03   -2.392344497608D-03     -.25 -1.196172248804D-03
1   -1.5      2.870813397129D-02    1.435406698565D-02      .25  1.121411483254D-03
2    -.5     -1.148325358852D-01   -5.741626794258D-02      .50  1.794258373206D-03
3     .5      4.306220095694D-01    2.153110047847D-01      .75  1.569976076555D-03
```

```
4   1,5  -6,076555023923D-01  -3,038277511962D-01     1.25 -2,915669856459D-03
5   2,5   2,679425837321D-01   1.339712918660D-01     1.50 -5,382775119617D-03
                                                      1.75 -5,158492822967D-03
      0   1,196172248804D-02  -1,078947368421D+00     2,25  1,054126794258D-02
      1   4,784688995215D-03   6,363636363636D-01     2,50  1,973684210526D-02
```

```
              N= 2        M= 6                       P=   3.0        Q= 3

I= 0      H= -3.0
0             1,000000000000D+00   4.961538461538D-01    -1.00  2,267948717949D+00
1            -1,267948717949D+00  -6,333333333333D-01     -.75  1,950961538462D+00
                                                          -.50  1,633974358974D+00
0   -3,0      2,679487179487D-01   1.339743589744D-01     -.25  1,316987179487D+00
1   -2,0     -6,076923076923D-01  -3,038461538462D-01      .25  6,871995192308D-01
2   -1,0      4,307692307692D-01   2,153846153846D-01      .50  3,995192307692D-01
3    0,0     -1,153846153846D-01  -5,769230769231D-02      .75  1,620793269231D-01
4    1,0      3,076923076923D-02   1.538461538462D-02     1.25 -7,109375000000D-02
5    2,0     -7,692307692308D-03  -3,846153846154D-03     1.50 -7,355769230769D-02
6    3,0      1,282051282051D-03   6,410256410256D-04     1,75 -3,924278846154D-02
                                                          2,25  1,905048076923D-02
          0  -2,803846153846D+00  -1.403846153846D+00     2.50  1,971153846154D-02
          1  -1,267948717949D+00  -6,333333333333D-01     2.75  1,051682692308D-02


I= 1      H= -2.0
0             0,                   2,307692307692D-02    -1.00 -1,607692307692D+00
1             1,607692307692D+00   8,000000000000D-01     -.75 -1,205769230769D+00
                                                          -.50 -8,038461538462D-01
0   -3,0     -6,076923076923D-01  -3,038461538462D-01     -.25 -4,019230769231D-01
1   -2,0      1,646153846154D+00   8,230769230769D-01      .25  3,924278846154D-01
2   -1,0     -1,584615384615D+00  -7,923076923077D-01      .50  7,278846153846D-01
3    0,0      6,923076923077D-01   3,461538461538D-01      .75  9,493990384615D-01
4    1,0     -1,846153846154D-01  -9,230769230769D-02     1.25  8,484375000000D-01
5    2,0      4.615384615385D-02   2,307692307692D-02     1.50  5,663461538462D-01
6    3,0     -7,692307692308D-03  -3,846153846154D-03     1.75  2,510817307692D-01
                                                          2,25 -1,143028846154D-01
          0   4.823076923077D+00   2.423076923077D+00     2.50 -1,182692307692D-01
          1   1.607692307692D+00   8,000000000000D-01     2.75 -6,310096153846D-02


I= 2      H= -1.0
0             0,                  -9.230769230769D-02    -1,00  4,307692307692D-01
1            -4,307692307692D-01  -2,000000000000D-01     -.75  3,230769230769D-01
                                                          -,50  2,153846153846D-01
0   -3,0      4,307692307692D-01   2,153846153846D-01     -.25  1,076923076923D-01
1   -2,0     -1,584615384615D+00  -7,923076923077D-01      ,25 -1,009615384615D-01
2   -1,0      2,338461538462D+00   1,169230769231D+00      ,50 -1,615384615385D-01
3    0.0     -1,769230769231D+00  -8,846153846154D-01      .75 -1,413461538462D-01
4    1.0      7,384615384615D-01   3,692307692308D-01     1.25  2,781250000000D-01
```

CARDINAL SPLINES N=2

```
5   2.0  -1.846153846154D-01  -9.230769230769D-02    1.50  6.096153846154D-01
6   3.0   3.076923076923D-02   1.538461538462D-02    1.75  8.862980769231D-01
                                                     2.25  8.790865384615D-01
       0 -1.292307692308D+00  -6.923076923077D-01    2.50  5.980769230769D-01
       1 -4.307692307692D-01  -2.000000000000D-01    2.75  2.680288461538D-01


I= 3     H=  0.0
0          0.                  3.461538461538D-01   -1.00 -1.153846153846D-01
1          1.153846153846D-01  0.                    -.75 -8.653846153846D-02
                                                     -.50 -5.769230769231D-02
0  -3.0  -1.153846153846D-01  -5.769230769231D-02    -.25 -2.884615384615D-02
1  -2.0   6.923076923077D-01   3.461538461538D-01     .25  2.704326923077D-02
2  -1.0  -1.769230769231D+00  -8.846153846154D-01     .50  4.326923076923D-02
3   0.0   2.384615384615D+00   1.192307692308D+00     .75  3.786057692308D-02
4   1.0  -1.769230769231D+00  -8.846153846154D-01    1.25 -7.031250000000D-02
5   2.0   6.923076923077D-01   3.461538461538D-01    1.50 -1.298076923077D-01
6   3.0  -1.153846153846D-01  -5.769230769231D-02    1.75 -1.243990384615D-01
                                                     2.25  2.698317307692D-01
       0  3.461538461538D-01   3.461538461538D-01    2.50  6.009615384615D-01
       1  1.153846153846D-01   0.                    2.75  8.816105769231D-01


I= 4     H=  1.0
0          0.                 -1.292307692308D+00   -1.00  3.076923076923D-02
1         -3.076923076923D-02  2.000000000000D-01    -.75  2.307692307692D-02
                                                     -.50  1.538461538462D-02
0  -3.0   3.076923076923D-02   1.538461538462D-02    -.25  7.692307692308D-03
1  -2.0  -1.846153846154D-01  -9.230769230769D-02     .25 -7.211538461538D-03
2  -1.0   7.384615384615D-01   3.692307692308D-01     .50 -1.153846153846D-02
3   0.0  -1.769230769231D+00  -8.846153846154D-01     .75 -1.009615384615D-02
4   1.0   2.338461538462D+00   1.169230769231D+00    1.25  1.875000000000D-02
5   2.0  -1.584615384615D+00  -7.923076923077D-01    1.50  3.461538461538D-02
6   3.0   4.307692307692D-01   2.153846153846D-01    1.75  3.317307692308D-02
                                                     2.25 -6.778846153846D-02
       0 -9.230769230769D-02  -6.923076923077D-01    2.50 -1.269230769231D-01
       1 -3.076923076923D-02   2.000000000000D-01    2.75 -1.225961538462D-01


I= 5     H=  2.0
0          0.                  4.823076923077D+00   -1.00 -7.692307692308D-03
1          7.692307692308D-03 -8.000000000000D-01    -.75 -5.769230769231D-03
                                                     -.50 -3.846153846154D-03
0  -3.0  -7.692307692308D-03  -3.846153846154D-03    -.25 -1.923076923077D-03
1  -2.0   4.615384615385D-02   2.307692307692D-02     .25  1.802884615385D-03
2  -1.0  -1.846153846154D-01  -9.230769230769D-02     .50  2.884615384615D-03
3   0.0   6.923076923077D-01   3.461538461538D-01     .75  2.524038461538D-03
4   1.0  -1.584615384615D+00  -7.923076923077D-01    1.25 -4.687500000000D-03
5   2.0   1.646153846154D+00   8.230769230769D-01    1.50 -8.653846153846D-03
6   3.0  -6.076923076923D-01  -3.038461538462D-01    1.75 -8.293269230769D-03
                                                     2.25  1.694711538462D-02
       0  2.307692307692D-02   2.423076923077D+00    2.50  3.173076923077D-02
       1  7.692307692308D-03  -8.000000000000D-01    2.75  3.064903846154D-02


I= 6     H=  3.0
0          0.                 -3.303846153846D+00   -1.00  1.282051282051D-03
```

```
                          CARDINAL SPLINES  N=2

1          -1.282051282051D-03   6.333333333333D-01      -.75  9.615384615385D-04
                                                         -.50  6.410256410256D-04
0  -3.0     1.282051282051D-03   6.410256410256D-04      -.25  3.205128205128D-04
1  -2.0    -7.692307692308D-03  -3.846153846154D-03       .25 -3.004807692308D-04
2  -1.0     3.076923076923D-02   1.538461538462D-02       .50 -4.807692307692D-04
3   0.0    -1.153846153846D-01  -5.769230769231D-02       .75 -4.206730769231D-04
4   1.0     4.307692307692D-01   2.153846153846D-01      1.25  7.812500000000D-04
5   2.0    -6.076923076923D-01  -3.038461538462D-01      1.50  1.442307692308D-03
6   3.0     2.679487179487D-01   1.339743589744D-01      1.75  1.382211538462D-03
                                                         2.25 -2.824519230769D-03
        0  -3.846153846154D-03  -1.403846153846D+00      2.50 -5.288461538462D-03
        1  -1.282051282051D-03   6.333333333333D-01      2.75 -5.108173076923D-03




                N= 2       M= 7                       P=  3.5      Q= 3

I= 0      H= -3.5
0           1.000000000000D+00   5.012023359670D-01     -1.00  2.267949158365D+00
1          -1.267949158365D+00  -6.341463414634D-01      -.75  1.950961868774D+00
                                                         -.50  1.633974579182D+00
0  -3.5     2.679491583648D-01   1.339745791824D-01      -.25  1.316987289591D+00
1  -2.5    -6.076949501889D-01  -3.038474750945D-01       .25  6.871994160082D-01
2  -1.5     4.307798007558D-01   2.153899003779D-01       .50  3.995190656132D-01
3   -.5    -1.154242528341D-01  -5.771212641704D-02       .75  1.620791824115D-01
4    .5     3.091721058056D-02   1.545860529028D-02      1.25 -7.109348162144D-02
5   1.5    -8.244589488148D-03  -4.122294744074D-03      1.50 -7.355719683957D-02
6   2.5     2.061147372037D-03   1.030573686019D-03      1.75 -3.924231363793D-02
7   3.5    -3.435245620062D-04  -1.717622810031D-04      2.25  1.904951047750D-02
                                                         2.50  1.970972174510D-02
        0  -3.437822054277D+00  -1.718309859155D+00      2.75  1.051507214016D-02
        1  -1.267949158365D+00  -6.341463414634D-01      3.25 -5.104560288561D-03
                                                         3.50 -5.281690140845D-03


I= 1      H= -2.5
0           0.                  -7.214015802130D-03     -1.00 -1.607694950189D+00
1           1.607694950189D+00   8.048780487805D-01      -.75 -1.205771212642D+00
                                                         -.50 -8.038474750945D-01
0  -3.5    -6.076949501889D-01  -3.038474750945D-01      -.25 -4.019237375472D-01
1  -2.5     1.646169701134D+00   8.230848505668D-01       .25  3.924285039505D-01
2  -1.5    -1.584678804535D+00  -7.923394022673D-01       .50  7.278856063209D-01
3   -.5     6.925455170045D-01   3.462727585022D-01       .75  9.493999055307D-01
4    .5    -1.855032634833D-01  -9.275163174167D-02      1.25  8.484358897286D-01
5   1.5     4.946753692889D-02   2.473376846445D-02      1.50  5.663431810374D-01
6   2.5    -1.236688423222D-02  -6.183442116111D-03      1.75  2.510788818276D-01
7   3.5     2.061147372037D-03   1.030573686019D-03      2.25 -1.142970628650D-01
                                                         2.50 -1.182583304706D-01
        0   5.626932325661D+00   2.809859154930D+00      2.75 -6.309043284095D-02
        1   1.607694950189D+00   8.048780487805D-01      3.25  3.062736173136D-02
                                                         3.50  3.169014084507D-02
```

CARDINAL SPLINES N=2

I= 2 H= -1.5

0		0.	2.885606320852D-02	-1.00	4.307798007558D-01
1		-4.307798007558D-01	-2.195121951220D-01	-.75	3.230848505668D-01
				-.50	2.153899003779D-01
0	-3.5	4.307798007558D-01	2.153899003779D-01	-.25	1.076949501889D-01
1	-2.5	-1.584678804535D+00	-7.923394022673D-01	.25	-1.009640158021D-01
2	-1.5	2.338715218138D+00	1.169357609069D+00	.50	-1.615424252834D-01
3	-.5	-1.770182068018D+00	-8.850910340089D-01	.75	-1.413496221230D-01
4	.5	7.420130539334D-01	3.710065269667D-01	1.25	2.781314410855D-01
5	1.5	-1.978701477156D-01	-9.893507385778D-02	1.50	6.096272758502D-01
6	2.5	4.946753692889D-02	2.473376846445D-02	1.75	8.863094726898D-01
7	3.5	-8.244589488148D-03	-4.122294744074D-03	2.25	8.790632514600D-01
				2.50	5.980333218825D-01
	0	-1.507729302645D+00	-7.394366197183D-01	2.75	2.679867313638D-01
	1	-4.307798007558D-01	-2.195121951220D-01	3.25	-1.225094469255D-01
				3.50	-1.267605633803D-01

I= 3 H= -.5

0		0.	-1.082102370319D-01	-1.00	-1.154242528341D-01
1		1.154242528341D-01	7.317073170732D-02	-.75	-8.656818962556D-02
				-.50	-5.771212641704D-02
0	-3.5	-1.154242528341D-01	-5.771212641704D-02	-.25	-2.885606320852D-02
1	-2.5	6.925455170045D-01	3.462727585022D-01	.25	2.705255925799D-02
2	-1.5	-1.770182068018D+00	-8.850910340089D-01	.50	4.328409481278D-02
3	-.5	2.388182755067D+00	1.194091377533D+00	.75	3.787358296118D-02
4	.5	-1.782548952250D+00	-8.912744761250D-01	1.25	-7.033665407077D-02
5	1.5	7.420130539334D-01	3.710065269667D-01	1.50	-1.298522844383D-01
6	2.5	-1.855032634833D-01	-9.275163174167D-02	1.75	-1.244417725867D-01
7	3.5	3.091721058056D-02	1.545860529028D-02	2.25	2.699190570251D-01
				2.50	6.011250429406D-01
	0	4.039848849193D-01	1.478873239437D-01	2.75	8.817685073858D-01
	1	1.154242528341D-01	7.317073170732D-02	3.25	8.812854259705D-01
				3.50	6.003521126761D-01

I= 4 H= .5

0		0.	4.039848849193D-01	-1.00	3.091721058056D-02
1		-3.091721058056D-02	-7.317073170732D-02	-.75	2.318790793542D-02
				-.50	1.545860529028D-02
0	-3.5	3.091721058056D-02	1.545860529028D-02	-.25	7.729302645139D-03
1	-2.5	-1.855032634833D-01	-9.275163174167D-02	.25	-7.246221229818D-03
2	-1.5	7.420130539334D-01	3.710065269667D-01	.50	-1.159395396771D-02
3	-.5	-1.782548952250D+00	-8.912744761250D-01	.75	-1.014470972175D-02
4	.5	2.388182755067D+00	1.194091377533D+00	1.25	1.884017519753D-02
5	1.5	-1.770182068018D+00	-8.850910340089D-01	1.50	3.478186190313D-02
6	2.5	6.925455170045D-01	3.462727585022D-01	1.75	3.333261765716D-02
7	3.5	-1.154242528341D-01	-5.771212641704D-02	2.25	-6.811447956029D-02
				2.50	-1.275334936448D-01
	0	-1.082102370319D-01	1.478873239437D-01	2.75	-1.231857609069D-01
	1	-3.091721058056D-02	-7.317073170732D-02	3.25	2.692427430436D-01
				3.50	6.003521126761D-01

I= 5 H= 1.5

CARDINAL SPLINES N=2

0		0.	-1.507729302645D+00	-1.00	-8,244589488148D-03
1		8,244589488148D-03	2.195121951220D-01	-.75	-6,183442116111D-03
				-.50	-4,122294744074D-03
0	-3,5	-8,244589488148D-03	-4.122294744074D-03	-.25	-2,061147372037D-03
1	-2,5	4,946753692889D-02	2.473376846445D-02	.25	1,932325661285D-03
2	-1,5	-1,978701477156D-01	-9,893507385778D-02	,50	3,091721058056D-03
3	-,5	7,420130539334D-01	3,710065269667D-01	,75	2,705255925799D-03
4	,5	-1,770182068018D+00	-8,850910340089D-01	1,25	-5,024046719340D-03
5	1,5	2,338715218138D+00	1,169357609069D+00	1,50	-9,275163174167D-03
6	2,5	-1,584678804535D+00	-7,923394022673D-01	1,75	-8,888698041910D-03
7	3,5	4,307798007558D-01	2,153899003779D-01	2,25	1,816386121608D-02
				2,50	3,400893163861D-02
	0	2,885606320852D-02	-7,394366197183D-01	2,75	3,284953624184D-02
	1	8,244589488148D-03	2,195121951220D-01	3,25	-6,763139814497D-02
				3,50	-1,267605633803D-01

I= 6 H= 2.5

0		0.	5.626932325661D+00	-1.00	2,061147372037D-03
1		-2,061147372037D-03	-8,048780487805D-01	-.75	1,545860529028D-03
				-.50	1,030573686019D-03
0	-3,5	2,061147372037D-03	1.030573686019D-03	-.25	5,152868430093D-04
1	-2,5	-1,236688423222D-02	-6,183442116111D-03	.25	-4,830814153212D-04
2	-1,5	4,946753692889D-02	2,473376846445D-02	,50	-7,729302645139D-04
3	-,5	-1,855032634833D-01	-9,275163174167D-02	,75	-6,763139814497D-04
4	,5	6,925455170045D-01	3,462727585022D-01	1,25	1,256011679835D-03
5	1,5	-1,584678804535D+00	-7,923394022673D-01	1,50	2,318790793542D-03
6	2,5	1,646169701134D+00	8,230848505668D-01	1,75	2,222174510477D-03
7	3,5	-6,076949501889D-01	-3,038474750945D-01	2,25	-4,540965304019D-03
				2,50	-8,502232909653D-03
	0	-7,214015802130D-03	2,809859154930D+00	2,75	-8,212840060460D-03
	1	-2,061147372037D-03	-8,048780487805D-01	3,25	1,690784953624D-02
				3,50	3,169014084507D-02

I= 7 H= 3.5

0		0.	-3.937822054277D+00	-1.00	-3,435245620062D-04
1		3,435245620062D-04	6,341463414634D-01	-.75	-2,576434215046D-04
				-.50	-1,717622810031D-04
0	-3,5	-3,435245620062D-04	-1,717622810031D-04	-.25	-8,588114050155D-05
1	-2,5	2,061147372037D-03	1,030573686019D-03	.25	8,051356922020D-05
2	-1,5	-8,244589488148D-03	-4,122294744074D-03	,50	1,288217107523D-04
3	-,5	3,091721058056D-02	1,545860529028D-02	,75	1,127189969083D-04
4	,5	-1,154242528341D-01	-5,771212641704D-02	1,25	-2,093352799725D-04
5	1,5	4,307798007558D-01	2,153899003779D-01	1,50	-3,864651322570D-04
6	2,5	-6,076949501889D-01	-3,038474750945D-01	1,75	-3,703624184129D-04
7	3,5	2,679491583648D-01	1,339745791824D-01	2,25	7,568275506699D-04
				2,50	1,417038818276D-03
	0	1,202335967022D-03	-1,718309859155D+00	2,75	1,368730676743D-03
	1	3,435245620062D-04	6,341463414634D-01	3,25	-2,817974922707D-03
				3,50	-5,281690140845D-03

CARDINAL SPLINES N=2

N= 2 M= 8 P= 4.0 Q= 3

I= 0	H= -4.0				
0		1.000000000000D+00	4.996318114875D-01	-1.00	2.267949189985D+00
1		-1.267949189985D+00	-6.339285714286D-01	-.75	1.950961892489D+00
				-.50	1.633974594993D+00
0	-4.0	2.679491899853D-01	1.339745949926D-01	-.25	1.316987297496D+00
1	-3.0	-6.076951399116D-01	-3.038475699558D-01	.25	6.871994085972D-01
2	-2.0	4.307805596465D-01	2.153902798233D-01	.50	3.995190537555D-01
3	-1.0	-1.154270986745D-01	-5.771354933726D-02	.75	1.620791720361D-01
4	0.0	3.092783505155D-02	1.546391752577D-02	1.25	-7.109346235272D-02
5	1.0	-8.284241531664D-03	-4.142120765832D-03	1.50	-7.355716126657D-02
6	2.0	2.209131075110D-03	1.104565537555D-03	1.75	-3.924227954713D-02
7	3.0	-5.522827687776D-04	-2.761413843888D-04	2.25	1.904944081370D-02
8	4.0	9.204712812960D-05	4.602356406480D-05	2.50	1.970959131075D-02
				2.75	1.051494615243D-02
	0	-4.071796759941D+00	-2.036082474227D+00	3.25	-5.104300902062D-03
	1	-1.267949189985D+00	-6.339285714286D-01	3.50	-5.281203976436D-03
				3.75	-2.817505062592D-03

I= 1	H= -3.0				
0		0.	2.209131075110D-03	-1.00	-1.607695139912D+00
1		1.607695139912D+00	8.035714285714D-01	-.75	-1.205771354934D+00
				-.50	-8.038475699558D-01
0	-4.0	-6.076951399116D-01	-3.038475699558D-01	-.25	-4.019237849779D-01
1	-3.0	1.646170839470D+00	8.230854197349D-01	.25	3.924285484168D-01
2	-2.0	-1.584683357879D+00	-7.923416789396D-01	.50	7.278856774669D-01
3	-1.0	6.925625920471D-01	3.462812960236D-01	.75	9.493999677835D-01
4	0.0	-1.855670103093D-01	-9.278350515464D-02	1.25	8.484357741163D-01
5	1.0	4.970544918999D-02	2.485272459499D-02	1.50	5.663429675994D-01
6	2.0	-1.325478645066D-02	-6.627393225331D-03	1.75	2.510786772828D-01
7	3.0	3.313696612666D-03	1.656848306333D-03	2.25	-1.142966448822D-01
8	4.0	-5.522827687776D-04	-2.761413843888D-04	2.50	-1.182575478645D-01
				2.75	-6.308967691458D-02
	0	6.430780559647D+00	3.216494845361D+00	3.25	3.062580541237D-02
	1	1.607695139912D+00	8.035714285714D-01	3.50	3.168722385862D-02
				3.75	1.690503037555D-02

I= 2	H= -2.0				
0		0.	-8.836524300442D-03	-1.00	4.307805596465D-01
1		-4.307805596465D-01	-2.142857142857D-01	-.75	3.230854197349D-01
				-.50	2.153902798233D-01
0	-4.0	4.307805596465D-01	2.153902798233D-01	-.25	1.076951399116D-01
1	-3.0	-1.584683357879D+00	-7.923416789396D-01	.25	-1.009641936672D-01
2	-2.0	2.338733431517D+00	1.169366715758D+00	.50	-1.615427098675D-01
3	-1.0	-1.770250368189D+00	-8.851251840943D-01	.75	-1.413498711340D-01
4	0.0	7.422680412371D-01	3.711340206186D-01	1.25	2.781319035346D-01
5	1.0	-1.988217967599D-01	-9.941089837997D-02	1.50	6.096281296024D-01

6	2.0	5.301914580265D-02	2.650957290133D-02	1.75	8.863102908689D-01
7	3.0	-1.325478645066D-02	-6.627393225331D-03	2.25	8.790615795287D-01
8	4.0	2.209131075110D-03	1.104565537555D-03	2.50	5.980301914580D-01
				2.75	2.679837076583D-01
	0	-1.723122238586D+00	-8.659793814433D-01	3.25	-1.225032216495D-01
	1	-4.307805596465D-01	-2.142857142857D-01	3.50	-1.267488954345D-01
				3.75	-6.762012150221D-02

I= 3 H= -1.0

0		0.	3.313696612666D-02	-1.00	-1.154270986745D-01
1		1.154270986745D-01	5.357142857143D-02	-.75	-8.657032400589D-02
				-.50	-5.771354933726D-02
0	-4.0	-1.154270986745D-01	-5.771354933726D-02	-.25	-2.885677466863D-02
1	-3.0	6.925625920471D-01	3.462812960236D-01	.25	2.705322625184D-02
2	-2.0	-1.770250368189D+00	-8.851251840943D-01	.50	4.328516200295D-02
3	-1.0	2.388438880707D+00	1.194219440353D+00	.75	3.787451675258D-02
4	0.0	-1.783505154639D+00	-8.917525773196D-01	1.25	-7.033838825479D-02
5	1.0	7.455817378498D-01	3.727908689249D-01	1.50	-1.298554860088D-01
6	2.0	-1.988217967599D-01	-9.941089837997D-02	1.75	-1.244448407585D-01
7	3.0	4.970544918999D-02	2.485272459499D-02	2.25	2.699253267673D-01
8	4.0	-8.284241531664D-03	-4.142120765832D-03	2.50	6.011367820324D-01
				2.75	8.817798462813D-01
	0	4.617083946981D-01	2.474226804124D-01	3.25	8.812620811856D-01
	1	1.154270986745D-01	5.357142857143D-02	3.50	6.003083578792D-01
				3.75	2.692004556333D-01

I= 4 H= 0.0

0		0.	-1.237113402062D-01	-1.00	3.092783505155D-02
1		-3.092783505155D-02	0.	-.75	2.319587628866D-02
				-.50	1.546391752577D-02
0	-4.0	3.092783505155D-02	1.546391752577D-02	-.25	7.731958762887D-03
1	-3.0	-1.855670103093D-01	-9.278350515464D-02	.25	-7.248711340206D-03
2	-2.0	7.422680412371D-01	3.711340206186D-01	.50	-1.159793814433D-02
3	-1.0	-1.783505154639D+00	-8.917525773196D-01	.75	-1.014819587629D-02
4	0.0	2.391752577320D+00	1.195876288660D+00	1.25	1.884664948454D-02
5	1.0	-1.783505154639D+00	-8.917525773196D-01	1.50	3.479381443299D-02
6	2.0	7.422680412371D-01	3.711340206186D-01	1.75	3.334407216495D-02
7	3.0	-1.855670103093D-01	-9.278350515464D-02	2.25	-6.813788659794D-02
8	4.0	3.092783505155D-02	1.546391752577D-02	2.50	-1.275773195876D-01
				2.75	-1.232280927835D-01
	0	-1.237113402062D-01	-1.237113402062D-01	3.25	2.693298969072D-01
	1	-3.092783505155D-02	0.	3.50	6.005154639175D-01
				3.75	8.814432989691D-01

I= 5 H= 1.0

0		0.	4.617083946981D-01	-1.00	-8.284241531664D-03
1		8.284241531664D-03	-5.357142857143D-02	-.75	-6.213181148748D-03
				-.50	-4.142120765832D-03
0	-4.0	-8.284241531664D-03	-4.142120765832D-03	-.25	-2.071060382916D-03
1	-3.0	4.970544918999D-02	2.485272459499D-02	.25	1.941619108984D-03
2	-2.0	-1.988217967599D-01	-9.941089837997D-02	.50	3.106590574374D-03
3	-1.0	7.455817378498D-01	3.727908689249D-01	.75	2.718266752577D-03
4	0.0	-1.783505154639D+00	-8.917525773196D-01	1.25	-5.048209683358D-03
5	1.0	2.388438880707D+00	1.194219440353D+00	1.50	-9.319771723122D-03

CARDINAL SPLINES N=2

6	2.0	-1.770250368189D+00	-8.851251840943D-01	1.75	-8.931447901325D-03
7	3.0	6.925625920471D-01	3.462812960236D-01	2.25	1.825121962445D-02
8	4.0	-1.154270986745D-01	-5.771354933726D-02	2.50	3.417249631811D-02
				2.75	3.300752485272D-02
	0	3.313696612666D-02	2.474226804124D-01	3.25	-6.795666881443D-02
	1	8.284241531664D-03	-5.357142857143D-02	3.50	-1.273702135493D-01
				3.75	-1.230986515096D-01

I= 6 H= 2.0

0		0.	-1.723122238586D+00	-1.00	2.209131075110D-03
1		-2.209131075110D-03	2.142857142857D-01	-.75	1.656848306333D-03
				-.50	1.104565537555D-03
0	-4.0	2.209131075110D-03	1.104565537555D-03	-.25	5.522827687776D-04
1	-3.0	-1.325478645066D-02	-6.627393225331D-03	.25	-5.177650957290D-04
2	-2.0	5.301914580265D-02	2.650957290133D-02	.50	-8.284241531664D-04
3	-1.0	-1.988217967599D-01	-9.941089837997D-02	.75	-7.248711340206D-04
4	0.0	7.422680412371D-01	3.711340206186D-01	1.25	1.346189248895D-03
5	1.0	-1.770250368189D+00	-8.851251840943D-01	1.50	2.485272459499D-03
6	2.0	2.338733431517D+00	1.169366715758D+00	1.75	2.381719440353D-03
7	3.0	-1.584683357879D+00	-7.923416789396D-01	2.25	-4.866991899853D-03
8	4.0	4.307805596465D-01	2.153902798233D-01	2.50	-9.112665684831D-03
				2.75	-8.802006627393D-03
	0	-8.836524300442D-03	-8.659793814433D-01	3.25	1.812177835052D-02
	1	-2.209131075110D-03	2.142857142857D-01	3.50	3.396539027982D-02
				3.75	3.282630706922D-02

I= 7 H= 3.0

0		0.	6.430780559647D+00	-1.00	-5.522827687776D-04
1		5.522827687776D-04	-8.035714285714D-01	-.75	-4.142120765832D-04
				-.50	-2.761413843888D-04
0	-4.0	-5.522827687776D-04	-2.761413843888D-04	-.25	-1.380706921944D-04
1	-3.0	3.313696612666D-03	1.656848306333D-03	.25	1.294412739323D-04
2	-2.0	-1.325478645066D-02	-6.627393225331D-03	.50	2.071060382916D-04
3	-1.0	4.970544918999D-02	2.485272459499D-02	.75	1.812177835052D-04
4	0.0	-1.855670103093D-01	-9.278350515464D-02	1.25	-3.365473122239D-04
5	1.0	6.925625920471D-01	3.462812960236D-01	1.50	-6.213181148748D-04
6	2.0	-1.584683357879D+00	-7.923416789396D-01	1.75	-5.954298600884D-04
7	3.0	1.646170839470D+00	8.230854197349D-01	2.25	1.216747974963D-03
8	4.0	-6.076951399116D-01	-3.038475699558D-01	2.50	2.278166421208D-03
				2.75	2.200501656848D-03
	0	2.209131075110D-03	3.216494845361D+00	3.25	-4.530444587629D-03
	1	5.522827687776D-04	-8.035714285714D-01	3.50	-8.491347569956D-03
				3.75	-8.206576767305D-03

I= 8 H= 4.0

0		0.	-4.571796759941D+00	-1.00	9.204712812960D-05
1		-9.204712812960D-05	6.339285714286D-01	-.75	6.903534609720D-05
				-.50	4.602356406480D-05
0	-4.0	9.204712812960D-05	4.602356406480D-05	-.25	2.301178203240D-05
1	-3.0	-5.522827687776D-04	-2.761413843888D-04	.25	-2.157354565538D-05
2	-2.0	2.209131075110D-03	1.104565537555D-03	.50	-3.451767304860D-05
3	-1.0	-8.284241531664D-03	-4.142120765832D-03	.75	-3.020296391753D-05
4	0.0	3.092783505155D-02	1.546391752577D-02	1.25	5.609121870398D-05
5	1.0	-1.154270986745D-01	-5.771354933726D-02	1.50	1.035530191458D-04

6	2,0	4.307805596465D-01	2.153902798233D-01	1.75	9,923831001473D-05
7	3.0	-6.076951399116D-01	-3.038475699558D-01	2.25	-2,027913291605D-04
8	4.0	2,679491899853D-01	1.339745949926D-01	2.50	-3,796944035346D-04
				2.75	-3,667502761414D-04
	0	-3,681885125184D-04	-2.036082474227D+00	3,25	7,550740979381D-04
	1	-9,204712812960D-05	6.339285714286D-01	3.50	1.415224594993D-03
				3.75	1.367762794551D-03

N= 2 M= 9 P= 4.5 Q= 3

I= 0 H= -4.5

0		1.000000000000D+00	5.001109877913D-01	-1.00	2,267949192256D+00
1		-1,267949192256D+00	-6.339869281046D-01	-.75	1,950961894192D+00
				-.50	1,633974596128D+00
0	-4,5	2,679491922555D-01	1.339745961278D-01	-,25	1,316987298064D+00
1	-3,5	-6,076951535331D-01	-3,038475767666D-01	.25	6,871994080651D-01
2	-2,5	4,307806141324D-01	2.153903070662D-01	.50	3,995190529042D-01
3	-1,5	-1,154273029967D-01	-5,771365149834D-02	.75	1,620791712912D-01
4	-,5	3.092859785424D-02	1,546429892712D-02	1.25	-7,109346096929D-02
5	,5	-8,287088420274D-03	-4.143544210137D-03	1.50	-7,355715871254D-02
6	1,5	2,219755826859D-03	1.109877913430D-03	1.75	-3,924227709952D-02
7	2,5	-5,919348871624D-04	-2,959674435812D-04	2.25	1,904943581206D-02
8	3,5	1.479837217906D-04	7.399186089530D-05	2.50	1,970958194599D-02
9	4,5	-2,466395363177D-05	-1,233197681588D-05	2.75	1,051493710692D-02
				3,25	-5,104282278949D-03
	0	-4,705771365150D+00	-2,352830188679D+00	3.50	-5,281169071402D-03
	1	-1,267949192256D+00	-6.339869281046D-01	3,75	-2,817471328154D-03
				4,25	1,367693303737D-03
				4.50	1.415094339623D-03

I= 1 H= -3.5

0		0.	-6.659267480577D-04	-1.00	-1,607695153533D+00
1		1.607695153533D+00	8.039215686275D-01	-.75	-1,205771365150D+00
				-.50	-8,038475767666D-01
0	-4,5	-6,076951535331D-01	-3.038475767666D-01	-,25	-4,019237883833D-01
1	-3.5	1.646170921199D+00	8,230854605993D-01	.25	3,924285516093D-01
2	-2,5	-1,584683684795D+00	-7,923418423973D-01	.50	7,278856825749D-01
3	-1.5	6,925638179800D-01	3.462819089900D-01	.75	9,493999722531D-01
4	-.5	-1,855715871254D-01	-9,278579356271D-02	1.25	8,484357658158D-01
5	,5	4.972253052164D-02	2.486126526082D-02	1.50	5,663429522752D-01
6	1,5	-1,331853496115D-02	-6,659267480577D-03	1.75	2,510786625971D-01
7	2.5	3,551609322974D-03	1.775804661487D-03	2,25	-1,142966148724D-01
8	3,5	-8,879023307436D-04	-4.439511653718D-04	2,50	-1,182574916759D-01
9	4,5	1.479837217906D-04	7.399186089530D-05	2.75	-6,308962264151D-02
				3,25	3,062569367370D-02
	0	7,234628190899D+00	3.616981132075D+00	3.50	3,168701442841D-02
	1	1,607695153533D+00	8.039215686275D-01	3,75	1,690482796892D-02
				4.25	-8,206159822420D-03

CARDINAL SPLINES N=2

```
                                                          4.50 -8.490566037736D-03

I= 2     H= -2.5
0            0.                   2.663706992231D-03     -1.00  4.307806141324D-01
1        -4.307806141324D-01     -2.156862745098D-01      -.75  3.230854605993D-01
                                                          -.50  2.153903070662D-01
0  -4.5   4.307806141324D-01      2.153903070662D-01      -.25  1.076951535331D-01
1  -3.5  -1.584683684795D+00     -7.923418423973D-01       .25 -1.009642064373D-01
2  -2.5   2.338734739179D+00      1.169367369589D+00       .50 -1.615427302997D-01
3  -1.5  -1.770255271920D+00     -8.851276359600D-01       .75 -1.413498890122D-01
4   -.5   7.422863485017D-01      3.711431742508D-01      1.25  2.781319367370D-01
5    .5  -1.988901220866D-01     -9.944506104329D-02      1.50  6.096281908990D-01
6   1.5   5.327413984462D-02      2.663706992231D-02      1.75  8.863103496115D-01
7   2.5  -1.420643729190D-02     -7.103218645949D-03      2.25  8.790614594895D-01
8   3.5   3.551609322974D-03      1.775804661487D-03      2.50  5.980299667037D-01
9   4.5  -5.919348871624D-04     -2.959674435812D-04      2.75  2.679834905660D-01
                                                          3.25 -1.225027746948D-01
      0  -1.938512763596D+00     -9.679245283019D-01      3.50 -1.267480577137D-01
      1  -4.307806141324D-01     -2.156862745098D-01      3.75 -6.761931187569D-02
                                                          4.25  3.282463928968D-02
                                                          4.50  3.396226415094D-02


I= 3     H= -1.5
0            0.                  -9.988901220866D-03     -1.00 -1.154273029967D-01
1         1.154273029967D-01      5.882352941176D-02      -.75 -8.657047724750D-02
                                                          -.50 -5.771365149834D-02
0  -4.5  -1.154273029967D-01     -5.771365149834D-02      -.25 -2.885682574917D-02
1  -3.5   6.925638179800D-01      3.462819089900D-01       .25  2.705327413984D-02
2  -2.5  -1.770255271920D+00     -8.851276359600D-01       .50  4.328523862375D-02
3  -1.5   2.388457269700D+00      1.194228634850D+00       .75  3.787458379578D-02
4   -.5  -1.783573806881D+00     -8.917869034406D-01      1.25 -7.033851276360D-02
5    .5   7.458379578246D-01      3.729189789123D-01      1.50 -1.298557158713D-01
6   1.5  -1.997780244173D-01     -9.988901220866D-02      1.75 -1.244450610433D-01
7   2.5   5.327413984462D-02      2.663706992231D-02      2.25  2.699257769145D-01
8   3.5  -1.331853496115D-02     -6.659267480577D-03      2.50  6.011376248613D-01
9   4.5   2.219755826859D-03      1.109877913430D-03      2.75  8.817806603774D-01
                                                          3.25  8.812604051054D-01
      0   5.194228634850D-01      2.547169811321D-01      3.50  6.003052164262D-01
      1   1.154273029967D-01      5.882352941176D-02      3.75  2.691974195339D-01
                                                          4.25 -1.230923973363D-01
                                                          4.50 -1.273584905660D-01


I= 4     H=  -.5
0            0.                   3.729189789123D-02     -1.00  3.092859785424D-02
1        -3.092859785424D-02     -1.960784313725D-02      -.75  2.319644839068D-02
                                                          -.50  1.546429892712D-02
0  -4.5   3.092859785424D-02      1.546429892712D-02      -.25  7.732149463559D-03
1  -3.5  -1.855715871254D-01     -9.278579356271D-02       .25 -7.248890122087D-03
2  -2.5   7.422863485017D-01      3.711431742508D-01       .50 -1.159822419534D-02
3  -1.5  -1.783573806881D+00     -8.917869034406D-01       .75 -1.014844617092D-02
4   -.5   2.392008879023D+00      1.196004439512D+00      1.25  1.884711431743D-02
5    .5  -1.784461709212D+00     -8.922308546060D-01      1.50  3.479467258602D-02
6   1.5   7.458379578246D-01      3.729189789123D-01      1.75  3.334489456160D-02
7   2.5  -1.988901220866D-01     -9.944506104329D-02      2.25 -6.813956714761D-02
```

8	3,5	4,972253052164D-02	2,486126526082D-02	2,50	-1,275804661487D-01
9	4,5	-8,287088420274D-03	-4,143544210137D-03	2,75	-1,232311320755D-01
				3,25	2,693361542730D-01
	0	-1,391786903441D-01	-5,094339622642D-02	3,50	6,005271920089D-01
	1	-3,092859785424D-02	-1,960784313725D-02	3,75	8,814546337403D-01
				4,25	8,814199500555D-01
				4,50	6,004716981132D-01

I= 5 H= .5

0		0,	-1,391786903441D-01	-1,00	-8,287088420274D-03
1		8,287088420274D-03	1,960784313725D-02	-,75	-6,215316315205D-03
				-,50	-4,143544210137D-03
0	-4,5	-8,287088420274D-03	-4,143544210137D-03	-,25	-2,071772105068D-03
1	-3,5	4,972253052164D-02	2,486126526082D-02	,25	1,942286348502D-03
2	-2,5	-1,988901220866D-01	-9,944506104329D-02	,50	3,107658157603D-03
3	-1,5	7,458379578246D-01	3,729189789123D-01	,75	2,719200887902D-03
4	-,5	-1,784461709212D+00	-8,922308546060D-01	1,25	-5,049944506104D-03
5	,5	2,392008879023D+00	1,196004439512D+00	1,50	-9,322974472808D-03
6	1,5	-1,783573806881D+00	-8,917869034406D-01	1,75	-8,934517203108D-03
7	2,5	7,422863485017D-01	3,711431742508D-01	2,25	1,825749167592D-02
8	3,5	-1,855715871254D-01	-9,278579356271D-02	2,50	3,418423973363D-02
9	4,5	3,092859785424D-02	1,546429892712D-02	2,75	3,301886792453D-02
				3,25	-6,798002219756D-02
	0	3,729189789123D-02	-5,094339622642D-02	3,50	-1,274139844617D-01
	1	8,287088420274D-03	1,960784313725D-02	3,75	-1,231409544950D-01
				4,25	2,692875971143D-01
				4,50	6,004716981132D-01

I= 6 H= 1.5

0		0,	5,194228634850D-01	-1,00	2,219755826859D-03
1		-2,219755826859D-03	-5,882352941176D-02	-,75	1,664816870144D-03
				-,50	1,109877913430D-03
0	-4,5	2,219755826859D-03	1,109877913430D-03	-,25	5,549389567148D-04
1	-3,5	-1,331853496115D-02	-6,659267480577D-03	,25	-5,202552719201D-04
2	-2,5	5,327413984462D-02	2,663706992231D-02	,50	-8,324084350721D-04
3	-1,5	-1,997780244173D-01	-9,988901220866D-02	,75	-7,283573806881D-04
4	-,5	7,458379578246D-01	3,729189789123D-01	1,25	1,352663706992D-03
5	,5	-1,783573806881D+00	-8,917869034406D-01	1,50	2,497225305216D-03
6	1,5	2,388457269700D+00	1,194228634850D+00	1,75	2,393174250832D-03
7	2,5	-1,770255271920D+00	-8,851276359600D-01	2,25	-4,890399556049D-03
8	3,5	6,925638179800D-01	3,462819089900D-01	2,50	-9,156492785794D-03
9	4,5	-1,154273029967D-01	-5,771365149834D-02	2,75	-8,844339622642D-03
				3,25	1,820893451720D-02
	0	-9,988901220866D-03	2,547169811321D-01	3,50	3,412874583796D-02
	1	-2,219755826859D-03	-5,882352941176D-02	3,75	3,298418423973D-02
				4,25	-6,794533851276D-02
				4,50	-1,273584905660D-01

I= 7 H= 2.5

0		0,	-1,938512763596D+00	-1,00	-5,919348871624D-04
1		5,919348871624D-04	2,156862745098D-01	-,75	-4,439511653718D-04
				-,50	-2,959674435812D-04
0	-4,5	-5,919348871624D-04	-2,959674435812D-04	-,25	-1,479837217906D-04
1	-3,5	3,551609322974D-03	1,775804661487D-03	,25	1,387347391787D-04

2	-2.5	-1.420643729190D-02	-7.103218645949D-03	.50	2.219755826859D-04
3	-1.5	5.327413984462D-02	2.663706992231D-02	.75	1.942286348502D-04
4	-.5	-1.988901220866D-01	-9.944506104329D-02	1.25	-3.607103218646D-04
5	.5	7.422863485017D-01	3.711431742508D-01	1.50	-6.659267480577D-04
6	1.5	-1.770255271920D+00	-8.851276359600D-01	1.75	-6.381798002220D-04
7	2.5	2.338734739179D+00	1.169367369589D+00	2.25	1.304106548280D-03
8	3.5	-1.584683684795D+00	-7.923418423973D-01	2.50	2.441731409545D-03
9	4.5	4.307806141324D-01	2.153903070662D-01	2.75	2.358490566038D-03
				3.25	-4.855715871254D-03
	0	2.663706992231D-03	-9.679245283019D-01	3.50	-9.100998890122D-03
	1	5.919348871624D-04	2.156862745098D-01	3.75	-8.795782463929D-03
				4.25	1.811875693674D-02
				4.50	3.396226415094D-02

I= 8 H= 3.5

0		0.	7.234628190899D+00	-1.00	1.479837217906D-04
1		-1.479837217906D-04	-8.039215686275D-01	-.75	1.109877913430D-04
				-.50	7.399186089530D-05
0	-4.5	1.479837217906D-04	7.399186089530D-05	-.25	3.699593044765D-05
1	-3.5	-8.879023307436D-04	-4.439511653718D-04	.25	-3.468368479467D-05
2	-2.5	3.551609322974D-03	1.775804661487D-03	.50	-5.549389567148D-05
3	-1.5	-1.331853496115D-02	-6.659267480577D-03	.75	-4.855715871254D-05
4	-.5	4.972253052164D-02	2.486126526082D-02	1.25	9.017758046615D-05
5	.5	-1.855715871254D-01	-9.278579356271D-02	1.50	1.664816870144D-04
6	1.5	6.925638179800D-01	3.462819089900D-01	1.75	1.595449500555D-04
7	2.5	-1.584683684795D+00	-7.923418423973D-01	2.25	-3.260266370699D-04
8	3.5	1.646170921199D+00	8.230854605993D-01	2.50	-6.104328523862D-04
9	4.5	-6.076951535331D-01	-3.038475767666D-01	2.75	-5.896226415094D-04
				3.25	1.213928967814D-03
	0	-6.659267480577D-04	3.616981132075D+00	3.50	2.275249722531D-03
	1	-1.479837217906D-04	-8.039215686275D-01	3.75	2.198945615982D-03
				4.25	-4.529689234184D-03
				4.50	-8.490566037736D-03

I= 9 H= 4.5

0		0.	-5.205771365150D+00	-1.00	-2.466395363177D-05
1		2.466395363177D-05	6.339869281046D-01	-.75	-1.849796522383D-05
				-.50	-1.233197681588D-05
0	-4.5	-2.466395363177D-05	-1.233197681588D-05	-.25	-6.165988407942D-06
1	-3.5	1.479837217906D-04	7.399186089530D-05	.25	5.780614132445D-06
2	-2.5	-5.919348871624D-04	-2.959674435812D-04	.50	9.248982611913D-06
3	-1.5	2.219755826859D-03	1.109877913430D-03	.75	8.092859785424D-06
4	-.5	-8.287088420274D-03	-4.143544210137D-03	1.25	-1.502959674436D-05
5	.5	3.092859785424D-02	1.546429892712D-02	1.50	-2.774694783574D-05
6	1.5	-1.154273029967D-01	-5.771365149834D-02	1.75	-2.659082500925D-05
7	2.5	4.307806141324D-01	2.153903070662D-01	2.25	5.433777284499D-05
8	3.5	-6.076951535331D-01	-3.038475767666D-01	2.50	1.017388087310D-04
9	4.5	2.679491922555D-01	1.339745961278D-01	2.75	9.827044025157D-05
				3.25	-2.023214946356D-04
	0	1.109877913430D-04	-2.352830188679D+00	3.50	-3.792082870884D-04
	1	2.466395363177D-05	6.339869281046D-01	3.75	-3.664909359970D-04
				4.25	7.549482056974D-04
				4.50	1.415094339623D-03

CARDINAL SPLINES N=2

N= 2 M=10 P= 5.0 Q= 3

I= 0 H= -5.0

0		1.000000000000D+00	4.999669565677D-01	-1.00	2.267949192419D+00
1		-1.267949192419D+00	-6.339712918660D-01	-.75	1.950961894314D+00
				-.50	1.633974596209D+00
0	-5.0	2.679491924185D-01	1.339745962093D-01	-.25	1.316987298105D+00
1	-4.0	-6.076951545111D-01	-3.038475772555D-01	.25	6.871994080269D-01
2	-3.0	4.307806180444D-01	2.153903090222D-01	.50	3.995190528431D-01
3	-2.0	-1.154273176663D-01	-5.771365883317D-02	.75	1.620791712377D-01
4	-1.0	3.092865262101D-02	1.546432631050D-02	1.25	-7.109346086997D-02
5	0.0	-8.287292817680D-03	-4.143646408840D-03	1.50	-7.355715852917D-02
6	1.0	2.220518649713D-03	1.110259324857D-03	1.75	-3.924227692379D-02
7	2.0	-5.947817811732D-04	-2.973908905866D-04	2.25	1.904943545296D-02
8	3.0	1.586084749795D-04	7.930423748976D-05	2.50	1.970958127363D-02
9	4.0	-3.965211874488D-05	-1.982605937244D-05	2.75	1.051493645748D-02
10	5.0	6.608686457480D-06	3.304343228740D-06	3.25	-5.104280941870D-03
				3.50	-5.281166565333D-03
	0	-5.339745962093D+00	-2.669889502762D+00	3.75	-2.817468906130D-03
	1	-1.267949192419D+00	-6.339712918660D-01	4.25	1.367688314521D-03
				4.50	1.415084987708D-03
				4.75	7.549391670412D-04

I= 1 H= -4.0

0		0.	1.982605937244D-04	-1.00	-1.607695154511D+00
1		1.607695154511D+00	8.038277511962D-01	-.75	-1.205771365883D+00
				-.50	-8.038475772555D-01
0	-5.0	-6.076951545111D-01	-3.038475772555D-01	-.25	-4.019237886278D-01
1	-4.0	1.646170927067D+00	8.230854635333D-01	.25	3.924285518385D-01
2	-3.0	-1.584683708266D+00	-7.923418541331D-01	.50	7.278856829417D-01
3	-2.0	6.925639059980D-01	3.462819529990D-01	.75	9.493999725740D-01
4	-1.0	-1.855719157260D-01	-9.278595786302D-02	1.25	8.484357652198D-01
5	0.0	4.972375690608D-02	2.486187845304D-02	1.50	5.663429511750D-01
6	1.0	-1.332311189828D-02	-6.661555949140D-03	1.75	2.510786615427D-01
7	2.0	3.568690687039D-03	1.784345343520D-03	2.25	-1.142966127178D-01
8	3.0	-9.516508498771D-04	-4.758254249385D-04	2.50	-1.182574876418D-01
9	4.0	2.379127124693D-04	1.189563562346D-04	2.75	-6.308961874488D-02
10	5.0	-3.965211874488D-05	-1.982605937244D-05	3.25	3.062568565122D-02
				3.50	3.168699939200D-02
	0	8.038475772555D+00	4.019337016575D+00	3.75	1.690481343678D-02
	1	1.607695154511D+00	8.038277511962D-01	4.25	-8.206129887124D-03
				4.50	-8.490509926247D-03
				4.75	-4.529635002247D-03

I= 2 H= -3.0

0		0.	-7.930423748976D-04	-1.00	4.307806180444D-01
1		-4.307806180444D-01	-2.153110047847D-01	-.75	3.230854635333D-01
				-.50	2.153903090222D-01

CARDINAL SPLINES N=2

0	-5,0	4.307806180444D-01	2.153903090222D-01	-.25	1,076951545111D-01
1	-4,0	-1.584683708266D+00	-7.923418541331D-01	.25	-1,009642073541D-01
2	-3,0	2.338734833065D+00	1.169367416532D+00	.50	-1,615427317666D-01
3	-2,0	-1,770255623992D+00	-8.851278119961D-01	.75	-1,413498902958D-01
4	-1,0	7.422876629041D-01	3.711438314521D-01	1.25	2,781319391208D-01
5	0,0	-1.988950276243D-01	-9.944751381215D-02	1.50	6,096281952999D-01
6	1,0	5.329244759312D-02	2.664622379656D-02	1.75	8,863103538291D-01
7	2,0	-1.427476274816D-02	-7.137381374078D-03	2.25	8,790614508710D-01
8	3,0	3.806603399508D-03	1.903301699754D-03	2.50	5,980299505670D-01
9	4,0	-9.516508498771D-04	-4.758254249385D-04	2.75	2,679834749795D-01
10	5,0	1.586084749795D-04	7.930423748976D-05	3.25	-1,225027426049D-01
				3.50	-1,267479975680D-01
	0	-2.153903090222D+00	-1.077348066298D+00	3.75	-6,761925374713D-02
	1	-4.307806180444D-01	-2.153110047847D-01	4.25	3,282451954849D-02
				4.50	3,396203970499D-02
				4.75	1,811854000899D-02

I= 3 H= -2.0

0		0.	2.973908905866D-03	-1.00	-1,154273176663D-01
1		1.154273176663D-01	5.741626794258D-02	-.75	-8,657048824976D-02
				-.50	-5,771365883317D-02
0	-5,0	-1.154273176663D-01	-5.771365883317D-02	-.25	-2,885682941659D-02
1	-4,0	6.925639059980D-01	3.462819529990D-01	.25	2,705327757805D-02
2	-3,0	-1,770255623992D+00	-8,851278119961D-01	.50	4,328524412488D-02
3	-2.0	2.388458589971D+00	1.194229294985D+00	.75	3,787458860927D-02
4	-1.0	-1.783578735890D+00	-8.917893679452D-01	1.25	-7,033852170293D-02
5	0.0	7.458563535912D-01	3.729281767956D-01	1.50	-1,298557323746D-01
6	1.0	-1.998466784742D-01	-9.992339923709D-02	1.75	-1,244450768590D-01
7	2.0	5.353036030559D-02	2.676518015279D-02	2.25	2,699258092337D-01
8	3,0	-1.427476274816D-02	-7.137381374078D-03	2.50	6,011376853737D-01
9	4.0	3.568690687039D-03	1.784345343520D-03	2.75	8,817807188268D-01
10	5.0	-5.947817811732D-04	-2.973908905866D-04	3.25	8,812602847683D-01
				3.50	6,003049908800D-01
	0	5.771365883317D-01	2.900552486188D-01	3.75	2,691972015517D-01
	1	1.154273176663D-01	5.741626794258D-02	4.25	-1,230919483069D-01
				4.50	-1,273576488937D-01
				4.75	-6,794452503370D-02

I= 4 H= -1.0

0		0.	-1.110259324857D-02	-1.00	3,092865262101D-02
1		-3.092865262101D-02	-1.435406698565D-02	-.75	2,319648946575D-02
				-.50	1,546432631050D-02
0	-5,0	3.092865262101D-02	1.546432631050D-02	-.25	7,732163155251D-03
1	-4,0	-1.855719157260D-01	-9.278595786302D-02	.25	-7,248902958048D-03
2	-3.0	7.422876629041D-01	3.711438314521D-01	.50	-1,159824473288D-02
3	-2.0	-1.783578735890D+00	-8.917893679452D-01	.75	-1,014846414127D-02
4	-1.0	2.392027280658D+00	1.196013640329D+00	1.25	1,884714769092D-02
5	0,0	-1,784530386740D+00	-8.922651933702D-01	1.50	3,479473419863D-02
6	1,0	7.460942663036D-01	3.730471331518D-01	1.75	3,334495360702D-02
7	2.0	-1.998466784742D-01	-9.992339923709D-02	2.25	-6,813968780565D-02
8	3.0	5.329244759312D-02	2.664622379656D-02	2.50	-1,275806920616D-01
9	4.0	-1.332311189828D-02	-6.661555949140D-03	2.75	-1,232313502868D-01
10	5.0	2.220518649713D-03	1.110259324857D-03	3.25	2,693366035317D-01
				3.50	6,005280340480D-01
	0	-1.546432631050D-01	-8.287292817680D-02	3.75	8,814554475402D-01

CARDINAL SPLINES N=2

```
        1  -3.092865262101D-02  -1.435406698565D-02     4.25   8.814182736789D-01
                                                         4.50   6.004685558698D-01
                                                         4.75   2.692845601258D-01


 I= 5     H=  0.0
 0            0.                 4.143646408840D-02    -1.00  -8.287292817680D-03
 1            8.287292817680D-03  0.                     -.75  -6.215469613260D-03
                                                          -.50  -4.143646408840D-03
 0   -5.0  -8.287292817680D-03  -4.143646408840D-03      -.25  -2.071823204420D-03
 1   -4.0   4.972375690608D-02   2.486187845304D-02       .25   1.942334254144D-03
 2   -3.0  -1.988950276243D-01  -9.944751381215D-02       .50   3.107734806630D-03
 3   -2.0   7.458563535912D-01   3.729281767956D-01       .75   2.719267955801D-03
 4   -1.0  -1.784530386740D+00  -8.922651933702D-01      1.25  -5.050069060773D-03
 5    0.0   2.392265193370D+00   1.196132596685D+00      1.50  -9.323204419890D-03
 6    1.0  -1.784530386740D+00  -8.922651933702D-01      1.75  -8.934737569061D-03
 7    2.0   7.458563535912D-01   3.729281767956D-01      2.25   1.825794198895D-02
 8    3.0  -1.988950276243D-01  -9.944751381215D-02      2.50   3.418508287293D-02
 9    4.0   4.972375690608D-02   2.486187845304D-02      2.75   3.301968232044D-02
10    5.0  -8.287292817680D-03  -4.143646408840D-03      3.25  -6.798169889503D-02
                                                         3.50  -1.274171270718D-01
        0   4.143646408840D-02   4.143646408840D-02      3.75  -1.231439917127D-01
        1   8.287292817680D-03   0.                      4.25   2.692938535912D-01
                                                         4.50   6.004834254144D-01
                                                         4.75   8.814312845304D-01


 I= 6     H=  1.0
 0            0.                -1.546432631050D-01    -1.00   2.220518649713D-03
 1           -2.220518649713D-03  1.435406698565D-02     -.75   1.665388987285D-03
                                                          -.50   1.110259324857D-03
 0   -5.0   2.220518649713D-03   1.110259324857D-03      -.25   5.551296624283D-04
 1   -4.0  -1.332311189828D-02  -6.661555949140D-03       .25  -5.204340585265D-04
 2   -3.0   5.329244759312D-02   2.664622379656D-02       .50  -8.326944936424D-04
 3   -2.0  -1.998466784742D-01  -9.992333923709D-02       .75  -7.286076819371D-04
 4   -1.0   7.460942663036D-01   3.730471331518D-01      1.25   1.353128552169D-03
 5    0.0  -1.784530386740D+00  -8.922651933702D-01      1.50   2.498083480927D-03
 6    1.0   2.392027280658D+00   1.196013640329D+00      1.75   2.393996669222D-03
 7    2.0  -1.783578735890D+00  -8.917893679452D-01      2.25  -4.892080150149D-03
 8    3.0   7.422876629041D-01   3.711438314521D-01      2.50  -9.159639430067D-03
 9    4.0  -1.855719157260D-01  -9.278595786302D-02      2.75  -8.847378994951D-03
10    5.0   3.092865262101D-02   1.546432631050D-02      3.25   1.821519204843D-02
                                                         3.50   3.414047423934D-02
        0  -1.110259324857D-02  -8.287292817680D-02      3.75   3.299551931058D-02
        1  -2.220518649713D-03   1.435406698565D-02      4.25  -6.796868804356D-02
                                                         4.50  -1.274022575273D-01
                                                         4.75  -1.231346982474D-01


 I= 7     H=  2.0
 0            0.                 5.771365883317D-01    -1.00  -5.947817811732D-04
 1            5.947817811732D-04 -5.741626794258D-02     -.75  -4.460863358799D-04
                                                          -.50  -2.973908905866D-04
 0   -5.0  -5.947817811732D-04  -2.973908905866D-04      -.25  -1.486954452933D-04
 1   -4.0   3.568690687039D-03   1.784345343520D-03       .25   1.394019799625D-04
 2   -3.0  -1.427476274816D-02  -7.137381374078D-03       .50   2.230431679399D-04
 3   -2.0   5.353036030559D-02   2.676518015279D-02       .75   1.951627719474D-04
```

4	-1.0	-1.998466784742D-01	-9.992333923709D-02	1.25	-3.624451479024D-04
5	0.0	7.458563535912D-01	3.729281767956D-01	1.50	-6.691295038198D-04
6	1.0	-1.783578735890D+00	-8.917893679452D-01	1.75	-6.412491078273D-04
7	2.0	2.388458589971D+00	1.194229294985D+00	2.25	1.310378611647D-03
8	3.0	-1.770255623992D+00	-8.851278119961D-01	2.50	2.453474847339D-03
9	4.0	6.925639059980D-01	3.462819529990D-01	2.75	2.369833659362D-03
10	5.0	-1.154273176663D-01	-5.771365883317D-02	3.25	-4.879069298686D-03
				3.50	-9.144769885538D-03
	0	2.973908905866D-03	2.900552486188D-01	3.75	-8.838085529620D-03
	1	5.947817811732D-04	-5.741626794258D-02	4.25	1.820589858310D-02
				4.50	3.412560469481D-02
				4.75	3.298250845912D-02

I= 8 H= 3.0

0		0.	-2.153903090222D+00	-1.00	1.586084749795D-04
1		-1.586084749795D-04	2.153110047847D-01	-.75	1.189563562346D-04
				-.50	7.930423748976D-05
0	-5.0	1.586084749795D-04	7.930423748976D-05	-.25	3.965211874488D-05
1	-4.0	-9.516508498771D-04	-4.758254249385D-04	.25	-3.717386132332D-05
2	-3.0	3.806603399508D-03	1.903301699754D-03	.50	-5.947817811732D-05
3	-2.0	-1.427476274816D-02	-7.137381374078D-03	.75	-5.204340585265D-05
4	-1.0	5.329244759312D-02	2.664622379656D-02	1.25	9.665203944064D-05
5	0.0	-1.988950276243D-01	-9.944751381215D-02	1.50	1.784345343520D-04
6	1.0	7.422876629041D-01	3.711438314521D-01	1.75	1.709997620873D-04
7	2.0	-1.770255623992D+00	-8.851278119961D-01	2.25	-3.494342964392D-04
8	3.0	2.338734833065D+00	1.169367416532D+00	2.50	-6.542599592905D-04
9	4.0	-1.584683708266D+00	-7.923418541331D-01	2.75	-6.319556424965D-04
10	5.0	4.307806180444D-01	2.153903090222D-01	3.25	1.301085146316D-03
				3.50	2.438605302810D-03
	0	-7.930423748976D-04	-1.077348066298D+00	3.75	2.356822807899D-03
	1	-1.586084749795D-04	2.153110047847D-01	4.25	-4.854906288826D-03
				4.50	-9.100161251950D-03
				4.75	-8.795335589098D-03

I= 9 H= 4.0

0		0.	8.038475772555D+00	-1.00	-3.965211874488D-05
1		3.965211874488D-05	-8.038277511962D-01	-.75	-2.973908905866D-05
				-.50	-1.982605937244D-05
0	-5.0	-3.965211874488D-05	-1.982605937244D-05	-.25	-9.913029686220D-06
1	-4.0	2.379127124693D-04	1.189563562346D-04	.25	9.293465330831D-06
2	-3.0	-9.516508498771D-04	-4.758254249385D-04	.50	1.486954452933D-05
3	-2.0	3.568690687039D-03	1.784345343520D-03	.75	1.301085146316D-05
4	-1.0	-1.332311189828D-02	-6.661555949140D-03	1.25	-2.416300986016D-05
5	0.0	4.972375690608D-02	2.486187845304D-02	1.50	-4.460863358799D-05
6	1.0	-1.855719157260D-01	-9.278595786302D-02	1.75	-4.274994052182D-05
7	2.0	6.925639059980D-01	3.462819529990D-01	2.25	8.735857410981D-05
8	3.0	-1.584683708266D+00	-7.923418541331D-01	2.50	1.635649898226D-04
9	4.0	1.646170927067D+00	8.230854635333D-01	2.75	1.579889106241D-04
10	5.0	-6.076951545111D-01	-3.038475772555D-01	3.25	-3.252712865791D-04
				3.50	-6.096513257025D-04
	0	1.982605937244D-04	4.019337016575D+00	3.75	-5.892057019747D-04
	1	3.965211874488D-05	-8.038277511962D-01	4.25	1.213726572207D-03
				4.50	2.275040312987D-03
				4.75	2.198833897275D-03

```
I=10    H=  5.0
 0              0.                     -5.839745962093D+00     -1.00   6.608686457480D-06
 1           -6.608686457480D-06        6.339712918660D-01      -.75   4.956514843110D-06
                                                                -.50   3.304343228740D-06
 0  -5.0      6.608686457480D-06        3.304343228740D-06      -.25   1.652171614370D-06
 1  -4.0     -3.965211874488D-05       -1.982605937244D-05       .25  -1.548910888472D-06
 2  -3.0      1.586084749795D-04        7.930423748976D-05       .50  -2.478257421555D-06
 3  -2.0     -5.947817811732D-04       -2.973908905866D-04       .75  -2.168475243861D-06
 4  -1.0      2.220518649713D-03        1.110259324857D-03      1.25   4.027168310027D-06
 5   0.0     -8.287292817680D-03       -4.143646408840D-03      1.50   7.434772264665D-06
 6   1.0      3.092865262101D-02        1.546432631050D-02      1.75   7.124990086970D-06
 7   2.0     -1.154273176663D-01       -5.771365883317D-02      2.25  -1.455976235163D-05
 8   3.0      4.307806180444D-01        2.153903090222D-01      2.50  -2.726083163710D-05
 9   4.0     -6.076951545111D-01       -3.038475772555D-01      2.75  -2.633148510402D-05
10   5.0      2.679491924185D-01        1.339745962093D-01      3.25   5.421188109651D-05
                                                                3.50   1.016085542838D-04
         0   -3.304343228740D-05       -2.669889502762D+00      3.75   9.820095032911D-05
         1   -6.608686457480D-06        6.339712918660D-01      4.25  -2.022877620344D-04
                                                                4.50  -3.791733854979D-04
                                                                4.75  -3.664723162124D-04
```

```
                   N= 2     M=11                        P=  5.5      Q= 3

I= 0    H= -5.5
 0            1.000000000000D+00        5.000097393571D-01     -1.00   2.267949192430D+00
 1           -1.267949192430D+00       -6.339754816112D-01      -.75   1.950961894323D+00
                                                                -.50   1.633974596215D+00
 0  -5.5      2.679491924302D-01        1.339745962151D-01      -.25   1.316987298108D+00
 1  -4.5     -6.076951545813D-01       -3.038475772907D-01       .25   6.871994080242D-01
 2  -3.5      4.307806183252D-01        2.153903091626D-01       .50   3.995190528387D-01
 3  -2.5     -1.154273187196D-01       -5.771365935979D-02       .75   1.620791712338D-01
 4  -1.5      3.092865655308D-02        1.546432827654D-02      1.25  -7.109346086284D-02
 5   -.5     -8.287307492753D-03       -4.143653746377D-03      1.50  -7.355715851601D-02
 6    .5      2.220573417930D-03        1.110286708965D-03      1.75  -3.924227691117D-02
 7   1.5     -5.949861789669D-04       -2.974930894834D-04      2.25   1.904943542718D-02
 8   2.5      1.593712979376D-04        7.968564896878D-05      2.50   1.970958122535D-02
 9   3.5     -4.249901278335D-05       -2.124950639167D-05      2.75   1.051493641085D-02
10   4.5      1.062475319584D-05        5.312376597919D-06      3.25  -5.104280845872D-03
11   5.5     -1.770792199306D-06       -8.853960996531D-07      3.50  -5.281166385406D-03
                                                                3.75  -2.817468732237D-03
         0   -5.973720558366D+00       -2.986855409505D+00      4.25   1.367687956311D-03
         1   -1.267949192430D+00       -6.339754816112D-01      4.50   1.415084316271D-03
                                                                4.75   7.549385180948D-04
                                                                5.25  -3.664709793720D-04
                                                                5.50  -3.791708796764D-04

I= 1    H= -4.5
```

CARDINAL SPLINES N=2

```
 0          0.                  -5.843614257710D-05   -1.00 -1.607695154581D+00
 1          1.607695154581D+00   8.038528896673D-01    -.75 -1.205771365936D+00
                                                       -.50 -8.038475772907D-01
 0  -5.5   -6.076951545813D-01  -3.038475772907D-01    -.25 -4.019237886453D-01
 1  -4.5    1.646170927488D+00   8.230854637439D-01     .25  3.924285518550D-01
 2  -3.5   -1.584683709951D+00  -7.923418549757D-01     .50  7.278856829680D-01
 3  -2.5    6.925639123175D-01   3.462819561587D-01     .75  9.493999725970D-01
 4  -1.5   -1.855719393185D-01  -9.278596965925D-02    1.25  8.484357651770D-01
 5   -.5    4.972384495652D-02   2.486192247826D-02    1.50  5.663429510960D-01
 6    .5   -1.332344050758D-02  -6.661720253790D-03    1.75  2.510786614670D-01
 7   1.5    3.569917073801D-03   1.784958536901D-03    2.25 -1.142966125631D-01
 8   2.5   -9.562277876253D-04  -4.781138938127D-04    2.50 -1.182574873521D-01
 9   3.5    2.549940767001D-04   1.274970383500D-04    2.75 -6.308961846511D-02
10   4.5   -6.374851917502D-05  -3.187425958751D-05    3.25  3.062568507523D-02
11   5.5    1.062475319584D-05   5.312376597919D-06    3.50  3.168699831244D-02
                                                       3.75  1.690481239342D-02
        0   8.842323350197D+00   4.421132457027D+00    4.25 -8.206127737866D-03
        1   1.607695154581D+00   8.038528896673D-01    4.50 -8.490505897623D-03
                                                       4.75 -4.529631108569D-03
                                                       5.25  2.198825876232D-03
                                                       5.50  2.275025278059D-03


I= 2     H= -3.5
 0          0.                   2.337445703084D-04   -1.00  4.307806183252D-01
 1         -4.307806183252D-01  -2.154115586690D-01    -.75  3.230854637439D-01
                                                       -.50  2.153903091626D-01
 0  -5.5    4.307806183252D-01   2.153903091626D-01    -.25  1.076951545813D-01
 1  -4.5   -1.584683709951D+00  -7.923418549757D-01     .25 -1.009642074200D-01
 2  -3.5    2.338734839805D+00   1.169367419903D+00     .50 -1.615427318720D-01
 3  -2.5   -1.770255649270D+00  -8.851278246349D-01     .75 -1.413498903880D-01
 4  -1.5    7.422877572740D-01   3.711438786370D-01    1.25  2.781319392919D-01
 5   -.5   -1.988953798261D-01  -9.944768991304D-02    1.50  6.096281956159D-01
 6    .5    5.329376203032D-02   2.664688101516D-02    1.75  8.863103541319D-01
 7   1.5   -1.427966829521D-02  -7.139834147603D-03    2.25  8.790614502522D-01
 8   2.5    3.824911150501D-03   1.912455575251D-03    2.50  5.980299494085D-01
 9   3.5   -1.019976306800D-03  -5.099881534002D-04    2.75  2.679834738605D-01
10   4.5    2.549940767001D-04   1.274970383500D-04    3.25 -1.225027403009D-01
11   5.5   -4.249901278335D-05  -2.124950639167D-05    3.50 -1.267479932497D-01
                                                       3.75 -6.761924957368D-02
        0  -2.369293400789D+00  -1.184529828109D+00    4.25  3.282451095146D-02
        1  -4.307806183252D-01  -2.154115586690D-01    4.50  3.396202359049D-02
                                                       4.75  1.811852443428D-02
                                                       5.25 -8.795303504929D-03
                                                       5.50 -9.100101112235D-03


I= 3     H= -2.5
 0          0.                  -8.765421386566D-04   -1.00 -1.154273187196D-01
 1          1.154273187196D-01   5.779334500876D-02    -.75 -8.657048903968D-02
                                                       -.50 -5.771365935979D-02
 0  -5.5   -1.154273187196D-01  -5.771365935979D-02    -.25 -2.885682967989D-02
 1  -4.5    6.925639123175D-01   3.462819561587D-01     .25  2.705327782490D-02
 2  -3.5   -1.770255649270D+00  -8.851278246349D-01     .50  4.328524451984D-02
 3  -2.5    2.388458684762D+00   1.194229342381D+00     .75  3.787458895486D-02
 4  -1.5   -1.783579089777D+00  -8.917895448887D-01    1.25 -7.033852234474D-02
 5   -.5    7.458576743478D-01   3.729288371739D-01    1.50 -1.298557335595D-01
```

6	.5	-1.998516076137D-01	-9.992580380685D-02	1.75	-1.244450779945D-01
7	1.5	5.354875610702D-02	2.677437805351D-02	2.25	2.699258115541D-01
8	2.5	-1.434341681438D-02	-7.171708407190D-03	2.50	6.011376897182D-01
9	3.5	3.824911150501D-03	1.912455575251D-03	2.75	8.817807230233D-01
10	4.5	-9.562277876253D-04	-4.781138938127D-04	3.25	8.812602761285D-01
11	5.5	1.593712979376D-04	7.968564896878D-05	3.50	6.003049746865D-01
				3.75	2.691971859013D-01
	0	6.348502529577D-01	3.169868554095D-01	4.25	-1.230919160680D-01
	1	1.154273187196D-01	5.779334500876D-02	4.50	-1.273575884644D-01
				4.75	-6.794446662854D-02
				5.25	3.298238814348D-02
				5.50	3.412537917088D-02

I= 4 H= -1.5

0		0.	3.272423984318D-03	-1.00	3.092865655308D-02
1		-3.092865655308D-02	-1.576182136602D-02	-.75	2.319649241481D-02
				-.50	1.546432827654D-02
0	-5.5	3.092865655308D-02	1.546432827654D-02	-.25	7.732164138271D-03
1	-4.5	-1.855719393185D-01	-9.278596965925D-02	.25	-7.248903879629D-03
2	-3.5	7.422877572740D-01	3.711438786370D-01	.50	-1.159824620741D-02
3	-2.5	-1.783579089777D+00	-8.917895448887D-01	.75	-1.014846543148D-02
4	-1.5	2.392028601836D+00	1.196014300918D+00	1.25	1.884715008703D-02
5	-.5	-1.784535317565D+00	-8.922676587825D-01	1.50	3.479473862222D-02
6	.5	7.461126684245D-01	3.730563342122D-01	1.75	3.334495784629D-02
7	1.5	-1.999153561329D-01	-9.995767806644D-02	2.25	-6.813969646851D-02
8	2.5	5.354875610702D-02	2.677437805351D-02	2.50	-1.275807082815D-01
9	3.5	-1.427966829521D-02	-7.139834147603D-03	2.75	-1.232313659537D-01
10	4.5	3.569917073801D-03	1.784958536901D-03	3.25	2.693366357870D-01
11	5.5	-5.949861789669D-04	-2.974930894834D-04	3.50	6.005280945036D-01
				3.75	8.814555059685D-01
	0	-1.701076110420D-01	-8.341759352882D-02	4.25	8.814181533205D-01
	1	-3.092865655308D-02	-1.576182136602D-02	4.50	6.004683302669D-01
				4.75	2.692843420799D-01
				5.25	-1.231342490690D-01
				5.50	-1.274014155713D-01

I= 5 H= -.5

0		0.	-1.221315379861D-02	-1.00	-8.287307492753D-03
1		8.287307492753D-03	5.253940455342D-03	-.75	-6.215480619565D-03
				-.50	-4.143653746377D-03
0	-5.5	-8.287307492753D-03	-4.143653746377D-03	-.25	-2.071826873188D-03
1	-4.5	4.972384495652D-02	2.486192247826D-02	.25	1.942337693614D-03
2	-3.5	-1.988953798261D-01	-9.944768991304D-02	.50	3.107740309782D-03
3	-2.5	7.458576743478D-01	3.729288371739D-01	.75	2.719272771060D-03
4	-1.5	-1.784535317565D+00	-8.922676587825D-01	1.25	-5.050078003396D-03
5	-.5	2.392283595912D+00	1.196141797956D+00	1.50	-9.323220929347D-03
6	.5	-1.784599066084D+00	-8.922995330421D-01	1.75	-8.934753390624D-03
7	1.5	7.461126684245D-01	3.730563342122D-01	2.25	1.825797431997D-02
8	2.5	-1.998516076137D-01	-9.992580380685D-02	2.50	3.418514340761D-02
9	3.5	5.329376203032D-02	2.664688101516D-02	2.75	3.301974079144D-02
10	4.5	-1.332344050758D-02	-6.661720253790D-03	3.25	-6.798181927649D-02
11	5.5	2.220573417930D-03	1.110286708965D-03	3.50	-1.274173527011D-01
				3.75	-1.231442097751D-01
	0	4.558019121014D-02	1.668351870576D-02	4.25	2.692943027860D-01
	1	8.287307492753D-03	5.253940455342D-03	4.50	6.004842673967D-01

CARDINAL SPLINES N=2

				4.75	8.814320983091D-01
				5.25	8.814296081325D-01
				5.50	6.004802831143D-01

I= 6 H= .5

0		0.	4.558019121014D-02	-1.00	2.220573417930D-03
1		-2.220573417930D-03	-5.253940455342D-03	-.75	1.665430063447D-03
				-.50	1.110286708965D-03
0	-5.5	2.220573417930D-03	1.110286708965D-03	-.25	5.551433544825D-04
1	-4.5	-1.332344050758D-02	-6.661720253790D-03	.25	-5.204468948273D-04
2	-3.5	5.329376203032D-02	2.664688101516D-02	.50	-8.327150317237D-04
3	-2.5	-1.998516076137D-01	-9.992580380685D-02	.75	-7.286256527583D-04
4	-1.5	7.461126684245D-01	3.730563342122D-01	1.25	1.353161926551D-03
5	-.5	-1.784599066084D+00	-8.922995330421D-01	1.50	2.498145095171D-03
6	.5	2.392283595912D+00	1.196141797956D+00	1.75	2.394055716206D-03
7	1.5	-1.784535317565D+00	-8.922676587825D-01	2.25	-4.892200811377D-03
8	2.5	7.458576743478D-01	3.729288371739D-01	2.50	-9.159865348961D-03
9	3.5	-1.988953798261D-01	-9.944768991304D-02	2.75	-8.847597212065D-03
10	4.5	4.972384495652D-02	2.486192247826D-02	3.25	1.821564131896D-02
11	5.5	-8.287307492753D-03	-4.143653746377D-03	3.50	3.414131630067D-02
				3.75	3.299633313205D-02
	0	-1.221315379861D-02	1.668351870576D-02	4.25	-6.797036446445D-02
	1	-2.220573417930D-03	-5.253940455342D-03	4.50	-1.274053998537D-01
				4.75	-1.231377353161D-01
				5.25	2.692908165388D-01
				5.50	6.004802831143D-01

I= 7 H= 1.5

0		0.	-1.701076110420D-01	-1.00	-5.949861789669D-04
1		5.949861789669D-04	1.576182136602D-02	-.75	-4.462396342252D-04
				-.50	-2.974930894834D-04
0	-5.5	-5.949861789669D-04	-2.974930894834D-04	-.25	-1.487465447417D-04
1	-4.5	3.569917073801D-03	1.784958536901D-03	.25	1.394498856954D-04
2	-3.5	-1.427966829521D-02	-7.139834147603D-03	.50	2.231198171126D-04
3	-2.5	5.354875610702D-02	2.677437805351D-02	.75	1.952298399735D-04
4	-1.5	-1.999153561329D-01	-9.995767806644D-02	1.25	-3.625697028079D-04
5	-.5	7.461126684245D-01	3.730563342122D-01	1.50	-6.693594513377D-04
6	.5	-1.784535317565D+00	-8.922676587825D-01	1.75	-6.414694741987D-04
7	1.5	2.392028601836D+00	1.196014300918D+00	2.25	1.310828925536D-03
8	2.5	-1.783579089777D+00	-8.917895448887D-01	2.50	2.454317988238D-03
9	3.5	7.422877572740D-01	3.711438786370D-01	2.75	2.370648056821D-03
10	4.5	-1.855719393185D-01	-9.278596965925D-02	3.25	-4.880745999338D-03
11	5.5	3.092865655308D-02	1.546432827654D-02	3.50	-9.147912501616D-03
				3.75	-8.841122753086D-03
	0	3.272423984318D-03	-8.341759352882D-02	4.25	1.821215507181D-02
	1	5.949861789669D-04	1.576182136602D-02	4.50	3.413733201822D-02
				4.75	3.299384295552D-02
				5.25	-6.796787428792D-02
				5.50	-1.274014155713D-01

I= 8 H= 2.5

0		0.	6.348502529577D-01	-1.00	1.593712979376D-04
1		-1.593712979376D-04	-5.779334500876D-02	-.75	1.195284734532D-04
				-.50	7.968564896878D-05

0	-5.5	1.593712979376D-04	7.968564896878D-05
1	-4.5	-9.562277876253D-04	-4.781138938127D-04
2	-3.5	3.824911150501D-03	1.912455575251D-03
3	-2.5	-1.434341681438D-02	-7.171708407190D-03
4	-1.5	5.354875610702D-02	2.677437805351D-02
5	-.5	-1.998516076137D-01	-9.992580380685D-02
6	.5	7.458576743478D-01	3.729288371739D-01
7	1.5	-1.783579089777D+00	-8.917895448887D-01
8	2.5	2.388458684762D+00	1.194229342381D+00
9	3.5	-1.770255649270D+00	-8.851278246349D-01
10	4.5	6.925639123175D-01	3.462819561587D-01
11	5.5	-1.154273187196D-01	-5.771365935979D-02
	0	-8.765421386566D-04	3.169868554095D-01
	1	-1.593712979376D-04	-5.779334500876D-02

-.25	3.984282448439D-05
.25	-3.735264795412D-05
.50	-5.976423672658D-05
.75	-5.229370713576D-05
1.25	9.711688468070D-05
1.50	1.792927101798D-04
1.75	1.718221805889D-04
2.25	-3.511148907687D-04
2.50	-6.574066039924D-04
2.75	-6.349950152200D-04
3.25	1.307342678394D-03
3.50	2.450333705790D-03
3.75	2.368157880291D-03
4.25	-4.878255822807D-03
4.50	-9.143928219167D-03
4.75	-8.837636505944D-03
5.25	1.820568061284D-02
5.50	3.412537917088D-02

I= 9   H= 3.5

0		0.	-2.369293400789D+00
1		4.249901278335D-05	2.154115586690D-01
0	-5.5	-4.249901278335D-05	-2.124950639167D-05
1	-4.5	2.549940767001D-04	1.274970383500D-04
2	-3.5	-1.019976306800D-03	-5.099881534002D-04
3	-2.5	3.824911150501D-03	1.912455575251D-03
4	-1.5	-1.427966829521D-02	-7.139834147603D-03
5	-.5	5.329376203032D-02	2.664688101516D-02
6	.5	-1.988953798261D-01	-9.944768991304D-02
7	1.5	7.422877572740D-01	3.711438786370D-01
8	2.5	-1.770255649270D+00	-8.851278246349D-01
9	3.5	2.338734839805D+00	1.169367419903D+00
10	4.5	-1.584683709951D+00	-7.923418549757D-01
11	5.5	4.307806183252D-01	2.153903091626D-01
	0	2.337445703084D-04	-1.184529828109D+00
	1	4.249901278335D-05	2.154115586690D-01

-1.00	-4.249901278335D-05
-.75	-3.187425958751D-05
-.50	-2.124950639167D-05
-.25	-1.062475319584D-05
.25	9.960706121097D-06
.50	1.593712979376D-05
.75	1.394498856954D-05
1.25	-2.589783591485D-05
1.50	-4.781138938127D-05
1.75	-4.581924815705D-05
2.25	9.363063753832D-05
2.50	1.753084277313D-04
2.75	1.693320040587D-04
3.25	-3.486247142384D-04
3.50	-6.534223215440D-04
3.75	-6.315087680776D-04
4.25	1.300868219415D-03
4.50	2.438380858445D-03
4.75	2.356703068252D-03
5.25	-4.854848163423D-03
5.50	-9.100101112235D-03

I=10   H= 4.5

0		0.	8.842323350197D+00
1		-1.062475319584D-05	-8.038528896673D-01
0	-5.5	1.062475319584D-05	5.312376597919D-06
1	-4.5	-6.374851917502D-05	-3.187425958751D-05
2	-3.5	2.549940767001D-04	1.274970383500D-04
3	-2.5	-9.562277876253D-04	-4.781138938127D-04
4	-1.5	3.569917073801D-03	1.784958536901D-03
5	-.5	-1.332344050758D-02	-6.661720253790D-03
6	.5	4.972384495652D-02	2.486192247826D-02
7	1.5	-1.855719393185D-01	-9.278596965925D-02
8	2.5	6.925639123175D-01	3.462819561587D-01

-1.00	1.062475319584D-05
-.75	7.968564896878D-06
-.50	5.312376597919D-06
-.25	2.656188298959D-06
.25	-2.490176530274D-06
.50	-3.984282448439D-06
.75	-3.486247142384D-06
1.25	6.474458978713D-06
1.50	1.195284734532D-05
1.75	1.145481203926D-05
2.25	-2.340765938458D-05
2.50	-4.382710693283D-05

9	3.5	-1.584683709951D+00	-7.923418549757D-01	2.75	-4.233300101466D-05
10	4.5	1.646170927488D+00	8.230854637439D-01	3.25	8.715617855960D-05
11	5.5	-6.076951545813D-01	-3.038475772907D-01	3.50	1.633555803860D-04
				3.75	1.578771920194D-04
	0	-5.843614257710D-05	4.421132457027D+00	4.25	-3.252170548538D-04
	1	-1.062475319584D-05	-8.038528896673D-01	4.50	-6.095952146112D-04
				4.75	-5.891757670629D-04
				5.25	1.213712040856D-03
				5.50	2.275025278059D-03

I=11 H= 5.5

0		0.	-6.473720558366D+00	-1.00	-1.770792199306D-06
1		1.770792199306D-06	6.339754816112D-01	-.75	-1.328094149480D-06
				-.50	-8.853960996531D-07
0	-5.5	-1.770792199306D-06	-8.853960996531D-07	-.25	-4.426980498266D-07
1	-4.5	1.062475319584D-05	5.312376597919D-06	.25	4.150294217124D-07
2	-3.5	-4.249901278335D-05	-2.124950639167D-05	.50	6.640470747398D-07
3	-2.5	1.593712979376D-04	7.968564896878D-05	.75	5.810411903973D-07
4	-1.5	-5.949861789669D-04	-2.974930894834D-04	1.25	-1.079076496452D-06
5	-.5	2.220573417930D-03	1.110286708965D-03	1.50	-1.992141224219D-06
6	.5	-8.287307492753D-03	-4.143653746377D-03	1.75	-1.909135339877D-06
7	1.5	3.092865655308D-02	1.546432827654D-02	2.25	3.901276564096D-06
8	2.5	-1.154273187196D-01	-5.771365935979D-02	2.50	7.304517822138D-06
9	3.5	4.307806183252D-01	2.153903091626D-01	2.75	7.055500169111D-06
10	4.5	-6.076951545813D-01	-3.038475772907D-01	3.25	-1.452602975993D-05
11	5.5	2.679491924302D-01	1.339745962151D-01	3.50	-2.722593006433D-05
				3.75	-2.631286533657D-05
	0	9.739357096184D-06	-2.986855409505D+00	4.25	5.420284247564D-05
	1	1.770792199306D-06	6.339754816112D-01	4.50	1.015992024352D-04
				4.75	9.819596117715D-05
				5.25	-2.022853401426D-04
				5.50	-3.791708796764D-04

N= 2 M=12 P= 6.0 Q= 3

I= 0 H= -6.0

0		1.000000000000D+00	4.999971531060D-01	-1.00	2.267949192431D+00
1		-1.267949192431D+00	-6.339743589744D-01	-.75	1.950961894323D+00
				-.50	1.633974596216D+00
0	-6.0	2.679491924311D-01	1.339745962155D-01	-.25	1.316987298108D+00
1	-5.0	-6.076951545863D-01	-3.038475772932D-01	.25	6.871994080240D-01
2	-4.0	4.307806183454D-01	2.153903091727D-01	.50	3.995190528384D-01
3	-3.0	-1.154273187952D-01	-5.771365939760D-02	.75	1.620791712336D-01
4	-2.0	3.092865683539D-02	1.546432841770D-02	1.25	-7.109346086232D-02
5	-1.0	-8.287308546376D-03	-4.143654273188D-03	1.50	-7.355715851506D-02
6	0.0	2.220577350111D-03	1.110288675056D-03	1.75	-3.924227691027D-02
7	1.0	-5.950008540682D-04	-2.975004270341D-04	2.25	1.904943542533D-02
8	2.0	1.594260661618D-04	7.971303308091D-05	2.50	1.970958122189D-02

9	3.0	-4.270341057906D-05	-2.135170528953D-05	2.75	1.051493640750D-02
10	4.0	1.138757615442D-05	5.693788077208D-06	3.25	-5.104280838980D-03
11	5.0	-2.846894038604D-06	-1.423447019302D-06	3.50	-5.281166372488D-03
12	6.0	4.744823397673D-07	2.372411698837D-07	3.75	-2.817468719752D-03
				4.25	1.367687930593D-03
	0	-6.607695154586D+00	-3.303849000740D+00	4.50	1.415084268064D-03
	1	-1.267949192431D+00	-6.339743589744D-01	4.75	7.549384715026D-04
				5.25	-3.664708833912D-04
				5.50	-3.791706997666D-04
				5.75	-2.022851662586D-04

I= 1		H= -5.0			
0		0.	1.708136423162D-05	-1.00	-1.607695154586D+00
1		1.607695154586D+00	8.038461538462D-01	-.75	-1.205771365940D+00
				-.50	-8.038475772932D-01
0	-6.0	-6.076951545863D-01	-3.038475772932D-01	-.25	-4.019237886466D-01
1	-5.0	1.646170927518D+00	8.230854637590D-01	.25	3.924285518562D-01
2	-4.0	-1.584683710072D+00	-7.923418550362D-01	.50	7.278856829699D-01
3	-3.0	6.925639127712D-01	3.462819563856D-01	.75	9.493999725986D-01
4	-2.0	-1.855719410124D-01	-9.278597050618D-02	1.25	8.484357651739D-01
5	-1.0	4.972385127826D-02	2.486192563913D-02	1.50	5.663429510904D-01
6	0.0	-1.332346410067D-02	-6.661732050333D-03	1.75	2.510786614616D-01
7	1.0	3.570005124409D-03	1.785002562205D-03	2.25	-1.142966125520D-01
8	2.0	-9.565563969709D-04	-4.782781984855D-04	2.50	-1.182574873313D-01
9	3.0	2.562204634743D-04	1.281102317372D-04	2.75	-6.308961844503D-02
10	4.0	-6.832545692649D-05	-3.416272846325D-05	3.25	3.062568503388D-02
11	5.0	1.708136423162D-05	8.540682115812D-06	3.50	3.168699823493D-02
12	6.0	-2.846894038604D-06	-1.423447019302D-06	3.75	1.690481231851D-02
				4.25	-8.206127583556D-03
	0	9.646170927518D+00	4.823094004441D+00	4.50	-8.490505608381D-03
	1	1.607695154586D+00	8.038461538462D-01	4.75	-4.529630829016D-03
				5.25	2.198825300347D-03
				5.50	2.275024198599D-03
				5.75	1.213710997552D-03

I= 2		H= -4.0			
0		0.	-6.832545692649D-05	-1.00	4.307806183454D-01
1		-4.307806183454D-01	-2.153846153846D-01	-.75	3.230854637590D-01
				-.50	2.153903091727D-01
0	-6.0	4.307806183454D-01	2.153903091727D-01	-.25	1.076951545863D-01
1	-5.0	-1.584683710072D+00	-7.923418550362D-01	.25	-1.009642074247D-01
2	-4.0	2.338734840289D+00	1.169367420145D+00	.50	-1.615427318795D-01
3	-3.0	-1.770255651085D+00	-8.851278255423D-01	.75	-1.413498903946D-01
4	-2.0	7.422877640494D-01	3.711438820247D-01	1.25	2.781319393042D-01
5	-1.0	-1.988954051130D-01	-9.944770255651D-02	1.50	6.096281956386D-01
6	0.0	5.329385640266D-02	2.664692820133D-02	1.75	8.863103541536D-01
7	1.0	-1.428002049764D-02	-7.140010248819D-03	2.25	8.790614502078D-01
8	2.0	3.826225587884D-03	1.913112793942D-03	2.50	5.980299493253D-01
9	3.0	-1.024881853897D-03	-5.124409269487D-04	2.75	2.679834737801D-01
10	4.0	2.733018277060D-04	1.366509138530D-04	3.25	-1.225027401355D-01
11	5.0	-6.832545692649D-05	-3.416272846325D-05	3.50	-1.267479929397D-01
12	6.0	1.138757615442D-05	5.693788077208D-06	3.75	-6.761924927404D-02
				4.25	3.282451033423D-02
	0	-2.584683710072D+00	-1.292376017765D+00	4.50	3.396202243353D-02
	1	-4.307806183454D-01	-2.153846153846D-01	4.75	1.811852331606D-02

CARDINAL SPLINES N=2

```
                                                                5.25 -8.795301201389D-03
                                                                5.50 -9.100096794397D-03
                                                                5.75 -4.854843990207D-03

 I= 3      H= -3.0
 0            0.                    2.562204634743D-04        -1.00 -1.154273187952D-01
 1            1.154273187952D-01    5.769230769231D-02         -.75 -8.657048909640D-02
                                                               -.50 -5.771365939760D-02
 0   -6.0   -1.154273187952D-01   -5.771365939760D-02          -.25 -2.885682969880D-02
 1   -5.0    6.925639127712D-01    3.462819563856D-01           .25  2.705327784262D-02
 2   -4.0   -1.770255651085D+00   -8.851278255423D-01           .50  4.328524454820D-02
 3   -3.0    2.388458691567D+00    1.194229345784D+00           .75  3.787458897967D-02
 4   -2.0   -1.783579115185D+00   -8.917895575927D-01          1.25 -7.033852239082D-02
 5   -1.0    7.458577691738D-01    3.729288845869D-01          1.50 -1.298557336446D-01
 6    0.0   -1.998519615100D-01   -9.992598075500D-02          1.75 -1.244450780761D-01
 7    1.0    5.355007686614D-02    2.677503843307D-02          2.25  2.699258117207D-01
 8    2.0   -1.434834595456D-02   -7.174172977282D-03          2.50  6.011376900302D-01
 9    3.0    3.843306952115D-03    1.921653476058D-03          2.75  8.817807233246D-01
10    4.0   -1.024881853897D-03   -5.124409269487D-04          3.25  8.812602755082D-01
11    5.0    2.562204634743D-04    1.281102317372D-04          3.50  6.003049735239D-01
12    6.0   -4.270341057906D-05   -2.135170528953D-05          3.75  2.691971847777D-01
                                                               4.25 -1.230919137533D-01
         0   6.925639127712D-01    3.464100666173D-01          4.50 -1.273575841257D-01
         1   1.154273187952D-01    5.769230769231D-02          4.75 -6.794446243523D-02
                                                               5.25  3.298237950521D-02
                                                               5.50  3.412536297899D-02
                                                               5.75  1.820566496328D-02


 I= 4      H= -2.0
 0            0.                   -9.565563969709D-04        -1.00  3.092865683539D-02
 1           -3.092865683539D-02   -1.538461538462D-02         -.75  2.319649262654D-02
                                                               -.50  1.546432841770D-02
 0   -6.0    3.092865683539D-02    1.546432841770D-02          -.25  7.732164208848D-03
 1   -5.0   -1.855719410124D-01   -9.278597050618D-02           .25 -7.248903945795D-03
 2   -4.0    7.422877640494D-01    3.711438820247D-01           .50 -1.159824631327D-02
 3   -3.0   -1.783579115185D+00   -8.917895575927D-01           .75 -1.014846552411D-02
 4   -2.0    2.392028696692D+00    1.196014348346D+00          1.25  1.884715025907D-02
 5   -1.0   -1.784535671582D+00   -8.922678357912D-01          1.50  3.479473893982D-02
 6    0.0    7.461139896373D-01    3.730569948187D-01          1.75  3.334495815066D-02
 7    1.0   -1.999202869669D-01   -9.996014348346D-02          2.25 -6.813969709047D-02
 8    2.0    5.356715823037D-02    2.678357911519D-02          2.50 -1.275807094460D-01
 9    3.0   -1.434834595456D-02   -7.174172977282D-03          2.75 -1.232313670785D-01
10    4.0    3.826225587884D-03    1.913112793942D-03          3.25  2.693366381028D-01
11    5.0   -9.565563969709D-04   -4.782781984855D-04          3.50  6.005280988442D-01
12    6.0    1.594260661618D-04    7.971303308091D-05          3.75  8.814555101634D-01
                                                               4.25  8.814181446792D-01
         0  -1.855719410124D-01   -9.326424870466D-02          4.50  6.004683140694D-01
         1  -3.092865683539D-02   -1.538461538462D-02          4.75  2.692843264249D-01
                                                               5.25 -1.231342168194D-01
                                                               5.50 -1.274013551216D-01
                                                               5.75 -6.796781586289D-02


 I= 5      H= -1.0
 0            0.                    3.570005124409D-03        -1.00 -8.287308546376D-03
```

CARDINAL SPLINES N=2

1		8.287308546376D-03	3.846153846154D-03	-.75	-6.215481409782D-03
				-.50	-4.143654273188D-03
0	-6.0	-8.287308546376D-03	-4.143654273188D-03	-.25	-2.071827136594D-03
1	-5.0	4.972385127826D-02	2.486192563913D-02	.25	1.942337940557D-03
2	-4.0	-1.988954051130D-01	-9.944770255651D-02	.50	3.107740704891D-03
3	-3.0	7.458577691738D-01	3.729288845869D-01	.75	2.719273116780D-03
4	-2.0	-1.784535671582D+00	-8.922678357912D-01	1.25	-5.050078645448D-03
5	-1.0	2.392284917155D+00	1.196142458578D+00	1.50	-9.323222114673D-03
6	0.0	-1.784603997039D+00	-8.923019985196D-01	1.75	-8.934754526562D-03
7	1.0	7.461310710015D-01	3.730655355008D-01	2.25	1.825797664123D-02
8	2.0	-1.999202869669D-01	-9.996014348346D-02	2.50	3.418514775380D-02
9	3.0	5.355007686614D-02	2.677503843307D-02	2.75	3.301974498947D-02
10	4.0	-1.428002049764D-02	-7.140010248819D-03	3.25	-6.798182791949D-02
11	5.0	3.570005124409D-03	1.785002562205D-03	3.50	-1.274173689005D-01
12	6.0	-5.950008540682D-04	-2.975004270341D-04	3.75	-1.231442254313D-01
				4.25	2.692943350367D-01
	0	4.972385127826D-02	2.664692820133D-02	4.50	6.004843278483D-01
	1	8.287308546376D-03	3.846153846154D-03	4.75	8.814321567358D-01
				5.25	8.814294877726D-01
				5.50	6.004800575073D-01
				5.75	2.692905984883D-01

I= 6 H= 0.0

0		0.	-1.332346410067D-02	-1.00	2.220577350111D-03
1		-2.220577350111D-03	0.	-.75	1.665433012583D-03
				-.50	1.110288675056D-03
0	-6.0	2.220577350111D-03	1.110288675056D-03	-.25	5.551443375278D-04
1	-5.0	-1.332346410067D-02	-6.661732050333D-03	.25	-5.204478164323D-04
2	-4.0	5.329385640266D-02	2.664692820133D-02	.50	-8.327165062916D-04
3	-3.0	-1.998519615100D-01	-9.992598075500D-02	.75	-7.286269430052D-04
4	-2.0	7.461139896373D-01	3.730569948187D-01	1.25	1.353164322724D-03
5	-1.0	-1.784603997039D+00	-8.923019985196D-01	1.50	2.498149518875D-03
6	0.0	2.392301998520D+00	1.196150999260D+00	1.75	2.394059955588D-03
7	1.0	-1.784603997039D+00	-8.923019985196D-01	2.25	-4.892209474463D-03
8	2.0	7.461139896373D-01	3.730569948187D-01	2.50	-9.159881569208D-03
9	3.0	-1.998519615100D-01	-9.992598075500D-02	2.75	-8.847612879349D-03
10	4.0	5.329385640266D-02	2.664692820133D-02	3.25	1.821567357513D-02
11	5.0	-1.332346410067D-02	-6.661732050333D-03	3.50	3.414137675796D-02
12	6.0	2.220577350111D-03	1.110288675056D-03	3.75	3.299639156181D-02
				4.25	-6.797048482605D-02
	0	-1.332346410067D-02	-1.332346410067D-02	4.50	-1.274056254626D-01
	1	-2.220577350111D-03	0.	4.75	-1.231379533679D-01
				5.25	2.692912657291D-01
				5.50	6.004811250925D-01
				5.75	8.814304219097D-01

I= 7 H= 1.0

0		0.	4.972385127826D-02	-1.00	-5.950008540682D-04
1		5.950008540682D-04	-3.846153846154D-03	-.75	-4.462506405512D-04
				-.50	-2.975004270341D-04
0	-6.0	-5.950008540682D-04	-2.975004270341D-04	-.25	-1.487502135171D-04
1	-5.0	3.570005124409D-03	1.785002562205D-03	.25	1.394533251722D-04
2	-4.0	-1.428002049764D-02	-7.140010248819D-03	.50	2.231253202756D-04
3	-3.0	5.355007686614D-02	2.677503843307D-02	.75	1.952346552411D-04
4	-2.0	-1.999202869669D-01	-9.996014348346D-02	1.25	-3.625786454478D-04

5	-1.0	7.461310710015D-01	3.730655355008D-01	1.50	-6.693759608267D-04
6	0.0	-1.784603997039D+00	-8.923019985196D-01	1.75	-6.414852957923D-04
7	1.0	2.392284917155D+00	1.196142458578D+00	2.25	1.310861256619D-03
8	2.0	-1.784535671582D+00	-8.922678357912D-01	2.50	2.454378523031D-03
9	3.0	7.458577691738D-01	3.729288845869D-01	2.75	2.370706527928D-03
10	4.0	-1.988954051130D-01	-9.944770255651D-02	3.25	-4.880866381028D-03
11	5.0	4.972385127826D-02	2.486192563913D-02	3.50	-9.148138131299D-03
12	6.0	-8.287308546376D-03	-4.143654273188D-03	3.75	-8.841340815920D-03
				4.25	1.821260426749D-02
	0	3.570005124409D-03	2.664692820133D-02	4.50	3.413817400216D-02
	1	5.950008540682D-04	-3.846153846154D-03	4.75	3.299465673575D-02
				5.25	-6.796955068895D-02
				5.50	-1.274045578774D-01
				5.75	-1.231372861271D-01

I= 8 H= 2.0

0		0.	-1.855719410124D-01	-1.00	1.594260661618D-04
1		-1.594260661618D-04	1.538461538462D-02	-.75	1.195695496214D-04
				-.50	7.971303308091D-05
0	-6.0	1.594260661618D-04	7.971303308091D-05	-.25	3.985651654045D-05
1	-5.0	-9.565563969709D-04	-4.782781984855D-04	.25	-3.736548425668D-05
2	-4.0	3.826225587884D-03	1.913112793942D-03	.50	-5.978477481068D-05
3	-3.0	-1.434834595456D-02	-7.174172977282D-03	.75	-5.231167795935D-05
4	-2.0	5.356715823037D-02	2.678357911519D-02	1.25	9.715025906736D-05
5	-1.0	-1.999202869669D-01	-9.996014348346D-02	1.50	1.793543244320D-04
6	0.0	7.461139896373D-01	3.730569948187D-01	1.75	1.718812275807D-04
7	1.0	-1.784535671582D+00	-8.922678357912D-01	2.25	-3.512355520128D-04
8	2.0	2.392028696692D+00	1.196014348346D+00	2.50	-6.576325229175D-04
9	3.0	-1.783579115185D+00	-8.917895575927D-01	2.75	-6.352132323635D-04
10	4.0	7.422877640494D-01	3.711438820247D-01	3.25	1.307791948984D-03
11	5.0	-1.855719410124D-01	-9.278597050618D-02	3.50	2.451175767238D-03
12	6.0	3.092865683539D-02	1.546432841770D-02	3.75	2.368971701873D-03
				4.25	-4.879932243922D-03
	0	-9.565563969709D-04	-9.326424870466D-02	4.50	-9.147070546034D-03
	1	-1.594260661618D-04	1.538461538462D-02	4.75	-8.840673575130D-03
				5.25	1.821193702670D-02
				5.50	3.413710641690D-02
				5.75	3.299372259864D-02

I= 9 H= 3.0

0		0.	6.925639127712D-01	-1.00	-4.270341057906D-05
1		4.270341057906D-05	-5.769230769231D-02	-.75	-3.202755793429D-05
				-.50	-2.135170528953D-05
0	-6.0	-4.270341057906D-05	-2.135170528953D-05	-.25	-1.067585264476D-05
1	-5.0	2.562204634743D-04	1.281102317372D-04	.25	1.000861185447D-05
2	-4.0	-1.024881853897D-03	-5.124409269487D-04	.50	1.601377896715D-05
3	-3.0	3.843306952115D-03	1.921653476058D-03	.75	1.401205659625D-05
4	-2.0	-1.434834595456D-02	-7.174172977282D-03	1.25	-2.602239082161D-05
5	-1.0	5.355007686614D-02	2.677503843307D-02	1.50	-4.804133690144D-05
6	0.0	-1.998519615100D-01	-9.992598075500D-02	1.75	-4.603961453055D-05
7	1.0	7.458577691738D-01	3.729288845869D-01	2.25	9.408095143199D-05
8	2.0	-1.783579115185D+00	-8.917895575927D-01	2.50	1.761515686386D-04
9	3.0	2.388458691567D+00	1.194229345784D+00	2.75	1.701464015259D-04
10	4.0	-1.770255651085D+00	-8.851278255423D-01	3.25	-3.503014149063D-04
11	5.0	6.925639127712D-01	3.462819563856D-01	3.50	-6.565649376530D-04

12	6.0	-1.154273187952D-01	-5.771365939760D-02	3.75	-6.345459915732D-04
				4.25	1.307124708193D-03
	0	2.562204634743D-04	3.464100666173D-01	4.50	2.450108181973D-03
	1	4.270341057906D-05	-5.769230769231D-02	4.75	2.368037564767D-03
				5.25	-4.878197417867D-03
				5.50	-9.143867790241D-03
				5.75	-8.837604267494D-03

I=10 H= 4.0

0		0.	-2.584683710072D+00	-1.00	1.138757615442D-05
1		-1.138757615442D-05	2.153846153846D-01	-.75	8.540682115812D-06
				-.50	5.693788077208D-06
0	-6.0	1.138757615442D-05	5.693788077208D-06	-.25	2.846894038604D-06
1	-5.0	-6.832545692649D-05	-3.416272846325D-05	.25	-2.668963161191D-06
2	-4.0	2.733018277060D-04	1.366509138530D-04	.50	-4.270341057906D-06
3	-3.0	-1.024881853897D-03	-5.124409269487D-04	.75	-3.736548425668D-06
4	-2.0	3.826225587884D-03	1.913112793942D-03	1.25	6.939304219097D-06
5	-1.0	-1.428002049764D-02	-7.140010248819D-03	1.50	1.281102317372D-05
6	0.0	5.329385640266D-02	2.664692820133D-02	1.75	1.227723054148D-05
7	1.0	-1.988954051130D-01	-9.944770255651D-02	2.25	-2.508825371520D-05
8	2.0	7.422877640494D-01	3.711438820247D-01	2.50	-4.697375163696D-05
9	3.0	-1.770255651085D+00	-8.851278255423D-01	2.75	-4.537237374025D-05
10	4.0	2.338734840289D+00	1.169367420145D+00	3.25	9.341371064169D-05
11	5.0	-1.584683710072D+00	-7.923418550362D-01	3.50	1.750839833741D-04
12	6.0	4.307806183454D-01	2.153903091727D-01	3.75	1.692122644195D-04
				4.25	-3.485665888516D-04
	0	-6.832545692649D-05	-1.292376017765D+00	4.50	-6.533621818596D-04
	1	-1.138757615442D-05	2.153846153846D-01	4.75	-6.314766839378D-04
				5.25	1.300852644765D-03
				5.50	2.438364744064D-03
				5.75	2.356694471332D-03

I=11 H= 5.0

0		0.	9.646170927518D+00	-1.00	-2.846894038604D-06
1		2.846894038604D-06	-8.038461538462D-01	-.75	-2.135170528953D-06
				-.50	-1.423447019302D-06
0	-6.0	-2.846894038604D-06	-1.423447019302D-06	-.25	-7.117235096510D-07
1	-5.0	1.708136423162D-05	8.540682115812D-06	.25	6.672407902978D-07
2	-4.0	-6.832545692649D-05	-3.416272846325D-05	.50	1.067585264476D-06
3	-3.0	2.562204634743D-04	1.281102317372D-04	.75	9.341371064169D-07
4	-2.0	-9.565563969709D-04	-4.782781984855D-04	1.25	-1.734826054774D-06
5	-1.0	3.570005124409D-03	1.785002562205D-03	1.50	-3.202755793429D-06
6	0.0	-1.332346410067D-02	-6.661732050333D-03	1.75	-3.069307635370D-06
7	1.0	4.972385127826D-02	2.486192563913D-02	2.25	6.272063428799D-06
8	2.0	-1.855719410124D-01	-9.278597050618D-02	2.50	1.174343790924D-05
9	3.0	6.925639127712D-01	3.462819563856D-01	2.75	1.134309343506D-05
10	4.0	-1.584683710072D+00	-7.923418550362D-01	3.25	-2.335342766042D-05
11	5.0	1.646170927518D+00	8.230854637590D-01	3.50	-4.377099584353D-05
12	6.0	-6.076951545863D-01	-3.038475772932D-01	3.75	-4.230306610488D-05
				4.25	8.714164721289D-05
	0	1.708136423162D-05	4.823094004441D+00	4.50	1.633405454649D-04
	1	2.846894038604D-06	-8.038461538462D-01	4.75	1.578691709845D-04
				5.25	-3.252131611911D-04
				5.50	-6.095911860161D-04
				5.75	-5.891736178329D-04

CARDINAL SPLINES N=2

```
I=12     H=  6.0
0            0.                    -7.107695154586D+00     -1.00   4.744823397673D-07
1           -4.744823397673D-07     6.339743589744D-01      -.75   3.558617548255D-07
                                                            -.50   2.372411698837D-07
0   -6.0     4.744823397673D-07     2.372411698837D-07      -.25   1.186205849418D-07
1   -5.0    -2.846894038604D-06    -1.423447019302D-06       .25  -1.112067983830D-07
2   -4.0     1.138757615442D-05     5.693788077208D-06       .50  -1.779308774127D-07
3   -3.0    -4.270341057906D-05    -2.135170528953D-05       .75  -1.556895177361D-07
4   -2.0     1.594260661618D-04     7.971303308091D-05      1.25   2.891376757957D-07
5   -1.0    -5.950008540682D-04    -2.975004270341D-04      1.50   5.337926322382D-07
6    0.0     2.220577350111D-03     1.110288675056D-03      1.75   5.115512725616D-07
7    1.0    -8.287308546376D-03    -4.143654273188D-03      2.25  -1.045343904800D-06
8    2.0     3.092865683539D-02     1.546432841770D-02      2.50  -1.957239651540D-06
9    3.0    -1.154273187952D-01    -5.771365939760D-02      2.75  -1.890515572510D-06
10   4.0     4.307806183454D-01     2.153903091727D-01      3.25   3.892237943404D-06
11   5.0    -6.076951545863D-01    -3.038475772932D-01      3.50   7.295165973922D-06
12   6.0     2.679491924311D-01     1.339745962155D-01      3.75   7.050511017480D-06
                                                            4.25  -1.452360786882D-05
       0    -2.846894038604D-06    -3.303849000740D+00      4.50  -2.722342424415D-05
       1    -4.744823397673D-07     6.339743589744D-01      4.75  -2.631152849741D-05
                                                            5.25   5.420219353186D-05
                                                            5.50   1.015985310027D-04
                                                            5.75   9.819560297216D-05
```

```
              N= 2      M=13                             P=  6.5       Q= 3

I= 0     H= -6.5
0            1.000000000000D+00     5.000008263915D-01     -1.00   2.267949192431D+00
1           -1.267949192431D+00    -6.339746597841D-01      -.75   1.950961894323D+00
                                                            -.50   1.633974596216D+00
0   -6.5     2.679491924311D-01     1.339745962156D-01      -.25   1.316987298108D+00
1   -5.5    -6.076951545867D-01    -3.038475772934D-01       .25   6.871994080240D-01
2   -4.5     4.307806183468D-01     2.153903091734D-01       .50   3.995190528383D-01
3   -3.5    -1.154273188006D-01    -5.771365940031D-02       .75   1.620791712335D-01
4   -2.5     3.092865685566D-02     1.546432842783D-02      1.25  -7.109346086229D-02
5   -1.5    -8.287308622023D-03    -4.143654311011D-03      1.50  -7.355715851499D-02
6    -.5     2.220577632429D-03     1.110288816214D-03      1.75  -3.924227691020D-02
7     .5    -5.950019076931D-04    -2.975009538465D-04      2.25   1.904943542519D-02
8    1.5     1.594299983434D-04     7.971499917170D-05      2.50   1.970958122164D-02
9    2.5    -4.271808568053D-05    -2.135904284026D-05      2.75   1.051493640726D-02
10   3.5     1.144234437871D-05     5.721172189357D-06      3.25  -5.104280838485D-03
11   4.5    -3.051291834323D-06    -1.525645917162D-06      3.50  -5.281166371560D-03
12   5.5     7.628229585809D-07     3.814114792904D-07      3.75  -2.817468718855D-03
13   6.5    -1.271371597635D-07    -6.356857988174D-08      4.25   1.367687928746D-03
                                                            4.50   1.415084264602D-03
       0    -7.241669750802D+00    -3.620834462205D+00      4.75   7.549384681574D-04
       1    -1.267949192431D+00    -6.339746597841D-01      5.25  -3.664708765001D-04
```

CARDINAL SPLINES N=2

5.50	-3.791706868496D-04
5.75	-2.022851537743D-04
6.25	9.819557725420D-05
6.50	1.015984827960D-04

I= 1 H= -5.5

0		0.	-4.958349230776D-06
1		1.607695154587D+00	8.038479587048D-01
0	-6.5	-6.076951545867D-01	-3.038475772934D-01
1	-5.5	1.646170927520D+00	8.230854637601D-01
2	-4.5	-1.584683710081D+00	-7.923418550405D-01
3	-3.5	6.925639128037D-01	3.462819564019D-01
4	-2.5	-1.855719411340D-01	-9.278597056698D-02
5	-1.5	4.972385173214D-02	2.486192586607D-02
6	-.5	-1.332346579457D-02	-6.661732897287D-03
7	.5	3.570011446158D-03	1.785005723079D-03
8	1.5	-9.565799900604D-04	-4.782899950302D-04
9	2.5	2.563085140832D-04	1.281542570416D-04
10	3.5	-6.865406627228D-05	-3.432703313614D-05
11	4.5	1.830775100594D-05	9.153875502970D-06
12	5.5	-4.576937751485D-06	-2.288468875743D-06
13	6.5	7.628229585809D-07	3.814114792904D-07
	0	1.045001850481D+01	5.225006773232D+00
	1	1.607695154587D+00	8.038479587048D-01

-1.00	-1.607695154587D+00
-.75	-1.205771365940D+00
-.50	-8.038475772934D-01
-.25	-4.019237886467D-01
.25	3.924285518563D-01
.50	7.278856829700D-01
.75	9.493999725988D-01
1.25	8.484357651737D-01
1.50	5.663429510900D-01
1.75	2.510786614612D-01
2.25	-1.142966125512D-01
2.50	-1.182574873298D-01
2.75	-6.308961844358D-02
3.25	3.062568503091D-02
3.50	3.168699822936D-02
3.75	1.690481231313D-02
4.25	-8.206127572477D-03
4.50	-8.490505587615D-03
4.75	-4.529630808945D-03
5.25	2.198825259001D-03
5.50	2.275024121098D-03
5.75	1.213710922646D-03
6.25	-5.891734635252D-04
6.50	-6.095908967759D-04

I= 2 H= -4.5

0		0.	1.983339692310D-05
1		-4.307806183468D-01	-2.153918348193D-01
0	-6.5	4.307806183468D-01	2.153903091734D-01
1	-5.5	-1.584683710081D+00	-7.923418550405D-01
2	-4.5	2.338734840324D+00	1.169367420162D+00
3	-3.5	-1.770255651215D+00	-8.851278256075D-01
4	-2.5	7.422877645359D-01	3.711438822679D-01
5	-1.5	-1.988954069285D-01	-9.944770346427D-02
6	-.5	5.329386317829D-02	2.664693158915D-02
7	.5	-1.428004578463D-02	-7.140022892317D-03
8	1.5	3.826319960242D-03	1.913159980121D-03
9	2.5	-1.025234056333D-03	-5.126170281663D-04
10	3.5	2.746162650891D-04	1.373081325446D-04
11	4.5	-7.323100402376D-05	-3.661550201188D-05
12	5.5	1.830775100594D-05	9.153875502970D-06
13	6.5	-3.051291834323D-06	-1.525645917162D-06
	0	-2.800074019254D+00	-1.400027092929D+00
	1	-4.307806183468D-01	-2.153918348193D-01

-1.00	4.307806183468D-01
-.75	3.230854637601D-01
-.50	2.153903091734D-01
-.25	1.076951545867D-01
.25	-1.009642074250D-01
.50	-1.615427318801D-01
.75	-1.413498903951D-01
1.25	2.781319393051D-01
1.50	6.096281956402D-01
1.75	8.863103541552D-01
2.25	8.790614502046D-01
2.50	5.980299493193D-01
2.75	2.679834737743D-01
3.25	-1.225027401236D-01
3.50	-1.267479929174D-01
3.75	-6.761924925253D-02
4.25	3.282451028991D-02
4.50	3.396202235046D-02
4.75	1.811852323578D-02
5.25	-8.795301036003D-03
5.50	-9.100096484391D-03
5.75	-4.854843690583D-03
6.25	2.356693854101D-03

6.50	2.438363587104D-03

I= 3 H= -3.5

0		0.	-7.437523846164D-05
1		1.154273188006D-01	5.771938057250D-02
0	-6.5	-1.154273188006D-01	-5.771365940031D-02
1	-5.5	6.925639128037D-01	3.462819564019D-01
2	-4.5	-1.770255651215D+00	-8.851278256075D-01
3	-3.5	2.388458692056D+00	1.194229346028D+00
4	-2.5	-1.783579117010D+00	-8.917895585048D-01
5	-1.5	7.458577759820D-01	3.729288879910D-01
6	-.5	-1.998519869186D-01	-9.992599345930D-02
7	.5	5.355017169238D-02	2.677508584619D-02
8	1.5	-1.434869985091D-02	-7.174349925453D-03
9	2.5	3.844627711248D-03	1.922313855624D-03
10	3.5	-1.029810994084D-03	-5.149054970421D-04
11	4.5	2.746162650891D-04	1.373081325446D-04
12	5.5	-6.865406627228D-05	-3.432703313614D-05
13	6.5	1.144234437871D-05	5.721172189357D-06
	0	7.502775722041D-01	3.751015984828D-01
	1	1.154273188006D-01	5.771938057250D-02

-1.00	-1.154273188006D-01
-.75	-8.657048910047D-02
-.50	-5.771365940031D-02
-.25	-2.885682970016D-02
.25	2.705327784390D-02
.50	4.328524455023D-02
.75	3.787458898145D-02
1.25	-7.033852239413D-02
1.50	-1.298557336507D-01
1.75	-1.244450780819D-01
2.25	2.699258117326D-01
2.50	6.011376900526D-01
2.75	8.817807233462D-01
3.25	8.812602754636D-01
3.50	6.003049734404D-01
3.75	2.691971846970D-01
4.25	-1.230919135872D-01
4.50	-1.273575838142D-01
4.75	-6.794446213417D-02
5.25	3.298237888501D-02
5.50	3.412536181646D-02
5.75	1.820566383969D-02
6.25	-8.837601952878D-03
6.50	-9.143863451639D-03

I= 4 H= -2.5

0		0.	2.776675569234D-04
1		-3.092865685566D-02	-1.548568747067D-02
0	-6.5	3.092865685566D-02	1.546432842783D-02
1	-5.5	-1.855719411340D-01	-9.278597056698D-02
2	-4.5	7.422877645359D-01	3.711438822679D-01
3	-3.5	-1.783579117010D+00	-8.917895585048D-01
4	-2.5	2.392028703502D+00	1.196014351751D+00
5	-1.5	-1.784535697000D+00	-8.922678484998D-01
6	-.5	7.461140844961D-01	3.730570422481D-01
7	.5	-1.999206409849D-01	-9.996032049244D-02
8	1.5	5.356847944338D-02	2.678423972169D-02
9	2.5	-1.435327678866D-02	-7.176638394329D-03
10	3.5	3.844627711248D-03	1.922313855624D-03
11	4.5	-1.025234056333D-03	-5.126170281663D-04
12	5.5	2.563085140832D-04	1.281542570416D-04
13	6.5	-4.271808568053D-05	-2.135904284026D-05
	0	-2.010362695618D-01	-1.003793010024D-01
	1	-3.092865685566D-02	-1.548568747067D-02

-1.00	3.092865685566D-02
-.75	2.319649264175D-02
-.50	1.546432842783D-02
-.25	7.732164213915D-03
.25	-7.248903950546D-03
.50	-1.159824632087D-02
.75	-1.014846553076D-02
1.25	1.884715027142D-02
1.50	3.479473896262D-02
1.75	3.334495817251D-02
2.25	-6.813969713513D-02
2.50	-1.275807095296D-01
2.75	-1.232313671593D-01
3.25	2.693366382691D-01
3.50	6.005280991558D-01
3.75	8.814555104646D-01
4.25	8.814181440587D-01
4.50	6.004683129064D-01
4.75	2.692843253009D-01
5.25	-1.231342145040D-01
5.50	-1.274013507815D-01
5.75	-6.796781166817D-02
6.25	3.299371395741D-02
6.50	3.413709021945D-02

CARDINAL SPLINES N=2

```
I= 5      H= -1.5
0          0.                    -1.036294989232D-03     -1.00  -8.287308622023D-03
1          8.287308622023D-03     4.223369310183D-03      -.75  -6.215481466517D-03
                                                          -.50  -4.143654311011D-03
0   -6.5  -8.287308622023D-03    -4.143654311011D-03      -.25  -2.071827155506D-03
1   -5.5   4.972385173214D-02     2.486192586607D-02       .25   1.942337958287D-03
2   -4.5  -1.988954069285D-01    -9.944770346427D-02       .50   3.107740733258D-03
3   -3.5   7.458577759820D-01     3.729288879910D-01       .75   2.719273141601D-03
4   -2.5  -1.784535697000D+00    -8.922678484998D-01      1.25  -5.050078691545D-03
5   -1.5   2.392285012016D+00     1.196142506008D+00      1.50  -9.323222199775D-03
6    -.5  -1.784604351066D+00    -8.923021755329D-01      1.75  -8.934754608118D-03
7     .5   7.461323922471D-01     3.730661961236D-01      2.25   1.825797680789D-02
8    1.5  -1.999252179226D-01    -9.996260896131D-02      2.50   3.418514806584D-02
9    2.5   5.356847944338D-02     2.678423972169D-02      2.75   3.301974529087D-02
10   3.5  -1.434869985091D-02    -7.174349925453D-03      3.25  -6.798182854003D-02
11   4.5   3.826319960242D-03     1.913159980121D-03      3.50  -1.274173700636D-01
12   5.5  -9.565799900604D-04    -4.782899950302D-04      3.75  -1.231442265554D-01
13   6.5   1.594299983434D-04     7.971499917170D-05      4.25   2.692943373522D-01
                                                          4.50   6.004843321885D-01
       0   5.386750604315D-02     2.641560552696D-02      4.75   8.814321609306D-01
       1   8.287308622023D-03     4.223369310183D-03      5.25   8.814294791311D-01
                                                          5.50   6.004800413094D-01
                                                          5.75   2.692905828330D-01
                                                          6.25  -1.231372538768D-01
                                                          6.50  -1.274044974262D-01

I= 6      H=  -.5
0          0.                     3.867512400005D-03     -1.00   2.220577632429D-03
1         -2.220577632429D-03    -1.407789770061D-03      -.75   1.665433224322D-03
                                                          -.50   1.110288816214D-03
0   -6.5   2.220577632429D-03     1.110288816214D-03      -.25   5.551444081072D-04
1   -5.5  -1.332346579457D-02    -6.661732897287D-03       .25  -5.204478826005D-04
2   -4.5   5.329386317829D-02     2.664693158915D-02       .50  -8.327166121608D-04
3   -3.5  -1.998519869186D-01    -9.992599345930D-02       .75  -7.286270356407D-04
4   -2.5   7.461140844961D-01     3.730570422481D-01      1.25   1.353164494761D-03
5   -1.5  -1.784604351066D+00    -8.923021755329D-01      1.50   2.498149836483D-03
6    -.5   2.392303319767D+00     1.196151659884D+00      1.75   2.394060259962D-03
7     .5  -1.784608928004D+00    -8.923044640018D-01      2.25  -4.892210096445D-03
8    1.5   7.461323922471D-01     3.730661961236D-01      2.50  -9.159882733769D-03
9    2.5  -1.999206409849D-01    -9.996032049244D-02      2.75  -8.847614004209D-03
10   3.5   5.355017169238D-02     2.677508584619D-02      3.25   1.821567589102D-02
11   4.5  -1.428004578463D-02    -7.140022892317D-03      3.50   3.414138109859D-02
12   5.5   3.570011446158D-03     1.785005723079D-03      3.75   3.299639575687D-02
13   6.5  -5.950019076931D-04    -2.975009538465D-04      4.25  -6.797049346763D-02
                                                          4.50  -1.274056416606D-01
       0  -1.443375461079D-02    -5.283121105391D-03      4.75  -1.231379690233D-01
       1  -2.220577632429D-03    -1.407789770061D-03      5.25   2.692912979795D-01
                                                          5.50   6.004811855438D-01
                                                          5.75   8.814304803363D-01
                                                          6.25   8.814303015496D-01
                                                          6.50   6.004808994852D-01

I= 7      H=   .5
0          0.                    -1.443375461079D-02     -1.00  -5.950019076931D-04
1          5.950019076931D-04     1.407789770061D-03      -.75  -4.462514307698D-04
```

				-.50	-2.975009538465D-04
0	-6.5	-5.950019076931D-04	-2.975009538465D-04	-.25	-1.487504769233D-04
1	-5.5	3.570011446158D-03	1.785005723079D-03	.25	1.394535721156D-04
2	-4.5	-1.428004578463D-02	-7.140022892317D-03	.50	2.231257153849D-04
3	-3.5	5.355017169238D-02	2.677508584619D-02	.75	1.952350009618D-04
4	-2.5	-1.999206409849D-01	-9.996032049244D-02	1.25	-3.625792875005D-04
5	-1.5	7.461323922471D-01	3.730661961236D-01	1.50	-6.693771461547D-04
6	-.5	-1.784608928004D+00	-8.923044640018D-01	1.75	-6.414864317316D-04
7	.5	2.392303319767D+00	1.196151659884D+00	2.25	1.310863577886D-03
8	1.5	-1.784604351066D+00	-8.923021755329D-01	2.50	2.454382869234D-03
9	2.5	7.461140844961D-01	3.730570422481D-01	2.75	2.370710725965D-03
10	3.5	-1.998519869186D-01	-9.992599345930D-02	3.25	-4.880875024045D-03
11	4.5	5.329386317829D-02	2.664693158915D-02	3.50	-9.148154330781D-03
12	5.5	-1.332346579457D-02	-6.661732897287D-03	3.75	-8.841356472127D-03
13	6.5	2.220577632429D-03	1.110288816214D-03	4.25	1.821263651829D-02
				4.50	3.413823445389D-02
	0	3.867512400005D-03	-5.283121105391D-03	4.75	3.299471516254D-02
	1	5.950019076931D-04	1.407789770061D-03	5.25	-6.796967104913D-02
				5.50	-1.274047834848D-01
				5.75	-1.231375041780D-01
				6.25	2.692910476782D-01
				6.50	6.004808994852D-01

I= 8 H= 1.5

0		0.	5.386750604315D-02	-1.00	1.594299983434D-04
1		-1.594299983434D-04	-4.223369310183D-03	-.75	1.195724987576D-04
				-.50	7.971499917170D-05
0	-6.5	1.594299983434D-04	7.971499917170D-05	-.25	3.985749958585D-05
1	-5.5	-9.565799900604D-04	-4.782899950302D-04	.25	-3.736640586174D-05
2	-4.5	3.826319960242D-03	1.913159980121D-03	.50	-5.978624937878D-05
3	-3.5	-1.434869985091D-02	-7.174349925453D-03	.75	-5.231296820643D-05
4	-2.5	5.356847944338D-02	2.678423972169D-02	1.25	9.715265524051D-05
5	-1.5	-1.999252179226D-01	-9.996260896131D-02	1.50	1.793587481363D-04
6	-.5	7.461323922471D-01	3.730661961236D-01	1.75	1.718854669640D-04
7	.5	-1.784604351066D+00	-8.923021755329D-01	2.25	-3.512442151003D-04
8	1.5	2.392285012016D+00	1.196142506008D+00	2.50	-6.576487431665D-04
9	2.5	-1.784535697000D+00	-8.922678484998D-01	2.75	-6.352288996495D-04
10	3.5	7.458577759820D-01	3.729288879910D-01	3.25	1.307824205161D-03
11	4.5	-1.988954069285D-01	-9.944770346427D-02	3.50	2.451236224530D-03
12	5.5	4.972385173214D-02	2.486192586607D-02	3.75	2.369030131634D-03
13	6.5	-8.287308622023D-03	-4.143654311011D-03	4.25	-4.880052605543D-03
				4.50	-9.147296154953D-03
	0	-1.036294989232D-03	2.641560552696D-02	4.75	-8.840891626887D-03
	1	-1.594299983434D-04	-4.223369310183D-03	5.25	1.821238621701D-02
				5.50	3.413794839528D-02
				5.75	3.299453637591D-02
				6.25	-6.796949226250D-02
				6.50	-1.274044974262D-01

I= 9 H= 2.5

0		0.	-2.010362695618D-01	-1.00	-4.271808568053D-05
1		4.271808568053D-05	1.548568747067D-02	-.75	-3.203856426040D-05
				-.50	-2.135904284026D-05
0	-6.5	-4.271808568053D-05	-2.135904284026D-05	-.25	-1.067952142013D-05
1	-5.5	2.563085140832D-04	1.281542570416D-04	.25	1.001205133137D-05

2	-4.5	-1.025234056333D-03	-5.126170281663D-04	.50	1.601928213020D-05
3	-3.5	3.844627711248D-03	1.922313855624D-03	.75	1.401687186392D-05
4	-2.5	-1.435327678866D-02	-7.176638394329D-03	1.25	-2.603133346157D-05
5	-1.5	5.356847944338D-02	2.678423972169D-02	1.50	-4.805784639060D-05
6	-.5	-1.999206409849D-01	-9.996032049244D-02	1.75	-4.605543612432D-05
7	.5	7.461140844961D-01	3.730570422481D-01	2.25	9.411328251492D-05
8	1.5	-1.784535697000D+00	-8.922678484998D-01	2.50	1.762121034322D-04
9	2.5	2.392028703502D+00	1.196014351751D+00	2.75	1.702048726334D-04
10	3.5	-1.783579117010D+00	-8.917895585048D-01	3.25	-3.504217965981D-04
11	4.5	7.422877645359D-01	3.711438822679D-01	3.50	-6.567905673381D-04
12	5.5	-1.855719411340D-01	-9.278597056698D-02	3.75	-6.347640544091D-04
13	6.5	3.092865685566D-02	1.546432842783D-02	4.25	1.307573903877D-03
				4.50	2.450950165920D-03
	0	2.776675569234D-04	-1.003793010024D-01	4.75	2.368851345003D-03
	1	4.271808568053D-05	1.548568747067D-02	5.25	-4.879873818912D-03
				5.50	-9.147010096343D-03
				5.75	-8.840641325603D-03
				6.25	1.821192137177D-02
				6.50	3.413709021945D-02

I=10 H= 3.5

0		0.	7.502775722041D-01	-1.00	1.144234437871D-05
1		-1.144234437871D-05	-5.771938057250D-02	-.75	8.581758284035D-06
				-.50	5.721172189357D-06
0	-6.5	1.144234437871D-05	5.721172189357D-06	-.25	2.860586094678D-06
1	-5.5	-6.865406627228D-05	-3.432703313614D-05	.25	-2.681799463761D-06
2	-4.5	2.746162650891D-04	1.373081325446D-04	.50	-4.290879142017D-06
3	-3.5	-1.029810994084D-03	-5.149054970421D-04	.75	-3.754519249265D-06
4	-2.5	3.844627711248D-03	1.922313855624D-03	1.25	6.972678605778D-06
5	-1.5	-1.434869985091D-02	-7.174349925453D-03	1.50	1.287263742605D-05
6	-.5	5.355017169238D-02	2.677508584619D-02	1.75	1.233627753330D-05
7	.5	-1.998519869186D-01	-9.992599345930D-02	2.25	-2.520891495935D-05
8	1.5	7.458577759820D-01	3.729288879910D-01	2.50	-4.719967056219D-05
9	2.5	-1.783579117010D+00	-8.917895585048D-01	2.75	-4.559059088394D-05
10	3.5	2.388458692056D+00	1.194229346028D+00	3.25	9.386298123163D-05
11	4.5	-1.770255651215D+00	-8.851278256075D-01	3.50	1.759260448227D-04
12	5.5	6.925639128037D-01	3.462819564019D-01	3.75	1.700260860024D-04
13	6.5	-1.154273188006D-01	-5.771365940031D-02	4.25	-3.502430099672D-04
				4.50	-6.565045087287D-04
	0	-7.437523846164D-05	3.751015984828D-01	4.75	-6.345137531258D-04
	1	-1.144234437871D-05	-5.771938057250D-02	5.25	1.307109058637D-03
				5.50	2.450091990092D-03
				5.75	2.368028926501D-03
				6.25	-4.878193224581D-03
				6.50	-9.143863451639D-03

I=11 H= 4.5

0		0.	-2.800074019254D+00	-1.00	-3.051291834323D-06
1		3.051291834323D-06	2.153918348193D-01	-.75	-2.288468875743D-06
				-.50	-1.525645917162D-06
0	-6.5	-3.051291834323D-06	-1.525645917162D-06	-.25	-7.628229585809D-07
1	-5.5	1.830775100594D-05	9.153875502970D-06	.25	7.151465236696D-07
2	-4.5	-7.323100402376D-05	-3.661550201188D-05	.50	1.144234437871D-06
3	-3.5	2.746162650891D-04	1.373081325446D-04	.75	1.001205133137D-06
4	-2.5	-1.025234056333D-03	-5.126170281663D-04	1.25	-1.859380961541D-06

5	-1.5	3.826319960242D-03	1.913159980121D-03	1.50	-3.432703313614D-06
6	-.5	-1.428004578463D-02	-7.140022892317D-03	1.75	-3.289674008880D-06
7	.5	5.329386317829D-02	2.664693158915D-02	2.25	6.722377322494D-06
8	1.5	-1.988954069285D-01	-9.944770346427D-02	2.50	1.258657881658D-05
9	2.5	7.422877645359D-01	3.711438822679D-01	2.75	1.215749090238D-05
10	3.5	-1.770255651215D+00	-8.851278256075D-01	3.25	-2.503012832843D-05
11	4.5	2.338734840324D+00	1.169367420162D+00	3.50	-4.691361195272D-05
12	5.5	-1.584683710081D+00	-7.923418550405D-01	3.75	-4.534028960065D-05
13	6.5	4.307806183468D-01	2.153903091734D-01	4.25	9.339813599125D-05
				4.50	1.750678689943D-04
	0	1.983339692310D-05	-1.400027092929D+00	4.75	1.692036675002D-04
	1	3.051291834323D-06	2.153918348193D-01	5.25	-3.485624156365D-04
				5.50	-6.533578640245D-04
				5.75	-6.314743804002D-04
				6.25	1.300851526555D-03
				6.50	2.438363587104D-03

I=12 H= 5.5

0		0.	1.045001850481D+01	-1.00	7.628229585809D-07
1		-7.628229585809D-07	-8.038479587048D-01	-.75	5.721172189357D-07
				-.50	3.814114792904D-07
0	-6.5	7.628229585809D-07	3.814114792904D-07	-.25	1.907057396452D-07
1	-5.5	-4.576937751485D-06	-2.288468875743D-06	.25	-1.787866309174D-07
2	-4.5	1.830775100594D-05	9.153875502970D-06	.50	-2.860586094678D-07
3	-3.5	-6.865406627228D-05	-3.432703313614D-05	.75	-2.503012832843D-07
4	-2.5	2.563085140832D-04	1.281542570416D-04	1.25	4.648452403852D-07
5	-1.5	-9.565799900604D-04	-4.782899950302D-04	1.50	8.581758284035D-07
6	-.5	3.570011446158D-03	1.785005723079D-03	1.75	8.224185022200D-07
7	.5	-1.332346579457D-02	-6.661732897287D-03	2.25	-1.680594330623D-06
8	1.5	4.972385173214D-02	2.486192586607D-02	2.50	-3.146644704146D-06
9	2.5	-1.855719411340D-01	-9.278597056698D-02	2.75	-3.039372725596D-06
10	3.5	6.925639128037D-01	3.462819564019D-01	3.25	6.257532082109D-06
11	4.5	-1.584683710081D+00	-7.923418550405D-01	3.50	1.172840298818D-05
12	5.5	1.646170927520D+00	8.230854637601D-01	3.75	1.133507240016D-05
13	6.5	-6.076951545867D-01	-3.038475772934D-01	4.25	-2.334953399781D-05
				4.50	-4.376696724858D-05
	0	-4.958349230776D-06	5.225006773232D+00	4.75	-4.230091687506D-05
	1	-7.628229585809D-07	-8.038479587048D-01	5.25	8.714060390914D-05
				5.50	1.633394660061D-04
				5.75	1.578685951001D-04
				6.25	-3.252128816387D-04
				6.50	-6.095908967759D-04

I=13 H= 6.5

0		0.	-7.741669750802D+00	-1.00	-1.271371597635D-07
1		1.271371597635D-07	6.339746597841D-01	-.75	-9.535286982261D-08
				-.50	-6.356857988174D-08
0	-6.5	-1.271371597635D-07	-6.356857988174D-08	-.25	-3.178428994087D-08
1	-5.5	7.628229585809D-07	3.814114792904D-07	.25	2.979777181957D-08
2	-4.5	-3.051291834323D-06	-1.525645917162D-06	.50	4.767643491130D-08
3	-3.5	1.144234437871D-05	5.721172189357D-06	.75	4.171688054739D-08
4	-2.5	-4.271808568053D-05	-2.135904284026D-05	1.25	-7.747420673087D-08
5	-1.5	1.594299983434D-04	7.971499917170D-05	1.50	-1.430293047339D-07
6	-.5	-5.950019076931D-04	-2.975009538465D-04	1.75	-1.370697503700D-07
7	.5	2.220577632429D-03	1.110288816214D-03	2.25	2.800990551039D-07

8	1.5	-8.287308622023D-03	-4.143654311011D-03	2.50	5.244407840244D-07
9	2.5	3.092865685566D-02	1.546432842783D-02	2.75	5.065621209326D-07
10	3.5	-1.154273188006D-01	-5.771365940031D-02	3.25	-1.042922013685D-06
11	4.5	4.307806183468D-01	2.153903091734D-01	3.50	-1.954733831363D-06
12	5.5	-6.076951545867D-01	-3.038475772934D-01	3.75	-1.889178733360D-06
13	6.5	2.679491924311D-01	1.339745962156D-01	4.25	3.891588999635D-06
				4.50	7.294494541430D-06
	0	8.263915384626D-07	-3.620834462205D+00	4.75	7.050152812509D-06
	1	1.271371597635D-07	6.339746597841D-01	5.25	-1.452343398486D-05
				5.50	-2.722324433435D-05
				5.75	-2.631143251668D-05
				6.25	5.420214693979D-05
				6.50	1.015984827960D-04

N= 2 M=14 P= 7.0 Q= 3

I= 0 H= -7.0

0		1.000000000000D+00	4.999997615359D-01	-1.00	2.267949192431D+00
1		-1.267949192431D+00	-6.339745791824D-01	-.75	1.950961894323D+00
				-.50	1.633974596216D+00
0	-7.0	2.679491924311D-01	1.339745962156D-01	-.25	1.316987298108D+00
1	-6.0	-6.076951545867D-01	-3.038475772934D-01	.25	6.871994080240D-01
2	-5.0	4.307806183469D-01	2.153903091735D-01	.50	3.995190528383D-01
3	-4.0	-1.154273188010D-01	-5.771365940051D-02	.75	1.620791712335D-01
4	-3.0	3.092865685712D-02	1.546432842856D-02	1.25	-7.109346086228D-02
5	-2.0	-8.287308627454D-03	-4.143654313727D-03	1.50	-7.355715851499D-02
6	-1.0	2.220577652698D-03	1.110288826349D-03	1.75	-3.924227691020D-02
7	0.0	-5.950019833399D-04	-2.975009916700D-04	2.25	1.904943542518D-02
8	1.0	1.594302806613D-04	7.971514033067D-05	2.50	1.970958122162D-02
9	2.0	-4.271913930541D-05	-2.135956965270D-05	2.75	1.051493640725D-02
10	3.0	1.144627656030D-05	5.723138280151D-06	3.25	-5.104280838449D-03
11	4.0	-3.065966935795D-06	-1.532983467897D-06	3.50	-5.281166371494D-03
12	5.0	8.175911828787D-07	4.087955914393D-07	3.75	-2.817468718791D-03
13	6.0	-2.043977957197D-07	-1.021988978598D-07	4.25	1.367687928614D-03
14	7.0	3.406629928661D-08	1.703314964331D-08	4.50	1.415084264354D-03
				4.75	7.549384679172D-04
	0	-7.875644347018D+00	-3.937822292741D+00	5.25	-3.664708760053D-04
	1	-1.267949192431D+00	-6.339745791824D-01	5.50	-3.791706859222D-04
				5.75	-2.022851528780D-04
				6.25	9.819557540773D-05
				6.50	1.015984793349D-04
				6.75	5.420214359463D-05

I= 1 H= -6.0

0		0.	1.430784570038D-06	-1.00	-1.607695154587D+00
1		1.607695154587D+00	8.038474750945D-01	-.75	-1.205771365940D+00
				-.50	-8.038475772934D-01
0	-7.0	-6.076951545867D-01	-3.038475772934D-01	-.25	-4.019237886467D-01

CARDINAL SPLINES N=2

1	-6.0	1.646170927520D+00	8.230854637602D-01	.25	3.924285518563D-01
2	-5.0	-1.584683710082D+00	-7.923418550408D-01	.50	7.278856829700D-01
3	-4.0	6.925639128061D-01	3.462819564030D-01	.75	9.493999725988D-01
4	-3.0	-1.855719411427D-01	-9.278597057135D-02	1.25	8.484357651737D-01
5	-2.0	4.972385176472D-02	2.486192588236D-02	1.50	5.663429510899D-01
6	-1.0	-1.332346591619D-02	-6.661732958095D-03	1.75	2.510786614612D-01
7	0.0	3.570011900040D-03	1.785005950020D-03	2.25	-1.142966125511D-01
8	1.0	-9.565816839680D-04	-4.782908419840D-04	2.50	-1.182574873297D-01
9	2.0	2.563148358325D-04	1.281574179162D-04	2.75	-6.308961844348D-02
10	3.0	-6.867765936181D-05	-3.433882968090D-05	3.25	3.062568503070D-02
11	4.0	1.839580161477D-05	9.197900807385D-06	3.50	3.168699822896D-02
12	5.0	-4.905547097272D-06	-2.452773548636D-06	3.75	1.690481231275D-02
13	6.0	1.226386774318D-06	6.131933871590D-07	4.25	-8.206127571682D-03
14	7.0	-2.043977957197D-07	-1.021988978598D-07	4.50	-8.490505586124D-03
				4.75	-4.529630807503D-03
	0	1.125386608211D+01	5.626933756446D+00	5.25	2.198825256032D-03
	1	1.607695154587D+00	8.038474750945D-01	5.50	2.275024115533D-03
				5.75	1.213710917268D-03
				6.25	-5.891734524464D-04
				6.50	-6.095908760094D-04
				6.75	-3.252128615678D-04

I= 2 H= -5.0

0		0.	-5.723138280151D-06	-1.00	4.307806183469D-01
1		-4.307806183469D-01	-2.153899003779D-01	-.75	3.230854637602D-01
				-.50	2.153903091735D-01
0	-7.0	4.307806183469D-01	2.153903091735D-01	-.25	1.076951545867D-01
1	-6.0	-1.584683710082D+00	-7.923418550408D-01	.25	-1.009642074251D-01
2	-5.0	2.338734840326D+00	1.169367420163D+00	.50	-1.615427318801D-01
3	-4.0	-1.770255651224D+00	-8.851278256122D-01	.75	-1.413498903951D-01
4	-3.0	7.422877645708D-01	3.711438822854D-01	1.25	2.781319393052D-01
5	-2.0	-1.988954070589D-01	-9.944770352945D-02	1.50	6.096281956403D-01
6	-1.0	5.329386366476D-02	2.664693183238D-02	1.75	8.863103541553D-01
7	0.0	-1.428004760016D-02	-7.140023800079D-03	2.25	8.790614502044D-01
8	1.0	3.826326735872D-03	1.913163367936D-03	2.50	5.980299493189D-01
9	2.0	-1.025259343330D-03	-5.126296716649D-04	2.75	2.679834737739D-01
10	3.0	2.747106374472D-04	1.373553187236D-04	3.25	-1.225027401228D-01
11	4.0	-7.358320645908D-05	-3.679160322954D-05	3.50	-1.267479929158D-01
12	5.0	1.962218838909D-05	9.811094194544D-06	3.75	-6.761924925098D-02
13	6.0	-4.905547097272D-06	-2.452773548636D-06	4.25	3.282451028673D-02
14	7.0	8.175911828787D-07	4.087955914393D-07	4.50	3.396202234449D-02
				4.75	1.811852323001D-02
	0	-3.015464328429D+00	-1.507735025783D+00	5.25	-8.795301024128D-03
	1	-4.307806183469D-01	-2.153899003779D-01	5.50	-9.100096462133D-03
				5.75	-4.854843669071D-03
				6.25	2.356693809786D-03
				6.50	2.438363504038D-03
				6.75	1.300851446271D-03

I= 3 H= -4.0

0		0.	2.146176855056D-05	-1.00	-1.154273188010D-01
1		1.154273188010D-01	5.771212641704D-02	-.75	-8.657048910076D-02
				-.50	-5.771365940051D-02
0	-7.0	-1.154273188010D-01	-5.771365940051D-02	-.25	-2.885682970025D-02
1	-6.0	6.925639128061D-01	3.462819564030D-01	.25	2.705327784399D-02

CARDINAL SPLINES N=2

2	-5.0	-1.770255651224D+00	-8.851278256122D-01	.50	4.328524455038D-02
3	-4.0	2.388458692091D+00	1.194229346046D+00	.75	3.787458898158D-02
4	-3.0	-1.783579117141D+00	-8.917895585703D-01	1.25	-7.033852239437D-02
5	-2.0	7.458577764708D-01	3.729288882354D-01	1.50	-1.298557336511D-01
6	-1.0	-1.998519887429D-01	-9.992599437143D-02	1.75	-1.244450780823D-01
7	0.0	5.355017850060D-02	2.677508925030D-02	2.25	2.699258117335D-01
8	1.0	-1.434872525952D-02	-7.174362629760D-03	2.50	6.011376900542D-01
9	2.0	3.844722537487D-03	1.922361268743D-03	2.75	8.817807233478D-01
10	3.0	-1.030164890427D-03	-5.150824452136D-04	3.25	8.812602754604D-01
11	4.0	2.759370242215D-04	1.379685121108D-04	3.50	6.003049734344D-01
12	5.0	-7.358320645908D-05	-3.679160322954D-05	3.75	2.691971846912D-01
13	6.0	1.839580161477D-05	9.197900807385D-06	4.25	-1.230919135752D-01
14	7.0	-3.065966935795D-06	-1.532983467897D-06	4.50	-1.273575837919D-01
				4.75	-6.794446211255D-02
	0	8.079912316071D-01	4.040063466878D-01	5.25	3.298237884048D-02
	1	1.154273188010D-01	5.771212641704D-02	5.50	3.412536173300D-02
				5.75	1.820566375902D-02
				6.25	-8.837601786696D-03
				6.50	-9.143863140142D-03
				6.75	-4.878192923517D-03

I= 4 H= -3.0

0		0.	-8.012393592211D-05	-1.00	3.092865685712D-02
1		-3.092865685712D-02	-1.545860529028D-02	-.75	2.319649264284D-02
				-.50	1.546432842856D-02
0	-7.0	3.092865685712D-02	1.546432842856D-02	-.25	7.732164214279D-03
1	-6.0	-1.855719411427D-01	-9.278597057135D-02	.25	-7.248903950887D-03
2	-5.0	7.422877645708D-01	3.711438822854D-01	.50	-1.159824632142D-02
3	-4.0	-1.783579117141D+00	-8.917895585703D-01	.75	-1.014846553124D-02
4	-3.0	2.392028703991D+00	1.196014351996D+00	1.25	1.884715027231D-02
5	-2.0	-1.784535698824D+00	-8.922678494122D-01	1.50	3.479473896426D-02
6	-1.0	7.461140913067D-01	3.730570456533D-01	1.75	3.334495817408D-02
7	0.0	-1.999206664022D-01	-9.996033320111D-02	2.25	-6.813969713834D-02
8	1.0	5.356857430221D-02	2.678428715110D-02	2.50	-1.275807095356D-01
9	2.0	-1.435363080662D-02	-7.176815403309D-03	2.75	-1.232313671651D-01
10	3.0	3.845948924261D-03	1.922974462131D-03	3.25	2.693366382810D-01
11	4.0	-1.030164890427D-03	-5.150824452136D-04	3.50	6.005280991782D-01
12	5.0	2.747106374472D-04	1.373553187236D-04	3.75	8.814555104862D-01
13	6.0	-6.867765936181D-05	-3.433882968090D-05	4.25	8.814181440142D-01
14	7.0	1.144627656030D-05	5.723138280151D-06	4.50	6.004683128229D-01
				4.75	2.692843252202D-01
	0	-2.165005979998D-01	-1.082903609679D-01	5.25	-1.231342143378D-01
	1	-3.092865685712D-02	-1.545860529028D-02	5.50	-1.274013504699D-01
				5.75	-6.796781136700D-02
				6.25	3.299371333700D-02
				6.50	3.413708905653D-02
				6.75	1.821192024780D-02

I= 5 H= -2.0

0		0.	2.990339751379D-04	-1.00	-8.287308627454D-03
1		8.287308627454D-03	4.122294744074D-03	-.75	-6.215481470590D-03
				-.50	-4.143654313727D-03
0	-7.0	-8.287308627454D-03	-4.143654313727D-03	-.25	-2.071827156863D-03
1	-6.0	4.972385176472D-02	2.486192588236D-02	.25	1.942337959559D-03
2	-5.0	-1.988954070589D-01	-9.944770352945D-02	.50	3.107740735295D-03

3	-4.0	7.458577764708D-01	3.729288882354D-01	.75	2.719273143383D-03
4	-3.0	-1.784535698824D+00	-8.922678494122D-01	1.25	-5.050078694855D-03
5	-2.0	2.392285018827D+00	1.196142509414D+00	1.50	-9.323222205886D-03
6	-1.0	-1.784604376484D+00	-8.923021882419D-01	1.75	-8.934754613974D-03
7	0.0	7.461324871083D-01	3.730662435541D-01	2.25	1.825797681986D-02
8	1.0	-1.999255719493D-01	-9.996278597466D-02	2.50	3.418514808825D-02
9	2.0	5.356980068898D-02	2.678490034449D-02	2.75	3.301974531251D-02
10	3.0	-1.435363080662D-02	-7.176815403309D-03	3.25	-6.798182858458D-02
11	4.0	3.844722537487D-03	1.922361268743D-03	3.50	-1.274173701471D-01
12	5.0	-1.025259343330D-03	-5.126296716649D-04	3.75	-1.231442266361D-01
13	6.0	2.563148358325D-04	1.281574179162D-04	4.25	2.692943375185D-01
14	7.0	-4.271913930541D-05	-2.135956965270D-05	4.50	6.004843325002D-01
				4.75	8.814321612318D-01
	0	5.801116039218D-02	2.915509718366D-02	5.25	8.814294785107D-01
	1	8.287308627454D-03	4.122294744074D-03	5.50	6.004800401465D-01
				5.75	2.692905817090D-01
				6.25	-1.231372515613D-01
				6.50	-1.274044930860D-01
				6.75	-6.796948806767D-02

I= 6 H= -1.0

0		0.	-1.116011964629D-03	-1.00	2.220577652698D-03
1		-2.220577652698D-03	-1.030573686019D-03	-.75	1.665433239524D-03
				-.50	1.110288826349D-03
0	-7.0	2.220577652698D-03	1.110288826349D-03	-.25	5.551444131746D-04
1	-6.0	-1.332346591619D-02	-6.661732958095D-03	.25	-5.204478873512D-04
2	-5.0	5.329386366476D-02	2.664693183238D-02	.50	-8.327166197619D-04
3	-4.0	-1.998519887429D-01	-9.992599437143D-02	.75	-7.286270422917D-04
4	-3.0	7.461140913067D-01	3.730570456533D-01	1.25	1.353164507113D-03
5	-2.0	-1.784604376484D+00	-8.923021882419D-01	1.50	2.498149859286D-03
6	-1.0	2.392303414629D+00	1.196151707314D+00	1.75	2.394060281816D-03
7	0.0	-1.784609282031D+00	-8.923046410155D-01	2.25	-4.892210141101D-03
8	1.0	7.461337134951D-01	3.730668567475D-01	2.50	-9.159882817381D-03
9	2.0	-1.999255719493D-01	-9.996278597466D-02	2.75	-8.847614084970D-03
10	3.0	5.356857430221D-02	2.678428715110D-02	3.25	1.821567605729D-02
11	4.0	-1.434872525952D-02	-7.174362629760D-03	3.50	3.414138141024D-02
12	5.0	3.826326735872D-03	1.913163367936D-03	3.75	3.299639605807D-02
13	6.0	-9.565816839680D-04	-4.782908419840D-04	4.25	-6.797049408807D-02
14	7.0	1.594302806613D-04	7.971514033067D-05	4.50	-1.274056428236D-01
				4.75	-1.231379701473D-01
	0	-1.554404356889D-02	-8.330027766759D-03	5.25	2.692913002950D-01
	1	-2.220577652698D-03	-1.030573686019D-03	5.50	6.004811898841D-01
				5.75	8.814304845311D-01
				6.25	8.814302929082D-01
				6.50	6.004808832874D-01
				6.75	2.692910320229D-01

I= 7 H= 0.0

0		0.	4.165013883380D-03	-1.00	-5.950019833399D-04
1		5.950019833399D-04	0.	-.75	-4.462514875050D-04
				-.50	-2.975009916700D-04
0	-7.0	-5.950019833399D-04	-2.975009916700D-04	-.25	-1.487504958350D-04
1	-6.0	3.570011900040D-03	1.785005950020D-03	.25	1.394535898453D-04
2	-5.0	-1.428004760016D-02	-7.140023800079D-03	.50	2.231257437525D-04
3	-4.0	5.355017850060D-02	2.677508925030D-02	.75	1.952350257834D-04

CARDINAL SPLINES N=2

4	-3.0	-1.999206664022D-01	-9.996033320111D-02	1.25	-3.625793335978D-04
5	-2.0	7.461324871083D-01	3.730662435541D-01	1.50	-6.693772312574D-04
6	-1.0	-1.784609282031D+00	-8.923046410155D-01	1.75	-6.414865132884D-04
7	0.0	2.392304641015D+00	1.196152320508D+00	2.25	1.310863744546D-03
8	1.0	-1.784609282031D+00	-8.923046410155D-01	2.50	2.454383181277D-03
9	2.0	7.461324871083D-01	3.730662435541D-01	2.75	2.370711027370D-03
10	3.0	-1.999206664022D-01	-9.996033320111D-02	3.25	-4.880875644585D-03
11	4.0	5.355017850060D-02	2.677508925030D-02	3.50	-9.148155493852D-03
12	5.0	-1.428004760016D-02	-7.140023800079D-03	3.75	-8.841357596192D-03
13	6.0	3.570011900040D-03	1.785005950020D-03	4.25	1.821263883380D-02
14	7.0	-5.950019833399D-04	-2.975009916700D-04	4.50	3.413823879413D-02
				4.75	3.299471935740D-02
	0	4.165013883380D-03	4.165013883380D-03	5.25	-6.796967969060D-02
	1	5.950019833399D-04	0.	5.50	-1.274047996827D-01
				5.75	-1.231375198334D-01
				6.25	2.692910799286D-01
				6.50	6.004809599365D-01
				6.75	8.814303599762D-01

I= 8 H= 1.0

0		0.	-1.554404356889D-02	-1.00	1.594302806613D-04
1		-1.594302806613D-04	1.030573686019D-03	-.75	1.195727104960D-04
				-.50	7.971514033067D-05
0	-7.0	1.594302806613D-04	7.971514033067D-05	-.25	3.985757016533D-05
1	-6.0	-9.565816839680D-04	-4.782908419840D-04	.25	-3.736647203000D-05
2	-5.0	3.826326735872D-03	1.913163367936D-03	.50	-5.978635524800D-05
3	-4.0	-1.434872525952D-02	-7.174362629760D-03	.75	-5.231306084200D-05
4	-3.0	5.356857430221D-02	2.678428715110D-02	1.25	9.715282727800D-05
5	-2.0	-1.999255719493D-01	-9.996278597466D-02	1.50	1.793590657440D-04
6	-1.0	7.461337134951D-01	3.730668567475D-01	1.75	1.718857713380D-04
7	0.0	-1.784609282031D+00	-8.923046410155D-01	2.25	-3.512448370820D-04
8	1.0	2.392303414629D+00	1.196151707314D+00	2.50	-6.576499077280D-04
9	2.0	-1.784604376484D+00	-8.923021882419D-01	2.75	-6.352300245100D-04
10	3.0	7.461140913067D-01	3.730570456533D-01	3.25	1.307826521050D-03
11	4.0	-1.998519887429D-01	-9.992599437143D-02	3.50	2.451240565168D-03
12	5.0	5.329386366476D-02	2.664693183238D-02	3.75	2.369034326702D-03
13	6.0	-1.332346591619D-02	-6.661732958095D-03	4.25	-4.880061247118D-03
14	7.0	2.220577652698D-03	1.110288826349D-03	4.50	-9.147312352944D-03
				4.75	-8.840907282298D-03
	0	-1.116011964629D-03	-8.330027766759D-03	5.25	1.821241846742D-02
	1	-1.594302806613D-04	1.030573686019D-03	5.50	3.413800884661D-02
				5.75	3.299459480249D-02
				6.25	-6.796961262257D-02
				6.50	-1.274047230335D-01
				6.75	-1.231374719277D-01

I= 9 H= 2.0

0		0.	5.801116039218D-02	-1.00	-4.271913930541D-05
1		4.271913930541D-05	-4.122294744074D-03	-.75	-3.203935447906D-05
				-.50	-2.135956965270D-05
0	-7.0	-4.271913930541D-05	-2.135956965270D-05	-.25	-1.067978482635D-05
1	-6.0	2.563148358325D-04	1.281574179162D-04	.25	1.001229827471D-05
2	-5.0	-1.025259343330D-03	-5.126296716649D-04	.50	1.601967723953D-05
3	-4.0	3.844722537487D-03	1.922361268743D-03	.75	1.401721758459D-05
4	-3.0	-1.435363080662D-02	-7.176815403309D-03	1.25	-2.603197551423D-05

5	-2.0	5.356980068898D-02	2.678490034449D-02	1.50	-4.805903171859D-05
6	-1.0	-1.999255719493D-01	-9.996278597466D-02	1.75	-4.605657206365D-05
7	0.0	7.461324871083D-01	3.730662435541D-01	2.25	9.411560378223D-05
8	1.0	-1.784604376484D+00	-8.923021882419D-01	2.50	1.762164496348D-04
9	2.0	2.392285018827D+00	1.196142509414D+00	2.75	1.702090706700D-04
10	3.0	-1.784535698824D+00	-8.922678494122D-01	3.25	-3.504304396147D-04
11	4.0	7.458577764708D-01	3.729288882354D-01	3.50	-6.568067668207D-04
12	5.0	-1.988954070589D-01	-9.944770352945D-02	3.75	-6.347797106163D-04
13	6.0	4.972385176472D-02	2.486192588236D-02	4.25	1.307606154677D-03
14	7.0	-8.287308627454D-03	-4.143654313727D-03	4.50	2.451010617648D-03
				4.75	2.368909771795D-03
	0	2.990339751379D-04	2.915509718366D-02	5.25	-4.879994179091D-03
	1	4.271913930541D-05	-4.122294744074D-03	5.50	-9.147235703771D-03
				5.75	-8.840859376565D-03
				6.25	1.821237056169D-02
				6.50	3.413793219744D-02
				6.75	3.299452773446D-02

I=10 H= 3.0

0		0.	-2.165005979998D-01	-1.00	1.144627656030D-05
1		-1.144627656030D-05	1.545860529028D-02	-.75	8.584707420226D-06
				-.50	5.723138280151D-06
0	-7.0	1.144627656030D-05	5.723138280151D-06	-.25	2.861569140075D-06
1	-6.0	-6.867765936181D-05	-3.433882968090D-05	.25	-2.682721068821D-06
2	-5.0	2.747106374472D-04	1.373553187236D-04	.50	-4.292353710113D-06
3	-4.0	-1.030164890427D-03	-5.150824452136D-04	.75	-3.755809496349D-06
4	-3.0	3.845948924261D-03	1.922974462131D-03	1.25	6.975074778934D-06
5	-2.0	-1.435363080662D-02	-7.176815403309D-03	1.50	1.287706113034D-05
6	-1.0	5.356857430221D-02	2.678428715110D-02	1.75	1.234051691657D-05
7	0.0	-1.999206664022D-01	-9.996033320111D-02	2.25	-2.521757804691D-05
8	1.0	7.461140913067D-01	3.730570456533D-01	2.50	-4.721589081124D-05
9	2.0	-1.784535698824D+00	-8.922678494122D-01	2.75	-4.560625816995D-05
10	3.0	2.392028703991D+00	1.196014351996D+00	3.25	9.389523740872D-05
11	4.0	-1.783579117141D+00	-8.917895585703D-01	3.50	1.759865021146D-04
12	5.0	7.422877645708D-01	3.711438822854D-01	3.75	1.700845157632D-04
13	6.0	-1.855719411427D-01	-9.278597057135D-02	4.25	-3.503633715880D-04
14	7.0	3.092865685712D-02	1.546432842856D-02	4.50	-6.567301176473D-04
				4.75	-6.347318048830D-04
	0	-8.012393592211D-05	-1.082903609679D-01	5.25	1.307558248943D-03
	1	-1.144627656030D-05	1.545860529028D-02	5.50	2.450933968475D-03
				5.75	2.368842703769D-03
				6.25	-4.879869624185D-03
				6.50	-9.147005756251D-03
				6.75	-8.840639010191D-03

I=11 H= 4.0

0		0.	8.079912316071D-01	-1.00	-3.065966935795D-06
1		3.065966935795D-06	-5.771212641704D-02	-.75	-2.299475201846D-06
				-.50	-1.532983467897D-06
0	-7.0	-3.065966935795D-06	-1.532983467897D-06	-.25	-7.664917339487D-07
1	-6.0	1.839580161477D-05	9.197900807385D-06	.25	7.185860005769D-07
2	-5.0	-7.358320645908D-05	-3.679160322954D-05	.50	1.149737600923D-06
3	-4.0	2.759370242215D-04	1.379685121108D-04	.75	1.006020400808D-06
4	-3.0	-1.030164890427D-03	-5.150824452136D-04	1.25	-1.868323601500D-06
5	-2.0	3.844722537487D-03	1.922361268743D-03	1.50	-3.449212802769D-06

6	-1.0	-1.434872525952D-02	-7.174362629760D-03	1.75	-3,305495602654D-06
7	0,0	5,355017850060D-02	2.677508925030D-02	2.25	6,754708405423D-06
8	1.0	-1,998519887429D-01	-9,992599437143D-02	2,50	1,264711361015D-05
9	2.0	7,458577764708D-01	3.729288882354D-01	2.75	1,221596200981D-05
10	3.0	-1,783579117141D+00	-8.917895585703D-01	3.25	-2,515051002019D-05
11	4,0	2,388458692091D+00	1.194229346046D+00	3.50	-4,713924163785D-05
12	5,0	-1,770255651224D+00	-8,851278256122D-01	3,75	-4,555835243658D-05
13	6.0	6,925639128061D-01	3.462819564030D-01	4.25	9,384733167535D-05
14	7,0	-1,154273188010D-01	-5,771365940051D-02	4.50	1,759098529412D-04
				4,75	1,700174477365D-04
	0	2,146176855056D-05	4,040063466878D-01	5,25	-3,502388166812D-04
	1	3,065966935795D-06	-5,771212641704D-02	5,50	-6,565001701271D-04
				5,75	-6,345114385094D-04
				6,25	1,307107935049D-03
				6,50	2,450090827567D-03
				6,75	2,368028306301D-03

I=12	H= 5.0				
0		0.	-3.015464328429D+00	-1.00	8,175911828787D-07
1		-8,175911828787D-07	2,153899003779D-01	-.75	6,131933871590D-07
				-.50	4,087955914393D-07
0	-7,0	8,175911828787D-07	4.087955914393D-07	-.25	2,043977957197D-07
1	-6,0	-4,905547097272D-06	-2.452773548636D-06	.25	-1,916229334872D-07
2	-5,0	1,962218838909D-05	9,811094194544D-06	.50	-3,065966935795D-07
3	-4,0	-7,358320645908D-05	-3,679160322954D-05	.75	-2,682721068821D-07
4	-3,0	2,747106374472D-04	1.373553187236D-04	1.25	4,982196270667D-07
5	-2.0	-1.025259343330D-03	-5.126296716649D-04	1.50	9,197900807385D-07
6	-1.0	3,826326735872D-03	1.913163367936D-03	1.75	8,814654940411D-07
7	0,0	-1,428004760016D-02	-7.140023800079D-03	2,25	-1,801255574780D-06
8	1,0	5,329386366476D-02	2.664693183238D-02	2,50	-3,372563629374D-06
9	2,0	-1,988954070589D-01	-9,944770352945D-02	2.75	-3,257589869282D-06
10	3,0	7,422877645708D-01	3.711438822854D-01	3,25	6,706802672052D-06
11	4,0	-1,770255651224D+00	-8,851278256122D-01	3,50	1,257046443676D-05
12	5,0	2,338734840326D+00	1.169367420163D+00	3,75	1,214889398309D-05
13	6,0	-1,584683710082D+00	-7,923418550408D-01	4.25	-2,502595511343D-05
14	7,0	4,307806183469D-01	2.153903091735D-01	4,50	-4,690929411766D-05
				4,75	-4,533798606307D-05
	0	-5,723138280151D-06	-1,507735025783D+00	5,25	9,339701778165D-05
	1	-8,175911828787D-07	2.153899003779D-01	5,50	1,750667120339D-04
				5,75	1,692030502692D-04
				6,25	-3,485621160132D-04
				6,50	-6,533575540179D-04
				6,75	-6,314742150137D-04

I=13	H= 6.0				
0		0.	1.125386608211D+01	-1.00	-2,043977957197D-07
1		2,043977957197D-07	-8,038474750945D-01	-.75	-1,532983467897D-07
				-.50	-1,021988978598D-07
0	-7,0	-2,043977957197D-07	-1.021988978598D-07	-.25	-5,109944892992D-08
1	-6,0	1.226386774318D-06	6.131933871590D-07	.25	4,790573337180D-08
2	-5,0	-4,905547097272D-06	-2.452773548636D-06	.50	7,664917339487D-08
3	-4,0	1,839580161477D-05	9.197900807385D-06	.75	6,706802672052D-08
4	-3,0	-6,867765936181D-05	-3.433882968090D-05	1.25	-1,245549067667D-07
5	-2.0	2,563148358325D-04	1.281574179162D-04	1.50	-2,299475201846D-07
6	-1,0	-9,565816839680D-04	-4.782908419840D-04	1.75	-2,203663735103D-07

```
 7   0.0   3.570011900040D-03   1.785005950020D-03     2.25  4.503138936949D-07
 8   1.0  -1.332346591619D-02  -6.661732958095D-03     2.50  8.431409073436D-07
 9   2.0   4.972385176472D-02   2.486192588236D-02     2.75  8.143974673205D-07
10   3.0  -1.855719411427D-01  -9.278597057135D-02     3.25 -1.676700668013D-06
11   4.0   6.925639128061D-01   3.462819564030D-01     3.50 -3.142616109190D-06
12   5.0  -1.584683710082D+00  -7.923418550408D-01     3.75 -3.037223495772D-06
13   6.0   1.646170927520D+00   8.230854637602D-01     4.25  6.256488778357D-06
14   7.0  -6.076951545867D-01  -3.038475772934D-01     4.50  1.172732352942D-05
                                                       4.75  1.133449651577D-05
       0   1.430784570038D-06   5.626933756446D+00     5.25 -2.334925444541D-05
       1   2.043977957197D-07  -8.038474750945D-01     5.50 -4.376667800847D-05
                                                       5.75 -4.230076256730D-05
                                                       6.25  8.714052900330D-05
                                                       6.50  1.633393885045D-04
                                                       6.75  1.578685537534D-04


I=14     H=  7.0
0          0.                  -8.375644347018D+00    -1.00  3.406629928661D-08
1         -3.406629928661D-08   6.339745791824D-01     -.75  2.554972446496D-08
                                                       -.50  1.703314964331D-08
0  -7.0    3.406629928661D-08   1.703314964331D-08     -.25  8.516574821653D-09
1  -6.0   -2.043977957197D-07  -1.021988978598D-07      .25 -7.984288895299D-09
2  -5.0    8.175911828787D-07   4.087955914393D-07      .50 -1.277486223248D-08
3  -4.0   -3.065966935795D-06  -1.532983467897D-06      .75 -1.117800445342D-08
4  -3.0    1.144627656030D-05   5.723138280151D-06     1.25  2.075915112778D-08
5  -2.0   -4.271913930541D-05  -2.135956965270D-05     1.50  3.832458669744D-08
6  -1.0    1.594302806613D-04   7.971514033067D-05     1.75  3.672772891838D-08
7   0.0   -5.950019833399D-04  -2.975009916700D-04     2.25 -7.505231561581D-08
8   1.0    2.220577652698D-03   1.110288826349D-03     2.50 -1.405234845573D-07
9   2.0   -8.287308627454D-03  -4.143654313727D-03     2.75 -1.357329112201D-07
10  3.0    3.092865685712D-02   1.546432842856D-02     3.25  2.794501113355D-07
11  4.0   -1.154273188010D-01  -5.771365940051D-02     3.50  5.237693515316D-07
12  5.0    4.307806183469D-01   2.153903091735D-01     3.75  5.062039159620D-07
13  6.0   -6.076951545867D-01  -3.038475772934D-01     4.25 -1.042748129726D-06
14  7.0    2.679491924311D-01   1.339745962156D-01     4.50 -1.954553921569D-06
                                                       4.75 -1.889082752628D-06
        0 -2.384640950063D-07  -3.937822292741D+00     5.25  3.891542407569D-06
        1 -3.406629928661D-08   6.339745791824D-01     5.50  7.294446334746D-06
                                                       5.75  7.050127094549D-06
                                                       6.25 -1.452342150055D-05
                                                       6.50 -2.722323141741D-05
                                                       6.75 -2.631142562557D-05



                N= 2      M=15                      P=  7.5       Q= 3

I= 0     H= -7.5
0           1.000000000000D+00   5.000000684603D-01    -1.00  2.267949192431D+00
1          -1.267949192431D+00  -6.339746007796D-01     -.75  1.950961894323D+00
```

CARDINAL SPLINES N=2

0	-7.5	2.679491924311D-01	1.339745962156D-01
1	-6.5	-6.076951545867D-01	-3.038475772934D-01
2	-5.5	4.307806183469D-01	2.153903091735D-01
3	-4.5	-1.154273188010D-01	-5.771365940052D-02
4	-3.5	3.092865685722D-02	1.546432842861D-02
5	-2.5	-8.287308627844D-03	-4.143654313922D-03
6	-1.5	2.220577654154D-03	1.110288827077D-03
7	-.5	-5.950019887711D-04	-2.975009943856D-04
8	.5	1.594303009309D-04	7.971515046543D-05
9	1.5	-4.271921495227D-05	-2.135960747614D-05
10	2.5	1.144655887824D-05	5.723279439119D-06
11	3.5	-3.067020560676D-06	-1.533510280338D-06
12	4.5	8.215233644668D-07	4.107616822334D-07
13	5.5	-2.190728971911D-07	-1.095364485956D-07
14	6.5	5.476822429779D-08	2.738411214889D-08
15	7.5	-9.128037382964D-09	-4.564018691482D-09
	0	-8.509618943233D+00	-4.254809437387D+00
	1	-1.267949192431D+00	-6.339746007796D-01

-.50	1.633974596216D+00
-.25	1.316987298108D+00
.25	6.871994080240D-01
.50	3.995190528383D-01
.75	1.620791712335D-01
1.25	-7.109346086228D-02
1.50	-7.355715851499D-02
1.75	-3.924227691020D-02
2.25	1.904943542518D-02
2.50	1.970958122162D-02
2.75	1.051493640725D-02
3.25	-5.104280838447D-03
3.50	-5.281166371489D-03
3.75	-2.817468718786D-03
4.25	1.367687928604D-03
4.50	1.415084264336D-03
4.75	7.549384679000D-04
5.25	-3.664708759698D-04
5.50	-3.791706858556D-04
5.75	-2.022851528136D-04
6.25	9.819557527516D-05
6.50	1.015984790864D-04
6.75	5.420214335446D-05
7.25	-2.631142513081D-05
7.50	-2.722323049002D-05

I= 1 H= -6.5

0		0.	-4.107616822334D-07
1		1.607695154587D+00	8.038476046775D-01
0	-7.5	-6.076951545867D-01	-3.038475772934D-01
1	-6.5	1.646170927520D+00	8.230854637602D-01
2	-5.5	-1.584683710082D+00	-7.923418550408D-01
3	-4.5	6.925639128062D-01	3.462819564031D-01
4	-3.5	-1.855719411433D-01	-9.278597057166D-02
5	-2.5	4.972385176706D-02	2.486192588353D-02
6	-1.5	-1.332346592492D-02	-6.661732962461D-03
7	-.5	3.570011932627D-03	1.785005966313D-03
8	.5	-9.565818055851D-04	-4.782909027926D-04
9	1.5	2.563152897136D-04	1.281576448568D-04
10	2.5	-6.867935326942D-05	-3.433967663471D-05
11	3.5	1.840212336406D-05	9.201061682028D-06
12	4.5	-4.929140186801D-06	-2.464570093400D-06
13	5.5	1.314437383147D-06	6.572186915734D-07
14	6.5	-3.286093457867D-07	-1.643046728934D-07
15	7.5	5.476822429779D-08	2.738411214889D-08
	0	1.205771365940D+01	6.028856624319D+00
	1	1.607695154587D+00	8.038476046775D-01

-1.00	-1.607695154587D+00
-.75	-1.205771365940D+00
-.50	-8.038475772934D-01
-.25	-4.019237886467D-01
.25	3.924285518563D-01
.50	7.278856829700D-01
.75	9.493999725988D-01
1.25	8.484357651737D-01
1.50	5.663429510899D-01
1.75	2.510786614612D-01
2.25	-1.142966125511D-01
2.50	-1.182574873297D-01
2.75	-6.308961844347D-02
3.25	3.062568503068D-02
3.50	3.168699822893D-02
3.75	1.690481231272D-02
4.25	-8.206127571625D-03
4.50	-8.490505586017D-03
4.75	-4.529630807400D-03
5.25	2.198825255819D-03
5.50	2.275024115134D-03
5.75	1.213710916882D-03
6.25	-5.891734516510D-04
6.50	-6.095908745185D-04
6.75	-3.252128601267D-04
7.25	1.578685507849D-04
7.50	1.633393829401D-04

CARDINAL SPLINES N=2

I= 2 H= -5.5

0		0.	1.643046728934D-06	-1.00	4.307806183469D-01
1		-4.307806183469D-01	-2.153904187099D-01	-.75	3.230854637602D-01
				-.50	2.153903091735D-01
0	-7.5	4.307806183469D-01	2.153903091735D-01	-.25	1.076951545867D-01
1	-6.5	-1.584683710082D+00	-7.923418550408D-01	.25	-1.009642074251D-01
2	-5.5	2.338734840327D+00	1.169367420163D+00	.50	-1.615427318801D-01
3	-4.5	-1.770255651225D+00	-8.851278256125D-01	.75	-1.413498903951D-01
4	-3.5	7.422877645733D-01	3.711438822867D-01	1.25	2.781319393052D-01
5	-2.5	-1.988954070683D-01	-9.944770353413D-02	1.50	6.096281956403D-01
6	-1.5	5.329386369969D-02	2.664693184984D-02	1.75	8.863103541553D-01
7	-.5	-1.428004773051D-02	-7.140023865254D-03	2.25	8.790614502044D-01
8	.5	3.826327222341D-03	1.913163611170D-03	2.50	5.980299493189D-01
9	1.5	-1.025261158855D-03	-5.126305794273D-04	2.75	2.679834737739D-01
10	2.5	2.747174130777D-04	1.373587065388D-04	3.25	-1.225027401227D-01
11	3.5	-7.360849345622D-05	-3.680424672811D-05	3.50	-1.267479929157D-01
12	4.5	1.971656074720D-05	9.858280373601D-06	3.75	-6.761924925087D-02
13	5.5	-5.257749532587D-06	-2.628874766294D-06	4.25	3.282451028650D-02
14	6.5	1.314437383147D-06	6.572186915734D-07	4.50	3.396202234407D-02
15	7.5	-2.190728971911D-07	-1.095364485956D-07	4.75	1.811852322960D-02
				5.25	-8.795301023276D-03
	0	-3.230854637602D+00	-1.615426497278D+00	5.50	-9.100096460535D-03
	1	-4.307806183469D-01	-2.153904187099D-01	5.75	-4.854843667527D-03
				6.25	2.356693806604D-03
				6.50	2.438363498074D-03
				6.75	1.300851440507D-03
				7.25	-6.314742031395D-04
				7.50	-6.533575317604D-04

I= 3 H= -4.5

0		0.	-6.161425233501D-06	-1.00	-1.154273188010D-01
1		1.154273188010D-01	5.771407016220D-02	-.75	-8.657048910078D-02
				-.50	-5.771365940052D-02
0	-7.5	-1.154273188010D-01	-5.771365940052D-02	-.25	-2.885682970026D-02
1	-6.5	6.925639128062D-01	3.462819564031D-01	.25	2.705327784399D-02
2	-5.5	-1.770255651225D+00	-8.851278256125D-01	.50	4.328524455039D-02
3	-4.5	2.388458692094D+00	1.194229346047D+00	.75	3.787458898159D-02
4	-3.5	-1.783579117150D+00	-8.917895585750D-01	1.25	-7.033852239438D-02
5	-2.5	7.458577765059D-01	3.729288882530D-01	1.50	-1.298557336512D-01
6	-1.5	-1.998519888738D-01	-9.992599443692D-02	1.75	-1.244450780824D-01
7	-.5	5.355017898940D-02	2.677508949470D-02	2.25	2.699258117335D-01
8	.5	-1.434872708378D-02	-7.174363541888D-03	2.50	6.011376900543D-01
9	1.5	3.844729345705D-03	1.922364672852D-03	2.75	8.817807233479D-01
10	2.5	-1.030190299041D-03	-5.150951495207D-04	3.25	8.812602754602D-01
11	3.5	2.760318504608D-04	1.380159252304D-04	3.50	6.003049734340D-01
12	4.5	-7.393710280201D-05	-3.696855140101D-05	3.75	2.691971846908D-01
13	5.5	1.971656074720D-05	9.858280373601D-06	4.25	-1.230919135744D-01
14	6.5	-4.929140186801D-06	-2.464570093400D-06	4.50	-1.273575837902D-01
15	7.5	8.215233644668D-07	4.107616822334D-07	4.75	-6.794446211100D-02
				5.25	3.298237883728D-02
	0	8.657048910078D-01	4.328493647913D-01	5.50	3.412536172701D-02
	1	1.154273188010D-01	5.771407016220D-02	5.75	1.820566375323D-02
				6.25	-8.837601774764D-03
				6.50	-9.143863117777D-03
				6.75	-4.878192901901D-03
				7.25	2.368028261773D-03

CARDINAL SPLINES N=2

7.50	2.450090744102D-03

I= 4 H= -3.5

0		0.	2.300265420507D-05
1		-3.092865685722D-02	-1.546586193889D-02
0	-7.5	3.092865685722D-02	1.546432842861D-02
1	-6.5	-1.855719411433D-01	-9.278597057166D-02
2	-5.5	7.422877645733D-01	3.711438822867D-01
3	-4.5	-1.783579117150D+00	-8.917895585750D-01
4	-3.5	2.392028704026D+00	1.196014352013D+00
5	-2.5	-1.784535698956D+00	-8.922678494778D-01
6	-1.5	7.461140917957D-01	3.730570458978D-01
7	-.5	-1.999206682271D-01	-9.996033411355D-02
8	.5	5.356858111277D-02	2.678429055638D-02
9	1.5	-1.435365622396D-02	-7.176828111982D-03
10	2.5	3.846043783088D-03	1.923021891544D-03
11	3.5	-1.030518908387D-03	-5.152594541936D-04
12	4.5	2.760318504608D-04	1.380159252304D-04
13	5.5	-7.360849345622D-05	-3.680424672811D-05
14	6.5	1.840212336406D-05	9.201061682028D-06
15	7.5	-3.067020560676D-06	-1.533510280338D-06
	0	-2.319649264292D-01	-1.159709618875D-01
	1	-3.092865685722D-02	-1.546586193889D-02

-1.00	3.092865685722D-02
-.75	2.319649264292D-02
-.50	1.546432842861D-02
-.25	7.732164214305D-03
.25	-7.248903950911D-03
.50	-1.159824632146D-02
.75	-1.014846553128D-02
1.25	1.884715027237D-02
1.50	3.479473896437D-02
1.75	3.334495817419D-02
2.25	-6.813969713857D-02
2.50	-1.275807095360D-01
2.75	-1.232313671655D-01
3.25	2.693366382819D-01
3.50	6.005280991798D-01
3.75	8.814555104878D-01
4.25	8.814181440110D-01
4.50	6.004683128169D-01
4.75	2.692843252144D-01
5.25	-1.231342143259D-01
5.50	-1.274013504475D-01
5.75	-6.796781134537D-02
6.25	3.299371329245D-02
6.50	3.413708897303D-02
6.75	1.821192016710D-02
7.25	-8.840638843952D-03
7.50	-9.147005444646D-03

I= 5 H= -2.5

0		0.	-8.584919158678D-05
1		8.287308627844D-03	4.149377593361D-03
0	-7.5	-8.287308627844D-03	-4.143654313922D-03
1	-6.5	4.972385176706D-02	2.486192588353D-02
2	-5.5	-1.988954070683D-01	-9.944770353413D-02
3	-4.5	7.458577765059D-01	3.729288882530D-01
4	-3.5	-1.784535698956D+00	-8.922678494778D-01
5	-2.5	2.392285019316D+00	1.196142509658D+00
6	-1.5	-1.784604378309D+00	-8.923021891544D-01
7	-.5	7.461324939190D-01	3.730662469595D-01
8	.5	-1.999255973673D-01	-9.996279868365D-02
9	1.5	5.356989555015D-02	2.678494777508D-02
10	2.5	-1.435398483331D-02	-7.176992416655D-03
11	3.5	3.846043783088D-03	1.923021891544D-03
12	4.5	-1.030190299041D-03	-5.150951495207D-04
13	5.5	2.747174130777D-04	1.373587065388D-04
14	6.5	-6.867935326942D-05	-3.433967663471D-05
15	7.5	1.144655887824D-05	5.723279439119D-06
	0	6.215481470883D-02	3.103448275862D-02
	1	8.287308627844D-03	4.149377593361D-03

-1.00	-8.287308627844D-03
-.75	-6.215481470883D-03
-.50	-4.143654313922D-03
-.25	-2.071827156961D-03
.25	1.942337959651D-03
.50	3.107740735441D-03
.75	2.719273143511D-03
1.25	-5.050078695092D-03
1.50	-9.323222206324D-03
1.75	-8.934754614394D-03
2.25	1.825797682072D-02
2.50	3.418514808986D-02
2.75	3.301974531406D-02
3.25	-6.798182858778D-02
3.50	-1.274173701531D-01
3.75	-1.231442266419D-01
4.25	2.692943375304D-01
4.50	6.004843325225D-01
4.75	8.814321612534D-01
5.25	8.814294784662D-01
5.50	6.004800400630D-01
5.75	2.692905816283D-01
6.25	-1.231372513950D-01

X	Value
6.50	-1.274044927744D-01
6.75	-6.796948776649D-02
7.25	3.299452711404D-02
7.50	3.413793103448D-02

I= 6 H= -1.5

0		0.	3.203941121420D-04
1		-2.220577654154D-03	-1.131648434553D-03
0	-7.5	2.220577654154D-03	1.110288827077D-03
1	-6.5	-1.332346592492D-02	-6.661732962461D-03
2	-5.5	5.329386369969D-02	2.664693184984D-02
3	-4.5	-1.998519888738D-01	-9.992599443692D-02
4	-3.5	7.461140917957D-01	3.730570458978D-01
5	-2.5	-1.784604378309D+00	-8.923021891544D-01
6	-1.5	2.392303421439D+00	1.196151710720D+00
7	-.5	-1.784609307449D+00	-8.923046537245D-01
8	.5	7.461338083564D-01	3.730669041782D-01
9	1.5	-1.999259259766D-01	-9.996296298832D-02
10	2.5	5.356989555015D-02	2.678494777508D-02
11	3.5	-1.435365622396D-02	-7.176828111982D-03
12	4.5	3.844729345705D-03	1.922364672852D-03
13	5.5	-1.025261158855D-03	-5.126305794273D-04
14	6.5	2.563152897136D-04	1.281576448568D-04
15	7.5	-4.271921495227D-05	-2.135960747614D-05
	0	-1.665433240615D-02	-8.166969147005D-03
	1	-2.220577654154D-03	-1.131648434553D-03

X	Value
-1.00	2.220577654154D-03
-.75	1.665433240615D-03
-.50	1.110288827077D-03
-.25	5.551444135384D-04
.25	-5.204478876923D-04
.50	-8.327166203076D-04
.75	-7.286270427692D-04
1.25	1.353164508000D-03
1.50	2.498149860923D-03
1.75	2.394060283384D-03
2.25	-4.892210144307D-03
2.50	-9.159882823384D-03
2.75	-8.847614090769D-03
3.25	1.821567606923D-02
3.50	3.414138143261D-02
3.75	3.299639607969D-02
4.25	-6.797049413261D-02
4.50	-1.274056429071D-01
4.75	-1.231379702280D-01
5.25	2.692913004612D-01
5.50	6.004811901957D-01
5.75	8.814304848323D-01
6.25	8.814302922877D-01
6.50	6.004808821244D-01
6.75	2.692910308989D-01
7.25	-1.231374696122D-01
7.50	-1.274047186933D-01

I= 7 H= -.5

0		0.	-1.195727256981D-03
1		5.950019887711D-04	3.772161448510D-04
0	-7.5	-5.950019887711D-04	-2.975009943856D-04
1	-6.5	3.570011932627D-03	1.785005966313D-03
2	-5.5	-1.428004773051D-02	-7.140023865254D-03
3	-4.5	5.355017898940D-02	2.677508949470D-02
4	-3.5	-1.999206682271D-01	-9.996033411355D-02
5	-2.5	7.461324939190D-01	3.730662469595D-01
6	-1.5	-1.784609307449D+00	-8.923046537245D-01
7	-.5	2.392304735877D+00	1.196152367938D+00
8	.5	-1.784609636058D+00	-8.923048180292D-01
9	1.5	7.461338083564D-01	3.730669041782D-01
10	2.5	-1.999259736730D-01	-9.996279868365D-02
11	3.5	5.356858111277D-02	2.678429055638D-02
12	4.5	-1.434872708378D-02	-7.174363541888D-03
13	5.5	3.826327222341D-03	1.913163611170D-03
14	6.5	-9.565818055851D-04	-4.782909027926D-04
15	7.5	1.594303009309D-04	7.971515046543D-05

X	Value
-1.00	-5.950019887711D-04
-.75	-4.462514915784D-04
-.50	-2.975009943856D-04
-.25	-1.487504971928D-04
.25	1.394535911182D-04
.50	2.231257457892D-04
.75	1.952350275655D-04
1.25	-3.625793369074D-04
1.50	-6.693772373675D-04
1.75	-6.414865191439D-04
2.25	1.310863756511D-03
2.50	2.454383203681D-03
2.75	2.370711049010D-03
3.25	-4.880875689138D-03
3.50	-9.148155577356D-03
3.75	-8.841357676896D-03
4.25	1.821263900004D-02
4.50	3.413823910574D-02
4.75	3.299471965857D-02
5.25	-6.796968031103D-02

CARDINAL SPLINES N=2

```
          0   4.462514915784D-03   1.633393829401D-03      5.50 -1.274048008456D-01
          1   5.950019887711D-04   3.772161448510D-04      5.75 -1.231375209574D-01
                                                           6.25  2.692910822441D-01
                                                           6.50  6.004809642767D-01
                                                           6.75  8.814303641710D-01
                                                           7.25  8.814303513347D-01
                                                           7.50  6.004809437387D-01


I= 8      H=   .5
0             0.                   4.462514915784D-03     -1.00  1.594303009309D-04
1            -1.594303009309D-04  -3.772161448510D-04      -.75  1.195727256981D-04
                                                           -.50  7.971515046543D-05
0   -7.5      1.594303009309D-04   7.971515046543D-05      -.25  3.985757523271D-05
1   -6.5     -9.565818055851D-04  -4.782909027926D-04       .25 -3.736647678067D-05
2   -5.5      3.826327222341D-03   1.913163611170D-03       .50 -5.978636284907D-05
3   -4.5     -1.434872708378D-02  -7.174363541888D-03       .75 -5.231306749294D-05
4   -3.5      5.356858111277D-02   2.678429055638D-02      1.25  9.715283962974D-05
5   -2.5     -1.999255973673D-01  -9.996279868365D-02      1.50  1.793590885472D-04
6   -1.5      7.461338083564D-01   3.730669041782D-01      1.75  1.718857931911D-04
7    -.5     -1.784609636058D+00  -8.923048180292D-01      2.25 -3.512448817383D-04
8     .5      2.392304735877D+00   1.196152367938D+00      2.50 -6.576499913398D-04
9    1.5     -1.784609307449D+00  -8.923046537245D-01      2.75 -6.352301052714D-04
10   2.5      7.461324939190D-01   3.730662469595D-01      3.25  1.307826687323D-03
11   3.5     -1.999206682271D-01  -9.996033411355D-02      3.50  2.451240876812D-03
12   4.5      5.355017898940D-02   2.677508949470D-02      3.75  2.369034627894D-03
13   5.5     -1.428004773051D-02  -7.140023865254D-03      4.25 -4.880061867555D-03
14   6.5      3.570011932627D-03   1.785005966313D-03      4.50 -9.147313515908D-03
15   7.5     -5.950019887711D-04  -2.975009943856D-04      4.75 -8.840908406306D-03
                                                           5.25  1.821242078290D-02
          0  -1.195727256981D-03   1.633393829401D-03      5.50  3.413801318682D-02
          1  -1.594303009309D-04  -3.772161448510D-04      5.75  3.299459899733D-02
                                                           6.25 -6.796962126404D-02
                                                           6.50 -1.274047392314D-01
                                                           6.75 -1.231374875830D-01
                                                           7.25  2.692910642732D-01
                                                           7.50  6.004809437387D-01


I= 9      H=  1.5
0             0.                  -1.665433240615D-02     -1.00 -4.271921495227D-05
1             4.271921495227D-05   1.131648434553D-03      -.75 -3.203941121420D-05
                                                           -.50 -2.135960747614D-05
0   -7.5     -4.271921495227D-05  -2.135960747614D-05      -.25 -1.067980373807D-05
1   -6.5      2.563152897136D-04   1.281576448568D-04       .25  1.001231600444D-05
2   -5.5     -1.025261158855D-03  -5.126305794273D-04       .50  1.601970560710D-05
3   -4.5      3.844729345705D-03   1.922364672852D-03       .75  1.401724240621D-05
4   -3.5     -1.435365622396D-02  -7.176828111982D-03      1.25 -2.603202161154D-05
5   -2.5      5.356989555015D-02   2.678494777508D-02      1.50 -4.805911682131D-05
6   -1.5     -1.999259259766D-01  -9.996296298832D-02      1.75 -4.605665362042D-05
7    -.5      7.461338083564D-01   3.730669041782D-01      2.25  9.411577044173D-05
8     .5     -1.784609307449D+00  -8.923046537245D-01      2.50  1.762167616781D-04
9    1.5      2.392303421439D+00   1.196151710720D+00      2.75  1.702093720755D-04
10   2.5     -1.784604378309D+00  -8.923021891544D-01      3.25 -3.504310601554D-04
11   3.5      7.461140917957D-01   3.730570458978D-01      3.50 -6.568079298912D-04
12   4.5     -1.998519888738D-01  -9.992599443692D-02      3.75 -6.347808346814D-04
13   5.5      5.329386369969D-02   2.664693184984D-02      4.25  1.307608470180D-03
```

14	6.5	-1.332346592492D-02	-6.661732962461D-03	4.50	2.451014957887D-03
15	7.5	2.220577654154D-03	1.110288827077D-03	4.75	2.368913966650D-03
				5.25	-4.880002820564D-03
	0	3.203941121420D-04	-8.166969147005D-03	5.50	-9.147251901655D-03
	1	4.271921495227D-05	1.131648434553D-03	5.75	-8.840875031920D-03
				6.25	1.821240281207D-02
				6.50	3.413799264874D-02
				6.75	3.299458616103D-02
				7.25	-6.796960842773D-02
				7.50	-1.274047186933D-01

I=10 H= 2.5

0		0.	6.215481470883D-02	-1.00	1.144655887824D-05
1		-1.144655887824D-05	-4.149377593361D-03	-.75	8.584919158678D-06
				-.50	5.723279439119D-06
0	-7.5	1.144655887824D-05	5.723279439119D-06	-.25	2.861639719559D-06
1	-6.5	-6.867935326942D-05	-3.433967663471D-05	.25	-2.682787237087D-06
2	-5.5	2.747174130777D-04	1.373587065388D-04	.50	-4.292459579339D-06
3	-4.5	-1.030190299041D-03	-5.150951495207D-04	.75	-3.755902131922D-06
4	-3.5	3.846043783088D-03	1.923021891544D-03	1.25	6.975246816426D-06
5	-2.5	-1.435398483331D-02	-7.176992416655D-03	1.50	1.287737873802D-05
6	-1.5	5.356989555015D-02	2.678494777508D-02	1.75	1.234082129060D-05
7	-.5	-1.999255973673D-01	-9.996279868365D-02	2.25	-2.521820002862D-05
8	.5	7.461324939190D-01	3.730662469595D-01	2.50	-4.721705537273D-05
9	1.5	-1.784604378309D+00	-8.923021891544D-01	2.75	-4.560738303048D-05
10	2.5	2.392285019316D+00	1.196142509658D+00	3.25	9.389755329804D-05
11	3.5	-1.784535698956D+00	-8.922678494778D-01	3.50	1.759908427529D-04
12	4.5	7.458577765059D-01	3.729288882530D-01	3.75	1.700887108313D-04
13	5.5	-1.988954070683D-01	-9.944770353413D-02	4.25	-3.503720131635D-04
14	6.5	4.972385176706D-02	2.486192588353D-02	4.50	-6.567463156389D-04
15	7.5	-8.287308627844D-03	-4.143654313922D-03	4.75	-6.347474602947D-04
				5.25	1.307590499356D-03
	0	-8.584919158678D-05	3.103448275862D-02	5.50	2.450994419803D-03
	1	-1.144655887824D-05	-4.149377593361D-03	5.75	2.368901130348D-03
				6.25	-4.879989984261D-03
				6.50	-9.147231363571D-03
				6.75	-8.840857061096D-03
				7.25	1.821236943769D-02
				7.50	3.413793103448D-02

I=11 H= 3.5

0		0.	-2.319649264292D-01	-1.00	-3.067020560676D-06
1		3.067020560676D-06	1.546586193889D-02	-.75	-2.300265420507D-06
				-.50	-1.533510280338D-06
0	-7.5	-3.067020560676D-06	-1.533510280338D-06	-.25	-7.667551401690D-07
1	-6.5	1.840212336406D-05	9.201061682028D-06	.25	7.188329439084D-07
2	-5.5	-7.360849345622D-05	-3.680424672811D-05	.50	1.150132710254D-06
3	-4.5	2.760318504608D-04	1.380159252304D-04	.75	1.006366121472D-06
4	-3.5	-1.030518908387D-03	-5.152594541936D-04	1.25	-1.868965654162D-06
5	-2.5	3.846043783088D-03	1.923021891544D-03	1.50	-3.450398130761D-06
6	-1.5	-1.435365622396D-02	-7.176828111982D-03	1.75	-3.306631541979D-06
7	-.5	5.356858111277D-02	2.678429055638D-02	2.25	6.757029672739D-06
8	.5	-1.999206682271D-01	-9.996033411355D-02	2.50	1.265145981279D-05
9	1.5	7.461140917957D-01	3.730570458978D-01	2.75	1.222016004644D-05
10	2.5	-1.784535698956D+00	-8.922678494778D-01	3.25	-2.515915303680D-05

CARDINAL SPLINES N=2

11	3,5	2.392028704026D+00	1.196014352013D+00	3.50	-4,715544112039D-05
12	4,5	-1.783579117150D+00	-8.917895585750D-01	3.75	-4,557400864379D-05
13	5,5	7.422877645733D-01	3.711438822867D-01	4.25	9,387958247444D-05
14	6,5	-1.855719411433D-01	-9.278597057166D-02	4.50	1,759703046688D-04
15	7,5	3.092865685722D-02	1.546432842861D-02	4.75	1,700758745287D-04
				5,25	-3,503591768610D-04
	0	2.300265420507D-05	-1.159709618875D-01	5,50	-6,567257775547D-04
	1	3.067020560676D-06	1.546586193889D-02	5,75	-6,347294894712D-04
				6,25	1,307557124969D-03
				6,50	2,450932805550D-03
				6.75	2,368842083356D-03
				7,25	-4,879869323017D-03
				7.50	-9,147005444646D-03

I=12 H= 4.5

0		0.	8,657048910078D-01	-1,00	8,215233644668D-07
1		-8,215233644668D-07	-5,771407016220D-02	-,75	6,161425233501D-07
				-,50	4,107616822334D-07
0	-7,5	8,215233644668D-07	4,107616822334D-07	-,25	2,053808411167D-07
1	-6,5	-4,929140186801D-06	-2,464570093400D-06	,25	-1,925445385469D-07
2	-5,5	1,971656074720D-05	9,858280373601D-06	,50	-3,080712616750D-07
3	-4,5	-7,393710280201D-05	-3,696855140101D-05	,75	-2,695623539657D-07
4	-3,5	2,760318504608D-04	1,380159252304D-04	1,25	5,006158002219D-07
5	-2,5	-1,030190299041D-03	-5,150951495207D-04	1,50	9,242137850251D-07
6	-1,5	3,844729345705D-03	1,922364672852D-03	1,75	8,857048773158D-07
7	-,5	-1,434872708378D-02	-7,174363541888D-03	2,25	-1,809918662341D-06
8	,5	5,355017898940D-02	2,677508949470D-02	2,50	-3,388783878425D-06
9	1,5	-1,998519888738D-01	-9,992599443692D-02	2,75	-3,273257155297D-06
10	2,5	7,458577765059D-01	3,729288882530D-01	3,25	6,739058849142D-06
11	3,5	-1,783579117150D+00	-8,917895585750D-01	3,50	1,263092172868D-05
12	4,5	2,388458692094D+00	1,194229346047D+00	3,75	1,220732374387D-05
13	5,5	-1,770255651225D+00	-8,851278256125D-01	4,25	-2,514631673423D-05
14	6,5	6,925639128062D-01	3,462819564031D-01	4,50	-4,713490303628D-05
15	7,5	-1,154273188010D-01	-5,771365940052D-02	4,75	-4,555603782020D-05
				5,25	9,384620808776D-05
	0	-6,161425233501D-06	4,328493647913D-01	5,50	1,759086904165D-04
	1	-8,215233644668D-07	-5,771407016220D-02	5,75	1,700168275369D-04
				6,25	-3,502385156168D-04
				6,50	-6,564998586295D-04
				6,75	-6,345112723275D-04
				7,25	1,307107854380D-03
				7,50	2,450090744102D-03

I=13 H= 5.5

0		0.	-3.230854637602D+00	-1.00	-2,190728971911D-07
1		2.190728971911D-07	2.153904187099D-01	-,75	-1,643046728934D-07
				-,50	-1,095364485956D-07
0	-7,5	-2.190728971911D-07	-1.095364485956D-07	-,25	-5,476822429779D-08
1	-6,5	1.314437383147D-06	6.572186915734D-07	.25	5,134521027917D-08
2	-5,5	-5,257749532587D-06	-2.628874766294D-06	.50	8,215233644668D-08
3	-4,5	1.971656074720D-05	9.858280373601D-06	.75	7,188329439084D-08
4	-3,5	-7.360849345622D-05	-3.680424672811D-05	1.25	-1,334975467259D-07
5	-2,5	2.747174130777D-04	1.373587065388D-04	1.50	-2,464570093400D-07
6	-1,5	-1.025261158855D-03	-5.126305794273D-04	1.75	-2,361879672842D-07
7	-,5	3.826327222341D-03	1.913163611170D-03	2.25	4,826449766242D-07

```
 8    .5  -1.428004773051D-02  -7.140023865254D-03    2.50  9.036757009135D-07
 9   1.5   5.329386369969D-02   2.664693184984D-02    2.75  8.728685747460D-07
10   2.5  -1.988954070683D-01  -9.944770353413D-02    3.25 -1.797082359771D-06
11   3.5   7.422877645733D-01   3.711438822867D-01    3.50 -3.368245794314D-06
12   4.5  -1.770255651225D+00  -8.851278256125D-01    3.75 -3.255286331700D-06
13   5.5   2.338734840327D+00   1.169367420163D+00    4.25  6.705684462460D-06
14   6.5  -1.584683710082D+00  -7.923418550408D-01    4.50  1.256930747634D-05
15   7.5   4.307806183469D-01   2.153903091735D-01    4.75  1.214827675205D-05
                                                      5.25 -2.502565549007D-05
       0   1.643046728934D-06  -1.615426497278D+00    5.50 -4.690898411105D-05
       1   2.190728971911D-07   2.153904187099D-01    5.75 -4.533782067651D-05
                                                      6.25  9.339693749782D-05
                                                      6.50  1.750666289679D-04
                                                      6.75  1.692030059540D-04
                                                      7.25 -3.485620945012D-04
                                                      7.50 -6.533575317604D-04


I=14      H=  6.5
0          0.                   1.205771365940D+01   -1.00  5.476822429779D-08
1         -5.476822429779D-08  -8.038476046775D-01    -.75  4.107616822334D-08
                                                      -.50  2.738411214889D-08
0  -7.5    5.476822429779D-08   2.738411214889D-08    -.25  1.369205607445D-08
1  -6.5   -3.286093457867D-07  -1.643046728934D-07     .25 -1.283630256979D-08
2  -5.5    1.314437383147D-06   6.572186915734D-07     .50 -2.053808411167D-08
3  -4.5   -4.929140186801D-06  -2.464570093400D-06     .75 -1.797082359771D-08
4  -3.5    1.840212336406D-05   9.201061682028D-06    1.25  3.337438668146D-08
5  -2.5   -6.867935326942D-05  -3.433967663471D-05    1.50  6.161425233501D-08
6  -1.5    2.563152897136D-04   1.281576448568D-04    1.75  5.904699182105D-08
7   -.5   -9.565818055851D-04  -4.782909027926D-04    2.25 -1.206612441561D-07
8    .5    3.570011932627D-03   1.785005966313D-03    2.50 -2.259189252284D-07
9   1.5   -1.332346592492D-02  -6.661732962461D-03    2.75 -2.182171436865D-07
10  2.5    4.972385176706D-02   2.486192588353D-02    3.25  4.492705899428D-07
11  3.5   -1.855719411433D-01  -9.278597057166D-02    3.50  8.420614485785D-07
12  4.5    6.925639128062D-01   3.462819564031D-01    3.75  8.138215829249D-07
13  5.5   -1.584683710082D+00  -7.923418550408D-01    4.25 -1.676421115615D-06
14  6.5    1.646170927520D+00   8.230854637602D-01    4.50 -3.142326869085D-06
15  7.5   -6.076951545867D-01  -3.038475772934D-01    4.75 -3.037069188013D-06
                                                      5.25  6.256413872517D-06
       0  -4.107616822334D-07   6.028856624319D+00    5.50  1.172724602776D-05
       1  -5.476822429779D-08  -8.038476046775D-01    5.75  1.133445516913D-05
                                                      6.25 -2.334923437445D-05
                                                      6.50 -4.376665724197D-05
                                                      6.75 -4.230075148850D-05
                                                      7.25  8.714052362530D-05
                                                      7.50  1.633393829401D-04


I=15      H=  7.5
0          0.                  -9.009618943233D+00   -1.00 -9.128037382964D-09
1          9.128037382964D-09   6.339746007796D-01    -.75 -6.846028037223D-09
                                                      -.50 -4.564018691482D-09
0  -7.5   -9.128037382964D-09  -4.564018691482D-09    -.25 -2.282009345741D-09
1  -6.5    5.476822429779D-08   2.738411214889D-08     .25  2.139383761632D-09
2  -5.5   -2.190728971911D-07  -1.095364485956D-07     .50  3.423014018612D-09
3  -4.5    8.215233644668D-07   4.107616822334D-07     .75  2.995137266285D-09
4  -3.5   -3.067020560676D-06  -1.533510280338D-06    1.25 -5.562397780244D-09
```

CARDINAL SPLINES N=2

5	-2.5	1.144655887824D-05	5.723279439119D-06
6	-1.5	-4.271921495227D-05	-2.135960747614D-05
7	-.5	1.594303009309D-04	7.971515046543D-05
8	.5	-5.950019887711D-04	-2.975009943856D-04
9	1.5	2.220577654154D-03	1.110288827077D-03
10	2.5	-8.287308627844D-03	-4.143654313922D-03
11	3.5	3.092865685722D-02	1.546432842861D-02
12	4.5	-1.154273188010D-01	-5.771365940052D-02
13	5.5	4.307806183469D-01	2.153903091735D-01
14	6.5	-6.076951545867D-01	-3.038475772934D-01
15	7.5	2.679491924311D-01	1.339745962156D-01
	0	6.846028037223D-08	-4.254809437387D+00
	1	9.128037382964D-09	6.339746007796D-01

1.50	-1.026904205583D-08
1.75	-9.841165303508D-09
2.25	2.011020735934D-08
2.50	3.765315420473D-08
2.75	3.636952394775D-08
3.25	-7.487843165713D-08
3.50	-1.403435747631D-07
3.75	-1.356369304875D-07
4.25	2.794035192692D-07
4.50	5.237211448476D-07
4.75	5.061781980022D-07
5.25	-1.042735645420D-06
5.50	-1.954541004627D-06
5.75	-1.889075861521D-06
6.25	3.891539062409D-06
6.50	7.294442873661D-06
6.75	7.050125248083D-06
7.25	-1.452342060422D-05
7.50	-2.722323049002D-05

N= 2 M=16 P= 8.0 Q= 3

I= 0 H= -8.0

0		1.000000000000D+00	4.999999804332D-01
1		-1.267949192431D+00	-6.339745949926D-01
0	-8.0	2.679491924311D-01	1.339745962156D-01
1	-7.0	-6.076951545867D-01	-3.038475772934D-01
2	-6.0	4.307806183469D-01	2.153903091735D-01
3	-5.0	-1.154273188010D-01	-5.771365940052D-02
4	-4.0	3.092865685723D-02	1.546432842861D-02
5	-3.0	-8.287308627872D-03	-4.143654313936D-03
6	-2.0	2.220577654258D-03	1.110288827129D-03
7	-1.0	-5.950019891611D-04	-2.975009945805D-04
8	0.0	1.594303023861D-04	7.971515119307D-05
9	1.0	-4.271922038347D-05	-2.135961019174D-05
10	2.0	1.144657914775D-05	5.723289573877D-06
11	3.0	-3.067096207539D-06	-1.533548103769D-06
12	4.0	8.218056824028D-07	4.109028412014D-07
13	5.0	-2.201265220722D-07	-1.100632610361D-07
14	6.0	5.870040588591D-08	2.935020294296D-08
15	7.0	-1.467510147148D-08	-7.337550735739D-09
16	8.0	2.445850245246D-09	1.222925122623D-09
	0	-9.143593539449D+00	-4.571796779508D+00
	1	-1.267949192431D+00	-6.339745949926D-01

-1.00	2.267949192431D+00
-.75	1.950961894323D+00
-.50	1.633974596216D+00
-.25	1.316987298108D+00
.25	6.871994080240D-01
.50	3.995190528383D-01
.75	1.620791712335D-01
1.25	-7.109346086228D-02
1.50	-7.355715851499D-02
1.75	-3.924227691020D-02
2.25	1.904943542518D-02
2.50	1.970958122162D-02
2.75	1.051493640725D-02
3.25	-5.104280838447D-03
3.50	-5.281166371488D-03
3.75	-2.817468718786D-03
4.25	1.367687928603D-03
4.50	1.415084264335D-03
4.75	7.549384678988D-04
5.25	-3.664708759673D-04
5.50	-3.791706858508D-04
5.75	-2.022851528090D-04
6.25	9.819557526564D-05
6.50	1.015984790686D-04
6.75	5.420214333721D-05
7.25	-2.631142509529D-05

CARDINAL SPLINES N=2

7.50	-2.722323042343D-05
7.75	-1.452342053986D-05

I= 1 H= -7.0

0		0.	1.174008117718D-07
1		1.607695154587D+00	8.038475699558D-01
0	-8.0	-6.076951545867D-01	-3.038475772934D-01
1	-7.0	1.646170927520D+00	8.230854637602D-01
2	-6.0	-1.584683710082D+00	-7.923418550408D-01
3	-5.0	6.925639128063D-01	3.462819564031D-01
4	-4.0	-1.855719411434D-01	-9.278597057169D-02
5	-3.0	4.972385176723D-02	2.486192588362D-02
6	-2.0	-1.332346592555D-02	-6.661732962775D-03
7	-1.0	3.570011934967D-03	1.785005967483D-03
8	0.0	-9.565818143168D-04	-4.782909071584D-04
9	1.0	2.563153223008D-04	1.281576611504D-04
10	2.0	-6.867947488652D-05	-3.433973744326D-05
11	3.0	1.840257724523D-05	9.201288622617D-06
12	4.0	-4.930834094417D-06	-2.465417047208D-06
13	5.0	1.320759132433D-06	6.603795662165D-07
14	6.0	-3.522024353155D-07	-1.761012176577D-07
15	7.0	8.805060882887D-08	4.402530441443D-08
16	8.0	-1.467510147148D-08	-7.337550735739D-09
	0	1.286156123669D+01	6.430780677047D+00
	1	1.607695154587D+00	8.038475699558D-01

-1.00	-1.607695154587D+00
-.75	-1.205771365940D+00
-.50	-8.038475772934D-01
-.25	-4.019237886467D-01
.25	3.924285518563D-01
.50	7.278856829700D-01
.75	9.493999725988D-01
1.25	8.484357651737D-01
1.50	5.663429510899D-01
1.75	2.510786614612D-01
2.25	-1.142966125511D-01
2.50	-1.182574873297D-01
2.75	-6.308961844347D-02
3.25	3.062568503068D-02
3.50	3.168699822893D-02
3.75	1.690481231272D-02
4.25	-8.206127571621D-03
4.50	-8.490505586009D-03
4.75	-4.529630807393D-03
5.25	2.198825255804D-03
5.50	2.275024115105D-03
5.75	1.213710916854D-03
6.25	-5.891734515938D-04
6.50	-6.095908744114D-04
6.75	-3.252128600233D-04
7.25	1.578685505717D-04
7.50	1.633393825406D-04
7.75	8.714052323918D-05

I= 2 H= -6.0

0		0.	-4.696032470873D-07
1		-4.307806183469D-01	-2.153902798233D-01
0	-8.0	4.307806183469D-01	2.153903091735D-01
1	-7.0	-1.584683710082D+00	-7.923418550408D-01
2	-6.0	2.338734840327D+00	1.169367420163D+00
3	-5.0	-1.770255651225D+00	-8.851278256125D-01
4	-4.0	7.422877645735D-01	3.711438822867D-01
5	-3.0	-1.988954070689D-01	-9.944770353446D-02
6	-2.0	5.329386370220D-02	2.664693185110D-02
7	-1.0	-1.428004773987D-02	-7.140023869933D-03
8	0.0	3.826327257267D-03	1.913163628634D-03
9	1.0	-1.025261289203D-03	-5.126306446017D-04
10	2.0	2.747178995461D-04	1.373589497730D-04
11	3.0	-7.361030898094D-05	-3.680515449047D-05
12	4.0	1.972333637767D-05	9.861668188833D-06
13	5.0	-5.283036529732D-06	-2.641518264866D-06
14	6.0	1.408809741262D-06	7.044048706310D-07
15	7.0	-3.522024353155D-07	-1.761012176577D-07
16	8.0	5.870040588591D-08	2.935020294296D-08

-1.00	4.307806183469D-01
-.75	3.230854637602D-01
-.50	2.153903091735D-01
-.25	1.076951545867D-01
.25	-1.009642074251D-01
.50	-1.615427318801D-01
.75	-1.413498903951D-01
1.25	2.781319393052D-01
1.50	6.096281956403D-01
1.75	8.863103541553D-01
2.25	8.790614502044D-01
2.50	5.980299493189D-01
2.75	2.679834737739D-01
3.25	-1.225027401227D-01
3.50	-1.267479929157D-01
3.75	-6.761924925087D-02
4.25	3.282451028648D-02
4.50	3.396202234404D-02
4.75	1.811852322957D-02
5.25	-8.795301023215D-03
5.50	-9.100096460420D-03

	0	-3.446244946776D+00	-1.723122708189D+00	5.75	-4.854843667416D-03
	1	-4.307806183469D-01	-2.153902798233D-01	6.25	2.356693806375D-03
				6.50	2.438363497646D-03
				6.75	1.300851440093D-03
				7.25	-6.314742022869D-04
				7.50	-6.533575301624D-04
				7.75	-3.485620929567D-04

I= 3 H= -5.0

0		0.	1.761012176577D-06	-1.00	-1.154273188010D-01
1		1.154273188010D-01	5.771354933726D-02	-.75	-8.657048910078D-02
				-.50	-5.771365940052D-02
0	-8.0	-1.154273188010D-01	-5.771365940052D-02	-.25	-2.885682970026D-02
1	-7.0	6.925639128063D-01	3.462819564031D-01	.25	2.705327784399D-02
2	-6.0	-1.770255651225D+00	-8.851278256125D-01	.50	4.328524455039D-02
3	-5.0	2.388458692094D+00	1.194229346047D+00	.75	3.787458898159D-02
4	-4.0	-1.783579117151D+00	-8.917895585753D-01	1.25	-7.033852239439D-02
5	-3.0	7.458577765085D-01	3.729288882542D-01	1.50	-1.298557336512D-01
6	-2.0	-1.998519888832D-01	-9.992599444162D-02	1.75	-1.244450780824D-01
7	-1.0	5.355017902450D-02	2.677508951225D-02	2.25	2.699258117335D-01
8	0.0	-1.434872721475D-02	-7.174363607376D-03	2.50	6.011376900543D-01
9	1.0	3.844729834513D-03	1.922364917256D-03	2.75	8.817807233479D-01
10	2.0	-1.030192123298D-03	-5.150960616489D-04	3.25	8.812602754602D-01
11	3.0	2.760386586785D-04	1.380193293393D-04	3.50	6.003049734340D-01
12	4.0	-7.396251141625D-05	-3.698125570813D-05	3.75	2.691971846907D-01
13	5.0	1.981138698650D-05	9.905693493248D-06	4.25	-1.230919135743D-01
14	6.0	-5.283036529732D-06	-2.641518264866D-06	4.50	-1.273575837901D-01
15	7.0	1.320759132433D-06	6.603795662165D-07	4.75	-6.794446211089D-02
16	8.0	-2.201265220722D-07	-1.100632610361D-07	5.25	3.298237883705D-02
				5.50	3.412536172658D-02
	0	9.234185504083D-01	4.617101557103D-01	5.75	1.820566375281D-02
	1	1.154273188010D-01	5.771354933726D-02	6.25	-8.837601773908D-03
				6.50	-9.143863116171D-03
				6.75	-4.878192900349D-03
				7.25	2.368028258576D-03
				7.50	2.450090738109D-03
				7.75	1.307107848588D-03

I= 4 H= -4.0

0		0.	-6.574445459222D-06	-1.00	3.092865685723D-02
1		-3.092865685723D-02	-1.546391752577D-02	-.75	2.319649264292D-02
				-.50	1.546432842861D-02
0	-8.0	3.092865685723D-02	1.546432842861D-02	-.25	7.732164214307D-03
1	-7.0	-1.855719411434D-01	-9.278597057169D-02	.25	-7.248903950913D-03
2	-6.0	7.422877645735D-01	3.711438822867D-01	.50	-1.159824632146D-02
3	-5.0	-1.783579117151D+00	-8.917895585753D-01	.75	-1.014846553128D-02
4	-4.0	2.392028704029D+00	1.196014352014D+00	1.25	1.884715027237D-02
5	-3.0	-1.784535698965D+00	-8.922678494825D-01	1.50	3.479473896438D-02
6	-2.0	7.461140918308D-01	3.730570459154D-01	1.75	3.334495817420D-02
7	-1.0	-1.999206683581D-01	-9.996033417906D-02	2.25	-6.813969713858D-02
8	0.0	5.356858160174D-02	2.678429080087D-02	2.50	-1.275807095361D-01
9	1.0	-1.435365804885D-02	-7.176829024424D-03	2.75	-1.232313671655D-01
10	2.0	3.846050593645D-03	1.923025296823D-03	3.25	2.693366382820D-01
11	3.0	-1.030544325733D-03	-5.152721628665D-04	3.50	6.005280991799D-01
12	4.0	2.761267092873D-04	1.380633546437D-04	3.75	8.814555104879D-01

```
13   5.0  -7.396251141625D-05   -3.698125570813D-05      4.25   8.814181440108D-01
14   6.0   1.972333637767D-05    9.861668188833D-06      4.50   6.004683128165D-01
15   7.0  -4.930034094417D-06   -2.465417047208D-06      4.75   2.692843252140D-01
16   8.0   8.218056824028D-07    4.109028412014D-07      5.25  -1.231342143250D-01
                                                         5.50  -1.274013504459D-01
      0   -2.474292548578D-01   -1.237179146516D-01      5.75  -6.796781134382D-02
      1   -3.092865685723D-02   -1.546391752577D-02      6.25   3.299371328926D-02
                                                         6.50   3.413708896704D-02
                                                         6.75   1.821192016130D-02
                                                         7.25  -8.840638832017D-03
                                                         7.50  -9.147005422274D-03
                                                         7.75  -4.879869301394D-03


I= 5      H= -3.0
0          0.                    2.453676966031D-05     -1.00  -8.287308627872D-03
1          8.287308627872D-03    4.142120765832D-03      -.75  -6.215481470904D-03
                                                         -.50  -4.143654313936D-03
0   -8.0  -8.287308627872D-03   -4.143654313936D-03      -.25  -2.071827156968D-03
1   -7.0   4.972385176723D-02    2.486192588362D-02       .25   1.942337959657D-03
2   -6.0  -1.988954070689D-01   -9.944770353446D-02       .50   3.107740735452D-03
3   -5.0   7.458577765085D-01    3.729288882542D-01       .75   2.719273143520D-03
4   -4.0  -1.784535698965D+00   -8.922678494825D-01      1.25  -5.050078695109D-03
5   -3.0   2.392285019351D+00    1.196142509676D+00      1.50  -9.323222206356D-03
6   -2.0  -1.784604378440D+00   -8.923021892199D-01      1.75  -8.934754614424D-03
7   -1.0   7.461324944080D-01    3.730662472040D-01      2.25   1.825797682078D-02
8    0.0  -1.999255991922D-01   -9.996279959611D-02      2.50   3.418514808997D-02
9    1.0   5.356990236088D-02    2.678495118044D-02      2.75   3.301974531418D-02
10   2.0  -1.435401025128D-02   -7.177005125641D-03      3.25  -6.798182858801D-02
11   3.0   3.846138644254D-03    1.923069322127D-03      3.50  -1.274173701535D-01
12   4.0  -1.030544325733D-03   -5.152721628665D-04      3.75  -1.231442266423D-01
13   5.0   2.760386586785D-04    1.380193293393D-04      4.25   2.692943375313D-01
14   6.0  -7.361030898094D-05   -3.680515449047D-05      4.50   6.004843325241D-01
15   7.0   1.840257724523D-05    9.201288622617D-06      4.75   8.814321612550D-01
16   8.0  -3.067096207539D-06   -1.533548103769D-06      5.25   8.814294784630D-01
                                                         5.50   6.004800400570D-01
      0    6.629846902297D-02    3.316150289632D-02      5.75   2.692905816225D-01
      1    8.287308627872D-03    4.142120765832D-03      6.25  -1.231372513831D-01
                                                         6.50  -1.274044927520D-01
                                                         6.75  -6.796948774487D-02
                                                         7.25   3.299452706949D-02
                                                         7.50   3.413793095099D-02
                                                         7.75   1.821236935699D-02


I= 6      H= -2.0
0          0.                   -9.157263318202D-05     -1.00   2.220577654258D-03
1         -2.220577654258D-03   -1.104565537555D-03      -.75   1.665433240694D-03
                                                         -.50   1.110288827129D-03
0   -8.0   2.220577654258D-03    1.110288827129D-03      -.25   5.551444135646D-04
1   -7.0  -1.332346592555D-02   -6.661732962775D-03       .25  -5.204478877168D-04
2   -6.0   5.329386370220D-02    2.664693185110D-02       .50  -8.327166203468D-04
3   -5.0  -1.998519888832D-01   -9.992599444162D-02       .75  -7.286270428035D-04
4   -4.0   7.461140918308D-01    3.730570459154D-01      1.25   1.353164508064D-03
5   -3.0  -1.784604378440D+00   -8.923021892199D-01      1.50   2.498149861040D-03
6   -2.0   2.392303421928D+00    1.196151710964D+00      1.75   2.394060283497D-03
7   -1.0  -1.784609309274D+00   -8.923046546369D-01      2.25  -4.892210144538D-03
```

8	0,0	7.461338151671D-01	3.730669075836D-01	2.50	-9,159882823815D-03
9	1,0	-1,999259513947D-01	-9,996297569733D-02	2.75	-8,847614091185D-03
10	2,0	5,356999041148D-02	2.678499520574D-02	3.25	1,821567607009D-02
11	3,0	-1.435401025128D-02	-7.177005125641D-03	3.50	3.414138143422D-02
12	4,0	3.846050593645D-03	1.923025296823D-03	3.75	3,299639608124D-02
13	5,0	-1.030192123298D-03	-5.150960616489D-04	4.25	-6,797049413581D-02
14	6,0	2.747178995461D-04	1.373589497730D-04	4.50	-1,274036429131D-01
15	7,0	-6,867947488652D-05	-3,433973744326D-05	4.75	-1,231379702338D-01
16	8,0	1,144657914775D-05	5.723289573877D-06	5,25	2,692913004732D-01
				5,50	6,004811902180D-01
	0	-1,776462123407D-02	-8,928096933624D-03	5,75	8,814304848539D-01
	1	-2,220577654258D-03	-1.104565537555D-03	6,25	8,814302922432D-01
				6,50	6,004808820409D-01
				6,75	2,692910308182D-01
				7,25	-1,231374694459D-01
				7,50	-1,274047183817D-01
				7,75	-6,796960812656D-02

I= 7 H= -1.0

0		0.	3,417537630678D-04	-1,00	-5,950019891611D-04
1		5,950019891611D-04	2,761413843888D-04	-.75	-4,462514918708D-04
				-.50	-2,975009945805D-04
0	-8,0	-5,950019891611D-04	-2.975009945805D-04	-.25	-1,487504972903D-04
1	-7,0	3,570011934967D-03	1,785005967483D-03	.25	1,394535912096D-04
2	-6,0	-1,428004773987D-02	-7,140023869933D-03	.50	2,231257459354D-04
3	-5,0	5,355017902450D-02	2.677508951225D-02	.75	1,952350276935D-04
4	-4,0	-1,999206683581D-01	-9,996033417906D-02	1.25	-3,625793371450D-04
5	-3,0	7,461324944080D-01	3,730662472040D-01	1,50	-6,693772378062D-04
6	-2,0	-1,784609309274D+00	-8,923046546369D-01	1,75	-6,414865195643D-04
7	-1,0	2,392304742688D+00	1,196152371344D+00	2,25	1,310863757371D-03
8	0,0	-1,784609661476D+00	-8,923048307382D-01	2,50	2,454383205289D-03
9	1,0	7.461339032177D-01	3,730669516089D-01	2,75	2,370711050564D-03
10	2,0	-1,999259513947D-01	-9,996297569733D-02	3,25	-4,880875692337D-03
11	3,0	5,356990236088D-02	2,678495118044D-02	3,50	-9,148155583352D-03
12	4,0	-1,435365804885D-02	-7,176829024424D-03	3,75	-8,841357682691D-03
13	5,0	3,844729834513D-03	1,922364917256D-03	4,25	1,821263901198D-02
14	6,0	-1,025261289203D-03	-5,126306446017D-04	4,50	3,413823912812D-02
15	7,0	2,563153223008D-04	1,281576611504D-04	4,75	3,299471968020D-02
16	8,0	-4,271922038347D-05	-2,135961019174D-05	5,25	-6,796968035557D-02
				5,50	-1,274048009291D-01
	0	4,760015913289D-03	2,550884838178D-03	5,75	-1,231375210381D-01
	1	5,950019891611D-04	2,761413843888D-04	6,25	2,692910824103D-01
				6,50	6,004809645884D-01
				6,75	8,814303644722D-01
				7,25	8,814303507143D-01
				7,50	6,004809425757D-01
				7,75	2,692910631492D-01

I= 8 H= 0.0

0		0,	-1.275442419089D-03	-1.00	1,594303023861D-04
1		-1,594303023861D-04	0.	-.75	1,195727267896D-04
				-.50	7,971515119307D-05
0	-8,0	1.594303023861D-04	7,971515119307D-05	-.25	3,985757559654D-05
1	-7,0	-9,565818143168D-04	-4,782909071584D-04	.25	-3,736647712175D-05
2	-6,0	3,826327257267D-03	1.913163628634D-03	.50	-5,978636339480D-05

3	-5.0	-1.434872721475D-02	-7.174363607376D-03	.75	-5.231306797045D-05
4	-4.0	5.356858160174D-02	2.678429080087D-02	1.25	9.715284051655D-05
5	-3.0	-1.999255991922D-01	-9.996279959611D-02	1.50	1.793590901844D-04
6	-2.0	7.461338151671D-01	3.730669075836D-01	1.75	1.718857947601D-04
7	-1.0	-1.784609661476D+00	-8.923048307382D-01	2.25	-3.512448849445D-04
8	0.0	2.392304830738D+00	1.196152415369D+00	2.50	-6.576499973428D-04
9	1.0	-1.784609661476D+00	-8.923048307382D-01	2.75	-6.352301110698D-04
10	2.0	7.461338151671D-01	3.730669075836D-01	3.25	1.307826699261D-03
11	3.0	-1.999255991922D-01	-9.996279959611D-02	3.50	2.451240899187D-03
12	4.0	5.356858160174D-02	2.678429080087D-02	3.75	2.369034649519D-03
13	5.0	-1.434872721475D-02	-7.174363607376D-03	4.25	-4.880061912101D-03
14	6.0	3.826327257267D-03	1.913163628634D-03	4.50	-9.147313599405D-03
15	7.0	-9.565818143168D-04	-4.782909071584D-04	4.75	-8.840908487006D-03
16	8.0	1.594303023861D-04	7.971515119307D-05	5.25	1.821242094914D-02
				5.50	3.413801349843D-02
	0	-1.275442419089D-03	-1.275442419089D-03	5.75	3.299459929851D-02
	1	-1.594303023861D-04	0.	6.25	-6.796962188447D-02
				6.50	-1.274047403943D-01
				6.75	-1.231374887070D-01
				7.25	2.692910665887D-01
				7.50	6.004809480789D-01
				7.75	8.814303555296D-01

I= 9 H= 1.0

0		0.	4.760015913289D-03	-1.00	-4.271922038347D-05
1		4.271922038347D-05	-2.761413843888D-04	-.75	-3.203941528760D-05
				-.50	-2.135961019174D-05
0	-8.0	-4.271922038347D-05	-2.135961019174D-05	-.25	-1.067980509587D-05
1	-7.0	2.563153223008D-04	1.281576611504D-04	.25	1.001231727738D-05
2	-6.0	-1.025261289203D-03	-5.126306446017D-04	.50	1.601970764380D-05
3	-5.0	3.844729834513D-03	1.922364917256D-03	.75	1.401724418833D-05
4	-4.0	-1.435365804885D-02	-7.176829024424D-03	1.25	-2.603202492118D-05
5	-3.0	5.356990236088D-02	2.678495118044D-02	1.50	-4.805912293141D-05
6	-2.0	-1.999259513947D-01	-9.996297569733D-02	1.75	-4.605665947593D-05
7	-1.0	7.461339032177D-01	3.730669516089D-01	2.25	9.411578240734D-05
8	0.0	-1.784609661476D+00	-8.923048307382D-01	2.50	1.762167840818D-04
9	1.0	2.392304742688D+00	1.196152371344D+00	2.75	1.702093937154D-04
10	2.0	-1.784609309274D+00	-8.923046546369D-01	3.25	-3.504311047082D-04
11	3.0	7.461324944080D-01	3.730662472040D-01	3.50	-6.568080133959D-04
12	4.0	-1.999206683581D-01	-9.996033417906D-02	3.75	-6.347809153857D-04
13	5.0	5.355017902450D-02	2.677508951225D-02	4.25	1.307608636425D-03
14	6.0	-1.428004773987D-02	-7.140023869933D-03	4.50	2.451015269502D-03
15	7.0	3.570011934967D-03	1.785005967483D-03	4.75	2.368914267827D-03
16	8.0	-5.950019891611D-04	-2.975009945805D-04	5.25	-4.880003440993D-03
				5.50	-9.147253064611D-03
	0	3.417537630678D-04	2.550884838178D-03	5.75	-8.840876155924D-03
	1	4.271922038347D-05	-2.761413843888D-04	6.25	1.821240512755D-02
				6.50	3.413799698894D-02
				6.75	3.299459035587D-02
				7.25	-6.796961706920D-02
				7.50	-1.274047348912D-01
				7.75	-1.231374852675D-01

I=10 H= 2.0

0		0.	-1.776462123407D-02	-1.00	1.144657914775D-05

CARDINAL SPLINES N=2

1		-1.144657914775D-05	1.104565537555D-03	-.75	8.584934360815D-06
				-.50	5.723289573877D-06
0	-8.0	1.144657914775D-05	5.723289573877D-06	-.25	2.861644786938D-06
1	-7.0	-6.867947488652D-05	-3.433973744326D-05	.25	-2.682791987755D-06
2	-6.0	2.747178995461D-04	1.373589497730D-04	.50	-4.292467180407D-06
3	-5.0	-1.030192123298D-03	-5.150960616489D-04	.75	-3.755908782856D-06
4	-4.0	3.846050593645D-03	1.923025296823D-03	1.25	6.975259168162D-06
5	-3.0	-1.435401025128D-02	-7.177005125641D-03	1.50	1.287740154122D-05
6	-2.0	5.356999041148D-02	2.678499520574D-02	1.75	1.234084314367D-05
7	-1.0	-1.999259513947D-01	-9.996297569733D-02	2.25	-2.521824468489D-05
8	0.0	7.461338151671D-01	3.730669075836D-01	2.50	-4.721713898448D-05
9	1.0	-1.784609309274D+00	-8.923046546369D-01	2.75	-4.560746379183D-05
10	2.0	2.392303421928D+00	1.196151710964D+00	3.25	9.389771957141D-05
11	3.0	-1.784604378440D+00	-8.923021892199D-01	3.50	1.759911543967D-04
12	4.0	7.461140918308D-01	3.730570459154D-01	3.75	1.700890120236D-04
13	5.0	-1.998519888832D-01	-9.992599444162D-02	4.25	-3.503726336008D-04
14	6.0	5.329386370220D-02	2.664693185110D-02	4.50	-6.567474786023D-04
15	7.0	-1.332346592555D-02	-6.661732962775D-03	4.75	-6.347485843027D-04
16	8.0	2.220577654258D-03	1.110288827129D-03	5.25	1.307592814832D-03
				5.50	2.450998760013D-03
	0	-9.157263318202D-05	-8.928096933624D-03	5.75	2.368905325187D-03
	1	-1.144657914775D-05	1.104565537555D-03	6.25	-4.879998625726D-03
				6.50	-9.147247561448D-03
				6.75	-8.840872716447D-03
				7.25	1.821240168807D-02
				7.50	3.413799148578D-02
				7.75	3.299458554060D-02

I=11 H= 3.0

0		0.	6.629846902297D-02	-1.00	-3.067096207539D-06
1		3.067096207539D-06	-4.142120765832D-03	-.75	-2.300322155654D-06
				-.50	-1.533548103769D-06
0	-8.0	-3.067096207539D-06	-1.533548103769D-06	-.25	-7.667740518847D-07
1	-7.0	1.840257724523D-05	9.201288622617D-06	.25	7.188506736419D-07
2	-6.0	-7.361030898094D-05	-3.680515449047D-05	.50	1.150161077827D-06
3	-5.0	2.760386586785D-04	1.380193293393D-04	.75	1.006390943099D-06
4	-4.0	-1.030544325733D-03	-5.152721628665D-04	1.25	-1.869011751469D-06
5	-3.0	3.846138644254D-03	1.923069322127D-03	1.50	-3.450483233481D-06
6	-2.0	-1.435401025128D-02	-7.177005125641D-03	1.75	-3.306713098753D-06
7	-1.0	5.356990236088D-02	2.678495118044D-02	2.25	6.757196332234D-06
8	0.0	-1.999255991922D-01	-9.996279959611D-02	2.50	1.265177185610D-05
9	1.0	7.461324944080D-01	3.730662472040D-01	2.75	1.222046145191D-05
10	2.0	-1.784604378440D+00	-8.923021892199D-01	3.25	-2.515977357747D-05
11	3.0	2.392285019351D+00	1.196142509676D+00	3.50	-4.715660419091D-05
12	4.0	-1.784535698965D+00	-8.922678494825D-01	3.75	-4.557513270890D-05
13	5.0	7.458577765085D-01	3.729288882542D-01	4.25	9.388189797764D-05
14	6.0	-1.988954070689D-01	-9.944770353446D-02	4.50	1.759746449075D-04
15	7.0	4.972385176723D-02	2.486192588362D-02	4.75	1.700800693837D-04
16	8.0	-8.287308627872D-03	-4.143654313936D-03	5.25	-3.503678183331D-04
				5.50	-6.567419754393D-04
	0	2.453676966031D-05	3.316150289632D-02	5.75	-6.347451448258D-04
	1	3.067096207539D-06	-4.142120765832D-03	6.25	1.307589375355D-03
				6.50	2.450993256850D-03
				6.75	2.368900509920D-03
				7.25	-4.879989683086D-03
				7.50	-9.147231051959D-03

CARDINAL SPLINES N=2

```
                                                                    7.75 -8.840856894853D-03


I=12      H=  4.0
0             0.                    -2.474292548578D-01     -1.00  8.218056824028D-07
1            -8.218056824028D-07     1.546391752577D-02      -.75  6.163542618021D-07
                                                             -.50  4.109028412014D-07
0   -8.0      8.218056824028D-07     4.109028412014D-07      -.25  2.054514206007D-07
1   -7.0     -4.930834094417D-06    -2.465417047208D-06       .25 -1.926107068132D-07
2   -6.0      1.972333637767D-05     9.861668188833D-06       .50 -3.081771309010D-07
3   -5.0     -7.396251141625D-05    -3.698125570813D-05       .75 -2.696549895384D-07
4   -4.0      2.761267092873D-04     1.380633546437D-04      1.25  5.007878377142D-07
5   -3.0     -1.030544325733D-03    -5.152721628665D-04      1.50  9.245313927031D-07
6   -2.0      3.846050593645D-03     1.923025296823D-03      1.75  8.860092513405D-07
7   -1.0     -1.435365804885D-02    -7.176829024424D-03      2.25 -1.810540644044D-06
8    0.0      5.356858160174D-02     2.678429080087D-02      2.50 -3.389948439911D-06
9    1.0     -1.999206683581D-01    -9.996033417906D-02      2.75 -3.274382015824D-06
10   2.0      7.461140918308D-01     3.730570459154D-01      3.25  6.741374738460D-06
11   3.0     -1.784535698965D+00    -8.922678494825D-01      3.50  1.263526236694D-05
12   4.0      2.392028704029D+00     1.196014352014D+00      3.75  1.221151881195D-05
13   5.0     -1.783579117151D+00    -8.917895585753D-01      4.25 -2.515495830980D-05
14   6.0      7.422877645735D-01     3.711438822867D-01      4.50 -4.715110102786D-05
15   7.0     -1.855719411434D-01    -9.278597057169D-02      4.75 -4.557169323199D-05
16   8.0      3.092865685723D-02     1.546432842861D-02      5.25  9.387845850073D-05
                                                             5.50  1.759691417445D-04
       0     -6.574445459222D-06    -1.237179146516D-01      5.75  1.700752541160D-04
       1     -8.218056824028D-07     1.546391752577D-02      6.25 -3.503588756931D-04
                                                             6.50 -6.567254659501D-04
                                                             6.75 -6.347293232321D-04
                                                             7.25  1.307557044272D-03
                                                             7.50  2.450932722056D-03
                                                             7.75  2.368842038812D-03


I=13      H=  5.0
0             0.                     9.234185504083D-01     -1.00 -2.201265220722D-07
1             2.201265220722D-07    -5.771354933726D-02      -.75 -1.650948915541D-07
                                                             -.50 -1.100632610361D-07
0   -8.0     -2.201265220722D-07    -1.100632610361D-07      -.25 -5.503163051804D-08
1   -7.0      1.320759132433D-06     6.603795662165D-07       .25  5.159215361067D-08
2   -6.0     -5.283036529732D-06    -2.641518264866D-06       .50  8.254744577707D-08
3   -5.0      1.981138698650D-05     9.905693493248D-06       .75  7.222901505493D-08
4   -4.0     -7.396251141625D-05    -3.698125570813D-05      1.25 -1.341395993877D-07
5   -3.0      2.760386586785D-04     1.380193293393D-04      1.50 -2.476423373312D-07
6   -2.0     -1.030192123298D-03    -5.150960616489D-04      1.75 -2.373239066091D-07
7   -1.0      3.844729834513D-03     1.922364917256D-03      2.25  4.849662439403D-07
8    0.0     -1.434872721475D-02    -7.174363607376D-03      2.50  9.080219035477D-07
9    1.0      5.355017902450D-02     2.677508951225D-02      2.75  8.770666113813D-07
10   2.0     -1.998519888832D-01    -9.992599444162D-02      3.25 -1.805725376373D-06
11   3.0      7.458577765085D-01     3.729288882542D-01      3.50 -3.384445276860D-06
12   4.0     -1.783579117151D+00    -8.917895585753D-01      3.75 -3.270942538916D-06
13   5.0      2.388458692094D+00     1.194229346047D+00      4.25  6.737935261553D-06
14   6.0     -1.770255651225D+00    -8.851278256125D-01      4.50  1.262975920389D-05
15   7.0      6.925639128063D-01     3.462819564031D-01      4.75  1.220670354428D-05
16   8.0     -1.154273188010D-01    -5.771365940052D-02      5.25 -2.514601566984D-05
                                                             5.50 -4.713459153870D-05
       0      1.761012176577D-06     4.617101557103D-01      5.75 -4.555587163822D-05
```

1		2.201265220722D-07	-5.771354933726D-02	6.25	9.384612741780D-05
				6.50	1.759086069509D-04
				6.75	1.700167830086D-04
				7.25	-3.502384940014D-04
				7.50	-6.564998362650D-04
				7.75	-6.345112603961D-04

I=14 H= 6.0

0		0.	-3.446244946776D+00	-1.00	5.870040588591D-08
1		-5.870040588591D-08	2.153902798233D-01	-.75	4.402530441443D-08
				-.50	2.935020294296D-08
0	-8.0	5.870040588591D-08	2.935020294296D-08	-.25	1.467510147148D-08
1	-7.0	-3.522024353155D-07	-1.761012176577D-07	.25	-1.375790762951D-08
2	-6.0	1.408809741262D-06	7.044048706310D-07	.50	-2.201265220722D-08
3	-5.0	-5.283036529732D-06	-2.641518264866D-06	.75	-1.926107068132D-08
4	-4.0	1.972333637767D-05	9.861668188833D-06	1.25	3.577055983673D-08
5	-3.0	-7.361030898094D-05	-3.680515449047D-05	1.50	6.603795662165D-08
6	-2.0	2.747178995461D-04	1.373589497730D-04	1.75	6.328637509575D-08
7	-1.0	-1.025261289203D-03	-5.126306446017D-04	2.25	-1.293243317174D-07
8	0.0	3.826327257267D-03	1.913163628634D-03	2.50	-2.421391742794D-07
9	1.0	-1.428004773987D-02	-7.140023869933D-03	2.75	-2.338844297017D-07
10	2.0	5.329386370220D-02	2.664693185110D-02	3.25	4.815267670329D-07
11	3.0	-1.988954070689D-01	-9.944770353446D-02	3.50	9.025187404959D-07
12	4.0	7.422877645735D-01	3.711438822867D-01	3.75	8.722513437110D-07
13	5.0	-1.770255651225D+00	-8.851278256125D-01	4.25	-1.796782736414D-06
14	6.0	2.338734840327D+00	1.169367420163D+00	4.50	-3.367935787704D-06
15	7.0	-1.584683710082D+00	-7.923418550408D-01	4.75	-3.255120945142D-06
16	8.0	4.307806183469D-01	2.153903091735D-01	5.25	6.705604178624D-06
				5.50	1.256922441032D-05
	0	-4.696032470873D-07	-1.723122708189D+00	5.75	1.214823243686D-05
	1	-5.870040588591D-08	2.153902798233D-01	6.25	-2.502563397808D-05
				6.50	-4.690896185358D-05
				6.75	-4.533780880229D-05
				7.25	9.339693173370D-05
				7.50	1.750666230040D-04
				7.75	1.692030027723D-04

I=15 H= 7.0

0		0.	1.286156123669D+01	-1.00	-1.467510147148D-08
1		1.467510147148D-08	-8.038475699558D-01	-.75	-1.100632610361D-08
				-.50	-7.337550735739D-09
0	-8.0	-1.467510147148D-08	-7.337550735739D-09	-.25	-3.668775367870D-09
1	-7.0	8.805060882887D-08	4.402530441443D-08	.25	3.439476907378D-09
2	-6.0	-3.522024353155D-07	-1.761012176577D-07	.50	5.503163051804D-09
3	-5.0	1.320759132433D-06	6.603795662165D-07	.75	4.815267670329D-09
4	-4.0	-4.930834094417D-06	-2.465417047208D-06	1.25	-8.942639959182D-09
5	-3.0	1.840257724523D-05	9.201288622617D-06	1.50	-1.650948915541D-08
6	-2.0	-6.867947488652D-05	-3.433973744326D-05	1.75	-1.582159377394D-08
7	-1.0	2.563153223008D-04	1.281576611504D-04	2.25	3.233108292935D-08
8	0.0	-9.565818143168D-04	-4.782909071584D-04	2.50	6.053479356985D-08
9	1.0	3.570011934967D-03	1.785005967483D-03	2.75	5.847110742542D-08
10	2.0	-1.332346592555D-02	-6.661732962775D-03	3.25	-1.203816917582D-07
11	3.0	4.972385176723D-02	2.486192588362D-02	3.50	-2.256296851240D-07
12	4.0	-1.855719411434D-01	-9.278597057169D-02	3.75	-2.180628359277D-07
13	5.0	6.925639128063D-01	3.462819564031D-01	4.25	4.491956841035D-07

14	6,0	-1,584683710082D+00	-7,923418550408D-01	4.50	8,419839469261D-07
15	7,0	1,646170927520D+00	8,230854637602D-01	4.75	8,137802362856D-07
16	8,0	-6,076951545867D-01	-3,038475772934D-01	5,25	-1,676401044656D-06
				5,50	-3,142306102580D-06
	0	1.174008117718D-07	6.430780677047D+00	5,75	-3,037058109215D-06
	1	1,467510147148D-08	-8,038475699558D-01	6,25	6,256408494520D-06
				6,50	1,172724046340D-05
				6,75	1,133445220057D-05
				7,25	-2,334923293342D-05
				7,50	-4,376665575100D-05
				7,75	-4,230075069308D-05

I=16 H= 8.0

0		0.	-9,643593539449D+00	-1,00	2,445850245246D-09
1		-2,445850245246D-09	6,339745949926D-01	-.75	1,834387683935D-09
				-.50	1,222925122623D-09
0	-8,0	2,445850245246D-09	1.222925122623D-09	-.25	6,114625613116D-10
1	-7,0	-1,467510147148D-08	-7,337550735739D-09	.25	-5,732461512296D-10
2	-6,0	5,870040588591D-08	2,935020294296D-08	.50	-9,171938419674D-10
3	-5,0	-2,201265220722D-07	-1.100632610361D-07	.75	-8,025446117215D-10
4	-4,0	8,218056824028D-07	4,109028412014D-07	1,25	1,490439993197D-09
5	-3,0	-3,067096207539D-06	-1.533548103769D-06	1,50	2,751581525902D-09
6	-2,0	1,144657914775D-05	5,723289573877D-06	1,75	2,636932295656D-09
7	-1,0	-4,271922038347D-05	-2,135961019174D-05	2,25	-5,388513821558D-09
8	0,0	1,594303023861D-04	7.971515119307D-05	2,50	-1,008913226164D-08
9	1,0	-5,950019891611D-04	-2.975009945805D-04	2,75	-9,745184570904D-09
10	2,0	2,220577654258D-03	1.110288827129D-03	3,25	2,006361529304D-08
11	3,0	-8,287308627872D-03	-4,143654313936D-03	3,50	3,760494752066D-08
12	4,0	3,092865685723D-02	1.546432842861D-02	3,75	3,634380598796D-08
13	5,0	-1,154273188010D-01	-5,771365940052D-02	4,25	-7,486594735059D-08
14	6,0	4,307806183469D-01	2,153903091735D-01	4,50	-1,403306578210D-07
15	7,0	-6,076951545867D-01	-3,038475772934D-01	4,75	-1,356300393809D-07
16	8,0	2,679491924311D-01	1.339745962156D-01	5,25	2,794001741093D-07
				5,50	5,237176837634D-07
	0	-1,956680196197D-08	-4,571796779508D+00	5,75	5,061763515358D-07
	1	-2,445850245246D-09	6,339745949926D-01	6,25	-1,042734749087D-06
				6,50	-1,954540077233D-06
				6,75	-1,889075366762D-06
				7,25	3,891538822237D-06
				7,50	7,294442625167D-06
				7,75	7,050125115513D-06

N= 2 M=17 P= 8.5 Q= 3

I= 0 H= -8.5

0		1,000000000000D+00	5,000000055706D-01	-1.00	2,267949192431D+00
1		-1.267949192431D+00	-6,339745965432D-01	-.75	1,950961894323D+00
				-.50	1,633974596216D+00

CARDINAL SPLINES N=2

0	-8.5	2.679491924311D-01	1.339745962156D-01
1	-7.5	-6.076951545867D-01	-3.038475772934D-01
2	-6.5	4.307806183469D-01	2.153903091735D-01
3	-5.5	-1.154273188010D-01	-5.771365940052D-02
4	-4.5	3.092865685723D-02	1.546432842861D-02
5	-3.5	-8.287308627874D-03	-4.143654313937D-03
6	-2.5	2.220577654266D-03	1.110288827133D-03
7	-1.5	-5.950019891891D-04	-2.975009945945D-04
8	-.5	1.594303024906D-04	7.971515124531D-05
9	.5	-4.271922077342D-05	-2.135961038671D-05
10	1.5	1.144658060304D-05	5.723290301519D-06
11	2.5	-3.067101638739D-06	-1.533550819370D-06
12	3.5	8.218259519186D-07	4.109129759593D-07
13	4.5	-2.202021689351D-07	-1.101010844676D-07
14	5.5	5.898272382191D-08	2.949136191096D-08
15	6.5	-1.572872635251D-08	-7.864363176255D-09
16	7.5	3.932181588127D-09	1.966090794064D-09
17	8.5	-6.553635980212D-10	-3.276817990106D-10
	0	-9.777568135665D+00	-4.888784065047D+00
	1	-1.267949192431D+00	-6.339745965432D-01

-.25	1.316987298108D+00
.25	6.871994080240D-01
.50	3.995190528383D-01
.75	1.620791712335D-01
1.25	-7.109346086228D-02
1.50	-7.355715851499D-02
1.75	-3.924227691020D-02
2.25	1.904943542518D-02
2.50	1.970958122162D-02
2.75	1.051493640725D-02
3.25	-5.104280838447D-03
3.50	-5.281166371488D-03
3.75	-2.817468718786D-03
4.25	1.367687928603D-03
4.50	1.415084264335D-03
4.75	7.549384678987D-04
5.25	-3.664708759671D-04
5.50	-3.791706858505D-04
5.75	-2.022851528087D-04
6.25	9.819557526496D-05
6.50	1.015984790673D-04
6.75	5.420214333598D-05
7.25	-2.631142509274D-05
7.50	-2.722323041865D-05
7.75	-1.452342053524D-05
8.25	7.050125105994D-06
8.50	7.294442607326D-06

I= 1 H= -7.5

0		0.	-3.342354349908D-08
1		1.607695154587D+00	8.038475792595D-01
0	-8.5	-6.076951545867D-01	-3.038475772934D-01
1	-7.5	1.646170927520D+00	8.230854637602D-01
2	-6.5	-1.584683710082D+00	-7.923418550408D-01
3	-5.5	6.925639128063D-01	3.462819564031D-01
4	-4.5	-1.855719411434D-01	-9.278597057169D-02
5	-3.5	4.972385176724D-02	2.486192588362D-02
6	-2.5	-1.332346592559D-02	-6.661732962797D-03
7	-1.5	3.570011935135D-03	1.785005967567D-03
8	-.5	-9.565818149438D-04	-4.782909074719D-04
9	.5	2.563153246405D-04	1.281576623202D-04
10	1.5	-6.867948361823D-05	-3.433974180912D-05
11	2.5	1.840260983244D-05	9.201304916218D-06
12	3.5	-4.930955711512D-06	-2.465477855756D-06
13	4.5	1.321213013611D-06	6.606065068054D-07
14	5.5	-3.538963429315D-07	-1.769481714657D-07
15	6.5	9.437235811506D-08	4.718617905753D-08
16	7.5	-2.359308952876D-08	-1.179654476438D-08
17	8.5	3.932181588127D-09	1.966090794064D-09
	0	1.366540881399D+01	6.832704390282D+00
	1	1.607695154587D+00	8.038475792595D-01

-1.00	-1.607695154587D+00
-.75	-1.205771365940D+00
-.50	-8.038475772934D-01
-.25	-4.019237886467D-01
.25	3.924285518563D-01
.50	7.278856829700D-01
.75	9.493999725988D-01
1.25	8.484357651737D-01
1.50	5.663429510899D-01
1.75	2.510786614612D-01
2.25	-1.142966125511D-01
2.50	-1.182574873297D-01
2.75	-6.308961844347D-02
3.25	3.062568503068D-02
3.50	3.168699822893D-02
3.75	1.690481231272D-02
4.25	-8.206127571620D-03
4.50	-8.490505586008D-03
4.75	-4.529630807392D-03
5.25	2.198825255803D-03
5.50	2.275024115103D-03
5.75	1.213710916852D-03
6.25	-5.891734515897D-04
6.50	-6.095908744037D-04
6.75	-3.252128600159D-04
7.25	1.578685505564D-04
7.50	1.633393825119D-04

				7.75	8.714052321146D-05
				8.25	-4.230075063597D-05
				8.50	-4.376665564395D-05

I= 2 H= -6.5

0		0.	1.336941739963D-07	-1.00	4.307806183469D-01
1		-4.307806183469D-01	-2.153903170378D-01	-.75	3.230854637602D-01
				-.50	2.153903091735D-01
0	-8.5	4.307806183469D-01	2.153903091735D-01	-.25	1.076951545867D-01
1	-7.5	-1.584683710082D+00	-7.923418550408D-01	.25	-1.009642074251D-01
2	-6.5	2.338734840327D+00	1.169367420163D+00	.50	-1.615427318801D-01
3	-5.5	-1.770255651225D+00	-8.851278256125D-01	.75	-1.413498903951D-01
4	-4.5	7.422877645735D-01	3.711438822868D-01	1.25	2.781319393052D-01
5	-3.5	-1.988954070690D-01	-9.944770353449D-02	1.50	6.096281956403D-01
6	-2.5	5.329386370238D-02	2.664693185119D-02	1.75	8.863103541553D-01
7	-1.5	-1.428004774054D-02	-7.140023870269D-03	2.25	8.790614502044D-01
8	-.5	3.826327259775D-03	1.913163629888D-03	2.50	5.980299493189D-01
9	.5	-1.025261298562D-03	-5.126306492810D-04	2.75	2.679834737739D-01
10	1.5	2.747179344729D-04	1.373589672365D-04	3.25	-1.225027401227D-01
11	2.5	-7.361043932974D-05	-3.680521966487D-05	3.50	-1.267479929157D-01
12	3.5	1.972382284605D-05	9.861911423023D-06	3.75	-6.761924925086D-02
13	4.5	-5.284852054443D-06	-2.642426027222D-06	4.25	3.282451028648D-02
14	5.5	1.415585371726D-06	7.077926858629D-07	4.50	3.396202234403D-02
15	6.5	-3.774894324602D-07	-1.887447162301D-07	4.75	1.811852322957D-02
16	7.5	9.437235811506D-08	4.718617905753D-08	5.25	-8.795301023210D-03
17	8.5	-1.572872635251D-08	-7.864363176255D-09	5.50	-9.100096460412D-03
				5.75	-4.854843667408D-03
	0	-3.661635255949D+00	-1.830817561127D+00	6.25	2.356693806359D-03
	1	-4.307806183469D-01	-2.153903170378D-01	6.50	2.438363497615D-03
				6.75	1.300851440063D-03
				7.25	-6.314742022257D-04
				7.50	-6.533575300477D-04
				7.75	-3.485620928458D-04
				8.25	1.692030025439D-04
				8.50	1.750666225758D-04

I= 3 H= -5.5

0		0.	-5.013531524862D-07	-1.00	-1.154273188010D-01
1		1.154273188010D-01	5.771368889188D-02	-.75	-8.657048910078D-02
				-.50	-5.771365940052D-02
0	-8.5	-1.154273188010D-01	-5.771365940052D-02	-.25	-2.885682970026D-02
1	-7.5	6.925639128063D-01	3.462819564031D-01	.25	2.705327784399D-02
2	-6.5	-1.770255651225D+00	-8.851278256125D-01	.50	4.328524455039D-02
3	-5.5	2.388458692094D+00	1.194229346047D+00	.75	3.787458898159D-02
4	-4.5	-1.783579117151D+00	-8.917895585753D-01	1.25	-7.033852239439D-02
5	-3.5	7.458577765086D-01	3.729288882543D-01	1.50	-1.298557336512D-01
6	-2.5	-1.998519888839D-01	-9.992599444196D-02	1.75	-1.244450780824D-01
7	-1.5	5.355017902702D-02	2.677508951351D-02	2.25	2.699258117335D-01
8	-.5	-1.434872722416D-02	-7.174363612078D-03	2.50	6.011376900543D-01
9	.5	3.844729869607D-03	1.922364934804D-03	2.75	8.817807233479D-01
10	1.5	-1.030192254273D-03	-5.150961271367D-04	3.25	8.812602754602D-01
11	2.5	2.760391474865D-04	1.380195737433D-04	3.50	6.003049734340D-01
12	3.5	-7.396433567268D-05	-3.698216783634D-05	3.75	2.691971846907D-01
13	4.5	1.981819520416D-05	9.909097602081D-06	4.25	-1.230919135743D-01
14	5.5	-5.308445143972D-06	-2.654222571986D-06	4.50	-1.273575837901D-01

15	6.5	1.415585371726D-06	7.077926858629D-07	4.75	-6.794446211088D-02
16	7.5	-3.538963429315D-07	-1.769481714657D-07	5.25	3.298237883704D-02
17	8.5	5.898272382191D-08	2.949136191096D-08	5.50	3.412536172655D-02
				5.75	1.820566375278D-02
	0	9.811322098089D-01	4.905658542279D-01	6.25	-8.837601773846D-03
	1	1.154273188010D-01	5.771368889188D-02	6.50	-9.143863116056D-03
				6.75	-4.878192900238D-03
				7.25	2.368028258346D-03
				7.50	2.450090737679D-03
				7.75	1.307107848172D-03
				8.25	-6.345112595395D-04
				8.50	-6.564998346593D-04

I= 4 H= -4.5

0		0.	1.871718435949D-06	-1.00	3.092865685723D-02
1		-3.092865685723D-02	-1.546443852970D-02	-.75	2.319649264292D-02
				-.50	1.546432842861D-02
0	-8.5	3.092865685723D-02	1.546432842861D-02	-.25	7.732164214307D-03
1	-7.5	-1.855719411434D-01	-9.278597057169D-02	.25	-7.248903950913D-03
2	-6.5	7.422877645735D-01	3.711438822868D-01	.50	-1.159824632146D-02
3	-5.5	-1.783579117151D+00	-8.917895585753D-01	.75	-1.014846553128D-02
4	-4.5	2.392028704029D+00	1.196014352015D+00	1.25	1.884715027237D-02
5	-3.5	-1.784535698966D+00	-8.922678494828D-01	1.50	3.479473896438D-02
6	-2.5	7.461140918333D-01	3.730570459166D-01	1.75	3.334495817420D-02
7	-1.5	-1.999206683675D-01	-9.996033418377D-02	2.25	-6.813969713858D-02
8	-.5	5.356858163685D-02	2.678429081843D-02	2.50	-1.275807095361D-01
9	.5	-1.435365817987D-02	-7.176829089934D-03	2.75	-1.232313671655D-01
10	1.5	3.846051082621D-03	1.923025541311D-03	3.25	2.693366382820D-01
11	2.5	-1.030546150616D-03	-5.152730753082D-04	3.50	6.005280991799D-01
12	3.5	2.761335198447D-04	1.380667599223D-04	3.75	8.814555104879D-01
13	4.5	-7.398792876220D-05	-3.699396438110D-05	4.25	8.814181440107D-01
14	5.5	1.981819520416D-05	9.909097602081D-06	4.50	6.004683128165D-01
15	6.5	-5.284852054443D-06	-2.642426027222D-06	4.75	2.692843252140D-01
16	7.5	1.321213013611D-06	6.606065068054D-07	5.25	-1.231342143249D-01
17	8.5	-2.202021689351D-07	-1.101010844676D-07	5.50	-1.274013504458D-01
				5.75	-6.796781134371D-02
	0	-2.628935832864D-01	-1.314458557840D-01	6.25	3.299371328903D-02
	1	-3.092865685723D-02	-1.546443852970D-02	6.50	3.413708896661D-02
				6.75	1.821192016089D-02
				7.25	-8.840638831160D-03
				7.50	-9.147005420668D-03
				7.75	-4.879869299841D-03
				8.25	2.368842035614D-03
				8.50	2.450932716061D-03

I= 5 H= -3.5

0		0.	-6.985520591308D-06	-1.00	-8.287308627874D-03
1		8.287308627874D-03	4.144065226913D-03	-.75	-6.215481470905D-03
				-.50	-4.143654313937D-03
0	-8.5	-8.287308627874D-03	-4.143654313937D-03	-.25	-2.071827156968D-03
1	-7.5	4.972385176724D-02	2.486192588362D-02	.25	1.942337959658D-03
2	-6.5	-1.988954070690D-01	-9.944770353449D-02	.50	3.107740735453D-03
3	-5.5	7.458577765086D-01	3.729288882543D-01	.75	2.719273143521D-03
4	-4.5	-1.784535698966D+00	-8.922678494828D-01	1.25	-5.050078695111D-03
5	-3.5	2.392285019354D+00	1.196142509677D+00	1.50	-9.323222206358D-03

CARDINAL SPLINES N=2

6	-2.5	-1.784604378449D+00	-8.923021892246D-01	1.75	-8.934754614426D-03
7	-1.5	7.461324944431D-01	3.730662472216D-01	2.25	1.825797682078D-02
8	-.5	-1.999255993232D-01	-9.996279966162D-02	2.50	3.418514808998D-02
9	.5	5.356990284986D-02	2.678495142493D-02	2.75	3.301974531418D-02
10	1.5	-1.435401207621D-02	-7.177006038105D-03	3.25	-6.798182858803D-02
11	2.5	3.846145454979D-03	1.923072727490D-03	3.50	-1.274173701536D-01
12	3.5	-1.030569743706D-03	-5.152848718530D-04	3.75	-1.231442266423D-01
13	4.5	2.761335198447D-04	1.380667599223D-04	4.25	2.692943375313D-01
14	5.5	-7.396433567268D-05	-3.698216783634D-05	4.50	6.004843325243D-01
15	6.5	1.972382284605D-05	9.861911423023D-06	4.75	8.814321612551D-01
16	7.5	-4.930955711512D-06	-2.465477855756D-06	5.25	8.814294784627D-01
17	8.5	8.218259519186D-07	4.109129759593D-07	5.50	6.004800400565D-01
				5.75	2.692905816221D-01
	0	7.044212333693D-02	3.521756890817D-02	6.25	-1.231372513823D-01
	1	8.287308627874D-03	4.144065226913D-03	6.50	-1.274044927504D-01
				6.75	-6.796948774331D-02
				7.25	3.299452706629D-02
				7.50	3.413793094499D-02
				7.75	1.821236935119D-02
				8.25	-8.840856882917D-03
				8.50	-9.147231029586D-03

I= 6 H= -2.5

0		0.	2.607036392928D-05	-1.00	2.220577654266D-03
1		-2.220577654266D-03	-1.111822377952D-03	-.75	1.665433240699D-03
				-.50	1.110288827133D-03
0	-8.5	2.220577654266D-03	1.110288827133D-03	-.25	5.551444135664D-04
1	-7.5	-1.332346592559D-02	-6.661732962797D-03	.25	-5.204478877185D-04
2	-6.5	5.329386370238D-02	2.664693185119D-02	.50	-8.327166203496D-04
3	-5.5	-1.998519888839D-01	-9.992599444196D-02	.75	-7.286270428059D-04
4	-4.5	7.461140918333D-01	3.730570459166D-01	1.25	1.353164508068D-03
5	-3.5	-1.784604378449D+00	-8.923021892246D-01	1.50	2.498149861049D-03
6	-2.5	2.392303421964D+00	1.196151710982D+00	1.75	2.394060283505D-03
7	-1.5	-1.784609309405D+00	-8.923046547025D-01	2.25	-4.892210144554D-03
8	-.5	7.461338156561D-01	3.730669078281D-01	2.50	-9.159882823846D-03
9	.5	-1.999259532196D-01	-9.996297660979D-02	2.75	-8.847614091215D-03
10	1.5	5.356999722222D-02	2.678499861111D-02	3.25	1.821567607015D-02
11	2.5	-1.435403566930D-02	-7.177017834650D-03	3.50	3.414138143434D-02
12	3.5	3.846145454979D-03	1.923072727490D-03	3.75	3.299639608135D-02
13	4.5	-1.030546150616D-03	-5.152730753082D-04	4.25	-6.797049413604D-02
14	5.5	2.760391474865D-04	1.380195737433D-04	4.50	-1.274056429135D-01
15	6.5	-7.361043932974D-05	-3.680521966487D-05	4.75	-1.231379702342D-01
16	7.5	1.840260983244D-05	9.201304916218D-06	5.25	2.692913004740D-01
17	8.5	-3.067101638739D-06	-1.533550819370D-06	5.50	6.004811902196D-01
				5.75	8.814304848555D-01
	0	-1.887491006126D-02	-9.424419848665D-03	6.25	8.814302922400D-01
	1	-2.220577654266D-03	-1.111822377952D-03	6.50	6.004808820349D-01
				6.75	2.692910308124D-01
				7.25	-1.231374694340D-01
				7.50	-1.274047183593D-01
				7.75	-6.796960810494D-02
				8.25	3.299458549605D-02
				8.50	3.413799140228D-02

I= 7 H= -1.5

CARDINAL SPLINES N=2

0		0.	-9.729593512583D-05	-1.00	-5.950019891891D-04
1		5.950019891891D-04	3.032242848961D-04	-.75	-4.462514918918D-04
				-.50	-2.975009945945D-04
0	-8.5	-5.950019891891D-04	-2.975009945945D-04	-.25	-1.487504972973D-04
1	-7.5	3.570011935135D-03	1.785005967567D-03	.25	1.394535912162D-04
2	-6.5	-1.428004774054D-02	-7.140023870269D-03	.50	2.231257459459D-04
3	-5.5	5.355017902702D-02	2.677508951351D-02	.75	1.952350277027D-04
4	-4.5	-1.999206683675D-01	-9.996033418377D-02	1.25	-3.625793371621D-04
5	-3.5	7.461324944431D-01	3.730662472216D-01	1.50	-6.693772378377D-04
6	-2.5	-1.784609309405D+00	-8.923046547025D-01	1.75	-6.414865195945D-04
7	-1.5	2.392304743177D+00	1.196152371588D+00	2.25	1.310863757432D-03
8	-.5	-1.784609663301D+00	-8.923048316506D-01	2.50	2.454383205405D-03
9	.5	7.461339100285D-01	3.730669550142D-01	2.75	2.370711050675D-03
10	1.5	-1.999259768127D-01	-9.996298840634D-02	3.25	-4.880875692567D-03
11	2.5	5.356999722222D-02	2.678499861111D-02	3.50	-9.148155583782D-03
12	3.5	-1.435401207621D-02	-7.177006038105D-03	3.75	-8.841357683107D-03
13	4.5	3.846051082621D-03	1.923025541311D-03	4.25	1.821263901283D-02
14	5.5	-1.030192254273D-03	-5.150961271367D-04	4.50	3.413823912972D-02
15	6.5	2.747179344729D-04	1.373589672365D-04	4.75	3.299471968175D-02
16	7.5	-6.867948361823D-05	-3.433974180912D-05	5.25	-6.796968035877D-02
17	8.5	1.144658060304D-05	5.723290301519D-06	5.50	-1.274048009351D-01
				5.75	-1.231375210439D-01
	0	5.057516908107D-03	2.480110486491D-03	6.25	2.692910824223D-01
	1	5.950019891891D-04	3.032242848961D-04	6.50	6.004809646107D-01
				6.75	8.814303644938D-01
				7.25	8.814303506698D-01
				7.50	6.004809424922D-01
				7.75	2.692910630685D-01
				8.25	-1.231374851013D-01
				8.50	-1.274047345795D-01

I= 8 H= -.5

0		0.	3.631133765740D-04	-1.00	1.594303024906D-04
1		-1.594303024906D-04	-1.010747616320D-04	-.75	1.195727268680D-04
				-.50	7.971515124531D-05
0	-8.5	1.594303024906D-04	7.971515124531D-05	-.25	3.985757562266D-05
1	-7.5	-9.565818149438D-04	-4.782909074719D-04	.25	-3.736647714624D-05
2	-6.5	3.826327259775D-03	1.913163629888D-03	.50	-5.978636343398D-05
3	-5.5	-1.434872722416D-02	-7.174363612078D-03	.75	-5.231306800474D-05
4	-4.5	5.356858163685D-02	2.678429081843D-02	1.25	9.715284058022D-05
5	-3.5	-1.999255993232D-01	-9.996279966162D-02	1.50	1.793590903020D-04
6	-2.5	7.461338156561D-01	3.730669078281D-01	1.75	1.718857948727D-04
7	-1.5	-1.784609663301D+00	-8.923048316506D-01	2.25	-3.512448851747D-04
8	-.5	2.392304837549D+00	1.196152418774D+00	2.50	-6.576499777738D-04
9	.5	-1.784609686894D+00	-8.923048434472D-01	2.75	-6.352301114861D-04
10	1.5	7.461339100285D-01	3.730669550142D-01	3.25	1.307826700118D-03
11	2.5	-1.999259532196D-01	-9.996297660979D-02	3.50	2.451240900793D-03
12	3.5	5.356990284986D-02	2.678495142493D-02	3.75	2.369034651072D-03
13	4.5	-1.435365817987D-02	-7.176829089934D-03	4.25	-4.880061915299D-03
14	5.5	3.844729869607D-03	1.922364934804D-03	4.50	-9.147313605400D-03
15	6.5	-1.025261298562D-03	-5.126306492810D-04	4.75	-8.840908492800D-03
16	7.5	2.563153246405D-04	1.281576623202D-04	5.25	1.821242096108D-02
17	8.5	-4.271922077342D-05	-2.135961038671D-05	5.50	3.413801352081D-02
				5.75	3.299459932013D-02
	0	-1.355157571170D-03	-4.960220972981D-04	6.25	-6.796962192901D-02
	1	-1.594303024906D-04	-1.010747616320D-04	6.50	-1.274047404778D-01

CARDINAL SPLINES N=2

				6.75	-1,231374887877D-01
				7.25	2,692910667550D-01
				7.50	6,004809483905D-01
				7.75	8,814303558308D-01
				8.25	8,814303549091D-01
				8.50	6,004809469159D-01

I= 9 H= .5

0		0.	-1,355157571170D-03	-1.00	-4,271922077342D-05
1		4,271922077342D-05	1.010747616320D-04	-.75	-3,203941558006D-05
				-.50	-2,135961038671D-05
0	-8,5	-4,271922077342D-05	-2.135961038671D-05	-.25	-1,067980519335D-05
1	-7,5	2,563153246405D-04	1.281576623202D-04	.25	1,001231736877D-05
2	-6,5	-1,025261298562D-03	-5.126306492810D-04	.50	1,601970779003D-05
3	-5,5	3,844729869607D-03	1,922364934804D-03	.75	1,401724431628D-05
4	-4,5	-1,435365817987D-02	-7.176829089934D-03	1.25	-2,603202515880D-05
5	-3,5	5,356990284986D-02	2,678495142493D-02	1.50	-4,805912337009D-05
6	-2,5	-1,999259532196D-01	-9,996297660979D-02	1.75	-4,605665989634D-05
7	-1,5	7,461339100285D-01	3,730669550142D-01	2.25	9,411578326643D-05
8	-,5	-1,784609686894D+00	-8,923048434472D-01	2,50	1,762167856903D-04
9	,5	2,392304837549D+00	1,196152418774D+00	2.75	1,702093952691D-04
10	1,5	-1,784609663301D+00	-8,923048316506D-01	3.25	-3,504311079069D-04
11	2,5	7,461338156561D-01	3,730669078281D-01	3.50	-6,568080193913D-04
12	3,5	-1,999255993232D-01	-9,996279966162D-02	3.75	-6,347809211800D-04
13	4,5	5,356858163685D-02	2,678429081843D-02	4.25	1,307608648361D-03
14	5,5	-1,434872722416D-02	-7.174363612078D-03	4.50	2,451015291875D-03
15	6,5	3,826327259775D-03	1,913163629888D-03	4,75	2,368914289451D-03
16	7,5	-9,565818149438D-04	-4,782909074719D-04	5,25	-4,880003485538D-03
17	8,5	1,594303024906D-04	7.971515124531D-05	5,50	-9,147253148108D-03
				5,75	-8,840876236623D-03
	0	3,631133765740D-04	-4,960220972981D-04	6.25	1,821240529379D-02
	1	4,271922077342D-05	1,010747616320D-04	6.50	3,413799730056D-02
				6.75	3,299459065704D-02
				7.25	-6,796961768963D-02
				7.50	-1,274047360541D-01
				7.75	-1,231374863915D-01
				8.25	2,692910654647D-01
				8.50	6,004809469159D-01

I=10 H= 1.5

0		0.	5,057516908107D-03	-1.00	1,144658060304D-05
1		-1.144658060304D-05	-3,032242848961D-04	-.75	8,584935452279D-06
				-,50	5,723290301519D-06
0	-8,5	1.144658060304D-05	5,723290301519D-06	-,25	2,861645150760D-06
1	-7,5	-6,867948361823D-05	-3,433974180912D-05	.25	-2,682792328837D-06
2	-6,5	2,747179344729D-04	1,373589672365D-04	,50	-4,292467726140D-06
3	-5,5	-1,030192254273D-03	-5,150961271367D-04	.75	-3,755909260372D-06
4	-4,5	3,846051082621D-03	1.923025541311D-03	1.25	6,975260054977D-06
5	-3,5	-1,435401207621D-02	-7.177006038105D-03	1.50	1,287740317842D-05
6	-2,5	5,356999722222D-02	2,678499861111D-02	1.75	1,234084471265D-05
7	-1,5	-1,999259768127D-01	-9,996298840634D-02	2.25	-2,521824789107D-05
8	-,5	7,461339100285D-01	3,730669550142D-01	2.50	-4,721714498754D-05
9	,5	-1,784609663301D+00	-8,923048316506D-01	2,75	-4,560746959023D-05
10	1,5	2,392304743177D+00	1,196152371588D+00	3.25	9,389773150930D-05
11	2,5	-1,784609309405D+00	-8,923046547025D-01	3.50	1,759911767717D-04

CARDINAL SPLINES N=2

12	3.5	7.461324944431D-01	3.730662472216D-01	3.75	1.700890336483D-04
13	4.5	-1.999206683675D-01	-9.996033418377D-02	4.25	-3.503726781461D-04
14	5.5	5.355017902702D-02	2.677508951351D-02	4.50	-6.567475620994D-04
15	6.5	-1.428004774054D-02	-7.140023870269D-03	4.75	-6.347486650029D-04
16	7.5	3.570011935135D-03	1.785005967567D-03	5.25	1.307592981075D-03
17	8.5	-5.950019891891D-04	-2.975009945945D-04	5.50	2.450999071626D-03
				5.75	2.368905626363D-03
	0	-9.729593512583D-05	2.480110486491D-03	6.25	-4.879999246155D-03
	1	-1.144658060304D-05	-3.032242848961D-04	6.50	-9.147248724403D-03
				6.75	-8.840873840450D-03
				7.25	1.821240400354D-02
				7.50	3.413799582599D-02
				7.75	3.299458973544D-02
				8.25	-6.796961676802D-02
				8.50	-1.274047345795D-01

I=11	H= 2.5				
0		0.	-1.887491006126D-02	-1.00	-3.067101638739D-06
1		3.067101638739D-06	1.111822377952D-03	-.75	-2.300326229055D-06
				-.50	-1.533550819370D-06
0	-8.5	-3.067101638739D-06	-1.533550819370D-06	-.25	-7.667754096848D-07
1	-7.5	1.840260983244D-05	9.201304916218D-06	.25	7.188519465795D-07
2	-6.5	-7.361043932974D-05	-3.680521966487D-05	.50	1.150163114527D-06
3	-5.5	2.760391474865D-04	1.380195737433D-04	.75	1.006392725211D-06
4	-4.5	-1.030546150616D-03	-5.152730753082D-04	1.25	-1.869015061107D-06
5	-3.5	3.846145454979D-03	1.923072727490D-03	1.50	-3.450489343582D-06
6	-2.5	-1.435403566930D-02	-7.177017834650D-03	1.75	-3.306718954266D-06
7	-1.5	5.356999722222D-02	2.678499861111D-02	2.25	6.757208297848D-06
8	-.5	-1.999259532196D-01	-9.996297660979D-02	2.50	1.265179425980D-05
9	.5	7.461338156561D-01	3.730669078281D-01	2.75	1.222048309185D-05
10	1.5	-1.784609309405D+00	-8.923046547025D-01	3.25	-2.515981813028D-05
11	2.5	2.392303421964D+00	1.196151710982D+00	3.50	-4.715668769562D-05
12	3.5	-1.784604378449D+00	-8.923021892246D-01	3.75	-4.557521341314D-05
13	4.5	7.461140918333D-01	3.730570459166D-01	4.25	9.388206422329D-05
14	5.5	-1.998519888839D-01	-9.992599444196D-02	4.50	1.759749565227D-04
15	6.5	5.329386370238D-02	2.664693185119D-02	4.75	1.700803705607D-04
16	7.5	-1.332346592559D-02	-6.661732962797D-03	5.25	-3.503684387629D-04
17	8.5	2.220577654266D-03	1.110288827133D-03	5.50	-6.567431383951D-04
				5.75	-6.347462688297D-04
	0	2.607036392928D-05	-9.424419848665D-03	6.25	1.307591690828D-03
	1	3.067101638739D-06	1.111822377952D-03	6.50	2.450997597058D-03
				6.75	2.368904704758D-03
				7.25	-4.879983245550D-03
				7.50	-9.147247249835D-03
				7.75	-8.840872550203D-03
				8.25	1.821240160737D-02
				8.50	3.413799140228D-02

I=12	H= 3.5				
0		0.	7.044212333693D-02	-1.00	8.218259519186D-07
1		-8.218259519186D-07	-4.144065226913D-03	-.75	6.163694639390D-07
				-.50	4.109129759593D-07
0	-8.5	8.218259519186D-07	4.109129759593D-07	-.25	2.054564879797D-07
1	-7.5	-4.930955711512D-06	-2.465477855756D-06	.25	-1.926154574809D-07
2	-6.5	1.972382284605D-05	9.861911423023D-06	.50	-3.081847319695D-07

CARDINAL SPLINES N=2

3	-5.5	-7.396433567268D-05	-3.698216783634D-05
4	-4.5	2.761335198447D-04	1.380667599223D-04
5	-3.5	-1.030569743706D-03	-5.152848718530D-04
6	-2.5	3.846145454979D-03	1.923072727490D-03
7	-1.5	-1.435401207621D-02	-7.177006038105D-03
8	-.5	5.356990284986D-02	2.678495142493D-02
9	.5	-1.999255993232D-01	-9.996279966162D-02
10	1.5	7.461324944431D-01	3.730662472216D-01
11	2.5	-1.784604378449D+00	-8.923021892246D-01
12	3.5	2.392285019354D+00	1.196142509677D+00
13	4.5	-1.784535698966D+00	-8.922678494828D-01
14	5.5	7.458577765086D-01	3.729288882543D-01
15	6.5	-1.988954070690D-01	-9.944770353449D-02
16	7.5	4.972385176724D-02	2.486192588362D-02
17	8.5	-8.287308627874D-03	-4.143654313937D-03
	0	-6.985520591308D-06	3.521756890817D-02
	1	-8.218259519186D-07	-4.144065226913D-03

.75	-2.696616404733D-07
1.25	5.008001894504D-07
1.50	9.245541959085D-07
1.75	8.860311044123D-07
2.25	-1.810585300321D-06
2.50	-3.390032051664D-06
2.75	-3.274462777176D-06
3.25	6.741541011832D-06
3.50	1.263557401075D-05
3.75	1.221182000429D-05
4.25	-2.515557874701D-05
4.50	-4.715226399133D-05
4.75	-4.557281723999D-05
5.25	9.388077397620D-05
5.50	1.759734819546D-04
5.75	1.700794489557D-04
6.25	-3.503675171578D-04
6.50	-6.567416638270D-04
6.75	-6.347449785826D-04
7.25	1.307589294655D-03
7.50	2.450993173353D-03
7.75	2.368900465375D-03
8.25	-4.879989661462D-03
8.50	-9.147231029586D-03

I=13 H= 4.5

0		0.	-2.628935832864D-01
1		2.202021689351D-07	1.546443852970D-02
0	-8.5	-2.202021689351D-07	-1.101010844676D-07
1	-7.5	1.321213013611D-06	6.606065068054D-07
2	-6.5	-5.284852054443D-06	-2.642426027222D-06
3	-5.5	1.981819520416D-05	9.909097602081D-06
4	-4.5	-7.398792876220D-05	-3.699396438110D-05
5	-3.5	2.761335198447D-04	1.380667599223D-04
6	-2.5	-1.030546150616D-03	-5.152730753082D-04
7	-1.5	3.846051082621D-03	1.923025541311D-03
8	-.5	-1.435365817987D-02	-7.176829089934D-03
9	.5	5.356858163685D-02	2.678429081843D-02
10	1.5	-1.999206683675D-01	-9.996033418377D-02
11	2.5	7.461140918333D-01	3.730570459166D-01
12	3.5	-1.784535698966D+00	-8.922678494828D-01
13	4.5	2.392028704029D+00	1.196014352015D+00
14	5.5	-1.783579117151D+00	-8.917895585753D-01
15	6.5	7.422877645735D-01	3.711438822868D-01
16	7.5	-1.855719411434D-01	-9.278597057169D-02
17	8.5	3.092865685723D-02	1.546432842861D-02
	0	1.871718435949D-06	-1.314458557840D-01
	1	2.202021689351D-07	1.546443852970D-02

-1.00	-2.202021689351D-07
-.75	-1.651516267014D-07
-.50	-1.101010844676D-07
-.25	-5.505054223378D-08
.25	5.160988334417D-08
.50	8.257581335068D-08
.75	7.225383668184D-08
1.25	-1.341856966948D-07
1.50	-2.477274400520D-07
1.75	-2.374054633832D-07
2.25	4.851329034352D-07
2.50	9.083339468574D-07
2.75	8.773680168509D-07
3.25	-1.806345917046D-06
3.50	-3.385608347378D-06
3.75	-3.272066604021D-06
4.25	6.740250764749D-06
4.50	1.263409944265D-05
4.75	1.221089839923D-05
5.25	-2.515465714195D-05
5.50	-4.715078942324D-05
5.75	-4.557152699290D-05
6.25	9.387837780305D-05
6.50	1.759690582503D-04
6.75	1.700752095724D-04
7.25	-3.503588540702D-04
7.50	-6.567254435779D-04
7.75	-6.347293112966D-04
8.25	1.307557038478D-03
8.50	2.450932716061D-03

CARDINAL SPLINES N=2

I=14 H= 5.5

0		0.	9.811322098089D-01
1		-5.898272382191D-08	-5.771368889188D-02

0	-8.5	5.898272382191D-08	2.949136191096D-08
1	-7.5	-3.538963429315D-07	-1.769481714657D-07
2	-6.5	1.415585371726D-06	7.077926858629D-07
3	-5.5	-5.308445143972D-06	-2.654222571986D-06
4	-4.5	1.981819520416D-05	9.909097602081D-06
5	-3.5	-7.396433567268D-05	-3.698216783634D-05
6	-2.5	2.760391474865D-04	1.380195737433D-04
7	-1.5	-1.030192254273D-03	-5.150961271367D-04
8	-.5	3.844729869607D-03	1.922364934804D-03
9	.5	-1.434872722416D-02	-7.174363612078D-03
10	1.5	5.355017902702D-02	2.677508951351D-02
11	2.5	-1.998519888839D-01	-9.992599444196D-02
12	3.5	7.458577765086D-01	3.729288882543D-01
13	4.5	-1.783579117151D+00	-8.917895585753D-01
14	5.5	2.388458692094D+00	1.194229346047D+00
15	6.5	-1.770255651225D+00	-8.851278256125D-01
16	7.5	6.925639128063D-01	3.462819564031D-01
17	8.5	-1.154273188010D-01	-5.771365940052D-02

0	-5.013531524862D-07	4.905658542279D-01
1	-5.898272382191D-08	-5.771368889188D-02

-1.00	5.898272382191D-08
-.75	4.423704286643D-08
-.50	2.949136191096D-08
-.25	1.474568095548D-08
.25	-1.382407589576D-08
.50	-2.211852143322D-08
.75	-1.935370625406D-08
1.25	3.594259732898D-08
1.50	6.635556429965D-08
1.75	6.359074912050D-08
2.25	-1.299463134201D-07
2.50	-2.433037357654D-07
2.75	-2.350092902279D-07
3.25	4.838426563516D-07
3.50	9.068593787619D-07
3.75	8.764464117912D-07
4.25	-1.805424311986D-06
4.50	-3.384133779282D-06
4.75	-3.270776356937D-06
5.25	6.737854591594D-06
5.50	1.262967573837D-05
5.75	1.220665901596D-05
6.25	-2.514599405439D-05
6.50	-4.713456917418D-05
6.75	-4.555585970689D-05
7.25	9.384612162596D-05
7.50	1.759086009584D-04
7.75	1.700167798116D-04
8.25	-3.502384924494D-04
8.50	-6.564998346593D-04

I=15 H= 6.5

0		0.	-3.661635255949D+00
1		1.572872635251D-08	2.153903170378D-01

0	-8.5	-1.572872635251D-08	-7.864363176255D-09
1	-7.5	9.437235811506D-08	4.718617905753D-08
2	-6.5	-3.774894324602D-07	-1.887447162301D-07
3	-5.5	1.415585371726D-06	7.077926858629D-07
4	-4.5	-5.284852054443D-06	-2.642426027222D-06
5	-3.5	1.972382284605D-05	9.861911423023D-06
6	-2.5	-7.361043932974D-05	-3.680521966487D-05
7	-1.5	2.747179344729D-04	1.373589672365D-04
8	-.5	-1.025261298562D-03	-5.126306492810D-04
9	.5	3.826327259775D-03	1.913163629888D-03
10	1.5	-1.428004774054D-02	-7.140023870269D-03
11	2.5	5.329386370238D-02	2.664693185119D-02
12	3.5	-1.988954070690D-01	-9.944770353449D-02
13	4.5	7.422877645735D-01	3.711438822868D-01
14	5.5	-1.770255651225D+00	-8.851278256125D-01
15	6.5	2.338734840327D+00	1.169367420163D+00
16	7.5	-1.584683710082D+00	-7.923418550408D-01
17	8.5	4.307806183469D-01	2.153903091735D-01

-1.00	-1.572872635251D-08
-.75	-1.179654476438D-08
-.50	-7.864363176255D-09
-.25	-3.932181588127D-09
.25	3.686420238869D-09
.50	5.898272382191D-09
.75	5.160988334417D-09
1.25	-9.584692621061D-09
1.50	-1.769481714657D-08
1.75	-1.695753309880D-08
2.25	3.465235024537D-08
2.50	6.488099620410D-08
2.75	6.266914406078D-08
3.25	-1.290247083604D-07
3.50	-2.418291676698D-07
3.75	-2.337190431443D-07
4.25	4.814464831963D-07
4.50	9.024356744752D-07
4.75	8.722070285165D-07
5.25	-1.796761224425D-06
5.50	-3.367913530231D-06

CARDINAL SPLINES N=2

				5.75	-3.255109070922D-06
	0	1.336941739963D-07	-1.830817561127D+00	6.25	6.705598414503D-06
	1	1.572872635251D-08	2.153903170378D-01	6.50	1.256921844645D-05
				6.75	1.214822925517D-05
				7.25	-2.502563243359D-05
				7.50	-4.690896025557D-05
				7.75	-4.533780794976D-05
				8.25	9.339693131985D-05
				8.50	1.750666225758D-04

I=16 H= 7.5

0		0.	1.366540881399D+01	-1.00	3.932181588127D-09
1		-3.932181588127D-09	-8.038475792595D-01	-.75	2.949136191096D-09
				-.50	1.966090794064D-09
0	-8.5	3.932181588127D-09	1.966090794064D-09	-.25	9.830453970318D-10
1	-7.5	-2.359308952876D-08	-1.179654476438D-08	.25	-9.216050597174D-10
2	-6.5	9.437235811506D-08	4.718617905753D-08	.50	-1.474568095548D-09
3	-5.5	-3.538963429315D-07	-1.769481714657D-07	.75	-1.290247083604D-09
4	-4.5	1.321213013611D-06	6.606065068054D-07	1.25	2.396173155265D-09
5	-3.5	-4.930955711512D-06	-2.465477855756D-06	1.50	4.423704286643D-09
6	-2.5	1.840260983244D-05	9.201304916218D-06	1.75	4.239383274700D-09
7	-1.5	-6.867948361823D-05	-3.433974180912D-05	2.25	-8.663087561343D-09
8	-.5	2.563153246405D-04	1.281576623202D-04	2.50	-1.622024905103D-08
9	.5	-9.565818149438D-04	-4.782909074719D-04	2.75	-1.566728601520D-08
10	1.5	3.570011935135D-03	1.785005967567D-03	3.25	3.225617709011D-08
11	2.5	-1.332346592559D-02	-6.661732962797D-03	3.50	6.045729191746D-08
12	3.5	4.972385176724D-02	2.486192588362D-02	3.75	5.842976078608D-08
13	4.5	-1.855719411434D-01	-9.278597057169D-02	4.25	-1.203616207991D-07
14	5.5	6.925639128063D-01	3.462819564031D-01	4.50	-2.256089186188D-07
15	6.5	-1.584683710082D+00	-7.923418550408D-01	4.75	-2.180517571291D-07
16	7.5	1.646170927520D+00	8.230854637602D-01	5.25	4.491903061062D-07
17	8.5	-6.076951545867D-01	-3.038475772934D-01	5.50	8.419783825578D-07
				5.75	8.137772677304D-07
	0	-3.342354349908D-08	6.832704390282D+00	6.25	-1.676399603626D-06
	1	-3.932181588127D-09	-8.038475792595D-01	6.50	-3.142304611612D-06
				6.75	-3.037057313793D-06
				7.25	6.256408108397D-06
				7.50	1.172724006389D-05
				7.75	1.133445198744D-05
				8.25	-2.334923282996D-05
				8.50	-4.376665564395D-05

I=17 H= 8.5

0		0.	-1.027756813566D+01	-1.00	-6.553635980212D-10
1		6.553635980212D-10	6.339745965432D-01	-.75	-4.915226985159D-10
				-.50	-3.276817990106D-10
0	-8.5	-6.553635980212D-10	-3.276817990106D-10	-.25	-1.638408995053D-10
1	-7.5	3.932181588127D-09	1.966090794064D-09	.25	1.536008432862D-10
2	-6.5	-1.572872635251D-08	-7.864363176255D-09	.50	2.457613492580D-10
3	-5.5	5.898272382191D-08	2.949136191096D-08	.75	2.150411806007D-10
4	-4.5	-2.202021689351D-07	-1.101010844676D-07	1.25	-3.993621925442D-10
5	-3.5	8.218259519186D-07	4.109129759593D-07	1.50	-7.372840477739D-10
6	-2.5	-3.067101638739D-06	-1.533550819370D-06	1.75	-7.065638791166D-10
7	-1.5	1.144658060304D-05	5.723290301519D-06	2.25	1.443847926891D-09
8	-.5	-4.271922077342D-05	-2.135961038671D-05	2.50	2.703374841838D-09

9	.5	1.594303024906D-04	7.971515124531D-05	2.75	2.611214335866D-09
10	1.5	-5.950019891891D-04	-2.975009945945D-04	3.25	-5.376029515018D-09
11	2.5	2.220577654266D-03	1.110288827133D-03	3.50	-1.007621531958D-08
12	3.5	-8.287308627874D-03	-4.143654313937D-03	3.75	-9.738293464347D-09
13	4.5	3.092865685723D-02	1.546432842861D-02	4.25	2.006027013318D-08
14	5.5	-1.154273188010D-01	-5.771365940052D-02	4.50	3.760148643647D-08
15	6.5	4.307806183469D-01	2.153903091735D-01	4.75	3.634195952152D-08
16	7.5	-6.076951545867D-01	-3.038475772934D-01	5.25	-7.486505101771D-08
17	8.5	2.679491924311D-01	1.339745962156D-01	5.50	-1.403297304263D-07
				5.75	-1.356295446217D-07
	0	5.570590583180D-09	-4.888784065047D+00	6.25	2.793999339376D-07
	1	6.553635980212D-10	6.339745965432D-01	6.50	5.237174352687D-07
				6.75	5.061762189654D-07
				7.25	-1.042734684733D-06
				7.50	-1.954540010649D-06
				7.75	-1.889075331240D-06
				8.25	3.891538804994D-06
				8.50	7.294442607326D-06

N= 2   M=18   P= 9.0   Q= 3

I= 0   H= -9.0

0		1.000000000000D+00	4.999999984196D-01	-1.00	2.267949192431D+00
1		-1.267949192431D+00	-6.339745961278D-01	-.75	1.950961894323D+00
				-.50	1.633974596216D+00
0	-9.0	2.679491924311D-01	1.339745962156D-01	-.25	1.316987298108D+00
1	-8.0	-6.076951545867D-01	-3.038475772934D-01	.25	6.871994080240D-01
2	-7.0	4.307806183469D-01	2.153903091735D-01	.50	3.995190528383D-01
3	-6.0	-1.154273188010D-01	-5.771365940052D-02	.75	1.620791712335D-01
4	-5.0	3.092865685723D-02	1.546432842861D-02	1.25	-7.109346086228D-02
5	-4.0	-8.287308627874D-03	-4.143654313937D-03	1.50	-7.355715851499D-02
6	-3.0	2.220577654266D-03	1.110288827133D-03	1.75	-3.924227691020D-02
7	-2.0	-5.950019891911D-04	-2.975009945955D-04	2.25	1.904943542518D-02
8	-1.0	1.594303024981D-04	7.971515124906D-05	2.50	1.970958122162D-02
9	0.0	-4.271922080141D-05	-2.135961040071D-05	2.75	1.051493640725D-02
10	1.0	1.144658070752D-05	5.723290353762D-06	3.25	-5.104280838447D-03
11	2.0	-3.067102028682D-06	-1.533551014341D-06	3.50	-5.281166371488D-03
12	3.0	8.218274072044D-07	4.109137036022D-07	3.75	-2.817468718786D-03
13	4.0	-2.202076001355D-07	-1.101038000678D-07	4.25	1.367687928603D-03
14	5.0	5.900299333775D-08	2.950149666888D-08	4.50	1.415084264335D-03
15	6.0	-1.580437321547D-08	-7.902186607734D-09	4.75	7.549384678987D-04
16	7.0	4.214499524125D-09	2.107249762063D-09	5.25	-3.664708759671D-04
17	8.0	-1.053624881031D-09	-5.268124405156D-10	5.50	-3.791706858505D-04
18	9.0	1.756041468385D-10	8.780207341927D-11	5.75	-2.022851528086D-04
				6.25	9.819557526491D-05
	0	-1.041154273188D+01	-5.205771366730D+00	6.50	1.015984790672D-04
	1	-1.267949192431D+00	-6.339745961278D-01	6.75	5.420214333589D-05
				7.25	-2.631142509255D-05
				7.50	-2.722323041831D-05

				7.75	-1,452342053491D-05
				8.25	7,050125105311D-06
				8.50	7,294442606045D-06
				8.75	3,891538803756D-06

I= 1 H= -8.0

0		0.	9,482623929281D-09	-1.00	-1,607695154587D+00
1		1.607695154587D+00	8.038475767666D-01	-.75	-1,205771365940D+00
				-.50	-8,038475772934D-01
0	-9.0	-6.076951545867D-01	-3.038475772934D-01	-.25	-4,019237886467D-01
1	-8.0	1.646170927520D+00	8.230854637602D-01	.25	3,924285518563D-01
2	-7.0	-1.584683710082D+00	-7.923418550408D-01	.50	7,278856829700D-01
3	-6.0	6.925639128063D-01	3.462819564031D-01	.75	9,493999725988D-01
4	-5.0	-1.855719411434D-01	-9.278597057169D-02	1.25	8,484357651737D-01
5	-4.0	4.972385176724D-02	2.486192588362D-02	1.50	5,663429510899D-01
6	-3.0	-1.332346592560D-02	-6.661732962799D-03	1.75	2,510786614612D-01
7	-2.0	3.570011935147D-03	1.785005967573D-03	2.25	-1,142966125511D-01
8	-1.0	-9.565818149888D-04	-4.782909074944D-04	2.50	-1,182574873297D-01
9	0.0	2.563153248085D-04	1.281576624042D-04	2.75	-6,308961844347D-02
10	1.0	-6.867948424514D-05	-3.433974212257D-05	3.25	3,062568503068D-02
11	2.0	1.840261217209D-05	9.201306086046D-06	3.50	3,168699822893D-02
12	3.0	-4.930964443226D-06	-2.465482221613D-06	3.75	1,690481231272D-02
13	4.0	1.321245600813D-06	6.606228004066D-07	4.25	-8,206127571620D-03
14	5.0	-3.540179600265D-07	-1.770089800133D-07	4.50	-8,490505586008D-03
15	6.0	9.482623929281D-08	4.741311964641D-08	4.75	-4,529630807392D-03
16	7.0	-2.528699714475D-08	-1.264349857238D-08	5.25	2,198825255802D-03
17	8.0	6.321749286188D-09	3.160874643094D-09	5.50	2,275024115103D-03
18	9.0	-1.053624881031D-09	-5.268124405156D-10	5.75	1,213710916852D-03
				6.25	-5,891734515895D-04
	0	1.446925639128D+01	7.234628200382D+00	6.50	-6,095908744032D-04
	1	1.607695154587D+00	8.038475767666D-01	6.75	-3,252128600153D-04
				7.25	1,578685505553D-04
				7.50	1,633393825099D-04
				7.75	8,714052320946D-05
				8.25	-4,230075063187D-05
				8.50	-4,376665563627D-05
				8.75	-2,334923282253D-05

I= 2 H= -7.0

0		0.	-3.793049571713D-08	-1.00	4,307806183469D-01
1		-4.307806183469D-01	-2,153903070662D-01	-.75	3,230854637602D-01
				-.50	2,153903091735D-01
0	-9.0	4.307806183469D-01	2.153903091735D-01	-.25	1,076951545867D-01
1	-8.0	-1.584683710082D+00	-7.923418550408D-01	.25	-1,009642074251D-01
2	-7.0	2.338734840327D+00	1.169367420163D+00	.50	-1,615427318801D-01
3	-6.0	-1.770255651225D+00	-8.851278256125D-01	.75	-1,413498903951D-01
4	-5.0	7.422877645735D-01	3.711438822868D-01	1.25	2,781319393052D-01
5	-4.0	-1.988954070690D-01	-9.944770353449D-02	1.50	6,096281956403D-01
6	-3.0	5.329386370239D-02	2.664693185119D-02	1.75	8,863103541553D-01
7	-2.0	-1.428004774059D-02	-7.140023870293D-03	2.25	8,790614502044D-01
8	-1.0	3.826327259955D-03	1.913163629978D-03	2.50	5,980299493189D-01
9	0.0	-1.025261299234D-03	-5.126306496170D-04	2.75	2,679834737739D-01
10	1.0	2.747179369806D-04	1.373589684903D-04	3.25	-1,225027401227D-01
11	2.0	-7.361044868837D-05	-3.680522434418D-05	3.50	-1,267479929157D-01
12	3.0	1.972385777291D-05	9.861928886453D-06	3.75	-6,761924925086D-02

13	4.0	-5.284982403253D-06	-2.642491201626D-06	4.25	3.282451028648D-02
14	5.0	1.416071840106D-06	7.080359200530D-07	4.50	3.396202234403D-02
15	6.0	-3.793049571713D-07	-1.896524785856D-07	4.75	1.811852322957D-02
16	7.0	1.011479885790D-07	5.057399428950D-08	5.25	-8.795301023210D-03
17	8.0	-2.528699714475D-08	-1.264349857238D-08	5.50	-9.100096460411D-03
18	9.0	4.214499524125D-09	2.107249762063D-09	5.75	-4.854843667407D-03
				6.25	2.356693806358D-03
	0	-3.877025565123D+00	-1.938512801527D+00	6.50	2.438363497613D-03
	1	-4.307806183469D-01	-2.153903070662D-01	6.75	1.300851440061D-03
				7.25	-6.314742022213D-04
				7.50	-6.533575300395D-04
				7.75	-3.485620928379D-04
				8.25	1.692030025275D-04
				8.50	1.750666225451D-04
				8.75	9.339693129014D-05

I= 3 H= -6.0

0		0.	1.422393589392D-07	-1.00	-1.154273188010D-01
1		1.154273188010D-01	5.771365149834D-02	-.75	-8.657048910078D-02
				-.50	-5.771365940052D-02
0	-9.0	-1.154273188010D-01	-5.771365940052D-02	-.25	-2.885682970026D-02
1	-8.0	6.925639128063D-01	3.462819564031D-01	.25	2.705327784399D-02
2	-7.0	-1.770255651225D+00	-8.851278256125D-01	.50	4.328524455039D-02
3	-6.0	2.388458692094D+00	1.194229346047D+00	.75	3.787458898159D-02
4	-5.0	-1.783579117151D+00	-8.917895585753D-01	1.25	-7.033852239439D-02
5	-4.0	7.458577765087D-01	3.729288882543D-01	1.50	-1.298557336512D-01
6	-3.0	-1.998519888840D-01	-9.992599444198D-02	1.75	-1.244450780824D-01
7	-2.0	5.355017902720D-02	2.677508951360D-02	2.25	2.699258117335D-01
8	-1.0	-1.434872722483D-02	-7.174363612416D-03	2.50	6.011376900543D-01
9	0.0	3.844729872127D-03	1.922364936064D-03	2.75	8.817807233479D-01
10	1.0	-1.030192263677D-03	-5.150961318386D-04	3.25	8.812602754602D-01
11	2.0	2.760391825814D-04	1.380195912907D-04	3.50	6.003049734340D-01
12	3.0	-7.396446664839D-05	-3.698223332420D-05	3.75	2.691971846907D-01
13	4.0	1.981868401220D-05	9.909342006099D-06	4.25	-1.230919135743D-01
14	5.0	-5.310269400398D-06	-2.655134700199D-06	4.50	-1.273575837901D-01
15	6.0	1.422393589392D-06	7.111967946961D-07	4.75	-6.794446211088D-02
16	7.0	-3.793049571713D-07	-1.896524785856D-07	5.25	3.298237883704D-02
17	8.0	9.482623929281D-08	4.741311964641D-08	5.50	3.412536172654D-02
18	9.0	-1.580437321547D-08	-7.902186607734D-09	5.75	1.820566375278D-02
				6.25	-8.837601773842D-03
	0	1.038845869209D+00	5.194230057244D-01	6.50	-9.143863116048D-03
	1	1.154273188010D-01	5.771365149834D-02	6.75	-4.878192900230D-03
				7.25	2.368028258330D-03
				7.50	2.450090737648D-03
				7.75	1.307107848142D-03
				8.25	-6.345112594780D-04
				8.50	-6.564998345440D-04
				8.75	-3.502384923380D-04

I= 4 H= -5.0

0		0.	-5.310269400398D-07	-1.00	3.092865685723D-02
1		-3.092865685723D-02	-1.546429892712D-02	-.75	2.319649264292D-02
				-.50	1.546432842861D-02
0	-9.0	3.092865685723D-02	1.546432842861D-02	-.25	7.732164214307D-03
1	-8.0	-1.855719411434D-01	-9.278597057169D-02	.25	-7.248903950913D-03

CARDINAL SPLINES N=2

2	-7.0	7.422877645735D-01	3.711438822868D-01
3	-6.0	-1.783579117151D+00	-8.917895585753D-01
4	-5.0	2.392028704029D+00	1.196014352015D+00
5	-4.0	-1.784535698966D+00	-8.922678494828D-01
6	-3.0	7.461140918335D-01	3.730570459167D-01
7	-2.0	-1.999206683682D-01	-9.996033418410D-02
8	-1.0	5.356858163937D-02	2.678429081969D-02
9	0.0	-1.435365818927D-02	-7.176829094637D-03
10	1.0	3.846051117728D-03	1.923025558864D-03
11	2.0	-1.030546281637D-03	-5.152731408186D-04
12	3.0	2.761340088207D-04	1.380670044103D-04
13	4.0	-7.398975364554D-05	-3.699487682277D-05
14	5.0	1.982500576148D-05	9.912502880742D-06
15	6.0	-5.310269400398D-06	-2.655134700199D-06
16	7.0	1.416071840106D-06	7.080359200530D-07
17	8.0	-3.540179600265D-07	-1.770089800133D-07
18	9.0	5.900299333775D-08	2.950149666888D-08
	0	-2.783579117151D-01	-1.391792213710D-01
	1	-3.092865685723D-02	-1.546429892712D-02

.50	-1.159824632146D-02
.75	-1.014846553128D-02
1.25	1.884715027237D-02
1.50	3.479473896438D-02
1.75	3.334495817420D-02
2.25	-6.813969713858D-02
2.50	-1.275807095361D-01
2.75	-1.232313671655D-01
3.25	2.693366382820D-01
3.50	6.005280991799D-01
3.75	8.814555104879D-01
4.25	8.814181440107D-01
4.50	6.004683128165D-01
4.75	2.692843252140D-01
5.25	-1.231342143249D-01
5.50	-1.274013504458D-01
5.75	-6.796781134370D-02
6.25	3.299371328901D-02
6.50	3.413708896658D-02
6.75	1.821192016086D-02
7.25	-8.840638831098D-03
7.50	-9.147005420552D-03
7.75	-4.879869299730D-03
8.25	2.368842035385D-03
8.50	2.450932715631D-03
8.75	1.307557038062D-03

I= 5 H= -4.0

0		0.	1.981868401220D-06
1		8.287308627874D-03	4.143544210137D-03
0	-9.0	-8.287308627874D-03	-4.143654313937D-03
1	-8.0	4.972385176724D-02	2.486192588362D-02
2	-7.0	-1.988954070690D-01	-9.944770353449D-02
3	-6.0	7.458577765087D-01	3.729288882543D-01
4	-5.0	-1.784535698966D+00	-8.922678494828D-01
5	-4.0	2.392285019354D+00	1.196142509677D+00
6	-3.0	-1.784604378450D+00	-8.923021892249D-01
7	-2.0	7.461324944456D-01	3.730662472228D-01
8	-1.0	-1.999255993327D-01	-9.996279966633D-02
9	0.0	5.356990288497D-02	2.678495144249D-02
10	1.0	-1.435401220723D-02	-7.177006103617D-03
11	2.0	3.846145943967D-03	1.923072971984D-03
12	3.0	-1.030571568634D-03	-5.152857843171D-04
13	4.0	2.761403305700D-04	1.380701652850D-04
14	5.0	-7.398975364554D-05	-3.699487682277D-05
15	6.0	1.981868401220D-05	9.909342006099D-06
16	7.0	-5.284982403253D-06	-2.642491201626D-06
17	8.0	1.321245600813D-06	6.606228004066D-07
18	9.0	-2.202076001355D-07	-1.101038000678D-07
	0	7.458577765087D-02	3.729387975963D-02
	1	8.287308627874D-03	4.143544210137D-03

-1.00	-8.287308627874D-03
-.75	-6.215481470905D-03
-.50	-4.143654313937D-03
-.25	-2.071827156968D-03
.25	1.942337959658D-03
.50	3.107740735453D-03
.75	2.719273143521D-03
1.25	-5.050078695111D-03
1.50	-9.323222206358D-03
1.75	-8.934754614427D-03
2.25	1.825797682078D-02
2.50	3.418514808998D-02
2.75	3.301974531419D-02
3.25	-6.798182858803D-02
3.50	-1.274173701536D-01
3.75	-1.231442266423D-01
4.25	2.692943375313D-01
4.50	6.004843325243D-01
4.75	8.814321612551D-01
5.25	8.814294784627D-01
5.50	6.004800400565D-01
5.75	2.692905816220D-01
6.25	-1.231372513822D-01
6.50	-1.274044927503D-01
6.75	-6.796948774320D-02
7.25	3.299452706606D-02
7.50	3.413793094456D-02
7.75	1.821236935078D-02

8.25	-8.840856882060D-03
8.50	-9.147231027980D-03
8.75	-4.879989659910D-03

I= 6   H= -3.0

0		0.	-7.396446664839D-06
1		-2.220577654266D-03	-1.109877913430D-03
0	-9.0	2.220577654266D-03	1.110288827133D-03
1	-8.0	-1.332346592560D-02	-6.661732962799D-03
2	-7.0	5.329386370239D-02	2.664693185119D-02
3	-6.0	-1.998519888840D-01	-9.992599444198D-02
4	-5.0	7.461140918335D-01	3.730570459167D-01
5	-4.0	-1.784604378450D+00	-8.923021892249D-01
6	-3.0	2.392303421966D+00	1.196151710983D+00
7	-2.0	-1.784609309414D+00	-8.923046547072D-01
8	-1.0	7.461338156912D-01	3.730669078456D-01
9	0.0	-1.999259533506D-01	-9.996297667531D-02
10	1.0	5.356999771121D-02	2.678499885561D-02
11	2.0	-1.435403749423D-02	-7.177018747116D-03
12	3.0	3.846152265717D-03	1.923076132858D-03
13	4.0	-1.030571568634D-03	-5.152857843171D-04
14	5.0	2.761340088207D-04	1.380670044103D-04
15	6.0	-7.396446664839D-05	-3.698223332420D-05
16	7.0	1.972385777291D-05	9.861928886453D-06
17	8.0	-4.930964443226D-06	-2.465482221613D-06
18	9.0	8.218274072044D-07	4.109137036022D-07
	0	-1.998519888840D-02	-9.996297667531D-03
	1	-2.220577654266D-03	-1.109877913430D-03

-1.00	2.220577654266D-03
-.75	1.665433240700D-03
-.50	1.110288827133D-03
-.25	5.551444135666D-04
.25	-5.204478877187D-04
.50	-8.327166203498D-04
.75	-7.286270428061D-04
1.25	1.353164508068D-03
1.50	2.498149861050D-03
1.75	2.394060283506D-03
2.25	-4.892210144555D-03
2.50	-9.159882823848D-03
2.75	-8.847614091217D-03
3.25	1.821567607015D-02
3.50	3.414138143434D-02
3.75	3.299639608136D-02
4.25	-6.797049413606D-02
4.50	-1.274056429135D-01
4.75	-1.231379702342D-01
5.25	2.692913004741D-01
5.50	6.004811902198D-01
5.75	8.814304848556D-01
6.25	8.814302922398D-01
6.50	6.004808820345D-01
6.75	2.692910308120D-01
7.25	-1.231374694332D-01
7.50	-1.274047183577D-01
7.75	-6.796960810338D-02
8.25	3.299458549286D-02
8.50	3.413799139629D-02
8.75	1.821240160158D-02

I= 7   H= -2.0

0		0.	2.760391825814D-05
1		5.950019891911D-04	2.959674435812D-04
0	-9.0	-5.950019891911D-04	-2.975009945955D-04
1	-8.0	3.570011935147D-03	1.785005967573D-03
2	-7.0	-1.428004774059D-02	-7.140023870293D-03
3	-6.0	5.355017902720D-02	2.677508951360D-02
4	-5.0	-1.999206683682D-01	-9.996033418410D-02
5	-4.0	7.461324944456D-01	3.730662472228D-01
6	-3.0	-1.784609309414D+00	-8.923046547072D-01
7	-2.0	2.392304743212D+00	1.196152371606D+00
8	-1.0	-1.784609663432D+00	-8.923048317161D-01
9	0.0	7.461339105175D-01	3.730669552587D-01
10	1.0	-1.999259786376D-01	-9.996298931880D-02
11	2.0	5.357000403296D-02	2.678500201648D-02
12	3.0	-1.435403749423D-02	-7.177018747116D-03
13	4.0	3.846145943967D-03	1.923072971984D-03

-1.00	-5.950019891911D-04
-.75	-4.462514918933D-04
-.50	-2.975009945955D-04
-.25	-1.487504972978D-04
.25	1.394535912167D-04
.50	2.231257459467D-04
.75	1.952350277033D-04
1.25	-3.625793371633D-04
1.50	-6.693772378400D-04
1.75	-6.414865195966D-04
2.25	1.310863757437D-03
2.50	2.454383205413D-03
2.75	2.370711050683D-03
3.25	-4.880875692583D-03
3.50	-9.148155583813D-03
3.75	-8.841357683136D-03
4.25	1.821263901290D-02

14	5.0	-1.030546281637D-03	-5.152731408186D-04	4.50	3.413823912984D-02
15	6.0	2.760391825814D-04	1.380195912907D-04	4.75	3.299471968186D-02
16	7.0	-7.361044868837D-05	-3.680522434418D-05	5.25	-6.796968035900D-02
17	8.0	1.840261217209D-05	9.201306086046D-06	5.50	-1.274048009355D-01
18	9.0	-3.067102028682D-06	-1.533551014341D-06	5.75	-1.231375210443D-01
				6.25	2.692910824231D-01
	0	5.355017902720D-03	2.691310910489D-03	6.50	6.004809646123D-01
	1	5.950019891911D-04	2.959674435812D-04	6.75	8.814303644954D-01
				7.25	8.814303506666D-01
				7.50	6.004809424862D-01
				7.75	2.692910630628D-01
				8.25	-1.231374850894D-01
				8.50	-1.274047345572D-01
				8.75	-6.796961674640D-02

I= 8 H= -1.0

0		0.	-1.030192263677D-04	-1.00	1.594303024981D-04
1		-1.594303024981D-04	-7.399186089530D-05	-.75	1.195727268736D-04
				-.50	7.971515124906D-05
0	-9.0	1.594303024981D-04	7.971515124906D-05	-.25	3.985757562453D-05
1	-8.0	-9.565818149888D-04	-4.782909074944D-04	.25	-3.736647714800D-05
2	-7.0	3.826327259955D-03	1.913163629978D-03	.50	-5.978636343680D-05
3	-6.0	-1.434872722483D-02	-7.174363612416D-03	.75	-5.231306800720D-05
4	-5.0	5.356858163937D-02	2.678429081969D-02	1.25	9.715284058480D-05
5	-4.0	-1.999255993327D-01	-9.996279966633D-02	1.50	1.793590903104D-04
6	-3.0	7.461338156912D-01	3.730669078456D-01	1.75	1.718857948808D-04
7	-2.0	-1.784609663432D+00	-8.923048317161D-01	2.25	-3.512448851912D-04
8	-1.0	2.392304838038D+00	1.196152419019D+00	2.50	-6.576499978048D-04
9	0.0	-1.784609688719D+00	-8.923048443596D-01	2.75	-6.352301115160D-04
10	1.0	7.461339168392D-01	3.730669584196D-01	3.25	1.307826700180D-03
11	2.0	-1.999259786376D-01	-9.996298931880D-02	3.50	2.451240900909D-03
12	3.0	5.356999771121D-02	2.678499885561D-02	3.75	2.369034651183D-03
13	4.0	-1.435401220723D-02	-7.177006103617D-03	4.25	-4.880061915529D-03
14	5.0	3.846051117728D-03	1.923025558864D-03	4.50	-9.147313605830D-03
15	6.0	-1.030192263677D-03	-5.150961318386D-04	4.75	-8.840908493216D-03
16	7.0	2.747179369806D-04	1.373589684903D-04	5.25	1.821242096193D-02
17	8.0	-6.867948424514D-05	-3.433974212257D-05	5.50	3.413801352241D-02
18	9.0	1.144658070752D-05	5.723290353762D-06	5.75	3.299459932168D-02
				6.25	-6.796962193221D-02
	0	-1.434872722483D-03	-7.689459744254D-04	6.50	-1.274047404838D-01
	1	-1.594303024981D-04	-7.399186089530D-05	6.75	-1.231374887935D-01
				7.25	2.692910667669D-01
				7.50	6.004809484129D-01
				7.75	8.814303558524D-01
				8.25	8.814303548646D-01
				8.50	6.004809468324D-01
				8.75	2.692910653840D-01

I= 9 H= 0.0

0		0.	3.844729872127D-04	-1.00	-4.271922080141D-05
1		4.271922080141D-05	0.	-.75	-3.203941560106D-05
				-.50	-2.135961040071D-05
0	-9.0	-4.271922080141D-05	-2.135961040071D-05	-.25	-1.067980520035D-05
1	-8.0	2.563153248085D-04	1.281576624042D-04	.25	1.001231737533D-05
2	-7.0	-1.025261299234D-03	-5.126306496170D-04	.50	1.601970780053D-05

CARDINAL SPLINES N=2

3	-6.0	3.844729872127D-03	1.922364936064D-03	.75	1.401724432546D-05
4	-5.0	-1.435365818927D-02	-7.176829094637D-03	1.25	-2.603202517586D-05
5	-4.0	5.356990288497D-02	2.678495144249D-02	1.50	-4.805912340159D-05
6	-3.0	-1.999259533506D-01	-9.996297667531D-02	1.75	-4.605665992652D-05
7	-2.0	7.461339105175D-01	3.730669552587D-01	2.25	9.411578332811D-05
8	-1.0	-1.784609688719D+00	-8.923048443596D-01	2.50	1.762167858058D-04
9	0.0	2.392304844360D+00	1.196152422180D+00	2.75	1.702093953806D-04
10	1.0	-1.784609688719D+00	-8.923048443596D-01	3.25	-3.504311081366D-04
11	2.0	7.461339105175D-01	3.730669552587D-01	3.50	-6.568080198217D-04
12	3.0	-1.999259533506D-01	-9.996297667531D-02	3.75	-6.347809215960D-04
13	4.0	5.356990288497D-02	2.678495144249D-02	4.25	1.307608649218D-03
14	5.0	-1.435365818927D-02	-7.176829094637D-03	4.50	2.451015293481D-03
15	6.0	3.844729872127D-03	1.922364936064D-03	4.75	2.368914291003D-03
16	7.0	-1.025261299234D-03	-5.126306496170D-04	5.25	-4.880003488736D-03
17	8.0	2.563153248085D-04	1.281576624042D-04	5.50	-9.147253154102D-03
18	9.0	-4.271922080141D-05	-2.135961040071D-05	5.75	-8.840876242417D-03
				6.25	1.821240530573D-02
	0	3.844729872127D-04	3.844729872127D-04	6.50	3.413799732293D-02
	1	4.271922080141D-05	0.	6.75	3.299459067867D-02
				7.25	-6.796961773417D-02
				7.50	-1.274047361376D-01
				7.75	-1.231374864722D-01
				8.25	2.692910656310D-01
				8.50	6.004809472275D-01
				8.75	8.814303552103D-01

I=10 H= 1.0

0		0.	-1.434872722483D-03	-1.00	1.144658070752D-05
1		-1.144658070752D-05	7.399186089530D-05	-.75	8.584935530643D-06
				-.50	5.723290353762D-06
0	-9.0	1.144658070752D-05	5.723290353762D-06	-.25	2.861645176881D-06
1	-8.0	-6.867948424514D-05	-3.433974212257D-05	.25	-2.682792353326D-06
2	-7.0	2.747179369806D-04	1.373589684903D-04	.50	-4.292467765321D-06
3	-6.0	-1.030192263677D-03	-5.150961318386D-04	.75	-3.755909294656D-06
4	-5.0	3.846051117728D-03	1.923025558864D-03	1.25	6.975260118647D-06
5	-4.0	-1.435401220723D-02	-7.177006103617D-03	1.50	1.287740329596D-05
6	-3.0	5.356999771121D-02	2.678499885561D-02	1.75	1.234084482530D-05
7	-2.0	-1.999259786376D-01	-9.996298931880D-02	2.25	-2.521824812126D-05
8	-1.0	7.461339168392D-01	3.730669584196D-01	2.50	-4.721714541854D-05
9	0.0	-1.784609688719D+00	-8.923048443596D-01	2.75	-4.560747000654D-05
10	1.0	2.392304838038D+00	1.196152419019D+00	3.25	9.389773236640D-05
11	2.0	-1.784609663432D+00	-8.923048317161D-01	3.50	1.759911783782D-04
12	3.0	7.461338156912D-01	3.730669078456D-01	3.75	1.700890352009D-04
13	4.0	-1.999255993327D-01	-9.996279966633D-02	4.25	-3.503726813444D-04
14	5.0	5.356858163937D-02	2.678429081969D-02	4.50	-6.567475680942D-04
15	6.0	-1.434872722483D-02	-7.174363612416D-03	4.75	-6.347486707969D-04
16	7.0	3.826327259955D-03	1.913163629978D-03	5.25	1.307592993011D-03
17	8.0	-9.565818149888D-04	-4.782909074944D-04	5.50	2.450999093999D-03
18	9.0	1.594303024981D-04	7.971515124906D-05	5.75	2.368905647987D-03
				6.25	-4.879999290700D-03
	0	-1.030192263677D-04	-7.689459744254D-04	6.50	-9.147248807900D-03
	1	-1.144658070752D-05	7.399186089530D-05	6.75	-8.840873921150D-03
				7.25	1.821240416979D-02
				7.50	3.413799613760D-02
				7.75	3.299459003661D-02
				8.25	-6.796961738845D-02

CARDINAL SPLINES N=2

8.50	-1.274047357425D-01
8.75	-1.231374862253D-01

I=11 H= 2.0

0		0.	5.355017902720D-03	-1.00	-3.067102028682D-06
1		3.067102028682D-06	-2.959674435812D-04	-.75	-2.300326521512D-06
				-.50	-1.533551014341D-06
0	-9.0	-3.067102028682D-06	-1.533551014341D-06	-.25	-7.667755071705D-07
1	-8.0	1.840261217209D-05	9.201306086046D-06	.25	7.188520379723D-07
2	-7.0	-7.361044868837D-05	-3.680522434418D-05	.50	1.150163260756D-06
3	-6.0	2.760391825814D-04	1.380195912907D-04	.75	1.006392853161D-06
4	-5.0	-1.030546281637D-03	-5.152731408186D-04	1.25	-1.869015298728D-06
5	-4.0	3.846145943967D-03	1.923072971984D-03	1.50	-3.450489782267D-06
6	-3.0	-1.435403749423D-02	-7.177018747116D-03	1.75	-3.306719374673D-06
7	-2.0	5.357000403296D-02	2.678500201648D-02	2.25	6.757209156940D-06
8	-1.0	-1.999259786376D-01	-9.996298931880D-02	2.50	1.265179586831D-05
9	0.0	7.461339105175D-01	3.730669552587D-01	2.75	1.222048464553D-05
10	1.0	-1.784609663432D+00	-8.923048317161D-01	3.25	-2.515982132903D-05
11	2.0	2.392304743212D+00	1.196152371606D+00	3.50	-4.715669369099D-05
12	3.0	-1.784609309414D+00	-8.923046547072D-01	3.75	-4.557521920745D-05
13	4.0	7.461324944456D-01	3.730662472228D-01	4.25	9.388207615919D-05
14	5.0	-1.999206683682D-01	-9.996033418410D-02	4.50	1.759749788956D-04
15	6.0	5.355017902720D-02	2.677508951360D-02	4.75	1.700803921843D-04
16	7.0	-1.428004774059D-02	-7.140023870293D-03	5.25	-3.503684833077D-04
17	8.0	3.570011935147D-03	1.785005967573D-03	5.50	-6.567432218915D-04
18	9.0	-5.950019891911D-04	-2.975009945955D-04	5.75	-6.347463495296D-04
				6.25	1.307591857072D-03
	0	2.760391825814D-05	2.691310910489D-03	6.50	2.450997908671D-03
	1	3.067102028682D-06	-2.959674435812D-04	6.75	2.368905005934D-03
				7.25	-4.879998944979D-03
				7.50	-9.147248412791D-03
				7.75	-8.840873674207D-03
				8.25	1.821240392284D-02
				8.50	3.413799574249D-02
				8.75	3.299458969089D-02

I=12 H= 3.0

0		0.	-1.998519888840D-02	-1.00	8.218274072044D-07
1		-8.218274072044D-07	1.109877913430D-03	-.75	6.163705554033D-07
				-.50	4.109137036022D-07
0	-9.0	8.218274072044D-07	4.109137036022D-07	-.25	2.054568518011D-07
1	-8.0	-4.930964443226D-06	-2.465482221613D-06	.25	-1.926157985635D-07
2	-7.0	1.972385777291D-05	9.861928886453D-06	.50	-3.081852777016D-07
3	-6.0	-7.396446664839D-05	-3.698223332420D-05	.75	-2.696621179889D-07
4	-5.0	2.761340088207D-04	1.380670044103D-04	1.25	5.008010762652D-07
5	-4.0	-1.030571568634D-03	-5.152857843171D-04	1.50	9.245558331049D-07
6	-3.0	3.846152265717D-03	1.923076132858D-03	1.75	8.860326733922D-07
7	-2.0	-1.435403749423D-02	-7.177018747116D-03	2.25	-1.810588506497D-06
8	-1.0	5.356999771121D-02	2.678499885561D-02	2.50	-3.390038054718D-06
9	0.0	-1.999259533506D-01	-9.996297667531D-02	2.75	-3.274468575580D-06
10	1.0	7.461338156912D-01	3.730669078456D-01	3.25	6.741552949723D-06
11	2.0	-1.784609309414D+00	-8.923046547072D-01	3.50	1.263559638577D-05
12	3.0	2.392303421966D+00	1.196151710983D+00	3.75	1.221184162893D-05
13	4.0	-1.784604378450D+00	-8.923021892249D-01	4.25	-2.515562329240D-05
14	5.0	7.461140918335D-01	3.730570459167D-01	4.50	-4.715234748835D-05

15	6,0	-1,998519888840D-01	-9,992599444198D-02	4.75	-4,557289794013D-05
16	7,0	5,329386370239D-02	2.664693185119D-02	5,25	9,388094021986D-05
17	8,0	-1,332346592560D-02	-6.661732962799D-03	5.50	1,759737935676D-04
18	9,0	2.220577654266D-03	1.110288827133D-03	5.75	1,700797501316D-04
				6.25	-3,503681375871D-04
	0	-7,396446664839D-06	-9,996297667531D-03	6.50	-6,567428267822D-04
	1	-8,218274072044D-07	1.109877913430D-03	6.75	-6,347461025862D-04
				7,25	1,307591610128D-03
				7.50	2,450997513561D-03
				7,75	2,368904660213D-03
				8,25	-4,879998302926D-03
				8,50	-9,147247227463D-03
				8,75	-8,840872538267D-03

I=13 H= 4.0

0		0.	7.458577765087D-02	-1.00	-2,202076001355D-07
1		2.202076001355D-07	-4.143544210137D-03	-.75	-1,651557001017D-07
				-.50	-1,101038000678D-07
0	-9.0	-2.202076001355D-07	-1.101038000678D-07	-.25	-5,505190003388D-08
1	-8,0	1.321245600813D-06	6.606228004066D-07	.25	5,161115628177D-08
2	-7,0	-5,284982403253D-06	-2,642491201626D-06	.50	8,257785005083D-08
3	-6.0	1.981868401220D-05	9.909342006099D-06	.75	7,225561879447D-08
4	-5,0	-7,398975364554D-05	-3,699487682277D-05	1,25	-1,341890063326D-07
5	-4.0	2.761403305700D-04	1.380701652850D-04	1,50	-2,477335501525D-07
6	-3,0	-1,030571568634D-03	-5,152857843171D-04	1,75	-2,374113188961D-07
7	-2,0	3,846145943967D-03	1.923072971984D-03	2,25	4,851448690486D-07
8	-1,0	-1,435401220723D-02	-7,177006103617D-03	2,50	9,083563505591D-07
9	0,0	5.356990288497D-02	2.678495144249D-02	2,75	8,773896567900D-07
10	1,0	-1,999255993327D-01	-9,996279966633D-02	3,25	-1,806390469862D-06
11	2,0	7,461324944456D-01	3,730662472228D-01	3,50	-3,385691852084D-06
12	3,0	-1,784604378450D+00	-8,923021892249D-01	3,75	-3,272147308264D-06
13	4,0	2,392285019354D+00	1.196142509677D+00	4,25	6,740417010399D-06
14	5,0	-1,784535698966D+00	-8,922678494828D-01	4,50	1,263441105778D-05
15	6,0	7,458577765087D-01	3,729288882543D-01	4,75	1,221119957627D-05
16	7,0	-1,988954070690D-01	-9,944770353449D-02	5,25	-2,515527757173D-05
17	8,0	4,972385176724D-02	2,486192588362D-02	5,50	-4,715195237902D-05
18	9,0	-8,287308627874D-03	-4,143654313937D-03	5,75	-4,557265099680D-05
				6,25	9,388069327653D-05
	0	1,981868401220D-06	3,729387975963D-02	6,50	1,759733984583D-04
	1	2,202076001355D-07	-4,143544210137D-03	6,75	1,700794044109D-04
				7,25	-3,503674955344D-04
				7,50	-6,567416414542D-04
				7,75	-6,347449666469D-04
				8,25	1,307589288861D-03
				8,50	2,450993167359D-03
				8,75	2,368900462177D-03

I=14 H= 5.0

0		0.	-2,783579117151D-01	-1.00	5,900299333775D-08
1		-5,900299333775D-08	1.546429892712D-02	-.75	4,425224500331D-08
				-.50	2,950149666888D-08
0	-9,0	5,900299333775D-08	2,950149666888D-08	-.25	1,475074833444D-08
1	-8,0	-3,540179600265D-07	-1,770089800133D-07	.25	-1,382882656354D-08
2	-7.0	1,416071840106D-06	7.080359200530D-07	.50	-2,212612250166D-08
3	-6,0	-5,310269400398D-06	-2,655134700199D-06	.75	-1,936035718895D-08

4	-5.0	1,982500576148D-05	9.912502880742D-06
5	-4.0	-7,398975364554D-05	-3.699487682277D-05
6	-3.0	2,761340088207D-04	1.380670044103D-04
7	-2.0	-1,030546281637D-03	-5.152731408186D-04
8	-1,0	3,846051117728D-03	1.923025558864D-03
9	0,0	-1,435365818927D-02	-7.176829094637D-03
10	1.0	5,356858163937D-02	2.678429081969D-02
11	2.0	-1,999206683682D-01	-9,996033418410D-02
12	3,0	7.461140918335D-01	3,730570459167D-01
13	4,0	-1,784535698966D+00	-8,922678494828D-01
14	5.0	2,392028704029D+00	1.196014352015D+00
15	6,0	-1,783579117151D+00	-8,917895585753D-01
16	7,0	7,422877645735D-01	3,711438822868D-01
17	8,0	-1,855719411434D-01	-9,278597057169D-02
18	9.0	3,092865685723D-02	1.546432842861D-02
	0	-5,310269400398D-07	-1,391792213710D-01
	1	-5,900299333775D-08	1.546429892712D-02

1.25	3,595494906519D-08
1.50	6,637836750497D-08
1.75	6,361260219226D-08
2.25	-1,299909696972D-07
2,50	-2,433873475182D-07
2.75	-2,350900515801D-07
3,25	4,840089297237D-07
3.50	9,071710225679D-07
3.75	8,767476041281D-07
4.25	-1,806044749198D-06
4.50	-3,385296742753D-06
4.75	-3,271900364932D-06
5,25	6,740170067067D-06
5.50	1,263401594845D-05
5.75	1,221085385560D-05
6,25	-2,515463551907D-05
6.50	-4,715076705103D-05
6.75	-4,557151505747D-05
7.25	9,387837200922D-05
7.50	1,759690522557D-04
7,75	1,700752063743D-04
8,25	-3,503588525178D-04
8,50	-6,567254419717D-04
8,75	-6,347293104397D-04

I=15 H= 6.0

0		0.	1.038845869209D+00
1		1,580437321547D-08	-5,771365149834D-02
0	-9,0	-1,580437321547D-08	-7,902186607734D-09
1	-8,0	9.482623929281D-08	4.741311964641D-08
2	-7.0	-3,793049571713D-07	-1.896524785856D-07
3	-6,0	1,422393589392D-06	7.111967946961D-07
4	-5,0	-5,310269400398D-06	-2.655134700199D-06
5	-4,0	1,981868401220D-05	9,909342006099D-06
6	-3,0	-7,396446664839D-05	-3,698223332420D-05
7	-2,0	2,760391825814D-04	1.380195912907D-04
8	-1,0	-1,030192263677D-03	-5,150961318386D-04
9	0.0	3,844729872127D-03	1,922364936064D-03
10	1,0	-1,434872722483D-02	-7,174363612416D-03
11	2,0	5,355017902720D-02	2,677508951360D-02
12	3,0	-1,998519888840D-01	-9,992599444198D-02
13	4,0	7,458577765087D-01	3,729288882543D-01
14	5,0	-1,783579117151D+00	-8,917895585753D-01
15	6,0	2,388458692094D+00	1.194229346047D+00
16	7,0	-1,770255651225D+00	-8,851278256125D-01
17	8,0	6,925639128063D-01	3,462819564031D-01
18	9,0	-1,154273188010D-01	-5,771365940052D-02
	0	1,422393589392D-07	5.194230057244D-01
	1	1,580437321547D-08	-5,771365149834D-02

-1.00	-1,580437321547D-08
-.75	-1,185327991160D-08
-.50	-7,902186607734D-09
-.25	-3,951093303867D-09
.25	3,704149972376D-09
.50	5,926639955801D-09
.75	5,185809961326D-09
1.25	-9,630789928176D-09
1.50	-1,777991986740D-08
1.75	-1,703908987293D-08
2.25	3,481900974033D-08
2,50	6,519303951381D-08
2.75	6,297054953038D-08
3,25	-1,296452490331D-07
3.50	-2,429922381878D-07
3,75	-2,348431082486D-07
4.25	4,837619863922D-07
4,50	9,067759132375D-07
4.75	8,764018834641D-07
5,25	-1,805402696536D-06
5,50	-3,384111414762D-06
5,75	-3,270764425608D-06
6,25	6,737848799751D-06
6.50	1,262966974581D-05
6.75	1,220665581897D-05
7.25	-2,514599250247D-05
7.50	-4,713456756848D-05
7,75	-4,555585885026D-05
8,25	9,384612121012D-05
8,50	1,759086005281D-04

				8.75	1.700167795821D-04

I=16 H= 7.0

0		0.	-3.877025565123D+00	-1.00	4.214499524125D-09
1		-4.214499524125D-09	2.153903070662D-01	-.75	3.160874643094D-09
				-.50	2.107249762063D-09
0	-9.0	4.214499524125D-09	2.107249762063D-09	-.25	1.053624881031D-09
1	-8.0	-2.528699714475D-08	-1.264349857238D-08	.25	-9.877733259668D-10
2	-7.0	1.011479885790D-07	5.057399428950D-08	.50	-1.580437321547D-09
3	-6.0	-3.793049571713D-07	-1.896524785856D-07	.75	-1.382882656354D-09
4	-5.0	1.416071840106D-06	7.080359200530D-07	1.25	2.568210647514D-09
5	-4.0	-5.284982403253D-06	-2.642491201626D-06	1.50	4.741311964641D-09
6	-3.0	1.972385777291D-05	9.861928886453D-06	1.75	4.543757299447D-09
7	-2.0	-7.361044868837D-05	-3.680522434418D-05	2.25	-9.285069264088D-09
8	-1.0	2.747179369806D-04	1.373589684903D-04	2.50	-1.738481053702D-08
9	0.0	-1.025261299234D-03	-5.126306496170D-04	2.75	-1.679214654144D-08
10	1.0	3.826327259955D-03	1.913163629978D-03	3.25	3.457206640884D-08
11	2.0	-1.428004774059D-02	-7.140023870293D-03	3.50	6.479793018342D-08
12	3.0	5.329386370239D-02	2.664693185119D-02	3.75	6.262482886630D-08
13	4.0	-1.988954070690D-01	-9.944770353449D-02	4.25	-1.290031963713D-07
14	5.0	7.422877645735D-01	3.711438822868D-01	4.50	-2.418069101967D-07
15	6.0	-1.770255651225D+00	-8.851278256125D-01	4.75	-2.337071689237D-07
16	7.0	2.338734840327D+00	1.169367420163D+00	5.25	4.814407190762D-07
17	8.0	-1.584683710082D+00	-7.923418550408D-01	5.50	9.024297106033D-07
18	9.0	4.307806183469D-01	2.153903091735D-01	5.75	8.722038468287D-07
				6.25	-1.796759679934D-06
	0	-3.793049571713D-08	-1.938512801527D+00	6.50	-3.367911932216D-06
	1	-4.214499524125D-09	2.153903070662D-01	6.75	-3.255108218391D-06
				7.25	6.705598000658D-06
				7.50	1.256921801826D-05
				7.75	1.214822902674D-05
				8.25	-2.502563232270D-05
				8.50	-4.690896014083D-05
				8.75	-4.533780788855D-05

I=17 H= 8.0

0		0.	1.446925639128D+01	-1.00	-1.053624881031D-09
1		1.053624881031D-09	-8.038475767666D-01	-.75	-7.902186607734D-10
				-.50	-5.268124405156D-10
0	-9.0	-1.053624881031D-09	-5.268124405156D-10	-.25	-2.634062202578D-10
1	-8.0	6.321749286188D-09	3.160874643094D-09	.25	2.469433314917D-10
2	-7.0	-2.528699714475D-08	-1.264349857238D-08	.50	3.951093303867D-10
3	-6.0	9.482623929281D-08	4.741311964641D-08	.75	3.457206640884D-10
4	-5.0	-3.540179600265D-07	-1.770089800133D-07	1.25	-6.420526618784D-10
5	-4.0	1.321245600813D-06	6.606228004066D-07	1.50	-1.185327991160D-09
6	-3.0	-4.930964443226D-06	-2.465482221613D-06	1.75	-1.135939324862D-09
7	-2.0	1.840261217209D-05	9.201306086046D-06	2.25	2.321267316022D-09
8	-1.0	-6.867948424514D-05	-3.433974212257D-05	2.50	4.346202634254D-09
9	0.0	2.563153248085D-04	1.281576624042D-04	2.75	4.198036635359D-09
10	1.0	-9.565818149888D-04	-4.782909074944D-04	3.25	-8.643016602210D-09
11	2.0	3.570011935147D-03	1.785005967573D-03	3.50	-1.619948254586D-08
12	3.0	-1.332346592560D-02	-6.661732962799D-03	3.75	-1.565620721657D-08
13	4.0	4.972385176724D-02	2.486192588362D-02	4.25	3.225079909282D-08
14	5.0	-1.855719411434D-01	-9.278597057169D-02	4.50	6.045172754917D-08
15	6.0	6.925639128063D-01	3.462819564031D-01	4.75	5.842679223094D-08

CARDINAL SPLINES N=2

```
16   7.0  -1.584683710082D+00  -7.923418550408D-01     5.25 -1.203601797691D-07
17   8.0   1.646170927520D+00   8.230854637602D-01     5.50 -2.256074276508D-07
18   9.0  -6.076951545867D-01  -3.038475772934D-01     5.75 -2.180509617072D-07
                                                      6.25  4.491899199834D-07
      0    9.482623929281D-09   7.234628200382D+00     6.50  8.419779830541D-07
      1    1.053624881031D-09  -8.038475767666D-01     6.75  8.137770545978D-07
                                                      7.25 -1.676399500165D-06
                                                      7.50 -3.142304504566D-06
                                                      7.75 -3.037057256684D-06
                                                      8.25  6.256408080675D-06
                                                      8.50  1.172724003521D-05
                                                      8.75  1.133445197214D-05


I=18     H=  9.0
 0            0.                 -1.091154273188D+01   -1.00  1.756041468385D-10
 1          -1.756041468385D-10   6.339745961278D-01    -.75  1.317031101289D-10
                                                        -.50  8.780207341927D-11
 0  -9.0   1.756041468385D-10   8.780207341927D-11    -.25  4.390103670964D-11
 1  -8.0  -1.053624881031D-09  -5.268124405156D-10     .25 -4.115722191528D-11
 2  -7.0   4.214499524125D-09   2.107249762063D-09     .50 -6.585155506445D-11
 3  -6.0  -1.580437321547D-08  -7.902186607734D-09     .75 -5.762011068140D-11
 4  -5.0   5.900299333775D-08   2.950149666888D-08    1.25  1.070087769797D-10
 5  -4.0  -2.202076001355D-07  -1.101038000678D-07    1.50  1.975546651934D-10
 6  -3.0   8.218274072044D-07   4.109137036022D-07    1.75  1.893232208103D-10
 7  -2.0  -3.067102028682D-06  -1.533551014341D-06    2.25 -3.868778860037D-10
 8  -1.0   1.144658070752D-05   5.723290353762D-06    2.50 -7.243671057090D-10
 9   0.0  -4.271922080141D-05  -2.135961040071D-05    2.75 -6.996727725598D-10
10   1.0   1.594303024981D-04   7.971515124906D-05    3.25  1.440502767035D-09
11   2.0  -5.950019891911D-04  -2.975009945955D-04    3.50  2.699913757643D-09
12   3.0   2.220577654266D-03   1.110288827133D-03    3.75  2.609367869429D-09
13   4.0  -8.287308627874D-03  -4.143654313937D-03    4.25 -5.375133182136D-09
14   5.0   3.092865685723D-02   1.546432842861D-02    4.50 -1.007528792486D-08
15   6.0  -1.154273188010D-01  -5.771365940052D-02    4.75 -9.737798705156D-09
16   7.0   4.307806183469D-01   2.153903091735D-01    5.25  2.006002996151D-08
17   8.0  -6.076951545867D-01  -3.038475772934D-01    5.50  3.760123794180D-08
18   9.0   2.679491924311D-01   1.339745962156D-01    5.75  3.634182695120D-08
                                                      6.25 -7.486498666390D-08
      0   -1.580437321547D-09  -5.205771366730D+00    6.50 -1.403296638424D-07
      1   -1.756041468385D-10   6.339745961278D-01    6.75 -1.356295090996D-07
                                                      7.25  2.793999166941D-07
                                                      7.50  5.237174174276D-07
                                                      7.75  5.061762094473D-07
                                                      8.25 -1.042734680112D-06
                                                      8.50 -1.954540005868D-06
                                                      8.75 -1.889075328690D-06
```

N= 2   M=19   P= 9.5   Q= 3

CARDINAL SPLINES N=2

I= 0 H= -9.5

0		1.000000000000D+00	5.000000004470D-01	-1.00	2,267949192431D+00
1		-1.267949192431D+00	-6,339745962391D-01	-.75	1,950961894323D+00
				-.50	1,633974596216D+00
0	-9,5	2,679491924311D-01	1.339745962156D-01	-.25	1,316987298108D+00
1	-8,5	-6,076951545867D-01	-3.038475772934D-01	.25	6,871994080240D-01
2	-7,5	4.307806183469D-01	2.153903091735D-01	.50	3,995190528383D-01
3	-6,5	-1.154273188010D-01	-5,771365940052D-02	.75	1,620791712335D-01
4	-5,5	3.092865685723D-02	1,546432842861D-02	1.25	-7.109346086228D-02
5	-4,5	-8,287308627874D-03	-4.143654313937D-03	1.50	-7,355715851499D-02
6	-3,5	2,220577654266D-03	1,110288827133D-03	1.75	-3,924227691020D-02
7	-2,5	-5,950019891912D-04	-2,975009945956D-04	2,25	1,904943542518D-02
8	-1,5	1,594303024987D-04	7,971515124933D-05	2,50	1,970958122162D-02
9	-,5	-4,271922080342D-05	-2,135961040171D-05	2,75	1,051493640725D-02
10	,5	1,144658071503D-05	5,723290357513D-06	3,25	-5,104280838447D-03
11	1,5	-3,067102056679D-06	-1,533551028339D-06	3,50	-5,281166371488D-03
12	2,5	8,218275116892D-07	4.109137558446D-07	3,75	-2,817468718786D-03
13	3,5	-2,202079900782D-07	-1.101039950391D-07	4.25	1,367687928603D-03
14	4,5	5,900444862351D-08	2,950222431176D-08	4,50	1,415084264335D-03
15	5,5	-1,580980441587D-08	-7,904902207935D-09	4,75	7,549384678987D-04
16	6,5	4,234769039965D-09	2,117384519982D-09	5,25	-3,664708759671D-04
17	7,5	-1,129271743991D-09	-5,646358719953D-10	5,50	-3,791706858505D-04
18	8,5	2,823179359977D-10	1,411589679988D-10	5,75	-2,022851528086D-04
19	9,5	-4,705298933294D-11	-2,352649466647D-11	6,25	9,819557526490D-05
				6,50	1,015984790672D-04
	0	-1,104551732810D+01	-5,522758663824D+00	6.75	5,420214333588D-05
	1	-1,267949192431D+00	-6,339745962391D-01	7,25	-2,631142509254D-05
				7.50	-2,722323041829D-05
				7.75	-1,452342053489D-05
				8,25	7,050125105262D-06
				8.50	7,294442605953D-06
				8,75	3,891538803667D-06
				9,25	-1,889075328506D-06
				9,50	-1,954540005525D-06

I= 1 H= -8.5

0		0.	-2,682020391978D-09	-1.00	-1,607695154587D+00
1		1,607695154587D+00	8.038475774345D-01	-.75	-1,205771365940D+00
				-.50	-8,038475772934D-01
0	-9,5	-6,076951545867D-01	-3,038475772934D-01	-.25	-4,019237886467D-01
1	-8,5	1,646170927520D+00	8,230854637602D-01	.25	3,924285518563D-01
2	-7,5	-1,584683710082D+00	-7,923418550408D-01	.50	7,278856829700D-01
3	-6,5	6,925639128063D-01	3,462819564031D-01	.75	9,493999725988D-01
4	-5,5	-1,855719411434D-01	-9,278597057169D-02	1,25	8,484357651737D-01
5	-4,5	4,972385176724D-02	2.486192588362D-02	1.50	5,663429510899D-01
6	-3,5	-1,332346592560D-02	-6,661732962799D-03	1.75	2,510786614612D-01
7	-2,5	3,570011935147D-03	1.785005967574D-03	2,25	-1,142966125511D-01
8	-1,5	-9,565818149920D-04	-4,782909074960D-04	2,50	-1,182574873297D-01
9	-,5	2,563153248205D-04	1.281576624103D-04	2,75	-6,308961844347D-02
10	,5	-6,867948429015D-05	-3,433974214508D-05	3,25	3,062568503068D-02
11	1,5	1,840261234007D-05	9,201306170036D-06	3,50	3,168699822893D-02
12	2,5	-4,930965070135D-06	-2,465482535068D-06	3,75	1,690481231272D-02
13	3,5	1,321247940469D-06	6,606239702345D-07	4,25	-8,206127571620D-03
14	4,5	-3,540266917411D-07	-1.770133458705D-07	4,50	-8,490505586008D-03
15	5,5	9,485882649522D-08	4,742941324761D-08	4,75	-4,529630807392D-03
16	6,5	-2,540861423979D-08	-1,270430711989D-08	5,25	2,198825255802D-03

CARDINAL SPLINES N=2

17	7,5	6,775630463944D-09	3,387815231972D-09	5,50	2,275024115103D-03
18	8,5	-1,693907615986D-09	-8,469538079930D-10	5,75	1,213710916852D-03
19	9,5	2,823179359977D-10	1,411589679988D-10	6,25	-5,891734515894D-04
				6,50	-6,095908744031D-04
	0	1,527310396857D+01	7,636551982946D+00	6,75	-3,252128600153D-04
	1	1,607695154587D+00	8,038475774345D-01	7,25	1,578685505553D-04
				7,50	1,633393825097D-04
				7,75	8,714052320932D-05
				8,25	-4,230075063157D-05
				8,50	-4,376665563572D-05
				8,75	-2,334923282200D-05
				9,25	1,133445197104D-05
				9,50	1,172724003315D-05
I= 2	H= -7.5				
0		0.	1,072808156791D-08	-1,00	4,307806183469D-01
1		-4,307806183469D-01	-2,153903097381D-01	-.75	3,230854637602D-01
				-.50	2,153903091735D-01
0	-9,5	4,307806183469D-01	2,153903091735D-01	-.25	1,076951545867D-01
1	-8,5	-1,584683710082D+00	-7,923418550408D-01	.25	-1,009642074251D-01
2	-7,5	2,338734840327D+00	1,169367420163D+00	.50	-1,615427318801D-01
3	-6,5	-1,770255651225D+00	-8,851278256125D-01	.75	-1,413498903951D-01
4	-5,5	7,422877645735D-01	3,711438822868D-01	1.25	2,781319393052D-01
5	-4,5	-1,988954070690D-01	-9,944770353449D-02	1,50	6,096281956403D-01
6	-3,5	5,329386370239D-02	2,664693185120D-02	1,75	8,863103541553D-01
7	-2,5	-1,428004774059D-02	-7,140023870295D-03	2,25	8,790614502044D-01
8	-1,5	3,826327259968D-03	1,913163629984D-03	2,50	5,980299493189D-01
9	-,5	-1,025261299282D-03	-5,126306496411D-04	2,75	2,679834737739D-01
10	,5	2,747179371606D-04	1,373589685803D-04	3,25	-1,225027401227D-01
11	1,5	-7,361044936029D-05	-3,680522468014D-05	3,50	-1,267479929157D-01
12	2,5	1,972386028054D-05	9,861930140270D-06	3,75	-6,761924925086D-02
13	3,5	-5,284991761876D-06	-2,642495880938D-06	4,25	3,282451028648D-02
14	4,5	1,416106766964D-06	7,080533834821D-07	4,50	3,396202234403D-02
15	5,5	-3,794353059809D-07	-1,897176529904D-07	4,75	1,811852322957D-02
16	6,5	1,016344569592D-07	5,081722847958D-08	5,25	-8,795301023210D-03
17	7,5	-2,710252185578D-08	-1,355126092789D-08	5,50	-9,100096460411D-03
18	8,5	6,775630463944D-09	3,387815231972D-09	5,75	-4,854843667407D-03
19	9,5	-1,129271743991D-09	-5,646358719953D-10	6,25	2,356693806358D-03
				6,50	2,438363497613D-03
	0	-4,092415874296D+00	-2,046207931784D+00	6,75	1,300851440061D-03
	1	-4,307806183469D-01	-2,153903097381D-01	7,25	-6,314742022210D-04
				7,50	-6,533575300389D-04
				7,75	-3,485620928373D-04
				8,25	1,692030025263D-04
				8,50	1,750666225429D-04
				8,75	9,339693128801D-05
				9,25	-4,533780788416D-05
				9,50	-4,690896013260D-05
I= 3	H= -6.5				
0		0.	-4,023030587967D-08	-1,00	-1,154273188010D-01
1		1,154273188010D-01	5,771366151791D-02	-.75	-8,657048910078D-02
				-.50	-5,771365940052D-02
0	-9,5	-1,154273188010D-01	-5,771365940052D-02	-.25	-2,885682970026D-02
1	-8,5	6,925639128063D-01	3,462819564031D-01	.25	2,705327784399D-02

CARDINAL SPLINES N=2

2	-7.5	-1.770255651225D+00	-8.851278256125D-01
3	-6.5	2.388458692094D+00	1.194229346047D+00
4	-5.5	-1.783579117151D+00	-8.917895585753D-01
5	-4.5	7.458577765087D-01	3.729288882543D-01
6	-3.5	-1.998519888840D-01	-9.992599444198D-02
7	-2.5	5.355017902721D-02	2.677508951361D-02
8	-1.5	-1.434872722488D-02	-7.174363612440D-03
9	-.5	3.844729872308D-03	1.922364936154D-03
10	.5	-1.030192264352D-03	-5.150961321761D-04
11	1.5	2.760391851011D-04	1.380195925505D-04
12	2.5	-7.396447605203D-05	-3.698223802601D-05
13	3.5	1.981871910704D-05	9.909359553518D-06
14	4.5	-5.310400376116D-06	-2.655200188058D-06
15	5.5	1.422882397428D-06	7.114411987141D-07
16	6.5	-3.811292135968D-07	-1.905646067984D-07
17	7.5	1.016344569592D-07	5.081722847958D-08
18	8.5	-2.540861423979D-08	-1.270430711989D-08
19	9.5	4.234769039965D-09	2.117384519982D-09
	0	1.096559528610D+00	5.482797441898D-01
	1	1.154273188010D-01	5.771366151791D-02

.50	4.328524455039D-02
.75	3.787458898159D-02
1.25	-7.033852239439D-02
1.50	-1.298557336512D-01
1.75	-1.244450780824D-01
2.25	2.699258117335D-01
2.50	6.011376900543D-01
2.75	8.817807233479D-01
3.25	8.812602754602D-01
3.50	6.003049734340D-01
3.75	2.691971846907D-01
4.25	-1.230919135743D-01
4.50	-1.273575837901D-01
4.75	-6.794446211088D-02
5.25	3.298237883704D-02
5.50	3.412536172654D-02
5.75	1.820566375278D-02
6.25	-8.837601773841D-03
6.50	-9.143863116047D-03
6.75	-4.878192900229D-03
7.25	2.368028258329D-03
7.50	2.450090737646D-03
7.75	1.307107848140D-03
8.25	-6.345112594736D-04
8.50	-6.564998345357D-04
8.75	-3.502384923300D-04
9.25	1.700167795656D-04
9.50	1.759086004972D-04

I= 4 H= -5.5

0		0.	1.501931419508D-07
1		-3.092865685723D-02	-1.546433633352D-02
0	-9.5	3.092865685723D-02	1.546432842861D-02
1	-8.5	-1.855719411434D-01	-9.278597057169D-02
2	-7.5	7.422877645735D-01	3.711438822868D-01
3	-6.5	-1.783579117151D+00	-8.917895585753D-01
4	-5.5	2.392028704029D+00	1.196014352015D+00
5	-4.5	-1.784535698966D+00	-8.922678494828D-01
6	-3.5	7.461140918335D-01	3.730570459167D-01
7	-2.5	-1.999206683683D-01	-9.996033418413D-02
8	-1.5	5.356858163955D-02	2.678429081978D-02
9	-.5	-1.435365818995D-02	-7.176829094975D-03
10	.5	3.846051120249D-03	1.923025560124D-03
11	1.5	-1.030546291044D-03	-5.152731455220D-04
12	2.5	2.761340439276D-04	1.380670219638D-04
13	3.5	-7.398988466627D-05	-3.699494233313D-05
14	4.5	1.982549473750D-05	9.912747368750D-06
15	5.5	-5.312094283732D-06	-2.656047141866D-06
16	6.5	1.422882397428D-06	7.114411987141D-07
17	7.5	-3.794353059809D-07	-1.897176529904D-07
18	8.5	9.485882649522D-08	4.742941324761D-08
19	9.5	-1.580980441587D-08	-7.904902207935D-09
	0	-2.938222401437D-01	-1.469110449753D-01
	1	-3.092865685723D-02	-1.546433633352D-02

-1.00	3.092865685723D-02
-.75	2.319649264292D-02
-.50	1.546432842861D-02
-.25	7.732164214307D-03
.25	-7.248903950913D-03
.50	-1.159824632146D-02
.75	-1.014846553128D-02
1.25	1.884715027237D-02
1.50	3.479473896438D-02
1.75	3.334495817420D-02
2.25	-6.813969713858D-02
2.50	-1.275807095361D-01
2.75	-1.232313671655D-01
3.25	2.693366382820D-01
3.50	6.005280991799D-01
3.75	8.814555104879D-01
4.25	8.814181440107D-01
4.50	6.004683128165D-01
4.75	2.692843252140D-01
5.25	-1.231342143249D-01
5.50	-1.274013504458D-01
5.75	-6.796781134370D-02
6.25	3.299371328901D-02
6.50	3.413708896658D-02
6.75	1.821192016086D-02
7.25	-8.840638831094D-03

CARDINAL SPLINES N=2

7.50	-9.147005420544D-03
7.75	-4.879869299722D-03
8.25	2.368842035368D-03
8.50	2.450932715600D-03
8.75	1.307557038032D-03
9.25	-6.347293103782D-04
9.50	-6.567254418563D-04

I= 5 H= -4.5

0		0.	-5.605422619234D-07
1		8.287308627874D-03	4.143683816161D-03
0	-9.5	-8.287308627874D-03	-4.143654313937D-03
1	-8.5	4.972385176724D-02	2.486192588362D-02
2	-7.5	-1.988954070690D-01	-9.944770353449D-02
3	-6.5	7.458577765087D-01	3.729288882543D-01
4	-5.5	-1.784535698966D+00	-8.922678494828D-01
5	-4.5	2.392285019354D+00	1.196142509677D+00
6	-3.5	-1.784604378450D+00	-8.923021892250D-01
7	-2.5	7.461324944458D-01	3.730662472229D-01
8	-1.5	-1.999255993333D-01	-9.996279966666D-02
9	-.5	5.356990288749D-02	2.678495144375D-02
10	.5	-1.435401221664D-02	-7.177006108321D-03
11	1.5	3.846145979075D-03	1.923072989538D-03
12	2.5	-1.030571699658D-03	-5.152858498291D-04
13	3.5	2.761408195580D-04	1.380704097790D-04
14	4.5	-7.399157857388D-05	-3.699578928694D-05
15	5.5	1.982549473750D-05	9.912747368750D-06
16	6.5	-5.310400376116D-06	-2.655200188058D-06
17	7.5	1.416106766964D-06	7.080533834821D-07
18	8.5	-3.540266917411D-07	-1.770133458705D-07
19	9.5	5.900444862351D-08	2.950222431176D-08
	0	7.872943196480D-02	3.936443571127D-02
	1	8.287308627874D-03	4.143683816161D-03

-1.00	-8.287308627874D-03
-.75	-6.215481470905D-03
-.50	-4.143654313937D-03
-.25	-2.071827156968D-03
.25	1.942337959658D-03
.50	3.107740735453D-03
.75	2.719273143521D-03
1.25	-5.050078695111D-03
1.50	-9.323222206358D-03
1.75	-8.934754614427D-03
2.25	1.825797682078D-02
2.50	3.418514808998D-02
2.75	3.301974531419D-02
3.25	-6.798182858803D-02
3.50	-1.274173701536D-01
3.75	-1.231442266423D-01
4.25	2.692943378313D-01
4.50	6.004843325243D-01
4.75	8.814321612551D-01
5.25	8.814294784627D-01
5.50	6.004800400565D-01
5.75	2.692905816220D-01
6.25	-1.231372513822D-01
6.50	-1.274044927503D-01
6.75	-6.796948774319D-02
7.25	3.299452706605D-02
7.50	3.413793094453D-02
7.75	1.821236935075D-02
8.25	-8.840856881999D-03
8.50	-9.147231027865D-03
8.75	-4.879989659798D-03
9.25	2.368900461947D-03
9.50	2.450993166928D-03

I= 6 H= -3.5

0		0.	2.091975905743D-06
1		-2.220577654266D-03	-1.110398931128D-03
0	-9.5	2.220577654266D-03	1.110288827133D-03
1	-8.5	-1.332346592560D-02	-6.661732962799D-03
2	-7.5	5.329386370239D-02	2.664693185120D-02
3	-6.5	-1.998519888840D-01	-9.992599444198D-02
4	-5.5	7.461140918335D-01	3.730570459167D-01
5	-4.5	-1.784604378450D+00	-8.923021892250D-01
6	-3.5	2.392303421966D+00	1.196151710983D+00
7	-2.5	-1.784609309415D+00	-8.923046547075D-01

-1.00	2.220577654266D-03
-.75	1.665433240700D-03
-.50	1.110288827133D-03
-.25	5.551444135666D-04
.25	-5.204478877187D-04
.50	-8.327166203499D-04
.75	-7.286270428061D-04
1.25	1.353164508069D-03
1.50	2.498149861050D-03
1.75	2.394060283506D-03
2.25	-4.892210144555D-03

CARDINAL SPLINES N=2

8	-1,5	7.461338156938D-01	3.730669078469D-01	2.50	-9,159882823848D-03
9	-,5	-1,999259533600D-01	-9,996297668001D-02	2,75	-8,847614091217D-03
10	,5	5,356999774632D-02	2.678499887316D-02	3,25	1,821567607015D-02
11	1,5	-1,435403762526D-02	-7.177018812628D-03	3.50	3,414138143434D-02
12	2,5	3,846152754705D-03	1,923076377353D-03	3,75	3,299639608136D-02
13	3,5	-1,030573393566D-03	-5,152866967829D-04	4,25	-6,797049413606D-02
14	4,5	2,761408195580D-04	1,380704097790D-04	4,50	-1,274056429135D-01
15	5,5	-7,398988466627D-05	-3,699494233313D-05	4,75	-1,231379702342D-01
16	6,5	1,981871910704D-05	9,909359553518D-06	5,25	2,692913004741D-01
17	7,5	-5,284991761876D-06	-2,642495880938D-06	5,50	6,004811902198D-01
18	8,5	1,321247940469D-06	6,606239702345D-07	5,75	8,814304848556D-01
19	9,5	-2,202079900782D-07	-1,101039950391D-07	6,25	8,814302922398D-01
				6,50	6,004808820345D-01
	0	-2,109548771553D-02	-1,054669786981D-02	6,75	2,692910308119D-01
	1	-2,220577654266D-03	-1,110398931128D-03	7,25	-1,231374694331D-01
				7,50	-1,274047183576D-01
				7,75	-6,796960810327D-02
				8,25	3,299458549263D-02
				8,50	3,413799139586D-02
				8,75	1,821240160116D-02
				9,25	-8,840872537410D-03
				9,50	-9,147247225856D-03

I= 7 H= -2.5

0		0,	-7,807361361047D-06	-1,00	-5,950019891912D-04
1		5,950019891912D-04	2.979119083515D-04	-,75	-4,462514918934D-04
				-,50	-2,975009945956D-04
0	-9,5	-5,950019891912D-04	-2,975009945956D-04	-,25	-1,487504972978D-04
1	-8,5	3,570011935147D-03	1.785005967574D-03	,25	1,394535912167D-04
2	-7,5	-1,428004774059D-02	-7,140023870295D-03	,50	2,231257459467D-04
3	-6,5	5,355017902721D-02	2,677508951361D-02	,75	1,952350277034D-04
4	-5,5	-1,999206683683D-01	-9,996033418413D-02	1,25	-3,625793371634D-04
5	-4,5	7,461324944458D-01	3,730662472229D-01	1,50	-6,693772378401D-04
6	-3,5	-1,784609309415D+00	-8,923046547075D-01	1,75	-6,414865195968D-04
7	-2,5	2,392304743214D+00	1.196152371607D+00	2,25	1,310863757437D-03
8	-1,5	-1,784609663442D+00	-8,923048317208D-01	2,50	2,454383205414D-03
9	-,5	7,461339105526D-01	3,730669552763D-01	2,75	2,370711050684D-03
10	,5	-1,999259787686D-01	-9,996298938432D-02	3,25	-4,880875692584D-03
11	1,5	5,357000452195D-02	2,678500226097D-02	3,50	-9,148155583815D-03
12	2,5	-1,435403931916D-02	-7,177019659582D-03	3,75	-8,841357683139D-03
13	3,5	3,846152754705D-03	1,923076377353D-03	4,25	1,821263901290D-02
14	4,5	-1,030571699658D-03	-5,152858498291D-04	4,50	3,413823912985D-02
15	5,5	2,761340439276D-04	1,380670219638D-04	4,75	3,299471968187D-02
16	6,5	-7,396447605203D-05	-3,698223802601D-05	5,25	-6,796968035902D-02
17	7,5	1,972386028054D-05	9,861930140270D-06	5,50	-1,274048009356D-01
18	8,5	-4,930965070135D-06	-2,465482535068D-06	5,75	-1,231375210443D-01
19	9,5	8,218275116892D-07	4,109137558446D-07	6,25	2,692910824232D-01
				6,50	6,004809646124D-01
	0	5,652518897317D-03	2,822355767978D-03	6,75	8,814303644955D-01
	1	5,950019891912D-04	2,979119083515D-04	7,25	8,814303506663D-01
				7,50	6,004809424858D-01
				7,75	2,692910630623D-01
				8,25	-1,231374850885D-01
				8,50	-1,274047345556D-01
				8,75	-6,796961674485D-02
				9,25	3,299458968769D-02

CARDINAL SPLINES N=2

				9.50	3.413799573650D-02

I= 8	H= -1.5				
0		0.	2.913746953845D-05	-1.00	1.594303024987D-04
1		-1.594303024987D-04	-8.124870227767D-05	-.75	1.195727268740D-04
				-.50	7.971515124933D-05
0	-9.5	1.594303024987D-04	7.971515124933D-05	-.25	3.985757562467D-05
1	-8.5	-9.565818149920D-04	-4.782909074960D-04	.25	-3.736647714812D-05
2	-7.5	3.826327259968D-03	1.913163629984D-03	.50	-5.978636343700D-05
3	-6.5	-1.434872722488D-02	-7.174363612440D-03	.75	-5.231306800737D-05
4	-5.5	5.356858163955D-02	2.678429081978D-02	1.25	9.715284058512D-05
5	-4.5	-1.999255993333D-01	-9.996279966666D-02	1.50	1.793590903110D-04
6	-3.5	7.461338156938D-01	3.730669078469D-01	1.75	1.718857948814D-04
7	-2.5	-1.784609663442D+00	-8.923048317208D-01	2.25	-3.512448851924D-04
8	-1.5	2.392304838073D+00	1.196152419036D+00	2.50	-6.576499978070D-04
9	-.5	-1.784609688850D+00	-8.923048444252D-01	2.75	-6.352301115181D-04
10	.5	7.461339173282D-01	3.730669586641D-01	3.25	1.307826700184D-03
11	1.5	-1.999259804625D-01	-9.996299023127D-02	3.50	2.451240900917D-03
12	2.5	5.357000452195D-02	2.678500226097D-02	3.75	2.369034651191D-03
13	3.5	-1.435403762526D-02	-7.177018812628D-03	4.25	-4.880061915545D-03
14	4.5	3.846145979075D-03	1.923072989538D-03	4.50	-9.147313605861D-03
15	5.5	-1.030546291044D-03	-5.152731455220D-04	4.75	-8.840908493246D-03
16	6.5	2.760391851011D-04	1.380195925505D-04	5.25	1.821242096200D-02
17	7.5	-7.361044936029D-05	-3.680522468014D-05	5.50	3.413801352253D-02
18	8.5	1.840261234007D-05	9.201306170036D-06	5.75	3.299459932179D-02
19	9.5	-3.067102056679D-06	-1.533551028339D-06	6.25	-6.796962193244D-02
				6.50	-1.274047404842D-01
	0	-1.514587873737D-03	-7.427252020994D-04	6.75	-1.231374887939D-01
	1	-1.594303024987D-04	-8.124870227767D-05	7.25	2.692910667678D-01
				7.50	6.004809484145D-01
				7.75	8.814303558539D-01
				8.25	8.814303548614D-01
				8.50	6.004809468264D-01
				8.75	2.692910653782D-01
				9.25	-1.231374862134D-01
				9.50	-1.274047357201D-01

I= 9	H= -.5				
0		0.	-1.087425167927D-04	-1.00	-4.271922080342D-05
1		4.271922080342D-05	2.708290075922D-05	-.75	-3.203941560257D-05
				-.50	-2.135961040171D-05
0	-9.5	-4.271922080342D-05	-2.135961040171D-05	-.25	-1.067980520086D-05
1	-8.5	2.563153248205D-04	1.281576624103D-04	.25	1.001231737580D-05
2	-7.5	-1.025261299282D-03	-5.126306496411D-04	.50	1.601970780128D-05
3	-6.5	3.844729872308D-03	1.922364936154D-03	.75	1.401724432612D-05
4	-5.5	-1.435365818995D-02	-7.176829094975D-03	1.25	-2.603202517709D-05
5	-4.5	5.356990288749D-02	2.678495144375D-02	1.50	-4.805912340385D-05
6	-3.5	-1.999259533600D-01	-9.996297668001D-02	1.75	-4.605665992869D-05
7	-2.5	7.461339105526D-01	3.730669552763D-01	2.25	9.411578333254D-05
8	-1.5	-1.784609688850D+00	-8.923048444252D-01	2.50	1.762167858141D-04
9	-.5	2.392304844849D+00	1.196152422424D+00	2.75	1.702093953886D-04
10	.5	-1.784609690544D+00	-8.923048452721D-01	3.25	-3.504311081531D-04
11	1.5	7.461339173282D-01	3.730669586641D-01	3.50	-6.568080198526D-04
12	2.5	-1.999259787686D-01	-9.996298938432D-02	3.75	-6.347809216259D-04
13	3.5	5.356999774632D-02	2.678499887316D-02	4.25	1.307608649280D-03

14	4,5	-1,435401221664D-02	-7.177006108321D-03	4.50	2,451015293596D-03
15	5,5	3,846051120249D-03	1.923025560124D-03	4.75	2,368914291115D-03
16	6,5	-1.030192264352D-03	-5.150961321761D-04	5.25	-4,880003488966D-03
17	7,5	2.747179371606D-04	1.373589685803D-04	5.50	-9,147253154533D-03
18	8,5	-6,867948429015D-05	-3.433974214508D-05	5,75	-8,840876242833D-03
19	9,5	1,144658071503D-05	5.723290357513D-06	6,25	1,821240530658D-02
				6,50	3,413799732454D-02
	0	4,058325976325D-04	1.485450404199D-04	6,75	3,299459068022D-02
	1	4.271922080342D-05	2.708290075922D-05	7,25	-6,796961773737D-02
				7.50	-1,274047361436D-01
				7.75	-1,231374864780D-01
				8,25	2,692910656429D-01
				8,50	6,004809472499D-01
				8,75	8,814303552319D-01
				9,25	8,814303551658D-01
				9,50	6,004809471440D-01

I=10 H= .5

0		0.	4.058325976325D-04	-1.00	1,144658071503D-05
1		-1,144658071503D-05	-2,708290075922D-05	-.75	8,584935536269D-06
				-.50	5,723290357513D-06
0	-9,5	1,144658071503D-05	5.723290357513D-06	-.25	2,861645178756D-06
1	-8,5	-6,867948429015D-05	-3.433974214508D-05	.25	-2,682792355084D-06
2	-7,5	2,747179371606D-04	1.373589685803D-04	.50	-4,292467768134D-06
3	-6,5	-1,030192264352D-03	-5.150961321761D-04	.75	-3,755909297118D-06
4	-5,5	3,846051120249D-03	1.923025560124D-03	1.25	6,975260123219D-06
5	-4,5	-1,435401221664D-02	-7,177006108321D-03	1,50	1,287740330440D-05
6	-3,5	5,356999774632D-02	2,678499887316D-02	1.75	1,234084483339D-05
7	-2,5	-1,999259787686D-01	-9,996298938432D-02	2,25	-2,521824813779D-05
8	-1,5	7,461339173282D-01	3,730669586641D-01	2,50	-4,721714544948D-05
9	-,5	-1,784609690544D+00	-8,923048452721D-01	2,75	-4,560747003643D-05
10	,5	2,392304844849D+00	1,196152422424D+00	3,25	9,389773242794D-05
11	1,5	-1,784609688850D+00	-8,923048444252D-01	3,50	1,759911784935D-04
12	2,5	7,461339105526D-01	3,730669552763D-01	3,75	1,700890353123D-04
13	3,5	-1,999259533600D-01	-9,996297668001D-02	4,25	-3,503726815740D-04
14	4,5	5,356990288749D-02	2,678495144375D-02	4,50	-6,567475685246D-04
15	5,5	-1,435365818995D-02	-7,176829094975D-03	4,75	-6,347486712129D-04
16	6,5	3,844729872308D-03	1,922364936154D-03	5,25	1,307592993868D-03
17	7,5	-1,025261299282D-03	-5,126306496411D-04	5,50	2,450999095605D-03
18	8,5	2,563153248205D-04	1,281576624103D-04	5,75	2,368905649539D-03
19	9,5	-4,271922080342D-05	-2,135961040171D-05	6,25	-4,879999293898D-03
				6,50	-9,147248813895D-03
	0	-1,087425167927D-04	1.485450404199D-04	6,75	-8,840873926944D-03
	1	-1,144658071503D-05	-2,708290075922D-05	7,25	1,821240418172D-02
				7,50	3,413799615997D-02
				7,75	3,299459005824D-02
				8,25	-6,796961743300D-02
				8,50	-1,274047358260D-01
				8,75	-1,231374863060D-01
				9,25	2,692910655503D-01
				9,50	6,004809471440D-01

I=11 H= 1.5

0		0.	-1,514587873737D-03	-1,00	-3,067102056679D-06
1		3.067102056679D-06	8.124870227767D-05	-.75	-2,300326542509D-06

CARDINAL SPLINES N=2

0	-9.5	-3.067102056679D-06	-1.533551028339D-06
1	-8.5	1.840261234007D-05	9.201306170036D-06
2	-7.5	-7.361044936029D-05	-3.680522468014D-05
3	-6.5	2.760391851011D-04	1.380195925505D-04
4	-5.5	-1.030546291044D-03	-5.152731455220D-04
5	-4.5	3.846145979075D-03	1.923072989538D-03
6	-3.5	-1.435403762526D-02	-7.177018812628D-03
7	-2.5	5.357000452195D-02	2.678500226097D-02
8	-1.5	-1.999259804625D-01	-9.996299023127D-02
9	-.5	7.461339173282D-01	3.730669586641D-01
10	.5	-1.784609688850D+00	-8.923048444252D-01
11	1.5	2.392304838073D+00	1.196152419036D+00
12	2.5	-1.784609663442D+00	-8.923048317208D-01
13	3.5	7.461338156938D-01	3.730669078469D-01
14	4.5	-1.999255993333D-01	-9.996279966666D-02
15	5.5	5.356858163955D-02	2.678429081978D-02
16	6.5	-1.434872722488D-02	-7.174363612440D-03
17	7.5	3.826327259968D-03	1.913163629984D-03
18	8.5	-9.565818149920D-04	-4.782909074960D-04
19	9.5	1.594303024987D-04	7.971515124933D-05
	0	2.913746953845D-05	-7.427252020994D-04
	1	3.067102056679D-06	8.124870227767D-05

-.50	-1.533551028339D-06
-.25	-7.667755141697D-07
.25	7.188520445341D-07
.50	1.150163271254D-06
.75	1.006392862348D-06
1.25	-1.869015315789D-06
1.50	-3.450489813763D-06
1.75	-3.306719404857D-06
2.25	6.757209218620D-06
2.50	1.265179598380D-05
2.75	1.222048475708D-05
3.25	-2.515982155869D-05
3.50	-4.715669412143D-05
3.75	-4.557521962346D-05
4.25	9.388207701615D-05
4.50	1.759749805019D-04
4.75	1.700803937368D-04
5.25	-3.503684865059D-04
5.50	-6.567432278863D-04
5.75	-6.347463553236D-04
6.25	1.307591869007D-03
6.50	2.450997931043D-03
6.75	2.368905027558D-03
7.25	-4.879998989524D-03
7.50	-9.147248496287D-03
7.75	-8.840873754907D-03
8.25	1.821240408909D-02
8.50	3.413799605410D-02
8.75	3.299458999207D-02
9.25	-6.796961736683D-02
9.50	-1.274047357201D-01

I=12 H= 2.5

0		0.	5.652518897317D-03
1		-8.218275116892D-07	-2.979119083515D-04
0	-9.5	8.218275116892D-07	4.109137558446D-07
1	-8.5	-4.930965070135D-06	-2.465482535068D-06
2	-7.5	1.972386028054D-05	9.861930140270D-06
3	-6.5	-7.396447605203D-05	-3.698223802601D-05
4	-5.5	2.761340439276D-04	1.380670219638D-04
5	-4.5	-1.030571699658D-03	-5.152858498291D-04
6	-3.5	3.846152754705D-03	1.923076377353D-03
7	-2.5	-1.435403931916D-02	-7.177019659582D-03
8	-1.5	5.357000452195D-02	2.678500226097D-02
9	-.5	-1.999259787686D-01	-9.996298938432D-02
10	.5	7.461339105526D-01	3.730669552763D-01
11	1.5	-1.784609663442D+00	-8.923048317208D-01
12	2.5	2.392304743214D+00	1.196152371607D+00
13	3.5	-1.784609309415D+00	-8.923046547075D-01
14	4.5	7.461324944458D-01	3.730662472229D-01
15	5.5	-1.999206683683D-01	-9.996033418413D-02
16	6.5	5.355017902721D-02	2.677508951361D-02
17	7.5	-1.428004774059D-02	-7.140023870295D-03
18	8.5	3.570011935147D-03	1.785005967574D-03
19	9.5	-5.950019891912D-04	-2.975009945956D-04

-1.00	8.218275116892D-07
-.75	6.163706337669D-07
-.50	4.109137558446D-07
-.25	2.054568779223D-07
.25	-1.926158230522D-07
.50	-3.081853168835D-07
.75	-2.696621522730D-07
1.25	5.008011399356D-07
1.50	9.245559506504D-07
1.75	8.860327860399D-07
2.25	-1.810588736690D-06
2.50	-3.390038485718D-06
2.75	-3.274468991887D-06
3.25	6.741553806825D-06
3.50	1.263559799222D-05
3.75	1.221184318151D-05
4.25	-2.515562649061D-05
4.50	-4.715235348317D-05
4.75	-4.557290373414D-05
5.25	9.388095215562D-05
5.50	1.759738159405D-04
5.75	1.700797717551D-04
6.25	-3.503681821319D-04

				6.50	-6.567429102786D-04
	0	-7.807361361047D-06	2.822355767978D-03	6.75	-6.347461832861D-04
	1	-8.218275116892D-07	-2.979119083515D-04	7.25	1.307591776372D-03
				7.50	2.450997825174D-03
				7.75	2.368904961389D-03
				8.25	-4.879998923356D-03
				8.50	-9.147248390418D-03
				8.75	-8.840873662271D-03
				9.25	1.821240391705D-02
				9.50	3.413799573650D-02

I=13 H= 3.5

0		0.	-2.109548771553D-02	-1.00	-2.202079900782D-07
1		2.202079900782D-07	1.110398931128D-03	-.75	-1.651559925586D-07
				-.50	-1.101039950391D-07
0	-9.5	-2.202079900782D-07	-1.101039950391D-07	-.25	-5.505199751954D-08
1	-8.5	1.321247940469D-06	6.606239702345D-07	.25	5.161124767457D-08
2	-7.5	-5.284991761876D-06	-2.642495880938D-06	.50	8.257799627932D-08
3	-6.5	1.981871910704D-05	9.909359553518D-06	.75	7.225574674440D-08
4	-5.5	-7.398988466627D-05	-3.699494233313D-05	1.25	-1.341892439539D-07
5	-4.5	2.761408195580D-04	1.380704097790D-04	1.50	-2.477339888380D-07
6	-3.5	-1.030573393566D-03	-5.152866967829D-04	1.75	-2.374117393030D-07
7	-2.5	3.846152754705D-03	1.923076377353D-03	2.25	4.851457281410D-07
8	-1.5	-1.435403762526D-02	-7.177018812628D-03	2.50	9.083579590725D-07
9	-.5	5.356999774632D-02	2.678499887316D-02	2.75	8.773912104677D-07
10	.5	-1.999259533600D-01	-9.996297668001D-02	3.25	-1.806393668610D-06
11	1.5	7.461338156938D-01	3.730669078469D-01	3.50	-3.385697847452D-06
12	2.5	-1.784609309415D+00	-8.923046547075D-01	3.75	-3.272153102568D-06
13	3.5	2.392303421966D+00	1.196151710983D+00	4.25	6.740428946299D-06
14	4.5	-1.784604378450D+00	-8.923021892250D-01	4.50	1.263443343074D-05
15	5.5	7.461140918335D-01	3.730570459167D-01	4.75	1.221122119980D-05
16	6.5	-1.998519888840D-01	-9.992599444198D-02	5.25	-2.515532211659D-05
17	7.5	5.329386370239D-02	2.664693185120D-02	5.50	-4.715203587549D-05
18	8.5	-1.332346592560D-02	-6.661732962799D-03	5.75	-4.557273169665D-05
19	9.5	2.220577654266D-03	1.110288827133D-03	6.25	9.388085952005D-05
				6.50	1.759737100712D-04
	0	2.091975905743D-06	-1.054669786981D-02	6.75	1.700797055868D-04
	1	2.202079900782D-07	1.110398931128D-03	7.25	-3.503681159636D-04
				7.50	-6.567428044094D-04
				7.75	-6.347460906505D-04
				8.25	1.307591604334D-03
				8.50	2.450997507566D-03
				8.75	2.368904657015D-03
				9.25	-4.879998301374D-03
				9.50	-9.147247225856D-03

I=14 H= 4.5

0		0.	7.872943196480D-02	-1.00	5.900444862351D-08
1		-5.900444862351D-08	-4.143683816161D-03	-.75	4.425333646763D-08
				-.50	2.950222431176D-08
0	-9.5	5.900444862351D-08	2.950222431176D-08	-.25	1.475111215588D-08
1	-8.5	-3.540266917411D-07	-1.770133458705D-07	.25	-1.382916764614D-08
2	-7.5	1.416106766964D-06	7.080533834821D-07	.50	-2.212666823382D-08
3	-6.5	-5.310400376116D-06	-2.655200188058D-06	.75	-1.936083470459D-08
4	-5.5	1.982549473750D-05	9.912747368750D-06	1.25	3.595583587995D-08

5	-4.5	-7.399157857388D-05	-3.699578928694D-05	1.50	6.638000470145D-08
6	-3.5	2.761408195580D-04	1.380704097790D-04	1.75	6.361417117222D-08
7	-2.5	-1.030571699658D-03	-5.152858498291D-04	2.25	-1.299941758737D-07
8	-1.5	3.846145979075D-03	1.923072989538D-03	2.50	-2.433933505720D-07
9	-.5	-1.435401221664D-02	-7.177006108321D-03	2.75	-2.350958499843D-07
10	.5	5.356990288749D-02	2.678495144375D-02	3.25	4.840208676147D-07
11	1.5	-1.999255993333D-01	-9.996279966666D-02	3.50	9.071933975865D-07
12	2.5	7.461324944458D-01	3.730662472229D-01	3.75	8.767692287650D-07
13	3.5	-1.784604378450D+00	-8.923021892250D-01	4.25	-1.806089294585D-06
14	4.5	2.392285019354D+00	1.196142509677D+00	4.50	-3.385380239774D-06
15	5.5	-1.784535698966D+00	-8.922678494828D-01	4.75	-3.271981065076D-06
16	6.5	7.458577765087D-01	3.729288882543D-01	5.25	6.740336310726D-06
17	7.5	-1.988954070690D-01	-9.944770353449D-02	5.50	1.263432756151D-05
18	8.5	4.972385176724D-02	2.486192588362D-02	5.75	1.221115503154D-05
19	9.5	-8.287308627874D-03	-4.143654313937D-03	6.25	-2.515525594832D-05
				6.50	-4.715193000626D-05
	0	-5.605422619234D-07	3.936443571127D-02	6.75	-4.557263906108D-05
	1	-5.900444862351D-08	-4.143683816161D-03	7.25	9.388068748256D-05
				7.50	1.759733924635D-04
				7.75	1.700794012128D-04
				8.25	-3.503674939819D-04
				8.50	-6.567416398479D-04
				8.75	-6.347449657900D-04
				9.25	1.307589288445D-03
				9.50	2.450993166928D-03

I=15 H= 5.5

0		0.	-2.938222401437D-01	-1.00	-1.580980441587D-08
1		1.580980441587D-08	1.546433633352D-02	-.75	-1.185735331190D-08
				-.50	-7.904902207935D-09
0	-9.5	-1.580980441587D-08	-7.904902207935D-09	-.25	-3.952451103967D-09
1	-8.5	9.485882649522D-08	4.742941324761D-08	.25	3.705422909969D-09
2	-7.5	-3.794353059809D-07	-1.897176529904D-07	.50	5.928676655951D-09
3	-6.5	1.422882397428D-06	7.114411987141D-07	.75	5.187592073957D-09
4	-5.5	-5.312094283732D-06	-2.656047141866D-06	1.25	-9.634099565920D-09
5	-4.5	1.982549473750D-05	9.912747368750D-06	1.50	-1.778602996785D-08
6	-3.5	-7.398988466627D-05	-3.699494233313D-05	1.75	-1.704494538586D-08
7	-2.5	2.761340439276D-04	1.380670219638D-04	2.25	3.483097535371D-08
8	-1.5	-1.030546291044D-03	-5.152731455220D-04	2.50	6.521544321546D-08
9	-.5	3.846051120249D-03	1.923025560124D-03	2.75	6.299218946948D-08
10	.5	-1.435365818995D-02	-7.176829094975D-03	3.25	-1.296898018489D-07
11	1.5	5.356858163955D-02	2.678429081978D-02	3.50	-2.430757428940D-07
12	2.5	-1.999206683683D-01	-9.996033418413D-02	3.75	-2.349238124921D-07
13	3.5	7.461140918335D-01	3.730570459167D-01	4.25	4.839282320420D-07
14	4.5	-1.784535698966D+00	-8.922678494828D-01	4.50	9.070875283605D-07
15	5.5	2.392028704029D+00	1.196014352015D+00	4.75	8.767030604987D-07
16	6.5	-1.783579117151D+00	-8.917895585753D-01	5.25	-1.806023126319D-06
17	7.5	7.422877645735D-01	3.711438822868D-01	5.50	-3.385274370548D-06
18	8.5	-1.855719411434D-01	-9.278597057169D-02	5.75	-3.271888429503D-06
19	9.5	3.092865685723D-02	1.546432842861D-02	6.25	6.740164273234D-06
				6.50	1.263400995383D-05
	0	1.501931419508D-07	-1.469110449753D-01	6.75	1.221085065751D-05
	1	1.580980441587D-08	1.546433633352D-02	7.25	-2.515463396662D-05
				7.50	-4.715076544478D-05
				7.75	-4.557151420055D-05
				8.25	9.387837159324D-05

				8.50	1,759690518253D-04
				8.75	1,700752061447D-04
				9.25	-3,503588524063D-04
				9.50	-6,567254418563D-04

I=16	H= 6.5				
0		0.	1.096559528610D+00	-1.00	4,234769039965D-09
1		-4.234769039965D-09	-5.771366151791D-02	-.75	3,176076779974D-09
				-.50	2,117384519982D-09
0	-9.5	4.234769039965D-09	2.117384519982D-09	-.25	1,058692259991D-09
1	-8,5	-2,540861423979D-08	-1,270430711989D-08	.25	-9,925239937418D-10
2	-7,5	1,016344569592D-07	5,081722847958D-08	.50	-1,588038389987D-09
3	-6,5	-3,811292135968D-07	-1,905646067984D-07	.75	-1,389533591239D-09
4	-5,5	1,422882397428D-06	7,114411987141D-07	1.25	2,580562383729D-09
5	-4,5	-5,310400376116D-06	-2,655200188058D-06	1.50	4,764115169961D-09
6	-3,5	1,981871910704D-05	9,909359553518D-06	1.75	4,565610371212D-09
7	-2,5	-7,396447605203D-05	-3,698223802601D-05	2.25	-9,329725541173D-09
8	-1,5	2,760391851011D-04	1.380195925505D-04	2.50	-1,746842228986D-08
9	-,5	-1.030192264352D-03	-5.150961321761D-04	2.75	-1,687290789361D-08
10	.5	3.844729872308D-03	1.922364936154D-03	3.25	3,473833978096D-08
11	1.5	-1.434872722488D-02	-7.174363612440D-03	3.50	6,510957398946D-08
12	2,5	5,355017902721D-02	2.677508951361D-02	3.75	6,292602120323D-08
13	3,5	-1,998519888840D-01	-9.992599444198D-02	4.25	-1,296236335827D-07
14	4.5	7.458577765087D-01	3.729288882543D-01	4.50	-2,429698736680D-07
15	5,5	-1.783579117151D+00	-8,917895585753D-01	4.75	-2,348311769193D-07
16	6,5	2,388458692094D+00	1,194229346047D+00	5,25	4,837561945497D-07
17	7.5	-1.770255651225D+00	-8.851278256125D-01	5.50	9,067699206825D-07
18	8,5	6,925639128063D-01	3.462819564031D-01	5.75	8,763986864740D-07
19	9,5	-1,154273188010D-01	-5,771365940052D-02	6.25	-1,805401144616D-06
				6.50	-3,384109809062D-06
	0	-4,023030587967D-08	5.482797441898D-01	6.75	-3,270763568977D-06
	1	-4.234769039965D-09	-5,771366151791D-02	7,25	6,737848383916D-06
				7.50	1,262966931557D-05
				7.75	1,220665558943D-05
				8.25	-2,514599239105D-05
				8.50	-4,713456745320D-05
				8.75	-4,555585878875D-05
				9,25	9,384612118027D-05
				9.50	1,759086004972D-04

I=17	H= 7.5				
0		0.	-4,092415874296D+00	-1.00	-1,129271743991D-09
1		1.129271743991D-09	2,153903097381D-01	-.75	-8,469538079930D-10
				-.50	-5,646358719953D-10
0	-9,5	-1,129271743991D-09	-5,646358719953D-10	-.25	-2,823179359977D-10
1	-8,5	6,775630463944D-09	3.387815231972D-09	.25	2,646730649978D-10
2	-7,5	-2,710252185578D-08	-1.355126092789D-08	.50	4,234769039965D-10
3	-6,5	1.016344569592D-07	5.081722847958D-08	.75	3,705422909969D-10
4	-5,5	-3,794353059809D-07	-1.897176529904D-07	1.25	-6,881499689943D-10
5	-4,5	1.416106766964D-06	7.080533834821D-07	1.50	-1,270430711989D-09
6	-3,5	-5,284991761876D-06	-2.642495880938D-06	1.75	-1,217496098990D-09
7	-2,5	1.972386028054D-05	9.861930140270D-06	2,25	2,487926810979D-09
8	-1.5	-7,361044936029D-05	-3,680522468014D-05	2,50	4,658245943961D-09
9	-.5	2.747179371606D-04	1.373589685803D-04	2.75	4,499442104963D-09
10	.5	-1.025261299282D-03	-5,126306496411D-04	3,25	-9,263557274923D-09

11	1.5	3.826327259968D-03	1.913163629984D-03
12	2.5	-1.428004774059D-02	-7.140023870295D-03
13	3.5	5.329386370239D-02	2.664693185120D-02
14	4.5	-1.988954070690D-01	-9.944770353449D-02
15	5.5	7.422877645735D-01	3.711438822868D-01
16	6.5	-1.770255651225D+00	-8.851278256125D-01
17	7.5	2.338734840327D+00	1.169367420163D+00
18	8.5	-1.584683710082D+00	-7.923418550408D-01
19	9.5	4.307806183469D-01	2.153903091735D-01
	0	1.072808156791D-08	-2.046207931784D+00
	1	1.129271743991D-09	2.153903097381D-01

3.50	-1.736255306386D-08
3.75	-1.678027232086D-08
4.25	3.456630228871D-08
4.50	6.479196631146D-08
4.75	6.262164717848D-08
5.25	-1.290016518799D-07
5.50	-2.418053121820D-07
5.75	-2.337063163931D-07
6.25	4.814403052310D-07
6.50	9.024292824165D-07
6.75	8.722036183938D-07
7.25	-1.796759569044D-06
7.50	-3.367911817484D-06
7.75	-3.255108157182D-06
8.25	6.705597970946D-06
8.50	1.256921798752D-05
8.75	1.214822901033D-05
9.25	-2.502563231474D-05
9.50	-4.690896013260D-05

I=18 H= 8.5

0		0.	1.527310396857D+01
1		-2.823179359977D-10	-8.038475774345D-01
0	-9.5	2.823179359977D-10	1.411589679988D-10
1	-8.5	-1.693907615986D-09	-8.469538079930D-10
2	-7.5	6.775630463944D-09	3.387815231972D-09
3	-6.5	-2.540861423979D-08	-1.270430711989D-08
4	-5.5	9.485882649522D-08	4.742941324761D-08
5	-4.5	-3.540266917411D-07	-1.770133458705D-07
6	-3.5	1.321247940469D-06	6.606239702345D-07
7	-2.5	-4.930965070135D-06	-2.465482535068D-06
8	-1.5	1.840261234007D-05	9.201306170036D-06
9	-.5	-6.867948429015D-05	-3.433974214508D-05
10	.5	2.563153248205D-04	1.281576624103D-04
11	1.5	-9.565818149920D-04	-4.782909074960D-04
12	2.5	3.570011935147D-03	1.785005967574D-03
13	3.5	-1.332346592560D-02	-6.661732962799D-03
14	4.5	4.972385176724D-02	2.486192588362D-02
15	5.5	-1.855719411434D-01	-9.278597057169D-02
16	6.5	6.925639128063D-01	3.462819564031D-01
17	7.5	-1.584683710082D+00	-7.923418550408D-01
18	8.5	1.646170927520D+00	8.230854637602D-01
19	9.5	-6.076951545867D-01	-3.038475772934D-01
	0	-2.682020391978D-09	7.636551982946D+00
	1	-2.823179359977D-10	-8.038475774345D-01

-1.00	2.823179359977D-10
-.75	2.117384519982D-10
-.50	1.411589679988D-10
-.25	7.057948399942D-11
.25	-6.616826624945D-11
.50	-1.058692259991D-10
.75	-9.263557274923D-11
1.25	1.720374922486D-10
1.50	3.176076779974D-10
1.75	3.043740247475D-10
2.25	-6.219817027449D-10
2.50	-1.164561485990D-09
2.75	-1.124860526241D-09
3.25	2.315889318731D-09
3.50	4.340638265964D-09
3.75	4.195068080215D-09
4.25	-8.641575572179D-09
4.50	-1.619799157787D-08
4.75	-1.565541179462D-08
5.25	3.225041296998D-08
5.50	6.045132804550D-08
5.75	5.842657909827D-08
6.25	-1.203600763078D-07
6.50	-2.256073206041D-07
6.75	-2.180509045984D-07
7.25	4.491898922610D-07
7.50	8.419779543710D-07
7.75	8.137770392955D-07
8.25	-1.676399492736D-06
8.50	-3.142304496880D-06
8.75	-3.037057252584D-06
9.25	6.256408078685D-06
9.50	1.172724003315D-05

CARDINAL SPLINES N=2

I=19 H= 9.5

0		0.	-1.154551732810D+01	-1.00	-4.705298933294D-11
1		4.705298933294D-11	6.339745962391D-01	-.75	-3.528974199971D-11
				-.50	-2.352649466647D-11
0	-9.5	-4.705298933294D-11	-2.352649466647D-11	-.25	-1.176324733324D-11
1	-8.5	2.823179359977D-10	1.411589679988D-10	.25	1.102804437491D-11
2	-7.5	-1.129271743991D-09	-5.646358719953D-10	.50	1.764487099985D-11
3	-6.5	4.234769039965D-09	2.117384519982D-09	.75	1.543926212487D-11
4	-5.5	-1.580980441587D-08	-7.904902207935D-09	1.25	-2.867291537476D-11
5	-4.5	5.900444862351D-08	2.950222431176D-08	1.50	-5.293461299956D-11
6	-3.5	-2.202079900782D-07	-1.101039950391D-07	1.75	-5.072900412458D-11
7	-2.5	8.218275116892D-07	4.109137558446D-07	2.25	1.036636171241D-10
8	-1.5	-3.067102056679D-06	-1.533551028339D-06	2.50	1.940935809984D-10
9	-.5	1.144658071503D-05	5.723290357513D-06	2.75	1.874767543734D-10
10	.5	-4.271922080342D-05	-2.135961040171D-05	3.25	-3.859815531218D-10
11	1.5	1.594303024987D-04	7.971515124933D-05	3.50	-7.234397109940D-10
12	2.5	-5.950019891912D-04	-2.975009945956D-04	3.75	-6.991780133692D-10
13	3.5	2.220577654266D-03	1.110288827133D-03	4.25	1.440262595363D-09
14	4.5	-8.287308627874D-03	-4.143654313937D-03	4.50	2.699665262978D-09
15	5.5	3.092865685723D-02	1.546432842861D-02	4.75	2.609235299103D-09
16	6.5	-1.154273188010D-01	-5.771365940052D-02	5.25	-5.375068828331D-09
17	7.5	4.307806183469D-01	2.153903091735D-01	5.50	-1.007522134092D-08
18	8.5	-6.076951545867D-01	-3.038475772934D-01	5.75	-9.737763183044D-09
19	9.5	2.679491924311D-01	1.339745962156D-01	6.25	2.006001271796D-08
				6.50	3.760122010069D-08
	0	4.470033986630D-10	-5.522758663824D+00	6.75	3.634181743307D-08
	1	4.705298933294D-11	6.339745962391D-01	7.25	-7.486498204351D-08
				7.50	-1.403296590618D-07
				7.75	-1.356295065493D-07
				8.25	2.793999154561D-07
				8.50	5.237174161467D-07
				8.75	5.061762087639D-07
				9.25	-1.042734679781D-06
				9.50	-1.954540005525D-06

N= 2 M=20 P= 10.0 Q= 3

I= 0 H=-10.0

0		1.000000000000D+00	4.999999998739D-01	-1.00	2.267949192431D+00
1		-1.267949192431D+00	-6.339745962093D-01	-.75	1.950961894323D+00
				-.50	1.633974596216D+00
0	-10.0	2.679491924311D-01	1.339745962156D-01	-.25	1.316987298108D+00
1	-9.0	-6.076951545867D-01	-3.038475772934D-01	.25	6.871994080240D-01
2	-8.0	4.307806183469D-01	2.153903091735D-01	.50	3.995190528383D-01
3	-7.0	-1.154273188010D-01	-5.771365940052D-02	.75	1.620791712335D-01
4	-6.0	3.092865685723D-02	1.546432842861D-02	1.25	-7.109346086228D-02
5	-5.0	-8.287308627874D-03	-4.143654313937D-03	1.50	-7.355715851499D-02
6	-4.0	2.220577654266D-03	1.110288827133D-03	1.75	-3.924227691020D-02
7	-3.0	-5.950019891912D-04	-2.975009945956D-04	2.25	1.904943542518D-02

CARDINAL SPLINES N=2

8	-2.0	1.594303024987D-04	7.971515124935D-05	2.50	1.970958122162D-02
9	-1.0	-4.271922080357D-05	-2.135961040178D-05	2.75	1.051493640725D-02
10	0.0	1.144658071556D-05	5.723290357782D-06	3.25	-5.104280838447D-03
11	1.0	-3.067102058689D-06	-1.533551029344D-06	3.50	-5.281166371488D-03
12	2.0	8.218275191909D-07	4.109137595954D-07	3.75	-2.817468718786D-03
13	3.0	-2.202080180748D-07	-1.101040090374D-07	4.25	1.367687928603D-03
14	4.0	5.900455310833D-08	2.950227655416D-08	4.50	1.415084264335D-03
15	5.0	-1.581019435851D-08	-7.905097179257D-09	4.75	7.549384678987D-04
16	6.0	4.236224325726D-09	2.118112162863D-09	5.25	-3.664708759671D-04
17	7.0	-1.134702944391D-09	-5.673514721955D-10	5.50	-3.791706858505D-04
18	8.0	3.025874518376D-10	1.512937259188D-10	5.75	-2.022851528086D-04
19	9.0	-7.564686295940D-11	-3.782343147970D-11	6.25	9.819557526490D-05
20	10.0	1.260781049323D-11	6.303905246616D-12	6.50	1.015984790672D-04
				6.75	5.420214333588D-05
	0	-1.167949192431D+01	-5.839745962219D+00	7.25	-2.631142509254D-05
	1	-1.267949192431D+00	-6.339745962093D-01	7.50	-2.722323041828D-05
				7.75	-1.452342053489D-05
				8.25	7.050125105259D-06
				8.50	7.294442605946D-06
				8.75	3.891538803661D-06
				9.25	-1.889075328493D-06
				9.50	-1.954540005500D-06
				9.75	-1.042734679757D-06

I= 1		H= -9.0			
0		0.	7.564686295940D-10	-1.00	-1.607695154587D+00
1		1.607695154587D+00	8.038475772555D-01	-.75	-1.205771365940D+00
				-.50	-8.038475772934D-01
0	-10.0	-6.076951545867D-01	-3.038475772934D-01	-.25	-4.019237886467D-01
1	-9.0	1.646170927520D+00	8.230854637602D-01	.25	3.924285518563D-01
2	-8.0	-1.584683710082D+00	-7.923418550408D-01	.50	7.278856829700D-01
3	-7.0	6.925639128063D-01	3.462819564031D-01	.75	9.493999725988D-01
4	-6.0	-1.855719411434D-01	-9.278597057169D-02	1.25	8.484357651737D-01
5	-5.0	4.972385176724D-02	2.486192588362D-02	1.50	5.663429510899D-01
6	-4.0	-1.332346592560D-02	-6.661732962799D-03	1.75	2.510786614612D-01
7	-3.0	3.570011935147D-03	1.785005967574D-03	2.25	-1.142966125511D-01
8	-2.0	-9.565818149922D-04	-4.782909074961D-04	2.50	-1.182574873297D-01
9	-1.0	2.563153248214D-04	1.281576624107D-04	2.75	-6.308961844347D-02
10	0.0	-6.867948429338D-05	-3.433974214669D-05	3.25	3.062568503068D-02
11	1.0	1.840261235213D-05	9.201306176066D-06	3.50	3.168699822893D-02
12	2.0	-4.930965115145D-06	-2.465482557573D-06	3.75	1.690481231272D-02
13	3.0	1.321248108449D-06	6.606240542244D-07	4.25	-8.206127571620D-03
14	4.0	-3.540273186500D-07	-1.770136593250D-07	4.50	-8.490505586008D-03
15	5.0	9.486116615108D-08	4.743058307554D-08	4.75	-4.529630807392D-03
16	6.0	-2.541734595436D-08	-1.270867297718D-08	5.25	2.198825255802D-03
17	7.0	6.808217666346D-09	3.404108833173D-09	5.50	2.275024115103D-03
18	8.0	-1.815524711025D-09	-9.077623555127D-10	5.75	1.213710916852D-03
19	9.0	4.538811777564D-10	2.269405888782D-10	6.25	-5.891734515894D-04
20	10.0	-7.564686295940D-11	-3.782343147970D-11	6.50	-6.095908744031D-04
				6.75	-3.252128600153D-04
	0	1.607695154587D+01	8.038475773312D+00	7.25	1.578685505552D-04
	1	1.607695154587D+00	8.038475772555D-01	7.50	1.633393825097D-04
				7.75	8.714052320931D-05
				8.25	-4.230075063155D-05
				8.50	-4.376655563568D-05
				8.75	-2.334923282196D-05

CARDINAL SPLINES N=2

```
                                                                9.25   1.133445197096D-05
                                                                9.50   1.172724003300D-05
                                                                9.75   6.256408078542D-06


I= 2     H= -8.0
0           0.                     -3.025874518376D-09    -1.00   4.307806183469D-01
1          -4.307806183469D-01     -2.153903090222D-01     -.75   3.230854637602D-01
                                                           -.50   2.153903091735D-01
0 -10.0     4.307806183469D-01      2.153903091735D-01     -.25   1.076951545867D-01
1  -9.0    -1.584683710082D+00     -7.923418550408D-01      .25  -1.009642074251D-01
2  -8.0     2.338734840327D+00      1.169367420163D+00      .50  -1.615427318801D-01
3  -7.0    -1.770255651225D+00     -8.851278256125D-01      .75  -1.413498903951D-01
4  -6.0     7.422877645735D-01      3.711438822868D-01     1.25   2.781319393052D-01
5  -5.0    -1.988954070690D-01     -9.944770353449D-02     1.50   6.096281956403D-01
6  -4.0     5.329386370239D-02      2.664693185120D-02     1.75   8.863103541553D-01
7  -3.0    -1.428004774059D-02     -7.140023870295D-03     2.25   8.790614502044D-01
8  -2.0     3.826327259969D-03      1.913163629984D-03     2.50   5.980299493189D-01
9  -1.0    -1.025261299286D-03     -5.126306496428D-04     2.75   2.679834737739D-01
10  0.0     2.747179371735D-04      1.373589685868D-04     3.25  -1.225027401227D-01
11  1.0    -7.361044940853D-05     -3.680522470426D-05     3.50  -1.267479929157D-01
12  2.0     1.972386046058D-05      9.861930230290D-06     3.75  -6.761924925086D-02
13  3.0    -5.284992433795D-06     -2.642496216898D-06     4.25   3.282451028648D-02
14  4.0     1.416109274600D-06      7.080546372999D-07     4.50   3.396202234403D-02
15  5.0    -3.794446646043D-07     -1.897223323022D-07     4.75   1.811852322957D-02
16  6.0     1.016693838174D-07      5.083469190871D-08     5.25  -8.795301023210D-03
17  7.0    -2.723287066538D-08     -1.361643533269D-08     5.50  -9.100096460411D-03
18  8.0     7.262098844102D-09      3.631049422051D-09     5.75  -4.854843667407D-03
19  9.0    -1.815524711025D-09     -9.077623555127D-10     6.25   2.356693806358D-03
20 10.0     3.025874518376D-10      1.512937259188D-10     6.50   2.438363497613D-03
                                                           6.75   1.300851440061D-03
        0  -4.307806183469D+00     -2.153903093248D+00     7.25  -6.314742022210D-04
        1  -4.307806183469D-01     -2.153903090222D-01     7.50  -6.533575300388D-04
                                                           7.75  -3.485620928372D-04
                                                           8.25   1.692030025262D-04
                                                           8.50   1.750666225427D-04
                                                           8.75   9.339693128785D-05
                                                           9.25  -4.533780788384D-05
                                                           9.50  -4.690896013200D-05
                                                           9.75  -2.502563231417D-05


I= 3     H= -7.0
0           0.                      1.134702944391D-08    -1.00  -1.154273188010D-01
1           1.154273188010D-01      5.771365883317D-02     -.75  -8.657048910078D-02
                                                           -.50  -5.771365940052D-02
0 -10.0    -1.154273188010D-01     -5.771365940052D-02     -.25  -2.885682970026D-02
1  -9.0     6.925639128063D-01      3.462819564031D-01      .25   2.705327784399D-02
2  -8.0    -1.770255651225D+00     -8.851278256125D-01      .50   4.328524455039D-02
3  -7.0     2.388458692094D+00      1.194229346047D+00      .75   3.787458898159D-02
4  -6.0    -1.783579117151D+00     -8.917895585753D-01     1.25  -7.033852239439D-02
5  -5.0     7.458577765087D-01      3.729288882543D-01     1.50  -1.298557336512D-01
6  -4.0    -1.998519888840D-01     -9.992599444198D-02     1.75  -1.244450780824D-01
7  -3.0     5.355017902721D-02      2.677508951361D-02     2.25   2.699258117335D-01
8  -2.0    -1.434872722488D-02     -7.174363612442D-03     2.50   6.011376900543D-01
9  -1.0     3.844729872321D-03      1.922364936161D-03     2.75   8.817807233479D-01
10  0.0    -1.030192264401D-03     -5.150961322004D-04     3.25   8.812602754602D-01
```

11	1.0	2,760391852820D-04	1.380195926410D-04
12	2.0	-7,396447672718D-05	-3,698223836359D-05
13	3.0	1.981872162673D-05	9.909360813366D-06
14	4.0	-5.310409779750D-06	-2.655204889875D-06
15	5.0	1.422917492266D-06	7.114587461331D-07
16	6.0	-3.812601893154D-07	-1.906300946577D-07
17	7.0	1.021232649952D-07	5.106163249759D-08
18	8.0	-2.723287066538D-08	-1.361643533269D-08
19	9.0	6.808217666346D-09	3.404108833173D-09
20	10.0	-1.134702944391D-09	-5.673514721955D-10
	0	1.154273188010D+00	5.771365996787D-01
	1	1.154273188010D-01	5.771365883317D-02

3.50	6,003049734340D-01
3.75	2,691971846907D-01
4.25	-1,230919135743D-01
4.50	-1,273575837901D-01
4.75	-6,794446211088D-02
5.25	3,298237883704D-02
5.50	3,412536172654D-02
5.75	1,820566375278D-02
6.25	-8,837601773841D-03
6.50	-9,143863116047D-03
6.75	-4,878192900229D-03
7.25	2,368028258329D-03
7.50	2,450090737646D-03
7.75	1,307107848140D-03
8.25	-6,345112594733D-04
8.50	-6,564998345351D-04
8.75	-3,502384923294D-04
9.25	1,700167795644D-04
9.50	1,759086004950D-04
9.75	9,384612117812D-05

I= 4 H= -6.0

0		0.	-4.236224325726D-08
1		-3.092865685723D-02	-1.546432631050D-02
0	-10.0	3.092865685723D-02	1.546432842861D-02
1	-9.0	-1.855719411434D-01	-9.278597057169D-02
2	-8.0	7.422877645735D-01	3.711438822868D-01
3	-7.0	-1.783579117151D+00	-8.917895585753D-01
4	-6.0	2.392028704029D+00	1.196014352015D+00
5	-5.0	-1.784535698966D+00	-8.922678494828D-01
6	-4.0	7.461140918335D-01	3.730570459167D-01
7	-3.0	-1.999206683683D-01	-9.996033418413D-02
8	-2.0	5.356858163956D-02	2.678429081978D-02
9	-1.0	-1.435365819000D-02	-7.176829094999D-03
10	0.0	3.846051120429D-03	1.923025560215D-03
11	1.0	-1.030546291719D-03	-5.152731458597D-04
12	2.0	2.761340464481D-04	1.380670232241D-04
13	3.0	-7.398989407313D-05	-3.699494703657D-05
14	4.0	1.982552984440D-05	9.912764922199D-06
15	5.0	-5.312225304461D-06	-2.656112652230D-06
16	6.0	1.423371373444D-06	7.116856867220D-07
17	7.0	-3.812601893154D-07	-1.906300946577D-07
18	8.0	1.016693838174D-07	5.083469190871D-08
19	9.0	-2.541734595436D-08	-1.270867297718D-08
20	10.0	4.236224325726D-09	2.118112162863D-09
	0	-3.092865685723D-01	-1.546433054673D-01
	1	-3.092865685723D-02	-1.546432631050D-02

-1.00	3,092865685723D-02
-.75	2,319649264292D-02
-.50	1,546432842861D-02
-.25	7,732164214307D-03
.25	-7,248903950913D-03
.50	-1,159824632146D-02
.75	-1,014846553128D-02
1.25	1,884715027237D-02
1.50	3,479473896438D-02
1.75	3,334495817420D-02
2.25	-6,813969713858D-02
2.50	-1,275807095361D-01
2.75	-1,232313671655D-01
3.25	2,693366382820D-01
3.50	6,005280991799D-01
3.75	8,814555104879D-01
4.25	8,814181440107D-01
4.50	6,004683128165D-01
4.75	2,692843252140D-01
5.25	-1,231342143249D-01
5.50	-1,274013504458D-01
5.75	-6,796781134370D-02
6.25	3,299371328901D-02
6.50	3,413708896658D-02
6.75	1,821192016086D-02
7.25	-8,840638831094D-03
7.50	-9,147005420543D-03
7.75	-4,879869299721D-03
8.25	2,368842035367D-03
8.50	2,450932715598D-03
8.75	1,307557038030D-03
9.25	-6,347293103738D-04
9.50	-6,567254418481D-04
9.75	-3,503588523983D-04

CARDINAL SPLINES N=2

I= 5 H= -5.0

0		0.	1.581019435851D-07	-1.00	-8.287308627874D-03
1		8.287308627874D-03	4.143646408840D-03	-.75	-6.215481470905D-03
				-.50	-4.143654313937D-03
0	-10.0	-8.287308627874D-03	-4.143654313937D-03	-.25	-2.071827156968D-03
1	-9.0	4.972385176724D-02	2.486192588362D-02	.25	1.942337959658D-03
2	-8.0	-1.988954070690D-01	-9.944770353449D-02	.50	3.107740735453D-03
3	-7.0	7.458577765087D-01	3.729288882543D-01	.75	2.719273143521D-03
4	-6.0	-1.784535698966D+00	-8.922678494828D-01	1.25	-5.050078695111D-03
5	-5.0	2.392285019354D+00	1.196142509677D+00	1.50	-9.323222206358D-03
6	-4.0	-1.784604378450D+00	-8.923021892250D-01	1.75	-8.934754614427D-03
7	-3.0	7.461324944458D-01	3.730662472229D-01	2.25	1.825797682078D-02
8	-2.0	-1.999255993334D-01	-9.996279966669D-02	2.50	3.418514808998D-02
9	-1.0	5.356990288767D-02	2.678495144384D-02	2.75	3.301974531419D-02
10	0.0	-1.435401221732D-02	-7.177006108659D-03	3.25	-6.798182858803D-02
11	1.0	3.846145981596D-03	1.923072990798D-03	3.50	-1.274173701536D-01
12	2.0	-1.030571709065D-03	-5.152858545327D-04	3.75	-1.231442266423D-01
13	3.0	2.761408546658D-04	1.380704273329D-04	4.25	2.692943375313D-01
14	4.0	-7.399170959784D-05	-3.699585479892D-05	4.50	6.004843325243D-01
15	5.0	1.982598372558D-05	9.912991862788D-06	4.75	8.814321612551D-01
16	6.0	-5.312225304461D-06	-2.656112652230D-06	5.25	8.814294784627D-01
17	7.0	1.422917492266D-06	7.114587461331D-07	5.50	6.004800400565D-01
18	8.0	-3.794446646043D-07	-1.897223323022D-07	5.75	2.692905816220D-01
19	9.0	9.486116615108D-08	4.743058307554D-08	6.25	-1.231372513822D-01
20	10.0	-1.581019435851D-08	-7.905097179257D-09	6.50	-1.274044927503D-01
				6.75	-6.796948774319D-02
	0	8.287308627874D-02	4.143662219034D-02	7.25	3.299452706605D-02
	1	8.287308627874D-03	4.143646408840D-03	7.50	3.413793094453D-02
				7.75	1.821236935075D-02
				8.25	-8.840856881994D-03
				8.50	-9.147231027856D-03
				8.75	-4.879989659790D-03
				9.25	2.368900461931D-03
				9.50	2.450993166897D-03
				9.75	1.307589288415D-03

I= 6 H= -4.0

0		0.	-5.900455310833D-07	-1.00	2.220577654266D-03
1		-2.220577654266D-03	-1.110259324857D-03	-.75	1.665433240700D-03
				-.50	1.110288827133D-03
0	-10.0	2.220577654266D-03	1.110288827133D-03	-.25	5.551444135666D-04
1	-9.0	-1.332346592560D-02	-6.661732962799D-03	.25	-5.204478877187D-04
2	-8.0	5.329386370239D-02	2.664693185120D-02	.50	-8.327166203499D-04
3	-7.0	-1.998519888840D-01	-9.992599444198D-02	.75	-7.286270428061D-04
4	-6.0	7.461140918335D-01	3.730570459167D-01	1.25	1.353164508069D-03
5	-5.0	-1.784604378450D+00	-8.923021892250D-01	1.50	2.498149861050D-03
6	-4.0	2.392303421966D+00	1.196151710983D+00	1.75	2.394060283506D-03
7	-3.0	-1.784609309415D+00	-8.923046547075D-01	2.25	-4.892210144555D-03
8	-2.0	7.461338156939D-01	3.730669078470D-01	2.50	-9.159882823848D-03
9	-1.0	-1.999259533607D-01	-9.996297668035D-02	2.75	-8.847614091217D-03
10	0.0	5.356999774884D-02	2.678499887442D-02	3.25	1.821567607015D-02
11	1.0	-1.435403763466D-02	-7.177018817332D-03	3.50	3.414138143434D-02
12	2.0	3.846152789813D-03	1.923076394907D-03	3.75	3.299639608136D-02
13	3.0	-1.030573524590D-03	-5.152867622950D-04	4.25	-6.797049413606D-02

14	4.0	2.761413085470D-04	1.380706542735D-04	4.50	-1.274056429135D-01
15	5.0	-7.399170959784D-05	-3.699585479892D-05	4.75	-1.231379702342D-01
16	6.0	1.982552984440D-05	9.912764922199D-06	5.25	2.692913004741D-01
17	7.0	-5.310409779750D-06	-2.655204889875D-06	5.50	6.004811902198D-01
18	8.0	1.416109274600D-06	7.080546372999D-07	5.75	8.814304848556D-01
19	9.0	-3.540273186500D-07	-1.770136593250D-07	6.25	8.814302922398D-01
20	10.0	5.900455310833D-08	2.950227655416D-08	6.50	6.004808820344D-01
				6.75	2.692910308119D-01
	0	-2.220577654266D-02	-1.110318329410D-02	7.25	-1.231374694331D-01
	1	-2.220577654266D-03	-1.110259324857D-03	7.50	-1.274047183576D-01
				7.75	-6.796960810326D-02
				8.25	3.299458549261D-02
				8.50	3.413799139583D-02
				8.75	1.821240160113D-02
				9.25	-8.840872537349D-03
				9.50	-9.147247225741D-03
				9.75	-4.879998301262D-03

I= 7 H= -3.0

0		0.	2.202080180748D-06	-1.00	-5.950019891912D-04
1		5.950019891912D-04	2.973908905866D-04	-.75	-4.462514918934D-04
				-.50	-2.975009945956D-04
0	-10.0	-5.950019891912D-04	-2.975009945956D-04	-.25	-1.487504972978D-04
1	-9.0	3.570011935147D-03	1.785005967574D-03	.25	1.394535912167D-04
2	-8.0	-1.428004774059D-02	-7.140023870295D-03	.50	2.231257459467D-04
3	-7.0	5.355017902721D-02	2.677508951361D-02	.75	1.952350277034D-04
4	-6.0	-1.999206683683D-01	-9.996033418413D-02	1.25	-3.625793371634D-04
5	-5.0	7.461324944458D-01	3.730662472229D-01	1.50	-6.693772378402D-04
6	-4.0	-1.784609309415D+00	-8.923046547075D-01	1.75	-6.414865195968D-04
7	-3.0	2.392304743214D+00	1.196152371607D+00	2.25	1.310863757437D-03
8	-2.0	-1.784609663442D+00	-8.923048317212D-01	2.50	2.454383205414D-03
9	-1.0	7.461339105551D-01	3.730669552776D-01	2.75	2.370711050684D-03
10	0.0	-1.999259787780D-01	-9.996298938902D-02	3.25	-4.880875692584D-03
11	1.0	5.357000455706D-02	2.678500227853D-02	3.50	-9.148155583815D-03
12	2.0	-1.435403945019D-02	-7.177019725094D-03	3.75	-8.841357683139D-03
13	3.0	3.846153243694D-03	1.923076621847D-03	4.25	1.821263901290D-02
14	4.0	-1.030573524590D-03	-5.152867622950D-04	4.50	3.413823912985D-02
15	5.0	2.761408546658D-04	1.380704273329D-04	4.75	3.299471968187D-02
16	6.0	-7.398989407313D-05	-3.699494703657D-05	5.25	-6.796968035902D-02
17	7.0	1.981872162673D-05	9.909360813366D-06	5.50	-1.274048009356D-01
18	8.0	-5.284992433795D-06	-2.642496216898D-06	5.75	-1.231375210443D-01
19	9.0	1.321248108449D-06	6.606240542244D-07	6.25	2.692910824232D-01
20	10.0	-2.202080180748D-07	-1.101040090374D-07	6.50	6.004809646125D-01
				6.75	8.814303644955D-01
	0	5.950019891912D-03	2.976110986047D-03	7.25	8.814303506663D-01
	1	5.950019891912D-04	2.973908905866D-04	7.50	6.004809424857D-01
				7.75	2.692910630623D-01
				8.25	-1.231374850884D-01
				8.50	-1.274047345555D-01
				8.75	-6.796961674474D-02
				9.25	3.299458968746D-02
				9.50	3.413799573607D-02
				9.75	1.821240391663D-02

I= 8 H= -2.0

CARDINAL SPLINES N=2

0		0.	-8.218275191909D-06
1		-1.594303024987D-04	-7.930423748976D-05
0	-10.0	1.594303024987D-04	7.971515124935D-05
1	-9.0	-9.565818149922D-04	-4.782909074961D-04
2	-8.0	3.826327259969D-03	1.913163629984D-03
3	-7.0	-1.434872722488D-02	-7.174363612442D-03
4	-6.0	5.356858163956D-02	2.678429081978D-02
5	-5.0	-1.999255993334D-01	-9.996279966669D-02
6	-4.0	7.461338156939D-01	3.730669078470D-01
7	-3.0	-1.784609663442D+00	-8.923048317212D-01
8	-2.0	2.392304838076D+00	1.196152419038D+00
9	-1.0	-1.784609688860D+00	-8.923048444299D-01
10	0.0	7.461339173633D-01	3.730669586817D-01
11	1.0	-1.999259805936D-01	-9.996299029678D-02
12	2.0	5.357000501094D-02	2.678500250547D-02
13	3.0	-1.435403945019D-02	-7.177019725094D-03
14	4.0	3.846152789813D-03	1.923076394907D-03
15	5.0	-1.030571709065D-03	-5.152858545327D-04
16	6.0	2.761340464481D-04	1.380670232241D-04
17	7.0	-7.396447672718D-05	-3.698223836359D-05
18	8.0	1.972386046058D-05	9.861930230290D-06
19	9.0	-4.930965115145D-06	-2.465482557573D-06
20	10.0	8.218275191909D-07	4.109137595954D-07
	0	-1.594303024987D-03	-8.012606500895D-04
	1	-1.594303024987D-04	-7.930423748976D-05

-1.00	1.594303024987D-04
-.75	1.195727268740D-04
-.50	7.971515124935D-05
-.25	3.985757562468D-05
.25	-3.736647714813D-05
.50	-5.978636343701D-05
.75	-5.231306800739D-05
1.25	9.715284058515D-05
1.50	1.793590903110D-04
1.75	1.718857948814D-04
2.25	-3.512448851925D-04
2.50	-6.576499978072D-04
2.75	-6.352301115183D-04
3.25	1.307826700185D-03
3.50	2.451240900918D-03
3.75	2.369034651192D-03
4.25	-4.880061915546D-03
4.50	-9.147313605863D-03
4.75	-8.840908493248D-03
5.25	1.821242096200D-02
5.50	3.413801352253D-02
5.75	3.299459932180D-02
6.25	-6.796962193246D-02
6.50	-1.274047404843D-01
6.75	-1.231374887940D-01
7.25	2.692910667678D-01
7.50	6.004809484146D-01
7.75	8.814303558540D-01
8.25	8.814303548612D-01
8.50	6.004809468260D-01
8.75	2.692910653778D-01
9.25	-1.231374862125D-01
9.50	-1.274047357185D-01
9.75	-6.796961736528D-02

I= 9 H= -1.0

0		0.	3.067102058689D-05
1		4.271922080357D-05	1.982605937244D-05
0	-10.0	-4.271922080357D-05	-2.135961040178D-05
1	-9.0	2.563153248214D-04	1.281576624107D-04
2	-8.0	-1.025261299286D-03	-5.126306496428D-04
3	-7.0	3.844729872321D-03	1.922364936161D-03
4	-6.0	-1.435365819000D-02	-7.176829094999D-03
5	-5.0	5.356990288767D-02	2.678495144384D-02
6	-4.0	-1.999259533607D-01	-9.996297668035D-02
7	-3.0	7.461339105551D-01	3.730669552776D-01
8	-2.0	-1.784609688860D+00	-8.923048444299D-01
9	-1.0	2.392304844884D+00	1.196152422442D+00
10	0.0	-1.784609690675D+00	-8.923048453376D-01
11	1.0	7.461339178172D-01	3.730669589086D-01
12	2.0	-1.999259805936D-01	-9.996299029678D-02
13	3.0	5.357000455706D-02	2.678500227853D-02
14	4.0	-1.435403763466D-02	-7.177018817332D-03
15	5.0	3.846145981596D-03	1.923072990798D-03
16	6.0	-1.030546291719D-03	-5.152731458597D-04

-1.00	-4.271922080357D-05
-.75	-3.203941560268D-05
-.50	-2.135961040178D-05
-.25	-1.067980520089D-05
.25	1.001231737584D-05
.50	1.601970780134D-05
.75	1.401724432617D-05
1.25	-2.603202517717D-05
1.50	-4.805912340401D-05
1.75	-4.605665992885D-05
2.25	9.411578333286D-05
2.50	1.762167858147D-04
2.75	1.702093953892D-04
3.25	-3.504311081543D-04
3.50	-6.568080198548D-04
3.75	-6.347809216280D-04
4.25	1.307608649284D-03
4.50	2.451015293605D-03
4.75	2.368914291123D-03
5.25	-4.880003488982D-03

17	7.0	2.760391852820D-04	1.380195926410D-04	5.50	-9.147253154564D-03
18	8.0	-7.361044940853D-05	-3.680522470426D-05	5.75	-8.840876242863D-03
19	9.0	1.840261235213D-05	9.201306176066D-06	6.25	1.821240530665D-02
20	10.0	-3.067102058689D-06	-1.533551029344D-06	6.50	3.413799732465D-02
				6.75	3.299459068033D-02
	0	4.271922080357D-04	2.289316143113D-04	7.25	-6.796961773760D-02
	1	4.271922080357D-05	1.982605937244D-05	7.50	-1.274047361440D-01
				7.75	-1.231374864785D-01
				8.25	2.692910656438D-01
				8.50	6.004809472515D-01
				8.75	8.814303552335D-01
				9.25	8.814303551626D-01
				9.50	6.004809471380D-01
				9.75	2.692910655445D-01

I=10 H= 0.0

0		0.	-1.144658071556D-04	-1.00	1.144658071556D-05
1		-1.144658071556D-05	-9.273151198994D-26	-.75	8.584935536673D-06
				-.50	5.723290357782D-06
0	-10.0	1.144658071556D-05	5.723290357782D-06	-.25	2.861645178891D-06
1	-9.0	-6.867948429338D-05	-3.433974214669D-05	.25	-2.682792355210D-06
2	-8.0	2.747179371735D-04	1.373589685868D-04	.50	-4.292467768336D-06
3	-7.0	-1.030192264401D-03	-5.150961322004D-04	.75	-3.755909297294D-06
4	-6.0	3.846051120429D-03	1.923025560215D-03	1.25	6.975260123547D-06
5	-5.0	-1.435401221732D-02	-7.177006108659D-03	1.50	1.287740330501D-05
6	-4.0	5.356999774884D-02	2.678499887442D-02	1.75	1.234084483397D-05
7	-3.0	-1.999259787780D-01	-9.996298938902D-02	2.25	-2.521824813898D-05
8	-2.0	7.461339173633D-01	3.730669586817D-01	2.50	-4.721714545170D-05
9	-1.0	-1.784609690675D+00	-8.923048453376D-01	2.75	-4.560747003857D-05
10	0.0	2.392304845338D+00	1.196152422669D+00	3.25	9.389773243236D-05
11	1.0	-1.784609690675D+00	-8.923048453376D-01	3.50	1.759911785018D-04
12	2.0	7.461339173633D-01	3.730669586817D-01	3.75	1.700890353203D-04
13	3.0	-1.999259787780D-01	-9.996298938902D-02	4.25	-3.503726815905D-04
14	4.0	5.356999774884D-02	2.678499887442D-02	4.50	-6.567475685555D-04
15	5.0	-1.435401221732D-02	-7.177006108659D-03	4.75	-6.347486712428D-04
16	6.0	3.846051120429D-03	1.923025560215D-03	5.25	1.307592993929D-03
17	7.0	-1.030192264401D-03	-5.150961322004D-04	5.50	2.450999095720D-03
18	8.0	2.747179371735D-04	1.373589685868D-04	5.75	2.368905649651D-03
19	9.0	-6.867948429338D-05	-3.433974214669D-05	6.25	-4.879999294128D-03
20	10.0	1.144658071556D-05	5.723290357782D-06	6.50	-9.147248814325D-03
				6.75	-8.840873927360D-03
	0	-1.144658071556D-04	-1.144658071556D-04	7.25	1.821240418258D-02
	1	-1.144658071556D-05	-9.273151198994D-26	7.50	3.413799616158D-02
				7.75	3.299459005979D-02
				8.25	-6.796961743619D-02
				8.50	-1.274047358320D-01
				8.75	-1.231374863118D-01
				9.25	2.692910655622D-01
				9.50	6.004809471664D-01
				9.75	8.814303551874D-01

I=11 H= 1.0

0		0.	4.271922080357D-04	-1.00	-3.067102058689D-06
1		3.067102058689D-06	-1.982605937244D-05	-.75	-2.300326544017D-06
				-.50	-1.533551029344D-06

CARDINAL SPLINES N=2

0	-10.0	-3.067102058689D-06	-1.533551029344D-06	-.25	-7.667755146722D-07
1	-9.0	1.840261235213D-05	9.201306176066D-06	.25	7.188520450052D-07
2	-8.0	-7.361044940853D-05	-3.680522470426D-05	.50	1.150163272008D-06
3	-7.0	2.760391852820D-04	1.380195926410D-04	.75	1.006392863007D-06
4	-6.0	-1.030546291719D-03	-5.152731458597D-04	1.25	-1.869015317013D-06
5	-5.0	3.846145981596D-03	1.923072990798D-03	1.50	-3.450489816025D-06
6	-4.0	-1.435403763466D-02	-7.177018817332D-03	1.75	-3.306719407024D-06
7	-3.0	5.357000455706D-02	2.678500227853D-02	2.25	6.757209223049D-06
8	-2.0	-1.999259805936D-01	-9.996299029678D-02	2.50	1.265179599209D-05
9	-1.0	7.461339178172D-01	3.730669589086D-01	2.75	1.222048476509D-05
10	0.0	-1.784609690675D+00	-8.923048453376D-01	3.25	-2.515982157518D-05
11	1.0	2.392304844884D+00	1.196152422442D+00	3.50	-4.715669415234D-05
12	2.0	-1.784609688860D+00	-8.923048444299D-01	3.75	-4.557521965333D-05
13	3.0	7.461339105551D-01	3.730669552776D-01	4.25	9.388207707767D-05
14	4.0	-1.999259533607D-01	-9.996297668035D-02	4.50	1.759749806173D-04
15	5.0	5.356990288767D-02	2.678495144384D-02	4.75	1.700803938482D-04
16	6.0	-1.435365819000D-02	-7.176829094999D-03	5.25	-3.503684867355D-04
17	7.0	3.844729872321D-03	1.922364936161D-03	5.50	-6.567432283167D-04
18	8.0	-1.025261299286D-03	-5.126306496428D-04	5.75	-6.347463557396D-04
19	9.0	2.563153248214D-04	1.281576624107D-04	6.25	1.307591869864D-03
20	10.0	-4.271922080357D-05	-2.135961040178D-05	6.50	2.450997932650D-03
				6.75	2.368905029110D-03
	0	3.067102058689D-05	2.289316143113D-04	7.25	-4.879998992722D-03
	1	3.067102058689D-06	-1.982605937244D-05	7.50	-9.147248502282D-03
				7.75	-8.840873760701D-03
				8.25	1.821240410102D-02
				8.50	3.413799607648D-02
				8.75	3.299459001369D-02
				9.25	-6.796961741137D-02
				9.50	-1.274047358036D-01
				9.75	-1.231374862941D-01

I=12 H= 2.0

0		0.	-1.594303024987D-03	-1.00	8.218275191909D-07
1		-8.218275191909D-07	7.930423748976D-05	-.75	6.163706393932D-07
				-.50	4.109137595954D-07
0	-10.0	8.218275191909D-07	4.109137595954D-07	-.25	2.054568797977D-07
1	-9.0	-4.930965115145D-06	-2.465482557573D-06	.25	-1.926158248104D-07
2	-8.0	1.972386046058D-05	9.861930230290D-06	.50	-3.081853196966D-07
3	-7.0	-7.396447672718D-05	-3.698223836359D-05	.75	-2.696621547345D-07
4	-6.0	2.761340464481D-04	1.380670232241D-04	1.25	5.008011445069D-07
5	-5.0	-1.030571709065D-03	-5.152858545327D-04	1.50	9.245559590897D-07
6	-4.0	3.846152789813D-03	1.923076394907D-03	1.75	8.860327941277D-07
7	-3.0	-1.435403945019D-02	-7.177019725094D-03	2.25	-1.810588753217D-06
8	-2.0	5.357000501094D-02	2.678500250547D-02	2.50	-3.390038516662D-06
9	-1.0	-1.999259805936D-01	-9.996299029678D-02	2.75	-3.274469021776D-06
10	0.0	7.461339173633D-01	3.730669586817D-01	3.25	6.741553868363D-06
11	1.0	-1.784609688860D+00	-8.923048444299D-01	3.50	1.263559810756D-05
12	2.0	2.392304838076D+00	1.196152419038D+00	3.75	1.221184329298D-05
13	3.0	-1.784609663442D+00	-8.923048317212D-01	4.25	-2.515562672023D-05
14	4.0	7.461338156939D-01	3.730669078470D-01	4.50	-4.715235391358D-05
15	5.0	-1.999255993334D-01	-9.996279966669D-02	4.75	-4.557290415013D-05
16	6.0	5.356858163956D-02	2.678429081978D-02	5.25	9.388095301257D-05
17	7.0	-1.434872722488D-02	-7.174363612442D-03	5.50	1.759738175467D-04
18	8.0	3.826327259969D-03	1.913163629984D-03	5.75	1.700797733075D-04
19	9.0	-9.565818149922D-04	-4.782909074961D-04	6.25	-3.503681853300D-04

20	10,0	1.594303024987D-04	7.971515124935D-05	6.50	-6,567429162734D-04
				6.75	-6,347461890801D-04
	0	-8,218275191909D-06	-8,012606500895D-04	7,25	1,307591788308D-03
	1	-8,218275191909D-07	7.930423748976D-05	7,50	2,450997847547D-03
				7,75	2,368904983013D-03
				8,25	-4,879998967900D-03
				8,50	-9,147248473914D-03
				8,75	-8,840873742971D-03
				9,25	1,821240408329D-02
				9,50	3,413799604811D-02
				9.75	3,299458998887D-02

I=13	H= 3.0				
0		0.	5,950019891912D-03	-1.00	-2,202080180748D-07
1		2,202080180748D-07	-2,973908905866D-04	-,75	-1,651560135561D-07
				-,50	-1,101040090374D-07
0	-10,0	-2,202080180748D-07	-1.101040090374D-07	-,25	-5,505200451870D-08
1	-9.0	1.321248108449D-06	6.606240542244D-07	.25	5,161125423628D-08
2	-8,0	-5,284992433795D-06	-2,642496216898D-06	.50	8,257800677805D-08
3	-7.0	1,981872162673D-05	9.909360813366D-06	.75	7,225575593079D-08
4	-6,0	-7,398989407313D-05	-3,699494703657D-05	1.25	-1,341892610143D-07
5	-5,0	2,761408546658D-04	1.380704273329D-04	1.50	-2,477340203342D-07
6	-4,0	-1,030573524590D-03	-5,152867622950D-04	1.75	-2,374117694869D-07
7	-3,0	3,846153243694D-03	1.923076621847D-03	2,25	4,851457898210D-07
8	-2,0	-1.435403945019D-02	-7.177019725094D-03	2,50	9,083580745586D-07
9	-1,0	5,357000455706D-02	2.678500227853D-02	2,75	8,773913220168D-07
10	0,0	-1,999259787780D-01	-9,996298938902D-02	3,25	-1,806393898270D-06
11	1,0	7,461339105551D-01	3.730669552776D-01	3,50	-3,385698277900D-06
12	2,0	-1,784609663442D+00	-8,923048317212D-01	3,75	-3,272153518580D-06
13	3,0	2,392304743214D+00	1.196152371607D+00	4,25	6,740429803258D-06
14	4,0	-1,784609309415D+00	-8,923046547075D-01	4,50	1,263443503704D-05
15	5,0	7,461324944458D-01	3.730662472229D-01	4.75	1,221122275230D-05
16	6,0	-1,999206683683D-01	-9,996033418413D-02	5,25	-2,515532531476D-05
17	7,0	5,355017902721D-02	2,677508951361D-02	5,50	-4,715204187027D-05
18	8,0	-1,428004774059D-02	-7,140023870295D-03	5,75	-4,557273749064D-05
19	9,0	3,570011935147D-03	1.785005967574D-03	6.25	9,388087145580D-05
20	10,0	-5,950019891912D-04	-2,975009945956D-04	6,50	1,759737324440D-04
				6,75	1,700797272102D-04
	0	2,202080180748D-06	2,976110986047D-03	7,25	-3,503681605084D-04
	1	2,202080180748D-07	-2,973908905866D-04	7,50	-6,567428879058D-04
				7,75	-6,347461713503D-04
				8,25	1,307591770578D-03
				8,50	2,450997819179D-03
				8,75	2,368904958191D-03
				9,25	-4,879998921803D-03
				9,50	-9,147248388811D-03
				9,75	-8,840873661414D-03

I=14	H= 4.0				
0		0.	-2,220577654266D-02	-1,00	5,900455310833D-08
1		-5,900455310833D-08	1.110259324857D-03	-,75	4,425341483125D-08
				-,50	2,950227655416D-08
0	-10.0	5,900455310833D-08	2,950227655416D-08	-,25	1,475113827708D-08
1	-9,0	-3,540273186500D-07	-1,770136593250D-07	.25	-1,382919213476D-08
2	-8,0	1.416109274600D-06	7,080546372999D-07	.50	-2.212670741562D-08

CARDINAL SPLINES N=2

3	-7,0	-5,310409779750D-06	-2,655204889875D-06	.75	-1,936086898867D-08
4	-6,0	1,982552984440D-05	9,912764922199D-06	1.25	3,595589955039D-08
5	-5,0	-7,399170959784D-05	-3,699585479892D-05	1,50	6,638012224687D-08
6	-4,0	2,761413085470D-04	1,380706542735D-04	1,75	6,361428381992D-08
7	-3,0	-1,030573524590D-03	-5,152867622950D-04	2,25	-1,299944060668D-07
8	-2,0	3,846152789813D-03	1,923076394907D-03	2,50	-2,433937815719D-07
9	-1,0	-1,435403763466D-02	-7,177018817332D-03	2,75	-2,350962662910D-07
10	0,0	5,356999774884D-02	2,678499887442D-02	3,25	4,840217247168D-07
11	1,0	-1,999259533607D-01	-9,996297668035D-02	3,50	9,071950040406D-07
12	2,0	7,461338156939D-01	3,730669078470D-01	3,75	8,767707813441D-07
13	3,0	-1,784609309415D+00	-8,923046547075D-01	4,25	-1,806092492800D-06
14	4,0	2,392303421966D+00	1,196151710983D+00	4,50	-3,385386234590D-06
15	5,0	-1,784604378450D+00	-8,923021892250D-01	4,75	-3,271986859085D-06
16	6,0	7,461140918335D-01	3,730570459167D-01	5,25	6,740348246484D-06
17	7,0	-1,998519888840D-01	-9,992599444198D-02	5,50	1,263434993432D-05
18	8,0	5,329386370239D-02	2,664693185120D-02	5,75	1,221117665500D-05
19	9,0	-1,332346592560D-02	-6,661732962799D-03	6,25	-2,515530049314D-05
20	10,0	2,220577654266D-03	1,110288827133D-03	6,50	-4,715201350269D-05
				6,75	-4,557271976090D-05
	0	-5,900455310833D-07	-1,110318329410D-02	7,25	9,388085372606D-05
	1	-5,900455310833D-08	1,110259324857D-03	7,50	1,759737040765D-04
				7,75	1,700797023886D-04
				8,25	-3,503681144111D-04
				8,50	-6,567428028031D-04
				8,75	-6,347460897936D-04
				9,25	1,307591603918D-03
				9,50	2,450997507136D-03
				9,75	2,368904656786D-03

I=15 H= 5.0

0		0.	8,287308627874D-02	-1.00	-1,581019435851D-08
1		1.581019435851D-08	-4.143646408840D-03	-.75	-1,185764576889D-08
				-.50	-7,905097179257D-09
0	-10.0	-1.581019435851D-08	-7,905097179257D-09	-.25	-3,952548589628D-09
1	-9,0	9.486116615108D-08	4.743058307554D-08	.25	3,705514302777D-09
2	-8,0	-3.794446646043D-07	-1.897223323022D-07	.50	5,928822884443D-09
3	-7,0	1,422917492266D-06	7,114587461331D-07	.75	5,187720023887D-09
4	-6,0	-5,312225304461D-06	-2,656112652230D-06	1.25	-9,634337187219D-09
5	-5,0	1,982598372558D-05	9,912991862788D-06	1,50	-1,778646865333D-08
6	-4,0	-7,399170959784D-05	-3,699585479892D-05	1,75	-1,704536579277D-08
7	-3,0	2,761408546658D-04	1,380704273329D-04	2,25	3,483183444610D-08
8	-2,0	-1,030571709065D-03	-5,152858545327D-04	2,50	6,521705172887D-08
9	-1,0	3,846145981596D-03	1,923072990798D-03	2,75	6,299374314720D-08
10	0,0	-1,435401221732D-02	-7,177006108659D-03	3,25	-1,296930005972D-07
11	1,0	5,356990288767D-02	2,678495144384D-02	3,50	-2,430817382621D-07
12	2,0	-1,999255993334D-01	-9,996279966669D-02	3,75	-2,349296067960D-07
13	3,0	7,461324944458D-01	3,730662472229D-01	4,25	4,839401679426D-07
14	4,0	-1,784604378450D+00	-8,923021892250D-01	4,50	9,071099013197D-07
15	5,0	2,392285019354D+00	1,196142509677D+00	4,75	8,767246840370D-07
16	6,0	-1,784535698966D+00	-8,922678494828D-01	5,25	-1,806067671173D-06
17	7,0	7,458577765087D-01	3,729288882543D-01	5,50	-3,385357867017D-06
18	8,0	-1,988954070690D-01	-9,944770353449D-02	5,75	-3,271969129352D-06
19	9,0	4,972385176724D-02	2,486192588362D-02	6,25	6,740330516751D-06
20	10,0	-8,287308627874D-03	-4,143654313937D-03	6,50	1,263432156675D-05
				6,75	1,221115183337D-05
	0	1.581019435851D-07	4.143662219034D-02	7,25	-2,515525439583D-05

1		1.581019435851D-08	-4.143646408840D-03	7.50	-4,715192839997D-05
				7.75	-4,557263820413D-05
				8,25	9,388068706657D-05
				8.50	1,759733920331D-04
				8,75	1,700794009831D-04
				9.25	-3,503674938704D-04
				9.50	-6,567416397326D-04
				9,75	-6,347449657285D-04

I=16 H= 6.0

0		0.	-3,092865685723D-01	-1.00	4,236224325726D-09
1		-4,236224325726D-09	1.546432631050D-02	-.75	3,177168244295D-09
				-.50	2,118112162863D-09
0	-10.0	4.236224325726D-09	2.118112162863D-09	-.25	1,059056081432D-09
1	-9.0	-2.541734595436D-08	-1,270867297718D-08	.25	-9,928650763421D-10
2	-8.0	1.016693838174D-07	5,083469190871D-08	.50	-1,588584122147D-09
3	-7,0	-3,812601893154D-07	-1,906300946577D-07	.75	-1,390011106879D-09
4	-6,0	1,423371373444D-06	7,116856867220D-07	1.25	2,581449198489D-09
5	-5,0	-5.312225304461D-06	-2.656112652230D-06	1.50	4,765752366442D-09
6	-4,0	1,982552984440D-05	9,912764922199D-06	1.75	4,567179351174D-09
7	-3,0	-7,398989407313D-05	-3,699494703657D-05	2,25	-9,332931717615D-09
8	-2,0	2,761340464481D-04	1,380670232241D-04	2,50	-1,747442534362D-08
9	-1,0	-1,030546291719D-03	-5,152731458597D-04	2,75	-1,687870629782D-08
10	0,0	3.846051120429D-03	1,923025560215D-03	3,25	3,475027767197D-08
11	1.0	-1.435365819000D-02	-7,176829094999D-03	3,50	6,513194900804D-08
12	2,0	5,356858163956D-02	2,678429081978D-02	3.75	6,294764584009D-08
13	3.0	-1,999206683683D-01	-9,996033418413D-02	4.25	-1,296681789703D-07
14	4.0	7,461140918335D-01	3,730570459167D-01	4.50	-2,430533706885D-07
15	5,0	-1,784535698966D+00	-8,922678494828D-01	4.75	-2,349118770625D-07
16	6.0	2.392028704029D+00	1.196014352015D+00	5,25	4,839224382091D-07
17	7.0	-1,783579117151D+00	-8,917895585753D-01	5,50	9,070815337461D-07
18	8,0	7,422877645735D-01	3,711438822868D-01	5,75	8,766998624100D-07
19	9,0	-1,855719411434D-01	-9,278597057169D-02	6,25	-1,806021573866D-06
20	10.0	3,092865685723D-02	1,546432842861D-02	6.50	-3,385272764296D-06
				6.75	-3,271887572578D-06
0		-4,236224325726D-08	-1,546433054673D-01	7.25	6,740163857256D-06
1		-4,236224325726D-09	1.546432631050D-02	7.50	1,263400952344D-05
				7.75	1,221085042790D-05
				8.25	-2,515463385516D-05
				8.50	-4,715076532945D-05
				8,75	-4,557151413902D-05
				9.25	9,387837156337D-05
				9.50	1,759690517944D-04
				9.75	1,700752061282D-04

I=17 H= 7.0

0		0.	1.154273188010D+00	-1.00	-1,134702944391D-09
1		1,134702944391D-09	-5,771365883317D-02	-.75	-8,510272082932D-10
				-.50	-5,673514721955D-10
0	-10.0	-1,134702944391D-09	-5,673514721955D-10	-.25	-2,836757360977D-10
1	-9,0	6,808217666346D-09	3,404108833173D-09	.25	2,659460025916D-10
2	-8,0	-2,723287066538D-08	-1,361643533269D-08	.50	4,255136041466D-10
3	-7,0	1,021232649952D-07	5,106163249759D-08	.75	3,723244036283D-10
4	-6,0	-3,812601893154D-07	-1,906300946577D-07	1.25	-6,914596067382D-10
5	-5,0	1,422917492266D-06	7,114587461331D-07	1.50	-1,276540812440D-09

CARDINAL SPLINES N=2

6	-4.0	-5.310409779750D-06	-2.655204889875D-06
7	-3.0	1.981872162673D-05	9.909360813366D-06
8	-2.0	-7.396447672718D-05	-3.698223836359D-05
9	-1.0	2.760391852820D-04	1.380195926410D-04
10	0.0	-1.030192264401D-03	-5.150961322004D-04
11	1.0	3.844729872321D-03	1.922364936161D-03
12	2.0	-1.434872722488D-02	-7.174363612442D-03
13	3.0	5.355017902721D-02	2.677508951361D-02
14	4.0	-1.998519888840D-01	-9.992599444198D-02
15	5.0	7.458577765087D-01	3.729288882543D-01
16	6.0	-1.783579117151D+00	-8.917895585753D-01
17	7.0	2.388458692094D+00	1.194229346047D+00
18	8.0	-1.770255651225D+00	-8.851278256125D-01
19	9.0	6.925639128063D-01	3.462819564031D-01
20	10.0	-1.154273188010D-01	-5.771365940052D-02
	0	1.134702944391D-08	5.771365996787D-01
	1	1.134702944391D-09	-5.771365883317D-02

1.75	-1.223351611921D-09
2.25	2.499892424361D-09
2.50	4.680649645613D-09
2.75	4.521082044058D-09
3.25	-9.308110090707D-09
3.50	-1.744605777001D-08
3.75	-1.686097656431D-08
4.25	3.473254793847D-08
4.50	6.510358143443D-08
4.75	6.292282421318D-08
5.25	-1.296220816632D-07
5.50	-2.429682679677D-07
5.75	-2.348303202884D-07
6.25	4.837557787142D-07
6.50	9.067694904364D-07
6.75	8.763984569404D-07
7.25	-1.805401033194D-06
7.50	-3.384109693778D-06
7.75	-3.270763507473D-06
8.25	6.737848354060D-06
8.50	1.262966928468D-05
8.75	1.220665557295D-05
9.25	-2.514599238305D-05
9.50	-4.713456744492D-05
9.75	-4.555585878434D-05

I=18 H= 8.0

0		0.	-4.307806183469D+00
1		-3.025874518376D-10	2.153903090222D-01
0	-10.0	3.025874518376D-10	1.512937259188D-10
1	-9.0	-1.815524711025D-09	-9.077623555127D-10
2	-8.0	7.262098844102D-09	3.631049422051D-09
3	-7.0	-2.723287066538D-08	-1.361643533269D-08
4	-6.0	1.016693838174D-07	5.083469190871D-08
5	-5.0	-3.794446646043D-07	-1.897223323022D-07
6	-4.0	1.416109274600D-06	7.080546372999D-07
7	-3.0	-5.284992433795D-06	-2.642496216898D-06
8	-2.0	1.972386046058D-05	9.861930230290D-06
9	-1.0	-7.361044940853D-05	-3.680522470426D-05
10	0.0	2.747179371735D-04	1.373589685868D-04
11	1.0	-1.025261299286D-03	-5.126306496428D-04
12	2.0	3.826327259969D-03	1.913163629984D-03
13	3.0	-1.428004774059D-02	-7.140023870295D-03
14	4.0	5.329386370239D-02	2.664693185120D-02
15	5.0	-1.988954070690D-01	-9.944770353449D-02
16	6.0	7.422877645735D-01	3.711438822868D-01
17	7.0	-1.770255651225D+00	-8.851278256125D-01
18	8.0	2.338734840327D+00	1.169367420163D+00
19	9.0	-1.584683710082D+00	-7.923418550408D-01
20	10.0	4.307806183469D-01	2.153903091735D-01
	0	-3.025874518376D-09	-2.153903093248D+00
	1	-3.025874518376D-10	2.153903090222D-01

-1.00	3.025874518376D-10
-.75	2.269405888782D-10
-.50	1.512937259188D-10
-.25	7.564686295940D-11
.25	-7.091893402443D-11
.50	-1.134702944391D-10
.75	-9.928650763421D-11
1.25	1.843892284635D-10
1.50	3.404108833173D-10
1.75	3.262270965124D-10
2.25	-6.666379798297D-10
2.50	-1.248173238830D-09
2.75	-1.205621878415D-09
3.25	2.482162690855D-09
3.50	4.652282072003D-09
3.75	4.496260417149D-09
4.25	-9.262012783591D-09
4.50	-1.736095504918D-08
4.75	-1.677941979018D-08
5.25	3.456588844351D-08
5.50	6.479153812472D-08
5.75	6.262141874357D-08
6.25	-1.290015409904D-07
6.50	-2.418051974497D-07
6.75	-2.337062551841D-07
7.25	4.814402755183D-07
7.50	9.024292516741D-07
7.75	8.722036019929D-07
8.25	-1.796759561083D-06

8.50	-3.367911809247D-06
8.75	-3.255108152787D-06
9.25	6.705597968812D-06
9.50	1.256921798531D-05
9.75	1.214822900916D-05

I=19 H= 9.0

0		0.	1.607695154587D+01
1		7.564686295940D-11	-8.038475772555D-01
0	-10.0	-7.564686295940D-11	-3.782343147970D-11
1	-9.0	4.538811777564D-10	2.269405888782D-10
2	-8.0	-1.815524711025D-09	-9.077623555127D-10
3	-7.0	6.808217666346D-09	3.404108833173D-09
4	-6.0	-2.541734595436D-08	-1.270867297718D-08
5	-5.0	9.486116615108D-08	4.743058307554D-08
6	-4.0	-3.540273186500D-07	-1.770136593250D-07
7	-3.0	1.321248108449D-06	6.606240542244D-07
8	-2.0	-4.930965115145D-06	-2.465482557573D-06
9	-1.0	1.840261235213D-05	9.201306176066D-06
10	0.0	-6.867948429338D-05	-3.433974214669D-05
11	1.0	2.563153248214D-04	1.281576624107D-04
12	2.0	-9.565818149922D-04	-4.782909074961D-04
13	3.0	3.570011935147D-03	1.785005967574D-03
14	4.0	-1.332346592560D-02	-6.661732962799D-03
15	5.0	4.972385176724D-02	2.486192588362D-02
16	6.0	-1.855719411434D-01	-9.278597057169D-02
17	7.0	6.925639128063D-01	3.462819564031D-01
18	8.0	-1.584683710082D+00	-7.923418550408D-01
19	9.0	1.646170927520D+00	8.230854637602D-01
20	10.0	-6.076951545867D-01	-3.038475772934D-01
	0	7.564686295940D-10	8.038475773312D+00
	1	7.564686295940D-11	-8.038475772555D-01

-1.00	-7.564686295940D-11
-.75	-5.673514721955D-11
-.50	-3.782343147970D-11
-.25	-1.891171573985D-11
.25	1.772973350611D-11
.50	2.836757360977D-11
.75	2.482162690855D-11
1.25	-4.609730711588D-11
1.50	-8.510272082932D-11
1.75	-8.155677412810D-11
2.25	1.666594949574D-10
2.50	3.120433097075D-10
2.75	3.014054696038D-10
3.25	-6.205406727138D-10
3.50	-1.163070518001D-09
3.75	-1.124065104287D-09
4.25	2.315503195898D-09
4.50	4.340238762295D-09
4.75	4.194854947545D-09
5.25	-8.641472110877D-09
5.50	-1.619788453118D-08
5.75	-1.565535468589D-08
6.25	3.225038524761D-08
6.50	6.045129936243D-08
6.75	5.842656379603D-08
7.25	-1.203600688796D-07
7.50	-2.256073129185D-07
7.75	-2.180509004982D-07
8.25	4.491898902707D-07
8.50	8.419779523117D-07
8.75	8.137770381969D-07
9.25	-1.676399492203D-06
9.50	-3.142304496328D-06
9.75	-3.037057252289D-06

I=20 H= 10.0

0		0.	-1.217949192431D+01
1		-1.260781049323D-11	6.339745962093D-01
0	-10.0	1.260781049323D-11	6.303905246616D-12
1	-9.0	-7.564686295940D-11	-3.782343147970D-11
2	-8.0	3.025874518376D-10	1.512937259188D-10
3	-7.0	-1.134702944391D-09	-5.673514721955D-10
4	-6.0	4.236224325726D-09	2.118112162863D-09
5	-5.0	-1.581019435851D-08	-7.905097179257D-09
6	-4.0	5.900455310833D-08	2.950227655416D-08
7	-3.0	-2.202080180748D-07	-1.101040090374D-07
8	-2.0	8.218275191909D-07	4.109137595954D-07

-1.00	1.260781049323D-11
-.75	9.455857869924D-12
-.50	6.303905246616D-12
-.25	3.151952623308D-12
.25	-2.954955584351D-12
.50	-4.727928934962D-12
.75	-4.136937818092D-12
1.25	7.682884519314D-12
1.50	1.418378680489D-11
1.75	1.359279568802D-11
2.25	-2.777658249290D-11
2.50	-5.200721828458D-11

CARDINAL SPLINES N=2

9	-1.0	-3.067102058689D-06	-1.533551029344D-06
10	0.0	1.144658071556D-05	5.723290357782D-06
11	1.0	-4.271922080357D-05	-2.135961040178D-05
12	2.0	1.594303024987D-04	7.971515124935D-05
13	3.0	-5.950019891912D-04	-2.975009945956D-04
14	4.0	2.220577654266D-03	1.110288827133D-03
15	5.0	-8.287308627874D-03	-4.143654313937D-03
16	6.0	3.092865685723D-02	1.546432842861D-02
17	7.0	-1.154273188010D-01	-5.771365940052D-02
18	8.0	4.307806183469D-01	2.153903091735D-01
19	9.0	-6.076951545867D-01	-3.038475772934D-01
20	10.0	2.679491924311D-01	1.339745962156D-01
	0	-1.260781049323D-10	-5.839745962219D+00
	1	-1.260781049323D-11	6.339745962093D-01

2.75	-5.023424493397D-11
3.25	1.034234454523D-10
3.50	1.938450863335D-10
3.75	1.873441840479D-10
4.25	-3.859171993163D-10
4.50	-7.233731270492D-10
4.75	-6.991424912575D-10
5.25	1.440245351813D-09
5.50	2.699647421863D-09
5.75	2.609225780982D-09
6.25	-5.375064207935D-09
6.50	-1.007521656040D-08
6.75	-9.737760632672D-09
7.25	2.006001147993D-08
7.50	3.760121881975D-08
7.75	3.634181674970D-08
8.25	-7.486498171178D-08
8.50	-1.403296587186D-07
8.75	-1.356295063661D-07
9.25	2.793999153672D-07
9.50	5.237174160547D-07
9.75	5.061762087149D-07

CARDINAL SPLINES N=3

```
                    N= 3      M= 3                      P=  1.5     Q= 5

I= 0     H= -1.5
0            1.000000000000D+00  -1.363636363636D-01  -1.00  3.393939393939D+00
1           -1.689393939394D+00  -3.257575757576D-01   -.75  2.663352272727D+00
2            7.045454545455D-01   2.500000000000D-01   -.50  2.020833333333D+00
                                                       -.25  1.466382575758D+00
0  -1.5     -1.515151515152D-02  -7.575757575758D-03    .25  6.216708096591D-01
1   -.5      4.545454545455D-02   2.272727272727D-02    .50  3.309659090909D-01
2    .5     -4.545454545455D-02  -2.272727272727D-02    .75  1.256658380682D-01
3   1.5      1.515151515152D-02   7.575757575758D-03   1.25 -5.708451704545D-02
                                                       1.50 -6.250000000000D-02
        0    5.113636363636D-02  -6.250000000000D-02   1.75 -3.666548295455D-02
        1    4.242424242424D-01   4.242424242424D-01   2.25  3.058416193182D-02
        2    7.045454545455D-01   2.500000000000D-01   2.50  4.403409090909D-02


I= 1     H=  -.5
0            0.                   3.409090909091D+00  -1.00 -4.181818181818D+00
1            2.568181818182D+00  -1.522727272727D+00   -.75 -2.833806818182D+00
2           -1.613636363636D+00  -2.500000000000D-01   -.50 -1.687500000000D+00
                                                       -.25 -7.428977272727D-01
0  -1.5      4.545454545455D-02   2.272727272727D-02    .25  5.412375710227D-01
1   -.5     -1.363636363636D-01  -6.818181818182D-02    .50  8.821022727273D-01
2    .5      1.363636363636D-01   6.818181818182D-02    .75  1.029252485795D+00
3   1.5     -4.545454545455D-02  -2.272727272727D-02   1.25  8.275035511364D-01
                                                       1.50  5.625000000000D-01
        0    2.215909090909D-01   5.625000000000D-01   1.75  2.662464488636D-01
        1   -2.272727272727D+00  -2.272727272727D+00   2.25 -1.855024857955D-01
        2   -1.613636363636D+00  -2.500000000000D-01   2.50 -2.571022727273D-01


I= 2     H=   .5
0            0.                  -3.409090909091D+00  -1.00  2.181818181818D+00
1           -1.068181818182D+00   3.022727272727D+00   -.75  1.427556818182D+00
2            1.113636363636D+00  -2.500000000000D-01   -.50  8.125000000000D-01
                                                       -.25  3.366477272727D-01
0  -1.5     -4.545454545455D-02  -2.272727272727D-02    .25 -1.974875710227D-01
1   -.5      1.363636363636D-01   6.818181818182D-02    .50 -2.571022727273D-01
2    .5     -1.363636363636D-01  -6.818181818182D-02    .75 -1.855024857955D-01
3   1.5      4.545454545455D-02   2.272727272727D-02   1.25  2.662464488636D-01
                                                       1.50  5.625000000000D-01
        0    9.034090909091D-01   5.625000000000D-01   1.75  8.275035511364D-01
        1    2.272727272727D+00   2.272727272727D+00   2.25  1.029252485795D+00
        2    1.113636363636D+00  -2.500000000000D-01   2.50  8.821022727273D-01


I= 3     H=  1.5
0            0.                   1.136363636364D+00  -1.00 -3.939393939394D-01
```

CARDINAL SPLINES N=3

```
1           1.893939393939D-01  -1.174242424242D+00     -.75 -2.571022727273D-01
2          -2.045454545455D-01   2.500000000000D-01     -.50 -1.458333333333D-01
                                                        -.25 -6.013257575758D-02
0   -1.5    1.515151515152D-02   7.575757575758D-03      .25  3.457919034091D-02
1    -.5   -4.545454545455D-02  -2.272727272727D-02      .50  4.403409090909D-02
2     .5    4.545454545455D-02   2.272727272727D-02      .75  3.058416193182D-02
3    1.5   -1.515151515152D-02  -7.575757575758D-03     1.25 -3.666548295455D-02
                                                        1.50 -6.250000000000D-02
       0   -1.761363636364D-01  -6.250000000000D-02     1.75 -5.708451704545D-02
       1   -4.242424242424D-01  -4.242424242424D-01     2.25  1.256658380682D-01
       2   -2.045454545455D-01   2.500000000000D-01     2.50  3.309659090909D-01
```

```
              N= 3       M= 4                        P=  2.0      Q= 5

I= 0     H= -2.0
0           1.000000000000D+00   1.032608695652D+00    -1.00  3.459239130435D+00
1          -1.720652173913D+00  -1.166304347826D+00     -.75  2.705944293478D+00
2           7.385869565217D-01   4.125000000000D-01     -.50  2.044972826087D+00
                                                        -.25  1.476324728261D+00
0   -2.0   -1.793478260870D-02  -8.967391304348D-03      .25  6.159811268682D-01
1   -1.0    6.086956521739D-02   3.043478260870D-02      .50  3.237601902174D-01
2    0.0   -7.500000000000D-02  -3.750000000000D-02      .75  1.207100246264D-01
3    1.0    3.913043478261D-02   1.956521739130D-02     1.25 -5.144626783288D-02
4    2.0   -7.065217391304D-03  -3.532608695652D-03     1.50 -5.344769021739D-02
                                                        1.75 -2.913924507473D-02
       0    5.130434782609D-01   3.500000000000D-01     2.25  1.910506538723D-02
       1    1.233695652174D+00   4.836956521739D-01     2.50  2.297894021739D-02
       2    7.385869565217D-01   4.125000000000D-01     2.75  1.431247877038D-02

I= 1     H= -1.0
0           0.                  -3.065217391304D+00    -1.00 -4.543478260870D+00
1           2.741304347826D+00   3.132608695652D+00     -.75 -3.069701086957D+00
2          -1.802173913043D+00  -1.150000000000D+00     -.50 -1.821195652174D+00
                                                        -.25 -7.979619565217D-01
0   -2.0    6.086956521739D-02   3.043478260870D-02      .25  5.727496603261D-01
1   -1.0   -2.217391304348D-01  -1.108695652174D-01      .50  9.220108695652D-01
2    0.0    3.000000000000D-01   1.500000000000D-01      .75  1.056700067935D+00
3    1.0   -1.782608695652D-01  -8.913043478261D-02     1.25  7.962763247283D-01
4    2.0    3.913043478261D-02   1.956521739130D-02     1.50  5.123641304348D-01
                                                        1.75  2.245626698370D-01
       0   -1.726086956522D+00  -1.400000000000D+00     2.25 -1.219259510870D-01
       1   -4.467391304348D+00  -1.467391304348D+00     2.50 -1.404891304348D-01
       2   -1.802173913043D+00  -1.150000000000D+00     2.75 -8.524116847826D-02

I= 2     H=  0.0
0           0.                   9.000000000000D+00    -1.00  2.875000000000D+00
1          -1.400000000000D+00  -5.900000000000D+00     -.75  1.879687500000D+00
```

2		1.475000000000D+00	1.475000000000D+00	-.50	1.068750000000D+00
				-.25	4.421875000000D-01
0	-2.0	-7.500000000000D-02	-3.750000000000D-02	.25	-2.578857421875D-01
1	-1.0	3.000000000000D-01	1.500000000000D-01	.50	-3.335937500000D-01
2	0.0	-4.500000000000D-01	-2.250000000000D-01	.75	-2.381103515625D-01
3	1.0	3.000000000000D-01	1.500000000000D-01	1.25	3.260986328125D-01
4	2.0	-7.500000000000D-02	-3.750000000000D-02	1.50	6.585937500000D-01
				1.75	9.073974609375D-01
	0	3.100000000000D+00	3.100000000000D+00	2.25	9.073974609375D-01
	1	4.500000000000D+00	8.991093614707D-28	2.50	6.585937500000D-01
	2	1.475000000000D+00	1.475000000000D+00	2.75	3.260986328125D-01

I= 3 H= 1.0

0		0.	-8.934782608696D+00	-1.00	-9.565217391304D-01
1		4.586956521739D-01	6.067391304348D+00	-.75	-6.240489130435D-01
2		-4.978260869565D-01	-1.150000000000D+00	-.50	-3.538043478261D-01
				-.25	-1.457880434783D-01
0	-2.0	3.913043478261D-02	1.956521739130D-02	.25	8.359799592391D-02
1	-1.0	-1.782608695652D-01	-8.913043478261D-02	.50	1.061141304348D-01
2	0.0	3.000000000000D-01	1.500000000000D-01	.75	7.328040081522D-02
3	1.0	-2.217391304348D-01	-1.108695652174D-01	1.25	-8.524116847826D-02
4	2.0	6.086956521739D-02	3.043478260870D-02	1.50	-1.404891304348D-01
				1.75	-1.219259510870D-01
	0	-1.073913043478D+00	-1.400000000000D+00	2.25	2.245626698370D-01
	1	-1.532608695652D+00	1.467391304348D+00	2.50	5.123641304348D-01
	2	-4.978260869565D-01	-1.150000000000D+00	2.75	7.962763247283D-01

I= 4 H= 2.0

0		0.	2.967391304348D+00	-1.00	1.657608695652D-01
1		-7.934782608696D-02	-2.133695652174D+00	-.75	1.081182065217D-01
2		8.641304347826D-02	4.125000000000D-01	-.50	6.127717391304D-02
				-.25	2.523777173913D-02
0	-2.0	-7.065217391304D-03	-3.532608695652D-03	.25	-1.444304093071D-02
1	-1.0	3.913043478261D-02	1.956521739130D-02	.50	-1.829144021739D-02
2	0.0	-7.500000000000D-02	-3.750000000000D-02	.75	-1.258014181386D-02
3	1.0	6.086956521739D-02	3.043478260870D-02	1.25	1.431247877038D-02
4	2.0	-1.793478260870D-02	-8.967391304348D-03	1.50	2.297894021739D-02
				1.75	1.910506538723D-02
	0	1.869565217391D-01	3.500000000000D-01	2.25	-2.913924507473D-02
	1	2.663043478261D-01	-4.836956521739D-01	2.50	-5.344769021739D-02
	2	8.641304347826D-02	4.125000000000D-01	2.75	-5.144626783288D-02

N= 3 M= 5 P= 2.5 Q= 5

I= 0 H= -2.5

0		1.000000000000D+00	1.220245552493D-01	-1.00	3.471009771987D+00
1		-1.726284139313D+00	-6.949636682536D-01	-.75	2.713621272864D+00

CARDINAL SPLINES N=3

```
2          7.447256326735D-01   3.538461538462D-01     -.50  2,049323477825D+00
                                                        -.25  1,478116386870D+00
0  -2,5   -1.844149336006D-02  -9,220746680030D-03      .25  6,149563079429D-01
1  -1,5    6,379353545477D-02   3.189676772739D-02      .50  3,224630418441D-01
2   -,5   -8.378852417940D-02  -4.189426208970D-02      .75  1,198188110749D-01
3    ,5    5,301929341017D-02   2.650964670509D-02     1.25 -5,043804615071D-02
4   1,5   -1,763968930093D-02  -8.819844650464D-03     1.50 -5,184007767477D-02
5   2,5    3.056877975445D-03   1.528438987722D-03     1.75 -2,781830407479D-02
                                                       2,25  1,720536245615D-02
       0   1,338824855926D+00   5.961538461538D-01     2.50  1,971153846154D-02
       1   1,997344024054D+00   1.074267100977D+00     2.75  1,146050292846D-02
       2   7,447256326735D-01   3.538461538462D-01     3,25 -7,968955540591D-03


I= 1      H= -1.5
0          0.                   2,189300927086D+00    -1.00 -4,611400651466D+00
1          2,773803558006D+00   4.127411676272D-01     -.75 -3,114001033576D+00
2         -1,837597093460D+00  -8.115384615385D-01     -.50 -1,846301052368D+00
                                                        -.25 -8,083007078426D-01
0  -2.5    6.379353545477D-02   3.189676772739D-02      .25  5,786633695346D-01
1  -1,5   -2,386118767226D-01  -1.193059383613D-01      ,50  9,294960536206D-01
2   -,5    3,507141067402D-01   1.753570533701D-01      ,75  1,061842808428D+00
3    ,5   -2,584064144325D-01  -1.292032072162D-01     1.25  7,904583897754D-01
4   1,5    1.001503382611D-01   5.007516913054D-02     1.50  5,030874154347D-01
5   2,5   -1.763968930093D-02  -8,819844650464D-03     1.75  2,169401906242D-01
                                                       2,25 -1,109637308945D-01
       0  -4,550472939113D+00  -1.850961538462D+00     2.50 -1,216346153846D-01
       1  -6,414181909296D+00  -3,644951140065D+00     2,75 -6,878386525933D-02
       2  -1,837597093460D+00  -8,115384615385D-01     3,25  4,667158822194D-02


I= 2      H=  -.5
0          0,                  -6,793410172889D+00    -1.00  3,079153094463D+00
1         -1.497682285142D+00   2,275056376848D+00     -.75  2,012839044099D+00
2          1.581470809321D+00   4,576923076923D-01     -.50  1,144208844901D+00
                                                        -.25  4,732624968680D-01
0  -2,5   -8,378852417940D-02  -4,189426208970D-02      .25 -2,756604704335D-01
1  -1,5    3,507141067402D-01   1.753570533701D-01      .50 -3,560918316211D-01
2   -,5   -6.024304685542D-01  -3,012152342771D-01      .75 -2,535677931596D-01
3    ,5    5,408920070158D-01   2,704460035079D-01     1,25  3,435854941196D-01
4   1,5   -2,584064144325D-01  -1,292032072162D-01     1,50  6,864766036081D-01
5   2,5    5,301929341017D-02   2,650964670509D-02     1.75  9,303082081167D-01
                                                       2,25  8,744485150182D-01
       0   6,139986845402D+00   1.754807692308D+00     2,50  6,019230769231D-01
       1   6,409671761463D+00   4,563517915309D+00     2,75  2,766332157511D-01
       2   1.581470809321D+00   4,576923076923D-01     3,25 -1,581327273475D-01


I= 3      H=   .5
0          0.                   1,602417940366D+01    -1.00 -1.279153094463D+00
1          6.130669005262D-01  -6,851979453771D+00     -,75 -8,344736594838D-01
2         -6,660861939364D-01   4,576923076923D-01     -,50 -4,730549987472D-01
                                                        -,25 -1,948971122526D-01
0  -2,5    5,301929341017D-02   2,650964670509D-02      .25  1,116881146642D-01
1  -1,5   -2,584064144325D-01  -1,292032072162D-01      ,50  1,416687546981D-01
2   -,5    5,408920070158D-01   2,704460035079D-01      ,75  9,770841815961D-02
3    ,5   -6,024304685542D-01  -3,012152342771D-01     1,25 -1,128763595042D-01
```

```
4   1,5   3,507141067402D-01   1,753570533701D-01   1.50 -1,845535266850D-01
5   2,5  -8,378852417940D-02  -4,189426208970D-02   1.75 -1,581327273475D-01
                                                    2,25  2,766332157511D-01
      0  -2,630371460787D+00   1,754807692308D+00   2,50  6,019230769231D-01
      1  -2,717364069156D+00  -4,563517915309D+00   2,75  8,744485150182D-01
      2  -6,660861939364D-01   4,576923076923D-01   3,25  9,303082081167D-01


I= 4     H=  1.5
0            0.               -1,603545477324D+01  -1.00  4,114006514658D-01
1            -1,968804810824D-01   7,702643447757D+00   -,75  2,683279566525D-01
2            2,145201703834D-01  -8,115384615385D-01    -,50  1,520702831371D-01
                                                     -,25  6,262763091957D-02
0  -2,5  -1,763968930093D-02  -8,819844650464D-03    .25 -3,582983588073D-02
1  -1,5   1,001503382611D-01   5,007516913054D-02    .50 -4,536143823603D-02
2   -,5  -2,584064144325D-01  -1,292032072162D-01    .75 -3,117874592834D-02
3    ,5   3,507141067402D-01   1,753570533701D-01   1,25  3,535290830149D-02
4   1,5  -2,386118767226D-01  -1,193059383613D-01   1,50  5,652796918066D-02
5   2,5   6,379353545477D-02   3,189676772739D-02   1,75  4,667158822194D-02
                                                    2.25 -6,878386525933D-02
      0   8,485498621899D-01  -1,850961538462D+00   2,50 -1,216346153846D-01
      1   8,757203708344D-01   3,644951140065D+00   2,75 -1,109637308945D-01
      2   2,145201703834D-01  -8,115384615385D-01   3,25  2,169401906242D-01


I= 5     H=  2.5
0            0.                5,493360060135D+00  -1,00 -7,100977198697D-02
1            3,397644700576D-02  -2,843497870208D+00   -,75 -4,631358055625D-02
2           -3,703332498121D-02   3,538461538462D-01   -,50 -2,624655474818D-02
                                                     -,25 -1,080869456277D-02
0  -2,5   3,056877975445D-03   1,528438987722D-03    .25  6,182514172513D-03
1  -1,5  -1,763968930093D-02  -8,819844650464D-03    .50  7,825419694312D-03
2   -,5   5,301929341017D-02   2,650964670509D-02    .75  5,376501425081D-03
3    ,5  -8,378852417940D-02  -4,189426208970D-02   1,25 -6,082386541594D-03
4   1,5   6,379353545477D-02   3,189676772739D-02   1,50 -9,698383863693D-03
5   2,5  -1,844149336006D-02  -9,220746680030D-03   1,75 -7,968955540591D-03
                                                    2.25  1,146050292846D-02
      0  -1,465171636181D-01   5,961538461538D-01   2,50  1,971153846154D-02
      1  -1,511901779003D-01  -1,074267100977D+00   2,75  1,720536245615D-02
      2  -3,703332498121D-02   3,538461538462D-01   3.25 -2,781830407479D-02




                   N= 3       M= 6                      P=  3.0      Q= 5

I= 0     H= -3.0
0            1.000000000000D+00   7.428954375034D-01  -1.00  3.473178654012D+00
1           -1.727321850114D+00  -9,519328877699D-01   -,75  2,715035839778D+00
2            7,458568038978D-01   3,808933002481D-01   -,50  2,050125126032D+00
                                                     -,25  1.478446512772D+00
0  -3.0  -1,853495378336D-02  -9,267476891682D-03    .25  6,147674871742D-01
```

CARDINAL SPLINES N=3

```
1  -2.0   6.433500570538D-02   3.216750285269D-02      .50  3.222240586115D-01
2  -1.0  -8.547209795150D-02  -4.273604897575D-02      .75  1.196546331913D-01
3   0.0   5.707196029777D-02   2.853598014888D-02     1.25 -5.025241941144D-02
4   1.0  -2.370904348771D-02  -1.185452174386D-02     1.50 -5.154430276575D-02
5   2.0   7.625292061365D-03   3.812646030682D-03     1.75 -2.757556663301D-02
6   3.0  -1.316162841946D-03  -6.580814209729D-04     2.25  1.685846046870D-02
                                                      2.50  1.911926518266D-02
      0   2.530745684737D+00   1.315136476427D+00     2.75  1.094984256421D-02
      1   2.747818973272D+00   1.333426913719D+00     3.25 -7.191409716814D-03
      2   7.458568038978D-01   3.808933002481D-01     3.50 -8.340729202514D-03


I= 1     H= -2.0
0          0.                 -1.407762945790D+00    -1.00 -4.623966238612D+00
1         2.779815616453D+00   1.901512379779D+00     -.75 -3.122196437304D+00
2        -1.844150622159D+00  -9.682382133995D-01     -.50 -1.850945463766D+00
                                                      -.25 -8.102133179982D-01
0  -3.0   6.433500570538D-02   3.216750285269D-02      .25  5.797573173824D-01
1  -2.0  -2.417489268443D-01  -1.208744634221D-01      .50  9.308806216153D-01
2  -1.0   3.604680226766D-01   1.802340113383D-01      .75  1.062793985800D+00
3   0.0  -2.818858560794D-01  -1.409429280397D-01     1.25  7.893829467860D-01
4   1.0   1.353136150405D-01   6.765680752024D-02     1.50  5.013738204343D-01
5   2.0  -4.410715256018D-02  -2.205357628009D-02     1.75  2.155338723262D-01
6   3.0   7.625292061365D-03   3.812646030682D-03     2.25 -1.089539271782D-01
                                                      2.50 -1.182032339570D-01
      0  -8.257908750068D+00  -4.417369727047D+00     2.75 -6.582531461688D-02
      1  -8.285088116499D+00  -3.907916900618D+00     3.25  4.216681585757D-02
      2  -1.844150622159D+00  -9.682382133995D-01     3.50  4.866229102896D-02


I= 2     H= -1.0
0          0.                  4.390809802394D+00    -1.00  3.118222817917D+00
1        -1.516375359983D+00  -2.353925848110D+00     -.75  2.038320715075D+00
2         1.601847457934D+00   9.449131513648D-01     -.50  1.158649544475D+00
                                                      -.25  4.792093061165D-01
0  -3.0  -8.547209795150D-02  -4.273604897575D-02      .25 -2.790618427204D-01
1  -2.0   3.604680226766D-01   1.802340113383D-01      .50 -3.603968185688D-01
2  -1.0  -6.327579648983D-01  -3.163789824491D-01      .75 -2.565252543934D-01
3   0.0   6.138957816377D-01   3.069478908189D-01     1.25  3.469293299229D-01
4   1.0  -3.677383130173D-01  -1.838691565087D-01     1.50  6.918046222673D-01
5   2.0   1.353136150405D-01   6.765680752024D-02     1.75  9.346808224923D-01
6   3.0  -2.370904348771D-02  -1.185452174386D-02     2.25  8.681995053635D-01
                                                      2.50  5.912540073536D-01
      0   9.867501041459D+00   5.833250620347D+00     2.75  2.674343017583D-01
      1   8.094709387622D+00   3.315553060079D+00     3.25 -1.441262023476D-01
      2   1.601847457934D+00   9.449131513648D-01     3.50 -1.600970594628D-01


I= 3     H=  0.0
0          0.                 -1.089826302730D+01    -1.00 -1.373200992556D+00
1         6.580645161290D-01   4.290818858561D+00     -.75 -8.958126550868D-01
2        -7.151364764268D-01  -7.151364764268D-01     -.50 -5.078163771712D-01
                                                      -.25 -2.092121588089D-01
0  -3.0   5.707196029777D-02   2.853598014888D-02      .25  1.198758335918D-01
1  -2.0  -2.818858560794D-01  -1.409429280397D-01      .50  1.520316377171D-01
2  -1.0   6.138957816377D-01   3.069478908189D-01      .75  1.048275628102D-01
3   0.0  -7.781637717122D-01  -3.890818858561D-01     1.25 -1.209255776985D-01
```

CARDINAL SPLINES N=3

```
4   1.0   6.138957816377D-01   3.069478908189D-01    1.50 -1.973790322581D-01
5   2.0  -2.818858560794D-01  -1.409429280397D-01    1.75 -1.686584018300D-01
6   3.0   5.707196029777D-02   2.853598014888D-02    2.25  2.916757133995D-01
                                                     2.50  6.276054590571D-01
      0  -4.462034739454D+00  -4.462034739454D+00    2.75  8.965919665012D-01
      1  -3.632754342432D+00   2.921055566306D-26    3.25  8.965919665012D-01
      2  -7.151364764268D-01  -7.151364764268D-01    3.50  6.276054590571D-01


I= 4    H=  1.0
0          0.                  2.428412816287D+01   -1.00  5.522486461031D-01
1         -2.642698013077D-01 -8.985031968267D+00    -.75  3.601904511782D-01
2          2.879788447954D-01  9.449131513648D-01    -.50  2.041296118527D-01
                                                     -.25  8.406612812664D-02
0  -3.0  -2.370904348771D-02  -1.185452174386D-02     .25 -4.809192588999D-02
1  -2.0   1.353136150405D-01   6.765680752024D-02     .50 -6.088109706399D-02
2  -1.0  -3.677383130173D-01  -1.838691565087D-01     .75 -4.184051793913D-02
3   0.0   6.138957816377D-01   3.069478908189D-01    1.25  4.740757686987D-02
4   1.0  -6.327579648983D-01  -3.163789824491D-01    1.50  7.573570031334D-02
5   2.0   3.604680226766D-01   1.802340113383D-01    1.75  6.243504676944D-02
6   3.0  -8.547209795150D-02  -4.273604897575D-02    2.25 -9.131180762778D-02
                                                     2.50 -1.600970594628D-01
      0   1.799000199236D+00   5.833250620347D+00    2.75 -1.441262023476D-01
      1   1.463603267465D+00  -3.315553060079D+00    3.25  2.674343017583D-01
      2   2.879788447954D-01   9.449131513648D-01    3.50  5.912540073536D-01


I= 5    H=  2.0
0          0.                 -2.485526434950D+01   -1.00 -1.770263172194D-01
1          8.470051257902D-02  9.717346181015D+00    -.75 -1.154586495445D-01
2         -9.232580464038D-02 -9.682382133995D-01    -.50 -6.543170744960D-02
                                                     -.25 -2.694549093478D-02
0  -3.0   7.625292061365D-03   3.812646030682D-03     .25  1.541221192901D-02
1  -2.0  -4.410715256018D-02  -2.205357628009D-02     .50  1.950709550633D-02
2  -1.0   1.353136150405D-01   6.765680752024D-02     .75  1.340163687377D-02
3   0.0  -2.818858560794D-01  -1.409429280397D-01    1.25 -1.515595779713D-02
4   1.0   3.604680226766D-01   1.802340113383D-01    1.50 -2.415607849885D-02
5   2.0  -2.417489268443D-01  -1.208744634221D-01    1.75 -1.983413985098D-02
6   3.0   6.433500570538D-02   3.216750285269D-02    2.25  2.841732668196D-02
                                                     2.50  4.866229102896D-02
      0  -5.768307040264D-01  -4.417369727047D+00    2.75  4.216681585757D-02
      1  -4.692543152633D-01   3.907916900618D+00    3.25 -6.582531461688D-02
      2  -9.232580464038D-02  -9.682382133995D-01    3.50 -1.182032339570D-01


I= 6    H=  3.0
0          0.                  8.743456919817D+00   -1.00  3.054343035506D-02
1         -1.461363375656D-02 -3.618786715208D+00    -.75  1.992073590408D-02
2          1.592979659850D-02  3.808933002481D-01    -.50  1.128926602791D-02
                                                     -.25  4.649020726546D-03
0  -3.0  -1.316162841946D-03  -6.580814209729D-04     .25 -2.659081467008D-03
1  -2.0   7.625292061365D-03   3.812646030682D-03     .50 -3.365497817464D-03
2  -1.0  -2.370904348771D-02  -1.185452174386D-02     .75 -2.312046342667D-03
3   0.0   5.707196029777D-02   2.853598014888D-02    1.25  2.614101328313D-03
4   1.0  -8.547209795150D-02  -4.273604897575D-02    1.50  4.165270507689D-03
5   2.0   6.433500570538D-02   3.216750285269D-02    1.75  3.418366726060D-03
6   3.0  -1.853495378336D-02  -9.267476891682D-03    2.25 -4.885271107660D-03
```

CARDINAL SPLINES N=3

```
                                                    2.50 -8.340729202514D-03
 0    9.952726811686D-02    1.315136476427D+00      2.75 -7.191409716814D-03
 1    8.096514583447D-02   -1.333426913719D+00      3.25  1.094984256421D-02
 2    1.592979659850D-02    3.808933002481D-01      3.50  1.911926518266D-02



              N= 3      M= 7                        P=  3.5      Q= 5

I= 0      H= -3.5
0           1.000000000000D+00    3.540065751526D-01   -1.00  3.473580257445D+00
1          -1.727513998118D+00   -8.188991036921D-01    -.75  2.715297769460D+00
2           7.460662593270D-01    3.696043165468D-01    -.50  2.050273563891D+00
                                                        -.25  1.478507640737D+00
0   -3.5   -1.855226120896D-02   -9.276130604478D-03     .25  6.147325242358D-01
1   -2.5    6.443531761460D-02    3.221765880730D-02     .50  3.221798076099D-01
2   -1.5   -8.578502948012D-02   -4.289251474006D-02     .75  1.196242337343D-01
3    -.5    5.785008422572D-02    2.892504211286D-02    1.25 -5.021805041822D-02
4     .5   -2.547598350629D-02   -1.273799175315D-02    1.50 -5.148954357134D-02
5    1.5    1.024546113480D-02    5.122730567399D-03    1.75 -2.753063216464D-02
6    2.5   -3.284238477914D-03   -1.642119238957D-03    2.25  1.679428425373D-02
7    3.5    5.666496981647D-04    2.833248490823D-04    2.50  1.900977777632D-02
                                                        2.75  1.085556108347D-02
       0    4.093012683343D+00    2.015512589928D+00    3.25 -7.048772586938D-03
       1    3.494949817171D+00    1.768331112135D+00    3.50 -8.093525179856D-03
       2    7.460662593270D-01    3.696043165468D-01    3.75 -4.669977413062D-03


I= 1      H= -2.5
0           0.                    8.461935453214D-01   -1.00 -4.626293887337D+00
1           2.780929284861D+00    1.130463398467D+00    -.75 -3.123714552538D+00
2          -1.845364602476D+00   -9.028085224128D-01    -.50 -1.851805793049D+00
                                                        -.25 -8.105676088700D-01
0   -3.5    6.443531761460D-02    3.221765880730D-02     .25  5.799599586754D-01
1   -2.5   -2.423303234869D-01   -1.211651617435D-01     .50  9.311370954871D-01
2   -1.5    3.622817389295D-01    1.811408694648D-01     .75  1.062970177664D+00
3    -.5   -2.863957756171D-01   -1.431978878085D-01    1.25  7.891837479329D-01
4     .5    1.455546024018D-01    7.277730120090D-02    1.50  5.010564422483D-01
5    1.5   -5.929336040160D-02   -2.964668020080D-02    1.75  2.152734371570D-01
6    2.5    1.903203903754D-02    9.516019518770D-03    2.25 -1.085819689909D-01
7    3.5   -3.284238477914D-03   -1.642119238957D-03    2.50 -1.175686571560D-01
                                                        2.75 -6.527886966544D-02
       0   -1.287246388331D+01   -6.256588959602D+00    3.25  4.134010695464D-02
       1   -1.013662293247D+01   -5.189196258423D+00    3.50  4.722952407305D-02
       2   -1.845364602476D+00   -9.028085224128D-01    3.75  2.716950842997D-02


I= 2      H= -1.5
0           0.                   -2.640599087216D+00   -1.00  3.125484115994D+00
1          -1.519849543257D+00    5.142699827650D-02    -.75  2.043056604607D+00
2           1.605634572737D+00    7.407996679579D-01    -.50  1.161333414813D+00
```

CARDINAL SPLINES N=3

				-.25	4.803145466103D-01
0	-3.5	-8.578502948012D-02	-4.289251474006D-02	.25	-2.796939994610D-01
1	-2.5	3.622817389295D-01	1.811408694648D-01	.50	-3.611969106155D-01
2	-1.5	-6.384160069387D-01	-3.192080034693D-01	.75	-2.570748998911D-01
3	-.5	6.279648590372D-01	3.139824295186D-01	1.25	3.475507476796D-01
4	.5	-3.996859437079D-01	-1.998429718540D-01	1.50	6.927947104998D-01
5	1.5	1.826882814268D-01	9.134414071340D-02	1.75	9.354932721430D-01
6	2.5	-5.929336040160D-02	-2.964668020080D-02	2.25	8.670391501769D-01
7	3.5	1.024546113480D-02	5.122730567399D-03	2.50	5.892743910755D-01
				2.75	2.657296202773D-01
	0	1.434955011463D+01	6.614191339236D+00	3.25	-1.415472136439D-01
	1	9.719592465902D+00	5.237024673982D+00	3.50	-1.556274211400D-01
	2	1.605634572737D+00	7.407996679579D-01	3.75	-8.741913250996D-02
I= 3	H= -.5				
0		0.	6.585777421536D+00	-1.00	-1.391256666929D+00
1		6.667032913515D-01	-1.690242280135D+00	-.75	-9.075887422758D-01
2		-7.245533755772D-01	-2.075954620919D-01	-.50	-5.144899895701D-01
				-.25	-2.119604088115D-01
0	-3.5	5.785008422572D-02	2.892504211286D-02	.25	1.214477310872D-01
1	-2.5	-2.863957756171D-01	-1.431978878085D-01	.50	1.540211169135D-01
2	-1.5	6.279648590372D-01	3.139824295186D-01	.75	1.061942909105D-01
3	-.5	-8.131474175493D-01	-4.065737087747D-01	1.25	-1.224707720350D-01
4	.5	6.933355747159D-01	3.466677873580D-01	1.50	-1.998409489205D-01
5	1.5	-3.996859437079D-01	-1.998429718540D-01	1.75	-1.706786089859D-01
6	2.5	1.455546024018D-01	7.277730120090D-02	2.25	2.945610095524D-01
7	3.5	-2.547598350629D-02	-1.273799175315D-02	2.50	6.325278994417D-01
				2.75	9.008307642206D-01
	0	-6.542317331091D+00	-1.873114969563D+00	3.25	8.901791488824D-01
	1	-4.405170337689D+00	-3.143410514778D+00	3.50	6.164914222468D-01
	2	-7.245533755772D-01	-2.075954620919D-01	3.75	2.819963318868D-01
I= 4	H= .5				
0		0.	-1.541809618191D+01	-1.00	5.932489192807D-01
1		-2.838864678872D-01	4.596578749421D+00	-.75	3.869312298243D-01
2		3.093624513935D-01	-2.075954620919D-01	-.50	2.192838467920D-01
				-.25	9.030677018390D-02
0	-3.5	-2.547598350629D-02	-1.273799175315D-02	.25	-5.166134264985D-02
1	-2.5	1.455546024018D-01	7.277730120090D-02	.50	-6.539874557980D-02
2	-1.5	-3.996859437079D-01	-1.998429718540D-01	.75	-4.494404231128D-02
3	-.5	6.933355747159D-01	3.466677873580D-01	1.25	5.091635691329D-02
4	.5	-8.131474175493D-01	-4.065737087747D-01	1.50	8.132614537871D-02
5	1.5	6.279648590372D-01	3.139824295186D-01	1.75	6.702247139601D-02
6	2.5	-2.863957756171D-01	-1.431978878085D-01	2.25	-9.786364989203D-02
7	3.5	5.785008422572D-02	2.892504211286D-02	2.50	-1.712747865474D-01
				2.75	-1.537515345755D-01
	0	2.796087391965D+00	-1.873114969563D+00	3.25	2.819963318868D-01
	1	1.881650691867D+00	3.143410514778D+00	3.50	6.164914222468D-01
	2	3.093624513935D-01	-2.075954620919D-01	3.75	8.901791488824D-01
I= 5	H= 1.5				
0		0.	3.401857363066D+01	-1.00	-2.378250125075D-01
1		1.137897756864D-01	-1.042262234969D+01	-.75	-1.551121524767D-01
2		-1.240352368212D-01	7.407996679579D-01	-.50	-8.790369704847D-02

				-.25	-3.619964622291D-02
0	-3.5	1.024546113480D-02	5.122730567399D-03	.25	2.070524695341D-02
1	-2.5	-5.929336040160D-02	-2.964668020080D-02	.50	2.620624929835D-02
2	-1.5	1.826882814268D-01	9.134414071340D-02	.75	1.800380700576D-02
3	-.5	-3.996859437079D-01	-1.998429718540D-01	1.25	-2.035907553659D-02
4	.5	6.279648590372D-01	3.139824295186D-01	1.50	-3.244606633825D-02
5	1.5	-6.384160069387D-01	-3.192080034693D-01	1.75	-2.663676359985D-02
6	2.5	3.622817389295D-01	1.811408694648D-01	2.25	3.813295612848D-02
7	3.5	-8.578502948012D-02	-4.289251474006D-02	2.50	6.523757627365D-02
				2.75	5.644007785840D-02
	0	-1.121167436157D+00	6.614191339236D+00	3.25	-8.741913250996D-02
	1	-7.544568820618D-01	-5.237024673982D+00	3.50	-1.556274211400D-01
	2	-1.240352368212D-01	7.407996679579D-01	3.75	-1.415472136439D-01

I= 6 H= 2.5

0		0.	-3.547818026364D+01	-1.00	7.621087682196D-02
1		-3.646331917202D-02	1.150885591531D+01	-.75	4.970549055711D-02
2		3.974755764994D-02	-9.028085224128D-01	-.50	2.816854899850D-02
				-.25	1.160005214613D-02
0	-3.5	-3.284238477914D-03	-1.642119238957D-03	.25	-6.634814704023D-03
1	-2.5	1.903203903754D-02	9.516019518770D-03	.50	-8.397402625962D-03
2	-1.5	-5.929336040160D-02	-2.964668020080D-02	.75	-5.768853386604D-03
3	-.5	1.455546024018D-01	7.277730120090D-02	1.25	6.522295405533D-03
4	.5	-2.863957756171D-01	-1.431978878085D-01	1.50	1.039209123259D-02
5	1.5	3.622817389295D-01	1.811408694648D-01	1.75	8.527990450939D-03
6	2.5	-2.423303234869D-01	-1.211651617435D-01	2.25	-1.218292750536D-02
7	3.5	6.443531761460D-02	3.221765880730D-02	2.50	-2.079084754795D-02
				2.75	-1.791241985102D-02
	0	3.592859641096D-01	-6.256588959602D+00	3.25	2.716950842997D-02
	1	2.417695843775D-01	5.189196258423D+00	3.50	4.722952407305D-02
	2	3.974755764994D-02	-9.028085224128D-01	3.75	4.134010695464D-02

I= 7 H= 3.5

0		0.	1.273232436010D+01	-1.00	-1.314860276886D-02
1		6.290976535349D-03	-4.355561327963D+00	-.75	-8.575647157863D-03
2		-6.857626233513D-03	3.696043165468D-01	-.50	-4.859894826053D-03
				-.25	-2.001345773432D-03
0	-3.5	5.666496981647D-04	2.833248490823D-04	.25	1.144695863088D-03
1	-2.5	-3.284238477914D-03	-1.642119238957D-03	.50	1.448789512364D-03
2	-1.5	1.024546113480D-02	5.122730567399D-03	.75	9.952862747052D-04
3	-.5	-2.547598350629D-02	-1.273799175315D-02	1.25	-1.125249941492D-03
4	.5	5.785008422572D-02	2.892504211286D-02	1.50	-1.792830529379D-03
5	1.5	-8.578502948012D-02	-4.289251474006D-02	1.75	-1.471166396516D-03
6	2.5	6.443531761460D-02	3.221765880730D-02	2.25	2.101146276846D-03
7	3.5	-1.855226120896D-02	-9.276130604478D-03	2.50	3.584646684112D-03
				2.75	3.086800652142D-03
	0	-6.198750348682D-02	2.015512589928D+00	3.25	-4.669977413062D-03
	1	-4.171240709924D-02	-1.768331112135D+00	3.50	-8.093525179856D-03
	2	-6.857626233513D-03	3.696043165468D-01	3.75	-7.048772586938D-03

CARDINAL SPLINES N=3

N= 3 M= 8 P= 4.0 Q= 5

I= 0 H= -4.0

0		1.000000000000D+00	5.836490811765D-01	-1.00	3.473654694943D+00
1		-1.727549612879D+00	-8.860413940439D-01	-.75	2.715346318320D+00
2		7.461050820640D-01	3.745288475892D-01	-.50	2.050301076955D+00
				-.25	1.478518970849D+00
0	-4.0	-1.855546918525D-02	-9.277734592626D-03	.25	6.147260438339D-01
1	-3.0	6.445391142952D-02	3.222695571476D-02	.50	3.221716056646D-01
2	-2.0	-8.584305366839D-02	-4.292152683420D-02	.75	1.196185991855D-01
3	-1.0	5.799482420332D-02	2.899741210166D-02	1.25	-5.021168015125D-02
4	0.0	-2.581542021197D-02	-1.290771010599D-02	1.50	-5.147939406745D-02
5	1.0	1.100801204090D-02	5.504006020449D-03	1.75	-2.752230375503D-02
6	2.0	-4.412954691065D-03	-2.206477345533D-03	2.25	1.678239022543D-02
7	3.0	1.414128576451D-03	7.070642882254D-04	2.50	1.898948757175D-02
8	4.0	-2.439784935067D-04	-1.219892467534D-04	2.75	1.083809105247D-02
				3.25	-7.022359515110D-03
	0	6.027482861509D+00	3.031945066428D+00	3.50	-8.047783431653D-03
	1	4.241291043633D+00	2.110189386670D+00	3.75	-4.630190791490D-03
	2	7.461050820640D-01	3.745288475892D-01	4.25	3.025920112681D-03

I= 1 H= -3.0

0		0.	-4.848419302593D-01	-1.00	-4.626725335990D+00
1		2.781135712280D+00	1.519628156845D+00	-.75	-3.123995947547D+00
2		-1.845589623710D+00	-9.313516942827D-01	-.50	-1.851965262068D+00
				-.25	-8.106332795520D-01
0	-4.0	6.445391142952D-02	3.222695571476D-02	.25	5.799975198611D-01
1	-3.0	-2.424380954657D-01	-1.212190477328D-01	.50	9.311846349449D-01
2	-2.0	3.626180540759D-01	1.813090270380D-01	.75	1.063002836183D+00
3	-1.0	-2.872347058668D-01	-1.436173529334D-01	1.25	7.891468251024D-01
4	0.0	1.475220182117D-01	7.376100910584D-02	1.50	5.009976145079D-01
5	1.0	-6.371319680527D-02	-3.185659840264D-02	1.75	2.152251646972D-01
6	2.0	2.557421342668D-02	1.278710671334D-02	2.25	-1.085130297780D-01
7	3.0	-8.196327582562D-03	-4.098163791281D-03	2.50	-1.174510526996D-01
8	4.0	1.414128576451D-03	7.070642882254D-04	2.75	-6.517761127288D-02
				3.25	4.118701362610D-02
	0	-1.840489113024D+01	-9.307956411405D+00	3.50	4.696439942070D-02
	1	-1.198358127740D+01	-5.931185397417D+00	3.75	2.693890040499D-02
	2	-1.845589623710D+00	-9.313516942827D-01	4.25	-1.759548950095D-02

I= 2 H= -2.0

0		0.	1.513053801038D+00	-1.00	3.126830502307D+00
1		-1.520493724320D+00	-1.163007441900D+00	-.75	2.043934730858D+00
2		1.606336777988D+00	8.298719958203D-01	-.50	1.161831056657D+00
				-.25	4.805194797041D-01
0	-4.0	-8.584305366839D-02	-4.292152683420D-02	.25	-2.798112135627D-01
1	-3.0	3.626180540759D-01	1.813090270380D-01	.50	-3.613452630899D-01
2	-2.0	-6.394655179433D-01	-3.197327589716D-01	.75	-2.571768146463D-01

CARDINAL SPLINES N=3

```
3  -1.0   6.305828396385D-01   3.152914198193D-01     1.25  3.476659697188D-01
4   0.0  -4.058254963427D-01  -2.029127481714D-01     1.50  6.929782893890D-01
5   1.0   1.964809012183D-01   9.824045060916D-02     1.75  9.356439120364D-01
6   2.0  -7.970898571397D-02  -3.985449285699D-02     2.25  8.668240172477D-01
7   3.0   2.557421342668D-02   1.278710671334D-02     2.50  5.889073925252D-01
8   4.0  -4.412954691065D-03  -2.206477345533D-03     2.75  2.654136315434D-01
                                                      3.25 -1.410694678903D-01
      0   1.961941355053D+01   1.013897596656D+01     3.50 -1.548000684524D-01
      1   1.133020049958D+01   5.475968524662D+00     3.75 -8.669949302131D-02
      2   1.606336777988D+00   8.298719958203D-01     4.25  5.533890267550D-02


I= 3     H= -1.0
0         0.                  -3.775412350067D+00    -1.00 -1.394615196029D+00
1         6.683101859131D-01   1.339135965558D+00     -.75 -9.097792076253D-01
2        -7.263050101164D-01  -4.297842961636D-01     -.50 -5.157313454856D-01
                                                      -.25 -2.124716096105D-01
0  -4.0   5.799482420332D-02   2.899741210166D-02      .25  1.217401189165D-01
1  -3.0  -2.872347058668D-01  -1.436173529334D-01      .50  1.543911786838D-01
2  -2.0   6.305828396385D-01   3.152914198193D-01      .75  1.064485148785D-01
3  -1.0  -8.196779087960D-01  -4.098389543980D-01     1.25 -1.227581907109D-01
4   0.0   7.086505448574D-01   3.543252724287D-01     1.50 -2.002988821567D-01
5   1.0  -4.340913104904D-01  -2.170456552452D-01     1.75 -1.710543766692D-01
6   2.0   1.964809012183D-01   9.824045060916D-02     2.25  2.950976536016D-01
7   3.0  -6.371319680527D-02  -3.185659840264D-02     2.50  6.334433687235D-01
8   4.0   1.100801204090D-02   5.504006020449D-03     2.75  9.016189907108D-01
                                                      3.25  8.889874232981D-01
      0  -8.947639418210D+00  -5.295417226452D+00     3.50  6.144276103601D-01
      1  -5.142129895018D+00  -2.099138403751D+00     3.75  2.802012080501D-01
      2  -7.263050101164D-01  -4.297842961636D-01     4.25 -1.510046788923D-01


I= 4     H=  0.0
0         0.                   8.880429914913D+00    -1.00  6.011251679355D-01
1        -2.876548738618D-01  -2.507762352590D+00     -.75  3.920681958128D-01
2         3.134702940737D-01   3.134702940737D-01     -.50  2.221950104493D-01
                                                      -.25  9.150561184505D-02
0  -4.0  -2.581542021197D-02  -1.290771010599D-02      .25 -5.234703545713D-02
1  -3.0   1.475220182117D-01   7.376100910584D-02      .50 -6.626659529407D-02
2  -2.0  -4.058254963427D-01  -2.029127481714D-01      .75 -4.554023520593D-02
3  -1.0   7.086505448574D-01   3.543252724287D-01     1.25  5.159039631906D-02
4   0.0  -8.490632930288D-01  -4.245316465144D-01     1.50  8.240006670772D-02
5   1.0   7.086505448574D-01   3.543252724287D-01     1.75  6.790370232206D-02
6   2.0  -4.058254963427D-01  -2.029127481714D-01     2.25 -9.912215974691D-02
7   3.0   1.475220182117D-01   7.376100910584D-02     2.50 -1.734216977347D-01
8   4.0  -2.581542021197D-02  -1.290771010599D-02     2.75 -1.556000425524D-01
                                                      3.25  2.847911050594D-01
      0   3.864905209733D+00   3.864905209733D+00     3.50  6.213313670884D-01
      1   2.220107478728D+00  -1.853832864698D-25     3.75  8.943889804128D-01
      2   3.134702940737D-01   3.134702940737D-01     4.25  8.943889804128D-01


I= 5     H=  1.0
0         0.                  -2.056851958007D+01    -1.00 -2.555191523807D-01
1         1.222555701699D-01   5.537412773060D+00     -.75 -1.666524426210D-01
2        -1.332635822108D-01  -4.297842961636D-01     -.50 -9.444368063767D-02
                                                      -.25 -3.889286643066D-02
```

CARDINAL SPLINES N=3

```
0  -4.0   1.100801204090D-02   5.504006020449D-03      .25  2.224566866606D-02
1  -3.0  -6.371319680527D-02  -3.185659840264D-02      .50  2.815588990853D-02
2  -2.0   1.964809012183D-01   9.824045060916D-02      .75  1.934316549122D-02
3  -1.0  -4.340913104904D-01  -2.170456552452D-01     1.25 -2.187331766486D-02
4   0.0   7.086505448574D-01   3.543252724287D-01     1.50 -3.485865068406D-02
5   1.0  -8.196779087960D-01  -4.098389543980D-01     1.75 -2.861646535249D-02
6   2.0   6.305828396385D-01   3.152914198193D-01     2.25  4.096022204911D-02
7   3.0  -2.872347058668D-01  -1.436173529334D-01     2.50  7.006065239908D-02
8   4.0   5.799482420332D-02   2.899741210166D-02     2.75  6.059278559207D-02
                                                      3.25 -9.369764269889D-02
        0 -1.643195034693D+00  -5.295417226452D+00     3.50 -1.665004477239D-01
        1 -9.438530875166D-01   2.099138403751D+00     3.75 -1.510046788923D-01
        2 -1.332635822108D-01  -4.297842961636D-01     4.25  2.802012080501D-01


I= 6     H=  2.0
0          0.                  4.532080199833D+01    -1.00  1.024014726142D-01
1         -4.899425896158D-02  -1.211494449122D+01     -.75  6.678725190079D-02
2          5.340721365264D-02   8.298719958203D-01     -.50  3.784893289395D-02
                                                       -.25  1.558651559368D-02
0  -4.0  -4.412954691065D-03  -2.206477345533D-03      .25 -8.914923413169D-03
1  -3.0   2.557421342668D-02   1.278710671334D-02      .50 -1.128323090172D-02
2  -2.0  -7.970898571397D-02  -3.985449285699D-02      .75 -7.751351375486D-03
3  -1.0   1.964809012183D-01   9.824045060916D-02     1.25  8.763653493414D-03
4   0.0  -4.058254963427D-01  -2.029127481714D-01     1.50  1.396316176038D-02
5   1.0   6.305828396385D-01   3.152914198193D-01     1.75  1.145831477098D-02
6   2.0  -6.394655179433D-01  -3.197327589716D-01     2.25 -1.636780337132D-02
7   3.0   3.626180540759D-01   1.813090270380D-01     2.50 -2.792989121899D-02
8   4.0  -8.584305366839D-02  -4.292152683420D-02     2.75 -2.405919445974D-02
                                                      3.25  3.646286338897D-02
        0  6.585383825960D-01   1.013897596656D+01     3.50  6.332361207836D-02
        1  3.782634502596D-01  -5.475968524662D+00     3.75  5.533890267550D-02
        2  5.340721365264D-02   8.298719958203D-01     4.25 -8.669949302131D-02


I= 7     H=  3.0
0          0.                 -4.793432510960D+01    -1.00 -3.281340113451D-02
1          1.569963627903D-02   1.338199895168D+01     -.75 -2.140121994048D-02
2         -1.711376485548D-02  -9.313516942827D-01     -.50 -1.212825935339D-02
                                                       -.25 -4.994519373225D-03
0  -4.0   1.414128576451D-03   7.070642882254D-04      .25  2.856679751228D-03
1  -3.0  -8.196327582562D-03  -4.098163791281D-03      .50  3.615568443659D-03
2  -2.0   2.557421342668D-02   1.278710671334D-02      .75  2.483813817984D-03
3  -1.0  -6.371319680527D-02  -3.185659840264D-02     1.25 -2.808138532993D-03
4   0.0   1.475220182117D-01   7.376100910584D-02     1.50 -4.474112865819D-03
5   1.0  -2.872347058668D-01  -1.436173529334D-01     1.75 -3.671354485274D-03
6   2.0   3.626180540759D-01   1.813090270380D-01     2.25  5.243294731285D-03
7   3.0  -2.424380954657D-01  -1.212190477328D-01     2.50  8.944885734346D-03
8   4.0   6.445391142952D-02   3.222695571476D-02     2.75  7.702010028754D-03
                                                      3.25 -1.164774732935D-02
        0 -2.110216925716D-01  -9.307956411405D+00     3.50 -2.017751891615D-02
        1 -1.212104825648D-01   5.931185397417D+00     3.75 -1.755954895095D-02
        2 -1.711376485548D-02  -9.313516942827D-01     4.25  2.693890040499D-02


I= 8     H=  4.0
0          0.                  1.746516417453D+01    -1.00  5.661247735244D-03
```

CARDINAL SPLINES N=3

```
1          -2.708634620869D-03  -5.106420167383D+00     -.75   3.692320842488D-03
2           2.952613114375D-03   3.745288475892D-01     -.50   2.092470589028D-03
                                                        -.25   8.616969748656D-04
0   -4.0   -2.439784935067D-04  -1.219892467534D-04      .25  -4.928585958163D-04
1   -3.0    1.414128576451D-03   7.070642882254D-04      .50  -6.237883597626D-04
2   -2.0   -4.412954691065D-03  -2.206477345533D-03      .75  -4.285283289737D-04
3   -1.0    1.100801204090D-02   5.504006020449D-03     1.25   4.844824264225D-04
4    0.0   -2.581542021197D-02  -1.290771010599D-02     1.50   7.719074089891D-04
5    1.0    5.799482420332D-02   2.899741210166D-02     1.75   6.334064353464D-04
6    2.0   -8.584305366839D-02  -4.292152683420D-02     2.25  -9.045849589062D-04
7    3.0    6.445391142952D-02   3.222695571476D-02     2.50  -1.543145300514D-03
8    4.0   -1.855546918525D-02  -9.277734592626D-03     2.75  -1.328660642494D-03
                                                        3.25   2.008812061082D-03
        0   3.640727134653D-02   3.031945066428D+00     3.50   3.478829576600D-03
        1   2.091227029413D-02  -2.110189386670D+00     3.75   3.025920112681D-03
        2   2.952613114375D-03   3.745288475892D-01     4.25  -4.630190791490D-03
```

```
               N= 3       M= 9                       P=  4.5       Q= 5

I= 0      H= -4.5
0           1.000000000000D+00   4.537599927332D-01    -1.00   3.473668494666D+00
1          -1.727556215382D+00  -8.529194190894D-01     -.75   2.715355318633D+00
2           7.461122792838D-01   3.724204829734D-01     -.50   2.050306177512D+00
                                                        -.25   1.478521071301D+00
0   -4.5   -1.855606390201D-02  -9.278031951007D-03      .25   6.147248424536D-01
1   -3.5    6.445735849289D-02   3.222867924644D-02      .50   3.221700851331D-01
2   -2.5   -8.585381099170D-02  -4.292690549585D-02      .75   1.196175546153D-01
3   -1.5    5.802166661158D-02   2.901083330579D-02     1.25  -5.021049918937D-02
4    -.5   -2.587856869624D-02  -1.293928434812D-02     1.50  -5.147751248717D-02
5     .5    1.115451702932D-02   5.577258514660D-03     1.75  -2.752075978280D-02
6    1.5   -4.741429719765D-03  -2.370714859883D-03     2.25   1.678018525073D-02
7    2.5    1.900168269490D-03   9.500841347448D-04     2.50   1.898572610000D-02
8    3.5   -6.088875682944D-04  -3.044437841472D-04     2.75   1.083485243577D-02
9    4.5    1.050504747403D-04   5.252523737016D-05     3.25  -7.017463343337D-03
                                                        3.50  -8.039304961336D-03
        0   8.334770686278D+00   4.157137387042D+00     3.75  -4.622817074457D-03
        1   4.987454298172D+00   2.498864927671D+00     4.25   3.014644348474D-03
        2   7.461122792838D-01   3.724204829734D-01     4.50   3.459245928422D-03

I= 1      H= -3.5
0           0.                   2.680137936748D-01    -1.00  -4.626805321156D+00
1           2.781173981332D+00   1.327648455305D+00     -.75  -3.124048114650D+00
2          -1.845631339824D+00  -9.191313115842D-01     -.50  -1.851994825622D+00
                                                        -.25  -8.106454540719D-01
0   -4.5    6.445735849289D-02   3.222867924644D-02      .25   5.800044832330D-01
1   -3.5   -2.424580751377D-01  -1.212290375688D-01      .50   9.311934481626D-01
2   -2.5    3.626804050596D-01   1.813402025298D-01      .75   1.063008890662D+00
3   -1.5   -2.873902883021D-01  -1.436951441511D-01     1.25   7.891399800785D-01
```

4	-.5	1.478880358534D-01	7.394401792671D-02	1.50	5.009867085996D-01
5	.5	-6.456236066069D-02	-3.228118033034D-02	1.75	2.152162156134D-01
6	1.5	2.747810164770D-02	1.373905082385D-02	2.25	-1.085002494295D-01
7	2.5	-1.101348312156D-02	-5.506741560780D-03	2.50	-1.174292506716D-01
8	3.5	3.529193736749D-03	1.764596868374D-03	2.75	-6.515883978755D-02
9	4.5	-6.088875682944D-04	-3.044437841472D-04	3.25	4.115863471460D-02
				3.50	4.691525699477D-02
	0	-2.485875171545D+01	-1.236997721703D+01	3.75	2.689616128702D-02
	1	-1.382950807709D+01	-6.944533348953D+00	4.25	-1.749419300977D-02
	2	-1.845631339824D+00	-9.191313115842D-01	4.50	-2.006400929188D-02

I= 2 H= -2.5

0		0.	-8.363989265491D-01	-1.00	3.127080113700D+00
1		-1.520613151354D+00	-5.638923404490D-01	-.75	2.044097529835D+00
2		1.606466962346D+00	7.917355899321D-01	-.50	1.161923316263D+00
				-.25	4.805574729851D-01
0	-4.5	-8.585381099170D-02	-4.292690549585D-02	.25	-2.798329443042D-01
1	-3.5	3.626804050596D-01	1.813402025298D-01	.50	-3.613727666841D-01
2	-2.5	-6.396600979723D-01	-3.198300489862D-01	.75	-2.571957089841D-01
3	-1.5	6.310683690241D-01	3.155341845121D-01	1.25	3.476873311289D-01
4	-.5	-4.069677353128D-01	-2.034838676564D-01	1.50	6.930123236867D-01
5	.5	1.991309047620D-01	9.956545238098D-02	1.75	9.356718396306D-01
6	1.5	-8.565048981853D-02	-4.282524490926D-02	2.25	8.667841333445D-01
7	2.5	3.436577010178D-02	1.718288505089D-02	2.50	5.888393544769D-01
8	3.5	-1.101348312156D-02	-5.506741560780D-03	2.75	2.653550509734D-01
9	4.5	1.900168269490D-03	9.500841347448D-04	3.25	-1.409809052230D-01
				3.50	-1.546467086479D-01
	0	2.568819680641D+01	1.265873123756D+01	3.75	-8.656611615492D-02
	1	1.293758950976D+01	6.561727968940D+00	4.25	5.513494501218D-02
	2	1.606466962346D+00	7.917355899321D-01	4.50	6.296938020115D-02

I= 3 H= -1.5

0		0.	2.087102652754D+00	-1.00	-1.395238043434D+00
1		6.686081884111D-01	-1.558169779217D-01	-.75	-9.101854347586D-01
2		-7.266298550227D-01	-3.346237300250D-01	-.50	-5.159615579612D-01
				-.25	-2.125664130417D-01
0	-4.5	5.802166661158D-02	2.901083330579D-02	.25	1.217943429477D-01
1	-3.5	-2.873902883021D-01	-1.436951441511D-01	.50	1.544598075315D-01
2	-2.5	6.310683690241D-01	3.155341845121D-01	.75	1.064956613216D-01
3	-1.5	-8.208894348991D-01	-4.104447174495D-01	1.25	-1.228114931613D-01
4	-.5	7.115007375940D-01	3.557503687970D-01	1.50	-2.003838068621D-01
5	.5	-4.407037804285D-01	-2.203518902143D-01	1.75	-1.711240635107D-01
6	1.5	2.113065482906D-01	1.056532741453D-01	2.25	2.951971746424D-01
7	2.5	-8.565048981853D-02	-4.282524490926D-02	2.50	6.336131419111D-01
8	3.5	2.747810164770D-02	1.373905082385D-02	2.75	9.017651649523D-01
9	4.5	-4.741429719765D-03	-2.370714859883D-03	3.25	8.887664356785D-01
				3.50	6.140449364956D-01
	0	-1.170551771636D+01	-5.390204280901D+00	3.75	2.798683969784D-01
	1	-5.871060506793D+00	-3.167430548147D+00	4.25	-1.504957497928D-01
	2	-7.266298550227D-01	-3.346237300250D-01	4.50	-1.656165441364D-01

I= 4 H= -.5

0		0.	-4.911512030572D+00	-1.00	6.025904563376D-01
1		-2.883559438207D-01	1.009210175035D+00	-.75	3.930238711563D-01

2		3.142345125169D-01	8.959896870373D-02	-.50	2.227366000396D-01
				-.25	9.172864298748D-02
0	-4.5	-2.587856869624D-02	-1.293928434812D-02	.25	-5.247460096260D-02
1	-3.5	1.478880358534D-01	7.394401792671D-02	.50	-6.642804905287D-02
2	-2.5	-4.069677353128D-01	-2.034838676564D-01	.75	-4.565115023215D-02
3	-1.5	7.115007375940D-01	3.557503687970D-01	1.25	5.171579374678D-02
4	-.5	-8.557685539513D-01	-4.278842769756D-01	1.50	8.259985751538D-02
5	.5	7.242068038109D-01	3.621034019055D-01	1.75	6.806764507915D-02
6	1.5	-4.407037804285D-01	-2.203518902143D-01	2.25	-9.935628936807D-02
7	2.5	1.991309047620D-01	9.956545238098D-02	2.50	-1.738211000300D-01
8	3.5	-6.456236066069D-02	-3.228118033034D-02	2.75	-1.559439268155D-01
9	4.5	1.115451702932D-02	5.577258514660D-03	3.25	2.853109925854D-01
				3.50	6.222316318589D-01
	0	5.065647131275D+00	1.444312873334D+00	3.75	8.951719397703D-01
	1	2.539754668832D+00	1.815600893368D+00	4.25	8.931916921171D-01
	2	3.142345125169D-01	8.959896870373D-02	4.50	6.192519272987D-01

I= 5 H= .5

0		0.	1.142889600974D+01	-1.00	-2.589186331896D-01
1		1.238820580801D-01	-2.621991611702D+00	-.75	-1.686961170592D-01
2		-1.350365751095D-01	8.959896870373D-02	-.50	-9.570017281743D-02
				-.25	-3.941030046437D-02
0	-4.5	1.115451702932D-02	5.577258514660D-03	.25	2.254162165873D-02
1	-3.5	-6.456236066069D-02	-3.228118033034D-02	.50	2.853046391987D-02
2	-2.5	1.991309047620D-01	9.956545238098D-02	.75	1.960048923889D-02
3	-1.5	-4.407037804285D-01	-2.203518902143D-01	1.25	-2.216424069980D-02
4	-.5	7.242068038109D-01	3.621034019055D-01	1.50	-3.532216695532D-02
5	.5	-8.557685539513D-01	-4.278842769756D-01	1.75	-2.899681385978D-02
6	1.5	7.115007375940D-01	3.557503687970D-01	2.25	4.150340464223D-02
7	2.5	-4.069677353128D-01	-2.034838676564D-01	2.50	7.098726891748D-02
8	3.5	1.478880358534D-01	7.394401792671D-02	2.75	6.139059983205D-02
9	4.5	-2.587856869624D-02	-1.293928434812D-02	3.25	-9.490378591616D-02
				3.50	-1.685890691898D-01
	0	-2.177021384606D+00	1.444312873334D+00	3.75	-1.528211508622D-01
	1	-1.091447117905D+00	-1.815600893368D+00	4.25	2.829789264692D-01
	2	-1.350365751095D-01	8.959896870373D-02	4.50	6.192519272987D-01

I= 6 H= 1.5

0		0.	-2.641977228057D+01	-1.00	1.100233602255D-01
1		-5.264096525286D-02	6.179044118372D+00	-.75	7.175832111175D-02
2		5.738239497263D-02	-3.346237300250D-01	-.50	4.066608136959D-02
				-.25	1.674664099901D-02
0	-4.5	-4.741429719765D-03	-2.370714859883D-03	.25	-9.578471929887D-03
1	-3.5	2.747810164770D-02	1.373905082385D-02	.50	-1.212305356202D-02
2	-2.5	-8.565048981853D-02	-4.282524490926D-02	.75	-8.328290265496D-03
3	-1.5	2.113065482906D-01	1.056532741453D-01	1.25	9.415924472580D-03
4	-.5	-4.407037804285D-01	-2.203518902143D-01	1.50	1.500239955114D-02
5	.5	7.115007375940D-01	3.557503687970D-01	1.75	1.231108427864D-02
6	1.5	-8.208894348991D-01	-4.104447174495D-01	2.25	-1.758565895280D-02
7	2.5	6.310683690241D-01	3.155341845121D-01	2.50	-3.000743404366D-02
8	3.5	-2.873902883021D-01	-1.436951441511D-01	2.75	-2.584795304194D-02
9	4.5	5.802166661158D-02	2.901083330579D-02	3.25	3.916712576376D-02
				3.50	6.800645600962D-02
	0	9.251091545579D-01	-5.390204280901D+00	3.75	5.941156729159D-02
	1	4.638005895008D-01	3.167430548147D+00	4.25	-9.292734333023D-02

	2	5.738239497263D-02	-3.346237300250D-01	4.50	-1.656165441364D-01

I= 7 H= 2.5

0		0.	5.821915279391D+01	-1.00	-4.409139669364D-02
1		2.109561421208D-02	-1.368734827833D+01	-.75	-2.875683830494D-02
2		-2.299578248157D-02	7.917355899321D-01	-.50	-1.629675272643D-02
				-.25	-6.711139958117D-03
0	-4.5	1.900168269490D-03	9.500841347448D-04	.25	3.838522780997D-03
1	-3.5	-1.101348312156D-02	-5.506741560780D-03	.50	4.858241744068D-03
2	-2.5	3.436577010178D-02	1.718288505089D-02	.75	3.337501850565D-03
3	-1.5	-8.565048981853D-02	-4.282524490926D-02	1.25	-3.773294355482D-03
4	-.5	1.991309047620D-01	9.956545238098D-02	1.50	-6.011857816520D-03
5	.5	-4.069677353128D-01	-2.034838676564D-01	1.75	-4.933185046265D-03
6	1.5	6.310683690241D-01	3.155341845121D-01	2.25	7.045337813000D-03
7	2.5	-6.396600979723D-01	-3.198300489862D-01	2.50	1.201899544909D-02
8	3.5	3.626804050596D-01	1.813402025298D-01	2.75	1.034880992600D-02
9	4.5	-8.585381099170D-02	-4.292690549585D-02	3.25	-1.564920479095D-02
				3.50	-2.710665455215D-02
	0	-3.707343312974D-01	1.265873123756D+01	3.75	-2.358581118369D-02
	1	-1.858664281220D-01	-6.561727968940D+00	4.25	3.615415936827D-02
	2	-2.299578248157D-02	7.917355899321D-01	4.50	6.296938020115D-02

I= 8 H= 3.5

0		0.	-6.223278634690D+01	-1.00	1.412854574379D-02
1		-6.759829087747D-03	1.521671515321D+01	-.75	9.214774934834D-03
2		7.368716656042D-03	-9.191313115842D-01	-.50	5.222093707884D-03
				-.25	2.150502062939D-03
0	-4.5	-6.088875682944D-04	-3.044437841472D-04	.25	-1.230007097700D-03
1	-3.5	3.529193736749D-03	1.764596868374D-03	.50	-1.556763116372D-03
2	-2.5	-1.101348312156D-02	-5.506741560780D-03	.75	-1.069460570904D-03
3	-1.5	2.747810164770D-02	1.373905082385D-02	1.25	1.209102500163D-03
4	-.5	-6.456236066069D-02	-3.228118033034D-02	1.50	1.926416177011D-03
5	.5	1.478880358534D-01	7.394401792671D-02	1.75	1.580764087704D-03
6	1.5	-2.873902883021D-01	-1.436951441511D-01	2.25	-2.257523583249D-03
7	2.5	3.626804050596D-01	1.813402025298D-01	2.50	-3.851126619470D-03
8	3.5	-2.424580751377D-01	-1.212290375688D-01	2.75	-3.315826126066D-03
9	4.5	6.445735849289D-02	3.222867924644D-02	3.25	5.013027891653D-03
				3.50	8.681088795807D-03
	0	1.187972813900D-01	-1.236997721703D+01	3.75	7.550319677719D-03
	1	5.955862081663D-02	6.944533348953D+00	4.25	-1.154882659507D-02
	2	7.368716656042D-03	-9.191313115842D-01	4.50	-2.006400929188D-02

I= 9 H= 4.5

0		0.	2.294354434177D+01	-1.00	-2.437576199175D-03
1		1.166262862217D-03	-5.850649274432D+00	-.75	-1.589810898702D-03
2		-1.271313336958D-03	3.724204829734D-01	-.50	-9.009597653480D-04
				-.25	-3.710227991142D-04
0	-4.5	1.050504747403D-04	5.252523737016D-05	.25	2.122112203487D-04
1	-3.5	-6.088875682944D-04	-3.044437841472D-04	.50	2.685859242049D-04
2	-2.5	1.900168269490D-03	9.500841347448D-04	.75	1.845123647043D-04
3	-1.5	-4.741429719765D-03	-2.370714859883D-03	1.25	-2.086045209950D-04
4	-.5	1.115451702932D-02	5.577258514660D-03	1.50	-3.323614087786D-04
5	.5	-2.587856869624D-02	-1.293928434812D-02	1.75	-2.727264899811D-04
6	1.5	5.802166661158D-02	2.901083330579D-02	2.25	3.894856406749D-04

7	2.5	-8.585381099170D-02	-4.292690549585D-02	2.50	6.644245101035D-04
8	3.5	6.445735849289D-02	3.222867924644D-02	2.75	5.720676515327D-04
9	4.5	-1.855606390201D-02	-9.278031951007D-03	3.25	-8.648573604466D-04
				3.50	-1.497632803508D-03
	0	-2.049591219341D-02	4.157137387042D+00	3.75	-1.302489729801D-03
	1	-1.027555717040D-02	-2.498864927671D+00	4.25	1.991745412592D-03
	2	-1.271313336958D-03	3.724204829734D-01	4.50	3.459245928422D-03

N= 3 M=10 P= 5.0 Q= 5

I= 0 H= -5.0

0		1.000000000000D+00	5.248589556671D-01	-1.00	3.473671053045D+00
1		-1.727557439443D+00	-8.690015925355D-01	-.75	2.715356987233D+00
2		7.461136136019D-01	3.733305045257D-01	-.50	2.050307123122D+00
				-.25	1.478521460711D+00
0	-5.0	-1.855617415868D-02	-9.278087079338D-03	.25	6.147246197255D-01
1	-4.0	6.445799755652D-02	3.222899877826D-02	.50	3.221698032364D-01
2	-3.0	-8.585580533766D-02	-4.292790266883D-02	.75	1.196173609587D-01
3	-2.0	5.802664319795D-02	2.901332159897D-02	1.25	-5.021028024668D-02
4	-1.0	-2.589028012677D-02	-1.294514006338D-02	1.50	-5.147716365438D-02
5	0.0	1.118177345416D-02	5.590886727078D-03	1.75	-2.752047354039D-02
6	1.0	-4.804539250392D-03	-2.402269625196D-03	2.25	1.677977646302D-02
7	2.0	2.041612633867D-03	1.020806316934D-03	2.50	1.898502874856D-02
8	3.0	-8.181676359465D-04	-4.090838179732D-04	2.75	1.083425201889D-02
9	4.0	2.621717554306D-04	1.310858777153D-04	3.25	-7.016555633223D-03
10	5.0	-4.523208847693D-05	-2.261604423847D-05	3.50	-8.037733134387D-03
				3.75	-4.621450075594D-03
	0	1.101505314283D+01	5.513113606131D+00	4.25	3.012554092684D-03
	1	5.733578696576D+00	2.864303452721D+00	4.50	3.455615866846D-03
	2	7.461136136019D-01	3.733305045257D-01	4.75	1.988582310341D-03

I= 1 H= -4.0

0		0.	-1.440861236480D-01	-1.00	-4.626820149896D+00
1		2.781181076170D+00	1.420863068769D+00	-.75	-3.124057786098D+00
2		-1.845639073726D+00	-9.244059286950D-01	-.50	-1.852000306516D+00
				-.25	-8.106477111503D-01
0	-5.0	6.445799755652D-02	3.222899877826D-02	.25	5.800057741978D-01
1	-4.0	-2.424617792433D-01	-1.212308896217D-01	.50	9.311950820769D-01
2	-3.0	3.626919645785D-01	1.813459822892D-01	.75	1.063010013123D+00
3	-2.0	-2.874191333196D-01	-1.437095666598D-01	1.25	7.891387110549D-01
4	-1.0	1.479559170060D-01	7.397795850300D-02	1.50	5.009846867141D-01
5	0.0	-6.472034285843D-02	-3.236017142922D-02	1.75	2.152145565108D-01
6	1.0	2.784389365479D-02	1.392194682740D-02	2.25	-1.084978800365D-01
7	2.0	-1.183331520434D-02	-5.916657602170D-03	2.50	-1.174252087214D-01
8	3.0	4.742211409832D-03	2.371105704916D-03	2.75	-6.515535968407D-02
9	4.0	-1.519585335325D-03	-7.597926676623D-04	3.25	4.115337349493D-02
10	5.0	2.621717554306D-04	1.310858777153D-04	3.50	4.690614645745D-02
				3.75	2.688823796303D-02

CARDINAL SPLINES N=3

```
          0  -3.223507146230D+01  -1.614991899718D+01    4.25 -1.748207758357D-02
          1  -1.567520966109D+01  -7.823196218181D+00    4.50 -2.004296892767D-02
          2  -1.845639073726D+00  -9.244059286950D-01    4.75 -1.153049279494D-02


I= 2      H= -3.0
0             0.                   4.496542357224D-01   -1.00  3.127126390214D+00
1            -1.520635292438D+00  -8.547901139626D-01    -.75  2.044127711827D+00
2             1.606491097776D+00   8.081962533050D-01    -.50  1.161940420663D+00
                                                         -.25  4.805645167205D-01
0   -5.0     -8.585580533766D-02  -4.292790266883D-02     .25 -2.798369730584D-01
1   -4.0      3.626919645785D-01   1.813459822892D-01     .50 -3.613778656919D-01
2   -3.0     -6.396961721261D-01  -3.198480860631D-01     .75 -2.571992118854D-01
3   -2.0      6.311583865767D-01   3.155791932883D-01    1.25  3.476912914109D-01
4   -1.0     -4.071795741589D-01  -2.035897870794D-01    1.50  6.930186334485D-01
5    0.0      1.996239247711D-01   9.981196238554D-02    1.75  9.356770172442D-01
6    1.0     -8.679202843011D-02  -4.339601421505D-02    2.25  8.667767391054D-01
7    2.0      3.692424584582D-02   1.846212292291D-02    2.50  5.888267406357D-01
8    3.0     -1.479898549315D-02  -7.399492746577D-03    2.75  2.653441905052D-01
9    4.0      4.742211409832D-03   2.371105704916D-03    3.25 -1.409644863694D-01
10   5.0     -8.181676359465D-04  -4.090838179732D-04    3.50 -1.546182771078D-01
                                                         3.75 -8.654138958877D-02
          0   3.255910098220D+01   1.638060999853D+01    4.25  5.509713602090D-02
          1   1.454427568532D+01   7.227172419087D+00    4.50  6.290371887536D-02
          2   1.606491097776D+00   8.081962533050D-01    4.75  3.609694450206D-02


I= 3      H= -2.0
0             0.                  -1.122046998932D+00   -1.00 -1.395353519421D+00
1             6.686634381114D-01   5.700740784781D-01    -.75 -9.102607493201D-01
2            -7.266900813094D-01  -3.756988066076D-01    -.50 -5.160042393831D-01
                                                         -.25 -2.125839896097D-01
0   -5.0      5.802664319795D-02   2.901332159897D-02     .25  1.218043960898D-01
1   -4.0     -2.874191333196D-01  -1.437095666598D-01     .50  1.544725313283D-01
2   -3.0      6.311583865767D-01   3.155791932883D-01     .75  1.065044022778D-01
3   -2.0     -8.211140599814D-01  -4.105570299907D-01    1.25 -1.228213754415D-01
4   -1.0      7.120293491467D-01   3.560146745733D-01    1.50 -2.003995519108D-01
5    0.0     -4.419340367172D-01  -2.209670183586D-01    1.75 -1.711369834564D-01
6    1.0      2.141550839074D-01   1.070775419537D-01    2.25  2.952156258391D-01
7    2.0     -9.203477608577D-02  -4.601738804289D-02    2.50  6.336446178291D-01
8    3.0      3.692424584582D-02   1.846212292291D-02    2.75  9.017922655955D-01
9    4.0     -1.183331520434D-02  -5.916657602170D-03    3.25  8.887254649318D-01
10   5.0      2.041612633867D-03   1.020806316934D-03    3.50  6.139739899206D-01
                                                         3.75  2.798066956013D-01
          0  -1.482393484218D+01  -7.664146771732D+00    4.25 -1.504014032177D-01
          1  -6.598237374982D+00  -3.186913987598D+00    4.50 -1.654526963057D-01
          2  -7.266900813094D-01  -3.756988066076D-01    4.75 -9.278457235204D-02


I= 4      H= -1.0
0             0.                   2.640599083034D+00   -1.00  6.028622066702D-01
1            -2.884859632717D-01  -6.990335995817D-01    -.75  3.932011093655D-01
2             3.143762433985D-01   1.862611926635D-01    -.50  2.228370424855D-01
                                                         -.25  9.177006030340D-02
0   -5.0     -2.589028012677D-02  -1.294514006338D-02     .25 -5.249825908221D-02
1   -4.0      1.479559170060D-01   7.397795850300D-02     .50 -6.645799204020D-02
2   -3.0     -4.071795741589D-01  -2.035897870794D-01     .75 -4.567172037691D-02
```

3	-2.0	7.120293491467D-01	3.560146745733D-01	1.25	5.173904977602D-02
4	-1.0	-8.570125386871D-01	-4.285062693436D-01	1.50	8.263691043283D-02
5	0.0	7.271019733103D-01	3.635509866551D-01	1.75	6.809804966469D-02
6	1.0	-4.474072563718D-01	-2.237036281859D-01	2.25	-9.939971067980D-02
7	2.0	2.141550839074D-01	1.070775419537D-01	2.50	-1.738951724964D-01
8	3.0	-8.679202843011D-02	-4.339601421505D-02	2.75	-1.560077029218D-01
9	4.0	2.784389365479D-02	1.392194682740D-02	3.25	2.854074092891D-01
10	5.0	-4.804539250392D-03	-2.402269625196D-03	3.50	6.223985908591D-01
				3.75	8.953171419910D-01
	0	6.416976268604D+00	3.801960901713D+00	4.25	8.929696657565D-01
	1	2.855276470713D+00	1.163578327053D+00	4.50	6.188663432165D-01
	2	3.143762433985D-01	1.862611926635D-01	4.75	2.826429426703D-01

I= 5 H= 0.0

0		0.	-6.147398234512D+00	-1.00	-2.595510873121D-01
1		1.241846569290D-01	1.353664303831D+00	-.75	-1.692821097872D-01
2		-1.353664303831D-01	-1.353664303831D-01	-.50	-9.593393606027D-02
				-.25	-3.950656613119D-02
0	-5.0	1.118177345416D-02	5.590886727078D-03	.25	2.259668203394D-02
1	-4.0	-6.472034285843D-02	-3.236017142922D-02	.50	2.860015128915D-02
2	-3.0	1.996239247711D-01	9.981196238554D-02	.75	1.964836286145D-02
3	-2.0	-4.419340367172D-01	-2.209670183586D-01	1.25	-2.221836527592D-02
4	-1.0	7.271019733103D-01	3.635509866551D-01	1.50	-3.540840151541D-02
5	0.0	-8.625065839197D-01	-4.312532919599D-01	1.75	-2.906757552282D-02
6	1.0	7.271019733103D-01	3.635509866551D-01	2.25	4.160446059068D-02
7	2.0	-4.419340367172D-01	-2.209670183586D-01	2.50	7.115966037821D-02
8	3.0	1.996239247711D-01	9.981196238554D-02	2.75	6.153902821101D-02
9	4.0	-6.472034285843D-02	-3.236017142922D-02	3.25	-9.512817990712D-02
10	5.0	1.118177345416D-02	5.590886727078D-03	3.50	-1.689776387526D-01
				3.75	-1.531590851055D-01
	0	-2.763237474933D+00	-2.763237474933D+00	4.25	2.834956562400D-01
	1	-1.229479646902D+00	2.084359492986D-24	4.50	6.201493107168D-01
	2	-1.353664303831D-01	-1.353664303831D-01	4.75	8.939736390485D-01

I= 6 H= 1.0

0		0.	1.427638235357D+01	-1.00	1.114877446066D-01
1		-5.334160267810D-02	-3.026190253688D+00	-.75	7.271340684335D-02
2		5.814614192849D-02	1.862611926635D-01	-.50	4.120733682117D-02
				-.25	1.696953454005D-02
0	-5.0	-4.804539250392D-03	-2.402269625196D-03	.25	-9.705958731855D-03
1	-4.0	2.784389365479D-02	1.392194682740D-02	.50	-1.228440770850D-02
2	-3.0	-8.679202843011D-02	-4.339601421505D-02	.75	-8.439136859193D-03
3	-2.0	2.141550839074D-01	1.070775419537D-01	1.25	9.541244525193D-03
4	-1.0	-4.474072563718D-01	-2.237036281859D-01	1.50	1.520206706600D-02
5	0.0	7.271019733103D-01	3.635509866551D-01	1.75	1.247492584508D-02
6	1.0	-8.570125386871D-01	-4.285062693436D-01	2.25	-1.781964391276D-02
7	2.0	7.120293491467D-01	3.560146745733D-01	2.50	-3.040658926068D-02
8	3.0	-4.071795741589D-01	-2.035897870794D-01	2.75	-2.619162413542D-02
9	4.0	1.479559170060D-01	7.397795850300D-02	3.25	3.968668764797D-02
10	5.0	-2.589028012677D-02	-1.294514006338D-02	3.50	6.890615004038D-02
				3.75	6.019402029512D-02
	0	1.186945534822D+00	3.801960901713D+00	4.25	-9.412377952808D-02
	1	5.281198166068D-01	-1.163578327053D+00	4.50	-1.676943458527D-01
	2	5.814614192849D-02	1.862611926635D-01	4.75	-1.523062698484D-01

I= 7 H= 2.0

0		0.	-3.299118687491D+01	-1.00	-4.737345117780D-02
1		2.266591927197D-02	6.943902053674D+00	-.75	-3.089742615101D-02
2		-2.470753190583D-02	-3.756988066076D-01	-.50	-1.750984261244D-02
				-.25	-7.210700562107D-03
0	-5.0	2.041612633867D-03	1.020806316934D-03	.25	4.124252836215D-03
1	-4.0	-1.183331520434D-02	-5.916657602170D-03	.50	5.219877054334D-03
2	-3.0	3.692424584582D-02	1.846212292291D-02	.75	3.585937005020D-03
3	-2.0	-9.203477608577D-02	-4.601738804289D-02	1.25	-4.054168179284D-03
4	-1.0	2.141550839074D-01	1.070775419537D-01	1.50	-6.459363041885D-03
5	0.0	-4.419340367172D-01	-2.209670183586D-01	1.75	-5.300395291586D-03
6	1.0	7.120293491467D-01	3.560146745733D-01	2.25	7.569757082286D-03
7	2.0	-8.211140599814D-01	-4.105570299907D-01	2.50	1.291360289425D-02
8	3.0	6.311583865767D-01	3.155791932883D-01	2.75	1.111906346599D-02
9	4.0	-2.874191333196D-01	-1.437095666598D-01	3.25	-1.681367392482D-02
10	5.0	5.802664319795D-02	2.901332159897D-02	3.50	-2.912309563562D-02
				3.75	-2.533948557550D-02
	0	-5.043587012860D-01	-7.664146771732D+00	4.25	3.883567443962D-02
	1	-2.244093997864D-01	3.186913987598D+00	4.50	6.762625754031D-02
	2	-2.470753190583D-02	-3.756988066076D-01	4.75	5.919277678372D-02

I= 8 H= 3.0

0		0.	7.272137842659D+01	-1.00	1.898465003262D-02
1		-9.083241198335D-03	-1.530913495214D+01	-.75	1.238197336803D-02
2		9.901408834281D-03	8.081962533050D-01	-.50	7.016972807738D-03
				-.25	2.889648351726D-03
0	-5.0	-8.181676359465D-04	-4.090838179732D-04	.25	-1.652771239273D-03
1	-4.0	4.742211409832D-03	2.371105704916D-03	.50	-2.091836129220D-03
2	-3.0	-1.479898549315D-02	-7.399492746577D-03	.75	-1.437043444639D-03
3	-2.0	3.692424584582D-02	1.846212292291D-02	1.25	1.624681396542D-03
4	-1.0	-8.679202843011D-02	-4.339601421505D-02	1.50	2.588541700719D-03
5	0.0	1.996239247711D-01	9.981196238554D-02	1.75	2.124085734499D-03
6	1.0	-4.071795741589D-01	-2.035897870794D-01	2.25	-3.033450575021D-03
7	2.0	6.311583865767D-01	3.155791932883D-01	2.50	-5.174781382727D-03
8	3.0	-6.396961721261D-01	-3.198480860631D-01	2.75	-4.455487742621D-03
9	4.0	3.626919645785D-01	1.813459822892D-01	3.25	6.735968075152D-03
10	5.0	-8.585580533766D-02	-4.292790266883D-02	3.50	1.166460056895D-02
				3.75	1.014504373742D-02
	0	2.021190148654D-01	1.638060999853D+01	4.25	-1.551637707339D-02
	1	8.993084714448D-02	-7.227172419087D+00	4.50	-2.695429162215D-02
	2	9.901408834281D-03	8.081962533050D-01	4.75	-2.349813181219D-02

I= 9 H= 4.0

0		0.	-7.837604830546D+01	-1.00	-6.083395572162D-03
1		2.910611908365D-03	1.706725550513D+01	-.75	-3.967649742159D-03
2		-3.172783663796D-03	-9.244059286950D-01	-.50	-2.248501870132D-03
				-.25	-9.259519560786D-04
0	-5.0	2.621717554306D-04	1.310858777153D-04	.25	5.296100252090D-04
1	-4.0	-1.519585335325D-03	-7.597926676623D-04	.50	6.703029055909D-04
2	-3.0	4.742211409832D-03	2.371105704916D-03	.75	4.604827068826D-04
3	-2.0	-1.183331520434D-02	-5.916657602170D-03	1.25	-5.206088564264D-04
4	-1.0	2.784389365479D-02	1.392194682740D-02	1.50	-8.294656549210D-04
5	0.0	-6.472034285843D-02	-3.236017142922D-02	1.75	-6.806363015976D-04
6	1.0	1.479559170060D-01	7.397795850300D-02	2.25	9.720286608346D-04

7	2.0	-2.874191333196D-01	-1.437095666598D-01	2.50	1.658185329889D-03
8	3.0	3.626919645785D-01	1.813459822892D-01	2.75	1.427691862456D-03
9	4.0	-2.424617792433D-01	-1.212308896217D-01	3.25	-2.158389856021D-03
10	5.0	6.445799755652D-02	3.222899877826D-02	3.50	-3.737565139204D-03
				3.75	-3.250531757270D-03
	0	-6.476653205307D-02	-1.614991899718D+01	4.25	4.970464883349D-03
	1	-2.881722472960D-02	7.823196218181D+00	4.50	8.632265968959D-03
	2	-3.172783663796D-03	-9.244059286950D-01	4.75	7.522223901582D-03

I=10 H= 5.0

0		0.	2.916789348288D+01	-1.00	1.049558810400D-03
1		-5.021633609618D-04	-6.597608497978D+00	-.75	6.845324610306D-04
2		5.473954494387D-04	3.733305045257D-01	-.50	3.879305428406D-04
				-.25	1.597530558304D-04
0	-5.0	-4.523208847693D-05	-2.261604423847D-05	.25	-9.137279661193D-05
1	-4.0	2.621717554306D-04	1.310858777153D-04	.50	-1.156463208861D-04
2	-3.0	-8.181676359465D-04	-4.090838179732D-04	.75	-7.944636703305D-05
3	-2.0	2.041612633867D-03	1.020806316934D-03	1.25	8.981983626550D-05
4	-1.0	-4.804539250392D-03	-2.402269625196D-03	1.50	1.431064152799D-04
5	0.0	1.118177345416D-02	5.590886727078D-03	1.75	1.174291134122D-04
6	1.0	-2.589028012677D-02	-1.294514006338D-02	2.25	-1.677025372612D-04
7	2.0	5.802664319795D-02	2.901332159897D-02	2.50	-2.860839545604D-04
8	3.0	-8.585580533766D-02	-4.292790266883D-02	2.75	-2.463171751240D-04
9	4.0	6.445799755652D-02	3.222899877826D-02	3.25	3.723822516631D-04
10	5.0	-1.855617415868D-02	-9.278087079338D-03	3.50	6.448319231198D-04
				3.75	5.608025148227D-04
	0	1.117406943116D-02	5.513113606131D+00	4.25	-8.575140302916D-04
	1	4.971791133425D-03	-2.864303452721D+00	4.50	-1.489209476518D-03
	2	5.473954494387D-04	3.733305045257D-01	4.75	-1.297642408891D-03

N= 3 M=11 P= 5.5 Q= 5

I= 0 H= -5.5

0		1.000000000000D+00	4.869296620234D-01	-1.00	3.473671527355D+00
1		-1.727557666378D+00	-8.612942985476D-01	-.75	2.715357296583D+00
2		7.461138609776D-01	3.729390830253D-01	-.50	2.050307298433D+00
				-.25	1.478521532906D+00
0	-5.5	-1.855619459969D-02	-9.278097299846D-03	.25	6.147245784328D-01
1	-4.5	6.445811603562D-02	3.222905801781D-02	.50	3.221697509742D-01
2	-3.5	-8.585617507924D-02	-4.292808753962D-02	.75	1.196173250558D-01
3	-2.5	5.802756583464D-02	2.901378291732D-02	1.25	-5.021023965583D-02
4	-1.5	-2.589245144174D-02	-1.294622572087D-02	1.50	-5.147709898258D-02
5	-.5	1.118682842377D-02	5.593414211886D-03	1.75	-2.752042047253D-02
6	.5	-4.816280484725D-03	-2.408140242362D-03	2.25	1.677970067590D-02
7	1.5	2.068788284401D-03	1.034394142200D-03	2.50	1.898489946322D-02
8	2.5	-8.790709789143D-04	-4.395354894572D-04	2.75	1.083414070473D-02
9	3.5	3.522828604123D-04	1.761414302061D-04	3.25	-7.016387348615D-03
10	4.5	-1.128846720639D-04	-5.644233603197D-05	3.50	-8.037441726287D-03

11	5.5	1.947581752961D-05	9.737908764806D-06	3.75	-4.621196641822D-03
				4.25	3.012166573734D-03
	0	1.406837712949D+01	7.031218281528D+00	4.50	3.454942883694D-03
	1	6.479694804375D+00	3.241035614731D+00	4.75	1.987995905044D-03
	2	7.461138609776D-01	3.729390830253D-01	5.25	-1.296743825945D-03
				5.50	-1.487648102345D-03

I= 1 H= -4.5

0		0.	7.575758088853D-02	-1.00	-4.626822899068D+00
1		2.781182391516D+00	1.376190471431D+00	-.75	-3.124059579135D+00
2		-1.845640507552D+00	-9.221371927286D-01	-.50	-1.852001322646D+00
				-.25	-8.106481296010D-01
0	-5.5	6.445811603562D-02	3.222905801781D-02	.25	5.800060135360D-01
1	-4.5	-2.424624659654D-01	-1.212312329827D-01	.50	9.311953849962D-01
2	-3.5	3.626941076544D-01	1.813470538272D-01	.75	1.063010221222D+00
3	-2.5	-2.874244810561D-01	-1.437122405281D-01	1.25	7.891384757844D-01
4	-1.5	1.479685022635D-01	7.398425113174D-02	1.50	5.009843118669D-01
5	-.5	-6.474964219781D-02	-3.237482109890D-02	1.75	2.152142489218D-01
6	.5	2.791194755729D-02	1.395597377865D-02	2.25	-1.084974407633D-01
7	1.5	-1.199082922879D-02	-5.995414614397D-03	2.50	-1.174244593648D-01
8	2.5	5.095216046894D-03	2.547608023447D-03	2.75	-6.515471449103D-02
9	3.5	-2.041882410644D-03	-1.020941205322D-03	3.25	4.115239809286D-02
10	4.5	6.542959731115D-04	3.271479865558D-04	3.50	4.690445741370D-02
11	5.5	-1.128846720639D-04	-5.644233603197D-05	3.75	2.688676902401D-02
				4.25	-1.747983146735D-02
	0	-4.053412220010D+01	-2.024984490628D+01	4.50	-2.003906821940D-02
	1	-1.752086319155D+01	-8.767318648584D+00	4.75	-1.152709390452D-02
	2	-1.845640507552D+00	-9.221371927286D-01	5.25	7.517015584082D-03
				5.50	8.623216017080D-03

I= 2 H= -3.5

0		0.	-2.364191492797D-01	-1.00	3.127134969644D+00
1		-1.520639397282D+00	-7.153788893518D-01	-.75	2.044133307415D+00
2		1.606495572361D+00	8.011161353303D-01	-.50	1.161943591731D+00
				-.25	4.805658225931D-01
0	-5.5	-8.585617507924D-02	-4.292808753962D-02	.25	-2.798377199689D-01
1	-4.5	3.626941076544D-01	1.813470538272D-01	.50	-3.613788110219D-01
2	-3.5	-6.397028600929D-01	-3.198514300465D-01	.75	-2.571998613055D-01
3	-2.5	6.311750754315D-01	3.155875377158D-01	1.25	3.476920256270D-01
4	-1.5	-4.072188493804D-01	-2.036094246902D-01	1.50	6.930198032461D-01
5	-.5	1.997153601705D-01	9.985768008527D-02	1.75	9.356779771474D-01
6	.5	-8.700440645271D-02	-4.350220322635D-02	2.25	8.667753682513D-01
7	1.5	3.741580497492D-02	1.870790248746D-02	2.50	5.888244020943D-01
8	2.5	-1.590061852277D-02	-7.950309261385D-03	2.75	2.653421770303D-01
9	3.5	6.372160846850D-03	3.186080423425D-03	3.25	-1.409614424005D-01
10	4.5	-2.041882410644D-03	-1.020941205322D-03	3.50	-1.546130060541D-01
11	5.5	3.522828604123D-04	1.761414302061D-04	3.75	-8.653680542309D-02
				4.25	5.509012649312D-02
	0	4.023297437888D+01	2.006276005303D+01	4.50	6.289154580846D-02
	1	1.615081189869D+01	8.096898599281D+00	4.75	3.608633747440D-02
	2	1.606495572361D+00	8.011161353303D-01	5.25	-2.348187804670D-02
				5.50	-2.692604914355D-02

I= 3 H= -2.5

CARDINAL SPLINES N=3

```
 0          0.                   5.899500066988D-01   -1.00 -1.395374928148D+00
 1          6.686736811569D-01   2.221935101562D-01    -.75 -9.102747123004D-01
 2         -7.267012469915D-01  -3.580313951656D-01    -.50 -5.160121523263D-01
                                                       -.25 -2.125872482262D-01
 0  -5.5    5.802756583464D-02   2.901378291732D-02     .25  1.218062598970D-01
 1  -4.5   -2.874244810561D-01  -1.437122405281D-01     .50  1.544748902629D-01
 2  -3.5    6.311750754315D-01   3.155875377158D-01     .75  1.065060228117D-01
 3  -2.5   -8.211557046075D-01  -4.105778523037D-01    1.25 -1.228232075717D-01
 4  -1.5    7.121273547874D-01   3.560636773937D-01    1.50 -2.004024709718D-01
 5   -.5   -4.421622005456D-01  -2.210811002728D-01    1.75 -1.711393787560D-01
 6    .5    2.146850425775D-01   1.073425212888D-01    2.25  2.952190466075D-01
 7   1.5   -9.326139087002D-02  -4.663069543501D-02    2.50  6.336504533215D-01
 8   2.5    3.967321190285D-02   1.983660595143D-02    2.75  9.017972899312D-01
 9   3.5   -1.590061852277D-02  -7.950309261385D-03    3.25  8.887178691472D-01
10   4.5    5.095216046894D-03   2.547608023447D-03    3.50  6.139608367678D-01
11   5.5   -8.790709789143D-04  -4.395354894572D-04    3.75  2.797952564782D-01
                                                       4.25 -1.503839119539D-01
       0   -1.830500747513D+01  -9.018435391202D+00    4.50 -1.654223201761D-01
       1   -7.325040035750D+00  -3.716151836666D+00    4.75 -9.275810404707D-02
       2   -7.267012469915D-01  -3.580313951656D-01    5.25  5.915221786036D-02
                                                       5.50  6.755578251556D-02


I= 4      H= -1.5
 0          0.                  -1.388380866403D+00   -1.00  6.029125895469D-01
 1         -2.885100690526D-01   1.196616609578D-01    -.75  3.932339695675D-01
 2          3.144025204943D-01   1.446830554214D-01    -.50  2.228556646499D-01
                                                       -.25  9.177767479404D-02
 0  -5.5   -2.589245144174D-02  -1.294622572087D-02     .25 -5.250264532936D-02
 1  -4.5    1.479685022635D-01   7.398425113174D-02     .50 -6.646354351027D-02
 2  -3.5   -4.072188493804D-01  -2.036094246902D-01     .75 -4.567553410937D-02
 3  -2.5    7.121273547874D-01   3.560636773937D-01    1.25  5.174336147504D-02
 4  -1.5   -8.572431832270D-01  -4.286215916135D-01    1.50  8.264378009339D-02
 5   -.5    7.276389295589D-01   3.638194647795D-01    1.75  6.810368671553D-02
 6    .5   -4.486544506421D-01  -2.243272253211D-01    2.25 -9.940776104858D-02
 7   1.5    2.170417749209D-01   1.085208874605D-01    2.50 -1.739089056289D-01
 8   2.5   -9.326139087002D-02  -4.663069543501D-02    2.75 -1.560195270941D-01
 9   3.5    3.741580497492D-02   1.870790248746D-02    3.25  2.854252850586D-01
10   4.5   -1.199082922879D-02  -5.995414614397D-03    3.50  6.224295452292D-01
11   5.5    2.068788284401D-03   1.034394142200D-03    3.75  8.953440625974D-01
                                                       4.25  8.929285021616D-01
       0    7.923870865164D+00   3.646420695361D+00    4.50  6.187948566328D-01
       1    3.169917656385D+00   1.711175270593D+00    4.75  2.825806526840D-01
       2    3.144025204943D-01   1.446830554214D-01    5.25 -1.522108192790D-01
                                                       5.50 -1.675284913210D-01


I= 5      H=  -.5
 0          0.                   3.232342384739D+00   -1.00 -2.596683820916D-01
 1          1.242407768339D-01  -5.523142034938D-01    -.75 -1.693586105829D-01
 2         -1.354276052577D-01  -3.856968588273D-02    -.50 -9.597728973138D-02
                                                       -.25 -3.952441953708D-02
 0  -5.5    1.118682842377D-02   5.593414211886D-03     .25  2.260689351701D-02
 1  -4.5   -6.474964219781D-02  -3.237482109890D-02     .50  2.861307549078D-02
 2  -3.5    1.997153601705D-01   9.985768008527D-02     .75  1.965724149121D-02
 3  -2.5   -4.421622005456D-01  -2.210811002728D-01    1.25 -2.222840320586D-02
 4  -1.5    7.276389295589D-01   3.638194647795D-01    1.50 -3.542439455458D-02
```

5	-.5	-8.637566547563D-01	-4.318783273782D-01	1.75	-2.908069896235D-02
6	.5	7.300055267913D-01	3.650027633956D-01	2.25	4.162320239928D-02
7	1.5	-4.486544506421D-01	-2.243272253211D-01	2.50	7.119163204902D-02
8	2.5	2.146850425775D-01	1.073425212888D-01	2.75	6.156655569206D-02
9	3.5	-8.700440645271D-02	-4.350220322635D-02	3.25	-9.516979591996D-02
10	4.5	2.791194755729D-02	1.395597377865D-02	3.50	-1.690497026414D-01
11	5.5	-4.816280484725D-03	-2.408140242362D-03	3.75	-1.532217581169D-01
				4.25	2.835914879016D-01
	0	-3.413360786458D+00	-9.721187324299D-01	4.50	6.203157363683D-01
	1	-1.365462881001D+00	-9.765807482038D-01	4.75	8.941186543928D-01
	2	-1.354276052577D-01	-3.856968588273D-02	5.25	8.937514236009D-01
				5.50	6.197631900343D-01

I= 6 H= .5

0		0.	-7.510045845503D+00	-1.00	1.117601864997D-01
1		-5.347195300750D-02	1.400847292914D+00	-.75	7.289109609501D-02
2		5.828823349223D-02	-3.856968588273D-02	-.50	4.130803487681D-02
				-.25	1.701100284514D-02
0	-5.5	-4.816280484725D-03	-2.408140242362D-03	.25	-9.729677057523D-03
1	-4.5	2.791194755729D-02	1.395597377865D-02	.50	-1.231442689584D-02
2	-3.5	-8.700440645271D-02	-4.350220322635D-02	.75	-8.459759351589D-03
3	-2.5	2.146850425775D-01	1.073425212888D-01	1.25	9.564559737063D-03
4	-1.5	-4.486544506421D-01	-2.243272253211D-01	1.50	1.523921427654D-02
5	-.5	7.300055267913D-01	3.650027633956D-01	1.75	1.250540780452D-02
6	.5	-8.637566547563D-01	-4.318783273782D-01	2.25	-1.786317572072D-02
7	1.5	7.276389295589D-01	3.638194647795D-01	2.50	-3.048085021722D-02
8	2.5	-4.421622005456D-01	-2.210811002728D-01	2.75	-2.625562522281D-02
9	3.5	1.997153601705D-01	9.985768008527D-02	3.25	3.978334962535D-02
10	4.5	-6.474964219781D-02	-3.237482109890D-02	3.50	6.907353363894D-02
11	5.5	1.118682842377D-02	5.593414211886D-03	3.75	6.033959159794D-02
				4.25	-9.434636879775D-02
	0	1.469123321599D+00	-9.721187324299D-01	4.50	-1.680809045705D-01
	1	5.876986154070D-01	9.765807482038D-01	4.75	-1.526430986068D-01
	2	5.828823349223D-02	-3.856968588273D-02	5.25	2.831590849700D-01
				5.50	6.197631900343D-01

I= 7 H= 1.5

0		0.	1.743454711012D+01	-1.00	-4.800403101882D-02
1		2.296762136721D-02	-3.302688880228D+00	-.75	-3.130869645444D-02
2		-2.503640965161D-02	1.446830554214D-01	-.50	-1.774291309651D-02
				-.25	-7.306680945028D-03
0	-5.5	2.068788284401D-03	1.034394142200D-03	.25	4.179150039636D-03
1	-4.5	-1.199082922879D-02	-5.995414614397D-03	.50	5.289357904590D-03
2	-3.5	3.741580497492D-02	1.870790248746D-02	.75	3.633668753710D-03
3	-2.5	-9.326139087002D-02	-4.663069543501D-02	1.25	-4.108132356470D-03
4	-1.5	2.170417749209D-01	1.085208874605D-01	1.50	-6.545342044043D-03
5	-.5	-4.486544506421D-01	-2.243272253211D-01	1.75	-5.370947250086D-03
6	.5	7.276389295589D-01	3.638194647795D-01	2.25	7.670513541664D-03
7	1.5	-8.572431832270D-01	-4.286215916135D-01	2.50	1.308548344326D-02
8	2.5	7.121273547874D-01	3.560636773937D-01	2.75	1.126705193316D-02
9	3.5	-4.072188493804D-01	-2.036094246902D-01	3.25	-1.703740271198D-02
10	4.5	1.479685022635D-01	7.398425113174D-02	3.50	-2.951051301021D-02
11	5.5	-2.589245144174D-02	-1.294622572087D-02	3.75	-2.567641734444D-02
				4.25	3.935086799029D-02
	0	-6.310294744416D-01	3.646420695361D+00	4.50	6.852096625053D-02

1		-2.524328848005D-01	-1.711175270593D+00	4.75	5.997238311538D-02
2		-2.503640965161D-02	1.446830554214D-01	5.25	-9.397920848359D-02
				5.50	-1.675284913210D-01

I= 8 H= 2.5

0		0.	-4.028772019662D+01	-1.00	2.039784234162D-02
1		-9.759385681353D-03	7.654497183487D+00	-.75	1.330367113242D-02
2		1.063845666027D-02	-3.580313951656D-01	-.50	7.539307005744D-03
				-.25	3.104749961605D-03
0	-5.5	-8.790709789143D-04	-4.395354894572D-04	.25	-1.775801346824D-03
1	-4.5	5.095216046894D-03	2.547608023447D-03	.50	-2.247549643701D-03
2	-3.5	-1.590061852277D-02	-7.950309261385D-03	.75	-1.544015053556D-03
3	-2.5	3.967321190285D-02	1.983660595143D-02	1.25	1.745620499254D-03
4	-1.5	-9.326139087002D-02	-4.663069543501D-02	1.50	2.781229218907D-03
5	-.5	2.146850425775D-01	1.073425212888D-01	1.75	2.282199709372D-03
6	.5	-4.421622005456D-01	-2.210811002728D-01	2.25	-3.259255848536D-03
7	1.5	7.121273547874D-01	3.560636773937D-01	2.50	-5.559982834048D-03
8	2.5	-8.211557046075D-01	-4.105778523037D-01	2.75	-4.787144655857D-03
9	3.5	6.311750754315D-01	3.155875377158D-01	3.25	7.237366598782D-03
10	4.5	-2.874244810561D-01	-1.437122405281D-01	3.50	1.253284154103D-02
11	5.5	5.802756583464D-02	2.901378291732D-02	3.75	1.090014143250D-02
				4.25	-1.667097719870D-02
	0	2.681366927257D-01	-9.018435391202D+00	4.50	-2.895942306760D-02
	1	1.072636375816D-01	3.716151836666D+00	4.75	-2.524530734855D-02
	2	1.063845666027D-02	-3.580313951656D-01	5.25	3.877424320929D-02
				5.50	6.755578251556D-02

I= 9 H= 3.5

0		0.	8.882946544281D+01	-1.00	-8.174320541311D-03
1		3.911018840449D-03	-1.690917608791D+01	-.75	-5.331371337072D-03
2		-4.263301700862D-03	8.011161353303D-01	-.50	-3.021334845440D-03
				-.25	-1.244211066416D-03
0	-5.5	3.522828604123D-04	1.761414302061D-04	.25	7.116423800394D-04
1	-4.5	-2.041882410644D-03	-1.020941205322D-03	.50	9.006928343972D-04
2	-3.5	6.372160846850D-03	3.186080423425D-03	.75	6.187552977041D-04
3	-2.5	-1.590061852277D-02	-7.950309261385D-03	1.25	-6.995474113533D-04
4	-1.5	3.741580497492D-02	1.870790248746D-02	1.50	-1.114561420342D-03
5	-.5	-8.700440645271D-02	-4.350220322635D-02	1.75	-9.145778932174D-04
6	.5	1.997153601705D-01	9.985768008527D-02	2.25	1.306124651558D-03
7	1.5	-4.072188493804D-01	-2.036094246902D-01	2.50	2.228120028506D-03
8	2.5	6.311750754315D-01	3.155875377158D-01	2.75	1.918403364815D-03
9	3.5	-6.397028600929D-01	-3.198514300465D-01	3.25	-2.900246916714D-03
10	4.5	3.626941076544D-01	1.813470538272D-01	3.50	-5.022193364396D-03
11	5.5	-8.585617507924D-02	-4.292808753962D-02	3.75	-4.367755941644D-03
				4.25	6.678783146910D-03
	0	-1.074542728286D-01	2.006276005303D+01	4.50	1.159900968742D-02
	1	-4.298529986903D-02	-8.096898599281D+00	4.75	1.010730233089D-02
	2	-4.263301700862D-03	8.011161353303D-01	5.25	-1.549175884567D-02
				5.50	-2.692604914355D-02

I=10 H= 4.5

0		0.	-9.636474755353D+01	-1.00	2.619359516743D-03
1		-1.253237422339D-03	1.891082776860D+01	-.75	1.708371744856D-03
2		1.366122094403D-03	-9.221371927286D-01	-.50	9.681492347705D-04

CARDINAL SPLINES N=3

```
 0  -5,5  -1,128846720639D-04  -5,644233603197D-05
 1  -4,5   6,542959731115D-04   3,271479865558D-04
 2  -3,5  -2,041882410644D-03  -1,020941205322D-03
 3  -2,5   5,095216046894D-03   2,547608023447D-03
 4  -1,5  -1,199082922879D-02  -5,995414614397D-03
 5   -,5   2,791194755729D-02   1,395597377865D-02
 6    ,5  -6,474964219781D-02  -3,237482109890D-02
 7   1,5   1,479685022635D-01   7,398425113174D-02
 8   2,5  -2,874244810561D-01  -1,437122405281D-01
 9   3,5   3,626941076544D-01   1,813470538272D-01
10   4,5  -2,424624659654D-01  -1,212312329827D-01
11   5,5   6,445811603562D-02   3,222905801781D-02

       0   3,443238753283D-02  -2,024984490628D+01
       1   1.377410561610D-02   8,767318648584D+00
       2   1.366122094403D-03  -9,221371927286D-01
```

```
 -,25   3,986919864850D-04
  ,25  -2,280369636222D-04
  .50  -2,886158335708D-04
  ,75  -1,982724504803D-04
 1,25   2,241612756096D-04
 1.50   3,571473495726D-04
 1.75   2,930651028360D-04
 2,25  -4,185312996325D-04
 2,50  -7,139730015885D-04
 2,75  -6,147279380063D-04
 3,25   9,293452011055D-04
 3,50   1,609290392480D-03
 3,75   1,399579295267D-03
 4,25  -2,140065655224D-03
 4,50  -3,716547354963D-03
 4,75  -3,238437991784D-03
 5,25   4,962576277700D-03
 5,50   8,623216017080D-03
```

```
I=11      H=  5,5
0            0.                    3,613832142406D+01
1            2.162191093826D-04   -7,343365528010D+00
2           -2,356949269122D-04    3,729390830253D-01

 0  -5,5   1,947581752961D-05   9,737908764806D-06
 1  -4,5  -1,128846720639D-04  -5,644233603197D-05
 2  -3,5   3,522828604123D-04   1,761414302061D-04
 3  -2,5  -8,790709789143D-04  -4,395354894572D-04
 4  -1,5   2,068788284401D-03   1.034394142200D-03
 5   -,5  -4,816280484725D-03  -2.408140242362D-03
 6    ,5   1,118682842377D-02   5,593414211886D-03
 7   1,5  -2,589245144174D-02  -1,294622572087D-02
 8   2,5   5,802756583464D-02   2,901378291732D-02
 9   3,5  -8,585617507924D-02  -4,292808753962D-02
10   4,5   6,445811603562D-02   3,222905801781D-02
11   5,5  -1,855619459969D-02  -9,278097299846D-03

       0  -5,940566437490D-03   7,031218281528D+00
       1  -2,376425086652D-03  -3,241035614731D+00
       2  -2,356949269122D-04   3,729390830253D-01
```

```
-1.00  -4,519140362948D-04
 -.75  -2,947427284251D-04
 -.50  -1,670332864194D-04
 -,25  -6,878571027766D-05
  ,25   3,934286376669D-05
  .50   4,979444226105D-05
  .75   3,420763844150D-05
 1,25  -3,867419720883D-05
 1.50  -6,161807811508D-05
 1,75  -5,056206726018D-05
 2,25   7,220855358523D-05
 2,50   1,231806467837D-04
 2,75   1,060580455824D-04
 3,25  -1,603384261023D-04
 3,50  -2,776481867835D-04
 3,75  -2,414669574740D-04
 4,25   3,692208056973D-04
 4,50   6,412057573768D-04
 4,75   5,587159962055D-04
 5,25  -8,561530213842D-04
 5,50  -1,487648102345D-03
```

```
          N= 3      M=12                      P=  6.0      Q= 5

I= 0      H= -6.0
0            1.000000000000D+00   5,067482865765D-01   -1.00   3,473671615290D+00
1           -1,727557708450D+00  -8,649501179615D-01    -,75   2,715357353935D+00
2            7.461139068398D-01   3,731076956300D-01    -.50   2,050307330935D+00
```

CARDINAL SPLINES N=3

0	-6,0	-1,855619838936D-02	-9,278099194678D-03
1	-5,0	6,445813800106D-02	3.222906900053D-02
2	-4,0	-8,585624362751D-02	-4,292812181376D-02
3	-3.0	5,802773688700D-02	2.901386844350D-02
4	-2,0	-2,589285399425D-02	-1,294642699713D-02
5	-1,0	1,118776562307D-02	5,593882811536D-03
6	0,0	-4,818458013535D-03	-2,409229006767D-03
7	1,0	2,073844192720D-03	1,036922096360D-03
8	2,0	-8,907723230740D-04	-4,453861615370D-04
9	3,0	3,785064092059D-04	1,892532046030D-04
10	4,0	-1,516843136628D-04	-7,584215683138D-05
11	5,0	4,860535484572D-05	2,430267742286D-05
12	6,0	-8,385806512421D-06	-4,192903256210D-06
	0	1,749475439553D+01	8,748924621489D+00
	1	7,225809173627D+00	3,612342229599D+00
	2	7,461139068398D-01	3,731076956300D-01

-,25	1,478521546290D+00
,25	6,147245707774D-01
,50	3,221697412851D-01
,75	1,196173183996D-01
1,25	-5,021023213049D-02
1,50	-5,147708699275D-02
1,75	-2,752041063401D-02
2,25	1,677968662534D-02
2,50	1,898487549436D-02
2,75	1,083412006763D-02
3,25	-7,016356149481D-03
3,50	-8,037387700669D-03
3,75	-4,621149656463D-03
4,25	3,012094729698D-03
4,50	3,454818116146D-03
4,75	1,987887188716D-03
5,25	-1,296577234675D-03
5,50	-1,487358636221D-03
5,75	-8,559007045807D-04

I= 1 H= -5.0

0		0.	-3.911405102048D-02
1		2.781182635375D+00	1.397380132747D+00
2		-1.845640773376D+00	-9,231144959219D-01
0	-6,0	6,445813800106D-02	3,222906900053D-02
1	-5,0	-2,424625932803D-01	-1,212312966402D-01
2	-4,0	3,626945049702D-01	1,813472524851D-01
3	-3,0	-2,874254725005D-01	-1,437127362502D-01
4	-2,0	1,479708355164D-01	7,398541775822D-02
5	-1.0	-6,475507434133D-02	-3,237753717066D-02
6	0,0	2,792456883115D-02	1,396228441557D-02
7	1.0	-1,202013400915D-02	-6,010067004575D-03
8	2.0	5,163038740419D-03	2,581519370210D-03
9	3,0	-2,193877915494D-03	-1,096938957747D-03
10	4,0	8,791843419290D-04	4,395921709645D-04
11	5,0	-2,817237092685D-04	-1,408618546343D-04
12	6,0	4,860535484572D-05	2,430267742286D-05
	0	-4,975597202928D+01	-2,488695510773D+01
	1	-1,936650664513D+01	-9,679993818316D+00
	2	-1.845640773376D+00	-9.231144959219D-01

-1,00	-4,626823408750D+00
-,75	-3,124059911555D+00
-,50	-1,852001511031D+00
-,25	-8,106482071797D-01
,25	5,800060579081D-01
,50	9,311954411559D-01
,75	1,063010259802D+00
1,25	7,891384321664D-01
1,50	5,009842423721D-01
1,75	2,152141918963D-01
2,25	-1,084973593242D-01
2,50	-1,174243204377D-01
2,75	-6,515459487542D-02
3,25	4,115221725813D-02
3,50	4,690414427336D-02
3,75	2,688649669003D-02
4,25	-1,747941504886D-02
4,50	-2,003834504854D-02
4,75	-1,152646376886D-02
5,25	7,516049996842D-03
5,50	8,621538229302D-03
5,75	4,961113812790D-03

I= 2 H= -4.0

0		0.	1.220644931244D-01
1		-1,520640158300D+00	-7,815061554476D-01
2		1.606496401928D+00	8,041660372308D-01
0	-6,0	-8,585624362751D-02	-4,292812181376D-02
1	-5,0	3,626945049702D-01	1,813472524851D-01
2	-4,0	-6,397041000091D-01	-3,198520500046D-01
3	-3,0	6,311781694643D-01	3,155890847321D-01
4	-2,0	-4,072261308389D-01	-2,036130654195D-01
5	-1,0	1,997323124377D-01	9,986615621886D-02

-1,00	3,127136560228D+00
-,75	2,044134344809D+00
-,50	1,161944179632D+00
-,25	4,805660646955D-01
,25	-2,798378584425D-01
,50	-3,613789862815D-01
,75	-2,571999817047D-01
1,25	3,476921617471D-01
1,50	6,930200201208D-01
1,75	9,356781551087D-01

CARDINAL SPLINES N=3

6	0.0	-8.704379407302D-02	-4.352189703651D-02	2.25	8.667751141017D-01
7	1.0	3.750725735727D-02	1.875362867863D-02	2.50	5.888239685402D-01
8	2.0	-1.611227501105D-02	-8.056137505524D-03	2.75	2.653418037420D-01
9	3.0	6.846498168451D-03	3.423249084226D-03	3.25	-1.409608780637D-01
10	4.0	-2.743698866529D-03	-1.371849433264D-03	3.50	-1.546120288269D-01
11	5.0	8.791843419290D-04	4.395921709645D-04	3.75	-8.653595554157D-02
12	6.0	-1.516843136628D-04	-7.584215683138D-05	4.25	5.508882696238D-02
				4.50	6.288928898560D-02
	0	4.871002951959D+01	2.438300490075D+01	4.75	3.608437098953D-02
	1	1.775731666483D+01	8.868486291322D+00	5.25	-2.347886470716D-02
	2	1.606496401928D+00	8.041660372308D-01	5.50	-2.692081321659D-02
				5.75	-1.548719488380D-02

I= 3 H= -3.0

0		0.	-3.045944123578D-01	-1.00	-1.395378897222D+00
1		6.686755801677D-01	3.872046008624D-01	-.75	-9.102773009690D-01
2		-7.267033170547D-01	-3.656419871175D-01	-.50	-5.160136193475D-01
				-.25	-2.125878523578D-01
0	-6.0	5.802773688700D-02	2.901386844350D-02	.25	1.218066054378D-01
1	-5.0	-2.874254725005D-01	-1.437127362502D-01	.50	1.544753275979D-01
2	-4.0	6.311781694643D-01	3.155890847321D-01	.75	1.065063232508D-01
3	-3.0	-8.211634253215D-01	-4.105817126608D-01	1.25	-1.228235472397D-01
4	-2.0	7.121455246208D-01	3.560727623104D-01	1.50	-2.004030121515D-01
5	-1.0	-4.422045024925D-01	-2.211022512463D-01	1.75	-1.711398228329D-01
6	0.0	2.147833287244D-01	1.073916643622D-01	2.25	2.952196808013D-01
7	1.0	-9.348959714936D-02	-4.674479857468D-02	2.50	6.336515351934D-01
8	2.0	4.020137025640D-02	2.010068512820D-02	2.75	9.017982214186D-01
9	3.0	-1.708425915123D-02	-8.542129575616D-03	3.25	8.887164609258D-01
10	4.0	6.846498168451D-03	3.423249084226D-03	3.50	6.139583982376D-01
11	5.0	-2.193877915494D-03	-1.096938957747D-03	3.75	2.797931357210D-01
12	6.0	3.785064092059D-04	1.892532046030D-04	4.25	-1.503806691617D-01
				4.50	-1.654166885989D-01
	0	-2.214926593296D+01	-1.114447834341D+01	4.75	-9.275319696663D-02
	1	-8.051764224489D+00	-4.000499244548D+00	5.25	5.914469850452D-02
	2	-7.267033170547D-01	-3.656419871175D-01	5.50	6.754271701219D-02
				5.75	3.876285449824D-02

I= 4 H= -2.0

0		0.	7.168289553479D-01	-1.00	6.029219303169D-01
1		-2.885145381613D-01	-2.686734082151D-01	-.75	3.932400617085D-01
2		3.144073921556D-01	1.625937288157D-01	-.50	2.228591171196D-01
				-.25	9.177909655005D-02
0	-6.0	-2.589285399425D-02	-1.294642699713D-02	.25	-5.250345852084D-02
1	-5.0	1.479708355164D-01	7.398541775822D-02	.50	-6.646457272909D-02
2	-4.0	-4.072261308389D-01	-2.036130654195D-01	.75	-4.567624115907D-02
3	-3.0	7.121455246208D-01	3.560727623104D-01	1.25	5.174416084561D-02
4	-2.0	-8.572859438892D-01	-4.286429719446D-01	1.50	8.264505369909D-02
5	-1.0	7.277384824399D-01	3.638692412199D-01	1.75	6.810473180066D-02
6	0.0	-4.488857560501D-01	-2.244428780250D-01	2.25	-9.940925355253D-02
7	1.0	2.175788327689D-01	1.087894163845D-01	2.50	-1.739114516930D-01
8	2.0	-9.450435223633D-02	-4.725217611816D-02	2.75	-1.560217192451D-01
9	3.0	4.020137025640D-02	2.010068512820D-02	3.25	2.854285991495D-01
10	4.0	-1.611227501105D-02	-8.056137505524D-03	3.50	6.224352840361D-01
11	5.0	5.163038740419D-03	2.581519370210D-03	3.75	8.953490535613D-01
12	6.0	-8.907723230740D-04	-4.453861615370D-04	4.25	8.929208706143D-01

CARDINAL SPLINES N=3

```
                                                          4.50  6,187816033484D-01
     0   9.587578888633D+00   4.958162743424D+00          4.75  2.825691044213D-01
     1   3.484374167706D+00   1.682451337574D+00          5,25 -1,521931233196D-01
     2   3.144073921556D-01   1.625937288157D-01          5,50 -1,674977431262D-01
                                                          5,75 -9,395240643122D-02


I= 5     H= -1.0
0             0.                 -1,668884393648D+00     -1.00 -2,596901287278D-01
1             1.242511815523D-01  3.517848655642D-01      -.75 -1,693727939504D-01
2            -1,354389471754D-01 -8.026827153083D-02      -.50 -9,598532757003D-02
                                                          -.25 -3,952772958655D-02
0   -6,0  1.118776562307D-02   5,593882811536D-03          .25  2,260878674199D-02
1   -5.0 -6.475507434133D-02  -3.237753717066D-02          .50  2.861547165804D-02
2   -4.0  1.997323124377D-01   9,986615621886D-02          .75  1,965888760309D-02
3   -3.0 -4,422045024925D-01  -2,211022512463D-01         1,25 -2,223026425382D-02
4   -2.0  7.277384824399D-01   3,638692412199D-01         1,50 -3,542735968913D-02
5   -1,0 -8,639884279701D-01  -4.319942139850D-01         1.75 -2,908313206863D-02
6    0.0  7.305440385575D-01   3,652720192787D-01         2,25  4,162667715997D-02
7    1.0 -4,499047974147D-01  -2.249523987073D-01         2,50  7,119755964686D-02
8    2,0  2,175788327689D-01   1.087894163845D-01         2,75  6,157165933009D-02
9    3,0 -9.348959714936D-02  -4,674479857468D-02         3,25 -9,517751159331D-02
10   4.0  3.750725735727D-02   1.875362867863D-02         3,50 -1,690630633971D-01
11   5,0 -1.202013400915D-02  -6,010067004575D-03         3.75 -1.532333777881D-01
12   6,0  2,073844192720D-03   1.036922096360D-03         4,25  2,836092552251D-01
                                                          4,50  6,203465918924D-01
     0  -4.130295009001D+00  -2.447832975372D+00          4,75  8,941455403848D-01
     1  -1,501016184553D+00  -6.114343928057D-01          5,25  8,937102248997D-01
     2  -1.354389471754D-01  -8,026827153083D-02          5,50  6,196916038798D-01
                                                          5,75  2,830966859942D-01


I= 6     H=  0.0
0             0.                  3,877673410053D+00     -1.00  1,118107135614D-01
1            -5,349612777391D-02 -6,997750294494D-01      -.75  7,292405033587D-02
2             5,831458578745D-02  5,831458578745D-02      -.50  4,132671033382D-02
                                                          -.25  1,701869355519D-02
0   -6,0 -4,818458013535D-03  -2,409229006767D-03          .25 -9,734075857166D-03
1   -5,0  2,792456883115D-02   1.396228441557D-02          .50 -1,231999425302D-02
2   -4.0 -8,704379407302D-02  -4.352189703651D-02          .75 -8,463583998129D-03
3   -3,0  2,147833287244D-01   1.073916643622D-01         1.25  9,568883775237D-03
4   -2,0 -4,488857560501D-01  -2,244428780250D-01         1.50  1,524610359658D-02
5   -1.0  7,305440385575D-01   3,652720192787D-01         1.75  1,251106098738D-02
6    0,0 -8,650078559529D-01  -4,325039279764D-01         2,25 -1,787124912785D-02
7    1,0  7,305440385575D-01   3,652720192787D-01         2,50 -3,049462265075D-02
8    2,0 -4,488857560501D-01  -2,244428780250D-01         2,75 -2,626742053300D-02
9    3,0  2,147833287244D-01   1,073916643622D-01         3,25  3,980127654977D-02
10   4,0 -8,704379407302D-02  -4,352189703651D-02         3,50  6,910457658854D-02
11   5,0  2,792456883115D-02   1,396228441557D-02         3,75  6,036658923736D-02
12   6,0 -4,818458013535D-03  -2,409229006767D-03         4,25 -9,438765015479D-02
                                                          4,50 -1,681525956107D-01
     0   1.778348321705D+00   1.778348321705D+00          4,75 -1,527055666668D-01
     1   6,462789016754D-01  -2,287453426811D-24          5,25  2,832548077888D-01
     2   5,831458578745D-02   5,831458578745D-02          5,50  6,199295163576D-01
                                                          5,75  8,938964040416D-01
```

CARDINAL SPLINES N=3

I= 7	H= 1.0				
0		0.	-9.006097107316D+00	-1.00	-4.812134757974D-02
1		2.302375169351D-02	1.574653651176D+00	-.75	-3.138521145614D-02
2		-2.509759588623D-02	-8.026827153083D-02	-.50	-1.778627481831D-02
				-.25	-7.324537666267D-03
0	-6.0	2.073844192720D-03	1.036922096360D-03	.25	4.189363418958D-03
1	-5.0	-1.202013400915D-02	-6.010067004575D-03	.50	5.302284506220D-03
2	-4.0	3.750725735727D-02	1.875362867863D-02	.75	3.642549032205D-03
3	-3.0	-9.348959714936D-02	-4.674479857468D-02	1.25	-4.118172150424D-03
4	-2.0	2.175788327689D-01	1.087894163845D-01	1.50	-6.561338053069D-03
5	-1.0	-4.499047974147D-01	-2.249523987073D-01	1.75	-5.384073126583D-03
6	0.0	7.305440385575D-01	3.652720192787D-01	2.25	7.689258830409D-03
7	1.0	-8.639884279701D-01	-4.319942139850D-01	2.50	1.311746105062D-02
8	2.0	7.277384824399D-01	3.638692412199D-01	2.75	1.129458452520D-02
9	3.0	-4.422045024925D-01	-2.211022512463D-01	3.25	-1.707902644890D-02
10	4.0	1.997323124377D-01	9.986615621886D-02	3.50	-2.958259026882D-02
11	5.0	-6.475507434133D-02	-3.237753717066D-02	3.75	-2.573910197537D-02
12	6.0	1.118776562307D-02	5.593882811536D-03	4.25	3.944671735660D-02
				4.50	6.868742252196D-02
	0	-7.653709417433D-01	-2.447832975372D+00	4.75	6.011742495499D-02
	1	-2.781473989413D-01	6.114343928057D-01	5.25	-9.420146308419D-02
	2	-2.509759588623D-02	-8.026827153083D-02	5.50	-1.679146770919D-01
				5.75	-1.525474429844D-01

I= 8	H= 2.0				
0		0.	2.090624500623D+01	-1.00	2.066935862873D-02
1		-9.889293152827D-03	-3.633576083363D+00	-.75	1.348075669481D-02
2		1.078006547590D-02	1.625937288157D-01	-.50	7.639662945389D-03
				-.25	3.146077380451D-03
0	-6.0	-8.907723230740D-04	-4.453861615370D-04	.25	-1.799439090810D-03
1	-5.0	5.163038740419D-03	2.581519370210D-03	.50	-2.277466842534D-03
2	-4.0	-1.611227501105D-02	-8.056137505524D-03	.75	-1.564567482187D-03
3	-3.0	4.020137025640D-02	2.010068512820D-02	1.25	1.768856498982D-03
4	-2.0	-9.450435223633D-02	-4.725217611816D-02	1.50	2.818250223831D-03
5	-1.0	2.175788327689D-01	1.087894163845D-01	1.75	2.312578107985D-03
6	0.0	-4.488857560501D-01	-2.244428780250D-01	2.25	-3.302639759221D-03
7	1.0	7.277384824399D-01	3.638692412199D-01	2.50	-5.633991491923D-03
8	2.0	-8.572859438892D-01	-4.286429719446D-01	2.75	-4.850865814298D-03
9	3.0	7.121455246208D-01	3.560727623104D-01	3.25	7.333700163425D-03
10	4.0	-4.072261308389D-01	-2.036130654195D-01	3.50	1.269965643477D-02
11	5.0	1.479708355164D-01	7.398541775822D-02	3.75	1.104521812171D-02
12	6.0	-2.589285399425D-02	-1.294642699713D-02	4.25	-1.689281002314D-02
				4.50	-2.934466781420D-02
	0	3.287465982155D-01	4.958162743424D+00	4.75	-2.558099074617D-02
	1	1.194714925580D-01	-1.682451337574D+00	5.25	3.928862705650D-02
	2	1.078006547590D-02	1.625937288157D-01	5.50	6.844956704043D-02
				5.75	5.993129643141D-02

I= 9	H= 3.0				
0		0.	-4.831058534693D+01	-1.00	-8.782807951477D-03
1		4.202150771136D-03	8.388203089958D+00	-.75	-5.728232742294D-03
2		-4.580657180342D-03	-3.656419871175D-01	-.50	-3.246239680653D-03
				-.25	-1.336828766555D-03
0	-6.0	3.785064092059D-04	1.892532046030D-04	.25	7.646162541778D-04
1	-5.0	-2.193877915494D-03	-1.096938957747D-03	.50	9.677394157701D-04

CARDINAL SPLINES N=3

2	-4.0	6.846498168451D-03	3.423249084226D-03	.75	6.648147595630D-04
3	-3.0	-1.708425915123D-02	-8.542129575616D-03	1.25	-7.516209478905D-04
4	-2.0	4.020137025640D-02	2.010068512820D-02	1.50	-1.197528139017D-03
5	-1.0	-9.348959714936D-02	-4.674479857468D-02	1.75	-9.826580562669D-04
6	0.0	2.147833287244D-01	1.073916643622D-01	2.25	1.403351096928D-03
7	1.0	-4.422045024925D-01	-2.211022512463D-01	2.50	2.393978721698D-03
8	2.0	7.121455246208D-01	3.560727623104D-01	2.75	2.061207036322D-03
9	3.0	-8.211634253215D-01	-4.105817126608D-01	3.25	-3.116137321987D-03
10	4.0	6.311781694643D-01	3.155890847321D-01	3.50	-5.396037466892D-03
11	5.0	-2.874254725005D-01	-1.437127362502D-01	3.75	-4.692883173727D-03
12	6.0	5.802773688700D-02	2.901386844350D-02	4.25	7.175926365368D-03
				4.50	1.246237070578D-02
	0	-1.396907538655D-01	-1.114447834341D+01	4.75	1.085959282303D-02
	1	-5.076573539296D-02	4.000499244548D+00	5.25	-1.664452982068D-02
	2	-4.580657180342D-03	-3.656419871175D-01	5.50	-2.892908416063D-02
				5.75	-2.522784882092D-02

I=10 H= 4.0

0		0.	1.065438999890D+02	-1.00	3.519660754291D-03
1		-1.683988220314D-03	-1.851847873809D+01	-.75	2.295556965597D-03
2		1.835672533977D-03	8.041660372308D-01	-.50	1.300912243651D-03
				-.25	5.357265884521D-04
0	-6.0	-1.516843136628D-04	-7.584215683138D-05	.25	-3.064156509175D-04
1	-5.0	8.791843419290D-04	4.395921709645D-04	.50	-3.878161114648D-04
2	-4.0	-2.743698866529D-03	-1.371849433264D-03	.75	-2.664207635260D-04
3	-3.0	6.846498168451D-03	3.423249084226D-03	1.25	3.012078481511D-04
4	-2.0	-1.611227501105D-02	-8.056137505524D-03	1.50	4.799026247855D-04
5	-1.0	3.750725735727D-02	1.875362867863D-02	1.75	3.937946397532D-04
6	0.0	-8.704379407302D-02	-4.352189703651D-02	2.25	-5.623848742950D-04
7	1.0	1.997323124377D-01	9.986615621886D-02	2.50	-9.593729626139D-04
8	2.0	-4.072261308389D-01	-2.036130654195D-01	2.75	-8.260163230015D-04
9	3.0	6.311781694643D-01	3.155890847321D-01	3.25	1.248770707360D-03
10	4.0	-6.397041000091D-01	-3.198520500046D-01	3.50	2.162419845126D-03
11	5.0	3.626945049702D-01	1.813472524851D-01	3.75	1.880628606254D-03
12	6.0	-8.585624362751D-02	-4.292812181376D-02	4.25	-2.875625088893D-03
				4.50	-4.993952573703D-03
	0	5.598028190128D-02	2.438300490075D+01	4.75	-4.351506314564D-03
	1	2.034408218741D-02	-8.868486291322D+00	5.25	6.668184508987D-03
	2	1.835672533977D-03	8.041660372308D-01	5.50	1.158685154117D-02
				5.75	1.010030592418D-02

I=11 H= 5.0

0		0.	-1.161990398708D+02	-1.00	-1.127831581257D-03
1		5.396131132058D-04	2.075736776938D+01	-.75	-7.355827231834D-04
2		-5.882184680515D-04	-9.231144959219D-01	-.50	-4.168611736158D-04
				-.25	-1.716669325547D-04
0	-6.0	4.860535484572D-05	2.430267742286D-05	.25	9.818709021508D-05
1	-5.0	-2.817237092685D-04	-1.408618546343D-04	.50	1.242708569290D-04
2	-4.0	8.791843419290D-04	4.395921709645D-04	.75	8.537122516786D-05
3	-3.0	-2.193877915494D-03	-1.096938957747D-03	1.25	-9.651831425333D-05
4	-2.0	5.163038740419D-03	2.581519370210D-03	1.50	-1.537788358622D-04
5	-1.0	-1.202013400915D-02	-6.010067004575D-03	1.75	-1.261866009815D-04
6	0.0	2.792456883115D-02	1.396228441557D-02	2.25	1.802092437595D-04
7	1.0	-6.475507434133D-02	-3.237753717066D-02	2.50	3.074191352722D-04
8	2.0	1.479708355164D-01	7.398541775822D-02	2.75	2.646866473313D-04

9	3.0	-2,874254725005D-01	-1.437127362502D-01	3.25	-4,001529403795D-04
10	4.0	3.626945049702D-01	1.813472524851D-01	3.50	-6,929201936250D-04
11	5.0	-2,424625932803D-01	-1.212312966402D-01	3.75	-6.026234856753D-04
12	6.0	6.445813800106D-02	3.222906900053D-02	4.25	9.214554421099D-04
				4.50	1.600240981769D-03
	0	-1,793818617062D-02	-2.488695510773D+01	4.75	1.394372304041D-03
	1	-6,519008503413D-03	9.679993818316D+00	5.25	-2,136669453305D-03
	2	-5,882184680515D-04	-9.231144959219D-01	5.50	-3,712651427965D-03
				5.75	-3,236196080128D-03

I=12 H= 6.0

0		0.	4.385485504176D+01	-1.00	1,945830340145D-04
1		-9.309861375103D-05	-8,089634577159D+00	-.75	1,269089467115D-04
2		1.014844202635D-04	3,731076956300D-01	-.50	7,192041194138D-05
				-.25	2,961742970422D-05
0	-6,0	-8,385806512421D-06	-4.192903256210D-06	.25	-1,694006643546D-05
1	-5,0	4.860535484572D-05	2,430267742286D-05	.50	-2,144025826317D-05
2	-4,0	-1,516843136628D-04	-7,584215683138D-05	,75	-1,472896510895D-05
3	-3.0	3,785064092059D-04	1.892532046030D-04	1.25	1,665215510130D-05
4	-2,0	-8,907723230740D-04	-4,453861615370D-04	1.50	2,653122410145D-05
5	-1,0	2,073844192720D-03	1.036922096360D-03	1.75	2,177077859140D-05
6	0.0	-4,818458013535D-03	-2,409229006767D-03	2.25	-3,109122129821D-05
7	1.0	1.118776562307D-02	5.593882811536D-03	2.50	-5,303854640187D-05
8	2,0	-2,589285399425D-02	-1,294642699713D-02	2.75	-4,566597633071D-05
9	3,0	5,802773688700D-02	2,901386844350D-02	3.25	6,903776375764D-05
10	4,0	-8,585624362751D-02	-4,292812181376D-02	3.50	1,195484384667D-04
11	5.0	6.445813800106D-02	3.222906900053D-02	3.75	1,039696831889D-04
12	6.0	-1,855619838936D-02	-9,278099194678D-03	4.25	-1,589772182153D-04
				4.50	-2,760869059840D-04
	0	3.094847446978D-03	8,748924621489D+00	4.75	-2,405686033508D-04
	1	1.124714429410D-03	-3.612342229599D+00	5.25	3,686348642458D-04
	2	1.014844202635D-04	3.731076956300D-01	5.50	6,405335991321D-04
				5.75	5,583292026635D-04

N= 3 M=13 P= 6.5 Q= 5

I= 0 H= -6.5

0		1.000000000000D+00	4.965681885600D-01	-1.00	3,473671631593D+00
1		-1.727557716250D+00	-d.632308441507D-01	-.75	2,715357364568D+00
2		7.461139153424D-01	3,730351093267D-01	-.50	2,050307336961D+00
				-.25	1,478521548772D+00
0	-6.5	-1.855619909194D-02	-9,278099545970D-03	.25	6,147245693581D-01
1	-5.5	6.445814207335D-02	3.222907103667D-02	.50	3,221697394887D-01
2	-4.5	-8.585625633602D-02	-4.292812816801D-02	.75	1.196173171655D-01
3	-3.5	5.802776859927D-02	2,901388429963D-02	1.25	-5,021023073533D-02
4	-2.5	-2.589292862555D-02	-1,294646431278D-02	1.50	-5,147708476990D-02
5	-1.5	1.118793937586D-02	5.593969687929D-03	1.75	-2,752040880999D-02
6	-.5	-4.818861730884D-03	-2.409430865442D-03	2.25	1.677968402043D-02

CARDINAL SPLINES N=3

7	,5	2,074781861456D-03	1.037390930728D-03	2,50	1,898487105065D-02
8	1,5	-8,929493062622D-04	-4.464746531311D-04	2,75	1,083411624162D-02
9	2,5	3,835447338174D-04	1.917723669087D-04	3,25	-7,016350365313D-03
10	3,5	-1,629755329339D-04	-8.148776646694D-05	3,50	-8,037377684584D-03
11	4,5	6,531152569271D-05	3,265576284635D-05	3,75	-4,621140945607D-03
12	5,5	-2,092826736399D-05	-1,046413368199D-05	4,25	3,012081410165D-03
13	6,5	3,610721519087D-06	1,805360759543D-06	4,50	3,454794984856D-03
				4,75	1,987867033245D-03
	0	2,129418776759D+01	1,064630107063D+01	5,25	-1,296546349496D-03
	1	7,971923183201D+00	3,986225577096D+00	5,50	-1,487304970706D-03
	2	7,461139153424D-01	3,730351093267D-01	5,75	-8,558539264167D-04
				6,25	5,582574937102D-04
				6,50	6,404089864085D-04

I= 1 H= -5.5

0		0.	1,989127851107D-02	-1,00	-4,626823503243D+00
1		2,781182680585D+00	1,387414971548D+00	-,75	-3,124059973184D+00
2		-1,845640822658D+00	-9,226937751447D-01	-,50	-1,852001545957D+00
				-,25	-8,106482215624D-01
0	-6,5	6,445814207335D-02	3,222907103667D-02	,25	5,800060661345D-01
1	-5,5	-2,424626168839D-01	-1,212313084419D-01	,50	9,311954515677D-01
2	-4,5	3,626945786306D-01	1,813472893153D-01	,75	1,063010266955D+00
3	-3,5	-2,874256563094D-01	-1,437128281547D-01	1,25	7,891384240799D-01
4	-2,5	1,479712680903D-01	7,398563404515D-02	1,50	5,009842294881D-01
5	-1,5	-6,475608143776D-02	-3,237804071888D-02	1,75	2,152141813240D-01
6	-,5	2,792690883565D-02	1,396345441783D-02	2,25	-1,084973442258D-01
7	,5	-1,202556887360D-02	-6,012784436798D-03	2,50	-1,174242946813D-01
8	1,5	5,175656851776D-03	2,587828425888D-03	2,75	-6,515457269927D-02
9	2,5	-2,223080778130D-03	-1,111540389065D-03	3,25	4,115218373226D-02
10	3,5	9,446298919452D-04	4,723149459726D-04	3,50	4,690408621867D-02
11	4,5	-3,785551066128D-04	-1,892775533064D-04	3,75	2,688644620064D-02
12	5,5	1,213032831452D-04	6,065164157261D-05	4,25	-1,747933784692D-02
13	6,5	-2,092826736399D-05	-1,046413368199D-05	4,50	-2,003821097621D-02
				4,75	-1,152634694482D-02
	0	-5,990063733351D+01	-2,994572340629D+01	5,25	7,515870981851D-03
	1	-2,121214801397D+01	-1,060760410533D+01	5,50	8,621227176169D-03
	2	-1,845640822658D+00	-9,226937751447D-01	5,75	4,960842679744D-03
				6,25	-3,235780444599D-03
				6,50	-3,711929154485D-03

I= 2 H= -4.5

0		0.	-6,207536081957D-02	-1,00	3,127136855114D+00
1		-1,520640299389D+00	-7,504075522806D-01	-,75	2,044134537137D+00
2		1,606496555725D+00	8,028530801965D-01	-,50	1,161944288626D+00
				-,25	4,805661095801D-01
0	-6,5	-8,585625633602D-02	-4,292812816801D-02	,25	-2,798378841148D-01
1	-5,5	3,626945786306D-01	1,813472893153D-01	,50	-3,613790187738D-01
2	-4,5	-6,397043298835D-01	-3,198521649418D-01	,75	-2,572000040262D-01
3	-3,5	6,311787430828D-01	3,155893715414D-01	1,25	3,476921869831D-01
4	-2,5	-4,072274807863D-01	-2,036137403931D-01	1,50	6,930200603284D-01
5	-1,5	1,997354553163D-01	9,986772765817D-02	1,75	9,356781881019D-01
6	-,5	-8,705109660132D-02	-4,352554830066D-02	2,25	8,667750669836D-01
7	,5	3,752421811578D-02	1,876210905789D-02	2,50	5,888238881614D-01
8	1,5	-1,615165276219D-02	-8,075826381095D-03	2,75	2,653417345361D-01
9	2,5	6,937632493148D-03	3,468816246574D-03	3,25	-1,409607734384D-01

CARDINAL SPLINES N=3

10	3.5	-2.947936926327D-03	-1.473968463163D-03	3.50	-1.546118476537D-01
11	4.5	1.181369237865D-03	5.906846189326D-04	3.75	-8.653579797769D-02
12	5.5	-3.785551066128D-04	-1.892775533064D-04	4.25	5.508858603575D-02
13	6.5	6.531152569271D-05	3.265576284635D-05	4.50	6.288887058172D-02
				4.75	3.608400641293D-02
	0	5.799031753336D+01	2.898081818766D+01	5.25	-2.347830604924D-02
	1	1.936381492504D+01	9.686682490274D+00	5.50	-2.691984250295D-02
	2	1.606496555725D+00	8.028530801965D-01	5.75	-1.548634875007D-02
				6.25	1.009900883685D-02
				6.50	1.158459751879D-02

I= 3 H= -3.5

0		0.	1.549001487109D-01	-1.00	-1.395379633069D+00
1		6.686759322351D-01	3.096025027506D-01	-.75	-9.102777808956D-01
2		-7.267037008343D-01	-3.623656913931D-01	-.50	-5.160138913261D-01
				-.25	-2.125879643609D-01
0	-6.5	5.802776859927D-02	2.901388429963D-02	.25	1.218066694994D-01
1	-5.5	-2.874256563094D-01	-1.437128281547D-01	.50	1.544754086777D-01
2	-4.5	6.311787430828D-01	3.155893715414D-01	.75	1.065063789508D-01
3	-3.5	-8.211648567043D-01	-4.105824283522D-01	1.25	-1.228236102125D-01
4	-2.5	7.121488932206D-01	3.560744466103D-01	1.50	-2.004031124836D-01
5	-1.5	-4.422123450948D-01	-2.211061725474D-01	1.75	-1.711399051626D-01
6	-.5	2.148015511350D-01	1.074007755675D-01	2.25	2.952197983777D-01
7	.5	-9.353192028648D-02	-4.676596014324D-02	2.50	6.336517357672D-01
8	1.5	4.029963177931D-02	2.014981588966D-02	2.75	9.017983941118D-01
9	2.5	-1.731167177065D-02	-8.655835885326D-03	3.25	8.887161998484D-01
10	3.5	7.356144915912D-03	3.678072457956D-03	3.50	6.139579461460D-01
11	4.5	-2.947936926327D-03	-1.473968463163D-03	3.75	2.797927425429D-01
12	5.5	9.446298919452D-04	4.723149459726D-04	4.25	-1.503800679638D-01
13	6.5	-1.629755329339D-04	-8.148776646694D-05	4.50	-1.654156445321D-01
				4.75	-9.275228721810D-02
	0	-2.635683780072D+01	-1.314263404477D+01	5.25	5.914330445392D-02
	1	-8.778472178611D+00	-4.401151485360D+00	5.50	6.754029473570D-02
	2	-7.267037008343D-01	-3.623656913931D-01	5.75	3.876074309304D-02
				6.25	-2.522461212576D-02
				6.50	-2.892345957149D-02

I= 4 H= -2.5

0		0.	-3.645402289891D-01	-1.00	6.029236620507D-01
1		-2.885153667126D-01	-8.604552826192D-02	-.75	3.932411911621D-01
2		3.144082953381D-01	1.548833323600D-01	-.50	2.228597571908D-01
				-.25	9.177936013677D-02
0	-6.5	-2.589292862555D-02	-1.294646431278D-02	.25	-5.250360928262D-02
1	-5.5	1.479712680903D-01	7.398563404515D-02	.50	-6.646476354130D-02
2	-4.5	-4.072274807863D-01	-2.036137403931D-01	.75	-4.567637224268D-02
3	-3.5	7.121488932206D-01	3.560744466103D-01	1.25	5.174430904506D-02
4	-2.5	-8.572938715129D-01	-4.286469357564D-01	1.50	8.264528981946D-02
5	-1.5	7.277569391293D-01	3.638784695647D-01	1.75	6.810492555441D-02
6	-.5	-4.489286404604D-01	-2.244643202302D-01	2.25	-9.940953025557D-02
7	.5	2.176784355502D-01	1.088392177751D-01	2.50	-1.739119237210D-01
8	1.5	-9.473559976760D-02	-4.736779988380D-02	2.75	-1.560221256593D-01
9	2.5	4.073656048411D-02	2.036828024205D-02	3.25	2.854292135660D-01
10	3.5	-1.731167177065D-02	-8.655835885326D-03	3.50	6.224363479833D-01
11	4.5	6.937632493148D-03	3.468816246574D-03	3.75	8.953499788620D-01
12	5.5	-2.223080778130D-03	-1.111540389065D-03	4.25	8.929194557623D-01

13	6.5	3.835447338174D-04	1.917723669087D-04	4.50	6.187791462537D-01
				4.75	2.825669634296D-01
	0	1.140840059440D+01	5.619984629517D+00	5.25	-1.521898425770D-01
	1	3.798792472683D+00	1.927437792418D+00	5.50	-1.674920425688D-01
	2	3.144082953381D-01	1.548833323600D-01	5.75	-9.394743747465D-02
				6.25	5.992367923046D-02
				6.50	6.843633019742D-02

I= 5 H= -1.5

0		0.	8.487042415982D-01	-1.00	-2.596941604621D-01
1		1.242531105431D-01	-7.340005898212D-02	-.75	-1.693754234868D-01
2		-1.354410499190D-01	-6.231731937547D-02	-.50	-9.598681775131D-02
				-.25	-3.952834325572D-02
0	-6.5	1.118793937586D-02	5.593969687929D-03	.25	2.260913773789D-02
1	-5.5	-6.475608143776D-02	-3.237804071888D-02	.50	2.861591589731D-02
2	-4.5	1.997354553163D-01	9.986772765817D-02	.75	1.965919278527D-02
3	-3.5	-4.422123450948D-01	-2.211061725474D-01	1.25	-2.223060928423D-02
4	-2.5	7.277569391293D-01	3.638784695647D-01	1.50	-3.542790941254D-02
5	-1.5	-8.640313978940D-01	-4.320156989470D-01	1.75	-2.908358315624D-02
6	-.5	7.306438798566D-01	3.653219399283D-01	2.25	4.162732136590D-02
7	.5	-4.501366875349D-01	-2.250683437675D-01	2.50	7.119865859842D-02
8	1.5	2.181172114883D-01	1.090586057442D-01	2.75	6.157260552299D-02
9	2.5	-9.473559976760D-02	-4.736779988380D-02	3.25	-9.517894204649D-02
10	3.5	4.029963177931D-02	2.014981588966D-02	3.50	-1.690655404247D-01
11	4.5	-1.615165276219D-02	-8.075826381095D-03	3.75	-1.532355320259D-01
12	5.5	5.175656851776D-03	2.587828425888D-03	4.25	2.836125492114D-01
13	6.5	-8.929493062622D-04	-4.464746531311D-04	4.50	6.203523123750D-01
				4.75	8.941505249324D-01
	0	-4.914739140547D+00	-2.261302885399D+00	5.25	8.937025868421D-01
	1	-1.636480538404D+00	-8.835252108632D-01	5.50	6.196783321323D-01
	2	-1.354410499190D-01	-6.231731937547D-02	5.75	2.830851175227D-01
				6.25	-1.525297090053D-01
				6.50	-1.678838597485D-01

I= 6 H= -.5

0		0.	-1.971983673933D+00	-1.00	1.118200813600D-01
1		-5.350060981455D-02	2.881488795866D-01	-.75	7.293016010522D-02
2		5.831947154544D-02	1.660526403019D-02	-.50	4.133017279363D-02
				-.25	1.702011942523D-02
0	-6.5	-4.818861730884D-03	-2.409430865442D-03	.25	-9.734891401707D-03
1	-5.5	2.792690883565D-02	1.396345441783D-02	.50	-1.232102645001D-02
2	-4.5	-8.705109660132D-02	-4.352554830066D-02	.75	-8.464293093759D-03
3	-3.5	2.148015511350D-01	1.074007755675D-01	1.25	9.569685458881D-03
4	-2.5	-4.489286404604D-01	-2.244643202302D-01	1.50	1.524738088762D-02
5	-1.5	7.306438798566D-01	3.653219399283D-01	1.75	1.251210909659D-02
6	-.5	-8.652398387933D-01	-4.326199193966D-01	2.25	-1.787274595054D-02
7	.5	7.310828389262D-01	3.655414194631D-01	2.50	-3.049717608211D-02
8	1.5	-4.501366875349D-01	-2.250683437675D-01	2.75	-2.626961902719D-02
9	2.5	2.176784355502D-01	1.088392177751D-01	3.25	3.980460023040D-02
10	3.5	-9.353192028648D-02	-4.676596014324D-02	3.50	6.911033200131D-02
11	4.5	3.752421811578D-02	1.876210905789D-02	3.75	6.037159464306D-02
12	5.5	-1.202556887360D-02	-6.012784436798D-03	4.25	-9.439530378417D-02
13	6.5	2.074781861456D-03	1.037390930728D-03	4.50	-1.681658872428D-01
				4.75	-1.527171483423D-01
	0	2.116243709000D+00	6.025564486550D-01	5.25	2.832725549367D-01

CARDINAL SPLINES N=3

```
      1   7.046525202761D-01   5.040173119791D-01     5.50  6.199603534731D-01
      2   5.831947154544D-02   1.660526403019D-02     5.75  8.939232835684D-01
                                                      6.25  8.938551988601D-01
                                                      6.50  6.198579117719D-01


I= 7      H=    .5
0             0.               4.580241381795D+00    -1.00 -4.814310510864D-02
1             2.303416162359D-02  -7.198857443716D-01   -.75 -3.139940192804D-02
2            -2.510894348505D-02   1.660526403019D-02   -.50 -1.779431668306D-02
                                                      -.25 -7.327849373714D-03
0   -6.5   2.074781861456D-03   1.037390930728D-03      .25  4.191257592244D-03
1   -5.5  -1.202556887360D-02  -6.012784436798D-03      .50  5.304681873705D-03
2   -4.5   3.752421811578D-02   1.876210905789D-02      .75  3.644195968618D-03
3   -3.5  -9.353192028648D-02  -4.676596014324D-02     1.25 -4.120034130569D-03
4   -2.5   2.176784355502D-01   1.088392177751D-01     1.50 -6.564304672837D-03
5   -1.5  -4.501366875349D-01  -2.250683437675D-01     1.75 -5.386507451583D-03
6    -.5   7.310828389262D-01   3.655414194631D-01     2.25  7.692735331578D-03
7     .5  -8.652398387933D-01  -4.326199193966D-01     2.50  1.312339161754D-02
8    1.5   7.306438798566D-01   3.653219399283D-01     2.75  1.129969071960D-02
9    2.5  -4.489286404604D-01  -2.244643202302D-01     3.25 -1.708674598691D-02
10   3.5   2.148015511350D-01   1.074007755675D-01     3.50 -2.959595771667D-02
11   4.5  -8.705109660132D-02  -4.352554830066D-02     3.75 -2.575072746641D-02
12   5.5   2.792690883565D-02   1.396345441783D-02     4.25  3.946449357799D-02
13   6.5  -4.818861730884D-03  -2.409430865442D-03     4.50  6.871829349608D-02
                                                      4.75  6.014432440591D-02
      0  -9.111308116900D-01   6.025564486550D-01     5.25 -9.424268238229D-02
      1  -3.033821036821D-01  -5.040173119791D-01     5.50 -1.679862989807D-01
      2  -2.510894348505D-02   1.660526403019D-02     5.75 -1.526098730286D-01
                                                      6.25  2.831923886201D-01
                                                      6.50  6.198579117719D-01


I= 8      H=   1.5
0             0.              -1.063712349962D+01    -1.00  2.071987302983D-02
1            -9.913461861782D-03   1.693650362744D+00   -.75  1.351370267836D-02
2             1.080641116804D-02  -6.231731937547D-02   -.50  7.658333722902D-03
                                                      -.25  3.153766163448D-03
0   -6.5  -8.929493062622D-04  -4.464746531311D-04      .25 -1.803836788250D-03
1   -5.5   5.175656851776D-03   2.587828425888D-03      .50 -2.283032804701D-03
2   -4.5  -1.615165276219D-02  -8.075826381095D-03      .75 -1.568391170388D-03
3   -3.5   4.029963177931D-02   2.014981588966D-02     1.25  1.773179453684D-03
4   -2.5  -9.473559976760D-02  -4.736779988380D-02     1.50  2.825137817616D-03
5   -1.5   2.181172114883D-01   1.090586057442D-01     1.75  2.318229874323D-03
6    -.5  -4.501366875349D-01  -2.250683437675D-01     2.25 -3.310711143398D-03
7     .5   7.306438798566D-01   3.653219399283D-01     2.50 -5.647760474485D-03
8    1.5  -8.640313978940D-01  -4.320156989470D-01     2.75 -4.862720853201D-03
9    2.5   7.277569391293D-01   3.638784695647D-01     3.25  7.351622595738D-03
10   3.5  -4.422123450948D-01  -2.211061725474D-01     3.50  1.273069160543D-02
11   4.5   1.997354553163D-01   9.986772765817D-02     3.75  1.107220899555D-02
12   5.5  -6.475608143776D-02  -3.237804071888D-02     4.25 -1.693408103243D-02
13   6.5   1.118793937586D-02   5.593969687929D-03     4.50 -2.941634087854D-02
                                                      4.75 -2.564344313492D-02
      0   3.921333697483D-01  -2.261302885399D+00     5.25  3.938432579907D-02
      1   1.305698833228D-01   8.835252108632D-01     5.50  6.861585140251D-02
      2   1.080641116804D-02  -6.231731937547D-02     5.75  6.007624011072D-02
                                                      6.25 -9.417459898129D-02
```

6.50	-1.678838597485D-01

I= 9 H= 2.5

0		0.	2.469215107244D+01	-1.00	-8.899716502575D-03
1		4.258085884379D-03	-3.940921113097D+00	-.75	-5.804481636019D-03
2		-4.641630618196D-03	1.548833323600D-01	-.50	-3.289450596738D-03
				-.25	-1.354623384732D-03
0	-6.5	3.835447338174D-04	1.917723669087D-04	.25	7.747941128615D-04
1	-5.5	-2.223080778130D-03	-1.111540389065D-03	.50	9.806210605721D-04
2	-4.5	6.937632493148D-03	3.468816246574D-03	.75	6.736641537495D-04
3	-3.5	-1.731167177065D-02	-8.655835885326D-03	1.25	-7.616258249100D-04
4	-2.5	4.073656048411D-02	2.036828024205D-02	1.50	-1.213468516264D-03
5	-1.5	-9.473559976760D-02	-4.736779988380D-02	1.75	-9.957382828837D-04
6	-.5	2.176784355502D-01	1.088392177751D-01	2.25	1.422031192287D-03
7	.5	-4.489286404604D-01	-2.244643202302D-01	2.50	2.425845115565D-03
8	1.5	7.277569391293D-01	3.638784695647D-01	2.75	2.088643873961D-03
9	2.5	-8.572938715129D-01	-4.286469357564D-01	3.25	-3.157616297256D-03
10	3.5	7.121488932206D-01	3.560744466103D-01	3.50	-5.467864050126D-03
11	4.5	-4.072274807863D-01	-2.036137403931D-01	3.75	-4.755349794819D-03
12	5.5	1.479712680903D-01	7.398563404515D-02	4.25	7.271442372539D-03
13	6.5	-2.589292862555D-02	-1.294646431278D-02	4.50	1.262824803824D-02
				4.75	1.100413018509D-02
	0	-1.684313353703D-01	5.619984629517D+00	5.25	-1.686601124167D-02
	1	-5.608311215217D-02	-1.927437792418D+00	5.50	-2.931392617437D-02
	2	-4.641630618196D-03	1.548833323600D-01	5.75	-2.556330079443D-02
				6.25	3.927708823215D-02
				6.50	6.843633019742D-02

I=10 H= 3.5

0		0.	-5.706006916097D+01	-1.00	3.781660563138D-03
1		-1.809342515102D-03	9.111905473471D+00	-.75	2.466435788346D-03
2		1.972318048036D-03	-3.623656913931D-01	-.50	1.397750769560D-03
				-.25	5.756055067777D-04
0	-6.5	-1.629755329339D-04	-8.148776646694D-05	.25	-3.292249065671D-04
1	-5.5	9.446298919452D-04	4.723149459726D-04	.50	-4.166847309462D-04
2	-4.5	-2.947936926327D-03	-1.473968463163D-03	.75	-2.862528422193D-04
3	-3.5	7.356144915912D-03	3.678072457956D-03	1.25	3.236294404300D-04
4	-2.5	-1.731167177065D-02	-8.655835885326D-03	1.50	5.156260663342D-04
5	-1.5	4.029963177931D-02	2.014981588966D-02	1.75	4.231082942387D-04
6	-.5	-9.353192028648D-02	-4.676596014324D-02	2.25	-6.042482056526D-04
7	.5	2.148015511350D-01	1.074007755675D-01	2.50	-1.030787662595D-03
8	1.5	-4.422123450948D-01	-2.211061725474D-01	2.75	-8.875040939928D-04
9	2.5	7.121488932206D-01	3.560744466103D-01	3.25	1.341727839149D-03
10	3.5	-8.211648567043D-01	-4.105824283522D-01	3.50	2.323387977061D-03
11	4.5	6.311787430828D-01	3.155893715414D-01	3.75	2.020620442880D-03
12	5.5	-2.874256563094D-01	-1.437128281547D-01	4.25	-3.089682789427D-03
13	6.5	5.802776859927D-02	2.901388429963D-02	4.50	-5.365694666028D-03
				4.75	-4.675424119341D-03
	0	7.156971118135D-02	-1.314263404477D+01	5.25	7.164539047780D-03
	1	2.383079210936D-02	4.401151485360D+00	5.50	1.244930799186D-02
	2	1.972318048036D-03	-3.623656913931D-01	5.75	1.085207602236D-02
				6.25	-1.663962675233D-02
				6.50	-2.892345957149D-02

CARDINAL SPLINES N=3

I=11	H= 4.5				
0		0.	1.258647970127D+02	-1.00	-1.515479138476D-03
1		7.250838063915D-04	-2.012377253283D+01	-.75	-9.884102290910D-04
2		-7.903953320842D-04	8.028530801965D-01	-.50	-5.601407362168D-04
				-.25	-2.306706598531D-04
0	-6.5	6.531152569271D-05	3.265576284635D-05	.25	1.319350241294D-04
1	-5.5	-3.785551066128D-04	-1.892775533064D-04	.50	1.669840553526D-04
2	-4.5	1.181369237865D-03	5.906846189326D-04	.75	1.147142116909D-04
3	-3.5	-2.947936926327D-03	-1.473968463163D-03	1.25	-1.296926723345D-04
4	-2.5	6.937632493148D-03	3.468816246574D-03	1.50	-2.066342364549D-04
5	-1.5	-1.615165276219D-02	-8.075826381095D-03	1.75	-1.695582606954D-04
6	-.5	3.752421811578D-02	1.876210905789D-02	2.25	2.421490528350D-04
7	.5	-8.705109660132D-02	-4.352554830066D-02	2.50	4.130823192232D-04
8	1.5	1.997354553163D-01	9.986772765817D-02	2.75	3.556622260102D-04
9	2.5	-4.072274807863D-01	-2.036137403931D-01	3.25	-5.376897012759D-04
10	3.5	6.311787430828D-01	3.155893715414D-01	3.50	-9.310841198101D-04
11	4.5	-6.397043298835D-01	-3.198521649418D-01	3.75	-8.097514745752D-04
12	5.5	3.626945786306D-01	1.813472893153D-01	4.25	1.238169202632D-03
13	6.5	-8.585625633602D-02	-4.292812816801D-02	4.50	2.150260138803D-03
				4.75	1.873631988672D-03
	0	-2.868115803901D-02	2.898081818766D+01	5.25	-2.871061685590D-03
	1	-9.550055510703D-03	-9.686682490274D+00	5.50	-4.988717764768D-03
	2	-7.903953320842D-04	8.028530801965D-01	5.75	-4.348493999088D-03
				6.25	6.666219631996D-03
				6.50	1.158459751879D-02

I=12	H= 5.5				
0		0.	-1.378789620908D+02	-1.00	4.856164704105D-04
1		-2.323441015232D-04	2.260262318221D+01	-.75	3.167237836415D-04
2		2.532723688872D-04	-9.226937751447D-01	-.50	1.794901429834D-04
				-.25	7.391554843626D-05
0	-6.5	-2.092826736399D-05	-1.046413368199D-05	.25	-4.227694008646D-05
1	-5.5	1.213032831452D-04	6.065164157261D-05	.50	-5.350796689494D-05
2	-4.5	-3.785551066128D-04	-1.892775533064D-04	.75	-3.675874459009D-05
3	-3.5	9.446298919452D-04	4.723149459726D-04	1.25	4.155840628955D-05
4	-2.5	-2.223080778130D-03	-1.111540389065D-03	1.50	6.621337501440D-05
5	-1.5	5.175656851776D-03	2.587828425888D-03	1.75	5.433283898284D-05
6	-.5	-1.202556887360D-02	-6.012784436798D-03	2.25	-7.759365669157D-05
7	.5	2.792690883565D-02	1.396345441783D-02	2.50	-1.323670987252D-04
8	1.5	-6.475608143776D-02	-3.237804071888D-02	2.75	-1.139675425292D-04
9	2.5	1.479712680903D-01	7.398563404515D-02	3.25	1.722959822324D-04
10	3.5	-2.874256563094D-01	-1.437128281547D-01	3.50	2.983543268891D-04
11	4.5	3.626945786306D-01	1.813472893153D-01	3.75	2.594747786314D-04
12	5.5	-2.424626168839D-01	-1.212313084419D-01	4.25	-3.967558198398D-04
13	6.5	6.445814207335D-02	3.222907103667D-02	4.50	-6.890237667343D-04
				4.75	-6.003815064559D-04
	0	9.190520925585D-03	-2.994572340629D+01	5.25	9.199931561904D-04
	1	3.060196694011D-03	1.060760410533D+01	5.50	1.598563552252D-03
	2	2.532723688872D-04	-9.226937751447D-01	5.75	1.393407044954D-03
				6.25	-2.136039832873D-03
				6.50	-3.711929154485D-03

I=13	H= 6.5				
0		0.	5.231750069081D+01	-1.00	-8.378265668952D-05
1		4.008596758522D-05	-8.835681998343D+00	-.75	-5.464386331008D-05

2		-4.369668910430D-05	3.730351093267D-01	-.50	-3,096715606869D-05
				-.25	-1,275253496532D-05
0	-6,5	3.610721519087D-06	1.805360759543D-06	.25	7,293974922519D-06
1	-5,5	-2,092826736399D-05	-1.046413368199D-05	.50	9,231646564004D-06
2	-4,5	6,531152569271D-05	3,265576284635D-05	.75	6,341929209479D-06
3	-3,5	-1,629755329339D-04	-8,148776646694D-05	1.25	-7,170007400417D-06
4	-2,5	3,835447338174D-04	1.917723669087D-04	1.50	-1,142369092642D-05
5	-1,5	-8.929493062622D-04	-4,464746531311D-04	1,75	-9,373960465380D-06
6	-,5	2.074781861456D-03	1,037390930728D-03	2.25	1,338711327480D-05
7	,5	-4,818861730884D-03	-2.409430865442D-03	2.50	2,283709029204D-05
8	1.5	1.118793937586D-02	5,593969687929D-03	2,75	1,966264338224D-05
9	2,5	-2,589292862555D-02	-1,294646431278D-02	3,25	-2,972595855642D-05
10	3,5	5.802776859927D-02	2,901388429963D-02	3,50	-5,147460906649D-05
11	4,5	-8,585625633602D-02	-4,292812816801D-02	3,75	-4,476678070491D-05
12	5,5	6.445814207335D-02	3,222907103667D-02	4,25	6,845166381815D-05
13	6,5	-1,855619909194D-02	-9,278099545970D-03	4,50	1,188761939766D-04
				4,75	1,035828779903D-04
	0	-1,585626325353D-03	1.064630107063D+01	5,25	-1,587249322929D-04
	1	-5,279709907707D-04	-3,986225577096D+00	5,50	-2,757975016703D-04
	2	-4,369668910430D-05	3,730351093267D-01	5,75	-2,404020686972D-04
				6,25	3,685262368089D-04
				6,50	6,404089864085D-04

N= 3 M=14 P= 7,0 Q= 5

I= 0 H= -7.0

0		1,000000000000D+00	5,017230220651D-01	-1,00	3,473671634615D+00
1		-1,727557717697D+00	-8,640336348723D-01	-,75	2,715357366539D+00
2		7.461139169188D-01	3,730663658179D-01	-,50	2,050307338078D+00
				-,25	1,478521549232D+00
0	-7,0	-1,855619922220D-02	-9,278099611098D-03	.25	6,147245690950D-01
1	-6,0	6,445814282833D-02	3,222907141417D-02	.50	3,221697391557D-01
2	-5,0	-8,585625869212D-02	-4,292812934606D-02	.75	1,196173169368D-01
3	-4,0	5,802777447857D-02	2,901388723929D-02	1.25	-5,021023047667D-02
4	-3,0	-2,589294246183D-02	-1,294647123092D-02	1.50	-5,147708435779D-02
5	-2,0	1,118797158879D-02	5,593985794394D-03	1,75	-2,752040847183D-02
6	-1,0	-4,818936578376D-03	-2,409468289188D-03	2,25	1,677968353749D-02
7	0,0	2,074955706758D-03	1.037477853379D-03	2,50	1,898487022681D-02
8	1,0	-8,933530495711D-04	-4,466765247856D-04	2.75	1,083411553229D-02
9	2,0	3,844820918001D-04	1.922410459000D-04	3.25	-7,016349292957D-03
10	3,0	-1,651449124428D-04	-8.257245622142D-05	3,50	-8,037375827650D-03
11	4,0	7,017324680030D-05	3,508662340015D-05	3,75	-4,621139330656D-03
12	5,0	-2,812153281064D-05	-1,406076640532D-05	4,25	3,012078940789D-03
13	6,0	9,011195970517D-06	4.505597985258D-06	4,50	3,454790696427D-03
14	7,0	-1,554687668896D-06	-7.773438344478D-07	4,75	1,987863296519D-03
				5,25	-1,296540623535D-03
	0	2,546667790514D+01	1,273373950304D+01	5,50	-1,487295021383D-03
	1	8,718037119166D+00	4,358895486579D+00	5,75	-8,558452539757D-04
	2	7,461139169188D-01	3,730663658179D-01	6,25	5,582441992323D-04

```
 6.50   6.403858838563D-04
 6.75   3.685060978716D-04
```

```
I= 1      H= -6.0
 0              0.                      -9.986886627155D-03
 1              2.781182688967D+00       1.392068063394D+00
 2             -1.845640831795D+00      -9.228749423165D-01

 0   -7.0       6.445814282833D-02       3.222907141417D-02
 1   -6.0      -2.424626212599D-01      -1.212313106299D-01
 2   -5.0       3.626945922869D-01       1.813472961434D-01
 3   -4.0      -2.874256903867D-01      -1.437128451934D-01
 4   -3.0       1.479713482874D-01       7.398567414370D-02
 5   -2.0      -6.475626814859D-02      -3.237813407430D-02
 6   -1.0       2.792734266261D-02       1.396367133130D-02
 7    0.0      -1.202657650626D-02      -6.013288253131D-03
 8    1.0       5.177997006749D-03       2.588998503374D-03
 9    2.0      -2.228513841404D-03      -1.114256920702D-03
10    3.0       9.572039312720D-04       4.786019656360D-04
11    4.0      -4.067343496409D-04      -2.033671748204D-04
12    5.0       1.629964964539D-04       8.149824822697D-05
13    6.0      -5.223020317350D-05      -2.611510158675D-05
14    7.0       9.011195970517D-06       4.505597985258D-06

        0      -7.096812193519D+01      -3.548638261638D+01
        1      -2.305778895616D+01      -1.152818112904D+01
        2      -1.845640831795D+00      -9.228749423165D-01
```

```
-1.00  -4.626823520762D+00
 -.75  -3.124059984610D+00
 -.50  -1.852001552432D+00
 -.25  -8.106482242289D-01
  .25   5.800060676596D-01
  .50   9.311954534980D-01
  .75   1.063010268281D+00
 1.25   7.891384225807D-01
 1.50   5.009842270995D-01
 1.75   2.152141793640D-01
 2.25  -1.084973414266D-01
 2.50  -1.174242899062D-01
 2.75  -6.515456858792D-02
 3.25   4.115217751672D-02
 3.50   4.690407545561D-02
 3.75   2.688643684016D-02
 4.25  -1.747932353405D-02
 4.50  -2.003818611985D-02
 4.75  -1.152632528621D-02
 5.25   7.515837793347D-03
 5.50   8.621169508444D-03
 5.75   4.960792413014D-03
 6.25  -3.235703387871D-03
 6.50  -3.711795248731D-03
 6.75  -2.135923104662D-03
```

```
I= 2      H= -5.0
 0              0.                       3.116640242754D-02
 1             -1.520640325546D+00      -7.649286075826D-01
 2              1.606496584238D+00       8.034184544912D-01

 0   -7.0      -8.585625869212D-02      -4.292812934606D-02
 1   -6.0       3.626945922869D-01       1.813472961434D-01
 2   -5.0      -6.397043725011D-01      -3.198521862506D-01
 3   -4.0       6.311788494289D-01       3.155894247145D-01
 4   -3.0      -4.072277310599D-01      -2.036138655300D-01
 5   -2.0       1.997360379909D-01       9.986801899547D-02
 6   -1.0      -8.705245045923D-02      -4.352622522961D-02
 7    0.0       3.752736266787D-02       1.876368133393D-02
 8    1.0      -1.615895576004D-02      -8.079477880022D-03
 9    2.0       6.954587630693D-03       3.477293815347D-03
10    3.0      -2.987177140461D-03      -1.493588570231D-03
11    4.0       1.269309119540D-03       6.346545597698D-04
12    5.0      -5.086684755683D-04      -2.543342377842D-04
13    6.0       1.629964964539D-04       8.149824822697D-05
14    7.0      -2.812153281064D-05      -1.406076640532D-05

        0       6.807385034886D+01       3.404417041942D+01
        1       2.097031185379D+01       1.048292975529D+01
        2       1.606496584238D+00       8.034184544912D-01
```

```
-1.00   3.127136909785D+00
 -.75   2.044134572794D+00
 -.50   1.161944308833D+00
 -.25   4.805661179015D-01
  .25  -2.798378888743D-01
  .50  -3.613790247977D-01
  .75  -2.572000081645D-01
 1.25   3.476921916617D-01
 1.50   6.930200677826D-01
 1.75   9.356781942186D-01
 2.25   8.667750582481D-01
 2.50   5.888238732595D-01
 2.75   2.653417217057D-01
 3.25  -1.409607540414D-01
 3.50  -1.546118140651D-01
 3.75  -8.653576876611D-02
 4.25   5.508854136912D-02
 4.50   6.288879301168D-02
 4.75   3.608393882221D-02
 5.25  -2.347820247680D-02
 5.50  -2.691966253740D-02
 5.75  -1.548619188105D-02
 6.25   1.009876836341D-02
 6.50   1.158417963474D-02
```

CARDINAL SPLINES N=3

6.75	6.665855354465D-03

I= 3 H= -4.0

0		0.	-7.777128170136D-02
1		6.686759975066D-01	3.458377109937D-01
2		-7.267037719852D-01	-3.637765017562D-01
0	-7.0	5.802777447857D-02	2.901388723929D-02
1	-6.0	-2.874256903867D-01	-1.437128451934D-01
2	-5.0	6.311788494289D-01	3.155894247145D-01
3	-4.0	-8.211651220757D-01	-4.105825610379D-01
4	-3.0	7.121495177426D-01	3.560747588713D-01
5	-2.0	-4.422137990755D-01	-2.211068995378D-01
6	-1.0	2.148049294930D-01	1.074024647465D-01
7	0.0	-9.353976706465D-02	-4.676988353232D-02
8	1.0	4.031785536173D-02	2.015892768086D-02
9	2.0	-1.735398088159D-02	-8.676990440795D-03
10	3.0	7.454063235283D-03	3.727031617642D-03
11	4.0	-3.167378272639D-03	-1.583689136319D-03
12	5.0	1.269309119540D-03	6.346545597698D-04
13	6.0	-4.067343496409D-04	-2.033671748204D-04
14	7.0	7.017324680030D-05	3.508662340015D-05
	0	-3.092775284473D+01	-1.548195589080D+01
	1	-9.505176810286D+00	-4.747033313593D+00
	2	-7.267037719852D-01	-3.637765017562D-01

-1.00	-1.395379769492D+00
-.75	-9.102778698717D-01
-.50	-5.160139417496D-01
-.25	-2.125879851257D-01
.25	1.218066813761D-01
.50	1.544754237095D-01
.75	1.065063892773D-01
1.25	-1.228236218874D-01
1.50	-2.004031310847D-01
1.75	-1.711399204261D-01
2.25	2.952198201758D-01
2.50	6.336517729526D-01
2.75	9.017984261283D-01
3.25	8.887161514460D-01
3.50	6.139578623304D-01
3.75	2.797926696496D-01
4.25	-1.503799565047D-01
4.50	-1.654154509672D-01
4.75	-9.275211855516D-02
5.25	5.914304600376D-02
5.50	6.753984565752D-02
5.75	3.876035164892D-02
6.25	-2.522401205882D-02
6.50	-2.892241680185D-02
6.75	-1.663871775006D-02

I= 4 H= -3.0

0		0.	1.830260409987D-01
1		-2.885155203219D-01	-1.713210508508D-01
2		3.144084627838D-01	1.582035172524D-01
0	-7.0	-2.589294246183D-02	-1.294647123092D-02
1	-6.0	1.479713482874D-01	7.398567414370D-02
2	-5.0	-4.072277310599D-01	-2.036138655300D-01
3	-4.0	7.121495177426D-01	3.560747588713D-01
4	-3.0	-8.572953412558D-01	-4.286476706279D-01
5	-2.0	7.277603609109D-01	3.638801804555D-01
6	-1.0	-4.489365910493D-01	-2.244682955247D-01
7	0.0	2.176969020678D-01	1.088484510339D-01
8	1.0	-9.477848693699D-02	-4.738924346850D-02
9	2.0	4.083613025949D-02	2.041806512974D-02
10	3.0	-1.754211161798D-02	-8.771055808989D-03
11	4.0	7.454063235283D-03	3.727031617642D-03
12	5.0	-2.987177140461D-03	-1.493588570231D-03
13	6.0	9.572039312720D-04	4.786019656360D-04
14	7.0	-1.651449124428D-04	-8.257245622142D-05
	0	1.338640603415D+01	6.735751030410D+00
	1	4.113202958651D+00	2.043528190683D+00
	2	3.144084627838D-01	1.582035172524D-01

-1.00	6.029239831057D-01
-.75	3.932414005573D-01
-.50	2.228598758569D-01
-.25	9.177940900447D-02
.25	-5.250363723312D-02
.50	-6.646479891696D-02
.75	-4.567639654494D-02
1.25	5.174433652052D-02
1.50	8.264533359502D-02
1.75	6.810496147542D-02
2.25	-9.940958155497D-02
2.50	-1.739120112327D-01
2.75	-1.560222010065D-01
3.25	2.854293274759D-01
3.50	6.224365452339D-01
3.75	8.953501504082D-01
4.25	8.929191934556D-01
4.50	6.187786907203D-01
4.75	2.825665665002D-01
5.25	-1.521892343434D-01
5.50	-1.674909857133D-01
5.75	-9.394651625456D-02
6.25	5.992226703981D-02
6.50	6.843387615534D-02
6.75	3.927494899667D-02

CARDINAL SPLINES N=3

I= 5 H= -2.0

0		0.	-4,261120501658D-01	-1.00	-2,596949079263D-01
1		1.242534681687D-01	1.251341302903D-01	-.75	-1,693759109902D-01
2		-1.354414397575D-01	-7.004720710692D-02	-.50	-9,598709402375D-02
				-.25	-3,952845702703D-02
0	-7.0	1.118797158879D-02	5.593985794394D-03	.25	2,260920281084D-02
1	-6,0	-6.475626814859D-02	-3.237813407430D-02	.50	2,861599825714D-02
2	-5,0	1.997360379909D-01	9.986801899547D-02	.75	1,965924936458D-02
3	-4,0	-4.422137990755D-01	-2,211068995378D-01	1.25	-2,223067325121D-02
4	-3,0	7.277603609109D-01	3,638801804555D-01	1,50	-3,542801132861D-02
5	-2,0	-8,640393643141D-01	-4.320196821571D-01	1,75	-2,908366678571D-02
6	-1,0	7.306623900183D-01	3,653311950092D-01	2,25	4,162744079859D-02
7	0,0	-4,501796803537D-01	-2,250898401769D-01	2,50	7,119886233875D-02
8	1,0	2,182170592424D-01	1,091085296212D-01	2,75	6,157278094260D-02
9	2,0	-9,496741311838D-02	-4,748370655919D-02	3,25	-9,517920724564D-02
10	3,0	4,083613025949D-02	2.041806512974D-02	3.50	-1,690659996537D-01
11	4,0	-1,735398088159D-02	-8,676990440795D-03	3,75	-1,532359314112D-01
12	5,0	6,954587630693D-03	3,477293815347D-03	4,25	2,836131599007D-01
13	6,0	-2,228513841404D-03	-1,114256920702D-03	4,50	6,203533729250D-01
14	7,0	3,844820918001D-04	1,922410459000D-04	4,75	8,941514490436D-01
				5,25	8,937011707830D-01
	0	-5,766856270938D+00	-2,982486286373D+00	5,50	6,196758716146D-01
	1	-1,771926688437D+00	-8,555267692065D-01	5,75	2,830829727843D-01
	2	-1,354414397575D-01	-7.004720710692D-02	6,25	-1,525264212139D-01
				6,50	-1,678781463710D-01
				6,75	-9,416961852067D-02

I= 6 H= -1.0

0		0.	9,900815202045D-01	-1.00	1,118218181103D-01
1		-5,350144076596D-02	-1,731499096504D-01	-.75	7,293129283066D-02
2		5,832037734434D-02	3,456583729226D-02	-.50	4,133081471906D-02
				-.25	1,702038377551D-02
0	-7,0	-4,818936578376D-03	-2,409468289188D-03	.25	-9,735042600221D-03
1	-6,0	2,792734266261D-02	1.396367133130D-02	.50	-1,232121781497D-02
2	-5,0	-8,705245045923D-02	-4,352622522961D-02	.75	-8,464424557095D-03
3	-4,0	2,148049294930D-01	1.074024647465D-01	1.25	9,569834087643D-03
4	-3,0	-4,489365910493D-01	-2,244682955247D-01	1.50	1,524761769198D-02
5	-2,0	7,306623900183D-01	3,653311950092D-01	1.75	1,251230341161D-02
6	-1,0	-8,652828475830D-01	-4,326414237915D-01	2.25	-1,787302345514D-02
7	0,0	7,311827337369D-01	3,655913668684D-01	2.50	-3,049764947750D-02
8	1,0	-4,503686861052D-01	-2,251843430526D-01	2,75	-2,627002661873D-02
9	2,0	2,182170592424D-01	1,091085296212D-01	3,25	3,980521642676D-02
10	3,0	-9,477848693699D-02	-4,738924346850D-02	3,50	6,911139903054D-02
11	4,0	4,031785536173D-02	2,015892768086D-02	3,75	6,037252262414D-02
12	5,0	-1,615895576004D-02	-8,079477880022D-03	4,25	-9,439672273472D-02
13	6,0	5.177997006749D-03	2.588998503374D-03	4,50	-1,681683514552D-01
14	7,0	-8,933530495711D-04	-4,466765247856D-04	4,75	-1,527192955360D-01
				5,25	2,832758451827D-01
	0	2,483188404511D+00	1.471758179973D+00	5,50	6,199660705431D-01
	1	7,629838420547D-01	3,107718124413D-01	5,75	8,939282669177D-01
	2	5,832037734434D-02	3,456583729226D-02	6,25	8,938475596007D-01
				6,50	6,198446366068D-01
				6,75	2,831808164044D-01

CARDINAL SPLINES N=3

I= 7 H= 0.0

0		0.	-2,299629998637D+00	-1.00	-4,814713898972D-02
1		2,303609164148D-02	3,515546628754D-01	-.75	-3,140203286450D-02
2		-2,511104734824D-02	-2,511104734824D-02	-.50	-1,779580765780D-02
				-.25	-7,328463369636D-03
0	-7,0	2,074955706758D-03	1,037477853379D-03	.25	4,191608775038D-03
1	-6,0	-1,202657650626D-02	-6,013288253131D-03	.50	5,305126349517D-03
2	-5,0	3,752736266787D-02	1,876368133393D-02	.75	3,644501313295D-03
3	-4,0	-9,353976706465D-02	-4,676988353232D-02	1.25	-4,120379344702D-03
4	-3,0	2,176969020678D-01	1,088484510339D-01	1.50	-6,564854688944D-03
5	-2,0	-4,501796803537D-01	-2,250898401769D-01	1.75	-5,386958779379D-03
6	-1,0	7,311827337369D-01	3,655913668684D-01	2.25	7,693379880523D-03
7	0,0	-8,654718605092D-01	-4,327359302546D-01	2.50	1,312449115425D-02
8	1,0	7,311827337369D-01	3,655913668684D-01	2.75	1,130063741630D-02
9	2,0	-4,501796803537D-01	-2,250898401769D-01	3,25	-1,708817720175D-02
10	3,0	2,176969020678D-01	1,088484510339D-01	3,50	-2,959843606314D-02
11	4,0	-9,353976706465D-02	-4,676988353232D-02	3.75	-2,575288285126D-02
12	5,0	3,752736266787D-02	1,876368133393D-02	4.25	3,946778931818D-02
13	6,0	-1,202657650626D-02	-6,013288253131D-03	4.50	6,872401702456D-02
14	7,0	2,074955706758D-03	1,037477853379D-03	4.75	6,014931160751D-02
				5,25	-9,425032450616D-02
	0	-1,069188678573D+00	-1,069188678573D+00	5,50	-1,679995777925D-01
	1	-3,285185712339D-01	-1,106852512649D-23	5,75	-1,526214476564D-01
	2	-2,511104734824D-02	-2,511104734824D-02	6,25	2,832101320249D-01
				6.50	6,198887454714D-01
				6.75	8,938820771866D-01

I= 8 H= 1.0

0		0.	5,340886894383D+00	-1.00	2,072924143081D-02
1		-9,917944190621D-03	-7,946935345330D-01	-.75	1,351981284057D-02
2		1.081129724019D-02	3.456583729226D-02	-.50	7,661796405359D-03
				-.25	3,155192125167D-03
0	-7.0	-8,933530495711D-04	-4,466765247856D-04	.25	-1,804652385231D-03
1	-6,0	5,177997006749D-03	2,588998503374D-03	.50	-2,284065068062D-03
2	-5,0	-1,615895576004D-02	-8,079477880022D-03	.75	-1,569100311613D-03
3	-4,0	4,031785536173D-02	2.015892768086D-02	1,25	1,773981188877D-03
4	-3,0	-9.477848693699D-02	-4,738924346850D-02	1.50	2,826415190781D-03
5	-2,0	2.182170592424D-01	1.091085296212D-01	1.75	2,319278050926D-03
6	-1,0	-4,503686861052D-01	-2,251843430526D-01	2,25	-3,312208062333D-03
7	0,0	7,311827337369D-01	3,655913668684D-01	2,50	-5,650314070034D-03
8	1,0	-8,652828475830D-01	-4,326414237915D-01	2,75	-4,864919488753D-03
9	2,0	7,306623900183D-01	3,653311950092D-01	3,25	7,354946490089D-03
10	3,0	-4,489365910493D-01	-2,244682955247D-01	3,50	1,273644738827D-02
11	4,0	2,148049294930D-01	1,074024647465D-01	3,75	1.107721472308D-02
12	5,0	-8,705245045923D-02	-4,352622522961D-02	4,25	-1,694173515387D-02
13	6,0	2,792734266261D-02	1,396367133130D-02	4,50	-2,942963336505D-02
14	7.0	-4,818936578376D-03	-2,409468289188D-03	4,75	-2,565502555478D-02
				5.25	3,940207408648D-02
	0	4.603279554351D-01	1,471758179973D+00	5,50	6,864669049554D-02
	1	1.414402171721D-01	-3,107718124413D-01	5,75	6,010312135789D-02
	2	1.081129724019D-02	3,456583729226D-02	6,25	-9,421580677313D-02
				6,50	-1,679554688157D-01
				6,75	-1,525921320008D-01

CARDINAL SPLINES N=3

I= 9	H= 2.0				
0		0,	-1,240348681906D+01	-1,00	-8,921466820799D-03
1		4,268492364499D-03	1,836187668703D+00	-,75	-5,818667405043D-03
2		-4,652974456299D-03	-7,004720710692D-02	-,50	-3,297489796324D-03
				-,25	-1,357933994644D-03
0	-7,0	3,844820918001D-04	1,922410459000D-04	,25	7,766876583989D-04
1	-6,0	-2,228513841404D-03	-1,114256920702D-03	,50	9,830176335435D-04
2	-5,0	6,954587630693D-03	3,477293815347D-03	,75	6,753105443500D-04
3	-4,0	-1,735398088159D-02	-8,676990440795D-03	1,25	-7,634871879747D-04
4	-3,0	4,083613025949D-02	2,041806512974D-02	1,50	-1,216434152862D-03
5	-2,0	-9,496741311838D-02	-4,748370655919D-02	1,75	-9,981718011223D-04
6	-1,0	2,182170592424D-01	1,091085296212D-01	2,25	1,425506541306D-03
7	0,0	-4,501796803537D-01	-2,250898401769D-01	2,50	2,431773717029D-03
8	1,0	7,306623900183D-01	3,653311950092D-01	2,75	2,093748376103D-03
9	2,0	-8,640393643141D-01	-4,320196821571D-01	3,25	-3,165333276915D-03
10	3,0	7,277603609109D-01	3,638801804555D-01	3,50	-5,481227067843D-03
11	4,0	-4,422137990755D-01	-2,211068995378D-01	3,75	-4,766971433017D-03
12	5,0	1,997360379909D-01	9,986801899547D-02	4,25	7,289212702529D-03
13	6,0	-6,475626814859D-02	-3,237813407430D-02	4,50	1,265910878086D-02
14	7,0	1,118797158879D-02	5,593985794394D-03	4,75	1,103102072044D-02
				5,25	-1,690721687533D-02
	0	-1,981163018072D-01	-2,982486286373D+00	5,50	-2,938552431454D-02
	1	-6,087315002369D-02	8,555267692065D-01	5,75	-2,562571012992D-02
	2	-4,652974456299D-03	-7,004720710692D-02	6,25	3,937275905014D-02
				6,50	6,860258268831D-02
				6,75	6,006860471116D-02

I=10	H= 3.0				
0		0,	2,879242071056D+01	-1,00	3,831998529571D-03
1		-1,833426808564D-03	-4,258377432216D+00	-,75	2,499266699489D-03
2		1,998571721007D-03	1,582035172524D-01	-,50	1,416356334534D-03
				-,25	5,832674347039D-04
0	-7,0	-1,651449124428D-04	-8,257245622142D-05	,25	-3,336072439066D-04
1	-6,0	9,572039312720D-04	4,786019656360D-04	,50	-4,222312525441D-04
2	-5,0	-2,987177140461D-03	-1,493588570231D-03	,75	-2,900631751961D-04
3	-4,0	7,454063235283D-03	3,727031617642D-03	1,25	3,279372960902D-04
4	-3,0	-1,754211161798D-02	-8,771055808989D-03	1,50	5,224896034087D-04
5	-2,0	4,083613025949D-02	2,041806512974D-02	1,75	4,287403203152D-04
6	-1,0	-9,477848693699D-02	-4,738924346850D-02	2,25	-6,122913984233D-04
7	0,0	2,176969020678D-01	1,088484510339D-01	2,50	-1,044508553407D-03
8	1,0	-4,489365910493D-01	-2,244682955247D-01	2,75	-8,993177260902D-04
9	2,0	7,277603609109D-01	3,638801804555D-01	3,25	1,359587672696D-03
10	3,0	-8,572953412558D-01	-4,286476706279D-01	3,50	2,354314749275D-03
11	4,0	7,121495177426D-01	3,560747588713D-01	3,75	2,047517043999D-03
12	5,0	-4,072277310599D-01	-2,036138655300D-01	4,25	-3,130809648600D-03
13	6,0	1,479713482874D-01	7,398567414370D-02	4,50	-5,437117392352D-03
14	7,0	-2,589294246183D-02	-1,294647123092D-02	4,75	-4,737658375034D-03
				5,25	7,259903528556D-03
	0	8,509602666938D-02	6,735751030410D+00	5,50	1,261501153419D-02
	1	2,614657728553D-02	-2,043528190683D+00	5,75	1,099651340496D-02
	2	1,998571721007D-03	1,582035172524D-01	6,25	-1,686104302656D-02
				6,50	-2,930822692155D-02
				6,75	-2,556002119463D-02

I=11 H= 4.0

CARDINAL SPLINES N=3

0		0.	-6.653623767200D+01	-1.00	-1.628289807568D-03
1		7.790582803838D-04	9.839904338180D+00	-.75	-1.061986444329D-03
2		-8.492315271841D-04	-3.637765017562D-01	-.50	-6.018370219879D-04
				-.25	-2.478415405450D-04
0	-7.0	7.017324680030D-05	3.508662340015D-05	.25	1.417561282083D-04
1	-6.0	-4.067343496409D-04	-2.033671748204D-04	.50	1.794141723584D-04
2	-5.0	1.269309119540D-03	6.346545597698D-04	.75	1.232534166496D-04
3	-4.0	-3.167378272639D-03	-1.583689136319D-03	1.25	-1.393468581074D-04
4	-3.0	7.454063235283D-03	3.727031617642D-03	1.50	-2.220158711250D-04
5	-2.0	-1.735398088159D-02	-8.676990440795D-03	1.75	-1.821799988200D-04
6	-1.0	4.031785536173D-02	2.015892768086D-02	2.25	2.601743729344D-04
7	0.0	-9.353976706465D-02	-4.676988353232D-02	2.50	4.438317313012D-04
8	1.0	2.148049294930D-01	1.074024647465D-01	2.75	3.821373467937D-04
9	2.0	-4.422137990755D-01	-2.211068995378D-01	3.25	-5.777147542342D-04
10	3.0	7.121495177426D-01	3.560747588713D-01	3.50	-1.000393035397D-03
11	4.0	-8.211651220757D-01	-4.105825610379D-01	3.75	-8.700285136293D-04
12	5.0	6.311788494289D-01	3.155894247145D-01	4.25	1.330337179006D-03
13	6.0	-2.874256903867D-01	-1.437128451934D-01	4.50	2.310323131330D-03
14	7.0	5.802777447857D-02	2.901388723929D-02	4.75	2.013103018647D-03
				5.25	-3.084779712882D-03
	0	-3.615893686934D-02	-1.548195589080D+01	5.50	-5.360070221100D-03
	1	-1.111018310019D-02	4.747033313593D+00	5.75	-4.672187603096D-03
	2	-8.492315271841D-04	-3.637765017562D-01	6.25	7.162427957707D-03
				6.50	1.244688626982D-02
				6.75	1.085068245738D-02

I=12 H= 5.0

0		0.	1.467921829765D+02	-1.00	6.525279550546D-04
1		-3.122032111220D-04	-2.173078811817D+01	-.75	4.255850768036D-04
2		3.403247439326D-04	8.034184544912D-01	-.50	2.411827915441D-04
				-.25	9.932109927628D-05
0	-7.0	-2.812153281064D-05	-1.406076640532D-05	.25	-5.680796871909D-05
1	-6.0	1.629964964539D-04	8.149824822697D-05	.50	-7.189921747817D-05
2	-5.0	-5.086684755683D-04	-2.543342377842D-04	.75	-4.939311143504D-05
3	-4.0	1.269309119540D-03	6.346545597698D-04	1.25	5.584246730397D-05
4	-3.0	-2.987177140461D-03	-1.493588570231D-03	1.50	8.897160789882D-05
5	-2.0	6.954587630693D-03	3.477293815347D-03	1.75	7.300760676953D-05
6	-1.0	-1.615895576004D-02	-8.079477880022D-03	2.25	-1.042634119832D-04
7	0.0	3.752736266787D-02	1.876368133393D-02	2.50	-1.778630616795D-04
8	1.0	-8.705245045923D-02	-4.352622522961D-02	2.75	-1.531393846224D-04
9	2.0	1.997360379909D-01	9.986801899547D-02	3.25	2.315159218532D-04
10	3.0	-4.072277310599D-01	-2.036138655300D-01	3.50	4.009018438283D-04
11	4.0	6.311788494289D-01	3.155894247145D-01	3.75	3.486589857586D-04
12	5.0	-6.397043725011D-01	-3.198521862506D-01	4.25	-5.331249584777D-04
13	6.0	3.626945922869D-01	1.813472961434D-01	4.50	-9.258484565818D-04
14	7.0	-8.585625869212D-02	-4.292812934606D-02	4.75	-8.067389092192D-04
				5.25	1.236204322365D-03
	0	1.449048997484D-02	3.404417041942D+01	5.50	2.148006174226D-03
	1	4.452343203935D-03	-1.048292975529D+01	5.75	1.872334973044D-03
	2	3.403247439326D-04	8.034184544912D-01	6.25	-2.870215678152D-03
				6.50	-4.987747273320D-03
				6.75	-4.347935535784D-03

I=13 H= 6.0

0		0.	-1.614045226931D+02	-1.00	-2.090944799715D-04

CARDINAL SPLINES N=3

1		1.000416420005D-04	2.444843032147D+01	-.75	-1.363734528591D-04
2		-1.090528379710D-04	-9.228749423165D-01	-.50	-7.728403049299D-05
				-.25	-3.182621287331D-05
0	-7.0	9.011195970517D-06	4.505597985258D-06	.25	1.820340812300D-05
1	-6.0	-5.223020317350D-05	-2.611510158675D-05	.50	2.303921138157D-05
2	-5.0	1.629964964539D-04	8.149824822697D-05	.75	1.582740918546D-05
3	-4.0	-4.067343496409D-04	-2.033671748204D-04	1.25	-1.789402518388D-05
4	-3.0	9.572039312720D-04	4.786019656360D-04	1.50	-2.850984688209D-05
5	-2.0	-2.228513841404D-03	-1.114256920702D-03	1.75	-2.339438096375D-05
6	-1.0	5.177997006749D-03	2.588998503374D-03	2.25	3.340991559547D-05
7	0.0	-1.202657650626D-02	-6.013288253131D-03	2.50	5.699400934366D-05
8	1.0	2.792734266261D-02	1.396367133130D-02	2.75	4.907161403803D-05
9	2.0	-6.475626814859D-02	-3.237813407430D-02	3.25	-7.418640191275D-05
10	3.0	1.479713482874D-01	7.398567414370D-02	3.50	-1.284640166444D-04
11	4.0	-2.874256903867D-01	-1.437128451934D-01	3.75	-1.117234411581D-04
12	5.0	3.626945922869D-01	1.813472961434D-01	4.25	1.708332670885D-04
13	6.0	-2.424626212599D-01	-1.212313106299D-01	4.50	2.966766236251D-04
14	7.0	6.445814282833D-02	3.222907141417D-02	4.75	2.585094394737D-04
				5.25	-3.961261983585D-04
	0	-4.643297566576D-03	-3.548638261638D+01	5.50	-6.883015117755D-04
	1	-1.426698089594D-03	1.152818112904D+01	5.75	-5.999658939044D-04
	2	-1.090528379710D-04	-9.228749423165D-01	6.25	9.197220636128D-04
				6.50	1.598252570320D-03
				6.75	1.393228092323D-03

I=14 H= 7.0

0		0.	6.152625983416D+01	-1.00	3.607474642746D-05
1		-1.726002937928D-05	-9.581824608029D+00	-.75	2.352830037406D-05
2		1.881471704818D-05	3.730663658179D-01	-.50	1.333369395168D-05
				-.25	5.490927160331D-06
0	-7.0	-1.554687668896D-06	-7.773438344478D-07	.25	-3.140605778986D-06
1	-6.0	9.011195970517D-06	4.505597985258D-06	.50	-3.974919417249D-06
2	-5.0	-2.812153281064D-05	-1.406076640532D-05	.75	-2.730678366289D-06
3	-4.0	7.017324680030D-05	3.508662340015D-05	1.25	3.087228420190D-06
4	-3.0	-1.651449124428D-04	-8.257245622142D-05	1.50	4.918759677878D-06
5	-2.0	3.844820918001D-04	1.922410459000D-04	1.75	4.036196274528D-06
6	-1.0	-8.933530495711D-04	-4.466765247856D-04	2.25	-5.764160935236D-06
7	0.0	2.074955706758D-03	1.037477853379D-03	2.50	-9.833088061911D-06
8	1.0	-4.818936578376D-03	-2.409468289188D-03	2.75	-8.466249483203D-06
9	2.0	1.118797158879D-02	5.593985794394D-03	3.25	1.279926489078D-05
10	3.0	-2.589294246183D-02	-1.294647123092D-02	3.50	2.216369759239D-05
11	4.0	5.802777447857D-02	2.901388723929D-02	3.75	1.927547205218D-05
12	5.0	-8.585625869212D-02	-4.292812934606D-02	4.25	-2.947359858060D-05
13	6.0	6.445814282833D-02	3.222907141417D-02	4.50	-5.118515752363D-05
14	7.0	-1.855619922220D-02	-9.278099611098D-03	4.75	-4.460023223679D-05
				5.25	6.834303620021D-05
	0	8.011009297057D-04	1.273373950304D+01	5.50	1.187515844484D-04
	1	2.461460092952D-04	-4.358895486579D+00	5.75	1.035111730012D-04
	2	1.881471704818D-05	3.730663658179D-01	6.25	-1.586781611108D-04
				6.50	-2.757438484396D-04
				6.75	-2.403711942773D-04

CARDINAL SPLINES N=3

N= 3 M=15 P= 7.5 Q= 5

I= 0 H= -7.5

0		1.000000000000D+00	4.991443583249D-01	-1.00	3,473671635176D+00
1		-1,727557717965D+00	-8,636610575541D-01	-.75	2,715357366905D+00
2		7,461139172110D-01	3,730529080289D-01	-.50	2,050307338285D+00
				-.25	1,478521549317D+00
0	-7,5	-1,855619924635D-02	-9,278099623173D-03	.25	6,147245690462D-01
1	-6,5	6,445814296830D-02	3,222907148415D-02	.50	3,221697390940D-01
2	-5,5	-8,585625912893D-02	-4,292812956447D-02	.75	1,196173168943D-01
3	-4,5	5,802777556856D-02	2,901388778428D-02	1,25	-5,021023042872D-02
4	-3,5	-2,589294502701D-02	-1,294647251350D-02	1,50	-5,147708428139D-02
5	-2,5	1,118797756091D-02	5,593988780455D-03	1,75	-2,752040840914D-02
6	-1,5	-4,818950454741D-03	-2,409475227371D-03	2,25	1,677968344796D-02
7	-,5	2,074987936943D-03	1,037493968472D-03	2,50	1,898487007407D-02
8	,5	-8,934279042348D-04	-4,467139521174D-04	2,75	1,083411540079D-02
9	1,5	3,846559342158D-04	1,923279671079D-04	3,25	-7,016349094147D-03
10	2,5	-1,655485158964D-04	-8,277425794820D-05	3,50	-8,037375483384D-03
11	3,5	7,110732822641D-05	3,555366411321D-05	3,75	-4,621139031252D-03
12	4,5	-3,021487010002D-05	-1,510743505001D-05	4,25	3,012078482979D-03
13	5,5	1,210843874780D-05	6,054219373902D-06	4,50	3,454789901374D-03
14	6,5	-3,879998831662D-06	-1,939999415831D-06	4,75	1,987862603748D-03
15	7,5	6,694101824693D-07	3,347050912346D-07	5,25	-1,296539561970D-03
				5,50	-1,487293176827D-03
	0	3,001222495839D+01	1,500591250329D+01	5,75	-8,558436461474D-04
	1	9,464151040201D+00	4,732132562879D+00	6,25	5,582417345011D-04
	2	7,461139172110D-01	3,730529080289D-01	6,50	6,403816007564D-04
				6,75	3,685023642119D-04
				7,25	-2,403654703157D-04
				7,50	-2,757339014091D-04

I= 1 H= -6.5

0		0.	4,959423663749D-03	-1.00	-4,626823524010D+00
1		2,781182690521D+00	1,389908551039D+00	-.75	-3,124059986728D+00
2		-1,845640833489D+00	-9,227969390127D-01	-.50	-1,852001553633D+00
				-.25	-8,106482247232D-01
0	-7,5	6,445814296830D-02	3,222907148415D-02	.25	5,800060679423D-01
1	-6,5	-2,424626220712D-01	-1,212313110356D-01	.50	9,311954538558D-01
2	-5,5	3,626945948187D-01	1,813472974093D-01	.75	1,063010268527D+00
3	-4,5	-2,874256967045D-01	-1,437128483522D-01	1,25	7,891384223027D-01
4	-3,5	1,479713631555D-01	7,398568157777D-02	1,50	5,009842266567D-01
5	-2,5	-6,475630276387D-02	-3,237815138194D-02	1,75	2,152141790006D-01
6	-1,5	2,792742309204D-02	1,396371154602D-02	2,25	-1,084973409077D-01
7	-,5	-1,202676331710D-02	-6,013381658552D-03	2,50	-1,174242890209D-01
8	,5	5,178430875272D-03	2,589215437636D-03	2,75	-6,515456782570D-02
9	1,5	-2,229521457343D-03	-1,114760728671D-03	3,25	4,115217636439D-02
10	2,5	9,595432756221D-04	4,797716378110D-04	3,50	4,690407346019D-02
11	3,5	-4,121484215162D-04	-2,060742107581D-04	3,75	2,688643510477D-02
12	4,5	1,751297840103D-04	8,756489200517D-05	4,25	-1,747932088051D-02

13	5.5	-7.018227302836D-05	-3.509113651418D-05	4.50	-2.003818151161D-02
14	6.5	2.248903785440D-05	1.124451892720D-05	4.75	-1.152632127081D-02
15	7.5	-3.879998831662D-06	-1.939999415831D-06	5.25	7.515831640360D-03
				5.50	8.621158817129D-03
	0	-8.295842670485D+01	-4.147805426300D+01	5.75	4.960783093807D-03
	1	-2.490342981181D+01	-1.245204553415D+01	6.25	-3.235689101930D-03
	2	-1.845640833489D+00	-9.227969390127D-01	6.50	-3.711770423260D-03
				6.75	-2.135901463827D-03
				7.25	1.393194915409D-03
				7.50	1.598194915883D-03

I= 2 H= -5.5

0		0.	-1.547703498441D-02	-1.00	3.127136919920D+00
1		-1.520640330396D+00	-7.581893470303D-01	-.75	2.044134579404D+00
2		1.606496589525D+00	8.031750270400D-01	-.50	1.161944312579D+00
				-.25	4.805661194442D-01
0	-7.5	-8.585625912893D-02	-4.292812956447D-02	.25	-2.798378897567D-01
1	-6.5	3.626945948187D-01	1.813472974093D-01	.50	-3.613790259145D-01
2	-5.5	-6.397043804022D-01	-3.198521902011D-01	.75	-2.572000089317D-01
3	-4.5	6.311788691450D-01	3.155894345725D-01	1.25	3.476921925291D-01
4	-3.5	-4.072277774595D-01	-2.036138887297D-01	1.50	6.930200691646D-01
5	-2.5	1.997361460160D-01	9.986807300798D-02	1.75	9.356781953527D-01
6	-1.5	-8.705270145798D-02	-4.352635072899D-02	2.25	8.667750566286D-01
7	-.5	3.752794565455D-02	1.876397282728D-02	2.50	5.888238704968D-01
8	.5	-1.616030974768D-02	-8.080154873840D-03	2.75	2.653417193270D-01
9	1.5	6.957732130579D-03	3.478866065289D-03	3.25	-1.409607504453D-01
10	2.5	-2.994477608580D-03	-1.497238804290D-03	3.50	-1.546118078379D-01
11	3.5	1.286204990009D-03	6.431024950043D-04	3.75	-8.653576335043D-02
12	4.5	-5.465332212964D-04	-2.732666106482D-04	4.25	5.508853308815D-02
13	5.5	2.190201056520D-04	1.095100528260D-04	4.50	6.288877863058D-02
14	6.5	-7.018227302836D-05	-3.509113651418D-05	4.75	3.608392629122D-02
15	7.5	1.210843874780D-05	6.054219373902D-06	5.25	-2.347818327497D-02
				5.50	-2.691962917267D-02
	0	7.896063068279D+01	3.947669813329D+01	5.75	-1.548616279830D-02
	1	2.257680851247D+01	1.128943605857D+01	6.25	1.009872378081D-02
	2	1.606496589525D+00	8.031750270400D-01	6.50	1.158410216109D-02
				6.75	6.665787819206D-03
				7.25	-4.347831999506D-03
				7.50	-4.987567349241D-03

I= 3 H= -4.5

0		0.	3.862071827274D-02	-1.00	-1.395379794784D+00
1		6.686760096077D-01	3.290208540561D-01	-.75	-9.102778863674D-01
2		-7.267037851762D-01	-3.631690635084D-01	-.50	-5.160139510979D-01
				-.25	-2.125879889754D-01
0	-7.5	5.802777556856D-02	2.901388778428D-02	.25	1.218066835780D-01
1	-6.5	-2.874256967045D-01	-1.437128483522D-01	.50	1.544754264963D-01
2	-5.5	6.311788691450D-01	3.155894345725D-01	.75	1.065063911917D-01
3	-4.5	-8.211651712743D-01	-4.105825856371D-01	1.25	-1.228236240519D-01
4	-3.5	7.121496335259D-01	3.560748167630D-01	1.50	-2.004031345332D-01
5	-2.5	-4.422140686365D-01	-2.211070343183D-01	1.75	-1.711399232559D-01
6	-1.5	2.148055558244D-01	1.074027779122D-01	2.25	2.952198242171D-01
7	-.5	-9.354122182420D-02	-4.677061091210D-02	2.50	6.336517798465D-01
8	.5	4.032123404343D-02	2.016061702172D-02	2.75	9.017984320640D-01
9	1.5	-1.736182752950D-02	-8.680913764749D-03	3.25	8.887161424724D-01

CARDINAL SPLINES N=3

10	2,5	7,472280505113D-03	3.736140252556D-03	3.50	6.139578467914D-01
11	3,5	-3.209539491114D-03	-1.604769745557D-03	3,75	2,797926561356D-01
12	4,5	1,363795155054D-03	6.818975775269D-04	4.25	-1,503799358407D-01
13	5,5	-5.465332212964D-04	-2,732666106482D-04	4,50	-1,654154150812D-01
14	6,5	1,751297840103D-04	8,756489200517D-05	4,75	-9,275208728587D-02
15	7,5	-3.021487010002D-05	-1.510743505001D-05	5,25	5.914299808835D-02
				5,50	6,753976240061D-02
	0	-3,586201784411D+01	-1,792198269865D+01	5,75	3,876027907709D-02
	1	-1,023188076804D+01	-5.118515098569D+00	6.25	-2,522390080934D-02
	2	-7,267037851762D-01	-3.631690635084D-01	6.50	-2,892222347747D-02
				6,75	-1,663854922552D-02
				7,25	1,085042409747D-02
				7,50	1,244643729512D-02

I= 4 H= -3.5

0		0.	-9,088955478870D-02	-1.00	6,029240426277D-01
1		-2,885155488004D-01	-1,317444529028D-01	-.75	3,932414393782D-01
2		3,144084938274D-01	1,567739790574D-01	-.50	2,228598978570D-01
				-.25	9.177941806430D-02
0	-7,5	-2,589294502701D-02	-1,294647251350D-02	,25	-5,250364241501D-02
1	-6,5	1,479713631555D-01	7,398568157777D-02	,50	-6,646480547543D-02
2	-5,5	-4,072277774595D-01	-2,036138887297D-01	,75	-4,567640105046D-02
3	-4,5	7,121496335259D-01	3,560748167630D-01	1,25	5,174434161434D-02
4	-3,5	-8,572956137390D-01	-4,286478068695D-01	1,50	8,264534171080D-02
5	-2,5	7,277609952926D-01	3,638804976463D-01	1,75	6,810496813499D-02
6	-1,5	-4,489380650503D-01	-2,244690325251D-01	2,25	-9,940959106563D-02
7	-,5	2,177003256822D-01	1,088501628411D-01	2,50	-1,739120274569D-01
8	,5	-9,478643828772D-02	-4,739321914386D-02	2,75	-1,560222149755D-01
9	1,5	4,085459647055D-02	2,042729823527D-02	3,25	2,854293485942D-01
10	2,5	-1,758498393154D-02	-8,792491965771D-03	3,50	6,224365818032D-01
11	3,5	7,553284962553D-03	3,776642481276D-03	3,75	8,953501822121D-01
12	4,5	-3,209539491114D-03	-1,604769745557D-03	4,25	8,929191448252D-01
13	5,5	1,286204990009D-03	6,431024950043D-04	4,50	6,187786062667D-01
14	6,5	-4,121484215162D-04	-2,060742107581D-04	4,75	2,825664929114D-01
15	7,5	7,110732822641D-05	3,555366411321D-05	5,25	-1,521891215798D-01
				5,50	-1,674907897775D-01
	0	1,552161116179D+01	7,739563370417D+00	5,75	-9,394634546484D-02
	1	4,427611858610D+00	2,219865232958D+00	6,25	5,992200522656D-02
	2	3,144084938274D-01	1,567739790574D-01	6,50	6,843342118797D-02
				6,75	3,927455239295D-02
				7,25	-2,555941317338D-02
				7,50	-2,930717030964D-02

I= 5 H= -2.5

0		0.	2,116045044359D-01	-1.00	-2,596950465025D-01
1		1,242535344708D-01	3,299387951824D-02	-.75	-1,693760013709D-01
2		-1,354415120317D-01	-6,671902796553D-02	-.50	-9,598714524332D-02
				-.25	-3,952847811968D-02
0	-7,5	1,118797756091D-02	5,593988780455D-03	,25	2,260921487505D-02
1	-6,5	-6,475630276387D-02	-3,237815138194D-02	.50	2,861601352625D-02
2	-5,5	1,997361460160D-01	9,986807300798D-02	,75	1,965925985411D-02
3	-4,5	-4,422140686365D-01	-2,211070343183D-01	1,25	-2,223068511038D-02
4	-3,5	7,277609952926D-01	3,638804976463D-01	1,50	-3,542803022336D-02
5	-2,5	-8,640408412497D-01	-4,320204206249D-01	1,75	-2,908368229021D-02
6	-1,5	7,306658217135D-01	3,653329108568D-01	2,25	4,162746294082D-02

CARDINAL SPLINES N=3

7	-,5	-4,501876510411D-01	-2,250938255206D-01	2.50	7,119890011122D-02
8	,5	2,182355711784D-01	1,091177855892D-01	2.75	6,157281346455D-02
9	1,5	-9,501040522498D-02	-4,750520261249D-02	3,25	-9,517925641227D-02
10	2,5	4,093594343009D-02	2,046797171504D-02	3,50	-1,690660847925D-01
11	3,5	-1.758498393154D-02	-8,792491965771D-03	3,75	-1,532360054553D-01
12	4,5	7,472280505113D-03	3,736140252556D-03	4,25	2,836132731195D-01
13	5,5	-2,994477608580D-03	-1.497238804290D-03	4,50	6,203535695458D-01
14	6,5	9,595432756221D-04	4,797716378110D-04	4,75	8,941516203693D-01
15	7,5	-1,655485158964D-04	-8,277425794820D-05	5,25	8,937009082526D-01
				5,50	6,196754154465D-01
	0	-6,686683543252D+00	-3,293886722239D+00	5,75	2,830825751603D-01
	1	-1.907369146005D+00	-9,677915399648D-01	6,25	-1,525258116735D-01
	2	-1,354415120317D-01	-6,671902796553D-02	6,50	-1,678770871388D-01
				6,75	-9,416869516777D-02
				7,25	6,006718914657D-02
				7,50	6,860012273745D-02

I= 6 H= -1.5

0		0.	-4,916681393860D-01	-1,00	1,118221400955D-01
1		-5,350159482036D-02	4,094016568312D-02	-.75	7,293150283251D-02
2		5,832054527510D-02	2,683273349527D-02	-.50	4,133093372895D-02
				-,25	1,702043278478D-02
0	-7,5	-4,818950454741D-03	-2,409475227371D-03	,25	-9,735070631699D-03
1	-6,5	2,792742309204D-02	1,396371154602D-02	,50	-1,232125329311D-02
2	-5,5	-8,705270145798D-02	-4,352635072899D-02	,75	-8,464448929765D-03
3	-4,5	2,148055558244D-01	1,074027779122D-01	1,25	9,569861642701D-03
4	-3,5	-4,489380650503D-01	-2,244690325251D-01	1,50	1,524766159437D-02
5	-2,5	7,306658217135D-01	3,653329108568D-01	1,75	1,251233943668D-02
6	-1,5	-8,652908212088D-01	-4,326454106044D-01	2,25	-1,787307490316D-02
7	-,5	7,312012538196D-01	3,656006269098D-01	2,50	-3,049773724273D-02
8	,5	-4,504116990311D-01	-2,252058495155D-01	2,75	-2,627010218424D-02
9	1,5	2,183169524298D-01	1,091584762149D-01	3,25	3,980533066660D-02
10	2,5	-9,501040522498D-02	-4,750520261249D-02	3,50	6,911159685263D-02
11	3,5	4,085459647055D-02	2,042729823527D-02	3,75	6,037269466737D-02
12	4,5	-1,736182752950D-02	-8,680913764749D-03	4,25	-9,439698580133D-02
13	5,5	6,957732130579D-03	3,478866065289D-03	4,50	-1,681688083083D-01
14	6,5	-2,229521457343D-03	-1.114760728671D-03	4,75	-1,527196936154D-01
15	7,5	3,846559342158D-04	1,923279671079D-04	5,25	2,832764551785D-01
				5,50	6,199671304604D-01
	0	2,879268710572D+00	1,324724362346D+00	5,75	8,939291908067D-01
	1	8.213065843061D-01	4,434311681121D-01	6,25	8,938461433189D-01
	2	5.832054527510D-02	2,683273349527D-02	6,50	6,198421754554D-01
				6,75	2,831786709719D-01
				7,25	-1,525888429026D-01
				7,50	-1,679497530611D-01

I= 7 H= -.5

0		0.	1,141982037500D+00	-1,00	-4,814788685425D-02
1		2,303644945865D-02	-1,457054345481D-01	-.75	-3,140252062901D-02
2		-2,511143739560D-02	-7.149617134855D-03	-.50	-1,779608407823D-02
				-.25	-7,328577201888D-03
0	-7,5	2,074987936943D-03	1,037493968472D-03	,25	4,191673882846D-03
1	-6,5	-1,202676331710D-02	-6,013381658552D-03	,50	5,305208753457D-03
2	-5,5	3,752794565455D-02	1,876397282728D-02	,75	3,644557922909D-03
3	-4,5	-9,354122182420D-02	-4,677061091210D-02	1,25	-4,120443345945D-03

4	-3.5	2.177003256822D-01	1.088501628411D-01	1.50	-6.564956659609D-03
5	-2.5	-4.501876510411D-01	-2.250938255206D-01	1.75	-5.387042453647D-03
6	-1.5	7.312012538196D-01	3.656006269098D-01	2.25	7.693499377177D-03
7	-.5	-8.655148765067D-01	-4.327574382534D-01	2.50	1.312469500371D-02
8	.5	7.312826384695D-01	3.656413192347D-01	2.75	1.130081292987D-02
9	1.5	-4.504116990311D-01	-2.252058495155D-01	3.25	-1.708844254294D-02
10	2.5	2.182355711784D-01	1.091177855892D-01	3.50	-2.959889553812D-02
11	3.5	-9.478643828772D-02	-4.739321914386D-02	3.75	-2.575328245052D-02
12	4.5	4.032123404343D-02	2.016061702172D-02	4.25	3.946840033450D-02
13	5.5	-1.616030974768D-02	-8.080154873840D-03	4.50	6.872507814257D-02
14	6.5	5.178430875272D-03	2.589215437636D-03	4.75	6.015023621363D-02
15	7.5	-8.934279042348D-04	-4.467139521174D-04	5.25	-9.425174132365D-02
				5.50	-1.680020396281D-01
	0	-1.239744982562D+00	-3.529746854468D-01	5.75	-1.526235935434D-01
	1	-3.536351114753D-01	-2.529496915710D-01	6.25	2.832134215770D-01
	2	-2.511143739560D-02	-7.149617134855D-03	6.50	6.198944619082D-01
				6.75	8.938870603134D-01
				7.25	8.938744377046D-01
				7.50	6.198754696730D-01

I= 8 H= .5

0		0.	-2.652263336065D+00	-1.00	2.073097834754D-02
1		-9.918775221651D-03	3.601939485938D-01	-.75	1.352094567455D-02
2		1.081220312589D-02	-7.149617134855D-03	-.50	7.662438392297D-03
				-.25	3.155456500781D-03
0	-7.5	-8.934279042348D-04	-4.467139521174D-04	.25	-1.804803598233D-03
1	-6.5	5.178430875272D-03	2.589215437636D-03	.50	-2.284256451361D-03
2	-5.5	-1.616030974768D-02	-8.080154873840D-03	.75	-1.569231787546D-03
3	-4.5	4.032123404343D-02	2.016061702172D-02	1.25	1.774129831880D-03
4	-3.5	-9.478643828772D-02	-4.739321914386D-02	1.50	2.826652017836D-03
5	-2.5	2.182355711784D-01	1.091177855892D-01	1.75	2.319472384567D-03
6	-1.5	-4.504116990311D-01	-2.252058495155D-01	2.25	-3.312485593529D-03
7	-.5	7.312826384695D-01	3.656413192347D-01	2.50	-5.650787510786D-03
8	.5	-8.655148765067D-01	-4.327574382534D-01	2.75	-4.865327119347D-03
9	1.5	7.312012538196D-01	3.656006269098D-01	3.25	7.355562745487D-03
10	2.5	-4.501876510411D-01	-2.250938255206D-01	3.50	1.273751451973D-02
11	3.5	2.177003256822D-01	1.088501628411D-01	3.75	1.107814279307D-02
12	4.5	-9.354122182420D-02	-4.677061091210D-02	4.25	-1.694315424038D-02
13	5.5	3.752794565455D-02	1.876397282728D-02	4.50	-2.943209781358D-02
14	6.5	-1.202676331710D-02	-6.013381658552D-03	4.75	-2.565717295419D-02
15	7.5	2.074987936943D-03	1.037493968472D-03	5.25	3.940536464769D-02
				5.50	6.865240811325D-02
	0	5.337956116687D-01	-3.529746854468D-01	5.75	6.010810518452D-02
	1	1.522642716666D-01	2.529496915710D-01	6.25	-9.422344676377D-02
	2	1.081220312589D-02	-7.149617134855D-03	6.50	-1.679687452506D-01
				6.75	-1.526037053218D-01
				7.25	2.831985591152D-01
				7.50	6.198754696730D-01

I= 9 H= 1.5

0		0.	6.159799382296D+00	-1.00	-8.925500634915D-03
1		4.270422350349D-03	-8.459221705411D-01	-.75	-5.821298297830D-03
2		-4.655078284565D-03	2.683273349527D-02	-.50	-3.298980746316D-03
				-.25	-1.358547980373D-03
0	-7.5	3.846559342158D-04	1.923279671079D-04	.25	7.770388353628D-04

1	-6.5	-2.229521457343D-03	-1.114760728671D-03	.50	9.834621019776D-04
2	-5.5	6.957732130579D-03	3.478866065289D-03	.75	6.756158839582D-04
3	-4.5	-1.736182752950D-02	-8.680913764749D-03	1.25	-7.638323963770D-04
4	-3.5	4.085459647055D-02	2.042729823527D-02	1.50	-1.216984159838D-03
5	-2.5	-9.501040522498D-02	-4.750520261249D-02	1.75	-9.986231214258D-04
6	-1.5	2.183169524298D-01	1.091584762149D-01	2.25	1.426151079551D-03
7	-.5	-4.504116990311D-01	-2.252058495155D-01	2.50	2.432873235483D-03
8	.5	7.312012538196D-01	3.656006269098D-01	2.75	2.094695057094D-03
9	1.5	-8.652908212088D-01	-4.326454106044D-01	3.25	-3.166764467998D-03
10	2.5	7.306658217135D-01	3.653329108568D-01	3.50	-5.483705373168D-03
11	3.5	-4.489380650503D-01	-2.244690325251D-01	3.75	-4.769126782089D-03
12	4.5	2.148055558244D-01	1.074027779122D-01	4.25	7.292508388006D-03
13	5.5	-8.705270145798D-02	-4.352635072899D-02	4.50	1.266483221431D-02
14	6.5	2.792742309204D-02	1.396371154602D-02	4.75	1.103600783923D-02
15	7.5	-4.818950454741D-03	-2.409475227371D-03	5.25	-1.691485887226D-02
				5.50	-2.939880290571D-02
	0	-2.298199858792D-01	1.324724362346D+00	5.75	-2.563728456522D-02
	1	-6.555575191813D-02	-4.434311681121D-01	6.25	3.939050215878D-02
	2	-4.655078284565D-03	2.683273349527D-02	6.50	6.863341587073D-02
				6.75	6.009548258342D-02
				7.25	-9.421082417926D-02
				7.50	-1.679497530611D-01

I=10 H= 2.5

0		0.	-1.430526859504D+01	-1.00	3.841363685373D-03
1		-1.837907584739D-03	1.968576959448D+00	-.75	2.505374745161D-03
2		2.003456100635D-03	-6.671902796553D-02	-.50	1.419817817528D-03
				-.25	5.846929024743D-04
0	-7.5	-1.655485158964D-04	-8.277425794820D-05	.25	-3.344225583675D-04
1	-6.5	9.595432756221D-04	4.797716378110D-04	.50	-4.232631583323D-04
2	-5.5	-2.994477608580D-03	-1.497238804290D-03	.75	-2.907720707776D-04
3	-4.5	7.472280505113D-03	3.736140252556D-03	1.25	3.287387535644D-04
4	-3.5	-1.758498393154D-02	-8.792491965771D-03	1.50	5.237665340957D-04
5	-2.5	4.093594343009D-02	2.046797171504D-02	1.75	4.297881338331D-04
6	-1.5	-9.501040522498D-02	-4.750520261249D-02	2.25	-6.137877988302D-04
7	-.5	2.182355711784D-01	1.091177855892D-01	2.50	-1.047061264399D-03
8	.5	-4.501876510411D-01	-2.250938255206D-01	2.75	-9.015156000415D-04
9	1.5	7.306658217135D-01	3.653329108568D-01	3.25	1.362910415660D-03
10	2.5	-8.640408412497D-01	-4.320204206249D-01	3.50	2.360068538328D-03
11	3.5	7.277609952926D-01	3.638804976463D-01	3.75	2.052521037562D-03
12	4.5	-4.422140686365D-01	-2.211070343183D-01	4.25	-3.138461118668D-03
13	5.5	1.997361460160D-01	9.986807300798D-02	4.50	-5.450405274365D-03
14	6.5	-6.475630276387D-02	-3.237815138194D-02	4.75	-4.749236782753D-03
15	7.5	1.118797756091D-02	5.593988780455D-03	5.25	7.277645667845D-03
				5.50	1.264583994413D-02
	0	9.891009877517D-02	-3.293886722239D+00	5.75	1.102338533975D-02
	1	2.821393392478D-02	9.677915399648D-01	6.25	-1.690223654022D-02
	2	2.003456100635D-03	-6.671902796553D-02	6.50	-2.937981117133D-02
				6.75	-2.562242254830D-02
				7.25	3.937061463757D-02
				7.50	6.860012273745D-02

I=11 H= 3.5

0		0.	3.320708893958D+01	-1.00	-1.649964097050D-03
1		7.894283844117D-04	-4.571474918818D+00	-.75	-1.076122626668D-03

2		-8.605357126381D-04	1.567739790574D-01	-.50	-6,098481203654D-04
				-.25	-2,511405781428D-04
0	-7,5	7.110732822641D-05	3.555366411321D-05	.25	1,436430548133D-04
1	-6,5	-4.121484215162D-04	-2,060742107581D-04	.50	1,818023680534D-04
2	-5,5	1,286204990009D-03	6,431024950043D-04	.75	1.248940522536D-04
3	-4,5	-3.209539491114D-03	-1,604769745557D-03	1.25	-1,412017147343D-04
4	-3,5	7.553284962553D-03	3,776642481276D-03	1.50	-2,249711412712D-04
5	-2,5	-1.758498393154D-02	-8,792491965771D-03	1.75	-1,846050106391D-04
6	-1,5	4,085459647055D-02	2,042729823527D-02	2.25	2,636375737910D-04
7	-,5	-9,478643828772D-02	-4,739321914386D-02	2.50	4,497396092008D-04
8	,5	2.177003256822D-01	1,088501628411D-01	2.75	3,872240060302D-04
9	1,5	-4,489380650503D-01	-2,244690325251D-01	3.25	-5,854047590116D-04
10	2,5	7,277609952926D-01	3,638804976463D-01	3.50	-1,013709342373D-03
11	3,5	-8,572956137390D-01	-4,286478068695D-01	3.75	-8,816095281175D-04
12	4,5	7,121496335259D-01	3,560748167630D-01	4.25	1,348045392378D-03
13	5,5	-4,072277774595D-01	-2,036138887297D-01	4.50	2,341075999453D-03
14	6.5	1.479713631555D-01	7,398568157777D-02	4.75	2,039899557425D-03
15	7,5	-2,589294502701D-02	-1.294647251350D-02	5.25	-3,125841310905D-03
				5.50	-5,431418087085D-03
	0	-4,248442095281D-02	7.739563370417D+00	5.75	-4,734378783441D-03
	1	-1,211860730516D-02	-2,219865232958D+00	6.25	7,257764347050D-03
	2	-8.605357126381D-04	1.567739790574D-01	6.50	1,261255759284D-02
				6.75	1,099510130444D-02
				7.25	-1,686012194643D-02
				7.50	-2,930717030964D-02

I=12 H= 4.5

0		0.	-7,673910576027D+01	-1.00	7,011014488904D-04
1		-3,354432893952D-04	1.056605105119D+01	-.75	4,572651817624D-04
2		3,656581594952D-04	-3.631690635084D-01	-.50	2,591361845714D-04
				-.25	1,067144573172D-04
0	-7,5	-3,021487010002D-05	-1,510743505001D-05	.25	-6,103669408942D-05
1	-6,5	1.751297840103D-04	8.756489200517D-05	.50	-7,725131951441D-05
2	-5,5	-5,465332212964D-04	-2,732666106482D-04	.75	-5,306988263727D-05
3	-4,5	1.363795155054D-03	6,818975775269D-04	1.25	5,999932176555D-05
4	-3,5	-3,209539491114D-03	-1,604769745557D-03	1.50	9,559456069969D-05
5	-2.5	7.472280505113D-03	3.736140252556D-03	1.75	7,844221613690D-05
6	-1,5	-1,736182752950D-02	-8,680913764749D-03	2.25	-1,120246705749D-04
7	-,5	4,032123404343D-02	2.016061702172D-02	2,50	-1,911030006854D-04
8	,5	-9,354122182420D-02	-4,677061091210D-02	2,75	-1,645389191388D-04
9	1,5	2,148055558244D-01	1.074027779122D-01	3,25	2,487497232180D-04
10	2,5	-4.422140686365D-01	-2.211070343183D-01	3,50	4,307445547220D-04
11	3,5	7,121496335259D-01	3.560748167630D-01	3,75	3,746127932098D-04
12	4,5	-8,211651712743D-01	-4.105825856371D-01	4,25	-5,728102175408D-04
13	5,5	6,311788691450D-01	3,155894345725D-01	4,50	-9,947676362131D-04
14	6,5	-2,874256967045D-01	-1,437128483522D-01	4,75	-8,667916972128D-04
15	7,5	5,802777556856D-02	2.901388778428D-02	5,25	1,328226036852D-03
				5.50	2,307901387009D-03
	0	1,805244680114D-02	-1,792198269865D+01	5.75	2,011709457161D-03
	1	5,149429103033D-03	5.118515098569D+00	6,25	-3,083870733556D-03
	2	3.656581594952D-04	-3,631690635084D-01	6,50	-5,359027494266D-03
				6,75	-4,671587574452D-03
				7.25	7,162036573062D-03
				7.50	1,244643729512D-02

CARDINAL SPLINES N=3

I=13	H= 5.5				
0		0.	1.693260638436D+02	-1.00	-2.809624506579D-04
1		1.344270059551D-04	-2.333706146417D+01	-.75	-1.832464421117D-04
2		-1.465354447029D-04	8.031750270400D-01	-.50	-1.038473641532D-04
				-.25	-4.276521678269D-05
0	-7.5	1.210843874780D-05	6.054219373902D-06	.25	2.446011084205D-05
1	-6.5	-7.018227302836D-05	-3.509113651418D-05	.50	3.095803051268D-05
2	-5.5	2.190201056520D-04	1.095100528260D-04	.75	2.126745609410D-05
3	-4.5	-5.465332212964D-04	-2.732666106482D-04	1.25	-2.404438973462D-05
4	-3.5	1.286204990009D-03	6.431024950043D-04	1.50	-3.830898093986D-05
5	-2.5	-2.994477608580D-03	-1.497238804290D-03	1.75	-3.143527561356D-05
6	-1.5	6.957732130579D-03	3.478866065289D-03	2.25	4.489325477721D-05
7	-.5	-1.616030974768D-02	-8.080154873840D-03	2.50	7.658344945241D-05
8	.5	3.752794565455D-02	1.876397282728D-02	2.75	6.593804360223D-05
9	1.5	-8.705270145798D-02	-4.352635072899D-02	3.25	-9.968504805684D-05
10	2.5	1.997361460160D-01	9.986807300798D-02	3.50	-1.726184494913D-04
11	3.5	-4.072277774595D-01	-2.036138887297D-01	3.75	-1.501239622203D-04
12	4.5	6.311788691450D-01	3.155894345725D-01	4.25	2.295504564403D-04
13	5.5	-6.397043804022D-01	-3.198521902011D-01	4.50	3.986474968388D-04
14	6.5	3.626945948187D-01	1.813472974093D-01	4.75	3.473618498388D-04
15	7.5	-8.585625912893D-02	-4.292812956447D-02	5.25	-5.322789301687D-04
				5.50	-9.248779562066D-04
	0	-7.234416219873D-03	3.947669813329D+01	5.75	-8.061804473149D-04
	1	-2.063604664588D-03	-1.128943605857D+01	6.25	1.235840054027D-03
	2	-1.465354447029D-04	8.031750270400D-01	6.50	2.147588307329D-03
				6.75	1.872094514950D-03
				7.25	-2.870058832963D-03
				7.50	-4.987567349241D-03

I=14	H= 6.5				
0		0.	-1.867757235886D+02	-1.00	9.003092826379D-05
1		-4.307546471607D-05	2.629399961934D+01	-.75	5.871904678265D-05
2		4.695546354773D-05	-9.227969390127D-01	-.50	3.327659824496D-05
				-.25	1.370358265075D-05
0	-7.5	-3.879998831662D-06	-1.939999415831D-06	.25	-7.837938768642D-06
1	-6.5	2.248903785440D-05	1.124451892720D-05	.50	-9.920116434590D-06
2	-5.5	-7.018227302836D-05	-3.509113651418D-05	.75	-6.814892201700D-06
3	-4.5	1.751297840103D-04	8.756489200517D-05	1.25	7.704726102257D-06
4	-3.5	-4.121484215162D-04	-2.060742107581D-04	1.50	1.227563721331D-05
5	-2.5	9.595432756221D-04	4.797716378110D-04	1.75	1.007304370055D-05
6	-1.5	-2.229521457343D-03	-1.114760728671D-03	2.25	-1.438548599924D-05
7	-.5	5.178430875272D-03	2.589215437636D-03	2.50	-2.454021534662D-05
8	.5	-1.202676331710D-02	-6.013381658552D-03	2.75	-2.112902723807D-05
9	1.5	2.792742309204D-02	1.396371154602D-02	3.25	3.194283573033D-05
10	2.5	-6.475630276387D-02	-3.237815138194D-02	3.50	5.531343849663D-05
11	3.5	1.479713631555D-01	7.398568157777D-02	3.75	4.810535937541D-05
12	4.5	-2.874256967045D-01	-1.437128483522D-01	4.25	-7.355659292872D-05
13	5.5	3.626945948187D-01	1.813472974093D-01	4.50	-1.277416391465D-04
14	6.5	-2.424626220712D-01	-1.212313110356D-01	4.75	-1.113077900608D-04
15	7.5	6.445814296830D-02	3.222907148415D-02	5.25	1.705621678231D-04
				5.50	2.963656388317D-04
	0	2.318178839189D-03	-4.147805426300D+01	5.75	2.583304872916D-04
	1	6.612564884998D-04	1.245204553415D+01	6.25	-3.960094730933D-04
	2	4.695546354773D-05	-9.227969390127D-01	6.50	-6.881676115177D-04
				6.75	-5.998884420942D-04
				7.25	9.196718045197D-04

CARDINAL SPLINES N=3

```
                                                                          7.50   1.598194915883D-03

I=15      H=  7.5
 0                0.                     7.148113280151D+01           -1.00  -1.553289645995D-05
 1                7.431743138739D-06    -1.032792618331D+01            -.75  -1.013070609723D-05
 2               -8.101153321208D-06     3.730529080289D-01            -.50  -5.741159899672D-06
                                                                       -.25  -2.364257867260D-06
 0   -7.5   6.694101824693D-07     3.347050912346D-07              .25   1.352267422991D-06
 1   -6.5  -3.879998831662D-06    -1.939999415831D-06              .50   1.711502307270D-06
 2   -5.5   1.210843874780D-05     6.054219373902D-06              .75   1.175762785035D-06
 3   -4.5  -3.021487010002D-05    -1.510743505001D-05             1.25  -1.329284448206D-06
 4   -3.5   7.110732822641D-05     3.555366411321D-05             1.50  -2.117896654973D-06
 5   -2.5  -1.655485158964D-04    -8.277425794820D-05             1.75  -1.737886611349D-06
 6   -1.5   3.846559342158D-04     1.923279671079D-04             2.25   2.481905594672D-06
 7    -.5  -8.934279042348D-04    -4.467139521174D-04             2.50   4.233885303998D-06
 8     .5   2.074987936943D-03     1.037493968472D-03             2.75   3.645358308679D-06
 9    1.5  -4.818950454741D-03    -2.409475227371D-03             3.25  -5.511047920818D-06
10    2.5   1.118797756091D-02     5.593988780455D-03             3.50  -9.543141779406D-06
11    3.5  -2.589294502701D-02    -1.294647251350D-02             3.75  -8.299543064445D-06
12    4.5   5.802777556856D-02     2.901388778428D-02             4.25   1.269060492328D-05
13    5.5  -8.585625912893D-02    -4.292812956447D-02             4.50   2.203906692274D-05
14    6.5   6.445814296830D-02     3.222907148415D-02             4.75   1.920376041282D-05
15    7.5  -1.855619924635D-02    -9.278099623173D-03             5.25  -2.942682624467D-05
                                                                      5.50  -5.113150379942D-05
        0  -3.999518007774D-04     1.500591250329D+01             5.75  -4.456935789428D-05
        1  -1.140855566794D-04    -4.732132562879D+00             6.25   6.832289777119D-05
        2  -8.101153321208D-06     3.730529080289D-01             6.50   1.187284828446D-04
                                                                      6.75   1.034978793719D-04
                                                                      7.25  -1.586694899874D-04
                                                                      7.50  -2.757339014091D-04
```

```
              N= 3        M=16                          P=  8.0        Q= 5

I= 0      H= -8.0
 0                1.000000000000D+00     5.004208846805D-01           -1.00   3.473671635280D+00
 1               -1.727557718014D+00    -8.638330695550D-01            -.75   2.715357366972D+00
 2                7.461139172652D-01     3.730587027111D-01            -.50   2.050307338323D+00
                                                                       -.25   1.478521549333D+00
 0   -8.0  -1.855619925082D-02    -9.278099625411D-03              .25   6.147245690371D-01
 1   -7.0   6.445814299425D-02     3.222907149713D-02              .50   3.221697390825D-01
 2   -6.0  -8.585625920991D-02    -4.292812960496D-02              .75   1.196173168865D-01
 3   -5.0   5.802777577064D-02     2.901388788532D-02             1.25  -5.021023041983D-02
 4   -4.0  -2.589294550258D-02    -1.294647275129D-02             1.50  -5.147708426722D-02
 5   -3.0   1.118797866811D-02     5.593989334056D-03             1.75  -2.752040839751D-02
 6   -2.0  -4.818953027352D-03    -2.409476513676D-03             2.25   1.677968343136D-02
 7   -1.0   2.074993912265D-03     1.037496956132D-03             2.50   1.896487004575D-02
 8    0.0  -8.934417819735D-04    -4.467208909867D-04             2.75   1.083411537641D-02
 9    1.0   3.846881648814D-04     1.923440824407D-04             3.25  -7.016349057289D-03
```

10	2.0	-1.656233681948D-04	-8.281168409741D-05	3.50	-8.037375419559D-03
11	3.0	7.128110994066D-05	3.564055497033D-05	3.75	-4.621138975744D-03
12	4.0	-3.061706254156D-05	-1.530853127078D-05	4.25	3.012078398103D-03
13	5.0	1.300977818065D-05	6.504889090325D-06	4.50	3.454789753975D-03
14	6.0	-5.213595216408D-06	-2.606797608204D-06	4.75	1.987862475312D-03
15	7.0	1.670631843599D-06	8.353159217997D-07	5.25	-1.296539365161D-03
16	8.0	-2.882315216532D-07	-1.441157608266D-07	5.50	-1.487292834855D-03
				5.75	-8.558433480639D-04
	0	3.493082896086D+01	1.746551330175D+01	6.25	5.582412775519D-04
	1	1.021026495823D+01	5.105106173822D+00	6.50	6.403808066905D-04
	2	7.461139172652D-01	3.730587027111D-01	6.75	3.685016720096D-04
				7.25	-2.403644091211D-04
				7.50	-2.757320572784D-04
				7.75	-1.586678824038D-04

I= 1 H= -7.0

0		0.	-2.439508856452D-03	-1.00	-4.626823524612D+00
1		2.781182690809D+00	1.390905557628D+00	-.75	-3.124059987121D+00
2		-1.845640833803D+00	-9.228305258340D-01	-.50	-1.852001553855D+00
				-.25	-8.106482248149D-01
0	-8.0	6.445814299425D-02	3.222907149713D-02	.25	5.800060679948D-01
1	-7.0	-2.424626222216D-01	-1.212313111108D-01	.50	9.311954539222D-01
2	-6.0	3.626945952881D-01	1.813472976440D-01	.75	1.063010268572D+00
3	-5.0	-2.874256978758D-01	-1.437128489379D-01	1.25	7.891384222512D-01
4	-4.0	1.479713659120D-01	7.398568295602D-02	1.50	5.009842265746D-01
5	-3.0	-6.475630918138D-02	-3.237815459069D-02	1.75	2.152141789332D-01
6	-2.0	2.792743800327D-02	1.396371900163D-02	2.25	-1.084973408115D-01
7	-1.0	-1.202679795094D-02	-6.013398975469D-03	2.50	-1.174242888568D-01
8	0.0	5.178511312665D-03	2.589255656333D-03	2.75	-6.515456768438D-02
9	1.0	-2.229708270971D-03	-1.114854135485D-03	3.25	4.115217615076D-02
10	2.0	9.599771304364D-04	4.799885652182D-04	3.50	4.690407309025D-02
11	3.0	-4.131556856211D-04	-2.065778428106D-04	3.75	2.688643478304D-02
12	4.0	1.774609499294D-04	8.873047496471D-05	4.25	-1.747932038856D-02
13	5.0	-7.540656754607D-05	-3.770328377304D-05	4.50	-2.003818065726D-02
14	6.0	3.021875656833D-05	1.510937828416D-05	4.75	-1.152632052637D-02
15	7.0	-9.683225279595D-06	-4.841612639797D-06	5.25	7.515830499626D-03
16	8.0	1.670631843599D-06	8.353159217997D-07	5.50	8.621156835011D-03
				5.75	4.960781366072D-03
	0	-9.587155183692D+01	-4.793634870121D+01	6.25	-3.235686453386D-03
	1	-2.674907065004D+01	-1.337438285572D+01	6.50	-3.711765820739D-03
	2	-1.845640833803D+00	-9.228305258340D-01	6.75	-2.135897451722D-03
				7.25	1.393188764571D-03
				7.50	1.598184227033D-03
				7.75	9.196624867315D-04

I= 2 H= -6.0

0		0.	7.613054757166D-03	-1.00	3.127136921799D+00
1		-1.520640331295D+00	-7.613007379694D-01	-.75	2.044134580630D+00
2		1.606496590505D+00	8.032798425273D-01	-.50	1.161944313274D+00
				-.25	4.805661197302D-01
0	-8.0	-8.585625920991D-02	-4.292812960496D-02	.25	-2.798378899203D-01
1	-7.0	3.626945952881D-01	1.813472976440D-01	.50	-3.613790261215D-01
2	-6.0	-6.397043818670D-01	-3.198521909335D-01	.75	-2.572000090739D-01
3	-5.0	6.311788728002D-01	3.155894364001D-01	1.25	3.476921926899D-01
4	-4.0	-4.072277860617D-01	-2.036138930309D-01	1.50	6.930200694208D-01

CARDINAL SPLINES N=3

5	-3.0	1.997361660433D-01	9.986808302164D-02	1.75	9.356781955629D-01
6	-2.0	-8.705274799192D-02	-4.352637399596D-02	2.25	8.667750563284D-01
7	-1.0	3.752805373748D-02	1.876402686874D-02	2.50	5.888238699846D-01
8	0.0	-1.616056077127D-02	-8.080280385636D-03	2.75	2.653417188860D-01
9	1.0	6.958315125954D-03	3.479157562977D-03	3.25	-1.409607497786D-01
10	2.0	-2.995831553432D-03	-1.497915776716D-03	3.50	-1.546118066835D-01
11	3.0	1.289348391913D-03	6.446741959565D-04	3.75	-8.653576234639D-02
12	4.0	-5.538081667196D-04	-2.769040833598D-04	4.25	5.508853155290D-02
13	5.0	2.353237317387D-04	1.176618658694D-04	4.50	6.288877596439D-02
14	6.0	-9.430465802030D-05	-4.715232901015D-05	4.75	3.608392396804D-02
15	7.0	3.021875656833D-05	1.510937828416D-05	5.25	-2.347817971505D-02
16	8.0	-5.213595216408D-06	-2.606797608204D-06	5.50	-2.691962298701D-02
				5.75	-1.548615740650D-02
	0	9.065065914194D+01	4.532711707275D+01	6.25	1.009871551541D-02
	1	2.418330511678D+01	1.209117674247D+01	6.50	1.158408779785D-02
	2	1.606496590505D+00	8.032798425273D-01	6.75	6.665775298498D-03
				7.25	-4.347812804386D-03
				7.50	-4.987533992199D-03
				7.75	-2.870029754638D-03
I= 3	H= -5.0				
0		0.	-1.899728489785D-02	-1.00	-1.395379799473D+00
1		6.686760118511D-01	3.367848833432D-01	-.75	-9.102778894256D-01
2		-7.267037876218D-01	-3.634306155054D-01	-.50	-5.160139528310D-01
				-.25	-2.125879896891D-01
0	-8.0	5.802777577064D-02	2.901388788532D-02	.25	1.218066839862D-01
1	-7.0	-2.874256978758D-01	-1.437128489379D-01	.50	1.544754270130D-01
2	-6.0	6.311788728002D-01	3.155894364001D-01	.75	1.065063915467D-01
3	-5.0	-8.211651803955D-01	-4.105825901977D-01	1.25	-1.228236244531D-01
4	-4.0	7.121496549916D-01	3.560748274958D-01	1.50	-2.004031351726D-01
5	-3.0	-4.422141186118D-01	-2.211070593059D-01	1.75	-1.711399237805D-01
6	-2.0	2.148056719432D-01	1.074028359716D-01	2.25	2.952198249663D-01
7	-1.0	-9.354149152963D-02	-4.677074576481D-02	2.50	6.336517811247D-01
8	0.0	4.032186043676D-02	2.016093021838D-02	2.75	9.017984331644D-01
9	1.0	-1.736328231074D-02	-8.681641155368D-03	3.25	8.887161408087D-01
10	2.0	7.475659080062D-03	3.737829540031D-03	3.50	6.139578439106D-01
11	3.0	-3.217383399169D-03	-1.608691699584D-03	3.75	2.797926536301D-01
12	4.0	1.381948736662D-03	6.909743683309D-04	4.25	-1.503799320097D-01
13	5.0	-5.872165789651D-04	-2.936082894826D-04	4.50	-1.654154084282D-01
14	6.0	2.353237317387D-04	1.176618658694D-04	4.75	-9.275208148870D-02
15	7.0	-7.540656754607D-05	-3.770328377304D-05	5.25	5.914298920507D-02
16	8.0	1.300977818065D-05	6.504889090325D-06	5.50	6.753974696518D-02
				5.75	3.876026562262D-02
	0	-4.115963431298D+01	-2.058427761050D+01	6.25	-2.522388018423D-02
	1	-1.095858459010D+01	-5.478104964743D+00	6.50	-2.892218763608D-02
	2	-7.267037876218D-01	-3.634306155054D-01	6.75	-1.663851798189D-02
				7.25	1.085037619881D-02
				7.50	1.244635405741D-02
				7.75	7.161964012276D-03
I= 4	H= -4.0				
0		0.	4.470799207937D-02	-1.00	6.029240536628D-01
1		-2.885155540801D-01	-1.500162302503D-01	-.75	3.932414465754D-01
2		3.144084995827D-01	1.573895125229D-01	-.50	2.228599019357D-01
				-.25	9.177941974395D-02

CARDINAL SPLINES N=3

0	-8.0	-2.589294550258D-02	-1.294647275129D-02	.25	-5.250364337570D-02
1	-7.0	1.479713659120D-01	7.398568295602D-02	.50	-6.646480669134D-02
2	-6.0	-4.072277860617D-01	-2.036138930309D-01	.75	-4.567640188577D-02
3	-5.0	7.121496549916D-01	3.560748274958D-01	1.25	5.174434255871D-02
4	-4.0	-8.572956642560D-01	-4.286478321280D-01	1.50	8.264534321542D-02
5	-3.0	7.277611129039D-01	3.638805564519D-01	1.75	6.810496936965D-02
6	-2.0	-4.489383383229D-01	-2.244691691614D-01	2.25	-9.940959282886D-02
7	-1.0	2.177009604039D-01	1.088504802019D-01	2.50	-1.739120304648D-01
8	0.0	-9.478791243458D-02	-4.739395621729D-02	2.75	-1.560222175653D-01
9	1.0	4.085802013600D-02	2.042901006800D-02	3.25	2.854293525094D-01
10	2.0	-1.759293503102D-02	-8.796467515511D-03	3.50	6.224365885830D-01
11	3.0	7.571744725674D-03	3.785872362837D-03	3.75	8.953501881083D-01
12	4.0	-3.252261921560D-03	-1.626130960780D-03	4.25	8.929191358094D-01
13	5.0	1.381948736662D-03	6.909743683309D-04	4.50	6.187785906094D-01
14	6.0	-5.538081667196D-04	-2.769040833598D-04	4.75	2.825664792684D-01
15	7.0	1.774609499294D-04	8.873047496471D-05	5.25	-1.521891006740D-01
16	8.0	-3.061706254156D-05	-1.530853127078D-05	5.50	-1.674907534519D-01
				5.75	-9.394631380126D-02
	0	1.781401954065D+01	8.917506951545D+00	6.25	5.992195668767D-02
	1	4.742020439243D+00	2.368215970117D+00	6.50	6.843333683923D-02
	2	3.144084995827D-01	1.573895125229D-01	6.75	3.927447886455D-02
				7.25	-2.555930044921D-02
				7.50	-2.930697441898D-02
				7.75	-1.685995118271D-02

I= 5 H= -3.0

0		0.	-1.040869056354D-01	-1.00	-2.596950721939D-01
1		1.242535467629D-01	7.553331543405D-02	-.75	-1.693760181271D-01
2		-1.354415254310D-01	-6.815208211864D-02	-.50	-9.598715473918D-02
				-.25	-3.952848203015D-02
0	-8.0	1.118797866811D-02	5.593989334056D-03	.25	2.260921711170D-02
1	-7.0	-6.475630918138D-02	-3.237815459069D-02	.50	2.861601635707D-02
2	-6.0	1.997361660433D-01	9.986808302164D-02	.75	1.965926179882D-02
3	-5.0	-4.422141186118D-01	-2.211070593059D-01	1.25	-2.223068730901D-02
4	-4.0	7.277611129039D-01	3.638805564519D-01	1.50	-3.542803372635D-02
5	-3.0	-8.640411150664D-01	-4.320205575332D-01	1.75	-2.908368516467D-02
6	-2.0	7.306664579331D-01	3.653332289665D-01	2.25	4.162746704589D-02
7	-1.0	-4.501891287683D-01	-2.250945643841D-01	2.50	7.119890711405D-02
8	0.0	2.182390032132D-01	1.091195016066D-01	2.75	6.157281949396D-02
9	1.0	-9.501837603124D-02	-4.750918801562D-02	3.25	-9.517926552752D-02
10	2.0	4.095445478125D-02	2.047722739063D-02	3.50	-1.690661005768D-01
11	3.0	-1.762796102651D-02	-8.813980513255D-03	3.75	-1.532360191827D-01
12	4.0	7.571744725674D-03	3.785872362837D-03	4.25	2.836132941097D-01
13	5.0	-3.217383399169D-03	-1.608691699584D-03	4.50	6.203536059983D-01
14	6.0	1.289348391913D-03	6.446741959565D-04	4.75	8.941516521323D-01
15	7.0	-4.131556856211D-04	-2.065778428106D-04	5.25	8.937008595807D-01
16	8.0	7.128110994066D-05	3.564055497033D-05	5.50	6.196753308752D-01
				5.75	2.830825014427D-01
	0	-7.674229253480D+00	-3.861553637756D+00	6.25	-1.525256986676D-01
	1	-2.042810860133D+00	-1.014899998464D+00	6.50	-1.678768907623D-01
	2	-1.354415254310D-01	-6.815208211864D-02	6.75	-9.416852398265D-02
				7.25	6.006692670782D-02
				7.50	6.859966667462D-02
				7.75	3.937021707404D-02

CARDINAL SPLINES N=3

I= 6 H= -2.0

0		0.	2,418484218700D-01	-1.00	1,118221997899D-01
1		-5,350162338129D-02	-5,790123371916D-02	-.75	7,293154176583D-02
2		5,832057640865D-02	3,016246918317D-02	-.50	4,133095579281D-02
				-.25	1,702044187086D-02
0	-8,0	-4,818953027352D-03	-2,409476513676D-03	.25	-9,735075828599D-03
1	-7,0	2,792743800327D-02	1,396371900163D-02	.50	-1,232125987059D-02
2	-6,0	-8,705274799192D-02	-4,352637399596D-02	.75	-8,464453448340D-03
3	-5,0	2,148056719432D-01	1,074028359716D-01	1.25	9,569866751275D-03
4	-4,0	-4,489383383229D-01	-2,244691691614D-01	1.50	1,524766973366D-02
5	-3,0	7,306664579331D-01	3,653332289665D-01	1.75	1,251234611555D-02
6	-2,0	-8,652922994802D-01	-4,326461497401D-01	2.25	-1,787308444138D-02
7	-1,0	7,312046873540D-01	3,656023436770D-01	2,50	-3,049775351398D-02
8	0,0	-4,504196734463D-01	-2,252098367231D-01	2,75	-2,627011619372D-02
9	1,0	2,183354727893D-01	1,091677363946D-01	3.25	3,980535184611D-02
10	2,0	-9,505341679328D-02	-4,752670839664D-02	3,50	6,911163352788D-02
11	3,0	4,095445478125D-02	2,047722739063D-02	3,75	6,037272656335D-02
12	4,0	-1,759293503102D-02	-8,796467515511D-03	4,25	-9,439703457260D-02
13	5,0	7,475659080062D-03	3,737829540031D-03	4,50	-1,681688930067D-01
14	6,0	-2,995831553432D-03	-1,497915776716D-03	4,75	-1,527197674173D-01
15	7,0	9,599771304364D-04	4,799885652182D-04	5,25	2,832765682687D-01
16	8,0	-1,656233681948D-04	-8,281168409741D-05	5,50	6,199673269639D-01
				5,75	8,939293620912D-01
	0	3,304503903103D+00	1,709036579840D+00	6,25	8,938458807471D-01
	1	8,796275991570D-01	4,246982732116D-01	6,50	6,198417191699D-01
	2	5,832057640865D-02	3,016246918317D-02	6,75	2,831782732191D-01
				7,25	-1,525882331199D-01
				7,50	-1,679486933882D-01
				7,75	-9,420990043087D-02

I= 7 H= -1.0

0		0.	-5,617337058473D-01	-1.00	-4,814802550474D-02
1		2,303651579624D-02	8,387036121545D-02	-.75	-3,140261105821D-02
2		-2,511150970850D-02	-1,488348902312D-02	-.50	-1,779613532524D-02
				-.25	-7,328598305840D-03
0	-8,0	2,074993912265D-03	1,037496956132D-03	.25	4,191685953520D-03
1	-7,0	-1,202679795094D-02	-6,013398975469D-03	.50	5,305224030751D-03
2	-6,0	3,752805373748D-02	1,876402686874D-02	.75	3,644568418060D-03
3	-5,0	-9,354149152963D-02	-4,677074576481D-02	1.25	-4,120455211468D-03
4	-4,0	2,177009604039D-01	1,088504802019D-01	1.50	-6,564975564480D-03
5	-3,0	-4,501891287683D-01	-2,250945643841D-01	1.75	-5,387057966455D-03
6	-2,0	7,312046873540D-01	3,656023436770D-01	2,25	7,693521531282D-03
7	-1,0	-8,655228514687D-01	-4,327614257344D-01	2,50	1,312473279642D-02
8	0,0	7,313011603917D-01	3,656505801958D-01	2,75	1,130084546925D-02
9	1,0	-4,504547156849D-01	-2,252273578425D-01	3,25	-1,708849173592D-02
10	2,0	2,183354727893D-01	1,091677363946D-01	3,50	-2,959898072257D-02
11	3,0	-9,501837603124D-02	-4,750918801562D-02	3,75	-2,575335653430D-02
12	4,0	4,085802013600D-02	2,042901006800D-02	4,25	3,946851361399D-02
13	5,0	-1,736328231074D-02	-8,681641155368D-03	4,50	6,872527486876D-02
14	6,0	6,958315125954D-03	3,479157562977D-03	4,75	6,015040763116D-02
15	7,0	-2,229708270971D-03	-1,114854135485D-03	5,25	-9,425200399480D-02
16	8,0	3,846881648814D-04	1,923440824407D-04	5,50	-1,680024960406D-01
				5,75	-1,526239913806D-01
	0	-1,422844494974D+00	-8,433141136032D-01	6,25	2,832140314441D-01
	1	-3,787476395398D-01	-1,542654631544D-01	6,50	6,198955217081D-01
	2	-2,511150970850D-02	-1,488348902312D-02	6,75	8,938879841611D-01

CARDINAL SPLINES N=3

				7.25	8.938730213815D-01
				7.50	6.198730084041D-01
				7.75	2.831964135540D-01

I= 8 H= 0.0

0		0.	1.304632063048D+00	-1.00	2.073130036457D-02
1		-9.918929291296D-03	-1.729979371723D-01	-.75	1.352115569719D-02
2		1.081237107327D-02	1.081237107327D-02	-.50	7.662557413965D-03
				-.25	3.155505514903D-03
0	-8.0	-8.934417819735D-04	-4.467208909867D-04	.25	-1.804831632485D-03
1	-7.0	5.178511312665D-03	2.589255656333D-03	.50	-2.284291933017D-03
2	-6.0	-1.616056077127D-02	-8.080280385636D-03	.75	-1.569256162629D-03
3	-5.0	4.032186043676D-02	2.016093021838D-02	1.25	1.774157389666D-03
4	-4.0	-9.478791243458D-02	-4.739395621729D-02	1.50	2.826695924572D-03
5	-3.0	2.182390032132D-01	1.091195016066D-01	1.75	2.319508413203D-03
6	-2.0	-4.504196734463D-01	-2.252098367231D-01	2.25	-3.312537046642D-03
7	-1.0	7.313011603917D-01	3.656505801958D-01	2.50	-5.650875284703D-03
8	0.0	-8.655578938405D-01	-4.327789469203D-01	2.75	-4.865402692338D-03
9	1.0	7.313011603917D-01	3.656505801958D-01	3.25	7.355676996636D-03
10	2.0	-4.504196734463D-01	-2.252098367231D-01	3.50	1.273771236140D-02
11	3.0	2.182390032132D-01	1.091195016066D-01	3.75	1.107831485333D-02
12	4.0	-9.478791243458D-02	-4.739395621729D-02	4.25	-1.694341733303D-02
13	5.0	4.032186043676D-02	2.016093021838D-02	4.50	-2.943255471194D-02
14	6.0	-1.616056077127D-02	-8.080280385636D-03	4.75	-2.565757107297D-02
15	7.0	5.178511312665D-03	2.589255656333D-03	5.25	3.940597470385D-02
16	8.0	-8.934417819735D-04	-4.467208909867D-04	5.50	6.865346813544D-02
				5.75	6.010902916494D-02
	0	6.126403143589D-01	6.126403143589D-01	6.25	-9.422486318578D-02
	1	1.630790078810D-01	-2.697613970278D-23	6.50	-1.679712066455D-01
	2	1.081237107327D-02	1.081237107327D-02	6.75	-1.526058509666D-01
				7.25	2.832018485386D-01
				7.50	6.198811959923D-01
				7.75	8.938794207901D-01

I= 9 H= 1.0

0		0.	-3.029981116318D+00	-1.00	-8.926248510589D-03
1		4.270780172854D-03	3.924012875243D-01	-.75	-5.821786069616D-03
2		-4.655468337735D-03	-1.488348902312D-02	-.50	-3.299257170861D-03
				-.25	-1.358661814322D-03
0	-8.0	3.846881648814D-04	1.923440824407D-04	.25	7.771039441410D-04
1	-7.0	-2.229708270971D-03	-1.114854135485D-03	.50	9.835445071456D-04
2	-6.0	6.958315125954D-03	3.479157562977D-03	.75	6.756724944164D-04
3	-5.0	-1.736328231074D-02	-8.681641155368D-03	1.25	-7.638963985743D-04
4	-4.0	4.085802013600D-02	2.042901006800D-02	1.50	-1.217086132023D-03
5	-3.0	-9.501837603124D-02	-4.750918801562D-02	1.75	-9.987067969415D-04
6	-2.0	2.183354727893D-01	1.091677363946D-01	2.25	1.426270577986D-03
7	-1.0	-4.504547156849D-01	-2.252273578425D-01	2.50	2.433077087986D-03
8	0.0	7.313011603917D-01	3.656505801958D-01	2.75	2.094870573284D-03
9	1.0	-8.655228514687D-01	-4.327614257344D-01	3.25	-3.167029813144D-03
10	2.0	7.312046873540D-01	3.656023436770D-01	3.50	-5.484164854992D-03
11	3.0	-4.501891287683D-01	-2.250945643841D-01	3.75	-4.769526387300D-03
12	4.0	2.177009604039D-01	1.088504802019D-01	4.25	7.293119413435D-03
13	5.0	-9.354149152963D-02	-4.677074576481D-02	4.50	1.266589334815D-02
14	6.0	3.752805373748D-02	1.876402686874D-02	4.75	1.103693245914D-02
15	7.0	-1.202679795094D-02	-6.013398975469D-03	5.25	-1.691627571088D-02

16	8,0	2,074993912265D-03	1.037496956132D-03	5.50	-2,940126477802D-02
				5.75	-2,563943048427D-02
	0	-2.637837322322D-01	-8,433141136032D-01	6,25	3,939379175985D-02
	1	-7,021671323091D-02	1,542654631544D-01	6,50	6,863913239262D-02
	2	-4,655468337735D-03	-1,488348902312D-02	6,75	6,010046578434D-02
				7,25	-9,421846377441D-02
				7,50	-1,679630290552D-01
				7,75	-1,526004159814D-01

I=10 H= 2.0

0		0.	7,037020793256D+00	-1.00	3,843100547213D-03
1		-1,838738589509D-03	-9,072977801424D-01	-.75	2,506507543340D-03
2		2,004361957704D-03	3,016246918317D-02	-.50	1,420459784181D-03
				-,25	5,849572697338D-04
0	-8,0	-1,656233681948D-04	-8,281168409741D-05	,25	-3,345737665913D-04
1	-7,0	9.599771304364D-04	4,799885652182D-04	.50	-4,234545355847D-04
2	-6,0	-2,995831553432D-03	-1.497915776716D-03	,75	-2,909035425556D-04
3	-5,0	7.475659080062D-03	3,737829540031D-03	1,25	3,288873918714D-04
4	-4,0	-1,759293503102D-02	-8,796467515511D-03	1.50	5,240033536670D-04
5	-3,0	4,095445478125D-02	2.047722739063D-02	1.75	4,299824613331D-04
6	-2,0	-9,505341679328D-02	-4,752670839664D-02	2,25	-6,140653212568D-04
7	-1,0	2,183354727893D-01	1,091677363946D-01	2,50	-1,047534690191D-03
8	0.0	-4,504196734463D-01	-2,252098367231D-01	2,75	-9,019232177550D-04
9	1,0	7,312046873540D-01	3,656023436770D-01	3,25	1,363526651586D-03
10	2.0	-8,652922994802D-01	-4,326461497401D-01	3,50	2,361135636073D-03
11	3.0	7,306664579331D-01	3,653332289665D-01	3,75	2,053449078232D-03
12	4,0	-4,489383383229D-01	-2.244691691614D-01	4,25	-3,139880160343D-03
13	5,0	2,148056719432D-01	1,074028359716D-01	4,50	-5,452869645026D-03
14	6,0	-8,705274799192D-02	-4,352637399596D-02	4,75	-4,751384114303D-03
15	7,0	2,792743800327D-02	1,396371900163D-02	5,25	7,280936125079D-03
16	8,0	-4,818953027352D-03	-2,409476513676D-03	5,50	1,265155738117D-02
				5,75	1,102836900889D-02
	0	1,135692565770D-01	1,709036579840D+00	6,25	-1,690987628938D-02
	1	3,023105273376D-02	-4,246982732116D-01	6,50	-2,939308718641D-02
	2	2,004361957704D-03	3,016246918317D-02	6,75	-2,563399550331D-02
				7,25	3,938835678610D-02
				7,50	6,863095482407D-02
				7,75	6,009406639313D-02

I=11 H= 3.0

0		0.	-1,634248688106D+01	-1,00	-1,653996502658D-03
1		7,913576963587D-04	2,105333312362D+00	-.75	-1,078752600812D-03
2		-8,626388062994D-04	-6,815208211864D-02	-,50	-6,113385497542D-04
				-,25	-2,517543494834D-04
0	-8,0	7,128110994066D-05	3,564055497033D-05	,25	1,439941091549D-04
1	-7,0	-4,131556856211D-04	-2,065778428106D-04	.50	1,822466812902D-04
2	-6,0	1,289348391913D-03	6,446741959565D-04	.75	1,251992852448D-04
3	-5,0	-3,217383399169D-03	-1,608691699584D-03	1,25	-1,415468025985D-04
4	-4,0	7.571744725674D-03	3,785872362837D-03	1,50	-2,255209561993D-04
5	-3,0	-1,762796102651D-02	-8,813980513255D-03	1,75	-1,850561733528D-04
6	-2,0	4,095445478125D-02	2,047722739063D-02	2,25	2,642818869791D-04
7	-1,0	-9,501837603124D-02	-4,750918801562D-02	2,50	4,508387437304D-04
8	0,0	2,182390032132D-01	1.091195016066D-01	2,75	3,881703564630D-04
9	1.0	-4,501891287683D-01	-2,250945643841D-01	3,25	-5,868354503576D-04
10	2,0	7,306664579331D-01	3,653332289665D-01	3,50	-1,016186782336D-03

11	3.0	-8.640411150664D-01	-4.320205575332D-01	3.75	-8.837641245950D-04
12	4.0	7.277611129039D-01	3.638805564519D-01	4.25	1.351339927084D-03
13	5.0	-4.422141186118D-01	-2.211070593059D-01	4.50	2.346797434426D-03
14	6.0	1.997361660433D-01	9.986808302164D-02	4.75	2.044884934838D-03
15	7.0	-6.475630918138D-02	-3.237815459069D-02	5.25	-3.133480639434D-03
16	8.0	1.118797866811D-02	5.593989334056D-03	5.50	-5.444692041679D-03
				5.75	-4.745949177196D-03
	0	-4.887802203229D-02	-3.861553637756D+00	6.25	7.275501260074D-03
	1	-1.301086320443D-02	1.014899998464D+00	6.50	1.264338000855D-02
	2	-8.626388062994D-04	-6.815208211864D-02	6.75	1.102196979081D-02
				7.25	-1.690131321311D-02
				7.50	-2.937875198422D-02
				7.75	-2.562181304726D-02

I=12	H= 4.0				
0		0.	3.793616351395D+01	-1.00	7.104338637758D-04
1		-3.399084006171D-04	-4.886448170484D+00	-.75	4.633518734896D-04
2		3.705254631587D-04	1.573895125229D-01	-.50	2.625855660982D-04
				-.25	1.081349416017D-04
0	-8.0	-3.061706254156D-05	-1.530853127078D-05	.25	-6.184915818200D-05
1	-7.0	1.774609499294D-04	8.873047496471D-05	.50	-7.827961772331D-05
2	-6.0	-5.538081667196D-04	-2.769040833598D-04	.75	-5.377629989467D-05
3	-5.0	1.381948736662D-03	6.909743683309D-04	1.25	6.079797731594D-05
4	-4.0	-3.252261921560D-03	-1.626130960780D-03	1.50	9.686702719167D-05
5	-3.0	7.571744725674D-03	3.785872362837D-03	1.75	7.948636646152D-05
6	-2.0	-1.759293503102D-02	-8.796467515511D-03	2.25	-1.135158395127D-04
7	-1.0	4.085802013600D-02	2.042901006800D-02	2.50	-1.936467873086D-04
8	0.0	-9.478791243458D-02	-4.739395621729D-02	2.75	-1.667291092447D-04
9	1.0	2.177009604039D-01	1.088504802019D-01	3.25	2.520608497541D-04
10	2.0	-4.489383383229D-01	-2.244691691614D-01	3.50	4.364782283240D-04
11	3.0	7.277611129039D-01	3.638805564519D-01	3.75	3.795992926327D-04
12	4.0	-8.572956642560D-01	-4.286478321280D-01	4.25	-5.804349377972D-04
13	5.0	7.121496549916D-01	3.560748274958D-01	4.50	-1.008009063316D-03
14	6.0	-4.072277860617D-01	-2.036138930309D-01	4.75	-8.783296264033D-04
15	7.0	1.479713659120D-01	7.398568295602D-02	5.25	1.345906148976D-03
16	8.0	-2.589294550258D-02	-1.294647275129D-02	5.50	2.338622019673D-03
				5.75	2.038487446628D-03
	0	2.099436243722D-02	8.917506951545D+00	6.25	-3.124920232839D-03
	1	5.588499009922D-03	-2.368215970117D+00	6.50	-5.430361481746D-03
	2	3.705254631587D-04	1.573895125229D-01	6.75	-4.733770768912D-03
				7.25	7.257367754405D-03
				7.50	1.261210264485D-02
				7.75	1.099483950813D-02

I=13	H= 5.0				
0		0.	-8.766867672078D+01	-1.00	-3.018769996928D-04
1		1.444336107561D-04	1.129299481283D+01	-.75	-1.968871143439D-04
2		-1.574433889367D-04	-3.634306155054D-01	-.50	-1.115776526122D-04
				-.25	-4.594861449756D-05
0	-8.0	1.300977818065D-05	6.504889090325D-06	.25	2.628089574197D-05
1	-7.0	-7.540656754607D-05	-3.770328377304D-05	.50	3.326251371200D-05
2	-6.0	2.353237317387D-04	1.176618658694D-04	.75	2.285058313575D-05
3	-5.0	-5.872165789651D-04	-2.936082894826D-04	1.25	-2.583422879296D-05
4	-4.0	1.381948736662D-03	6.909743683309D-04	1.50	-4.116066115001D-05
5	-3.0	-3.217383399169D-03	-1.608691699584D-03	1.75	-3.377528443572D-05

6	-2.0	7.475659080062D-03	3.737829540031D-03	2.25	4.823506139990D-05
7	-1.0	-1.736328231074D-02	-8.681641155368D-03	2.50	8.228424080362D-05
8	0.0	4.032186043676D-02	2.016093021838D-02	2.75	7.084640214906D-05
9	1.0	-9.354149152963D-02	-4.677074576481D-02	3.25	-1.071054980849D-04
10	2.0	2.148056719432D-01	1.074028359716D-01	3.50	-1.854679851335D-04
11	3.0	-4.422141186118D-01	-2.211070593059D-01	3.75	-1.612990319074D-04
12	4.0	7.121496549916D-01	3.560748274958D-01	4.25	2.466379506889D-04
13	5.0	-8.211651803955D-01	-4.105825901977D-01	4.50	4.283223966199D-04
14	6.0	6.311788728002D-01	3.155894364001D-01	4.75	3.732190999481D-04
15	7.0	-2.874256978758D-01	-1.437128489379D-01	5.25	-5.719012119133D-04
16	8.0	5.802777577064D-02	2.901388788532D-02	5.50	-9.937248930517D-04
				5.75	-8.661916642014D-04
	0	-8.920908005901D-03	-2.058427761050D+01	6.25	1.327834653084D-03
	1	-2.374660612231D-03	5.478104964743D+00	6.50	2.307452415098D-03
	2	-1.574433889367D-04	-3.634306155054D-01	6.75	2.011451100105D-03
				7.25	-3.083702213720D-03
				7.50	-5.358834178121D-03
				7.75	-4.671476332170D-03

I=14 H= 6.0

0		0.	1.934664409342D+02	-1.00	1.209755047075D-04
1		-5.788095474555D-05	-2.494365422290D+01	-.75	7.890140041276D-05
2		6.309454996195D-05	8.032798425273D-01	-.50	4.471411486326D-05
				-.25	1.841364805901D-05
0	-8.0	-5.213595216408D-06	-2.606797608204D-06	.25	-1.053192071534D-05
1	-7.0	3.021875656833D-05	1.510937828416D-05	.50	-1.332976473280D-05
2	-6.0	-9.430465802030D-05	-4.715232901015D-05	.75	-9.157242289142D-06
3	-5.0	2.353237317387D-04	1.176618658694D-04	1.25	1.035292145520D-05
4	-4.0	-5.538081667196D-04	-2.769040833598D-04	1.50	1.649490276424D-05
5	-3.0	1.289348391913D-03	6.446741959565D-04	1.75	1.353525470762D-05
6	-2.0	-2.995831553432D-03	-1.497915776716D-03	2.25	-1.932992875648D-05
7	-1.0	6.958315125954D-03	3.479157562977D-03	2.50	-3.297494532638D-05
8	0.0	-1.616056077127D-02	-8.080280385636D-03	2.75	-2.839129600669D-05
9	1.0	3.752805373748D-02	1.876402686874D-02	3.25	4.292192415140D-05
10	2.0	-8.705274799192D-02	-4.352637399596D-02	3.50	7.432524875767D-05
11	3.0	1.997361660433D-01	9.986808302164D-02	3.75	6.463967707077D-05
12	4.0	-4.072277860617D-01	-2.036138930309D-01	4.25	-9.883876712853D-05
13	5.0	6.311788728002D-01	3.155894364001D-01	4.50	-1.716477832955D-04
14	6.0	-6.397043818670D-01	-3.198521909335D-01	4.75	-1.495654475077D-04
15	7.0	3.626945952881D-01	1.813472976440D-01	5.25	2.291861775629D-04
16	8.0	-8.585625920991D-02	-4.292812960496D-02	5.50	3.982296233991D-04
				5.75	3.471213899944D-04
	0	3.575003559601D-03	4.532711707275D+01	6.25	-5.321220853313D-04
	1	9.516318446457D-04	-1.209117674247D+01	6.50	-9.246980332458D-04
	2	6.309454996195D-05	8.032798425273D-01	6.75	-8.060769121821D-04
				7.25	1.235772520654D-03
				7.50	2.147510836975D-03
				7.75	1.872049935233D-03

I=15 H= 7.0

0		0.	-2.139925652003D+02	-1.00	-3.876509818479D-05
1		1.854723317059D-05	2.813967126906D+01	-.75	-2.528297394843D-05
2		-2.021786501419D-05	-9.228305258340D-01	-.50	-1.432808283885D-05
				-.25	-5.900424856036D-06
0	-8.0	1.670631843599D-06	8.353159217997D-07	.25	3.374823205671D-06

CARDINAL SPLINES N=3

1	-7,0	-9,683225279595D-06	-4,841612639797D-06	.50	4,271337576861D-06
2	-6,0	3,021875656833D-05	1.510937828416D-05	.75	2,934324575035D-06
3	-5,0	-7,540656754607D-05	-3,770328377304D-05	1,25	-3,317465115606D-06
4	-4,0	1,774609499294D-04	8,873047496471D-05	1,50	-5,285586753699D-06
5	-3,0	-4,131556856211D-04	-2,065778428106D-04	1,75	-4,337204287474D-06
6	-2,0	9,599771304364D-04	4,799885652182D-04	2,25	6,194035626959D-06
7	-1,0	-2,229708270971D-03	-1,114854135485D-03	2,50	1,056641174012D-05
8	0,0	5.178511312665D-03	2.589255656333D-03	2,75	9,097638236294D-06
9	1,0	-1,202679795094D-02	-6,013398975469D-03	3,25	-1,375379758059D-05
10	2,0	2,792743800327D-02	1,396371900163D-02	3,50	-2,381660297680D-05
11	3,0	-6,475630918138D-02	-3,237815459069D-02	3,75	-2,071298180645D-05
12	4,0	1,479713659120D-01	7,398568295602D-02	4,25	3,167165551614D-05
13	5,0	-2,874256978758D-01	-1,437128489379D-01	4,50	5,500240056812D-05
14	6,0	3,626945952881D-01	1,813472976440D-01	4,75	4,792639027150D-05
15	7,0	-2,424626222216D-01	-1.212313111108D-01	5,25	-7,343986428615D-05
16	8,0	6,445814299425D-02	3,222907149713D-02	5,50	-1,276077367920D-04
				5,75	-1,112307376897D-04
	0	-1,145565495544D-03	-4,793634870121D+01	6,25	1,705119088426D-04
	1	-3,049386070565D-04	1,337438285572D+01	6,50	2,963079847532D-04
	2	-2,021786501419D-05	-9,228305258340D-01	6,75	2,582973107440D-04
				7,25	-3,959878328624D-04
				7,50	-6,881427871047D-04
				7,75	-5,998745570775D-04

I=16 H= 8.0

0		0.	8.218211966583D+01	-1,00	6,688082284343D-06
1		-3,199925381345D-06	-1,107404541720D+01	-.75	4,362032293945D-06
2		3,488156902998D-06	3,730587027111D-01	-.50	2,472001916422D-06
				-.25	1,017991151774D-06
0	-8,0	-2,882315216532D-07	-1,441157608266D-07	.25	-5,822530149942D-07
1	-7,0	1,670631843599D-06	8,353159217997D-07	.50	-7,369306999745D-07
2	-6,0	-5,213595216408D-06	-2,606797608204D-06	.75	-5,062544692458D-07
3	-5,0	1.300977818065D-05	6,504889090325D-06	1,25	5,723571126492D-07
4	-4,0	-3,061706254156D-05	-1.530853127078D-05	1,50	9,119140872869D-07
5	-3,0	7,128110994066D-05	3,564055497033D-05	1,75	7,482911308608D-07
6	-2,0	-1,656233681948D-04	-8,281168409741D-05	2,25	-1,068647362836D-06
7	-1,0	3,846881648814D-04	1,923440824407D-04	2,50	-1,823006634247D-06
8	0,0	-8,934417819735D-04	-4,467208909867D-04	2,75	-1,569601418972D-06
9	1,0	2,074993912265D-03	1,037496956132D-03	3,25	2,372921371238D-06
10	2,0	-4,818953027352D-03	-2,409476513676D-03	3,50	4,109041583796D-06
11	3,0	1,118797866811D-02	5,593989334056D-03	3,75	3,573578635441D-06
12	4,0	-2,589294550258D-02	-1,294647275129D-02	4,25	-5,464261618931D-06
13	5,0	5,802777577064D-02	2,901388788532D-02	4,50	-9,489478887867D-06
14	6,0	-8,585625920991D-02	-4,292812960496D-02	4,75	-8,268665802439D-06
15	7,0	6,445814299425D-02	3,222907149713D-02	5,25	1,267046591157D-05
16	8,0	-1,855619925082D-02	-9,278099625411D-03	5,50	2,201596495713D-05
				5,75	1,919046668671D-05
	0	1,976426387411D-04	1,746551330175D+01	6,25	-2,941815514070D-05
	1	5,261058506662D-05	-5,105106173822D+00	6,50	-5,112155683065D-05
	2	3,488156902998D-06	3,730587027111D-01	6,75	-4,456363399596D-05
				7,25	6,831916421571D-05
				7,50	1,187241999271D-04
				7,75	1,034954148001D-04

CARDINAL SPLINES N=3

```
               N= 3      M=17                       P=  8.5      Q= 5

I= 0     H= -8.5
0           1.000000000000D+00   4.997946847500D-01   -1.00   3.473671635299D+00
1          -1.727557718024D+00  -8.637540153720D-01    -.75   2.715357366985D+00
2           7.461139172753D-01   3.730562076804D-01    -.50   2.050307338331D+00
                                                       -.25   1.478521549336D+00
0   -8.5   -1.855619925165D-02  -9.278099625826D-03     .25   6.147245690355D-01
1   -7.5    6.445814299906D-02   3.222907149953D-02     .50   3.221697390804D-01
2   -6.5   -8.585625922493D-02  -4.292812961246D-02     .75   1.196173168850D-01
3   -5.5    5.802777580811D-02   2.901388790405D-02    1.25  -5.021023041818D-02
4   -4.5   -2.589294559075D-02  -1.294647279538D-02    1.50  -5.147708426460D-02
5   -3.5    1.118797887338D-02   5.593989436691D-03    1.75  -2.752040839536D-02
6   -2.5   -4.818953504301D-03  -2.409476752151D-03    2.25   1.677968342828D-02
7   -1.5    2.074995020060D-03   1.037497510030D-03    2.50   1.898487004050D-02
8    -.5   -8.934443548395D-04  -4.467221774198D-04    2.75   1.083411537189D-02
9     .5    3.846941403107D-04   1.923470701554D-04    3.25  -7.016349050456D-03
10   1.5   -1.656372459322D-04  -8.281862296612D-05    3.50  -8.037375407726D-03
11   2.5    7.131333949822D-05   3.565666974911D-05    3.75  -4.621138965453D-03
12   3.5   -3.069188866486D-05  -1.534594433243D-05    4.25   3.012078382367D-03
13   4.5    1.318295233134D-05   6.591476165672D-06    4.50   3.454789726647D-03
14   5.5   -5.601689755790D-06  -2.800844877895D-06    4.75   1.987862451501D-03
15   6.5    2.244845569922D-06   1.122422784961D-06    5.25  -1.296539328673D-03
16   7.5   -7.193328859268D-07  -3.596664429634D-07    5.50  -1.487292771456D-03
17   8.5    1.241053874797D-07   6.205269373985D-08    5.75  -8.558432928006D-04
                                                       6.25   5.582411928358D-04
       0    4.022248991994D+01   2.011119655900D+01    6.50   6.403806594746D-04
       1    1.095637887566D+01   5.478201515196D+00    6.75   3.685015436786D-04
       2    7.461139172753D-01   3.730562076804D-01    7.25  -2.403642123809D-04
                                                       7.50  -2.757317153855D-04
                                                       7.75  -1.586675843657D-04
                                                       8.25   1.034949578805D-04
                                                       8.50   1.187234058952D-04

I= 1     H= -7.5
0           0.                   1.190037066989D-03   -1.00  -4.626823524723D+00
1           2.781182690862D+00   1.390447348080D+00    -.75  -3.124059987194D+00
2          -1.845640833861D+00  -9.228160642735D-01    -.50  -1.852001553896D+00
                                                       -.25  -8.106482248319D-01
0   -8.5    6.445814299906D-02   3.222907149953D-02     .25   5.800060680045D-01
1   -7.5   -2.424626222495D-01  -1.212313111247D-01     .50   9.311954539345D-01
2   -6.5    3.626945953751D-01   1.813472976875D-01     .75   1.063010268581D+00
3   -5.5   -2.874256980929D-01  -1.437128490465D-01    1.25   7.891384222416D-01
4   -4.5    1.479713664231D-01   7.398568321153D-02    1.50   5.009842265593D-01
5   -3.5   -6.475631037115D-02  -3.237815518558D-02    1.75   2.152141789208D-01
6   -2.5    2.792744076773D-02   1.396372038387D-02    2.25  -1.084973407936D-01
7   -1.5   -1.202680437188D-02  -6.013402185941D-03    2.50  -1.174242888264D-01
8    -.5    5.178526225371D-03   2.589263112686D-03    2.75  -6.515456765818D-02
9     .5   -2.229742905429D-03  -1.114871452715D-03    3.25   4.115217611115D-02
```

CARDINAL SPLINES N=3

```
10   1.5   9.600575678220D-04   4.800287839110D-04    3.50   4.690407302167D-02
11   2.5  -4.133424928269D-04  -2.066712464134D-04    3.75   2.688643472339D-02
12   3.5   1.778946530289D-04   8.894732651447D-05    4.25  -1.747932029736D-02
13   4.5  -7.641031012419D-05  -3.820515506210D-05    4.50  -2.003818049887D-02
14   5.5   3.246820899497D-05   1.623410449749D-05    4.75  -1.152632038836D-02
15   6.5  -1.301145159823D-05  -6.505725799114D-06    5.25   7.515830288140D-03
16   7.5   4.169358085767D-06   2.084679042884D-06    5.50   8.621156467536D-03
17   8.5  -7.193328859268D-07  -3.596664429634D-07    5.75   4.960781045758D-03
                                                      6.25  -3.235685962359D-03
       0  -1.097074973741D+02  -5.485346814801D+01    6.50  -3.711764967454D-03
       1  -2.859471148478D+01  -1.429742574457D+01    6.75  -2.135896707897D-03
       2  -1.845640833861D+00  -9.228160642735D-01    7.25   1.393187624236D-03
                                                      7.50   1.598182245372D-03
                                                      7.75   9.196607592590D-04
                                                      8.25  -5.998719087052D-04
                                                      8.50  -6.881381847802D-04


I= 2     H= -6.5
0               0.                 -3.713787441305D-03   -1.00   3.127136922148D+00
1         -1.520640331461D+00      -7.598707885051D-01    -.75   2.044134580857D+00
2          1.606496590686D+00       8.032347118643D-01    -.50   1.161944313402D+00
                                                          -.25   4.805661197833D-01
0   -8.5  -8.585625922493D-02  -4.292812961246D-02         .25  -2.798378899506D-01
1   -7.5   3.626945953751D-01   1.813472976875D-01         .50  -3.613790261599D-01
2   -6.5  -6.397043821386D-01  -3.198521910693D-01         .75  -2.572000091003D-01
3   -5.5   6.311788734779D-01   3.155894367390D-01        1.25   3.476921927197D-01
4   -4.5  -4.072277876565D-01  -2.036138938283D-01        1.50   6.930200694683D-01
5   -3.5   1.997361697563D-01   9.986808487813D-02        1.75   9.356781956019D-01
6   -2.5  -8.705275661909D-02  -4.352637830955D-02        2.25   8.667750562727D-01
7   -1.5   3.752807377553D-02   1.876403688777D-02        2.50   5.888238698897D-01
8    -.5  -1.616060730984D-02  -8.080303654920D-03        2.75   2.653417188042D-01
9     .5   6.958423210835D-03   3.479211605418D-03        3.25  -1.409607496550D-01
10   1.5  -2.996082577002D-03  -1.498041288501D-03        3.50  -1.546118064694D-01
11   2.5   1.289931367245D-03   6.449656836227D-04        3.75  -8.653576216024D-02
12   3.5  -5.551616381104D-04  -2.775808190552D-04        4.25   5.508853126827D-02
13   4.5   2.384561438999D-04   1.192280719500D-04        4.50   6.288877547009D-02
14   5.5  -1.013245974857D-04  -5.066229874287D-05        4.75   3.608392353733D-02
15   6.5   4.060526086036D-05   2.030263043018D-05        5.25  -2.347817905505D-02
16   7.5  -1.301145159823D-05  -6.505725799114D-06        5.50  -2.691962184022D-02
17   8.5   2.244845569922D-06   1.122422784961D-06        5.75  -1.548615640689D-02
                                                          6.25   1.009871398305D-02
       0   1.031439358597D+02   5.157109244246D+01        6.50   1.158408513497D-02
       1   2.578980171021D+01   1.289511931319D+01        6.75   6.665772977220D-03
       2   1.606496590686D+00   8.032347118643D-01        7.25  -4.347809245705D-03
                                                          7.50  -4.987527807966D-03
                                                          7.75  -2.870024363658D-03
                                                          8.25   1.872041670371D-03
                                                          8.50   2.147496474351D-03


I= 3     H= -5.5
0               0.                  9.267223253077D-03   -1.00  -1.395379800342D+00
1          6.686760122671D-01       3.332166497983D-01    -.75  -9.102778899926D-01
2         -7.267037880752D-01      -3.633179984167D-01    -.50  -5.160139531523D-01
                                                          -.25  -2.125879898215D-01
0   -8.5   5.802777580811D-02   2.901388790405D-02         .25   1.218066840619D-01
```

CARDINAL SPLINES N=3

1	-7.5	-2.874256980929D-01	-1.437128490465D-01
2	-6.5	6.311788734779D-01	3.155894367390D-01
3	-5.5	-8.211651820865D-01	-4.105825910432D-01
4	-4.5	7.121496589712D-01	3.560748294856D-01
5	-3.5	-4.422141278770D-01	-2.211070639385D-01
6	-2.5	2.148056934710D-01	1.074028467355D-01
7	-1.5	-9.354154153170D-02	-4.677077076585D-02
8	-.5	4.032197656707D-02	2.016098828354D-02
9	.5	-1.736355202104D-02	-8.681776010519D-03
10	1.5	7.476285473333D-03	3.738142736667D-03
11	2.5	-3.218838130394D-03	-1.609419065197D-03
12	3.5	1.385326130157D-03	6.926630650783D-04
13	4.5	-5.950330636933D-04	-2.975165318466D-04
14	5.5	2.528409825985D-04	1.264204912993D-04
15	6.5	-1.013245974857D-04	-5.066229874287D-05
16	7.5	3.246820899497D-05	1.623410449749D-05
17	8.5	-5.601689755790D-06	-2.800844877895D-06
	0	-4.682060258416D+01	-2.340811663907D+01
	1	-1.168528838501D+01	-5.843189323285D+00
	2	-7.267037880752D-01	-3.633179984167D-01

.50	1.544754271087D-01
.75	1.065063916125D-01
1.25	-1.228236245275D-01
1.50	-2.004031352911D-01
1.75	-1.711399238778D-01
2.25	2.952198251052D-01
2.50	6.336517813616D-01
2.75	9.017984333684D-01
3.25	8.887161405003D-01
3.50	6.139578433765D-01
3.75	2.797926531656D-01
4.25	-1.503799312995D-01
4.50	-1.654154071947D-01
4.75	-9.275208041393D-02
5.25	5.914298755816D-02
5.50	6.753974410353D-02
5.75	3.876026312823D-02
6.25	-2.522387636043D-02
6.50	-2.892218099126D-02
6.75	-1.663851218948D-02
7.25	1.085036731863D-02
7.50	1.244633862554D-02
7.75	7.161950559861D-03
8.25	-4.671455708395D-03
8.50	-5.358798338256D-03

I= 4 H= -4.5

0		0.	-2.180937676231D-02
1		-2.885155550590D-01	-1.416187904098D-01
2		3.144085006497D-01	1.571244807599D-01
0	-8.5	-2.589294559075D-02	-1.294647279538D-02
1	-7.5	1.479713664231D-01	7.398568321153D-02
2	-6.5	-4.072277876565D-01	-2.036138938283D-01
3	-5.5	7.121496589712D-01	3.560748294856D-01
4	-4.5	-8.572956736216D-01	-4.286478368108D-01
5	-3.5	7.277611347084D-01	3.638805673542D-01
6	-2.5	-4.489383889863D-01	-2.244691944931D-01
7	-1.5	2.177010780782D-01	1.088505390391D-01
8	-.5	-9.478818573432D-02	-4.739409286716D-02
9	.5	4.085865486913D-02	2.042932743456D-02
10	1.5	-1.759440917775D-02	-8.797204588876D-03
11	2.5	7.575168273431D-03	3.787584136716D-03
12	3.5	-3.260210240619D-03	-1.630105120310D-03
13	4.5	1.400343962042D-03	7.001719810212D-04
14	5.5	-5.950330636933D-04	-2.975165318466D-04
15	6.5	2.384561438999D-04	1.192280719500D-04
16	7.5	-7.641031012419D-05	-3.820515506210D-05
17	8.5	1.318295233134D-05	6.591476165672D-06
	0	2.026363195394D+01	1.012667463966D+01
	1	5.056428955986D+00	2.529497382509D+00
	2	3.144085006497D-01	1.571244807599D-01

-1.00	6.029240557087D-01
-.75	3.932414479097D-01
-.50	2.228599026919D-01
-.25	9.177942005535D-02
.25	-5.250364355381D-02
.50	-6.646480691676D-02
.75	-4.567640204063D-02
1.25	5.174434273379D-02
1.50	8.264534349437D-02
1.75	6.810496959855D-02
2.25	-9.940959315575D-02
2.50	-1.739120310225D-01
2.75	-1.560222180454D-01
3.25	2.854293532353D-01
3.50	6.224365898399D-01
3.75	8.953501892015D-01
4.25	8.929191341379D-01
4.50	6.187785877066D-01
4.75	2.825664767391D-01
5.25	-1.521890967982D-01
5.50	-1.674907467174D-01
5.75	-9.394630793099D-02
6.25	5.992194768879D-02
6.50	6.843332120139D-02
6.75	3.927446523275D-02
7.25	-2.555927955070D-02
7.50	-2.930693810180D-02
7.75	-1.685991952395D-02
8.25	1.099479097238D-02

				8.50	1,261201829972D-02

I= 5	H= -3.5				
0		0.	5.077549752400D-02	-1.00	-2.596950769569D-01
1		1.242535490418D-01	5.598281701948D-02	-.75	-1.693760212336D-01
2		-1.354415279151D-01	-6.753504845584D-02	-.50	-9.598715649967D-02
				-.25	-3.952848275514D-02
0	-8.5	1.118797887338D-02	5.593989436691D-03	.25	2.260921752636D-02
1	-7.5	-6.475631037115D-02	-3.237815518558D-02	.50	2.861601688189D-02
2	-6.5	1.997361697563D-01	9.986808487813D-02	.75	1.965926215936D-02
3	-5.5	-4.422141278770D-01	-2.211070639385D-01	1.25	-2.223068771663D-02
4	-4.5	7.277611347084D-01	3.638805673542D-01	1.50	-3.542803437579D-02
5	-3.5	-8.640411658307D-01	-4.320205829153D-01	1.75	-2.908368569758D-02
6	-2.5	7.306665758851D-01	3.653332879426D-01	2.25	4.162746780695D-02
7	-1.5	-4.501894027317D-01	-2.250947013659D-01	2.50	7.119890841234D-02
8	-.5	2.182396394960D-01	1.091198197480D-01	2.75	6.157282061178D-02
9	.5	-9.501985378508D-02	-4.750992689254D-02	3.25	-9.517926721745D-02
10	1.5	4.095788681576D-02	2.047894340788D-02	3.50	-1.690661035032D-01
11	2.5	-1.763593155875D-02	-8.817965779376D-03	3.75	-1.532360217277D-01
12	3.5	7.590249603602D-03	3.795124801801D-03	4.25	2.836132980012D-01
13	4.5	-3.260210240619D-03	-1.630105120310D-03	4.50	6.203536127565D-01
14	5.5	1.385326130157D-03	6.926630650783D-04	4.75	8.941516580210D-01
15	6.5	-5.551616381104D-04	-2.775808190552D-04	5.25	8.937008505572D-01
16	7.5	1.778946530289D-04	8.894732651447D-05	5.50	6.196753151961D-01
17	8.5	-3.069188866486D-05	-1.534594433243D-05	5.75	2.830824877758D-01
				6.25	-1.525256777169D-01
	0	-8.729495225014D+00	-4.352777808745D+00	6.50	-1.678768543550D-01
	1	-2.178252425516D+00	-1.092113006730D+00	6.75	-9.416849224576D-02
	2	-1.354415279151D-01	-6.753504845584D-02	7.25	6.006687805295D-02
				7.50	6.859958212280D-02
				7.75	3.937014336770D-02
				8.25	-2.562170004875D-02
				8.50	-2.937855561612D-02

I= 6	H= -2.5				
0		0.	-1.179780864499D-01	-1.00	1.118222108570D-01
1		-5.350162867635D-02	-1.247518170844D-02	-.75	7.293154898388D-02
2		5.832058218065D-02	2.872877666730D-02	-.50	4.133095988334D-02
				-.25	1.702044355538D-02
0	-8.5	-4.818953504301D-03	-2.409476752151D-03	.25	-9.735076792079D-03
1	-7.5	2.792744076773D-02	1.396372038387D-02	.50	-1.232126109002D-02
2	-6.5	-8.705275661909D-02	-4.352637830955D-02	.75	-8.464454286062D-03
3	-5.5	2.148056934710D-01	1.074028467355D-01	1.25	9.569867698380D-03
4	-4.5	-4.489383889863D-01	-2.244691944931D-01	1.50	1.524767124265D-02
5	-3.5	7.306665758851D-01	3.653332879426D-01	1.75	1.251234735378D-02
6	-2.5	-8.652925735446D-01	-4.326462867723D-01	2.25	-1.787308620972D-02
7	-1.5	7.312053239146D-01	3.656026619573D-01	2.50	-3.049775653059D-02
8	-.5	-4.504211518645D-01	-2.252105759323D-01	2.75	-2.627011879101D-02
9	.5	2.183389063858D-01	1.091694531929D-01	3.25	3.980535577269D-02
10	1.5	-9.506139120778D-02	-4.753069560389D-02	3.50	6.911164032729D-02
11	2.5	4.097297450410D-02	2.048648725205D-02	3.75	6.037273247670D-02
12	3.5	-1.763593155875D-02	-8.817965779376D-03	4.25	-9.439704361456D-02
13	4.5	7.575168273431D-03	3.787584136716D-03	4.50	-1.681689087094D-01
14	5.5	-3.218838130394D-03	-1.609419065197D-03	4.75	-1.527197810999D-01
15	6.5	1.289931367245D-03	6.449656836227D-04	5.25	2.832765892351D-01

```
16   7.5  -4.133424928269D-04  -2.066712464134D-04      5.50  6.199673633947D-01
17   8.5   7.131333949822D-05   3.565666974911D-05      5.75  8.939293938465D-01
                                                        6.25  8.938458320676D-01
       0   3.758898218803D+00   1.851636983241D+00      6.50  6.198416345768D-01
       1   9.379482683948D-01   4.759140216356D-01      6.75  2.831781994777D-01
       2   5.832058218065D-02   2.872877666730D-02      7.25 -1.525881200691D-01
                                                        7.50 -1.679484969300D-01
                                                        7.75 -9.420972917242D-02
                                                        8.25  6.009380383837D-02
                                                        8.50  6.863049855809D-02


I= 7     H= -1.5
0              0.                 2.740239836803D-01   -1.00 -4.814805120986D-02
1              2.303652809490D-02 -2.163929800921D-02    -.75 -3.140262782334D-02
2             -2.511152311496D-02 -1.155349660494D-02    -.50 -1.779614482619D-02
                                                         -.25 -7.328602218411D-03
0   -8.5   2.074995020060D-03   1.037497510030D-03       .25  4.191688191365D-03
1   -7.5  -1.202680437188D-02  -6.013402185941D-03       .50  5.305226863087D-03
2   -6.5   3.752807377553D-02   1.876403688777D-02       .75  3.644570363810D-03
3   -5.5  -9.354154153170D-02  -4.677077076585D-02      1.25 -4.120457411278D-03
4   -4.5   2.177010780782D-01   1.088505390391D-01      1.50 -6.564979069351D-03
5   -3.5  -4.501894027317D-01  -2.250947013659D-01      1.75 -5.387060842454D-03
6   -2.5   7.312053239146D-01   3.656026619573D-01      2.25  7.693525638545D-03
7   -1.5  -8.655243299879D-01  -4.327621649940D-01      2.50  1.312473980301D-02
8    -.5   7.313045942672D-01   3.656522971336D-01      2.75  1.130085150189D-02
9     .5  -4.504626907913D-01  -2.252313453956D-01      3.25 -1.708850085606D-02
10   1.5   2.183539947105D-01   1.091769973552D-01      3.50 -2.959899651535D-02
11   2.5  -9.506139120778D-02  -4.753069560389D-02      3.75 -2.575337026907D-02
12   3.5   4.095788681576D-02   2.047894340788D-02      4.25  3.946853461546D-02
13   4.5  -1.759440917775D-02  -8.797204588876D-03      4.50  6.872531134083D-02
14   5.5   7.476285473333D-03   3.738142736667D-03      4.75  6.015043941114D-02
15   6.5  -2.996082577002D-03  -1.498041288501D-03      5.25 -9.425205269276D-02
16   7.5   9.600575678220D-04   4.800287839110D-04      5.50 -1.680025806573D-01
17   8.5  -1.656372459322D-04  -8.281862296612D-05      5.75 -1.526240651377D-01
                                                        6.25  2.832141445105D-01
       0  -1.618497056249D+00  -7.446501791053D-01      6.50  6.198957181898D-01
       1  -4.038593648595D-01  -2.180487402933D-01      6.75  8.938881554380D-01
       2  -2.511152311496D-02  -1.155349660494D-02      7.25  8.938727588021D-01
                                                        7.50  6.198725520969D-01
                                                        7.75  2.831960157774D-01
                                                        8.25 -1.525998061537D-01
                                                        8.50 -1.679619693006D-01


I= 8     H=  -.5
0              0.                -6.364234288573D-01   -1.00  2.073136006498D-02
1             -9.918957855068D-03  7.204928889974D-02    -.75  1.352119463437D-02
2              1.081240220991D-02  3.078430778981D-03    -.50  7.662579480011D-03
                                                         -.25  3.155514601886D-03
0   -8.5  -8.934443548395D-04  -4.467221774198D-04       .25 -1.804836829901D-03
1   -7.5   5.178526225371D-03   2.589263112686D-03       .50 -2.284298511146D-03
2   -6.5  -1.616060730984D-02  -8.080303654920D-03       .75 -1.569260681652D-03
3   -5.5   4.032197656707D-02   2.016098828354D-02      1.25  1.774162498747D-03
4   -4.5  -9.478818573432D-02  -4.739409286716D-02      1.50  2.826704064670D-03
5   -3.5   2.182396394960D-01   1.091198197480D-01      1.75  2.319515092739D-03
6   -2.5  -4.504211518645D-01  -2.252105759323D-01      2.25 -3.312546585801D-03
```

7	-1.5	7.313045942672D-01	3.656522971336D-01	2.50	-5.650891557566D-03
8	-.5	-8.655658690503D-01	-4.327829345252D-01	2.75	-4.865416703208D-03
9	.5	7.313196826550D-01	3.656598413275D-01	3.25	7.355698178251D-03
10	1.5	-4.504626907913D-01	-2.252313453956D-01	3.50	1.273774904030D-02
11	2.5	2.183389063858D-01	1.091694531929D-01	3.75	1.107834675248D-02
12	3.5	-9.501985378508D-02	-4.750992689254D-02	4.25	-1.694346610915D-02
13	4.5	4.085865486913D-02	2.042932743456D-02	4.50	-2.943263941870D-02
14	5.5	-1.736355202104D-02	-8.681776010519D-03	4.75	-2.565764488228D-02
15	6.5	6.958423210835D-03	3.479211605418D-03	5.25	3.940608780533D-02
16	7.5	-2.229742905429D-03	-1.114871452715D-03	5.50	6.865366465846D-02
17	8.5	3.846941403107D-04	1.923470701554D-04	5.75	6.010920046647D-02
				6.25	-9.422512578361D-02
	0	6.968849178977D-01	1.984121505718D-01	6.50	-1.679716629763D-01
	1	1.738918797134D-01	1.243826121424D-01	6.75	-1.526062487589D-01
	2	1.081240220991D-02	3.078430778981D-03	7.25	2.832024583819D-01
				7.50	6.198822457704D-01
				7.75	8.938803446302D-01
				8.25	8.938780044594D-01
				8.50	6.198787247017D-01

I= 9 H= .5

0		0.	1.478080977564D+00	-1.00	-8.926387163582D-03
1		4.270846511636D-03	-1.767159353851D-01	-.75	-5.821876500447D-03
2		-4.655540651946D-03	3.078430778981D-03	-.50	-3.299308418804D-03
				-.25	-1.358682918656D-03
0	-8.5	3.846941403107D-04	1.923470701554D-04	.25	7.771160150337D-04
1	-7.5	-2.229742905429D-03	-1.114871452715D-03	.50	9.835597847159D-04
2	-6.5	6.958423210835D-03	3.479211605418D-03	.75	6.756829897564D-04
3	-5.5	-1.736355202104D-02	-8.681776010519D-03	1.25	-7.639082643115D-04
4	-4.5	4.085865486913D-02	2.042932743456D-02	1.50	-1.217105037236D-03
5	-3.5	-9.501985378508D-02	-4.750992689254D-02	1.75	-9.987223100289D-04
6	-2.5	2.183389063858D-01	1.091694531929D-01	2.25	1.426292732492D-03
7	-1.5	-4.504626907913D-01	-2.252313453956D-01	2.50	2.433114881380D-03
8	-.5	7.313196826550D-01	3.656598413275D-01	2.75	2.094903113244D-03
9	.5	-8.655658690503D-01	-4.327829345252D-01	3.25	-3.167079007014D-03
10	1.5	7.313045942672D-01	3.656522971336D-01	3.50	-5.484250040983D-03
11	2.5	-4.504211518645D-01	-2.252105759323D-01	3.75	-4.769600472419D-03
12	3.5	2.182396394960D-01	1.091198197480D-01	4.25	7.293232694970D-03
13	4.5	-9.478818573432D-02	-4.739409286716D-02	4.50	1.266609007788D-02
14	5.5	4.032197656707D-02	2.016098828354D-02	4.75	1.103710387977D-02
15	6.5	-1.616060730984D-02	-8.080303654920D-03	5.25	-1.691653838678D-02
16	7.5	5.178526225371D-03	2.589263112686D-03	5.50	-2.940172119876D-02
17	8.5	-8.934443548395D-04	-4.467221774198D-04	5.75	-2.563982832860D-02
				6.25	3.939440163800D-02
	0	-3.000606167542D-01	1.984121505718D-01	6.50	6.864019221164D-02
	1	-7.487334457145D-02	-1.243826121424D-01	6.75	6.010138964876D-02
	2	-4.655540651946D-03	3.078430778981D-03	7.25	-9.421988012310D-02
				7.50	-1.679654903685D-01
				7.75	-1.526025615813D-01
				8.25	2.831997029536D-01
				8.50	6.198787247017D-01

I=10 H= 1.5

0		0.	-3.432804601305D+00	-1.00	3.843422564213D-03
1		-1.838892659140D-03	4.144581825773D-01	-.75	2.506717565959D-03

2		2.004529905073D-03	-1.155349660494D-02
0	-8.5	-1.656372459322D-04	-8.281862296612D-05
1	-7.5	9.600575678220D-04	4.800287839110D-04
2	-6.5	-2.996082577002D-03	-1.498041288501D-03
3	-5.5	7.476285473333D-03	3.738142736667D-03
4	-4.5	-1.759440917775D-02	-8.797204588876D-03
5	-3.5	4.095788681576D-02	2.047894340788D-02
6	-2.5	-9.506139120778D-02	-4.753069560389D-02
7	-1.5	2.183539947105D-01	1.091769973552D-01
8	-.5	-4.504626907913D-01	-2.252313453956D-01
9	.5	7.313045942672D-01	3.656522971336D-01
10	1.5	-8.655243299879D-01	-4.327621649940D-01
11	2.5	7.312053239146D-01	3.656026619573D-01
12	3.5	-4.501894027317D-01	-2.250947013659D-01
13	4.5	2.177010780782D-01	1.088505390391D-01
14	5.5	-9.354154153170D-02	-4.677077076585D-02
15	6.5	3.752807377553D-02	1.876403688777D-02
16	7.5	-1.202680437188D-02	-6.013402185941D-03
17	8.5	2.074995020060D-03	1.037497510030D-03
	0	1.291966980388D-01	-7.446501791053D-01
	1	3.223811572710D-02	2.180487402933D-01
	2	2.004529905073D-03	-1.155349660494D-02

-.50	1.420578805838D-03
-.25	5.850062838522D-04
.25	-3.346018008410D-04
.50	-4.234900172374D-04
.75	-2.909279176363D-04
1.25	3.289149496544D-04
1.50	5.240472603993D-04
1.75	4.300184899666D-04
2.25	-6.141167743644D-04
2.50	-1.047622464101D-03
2.75	-9.019987907396D-04
3.25	1.363640902725D-03
3.50	2.361333477724D-03
3.75	2.053621138476D-03
4.25	-3.140143252969D-03
4.50	-5.453326543340D-03
4.75	-4.751782233053D-03
5.25	7.281546181183D-03
5.50	1.265261740326D-02
5.75	1.102929298922D-02
6.25	-1.691129271126D-02
6.50	-2.939554858112D-02
6.75	-2.563614114791D-02
7.25	3.939164620916D-02
7.50	6.863667114280D-02
7.75	6.009904947805D-02
8.25	-9.421753995270D-02
8.50	-1.679619693006D-01

I=11 H= 2.5

0		0.	7.972560281355D+00
1		7.917155065617D-04	-9.643032249796D-01
2		-8.630288460599D-04	2.872877666730D-02
0	-8.5	7.131333949822D-05	3.565666974911D-05
1	-7.5	-4.133424928269D-04	-2.066712464134D-04
2	-6.5	1.289931367245D-03	6.449656836227D-04
3	-5.5	-3.218838130394D-03	-1.609419065197D-03
4	-4.5	7.575168273431D-03	3.787584136716D-03
5	-3.5	-1.763593155875D-02	-8.817965779376D-03
6	-2.5	4.097297450410D-02	2.048648725205D-02
7	-1.5	-9.506139120778D-02	-4.753069560389D-02
8	-.5	2.183389063858D-01	1.091694531929D-01
9	.5	-4.504211518645D-01	-2.252105759323D-01
10	1.5	7.312053239146D-01	3.656026619573D-01
11	2.5	-8.652925735446D-01	-4.326462867723D-01
12	3.5	7.306665758851D-01	3.653332879426D-01
13	4.5	-4.489383889863D-01	-2.244691944931D-01
14	5.5	2.148056934710D-01	1.074028467355D-01
15	6.5	-8.705275661909D-02	-4.352637830955D-02
16	7.5	2.792744076773D-02	1.396372038387D-02
17	8.5	-4.818953504301D-03	-2.409476752151D-03
	0	-5.562425232205D-02	1.851636983241D+00
	1	-1.387977487646D-02	-4.759140216356D-01
	2	-8.630288460599D-04	2.872877666730D-02

-1.00	-1.654744352622D-03
-.75	-1.079240355830D-03
-.50	-6.116149647958D-04
-.25	-2.518681795192D-04
.25	1.440592156948D-04
.50	1.823290836252D-04
.75	1.252558937568D-04
1.25	-1.416108025955D-04
1.50	-2.256229248785D-04
1.75	-1.851398459918D-04
2.25	2.644013813061D-04
2.50	4.510425892256D-04
2.75	3.883458666189D-04
3.25	-5.871007863815D-04
3.50	-1.016646248364D-03
3.75	-8.841637160681D-04
4.25	1.351950931506D-03
4.50	2.347858531784D-03
4.75	2.045809522957D-03
5.25	-3.134897429340D-03
5.50	-5.447153829362D-03
5.75	-4.748095022476D-03
6.25	7.278790748046D-03
6.50	1.264909633391D-02
6.75	1.102695282040D-02
7.25	-1.690895254555D-02

CARDINAL SPLINES N=3

7.50	-2,939202752170D-02
7.75	-2,563338572783D-02
8.25	3,938795904457D-02
8,50	6,863049855809D-02

I=12 H= 3.5

0		0.	-1,851514561688D+01	-1.00	7,121701182520D-04
1		-3,407391147936D-04	2.240208830479D+00	-.75	4,644842755405D-04
2		3,714310034584D-04	-6,753504845584D-02	-.50	2,632273082614D-04
				-.25	1,083992164145D-04
0	-8,5	-3,069188866486D-05	-1,534594433243D-05	.25	-6,200031352976D-05
1	-7,5	1.778946530289D-04	8.894732651447D-05	.50	-7,847092805295D-05
2	-6,5	-5,551616381104D-04	-2,775808190552D-04	.75	-5,390772569821D-05
3	-5,5	1,385326130157D-03	6,926630650783D-04	1.25	6,094656364556D-05
4	-4,5	-3,260210240619D-03	-1,630105120310D-03	1.50	9,710376394948D-05
5	-3,5	7,590249603602D-03	3,795124801801D-03	1.75	7,968062600711D-05
6	-2,5	-1,763593155875D-02	-8,817965779376D-03	2.25	-1,137932648924D-04
7	-1,5	4,095788681576D-02	2.047894340788D-02	2.50	-1,941200475489D-04
8	-,5	-9,501985378508D-02	-4,750992689254D-02	2.75	-1,671365844183D-04
9	,5	2,182396394960D-01	1.091198197480D-01	3.25	2,526768701882D-04
10	1,5	-4,501894027317D-01	-2.250947013659D-01	3,50	4,375449529155D-04
11	2,5	7,306665758851D-01	3,653332879426D-01	3,75	3,805270087759D-04
12	3,5	-8,640411658307D-01	-4,320205829153D-01	4.25	-5,818534832474D-04
13	4,5	7,277611347084D-01	3,638805673542D-01	4,50	-1,010472572211D-03
14	5,5	-4,422141278770D-01	-2,211070639385D-01	4,75	-8,804762070524D-04
15	6,5	1.997361697563D-01	9,986808487813D-02	5,25	1,349195455569D-03
16	7,5	-6,475631037115D-02	-3,237815518558D-02	5,50	2,344337457384D-03
17	8,5	1.118797887338D-02	5,593989436691D-03	5,75	2,043469373025D-03
				6,25	-3,132557310453D-03
	0	2,393960752413D-02	-4,352777808745D+00	6,50	-5,443632854308D-03
	1	5,973587944000D-03	1.092113006730D+00	6,75	-4,745339676931D-03
	2	3.714310034584D-04	-6,753504845584D-02	7,25	7,275103698507D-03
				7.50	1,264292394926D-02
				7.75	1,102170735518D-02
				8,25	-1,690114203281D-02
				8,50	-2,937855561612D-02

I=13 H= 4.5

0		0.	4.297964612588D+01	-1.00	-3,058953074848D-04
1		1.463561775767D-04	-5,200613555427D+00	-.75	-1,995078937558D-04
2		-1,595391299081D-04	1.571244807599D-01	-.50	-1,130628712654D-04
				-.25	-4,656024001344D-05
0	-8,5	1,318295233134D-05	6,591476165672D-06	.25	2,663072275181D-05
1	-7,5	-7,641031012419D-05	-3,820515506210D-05	.50	3,370527357170D-05
2	-6,5	2,384561438999D-04	1.192280719500D-04	.75	2,315474899257D-05
3	-5,5	-5,950330636933D-04	-2,975165318466D-04	1.25	-2,617811018494D-05
4	-4,5	1,400343962042D-03	7,001719810212D-04	1.50	-4,170855385330D-05
5	-3,5	-3,260210240619D-03	-1,630105120310D-03	1.75	-3,422486982567D-05
6	-2,5	7,575168273431D-03	3,787584136716D-03	2,25	4,887712198505D-05
7	-1,5	-1.759440917775D-02	-8,797204588876D-03	2,50	8,337953261551D-05
8	-,5	4,085865486913D-02	2,042932743456D-02	2,75	7,178944401728D-05
9	,5	-9,478818573432D-02	-4,739409286716D-02	3,25	-1,085311875472D-04
10	1,5	2.177010780782D-01	1.088505390391D-01	3,50	-1,879367636431D-04
11	2,5	-4,489383889863D-01	-2,244691944931D-01	3,75	-1,634460956345D-04
12	3,5	7,277611347084D-01	3,638805673542D-01	4,25	2,499209672709D-04

13	4.5	-8.572956736216D-01	-4.286478368108D-01	4.50	4.340238287099D-04
14	5.5	7.121496589712D-01	3.560748294856D-01	4.75	3.781870478307D-04
15	6.5	-4.072277876565D-01	-2.036138938283D-01	5.25	-5.795138323524D-04
16	7.5	1.479713664231D-01	7.398568321153D-02	5.50	-1.006952440166D-03
17	8.5	-2.589294559075D-02	-1.294647279538D-02	5.75	-8.777216063460D-04
				6.25	1.345509555528D-03
	0	-1.028267462646D-02	1.012667463966D+01	6.50	2.338167071567D-03
	1	-2.565809030860D-03	-2.529497382509D+00	6.75	2.038225650652D-03
	2	-1.595391299081D-04	1.571244807599D-01	7.25	-3.124749469963D-03
				7.50	-5.430165592598D-03
				7.75	-4.733658046087D-03
				8.25	7.257294228082D-03
				8.50	1.261201829972D-02

I=14 H= 5.5

0		0.	-9.932495127259D+01	-1.00	1.299807939229D-04
1		-6.218955208357D-05	1.201959529637D+01	-.75	8.477473759732D-05
2		6.779124183936D-05	-3.633179984167D-01	-.50	4.804258650163D-05
				-.25	1.978434063585D-05
0	-8.5	-5.601689755790D-06	-2.800844877895D-06	.25	-1.131590580608D-05
1	-7.5	3.246820899497D-05	1.623410449749D-05	.50	-1.432201838681D-05
2	-6.5	-1.013245974857D-04	-5.066229874287D-05	.75	-9.838897765008D-06
3	-5.5	2.528409825985D-04	1.264204912993D-04	1.25	1.112358202945D-05
4	-4.5	-5.950330636933D-04	-2.975165318466D-04	1.50	1.772276596127D-05
5	-3.5	1.385326130157D-03	6.926630650783D-04	1.75	1.454280481904D-05
6	-2.5	-3.218838130394D-03	-1.609419065197D-03	2.25	-2.076882830384D-05
7	-1.5	7.476285473333D-03	3.738142736667D-03	2.50	-3.542956554263D-05
8	-.5	-1.736355202104D-02	-8.681776010519D-03	2.75	-3.050471419293D-05
9	.5	4.032197656707D-02	2.016098828354D-02	3.25	4.611698699976D-05
10	1.5	-9.354154153170D-02	-4.677077076585D-02	3.50	7.985794203013D-05
11	2.5	2.148056934710D-01	1.074028467355D-01	3.75	6.945138658300D-05
12	3.5	-4.422141278770D-01	-2.211070639385D-01	4.25	-1.061962209000D-04
13	4.5	7.121496589712D-01	3.560748294856D-01	4.50	-1.844250635762D-04
14	5.5	-8.211651820865D-01	-4.105825910432D-01	4.75	-1.606989419548D-04
15	6.5	6.311788734779D-01	3.155894367390D-01	5.25	2.462465552864D-04
16	7.5	-2.874256980929D-01	-1.437128490465D-01	5.50	4.278734171404D-04
17	8.5	5.802777580811D-02	2.901388790405D-02	5.75	3.729607405427D-04
				6.25	-5.717326917369D-04
	0	4.369306030184D-03	-2.340811663907D+01	6.50	-9.935315768596D-04
	1	1.090261559186D-03	5.843189323285D+00	6.75	-8.660804220610D-04
	2	6.779124183936D-05	-3.633179984167D-01	7.25	1.327762092656D-03
				7.50	2.307369178035D-03
				7.75	2.011403202010D-03
				8.25	-3.083670970973D-03
				8.50	-5.358798338256D-03

I=15 H= 6.5

0		0.	2.192133145368D+02	-1.00	-5.208906992950D-05
1		2.492211217979D-05	-2.655010941488D+01	-.75	-3.397299786905D-05
2		-2.716695774971D-05	8.032347118643D-01	-.50	-1.925279552732D-05
				-.25	-7.928462904304D-06
0	-8.5	2.244845569922D-06	1.122422784961D-06	.25	4.534785417592D-06
1	-7.5	-1.301145159823D-05	-6.505725799114D-06	.50	5.739468076527D-06
2	-6.5	4.060526086036D-05	2.030263043018D-05	.75	3.942882777085D-06
3	-5.5	-1.013245974857D-04	-5.066229874287D-05	1.25	-4.457712749024D-06

CARDINAL SPLINES N=3

4	-4,5	2,384561438999D-04	1.192280719500D-04
5	-3,5	-5,551616381104D-04	-2,775808190552D-04
6	-2,5	1,289931367245D-03	6,449656836227D-04
7	-1,5	-2,996082577002D-03	-1.498041288501D-03
8	-,5	6,958423210835D-03	3,479211605418D-03
9	,5	-1,616060730984D-02	-8,080303654920D-03
10	1,5	3,752807377553D-02	1,876403688777D-02
11	2,5	-8,705275661909D-02	-4,352637830955D-02
12	3,5	1,997361697563D-01	9,986808487813D-02
13	4,5	-4,072277876565D-01	-2,036138938283D-01
14	5,5	6,311788734779D-01	3,155894367390D-01
15	6,5	-6,397043821386D-01	-3,198521910693D-01
16	7,5	3,626945953751D-01	1,813472976875D-01
17	8,5	-8,585625922493D-02	-4,292812961246D-02
	0	-1.750974743888D-03	5.157109244246D+01
	1	-4,369161695653D-04	-1,289511931319D+01
	2	-2,716695774971D-05	8,032347118643D-01

1,50	-7,102298483017D-06
1.75	-5,827946993757D-06
2,25	8,322990783628D-06
2,50	1,419819852928D-05
2,75	1,222459213245D-05
3,25	-1,848112238890D-05
3,50	-3,200261977991D-05
3,75	-2,783225138802D-05
4,25	4,255753644887D-05
4,50	7,390730382683D-05
4,75	6,439919440741D-05
5,25	-9,868191762888D-05
5,50	-1,714678573018D-04
5,75	-1,494619114334D-04
6.25	2,291186440528D-04
6,50	3,981521530256D-04
6,75	3,470768103341D-04
7,25	-5,320930071494D-04
7,50	-9,246646764624D-04
7,75	-8,060577172910D-04
8,25	1,235760000299D-03
8,50	2,147496474351D-03

I=16 H= 7.5

0		0,	-2,430550476206D+02
1		-7,985981360092D-06	2,998529883722D+01
2		8,705314246019D-06	-9,228160642735D-01
0	-8,5	-7,193328859268D-07	-3,596664429634D-07
1	-7,5	4,169358085767D-06	2,084679042884D-06
2	-6,5	-1,301145159823D-05	-6,505725799114D-06
3	-5,5	3,246820899497D-05	1.623410449749D-05
4	-4,5	-7,641031012419D-05	-3,820515506210D-05
5	-3,5	1,778946530289D-04	8,894732651447D-05
6	-2,5	-4,133424928269D-04	-2,066712464134D-04
7	-1,5	9,600575678220D-04	4,800287839110D-04
8	-,5	-2,229742905429D-03	-1,114871452715D-03
9	,5	5,178526225371D-03	2,589263112686D-03
10	1,5	-1,202680437188D-02	-6,013402185941D-03
11	2,5	2,792744076773D-02	1,396372038387D-02
12	3,5	-6,475631037115D-02	-3,237815518558D-02
13	4,5	1,479713664231D-01	7,398568321153D-02
14	5,5	-2,874256980929D-01	-1,437128490465D-01
15	6,5	3,626945953751D-01	1,813472976875D-01
16	7,5	-2,424626222495D-01	-1,212313111247D-01
17	8,5	6,445814299906D-02	3.222907149953D-02
	0	5,610781127141D-04	-5,485346814801D+01
	1	1.400043608222D-04	1.429742574457D+01
	2	8.705314246019D-06	-9,228160642735D-01

-1.00	1,669129560611D-05
-,75	1,088622528345D-05
-,50	6,169319241550D-06
-,25	2,540577480399D-06
,25	-1,453115673168D-06
.50	-1,839141271226D-06
,75	-1,263447822387D-06
1,25	1,428418693630D-06
1,50	2,275843351078D-06
1,75	1,867493241503D-06
2,25	-2,666999039999D-06
2,50	-4,549636402553D-06
2,75	-3,917218741337D-06
3,25	5,922046167119D-06
3,50	1,025484209329D-05
3,75	8,918499330573D-06
4,25	-1,363703406636D-05
4.50	-2,368267771383D-05
4,75	-2,063592212334D-05
5,25	3,162139504177D-05
5,50	5,494474551781D-05
5,75	4,789321342222D-05
6,25	-7,341822401150D-05
6,50	-1,275829123729D-04
6,75	-1,112164526912D-04
7,25	1,705025911004D-04
7,50	2,962972959862D-04
7,75	2,582911599795D-04
8,25	-3,959838208702D-04
8,50	-6,881381847802D-04

CARDINAL SPLINES N=3

```
I=17     H=  8.5
0                 C.                   9.362922044307D+01     -1.00  -2.879723350985D-06
1                 1.377808981752D-06  -1.182015704576D+01      -.75  -1.878183569007D-06
2                -1.501914369232D-06   3.730562076804D-01      -.50  -1.064383083184D-06
                                                               -.25  -4.383218935151D-07
0   -8.5          1.241053874797D-07   6.205269373985D-08       .25   2.507037940286D-07
1   -7.5         -7.193328859268D-07  -3.596664429634D-07       .50   3.173041919269D-07
2   -6.5          2.244845569922D-06   1.122422784961D-06       .75   2.179806938142D-07
3   -5.5         -5.601689755790D-06  -2.800844877895D-06      1.25  -2.464428624416D-07
4   -4.5          1.318295233134D-05   6.591476165672D-06      1.50  -3.926477246549D-07
5   -3.5         -3.069188866486D-05  -1.534594433243D-05      1.75  -3.221957133989D-07
6   -2.5          7.131333949822D-05   3.565666974911D-05      2.25   4.601332091756D-07
7   -1.5         -1.656372459322D-04  -8.281862296612D-05      2.50   7.849417142984D-07
8    -.5          3.846941403107D-04   1.923470701554D-04      2.75   6.758316757755D-07
9     .5         -8.934443548395D-04  -4.467221774198D-04      3.25  -1.021721443050D-06
10   1.5          2.074995020060D-03   1.037497510030D-03      3.50  -1.769252006170D-06
11   2.5         -4.818953504301D-03  -2.409476752151D-03      3.75  -1.538694861324D-06
12   3.5          1.118797887338D-02   5.593989436691D-03      4.25   2.352776343170D-06
13   4.5         -2.589294559075D-02  -1.294647279538D-02      4.50   4.085935665829D-06
14   5.5          5.802777580811D-02   2.901388790405D-02      4.75   3.560283647796D-06
15   6.5         -8.585625922493D-02  -4.292812961246D-02      5.25  -5.455590257216D-06
16   7.5          6.445814299906D-02   3.222907149953D-02      5.50  -9.479531751430D-06
17   8.5         -1.855619925165D-02  -9.278099625826D-03      5.75  -8.262941852074D-06
                                                               6.25   1.266673234854D-05
          0      -9.680193683213D-05   2.011119655900D+01      6.50   2.201168203859D-05
          1      -2.415473529519D-05  -5.478201515196D+00      6.75   1.918800211806D-05
          2      -1.501914369232D-06   3.730562076804D-01      7.25  -2.941654756507D-05
                                                               7.50  -5.111971271425D-05
                                                               7.75  -4.456257281410D-05
                                                               8.25   6.831847203291D-05
                                                               8.50   1.187234058952D-04
```

```
                 N= 3       M=18                       P=  9.0       Q= 5

I= 0     H= -9.0
0                 1.000000000000D+00   5.000994240875D-01     -1.00   3.473671635302D+00
1                -1.727557718025D+00  -8.637902027588D-01      -.75   2.715357366987D+00
2                 7.461139172771D-01   3.730572819822D-01      -.50   2.050307338332D+00
                                                               -.25   1.478521549336D+00
0   -9.0         -1.855619925181D-02  -9.278099625903D-03       .25   6.147245690352D-01
1   -8.0          6.445814299995D-02   3.222907149998D-02       .50   3.221697390800D-01
2   -7.0         -8.585625922771D-02  -4.292812961385D-02       .75   1.196173168848D-01
3   -6.0          5.802777581505D-02   2.901388790753D-02      1.25  -5.021023041788D-02
4   -5.0         -2.589294560710D-02  -1.294647280355D-02      1.50  -5.147708426411D-02
5   -4.0          1.118797891144D-02   5.593989455719D-03      1.75  -2.752040839496D-02
6   -3.0         -4.818953592725D-03  -2.409476796363D-03      2.25   1.677968342771D-02
7   -2.0          2.074995225440D-03   1.037497612720D-03      2.50   1.898487003953D-02
8   -1.0         -8.934448318364D-04  -4.467224159182D-04      2.75   1.083411537105D-02
9    0.0          3.846952481266D-04   1.923476240633D-04      3.25  -7.016349049189D-03
```

CARDINAL SPLINES N=3

```
10    1.0   -1.656398188062D-04   -8.281990940310D-05      3.50  -8.037375405532D-03
11    2.0    7.131931491043D-05    3.565965745521D-05      3.75  -4.621138963546D-03
12    3.0   -3.070576591805D-05   -1.535288295902D-05      4.25   3.012078379450D-03
13    4.0    1.321517061547D-05    6.607585307733D-06      4.50   3.454789721581D-03
14    5.0   -5.676254275880D-06   -2.838127137940D-06      4.75   1.987862447086D-03
15    6.0    2.411949510942D-06    1.205974755471D-06      5.25  -1.296539321909D-03
16    7.0   -9.665751604530D-07   -4.832875802265D-07      5.50  -1.487292759701D-03
17    8.0    3.097270070378D-07    1.548635035189D-07      5.75  -8.558432825551D-04
18    9.0   -5.343672029095D-08   -2.671836014548D-08      6.25   5.582411771298D-04
                                                           6.50   6.403806321815D-04
          0  4.588720783722D+01    2.294362743982D+01      6.75   3.685015198867D-04
          1  1.170249279296D+01    5.851240872921D+00      7.25  -2.403641759062D-04
          2  7.461139172771D-01    3.730572819822D-01      7.50  -2.757316520003D-04
                                                           7.75  -1.586675291109D-04
                                                           8.25   1.034948731699D-04
                                                           8.50   1.187232586855D-04
                                                           8.75   6.831834370560D-05


I= 1      H= -8.0
0              0.                 -5.762764796356D-04     -1.00  -4.626823524744D+00
1              2.781182690872D+00  1.390657095439D+00      -.75  -3.124059987207D+00
2             -1.845640833872D+00 -9.228222910828D-01      -.50  -1.852001553904D+00
                                                           -.25  -8.106482248350D-01
0    -9.0    6.445814299995D-02    3.222907149998D-02       .25   5.800060680063D-01
1    -8.0   -2.424626222546D-01   -1.212313111273D-01       .50   9.311954539368D-01
2    -7.0    3.626945953912D-01    1.813472976956D-01       .75   1.063010268582D+00
3    -6.0   -2.874256981332D-01   -1.437128490666D-01      1.25   7.891384222399D-01
4    -5.0    1.479713665178D-01    7.398568325891D-02      1.50   5.009842265565D-01
5    -4.0   -6.475631059173D-02   -3.237815529587D-02      1.75   2.152141789184D-01
6    -3.0    2.792744128025D-02    1.396372064013D-02      2.25  -1.084973407903D-01
7    -2.0   -1.202680556229D-02   -6.013402781147D-03      2.50  -1.174242888207D-01
8    -1.0    5.178528990115D-03    2.589264495057D-03      2.75  -6.515456765333D-02
9     0.0   -2.229749326491D-03   -1.114874663246D-03      3.25   4.115217610381D-02
10    1.0    9.600724805738D-04    4.800362402869D-04      3.50   4.690407300895D-02
11    2.0   -4.133771271855D-04   -2.066885635928D-04      3.75   2.688643471233D-02
12    3.0    1.779750876079D-04    8.898754380394D-05      4.25  -1.747932028045D-02
13    4.0   -7.659705198744D-05   -3.829852599372D-05      4.50  -2.003818046950D-02
14    5.0    3.290039580421D-05    1.645019790211D-05      4.75  -1.152632036277D-02
15    6.0   -1.398001035772D-05   -6.990005178861D-06      5.25   7.515830248931D-03
16    7.0    5.602410288172D-06    2.801205144086D-06      5.50   8.621156399408D-03
17    8.0   -1.795222804960D-06   -8.976114024800D-07      5.75   4.960780986373D-03
18    9.0    3.097270070378D-07    1.548635035189D-07      6.25  -3.235685871325D-03
                                                           6.50  -3.711764809259D-03
          0 -1.244662633258D+02   -6.223326799524D+01      6.75  -2.135896569996D-03
          1 -3.044035231882D+01   -1.522014414405D+01      7.25   1.393187412823D-03
          2 -1.845640833872D+00   -9.228222910828D-01      7.50   1.598181877982D-03
                                                           7.75   9.196604389940D-04
                                                           8.25  -5.998714177099D-04
                                                           8.50  -6.881373315317D-04
                                                           8.75  -3.959830770664D-04


I= 2      H= -7.0
0              0.                  1.798404782639D-03     -1.00   3.127136922212D+00
1             -1.520640331492D+00 -7.605253539233D-01      -.75   2.044134580899D+00
2              1.606496590720D+00  8.032541440711D-01      -.50   1.161944313426D+00
```

CARDINAL SPLINES N=3

0	-9.0	-8.585625922771D-02	-4.292812961385D-02
1	-8.0	3.626945953912D-01	1.813472976956D-01
2	-7.0	-6.397043821890D-01	-3.198521910945D-01
3	-6.0	6.311788736036D-01	3.155894368018D-01
4	-5.0	-4.072277879522D-01	-2.036138939761D-01
5	-4.0	1.997361704446D-01	9.986808522231D-02
6	-3.0	-8.705275821853D-02	-4.352637910926D-02
7	-2.0	3.752807749049D-02	1.876403874524D-02
8	-1.0	-1.616061593786D-02	-8.080307968932D-03
9	0.0	6.958443249253D-03	3.479221624626D-03
10	1.0	-2.996129115713D-03	-1.498064557856D-03
11	2.0	1.290039451817D-03	6.450197259084D-04
12	3.0	-5.554126529211D-04	-2.777063264606D-04
13	4.0	2.390389153158D-04	1.195194576579D-04
14	5.0	-1.026733369401D-04	-5.133666847007D-05
15	6.0	4.362787373264D-05	2.181393686632D-05
16	7.0	-1.748362428902D-05	-8.741812144512D-06
17	8.0	5.602410288172D-06	2.801205144086D-06
18	9.0	-9.665751604530D-07	-4.832875802265D-07

0.	1.164404608649D+02	5.822065588924D+01
1	2.739629830147D+01	1.369804923936D+01
2	1.606496590720D+00	8.032541440711D-01

-.25	4.805661197931D-01
.25	-2.798378899562D-01
.50	-3.613790261670D-01
.75	-2.572000091052D-01
1.25	3.476921927253D-01
1.50	6.930200694771D-01
1.75	9.356781956091D-01
2.25	8.667750562624D-01
2.50	5.888238698721D-01
2.75	2.653417187891D-01
3.25	-1.409607496321D-01
3.50	-1.546118064297D-01
3.75	-8.653576212573D-02
4.25	5.508853121550D-02
4.50	6.288877537845D-02
4.75	3.608392345748D-02
5.25	-2.347817893269D-02
5.50	-2.691962162761D-02
5.75	-1.548615622156D-02
6.25	1.009871369895D-02
6.50	1.158408464129D-02
6.75	6.665772546866D-03
7.25	-4.347808585942D-03
7.50	-4.987526661440D-03
7.75	-2.870023364197D-03
8.25	1.872040138106D-03
8.50	2.147493811590D-03
8.75	1.235757679086D-03

I= 3 H= -6.0

0	0.	-4.487660880844D-03
1	6.686760123442D-01	3.348500238028D-01
2	-7.267037881592D-01	-3.633664886984D-01

0	-9.0	5.802777581505D-02	2.901388790753D-02
1	-8.0	-2.874256981332D-01	-1.437128490666D-01
2	-7.0	6.311788736036D-01	3.155894368018D-01
3	-6.0	-8.211651824000D-01	-4.105825912000D-01
4	-5.0	7.121496597090D-01	3.560748298545D-01
5	-4.0	-4.422141295947D-01	-2.211070647973D-01
6	-3.0	2.148056974622D-01	1.074028487311D-01
7	-2.0	-9.354155080184D-02	-4.677077540092D-02
8	-1.0	4.032199809707D-02	2.016099904853D-02
9	0.0	-1.736360202403D-02	-8.681801012016D-03
10	1.0	7.476401604002D-03	3.738200802001D-03
11	2.0	-3.219107839921D-03	-1.609553919960D-03
12	3.0	1.385952501571D-03	6.929762507853D-04
13	4.0	-5.964872860746D-04	-2.982436430373D-04
14	5.0	2.562065682255D-04	1.281032841128D-04
15	6.0	-1.088670938558D-04	-5.443354692788D-05
16	7.0	4.362787373264D-05	2.181393686632D-05
17	8.0	-1.398001035772D-05	-6.990005178861D-06
18	9.0	2.411949510942D-06	1.205974755471D-06

0	-5.284492272980D+01	-2.642352303122D+01
1	-1.241199217452D+01	-6.205746772768D+00

-1.00	-1.395379800503D+00
-.75	-9.102778900977D-01
-.50	-5.160139532119D-01
-.25	-2.125879898460D-01
.25	1.218066840759D-01
.50	1.544754271265D-01
.75	1.065063916247D-01
1.25	-1.228236245413D-01
1.50	-2.004031353131D-01
1.75	-1.711399238958D-01
2.25	2.952198251310D-01
2.50	6.336517814055D-01
2.75	9.017984334063D-01
3.25	8.887161404431D-01
3.50	6.139578432775D-01
3.75	2.797926530795D-01
4.25	-1.503799311678D-01
4.50	-1.654154069660D-01
4.75	-9.275208021468D-02
5.25	5.914298725283D-02
5.50	6.753974357300D-02
5.75	3.876026266578D-02
6.25	-2.522387565152D-02
6.50	-2.892217975934D-02
6.75	-1.663851111559D-02
7.25	1.085036567228D-02

	2	-7.267037881592D-01	-3.633664886984D-01	7.50	1,244633576455D-02
				7.75	7,161948065848D-03
				8.25	-4,671451884847D-03
				8,50	-5,358791693719D-03
				8.75	-3,083665178720D-03

I= 4 H= -5.0

0		0.	1.056120957261D-02	-1.00	6,029240560880D-01
1		-2,885155552404D-01	-1,454627540315D-01	-.75	3,932414481571D-01
2		3.144085008475D-01	1.572385972319D-01	-.50	2,228599028321D-01
				-.25	9,177942011308D-02
0	-9,0	-2,589294560710D-02	-1.294647280355D-02	.25	-5,250364358683D-02
1	-8,0	1.479713665178D-01	7.398568325891D-02	.50	-6,646480695856D-02
2	-7,0	-4,072277879522D-01	-2.036138939761D-01	.75	-4,567640206934D-02
3	-6,0	7,121496597090D-01	3.560748298545D-01	1.25	5,174434276625D-02
4	-5,0	-8,572956753580D-01	-4,286478376790D-01	1.50	8,264534354609D-02
5	-4,0	7,277611387509D-01	3,638805693754D-01	1.75	6,810496964098D-02
6	-3,0	-4,489383983790D-01	-2,244691991895D-01	2,25	-9,940959321636D-02
7	-2,0	2,177010998944D-01	1,088505499472D-01	2.50	-1,739120311258D-01
8	-1,0	-9,478823640277D-02	-4.739411820138D-02	2.75	-1.560222181345D-01
9	0,0	4.085877254560D-02	2.042938627280D-02	3.25	2,854293533699D-01
10	1,0	-1.759468247833D-02	-8.797341239166D-03	3.50	6,224365900730D-01
11	2.0	7,575803004734D-03	3,787901502367D-03	3.75	8,953501894041D-01
12	3,0	-3,261684335910D-03	-1,630842167955D-03	4.25	8,929191338280D-01
13	4,0	1.403766312292D-03	7.018831561458D-04	4.50	6,187785871684D-01
14	5,0	-6,029535942514D-04	-3,014767971257D-04	4,75	2,825664762702D-01
15	6,0	2,562065682255D-04	1,281032841128D-04	5,25	-1,521890960796D-01
16	7,0	-1,026733369401D-04	-5,133666847007D-05	5,50	-1,674907454688D-01
17	8,0	3,290039580421D-05	1,645019790211D-05	5,75	-9,394630684267D-02
18	9,0	-5,676254275880D-06	-2,838127137940D-06	6,25	5,992194602044D-02
				6,50	6,843331830221D-02
	0	2,287044857149D+01	1.143772279907D+01	6,75	3,927446270547D-02
	1	5,370837460015D+00	2,684831996142D+00	7,25	-2,555927567622D-02
	2	3,144085008475D-01	1.572385972319D-01	7,50	-2,930693136877D-02
				7.75	-1,685991365457D-02
				8,25	1,099478197409D-02
				8,50	1,261200266254D-02
				8,75	7,257280596660D-03

I= 5 H= -4.0

0		0.	-2,458807862099D-02	-1.00	-2,596950778400D-01
1		1.242535494643D-01	6.493214039911D-02	-.75	-1,693760218095D-01
2		-1,354415283757D-01	-6.780072870447D-02	-.50	-9,598715682605D-02
				-.25	-3,952848288955D-02
0	-9,0	1,118797891144D-02	5.593989455719D-03	.25	2,260921760324D-02
1	-8.0	-6,475631059173D-02	-3,237815529587D-02	.50	2,861601697919D-02
2	-7.0	1,997361704446D-01	9.986808522231D-02	.75	1,965926222620D-02
3	-6,0	-4,422141295947D-01	-2.211070647973D-01	1.25	-2,223068779220D-02
4	-5,0	7,277611387509D-01	3.638805693754D-01	1.50	-3,542803449619D-02
5	-4,0	-8,640411752421D-01	-4,320205876211D-01	1.75	-2,908368579638D-02
6	-3,0	7,306665977528D-01	3,653332988764D-01	2.25	4,162746794804D-02
7	-2,0	-4.501894535232D-01	-2,250947267616D-01	2.50	7,119890865304D-02
8	-1,0	2,182397574597D-01	1,091198787299D-01	2.75	6,157282081902D-02
9	0,0	-9,502012775354D-02	-4,751006387677D-02	3,25	-9,517926753075D-02
10	1.0	4,095852310047D-02	2,047926155023D-02	3,50	-1,690661040457D-01

11	2.0	-1.763740930835D-02	-8.818704654174D-03	3.75	-1.532360221995D-01
12	3.0	7.593681518352D-03	3.796840759176D-03	4.25	2.836132987226D-01
13	4.0	-3.268177984884D-03	-1.634088992442D-03	4.50	6.203536140094D-01
14	5.0	1.403766312292D-03	7.018831561458D-04	4.75	8.941516591127D-01
15	6.0	-5.964872860746D-04	-2.982436430373D-04	5.25	8.937008488843D-01
16	7.0	2.390389153158D-04	1.195194576579D-04	5.50	6.196753122893D-01
17	8.0	-7.659705198744D-05	-3.829852599372D-05	5.75	2.830824852421D-01
18	9.0	1.321517061547D-05	6.607585307733D-06	6.25	-1.525256738327D-01
				6.50	-1.678768476053D-01
	0	-9.852481853253D+00	-4.932057840091D+00	6.75	-9.416848636189D-02
	1	-2.313693961298D+00	-1.155480976281D+00	7.25	6.006686903257D-02
	2	-1.354415283757D-01	-6.780072870447D-02	7.50	6.859956644730D-02
				7.75	3.937012970291D-02
				8.25	-2.562167909938D-02
				8.50	-2.937851921043D-02
				8.75	-1.690111029681D-02

I= 6	H= -3.0				
0		0.	5.713099047318D-02	-1.00	1.118222129088D-01
1		-5.350162965803D-02	-3.326914924889D-02	-.75	7.293155032207D-02
2		5.832058325076D-02	2.934609109786D-02	-.50	4.133096064170D-02
				-.25	1.702044386768D-02
0	-9.0	-4.818953592725D-03	-2.409476796363D-03	.25	-9.735076970703D-03
1	-8.0	2.792744128025D-02	1.396372064013D-02	.50	-1.232126131610D-02
2	-7.0	-8.705275821853D-02	-4.352637910926D-02	.75	-8.464454441371D-03
3	-6.0	2.148056974622D-01	1.074028487311D-01	1.25	9.569867873968D-03
4	-5.0	-4.489383983790D-01	-2.244691991895D-01	1.50	1.524767152240D-02
5	-4.0	7.306665977528D-01	3.653332988764D-01	1.75	1.251234758334D-02
6	-3.0	-8.652926243548D-01	-4.326463121774D-01	2.25	-1.787308653756D-02
7	-2.0	7.312054419299D-01	3.656027209649D-01	2.50	-3.049775708986D-02
8	-1.0	-4.504214259561D-01	-2.252107129780D-01	2.75	-2.627011927254D-02
9	0.0	2.183395429581D-01	1.091697714790D-01	3.25	3.980535650066D-02
10	1.0	-9.506286963057D-02	-4.753143481528D-02	3.50	6.911164158787D-02
11	2.0	4.097640809073D-02	2.048820404537D-02	3.75	6.037273357301D-02
12	3.0	-1.764390569500D-02	-8.821952847498D-03	4.25	-9.439704529089D-02
13	4.0	7.593681518352D-03	3.796840759176D-03	4.50	-1.681689116206D-01
14	5.0	-3.261684335910D-03	-1.630842167955D-03	4.75	-1.527197836365D-01
15	6.0	1.385952501571D-03	6.929762507853D-04	5.25	2.832765931222D-01
16	7.0	-5.554126529211D-04	-2.777063264606D-04	5.50	6.199673701488D-01
17	8.0	1.779750876079D-04	8.898754380394D-05	5.75	8.939293997338D-01
18	9.0	-3.070576591805D-05	-1.535288295902D-05	6.25	8.938458230426D-01
				6.50	6.198416188937D-01
	0	4.242452576389D+00	2.134742026160D+00	6.75	2.831781858064D-01
	1	9.962688688556D-01	4.949604905126D-01	7.25	-1.525880991101D-01
	2	5.832058325076D-02	2.934609109786D-02	7.50	-1.679484605076D-01
				7.75	-9.420969742194D-02
				8.25	6.009375516200D-02
				8.50	6.863041396860D-02
				8.75	3.938788530523D-02

I= 7	H= -2.0				
0		0.	-1.326963513239D-01	-1.00	-4.814805597547D-02
1		2.303653037501D-02	2.665819514608D-02	-.75	-3.140263093152D-02
2		-2.511152560045D-02	-1.298731338026D-02	-.50	-1.779614658762D-02
				-.25	-7.328602943782D-03

CARDINAL SPLINES N=3

0	-9.0	2.074995225440D-03	1.037497612720D-03	.25	4.191688606250D-03
1	-8.0	-1.202680556229D-02	-6.013402781147D-03	.50	5.305227388189D-03
2	-7.0	3.752807749049D-02	1.876403874524D-02	.75	3.644570724542D-03
3	-6.0	-9.354155080184D-02	-4.677077540092D-02	1.25	-4.120457819112D-03
4	-5.0	2.177010998944D-01	1.088505499472D-01	1.50	-6.564979719137D-03
5	-4.0	-4.501894535232D-01	-2.250947267616D-01	1.75	-5.387061375650D-03
6	-3.0	7.312054419299D-01	3.656027209649D-01	2.25	7.693526400012D-03
7	-2.0	-8.655246040982D-01	-4.327623020491D-01	2.50	1.312474110199D-02
8	-1.0	7.313052308910D-01	3.656526154455D-01	2.75	1.130085262031D-02
9	0.0	-4.504641693377D-01	-2.252320846688D-01	3.25	-1.708850254689D-02
10	1.0	2.183574285965D-01	1.091787142983D-01	3.50	-2.959899944326D-02
11	2.0	-9.506936629125D-02	-4.753468314562D-02	3.75	-2.575337281543D-02
12	3.0	4.097640809073D-02	2.048820404537D-02	4.25	3.946853850903D-02
13	4.0	-1.763740930835D-02	-8.818704654174D-03	4.50	6.872531810258D-02
14	5.0	7.575803004734D-03	3.787901502367D-03	4.75	6.015044530299D-02
15	6.0	-3.219107839921D-03	-1.609553919960D-03	5.25	-9.425206172112D-02
16	7.0	1.290039451817D-03	6.450197259084D-04	5.50	-1.680025963448D-01
17	8.0	-4.133771271855D-04	-2.066885635928D-04	5.75	-1.526240788119D-01
18	9.0	7.131931491043D-05	3.565965745521D-05	6.25	2.832141654725D-01
				6.50	6.198957546166D-01
	0	-1.826704800262D+00	-9.447449788102D-01	6.75	8.938881871919D-01
	1	-4.289709304332D-01	-2.071134456986D-01	7.25	8.938727101211D-01
	2	-2.511152560045D-02	-1.298731338026D-02	7.50	6.198724674997D-01
				7.75	2.831959420316D-01
				8.25	-1.525996930947D-01
				8.50	-1.679617728273D-01
				8.75	-9.421736868066D-02

I= 8 H= -1.0

0		0.	3.081885977656D-01	-1.00	2.073137113314D-02
1		-9.918963150652D-03	-4.012211971633D-02	-.75	1.352120185314D-02
2		1.081240798249D-02	6.408484512147D-03	-.50	7.662583570948D-03
				-.25	3.155516286569D-03
0	-9.0	-8.934448318364D-04	-4.467224159182D-04	.25	-1.804837793476D-03
1	-8.0	5.178528990115D-03	2.589264495057D-03	.50	-2.284299730699D-03
2	-7.0	-1.616061593786D-02	-8.080307968932D-03	.75	-1.569261519457D-03
3	-6.0	4.032199809707D-02	2.016099904853D-02	1.25	1.774163445946D-03
4	-5.0	-9.478823640277D-02	-4.739411820138D-02	1.50	2.826705573805D-03
5	-4.0	2.182397574597D-01	1.091198787299D-01	1.75	2.319516331093D-03
6	-3.0	-4.504214259561D-01	-2.252107129780D-01	2.25	-3.312548354315D-03
7	-2.0	7.313052308910D-01	3.656526154455D-01	2.50	-5.650894574475D-03
8	-1.0	-8.655673476155D-01	-4.327836738077D-01	2.75	-4.865419300755D-03
9	0.0	7.313231165937D-01	3.656615582968D-01	3.25	7.355702105219D-03
10	1.0	-4.504706660258D-01	-2.252353330129D-01	3.50	1.273775584039D-02
11	2.0	2.183574285965D-01	1.091787142983D-01	3.75	1.107835266642D-02
12	3.0	-9.506286963057D-02	-4.753143481528D-02	4.25	-1.694347515200D-02
13	4.0	4.095852310047D-02	2.047926155023D-02	4.50	-2.943265512292D-02
14	5.0	-1.759468247833D-02	-8.797341239166D-03	4.75	-2.565765856617D-02
15	6.0	7.476401604002D-03	3.738200802001D-03	5.25	3.940610877379D-02
16	7.0	-2.996129115713D-03	-1.498064557856D-03	5.50	6.865370109288D-02
17	8.0	9.600724805738D-04	4.800362402869D-04	5.75	6.010923222494D-02
18	9.0	-1.656398188062D-04	-8.281990940310D-05	6.25	-9.422517446797D-02
				6.50	-1.679717475778D-01
	0	7.865343782257D-01	4.661767658025D-01	6.75	-1.526063225076D-01
	1	1.847043805341D-01	7.523060150232D-02	7.25	2.832025714439D-01
	2	1.081240798249D-02	6.408484512147D-03	7.50	6.198824422481D-01

CARDINAL SPLINES N=3

				7.75	8.938805159057D-01
				8.25	8.938777418785D-01
				8.50	6.198782683904D-01
				8.75	2.831993051725D-01

I= 9 H= 0.0

0		0.	-7.157620282071D-01	-1.00	-8.926412869181D-03
1		4.270858810527D-03	8.379997305577D-02	-.75	-5.821893265888D-03
2		-4.655554058654D-03	-4.655554058654D-03	-.50	-3.299317919927D-03
				-.25	-1.358686831298D-03
0	-9.0	3.846952481266D-04	1.923476240633D-04	.25	7.771182529192D-04
1	-8.0	-2.229749326491D-03	-1.114874663246D-03	.50	9.835626171041D-04
2	-7.0	6.958443249253D-03	3.479221624626D-03	.75	6.756849355420D-04
3	-6.0	-1.736360202403D-02	-8.681801012016D-03	1.25	-7.639104641622D-04
4	-5.0	4.085877254560D-02	2.042938627280D-02	1.50	-1.217108542172D-03
5	-4.0	-9.502012775354D-02	-4.751006387677D-02	1.75	-9.987251860808D-04
6	-3.0	2.183395429581D-01	1.091697714790D-01	2.25	1.426296839831D-03
7	-2.0	-4.504641693377D-01	-2.252320846688D-01	2.50	2.433121888094D-03
8	-1.0	7.313231165937D-01	3.656615582968D-01	2.75	2.094909145997D-03
9	0.0	-8.655738443060D-01	-4.327869221530D-01	3.25	-3.167088127321D-03
10	1.0	7.313231165937D-01	3.656615582968D-01	3.50	-5.484265834056D-03
11	2.0	-4.504641693377D-01	-2.252320846688D-01	3.75	-4.769614207444D-03
12	3.0	2.183395429581D-01	1.091697714790D-01	4.25	7.293253696822D-03
13	4.0	-9.502012775354D-02	-4.751006387677D-02	4.50	1.266612655063D-02
14	5.0	4.085877254560D-02	2.042938627280D-02	4.75	1.103713566033D-02
15	6.0	-1.736360202403D-02	-8.681801012016D-03	5.25	-1.691658708562D-02
16	7.0	6.958443249253D-03	3.479221624626D-03	5.50	-2.940180581697D-02
17	8.0	-2.229749326491D-03	-1.114874663246D-03	5.75	-2.563990208703D-02
18	9.0	3.846952481266D-04	1.923476240633D-04	6.25	3.939451470648D-02
				6.50	6.864038869700D-02
	0	-3.386621494562D-01	-3.386621494562D-01	6.75	6.010156092878D-02
	1	-7.952911424524D-02	2.378865867675D-23	7.25	-9.422014270734D-02
	2	-4.655554058654D-03	-4.655554058654D-03	7.50	-1.679659466841D-01
				7.75	-1.526029593652D-01
				8.25	2.832003127924D-01
				8.50	6.198797844758D-01
				8.75	8.938789282980D-01

I=10 H= 1.0

0		0.	1.662339424807D+00	-1.00	3.843482264806D-03
1		-1.838921223000D-03	-1.905833227210D-01	-.75	2.506756503266D-03
2		2.004561041806D-03	6.408484512147D-03	-.50	1.420600871951D-03
				-.25	5.850153708628D-04
0	-9.0	-1.656398188062D-04	-8.281990940310D-05	.25	-3.346069982726D-04
1	-8.0	9.600724805738D-04	4.800362402869D-04	.50	-4.234965953861D-04
2	-7.0	-2.996129115713D-03	-1.498064557856D-03	.75	-2.909324366733D-04
3	-6.0	7.476401604002D-03	3.738200802001D-03	1.25	3.289200587510D-04
4	-5.0	-1.759468247833D-02	-8.797341239166D-03	1.50	5.240554005222D-04
5	-4.0	4.095852310047D-02	2.047926155023D-02	1.75	4.300251695231D-04
6	-3.0	-9.506286963057D-02	-4.753143481528D-02	2.25	-6.141263135534D-04
7	-2.0	2.183574285965D-01	1.091787142983D-01	2.50	-1.047638737013D-03
8	-1.0	-4.504706660258D-01	-2.252353330129D-01	2.75	-9.020128016520D-04
9	0.0	7.313231165937D-01	3.656615582968D-01	3.25	1.363662084404D-03
10	1.0	-8.655673476155D-01	-4.327836738077D-01	3.50	2.361370156731D-03
11	2.0	7.313052308910D-01	3.656526154455D-01	3.75	2.053653037719D-03

12	3.0	-4.504214259561D-01	-2.252107129780D-01	4.25	-3,140192029232D-03
13	4,0	2,182397574597D-01	1,091198787299D-01	4.50	-5,453411250359D-03
14	5,0	-9,478823640277D-02	-4,739411820138D-02	4,75	-4,751856042588D-03
15	6,0	4,032199809707D-02	2,016099904853D-02	5,25	7,281659283010D-03
16	7,0	-1,616061593786D-02	-8,080307968932D-03	5,50	1,265281392689D-02
17	8,0	5,178528990115D-03	2,589264495057D-03	5,75	1,102946429128D-02
18	9,0	-8,934448318364D-04	-4,467224159182D-04	6,25	-1,691155530990D-02
				6.50	-2,939600491331D-02
	0	1.458191533793D-01	4.661767658025D-01	6,75	-2,563653894137D-02
	1	3.424317752951D-02	-7.523060150232D-02	7,25	3,939225605431D-02
	2	2.004561041806D-03	6.408484512147D-03	7,50	6,863773092416D-02
				7,75	6,009997332096D-02
				8,25	-9,421895628779D-02
				8,50	-1,679644305987D-01
				8,75	-1,526019517453D-01

I=11 H= 2.0

0		0.	-3.860738373898D+00	-1,00	-1,654883005217D-03
1		7,917818451532D-04	4,408850865432D-01	-.75	-1,079330786401D-03
2		-8,631011600636D-04	-1,298731338026D-02	-.50	-6,116662125925D-04
				-.25	-2,518892837923D-04
0	-9,0	7,131931491043D-05	3,565965745521D-05	.25	1,440712865528D-04
1	-8,0	-4,133771271855D-04	-2,066885635928D-04	.50	1,823443611516D-04
2	-7,0	1,290039451817D-03	6,450197259084D-04	.75	1,252663890666D-04
3	-6,0	-3,219107839921D-03	-1,609553919960D-03	1,25	-1,416226682986D-04
4	-5,0	7,575803004734D-03	3,787901502367D-03	1,50	-2,256418300369D-04
5	-4,0	-1,763740930835D-02	-8,818704654174D-03	1,75	-1,851553590347D-04
6	-3,0	4,097640809073D-02	2,048820404537D-02	2,25	2,644235357480D-04
7	-2,0	-9,506936629125D-02	-4,753468314562D-02	2,50	4,510803825107D-04
8	-1,0	2,183574285965D-01	1,091787142983D-01	2,75	3,883784064862D-04
9	0,0	-4,504641693377D-01	-2,252320846688D-01	3,25	-5,871499801104D-04
10	1,0	7,313052308910D-01	3,656526154455D-01	3,50	-1,016731434110D-03
11	2,0	-8,655246040982D-01	-4,327623020491D-01	3,75	-8,842378009748D-04
12	3,0	7,312054419299D-01	3,656027209649D-01	4,25	1,352064212716D-03
13	4,0	-4,501894535232D-01	-2,250947267616D-01	4,50	2,348055260954D-03
14	5,0	2,177010998944D-01	1,088505499472D-01	4,75	2,045980943092D-03
15	6,0	-9,354155080184D-02	-4,677077540092D-02	5,25	-3,135160104484D-03
16	7,0	3,752807749049D-02	1,876403874524D-02	5,50	-5,447610248788D-03
17	8,0	-1,202680556229D-02	-6,013402781147D-03	5,75	-4,748492865669D-03
18	9,0	2,074995225440D-03	1,037497612720D-03	6,25	7,279400624447D-03
				6,50	1,265015614989D-02
	0	-6,278515735878D-02	-9,447449788102D-01	6,75	1,102787668216D-02
	1	-1,474403903599D-02	2,071134456986D-01	7,25	-1,691036889016D-02
	2	-8,631011600636D-04	-1,298731338026D-02	7,50	-2,939448882786D-02
				7,75	-2,563553132155D-02
				8,25	3,939124843463D-02
				8,50	6,863621483915D-02
				8,75	6,009878690179D-02

I=12 H= 3.0

0		0.	8.966419819700D+00	-1.00	7,124921240155D-04
1		-3.408931790487D-04	-1.023190130274D+00	-.75	4,646942908304D-04
2		3.715989449668D-04	2.934609109786D-02	-,50	2,633463257661D-04
				-,25	1,084482288226D-04
0	-9,0	-3,070576591805D-05	-1.535288295902D-05	.25	-6.202834680129D-05

CARDINAL SPLINES N=3

1	-8.0	1.779750876079D-04	8.898754380394D-05	.50	-7.850640846761D-05
2	-7.0	-5.554126529211D-04	-2.777063264606D-04	.75	-5.393209992837D-05
3	-6.0	1.385952501571D-03	6.929762507853D-04	1.25	6.097412046697D-05
4	-5.0	-3.261684335910D-03	-1.630842167955D-03	1.50	9.714766914973D-05
5	-4.0	7.593681518352D-03	3.796840759176D-03	1.75	7.971665338338D-05
6	-3.0	-1.764390569500D-02	-8.821952847498D-03	2.25	-1.138447162047D-04
7	-2.0	4.097640809073D-02	2.048820404537D-02	2.50	-1.942078183953D-04
8	-1.0	-9.506286963057D-02	-4.753143481528D-02	2.75	-1.672121547659D-04
9	0.0	2.183395429581D-01	1.091697714790D-01	3.25	2.527911173409D-04
10	1.0	-4.504214259561D-01	-2.252107129780D-01	3.50	4.377427876634D-04
11	2.0	7.312054419299D-01	3.656027209649D-01	3.75	3.806990630162D-04
12	3.0	-8.652926243548D-01	-4.326463121774D-01	4.25	-5.821165666935D-04
13	4.0	7.306665977528D-01	3.653332988764D-01	4.50	-1.010929454582D-03
14	5.0	-4.489383983790D-01	-2.244691991895D-01	4.75	-8.808743119100D-04
15	6.0	2.148056974622D-01	1.074028487311D-01	5.25	1.349805490385D-03
16	7.0	-8.705275821853D-02	-4.352637910926D-02	5.50	2.345397442489D-03
17	8.0	2.792744128025D-02	1.396372064013D-02	5.75	2.044393321118D-03
18	9.0	-4.818953592725D-03	-2.409476796363D-03	6.25	-3.133973682904D-03
				6.50	-5.446094163127D-03
	0	2.703147593087D-02	2.134742026160D+00	6.75	-4.747485246667D-03
	1	6.347887830353D-03	-4.949604905126D-01	7.25	7.278393006783D-03
	2	3.715989449668D-04	2.934609109786D-02	7.50	1.264864006851D-02
				7.75	1.102669026620D-02
				8.25	-1.690878128799D-02
				8.50	-2.939183106506D-02
				8.75	-2.563327267843D-02

I=13	H=	4.0			
0		0.	-2.082324565168D+01	-1.00	-3.066428958611D-04
1		1.467138626228D-04	2.375894092962D+00	-.75	-1.999954781636D-04
2		-1.599290332383D-04	-6.780072870447D-02	-.50	-1.133391896210D-04
				-.25	-4.667403023309D-05
0	-9.0	1.321517061547D-05	6.607585307733D-06	.25	2.669580651836D-05
1	-8.0	-7.659705198744D-05	-3.829852599372D-05	.50	3.378764708357D-05
2	-7.0	2.390389153158D-04	1.195194576579D-04	.75	2.321133770374D-05
3	-6.0	-5.964872860746D-04	-2.982436430373D-04	1.25	-2.624208779560D-05
4	-5.0	1.403766312292D-03	7.018831561458D-04	1.50	-4.181048686531D-05
5	-4.0	-3.268177984884D-03	-1.634088992442D-03	1.75	-3.430851319723D-05
6	-3.0	7.593681518352D-03	3.796840759176D-03	2.25	4.899657451462D-05
7	-2.0	-1.763740930835D-02	-8.818704654174D-03	2.50	8.358330680843D-05
8	-1.0	4.095852310047D-02	2.047926155023D-02	2.75	7.196489278216D-05
9	0.0	-9.502012775354D-02	-4.751006387677D-02	3.25	-1.087964307604D-04
10	1.0	2.182397574597D-01	1.091198787299D-01	3.50	-1.883960689559D-04
11	2.0	-4.501894535232D-01	-2.250947267616D-01	3.75	-1.638455473362D-04
12	3.0	7.306665977528D-01	3.653332988764D-01	4.25	2.505317579725D-04
13	4.0	-8.640411752421D-01	-4.320205876211D-01	4.50	4.350845549109D-04
14	5.0	7.277611387509D-01	3.638805693754D-01	4.75	3.791113125419D-04
15	6.0	-4.422141295947D-01	-2.211070647973D-01	5.25	-5.809301266852D-04
16	7.0	1.997361704446D-01	9.986808522231D-02	5.50	-1.009413366750D-03
17	8.0	-6.475631059173D-02	-3.237815529587D-02	5.75	-8.798667010407D-04
18	9.0	1.118797891144D-02	5.593989455719D-03	6.25	1.348797892883D-03
				6.50	2.343881397433D-03
	0	-1.163382692869D-02	-4.932057840091D+00	6.75	2.043206937254D-03
	1	-2.732008735666D-03	1.155480976281D+00	7.25	-3.132386130270D-03
	2	-1.599290332383D-04	-6.780072870447D-02	7.50	-5.443436486465D-03
				7.75	-4.745226678658D-03

CARDINAL SPLINES N=3

				8.25	7.275029992551D-03
				8.50	1.264283939810D-02
				8.75	1.102165870089D-02

I=14 H= 5.0

0		0.	4.833753714014D+01	-1.00	1.317109781963D-04
1		-6.301736196022D-05	-5.515126746316D+00	-.75	8.590318060297D-05
2		6.869361623610D-05	1.572385972319D-01	-.50	4.868208503913D-05
				-.25	2.004769150481D-05
0	-9.0	-5.676254275880D-06	-2.838127137940D-06	.25	-1.146653269236D-05
1	-8.0	3.290039580421D-05	1.645019790211D-05	.50	-1.451265986721D-05
2	-7.0	-1.026733369401D-04	-5.133666847007D-05	.75	-9.969864084467D-06
3	-6.0	2.562065682255D-04	1.281032841128D-04	1.25	1.127164887925D-05
4	-5.0	-6.029535942514D-04	-3.014767971257D-04	1.50	1.795867505231D-05
5	-4.0	1.403766312292D-03	7.018831561458D-04	1.75	1.473638520449D-05
6	-3.0	-3.261684335910D-03	-1.630842167955D-03	2.25	-2.104528376332D-05
7	-2.0	7.575803004734D-03	3.787901502367D-03	2.50	-3.590117119500D-05
8	-1.0	-1.759468247833D-02	-8.797341239166D-03	2.75	-3.091076477292D-05
9	0.0	4.085877254560D-02	2.042938627280D-02	3.25	4.673085373523D-05
10	1.0	-9.478823640277D-02	-4.739411820138D-02	3.50	8.092093719447D-05
11	2.0	2.177010998944D-01	1.088505499472D-01	3.75	7.037585929309D-05
12	3.0	-4.489383983790D-01	-2.244691991895D-01	4.25	-1.076098069052D-04
13	4.0	7.277611387509D-01	3.638805693754D-01	4.50	-1.868799596789D-04
14	5.0	-8.572956753580D-01	-4.286478376790D-01	4.75	-1.628380178400D-04
15	6.0	7.121496597090D-01	3.560748298545D-01	5.25	2.495243619763D-04
16	7.0	-4.072277879522D-01	-2.036138939761D-01	5.50	4.335688728339D-04
17	8.0	1.479713665178D-01	7.398568325891D-02	5.75	3.779252493866D-04
18	9.0	-2.589294560710D-02	-1.294647280355D-02	6.25	-5.793430689959D-04
				6.50	-1.006756550736D-03
	0	4.997026657482D-03	1.143772279907D+01	6.75	-8.776088834603D-04
	1	1.173467730290D-03	-2.684831996142D+00	7.25	1.345436029255D-03
	2	6.869361623610D-05	1.572385972319D-01	7.50	2.338082726549D-03
				7.75	2.038177115000D-03
				8.25	-3.124717811367D-03
				8.50	-5.430129275710D-03
				8.75	-4.733637147827D-03

I=15 H= 6.0

0		0.	-1.117079295707D+02	-1.00	-5.596652545959D-05
1		2.677728797433D-05	1.274634356934D+01	-.75	-3.650191206621D-05
2		-2.918923748527D-05	-3.633664886984D-01	-.50	-2.068595335848D-05
				-.25	-8.518649336411D-06
0	-9.0	2.411949510942D-06	1.205974755471D-06	.25	4.872350070196D-06
1	-8.0	-1.398001035772D-05	-6.990005178861D-06	.50	6.166708038063D-06
2	-7.0	4.362787373264D-05	2.181393686632D-05	.75	4.236386820241D-06
3	-6.0	-1.088670938558D-04	-5.443354692788D-05	1.25	-4.789540193316D-06
4	-5.0	2.562065682255D-04	1.281032841128D-04	1.50	-7.630986105327D-06
5	-4.0	-5.964872860746D-04	-2.982436430373D-04	1.75	-6.261773232748D-06
6	-3.0	1.385952501571D-03	6.929762507853D-04	2.25	8.942545455742D-06
7	-2.0	-3.219107839921D-03	-1.609553919960D-03	2.50	1.525509748101D-05
8	-1.0	7.476401604002D-03	3.738200802001D-03	2.75	1.313457790166D-05
9	0.0	-1.736360202403D-02	-8.681801012016D-03	3.25	-1.985683768398D-05
10	1.0	4.032199809707D-02	2.016099904853D-02	3.50	-3.438486110636D-05
11	2.0	-9.354155080184D-02	-4.677077540092D-02	3.75	-2.990405488162D-05
12	3.0	2.148056974622D-01	1.074028487311D-01	4.25	4.572547466064D-05

13	4,0	-4,422141295947D-01	-2,211070647973D-01	4.50	7,940888571928D-05
14	5,0	7.121496597090D-01	3,560748298545D-01	4.75	6,919300264392D-05
15	6,0	-8,211651824000D-01	-4,105825912000D-01	5.25	-1,060276956896D-04
16	7,0	6,311788736036D-01	3,155894368018D-01	5,50	-1,842317440825D-04
17	8,0	-2,874256981332D-01	-1.437128490666D-01	5,75	-1,605876987654D-04
18	9,0	5,802777581505D-02	2,901388790753D-02	6,25	2,461739946550D-04
				6,50	4,277901799574D-04
	0	-2,123332644538D-03	-2,642352303122D+01	6,75	3,729128424226D-04
	1	-4,986289867605D-04	6,205746772768D+00	7,25	-5,717014490105D-04
	2	-2,918923748527D-05	-3,633664886984D-01	7.50	-9,934957370405D-04
				7.75	-8,660597983296D-04
				8,25	1,327748640310D-03
				8,50	2,307353746292D-03
				8,75	2,011394321938D-03

I=16 H= 7.0

0		0.	2.465666847132D+02	-1.00	2,242826936497D-05
1		-1,073084710226D-05	-2,815662383264D+01	-.75	1,462793534947D-05
2		1,169742226271D-05	8,032541440711D-01	-.50	8,289779116806D-06
				-,25	3,413800666984D-06
0	-9,0	-9,665751604530D-07	-4,832875802265D-07	.25	-1,952566805200D-06
1	-8,0	5,602410288172D-06	2,801205144086D-06	.50	-2,471273459215D-06
2	-7,0	-1.748362428902D-05	-8.741812144512D-06	.75	-1,697708120315D-06
3	-6,0	4,362787373264D-05	2,181393686632D-05	1,25	1,919381214179D-06
4	-5,0	-1,026733369401D-04	-5,133666847007D-05	1,50	3,058074634527D-06
5	-4,0	2,390389153158D-04	1,195194576579D-04	1,75	2,509370299712D-06
6	-3,0	-5,554126529211D-04	-2,777063264606D-04	2,25	-3,583674645563D-06
7	-2,0	1,290039451817D-03	6,450197259084D-04	2,50	-6,113394259926D-06
8	-1,0	-2,996129115713D-03	-1,498064557856D-03	2,75	-5,263608000570D-06
9	0,0	6,958443249253D-03	3,479221624626D-03	3,25	7,957515687355D-06
10	1,0	-1,616061593786D-02	-8,080307968932D-03	3,50	1,377953911974D-05
11	2,0	3,752807749049D-02	1,876403874524D-02	3,75	1,198388130182D-05
12	3,0	-8,705275821853D-02	-4,352637910926D-02	4,25	-1,832422602756D-05
13	4,0	1,997361704446D-01	9,986808522231D-02	4,50	-3,182266299651D-05
14	5,0	-4,072277879522D-01	-2,036138939761D-01	4,75	-2,772870548200D-05
15	6,0	6,311788736036D-01	3,155894368018D-01	5,25	4,249000092147D-05
16	7,0	-6,397043821890D-01	-3,198521910945D-01	5,50	7,382983213026D-05
17	8,0	3,626945953912D-01	1,813472976956D-01	5,75	6,435461432682D-05
18	9,0	-8,585625922771D-02	-4,292812961385D-02	6,25	-9,865283936516D-05
				6,50	-1,714345004703D-04
	0	8,509135793592D-04	5,822065588924D+01	6,75	-1,494427165319D-04
	1	1,998227536265D-04	-1,369804923936D+01	7,25	2,291061237065D-04
	2	1,169742226271D-05	8,032541440711D-01	7,50	3,981377904203D-04
				7,75	3,470685454901D-04
				8,25	-5,320876161977D-04
				8,50	-9,246584922782D-04
				8,75	-8,060541586520D-04

I=17 H= 8.0

0		0.	-2,739631708694D+02	-1.00	-7,186860399136D-06
1		3,438566696049D-06	3,183094538354D+01	-.75	-4,687340230023D-06
2		-3,748293703087D-06	-9,228222910828D-01	-,50	-2,656356773796D-06
				-,25	-1,093910030455D-06
0	-9,0	3,097270070378D-07	1.548635035189D-07	.25	6,256757853496D-07
1	-8,0	-1,795222804960D-06	-8,976114024800D-07	.50	7,918888912227D-07

2	-7.0	5,602410288172D-06	2.801205144086D-06
3	-6.0	-1,398001035772D-05	-6,990005178861D-06
4	-5.0	3,290039580421D-05	1.645019790211D-05
5	-4.0	-7,659705198744D-05	-3.829852599372D-05
6	-3.0	1,779750876079D-04	8,898754380394D-05
7	-2.0	-4,133771271855D-04	-2.066885635928D-04
8	-1.0	9,600724805738D-04	4.800362402869D-04
9	0.0	-2,229749326491D-03	-1.114874663246D-03
10	1.0	5,178528990115D-03	2,589264495057D-03
11	2.0	-1,202680556229D-02	-6.013402781147D-03
12	3.0	2,792744128025D-02	1.396372064013D-02
13	4.0	-6,475631059173D-02	-3,237815529587D-02
14	5.0	1,479713665178D-01	7.398568325891D-02
15	6.0	-2,874256981332D-01	-1.437128490666D-01
16	7.0	3,626945953912D-01	1.813472976956D-01
17	8.0	-2.424626222546D-01	-1.212313111273D-01
18	9.0	6,445814299995D-02	3,222907149998D-02
	0	-2,726646896856D-04	-6.223326799524D+01
	1	-6,403071995951D-05	1.522014414405D+01
	2	-3,748293703087D-06	-9.228222910828D-01

.75	5,440094846658D-07
1.25	-6,150418748125D-07
1.50	-9,799220408336D-07
1,75	-8,040965506656D-07
2.25	1,148344037361D-06
2.50	1,958961273204D-06
2,75	1,686657819214D-06
3.25	-2,549887083943D-06
3.50	-4,415482193762D-06
3,75	-3,840085944860D-06
4,25	5,871770676499D-06
4.50	1,019717717679D-05
4,75	8,885319330844D-06
5,25	-1,361539314527D-05
5.50	-2,365785287073D-05
5,75	-2,062163699012D-05
6.25	3,161207727326D-05
6.50	5,493405673537D-05
6,75	4,788706265442D-05
7,25	-7,341421202198D-05
7.50	-1,275783100544D-04
7.75	-1,112138043245D-04
8.25	1,705008636368D-04
8.50	2,962953143412D-04
8.75	2,582900196578D-04

I=18 H= 9.0

0		0.	1.058224351367D+02
1		-5,932505805544D-07	-1.256627194860D+01
2		6.466873008454D-07	3.730572819822D-01
0	-9.0	-5,343672029095D-08	-2.671836014548D-08
1	-8.0	3,097270070378D-07	1.548635035189D-07
2	-7.0	-9,665751604530D-07	-4.832875802265D-07
3	-6.0	2,411949510942D-06	1.205974755471D-06
4	-5.0	-5,676254275880D-06	-2.838127137940D-06
5	-4.0	1,321517061547D-05	6.607585307733D-06
6	-3.0	-3,070576591805D-05	-1.535288295902D-05
7	-2.0	7,131931491043D-05	3.565965745521D-05
8	-1.0	-1,656398188062D-04	-8.281990940310D-05
9	0.0	3,846952481266D-04	1.923476240633D-04
10	1.0	-8,934448318364D-04	-4.467224159182D-04
11	2.0	2,074995225440D-03	1.037497612720D-03
12	3.0	-4,818953592725D-03	-2,409476796363D-03
13	4.0	1,118797891144D-02	5,593989455719D-03
14	5.0	-2,589294560710D-02	-1.294647280355D-02
15	6.0	5,802777581505D-02	2.901388790753D-02
16	7.0	-8,585625922771D-02	-4.292812961385D-02
17	8.0	6,445814299995D-02	3.222907149998D-02
18	9.0	-1,855619925181D-02	-9.278099625903D-03
	0	4,704241614348D-05	2.294362743982D+01
	1	1,104712083466D-05	-5,851240872921D+00
	2	6,466873008454D-07	3.730572819822D-01

-1.00	1,239937881400D-06
-.75	8,086995421413D-07
-.50	4,582971154885D-07
-.25	1,887306014414D-07
.25	-1,079468731329D-07
.50	-1,366233625750D-07
.75	-9,385711289996D-08
1.25	1,061122210359D-07
1.50	1,690644303310D-07
1.75	1,387295311306D-07
2.25	-1,981220162527D-07
2.50	-3,379765510866D-07
2.75	-2,909964583776D-07
3.25	4,399280649800D-07
3.50	7,617962966625D-07
3.75	6,625240739940D-07
4.25	-1,013047490606D-06
4.50	-1,759303167536D-06
4.75	-1,532970367416D-06
5.25	2,349042668609D-06
5.50	4,081652674139D-06
5.75	3,557819055910D-06
6.25	-5,453982677063D-06
6.50	-9,477687632370D-06
6.75	-8,261880669643D-06
7.25	1,266604016621D-05
7.50	2,201088800771D-05
7.75	1,918754519947D-05
8.25	-2,941624952847D-05

CARDINAL SPLINES N=3

8.50	-5.111937082411D-05
8.75	-4.456237607617D-05

N= 3 M=19 P= 9.5 Q= 5

I= 0 H= -9.5

0		1.000000000000D+00	4.999521668778D-01
1		-1.727557718026D+00	-8.637736962278D-01
2		7.461139172775D-01	3.730568194149D-01
0	-9.5	-1.855619925183D-02	-9.278099625917D-03
1	-8.5	6.445814300012D-02	3.222907150006D-02
2	-7.5	-8.585625922823D-02	-4.292812961411D-02
3	-6.5	5.802777581634D-02	2.901388790817D-02
4	-5.5	-2.589294561013D-02	-1.294647280506D-02
5	-4.5	1.118797891849D-02	5.593989459246D-03
6	-3.5	-4.818953609119D-03	-2.409476804559D-03
7	-2.5	2.074995263516D-03	1.037497631758D-03
8	-1.5	-8.934449202693D-04	-4.467224601346D-04
9	-.5	3.846954535103D-04	1.923477267551D-04
10	.5	-1.656402958047D-04	-8.282014790233D-05
11	1.5	7.132042272663D-05	3.566021136331D-05
12	2.5	-3.070833878328D-05	-1.535416939164D-05
13	3.5	1.322114581859D-05	6.610572909297D-06
14	4.5	-5.690126674758D-06	-2.845063337379D-06
15	5.5	2.444055155065D-06	1.222027577532D-06
16	6.5	-1.038525997861D-06	-5.192629989305D-07
17	7.5	4.161834352101D-07	2.080917176051D-07
18	8.5	-1.333608135615D-07	-6.668040678074D-08
19	9.5	2.300853438715D-08	1.150426719358D-08
	0	5.192498271305D+01	2.596248000491D+01
	1	1.244860671025D+01	6.224305872656D+00
	2	7.461139172775D-01	3.730568194149D-01

-1.00	3.473671635303D+00
-.75	2.715357366988D+00
-.50	2.050307338332D+00
-.25	1.478521549336D+00
.25	6.147245690351D-01
.50	3.221697390799D-01
.75	1.196173168847D-01
1.25	-5.021023041782D-02
1.50	-5.147708426402D-02
1.75	-2.752040839488D-02
2.25	1.677968342761D-02
2.50	1.898487003935D-02
2.75	1.083411537089D-02
3.25	-7.016349048954D-03
3.50	-8.037375405125D-03
3.75	-4.621138963192D-03
4.25	3.012078378909D-03
4.50	3.454789720642D-03
4.75	1.987862446268D-03
5.25	-1.296539320655D-03
5.50	-1.487292757522D-03
5.75	-8.558432806556D-04
6.25	5.582411742180D-04
6.50	6.403806271215D-04
6.75	3.685015154758D-04
7.25	-2.403641691439D-04
7.50	-2.757316402489D-04
7.75	-1.586675188669D-04
8.25	1.034948574649D-04
8.50	1.187232313936D-04
8.75	6.831831991434D-05
9.25	-4.456233960192D-05
9.50	-5.111930743934D-05

I= 1 H= -8.5

0		0.	2.772477371716D-04
1		2.781182690874D+00	1.390561421184D+00
2		-1.845640833874D+00	-9.228196099756D-01
0	-9.5	6.445814300012D-02	3.222907150006D-02
1	-8.5	-2.424626222556D-01	-1.212313111278D-01
2	-7.5	3.626945953942D-01	1.813472976971D-01

-1.00	-4.626823524748D+00
-.75	-3.124059987210D+00
-.50	-1.852001553905D+00
-.25	-8.106482248356D-01
.25	5.800060680066D-01
.50	9.311954539372D-01
.75	1.063010268583D+00

CARDINAL SPLINES N=3

3	-6.5	-2.874256981406D-01	-1.437128490703D-01	1.25	7.891384222395D-01
4	-5.5	1.479713665354D-01	7.398568326769D-02	1.50	5.009842265560D-01
5	-4.5	-6.475631063263D-02	-3.237815531631D-02	1.75	2.152141789180D-01
6	-3.5	2.792744137527D-02	1.396372068764D-02	2.25	-1.084973407897D-01
7	-2.5	-1.202680578299D-02	-6.013402891495D-03	2.50	-1.174242888197D-01
8	-1.5	5.178529502685D-03	2.589264751342D-03	2.75	-6.515456765243D-02
9	-.5	-2.229750516925D-03	-1.114875258462D-03	3.25	4.115217610245D-02
10	.5	9.600752453264D-04	4.800376226632D-04	3.50	4.690407300659D-02
11	1.5	-4.133835482495D-04	-2.066917741247D-04	3.75	2.688643471028D-02
12	2.5	1.779900003092D-04	8.899500015458D-05	4.25	-1.747932027731D-02
13	3.5	-7.663168513430D-05	-3.831584256715D-05	4.50	-2.003818046406D-02
14	4.5	3.298080224685D-05	1.649040112342D-05	4.75	-1.152632035803D-02
15	5.5	-1.416609934314D-05	-7.083049671572D-06	5.25	7.515830241662D-03
16	6.5	6.019447812236D-06	3.009723906118D-06	5.50	8.621156386777D-03
17	7.5	-2.412259754425D-06	-1.206129877213D-06	5.75	4.960780975364D-03
18	8.5	7.729786823672D-07	3.864893411836D-07	6.25	-3.235685854448D-03
19	9.5	-1.333608135615D-07	-6.668040678074D-08	6.50	-3.711764779931D-03
				6.75	-2.135896544429D-03
	0	-1.401478496938D+02	-7.007385905131D+01	7.25	1.393187373628D-03
	1	-3.228599315273D+01	-1.614301116835D+01	7.50	1.598181809870D-03
	2	-1.845640833874D+00	-9.228196099756D-01	7.75	9.196603796184D-04
				8.25	-5.998713266817D-04
				8.50	-6.881371733435D-04
				8.75	-3.959829391688D-04
				9.25	2.582898082478D-04
				9.50	2.962949469539D-04

I= 2 H= -7.5

0		0.	-8.652160449451D-04	-1.00	3.127136922224D+00
1		-1.520640331498D+00	-7.602267801568D-01	-.75	2.044134580907D+00
2		1.606496590726D+00	8.032457770523D-01	-.50	1.161944313431D+00
				-.25	4.805661197949D-01
0	-9.5	-8.585625922823D-02	-4.292812961411D-02	.25	-2.798378899573D-01
1	-8.5	3.626945953942D-01	1.813472976971D-01	.50	-3.613790261683D-01
2	-7.5	-6.397043821983D-01	-3.198521910992D-01	.75	-2.572000091061D-01
3	-6.5	6.311788736268D-01	3.155894368134D-01	1.25	3.476921927263D-01
4	-5.5	-4.072277880070D-01	-2.036138940035D-01	1.50	6.930200694788D-01
5	-4.5	1.997361705722D-01	9.986808528612D-02	1.75	9.356781956105D-01
6	-3.5	-8.705275851505D-02	-4.352637925753D-02	2.25	8.667750562605D-01
7	-2.5	3.752807817922D-02	1.876403908961D-02	2.50	5.888238698688D-01
8	-1.5	-1.616061753746D-02	-8.080308768729D-03	2.75	2.653417187863D-01
9	-.5	6.958446964278D-03	3.479223482139D-03	3.25	-1.409607496278D-01
10	.5	-2.996137743766D-03	-1.498068871883D-03	3.50	-1.546118064224D-01
11	1.5	1.290059490240D-03	6.450297451200D-04	3.75	-8.653576211933D-02
12	2.5	-5.554591914738D-04	-2.777295957369D-04	4.25	5.508853120572D-02
13	3.5	2.391469961054D-04	1.195734980527D-04	4.50	6.288877536146D-02
14	4.5	-1.029242639450D-04	-5.146213197249D-05	4.75	3.608392344268D-02
15	5.5	4.420860769144D-05	2.210430384572D-05	5.25	-2.347817891001D-02
16	6.5	-1.878508687568D-05	-9.392543437842D-06	5.50	-2.691962158819D-02
17	7.5	7.528017596811D-06	3.764008798405D-06	5.75	-1.548615618720D-02
18	8.5	-2.412259754425D-06	-1.206129877213D-06	6.25	1.009871364628D-02
19	9.5	4.161834352101D-07	2.080917176051D-07	6.50	1.158408454976D-02
				6.75	6.665772467081D-03
	0	1.305402341638D+02	6.526991175144D+01	7.25	-4.347808463626D-03
	1	2.900279489230D+01	1.450144298384D+01	7.50	-4.987526448879D-03
	2	1.606496590726D+00	8.032457770523D-01	7.75	-2.870023178902D-03

8.25	1.872039854032D-03
8.50	2.147493317927D-03
8.75	1.235757248745D-03
9.25	-8.060534988977D-04
9.50	-9.246573457606D-04

I= 3 H= -6.5

0		0.	2.159022393547D-03	-1.00	-1.395379800533D+00
1		6.686760123585D-01	3.341049758441D-01	-.75	-9.102778901172D-01
2		-7.267037881748D-01	-3.633456100044D-01	-.50	-5.160139532229D-01
				-.25	-2.125879898505D-01
0	-9.5	5.802777581634D-02	2.901388790817D-02	.25	1.218066840785D-01
1	-8.5	-2.874256981406D-01	-1.437128490703D-01	.50	1.544754271298D-01
2	-7.5	6.311788736268D-01	3.155894368134D-01	.75	1.065063916269D-01
3	-6.5	-8.211651824581D-01	-4.105825912291D-01	1.25	-1.228236245439D-01
4	-5.5	7.121496598458D-01	3.560748299229D-01	1.50	-2.004031353172D-01
5	-4.5	-4.422141299131D-01	-2.211070649566D-01	1.75	-1.711399238992D-01
6	-3.5	2.148056982021D-01	1.074028491011D-01	2.25	2.952198251358D-01
7	-2.5	-9.354155252048D-02	-4.677077626024D-02	2.50	6.336517814137D-01
8	-1.5	4.032200208863D-02	2.016100104431D-02	2.75	9.017984334133D-01
9	-.5	-1.736361129434D-02	-8.681805647172D-03	3.25	8.887161404325D-01
10	.5	7.476423134070D-03	3.738211567035D-03	3.50	6.139578432591D-01
11	1.5	-3.219157842929D-03	-1.609578921464D-03	3.75	2.797926530636D-01
12	2.5	1.386068631846D-03	6.930343159229D-04	4.25	-1.503799311434D-01
13	3.5	-5.967569861644D-04	-2.983784930822D-04	4.50	-1.654154069236D-01
14	4.5	2.568327205325D-04	1.284163602663D-04	4.75	-9.275208017773D-02
15	5.5	-1.103162320555D-04	-5.515811602774D-05	5.25	5.914298719622D-02
16	6.5	4.687548672523D-05	2.343774336262D-05	5.50	6.753974347464D-02
17	7.5	-1.878508687568D-05	-9.392543437842D-06	5.75	3.876026258005D-02
18	8.5	6.019447812236D-06	3.009723906118D-06	6.25	-2.522387552009D-02
19	9.5	-1.038525997861D-06	-5.192629989305D-07	6.50	-2.892217953095D-02
				6.75	-1.663851091650D-02
	0	-5.923259476537D+01	-2.961578500998D+01	7.25	1.085036536706D-02
	1	-1.313869596296D+01	-6.569461614239D+00	7.50	1.244633523414D-02
	2	-7.267037881748D-01	-3.633456100044D-01	7.75	7.161947603470D-03
				8.25	-4.671451175980D-03
				8.50	-5.358790461855D-03
				8.75	-3.083664104865D-03
				9.25	2.011392675616D-03
				9.50	2.307350885322D-03

I= 4 H= -5.5

0		0.	-5.081018502876D-03	-1.00	6.029240561583D-01
1		-2.885155552741D-01	-1.437093667229D-01	-.75	3.932414482029D-01
2		3.144085008842D-01	1.571894615533D-01	-.50	2.228599028581D-01
				-.25	9.177942012378D-02
0	-9.5	-2.589294561013D-02	-1.294647280506D-02	.25	-5.250364359295D-02
1	-8.5	1.479713665354D-01	7.398568326769D-02	.50	-6.646480696631D-02
2	-7.5	-4.072277880070D-01	-2.036138940035D-01	.75	-4.567640207466D-02
3	-6.5	7.121496598458D-01	3.560748299229D-01	1.25	5.174434277226D-02
4	-5.5	-8.572956756799D-01	-4.286478378399D-01	1.50	8.264534355568D-02
5	-4.5	7.277611395004D-01	3.638805697502D-01	1.75	6.810496964885D-02
6	-3.5	-4.489384001204D-01	-2.244692000602D-01	2.25	-9.940959322760D-02
7	-2.5	2.177011039390D-01	1.088505519695D-01	2.50	-1.739120311450D-01
8	-1.5	-9.478824579645D-02	-4.739412289823D-02	2.75	-1.560222181510D-01

CARDINAL SPLINES N=3

9	-,5	4,085879436224D-02	2,042939718112D-02	3,25	2,854293533948D-01
10	,5	-1,759473314695D-02	-8,797366573474D-03	3,50	6,224365901162D-01
11	1,5	7,575920681238D-03	3,787960340619D-03	3,75	8,953501894417D-01
12	2,5	-3,261957635566D-03	-1,630978817783D-03	4,25	8,929191337706D-01
13	3,5	1,404401021385D-03	7,022005106926D-04	4,50	6,187785870686D-01
14	4,5	-6,044271738988D-04	-3,022135869494D-04	4,75	2,825664761832D-01
15	5,5	2,596169534202D-04	1,298084767101D-04	5,25	-1,521890959464D-01
16	6,5	-1,103162320555D-04	-5,515811602774D-05	5,50	-1,674907452373D-01
17	7,5	4,420860769144D-05	2,210430384572D-05	5,75	-9,394630664090D-02
18	8,5	-1,416609934314D-05	-7,083049671572D-06	6,25	5,992194571114D-02
19	9,5	2,444055155065D-06	1,222027577532D-06	6,50	6,843331776471D-02
				6,75	3,927446223693D-02
	0	2,563446942970D+01	1,281602890281D+01	7,25	-2,555927495791D-02
	1	5,685245961526D+00	2,842890402789D+00	7,50	-2,930693012050D-02
	2	3,144085008842D-01	1,571894615533D-01	7,75	-1,685991256641D-02
				8,25	1,099478030585D-02
				8,50	1,261199976349D-02
				8,75	7,257278069461D-03
				9,25	-4,733633273391D-03
				9,50	-5,430122542736D-03

I= 5 H= -4.5

0		0,	1,182937253206D-02	-1,00	-2,596950780037D-01
1		1,242535495426D-01	6,084999197649D-02	-,75	-1,693760219163D-01
2		-1,354415284611D-01	-6,768633348104D-02	-,50	-9,598715688656D-02
				-,25	-3,952848291446D-02
0	-9,5	1,118797891849D-02	5,593989459246D-03	,25	2,260921761749D-02
1	-8,5	-6,475631063263D-02	-3,237815531631D-02	,50	2,861601699723D-02
2	-7,5	1,997361705722D-01	9,986808528612D-02	,75	1,965926223859D-02
3	-6,5	-4,422141299131D-01	-2,211070649566D-01	1,25	-2,223068780621D-02
4	-5,5	7,277611395004D-01	3,638805697502D-01	1,50	-3,542803451851D-02
5	-4,5	-8,640411769869D-01	-4,320205884935D-01	1,75	-2,908368581470D-02
6	-3,5	7,306666018070D-01	3,653333009035D-01	2,25	4,162746797420D-02
7	-2,5	-4,501894629397D-01	-2,250947314698D-01	2,50	7,119890869766D-02
8	-1,5	2,182397793296D-01	1,091198896648D-01	2,75	6,157282085744D-02
9	-,5	-9,502017854596D-02	-4,751008927298D-02	3,25	-9,517926758883D-02
10	,5	4,095864106460D-02	2,047932053230D-02	3,50	-1,690661041463D-01
11	1,5	-1,763768327688D-02	-8,818841638442D-03	3,75	-1,532360222870D-01
12	2,5	7,594317800909D-03	3,797158900454D-03	4,25	2,836132988564D-01
13	3,5	-3,269655682773D-03	-1,634827841387D-03	4,50	6,203536142417D-01
14	4,5	1,407197026546D-03	7,035985132730D-04	4,75	8,941516593151D-01
15	5,5	-6,044271738988D-04	-3,022135869494D-04	5,25	8,937008485741D-01
16	6,5	2,568327205325D-04	1,284163602663D-04	5,50	6,196753117504D-01
17	7,5	-1,029242639450D-04	-5,146213197249D-05	5,75	2,830824847723D-01
18	8,5	3,298080224685D-05	1,649040112342D-05	6,25	-1,525256731126D-01
19	9,5	-5,690126674758D-06	-2,845063337379D-06	6,50	-1,678768463539D-01
				6,75	-9,416848527105D-02
	0	-1,104318922296D+01	-5,518787300355D+00	7,25	6,006686736024D-02
	1	-2,449135491218D+00	-1,225190344163D+00	7,50	6,859956354114D-02
	2	-1,354415284611D-01	-6,768633348104D-02	7,75	3,937012716952D-02
				8,25	-2,562167521546D-02
				8,50	-2,937851246100D-02
				8,75	-1,690110441312D-02
				9,25	1,102164968063D-02
				9,50	1,264282372273D-02

CARDINAL SPLINES N=3

I= 6 H= -3.5

0		0.	-2.748583083099D-02	-1.00	1,118222132892D-01
1		-5,350162984003D-02	-2.378417946994D-02	-.75	7,293155057017D-02
2		5,832058344915D-02	2.908029105222D-02	-.50	4,133096078230D-02
				-.25	1,702044392558D-02
0	-9.5	-4,818953609119D-03	-2.409476804559D-03	.25	-9,735077003819D-03
1	-8.5	2,792744137527D-02	1.396372068764D-02	.50	-1,232126135801D-02
2	-7.5	-8,705275851505D-02	-4.352637925753D-02	.75	-8,464454470165D-03
3	-6.5	2,148056982021D-01	1.074028491011D-01	1,25	9.569867906522D-03
4	-5.5	-4,489384001204D-01	-2.244692000602D-01	1.50	1,524767157427D-02
5	-4.5	7,306666018070D-01	3.653333009035D-01	1.75	1,251234762590D-02
6	-3.5	-8,652926337747D-01	-4.326463168874D-01	2.25	-1,787308659834D-02
7	-2.5	7,312054638093D-01	3.656027319047D-01	2,50	-3,049775719354D-02
8	-1.5	-4,504214767713D-01	-2,252107383857D-01	2.75	-2,627011936181D-02
9	-.5	2,183396609755D-01	1.091698304877D-01	3,25	3,980535663562D-02
10	.5	-9,506314372305D-02	-4.753157186152D-02	3.50	6,911164182158D-02
11	1.5	4,097704466320D-02	2.048852233160D-02	3.75	6,037273377626D-02
12	2.5	-1,764538411278D-02	-8.822692056389D-03	4.25	-9,439704560168D-02
13	3.5	7,597114984845D-03	3.798557492422D-03	4.50	-1,681689121603D-01
14	4.5	-3,269655682773D-03	-1.634827841387D-03	4.75	-1,527197841068D-01
15	5.5	1,404401021385D-03	7.022005106926D-04	5,25	2,832765938428D-01
16	6.5	-5,967569861644D-04	-2,983784930822D-04	5,50	6,199673714009D-01
17	7.5	2,391469961054D-04	1,195734980527D-04	5,75	8,939294008253D-01
18	8.5	-7,663168513430D-05	-3,831584256715D-05	6,25	8,938458213695D-01
19	9.5	1,322114581859D-05	6,610572909297D-06	6,50	6,198416159861D-01
				6.75	2,831781832718D-01
	0	4.755167172805D+00	2.371060731667D+00	7.25	-1,525880952244D-01
	1	1,054589455694D+00	5,287413505222D-01	7.50	-1,679484537551D-01
	2	5,832058344915D-02	2.908029105222D-02	7.75	-9,420969153556D-02
				8.25	6,009374613764D-02
				8.50	6,863039828612D-02
				8.75	3,938787163431D-02
				9.25	-2,563325171963D-02
				9.50	-2,939179464296D-02

I= 7 H= -2.5

0		0.	6.384047317112D-02	-1.00	-4,814805685899D-02
1		2.303653079774D-02	4.627754719976D-03	-.75	-3,140263150776D-02
2		-2.511152606125D-02	-1.236994798983D-02	-.50	-1,779614691418D-02
				-.25	-7,328603078262D-03
0	-9.5	2.074995263516D-03	1.037497631758D-03	.25	4,191688683168D-03
1	-8.5	-1.202680578299D-02	-6.013402891495D-03	.50	5,305227485540D-03
2	-7.5	3.752807817922D-02	1.876403908961D-02	.75	3.644570791420D-03
3	-6.5	-9.354155252048D-02	-4.677077626024D-02	1.25	-4,120457894723D-03
4	-5.5	2.177011039390D-01	1.088505519695D-01	1.50	-6,564979839604D-03
5	-4.5	-4.501894629397D-01	-2.250947314698D-01	1.75	-5,387061474502D-03
6	-3.5	7.312054638093D-01	3.656027319047D-01	2.25	7.693526541184D-03
7	-2.5	-8.655246549169D-01	-4.327623274585D-01	2.50	1.312474134282D-02
8	-1.5	7.313053489180D-01	3.656526744590D-01	2.75	1,130085282766D-02
9	-.5	-4.504644434530D-01	-2.252322217265D-01	3,25	-1,708850286036D-02
10	.5	2.183580652225D-01	1.091790326112D-01	3,50	-2,959899998608D-02
11	1.5	-9.507084483806D-02	-4.753542241903D-02	3,75	-2,575337328751D-02
12	2.5	4.097984196513D-02	2.048992098256D-02	4.25	3.946853923088D-02
13	3.5	-1.764538411278D-02	-8.822692056389D-03	4.50	6,872531935617D-02
14	4.5	7.594317800909D-03	3.797158900454D-03	4.75	6,015044639531D-02

15	5.5	-3.261957635566D-03	-1.630978817783D-03
16	6.5	1.386068631846D-03	6.930343159229D-04
17	7.5	-5.554591914738D-04	-2.777295957369D-04
18	8.5	1.779900003092D-04	8.899500015458D-05
19	9.5	-3.070833878328D-05	-1.535416939164D-05
	0	-2.047468184450D+00	-1.008583663071D+00
	1	-4.540824643661D-01	-2.304012570868D-01
	2	-2.511152606125D-02	-1.236994798983D-02

5.25	-9.425206339494D-02
5.50	-1.680025992532D-01
5.75	-1.526240813470D-01
6.25	2.832141693587D-01
6.50	6.198957613699D-01
6.75	8.938881930789D-01
7.25	8.938727010959D-01
7.50	6.198724518158D-01
7.75	2.831959283595D-01
8.25	-1.525996721341D-01
8.50	-1.679617364021D-01
8.75	-9.421733692766D-02
9.25	6.009873822143D-02
9.50	6.863613024268D-02

I= 8 H= -1.5

0		0.	-1.482701349344D-01
1		-9.918964132428D-03	1.104379763048D-02
2		1.081240905270D-02	4.974647242961D-03
0	-9.5	-8.934449202693D-04	-4.467224601346D-04
1	-8.5	5.178529502685D-03	2.589264751342D-03
2	-7.5	-1.616061753746D-02	-8.080308768729D-03
3	-6.5	4.032200208863D-02	2.016100104431D-02
4	-5.5	-9.478824579645D-02	-4.739412289823D-02
5	-4.5	2.182397793296D-01	1.091198896648D-01
6	-3.5	-4.504214767713D-01	-2.252107383857D-01
7	-2.5	7.313053489180D-01	3.656526744590D-01
8	-1.5	-8.655676217342D-01	-4.327838108671D-01
9	-.5	7.313237532292D-01	3.656618766146D-01
10	.5	-4.504721445959D-01	-2.252360722980D-01
11	1.5	2.183608625362D-01	1.091804312681D-01
12	2.5	-9.507084483806D-02	-4.753542241903D-02
13	3.5	4.097704466320D-02	2.048852233160D-02
14	4.5	-1.763768327688D-02	-8.818841638442D-03
15	5.5	7.575920681238D-03	3.787960340619D-03
16	6.5	-3.219157842929D-03	-1.609578921464D-03
17	7.5	1.290059490240D-03	6.450297451200D-04
18	8.5	-4.133835482495D-04	-2.066917741247D-04
19	9.5	7.132042272663D-05	3.566021136331D-05
	0	8.815897577478D-01	4.056078562324D-01
	1	1.955168078688D-01	1.055620952467D-01
	2	1.081240905270D-02	4.974647242961D-03

-1.00	2.073137318512D-02
-.75	1.352120319146D-02
-.50	7.662584329388D-03
-.25	3.155516598901D-03
.25	-1.804837972118D-03
.50	-2.284299956798D-03
.75	-1.569261674782D-03
1.25	1.774163621552D-03
1.50	2.826705853591D-03
1.75	2.319516560678D-03
2.25	-3.312548682189D-03
2.50	-5.650895133795D-03
2.75	-4.865419782327D-03
3.25	7.355702833260D-03
3.50	1.273775710109D-02
3.75	1.107835376284D-02
4.25	-1.694347682850D-02
4.50	-2.943265803440D-02
4.75	-2.565766110310D-02
5.25	3.940611266124D-02
5.50	6.865370784764D-02
5.75	6.010923811281D-02
6.25	-9.422518349382D-02
6.50	-1.679717632625D-01
6.75	-1.526063361803D-01
7.25	2.832025924050D-01
7.50	6.198824786741D-01
7.75	8.938805476593D-01
8.25	8.938776931973D-01
8.50	6.198781837925D-01
8.75	2.831992314259D-01
9.25	-1.526018386847D-01
9.50	-1.679642341225D-01

I= 9 H= -.5

0		0.	3.443545080737D-01
1		4.270861090682D-03	-3.503187566909D-02
2		-4.655556544192D-03	-1.325494864893D-03

-1.00	-8.926417634874D-03
-.75	-5.821896374119D-03
-.50	-3.299319681389D-03
-.25	-1.358687556682D-03

CARDINAL SPLINES N=3

0	-9.5	3.846954535103D-04	1.923477267551D-04
1	-8.5	-2.229750516925D-03	-1.114875258462D-03
2	-7.5	6.958446964278D-03	3.479223482139D-03
3	-6.5	-1.736361129434D-02	-8.681805647172D-03
4	-5.5	4.085879436224D-02	2.042939718112D-02
5	-4.5	-9.502017854596D-02	-4.751008927298D-02
6	-3.5	2.183396609755D-01	1.091698304877D-01
7	-2.5	-4.504644434530D-01	-2.252322217265D-01
8	-1.5	7.313237532292D-01	3.656618766146D-01
9	-.5	-8.655753228797D-01	-4.327876614398D-01
10	.5	7.313265505441D-01	3.656632752721D-01
11	1.5	-4.504721445959D-01	-2.252360722980D-01
12	2.5	2.183580652225D-01	1.091790326112D-01
13	3.5	-9.506314372305D-02	-4.753157186152D-02
14	4.5	4.095864106460D-02	2.047932053230D-02
15	5.5	-1.759473314695D-02	-8.797366573474D-03
16	6.5	7.476423134070D-03	3.738211567035D-03
17	7.5	-2.996137743766D-03	-1.498068871883D-03
18	8.5	9.600752453264D-04	4.800376226632D-04
19	9.5	-1.656402958047D-04	-8.282014790233D-05

0	-3.795907977519D-01	-1.080742223392D-01
1	-8.418471324897D-02	-6.021627810205D-02
2	-4.655556544192D-03	-1.325494864893D-03

.25	7.771186678122D-04
.50	9.835631422150D-04
.75	6.756852962812D-04
1.25	-7.639108720038D-04
1.50	-1.217109191970D-03
1.75	-9.987257192868D-04
2.25	1.426297601311D-03
2.50	2.433123187104D-03
2.75	2.094910264440D-03
3.25	-3.167089818182D-03
3.50	-5.484268762015D-03
3.75	-4.769616753851D-03
4.25	7.293257590464D-03
4.50	1.266613331250D-02
4.75	1.103714155229D-02
5.25	-1.691659611415D-02
5.50	-2.940182150478D-02
5.75	-2.563991576149D-02
6.25	3.939453566883D-02
6.50	6.864042512443D-02
6.75	6.010159268326D-02
7.25	-9.422019138918D-02
7.50	-1.679660312828D-01
7.75	-1.526030331124D-01
8.25	2.832004258536D-01
8.50	6.198799809528D-01
8.75	8.938790995732D-01
9.25	8.938786657169D-01
9.50	6.198793281638D-01

I=10 H= .5

0		0.	-7.997547758652D-01
1		-1.838926518602D-03	8.540068053501D-02
2		2.004566814407D-03	-1.325494864893D-03

0	-9.5	-1.656402958047D-04	-8.282014790233D-05
1	-8.5	9.600752453264D-04	4.800376226632D-04
2	-7.5	-2.996137743766D-03	-1.498068871883D-03
3	-6.5	7.476423134070D-03	3.738211567035D-03
4	-5.5	-1.759473314695D-02	-8.797366573474D-03
5	-4.5	4.095864106460D-02	2.047932053230D-02
6	-3.5	-9.506314372305D-02	-4.753157186152D-02
7	-2.5	2.183580652225D-01	1.091790326112D-01
8	-1.5	-4.504721445959D-01	-2.252360722980D-01
9	-.5	7.313265505441D-01	3.656632752721D-01
10	.5	-8.655753228797D-01	-4.327876614398D-01
11	1.5	7.313237532292D-01	3.656618766146D-01
12	2.5	-4.504644434530D-01	-2.252322217265D-01
13	3.5	2.183396609755D-01	1.091698304877D-01
14	4.5	-9.502017854596D-02	-4.751008927298D-02
15	5.5	4.085879436224D-02	2.042939718112D-02
16	6.5	-1.736361129434D-02	-8.681805647172D-03
17	7.5	6.958446964278D-03	3.479223482139D-03
18	8.5	-2.229750516925D-03	-1.114875258462D-03
19	9.5	3.846954535103D-04	1.923477267551D-04

-1.00	3.843493333009D-03
-.75	2.506763722055D-03
-.50	1.420604962903D-03
-.25	5.850170555509D-04
.25	-3.346079618515D-04
.50	-4.234978149432D-04
.75	-2.909332744811D-04
1.25	3.289210059530D-04
1.50	5.240569096618D-04
1.75	4.300264078808D-04
2.25	-6.141280820732D-04
2.50	-1.047641753933D-03
2.75	-9.020153992078D-04
3.25	1.363666011385D-03
3.50	2.361376956843D-03
3.75	2.053658951685D-03
4.25	-3.140201072117D-03
4.50	-5.453426954633D-03
4.75	-4.751869726522D-03
5.25	7.281680251545D-03
5.50	1.265285036142D-02
5.75	1.102949604985D-02
6.25	-1.691160399442D-02
6.50	-2.939608951510D-02
6.75	-2.563661269036D-02

	0	1.634423530735D-01	-1.080742223392D-01	7.25	3,939236911667D-02
	1	3.624784295513D-02	6.021627810205D-02	7.50	6,863792740254D-02
	2	2.004566814407D-03	-1.325494864893D-03	7.75	6,010014459700D-02
				8,25	-9,421921886951D-02
				8.50	-1,679648869115D-01
				8,75	-1,526023495277D-01
				9,25	2,831999150106D-01
				9.50	6,198793281638D-01

I=11 H= 1.5

0		0.	1.857409674754D+00	-1.00	-1,654908710823D-03
1		7.917941440481D-04	-2,000803928630D-01	-.75	-1,079347551847D-03
2		-8,631145667747D-04	4.974647242961D-03	-.50	-6,116757137177D-04
				-.25	-2,518931964354D-04
0	-9,5	7.132042272663D-05	3.566021136331D-05	.25	1,440735244389D-04
1	-8,5	-4.133835482495D-04	-2.066917741247D-04	.50	1,823471935406D-04
2	-7,5	1.290059490240D-03	6.450297451200D-04	.75	1,252683348528D-04
3	-6,5	-3,219157842929D-03	-1.609578921464D-03	1.25	-1,416248681499D-04
4	-5,5	7.575920681238D-03	3.787960340619D-03	1.50	-2,256453349734D-04
5	-4,5	-1,763768327688D-02	-8,818841638442D-03	1.75	-1,851582350874D-04
6	-3,5	4.097704466320D-02	2.048852233160D-02	2.25	2,644276430880D-04
7	-2,5	-9,507084483806D-02	-4.753542241903D-02	2.50	4,510873892263D-04
8	-1.5	2,183608625362D-01	1.091804312681D-01	2.75	3,883844392402D-04
9	-.5	-4,504721445959D-01	-2,252360722980D-01	3,25	-5,871591004200D-04
10	.5	7.313237532292D-01	3,656618766146D-01	3.50	-1,016747227188D-03
11	1.5	-8,655676217342D-01	-4,327838108671D-01	3,75	-8,842515360038D-04
12	2,5	7,313053489180D-01	3,656526744590D-01	4,25	1,352085214574D-03
13	3,5	-4,504214767713D-01	-2,252107383857D-01	4.50	2,348091733711D-03
14	4,5	2,182397793296D-01	1.091198896648D-01	4,75	2,046012723662D-03
15	5.5	-9.478824579645D-02	-4.739412289823D-02	5.25	-3,135208803347D-03
16	6,5	4.032200208863D-02	2.016100104431D-02	5,50	-5,447694867023D-03
17	7,5	-1.616061753746D-02	-8,080308768729D-03	5,75	-4,748566624118D-03
18	8,5	5.178529502685D-03	2.589264751342D-03	6,25	7,279513692957D-03
19	9,5	-8,934449202693D-04	-4.467224601346D-04	6.50	1,265035263530D-02
				6,75	1,102804796224D-02
	0	-7,037404528296D-02	4,056078562324D-01	7,25	-1,691063147448D-02
	1	-1,560738262467D-02	-1,055620952467D-01	7.50	-2,939494514364D-02
	2	-8,631145667747D-04	4.974647242961D-03	7.75	-2,563592910557D-02
				8,25	3,939185827366D-02
				8,50	6,863727461353D-02
				8,75	6,009971074071D-02
				9.25	-9,421878501323D-02
				9,50	-1,679642341225D-01

I=12 H= 2.5

0		0.	-4,313783411478D+00	-1.00	7,125518244059D-04
1		-3,409217428113D-04	4,654302688935D-01	-.75	4,647332280055D-04
2		3.716300815946D-04	-1.236994798983D-02	-.50	2,633683918043D-04
				-.25	1,084573158025D-04
0	-9,5	-3.070833878328D-05	-1,535416939164D-05	.25	-6,203354421526D-05
1	-8,5	1.779900003092D-04	8.899500015458D-05	.50	-7,851298659399D-05
2	-7,5	-5,554591914738D-04	-2,777295957369D-04	.75	-5,393661895014D-05
3	-6,5	1.386068631846D-03	6,930343159229D-04	1.25	6,097922954632D-05
4	-5,5	-3.261957635566D-03	-1.630978817783D-03	1.50	9,715580924501D-05
5	-4,5	7,594317800909D-03	3.797158900454D-03	1.75	7,972333291732D-05

6	-3.5	-1.764538411278D-02	-8.822692056389D-03
7	-2.5	4.097984196513D-02	2.048992098256D-02
8	-1.5	-9.507084483806D-02	-4.753542241903D-02
9	-.5	2.183580652225D-01	1.091790326112D-01
10	.5	-4.504644434530D-01	-2.252322217265D-01
11	1.5	7.313053489180D-01	3.656526744590D-01
12	2.5	-8.655246549169D-01	-4.327623274585D-01
13	3.5	7.312054638093D-01	3.656027319047D-01
14	4.5	-4.501894629397D-01	-2.250947314698D-01
15	5.5	2.177011039390D-01	1.088505519695D-01
16	6.5	-9.354155252048D-02	-4.677077626024D-02
17	7.5	3.752807817922D-02	1.876403908961D-02
18	8.5	-1.202680578299D-02	-6.013402891495D-03
19	9.5	2.074995263516D-03	1.037497631758D-03
	0	3.030085830720D-02	-1.008583663071D+00
	1	6.720049807486D-03	2.304012570868D-01
	2	3.716300815946D-04	-1.236994798983D-02

2.25	-1.138542553614D-04
2.50	-1.942240912525D-04
2.75	-1.672261656309D-04
3.25	2.528122989482D-04
3.50	4.377794665462D-04
3.75	3.807309621512D-04
4.25	-5.821653427907D-04
4.50	-1.011014161315D-03
4.75	-8.809481211957D-04
5.25	1.349918591829D-03
5.50	2.345593965445D-03
5.75	2.044564622592D-03
6.25	-3.134236280652D-03
6.50	-5.446550493771D-03
6.75	-4.747883038774D-03
7.25	7.279002849868D-03
7.50	1.264969984629D-02
7.75	1.102761410598D-02
8.25	-1.691019761828D-02
8.50	-2.939429235481D-02
8.75	-2.563541826273D-02
9.25	3.939117468917D-02
9.50	6.863613024268D-02

I=13 H= 3.5

0		0.	1.001859982909D+01
1		1.467801988932D-04	-1.081266880514D+00
2		-1.600013447118D-04	2.908029105222D-02
0	-9.5	1.322114581859D-05	6.610572909297D-06
1	-8.5	-7.663168513430D-05	-3.831584256715D-05
2	-7.5	2.391469961054D-04	1.195734980527D-04
3	-6.5	-5.967569861644D-04	-2.983784930822D-04
4	-5.5	1.404401021385D-03	7.022005106926D-04
5	-4.5	-3.269655682773D-03	-1.634827841387D-03
6	-3.5	7.597114984845D-03	3.798557492422D-03
7	-2.5	-1.764538411278D-02	-8.822692056389D-03
8	-1.5	4.097704466320D-02	2.048852233160D-02
9	-.5	-9.506314372305D-02	-4.753157186152D-02
10	.5	2.183396609755D-01	1.091698304877D-01
11	1.5	-4.504214767713D-01	-2.252107383857D-01
12	2.5	7.312054638093D-01	3.656027319047D-01
13	3.5	-8.652926337747D-01	-4.326463168874D-01
14	4.5	7.306666018070D-01	3.653333009035D-01
15	5.5	-4.489384001204D-01	-2.244692000602D-01
16	6.5	2.148056982021D-01	1.074028491011D-01
17	7.5	-8.705275851505D-02	-4.352637925753D-02
18	8.5	2.792744137527D-02	1.396372068764D-02
19	9.5	-4.818953609119D-03	-2.409476804559D-03
	0	-1.304570947075D-02	2.371060731667D+00
	1	-2.893245350630D-03	-5.287413505222D-01
	2	-1.600013447118D-04	2.908029105222D-02

-1.00	-3.067815436049D-04
-.75	-2.000859055703D-04
-.50	-1.133904356245D-04
-.25	-4.669513376778D-05
.25	2.670787695402D-05
.50	3.380292407548D-05
.75	2.322183264638D-05
1.25	-2.625395308359D-05
1.50	-4.182939136221D-05
1.75	-3.432402569732D-05
2.25	4.901872818133D-05
2.50	8.362109877123D-05
2.75	7.199743151094D-05
3.25	-1.088456227680D-04
3.50	-1.884812517214D-04
3.75	-1.639196296507D-04
4.25	2.506450352190D-04
4.50	4.352812771973D-04
4.75	3.792827266797D-04
5.25	-5.811927926390D-04
5.50	-1.009869770207D-03
5.75	-8.802645303131D-04
6.25	1.349407747945D-03
6.50	2.344941176336D-03
6.75	2.044130766691D-03
7.25	-3.133802425327D-03
7.50	-5.445897706505D-03
7.75	-4.747372197310D-03
8.25	7.278319267512D-03
8.50	1.264855547915D-02
8.75	1.102664158994D-02

9.25	-1.690874953768D-02
9.50	-2.939179464296D-02

I=14 H= 4.5

0		0.	-2.326678716657D+01	-1.00	1.320328713211D-04
1		-6.317137232318D-05	2.511230680303D+00	-.75	8.611312242873D-05
2		6.886149899794D-05	-6.768633348104D-02	-.50	4.880106091108D-05
				-.25	2.009668676817D-05
0	-9.5	-5.690126674758D-06	-2.845063337379D-06	.25	-1.149455615776D-05
1	-8.5	3.298080224685D-05	1.649040112342D-05	.50	-1.454812787069D-05
2	-7.5	-1.029242639450D-04	-5.146213197249D-05	.75	-9.994229788434D-06
3	-6.5	2.568327205325D-04	1.284163602663D-04	1.25	1.129919606119D-05
4	-5.5	-6.044271738988D-04	-3.022135869494D-04	1.50	1.800256489436D-05
5	-4.5	1.407197026546D-03	7.035985132730D-04	1.75	1.477239997825D-05
6	-3.5	-3.269655682773D-03	-1.634827841387D-03	2.25	-2.109671707773D-05
7	-2.5	7.594317800909D-03	3.797158900454D-03	2.50	-3.598891133890D-05
8	-1.5	-1.763768327688D-02	-8.818841638442D-03	2.75	-3.098630868580D-05
9	-.5	4.095864106460D-02	2.047932053230D-02	3.25	4.684506092388D-05
10	.5	-9.502017854596D-02	-4.751008927298D-02	3.50	8.111870273908D-05
11	1.5	2.182397793296D-01	1.091198896648D-01	3.75	7.054785334831D-05
12	2.5	-4.501894629397D-01	-2.250947314698D-01	4.25	-1.078727983237D-04
13	3.5	7.306666018070D-01	3.653333009035D-01	4.50	-1.873366822313D-04
14	4.5	-8.640411769869D-01	-4.320205884935D-01	4.75	-1.632359834391D-04
15	5.5	7.277611395004D-01	3.638805697502D-01	5.25	2.501341834007D-04
16	6.5	-4.422141299131D-01	-2.211070649566D-01	5.50	4.346284871528D-04
17	7.5	1.997361705722D-01	9.986808528612D-02	5.75	3.788488742799D-04
18	8.5	-6.475631063263D-02	-3.237815531631D-02	6.25	-5.807589459950D-04
19	9.5	1.118797891849D-02	5.593989459246D-03	6.50	-1.009216998580D-03
				6.75	-8.797537026689D-04
	0	5.614622247494D-03	-5.518787300355D+00	7.25	1.348724186919D-03
	1	1.245197108638D-03	1.225190344163D+00	7.50	2.343796846285D-03
	2	6.886149899794D-05	-6.768633348104D-02	7.75	2.043158282986D-03
				8.25	-3.132354394307D-03
				8.50	-5.443400080830D-03
				8.75	-4.745205729331D-03
				9.25	7.275016327826D-03
				9.50	1.264282372273D-02

I=15 H= 5.5

0		0.	5.400983663450D+01	-1.00	-5.671150015374D-05
1		2.713372249934D-05	-5.829490172302D+00	-.75	-3.698779180510D-05
2		-2.957777765440D-05	1.571894615533D-01	-.50	-2.096130566327D-05
				-.25	-8.632041728234D-06
0	-9.5	2.444055155065D-06	1.222027577532D-06	.25	4.937206294046D-06
1	-8.5	-1.416609934314D-05	-7.083049671572D-06	.50	6.248793559664D-06
2	-7.5	4.420860769144D-05	2.210430384572D-05	.75	4.292777688707D-06
3	-6.5	-1.103162320555D-04	-5.515811602774D-05	1.25	-4.853294128571D-06
4	-5.5	2.596169534202D-04	1.298084767101D-04	1.50	-7.732562744098D-06
5	-4.5	-6.044271738988D-04	-3.022135869494D-04	1.75	-6.345124174414D-06
6	-3.5	1.404401021385D-03	7.022005106926D-04	2.25	9.061580361180D-06
7	-2.5	-3.261957635566D-03	-1.630978817783D-03	2.50	1.545815924850D-05
8	-1.5	7.575920681238D-03	3.787960340619D-03	2.75	1.330941327110D-05
9	-.5	-1.759473314695D-02	-8.797366573474D-03	3.25	-2.012115356672D-05
10	.5	4.085879436224D-02	2.042939718112D-02	3.50	-3.484256061826D-05
11	1.5	-9.478824579645D-02	-4.739412289823D-02	3.75	-3.030211003970D-05

CARDINAL SPLINES N=3

```
12     2.5   2.177011039390D-01    1.088505519695D-01     4.25   4.633412994554D-05
13     3.5  -4.489384001204D-01   -2.244692000602D-01     4.50   8.046590346090D-05
14     4.5   7.277611395004D-01    3.638805697502D-01     4.75   7.011403598577D-05
15     5.5  -8.572956756799D-01   -4.286478378399D-01     5.25  -1.074390384430D-04
16     6.5   7.121496598458D-01    3.560748299229D-01     5.50  -1.866840668949D-04
17     7.5  -4.072277880070D-01   -2.036138940035D-01     5.75  -1.627252938843D-04
18     8.5   1.479713665354D-01    7.398568326769D-02     6.25   2.494508354852D-04
19     9.5  -2.589294561013D-02   -1.294647280506D-02     6.50   4.334845276751D-04
                                                          6.75   3.778767136915D-04
         0  -2.411624069566D-03    1.281602890281D+01     7.25  -5.793114103961D-04
         1  -5.348440529343D-04   -2.842890402789D+00     7.50  -1.006720233852D-03
         2  -2.957777765440D-05    1.571894615533D-01     7.75  -8.775879852062D-04
                                                          8.25   1.345422397846D-03
                                                          8.50   2.338067089396D-03
                                                          8.75   2.038168116727D-03
                                                          9.25  -3.124711942018D-03
                                                          9.50  -5.430122542736D-03


I=16      H=  6.5
 0            0.                  -1.248176116481D+02    -1.00   2.409780612574D-05
 1           -1.152964006394D-05   1.347302820432D+01     -.75   1.571682345771D-05
 2            1.256816606180D-05  -3.633456100044D-01     -.50   8.906861547418D-06
                                                          -.25   3.667920394847D-06
 0    -9.5  -1.038525997861D-06   -5.192629989305D-07      .25  -2.097913822667D-06
 1    -8.5   6.019447812236D-06    3.009723906118D-06      .50  -2.655232453952D-06
 2    -7.5  -1.878508687568D-05   -9.392543437842D-06      .75  -1.824083725574D-06
 3    -6.5   4.687548672523D-05    2.343774336262D-05     1.25   2.062257931186D-06
 4    -5.5  -1.103162320555D-04   -5.515811602774D-05     1.50   3.285714491015D-06
 5    -4.5   2.568327205325D-04    1.284163602663D-04     1.75   2.696165183150D-06
 6    -3.5  -5.967569861644D-04   -2.983784930822D-04     2.25  -3.850439613561D-06
 7    -2.5   1.386068631846D-03    6.930343159229D-04     2.50  -6.568468892923D-06
 8    -1.5  -3.219157842929D-03   -1.609578921464D-03     2.75  -5.655425439010D-06
 9     -.5   7.476423134070D-03    3.738211567035D-03     3.25   8.549864778060D-06
10      .5  -1.736361129434D-02   -8.681805647172D-03     3.50   1.480527350577D-05
11     1.5   4.032200208863D-02    2.016100104431D-02     3.75   1.287594881022D-05
12     2.5  -9.354155252048D-02   -4.677077626024D-02     4.25  -1.968826212270D-05
13     3.5   2.148056982021D-01    1.074028491011D-01     4.50  -3.419150852950D-05
14     4.5  -4.422141299131D-01   -2.211070649566D-01     4.75  -2.979280112741D-05
15     5.5   7.121496598458D-01    3.560748299229D-01     5.25   4.565291185974D-05
16     6.5  -8.211651824581D-01   -4.105825912291D-01     5.50   7.932564711130D-05
17     7.5   6.311788736268D-01    3.155894368134D-01     5.75   6.914510406916D-05
18     8.5  -2.874256981406D-01   -1.437128490703D-01     6.25  -1.059964528707D-04
19     9.5   5.802777581634D-02    2.901388790817D-02     6.50  -1.841959042039D-04
                                                          6.75  -1.605670750160D-04
         0   1.024745406470D-03   -2.961578500998D+01     7.25   2.461605423076D-04
         1   2.272655151102D-04    6.569461614239D+00     7.50   4.277747482155D-04
         2   1.256816606180D-05   -3.633456100044D-01     7.75   3.729039623528D-04
                                                          8.25  -5.716956567629D-04
                                                          8.50  -9.934890925138D-04
                                                          8.75  -8.660559747906D-04
                                                          9.25   1.327746146311D-03
                                                          9.50   2.307350885322D-03


I=17      H=  7.5
 0            0.                   2.755265514769D+02    -1.00  -9.657059866669D-06
```

CARDINAL SPLINES N=3

1		4.620438215729D-06	-2.976311274783D+01	-.75	-6.298428340450D-06
2		-5.036621650939D-06	8.032457770523D-01	-.50	-3.569374520600D-06
				-.25	-1.469898407116D-06
0	-9.5	4.161834352101D-07	2.080917176051D-07	.25	8.407271298846D-07
1	-8.5	-2.412259754425D-06	-1.206129877213D-06	.50	1.064069427480D-06
2	-7.5	7.528017596811D-06	3.764008798405D-06	.75	7.309912631788D-07
3	-6.5	-1.878508687568D-05	-9.392543437842D-06	1.25	-8.264382325121D-07
4	-5.5	4.420860769144D-05	2.210430384572D-05	1.50	-1.316731547219D-06
5	-4.5	-1.029242639450D-04	-5.146213197249D-05	1.75	-1.080472987801D-06
6	-3.5	2.391469961054D-04	1.195734980527D-04	2.25	1.543041954406D-06
7	-2.5	-5.554591914738D-04	-2.777295957369D-04	2.50	2.632276855426D-06
8	-1.5	1.290059490240D-03	6.450297451200D-04	2.75	2.266379841843D-06
9	-.5	-2.996137743766D-03	-1.498068871883D-03	3.25	-3.426310079134D-06
10	.5	6.958446964278D-03	3.479223482139D-03	3.50	-5.933129839352D-06
11	1.5	-1.616061753746D-02	-8.080308768729D-03	3.75	-5.159963851131D-06
12	2.5	3.752807817922D-02	1.876403908961D-02	4.25	7.889959982124D-06
13	3.5	-8.705275851505D-02	-4.352637925753D-02	4.50	1.370205416528D-05
14	4.5	1.997361705722D-01	9.986808528612D-02	4.75	1.193929698742D-05
15	5.5	-4.072277880070D-01	-2.036138940035D-01	5.25	-1.829514689443D-05
16	6.5	6.311788736268D-01	3.155894368134D-01	5.50	-3.178930559392D-05
17	7.5	-6.397043821983D-01	-3.198521910992D-01	5.75	-2.770951039837D-05
18	8.5	3.626945953942D-01	1.813472976971D-01	6.25	4.247748053804D-05
19	9.5	-8.585625922823D-02	-4.292812961411D-02	6.50	7.381546950106D-05
				6.75	6.434634947557D-05
	0	-4.106609409479D-04	6.526991175144D+01	7.25	-9.864744841283D-05
	1	-9.107537315212D-05	-1.450144298384D+01	7.50	-1.714283162868D-04
	2	-5.036621650939D-06	8.032457770523D-01	7.75	-1.494391578941D-04
				8.25	2.291038024959D-04
				8.50	3.981351276637D-04
				8.75	3.470670132284D-04
				9.25	-5.320866167415D-04
				9.50	-9.246573457606D-04

I=18 H= 8.5

0		0.	-3.067169349510D+02	-1.00	3.094484910916D-06
1		-1.480562048677D-06	3.367658375789D+01	-.75	2.018253146517D-06
2		1.613922862239D-06	-9.228196099756D-01	-.50	1.143761739898D-06
				-.25	4.710106910592D-07
0	-9.5	-1.333608135615D-07	-6.668040678074D-08	.25	-2.694005684489D-07
1	-8.5	7.729786823672D-07	3.864893411836D-07	.50	-3.409678342027D-07
2	-7.5	-2.412259754425D-06	-1.206129877213D-06	.75	-2.342370726856D-07
3	-6.5	6.019447812236D-06	3.009723906118D-06	1.25	2.648218687283D-07
4	-5.5	-1.416609934314D-05	-7.083049671572D-06	1.50	4.219302728627D-07
5	-4.5	3.298080224685D-05	1.649040112342D-05	1.75	3.462241514046D-07
6	-3.5	-7.663168513430D-05	-3.831584256715D-05	2.25	-4.944486324766D-07
7	-2.5	1.779900003092D-04	8.899500015458D-05	2.50	-8.434804301650D-07
8	-1.5	-4.133835482495D-04	-2.066917741247D-04	2.75	-7.262332759468D-07
9	-.5	9.600752453264D-04	4.800376226632D-04	3.25	1.097918516234D-06
10	.5	-2.229750516925D-03	-1.114875258462D-03	3.50	1.901197778192D-06
11	1.5	5.178529502685D-03	2.589264751342D-03	3.75	1.653446338601D-06
12	2.5	-1.202680578299D-02	-6.013402891495D-03	4.25	-2.528239697123D-06
13	3.5	2.792744137527D-02	1.396372068764D-02	4.50	-4.390653102332D-06
14	4.5	-6.475631063263D-02	-3.237815531631D-02	4.75	-3.825799454966D-06
15	5.5	1.479713665354D-01	7.398568326769D-02	5.25	5.862452629399D-06
16	6.5	-2.874256981406D-01	-1.437128490703D-01	5.50	1.018648821137D-05
17	7.5	3.626945953942D-01	1.813472976971D-01	5.75	8.879168504662D-06

18	8.5	-2.424626222556D-01	-1.212313111278D-01
19	9.5	6.445814300012D-02	3.222907150006D-02
	0	1.315911988546D-04	-7.007385905131D+01
	1	2.918397233386D-05	1.614301116835D+01
	2	1.613922862239D-06	-9.228196099756D-01

6.25	-1,361138114388D-05
6.50	-2,365325054456D-05
6,75	-2,061898862117D-05
7.25	3,161034980953D-05
7,50	5,493207509057D-05
7,75	4,788592233312D-05
8,25	-7,341346821889D-05
8.50	-1,275774568071D-04
8.75	-1,112133133303D-04
9.25	1,705005433735D-04
9,50	2,962949469539D-04

I=19 H= 9.5

0		0.	1.187617637473D+02
1		2.554390746394D-07	-1.331238544154D+01
2		-2.784476090265D-07	3.730568194149D-01
0	-9.5	2.300853438715D-08	1.150426719358D-08
1	-8,5	-1.333608135615D-07	-6.668040678074D-08
2	-7.5	4.161834352101D-07	2.080917176051D-07
3	-6,5	-1,038525997861D-06	-5.192629989305D-07
4	-5,5	2.444055155065D-06	1.222027577532D-06
5	-4,5	-5,690126674758D-06	-2.845063337379D-06
6	-3,5	1,322114581859D-05	6.610572909297D-06
7	-2,5	-3,070833878328D-05	-1.535416939164D-05
8	-1.5	7.132042272663D-05	3.566021136331D-05
9	-,5	-1,656402958047D-04	-8,282014790233D-05
10	,5	3,846954535103D-04	1,923477267551D-04
11	1,5	-8,934449202693D-04	-4.467224601346D-04
12	2,5	2,074995263516D-03	1,037497631758D-03
13	3,5	-4,818953609119D-03	-2.409476804559D-03
14	4,5	1,118797891849D-02	5,593989459246D-03
15	5,5	-2,589294561013D-02	-1.294647280506D-02
16	6,5	5,802777581634D-02	2,901388790817D-02
17	7,5	-8,585625922823D-02	-4,292812961411D-02
18	8,5	6,445814300012D-02	3,222907150006D-02
19	9,5	-1,855619925183D-02	-9,278099625917D-03
	0	-2,270322550557D-05	2,596248000491D+01
	1	-5,035065496865D-06	-6,224305872656D+00
	2	-2,784476090265D-07	3,730568194149D-01

-1.00	-5,338866836659D-07
-.75	-3,482060860570D-07
-.50	-1,973314395763D-07
-.25	-8,126274422401D-08
.25	4,647926236755D-08
.50	5,882665176266D-08
.75	4,041255896469D-08
1.25	-4,568930640406D-08
1.50	-7,279497577199D-08
1.75	-5,973351601955D-08
2.25	8,530645591610D-08
2.50	1,455243707957D-07
2.75	1,252959010708D-07
3.25	-1,894221792777D-07
3.50	-3,280107048548D-07
3.75	-2,852665331220D-07
4.25	4,361932748963D-07
4.50	7,575125720156D-07
4,75	6,600592480439D-07
5,25	-1,011439862388D-06
5.50	-1,757459016905D-06
5,75	-1,531909174912D-06
6.25	2,348350484230D-06
6.50	4,080858641941D-06
6.75	3,557362136922D-06
7.25	-5,453684640424D-06
7.50	-9,477345742264D-06
7.75	-8,261683931776D-06
8.25	1,266591183901D-05
8.50	2,201074079828D-05
8.75	1,918746048900D-05
9.25	-2,941619427394D-05
9,50	-5,111930743934D-05

CARDINAL SPLINES N=3

N= 3 M=20 P= 10.0 Q= 5

I= 0	H=-10.0				
0		1.000000000000D+00	5,000228786775D-01	-1.00	3,473671635303D+00
1		-1,727557718026D+00	-8,637812018735D-01	-.75	2,715357366988D+00
2		7.461139172775D-01	3.730570185851D-01	-.50	2,050307338332D+00
				-.25	1,478521549336D+00
0	-10.0	-1.855619925184D-02	-9.278099625920D-03	.25	6,147245690351D-01
1	-9,0	6,445814300015D-02	3.222907150007D-02	.50	3,221697390799D-01
2	-8,0	-8,585625922832D-02	-4,292812961416D-02	.75	1,196173168847D-01
3	-7,0	5,802777581658D-02	2,901388790829D-02	1,25	-5,021023041781D-02
4	-6,0	-2,589294561069D-02	-1,294647280534D-02	1,50	-5,147708426400D-02
5	-5,0	1,118797891980D-02	5,593989459900D-03	1,75	-2,752040839487D-02
6	-4,0	-4,818953612158D-03	-2,409476806079D-03	2,25	1,677968342759D-02
7	-3,0	2,074995270576D-03	1,037497635288D-03	2,50	1,898487003932D-02
8	-2,0	-8,934449366643D-04	-4,467224683322D-04	2,75	1,083411537087D-02
9	-1,0	3,846954915874D-04	1,923477457937D-04	3,25	-7,016349048910D-03
10	0,0	-1,656403842379D-04	-8,282019211893D-05	3,50	-8,037375405050D-03
11	1,0	7,132062811043D-05	3,566031405521D-05	3,75	-4,621138963126D-03
12	2,0	-3,070881578163D-05	-1,535440789082D-05	4,25	3,012078378809D-03
13	3,0	1,322225363094D-05	6,611126815469D-06	4,50	3,454789720468D-03
14	4,0	-5,692699449920D-06	-2,846349724960D-06	4,75	1,987862446116D-03
15	5,0	2,450028268027D-06	1,225014134014D-06	5,25	-1,296539320422D-03
16	6,0	-1,052349896724D-06	-5,261749483618D-07	5,50	-1,487292757118D-03
17	7,0	4,471636920014D-07	2,235818460007D-07	5,75	-8,558432803035D-04
18	8,0	-1,791983270729D-07	-8,959916353643D-08	6,25	5,582411736782D-04
19	9,0	5,742187858877D-08	2,871093929438D-08	6,50	6,403806261834D-04
20	10,0	-9,906907680009D-09	-4,953453840005D-09	6,75	3,685015146580D-04
				7,25	-2,403641678903D-04
	0	5,833581454750D+01	2,916791271845D+01	7,50	-2,757316380703D-04
	1	1,319472062752D+01	6,597359169829D+00	7,75	-1,586675169677D-04
	2	7,461139172775D-01	3,730570185851D-01	8,25	1,034948545533D-04
				8,50	1,187232263338D-04
				8,75	6,831831550356D-05
				9,25	-4,456233283977D-05
				9,50	-5,111929568812D-05
				9,75	-2,941618403002D-05

I= 1	H= -9.0				
0		0.	-1,326081441401D-04	-1,00	-4,626823524749D+00
1		2.781182690874D+00	1.390604924998D+00	-.75	-3,124059987210D+00
2		-1.845640833874D+00	-9.228207643949D-01	-.50	-1,852001553906D+00
				-.25	-8,106482248357D-01
0	-10.0	6.445814300015D-02	3.222907150007D-02	.25	5,800060680067D-01
1	-9,0	-2.424626222558D-01	-1.212313111279D-01	.50	9,311954539373D-01
2	-8,0	3.626945953948D-01	1.813472976974D-01	.75	1,063010268583D+00
3	-7,0	-2,874256981420D-01	-1.437128490710D-01	1,25	7,891384222395D-01
4	-6,0	1.479713665386D-01	7.398568326932D-02	1,50	5,009842265559D-01
5	-5,0	-6.475631064021D-02	-3.237815532010D-02	1,75	2,152141789179D-01

CARDINAL SPLINES N=3

6	-4.0	2.792744139289D-02	1.396372069644D-02	2.25	-1.084973407896D-01
7	-3.0	-1.202680582391D-02	-6.013402911953D-03	2.50	-1.174242888195D-01
8	-2.0	5.178529597713D-03	2.589264798856D-03	2.75	-6.515456765226D-02
9	-1.0	-2.229750737626D-03	-1.114875368813D-03	3.25	4.115217610219D-02
10	0.0	9.600757578980D-04	4.800378789490D-04	3.50	4.690407300616D-02
11	1.0	-4.133847386839D-04	-2.066923693419D-04	3.75	2.688643470990D-02
12	2.0	1.779927650610D-04	8.899638253052D-05	4.25	-1.747932027673D-02
13	3.0	-7.663810617585D-05	-3.831905308793D-05	4.50	-2.003818046305D-02
14	4.0	3.299571442609D-05	1.649785721304D-05	4.75	-1.152632035715D-02
15	5.0	-1.420072037510D-05	-7.100360187548D-06	5.25	7.515830240314D-03
16	6.0	6.099573141728D-06	3.049786570864D-06	5.50	8.621156384436D-03
17	7.0	-2.591825831104D-06	-1.295912915552D-06	5.75	4.960780973323D-03
18	8.0	1.038659581057D-06	5.193297905284D-07	6.25	-3.235685851319D-03
19	9.0	-3.328255644611D-07	-1.664127822305D-07	6.50	-3.711764774493D-03
20	10.0	5.742187858877D-08	2.871093929438D-08	6.75	-2.135896539689D-03
				7.25	1.393187366362D-03
	0	-1.567522564787D+02	-7.837615979765D+01	7.50	1.598181797242D-03
	1	-3.413163398661D+01	-1.706581036290D+01	7.75	9.196603686105D-04
	2	-1.845640833874D+00	-9.228207643949D-01	8.25	-5.998713098056D-04
				8.50	-6.881371440162D-04
				8.75	-3.959829136033D-04
				9.25	2.582897690534D-04
				9.50	2.962948788421D-04
				9.75	1.705004839983D-04

I= 2 H= -8.0

0		0.	4.138345552280D-04	-1.00	3.127136922227D+00
1		-1.520640331499D+00	-7.603625439285D-01	-.75	2.044134580909D+00
2		1.606496590727D+00	8.032493796863D-01	-.50	1.161944313431D+00
				-.25	4.805661197952D-01
0	-10.0	-8.585625922832D-02	-4.292812961416D-02	.25	-2.798378899575D-01
1	-9.0	3.626945953948D-01	1.813472976974D-01	.50	-3.613790261686D-01
2	-8.0	-6.397043822000D-01	-3.198521911000D-01	.75	-2.572000091062D-01
3	-7.0	6.311788736312D-01	3.155894368156D-01	1.25	3.476921927265D-01
4	-6.0	-4.072277880172D-01	-2.036138940086D-01	1.50	6.930200694791D-01
5	-5.0	1.997361705959D-01	9.986808529795D-02	1.75	9.356781956107D-01
6	-4.0	-8.705275857003D-02	-4.352637928501D-02	2.25	8.667750562601D-01
7	-3.0	3.752807830691D-02	1.876403915346D-02	2.50	5.888238698682D-01
8	-2.0	-1.616061783401D-02	-8.080308917007D-03	2.75	2.653417187858D-01
9	-1.0	6.958447653026D-03	3.479223826513D-03	3.25	-1.409607496271D-01
10	0.0	-2.996139343365D-03	-1.498069671682D-03	3.50	-1.546118064210D-01
11	1.0	1.290063205267D-03	6.450316026337D-04	3.75	-8.653576211815D-02
12	2.0	-5.554678195251D-04	-2.777339097625D-04	4.25	5.508853120390D-02
13	3.0	2.391670344590D-04	1.195835172295D-04	4.50	6.288877535831D-02
14	4.0	-1.029708008685D-04	-5.148540043425D-05	4.75	3.608392343993D-02
15	5.0	4.431665067366D-05	2.215832533683D-05	5.25	-2.347817890580D-02
16	6.0	-1.903513660157D-05	-9.517568300785D-06	5.50	-2.691962158088D-02
17	7.0	8.088395301801D-06	4.044197650901D-06	5.75	-1.548615618083D-02
18	8.0	-3.241378789721D-06	-1.620689394861D-06	6.25	1.009871363652D-02
19	9.0	1.038659581057D-06	5.193297905284D-07	6.50	1.158408453280D-02
20	10.0	-1.791983270729D-07	-8.959916353643D-08	6.75	6.665772452289D-03
				7.25	-4.347808440949D-03
	0	1.454432557578D+02	7.272172636390D+01	7.50	-4.987526409471D-03
	1	3.060929148305D+01	1.530462504980D+01	7.75	-2.870023144549D-03
	2	1.606496590727D+00	8.032493796863D-01	8.25	1.872039801366D-03
				8.50	2.147493226405D-03

CARDINAL SPLINES N=3

8,75	1,235757168961D-03
9,25	-8,060533765824D-04
9,50	-9,246571332019D-04
9,75	-5,320864314472D-04

I= 3 H= -7.0

0		0.	-1.032664705169D-03
1		6.686760123611D-01	3,344437548405D-01
2		-7,267037881777D-01	-3,633545998601D-01
0	-10,0	5,802777581658D-02	2,901388790829D-02
1	-9,0	-2,874256981420D-01	-1,437128490710D-01
2	-8,0	6,311788736312D-01	3.155894368156D-01
3	-7,0	-8,211651824689D-01	-4,105825912344D-01
4	-6,0	7,121496598712D-01	3,560748299356D-01
5	-5,0	-4,422141299722D-01	-2,211070649861D-01
6	-4,0	2,148056983393D-01	1.074028491697D-01
7	-3,0	-9,354155283911D-02	-4,677077641956D-02
8	-2,0	4,032200282864D-02	2,016100141432D-02
9	-1,0	-1,736361301301D-02	-8,681806506507D-03
10	0,0	7,476427125640D-03	3,738213562820D-03
11	1,0	-3,219167113246D-03	-1,609583556623D-03
12	2,0	1,386090161909D-03	6,930450809543D-04
13	3,0	-5,968069889981D-04	-2,984034944991D-04
14	4,0	2,569488467424D-04	1,284744233712D-04
15	5,0	-1,105858378024D-04	-5,529291890121D-05
16	6,0	4,749944990858D-05	2,374972495429D-05
17	7,0	-2,018342896722D-05	-1,009171448361D-05
18	8,0	8,088395301801D-06	4,044197650901D-06
19	9,0	-2,591825831104D-06	-1,295912915552D-06
20	10,0	4,471636920014D-07	2,235818460007D-07
	0	-6,598361869416D+01	-3,299205510231D+01
	1	-1,386539975119D+01	-6,932648242361D+00
	2	-7,267037881777D-01	-3,633545998601D-01

-1,00	-1,395379800539D+00
-,75	-9,102778901208D-01
-,50	-5,160139532250D-01
-,25	-2,125879898514D-01
,25	1,218066840790D-01
,50	1,544754271304D-01
,75	1,065063916274D-01
1,25	-1,228236245444D-01
1,50	-2,004031353179D-01
1,75	-1,711399238998D-01
2,25	2,952198251367D-01
2,50	6,336517814152D-01
2,75	9,017984334146D-01
3,25	8,887161404305D-01
3,50	6,139578432557D-01
3,75	2,797926530606D-01
4,25	-1,503799311388D-01
4,50	-1,654154069158D-01
4,75	-9,275208017089D-02
5,25	5,914298718572D-02
5,50	6,753974345640D-02
5,75	3,876026256415D-02
6,25	-2,522387549572D-02
6,50	-2,892217948861D-02
6,75	-1,663851087959D-02
7,25	1,085036531047D-02
7,50	1,244633513580D-02
7,75	7,161947517748D-03
8,25	-4,671451044560D-03
8,50	-5,358790233473D-03
8,75	-3,083663905777D-03
9,25	2,011392370395D-03
9,50	2,307350354912D-03
9,75	1,327745683936D-03

I= 4 H= -6.0

0		0.	2.430261256165D-03
1		-2,885155552803D-01	-1.445066453218D-01
2		3,144085008910D-01	1.572106181764D-01
0	-10.0	-2,589294561069D-02	-1.294647280534D-02
1	-9,0	1.479713665386D-01	7.398568326932D-02
2	-8,0	-4,072277880172D-01	-2.036138940086D-01
3	-7,0	7,121496598712D-01	3.560748299356D-01
4	-6,0	-8,572956757396D-01	-4,286478378698D-01
5	-5,0	7,277611396393D-01	3.638805698196D-01
6	-4,0	-4.489384004433D-01	-2,244692002216D-01
7	-3,0	2,177011046889D-01	1.088505523444D-01
8	-2.0	-9.478824753799D-02	-4.739412376900D-02

-1,00	6,029240561713D-01
-,75	3,932414482114D-01
-,50	2,228599028629D-01
-,25	9,177942012577D-02
,25	-5,250364359409D-02
,50	-6,646480696774D-02
,75	-4,567640207565D-02
1,25	5,174434277338D-02
1,50	8,264534355746D-02
1.75	6,810496965031D-02
2,25	-9,940959322968D-02
2,50	-1,739120311486D-01
2,75	-1,560222181540D-01

CARDINAL SPLINES N=3

9	-1.0	4.085879840694D-02	2.042939920347D-02
10	0.0	-1.759474254066D-02	-8.797371270331D-03
11	1.0	7.575942497897D-03	3.787971248948D-03
12	2.0	-3.262008304169D-03	-1.631004152084D-03
13	3.0	1.404518697480D-03	7.022593487399D-04
14	4.0	-6.047004639873D-04	-3.023502319937D-04
15	5.0	2.602514404885D-04	1.301257202442D-04
16	6.0	-1.117846598445D-04	-5.589232992223D-05
17	7.0	4.749944990858D-05	2.374972495429D-05
18	8.0	-1.903513660157D-05	-9.517568300785D-06
19	9.0	6.099573141728D-06	3.049786570864D-06
20	10.0	-1.052349896724D-06	-5.261749483618D-07
	0	2.855569453630D+01	1.427842562568D+01
	1	5.999654462540D+00	2.999705718207D+00
	2	3.144085008910D-01	1.572106181764D-01

3.25	2.854293533994D-01
3.50	6.224365901242D-01
3.75	8.953501894487D-01
4.25	8.929191337599D-01
4.50	6.187785870501D-01
4.75	2.825664761671D-01
5.25	-1.521890959217D-01
5.50	-1.674907451944D-01
5.75	-9.394630660349D-02
6.25	5.992194565379D-02
6.50	6.843331766507D-02
6.75	3.927446215007D-02
7.25	-2.555927482474D-02
7.50	-2.930692988908D-02
7.75	-1.685991236467D-02
8.25	1.099477999656D-02
8.50	1.261199922602D-02
8.75	7.257277600931D-03
9.25	-4.733632555089D-03
9.50	-5.430121294475D-03
9.75	-3.124710853869D-03

I= 5 H= -5.0

0		0.	-5.658012410924D-03
1		1.242535495571D-01	6.270617603743D-02
2		-1.354415284769D-01	-6.773558927037D-02
0	-10.0	1.118797891980D-02	5.593989459900D-03
1	-9.0	-6.475631064021D-02	-3.237815532010D-02
2	-8.0	1.997361705959D-01	9.986808529795D-02
3	-7.0	-4.422141299722D-01	-2.211070649861D-01
4	-6.0	7.277611396393D-01	3.638805698196D-01
5	-5.0	-8.640411773104D-01	-4.320205886552D-01
6	-4.0	7.306666025586D-01	3.653333012793D-01
7	-3.0	-4.501894646855D-01	-2.250947323427D-01
8	-2.0	2.182397833842D-01	1.091198916921D-01
9	-1.0	-9.502018796262D-02	-4.751009398131D-02
10	0.0	4.095866293457D-02	2.047933146729D-02
11	1.0	-1.763773406934D-02	-8.818867034668D-03
12	2.0	7.594435765009D-03	3.797217882504D-03
13	3.0	-3.269929650357D-03	-1.634964825178D-03
14	4.0	1.407833286828D-03	7.039166434141D-04
15	5.0	-6.059043548799D-04	-3.029521774399D-04
16	6.0	2.602514404885D-04	1.301257202442D-04
17	7.0	-1.105858378024D-04	-5.529291890121D-05
18	8.0	4.431665067366D-05	2.215832533683D-05
19	9.0	-1.420072037510D-05	-7.100360187548D-06
20	10.0	2.450028268027D-06	1.225014134014D-06
	0	-1.230161735212D+01	-6.152155179074D+00
	1	-2.584577019981D+00	-1.292005609370D+00
	2	-1.354415284769D-01	-6.773558927037D-02

-1.00	-2.596950780340D-01
-.75	-1.693760219361D-01
-.50	-9.598715689778D-02
-.25	-3.952848291908D-02
.25	2.260921762013D-02
.50	2.861601700057D-02
.75	1.965926224089D-02
1.25	-2.223068780881D-02
1.50	-3.542803452265D-02
1.75	-2.908368581809D-02
2.25	4.162746797905D-02
2.50	7.119890870594D-02
2.75	6.157282086456D-02
3.25	-9.517926759960D-02
3.50	-1.690661041649D-01
3.75	-1.532360223032D-01
4.25	2.836132988812D-01
4.50	6.203536142847D-01
4.75	8.941516593526D-01
5.25	8.937008485166D-01
5.50	6.196753116505D-01
5.75	2.830824846852D-01
6.25	-1.525256729791D-01
6.50	-1.678768461219D-01
6.75	-9.416848506882D-02
7.25	6.006686705020D-02
7.50	6.859956300235D-02
7.75	3.937012669984D-02
8.25	-2.562167449541D-02
8.50	-2.937851120968D-02
8.75	-1.690110332231D-02
9.25	1.102164800831D-02
9.50	1.264282081659D-02

CARDINAL SPLINES N=3

9.75	7.275013794453D-03

I= 6 H= -4.0

0		0.	1.314652755629D-02
1		-5.350162987377D-02	-2.809706766006D-02
2		5.832058348593D-02	2.919473806023D-02
0	-10.0	-4.818953612158D-03	-2.409476806079D-03
1	-9.0	2.792744139289D-02	1.396372069644D-02
2	-8.0	-8.705275857003D-02	-4.352637928501D-02
3	-7.0	2.148056983393D-01	1.074028491697D-01
4	-6.0	-4.489384004433D-01	-2.244692002216D-01
5	-5.0	7.306666025586D-01	3.653333012793D-01
6	-4.0	-8.652926355212D-01	-4.326463177606D-01
7	-3.0	7.312054678657D-01	3.656027339328D-01
8	-2.0	-4.504214861922D-01	-2.252107430961D-01
9	-1.0	2.183396828553D-01	1.091698414277D-01
10	0.0	-9.506319453846D-02	-4.753159726923D-02
11	1.0	4.097716268068D-02	2.048858134034D-02
12	2.0	-1.764565820519D-02	-8.822829102597D-03
13	3.0	7.597751555097D-03	3.798875777548D-03
14	4.0	-3.271134048801D-03	-1.635567024400D-03
15	5.0	1.407833286828D-03	7.039166434141D-04
16	6.0	-6.047004639873D-04	-3.023502319937D-04
17	7.0	2.569488467424D-04	1.284744233712D-04
18	8.0	-1.029708008685D-04	-5.148540043425D-05
19	9.0	3.299571442609D-05	1.649785721304D-05
20	10.0	-5.692699449920D-06	-2.846349724960D-06
	0	5.297042049855D+00	2.651649656979D+00
	1	1.112910039845D+00	5.557976935446D-01
	2	5.832058348593D-02	2.919473806023D-02

-1.00	1.118222133597D-01
-.75	7.293155061616D-02
-.50	4.133096080837D-02
-.25	1.702044393631D-02
.25	-9.735077009959D-03
.50	-1.232126136578D-02
.75	-8.464454475503D-03
1.25	9.569867912557D-03
1.50	1.524767158389D-02
1.75	1.251234763379D-02
2.25	-1.787308660961D-02
2.50	-3.049775721276D-02
2.75	-2.627011937836D-02
3.25	3.980535666064D-02
3.50	6.911164186490D-02
3.75	6.037273381394D-02
4.25	-9.439704565930D-02
4.50	-1.681689122603D-01
4.75	-1.527197841940D-01
5.25	2.832765939764D-01
5.50	6.199673716331D-01
5.75	8.939294010277D-01
6.25	8.938458210593D-01
6.50	6.198416154470D-01
6.75	2.831781828019D-01
7.25	-1.525880945040D-01
7.50	-1.679484525032D-01
7.75	-9.420969044425D-02
8.25	6.009374444657D-02
8.50	6.863039537866D-02
8.75	3.938786909979D-02
9.25	-2.563324783397D-02
9.50	-2.939178789048D-02
9.75	-1.690874365132D-02

I= 7 H= -3.0

0		0.	-3.053502529778D-02
1		2.303653087611D-02	1.464516421670D-02
2		-2.511152614668D-02	-1.263577044903D-02
0	-10.0	2.074995270576D-03	1.037497635288D-03
1	-9.0	-1.202680582391D-02	-6.013402911953D-03
2	-8.0	3.752807830691D-02	1.876403915346D-02
3	-7.0	-9.354155283911D-02	-4.677077641956D-02
4	-6.0	2.177011046889D-01	1.088505523444D-01
5	-5.0	-4.501894646855D-01	-2.250947323427D-01
6	-4.0	7.312054678657D-01	3.656027339328D-01
7	-3.0	-8.655246643385D-01	-4.327623321692D-01
8	-2.0	7.313053707996D-01	3.656526853998D-01
9	-1.0	-4.504644942726D-01	-2.252322471363D-01
10	0.0	2.183581832498D-01	1.091790916249D-01
11	1.0	-9.507111895354D-02	-4.753555947677D-02

-1.00	-4.814805702279D-02
-.75	-3.140263161459D-02
-.50	-1.779614697472D-02
-.25	-7.328603103194D-03
.25	4.191688697428D-03
.50	5.305227503588D-03
.75	3.644570803819D-03
1.25	-4.120457908740D-03
1.50	-6.564979861938D-03
1.75	-5.387061492828D-03
2.25	7.693526567357D-03
2.50	1.312474138747D-02
2.75	1.130085286610D-02
3.25	-1.708850291848D-02
3.50	-2.959900008671D-02
3.75	-2.575337337503D-02

12	2.0	4.098047859095D-02	2.049023929547D-02	4.25	3.946853936470D-02
13	3.0	-1.764686265444D-02	-8.823431327222D-03	4.50	6.872531958858D-02
14	4.0	7.597751555097D-03	3.798875777548D-03	4.75	6.015044659782D-02
15	5.0	-3.269929650357D-03	-1.634964825178D-03	5.25	-9.425206370525D-02
16	6.0	1.404518697480D-03	7.022593487399D-04	5.50	-1.680025997924D-01
17	7.0	-5.968069889981D-04	-2.984034944991D-04	5.75	-1.526240818170D-01
18	8.0	2.391670344590D-04	1.195835172295D-04	6.25	2.832141700792D-01
19	9.0	-7.663810617585D-05	-3.831905308793D-05	6.50	6.198957626220D-01
20	10.0	1.322225363094D-05	6.611126815469D-06	6.75	8.938881941703D-01
				7.25	8.938726994227D-01
	0	-2.280787305907D+00	-1.147660428034D+00	7.50	6.198724489081D-01
	1	-4.791939920575D-01	-2.380702447639D-01	7.75	2.831959258247D-01
	2	-2.511152614668D-02	-1.263577044903D-02	8.25	-1.525996682481D-01
				8.50	-1.679617296490D-01
				8.75	-9.421733104080D-02
				9.25	6.009872919633D-02
				9.50	6.863611455890D-02
				9.75	3.939116101712D-02

I= 8 H= -2.0

0		0.	7.091790045467D-02	-1.00	2.073137356555D-02
1		-9.918964314444D-03	-1.222173572190D-02	-.75	1.352120343958D-02
2		1.081240925111D-02	5.592022552651D-03	-.50	7.662584469999D-03
				-.25	3.155516656805D-03
0	-10.0	-8.934449366643D-04	-4.467224683322D-04	.25	-1.804838005238D-03
1	-9.0	5.178529597713D-03	2.589264798856D-03	.50	-2.284299998716D-03
2	-8.0	-1.616061783401D-02	-8.080308917007D-03	.75	-1.569261703578D-03
3	-7.0	4.032200282864D-02	2.016100141432D-02	1.25	1.774163654108D-03
4	-6.0	-9.478824753799D-02	-4.739412376900D-02	1.50	2.826705905462D-03
5	-5.0	2.182397833842D-01	1.091198916921D-01	1.75	2.319516603241D-03
6	-4.0	-4.504214861922D-01	-2.252107430961D-01	2.25	-3.312548742975D-03
7	-3.0	7.313053707996D-01	3.656526853998D-01	2.50	-5.650895237491D-03
8	-2.0	-8.655676725545D-01	-4.327838362773D-01	2.75	-4.865419871609D-03
9	-1.0	7.313238712584D-01	3.656619356292D-01	3.25	7.355702968235D-03
10	0.0	-4.504724187156D-01	-2.252362093578D-01	3.50	1.273775733482D-02
11	1.0	2.183614991721D-01	1.091807495861D-01	3.75	1.107835396611D-02
12	2.0	-9.507232340787D-02	-4.753616170394D-02	4.25	-1.694347713932D-02
13	3.0	4.098047859095D-02	2.049023929547D-02	4.50	-2.943265857418D-02
14	4.0	-1.764565820519D-02	-8.822829102597D-03	4.75	-2.565766157343D-02
15	5.0	7.594435765009D-03	3.797217882504D-03	5.25	3.940611338196D-02
16	6.0	-3.262008304169D-03	-1.631004152084D-03	5.50	6.865370909994D-02
17	7.0	1.386090161909D-03	6.930450809543D-04	5.75	6.010923920439D-02
18	8.0	-5.554678195251D-04	-2.777339097625D-04	6.25	-9.422518516716D-02
19	9.0	1.779927650610D-04	8.899638253052D-05	6.50	-1.679717661704D-01
20	10.0	-3.070881578163D-05	-1.535440789082D-05	6.75	-1.526063387151D-01
				7.25	2.832025962911D-01
	0	9.820512819664D-01	5.079027985008D-01	7.50	6.198824854274D-01
	1	2.063292207077D-01	9.961871533113D-02	7.75	8.938805535463D-01
	2	1.081240925111D-02	5.592022552651D-03	8.25	8.938776841721D-01
				8.50	6.198781681085D-01
				8.75	2.831992177536D-01
				9.25	-1.526018177238D-01
				9.50	-1.679641976968D-01
				9.75	-9.421875325976D-02

CARDINAL SPLINES N=3

I= 9	H= -1.0				
0		0.	-1.647054462209D-01	-1.00	-8.926418518410D-03
1		4.270861513411D-03	1.900187359089D-02	-.75	-5.821896950370D-03
2		-4.655557004999D-03	-2.759337028657D-03	-.50	-3.299320007955D-03
				-.25	-1.358687691165D-03
0	-10.0	3.846954915874D-04	1.923477457937D-04	.25	7.771187447314D-04
1	-9.0	-2.229750737626D-03	-1.114875368813D-03	.50	9.835632395681D-04
2	-8.0	6.958447653026D-03	3.479223826513D-03	.75	6.756853631605D-04
3	-7.0	-1.736361301301D-02	-8.681806506507D-03	1.25	-7.639109476156D-04
4	-6.0	4.085879840694D-02	2.042939920347D-02	1.50	-1.217109312439D-03
5	-5.0	-9.502018796262D-02	-4.751009398131D-02	1.75	-9.987258181406D-04
6	-4.0	2.183396828553D-01	1.091698414277D-01	2.25	1.426297742486D-03
7	-3.0	-4.504644942726D-01	-2.252322471363D-01	2.50	2.433123427935D-03
8	-2.0	7.313238712584D-01	3.656619356292D-01	2.75	2.094910471793D-03
9	-1.0	-8.655755970000D-01	-4.327877985000D-01	3.25	-3.167090131659D-03
10	0.0	7.313271871819D-01	3.656635935909D-01	3.50	-5.484269304844D-03
11	1.0	-4.504736231705D-01	-2.252368115852D-01	3.75	-4.769617225942D-03
12	2.0	2.183614991721D-01	1.091807495861D-01	4.25	7.293258312326D-03
13	3.0	-9.507111895354D-02	-4.753555947677D-02	4.50	1.266613456612D-02
14	4.0	4.097716268068D-02	2.048858134034D-02	4.75	1.103714264463D-02
15	5.0	-1.763773406934D-02	-8.818867034668D-03	5.25	-1.691659778800D-02
16	6.0	7.575942497897D-03	3.787971248948D-03	5.50	-2.940182441322D-02
17	7.0	-3.219167113246D-03	-1.609583556623D-03	5.75	-2.563991829666D-02
18	8.0	1.290063205267D-03	6.450316026337D-04	6.25	3.939453955514D-02
19	9.0	-4.133847386839D-04	-2.066923693419D-04	6.50	6.864043187790D-02
20	10.0	7.132062811043D-05	3.566031405521D-05	6.75	6.010159857039D-02
				7.25	-9.422020041455D-02
	0	-4.228470853658D-01	-2.506204131776D-01	7.50	-1.679660469670D-01
	1	-8.884027858656D-02	-3.618486698224D-02	7.75	-1.526030467848D-01
	2	-4.655557004999D-03	-2.759337028657D-03	8.25	2.832004468146D-01
				8.50	6.198800173786D-01
				8.75	8.938791313268D-01
				9.25	8.938786170357D-01
				9.50	6.198792435657D-01
				9.75	2.831998412638D-01

I=10	H= 0.0				
0		0.	3.825243019199D-01	-1.00	3.843495385000D-03
1		-1.838927500381D-03	-4.009135769238D-02	-.75	2.506765060384D-03
2		2.004567884619D-03	2.004567884619D-03	-.50	1.420605721345D-03
				-.25	5.850173678839D-04
0	-10.0	-1.656403842379D-04	-8.282019211893D-05	.25	-3.346081404943D-04
1	-9.0	9.600757578980D-04	4.800378789490D-04	.50	-4.234980410432D-04
2	-8.0	-2.996139343365D-03	-1.498069671682D-03	.75	-2.909334298066D-04
3	-7.0	7.476427125640D-03	3.738213562820D-03	1.25	3.289211815596D-04
4	-6.0	-1.759474254066D-02	-8.797371270331D-03	1.50	5.240571894490D-04
5	-5.0	4.095866293457D-02	2.047933146729D-02	1.75	4.300266374663D-04
6	-4.0	-9.506319453846D-02	-4.753159726923D-02	2.25	-6.141284099482D-04
7	-3.0	2.183581832498D-01	1.091790916249D-01	2.50	-1.047642313255D-03
8	-2.0	-4.504724187156D-01	-2.252362093578D-01	2.75	-9.020158807821D-04
9	-1.0	7.313271871819D-01	3.656635935909D-01	3.25	1.363666739429D-03
10	0.0	-8.655768014549D-01	-4.327884007275D-01	3.50	2.361378217551D-03
11	1.0	7.313271871819D-01	3.656635935909D-01	3.75	2.053660048106D-03
12	2.0	-4.504724187156D-01	-2.252362093578D-01	4.25	-3.140202748623D-03
13	3.0	2.183581832498D-01	1.091790916249D-01	4.50	-5.453429866130D-03
14	4.0	-9.506319453846D-02	-4.753159726923D-02	4.75	-4.751872263456D-03

15	5,0	4,095866293457D-02	2,047933146729D-02	5,25	7,281684139009D-03
16	6,0	-1,759474254066D-02	-8,797371270331D-03	5,50	1,265285711621D-02
17	7,0	7,476427125640D-03	3,738213562820D-03	5,75	1,102950193774D-02
18	8,0	-2,996139343365D-03	-1,498069671682D-03	6,25	-1,691161302030D-02
19	9,0	9,600757578980D-04	4,800378789490D-04	6,50	-2,939610519986D-02
20	10,0	-1,656403842379D-04	-8,282019211893D-05	6,75	-2,563662636307D-02
				7,25	3,939239007788D-02
	0	1,820675134581D-01	1,820675134581D-01	7,50	6,863796382867D-02
	1	3,825243019199D-02	-8,015658085331D-24	7,75	6,010017635074D-02
	2	2,004567884619D-03	2,004567884619D-03	8,25	-9,421926755088D-02
				8,50	-1,679649715097D-01
				8,75	-1,526024232746D-01
				9,25	2,832000280716D-01
				9,50	6,198795246406D-01
				9,75	8,938788369921D-01

I=11	H= 1.0				
0		0.	-8,884027858656D-01	-1,00	-1,654913476519D-03
1		7,917964242041D-04	9,137160755537D-02	-,75	-1,079350660080D-03
2		-8,631170523145D-04	-2,759337028657D-03	-,50	-6,116774751807D-04
				-,25	-2,518939218207D-04
0	-10,0	7,132062811043D-05	3,566031405521D-05	,25	1,440739393323D-04
1	-9,0	-4,133847386839D-04	-2,066923693419D-04	,50	1,823477186519D-04
2	-8,0	1,290063205267D-03	6,450316026337D-04	,75	1,252686955922D-04
3	-7,0	-3,219167113246D-03	-1,609583556623D-03	1,25	-1,416252759918D-04
4	-6,0	7,575942497897D-03	3,787971248948D-03	1,50	-2,256459847718D-04
5	-5,0	-1,763773406934D-02	-8,818867034668D-03	1,75	-1,851587682937D-04
6	-4,0	4,097716268068D-02	2,048858134034D-02	2,25	2,644284045691D-04
7	-3,0	-9,507111895354D-02	-4,753555947677D-02	2,50	4,510886882377D-04
8	-2,0	2,183614991721D-01	1,091807495861D-01	2,75	3,883855576839D-04
9	-1,0	-4,504736231705D-01	-2,252368115852D-01	3,25	-5,871607912816D-04
10	0,0	7,313271871819D-01	3,656635935909D-01	3,50	-1,016730155148D-03
11	1,0	-8,655755970000D-01	-4,327877985000D-01	3,75	-8,842540824122D-04
12	2,0	7,313238712584D-01	3,656619356292D-01	4,25	1,352089108218D-03
13	3,0	-4,504644942726D-01	-2,252322471363D-01	4,50	2,348098495585D-03
14	4,0	2,183396828553D-01	1,091698414277D-01	4,75	2,046018615627D-03
15	5,0	-9,502018796262D-02	-4,751009398131D-02	5,25	-3,135217831883D-03
16	6,0	4,085879840694D-02	2,042939920347D-02	5,50	-5,447710554837D-03
17	7,0	-1,736361301301D-02	-8,681806506507D-03	5,75	-4,748580298580D-03
18	8,0	6,958447653026D-03	3,479223826513D-03	6,25	7,279534655316D-03
19	9,0	-2,229750737626D-03	-1,114875368813D-03	6,50	1,265038906275D-02
20	10,0	3,846954915874D-04	1,923477457937D-04	6,75	1,102807971674D-02
				7,25	-1,691068015635D-02
	0	-7,839374098941D-02	-2,506204131776D-01	7,50	-2,939502974239D-02
	1	-1,647054462209D-02	3,618486698224D-02	7,75	-2,563600285282D-02
	2	-8,631170523145D-04	-2,759337028657D-03	8,25	3,939197133488D-02
				8,50	6,863747109061D-02
				8,75	6,009988201601D-02
				9,25	-9,421904759448D-02
				9,50	-1,679646904349D-01
				9,75	-1,526022364668D-01

I=12	H= 2.0				
0		0.	2,063292207077D+00	-1,00	7,125628926063D-04
1		-3,409270384123D-04	-2,114591663842D-01	-,75	4,647404467934D-04

2		3.716358541940D-04	5.592022552651D-03
0	-10.0	-3.070881578163D-05	-1.535440789082D-05
1	-9.0	1.779927650610D-04	8.899638253052D-05
2	-8.0	-5.554678195251D-04	-2.777339097625D-04
3	-7.0	1.386090161909D-03	6.930450809543D-04
4	-6.0	-3.262008304169D-03	-1.631004152084D-03
5	-5.0	7.594435765009D-03	3.797217882504D-03
6	-4.0	-1.764565820519D-02	-8.822829102597D-03
7	-3.0	4.098047859095D-02	2.049023929547D-02
8	-2.0	-9.507232340787D-02	-4.753616170394D-02
9	-1.0	2.183614991721D-01	1.091807495861D-01
10	0.0	-4.504724187156D-01	-2.252362093578D-01
11	1.0	7.313238712584D-01	3.656619356292D-01
12	2.0	-8.655676725545D-01	-4.327838362773D-01
13	3.0	7.313053707996D-01	3.656526853998D-01
14	4.0	-4.504214861922D-01	-2.252107430961D-01
15	5.0	2.182397833842D-01	1.091198916921D-01
16	6.0	-9.478824753799D-02	-4.739412376900D-02
17	7.0	4.032200282864D-02	2.016100141432D-02
18	8.0	-1.616061783401D-02	-8.080308917007D-03
19	9.0	5.178529597713D-03	2.589264798856D-03
20	10.0	-8.934449366643D-04	-4.467224683322D-04
	0	3.375431503527D-02	5.079027985008D-01
	1	7.091790045467D-03	-9.961871533113D-02
	2	3.716358541940D-04	5.592022552651D-03

-.50	2.633724827547D-04
-.25	1.084590004902D-04
.25	-6.203450779387D-05
.50	-7.851420615085D-05
.75	-5.393745675770D-05
1.25	6.098017674807D-05
1.50	9.715731838431D-05
1.75	7.972457127472D-05
2.25	-1.138560238808D-04
2.50	-1.942271081715D-04
2.75	-1.672287631861D-04
3.25	2.528162259288D-04
3.50	4.377862666563D-04
3.75	3.807368761162D-04
4.25	-5.821743856733D-04
4.50	-1.011029865585D-03
4.75	-8.809618051258D-04
5.25	1.349939560359D-03
5.50	2.345630399972D-03
5.75	2.044596381162D-03
6.25	-3.134284965166D-03
6.50	-5.446635095547D-03
6.75	-4.747956787751D-03
7.25	7.279115912202D-03
7.50	1.264989632461D-02
7.75	1.102778538198D-02
8.25	-1.691046019994D-02
8.50	-2.939474866754D-02
8.75	-2.563581604500D-02
9.25	3.939178452707D-02
9.50	6.863719001576D-02
9.75	6.009966205961D-02

I=13 H= 3.0

0		0.	-4.791939920575D+00
1		1.467924977452D-04	4.907856537445D-01
2		-1.600147513761D-04	-1.263577044903D-02
0	-10.0	1.322225363094D-05	6.611126815469D-06
1	-9.0	-7.663810617585D-05	-3.831905308793D-05
2	-8.0	2.391670344590D-04	1.195835172295D-04
3	-7.0	-5.968069889981D-04	-2.984034944991D-04
4	-6.0	1.404518697480D-03	7.022593487399D-04
5	-5.0	-3.269929650357D-03	-1.634964825178D-03
6	-4.0	7.597751555097D-03	3.798875777548D-03
7	-3.0	-1.764686265444D-02	-8.823431327222D-03
8	-2.0	4.098047859095D-02	2.049023929547D-02
9	-1.0	-9.507111895354D-02	-4.753555947677D-02
10	0.0	2.183581832498D-01	1.091790916249D-01
11	1.0	-4.504644942726D-01	-2.252322471363D-01
12	2.0	7.313053707996D-01	3.656526853998D-01
13	3.0	-8.655246643385D-01	-4.327623321692D-01
14	4.0	7.312054678657D-01	3.656027339328D-01
15	5.0	-4.501894646855D-01	-2.250947323427D-01
16	6.0	2.177011046889D-01	1.088505523444D-01
17	7.0	-9.354155283911D-02	-4.677077641956D-02

-1.00	-3.068072491214D-04
-.75	-2.001026709580D-04
-.50	-1.133999367166D-04
-.25	-4.669904639731D-05
.25	2.671011483235D-05
.50	3.380575645453D-05
.75	2.322377842576D-05
1.25	-2.625615292721D-05
1.50	-4.183289628658D-05
1.75	-3.432690173998D-05
2.25	4.902283550704D-05
2.50	8.362810546238D-05
2.75	7.200346424399D-05
3.25	-1.088547430459D-04
3.50	-1.884970447442D-04
3.75	-1.639333646319D-04
4.25	2.506660370041D-04
4.50	4.353177498280D-04
4.75	3.793145071386D-04
5.25	-5.812414913325D-04
5.50	-1.009954388147D-03
5.75	-8.803382885050D-04

18	8,0	3,752807830691D-02	1,876403915346D-02	6.25	1,349520816062D-03
19	9,0	-1,202680582391D-02	-6,013402911953D-03	6.50	2,345137661062D-03
20	10,0	2.074995270576D-03	1,037497635288D-03	6,75	2,044302046165D-03
				7,25	-3,134065008726D-03
	0	-1,453355016016D-02	-1,147660428034D+00	7.50	-5,446354020689D-03
	1	-3,053502529778D-03	2,380702447639D-01	7.75	-4,747769979945D-03
	2	-1,600147513761D-04	-1,263577044903D-02	8,25	7,278929104421D-03
				8,50	1,264961524983D-02
				8,75	1,102756542564D-02
				9,25	-1,691016586531D-02
				9,50	-2,939425592966D-02
				9,75	-2,563539730217D-02

I=14 H= 4.0

0		0.	1.112910039845D+01	-1.00	1,320925696216D-04
1		-6,319993508584D-05	-1,139692454749D+00	-,75	8,615205824075D-05
2		6.889263453576D-05	2.919473806023D-02	-,50	4,882312617686D-05
				-,25	2,010577342995D-05
0	-10.0	-5,692699449920D-06	-2,846349724960D-06	.25	-1,149975338978D-05
1	-9.0	3.299571442609D-05	1.649785721304D-05	.50	-1,455470576679D-05
2	-8,0	-1.029708008685D-04	-5,148540043425D-05	.75	-9,998748652010D-06
3	-7,0	2,569488467424D-04	1,284744233712D-04	1.25	1,130430496169D-05
4	-6,0	-6,047004639873D-04	-3,023502319937D-04	1.50	1,801070470468D-05
5	-5,0	1.407833286828D-03	7.039166434141D-04	1,75	1,477907927837D-05
6	-4,0	-3,271134048801D-03	-1,635567024400D-03	2,25	-2,110625590047D-05
7	-3,0	7,597751555097D-03	3,798875777548D-03	2,50	-3,600518362641D-05
8	-2.0	-1,764565820519D-02	-8,822829102597D-03	2,75	-3,100031906025D-05
9	-1,0	4,097716268068D-02	2,048858134034D-02	3,25	4,686624178963D-05
10	0,0	-9,506319453846D-02	-4,753159726923D-02	3,50	8,115538033794D-05
11	1,0	2,183396828553D-01	1,091698414277D-01	3,75	7,057975136662D-05
12	2,0	-4,504214861922D-01	-2,252107430961D-01	4,25	-1,079215727135D-04
13	3,0	7.312054678657D-01	3,656027339328D-01	4,50	-1,874213859981D-04
14	4,0	-8,652926355212D-01	-4,326463177606D-01	4,75	-1,633097901410D-04
15	5,0	7,306666025586D-01	3,653333012793D-01	5,25	2,502472808850D-04
16	6,0	-4,489384004433D-01	-2,244692002216D-01	5,50	4,348250032295D-04
17	7,0	2,148056983393D-01	1,074028491697D-01	5,75	3,790201697570D-04
18	8,0	-8,705275857003D-02	-4,352637928501D-02	6,25	-5,810215345501D-04
19	9,0	2,792744139289D-02	1,396372069644D-02	6,50	-1,009673313250D-03
20	10,0	-4,818953612158D-03	-2,409476806079D-03	6,75	-8,801514808496D-04
				7,25	1,349334008655D-03
	0	6,257264102718D-03	2,651649656979D+00	7,50	2,344856586958D-03
	1	1.314652755629D-03	-5,557976935446D-01	7,75	2,044082090425D-03
	2	6,889263453576D-05	2.919473806023D-02	8,25	-3,133770675016D-03
				8,50	-5,445861284411D-03
				8,75	-4,747351238512D-03
				9,25	7,278305596611D-03
				9,50	1,264853979669D-02
				9,75	1,102663256560D-02

I=15 H= 5.0

0		0.	-2,584577019981D+01	-1,00	-5,685009939770D-05
1		2,720003556484D-05	2,646717394777D+00	-,75	-3,707818757961D-05
2		-2.965006383286D-05	-6,773558927037D-02	-,50	-2,101253374063D-05
				-,25	-8,653137880763D-06
0	-10.0	2.450028268027D-06	1.225014134014D-06	.25	4,949272507385D-06

CARDINAL SPLINES N=3

1	-9.0	-1.420072037510D-05	-7.100360187548D-06	.50	6.264065207578D-06
2	-8.0	4.431665067366D-05	2.215832533683D-05	.75	4.303268960152D-06
3	-7.0	-1.105858378024D-04	-5.529291890121D-05	1.25	-4.865155265997D-06
4	-6.0	2.602514404885D-04	1.301257202442D-04	1.50	-7.751460628077D-06
5	-5.0	-6.059043548799D-04	-3.029521774399D-04	1.75	-6.360631248130D-06
6	-4.0	1.407833286828D-03	7.039166434141D-04	2.25	9.083726278387D-06
7	-3.0	-3.269929650357D-03	-1.634964825178D-03	2.50	1.549593799142D-05
8	-2.0	7.594435765009D-03	3.797217882504D-03	2.75	1.334194061761D-05
9	-1.0	-1.763773406934D-02	-8.818867034668D-03	3.25	-2.017032836661D-05
10	0.0	4.095866293457D-02	2.047933146729D-02	3.50	-3.492771358629D-05
11	1.0	-9.502018796262D-02	-4.751009398131D-02	3.75	-3.037616643973D-05
12	2.0	2.182397833842D-01	1.091198916921D-01	4.25	4.644736756688D-05
13	3.0	-4.501894646855D-01	-2.250947323427D-01	4.50	8.066255693248D-05
14	4.0	7.306666025586D-01	3.653333012793D-01	4.75	7.028539016176D-05
15	5.0	-8.640411773104D-01	-4.320205886552D-01	5.25	-1.077016125145D-04
16	6.0	7.277611396393D-01	3.638805698196D-01	5.50	-1.871403106981D-04
17	7.0	-4.422141299722D-01	-2.211070649861D-01	5.75	-1.631229839934D-04
18	8.0	1.997361705959D-01	9.986808529795D-02	6.25	2.500604772156D-04
19	9.0	-6.475631064021D-02	-3.237815532010D-02	6.50	4.345439358600D-04
20	10.0	1.118797891980D-02	5.593989459900D-03	6.75	3.788002199668D-04
				7.25	-5.807272100237D-04
	0	-2.693006027638D-03	-6.152155179074D+00	7.50	-1.009180592941D-03
	1	-5.658012410924D-04	1.292005609370D+00	7.75	-8.797327533410D-04
	2	-2.965006383286D-05	-6.773558927037D-02	8.25	1.348710522196D-03
				8.50	2.343781170916D-03
				8.75	2.043149262723D-03
				9.25	-3.132348510614D-03
				9.50	-5.443393331402D-03
				9.75	-4.745201845427D-03

I=16		H= 6.0			
0		0.	5.999654462540D+01	-1.00	2.441857386325D-05
1		-1.168311198327D-05	-6.143918081736D+00	-.75	1.592603129494D-05
2		1.273546187999D-05	1.572106181764D-01	-.50	9.025421461630D-06
				-.25	3.716744363316D-06
0	-10.0	-1.052349896724D-06	-5.261749483618D-07	.25	-2.125839313763D-06
1	-9.0	6.099573141728D-06	3.049786570864D-06	.50	-2.690576455908D-06
2	-8.0	-1.903513660157D-05	-9.517568300785D-06	.75	-1.848364243338D-06
3	-7.0	4.749944990858D-05	2.374972495429D-05	1.25	2.089708803988D-06
4	-6.0	-1.117846598445D-04	-5.589232992223D-05	1.50	3.329450887512D-06
5	-5.0	2.602514404885D-04	1.301257202442D-04	1.75	2.732054043790D-06
6	-4.0	-6.047004639873D-04	-3.023502319937D-04	2.25	-3.901693109288D-06
7	-3.0	1.404518697480D-03	7.022593487399D-04	2.50	-6.655902283944D-06
8	-2.0	-3.262008304169D-03	-1.631004152084D-03	2.75	-5.730705238894D-06
9	-1.0	7.575942497897D-03	3.787971248948D-03	3.25	8.663672681014D-06
10	0.0	-1.759474254066D-02	-8.797371270331D-03	3.50	1.500234763198D-05
11	1.0	4.085879840694D-02	2.042939920347D-02	3.75	1.304734154808D-05
12	2.0	-9.478824753799D-02	-4.739412376900D-02	4.25	-1.995033408327D-05
13	3.0	2.177011046889D-01	1.088505523444D-01	4.50	-3.464663431050D-05
14	4.0	-4.489384004433D-01	-2.244692002216D-01	4.75	-3.018937537828D-05
15	5.0	7.277611396393D-01	3.638805698196D-01	5.25	4.626060125567D-05
16	6.0	-8.572956757396D-01	-4.286478378698D-01	5.50	8.038155685748D-05
17	7.0	7.121496598712D-01	3.560748299356D-01	5.75	7.006549982945D-05
18	8.0	-4.072277880172D-01	-2.036138940086D-01	6.25	-1.074073797486D-04
19	9.0	1.479713665386D-01	7.398568326932D-02	6.50	-1.866477499489D-04
20	10.0	-2.589294561069D-02	-1.294647280534D-02	6.75	-1.627043956106D-04

```
                                                                   7,25   2,494372040727D-04
        0    1.156715068166D-03    1.427842562568D+01              7,50   4,334688905202D-04
        1    2,430261256165D-04   -2,999705718207D+00              7,75   3,778677154186D-04
        2    1.273546187999D-05    1.572106181764D-01              8.25  -5,793055410476D-04
                                                                   8.50  -1,006713500879D-03
                                                                   8,75  -8,775841107720D-04
                                                                   9,25   1,345419870649D-03
                                                                   9,50   2,338064190344D-03
                                                                   9,75   2,038166448491D-03


I=17      H=   7.0
0             0.                   -1,386539975119D+02            -1.00  -1,037592123694D-05
1             4,964378772467D-06    1.419974023956D+01             -,75  -6,767276715614D-06
2            -5,411542464469D-06   -3.633545998601D-01             -,50  -3,835075002351D-06
                                                                   -,25  -1,579316097146D-06
0  -10,0      4,471636920014D-07    2.235818460007D-07              .25   9,033099723805D-07
1   -9,0     -2,591825831104D-06   -1.295912915552D-06              .50   1,143277635492D-06
2   -8,0      8,088395301801D-06    4,044197650901D-06              .75   7,854054832786D-07
3   -7,0     -2,018342896722D-05   -1.009171448361D-05             1.25  -8,879574245298D-07
4   -6,0      4.749944990858D-05    2.374972495429D-05             1.50  -1,414747657440D-06
5   -5.0     -1.105858378024D-04   -5,529291890121D-05             1.75  -1,160902259574D-06
6   -4,0      2,569488467424D-04    1.284744233712D-04             2.25   1,657904373096D-06
7   -3.0     -5.968069889981D-04   -2.984034944991D-04             2,50   2,828220773485D-06
8   -2.0      1.386090161909D-03    6,930450809543D-04             2,75   2,435086771399D-06
9   -1.0     -3.219167113246D-03   -1.609583556623D-03             3.25  -3,681360994470D-06
10   0.0      7.476427125640D-03    3.738213562820D-03             3.50  -6,374785778652D-06
11   1.0     -1.736361301301D-02   -8,681806506507D-03             3.75  -5,544066128197D-06
12   2.0      4.032200282864D-02    2.016100141432D-02             4,25   8,477280297252D-06
13   3,0     -9,354155283911D-02   -4.677077641956D-02             4.50   1,472202065288D-05
14   4,0      2.148056983393D-01    1.074028491697D-01             4.75   1,282804568639D-05
15   5,0     -4.422141299722D-01   -2.211070649861D-01             5,25  -1,965701836953D-05
16   6.0      7.121496598712D-01    3.560748299356D-01             5,50  -3,415566803699D-05
17   7,0     -8.211651824689D-01   -4,105825912344D-01             5.75  -2,977217718197D-05
18   8,0      6,311788736312D-01    3,155894368156D-01             6.25   4,563945947216D-05
19   9,0     -2,874256981420D-01   -1,437128490710D-01             6,50   7,931021534313D-05
20  10,0      5,802777581658D-02    2.901388790829D-02             6,75   6,913622399111D-05
                                                                   7,25  -1,059906606216D-04
          0  -4,915104587222D-04   -3,299205510231D+01             7,50  -1.841892596763D-04
          1  -1,032664705169D-04    6,932648242361D+00             7,75  -1,605632514769D-04
          2  -5.411542464469D-06   -3,633545998601D-01             8,25   2,461580483087D-04
                                                                   8,50   4,277718872462D-04
                                                                   8,75   3,729023160313D-04
                                                                   9,25  -5,716945829085D-04
                                                                   9,50  -9,934878606509D-04
                                                                   9,75  -8,660552659251D-04


I=18      H=   8.0
0             0.                    3,060929148305D+02            -1.00   4,158091904056D-06
1            -1,989446788491D-06   -3,136961264352D+01             -,75   2,711947968874D-06
2             2,168645115564D-06    8,032493796863D-01             -,50   1,536884673137D-06
                                                                   -,25   6,329020168456D-07
0  -10,0     -1.791983270729D-07   -8,959916353643D-08              .25  -3,619963757664D-07
1   -9,0      1,038659581057D-06    5,193297905284D-07              .50  -4,581620630757D-07
2   -8,0     -3,241378789721D-06   -1,620689394861D-06              .75  -3,147468168702D-07
3   -7,0      8,088395301801D-06    4,044197650901D-06             1,25   3,558439288205D-07
```

CARDINAL SPLINES N=3

```
 4  -6.0  -1.903513660157D-05  -9.517568300785D-06     1.50   5.669521429811D-07
 5  -5.0   4.431665067366D-05   2.215832533683D-05     1.75   4.652250317543D-07
 6  -4.0  -1.029708008685D-04  -5.148540043425D-05     2.25  -6.643958251081D-07
 7  -3.0   2.391670344590D-04   1.195835172295D-04     2.50  -1.133393520688D-06
 8  -2.0  -5.554678195251D-04  -2.777339097625D-04     2.75  -9.758472870617D-07
 9  -1.0   1.290063205267D-03   6.450316026337D-04     3.25   1.475284651595D-06
10   0.0  -2.996139343365D-03  -1.498069671682D-03     3.50   2.554659439968D-06
11   1.0   6.958447653026D-03   3.479223826513D-03     3.75   2.221753226223D-06
12   2.0  -1.616061783401D-02  -8.080308917007D-03     4.25  -3.397222257907D-06
13   3.0   3.752807830691D-02   1.876403915346D-02     4.50  -5.899766728195D-06
14   4.0  -8.705275857003D-02  -4.352637928501D-02     4.75  -5.140766944482D-06
15   5.0   1.997361705959D-01   9.986808529795D-02     5.25   7.877439224287D-06
16   6.0  -4.072277880172D-01  -2.036138940086D-01     5.50   1.368769129009D-05
17   7.0   6.311788736312D-01   3.155894368156D-01     5.75   1.193103205762D-05
18   8.0  -6.397043822000D-01  -3.198521911000D-01     6.25  -1.828975592600D-05
19   9.0   3.626945953948D-01   1.813472976974D-01     6.50  -3.178312139988D-05
20  10.0  -8.585625922832D-02  -4.292812961416D-02     6.75  -2.770595175718D-05
                                                       7.25   4.247515932688D-05
        0  1.969700436715D-04   7.272172636390D+01     7.50   7.381280674413D-05
        1  4.138345552280D-05  -1.530462504980D+01     7.75   6.434481721386D-05
        2  2.168645115564D-06   8.032493796863D-01     8.25  -9.864644895682D-05
                                                       8.50  -1.714271697695D-04
                                                       8.75  -1.494384981400D-04
                                                       9.25   2.291033721548D-04
                                                       9.50   3.981346340016D-04
                                                       9.75   3.470667291546D-04


I=19      H=  9.0
0              0.                  -3.413163398661D+02    -1.00  -1.332408914613D-06
1              6.374935180122D-07   3.552222565080D+01     -.75  -8.690100490972D-07
2             -6.949153966009D-07  -9.228207643949D-01     -.50  -4.924756081563D-07
                                                           -.25  -2.028055917906D-07
 0 -10.0   5.742187858877D-08   2.871093929438D-08          .25   1.159972432688D-07
 1  -9.0  -3.328255644611D-07  -1.664127822305D-07          .50   1.468123435618D-07
 2  -8.0   1.038659581057D-06   5.193297905284D-07          .75   1.008567088753D-07
 3  -7.0  -2.591825831104D-06  -1.295912915552D-06         1.25  -1.140257680474D-07
 4  -6.0   6.099573141728D-06   3.049786570864D-06         1.50  -1.816727736898D-07
 5  -5.0  -1.420072037510D-05  -7.100360187548D-06         1.75  -1.490755841654D-07
 6  -4.0   3.299571442609D-05   1.649785721304D-05         2.25   2.128973915517D-07
 7  -3.0  -7.663810617585D-05  -3.831905308793D-05         2.50   3.631818789903D-07
 8  -2.0   1.779927650610D-04   8.899638253052D-05         2.75   3.126981448663D-07
 9  -1.0  -4.133847386839D-04  -2.066923693419D-04         3.25  -4.727366462150D-07
10   0.0   9.600757578980D-04   4.800378789490D-04         3.50  -8.186088932507D-07
11   1.0  -2.229750737626D-03  -1.114875368813D-03         3.75  -7.119332311542D-07
12   2.0   5.178529597713D-03   2.589264798856D-03         4.25   1.088597685141D-06
13   3.0  -1.202680582391D-02  -6.013402911953D-03         4.50   1.890506983532D-06
14   4.0   2.792744139289D-02   1.396372069644D-02         4.75   1.647294928256D-06
15   5.0  -6.475631064021D-02  -3.237815532010D-02         5.25  -2.524227575760D-06
16   6.0   1.479713665386D-01   7.398568326932D-02         5.50  -4.386050697337D-06
17   7.0  -2.874256981420D-01  -1.437128490710D-01         5.75  -3.823151060849D-06
18   8.0   3.626945953948D-01   1.813472976974D-01         6.25   5.860725160511D-06
19   9.0  -2.424626222558D-01  -1.212313111279D-01         6.50   1.018450656319D-05
20  10.0   6.445814300015D-02   3.222907150007D-02         6.75   8.878028182286D-06
                                                           7.25  -1.361063734059D-05
        0 -6.311660447997D-05  -7.837615979765D+01         7.50  -2.365239729717D-05
        1 -1.326081441401D-05   1.706581036290D+01         7.75  -2.061849762693D-05
```

CARDINAL SPLINES N=3

```
       2  -6.949153966009D-07  -9.228207643949D-01     8.25  3,161002954627D-05
                                                        8.50  5,493170770333D-05
                                                        8.75  4,788571092312D-05
                                                        9.25 -7,341333032142D-05
                                                        9.50 -1,275772986192D-04
                                                        9.75 -1,112132223024D-04


I=20     H= 10.0
0            0.                    1,324472062752D+02   -1.00  2,298784441315D-07
1           -1.099857682257D-07   -1.405849954153D+01    -.75  1,499289563663D-07
2            1,198926759058D-07    3.730570185851D-01    -.50  8,496605308931D-08
                                                         -.25  3,498973430055D-08
0  -10.0    -9.906907680009D-09   -4.953453840005D-09     .25 -2,001282452686D-08
1   -9.0     5.742187858877D-08    2.871093929438D-08     .50 -2,532930600143D-08
2   -8.0    -1.791983270729D-07   -8.959916353643D-08     .75 -1,740065160342D-08
3   -7.0     4.471636920014D-07    2.235818460007D-07    1.25  1.967268896368D-08
4   -6.0    -1.052349896724D-06   -5.261749483618D-07    1.50  3.134372196016D-08
5   -5.0     2.450028268027D-06    1.225014134014D-06    1.75  2,571977939362D-08
6   -4.0    -5.692699449920D-06   -2.846349724960D-06    2.25 -3,673085686594D-08
7   -3.0     1.322225363094D-05    6.611126815469D-06    2.50 -6,265920646687D-08
8   -2.0    -3.070881578163D-05   -1.535440789082D-05    2.75 -5,394932609375D-08
9   -1.0     7.132062811043D-05    3.566031405521D-05    3.25  8,156052059094D-08
10   0.0    -1.656403842379D-04   -8.282019211893D-05    3.50  1,412333230954D-07
11   1.0     3.846954915874D-04    1.923477457937D-04    3.75  1,228287365150D-07
12   2.0    -8.934449366643D-04   -4.467224683322D-04    4.25 -1,878140707411D-07
13   3.0     2.074995270576D-03    1.037497635288D-03    4.50 -3,261662386282D-07
14   4.0    -4.818953612158D-03   -2.409476806079D-03    4.75 -2,842052398330D-07
15   5.0     1.118797891980D-02    5.593989459900D-03    5.25  4,355010698183D-07
16   6.0    -2.589294561069D-02   -1.294647280534D-02    5.50  7,567185262178D-07
17   7.0     5.802777581658D-02    2.901388790829D-02    5.75  6,596023247132D-07
18   8.0    -8.585625922832D-02   -4.292812961416D-02    6.25 -1,011141824859D-06
19   9.0     6.445814300015D-02    3.222907150007D-02    6.50 -1,757117126217D-06
20  10.0    -1.855619925184D-02   -9.278099625920D-03    6.75 -1,531712436861D-06
                                                         7.25  2,348222157002D-06
       0     1.088940990832D-05    2.916791271845D+01    7.50  4,080711432491D-06
       1     2.287867749889D-06   -6.597359169829D+00    7.75  3,557277426447D-06
       2     1.198926759058D-07    3.730570185851D-01    8.25 -5,453629385899D-06
                                                         8.50 -9,477282357515D-06
                                                         8.75 -8,261647457537D-06
                                                         9.25  1,266588804778D-05
                                                         9.50  2,201071350637D-05
                                                         9.75  1,918744478409D-05
```

CARDINAL SPLINES N=4

```
             N= 4        M= 4                          P=  2.0      Q= 7

I= 0     H= -2.0
0             1.000000000000D+00  -7.384105960265D-01   -1.00  4.492963576159D+00
1            -1.989514348786D+00   2.704194260486D-01    -.75  3.301660802980D+00
2             1.246274834437D+00   2.032284768212D-01    -.50  2.338472682119D+00
3            -2.571743929360D-01  -8.333333333333D-02    -.25  1.579289114238D+00
                                                          .25  5.764952653291D-01
0   -2.0      4.139072847682D-04   2.069536423841D-04     .50  2.846679687500D-01
1   -1.0     -1.655629139073D-03  -8.278145695364D-04     .75  1.004536357147D-01
2    0.0      2.483443708609D-03   1.241721854305D-03    1.25 -3.990868069478D-02
3    1.0     -1.655629139073D-03  -8.278145695364D-04    1.50 -4.105766245861D-02
4    2.0      4.139072847682D-04   2.069536423841D-04    1.75 -2.264336087056D-02
                                                         2.25  1.641913912944D-02
        0    -5.132450331126D-02  -5.132450331126D-02    2.50  2.144233754139D-02
        1    -9.050772626932D-02   8.333333333333D-02    2.75  1.477881930522D-02
        2    -2.967715231788D-01  -2.967715231788D-01    3.25 -1.673386428530D-02
        3    -2.571743929360D-01  -8.333333333333D-02    3.50 -2.783203125000D-02

I= 1     H= -1.0
0             0.                   6.953642384106D+00   -1.00 -7.971854304636D+00
1             3.624724061810D+00  -5.415011037528D+00    -.75 -5.042580711921D+00
2            -3.485099337748D+00   6.870860927152D-01    -.50 -2.791390728477D+00
3             8.620309050773D-01   1.666666666667D-01    -.25 -1.137468956954D+00
                                                          .25  7.018314386835D-01
0   -2.0     -1.655629139073D-03  -8.278145695364D-04     .50  1.048828125000D+00
1   -1.0      6.622516556291D-03   3.311258278146D-03     .75  1.121622957141D+00
2    0.0     -9.933774834437D-03  -4.966887417219D-03    1.25  7.611972227791D-01
3    1.0      6.622516556291D-03   3.311258278146D-03    1.50  4.767306498344D-01
4    2.0     -1.655629139073D-03  -8.278145695364D-04    1.75  2.077609434823D-01
                                                         2.25 -1.203640565177D-01
        0     2.052980132450D-01   2.052980132450D-01    2.50 -1.482693501656D-01
        1     2.869757174393D-02  -6.666666666667D-01    2.75 -9.817777722087D-02
        2     1.687086092715D+00   1.687086092715D+00    3.25  1.059979571412D-01
        3     8.620309050773D-01   1.666666666667D-01    3.50  1.738281250000D-01

I= 2     H=  0.0
0             0.                  -1.043046357616D+01   -1.00  6.957781456954D+00
1            -2.437086092715D+00   1.112251655629D+01    -.75  4.224027317881D+00
2             3.477649006623D+00  -2.780629139073D+00    -.50  2.218336092715D+00
3            -1.043046357616D+00   3.887506540929D-27    -.25  8.429221854305D-01
                                                          .25 -4.082159080253D-01
0   -2.0      2.483443708609D-03   1.241721854305D-03     .50 -4.794921875000D-01
1   -1.0     -9.933774834437D-03  -4.966887417219D-03     .75 -3.113406857118D-01
2    0.0      1.490066225166D-02   7.450331125828D-03    1.25  3.621104158313D-01
3    1.0     -9.933774834437D-03  -4.966887417219D-03    1.50  6.911540252483D-01
4    2.0      2.483443708609D-03   1.241721854305D-03    1.75  9.188273347766D-01
```

```
                                                              2.25   9.188273347766D-01
       0   6.920529801325D-01   6.920529801325D-01           2.50   6.911540252483D-01
       1  -1.043046357616D+00  -5.391085874769D-27           2.75   3.621104158313D-01
       2  -2.780629139073D+00  -2.780629139073D+00           3.25  -3.113406857118D-01
       3  -1.043046357616D+00   3.887506540929D-27           3.50  -4.794921875000D-01


I= 3      H=  1.0
0              0.                6.953642384106D+00         -1.00  -2.971854304636D+00
1              9.580573951435D-01  -8.081677704194D+00       -.75  -1.776955711921D+00
2             -1.485099337748D+00   2.687086092715D+00       -.50  -9.163907284768D-01
3              5.286975717439D-01  -1.666666666667D-01       -.25  -3.405939569536D-01
                                                               .25   1.549564386835D-01
0   -2.0  -1.655629139073D-03  -8.278145695364D-04           .50   1.738281250000D-01
1   -1.0   6.622516556291D-03   3.311258278146D-03           .75   1.059979571412D-01
2    0.0  -9.933774834437D-03  -4.966887417219D-03          1.25  -9.817777722087D-02
3    1.0   6.622516556291D-03   3.311258278146D-03          1.50  -1.482693501656D-01
4    2.0  -1.655629139073D-03  -8.278145695364D-04          1.75  -1.203640565177D-01
                                                              2.25   2.077609434823D-01
       0   2.052980132450D-01   2.052980132450D-01           2.50   4.767306498344D-01
       1   1.362030905077D+00   6.666666666667D-01           2.75   7.611972227791D-01
       2   1.687086092715D+00   1.687086092715D+00           3.25   1.121622957141D+00
       3   5.286975717439D-01  -1.666666666667D-01           3.50   1.048828125000D+00


I= 4      H=  2.0
0              0.               -1.738410596026D+00         -1.00   4.929635761589D-01
1             -1.561810154525D-01   2.103752759382D+00       -.75   2.938483029801D-01
2              2.462748344371D-01  -7.967715231788D-01       -.50   1.509726821192D-01
3             -9.050772626932D-02   8.333333333333D-02       -.25   5.585161423841D-02
                                                               .25  -2.506723467088D-02
0   -2.0   4.139072847682D-04   2.069536423841D-04           .50  -2.783203125000D-02
1   -1.0  -1.655629139073D-03  -8.278145695364D-04           .75  -1.673386428530D-02
2    0.0   2.483443708609D-03   1.241721854305D-03          1.25   1.477881930522D-02
3    1.0  -1.655629139073D-03  -8.278145695364D-04          1.50   2.144233754139D-02
4    2.0   4.139072847682D-04   2.069536423841D-04          1.75   1.641913912944D-02
                                                              2.25  -2.264336087056D-02
       0  -5.132450331126D-02  -5.132450331126D-02           2.50  -4.105766245861D-02
       1  -2.571743929360D-01  -8.333333333333D-02           2.75  -3.990868069478D-02
       2  -2.967715231788D-01  -2.967715231788D-01           3.25   1.004536357147D-01
       3  -9.050772626932D-02   8.333333333333D-02           3.50   2.846679687500D-01




               N= 4        M= 5                    P=  2.5       Q= 7

I= 0      H= -2.5
0              1.000000000000D+00   2.099766327379D+00        -1.00   4.618872826647D+00
1             -2.028915808075D+00  -2.213319150782D+00         -.75   3.376551023237D+00
2              1.309163022015D+00   9.501468233537D-01         -.50   2.376847909111D+00
3             -2.807939965562D-01  -1.643537414966D-01         -.25   1.593439047091D+00
```

CARDINAL SPLINES N=4

				.25	5.702063640339D-01
0	-2.5	5.467826165676D-04	2.733913082838D-04	.50	2.777378736357D-01
1	-1.5	-2.456674380315D-03	-1.228337190157D-03	.75	9.633036319861D-02
2	-.5	4.358871355584D-03	2.179435677792D-03	1.25	-3.639619997271D-02
3	.5	-3.804393950538D-03	-1.902196975269D-03	1.50	-3.611355011231D-02
4	1.5	1.624958272746D-03	8.124791363732D-04	1.75	-1.901509134478D-02
5	2.5	-2.695439140447D-04	-1.347719570223D-04	2.25	1.206588827565D-02
				2.50	1.436226434710D-02
	0	-2.774268287853D-01	-6.314111449958D-02	2.75	8.830775871317D-03
	1	-7.479881334291D-01	-5.442176870748D-01	3.25	-7.360116854986D-03
	2	-7.967919521565D-01	-2.825062378708D-01	3.50	-1.002936643884D-02
	3	-2.807939965562D-01	-1.643537414966D-01	3.75	-7.125233773734D-03
I= 1	H= -1.5				
0		0.	-1.015644183928D+01	-1.00	-8.730904194226D+00
1		3.862257628308D+00	9.558324799285D+00	-.75	-5.494059957466D+00
2		-3.864223759923D+00	-3.815746592510D+00	-.50	-3.022737604884D+00
3		1.004422805995D+00	6.551020408163D-01	-.25	-1.222772498416D+00
				.25	7.397443784819D-01
0	-2.5	-2.456674380315D-03	-1.228337190157D-03	.50	1.090606532154D+00
1	-1.5	1.145165579401D-02	5.725827897003D-03	.75	1.146480301109D+00
2	-.5	-2.123987937287D-02	-1.061993968644D-02	1.25	7.400220661171D-01
3	.5	1.957644715774D-02	9.788223578869D-03	1.50	4.469248340104D-01
4	1.5	-8.956507471300D-03	-4.478253735650D-03	1.75	1.858877485241D-01
5	2.5	1.624958272746D-03	8.124791363732D-04	2.25	-9.412027723130D-02
				2.50	-1.055867930413D-01
	0	1.198351914927D+00	1.269233434988D-01	2.75	-6.231971520961D-02
	1	3.374066441104D+00	2.762755102041D+00	3.25	4.948787607516D-02
	2	3.668947285041D+00	1.097518713612D+00	3.50	6.650391564305D-02
	3	1.004422805995D+00	6.551020408163D-01	3.75	4.679223512222D-02
I= 2	H= -.5				
0		0.	2.962810408333D+01	-1.00	8.734888510436D+00
1		-2.993205765811D+00	-2.393344102265D+01	-.75	5.281042097495D+00
2		4.365264819540D+00	7.761518136503D+00	-.50	2.759971328426D+00
3		-1.376417925085D+00	-1.143537414966D+00	-.25	1.042637022753D+00
				.25	-4.969786542665D-01
0	-2.5	4.358871355584D-03	2.179435677792D-03	.50	-5.773048649734D-01
1	-1.5	-2.123987937287D-02	-1.061993968644D-02	.75	-3.695373364217D-01
2	-.5	4.137080393566D-02	2.068540196783D-02	1.25	4.116862352587D-01
3	.5	-4.026184912557D-02	-2.013092456278D-02	1.50	7.609361650815D-01
4	1.5	1.957644715774D-02	9.788223578869D-03	1.75	9.700374193513D-01
5	2.5	-3.804393950538D-03	-1.902196975269D-03	2.25	8.573847261687D-01
				2.50	5.912245286942D-01
	0	-1.706639371856D+00	4.362177710008D-01	2.75	2.781586021253D-01
	1	-6.974717763457D+00	-6.567176870748D+00	3.25	-1.790378357508D-01
	2	-5.957869618599D+00	-8.150124757416D-01	3.50	-2.282219981838D-01
	3	-1.376417925085D+00	-1.143537414966D+00	3.75	-1.549791027515D-01
I= 3	H= .5				
0		0.	-3.894332448810D+01	-1.00	-5.007968632421D+00
1		1.595229608339D+00	3.208356578007D+01	-.75	-2.988026780056D+00
2		-2.502082119235D+00	-9.391543087987D+00	-.50	-1.536967447084D+00
3		9.106569048468D-01	1.143537414966D+00	-.25	-5.694165486752D-01

CARDINAL SPLINES N=4

```
                                                          .25  2.566560515690D-01
0  -2.5  -3.804393950538D-03  -1.902196975269D-03        .50  2.858966656388D-01
1  -1.5   1.957644715774D-02   9.788223578869D-03        .75  1.726765706255D-01
2   -.5  -4.026184912557D-02  -2.013092456278D-02       1.25 -1.549791027515D-01
3    .5   4.137080393566D-02   2.068540196783D-02       1.50 -2.282219981838D-01
4   1.5  -2.123987937287D-02  -1.061993968644D-02       1.75 -1.790378357508D-01
5   2.5   4.358871355584D-03   2.179435677792D-03       2.25  2.781586021253D-01
                                                         2.50  5.912245286942D-01
      0   2.579074913858D+00   4.362177710008D-01       2.75  8.573847261687D-01
      1   6.159635978040D+00   6.567176870748D+00       3.25  9.700374193513D-01
      2   4.327844667116D+00  -8.150124757416D-01       3.50  7.609361650815D-01
      3   9.106569048468D-01   1.143537414966D+00       3.75  4.116862352587D-01


I= 4     H=  1.5
0             0.               2.412927244643D+01      -1.00  1.640524377203D+00
1        -5.152933921004D-01  -2.053351193541D+01       -.75  9.764119813090D-01
2         8.194497094652D-01   6.010784019735D+00       -.50  5.007317828712D-01
3        -3.057812756375D-01  -6.551020408163D-01       -.25  1.848167872985D-01
                                                          .25 -8.238547443578D-02
0  -2.5   1.624958272746D-03   8.124791363732D-04        .50 -9.099423315209D-02
1  -1.5  -8.956507471300D-03  -4.478253735650D-03        .75 -5.431415241472D-02
2   -.5   1.957644715774D-02   9.788223578869D-03       1.25  4.679223512222D-02
3    .5  -2.123987937287D-02  -1.061993968644D-02       1.50  6.650391564305D-02
4   1.5   1.145165579401D-02   5.725827897003D-03       1.75  4.948787607516D-02
5   2.5  -2.456674380315D-03  -1.228337190157D-03       2.25 -6.231971520961D-02
                                                         2.50 -1.055867930413D-01
      0  -9.445052279298D-01   1.269233434988D-01       2.75 -9.412027723130D-02
      1  -2.151443762978D+00  -2.762755102041D+00       3.25  1.858877485241D-01
      2  -1.473909857816D+00   1.097518713612D+00       3.50  4.469248340104D-01
      3  -3.057812756375D-01  -6.551020408163D-01       3.75  7.400220661171D-01


I= 5     H=  2.5
0             0.              -5.757376529764D+00      -1.00 -2.554128876391D-01
1         7.992772933959D-02   5.038381529490D+00       -.75 -1.519183645180D-01
2        -1.275716718625D-01  -1.515159299095D+00       -.50 -7.784596844005D-02
3         4.791348643699D-02   1.643537414966D-01       -.25 -2.870381005188D-02
                                                          .25  1.275733461741D-02
0  -2.5  -2.695439140447D-04  -1.347719570223D-04        .50  1.405802669696D-02
1  -1.5   1.624958272746D-03   8.124791363732D-04        .75  8.364253903333D-03
2   -.5  -3.804393950538D-03  -1.902196975269D-03       1.25 -7.125233773734D-03
3    .5   4.358871355584D-03   2.179435677792D-03       1.50 -1.002936643884D-02
4   1.5  -2.456674380315D-03  -1.228337190157D-03       1.75 -7.360116854986D-03
5   2.5   5.467826165676D-04   2.733913082838D-04       2.25  8.830775871317D-03
                                                         2.50  1.436226434710D-02
      0   1.511445997861D-01  -6.314111449958D-02       2.75  1.206588827565D-02
      1   3.404472407205D-01   5.442176870748D-01       3.25 -1.901509134478D-02
      2   2.317794764149D-01  -2.825062378708D-01       3.50 -3.611355011231D-02
      3   4.791348643699D-02   1.643537414966D-01       3.75 -3.639619997271D-02
```

CARDINAL SPLINES N=4

```
                    N= 4        M= 6                        P=   3.0       Q= 7

I= 0     H= -3.0
0            1.000000000000D+00   -1.162409324465D+00    -1.00   4.653374451131D+00
1           -2.039681297215D+00   -2.838679416053D-02     -.75   3.397062034600D+00
2            1.326394754703D+00    4.677344231073D-01     -.50   2.387351637185D+00
3           -2.872983992133D-01   -1.309160859466D-01     -.25   1.597309033960D+00
                                                           .25   5.684903460796D-01
0   -3.0     5.849417246901D-04    2.924708623451D-04      .50   2.758503100241D-01
1   -2.0    -2.701771437343D-03   -1.350885718671D-03      .75   9.521014473643D-02
2   -1.0     5.107070157474D-03    2.553535078737D-03     1.25  -3.545044866388D-02
3    0.0    -5.109400750270D-03   -2.554700375135D-03     1.50  -3.479267492480D-02
4    1.0     2.929369809042D-03    1.464684904521D-03     1.75  -1.805669502133D-02
5    2.0    -9.596111585973D-04   -4.798055792987D-04     2.25   1.096106496110D-02
6    3.0     1.494016550037D-04    7.470082750185D-05     2.50   1.262418285420D-02
                                                          2.75   7.436000169296D-03
        0   -9.385478780749D-01   -5.726942195383D-01     3.25  -5.460866579261D-03
        1   -1.838369547755D+00   -7.567145760743D-01     3.50  -6.804987441964D-03
        2   -1.259290838217D+00   -7.105103504119D-01     3.75  -4.325636520421D-03
        3   -2.872983992133D-01   -1.309160859466D-01     4.25   3.752341286715D-03


I= 1     H= -2.0
0            0.                    1.079660688522D+01    -1.00  -8.952509127880D+00
1            3.931404676949D+00   -4.475559029796D+00     -.75  -5.625802777698D+00
2           -3.974903678221D+00   -7.171977691340D-01     -.50  -3.090203354619D+00
3            1.046200772709D+00    4.403310104530D-01     -.25  -1.247629536200D+00
                                                           .25   7.507664115191D-01
0   -3.0    -2.701771437343D-03   -1.350885718671D-03      .50   1.102730407919D+00
1   -2.0     1.302592129232D-02    6.512960646160D-03      .75   1.153675496747D+00
2   -1.0    -2.604558195445D-02   -1.302279097723D-02     1.25   7.339474780599D-01
3    0.0     2.795854364146D-02    1.397927182073D-02     1.50   4.384408144122D-01
4    1.0    -1.733478056072D-02   -8.667390280362D-03     1.75   1.797319411717D-01
5    2.0     6.057280177338D-03    3.028640088669D-03     2.25  -8.702396507263D-02
6    3.0    -9.596111585973D-04   -4.798055792987D-04     2.50  -9.442304525621D-02
                                                          2.75  -5.336103046568D-02
        0    4.267501790006D+00    2.804087155860D+00     3.25   3.728893652855D-02
        1    8.329403470772D+00    3.110191637631D+00     3.50   4.579363592846D-02
        2    5.440903276162D+00    3.245781324943D+00     3.75   2.881034085344D-02
        3    1.046200772709D+00    4.403310104530D-01     4.25  -2.469170406050D-02


I= 2     H= -1.0
0            0.                   -3.433449971827D+01    -1.00   9.411373768415D+00
1           -3.204288426595D+00    1.890728172656D+01     -.75   5.683208575748D+00
2            4.703133349129D+00   -1.697308278468D+00     -.50   2.965921549666D+00
3           -1.503951992691D+00   -4.879137630662D-01     -.25   1.118517190855D+00
                                                           .25  -5.306252105033D-01
0   -3.0     5.107070157474D-03    2.553535078737D-03      .50  -6.143149761162D-01
1   -2.0    -2.604558195445D-02   -1.302279097723D-02      .75  -3.915018464037D-01
```

CARDINAL SPLINES N=4

```
2  -1.0   5.604099605210D-02   2.802049802605D-02    1.25  4.302299075553D-01
3   0.0  -6.584956781339D-02  -3.292478390669D-02    1.50  7.868350207248D-01
4   1.0   4.515249430994D-02   2.257624715497D-02    1.75  9.888290268133D-01
5   2.0  -1.733478056072D-02  -8.667390280362D-03    2.25  8.357220741591D-01
6   3.0   2.929369809042D-03   1.464684904521D-03    2.50  5.571453650408D-01
                                                     2.75  2.508107535905D-01
      0  -7.891368940279D+00  -6.062100647596D+00    3.25 -1.417985801523D-01
      1  -1.559219213447D+01  -4.450239547038D+00    3.50 -1.650004864400D-01
      2  -8.832434585088D+00  -6.088532146064D+00    3.75 -1.000864332485D-01
      3  -1.503951992691D+00  -4.879137630662D-01    4.25  8.224920835351D-02


I= 3     H=  0.0
0           0.                 7.262011651014D+01   -1.00 -6.187892817477D+00
1           1.963399547844D+00  -4.263913405840D+01   -.75 -3.689484717998D+00
2          -3.091391708364D+00   7.106522343066D+00   -.50 -1.896185396172D+00
3           1.133101561270D+00   1.435535197123D-26   -.25 -7.017665806285D-01
                                                      .25  3.153423052300D-01
0  -3.0  -5.109400750270D-03  -2.554700375135D-03     .50  3.504496247964D-01
1  -2.0   2.795854364146D-02   1.397927182073D-02     .75  2.109870234182D-01
2  -1.0  -6.584956781339D-02  -3.292478390669D-02    1.25 -1.873229375426D-01
3   0.0   8.600084984441D-02   4.300042492220D-02    1.50 -2.733947285255D-01
4   1.0  -6.584956781339D-02  -3.292478390669D-02    1.75 -2.118141185433D-01
5   2.0   2.795854364146D-02   1.397927182073D-02    2.25  3.159425548678D-01
6   3.0  -5.109400750270D-03  -2.554700375135D-03    2.50  6.506653353147D-01
                                                     2.75  9.050847869089D-01
      0   8.661415422548D+00   8.661415422548D+00    3.25  9.050847869089D-01
      1   1.400879145195D+01   1.221475222939D-25    3.50  6.506653353147D-01
      2   7.106522343066D+00   7.106522343066D+00    3.75  3.159425548678D-01
      3   1.133101561270D+00   1.435535197123D-26    4.25 -2.118141185433D-01


I= 4     H=  1.0
0           0.                -8.738328020607D+01   -1.00  2.819910353781D+00
1          -8.832953952364D-01   5.415510402621D+01   -.75  1.677549957499D+00
2           1.408490491986D+00  -1.047975601366D+01   -.50  8.597858789345D-01
3          -5.281244665584D-01   4.879137630662D-01   -.25  3.171064493482D-01
                                                      .25 -1.410449590554D-01
0  -3.0   2.929369809042D-03   1.464684904521D-03     .50 -1.555177472399D-01
1  -2.0  -1.733478056072D-02  -8.667390280362D-03     .75 -9.260713034690D-02
2  -1.0   4.515249430994D-02   2.257624715497D-02    1.25  7.912131665506D-02
3   0.0  -6.584956781339D-02  -3.292478390669D-02    1.50  1.116560409774D-01
4   1.0   5.604099605210D-02   2.802049802605D-02    1.75  8.224920835351D-02
5   2.0  -2.604558195445D-02  -1.302279097723D-02    2.25 -1.000864332485D-01
6   3.0   5.107070157474D-03   2.553535078737D-03    2.50 -1.650004864400D-01
                                                     2.75 -1.417985801523D-01
      0  -4.232832354913D+00  -6.062100647596D+00    3.25  2.508107535905D-01
      1  -6.691713040397D+00   4.450239547038D+00    3.50  5.571453650408D-01
      2  -3.344629707039D+00  -6.088532146064D+00    3.75  8.357220741591D-01
      3  -5.281244665584D-01   4.879137630662D-01    4.25  9.888290268133D-01


I= 5     H=  2.0
0           0.                 5.323563127547D+01   -1.00 -8.793383961729D-01
1           2.746102518624D-01  -3.447381686952D+01   -.75 -5.228383830991D-01
2          -4.391893925071D-01   7.208760419019D+00   -.50 -2.677948180334D-01
3           1.655387518034D-01  -4.403310104530D-01   -.25 -9.868844299422D-02
```

CARDINAL SPLINES N=4

```
                                                              .25   4,378971036081D-02
0  -3,0  -9,596111585973D-04   -4,798055792987D-04           .50   4,819262481764D-02
1  -2,0   6,057280177338D-03    3,028640088669D-03           .75   2,862222390137D-02
2  -1,0  -1,733478056072D-02   -8,667390280362D-03          1,25  -2,422814921991D-02
3   0,0   2,795854364146D-02    1,397927182073D-02          1,50  -3,391600178986D-02
4   1,0  -2,604558195445D-02   -1,302279097723D-02          1,75  -2,469170406050D-02
5   2,0   1,302592129232D-02    6,512960646160D-03          2,25   2,881034085344D-02
6   3,0  -2,701771437343D-03   -1,350885718671D-03          2,50   4,579363592846D-02
                                                             2,75   3,728893652855D-02
      0   1,340672521714D+00    2,804087155860D+00          3,25  -5,336103046568D-02
      1   2,109020195510D+00   -3,110191637631D+00          3,50  -9,442304525621D-02
      2   1,050659373723D+00    3,245781324943D+00          3,75  -8,702396507263D-02
      3   1,655387518034D-01   -4,403310104530D-01          4,25   1,797319411717D-01


I= 6     H=  3,0
0           0.                 -1,277216542203D+01         -1,00   1,350817682042D-01
1          -4,214935760943D-02  8,554510999103D+00          -,75   8,030531094973D-02
2           6,746618327460D-02 -1,888755123931D+00          -,50   4,112450303839D-02
3          -2,546622732018D-02  1,309160859466D-01          -,25   1,515188565890D-02
                                                              ,25  -6,718603630818D-03
0  -3,0   1,494016550037D-04    7,470082750185D-05           .50  -7,390244200655D-03
1  -2,0  -9,596111585973D-04   -4,798055792987D-04           ,75  -4,385912052215D-03
2  -1,0   2,929369809042D-03    1,464684904521D-03          1,25   3,702833156069D-03
3   0,0  -5,109400750270D-03   -2,554700375135D-03          1,50   5,171529125724D-03
4   1,0   5,107070157474D-03    2,553535078737D-03          1,75   3,752341286715D-03
5   2,0  -2,701771437343D-03   -1,350885718671D-03          2,25  -4,325636520421D-03
6   3,0   5,849417246901D-04    2,924708623451D-04          2,50  -6,804987441964D-03
                                                             2,75  -5,460866579261D-03
      0  -2,068405610017D-01   -5,726942195383D-01          3,25   7,436000169296D-03
      1  -3,249403956066D-01    7,567145760743D-01          3,50   1,262418285420D-02
      2  -1,617298626070D-01   -7,105103504119D-01          3,75   1,096106496110D-02
      3  -2,546622732018D-02    1,309160859466D-01          4,25  -1,805669502133D-02
```

```
                N= 4      M= 7                    P=  3,5       Q= 7

I= 0     H= -3.5
0           1,000000000000D+00  2,028266544865D+00         -1,00   4,663101242720D+00
1          -2,042714295292D+00 -1,779746340005D+00          -,75   3,402843877901D+00
2           1,331252714567D+00  7,903454336780D-01          -,50   2,390312105395D+00
3          -2,891342328610D-01 -1,513623844820D-01          -,25   1,598399590872D+00
                                                              ,25   5,680070348147D-01
0  -3,5   5,958135852895D-04    2,979067926448D-04           .50   2,753189066820D-01
1  -2,5  -2,772543715423D-03   -1,386271857712D-03           ,75   9,489495750028D-02
2  -1,5   5,334026560819D-03    2,667013280410D-03          1,25  -3,518490981403D-02
3   -,5  -5,607424347459D-03   -2,803712173729D-03          1,50  -3,442249788459D-02
4    ,5   3,686830604968D-03    1,843415302484D-03          1,75  -1,778883259047D-02
5   1,5  -1,680140898081D-03   -8,400704490405D-04          2,25   1,065530801244D-02
6   2,5   5,239217625065D-04    2,619608812533D-04          2,50   1,214725554918D-02
```

CARDINAL SPLINES N=4

7	3,5	-8,048355262099D-05	-4,024177631050D-05	2,75	7,057913616724D-03
				3,25	-4,967188658675D-03
	0	-2,238284513985D+00	-1,008776317263D+00	3,50	-5,995151743764D-03
	1	-3,349628350961D+00	-1,809895933973D+00	3,75	-3,654659893403D-03
	2	-1,704656730473D+00	-7,989596033830D-01	4,25	2,798945835535D-03
	3	-2,891342328610D-01	-1,513623844820D-01	4,50	3,530656367311D-03

I= 1 H= -2.5

0		0.	-9,973655741211D+00	-1.00	-9,015827378962D+00
1		3,951148508410D+00	6,925222249860D+00	-,75	-5,663440697925D+00
2		-4,006527417624D+00	-2,817290567699D+00	-,50	-3,109475040227D+00
3		1,058151452929D+00	5,734297668643D-01	-,25	-1,254728707156D+00
				,25	7,539126107303D-01
0	-3,5	-2.772543715423D-03	-1,386271857712D-03	,50	1,106189670917D+00
1	-2,5	1,348662581468D-02	6,743312907342D-03	,75	1,155727263187D+00
2	-1,5	-2,752299435711D-02	-1,376149717856D-02	1,25	7,322189064736D-01
3	-,5	3.120051535259D-02	1,560025767629D-02	1,50	4,360310822243D-01
4	,5	-2,226560409308D-02	-1,113280204654D-02	1,75	1,779882438284D-01
5	1,5	1,074769453847D-02	5,373847269236D-03	2,25	-8,503358673788D-02
6	2,5	-3,397615302637D-03	-1,698807651318D-03	2,50	-9,131840354874D-02
7	3,5	5.239217625065D-04	2,619608812533D-04	2,75	-5,089980992369D-02
				3,25	3,407525375932D-02
	0	1,011730245789D+01	4,338613933298D+00	3,50	4,052186880929D-02
	1	1,479252248020D+01	8,277732208233D+00	3,75	2,444250114483D-02
	2	7,104062838134D+00	3,203721984376D+00	4,25	-1,848540962044D-02
	3	1.058151452929D+00	5,734297668643D-01	4,50	-2,323445357733D-02

I= 2 H= -1.5

0		0.	3,227271189211D+01	-1,00	9,614426191222D+00
1		-3,267604022730D+00	-1,765336581438D+01	-,75	5,803907912208D+00
2		4,804546082330D+00	5,037383908688D+00	-,50	3,027723042718D+00
3		-1,542276086161D+00	-9,147421182399D-01	-,25	1,141283199674D+00
				,25	-5,407146138200D-01
0	-3,5	5,334026560819D-03	2,667013280410D-03	,50	-6,254083294701D-01
1	-2,5	-2.752299435711D-02	-1,376149717856D-02	,75	-3,980815629887D-01
2	-1,5	6.077884276641D-02	3,038942138321D-02	1,25	4,357731852453D-01
3	-,5	-7,624609960897D-02	-3,812304980449D-02	1,50	7,945626814025D-01
4	,5	6.096492828538D-02	3,048246414269D-02	1.75	9,944208106063D-01
5	1,5	-3,237625728692D-02	-1,618812864346D-02	2.25	8,293392203016D-01
6	2.5	1,074769453847D-02	5,373847269236D-03	2,50	5,471892305950D-01
7	3,5	-1,680140898081D-03	-8,400704490405D-04	2,75	2,429179772492D-01
				3,25	-1,314927668793D-01
	0	-1,870601176517D+01	-7,025683896315D+00	3,50	-1,480946959653D-01
	1	-2,631442761284D+01	-1,600845129888D+01	3,75	-8,607940652067D-02
	2	-1,138935282236D+01	-4,567408332831D+00	4,25	6,234652466962D-02
	3	-1.542276086161D+00	-9,147421182399D-01	4,50	7,740186928423D-02

I= 3 H= -.5

0		0.	-7,353995051593D+01	-1,00	-6,633462523966D+00
1		2.102336632149D+00	3,758801501305D+01	-,75	-3,954342271414D+00
2		-3,313927549809D+00	-7,671803271773D+00	-,50	-2,031799996278D+00
3		1,217198342008D+00	9,366142117725D-01	-,25	-7,517233539943D-01
				,25	3,374820680181D-01
0	-3,5	-5,607424347459D-03	-2,803712173729D-03	,50	3,747924133705D-01

CARDINAL SPLINES N=4

```
1  -2,5   3,120051535259D-02    1.560025767629D-02        .75  2,254252770822D-01
2  -1,5  -7,624609960897D-02   -3.812304980449D-02       1.25 -1,994868732355D-01
3   -,5   1.088145624641D-01    5.440728123207D-02       1.50 -2,903519824819D-01
4    ,5  -1.005477086576D-01   -5,027385432879D-02       1,75 -2,240844939910D-01
5   1,5   6,096492828538D-02    3,048246414269D-02       2,25  3,299488212327D-01
6   2,5  -2,226560409308D-02   -1,113280204654D-02       2,50  6,725126584909D-01
7   3.5   3.686830604968D-03    1.843415302484D-03       2,75  9,224043637411D-01
                                                         3,25  8,824701432042D-01
      0   1,894994464094D+01    4,195846280280D+00       3.50  6,135679788998D-01
      1   2,363688285227D+01    1.830596439328D+01       3.75  2,852061238438D-01
      2   9,466655041271D+00    2.162645951838D+00       4.25 -1,681405055680D-01
      3   1.217198342008D+00    9,366142117725D-01       4,50 -1,982288131606D-01


I=  4     H=    .5
0             0.                1.349164688965D+02      -1,00  3,497592267696D+00
1            -1,094609467613D+00  -6,786505833878D+01    -,75  2,080380695540D+00
2             1.746952718545D+00  1,199709517545D+01      -,50  1,066046673635D+00
3            -6,560300815373D-01  -9,366142117725D-01    -,25  3,930873818364D-01
                                                          .25 -1,747180669919D-01
0  -3,5   3.686830604968D-03    1.843415302484D-03        .50 -1,925415109982D-01
1  -2,5  -2,226560409308D-02   -1,113280204654D-02        .75 -1,145667546977D-01
2  -1,5   6,096492828538D-02    3,048246414269D-02       1,25  9,762185450818D-02
3   -,5  -1,005477086576D-01   -5,027385432879D-02       1.50  1,374468973768D-01
4    ,5   1,088145624641D-01    5,440728123207D-02       1,75  1,009116340003D-01
5   1,5  -7,624609960897D-02   -3,812304980449D-02       2,25 -1,213890336244D-01
6   2,5   3,120051535259D-02    1,560025767629D-02       2,50 -1,982288131606D-01
7   3,5  -5,607424347459D-03   -2,803712173729D-03       2,75 -1,681405055680D-01
                                                         3,25  2,852061238438D-01
      0  -1,055825208038D+01    4,195846280280D+00       3,50  6,135679788998D-01
      1  -1,297504593429D+01   -1,830596439328D+01       3,75  8,824701432042D-01
      2  -5,141363137596D+00    2,162645951838D+00       4,25  9,224043637411D-01
      3  -6,560300815373D-01   -9,366142117725D-01       4,50  6,725126584909D-01


I=  5     H=   1.5
0             0.               -1,582255838391D+02      -1.00 -1,523978984190D+00
1             4.756214122256D-01  8,159708247401D+01     -,75 -9,060285472609D-01
2            -7.611494216462D-01  -1,417220057435D+01    -,50 -4,639990803142D-01
3             2,872081503187D-01  9,147421182399D-01     -,25 -1,709648192580D-01
                                                          .25  7,582103900458D-02
0  -3,5  -1,680140898081D-03   -8,400704490405D-04        .50  8,341124339031D-02
1  -2,5   1,074769453847D-02    5,373847269236D-03        .75  4,951117617159D-02
2  -1,5  -3.237625728692D-02   -1,618812864346D-02       1,25 -4,182666757401D-02
3   -,5   6,096492828538D-02    3,048246414269D-02       1,50 -5,844938905221D-02
4    .5  -7,624609960897D-02   -3,812304980449D-02       1,75 -4,244421714336D-02
5   1,5   6,077884276641D-02    3,038942138321D-02       2,25  4,907430326807D-02
6   2,5  -2,752299435711D-02   -1,376149717856D-02       2,50  7,740186928423D-02
7   3,5   5,334026560819D-03    2,667013280410D-03       2,75  6,234652466962D-02
                                                         3,25 -8,607940652067D-02
      0   4,654643972537D+00   -7,025683896315D+00       3,50 -1,480946959653D-01
      1   5,702474984914D+00    1,600845129888D+01       3,75 -1,314927668793D-01
      2   2,254536156700D+00   -4,567408332831D+00       4,25  2,429179772492D-01
      3   2,872081503187D-01    9,147421182399D-01       4,50  5,471892305950D-01


I=  6     H=   2.5
```

0		0.	9,714207222503D+01	-1.00	4,701558797850D-01
1		-1,466318199745D-01	-5,177733003113D+01	-,75	2,794815747619D-01
2		2,348159790112D-01	9,224734536451D+00	-,50	1,431084148399D-01
3		-8,870808079927D-02	-5,734297668643D-01	-,25	5,272001744431D-02
				,25	-2,336798809026D-02
0	-3,5	5,239217625065D-04	2.619608812533D-04	,50	-2,569633219556D-02
1	-2,5	-3,397615302637D-03	-1,698807651318D-03	,75	-1,524366326103D-02
2	-1,5	1,074769453847D-02	5,373847269236D-03	1,25	1,285026819545D-02
3	-,5	-2,226560409308D-02	-1.113280204654D-02	1,50	1,792360073555D-02
4	,5	3,120051535259D-02	1,560025767629D-02	1,75	1,297982038147D-02
5	1,5	-2,752299435711D-02	-1,376149717856D-02	2,25	-1,485852776152D-02
6	2,5	1,348662581468D-02	6,743312907342D-03	2,50	-2,323445357733D-02
7	3,5	-2,772543715423D-03	-1,386271857712D-03	2,75	-1,848540962044D-02
				3,25	2,444250114483D-02
	0	-1,440074591292D+00	4.338613933298D+00	3,50	4,052186880929D-02
	1	-1,762941936269D+00	-8,277732208233D+00	3,75	3,407525375932D-02
	2	-6,966188693811D-01	3,203721984376D+00	4,25	-5,089980992369D-02
	3	-8,870808079927D-02	-5,734297668643D-01	4,50	-9,131840354874D-02

I= 7 H= 3.5

0		0,	-2,362032946228D+01	-1.00	-7,200669430317D-02
1		2,245305282485D-02	1.296518078737D+01	-,75	-4,280254381070D-02
2		-3,596310537527D-02	-2,388264640444D+00	-,50	-2,191611976912D-02
3		1,359053610305D-02	1.513623844820D-01	-,25	-8,073309418776D-03
				,25	3,577916334541D-03
0	-3,5	-8,048355262099D-05	-4,024177631050D-05	,50	3,933938303731D-03
1	-2,5	5,239217625065D-04	2,619608812533D-04	,75	2,333307006096D-03
2	-1,5	-1,680140898081D-03	-8,400704490405D-04	1,25	-1,965763799034D-03
3	-,5	3,686830604968D-03	1,843415302484D-03	1,50	-2,740392320400D-03
4	,5	-5,607424347459D-03	-2,803712173729D-03	1,75	-1,982965091605D-03
5	1,5	5,334026560819D-03	2,667013280410D-03	2,25	2,263495308988D-03
6	2,5	-2,772543715423D-03	-1,386271857712D-03	2,50	3,530656367311D-03
7	3,5	5,958135852895D-04	2,979067926448D-04	2,75	2,798945835535D-03
				3,25	-3,654659893403D-03
	0	2,207318794580D-01	-1,008776317263D+00	3,50	-5,995151743764D-03
	1	2,701635169849D-01	1,809895933973D+00	3,75	-4,967188658675D-03
	2	1,067375237067D-01	-7,989596033830D-01	4,25	7,057913616724D-03
	3	1,359053610305D-02	1,513623844820D-01	4,50	1,214725554918D-02

N= 4 M= 8 P= 4.0 Q= 7

I= 0 H= -4.0

0		1,000000000000D+00	-7,931175403479D-01	-1,00	4,665872871511D+00
1		-2,043578408303D+00	-4,719743016343D-01	-,75	3,404491357682D+00
2		1,332636976276D+00	5,887296188472D-01	-,50	2,391155634087D+00
3		-2,896574869315D-01	-1,411953724602D-01	-,25	1,598710311326D+00
				,25	5,678693472632D-01
0	-4,0	5,989189585431D-04	2,994594792716D-04	,50	2,751675331053D-01

CARDINAL SPLINES N=4

1	-3.0	-2.792819739871D-03	-1.396409869936D-03	.75	9.480518665911D-02
2	-2.0	5.399749944213D-03	2.699874972106D-03	1.25	-3.510931642890D-02
3	-1.0	-5.758262121436D-03	-2.879131060718D-03	1.50	-3.431716123818D-02
4	0.0	3.972906466389D-03	1.986453233194D-03	1.75	-1.771265836256D-02
5	1.0	-2.094774318377D-03	-1.047387159188D-03	2.25	1.056855744965D-02
6	2.0	9.127399060530D-04	4.563699530265D-04	2.50	1.201221184336D-02
7	3.0	-2.815852698737D-04	-1.407926349369D-04	2.75	6.951168413712D-03
8	4.0	4.312617436019D-05	2.156308718009D-05	3.25	-4.829241286239D-03
				3.50	-5.770842460583D-03
	0	-4.390201176410D+00	-2.297844682785D+00	3.75	-3.471114660894D-03
	1	-5.286041970807D+00	-2.539515228948D+00	4.25	2.549002343257D-03
	2	-2.143252866902D+00	-1.105614850676D+00	4.50	3.115117078380D-03
	3	-2.896574869315D-01	-1.411953724602D-01	4.75	1.915757607974D-03

I= 1 H= -3.0

0		0.	8.448108091050D+00	-1.00	-9.033924273777D+00
1		3.956790592346D+00	-1.613660408325D+00	-.75	-5.674197646475D+00
2		-4.015565727019D+00	-1.500873230148D+00	-.50	-3.114982722229D+00
3		1.061567954413D+00	5.070459323037D-01	-.25	-1.256757505313D+00
				.25	7.548116189753D-01
0	-4.0	-2.792819739871D-03	-1.396409869936D-03	.50	1.107178039816D+00
1	-3.0	1.361901478498D-02	6.809507392488D-03	.75	1.156313407145D+00
2	-2.0	-2.795212439066D-02	-1.397606219533D-02	1.25	7.317253318853D-01
3	-1.0	3.218538581520D-02	1.609269290760D-02	1.50	4.353433038984D-01
4	0.0	-2.413348942809D-02	-1.206674471404D-02	1.75	1.774908767226D-01
5	1.0	1.345497533221D-02	6.727487666105D-03	2.25	-8.446716319127D-02
6	2.0	-5.936339488277D-03	-2.968169744138D-03	2.50	-9.043665785918D-02
7	3.0	1.836982384386D-03	9.184911921932D-04	2.75	-5.020283465881D-02
8	4.0	-2.815852698737D-04	-1.407926349369D-04	3.25	3.317454905569D-02
				3.50	3.905727819867D-02
	0	1.951845981950D+01	1.043043444281D+01	3.75	2.324407270556D-02
	1	2.278752658800D+01	1.071755850106D+01	4.25	-1.685344465321D-02
	2	8.723249725933D+00	4.583677957496D+00	4.50	-2.052125806318D-02
	3	1.061567954413D+00	5.070459323037D-01	4.75	-1.258803157590D-02

I= 2 H= -2.0

0		0.	-2.744020916064D+01	-1.00	9.673086070981D+00
1		-3.285892462193D+00	1.002485385501D+01	-.75	5.838775844800D+00
2		4.833843160519D+00	7.703047481175D-01	-.50	3.045575827260D+00
3		-1.553350448270D+00	-6.995633379456D-01	-.25	1.147859413835D+00
				.25	-5.436286891955D-01
0	-4.0	5.399749944213D-03	2.699874972106D-03	.50	-6.286120614542D-01
1	-3.0	-2.795212439066D-02	-1.397606219533D-02	.75	-3.999815096071D-01
2	-2.0	6.216983920127D-02	3.108491960063D-02	1.25	4.373730744140D-01
3	-1.0	-7.943849164680D-02	-3.971924582340D-02	1.50	7.967920690578D-01
4	0.0	6.701955424631D-02	3.350977712316D-02	1.75	9.960329929680D-01
5	1.0	-4.115172774545D-02	-2.057586387272D-02	2.25	8.275031960843D-01
6	2.0	1.897679997334D-02	9.488399986669D-03	2.50	5.443311106436D-01
7	3.0	-5.936339488277D-03	-2.968169744138D-03	2.75	2.406587783285D-01
8	4.0	9.127399060530D-04	4.563699530265D-04	3.25	-1.285731925460D-01
				3.50	-1.433473230045D-01
	0	-3.521650796973D+01	-1.978797139924D+01	3.75	-8.219478051890D-02
	1	-3.917596869498D+01	-1.739174838144D+01	4.25	5.705661887470D-02
	2	-1.380636221872D+01	-7.624455307229D+00	4.50	6.860722665322D-02
	3	-1.553350448270D+00	-6.995633379456D-01	4.75	4.171464103826D-02

CARDINAL SPLINES N=4

I= 3 H= -1.0

0		0.	6.350357378606D+01	-1.00	-6.768089276697D+00
1		2.144309326732D+00	-2.593459789926D+01	-.75	-4.034365560903D+00
2		-3.381165507288D+00	2.121312796416D+00	-.50	-2.072772845523D+00
3		1.242614442677D+00	4.427703688211D-01	-.25	-7.668160265554D-01
				.25	3.441699866880D-01
0	-4.0	-5.758262121436D-03	-2.879131060718D-03	.50	3.821451054560D-01
1	-3.0	3.218538581520D-02	1.609269290760D-02	.75	2.297857299955D-01
2	-2.0	-7.943849164680D-02	-3.971924582340D-02	1.25	-2.031586823901D-01
3	-1.0	1.161412289227D-01	5.807061446133D-02	1.50	-2.954685156533D-01
4	0.0	-1.144433155947D-01	-5.722165779735D-02	1.75	-2.277845165116D-01
5	1.0	8.110498135668D-02	4.055249067834D-02	2.25	3.341625696979D-01
6	2.0	-4.115172774545D-02	-2.057586387272D-02	2.50	6.790721572412D-01
7	3.0	1.345497533221D-02	6.727487666105D-03	2.75	9.275893149499D-01
8	4.0	-2.094774318377D-03	-1.047387159188D-03	3.25	8.757696042077D-01
				3.50	6.026725694934D-01
	0	3.400591352165D+01	2.204349053622D+01	3.75	2.762907529576D-01
	1	3.474047851692D+01	1.228888217548D+01	4.25	-1.559999617698D-01
	2	1.153020780483D+01	7.434557222269D+00	4.50	-1.780447591924D-01
	3	1.242614442677D+00	4.427703688211D-01	4.75	-1.044983166428D-01

I= 4 H= 0.0

0		0.	-1.249974984935D+02	-1.00	3.752922635163D+00
1		-1.174214028476D+00	5.261064034975D+01	-.75	2.232151242602D+00
2		1.874474864348D+00	-6.576330043719D+00	-.50	1.143754948117D+00
3		-7.042337423389D-01	3.453106931948D-26	-.25	4.217118383648D-01
				.25	-1.874022378343D-01
0	-4.0	3.972906466389D-03	1.986453233194D-03	.50	-2.064864776114D-01
1	-3.0	-2.413348942809D-02	-1.206674471404D-02	.75	-1.228367012134D-01
2	-2.0	6.701955424631D-02	3.350977712316D-02	1.25	1.045857332449D-01
3	-1.0	-1.144433155947D-01	-5.722165779735D-02	1.50	1.471508103681D-01
4	0.0	1.351686886202D-01	6.758434431008D-02	1.75	1.079290216344D-01
5	1.0	-1.144433155947D-01	-5.722165779735D-02	2.25	-1.293807434608D-01
6	2.0	6.701955424631D-02	3.350977712316D-02	2.50	-2.106694253891D-01
7	3.0	-2.413348942809D-02	-1.206674471404D-02	2.75	-1.779741788209D-01
8	4.0	3.972906466389D-03	1.986453233194D-03	3.25	2.979142301413D-01
				3.50	6.342319912970D-01
	0	-1.977621779402D+01	-1.977621779402D+01	3.75	8.993788547217D-01
	1	-1.998163474596D+01	-5.517597181372D-26	4.25	8.993788547217D-01
	2	-6.576330043719D+00	-6.576330043719D+00	4.50	6.342319912970D-01
	3	-7.042337423389D-01	3.453106931948D-26	4.75	2.979142301413D-01

I= 5 H= 1.0

0		0.	2.184892383990D+02	-1.00	-1.894050415743D+00
1		5.909988899960D-01	-9.301831765705D+01	-.75	-1.126002160959D+00
2		-9.459778207122D-01	1.274780164812D+01	-.50	-5.766281133054D-01
3		3.570737050346D-01	-4.427703688211D-01	-.25	-2.124526129347D-01
				.25	9.420525749077D-02
0	-4.0	-2.094774318377D-03	-1.047387159188D-03	.50	1.036228375249D-01
1	-3.0	1.345497533221D-02	6.727487666105D-03	.75	6.149749404003D-02
2	-2.0	-4.115172774545D-02	-2.057586387272D-02	1.25	-5.191999332613D-02
3	-1.0	8.110498135668D-02	4.055249067834D-02	1.50	-7.251407320230D-02
4	0.0	-1.144433155947D-01	-5.722165779735D-02	1.75	-5.261509776852D-02

5	1.0	1.161412289227D-01	5.807061446133D-02	2.25	6.065734969158D-02
6	2.0	-7.943849164680D-02	-3.971924582340D-02	2.50	9.543307811551D-02
7	3.0	3.218538581520D-02	1.609269290760D-02	2.75	7.659928110507D-02
8	4.0	-5.758262121436D-03	-2.879131060718D-03	3.25	-1.044983166428D-01
				3.50	-1.780447591924D-01
	0	1.008106755080D+01	2.204349053622D+01	3.75	-1.559999617698D-01
	1	1.016271416596D+01	-1.228888217548D+01	4.25	2.762907529576D-01
	2	3.338906639703D+00	7.434557222269D+00	4.50	6.026725694934D-01
	3	3.570737050346D-01	-4.427703688211D-01	4.75	8.757696042077D-01

I= 6 H= 2.0

0		0.	-2.561183034692D+02	-1.00	8.171864855021D-01
1		-2.548258403257D-01	1.119664310607D+02	-.75	4.857595251653D-01
2		4.081368727980D-01	-1.601921536258D+01	-.50	2.487251099096D-01
3		-1.542237723784D-01	6.995633379456D-01	-.25	9.162476107470D-02
				.25	-4.060759626573D-02
0	-4.0	9.127399060530D-04	4.563699530265D-04	.50	-4.464954273011D-02
1	-3.0	-5.936339488277D-03	-2.968169744138D-03	.75	-2.648370719728D-02
2	-2.0	1.897679997334D-02	9.488399986669D-03	1.25	2.231517861765D-02
3	-1.0	-4.115172774545D-02	-2.057586387272D-02	1.50	3.111261086625D-02
4	0.0	6.701955424631D-02	3.350977712316D-02	1.75	2.251745707291D-02
5	1.0	-7.943849164680D-02	-3.971924582340D-02	2.25	-2.572040812753D-02
6	2.0	6.216983920127D-02	3.108491960063D-02	2.50	-4.014303064928D-02
7	3.0	-2.795212439066D-02	-1.397606219533D-02	2.75	-3.185078256529D-02
8	4.0	5.399749944213D-03	2.699874972106D-03	3.25	4.171464103826D-02
				3.50	6.860722665322D-02
	0	-4.359434828752D+00	-1.978797139924D+01	3.75	5.705661887470D-02
	1	-4.392471932105D+00	1.739174838144D+01	4.25	-8.219478051890D-02
	2	-1.442548395743D+00	-7.624455307229D+00	4.50	-1.433473230045D-01
	3	-1.542237723784D-01	6.995633379456D-01	4.75	-1.285731925460D-01

I= 7 H= 3.0

0		0.	1.590904554344D+02	-1.00	-2.514953624821D-01
1		7.841238407067D-02	-7.172518691160D+01	-.75	-1.494921382806D-01
2		-1.256068886061D-01	1.066822914514D+01	-.50	-7.654242541253D-02
3		4.747608980533D-02	-5.070459323037D-01	-.25	-2.819534045876D-02
				.25	1.249446219639D-02
0	-4.0	-2.815852698737D-04	-1.407926349369D-04	.50	1.373678122455D-02
1	-3.0	1.836982384386D-03	9.184911921932D-04	.75	8.146801502416D-03
2	-2.0	-5.936339488277D-03	-2.968169744138D-03	1.25	-6.861136535980D-03
3	-1.0	1.345497533221D-02	6.727487666105D-03	1.50	-9.561917061902D-03
4	0.0	-2.413348942809D-02	-1.206674471404D-02	1.75	-6.915952785249D-03
5	1.0	3.218538581520D-02	1.609269290760D-02	2.25	7.881398769650D-03
6	2.0	-2.795212439066D-02	-1.397606219533D-02	2.50	1.227598870817D-02
7	3.0	1.361901478498D-02	6.809507392488D-03	2.75	9.711687680392D-03
8	4.0	-2.792819739871D-03	-1.396409869936D-03	3.25	-1.258803157590D-02
				3.50	-2.052125806318D-02
	0	1.342409066126D+00	1.043043444281D+01	3.75	-1.685344465321D-02
	1	1.352409585877D+00	-1.071755850106D+01	4.25	2.324407270556D-02
	2	4.441061890578D-01	4.583677957496D+00	4.50	3.905727819867D-02
	3	4.747608980533D-02	-5.070459323037D-01	4.75	3.317454905569D-02

I= 8 H= 4.0

0		0.	-3.918224704684D+01	-1.00	3.849126554118D-02

CARDINAL SPLINES N=4

1		-1.200045384672D-02	1.816181191245D+01	-.75	2.287953636787D-02
2		1.922406968341D-02	-2.799959320199D+00	-.50	1.171458709559D-02
3		-7.266742011055D-03	1.411953724602D-01	-.25	4.315160660815D-03
				.25	-1.912149318176D-03
0	-4.0	4.312617436019D-05	2.156308718009D-05	.50	-2.102215330650D-03
1	-3.0	-2.815852698737D-04	-1.407926349369D-04	.75	-1.246701324659D-03
2	-2.0	9.127399060530D-04	4.563699530265D-04	1.25	1.049810519252D-03
3	-1.0	-2.094774318377D-03	-1.047387159188D-03	1.50	1.462872965101D-03
4	0.0	3.972906466389D-03	1.986453233194D-03	1.75	1.057877030087D-03
5	1.0	-5.758262121436D-03	-2.879131060718D-03	2.25	-1.204756913531D-03
6	2.0	5.399749944213D-03	2.699874972106D-03	2.50	-1.875432654290D-03
7	3.0	-2.792819739871D-03	-1.396409869936D-03	2.75	-1.482434432589D-03
8	4.0	5.989189585431D-04	2.994594792716D-04	3.25	1.915757607974D-03
				3.50	3.115117078380D-03
	0	-2.054881891598D-01	-2.297844682785D+00	3.75	2.549002343257D-03
	1	-2.070115129101D-01	2.539515228948D+00	4.25	-3.471114660894D-03
	2	-6.797683444925D-02	-1.105614850676D+00	4.50	-5.770842460583D-03
	3	-7.266742011055D-03	1.411953724602D-01	4.75	-4.829241286239D-03

N= 4 M= 9 P= 4.5 Q= 7

I= 0 H= -4.5

0		1.000000000000D+00	1.529602851740D+00	-1.00	4.666665589189D+00
1		-2.043825545806D+00	-1.404979333586D+00	-.75	3.404962552923D+00
2		1.333032890785D+00	7.138656095106D-01	-.50	2.391396889674D+00
3		-2.898071525985D-01	-1.468477273269D-01	-.25	1.598799178885D+00
				.25	5.678299690725D-01
0	-4.5	5.998076200311D-04	2.999038100155D-04	.50	2.751242417153D-01
1	-3.5	-2.798626102429D-03	-1.399313051214D-03	.75	9.477951386163D-02
2	-2.5	5.418616618195D-03	2.709308309098D-03	1.25	-3.508770053017D-02
3	-1.5	-5.801990734591D-03	-2.900995367296D-03	1.50	-3.428704322223D-02
4	-.5	4.059551521213D-03	2.029775760607D-03	1.75	-1.769088166345D-02
5	.5	-2.251031593958D-03	-1.125515796979D-03	2.25	1.054377032731D-02
6	1.5	1.136065872086D-03	5.680329360428D-04	2.50	1.197364389327D-02
7	2.5	-4.901546809906D-04	-2.450773404953D-04	2.75	6.920702945965D-03
8	3.5	1.508488348193D-04	7.542441740963D-05	3.25	-4.789965373000D-03
9	4.5	-2.308735437624D-05	-1.154367718812D-05	3.50	-5.707110210491D-03
				3.75	-3.419119574211D-03
	0	-7.611975698274D+00	-3.718524709472D+00	4.25	2.478935143045D-03
	1	-7.652314049101D+00	-3.901188283100D+00	4.50	2.999656503377D-03
	2	-2.579363669295D+00	-1.268578709403D+00	4.75	1.820340257124D-03
	3	-2.898071525985D-01	-1.468477273269D-01	5.25	-1.350828193141D-03

I= 1 H= -3.5

0		0.	-6.728150697528D+00	-1.00	-9.039103755849D+00
1		3.958405346628D+00	4.482435300905D+00	-.75	-5.677276355796D+00
2		-4.018152564873D+00	-2.318490445092D+00	-.50	-3.116559045076D+00
3		1.062545844348D+00	5.439774547174D-01	-.25	-1.257338150780D+00

CARDINAL SPLINES N=4

				.25	7.550689093558D-01
0	-4.5	-2.798626102429D-03	-1.399313051214D-03	.50	1.107460898373D+00
1	-3.5	1.365695256737D-02	6.828476283684D-03	.75	1.156481148824D+00
2	-2.5	-2.807539601983D-02	-1.403769800991D-02	1.25	7.315840972907D-01
3	-1.5	3.247110111350D-02	1.623555055675D-02	1.50	4.351465179203D-01
4	-.5	-2.469961343477D-02	-1.234980671738D-02	1.75	1.773485914857D-01
5	.5	1.447593371355D-02	7.237966856776D-03	2.25	-8.430520836255D-02
6	1.5	-7.395513251191D-03	-3.697756625595D-03	2.50	-9.018466145292D-02
7	2.5	3.199739334931D-03	1.599869667466D-03	2.75	-5.000377848992D-02
8	3.5	-9.854267559560D-04	-4.927133779780D-04	3.25	3.291792693901D-02
9	4.5	1.508488348193D-04	7.542441740963D-05	3.50	3.864086254921D-02
				3.75	2.290434567948D-02
	0	3.326972468731D+01	1.606332220455D+01	4.25	-1.639563752041D-02
	1	3.234469230688D+01	1.666265166916D+01	4.50	-1.976685836354D-02
	2	1.032621633382D+01	5.025205193593D+00	4.75	-1.196459088353D-02
	3	1.062545844348D+00	5.439774547174D-01	5.25	8.851794963706D-03

I= 2 H= -2.5

0		0.	2.187216566850D+01	-1.00	9.689915815710D+00
1		-3.291139300003D+00	-9.783253056957D+00	-.75	5.848779526877D+00
2		4.842248599546D+00	3.426996830913D+00	-.50	3.050697789408D+00
3		-1.556527916161D+00	-8.195653161438D-01	-.25	1.149746111162D+00
				.25	-5.444647054931D-01
0	-4.5	5.418616618195D-03	2.709308309098D-03	.50	-6.295311566929D-01
1	-3.5	-2.807539601983D-02	-1.403769800991D-02	.75	-4.005265543814D-01
2	-2.5	6.257038698238D-02	3.128519349119D-02	1.25	4.378319894447D-01
3	-1.5	-8.036686932411D-02	-4.018343466206D-02	1.50	7.974314877742D-01
4	-.5	6.885906684334D-02	3.442953342167D-02	1.75	9.964953218615D-01
5	.5	-4.446913838578D-02	-2.223456919289D-02	2.25	8.269769545823D-01
6	1.5	2.371810839065D-02	1.185905419533D-02	2.50	5.435122961114D-01
7	2.5	-1.036435975878D-02	-5.182179879391D-03	2.75	2.400119830578D-01
8	3.5	3.199739334931D-03	1.599869667466D-03	3.25	-1.277393476450D-01
9	4.5	-4.901546809906D-04	-2.450773404953D-04	3.50	-1.419942593229D-01
				3.75	-8.109090197523D-02
	0	-5.859319906936D+01	-2.743867669542D+01	4.25	5.556906141756D-02
	1	-5.426997281086D+01	-2.872887453448D+01	4.50	6.615594789752D-02
	2	-1.617087826862D+01	-7.637134937028D+00	4.75	3.968888874093D-02
	3	-1.556527916161D+00	-8.195653161438D-01	5.25	-2.905672425693D-02

I= 3 H= -1.5

0		0.	-5.079117496889D+01	-1.00	-6.807096756098D+00
1		2.156470290615D+00	1.997604036667D+01	-.75	-4.057551796276D+00
2		-3.400647382682D+00	-4.036288721955D+00	-.50	-2.084644376328D+00
3		1.249979082801D+00	7.209073678155D-01	-.25	-7.711889572401D-01
				.25	3.461076802795D-01
0	-4.5	-5.801990734591D-03	-2.900995367296D-03	.50	3.842753569346D-01
1	-3.5	3.247110111350D-02	1.623555055675D-02	.75	2.310490184856D-01
2	-2.5	-8.036686932411D-02	-4.018343466206D-02	1.25	-2.042223419343D-01
3	-1.5	1.182929949600D-01	5.914649748000D-02	1.50	-2.969505412461D-01
4	-.5	-1.187068829211D-01	-5.935344146054D-02	1.75	-2.288560886050D-01
5	.5	8.879397660317D-02	4.439698830158D-02	2.25	3.353822765253D-01
6	1.5	-5.214099070844D-02	-2.607049535422D-02	2.50	6.809699810950D-01
7	2.5	2.371810839065D-02	1.185905419533D-02	2.75	9.290884377000D-01
8	3.5	-7.395513251191D-03	-3.697756625595D-03	3.25	8.738369434144D-01
9	4.5	1.136065872086D-03	5.680329360428D-04	3.50	5.995364789263D-01

```
                                                      3.75  2.737322162568D-01
       0  5.474535072874D+01  2.305884395371D+01      4.25 -1.525521455490D-01
       1  4.748687312666D+01  2.744456446386D+01      4.50 -1.723632585451D-01
       2  1.347407023514D+01  5.695960743554D+00      4.75 -9.980308850452D-02
       3  1.249979082801D+00  7.209073678155D-01      5.25  7.012329625406D-02


I= 4     H=  -.5
0            0.               1.014692027714D+02     -1.00  3.830213108092D+00
1           -1.198310090672D+00 -3.835793794725D+01   -.75  2.278093075424D+00
2            1.913076778285D+00  5.624507866684D+00   -.50  1.167277519799D+00
3           -7.188262391348D-01 -5.511081598074D-01   -.25  4.303764812972D-01
                                                       .25 -1.912416362362D-01
0  -4.5   4.059551521213D-03  2.029775760607D-03      .50 -2.107074154101D-01
1  -3.5  -2.469961343477D-02 -1.234980671738D-02      .75 -1.253398151379D-01
2  -2.5   6.885906684334D-02  3.442953342167D-02     1.25  1.066932969858D-01
3  -1.5  -1.187068829211D-01 -5.935344146054D-02     1.50  1.500873358940D-01
4   -.5   1.436166360416D-01  7.180831802080D-02     1.75  1.100522635137D-01
5    .5  -1.296784983873D-01 -6.483924919363D-02     2.25 -1.317975035019D-01
6   1.5   8.879397660317D-02  4.439698830158D-02     2.50 -2.144298248163D-01
7   2.5  -4.446913838578D-02 -2.223456919289D-02     2.75 -1.809445811158D-01
8   3.5   1.447593371355D-02  7.237966856776D-03     3.25  3.017436564212D-01
9   4.5  -2.251031593958D-03 -1.125515796979D-03     3.50  6.404459258246D-01
                                                      3.75  9.044484084239D-01
       0 -3.215563168890D+01 -7.464964753360D+00      4.25  8.925472585670D-01
       1 -2.764931311354D+01 -2.121718785540D+01      4.50  6.229745125077D-01
       2 -7.791077450034D+00 -1.815452290716D+00      4.75  2.886109783318D-01
       3 -7.188262391348D-01 -5.511081598074D-01      5.25 -1.651424875785D-01


I= 5     H=   .5
0            0.              -1.899249500521D+02     -1.00 -2.033437467349D+00
1            6.344541689913D-01  7.103607918014D+01   -.75 -1.208854502002D+00
2           -1.015593217877D+00 -9.255412448116D+00   -.50 -6.190491490250D-01
3            3.833900804800D-01  5.511081598074D-01   -.25 -2.280785883727D-01
                                                       .25  1.011292987459D-01
0  -4.5  -2.251031593958D-03 -1.125515796979D-03      .50  1.112349539020D-01
1  -3.5   1.447593371355D-02  7.237966856776D-03      .75  6.601165544369D-02
2  -2.5  -4.446913838578D-02 -2.223456919289D-02     1.25 -5.572081229875D-02
3  -1.5   8.879397660317D-02  4.439698830158D-02     1.50 -7.780985700988D-02
4   -.5  -1.296784983873D-01 -6.483924919363D-02     1.75 -5.644419098267D-02
5    .5   1.436166360416D-01  7.180831802080D-02     2.25  6.501577892357D-02
6   1.5  -1.187068829211D-01 -5.935344146054D-02     2.50  1.022146511757D-01
7   2.5   6.885906684334D-02  3.442953342167D-02     2.75  8.195615889721D-02
8   3.5  -2.469961343477D-02 -1.234980671738D-02     3.25 -1.114043739908D-01
9   4.5   4.059551521213D-03  2.029775760607D-03     3.50 -1.892510822987D-01
                                                      3.75 -1.651424875785D-01
       0  1.722570218218D+01 -7.464964753360D+00      4.25  2.886109783318D-01
       1  1.478506259725D+01  2.121718785540D+01      4.50  6.229745125077D-01
       2  4.160172868602D+00 -1.815452290716D+00      4.75  8.925472585670D-01
       3  3.833900804800D-01  5.511081598074D-01      5.25  9.044484084239D-01


I= 6     H=  1.5
0            0.               3.275952729903D+02     -1.00  1.016401196219D+00
1           -3.169329782158D-01 -1.225033337506D+02   -.75  6.041737176094D-01
2            5.076325651734D-01  1.542821020906D+01   -.50  3.093540870049D-01
```

3		-1,918356528297D-01	-7,209073678155D-01	-,25	1,139577119527D-01
				,25	-5,050357196611D-02
0	-4,5	1,136065872086D-03	5,680329360428D-04	,50	-5,552892890362D-02
1	-3,5	-7,395513251191D-03	-3,697756625595D-03	,75	-3,293543530115D-02
2	-2,5	2,371810839065D-02	1,185905419533D-02	1,25	2,774738371414D-02
3	-1,5	-5,214099070844D-02	-2,607049535422D-02	1,50	3,868144905524D-02
4	-,5	8,879397660317D-02	4.439698830158D-02	1,75	2,799007227050D-02
5	,5	-1,187068829211D-01	-5,935344146054D-02	2,25	-3,194956075418D-02
6	1,5	1,182929949600D-01	5,914649748000D-02	2,50	-4,983538800830D-02
7	2,5	-8,036686932411D-02	-4.018343466206D-02	2,75	-3,950693747648D-02
8	3,5	3,247110111350D-02	1.623555055675D-02	3,25	5,158491378545D-02
9	4,5	-5.801990734591D-03	-2.900995367296D-03	3,50	8,462352364864D-02
				3,75	7,012329625406D-02
	0	-8,627662821319D+00	2,305884395371D+01	4,25	-9,980308850452D-02
	1	-7.402255801062D+00	-2,744456446386D+01	4,50	-1,723632585451D-01
	2	-2.082148748028D+00	5.695960743554D+00	4,75	-1,525521455490D-01
	3	-1.918356528297D-01	-7.209073678155D-01	5,25	2,737322162568D-01

I= 7 H= 2.5

0		0.	-3,860534840090D+02	-1,00	-4,375467009725D-01
1		1,364157116998D-01	1.472516819235D+02	-,75	-2,600819582854D-01
2		-2.185282731457D-01	-1,870126670497D+01	-,50	-1,331652636522D-01
3		8,260271612689D-02	8,195653161438D-01	-,25	-4,905261243605D-02
				,25	2,173654837616D-02
0	-4,5	-4,901546809906D-04	-2,450773404953D-04	,50	2,389729774590D-02
1	-3,5	3,199739334931D-03	1,599869667466D-03	,75	1,417222324452D-02
2	-2,5	-1,036435975878D-02	-5,182179879391D-03	1,25	-1,193440158513D-02
3	-1,5	2,371810839065D-02	1,185905419533D-02	1,50	-1,663063438108D-02
4	-,5	-4.446913838578D-02	-2,223456919289D-02	1,75	-1,202695778529D-02
5	,5	6,885906684334D-02	3,442953342167D-02	2,25	1,369895199993D-02
6	1,5	-8,036686932411D-02	-4,018343466206D-02	2,50	2,132791089445D-02
7	2,5	6,257038698238D-02	3,128519349119D-02	2,75	1,686195215529D-02
8	3,5	-2,807539601983D-02	-1,403769800991D-02	3,25	-2,180611315869D-02
9	4,5	5,418616618195D-03	2.709308309098D-03	3,50	-3,547925731400D-02
				3,75	-2,905672425693D-02
	0	3,715845678511D+00	-2.743867669542D+01	4,25	3,968888874093D-02
	1	3,187776258097D+00	2.872887453448D+01	4,50	6,615594789752D-02
	2	8,966083945673D-01	-7,637134937028D+00	4,75	5,556906141756D-02
	3	8,260271612689D-02	8,195653161438D-01	5,25	-8,109090197523D-02

I= 8 H= 3.5

0		0.	2,423756054471D+02	-1,00	1,345837129276D-01
1		-4,195821579385D-02	-9,493612878557D+01	-,75	7,999735420519D-02
2		6,721643204637D-02	1,236890083228D+01	-,50	4,095934904444D-02
3		-2,540906508733D-02	-5,439774547174D-01	-,25	1,508759759335D-02
				,25	-6,685534380473D-03
0	-4,5	1,508488348193D-04	7,542441740963D-05	,50	-7,349954514729D-03
1	-3,5	-9,854267559560D-04	-4,927133779780D-04	,75	-4,358732264248D-03
2	-2,5	3,199739334931D-03	1,599869667466D-03	1,25	3,670068232893D-03
3	-1,5	-7,395513251191D-03	-3,697756625595D-03	1,50	5,113748860934D-03
4	-,5	1,447593371355D-02	7,237966856776D-03	1,75	3,697626805837D-03
5	,5	-2,469961343477D-02	-1,234980671738D-02	2,25	-4,209427195584D-03
6	1,5	3,247110111350D-02	1,623555055675D-02	2,50	-6,550601150733D-03
7	2,5	-2,807539601983D-02	-1,403769800991D-02	2,75	-5,175428089222D-03
8	3,5	1,365695256737D-02	6,828476283684D-03	3,25	6,676744691890D-03

9	4.5	-2.798626102429D-03	-1.399313051214D-03	3.50	1.084067695706D-02
				3.75	8.851794963706D-03
	0	-1.143080278217D+00	1.606332220455D+01	4.25	-1.196459088353D-02
	1	-9.806110314321D-01	-1.666265166916D+01	4.50	-1.976685836354D-02
	2	-2.758059466326D-01	5.025205193593D+00	4.75	-1.639563752041D-02
	3	-2.540906508733D-02	-5.439774547174D-01	5.25	2.290434567948D-02

I= 9 H= 4.5

0		0.	-6.034409000149D+01	-1.00	-2.059474186986D-02
1		6.420612556774D-03	2.423939610283D+01	-.75	-1.224161467966D-02
2		-1.028582725774D-02	-3.251023028316D+00	-.50	-6.267800849741D-03
3		3.888302055346D-03	1.468477273269D-01	-.25	-2.308772062417D-03
				.25	1.023042246059D-03
0	-4.5	-2.308735437624D-05	-1.154367718812D-05	.50	1.124706850913D-03
1	-3.5	1.508488348193D-04	7.542441740963D-05	.75	6.669772249026D-04
2	-2.5	-4.901546809906D-04	-2.450773404953D-04	1.25	-5.615793198394D-04
3	-1.5	1.136065872086D-03	5.680329360428D-04	1.50	-7.824636453609D-04
4	-.5	-2.251031593958D-03	-1.125515796979D-03	1.75	-5.657569008937D-04
5	.5	4.059551521213D-03	2.029775760607D-03	2.25	6.439674557785D-04
6	1.5	-5.801990734591D-03	-2.900995367296D-03	2.50	1.001992258509D-03
7	2.5	5.418616618195D-03	2.709308309098D-03	2.75	7.914904152623D-04
8	3.5	-2.798626102429D-03	-1.399313051214D-03	3.25	-1.020385084426D-03
9	4.5	5.998076200311D-04	2.999038100155D-04	3.50	-1.655758759705D-03
				3.75	-1.350828193141D-03
	0	1.749262793296D-01	-3.718524709472D+00	4.25	1.820340257124D-03
	1	1.500625170993D-01	3.901188283100D+00	4.50	2.999656503377D-03
	2	4.220625048942D-02	-1.268578709403D+00	4.75	2.478935143045D-03
	3	3.888302055346D-03	1.468477273269D-01	5.25	-3.419119574211D-03

N= 4 M=10 P= 5.0 Q= 7

I= 0 H= -5.0

0		1.000000000000D+00	-2.824702900581D-01	-1.00	4.666892594685D+00
1		-2.043896316420D+00	-7.631180659749D-01	-.75	3.405097485903D+00
2		1.333146266276D+00	6.381094062025D-01	-.50	2.391465976278D+00
3		-2.898500119888D-01	-1.438844639331D-01	-.25	1.598824627185D+00
				.25	5.678186927249D-01
0	-5.0	6.000621324197D-04	3.000310662099D-04	.50	2.751118448459D-01
1	-4.0	-2.800289304739D-03	-1.400144652369D-03	.75	9.477216228341D-02
2	-3.0	5.424023874898D-03	2.712011937449D-03	1.25	-3.508151083022D-02
3	-2.0	-5.814551547343D-03	-2.907275773671D-03	1.50	-3.427841913995D-02
4	-1.0	4.084680995051D-03	2.042340497526D-03	1.75	-1.768464626554D-02
5	0.0	-2.298365845264D-03	-1.149182922632D-03	2.25	1.053667380676D-02
6	1.0	1.220188822799D-03	6.100944113997D-04	2.50	1.196260311554D-02
7	2.0	-6.099038691971D-04	-3.049519345985D-04	2.75	6.911983000550D-03
8	3.0	2.625552064188D-04	1.312776032094D-04	3.25	-4.778729896187D-03
9	4.0	-8.075858976828D-05	-4.037929488414D-05	3.50	-5.688887342549D-03
10	5.0	1.235812472306D-05	6.179062361532D-06	3.75	-3.404262929045D-03

CARDINAL SPLINES N=4

				4.25	2.458963528757D-03
	0	-1.212207642379D+01	-6.130883456510D+00	4.50	2.966814934073D-03
	1	-1.045118455282D+01	-5.173358798934D+00	4.75	1.793280907423D-03
	2	-3.014603913555D+00	-1.520157552794D+00	5.25	-1.313896035752D-03
	3	-2.898500119888D-01	-1.438844639331D-01	5.50	-1.594657971962D-03
I= 1	H= -4.0				
0		0.	5.113489904746D+00	-1.00	-9.040587204520D+00
1		3.958867822537D+00	2.879630843855D-01	-.75	-5.678158123653D+00
2		-4.018893457607D+00	-1.823434416572D+00	-.50	-3.117010516217D+00
3		1.062825924376D+00	5.246129493287D-01	-.25	-1.257504451803D+00
				.25	7.551425986860D-01
0	-5.0	-2.800289304739D-03	-1.400144652369D-03	.50	1.107541910153D+00
1	-4.0	1.366782135823D-02	6.833910679115D-03	.75	1.156529190344D+00
2	-3.0	-2.811073167436D-02	-1.405536583718D-02	1.25	7.315436484806D-01
3	-2.0	3.255318424075D-02	1.627659212037D-02	1.50	4.350901607675D-01
4	-1.0	-2.486383097701D-02	-1.243191548851D-02	1.75	1.773078440460D-01
5	0.0	1.478525632267D-02	7.392628161337D-03	2.25	-8.425883360868D-02
6	1.0	-7.945244785629D-03	-3.972622392815D-03	2.50	-9.011251153866D-02
7	2.0	3.982283271181D-03	1.991141635591D-03	2.75	-4.994679488518D-02
8	3.0	-1.715412028047D-03	-8.577060140237D-04	3.25	3.284450469388D-02
9	4.0	5.277221667223D-04	2.638610833612D-04	3.50	3.852177869552D-02
10	5.0	-8.075858976828D-05	-4.037929488414D-05	3.75	2.280725961349D-02
				4.25	-1.626512585772D-02
	0	5.217524321945D+01	2.654406357847D+01	4.50	-1.955224337282D-02
	1	4.348187757463D+01	2.139959011832D+01	4.75	-1.178776187663D-02
	2	1.192349540803D+01	6.045759823359D+00	5.25	8.610448560862D-03
	3	1.062825924376D+00	5.246129493287D-01	5.50	1.044139198766D-02
I= 2	H= -3.0				
0		0.	-1.662633215802D+01	-1.00	9.694738673230D+00
1		-3.292642860899D+00	3.853445178932D+00	-.75	5.851646252784D+00
2		4.844657324678D+00	1.817514321003D+00	-.50	3.052165572576D+00
3		-1.557438487653D+00	-7.566091436287D-01	-.25	1.150286774387D+00
				.25	-5.447042777460D-01
0	-5.0	5.424023874898D-03	2.712011937449D-03	.50	-6.297945350505D-01
1	-4.0	-2.811073167436D-02	-1.405536583718D-02	.75	-4.006827427336D-01
2	-3.0	6.268526714677D-02	3.134263357339D-02	1.25	4.379634930495D-01
3	-2.0	-8.063373074999D-02	-4.031686537499D-02	1.50	7.976147111803D-01
4	-1.0	6.939295643731D-02	3.469647821866D-02	1.75	9.966277963447D-01
5	0.0	-4.547478077901D-02	-2.273739038950D-02	2.25	8.268261850715D-01
6	1.0	2.550534708456D-02	1.275267354228D-02	2.50	5.432777286717D-01
7	2.0	-1.290849760596D-02	-6.454248802978D-03	2.75	2.398267229864D-01
8	3.0	5.573003087388D-03	2.786501543694D-03	3.25	-1.275006437154D-01
9	4.0	-1.715412028047D-03	-8.577060140237D-04	3.50	-1.416071043976D-01
10	5.0	2.625552064188D-04	1.312776032094D-04	3.75	-8.077526431775D-02
				4.25	5.514475340729D-02
	0	-9.002659214418D+01	-4.649739119188D+01	4.50	6.545821056270D-02
	1	-7.165395618810D+01	-3.471709738319D+01	4.75	3.911399786506D-02
	2	-1.851691999012D+01	-9.531622833428D+00	5.25	-2.827208011197D-02
	3	-1.557438487653D+00	-7.566091436287D-01	5.50	-3.418113722720D-02
I= 3	H= -2.0				
0		0.	3.863909420361D+01	-1.00	-6.818300035980D+00

CARDINAL SPLINES N=4

1		2.159962994571D+00	-1.170139285047D+01	-.75	-4.064211070897D+00
2		-3.406242742217D+00	-2.975336479702D-01	-.50	-2.088053970239D+00
3		1.252094299193D+00	5.746630265566D-01	-.25	-7.724448934562D-01
				.25	3.466641957870D-01
0	-5.0	-5.814551547343D-03	-2.907275773671D-03	.50	3.848871729465D-01
1	-4.0	3.255318424075D-02	1.627659212037D-02	.75	2.314118369634D-01
2	-3.0	-8.063373074999D-02	-4.031686537499D-02	1.25	-2.045278188604D-01
3	-2.0	1.189129019822D-01	5.945645099112D-02	1.50	-2.973761609312D-01
4	-1.0	-1.199470843002D-01	-5.997354215009D-02	1.75	-2.291638208377D-01
5	0.0	9.113003839764D-02	4.556501919882D-02	2.25	3.357325072797D-01
6	1.0	-5.629266535798D-02	-2.814633267899D-02	2.50	6.815148706479D-01
7	2.0	2.962802553885D-02	1.481401276942D-02	2.75	9.295187884665D-01
8	3.0	-1.290849760596D-02	-6.454248802978D-03	3.25	8.732824449824D-01
9	4.0	3.982283271181D-03	1.991141635591D-03	3.50	5.986371355457D-01
10	5.0	-6.099038691971D-04	-3.049519345985D-04	3.75	2.729990042547D-01
				4.25	-1.515664972325D-01
	0	8.215553381656D+01	4.452666707158D+01	4.50	-1.707424462748D-01
	1	6.200460801187D+01	2.842299766157D+01	4.75	-9.846764300208D-02
	2	1.537517174568D+01	8.322411750379D+00	5.25	6.830060339460D-02
	3	1.252094299193D+00	5.746630265566D-01	5.50	8.160804940068D-02

I= 4 H= -1.0

0		0.	-7.744721402047D+01	-1.00	3.852626667969D+00
1		-1.205297680920D+00	2.501672140261D+01	-.75	2.291415785560D+00
2		1.924270993487D+00	-1.855338356406D+00	-.50	1.174098838027D+00
3		-7.230579935623D-01	-2.585280998764D-01	-.25	4.328891384723D-01
				.25	-1.923550149773D-01
0	-5.0	4.084680995051D-03	2.042340497526D-03	.50	-2.119314297132D-01
1	-4.0	-2.486383097701D-02	-1.243191548851D-02	.75	-1.260656787923D-01
2	-3.0	6.939295643731D-02	3.469647821866D-02	1.25	1.073044417081D-01
3	-2.0	-1.199470843002D-01	-5.997354215009D-02	1.50	1.509388412029D-01
4	-1.0	1.460978137111D-01	7.304890685556D-02	1.75	1.106679202587D-01
5	0.0	-1.343520816107D-01	-6.717604080535D-02	2.25	-1.324981838443D-01
6	1.0	9.709991998067D-02	4.854995999034D-02	2.50	-2.155199443838D-01
7	2.0	-5.629266535798D-02	-2.814633267899D-02	2.75	-1.818055515437D-01
8	3.0	2.550534708456D-02	1.275267354228D-02	3.25	3.028529997507D-01
9	4.0	-7.945244785629D-03	-3.972622392815D-03	3.50	6.422451745197D-01
10	5.0	1.220188822799D-03	6.100944113997D-04	3.75	9.059152905588D-01
				4.25	8.905753460743D-01
	0	-4.830196276271D+01	-3.106307840213D+01	4.50	6.197318752399D-01
	1	-3.619193726322D+01	-1.292626965218D+01	4.75	2.859392529045D-01
	2	-8.921598909947D+00	-5.733259854552D+00	5.25	-1.614959629918D-01
	3	-7.230579935623D-01	-2.585280998764D-01	5.50	-1.832182496531D-01

I= 5 H= 0.0

0		0.	1.470846781527D+02	-1.00	-2.075655982656D+00
1		6.476160980853D-01	-4.833737334072D+01	-.75	-1.233949357299D+00
2		-1.036678808405D+00	4.833737334072D+00	-.50	-6.318978856646D-01
3		3.913610761652D-01	1.771421783493D-27	-.25	-2.328114668617D-01
				.25	1.032264755300D-01
0	-5.0	-2.298365845264D-03	-1.149182922632D-03	.50	1.135405254788D-01
1	-4.0	1.478525632267D-02	7.392628161337D-03	.75	6.737890303710D-02
2	-3.0	-4.547478077901D-02	-2.273739038950D-02	1.25	-5.687197360266D-02
3	-2.0	9.113003839764D-02	4.556501919882D-02	1.50	-7.941376509563D-02
4	-1.0	-1.343520816107D-01	-6.717604080535D-02	1.75	-5.760385119968D-02

CARDINAL SPLINES N=4

5	0.0	1.524198670293D-01	7.620993351466D-02	2.25	6.633559085544D-02
6	1.0	-1.343520816107D-01	-6.717604080535D-02	2.50	1.042680166339D-01
7	2.0	9.113003839764D-02	4.556501919882D-02	2.75	8.357789562211D-02
8	3.0	-4.547478077901D-02	-2.273739038950D-02	3.25	-1.134939496135D-01
9	4.0	1.478525632267D-02	7.392628161337D-03	3.50	-1.926401739464D-01
10	5.0	-2.298365845264D-03	-1.149182922632D-03	3.75	-1.679055286216D-01
				4.25	2.923253020810D-01
	0	2.624124480094D+01	2.624124480094D+01	4.50	6.290823923749D-01
	1	1.963290872642D+01	2.642130165783D-25	4.75	8.975797603858D-01
	2	4.833737334072D+00	4.833737334072D+00	5.25	8.975797603858D-01
	3	3.913610761652D-01	1.771421783493D-27	5.50	6.290823923749D-01

I= 6 H= 1.0

0		0.	-2.713419355114D+02	-1.00	1.091432404792D+00
1		-3.403245029981D-01	8.964847568843D+01	-.75	6.487725697534D-01
2		5.451061079848D-01	-9.611181352698D+00	-.50	3.321890027214D-01
3		-2.060017938094D-01	2.585280998764D-01	-.25	1.223690355269D-01
				.25	-5.423069755434D-02
0	-5.0	1.220188822799D-03	6.100944113997D-04	.50	-5.962641600388D-02
1	-4.0	-7.945244785629D-03	-3.972622392815D-03	.75	-3.536532272392D-02
2	-3.0	2.550534708456D-02	1.275267354228D-02	1.25	2.979324021257D-02
3	-2.0	-5.629266535798D-02	-2.814633267899D-02	1.50	4.153193210134D-02
4	-1.0	9.709991998067D-02	4.854995999034D-02	1.75	3.005103313054D-02
5	0.0	-1.343520816107D-01	-6.717604080535D-02	2.25	-3.429514499886D-02
6	1.0	1.460978137111D-01	7.304890685556D-02	2.50	-5.348465162771D-02
7	2.0	-1.199470843002D-01	-5.997354215009D-02	2.75	-4.238910577568D-02
8	3.0	6.939295643731D-02	3.469647821866D-02	3.25	5.529853050633D-02
9	4.0	-2.486383097701D-02	-1.243191548851D-02	3.50	9.064665450579D-02
10	5.0	4.084680995051D-03	2.042340497526D-03	3.75	7.503380310859D-02
				4.25	-1.064042254206D-01
	0	-1.382419404155D+01	-3.106307840213D+01	4.50	-1.832182496531D-01
	1	-1.033939795886D+01	1.292626965218D+01	4.75	-1.614959629918D-01
	2	-2.544920799157D+00	-5.733259854552D+00	5.25	2.859392529045D-01
	3	-2.060017938094D-01	2.585280998764D-01	5.50	6.197318752399D-01

I= 7 H= 2.0

0		0.	4.665348274585D+02	-1.00	-5.443537760998D-01
1		1.697135939048D-01	-1.547468421571D+02	-.75	-3.235685133083D-01
2		-2.718719361153D-01	1.694235714873D+01	-.50	-1.656708117412D-01
3		1.027682460797D-01	-5.746630265566D-01	-.25	-6.102614832840D-02
				.25	2.704211908841D-02
0	-5.0	-6.099038691971D-04	-3.049519345985D-04	.50	2.973007880956D-02
1	-4.0	3.982283271181D-03	1.991141635591D-03	.75	1.763117283963D-02
2	-3.0	-1.290849760596D-02	-6.454248802978D-03	1.25	-1.484668239311D-02
3	-2.0	2.962802553885D-02	1.481401276942D-02	1.50	-2.068830268528D-02
4	-1.0	-5.629266535798D-02	-2.814633267899D-02	1.75	-1.496073968214D-02
5	0.0	9.113003839764D-02	4.556501919882D-02	2.25	1.703789585278D-02
6	1.0	-1.199470843002D-01	-5.997354215009D-02	2.50	2.652264493871D-02
7	2.0	1.189129019822D-01	5.945645099112D-02	2.75	2.096472448050D-02
8	3.0	-8.063373074999D-02	-4.031686537499D-02	3.25	-2.709245397129D-02
9	4.0	3.255318424075D-02	1.627659212037D-02	3.50	-4.405319610818D-02
10	5.0	-5.814551547343D-03	-2.907275773671D-03	3.75	-3.604684056316D-02
				4.25	4.908562031337D-02
	0	6.897800326605D+00	4.452666707158D+01	4.50	8.160804940068D-02
	1	5.158612688730D+00	-2.842299766157D+01	4.75	6.830060339460D-02

	2	1.269651755080D+00	8.322411750379D+00	5.25	-9.846764300208D-02
	3	1.027682460797D-01	-5.746630265566D-01	5.50	-1.707424462748D-01

I= 8 H= 3.0

0		0.	-5.529495918971D+02	-1.00	2.342172135994D-01
1		-7.301968400730D-02	1.867790114896D+02	-.75	1.392199077204D-01
2		1.169773291965D-01	-2.088075998786D+01	-.50	7.128169935223D-02
3		-4.422020039562D-02	7.566091436287D-01	-.25	2.625694470779D-02
				.25	-1.163476253313D-02
0	-5.0	2.625552064188D-04	1.312776032094D-04	.50	-1.279098354143D-02
1	-4.0	-1.715412028047D-03	-8.577060140237D-04	.75	-7.585365484915D-03
2	-3.0	5.573003087388D-03	2.786501543694D-03	1.25	6.386749047274D-03
3	-2.0	-1.290849760596D-02	-6.454248802978D-03	1.50	8.898888536663D-03
4	-1.0	2.550534708456D-02	1.275267354228D-02	1.75	6.434364613168D-03
5	0.0	-4.547478077901D-02	-2.273739038950D-02	2.25	-7.324114721758D-03
6	1.0	6.939295643731D-02	3.469647821866D-02	2.50	-1.139643685283D-02
7	2.0	-8.063373074999D-02	-4.031686537499D-02	2.75	-9.002642426221D-03
8	3.0	6.268526714677D-02	3.134263357339D-02	3.25	1.160803449901D-02
9	4.0	-2.811073167436D-02	-1.405536583718D-02	3.50	1.883875700552D-02
10	5.0	5.424023874898D-03	2.712011937449D-03	3.75	1.537242814144D-02
				4.25	-2.073020177887D-02
	0	-2.968190239577D+00	-4.649739119168D+01	4.50	-3.418113722720D-02
	1	-2.219761421714D+00	3.471709738319D+01	4.75	-2.827208011197D-02
	2	-5.463256767378D-01	-9.531622833428D+00	5.25	3.911399786506D-02
	3	-4.422020039562D-02	7.566091436287D-01	5.50	6.545821056270D-02

I= 9 H= 4.0

0		0.	3.502626284201D+02	-1.00	-7.203305275495D-02
1		2.245687995415D-02	-1.212031595516D+02	-.75	-4.281675354944D-02
2		-3.597614708259D-02	1.391495406329D+01	-.50	-2.192247996250D-02
3		1.360002571821D-02	-5.246129493287D-01	-.25	-8.075229583046D-03
				.25	3.578206268609D-03
0	-5.0	-8.075858976828D-05	-4.037929488414D-05	.50	3.933775494721D-03
1	-4.0	5.277221667223D-04	2.638610833612D-04	.75	2.332808111076D-03
2	-3.0	-1.715412028047D-03	-8.577060140237D-04	1.25	-1.964134397834D-03
3	-2.0	3.982283271181D-03	1.991141635591D-03	1.50	-2.736637593588D-03
4	-1.0	-7.945244785629D-03	-3.972622392815D-03	1.75	-1.978666908022D-03
5	0.0	1.478525632267D-02	7.392628161337D-03	2.25	2.252003550073D-03
6	1.0	-2.486383097701D-02	-1.243191548851D-02	2.50	3.503777317366D-03
7	2.0	3.255318424075D-02	1.627659212037D-02	2.75	2.767386469931D-03
8	3.0	-2.811073167436D-02	-1.405536583718D-02	3.25	-3.566287983077D-03
9	4.0	1.366782135823D-02	6.833910679115D-03	3.50	-5.784969569746D-03
10	5.0	-2.800289304739D-03	-1.400144652369D-03	3.75	-4.717269744450D-03
				4.25	6.345808250237D-03
	0	9.128839374823D-01	2.654406357847D+01	4.50	1.044139198766D-02
	1	6.826973379940D-01	-2.139959011832D+01	4.75	8.610448560862D-03
	2	1.680242386906D-01	6.045759823359D+00	5.25	-1.178776187663D-02
	3	1.360002571821D-02	-5.246129493287D-01	5.50	-1.955224337282D-02

I=10 H= 5.0

0		0.	-8.798717426267D+01	-1.00	1.102249773462D-02
1		-3.436343807134D-03	3.116626912186D+01	-.75	6.551816984832D-03
2		5.505069804951D-03	-3.678424511791D+00	-.50	3.354574870122D-03
3		-2.081084122540D-03	1.438844639331D-01	-.25	1.235669754008D-03

CARDINAL SPLINES N=4

				.25	-5.475352741087D-04
0	-5.0	1.235812472306D-05	6.179062361532D-06	.50	-6.019434197973D-04
1	-4.0	-8.075858976828D-05	-4.037929488414D-05	.75	-3.569638437416D-04
2	-3.0	2.625552064188D-04	1.312776032094D-04	1.25	3.005475861059D-04
3	-2.0	-6.099038691971D-04	-3.049519345985D-04	1.50	4.187516570191D-04
4	-1.0	1.220188822799D-03	6.100944113997D-04	1.75	3.027665000040D-04
5	0.0	-2.298365845264D-03	-1.149182922632D-03	2.25	-3.445792426222D-04
6	1.0	4.084680995051D-03	2.042340497526D-03	2.50	-5.360969221047D-04
7	2.0	-5.814551547343D-03	-2.907275773671D-03	2.75	-4.234063951884D-04
8	3.0	5.424023874898D-03	2.712011937449D-03	3.25	5.455507471007D-04
9	4.0	-2.800289304739D-03	-1.400144652369D-03	3.50	8.848310921628D-04
10	5.0	6.000621324197D-04	3.000310662099D-04	3.75	7.213804989815D-04
				4.25	-9.697433652835D-04
	0	-1.396904892294D-01	-6.130883456510D+00	4.50	-1.594657971962D-03
	1	-1.044669549481D-01	5.173358798934D+00	4.75	-1.313896035752D-03
	2	-2.571119203315D-02	-1.520157552794D+00	5.25	1.793280907423D-03
	3	-2.081084122540D-03	1.438844639331D-01	5.50	2.966814934073D-03

N= 4 M=11 P= 5.5 Q= 7

I= 0 H= -5.5

0		1.000000000000D+00	1.073160045816D+00	-1.00	4.666957626210D+00
1		-2.043916590433D+00	-1.192641847562D+00	-.75	3.405136140890D+00
2		1.333178745582D+00	6.834957722147D-01	-.50	2.391485767886D+00
3		-2.898622901950D-01	-1.454881062547D-01	-.25	1.598831917491D+00
				.25	5.678154623357D-01
0	-5.5	6.001350461010D-04	3.000675230505D-04	.50	2.751082934597D-01
1	-4.5	-2.800765802460D-03	-1.400382901230D-03	.75	9.477005624758D-02
2	-3.5	5.425573218765D-03	2.712786609382D-03	1.25	-3.507973765263D-02
3	-2.5	-5.818152440596D-03	-2.909076220298D-03	1.50	-3.427594859198D-02
4	-1.5	4.091900806424D-03	2.045950403212D-03	1.75	-1.768286001953D-02
5	-.5	-2.312096282992D-03	-1.156048141496D-03	2.25	1.053464093268D-02
6	.5	1.245673898841D-03	6.228369494205D-04	2.50	1.195944044402D-02
7	1.5	-6.550068365662D-04	-3.275034182831D-04	2.75	6.909485228551D-03
8	2.5	3.266857785149D-04	1.633428892575D-04	3.25	-4.775511975272D-03
9	3.5	-1.405619582811D-04	-7.028097914057D-05	3.50	-5.683668755369D-03
10	4.5	4.322955943589D-05	2.161477971794D-05	3.75	-3.400009018660D-03
11	5.5	-6.614987186117D-06	-3.307493593059D-06	4.25	2.453248255399D-03
				4.50	2.957421198602D-03
	0	-1.813872272472D+01	-9.016206684413D+00	4.75	1.785546428286D-03
	1	-1.368395322423D+01	-6.877233995817D+00	5.25	-1.303365394679D-03
	2	-3.449549042635D+00	-1.717057980988D+00	5.50	-1.577272489520D-03
	3	-2.898622901950D-01	-1.454881062547D-01	5.75	-9.553770231733D-04

I= 1 H= -4.5

0		0.	-3.745681024678D+00	-1.00	-9.041012191611D+00
1		3.959000315106D+00	3.094941136321D+00	-.75	-5.678410737607D+00
2		-4.019105712904D+00	-2.120038559845D+00	-.50	-3.117139856229D+00

3		1.062906163600D+00	5.350929017618D-01	-.25	-1,257552094639D+00
				.25	7,551637095811D-01
0	-5,5	-2,800765802460D-03	-1.400382901230D-03	.50	1,107565118794D+00
1	-4,5	1,367093531537D-02	6,835467657683D-03	.75	1,156542953485D+00
2	-3,5	-2,812085678131D-02	-1,406042839066D-02	1.25	7,315320605994D-01
3	-2,5	3.257671641466D-02	1.628835820733D-02	1,50	4,350740155057D-01
4	-1,5	-2,491101311903D-02	-1,245550655951D-02	1.75	1,772961707614D-01
5	-,5	1,487498601634D-02	7,437493008172D-03	2,25	-8,424554858660D-02
6	,5	-8.111792139157D-03	-4,055896069578D-03	2,50	-9,009184318474D-02
7	1,5	4.277035377838D-03	2,138517688919D-03	2,75	-4,993047171175D-02
8	2,5	-2,134511309737D-03	-1,067255654869D-03	3,25	3,282347527998D-02
9	3,5	9,185427686154D-04	4.592713843077D-04	3.50	3,848767474061D-02
10	4,5	-2,825063005711D-04	-1.412531502856D-04	3.75	2,277945991160D-02
11	5,5	4.322955943589D-05	2,161477971794D-05	4,25	-1,622777601223D-02
				4.50	-1,949085443367D-02
	0	7.703756688676D+01	3.817141032038D+01	4,75	-1,173721633268D-02
	1	5,620757181990D+01	2.833419781290D+01	5,25	8,541629837351D-03
	2	1,351884598650D+01	6.708994319224D+00	5,50	1,032777623521D-02
	3	1.062906163600D+00	5.350929017618D-01	5,75	6,251922862003D-03

I= 2 H= -3.5

0		0.	1.217947643307D+01	-1.00	9,696120529038D+00
1		-3,293073663695D+00	-5,273511861132D+00	-,75	5,852467633169D+00
2		4,845347477909D+00	2,781929868069D+00	-,50	3,052586124754D+00
3		-1,557699387434D+00	-7,906849624875D-01	-,25	1,150441686222D+00
				.25	-5,447729203322D-01
0	-5,5	5,425573218765D-03	2.712786609382D-03	.50	-6,298699985084D-01
1	-4,5	-2,812085678131D-02	-1,406042839066D-02	.75	-4,007274939195D-01
2	-3,5	6.271818917698D-02	3,135909458849D-02	1.25	4,380011713257D-01
3	-2,5	-8,071024618276D-02	-4,035512309138D-02	1.50	7,976672078897D-01
4	-1,5	6,954637031346D-02	3,477318515673D-02	1.75	9,966657523120D-01
5	-,5	-4,576653904865D-02	-2.288326952432D-02	2,25	8,267829885008D-01
6	,5	2,604687983354D-02	1,302343991677D-02	2,50	5,432105250190D-01
7	1,5	-1,386689121116D-02	-6,933445605581D-03	2,75	2,397736477931D-01
8	2,5	6,935714540322D-03	3,467857270161D-03	3,25	-1,274322660674D-01
9	3,5	-2,986174669526D-03	-1,493087334763D-03	3,50	-1,414962145632D-01
10	4,5	9.185427686154D-04	4.592713843077D-04	3,75	-8,068487291453D-02
11	5.5	-1,405619582811D-04	-7,028097914057D-05	4,25	5,502330948115D-02
				4,50	6,525860294198D-02
	0	-1.307023795278D+02	-6,422167092793D+01	4,75	3,894964780613D-02
	1	-9.135547081630D+01	-4,642694365811D+01	5,25	-2,804831436767D-02
	2	-2,085669241475D+01	-1,026437201297D+01	5,50	-3,381171285962D-02
	3	-1,557699387434D+00	-7,906849624875D-01	5,75	-2,042493116783D-02

I= 3 H= -2.5

0		0.	-2,830966007106D+01	-1,00	-6,821511663407D+00
1		2,160964240913D+00	9,510940148917D+00	-,75	-4,066120074539D+00
2		-3,407846755483D+00	-2,538971143990D+00	-,50	-2,089031392703D+00
3		1,252700667011D+00	6,538600265740D-01	-,25	-7,728049303679D-01
				,25	3,468237308207D-01
0	-5,5	-5,818152440596D-03	-2,909076220298D-03	.50	3,850625606460D-01
1	-4,5	3,257671641466D-02	1,628835820733D-02	.75	2,315158450192D-01
2	-3,5	-8,071024618276D-02	-4,035512309138D-02	1,25	-2,046153884782D-01
3	-2,5	1,190907346194D-01	5,954536730968D-02	1,50	-2,974981706741D-01
4	-1,5	-1,203036397450D-01	-6,015181987252D-02	1,75	-2,292520358490D-01

CARDINAL SPLINES N=4

5	-.5	9.180812568827D-02	4.590406284413D-02	2.25	3.358329021909D-01
6	.5	-5.755126370033D-02	-2.877563185017D-02	2.50	6.816710613974D-01
7	1.5	3.185546719425D-02	1.592773359713D-02	2.75	9.296421426874D-01
8	2.5	-1.607563085692D-02	-8.037815428460D-03	3.25	8.731235257026D-01
9	3.5	6.935714540322D-03	3.467857270161D-03	3.50	5.983794119602D-01
10	4.5	-2.134511309737D-03	-1.067255654869D-03	3.75	2.727889219089D-01
11	5.5	3.266857785149D-04	1.633428892575D-04	4.25	-1.512842444520D-01
				4.50	-1.702785300647D-01
	0	1.172160124456D+02	5.598259556352D+01	4.75	-9.808567032890D-02
	1	7.835723546186D+01	4.092005497661D+01	5.25	6.778054030443D-02
	2	1.726171425020D+01	8.249719294480D+00	5.50	8.074945516507D-02
	3	1.252700667011D+00	6.538600265740D-01	5.75	4.837612843913D-02

I= 4 H= -1.5

0		0.	5.678541755856D+01	-1.00	3.859065999152D+00
1		-1.207305185507D+00	-1.751412642701D+01	-.75	2.295243348677D+00
2		1.927487049173D+00	2.638756520644D+00	-.50	1.176058575606D+00
3		-7.242737644727D-01	-4.173185335547D-01	-.25	4.336110145199D-01
				.25	-1.926748836234D-01
0	-5.5	4.091900806424D-03	2.045950403212D-03	.50	-2.122830830441D-01
1	-4.5	-2.491101311903D-02	-1.245550655951D-02	.75	-1.262742155334D-01
2	-3.5	6.954637031346D-02	3.477318515673D-02	1.25	1.074800192879D-01
3	-2.5	-1.203036397450D-01	-6.015181987252D-02	1.50	1.511834714285D-01
4	-1.5	1.468127093979D-01	7.340635469895D-02	1.75	1.108447918560D-01
5	-.5	-1.357116504697D-01	-6.785582523485D-02	2.25	-1.326994762101D-01
6	.5	9.962341678693D-02	4.981170839346D-02	2.50	-2.158331077214D-01
7	1.5	-6.075869849531D-02	-3.037934924765D-02	2.75	-1.820528774553D-01
8	2.5	3.185546719425D-02	1.592773359713D-02	3.25	3.031716338068D-01
9	3.5	-1.386689121116D-02	-6.933445605581D-03	3.50	6.427619117687D-01
10	4.5	4.277035377838D-03	2.138517688919D-03	3.75	9.063365068522D-01
11	5.5	-6.550068365662D-04	-3.275034182831D-04	4.25	8.900094276505D-01
				4.50	6.188017206106D-01
	0	-6.883474284695D+01	-2.915126406067D+01	4.75	2.851733955310D-01
	1	-4.573279177050D+01	-2.635946162001D+01	5.25	-1.604532335500D-01
	2	-1.002303006463D+01	-4.246999283009D+00	5.50	-1.814967633469D-01
	3	-7.242737644727D-01	-4.173185335547D-01	5.75	-1.049816901936D-01

I= 5 H= -.5

0		0.	-1.081952306911D+02	-1.00	-2.087902124731D+00
1		6.514339148256D-01	3.254661760304D+01	-.75	-1.241228511048D+00
2		-1.042795014224D+00	-3.713008096255D+00	-.50	-6.356248604290D-01
3		3.936731956814D-01	3.019832581529D-01	-.25	-2.341843107779D-01
				.25	1.038347928808D-01
0	-5.5	-2.312096282992D-03	-1.156048141496D-03	.50	1.142092900648D-01
1	-4.5	1.487498601634D-02	7.437493008172D-03	.75	6.777549244795D-02
2	-3.5	-4.576653904865D-02	-2.288326952432D-02	1.25	-5.720588219140D-02
3	-2.5	9.180812568827D-02	4.590406284413D-02	1.50	-7.987899609775D-02
4	-1.5	-1.357116504697D-01	-6.785582523485D-02	1.75	-5.794022071507D-02
5	-.5	1.550054575099D-01	7.750272875494D-02	2.25	6.671840311337D-02
6	.5	-1.391511980930D-01	-6.957559904649D-02	2.50	1.048635820142D-01
7	1.5	9.962341678693D-02	4.981170839346D-02	2.75	8.404825320524D-02
8	2.5	-5.755126370033D-02	-2.877563185017D-02	3.25	-1.140999190552D-01
9	3.5	2.604687983354D-02	1.302343991677D-02	3.50	-1.936228905650D-01
10	4.5	-8.111792139157D-03	-4.055896069578D-03	3.75	-1.687065861271D-01
11	5.5	1.245673898841D-03	6.228369494205D-04	4.25	2.934015500988D-01

CARDINAL SPLINES N=4

				4.50	6.308513347416D-01
	0	3.753571528276D+01	8.735135789098D+00	4.75	8.990362467722D-01
	1	2.490653126645D+01	1.910850922161D+01	5.25	8.955967263520D-01
	2	5.452812714520D+00	1.269715663267D+00	5.50	6.258085172957D-01
	3	3.936731956814D-01	3.019832581529D-01	5.75	2.896199639020D-01
I= 6	H= .5				
0		0.	2.024832998970D+02	-1.00	1.114162478530D+00
1		-3.474107555900D-01	-6.048036219493D+01	-.75	6.622834126333D-01
2		5.564584023154D-01	6.252439422790D+00	-.50	3.391066434519D-01
3		-2.102933206243D-01	-3.019832581529D-01	-.25	1.249171721770D-01
				.25	-5.535979585764D-02
0	-5.5	1.245673898841D-03	6.228369494205D-04	.50	-6.086771046684D-02
1	-4.5	-8.111792139157D-03	-4.055896069578D-03	.75	-3.610143263483D-02
2	-3.5	2.604687983354D-02	1.302343991677D-02	1.25	3.041300820129D-02
3	-2.5	-5.755126370033D-02	-2.877563185017D-02	1.50	4.239544772268D-02
4	-1.5	9.962341678693D-02	4.981170839346D-02	1.75	3.067536884711D-02
5	-.5	-1.391511980930D-01	-6.957559904649D-02	2.25	-3.500568312561D-02
6	.5	1.550054575099D-01	7.750272875494D-02	2.50	-5.459008098877D-02
7	1.5	-1.357116504697D-01	-6.785582523485D-02	2.75	-4.326213684797D-02
8	2.5	9.180812568827D-02	4.590406284413D-02	3.25	5.642327085372D-02
9	3.5	-4.576653904865D-02	-2.288326952432D-02	3.50	9.247067557839D-02
10	4.5	1.487498601634D-02	7.437493008172D-03	3.75	7.652064656145D-02
11	5.5	-2.312096282992D-03	-1.156048141496D-03	4.25	-1.084018501952D-01
				4.50	-1.865015849423D-01
	0	-2.006544370457D+01	8.735135789098D+00	4.75	-1.641993484954D-01
	1	-1.331048717677D+01	-1.910850922161D+01	5.25	2.896199639020D-01
	2	-2.913381387985D+00	1.269715663267D+00	5.50	6.258085172957D-01
	3	-2.102933206243D-01	-3.019832581529D-01	5.75	8.955967263520D-01
I= 7	H= 1.5				
0		0.	-3.720314023019D+02	-1.00	-5.845809966270D-01
1		1.822546990951D-01	1.109481106532D+02	-.75	-3.474797272498D-01
2		-2.919629948952D-01	-1.113275508666D+01	-.50	-1.779135111010D-01
3		1.103633026367D-01	4.173185335547D-01	-.25	-6.553578855843D-02
				.25	2.904037421808D-02
0	-5.5	-6.550068365662D-04	-3.275034182831D-04	.50	3.192689641243D-02
1	-4.5	4.277035377838D-03	2.138517688919D-03	.75	1.893392513425D-02
2	-3.5	-1.386689121116D-02	-6.933445605581D-03	1.25	-1.594353512865D-02
3	-2.5	3.185546719425D-02	1.592773359713D-02	1.50	-2.221653500559D-02
4	-1.5	-6.075869849531D-02	-3.037934924765D-02	1.75	-1.606567628921D-02
5	-.5	9.962341678693D-02	4.981170839346D-02	2.25	1.829539172579D-02
6	.5	-1.357116504697D-01	-6.785582523485D-02	2.50	2.847901125755D-02
7	1.5	1.468127093979D-01	7.340635469895D-02	2.75	2.250979705166D-02
8	2.5	-1.203036397450D-01	-6.015181987252D-02	3.25	-2.908299647957D-02
9	3.5	6.954637031346D-02	3.477318515673D-02	3.50	-4.728131137000D-02
10	4.5	-2.491101311903D-02	-1.245550655951D-02	3.75	-3.867822575613D-02
11	5.5	4.091900806424D-03	2.045950403212D-03	4.25	5.262097586339D-02
				4.50	8.741882917502D-02
	0	1.053221472562D+01	-2.915126406067D+01	4.75	7.308499987649D-02
	1	6.986131469524D+00	2.635946162001D+01	5.25	-1.049816901936D-01
	2	1.529031498610D+00	-4.246999283009D+00	5.50	-1.814967633469D-01
	3	1.103633026367D-01	4.173185335547D-01	5.75	-1.604532335500D-01

CARDINAL SPLINES N=4

I= 8		H= 2.5			
0		0.	6,393828685142D+02	-1.00	2,914151058165D-01
1		-9,085150966065D-02	-1.910047646275D+02	-.75	1,732185539077D-01
2		1.455442100190D-01	1.903840973295D+01	-.50	8,868923060218D-02
3		-5,501938613685D-02	-6.538600265740D-01	-.25	3,266906844974D-02
				.25	-1,447602225805D-02
0	-5,5	3,266857785149D-04	1.633428892575D-04	.50	-1,591457336004D-02
1	-4,5	-2.134511309737D-03	-1,067255654869D-03	.75	-9,437710348840D-03
2	-3,5	6,935714540322D-03	3.467857270161D-03	1.25	7,946331432608D-03
3	-2,5	-1.607563085692D-02	-8,037815428460D-03	1.50	1,107183676411D-02
4	-1,5	3,185546719425D-02	1.592773359713D-02	1.75	8,005441215754D-03
5	-,5	-5,755126370033D-02	-2,877563185017D-02	2.25	-9,112110821550D-03
6	,5	9.180812568827D-02	4.590406284413D-02	2,50	-1,417813611953D-02
7	1,5	-1.203036397450D-01	-6.015181987252D-02	2,75	-1,119953530708D-02
8	2.5	1,190907346194D-01	5,954536730968D-02	3,25	1,443832787623D-02
9	3,5	-8.071024618276D-02	-4.035512309138D-02	3,50	2,342871838962D-02
10	4,5	3.257671641466D-02	1.628835820733D-02	3.75	1,911391673242D-02
11	5,5	-5,818152440596D-03	-2,909076220298D-03	4.25	-2,575701901976D-02
				4,50	-4,244331271120D-02
	0	-5,250821318578D+00	5,598259556352D+01	4.75	-3,507487162252D-02
	1	-3,482874491371D+00	-4,092005497661D+01	5,25	4,837612843913D-02
	2	-7.622756612391D-01	8,249719294480D+00	5,50	8,074945516507D-02
	3	-5,501938613685D-02	-6.538600265740D-01	5,75	6,778054030443D-02

I= 9		H= 3.5			
0		0,	-7,616173250739D+02	-1.00	-1,253715232318D-01
1		3,908550505366D-02	2,310896961466D+02	-.75	-7,452134467370D-02
2		-6,261548063677D-02	-2,331067389402D+01	-.50	-3,815543987870D-02
3		2.367053754139D-02	7.906849624875D-01	-.25	-1,405469595230D-02
				.25	6,227752293480D-03
0	-5,5	-1.405619582811D-04	-7,028097914057D-05	.50	6,846601420012D-03
1	-4,5	9,185427686154D-04	4.592713843077D-04	.75	4,060166200820D-03
2	-3,5	-2,986174669526D-03	-1.493087334763D-03	1.25	-3,418484126329D-03
3	-2,5	6,935714540322D-03	3,467857270161D-03	1.50	-4,762966368736D-03
4	-1,5	-1,386689121116D-02	-6,933445605581D-03	1.75	-3,443735282672D-03
5	-,5	2,604687983354D-02	1.302343991677D-02	2,25	3,919354801752D-03
6	,5	-4,576653904865D-02	-2,288326952432D-02	2,50	6,097781719655D-03
7	1,5	6.954637031346D-02	3.477318515673D-02	2.75	4,816044235531D-03
8	2,5	-8,071024618276D-02	-4,035512309138D-02	3,25	-6,205607613399D-03
9	3,5	6,271818917698D-02	3,135909458849D-02	3,50	-1,006522379392D-02
10	4.5	-2,812085678131D-02	-1,406042839066D-02	3,75	-8,206301787657D-03
11	5,5	5,425573218765D-03	2.712786609382D-03	4,25	1,103344152837D-02
				4.50	1,814607805294D-02
	0	2.259037671982D+00	-6.422167092793D+01	4.75	1,495422243226D-02
	1	1.498416499930D+00	4.642694365811D+01	5.25	-2,042493116783D-02
	2	3,279483887962D-01	-1,026437201297D+01	5.50	-3,381171285962D-02
	3	2.367053754139D-02	7.906849624875D-01	5,75	-2,804831436767D-02

I=10		H= 4.5			
0		0.	4,859826579785D+02	-1.00	3,855665090953D-02
1		-1,202030031141D-02	-1.506928161592D+02	-.75	2,291822308082D-02
2		1,925671067505D-02	1.553802719829D+01	-.50	1,173428281485D-02
3		-7,279639923074D-03	-5,350929017618D-01	-.25	4,322363868840D-03
				.25	-1,915272395937D-03
0	-5,5	4,322955943589D-05	2.161477971794D-05	.50	-2,105589746394D-03

CARDINAL SPLINES N=4

```
 1  -4,5  -2,825063005711D-04  -1.412531502856D-04      .75 -1,248653121775D-03
 2  -3,5   9,185427686154D-04   4.592713843077D-04     1.25  1,051305675647D-03
 3  -2,5  -2.134511309737D-03  -1.067255654869D-03     1.50  1.464774281568D-03
 4  -1,5   4.277035377838D-03   2.138517688919D-03     1.75  1,059057722431D-03
 5   -,5  -8,111792139157D-03  -4,055896069578D-03     2.25 -1,205292051776D-03
 6    ,5   1,487498601634D-02   7.437493008172D-03     2.50 -1,875162589164D-03
 7   1.5  -2.491101311903D-02  -1.245550655951D-02     2.75 -1,480955594044D-03
 8   2,5   3,257671641466D-02   1.628835820733D-02     3,25  1,908008844026D-03
 9   3,5  -2,812085678131D-02  -1,406042839066D-02     3.50  3,094365306069D-03
10   4.5   1,367093531537D-02   6.835467657683D-03     3,75  2,522473440922D-03
11   5.5  -2,800765802460D-03  -1.400382901230D-03     4,25 -3,389573022075D-03
                                                       4,50 -5,571937492529D-03
       0  -6,947462459940D-01   3.817141032038D+01     4,75 -4,588651384385D-03
       1  -4,608238059048D-01  -2,833419781290D+01     5,25  6,251922862003D-03
       2  -1,008573480557D-01   6.708994319224D+00     5,50  1,032777623521D-02
       3  -7,279639923074D-03  -5.350929017618D-01     5,75  8,541629837351D-03


I=11     H=  5.5
0           0.                 -1,229875812644D+02    -1.00 -5,899890048094D-03
1           1,839330203153D-03  3.896791742930D+01     -,75 -3,506917239669D-03
2          -2,946637530454D-03 -4.117611734191D+00     -,50 -1,795564773501D-03
3           1,113922314487D-03  1.454881062547D-01     -,25 -6,614024326054D-04
                                                        .25  2,930723375519D-04
0   -5,5  -6,614987186117D-06  -3,307493593059D-06      .50  3,221943286865D-04
1   -4.5   4,322955943589D-05   2.161477971794D-05      .75  1,910670237084D-04
2   -3,5  -1,405619582811D-04  -7,028097914057D-05     1.25 -1,608689453013D-04
3   -2,5   3,266857785149D-04   1.633428892575D-04     1,50 -2,241368540901D-04
4   -1,5  -6,550068365662D-04  -3,275034182831D-04     1,75 -1,620545592346D-04
5    -,5   1,245673898841D-03   6,228369494205D-04     2,25  1,844295302904D-04
6     ,5  -2,312096282992D-03  -1.156048141496D-03     2,50  2,869287518174D-04
7    1,5   4,091900806424D-03   2.045950403212D-03     2,75  2,266067147249D-04
8    2,5  -5,818152440596D-03  -2,909076220298D-03     3,25 -2,919411726931D-04
9    3,5   5,425573218765D-03   2.712786609382D-03     3,50 -4,734486961243D-04
10   4,5  -2,800765802460D-03  -1,400382901230D-03     3,75 -3,859298034826D-04
11   5,5   6,001350461010D-04   3.000675230505D-04     4,25  5,185098237631D-04
                                                       4,50  8,522329237008D-04
       0   1,063093558939D-01  -9.016206684413D+00     4,75  7,016993175206D-04
       1   7,051476740786D-02   6,877233995817D+00     5,25 -9,553770231733D-04
       2   1,543308065858D-02  -1,717057980988D+00     5,50 -1,577272489520D-03
       3   1,113922314487D-03   1.454881062547D-01     5,75 -1,303365394679D-03
```

```
              N= 4       M=12                      P=  6,0       Q= 7

I= 0     H= -6.0
0           1,000000000000D+00   9,249181003606D-02   -1.00  4,666976258403D+00
1          -2,043922399141D+00  -9,116051466432D-01    -,75  3,405147215937D+00
2           1,333188051233D+00   6,566500813690D-01    -,50  2,391491438383D+00
3          -2,898658080285D-01  -1,446347769734D-01    -,25  1,598834006238D+00
```

CARDINAL SPLINES N=4

				.25	5.678145367969D-01
0	-6.0	6.001559367505D-04	3.000779683753D-04	.50	2.751072759523D-01
1	-5.0	-2.800902325933D-03	-1.400451162966D-03	.75	9.476945284765D-02
2	-4.0	5.426017141029D-03	2.713008570515D-03	1.25	-3.507922962053D-02
3	-3.0	-5.819184298951D-03	-2.909592149475D-03	1.50	-3.427524075753D-02
4	-2.0	4.093970722406D-03	2.046985361203D-03	1.75	-1.768234824470D-02
5	-1.0	-2.316041369703D-03	-1.158020684852D-03	2.25	1.053405850050D-02
6	0.0	1.253066855511D-03	6.265334277557D-04	2.50	1.195853432229D-02
7	1.0	-6.686712259943D-04	-3.343356129971D-04	2.75	6.908769609776D-03
8	2.0	3.508397556372D-04	1.754198778186D-04	3.25	-4.774590058607D-03
9	3.0	-1.748945058145D-04	-8.744725290727D-05	3.50	-5.682173696376D-03
10	4.0	7.524246396703D-05	3.762123198352D-05	3.75	-3.398790371442D-03
11	5.0	-2.314000294457D-05	-1.157000147229D-05	4.25	2.451611171983D-03
12	6.0	3.540854038539D-06	1.770427019269D-06	4.50	2.954730753696D-03
				4.75	1.783331560337D-03
	0	-2.587977908461D+01	-1.297884796681D+01	5.25	-1.300351507279D-03
	1	-1.735117305142D+01	-8.652360083348D+00	5.50	-1.572299138668D-03
	2	-3.884396493279D+00	-1.946775904153D+00	5.75	-9.512701868341D-04
	3	-2.898658080285D-01	-1.446347769734D-01	6.25	6.960869280123D-04

I= 1 H= -5.0

0		0.	2.663130413638D+00	-1.00	-9.041133955740D+00
1		3.959038275872D+00	1.258324880462D+00	-.75	-5.678483114667D+00
2		-4.019166526707D+00	-1.944598014017D+00	-.50	-3.117176913758D+00
3		1.062929153161D+00	5.295162691566D-01	-.25	-1.257565744905D+00
				.25	7.551697581135D-01
0	-6.0	-2.800902325933D-03	-1.400451162966D-03	.50	1.107571768355D+00
1	-5.0	1.367182751638D-02	6.835913758191D-03	.75	1.156546896792D+00
2	-4.0	-2.812375787873D-02	-1.406187893936D-02	1.25	7.315287405348D-01
3	-3.0	3.258345976107D-02	1.629172988054D-02	1.50	4.350693897032D-01
4	-2.0	-2.492454032501D-02	-1.246227016250D-02	1.75	1.772928262375D-01
5	-1.0	1.490076773938D-02	7.450383869689D-03	2.25	-8.424174230647D-02
6	0.0	-8.160106200471D-03	-4.080053100235D-03	2.50	-9.008592154570D-02
7	1.0	4.366334176422D-03	2.183167088211D-03	2.75	-4.992579503760D-02
8	2.0	-2.292361109715D-03	-1.146180554857D-03	3.25	3.281745041872D-02
9	3.0	1.142911026208D-03	5.714555131041D-04	3.50	3.847790430977D-02
10	4.0	-4.917153493069D-04	-2.458576746534D-04	3.75	2.277149587244D-02
11	5.0	1.512229726486D-04	7.561148632428D-05	4.25	-1.621707743086D-02
12	6.0	-2.314000294457D-05	-1.157000147229D-05	4.50	-1.947327197973D-02
				4.75	-1.172274184423D-02
	0	1.086569317765D+02	5.458306532963D+01	5.25	8.521933639046D-03
	1	7.052538849677D+01	3.511090578117D+01	5.50	1.029527465449D-02
	2	1.511355823019D+01	7.586694830802D+00	5.75	6.225084081317D-03
	3	1.062929153161D+00	5.295162691566D-01	6.25	-4.551973591870D-03

I= 2 H= -4.0

0		0.	-8.659534040446D+00	-1.00	9.696516459547D+00
1		-3.293197097628D+00	6.984637165607D-01	-.75	5.852702975762D+00
2		4.845545221203D+00	2.211464126148D+00	-.50	3.052706621704D+00
3		-1.557774140716D+00	-7.725518801861D-01	-.25	1.150486071681D+00
				.25	-5.447925878526D-01
0	-6.0	5.426017141029D-03	2.713008570515D-03	.50	-6.298916203438D-01
1	-5.0	-2.812375787873D-02	-1.406187893936D-02	.75	-4.007403160514D-01
2	-4.0	6.272762243930D-02	3.136381121965D-02	1.25	4.380119669095D-01
3	-3.0	-8.073217297335D-02	-4.036608648668D-02	1.50	7.976822492350D-01

CARDINAL SPLINES N=4

```
 4  -2,0   6.959035562804D-02   3.479517781402D-02     1.75  9,966766274283D-01
 5  -1,0  -4.585037137879D-02  -2,292518568939D-02     2,25  8,267706119296D-01
 6   0,0   2.620397873393D-02   1.310198936696D-02     2.50  5,431912701073D-01
 7   1,0  -1,415725684555D-02  -7,078628422775D-03     2.75  2,397584410324D-01
 8   2,0   7.448981885856D-03   3,724490942928D-03     3.25 -1,274126755164D-01
 9   3,0  -3.715734669052D-03  -1,857867334526D-03     3.50 -1,414644448485D-01
10   4.0   1.598810802668D-03   7.994054013341D-04     3.75 -8,065897689669D-02
11   5,0  -4,917153493069D-04  -2,458576746534D-04     4,25  5,498852177498D-02
12   6,0   7,524246396703D-05   3,762123198352D-05     4,50  6,520143150770D-02
                                                       4,75  3,890258229029D-02
      0   -1,817987690171D+02  -9,172724931993D+01     5,25 -2,798426984292D-02
      1   -1,133862616405D+02  -5,619956982975D+01     5,50 -3,370603011522D-02
      2   -2,319438931169D+01  -1,169446971720D+01     5,75 -2,033766168936D-02
      3   -1,557774140716D+00  -7,725518801861D-01     6,25  1,483496023993D-02


I= 3     H= -3.0
 0         0.                   2,012879138670D+01    -1.00 -6,822431969136D+00
 1         2.161251152260D+00  -4,370393107990D+00     -.75 -4,066667107749D+00
 2        -3,408306392419D+00  -1,212973578824D+00     -,50 -2,089311477292D+00
 3         1,252874424457D+00   6,117112677375D-01     -.25 -7,729081004734D-01
                                                        ,25  3,468694462462D-01
 0  -6,0  -5,819184298951D-03  -2,909592149475D-03      .50  3,851128187053D-01
 1  -5,0   3.258345976107D-02   1.629172988054D-02      .75  2,315456489399D-01
 2  -4,0  -8,073217297335D-02  -4,036608648668D-02     1.25 -2,046404818657D-01
 3  -3,0   1,191417015203D-01   5,957085076014D-02     1.50 -2,975331329615D-01
 4  -2,0  -1,204058797496D-01  -6,020293987478D-02     1,75 -2,292773141025D-01
 5  -1,0   9,200298658435D-02   4,600149329217D-02     2,25  3,358616704449D-01
 6   0,0  -5,791642631063D-02  -2,895821315531D-02     2,50  6,817158177501D-01
 7   1,0   3,253039663508D-02   1.626519831754D-02     2,75  9,296774894683D-01
 8   2,0  -1,726867576820D-02  -8,634337884100D-03     3,25  8,730779891858D-01
 9   3,0   8,631512748566D-03   4,315756374283D-03     3,50  5,983055660459D-01
10   4,0  -3,715734669052D-03  -1,857867334526D-03     3,75  2,727287288875D-01
11   5,0   1,142911026208D-03   5,714555131041D-04     4,25 -1,512033834812D-01
12   6,0  -1,748945058145D-04  -8,744725290727D-05     4,50 -1,701456400819D-01
                                                       4,75 -9,797627066680D-02
      0    1,608893524693D+02   8,236901773240D+01     5,25  6,763167442583D-02
      1    9,657201228463D+01   4,713874086177D+01     5,50  8,050380489820D-02
      2    1,914343324781D+01   9,797829240451D+00     5,75  4,817327819335D-02
      3    1,252874424457D+00   6,117112677375D-01     6,25 -3,479765712036D-02


I= 4     H= -2.0
 0         0.                  -4,038249742961D+01    -1,00  3,860912139692D+00
 1        -1.207880731929D+00   1,033193737637D+01     -.75  2,296340701603D+00
 2         1.928409084485D+00  -2,120503487668D-02     -,50  1,176620427496D+00
 3        -7.246223232780D-01  -3,327677914846D-01     -.25  4,338179745638D-01
                                                        .25 -1,927665891271D-01
 0  -6,0   4,093970722406D-03   2,046985361203D-03      .50 -2,123839011069D-01
 1  -5,0  -2,492454032501D-02  -1,246227016250D-02      .75 -1,263340024330D-01
 2  -4,0   6,959035562804D-02   3,479517781402D-02     1,25  1,075303568209D-01
 3  -3,0  -1,204058797496D-01  -6,020293987478D-02     1,50  1,512536060524D-01
 4  -2,0   1,470178036519D-01   7,350890182595D-02     1,75  1,108955002316D-01
 5  -1,0  -1,361025429614D-01  -6,805127148070D-02     2,25 -1,327571855541D-01
 6   0,0   1,003559358582D-01   5,017796792908D-02     2,50 -2,159228893173D-01
 7   1.0  -6,211261226744D-02  -3,105630613372D-02     2.75 -1,821237833761D-01
 8   2,0   3,424872467932D-02   1,712436233966D-02     3,25  3,032629804181D-01
```

9	3.0	-1.726867576820D-02	-8.634337884100D-03	3.50	6.429100472541D-01
10	4.0	7.448981885856D-03	3.724490942928D-03	3.75	9.064572545271D-01
11	5.0	-2.292361109715D-03	-1.146180554857D-03	4.25	8.898472199065D-01
12	6.0	3.508397556372D-04	1.754198778186D-04	4.50	6.185351422588D-01
				4.75	2.849539389475D-01
	0	-9.434297917816D+01	-5.103209738763D+01	5.25	-1.601546074254D-01
	1	-5.632618263213D+01	-2.586144452249D+01	5.50	-1.810039869775D-01
	2	-1.111479273452D+01	-6.011025281600D+00	5.75	-1.045747710034D-01
	3	-7.246223232780D-01	-3.327677914846D-01	6.25	7.252890540261D-02
I= 5		H= -1.0			
0		0.	7.699868704045D+01	-1.00	-2.091420714167D+00
1		6.525308581617D-01	-2.052565064290D+01	-.75	-1.243319973930D+00
2		-1.044552336399D+00	1.356656118956D+00	-.50	-6.366957031314D-01
3		3.943375196068D-01	1.408366165585D-01	-.25	-2.345787593092D-01
				.25	1.040095758994D-01
0	-6.0	-2.316041369703D-03	-1.158020684852D-03	.50	1.144014408588D-01
1	-5.0	1.490076773938D-02	7.450383869689D-03	.75	6.788944127724D-02
2	-4.0	-4.585037137879D-02	-2.292518568939D-02	1.25	-5.730182131838D-02
3	-3.0	9.200298658435D-02	4.600149329217D-02	1.50	-8.001266682398D-02
4	-2.0	-1.361025429614D-01	-6.805127148070D-02	1.75	-5.803686663715D-02
5	-1.0	1.557504659004D-01	7.787523295018D-02	2.25	6.682839229555D-02
6	0.0	-1.405473181665D-01	-7.027365908324D-02	2.50	1.050346982257D-01
7	1.0	1.022038631798D-01	5.110193158989D-02	2.75	8.418339395892D-02
8	2.0	-6.211261226744D-02	-3.105630613372D-02	3.25	-1.142740180550D-01
9	3.0	3.253039663508D-02	1.626519831754D-02	3.50	-1.939052244085D-01
10	4.0	-1.415725684555D-02	-7.078628422775D-03	3.75	-1.689367210946D-01
11	5.0	4.366334176422D-03	2.183167088211D-03	4.25	2.937107044899D-01
12	6.0	-6.686712259943D-04	-3.343356129971D-04	4.50	6.313594107804D-01
				4.75	8.994545126625D-01
	0	5.148820527368D+01	3.310511264215D+01	5.25	8.950275699391D-01
	1	3.070635493891D+01	1.096457737290D+01	5.50	6.248693267571D-01
	2	6.053523016524D+00	3.891715217010D+00	5.75	2.888444099650D-01
	3	3.943375196068D-01	1.408366165585D-01	6.25	-1.631394789463D-01
I= 6		H= 0.0			
0		0.	-1.445637258019D+02	-1.00	1.120756193990D+00
1		-3.494663896347D-01	3.897524124742D+01	-.75	6.662027420651D-01
2		5.597515635673D-01	-3.247936770618D+00	-.50	3.411133658077D-01
3		-2.115382407881D-01	1.139722976452D-24	-.25	1.256563551440D-01
				.25	-5.568733321691D-02
0	-6.0	1.253066855511D-03	6.265334277557D-04	.50	-6.122779443924D-02
1	-5.0	-8.160106200471D-03	-4.080053100235D-03	.75	-3.631496881772D-02
2	-4.0	2.620397873393D-02	1.310198936696D-02	1.25	3.059279482211D-02
3	-3.0	-5.791642631063D-02	-2.895821315531D-02	1.50	4.264594206143D-02
4	-2.0	1.003559358582D-01	5.017796792908D-02	1.75	3.085647997762D-02
5	-1.0	-1.405473181665D-01	-7.027365908324D-02	2.25	-3.521179906616D-02
6	0.0	1.576217384599D-01	7.881086922997D-02	2.50	-5.491074687757D-02
7	1.0	-1.405473181665D-01	-7.027365908324D-02	2.75	-4.351538596754D-02
8	2.0	1.003559358582D-01	5.017796792908D-02	3.25	5.674952638528D-02
9	3.0	-5.791642631063D-02	-2.895821315531D-02	3.50	9.299975948046D-02
10	4.0	2.620397873393D-02	1.310198936696D-02	3.75	7.695191156719D-02
11	5.0	-8.160106200471D-03	-4.080053100235D-03	4.25	-1.089811948802D-01
12	6.0	1.253066855511D-03	6.265334277557D-04	4.50	-1.874537019463D-01
				4.75	-1.649831643731D-01

CARDINAL SPLINES N=4

	0	-2,763800205961D+01	-2,763800205961D+01	5,25	2,906865434238D-01
	1	-1.647857763194D+01	8.760756642647D-23	5,50	6,275685280292D-01
	2	-3,247936770618D+00	-3,247936770618D+00	5,75	8,970500877280D-01
	3	-2,115382407881D-01	1.139722976452D-24	6,25	8,970500877280D-01

I= 7 H= 1.0

0		0.	2,694150338685D+02	-1,00	-5,967681498656D-01
1		1,860541240561D-01	-7,287551456534D+01	-,75	-3,547238172723D-01
2		-2,980497393198D-01	6,426774315063D+00	-,50	-1,816225326692D-01
3		1,126642864898D-01	-1,408366165585D-01	-,25	-6,690201919791D-02
				,25	2,964576097048D-02
0	-6,0	-6,686712259943D-04	-3,343356129971D-04	,50	3,259243901534D-02
1	-5,0	4,366334176422D-03	2.183167088211D-03	,75	1,932860370333D-02
2	-4.0	-1,415725684555D-02	-7,078628422775D-03	1.25	-1,627583443871D-02
3	-3,0	3,253039663508D-02	1,626519831754D-02	1,50	-2,267952327223D-02
4	-2,0	-6,211261226744D-02	-3,105630613372D-02	1,75	-1,640042368830D-02
5	-1,0	1.022038631798D-01	5.110193158989D-02	2.25	1,867635547340D-02
6	0.0	-1,405473181665D-01	-7,027365908324D-02	2,50	2,907169748257D-02
7	1.0	1,557504659004D-01	7,787523295018D-02	2,75	2,297787697509D-02
8	2,0	-1,361025429614D-01	-6,805127148070D-02	3,25	-2,968601403195D-02
9	3,0	9,200298658435D-02	4,600149329217D-02	3,50	-4,825921625978D-02
10	4,0	-4,585037137879D-02	-2,292518568939D-02	3,75	-3,947533214372D-02
11	5,0	1,490076773938D-02	7,450383869689D-03	4,25	5,369177767396D-02
12	6,0	-2,316041369703D-03	-1,158020684852D-03	4,50	8,917862543847D-02
				4,75	7,453372545014D-02
	0	1,472202001061D+01	3,310511264215D+01	5,25	-1,069530473369D-01
	1	8,777200193111D+00	-1,096457737290D+01	5,50	-1,847497882370D-01
	2	1,729907417496D+00	3,891715217010D+00	5,75	-1,631394789463D-01
	3	1,126642864898D-01	-1,408366165585D-01	6,25	2,888444099650D-01

I= 8 H= 2.0

0		0,	-4,944755176208D+02	-1,00	3,129578335074D-01
1		-9,756759632280D-02	1,339326693820D+02	-,75	1,860236328025D-01
2		1,563034968759D-01	-1,200084552832D+01	-,50	9,524551491896D-02
3		-5,908674030872D-02	3,327677914846D-01	-,25	3,508409795277D-02
				,25	-1,554613942972D-02
0	-6,0	3,508397556372D-04	1,754198778186D-04	,50	-1,709102554543D-02
1	-5,0	-2,292361109715D-03	-1,146180554857D-03	,75	-1,013536736296D-02
2	-4,0	7,448981885856D-03	3,724490942928D-03	1,25	8,533723205604D-03
3	-3,0	-1,726867576820D-02	-8,634337884100D-03	1,50	1,189024201076D-02
4	-2,0	3,424872467932D-02	1,712436233966D-02	1,75	8,597160374663D-03
5	-1,0	-6,211261226744D-02	-3,105630613372D-02	2,25	-9,785524711897D-03
6	0,0	1,003559358582D-01	5,017796792908D-02	2,50	-1,522580309326D-02
7	1,0	-1,361025429614D-01	-6,805127148070D-02	2,75	-1,202694086526D-02
8	2,0	1,470178036519D-01	7,350890182595D-02	3,25	1,550425711085D-02
9	3,0	-1,204058797496D-01	-6,020293987478D-02	3,50	2,515732049511D-02
10	4,0	6,959035562804D-02	3,479517781402D-02	3,75	2,052292878900D-02
11	5,0	-2,492454032501D-02	-1,246227016250D-02	4,25	-2,764983117731D-02
12	6,0	4,093970722406D-03	2,046985361203D-03	4,50	-4,555403192124D-02
				4,75	-3,763572401527D-02
	0	-7,721215597089D+00	-5,103209738763D+01	5,25	5,186081508330D-02
	1	-4,603293587154D+00	2,586144452249D+01	5,50	8,649969297754D-02
	2	-9,072578286811D-01	-6,011025281600D+00	5,75	7,252890540261D-02
	3	-5,908674030872D-02	3,327677914846D-01	6,25	-1,045747710034D-01

CARDINAL SPLINES N=4

```
I= 9      H=  3.0
0             0.                    8.500529493906D+02   -1.00 -1.559924316935D-01
1             4.863177411730D-02   -2.307775086628D+02    -.75 -9.272252858643D-02
2            -7.790876859383D-02    2.080863205973D+01    -.50 -4.747456532990D-02
3             2.945188898235D-02   -6.117112677375D-01    -.25 -1.748742733179D-02
                                                           .25  7.748820582846D-03
0   -6.0     -1.748945058145D-04   -8.744725290727D-05     .50  8.518814669658D-03
1   -5.0      1.142911026208D-03    5.714555131041D-04     .75  5.051818319970D-03
2   -4.0     -3.715734669052D-03   -1.857867334526D-03    1.25 -4.253404846217D-03
3   -3.0      8.631512748566D-03    4.315756374283D-03    1.50 -5.926250447798D-03
4   -2.0     -1.726867576820D-02   -8.634337884100D-03    1.75 -4.284806964027D-03
5   -1.0      3.253039663508D-02    1.626519831754D-02    2.25  4.876547660008D-03
6    0.0     -5.791642631063D-02   -2.895821315531D-02    2.50  7.586939219165D-03
7    1.0      9.200298658435D-02    4.600149329217D-02    2.75  5.992121388859D-03
8    2.0     -1.204058797496D-01   -6.020293987478D-02    3.25 -7.720723154391D-03
9    3.0      1.191417015203D-01    5.957085076014D-02    3.50 -1.252226487054D-02
10   4.0     -8.073217297335D-02   -4.036608648668D-02    3.75 -1.020907645134D-02
11   5.0      3.258345976107D-02    1.629172988054D-02    4.25  1.372389134713D-02
12   6.0     -5.819184298951D-03   -2.909592149475D-03    4.50  2.256766521438D-02
                                                          4.75  1.859422697706D-02
         0    3.848682995513D+00    8.236901773240D+01    5.25 -2.537807694766D-02
         1    2.294530561085D+00   -4.713874086177D+01    5.50 -4.198512099133D-02
         2    4.522252330884D-01    9.797829240451D+00    5.75 -3.479765712036D-02
         3    2.945188898235D-02   -6.117112677375D-01    6.25  4.817327819335D-02


I=10      H=  4.0
0             0.                   -1.016796784238D+03   -1.00  6.710868941720D-02
1            -2.092158559662D-02    2.799688094963D+02    -.75  3.988966286067D-02
2             3.351672347661D-02   -2.560040356055D+01    -.50  2.042377121046D-02
3            -1.267038034396D-02    7.725518801861D-01    -.25  7.523166309318D-03
                                                           .25 -3.333571282305D-03
0   -6.0      7.524246396703D-05    3.762123198352D-05     .50 -3.664821640402D-03
1   -5.0     -4.917153493069D-04   -2.458576746534D-04     .75 -2.173305292942D-03
2   -4.0      1.598810802668D-03    7.994054013341D-04    1.25  1.829815842155D-03
3   -3.0     -3.715734669052D-03   -1.857867334526D-03    1.50  2.549462277103D-03
4   -2.0      7.448981885856D-03    3.724490942928D-03    1.75  1.843303267067D-03
5   -1.0     -1.415725684555D-02   -7.078628422775D-03    2.25 -2.097813166309D-03
6    0.0      2.620397873393D-02    1.310198936696D-02    2.50 -3.263706705130D-03
7    1.0     -4.585037137879D-02   -2.292518568939D-02    2.75 -2.577572313037D-03
8    2.0      6.959035562804D-02    3.479517781402D-02    3.25  3.320757173296D-03
9    3.0     -8.073217297335D-02   -4.036608648668D-02    3.50  5.385398948073D-03
10   4.0      6.272762243930D-02    3.136381121965D-02    3.75  4.389932709517D-03
11   5.0     -2.812375787873D-02   -1.406187893936D-02    4.25 -5.898245386395D-03
12   6.0      5.426017141029D-03    2.713008570515D-03    4.50 -9.694784698593D-03
                                                          4.75 -7.982722793510D-03
         0   -1.655729622717D+00   -9.172724931993D+01    5.25  1.087041447916D-02
         1   -9.871219810251D-01    5.619956982975D+01    5.50  1.794895651495D-02
         2   -1.945501227147D-01   -1.169446971720D+01    5.75  1.483496023993D-02
         3   -1.267038034396D-02    7.725518801861D-01    6.25 -2.033766168936D-02


I=11      H=  5.0
0             0.                    6.527450280634D+02   -1.00 -2.063841169071D-02
1             6.434160999138D-03   -1.833390008197D+02    -.75 -1.226755030041D-02
2            -1.030763584388D-02    1.711798767562D+01    -.50 -6.281066316500D-03
```

CARDINAL SPLINES N=4

3		3.896614847688D-03	-5.295162691566D-01	-.25	-2.313652097022D-03
				.25	1.025196204183D-03
0	-6.0	-2.314000294457D-05	-1.157000147229D-05	.50	1.127067613287D-03
1	-5.0	1.512229726486D-04	7.561148632428D-05	.75	6.683711585067D-04
2	-4.0	-4.917153493069D-04	-2.458576746534D-04	1.25	-5.627347852826D-04
3	-3.0	1.142911026208D-03	5.714555131041D-04	1.50	-7.840512536655D-04
4	-2.0	-2.292361109715D-03	-1.146180554857D-03	1.75	-5.668809932433D-04
5	-1.0	4.366334176422D-03	2.183167088211D-03	2.25	6.451476593831D-04
6	0.0	-8.160106200471D-03	-4.080053100235D-03	2.50	1.003693248314D-03
7	1.0	1.490076773938D-02	7.450383869689D-03	2.75	7.926787043726D-04
8	2.0	-2.492454032501D-02	-1.246227016250D-02	3.25	-1.021199849650D-03
9	3.0	3.258345976107D-02	1.629172988054D-02	3.50	-1.656077001545D-03
10	4.0	-2.812375787873D-02	-1.406187893936D-02	3.75	-1.349909620177D-03
11	5.0	1.367182751638D-02	6.835913758191D-03	4.25	1.813482927902D-03
12	6.0	-2.800902325933D-03	-1.400451162966D-03	4.50	2.980440783867D-03
				4.75	2.453714148495D-03
	0	5.091988827158D-01	5.458306532963D+01	5.25	-3.339435822949D-03
	1	3.035769344229D-01	-3.511090578117D+01	5.50	-5.511314903514D-03
	2	5.983143141451D-02	7.586694830802D+00	5.75	-4.551973591870D-03
	3	3.896614847688D-03	-5.295162691566D-01	6.25	6.225084081317D-03

I=12	H= 6.0				
0		0.	-1.662180528427D+02	-1.00	3.158057737624D-03
1		-9.845452142484D-04	4.763422684632D+01	-.75	1.877161474862D-03
2		1.577258441793D-03	-4.550201889675D+00	-.50	9.611189777702D-04
3		-5.962540815828D-04	1.446347769734D-01	-.25	3.540314261989D-04
				.25	-1.568739048582D-04
0	-6.0	3.540854038539D-06	1.770427019269D-06	.50	-1.724620939517D-04
1	-5.0	-2.314000294457D-05	-1.157000147229D-05	.75	-1.022730808863D-04
2	-4.0	7.524246396703D-05	3.762123198352D-05	1.25	8.610873981894D-05
3	-3.0	-1.748945058145D-04	-8.744725290727D-05	1.50	1.199741768451D-04
4	-2.0	3.508397556372D-04	1.754198778186D-04	1.75	8.674311327259D-05
5	-1.0	-6.686712259943D-04	-3.343356129971D-04	2.25	-9.871915838818D-05
6	0.0	1.253066855511D-03	6.265334277557D-04	2.50	-1.535828164522D-04
7	1.0	-2.316041369703D-03	-1.158020684852D-03	2.75	-1.212935781694D-04
8	2.0	4.093970722406D-03	2.046985361203D-03	3.25	1.562599739199D-04
9	3.0	-5.819184298951D-03	-2.909592149475D-03	3.50	2.534045517747D-04
10	4.0	5.426017141029D-03	2.713008570515D-03	3.75	2.065542252368D-04
11	5.0	-2.800902325933D-03	-1.400451162966D-03	4.25	-2.774769363378D-04
12	6.0	6.001559367505D-04	3.000779683753D-04	4.50	-4.560161095558D-04
				4.75	-3.754083433573D-04
	0	-7.791684900283D-02	-1.297884796681D+01	5.25	5.108378928463D-04
	1	-4.645288472368D-02	8.652360083348D+00	5.50	8.429565316600D-04
	2	-9.155315026697D-03	-1.946775904153D+00	5.75	6.960869280123D-04
	3	-5.962540815828D-04	1.446347769734D-01	6.25	-9.512701868341D-04

CARDINAL SPLINES N=4

```
I= 0     H= -6.5
0           1.000000000000D+00    7.826840278434D-01    -1.00  4,666981596912D+00
1          -2.043924063456D+00   -1,092157573258D+00     -.75  3,405150389166D+00
2           1.333190717495D+00    6.723968826192D-01     -.50  2,391493063097D+00
3          -2.898668159607D-01   -1.450929890101D-01     -.25  1,598834604707D+00
                                                          .25  5,678142716109D-01
0   -6,5    6,001619223626D-04    3.000809611813D-04      .50  2,751069844154D-01
1   -5,5   -2.800941442864D-03   -1.400470721432D-03      .75  9,476927996114D-02
2   -4,5    5,426144335215D-03    2.713072167608D-03     1.25 -3,507908405892D-02
3   -3,5   -5.819479958582D-03   -2.909739979291D-03     1.50 -3,427503794850D-02
4   -2,5    4.094563885785D-03    2.047281942893D-03     1.75 -1,768220161081D-02
5   -1,5   -2.317172451831D-03   -1.158586225915D-03     2.25  1,053389162209D-02
6    -,5    1.255191076284D-03    6,275955381418D-04     2.50  1,195827470072D-02
7     ,5   -6,726351918849D-04   -3,363175959425D-04     2.75  6,908564571401D-03
8    1,5    3,581575200282D-04    1.790787600141D-04     3.25 -4,774325913726D-03
9    2,5   -1.878253749762D-04   -9.391268748812D-05     3,50 -5,681745338887D-03
10   3,5    9,362071767918D-05    4,681035883959D-05     3.75 -3,398441213090D-03
11   4,5   -4.027608004786D-05   -2.013804002393D-05     4,25  2,451142140010D-03
12   5,5    1,238638883481D-05    6.193194417403D-06     4.50  2,953959948304D-03
13   6,5   -1,895346002745D-06   -9.476730013723D-07     4.75  1,782697029591D-03
                                                         5,25 -1,299488179315D-03
       0   -3.556287293152D+01   -1,775373401456D+01     5,50 -1,570874680314D-03
       1   -2.145306365904D+01   -1,074153445623D+01     5,75 -9,500941012113D-04
       2   -4,319212193739D+00   -2.156916403077D+00     6,25  6,944805987353D-04
       3   -2,898668159607D-01   -1,450929890101D-01     6,50  8,403028068788D-04


I= 1     H= -5.5
0           0.                   -1,847385960755D+00    -1.00 -9,041168843747D+00
1           3,959049152436D+00    2,438263824528D+00     -.75 -5,678503852230D+00
2          -4,019183951152D+00   -2.047505875119D+00     -.50 -3,117187531526D+00
3           1,062935740159D+00    5,325107579973D-01     -.25 -1,257569655996D+00
                                                          .25  7,551714911461D-01
0   -6,5   -2,800941442864D-03   -1,400470721432D-03      .50  1,107573673595D+00
1   -5,5    1.367208315177D-02    6.836041575887D-03      .75  1,156548026633D+00
2   -4,5   -2,812458911305D-02   -1,406229455653D-02     1.25  7,315277892664D-01
3   -3,5    3,258539194399D-02    1.629269597199D-02     1.50  4,350680643138D-01
4   -2,5   -2,492841674242D-02   -1,246420837121D-02     1.75  1,772918679616D-01
5   -1,5    1,490815954173D-02    7.454079770863D-03     2,25 -8,424065172942D-02
6    -,5   -8,173988322479D-03   -4,086994161239D-03     2,50 -9,008422487725D-02
7     ,5    4,392239326605D-03    2.196119663302D-03     2,75 -4,992445507907D-02
8    1,5   -2,340183868734D-03   -1.170091934367D-03     3,25  3,281572418974D-02
9    2.5    1.227416321377D-03    6,137081606885D-04     3,50  3,847510492517D-02
10   3,5   -6,118201725179D-04   -3.059100862590D-04     3.75  2,276921406685D-02
11   4,5    2,632099733319D-04    1.316049866660D-04     4,25 -1,621401223200D-02
12   5,5   -8.094698556396D-05   -4,047349278198D-05     4,50 -1,946823464338D-02
13   6,5    1,238638883481D-05    6,193194417403D-06     4,75 -1,171859508447D-02
                                                         5,25  8,516291652924D-03
       0    1,478320251958D+02    7,373497258992D+01     5,50  1,028596559165D-02
       1    8,643676285260D+01    4,331642602414D+01     5,75  6,217398173939D-03
       2    1,670806298195D+01    8,336453905828D+00     6,25 -4,541475973498D-03
       3    1,062935740159D+00    5,325107579973D-01     6,50 -5,493972388240D-03


I= 2     H= -4.5
0           0.                    6,007042390753D+00    -1,00  9,696629902793D+00
1          -3,293232464300D+00   -3,138273204635D+00     -.75  5,852770406856D+00
```

CARDINAL SPLINES N=4

2		4.845601879229D+00	2,546083465695D+00	-.50	3,052741146865D+00
3		-1.557795559264D+00	-7,822888805222D-01	-.25	1,150498789140D+00
				.25	-5,447982230512D-01
0	-6,5	5,426144335215D-03	2,713072167608D-03	.50	-6,298978154983D-01
1	-5,5	-2,812458911305D-02	-1,406229455653D-02	.75	-4,007439898877D-01
2	-4,5	6,273032531424D-02	3,136516265712D-02	1.25	4,380150600921D-01
3	-3,5	-8,073845573702D-02	-4.036922786851D-02	1.50	7,976865589245D-01
4	-2,5	6,960296034279D-02	3,480148017140D-02	1.75	9,966797433966D-01
5	-1.5	-4,587440686056D-02	-2.293720343028D-02	2.25	8,267670657654D-01
6	-.5	2,624911840011D-02	1.312455920005D-02	2,50	5,431857531520D-01
7	.5	-1.424149107365D-02	-7.120745536824D-03	2.75	2,397540839694D-01
8	1,5	7,604484292252D-03	3.802242146126D-03	3,25	-1.274070624408D-01
9	2,5	-3,990515485162D-03	-1,995257742581D-03	3,50	-1.414553422570D-01
10	3,5	1,989348475778D-03	9,946742378892D-04	3,75	-8,065155728586D-02
11	4,5	-8,558567842317D-04	-4,279283921159D-04	4,25	5,497855485111D-02
12	5,5	2,632099733319D-04	1.316049866660D-04	4,50	6,518505190228D-02
13	6,5	-4,027608004786D-05	-2,013804002393D-05	4,75	3,888909851947D-02
				5,25	-2,796592413395D-02
	0	-2,444889370833D+02	-1,216557908272D+02	5,50	-3,367576039212D-02
	1	-1,377509951710D+02	-6,919430375678D+01	5,75	-2,031266988384D-02
	2	-2,553141152641D+01	-1,270854970449D+01	6.25	1,480082576187D-02
	3	-1,557795559264D+00	-7,822888805222D-01	6,50	1,789256489487D-02

I= 3 H= -3.5

0		0.	-1.396329030928D+01	-1.00	-6,822695665060D+00
1		2.161333361186D+00	4.548003709220D+00	-.75	-4,066823849601D+00
2		-3.408438092551D+00	-1,990787672958D+00	-.50	-2,089391730146D+00
3		1.252924211323D+00	6.343446759149D-01	-.25	-7,729376618829D-01
				.25	3,468825451212D-01
0	-6,5	-5,819479958582D-03	-2,909739979291D-03	.50	3,851272191836D-01
1	-5,5	3,258539194399D-02	1.629269597199D-02	.75	2,315541886782D-01
2	-4,5	-8,073845573702D-02	-4,036922786851D-02	1.25	-2,046476718895D-01
3	-3,5	1,191563056437D-01	5,957815282185D-02	1,50	-2,975431507246D-01
4	-2,5	-1,204351790865D-01	-6,021758954327D-02	1,75	-2,292845570912D-01
5	-1.5	9,205885644682D-02	4,602942822341D-02	2,25	3,358699134129D-01
6	-.5	-5,802135230985D-02	-2,901067615493D-02	2,50	6,817286417717D-01
7	.5	3,272619694331D-02	1,636309847165D-02	2,75	9,296876173498D-01
8	1,5	-1,763013713361D-02	-8,815068566807D-03	3.25	8,730649417352D-01
9	2,5	9,270233725834D-03	4,635116862917D-03	3,50	5,982844073048D-01
10	3,5	-4,623529498983D-03	-2,311764749492D-03	3,75	2,727114821919D-01
11	4,5	1,989348475778D-03	9,946742378892D-04	4,25	-1,511802156217D-01
12	5,5	-6,118201725179D-04	-3,059100862590D-04	4,50	-1,701075661081D-01
13	6,5	9,362071767918D-05	4,681035883959D-05	4,75	-9,794492798650D-02
				5,25	6,758903029469D-02
	0	2,141264689721D+02	1,056948612413D+02	5,50	8,043344370125D-02
	1	1,166597819432D+02	5,907095163297D+01	5,75	4,811518538080D-02
	2	2,102358402825D+01	1,037893350738D+01	6,25	-3,471831239939D-02
	3	1,252924211323D+00	6.343446759149D-01	6,50	-4,185404011313D-02

I= 4 H= -2.5

0		0.	2,801430818142D+01	-1.00	3,861441176287D+00
1		-1.208045662542D+00	-7,560482890259D+00	-.75	2,296655162952D+00
2		1.928673306201D+00	1.539274605892D+00	-.50	1,176781433764D+00
3		-7.247222075446D-01	-3,781757789823D-01	-.25	4,338772817659D-01
				.25	-1,927928685784D-01

CARDINAL SPLINES N=4

0	-6.5	4.094563885785D-03	2.047281942893D-03	.50	-2.124127918835D-01
1	-5.5	-2.492841674242D-02	-1.246420837121D-02	.75	-1.263511351745D-01
2	-4.5	6.960296034279D-02	3.480148017140D-02	1.25	1.075447817146D-01
3	-3.5	-1.204351790865D-01	-6.021758954327D-02	1.50	1.512737040621D-01
4	-2.5	1.470765850743D-01	7.353829253717D-02	1.75	1.109100313856D-01
5	-1.5	-1.362146311629D-01	-6.810731558147D-02	2.25	-1.327737229038D-01
6	-.5	1.005664423106D-01	5.028322115528D-02	2.50	-2.159486173476D-01
7	.5	-6.250543414731D-02	-3.125271707365D-02	2.75	-1.821441023097D-01
8	1.5	3.497390194971D-02	1.748695097485D-02	3.25	3.032891567000D-01
9	2.5	-1.855010161105D-02	-9.275050805526D-03	3.50	6.429524967096D-01
10	3.5	9.270233725834D-03	4.635116862917D-03	3.75	9.064918554909D-01
11	4.5	-3.990515485162D-03	-1.995257742581D-03	4.25	8.898007396827D-01
12	5.5	1.227416321377D-03	6.137081606885D-04	4.50	6.184587568328D-01
13	6.5	-1.878253749762D-04	-9.391268748812D-05	4.75	2.848910580935D-01
				5.25	-1.600690531793D-01
	0	-1.253926858665D+02	-5.995100180934D+01	5.50	-1.808628257203D-01
	1	-6.799383248822D+01	-3.548369299967D+01	5.75	-1.044582230366D-01
	2	-1.220340974092D+01	-5.835153084263D+00	6.25	7.236972106459D-02
	3	-7.247222075446D-01	-3.781757789823D-01	6.50	8.623671363161D-02

I= 5 H= -1.5

0		0.	-5.342474631402D+01	-1.00	-2.092429515214D+00
1		6.528453584744D-01	1.359276850472D+01	-.75	-1.243919609081D+00
2		-1.045056171381D+00	-1.618966947218D+00	-.50	-6.370027202523D-01
3		3.945279853585D-01	2.274234911685D-01	-.25	-2.346918501012D-01
				.25	1.040596872495D-01
0	-6.5	-2.317172451831D-03	-1.158586225915D-03	.50	1.144565316520D-01
1	-5.5	1.490815954173D-02	7.454079770863D-03	.75	6.792211109200D-02
2	-4.5	-4.587440686056D-02	-2.293720343028D-02	1.25	-5.732932763472D-02
3	-3.5	9.205885644682D-02	4.602942822341D-02	1.50	-8.005099100313D-02
4	-2.5	-1.362146311629D-01	-6.810731558147D-02	1.75	-5.806457557732D-02
5	-1.5	1.559642029009D-01	7.798210145047D-02	2.25	6.685992677893D-02
6	-.5	-1.409487254255D-01	-7.047436271273D-02	2.50	1.050837580906D-01
7	.5	1.029529212361D-01	5.147646061805D-02	2.75	8.422213941003D-02
8	1.5	-6.349542698443D-02	-3.174771349222D-02	3.25	-1.143239326749D-01
9	2.5	3.497390194971D-02	1.748695097485D-02	3.50	-1.939861697638D-01
10	3.5	-1.763013713361D-02	-8.815068566807D-03	3.75	-1.690027004405D-01
11	4.5	7.604484292252D-03	3.802242146126D-03	4.25	2.937993359737D-01
12	5.5	-2.340183868734D-03	-1.170091934367D-03	4.50	6.315050674295D-01
13	6.5	3.581575200282D-04	1.790787600141D-04	4.75	8.995744179242D-01
				5.25	8.948644295934D-01
	0	6.843711956832D+01	2.898307170884D+01	5.50	6.246001513819D-01
	1	3.707353727471D+01	2.137212569649D+01	5.75	2.886221687967D-01
	2	6.648239543110D+00	2.815791130568D+00	6.25	-1.628359359999D-01
	3	3.945279853585D-01	2.274234911685D-01	6.50	-1.842483222856D-01

I= 6 H= -.5

0		0.	1.003770787742D+02	-1.00	1.122650765646D+00
1		-3.500570347179D-01	-2.510061400298D+01	-.75	6.673288826105D-01
2		5.606977872847D-01	2.340411037732D+00	-.50	3.416899571355D-01
3		-2.118959436431D-01	-1.626138661376D-01	-.25	1.258687445042D-01
				.25	-5.578144448281D-02
0	-6.5	1.255191076284D-03	6.275955381418D-04	.50	-6.133125731286D-02
1	-5.5	-8.173988322479D-03	-4.086994161239D-03	.75	-3.637632413176D-02
2	-4.5	2.624911840011D-02	1.312455920005D-02	1.25	3.064445286457D-02

3	-3.5	-5.802135230985D-02	-2.901067615493D-02	1.50	4.271791651444D-02
4	-2.5	1.005664423106D-01	5.028322115528D-02	1.75	3.090851855631D-02
5	-1.5	-1.409487254255D-01	-7.047436271273D-02	2.25	-3.527102217914D-02
6	-.5	1.583755985175D-01	7.918779925877D-02	2.50	-5.500288340920D-02
7	.5	-1.419540813399D-01	-7.097704066993D-02	2.75	-4.358815158736D-02
8	1.5	1.029529212361D-01	5.147646061805D-02	3.25	5.684326818340D-02
9	2.5	-6.250543414731D-02	-3.125271707365D-02	3.50	9.315177833128D-02
10	3.5	3.272619694331D-02	1.636309847165D-02	3.75	7.707582360999D-02
11	4.5	-1.424149107365D-02	-7.120745536824D-03	4.25	-1.091476486100D-01
12	5.5	4.392239326605D-03	2.196119663302D-03	4.50	-1.877272513838D-01
13	6.5	-6.726351918849D-04	-3.363175959425D-04	4.75	-1.652083515998D-01
				5.25	2.909929279939D-01
	0	-3.677781273588D+01	-8.552378888994D+00	5.50	6.280740509347D-01
	1	-1.991879665678D+01	-1.528657804540D+01	5.75	8.974674661789D-01
	2	-3.571273113756D+00	-8.305593519508D-01	6.25	8.964800210489D-01
	3	-2.118959436431D-01	-1.626138661376D-01	6.50	6.266267534536D-01

I= 7 H= .5

0		0.	-1.876641017921D+02	-1.00	-6.003035721333D-01
1		1.871563150305D-01	4.669515715370D+01	-.75	-3.568252856168D-01
2		-2.998154684707D-01	-4.001529741634D+00	-.50	-1.826984982119D-01
3		1.133317886320D-01	1.626138661376D-01	-.25	-6.729835473442D-02
				.25	2.982138012120D-02
0	-6.5	-6.726351918849D-04	-3.363175959425D-04	.50	3.278550901416D-02
1	-5.5	4.392239326605D-03	2.196119663302D-03	.75	1.944309762752D-02
2	-4.5	-1.424149107365D-02	-7.120745536824D-03	1.25	-1.637223247837D-02
3	-3.5	3.272619694331D-02	1.636309847165D-02	1.50	-2.281383335921D-02
4	-2.5	-6.250543414731D-02	-3.125271707365D-02	1.75	-1.649753183902D-02
5	-1.5	1.029529212361D-01	5.147646061805D-02	2.25	1.878687053984D-02
6	-.5	-1.419540813399D-01	-7.097704066993D-02	2.50	2.924363162116D-02
7	.5	1.583755985175D-01	7.918779925877D-02	2.75	2.311366344264D-02
8	1.5	-1.409487254255D-01	-7.047436271273D-02	3.25	-2.986094372623D-02
9	2.5	1.005664423106D-01	5.028322115528D-02	3.50	-4.854289559633D-02
10	3.5	-5.802135230985D-02	-2.901067615493D-02	3.75	-3.970656192806D-02
11	4.5	2.624911840011D-02	1.312455920005D-02	4.25	5.400239364952D-02
12	5.5	-8.173988322479D-03	-4.086994161239D-03	4.50	8.968909057160D-02
13	6.5	1.255191076284D-03	6.275955381418D-04	4.75	7.495394282927D-02
				5.25	-1.075247854542D-01
	0	1.967305495789D+01	-8.552378888994D+00	5.50	-1.856931344184D-01
	1	1.065435943402D+01	1.528657804540D+01	5.75	-1.639183405243D-01
	2	1.910154409854D+00	-8.305593519508D-01	6.25	2.899081999912D-01
	3	1.133317886320D-01	1.626138661376D-01	6.50	6.266267534536D-01

I= 8 H= 1.5

0		0.	3.493252402647D+02	-1.00	3.194844757365D-01
1		-9.960231964973D-02	-8.680333789950D+01	-.75	1.899030935860D-01
2		1.595631591083D-01	7.250549208355D+00	-.50	9.723182422425D-02
3		-6.031899697855D-02	-2.274234911685D-01	-.25	3.581576168449D-02
				.25	-1.587034493576D-02
0	-6.5	3.581575200282D-04	1.790787600141D-04	.50	-1.744744656449D-02
1	-5.5	-2.340183868734D-03	-1.170091934367D-03	.75	-1.034673133140D-02
2	-4.5	7.604484292252D-03	3.802242146126D-03	1.25	8.711680877654D-03
3	-3.5	-1.763013713361D-02	-8.815068566807D-03	1.50	1.213818803224D-02
4	-2.5	3.497390194971D-02	1.748695097485D-02	1.75	8.776428962496D-03
5	-1.5	-6.349542698443D-02	-3.174771349222D-02	2.25	-9.989543424511D-03

CARDINAL SPLINES N=4

6	-.5	1.029529212361D-01	5.147646061805D-02	2.50	-1.554320580349D-02
7	.5	-1.409487254255D-01	-7.047436271273D-02	2.75	-1.227761239079D-02
8	1.5	1.559642029009D-01	7.798210145047D-02	3.25	1.582718983101D-02
9	2.5	-1.362146311629D-01	-6.810731558147D-02	3.50	2.568101282868D-02
10	3.5	9.205885644682D-02	4.602942822341D-02	3.75	2.094979549993D-02
11	4.5	-4.587440686056D-02	-2.293720343028D-02	4.25	-2.822325047200D-02
12	5.5	1.490815954173D-02	7.454079770863D-03	4.50	-4.649638704666D-02
13	6.5	-2.317172451831D-03	-1.158586225915D-03	4.75	-3.841147533560D-02
				5.25	5.291628451741D-02
	0	-1.047097615063D+01	2.898307170884D+01	5.50	8.824117746183D-02
	1	-5.670714118274D+00	-2.137212569649D+01	5.75	7.396673955116D-02
	2	-1.016657281974D+00	2.815791130568D+00	6.25	-1.065386034349D-01
	3	-6.031899697855D-02	-2.274234911685D-01	6.50	-1.842483222856D-01
I= 9	H= 2.5				
0		0.	-6.409867474203D+02	-1.00	-1.675253469695D-01
1		5.222723575221D-02	1.592744630811D+02	-.75	-9.957773697106D-02
2		-8.366876079725D-02	-1.320958077442D+01	-.50	-5.098447687792D-02
3		3.162935042002D-02	3.781757789823D-01	-.25	-1.878031508819D-02
				.25	8.321708524585D-03
0	-6.5	-1.878253749762D-04	-9.391268748812D-05	.50	9.148629093551D-03
1	-5.5	1.227416321377D-03	6.137081606885D-04	.75	5.425309413071D-03
2	-4.5	-3.990515485162D-03	-1.995257742581D-03	1.25	-4.567865281217D-03
3	-3.5	9.270233725834D-03	4.635116862917D-03	1.50	-6.364383925680D-03
4	-2.5	-1.855010161105D-02	-9.275050805526D-03	1.75	-4.601583855198D-03
5	-1.5	3.497390194971D-02	1.748695097485D-02	2.25	5.237059305153D-03
6	-.5	-6.250543414731D-02	-3.125271707365D-02	2.50	8.147806267939D-03
7	.5	1.005664423106D-01	5.028322115528D-02	2.75	6.435070973751D-03
8	1.5	-1.362146311629D-01	-6.810731558147D-02	3.25	-8.291362017479D-03
9	2.5	1.470765850743D-01	7.353829253717D-02	3.50	-1.344765638146D-02
10	3.5	-1.204351790865D-01	-6.021758954327D-02	3.75	-1.096337206590D-02
11	4.5	6.960296034279D-02	3.480148017140D-02	4.25	1.473715297709D-02
12	5.5	-2.492841674242D-02	-1.246420837121D-02	4.50	2.423285558943D-02
13	6.5	4.094563885785D-03	2.047281942893D-03	4.75	1.996501978882D-02
				5.25	-2.724314617204D-02
	0	5.490682247804D+00	-5.995100180934D+01	5.50	-4.506241438458D-02
	1	2.973553511126D+00	3.548369299967D+01	5.75	-3.733838462373D-02
	2	5.331035723932D-01	-5.835153084263D+00	6.25	5.164347193970D-02
	3	3.162935042002D-02	3.781757789823D-01	6.50	8.623671363161D-02
I=10	H= 3.5				
0		0.	1.102372894166D+03	-1.00	8.350007350226D-02
1		-2.603170660336D-02	-2.744002749011D+02	-.75	4.963276344941D-02
2		4.170322639229D-02	2.274865468772D+01	-.50	2.541230246308D-02
3		-1.576514050661D-02	-6.343446759149D-01	-.25	9.360708620774D-03
				.25	-4.147799607583D-03
0	-6.5	9.362071767918D-05	4.681035883959D-05	.50	-4.559957855078D-03
1	-5.5	-6.118201725179D-04	-3.059100862590D-04	.75	-2.704136900810D-03
2	-4.5	1.989348475778D-03	9.946742378892D-04	1.25	2.276748958915D-03
3	-3.5	-4.623529498983D-03	-2.311764749492D-03	1.50	3.172168153685D-03
4	-2.5	9.270233725834D-03	4.635116862917D-03	1.75	2.293528693005D-03
5	-1.5	-1.763013713361D-02	-8.815068566807D-03	2.25	-2.610197476892D-03
6	-.5	3.272619694331D-02	1.636309847165D-02	2.50	-4.060850098374D-03
7	.5	-5.802135230985D-02	-2.901067615493D-02	2.75	-3.207123177023D-03
8	1.5	9.205885644682D-02	4.602942822341D-02	3.25	4.131788951029D-03

9	2,5	-1,204351790865D-01	-6,021758954327D-02	3,50	6,700629947489D-03
10	3,5	1,191563056437D-01	5,957815282185D-02	3,75	5,461990280062D-03
11	4,5	-8,073845573702D-02	-4,036922786851D-02	4,25	-7,338363531544D-03
12	5,5	3.258539194399D-02	1.629269597199D-02	4.50	-1,206146947508D-02
13	6,5	-5,819479958582D-03	-2,909739979291D-03	4,75	-9,930989206717D-03
				5,25	1,352118102608D-02
	0	-2,736746489474D+00	1.056948612413D+02	5,50	2,232262065311D-02
	1	-1,482121322716D+00	-5,907095163297D+01	5,75	1,844601948787D-02
	2	-2,657170134865D-01	1.037893350738D+01	6.25	-2,526974323030D-02
	3	-1,576514050661D-02	-6,343446759149D-01	6,50	-4,185404011313D-02

I=11 H= 4.5

0		0.	-1,323191074074D+03	-1,00	-3,592191065153D-02
1		1,119889158509D-02	3,335605655213D+02	-,75	-2,135211978828D-02
2		-1,794081728574D-02	-2,796318287467D+01	-,50	-1,093242533657D-02
3		6,782201780696D-03	7,822888805222D-01	-,25	-4,026995879455D-03
				,25	1,784391260481D-03
0	-6,5	-4,027608004786D-05	-2,013804002393D-05	,50	1,961702036822D-03
1	-5,5	2,632099733319D-04	1.316049866660D-04	,75	1,163324134114D-03
2	-4,5	-8,558567842317D-04	-4,279283921159D-04	1.25	-9,794599251556D-04
3	-3.5	1,989348475778D-03	9,946742378892D-04	1,50	-1,364668763953D-03
4	-2,5	-3,990515485162D-03	-1,995257742581D-03	1,75	-9,866759168294D-04
5	-1,5	7,604484292252D-03	3.802242146126D-03	2,25	1,122900180933D-03
6	-,5	-1,424149107365D-02	-7,120745536824D-03	2,50	1,746958134247D-03
7	,5	2,624911840011D-02	1,312455920005D-02	2,75	1,379678552926D-03
8	1,5	-4,587440686056D-02	-2,293720343028D-02	3,25	-1,777414411388D-03
9	2,5	6,960296034279D-02	3,480148017140D-02	3,50	-2,882412221835D-03
10	3,5	-8,073845573702D-02	-4,036922786851D-02	3,75	-2,349507377165D-03
11	4,5	6,273032531424D-02	3,136516265712D-02	4,25	3,156264236705D-03
12	5,5	-2,812458911305D-02	-1,406229455653D-02	4,50	5,187162567351D-03
13	6,5	5,426144335215D-03	2,713072167608D-03	4,75	4,270298248234D-03
				5,25	-5,811038437768D-03
	0	1,177355429004D+00	-1,216557908272D+02	5,50	-9,589365357559D-03
	1	6,376123425737D-01	6,919430375678D+01	5,75	-7,918963223555D-03
	2	1,143121174378D-01	-1,270854970449D+01	6,25	1,082380879636D-02
	3	6,782201780696D-03	7,822888805222D-01	6,50	1,789256489487D-02

I=12 H= 5.5

0		0.	8,537476861831D+02	-1,00	1,104730339471D-02
1		-3,444069056190D-03	-2,191860653761D+02	-,75	6,566558880681D-03
2		5,517458502938D-03	1,872041368677D+01	-,50	3,362121133277D-03
3		-2,085775835583D-03	-5,325107579973D-01	-,25	1,238448667912D-03
				,25	-5,487655990396D-04
0	-6,5	1,238638883481D-05	6,193194417403D-06	,50	-6,032951131455D-04
1	-5,5	-8,094698556396D-05	-4,047349278198D-05	,75	-3,577646815522D-04
2	-4,5	2,632099733319D-04	1,316049866660D-04	1,25	3,012196256423D-04
3	-3,5	-6,118201725179D-04	-3,059100862590D-04	1,50	4,196852693907D-04
4	-2,5	1,227416321377D-03	6,137081606885D-04	1,75	3,034386020647D-04
5	-1,5	-2,340183868734D-03	-1,170091934367D-03	2,25	-3,453319932806D-04
6	-,5	4,392239326605D-03	2,196119663302D-03	2,50	-5,372514572814D-04
7	,5	-8,173988322479D-03	-4,086994161239D-03	2,75	-4,242991934269D-04
8	1,5	1,490815954173D-02	7,454079770863D-03	3,25	5,466131556300D-04
9	2,5	-2,492841674242D-02	-1,246420837121D-02	3,50	8,864310488881D-04
10	3,5	3,258539194399D-02	1,629269597199D-02	3,75	7,225402755102D-04
11	4,5	-2,812458911305D-02	-1,406229455653D-02	4,25	-9,706121690519D-04

12	5.5	1.367208315177D-02	6.836041575887D-03	4.50	-1.595111960014D-03
13	6.5	-2.800941442864D-03	-1.400470721432D-03	4.75	-1.313117584961D-03
				5.25	1.786663555870D-03
	0	-3.620800159631D-01	7.373497258992D+01	5.50	2.948020425500D-03
	1	-1.960891956781D-01	-4.331642602414D+01	5.75	2.434105715289D-03
	2	-3.515517029093D-02	8.336453905828D+00	6.25	-3.325102846441D-03
	3	-2.085775835583D-03	-5.325107579973D-01	6.50	-5.493972388240D-03

I=13 H= 6.5

0		0.	-2.185495881170D+02	-1.00	-1.690440495145D-03
1		5.270058611318D-04	5.717198405326D+01	-.75	-1.004804212872D-03
2		-8.442725745713D-04	-4.986229688773D+00	-.50	-5.144663316390D-04
3		3.191620594423D-04	1.450929890101D-01	-.25	-1.895054083724D-04
				.25	8.397122086829D-05
0	-6.5	-1.895346002745D-06	-9.476730013723D-07	.50	9.231523696270D-05
1	-5.5	1.238638883481D-05	6.193194417403D-06	.75	5.474456832856D-05
2	-4.5	-4.027608004786D-05	-2.013804002393D-05	1.25	-4.609213201718D-05
3	-3.5	9.362071767918D-05	4.681035883959D-05	1.50	-6.421954515111D-05
4	-2.5	-1.878253749762D-04	-9.391268748812D-05	1.75	-4.643166727107D-05
5	-1.5	3.581575200282D-04	1.790787600141D-04	2.25	5.284210187451D-05
6	-.5	-6.726351918849D-04	-3.363175959425D-04	2.50	8.220925489784D-05
7	.5	1.255191076284D-03	6.275955381418D-04	2.75	6.492546747201D-05
8	1.5	-2.317172451831D-03	-1.158586225915D-03	3.25	-8.364156149440D-05
9	2.5	4.094563885785D-03	2.047281942893D-03	3.50	-1.356395366215D-04
10	3.5	-5.819479958582D-03	-2.909739979291D-03	3.75	-1.105611046005D-04
11	4.5	5.426144335215D-03	2.713072167608D-03	4.25	1.485191255186D-04
12	5.5	-2.800941442864D-03	-1.400470721432D-03	4.50	2.440757757033D-04
13	6.5	6.001619223626D-04	3.000809611813D-04	4.75	2.009243650887D-04
				5.25	-2.733730776381D-04
	0	5.540490239605D-02	-1.775373401456D+01	5.50	-4.510551966168D-04
	1	3.000525342601D-02	1.074153445623D+01	5.75	-3.724078913295D-04
	2	5.379387584553D-03	-2.156916403077D+00	6.25	5.086446830178D-04
	3	3.191620594423D-04	1.450929890101D-01	6.50	8.403028068788D-04

N= 4 M=14 P= 7.0 Q= 7

I= 0 H= -7.0

0		1.000000000000D+00	3.079198011217D-01	-1.00	4.666983126522D+00
1		-2.043924540322D+00	-9.779208683619D-01	-.75	3.405151298372D+00
2		1.333191481442D+00	6.632340400930D-01	-.50	2.391493528616D+00
3		-2.898671047573D-01	-1.448481318623D-01	-.25	1.598834776183D+00
				.25	5.678141956288D-01
0	-7.0	6.001636373837D-04	3.000818186918D-04	.50	2.751069008832D-01
1	-6.0	-2.800952650805D-03	-1.400476325403D-03	.75	9.476923042504D-02
2	-5.0	5.426180779463D-03	2.713090389732D-03	1.25	-3.507904235206D-02
3	-4.0	-5.819564672824D-03	-2.909782336412D-03	1.50	-3.427497983889D-02
4	-3.0	4.094733847104D-03	2.047366923552D-03	1.75	-1.768215959672D-02
5	-2.0	-2.317496581904D-03	-1.158748290952D-03	2.25	1.053384380748D-02

CARDINAL SPLINES N=4

```
 6  -1.0   1.255800108820D-03   6.279000544099D-04    2.50   1.195820031302D-02
 7   0.0  -6.737741670215D-04  -3.368870835107D-04    2.75   6.908505823101D-03
 8   1.0   3.602803753859D-04   1.801401876930D-04    3.25  -4.774250230143D-03
 9   2.0  -1.917429650457D-04  -9.587148252283D-05    3.50  -5.681622604777D-03
10   3.0   1.005426215912D-04   5.027131079558D-05    3.75  -3.398341171514D-03
11   4.0  -5.011370783492D-05  -2.505685391746D-05    4.25   2.451007752900D-03
12   5.0   2.155902455839D-05   1.077951227920D-05    4.50   2.953739098311D-03
13   6.0  -6.630189788651D-06  -3.315094894325D-06    4.75   1.782515226319D-03
14   7.0   1.014540916347D-06   5.072704581736D-07    5.25  -1.299240829268D-03
                                                      5.50  -1.570466572487D-03
         0  -4.740550612334D+01  -2.372196754163D+01    5.75  -9.497571645438D-04
         1  -2.598970819945D+01  -1.298531969082D+01    6.25   6.940204610399D-04
         2  -4.754017718461D+00  -2.378576729016D+00    6.50   8.395427239770D-04
         3  -2.898671047573D-01  -1.448481318623D-01    6.75   5.080166013403D-04


I= 1     H= -6.0
 0          0.                   1.255275427571D+00   -1.00  -9.041178839997D+00
 1          3.959052268832D+00   1.691708390354D+00    -.75  -5.678509794038D+00
 2         -4.019188943673D+00  -1.987625209910D+00    -.50  -3.117190573771D+00
 3          1.062937627492D+00   5.309105767384D-01    -.25  -1.257570776617D+00
                                                        .25   7.551719877014D-01
 0  -7.0  -2.800952650805D-03  -1.400476325403D-03     .50   1.107574219492D+00
 1  -6.0   1.367215639749D-02   6.836078198746D-03     .75   1.156548350360D+00
 2  -5.0  -2.812482728214D-02  -1.406241364107D-02    1.25   7.315275167053D-01
 3  -4.0   3.258594556532D-02   1.629297278266D-02    1.50   4.350676845581D-01
 4  -3.0  -2.492952746727D-02  -1.246476373363D-02    1.75   1.772915933927D-01
 5  -2.0   1.491027778443D-02   7.455138892217D-03    2.25  -8.424033925317D-02
 6  -1.0  -8.177968449093D-03  -4.088984224546D-03    2.50  -9.008373874156D-02
 7   0.0   4.399682714247D-03   2.199841357124D-03    2.75  -4.992407114938D-02
 8   1.0  -2.354057073539D-03  -1.177028536770D-03    3.25   3.281522958522D-02
 9   2.0   1.253018409234D-03   6.265092046168D-04    3.50   3.847430283781D-02
10   3.0  -6.570559396475D-04  -3.285279698237D-04    3.75   2.276856027888D-02
11   4.0   3.275004698978D-04   1.637502349489D-04    4.25  -1.621313399040D-02
12   5.0  -1.408916509092D-04  -7.044582545462D-05    4.50  -1.946679135278D-02
13   6.0   4.332936256247D-05   2.166468128124D-05    4.75  -1.171740697053D-02
14   7.0  -6.630189788651D-06  -3.315094894325D-06    5.25   8.514675180159D-03
                                                      5.50   1.028329854067D-02
         0   1.953607138715D+02   9.780592669575D+01    5.75   6.215196238052D-03
         1   1.039422382987D+02   5.190881023216D+01    6.25  -4.538468898869D-03
         2   1.830250123365D+01   9.161496901597D+00    6.50  -5.489005122936D-03
         3   1.062937627492D+00   5.309105767384D-01    6.75  -3.320998230592D-03


I= 2     H= -5.0
 0          0.                  -4.081711506124D+00   -1.00   9.696662407049D+00
 1         -3.293242597714D+00  -7.107398379270D-01    -.75   5.852789727508D+00
 2          4.845618113135D+00   2.351372790291D+00    -.50   3.052751039166D+00
 3         -1.557801696200D+00  -7.770856588714D-01    -.25   1.150502433003D+00
                                                        .25  -5.447998376730D-01
 0  -7.0   5.426180779463D-03   2.713090389732D-03     .50  -6.298995905611D-01
 1  -6.0  -2.812482728214D-02  -1.406241364107D-02     .75  -4.007450425313D-01
 2  -5.0   6.273109975560D-02   3.136554987780D-02    1.25   4.380159463642D-01
 3  -4.0  -8.074025591716D-02  -4.037012795858D-02    1.50   7.976877937554D-01
 4  -3.0   6.960657202586D-02   3.480328601293D-02    1.75   9.966806361972D-01
 5  -2.0  -4.588129463422D-02  -2.294064731711D-02    2.25   8.267660497035D-01
 6  -1.0   2.626206035956D-02   1.313103017978D-02    2.50   5.431841724112D-01
```

CARDINAL SPLINES N=4

7	0.0	-1.426569432907D-02	-7.132847164537D-03	2.75	2.397528355663D-01
8	1.0	7.649595031557D-03	3.824797515779D-03	3.25	-1.274054541624D-01
9	2.0	-4.073764390313D-03	-2.036882195156D-03	3.50	-1.414527341535D-01
10	3.0	2.136439139176D-03	1.068219569588D-03	3.75	-8.064943139938D-02
11	4.0	-1.064906666440D-03	-5.324533332198D-04	4.25	5.497569912104D-02
12	5.0	4.581287544708D-04	2.290643772354D-04	4.50	6.518035883338D-02
13	6.0	-1.408916509092D-04	-7.044582545462D-05	4.75	3.888523519454D-02
14	7.0	2.155902455839D-05	1.077951227920D-05	5.25	-2.796066793810D-02
				5.50	-3.366708808870D-02
	0	-3.199433924370D+02	-1.603800046402D+02	5.75	-2.030550996967D-02
	1	-1.644514383552D+02	-8.202311262795D+01	6.25	1.479104782220D-02
	2	-2.786821750706D+01	-1.396742604601D+01	6.50	1.787641311066D-02
	3	-1.557801696200D+00	-7.770856588714D-01	6.75	1.081046204208D-02

I= 3 H= -4.0

0		0.	9.487903590385D+00	-1.00	-6.822771220834D+00
1		2.161356916191D+00	-1.094770042868D+00	-.75	-4.066868760238D+00
2		-3.408475828080D+00	-1.538184918454D+00	-.50	-2.089414724686D+00
3		1.252938476562D+00	6.222498462577D-01	-.25	-7.729461319992D-01
				.25	3.468862982911D-01
0	-7.0	-5.819564672824D-03	-2.909782336412D-03	.50	3.851313452968D-01
1	-6.0	3.258594556532D-02	1.629297278266D-02	.75	2.315566355363D-01
2	-5.0	-8.074025591716D-02	-4.037012795858D-02	1.25	-2.046497320189D-01
3	-4.0	1.191604901420D-01	5.958024507099D-02	1.50	-2.975460210749D-01
4	-3.0	-1.204435744028D-01	-6.022178720141D-02	1.75	-2.292866323962D-01
5	-2.0	9.207486699850D-02	4.603743349925D-02	2.25	3.358722752371D-01
6	-1.0	-5.805143574768D-02	-2.902571787384D-02	2.50	6.817323161857D-01
7	0.0	3.278245713578D-02	1.639122856789D-02	2.75	9.296905192487D-01
8	1.0	-1.773499653500D-02	-8.867498267502D-03	3.25	8.730612033104D-01
9	2.0	9.463744860764D-03	4.731872430382D-03	3.50	5.982783447978D-01
10	3.0	-4.965440076188D-03	-2.482720038094D-03	3.75	2.727065405930D-01
11	4.0	2.475282553710D-03	1.237641276855D-03	4.25	-1.511735775096D-01
12	5.0	-1.064906666440D-03	-5.324533332198D-04	4.50	-1.700966571227D-01
13	6.0	3.275004698978D-04	1.637502349489D-04	4.75	-9.793594773156D-02
14	7.0	-5.011370783492D-05	-2.505685391746D-05	5.25	6.757681232712D-02
				5.50	8.041328503039D-02
	0	2.778720802981D+02	1.398851495525D+02	5.75	4.809854224158D-02
	1	1.386246513777D+02	6.884136849866D+01	6.25	-3.469558368962D-02
	2	2.290323217972D+01	1.152906185296D+01	6.50	-4.181649547411D-02
	3	1.252938476562D+00	6.222498462577D-01	6.75	-2.523871885152D-02

I= 4 H= -3.0

0		0.	-1.903558327507D+01	-1.00	3.861592763067D+00
1		-1.208092920709D+00	3.760556613536D+00	-.75	2.296745266706D+00
2		1.928749014610D+00	6.312223095598D-01	-.50	1.176827567475D+00
3		-7.247508277479D-01	-3.539100454342D-01	-.25	4.338942752739D-01
				.25	-1.928003985250D-01
0	-7.0	4.094733847104D-03	2.047366923552D-03	.50	-2.124210700624D-01
1	-6.0	-2.492952746727D-02	-1.246476373363D-02	.75	-1.263560442809D-01
2	-5.0	6.960657202586D-02	3.480328601293D-02	1.25	1.075489149313D-01
3	-4.0	-1.204435744028D-01	-6.022178720141D-02	1.50	1.512794628168D-01
4	-3.0	1.470934285119D-01	7.354671425595D-02	1.75	1.109141950488D-01
5	-2.0	-1.362467529697D-01	-6.812337648484D-02	2.25	-1.327784614078D-01
6	-1.0	1.006267984055D-01	5.031339920273D-02	2.50	-2.159559892870D-01
7	0.0	-6.261830839842D-02	-3.130915419921D-02	2.75	-1.821499243599D-01

8	1.0	3.518428029866D-02	1.759214014933D-02	3.25	3.032966570638D-01
9	2.0	-1.893834103004D-02	-9.469170515018D-03	3.50	6.429646598556D-01
10	3.0	9.956205434318D-03	4.978102717159D-03	3.75	9.065017697705D-01
11	4.0	-4.965440076188D-03	-2.482720038094D-03	4.25	8.897874217060D-01
12	5.0	2.136439139176D-03	1.068219569588D-03	4.50	6.184368702465D-01
13	6.0	-6.570559396475D-04	-3.285279698237D-04	4.75	2.848730410994D-01
14	7.0	1.005426215912D-04	5.027131079558D-05	5.25	-1.600445403954D-01
				5.50	-1.808223815843D-01
	0	-1.625374826466D+02	-8.317293939583D+01	5.75	-1.044248320759D-01
	1	-8.074397839511D+01	-3.942710773146D+01	6.25	7.232412068569D-02
	2	-1.329101836810D+01	-6.800888644559D+00	6.50	8.616138820476D-02
	3	-7.247508277479D-01	-3.539100454342D-01	6.75	5.158122804454D-02

I= 5 H= -2.0

0		0.	3.630322725750D+01	-1.00	-2.092718603544D+00
1		6.529354836468D-01	-7.997375530747D+00	-.75	-1.244091444275D+00
2		-1.045200553481D+00	1.127627109089D-01	-.50	-6.370907009957D-01
3		3.945825664163D-01	1.811467616991D-01	-.25	-2.347242581045D-01
				.25	1.040740474706D-01
0	-7.0	-2.317496581904D-03	-1.158748290952D-03	.50	1.144723188131D-01
1	-6.0	1.491027778443D-02	7.455138892217D-03	.75	6.793147315759D-02
2	-5.0	-4.588129463422D-02	-2.294064731711D-02	1.25	-5.733721001550D-02
3	-4.0	9.207486699850D-02	4.603743349925D-02	1.50	-8.006197341715D-02
4	-3.0	-1.362467529697D-01	-6.812337648484D-02	1.75	-5.807251602203D-02
5	-2.0	1.560254618016D-01	7.801273090080D-02	2.25	6.686896349185D-02
6	-1.0	-1.410638294123D-01	-7.053191470616D-02	2.50	1.050978169791D-01
7	0.0	1.031681816226D-01	5.158409081131D-02	2.75	8.423324253365D-02
8	1.0	-6.389663562383D-02	-3.194831781192D-02	3.25	-1.143382364792D-01
9	2.0	3.571430613020D-02	1.785715306510D-02	3.50	-1.940093658734D-01
10	3.0	-1.893834103004D-02	-9.469170515018D-03	3.75	-1.690216077790D-01
11	4.0	9.463744860764D-03	4.731872430382D-03	4.25	2.938247344396D-01
12	5.0	-4.073764390313D-03	-2.036882195156D-03	4.50	6.315468069314D-01
13	6.0	1.253018409234D-03	6.265092046168D-04	4.75	8.996087777986D-01
14	7.0	-1.917429650457D-04	-9.587148252283D-05	5.25	8.948176817188D-01
				5.50	6.245230211218D-01
	0	8.869754154574D+01	4.798031063960D+01	5.75	2.885584895161D-01
	1	4.402376499810D+01	2.020987639175D+01	6.25	-1.627489723648D-01
	2	7.241033341261D+00	3.916844706590D+00	6.50	-1.841046705640D-01
	3	3.945825664163D-01	1.811467616991D-01	6.75	-1.064198992592D-01

I= 6 H= -1.0

0		0.	-6.821960477694D+01	-1.00	1.123193955598D+00
1		-3.502263777283D-01	1.546674238416D+01	-.75	6.676517567679D-01
2		5.609690777447D-01	-9.134669397387D-01	-.50	3.418552708160D-01
3		-2.119985001253D-01	-7.566101516380D-02	-.25	1.259296383556D-01
				.25	-5.580842698952D-02
0	-7.0	1.255800108820D-03	6.279000544099D-04	.50	-6.136092100526D-02
1	-6.0	-8.177968449093D-03	-4.088984224546D-03	.75	-3.639391522568D-02
2	-5.0	2.626206035956D-02	1.313103017978D-02	1.25	3.065926366664D-02
3	-4.0	-5.805143574768D-02	-2.902571787384D-02	1.50	4.273855220370D-02
4	-3.0	1.006267984055D-01	5.031339920273D-02	1.75	3.092343845909D-02
5	-2.0	-1.410638294123D-01	-7.053191470616D-02	2.25	-3.528800194330D-02
6	-1.0	1.585918761138D-01	7.929593805692D-02	2.50	-5.502929971932D-02
7	0.0	-1.423585503791D-01	-7.117927518957D-02	2.75	-4.360901408704D-02
8	1.0	1.037067825273D-01	5.185339126365D-02	3.25	5.687014468436D-02

CARDINAL SPLINES N=4

9	2.0	-6.389663562383D-02	-3.194831781192D-02	3.50	9.319536325806D-02
10	3.0	3.518428029866D-02	1.759214014933D-02	3.75	7.711135003998D-02
11	4.0	-1.773499653500D-02	-8.867498267502D-03	4.25	-1.091953717107D-01
12	5.0	7.649595031557D-03	3.824797515779D-03	4.50	-1.878056788942D-01
13	6.0	-2.354057073539D-03	-1.177028536770D-03	4.75	-1.652729129696D-01
14	7.0	3.602803753859D-04	1.801401876930D-04	5.25	2.910807661151D-01
				5.50	6.282189768206D-01
	0	-4.767958537758D+01	-3.066401633622D+01	5.75	8.975871180008D-01
	1	-2.366043880772D+01	-8.443964001263D+00	6.25	8.963166184902D-01
	2	-3.890999424886D+00	-2.502348258178D+00	6.50	6.263568353569D-01
	3	-2.119985001253D-01	-7.566101516380D-02	6.75	2.896851577275D-01

I= 7 H= 0.0

0		0.	1.276350312806D+02	-1.00	-6.013194125255D-01
1		1.874730098981D-01	-2.917141406525D+01	-.75	-3.574291049791D-01
2		-3.003228191792D-01	2.083672433232D+00	-.50	-1.830076576749D-01
3		1.135235834482D-01	4.599005999256D-24	-.25	-6.741223466460D-02
				.25	2.987184114329D-02
0	-7.0	-6.737741670215D-04	-3.368870835107D-04	.50	3.284098422458D-02
1	-6.0	4.399682714247D-03	2.199841357124D-03	.75	1.947599540786D-02
2	-5.0	-1.426569432907D-02	-7.132847164537D-03	1.25	-1.639993072816D-02
3	-4.0	3.278245713578D-02	1.639122856789D-02	1.50	-2.285242495424D-02
4	-3.0	-6.261830839842D-02	-3.130915419921D-02	1.75	-1.652543412223D-02
5	-2.0	1.031681816226D-01	5.158409081131D-02	2.25	1.881862504891D-02
6	-1.0	-1.423585503791D-01	-7.117927518957D-02	2.50	2.929303377728D-02
7	0.0	1.591320116020D-01	7.956600580101D-02	2.75	2.315267920454D-02
8	1.0	-1.423585503791D-01	-7.117927518957D-02	3.25	-2.991120650289D-02
9	2.0	1.031681816226D-01	5.158409081131D-02	3.50	-4.862440544183D-02
10	3.0	-6.261830839842D-02	-3.130915419921D-02	3.75	-3.977300126924D-02
11	4.0	3.278245713578D-02	1.639122856789D-02	4.25	5.409164245300D-02
12	5.0	-1.426569432907D-02	-7.132847164537D-03	4.50	8.983576087104D-02
13	6.0	4.399682714247D-03	2.199841357124D-03	4.75	7.507468152639D-02
14	7.0	-6.737741670215D-04	-3.368870835107D-04	5.25	-1.076890548983D-01
				5.50	-1.859641658837D-01
	0	2.553508205222D+01	2.553508205222D+01	5.75	-1.641421059919D-01
	1	1.267092030827D+01	4.126629591578D-22	6.25	2.902137853919D-01
	2	2.083672433232D+00	2.083672433232D+00	6.50	6.271315376528D-01
	3	1.135235834482D-01	4.599005999256D-24	6.75	8.968971409524D-01

I= 8 H= 1.0

0		0.	-2.383385571970D+02	-1.00	3.213778289364D-01
1		-1.001925848582D-01	5.459900884484D+01	-.75	1.910285098723D-01
2		1.605087742805D-01	-4.091229576618D+00	-.50	9.780804472394D-02
3		-6.067646979768D-02	7.566101516380D-02	-.25	3.602801444767D-02
				.25	-1.596439567284D-02
0	-7.0	3.602803753859D-04	1.801401876930D-04	.50	-1.755084289325D-02
1	-6.0	-2.354057073539D-03	-1.177028536770D-03	.75	-1.040804718195D-02
2	-5.0	7.649595031557D-03	3.824797515779D-03	1.25	8.763305690242D-03
3	-4.0	-1.773499653500D-02	-8.867498267502D-03	1.50	1.221011618208D-02
4	-3.0	3.518428029866D-02	1.759214014933D-02	1.75	8.828434058579D-03
5	-2.0	-6.389663562383D-02	-3.194831781192D-02	2.25	-1.004872841245D-02
6	-1.0	1.037067825273D-01	5.185339126365D-02	2.50	-1.563528299502D-02
7	0.0	-1.423585503791D-01	-7.117927518957D-02	2.75	-1.235033111531D-02
8	1.0	1.585918761138D-01	7.929593805692D-02	3.25	1.592087107242D-02
9	2.0	-1.410638294123D-01	-7.053191470616D-02	3.50	2.583293327600D-02

CARDINAL SPLINES N=4

10	3.0	1.006267984055D-01	5.031339920273D-02	3.75	2.107362709805D-02
11	4.0	-5.805143574768D-02	-2.902571787384D-02	4.25	-2.838959501684D-02
12	5.0	2.626206035956D-02	1.313103017978D-02	4.50	-4.676975546447D-02
13	6.0	-8.177968449093D-03	-4.088984224546D-03	4.75	-3.863651167032D-02
14	7.0	1.255800108820D-03	6.279000544099D-04	5.25	5.322245473864D-02
				5.50	8.874633387822D-02
	0	-1.364844729487D+01	-3.066401633622D+01	5.75	7.438380021074D-02
	1	-6.772510805190D+00	8.443964001263D+00	6.25	-1.071081624923D-01
	2	-1.113697091471D+00	-2.502348258178D+00	6.50	-1.851891539222D-01
	3	-6.067646979768D-02	7.566101516380D-02	6.75	-1.636133763439D-01
I= 9	H= 2.0				
0		0.	4.435081752675D+02	-1.00	-1.710194058871D-01
1		5.331653140798D-02	-1.016742762538D+02	-.75	-1.016546187761D-01
2		-8.541383146103D-02	7.720926702271D+00	-.50	-5.204785394651D-02
3		3.228904301810D-02	-1.811467616991D-01	-.25	-1.917201361547D-02
				.25	8.495272979776D-03
0	-7.0	-1.917429650457D-04	-9.587148252283D-05	.50	9.339440224080D-03
1	-6.0	1.253018409234D-03	6.265092046168D-04	.75	5.538463786189D-03
2	-5.0	-4.073764390313D-03	-2.036882195156D-03	1.25	-4.663135480569D-03
3	-4.0	9.463744860764D-03	4.731872430382D-03	1.50	-6.497122599786D-03
4	-3.0	-1.893834103004D-02	-9.469170515018D-03	1.75	-4.697555842752D-03
5	-2.0	3.571430613020D-02	1.785715306510D-02	2.25	5.346281311331D-03
6	-1.0	-6.389663562383D-02	-3.194831781192D-02	2.50	8.317728674484D-03
7	0.0	1.031681816226D-01	5.158409081131D-02	2.75	6.569268601480D-03
8	1.0	-1.410638294123D-01	-7.053191470616D-02	3.25	-8.464244592699D-03
9	2.0	1.560254618016D-01	7.801273090080D-02	3.50	-1.372801558427D-02
10	3.0	-1.362467529697D-01	-6.812337648484D-02	3.75	-1.119189514324D-02
11	4.0	9.207486699850D-02	4.603743349925D-02	4.25	1.504413090404D-02
12	5.0	-4.588129463422D-02	-2.294064731711D-02	4.50	2.473733904139D-02
13	6.0	1.491027778443D-02	7.455138892217D-03	4.75	2.038030955736D-02
14	7.0	-2.317496581904D-03	-1.158748290952D-03	5.25	-2.780816319681D-02
				5.50	-4.599464735708D-02
	0	7.263079733473D+00	4.798031063960D+01	5.75	-3.810804266581D-02
	1	3.604012214614D+00	-2.020987639175D+01	6.25	5.269455575427D-02
	2	5.926560719190D-01	3.916844706590D+00	6.50	8.797295659480D-02
	3	3.228904301810D-02	-1.811467616991D-01	6.75	7.380443613362D-02
I=10	H= 3.0				
0		0.	-8.137973826833D+02	-1.00	8.967364951764D-02
1		-2.795635919013D-02	1.866643254341D+02	-.75	5.330236032815D-02
2		4.478655344803D-02	-1.423299959868D+01	-.50	2.729116006701D-02
3		-1.693073687948D-02	3.539100454342D-01	-.25	1.005279215178D-02
				.25	-4.454466834139D-03
0	-7.0	1.005426215912D-04	5.027131079558D-05	.50	-4.897097853765D-03
1	-6.0	-6.570559396475D-04	-3.285279698237D-04	.75	-2.904066879207D-03
2	-5.0	2.136439139176D-03	1.068219569588D-03	1.25	2.445079783200D-03
3	-4.0	-4.965440076188D-03	-2.482720038094D-03	1.50	3.406701203668D-03
4	-3.0	9.956205434318D-03	4.978102717159D-03	1.75	2.463099491461D-03
5	-2.0	-1.893834103004D-02	-9.469170515018D-03	2.25	-2.803179442447D-03
6	-1.0	3.518428029866D-02	1.759214014933D-02	2.50	-4.361082269111D-03
7	0.0	-6.261830839842D-02	-3.130915419921D-02	2.75	-3.444234020142D-03
8	1.0	1.006267984055D-01	5.031339920273D-02	3.25	4.437251378830D-03
9	2.0	-1.362467529697D-01	-6.812337648484D-02	3.50	7.195990469724D-03
10	3.0	1.470934285119D-01	7.354671425595D-02	3.75	5.865762689873D-03

11	4,0	-1,204435744028D-01	-6,022178720141D-02	4,25	-7,880756093249D-03
12	5,0	6,960657202586D-02	3,480328601293D-02	4,50	-1,295283021416D-02
13	6,0	-2,492952746727D-02	-1,246476373363D-02	4,75	-1,066475558414D-02
14	7,0	4,094733847104D-03	2,047366923552D-03	5,25	1,451949720637D-02
				5,50	2,396976263197D-02
	0	-3,808396145040D+00	-8,317293939583D+01	5,75	1,980591139029D-02
	1	-1,889762932202D+00	3,942710773146D+01	6,25	-2,712688015350D-02
	2	-3,107589210211D-01	-6,800888644559D+00	6,50	-4,492176969845D-02
	3	-1,693073687948D-02	3,539100454342D-01	6,75	-3,725327896632D-02

I=11 H= 4.0

0		0,	1,400130457104D+03	-1,00	-4,469599126962D-02
1		1,393426844221D-02	-3,217189618399D+02	-,75	-2,656746641553D-02
2		-2,232293878089D-02	2,459630862437D+01	-,50	-1,360271692214D-02
3		8,438784046514D-03	-6,222498462577D-01	-,25	-5,010606785085D-03
				,25	2,220236378776D-03
0	-7,0	-5,011370783492D-05	-2,505685391746D-05	,50	2,440856018355D-03
1	-6,0	3,275004698978D-04	1,637502349489D-04	,75	1,447470914531D-03
2	-5,0	-1,064906666440D-03	-5,324533332198D-04	1,25	-1,218696992798D-03
3	-4,0	2,475282553710D-03	1,237641276855D-03	1,50	-1,697994519242D-03
4	-3,0	-4,965440076188D-03	-2,482720038094D-03	1,75	-1,227675274805D-03
5	-2,0	9,463744860764D-03	4,731872430382D-03	2,25	1,397172226729D-03
6	-1,0	-1,773499653500D-02	-8,867498267502D-03	2,50	2,173657548593D-03
7	0,0	3,278245713578D-02	1,639122856789D-02	2,75	1,716667948985D-03
8	1,0	-5,805143574768D-02	-2,902571787384D-02	3,25	-2,211547231757D-03
9	2,0	9,207486699850D-02	4,603743349925D-02	3,50	-3,586434193263D-03
10	3,0	-1,204435744028D-01	-6,022178720141D-02	3,75	-2,923361439416D-03
11	4,0	1,191604901420D-01	5,958024507099D-02	4,25	3,927129623507D-03
12	5,0	-8,074025591716D-02	-4,037012795858D-02	4,50	6,453992515673D-03
13	6,0	3,258594556532D-02	1,629297278266D-02	4,75	5,313150127645D-03
14	7,0	-5,819564672824D-03	-2,909782336412D-03	5,25	-7,229876826803D-03
				5,50	-1,193033540084D-02
	0	1,898218806786D+00	1,398851495525D+02	5,75	-9,851684413668D-03
	1	9,419143803473D-01	-6,884136849866D+01	6,25	1,346323026657D-02
	2	1,548915261959D-01	1,152906185296D+01	6,50	2,225251877395D-02
	3	8,438784046514D-03	-6,222498462577D-01	6,75	1,840360006432D-02

I=12 H= 5.0

0		0,	-1,685486050283D+03	-1,00	1,922828432438D-02
1		-5,994543217258D-03	3,917986691261D+02	-,75	1,142936481515D-02
2		9,603362649911D-03	-3,028622488231D+01	-,50	5,851909578258D-03
3		-3,630378457212D-03	7,770856588714D-01	-,25	2,155570633328D-03
				,25	-9,551489862305D-04
0	-7,0	2,155902455839D-05	1,077951227920D-05	,50	-1,050059823423D-03
1	-6,0	-1,408916509092D-04	-7,044582545462D-05	,75	-6,227040516126D-04
2	-5,0	4,581287544708D-04	2,290643772354D-04	1,25	5,242850353689D-04
3	-4,0	-1,064906666440D-03	-5,324533332198D-04	1,50	7,304792734667D-04
4	-3,0	2,136439139176D-03	1,068219569588D-03	1,75	5,281471769550D-04
5	-2,0	-4,073764390313D-03	-2,036882195156D-03	2,25	-6,010641290862D-04
6	-1,0	7,649595031557D-03	3,824797515779D-03	2,50	-9,351073607559D-04
7	0,0	-1,426569432907D-02	-7,132847164537D-03	2,75	-7,385091879165D-04
8	1,0	2,626206035956D-02	1,313103017978D-02	3,25	9,513999907382D-04
9	2,0	-4,588129463422D-02	-2,294064731711D-02	3,50	1,542863394682D-03
10	3,0	6,960657202586D-02	3,480328601293D-02	3,75	1,257603645000D-03
11	4,0	-8,074025591716D-02	-4,037012795858D-02	4,25	-1,689369529882D-03

12	5.0	6,273109975560D-02	3.136554987780D-02	4.50	-2,776308269858D-03
13	6.0	-2,812482728214D-02	-1.406241364107D-02	4.75	-2,285476024470D-03
14	7.0	5,426180779463D-03	2,713090389732D-03	5,25	3,109593023721D-03
				5,50	5,130748407453D-03
	0	-8,166168434988D-01	-1,603800046402D+02	5,75	4,236181156228D-03
	1	-4,052130993286D-01	8,202311262795D+01	6,25	-5,786107896777D-03
	2	-6,663458495153D-02	-1,396742604601D+01	6,50	-9,559207378840D-03
	3	-3,630378457212D-03	7.770856588714D-01	6,75	-7,900714296592D-03

I=13 H= 6.0

0		0.	1,092183274320D+03	-1,00	-5,913399727884D-03
1		1,843540944019D-03	-2,582136216351D+02	-,75	-3,514947115605D-03
2		-2,953384769048D-03	2,031061901310D+01	-,50	-1,799675916124D-03
3		1,116474014817D-03	-5,309105767384D-01	-,25	-6,629166905519D-04
				.25	2,937431897462D-04
0	-7.0	-6,630189788651D-06	-3,315094894325D-06	,50	3,229317332421D-04
1	-6,0	4,332936256247D-05	2,166468128124D-05	,75	1,915042269234D-04
2	-5,0	-1,408916509092D-04	-7,044582545462D-05	1,25	-1,612367753806D-04
3	-4,0	3,275004698978D-04	1,637502349489D-04	1,50	-2,246490120177D-04
4	-3,0	-6,570559396475D-04	-3,285279698237D-04	1,75	-1,624244998136D-04
5	-2,0	1,253018409234D-03	6,265092046168D-04	2,25	1,848490571886D-04
6	-1,0	-2,354057073539D-03	-1.177028536770D-03	2,50	2,875794024781D-04
7	0,0	4,399682714247D-03	2,199841357124D-03	2,75	2,271182422328D-04
8	1,0	-8,177968449093D-03	-4,088984224546D-03	3,25	-2,925894025060D-04
9	2,0	1,491027778443D-02	7.455138892217D-03	3,50	-4,744848485922D-04
10	3,0	-2,492952746727D-02	-1,246476373363D-02	3,75	-3,867566874665D-04
11	4,0	3,258594556532D-02	1,629297278266D-02	4,25	5,195361293609D-04
12	5,0	-2,812482728214D-02	-1.406241364107D-02	4,50	8,538001969203D-04
13	6,0	1,367215639749D-02	6,836078198746D-03	4,75	7,028482889993D-04
14	7,0	-2,800952650805D-03	-1.400476325403D-03	5,25	-9,562590614282D-04
				5,50	-1,577762539317D-03
	0	2.511395200071D-01	9,780592669575D+01	5,75	-1,302625329413D-03
	1	1,246178343555D-01	-5,190881023216D+01	6,25	1,778996501275D-03
	2	2,049256954211D-02	9,161496901597D+00	6,50	2,938745742315D-03
	3	1,116474014817D-03	-5,309105767384D-01	6,75	2,428493501764D-03

I=14 H= 7.0

0		0.	-2,808523743279D+02	-1,00	9,048587711453D-04
1		-2,820956233808D-04	6,757806928080D+01	-,75	5,378514679367D-04
2		4,519221151145D-04	-5,420387498124D+00	-,50	2,753834695503D-04
3		-1,708410326500D-04	1,448481318623D-01	-,25	1,014384291750D-04
				.25	-4,494810286304D-05
0	-7,0	1,014540916347D-06	5,072704581736D-07	.50	-4,941448589213D-05
1	-6,0	-6,630189788651D-06	-3,315094894325D-06	.75	-2,930366356819D-05
2	-5,0	2,155902455839D-05	1.077951227920D-05	1,25	2,467218714545D-05
3	-4,0	-5,011370783492D-05	-2,505685391746D-05	1.50	3,437542307275D-05
4	-3,0	1,005426215912D-04	5,027131079558D-05	1.75	2,485392970464D-05
5	-2,0	-1,917429650457D-04	-9,587148252283D-05	2,25	-2,828529587548D-05
6	-1,0	3,602803753859D-04	1,801401876930D-04	2,50	-4,400491909291D-05
7	0,0	-6,737741670215D-04	-3,368870835107D-04	2,75	-3,475324926675D-05
8	1.0	1,255800108820D-03	6,279000544099D-04	3,25	4,477151595152D-05
9	2,0	-2,317496581904D-03	-1,158748290952D-03	3,50	7,260480996859D-05
10	3,0	4,094733847104D-03	2,047366923552D-03	3,75	5,918077393936D-05
11	4,0	-5,819564672824D-03	-2,909782336412D-03	4,25	-7,949827873498D-05
12	5,0	5,426180779463D-03	2,713090389732D-03	4,50	-1,306464164377D-04

CARDINAL SPLINES N=4

```
13   6.0  -2.800952650805D-03  -1.400476325403D-03     4.75 -1.075478686787D-04
14   7.0   6.001636373837D-04   3.000818186918D-04     5.25  1.463228362079D-04
                                                       5.50  2.414209954129D-04
       0  -3.842895992201D-02  -2.372196754163D+01     5.75  1.993188571172D-04
       1  -1.906881781133D-02   1.298531969082D+01     6.25 -2.721998773463D-04
       2  -3.135739570536D-03  -2.378576729016D+00     6.50 -4.496359995988D-04
       3  -1.708410326500D-04   1.448481318623D-01     6.75 -3.715491194944D-04
```

```
              N= 4      M=15                        P=  7.5        Q= 7

I= 0      H= -7.5
0               1.000000000000D+00   6.282364108092D-01    -1.00  4.666983564793D+00
1              -2.043924676956D+00  -1.049272907849D+00     -.75  3.405151558882D+00
2               1.333191700332D+00   6.685324739454D-01     -.50  2.391493661999D+00
3              -2.898671875047D-01  -1.449793176695D-01     -.25  1.598834825315D+00
                                                             .25  5.678141738581D-01
0    -7.5   6.001641287797D-04   3.000820643899D-04          .50  2.751068769491D-01
1    -6.5  -2.800955862158D-03  -1.400477931079D-03          .75  9.476921623172D-02
2    -5.5   5.426191221648D-03   2.713095610824D-03         1.25 -3.507903040201D-02
3    -4.5  -5.819588945597D-03  -2.909794472799D-03         1.50 -3.427496318905D-02
4    -3.5   4.094782545615D-03   2.047391272807D-03         1.75 -1.768214755865D-02
5    -2.5  -2.317589456348D-03  -1.158794728174D-03         2.25  1.053383010741D-02
6    -1.5   1.255974637477D-03   6.279873187386D-04         2.50  1.195817899911D-02
7     -.5  -6.741007218122D-04  -3.370503609061D-04         2.75  6.908488990264D-03
8      .5   3.608903413367D-04   1.804451706683D-04         3.25 -4.774228544937D-03
9     1.5  -1.928794447127D-04  -9.643972235636D-05         3.50 -5.681587438449D-03
10    2.5   1.026397127387D-04   5.131985636933D-05         3.75 -3.398312507166D-03
11    3.5  -5.381890773725D-05  -2.690945386863D-05         4.25  2.450969247780D-03
12    4.5   2.682493066027D-05   1.341246533014D-05         4.50  2.953675819591D-03
13    5.5  -1.154012650947D-05  -5.770063254735D-06         4.75  1.782463135502D-03
14    6.5   3.549010395812D-06   1.774505197906D-06         5.25 -1.299169958274D-03
15    7.5  -5.430637760531D-07  -2.715318880265D-07         5.50 -1.570349641685D-03
                                                            5.75 -9.496062264645D-04
        0  -6.162512166203D+01  -3.079950838047D+01         6.25  6.938886276729D-04
        1  -3.096113706339D+01  -1.548654565541D+01         6.50  8.393249592853D-04
        2  -5.188820018523D+00  -2.593502173619D+00         6.75  5.078366617253D-04
        3  -2.898671875047D-01  -1.449793176695D-01         7.25 -3.713031210595D-04
                                                            7.50 -4.492295103361D-04

I= 1      H= -6.5
0               0.                  -8.380458872740D-01    -1.00 -9.041181704170D+00
1               3.959053161757D+00   2.158005611915D+00     -.75 -5.678511496514D+00
2              -4.019190374154D+00  -2.022251339755D+00     -.50 -3.117191445449D+00
3               1.062938168259D+00   5.317678973961D-01     -.25 -1.257571097703D+00
                                                             .25  7.551721299769D-01
0    -7.5  -2.800955862158D-03  -1.400477931079D-03          .50  1.107574375905D+00
1    -6.5   1.367217738421D-02   6.836088692104D-03          .75  1.156548443115D+00
2    -5.5  -2.812489552352D-02  -1.406244776176D-02         1.25  7.315274386097D-01
```

CARDINAL SPLINES N=4

3	-4.5	3.258610419186D-02	1.629305209593D-02	1.50	4.350675757486D-01
4	-3.5	-2.492984571998D-02	-1.246492285999D-02	1.75	1.772915147219D-01
5	-2.5	1.491088473408D-02	7.455442367042D-03	2.25	-8.424024972100D-02
6	-1.5	-8.179109022349D-03	-4.089554511174D-03	2.50	-9.008359945167D-02
7	-.5	4.401816803094D-03	2.200908401547D-03	2.75	-4.992396114404D-02
8	.5	-2.358043300564D-03	-1.179021650282D-03	3.25	3.281508786886D-02
9	1.5	1.260445489313D-03	6.302227446565D-04	3.50	3.847407302012D-02
10	2.5	-6.707607725080D-04	-3.353803862540D-04	3.75	2.276837295269D-02
11	3.5	3.517145558887D-04	1.758572779444D-04	4.25	-1.621288235316D-02
12	4.5	-1.753052065965D-04	-8.765260329827D-05	4.50	-1.946637781602D-02
13	5.5	7.541660025601D-05	3.770830012801D-05	4.75	-1.171706654855D-02
14	6.5	-2.319336142175D-05	-1.159668071088D-05	5.25	8.514212026647D-03
15	7.5	3.549010395812D-06	1.774505197906D-06	5.50	1.028253437880D-02
				5.75	6.214565345950D-03
	0	2.520404799014D+02	1.259349400548D+02	6.25	-4.537607346303D-03
	1	1.230420134432D+02	6.156006820118D+01	6.50	-5.487581995140D-03
	2	1.989691841168D+01	9.942526351657D+00	6.75	-3.319822295845D-03
	3	1.062938168259D+00	5.317678973961D-01	7.25	2.426885861849D-03
				7.50	2.936089268651D-03

I= 2	H= -5.5				
0		0.	2.725028994491D+00	-1.00	9.696671720327D+00
1		-3.293245501189D+00	-2.226973471441D+00	-.75	5.852795263356D+00
2		4.845622764553D+00	2.463964717160D+00	-.50	3.052753873556D+00
3		-1.557803454585D+00	-7.798733624127D-01	-.25	1.150503477060D+00
				.25	-5.448003003023D-01
0	-7.5	5.426191221648D-03	2.713095610824D-03	.50	-6.299000991607D-01
1	-6.5	-2.812489552352D-02	-1.406244776176D-02	.75	-4.007453441398D-01
2	-5.5	6.273132165243D-02	3.136566082622D-02	1.25	4.380162003032D-01
3	-4.5	-8.074077171455D-02	-4.037038585727D-02	1.50	7.976881475652D-01
4	-3.5	6.960760687112D-02	3.480380343556D-02	1.75	9.966808920068D-01
5	-2.5	-4.588326821980D-02	-2.294163410990D-02	2.25	8.267657585765D-01
6	-1.5	2.626576910037D-02	1.313288455018D-02	2.50	5.431837194897D-01
7	-.5	-1.427263363117D-02	-7.136316815584D-03	2.75	2.397524778678D-01
8	.5	7.662556831900D-03	3.831278415950D-03	3.25	-1.274049933510D-01
9	1.5	-4.097914627746D-03	-2.048957313873D-03	3.50	-1.414519868677D-01
10	2.5	2.181002408190D-03	1.090501204095D-03	3.75	-8.064882228086D-02
11	3.5	-1.143642309473D-03	-5.718211547366D-04	4.25	5.497488088573D-02
12	4.5	5.700294653678D-04	2.850147326839D-04	4.50	6.517901415811D-02
13	5.5	-2.452279985175D-04	-1.226139992588D-04	4.75	3.888412826266D-02
14	6.5	7.541660025601D-05	3.770830012801D-05	5.25	-2.795916192670D-02
15	7.5	-1.154012650947D-05	-5.770063254735D-06	5.50	-3.366460330458D-02
				5.75	-2.030345853170D-02
	0	-4.093313931558D+02	-2.043883314689D+02	6.25	1.478824635799D-02
	1	-1.934882369941D+02	-9.687113262117D+01	6.50	1.787178560244D-02
	2	-3.020495496361D+01	-1.508318593712D+01	6.75	1.080663831822D-02
	3	-1.557803454585D+00	-7.798733624127D-01	7.25	-7.895486820217D-03
				7.50	-9.550569466088D-03

I= 3	H= -4.5				
0		0.	-6.334309606467D+00	-1.00	-6.822792869473D+00
1		2.161363665296D+00	2.429702765155D+00	-.75	-4.066881628271D+00
2		-3.408486640264D+00	-1.799903941092D+00	-.50	-2.089421313203D+00
3		1.252942563914D+00	6.287298405745D-01	-.25	-7.729485589016D-01
				.25	3.468873736691D-01

0	-7.5	-5.819588945597D-03	-2.909794472799D-03	.50	3.851325275326D-01
1	-6.5	3.258610419186D-02	1.629305209593D-02	.75	2.315573366230D-01
2	-5.5	-8.074077171455D-02	-4.037038585727D-02	1.25	-2.046503222981D-01
3	-4.5	1.191616891088D-01	5.958084455442D-02	1.50	-2.975468435028D-01
4	-3.5	-1.204459798923D-01	-6.022298994613D-02	1.75	-2.292872270235D-01
5	-2.5	9.207945458220D-02	4.603972729110D-02	2.25	3.358729519594D-01
6	-1.5	-5.806005668565D-02	-2.903002834283D-02	2.50	6.817333689980D-01
7	-.5	3.279858748732D-02	1.639929374366D-02	2.75	9.296913507158D-01
8	.5	-1.776512613498D-02	-8.882563067491D-03	3.25	8.730601321578D-01
9	1.5	9.519881892196D-03	4.759940946098D-03	3.50	5.982766077380D-01
10	2.5	-5.069027034296D-03	-2.534513517148D-03	3.75	2.727051247019D-01
11	3.5	2.658302918620D-03	1.329151459310D-03	4.25	-1.511716755282D-01
12	4.5	-1.325018963145D-03	-6.625094815724D-04	4.50	-1.700935314359D-01
13	5.5	5.700294653678D-04	2.850147326839D-04	4.75	-9.793337467736D-02
14	6.5	-1.753052065965D-04	-8.765260329827D-05	5.25	6.757331161580D-02
15	7.5	2.682493066027D-05	1.341246533014D-05	5.50	8.040750916983D-02
				5.75	4.809377369064D-02
	0	3.530679981259D+02	1.758892659381D+02	6.25	-3.468907170874D-02
	1	1.624681217217D+02	8.152930424572D+01	6.50	-4.180573886870D-02
	2	2.478272104779D+01	1.234651747183D+01	6.75	-2.522983063634D-02
	3	1.252942563914D+00	6.287298405745D-01	7.25	1.839144883780D-02
				7.50	2.223244001568D-02

I= 4 H= -3.5

0		0.	1.270855217209D+01	-1.00	3.861636196772D+00
1		-1.208106461450D+00	-3.310599665663D+00	-.75	2.296771083864D+00
2		1.928770707113D+00	1.156309666200D+00	-.50	1.176840786029D+00
3		-7.247590282090D-01	-3.669108699542D-01	-.25	4.338991443728D-01
				.25	-1.928025560579D-01
0	-7.5	4.094782545615D-03	2.047391272807D-03	.50	-2.124234419841D-01
1	-6.5	-2.492984571998D-02	-1.246492285999D-02	.75	-1.263574508724D-01
2	-5.5	6.960760687112D-02	3.480380343556D-02	1.25	1.075500992095D-01
3	-4.5	-1.204459798923D-01	-6.022298994613D-02	1.50	1.512811128554D-01
4	-3.5	1.470982546498D-01	7.354912732489D-02	1.75	1.109153880506D-01
5	-2.5	-1.362559570473D-01	-6.812797852365D-02	2.25	-1.327798191169D-01
6	-1.5	1.006440946092D-01	5.032204730459D-02	2.50	-2.159581015462D-01
7	-.5	-6.265067075240D-02	-3.132533537620D-02	2.75	-1.821515925339D-01
8	.5	3.524472937053D-02	1.762236468526D-02	3.25	3.032988061191D-01
9	1.5	-1.905096885948D-02	-9.525484429740D-03	3.50	6.429681449220D-01
10	2.5	1.016403213895D-02	5.082016069476D-03	3.75	9.065046104754D-01
11	3.5	-5.332634171035D-03	-2.666317085517D-03	4.25	8.897836057573D-01
12	4.5	2.658302918620D-03	1.329151459310D-03	4.50	6.184305991754D-01
13	5.5	-1.143642309473D-03	-5.718211547366D-04	4.75	2.848678787761D-01
14	6.5	3.517145558887D-04	1.758572779444D-04	5.25	-1.600375169119D-01
15	7.5	-5.381890773725D-05	-2.690945386863D-05	5.50	-1.808107934647D-01
				5.75	-1.044152649234D-01
	0	-2.063251612114D+02	-1.018690498585D+02	6.25	7.231105568647D-02
	1	-9.457963186501D+01	-4.788216397743D+01	6.50	8.613980720753D-02
	2	-1.437830742759D+01	-7.099184907770D+00	6.75	5.156339560208D-02
	3	-7.247590282090D-01	-3.669108699542D-01	7.25	-3.722889993819D-02
				7.50	-4.488148564878D-02

I= 5 H= -2.5

0		0.	-2.423700071109D+01	-1.00	-2.092801437314D+00
1		6.529613076147D-01	5.488246525547D+00	-.75	-1.244140680980D+00

CARDINAL SPLINES N=4

2		-1.045241923929D+00	-8.886477174001D-01	-.50	-6.371159105109D-01
3		3.945982057706D-01	2.059410384729D-01	-.25	-2.347335441144D-01
				.25	1.040781621688D-01
0	-7.5	-2.317589456348D-03	-1.158794728174D-03	.50	1.144768423788D-01
1	-6.5	1.491088473408D-02	7.455442367042D-03	.75	6.793415571200D-02
2	-5.5	-4.588326821980D-02	-2.294163410990D-02	1.25	-5.733946858919D-02
3	-4.5	9.207945458220D-02	4.603972729110D-02	1.50	-8.006512025713D-02
4	-3.5	-1.362559570473D-01	-6.812797852365D-02	1.75	-5.807479123295D-02
5	-2.5	1.560430151843D-01	7.802150759215D-02	2.25	6.687155282118D-02
6	-1.5	-1.410968155404D-01	-7.054840777018D-02	2.50	1.051018453344D-01
7	-.5	1.032299008744D-01	5.161495043719D-02	2.75	8.423642396026D-02
8	.5	-6.401191992878D-02	-3.200595996439D-02	3.25	-1.143423350094D-01
9	1.5	3.592910216651D-02	1.796455108325D-02	3.50	-1.940160123519D-01
10	2.5	-1.933469380195D-02	-9.667346900976D-03	3.75	-1.690270253756D-01
11	3.5	1.016403213895D-02	5.082016069476D-03	4.25	2.938320119540D-01
12	4.5	-5.069027034296D-03	-2.534513517148D-03	4.50	6.315587666862D-01
13	5.5	2.181002408190D-03	1.090501204095D-03	4.75	8.996186230261D-01
14	6.5	-6.707607725080D-04	-3.353803862540D-04	5.25	8.948042870148D-01
15	7.5	1.026397127387D-04	5.131985636933D-05	5.50	6.245009210584D-01
				5.75	2.885402437020D-01
	0	1.125734696456D+02	5.381978973249D+01	6.25	-1.627240556984D-01
	1	5.156277967247D+01	2.691108100684D+01	6.50	-1.840635127730D-01
	2	7.833217705910D+00	3.745025648239D+00	6.75	-1.063858904536D-01
	3	3.945982057706D-01	2.059410384729D-01	7.25	7.375794213088D-02
				7.50	8.789612963083D-02

I= 6 H= -1.5

0		0.	4.554694076178D+01	-1.00	1.123349615928D+00
1		-3.502749058511D-01	-9.875293601919D+00	-.75	6.677442818554D-01
2		5.610468206454D-01	9.683727711106D-01	-.50	3.419026442659D-01
3		-2.120278894318D-01	-1.222541535171D-01	-.25	1.259470885255D-01
				.25	-5.581615928620D-02
0	-7.5	1.255974637477D-03	6.279873187386D-04	.50	-6.136942164132D-02
1	-6.5	-8.179109022349D-03	-4.089554511174D-03	.75	-3.639895625309D-02
2	-5.5	2.626576910037D-02	1.313288455018D-02	1.25	3.066350795403D-02
3	-4.5	-5.806005668565D-02	-2.903002834283D-02	1.50	4.274446571156D-02
4	-3.5	1.006440946092D-01	5.032204730459D-02	1.75	3.092771401100D-02
5	-2.5	-1.410968155404D-01	-7.054840777018D-02	2.25	-3.529286778314D-02
6	-1.5	1.586538632910D-01	7.932693164552D-02	2.50	-5.503686976120D-02
7	-.5	-1.424745325335D-01	-7.123726626676D-02	2.75	-4.361499258972D-02
8	.5	1.039234235538D-01	5.196171177688D-02	3.25	5.687784659829D-02
9	1.5	-6.430027802602D-02	-3.215013901301D-02	3.50	9.320785324900D-02
10	2.5	3.592910216651D-02	1.796455108325D-02	3.75	7.712153072925D-02
11	3.5	-1.905096885948D-02	-9.525484429740D-03	4.25	-1.092090475377D-01
12	4.5	9.519881892196D-03	4.759940946098D-03	4.50	-1.878281535369D-01
13	5.5	-4.097914627746D-03	-2.048957313873D-03	4.75	-1.652914140152D-01
14	6.5	1.260445489313D-03	6.302227446565D-04	5.25	2.911059372987D-01
15	7.5	-1.928794447127D-04	-9.643972235636D-05	5.50	6.282605070222D-01
				5.75	8.976214053385D-01
	0	-6.051744398660D+01	-2.562276389267D+01	6.25	8.962697953587D-01
	1	-2.771427893778D+01	-1.598009044127D+01	6.50	6.262794920785D-01
	2	-4.209580691569D+00	-1.782345683024D+00	6.75	2.896212487457D-01
	3	-2.120278894318D-01	-1.222541535171D-01	7.25	-1.635260053138D-01
				7.50	-1.850447815175D-01

CARDINAL SPLINES N=4

I= 7		H= -.5			
0		0.	-8.522978442497D+01	-1.00	-6.016106634139D-01
1		1.875638092616D-01	1.824523020601D+01	-.75	-3.576022256060D-01
2		-3.004682813461D-01	-1.437375528104D+00	-.50	-1.830962965681D-01
3		1.135785728062D-01	8.717887812940D-02	-.25	-6.744488509964D-02
				.25	2.988630878754D-02
0	-7.5	-6.741007218122D-04	-3.370503609061D-04	.50	3.285688948320D-02
1	-6.5	4.401816803094D-03	2.200908401547D-03	.75	1.948542750731D-02
2	-5.5	-1.427263363117D-02	-7.136316815584D-03	1.25	-1.640787207374D-02
3	-4.5	3.279858748732D-02	1.639929374366D-02	1.50	-2.286348952296D-02
4	-3.5	-6.265067075240D-02	-3.132533537620D-02	1.75	-1.653343396581D-02
5	-2.5	1.032299008744D-01	5.161495043719D-02	2.25	1.882772936070D-02
6	-1.5	-1.424745325335D-01	-7.123726626676D-02	2.50	2.930719783194D-02
7	-.5	1.593490219701D-01	7.967451098503D-02	2.75	2.316386538291D-02
8	.5	-1.427639002292D-01	-7.138195011458D-02	3.25	-2.992561729898D-02
9	1.5	1.039234235538D-01	5.196171177688D-02	3.50	-4.864777505025D-02
10	2.5	-6.401191992878D-02	-3.200595996439D-02	3.75	-3.979205001976D-02
11	3.5	3.524472937053D-02	1.762236468526D-02	4.25	5.411723083989D-02
12	4.5	-1.776512613498D-02	-8.882563067491D-03	4.50	8.987781243070D-02
13	5.5	7.662556831900D-03	3.831278415950D-03	4.75	7.510929822011D-02
14	6.5	-2.358043300564D-03	-1.179021650282D-03	5.25	-1.077361518663D-01
15	7.5	3.608903413367D-04	1.804451706683D-04	5.50	-1.860418716685D-01
				5.75	-1.642062598941D-01
	0	3.242134814637D+01	7.535657875108D+00	6.25	2.903013946027D-01
	1	1.484692375012D+01	1.139603296879D+01	6.50	6.272762520995D-01
	2	2.255049606794D+00	5.241492298076D-01	6.75	8.970167189319D-01
	3	1.135785728062D-01	8.717887812940D-02	7.25	8.967336639115D-01
				7.50	6.268614072275D-01

I= 8		H= .5			
0		0.	1.592678885285D+02	-1.00	3.219218513489D-01
1		-1.003621873925D-01	-3.396970710024D+01	-.75	1.913518788468D-01
2		1.607804805038D-01	2.485673987719D+00	-.50	9.797361175378D-02
3		-6.077918345257D-02	-8.717887812940D-02	-.25	3.608900162107D-02
				.25	-1.599141953110D-02
0	-7.5	3.608903413367D-04	1.804451706683D-04	.50	-1.758055204611D-02
1	-6.5	-2.358043300564D-03	-1.179021650282D-03	.75	-1.042566523465D-02
2	-5.5	7.662556831900D-03	3.831278415950D-03	1.25	8.778139189990D-03
3	-4.5	-1.776512613498D-02	-8.882563067491D-03	1.50	1.223078349541D-02
4	-3.5	3.524472937053D-02	1.762236468526D-02	1.75	8.843376825708D-03
5	-2.5	-6.401191992878D-02	-3.200595996439D-02	2.25	-1.006573419633D-02
6	-1.5	1.039234235538D-01	5.196171177688D-02	2.50	-1.566173978362D-02
7	-.5	-1.427639002292D-01	-7.138195011458D-02	2.75	-1.237122558117D-02
8	.5	1.593490219701D-01	7.967451098503D-02	3.25	1.594778874500D-02
9	1.5	-1.424745325335D-01	-7.123726626676D-02	3.50	2.587658495658D-02
10	2.5	1.032299008744D-01	5.161495043719D-02	3.75	2.110920792415D-02
11	3.5	-6.265067075240D-02	-3.132533537620D-02	4.25	-2.843739111470D-02
12	4.5	3.279858748732D-02	1.639929374366D-02	4.50	-4.684830283321D-02
13	5.5	-1.427263363117D-02	-7.136316815584D-03	4.75	-3.870117158375D-02
14	6.5	4.401816803094D-03	2.200908401547D-03	5.25	5.331042633693D-02
15	7.5	-6.741007218122D-04	-3.370503609061D-04	5.50	8.889147914705D-02
				5.75	7.450363215638D-02
	0	-1.735003239616D+01	7.535657875108D+00	6.25	-1.072718061963D-01
	1	-7.945142187456D+00	-1.139603296879D+01	6.50	-1.854594634923D-01
	2	-1.206751147179D+00	5.241492298076D-01	6.75	-1.638367339488D-01
	3	-6.077918345257D-02	-8.717887812940D-02	7.25	2.899905136173D-01

CARDINAL SPLINES N=4

				7,50	6,268614072275D-01

I= 9 H= 1.5

0		0.	-2,973063578874D+02	-1,00	-1,720330205150D-01
1		5,363253237740D-02	6,334566409265D+01	-,75	-1,022571151351D-01
2		-8,592007053515D-02	-4,533064137160D+00	-,50	-5,235633602279D-02
3		3,248041760246D-02	1,222541535171D-01	-,25	-1,928564402783D-02
				,25	8,545623438514D-03
0	-7,5	-1,928794447127D-04	-9,643972235636D-05	,50	9,394793884557D-03
1	-6,5	1,260445489313D-03	6,302227446565D-04	,75	5,571289485118D-03
2	-5,5	-4,097914627746D-03	-2,048957313873D-03	1,25	-4,690773041049D-03
3	-4,5	9,519881892196D-03	4,759940946098D-03	1.50	-6,535629636682D-03
4	-3,5	-1,905096885948D-02	-9,525484429740D-03	1,75	-4,725396988624D-03
5	-2,5	3,592910216651D-02	1,796455108325D-02	2,25	5,377966239937D-03
6	-1,5	-6,430027802602D-02	-3,215013901301D-02	2,50	8,367022577311D-03
7	-,5	1,039234235538D-01	5,196171177688D-02	2,75	6,608198865657D-03
8	,5	-1,424745325335D-01	-7,123726626676D-02	3,25	-8,514397207829D-03
9	1,5	1,586538632910D-01	7,932693164552D-02	3,50	-1,380934675914D-02
10	2,5	-1,410968155404D-01	-7,054840777018D-02	3,75	-1,125818881971D-02
11	3,5	1,006440946092D-01	5,032204730459D-02	4,25	1,513318389678D-02
12	4,5	-5,806005668565D-02	-2,903002834283D-02	4,50	2,488368735372D-02
13	5,5	2,626576910037D-02	1,313288455018D-02	4.75	2,050078296541D-02
14	6,5	-8,179109022349D-03	-4,089554511174D-03	5,25	-2,797207059623D-02
15	7,5	1,255974637477D-03	6,279873187386D-04	5,50	-4,626507990595D-02
				5,75	-3,833131179996D-02
	0	9,271916201268D+00	-2,562276389267D+01	6,25	5,299945431796D-02
	1	4,245901944766D+00	1,598009044127D+01	6,50	8,847659344503D-02
	2	6,448893255203D-01	-1,782345683024D+00	6,75	7,422059276694D-02
	3	3,248041760246D-02	1,222541535171D-01	7,25	-1,069888339014D-01
				7,50	-1,850447815175D-01

I=10 H= 2.5

0		0.	5,531919656030D+02	-1,00	9,154402373088D-02
1		-2,853946054669D-02	-1,178390159727D+02	-,75	5,441411781713D-02
2		4,572069200907D-02	8,378699013879D+00	-,50	2,786038717250D-02
3		-1,728387117512D-02	-2,059410384729D-01	-,25	1,026246887435D-02
				,25	-4,547376108587D-03
0	-7,5	1,026397127387D-04	5,131985636933D-05	,50	-4,999239295212D-03
1	-6,5	-6,707607725080D-04	-3,353803862540D-04	,75	-2,964638559131D-03
2	-5,5	2,181002408190D-03	1,090501204095D-03	1.25	2,496078041329D-03
3	-4,5	-5,069027034296D-03	-2,534513517148D-03	1.50	3,477756382686D-03
4	-3,5	1,016403213895D-02	5,082016069476D-03	1,75	2,514473415902D-03
5	-2,5	-1,933469380195D-02	-9,667346900976D-03	2,25	-2,861646113884D-03
6	-1,5	3,592910216651D-02	1,796455108325D-02	2,50	-4,452041931866D-03
7	-,5	-6,401191992878D-02	-3,200595996439D-02	2,75	-3,516070160059D-03
8	,5	1,032299008744D-01	5,161495043719D-02	3,25	4,529795582006D-03
9	1,5	-1,410968155404D-01	-7,054840777018D-02	3,50	7,346066966264D-03
10	2,5	1,560430151843D-01	7,802150759215D-02	3,75	5,988091215485D-03
11	3,5	-1,362559570473D-01	-6,812797852365D-02	4,25	-8,045081288096D-03
12	4,5	9,207945458220D-02	4,603972729110D-02	4,50	-1,322287970042D-02
13	5,5	-4,588326821980D-02	-2,294163410990D-02	4,75	-1,088705935679D-02
14	6,5	1,491088473408D-02	7,455442367042D-03	5,25	1,482194762963D-02
15	7,5	-2,317589456348D-03	-1.158794728174D-03	5,50	2,446877877864D-02
				5,75	2,021789916122D-02
	0	-4,933890180593D+00	5,381978973249D+01	6,25	-2,768949477595D-02

CARDINAL SPLINES N=4

	1	-2,259382341212D+00	-2,691108100684D+01	6,50	-4,585110650120D-02
	2	-3,431664094311D-01	3.745025648239D+00	6.75	-3,802119274163D-02
	3	-1,728387117512D-02	-2.059410384729D-01	7.25	5,263105571510D-02
				7.50	8,789612963083D-02

I=11	H=	3.5			
0		0.	-1.015104954013D+03	-1.00	-4,800062128559D-02
1		1.496450839719D-02	2.162861468988D+02	-.75	-2,853175096488D-02
2		-2,397340118893D-02	-1,535467948174D+01	-.50	-1,460844345826D-02
3		9,062711699476D-03	3,669108699542D-01	-,25	-5,381069543909D-03
				.25	2,384391110448D-03
0	-7,5	-5,381890773725D-05	-2.690945386863D-05	.50	2,621322403580D-03
1	-6,5	3,517145558887D-04	1,758572779444D-04	.75	1,554490670352D-03
2	-5,5	-1.143642309473D-03	-5.718211547366D-04	1.25	-1,308802156451D-03
3	-4,5	2,658302918620D-03	1.329151459310D-03	1.50	-1,823536817520D-03
4	-3,5	-5,332634171035D-03	-2,666317085517D-03	1.75	-1,318444176314D-03
5	-2,5	1.016403213895D-02	5,082016069476D-03	2.25	1,500472794069D-03
6	-1,5	-1,905096885948D-02	-9.525484429740D-03	2.50	2.334367651233D-03
7	-,5	3,524472937053D-02	1,762236468526D-02	2,75	1,843590071051D-03
8	,5	-6,265067075240D-02	-3,132533537620D-02	3,25	-2,375056945191D-03
9	1,5	1.006440946092D-01	5,032204730459D-02	3.50	-3,851593588451D-03
10	2,5	-1,362559570473D-01	-6,812797852365D-02	3,75	-3,139494935423D-03
11	3,5	1,470982546498D-01	7,354912732489D-02	4,25	4,217464021526D-03
12	4,5	-1,204459798923D-01	-6,022298994613D-02	4.50	6,931123579489D-03
13	5,5	6,960760687112D-02	3.480380343556D-02	4.75	5,705922715382D-03
14	6,5	-2,492984571998D-02	-1,246492285999D-02	5,25	-7,764254781747D-03
15	7,5	4,094782545615D-03	2,047391272807D-03	5,50	-1,281201123077D-02
				5,75	-1,057959604996D-02
	0	2,587061494318D+00	-1,018690498585D+02	6,25	1,445727368077D-02
	1	1,184696089850D+00	4,788216397743D+01	6,50	2,389449729974D-02
	2	1,799376120493D-01	-7,099184907770D+00	6,75	1,976037182347D-02
	3	9,062711699476D-03	3,669108699542D-01	7,25	-2,709358404756D-02
				7,50	-4,488148564878D-02

I=12	H=	4.5			
0		0,	1,747096057064D+03	-1.00	2,392489228732D-02
1		-7,458741373447D-03	-3,728252269202D+02	-.75	1,422104630789D-02
2		1,194903367833D-02	2,649293888476D+01	-.50	7,281268760749D-03
3		-4,517117235543D-03	-6,287298405745D-01	-.25	2,682079905063D-03
				.25	-1,188449058008D-03
0	-7,5	2,682493066027D-05	1,341246533014D-05	.50	-1,306542351813D-03
1	-6,5	-1,753052065965D-04	-8,765260329827D-05	.75	-7,748027237640D-04
2	-5,5	5,700294653678D-04	2,850147326839D-04	1.25	6,523443421813D-04
3	-4,5	-1,325018963145D-03	-6,625094815724D-04	1.50	9,089025628906D-04
4	-3,5	2,658302918620D-03	1,329151459310D-03	1.75	6,571498016219D-04
5	-2,5	-5,069027034296D-03	-2,534513517148D-03	2.25	-7,478770142810D-04
6	-1,5	9,519881892196D-03	4,759940946098D-03	2,50	-1,163511855097D-03
7	-,5	-1,776512613498D-02	-8,882563067491D-03	2,75	-9,188935112103D-04
8	,5	3,279858748732D-02	1.639929374366D-02	3,25	1,183783349365D-03
9	1,5	-5,806005668565D-02	-2,903002834283D-02	3,50	1,919713366047D-03
10	2,5	9,207945458220D-02	4,603972729110D-02	3,75	1,564777002834D-03
11	3,5	-1.204459798923D-01	-6,022298994613D-02	4,25	-2,101998731656D-03
12	4,5	1.191616891088D-01	5,958084455442D-02	4,50	-3,454416724617D-03
13	5,5	-8,074077171455D-02	-4.037038585727D-02	4.75	-2,843692482673D-03
14	6,5	3,258610419186D-02	1.629305209593D-02	5,25	3,869061934395D-03

15	7.5	-5.819588945597D-03	-2.909794472799D-03	5.50	6.383804181798D-03
				5.75	5.270704104576D-03
	0	-1.289466249639D+00	1.758892659381D+02	6.25	-7.198862786759D-03
	1	-5.904867696964D-01	-8.152930424572D+01	6.50	-1.189282093979D-02
	2	-8.968610412139D-02	1.234651747183D+01	6.75	-9.828986150669D-03
	3	-4.517117235543D-03	-6.287298405745D-01	7.25	1.344663450551D-02
				7.50	2.223244001568D-02

I=13 H= 5.5

0		0.	-2.108360109859D+03	-1.00	-1.029252222919D-02
1		3.208760937414D-03	4.547225515852D+02	-.75	-6.117914052124D-03
2		-5.140491051341D-03	-3.263033659141D+01	-.50	-3.132412011597D-03
3		1.943270240437D-03	7.798733624127D-01	-.25	-1.153834522569D-03
				.25	5.112724367981D-04
0	-7.5	-1.154012650947D-05	-5.770063254735D-06	.50	5.620763286879D-04
1	-6.5	7.541660025601D-05	3.770830012801D-05	.75	3.333211984256D-04
2	-5.5	-2.452279985175D-04	-1.226139992588D-04	1.25	-2.806394158371D-04
3	-4.5	5.700294653678D-04	2.850147326839D-04	1.50	-3.910110860164D-04
4	-3.5	-1.143642309473D-03	-5.718211547366D-04	1.75	-2.827066908893D-04
5	-2.5	2.181002408190D-03	1.090501204095D-03	2.25	3.217375550388D-04
6	-1.5	-4.097914627746D-03	-2.048957313873D-03	2.50	5.005440076700D-04
7	-.5	7.662556831900D-03	3.831278415950D-03	2.75	3.953087863984D-04
8	.5	-1.427263363117D-02	-7.136316815584D-03	3.25	-5.092639051946D-04
9	1.5	2.626576910037D-02	1.313288455018D-02	3.50	-8.258601820756D-04
10	2.5	-4.588326821980D-02	-2.294163410990D-02	3.75	-6.731654688056D-04
11	3.5	6.960760687112D-02	3.480380343556D-02	4.25	9.042720571716D-04
12	4.5	-8.074077171455D-02	-4.037038585727D-02	4.50	1.486069276221D-03
13	5.5	6.273132165243D-02	3.136566082622D-02	4.75	1.223329937032D-03
14	6.5	-2.812489552352D-02	-1.406244776176D-02	5.25	-1.664388720539D-03
15	7.5	5.426191221648D-03	2.713095610824D-03	5.50	-2.746113140459D-03
				5.75	-2.267215679933D-03
	0	5.547302180768D-01	-2.043883314689D+02	6.25	3.096250735899D-03
	1	2.540282482410D-01	9.687113262117D+01	6.50	5.114609628496D-03
	2	3.858308935848D-02	-1.508318593712D+01	6.75	4.226416328506D-03
	3	1.943270240437D-03	7.798733624127D-01	7.25	-5.778968374794D-03
				7.50	-9.550569466088D-03

I=14 H= 6.5

0		0.	1.371242140558D+03	-1.00	3.165326259318D-03
1		-9.868111019615D-04	-3.004337961616D+02	-.75	1.881481871296D-03
2		1.580888624461D-03	2.190730404307D+01	-.50	9.633310237080D-04
3		-5.976265328955D-04	-5.317678973961D-01	-.25	3.548462290957D-04
				.25	-1.572349344236D-04
0	-7.5	3.549010395812D-06	1.774505197906D-06	.50	-1.728589848337D-04
1	-6.5	-2.319336142175D-05	-1.159668071088D-05	.75	-1.025084330754D-04
2	-5.5	7.541660025601D-05	3.770830012801D-05	1.25	8.630686091251D-05
3	-4.5	-1.753052065965D-04	-8.765260329827D-05	1.50	1.202501752467D-04
4	-3.5	3.517145558887D-04	1.758572779444D-04	1.75	8.694262036169D-05
5	-2.5	-6.707607725080D-04	-3.353803862540D-04	2.25	-9.894602728981D-05
6	-1.5	1.260445489313D-03	6.302227446565D-04	2.50	-1.539355195783D-04
7	-.5	-2.358043300564D-03	-1.179021650282D-03	2.75	-1.215718428984D-04
8	.5	4.401816803094D-03	2.200908401547D-03	3.25	1.566171354101D-04
9	1.5	-8.179109022349D-03	-4.089554511174D-03	3.50	2.539819048656D-04
10	2.5	1.491088473408D-02	7.455442367042D-03	3.75	2.070226619777D-04
11	3.5	-2.492984571998D-02	-1.246492285999D-02	4.25	-2.780958416231D-04

12	4.5	3,258610419186D-02	1.629305209593D-02	4.50	-4,570185828044D-04
13	5,5	-2,812489552352D-02	-1,406244776176D-02	4,75	-3,762162704460D-04
14	6.5	1,367217738421D-02	6.836088692104D-03	5,25	5,118536180832D-04
15	7,5	-2,800955862158D-03	-1,400477931079D-03	5,50	8,445140917765D-04
				5,75	6,972325661092D-04
	0	-1,705997917041D-01	1.259349400548D+02	6,25	-9,521558206061D-04
	1	-7,812295916116D-02	-6,156006820118D+01	6,50	-1,572799275331D-03
	2	-1,186570836569D-02	9,942526351657D+00	6,75	-1,299622288236D-03
	3	-5,976265328955D-04	-5,317678973961D-01	7,25	1,776800837500D-03
				7,50	2,936089268651D-03

I=15 H= 7.5

0		0.	-3,539962477039D+02	-1.00	-4,843530449740D-04
1		1,510002199442D-04	7,885433811643D+01	-.75	-2.879012773206D-04
2		-2,419049905990D-04	-5,855536821184D+00	-,50	-1,474073369257D-04
3		9,144783443080D-05	1,449793176695D-01	-,25	-5,429798931147D-05
				.25	2,405983234062D-05
0	-7,5	-5,430637760531D-07	-2,715318880265D-07	.50	2,645059894047D-05
1	-6,5	3,549010395812D-06	1,774505197906D-06	.75	1,568567263318D-05
2	-5,5	-1,154012650947D-05	-5,770063254735D-06	1.25	-1,320653452051D-05
3	-4,5	2,682493066027D-05	1,341246533014D-05	1,50	-1,840048481983D-05
4	-3,5	-5,381890773725D-05	-2,690945386863D-05	1,75	-1,330381725707D-05
5	-2,5	1,026397127387D-04	5,131985636933D-05	2,25	1,514055857900D-05
6	-1,5	-1,928794447127D-04	-9.643972235636D-05	2,50	2,355495991848D-05
7	-,5	3,608903413367D-04	1,804451706683D-04	2.75	1,860272284644D-05
8	,5	-6.741007218122D-04	-3,370503609061D-04	3,25	-2,396529330115D-05
9	1,5	1,255974637477D-03	6,279873187386D-04	3,50	-3,886388486520D-05
10	2,5	-2,317589456348D-03	-1,158794728174D-03	3,75	-3,167825638755D-05
11	3,5	4,094782545615D-03	2,047391272807D-03	4,25	4,255373492986D-05
12	4,5	-5,819588945597D-03	-2,909794472799D-03	4,50	6,993215049612D-05
13	5,5	5,426191221648D-03	2,713095610824D-03	4,75	5,756789653696D-05
14	6,5	-2,800955862158D-03	-1,400477931079D-03	5,25	-7,832271340303D-05
15	7.5	6,001641287797D-04	3,000820643899D-04	5,50	-1,292254717906D-04
				5,75	-1,066885599312D-04
	0	2,610490107888D-02	-3,079950838047D+01	6,25	1,456949649969D-04
	1	1,195424742116D-02	1,548654565541D+01	6,50	2,406615248616D-04
	2	1,815671284094D-03	-2,593502173619D+00	6,75	1,988593366595D-04
	3	9,144783443080D-05	1,449793176695D-01	7,25	-2,718639004605D-04
				7,50	-4,492295103361D-04

N= 4 M=16 P= 8,0 Q= 7

I= 0 H= -8.0

0		1,000000000000D+00	4,156776408021D-01	-1,00	4,666983690368D+00
1		-2,043924716105D+00	-1,005196847190D+00	-,75	3,405151633525D+00
2		1,333191763049D+00	6,654858194557D-01	-,50	2,391493700217D+00
3		-2,898672112139D-01	-1.449091304909D-01	-,25	1,598834839392D+00
				,25	5,678141676202D-01

CARDINAL SPLINES N=4

0	-8.0	6.001642695770D-04	3.000821347885D-04	.50	2.751068700914D-01
1	-7.0	-2.800956782291D-03	-1.400478391146D-03	.75	9.476921216498D-02
2	-6.0	5.426194213596D-03	2.713097106798D-03	1.25	-3.507902697802D-02
3	-5.0	-5.819595900358D-03	-2.909797950179D-03	1.50	-3.427495841846D-02
4	-4.0	4.094796498982D-03	2.047398249491D-03	1.75	-1.768214410944D-02
5	-3.0	-2.317616067408D-03	-1.158808033704D-03	2.25	1.053382618200D-02
6	-2.0	1.256024645995D-03	6.280123229976D-04	2.50	1.195817289214D-02
7	-1.0	-6.741943017231D-04	-3.370971508616D-04	2.75	6.908484167235D-03
8	0.0	3.610652243775D-04	1.805326121888D-04	3.25	-4.774222331584D-03
9	1.0	-1.932059927757D-04	-9.660299638783D-05	3.50	-5.681577362419D-03
10	2.0	1.032480721005D-04	5.162403605025D-05	3.75	-3.398304294115D-03
11	3.0	-5.494145213658D-05	-2.747072606829D-05	4.25	2.450958215107D-03
12	4.0	2.880825781311D-05	1.440412890655D-05	4.50	2.953657688672D-03
13	5.0	-1.435886550433D-05	-7.179432752163D-06	4.75	1.782448210201D-03
14	6.0	6.177204085787D-06	3.088602042894D-06	5.25	-1.299149652020D-03
15	7.0	-1.899715722397D-06	-9.498578611983D-07	5.50	-1.570316138228D-03
16	8.0	2.906913914542D-07	1.453456957271D-07	5.75	-9.496329660604D-04
				6.25	6.938508546041D-04
	0	-7.843913703518D+01	-3.922827950292D+01	6.50	8.392625654238D-04
	1	-3.636736106038D+01	-1.817997679016D+01	6.75	5.077851059218D-04
	2	-5.623621306083D+00	-2.812333312327D+00	7.25	-3.712326403499D-04
	3	-2.898672112139D-01	-1.449091304909D-01	7.50	-4.491130505547D-04
				7.75	-2.717676460914D-04

I= 1 H= -7.0

0		0.	5.510603923982D-01	-1.00	-9.041182524827D+00
1		3.959053417602D+00	1.869961356468D+00	-.75	-5.678511984315D+00
2		-4.019190784022D+00	-2.002340955048D+00	-.50	-3.117191695207D+00
3		1.062938323203D+00	5.313092127201D-01	-.25	-1.257571189702D+00
				.25	7.551721707423D-01
0	-8.0	-2.800956782291D-03	-1.400478391146D-03	.50	1.107574420721D+00
1	-7.0	1.367218339743D-02	6.836091698714D-03	.75	1.156548469692D+00
2	-6.0	-2.812491507638D-02	-1.406245753819D-02	1.25	7.315274162334D-01
3	-5.0	3.258614964235D-02	1.629307482118D-02	1.50	4.350675445720D-01
4	-4.0	-2.492993690752D-02	-1.246496845376D-02	1.75	1.772914921808D-01
5	-3.0	1.491105864171D-02	7.455529320853D-03	2.25	-8.424022406779D-02
6	-2.0	-8.179435836182D-03	-4.089717918091D-03	2.50	-9.008355954163D-02
7	-1.0	4.402428363096D-03	2.201214181548D-03	2.75	-4.992392962476D-02
8	0.0	-2.359186189800D-03	-1.179593094900D-03	3.25	3.281504726358D-02
9	1.0	1.262579534240D-03	6.312897671201D-04	3.50	3.847400717162D-02
10	2.0	-6.747365003023D-04	-3.373682501512D-04	3.75	2.276831927906D-02
11	3.0	3.590505668859D-04	1.795252834430D-04	4.25	-1.621281025283D-02
12	4.0	-1.882665734813D-04	-9.413328674066D-05	4.50	-1.946625932751D-02
13	5.0	9.383751997804D-05	4.691875998902D-05	4.75	-1.171696900927D-02
14	6.0	-4.036903645421D-05	-2.018451822710D-05	5.25	8.514079321965D-03
15	7.0	1.241495245001D-05	6.207476225007D-06	5.50	1.028231542823D-02
16	8.0	-1.899715722397D-06	-9.498578611983D-07	5.75	6.214384580692D-03
				6.25	-4.537360493128D-03
	0	3.186686386432D+02	1.593912470337D+02	6.50	-5.487174241064D-03
	1	1.437361589282D+02	7.184387491796D+01	6.75	-3.319485370248D-03
	2	2.149132897284D+01	1.074908015023D+01	7.25	2.426425258901D-03
	3	1.062938323203D+00	5.313092127201D-01	7.50	2.935328184958D-03
				7.75	1.776171799476D-03

I= 2 H= -6.0

0		0.	-1.791853655487D+00	-1.00	9,696674388814D+00
1		-3.293246333108D+00	-1,290355360289D+00	-.75	5,852796849515D+00
2		4.845624097300D+00	2,399223183304D+00	-.50	3,052754685680D+00
3		-1,557803958406D+00	-7,783818819659D-01	-.25	1,150503776208D+00
				.25	-5,448004328571D-01
0	-8,0	5,426194213596D-03	2.713097106798D-03	.50	-6,299002448873D-01
1	-7.0	-2.812491507638D-02	-1.406245753819D-02	.75	-4,007454305582D-01
2	-6,0	6.273138523144D-02	3.136569261572D-02	1,25	4,380162730631D-01
3	-5,0	-8,074091950349D-02	-4.037045975175D-02	1.50	7,976882489405D-01
4	-4,0	6.960790338076D-02	3,480395169038D-02	1.75	9,966809653027D-01
5	-3,0	-4,588383370592D-02	-2,294191685296D-02	2,25	8,267656751613D-01
6	-2,0	2,626683178344D-02	1.313341589172D-02	2,50	5,431835897164D-01
7	-1,0	-1,427462220814D-02	-7.137311104072D-03	2,75	2,397523753783D-01
8	0,0	7.666273103748D-03	3.833136551874D-03	3,25	-1,274048613169D-01
9	1.0	-4.104853787584D-03	-2.052426893792D-03	3,50	-1,414517727516D-01
10	2,0	2,193930069779D-03	1.096965034889D-03	3,75	-8,064864775320D-02
11	3.0	-1,167496424366D-03	-5,837482121829D-04	4,25	5,497464644097D-02
12	4.0	6,121752493576D-04	3.060876246788D-04	4,50	6,517862887533D-02
13	5,0	-3,051263186869D-04	-1,525631593434D-04	4,75	3,888381109939D-02
14	6.0	1,312658248194D-04	6,563291240968D-05	5,25	-2,795873041798D-02
15	7.0	-4,036903645421D-05	-2.018451822710D-05	5,50	-3,366389135473D-02
16	8,0	6.177204085787D-06	3,088602042894D-06	5,75	-2,030287074697D-02
				6,25	1,478744367872D-02
	0	-5,138216551416D+02	-2,570959363728D+02	6,50	1,787045973030D-02
	1	-2,248616207903D+02	-1,123521057649D+02	6,75	1,080554275526D-02
	2	-3,254167090445D+01	-1,628194198388D+01	7,25	-7,893989102225D-03
	3	-1,557803958406D+00	-7,783818819659D-01	7,50	-9,548094690919D-03
				7,75	-5,776922965471D-03

I= 3 H= -5.0

0		0,	4,165150921167D+00	-1.00	-6,822799072353D+00
1		2.161365599084D+00	2.525408485080D-01	-.75	-4,066885315286D+00
2		-3,408489738226D+00	-1,649412721396D+00	-,50	-2,089423200979D+00
3		1,252943735043D+00	6,252629052361D-01	-,25	-7,729492542702D-01
				.25	3,468876817920D-01
0	-8,0	-5,819595900358D-03	-2.909797950179D-03	.50	3,851328662728D-01
1	-7,0	3,258614964235D-02	1.629307482118D-02	.75	2,315575375019D-01
2	-6,0	-8,074091950349D-02	-4,037045975175D-02	1.25	-2,046504914279D-01
3	-5.0	1,191620326432D-01	5,958101632158D-02	1.50	-2,975470791491D-01
4	-4,0	-1,204466691267D-01	-6,022333456336D-02	1.75	-2,292873973991D-01
5	-3,0	9,208076905053D-02	4.604038452526D-02	2.25	3,358731458574D-01
6	-2,0	-5,806252688466D-02	-2,903126344233D-02	2.50	6,817336706551D-01
7	-1,0	3,280320991989D-02	1.640160495995D-02	2,75	9,296915889521D-01
8	0,0	-1,777376458158D-02	-8,886882290792D-03	3,25	8,730598252457D-01
9	1.0	9,536011918245D-03	4,768005959123D-03	3,50	5,982761100267D-01
10	2,0	-5,099077288662D-03	-2,549538644331D-03	3,75	2,727047190136D-01
11	3,0	2,713751633396D-03	1,356875816698D-03	4,25	-1,511711305631D-01
12	4.0	-1,422986528321D-03	-7,114932641607D-04	4,50	-1,700926358489D-01
13	5,0	7,092626704396D-04	3,546313352198D-04	4,75	-9,793263743366D-02
14	6,0	-3,051263186869D-04	-1,525631593434D-04	5,25	6,757230857695D-02
15	7,0	9,383751997804D-05	4,691875998902D-05	5,50	8,040585424764D-02
16	8,0	-1,435886550433D-05	-7,179432752163D-06	5,75	4,809240738935D-02
				6,25	-3,468720588667D-02
	0	4,406547738880D+02	2,207576710207D+02	6,50	-4,180265688866D-02
	1	1,881907269156D+02	9,391241511149D+01	6,75	-2,522728400830D-02
	2	2,666215990279D+01	1,335689700427D+01	7,25	1,838796740333D-02

	3	1.252943735043D+00	6.252629052361D-01	7.50	2.222668741898D-02
				7.75	1.344187996657D-02

I= 4 H= -4.0

0		0.	-8.356563709626D+00	-1.00	3.861648641638D+00
1		-1.208110341218D+00	1.057450287640D+00	-.75	2.296778481139D+00
2		1.928776922569D+00	8.543784409020D-01	-.50	1.176844573483D+00
3		-7.247613778505D-01	-3.599551416607D-01	-.25	4.339005394940D-01
				.25	-1.928031742462D-01
0	-8.0	4.094796498982D-03	2.047398249491D-03	.50	-2.124241216003D-01
1	-7.0	-2.492993690752D-02	-1.246496845376D-02	.75	-1.263578538968D-01
2	-6.0	6.960790338076D-02	3.480395169038D-02	1.25	1.075504385354D-01
3	-5.0	-1.204466691267D-01	-6.022333456336D-02	1.50	1.512815856336D-01
4	-4.0	1.470996374640D-01	7.354981873202D-02	1.75	1.109157298761D-01
5	-3.0	-1.362585942711D-01	-6.812929713557D-02	2.25	-1.327802081353D-01
6	-2.0	1.006490505808D-01	5.032452529039D-02	2.50	-2.159587067624D-01
7	-1.0	-6.265994476009D-02	-3.132997238004D-02	2.75	-1.821520705084D-01
8	0.0	3.526206072562D-02	1.763103036281D-02	3.25	3.032994218782D-01
9	1.0	-1.908333060483D-02	-9.541665302415D-03	3.50	6.429691434823D-01
10	2.0	1.022432210231D-02	5.112161051155D-03	3.75	9.065054244096D-01
11	3.0	-5.443881182172D-03	-2.721940591086D-03	4.25	8.897825123912D-01
12	4.0	2.854855694609D-03	1.427427847305D-03	4.50	6.184288023551D-01
13	5.0	-1.422986528321D-03	-7.114932641607D-04	4.75	2.848663996407D-01
14	6.0	6.121752493576D-04	3.060876246788D-04	5.25	-1.600355045104D-01
15	7.0	-1.882665734813D-04	-9.413328674066D-05	5.50	-1.808074731867D-01
16	8.0	2.880825781311D-05	1.440412890655D-05	5.75	-1.044125237068D-01
				6.25	7.230731227907D-02
	0	-2.573009851448D+02	-1.295137737211D+02	6.50	8.613362381682D-02
	1	-1.095018641274D+02	-5.438388185678D+01	6.75	5.155828629054D-02
	2	-1.546549614584D+01	-7.784544958955D+00	7.25	-3.722191512023D-02
	3	-7.247613778505D-01	-3.599551416607D-01	7.50	-4.486994418761D-02
				7.75	-2.708404499425D-02

I= 5 H= -3.0

0		0.	1.593717655000D+01	-1.00	-2.092825171447D+00
1		6.529687068852D-01	-2.842247077937D+00	-.75	-1.244154788638D+00
2		-1.045253777690D+00	-3.128218527856D-01	-.50	-6.371231337240D-01
3		3.946026868718D-01	1.926754735924D-01	-.25	-2.347362048093D-01
				.25	1.040793411420D-01
0	-8.0	-2.317616067408D-03	-1.158808033704D-03	.50	1.144781385036D-01
1	-7.0	1.491105864171D-02	7.455529320853D-03	.75	6.793492433694D-02
2	-6.0	-4.588383370592D-02	-2.294191685296D-02	1.25	-5.734011573209D-02
3	-5.0	9.208076905053D-02	4.604038452526D-02	1.50	-8.006602191253D-02
4	-4.0	-1.362585942711D-01	-6.812929713557D-02	1.75	-5.807544314287D-02
5	-3.0	1.560480447459D-01	7.802402237295D-02	2.25	6.687229473457D-02
6	-2.0	-1.411062672843D-01	-7.055313364216D-02	2.50	1.051029995679D-01
7	-1.0	1.032475877284D-01	5.162379386418D-02	2.75	8.423733552561D-02
8	0.0	-6.404497329225D-02	-3.202248664613D-02	3.25	-1.143435093499D-01
9	1.0	3.599082062566D-02	1.799541031283D-02	3.50	-1.940179167487D-01
10	2.0	-1.944967535195D-02	-9.724837675975D-03	3.75	-1.690285776641D-01
11	3.0	1.037619603923D-02	5.188098019613D-03	4.25	2.938340971588D-01
12	4.0	-5.443881182172D-03	-2.721940591086D-03	4.50	6.315621934785D-01
13	5.0	2.713751633396D-03	1.356875816698D-03	4.75	8.996214439480D-01
14	6.0	-1.167496424366D-03	-5.837482121829D-04	5.25	8.948004490785D-01
15	7.0	3.590505668859D-04	1.795252834430D-04	5.50	6.244945888152D-01

16	8,0	-5,494145213658D-05	-2,747072606829D-05	5,75	2,885350158115D-01
				6,25	-1,627169164872D-01
	0	1,403640835613D+02	7,182844382752D+01	6,50	-1,840517201659D-01
	1	5,969262414324D+01	2,914629420723D+01	6,75	-1,063761462686D-01
	2	8,425210707234D+00	4,311389513431D+00	7,25	7,374462108786D-02
	3	3,946026868718D-01	1,926754735924D-01	7,50	8,787411842000D-02
				7,75	5,261286338134D-02

I= 6 H= -2.0

0		0,	-2,994990522838D+01	-1,00	1,123394218015D+00
1		-3,502888108424D-01	5,779687669183D+00	-,75	6,677707935040D-01
2		5,610690966844D-01	-1,137411533495D-01	-,50	3,419162184033D-01
3		-2,120363104880D-01	-9,732499809855D-02	-,25	1,259520886048D-01
				,25	-5,581837485754D-02
0	-8,0	1,256024645995D-03	6,280123229976D-04	,50	-6,137185736856D-02
1	-7,0	-8,179435836182D-03	-4,089717918091D-03	,75	-3,640040068239D-02
2	-6,0	2,626683178344D-02	1,313341589172D-02	1,25	3,066472408966D-02
3	-5,0	-5,806252688466D-02	-2,903126344233D-02	1,50	4,274616013677D-02
4	-4,0	1,006490505808D-01	5,032452529039D-02	1,75	3,092893910498D-02
5	-3,0	-1,411062672843D-01	-7,055313364216D-02	2,25	-3,529426201508D-02
6	-2,0	1,586716253687D-01	7,933581268437D-02	2,50	-5,503903884080D-02
7	-1,0	-1,425077703440D-01	-7,125388517202D-02	2,75	-4,361670563810D-02
8	0,0	1,039855386949D-01	5,199276934746D-02	3,25	5,688005346359D-02
9	1,0	-6,441626170816D-02	-3,220813085408D-02	3,50	9,321143206408D-02
10	2,0	3,614517988020D-02	1,807258994010D-02	3,75	7,712444784910D-02
11	3,0	-1,944967535195D-02	-9,724837675975D-03	4,25	-1,092129661341D-01
12	4,0	1,022432210231D-02	5,112161051155D-03	4,50	-1,878345932957D-01
13	5,0	-5,099077288662D-03	-2,549538644331D-03	4,75	-1,652967151993D-01
14	6,0	2,193930069779D-03	1,096965034889D-03	5,25	2,911131496950D-01
15	7,0	-6,747365003023D-04	-3,373682501512D-04	5,50	6,282724068151D-01
16	8,0	1,032480721005D-04	5,162403605025D-05	5,75	8,976312297898D-01
				6,25	8,962563790806D-01
	0	-7,545647926880D+01	-4,082223671574D+01	6,50	6,262573309616D-01
	1	-3,208415487759D+01	-1,472657041933D+01	6,75	2,896029371015D-01
	2	-4,527802355028D+00	-2,449541107715D+00	7,25	-1,635009719017D-01
	3	-2,120363104880D-01	-9,732499809855D-02	7,50	-1,850034172107D-01
				7,75	-1,069546461740D-01

I= 7 H= -1.0

0		0,	5,604590997019D+01	-1,00	-6,016941263809D-01
1		1,875898293858D-01	-1,104961411247D+01	-,75	-3,576518363084D-01
2		-3,005099660396D-01	5,875619919706D-01	-,50	-1,831216975722D-01
3		1,135943309555D-01	4,052946256440D-02	-,25	-6,745424164509D-02
				,25	2,989045474059D-02
0	-8,0	-6,741943017231D-04	-3,370971508616D-04	,50	3,286144740945D-02
1	-7,0	4,402428363096D-03	2,201214181548D-03	,75	1,948813043813D-02
2	-6,0	-1,427462220814D-02	-7,137311104072D-03	1,25	-1,641014780331D-02
3	-5,0	3,280320991989D-02	1,640160495995D-02	1,50	-2,286660026599D-02
4	-4,0	-6,265994476009D-02	-3,132997238004D-02	1,75	-1,653572645897D-02
5	-3,0	1,032475877284D-01	5,162379386418D-02	2,25	1,883033835823D-02
6	-2,0	-1,425077703440D-01	-7,125388517202D-02	2,50	2,931125678597D-02
7	-1,0	1,594112192009D-01	7,970560960047D-02	2,75	2,316707097509D-02
8	0,0	-1,428801350146D-01	-7,144006750732D-02	3,25	-2,992974696060D-02
9	1,0	1,041404614315D-01	5,207023071577D-02	3,50	-4,865447201326D-02
10	2,0	-6,441626170816D-02	-3,220813085408D-02	3,75	-3,979750876612D-02

11	3,0	3,599082062566D-02	1,799541031283D-02	4,25	5,412456362869D-02
12	4,0	-1,908333060483D-02	-9,541665302415D-03	4,50	8,988986301885D-02
13	5,0	9,536011918245D-03	4,768005959123D-03	4,75	7,511921821691D-02
14	6,0	-4,104853787584D-03	-2,052426893792D-03	5,25	-1,077496482750D-01
15	7,0	1,262579534240D-03	6,312897671201D-04	5,50	-1,860641395062D-01
16	8,0	-1,932059927757D-04	-9,660299638783D-05	5,75	-1,642246441875D-01
				6,25	2,903265002077D-01
	0	4,042837825778D+01	2,600404938953D+01	6,50	6,273177217416D-01
	1	1,718954191621D+01	6,133034571426D+00	6,75	8,970509851348D-01
	2	2,425753976893D+00	1,560269093516D+00	7,25	8,966868194026D-01
	3	1,135943309555D-01	4,052946256440D-02	7,50	6,267840030513D-01
				7,75	2,899265388265D-01

I= 8 H= 0.0

0		0.	-1,047494929360D+02	-1,00	3,220778277471D-01
1		-1,004108140517D-01	2,077678075451D+01	-,75	1,914445918064D-01
2		1,608583812614D-01	-1,298548797157D+00	-,50	9,802108139544D-02
3		-6,080863243409D-02	1.714135798076D-23	-,25	3,610648722354D-02
				,25	-1,599916752819D-02
0	-8,0	3,610652243775D-04	1,805326121888D-04	,50	-1,758906994269D-02
1	-7,0	-2,359186189800D-03	-1,179593094900D-03	,75	-1,043071649785D-02
2	-6,0	7,666273103748D-03	3,833136551874D-03	1,25	8,782392095375D-03
3	-5,0	-1,777376458158D-02	-8,886882290792D-03	1,50	1,223670901058D-02
4	-4,0	3,526206072562D-02	1,763103036281D-02	1,75	8,847661059056D-03
5	-3,0	-6,404497329225D-02	-3,202248664613D-02	2,25	-1,007060991610D-02
6	-2,0	1,039855386949D-01	5,199276934746D-02	2,50	-1,566932519610D-02
7	-1,0	-1,428801350146D-01	-7,144006750732D-02	2,75	-1,237721622277D-02
8	0,0	1,595662426592D-01	7,978312132961D-02	3,25	1,595550629651D-02
9	1,0	-1,428801350146D-01	-7,144006750732D-02	3,50	2,588910030569D-02
10	2,0	1,039855386949D-01	5,199276934746D-02	3,75	2,111940928205D-02
11	3,0	-6,404497329225D-02	-3,202248664613D-02	4,25	-2,845109470119D-02
12	4,0	3,526206072562D-02	1,763103036281D-02	4,50	-4,687082308867D-02
13	5,0	-1,777376458158D-02	-8,886882290792D-03	4,75	-3,871971016952D-02
14	6,0	7,666273103748D-03	3,833136551874D-03	5,25	5,333564855587D-02
15	7,0	-2,359186189800D-03	-1,179593094900D-03	5,50	8,893309349701D-02
16	8,0	3.610652243775D-04	1,805326121888D-04	5,75	7,453798890140D-02
				6,25	-1,073187237933D-01
	0	-2,164236991794D+01	-2,164236991794D+01	6,50	-1,855369623597D-01
	1	-9,201934141215D+00	2,350462493393D-21	6,75	-1,639007709596D-01
	2	-1,298548797157D+00	-1,298548797157D+00	7,25	2,900780570876D-01
	3	-6,080863243409D-02	1.714135798076D-23	7,50	6,270060608992D-01
				7,75	8,968532206180D-01

I= 9 H= 1.0

0		0.	1,956766327789D+02	-1,00	-1,723242653977D-01
1		5,372332986850D-02	-3,887899688005D+01	-,75	-1,024302321922D-01
2		-8,606552970246D-02	2,532976195062D+00	-,50	-5,244497308821D-02
3		3,253540582673D-02	-4,052946256440D-02	-,25	-1,931829378957D-02
				,25	8,560090784406D-03
0	-8,0	-1,932059927757D-04	-9,660299638783D-05	,50	9,410698815158D-03
1	-7,0	1,262579534240D-03	6,312897671201D-04	,75	5,580721390042D-03
2	-6,0	-4,104853787584D-03	-2,052426893792D-03	1,25	-4,698714222816D-03
3	-5,0	9,536011918245D-03	4,768005959123D-03	1,50	-6,546693977131D-03
4	-4,0	-1,908333060483D-02	-9,541665302415D-03	1,75	-4,733396667118D-03
5	-3,0	3,599082062566D-02	1,799541031283D-02	2,25	5,387070363693D-03

```
 6  -2.0  -6.441626170816D-02  -3.220813085408D-02    2.50   8.381186339219D-03
 7  -1.0   1.041404614315D-01   5.207023071577D-02    2.75   6.619384812578D-03
 8   0.0  -1.428801350146D-01  -7.144006750732D-02    3.25  -8.528807704593D-03
 9   1.0   1.594112192009D-01   7.970560960047D-02    3.50  -1.383271588053D-02
10   2.0  -1.425077703440D-01  -7.125388517202D-02    3.75  -1.127723717136D-02
11   3.0   1.032475877284D-01   5.162379386418D-02    4.25   1.515877173882D-02
12   4.0  -6.265994476009D-02  -3.132997238004D-02    4.50   2.492573800519D-02
13   5.0   3.280320991989D-02   1.640160495995D-02    4.75   2.053539889639D-02
14   6.0  -1.427462220814D-02  -7.137311104072D-03    5.25  -2.801916645354D-02
15   7.0   4.402428363096D-03   2.201214181548D-03    5.50  -4.634278375478D-02
16   8.0  -6.741943017231D-04  -3.370971508616D-04    5.75  -3.839546398121D-02
                                                       6.25   5.308706058525D-02
        0  1.157972052128D+01   2.600404938953D+01    6.50   8.862130218822D-02
        1  4.923472773362D+00  -6.133034571426D+00    6.75   7.434016503500D-02
        2  6.947842101392D-01   1.560269093516D+00    7.25  -1.071522982975D-01
        3  3.253540582673D-02  -4.052946256440D-02    7.50  -1.853148841876D-01
                                                       7.75  -1.637492460050D-01


I=10     H=  2.0
0            0.                -3.652358299906D+02   -1.00   9.208661323325D-02
1           -2.870861636176D-02   7.260562777769D+01    -.75   5.473663506437D-02
2            4.599168258057D-02  -4.785341062080D+00    -.50   2.802551811239D-02
3           -1.738631429092D-02   9.732499809855D-02    -.25   1.032329541252D-02
                                                         .25  -4.574328788186D-03
0  -8.0   1.032480721005D-04   5.162403605025D-05     .50  -5.028870196536D-03
1  -7.0  -6.747365003023D-04  -3.373682501512D-04     .75  -2.982210207265D-03
2  -6.0   2.193930069779D-03   1.096965034889D-03    1.25   2.510872470758D-03
3  -5.0  -5.099077288662D-03  -2.549538644331D-03    1.50   3.498369259804D-03
4  -4.0   1.022432210231D-02   5.112161051155D-03    1.75   2.529376824790D-03
5  -3.0  -1.944967535195D-02  -9.724837675975D-03    2.25  -2.878607105379D-03
6  -2.0   3.614517988020D-02   1.807258994010D-02    2.50  -4.478429034207D-03
7  -1.0  -6.441626170816D-02  -3.220813085408D-02    2.75  -3.536909590154D-03
8   0.0   1.039855386949D-01   5.199276934746D-02    3.25   4.556642351624D-03
9   1.0  -1.425077703440D-01  -7.125388517202D-02    3.50   7.389603662364D-03
10  2.0   1.586716253687D-01   7.933581268437D-02    3.75   6.023578313311D-03
11  3.0  -1.411062672843D-01  -7.055313364216D-02    4.25  -8.092751462928D-03
12  4.0   1.006490505808D-01   5.032452529039D-02    4.50  -1.330122010539D-02
13  5.0  -5.806252688466D-02  -2.903126344233D-02    4.75  -1.095154887020D-02
14  6.0   2.626683178344D-02   1.313341589172D-02    5.25   1.490968725804D-02
15  7.0  -8.179435836182D-03  -4.089717918091D-03    5.50   2.461354112530D-02
16  8.0   1.256024645995D-03   6.280123229976D-04    5.75   2.033741473787D-02
                                                       6.25  -2.785270533792D-02
        0 -6.187994162687D+00  -4.082223671574D+01    6.50  -4.612069903182D-02
        1 -2.631014038929D+00   1.472657041933D+01    6.75  -3.824395599442D-02
        2 -3.712798604015D-01  -2.449541107715D+00    7.25   5.293559004553D-02
        3 -1.738631429092D-02   9.732499809855D-02    7.50   8.839933117039D-02
                                                       7.75   7.417383975790D-02


I=11     H=  3.0
0            0.                 6.795775688243D+02   -1.00  -4.900180717210D-02
1            1.527663462507D-02  -1.351222173519D+02    -.75  -2.912685963302D-02
2           -2.447343285998D-02   8.935600879648D+00    -.50  -1.491314298841D-02
3            9.251739687048D-03  -1.926754735924D-01    -.25  -5.493306642627D-03
                                                         .25   2.434124181769D-03
0  -8.0  -5.494145213658D-05  -2.747072606829D-05     .50   2.675997328326D-03
```

CARDINAL SPLINES N=4

1	-7.0	3.590505668859D-04	1.795252834430D-04	.75	1.586913866963D-03
2	-6.0	-1.167496424366D-03	-5.837482121829D-04	1.25	-1.336100830454D-03
3	-5.0	2.713751633396D-03	1.356875816698D-03	1.50	-1.861571688378D-03
4	-4.0	-5.443881182172D-03	-2.721940591086D-03	1.75	-1.345943939164D-03
5	-3.0	1.037619603923D-02	5.188098019613D-03	2.25	1.531769207347D-03
6	-2.0	-1.944967535195D-02	-9.724837675975D-03	2.50	2.383057119669D-03
7	-1.0	3.599082062566D-02	1.799541031283D-02	2.75	1.882042977352D-03
8	0.0	-6.404497329225D-02	-3.202248664613D-02	3.25	-2.424594591465D-03
9	1.0	1.032475877284D-01	5.162379386418D-02	3.50	-3.931927478972D-03
10	2.0	-1.411062672843D-01	-7.055313364216D-02	3.75	-3.204975711114D-03
11	3.0	1.560480447459D-01	7.802402237295D-02	4.25	4.305425004999D-03
12	4.0	-1.362585942711D-01	-6.812929713557D-02	4.50	7.075677256720D-03
13	5.0	9.208076905053D-02	4.604038452526D-02	4.75	5.824918734232D-03
14	6.0	-4.588383370592D-02	-2.294191685296D-02	5.25	-7.926151901521D-03
15	7.0	1.491105864171D-02	7.455529320853D-03	5.50	-1.307912664260D-02
16	8.0	-2.317616067408D-03	-1.158808033704D-03	5.75	-1.080012613401D-02
				6.25	1.475842973053D-02
	0	3.292804093730D+00	7.182844382752D+01	6.50	2.439194930643D-02
	1	1.400035728779D+00	-2.914629420723D+01	6.75	2.017141446944D-02
	2	1.975683196292D-01	4.311389513431D+00	7.25	-2.765551064631D-02
	3	9.251739687048D-03	-1.926754735924D-01	7.50	-4.580999288544D-02
				7.75	-3.799631403779D-02

I=12 H= 4.0

0		0.	-1.247092738479D+03	-1.00	2.569380152150D-02
1		-8.010210360548D-03	2.480479883989D+02	-.75	1.527249263029D-02
2		1.283249663185D-02	-1.642346835881D+01	-.50	7.819616154374D-03
3		-4.851094529110D-03	3.599551416607D-01	-.25	2.880381981645D-03
				.25	-1.276318144348D-03
0	-8.0	2.880825781311D-05	1.440412890655D-05	.50	-1.403142773937D-03
1	-7.0	-1.882665734813D-04	-9.413328674066D-05	.75	-8.320884812441D-04
2	-6.0	6.121752493576D-04	3.060876246788D-04	1.25	7.005760214089D-04
3	-5.0	-1.422986528321D-03	-7.114932641607D-04	1.50	9.761031049562D-04
4	-4.0	2.854855694609D-03	1.427427847305D-03	1.75	7.057367674380D-04
5	-3.0	-5.443881182172D-03	-2.721940591086D-03	2.25	-8.031719552006D-04
6	-2.0	1.022432210231D-02	5.112161051155D-03	2.50	-1.249537089257D-03
7	-1.0	-1.908333060483D-02	-9.541665302415D-03	2.75	-9.868326441452D-04
8	0.0	3.526206072562D-02	1.763103036281D-02	3.25	1.271307155998D-03
9	1.0	-6.265994476009D-02	-3.132997238004D-02	3.50	2.061648407955D-03
10	2.0	1.006490505808D-01	5.032452529039D-02	3.75	1.680469353616D-03
11	3.0	-1.362585942711D-01	-6.812929713557D-02	4.25	-2.257409428121D-03
12	4.0	1.470996374640D-01	7.354981873202D-02	4.50	-3.709816184330D-03
13	5.0	-1.204466691267D-01	-6.022333456336D-02	4.75	-3.053936313924D-03
14	6.0	6.960790338076D-02	3.480395169038D-02	5.25	4.155104031086D-03
15	7.0	-2.492993690752D-02	-1.246496845376D-02	5.50	6.855747429240D-03
16	8.0	4.094796498982D-03	2.047398249491D-03	5.75	5.660339743017D-03
				6.25	-7.730949509646D-03
	0	-1.726562297351D+00	-1.295137737211D+02	6.50	-1.277172610621D-02
	1	-7.341004138402D-01	5.438388185678D+01	6.75	-1.055522204940D-02
	2	-1.035937720668D-01	-7.784544958955D+00	7.25	1.443945428339D-02
	3	-4.851094529110D-03	3.599551416607D-01	7.50	2.387293959428D-02
				7.75	1.974732680762D-02

I=13 H= 5.0

0		0.	2.147033007667D+03	-1.00	-1.280652675418D-02

CARDINAL SPLINES N=4

1		3.992518239470D-03	-4,276732449851D+02	-.75	-7,612247826420D-03
2		-6,396083944340D-03	2,836320672994D+01	-.50	-3,897520677116D-03
3		2.417924570374D-03	-6,252629052361D-01	-.25	-1,435664877801D-03
				.25	6,361535083627D-04
0	-8.0	-1.435886550433D-05	-7,179432752163D-06	.50	6,993665263099D-04
1	-7,0	9,383751997804D-05	4,691875998902D-05	.75	4,147367118614D-04
2	-6,0	-3,051263186869D-04	-1.525631593434D-04	1.25	-3,491871160146D-04
3	-5,0	7.092626704396D-04	3.546313352198D-04	1.50	-4,865176634992D-04
4	-4,0	-1,422986528321D-03	-7.114932641607D-04	1.75	-3,517593305369D-04
5	-3,0	2,713751633396D-03	1.356875816698D-03	2.25	4,003236854036D-04
6	-2,0	-5,099077288662D-03	-2,549538644331D-03	2,50	6,228045644930D-04
7	-1,0	9,536011918245D-03	4,768005959123D-03	2.75	4,918650620521D-04
8	0,0	-1,777376458158D-02	-8,886882290792D-03	3,25	-6,336542592421D-04
9	1.0	3,280320991989D-02	1,640160495995D-02	3.50	-1,027580728679D-03
10	2,0	-5,806252688466D-02	-2,903126344233D-02	3,75	-8,375894470140D-04
11	3,0	9,208076905053D-02	4.604038452526D-02	4,25	1,125144438068D-03
12	4,0	-1,204466691267D-01	-6,022333456336D-02	4,50	1,849047423992D-03
13	5,0	1.191620326432D-01	5,958101632158D-02	4,75	1,522132121397D-03
14	6,0	-8,074091950349D-02	-4,037045975175D-02	5,25	-2,070916716166D-03
15	7,0	3,258614964235D-02	1.629307482118D-02	5,50	-3,416847080153D-03
16	8,0	-5.819595900358D-03	-2.909797950179D-03	5,75	-2,820972616248D-03
				6,25	3,852461627435D-03
	0	8,605681535097D-01	2,207576710207D+02	6,50	6,363724915320D-03
	1	3,658966926419D-01	-9,391241511149D+01	6,75	5,258555400611D-03
	2	5,163410574465D-02	1,335689700427D+01	7,25	-7,189981085825D-03
	3	2,417924570374D-03	-6,252629052361D-01	7,50	-1,188207595843D-02
				7,75	-9,822484139009D-03

I=14 H= 6.0

0		0.	-2.596488593026D+03	-1,00	5,509385372706D-03
1		-1,717586813900D-03	5,223124988443D+02	-.75	3,274799451777D-03
2		2,751604084310D-03	-3,496310715106D+01	-.50	1,676718737339D-03
3		-1.040194474496D-03	7,783818819659D-01	-.25	6,176249974084D-04
				.25	-2,736741098430D-04
0	-8,0	6.177204085787D-06	3,088602042894D-06	.50	-3,008684357775D-04
1	-7,0	-4,036903645421D-05	-2,018451822710D-05	.75	-1,784202997181D-04
2	-6,0	1.312658248194D-04	6,563291240968D-05	1.25	1,502207718484D-04
3	-5,0	-3,051263186869D-04	-1,525631593434D-04	1.50	2,093005572613D-04
4	-4,0	6,121752493576D-04	3.060876246788D-04	1.75	1,513273371195D-04
5	-3,0	-1,167496424366D-03	-5,837482121829D-04	2,25	-1,722197757004D-04
6	-2.0	2.193930069779D-03	1.096965034889D-03	2,50	-2,679313257035D-04
7	-1.0	-4,104853787584D-03	-2,052426893792D-03	2,75	-2,116009640280D-04
8	0,0	7,666273103748D-03	3,833136551874D-03	3,25	2,725987655639D-04
9	1,0	-1,427462220814D-02	-7,137311104072D-03	3,50	4,420662463704D-04
10	2,0	2,626683178344D-02	1,313341589172D-02	3,75	3,603316639096D-04
11	3,0	-4,588383370592D-02	-2,294191685296D-02	4,25	-4,840373632828D-04
12	4,0	6,960790338076D-02	3,480395169038D-02	4,50	-7,954595946575D-04
13	5,0	-8,074091950349D-02	-4,037045975175D-02	4,75	-6,548195809797D-04
14	6,0	6,273138523144D-02	3,136569261572D-02	5,25	8,909005304991D-04
15	7,0	-2,812491507638D-02	-1,406245753819D-02	5,50	1,469906762462D-03
16	8,0	5,426194213596D-03	2,713097106798D-03	5,75	1,213555818620D-03
				6,25	-1,657247245194D-03
	0	-3,702176040572D-01	-2,570959363728D+02	6,50	-2,737475011848D-03
	1	-1,574092605681D-01	1,123521057649D+02	6,75	-2,261989290443D-03
	2	-2,221306330359D-02	-1,628194198388D+01	7,25	3,092429815696D-03
	3	-1,040194474496D-03	7,783818819659D-01	7,50	5,109987122439D-03

				7.75	4.223619153968D-03

I=15	H= 7.0				
0		0.	1.694113692905D+03	-1.00	-1.694337062001D-03
1		5.282206263866D-04	-3.458405261640D+02	-.75	-1.007120341975D-03
2		-8.462186731393D-04	2.350050125552D+01	-.50	-5.156522017875D-04
3		3.198977624751D-04	-5.313092127201D-01	-.25	-1.899422262065D-04
				.25	8.416477611467D-05
0	-8.0	-1.899715722397D-06	-9.498578611983D-07	.50	9.252802368878D-05
1	-7.0	1.241495245001D-05	6.207476225007D-06	.75	5.487075325499D-05
2	-6.0	-4.036903645421D-05	-2.018451822710D-05	1.25	-4.619836917879D-05
3	-5.0	9.383751997804D-05	4.691875998902D-05	1.50	-6.436755891470D-05
4	-4.0	-1.882665734813D-04	-9.413328674066D-05	1.75	-4.653867800472D-05
5	-3.0	3.590505668859D-04	1.795252834430D-04	2.25	5.296386428752D-05
6	-2.0	-6.747365003023D-04	-3.373682501512D-04	2.50	8.239865657776D-05
7	-1.0	1.262579534240D-03	6.312897671201D-04	2.75	6.507501401506D-05
8	0.0	-2.359186189800D-03	-1.179593094900D-03	3.25	-8.383405597582D-05
9	1.0	4.402428363096D-03	2.201214181548D-03	3.50	-1.359514735587D-04
10	2.0	-8.179435836182D-03	-4.089717918091D-03	3.75	-1.108151010791D-04
11	3.0	1.491105864171D-02	7.455529320853D-03	4.25	1.488590530113D-04
12	4.0	-2.492993690752D-02	-1.246496845376D-02	4.50	2.446326146226D-04
13	5.0	3.258614964235D-02	1.629307482118D-02	4.75	2.013806318521D-04
14	6.0	-2.812491507638D-02	-1.406245753819D-02	5.25	-2.739836091148D-04
15	7.0	1.367218339743D-02	6.836091698714D-03	5.50	-4.520480198966D-04
16	8.0	-2.800956782291D-03	-1.400478391146D-03	5.75	-3.732103722818D-04
				6.25	5.096573537531D-04
	0	1.138554243174D-01	1.593912470337D+02	6.50	8.418575519397D-04
	1	4.840909225137D-02	-7.184387491796D+01	6.75	6.956252605793D-04
	2	6.831327626263D-03	1.074908015023D+01	7.25	-9.509807481813D-04
	3	3.198977624751D-04	-5.313092127201D-01	7.50	-1.571377685999D-03
				7.75	-1.298762054969D-03

I=16	H= 8.0				
0		0.	-4.388509006245D+02	-1.00	2.592646835431D-04
1		-8.082745550854D-05	9.099986284166D+01	-.75	1.541079060317D-04
2		1.294869960758D-04	-6.290152444110D+00	-.50	7.890425576807D-05
3		-4.895023195875D-05	1.449091304909D-01	-.25	2.906464850623D-05
				.25	-1.287875625436D-05
0	-8.0	2.906913914542D-07	1.453456957271D-07	.50	-1.415848670366D-05
1	-7.0	-1.899715722397D-06	-9.498578611983D-07	.75	-8.396232829578D-06
2	-6.0	6.177204085787D-06	3.088602042894D-06	1.25	7.069198809090D-06
3	-5.0	-1.435886550433D-05	-7.179432752163D-06	1.50	9.849418464131D-06
4	-4.0	2.880825781311D-05	1.440412890655D-05	1.75	7.121272292102D-06
5	-3.0	-5.494145213658D-05	-2.747072606829D-05	2.25	-8.104443734033D-06
6	-2.0	1.032480721005D-04	5.162403605025D-05	2.50	-1.260850736720D-05
7	-1.0	-1.932059927757D-04	-9.660299638783D-05	2.75	-9.957671914510D-06
8	0.0	3.610652243775D-04	1.805326121888D-04	3.25	1.282814945532D-05
9	1.0	-6.741943017231D-04	-3.370971508616D-04	3.50	2.080307062273D-05
10	2.0	1.256024645995D-03	6.280123229976D-04	3.75	1.695674381384D-05
11	3.0	-2.317616067408D-03	-1.158808033704D-03	4.25	-2.277816422655D-05
12	4.0	4.094796498982D-03	2.047398249491D-03	4.50	-3.743327176988D-05
13	5.0	-5.819595900358D-03	-2.909797950179D-03	4.75	-3.081492216943D-05
14	6.0	5.426194213596D-03	2.713097106798D-03	5.25	4.192448784003D-05
15	7.0	-2.800956782291D-03	-1.400478391146D-03	5.50	6.917156309299D-05
16	8.0	6.001642695770D-04	3.000821347885D-04	5.75	5.710793891824D-05

CARDINAL SPLINES N=4

			x	value
			6.25	-7.798664462884D-05
0	-1.742197065810D-02	-3.922827950292D+01	6.50	-1.288189724115D-04
1	-7.407480054375D-03	1.817997679016D+01	6.75	-1.064426126690D-04
2	-1.045318570934D-03	-2.812333312327D+00	7.25	1.455151573372D-04
3	-4.895023195875D-05	1.449091304909D-01	7.50	2.404439955881D-04
			7.75	1.987277051726D-04

N= 4 M=17 P= 8.5 Q= 7

I= 0 H= -8.5

				x	value
0		1.000000000000D+00	5.547176528071D-01	-1.00	4.666983726349D+00
1		-2.043924727322D+00	-1.032164305437D+00	-.75	3.405151654912D+00
2		1.333191781019D+00	6.672293881481D-01	-.50	2.391493711167D+00
3		-2.898672180071D-01	-1.449467100539D-01	-.25	1.598834843426D+00
				.25	5.678141658329D-01
0	-8.5	6.001643099189D-04	3.000821549595D-04	.50	2.751068681265D-01
1	-7.5	-2.800957045933D-03	-1.400478522966D-03	.75	9.476921099976D-02
2	-6.5	5.426195070864D-03	2.713097535432D-03	1.25	-3.507902599696D-02
3	-5.5	-5.819597893071D-03	-2.909798946536D-03	1.50	-3.427495705156D-02
4	-4.5	4.094800496974D-03	2.047400248487D-03	1.75	-1.768214312115D-02
5	-3.5	-2.317623692158D-03	-1.158811846079D-03	2.25	1.053382505727D-02
6	-2.5	1.256038974802D-03	6.280194874011D-04	2.50	1.195817114234D-02
7	-1.5	-6.742211156218D-04	-3.371105578109D-04	2.75	6.908482785317D-03
8	-.5	3.611153401416D-04	1.805576700708D-04	3.25	-4.774220551302D-03
9	.5	-1.932996172300D-04	-9.664980861502D-05	3.50	-5.681574475383D-03
10	1.5	1.034228738247D-04	5.171143691234D-05	3.75	-3.398301940870D-03
11	2.5	-5.526709868390D-05	-2.763354934195D-05	4.25	2.450955053969D-03
12	3.5	2.940913578623D-05	1.470456789311D-05	4.50	2.953652493710D-03
13	4.5	-1.542050265935D-05	-7.710251329673D-06	4.75	1.782443933729D-03
14	5.5	7.686020291883D-06	3.843010145941D-06	5.25	-1.299143833773D-03
15	6.5	-3.306536470287D-06	-1.653268235144D-06	5.50	-1.570306538656D-03
16	7.5	1.016880633586D-06	5.084403167929D-07	5.75	-9.496250406715D-04
17	8.5	-1.556014098850D-07	-7.780070494250D-08	6.25	6.938400317037D-04
				6.50	8.392446880929D-04
	0	-9.806495926221D+01	-4.902675396154D+01	6.75	5.077703339814D-04
	1	-4.220838395304D+01	-2.110646411109D+01	7.25	-3.712124461194D-04
	2	-6.058422278163D+00	-3.028911718225D+00	7.50	-4.490796825309D-04
	3	-2.898672180071D-01	-1.449467100539D-01	7.75	-2.717400675469D-04
				8.25	1.986899915870D-04
				8.50	2.403816731096D-04

I= 1 H= -7.5

				x	value
0		0.	-3.575888004714D-01	-1.00	-9.041182759966D+00
1		3.959053490908D+00	2.046198101683D+00	-.75	-5.678512124083D+00
2		-4.019190901460D+00	-2.013735461293D+00	-.50	-3.117191766769D+00
3		1.062938367598D+00	5.315548013030D-01	-.25	-1.257571216062D+00
				.25	7.551721824226D-01
0	-8.5	-2.800957045933D-03	-1.400478522966D-03	.50	1.107574433562D+00

CARDINAL SPLINES N=4

1	-7.5	1.367218512037D-02	6.836092560184D-03	.75	1.156548477307D+00
2	-6.5	-2.812492067877D-02	-1.406246033939D-02	1.25	7.315274098220D-01
3	-5.5	3.258616266506D-02	1.629308133253D-02	1.50	4.350675356391D-01
4	-4.5	-2.492996303505D-02	-1.246498151752D-02	1.75	1.772914857221D-01
5	-3.5	1.491110847069D-02	7.455554235347D-03	2.25	-8.424021671750D-02
6	-2.5	-8.179529477276D-03	-4.089764738638D-03	2.50	-9.008354810640D-02
7	-1.5	4.402603596304D-03	2.201301798152D-03	2.75	-4.992392059370D-02
8	-.5	-2.359513704506D-03	-1.179756852253D-03	3.25	3.281503562915D-02
9	.5	1.263191385345D-03	6.315956926723D-04	3.50	3.847398830437D-02
10	1.5	-6.758788581265D-04	-3.379394290633D-04	3.75	2.276830390022D-02
11	2.5	3.611787202723D-04	1.805893601361D-04	4.25	-1.621278959428D-02
12	3.5	-1.921934091987D-04	-9.609670459936D-05	4.50	-1.946622537758D-02
13	4.5	1.007754922216D-04	5.038774611078D-05	4.75	-1.171694106182D-02
14	5.5	-5.022939685038D-05	-2.511469842519D-05	5.25	8.514041298767D-03
15	6.5	2.160875586423D-05	1.080437793211D-05	5.50	1.028225269346D-02
16	7.5	-6.645481054393D-06	-3.322740527197D-06	5.75	6.214332786980D-03
17	8.5	1.016880633586D-06	5.084403167929D-07	6.25	-4.537289763706D-03
				6.50	-5.487057409786D-03
	0	3.960424370433D+02	1.979838003356D+02	6.75	-3.319388833205D-03
	1	1.660246993429D+02	8.302719844212D+01	7.25	2.426293286305D-03
	2	2.308573747229D+01	1.154091197193D+01	7.50	2.935110119471D-03
	3	1.062938367598D+00	5.315548013030D-01	7.75	1.775991569183D-03
				8.25	-1.298515590526D-03
				8.50	-1.570970398421D-03

I= 2 H= -6.5

0		0.	1.162752416935D+00	-1.00	9.696675153403D+00
1		-3.293246571473D+00	-1.863414976347D+00	-.75	5.852797303989D+00
2		4.845624479166D+00	2.436274090967D+00	-.50	3.052754918374D+00
3		-1.557804102764D+00	-7.791804492501D-01	-.25	1.150503861922D+00
				.25	-5.448004708375D-01
0	-8.5	5.426195070864D-03	2.713097535432D-03	.50	-6.299002866416D-01
1	-7.5	-2.812492067877D-02	-1.406246033939D-02	.75	-4.007454553193D-01
2	-6.5	6.273140344843D-02	3.136570172421D-02	1.25	4.380162939106D-01
3	-5.5	-8.074096184874D-02	-4.037048092437D-02	1.50	7.976882779871D-01
4	-4.5	6.960798833825D-02	3.480399416912D-02	1.75	9.966809863038D-01
5	-3.5	-4.588399573218D-02	-2.294199786609D-02	2.25	8.267656512608D-01
6	-2.5	2.626713627119D-02	1.313356813559D-02	2.50	5.431835525331D-01
7	-1.5	-1.427519200461D-02	-7.137596002306D-03	2.75	2.397523460124D-01
8	-.5	7.667338065832D-03	3.833669032916D-03	3.25	-1.274048234859D-01
9	.5	-4.106843311156D-03	-2.053421655578D-03	3.50	-1.414517114020D-01
10	1.5	2.197644613731D-03	1.098822306866D-03	3.75	-8.064859774664D-02
11	2.5	-1.174416427121D-03	-5.872082135606D-04	4.25	5.497457926667D-02
12	3.5	6.249439314266D-04	3.124719657133D-04	4.50	6.517851848216D-02
13	4.5	-3.276861526491D-04	-1.638430763245D-04	4.75	3.888372022418D-02
14	5.5	1.633282323265D-04	8.166411616324D-05	5.25	-2.795860677998D-02
15	6.5	-7.026403620708D-05	-3.513201810354D-05	5.50	-3.366368736342D-02
16	7.5	2.160875586423D-05	1.080437793211D-05	5.75	-2.030270233212D-02
17	8.5	-3.306536470287D-06	-1.653268235144D-06	6.25	1.478721369163D-02
				6.50	1.787007983627D-02
	0	-6.345826718475D+02	-3.171696652053D+02	6.75	1.080522885091D-02
	1	-2.585716696997D+02	-1.293341178049D+02	7.25	-7.893559973982D-03
	2	-3.487838014131D+01	-1.743282736491D+01	7.50	-9.547385619017D-03
	3	-1.557804102764D+00	-7.791804492501D-01	7.75	-5.776336920249D-03
				8.25	4.222817738695D-03
				8.50	5.108662767166D-03

CARDINAL SPLINES N=4

```
I= 3     H= -5.5
 0            0.                 -2,702809686925D+00    -1,00 -6,822800849634D+00
 1            2,161366153163D+00  1,584613812788D+00     -,75 -4,066886371709D+00
 2           -3,408490625870D+00 -1,735537290318D+00     -,50 -2,089423741874D+00
 3            1,252944070600D+00  6,271191691563D-01     -,25 -7,729494535108D-01
                                                          ,25  3,468877700769D-01
 0    -8,5   -5,819597893071D-03 -2,909798946536D-03      ,50  3,851329633304D-01
 1    -7,5    3,258616266506D-02  1,629308133253D-02      ,75  2,315575950588D-01
 2    -6,5   -8,074096184874D-02 -4,037048092437D-02     1,25 -2,046505398879D-01
 3    -5,5    1,191621310744D-01  5,958106553719D-02     1,50 -2,975471466677D-01
 4    -4,5   -1,204468666098D-01 -6,022343330490D-02     1,75 -2,292874462160D-01
 5    -3,5    9,208114567941D-02  4,604057283971D-02     2,25  3,358732014140D-01
 6    -2,5   -5,806323466424D-02 -2,903161733212D-02     2,50  6,817337570875D-01
 7    -1,5    3,280453440769D-02  1,640226720385D-02     2,75  9,296916572127D-01
 8     -,5   -1,777624007830D-02 -8,888120039151D-03     3,25  8,730597373077D-01
 9      ,5    9,540636551510D-03  4,770318275755D-03     3,50  5,982759674200D-01
10     1,5   -5,107711719417D-03 -2,553855859709D-03     3,75  2,727046027738D-01
11     2,5    2,729837130131D-03  1,364918565066D-03     4,25 -1,511709744169D-01
12     3,5   -1,452667238141D-03 -7,263336190706D-04     4,50 -1,700923792408D-01
13     4,5    7,617028425571D-04  3,808514212786D-04     4,75 -9,793242619488D-02
14     5,5   -3,796551548031D-04 -1,898275774015D-04     5,25  6,757202118130D-02
15     6,5    1,633282323265D-04  8,166411616324D-05     5,50  8,040538007130D-02
16     7,5   -5,022939685038D-05 -2,511469842519D-05     5,75  4,809201591025D-02
17     8,5    7,686020291883D-06  3,843010145941D-06     6,25 -3,468667128333D-02
                                                         6,50 -4,180177382771D-02
          0   5,415724419402D+02  2,705033982544D+02     6,75 -2,522655433989D-02
          1   2,157926528160D+02  1,080085597920D+02     7,25  1,838696989781D-02
          2   2,854158317444D+01  1,425600152317D+01     7,50  2,222503918643D-02
          3   1,252944070600D+00  6,271191691563D-01     7,75  1,344051770866D-02
                                                         8,25 -9,820621254960D-03
                                                         8,50 -1,187899750410D-02


I= 4     H= -4.5
 0            0.                  5,422661533017D+00    -1,00  3,861652207406D+00
 1           -1,208111452869D+00 -1,615094658646D+00     -,75  2,296780600645D+00
 2            1,928778703454D+00  1,027170613399D+00     -,50  1,176845658684D+00
 3           -7,247620510821D-01 -3,636793736600D-01     -,25  4,339009392314D-01
                                                          ,25 -1,928033513728D-01
 0    -8,5    4,094800496974D-03  2,047400248487D-03      ,50 -2,124243163274D-01
 1    -7,5   -2,492996303505D-02 -1,246498151752D-02      ,75 -1,263579693734D-01
 2    -6,5    6,960798833825D-02  3,480399416912D-02     1,25  1,075505357609D-01
 3    -5,5   -1,204468666098D-01 -6,022343330490D-02     1,50  1,512817210965D-01
 4    -4,5    1,471000336754D-01  7,355001683769D-02     1,75  1,109158278178D-01
 5    -3,5   -1,362593499036D-01 -6,812967495182D-02     2,25 -1,327803195988D-01
 6    -2,5    1,006504706026D-01  5,032523530132D-02     2,50 -2,159588801721D-01
 7    -1,5   -6,266260208688D-02 -3,133130104344D-02     2,75 -1,821522074601D-01
 8     -,5    3,526702732762D-02  1,763351366381D-02     3,25  3,032995983087D-01
 9      ,5   -1,909260903071D-02 -9,546304515353D-03     3,50  6,429694295950D-01
10     1,5    1,024164540584D-02  5,120822702918D-03     3,75  9,065056576223D-01
11     2,5   -5,476153598294D-03 -2,738076799147D-03     4,25  8,897821991143D-01
12     3,5    2,914404257757D-03  1,457202128879D-03     4,50  6,184282875209D-01
13     4,5   -1,528197519561D-03 -7,640987597804D-04     4,75  2,848659758312D-01
14     5,5    7,617028425571D-04  3,808514212786D-04     5,25 -1,600349279070D-01
15     6,5   -3,276861526491D-04 -1,638430763245D-04     5,50 -1,808065218442D-01
```

16	7.5	1,007754922216D-04	5,038774611078D-05	5.75	-1,044117382802D-01
17	8,5	-1,542050265935D-05	-7,710251329673D-06	6,25	7,230623970161D-02
				6,50	8,613185212703D-02
	0	-3,160091806456D+02	-1,574371615964D+02	6,75	5,155682235301D-02
	1	-1,255110480662D+02	-6,298069847167D+01	7,25	-3,721991381970D-02
	2	-1,655265359914D+01	-8,246653414932D+00	7,50	-4,486663733004D-02
	3	-7,247620510821D-01	-3,636793736600D-01	7,75	-2,708131188908D-02
				8,25	1,974358929362D-02
				8,50	2,386676327528D-02

I= 5 H= -3.5

0		0.	-1,034180508516D+01	-1,00	-2,092831971883D+00
1		6,529708269657D-01	2,254683868894D+00	-,75	-1,244158830844D+00
2		-1,045257174096D+00	-6,423615945188D-01	-,50	-6,371252033595D-01
3		3,946039708221D-01	1,997781243287D-01	-,25	-2,347369671665D-01
				,25	1,040796789480D-01
0	-8,5	-2,317623692158D-03	-1,158811846079D-03	,50	1,144785098766D-01
1	-7,5	1,491110847069D-02	7,455554235347D-03	,75	6,793514456767D-02
2	-6,5	-4,588399573218D-02	-2,294199786609D-02	1,25	-5,734030115511D-02
3	-5,5	9,208114567941D-02	4,604057283971D-02	1,50	-8,006628025990D-02
4	-4,5	-1,362593499036D-01	-6,812967495182D-02	1,75	-5,807562993176D-02
5	-3,5	1,560494858467D-01	7,802474292335D-02	2,25	6,687250731176D-02
6	-2,5	-1,411089754720D-01	-7,055448773602D-02	2,50	1,051033302853D-01
7	-1,5	1,032526556361D-01	5,162632781804D-02	2,75	8,423759671243D-02
8	-,5	-6,405444532282D-02	-3,202722266141D-02	3,25	-1,143438458285D-01
9	,5	3,600851592993D-02	1,800425796496D-02	3,50	-1,940184624072D-01
10	1,5	-1,948271340505D-02	-9,741356702525D-03	3,75	-1,690290224345D-01
11	2,5	1,043774421887D-02	5,218872109435D-03	4,25	2,938346946233D-01
12	3,5	-5,557448931433D-03	-2,778724465716D-03	4,50	6,315631753421D-01
13	4,5	2,914404257757D-03	1,457202128879D-03	4,75	8,996222522142D-01
14	5,5	-1,452667238141D-03	-7,263336190706D-04	5,25	8,947993494122D-01
15	6,5	6,249439314266D-04	3,124719657133D-04	5,50	6,244927744670D-01
16	7,5	-1,921934091987D-04	-9,609670459936D-05	5,75	2,885335178890D-01
17	8,5	2,940913578623D-05	1,470456789311D-05	6,25	-1,627148709263D-01
				6,50	-1,840483412964D-01
	0	1,723665847819D+02	8,510112319979D+01	6,75	-1,063733543273D-01
	1	6,841400954302D+01	3,463644521031D+01	7,25	7,374080431739D-02
	2	9,017144081867D+00	4,451980575862D+00	7,50	8,786781176285D-02
	3	3,946039708221D-01	1,997781243287D-01	7,75	5,260765095327D-02
				8,25	-3,798918605635D-02
				8,50	-4,579821374905D-02

I= 6 H= -2.5

0		0.	1,943485336840D+01	-1,00	1,123406997731D+00
1		-3,502927950024D-01	-3,798716391077D+00	-,75	6,677783898156D-01
2		5,610754793783D-01	5,055461850825D-01	-,50	3,419201077646D-01
3		-2,120387233507D-01	-1,106726497267D-01	-,25	1,259535212641D-01
				,25	-5,581900967927D-02
0	-8,5	1,256038974802D-03	6,280194874011D-04	,50	-6,137255527096D-02
1	-7,5	-8,179529477276D-03	-4,089764738638D-03	,75	-3,640081455084D-02
2	-6,5	2,626713627119D-02	1,313356813559D-02	1,25	3,066507254574D-02
3	-5,5	-5,806323466424D-02	-2,903161733212D-02	1,50	4,274664563589D-02
4	-4,5	1,006504706026D-01	5,032523530132D-02	1,75	3,092929012788D-02
5	-3,5	-1,411089754720D-01	-7,055448773602D-02	2,25	-3,529466150062D-02
6	-2,5	1,586767147287D-01	7,933835736434D-02	2,50	-5,503966034137D-02

7	-1,5	-1,425172942062D-01	-7,125864710311D-02	2.75	-4,361719647325D-02
8	-,5	1,040033390020D-01	5,200166950101D-02	3.25	5,688068579077D-02
9	,5	-6,444951559682D-02	-3,222475779841D-02	3.50	9,321245749232D-02
10	1,5	3,620726661931D-02	1,810363330965D-02	3.75	7,712528368358D-02
11	2,5	-1,956533972723D-02	-9,782669863615D-03	4.25	-1,092140889188D-01
12	3,5	1,043774421887D-02	5,218872109435D-03	4.50	-1,878364384622D-01
13	4,5	-5,476153598294D-03	-2,738076799147D-03	4.75	-1,652982341329D-01
14	5,5	2,729837130131D-03	1,364918565066D-03	5.25	2,911152162422D-01
15	6,5	-1,174416427121D-03	-5,872082135606D-04	5.50	6,282758164275D-01
16	7,5	3,611787202723D-04	1,805893601361D-04	5.75	8,976340447595D-01
17	8,5	-5,526709868390D-05	-2,763354934195D-05	6.25	8,962525349618D-01
				6.50	6,262509812234D-01
	0	-9,265806635020D+01	-4,429536509696D+01	6.75	2,895976903477D-01
	1	-3,677140293184D+01	-1,919272807294D+01	7.25	-1,634937992384D-01
	2	-4,845911966065D+00	-2,316606382949D+00	7.50	-1,849915654295D-01
	3	-2,120387233507D-01	-1,106726497267D-01	7.75	-1,069448507220D-01
				8.25	7,416044450357D-02
				8.50	8,837719523623D-02

I= 7 H= -1.5

0		0,	-3,636917565661D+01	-1.00	-6,017180414231D-01
1		1,875972850561D-01	6,874722394356D+00	-.75	-3,576660514998D-01
2		-3,005219101538D-01	-5,713277957891D-01	-.50	-1,831289758432D-01
3		1,135988462133D-01	6,550729789624D-02	-.25	-6,745692262072D-02
				.25	2,989164270031D-02
0	-8,5	-6,742211156218D-04	-3,371105578109D-04	.50	3,286275341381D-02
1	-7,5	4,402603596304D-03	2,201301798152D-03	.75	1,948890492180D-02
2	-6,5	-1,427519200461D-02	-7,137596002306D-03	1,25	-1,641079987896D-02
3	-5,5	3,280453440769D-02	1,640226720385D-02	1,50	-2,286756879412D-02
4	-4,5	-6,266260208688D-02	-3,133130104344D-02	1,75	-1,653638333798D-02
5	-3,5	1,032526556361D-01	5,162632781804D-02	2,25	1,883108592674D-02
6	-2,5	-1,425172942062D-01	-7,125864710311D-02	2,50	2,931241981743D-02
7	-1,5	1,594290414717D-01	7,971452073585D-02	2,75	2,316798948870D-02
8	-,5	-1,429134452292D-01	-7,145672261459D-02	3,25	-2,993093025222D-02
9	,5	1,042026903673D-01	5,210134518363D-02	3,50	-4,865639092594D-02
10	1,5	-6,453244636663D-02	-3,226622318331D-02	3,75	-3,979907288665D-02
11	2,5	3,620726661931D-02	1,810363330965D-02	4,25	5,412666472734D-02
12	3,5	-1,948271340505D-02	-9,741356702525D-03	4,50	8,989331593062D-02
13	4,5	1,024164540584D-02	5,120822702918D-03	4,75	7,512206064002D-02
14	5,5	-5,107711719417D-03	-2,553855859709D-03	5,25	-1,077351154628D-01
15	6,5	2,197644613731D-03	1,098822306866D-03	5,50	-1,860705200097D-01
16	7,5	-6,758788581265D-04	-3,379394290633D-04	5,75	-1,642299119194D-01
17	8,5	1,034228738247D-04	5,171143691234D-05	6,25	2,903336938152D-01
				6,50	6,273296041850D-01
	0	4,964576034509D+01	2,101720077019D+01	6,75	8,970608035325D-01
	1	1,970127472917D+01	1,136085668495D+01	7,25	8,966733969963D-01
	2	2,596248668284D+00	1,099108300565D+00	7,50	6,267618244804D-01
	3	1,135988462133D-01	6,550729789624D-02	7,75	2,899082083222D-01
				8,25	-1,637241790894D-01
				8,50	-1,852734605924D-01

I= 8 H= -.5

0		0,	6,797632696555D+01	-1,00	3,221225254852D-01
1		-1,004247488629D-01	-1,272419690876D+01	-,75	1,914711603105D-01
2		1,608807050725D-01	8,674416080748D-01	-,50	9,803468464331D-02

3		-6.081707154973D-02	-4.668412151324D-02
0	-8.5	3.611153401416D-04	1.805576700708D-04
1	-7.5	-2.359513704506D-03	-1.179756852253D-03
2	-6.5	7.667338065832D-03	3.833669032916D-03
3	-5.5	-1.777624007830D-02	-8.888120039151D-03
4	-4.5	3.526702732762D-02	1.763351366381D-02
5	-3.5	-6.405444532282D-02	-3.202722266141D-02
6	-2.5	1.040033390020D-01	5.200166950101D-02
7	-1.5	-1.429134452292D-01	-7.145672261459D-02
8	-.5	1.596285001800D-01	7.981425009002D-02
9	.5	-1.429964422590D-01	-7.149822112948D-02
10	1.5	1.042026903673D-01	5.210134518363D-02
11	2.5	-6.444951559682D-02	-3.222475779841D-02
12	3.5	3.600851592993D-02	1.800425796496D-02
13	4.5	-1.909260903071D-02	-9.546304515353D-03
14	5.5	9.540636551510D-03	4.770318275755D-03
15	6.5	-4.106843311156D-03	-2.053421655578D-03
16	7.5	1.263191385345D-03	6.315956926723D-04
17	8.5	-1.932996172300D-04	-9.664980861502D-05
	0	-2.657926348932D+01	-6.176576699801D+00
	1	-1.054755302103D+01	-8.096472909481D+00
	2	-1.389954619446D+00	-3.230034905129D-01
	3	-6.081707154973D-02	-4.668412151324D-02

-.25	3.611149802573D-02
.25	-1.600138785093D-02
.50	-1.759151089346D-02
.75	-1.043216402479D-02
1.25	8.783610839061D-03
1.50	1.223840706957D-02
1.75	8.848888780310D-03
2.25	-1.007200713802D-02
2.50	-1.567149892739D-02
2.75	-1.237893294484D-02
3.25	1.595771789446D-02
3.50	2.589268679554D-02
3.75	2.112233265760D-02
4.25	-2.845502170060D-02
4.50	-4.687727665642D-02
4.75	-3.872502272059D-02
5.25	5.334287641459D-02
5.50	8.894501879792D-02
5.75	7.454783440767D-02
6.25	-1.073321688033D-01
6.50	-1.855591709067D-01
6.75	-1.639191217600D-01
7.25	2.901031438592D-01
7.50	6.270475131344D-01
7.75	8.968874807323D-01
8.25	8.968063700035D-01
8.50	6.269286392921D-01

I= 9 H= .5

0		0.	-1.270034847443D+02
1		5.374936237792D-02	2.370631558619D+01
2		-8.610723423745D-02	-1.513448589101D+00
3		3.255117147676D-02	4.668412151324D-02
0	-8.5	-1.932996172300D-04	-9.664980861502D-05
1	-7.5	1.263191385345D-03	6.315956926723D-04
2	-6.5	-4.106843311156D-03	-2.053421655578D-03
3	-5.5	9.540636551510D-03	4.770318275755D-03
4	-4.5	-1.909260903071D-02	-9.546304515353D-03
5	-3.5	3.600851592993D-02	1.800425796496D-02
6	-2.5	-6.444951559682D-02	-3.222475779841D-02
7	-1.5	1.042026903673D-01	5.210134518363D-02
8	-.5	-1.429964422590D-01	-7.149822112948D-02
9	.5	1.596285001800D-01	7.981425009002D-02
10	1.5	-1.429134452292D-01	-7.145672261459D-02
11	2.5	1.040033390020D-01	5.200166950101D-02
12	3.5	-6.405444532282D-02	-3.202722266141D-02
13	4.5	3.526702732762D-02	1.763351366381D-02
14	5.5	-1.777624007830D-02	-8.888120039151D-03
15	6.5	7.667338065832D-03	3.833669032916D-03
16	7.5	-2.359513704506D-03	-1.179756852253D-03
17	8.5	3.611153401416D-04	1.805576700708D-04
	0	1.422611008972D+01	-6.176576699801D+00
	1	5.645392797928D+00	8.096472909481D+00
	2	7.439476384199D-01	-3.230034905129D-01

-1.00	-1.724077680921D-01
-.75	-1.024798665088D-01
-.50	-5.247038618292D-02
-.25	-1.932765478865D-02
.25	8.564238710892D-03
.50	9.415258910934D-03
.75	5.583425607424D-03
1.25	-4.700991035605D-03
1.50	-6.549866229392D-03
1.75	-4.735690251474D-03
2.25	5.389680603061D-03
2.50	8.385247225230D-03
2.75	6.622591930533D-03
3.25	-8.532939331750D-03
3.50	-1.383941603089D-02
3.75	-1.128269851563D-02
4.25	1.516610801682D-02
4.50	2.493779432660D-02
4.75	2.054532361179D-02
5.25	-2.803266927728D-02
5.50	-4.636506216980D-02
5.75	-3.841385699924D-02
6.25	5.311217806565D-02
6.50	8.866279139092D-02
6.75	7.437444733524D-02
7.25	-1.071991644951D-01
7.50	-1.853923237511D-01

CARDINAL SPLINES N=4

```
      3   3.255117147676D-02   4.668412151324D-02     7.75 -1.638132495090D-01
                                                      8.25  2.900140634467D-01
                                                      8.50  6.269286392921D-01


I=10      H=  1.5
0           0.                 2.372247266286D+02    -1.00  9.224251710535D-02
1          -2.875722041037D-02 -4.424440461357D+01    -.75  5.482930491412D-02
2           4.606954711576D-02  2.769544396919D+00    -.50  2.807296568153D-02
3          -1.741574957922D-02 -6.550729789624D-02    -.25  1.034077288450D-02
                                                       .25 -4.582073182601D-03
0   -8.5    1.034228738247D-04  5.171143691234D-05     .50 -5.037384132444D-03
1   -7.5   -6.758788581265D-04 -3.379394290633D-04     .75 -2.987259121719D-03
2   -6.5    2.197644613731D-03  1.098822306866D-03    1.25  2.515123398613D-03
3   -5.5   -5.107711719417D-03 -2.553855859709D-03    1.50  3.504292019710D-03
4   -4.5    1.024164540584D-02  5.120822702918D-03    1.75  2.533659066033D-03
5   -3.5   -1.948271340505D-02 -9.741356702525D-03    2.25 -2.883480557994D-03
6   -2.5    3.620726661931D-02  1.810363330965D-02    2.50 -4.486010919527D-03
7   -1.5   -6.453244636663D-02 -3.226622318331D-02    2.75 -3.542897446126D-03
8    -.5    1.042026903673D-01  5.210134518363D-02    3.25  4.564356314343D-03
9     .5   -1.429134452292D-01 -7.145672261459D-02    3.50  7.402113191446D-03
10   1.5    1.594290414717D-01  7.971452073585D-02    3.75  6.033774927018D-03
11   2.5   -1.425172942062D-01 -7.125864710311D-02    4.25 -8.106448675395D-03
12   3.5    1.032526556361D-01  5.162632781804D-02    4.50 -1.332372988431D-02
13   4.5   -6.266260208688D-02 -3.133130104344D-02    4.75 -1.097007882987D-02
14   5.5    3.280453440769D-02  1.640226720385D-02    5.25  1.493489773202D-02
15   6.5   -1.427519200461D-02 -7.137596002306D-03    5.50  2.465513608446D-02
16   7.5    4.402603596304D-03  2.201301798152D-03    5.75  2.037175545885D-02
17   8.5   -6.742211156218D-04 -3.371105578109D-04    6.25 -2.789960097977D-02
                                                      6.50 -4.619816153042D-02
       0   -7.611358804711D+00  2.101720077019D+01    6.75 -3.830796283142D-02
       1   -3.020438640738D+00 -1.136085668495D+01    7.25  5.302309167317D-02
       2   -3.980320671543D-01  1.099108300565D+00    7.50  8.854391486182D-02
       3   -1.741574957922D-02 -6.550729789624D-02    7.75  7.429333762748D-02
                                                      8.25 -1.071180591906D-01
                                                      8.50 -1.852734605924D-01


I=11      H=  2.5
0           0.                -4.427752058984D+02    -1.00 -4.929224800100D-02
1           1.536718144714D-02  8.256333341133D+01    -.75 -2.929949875746D-02
2          -2.461849045116D-02 -5.138758950980D+00    -.50 -1.500153534920D-02
3           9.306576102702D-03  1.106726497267D-01    -.25 -5.525866266587D-03
                                                       .25  2.448551586956D-03
0   -8.5   -5.526709868390D-05 -2.763354934195D-05     .50  2.691858349410D-03
1   -7.5    3.611787202723D-04  1.805893601361D-04     .75  1.596319732766D-03
2   -6.5   -1.174416427121D-03 -5.872082135606D-04    1.25 -1.344020088605D-03
3   -5.5    2.729837130131D-03  1.364918565066D-03    1.50 -1.872605482943D-03
4   -4.5   -5.476153598294D-03 -2.738076799147D-03    1.75 -1.353921532533D-03
5   -3.5    1.043774421887D-02  5.218872109435D-03    2.25  1.540848196842D-03
6   -2.5   -1.956533972723D-02 -9.782669863615D-03    2.50  2.397181778842D-03
7   -1.5    3.620726661931D-02  1.810363330965D-02    2.75  1.893198042510D-03
8    -.5   -6.444951559682D-02 -3.222475779841D-02    3.25 -2.438965303981D-03
9     .5    1.040033390020D-01  5.200166950101D-02    3.50 -3.955232082936D-03
10   1.5   -1.425172942062D-01 -7.125864710311D-02    3.75 -3.223971473676D-03
11   2.5    1.586767147287D-01  7.933835736434D-02    4.25  4.330942201517D-03
12   3.5   -1.411089754720D-01 -7.055448773602D-02    4.50  7.117611807550D-03
```

CARDINAL SPLINES N=4

13	4.5	1.006504706026D-01	5.032523530132D-02	4.75	5.859439088176D-03
14	5.5	-5.806323466424D-02	-2.903161733212D-02	5.25	-7.973117707173D-03
15	6.5	2.626713627119D-02	1.313356813559D-02	5.50	-1.315661589450D-02
16	7.5	-8.179529477276D-03	-4.089764738638D-03	5.75	-1.086410111636D-02
17	8.5	1.256038974802D-03	6.280194874011D-04	6.25	1.484579387940D-02
				6.50	2.453625792430D-02
	0	4.067336156277D+00	-4.429536509696D+01	6.75	2.029065588752D-02
	1	1.614053214038D+00	1.919272807294D+01	7.25	-2.781852164097D-02
	2	2.126992001678D-01	-2.316606382949D+00	7.50	-4.607934480478D-02
	3	9.306576102702D-03	1.106726497267D-01	7.75	-3.821893237192D-02
				8.25	5.291729335712D-02
				8.50	8.837719523623D-02

I=12 H= 3.5

0		0.	8.238552446968D+02	-1.00	2.622971844743D-02
1		-8.177285956356D-03	-1.536220234482D+02	-.75	1.559104367291D-02
2		1.310015465582D-02	9.546322746243D+00	-.50	7.982716371540D-03
3		-4.952277835252D-03	-1.997781243287D-01	-.25	2.940460496254D-03
				.25	-1.302939369285D-03
0	-8.5	2.940913578623D-05	1.470456789311D-05	.50	-1.432409284756D-03
1	-7.5	-1.921934091987D-04	-9.609670459936D-05	.75	-8.494440393646D-04
2	-6.5	6.249439314266D-04	3.124719657133D-04	1.25	7.151885140807D-04
3	-5.5	-1.452667238141D-03	-7.263336190706D-04	1.50	9.964624920535D-04
4	-4.5	2.914404257757D-03	1.457202128879D-03	1.75	7.204568993513D-04
5	-3.5	-5.557448931433D-03	-2.778724465716D-03	2.25	-8.199243662133D-04
6	-2.5	1.043774421887D-02	5.218872109435D-03	2.50	-1.275599691930D-03
7	-1.5	-1.948271340505D-02	-9.741356702525D-03	2.75	-1.007415797866D-03
8	-.5	3.600851592993D-02	1.800425796496D-02	3.25	1.297823772495D-03
9	.5	-6.405444532282D-02	-3.202722266141D-02	3.50	2.104649702793D-03
10	1.5	1.032526556361D-01	5.162632781804D-02	3.75	1.715520040933D-03
11	2.5	-1.411089754720D-01	-7.055448773602D-02	4.25	-2.304493364523D-03
12	3.5	1.560494858467D-01	7.802474292335D-02	4.50	-3.787193169022D-03
13	4.5	-1.362593499036D-01	-6.812967495182D-02	4.75	-3.117632737178D-03
14	5.5	9.208114567941D-02	4.604057283971D-02	5.25	4.241764609368D-03
15	6.5	-4.588399573218D-02	-2.294199786609D-02	5.50	6.998729398511D-03
16	7.5	1.491110847069D-02	7.455554235347D-03	5.75	5.778385391103D-03
17	8.5	-2.317623692158D-03	-1.158811846079D-03	6.25	-7.892152486079D-03
				6.50	-1.303800213622D-02
	0	-2.164338382320D+00	8.510112319979D+01	6.75	-1.077524447002D-02
	1	-8.588808775982D-01	-3.463644521031D+01	7.25	1.474023964983D-02
	2	-1.131829301431D-01	4.451980575862D+00	7.50	2.436994359042D-02
	3	-4.952277835252D-03	-1.997781243287D-01	7.75	2.015809871505D-02
				8.25	-2.764577447766D-02
				8.50	-4.579821374905D-02

I=13 H= 4.5

0		0.	-1.511938403183D+03	-1.00	-1.375339008754D-02
1		4.287709057143D-03	2.820013107663D+02	-.75	-8.175066963992D-03
2		-6.868984792439D-03	-1.752047744326D+01	-.50	-4.185687756426D-03
3		2.596696237956D-03	3.636793736600D-01	-.25	-1.541812192531D-03
				.25	6.831881522835D-04
0	-8.5	-1.542050265935D-05	-7.710251329673D-06	.50	7.510748875290D-04
1	-7.5	1.007754922216D-04	5.038774611078D-05	.75	4.454006838675D-04
2	-6.5	-3.276861526491D-04	-1.638430763245D-04	1.25	-3.750046128140D-04
3	-5.5	7.617028425571D-04	3.808514212786D-04	1.50	-5.224888302965D-04

4	-4,5	-1.528197519561D-03	-7.640987597804D-04	1.75	-3,777670054144D-04
5	-3,5	2,914404257757D-03	1.457202128879D-03	2.25	4,299220111251D-04
6	-2,5	-5,476153598294D-03	-2,738076799147D-03	2,50	6,688522290647D-04
7	-1,5	1.024164540584D-02	5,120822702918D-03	2,75	5,282315819373D-04
8	-,5	-1,909260903071D-02	-9,546304515353D-03	3,25	-6,805040799356D-04
9	,5	3,526702732762D-02	1,763351366381D-02	3,50	-1,103555842356D-03
10	1,5	-6,266260208688D-02	-3,133130104344D-02	3,75	-8,995173485663D-04
11	2,5	1,006504706026D-01	5,032523530132D-02	4,25	1,208332803303D-03
12	3,5	-1,362593499036D-01	-6,812967495182D-02	4,50	1,985757847018D-03
13	4,5	1.471000336754D-01	7,355001683769D-02	4,75	1,634671592960D-03
14	5,5	-1,204468666098D-01	-6,022343330490D-02	5,25	-2,224029484308D-03
15	6,5	6,960798833825D-02	3,480399416912D-02	5,50	-3,669469039798D-03
16	7,5	-2,492996303505D-02	-1,246498151752D-02	5,75	-3,029536836609D-03
17	8,5	4,094800496974D-03	2,047400248487D-03	6,25	4,137276643024D-03
				6,50	6,834184042796D-03
	0	1,134857452867D+00	-1,574371615964D+02	6,75	5,647293192838D-03
	1	4,503488771626D-01	6,298069847167D+01	7,25	-7,721411650734D-03
	2	5,934676927543D-02	-8,246653414932D+00	7,50	-1,276018754198D-02
	3	2,596696237956D-03	3,636793736600D-01	7,75	-1,054824001403D-02
				8,25	1,443434918218D-02
				8,50	2,386676327528D-02

I=14 H= 5.5

0		0.	2,603701826294D+03	-1,00	6,855083133534D-03
1		-2,137116864718D-03	-4,862886656005D+02	-,75	4,074687277720D-03
2		3,423698556621D-03	3,024754033665D+01	-,50	2,086266535539D-03
3		-1,294267712195D-03	-6,271191691563D-01	-,25	7,684833089714D-04
				,25	-3,405205202763D-04
0	-8,5	7,686020291883D-06	3,843010145941D-06	,50	-3,743572101946D-04
1	-7,5	-5,022939685038D-05	-2,511469842519D-05	,75	-2,220004416595D-04
2	-6,5	1,633282323265D-04	8,166411616324D-05	1,25	1,869130234219D-04
3	-5,5	-3,796551548031D-04	-1,898275774015D-04	1,50	2,604233718306D-04
4	-4,5	7,617028425571D-04	3,808514212786D-04	1,75	1,882898729091D-04
5	-3,5	-1,452667238141D-03	-7,263336190706D-04	2,25	-2,142854035947D-04
6	-2,5	2,729837130131D-03	1,364918565066D-03	2,50	-3,333750248337D-04
7	-1,5	-5,107711719417D-03	-2,553855859709D-03	2,75	-2,632856609159D-04
8	-,5	9,540636551510D-03	4,770318275755D-03	3,25	3,391825021830D-04
9	,5	-1,777624007830D-02	-8,888120039151D-03	3,50	5,500433289681D-04
10	1,5	3,280453440769D-02	1,640226720385D-02	3,75	4,483446173276D-04
11	2,5	-5,806323466424D-02	-2,903161733212D-02	4,25	-6,022660377702D-04
12	3,5	9,208114567941D-02	4,604057283971D-02	4,50	-9,897546989958D-04
13	4,5	-1,204468666098D-01	-6,022343330490D-02	4,75	-8,147625310889D-04
14	5,5	1,191621310744D-01	5,958106553719D-02	5,25	1,108506917803D-03
15	6,5	-8,074096184874D-02	-4,037048092437D-02	5,50	1,828937251292D-03
16	7,5	3,258616266506D-02	1,629308133253D-02	5,75	1,509970722907D-03
17	8,5	-5,819597893071D-03	-2,909798946536D-03	6,25	-2,062031035237D-03
				6,50	-3,406099270602D-03
	0	-5,656454313859D-01	2,705033982544D+02	6,75	-2,814469841881D-03
	1	-2,244667680204D-01	-1,080085597920D+02	7,25	3,847707685312D-03
	2	-2,958012810435D-02	1,425600152317D+01	7,50	6,357973764351D-03
	3	-1,294267712195D-03	-6,271191691563D-01	7,75	5,255075354166D-03
				8,25	-7,187436557132D-03
				8,50	-1,187899750410D-02

I=15 H= 6.5

CARDINAL SPLINES N=4

0		0.	-3.154545637057D+03	-1.00	-2.949066115596D-03
1		9.193905896135D-04	5.945795453833D+02	-.75	-1.752936025141D-03
2		-1.472879789563D-03	-3.730192882079D+01	-.50	-8.975147092499D-04
3		5.567957364196D-04	7.791804492501D-01	-.25	-3.306025676326D-04
				.25	1.464923921223D-04
0	-8.5	-3.306536470287D-06	-1.653268235144D-06	.50	1.610489821523D-04
1	-7.5	2.160875586423D-05	1.080437793211D-05	.75	9.550489254343D-05
2	-6.5	-7.026403620708D-05	-3.513201810354D-05	1.25	-8.041023712657D-05
3	-5.5	1.633282323265D-04	8.166411616324D-05	1.50	-1.120344887475D-04
4	-4.5	-3.276861526491D-04	-1.638430763245D-04	1.75	-8.100255904834D-05
5	-3.5	6.249439314266D-04	3.124719657133D-04	2.25	9.218587015921D-05
6	-2.5	-1.174416427121D-03	-5.872082135606D-04	2.50	1.434183840854D-04
7	-1.5	2.197644613731D-03	1.098822306866D-03	2.75	1.132658431144D-04
8	-.5	-4.106843311156D-03	-2.053421655578D-03	3.25	-1.459167548290D-04
9	.5	7.667338065832D-03	3.833669032916D-03	3.50	-2.366293423949D-04
10	1.5	-1.427519200461D-02	-7.137596002306D-03	3.75	-1.928784090230D-04
11	2.5	2.626713627119D-02	1.313356813559D-02	4.25	2.590955096404D-04
12	3.5	-4.588399573218D-02	-2.294199786609D-02	4.50	4.257934365255D-04
13	4.5	6.960798833825D-02	3.480399416912D-02	4.75	3.505114933735D-04
14	5.5	-8.074096184874D-02	-4.037048092437D-02	5.25	-4.768798792480D-04
15	6.5	6.273140344843D-02	3.136570172421D-02	5.50	-7.868081707828D-04
16	7.5	-2.812492067877D-02	-1.406246033939D-02	5.75	-6.495877306711D-04
17	8.5	5.426195070864D-03	2.713097535432D-03	6.25	8.870778983775D-04
				6.50	1.465283040053D-03
	0	2.433414368445D-01	-3.171696652053D+02	6.75	1.210758316169D-03
	1	9.656591003600D-02	1.293341178049D+02	7.25	-1.655202092773D-03
	2	1.272541148914D-02	-1.743282736491D+01	7.50	-2.735000858814D-03
	3	5.567957364196D-04	7.791804492501D-01	7.75	-2.260492169689D-03
				8.25	3.091335155963D-03
				8.50	5.108662767166D-03

I=16 H= 7.5

0		0.	2.063986969416D+03	-1.00	9.069454501773D-04
1		-2.827461573519D-04	-3.944372051474D+02	-.75	5.390917972207D-04
2		4.529642847718D-04	2.509555940516D+01	-.50	2.760185258756D-04
3		-1.712350080535D-04	-5.315548013030D-01	-.25	1.016723541371D-04
				.25	-4.505175647511D-05
0	-8.5	1.016880633586D-06	5.084403167929D-07	.50	-4.952843910973D-05
1	-7.5	-6.645481054393D-06	-3.322740527197D-06	.75	-2.937123967305D-05
2	-6.5	2.160875586423D-05	1.080437793211D-05	1.25	2.472908226912D-05
3	-5.5	-5.022939685038D-05	-2.511469842519D-05	1.50	3.445469365783D-05
4	-4.5	1.007754922216D-04	5.038774611078D-05	1.75	2.491124282922D-05
5	-3.5	-1.921934091987D-04	-9.609670459936D-05	2.25	-2.835051897962D-05
6	-2.5	3.611787202723D-04	1.805893601361D-04	2.50	-4.410638635448D-05
7	-1.5	-6.758788581265D-04	-3.379394290633D-04	2.75	-3.483337957936D-05
8	-.5	1.263191385345D-03	6.315956926723D-04	3.25	4.487472546529D-05
9	.5	-2.359513704506D-03	-1.179756852253D-03	3.50	7.277215437360D-05
10	1.5	4.402603596304D-03	2.201301798152D-03	3.75	5.931714508877D-05
11	2.5	-8.179529477276D-03	-4.089764738638D-03	4.25	-7.968131199058D-05
12	3.5	1.491110847069D-02	7.455554235347D-03	4.50	-1.309469906704D-04
13	4.5	-2.492996303505D-02	-1.246498151752D-02	4.75	-1.077950402613D-04
14	5.5	3.258616266506D-02	1.629308133253D-02	5.25	1.466578654649D-04
15	6.5	-2.812492067877D-02	-1.406246033939D-02	5.50	2.419719860554D-04
16	7.5	1.367218512037D-02	6.836092560184D-03	5.75	1.997716469506D-04
17	8.5	-2.800957045933D-03	-1.400478522966D-03	6.25	-2.728080102340D-04
				6.50	-4.506260565175D-04

	0	-7,483637208359D-02	1,979838003356D+02	6,75	-3,723500381860D-04
	1	-2,969754131183D-02	-8,302719844212D+01	7,25	5,090283947477D-04
	2	-3,913528420593D-03	1,154091197193D+01	7,50	8,410966595898D-04
	3	-1,712350080535D-04	-5,315548013030D-01	7,75	6,951648413117D-04
				8,25	-9,506441003647D-04
				8,50	-1,570970398421D-03

I=17 H= 8.5

0		0.	-5,362859688595D+02	-1,00	-1,387793079223D-04
1		4,326535407871D-05	1.040151627251D+02	-,75	-8,249094420097D-05
2		-6,931185325623D-05	-6,725052824599D+00	-,50	-4,223590292684D-05
3		2,620210058740D-05	1,449467100539D-01	-,25	-1,555773716987D-05
				,25	6,893746015686D-06
0	-8,5	-1,556014098850D-07	-7,780070494250D-08	,50	7,578760662711D-06
1	-7,5	1,016880633586D-06	5,084403167929D-07	,75	4,494339006746D-06
2	-6,5	-3,306536470287D-06	-1,653268235144D-06	1,25	-3,784003680253D-06
3	-5,5	7,686020291883D-06	3,843010145941D-06	1,50	-5,272200810231D-06
4	-4,5	-1,542050265935D-05	-7,710251329673D-06	1,75	-3,811877588909D-06
5	-3,5	2,940913578623D-05	1,470456789311D-05	2,25	4,338149989118D-06
6	-2,5	-5,526709868390D-05	-2,763354934195D-05	2,50	6,749087007157D-06
7	-1,5	1,034228738247D-04	5,171143691234D-05	2,75	5,330146689004D-06
8	-,5	-1,932996172300D-04	-9,664980861502D-05	3,25	-6,866656953635D-06
9	,5	3,611153401416D-04	1.805576700708D-04	3,50	-1,113547575881D-05
10	1,5	-6,742211156218D-04	-3,371105578109D-04	3,75	-9,076612237227D-06
11	2,5	1,256038974802D-03	6,280194874011D-04	4,25	1,219270336887D-05
12	3,5	-2,317623692158D-03	-1,158811846079D-03	4,50	2,003729284366D-05
13	4,5	4,094800496974D-03	2,047400248487D-03	4,75	1,649461905143D-05
14	5,5	-5,819597893071D-03	-2,909798946536D-03	5,25	-2,244134211161D-05
15	6,5	5,426195070864D-03	2,713097535432D-03	5,50	-3,702614674207D-05
16	7,5	-2,800957045933D-03	-1,400478522966D-03	5,75	-3,056871793378D-05
17	8,5	6,001643099189D-04	3,000821549595D-04	6,25	4,174459962378D-05
				6,50	6,895397658452D-05
	0	1,145133913514D-02	-4,902675396154D+01	6,75	5,697629200328D-05
	1	4,544269151041D-03	2,110646411109D+01	7,25	-7,789040235147D-05
	2	5,988417117224D-04	-3,028911718225D+00	7,50	-1,287025419097D-04
	3	2,620210058740D-05	1,449467100539D-01	7,75	-1,063721600664D-04
				8,25	1,454636440394D-04
				8,50	2,403816731096D-04

N= 4 M=18 P= 9,0 Q= 7

I= 0 H= -9,0

0		1,000000000000D+00	4,649037000769D-01	-1,00	4,666983736658D+00
1		-2,043924730536D+00	-1,015802229539D+00	-,75	3,405151661040D+00
2		1,333191786168D+00	6,662357559905D-01	-,50	2,391493714304D+00
3		-2,898672199536D-01	-1,449265972410D-01	-,25	1,598834844581D+00
				,25	5,678141653208D-01
0	-9,0	6,001643214779D-04	3,000821607389D-04	,50	2,751068675635D-01

CARDINAL SPLINES N=4

1	-8.0	-2.800957121473D-03	-1.400478560736D-03	.75	9.476921066589D-02
2	-7.0	5.426195316493D-03	2.713097658246D-03	1.25	-3.507902571586D-02
3	-6.0	-5.819598464034D-03	-2.909799232017D-03	1.50	-3.427495665991D-02
4	-5.0	4.094801642499D-03	2.047400821250D-03	1.75	-1.768214283798D-02
5	-4.0	-2.317625876841D-03	-1.158812938421D-03	2.25	1.053382473501D-02
6	-3.0	1.256043080373D-03	6.280215401863D-04	2.50	1.195817064098D-02
7	-2.0	-6.742287985370D-04	-3.371143992685D-04	2.75	6.908482389362D-03
8	-1.0	3.611297000517D-04	1.805648500259D-04	3.25	-4.774220041207D-03
9	0.0	-1.933264469406D-04	-9.666322347028D-05	3.50	-5.681573648175D-03
10	1.0	1.034729911545D-04	5.173649557727D-05	3.75	-3.398301266606D-03
11	2.0	-5.536066768994D-05	-2.768033384497D-05	4.25	2.450954148224D-03
12	3.0	2.958344857564D-05	1.479172428782D-05	4.50	2.953651005222D-03
13	4.0	-1.574214117456D-05	-7.871070587279D-06	4.75	1.782442708412D-03
14	5.0	8.254294029908D-06	4.127147014954D-06	5.25	-1.299142166699D-03
15	6.0	-4.114176317512D-06	-2.057088158756D-06	5.50	-1.570303788138D-03
16	7.0	1.769924264260D-06	8.849621321301D-07	5.75	-9.496227698491D-04
17	8.0	-5.443163007231D-07	-2.721581503616D-07	6.25	6.938369306729D-04
18	9.0	8.329038917438D-08	4.164519458719D-08	6.50	8.392395657934D-04
				6.75	5.077661014553D-04
	0	-1.207199912413D+02	-6.036370951923D+01	7.25	-3.712066599821D-04
	1	-4.848420702823D+01	-2.424072175127D+01	7.50	-4.490701217947D-04
	2	-6.493223152578D+00	-3.246782369516D+00	7.75	-2.717321656500D-04
	3	-2.898672199536D-01	-1.449265972410D-01	8.25	1.986791858373D-04
				8.50	2.403638164893D-04
				8.75	1.454488845859D-04

I= 1 H= -8.0

0		0.	2.293600538680D-01	-1.00	-9.041182827339D+00
1		3.959053511912D+00	1.939269262554D+00	-.75	-5.678512164130D+00
2		-4.019190935109D+00	-2.007241912810D+00	-.50	-3.117191787273D+00
3		1.062938380318D+00	5.314233607850D-01	-.25	-1.257571223615D+00
				.25	7.551721857694D-01
0	-9.0	-2.800957121473D-03	-1.400478560736D-03	.50	1.107574437241D+00
1	-8.0	1.367218561403D-02	6.836092807016D-03	.75	1.156548479489D+00
2	-7.0	-2.812492228399D-02	-1.406246114200D-02	1.25	7.315274079850D-01
3	-6.0	3.258616639639D-02	1.629308319820D-02	1.50	4.350675330796D-01
4	-5.0	-2.492997052124D-02	-1.246498526062D-02	1.75	1.772914838716D-01
5	-4.0	1.491112274796D-02	7.455561373979D-03	2.25	-8.424021461146D-02
6	-3.0	-8.179556307850D-03	-4.089778153925D-03	2.50	-9.008354482992D-02
7	-2.0	4.402653805410D-03	2.201326902705D-03	2.75	-4.992391800607D-02
8	-1.0	-2.359607548864D-03	-1.179803774432D-03	3.25	3.281503229559D-02
9	0.0	1.263366721886D-03	6.316833609430D-04	3.50	3.847398289843D-02
10	1.0	-6.762063830649D-04	-3.381031915324D-04	3.75	2.276829949380D-02
11	2.0	3.617902090135D-04	1.808951045067D-04	4.25	-1.621278367509D-02
12	3.0	-1.933325717557D-04	-9.666628587785D-05	4.50	-1.946621565007D-02
13	4.0	1.028774524658D-04	5.143872623289D-05	4.75	-1.171693305418D-02
14	5.0	-5.394315856943D-05	-2.697157928471D-05	5.25	8.514030404167D-03
15	6.0	2.688681421304D-05	1.344340710652D-05	5.50	1.028223471837D-02
16	7.0	-1.156674420846D-05	-5.783372104230D-06	5.75	6.214317946785D-03
17	8.0	3.557195951047D-06	1.778597975524D-06	6.25	-4.537269497963D-03
18	9.0	-5.443163007231D-07	-2.721581503616D-07	6.50	-5.487023934722D-03
				6.75	-3.319361172955D-03
	0	4.849590951154D+02	2.425038184915D+02	7.25	2.426255472952D-03
	1	1.899076430973D+02	9.494479150273D+01	7.50	2.935047638498D-03
	2	2.468014533348D+01	1.234118882838D+01	7.75	1.775939928996D-03
	3	1.062938380318D+00	5.314233607850D-01	8.25	-1.298444973187D-03

				8,50	-1,570853702489D-03
				8,75	-9,505476449251D-04

I= 2	H= -7.0				
0		0.	-7,457978465701D-01	-1,00	9,696675372477D+00
1		-3,293246639771D+00	-1,515720177068D+00	-.75	5,852797434208D+00
2		4,845624588580D+00	2,415159365933D+00	-.50	3,052754985046D+00
3		-1,557804144126D+00	-7,787530511329D-01	-,25	1,150503886481D+00
				,25	-5,448004817198D-01
0	-9,0	5,426195316493D-03	2,713097658246D-03	.50	-6,299002986052D-01
1	-8,0	-2,812492228399D-02	-1,406246114200D-02	.75	-4,007454624140D-01
2	-7,0	6,273140866805D-02	3,136570433403D-02	1,25	4,380162998840D-01
3	-6,0	-8,074097398172D-02	-4,037048699086D-02	1,50	7,976882863097D-01
4	-5,0	6,960801268071D-02	3,480400634035D-02	1,75	9,966809923211D-01
5	-4,0	-4,588404215680D-02	-2,294202107840D-02	2,25	8,267656444127D-01
6	-3,0	2,626722351473D-02	1,313361175737D-02	2,50	5,431835418791D-01
7	-2,0	-1,427535526688D-02	-7,137677633441D-03	2,75	2,397523375984D-01
8	-1,0	7,667643214524D-03	3,833821607262D-03	3,25	-1,274048126463D-01
9	0,0	-4,107413443630D-03	-2,053706721815D-03	3,50	-1,414516938238D-01
10	1,0	2,198709609093D-03	1,099354804547D-03	3,75	-8,064858341851D-02
11	2,0	-1,176404772425D-03	-5,882023862125D-04	4,25	5,497456001954D-02
12	3,0	6,286480855147D-04	3,143240427573D-04	4,50	6,517848685174D-02
13	4,0	-3,345209845911D-04	-1,672604922955D-04	4,75	3,888369418615D-02
14	5,0	1,754040731005D-04	8,770203655023D-05	5,25	-2,795857135459D-02
15	6,0	-8,742641684373D-05	-4,371320842187D-05	5,50	-3,366362891480D-02
16	7,0	3,761096460870D-05	1,880548230435D-05	5,75	-2,030265407705D-02
17	8,0	-1,156674420846D-05	-5,783372104230D-06	6,25	1,478714779459D-02
18	9,0	1,769924264260D-06	8,849621321301D-07	6,50	1,786997098719D-02
				6,75	1,080513890955D-02
	0	-7,727828491506D+02	-3,864703450755D+02	7,25	-7,893437018321D-03
	1	-2,946184110679D+02	-1,472798430156D+02	7,50	-9,547182452972D-03
	2	-3,721508730282D+01	-1,861117301466D+01	7,75	-5,776169004608D-03
	3	-1,557804144126D+00	-7,787530511329D-01	8,25	4,222588116062D-03
				8,50	5,108283313234D-03
				8,75	3,091021516958D-03

I= 3	H= -6.0				
0		0.	1,733601776237D+00	-1,00	-6,822801358869D+00
1		2,161366311921D+00	7,763997339141D-01	-.75	-4,066886674401D+00
2		-3,408490880203D+00	-1,686456262705D+00	-,50	-2,089423896854D+00
3		1,252944166746D+00	6,261256852840D-01	-,25	-7,729495105982D-01
				,25	3,468877953728D-01
0	-9,0	-5,819598464034D-03	-2,909799232017D-03	.50	3,851329911399D-01
1	-8,0	3,258616639639D-02	1,629308319820D-02	.75	2,315576115503D-01
2	-7,0	-8,074097398172D-02	-4,037048699086D-02	1,25	-2,046505537728D-01
3	-6,0	1,191621592774D-01	5,958107963870D-02	1,50	-2,975471660134D-01
4	-5,0	-1,204469231937D-01	-6,022346159684D-02	1,75	-2,292874602033D-01
5	-4,0	9,208125359310D-02	4,604062679655D-02	2,25	3,358732173324D-01
6	-3,0	-5,806343746123D-02	-2,903171873062D-02	2,50	6,817337818526D-01
7	-2,0	3,280491390967D-02	1,640245695484D-02	2,75	9,296916767711D-01
8	-1,0	-1,777694939426D-02	-8,888474697130D-03	3,25	8,730597121113D-01
9	0,0	9,541961820376D-03	4,770980910188D-03	3,50	5,982759265595D-01
10	1,0	-5,110187293533D-03	-2,555093646766D-03	3,75	2,727045694681D-01
11	2,0	2,734459024596D-03	1,367229512298D-03	4,25	-1,511709296771D-01
12	3,0	-1,461277517876D-03	-7,306387589382D-04	4,50	-1,700923057161D-01

13	4.0	7.775903606153D-04	3.887951803077D-04	4.75	-9.793236566967D-02
14	5.0	-4.077253604778D-04	-2.038626802389D-04	5.25	6.757193883523D-02
15	6.0	2.032220636012D-04	1.016110318006D-04	5.50	8.040524420789D-02
16	7.0	-8.742641684373D-05	-4.371320842187D-05	5.75	4.809190374168D-02
17	8.0	2.688681421304D-05	1.344340710652D-05	6.25	-3.468651810615D-02
18	9.0	-4.114176317512D-06	-2.057088158756D-06	6.50	-4.180152080880D-02
				6.75	-2.522634527185D-02
	0	6.567608330689D+02	3.285638666744D+02	7.25	1.838668408826D-02
	1	2.452739629876D+02	1.225687285292D+02	7.50	2.222456692841D-02
	2	3.042100162195D+01	1.521893723996D+01	7.75	1.344012738995D-02
	3	1.252944166746D+00	6.261256852840D-01	8.25	-9.820087498789D-03
				8.50	-1.187811546615D-02
				8.75	-7.186707505503D-03

I= 4 H= -5.0

0		0.	-3.478134599304D+00	-1.00	3.861653229088D+00
1		-1.208111771385D+00	6.429586131857D-03	-.75	2.296781207937D+00
2		1.928779213723D+00	9.286990842418D-01	-.50	1.176845969621D+00
3		-7.247622439799D-01	-3.616861416341D-01	-.25	4.339010537662D-01
				.25	-1.928034021240D-01
0	-9.0	4.094801642499D-03	2.047400821250D-03	.50	-2.124243721217D-01
1	-8.0	-2.492997052124D-02	-1.246498526062D-02	.75	-1.263580024604D-01
2	-7.0	6.960801268071D-02	3.480400634035D-02	1.25	1.075505636184D-01
3	-6.0	-1.204469231937D-01	-6.022346159684D-02	1.50	1.512817599100D-01
4	-5.0	1.471001471999D-01	7.355007359995D-02	1.75	1.109158558805D-01
5	-4.0	-1.362595664115D-01	-6.812978320573D-02	2.25	-1.327803515359D-01
6	-3.0	1.006508774754D-01	5.032543873768D-02	2.50	-2.159589298583D-01
7	-2.0	-6.266336348374D-02	-3.133168174187D-02	2.75	-1.821522467002D-01
8	-1.0	3.526845043203D-02	1.763422521602D-02	3.25	3.032996488605D-01
9	0.0	-1.909526792475D-02	-9.547633962373D-03	3.50	6.429695115735D-01
10	1.0	1.024661216347D-02	5.123306081737D-03	3.75	9.065057244436D-01
11	2.0	-5.485426529968D-03	-2.742713264984D-03	4.25	8.897821093526D-01
12	3.0	2.931679108208D-03	1.465839554104D-03	4.50	6.184281400079D-01
13	4.0	-1.560072732322D-03	-7.800363661610D-04	4.75	2.848658543991D-01
14	5.0	8.180202468775D-04	4.090101234388D-04	5.25	-1.600347626957D-01
15	6.0	-4.077253604778D-04	-2.038626802389D-04	5.50	-1.808062492607D-01
16	7.0	1.754040731005D-04	8.770203655023D-05	5.75	-1.044115132358D-01
17	8.0	-5.394315856943D-05	-2.697157928471D-05	6.25	7.230593238140D-02
18	9.0	8.254294029908D-06	4.127147014954D-06	6.50	8.613134449384D-02
				6.75	5.155640289867D-02
	0	-3.829935654922D+02	-1.918648397518D+02	7.25	-3.721934039845D-02
	1	-1.426073112115D+02	-7.116671931460D+01	7.50	-4.486568983624D-02
	2	-1.763980137373D+01	-8.836826739879D+00	7.75	-2.708052879056D-02
	3	-7.247622439799D-01	-3.616861416341D-01	8.25	1.974251841575D-02
				8.50	2.386499363780D-02
				8.75	1.443288648199D-02

I= 5 H= -4.0

0		0.	6.633309779088D+00	-1.00	-2.092833920380D+00
1		6.529714344223D-01	-8.377990844353D-01	-.75	-1.244159989038D+00
2		-1.045258147252D+00	-4.545620433946D-01	-.50	-6.371257963624D-01
3		3.946043387062D-01	1.959767404875D-01	-.25	-2.347371856011D-01
				.25	1.040797757380D-01
0	-9.0	-2.317625876841D-03	-1.158812938421D-03	.50	1.144786162843D-01
1	-8.0	1.491112274796D-02	7.455561373979D-03	.75	6.793520766934D-02

CARDINAL SPLINES N=4

2	-7,0	-4,588404215680D-02	-2,294202107840D-02	1,25	-5,734035428348D-02
3	-6,0	9,208125359310D-02	4,604062679655D-02	1,50	-8,006635428294D-02
4	-5,0	-1,362595664115D-01	-6,812978320573D-02	1,75	-5,807568345149D-02
5	-4,0	1,560498987586D-01	7,802494937931D-02	2,25	6,687256822049D-02
6	-3,0	-1,411097514376D-01	-7,055487571878D-02	2,50	1,051034250442D-01
7	-2,0	1,032541077308D-01	5,162705386541D-02	2,75	8,423767154905D-02
8	-1,0	-6,405715939025D-02	-3,202857969512D-02	3,25	-1,143439422381D-01
9	0,0	3,601358682820D-02	1,800679341410D-02	3,50	-1,940186187521D-01
10	1,0	-1,949218573541D-02	-9,746092867703D-03	3,75	-1,690291498725D-01
11	2,0	1,045542905042D-02	5,227714525212D-03	4,25	2,938348658120D-01
12	3,0	-5,590394587789D-03	-2,795197293894D-03	4,50	6,315634566708D-01
13	4,0	2,975194931973D-03	1,487597465987D-03	4,75	8,996224838028D-01
14	5,0	-1,560072732322D-03	-7,800363661610D-04	5,25	8,947990343301D-01
15	6,0	7,775903606153D-04	3,887951803077D-04	5,50	6,244922546106D-01
16	7,0	-3,345209845911D-04	-1,672604922955D-04	5,75	2,885330886966D-01
17	8,0	1,028774524658D-04	5,143872623289D-05	6,25	-1,627142848219D-01
18	9,0	-1,574214117456D-05	-7,871070587279D-06	6,50	-1,840473731660D-01
				6,75	-1,063725543668D-01
	0	2,088773958993D+02	1,051406363196D+02	7,25	7,373971071954D-02
	1	7,772717908951D+01	3,860243207292D+01	7,50	8,786600475414D-02
	2	9,609058997817D+00	4,836809949768D+00	7,75	5,260615747031D-02
	3	3,946043387062D-01	1,959767404875D-01	8,25	-3,798714373625D-02
				8,50	-4,579483879257D-02
				8,75	-2,764298489530D-02

I= 6 H= -3,0

0		0,	-1,246566334553D+01	-1,00	1,123410659448D+00
1		-3,502939365663D-01	2,012838047827D+00	-,75	6,677805663537D-01
2		5,610773081837D-01	1,526235234441D-01	-,50	3,419212221663D-01
3		-2,120394146977D-01	-1,035288919536D-01	-,25	1,259539317577D-01
				,25	-5,581919157196D-02
0	-9,0	1,256043080373D-03	6,280215401863D-04	,50	-6,137275523789D-02
1	-8,0	-8,179556307850D-03	-4,089778153925D-03	,75	-3,640093313477D-02
2	-7,0	2,626722351473D-02	1,313361175737D-02	1,25	3,066517238734D-02
3	-6,0	-5,806343746123D-02	-2,903171873062D-02	1,50	4,274678474383D-02
4	-5,0	1,006508774754D-01	5,032543873768D-02	1,75	3,092939070494D-02
5	-4,0	-1,411097514376D-01	-7,055487571878D-02	2,25	-3,529477596347D-02
6	-3,0	1,586781729634D-01	7,933908648168D-02	2,50	-5,503983841721D-02
7	-2,0	-1,425200230580D-01	-7,126001152899D-02	2,75	-4,361733711011D-02
8	-1,0	1,040084394184D-01	5,200421970918D-02	3,25	5,688086696872D-02
9	0,0	-6,445904509199D-02	-3,222952254600D-02	3,50	9,321275130380D-02
10	1,0	3,622506751404D-02	1,811253375702D-02	3,75	7,712552317159D-02
11	2,0	-1,959857398038D-02	-9,799286990188D-03	4,25	-1,092144106254D-01
12	3,0	1,049965740589D-02	5,249828702944D-03	4,50	-1,878369671496D-01
13	4,0	-5,590394587789D-03	-2,795197293894D-03	4,75	-1,652986693463D-01
14	5,0	2,931679108208D-03	1,465839554104D-03	5,25	2,911158083608D-01
15	6,0	-1,461277517876D-03	-7,306387589382D-04	5,50	6,282767933687D-01
16	7,0	6,286480855147D-04	3,143240427573D-04	5,75	8,976348513202D-01
17	8,0	-1,933325717557D-04	-9,666628587785D-05	6,25	8,962514335240D-01
18	9,0	2,958344857564D-05	1,479172428782D-05	6,50	6,262491618624D-01
				6,75	2,895961870204D-01
	0	-1,122821167809D+02	-5,746017775027D+01	7,25	-1,634917440925D-01
	1	-4,177648016081D+01	-2,039745927490D+01	7,50	-1,849881696049D-01
	2	-5,163986888655D+00	-2,642656559302D+00	7,75	-1,069420440914D-01
	3	-2,120394146977D-01	-1,035288919536D-01	8,25	7,415660646966D-02
				8,50	8,837085284289D-02

CARDINAL SPLINES N=4

				8.75	5,291205102920D-02

I= 7	H= -2.0				
0		0.	2.332751252435D+01	-1.00	-6,017248937367D-01
1		1,875994213095D-01	-4,000667377446D+00	-.75	-3,576701245408D-01
2		-3,005253324691D-01	8,911022384401D-02	-.50	-1,831310612668D-01
3		1.136001399581D-01	5,213890377122D-02	-.25	-6,745769079355D-02
				.25	2,989198308325D-02
0	-9,0	-6,742287985370D-04	-3,371143992685D-04	.50	3,286312761977D-02
1	-8,0	4,402653805410D-03	2,201326902705D-03	.75	1,948912683255D-02
2	-7,0	-1,427535526688D-02	-7.137677633441D-03	1.25	-1,641098671647D-02
3	-6,0	3,280491390967D-02	1,640245695484D-02	1.50	-2,286782911227D-02
4	-5,0	-6,266336348374D-02	-3,133168174187D-02	1.75	-1,653657155177D-02
5	-4,0	1,032541077308D-01	5.162705386541D-02	2.25	1,883130012556D-02
6	-3,0	-1,425200230580D-01	-7,126001152899D-02	2,50	2,931275305774D-02
7	-2,0	1,594341480790D-01	7,971707403951D-02	2,75	2,316825266797D-02
8	-1,0	-1,429229898383D-01	-7,146149491917D-02	3.25	-2,993126929762D-02
9	0,0	1,042205232853D-01	5,211026164267D-02	3,50	-4,865694074688D-02
10	1,0	-6,456575787916D-02	-3,228287893958D-02	3,75	-3,979952104995D-02
11	2,0	3,626945918079D-02	1,813472959039D-02	4,25	5,412726674954D-02
12	3,0	-1,959857398038D-02	-9,799286990188D-03	4,50	8,989430528420D-02
13	4,0	1,045542905042D-02	5,227714525212D-03	4,75	7,512287507190D-02
14	5,0	-5,485426529968D-03	-2,742713264984D-03	5,25	-1,077546235177D-01
15	6,0	2,734459024596D-03	1,367229512298D-03	5,50	-1,860723481980D-01
16	7,0	-1.176404772425D-03	-5,882023862125D-04	5,75	-1,642314212681D-01
17	8,0	3,617902090135D-04	1,808951045067D-04	6,25	2,903357549790D-01
18	9,0	-5,536066768994D-05	-2,768033384497D-05	6,50	6,273330088266D-01
				6,75	8,970636167679D-01
	0	6,016034489122D+01	3,254869510791D+01	7,25	8,966695511213D-01
	1	2,238297744668D+01	1.027307026815D+01	7,50	6,267554697407D-01
	2	2,766678446399D+00	1,496860625667D+00	7,75	2,899029561641D-01
	3	1,136001399581D-01	5,213890377122D-02	8,25	-1,637169968260D-01
				8,50	-1,852615918229D-01
				8,75	-1,071082490168D-01

I= 8	H= -1.0				
0		0.	-4,360099220079D+01	-1,00	3,221353329423D-01
1		-1,004287416712D-01	7,602673351591D+00	-.75	1,914787731114D-01
2		1,608871016211D-01	-3,669636152571D-01	-.50	9,803858244712D-02
3		-6,081948964998D-02	-2,169765034573D-02	-.25	3,611293379490D-02
				.25	-1,600202405065D-02
0	-9,0	3,611297000517D-04	1,805648500259D-04	.50	-1,759221031078D-02
1	-8,0	-2,359607548864D-03	-1,179803774432D-03	.75	-1,043257879163D-02
2	-7,0	7,667643214524D-03	3,833821607262D-03	1.25	8,783960051530D-03
3	-6,0	-1,777694939426D-02	-8,888474697130D-03	1,50	1,223889362255D-02
4	-5,0	3,526845043203D-02	1,763422521602D-02	1,75	8,849240565165D-03
5	-4.0	-6,405715939025D-02	-3,202857969512D-02	2,25	-1,007240749072D-02
6	-3.0	1,040084394184D-01	5,200421970918D-02	2,50	-1,567212177703D-02
7	-2.0	-1,429229898383D-01	-7,146149491917D-02	2,75	-1,237942484544D-02
8	-1.0	1,596463397275D-01	7,982316986375D-02	3,25	1,595835159422D-02
9	0.0	-1,430297732402D-01	-7,151488662010D-02	3,50	2,589371444965D-02
10	1.0	1,042649519328D-01	5,213247596642D-02	3,75	2,112317030639D-02
11	2,0	-6,456575787916D-02	-3,228287893958D-02	4,25	-2,845614692253D-02
12	3,0	3,622506751404D-02	1,811253375702D-02	4,50	-4,687912582803D-02
13	4,0	-1,949218573541D-02	-9,746092867703D-03	4,75	-3,872654495120D-02

14	5.0	1.024661216347D-02	5.123306081737D-03	5.25	5.334494744727D-02
15	6.0	-5.110187293533D-03	-2.555093646766D-03	5.50	8.894843581078D-02
16	7.0	2.198709609093D-03	1.099354804547D-03	5.75	7.455065548685D-02
17	8.0	-6.762063830649D-04	-3.381031915324D-04	6.25	-1.073360217638D-01
18	9.0	1.034729911545D-04	5.173649557727D-05	6.50	-1.855655344219D-01
				6.75	-1.639243798953D-01
	0	-3.220941139857D+01	-2.071857197433D+01	7.25	2.901103320707D-01
	1	-1.198359689744D+01	-4.275200757050D+00	7.50	6.270593905908D-01
	2	-1.481239118928D+00	-9.528001745919D-01	7.75	8.968972973860D-01
	3	-6.081948964998D-02	-2.169765034573D-02	8.25	8.967929458468D-01
				8.50	6.269064557252D-01
				8.75	2.899957275401D-01

I= 9 H= 0.0

0		0.	8.146489364992D+01	-1.00	-1.724316972368D-01
1		5.375682244480D-02	-1.427191970987D+01	-.75	-1.024940900827D-01
2		-8.611918539495D-02	7.928844283264D-01	-.50	-5.247766874577D-02
3		3.255568939709D-02	1.629156614018D-23	-.25	-1.933033734521D-02
				.25	8.565427371136D-03
0	-9.0	-1.933264469406D-04	-9.666322347028D-05	.50	9.416565685433D-03
1	-8.0	1.263366721886D-03	6.316833609430D-04	.75	5.584200547796D-03
2	-7.0	-4.107413443630D-03	-2.053706721815D-03	1.25	-4.701643495780D-03
3	-6.0	9.541961820376D-03	4.770980910188D-03	1.50	-6.550775293271D-03
4	-5.0	-1.909526792475D-02	-9.547633962373D-03	1.75	-4.736347517832D-03
5	-4.0	3.601358682820D-02	1.800679341410D-02	2.25	5.390428612401D-03
6	-3.0	-6.445904509199D-02	-3.222952254600D-02	2.50	8.386410942513D-03
7	-2.0	1.042205232853D-01	5.211026164267D-02	2.75	6.623510985779D-03
8	-1.0	-1.430297732402D-01	-7.151488662010D-02	3.25	-8.534123321128D-03
9	0.0	1.596907749834D-01	7.984538749169D-02	3.50	-1.384133607511D-02
10	1.0	-1.430297732402D-01	-7.151488662010D-02	3.75	-1.128426355848D-02
11	2.0	1.042205232853D-01	5.211026164267D-02	4.25	1.516821035439D-02
12	3.0	-6.445904509199D-02	-3.222952254600D-02	4.50	2.494124927433D-02
13	4.0	3.601358682820D-02	1.800679341410D-02	4.75	2.054816771083D-02
14	5.0	-1.909526792475D-02	-9.547633962373D-03	5.25	-2.803653874490D-02
15	6.0	9.541961820376D-03	4.770980910188D-03	5.50	-4.637144643441D-02
16	7.0	-4.107413443630D-03	-2.053706721815D-03	5.75	-3.841912783571D-02
17	8.0	1.263366721886D-03	6.316833609430D-04	6.25	5.311937591020D-02
18	9.0	-1.933264469406D-04	-9.666322347028D-05	6.50	8.867468082950D-02
				6.75	7.438427150871D-02
	0	1.724125495549D+01	1.724125495549D+01	7.25	-1.072125947765D-01
	1	6.414644008828D+00	2.816470223318D-21	7.50	-1.854145153050D-01
	2	7.928844283264D-01	7.928844283264D-01	7.75	-1.638315907085D-01
	3	3.255568939709D-02	1.629156614018D-23	8.25	2.900391448181D-01
				8.50	6.269700865348D-01
				8.75	8.968406283702D-01

I=10 H= 1.0

0		0.	-1.521897800318D+02	-1.00	9.228721623982D-02
1		-2.877115565697D-02	2.669813293372D+01	-.75	5.485587424829D-02
2		4.609187162433D-02	-1.538636733927D+00	-.50	2.808656935438D-02
3		-1.742418895851D-02	2.169765034573D-02	-.25	1.034578384324D-02
				.25	-4.584293574709D-03
0	-9.0	1.034729911545D-04	5.173649557727D-05	.50	-5.039825159475D-03
1	-8.0	-6.762063830649D-04	-3.381031915324D-04	.75	-2.988706693887D-03
2	-7.0	2.198709609093D-03	1.099354804547D-03	1.25	2.516342180374D-03

CARDINAL SPLINES N=4

3	-6.0	-5.110187293533D-03	-2.555093646766D-03	1.50	3.505990131744D-03
4	-5.0	1.024661216347D-02	5.123306081737D-03	1.75	2.534886825642D-03
5	-4.0	-1.949218573541D-02	-9.746092867703D-03	2.25	-2.884877823568D-03
6	-3.0	3.622506751404D-02	1.811253375702D-02	2.50	-4.488184718716D-03
7	-2.0	-6.456575787916D-02	-3.228287893958D-02	2.75	-3.544614221820D-03
8	-1.0	1.042649519328D-01	5.213247596642D-02	3.25	4.566567981375D-03
9	0.0	-1.430297732402D-01	-7.151488662010D-02	3.50	7.405699793297D-03
10	1.0	1.596463397275D-01	7.982316986375D-02	3.75	6.036698393853D-03
11	2.0	-1.429229898383D-01	-7.146149491917D-02	4.25	-8.110375797358D-03
12	3.0	1.040084394184D-01	5.200421970918D-02	4.50	-1.333018365336D-02
13	4.0	-6.405715939025D-02	-3.202857969512D-02	4.75	-1.097539154652D-02
14	5.0	3.526845043203D-02	1.763422521602D-02	5.25	1.494212581542D-02
15	6.0	-1.777694939426D-02	-8.888474697130D-03	5.50	2.466706175526D-02
16	7.0	7.667643214524D-03	3.833821607262D-03	5.75	2.038160126951D-02
17	8.0	-2.359607548864D-03	-1.179803774432D-03	6.25	-2.791304640064D-02
18	9.0	3.611297000517D-04	1.805648500259D-04	6.50	-4.622037074937D-02
				6.75	-3.832631417898D-02
	0	-9.227732550097D+00	-2.071857197433D+01	7.25	5.304817915390D-02
	1	-3.433195383337D+00	4.275200757050D+00	7.50	8.858536821350D-02
	2	-4.243612302555D-01	-9.528001745919D-01	7.75	7.432759859904D-02
	3	-1.742418895851D-02	2.169765034573D-02	8.25	-1.071649106601D-01
				8.50	-1.853508831620D-01
				8.75	-1.637881729914D-01

I=11 H= 2.0

0		0.	2.842612990495D+02	-1.00	-4.937570124155D-02
1		1.539319853899D-02	-4.988631514656D+01	-.75	-2.934910367848D-02
2		-2.466017028693D-02	2.904611027490D+00	-.50	-1.502693339318D-02
3		9.322332415630D-03	-5.213890377122D-02	-.25	-5.535221721675D-03
				.25	2.452697056862D-03
0	-9.0	-5.536066768994D-05	-2.768033384497D-05	.50	2.696415744500D-03
1	-8.0	3.617902090135D-04	1.808951045067D-04	.75	1.599022348594D-03
2	-7.0	-1.176404772425D-03	-5.882023862125D-04	1.25	-1.346295552966D-03
3	-6.0	2.734459024596D-03	1.367229512298D-03	1.50	-1.875775856457D-03
4	-5.0	-5.485426529968D-03	-2.742713264984D-03	1.75	-1.356213758527D-03
5	-4.0	1.045542905042D-02	5.227714525212D-03	2.25	1.543456890309D-03
6	-3.0	-1.959857398038D-02	-9.799286990188D-03	2.50	2.401240259811D-03
7	-2.0	3.626945918079D-02	1.813472959039D-02	2.75	1.896403261060D-03
8	-1.0	-6.456575787916D-02	-3.228287893958D-02	3.25	-2.443094484178D-03
9	0.0	1.042205232853D-01	5.211026164267D-02	3.50	-3.961928265100D-03
10	1.0	-1.429229898383D-01	-7.146149491917D-02	3.75	-3.229429583412D-03
11	2.0	1.594341480790D-01	7.971707403951D-02	4.25	4.338274134401D-03
12	3.0	-1.425200230580D-01	-7.126001152899D-02	4.50	7.129660988074D-03
13	4.0	1.032541077308D-01	5.162705386541D-02	4.75	5.869357924998D-03
14	5.0	-6.266336348374D-02	-3.133168174187D-02	5.25	-7.986612531861D-03
15	6.0	3.280491390967D-02	1.640245695484D-02	5.50	-1.317888111027D-02
16	7.0	-1.427535526688D-02	-7.137677633441D-03	5.75	-1.088248323528D-02
17	8.0	4.402653805410D-03	2.201326902705D-03	6.25	1.487089646707D-02
18	9.0	-6.742287985370D-04	-3.371143992685D-04	6.50	2.457772251446D-02
				6.75	2.032491783555D-02
	0	4.937045324604D+00	3.254869510791D+01	7.25	-2.786535994309D-02
	1	1.836836910372D+00	-1.027307026815D+01	7.50	-4.615673817203D-02
	2	2.270428049351D-01	1.496860625667D+00	7.75	-3.828289757248D-02
	3	9.322332415630D-03	-5.213890377122D-02	8.25	5.300476500497D-02
				8.50	8.852174309837D-02
				8.75	7.427992105625D-02

CARDINAL SPLINES N=4

I=12	H= 3.0				
0		0.	-5,305650547620D+02	-1.00	2,638518624346D-02
1		-8,225754055408D-03	9,312279808705D+01	-.75	1,568345431741D-02
2		1,317780139744D-02	-5,437936642049D+00	-.50	8,030031225891D-03
3		-4,981630790610D-03	1,035288919536D-01	-.25	2,957889082295D-03
				.25	-1,310662101985D-03
0	-9,0	2,958344857564D-05	1,479172428782D-05	.50	-1,440899406478D-03
1	-8,0	-1,933325717557D-04	-9,666628587785D-05	.75	-8,544788315848D-04
2	-7,0	6,286480855147D-04	3,143240427573D-04	1.25	7,194275517364D-04
3	-6,0	-1,461277517876D-03	-7,306387589382D-04	1.50	1,002368685506D-03
4	-5,0	2,931679108208D-03	1,465839554104D-03	1.75	7,247271628069D-04
5	-4,0	-5,590394587789D-03	-2,795197293894D-03	2.25	-8,247841873682D-04
6	-3,0	1,049965740589D-02	5,249828702944D-03	2.50	-1,283160370069D-03
7	-2,0	-1,959857398038D-02	-9,799286990188D-03	2,75	-1,013386905286D-03
8	-1.0	3,622506751404D-02	1,811253375702D-02	3.25	1,305516158559D-03
9	0,0	-6,445904509199D-02	-3,222952254600D-02	3,50	2,117124241540D-03
10	1.0	1,040084394184D-01	5,200421970918D-02	3.75	1,725688133694D-03
11	2,0	-1,425200230580D-01	-7.126001152899D-02	4.25	-2,318152264268D-03
12	3,0	1,586781729634D-01	7,933908648168D-02	4,50	-3,809639985078D-03
13	4,0	-1,411097514376D-01	-7,055487571878D-02	4.75	-3,136110865645D-03
14	5,0	1,006508774754D-01	5,032543873768D-02	5,25	4,266904563648D-03
15	6,0	-5,806343746123D-02	-2,903171873062D-02	5,50	7,040208003789D-03
16	7,0	2,626722351473D-02	1,313361175737D-02	5.75	5,812630047097D-03
17	8,0	-8.179556307850D-03	-4,089778153925D-03	6,25	-7,938916925114D-03
18	9,0	1.256043080373D-03	6,280215401863D-04	6.50	-1,311524788969D-02
				6,75	-1,083907218376D-02
	0	-2,638238719660D+00	-5,746017775027D+01	7,25	1,482749626866D-02
	1	-9,815616110196D-01	2,039745927490D+01	7.50	2,451412224849D-02
	2	-1,213262299490D-01	-2,642656559302D+00	7.75	2,027726159822D-02
	3	-4,981630790610D-03	1,035288919536D-01	8,25	-2,780872829727D-02
				8.50	-4,606749673193D-02
				8,75	-3,821176286943D-02

I=13	H= 4.0				
0		0.	9,872111747224D+02	-1.00	-1,404025619327D-02
1		4,377141435951D-03	-1,732873591072D+02	-.75	-8,345581259493D-03
2		-7,012257026046D-03	1,012818194293D+01	-.50	-4,272992190896D-03
3		2,650857731270D-03	-1,959767404875D-01	-.25	-1,573971075167D-03
				.25	6,974379860868D-04
0	-9,0	-1,574214117456D-05	-7,871070587279D-06	.50	7,667406923946D-04
1	-8,0	1,028774524658D-04	5,143872623289D-05	.75	4,546907830023D-04
2	-7,0	-3,345209845911D-04	-1,672604922955D-04	1.25	-3,828264013476D-04
3	-6,0	7,775903606153D-04	3,887951803077D-04	1,50	-5,333868217686D-04
4	-5,0	-1,560072732322D-03	-7,800363661610D-04	1,75	-3,856464111771D-04
5	-4,0	2,975194931973D-03	1,487597465987D-03	2.25	4,388892571068D-04
6	-3,0	-5,590394587789D-03	-2,795197293894D-03	2,50	6,828030429942D-04
7	-2,0	1,045542905042D-02	5,227714525212D-03	2,75	5,392493513793D-04
8	-1.0	-1,949218573541D-02	-9,746092867703D-03	3,25	-6,946979187264D-04
9	0,0	3,601358682820D-02	1,800679341410D-02	3,50	-1,126573614800D-03
10	1,0	-6,405715939025D-02	-3,202857969512D-02	3,75	-9,182793124439D-04
11	2,0	1,032541077308D-01	5,162705386541D-02	4,25	1,233535935260D-03
12	3,0	-1,411097514376D-01	-7,055487571878D-02	4,50	2,027176268534D-03
13	4,0	1,560498987586D-01	7,802494937931D-02	4,75	1,668767070542D-03
14	5,0	-1,362595664115D-01	-6,812978320573D-02	5,25	-2,270417231267D-03

15	6,0	9,208125359310D-02	4,604062679655D-02	5,50	-3,746004542782D-03
16	7,0	-4,588404215680D-02	-2,294202107840D-02	5,75	-3,092724399294D-03
17	8,0	1,491112274796D-02	7,455561373979D-03	6,25	4,223565462028D-03
18	9,0	-2,317625876841D-03	-1,158812938421D-03	6,50	6,976716380991D-03
				6,75	5,765066829225D-03
	0	1,403876739910D+00	1,051406363196D+02	7,25	-7,882415839902D-03
	1	5,223149436658D-01	-3,860243207292D+01	7,50	-1,302622315337D-02
	2	6,456090171825D-02	4,836809949768D+00	7,75	-1,076811701182D-02
	3	2.650857731270D-03	-1.959767404875D-01	8,25	1,473502835072D-02
				8,50	2,436363891519D-02
				8,75	2,015428363175D-02

I=14 H= 5.0

0		0.	-1.811817476765D+03	-1.00	7,361920674767D-03
1		-2,295126772742D-03	3,181193330495D+02	-.75	4,375953424369D-03
2		3,676833190368D-03	-1,860235256400D+01	-.50	2,240516772920D-03
3		-1.389960711656D-03	3.616861416341D-01	-.25	8,253019037033D-04
				.25	-3,656972511051D-04
0	-9,0	8,254294029908D-06	4.127147014954D-06	.50	-4,020356910640D-04
1	-8,0	-5,394315856943D-05	-2,697157928471D-05	.75	-2,384142699552D-04
2	-7,0	1,754040731005D-04	8,770203655023D-05	1.25	2,007326277225D-04
3	-6,0	-4,077253604778D-04	-2,038626802389D-04	1.50	2,796780383518D-04
4	-5,0	8,180202468775D-04	4,090101234388D-04	1.75	2,022112758332D-04
5	-4.0	-1,560072732322D-03	-7.800363661610D-04	2.25	-2,301288121197D-04
6	-3,0	2.931679108208D-03	1.465839554104D-03	2.50	-3,580234447378D-04
7	-2,0	-5,485426529968D-03	-2,742713264984D-03	2.75	-2,827519522007D-04
8	-1,0	1,024661216347D-02	5.123306081737D-03	3,25	3,642603005346D-04
9	0,0	-1,909526792475D-02	-9,547633962373D-03	3.50	5,907113303702D-04
10	1.0	3,526845043203D-02	1,763422521602D-02	3.75	4,814934189827D-04
11	2.0	-6,266336348374D-02	-3,133168174187D-02	4,25	-6,467951514484D-04
12	3,0	1,006508774754D-01	5,032543873768D-02	4.50	-1,062933127899D-03
13	4,0	-1,362595664115D-01	-6,812978320573D-02	4,75	-8,750027186888D-04
14	5,0	1.471001471999D-01	7.355007359995D-02	5,25	1,190465194305D-03
15	6,0	-1.204469231937D-01	-6,022346159684D-02	5.50	1,964160846464D-03
16	7,0	6.960801268071D-02	3,480400634035D-02	5,75	1,621611060570D-03
17	8,0	-2,492997052124D-02	-1,246498526062D-02	6,25	-2,214486874343D-03
18	9,0	4,094801642499D-03	2,047400821250D-03	6.50	-3,657926650673D-03
				6,75	-3,022553331343D-03
	0	-7,361140113321D-01	-1,918648397518D+02	7.25	4,132171293026D-03
	1	-2,738725822785D-01	7.116671931460D+01	7.50	6,828007806234D-03
	2	-3,385210602434D-02	-8,836826739879D+00	7,75	5,643555954364D-03
	3	-1,389960711656D-03	3,616861416341D-01	8,25	-7,718679138388D-03
				8,50	-1,275688172465D-02
				8,75	-1,054623959879D-02

I=15 H= 6.0

0		0.	3,120861964446D+03	-1.00	-3,669391844164D-03
1		1.143956831967D-03	-5,486581403726D+02	-.75	-2,181100355767D-03
2		-1,832638833923D-03	3,212433074263D+01	-.50	-1,116737646749D-03
3		6,927961782733D-04	-6,261256852840D-01	-.25	-4,113540753976D-04
				.25	1,822739700478D-04
0	-9,0	-4,114176317512D-06	-2,057088158756D-06	.50	2,003860877846D-04
1	-8,0	2,688681421304D-05	1,344340710652D-05	.75	1,188324913573D-04
2	-7,0	-8,742641684373D-05	-4,371320842187D-05	1.25	-1,000508827507D-04
3	-6,0	2,032220636012D-04	1.016110318006D-04	1.50	-1.393995328926D-04

CARDINAL SPLINES N=4

4	-5.0	-4.077253604778D-04	-2.038626802389D-04	1.75	-1.007878825261D-04
5	-4.0	7.775903606153D-04	3.887951803077D-04	2.25	1.147027791493D-04
6	-3.0	-1.461277517876D-03	-7.306387589382D-04	2.50	1.784491179617D-04
7	-2.0	2.734459024596D-03	1.367229512298D-03	2.75	1.409316518035D-04
8	-1.0	-5.110187293533D-03	-2.555093646766D-03	3.25	-1.815577283145D-04
9	0.0	9.541961820376D-03	4.770980910188D-03	3.50	-2.944273651462D-04
10	1.0	-1.777694939426D-02	-8.888474697130D-03	3.75	-2.399900232206D-04
11	2.0	3.280491390967D-02	1.640245695484D-02	4.25	3.223810077404D-04
12	3.0	-5.806343746123D-02	-2.903171873062D-02	4.50	5.297958057799D-04
13	4.0	9.208125359310D-02	4.604062679655D-02	4.75	4.361258243806D-04
14	5.0	-1.204469231937D-01	-6.022346159684D-02	5.25	-5.933603102152D-04
15	6.0	1.191621592774D-01	5.958107963870D-02	5.50	-9.789901359832D-04
16	7.0	-8.074097398172D-02	-4.037048699086D-02	5.75	-8.082527880561D-04
17	8.0	3.258616639639D-02	1.629308319820D-02	6.25	1.103750607963D-03
18	9.0	-5.819598464034D-03	-2.909799232017D-03	6.50	1.823184194300D-03
				6.75	1.506489944069D-03
	0	3.669002799012D-01	3.285638666744D+02	7.25	-2.059486382723D-03
	1	1.365059291418D-01	-1.225687285292D+02	7.50	-3.403020857584D-03
	2	1.687285797946D-02	1.521893723996D+01	7.75	-2.812607095295D-03
	3	6.927961782733D-04	-6.261256852840D-01	8.25	3.846345722976D-03
				8.50	6.356326050406D-03
				8.75	5.254078289971D-03
I=16	H= 7.0				
0		0.	-3.787204920679D+03	-1.00	1.578577370981D-03
1		-4.921317876551D-04	6.715179487047D+02	-.75	9.383123448042D-04
2		7.884037233584D-04	-3.963750539524D+01	-.50	4.804220571631D-04
3		-2.980418599675D-04	7.787530511329D-01	-.25	1.769650836857D-04
				.25	-7.841451023827D-05
0	-9.0	1.769924264260D-06	8.849621321301D-07	.50	-8.620636795059D-05
1	-8.0	-1.156674420846D-05	-5.783372104230D-06	.75	-5.112189965606D-05
2	-7.0	3.761096460870D-05	1.880548230435D-05	1.25	4.304202601552D-05
3	-6.0	-8.742641684373D-05	-4.371320842187D-05	1.50	5.996986885306D-05
4	-5.0	1.754040731005D-04	8.770203655023D-05	1.75	4.335908424881D-05
5	-4.0	-3.345209845911D-04	-1.672604922955D-04	2.25	-4.934529153293D-05
6	-3.0	6.286480855147D-04	3.143240427573D-04	2.50	-7.676905290571D-05
7	-2.0	-1.176404772425D-03	-5.882023862125D-04	2.75	-6.062898772368D-05
8	-1.0	2.198709609093D-03	1.099354804547D-03	3.25	7.810637960729D-05
9	0.0	-4.107413443630D-03	-2.053706721815D-03	3.50	1.266630477563D-04
10	1.0	7.667643214524D-03	3.833821607262D-03	3.75	1.032440279985D-04
11	2.0	-1.427535526688D-02	-7.137677633441D-03	4.25	-1.386887289996D-04
12	3.0	2.626722351473D-02	1.313361175737D-02	4.50	-2.279188302981D-04
13	4.0	-4.588404215680D-02	-2.294202107840D-02	4.75	-1.876218676286D-04
14	5.0	6.960801268071D-02	3.480400634035D-02	5.25	2.552642534972D-04
15	6.0	-8.074097398172D-02	-4.037048699086D-02	5.50	4.211625068327D-04
16	7.0	6.273140866805D-02	3.136570433403D-02	5.75	3.477109930403D-04
17	8.0	-2.812492228399D-02	-1.406246114200D-02	6.25	-4.748337082660D-04
18	9.0	5.426195316493D-03	2.713097658246D-03	6.50	-7.843331978164D-04
				6.75	-6.480902948666D-04
	0	-1.578410004132D-01	-3.864703450755D+02	7.25	8.859831853946D-04
	1	-5.872503673931D-02	1.472798430156D+02	7.50	1.463958702573D-03
	2	-7.258726495764D-03	-1.861117301466D+01	7.75	1.209956960047D-03
	3	-2.980418599675D-04	7.787530511329D-01	8.25	-1.654616174718D-03
				8.50	-2.734292010027D-03
				8.75	-2.260063231312D-03

CARDINAL SPLINES N=4

I=17	H= 8.0				
0		0.	2,484050867128D+03	-1.00	-4,854701477942D-04
1		1,513484838085D-04	-4,462220670844D+02	-.75	-2,885652873772D-04
2		-2,424629157467D-04	2,668961956958D+01	-.50	-1,477473143708D-04
3		9,165874823892D-05	-5,314233607850D-01	-.25	-5,442322112754D-05
				.25	2,411532343676D-05
0	-9,0	-5,443163007231D-07	-2,721581503616D-07	.50	2,651160402635D-05
1	-8,0	3,557195951047D-06	1,778597975524D-06	.75	1,572184970673D-05
2	-7,0	-1,156674420846D-05	-5,783372104230D-06	1,25	-1,323699370545D-05
3	-6,0	2,688681421304D-05	1,344340710652D-05	1,50	-1,844292311200D-05
4	-5,0	-5,394315856943D-05	-2,697157928471D-05	1,75	-1,333450067676D-05
5	-4,0	1,028774524658D-04	5,143872623289D-05	2,25	1,517547785274D-05
6	-3,0	-1,933325717557D-04	-9,666628587785D-05	2,50	2,360928520255D-05
7	-2,0	3,617902090135D-04	1,808951045067D-04	2,75	1,864562615856D-05
8	-1,0	-6,762063830649D-04	-3,381031915324D-04	3,25	-2,402056183510D-05
9	0,0	1,263366721886D-03	6,316833609430D-04	3,50	-3,895350898442D-05
10	1,0	-2,359607548864D-03	-1,179803774432D-03	3,75	-3,175130570728D-05
11	2,0	4,402653805410D-03	2,201326902705D-03	4,25	4,265184375196D-05
12	3,0	-8,179556307850D-03	-4,089778153925D-03	4,50	7,009335403300D-05
13	4,0	1,491112274796D-02	7,455561373979D-03	4,75	5,770056678418D-05
14	5,0	-2,492997052124D-02	-1,246498526062D-02	5,25	-7,850306091336D-05
15	6,0	3,258616639639D-02	1,629308319820D-02	5,50	-1,295228107943D-04
16	7,0	-2,812492228399D-02	-1,406246114200D-02	5,75	-1,069337842108D-04
17	8,0	1,367218561403D-02	6,836092807016D-03	6,25	1,460285932175D-04
18	9,0	-2,800957121473D-03	-1,400478560736D-03	6,50	2,412108415496D-04
				6,75	1,993111307952D-04
	0	4,854186764496D-02	2,425038184915D+02	7,25	-2,724713460426D-04
	1	1,806009182242D-02	-9,494479150273D+01	7,50	-4,502187744137D-04
	2	2,232323286704D-03	1,234118882838D+01	7,75	-3,721035919347D-04
	3	9,165874823892D-05	-5,314233607850D-01	8,25	5,088482035641D-04
				8,50	8,408786627200D-04
				8,75	6,950329271076D-04

I=18	H= 9.0				
0		0.	-6,471710666002D+02	-1.00	7,428584717673D-05
1		-2,315909719582D-05	1,178999675321D+02	-.75	4,415578781919D-05
2		3,710127839378D-05	-7,159800495023D+00	-.50	2,260805214475D-05
3		-1,402547158713D-05	1,449265972410D-01	-.25	8,327752192115D-06
				.25	-3,690087309250D-06
0	-9,0	8,329038917438D-08	4,164519458719D-08	.50	-4,056762241691D-06
1	-8,0	-5,443163007231D-07	-2,721581503616D-07	.75	-2,405731701312D-06
2	-7,0	1,769924264260D-06	8,849621321301D-07	1,25	2,025503104249D-06
3	-6,0	-4,114176317512D-06	-2,057088158756D-06	1,50	2,822105898380D-06
4	-5,0	8,254294029908D-06	4,127147014954D-06	1,75	2,040423461107D-06
5	-4,0	-1,574214117456D-05	-7,871070587279D-06	2,25	-2,322126773753D-06
6	-3,0	2,958344857564D-05	1,479172428782D-05	2,50	-3,612654164659D-06
7	-2,0	-5,536066768994D-05	-2,768033384497D-05	2,75	-2,853123184119D-06
8	-1,0	1,034729911545D-04	5,173649557727D-05	3,25	3,675587035930D-06
9	0,0	-1,933264469406D-04	-9,666322347028D-05	3,50	5,960602148354D-06
10	1,0	3,611297000517D-04	1,805648500259D-04	3,75	4,858532791113D-06
11	2,0	-6,742287985370D-04	-3,371143992685D-04	4,25	-6,526515233195D-06
12	3,0	1,256043080373D-03	6,280215401863D-04	4,50	-1,072557011646D-05
13	4,0	-2,317625876841D-03	-1,158812938421D-03	4,75	-8,829246071742D-06
14	5,0	4,094801642499D-03	2,047400821250D-03	5,25	1,201240931809D-05
15	6,0	-5,819598464034D-03	-2,909799232017D-03	5,50	1,981936717071D-05

16	7.0	5.426195316493D-03	2.713097658246D-03	5.75	1.636283106064D-05
17	8.0	-2.800957121473D-03	-1.400478560736D-03	6.25	-2.234505190256D-05
18	9.0	6.001643214779D-04	3.000821607389D-04	6.50	-3.690967765595D-05
				6.75	-3.049825050563D-05
	0	-7.427797111887D-03	-6.036370951923D+01	7.25	4.169308382041D-05
	1	-2.763525681781D-03	2.424072175127D+01	7.50	6.889165494381D-05
	2	-3.415864544588D-04	-3.246782369516D+00	7.75	5.693858120140D-05
	3	-1.402547158713D-05	1.449265972410D-01	8.25	-7.786282979156D-05
				8.50	-1.286691843856D-04
				8.75	-1.063519747708D-04

N= 4 M=19 P= 9.5 Q= 7

I= 0 H= -9.5

0		1.000000000000D+00	5.222802907204D-01	-1.00	4.666983739612D+00
1		-2.043924731457D+00	-1.025656575845D+00	-.75	3.405151662796D+00
2		1.333191787644D+00	6.667999295007D-01	-.50	2.391493715203D+00
3		-2.898672205113D-01	-1.449373640359D-01	-.25	1.598834844913D+00
				.25	5.678141651741D-01
0	-9.5	6.001643247898D-04	3.000821623949D-04	.50	2.751068674022D-01
1	-8.5	-2.800957143117D-03	-1.400478571558D-03	.75	9.476921057023D-02
2	-7.5	5.426195386872D-03	2.713097693436D-03	1.25	-3.507902563532D-02
3	-6.5	-5.819598627629D-03	-2.909799313814D-03	1.50	-3.427495654769D-02
4	-5.5	4.094801970721D-03	2.047400985361D-03	1.75	-1.768214275685D-02
5	-4.5	-2.317626502808D-03	-1.158813251404D-03	2.25	1.053382464267D-02
6	-3.5	1.256044256722D-03	6.280221283611D-04	2.50	1.195817049733D-02
7	-2.5	-6.742309998892D-04	-3.371154999446D-04	2.75	6.908482275911D-03
8	-1.5	3.611338145587D-04	1.805669072794D-04	3.25	-4.774218950051D-03
9	-.5	-1.933341345863D-04	-9.666706729313D-05	3.50	-5.681573411159D-03
10	.5	1.034873531445D-04	5.174367657225D-05	3.75	-3.398301073413D-03
11	1.5	-5.538749482130D-05	-2.769374741065D-05	4.25	2.450953888705D-03
12	2.5	2.963353439666D-05	1.481676719833D-05	4.50	2.953650578733D-03
13	3.5	-1.583544748626D-05	-7.917723743130D-06	4.75	1.782442357328D-03
14	4.5	8.426460878108D-06	4.213230439054D-06	5.25	-1.299141689040D-03
15	5.5	-4.418362150928D-06	-2.209181075464D-06	5.50	-1.570303000045D-03
16	6.5	2.202238078542D-06	1.101119039271D-06	5.75	-9.496221192014D-04
17	7.5	-9.474058219833D-07	-4.737029109917D-07	6.25	6.938360421497D-04
18	8.5	2.913618634565D-07	1.456809317283D-07	6.50	8.392380981261D-04
19	9.5	-4.458371531226D-08	-2.229185765613D-08	6.75	5.077648887306D-04
				7.25	-3.712050021094D-04
	0	-1.466216342999D+02	-7.330843603263D+01	7.50	-4.490673824061D-04
	1	-5.519483071966D+01	-2.759824922804D+01	7.75	-2.717299015612D-04
	2	-6.928023996928D+00	-3.463914945522D+00	8.25	1.986760897277D-04
	3	-2.898672205113D-01	-1.449373640359D-01	8.50	2.403587001407D-04
				8.75	1.454446556578D-04
				9.25	-1.063461912514D-04
				9.50	-1.286596267963D-04

CARDINAL SPLINES N=4

I= 1	H= -8.5				
0		0.	-1.456053399216D-01	-1.00	-9.041182846643D+00
1		3.959053517930D+00	2.003669025624D+00	-.75	-5.678512175604D+00
2		-4.019190944750D+00	-2.010928878871D+00	-.50	-3.117191793148D+00
3		1.062938383963D+00	5.314937235485D-01	-.25	-1.257571225779D+00
				.25	7.551721867283D-01
0	-9.5	-2.800957143117D-03	-1.400478571558D-03	.50	1.107574438295D+00
1	-8.5	1.367218575548D-02	6.836092877740D-03	.75	1.156548480114D+00
2	-7.5	-2.812492274393D-02	-1.406246137197D-02	1.25	7.315274074587D-01
3	-6.5	3.258616746551D-02	1.629308373276D-02	1.50	4.350675323463D-01
4	-5.5	-2.492997266622D-02	-1.246498633311D-02	1.75	1.772914833414D-01
5	-4.5	1.491112683875D-02	7.455563419376D-03	2.25	-8.424021400802D-02
6	-3.5	-8.179563995487D-03	-4.089781997744D-03	2.50	-9.008354389113D-02
7	-2.5	4.402668191606D-03	2.201334095803D-03	2.75	-4.992391726465D-02
8	-1.5	-2.359634437840D-03	-1.179817218920D-03	3.25	3.281503134044D-02
9	-.5	1.263416961907D-03	6.317084809533D-04	3.50	3.847398134949D-02
10	.5	-6.763002410158D-04	-3.381501205079D-04	3.75	2.276829823125D-02
11	1.5	3.619655286998D-04	1.809827643499D-04	4.25	-1.621278197909D-02
12	2.5	-1.936598907789D-04	-9.682994538943D-05	4.50	-1.946621286290D-02
13	3.5	1.034872244567D-04	5.174361222836D-05	4.75	-1.171693075979D-02
14	4.5	-5.506829705028D-05	-2.753414852514D-05	5.25	8.514027282590D-03
15	5.5	2.887471833095D-05	1.443735916548D-05	5.50	1.028222956806D-02
16	6.5	-1.439198561432D-05	-7.195992807159D-06	5.75	6.214313694696D-03
17	7.5	6.191451827472D-06	3.095725913736D-06	6.25	-4.537263691319D-03
18	8.5	-1.904097378957D-06	-9.520486894785D-07	6.50	-5.487014343276D-03
19	9.5	2.913618634565D-07	1.456809317283D-07	6.75	-3.319353247601D-03
				7.25	2.426244638484D-03
	0	5.862158226068D+02	2.930923503128D+02	7.50	2.935029736147D-03
	1	2.153849930256D+02	1.076979459778D+02	7.75	1.775925132805D-03
	2	2.627455299819D+01	1.313664224226D+01	8.25	-1.298424739604D-03
	3	1.062938383963D+00	5.314937235485D-01	8.50	-1.570820266316D-03
				8.75	-9.505200081889D-04
				9.25	6.949951308633D-04
				9.50	8.408162023120D-04

I= 2	H= -7.5				
0		0.	4.734571132390D-01	-1.00	9.696675435247D+00
1		-3.293246659340D+00	-1.725125449753D+00	-.75	5.852797471518D+00
2		4.845624619930D+00	2.427148076708D+00	-.50	3.052755004150D+00
3		-1.557804155977D+00	-7.789818459760D-01	-.25	1.150503893518D+00
				.25	-5.448004848378D-01
0	-9.5	5.426195386872D-03	2.713097693436D-03	.50	-6.299003020331D-01
1	-8.5	-2.812492274393D-02	-1.406246137197D-02	.75	-4.007454644468D-01
2	-7.5	6.273141016360D-02	3.136570508180D-02	1.25	4.380163015955D-01
3	-6.5	-8.074097745812D-02	-4.037048872906D-02	1.50	7.976882886943D-01
4	-5.5	6.960801965543D-02	3.480400982772D-02	1.75	9.966809940452D-01
5	-4.5	-4.588405545862D-02	-2.294202772931D-02	2.25	8.267656424505D-01
6	-3.5	2.626724851221D-02	1.313362425611D-02	2.50	5.431835388265D-01
7	-2.5	-1.427540204571D-02	-7.137701022854D-03	2.75	2.397523351875D-01
8	-1.5	7.667730647970D-03	3.833865323985D-03	3.25	-1.274048095405D-01
9	-.5	-4.107576806424D-03	-2.053788403212D-03	3.50	-1.414516887872D-01
10	.5	2.199014801983D-03	1.099507400991D-03	3.75	-8.064857931314D-02
11	1.5	-1.176974850093D-03	-5.884874250463D-04	4.25	5.497455450476D-02
12	2.5	6.297124113139D-04	3.148562056570D-04	4.50	6.517847778883D-02
13	3.5	-3.365037476317D-04	-1.682518738158D-04	4.75	3.888368672560D-02
14	4.5	1.790626258522D-04	8.953131292609D-05	5.25	-2.795856120432D-02

CARDINAL SPLINES N=4

15	5.5	-9.389037857336D-05	-4.694518928668D-05	5.50	-3.366361216780D-02
16	6.5	4.679765131047D-05	2.339882565523D-05	5.75	-2.030264025076D-02
17	7.5	-2.013241345670D-05	-1.006620672835D-05	6.25	1.478712891344D-02
18	8.5	6.191451827472D-06	3.095725913736D-06	6.50	1.786993979920D-02
19	9.5	-9.474058219833D-07	-4.737029109917D-07	6.75	1.080511313910D-02
				7.25	-7.893401788457D-03
	0	-9.295905595458D+02	-4.647446809301D+02	7.50	-9.547124240850D-03
	1	-3.330018541114D+02	-1.665186467903D+02	7.75	-5.776120892625D-03
	2	-3.955179382541D+01	-1.977383453361D+01	8.25	4.222522323601D-03
	3	-1.557804155977D+00	-7.789818459760D-01	8.50	5.108174590612D-03
				8.75	3.090931652057D-03
				9.25	-2.259940331282D-03
				9.50	-2.734088910852D-03

I= 3 H= -6.5

0		0.	-1.100547684996D+00	-1.00	-6.822801504778D+00
1		2.161366357409D+00	1.263160797287D+00	-.75	-4.066886761129D+00
2		-3.408490953075D+00	-1.714323935428D+00	-.50	-2.089423941260D+00
3		1.252944194294D+00	6.266575172655D-01	-.25	-7.729495269552D-01
				.25	3.468878026207D-01
0	-9.5	-5.819598627629D-03	-2.909799313814D-03	.50	3.851329991080D-01
1	-8.5	3.258616746551D-02	1.629308373276D-02	.75	2.315576162755D-01
2	-7.5	-8.074097745812D-02	-4.037048872906D-02	1.25	-2.046505577512D-01
3	-6.5	1.191621673583D-01	5.958108367914D-02	1.50	-2.975471715565D-01
4	-5.5	-1.204469394064D-01	-6.022346970319D-02	1.75	-2.292874642110D-01
5	-4.5	9.208128451309D-02	4.604064225654D-02	2.25	3.358732218934D-01
6	-3.5	-5.806349556770D-02	-2.903174778385D-02	2.50	6.817337889484D-01
7	-2.5	3.280502264672D-02	1.640251132336D-02	2.75	9.296916823751D-01
8	-1.5	-1.777715263268D-02	-8.888576316341D-03	3.25	8.730597048918D-01
9	-.5	9.542341556025D-03	4.771170778013D-03	3.50	5.982759148520D-01
10	.5	-5.110896712230D-03	-2.555448356115D-03	3.75	2.727045599252D-01
11	1.5	2.735784166070D-03	1.367892083035D-03	4.25	-1.511709168580D-01
12	2.5	-1.463751535609D-03	-7.318757678043D-04	4.50	-1.700922846495D-01
13	3.5	7.821992791668D-04	3.910996395834D-04	4.75	-9.793234832767D-02
14	4.5	-4.162296402650D-04	-2.081148201325D-04	5.25	6.757191524102D-02
15	5.5	2.182474965560D-04	1.091237482780D-04	5.50	8.040520527961D-02
16	6.5	-1.087808045843D-04	-5.439040229213D-05	5.75	4.809187160257D-02
17	7.5	4.679765131047D-05	2.339882565523D-05	6.25	-3.468647421704D-02
18	8.5	-1.439198561432D-05	-7.195992807159D-06	6.50	-4.180144831254D-02
19	9.5	2.202238078542D-06	1.101119039271D-06	6.75	-2.522628536862D-02
				7.25	1.838660219669D-02
	0	7.871597004635D+02	3.934622335824D+02	7.50	2.222443161475D-02
	1	2.766346788542D+02	1.383585288238D+02	7.75	1.344001555399D-02
	2	3.230041858432D+01	1.614541530664D+01	8.25	-9.819934564683D-03
	3	1.252944194294D+00	6.266575172655D-01	8.50	-1.187786274119D-02
				8.75	-7.186498615184D-03
				9.25	5.253792609726D-03
				9.50	6.355853947821D-03

I= 4 H= -5.5

0		0.	2.208034771572D+00	-1.00	3.861653521826D+00
1		-1.208111862648D+00	-9.701617642357D-01	-.75	2.296781381941D+00
2		1.928779359927D+00	9.846101465349D-01	-.50	1.176846058712D+00
3		-7.247622992499D-01	-3.627531590109D-01	-.25	4.339010865833D-01
				.25	-1.928034166655D-01

CARDINAL SPLINES N=4

0	-9.5	4.094801970721D-03	2.047400985361D-03
1	-8.5	-2.492997266622D-02	-1.246498633311D-02
2	-7.5	6.960801965543D-02	3.480400982772D-02
3	-6.5	-1.204469394064D-01	-6.022346970319D-02
4	-5.5	1.471001797275D-01	7.355008986377D-02
5	-4.5	-1.362596284464D-01	-6.812981422319D-02
6	-3.5	1.006509940547D-01	5.032549702733D-02
7	-2.5	-6.266358164347D-02	-3.133179082174D-02
8	-1.5	3.526885819038D-02	1.763442909519D-02
9	-.5	-1.909602979043D-02	-9.548014895215D-03
10	.5	1.024803547404D-02	5.124017737018D-03
11	1.5	-5.488085168464D-03	-2.744042584232D-03
12	2.5	2.936642743346D-03	1.468321371673D-03
13	3.5	-1.569319630504D-03	-7.846598152520D-04
14	4.5	8.350824293449D-04	4.175412146725D-04
15	5.5	-4.378709677651D-04	-2.189354838825D-04
16	6.5	2.182474965560D-04	1.091237482780D-04
17	7.5	-9.389037857336D-05	-4.694518928668D-05
18	8.5	2.887471833095D-05	1.443735916548D-05
19	9.5	-4.418362150928D-06	-2.209181075464D-06
	0	-4.587978017811D+02	-2.291629259709D+02
	1	-1.607906965459D+02	-8.047798678229D+01
	2	-1.872694616870D+01	-9.353854885277D+00
	3	-7.247622992499D-01	-3.627531590109D-01

.50	-2.124243881081D-01
.75	-1.263580119407D-01
1.25	1.075505716003D-01
1.50	1.512817710311D-01
1.75	1.109158639212D-01
2.25	-1.327803606867D-01
2.50	-2.159589440947D-01
2.75	-1.821522579435D-01
3.25	3.032996633449D-01
3.50	6.429695350624D-01
3.75	9.065057435896D-01
4.25	8.897820836336D-01
4.50	6.184280977418D-01
4.75	2.848658196058D-01
5.25	-1.600347153585D-01
5.50	-1.808061711587D-01
5.75	-1.044114487550D-01
6.25	7.230584432644D-02
6.50	8.613119904420D-02
6.75	5.155628271450D-02
7.25	-3.721917609896D-02
7.50	-4.486541835571D-02
7.75	-2.708030441346D-02
8.25	1.974221158322D-02
8.50	2.386448659435D-02
8.75	1.443246738421D-02
9.25	-1.054566643697D-02
9.50	-1.275593454268D-02

I= 5   H= -4.5

0		0.	-4.211044330709D+00
1		6.529716084739D-01	1.024703020795D+00
2		-1.045258426085D+00	-5.611925917680D-01
3		3.946044441143D-01	1.980116982105D-01
0	-9.5	-2.317626502808D-03	-1.158813251404D-03
1	-8.5	1.491112683875D-02	7.455563419376D-03
2	-7.5	-4.588405545862D-02	-2.294202772931D-02
3	-6.5	9.208128451309D-02	4.604064225654D-02
4	-5.5	-1.362596284464D-01	-6.812981422319D-02
5	-4.5	1.560500170683D-01	7.802500853415D-02
6	-3.5	-1.411099737713D-01	-7.055498688565D-02
7	-2.5	1.032545237932D-01	5.162726189661D-02
8	-1.5	-6.405793704486D-02	-3.202896852243D-02
9	-.5	3.601503981715D-02	1.800751990857D-02
10	.5	-1.949490019619D-02	-9.747450098093D-03
11	1.5	1.046049946170D-02	5.230249730851D-03
12	2.5	-5.599860963760D-03	-2.799930481880D-03
13	3.5	2.992830115128D-03	1.496415057564D-03
14	4.5	-1.592612802182D-03	-7.963064010909D-04
15	5.5	8.350824293449D-04	4.175412146725D-04
16	6.5	-4.162296402650D-04	-2.081148201325D-04
17	7.5	1.790626258522D-04	8.953131292609D-05
18	8.5	-5.506829705028D-05	-2.753414852514D-05
19	9.5	8.426460878108D-06	4.213230439054D-06

-1.00	-2.092834478674D+00
-.75	-1.244160320889D+00
-.50	-6.371259662726D-01
-.25	-2.347372481881D-01
.25	1.040798034707D-01
.50	1.144786467728D-01
.75	6.793522574955D-02
1.25	-5.734036950610D-02
1.50	-8.006637549240D-02
1.75	-5.807569878624D-02
2.25	6.687258567238D-02
2.50	1.051034521950D-01
2.75	8.423769299162D-02
3.25	-1.143439698618D-01
3.50	-1.940186635488D-01
3.75	-1.690291863867D-01
4.25	2.938349148619D-01
4.50	6.315635372785D-01
4.75	8.996225501588D-01
5.25	8.947989440512D-01
5.50	6.244921056587D-01
5.75	2.885329657222D-01
6.25	-1.627141168883D-01
6.50	-1.840470957723D-01
6.75	-1.063723251580D-01
7.25	7.373939737645D-02

```
     0   2.501926425988D+02   1.246462827130D+02     7.50  8.786548700119D-02
     1   8.763221475680D+01   4.397371106770D+01     7.75  5.260572955047D-02
     2   1.020096823117D+01   5.082140807232D+00     8.25 -3.798655856189D-02
     3   3.946044441143D-01   1.980116982105D-01     8.50 -4.579387178678D-02
                                                     8.75 -2.764218561473D-02
                                                     9.25  2.015319052859D-02
                                                     9.50  2.436183250107D-02


I= 6     H= -3.5
0          0.                 7.913615988007D+00    -1.00  1.123411708622D+00
1         -3.502942636532D-01 -1.487273648774D+00    -.75  6.677811899868D-01
2          5.610778321826D-01  3.530092457093D-01    -.50  3.419215414705D-01
3         -2.120396127861D-01 -1.073530912973D-01    -.25  1.259540493745D-01
                                                      .25 -5.581924368882D-02
0   -9.5   1.256044256722D-03  6.280221283611D-04     .50 -6.137281253346D-02
1   -8.5  -8.179563995487D-03 -4.089781997744D-03     .75 -3.640096711206D-02
2   -7.5   2.626724851221D-02  1.313362425611D-02    1.25  3.066520099448D-02
3   -6.5  -5.806349556770D-02 -2.903174778385D-02    1.50  4.274682460177D-02
4   -5.5   1.006509940547D-01  5.032549702733D-02    1.75  3.092941952281D-02
5   -4.5  -1.411099737713D-01 -7.055498688565D-02    2.25 -3.529480875996D-02
6   -3.5   1.586785907846D-01  7.933929539230D-02    2.50 -5.503988944044D-02
7   -2.5  -1.425208049443D-01 -7.126040247215D-02    2.75 -4.361737740612D-02
8   -1.5   1.040099008277D-01  5.200495041386D-02    3.25  5.688091888077D-02
9    -.5  -6.446177562508D-02 -3.223088781254D-02    3.50  9.321283548821D-02
10    .5   3.623016867119D-02  1.811508433560D-02    3.75  7.712559179095D-02
11   1.5  -1.960810256333D-02 -9.804051281666D-03    4.25 -1.092145028025D-01
12   2.5   1.051744711639D-02  5.258723558195D-03    4.50 -1.878371186319D-01
13   3.5  -5.623535549432D-03 -2.811767774716D-03    4.75 -1.652987940459D-01
14   4.5   2.992830115128D-03  1.496415057564D-03    5.25  2.911159780178D-01
15   5.5  -1.569319630504D-03 -7.846598152520D-04    5.50  6.282770732870D-01
16   6.5   7.821992791668D-04  3.910996395834D-04    5.75  8.976350824202D-01
17   7.5  -3.365037476317D-04 -1.682518738158D-04    6.25  8.962511179343D-01
18   8.5   1.034872244567D-04  5.174361222836D-05    6.50  6.262486405697D-01
19   9.5  -1.583544748626D-05 -7.917723743130D-06    6.75  2.895957562793D-01
                                                     7.25 -1.634911552418D-01
     0  -1.344879841627D+02  -6.639825590110D+01     7.50 -1.849871966164D-01
     1  -4.709954061403D+01  -2.384594744904D+01     7.75 -1.069412399219D-01
     2  -5.482051132222D+00  -2.706553856264D+00     8.25  7.415550677929D-02
     3  -2.120396127861D-01  -1.073530912973D-01     8.50  8.836903559486D-02
                                                     8.75  5.291054897921D-02
                                                     9.25 -3.820970865260D-02
                                                     9.50 -4.606410202371D-02


I= 7     H= -2.5
0          0.                -1.480908422378D+01    -1.00 -6.017268571002D-01
1          1.876000334009D-01  2.549237855315D+00    -.75 -3.576712915715D-01
2         -3.005263130502D-01 -2.858799492153D-01    -.50 -1.831316587941D-01
3          1.136005106491D-01  5.929528775659D-02    -.25 -6.745791089476D-02
                                                      .25  2.989208061169D-02
0   -9.5  -6.742309998892D-04 -3.371154999446D-04     .50  3.286323483938D-02
1   -8.5   4.402668191606D-03  2.201334095803D-03     .75  1.948919041567D-02
2   -7.5  -1.427540204571D-02 -7.137701022854D-03    1.25 -1.641104025020D-02
3   -6.5   3.280502264672D-02  1.640251132336D-02    1.50 -2.286790370010D-02
4   -5.5  -6.266358164347D-02 -3.133179082174D-02    1.75 -1.653662547985D-02
5   -4.5   1.032545237932D-01  5.162726189661D-02    2.25  1.883136149901D-02
```

CARDINAL SPLINES N=4

6	-3.5	-1.425208049443D-01	-7.126040247215D-02	2.50	2.931284853964D-02
7	-2.5	1.594356112556D-01	7.971780562779D-02	2.75	2.316832807557D-02
8	-1.5	-1.429257246347D-01	-7.146286231737D-02	3.25	-2.993136644282D-02
9	-.5	1.042256330460D-01	5.211281652299D-02	3.50	-4.865709828468D-02
10	.5	-6.457530388763D-02	-3.228765194382D-02	3.75	-3.979964946022D-02
11	1.5	3.628729041612D-02	1.814364520806D-02	4.25	5.412743924433D-02
12	2.5	-1.963186460879D-02	-9.815932304397D-03	4.50	8.989458875937D-02
13	3.5	1.051744711639D-02	5.258723558195D-03	4.75	7.512310842751D-02
14	4.5	-5.599860963760D-03	-2.799930481880D-03	5.25	-1.077549410038D-01
15	5.5	2.936642743346D-03	1.468321371673D-03	5.50	-1.860728720208D-01
16	6.5	-1.463751535609D-03	-7.318757678043D-04	5.75	-1.642318537352D-01
17	7.5	6.297124113139D-04	3.148562056570D-04	6.25	2.903363455552D-01
18	8.5	-1.936598907789D-04	-9.682994538943D-05	6.50	6.273339843435D-01
19	9.5	2.963353439666D-05	1.481676719833D-05	6.75	8.970644228317D-01
				7.25	8.966684491802D-01
	0	7.205793838233D+01	3.444630732534D+01	7.50	6.267536489466D-01
	1	2.523493834370D+01	1.317171798032D+01	7.75	2.899014512883D-01
	2	2.937088240450D+00	1.404035751847D+00	8.25	-1.637149389295D-01
	3	1.136005106491D-01	5.929528775659D-02	8.50	-1.852581911308D-01
				8.75	-1.071054381679D-01
				9.25	7.427607691450D-02
				9.50	8.851539043907D-02

I= 8 H= -1.5

0		0.	2.767942027430D+01	-1.00	3.221390026290D-01
1		-1.004298857200D-01	-4.639634207931D+00	-.75	1.914809543870D-01
2		1.608889344072D-01	3.339236292528D-01	-.50	9.803969927451D-02
3		-6.082018250184D-02	-3.507351620174D-02	-.25	3.611334518203D-02
				.25	-1.600220633927D-02
0	-9.5	3.611338145587D-04	1.805669072794D-04	.50	-1.759241071297D-02
1	-8.5	-2.359634437840D-03	-1.179817218920D-03	.75	-1.043269763367D-02
2	-7.5	7.667730647970D-03	3.833865323985D-03	1.25	8.784060110453D-03
3	-6.5	-1.777715263268D-02	-8.888576316341D-03	1.50	1.223903303328D-02
4	-5.5	3.526885819038D-02	1.763442909519D-02	1.75	8.849341361147D-03
5	-4.5	-6.405793704486D-02	-3.202896852243D-02	2.25	-1.007252220271D-02
6	-3.5	1.040099008277D-01	5.200495041386D-02	2.50	-1.567230024048D-02
7	-2.5	-1.429257246347D-01	-7.146286231737D-02	2.75	-1.237956578842D-02
8	-1.5	1.596514512852D-01	7.982572564261D-02	3.25	1.595853316653D-02
9	-.5	-1.430393237999D-01	-7.151966189997D-02	3.50	2.589400890066D-02
10	.5	1.042827942019D-01	5.214139710096D-02	3.75	2.112341031568D-02
11	1.5	-6.459908591360D-02	-3.229954295680D-02	4.25	-2.845646932938D-02
12	2.5	3.628729041612D-02	1.814364520806D-02	4.50	-4.687965566626D-02
13	3.5	-1.960810256333D-02	-9.804051281666D-03	4.75	-3.872698111189D-02
14	4.5	1.046049946170D-02	5.230249730851D-03	5.25	5.334554085475D-02
15	5.5	-5.488085168464D-03	-2.744042584232D-03	5.50	8.894941487836D-02
16	6.5	2.735784166070D-03	1.367892083035D-03	5.75	7.455146380313D-02
17	7.5	-1.176974850093D-03	-5.884874250463D-04	6.25	-1.073371250990D-01
18	8.5	3.619655286998D-04	1.809827643499D-04	6.50	-1.855673577426D-01
19	9.5	-5.538749482130D-05	-2.769374741065D-05	6.75	-1.639258864944D-01
				7.25	2.901123916885D-01
	0	-3.857956155660D+01	-1.633165311445D+01	7.50	6.270627938036D-01
	1	-1.351060454435D+01	-7.791239763750D+00	7.75	8.969001101217D-01
	2	-1.572486266895D+00	-6.656715824968D-01	8.25	8.967890994703D-01
	3	-6.082018250184D-02	-3.507351620174D-02	8.50	6.269000995540D-01
				8.75	2.899904738341D-01
				9.25	-1.637809879767D-01

				9.50	-1.853390095232D-01

I= 9 H= -.5

0		0.	-5.171717645263D+01	-1.00	-1.724385537695D-01
1		5.375896001356D-02	8.601907853613D+00	-.75	-1.024981656315D-01
2		-8.612260981745D-02	-5.166704815155D-01	-.50	-5.247975545345D-02
3		3.255698393848D-02	2.499179567374D-02	-.25	-1.933110599102D-02
				.25	8.565767963659D-03
0	-9.5	-1.933341345863D-04	-9.666706729313D-05	.50	9.416940121801D-03
1	-8.5	1.263416961907D-03	6.317084809533D-04	.75	5.584422595179D-03
2	-7.5	-4.107576806424D-03	-2.053788403212D-03	1.25	-4.701830448325D-03
3	-6.5	9.542341556025D-03	4.771170778013D-03	1.50	-6.551035771707D-03
4	-5.5	-1.909602979043D-02	-9.548014895215D-03	1.75	-4.736535847516D-03
5	-4.5	3.601503981715D-02	1.800751990857D-02	2.25	5.390642943107D-03
6	-3.5	-6.446177562508D-02	-3.223088781254D-02	2.50	8.386744388010D-03
7	-2.5	1.042256330460D-01	5.211281652299D-02	2.75	6.623774327088D-03
8	-1.5	-1.430393237999D-01	-7.151966189997D-02	3.25	-8.534462575286D-03
9	-.5	1.597086194833D-01	7.985430974166D-02	3.50	-1.384188623459D-02
10	.5	-1.430631101732D-01	-7.153155508658D-02	3.75	-1.128471199772D-02
11	1.5	1.042827942019D-01	5.214139710096D-02	4.25	1.516881274726D-02
12	2.5	-6.457530388763D-02	-3.228765194382D-02	4.50	2.494223923707D-02
13	3.5	3.623016867119D-02	1.811508433560D-02	4.75	2.054898264416D-02
14	4.5	-1.949490019619D-02	-9.747450098093D-03	5.25	-2.803764748199D-02
15	5.5	1.024803547404D-02	5.124017737018D-03	5.50	-4.637327574829D-02
16	6.5	-5.110896712230D-03	-2.555448356115D-03	5.75	-3.842063811366D-02
17	7.5	2.199014801983D-03	1.099507400991D-03	6.25	5.312143834282D-02
18	8.5	-6.763002410158D-04	-3.381501205079D-04	6.50	8.867808756662D-02
19	9.5	1.034873531445D-04	5.174367657225D-05	6.75	7.438708647536D-02
				7.25	-1.072164430169D-01
	0	2.065168868836D+01	4.798778015686D+00	7.50	-1.854208739511D-01
	1	7.232232774825D+00	5.551697383483D+00	7.75	-1.638368460930D-01
	2	8.417514324292D-01	1.955956951860D-01	8.25	2.900463314824D-01
	3	3.255698393848D-02	2.499179567374D-02	8.50	6.269819625609D-01
				8.75	8.968504445234D-01
				9.25	8.968272037126D-01
				9.50	6.269479015370D-01

I=10 H= .5

0		0.	9.661975546509D+01	-1.00	9.230002555194D-02
1		-2.877514904355D-02	-1.603454427068D+01	-.75	5.486348815174D-02
2		4.609826909940D-02	9.078618718876D-01	-.50	2.809046772275D-02
3		-1.742660740900D-02	-2.499179567374D-02	-.25	1.034721982036D-02
				.25	-4.584929866573D-03
0	-9.5	1.034873531445D-04	5.174367657225D-05	.50	-5.040524678101D-03
1	-8.5	-6.763002410158D-04	-3.381501205079D-04	.75	-2.989121520794D-03
2	-7.5	2.199014801983D-03	1.099507400991D-03	1.25	2.516691443423D-03
3	-6.5	-5.110896712230D-03	-2.555448356115D-03	1.50	3.506476755195D-03
4	-5.5	1.024803547404D-02	5.124017737018D-03	1.75	2.535238661449D-03
5	-4.5	-1.949490019619D-02	-9.747450098093D-03	2.25	-2.885278234249D-03
6	-3.5	3.623016867119D-02	1.811508433560D-02	2.50	-4.488807658569D-03
7	-2.5	-6.457530388763D-02	-3.228765194382D-02	2.75	-3.545106193666D-03
8	-1.5	1.042827942019D-01	5.214139710096D-02	3.25	4.567201772916D-03
9	-.5	-1.430631101732D-01	-7.153155508658D-02	3.50	7.406727596251D-03
10	.5	1.597086194833D-01	7.985430974166D-02	3.75	6.037536163968D-03
11	1.5	-1.430393237999D-01	-7.151966189997D-02	4.25	-8.111501182250D-03

12	2.5	1.042256330460D-01	5.211281652299D-02	4.50	-1.333203309277D-02
13	3.5	-6.446177562508D-02	-3.223088781254D-02	4.75	-1.097691399758D-02
14	4.5	3.601503981715D-02	1.800751990857D-02	5.25	1.494419714800D-02
15	5.5	-1.909602979043D-02	-9.548014895215D-03	5.50	2.467047926286D-02
16	6.5	9.542341556025D-03	4.771170778013D-03	5.75	2.038442275708D-02
17	7.5	-4.107576806424D-03	-2.053788403212D-03	6.25	-2.791689941855D-02
18	8.5	1.263416961907D-03	6.317084809533D-04	6.50	-4.622673518476D-02
19	9.5	-1.933341345863D-04	-9.666706729313D-05	6.75	-3.833157307415D-02
				7.25	5.305536840167D-02
	0	-1.105413265698D+01	4.798778015686D+00	7.50	8.859724737858D-02
	1	-3.871161992141D+00	-5.551697383483D+00	7.75	7.433741666060D-02
	2	-4.505600420571D-01	1.955956951860D-01	8.25	-1.071783367212D-01
	3	-1.742660740900D-02	-2.499179567374D-02	8.50	-1.853730698464D-01
				8.75	-1.638065114397D-01
				9.25	2.900208073639D-01
				9.50	6.269479015370D-01

I=11 H= 1.5

0		0.	-1.804964471439D+02	-1.00	-4.939962808593D-02
1		1.540065788873D-02	2.993515434281D+01	-.75	-2.936332588505D-02
2		-2.467212029555D-02	-1.665266794246D+00	-.50	-1.503421525596D-02
3		9.326849901648D-03	3.507351620174D-02	-.25	-5.537904020367D-03
				.25	2.453885602838D-03
0	-9.5	-5.538749482130D-05	-2.769374741065D-05	.50	2.697722393377D-03
1	-8.5	3.619655286998D-04	1.809827643499D-04	.75	1.599797214469D-03
2	-7.5	-1.176974850093D-03	-5.884874250463D-04	1.25	-1.346947950420D-03
3	-6.5	2.735784166070D-03	1.367892083035D-03	1.50	-1.876684832947D-03
4	-5.5	-5.488085168464D-03	-2.744042584232D-03	1.75	-1.356870961702D-03
5	-4.5	1.046049946170D-02	5.230249730851D-03	2.25	1.544204827743D-03
6	-3.5	-1.960810256333D-02	-9.804051281666D-03	2.50	2.402403865225D-03
7	-2.5	3.628729041612D-02	1.814364520806D-02	2.75	1.897322227956D-03
8	-1.5	-6.459908591360D-02	-3.229954295680D-02	3.25	-2.444278359737D-03
9	-.5	1.042827942019D-01	5.214139710096D-02	3.50	-3.963848124738D-03
10	.5	-1.430393237999D-01	-7.151966189997D-02	3.75	-3.230994475810D-03
11	1.5	1.596514512852D-01	7.982572564261D-02	4.25	4.340376269855D-03
12	2.5	-1.429257246347D-01	-7.146286231737D-02	4.50	7.133115603640D-03
13	3.5	1.040099008277D-01	5.200495041386D-02	4.75	5.872201750587D-03
14	4.5	-6.405793704486D-02	-3.202896852243D-02	5.25	-7.990481627376D-03
15	5.5	3.526885819038D-02	1.763442909519D-02	5.50	-1.318526476083D-02
16	6.5	-1.777152632680D-02	-8.885763163410D-03	5.75	-1.088775356466D-02
17	7.5	7.667730647970D-03	3.833865323985D-03	6.25	1.487809361857D-02
18	8.5	-2.359634437840D-03	-1.179817218920D-03	6.50	2.458961080742D-02
19	9.5	3.611338145587D-04	1.805669072794D-04	6.75	2.033474106142D-02
				7.25	-2.787878892434D-02
	0	5.916255327695D+00	-1.633165311445D+01	7.50	-4.617892757074D-02
	1	2.071874983145D+00	7.791239763750D+00	7.75	-3.830123698279D-02
	2	2.411431019014D-01	-6.656715824968D-01	8.25	5.302984389065D-02
	3	9.326849901648D-03	3.507351620174D-02	8.50	8.856318617817D-02
				8.75	7.431417591665D-02
				9.25	-1.071550962661D-01
				9.50	-1.853390095232D-01

I=12 H= 2.5

0		0.	3.371301520829D+02	-1.00	2.642985727552D-02
1		-8.239680540905D-03	-5.590259642552D+01	-.75	1.571000694739D-02

2		1.320011187056D-02	3.093951452910D+00
3		-4.990064864054D-03	-5.929528775659D-02
0	-9.5	2.963353439666D-05	1.481676719833D-05
1	-8.5	-1.936598907789D-04	-9.682994538943D-05
2	-7.5	6.297124113139D-04	3.148562056570D-04
3	-6.5	-1.463751535609D-03	-7.318757678043D-04
4	-5.5	2.936642743346D-03	1.468321371673D-03
5	-4.5	-5.599860963760D-03	-2.799930481880D-03
6	-3.5	1.051744711639D-02	5.258723558195D-03
7	-2.5	-1.963186460879D-02	-9.815932304397D-03
8	-1.5	3.628729041612D-02	1.814364520806D-02
9	-.5	-6.457530388763D-02	-3.228765194382D-02
10	.5	1.042256330460D-01	5.211281652299D-02
11	1.5	-1.429257246347D-01	-7.146286231737D-02
12	2.5	1.594356112556D-01	7.971780562779D-02
13	3.5	-1.425208049443D-01	-7.126040247215D-02
14	4.5	1.032545237932D-01	5.162726189661D-02
15	5.5	-6.266358164347D-02	-3.133179082174D-02
16	6.5	3.280502264672D-02	1.640251132336D-02
17	7.5	-1.427540204571D-02	-7.137701022854D-03
18	8.5	4.402668191606D-03	2.201334095803D-03
19	9.5	-6.742309998892D-04	-3.371154999446D-04
	0	-3.165323731639D+00	3.444630732534D+01
	1	-1.108497616943D+00	-1.317171798032D+01
	2	-1.290167367550D-01	1.404035751847D+00
	3	-4.990064864054D-03	-5.929528775659D-02

-.50	8.043626346099D-03
-.25	2.962896890637D-03
.25	-1.312881098129D-03
.50	-1.443338898831D-03
.75	-8.559254936616D-04
1.25	7.206455672473D-04
1.50	1.004065729934D-03
1.75	7.259541505212D-04
2.25	-8.261805744791D-04
2.50	-1.285332802586D-03
2.75	-1.015102601640D-03
3.25	1.307726435110D-03
3.50	2.120708588480D-03
3.75	1.728609762531D-03
4.25	-2.322076917213D-03
4.50	-3.816089696553D-03
4.75	-3.141420242096D-03
5.25	4.274128102493D-03
5.50	7.052126176285D-03
5.75	5.822469666948D-03
6.25	-7.952353890733D-03
6.50	-1.313744314067D-02
6.75	-1.085741198783D-02
7.25	1.485256795979D-02
7.50	2.455554949755D-02
7.75	2.031150098112D-02
8.25	-2.785555017184D-02
8.50	-4.614487029266D-02
8.75	-3.827571614061D-02
9.25	5.299951408720D-02
9.50	8.851539043907D-02

I=13 H= 3.5

0		0.	-6.292430988458D+02
1		4.403085484989D-03	1.043363201868D+02
2		-7.053819845957D-03	-5.766116958236D+00
3		2.666569808455D-03	1.073530912973D-01
0	-9.5	-1.583544748626D-05	-7.917723743130D-06
1	-8.5	1.034872244567D-04	5.174361222836D-05
2	-7.5	-3.365037476317D-04	-1.682518738158D-04
3	-6.5	7.821992791668D-04	3.910996395834D-04
4	-5.5	-1.569319630504D-03	-7.846598152520D-04
5	-4.5	2.992830115128D-03	1.496415057564D-03
6	-3.5	-5.623535549432D-03	-2.811767774716D-03
7	-2.5	1.051744711639D-02	5.258723558195D-03
8	-1.5	-1.960810256333D-02	-9.804051281666D-03
9	-.5	3.623016867119D-02	1.811508433560D-02
10	.5	-6.446177562508D-02	-3.223088781254D-02
11	1.5	1.040099008277D-01	5.200495041386D-02
12	2.5	-1.425208049443D-01	-7.126040247215D-02
13	3.5	1.586785907846D-01	7.933929539230D-02
14	4.5	-1.411099737713D-01	-7.055498688565D-02
15	5.5	1.006509940547D-01	5.032549702733D-02
16	6.5	-5.806349556770D-02	-2.903174778385D-02
17	7.5	2.626724851221D-02	1.313362425611D-02

-1.00	-1.412347513940D-02
-.75	-8.395046915035D-03
-.50	-4.298318930041D-03
-.25	-1.583300264877D-03
.25	7.015718176131D-04
.50	7.712852926286D-04
.75	4.573858112542D-04
1.25	-3.850954773576D-04
1.50	-5.365482944802D-04
1.75	-3.879322017620D-04
2.25	4.414906266835D-04
2.50	6.868501298026D-04
2.75	5.424455712973D-04
3.25	-6.988155062700D-04
3.50	-1.133250997460D-03
3.75	-9.237220985526D-04
4.25	1.240847283732D-03
4.50	2.039191620922D-03
4.75	1.678658060128D-03
5.25	-2.283874168846D-03
5.50	-3.768207247963D-03
5.75	-3.111054909169D-03
6.25	4.248597570453D-03

```
18    8.5  -8.179563995487D-03  -4.089781997744D-03     6.50  7.018064550267D-03
19    9.5   1.256044256722D-03   6.280221283611D-04     6.75  5.799232575940D-03
                                                        7.25 -7.929122611999D-03
        0   1.691472360533D+00  -6.639825590110D+01     7.50 -1.310339916317D-02
        1   5.923542840509D-01   2.384594744904D+01     7.75 -1.083190253970D-02
        2   6.894341969500D-02  -2.706553856264D+00     8.25  1.482225416303D-02
        3   2.666569808455D-03   1.073530912973D-01     8.50  2.450778033942D-02
                                                        8.75  2.027342401389D-02
                                                        9.25 -2.780592233321D-02
                                                        9.50 -4.606410202371D-02


I=14      H=  4.5
0           0.                   1.170830025462D+03    -1.00  7.515474527718D-03
1          -2.342998187559D-03  -1.941460536956D+02     -.75  4.467226413874D-03
2           3.753524033420D-03   1.072547420623D+01     -.50  2.287249140477D-03
3          -1.418952306739D-03  -1.980116982105D-01     -.25  8.425159287714D-04
                                                         .25 -3.733249102836D-04
0    -9.5   8.426460878108D-06   4.213230439054D-06      .50 -4.104212920415D-04
1    -8.5  -5.506829705028D-05  -2.753414852514D-05      .75 -2.433870795269D-04
2    -7.5   1.790626258522D-04   8.953131292609D-05     1.25  2.049194791567D-04
3    -6.5  -4.162296402650D-04  -2.081148201325D-04     1.50  2.855115214519D-04
4    -5.5   8.350824293449D-04   4.175412146725D-04     1.75  2.064289686503D-04
5    -4.5  -1.592612802182D-03  -7.963064010909D-04     2.25 -2.349288046696D-04
6    -3.5   2.992830115128D-03   1.496415057564D-03     2.50 -3.654910443833D-04
7    -2.5  -5.599860963760D-03  -2.799930481880D-03     2.75 -2.886495501166D-04
8    -1.5   1.046049946170D-02   5.230249730851D-03     3.25  3.718579866573D-04
9     -.5  -1.949490019619D-02  -9.747450098093D-03     3.50  6.030322967207D-04
10     .5   3.601503981715D-02   1.800751990857D-02     3.75  4.915363336867D-04
11    1.5  -6.405793704486D-02  -3.202896852243D-02     4.25 -6.602858982111D-04
12    2.5   1.032545237932D-01   5.162726189661D-02     4.50 -1.085103603969D-03
13    3.5  -1.411099737713D-01  -7.055498688565D-02     4.75 -8.932533650972D-04
14    4.5   1.560500170683D-01   7.802500853415D-02     5.25  1.215295653114D-03
15    5.5  -1.362596284464D-01  -6.812981422319D-02     5.50  2.005128812112D-03
16    6.5   9.208128451309D-02   4.604064225654D-02     5.75  1.655434132702D-03
17    7.5  -4.588405545862D-02  -2.294202772931D-02     6.25 -2.260675595479D-03
18    8.5   1.491112683875D-02   7.455563419376D-03     6.50 -3.734221424877D-03
19    9.5  -2.317626502808D-03  -1.158813251404D-03     6.75 -3.085595250170D-03
                                                        7.25  4.218353648698D-03
        0  -9.000771727556D-01   1.246462827130D+02     7.50  6.970411359894D-03
        1  -3.152073786020D-01  -4.397371106770D+01     7.75  5.761251671791D-03
        2  -3.668661670863D-02   5.082140807232D+00     8.25 -7.879626376703D-03
        3  -1.418952306739D-03  -1.980116982105D-01     8.50 -1.302284845564D-02
                                                        8.75 -1.076607493159D-02
                                                        9.25  1.473353522295D-02
                                                        9.50  2.436183250107D-02


I=15      H=  5.5
0           0.                  -2.148904693506D+03    -1.00 -3.940692081245D-03
1           1.228536449735D-03   3.564166474048D+02     -.75 -2.342362240218D-03
2          -1.968136859547D-03  -1.969231991709D+01     -.50 -1.199304786249D-03
3           7.440187719633D-04   3.627531590109D-01     -.25 -4.417679594672D-04
                                                         .25  1.957505823484D-04
0    -9.5  -4.418362150928D-06  -2.209181075464D-06      .50  2.152018380216D-04
1    -8.5   2.887471833095D-05   1.443735916548D-05      .75  1.276184930749D-04
2    -7.5  -9.389037857336D-05  -4.694518928668D-05     1.25 -1.074482470297D-04
```

CARDINAL SPLINES N=4

3	-6.5	2.182474965560D-04	1.091237482780D-04	1.50	-1.497061798288D-04
4	-5.5	-4.378709677651D-04	-2.189354838825D-04	1.75	-1.082397376194D-04
5	-4.5	8.350824293449D-04	4.175412146725D-04	2.25	1.231834463012D-04
6	-3.5	-1.569319630504D-03	-7.846598152520D-04	2.50	1.916429357352D-04
7	-2.5	2.936642743346D-03	1.468321371673D-03	2.75	1.513515773349D-04
8	-1.5	-5.488085168464D-03	-2.744042584232D-03	3.25	-1.949813839684D-04
9	-.5	1.024803547404D-02	5.124017737018D-03	3.50	-3.161961519770D-04
10	.5	-1.909602979043D-02	-9.548014895215D-03	3.75	-2.577339292692D-04
11	1.5	3.526885819038D-02	1.763442909519D-02	4.25	3.462165724285D-04
12	2.5	-6.266358164347D-02	-3.133179082174D-02	4.50	5.689667891086D-04
13	3.5	1.006509940547D-01	5.032549702733D-02	4.75	4.683712197407D-04
14	4.5	-1.362596284464D-01	-6.812981422319D-02	5.25	-6.372309733884D-04
15	5.5	1.471001797275D-01	7.355008986377D-02	5.50	-1.051372681777D-03
16	6.5	-1.204469394064D-01	-6.022346970319D-02	5.75	-8.680116759763D-04
17	7.5	6.960801965543D-02	3.480400982772D-02	6.25	1.185357228401D-03
18	8.5	-2.492997266622D-02	-1.246498633311D-02	6.50	1.957982442158D-03
19	9.5	4.094801970721D-03	2.047400985361D-03	6.75	1.617872935738D-03
				7.25	-2.211754092564D-03
	0	4.719498393104D-01	-2.291629259709D+02	7.50	-3.654620652226D-03
	1	1.652770186274D-01	8.047798678229D+01	7.75	-3.020552877362D-03
	2	1.923639814141D-02	-9.353854885277D+00	8.25	4.130708655413D-03
	3	7.440187719633D-04	3.627531590109D-01	8.50	6.826238304550D-03
				8.75	5.642485202346D-03
				9.25	-7.717896226164D-03
				9.50	-1.275593454268D-02

I=16 H= 6.5

0		0.	3.702272477698D+03	-1.00	1.964153645164D-03
1		-6.123377050601D-04	-6.147889424495D+02	-.75	1.167500336837D-03
2		9.809757035429D-04	3.400515454870D+01	-.50	5.977678079860D-04
3		-3.708402365613D-04	-6.266575172655D-01	-.25	2.201897864327D-04
				.25	-9.756768907593D-05
0	-9.5	2.202238078542D-06	1.101119039271D-06	.50	-1.072627512295D-04
1	-8.5	-1.439198561432D-05	-7.195992807159D-06	.75	-6.360870705332D-05
2	-7.5	4.679765131047D-05	2.339882565523D-05	1.25	5.355527948031D-05
3	-6.5	-1.087808045843D-04	-5.439040229213D-05	1.50	7.461784176941D-05
4	-5.5	2.182474965560D-04	1.091237482780D-04	1.75	5.394978093799D-05
5	-4.5	-4.162296402650D-04	-2.081148201325D-04	2.25	-6.139815251901D-05
6	-3.5	7.821992791668D-04	3.910996395834D-04	2.50	-9.552031959862D-05
7	-2.5	-1.463751535609D-03	-7.318757678043D-04	2.75	-7.543795400730D-05
8	-1.5	2.735784166070D-03	1.367892083035D-03	3.25	9.718429560417D-05
9	-.5	-5.110896712230D-03	-2.555448356115D-03	3.50	1.576011988248D-04
10	.5	9.542341556025D-03	4.771170778013D-03	3.75	1.284619535607D-04
11	1.5	-1.777715263268D-02	-8.888576316341D-03	4.25	-1.725642181048D-04
12	2.5	3.280502264672D-02	1.640251132336D-02	4.50	-2.835892638021D-04
13	3.5	-5.806349556770D-02	-2.903174778385D-02	4.75	-2.334495437719D-04
14	4.5	9.208128451309D-02	4.604064225654D-02	5.25	3.176139478222D-04
15	5.5	-1.204469394064D-01	-6.022346970319D-02	5.50	5.240337487586D-04
16	6.5	1.191621673583D-01	5.958108367914D-02	5.75	4.326412887214D-04
17	7.5	-8.074097745812D-02	-4.037048872906D-02	6.25	-5.908143539009D-04
18	8.5	3.258616746551D-02	1.629308373276D-02	6.50	-9.759106425506D-04
19	9.5	-5.819598627629D-03	-2.909799313814D-03	6.75	-8.063895997146D-04
				7.25	1.102388511355D-03
	0	-2.352332987751D-01	3.934622335824D+02	7.50	1.821536390109D-03
	1	-8.237879338672D-02	-1.383585288238D+02	7.75	1.505492860585D-03
	2	-9.587971038454D-03	1.614541530664D+01	8.25	-2.058757362300D-03

	3	-3.708402365613D-04	-6.266575172655D-01	8.50	-3.402138887330D-03
				8.75	-2.812073401862D-03
				9.25	3.845955497129D-03
				9.50	6.355853947821D-03

I=17 H= 7.5

0		0.	-4.499139952290D+03	-1.00	-8.449815736042D-04
1		2.634285147923D-04	7.531308377268D+02	-.75	-5.022602352027D-04
2		-4.220170838911D-04	-4.197481714392D+01	-.50	-2.571605252340D-04
3		1.595359749208D-04	7.789818459760D-01	-.25	-9.472594604941D-05
				.25	4.197375273796D-05
0	-9.5	-9.474058219833D-07	-4.737029109917D-07	.50	4.614458168049D-05
1	-8.5	6.191451827472D-06	3.095725913736D-06	.75	2.736455241377D-05
2	-7.5	-2.013241345670D-05	-1.006620672835D-05	1.25	-2.303955417429D-05
3	-6.5	4.679765131047D-05	2.339882565523D-05	1.50	-3.210069715305D-05
4	-5.5	-9.389037857336D-05	-4.694518928668D-05	1.75	-2.320926922965D-05
5	-4.5	1.790626258522D-04	8.953131292609D-05	2.25	2.641356880471D-05
6	-3.5	-3.365037476317D-04	-1.682518738158D-04	2.50	4.109297150765D-05
7	-2.5	6.297124113139D-04	3.148562056570D-04	2.75	3.245351046024D-05
8	-1.5	-1.176974850093D-03	-5.884874250463D-04	3.25	-4.180881604251D-05
9	-.5	2.199014801983D-03	1.099507400991D-03	3.50	-6.780024968729D-05
10	.5	-4.107576806424D-03	-2.053788403212D-03	3.75	-5.526450654941D-05
11	1.5	7.667730647970D-03	3.833865323985D-03	4.25	7.423735926126D-05
12	2.5	-1.427540204571D-02	-7.137701022854D-03	4.50	1.220004817909D-04
13	3.5	2.626724851221D-02	1.313362425611D-02	4.75	1.004303048282D-04
14	4.5	-4.588405545862D-02	-2.294202772931D-02	5.25	-1.366379333276D-04
15	5.5	6.960801965543D-02	3.480400982772D-02	5.50	-2.254399855167D-04
16	6.5	-8.074097745812D-02	-4.037048872906D-02	5.75	-1.861228156403D-04
17	7.5	6.273141016360D-02	3.136570508180D-02	6.25	2.541689796219D-04
18	8.5	-2.812492274393D-02	-1.406246137197D-02	6.50	4.198377045621D-04
19	9.5	5.426195386872D-03	2.713097693436D-03	6.75	3.469094468772D-04
				7.25	-4.742477324497D-04
	0	1.011976855671D-01	-4.647446809301D+02	7.50	-7.836243102085D-04
	1	3.543946913066D-02	1.665186467903D+02	7.75	-6.476613481930D-04
	2	4.124758201351D-03	-1.977383453361D+01	8.25	8.856695598152D-04
	3	1.595359749208D-04	7.789818459760D-01	8.50	1.463579277768D-03
				8.75	1.209727364405D-03
				9.25	-1.654448299027D-03
				9.50	-2.734088910852D-03

I=18 H= 8.5

0		0.	2.957494230694D+03	-1.00	2.598626693072D-04
1		-8.101388144194D-05	-5.011960742316D+02	-.75	1.544633510167D-04
2		1.297856537219D-04	2.828421336340D+01	-.50	7.908624591936D-05
3		-4.906313414338D-05	-5.314937235485D-01	-.25	2.913168518909D-05
				.25	-1.290846069054D-05
0	-9.5	2.913618634565D-07	1.456809317283D-07	.50	-1.419114279387D-05
1	-8.5	-1.904097378957D-06	-9.520486894785D-07	.75	-8.415598465669D-06
2	-7.5	6.191451827472D-06	3.095725913736D-06	1.25	7.085503677109D-06
3	-6.5	-1.439198561432D-05	-7.195992807159D-06	1.50	9.872135805501D-06
4	-5.5	2.887471833095D-05	1.443735916548D-05	1.75	7.137697249249D-06
5	-4.5	-5.506829705028D-05	-2.753414852514D-05	2.25	-8.123136299336D-06
6	-3.5	1.034872244567D-04	5.174361222836D-05	2.50	-1.263758831332D-05
7	-2.5	-1.936598907789D-04	-9.682994538943D-05	2.75	-9.980638765237D-06
8	-1.5	3.619655286998D-04	1.809827643499D-04	3.25	1.285773661469D-05

CARDINAL SPLINES N=4

```
 9   -.5  -6.763002410158D-04  -3.381501205079D-04     3.50  2,085105091893D-05
10    .5   1,263416961907D-03   6.317084809533D-04     3.75  1,699585243420D-05
11   1.5  -2,359634437840D-03  -1.179817218920D-03     4.25 -2,283069688851D-05
12   2,5   4.402668191606D-03   2,201334095803D-03     4.50 -3,751959980143D-05
13   3,5  -8.179563995487D-03  -4,089781997744D-03     4,75 -3,088598315264D-05
14   4,5   1.491112683875D-02   7.455563419376D-03     5,25  4,202114924418D-05
15   5,5  -2,492997266622D-02  -1.246498633311D-02     5,50  6,933101880349D-05
16   6,5   3,258616746551D-02   1.629308373276D-02     5,75  5,723955359362D-05
17   7,5  -2,812492274393D-02  -1,406246137197D-02     6,25 -7,816622422743D-05
18   8,5   1,367218575548D-02   6,836092877740D-03     6,50 -1,291153857508D-04
19   9,5  -2,800957143117D-03  -1,400478571558D-03     6,75 -1,066872795152D-04
                                                       7,25  1,458483842702D-04
       0  -3,112198126148D-02   2,930923503128D+02     7,50  2,409928327411D-04
       1  -1.089893003005D-02  -1.076979459778D+02     7,75  1,991792140397D-04
       2  -1.268513669365D-03   1.313664224226D+01     8.25 -2,723748947320D-04
       3  -4,906313414338D-05  -5.314937235485D-01     8,50 -4,501020874401D-04
                                                       8,75 -3,720329828962D-04
                                                       9,25  5,087965756629D-04
                                                       9,50  8,408162023120D-04


I=19     H=   9.5
 0                 0.                  -7.723758000226D+02    -1.00 -3,976376019329D-05
 1                 1.239661149699D-05   1.326544245057D+02     -.75 -2,363572907510D-05
 2                -1,985958823899D-05  -7.594629820544D+00     -.50 -1,210164786541D-05
 3                 7.507560457311D-06   1.449373640359D-01     -.25 -4,457682771330D-06
                                                                .25  1,975231520282D-06
 0  -9,5  -4.458371531226D-08  -2.229185765613D-08              .50  2,171505435636D-06
 1  -8,5   2.913618634565D-07   1,456809317283D-07              .75  1,287741098635D-06
 2  -7,5  -9.474058219833D-07  -4.737029109917D-07             1.25 -1.084212171796D-06
 3  -6,5   2.202238078542D-06   1.101119039271D-06             1.50 -1.510618057600D-06
 4  -5,5  -4.418362150928D-06  -2.209181075464D-06             1.75 -1,092198746720D-06
 5  -4.5   8.426460878108D-06   4.213230439054D-06             2.25  1.242989016657D-06
 6  -3,5  -1,583544748626D-05  -7.917723743130D-06             2,50  1,933783072302D-06
 7  -2.5   2.963353439666D-05   1.481676719833D-05             2.75  1,527221002536D-06
 8  -1.5  -5.538749482130D-05  -2.769374741065D-05             3.25 -1,967469804241D-06
 9   -.5   1.034873531445D-04   5.174367657225D-05             3.50 -3,190593670514D-06
10    .5  -1.933341345863D-04  -9.666706729313D-05             3.75 -2,600677509919D-06
11   1,5   3.611338145587D-04   1.805669072794D-04             4.25  3,493515849916D-06
12   2,5  -6.742309998892D-04  -3.371154999446D-04             4,50  5,741187712849D-06
13   3,5   1.256044256722D-03   6,280221283611D-04             4.75  4,726122550555D-06
14   4,5  -2.317626502808D-03  -1.158813251404D-03             5.25 -6,430007392222D-06
15   5,5   4.094801970721D-03   2,047400985361D-03             5,50 -1,060891882778D-05
16   6,5  -5,819598627629D-03  -2,909799313814D-03             5,75 -8,758702588120D-06
17   7,5   5,426195386872D-03   2.713097693436D-03             6.25  1,196086711989D-05
18   8,5  -2.800957143117D-03  -1,400478571558D-03             6,50  1,975702365761D-05
19   9,5   6,001643247898D-04   3.000821623949D-04             6,75  1,632511131571D-05
                                                               7.25 -2,231747662451D-05
       0   4.762234617740D-03  -7.330843603263D+01             7,50 -3,687631830501D-05
       1   1.667736428773D-03   2.759824922804D+01             7,75 -3,047806481969D-05
       2   1.941058847944D-04  -3.463914945522D+00             8,25  4.167832499877D-05
       3   7.507560457311D-06   1.449373640359D-01             8,50  6,887379969414D-05
                                                               8,75  5,692777672193D-05
                                                               9,25 -7,785492977465D-05
                                                               9,50 -1,286596267963D-04
```

CARDINAL SPLINES N=4

```
                 N= 4      M=20                        P= 10.0      Q= 7

I= 0      H=-10.0
0           1.000000000000D+00   4.859851323795D-01   -1.00  4.666983740458D+00
1          -2.043924731721D+00  -1.019760466892D+00    -.75  3.405151663299D+00
2           1.333191788066D+00   6.664806497519D-01    -.50  2.391493715461D+00
3          -2.898672206711D-01  -1.449316010104D-01    -.25  1.598834845007D+00
                                                        .25  5.678141651320D-01
0  -10.0    6.001643257388D-04   3.000821628694D-04     .50  2.751068673560D-01
1   -9.0   -2.800957149318D-03  -1.400478574659D-03     .75  9.476921054282D-02
2   -8.0    5.426195407037D-03   2.713097703518D-03    1.25 -3.507902561224D-02
3   -7.0   -5.819598674503D-03  -2.909799337251D-03    1.50 -3.427495651554D-02
4   -6.0    4.094802064765D-03   2.047401032382D-03    1.75 -1.768214273360D-02
5   -5.0   -2.317626682163D-03  -1.158813341082D-03    2.25  1.053382461622D-02
6   -4.0    1.256044593776D-03   6.280222968880D-04    2.50  1.195817045617D-02
7   -3.0   -6.742316306322D-04  -3.371158153161D-04    2.75  6.908482243405D-03
8   -2.0    3.611349934705D-04   1.805674967352D-04    3.25 -4.774219853174D-03
9   -1.0   -1.933363373070D-04  -9.666816865351D-05    3.50 -5.681573343248D-03
10   0.0    1.034914683544D-04   5.174573417722D-05    3.75 -3.398301018058D-03
11   1.0   -5.539518260106D-05  -2.769759130053D-05    4.25  2.450953814347D-03
12   2.0    2.964789448194D-05   1.482394724097D-05    4.50  2.953650456534D-03
13   3.0   -1.586225747114D-05  -7.931128735568D-06    4.75  1.782442256734D-03
14   4.0    8.476405934127D-06   4.238202967064D-06    5.25 -1.299141552179D-03
15   5.0   -4.510519702535D-06  -2.255259851267D-06    5.50 -1.570302774237D-03
16   6.0    2.365062805884D-06   1.182531402942D-06    5.75 -9.496219327745D-04
17   7.0   -1.178814948097D-06  -5.894074740487D-07    6.25  6.938357875655D-04
18   8.0    5.071277963839D-07   2.535638981919D-07    6.50  8.392376776026D-04
19   9.0   -1.559603037094D-07  -7.798015185472D-08    6.75  5.077645412545D-04
20  10.0    2.386479032524D-08   1.193239516262D-08    7.25 -3.712045270873D-04
                                                       7.50 -4.490665975026D-04
        0  -1.759872891817D+02  -8.799515557180D+01    7.75 -2.717292528430D-04
        1  -6.234025517172D+01  -3.116962777499D+01    8.25  1.986752026150D-04
        2  -7.362824832066D+00  -3.681467380562D+00    8.50  2.403572341797D-04
        3  -2.898672206711D-01  -1.449316010104D-01    8.75  1.454434439653D-04
                                                       9.25 -1.063445341327D-04
                                                       9.50 -1.286568883184D-04
                                                       9.75 -7.785266623529D-05


I= 1      H= -9.0
0           0.                   9.158944959410D-02   -1.00 -9.041182852175D+00
1           3.959053519655D+00   1.965136990328D+00    -.75 -5.678512178892D+00
2          -4.019190947513D+00  -2.008842333550D+00    -.50 -3.117191794831D+00
3           1.062938385007D+00   5.314560612358D-01    -.25 -1.257571226399D+00
                                                        .25  7.551721870030D-01
0  -10.0   -2.800957149318D-03  -1.400478574659D-03     .50  1.107574438597D+00
1   -9.0    1.367218579601D-02   6.836092898004D-03     .75  1.156548480293D+00
2   -8.0   -2.812492287572D-02  -1.406246143786D-02    1.25  7.315274073079D-01
3   -7.0    3.258616777184D-02   1.629308388592D-02    1.50  4.350675321362D-01
4   -6.0   -2.492997328082D-02  -1.246498664041D-02    1.75  1.772914831894D-01
```

CARDINAL SPLINES N=4

5	-5.0	1.491112801087D-02	7.455564005434D-03	2.25	-8,424021383512D-02
6	-4.0	-8.179566198189D-03	-4.089783099094D-03	2.50	-9,008354362214D-02
7	-3.0	4.402672313615D-03	2.201336156807D-03	2.75	-4,992391705222D-02
8	-2.0	-2.359642142221D-03	-1.179821071110D-03	3.25	3,281503106677D-02
9	-1.0	1.263431357047D-03	6.317156785233D-04	3.50	3,847398090568D-02
10	0.0	-6.763271345850D-04	-3.381635672925D-04	3.75	2,276829786949D-02
11	1.0	3.620157695965D-04	1.810078847982D-04	4.25	-1,621278149315D-02
12	2.0	-1.937537362822D-04	-9.687686814112D-05	4.50	-1,946621206430D-02
13	3.0	1.036624320880D-04	5.183121604399D-05	4.75	-1,171693010239D-02
14	4.0	-5.539469615118D-05	-2.769734807559D-05	5.25	8.514026388180D-03
15	5.0	2.947698298704D-05	1.473849149352D-05	5.50	1.028222809237D-02
16	6.0	-1.545607180915D-05	-7.728035904576D-06	5.75	6,214312476366D-03
17	7.0	7.703748273325D-06	3.851874136662D-06	6.25	-4,537262027570D-03
18	8.0	-3.314163001612D-06	-1.657081500806D-06	6.50	-5,487011595086D-03
19	9.0	1.019226064893D-06	5.096130324467D-07	6.75	-3,319350976788D-03
20	10.0	-1.559603037094D-07	-7.798015185472D-08	7.25	2.426241534137D-03
				7.50	2,935024606674D-03
	0	7.006098254525D+02	3.503147872337D+02	7.75	1.775920893326D-03
	1	2.424567500716D+02	1.212251086901D+02	8.25	-1,298418942178D-03
	2	2.786896060270D+01	1.393483950353D+01	8.50	-1,570810686021D-03
	3	1.062938385007D+00	5.314560612358D-01	8.75	-9,505120895805D-04
				9.25	6,949843013214D-04
				9.50	8,407983059120D-04
				9.75	5,087817830633D-04

I= 2 H= -8.0

0		0.	-2.978165253720D-01	-1.00	9,696675453232D+00
1		-3.293246664947D+00	-1.599832886933D+00	-.75	5,852797482209D+00
2		4.845624628913D+00	2.420363368639D+00	-.50	3,052755009623D+00
3		-1.557804159373D+00	-7.788593814437D-01	-.25	1,150503895534D+00
				.25	-5,448004857312D-01
0	-10.0	5.426195407037D-03	2.713097703518D-03	.50	-6,299003030153D-01
1	-9.0	-2.812492287572D-02	-1.406246143786D-02	.75	-4,007454650292D-01
2	-8.0	6.273141059212D-02	3.136570529606D-02	1.25	4,380163020859D-01
3	-7.0	-8.074097845420D-02	-4.037048922710D-02	1.50	7.976882893776D-01
4	-6.0	6.960802165387D-02	3.480401082693D-02	1.75	9,966809945392D-01
5	-5.0	-4.588405926992D-02	-2.294202963496D-02	2.25	8,267656418883D-01
6	-4.0	2.626725567462D-02	1.313362783731D-02	2.50	5,431835379518D-01
7	-3.0	-1.427541544902D-02	-7.137707724512D-03	2.75	2,397523344968D-01
8	-2.0	7.667755699894D-03	3.833877849947D-03	3.25	-1,274048086506D-01
9	-1.0	-4.107623614333D-03	-2.053811807166D-03	3.50	-1,414516873440D-01
10	0.0	2.199102250367D-03	1.099551125183D-03	3.75	-8,064857813685D-02
11	1.0	-1.177138215735D-03	-5.885691078676D-04	4.25	5,497455292463D-02
12	2.0	6.300175637290D-04	3.150087818645D-04	4.50	6,517847519208D-02
13	3.0	-3.370734609359D-04	-1.685367304680D-04	4.75	3,888368458796D-02
14	4.0	1.801239603895D-04	9.006198019477D-05	5.25	-2,795855829602D-02
15	5.0	-9.584873041428D-05	-4.792436520714D-05	5.50	-3,366360736936D-02
16	6.0	5.025768360275D-05	2.512884180138D-05	5.75	-2,030263628919D-02
17	7.0	-2.504986710242D-05	-1.252493355121D-05	6.25	1,478712350351D-02
18	8.0	1.077648696119D-05	5.388243480597D-06	6.50	1,786993086306D-02
19	9.0	-3.314163001612D-06	-1.657081500806D-06	6.75	1,080510575522D-02
20	10.0	5.071277963839D-07	2.535638981919D-07	7.25	-7,893391694217D-03
				7.50	-9,547107561618D-03
	0	-1.106174163131D+03	-5.531191899745D+02	7.75	-5,776107107336D-03
	1	-3.737220018985D+02	-1.868503799473D+02	8.25	4,222503472420D-03
	2	-4.188850015227D+01	-2.094541807467D+01	8.50	5,108143438879D-03

	3	-1.557804159373D+00	-7.788593814437D-01	8.75	3,090905903541D-03
				9.25	-2,259905117438D-03
				9.50	-2,734030718083D-03
				9.75	-1,654400198721D-03

I= 3 H= -7.0

0		0.	6.922723906367D-01	-1,00	-6,822801546585D+00
1		2.161366370442D+00	9.719191280459D-01	-.75	-4,066886785979D+00
2		-3.408490973955D+00	-1.698552929880D+00	-.50	-2,089423953983D+00
3		1.252944202188D+00	6.263728493331D-01	-.25	-7,729495316419D-01
				.25	3,468878046974D-01
0	-10.0	-5.819598674503D-03	-2.909799337251D-03	.50	3,851330013910D-01
1	-9.0	3.258616777184D-02	1.629308388592D-02	.75	2,315576176294D-01
2	-8.0	-8.074097845420D-02	-4.037048922710D-02	1.25	-2,046505588912D-01
3	-7.0	1.191621696737D-01	5.958108483683D-02	1.50	-2,975471731447D-01
4	-6.0	-1.204469440517D-01	-6.022347202586D-02	1.75	-2,292874653593D-01
5	-5.0	9.208129337244D-02	4.604064668622D-02	2.25	3,358732232002D-01
6	-4.0	-5.806351221666D-02	-2.903175610833D-02	2.50	6,817337909815D-01
7	-3.0	3.280505380263D-02	1.640252690132D-02	2.75	9,296916839808D-01
8	-2.0	-1.777721086570D-02	-8.888605432851D-03	3.25	8,730597028233D-01
9	-1.0	9.542450360675D-03	4.771225180338D-03	3.50	5,982759114975D-01
10	0.0	-5.111099985374D-03	-2.555549992687D-03	3.75	2,727045571909D-01
11	1.0	2.736163908342D-03	1.368081954171D-03	4.25	-1,511709131850D-01
12	2.0	-1.464460860222D-03	-7.322304301112D-04	4.50	-1,700922786133D-01
13	3.0	7.835235736818D-04	3.917617868409D-04	4.75	-9,793234335876D-02
14	4.0	-4.186967048323D-04	-2.093483524162D-04	5.25	6,757190848068D-02
15	5.0	2.227996714519D-04	1.113998357260D-04	5.50	8,040519412568D-02
16	6.0	-1.168236249860D-04	-5.841181249300D-05	5.75	4,809186239390D-02
17	7.0	5.822823726039D-05	2.911411863019D-05	6.25	-3,468646164171D-02
18	8.0	-2.504986710242D-05	-1.252493355121D-05	6.50	-4,180142754054D-02
19	9.0	7.703748273325D-06	3.851874136662D-06	6.75	-2,522626820484D-02
20	10.0	-1.178814948097D-06	-5.894074740487D-07	7.25	1,838657873270D-02
				7.50	2,222439284399D-02
	0	9.337087684965D+02	4.669290200161D+02	7.75	1,343998351018D-02
	1	3.098748075476D+02	1.549127153304D+02	8.25	-9,819890745247D-03
	2	3.417983509167D+01	1.709263255011D+01	8.50	-1,187779032920D-02
	3	1.252944202188D+00	6.263728493331D-01	8.75	-7,186438762941D-03
				9.25	5,253710755398D-03
				9.50	6,355718679145D-03
				9.75	3,845843688310D-03

I= 4 H= -6.0

0		0.	-1.388909842478D+00	-1,00	3,861653605702D+00
1		-1.208111888797D+00	-3.858420280815D-01	-.75	2,296781431798D+00
2		1.928779401819D+00	9.529686928083D-01	-.50	1,176846084239D+00
3		-7.247623150862D-01	-3.621820281985D-01	-.25	4,339010959862D-01
				.25	-1,928034208320D-01
0	-10.0	4.094802064765D-03	2.047401032382D-03	.50	-2,124243926886D-01
1	-9.0	-2.492997328082D-02	-1.246498664041D-02	.75	-1,263580146570D-01
2	-8.0	6.960802165387D-02	3.480401082693D-02	1.25	1,075505738873D-01
3	-7.0	-1.204469440517D-01	-6.022347202586D-02	1.50	1,512817742175D-01
4	-6.0	1.471001890475D-01	7.355009452376D-02	1.75	1,109158662250D-01
5	-5.0	-1.362596462210D-01	-6.812982311048D-02	2.25	-1,327803633087D-01
6	-4.0	1.006510274576D-01	5.032551372878D-02	2.50	-2,159589481738D-01
7	-3.0	-6.266364415174D-02	-3.133182207587D-02	2.75	-1,821522611650D-01

8	-2.0	3,526897502360D-02	1,763448751180D-02
9	-1.0	-1,909624808579D-02	-9,548124042894D-03
10	0.0	1,024844330205D-02	5,124221651023D-03
11	1.0	-5,488847047437D-03	-2,744423523719D-03
12	2.0	2,938065865155D-03	1,469032932577D-03
13	3,0	-1,571976569752D-03	-7,859882848759D-04
14	4.0	8,400321143293D-04	4,200160571647D-04
15	5.0	-4,470040208788D-04	-2,235020104394D-04
16	6.0	2,343838506269D-04	1,171919253135D-04
17	7,0	-1,168236249860D-04	-5,841181249300D-05
18	8.0	5,025768360275D-05	2,512884180138D-05
19	9.0	-1,545607180915D-05	-7,728035904576D-06
20	10,0	2,365062805884D-06	1.182531402942D-06
	0	-5,439654937923D+02	-2,721324890410D+02
	1	-1,800612183783D+02	-8,998107663147D+01
	2	-1,981409005077D+01	-9,912492153147D+00
	3	-7,247623150862D-01	-3.621820281985D-01

3,25	3,032996674950D-01
3,50	6,429695417925D-01
3,75	9,065057490754D-01
4,25	8,897820762645D-01
4,50	6,184280856314D-01
4,75	2,848658096366D-01
5,25	-1,600347017952D-01
5,50	-1,808061487805D-01
5,75	-1,044114302796D-01
6,25	7,230581909649D-02
6,50	8,613115736922D-02
6,75	5,155624827871D-02
7,25	-3,721912902303D-02
7,50	-4,486534056973D-02
7,75	-2,708024012381D-02
8,25	1,974212366805D-02
8,50	2,386434131380D-02
8,75	1,443234730233D-02
9,25	-1,054550221219D-02
9,50	-1,275566315241D-02
9,75	-7,717671903525D-03

I= 5 H= -5.0

0		0.	2,648853643346D+00
1		6,529716583441D-01	-8,967993594801D-02
2		-1,045258505978D+00	-5,008477247752D-01
3		3,946044743164D-01	1,969224684549D-01
0	-10,0	-2,317626682163D-03	-1,158813341082D-03
1	-9,0	1,491112801087D-02	7,455564005434D-03
2	-8,0	-4,588405926992D-02	-2,294202963496D-02
3	-7,0	9,208129337244D-02	4,604064668622D-02
4	-6,0	-1,362596462210D-01	-6,812982311048D-02
5	-5,0	1,560500509670D-01	7,802502548350D-02
6	-4,0	-1,411100374755D-01	-7,055501873776D-02
7	-3,0	1,032546430056D-01	5,162732150281D-02
8	-2,0	-6,405815986286D-02	-3,202907993143D-02
9	-1,0	3.601545613823D-02	1,800772806912D-02
10	0.0	-1,949567798368D-02	-9,747838991840D-03
11	1.0	1.046195247602D-02	5,230976238010D-03
12	2,0	-5,602575064605D-03	-2,801287532302D-03
13	3.0	2,997897285773D-03	1.498948642887D-03
14	4,0	-1.602052573332D-03	-8,010262866660D-04
15	5,0	8,525004936395D-04	4,262502468197D-04
16	6,0	-4,470040208788D-04	-2,235020104394D-04
17	7,0	2,227996714519D-04	1,113998357260D-04
18	8,0	-9,584873041428D-05	-4,792436520714D-05
19	9,0	2,947698298704D-05	1.473849149352D-05
20	10,0	-4,510519702535D-06	-2,255259851267D-06
	0	2.966083403020D+02	1,485897502612D+02
	1	9,812914383369D+01	4,897010610501D+01
	2	1.079287572351D+01	5,406826328871D+00
	3	3,946044743164D-01	1,969224684549D-01

-1.00	-2,092834638639D+00
-.75	-1,244160415973D+00
-.50	-6,371260149562D-01
-.25	-2,347372661209D-01
.25	1,040798114169D-01
.50	1,144786555086D-01
,75	6,793523092999D-02
1.25	-5,734037386776D-02
1.50	-8,006638156945D-02
1.75	-5,807570318003D-02
2.25	6,687259067278D-02
2.50	1,051034599744D-01
2.75	8,423769913546D-02
3,25	-1,143439777768D-01
3,50	-1,940186763842D-01
3,75	-1,690291968489D-01
4,25	2,938349289159D-01
4,50	6,315635603747D-01
4,75	8,996225691714D-01
5,25	8,947989181840D-01
5,50	6,244920629802D-01
5,75	2,885329304869D-01
6,25	-1,627140687711D-01
6,50	-1,840470162920D-01
6,75	-1,063722594840D-01
7,25	7,373930759579D-02
7,50	8,786533865198D-02
7,75	5,260560694072D-02
8,25	-3,798639089484D-02
8,50	-4,579359471557D-02
8,75	-2,764195660108D-02
9,25	2,015287732798D-02
9,50	2,436131492023D-02

9,75	1,473310740696D-02

I= 6 H= -4.0

0	0.	-4,977865200316D+00
1	-3,502943573718D-01	6.069334338634D-01
2	5,610779823215D-01	2,396059967699D-01
3	-2,120396695434D-01	-1,053061534335D-01

0	-10,0	1.256044593776D-03	6.280222968880D-04
1	-9,0	-8,179566198189D-03	-4,089783099094D-03
2	-8,0	2,626725567462D-02	1,313362783731D-02
3	-7,0	-5,806351221666D-02	-2,903175610833D-02
4	-6,0	1.006510274576D-01	5.032551372878D-02
5	-5,0	-1,411100374755D-01	-7,055501873776D-02
6	-4,0	1,586787105009D-01	7.933935525045D-02
7	-3,0	-1,425210289745D-01	-7,126051448725D-02
8	-2,0	1,040103195590D-01	5,200515977951D-02
9	-1,0	-6,446255799756D-02	-3,223127899878D-02
10	0,0	3,623163033037D-02	1,811581516519D-02
11	1,0	-1,961083314429D-02	-9,805416572143D-03
12	2,0	1,052254759749D-02	5,261273798746D-03
13	3.0	-5,633058043106D-03	-2,816529021553D-03
14	4,0	3,010569829722D-03	1,505284914861D-03
15	5,0	-1,602052573332D-03	-8,010262866660D-04
16	6,0	8,400321143293D-04	4,200160571647D-04
17	7,0	-4,186967048323D-04	-2,093483524162D-04
18	8,0	1,801239603895D-04	9,006198019477D-05
19	9,0	-5,539469615118D-05	-2,769734807559D-05
20	10,0	8,476405934127D-06	4,238202967064D-06

0	-1,594348148850D+02	-8,025408461820D+01
1	-5,274063557397D+01	-2,619279266079D+01
2	-5,800112103982D+00	-2,919578606235D+00
3	-2,120396695434D-01	-1,053061534335D-01

-1,00	1,123412009237D+00
-.75	6,677813686734D-01
-.50	3,419216329592D-01
-.25	1,259540830747D-01
.25	-5,581925862161D-02
.50	-6,137282895009D-02
.75	-3,640097684741D-02
1,25	3,066520919115D-02
1,50	4,274683602207D-02
1,75	3,092942777985D-02
2,25	-3,529481815698D-02
2,50	-5,503990405988D-02
2,75	-4,361738895194D-02
3,25	5,688093375489D-02
3,50	9,321285960916D-02
3,75	7,712561145212D-02
4,25	-1,092145292135D-01
4,50	-1,878371620354D-01
4,75	-1,652988297755D-01
5,25	2,911160266288D-01
5,50	6,282771534906D-01
5,75	8,976351486362D-01
6,25	8,962510275099D-01
6,50	6,262484912062D-01
6,75	2,895956328611D-01
7,25	-1,634909865212D-01
7,50	-1,849869178308D-01
7,75	-1,069410095072D-01
8,25	7,415519169054D-02
8,50	8,836851490808D-02
8,75	5,291011860470D-02
9,25	-3,820912006954D-02
9,50	-4,606312935857D-02
9,75	-2,780511835889D-02

I= 7 H= -3.0

0	0.	9,315290585856D+00
1	1,876002087806D-01	-1,369740565589D+00
2	-3,005265940114D-01	-7,366364590005D-02
3	1,136006168615D-01	5,546476622717D-02

0	-10,0	-6,742316306322D-04	-3,371158153161D-04
1	-9,0	4.402672313615D-03	2.201336156807D-03
2	-8,0	-1,427541544902D-02	-7,137707724512D-03
3	-7,0	3.280505380263D-02	1,640252690132D-02
4	-6,0	-6,266364415174D-02	-3,133182207587D-02
5	-5,0	1.032546430056D-01	5,162732150281D-02
6	-4,0	-1,425210289745D-01	-7,126051448725D-02
7	-3,0	1,594360304928D-01	7.971801524639D-02
8	-2,0	-1,429265082244D-01	-7,146325411218D-02
9	-1,0	1,042270971328D-01	5,211354856638D-02
10	0,0	-6,457803915237D-02	-3,228901957618D-02

-1,00	-6,017274196535D-01
-.75	-3,576716259553D-01
-.50	-1,831318300008D-01
-.25	-6,745797395932D-02
.25	2,989210855606D-02
.50	3,286326556050D-02
.75	1,948920863384D-02
1,25	-1,641105558897D-02
1,50	-2,286792507140D-02
1,75	-1,653664093161D-02
2,25	1,883137908406D-02
2,50	2,931287589761D-02
2,75	2,316834968176D-02
3,25	-2,993139427737D-02
3,50	-4,865714342324D-02
3,75	-3,979968625301D-02

CARDINAL SPLINES N=4

11	1.0	3.629240026793D-02	1.814620013396D-02
12	2.0	-1.964140935513D-02	-9.820704677565D-03
13	3.0	1.053526696216D-02	5.267633481079D-03
14	4.0	-5.633058043106D-03	-2.816529021553D-03
15	5.0	2.997897285773D-03	1.498948642887D-03
16	6.0	-1.571976569752D-03	-7.859882848759D-04
17	7.0	7.835235736818D-04	3.917617868409D-04
18	8.0	-3.370734609359D-04	-1.685367304680D-04
19	9.0	1.036624320880D-04	5.183121604399D-05
20	10.0	-1.586225747114D-05	-7.931128735568D-06
	0	8.542395954814D+01	4.371628656714D+01
	1	2.825725338700D+01	1.379641638456D+01
	2	3.107491911833D+00	1.590279340915D+00
	3	1.136006168615D-01	5.546476622717D-02

4.25	5.412748866845D-02
4.50	8.989466998217D-02
4.75	7.512317528979D-02
5.25	-1.077550319716D-01
5.50	-1.860730221093D-01
5.75	-1.642319776479D-01
6.25	2.903365147702D-01
6.50	6.273342638537D-01
6.75	8.970646537893D-01
7.25	8.966681334463D-01
7.50	6.267531272432D-01
7.75	2.899010201035D-01
8.25	-1.637143492906D-01
8.50	-1.852572167475D-01
8.75	-1.071046327898D-01
9.25	7.427497547408D-02
9.50	8.851357024959D-02
9.75	5.299800957601D-02

I= 8 H= -2.0

0		0.	-1.741106700976D+01
1		-1.004302135193D-01	2.685266356278D+00
2		1.608894595475D-01	-6.272650579182D-02
3		-6.082038102166D-02	-2.791394901393D-02
0	-10.0	3.611349934705D-04	1.805674967352D-04
1	-9.0	-2.359642142221D-03	-1.179821071110D-03
2	-8.0	7.667755699894D-03	3.833877849947D-03
3	-7.0	-1.777721086570D-02	-8.888605432851D-03
4	-6.0	3.526897502360D-02	1.763448751180D-02
5	-5.0	-6.405815986286D-02	-3.202907993143D-02
6	-4.0	1.040103195590D-01	5.200515977951D-02
7	-3.0	-1.429265082244D-01	-7.146325411218D-02
8	-2.0	1.596529158802D-01	7.982645794009D-02
9	-1.0	-1.430420603013D-01	-7.152103015067D-02
10	0.0	1.042879066420D-01	5.214395332099D-02
11	1.0	-6.460863665676D-02	-3.230431832838D-02
12	2.0	3.630513035049D-02	1.815256517525D-02
13	3.0	-1.964140935513D-02	-9.820704677565D-03
14	4.0	1.052254759749D-02	5.261273798746D-03
15	5.0	-5.602575064605D-03	-2.801287532302D-03
16	6.0	2.938065865155D-03	1.469032932577D-03
17	7.0	-1.464460860222D-03	-7.322304301112D-04
18	8.0	6.300175637290D-04	3.150087818645D-04
19	9.0	-1.937537362822D-04	-9.687686814112D-05
20	10.0	2.964789448194D-05	1.482394724097D-05
	0	-4.573573720211D+01	-2.474500304009D+01
	1	-1.512875532907D+01	-6.943448463737D+00
	2	-1.663721971102D+00	-9.001449762097D-01
	3	-6.082038102166D-02	-2.791394901393D-02

-1.00	3.221400540884D-01
-.75	1.914815793784D-01
-.50	9.804001927420D-02
-.25	3.611346305500D-02
.25	-1.600225856963D-02
.50	-1.759246813334D-02
.75	-1.043273168496D-02
1.25	8.784088779901D-03
1.50	1.223907297803D-02
1.75	8.849370241780D-03
2.25	-1.007255507064D-02
2.50	-1.567235137484D-02
2.75	-1.237960617219D-02
3.25	1.595858519166D-02
3.50	2.589409326842D-02
3.75	2.112347908450D-02
4.25	-2.845656170721D-02
4.50	-4.687980747850D-02
4.75	-3.872710608312D-02
5.25	5.334571088121D-02
5.50	8.894969540633D-02
5.75	7.455169540647D-02
6.25	-1.073374413760D-01
6.50	-1.855678801707D-01
6.75	-1.639263181736D-01
7.25	2.901129818218D-01
7.50	6.270637689111D-01
7.75	8.969009160424D-01
8.25	8.967879973855D-01
8.50	6.268982783497D-01
8.75	2.899889685147D-01
9.25	-1.637789292918D-01
9.50	-1.853356074358D-01
9.75	-1.071522842081D-01

CARDINAL SPLINES N=4

I= 9 H= -1.0

0		0.	3,253149686185D+01	-1,00	-1,724405183536D-01
1		5,375957248547D-02	-5,084197395317D+00	-.75	-1,024993333878D-01
2		-8,612359100816D-02	2.244447660754D-01	-.50	-5,248035335228D-02
3		3,255735485999D-02	1.161460519719D-02	-.25	-1,933132622907D-02
				.25	8,565865552732D-03
0	-10,0	-1,933363373070D-04	-9,666816865351D-05	.50	9,417047408061D-03
1	-9,0	1,263431357047D-03	6,317156785233D-04	.75	5,584486217827D-03
2	-8,0	-4,107623614333D-03	-2,053811807166D-03	1.25	-4,701884015342D-03
3	-7,0	9,542450360675D-03	4,771225180338D-03	1,50	-6,551110405903D-03
4	-6,0	-1,909624808579D-02	-9,548124042894D-03	1,75	-4,736589809121D-03
5	-5,0	3,601545613823D-02	1,800772806912D-02	2,25	5,390704354717D-03
6	-4,0	-6,446255799756D-02	-3,223127899878D-02	2,50	8,386839929265D-03
7	-3,0	1,042270971328D-01	5,211354856638D-02	2,75	6,623849781576D-03
8	-2,0	-1,430420603013D-01	-7,152103015067D-02	3,25	-8,534559780881D-03
9	-1,0	1,597137324601D-01	7.985686623005D-02	3,50	-1,384204387033D-02
10	0,0	-1,430726624383D-01	-7,153633121916D-02	3,75	-1,128484048783D-02
11	1,0	1,043006391508D-01	5,215031957540D-02	4,25	1,516898534929D-02
12	2,0	-6,460863665676D-02	-3,230431832838D-02	4,50	2,494252288848D-02
13	3,0	3,629240026793D-02	1,814620013396D-02	4,75	2,054921614485D-02
14	4,0	-1,961083314429D-02	-9,805416572143D-03	5,25	-2,803796516549D-02
15	5,0	1,046195247602D-02	5,230976238010D-03	5,50	-4,637379989674D-02
16	6,0	-5,488847047437D-03	-2,744423523719D-03	5,75	-3,842107084960D-02
17	7,0	2,736163908342D-03	1,368081954171D-03	6,25	5,312202928615D-02
18	8,0	-1,177138215735D-03	-5,885691078676D-04	6,50	8,867906368993D-02
19	9,0	3,620157695965D-04	1,810078847982D-04	6,75	7,438789304024D-02
20	10,0	-5,539518260106D-05	-2,769759130053D-05	7,25	-1,072175456428D-01
				7,50	-1,854226958768D-01
	0	2,448259148403D+01	1.574860471341D+01	7,75	-1,638383519039D-01
	1	8,098494210320D+00	2,889079485348D+00	8,25	2,900483906569D-01
	2	8,905970547916D-01	5,728829219911D-01	8,50	6,269853653639D-01
	3	3,255735485999D-02	1,161460519719D-02	8,75	8,968532571158D-01
				9,25	8,968233571924D-01
				9,50	6,269415449557D-01
				9,75	2,900155532144D-01

I=10 H= 0.0

0		0,	-6.077694401005D+01	-1,00	9,230369586564D-02
1		-2,877629328777D-02	9,534378183563D+00	-,75	5,486566980005D-02
2		4,610010219864D-02	-4,767189091781D-01	-,50	2,809158474095D-02
3		-1,742730037923D-02	-1,854891945371D-23	-,25	1,034763127778D-02
				,25	-4,585112186335D-03
0	-10,0	1,034914683544D-04	5,174573417722D-05	,50	-5,040725114531D-03
1	-9,0	-6,763271345850D-04	-3,381635672925D-04	,75	-2,989240383139D-03
2	-8,0	2,199102250367D-03	1,099551125183D-03	1,25	2,516791519441D-03
3	-7,0	-5,111099985374D-03	-2,555549992687D-03	1,50	3,506616189749D-03
4	-6,0	1,024844330205D-02	5,124221651023D-03	1,75	2,535339474651D-03
5	-5,0	-1,949567798368D-02	-9,747838991840D-03	2,25	-2,885392965843D-03
6	-4,0	3,623163033037D-02	1,811581516519D-02	2,50	-4,488986152516D-03
7	-3,0	-6,457803915237D-02	-3,228901957618D-02	2,75	-3,545247160721D-03
8	-2,0	1,042879066420D-01	5,214395332099D-02	3,25	4,567383376248D-03
9	-1,0	-1,430726624383D-01	-7,153633121916D-02	3,50	7,407022097565D-03
10	0,0	1,597264654025D-01	7,986323270123D-02	3,75	6,037776214261D-03
11	1,0	-1,430726624383D-01	-7,153633121916D-02	4,25	-8,111823644184D-03
12	2,0	1,042879066420D-01	5,214395332099D-02	4,50	-1,333256302152D-02
13	3,0	-6,457803915237D-02	-3,228901957618D-02	4,75	-1,097735023279D-02

14	4.0	3.623163033037D-02	1.811581516519D-02	5.25	1.494479065686D-02
15	5.0	-1.949567798368D-02	-9.747838991840D-03	5.50	2.467145849770D-02
16	6.0	1.024844330205D-02	5.124221651023D-03	5.75	2.038523121145D-02
17	7.0	-5.111099985374D-03	-2.555549992687D-03	6.25	-2.791800344223D-02
18	8.0	2.199102250367D-03	1.099551125183D-03	6.50	-4.622855881689D-02
19	9.0	-6.763271345850D-04	-3.381635672925D-04	6.75	-3.833307993051D-02
20	10.0	1.034914683544D-04	5.174573417722D-05	7.25	5.305742837102D-02
				7.50	8.860065117200D-02
	0	-1.310505309224D+01	-1.310505309224D+01	7.75	7.434022987599D-02
	1	-4.334964363083D+00	-4.499600743675D-21	8.25	-1.071821837523D-01
	2	-4.767189091781D-01	-4.767189091781D-01	8.50	-1.853794270973D-01
	3	-1.742730037923D-02	-1.854891945371D-23	8.75	-1.638117660358D-01
				9.25	2.900279935848D-01
				9.50	6.269597771532D-01
				9.75	8.968370197222D-01

I=11 H= 1.0

0		0.	1.135422969632D+02	-1.00	-4.940648473815D-02
1		1.540279549476D-02	-1.783111948433D+01	-.75	-2.936740150500D-02
2		-2.467554477777D-02	9.213210779068D-01	-.50	-1.503630200002D-02
3		9.328144465614D-03	-1.161460519719D-02	-.25	-5.538672679576D-03
				.25	2.454226201301D-03
0	-10.0	-5.539518260106D-05	-2.769759130053D-05	.50	2.698096836274D-03
1	-9.0	3.620157695965D-04	1.810078847982D-04	.75	1.600019265725D-03
2	-8.0	-1.177138215735D-03	-5.885691078676D-04	1.25	-1.347134906225D-03
3	-7.0	2.736163908342D-03	1.368081954171D-03	1.50	-1.876945315925D-03
4	-6.0	-5.488847047437D-03	-2.744423523719D-03	1.75	-1.357059294670D-03
5	-5.0	1.046195247602D-02	5.230976238010D-03	2.25	1.544419162186D-03
6	-4.0	-1.961083314429D-02	-9.805416572143D-03	2.50	2.402737316536D-03
7	-3.0	3.629240026793D-02	1.814620013396D-02	2.75	1.897585573857D-03
8	-2.0	-6.460863665676D-02	-3.230431832838D-02	3.25	-2.444617619810D-03
9	-1.0	1.043006391508D-01	5.215031957540D-02	3.50	-3.964398293812D-03
10	0.0	-1.430726624383D-01	-7.153633121916D-02	3.75	-3.231442922866D-03
11	1.0	1.597137324601D-01	7.985686623005D-02	4.25	4.340978673225D-03
12	2.0	-1.430420603013D-01	-7.152103015067D-02	4.50	7.134105583642D-03
13	3.0	1.042270971328D-01	5.211354856638D-02	4.75	5.873016698127D-03
14	4.0	-6.446255799756D-02	-3.223127899878D-02	5.25	-7.991590383788D-03
15	5.0	3.601545613823D-02	1.800772806912D-02	5.50	-1.318709410658D-02
16	6.0	-1.909624808579D-02	-9.548124042894D-03	5.75	-1.088926386892D-02
17	7.0	9.542450360675D-03	4.771225180338D-03	6.25	1.488015608708D-02
18	8.0	-4.107623614333D-03	-2.053811807166D-03	6.50	2.459301760377D-02
19	9.0	1.263431357047D-03	6.317156785233D-04	6.75	2.033755607695D-02
20	10.0	-1.933363373070D-04	-9.666816865351D-05	7.25	-2.788263723122D-02
				7.50	-4.618528632634D-02
	0	7.014617942784D+00	1.574860471341D+01	7.75	-3.830649245717D-02
	1	2.320335239623D+00	-2.889079485348D+00	8.25	5.303703067539D-02
	2	2.551687891906D-01	5.728829219911D-01	8.50	8.857506239972D-02
	3	9.328144465614D-03	-1.161460519719D-02	8.75	7.432399222700D-02
				9.25	-1.071685211180D-01
				9.50	-1.853611948123D-01
				9.75	-1.637993256366D-01

I=12 H= 2.0

0		0.	-2.121079343123D+02	-1.00	2.644266488887D-02
1		-8.243673397870D-03	3.332053269211D+01	-.75	1.571761984109D-02

2		1.320650849719D-02	-1.737563446628D+00	-.50	8.047524197459D-03
3		-4.992482993805D-03	2.791394901393D-02	-.25	2.964332677320D-03
				.25	-1.313517305607D-03
0	-10.0	2.964789448194D-05	1.482394724097D-05	.50	-1.444038324687D-03
1	-9.0	-1.937537362822D-04	-9.687686814112D-05	.75	-8.563402655545D-04
2	-8.0	6.300175637290D-04	3.150087818645D-04	1.25	7.209947839767D-04
3	-7.0	-1.464460860222D-03	-7.322304301112D-04	1.50	1.004552288848D-03
4	-6.0	2.938065865155D-03	1.469032932577D-03	1.75	7.263059396673D-04
5	-5.0	-5.602575064605D-03	-2.801287532302D-03	2.25	-8.265809320570D-04
6	-4.0	1.052254759749D-02	5.261273798746D-03	2.50	-1.285955659825D-03
7	-3.0	-1.964140935513D-02	-9.820704677565D-03	2.75	-1.015594508240D-03
8	-2.0	3.630513035049D-02	1.815256517525D-02	3.25	1.308360142596D-03
9	-1.0	-6.460863665676D-02	-3.230431832838D-02	3.50	2.121736255125D-03
10	0.0	1.042879066420D-01	5.214395332099D-02	3.75	1.729447421540D-03
11	1.0	-1.430420603013D-01	-7.152103015067D-02	4.25	-2.323202152853D-03
12	2.0	1.596529158802D-01	7.982645794009D-02	4.50	-3.817938890689D-03
13	3.0	-1.429265082244D-01	-7.146325411218D-02	4.75	-3.142942491243D-03
14	4.0	1.040103195590D-01	5.200515977951D-02	5.25	4.276199160346D-03
15	5.0	-6.405815986286D-02	-3.202907993143D-02	5.50	7.055543230611D-03
16	6.0	3.526897502360D-02	1.763448751180D-02	5.75	5.825290780286D-03
17	7.0	-1.777721086570D-02	-8.888605432851D-03	6.25	-7.956206397517D-03
18	8.0	7.667755699894D-03	3.833877849947D-03	6.50	-1.314380673168D-02
19	9.0	-2.359642142221D-03	-1.179821071110D-03	6.75	-1.086267018517D-02
20	10.0	3.611349934705D-04	1.805674967352D-04	7.25	1.485975625299D-02
				7.50	2.456742708452D-02
	0	-3.754268878064D+00	-2.474500304009D+01	7.75	2.032131773739D-02
	1	-1.241858401595D+00	6.943448463737D+00	8.25	-2.786897444322D-02
	2	-1.365679813169D-01	-9.001449762097D-01	8.50	-4.616705401277D-02
	3	-4.992482993805D-03	2.791394901393D-02	8.75	-3.829405213076D-02
				9.25	5.302459051021D-02
				9.50	8.855683057573D-02
				9.75	7.431033002389D-02

I=13 H= 3.0

0		0.	3.961731507314D+02	-1.00	-1.414738669100D-02
1		4.410540067103D-03	-6.224143307101D+01	-.75	-8.409260031518D-03
2		-7.065762216766D-03	3.254222327730D+00	-.50	-4.305596138635D-03
3		2.671084407135D-03	-5.546476622717D-02	-.25	-1.585980849185D-03
				.25	7.027596039339D-04
0	-10.0	-1.586225747114D-05	-7.931128735568D-06	.50	7.725911063652D-04
1	-9.0	1.036624320880D-04	5.183121604399D-05	.75	4.581601818766D-04
2	-8.0	-3.370734609359D-04	-1.685367304680D-04	1.25	-3.857474578335D-04
3	-7.0	7.835235736818D-04	3.917617868409D-04	1.50	-5.374566900007D-04
4	-6.0	-1.571976569752D-03	-7.859882848759D-04	1.75	-3.885889848872D-04
5	-5.0	2.997897285773D-03	1.498948642887D-03	2.25	4.422380860753D-04
6	-4.0	-5.633058043106D-03	-2.816529021553D-03	2.50	6.880129915010D-04
7	-3.0	1.053526696216D-02	5.267633481079D-03	2.75	5.433639508375D-04
8	-2.0	-1.964140935513D-02	-9.820704677565D-03	3.25	-6.999986251577D-04
9	-1.0	3.629240026793D-02	1.814620013396D-02	3.50	-1.135169630023D-03
10	0.0	-6.457803915237D-02	-3.228901957618D-02	3.75	-9.252859907530D-04
11	1.0	1.042270971328D-01	5.211354856638D-02	4.25	1.242948075608D-03
12	2.0	-1.429265082244D-01	-7.146325411218D-02	4.50	2.042644028472D-03
13	3.0	1.594360304928D-01	7.971801524639D-02	4.75	1.681500068083D-03
14	4.0	-1.425210289745D-01	-7.126051448725D-02	5.25	-2.287740791408D-03
15	5.0	1.032546430056D-01	5.162732150281D-02	5.50	-3.774586818350D-03
16	6.0	-6.266364415174D-02	-3.133182207587D-02	5.75	-3.116321869947D-03

17	7.0	3.280505380263D-02	1.640252690132D-02	6.25	4.255790121655D-03
18	8.0	-1.427541544902D-02	-7.137707724512D-03	6.50	7.029945244246D-03
19	9.0	4.402672313615D-03	2.201336156807D-03	6.75	5.809049522593D-03
20	10.0	-6.742316306322D-04	-3.371158153161D-04	7.25	-7.942543008062D-03
				7.50	-1.312557437459D-02
	0	2.008613586129D+00	4.371628656714D+01	7.75	-1.085023022246D-02
	1	6.644206178721D-01	-1.379641638456D+01	8.25	1.484731700254D-02
	2	7.306676999727D-02	1.590279340915D+00	8.50	2.454919689016D-02
	3	2.671084407135D-03	-5.546476622717D-02	8.75	2.030765693165D-02
				9.25	-2.785273950098D-02
				9.50	-4.614146990946D-02
				9.75	-3.827365850576D-02

I=14 H= 4.0

0		0.	-7.394460252832D+02	-1.00	7.560020012911D-03
1		-2.356885533002D-03	1.161762108155D+02	-.75	4.493704418328D-03
2		3.775771803488D-03	-6.078763209241D+00	-.50	2.300806051925D-03
3		-1.427362676421D-03	1.053061534335D-01	-.25	8.475096627875D-04
				.25	-3.755376699927D-04
0	-10.0	8.476405934127D-06	4.238202967064D-06	.50	-4.128539282599D-04
1	-9.0	-5.539469615118D-05	-2.769734807559D-05	.75	-2.448296757945D-04
2	-8.0	1.801239603895D-04	9.006198019477D-05	1.25	2.061340714641D-04
3	-7.0	-4.186967048323D-04	-2.093483524162D-04	1.50	2.872037963773D-04
4	-6.0	8.400321143293D-04	4.200160571647D-04	1.75	2.076525079449D-04
5	-5.0	-1.602052573332D-03	-8.010262866660D-04	2.25	-2.363212672677D-04
6	-4.0	3.010569829722D-03	1.505284914861D-03	2.50	-3.676573713298D-04
7	-3.0	-5.633058043106D-03	-2.816529021553D-03	2.75	-2.903604245463D-04
8	-2.0	1.052254759749D-02	5.261273798746D-03	3.25	3.740620512772D-04
9	-1.0	-1.961083314429D-02	-9.805416572143D-03	3.50	6.066065699365D-04
10	0.0	3.623163033037D-02	1.811581516519D-02	3.75	4.944497513526D-04
11	1.0	-6.446255799756D-02	-3.223127899878D-02	4.25	-6.641995210053D-04
12	2.0	1.040103195590D-01	5.200515977951D-02	4.50	-1.091535188667D-03
13	3.0	-1.425210289745D-01	-7.126051448725D-02	4.75	-8.985478196522D-04
14	4.0	1.586787105009D-01	7.933935525045D-02	5.25	1.222498890317D-03
15	5.0	-1.411100374755D-01	-7.055501873776D-02	5.50	2.017013488729D-03
16	6.0	1.006510274576D-01	5.032551372878D-02	5.75	1.665246098369D-03
17	7.0	-5.806351221666D-02	-2.903175610833D-02	6.25	-2.274074796349D-03
18	8.0	2.626725567462D-02	1.313362783731D-02	6.50	-3.756354295488D-03
19	9.0	-8.179566198189D-03	-4.089783099094D-03	6.75	-3.103883509294D-03
20	10.0	1.256044593776D-03	6.280222968880D-04	7.25	4.243354872543D-03
				7.50	7.011722169341D-03
	0	-1.073354351402D+00	-8.025408461820D+01	7.75	5.795394814593D-03
	1	-3.550502523895D-01	2.619279266079D+01	8.25	-7.926316627714D-03
	2	-3.904510848914D-02	-2.919578606235D+00	8.50	-1.310000448375D-02
	3	-1.427362676421D-03	1.053061534335D-01	8.75	-1.082984837283D-02
				9.25	1.482075220864D-02
				9.50	2.450596325705D-02
				9.75	2.027232446369D-02

I=15 H= 5.0

0		0.	1.375895912653D+03	-1.00	-4.022886459983D-03
1		1.254161083207D-03	-2.161833732189D+02	-.75	-2.391218889033D-03
2		-2.009187970140D-03	1.131450038252D+01	-.50	-1.224319709968D-03
3		7.595374066354D-04	-1.969224684549D-01	-.25	-4.509822909142D-04
				.25	1.998335193464D-04

CARDINAL SPLINES N=4

0	-10.0	-4,510519702535D-06	-2.255259851267D-06	.50	2,196904864628D-04
1	-9.0	2.947698298704D-05	1.473849149352D-05	.75	1,302803409249D-04
2	-8.0	-9,584873041428D-05	-4.792436520714D-05	1.25	-1,096893868376D-04
3	-7.0	2.227996714519D-04	1.113998357260D-04	1.50	-1,528287294103D-04
4	-6.0	-4,470040208788D-04	-2.235020104394D-04	1.75	-1,104973862171D-04
5	-5.0	8,525004936395D-04	4.262502468197D-04	2,25	1,257527885741D-04
6	-4.0	-1.602052573332D-03	-8.010262866660D-04	2,50	1,956401959904D-04
7	-3.0	2.997897285773D-03	1.498948642887D-03	2.75	1,545084463248D-04
8	-2.0	-5,602575064605D-03	-2.801287532302D-03	3.25	-1,990482769733D-04
9	-1.0	1.046195247602D-02	5.230976238010D-03	3,50	-3,227913246448D-04
10	0.0	-1.949567798368D-02	-9.747838991840D-03	3,75	-2,631097053779D-04
11	1.0	3,601545613823D-02	1.800772806912D-02	4,25	3,534379056970D-04
12	2.0	-6.405815986286D-02	-3.202907993143D-02	4.50	5,808342119508D-04
13	3.0	1.032546430056D-01	5.162732150281D-02	4,75	4,781404343091D-04
14	4.0	-1.411100374755D-01	-7.055501873776D-02	5.25	-6,505222329678D-04
15	5.0	1.560500509670D-01	7.802502548350D-02	5,50	-1,073302033469D-03
16	6.0	-1,362596462210D-01	-6.812982311048D-02	5.75	-8,861165056505D-04
17	7.0	9.208129337244D-02	4.604064668622D-02	6.25	1,210081147926D-03
18	8.0	-4,588405926992D-02	-2.294202963496D-02	6,50	1,998821542723D-03
19	9.0	1.491112801087D-02	7.455564005434D-03	6,75	1,651618041226D-03
20	10.0	-2.317626682163D-03	-1.158813341082D-03	7,25	-2,257885817311D-03
				7,50	-3,730846476302D-03
	0	5,711602204535D-01	1.485897502612D+02	7,75	-3,083553075892D-03
	1	1,889316236710D-01	-4.897010610501D+01	8,25	4,216860510273D-03
	2	2,077693422892D-02	5.406826328871D+00	8,50	6,968604961222D-03
	3	7,595374066354D-04	-1.969224684549D-01	8,75	5,760158595293D-03
				9,25	-7,878827147065D-03
				9,50	-1,302188153807D-02
				9,75	-1,076548983186D-02
I=16	H= 6.0				
0		0.	-2,525374498869D+03	-1.00	2,109375355960D-03
1		-6,576115202351D-04	3,968855281540D+02	-.75	1,253820669610D-03
2		1,053505146577D-03	-2.077795299910D+01	-.50	6,419643829052D-04
3		-3,982586891477D-04	3,621820281985D-01	-.25	2,364697437378D-04
				.25	-1,047814560637D-04
0	-10.0	2,365062805884D-06	1.182531402942D-06	.50	-1,151933325636D-04
1	-9.0	-1,545607180915D-05	-7.728035904576D-06	.75	-6,831168193533D-05
2	-8.0	5,025768360275D-05	2,512884180138D-05	1,25	5,751494390050D-05
3	-7.0	-1.168236249860D-04	-5,841181249300D-05	1,50	8,013478829585D-05
4	-6.0	2,343838506269D-04	1.171919253135D-04	1,75	5,793861322521D-05
5	-5.0	-4.470040208788D-04	-2,235020104394D-04	2,25	-6,593768777938D-05
6	-4.0	8,400321143293D-04	4.200160571647D-04	2,50	-1,025827122060D-04
7	-3.0	-1.571976569752D-03	-7.859882848759D-04	2.75	-8,101553634936D-05
8	-2.0	2,938065865155D-03	1.469032932577D-03	3,25	1,043697159317D-04
9	-1.0	-5,488847047437D-03	-2.744423523719D-03	3,50	1,692536046524D-04
10	0.0	1,024844330205D-02	5.124221651023D-03	3.75	1,379599194233D-04
11	1.0	-1.909624808579D-02	-9,548124042894D-03	4,25	-1,853229296807D-04
12	2.0	3,526897502360D-02	1,763448751180D-02	4,50	-3,045567250122D-04
13	3.0	-6,266364415174D-02	-3.133182207587D-02	4,75	-2,507098731828D-04
14	4.0	1,006510274576D-01	5,032551372878D-02	5,25	3,410970556014D-04
15	5.0	-1.362596462210D-01	-6.812982311048D-02	5,50	5,627787094010D-04
16	6.0	1,471001890475D-01	7,355009452376D-02	5.75	4,646290521101D-04
17	7.0	-1,204469440517D-01	-6.022347202586D-02	6,25	-6,344967801946D-04
18	8.0	6.960802165387D-02	3,480401082693D-02	6,50	-1,048065504507D-03
19	9.0	-2,492997328082D-02	-1,246498664041D-02	6,75	-8,660107322694D-04

CARDINAL SPLINES N=4

```
20  10.0   4.094802064765D-03   2.047401032382D-03       7.25  1.183894425661D-03
                                                          7.50  1.956212808973D-03
       0  -2.994842896923D-01  -2.721324890410D+02       7.75  1.616802134423D-03
       1  -9.906511533300D-02   8.998107663147D+01       8.25 -2.210971174766D-03
       2  -1.089425552785D-02  -9.912492153147D+00       8.50 -3.653673478376D-03
       3  -3.982586891477D-04   3.621820281985D-01       8.75 -3.019979729535D-03
                                                          9.25  4.130289584350D-03
                                                          9.50  6.825731307369D-03
                                                          9.75  5.642178408961D-03


I=17      H=  7.0
0              0.                4.351692277664D+03     -1.00 -1.051373008962D-03
1              3.277723904216D-04  -6.846772211325D+02    -.75 -6.249400830295D-04
2             -5.250970970069D-04   3.588381803010D+01    -.50 -3.199734096542D-04
3              1.985035215335D-04  -6.263728493331D-01    -.25 -1.178632836923D-04
                                                           .25  5.222607461726D-05
0  -10.0  -1.178814948097D-06  -5.894074740487D-07       .50  5.741565165895D-05
1   -9.0   7.703748273325D-06   3.851874136662D-06       .75  3.404849609571D-05
2   -8.0  -2.504986710242D-05  -1.252493355121D-05      1.25 -2.866709305083D-05
3   -7.0   5.822823726039D-05   2.911411863019D-05      1.50 -3.994147045167D-05
4   -6.0  -1.168236249860D-04  -5.841181249300D-05      1.75 -2.887826194881D-05
5   -5.0   2.227996714519D-04   1.113998357260D-04      2.25  3.286522946285D-05
6   -4.0  -4.186967048323D-04  -2.093483524162D-04      2.50  5.113015767843D-05
7   -3.0   7.835235736818D-04   3.917617868409D-04      2.75  4.038046035705D-05
8   -2.0  -1.464460860222D-03  -7.322304301112D-04      3.25 -5.202085119158D-05
9   -1.0   2.736163908342D-03   1.368081954171D-03      3.50 -8.436083660621D-05
10   0.0  -5.111099985374D-03  -2.555549992687D-03      3.75 -6.876316867532D-05
11   1.0   9.542450360675D-03   4.771225180338D-03      4.25  9.237024573559D-05
12   2.0  -1.777721086570D-02  -8.888605432851D-03      4.50  1.517997755140D-04
13   3.0   3.280505380263D-02   1.640252690132D-02      4.75  1.249609631048D-04
14   4.0  -5.806351221666D-02  -2.903175610833D-02      5.25 -1.700125044895D-04
15   5.0   9.208129337244D-02   4.604064668622D-02      5.50 -2.805049478841D-04
16   6.0  -1.204469440517D-01  -6.022347202586D-02      5.75 -2.315843403911D-04
17   7.0   1.191621696737D-01   5.958108483683D-02      6.25  3.162511477310D-04
18   8.0  -8.074097845420D-02  -4.037048922710D-02      6.50  5.223853570150D-04
19   9.0   3.258616777184D-02   1.629308388592D-02      6.75  4.316439611468D-04
20  10.0  -5.819598674503D-03  -2.909799337251D-03      7.25 -5.900852511759D-04
                                                          7.50 -9.750286067562D-04
       0   1.492715357370D-01   4.669290200161D+02       7.75 -8.058558817122D-04
       1   4.937688691032D-02  -1.549127153304D+02       8.25  1.101998282732D-03
       2   5.430008548997D-03   1.709263255011D+01       8.50  1.821064291571D-03
       3   1.985035215335D-04  -6.263728493331D-01       8.75  1.505207187314D-03
                                                          9.25 -2.058548485296D-03
                                                          9.50 -3.401886185434D-03
                                                          9.75 -2.811920487264D-03


I=18      H=  8.0
0              0.               -5.295024178358D+03     -1.00  4.523021004826D-04
1             -1.410081288013D-04   8.394165558738D+02    -.75  2.688500749210D-04
2              2.258974863431D-04  -4.431119951798D+01    -.50  1.376529966537D-04
3             -8.539648533815D-05   7.788593814437D-01    -.25  5.070494518018D-05
                                                           .25 -2.246772843467D-05
0  -10.0   5.071277963839D-07   2.535638981919D-07       .50 -2.470029154625D-05
1   -9.0  -3.314163001612D-06  -1.657081500806D-06       .75 -1.464770939571D-05
2   -8.0   1.077648696119D-05   5.388243480597D-06      1.25  1.233262247562D-05
```

CARDINAL SPLINES N=4

3	-7,0	-2,504986710242D-05	-1,252493355121D-05	1.50	1,718287498878D-05
4	-6,0	5,025768360275D-05	2,512884180138D-05	1.75	1,242346762204D-05
5	-5,0	-9,584873041428D-05	-4,792436520714D-05	2.25	-1,413866647401D-05
6	-4,0	1,801239603895D-04	9,006198019477D-05	2.50	-2,199626346357D-05
7	-3,0	-3,370734609359D-04	-1,685367304680D-04	2.75	-1,737172903240D-05
8	-2,0	6,300175637290D-04	3,150087818645D-04	3.25	2,237944100044D-05
9	-1,0	-1,177138215735D-03	-5,885691078676D-04	3.50	3,629214672352D-05
10	0,0	2,199102250367D-03	1,099551125183D-03	3.75	2,958200871842D-05
11	1,0	-4,107623614333D-03	-2,053811807166D-03	4.25	-3,973780523164D-05
12	2.0	7,667755699894D-03	3,833877849947D-03	4.50	-6,530446947665D-05
13	3,0	-1,427541544902D-02	-7,137707724512D-03	4.75	-5,375837568646D-05
14	4,0	2,626725567462D-02	1,313362783731D-02	5.25	7,313960858817D-05
15	5,0	-4,588405926992D-02	-2,294202963496D-02	5.50	1,206736048817D-04
16	6,0	6,960802165387D-02	3,480401082693D-02	5.75	9,962789179307D-05
17	7,0	-8,074097845420D-02	-4,037048922710D-02	6.25	-1,360516548720D-04
18	8,0	6,273141059212D-02	3,136570529606D-02	6.50	-2,247308451432D-04
19	9,0	-2,812492287572D-02	-1,406246143786D-02	6.75	-1,856937639592D-04
20	10,0	5,426195407037D-03	2,713097703518D-03	7.25	2,538553186365D-04
				7.50	4,194582515619D-04
	0	-6,421681799186D-02	-5,531191899745D+02	7.75	3,466798406655D-04
	1	-2,124200400339D-02	1.868503799473D+02	8.25	-4,740798555643D-04
	2	-2,335997073802D-03	-2,094541807467D+01	8.50	-7,834212127745D-04
	3	-8,539648533815D-05	7,788593814437D-01	8.75	-6,475384511628D-04
				9.25	8,855797006434D-04
				9.50	1,463470565068D-03
				9.75	1,209661580336D-03

I=19 H= 9.0

0		0.	3,487505885723D+03	-1.00	-1,390994014904D-04
1		4,336514533219D-05	-5,593587171313D+02	-.75	-8,268120902432D-05
2		-6,947172059334D-05	2,987852134060D+01	-.50	-4,233331976003D-05
3		2,626253556486D-05	-5,314560612358D-01	-.25	-1,559362098833D-05
				.25	6,909646395102D-06
0	-10.0	-1,559603037094D-07	-7,798015185472D-08	.50	7,596241023493D-06
1	-9,0	1,019226064893D-06	5,096130324467D-07	.75	4,504705169300D-06
2	-8,0	-3,314163001612D-06	-1,657081500806D-06	1.25	-3,792731458249D-06
3	-7,0	7,703748273325D-06	3,851874136662D-06	1.50	-5,284361103352D-06
4	-6,0	-1,545607180915D-05	-7,728035904576D-06	1.75	-3,820669655845D-06
5	-5,0	2,947698298704D-05	1,473849149352D-05	2,25	4,348155894326D-06
6	-4,0	-5,539469615118D-05	-2,769734807559D-05	2.50	6,764653710731D-06
7	-3,0	1,036624320880D-04	5,183121604399D-05	2.75	5,342440612617D-06
8	-2,0	-1,937537362822D-04	-9,687686814112D-05	3.25	-6,882494784430D-06
9	-1,0	3,620157695965D-04	1,810078847982D-04	3,50	-1,116115949848D-05
10	0,0	-6,763271345850D-04	-3,381635672925D-04	3.75	-9,097547191576D-06
11	1,0	1,263431357047D-03	6,317156785233D-04	4.25	1,222082521415D-05
12	2,0	-2,359642142221D-03	-1,179821071110D-03	4,50	2,008350742709D-05
13	3,0	4,402672313615D-03	2,201336156807D-03	4.75	1,653266223148D-05
14	4,0	-8,179566198189D-03	-4,089783099094D-03	5,25	-2,249309849696D-05
15	5,0	1,491112801087D-02	7,455564005434D-03	5,50	-3,711153673090D-05
16	6,0	-2,492997328082D-02	-1,246498664041D-02	5,75	-3,063921186235D-05
17	7,0	3,258616777184D-02	1,629308388592D-02	6.25	4,184084722436D-05
18	8,0	-2,812492287572D-02	-1,406246143786D-02	6.50	6,911293226040D-05
19	9,0	1,367218579601D-02	6,836092898004D-03	6,75	5,710760454455D-05
20	10,0	-2,800957149318D-03	-1,400478574659D-03	7.25	-7,806976202817D-05
				7.50	-1,289986901061D-04
	0	1,974901495885D-02	3,503147872337D+02	7.75	-1,066166672262D-04

CARDINAL SPLINES N=4

	1	6,532691402924D-03	-1,212251086901D+02	8,25	1,457967560018D-04
	2	7,184043463525D-04	1,393483950353D+01	8,50	2,409303728687D-04
	3	2,626253556486D-05	-5,314560612358D-01	8,75	1,991414187181D-04
				9,25	-2,723472597575D-04
				9,50	-4,500686543176D-04
				9,75	-3,720127518946D-04
I=20	H= 10.0				
0		0.	-9,127697723883D+02	-1,00	2,128476267315D-05
1		-6,635663538388D-06	1,482784556894D+02	-,75	1,265174323366D-05
2		1,063044894141D-05	-8,029415410875D+00	-,50	6,477775278716D-06
3		-4,018650193349D-06	1.449316010104D-01	-,25	2,386110352706D-06
				,25	-1,057302778439D-06
0	-10.0	2,386479032524D-08	1,193239516262D-08	,50	-1.162364364335D-06
1	-9,0	-1,559603037094D-07	-7,798015185472D-08	,75	-6,893026095977D-07
2	-8,0	5,071277963839D-07	2,535638981919D-07	1,25	5,803575580226D-07
3	-7,0	-1,178814948097D-06	-5,894074740487D-07	1,50	8,086042841164D-07
4	-6,0	2,365062805884D-06	1,182531402942D-06	1,75	5,846326152728D-07
5	-5,0	-4,510519702535D-06	-2,255259851267D-06	2,25	-6,653476958390D-07
6	-4,0	8,476405934127D-06	4,238202967064D-06	2,50	-1,035116235230D-06
7	-3,0	-1,586225747114D-05	-7,931128735568D-06	2,75	-8,174915154939D-07
8	-2,0	2,964789448194D-05	1,482394724097D-05	3,25	1,053148083305D-06
9	-1,0	-5,539518260106D-05	-2,769759130053D-05	3,50	1,707862352361D-06
10	0,0	1,034914683544D-04	5,174573417722D-05	3,75	1,392091775789D-06
11	1,0	-1,933363373070D-04	-9,666816865351D-05	4,25	-1,870010664523D-06
12	2,0	3,611349934705D-04	1.805674967352D-04	4,50	-3,073145422066D-06
13	3,0	-6,742316306322D-04	-3,371158153161D-04	4,75	-2,529800903464D-06
14	4,0	1,256044593776D-03	6,280222968880D-04	5,25	3,441857096778D-06
15	5,0	-2.317626682163D-03	-1,158813341082D-03	5,50	5,678746569903D-06
16	6,0	4,094802064765D-03	2,047401032382D-03	5,75	4,688362011746D-06
17	7,0	-5,819598674503D-03	-2,909799337251D-03	6,25	-6,402417872352D-06
18	8,0	5,426195407037D-03	2,713097703518D-03	6,50	-1.057554758202D-05
19	9,0	-2,800957149318D-03	-1,400478574659D-03	6,75	-8,738511960666D-06
20	10,0	6,001643257388D-04	3,000821628694D-04	7,25	1,194610663210D-05
				7,50	1,973916708111D-05
	0	-3,021961934592D-03	-8,799515557180D+01	7,75	1,631430633887D-05
	1	-9,996217427150D-04	3,116962777499D+01	8,25	-2,230957655139D-05
	2	-1,099290568591D-04	-3,681467380562D+00	8,50	-3,686676079759D-05
	3	-4,018650193349D-06	1,449316010104D-01	8,75	-3,047228144146D-05
				9,25	4,167409634022D-05
				9,50	6,886868381249D-05
				9,75	5,692468100718D-05

CARDINAL SPLINES N=5

```
                  N= 5       M= 5                        P=  2,5       Q= 9

I= 0      H= -2.5
0            1.000000000000D+00  -1.670209360394D+00     -1.00  5,564914527178D+00
1           -2.217034274495D+00   1.413159719999D+00      -.75  3,886907663168D+00
2            1.713224544892D+00  -2.228789188382D-01      -.50  2,611827991949D+00
3           -5.654197878652D-01  -8.139392193269D-02      -.25  1,670440239677D+00
4            6.923591992659D-02   2.083333333333D-02       .25  5,442537340345D-01
                                                           .50  2,534387579831D-01
0   -2,5    -6.402458544081D-06  -3.201229272040D-06       .75  8,428282468769D-02
1   -1,5     3.201229272040D-05   1.600614636020D-05      1.25 -2,972970917707D-02
2    -,5    -6.402458544081D-05  -3.201229272040D-05      1.50 -2,882719464274D-02
3     ,5     6.402458544081D-05   3.201229272040D-05      1.75 -1,498245541132D-02
4    1,5    -3.201229272040D-05  -1.600614636020D-05      2.25  9,604639109152D-03
5    2,5     6.402458544081D-06   3.201229272040D-06      2.50  1,171875000000D-02
                                                          2.75  7,485204640848D-03
       0     3.491165607593D-02   1.171875000000D-02      3.25 -6,990200838685D-03
       1     7.471242290373D-02   7.471242290373D-02      3.50 -1,023530535726D-02
       2     6.892313314980D-02  -5.208333333333D-02      3.75 -7,867947072926D-03
       3     1.269394114006D-01   1.269394114006D-01      4.25  1,093201906231D-02
       4     6.923591992659D-02   2.083333333333D-02      4.50  1,999874201693D-02


I= 1      H= -1.5
0            0.                   1.335104680197D+01     -1.00 -1,282457263589D+01
1            4.668504705807D+00  -1.348246526666D+01      -.75 -7,698698472088D+00
2           -5.607789391126D+00   4.072727927524D+00      -.50 -4,035702459745D+00
3            2.243765605993D+00  -1.763637236699D-01      -.25 -1,553861354635D+00
4           -3.045129329663D-01  -6.250000000000D-02       .25  8,505086735777D-01
                                                           .50  1,193743710085D+00
0   -2,5     3.201229272040D-05   1.600614636020D-05       .75  1,197238220312D+00
1   -1,5    -1.600614636020D-04  -8.003073180101D-05      1.25  7,126133896354D-01
2    -,5     3.201229272040D-04   1.600614636020D-04      1.50  4,175734732137D-01
3     ,5    -3.201229272040D-04  -1.600614636020D-04      1.75  1,701271208066D-01
4    1,5     1.600614636020D-04   8.003073180101D-05      2.25 -8,562085179576D-02
5    2,5    -3.201229272040D-05  -1.600614636020D-05      2.50 -9,765625000000D-02
                                                          2.75 -5,939867945424D-02
       0    -2.136207803797D-01  -9.765625000000D-02      3.25  5,204084794342D-02
       1    -3.318954478520D-01  -3.318954478520D-01      3.50  7,461402678629D-02
       2    -1.987823324156D-01   4.062500000000D-01      3.75  5,642957911463D-02
       3    -8.013637236699D-01  -8.013637236699D-01      4.25 -7,663275156156D-02
       4    -3.045129329663D-01  -6.250000000000D-02      4.50 -1,390562100847D-01


I= 2      H=  -.5
0            0.                  -2.670209360394D+01     -1.00  1,564914527178D+01
1           -4.337009411614D+00   3.196493053332D+01      -.75  8,996029756677D+00
2            7.298912115586D+00  -1.206212252172D+01      -.50  4,462029919489D+00
3           -3.487531211985D+00   1.352727447340D+00      -.25  1,596980521770D+00
```

4		5.256925325992D-01	4.166666666667D-02	.25	-6.805095346554D-01
				.50	-7.468624201693D-01
0	-2.5	-6.402458544081D-05	-3.201229272040D-05	.75	-4.520936281231D-01
1	-1.5	3.201229272040D-04	1.600614636020D-04	1.25	4.546560332293D-01
2	-.5	-6.402458544081D-04	-3.201229272040D-04	1.50	8.054780535726D-01
3	.5	6.402458544081D-04	3.201229272040D-04	1.75	9.927535708868D-01
4	1.5	-3.201229272040D-04	-1.600614636020D-04	2.25	8.479995160915D-01
5	2.5	6.402458544081D-05	3.201229272040D-05	2.50	5.859375000000D-01
				2.75	2.799301714085D-01
	0	8.178665607593D-01	5.859375000000D-01	3.25	-1.929488833868D-01
	1	-3.778757709627D-01	-3.778757709627D-01	3.50	-2.586030535726D-01
	2	8.558979981646D-01	-3.541666666667D-01	3.75	-1.861013457293D-01
	3	1.769394114006D+00	1.769394114006D+00	4.25	2.362733156231D-01
	4	5.256925325992D-01	4.166666666667D-02	4.50	4.187374201693D-01

I= 3 H= .5

0		0.	2.670209360394D+01	-1.00	-1.064914527178D+01
1		2.670342744947D+00	-3.363159719999D+01	-.75	-6.008725069177D+00
2		-4.882245448919D+00	1.447878918838D+01	-.50	-2.915154919489D+00
3		2.654197878652D+00	-2.186060780673D+00	-.25	-1.015925834270D+00
4		-4.423591992659D-01	4.166666666667D-02	.25	4.021892221554D-01
				.50	4.187374201693D-01
0	-2.5	6.402458544081D-05	3.201229272040D-05	.75	2.362733156231D-01
1	-1.5	-3.201229272040D-04	-1.600614636020D-04	1.25	-1.861013457293D-01
2	-.5	6.402458544081D-04	3.201229272040D-04	1.50	-2.586030535726D-01
3	.5	-6.402458544081D-04	-3.201229272040D-04	1.75	-1.929488833868D-01
4	1.5	3.201229272040D-04	1.600614636020D-04	2.25	2.799301714085D-01
5	2.5	-6.402458544081D-05	-3.201229272040D-05	2.50	5.859375000000D-01
				2.75	8.479995160915D-01
	0	3.540084392407D-01	5.859375000000D-01	3.25	9.927535708868D-01
	1	3.778757709627D-01	3.778757709627D-01	3.50	8.054780535726D-01
	2	-1.564231331498D+00	-3.541666666667D-01	3.75	4.546560332293D-01
	3	-1.769394114006D+00	-1.769394114006D+00	4.25	-4.520936281231D-01
	4	-4.423591992659D-01	4.166666666667D-02	4.50	-7.468624201693D-01

I= 4 H= 1.5

0		0.	-1.335104680197D+01	-1.00	3.824572635892D+00
1		-9.185047058070D-01	1.723246526666D+01	-.75	2.139616440838D+00
2		1.732789391126D+00	-7.947727927524D+00	-.50	1.027889959745D+00
3		-9.937656059927D-01	1.426363723670D+00	-.25	3.541543233852D-01
4		1.795129329663D-01	-6.250000000000D-02	.25	-1.361532048277D-01
				.50	-1.390562100847D-01
0	-2.5	-3.201229272040D-05	-1.600614636020D-05	.75	-7.663275156156D-02
1	-1.5	1.600614636020D-04	8.003073180101D-05	1.25	5.642957911463D-02
2	-.5	-3.201229272040D-04	-1.600614636020D-04	1.50	7.461402678629D-02
3	.5	3.201229272040D-04	1.600614636020D-04	1.75	5.204084794342D-02
4	1.5	-1.600614636020D-04	-8.003073180101D-05	2.25	-5.939867945424D-02
5	2.5	3.201229272040D-05	1.600614636020D-05	2.50	-9.765625000000D-02
				2.75	-8.562085179576D-02
	0	1.830828037967D-02	-9.765625000000D-02	3.25	1.701271208066D-01
	1	3.318954478520D-01	3.318954478520D-01	3.50	4.175734732137D-01
	2	1.011282332416D+00	4.062500000000D-01	3.75	7.126133896354D-01
	3	8.013637236699D-01	8.013637236699D-01	4.25	1.197238220312D+00
	4	1.795129329663D-01	-6.250000000000D-02	4.50	1.193743710085D+00

CARDINAL SPLINES N=5

I= 5 H= 2.5

0		0.	2.670209360394D+00	-1.00	-5,649145271784D-01
1		1,337009411614D-01	-3.496493053332D+00	-.75	-3,151303194177D-01
2		-2.548912115586D-01	1.681212252172D+00	-.50	-1.508904919489D-01
3		1.487531211985D-01	-3.352727447340D-01	-.25	-5,178789592704D-02
4		-2,756925325992D-02	2.083333333333D-02	.25	1,971110971554D-02
				.50	1,999874201693D-02
0	-2,5	6.402458544081D-06	3.201229272040D-06	.75	1,093201906231D-02
1	-1,5	-3.201229272040D-05	-1.600614636020D-05	1.25	-7,867947072926D-03
2	-,5	6.402458544081D-05	3.201229272040D-05	1.50	-1,023530535726D-02
3	,5	-6.402458544081D-05	-3.201229272040D-05	1.75	-6,990200838685D-03
4	1,5	3,201229272040D-05	1.600614636020D-05	2,25	7,485204640848D-03
5	2,5	-6.402458544081D-06	-3.201229272040D-06	2.50	1,171875000000D-02
				2.75	9,604639109152D-03
	0	-1.147415607593D-02	1.171875000000D-02	3,25	-1,498245541132D-02
	1	-7.471242290373D-02	-7.471242290373D-02	3.50	-2,882719464274D-02
	2	-1,730897998165D-01	-5,208333333333D-02	3.75	-2,972970917707D-02
	3	-1.269394114006D-01	-1.269394114006D-01	4.25	8,428282468769D-02
	4	-2,756925325992D-02	2,083333333333D-02	4,50	2,534387579831D-01

N= 5 M= 6 P= 3,0 Q= 9

I= 0 H= -3.0

0		1.000000000000D+00	4,353774249590D+00	-1.00	5,752094402048D+00
1		-2.260522750519D+00	-4,518943213176D+00	-.75	3,990948223309D+00
2		1.797238731349D+00	2,005779063116D+00	-.50	2,661436849911D+00
3		-6,155197488405D-01	-4,668583435466D-01	-.25	1,687383468365D+00
4		7.881317133900D-02	4,788331175074D-02	.25	5,378871009185D-01
				.50	2,470341438152D-01
0	-3,0	-9,403327676642D-06	-4,701663838321D-06	.75	8,081860340000D-02
1	-2,0	5.232871479156D-05	2.616435739578D-05	1,25	-2,731002083294D-02
2	-1.0	-1.205936588082D-04	-6,029682940410D-05	1.50	-2,574584597842D-02
3	0.0	1.471540408500D-04	7.357702042498D-05	1.75	-1,293459637521D-02
4	1.0	-1.001374024668D-04	-5,006870123338D-05	2,25	7.578406388252D-03
5	2,0	3,596370971841D-05	1.798185485921D-05	2.50	8,719232105247D-03
6	3,0	-5,312076408354D-06	-2,656038204177D-06	2,75	5,193840892656D-03
				3,25	-4,056124666570D-03
	0	1,584139903449D-01	1.223291541586D-01	3,50	-5,309684475163D-03
	1	4.156989234908D-01	8.195355884310D-02	3,75	-3,586795211341D-03
	2	5,134722440901D-01	3.897528057370D-01	4,25	3,625077666767D-03
	3	3,302383072275D-01	1.077413974623D-01	4,50	5,454539609226D-03
	4	7,881317133900D-02	4.788331175074D-02	4,75	4,283282209472D-03

I= 1 H= -2.0

0		0.	-2.743240231356D+01	-1.00	-1,409181394935D+01
1		4.962929485803D+00	2.667893519304D+01	-.75	-8,403071718299D+00
2		-6.176580496843D+00	-1.101568619968D+01	-.50	-4,371563322813D+00

CARDINAL SPLINES N=5

```
3            2.582951324516D+00   2.433299522871D+00    -.25 -1.668570050707D+00
4           -3.693526421908D-01  -2.456332038378D-01     .25  8.936119212846D-01
                                                         .50  1.237104096323D+00
0  -3.0      5.232871479156D-05   2.616435739578D-05     .75  1.220691619550D+00
1  -2.0     -2.976072836762D-04  -1.488036418381D-04    1.25  6.962316657052D-01
2  -1.0      7.031056965077D-04   3.515528482539D-04    1.50  3.967121902829D-01
3   0.0     -8.829242450998D-04  -4.414621225499D-04    1.75  1.562627479593D-01
4   1.0      6.212806711419D-04   3.106403355710D-04    2.25 -7.190289297992D-02
5   2.0     -2.321472633836D-04  -1.160736316918D-04    2.50 -7.734897605107D-02
6   3.0      3.596370971841D-05   1.798185485921D-05    2.75 -4.388573604671D-02
                                                        3.25  3.217662619019D-02
        0   -8.783142696966D-01  -7.339749249514D-01    3.50  4.126669027057D-02
        1   -2.246953089925D+00  -2.444809020391D-01    3.75  2.744541341845D-02
        2   -2.875061254501D+00  -2.380183501089D+00    4.25 -2.716344820863D-02
        3   -1.849280381774D+00  -5.142989231827D-01    4.50 -4.058935206771D-02
        4   -3.693526421908D-01  -2.456332038378D-01    4.75 -3.170404646442D-02


I= 2    H= -1.0
0            0.                   8.685539782397D+01   -1.00  1.917765371605D+01
1           -5.156806171221D+00  -7.986052776757D+01    -.75  1.095728758561D+01
2            8.882654847317D+00   2.995007838502D+01    -.50  5.397201365408D+00
3           -4.431960389973D+00  -5.913622461155D+00    -.25  1.916375571807D+00
4            7.062323075357D-01   5.515830095944D-01     .25 -8.005262764500D-01
                                                         .50 -8.675951388414D-01
0  -3.0     -1.205936588082D-04  -6.029682940410D-05     .75 -5.173973050006D-01
1  -2.0      7.031056965077D-04   3.515528482539D-04    1.25  5.002693277180D-01
2  -1.0     -1.706623600416D-03  -8.533118002078D-04    1.50  8.635642382619D-01
3   0.0      2.207310612749D-03   1.103655306375D-03    1.75  1.031357549582D+00
4   1.0     -1.604342318708D-03  -8.021711593542D-04    2.25  8.098032128258D-01
5   2.0      6.212806711419D-04   3.106403355710D-04    2.50  5.293938986767D-01
6   3.0     -1.001374024668D-04  -5.006870123338D-05    2.75  2.367359105937D-01
                                                        3.25 -1.376389172024D-01
        0    2.015361493310D+00   1.834937312379D+00    3.50 -1.657506842254D-01
        1    4.749281597265D+00  -2.568988724512D-01    3.75 -1.053977951721D-01
        2    7.131555944487D+00   6.512958752722D+00    4.25  9.853091979163D-02
        3    4.042827300455D+00   7.053736539783D-01    4.50  1.445661662002D-01
        4    7.062323075357D-01   5.515830095944D-01    4.75  1.113608429300D-01


I= 3    H=  0.0
0            0.                  -1.401730531520D+02   -1.00 -1.583436341162D+01
1            3.875051503914D+00   1.306982900700D+02    -.75 -8.890835125624D+00
2           -7.209589047815D+00  -4.725903280554D+01    -.50 -4.289409231709D+00
3            4.042056624875D+00   8.491994820178D+00    -.25 -1.485283647461D+00
4           -7.076662350148D-01  -7.076662350148D-01     .25  5.785563745846D-01
                                                         .50  5.961567158345D-01
0  -3.0      1.471540408500D-04   7.357702042498D-05     .75  3.322384565011D-01
1  -2.0     -8.829242450998D-04  -4.414621225499D-04    1.25 -2.531310512653D-01
2  -1.0      2.207310612749D-03   1.103655306375D-03    1.50 -3.439619363081D-01
3   0.0     -2.943080816999D-03  -1.471540408500D-03    1.75 -2.496782504156D-01
4   1.0      2.207310612749D-03   1.103655306375D-03    2.25  3.360604507309D-01
5   2.0     -8.829242450998D-04  -4.414621225499D-04    2.50  6.690295236992D-01
6   3.0      1.471540408500D-04   7.357702042498D-05    2.75  9.114744002391D-01
                                                        3.25  9.114744002391D-01
        0   -1.446583083171D+00  -1.446583083171D+00    3.50  6.690295236992D-01
        1   -6.674907292954D+00  -3.007814778549D-26    3.75  3.360604507309D-01
```

CARDINAL SPLINES N=5

```
       2  -9.045056114741D+00  -9.045056114741D+00    4.25 -2.496782504156D-01
       3  -4.449938195303D+00  -1.357694170873D-26    4.50 -3.439619363081D-01
       4  -7.076662350148D-01  -7.076662350148D-01    4.75 -2.531310512653D-01


I= 4     H=  1.0
0             0.               1.234041819041D+02   -1.00  8.073891401381D+00
1          -1.905771084650D+00 -1.174369073375D+02    -.75  4.501523696577D+00
2           3.640062057739D+00  4.264680415663D+01    -.50  2.154099982156D+00
3          -2.131124547339D+00 -7.324369769112D+00    -.25  7.387959931346D-01
4           3.969337116532D-01  5.515830095944D-01     .25 -2.806871916769D-01
                                                       .50 -2.844524349104D-01
0  -3.0   -1.001374024668D-04  -5.006870123338D-05     .75 -1.552767860011D-01
1  -2.0    6.212806711419D-04   3.106403355710D-04    1.25  1.113608429300D-01
2  -1.0   -1.604342318708D-03  -8.021711593542D-04    1.50  1.445661662002D-01
3   0.0    2.207310612749D-03   1.103655306375D-03    1.75  9.853091979163D-02
4   1.0   -1.706623600416D-03  -8.533118002078D-04    2.25 -1.053977951721D-01
5   2.0    7.031056965077D-04   3.515528482539D-04    2.50 -1.657506842254D-01
6   3.0   -1.205936588082D-04  -6.029682940410D-05    2.75 -1.376389172024D-01
                                                      3.25  2.367359105937D-01
       0   1.654513131447D+00   1.834937312379D+00    3.50  5.293938986767D-01
       1   5.263079342167D+00   2.568988724512D-01    3.75  8.098032128258D-01
       2   5.894361560957D+00   6.512958752722D+00    4.25  1.031357549582D+00
       3   2.632079992499D+00  -7.053736539783D-01    4.50  8.635642382619D-01
       4   3.969337116532D-01   5.515830095944D-01    4.75  5.002693277180D-01


I= 5     H=  2.0
0             0.              -5.667142957766D+01   -1.00 -2.408804097619D+00
1           5.621014165459D-01  5.494003884898D+01    -.75 -1.340023107075D+00
2          -1.082506265180D+00 -2.027306681697D+01    -.50 -6.395822162118D-01
3           6.422826504088D-01  3.461897369236D+00    -.25 -2.186938877693D-01
4          -1.219137654848D-01 -2.456332038378D-01     .25  8.242815346615D-02
                                                       .50  8.308993317796D-02
0  -3.0    3.596370971841D-05   1.798185485921D-05     .75  4.505770435053D-02
1  -2.0   -2.321472633836D-04  -1.160736316918D-04    1.25 -3.170404646442D-02
2  -1.0    6.212806711419D-04   3.106403355710D-04    1.50 -4.058935206771D-02
3   0.0   -8.829242450998D-04  -4.414621225499D-04    1.75 -2.716344820863D-02
4   1.0    7.031056965077D-04   3.515528482539D-04    2.25  2.744541341845D-02
5   2.0   -2.976072836762D-04  -1.488036418381D-04    2.50  4.126669027057D-02
6   3.0    5.232871479156D-05   2.616435739578D-05    2.75  3.217662619019D-02
                                                      3.25 -4.388573604671D-02
       0  -5.896355802062D-01  -7.339749249514D-01    3.50 -7.734897605107D-02
       1  -1.757991285847D+00   2.444809020391D-01    3.75 -7.190289297992D-02
       2  -1.885305747677D+00  -2.380183501089D+00    4.25  1.562627479593D-01
       3  -8.206825354082D-01   5.142989231827D-01    4.50  3.967121902829D-01
       4  -1.219137654848D-01  -2.456332038378D-01    4.75  6.962316657052D-01


I= 6     H=  3.0
0             0.               1.066353106561D+01   -1.00  3.313419391143D-01
1          -7.698239987196D-02 -1.050088579383D+01    -.75  1.841704455031D-01
2           1.487201734329D-01  3.945124217438D+00    -.50  8.781657326023D-02
3          -8.868591364700D-02 -6.823411384712D-01    -.25  2.999255263079D-02
4           1.695345216248D-02  4.788331175074D-02     .25 -1.127008212692D-02
                                                       .50 -1.133731539863D-02
0  -3.0   -5.312076408354D-06  -2.656038204177D-06     .75 -6.132292800105D-03
```

CARDINAL SPLINES N=5

1	-2.0	3.596370971841D-05	1.798185485921D-05	1.25	4.283282209472D-03
2	-1.0	-1.001374024668D-04	-5.006870123338D-05	1.50	5.454539609226D-03
3	0.0	1.471540408500D-04	7.357702042498D-05	1.75	3.625077666767D-03
4	1.0	-1.205936588082D-04	-6.029682940410D-05	2.25	-3.586795211341D-03
5	2.0	5.232871479156D-05	2.616435739578D-05	2.50	-5.309684475163D-03
6	3.0	-9.403327676642D-06	-4.701663838321D-06	2.75	-4.056124666570D-03
				3.25	5.193840892656D-03
	0	8.624431797227D-02	1.223291541586D-01	3.50	8.719232105247D-03
	1	2.517918058046D-01	-8.195355884310D-02	3.75	7.578406388252D-03
	2	2.660333673840D-01	3.897528057370D-01	4.25	-1.293459637521D-02
	3	1.147555123028D-01	-1.077413974623D-01	4.50	-2.574584597842D-02
	4	1.695345216248D-02	4.788331175074D-02	4.75	-2.731002083294D-02

N= 5 M= 7 P= 3.5 Q= 9

I= 0 H= -3.5

0		1.000000000000D+00	-4.726987853791D+00	-1.00	5.817722437222D+00
1		-2.275675864806D+00	2.602230949999D+00	-.75	4.027382626398D+00
2		1.826643682498D+00	-8.409817608996D-02	-.50	2.678785288166D+00
3		-6.331800722976D-01	-1.991739121394D-01	-.25	1.693298817869D+00
4		8.222281762051D-02	3.596620097658D-02	.25	5.356740081660D-01
				.50	2.448143846545D-01
0	-3.5	-1.056301441722D-05	-5.281507208610D-06	.75	7.962235006690D-02
1	-2.5	6.075945609804D-05	3.037972804902D-05	1.25	-2.648330533853D-02
2	-1.5	-1.488520339118D-04	-7.442601695589D-05	1.50	-2.470120047863D-02
3	-.5	2.025738326911D-04	1.012869163455D-04	1.75	-1.224731213076D-02
4	.5	-1.672586090044D-04	-8.362930450221D-05	2.25	6.918769643291D-03
5	1.5	8.528463127586D-05	4.264231563793D-05	2.50	7.764341911225D-03
6	2.5	-2.544423241141D-05	-1.272211620570D-05	2.75	4.485109501574D-03
7	3.5	3.499969679894D-06	1.749984839947D-06	3.25	-3.221330776187D-03
				3.50	-3.988219944979D-03
	0	6.024855531917D-01	2.082143651735D-01	3.75	-2.519433225430D-03
	1	1.342675477657D+00	8.621059137293D-01	4.25	2.138972787376D-03
	2	1.221630018480D+00	4.680915182250D-01	4.50	2.881887504414D-03
	3	5.179393743895D-01	3.043529015327D-01	4.75	1.990803659579D-03
	4	8.222281762051D-02	3.596620097658D-02	5.25	-2.074244649761D-03

I= 1 H= -2.5

0		0.	3.858331855351D+01	-1.00	-1.456891950523D+01
1		5.073090257732D+00	-2.509088769737D+01	-.75	-8.667944132847D+00
2		-6.390349894726D+00	4.177394612170D+00	-.50	-4.497683756793D+00
3		2.711339115155D+00	4.872755545016D-01	-.25	-1.711573716831D+00
4		-3.941402376165D-01	-1.589976715496D-01	.25	9.097007596155D-01
				.50	1.253241398399D+00
0	-3.5	6.075945609804D-05	3.037972804902D-05	.75	1.229388194268D+00
1	-2.5	-3.588974549696D-04	-1.794487274848D-04	1.25	6.902215733039D-01
2	-1.5	9.085396657889D-04	4.542698328945D-04	1.50	3.891177808122D-01
3	-.5	-1.285817467832D-03	-6.429087339159D-04	1.75	1.512662989475D-01

CARDINAL SPLINES N=5

4	,5	1,109241349399D-03	5,546206746993D-04	2.25	-6,710743696717D-02
5	1,5	-5,907026526093D-04	-2,953513263046D-04	2.50	-7,040707362156D-02
6	2,5	1,823213365365D-04	9.116066826824D-05	2.75	-3,873336963178D-02
7	3,5	-2,544423241141D-05	-1,272211620570D-05	3,25	2,610780511310D-02
				3.50	3,165984972489D-02
	0	-3,422975153401D+00	-1,029353075868D+00	3,75	1,968585864499D-02
	1	-7,612697274645D+00	-5,209849455007D+00	4,25	-1,635969806297D-02
	2	-6,890596650414D+00	-2,392540924458D+00	4,50	-2,188657345663D-02
	3	-2,806624211476D+00	-1,738691847193D+00	4,75	-1,503808579718D-02
	4	-3,941402376165D-01	-1,589976715496D-01	5,25	1,555632870514D-02

I= 2 H= -1.5

0		0.	-1,344177898095D+02	-1.00	2.077682836414D+01
1		-5,526045874079D+00	9.366289337700D+01	-.75	1,184509374319D+01
2		9,599172601980D+00	-2,097447984226D+01	-.50	5,819935073164D+00
3		-4,862293881976D+00	6,091106247884D-01	-.25	2,060516363698D+00
4		7,893160061089D-01	2,611958087893D-01	.25	-8,544532577209D-01
				.50	-9,216845621364D-01
0	-3,5	-1,488520339118D-04	-7,442601695589D-05	.75	-5,465467073546D-01
1	-2,5	9,085396657889D-04	4,542698328945D-04	1.25	5,204141096902D-01
2	-1,5	-2,395202451544D-03	-1,197601225772D-03	1.50	8,890193748522D-01
3	-,5	3,557738457797D-03	1.778869228899D-03	1.75	1,048104775624D+00
4	.5	-3.239901444618D-03	-1,619950722309D-03	2,25	7,937296802769D-01
5	1,5	1.823095827820D-03	9.115479139102D-04	2,50	5,061258519350D-01
6	2,5	-5,907026526093D-04	-2,953513263046D-04	2,75	2,194660765979D-01
7	3,5	8.528463127586D-05	4,264231563793D-05	3,25	-1,172972869473D-01
				3,50	-1,335502295048D-01
	0	8.224586791976D+00	1.776273036562D+00	3,75	-7,938912031729D-02
	1	1,834655722484D+01	1.402143114949D+01	4.25	6,231863275262D-02
	2	1,655981329024D+01	4,619073664025D+00	4.50	8,187771280573D-02
	3	6,188130203548D+00	4,265851947838D+00	4.75	5,549944819893D-02
	4	7,893160061089D-01	2,611958087893D-01	5,25	-5,639331789850D-02

I= 3 H= -.5

0		0.	2.937837456273D+02	-1.00	-1,897063454505D+01
1		4.599197439300D+00	-2,096125680077D+02	-.75	-1,063198380342D+01
2		-8,614810143094D+00	5,261324952525D+01	-.50	-5,118466599734D+00
3		4,886018546309D+00	-4,300266183252D+00	-.25	-1,767969847681D+00
4		-8,706084163481D-01	-1,381643382163D-01	.25	6,843169523141D-01
				.50	7,022358717954D-01
0	-3,5	2,025738326911D-04	1,012869163455D-04	.75	3,894057141664D-01
1	-2,5	-1,285817467832D-03	-6.429087339159D-04	1,25	-2,926386174114D-01
2	-1,5	3,557738457797D-03	1,778869228899D-03	1.50	-3,938840696915D-01
3	-,5	-5,591514415697D-03	-2,795757207849D-03	1,75	-2,825225940045D-01
4	,5	5,414938297264D-03	2,707469148632D-03	2,25	3,675835593578D-01
5	1,5	-3,239901444618D-03	-1,619950722309D-03	2,50	7,146623777555D-01
6	2,5	1,109241349399D-03	5,546206746993D-04	2,75	9,453436726820D-01
7	3,5	-1,672586090044D-04	-8,362930450221D-05	3,25	8,715807789564D-01
				3,50	6,058785997249D-01
	0	-1,059186352057D+01	-4,551343258677D-01	3,75	2,850527285517D-01
	1	-2,545263538918D+01	-2,304978756954D+01	4,25	-1,786593966268D-01
	2	-2,130133400843D+01	-2,694624257792D+00	4,50	-2,210185248337D-01
	3	-7,302499282564D+00	-6,234566918280D+00	4,75	-1,435767383722D-01
	4	-8,706084163481D-01	-1,381643382163D-01	5,25	1,383100654071D-01

CARDINAL SPLINES N=5

I= 4 H= .5

0		0.	-4.021788806020D+02	-1.00	1.187235934058D+01
1		-2.782814236162D+00	2.947276756250D+02	-.75	6.610301014807D+00
2		5.341983559677D+00	-7.831265575863D+01	-.50	3.158205717348D+00
3		-3.153281804822D+00	8.168867653307D+00	-.25	1.081168964955D+00
4		5.942797399156D-01	-1.381643382163D-01	.25	-4.087782101650D-01
				.50	-4.129292966967D-01
0	-3.5	-1.672586090044D-04	-8.362930450221D-05	.75	-2.245144208148D-01
1	-2.5	1.109241349399D-03	5.546206746993D-04	1.25	1.592100935135D-01
2	-1.5	-3.239901444618D-03	-1.619950722309D-03	1.50	2.050289307526D-01
3	-.5	5.414938297264D-03	2.707469148632D-03	1.75	1.383100654071D-01
4	.5	-5.591514415697D-03	-2.795757207849D-03	2.25	-1.435767383722D-01
5	1.5	3.557738457797D-03	1.778869228899D-03	2.50	-2.210185248337D-01
6	2.5	-1.285817467832D-03	-6.429087339159D-04	2.75	-1.786593966268D-01
7	3.5	2.025738326911D-04	1.012869163455D-04	3.25	2.850527285517D-01
				3.50	6.058785997249D-01
	0	9.681594868833D+00	-4.551343258677D-01	3.75	8.715807789564D-01
	1	2.064693974991D+01	2.304978756954D+01	4.25	9.453436726820D-01
	2	1.591208549284D+01	-2.694624257792D+00	4.50	7.146623777555D-01
	3	5.166634553996D+00	6.234566918280D+00	4.75	3.675835593578D-01
	4	5.942797399156D-01	-1.381643382163D-01	5.25	-2.825225940045D-01

I= 5 H= 1.5

0		0.	3.295290327641D+02	-1.00	-5.199932996089D+00
1		1.206556108431D+00	-2.479200870881D+02	-.75	-2.889560817428D+00
2		-2.333084751830D+00	6.860841106233D+01	-.50	-1.377402984868D+00
3		1.393367747298D+00	-7.922593270887D+00	-.25	-4.702708685414D-01
4		-2.669243885304D-01	2.611958087893D-01	.25	1.765499281026D-01
				.50	1.774952269587D-01
0	-3.5	8.528463127586D-05	4.264231563793D-05	.75	9.593378557168D-02
1	-2.5	-5.907026526093D-04	-2.953513263046D-04	1.25	-6.686386042403D-02
2	-1.5	1.823095827820D-03	9.115479139102D-04	1.50	-8.501762476219D-02
3	-.5	-3.239901444618D-03	-1.619950722309D-03	1.75	-5.639331789850D-02
4	.5	3.557738457797D-03	1.778869228899D-03	2.25	5.549944819893D-02
5	1.5	-2.395202451544D-03	-1.197601225772D-03	2.50	8.187771280573D-02
6	2.5	9.085396657889D-04	4.542698328945D-04	2.75	6.231863275262D-02
7	3.5	-1.488520339118D-04	-7.442601695589D-05	3.25	-7.938912031729D-02
				3.50	-1.335502295048D-01
	0	-4.672040718852D+00	1.776273036562D+00	3.75	-1.172972869473D-01
	1	-9.696305074142D+00	-1.402143114949D+01	4.25	2.194660765979D-01
	2	-7.321665962185D+00	4.619073664025D+00	4.50	5.061258519350D-01
	3	-2.343573692127D+00	-4.265851947838D+00	4.75	7.937296802769D-01
	4	-2.669243885304D-01	2.611958087893D-01	5.25	1.048104775624D+00

I= 6 H= 2.5

0		0.	-1.469784535283D+02	-1.00	1.470644300754D+00
1		-3.400403879273D-01	1.131226619813D+02	-.75	8.166715004785D-01
2		6.591899779759D-01	-3.233513417888D+01	-.50	3.889853744066D-01
3		-3.952690403335D-01	3.964659248887D+00	-.25	1.326829903547D-01
4		7.614489451732D-02	-1.589976715496D-01	.25	-4.968936121640D-02
				.50	-4.987232329980D-02
0	-3.5	-2.544423241141D-05	-1.272211620570D-05	.75	-2.689924466607D-02
1	-2.5	1.823213365365D-04	9.116066826824D-05	1.25	1.863505904821D-02
2	-1.5	-5.907026526093D-04	-2.953513263046D-04	1.50	2.358958024893D-02

3	-.5	1.109241349399D-03	5.546206746993D-04	1.75	1.555632870514D-02
4	.5	-1.285817467832D-03	-6.429087339159D-04	2.25	-1.503808579718D-02
5	1.5	9.085396657889D-04	4.542698328945D-04	2.50	-2.188657345663D-02
6	2.5	-3.588974549696D-04	-1.794487274848D-04	2.75	-1.635969806297D-02
7	3.5	6.075945609804D-05	3.037972804902D-05	3.25	1.968585864499D-02
				3.50	3.165984972489D-02
	0	1.364269001666D+00	-1.029353075868D+00	3.75	2.610780511310D-02
	1	2.807001635369D+00	5.209849455007D+00	4.25	-3.873336963178D-02
	2	2.105514801497D+00	-2.392540924458D+00	4.50	-7.040707362156D-02
	3	6.707594829090D-01	1.738691847193D+00	4.75	-6.710743696717D-02
	4	7.614489451732D-02	-1.589976715496D-01	5.25	1.512662989475D-01

I= 7 H= 3.5

0		0.	2.740601484875D+01	-1.00	-1.980673963268D-01
1		4.573255751187D-02	-2.149191914012D+01	-.75	-1.099601311742D-01
2		-8.874503248091D-02	6.307312756098D+00	-.50	-5.235811168871D-02
3		5.329939066670D-02	-8.078797152049D-01	-.25	-1.785270382339D-02
4		-1.029041566734D-02	3.596620097658D-02	.25	6.679180904228D-03
				.50	6.699300325713D-03
0	-3.5	3.499969679894D-06	1.749984839947D-06	.75	3.610328762784D-03
1	-2.5	-2.544423241141D-05	-1.272211620570D-05	1.25	-2.495052381899D-03
2	-1.5	8.528463127586D-05	4.264231563793D-05	1.50	-3.152771733587D-03
3	-.5	-1.672586090044D-04	-8.362930450221D-05	1.75	-2.074244649761D-03
4	.5	2.025738326911D-04	1.012869163455D-04	2.25	1.990803659579D-03
5	1.5	-1.488520339118D-04	-7.442601695589D-05	2.50	2.881887504414D-03
6	2.5	6.075945609804D-05	3.037972804902D-05	2.75	2.138972787376D-03
7	3.5	-1.056301441722D-05	-5.281507208610D-06	3.25	-2.519433225430D-03
				3.50	-3.988219944979D-03
	0	-1.860568228446D-01	2.082143651735D-01	3.75	-3.221330776187D-03
	1	-3.815363498016D-01	-8.621059137293D-01	4.25	4.485109501574D-03
	2	-2.854469820299D-01	4.680915182250D-01	4.50	7.764341911225D-03
	3	-9.076642867603D-02	-3.043529015327D-01	4.75	6.918769643291D-03
	4	-1.029041566734D-02	3.596620097658D-02	5.25	-1.224731213076D-02

N= 5 M= 8 P= 4.0 Q= 9

I= 0 H= -4.0

0		1.000000000000D+00	6.634797711454D+00	-1.00	5.841391941038D+00
1		-2.281129640963D+00	-4.852051723272D+00	-.75	4.040517851495D+00
2		1.837242700591D+00	1.756884939803D+00	-.50	2.685036772621D+00
3		-6.395608323913D-01	-4.035135889337D-01	-.25	1.695429227843D+00
4		8.345876709283D-02	4.484940606211D-02	.25	5.348781313071D-01
				.50	2.440169020874D-01
0	-4.0	-1.099432959153D-05	-5.497164795764D-06	.75	7.919311221484D-02
1	-3.0	6.396195794684D-05	3.198097897342D-05	1.25	-2.618774059151D-02
2	-2.0	-1.601755028329D-04	-8.008775141646D-05	1.50	-2.432872006575D-02
3	-1.0	2.284272989020D-04	1.142136494510D-04	1.75	-1.200311420719D-02
4	0.0	-2.085694087487D-04	-1.042847043744D-04	2.25	6.686913362825D-03

CARDINAL SPLINES N=5

5	1.0	1.305286316193D-04	6.526431580967D-05	2.50	7.431420398575D-03
6	2.0	-5.716977721287D-05	-2.858488860643D-05	2.75	4.240632206748D-03
7	3.0	1.616874815439D-05	8.084374077195D-06	3.25	-2.942451907627D-03
8	4.0	-2.177618236553D-06	-1.088809118276D-06	3.50	-3.556953752903D-03
				3.75	-2.181348259582D-03
	0	1.704915748325D+00	9.933281153533D-01	4.25	1.706918817118D-03
	1	3.083336384747D+00	1.315823478233D+00	4.50	2.177370935722D-03
	2	2.174554352807D+00	1.220264854561D+00	4.75	1.407999871418D-03
	3	6.957794410940D-01	3.140769080601D-01	5.25	-1.234313669791D-03
	4	8.345876709283D-02	4.484940606211D-02	5.50	-1.680849644231D-03
I= 1	H= -3.0				
0		0.	-4.577758837912D+01	-1.00	-1.474466484007D+01
1		5.113584377266D+00	3.025693093889D+01	-.75	-8.765472773536D+00
2		-6.469047276737D+00	-9.491848167720D+00	-.50	-4.544100835808D+00
3		2.758716061792D+00	2.004491344018D+00	-.25	-1.727391945095D+00
4		-4.033171242783D-01	-2.249551949670D-01	.25	9.156101207131D-01
				.50	1.259162681831D+00
0	-4.0	6.396195794684D-05	3.198097897342D-05	.75	1.232575272063D+00
1	-3.0	-3.826759322932D-04	-1.913379661466D-04	1.25	6.880270141853D-01
2	-2.0	9.926160728224D-04	4.963080364112D-04	1.50	3.863521252500D-01
3	-1.0	-1.477778656008D-03	-7.388893280040D-04	1.75	1.494531369066D-01
4	0.0	1.415972761687D-03	7.079863808433D-04	2.25	-6.538591124519D-02
5	1.0	-9.266379578762D-04	-4.633189789381D-04	2.50	-6.793514167185D-02
6	2.0	4.178825268981D-04	2.089412634491D-04	2.75	-3.691813326796D-02
7	3.0	-1.195095213310D-04	-5.975476066550D-05	3.25	2.403713812674D-02
8	4.0	1.616874815439D-05	8.084374077195D-06	3.50	2.845771156765D-02
				3.75	1.717558821474D-02
	0	-9.741774779319D+00	-5.920519201480D+00	4.25	-1.315171067705D-02
	1	-1.746960668589D+01	-7.050799801573D+00	4.50	-1.66555969188D-02
	2	-1.208289846596D+01	-7.033650756342D+00	4.75	-1.071078566899D-02
	3	-3.694357926662D+00	-1.594791775455D+00	5.25	9.319867118688D-03
	4	-4.033171242783D-01	-2.249551949670D-01	5.50	1.266060419325D-02
I= 2	H= -2.0				
0		0.	1.638671614287D+02	-1.00	2.139823217001D+01
1		-5.669225777067D+00	-1.020370105636D+02	-.75	1.218993742601D+01
2		9.877432350591D+00	2.735749718730D+01	-.50	5.984057492590D+00
3		-5.029810220186D+00	-4.755490755741D+00	-.25	2.116446765799D+00
4		8.217638221647D-01	4.944097152266D-01	.25	-8.753476927259D-01
				.50	-9.426211523664D-01
0	-4.0	-1.601755028329D-04	-8.008775141646D-05	.75	-5.578156396043D-01
1	-3.0	9.926160728224D-04	4.963080364112D-04	1.25	5.281736749139D-01
2	-2.0	-2.692481462452D-03	-1.346240731226D-03	1.50	8.987982340953D-01
3	-1.0	4.236478596289D-03	2.118239298145D-03	1.75	1.054515789571D+00
4	0.0	-4.324448395145D-03	-2.162224197572D-03	2.25	7.876426758254D-01
5	1.0	3.010902476589D-03	1.505451238294D-03	2.50	4.973855464085D-01
6	2.0	-1.423604534955D-03	-7.118022674777D-04	2.75	2.130477282189D-01
7	3.0	4.178825268981D-04	2.089412634491D-04	3.25	-1.099757819574D-01
8	4.0	-5.716977721287D-05	-2.858488860643D-05	3.50	-1.222280465314D-01
				3.75	-7.051325683851D-02
	0	2.382569888345D+01	1.565655290181D+01	4.25	5.097576796166D-02
	1	4.229088093290D+01	1.512829775730D+01	4.50	6.338179162045D-02
	2	2.840903663617D+01	1.775494078017D+01	4.75	4.019889543505D-02
	3	8.118410934449D+00	3.155064687885D+00	5.25	-3.434231456235D-02

	4	8.217638221647D-01	4.944097152266D-01	5.50	-4.637473662383D-02
I= 3		H= -1.0			
0		0.	-3.872531311264D+02	-1.00	-2.038940845702D+01
1		4.926102278246D+00	2.372046475691D+02	-.75	-1.141932243294D+01
2		-9.250125968895D+00	-5.773713371372D+01	-.50	-5.493186878002D+00
3		5.268487736612D+00	7.948059969507D+00	-.25	-1.895668768476D+00
4		-9.446924732621D-01	-6.706326031970D-01	.25	7.320226132877D-01
				.50	7.500377805438D-01
0	-4.0	2.284272989020D-04	1.142136494510D-04	.75	4.151346635882D-01
1	-3.0	-1.477778656008D-03	-7.388893280040D-04	1.25	-3.103550662063D-01
2	-2.0	4.236478596289D-03	2.118239298145D-03	1.50	-4.162109210110D-01
3	-1.0	-7.141197234516D-03	-3.570598617258D-03	1.75	-2.971600636365D-01
4	0.0	7.891149265380D-03	3.945574632690D-03	2.25	3.814812584649D-01
5	1.0	-5.951872420379D-03	-2.975936210190D-03	2.50	7.346180287292D-01
6	2.0	3.010902476589D-03	1.505451238294D-03	2.75	9.599978881088D-01
7	3.0	-9.266379578762D-04	-4.633189789381D-04	3.25	8.548644985314D-01
8	4.0	1.305286316193D-04	6.526431580967D-05	3.50	5.800280695583D-01
				3.75	2.647875749897D-01
	0	-3.295566440123D+01	-2.523478863958D+01	4.25	-1.527616462421D-01
	1	-5.802876727060D+01	-1.486749002281D+01	4.50	-1.787890917214D-01
	2	-3.671875056270D+01	-2.674114398656D+01	4.75	-1.086428919825D-01
	3	-9.846591835581D+00	-2.782061681646D+00	5.25	8.796375601738D-02
	4	-9.446924732621D-01	-6.706326031970D-01	5.50	1.168004373699D-01
I= 4		H= 0.0			
0		0.	6.860379876696D+02	-1.00	1.413939326482D+01
1		-3.305169738783D+00	-4.192337082910D+02	-.75	7.868375587455D+00
2		6.357143563364D+00	9.801428417641D+01	-.50	3.756964670958D+00
3		-3.764422608923D+00	-1.140251766001D+01	-.25	1.285216828459D+00
4		7.126573537507D-01	7.126573537507D-01	.25	-4.850062482575D-01
				.50	-4.893111274187D-01
0	-4.0	-2.085694087487D-04	-1.042847043744D-04	.75	-2.656262573679D-01
1	-3.0	1.415972761687D-03	7.079863808433D-04	1.25	1.875188959919D-01
2	-2.0	-4.324448395145D-03	-2.162224197572D-03	1.50	2.407046144792D-01
3	-1.0	7.891149265380D-03	3.945574632690D-03	1.75	1.616990211393D-01
4	0.0	-9.548208446347D-03	-4.774104223173D-03	2.25	-1.657836277708D-01
5	1.0	7.891149265380D-03	3.945574632690D-03	2.50	-2.529053091595D-01
6	2.0	-4.324448395145D-03	-2.162224197572D-03	2.75	-2.020751101703D-01
7	3.0	1.415972761687D-03	7.079863808433D-04	3.25	3.117633795519D-01
8	4.0	-2.085694087487D-04	-1.042847043744D-04	3.50	6.471847080155D-01
				3.75	9.039621120340D-01
	0	3.001085364779D+01	3.001085364779D+01	4.25	9.039621120340D-01
	1	4.929997610000D+01	4.682579662589D-24	4.50	6.471847080155D-01
	2	2.959917821635D+01	2.959917821635D+01	4.75	3.117633795519D-01
	3	7.638095051088D+00	1.054571296442D-24	5.25	-2.020751101703D-01
	4	7.126573537507D-01	7.126573537507D-01	5.50	-2.529053091595D-01
I= 5		H= 1.0			
0		0.	-8.622969465596D+02	-1.00	-7.682811184896D+00
1		1.778645070308D+00	5.340175490527D+02	-.75	-4.267416650511D+00
2		-3.444898123632D+00	-1.245066140732D+02	-.50	-2.033169769111D+00
3		2.062695257824D+00	1.351218333280D+01	-.25	-6.937461259464D-01
4		-3.965727331320D-01	-6.706326031970D-01	.25	2.600356365127D-01

CARDINAL SPLINES N=5

```
                                                              .50   2.611493705922D-01
0  -4.0   1.305286316193D-04   6.526431580967D-05            .75   1.409598794416D-01
1  -3.0  -9.266379578762D-04  -4.633189789381D-04           1.25  -9.786794354481D-02
2  -2.0   3.010902476589D-03   1.505451238294D-03           1.50  -1.240899897798D-01
3  -1.0  -5.951872420379D-03  -2.975936210190D-03           1.75  -8.200913574109D-02
4   0.0   7.891149265380D-03   3.945574632690D-03           2.25   7.982065519072D-02
5   1.0  -7.141197234516D-03  -3.570598617258D-03           2.50   1.168004373699D-01
6   2.0   4.236478596289D-03   2.118239298145D-03           2.75   8.796375601738D-02
7   3.0  -1.477778656008D-03  -7.388893280040D-04           3.25  -1.086428919825D-01
8   4.0   2.284272989020D-04   1.142136494510D-04           3.50  -1.787890917214D-01
                                                             3.75  -1.527616462421D-01
      0  -1.751391287794D+01  -2.523478863958D+01           4.25   2.647875749897D-01
      1  -2.829378722499D+01   1.486749002281D+01           4.50   5.800280695583D-01
      2  -1.676353741042D+01  -2.674114398656D+01           4.75   8.548644985314D-01
      3  -4.282468472288D+00   2.782061681646D+00           5.25   9.599978881088D-01
      4  -3.965727331320D-01  -6.706326031970D-01           5.50   7.346180287292D-01


I= 6     H=  2.0
0          0.                  6.887418235363D+02          -1.00   3.211663278243D+00
1        -7.411948802944D-01  -4.351798161151D+02           -.75   1.782837768505D+00
2         1.438804615722D+00   1.030790496965D+02           -.50   8.488155913380D-01
3        -8.646081739387D-01  -1.106562013151D+01           -.25   2.893860722439D-01
4         1.670556082886D-01   4.944097152266D-01            .25  -1.082303735570D-01
                                                              .50  -1.085314441008D-01
0  -4.0  -5.716977721287D-05  -2.858488860643D-05            .75  -5.847199127245D-02
1  -3.0   4.178825268981D-04   2.089412634491D-04           1.25   4.037543170776D-02
2  -2.0  -1.423604534955D-03  -7.118022674777D-04           1.50   5.098751273622D-02
3  -1.0   3.010902476589D-03   1.505451238294D-03           1.75   3.351839594988D-02
4   0.0  -4.324448395145D-03  -2.162224197572D-03           2.25  -3.209235898668D-02
5   1.0   4.236478596289D-03   2.118239298145D-03           2.50  -4.637473662383D-02
6   2.0  -2.692481462452D-03  -1.346240731226D-03           2.75  -3.434231456235D-02
7   3.0   9.926160728224D-04   4.963080364112D-04           3.25   4.019889543505D-02
8   4.0  -1.601755028329D-04  -8.008775141646D-05           3.50   6.338179162045D-02
                                                             3.75   5.097576796166D-02
      0   7.487406920174D+00   1.565655290181D+01           4.25  -7.051325683851D-02
      1   1.203428541830D+01  -1.512829775730D+01           4.50  -1.222280465314D-01
      2   7.100844924163D+00   1.775494078017D+01           4.75  -1.099757819574D-01
      3   1.808281558679D+00  -3.155064687885D+00           5.25   2.130477282189D-01
      4   1.670556082886D-01   4.944097152266D-01           5.50   4.973855464085D-01


I= 7     H=  3.0
0          0.                 -3.063173340499D+02          -1.00  -8.932984438547D-01
1         2.059232628143D-01   1.974585409857D+02           -.75  -4.957737238408D-01
2        -4.000640406457D-01  -4.776685077863D+01           -.50  -2.359794550144D-01
3         2.407178747390D-01   5.194074894927D+00           -.25  -8.042803998069D-02
4        -4.659326565571D-02  -2.249551949670D-01            .25   3.005602507372D-02
                                                              .50   3.012330806419D-02
0  -4.0   1.616874815439D-05   8.084374077195D-06            .75   1.621809121701D-02
1  -3.0  -1.195095213310D-04  -5.975476066550D-05           1.25  -1.117650976586D-02
2  -2.0   4.178825268981D-04   2.089412634491D-04           1.50  -1.409342979046D-02
3  -1.0  -9.266379578762D-04  -4.633189789381D-04           1.75  -9.246933165964D-03
4   0.0   1.415972761687D-03   7.079863808433D-04           2.25   8.800988021532D-03
5   1.0  -1.477778656008D-03  -7.388893280040D-04           2.50   1.266060419325D-02
6   2.0   9.926160728224D-04   4.963080364112D-04           2.75   9.319867118688D-03
7   3.0  -3.826759322932D-04  -1.913379661466D-04           3.25  -1.071078566899D-02
```

8	4.0	6.396195794684D-05	3.198097897342D-05	3.50	-1.665555969188D-02
				3.75	-1.315171067705D-02
	0	-2.099263623641D+00	-5.920519201480D+00	4.25	1.717558821474D-02
	1	-3.368007082742D+00	7.050799801573D+00	4.50	2.845771156765D-02
	2	-1.984403046726D+00	-7.033650756342D+00	4.75	2.403713812674D-02
	3	-5.047743757524D-01	1.594791775455D+00	5.25	-3.691813326796D-02
	4	-4.659326565571D-02	-2.249551949670D-01	5.50	-6.793514167185D-02

I= 8 H= 4.0

0		0.	5.736322976901D+01	-1.00	1.195022717275D-01
1		-2.753495152630D-02	-3.763508185351D+01	-.75	6.631694736727D-02
2		5.351217964146D-02	9.294730733245D+00	-.50	3.156241042902D-02
3		-3.221509552831D-02	-1.031667405054D+00	-.25	1.075598515270D-02
4		6.240045031386D-03	4.484940606211D-02	.25	-4.018212354018D-03
				.50	-4.026319232524D-03
0	-4.0	-2.177618236553D-06	-1.088809118276D-06	.75	-2.167130279981D-03
1	-3.0	1.616874815439D-05	8.084374077195D-06	1.25	1.497243309585D-03
2	-2.0	-5.716977721287D-05	-2.858488860643D-05	1.50	1.880574086326D-03
3	-1.0	1.305286316193D-04	6.526431580967D-05	1.75	1.232903183953D-03
4	0.0	-2.085694087487D-04	-1.042847043744D-04	2.25	-1.170592862661D-03
5	1.0	2.284272989020D-04	1.142136494510D-04	2.50	-1.680849644231D-03
6	2.0	-1.601755028329D-04	-8.008775141646D-05	2.75	-1.234313669791D-03
7	3.0	6.396195794684D-05	3.198097897342D-05	3.25	1.407999871418D-03
8	4.0	-1.099432959153D-05	-5.497164795764D-06	3.50	2.177370935722D-03
				3.75	1.706918817118D-03
	0	2.817404823814D-01	9.933281153533D-01	4.25	-2.181348259582D-03
	1	4.516894282815D-01	-1.315823478233D+00	4.50	-3.556953752903D-03
	2	2.659753563149D-01	1.220264854561D+00	4.75	-2.942451907627D-03
	3	6.762562497387D-02	-3.140769080601D-01	5.25	4.240632206748D-03
	4	6.240045031386D-03	4.484940606211D-02	5.50	7.431420398575D-03

N= 5 M= 9 P= 4.5 Q= 9

I= 0 H= -4.5

0		1.000000000000D+00	-6.043408647504D+00	-1.00	5.850045533686D+00
1		-2.283122162998D+00	2.290923351705D+00	-.75	4.045319461769D+00
2		1.841116954945D+00	2.476628934130D-01	-.50	2.687321660409D+00
3		-6.418950270396D-01	-2.625378148980D-01	-.25	1.696207739093D+00
4		8.391138870396D-02	4.006191608071D-02	.25	5.345874379568D-01
				.50	2.437257188672D-01
0	-4.5	-1.115361201534D-05	-5.576806007671D-06	.75	7.903645074283D-02
1	-3.5	6.515261933988D-05	3.257630966994D-05	1.25	-2.607998814045D-02
2	-2.5	-1.644562709224D-04	-8.222813546120D-05	1.50	-2.419306187823D-02
3	-1.5	2.386675794385D-04	1.193337897192D-04	1.75	-1.191428261881D-02
4	-.5	-2.274155511519D-04	-1.137077755759D-04	2.25	6.602881277748D-03
5	.5	1.577053066081D-04	7.885265330407D-05	2.50	7.311095767993D-03
6	1.5	-8.569109514346D-05	-4.284554757173D-05	2.75	4.152599894738D-03
7	2.5	3.582164612945D-05	1.791082306472D-05	3.25	-2.843166067193D-03

8	3.5	-9.963504256663D-06	-4.981752128331D-06	3.50	-3.404710365766D-03
9	4.5	1.332881973800D-06	6.664409869000D-07	3.75	-2.063310684694D-03
				4.25	1.560951255575D-03
	0	3.924798095590D+00	1.785051107054D+00	4.50	1.944945526874D-03
	1	5.877508721451D+00	3.173285548790D+00	4.75	1.221462650321D-03
	2	3.370767817443D+00	1.570925196097D+00	5.25	-9.877433556905D-04
	3	8.685099696318D-01	4.585766745548D-01	5.50	-1.273979020948D-03
	4	8.391138870396D-02	4.006191608071D-02	5.75	-8.308280587327D-04

I= 1 H= -3.5

0		0.	4.899401593945D+01	-1.00	-1.480935181827D+01
1		5.128478795649D+00	-2.313794119811D+01	-.75	-8.801365572257D+00
2		-6.498007942803D+00	1.789826113299D+00	-.50	-4.561180734588D+00
3		2.776164537179D+00	9.506750629806D-01	-.25	-1.733211440226D+00
4		-4.067005426438D-01	-1.891679476743D-01	.25	9.177830996343D-01
				.50	1.261339322607D+00
0	-4.5	6.515261933988D-05	3.257630966994D-05	.75	1.233746341425D+00
1	-3.5	-3.915763151410D-04	-1.957881575705D-04	1.25	6.872216222670D-01
2	-2.5	1.024615493566D-03	5.123077467831D-04	1.50	3.853380587719D-01
3	-1.5	-1.554326377952D-03	-7.771631889759D-04	1.75	1.487891066876D-01
4	-.5	1.556850666667D-03	7.784253333336D-04	2.25	-6.475775807498D-02
5	.5	-1.129787915828D-03	-5.648939579142D-04	2.50	-6.703569596506D-02
6	1.5	6.310839042091D-04	3.155419521045D-04	2.75	-3.626007776667D-02
7	2.5	-2.664180522281D-04	-1.332090261141D-04	3.25	2.329496071476D-02
8	3.5	7.436948162376D-05	3.718474081188D-05	3.50	2.731966809073D-02
9	4.5	-9.963504256663D-06	-4.981752128331D-06	3.75	1.629323861217D-02
				4.25	-1.206057999229D-02
	0	-2.230115407882D+01	-9.823157086837D+00	4.50	-1.491814287934D-02
	1	-3.294394484965D+01	-1.822771302963D+01	4.75	-9.316390341449D-03
	2	-1.843390262212D+01	-8.359966178890D+00	5.25	7.476714894862D-03
	3	-4.544445230410D+00	-2.454347995157D+00	5.50	9.619181721739D-03
	4	-4.067005426438D-01	-1.891679476743D-01	5.75	6.261192198088D-03

I= 2 H= -2.5

0		0.	-1.768638490238D+02	-1.00	2.163080034910D+01
1		-5.722775469113D+00	8.993281908145D+01	-.75	1.231898229830D+01
2		9.981554224259D+00	-1.320334701288D+01	-.50	6.045464611436D+00
3		-5.092542477302D+00	-9.667199646296D-01	-.25	2.137369514449D+00
4		8.339281784266D-01	3.657443304941D-01	.25	-8.831601731501D-01
				.50	-9.504467982062D-01
0	-4.5	-1.644562709224D-04	-8.222813546120D-05	.75	-5.620259687427D-01
1	-3.5	1.024615493566D-03	5.123077467831D-04	1.25	5.310692891024D-01
2	-2.5	-2.807528527892D-03	-1.403764263946D-03	1.50	9.024440929813D-01
3	-1.5	4.511689539834D-03	2.255844769917D-03	1.75	1.056903168039D+00
4	-.5	-4.830944737162D-03	-2.415472368581D-03	2.25	7.853842856045D-01
5	.5	3.741284651124D-03	1.870642325562D-03	2.50	4.941517819627D-01
6	1.5	-2.190124435545D-03	-1.095062217773D-03	2.75	2.106818305694D-01
7	2.5	9.460606930965D-04	4.730303465482D-04	3.25	-1.073074419906D-01
8	3.5	-2.664180522281D-04	-1.332090261141D-04	3.50	-1.181364548926D-01
9	4.5	3.582164612945D-05	1.791082306472D-05	3.75	-6.734095777471D-02
				4.25	4.705284122187D-02
	0	5.427872385213D+01	2.235173757838D+01	4.50	5.713528149360D-02
	1	7.870607808961D+01	4.568826657942D+01	4.75	3.518564560840D-02
	2	4.255450445951D+01	1.818386961966D+01	5.25	-2.771565540472D-02
	3	9.918164734377D+00	5.616677984265D+00	5.50	-3.543995336904D-02

```
      4   8.339281784266D-01   3.657443304941D-01     5.75 -2.296106693473D-02

I= 3     H= -1.5
0           0.                 4.278298309880D+02    -1.00 -2.094574866232D+01
1           5.054201688848D+00 -2.220178258298D+02     -.75 -1.172801836471D+01
2          -9.499202109919D+00  3.929087965274D+01     -.50 -5.640082507658D+00
3           5.418553308522D+00 -1.115284500054D+00     -.25 -1.945719322789D+00
4          -9.737915550305D-01 -3.628444662438D-01      .25  7.507113134364D-01
                                                        .50  7.687579744678D-01
0   -4.5    2.386675794385D-04  1.193337897192D-04      .75  4.252064429109D-01
1   -3.5   -1.554326377952D-03 -7.771631889759D-04     1.25 -3.172818375231D-01
2   -2.5    4.511689539834D-03  2.255844769917D-03     1.50 -4.249323973928D-01
3   -1.5   -7.799545738730D-03 -3.899772869365D-03     1.75 -3.028710539879D-01
4    -.5    9.102769358907D-03  4.551384679453D-03     2.25  3.868836883416D-01
5     .5   -7.699063160388D-03 -3.849531580194D-03     2.50  7.423537094346D-01
6    1.5    4.844540425370D-03  2.422270212685D-03     2.75  9.656574929143D-01
7    2.5   -2.190124435545D-03 -1.095062217773D-03     3.25  8.484814034291D-01
8    3.5    6.310839042091D-04  3.155419521045D-04     3.50  5.702403292868D-01
9    4.5   -8.569109514346D-05 -4.284554757173D-05     3.75  2.571989292888D-01
                                                       4.25 -1.433773793302D-01
      0   -7.516466442172D+01 -2.602928128465D+01     4.50 -1.638464424735D-01
      1   -1.062085256164D+02 -6.841025027932D+01     4.75 -9.665039821708D-02
      2   -5.466440638109D+01 -1.985106374662D+01     5.25  7.211172961110D-02
      3   -1.210969468203D+01 -7.646484892443D+00     5.50  9.064269049042D-02
      4   -9.737915550305D-01 -3.628444662438D-01     5.75  5.797714712917D-02

I= 4     H=  -.5
0           0.                -8.140351323686D+02    -1.00  1.516327798130D+01
1          -3.540923019700D+00  4.259161972795D+02     -.75  8.436497473242D+00
2           6.815541593208D+00 -8.055540934798D+01     -.50  4.027310385125D+00
3          -4.040602263176D+00  5.277599638928D+00     -.25  1.377329526992D+00
4           7.662111052184D-01  1.462061673432D-01      .25 -5.194008044493D-01
                                                        .50 -5.237636445397D-01
0   -4.5   -2.274155511519D-04 -1.137077755759D-04      .75 -2.841622912873D-01
1   -3.5    1.556850666667D-03  7.784253333336D-04     1.25  2.002668786439D-01
2   -2.5   -4.830944737162D-03 -2.415472368581D-03     1.50  2.567555600828D-01
3   -1.5    9.102769358907D-03  4.551384679453D-03     1.75  1.722094888010D-01
4    -.5   -1.177806579643D-02 -5.889032898214D-03     2.25 -1.757262228411D-01
5     .5    1.110666717765D-02  5.553333588826D-03     2.50 -2.671420031290D-01
6    1.5   -7.699063160388D-03 -3.849531580194D-03     2.75 -2.124910082242D-01
7    2.5    3.741284651124D-03  1.870642325562D-03     3.25  3.235107842097D-01
8    3.5   -1.129787915828D-03 -5.648939579142D-04     3.50  6.651979984038D-01
9    4.5    1.577053066081D-04  7.885265330407D-05     3.75  9.179282037848D-01
                                                       4.25  8.866913712628D-01
      0    6.807512377557D+01  1.221564968605D+01     4.50  6.196843583324D-01
      1    9.161631168337D+01  7.482383920915D+01     4.75  2.896924758821D-01
      2    4.536206032438D+01  8.456235109748D+00     5.25 -1.729011489719D-01
      3    9.751197630756D+00  7.909310651106D+00     5.50 -2.047647703452D-01
      4    7.662111052184D-01  1.462061673432D-01     5.75 -1.255829846295D-01

I= 5     H=   .5
0           0.                 1.300851286678D+03    -1.00 -9.159282155483D+00
1           2.118608026968D+00 -6.847127252482D+02     -.75 -5.086664617220D+00
2          -4.105921159863D+00  1.329959782319D+02     -.50 -2.423016001373D+00
```

3		2.460954198120D+00	-1.054102166328D+01	-.25	-8.265752650265D-01
4		-4.737987705320D-01	1.462061673432D-01	.25	3.096335677505D-01
				.50	3.108308831434D-01
0	-4.5	1.577053066081D-04	7.885265330407D-05	.75	1.676893691501D-01
1	-3.5	-1.129787915828D-03	-5.648939579142D-04	1.25	-1.162508982039D-01
2	-2.5	3.741284651124D-03	1.870642325562D-03	1.50	-1.472359112423D-01
3	-1.5	-7.699063160388D-03	-3.849531580194D-03	1.75	-9.716552995232D-02
4	-.5	1.110666717765D-02	5.553333588826D-03	2.25	9.415816092491D-02
5	.5	-1.177806579643D-02	-5.889032898214D-03	2.50	1.373301562188D-01
6	1.5	9.102769358907D-03	4.551384679453D-03	2.75	1.029837781562D-01
7	2.5	-4.830944737162D-03	-2.415472368581D-03	3.25	-1.255829846295D-01
8	3.5	1.556850666667D-03	7.784253333336D-04	3.50	-2.047647703452D-01
9	4.5	-2.274155511519D-04	-1.137077755759D-04	3.75	-1.729011489719D-01
				4.25	2.896924758821D-01
	0	-4.364382440346D+01	1.221564968605D+01	4.50	6.196843583324D-01
	1	-5.803136673492D+01	-7.482383920915D+01	4.75	8.866913712628D-01
	2	-2.844959010488D+01	8.456235109748D+00	5.25	9.179282037848D-01
	3	-6.067423671456D+00	-7.909310651106D+00	5.50	6.651979984038D-01
	4	-4.737987705320D-01	1.462061673432D-01	5.75	3.235107842097D-01

I= 6 H= 1.5

0		0.	-1.581434293173D+03	-1.00	4.761186830508D+00
1		-1.097978466790D+00	8.438505891606D+02	-.75	2.642620347945D+00
2		2.132533638259D+00	-1.671642124432D+02	-.50	1.257950569733D+00
3		-1.282572102917D+00	1.417768528483D+01	-.25	4.287873090660D-01
4		2.481026225430D-01	-3.628444662438D-01	.25	-1.602823028719D-01
				.50	-1.606710901513D-01
0	-4.5	-8.569109514346D-05	-4.284554757173D-05	.75	-8.652399816486D-02
1	-3.5	6.310839042091D-04	3.155419521045D-04	1.25	5.966793509717D-02
2	-2.5	-2.190124435545D-03	-1.095062217773D-03	1.50	7.527864419723D-02
3	-1.5	4.844540425370D-03	2.422270212685D-03	1.75	4.942469567779D-02
4	-.5	-7.699063160388D-03	-3.849531580194D-03	2.25	-4.713925341159D-02
5	.5	9.102769358907D-03	4.551384679453D-03	2.50	-6.792022143965D-02
6	1.5	-7.799545738730D-03	-3.899772869365D-03	2.75	-5.010549482340D-02
7	2.5	4.511689539834D-03	2.255844769917D-03	3.25	5.797714712917D-02
8	3.5	-1.554326377952D-03	-7.771631889759D-04	3.50	9.064269049042D-02
9	4.5	2.386675794385D-04	1.193337897192D-04	3.75	7.211172961110D-02
				4.25	-9.665039821708D-02
	0	2.310610185243D+01	-2.602928128465D+01	4.50	-1.638464424735D-01
	1	3.061197494226D+01	6.841025027932D+01	4.75	-1.433773793302D-01
	2	1.496227888785D+01	-1.985106374662D+01	5.25	2.571989292888D-01
	3	3.183275102857D+00	7.646484892443D+00	5.50	5.702403292868D-01
	4	2.481026225430D-01	-3.628444662438D-01	5.75	8.484814034291D-01

I= 7 H= 2.5

0		0.	1.257970112823D+03	-1.00	-1.961013068989D+00
1		4.517685439346D-01	-6.838700891656D+02	-.75	-1.088215467087D+00
2		-8.780849278794D-01	1.384469585623D+02	-.50	-5.178979837442D-01
3		5.287200797369D-01	-1.220007593316D+01	-.25	-1.764838495870D-01
4		-1.024395174383D-01	3.657443304941D-01	.25	6.592292500874D-02
				.50	6.605065008885D-02
0	-4.5	3.582164612945D-05	1.791082306472D-05	.75	3.554760575044D-02
1	-3.5	-2.664180522281D-04	-1.332090261141D-04	1.25	-2.447020123265D-02
2	-2.5	9.460606930965D-04	4.730303465482D-04	1.50	-3.083147567823D-02
3	-1.5	-2.190124435545D-03	-1.095062217773D-03	1.75	-2.020732729902D-02

```
4   -.5    3.741284651124D-03    1.870642325562D-03     2.25  1.916920037742D-02
5    .5   -4.830944737162D-03   -2.415472368581D-03     2.50  2.750673493444D-02
6   1.5    4.511689539834D-03    2.255844769917D-03     2.75  2.018164335871D-02
7   2.5   -2.807528527892D-03   -1.403764263946D-03     3.25 -2.296106693473D-02
8   3.5    1.024615493566D-03    5.123077467831D-04     3.50 -3.543995336904D-02
9   4.5   -1.644562709224D-04   -8.222813546120D-05     3.75 -2.771565540472D-02
                                                        4.25  3.518564560840D-02
      0   -9.575248695380D+00    2.235173757838D+01     4.50  5.713528149360D-02
      1   -1.267045506923D+01   -4.568826657942D+01     4.75  4.705284122187D-02
      2   -6.186765220188D+00    1.818386961966D+01     5.25 -6.734095777471D-02
      3   -1.315191234153D+00   -5.616677984265D+00     5.50 -1.181364548926D-01
      4   -1.024395174383D-01    3.657443304941D-01     5.75 -1.073074419906D-01


I= 8      H=  3.5
0              0.               -5.623603234445D+02    -1.00  5.424986351063D-01
1         -1.249314443903D-01    3.115207662727D+02     -.75  3.010245152955D-01
2          2.428896520100D-01   -6.447756975593D+01     -.50  1.432500370800D-01
3         -1.463128914107D-01    5.859371053294D+00     -.25  4.881040268000D-02
4          2.836464729523D-02   -1.891679476743D-01      .25 -1.822759740976D-02
                                                         .50 -1.825964962301D-02
0  -4.5   -9.963504256663D-06   -4.981752128331D-06      .75 -9.824901524322D-03
1  -3.5    7.436948162376D-05    3.718474081188D-05     1.25  6.758803319120D-03
2  -2.5   -2.664180522281D-04   -1.332090261141D-04     1.50  8.511683536288D-03
3  -1.5    6.310839042091D-04    3.155419521045D-04     1.75  5.575081093986D-03
4   -.5   -1.129787915828D-03   -5.648939579142D-04     2.25 -5.278166195494D-03
5    .5    1.556850666667D-03    7.784253333336D-04     2.50 -7.562439318040D-03
6   1.5   -1.554326377952D-03   -7.771631889759D-04     2.75 -5.537422142044D-03
7   2.5    1.024615493566D-03    5.123077467831D-04     3.25  6.261192198088D-03
8   3.5   -3.915763151410D-04   -1.957881575705D-04     3.50  9.619181721739D-03
9   4.5    6.515261933988D-05    3.257630966994D-05     3.75  7.476714894862D-03
                                                        4.25 -9.316390341449D-03
      0    2.654839905149D+00   -9.823157086837D+00     4.50 -1.491814287934D-02
      1    3.511481209612D+00    1.822771302963D+01     4.75 -1.206057999229D-02
      2    1.713970264336D+00   -8.359966178890D+00     5.25  1.629323861217D-02
      3    3.642507599034D-01    2.454347995157D+00     5.50  2.731966809073D-02
      4    2.836464729523D-02   -1.891679476743D-01     5.75  2.329496071476D-02


I= 9      H=  4.5
0              0.                1.060917602292D+02    -1.00 -7.241362463749D-02
1          1.667350758983D-02   -5.977271370429D+01     -.75 -4.018007527955D-02
2         -3.241992221719D-02    1.262923310639D+01     -.50 -1.912003641911D-02
3          1.953263828793D-02   -1.179691164008D+00     -.25 -6.514614652024D-03
4         -3.787556542543D-03    4.006191608071D-02      .25  2.432534094471D-03
                                                         .50  2.436633345983D-03
0  -4.5    1.332881973800D-06    6.664409869000D-07      .75  1.310949739658D-03
1  -3.5   -9.963504256663D-06   -4.981752128331D-06     1.25 -9.015933295856D-04
2  -2.5    3.582164612945D-05    1.791082306472D-05     1.50 -1.135193378033D-03
3  -1.5   -8.569109514346D-05   -4.284554757173D-05     1.75 -7.433464410302D-04
4   -.5    1.577053066081D-04    7.885265330407D-05     2.25  7.031839969471D-04
5    .5   -2.274155511519D-04   -1.137077755759D-04     2.50  1.006881533233D-03
6   1.5    2.386675794385D-04    1.193337897192D-04     2.75  7.366580629561D-04
7   2.5   -1.644562709224D-04   -8.222813546120D-05     3.25 -8.308280587327D-04
8   3.5    6.515261933988D-05    3.257630966994D-05     3.50 -1.273979020948D-03
9   4.5   -1.115361201534D-05   -5.576806007671D-06     3.75 -9.877433556905D-04
                                                        4.25  1.221462650321D-03
```

CARDINAL SPLINES N=5

0	-3.546958814826D-01	1.785051107054D+00	4.50	1.944945526874D-03
1	-4.690623761300D-01	-3.173285548790D+00	4.75	1.560951255575D-03
2	-2.289174252491D-01	1.570925196097D+00	5.25	-2.063310684694D-03
3	-4.864337947785D-02	-4.585766745548D-01	5.50	-3.404710365766D-03
4	-3.787556542543D-03	4.006191608071D-02	5.75	-2.843166067193D-03

N= 5 M=10 P= 5.0 Q= 9

I= 0 H= -5.0

0		1.000000000000D+00	7.007037429565D+00	-1.00	5.853229186106D+00
1		-2.283855041643D+00	-4.156167680680D+00	-.75	4.047085893489D+00
2		1.842542198661D+00	1.443443783624D+00	-.50	2.688162188675D+00
3		-6.427539452060D-01	-3.614630580275D-01	-.25	1.696494107911D+00
4		8.407800059528D-02	4.318941841406D-02	.25	5.344805262587D-01
				.50	2.436186388308D-01
0	-5.0	-1.121240688370D-05	-5.606203441849D-06	.75	7.897884787629D-02
1	-4.0	6.559309081185D-05	3.279654540593D-05	1.25	-2.604039842705D-02
2	-3.0	-1.660484782703D-04	-8.302423913517D-05	1.50	-2.414321701029D-02
3	-2.0	2.425333347342D-04	1.212666673671D-04	1.75	-1.188165622970D-02
4	-1.0	-2.348522564856D-04	-1.174261282428D-04	2.25	6.572055742596D-03
5	0.0	1.700168424170D-04	8.500842120849D-05	2.50	7.266998500917D-03
6	1.0	-1.026975982899D-04	-5.134879914494D-05	2.75	4.120377644947D-03
7	2.0	5.338436399296D-05	2.669218199648D-05	3.25	-2.806964939966D-03
8	3.0	-2.198767643376D-05	-1.099383821688D-05	3.50	-3.349361582334D-03
9	4.0	6.082715430662D-06	3.041357715331D-06	3.75	-2.020561891372D-03
10	5.0	-8.119310233590D-07	-4.059655116795D-07	4.25	1.508699200225D-03
				4.50	1.862458322484D-03
	0	7.848786979608D+00	4.122797872115D+00	4.75	1.155998320303D-03
	1	9.974021352157D+00	4.763250010527D+00	5.25	-9.040333272425D-04
	2	4.812933109863D+00	2.499910675320D+00	5.50	-1.139111663081D-03
	3	1.038806066700D+00	5.023253102536D-01	5.75	-7.215776811568D-04
	4	8.407800059528D-02	4.318941841406D-02	6.25	5.903418525975D-04

I= 1 H= -4.0

0		0.	-4.877555739401D+01	-1.00	-1.483320267646D+01
1		5.133969277130D+00	2.516150663348D+01	-.75	-8.814599087382D+00
2		-6.508685391617D+00	-7.168564024927D+00	-.50	-4.567477690986D+00
3		2.782599264556D+00	1.691789868986D+00	-.25	-1.735356819545D+00
4		-4.079487431597D-01	-2.125981482150D-01	.25	9.185840462874D-01
				.50	1.262141530394D+00
0	-5.0	6.559309081185D-05	3.279654540593D-05	.75	1.234177882819D+00
1	-4.0	-3.948761799599D-04	-1.974380899799D-04	1.25	6.869249544703D-01
2	-3.0	1.036543778021D-03	5.182718890103D-04	1.50	3.849646377071D-01
3	-2.0	-1.583287322554D-03	-7.916436612770D-04	1.75	1.485446807022D-01
4	-1.0	1.612563973568D-03	8.062819867841D-04	2.25	-6.452682348457D-02
5	0.0	-1.222021820785D-03	-6.110109103924D-04	2.50	-6.670533400182D-02
6	1.0	7.584909328041D-04	3.792454664021D-04	2.75	-3.601867945773D-02
7	2.0	-3.979920552892D-04	-1.989960276446D-04	3.25	2.302375398801D-02

CARDINAL SPLINES N=5

8	3.0	1.644505544469D-04	8.222527722344D-05	3.50	2.690501353342D-02
9	4.0	-4.554766649449D-05	-2.277383324725D-05	3.75	1.597297896289D-02
10	5.0	6.082715430662D-06	3.041357715331D-06	4.25	-1.166912508514D-02
				4.50	-1.430017635689D-02
	0	-4.419034481010D+01	-2.358223386094D+01	4.75	-8.825953495329D-03
	1	-5.523231137720D+01	-2.593896754934D+01	5.25	6.849587383913D-03
	2	-2.596200789723D+01	-1.368143822238D+01	5.50	8.608800623850D-03
	3	-5.376375598638D+00	-2.560173095313D+00	5.75	5.442724941180D-03
	4	-4.079487431597D-01	-2.125981482150D-01	6.25	-4.441270073827D-03

I= 2 H= -3.0

0		0.	1.765516227385D+02	-1.00	2.171701594779D+01
1		-5.742622350270D+00	-8.465904584401D+01	-.75	1.236681854051D+01
2		1.002015084979D+01	1.917926005991D+01	-.50	6.068226721777D+00
3		-5.115802599386D+00	-3.645686741961D+00	-.25	2.145124588124D+00
4		8.384401483488D-01	4.504393444462D-01	.25	-8.860554188753D-01
				.50	-9.533466026536D-01
0	-5.0	-1.660484782703D-04	-8.302423913517D-05	.75	-5.635858958236D-01
1	-4.0	1.036543778021D-03	5.182718890103D-04	1.25	5.321416778402D-01
2	-3.0	-2.850646648736D-03	-1.425323324368D-03	1.50	9.037939278766D-01
3	-2.0	4.616376975304D-03	2.308188487652D-03	1.75	1.057786714136D+00
4	-1.0	-5.032336069563D-03	-2.516168034781D-03	2.25	7.845495079285D-01
5	0.0	4.074689901052D-03	2.037344950526D-03	2.50	4.929575962348D-01
6	1.0	-2.650672779795D-03	-1.325336389898D-03	2.75	2.098092288558D-01
7	2.0	1.421671732176D-03	7.108358660879D-04	3.25	-1.063270894125D-01
8	3.0	-5.920412882020D-04	-2.960206441010D-04	3.50	-1.166375700045D-01
9	4.0	1.644505544469D-04	8.222527722344D-05	3.75	-6.618328968659D-02
10	5.0	-2.198767643376D-05	-1.099383821688D-05	4.25	4.563781795883D-02
				4.50	5.490146863527D-02
	0	1.063404272881D+02	5.855164254971D+01	4.75	3.341282443887D-02
	1	1.299937653681D+02	5.892672133102D+01	5.25	-2.544872759087D-02
	2	5.904913411132D+01	3.205986059741D+01	5.50	-3.178764825821D-02
	3	1.165300036759D+01	5.363100146962D+00	5.75	-2.000248808591D-02
	4	8.384401483488D-01	4.504393444462D-01	6.25	1.621929508303D-02

I= 3 H= -2.0

0		0.	-4.302353785005D+02	-1.00	-2.115507341367D+01
1		5.102388368830D+00	2.018776092095D+02	-.75	-1.184416103099D+01
2		-9.592911707302D+00	-3.933156543475D+01	-.50	-5.695347136795D+00
3		5.475027071339D+00	5.389037955163D+00	-.25	-1.964548037006D+00
4		-9.847462662015D-01	-5.684773564141D-01	.25	7.577407443136D-01
				.50	7.757984735671D-01
0	-5.0	2.425333347342D-04	1.212666673671D-04	.75	4.289938242352D-01
1	-4.0	-1.583287322554D-03	-7.916436612770D-04	1.25	-3.198855137637D-01
2	-3.0	4.616376975304D-03	2.308188487652D-03	1.50	-4.282096912716D-01
3	-2.0	-8.053718669528D-03	-4.026859334764D-03	1.75	-3.050162350214D-01
4	-1.0	9.591731811409D-03	4.795865905705D-03	2.25	3.889104634512D-01
5	0.0	-8.508545111469D-03	-4.254272555734D-03	2.50	7.452530992707D-01
6	1.0	5.962715896065D-03	2.981357948032D-03	2.75	9.677761018380D-01
7	2.0	-3.344870954841D-03	-1.672435477420D-03	3.25	8.461011838078D-01
8	3.0	1.421671732176D-03	7.108358660879D-04	3.50	5.666011536471D-01
9	4.0	-3.979920552892D-04	-1.989960276446D-04	3.75	2.543882014347D-01
10	5.0	5.338436399296D-05	2.669218199648D-05	4.25	-1.399418131775D-01
				4.50	-1.584229190248D-01
	0	-1.453988832970D+02	-8.580507168522D+01	4.75	-9.234612664507D-02

CARDINAL SPLINES N=5

	1	-1,725728314545D+02	-7,149887670785D+01	5.25	6,660780563239D-02
	2	-7,517944556744D+01	-4,376759956943D+01	5,50	8,177517845304D-02
	3	-1,421989825269D+01	-5,980509173120D+00	5.75	5,079394836587D-02
	4	-9,847462662015D-01	-5,684773564141D-01	6.25	-4,048522215597D-02
I= 4	H= -1.0				
0		0.	8.366586213715D+02	-1.00	1.556596422909D+01
1		-3,633621629059D+00	-3.895481044084D+02	-.75	8,659925683195D+00
2		6,995814427202D+00	7.069367705542D+01	-.50	4,133625123341D+00
3		-4,149243059355D+00	-7.235020999237D+00	-.25	1,413551064242D+00
4		7,872851134685D-01	5.417902367288D-01	.25	-5,329235967885D-01
				.50	-5,373077292525D-01
0	-5,0	-2,348522564856D-04	-1.174261282428D-04	.75	-2,914482255332D-01
1	-4,0	1,612563973568D-03	8.062819867841D-04	1.25	2,052756729168D-01
2	-3,0	-5,032336069563D-03	-2.516168034781D-03	1,50	2,630602190704D-01
3	-2,0	9,591731811409D-03	4.795865905705D-03	1,75	1,763362578315D-01
4	-1,0	-1,271870207586D-02	-6.359351037928D-03	2,25	-1,796252097439D-01
5	0,0	1,266389940853D-02	6.331949704267D-03	2,50	-2,727196732631D-01
6	1,0	-9,850141242392D-03	-4.925070621196D-03	2.75	-2,165666594430D-01
7	2,0	5,962715896065D-03	2.981357948032D-03	3,25	3,280897062052D-01
8	3,0	-2,650672779795D-03	-1.325336389898D-03	3,50	6,721988235519D-01
9	4,0	7,584909328041D-04	3.792454664021D-04	3.75	9,233353113810D-01
10	5,0	-1,026975982899D-04	-5.134879914494D-05	4,25	8,800822376840D-01
				4,50	6,092509132036D-01
	0	1,301250660332D+02	9,050129876564D+01	4,75	2,814121795599D-01
	1	1.487738499256D+02	4.565720956737D+01	5,25	-1,623130341850D-01
	2	6,284993555716D+01	4.343689757618D+01	5,50	-1,877059890667D-01
	3	1,159645921002D+01	3,600783735339D+00	5,75	-1,117643829669D-01
	4	7,872851134685D-01	5,417902367288D-01	6,25	8,447688168046D-02
I= 5	H= 0.0				
0		0.	-1.431887788838D+03	-1,00	-9,825933036057D+00
1		2,272071448920D+00	6,652961161019D+02	-.75	-5,456552125745D+00
2		-4,404364536530D+00	-1,173983106022D+02	-,50	-2,599021053048D+00
3		2,640810060688D+00	1.017373979840D+01	-.25	-8,865403615157D-01
4		-5,086869899199D-01	-5,086869899199D-01	.25	3,320206779892D-01
				.50	3,332532431075D-01
0	-5,0	1,700168424170D-04	8,500842120849D-05	.75	1,797513020012D-01
1	-4,0	-1,222021820785D-03	-6,110109103924D-04	1.25	-1,245430043562D-01
2	-3,0	4,074689901052D-03	2.037344950526D-03	1,50	-1,576733336497D-01
3	-2,0	-8,508545111469D-03	-4.254272555734D-03	1,75	-1,039974349981D-01
4	-1,0	1,266389940853D-02	6,331949704267D-03	2,25	1,006129705103D-01
5	0,0	-1,435607843950D-02	-7,178039219750D-03	2,50	1,465640417237D-01
6	1,0	1,266389940853D-02	6,331949704267D-03	2.75	1,097310571828D-01
7	2,0	-8,508545111469D-03	-4,254272555734D-03	3,25	-1,331634331727D-01
8	3,0	4,074689901052D-03	2,037344950526D-03	3,50	-2,163547023962D-01
9	4,0	-1,222021820785D-03	-6,110109103924D-04	3,75	-1,818526665869D-01
10	5,0	1,700168424170D-04	8,500842120849D-05	4,25	3,006339588466D-01
				4,50	6,369570251314D-01
	0	-8,657686728262D+01	-8,657686728262D+01	4,75	9,003994799082D-01
	1	-9,805431432475D+01	8,905745349891D-23	5,25	9,003994799082D-01
	2	-4,109526211420D+01	-4,109526211420D+01	5,50	6,369570251314D-01
	3	-7,532929737710D+00	8,739349530711D-24	5,75	3,006339588466D-01
	4	-5,086869899199D-01	-5,086869899199D-01	6,25	-1,818526665869D-01

CARDINAL SPLINES N=5

```
 I= 6     H=  1.0
 0            0.                    2.193426650880D+03    -1.00  5.682063029391D+00
 1           -1.309964705452D+00   -1.020980083844D+03     -.75  3.153563350932D+00
 2            2.544787503505D+00    1.787171891156D+02     -.50  1.501074621157D+00
 3           -1.531015460444D+00   -1.443658846991D+01     -.25  5.116199156516D-01
 4            2.962953599891D-01    5.417902367288D-01      .25 -1.912066706053D-01
                                                            .50 -1.916441499873D-01
 0   -5.0    -1.026975982899D-04   -5.134879914494D-05      .75 -1.031857130126D-01
 1   -4.0     7.584909328041D-04    3.792454664021D-04     1.25  7.112221111599D-02
 2   -3.0    -2.650672779795D-03   -1.325336389898D-03     1.50  8.969634640504D-02
 3   -2.0     5.962715896065D-03    2.981357948032D-03     1.75  5.886192726779D-02
 4   -1.0    -9.850141242392D-03   -4.925070621196D-03     2.25 -5.605558531585D-02
 5    0.0     1.266389940853D-02    6.331949704267D-03     2.50 -8.067542154889D-02
 6    1.0    -1.271870207586D-02   -6.359351037928D-03     2.75 -5.942582851510D-02
 7    2.0     9.591731811409D-03    4.795865905705D-03     3.25  6.844837703922D-02
 8    3.0    -5.032336069563D-03   -2.516168034781D-03     3.50  1.066524084454D-01
 9    4.0     1.612563973568D-03    8.062819867841D-04     3.75  8.447688168046D-02
10    5.0    -2.348522564856D-04   -1.174261282428D-04     4.25 -1.117643829669D-01
                                                           4.50 -1.877059890667D-01
          0   5.087753149807D+01    9.050129876564D+01     4.75 -1.623130341850D-01
          1   5.745943079085D+01   -4.565720956737D+01     5.25  2.814121795599D-01
          2   2.402385959521D+01    4.343689757618D+01     5.50  6.092509132036D-01
          3   4.394891739338D+00   -3.600783735339D+00     5.75  8.800822376840D-01
          4   2.962953599891D-01    5.417902367288D-01     6.25  9.233353113810D-01


 I= 7     H=  2.0
 0            0.                   -2.640351438859D+03    -1.00 -2.912007446197D+00
 1            6.706880048330D-01    1.241951738593D+03     -.75 -1.615869372687D+00
 2           -1.303821968654D+00   -2.187468406283D+02     -.50 -7.689736507545D-01
 3            7.852890260834D-01    1.735005630140D+01     -.25 -2.620255795263D-01
 4           -1.522084466267D-01   -5.684773564141D-01      .25  9.785870515895D-02
                                                            .50  9.803671486566D-02
 0   -5.0     5.338436399296D-05    2.669218199648D-05      .75  5.275425866873D-02
 1   -4.0    -3.979920552892D-04   -1.989960276446D-04     1.25 -3.629910081991D-02
 2   -3.0     1.421671732176D-03    7.108358660879D-04     1.50 -4.572072327581D-02
 3   -2.0    -3.344870954841D-03   -1.672435477420D-03     1.75 -2.995321299938D-02
 4   -1.0     5.962715896065D-03    2.981357948032D-03     2.25  2.837714984615D-02
 5    0.0    -8.508545111469D-03   -4.254272555734D-03     2.50  4.067910663734D-02
 6    1.0     9.591731811409D-03    4.795865905705D-03     2.75  2.980680789795D-02
 7    2.0    -8.053718669528D-03   -4.026859334764D-03     3.25 -3.377476890573D-02
 8    3.0     4.616376975304D-03    2.308188487652D-03     3.50 -5.197328518323D-02
 9    4.0    -1.583287322554D-03   -7.916436612770D-04     3.75 -4.048522215597D-02
10    5.0     2.425333347342D-04    1.212666673671D-04     4.25  5.079394836587D-02
                                                           4.50  8.177517845304D-02
          0  -2.621126007346D+01   -8.580507168522D+01     4.75  6.660780563239D-02
          1  -2.957507803882D+01    7.149887670785D+01     5.25 -9.234612664507D-02
          2  -1.235575357141D+01   -4.376759956943D+01     5.50 -1.584229190248D-01
          3  -2.258879906452D+00    5.980509173120D+00     5.75 -1.399418131775D-01
          4  -1.522084466267D-01   -5.684773564141D-01     6.25  2.543882014347D-01


 I= 8     H=  3.0
 0            0.                    2.106593872789D+03    -1.00  1.193589213447D+00
 1           -2.748128756821D-01   -1.006977510550D+03     -.75  6.622784772482D-01
 2            5.343670600182D-01    1.800722644688D+02     -.50  3.151469537800D-01
```

3		-3.219707372031D-01	-1.437188703589D+01	-.25	1.073758534895D-01
4		6.243854054350D-02	4.504393444462D-01	.25	-4.009216997308D-02
				.50	-4.015864914763D-02
0	-5.0	-2.198767643376D-05	-1.099383821688D-05	.75	-2.160529672789D-02
1	-4.0	1.644505544469D-04	8.222527722344D-05	1.25	1.485736337162D-02
2	-3.0	-5.920412882020D-04	-2.960206441010D-04	1.50	1.870548600422D-02
3	-2.0	1.421671732176D-03	7.108358660879D-04	1.75	1.224752260276D-02
4	-1.0	-2.650672779795D-03	-1.325336389898D-03	2.25	-1.158231403129D-02
5	0.0	4.074689901052D-03	2.037344950526D-03	2.50	-1.658079664561D-02
6	1.0	-5.032336069563D-03	-2.516168034781D-03	2.75	-1.212721288810D-02
7	2.0	4.616376975304D-03	2.308188487652D-03	3.25	1.366470544595D-02
8	3.0	-2.850646648736D-03	-1.425323324368D-03	3.50	2.093859309871D-02
9	4.0	1.036543778021D-03	5.182718890103D-04	3.75	1.621929508303D-02
10	5.0	-1.660484782703D-04	-8.302423913517D-05	4.25	-2.000248808591D-02
				4.50	-3.178764825821D-02
	0	1.076285781134D+01	5.855164254971D+01	4.75	-2.544872759087D-02
	1	1.214032270602D+01	-5.892672133102D+01	5.25	3.341282443887D-02
	2	5.070587083497D+00	3.205986059741D+01	5.50	5.490146863527D-02
	3	9.268000736669D-01	-5.363100146962D+00	5.75	4.563781795883D-02
	4	6.243854054350D-02	4.504393444462D-01	6.25	-6.618328968659D-02

I= 9 H= 4.0

0		0.	-9.482085067158D+02	-1.00	-3.296098595967D-01
1		7.588023104879D-02	4.610654060292D+02	-.75	-1.828839662115D-01
2		-1.475604178859D-01	-8.397375688433D+01	-.50	-8.702339924924D-02
3		8.892165739186D-02	6.812136059612D+00	-.25	-2.964935803178D-02
4		-1.724755327018D-02	-2.125981482150D-01	.25	1.106955930932D-02
				.50	1.108725802782D-02
0	-5.0	6.082715430662D-06	3.041357715331D-06	.75	5.964485504851D-03
1	-4.0	-4.554766649449D-05	-2.277383324725D-05	1.25	-4.100716805466D-03
2	-3.0	1.644505544469D-04	8.222527722344D-05	1.50	-5.161987347855D-03
3	-2.0	-3.979920552892D-04	-1.989960276446D-04	1.75	-3.379119236908D-03
4	-1.0	7.584909328041D-04	3.792454664021D-04	2.25	3.193472048919D-03
5	0.0	-1.222021820785D-03	-6.110109103924D-04	2.50	4.569346768221D-03
6	1.0	1.612563973568D-03	8.062819867841D-04	2.75	3.339781845479D-03
7	2.0	-1.583287322554D-03	-7.916436612770D-04	3.25	-3.755391546772D-03
8	3.0	1.036543778021D-03	5.182718890103D-04	3.50	-5.745415231147D-03
9	4.0	-3.948761799599D-04	-1.974380899799D-04	3.75	-4.441270073827D-03
10	5.0	6.559309081185D-05	3.279654540593D-05	4.25	5.442724941180D-03
				4.50	8.608800623850D-03
	0	-2.974122911783D+00	-2.358223386094D+01	4.75	6.849587383913D-03
	1	-3.354376278510D+00	2.593896754934D+01	5.25	-8.825953495329D-03
	2	-1.400868547535D+00	-1.368143822238D+01	5.50	-1.430017635689D-02
	3	-2.560294080117D-01	2.560173095313D+00	5.75	-1.166912508514D-02
	4	-1.724755327018D-02	-2.125981482150D-01	6.25	1.597297896289D-02

I=10 H= 5.0

0		0.	1.802208650982D+02	-1.00	4.396482616729D-02
1		-1.012072865383D-02	-8.903146423978D+01	-.75	2.439363764097D-02
2		1.968198281632D-02	1.651320309123D+01	-.50	1.160732210358D-02
3		-1.186127846431D-02	-1.366113678535D+00	-.25	3.954626207016D-03
4		2.300836232834D-03	4.318941841406D-02	.25	-1.476403075004D-03
				.50	-1.478727752121D-03
0	-5.0	-8.119310233590D-07	-4.059655116795D-07	.75	-7.954700076088D-04
1	-4.0	6.082715430662D-06	3.041357715331D-06	1.25	5.468544573643D-04

CARDINAL SPLINES N=5

2	-3.0	-2.198767643376D-05	-1.099383821688D-05	1.50	6.883354918581D-04
3	-2.0	5.338436399296D-05	2.669218199648D-05	1.75	4.505559453683D-04
4	-1.0	-1.026975982899D-04	-5.134879914494D-05	2.25	-4.256869519446D-04
5	0.0	1.700168424170D-04	8.500842120849D-05	2.50	-6.089636763465D-04
6	1.0	-2.348522564856D-04	-1.174261282428D-04	2.75	-4.449749608843D-04
7	2.0	2.425333347342D-04	1.212666673671D-04	3.25	4.999214914138D-04
8	3.0	-1.660484782703D-04	-8.302423913517D-05	3.50	7.643421208349D-04
9	4.0	6.559309081185D-05	3.279654540593D-05	3.75	5.903418525975D-04
10	5.0	-1.121240688370D-05	-5.606203441849D-06	4.25	-7.215776811568D-04
				4.50	-1.139111663081D-03
	0	3.968087646220D-01	4.122797872115D+00	4.75	-9.040333272425D-04
	1	4.475213311035D-01	-4.763250010527D+00	5.25	1.155998320303D-03
	2	1.868882407769D-01	2.499910675320D+00	5.50	1.862458322484D-03
	3	3.415544619238D-02	-5.023253102536D-01	5.75	1.508699200225D-03
	4	2.300836232834D-03	4.318941841406D-02	6.25	-2.020561891372D-03

N= 5 M=11 P= 5.5 Q= 9

I= 0 H= -5.5

0		1.000000000000D+00	-5.629435463209D+00	-1.00	5.854403699620D+00
1		-2.284125394606D+00	1.400570658424D+00	-.75	4.047737556142D+00
2		1.843067988443D+00	5.268863591848D-01	-.50	2.688472266584D+00
3		-6.430708381430D-01	-2.943758473586D-01	-.25	1.696599749613D+00
4		8.413947842790D-02	4.137056855892D-02	.25	5.344410885749D-01
				.50	2.435791404997D-01
0	-5.5	-1.123412138365D-05	-5.617060691825D-06	.75	7.895760103859D-02
1	-4.5	6.575588716303D-05	3.287794358152D-05	1.25	-2.602579403438D-02
2	-3.5	-1.666379993757D-04	-8.331899968784D-05	1.50	-2.412483604061D-02
3	-2.5	2.439715932482D-04	1.219857966241D-04	1.75	-1.186962640047D-02
4	-1.5	-2.376586626092D-04	-1.188293313046D-04	2.25	6.560694586977D-03
5	-.5	1.748699316031D-04	8.743496580155D-05	2.50	7.250750970292D-03
6	.5	-1.103857244347D-04	-5.519286221734D-05	2.75	4.108510395278D-03
7	1.5	6.383298578173D-05	3.191649289086D-05	3.25	-2.793649548929D-03
8	2.5	-3.271384512067D-05	-1.635692256033D-05	3.50	-3.329023289050D-03
9	3.5	1.340892610608D-05	6.704463053039D-06	3.75	-2.004873877194D-03
10	4.5	-3.702877171958D-06	-1.851438585979D-06	4.25	1.489599494468D-03
11	5.5	4.939061937744D-07	2.469530968872D-07	4.50	1.832396003127D-03
				4.75	1.132232542380D-03
	0	1.419208776296D+01	6.891889811131D+00	5.25	-8.739986471007D-04
	1	1.562576681056D+01	8.013825837630D+00	5.50	-1.091139887598D-03
	2	6.503714493748D+00	3.178443071213D+00	5.75	-6.831524825666D-04
	3	1.207997687271D+00	6.157766609377D-01	6.25	5.405634456908D-04
	4	8.413947842790D-02	4.137056855892D-02	6.50	6.837811490878D-04

I= 1 H= -4.5

0		0.	4.596167732487D+01	-1.00	-1.484200815276D+01
1		5.135996147471D+00	-1.649806315076D+01	-.75	-8.819484684671D+00
2		-6.512627304080D+00	-2.970171471501D-01	-.50	-4.569802384266D+00

CARDINAL SPLINES N=5

3		2.784975050965D+00	1.188828569760D+00	-.25	-1.736148828740D+00
4		-4.084096502426D-01	-1.989620009208D-01	.25	9.188797155886D-01
				.50	1.262437654375D+00
0	-5.5	6.575588716303D-05	3.287794358152D-05	.75	1.234337173047D+00
1	-4.5	-3.960966847846D-04	-1.980483423923D-04	1.25	6.868154634923D-01
2	-3.5	1.040963492272D-03	5.204817461359D-04	1.50	3.848268332537D-01
3	-2.5	-1.594070128451D-03	-7.970350642255D-04	1.75	1.484544915517D-01
4	-1.5	1.633603955049D-03	8.168019775247D-04	2.25	-6.444164746466D-02
5	-.5	-1.258406043917D-03	-6.292030219586D-04	2.50	-6.658352421079D-02
6	.5	8.161297864075D-04	4.080648932038D-04	2.75	-3.592970918676D-02
7	1.5	-4.763266937037D-04	-2.381633468519D-04	3.25	2.292392681814D-02
8	2.5	2.448659970430D-04	1.224329985215D-04	3.50	2.675253477761D-02
9	3.5	-1.004731940722D-04	-5.023659703609D-05	3.75	1.585536393740D-02
10	4.5	2.775650416613D-05	1.387825208306D-05	4.25	-1.152593217659D-02
11	5.5	-3.702877171958D-06	-1.851438585979D-06	4.50	-1.407479534834D-02
				4.75	-8.647778453892D-03
	0	-7.912912860821D+01	-3.803375137938D+01	5.25	6.624413590051D-03
	1	-8.556304055882D+01	-4.428827067650D+01	5.50	8.249150158698D-03
	2	-3.468689048220D+01	-1.679294891324D+01	5.75	5.154646369822D-03
	3	-6.200037254373D+00	-3.188335450498D+00	6.25	-4.068075064502D-03
	4	-4.084096502426D-01	-1.989620009208D-01	6.50	-5.141439441830D-03
I= 2	H= -3.5				
0		0.	-1.665125732495D+02	-1.00	2.174890249857D+01
1		-5.749962089981D+00	6.619934532226D+01	-.75	1.238451035420D+01
2		1.003442537483D+01	-5.704110367922D+00	-.50	6.076644943327D+00
3		-5.124405840303D+00	-1.824354145080D+00	-.25	2.147992626214D+00
4		8.401091934532D-01	4.010598800601D-01	.25	-8.871261019219D-01
				.50	-9.544189321949D-01
0	-5.5	-1.666379993757D-04	-8.331899968784D-05	.75	-5.641627204925D-01
1	-4.5	1.040963492272D-03	5.204817461359D-04	1.25	5.325381685595D-01
2	-3.5	-2.866651398300D-03	-1.433325699150D-03	1.50	9.042929478675D-01
3	-2.5	4.655423867970D-03	2.327711933985D-03	1.75	1.058113308727D+00
4	-1.5	-5.108526432854D-03	-2.554263216427D-03	2.25	7.842410669773D-01
5	-.5	4.206445111056D-03	2.103222555528D-03	2.50	4.925164964060D-01
6	.5	-2.859395643913D-03	-1.429697821956D-03	2.75	2.094870480937D-01
7	1.5	1.705338548348D-03	8.526692741741D-04	3.25	-1.059655934508D-01
8	2.5	-8.832431502905D-04	-4.416215751452D-04	3.50	-1.160854111630D-01
9	3.5	3.633478730521D-04	1.816739365260D-04	3.75	-6.575738001889D-02
10	4.5	-1.004731940722D-04	-5.023659703609D-05	4.25	4.511928519621D-02
11	5.5	1.340892610608D-05	6.704463053039D-06	4.50	5.408531483136D-02
				4.75	3.276761374091D-02
	0	1.880959732477D+02	8.850242300314D+01	5.25	-2.463332415654D-02
	1	1.986815552688D+02	1.047993427891D+02	5.50	-3.048527545801D-02
	2	7.796154762160D+01	3.698641446917D+01	5.75	-1.895929272817D-02
	3	1.335799641567D+01	6.998963216242D+00	6.25	1.486787453333D-02
	4	8.401091934532D-01	4.010598800601D-01	6.50	1.875146499461D-02
I= 3	H= -2.5				
0		0.	4.067405944604D+02	-1.00	-2.123286724110D+01
1		5.120295180026D+00	-1.661725988925D+02	-.75	-1.188732386508D+01
2		-9.627737359839D+00	2.137655688738D+01	-.50	-5.715885127221D+00
3		5.496016454729D+00	9.455208519842D-01	-.25	-1.971545208377D+00
4		-9.888182465087D-01	-4.480059527171D-01	.25	7.603528967770D-01
				.50	7.784146429945D-01

0	-5.5	2.439715932482D-04	1.219857966241D-04	.75	4.304011071698D-01
1	-4.5	-1.594070128451D-03	-7.970350642255D-04	1.25	-3.208528347768D-01
2	-3.5	4.655423867970D-03	2.327711933985D-03	1.50	-4.294271536220D-01
3	-2.5	-8.148981630113D-03	-4.074490815057D-03	1.75	-3.058130299903D-01
4	-1.5	9.777613941319D-03	4.888806970659D-03	2.25	3.896629688662D-01
5	-.5	-8.829989162976D-03	-4.414994581488D-03	2.50	7.463292534214D-01
6	.5	6.471938439009D-03	3.235969219504D-03	2.75	9.685621283600D-01
7	1.5	-4.036934750012D-03	-2.018467375006D-03	3.25	8.452192397351D-01
8	2.5	2.132118828375D-03	1.066059414187D-03	3.50	5.652540480937D-01
9	3.5	-8.832431502905D-04	-4.416215751452D-04	3.75	2.533491068201D-01
10	4.5	2.448659970430D-04	1.224329985215D-04	4.25	-1.386767453897D-01
11	5.5	-3.271384512067D-05	-1.635692256033D-05	4.50	-1.564317432270D-01
				4.75	-9.077200186608D-02
	0	-2.535081906854D+02	-1.132102689645D+02	5.25	6.461846051627D-02
	1	-2.600798655631D+02	-1.433724173470D+02	5.50	7.859777097551D-02
	2	-9.841397759814D+01	-4.433542947304D+01	5.75	4.824885779754D-02
	3	-1.625798496846D+01	-8.910610107793D+00	6.25	-3.718815256173D-02
	4	-9.888182465087D-01	-4.480059527171D-01	6.50	-4.663733437774D-02

I= 4 H= -1.5

0		0.	-7.964931889581D+02	-1.00	1.571775966194D+01
1		-3.668562346675D+00	3.286109384702D+02	-.75	8.744147289195D+00
2		7.063768083370D+00	-4.776322382951D+01	-.50	4.173699937108D+00
3		-4.190198654961D+00	1.435404718737D+00	-.25	1.427204315304D+00
4		7.952305769288D-01	3.067200586153D-01	.25	-5.380205669073D-01
				.50	-5.424125374841D-01
0	-5.5	-2.376586626092D-04	-1.188293313046D-04	.75	-2.941941904671D-01
1	-4.5	1.633603955049D-03	8.168019775247D-04	1.25	2.071631608672D-01
2	-3.5	-5.108526432854D-03	-2.554263216427D-03	1.50	2.654357960353D-01
3	-2.5	9.777613941319D-03	4.888806970659D-03	1.75	1.778910063404D-01
4	-1.5	-1.308140512442D-02	-6.540702562211D-03	2.25	-1.810935381315D-01
5	-.5	1.329111805232D-02	6.645559026158D-03	2.50	-2.748195221908D-01
6	.5	-1.084376321571D-02	-5.421881607853D-03	2.75	-2.181003959885D-01
7	1.5	7.313107396491D-03	3.656553698246D-03	3.25	3.298106021515D-01
8	2.5	-4.036934750012D-03	-2.018467375006D-03	3.50	6.748273671950D-01
9	3.5	1.705338548348D-03	8.526692741741D-04	3.75	9.253628477808D-01
10	4.5	-4.763266937037D-04	-2.381633468519D-04	4.25	8.776137704703D-01
11	5.5	6.383298578173D-05	3.191649289086D-05	4.50	6.053656256162D-01
				4.75	2.783406639901D-01
	0	2.240432701970D+02	8.551293550162D+01	5.25	-1.584313187204D-01
	1	2.229983075788D+02	1.376006535794D+02	5.50	-1.815060634515D-01
	2	8.225983998909D+01	3.159064466832D+01	5.75	-1.067982675665D-01
	3	1.330487403747D+01	8.183246008273D+00	6.25	7.804346501073D-02
	4	7.952305769288D-01	3.067200586153D-01	6.50	9.624061886087D-02

I= 5 H= -.5

0		0.	1.392304805122D+03	-1.00	-1.008843131365D+01
1		2.332494071785D+00	-5.766089152332D+02	-.75	-5.602195680994D+00
2		-4.521876093795D+00	8.744798725364D+01	-.50	-2.668322015475D+00
3		2.711634150072D+00	-4.819938036262D+00	-.25	-9.101507878642D-01
4		-5.224269979941D-01	-1.021825535963D-01	.25	3.408348158851D-01
				.50	3.420809353711D-01
0	-5.5	1.748699316031D-04	8.743496580155D-05	.75	1.844998708608D-01
1	-4.5	-1.258406043917D-03	-6.292030219586D-04	1.25	-1.278070178158D-01
2	-3.5	4.206445111056D-03	2.103222555528D-03	1.50	-1.617813944119D-01

3	-2.5	-8.829989162976D-03	-4.414994581488D-03	1.75	-1.066860455709D-01
4	-1.5	1.329111805232D-02	6.645559026158D-03	2.25	1.031521356580D-01
5	-.5	-1.544072117593D-02	-7.720360587966D-03	2.50	1.501952888328D-01
6	.5	1.438215964649D-02	7.191079823243D-03	2.75	1.123833319838D-01
7	1.5	-1.084376321571D-02	-5.421881607853D-03	3.25	-1.361393607967D-01
8	2.5	6.471938439009D-03	3.235969219504D-03	3.50	-2.209002158280D-01
9	3.5	-2.859395643913D-03	-1.429697821956D-03	3.75	-1.853588643868D-01
10	4.5	8.161297864075D-04	4.080648932038D-04	4.25	3.049026537586D-01
11	5.5	-1.103857244347D-04	-5.519286221734D-05	4.50	6.436758129137D-01
				4.75	9.057110201273D-01
	0	-1.508632575762D+02	-2.916322797198D+01	5.25	8.936868693610D-01
	1	-1.490025110060D+02	-1.200929216523D+02	5.50	6.262355576629D-01
	2	-5.460041275353D+01	-1.062712382242D+01	5.75	2.920461066665D-01
	3	-8.781759805797D+00	-7.067954215381D+00	6.25	-1.707274254442D-01
	4	-5.224269979941D-01	-1.021825535963D-01	6.50	-1.983497019739D-01

I= 6 H= .5

0		0.	-2.280579098222D+03	-1.00	6.097905350333D+00
1		-1.405684509638D+00	9.464106181631D+02	-.75	3.384287737399D+00
2		2.730945977227D+00	-1.457945018539D+02	-.50	1.610859238884D+00
3		-1.643212972667D+00	9.315970394500D+00	-.25	5.490228829450D-01
4		3.180618908014D-01	-1.021825535963D-01	.25	-2.051697776908D-01
				.50	-2.056287295174D-01
0	-5.5	-1.103857244347D-04	-5.519286221734D-05	.75	-1.107082610228D-01
1	-4.5	8.161297864075D-04	4.080648932038D-04	1.25	7.629296859719D-02
2	-3.5	-2.859395643913D-03	-1.429697821956D-03	1.50	9.620421966661D-02
3	-2.5	6.471938439009D-03	3.235969219504D-03	1.75	6.312114782721D-02
4	-1.5	-1.084376321571D-02	-5.421881607853D-03	2.25	-6.007805867452D-02
5	-.5	1.438215964649D-02	7.191079823243D-03	2.50	-8.642794015448D-02
6	.5	-1.544072117593D-02	-7.720360587966D-03	2.75	-6.362748697612D-02
7	1.5	1.329111805232D-02	6.645559026158D-03	3.25	7.316275711627D-02
8	2.5	-8.829989162976D-03	-4.414994581488D-03	3.50	1.138532818101D-01
9	3.5	4.206445111056D-03	2.103222555528D-03	3.75	9.003130071971D-02
10	4.5	-1.258406043917D-03	-6.292030219586D-04	4.25	-1.185267282895D-01
11	5.5	1.748699316031D-04	8.743496580155D-05	4.50	-1.983497019739D-01
				4.75	-1.707274254442D-01
	0	9.253680163220D+01	-2.916322797198D+01	5.25	2.920461066665D-01
	1	9.118333229873D+01	1.200929216523D+02	5.50	6.262355576629D-01
	2	3.334616510869D+01	-1.062712382242D+01	5.75	8.936868693610D-01
	3	5.354148624965D+00	7.067954215381D+00	6.25	9.057110201273D-01
	4	3.180618908014D-01	-1.021825535963D-01	6.50	6.436758129137D-01

I= 7 H= 1.5

0		0.	3.440089109668D+03	-1.00	-3.477161952438D+00
1		8.007769254345D-01	-1.431849519190D+03	-.75	-1.929437569439D+00
2		-1.556822433013D+00	2.222838944435D+02	-.50	-9.181774914881D-01
3		9.377721342914D-01	-1.493108729781D+01	-.25	-3.128584420035D-01
4		-1.817904596983D-01	3.067200586153D-01	.25	1.168354001539D-01
				.50	1.170425921930D-01
0	-5.5	6.383298578173D-05	3.191649289086D-05	.75	6.297784990084D-02
1	-4.5	-4.763266937037D-04	-2.381633468519D-04	1.25	-4.332646851196D-02
2	-3.5	1.705338548348D-03	8.526692741741D-04	1.50	-5.456531103082D-02
3	-2.5	-4.036934750012D-03	-2.018467375006D-03	1.75	-3.574174766985D-02
4	-1.5	7.313107396491D-03	3.656553698246D-03	2.25	3.384393095620D-02
5	-.5	-1.084376321571D-02	-5.421881607853D-03	2.50	4.849712238143D-02

CARDINAL SPLINES N=5

```
 6     .5   1.329111805232D-02    6.645559026158D-03     2.75  3.551711224491D-02
 7    1.5  -1.308140512442D-02   -6.540702562211D-03     3.25 -4.018189249680D-02
 8    2.5   9.777613941319D-03    4.888806970659D-03     3.50 -6.175970139419D-02
 9    3.5  -5.108526432854D-03   -2.554263216427D-03     3.75 -4.803400877891D-02
10    4.5   1.633603955049D-03    8.168019775247D-04     4.25  5.998437882310D-02
11    5.5  -2.376586626092D-04   -1.188293313046D-04     4.50  9.624061886087D-02
                                                         4.75  7.804346501073D-02
        0  -5.301739919374D+01    8.551293550162D+01     5.25 -1.067982675665D-01
        1  -5.220299958001D+01   -1.376006535794D+02     5.50 -1.815060634515D-01
        2  -1.907855065245D+01    3.159064466832D+01     5.75 -1.584313187204D-01
        3  -3.061617979072D+00   -8.183246008273D+00     6.25  2.783406639901D-01
        4  -1.817904596983D-01    3.067200586153D-01     6.50  6.053656256162D-01


I= 8     H=  2.5
0               0.               -4.135361509725D+03    -1.00  1.773755927360D+00
1              -4.083573502434D-01    1.737847970366D+03     -.75  9.841759907395D-01
2               7.940879795280D-01   -2.726735766698D+02     -.50  4.683140983851D-01
3              -4.785042565139D-01    1.876674106757D+01     -.25  1.595589900592D-01
4               9.280634107441D-02   -4.480059527171D-01      .25 -5.957294320336D-02
                                                          .50 -5.966937988102D-02
0    -5.5  -3.271384512067D-05   -1.635692256033D-05      .75 -3.210045736808D-02
1    -4.5   2.448659970430D-04    1.224329985215D-04     1.25  2.207139913765D-02
2    -3.5  -8.832431502905D-04   -4.416215751452D-04     1.50  2.778501267714D-02
3    -2.5   2.132118828375D-03    1.066059414187D-03     1.75  1.818981820698D-02
4    -1.5  -4.036934750012D-03   -2.018467375006D-03     2.25 -1.719430933041D-02
5     -.5   6.471938439009D-03    3.235969219504D-03     2.50 -2.460648245980D-02
6      .5  -8.829989162976D-03   -4.414994581488D-03     2.75 -1.798920013555D-02
7     1.5   9.777613941319D-03    4.888806970659D-03     3.25  2.024202155760D-02
8     2.5  -8.148981630113D-03   -4.074490815057D-03     3.50  3.098496603106D-02
9     3.5   4.655423867970D-03    2.327711933985D-03     3.75  2.396860023888D-02
10    4.5  -1.594070128451D-03   -7.970350642255D-04     4.25 -2.943704408598D-02
11    5.5   2.439715932482D-04    1.219857966241D-04     4.50 -4.663733437774D-02
                                                         4.75 -3.718815256173D-02
        0   2.708765275629D+01   -1.132102689645D+02     5.25  4.824885779754D-02
        1   2.666496913095D+01    1.433724173470D+02     5.50  7.859777097551D-02
        2   9.743118652055D+00   -4.433542947304D+01     5.75  6.461846051627D-02
        3   1.563235247123D+00    8.910610107793D+00     6.25 -9.077200186608D-02
        4   9.280634107441D-02   -4.480059527171D-01     6.50 -1.564317432270D-01


I= 9     H=  3.5
0               0.                3.315185207635D+03    -1.00 -7.258765692725D-01
1               1.670940636606D-01   -1.413711164004D+03     -.75 -4.027470942832D-01
2              -3.249555557662D-01    2.252616757681D+02     -.50 -1.916399499192D-01
3               1.958375165126D-01   -1.582228057756D+01     -.25 -6.529159557000D-02
4              -3.798943333309D-02    4.010598800601D-01      .25  2.437535870246D-02
                                                          .50  2.441351905881D-02
0    -5.5   1.340892610608D-05    6.704463053039D-06      .75  1.313291257016D-02
1    -4.5  -1.004731940722D-04   -5.023659703609D-05     1.25 -9.028062942249D-03
2    -3.5   3.633478730521D-04    1.816739365260D-04     1.50 -1.136350507338D-02
3    -2.5  -8.832431502905D-04   -4.416215751452D-04     1.75 -7.437838688642D-03
4    -1.5   1.705338548348D-03    8.526692741741D-04     2.25  7.026589079884D-03
5     -.5  -2.859395643913D-03   -1.429697821956D-03     2.50  1.005106785669D-02
6      .5   4.206445111056D-03    2.103222555528D-03     2.75  7.343648911622D-03
7     1.5  -5.108526432854D-03   -2.554263216427D-03     3.25 -8.247844063538D-03
8     2.5   4.655423867970D-03    2.327711933985D-03     3.50 -1.260731035648D-02
```

9	3.5	-2.866651398300D-03	-1.433325699150D-03	3.75	-9.734217053962D-03
10	4.5	1.040963492272D-03	5.204817461359D-04	4.25	1.188673555700D-02
11	5.5	-1.666379993757D-04	-8.331899968784D-05	4.50	1.875146499461D-02
				4.75	1.486787453333D-02
	0	-1.109112724137D+01	8.850242300314D+01	5.25	-1.895929272817D-02
	1	-1.091713030942D+01	-1.047993427891D+02	5.50	-3.048527545801D-02
	2	-3.988718683264D+00	3.698641446917D+01	5.75	-2.463332415654D-02
	3	-6.399300168154D-01	-6.998963216242D+00	6.25	3.276761374091D-02
	4	-3.798943333309D-02	4.010598800601D-01	6.50	5.408531483136D-02

I=10 H= 4.5

0		0.	-1.502127921270D+03	-1.00	2.003329363554D-01
1		-4.611399854065D-02	6.507613624676D+02	-.75	1.111523307060D-01
2		8.968267121531D-02	-1.055120870136D+02	-.50	5.288934737402D-02
3		-5.405061819848D-02	7.565499470756D+00	-.25	1.801916705954D-02
4		1.048564840098D-02	-1.989620009208D-01	.25	-6.726914043615D-03
				.50	-6.737312948426D-03
0	-5.5	-3.702877171958D-06	-1.851438585979D-06	.75	-3.624154239352D-03
1	-4.5	2.775650416613D-05	1.387825208306D-05	1.25	2.491200991753D-03
2	-3.5	-1.004731940722D-04	-5.023659703609D-05	1.50	3.135474207681D-03
3	-2.5	2.448659970430D-04	1.224329985215D-04	1.75	2.052139613526D-03
4	-1.5	-4.763266937037D-04	-2.381633468519D-04	2.25	-1.938247183795D-03
5	-.5	8.161297864075D-04	4.080648932038D-04	2.50	-2.772068338566D-03
6	.5	-1.258406043917D-03	-6.292030219586D-04	2.75	-2.024913702368D-03
7	1.5	1.633603955049D-03	8.168019775247D-04	3.25	2.272657613334D-03
8	2.5	-1.594070128451D-03	-7.970350642255D-04	3.50	3.472067869323D-03
9	3.5	1.040963492272D-03	5.204817461359D-04	3.75	2.678955648308D-03
10	4.5	-3.960966847846D-04	-1.980483423923D-04	4.25	-3.264404866545D-03
11	5.5	6.575588716303D-05	3.287794358152D-05	4.50	-5.141439441830D-03
				4.75	-4.068075064502D-03
	0	3.061625849443D+00	-3.803375137938D+01	5.25	5.154646369822D-03
	1	3.013500794171D+00	4.428827067650D+01	5.50	8.249150158698D-03
	2	1.100992655719D+00	-1.679294891324D+01	5.75	6.624413590051D-03
	3	1.766336466232D-01	3.188335450498D+00	6.25	-8.647778453892D-03
	4	1.048564840098D-02	-1.989620009208D-01	6.50	-1.407479534834D-02

I=11 H= 5.5

0		0.	2.874223326778D+02	-1.00	-2.671484495419D-02
1		6.149301307433D-03	-1.263905449770D+02	-.75	-1.482236391340D-02
2		-1.195932812014D-02	2.084751617013D+01	-.50	-7.052863292699D-03
3		7.207874216564D-03	-1.525929169234D+00	-.25	-2.402868639743D-03
4		-1.398341310055D-03	4.137056855892D-02	.25	8.970280851252D-04
				.50	8.984075335350D-04
0	-5.5	4.939061937744D-07	2.469530968872D-07	.75	4.832690027682D-04
1	-4.5	-3.702877171958D-06	-1.851438585979D-06	1.25	-3.321835644167D-04
2	-3.5	1.340892610608D-05	6.704463053039D-06	1.50	-4.180835292819D-04
3	-2.5	-3.271384512067D-05	-1.635692256033D-05	1.75	-2.736239462415D-04
4	-1.5	6.383298578173D-05	3.191649289086D-05	2.25	2.584146603112D-04
5	-.5	-1.103857244347D-04	-5.519286221734D-05	2.50	3.695574858669D-04
6	.5	1.748699316031D-04	8.743496580155D-05	2.75	2.699260000334D-04
7	1.5	-2.376586626092D-04	-1.188293313046D-04	3.25	-3.028646351536D-04
8	2.5	2.439715932482D-04	1.219857966241D-04	3.50	-4.626037461365D-04
9	3.5	-1.666379993757D-04	-8.331899968784D-05	3.75	-3.568310293861D-04
10	4.5	6.575588716303D-05	3.287794358152D-05	4.25	4.344315085806D-04
11	5.5	-1.123412138365D-05	-5.617060691825D-06	4.50	6.837811490878D-04

CARDINAL SPLINES N=5

			4.75	5.405634456908D-04
0	-4.083081406943D-01	6.891889811131D+00	5.25	-6.831524825666D-04
1	-4.018848647023D-01	-8.013825837630D+00	5.50	-1.091139887598D-03
2	-1.468283513218D-01	3.178443071213D+00	5.75	-8.739986471007D-04
3	-2.355563460464D-02	-6.157766609377D-01	6.25	1.132232542380D-03
4	-1.398341310055D-03	4.137056855892D-02	6.50	1.832396003127D-03

N= 5 M=12 P= 6.0 Q= 9

I= 0 H= -6.0

0		1.000000000000D+00	6.028622259906D+00	-1.00	5.854837514150D+00
1		-2.284225248764D+00	-3.218822414090D+00	-.75	4.047978250985D+00
2		1.843262191074D+00	1.213616792401D+00	-.50	2.688586794747D+00
3		-6.431878872386D-01	-3.398041542527D-01	-.25	1.696638768414D+00
4		8.416218707379D-02	4.250611277121D-02	.25	5.344265225133D-01
				.50	2.435645522163D-01
0	-6.0	-1.124214471055D-05	-5.621072355275D-06	.75	7.894975386796D-02
1	-5.0	6.581605358511D-05	3.290802679255D-05	1.25	-2.602040038404D-02
2	-4.0	-1.668560034485D-04	-8.342800172423D-05	1.50	-2.411804786197D-02
3	-3.0	2.445043091845D-04	1.222521545923D-04	1.75	-1.186518392335D-02
4	-2.0	-2.387029661444D-04	-1.193514830722D-04	2.25	6.556499616485D-03
5	-1.0	1.767013286451D-04	8.835066432256D-05	2.50	7.244752389821D-03
6	0.0	-1.134156047304D-04	-5.670780236522D-05	2.75	4.104129621276D-03
7	1.0	6.855353437968D-05	3.427676718984D-05	3.25	-2.788736314879D-03
8	2.0	-3.909075785132D-05	-1.954537892566D-05	3.50	-3.321521131083D-03
9	3.0	1.994082081657D-05	9.970410408287D-06	3.75	-1.999089562072D-03
10	4.0	-8.160409433759D-06	-4.080204716879D-06	4.25	1.482566595186D-03
11	5.0	2.252171860797D-06	1.126085930398D-06	4.50	1.821337473259D-03
12	6.0	-3.003321528949D-07	-1.501660764474D-07	4.75	1.123501680739D-03
				5.25	-8.630087913580D-04
	0	2.379769819017D+01	1.209611713470D+01	5.50	-1.073639018070D-03
	1	2.308675885411D+01	1.137101186975D+01	5.75	-6.691890027025D-04
	2	8.444912628719D+00	4.278462374433D+00	6.25	5.226861652386D-04
	3	1.376704602532D+00	6.803425522563D-01	6.50	6.550990491833D-04
	4	8.416218707379D-02	4.250611277121D-02	6.75	4.113763625254D-04

I= 1 H= -5.0

0		0.	-4.146136180853D+01	-1.00	-1.484526130054D+01
1		5.136744947498D+00	1.814247397321D+01	-.75	-8.821289640107D+00
2		-6.514083617352D+00	-5.446765326740D+00	-.50	-4.570661223739D+00
3		2.785852794746D+00	1.529492576209D+00	-.25	-1.736441428271D+00
4		-4.085799409450D-01	-2.074773753053D-01	.25	9.189889455645D-01
				.50	1.262547050992D+00
0	-6.0	6.581605358511D-05	3.290802679255D-05	.75	1.234396018484D+00
1	-5.0	-3.965478689867D-04	-1.982739344933D-04	1.25	6.867750168488D-01
2	-4.0	1.042598291056D-03	5.212991455278D-04	1.50	3.847759291305D-01
3	-3.0	-1.598064931639D-03	-7.990324658193D-04	1.75	1.484211776961D-01
4	-2.0	1.641435121354D-03	8.207175606770D-04	2.25	-6.441018964569D-02

CARDINAL SPLINES N=5

5	-1.0	-1.272139574757D-03	-6.360697873785D-04	2.50	-6.653854123440D-02
6	0.0	8.388506675317D-04	4.194253337658D-04	2.75	-3.589685803899D-02
7	1.0	-5.117257897291D-04	-2.558628948646D-04	3.25	2.288708278607D-02
8	2.0	2.926860631674D-04	1.463430315837D-04	3.50	2.669627656847D-02
9	3.0	-1.494554602827D-04	-7.472773014134D-05	3.75	1.581198772334D-02
10	4.0	6.118325728549D-05	3.059162864274D-05	4.25	-1.147319290872D-02
11	5.0	-1.688800044599D-05	-8.444000222994D-06	4.50	-1.399186813075D-02
12	6.0	2.252171860797D-06	1.126085930398D-06	4.75	-8.582306273443D-03
				5.25	6.542001355399D-03
	0	-1.314619403394D+02	-6.721035166640D+01	5.50	8.117912242571D-03
	1	-1.251732256047D+02	-6.129396398087D+01	5.75	5.049935115547D-03
	2	-4.462200055606D+01	-2.273101202092D+01	6.25	-3.934014466614D-03
	3	-7.020065787935D+00	-3.449964431118D+00	6.50	-4.926354184755D-03
	4	-4.085799409450D-01	-2.074773753053D-01	6.75	-3.091515781272D-03

I= 2 H= -4.0

0		0.	1.502517934892D+02	-1.00	2.176068979523D+01
1		-5.752675255397D+00	-5.931548365570D+01	-.75	1.239105034150D+01
2		1.003970210915D+01	1.295523546519D+01	-.50	6.079756820293D+00
3		-5.127586214215D+00	-3.058699454317D+00	-.25	2.149052817052D+00
4		8.407262164606D-01	4.319140713715D-01	.25	-8.875218804776D-01
				.50	-9.548153145485D-01
0	-6.0	-1.668560034485D-04	-8.342800172423D-05	.75	-5.643759381738D-01
1	-5.0	1.042598291056D-03	5.212991455278D-04	1.25	5.326847209507D-01
2	-4.0	-2.872574848303D-03	-1.436287424152D-03	1.50	9.044773913793D-01
3	-3.0	4.669898442573D-03	2.334949221286D-03	1.75	1.058234016523D+00
4	-2.0	-5.136901498053D-03	-2.568450749027D-03	2.25	7.841270842537D-01
5	-1.0	4.256206515403D-03	2.128103257702D-03	2.50	4.923535072881D-01
6	0.0	-2.941721373976D-03	-1.470860686988D-03	2.75	2.093680168506D-01
7	1.0	1.833601902571D-03	9.168009512856D-04	3.25	-1.058320945860D-01
8	2.0	-1.056512040431D-03	-5.282560202153D-04	3.50	-1.158815679179D-01
9	3.0	5.408278221753D-04	2.704139110876D-04	3.75	-6.560021276512D-02
10	4.0	-2.215900574179D-04	-1.107950287089D-04	4.25	4.492819231089D-02
11	5.0	6.118325728549D-05	3.059162864274D-05	4.50	5.378484040464D-02
12	6.0	-8.160409433759D-06	-4.080204716879D-06	4.75	3.253038504257D-02
				5.25	-2.433471569383D-02
	0	3.089357786597D+02	1.598289226668D+02	5.50	-3.000975440765D-02
	1	2.873318899412D+02	1.389815585253D+02	5.75	-1.857988708774D-02
	2	9.934001300877D+01	5.119208470373D+01	6.25	1.438212591586D-02
	3	1.504984298084D+01	7.307238258600D+00	6.50	1.797213558754D-02
	4	8.407262164606D-01	4.319140713715D-01	6.75	1.126029769628D-02

I= 3 H= -3.0

0		0.	-3.673065326459D+02	-1.00	-2.126167074325D+01
1		5.126925085773D+00	1.405361189469D+02	-.75	-1.190330501357D+01
2		-9.640631616497D+00	-2.421952194088D+01	-.50	-5.723489326678D+00
3		5.503788033696D+00	3.961773754131D+00	-.25	-1.974135896467D+00
4		-9.903260072809D-01	-5.234014221941D-01	.25	7.613200234055D-01
				.50	7.793832450667D-01
0	-6.0	2.445043091845D-04	1.222521545923D-04	.75	4.309221270537D-01
1	-5.0	-1.598064931639D-03	-7.990324658193D-04	1.25	-3.212109509958D-01
2	-4.0	4.669898442573D-03	2.334949221286D-03	1.50	-4.298778607892D-01
3	-3.0	-8.184351779233D-03	-4.092175889617D-03	1.75	-3.061079922161D-01
4	-2.0	9.846951404763D-03	4.923475702381D-03	2.25	3.899414976687D-01
5	-1.0	-8.951586400924D-03	-4.475793200462D-03	2.50	7.467275345150D-01

CARDINAL SPLINES N=5

6	0.0	6.673110040155D-03	3.336555020077D-03	2.75	9.688529937520D-01
7	1.0	-4.350359779069D-03	-2.175179889534D-03	3.25	8.448930211838D-01
8	2.0	2.555519629452D-03	1.277759814726D-03	3.50	5.647559356328D-01
9	3.0	-1.316934117972D-03	-6.584670589858D-04	3.75	2.529650520651D-01
10	4.0	5.408278221753D-04	2.704139110876D-04	4.25	-1.382097897772D-01
11	5.0	-1.494554602827D-04	-7.472773014134D-05	4.50	-1.556975022859D-01
12	6.0	1.994082081657D-05	9.970410408287D-06	4.75	-9.019230853027D-02
				5.25	6.388877925486D-02
	0	-4.109454778370D+02	-2.185777211075D+02	5.50	7.743578515401D-02
	1	-3.717932169637D+02	-1.744454076732D+02	5.75	4.732174011380D-02
	2	-1.244828645826D+02	-6.596230156044D+01	6.25	-3.600117460167D-02
	3	-1.826403614105D+01	-8.599860378527D+00	6.50	-4.473296080560D-02
	4	-9.903260072809D-01	-5.234014221941D-01	6.75	-2.790627712695D-02

I= 4 H= -2.0

0		0.	7.209011595269D+02	-1.00	1.577422427480D+01
1		-3.681559205857D+00	-2.726419552915D+02	-.75	8.775475749117D+00
2		7.089045190532D+00	4.162052284975D+01	-.50	4.188606741716D+00
3		-4.205433580059D+00	-4.477471832371D+00	-.25	1.432282940789D+00
4		7.981862983501D-01	4.545206966893D-01	.25	-5.399164624271D-01
				.50	-5.443113253728D-01
0	-6.0	-2.387029661444D-04	-1.193514830722D-04	.75	-2.952155657809D-01
1	-5.0	1.641435121354D-03	8.207175606770D-04	1.25	2.078651898624D-01
2	-4.0	-5.136901498053D-03	-2.568450749027D-03	1.50	2.663193346268D-01
3	-3.0	9.846951404763D-03	4.923475702381D-03	1.75	1.784692321341D-01
4	-2.0	-1.321733002031D-02	-6.608665010157D-03	2.25	-1.816395488536D-01
5	-1.0	1.352948979377D-02	6.764744896884D-03	2.50	-2.756002879342D-01
6	0.0	-1.123812764129D-02	-5.619063820644D-03	2.75	-2.186705906015D-01
7	1.0	7.927526533401D-03	3.963763266701D-03	3.25	3.304501009219D-01
8	2.0	-4.866943621823D-03	-2.433471810912D-03	3.50	6.758038362066D-01
9	3.0	2.555519629452D-03	1.277759814726D-03	3.75	9.261157250854D-01
10	4.0	-1.056512040431D-03	-5.282560202153D-04	4.25	8.766983794268D-01
11	5.0	2.926860631674D-04	1.463430315837D-04	4.50	6.039262648647D-01
12	6.0	-3.909075785132D-05	-1.954537892566D-05	4.75	2.772042688524D-01
				5.25	-1.570008964845D-01
	0	3.591920609930D+02	2.053131574861D+02	5.50	-1.792281779626D-01
	1	3.168331182086D+02	1.359432429490D+02	5.75	-1.049808031197D-01
	2	1.037994811931D+02	5.920250035197D+01	6.25	7.571658647154D-02
	3	1.495103758034D+01	6.431024888173D+00	6.50	9.250740211659D-02
	4	7.981862983501D-01	4.545206966893D-01	6.75	5.698355748917D-02

I= 5 H= -1.0

0		0.	-1.268752471692D+03	-1.00	-1.018745340971D+01
1		2.355286687783D+00	4.778093971500D+02	-.75	-5.657136459131D+00
2		-4.566204603399D+00	-6.930445469867D+01	-.50	-2.694464106300D+00
3		2.738351666407D+00	5.549484512947D+00	-.25	-9.190571827745D-01
4		-5.276104521199D-01	-3.613808060857D-01	.25	3.441596513665D-01
				.50	3.454108432051D-01
0	-6.0	1.767013286451D-04	8.835066432256D-05	.75	1.862910586387D-01
1	-5.0	-1.272139574757D-03	-6.360697873785D-04	1.25	-1.290381673637D-01
2	-4.0	4.256206515403D-03	2.128103257702D-03	1.50	-1.633308576747D-01
3	-3.0	-8.951586400924D-03	-4.475793200462D-03	1.75	-1.077000812153D-01
4	-2.0	1.352948979377D-02	6.764744896884D-03	2.25	1.041096756717D-01
5	-1.0	-1.585875414001D-02	-7.929377070005D-03	2.50	1.515645191619D-01
6	0.0	1.507375726938D-02	7.536878634692D-03	2.75	1.133832833317D-01

CARDINAL SPLINES N=5

7	1.0	-1.192127119306D-02	-5.960635596531D-03	3.25	-1.372608509737D-01
8	2.0	7.927526533401D-03	3.963763266701D-03	3.50	-2.226126513404D-01
9	3.0	-4.350359779069D-03	-2.175179889534D-03	3.75	-1.866791867092D-01
10	4.0	1.833601902571D-03	9.168009512856D-04	4.25	3.065079767256D-01
11	5.0	-5.117257897291D-04	-2.558628948646D-04	4.50	6.462000225271D-01
12	6.0	6.855353437968D-05	3.427676718984D-05	4.75	9.077039183831D-01
				5.25	8.911783352390D-01
	0	-2.425508315991D+02	-1.665173278342D+02	5.50	6.222408256659D-01
	1	-2.125526192126D+02	-6.673274829382D+01	5.75	2.888588159030D-01
	2	-6.923973226597D+01	-4.747198758014D+01	6.25	-1.666467742442D-01
	3	-9.924299184470D+00	-3.123654833110D+00	6.50	-1.918027528613D-01
	4	-5.276104521199D-01	-3.613808060857D-01	6.75	-1.132641823633D-01

I= 6 H= 0.0

0		0.	2.121898804676D+03	-1.00	6.261728426830D+00
1		-1.443392825485D+00	-7.980288424703D+02	-.75	3.475182271622D+00
2		2.804283475711D+00	1.135381956921D+02	-.50	1.654108957030D+00
3		-1.687414680127D+00	-7.839298692147D+00	-.25	5.637577055018D-01
4		3.266374455061D-01	3.266374455061D-01	.25	-2.106704164275D-01
				.50	-2.111377600015D-01
0	-6.0	-1.134156047304D-04	-5.670780236522D-05	.75	-1.136716187381D-01
1	-5.0	8.388506675317D-04	4.194253337658D-04	1.25	7.832979385721D-02
2	-4.0	-2.941721373976D-03	-1.470860686988D-03	1.50	9.876766609637D-02
3	-3.0	6.673110040155D-03	3.336555020077D-03	1.75	6.479877782639D-02
4	-2.0	-1.123812764129D-02	-5.619063820644D-03	2.25	-6.166222177696D-02
5	-1.0	1.507375726938D-02	7.536878634692D-03	2.50	-8.869320754282D-02
6	0.0	-1.658490671415D-02	-8.292453357075D-03	2.75	-6.528181578436D-02
7	1.0	1.507375726938D-02	7.536878634692D-03	3.25	7.501816089348D-02
8	2.0	-1.123812764129D-02	-5.619063820644D-03	3.50	1.166863510451D-01
9	3.0	6.673110040155D-03	3.336555020077D-03	3.75	9.221565424699D-02
10	4.0	-2.941721373976D-03	-1.470860686988D-03	4.25	-1.211825895332D-01
11	5.0	8.388506675317D-04	4.194253337658D-04	4.50	-2.025257778300D-01
12	6.0	-1.134156047304D-04	-5.670780236522D-05	4.75	-1.740244948499D-01
				5.25	2.961962488439D-01
	0	1.511344066412D+02	1.511344066412D+02	5.50	6.328444794242D-01
	1	1.321819763466D+02	-3.086516846051D-21	5.75	8.989599528376D-01
	2	4.298450746275D+01	4.298450746275D+01	6.25	8.989599528376D-01
	3	6.151884012020D+00	-1.672655262580D-22	6.50	6.328444794242D-01
	4	3.266374455061D-01	3.266374455061D-01	6.75	2.961962488439D-01

I= 7 H= 1.0

0		0.	-3.418964339121D+03	-1.00	-3.732398040526D+00
1		8.595264205067D-01	1.285984337690D+03	-.75	-2.071051106677D+00
2		-1.671082136979D+00	-1.817560286907D+02	-.50	-9.855604823749D-01
3		1.006638322989D+00	1.179679417917D+01	-.25	-3.358152717035D-01
4		-1.951511600516D-01	-3.613808060857D-01	.25	1.254053864048D-01
				.50	1.256256527727D-01
0	-6.0	6.855353437968D-05	3.427676718984D-05	.75	6.759475642610D-02
1	-5.0	-5.117257897291D-04	-2.558628948646D-04	1.25	-4.649983897117D-02
2	-4.0	1.833601902571D-03	9.168009512856D-04	1.50	-5.855915642031D-02
3	-3.0	-4.350359779069D-03	-2.175179889534D-03	1.75	-3.835549249390D-02
4	-2.0	7.927526533401D-03	3.963763266701D-03	2.25	3.631205450636D-02
5	-1.0	-1.192127119306D-02	-5.960635596531D-03	2.50	5.202640530689D-02
6	0.0	1.507375726938D-02	7.536878634692D-03	2.75	3.809455385117D-02
7	1.0	-1.585875414001D-02	-7.929377070005D-03	3.25	-4.307260857935D-02

8	2.0	1.352948979377D-02	6.764744896884D-03	3.50	-6,617361867800D-02
9	3.0	-8.951586400924D-03	-4.475793200462D-03	3.75	-5,143722803977D-02
10	4.0	4.256206515403D-03	2.128103257702D-03	4.25	6,412220634326D-02
11	5.0	-1.272139574757D-03	-6.360697873785D-04	4.50	1,027469382871D-01
12	6.0	1.767013286451D-04	8.835066432256D-05	4.75	8,318029380962D-02
				5.25	-1,132641823633D-01
	0	-9,048382406938D+01	-1.665173278342D+02	5.50	-1,918027528613D-01
	1	-7,908712262498D+01	6.673274829382D+01	5.75	-1,666467742442D-01
	2	-2.570424289432D+01	-4.747198758014D+01	6.25	2,888588159030D-01
	3	-3,676989518249D+00	3.123654833110D+00	6.50	6,222408256659D-01
	4	-1.951511600516D-01	-3.613808060857D-01	6.75	8,911783352390D-01

I= 8   H= 2.0

0		0.	5.130422826606D+03	-1.00	2.118550229480D+00
1		-4,877210926710D-01	-1.933629817035D+03	-.75	1,175479427132D+00
2		9.484395650903D-01	2.731374188240D+02	-.50	5,593406906336D-01
3		-5,715344766900D-01	-1.733952160872D+01	-.25	1,905709998991D-01
4		1.108550950286D-01	4.545206966893D-01	.25	-7,114999898206D-02
				.50	-7,126409755905D-02
0	-6.0	-3.909075785132D-05	-1.954537892566D-05	.75	-3,833736169948D-02
1	-5.0	2.926860631674D-04	1.463430315837D-04	1.25	2,635825395134D-02
2	-4.0	-1.056512040431D-03	-5.282560202153D-04	1.50	3,318023380029D-02
3	-3.0	2.555519629452D-03	1.277759814726D-03	1.75	2,172068380391D-02
4	-2.0	-4.866943621823D-03	-2.433471810912D-03	2.25	-2,052845750435D-02
5	-1.0	7.927526533401D-03	3.963763266701D-03	2.50	-2,937413366610D-02
6	0.0	-1,123812764129D-02	-5.619063820644D-03	2.75	-2,147102430288D-02
7	1.0	1.352948979377D-02	6.764744896884D-03	3.25	2,414704314672D-02
8	2.0	-1.321733002031D-02	-6.608665010157D-03	3.50	3,694765548021D-02
9	3.0	9.846951404763D-03	4.923475702381D-03	3.75	2,856595408016D-02
10	4.0	-5.136901498053D-03	-2.568450749027D-03	4.25	-3,502676834007D-02
11	5.0	1.641435121354D-03	8.207175606770D-04	4.50	-5,542661602963D-02
12	6.0	-2,387029661444D-04	-1.193514830722D-04	4.75	-4,412741148114D-02
				5.25	5,698355748917D-02
	0	5.143425397925D+01	2.053131574861D+02	5.50	9,250740211659D-02
	1	4.494663231060D+01	-1.359432429490D+02	5.75	7,571658647154D-02
	2	1.460551951085D+01	5.920250035197D+01	6.25	-1,049808031197D-01
	3	2.088987803996D+00	-6.431024888173D+00	6.50	-1,792281779626D-01
	4	1.108550950286D-01	4.545206966893D-01	6.75	-1,570008964845D-01

I= 9   H= 3.0

0		0.	-6.175791108248D+03	-1.00	-1,079050616551D+00
1		2.483866312813D-01	2.346996776055D+03	-.75	-5,986998941801D-01
2		-4,830584415787D-01	-3.338144955679D+02	-.50	-2,848788166775D-01
3		2.911287065839D-01	2.116149451119D+01	-.25	-9,705730910432D-02
4		-5,647683710729D-02	-5.234014221941D-01	.25	3,623377869315D-02
				.50	3,629003019667D-02
0	-6.0	1.994082081657D-05	9.970410408287D-06	.75	1,952139617232D-02
1	-5.0	-1.494554602827D-04	-7.472773014134D-05	1.25	-1,341910378990D-02
2	-4.0	5.408278221753D-04	2.704139110876D-04	1.50	-1,688984952951D-02
3	-3.0	-1.316934117972D-03	-6.584670589858D-04	1.75	-1,105451705483D-02
4	-2.0	2.555519629452D-03	1.277759814726D-03	2.25	1,044176908018D-02
5	-1.0	-4,350359779069D-03	-2.175179889534D-03	2.50	1,493459018344D-02
6	0.0	6.673110040155D-03	3.336555020077D-03	2.75	1,091009396459D-02
7	1.0	-8.951586400924D-03	-4.475793200462D-03	3.25	-1,224777176205D-02
8	2.0	9.846951404763D-03	4.923475702381D-03	3.50	-1,871491466569D-02

9	3.0	-8.184351779233D-03	-4.092175889617D-03	3.75	-1.444330317550D-02
10	4.0	4.669898442573D-03	2.334949221286D-03	4.25	1.761231027169D-02
11	5.0	-1.598064931639D-03	-7.990324658193D-04	4.50	2.775435789215D-02
12	6.0	2.445043091845D-04	1.222521545923D-04	4.75	2.197578247327D-02
				5.25	-2.790627712695D-02
	0	-2.620996437807D+01	-2.185777211075D+02	5.50	-4.473296080560D-02
	1	-2.290240161730D+01	1.744454076732D+02	5.75	-3.600117460167D-02
	2	-7.441738538244D+00	-6.596230156044D+01	6.25	4.732174011380D-02
	3	-1.064315383991D+00	8.599860378527D+00	6.50	7.743578515401D-02
	4	-5.647683710729D-02	-5.234014221941D-01	6.75	6.388877925486D-02

I=10 H= 4.0

0		0.	4.974757423508D+03	-1.00	4.413479526028D-01
1		-1.015901289553D-01	-1.915642064564D+03	-.75	2.448755405928D-01
2		1.975761302236D-01	2.760158127748D+02	-.50	1.165179383189D-01
3		-1.190797671414D-01	-1.767317597152D+01	-.25	3.969690363892D-02
4		2.310192628250D-02	4.319140713715D-01	.25	-1.481940359302D-02
				.50	-1.484214836006D-02
0	-6.0	-8.160409433759D-06	-4.080204716879D-06	.75	-7.983819087418D-03
1	-5.0	6.118325728549D-05	3.059162864274D-05	1.25	5.487759740599D-03
2	-4.0	-2.215900574179D-04	-1.107950287089D-04	1.50	6.906793219210D-03
3	-3.0	5.408278221753D-04	2.704139110876D-04	1.75	4.520253570098D-03
4	-2.0	-1.056512040431D-03	-5.282560202153D-04	2.25	-4.268853494198D-03
5	-1.0	1.833601902571D-03	9.168009512856D-04	2.50	-6.104709063433D-03
6	0.0	-2.941721373976D-03	-1.470860686988D-03	2.75	-4.458747159300D-03
7	1.0	4.256206515403D-03	2.128103257702D-03	3.25	5.002310731034D-03
8	2.0	-5.136901498053D-03	-2.568450749027D-03	3.50	7.640053493293D-03
9	3.0	4.669898442573D-03	2.334949221286D-03	3.75	5.892556638637D-03
10	4.0	-2.872574848303D-03	-1.436287424152D-03	4.25	-7.171683709706D-03
11	5.0	1.042598291056D-03	5.212991455278D-04	4.50	-1.128524415990D-02
12	6.0	-1.668560034485D-04	-8.342800172423D-05	4.75	-8.918693508375D-03
				5.25	1.126029769628D-02
	0	1.072206667390D+01	1.598289226668D+02	5.50	1.797213558754D-02
	1	9.368772890541D+00	-1.389815585253D+02	5.75	1.438212591586D-02
	2	3.044156398699D+00	5.119208470373D+01	6.25	-1.857988708774D-02
	3	4.353664636387D-01	-7.307238258600D+00	6.50	-3.000975440765D-02
	4	2.310192628250D-02	4.319140713715D-01	6.75	-2.433471569383D-02

I=11 H= 5.0

0		0.	-2.267373563822D+03	-1.00	-1.217827891475D-01
1		2.803176221146D-02	8.859227190565D+02	-.75	-6.756924604884D-02
2		-5.451771099405D-02	-1.296454848470D+02	-.50	-3.215104774289D-02
3		3.285850627633D-02	8.429421438446D+00	-.25	-1.095361325082D-02
4		-6.374809665604D-03	-2.074773753053D-01	.25	4.089096184640D-03
				.50	4.095345436433D-03
0	-6.0	2.252171860797D-06	1.126085930398D-06	.75	2.202931042631D-03
1	-5.0	-1.688800044599D-05	-8.444000222994D-06	1.25	-1.514170829537D-03
2	-4.0	6.118325728549D-05	3.059162864274D-05	1.50	-1.905673601077D-03
3	-3.0	-1.494554602827D-04	-7.472773014134D-05	1.75	-1.247167103591D-03
4	-2.0	2.926860631674D-04	1.463430315837D-04	2.25	1.177718171369D-03
5	-1.0	-5.117257897291D-04	-2.558628948646D-04	2.50	1.684111688260D-03
6	0.0	8.388506675317D-04	4.194253337658D-04	2.75	1.229947275313D-03
7	1.0	-1.272139574757D-03	-6.360697873785D-04	3.25	-1.379571449115D-03
8	2.0	1.641435121354D-03	8.207175606770D-04	3.50	-2.106658257210D-03
9	3.0	-1.598064931639D-03	-7.990324658193D-04	3.75	-1.624430220361D-03

10	4,0	1.042598291056D-03	5,212991455278D-04	4,25	1,975650693978D-03
11	5,0	-3,965478689867D-04	-1,982739344933D-04	4,50	3,107193985211D-03
12	6,0	6,581605358511D-05	3,290802679255D-05	4,75	2,453881257606D-03
				5,25	-3,091515781272D-03
	0	-2,958762993452D+00	-6,721035166640D+01	5,50	-4,926354184755D-03
	1	-2,585297642955D+00	6,129396398087D+01	5,75	-3,934014466614D-03
	2	-8,400234857905D-01	-2.273101202092D+01	6,25	5,049935115547D-03
	3	-1.201369256982D-01	3.449964431118D+00	6.50	8.117912242571D-03
	4	-6,374809665604D-03	-2,074773753053D-01	6,75	6,542001355399D-03

I=12 H= 6.0

0		0.	4.363887472717D+02	-1.00	1,623870663368D-02
1		-3,737777923557D-03	-1,729148374410D+02	-,75	9,009778765148D-03
2		7.269465014290D-03	2.570594867363D+01	-,50	4,287060773041D-03
3		-4,381425227207D-03	-1,700489258765D+00	-,25	1,460566276226D-03
4		8,500384686270D-04	4.250611277121D-02	,25	-5,452422250488D-04
				,50	-5,460740439039D-04
0	-6,0	-3,003321528949D-07	-1,501660764474D-07	,75	-2,937382059891D-04
1	-5,0	2.252171860797D-06	1.126085930398D-06	1.25	2.018971232198D-04
2	-4,0	-8,160409433759D-06	-4,080204716879D-06	1.50	2,540976243048D-04
3	-3,0	1,994082081657D-05	9,970410408287D-06	1,75	1,662924539031D-04
4	-2,0	-3.909075785132D-05	-1,954537892566D-05	2,25	-1,570276937201D-04
5	-1,0	6.855353437968D-05	3.427676718984D-05	2.50	-2,245410924564D-04
6	0,0	-1.134156047304D-04	-5,670780236522D-05	2,75	-1,639827597372D-04
7	1,0	1.767013286451D-04	8,835066432256D-05	3,25	1,839140020411D-04
8	2,0	-2,387029661444D-04	-1,193514830722D-04	3,50	2,808235638380D-04
9	3,0	2.445043091845D-04	1.222521545923D-04	3,75	2,165206323331D-04
10	4,0	-1,668560034485D-04	-8,342800172423D-05	4,25	-2,632580984289D-04
11	5,0	6,581605358511D-05	3,290802679255D-05	4,50	-4,139469979292D-04
12	6,0	-1.124214471055D-05	-5,621072355275D-06	4,75	-3,268168561794D-04
				5,25	4,113763625254D-04
	0	3,945360792370D-01	1.209611713470D+01	5,50	6,550990491833D-04
	1	3,447351146033D-01	-1,137101186975D+01	5,75	5,226861652386D-04
	2	1.120121201480D-01	4.278462374433D+00	6,25	-6,691890027025D-04
	3	1.601949801984D-02	-6,803425522563D-01	6,50	-1,073639018070D-03
	4	8.500384686270D-04	4.250611277121D-02	6,75	-8,630087913580D-04

N= 5 M=13 P= 6,5 Q= 9

I= 0 H= -6.5

0		1.000000000000D+00	-4.312552453924D+00	-1.00	5,854997825241D+00
1		-2.284262148390D+00	5,096902411002D-01	-,75	4,048067196807D+00
2		1.843333956216D+00	7,093897917184D-01	-,50	2,688629117143D+00
3		-6.432311416758D-01	-3,095110475635D-01	-,25	1,696653187274D+00
4		8.417057895922D-02	4,182689804447D-02	,25	5,344211398586D-01
				.50	2,435591613716D-01
0	-6,5	-1.124510999650D-05	-5,622554998250D-06	.75	7,894685409820D-02
1	-5,5	6.583829185825D-05	3.291914592913D-05	1.25	-2.601840729484D-02

2	-4.5	-1.669365962514D-04	-8.346829812568D-05	1.50	-2.411553948726D-02
3	-3.5	2.447013503471D-04	1.223506751735D-04	1.75	-1.186354235863D-02
4	-2.5	-2.390898262172D-04	-1.195449131086D-04	2.25	6.554949579649D-03
5	-1.5	1.773828870796D-04	8.869144353979D-05	2.50	7.242535997869D-03
6	-.5	-1.145590341764D-04	-5.727951708820D-05	2.75	4.102511060990D-03
7	.5	7.041382492791D-05	3.520691246396D-05	3.25	-2.786921287090D-03
8	1.5	-4.197122651644D-05	-2.098561325822D-05	3.50	-3.318750013909D-03
9	2.5	2.382333036361D-05	1.191166518181D-05	3.75	-1.996953281923D-03
10	3.5	-1.213404287491D-05	-6.067021437455D-06	4.25	1.479970331315D-03
11	4.5	4.963006273748D-06	2.481503136874D-06	4.50	1.817256464311D-03
12	5.5	-1.369460859644D-06	-6.847304298219D-07	4.75	1.120281078894D-03
13	6.5	1.826060422966D-07	9.130302114832D-08	5.25	-8.589603243809D-04
				5.50	-1.067198469742D-03
	0	3.763554750626D+01	1.863630856672D+01	5.75	-6.640570285641D-04
	1	3.261091306171D+01	1.644807975662D+01	6.25	5.161420495554D-04
	2	1.063756845970D+01	5.277043018504D+00	6.50	6.446310742811D-04
	3	1.545203911264D+00	7.779883015927D-01	6.75	4.029948832014D-04
	4	8.417057895922D-02	4.182689804447D-02	7.25	-3.160334289045D-04

I= 1 H= -5.5

0		0.	3.609266562240D+01	-1.00	-1.484646355957D+01
1		5.137021677622D+00	-9.819646456925D+00	-.75	-8.821956692637D+00
2		-6.514621822729D+00	-1.665296032922D+00	-.50	-4.570978622133D+00
3		2.786177183019D+00	1.302308286981D+00	-.25	-1.736549563046D+00
4		-4.086428762044D-01	-2.023835790001D-01	.25	9.190293129856D-01
				.50	1.262587479834D+00
0	-6.5	6.583829185825D-05	3.291914592913D-05	.75	1.234417765416D+00
1	-5.5	-3.967146457498D-04	-1.983573228749D-04	1.25	6.867600696016D-01
2	-4.5	1.043202699802D-03	5.216013499010D-04	1.50	3.847571174803D-01
3	-3.5	-1.599542649224D-03	-7.997713246118D-04	1.75	1.484088667200D-01
4	-2.5	1.644336392972D-03	8.221681964862D-04	2.25	-6.439856508633D-02
5	-1.5	-1.277250947786D-03	-6.386254738930D-04	2.50	-6.652191931988D-02
6	-.5	8.474258595259D-04	4.237129297630D-04	2.75	-3.588471958544D-02
7	.5	-5.256771084332D-04	-2.628385542166D-04	3.25	2.287347091712D-02
8	1.5	3.142882459030D-04	1.571441229515D-04	3.50	2.667549447134D-02
9	2.5	-1.785724861762D-04	-8.928624308809D-05	3.75	1.579596661029D-02
10	3.5	9.098367023890D-05	4.549183511945D-05	4.25	-1.145372213049D-02
11	4.5	-3.721800474826D-05	-1.860900237413D-05	4.50	-1.396126245116D-02
12	5.5	1.027014267600D-05	5.135071338000D-06	4.75	-8.558153249174D-03
13	6.5	-1.369460859644D-06	-6.847304298219D-07	5.25	6.511639725538D-03
				5.50	8.069611108947D-03
	0	-2.061512964236D+02	-1.017147179152D+02	5.75	5.011447683111D-03
	1	-1.752993035806D+02	-8.871928104166D+01	6.25	-3.884936625725D-03
	2	-5.577513587166D+01	-2.757452171331D+01	6.50	-4.847849214558D-03
	3	-7.838537598294D+00	-3.959664767021D+00	6.75	-3.028658562880D-03
	4	-4.086428762044D-01	-2.023835790001D-01	7.25	2.373010540514D-03

I= 2 H= -4.5

0		0.	-1.308085391108D+02	-1.00	2.176504685230D+01
1		-5.753678141557D+00	4.202087666801D+01	-.75	1.239346777889D+01
2		1.004165259693D+01	-7.490304852038D-01	-.50	6.080907090643D+00
3		-5.128761816297D+00	-2.235370299053D+00	-.25	2.149444703802D+00
4		8.409542975223D-01	4.134538555559D-01	.25	-8.876681743731D-01
				.50	-9.549618310366D-01
0	-6.5	-1.669365962514D-04	-8.346829812568D-05	.75	-5.644547503277D-01

CARDINAL SPLINES N=5

```
 1   -5.5   1.043202699802D-03   5.216013499010D-04    1.25   5.327388906487D-01
 2   -4.5  -2.874765260961D-03  -1.437382630480D-03    1.50   9.045455659001D-01
 3   -3.5   4.675253777553D-03   2.337626888776D-03    1.75   1.058278632220D+00
 4   -2.5  -5.147415876133D-03  -2.573707938067D-03    2.25   7.840849561705D-01
 5   -1.5   4.274730430294D-03   2.137365215147D-03    2.50   4.922932684974D-01
 6    -.5  -2.972798371443D-03  -1.486399185722D-03    2.75   2.093240263848D-01
 7     .5   1.884162297757D-03   9.420811488785D-04    3.25  -1.057827643766D-01
 8    1.5  -1.134799614470D-03  -5.673998072349D-04    3.50  -1.158062523821D-01
 9    2.5   6.463496274644D-04   3.231748137322D-04    3.75  -6.554215131411D-02
10    3.5  -3.295884973927D-04  -1.647942486963D-04    4.25   4.485762907149D-02
11    4.5   1.348603822555D-04   6.743019112775D-05    4.50   5.367392363126D-02
12    5.5  -3.721800474826D-05  -1.860900237413D-05    4.75   3.244285306958D-02
13    6.5   4.963006273748D-06   2.481503136874D-06    5.25  -2.422468337095D-02
                                                       5.50  -2.983470827361D-02
        0   4.795306812206D+02   2.348330258872D+02    5.75  -1.844040638094D-02
        1   3.985055412312D+02   2.031293552835D+02    6.25   1.420426494925D-02
        2   1.232127116011D+02   6.047180106668D+01    6.50   1.768762898272D-02
        3   1.673604991928D+01   8.514429945400D+00    6.75   1.103249945945D-02
        4   8.409542975223D-01   4.134538555559D-01    7.25  -8.625613298407D-03


I= 3     H= -3.5
0           0.                   3.198572423113D+02   -1.00  -2.127232330229D+01
1           5.129377039890D+00  -1.072209204627D+02    -.75  -1.190921540081D+01
2          -9.645400359666D+00   9.286007054880D+00    -.50  -5.726301620069D+00
3           5.506662260580D+00   1.948818154421D+00    -.25  -1.975094019500D+00
4          -9.908836421557D-01  -4.782680820708D-01     .25   7.616776970214D-01
                                                        .50   7.797414628990D-01
 0   -6.5   2.447013503471D-04   1.223506751735D-04     .75   4.311148147116D-01
 1   -5.5  -1.599542649224D-03  -7.997713246118D-04    1.25  -3.213433903686D-01
 2   -4.5   4.675253777553D-03   2.337626888776D-03    1.50  -4.300445405217D-01
 3   -3.5  -8.197445025679D-03  -4.098722512839D-03    1.75  -3.062170730351D-01
 4   -2.5   9.872657984080D-03   4.936328992040D-03    2.25   3.900444965246D-01
 5   -1.5  -8.996875478774D-03  -4.498437739387D-03    2.50   7.468748121989D-01
 6    -.5   6.749090121357D-03   3.374545060678D-03    2.75   9.689605459429D-01
 7     .5  -4.473974775528D-03  -2.236987387764D-03    3.25   8.447724138658D-01
 8    1.5   2.746924743121D-03   1.373462371560D-03    3.50   5.645717968487D-01
 9    2.5  -1.574924142362D-03  -7.874620711812D-04    3.75   2.528230977541D-01
10    3.5   8.048729651382D-04   4.024364825691D-04    4.25  -1.380372698719D-01
11    4.5  -3.295884973927D-04  -1.647942486963D-04    4.50  -1.554263221161D-01
12    5.5   9.098367023890D-05   4.549183511945D-05    4.75  -8.997830180640D-02
13    6.5  -1.213404287491D-05  -6.067021437455D-06    5.25   6.361976147615D-02
                                                       5.50   7.700781525463D-02
        0  -6.306993226002D+02  -3.032891752213D+02    5.75   4.698072404749D-02
        1  -5.107770670152D+02  -2.648676158311D+02    6.25  -3.556632272448D-02
        2  -1.534544895648D+02  -7.395299773884D+01    6.50  -4.403737124745D-02
        3  -2.025631243547D+01  -1.048615197942D+01    6.75  -2.734933373015D-02
        4  -9.908836421557D-01  -4.782680820708D-01    7.25   2.125923132717D-02


I= 4     H= -2.5
0           0.                  -6.282394190483D+02   -1.00   1.579513893910D+01
1          -3.686373241342D+00   2.137909558671D+02    -.75   8.787079887112D+00
2           7.098407885645D+00  -2.416243955630D+01    -.50   4.194128248045D+00
3          -4.211076683295D+00  -5.253425417847D-01    -.25   1.434164068274D+00
4           7.992811288185D-01   3.659083122388D-01     .25  -5.406186996618D-01
                                                        .50  -5.450146310928D-01
```

CARDINAL SPLINES N=5

0	-6,5	-2,390898262172D-04	-1.195449131086D-04	.75	-2,955938784158D-01
1	-5,5	1,644336392972D-03	8.221681964862D-04	1.25	2,081252142388D-01
2	-4,5	-5,147415876133D-03	-2.573707938067D-03	1.50	2,666465846928D-01
3	-3,5	9,872657984080D-03	4.936328992040D-03	1.75	1,786833955759D-01
4	-2,5	-1,326780094232D-02	-6.633900471162D-03	2.25	-1,818417712947D-01
5	-1,5	1,361840794534D-02	6.809203972668D-03	2,50	-2,758894450564D-01
6	-,5	-1.138730286625D-02	-5.693651433123D-03	2,75	-2,188817528168D-01
7	,5	8.170225601545D-03	4.085112800772D-03	3.25	3,306868948752D-01
8	1.5	-5.242738177951D-03	-2.621369088976D-03	3.50	6,761653644402D-01
9	2.5	3.062043435650D-03	1.531021717825D-03	3.75	9.263944305821D-01
10	3.5	-1,574924142362D-03	-7,874620711812D-04	4.25	8,763596630780D-01
11	4.5	6.463496274644D-04	3.231748137322D-04	4.50	6,033938442330D-01
12	5.5	-1.785724861762D-04	-8.928624308809D-05	4.75	2,767840995044D-01
13	6,5	2.382333036361D-05	1.191166518181D-05	5.25	-1,564727213757D-01
				5.50	-1,783879247672D-01
	0	5,462461429614D+02	2.494357339125D+02	5.75	-1,043112704263D-01
	1	4,328492796715D+02	2.350423554583D+02	6.25	7,486282156925D-02
	2	1.276001787169D+02	5.835113803143D+01	6.50	9,114171880624D-02
	3	1,657023266599D+01	8.988273576424D+00	6.75	5,589008463164D-02
	4	7,992811288185D-01	3.659083122388D-01	7.25	-4,272057331089D-02

I= 5 H= -1.5

0		0.	1.108122799802D+03	-1.00	-1,022430023589D+01
1		2,363767910631D+00	-3.791734674486D+02	-.75	-5,677580280246D+00
2		-4,582699517896D+00	4.658998885418D+01	-.50	-2,704191729991D+00
3		2,748293515870D+00	-1.413257939334D+00	-,25	-9,223712965691D-01
4		-5,295392914923D-01	-2.052661737833D-01	.25	3,453968317940D-01
				.50	3,466499060579D-01
0	-6,5	1,773828870796D-04	8.869144353979D-05	.75	1,869575584542D-01
1	-5,5	-1,277250947786D-03	-6,386254738930D-04	1.25	-1,294962704852D-01
2	-4,5	4,274730430294D-03	2.137365215147D-03	1,50	-1,639073969930D-01
3	-3,5	-8,996875478774D-03	-4.498437739387D-03	1.75	-1,080773879223D-01
4	-2.5	1,361840794534D-02	6.809203972668D-03	2,25	1,044659450830D-01
5	-1.5	-1,601540746362D-02	-8.007703731809D-03	2,50	1,520739474781D-01
6	-,5	1,533656969157D-02	7,668284845785D-03	2.75	1,137553025682D-01
7	,5	-1,234885110775D-02	-6.174425553873D-03	3,25	-1,376780274348D-01
8	1,5	8,589590073560D-03	4,294795036780D-03	3,50	-2,232495809028D-01
9	2,5	-5,242738177951D-03	-2,621369088976D-03	3,75	-1,871702016662D-01
10	3,5	2,746924743121D-03	1.373462371560D-03	4,25	3,071047169957D-01
11	4,5	-1,134799614470D-03	-5,673998072349D-04	4,50	6,471380251737D-01
12	5,5	3,142882459030D-04	1,571441229515D-04	4,75	9,084441600961D-01
13	6,5	-4,197122651644D-05	-2,098561325822D-05	5,25	8,902478122267D-01
				5,50	6,207604928721D-01
	0	-3,687651879356D+02	-1.426066204536D+02	5,75	2,876792533522D-01
	1	-2,905640343898D+02	-1,781189480559D+02	6,25	-1,651426369318D-01
	2	-8,522918635172D+01	-3,300351601690D+01	6,50	-1,893967331028D-01
	3	-1,101972806293D+01	-6,750178457699D+00	6,75	-1,113377347889D-01
	4	-5,295392914923D-01	-2,052661737833D-01	7,25	8,070176865741D-02

I= 6 H= -.5

0		0.	-1,865711116061D+03	-1.00	6,323545204253D+00
1		-1.457621509999D+00	6,397047288952D+02	-,75	3,509480236212D+00
2		2,831956482148D+00	-8,089431762158D+01	-,50	1,670428689581D+00
3		-1,704093812610D+00	3.841878360447D+00	-,25	5,693176914228D-01
4		3,298733994955D-01	6,472876901498D-02	.25	-2,127459956578D-01

CARDINAL SPLINES N=5

				.50	-2.132164973184D-01
0	-6.5	-1.145590341764D-04	-5.727951708820D-05	.75	-1.147897848060D-01
1	-5.5	8.474258595259D-04	4.237129297630D-04	1.25	7.909833925541D-02
2	-4.5	-2.972798371443D-03	-1.486399185722D-03	1.50	9.973490822861D-02
3	-3.5	6.749090121357D-03	3.374545060678D-03	1.75	6.543177358607D-02
4	-2.5	-1.138730286625D-02	-5.693651433123D-03	2.25	-6.225992391805D-02
5	-1.5	1.533656969157D-02	7.668284845785D-03	2.50	-8.954785962821D-02
6	-.5	-1.702581896334D-02	-8.512909481672D-03	2.75	-6.590594091863D-02
7	.5	1.579109483024D-02	7.895547415118D-03	3.25	7.571804490295D-02
8	1.5	-1.234885110775D-02	-6.174425553873D-03	3.50	1.177549080167D-01
9	2.5	8.170225601545D-03	4.085112800772D-03	3.75	9.303941481078D-02
10	3.5	-4.473974775528D-03	-2.236987387764D-03	4.25	-1.221837221757D-01
11	4.5	1.884162297757D-03	9.420811488785D-04	4.50	-2.040994357767D-01
12	5.5	-5.256771084332D-04	-2.628385542166D-04	4.75	-1.752663754122D-01
13	6.5	7.041382492791D-05	3.520691246396D-05	5.25	2.977573584401D-01
				5.50	6.353279878377D-01
	0	2.310334934545D+02	4.520544522381D+01	5.75	9.009388683432D-01
	1	1.817298513553D+02	1.461412347643D+02	6.25	8.964365083216D-01
	2	5.322503390835D+01	1.043105335244D+01	6.50	6.288079747016D-01
	3	6.872614574272D+00	5.524826354836D+00	6.75	2.929643075000D-01
	4	3.298733994955D-01	6.472876901498D-02	7.25	-1.698663500771D-01

I= 7 H= .5

0		0.	3.068635811269D+03	-1.00	-3.832970198405D+00
1		8.826756295882D-01	-1.053121221584D+03	-.75	-2.126851820551D+00
2		-1.716104444650D+00	1.345739102170D+02	-.50	-1.012111700136D+00
3		1.033774262702D+00	-7.207774349226D+00	-.25	-3.448610325012D-01
4		-2.004158614655D-01	6.472876901498D-02	.25	1.287822282709D-01
				.50	1.290076326549D-01
0	-6.5	7.041382492791D-05	3.520691246396D-05	.75	6.941394498630D-02
1	-5.5	-5.256771084332D-04	-2.628385542166D-04	1.25	-4.775021584855D-02
2	-4.5	1.884162297757D-03	9.420811488785D-04	1.50	-6.013280086579D-02
3	-3.5	-4.473974775528D-03	-2.236987387764D-03	1.75	-3.938533830814D-02
4	-2.5	8.170225601545D-03	4.085112800772D-03	2.25	3.728447973789D-02
5	-1.5	-1.234885110775D-02	-6.174425553873D-03	2.50	5.341687255618D-02
6	-.5	1.579109483024D-02	7.895547415118D-03	2.75	3.910996769441D-02
7	.5	-1.702581896334D-02	-8.512909481672D-03	3.25	-4.421127753025D-02
8	1.5	1.533656969157D-02	7.668284845785D-03	3.50	-6.791209623918D-02
9	2.5	-1.138730286625D-02	-5.693651433123D-03	3.75	-5.277743665189D-02
10	3.5	6.749090121357D-03	3.374545060678D-03	4.25	6.575098859902D-02
11	4.5	-2.972798371443D-03	-1.486399185722D-03	4.50	1.053071845837D-01
12	5.5	8.474258595259D-04	4.237129297630D-04	4.75	8.520075836924D-02
13	6.5	-1.145590341764D-04	-5.727951708820D-05	5.25	-1.158040132525D-01
				5.50	-1.958432708350D-01
	0	-1.406226030069D+02	4.520544522381D+01	5.75	-1.698663500771D-01
	1	-1.105526181732D+02	-1.461412347643D+02	6.25	2.929643075000D-01
	2	-3.236292720347D+01	1.043105335244D+01	6.50	6.288079747016D-01
	3	-4.177038135401D+00	-5.524826354836D+00	6.75	8.964365083216D-01
	4	-2.004158614655D-01	6.472876901498D-02	7.25	9.009388683432D-01

I= 8 H= 1.5

0		0.	-4.914959042815D+03	-1.00	2.274275876546D+00
1		-5.235652630590D-01	1.688234667690D+03	-.75	1.261881095924D+00
2		1.018151979960D+00	-2.166669709961D+02	-.50	6.004525217151D-01
3		-6.135516896007D-01	1.208709897606D+01	-.25	2.045774305370D-01

CARDINAL SPLINES N=5

```
4         1.190069439258D-01  -2.052661737833D-01      .25 -7.637869145263D-02
                                                       .50 -7.650074571917D-02
0  -6.5  -4.197122651644D-05  -2.098561325822D-05      .75 -4.115418816166D-02
1  -5.5   3.142882459030D-04   1.571441229515D-04     1.25  2.829433399986D-02
2  -4.5  -1.134799614470D-03  -5.673998072349D-04     1.50  3.561686044421D-02
3  -3.5   2.746924743121D-03   1.373462371560D-03     1.75  2.331529417305D-02
4  -2.5  -5.242738177951D-03  -2.621369088976D-03     2.25 -2.203415800434D-02
5  -1.5   8.589590073560D-03   4.294795036780D-03     2.50 -3.152712925342D-02
6   -.5  -1.234885110775D-02  -6.174425553873D-03     2.75 -2.304328824862D-02
7    .5   1.533656969157D-02   7.668284845785D-03     3.25  2.591015495533D-02
8   1.5  -1.601540746362D-02  -8.007703731809D-03     3.50  3.963950925798D-02
9   2.5   1.361840794534D-02   6.809203972668D-03     3.75  3.064112933469D-02
10  3.5  -8.996875478774D-03  -4.498437739387D-03     4.25 -3.754877021471D-02
11  4.5   4.274730430294D-03   2.137365215147D-03     4.50 -5.939089421148D-02
12  5.5  -1.277250947786D-03  -6.386254738930D-04     4.75 -4.725589313607D-02
13  6.5   1.773828870796D-04   8.869144353979D-05     5.25  6.091622451166D-02
                                                      5.50  9.876372880979D-02
      0   8.355194702833D+01  -1.426066204536D+02     5.75  8.070176865741D-02
      1   6.567386172198D+01   1.781189480559D+02     6.25 -1.113377347889D-01
      2   1.922215431793D+01  -3.300351601690D+01     6.50 -1.893967331028D-01
      3   2.480628852469D+00   6.750178457699D+00     6.75 -1.651426369318D-01
      4   1.190069439258D-01  -2.052661737833D-01     7.25  2.876792533522D-01


I= 9     H=  2.5
0         0.                   7.364120463761D+03    -1.00 -1.288949206587D+00
1         2.967000656269D-01  -2.534821106673D+03     -.75 -7.151584788638D-01
2        -5.770220106180D-01   3.263802299242D+02     -.50 -3.402923952395D-01
3         3.477626260016D-01  -1.850188969463D+01     -.25 -1.159362163217D-01
4        -6.746450434091D-02   3.659083122388D-01      .25  4.328139864519D-02
                                                       .50  4.334837341781D-02
0  -6.5   2.382333036361D-05   1.191166518181D-05      .75  2.331812403268D-02
1  -5.5  -1.785724861762D-04  -8.928624308809D-05     1.25 -1.602869623687D-02
2  -4.5   6.463496274644D-04   3.231748137322D-04     1.50 -2.017411584428D-02
3  -3.5  -1.574924142362D-03  -7.874620711812D-04     1.75 -1.320385131316D-02
4  -2.5   3.062043435650D-03   1.531021717825D-03     2.25  1.247126401514D-02
5  -1.5  -5.242738177951D-03  -2.621369088976D-03     2.50  1.783655749998D-02
6   -.5   8.170225601545D-03   4.085112800772D-03     2.75  1.302930805420D-02
7    .5  -1.138730286625D-02  -5.693651433123D-03     3.25 -1.462422477204D-02
8   1.5   1.361840794534D-02   6.809203972668D-03     3.50 -2.234319506255D-02
9   2.5  -1.326780094232D-02  -6.633900471162D-03     3.75 -1.724037846920D-02
10  3.5   9.872657984080D-03   4.936328992040D-03     4.25  2.101165166169D-02
11  4.5  -5.147415876133D-03  -2.573707938067D-03     4.50  3.309770638072D-02
12  5.5   1.644336392972D-03   8.221681964862D-04     4.75  2.619258234339D-02
13  6.5  -2.390898262172D-04  -1.195449131086D-04     5.25 -3.320701774774D-02
                                                      5.50 -5.316570246666D-02
      0  -4.737467513640D+01   2.494357339125D+02     5.75 -4.272057331089D-02
      1  -3.723543124520D+01  -2.350423554583D+02     6.25  5.589008463164D-02
      2  -1.089790265401D+01   5.835113803143D+01     6.50  9.114171880624D-02
      3  -1.406314486862D+00  -8.988273576424D+00     6.75  7.486282156925D-02
      4  -6.746450434091D-02   3.659083122388D-01     7.25 -1.043112704263D-01


I=10     H=  3.5
0         0.                  -8.882940738189D+03    -1.00  6.561729379668D-01
1        -1.510374969647D-01   3.080753837982D+03     -.75  3.640674501044D-01
2         2.937450579907D-01  -3.996739201424D+02     -.50  1.732320934806D-01
```

3		-1.770429049972D-01	2.292112211326D+01	-.25	5.901890559218D-02
4		3.434747801413D-02	-4.782680820708D-01	.25	-2.203243371764D-02
				.50	-2.206615343277D-02
0	-6.5	-1.213404287491D-05	-6.067021437455D-06	.75	-1.186965751488D-02
1	-5.5	9.098367023890D-05	4.549183511945D-05	1.25	8.158600259966D-03
2	-4.5	-3.295884973927D-04	-1.647942486963D-04	1.50	1.026814244448D-02
3	-3.5	8.048729651382D-04	4.024364825691D-04	1.75	6.720033475762D-03
4	-2.5	-1.574924142362D-03	-7.874620711812D-04	2.25	-6.345981404515D-03
5	-1.5	2.746924743121D-03	1.373462371560D-03	2.50	-9.074786595613D-03
6	-.5	-4.473974775528D-03	-2.236987387764D-03	2.75	-6.627699965058D-03
7	.5	6.749090121357D-03	3.374545060678D-03	3.25	7.434539947064D-03
8	1.5	-8.996875478774D-03	-4.498437739387D-03	3.50	1.135349093476D-02
9	2.5	9.872657984080D-03	4.936328992040D-03	3.75	8.755280294509D-03
10	3.5	-8.197445025679D-03	-4.098722512839D-03	4.25	-1.065080886357D-02
11	4.5	4.675253777553D-03	2.337626888776D-03	4.50	-1.675400295745D-02
12	5.5	-1.599542649224D-03	-7.997713246118D-04	4.75	-1.323446318661D-02
13	6.5	2.447013503471D-04	1.223506751735D-04	5.25	1.668544860407D-02
				5.50	2.660279672539D-02
	0	2.412097215757D+01	-3.032891752213D+02	5.75	2.125923132717D-02
	1	1.895816464704D+01	2.648676158311D+02	6.25	-2.734933373015D-02
	2	5.548494087127D+00	-7.395299773884D+01	6.50	-4.403737124745D-02
	3	7.159915233701D-01	1.048615197942D+01	6.75	-3.556632272448D-02
	4	3.434747801413D-02	-4.782680820708D-01	7.25	4.698072404749D-02

I=11 H= 4.5

0		0.	7.186423727086D+03	-1.00	-2.683375662075D-01
1		6.176502644307D-02	-2.522645825058D+03	-.75	-1.488826170063D-01
2		-1.201246781964D-01	3.313137373854D+02	-.50	-7.084175381598D-02
3		7.240127515757D-02	-1.926423018985D+01	-.25	-2.413518840054D-02
4		-1.404658641054D-02	4.134538555559D-01	.25	9.009864688598D-03
				.50	9.023601109851D-03
0	-6.5	4.963006273748D-06	2.481503136874D-06	.75	4.853871218251D-03
1	-5.5	-3.721800474826D-05	-1.860900237413D-05	1.25	-3.336232803454D-03
2	-4.5	1.348603822555D-04	6.743019112775D-05	1.50	-4.198804381445D-03
3	-3.5	-3.295884973927D-04	-1.647942486963D-04	1.75	-2.747868956820D-03
4	-2.5	6.463496274644D-04	3.231748137322D-04	2.25	2.594746164207D-03
5	-1.5	-1.134799614470D-03	-5.673998072349D-04	2.50	3.710314784898D-03
6	-.5	1.884162297757D-03	9.420811488785D-04	2.75	2.709618712800D-03
7	.5	-2.972798371443D-03	-1.486399185722D-03	3.25	-3.038851490076D-03
8	1.5	4.274730430294D-03	2.137365215147D-03	3.50	-4.639985539594D-03
9	2.5	-5.147415876133D-03	-2.573707938067D-03	3.75	-3.577395916880D-03
10	3.5	4.675253777553D-03	2.337626888776D-03	4.25	4.349128855395D-03
11	4.5	-2.874765260961D-03	-1.437382630480D-03	4.50	6.838011073415D-03
12	5.5	1.043202699802D-03	5.216013499010D-04	4.75	5.398122878127D-03
13	6.5	-1.669365962514D-04	-8.346829812568D-05	5.25	-6.792583329174D-03
				5.50	-1.081423830026D-02
	0	-9.864629446229D+00	2.348330258872D+02	5.75	-8.625613298407D-03
	1	-7.753169335863D+00	-2.031293552835D+02	6.25	1.103249945945D-02
	2	-2.269109467695D+00	6.047180106668D+01	6.50	1.768762898272D-02
	3	-2.928099715164D-01	-8.514429945400D+00	6.75	1.420426494925D-02
	4	-1.404658641054D-02	4.134538555559D-01	7.25	-1.844040638094D-02

I=12 H= 5.5

0		0.	-3.292103861206D+03	-1.00	7.403859235062D-02
1		-1.704187388385D-02	1.171393934066D+03	-.75	4.107904080411D-02

2		3.314426270153D-02	-1.560922219467D+02	-.50	1,954632720020D-02
3		-1.997673756103D-02	9.221637821023D+00	-.25	6,659260938434D-03
4		3,875718204206D-03	-2.023835790001D-01	.25	-2,485949057496D-03
				.50	-2,489733748634D-03
0	-6,5	-1,369460859644D-06	-6,847304298219D-07	.75	-1,339245164322D-03
1	-5,5	1,027014267600D-05	5,135071338000D-06	1.25	9,205019750996D-04
2	-4,5	-3,721800474826D-05	-1,860900237413D-05	1.50	1,158487775879D-03
3	-3,5	9,098367023890D-05	4,549183511945D-05	1.75	7,581557650275D-04
4	-2,5	-1,785724861762D-04	-8,928624308809D-05	2.25	-7,158907870066D-04
5	-1,5	3,142882459030D-04	1,571441229515D-04	2.50	-1,023657366755D-03
6	-,5	-5,256771084332D-04	-2,628385542166D-04	2.75	-7,475518640750D-04
7	,5	8,474258595259D-04	4,237129297630D-04	3.25	8,383191238929D-04
8	1,5	-1,277250947786D-03	-6,386254738930D-04	3.50	1,279945055083D-03
9	2,5	1,644336392972D-03	8,221681964862D-04	3.75	9,867527348431D-04
10	3,5	-1,599542649224D-03	-7,997713246118D-04	4.25	-1,199336534312D-03
11	4,5	1,043202699802D-03	5,216013499010D-04	4.50	-1,885347696390D-03
12	5,5	-3,967146457498D-04	-1,983573228749D-04	4.75	-1,487999265944D-03
13	6,5	6,583829185825D-05	3,291914592913D-05	5,25	1,871044109908D-03
				5,50	2,977227578992D-03
	0	2,721860593091D+00	-1,017147179152D+02	5.75	2,373010540514D-03
	1	2.139258502695D+00	8.871928104166D+01	6.25	-3,028658562880D-03
	2	6,260924450275D-01	-2,757452171331D+01	6.50	-4,847849214558D-03
	3	8,079193574831D-02	3,959664767021D+00	6.75	-3,884936625725D-03
	4	3,875718204206D-03	-2,023835790001D-01	7.25	5,011447683111D-03

I=13 H= 6.5

0		0.	6,368225590320D+02	-1.00	-9,872158804962D-03
1		2.272325393853D-03	-2.296065037259D+02	-.75	-5,477395736680D-03
2		-4,419387835223D-03	3,105093355384D+01	-.50	-2,606266423326D-03
3		2.663662705606D-03	-1.865487650749D+00	-.25	-8,879315010270D-04
4		-5,167828702788D-04	4.182689804447D-02	.25	3,314706561466D-04
				.50	3,319750035817D-04
0	-6.5	1,826060422966D-07	9,130302114832D-08	.75	1,785714728729D-04
1	-5,5	-1,369460859644D-06	-6,847304298219D-07	1.25	-1,227369419122D-04
2	-4.5	4,963006273748D-06	2,481503136874D-06	1.50	-1,544688728574D-04
3	-3,5	-1,213404287491D-05	-6,067021437455D-06	1.75	-1,010896219910D-04
4	-2,5	2,382333036361D-05	1,191166518181D-05	2.25	9,545321999728D-05
5	-1,5	-4,197122651644D-05	-2,098561325822D-05	2.50	1,364882069516D-04
6	-,5	7,041382492791D-05	3,520691246396D-05	2.75	9,967298032224D-05
7	,5	-1,145590341764D-04	-5,727951708820D-05	3.25	-1,117716965279D-04
8	1,5	1,773828870796D-04	8,869144353979D-05	3.50	-1,706488845900D-04
9	2,5	-2,390898262172D-04	-1,195449131086D-04	3.75	-1,315548212309D-04
10	3,5	2,447013503471D-04	1,223506751735D-04	4.25	1,598811980043D-04
11	4,5	-1,669365962514D-04	-8,346829812568D-05	4.50	2,513136692359D-04
12	5,5	6,583829185825D-05	3,291914592913D-05	4.75	1,983287165457D-04
13	6,5	-1,124510999650D-05	-5,622554998250D-06	5,25	-2,493096937113D-04
				5,50	-3,966170750591D-04
	0	-3,629303728281D-01	1,863630856672D+01	5.75	-3,160334289045D-04
	1	-2,852464515297D-01	-1,644807975662D+01	6.25	4,029948832014D-04
	2	-8,348242269158D-02	5,277043018504D+00	6.50	6,446310742811D-04
	3	-1,077269192164D-02	-7,779883015927D-01	6.75	5,161420495554D-04
	4	-5,167828702788D-04	4,182689804447D-02	7.25	-6,640570285641D-04

CARDINAL SPLINES N=5

N= 5 M=14 P= 7.0 Q= 9

I= 0 H= -7.0

0		1.000000000000D+00	4.566786982923D+00	-1.00	5.855037078171D+00
1		-2.284275786902D+00	-2.427238536137D+00	-.75	4.048100072248D+00
2		1.843360481465D+00	1.073760888714D+00	-.50	2.688644759998D+00
3		-6.432471290804D-01	-3.296112487735D-01	-.25	1.696658516649D+00
4		8.417368072313D-02	4.224393689807D-02	.25	5.344191503716D-01
				.50	2.435571688603D-01
0	-7.0	-1.124620604634D-05	-5.623103023169D-06	.75	7.894578231542D-02
1	-6.0	6.584651194351D-05	3.292325597176D-05	1.25	-2.601767063357D-02
2	-5.0	-1.669663882504D-04	-8.348319412521D-05	1.50	-2.411461237584D-02
3	-4.0	2.447742015092D-04	1.223871007546D-04	1.75	-1.186293562870D-02
4	-3.0	-2.392329309419D-04	-1.196164654709D-04	2.25	6.554376687469D-03
5	-2.0	1.776353870396D-04	8.881769351982D-05	2.50	7.241716830868D-03
6	-1.0	-1.149845871324D-04	-5.749229356621D-05	2.75	4.101912858644D-03
7	0.0	7.111589231035D-05	3.555794615518D-05	3.25	-2.786250504379D-03
8	1.0	-4.310637262073D-05	-2.155318631037D-05	3.50	-3.317725924629D-03
9	2.0	2.557697511385D-05	1.278848755692D-05	3.75	-1.996163839602D-03
10	3.0	-1.449578836054D-05	-7.247894180272D-06	4.25	1.479011047083D-03
11	4.0	7.379456062405D-06	3.689728031202D-06	4.50	1.815748753423D-03
12	5.0	-3.017779559543D-06	-1.508889779771D-06	4.75	1.119091414919D-03
13	6.0	8.326532649558D-07	4.163266324779D-07	5.25	-8.574655199818D-04
14	7.0	-1.110243320428D-07	-5.551216602138D-08	5.50	-1.064821238256D-03
				5.75	-6.621636301280D-04
	0	5.680197522513D+01	2.856143493991D+01	6.25	5.137308926334D-04
	1	4.445173293092D+01	2.211124176031D+01	6.50	6.407780574529D-04
	2	1.308223290338D+01	6.571642112504D+00	6.75	3.999139264226D-04
	3	1.713615931167D+00	8.532189843726D-01	7.25	-3.120854151146D-04
	4	8.417368072313D-02	4.224393689807D-02	7.50	-3.902912123172D-04

I= 1 H= -6.0

0		0.	-3.050004281105D+01	-1.00	-1.484690794095D+01
1		5.137123962875D+00	1.220654627322D+01	-.75	-8.822203249788D+00
2		-6.514820755115D+00	-4.397983345573D+00	-.50	-4.571095939423D+00
3		2.786297084346D+00	1.453054505003D+00	-.25	-1.736589531960D+00
4		-4.086661386172D-01	-2.055112606234D-01	.25	9.190442336141D-01
				.50	1.262602423145D+00
0	-7.0	6.584651194351D-05	3.292325597176D-05	.75	1.234425803505D+00
1	-6.0	-3.967762942255D-04	-1.983881471127D-04	1.25	6.867545448360D-01
2	-5.0	1.043426131953D-03	5.217130659764D-04	1.50	3.847501643887D-01
3	-4.0	-1.600089013763D-03	-8.000445068814D-04	1.75	1.484043164052D-01
4	-3.0	1.645409640738D-03	8.227048203689D-04	2.25	-6.439426854579D-02
5	-2.0	-1.279144631020D-03	-6.395723155100D-04	2.50	-6.651577578300D-02
6	-1.0	8.506173946761D-04	4.253086973381D-04	2.75	-3.588023322532D-02
7	0.0	-5.309424289936D-04	-2.654712144968D-04	3.25	2.286844022336D-02
8	1.0	3.228015428763D-04	1.614007714382D-04	3.50	2.666781407132D-02
9	2.0	-1.917243601843D-04	-9.586218009213D-05	3.75	1.579004600068D-02
10	3.0	1.086961396898D-04	5.434806984488D-05	4.25	-1.144652775127D-02

CARDINAL SPLINES N=5

11	4,0	-5.534074207175D-05	-2.767037103587D-05	4.50	-1.394995501639D-02
12	5,0	2.263209903193D-05	1.131604951596D-05	4.75	-8,549231082520D-03
13	6,0	-6.244643915140D-06	-3.122321957570D-06	5.25	6,500429086028D-03
14	7,0	8.326532649558D-07	4.163266324779D-07	5.50	8,051782498567D-03
				5,75	4.997247693247D-03
	0	-3,087738481499D+02	-1.555902443724D+02	6,25	-3,866853583508D-03
	1	-2.371746373927D+02	-1.177276579047D+02	6.50	-4.818952602591D-03
	2	-6.815042673732D+01	-3.430414936379D+01	6.75	-3,005552198052D-03
	3	-8.656354796937D+00	-4.301260792452D+00	7.25	2,343401476340D-03
	4	-4,086661386172D-01	-2.055112606234D-01	7.50	2,929785273609D-03

I= 2 H= -5.0

0		0.	1.105429607067D+02	-1.00	2.176665742071D+01
1		-5.754048853302D+00	-3.780850369129D+01	-.75	1.239436137430D+01
2		1.004237358622D+01	9.155029538124D+00	-.50	6,081332282896D+00
3		-5.129196373857D+00	-2,781718790761D+00	-.25	2,149589562866D+00
4		8,410386073269D-01	4,247894907701D-01	.25	-8,877222511053D-01
				.50	-9,550159899764D-01
0	-7,0	-1,669663882504D-04	-8,348319412521D-05	.75	-5,644838827179D-01
1	-6,0	1.043426131953D-03	5,217130659764D-04	1.25	5,327589140193D-01
2	-5,0	-2,875575044583D-03	-1.437787522292D-03	1.50	9,045707659426D-01
3	-4,0	4.677233962832D-03	2.338616981416D-03	1.75	1,058295123895D+00
4	-3,0	-5,151305640702D-03	-2,575652820351D-03	2.25	7,840693842479D-01
5	-2,0	4.281593693547D-03	2,140796846774D-03	2,50	4,922710025185D-01
6	-1,0	-2,984365430560D-03	-1.492182715280D-03	2,75	2,093077665004D-01
7	0.0	1.903245363218D-03	9.516226816091D-04	3.25	-1,057645316674D-01
8	1,0	-1,165654298732D-03	-5,828271493662D-04	3,50	-1.157784163605D-01
9	2,0	6,940158748949D-04	3.470079374475D-04	3.75	-6,552069328903D-02
10	3,0	-3,937836798383D-04	-1,968918399192D-04	4,25	4,483155453185D-02
11	4,0	2,005424957589D-04	1.002712478794D-04	4,50	5,363294217238D-02
12	5,0	-8,202135900984D-05	-4.101067950492D-05	4.75	3.241051652171D-02
13	6,0	2.263209903193D-05	1.131604951596D-05	5.25	-2.418405272649D-02
14	7,0	-3,017779559543D-06	-1.508889779771D-06	5.50	-2,977009216238D-02
				5.75	-1,838894145434D-02
	0	7.118173037106D+02	3.602699043439D+02	6.25	1.413872670245D-02
	1	5.347522836493D+02	2.642604289372D+02	6.50	1.758289918870D-02
	2	1.495946002893D+02	7.562704521857D+01	6.75	1.094875521871D-02
	3	1.841988463130D+01	9.112386950803D+00	7.25	-8,518301368698D-03
	4	8,410386073269D-01	4.247894907701D-01	7.50	-1,064229347585D-02

I= 3 H= -4.0

0		0.	-2.703259547572D+02	-1.00	-2,127626166774D+01
1		5.130283551112D+00	8.798797177977D+01	-.75	-1.191140053326D+01
2		-9.647163413826D+00	-1.493265250574D+01	-.50	-5.727341353916D+00
3		5.507724895660D+00	3.284818551326D+00	-.25	-1,975448247196D+00
4		-9,910898071475D-01	-5,059874102565D-01	.25	7,618099322831D-01
				.50	7,798738991848D-01
0	-7,0	2.447742015092D-04	1.223871007546D-04	.75	4,311860529147D-01
1	-6,0	-1,600089013763D-03	-8,000445068814D-04	1.25	-3,213923540457D-01
2	-5,0	4.677233962832D-03	2.338616981416D-03	1,50	-4,301061628514D-01
3	-4,0	-8,202287225044D-03	-4.101143612522D-03	1.75	-3,062574005632D-01
4	-3,0	9,882169728116D-03	4.941084864058D-03	2,25	3,900825749583D-01
5	-2,0	-9,013658397678D-03	-4,506829198839D-03	2.50	7,469292597851D-01
6	-1,0	6,777375356525D-03	3,388687678262D-03	2.75	9,690003066678D-01
7	0,0	-4,520639099522D-03	-2,260319549761D-03	3,25	8,447278289406D-01

8	1.0	2.822374517480D-03	1.411187258740D-03	3.50	5.645037286901D-01
9	2.0	-1.691483676163D-03	-8.457418380815D-04	3.75	2.527706258787D-01
10	3.0	9.618511402927D-04	4.809255701464D-04	4.25	-1.379735091114D-01
11	4.0	-4.902027043354D-04	-2.451013521677D-04	4.50	-1.553261090723D-01
12	5.0	2.005424957589D-04	1.002712478794D-04	4.75	-8.989922839184D-02
13	6.0	-5.534074207175D-05	-2.767037103587D-05	5.25	6.352040628813D-02
14	7.0	7.379456062405D-06	3.689728031202D-06	5.50	7.684980777122D-02
				5.75	4.685487550320D-02
	0	-9.272560101694D+02	-4.742931340012D+02	6.25	-3.540606031859D-02
	1	-6.800696599868D+02	-3.324155631276D+02	6.50	-4.378127271600D-02
	2	-1.853653439063D+02	-9.471176154330D+01	6.75	-2.714455172641D-02
	3	-2.224278970447D+01	-1.088282893586D+01	7.25	2.099681863113D-02
	4	-9.910898071475D-01	-5.059874102565D-01	7.50	2.618233550485D-02

I= 4 H= -3.0

0		0.	5.310832527458D+02	-1.00	1.580287524264D+01
1		-3.688153940989D+00	-1.696664106534D+02	-.75	8.791372238367D+00
2		7.101871130057D+00	2.341133287327D+01	-.50	4.196170642726D+00
3		-4.213164063867D+00	-3.149706565278D+00	-.25	1.434859893232D+00
4		7.996861077307D-01	4.203586011867D-01	.25	-5.408784551710D-01
				.50	-5.452747814825D-01
0	-7.0	-2.392329309419D-04	-1.196164654709D-04	.75	-2.957338147344D-01
1	-6.0	1.645409640738D-03	8.227048203689D-04	1.25	2.082213957316D-01
2	-5.0	-5.151305640702D-03	-2.575652820351D-03	1.50	2.667676321307D-01
3	-4.0	9.882169728116D-03	4.941084864058D-03	1.75	1.787626127037D-01
4	-3.0	-1.328648527723D-02	-6.643242638615D-03	2.25	-1.819165704278D-01
5	-2.0	1.365137536711D-02	6.825687683557D-03	2.50	-2.759963988318D-01
6	-1.0	-1.144286479104D-02	-5.721432395518D-03	2.75	-2.189598565482D-01
7	0.0	8.261890374514D-03	4.130945187257D-03	3.25	3.307744749943D-01
8	1.0	-5.390947470305D-03	-2.695473735153D-03	3.50	6.762990737018D-01
9	2.0	3.291006426837D-03	1.645503213419D-03	3.75	9.264975033826D-01
10	3.0	-1.883283238889D-03	-9.416416194447D-04	4.25	8.762344150270D-01
11	4.0	9.618511402927D-04	4.809255701464D-04	4.50	6.031969913672D-01
12	5.0	-3.937836798383D-04	-1.968918399192D-04	4.75	2.766287721371D-01
13	6.0	1.086961396898D-04	5.434806984488D-05	5.25	-1.562775536335D-01
14	7.0	-1.449578836054D-05	-7.247894180272D-06	5.50	-1.780775437554D-01
				5.75	-1.040640606255D-01
	0	7.971056785407D+02	4.195053385209D+02	6.25	7.454801111462D-02
	1	5.735722642978D+02	2.718173853046D+02	6.50	9.063865325779D-02
	2	1.537331414617D+02	8.085292375132D+01	6.75	5.548782238344D-02
	3	1.817804695259D+01	8.620334267949D+00	7.25	-4.220510457273D-02
	4	7.996861077307D-01	4.203586011867D-01	7.50	-5.233977209790D-02

I= 5 H= -2.0

0		0.	-9.374345985182D+02	-1.00	-1.023795049335D+01
1		2.366909851548D+00	2.974147440220D+02	-.75	-5.685153884282D+00
2		-4.588810211150D+00	-3.735116448212D+01	-.50	-2.707795416567D+00
3		2.751976577434D+00	3.217279679528D+00	-.25	-9.235990392203D-01
4		-5.302538532188D-01	-3.013405404009D-01	.25	3.458551552759D-01
				.50	3.471089262835D-01
0	-7.0	1.776353870396D-04	8.881769351982D-05	.75	1.872044679462D-01
1	-6.0	-1.279144631020D-03	-6.395723155100D-04	1.25	-1.296659771329D-01
2	-5.0	4.281593693547D-03	2.140796846774D-03	1.50	-1.641209781539D-01
3	-4.0	-9.013658397678D-03	-4.506829198839D-03	1.75	-1.082171619349D-01
4	-3.0	1.365137536711D-02	6.825687683557D-03	2.25	1.045979238010D-01

5	-2.0	-1.607357656112D-02	-8.036788280558D-03	2.50	1.522626611916D-01
6	-1.0	1.543460547495D-02	7.717302737473D-03	2.75	1.138931120576D-01
7	0.0	-1.251058826607D-02	-6.255294133033D-03	3.25	-1.378325574589D-01
8	1.0	8.851096746250D-03	4.425548373125D-03	3.50	-2.234855031150D-01
9	2.0	-5.646730053516D-03	-2.823365026758D-03	3.75	-1.873520676251D-01
10	3.0	3.291006426837D-03	1.645503213419D-03	4.25	3.073257098941D-01
11	4.0	-1.691483676163D-03	-8.457418380815D-04	4.50	6.474853606013D-01
12	5.0	6.940158748949D-04	3.470079374475D-04	4.75	9.087182261979D-01
13	6.0	-1.917243601843D-04	-9.586218009213D-05	5.25	8.899034501016D-01
14	7.0	2.557697511385D-05	1.278848755692D-05	5.50	6.202128436366D-01
				5.75	2.872430660427D-01
	0	-5.374948669041D+02	-3.057301574122D+02	6.25	-1.645871722021D-01
	1	-3.848441628380D+02	-1.660006672670D+02	6.50	-1.885091032138D-01
	2	-1.026919349314D+02	-5.838241008988D+01	6.75	-1.106279664579D-01
	3	-1.209513131269D+01	-5.220255451696D+00	7.25	7.979225406186D-02
	4	-5.302538532188D-01	-3.013405404009D-01	7.50	9.730642273150D-02

I= 6 H= -1.0

0		0.	1.581786449097D+03	-1.00	6.346550781803D+00
1		-1.462916806855D+00	-5.005889698174D+02	-.75	3.522244474212D+00
2		2.842255191014D+00	6.057661933725D+01	-.50	1.676502193411D+00
3		-1.710301091753D+00	-3.962237612788D+00	-.25	5.713868779459D-01
4		3.310776921809D-01	2.266485202086D-01	.25	-2.135184350375D-01
				.50	-2.139901109610D-01
0	-7.0	-1.149845871324D-04	-5.749229356621D-05	.75	-1.152059158191D-01
1	-6.0	8.506173946761D-04	4.253086973381D-04	1.25	7.938435579793D-02
2	-5.0	-2.984365430560D-03	-1.492182715280D-03	1.50	1.000948690553D-01
3	-4.0	6.777375356525D-03	3.388687678262D-03	1.75	6.566734290734D-02
4	-3.0	-1.144286479104D-02	-5.721432395518D-03	2.25	-6.248235537341D-02
5	-2.0	1.543460547495D-02	7.717302737473D-03	2.50	-8.986590989094D-02
6	-1.0	-1.719104440496D-02	-8.595522202480D-03	2.75	-6.613819931408D-02
7	0.0	1.606367992609D-02	8.031839963047D-03	3.25	7.597848339381D-02
8	1.0	-1.278958360350D-02	-6.394791801751D-03	3.50	1.181525215243D-01
9	2.0	8.851096746250D-03	4.425548373125D-03	3.75	9.334592415208D-02
10	3.0	-5.390947470305D-03	-2.695473735153D-03	4.25	-1.225561744373D-01
11	4.0	2.822374517480D-03	1.411187258740D-03	4.50	-2.046848204949D-01
12	5.0	-1.165654298732D-03	-5.828271493662D-04	4.75	-1.757282750486D-01
13	6.0	3.228015428763D-04	1.614007714382D-04	5.25	2.983377320775D-01
14	7.0	-4.310637262073D-05	-2.155318631037D-05	5.50	6.362509731362D-01
				5.75	9.016740003364D-01
	0	3.373143511667D+02	2.310536037348D+02	6.25	8.955003511134D-01
	1	2.411529890518D+02	7.599654155037D+01	6.50	6.273120001161D-01
	2	6.426277376538D+01	4.400429441003D+01	6.75	2.917680934045D-01
	3	7.559874289312D+00	2.383920953053D+00	7.25	-1.683334919152D-01
	4	3.310776921809D-01	2.266485202086D-01	7.50	-1.933871876370D-01

I= 7 H= 0.0

0		0.	-2.618965596094D+03	-1.00	-3.870924265893D+00
1		8.914116873465D-01	8.281088367484D+02	-.75	-2.147909963926D+00
2		-1.733095015327D+00	-9.882155560717D+01	-.50	-1.022131625685D+00
3		1.044014887654D+00	5.667274915837D+00	-.25	-3.482747283656D-01
4		-2.024026755656D-01	-2.024026755656D-01	.25	1.300565808181D-01
				.50	1.302839224736D-01
0	-7.0	7.111589231035D-05	3.555794615518D-05	.75	7.010046826615D-02
1	-6.0	-5.309424289936D-04	-2.654712144968D-04	1.25	-4.822207929705D-02

2	-5.0	1.903245363218D-03	9.516226816091D-04	1.50	-6.072665587677D-02
3	-4.0	-4.520639099522D-03	-2.260319549761D-03	1.75	-3.977397510276D-02
4	-3.0	8.261890374514D-03	4.130945187257D-03	2.25	3.765144198649D-02
5	-2.0	-1.251058826607D-02	-6.255294133033D-03	2.50	5.394158449221D-02
6	-1.0	1.606367992609D-02	8.031839963047D-03	2.75	3.949314220053D-02
7	0.0	-1.747552352311D-02	-8.737761761555D-03	3.25	-4.464094290324D-02
8	1.0	1.606367992609D-02	8.031839963047D-03	3.50	-6.856806976850D-02
9	2.0	-1.251058826607D-02	-6.255294133033D-03	3.75	-5.328310864381D-02
10	3.0	8.261890374514D-03	4.130945187257D-03	4.25	6.636545168913D-02
11	4.0	-4.520639099522D-03	-2.260319549761D-03	4.50	1.062729386949D-01
12	5.0	1.903245363218D-03	9.516226816091D-04	4.75	8.596278966085D-02
13	6.0	-5.309424289936D-04	-2.654712144968D-04	5.25	-1.167615001990D-01
14	7.0	7.111589231035D-05	3.555794615518D-05	5.50	-1.973659905422D-01
				5.75	-1.710791537644D-01
	0	-2.065534915073D+02	-2.065534915073D+02	6.25	2.945087579178D-01
	1	-1.475982009181D+02	-8.595999619562D-21	6.50	6.312759988255D-01
	2	-3.931516899088D+01	-3.931516899088D+01	6.75	8.984099945517D-01
	3	-4.623260028183D+00	-2.678758601526D-22	7.25	8.984099945517D-01
	4	-2.024026755656D-01	-2.024026755656D-01	7.50	6.312759988255D-01

I= 8 H= 1.0

0		0.	4.281107804597D+03	-1.00	2.335642367752D+00
1		-5.376902632785D-01	-1.353454813116D+03	-.75	1.295929208703D+00
2		1.045623388826D+00	1.607012993655D+02	-.50	6.166533590369D-01
3		-6.301093674111D-01	-8.730079518894D+00	-.25	2.100969058161D-01
4		1.222193482363D-01	2.266485202086D-01	.25	-7.843914371920D-02
				.50	-7.856433028653D-02
0	-7.0	-4.310637262073D-05	-2.155318631037D-05	.75	-4.226420159643D-02
1	-6.0	3.228015428763D-04	1.614007714382D-04	1.25	2.905727210056D-02
2	-5.0	-1.165654298732D-03	-5.828271493662D-04	1.50	3.657704207219D-02
3	-4.0	2.822374517480D-03	1.411187258740D-03	1.75	2.394366625590D-02
4	-3.0	-5.390947470305D-03	-2.695473735153D-03	2.25	-2.262748533636D-02
5	-2.0	8.851096746250D-03	4.425548373125D-03	2.50	-3.237551606779D-02
6	-1.0	-1.278958360350D-02	-6.394791801751D-03	2.75	-2.366282856450D-02
7	0.0	1.606367992609D-02	8.031839963047D-03	3.25	2.660486458932D-02
8	1.0	-1.719104440496D-02	-8.595522202480D-03	3.50	4.070012796060D-02
9	2.0	1.543460547495D-02	7.717302737473D-03	3.75	3.145873118608D-02
10	3.0	-1.144286479104D-02	-5.721432395518D-03	4.25	-3.854227226367D-02
11	4.0	6.777375356525D-03	3.388687678262D-03	4.50	-6.095238537952D-02
12	5.0	-2.984365430560D-03	-1.492182715280D-03	4.75	-4.848799260921D-02
13	6.0	8.506173946761D-04	4.253086973381D-04	5.25	6.246434881477D-02
14	7.0	-1.149845871324D-04	-5.749229356621D-05	5.50	1.012257565108D-01
				5.75	8.266270490849D-02
	0	1.247928563029D+02	2.310536037348D+02	6.25	-1.138348980511D-01
	1	8.915990595109D+01	-7.599654155037D+01	6.50	-1.933871876370D-01
	2	2.374581505467D+01	4.400429441003D+01	6.75	-1.683334919152D-01
	3	2.792032383206D+00	-2.383920953053D+00	7.25	2.917680934045D-01
	4	1.222193482363D-01	2.266485202086D-01	7.50	6.273120001161D-01

I= 9 H= 2.0

0		0.	-6.842539180120D+03	-1.00	-1.383752003106D+00
1		3.185212500439D-01	2.164171181355D+03	-.75	-7.677581345423D-01
2		-6.194615624575D-01	-2.566018934534D+02	-.50	-3.653204627379D-01
3		3.733419630214D-01	1.365779058292D+01	-.25	-1.244630471945D-01
4		-7.242722758293D-02	-3.013405404009D-01	.25	4.646451426942D-02

CARDINAL SPLINES N=5

				.50	4.653632801637D-02
0	-7.0	2.557697511385D-05	1.278848755692D-05	.75	2.503294226200D-02
1	-6.0	-1.917243601843D-04	-9.586218009213D-05	1.25	-1.720733076680D-02
2	-5.0	6.940158748949D-04	3.470079374475D-04	1.50	-2.165746452055D-02
3	-4.0	-1.691483676163D-03	-8.457418380815D-04	1.75	-1.417459983630D-02
4	-3.0	3.291006426837D-03	1.645503213419D-03	2.25	1.338787321376D-02
5	-2.0	-5.646730053516D-03	-2.823365026758D-03	2.50	1.914719855099D-02
6	-1.0	8.851096746250D-03	4.425548373125D-03	2.75	1.398641271094D-02
7	0.0	-1.251058826607D-02	-6.255294133033D-03	3.25	-1.569745569621D-02
8	1.0	1.543460547495D-02	7.717302737473D-03	3.50	-2.398170521804D-02
9	2.0	-1.607357656112D-02	-8.036788280558D-03	3.75	-1.850346098258D-02
10	3.0	1.365137536711D-02	6.825687683557D-03	4.25	2.254647580980D-02
11	4.0	-9.013658397678D-03	-4.506829198839D-03	4.50	3.550999567020D-02
12	5.0	4.281593693547D-03	2.140796846774D-03	4.75	2.809600672609D-02
13	6.0	-1.279144631020D-03	-6.395723155100D-04	5.25	-3.559865706765D-02
14	7.0	1.776353870396D-04	8.881769351982D-05	5.50	-5.696919695648D-02
				5.75	-4.574995036574D-02
	0	-7.396544792041D+01	-3.057301574122D+02	6.25	5.974785873550D-02
	1	-5.284282830401D+01	1.660006672670D+02	6.50	9.730642273150D-02
	2	-1.407288524839D+01	-5.838241008988D+01	6.75	7.979225406186D-02
	3	-1.654620409301D+00	5.220255451696D+00	7.25	-1.106279664579D-01
	4	-7.242722758293D-02	-3.013405404009D-01	7.50	-1.885091032138D-01
I=10	H= 3.0				
0		0.	1.025007595482D+04	-1.00	7.838499242392D-01
1		-1.804254809228D-01	-3.247679456040D+03	-.75	4.349067688022D-01
2		3.509011153711D-01	3.854653721271D+02	-.50	2.069389918822D-01
3		-2.114922333027D-01	-2.039037510118D+01	-.25	7.050253380019D-02
4		4.103109464271D-02	4.203586011867D-01	.25	-2.631933900721D-02
				.50	-2.635957567837D-02
0	-7.0	-1.449578836054D-05	-7.247894180272D-06	.75	-1.417911284283D-02
1	-6.0	1.086961396898D-04	5.434806984488D-05	1.25	9.745942724799D-03
2	-5.0	-3.937836798383D-04	-1.968918399192D-04	1.50	1.226586292881D-02
3	-4.0	9.618511402927D-04	4.809255701464D-04	1.75	8.027402566276D-03
4	-3.0	-1.883283238889D-03	-9.416416194447D-04	2.25	-7.580437605879D-03
5	-2.0	3.291006426837D-03	1.645503213419D-03	2.50	-1.083991068080D-02
6	-1.0	-5.390947470305D-03	-2.695473735153D-03	2.75	-7.916693988520D-03
7	0.0	8.261890374514D-03	4.130945187257D-03	3.25	8.879928638170D-03
8	1.0	-1.144286479104D-02	-5.721432395518D-03	3.50	1.356017730176D-02
9	2.0	1.365137536711D-02	6.825687683557D-03	3.75	1.045635425693D-02
10	3.0	-1.328648527723D-02	-6.643242638615D-03	4.25	-1.271785465637D-02
11	4.0	9.882169728116D-03	4.941084864058D-03	4.50	-2.000278714746D-02
12	5.0	-5.151305640702D-03	-2.575652820351D-03	4.75	-1.579792647784D-02
13	6.0	1.645409640738D-03	8.227048203689D-04	5.25	1.990642216003D-02
14	7.0	-2.392329309419D-04	-1.196164654709D-04	5.50	3.172520589051D-02
				5.75	2.533908705771D-02
	0	4.190499850106D+01	4.195053385209D+02	6.25	-3.254484472128D-02
	1	2.993749368858D+01	-2.718173853046D+02	6.50	-5.233977209790D-02
	2	7.972706040972D+00	8.085292375132D+01	6.75	-4.220510457273D-02
	3	9.373784166932D-01	-8.620334267949D+00	7.25	5.548782238344D-02
	4	4.103109464271D-02	4.203586011867D-01	7.50	9.063865325779D-02
I=11	H= 4.0				
0		0.	-1.238976448854D+04	-1.00	-3.989718908505D-01
1		9.183371422352D-02	3.952370805177D+03	-.75	-2.213627626177D-01

2		-1.786046217878D-01	-4.720114678117D+02	-.50	-1.053293935783D-01
3		1.076485414737D-01	2.505047642304D+01	-.25	-3.588480796161D-02
4		-2.088501336544D-02	-5.059874102565D-01	.25	1.339606609936D-02
				.50	1.341647042668D-02
0	-7.0	7.379456062405D-06	3.689728031202D-06	.75	7.216819670184D-03
1	-6.0	-5.534074207175D-05	-2.767037103587D-05	1.25	-4.960342338966D-03
2	-5.0	2.005424957589D-04	1.002712478794D-04	1.50	-6.242797382184D-03
3	-4.0	-4.902027043354D-04	-2.451013521677D-04	1.75	-4.085520190489D-03
4	-3.0	9.618511402927D-04	4.809255701464D-04	2.25	3.857795652408D-03
5	-2.0	-1.691483676163D-03	-8.457418380815D-04	2.50	5.516323835644D-03
6	-1.0	2.822374517480D-03	1.411187258740D-03	2.75	4.028469264022D-03
7	0.0	-4.520639099522D-03	-2.260319549761D-03	3.25	-4.517719225145D-03
8	1.0	6.777375356525D-03	3.388687678262D-03	3.50	-6.897784628280D-03
9	2.0	-9.013658397678D-03	-4.506829198839D-03	3.75	-5.317871273764D-03
10	3.0	9.882169728116D-03	4.941084864058D-03	4.25	6.464052923014D-03
11	4.0	-8.202287225044D-03	-4.101143612522D-03	4.50	1.016204574061D-02
12	5.0	4.677233962832D-03	2.338616981416D-03	4.75	8.020962792928D-03
13	6.0	-1.600089013763D-03	-8.000445068814D-04	5.25	-1.008816319412D-02
14	7.0	2.447742015092D-04	1.223871007546D-04	5.50	-1.605529607713D-02
				5.75	-1.279996933202D-02
	0	-2.133025783300D+01	-4.742931340012D+02	6.25	1.634835229493D-02
	1	-1.523853373156D+01	3.324155631276D+02	6.50	2.618233550485D-02
	2	-4.058179180281D+00	-9.471176154330D+01	6.75	2.099681863113D-02
	3	-4.771318327588D-01	1.088282893586D+01	7.25	-2.714455172641D-02
	4	-2.088501336544D-02	-5.059874102565D-01	7.50	-4.378127271600D-02

I=12 H= 5.0

0		0.	1.006128641408D+04	-1.00	1.631474168619D-01
1		-3.755245125343D-02	-3.245371125102D+03	-.75	9.051949677527D-02
2		7.303484310727D-02	3.918752814719D+02	-.50	4.307117832784D-02
3		-4.401974828773D-02	-2.100649269237D+01	-.25	1.467395991133D-02
4		8.540374213440D-03	4.247894907701D-01	.25	-5.477882860888D-03
				.50	-5.486215891621D-03
0	-7.0	-3.017779559543D-06	-1.508889779771D-06	.75	-2.951069311673D-03
1	-6.0	2.263209903193D-05	1.131604951596D-05	1.25	2.028346272397D-03
2	-5.0	-8.202135900984D-05	-4.101067950492D-05	1.50	2.552744714880D-03
3	-4.0	2.005424957589D-04	1.002712478794D-04	1.75	1.670599943569D-03
4	-3.0	-3.937836798383D-04	-1.968918399192D-04	2.25	-1.577447315131D-03
5	-2.0	6.940158748949D-04	3.470079374475D-04	2.50	-2.255579711885D-03
6	-1.0	-1.165654298732D-03	-5.828271493662D-04	2.75	-1.647171643537D-03
7	0.0	1.903245363218D-03	9.516226816091D-04	3.25	1.847090485017D-03
8	1.0	-2.984365430560D-03	-1.492182715280D-03	3.50	2.820044284940D-03
9	2.0	4.281593693547D-03	2.140796846774D-03	3.75	2.173972966330D-03
10	3.0	-5.151305640702D-03	-2.575652820351D-03	4.25	-2.641977265308D-03
11	4.0	4.677233962832D-03	2.338616981416D-03	4.50	-4.152751938256D-03
12	5.0	-2.875575044583D-03	-1.437787522292D-03	4.75	-3.277101650237D-03
13	6.0	1.043426131953D-03	5.217130659764D-04	5.25	4.119038644656D-03
14	7.0	-1.669663882504D-04	-8.348319412521D-05	5.50	6.552279533895D-03
				5.75	5.220439501188D-03
	0	8.722504977260D+00	3.602699043439D+02	6.25	-6.654730031759D-03
	1	6.231425774792D+00	-2.642604289372D+02	6.50	-1.064229347585D-02
	2	1.659490147816D+00	7.562704521857D+01	6.75	-8.518301368698D-03
	3	1.951107296886D-01	-9.112386950803D+00	7.25	1.094875521871D-02
	4	8.540374213440D-03	4.247894907701D-01	7.50	1.758289918870D-02

CARDINAL SPLINES N=5

I=13	H= 6.0				
0		0.	-4.629352157098D+03	-1.00	-4.501399387091D-02
1		1.036107901598D-02	1.512232535063D+03	-.75	-2.497522166866D-02
2		-2.015103063258D-02	-1.850509366285D+02	-.50	-1.188375877959D-02
3		1.214550159284D-02	1.005557608991D+01	-.25	-4.048687250567D-03
4		-2.356382629506D-03	-2.055112606234D-01	.25	1.511399185377D-03
				.50	1.513697260883D-03
0	-7.0	8.326532649558D-07	4.163266324779D-07	.75	8.142263437815D-04
1	-6.0	-6.244643915140D-06	-3.122321957570D-06	1.25	-5.596373475444D-04
2	-5.0	2.263209903193D-05	1.131604951596D-05	1.50	-7.043217932462D-04
3	-4.0	-5.534074207175D-05	-2.767037103587D-05	1.75	-4.609301092275D-04
4	-3.0	1.086961396898D-04	5.434806984488D-05	2.25	4.352251514236D-04
5	-2.0	-1.917243601843D-04	-9.586218009213D-05	2.50	6.223212555195D-04
6	-1.0	3.228015428763D-04	1.614007714382D-04	2.75	4.544559110286D-04
7	0.0	-5.309424289936D-04	-2.654712144968D-04	3.25	-5.096007217262D-04
8	1.0	8.506173946761D-04	4.253086973381D-04	3.50	-7.780176094907D-04
9	2.0	-1.279144631020D-03	-6.395723155100D-04	3.75	-5.997587032897D-04
10	3.0	1.645409640738D-03	8.227048203689D-04	4.25	7.288152278958D-04
11	4.0	-1.600089013763D-03	-8.000445068814D-04	4.50	1.145509573317D-03
12	5.0	1.043426131953D-03	5.217130659764D-04	4.75	9.038967817709D-04
13	6.0	-3.967762942255D-04	-1.983881471127D-04	5.25	-1.135850994223D-03
14	7.0	6.584651194351D-05	3.292325597176D-05	5.50	-1.806509830804D-03
				5.75	-1.438973602864D-03
	0	-2.406640594984D+00	-1.555902443724D+02	6.25	1.833008197770D-03
	1	-1.719321583375D+00	1.177276579047D+02	6.50	2.929785273609D-03
	2	-4.578719902576D-01	-3.430414936379D+01	6.75	2.343401476340D-03
	3	-5.383321203332D-02	4.301260792452D+00	7.25	-3.005552198052D-03
	4	-2.356382629506D-03	-2.055112606234D-01	7.50	-4.818952602591D-03

I=14	H= 7.0				
0		0.	8.994323949068D+02	-1.00	6.002023589906D-03
1		-1.381512660391D-03	-2.974961034623D+02	-.75	3.330116682183D-03
2		2.686874234101D-03	3.690895823236D+01	-.50	1.584542408584D-03
3		-1.619443622396D-03	-2.036049217519D+00	-.25	5.398389280204D-04
4		3.141930730181D-04	4.224393689807D-02	.25	-2.015250157983D-04
				.50	-2.018313742504D-04
0	-7.0	-1.110243320428D-07	-5.551216602138D-08	.75	-1.085662010272D-04
1	-6.0	8.326532649558D-07	4.163266324779D-07	1.25	7.462007991681D-05
2	-5.0	-3.017779559543D-06	-1.508889779771D-06	1.50	9.391172068731D-05
3	-4.0	7.379456062405D-06	3.689728031202D-06	1.75	6.145868845335D-05
4	-3.0	-1.449578836054D-05	-7.247894180272D-06	2.25	-5.803109438600D-05
5	-2.0	2.557697511385D-05	1.278848755692D-05	2.50	-8.297749428249D-05
6	-1.0	-4.310637262073D-05	-2.155318631037D-05	2.75	-6.059488684639D-05
7	0.0	7.111589231035D-05	3.555794615518D-05	3.25	6.794691254765D-05
8	1.0	-1.149845871324D-04	-5.749229356621D-05	3.50	1.037350897523D-04
9	2.0	1.776353870396D-04	8.881769351982D-05	3.75	7.996653389697D-05
10	3.0	-2.392329309419D-04	-1.196164654709D-04	4.25	-9.717066459681D-05
11	4.0	2.447742015092D-04	1.223871007546D-04	4.50	-1.527235245463D-04
12	5.0	-1.669663882504D-04	-8.348319412521D-05	4.75	-1.205069730656D-04
13	6.0	6.584651194351D-05	3.292325597176D-05	5.25	1.514161622212D-04
14	7.0	-1.124620604634D-05	-5.623103023169D-06	5.50	2.408015848729D-04
				5.75	1.917917320342D-04
	0	3.208946546827D-01	2.856143493991D+01	6.25	-2.442380607562D-04
	1	2.292494103056D-01	-2.211124176031D+01	6.50	-3.902912123172D-04
	2	6.105132163110D-02	6.571642112504D+00	6.75	-3.120854151146D-04
	3	7.177962422110D-03	-8.532189843726D-01	7.25	3.999139264226D-04

	4	3,141930730181D-04	4,224393689807D-02	7.50	6,407780574529D-04

N= 5 M=15 P= 7.5 Q= 9

I= 0 H= -7.5

0		1.000000000000D+00	-2,851311218702D+00	-1.00	5,855078980526D+00
1		-2,284280828260D+00	-1,609333840583D-01	-.75	4,048112224380D+00
2		1,843370286301D+00	8,140792364127D-01	-.50	2,688650542247D+00
3		-6,432530386969D-01	-3,163880345700D-01	-,25	1,696660486608D+00
4		8,417482726705D-02	4,199190052057D-02	.25	5,344184149752D-01
				.50	2,435564323462D-01
0	-7,5	-1,124661119749D-05	-5,623305598744D-06	.75	7,894538614070D-02
1	-6,5	6,584955049729D-05	3,292477524865D-05	1.25	-2,601739833390D-02
2	-5.5	-1,669774010968D-04	-8,348870054840D-05	1.50	-2,411426967855D-02
3	-4,5	2,448011330796D-04	1.224005665398D-04	1.75	-1,186271135752D-02
4	-3,5	-2,392858427132D-04	-1,196429213566D-04	2,25	6,554164925109D-03
5	-2,5	1,777287936564D-04	8,886439682822D-05	2,50	7,241414037284D-03
6	-1,5	-1,151422486578D-04	-5,757112432892D-05	2,75	4,101691742667D-03
7	-,5	7,137718908806D-05	3,568859454403D-05	3,25	-2,786002564138D-03
8	,5	-4,353478019342D-05	-2,176739009671D-05	3,50	-3,317347396826D-03
9	1,5	2,626806160649D-05	1,313403080324D-05	3,75	-1,995872047529D-03
10	2,5	-1,556252077268D-05	-7,781260386338D-06	4,25	1,478656495657D-03
11	3,5	8,815655335100D-06	4,407827667550D-06	4,50	1,815191523948D-03
12	4,5	-4,487081735122D-06	-2,243540867561D-06	4,75	1,118651752528D-03
13	5,5	1,834855565320D-06	9,174277826600D-07	5,25	-8,569131690468D-04
14	6,5	-5,062548875122D-07	-2,531274437561D-07	5,50	-1,063942915844D-03
15	7,5	6,750242571383D-08	3,375121285691D-08	5,75	-6,614641732082D-04
				6,25	5,128405639716D-04
	0	8,251951109189D+01	4,112244110570D+01	6,50	6,393557929967D-04
	1	5,886234419930D+01	2,952110645689D+01	6,75	3,987771535853D-04
	2	1,577918111825D+01	7,867614884278D+00	7,25	-3,106306827613D-04
	3	1,881991779314D+00	9,433689810469D-01	7,50	-3,879626513319D-04
	4	8,417482726705D-02	4,199190052057D-02	7,75	-2,423736480706D-04

I= 1 H= -6.5

0		0.	2,513423152658D+01	-1.00	-1,484707220430D+01
1		5,137161772069D+00	-4,790295658034D+00	-.75	-8,822294388385D+00
2		-6,514894289453D+00	-2,450422138927D+00	-.50	-4,571139305153D+00
3		2,786341405304D+00	1,353883004892D+00	-,25	-1,736604306259D+00
4		-4,086747374706D-01	-2,036210374707D-01	.25	9,190497489423D-01
				.50	1,262607946855D+00
0	-7.5	6,584955049729D-05	3,292477524865D-05	.75	1,234428774737D+00
1	-6,5	-3,967990827799D-04	-1,983995413899D-04	1.25	6,867525026421D-01
2	-5,5	1,043508726130D-03	5,217543630649D-04	1.50	3,847475942265D-01
3	-4,5	-1,600290995232D-03	-8,001454976162D-04	1.75	1,484026344156D-01
4	-3,5	1,645806468593D-03	8,229032342965D-04	2,25	-6,439268036982D-02
5	-2,5	-1,279845162235D-03	-6,399225811174D-04	2,50	-6,651350489080D-02
6	-1,5	8,517998250403D-04	4,258999125201D-04	2.75	-3,587857489908D-02

CARDINAL SPLINES N=5

7	-,5	-5,329021033224D-04	-2,664510516612D-04	3.25	2,286658072043D-02
8	,5	3.260145152285D-04	1.630072576142D-04	3.50	2,666497518741D-02
9	1,5	-1.969073726596D-04	-9,845368632981D-05	3,75	1,578785761765D-02
10	2,5	1.166964225181D-04	5.834821125904D-05	4.25	-1,144386868546D-02
11	3,5	-6,611195352903D-05	-3,305597676452D-05	4,50	-1,394577590516D-02
12	4,5	3,365157573596D-05	1.682578786798D-05	4.75	-8,545933701250D-03
13	5,5	-1.376096362469D-05	-6.880481812345D-06	5.25	6,496286562888D-03
14	6,5	3.796804527443D-06	1.898402263721D-06	5.50	8,045195253603D-03
15	7,5	-5.062548875122D-07	-2,531274437561D-07	5,75	4,992001904218D-03
				6,25	-3,860176294037D-03
	0	-4.455187216566D+02	-2.217295274067D+02	6,50	-4,808285899511D-03
	1	-3,120297599063D+02	-1.566893713982D+02	6.75	-2,997026625840D-03
	2	-8,174993656644D+01	-4.071015467522D+01	7,25	2,332491270432D-03
	3	-9.473900718813D+00	-4.754748119229D+00	7,50	2,912321525200D-03
	4	-4,086747374706D-01	-2,036210374707D-01	7,75	1,819025470120D-03

I= 2 H= -5.5

0		0.	-9.109629337054D+01	-1.00	2,176725277202D+01
1		-5,754185887854D+00	2.379435838413D+01	-.75	1,239469169439D+01
2		1.004264010195D+01	2.096345055339D+00	-.50	6,081489456394D+00
3		-5,129357009456D+00	-2.422284475384D+00	-,25	2,149643110408D+00
4		8,410697727622D-01	4.179386206711D-01	.25	-8,877422407016D-01
				.50	-9,550360099519D-01
0	-7,5	-1,669774010968D-04	-8,348870054840D-05	.75	-5,644946515663D-01
1	-6,5	1.043508726130D-03	5.217543630649D-04	1.25	5,327663156876D-01
2	-5,5	-2,875874396529D-03	-1.437937198265D-03	1,50	9,045800811640D-01
3	-4,5	4,677966018627D-03	2,338983009313D-03	1.75	1,058301220050D+00
4	-3,5	-5,152743892098D-03	-2,576371946049D-03	2,25	7,840636281089D-01
5	-2,5	4,284132678611D-03	2,142066339306D-03	2,50	4,922627719626D-01
6	-1,5	-2,988650996957D-03	-1.494325498479D-03	2,75	2,093017561107D-01
7	-,5	1,910347950150D-03	9,551739750752D-04	3,25	-1,057577921388D-01
8	,5	-1.177299302755D-03	-5.886496513775D-04	3,50	-1,157681271918D-01
9	1,5	7,128010353351D-04	3,564005176676D-04	3,75	-6,551276177690D-02
10	2,5	-4,227796734253D-04	-2,113898367127D-04	4,25	4,482191709068D-02
11	3,5	2,395813628891D-04	1,197906814446D-04	4,50	5,361779552257D-02
12	4,5	-1,219600315166D-04	-6,098001575832D-05	4,75	3,239856558845D-02
13	5,5	4,987403069518D-05	2,493701534759D-05	5,25	-2,416903868549D-02
14	6,5	-1,376096362469D-05	-6,880481812345D-06	5,50	-2,974621754234D-02
15	7,5	1,834855565320D-06	9,174277826600D-07	5,75	-1,836992876838D-02
				6,25	1,411452572770D-02
	0	1,018991951092D+03	5,057634577877D+02	6,50	1,754423909870D-02
	1	6,986116618319D+02	3,517504513756D+02	6,75	1,091785538148D-02
	2	1,784931556964D+02	8,864922883569D+01	7,25	-8,478758734100D-03
	3	2,010273617341D+01	1,011587414475D+01	7,50	-1,057899837144D-02
	4	8,410697727622D-01	4,179386206711D-01	7,75	-6,604051438269D-03

I= 3 H= -4.5

0		0.	2,227765167150D+02	-1.00	-2,127771758071D+01
1		5,130618664810D+00	-6,265989526455D+01	-,75	-1,191220832069D+01
2		-9,647815169680D+00	2,329139002722D+00	-,50	-5,727725716775D+00
3		5,508117724979D+00	2,405833206655D+00	-,25	-1,975579196031D+00
4		-9,911660212420D-01	-4,892338223054D-01	,25	7,618588162137D-01
				,50	7,799228574070D-01
0	-7,5	2,448011330796D-04	1,224005665398D-04	,75	4,312123877957D-01
1	-6,5	-1,600290995232D-03	-8,001454976162D-04	1,25	-3,214104545934D-01

2	-5.5	4.677966018627D-03	2.338983009313D-03	1.50	-4.301289429330D-01
3	-4.5	-8.204077444528D-03	-4.102038722264D-03	1.75	-3.062723085180D-01
4	-3.5	9.885686926795D-03	4.942843463397D-03	2.25	3.900966514157D-01
5	-2.5	-9.019867406017D-03	-4.509933703009D-03	2.50	7.469493873516D-01
6	-1.5	6.787855574733D-03	3.393927787366D-03	2.75	9.690150048872D-01
7	-.5	-4.538008253093D-03	-2.269004126547D-03	3.25	8.447113476347D-01
8	.5	2.850852009574D-03	1.425426004787D-03	3.50	5.644785668507D-01
9	1.5	-1.737422196827D-03	-8.687110984134D-04	3.75	2.527512296145D-01
10	2.5	1.032759933156D-03	5.163799665782D-04	4.25	-1.379499410513D-01
11	3.5	-5.856710294565D-04	-2.928355147282D-04	4.50	-1.552390684153D-01
12	4.5	2.982112667060D-04	1.491056333530D-04	4.75	-8.987000275938D-02
13	5.5	-1.219600315166D-04	-6.098001575832D-05	5.25	6.348368992178D-02
14	6.5	3.365157573596D-05	1.682578786798D-05	5.50	7.679142313674D-02
15	7.5	-4.487081735122D-06	-2.243540867561D-06	5.75	4.680838057601D-02
				6.25	-3.534687759391D-02
	0	-1.316584037169D+03	-6.491641356966D+02	6.50	-4.368673067849D-02
	1	-8.826844036362D+02	-4.473205317410D+02	6.75	-2.706898714364D-02
	2	-2.202336985268D+02	-1.086560288756D+02	7.25	2.090011835847D-02
	3	-2.422686291228D+01	-1.227118146251D+01	7.50	2.602754931227D-02
	4	-9.911660212420D-01	-4.892338223054D-01	7.75	1.622441938407D-02

I= 4 H= -3.5

0		0.	-4.377026285896D+02	-1.00	1.580573563781D+01
1		-3.688812330348D+00	1.263076212960D+02	-.75	8.792959277775D+00
2		7.103151618204D+00	-1.050246976906D+01	-.50	4.196925790656D+00
3		-4.213935845637D+00	-1.422786419704D+00	-.25	1.435117165077D+00
4		7.998358436234D-01	3.874432531450D-01	.25	-5.409744961859D-01
				.50	-5.453709684562D-01
0	-7.5	-2.392858427132D-04	-1.196429213566D-04	.75	-2.957855542052D-01
1	-6.5	1.645806468593D-03	8.229032342965D-04	1.25	2.082569574171D-01
2	-5.5	-5.152743892098D-03	-2.576371946049D-03	1.50	2.668123875775D-01
3	-4.5	9.885686926795D-03	4.942843463397D-03	1.75	1.787919019832D-01
4	-3.5	-1.329339542800D-02	-6.646697713998D-03	2.25	-1.819442260855D-01
5	-2.5	1.366357404784D-02	6.831787023922D-03	2.50	-2.760359429496D-01
6	-1.5	-1.146345500914D-02	-5.731727504568D-03	2.75	-2.189887337658D-01
7	-.5	8.296015108670D-03	4.148007554335D-03	3.25	3.308068553967D-01
8	.5	-5.446896474565D-03	-2.723448237282D-03	3.50	6.763485085274D-01
9	1.5	3.381260669724D-03	1.690630334862D-03	3.75	9.265356107293D-01
10	2.5	-2.022595940332D-03	-1.011297970166D-03	4.25	8.761881114589D-01
11	3.5	1.149415329975D-03	5.747076649875D-04	4.50	6.031242185307D-01
12	4.5	-5.856710294565D-04	-2.928355147282D-04	4.75	2.765713532805D-01
13	5.5	2.395813628891D-04	1.197906814446D-04	5.25	-1.562054179226D-01
14	6.5	-6.611195352903D-05	-3.305597676452D-05	5.50	-1.779628369494D-01
15	7.5	8.815655335100D-06	4.407827667550D-06	5.75	-1.039727132249D-01
				6.25	7.443173632102D-02
	0	1.124862600133D+03	5.444972539619D+02	6.50	9.045290892181D-02
	1	7.414797741060D+02	3.824858561172D+02	6.75	5.533936257000D-02
	2	1.822341923143D+02	8.824693372403D+01	7.25	-4.201512000705D-02
	3	1.978113946306D+01	1.020051117465D+01	7.50	-5.203566760305D-02
	4	7.998358436234D-01	3.874432531450D-01	7.75	-3.230135688413D-02

I= 5 H= -2.5

0		0.	7.727899769061D+02	-1.00	-1.024300002824D+01
1		2.368072124535D+00	-2.250764203099D+02	-.75	-5.687955528948D+00
2		-4.591070692376D+00	2.251781156557D+01	-.50	-2.709128500144D+00

CARDINAL SPLINES N=5

3		2.753339025189D+00	1.686996685505D-01	-.25	-9.240532083405D-01
4		-5.305181861416D-01	-2.432341639413D-01	.25	3.460246991423D-01
				.50	3.472787278149D-01
0	-7.5	1.777287936564D-04	8.886439682822D-05	.75	1.872958050683D-01
1	-6.5	-1.279845162235D-03	-6.399225811174D-04	1.25	-1.297287551623D-01
2	-5.5	4.284132678611D-03	2.142066339306D-03	1.50	-1.641999861850D-01
3	-4.5	-9.019867406017D-03	-4.509933703009D-03	1.75	-1.082688671121D-01
4	-3.5	1.366357404784D-02	6.831787023922D-03	2.25	1.046467451004D-01
5	-2.5	-1.609511123041D-02	-8.047555615204D-03	2.50	1.523324695191D-01
6	-1.5	1.547095395781D-02	7.735476978906D-03	2.75	1.139440898105D-01
7	-.5	-1.257082960536D-02	-6.285414802682D-03	3.25	-1.378897194810D-01
8	.5	8.949865074626D-03	4.474932537313D-03	3.50	-2.235727717833D-01
9	1.5	-5.806058370739D-03	-2.903029185369D-03	3.75	-1.874193395808D-01
10	2.5	3.536939001069D-03	1.768469500534D-03	4.25	3.074074508667D-01
11	3.5	-2.022595940332D-03	-1.011297970166D-03	4.50	6.476138285073D-01
12	4.5	1.032759933156D-03	5.163799665782D-04	4.75	9.088195892974D-01
13	5.5	-4.227796734253D-04	-2.113898367127D-04	5.25	8.897761069312D-01
14	6.5	1.166964225181D-04	5.834821125904D-05	5.50	6.200103485316D-01
15	7.5	-1.556252077268D-05	-7.781260386338D-06	5.75	2.870818079438D-01
				6.25	-1.643819090853D-01
	0	-7.575149825993D+02	-3.470941990308D+02	6.50	-1.881812035878D-01
	1	-4.971214668745D+02	-2.692988294095D+02	6.75	-1.103658862332D-01
	2	-1.216908304484D+02	-5.577797622224D+01	7.25	7.945686904084D-02
	3	-1.316220655906D+01	-7.128325249690D+00	7.50	9.676957863559D-02
	4	-5.305181861416D-01	-2.432341639413D-01	7.75	5.931802290601D-02

I= 6 H= -1.5

0		0.	-1.304910710340D+03	-1.00	6.355073918610D+00
1		-1.464878613614D+00	3.813266461677D+02	-.75	3.526973385148D+00
2		2.846070669302D+00	-4.047654384305D+01	-.50	1.678752312274D+00
3		-1.712600774566D+00	1.183476806516D+00	-.25	5.721534724201D-01
4		3.315238611275D-01	1.285704582504D-01	.25	-2.138046090315D-01
				.50	-2.142767198692D-01
0	-7.5	-1.151422486578D-04	-5.757112432892D-05	.75	-1.153600842334D-01
1	-6.5	8.517998250403D-04	4.258999125201D-04	1.25	7.949031916774D-02
2	-5.5	-2.988650996957D-03	-1.494325498479D-03	1.50	1.002282271313D-01
3	-4.5	6.787855574733D-03	3.393927787366D-03	1.75	6.575461635100D-02
4	-3.5	-1.146345500914D-02	-5.731727504568D-03	2.25	-6.256476110435D-02
5	-2.5	1.547095395781D-02	7.735476978906D-03	2.50	-8.998373973834D-02
6	-1.5	-1.725239720257D-02	-8.626198601284D-03	2.75	-6.622424493446D-02
7	-.5	1.616536160318D-02	8.082680801592D-03	3.25	7.607496747504D-02
8	.5	-1.295629519043D-02	-6.478147595213D-03	3.50	1.182998227737D-01
9	1.5	9.120027860533D-03	4.560013930267D-03	3.75	9.345947284392D-02
10	2.5	-5.806058370739D-03	-2.903029185369D-03	4.25	-1.226941454597D-01
11	3.5	3.381260669724D-03	1.690630334862D-03	4.50	-2.049016621569D-01
12	4.5	-1.737422196827D-03	-8.687110984134D-04	4.75	-1.758993663634D-01
13	5.5	7.128010353351D-04	3.564005176676D-04	5.25	2.985526752923D-01
14	6.5	-1.969073726596D-04	-9.845368632981D-05	5.50	6.365927657012D-01
15	7.5	2.626806160649D-05	1.313403080324D-05	5.75	9.019461887395D-01
				6.25	8.951538864066D-01
	0	4.755636506246D+02	1.843177880523D+02	6.50	6.267585365929D-01
	1	3.116713163705D+02	1.908528479189D+02	6.75	2.913257267990D-01
	2	7.620185637209D+01	2.954421396304D+01	7.25	-1.677673937539D-01
	3	8.233115059258D+00	5.040590554026D+00	7.50	-1.924810456314D-01
	4	3.315238611275D-01	1.285704582504D-01	7.75	-1.131093758677D-01

CARDINAL SPLINES N=5

I= 7 H= -.5

0		0.	2.165236915829D+03	-1.00	-3.885049894689D+00
1		8.946630435878D-01	-6.335141435167D+02	-.75	-2.155747317944D+00
2		-1.739418512320D+00	6.865662808256D+01	-.50	-1.025860809495D+00
3		1.047826215162D+00	-2.860858965469D+00	-.25	-3.495452264493D-01
4		-2.031421236189D-01	-3.985520886958D-02	.25	1.305308648407D-01
				.50	1.307589272915D-01
0	-7.5	7.137718908806D-05	3.568859454403D-05	.75	7.035597581484D-02
1	-6.5	-5.329021033224D-04	-2.664510516612D-04	1.25	-4.839769530108D-02
2	-5.5	1.910347950150D-03	9.551739750752D-04	1.50	-6.094767388004D-02
3	-4.5	-4.538008253093D-03	-2.269004126547D-03	1.75	-3.991861577870D-02
4	-3.5	8.296015108670D-03	4.148007554335D-03	2.25	3.778801526841D-02
5	-2.5	-1.257082960536D-02	-6.285414802682D-03	2.50	5.413686713559D-02
6	-1.5	1.616536160318D-02	8.082680801592D-03	2.75	3.963574797120D-02
7	-.5	-1.764404336017D-02	-8.822021680085D-03	3.25	-4.480084862188D-02
8	.5	1.633997561937D-02	8.169987809683D-03	3.50	-6.881219617610D-02
9	1.5	-1.295629519043D-02	-6.478147595213D-03	3.75	-5.347129601046D-02
10	2.5	8.949865074626D-03	4.474932537313D-03	4.25	6.659411485528D-02
11	3.5	-5.446896474565D-03	-2.723448237282D-03	4.50	1.066323163388D-01
12	4.5	2.850852009574D-03	1.425426004787D-03	4.75	8.624634399750D-02
13	5.5	-1.177299302755D-03	-5.886496513775D-04	5.25	-1.171177314943D-01
14	6.5	3.260145152285D-04	1.630072576142D-04	5.50	-1.979324527506D-01
15	7.5	-4.353478019342D-05	-2.176739009671D-05	5.75	-1.715302590830D-01
				6.25	2.950829633953D-01
	0	-2.918350094825D+02	-5.721307877348D+01	6.50	6.321932691283D-01
	1	-1.911782744395D+02	-1.536903376687D+02	6.75	8.991431408806D-01
	2	-4.672379539255D+01	-9.163831633979D+00	7.25	8.974717845160D-01
	3	-5.046437493405D+00	-4.056515231556D+00	7.50	6.297742247842D-01
	4	-2.031421236189D-01	-3.985520886958D-02	7.75	2.933063300701D-01

I= 8 H= .5

0		0.	-3.562802875827D+03	-1.00	2.358801958155D+00
1		-5.430210050027D-01	1.042940422471D+03	-.75	1.308778895716D+00
2		1.055991040588D+00	-1.138865573375D+02	-.50	6.227675201014D-01
3		-6.363582066848D-01	5.252171497644D+00	-.25	2.121799433680D-01
4		1.234317058798D-01	-3.985520886958D-02	.25	-7.921675325835D-02
				.50	-7.934312160134D-02
0	-7.5	-4.353478019342D-05	-2.176739009671D-05	.75	-4.268311747325D-02
1	-6.5	3.260145152285D-04	1.630072576142D-04	1.25	2.934520227205D-02
2	-5.5	-1.177299302755D-03	-5.886496513775D-04	1.50	3.693941082062D-02
3	-4.5	2.850852009574D-03	1.425426004787D-03	1.75	2.418081101304D-02
4	-3.5	-5.446896474565D-03	-2.723448237282D-03	2.25	-2.285140324578D-02
5	-2.5	8.949865074626D-03	4.474932537313D-03	2.50	-3.269569056047D-02
6	-1.5	-1.295629519043D-02	-6.478147595213D-03	2.75	-2.389663700486D-02
7	-.5	1.633997561937D-02	8.169987809683D-03	3.25	2.686703705613D-02
8	.5	-1.764404336017D-02	-8.822021680085D-03	3.50	4.110038395544D-02
9	1.5	1.616536160318D-02	8.082680801592D-03	3.75	3.176727266013D-02
10	2.5	-1.257082960536D-02	-6.285414802682D-03	4.25	-3.891717559328D-02
11	3.5	8.296015108670D-03	4.148007554335D-03	4.50	-6.154160085121D-02
12	4.5	-4.538008253093D-03	-2.269004126547D-03	4.75	-4.895289242987D-02
13	5.5	1.910347950150D-03	9.551739750752D-04	5.25	6.304840570930D-02
14	6.5	-5.329021033224D-04	-2.664510516612D-04	5.50	1.021544962444D-01
15	7.5	7.137718908806D-05	3.568859454403D-05	5.75	8.340231194176D-02
				6.25	-1.147763332154D-01
	0	1.774088519355D+02	-5.721307877348D+01	6.50	-1.948910926885D-01

CARDINAL SPLINES N=5

	1	1.162024008978D+02	1.536903376687D+02	6.75	-1.695355176039D-01
	2	2.839613212460D+01	-9.163831633979D+00	7.25	2.933063300701D-01
	3	3.066592969708D+00	4.056515231556D+00	7.50	6.297742247842D-01
	4	1.234317058798D-01	-3.985520886958D-02	7.75	8.974717845160D-01
I= 9	H= 1.5				
0		0.	5.810880288403D+03	-1.00	-1.421111940518D+00
1		3.271205462193D-01	-1.701578361654D+03	-.75	-7.884866316460D-01
2		-6.361861596633D-01	1.863500310881D+02	-.50	-3.751835333158D-01
3		3.834222900091D-01	-8.897704301537D+00	-.25	-1.278233031926D-01
4		-7.438294462678D-02	1.285704582504D-01	.25	4.771891658002D-02
				.50	4.779263671061D-02
0	-7.5	2.626806160649D-05	1.313403080324D-05	.75	2.570871720734D-02
1	-6.5	-1.969073726596D-04	-9.845368632981D-05	1.25	-1.767180587187D-02
2	-5.5	7.128010353351D-04	3.564005176676D-04	1.50	-2.224202035792D-02
3	-4.5	-1.737422196827D-03	-8.687110984134D-04	1.75	-1.455715035340D-02
4	-3.5	3.381260669724D-03	1.690630334862D-03	2.25	1.374908681624D-02
5	-2.5	-5.806058370739D-03	-2.903029185369D-03	2.50	1.966368864428D-02
6	-1.5	9.120027860533D-03	4.560013930267D-03	2.75	1.436358124227D-02
7	-.5	-1.295629519043D-02	-6.478147595213D-03	3.25	-1.612037971089D-02
8	.5	1.616536160318D-02	8.082680801592D-03	3.50	-2.462737893921D-02
9	1.5	-1.725239720257D-02	-8.626198601284D-03	3.75	-1.900118524968D-02
10	2.5	1.547095395781D-02	7.735476978906D-03	4.25	2.315125183071D-02
11	3.5	-1.146345500914D-02	-5.731727504568D-03	4.50	3.646048973168D-02
12	4.5	6.787855574733D-03	3.393927787366D-03	4.75	2.884596075811D-02
13	5.5	-2.988650996957D-03	-1.494325498479D-03	5.25	-3.654082956056D-02
14	6.5	8.517998250403D-04	4.258999125201D-04	5.50	-5.846739522841D-02
15	7.5	-1.151422486578D-04	-5.757112432892D-05	5.75	-4.694304886151D-02
				6.25	6.126653666566D-02
	0	-1.069280745200D+02	1.843177880523D+02	6.50	9.973245003261D-02
	1	-7.003437946738D+01	-1.908528479189D+02	6.75	8.173130409329D-02
	2	-1.711342844600D+01	2.954421396304D+01	7.25	-1.131093758677D-01
	3	-1.848066048794D+00	-5.040590554026D+00	7.50	-1.924810456314D-01
	4	-7.438294462678D-02	1.285704582504D-01	7.75	-1.677673937539D-01
I=10	H= 2.5				
0		0.	-9.281216893661D+03	-1.00	8.415171722290D-01
1		-1.936989974066D-01	2.719331010279D+03	-.75	4.669024161643D-01
2		3.767165091160D-01	-2.982568246705D+02	-.50	2.221632180544D-01
3		-2.270518074475D-01	1.442535016793D+01	-.25	7.568928542158D-02
4		4.404985825892D-02	-2.432341639413D-01	.25	-2.825558232381D-02
				.50	-2.829876160960D-02
0	-7.5	-1.556252077268D-05	-7.781260386338D-06	.75	-1.522221098638D-02
1	-6.5	1.166964225181D-04	5.834821125904D-05	1.25	1.046288720930D-02
2	-5.5	-4.227796734253D-04	-2.113898367127D-04	1.50	1.316815904525D-02
3	-4.5	1.032759933156D-03	5.163799665782D-04	1.75	8.617891658766D-03
4	-3.5	-2.022595940332D-03	-1.011297970166D-03	2.25	-8.137991922996D-03
5	-2.5	3.536939001069D-03	1.768469500534D-03	2.50	-1.163714334114D-02
6	-1.5	-5.806058370739D-03	-2.903029185369D-03	2.75	-8.498875671258D-03
7	-.5	8.949865074626D-03	4.474932537313D-03	3.25	9.532736584974D-03
8	.5	-1.257082960536D-02	-6.285414802682D-03	3.50	1.455681244564D-02
9	1.5	1.547095395781D-02	7.735476978906D-03	3.75	1.122462076723D-02
10	2.5	-1.609511123041D-02	-8.047555615204D-03	4.25	-1.365136180844D-02
11	3.5	1.366357404784D-02	6.831787023922D-03	4.50	-2.146993031076D-02
12	4.5	-9.019867406017D-03	-4.509933703009D-03	4.75	-1.695552437924D-02

13	5.5	4.284132678611D-03	2.142066339306D-03	5.25	2.136072049565D-02
14	6.5	-1.279845162235D-03	-6.399225811174D-04	5.50	3.403776252921D-02
15	7.5	1.777287936564D-04	8.886439682822D-05	5.75	2.718070435143D-02
				6.25	-3.488901290998D-02
	0	6.332658453765D+01	-3.470941990308D+02	6.50	-5.608448709722D-02
	1	4.147619194448D+01	2.692988294095D+02	6.75	-4.519814168068D-02
	2	1.013487800393D+01	-5.577797622224D+01	7.25	5.931802290601D-02
	3	1.094443940320D+00	7.128325249690D+00	7.50	9.676957863559D-02
	4	4.404985825892D-02	-2.432341639413D-01	7.75	7.945686904084D-02

I=11 H= 3.5

0		0.	1.390626651678D+04	-1.00	-4.766124112653D-01
1		1.097045640293D-01	-4.081336612381D+03	-.75	-2.644402251081D-01
2		-2.133612761269D-01	4.485205330900D+02	-.50	-1.258265888517D-01
3		1.285972337757D-01	-2.182380876900D+01	-.25	-4.286801089195D-02
4		-2.494933733341D-02	3.874432531450D-01	.25	1.600293471180D-02
				.50	1.602730083961D-02
0	-7.5	8.815655335100D-06	4.407827667550D-06	.75	8.621198854765D-03
1	-6.5	-6.611195352903D-05	-3.305597676452D-05	1.25	-5.925603294279D-03
2	-5.5	2.395813628891D-04	1.197906814446D-04	1.50	-7.457607217756D-03
3	-4.5	-5.856710294565D-04	-2.928355147282D-04	1.75	-4.880527443970D-03
4	-3.5	1.149415329975D-03	5.747076649875D-04	2.25	4.608461044801D-03
5	-2.5	-2.022595940332D-03	-1.011297970166D-03	2.50	6.589681084303D-03
6	-1.5	3.381260669724D-03	1.690630334862D-03	2.75	4.812291804534D-03
7	-.5	-5.446896474565D-03	-2.723448237282D-03	3.25	-5.396629707050D-03
8	.5	8.296015108670D-03	4.148007554335D-03	3.50	-8.239608175126D-03
9	1.5	-1.146345500914D-02	-5.731727504568D-03	3.75	-6.352229835273D-03
10	2.5	1.366357404784D-02	6.831787023922D-03	4.25	7.720883856073D-03
11	3.5	-1.329339542800D-02	-6.646697713998D-03	4.50	1.213733956323D-02
12	4.5	9.885686926795D-03	4.942843463397D-03	4.75	9.579499165460D-03
13	5.5	-5.152743892098D-03	-2.576371946049D-03	5.25	-1.204616333373D-02
14	6.5	1.645806468593D-03	8.229032342965D-04	5.50	-1.916881557348D-02
15	7.5	-2.392858427132D-04	-1.196429213566D-04	5.75	-1.527943783579D-02
				6.25	1.950443212282D-02
	0	-3.586809220930D+01	5.444972539619D+02	6.50	3.122404690061D-02
	1	-2.349193812836D+01	-3.824858561172D+02	6.75	2.502650561608D-02
	2	-5.740324866201D+00	8.824693372403D+01	7.25	-3.230135688413D-02
	3	-6.198828862267D-01	-1.020051117465D+01	7.50	-5.203567760305D-02
	4	-2.494933733341D-02	3.874432531450D-01	7.75	-4.201512000705D-02

I=12 H= 4.5

0		0.	-1.684084081839D+04	-1.00	2.425774707043D-01
1		-5.583520560253D-02	4.973504911813D+03	-.75	1.345898500744D-01
2		1.085926022617D-01	-5.498740268101D+02	-.50	6.404081268224D-02
3		-6.545128620874D-02	2.694819613167D+01	-.25	2.181811842272D-02
4		1.269837663130D-02	-4.892338223054D-01	.25	-8.144837089689D-03
				.50	-8.157223236307D-03
0	-7.5	-4.487081735122D-06	-2.243540867561D-06	.75	-4.387817979128D-03
1	-6.5	3.365157573596D-05	1.682578786798D-05	1.25	3.015855491075D-03
2	-5.5	-1.219600315166D-04	-6.098001575832D-05	1.50	3.795554656547D-03
3	-4.5	2.982112667060D-04	1.491056333530D-04	1.75	2.483931289261D-03
4	-3.5	-5.856710294565D-04	-2.928355147282D-04	2.25	-2.345414765849D-03
5	-2.5	1.032759933156D-03	5.163799665782D-04	2.50	-3.353676731298D-03
6	-1.5	-1.737422196827D-03	-8.687110984134D-04	2.75	-2.449060480231D-03
7	-.5	2.850852009574D-03	1.425426004787D-03	3.25	2.746258941480D-03

8	.5	-4.538008253093D-03	-2.269004126547D-03	3.50	4.192795472124D-03
9	1.5	6.787855574733D-03	3.393927787366D-03	3.75	3.232172419324D-03
10	2.5	-9.019867406017D-03	-4.509933703009D-03	4.25	-3.927776847154D-03
11	3.5	9.885686926795D-03	4.942843463397D-03	4.50	-6.173574233715D-03
12	4.5	-8.204077444528D-03	-4.102038722264D-03	4.75	-4.871560668854D-03
13	5.5	4.677966018627D-03	2.338983009313D-03	5.25	6.122168655665D-03
14	6.5	-1.600290995232D-03	-8.001454976162D-04	5.50	9.737562423722D-03
15	7.5	2.448011330796D-04	1.224005665398D-04	5.75	7.757057181014D-03
				6.25	-9.883554224603D-03
	0	1.825576577586D+01	-6.491641356966D+02	6.50	-1.580021108747D-02
	1	1.195665984592D+01	4.473205317410D+02	6.75	-1.264086845702D-02
	2	2.921640775629D+00	-1.086560288756D+02	7.25	1.622441938407D-02
	3	3.155000127303D-01	1.227118146251D+01	7.50	2.602754931227D-02
	4	1.269837663130D-02	-4.892338223054D-01	7.75	2.090011835847D-02

I=13 H= 5.5

0		0.	1.372042928690D+04	-1.00	-9.919276254218D-02
1		2.283163733285D-02	-4.093814068220D+03	-.75	-5.503534829772D-02
2		-4.440476727890D-02	4.573106815690D+02	-.50	-2.618702201371D-02
3		2.676382651045D-02	-2.265403276488D+01	-.25	-8.921675403229D-03
4		-5.192531419973D-03	4.179386206711D-01	.25	3.330512848648D-03
				.50	3.335575530462D-03
0	-7.5	1.834855565320D-06	9.174277826600D-07	.75	1.794224089310D-03
1	-6.5	-1.376096362469D-05	-6.880481812345D-06	1.25	-1.233211529247D-03
2	-5.5	4.987403069518D-05	2.493701534759D-05	1.50	-1.552035084056D-03
3	-4.5	-1.219600315166D-04	-6.098001575832D-05	1.75	-1.015698596716D-03
4	-3.5	2.395813628891D-04	1.197906814446D-04	2.25	9.590511951789D-04
5	-2.5	-4.227796734253D-04	-2.113898367127D-04	2.50	1.371326720270D-03
6	-1.5	7.128010353351D-04	3.564005176676D-04	2.75	1.001419530990D-03
7	-.5	-1.177299302755D-03	-5.886496513775D-04	3.25	-1.122918193421D-03
8	.5	1.910347950150D-03	9.551739750752D-04	3.50	-1.714363053260D-03
9	1.5	-2.988650996957D-03	-1.494325498479D-03	3.75	-1.321550274595D-03
10	2.5	4.284132678611D-03	2.142066339306D-03	4.25	1.605851497325D-03
11	3.5	-5.152743892098D-03	-2.576371946049D-03	4.50	2.523900476698D-03
12	4.5	4.677966018627D-03	2.338983009313D-03	4.75	1.991467822537D-03
13	5.5	-2.875874396529D-03	-1.437937198265D-03	5.25	-2.502174095303D-03
14	6.5	1.043508726130D-03	5.217543630649D-04	5.50	-3.979172399848D-03
15	7.5	-1.669774010968D-04	-8.348870054840D-05	5.75	-3.169185479031D-03
				6.25	4.035370031382D-03
	0	-7.465035516351D+00	5.057634577877D+02	6.50	6.447970287812D-03
	1	-4.889240919415D+00	-3.517504513756D+02	6.75	5.155380038677D-03
	2	-1.194698025034D+00	8.864922883569D+01	7.25	-6.604051438269D-03
	3	-1.290121160887D-01	-1.011587414475D+01	7.50	-1.057899837144D-02
	4	-5.192531419973D-03	4.179386206711D-01	7.75	-8.478758734100D-03

I=14 H= 6.5

0		0.	-6.337025065047D+03	-1.00	2.736807396598D-02
1		-6.299427625063D-03	1.913315937378D+03	-.75	1.518468989636D-02
2		1.225162758126D-02	-2.164140875042D+02	-.50	7.225206644730D-03
3		-7.384356230483D-03	1.086337924335D+01	-.25	2.461560534200D-03
4		1.432662529178D-03	-2.036210374707D-01	.25	-9.189144124652D-04
				.50	-9.203110267335D-04
0	-7.5	-5.062548875122D-07	-2.531274437561D-07	.75	-4.950401226928D-04
1	-6.5	3.796804527443D-06	1.898402263721D-06	1.25	3.402521762385D-04
2	-5.5	-1.376096362469D-05	-6.880481812345D-06	1.50	4.282176733716D-04

CARDINAL SPLINES N=5

3	-4.5	3.365157573596D-05	1.682578786798D-05	1.75	2.802383274418D-04
4	-3.5	-6.611195352903D-05	-3.305597676452D-05	2.25	-2.646082077767D-04
5	-2.5	1.166964225181D-04	5.834821125904D-05	2.50	-3.783568261552D-04
6	-1.5	-1.969073726596D-04	-9.845368632981D-05	2.75	-2.762965822628D-04
7	-.5	3.260145152285D-04	1.630072576142D-04	3.25	3.098160490772D-04
8	.5	-5.329021033224D-04	-2.664510516612D-04	3.50	4.729942001311D-04
9	1.5	8.517998250403D-04	4.258999125201D-04	3.75	3.646137389140D-04
10	2.5	-1.279845162235D-03	-6.399225811174D-04	4.25	-4.430404872935D-04
11	3.5	1.645806468593D-03	8.229032342965D-04	4.50	-6.963085893572D-04
12	4.5	-1.600290995232D-03	-8.001454976162D-04	4.75	-5.494037201577D-04
13	5.5	1.043508726130D-03	5.217543630649D-04	5.25	6.902421437564D-04
14	6.5	-3.967990827799D-04	-1.983995413899D-04	5.50	1.097617253425D-03
15	7.5	6.584955049729D-05	3.292477524865D-05	5.75	8.741216375322D-04
				6.25	-1.112765992220D-03
	0	2.059666843249D+00	-2.217295274067D+02	6.50	-1.777729873057D-03
	1	1.348982890187D+00	1.566893713982D+02	6.75	-1.421023050517D-03
	2	3.296272159929D-01	-4.071015467522D+01	7.25	1.819025470120D-03
	3	3.559551964485D-02	4.754748119229D+00	7.50	2.912321525200D-03
	4	1.432662529178D-03	-2.036210374707D-01	7.75	2.332491270432D-03

I=15 H= 7.5

0		0.	1.235932863393D+03	-1.00	-3.649161757232D-03
1		8.399431293348D-04	-3.775901774012D+02	-.75	-2.024672527681D-03
2		-1.633588403914D-03	4.326568338352D+01	-.50	-9.633833045242D-04
3		9.846039980685D-04	-2.203125996664D+00	-.25	-3.282156912431D-04
4		-1.910262259150D-04	4.199190052057D-02	.25	1.225247486214D-04
				.50	1.227109561684D-04
0	-7.5	6.750242571383D-08	3.375121285691D-08	.75	6.600685808965D-05
1	-6.5	-5.062548875122D-07	-2.531274437561D-07	1.25	-4.536797725031D-05
2	-5.5	1.834855565320D-06	9.174277826600D-07	1.50	-5.709695878639D-05
3	-4.5	-4.487081735122D-06	-2.243540867561D-06	1.75	-3.736592761908D-05
4	-3.5	8.815655335100D-06	4.407827667550D-06	2.25	3.528182742417D-05
5	-2.5	-1.556252077268D-05	-7.781260386338D-06	2.50	5.044858281125D-05
6	-1.5	2.626806160649D-05	1.313403080324D-05	2.75	3.684023785534D-05
7	-.5	-4.353478019342D-05	-2.176739009671D-05	3.25	-4.130944127451D-05
8	.5	7.137718908806D-05	3.568859454403D-05	3.50	-6.306696950083D-05
9	1.5	-1.151422486578D-04	-5.757112432892D-05	3.75	-4.861561573469D-05
10	2.5	1.777287936564D-04	8.886439682822D-05	4.25	5.907198137934D-05
11	3.5	-2.392858427132D-04	-1.196429213566D-04	4.50	9.284026740534D-05
12	4.5	2.448011330796D-04	1.224005665398D-04	4.75	7.325235968940D-05
13	5.5	-1.669774010968D-04	-8.348870054840D-05	5.25	-9.202745154169D-05
14	6.5	6.584955049729D-05	3.292477524865D-05	5.50	-1.463377140671D-04
15	7.5	-1.124661119749D-05	-5.623305598744D-06	5.75	-1.165368494941D-04
				6.25	1.483380810584D-04
	0	-2.746288804991D-01	4.112244110570D+01	6.50	2.369641563210D-04
	1	-1.798687144868D-01	-2.952110645689D+01	6.75	1.893982621513D-04
	2	-4.395134969367D-02	7.867614884278D+00	7.25	-2.423736480706D-04
	3	-4.746182779380D-03	-9.433689810469D-01	7.50	-3.879626513319D-04
	4	-1.910262259150D-04	4.199190052057D-02	7.75	-3.106306827613D-04

CARDINAL SPLINES N=5

N= 5 M=16 P= 8.0 Q= 9

I= 0 H= -8.0

0		1.000000000000D+00	3.202814281649D+00	-1.00	5.855087076805D+00
1		-2.284282691815D+00	-1.879344630165D+00	-.75	4.048116716457D+00
2		1.843373910692D+00	9.970140910615D-01	-.50	2.688652679675D+00
3		-6.432552232072D-01	-3.250448670748D-01	-.25	1.696661214809D+00
4		8.417525109121D-02	4.214569698624D-02	.25	5.344181431335D-01
				.50	2.435561600914D-01
0	-8.0	-1.124676096365D-05	-5.623380481824D-06	.75	7.894523969348D-02
1	-7.0	6.585067371726D-05	3.292533685863D-05	1.25	-2.601729767753D-02
2	-6.0	-1.669814720918D-04	-8.349073604588D-05	1.50	-2.411414299955D-02
3	-5.0	2.448110887622D-04	1.224055443811D-04	1.75	-1.186262845510D-02
4	-4.0	-2.393054034823D-04	-1.196527017411D-04	2.25	6.554086646740D-03
5	-3.0	1.777633305731D-04	8.888166528654D-05	2.50	7.241302109187D-03
6	-2.0	-1.152005728683D-04	-5.760028643414D-05	2.75	4.101610006957D-03
7	-1.0	7.147399726893D-05	3.573699863446D-05	3.25	-2.785910913295D-03
8	0.0	-4.369422749742D-05	-2.184711374871D-05	3.50	-3.317207474987D-03
9	1.0	2.652888178396D-05	1.326444089198D-05	3.75	-1.995764187899D-03
10	2.0	-1.598290629760D-05	-7.991453148799D-06	4.25	1.478525439433D-03
11	3.0	9.464342482865D-06	4.732171241433D-06	4.50	1.814985552410D-03
12	4.0	-5.360345861809D-06	-2.680172930904D-06	4.75	1.118489240512D-03
13	5.0	2.728210828178D-06	1.364105414089D-06	5.25	-8.567090140955D-04
14	6.0	-1.115597289636D-06	-5.577986448179D-07	5.50	-1.063618290237D-03
15	7.0	3.078021859308D-07	1.539010929654D-07	5.75	-6.612056683656D-04
16	8.0	-4.104125000128D-08	-2.052062500064D-08	6.25	5.125115657143D-04
				6.50	6.388302899148D-04
	0	1.161368229373D+02	5.818276198158D+01	6.75	3.983571964472D-04
	1	7.609561125827D+01	3.797865377627D+01	7.25	-3.100935024321D-04
	2	1.872854497274D+01	9.379884923981D+00	7.50	-3.871030856072D-04
	3	2.050352811711D+00	1.023617436485D+00	7.75	-2.416857215547D-04
	4	8.417525109121D-02	4.214569698624D-02	8.25	1.885163118589D-04

I= 1 H= -7.0

0		0.	-2.027064924215D+01	-1.00	-1.484713292498D+01
1		5.137175748403D+00	8.097487678566D+00	-.75	-8.822328078173D+00
2		-6.514921471748D+00	-3.822401504616D+00	-.50	-4.571155335487D+00
3		2.786357788749D+00	1.418807732285D+00	-.25	-1.736609767644D+00
4		-4.086779160775D-01	-2.047744840232D-01	.25	9.190517877072D-01
				.50	1.262609988718D+00
0	-8.0	6.585067371726D-05	3.292533685863D-05	.75	1.234429873066D+00
1	-7.0	-3.968075067329D-04	-1.984037533664D-04	1.25	6.867517477370D-01
2	-6.0	1.043539257879D-03	5.217696289395D-04	1.50	3.847466441562D-01
3	-5.0	-1.600365661108D-03	-8.001828305541D-04	1.75	1.484020126620D-01
4	-4.0	1.645953170935D-03	8.229765854675D-04	2.25	-6.439209329576D-02
5	-3.0	-1.280104183060D-03	-6.400520915299D-04	2.50	-6.651266544968D-02
6	-2.0	8.522372464020D-04	4.261186232010D-04	2.75	-3.587796189557D-02
7	-1.0	-5.336281477213D-04	-2.668140738606D-04	3.25	2.286589335516D-02
8	0.0	3.272103420785D-04	1.636051710392D-04	3.50	2.666392579813D-02

CARDINAL SPLINES N=5

9	1.0	-1.988634783035D-04	-9.943173915176D-05	3.75	1.578704868932D-02
10	2.0	1.198492403173D-04	5.992462015863D-05	4.25	-1.144288578673D-02
11	3.0	-7.097699350860D-05	-3.548849675430D-05	4.50	-1.394423115470D-02
12	4.0	4.020090371893D-05	2.010045185946D-05	4.75	-8.544714889594D-03
13	5.0	-2.046097160963D-05	-1.023048580481D-05	5.25	6.494755436515D-03
14	6.0	8.366765806598D-06	4.183382903299D-06	5.50	8.042760618419D-03
15	7.0	-2.308460996516D-06	-1.154230498258D-06	5.75	4.990063163180D-03
16	8.0	3.078021859308D-07	1.539010929654D-07	6.25	-3.857708864737D-03
				6.50	-4.804344718447D-03
	0	-6.231871246189D+02	-3.124511717378D+02	6.75	-2.993877020868D-03
	1	-4.010932444866D+02	-2.000279950759D+02	7.25	2.328462512059D-03
	2	-9.657465431556D+01	-4.840441779466D+01	7.50	2.905874932832D-03
	3	-1.029133552573D+01	-5.133975756456D+00	7.75	1.813866141754D-03
	4	-4.086779160775D-01	-2.047744840232D-01	8.25	-1.414408577813D-03

I= 2 H= -6.0

0		0.	7.346901589992D+01	-1.00	2.176747284787D+01
1		-5.754236543635D+00	-2.291608363644D+01	-.75	1.239481379955D+01
2		1.004273862137D+01	7.068942820266D+00	-.50	6.081547556687D+00
3		-5.129416389562D+00	-2.657597456322D+00	-.25	2.149662904633D+00
4		8.410812932954D-01	4.221191687844D-01	.25	-8.877496299947D-01
				.50	-9.550434104739D-01
0	-8.0	-1.669814720918D-04	-8.349073604588D-05	.75	-5.644986323446D-01
1	-7.0	1.043539257879D-03	5.217696289395D-04	1.25	5.327690517634D-01
2	-6.0	-2.875985055706D-03	-1.437992527853D-03	1.50	9.045835245959D-01
3	-5.0	4.678236637395D-03	2.339118318698D-03	1.75	1.058303473532D+00
4	-4.0	-5.153275599612D-03	-2.576637799806D-03	2.25	7.840615003195D-01
5	-3.0	4.285071472888D-03	2.142535736444D-03	2.50	4.922597294948D-01
6	-2.0	-2.990236385587D-03	-1.495118192793D-03	2.75	2.092995343427D-01
7	-1.0	1.912979423222D-03	9.564897116109D-04	3.25	-1.057553008543D-01
8	0.0	-1.181633453888D-03	-5.908167269438D-04	3.50	-1.157643237886D-01
9	1.0	7.198907385742D-04	3.599453692871D-04	3.75	-6.550982989957D-02
10	2.0	-4.342067364946D-04	-2.171033682473D-04	4.25	4.481835467557D-02
11	3.0	2.572141987274D-04	1.286070993637D-04	4.50	5.361219673379D-02
12	4.0	-1.456973955793D-04	-7.284869778966D-05	4.75	3.239414813126D-02
13	5.0	7.415751889046D-05	3.707875944523D-05	5.25	-2.416348927581D-02
14	6.0	-3.032431713483D-05	-1.516215856742D-05	5.50	-2.973739345812D-02
15	7.0	8.366765806598D-06	4.183382903299D-06	5.75	-1.836290200136D-02
16	8.0	-1.115597289636D-06	-5.577986448179D-07	6.25	1.410558278459D-02
				6.50	1.752995469426D-02
	0	1.415509165301D+03	7.108629050097D+02	6.75	1.090643996297D-02
	1	8.926161232714D+02	4.444283475445D+02	7.25	-8.464156914730D-03
	2	2.099119618973D+02	1.053803646818D+02	7.50	-1.055563336203D-02
	3	2.178518499589D+01	1.085021594478D+01	7.75	-6.585351984511D-03
	4	8.410812932954D-01	4.221191687844D-01	8.25	5.131406564289D-03

I= 3 H= -5.0

0		0.	-1.796705516538D+02	-1.00	-2.127825577967D+01
1		5.130742544323D+00	5.157122997515D+01	-.75	-1.191250693080D+01
2		-9.648056100493D+00	-9.831427531886D+00	-.50	-5.727867801961D+00
3		5.508262939969D+00	2.981294825798D+00	-.25	-1.975627603122D+00
4		-9.911941948884D-01	-4.994574188814D-01	.25	7.618768868471D-01
				.50	7.799409555006D-01
0	-8.0	2.448110887622D-04	1.224055443811D-04	.75	4.312221228521D-01
1	-7.0	-1.600365661108D-03	-8.001828305541D-04	1.25	-3.214171457099D-01

CARDINAL SPLINES N=5

2	-6.0	4.678236637395D-03	2.339118318698D-03	1.50	-4.301373639002D-01
3	-5.0	-8.204739247010D-03	-4.102369623505D-03	1.75	-3.062778194435D-01
4	-4.0	9.886987225936D-03	4.943493612968D-03	2.25	3.901018549583D-01
5	-3.0	-9.022163242279D-03	-4.511081621139D-03	2.50	7.469568277547D-01
6	-2.0	6.791732667717D-03	3.395866333858D-03	2.75	9.690204382561D-01
7	-1.0	-4.544443562137D-03	-2.272221781068D-03	3.25	8.447052551589D-01
8	0.0	2.861451244765D-03	1.430725622382D-03	3.50	5.644692655679D-01
9	1.0	-1.754760178305D-03	-8.773800891527D-04	3.75	2.527440596621D-01
10	2.0	1.060704996746D-03	5.303524983732D-04	4.25	-1.379412291088D-01
11	3.0	-6.287924085708D-04	-3.143962042854D-04	4.50	-1.552753764887D-01
12	4.0	3.562613662086D-04	1.781306831043D-04	4.75	-8.985919979791D-02
13	5.0	-1.813456862289D-04	-9.067284311447D-05	5.25	6.347011875256D-02
14	6.0	7.415751889046D-05	3.707875944523D-05	5.50	7.676984369911D-02
15	7.0	-2.046097160963D-05	-1.023048580481D-05	5.75	4.679119650606D-02
16	8.0	2.728210828178D-06	1.364105414089D-06	6.25	-3.532500748498D-02
				6.50	-4.365179794134D-02
	0	-1.816130447076D+03	-9.156667108228D+02	6.75	-2.704107055700D-02
	1	-1.121617381721D+03	-5.562117978509D+02	7.25	2.086440937810D-02
	2	-2.580683163784D+02	-1.300720005632D+02	7.50	2.597040981195D-02
	3	-2.620995129646D+01	-1.300134257841D+01	7.75	1.617868957395D-02
	4	-9.911941948884D-01	-4.994574188814D-01	8.25	-1.258224094598D-02

I= 4   H= -4.0

0		0.	3.530190617017D+02	-1.00	1.580679308269D+01
1		-3.689055726872D+00	-9.813190263828D+01	-.75	8.793545982234D+00
2		7.103624995287D+00	1.339042070252D+01	-.50	4.197204957402D+00
3		-4.214221161773D+00	-2.553444386802D+00	-.25	1.435212274571D+00
4		7.998911987621D-01	4.075304154100D-01	.25	-5.410100010831D-01
				.50	-5.454065273069D-01
0	-8.0	-2.393054034823D-04	-1.196527017411D-04	.75	-2.958046814913D-01
1	-7.0	1.645953170935D-03	8.229765854675D-04	1.25	2.082701040179D-01
2	-6.0	-5.153275599612D-03	-2.576637799806D-03	1.50	2.668289329618D-01
3	-5.0	9.886987225936D-03	4.943493612968D-03	1.75	1.788027297631D-01
4	-4.0	-1.329595023536D-02	-6.647975117681D-03	2.25	-1.819544499245D-01
5	-3.0	1.366808487094D-02	6.834042435468D-03	2.50	-2.760505617371D-01
6	-2.0	-1.147107266064D-02	-5.735536330322D-03	2.75	-2.189994091638D-01
7	-1.0	8.308659102985D-03	4.154329551493D-03	3.25	3.308188257976D-01
8	0.0	-5.467721685690D-03	-2.733860842845D-03	3.50	6.763667835422D-01
9	1.0	3.415326063905D-03	1.707663031953D-03	3.75	9.265496981394D-01
10	2.0	-2.077501963146D-03	-1.038750981573D-03	4.25	8.761709943708D-01
11	3.0	1.234139539594D-03	6.170697697970D-04	4.50	6.030973168480D-01
12	4.0	-6.997269548390D-04	-3.498634774195D-04	4.75	2.765501277912D-01
13	5.0	3.562613662086D-04	1.781306831043D-04	5.25	-1.561787535021D-01
14	6.0	-1.456973955793D-04	-7.284869778966D-05	5.50	-1.779204380086D-01
15	7.0	4.020090371893D-05	2.010045185946D-05	5.75	-1.039389502336D-01
16	8.0	-5.360345861809D-06	-2.680172930904D-06	6.25	7.438876627390D-02
				6.50	9.038427362755D-02
	0	1.543792669185D+03	7.868318210337D+02	6.75	5.528451249826D-02
	1	9.390156562020D+02	4.604757970958D+02	7.25	-4.194495956227D-02
	2	2.131205374374D+02	1.085994349367D+02	7.50	-5.192340080866D-02
	3	2.138229719861D+01	1.048752890632D+01	7.75	-3.221150767017D-02
	4	7.998911987621D-01	4.075304154100D-01	8.25	2.491131520156D-02

I= 5   H= -3.0

0		0.	-6.233253258145D+02	-1.00	-1.024486707578D+01

CARDINAL SPLINES N=5

1		2.368501870699D+00	1.711988529591D+02	-.75	-5.688991427020D+00
2		-4.591906497183D+00	-1.966799214157D+01	-.50	-2.709621402984D+00
3		2.753842785524D+00	2.165013807129D+00	-.25	-9.242211357192D-01
4		-5.306159223706D-01	-2.787004916908D-01	.25	3.460873873559D-01
				.50	3.473415112900D-01
0	-8.0	1.777633305731D-04	8.888166528654D-05	.75	1.873295766180D-01
1	-7.0	-1.280104183060D-03	-6.400520915299D-04	1.25	-1.297519670846D-01
2	-6.0	4.285071472888D-03	2.142535736444D-03	1.50	-1.642291990730D-01
3	-5.0	-9.022163242279D-03	-4.511081621139D-03	1.75	-1.082879848742D-01
4	-4.0	1.366808487094D-02	6.834042435468D-03	2.25	1.046647965315D-01
5	-3.0	-1.610307563713D-02	-8.051537818565D-03	2.50	1.523582807662D-01
6	-2.0	1.548440384792D-02	7.742201923962D-03	2.75	1.139629385238D-01
7	-1.0	-1.259315411495D-02	-6.296577057473D-03	3.25	-1.379108546795D-01
8	0.0	8.986634517393D-03	4.493317258696D-03	3.50	-2.236050385442D-01
9	1.0	-5.866204966809D-03	-2.933102483404D-03	3.75	-1.874442126165D-01
10	2.0	3.633882261388D-03	1.816941130694D-03	4.25	3.074376731672D-01
11	3.0	-2.172186835602D-03	-1.086093417801D-03	4.50	6.476613266996D-01
12	4.0	1.234139539594D-03	6.170697697970D-04	4.75	9.088570654800D-01
13	5.0	-6.287924085708D-04	-3.143962042854D-04	5.25	8.897290276542D-01
14	6.0	2.572141987274D-04	1.286070993637D-04	5.50	6.199354880457D-01
15	7.0	-7.097699350860D-05	-3.548849675430D-05	5.75	2.870221952769D-01
16	8.0	9.464342482865D-06	4.732171241433D-06	6.25	-1.643060402416D-01
				6.50	-1.880600196292D-01
	0	-1.038369312696D+03	-5.455561439178D+02	6.75	-1.102690417613D-01
	1	-6.290655962786D+02	-2.985849773201D+02	7.25	7.933299224347D-02
	2	-1.422561938349D+02	-7.472864957975D+01	7.50	9.657135795750D-02
	3	-1.422586673034D+01	-6.753401926977D+00	7.75	5.915938319233D-02
	4	-5.306159223706D-01	-2.787004916908D-01	8.25	-4.499475899571D-02

I= 6 H= -2.0

0		0.	1.052778701727D+03	-1.00	6.358226894726D+00
1		-1.465604347354D+00	-2.878831348795D+02	-.75	3.528722757790D+00
2		2.847482134487D+00	3.076472293917D+01	-.50	1.679584701837D+00
3		-1.713451499723D+00	-2.187798258312D+00	-.25	5.724370597441D-01
4		3.316889131629D-01	1.884642114790D-01	.25	-2.139104737386D-01
				.50	-2.143827454490D-01
0	-8.0	-1.152005728683D-04	-5.760028643414D-05	.75	-1.154171159306D-01
1	-7.0	8.522372464020D-04	4.261186232010D-04	1.25	7.952951829655D-02
2	-6.0	-2.990236385587D-03	-1.495118192793D-03	1.50	1.002775603896D-01
3	-5.0	6.791732667717D-03	3.395866333858D-03	1.75	6.578690146764D-02
4	-4.0	-1.147107266064D-02	-5.735536330322D-03	2.25	-6.259524545462D-02
5	-3.0	1.548440384792D-02	7.742201923962D-03	2.50	-9.002732847609D-02
6	-2.0	-1.727511070155D-02	-8.637555350775D-03	2.75	-6.625607569496D-02
7	-1.0	1.620306211386D-02	8.101531056928D-03	3.25	7.611065953685D-02
8	0.0	-1.301838957885D-02	-6.509194789427D-03	3.50	1.183543132592D-01
9	1.0	9.221600411489D-03	4.610800205744D-03	3.75	9.350147717750D-02
10	2.0	-5.969771280397D-03	-2.984885640198D-03	4.25	-1.227451833632D-01
11	3.0	3.633882261388D-03	1.816941130694D-03	4.50	-2.049818747187D-01
12	4.0	-2.077501963146D-03	-1.038750981573D-03	4.75	-1.759626542586D-01
13	5.0	1.060704996746D-03	5.303524983732D-04	5.25	2.986321804111D-01
14	6.0	-4.342067364946D-04	-2.171033682473D-04	5.50	6.367191863304D-01
15	7.0	1.198492403173D-04	5.992462015863D-05	5.75	9.020468596177D-01
16	8.0	-1.598290629760D-05	-7.991453148799D-06	6.25	8.950257629133D-01
				6.50	6.265538872092D-01
	0	6.518246422853D+02	3.704525927598D+02	6.75	2.911621807183D-01
	1	3.944103160152D+02	1.702698716603D+02	7.25	-1.675581968395D-01

	2	8.909318879568D+01	5.062782194762D+01	7.50	-1,921463005046D-01
	3	8,900593721489D+00	3.843056509015D+00	7.75	-1,128414730859D-01
	4	3,316889131629D-01	1.884642114790D-01	8,25	8,138784162853D-02

I= 7 H= -1.0

0		0.	-1.748122928207D+03	-1.00	-3,890283293991D+00
1		8.958676370592D-01	4.772592564836D+02	-.75	-2,158650976337D+00
2		-1,741761302319D+00	-4.959164975969D+01	-.50	-1,027242433331D+00
3		1.049238272937D+00	2,734879130640D+00	-.25	-3,500159327434D-01
4		-2,034160816747D-01	-1.392685614485D-01	.25	1,307065820881D-01
				.50	1,309349115600D-01
0	-8,0	7,147399726893D-05	3.573699863446D-05	.75	7,045063864560D-02
1	-7.0	-5,336281477213D-04	-2,668140738606D-04	1.25	-4,846275912833D-02
2	-6.0	1,912979423222D-03	9,564897116109D-04	1.50	-6,102955862320D-02
3	-5,0	-4.544443562137D-03	-2.272221781068D-03	1.75	-3,997220353220D-02
4	-4.0	8,308659102985D-03	4.154329551493D-03	2.25	3,783861405873D-02
5	-3,0	-1,259315411495D-02	-6.296577057473D-03	2,50	5,420921696042D-02
6	-2.0	1.620306211386D-02	8,101531056928D-03	2.75	3,968858157148D-02
7	-1.0	-1,770661973847D-02	-8,853309869235D-03	3.25	-4,486009131970D-02
8	0,0	1,644304164875D-02	8.221520824374D-03	3,50	-6,890264103143D-02
9	1.0	-1,312488819514D-02	-6,562444097571D-03	3,75	-5,354101599624D-02
10	2.0	9,221600411489D-03	4,610800205744D-03	4.25	6,667882901690D-02
11	3,0	-5,866204966809D-03	-2,933102483404D-03	4,50	1,067654554269D-01
12	4.0	3,415326063905D-03	1.707663031953D-03	4.75	8,635139104330D-02
13	5.0	-1,754760178305D-03	-8.773800891527D-04	5.25	-1,172496963484D-01
14	6,0	7.198907385742D-04	3.599453692871D-04	5.50	-1,981422893013D-01
15	7.0	-1,988634783035D-04	-9,943173915176D-05	5,75	-1,716973554656D-01
16	8,0	2.652888178396D-05	1.326444089198D-05	6,25	2,952956264081D-01
				6.50	6,325329519873D-01
	0	-4,002880570475D+02	-2,741003737638D+02	6.75	8,994145993101D-01
	1	-2,421147000658D+02	-7,633236043519D+01	7.25	8,971245535362D-01
	2	-5,467181811490D+01	-3,743367822057D+01	7,50	6,292186053161D-01
	3	-5,460076340653D+00	-1,721714835713D+00	7.75	2,928616574246D-01
	4	-2.034160816747D-01	-1.392685614485D-01	8.25	-1,689654288507D-01

I= 8 H= 0.0

0		0.	2,882671784473D+03	-1.00	2,367421595482D+00
1		-5,450050230857D-01	-7,865519034003D+02	-.75	1,313561347967D+00
2		1,059849718086D+00	8,087352321523D+01	-.50	6,250431149301D-01
3		-6,386839275414D-01	-3,964253656589D+00	-.25	2,129552172024D-01
4		1.238829267684D-01	1.238829267684D-01	.25	-7,950616724287D-02
				.50	-7,963297538125D-02
0	-8,0	-4,369422749742D-05	-2,184711374871D-05	.75	-4,283903129187D-02
1	-7.0	3.272103420785D-04	1.636051710392D-04	1.25	2,945236523855D-02
2	-6,0	-1,181633453888D-03	-5,908167269438D-04	1,50	3,707427857060D-02
3	-5.0	2,861451244765D-03	1.430725622382D-03	1.75	2,426907238460D-02
4	-4,0	-5,467721685690D-03	-2,733860842845D-03	2,25	-2,293474166438D-02
5	-3,0	8,986634517393D-03	4,493317258696D-03	2.50	-3,281485388333D-02
6	-2.0	-1,301838957885D-02	-6,509194789427D-03	2.75	-2,398365625312D-02
7	-1,0	1,644304164875D-02	8,221520824374D-03	3.25	2,696461236810D-02
8	0,0	-1,781379761411D-02	-8,906898807054D-03	3.50	4,124935058423D-02
9	1,0	1,644304164875D-02	8,221520824374D-03	3.75	3,188210452318D-02
10	2.0	-1,301838957885D-02	-6,509194789427D-03	4.25	-3,905670351911D-02
11	3.0	8,986634517393D-03	4,493317258696D-03	4,50	-6,176088674684D-02
12	4,0	-5,467721685690D-03	-2,733860842845D-03	4.75	-4,912590950440D-02

13	5.0	2.861451244765D-03	1.430725622382D-03	5.25	6.326575759029D-02
14	6.0	-1.181633453888D-03	-5.908167269438D-04	5.50	1.025001062124D-01
15	7.0	3.272103420785D-04	1.636051710392D-04	5.75	8.367752698525D-02
16	8.0	-4.369422749742D-05	-2.184711374871D-05	6.25	-1.151265984899D-01
				6.50	-1.954505652000D-01
	0	2.438886389150D+02	2.438886389150D+02	6.75	-1.699826214989D-01
	1	1.474975104000D+02	-8.064128181864D-20	7.25	2.938782346663D-01
	2	3.330247933616D+01	3.330247933616D+01	7.50	6.306893543263D-01
	3	3.325569729048D+00	-2.065749953842D-21	7.75	8.982041797935D-01
	4	1.238829267684D-01	1.238829267684D-01	8.25	8.982041797935D-01

I= 9 H= 1.0

0		0.	-4.732476686941D+03	-1.00	-1.435211742026D+00
1		3.303659566813D-01	1.291062474268D+03	-.75	-7.963096553882D-01
2		-6.424980942316D-01	-1.322339618739D+02	-.50	-3.789058982113D-01
3		3.872266498909D-01	6.178308802067D+00	-.25	-1.290914780316D-01
4		-7.512104122239D-02	-1.392685614485D-01	.25	4.819233321933D-02
				.50	4.826677275695D-02
0	-8.0	2.652888178396D-05	1.326444089198D-05	.75	2.596375739194D-02
1	-7.0	-1.988634783035D-04	-9.943173915176D-05	1.25	-1.784710055068D-02
2	-6.0	7.198907385742D-04	3.599453692871D-04	1.50	-2.246263387477D-02
3	-5.0	-1.754760178305D-03	-8.773800891527D-04	1.75	-1.470152624448D-02
4	-4.0	3.415326063905D-03	1.707663031953D-03	2.25	1.388540985580D-02
5	-3.0	-5.866204966809D-03	-2.933102483404D-03	2.50	1.985861322587D-02
6	-2.0	9.221600411489D-03	4.610800205744D-03	2.75	1.450592529704D-02
7	-1.0	-1.312488819514D-02	-6.562444097571D-03	3.25	-1.627999112825D-02
8	0.0	1.644304164875D-02	8.221520824374D-03	3.50	-2.487105507286D-02
9	1.0	-1.770661973847D-02	-8.853309869235D-03	3.75	-1.918902453043D-02
10	2.0	1.620306211386D-02	8.101531056928D-03	4.25	2.337948835442D-02
11	3.0	-1.259315411495D-02	-6.296577057473D-03	4.50	3.681919248039D-02
12	4.0	8.308659102985D-03	4.154329551493D-03	4.75	2.912897804943D-02
13	5.0	-4.544443562137D-03	-2.272221781068D-03	5.25	-3.689636869326D-02
14	6.0	1.912979423222D-03	9.564897116109D-04	5.50	-5.903273594546D-02
15	7.0	-5.336281477213D-04	-2.668140738606D-04	5.75	-4.739323920458D-02
16	8.0	7.147399726893D-05	3.573699863446D-05	6.25	6.183949242452D-02
				6.50	1.006476221056D-01
	0	-1.479126904802D+02	-2.741003737638D+02	6.75	8.246266620673D-02
	1	-8.944997919543D+01	7.633236043519D+01	7.25	-1.140448840546D-01
	2	-2.019553832625D+01	-3.743367822057D+01	7.50	-1.939779931722D-01
	3	-2.016646669226D+00	1.721714835713D+00	7.75	-1.689654288507D-01
	4	-7.512104122239D-02	-1.392685614485D-01	8.25	2.928616574246D-01

I=10 H= 2.0

0		0.	7.712386513524D+03	-1.00	8.642429924305D-01
1		-1.989298947777D-01	-2.104156577662D+03	-.75	4.795114327453D-01
2		3.868899778733D-01	2.152314353719D+02	-.50	2.281628624674D-01
3		-2.331836099844D-01	-9.873911276343D+00	-.25	7.773330805266D-02
4		4.523950979507D-02	1.884642114790D-01	.25	-2.901862720919D-02
				.50	-2.906296602300D-02
0	-8.0	-1.598290629760D-05	-7.991453148799D-06	.75	-1.563328041712D-02
1	-7.0	1.198492403173D-04	5.992462015863D-05	1.25	1.074542419046D-02
2	-6.0	-4.342067364946D-04	-2.171033682473D-04	1.50	1.352374015859D-02
3	-5.0	1.060704996746D-03	5.303524983732D-04	1.75	8.850594263860D-03
4	-4.0	-2.077501963146D-03	-1.038750981573D-03	2.25	-8.357715078998D-03
5	-3.0	3.633882261388D-03	1.816941130694D-03	2.50	-1.195131946566D-02

6	-2.0	-5.969771280397D-03	-2.984885640198D-03	2.75	-8.728303397161D-03
7	-1.0	9.221600411489D-03	4.610800205744D-03	3.25	9.789995552137D-03
8	0.0	-1.301838957885D-02	-6.509194789427D-03	3.50	1.494956549239D-02
9	1.0	1.620306211386D-02	8.101531056928D-03	3.75	1.152737692438D-02
10	2.0	-1.727511070155D-02	-8.637555350775D-03	4.25	-1.401922955455D-02
11	3.0	1.548440384792D-02	7.742201923962D-03	4.50	-2.204808129881D-02
12	4.0	-1.147107266064D-02	-5.735536330322D-03	4.75	-1.741168683464D-02
13	5.0	6.791732667717D-03	3.395866333858D-03	5.25	2.193377242123D-02
14	6.0	-2.990236385587D-03	-1.495118192793D-03	5.50	3.494896907927D-02
15	7.0	8.522372464020D-04	4.261186232010D-04	5.75	2.790631348766D-02
16	8.0	-1.152005728683D-04	-5.760028643414D-05	6.25	-3.581249325487D-02
				6.50	-5.755954587544D-02
	0	8.908054323429D+01	3.704525927598D+02	6.75	-4.637693869305D-02
	1	5.387057269450D+01	-1.702698716603D+02	7.25	6.082585908667D-02
	2	1.216245509956D+01	5.062782194762D+01	7.50	9.918233321545D-02
	3	1.214480703458D+00	-3.843056509015D+00	7.75	8.138784162853D-02
	4	4.523950979507D-02	1.884642114790D-01	8.25	-1.128414730859D-01

I=11 H= 3.0

0		0.	-1.231616853616D+04	-1.00	-5.116800970279D-01
1		1.177762403711D-01	3.361675147559D+03	-.75	-2.838969059811D-01
2		-2.290597196742D-01	-3.438312846365D+02	-.50	-1.350845009137D-01
3		1.380590759716D-01	1.567181766108D+01	-.25	-4.602209477903D-02
4		-2.678506101102D-02	-2.787004916908D-01	.25	1.718037156672D-02
				.50	1.720652693530D-02
0	-8.0	9.464342482865D-06	4.732171241433D-06	.75	9.255510553782D-03
1	-7.0	-7.097699350860D-05	-3.548849675430D-05	1.25	-6.361579553170D-03
2	-6.0	2.572141987274D-04	1.286070993637D-04	1.50	-8.006296181901D-03
3	-5.0	-6.287924085708D-04	-3.143962042854D-04	1.75	-5.239605433641D-03
4	-4.0	1.234139539594D-03	6.170697697970D-04	2.25	4.947510747789D-03
5	-3.0	-2.172186835602D-03	-1.086093417801D-03	2.50	7.074478971069D-03
6	-2.0	3.633882261388D-03	1.816941130694D-03	2.75	5.166316404126D-03
7	-1.0	-5.866204966809D-03	-2.933102483404D-03	3.25	-5.793600049158D-03
8	0.0	8.986634517393D-03	4.493317258696D-03	3.50	-8.845656287090D-03
9	1.0	-1.259315411495D-02	-6.296577057473D-03	3.75	-6.819405835421D-03
10	2.0	1.548440384792D-02	7.742201923962D-03	4.25	8.288532038114D-03
11	3.0	-1.610307563713D-02	-8.051537818565D-03	4.50	1.302947093594D-02
12	4.0	1.366808487094D-02	6.834042435468D-03	4.75	1.028339286438D-02
13	5.0	-9.022163242279D-03	-4.511081621139D-03	5.25	-1.293042646148D-02
14	6.0	4.285071472888D-03	2.142535736444D-03	5.50	-2.057487725483D-02
15	7.0	-1.280104183060D-03	-6.400520915299D-04	5.75	-1.639910845494D-02
16	8.0	1.777633305731D-04	8.888166528654D-05	6.25	2.092943321213D-02
				6.50	3.350017587011D-02
	0	-5.274297513983D+01	-5.455561439178D+02	6.75	2.684547992994D-02
	1	-3.189564163843D+01	2.985849773201D+02	7.25	-3.462806390269D-02
	2	-7.201105324586D+00	-7.472864957975D+01	7.50	-5.575873320326D-02
	3	-7.190628763810D-01	6.753401926977D+00	7.75	-4.499475899571D-02
	4	-2.678506101102D-02	-2.787004916908D-01	8.25	5.915938319233D-02

I=12 H= 4.0

0		0.	1.845986141531D+04	-1.00	2.897856697381D-01
1		-6.670131481371D-02	-5.046294596857D+03	-.75	1.607824785075D-01
2		1.297258829840D-01	5.167918082059D+02	-.50	7.650383514178D-02
3		-7.818883988242D-02	-2.352850219944D+01	-.25	2.606415338832D-02
4		1.516963205803D-02	4.075304154100D-01	.25	-9.729905285314D-03

CARDINAL SPLINES N=5

				.50	-9.744700111966D-03
0	-8.0	-5.360345861809D-06	-2.680172930904D-06	.75	-5.241729843726D-03
1	-7.0	4.020090371893D-05	2.010045185946D-05	1.25	3.602767731109D-03
2	-6.0	-1.456973955793D-04	-7.284869778966D-05	1.50	4.534200999005D-03
3	-5.0	3.562613662086D-04	1.781306831043D-04	1.75	2.967322902834D-03
4	-4.0	-6.997269548390D-04	-3.498634774195D-04	2.25	-2.801844253935D-03
5	-3.0	1.234139539594D-03	6.170697697970D-04	2.50	-4.006312758796D-03
6	-2.0	-2.077501963146D-03	-1.038750981573D-03	2.75	-2.925649208485D-03
7	-1.0	3.415326063905D-03	1.707663031953D-03	3.25	3.280661326845D-03
8	0.0	-5.467721685690D-03	-2.733860842845D-03	3.50	5.008658830982D-03
9	1.0	8.308659102985D-03	4.154329551493D-03	3.75	3.861085822622D-03
10	2.0	-1.147107266064D-02	-5.735536330322D-03	4.25	-4.691946132245D-03
11	3.0	1.366808487094D-02	6.834042435468D-03	4.50	-7.374563534581D-03
12	4.0	-1.329595023536D-02	-6.647975117681D-03	4.75	-5.819143981379D-03
13	5.0	9.886987225936D-03	4.943493612968D-03	5.25	7.312565707112D-03
14	6.0	-5.153275599612D-03	-2.576637799806D-03	5.50	1.163040588526D-02
15	7.0	1.645953170935D-03	8.229765854675D-04	5.75	9.264360337518D-03
16	8.0	-2.393054034823D-04	-1.196527017411D-04	6.25	-1.180189396060D-02
				6.50	-1.886434112862D-02
	0	2.987097288235D+01	7.868318210337D+02	6.75	-1.508957585184D-02
	1	1.806406201035D+01	-4.604757970958D+02	7.25	1.935663771091D-02
	2	4.078332436088D+00	1.085994349367D+02	7.50	3.103954876557D-02
	3	4.072393859745D-01	-1.048752890632D+01	7.75	2.491131520156D-02
	4	1.516963205803D-02	4.075304154100D-01	8.25	-3.221150767017D-02

I=13 H= 5.0

0		0.	-2.239243411756D+04	-1.00	-1.474870778222D-01
1		3.394774247669D-02	6.156510375785D+03	-.75	-8.183058898574D-02
2		-6.602426014219D-02	-6.338958712954D+02	-.50	-3.893678049466D-02
3		3.979443232898D-02	2.898397998261D+01	-.25	-1.326539864443D-02
4		-7.720642874302D-03	-4.994574188814D-01	.25	4.952048614606D-03
				.50	4.959575392812D-03
0	-8.0	2.728210828178D-06	1.364105414089D-06	.75	2.667781853729D-03
1	-7.0	-2.046097160963D-05	-1.023048580481D-05	1.25	-1.833626824513D-03
2	-6.0	7.415751889046D-05	3.707875944523D-05	1.50	-2.307675419276D-03
3	-5.0	-1.813456862289D-04	-9.067284311447D-05	1.75	-1.510211572584D-03
4	-4.0	3.562613662086D-04	1.781306831043D-04	2.25	1.425981729532D-03
5	-3.0	-6.287924085708D-04	-3.143962042854D-04	2.50	2.038977906295D-03
6	-2.0	1.060704996746D-03	5.303524983732D-04	2.75	1.488973107958D-03
7	-1.0	-1.754760178305D-03	-8.773800891527D-04	3.25	-1.669615542894D-03
8	0.0	2.861451244765D-03	1.430725622382D-03	3.50	-2.548996932214D-03
9	1.0	-4.544443562137D-03	-2.272221781068D-03	3.75	-1.964933051408D-03
10	2.0	6.791732667717D-03	3.395866333858D-03	4.25	2.387601980011D-03
11	3.0	-9.022163242279D-03	-4.511081621139D-03	4.50	3.752520867455D-03
12	4.0	9.886987225936D-03	4.943493612968D-03	4.75	2.960852128350D-03
13	5.0	-8.204739247010D-03	-4.102369623505D-03	5.25	-3.719958541386D-03
14	6.0	4.678236637395D-03	2.339118318698D-03	5.50	-5.915564399074D-03
15	7.0	-1.600365661108D-03	-8.001828305541D-04	5.75	-4.711167070207D-03
16	8.0	2.448110887622D-04	1.224055443811D-04	6.25	5.997844896347D-03
				6.50	9.582596587786D-03
	0	-1.520297456999D+01	-9.156667108228D+02	6.75	7.660424699863D-03
	1	-9.193786019205D+00	5.562117978509D+02	7.25	-9.808332527101D-03
	2	-2.075684747979D+00	-1.300720005632D+02	7.50	-1.570630860006D-02
	3	-2.072661396487D-01	1.300134257841D+01	7.75	-1.258224094598D-02
	4	-7.720642874302D-03	-4.994574188814D-01	8.25	1.617868957395D-02

CARDINAL SPLINES N=5

I=14 H= 6.0

0		0.	1.829494370407D+04	-1.00	6.030880348885D-02
1		-1.388153352652D-02	-5.078255701521D+03	-.75	3.346126698182D-02
2		2.699791526962D-02	5.278793081697D+02	-.50	1.592159965016D-02
3		-1.627231041926D-02	-2.435802934588D+01	-.25	5.424340140475D-03
4		3.157044273446D-03	4.221191687844D-01	.25	-2.024936327641D-03
				.50	-2.028013660071D-03
0	-8.0	-1.115597289636D-06	-5.577986448179D-07	.75	-1.090878988478D-03
1	-7.0	8.366765806598D-06	4.183382903299D-06	1.25	7.497852181314D-04
2	-6.0	-3.032431713483D-05	-1.516215856742D-05	1.50	9.436270700001D-04
3	-5.0	7.415751889046D-05	3.707875944523D-05	1.75	6.175372018368D-04
4	-4.0	-1.456973955793D-04	-7.284869778966D-05	2.25	-5.830935688161D-04
5	-3.0	2.572141987274D-04	1.286070993637D-04	2.50	-8.337503234433D-04
6	-2.0	-4.342067364946D-04	-2.171033682473D-04	2.75	-6.088485692556D-04
7	-1.0	7.198907385742D-04	3.599453692871D-04	3.25	6.827089954928D-04
8	0.0	-1.181633453888D-03	-5.908167269438D-04	3.50	1.042283746900D-03
9	1.0	1.912979423222D-03	9.564897116109D-04	3.75	8.034541678091D-04
10	2.0	-2.990236385587D-03	-1.495118192793D-03	4.25	-9.762591705948D-04
11	3.0	4.285071472888D-03	2.142535736444D-03	4.50	-1.534329622342D-03
12	4.0	-5.153275599612D-03	-2.576637799806D-03	4.75	-1.210604236955D-03
13	5.0	4.678236637395D-03	2.339118318698D-03	5.25	1.520872163522D-03
14	6.0	-2.875985055706D-03	-1.437992527853D-03	5.50	2.418397240199D-03
15	7.0	1.043539257879D-03	5.217696289395D-04	5.75	1.925881005683D-03
16	8.0	-1.669814720918D-04	-8.349073604588D-05	6.25	-2.451336657265D-03
				6.50	-3.915805058595D-03
	0	6.216644718415D+00	7.108629050097D+02	6.75	-3.129671328237D-03
	1	3.759428182306D+00	-4.444283475445D+02	7.25	4.004611000229D-03
	2	8.487674662105D-01	1.053803646818D+02	7.50	6.409572463362D-03
	3	8.475310633100D-02	-1.085021594478D+01	7.75	5.131406564289D-03
	4	3.157044273446D-03	4.221191687844D-01	8.25	-6.585351984511D-03

I=15 H= 7.0

0		0.	-8.477909745068D+03	-1.00	-1.663964030941D-02
1		3.830016073068D-03	2.379600168310D+03	-.75	-9.232208185658D-03
2		-7.448922087006D-03	-2.502532378145D+02	-.50	-4.392885578903D-03
3		4.489650180543D-03	1.168675924520D+01	-.25	-1.496614979529D-03
4		-8.710519687913D-04	-2.047744840232D-01	.25	5.586946263211D-04
				.50	5.595436404770D-04
0	-8.0	3.078021859308D-07	1.539010929654D-07	.75	3.009813750069D-04
1	-7.0	-2.308460996516D-06	-1.154230498258D-06	1.25	-2.068710852317D-04
2	-6.0	8.366765806598D-06	4.183382903299D-06	1.50	-2.603533774958D-04
3	-5.0	-2.046097160963D-05	-1.023048580481D-05	1.75	-1.703828297025D-04
4	-4.0	4.020090371893D-05	2.010045185946D-05	2.25	1.608794481111D-04
5	-3.0	-7.097699350860D-05	-3.548849675430D-05	2.50	2.300371816374D-04
6	-2.0	1.198492403173D-04	5.992462015863D-05	2.75	1.679851613918D-04
7	-1.0	-1.988634783035D-04	-9.943173915176D-05	3.25	-1.883631918390D-04
8	0.0	3.272103420785D-04	1.636051710392D-04	3.50	-2.875712466192D-04
9	1.0	-5.336281477213D-04	-2.668140738606D-04	3.75	-2.216763603497D-04
10	2.0	8.522372464020D-04	4.261186232010D-04	4.25	2.693516554902D-04
11	3.0	-1.280104183060D-03	-6.400520915299D-04	4.50	4.233215412856D-04
12	4.0	1.645953170935D-03	8.229765854675D-04	4.75	3.340028426495D-04
13	5.0	-1.600365661108D-03	-8.001828305541D-04	5.25	-4.195940116284D-04
14	6.0	1.043539257879D-03	5.217696289395D-04	5.50	-6.671994053184D-04
15	7.0	-3.968075067329D-04	-1.984037533664D-04	5.75	-5.313077945192D-04
16	8.0	6.585067371726D-05	3.292533685863D-05	6.25	6.762156893176D-04

```
                                                        6.50   1,080133698799D-03
      0  -1.715218856715D+00  -3.124511717378D+02      6.75   8,632193683150D-04
      1  -1.037254334739D+00   2.000279950759D+02      7.25  -1,104279330309D-03
      2  -2,341812737698D-01  -4.840441779466D+01      7.50  -1,767135607178D-03
      3  -2,338401282078D-02   5.133975756456D+00      7.75  -1,414408577813D-03
      4  -8,710519687913D-04  -2.047744840232D-01      8.25   1,813866141754D-03


I=16       H=  8.0
0             0.                   1.659045529662D+03  -1.00   2,218668372431D-03
1            -5,106802066510D-04  -4.709057477929D+02   -.75   1,230988630888D-03
2             9,932118255678D-04   5.013065104233D+01   -.50   5,857311721644D-04
3            -5,986334589393D-04  -2.372279740044D+00   -.25   1,995531216866D-04
4             1.161428812725D-04   4.214569698624D-02    .25  -7,449427738728D-05
                                                          .50  -7,460747938012D-05
0    -8,0    -4.104125000128D-08  -2,052062500064D-08    .75  -4,013174163813D-05
1    -7,0     3.078021859308D-07   1.539010929654D-07   1.25   2,758342091004D-05
2    -6,0    -1.115597289636D-06  -5,577986448179D-07   1.50   3,471454772607D-05
3    -5,0     2,728210828178D-06   1.364105414089D-06   1.75   2,271820799103D-05
4    -4,0    -5,360345861809D-06  -2,680172930904D-06   2,25  -2.145105490544D-05
5    -3,0     9,464342482865D-06   4.732171241433D-06   2,50  -3,067227605698D-05
6    -2,0    -1,598290629760D-05  -7,991453148799D-06   2.75  -2,239848923861D-05
7    -1,0     2,652888178396D-05   1.326444089198D-05   3.25   2,511558786032D-05
8     0,0    -4,369422749742D-05  -2.184711374871D-05   3.50   3,834355615272D-05
9     1.0     7.147399726893D-05   3,573699863446D-05   3.75   2,955737150440D-05
10    2,0    -1,152005728683D-04  -5,760028643414D-05   4.25  -3,591406278383D-05
11    3,0     1.777633305731D-04   8,888166528654D-05   4.50  -5,644352108305D-05
12    4,0    -2,393054034823D-04  -1.196527017411D-04   4.75  -4,453406757234D-05
13    5,0     2.448110887622D-04   1.224055443811D-04   5.25   5,594571160803D-05
14    6,0    -1,669814720918D-04  -8,349073604588D-05   5.50   8,895895221631D-05
15    7,0     6,585067371726D-05   3.292533685863D-05   5,75   7.083951325074D-05
16    8,0    -1.124676096365D-05  -5,623380481824D-06   6.25  -9,015721393595D-05
                                                        6,50  -1,440065190038D-04
        0     2.287010258984D-01   5.818276198158D+01   6.75  -1,150831796471D-04
        1     1.383037057322D-01  -3,797865377627D+01   7.25   1,472064997082D-04
        2     3.122487521966D-02   9,379884923981D+00   7.50   2,355155545449D-04
        3     3.117938741781D-03  -1.023617436485D+00   7.75   1,885163118589D-04
        4     1.161428812725D-04   4.214569698624D-02   8.25  -2,416857215547D-04




                N= 5       M=17                         P=   8,5       Q= 9

I= 0       H= -8.5
0              1.000000000000D+00  -1.639429421845D+00   -1.00   5,855090069660D+00
1             -2,284283380693D+00  -5.959533594121D-01    -.75   4.048118376988D+00
2              1.843375250477D+00   8.694441308950D-01    -.50   2.688653469792D+00
3             -6,432560307289D-01  -3.194091080332D-01    -.25   1.696661483995D+00
4              8,417540776122D-02   4.205239666378D-02     .25   5,344180426451D-01
                                                           .50   2,435560594503D-01
0    -8,5     -1.124681632598D-05  -5,623408162991D-06     .75   7,894518555812D-02
```

CARDINAL SPLINES N=5

1	-7.5	6.585108892546D-05	3.292554446273D-05	1.25	-2.601726046912D-02
2	-6.5	-1.669829769747D-04	-8.349148848733D-05	1.50	-2.411409617167D-02
3	-5.5	2.448147690008D-04	1.224073845004D-04	1.75	-1.186259780959D-02
4	-4.5	-2.393126344913D-04	-1.196563172456D-04	2.25	6.554057710555D-03
5	-3.5	1.777760985039D-04	8.888804925193D-05	2.50	7.241260734147D-03
6	-2.5	-1.152221382854D-04	-5.761106914268D-05	2.75	4.101579792770D-03
7	-1.5	7.150981004391D-05	3.575490502195D-05	3.25	-2.785877033968D-03
8	-.5	-4.375330166355D-05	-2.187665083177D-05	3.50	-3.317155752043D-03
9	.5	2.662595585055D-05	1.331297792527D-05	3.75	-1.995724317011D-03
10	1.5	-1.614156285146D-05	-8.070781425731D-06	4.25	1.478476994060D-03
11	2.5	9.719982196883D-06	4.859991098441D-06	4.50	1.814909414647D-03
12	3.5	-5.754772068934D-06	-2.877386034467D-06	4.75	1.118429167962D-03
13	4.5	3.259166201446D-06	1.629583100723D-06	5.25	-8.566335494698D-04
14	5.5	-1.658760074722D-06	-8.293800373609D-07	5.50	-1.063498295841D-03
15	6.5	6.782820356788D-07	3.391410178394D-07	5.75	-6.611101163459D-04
16	7.5	-1.871429782903D-07	-9.357148914516D-08	6.25	5.123899629676D-04
17	8.5	2.495295567299D-08	1.247647783650D-08	6.50	6.386360631803D-04
				6.75	3.982019872047D-04
	0	1.591287327162D+02	7.947132586911D+01	7.25	-3.098949989278D-04
	1	9.640424038236D+01	4.825438510417D+01	7.50	-3.867854868455D-04
	2	2.193038573138D+01	1.095422582979D+01	7.75	-2.414315785882D-04
	3	2.218707833152D+00	1.110372378535D+00	8.25	1.881906358202D-04
	4	8.417540776122D-02	4.205239666378D-02	8.50	2.350301006939D-04

I= 1 H= -7.5

0		0.	1.604536558811D+01	-1.00	-1.484715537088D+01
1		5.137180914870D+00	-1.527731388385D+00	-.75	-8.822340531883D+00
2		-6.514931519909D+00	-2.865648220604D+00	-.50	-4.571161261233D+00
3		2.786363845026D+00	1.376540485640D+00	-.25	-1.736611786490D+00
4		-4.086790910763D-01	-2.040747472685D-01	.25	9.190525413535D-01
				.50	1.262610743509D+00
0	-8.5	6.585108892546D-05	3.292554446273D-05	.75	1.234430279072D+00
1	-7.5	-3.968106207247D-04	-1.984053103623D-04	1.25	6.867514686802D-01
2	-6.5	1.043550544248D-03	5.217752721241D-04	1.50	3.847462929550D-01
3	-5.5	-1.600393262279D-03	-8.001966311396D-04	1.75	1.484017828258D-01
4	-4.5	1.646007402288D-03	8.230037011441D-04	2.25	-6.439187627924D-02
5	-3.5	-1.280199940397D-03	-6.400999701986D-04	2.50	-6.651235514383D-02
6	-2.5	8.523989834096D-04	4.261994917048D-04	2.75	-3.587773529424D-02
7	-1.5	-5.338967375212D-04	-2.669483687606D-04	3.25	2.286563926589D-02
8	-.5	3.276533884067D-04	1.638266942034D-04	3.50	2.666353788472D-02
9	.5	-1.995915175056D-04	-9.979575875278D-05	3.75	1.578674966436D-02
10	1.5	1.210391378350D-04	6.051956891750D-05	4.25	-1.144252245457D-02
11	2.5	-7.289424844535D-05	-3.644712422267D-05	4.50	-1.394366013426D-02
12	3.5	4.315903405364D-05	2.157951702682D-05	4.75	-8.544264355557D-03
13	4.5	-2.444304776906D-05	-1.222152388453D-05	5.25	6.494189464491D-03
14	5.5	1.244039550520D-05	6.220197752601D-06	5.50	8.041860680594D-03
15	6.5	-5.086997671168D-06	-2.543498835584D-06	5.75	4.989346539074D-03
16	7.5	1.403540619633D-06	7.017703098167D-07	6.25	-3.856796864552D-03
17	8.5	-1.871429782903D-07	-9.357148914516D-08	6.50	-4.802888050546D-03
				6.75	-2.992712977606D-03
	0	-8.491924660721D+02	-4.238984448219D+02	7.25	2.326973769103D-03
	1	-5.055924787432D+02	-2.531882175414D+02	7.50	2.903492995440D-03
	2	-1.126250394533D+02	-5.623026877771D+01	7.75	1.811960112172D-03
	3	-1.110872525157D+01	-5.562000921491D+00	8.25	-1.411966062199D-03
	4	-4.086790910763D-01	-2.040747472685D-01	8.50	-1.763224790840D-03

CARDINAL SPLINES N=5

I= 2 H= -6.5

0		0.	-5.815494718889D+01	-1.00	2.176755420092D+01
1		-5.754255269007D+00	1.196961307647D+01	-.75	1.239485893685D+01
2		1.004277503998D+01	3.601280976628D+00	-.50	6.081569033993D+00
3		-5.129438339966D+00	-2.504403816040D+00	-.25	2.149670221750D+00
4		8.410855519674D-01	4.195830390578D-01	.25	-8.877523615145D-01
				.50	-9.550461461443D-01
0	-8.5	-1.669829769747D-04	-8.349148848733D-05	.75	-5.645001038755D-01
1	-7.5	1.043550544248D-03	5.217752721241D-04	1.25	5.327700631784D-01
2	-6.5	-2.876025962084D-03	-1.438012981042D-03	1.50	9.045847974914D-01
3	-5.5	4.678336675231D-03	2.339168337615D-03	1.75	1.058304306551D+00
4	-4.5	-5.153472156026D-03	-2.576736078013D-03	2.25	7.840607137636D-01
5	-3.5	4.285418536306D-03	2.142709268153D-03	2.50	4.922586048205D-01
6	-2.5	-2.990822586083D-03	-1.495411293041D-03	2.75	2.092987130476D-01
7	-1.5	1.913952901564D-03	9.569764507819D-04	3.25	-1.057543799317D-01
8	-.5	-1.183239233427D-03	-5.916196167133D-04	3.50	-1.157629178331D-01
9	.5	7.225294478462D-04	3.612647239231D-04	3.75	-6.550874611185D-02
10	1.5	-4.385194075761D-04	-2.192597037880D-04	4.25	4.481703781248D-02
11	2.5	2.641631080228D-04	1.320815540114D-04	4.50	5.361012712418D-02
12	3.5	-1.564188593690D-04	-7.820942968450D-05	4.75	3.239251521323D-02
13	4.5	8.859017753639D-05	4.429508876820D-05	5.25	-2.416143796371D-02
14	5.5	-4.508880292967D-05	-2.254440146483D-05	5.50	-2.973413171854D-02
15	6.5	1.843730935095D-05	9.218654675473D-06	5.75	-1.836030466503D-02
16	7.5	-5.086997671168D-06	-2.543498835584D-06	6.25	1.410227732612D-02
17	8.5	6.782820356788D-07	3.391410178394D-07	6.50	1.752467513914D-02
				6.75	1.090222099820D-02
	0	1.917082155437D+03	9.560120088187D+02	7.25	-8.458761106605D-03
	1	1.119293818631D+03	5.610675979980D+02	7.50	-1.054700025509D-02
	2	2.438526841487D+02	1.216282310992D+02	7.75	-6.578443760454D-03
	3	2.346747042693D+01	1.176141951193D+01	8.25	5.122553897318D-03
	4	8.410855519674D-01	4.195830390578D-01	8.50	6.395398079205D-03

I= 3 H= -5.5

0		0.	1.422199944077D+02	-1.00	-2.127845473114D+01
1		5.130788337812D+00	-3.374283975148D+01	-.75	-1.191261731547D+01
2		-9.648145163349D+00	-1.351151310056D+00	-.50	-5.727920325390D+00
3		5.508316620373D+00	2.606654955015D+00	-.25	-1.975645497362D+00
4		-9.912046096049D-01	-4.932552337820D-01	.25	7.618835668645D-01
				.50	7.799476456688D-01
0	-8.5	2.448147690008D-04	1.224073845004D-04	.75	4.312257215272D-01
1	-7.5	-1.600393262279D-03	-8.001966311396D-04	1.25	-3.214196191573D-01
2	-6.5	4.678336675231D-03	2.339168337615D-03	1.50	-4.301404768062D-01
3	-5.5	-8.204983892687D-03	-4.102491946343D-03	1.75	-3.062798566192D-01
4	-4.5	9.887467910835D-03	4.943733955418D-03	2.25	3.901037785055D-01
5	-3.5	-9.023011996795D-03	-4.511505998397D-03	2.50	7.469595781810D-01
6	-2.5	6.793166239488D-03	3.396583119744D-03	2.75	9.690224467591D-01
7	-1.5	-4.546824234073D-03	-2.273412117037D-03	3.25	8.447030030139D-01
8	-.5	2.865378229188D-03	1.432689114594D-03	3.50	5.644658272595D-01
9	.5	-1.761213224913D-03	-8.806066124564D-04	3.75	2.527414092251D-01
10	1.5	1.071251769665D-03	5.356258848327D-04	4.25	-1.379380086786D-01
11	2.5	-6.457861850470D-04	-3.228930925235D-04	4.50	-1.552703151932D-01
12	3.5	3.824810434836D-04	1.912405217418D-04	4.75	-8.985520644545D-02
13	4.5	-2.166412073760D-04	-1.083206036880D-04	5.25	6.346510220423D-02
14	5.5	1.102645337781D-04	5.513226688905D-05	5.50	7.676186701226D-02
15	6.5	-4.508880292967D-05	-2.254440146483D-05	5.75	4.678484463827D-02

```
16   7,5   1.244039550520D-05   6.220197752601D-06     6.25 -3,531692388222D-02
17   8,5  -1.658760074722D-06  -8.293800373609D-07     6.50 -4,363888662307D-02
                                                       6.75 -2,703075294581D-02
        0  -2,444821855120D+03  -1.216226000177D+03    7.25  2,085121375944D-02
        1  -1.399854175468D+03  -7.034014323084D+02    7.50  2,594929727712D-02
        2  -2,988732696076D+02  -1.487075938017D+02    7.75  1.616179529451D-02
        3  -2,819264010619D+01  -1.416402299357D+01    8.25 -1,256059147017D-02
        4  -9.912046096049D-01  -4.932552337820D-01    8.50 -1.567164469731D-02


I= 4      H= -4.5
0           0.                  -2.794381599762D+02   -1.00  1.580718398666D+01
1          -3.689145702869D+00   6.949495117752D+01    -.75  8.793762868333D+00
2           7.103799987834D+00  -3.271801919248D+00    -.50  4.197308156524D+00
3          -4.214326634144D+00  -1.817344197765D+00    -.25  1.435247433544D+00
4           7.999116618130D-01   3.953442334787D-01     .25 -5,410231261200D-01
                                                        .50 -5,454196722880D-01
0   -8,5   -2,393126344913D-04  -1.196563172456D-04     .75 -2,958117522429D-01
1   -7,5    1.646007402288D-03   8.230037011441D-04    1.25  2,082749638987D-01
2   -6,5   -5.153472156026D-03  -2.576736078013D-03    1.50  2,668350492642D-01
3   -5,5    9.887467910835D-03   4.943733955418D-03    1.75  1,788067324482D-01
4   -4,5   -1.329689469502D-02  -6.648447347511D-03    2.25 -1,819582293499D-01
5   -3,5    1.366975252141D-02   6.834876260707D-03    2.50 -2,760559658317D-01
6   -2,5   -1.147388937220D-02  -5.736944686098D-03    2.75 -2,190033555122D-01
7   -1,5    8.313336696587D-03   4.156668348293D-03    3.25  3,308232508588D-01
8    -,5   -5,475437506139D-03  -2,737718753070D-03    3.50  6,763735392018D-01
9     ,5    3,428005143731D-03   1.714002571866D-03    3.75  9,265549057730D-01
10   1,5   -2.098224481342D-03  -1,049112240671D-03    4.25  8,761646668029D-01
11   2,5    1.267529264106D-03   6.337646320528D-04    4.50  6,030873723096D-01
12   3,5   -7.512439206073D-04  -3,756219603037D-04    4.75  2,765422815694D-01
13   4,5    4.256107382714D-04   2.128053691357D-04    5.25 -1,561688968839D-01
14   5,5   -2,166412073760D-04  -1.083206036880D-04    5.50 -1.779047652489D-01
15   6,5    8,859017753639D-05   4.429508876820D-05    5.75 -1,039264699520D-01
16   7,5   -2,444304776906D-05  -1.222152388453D-05    6.25  7.437288344352D-02
17   8,5    3.259166201446D-06   1.629583100723D-06    6.50  9.035890520159D-02
                                                       6.75  5,526424024177D-02
        0   2.069357335596D+03   1.022526338688D+03    7.25 -4,191903253701D-02
        1   1.168603153383D+03   5.911280732252D+02    7.50 -5,188191846231D-02
        2   2.464001762131D+02   1.217676462508D+02    7.75 -3,217831343949D-02
        3   2.298266986750D+01   1.162435974051D+01    8.25  2,486877786242D-02
        4   7.999116618130D-01   3,953442334787D-01    8.50  3,097144041043D-02


I= 5      H= -3.5
0           0.                   4.934164814895D+02   -1.00 -1.024555730238D+01
1           2.368660743035D+00  -1.247830874262D+02    -.75 -5.689374386950D+00
2          -4,592215484868D+00   9,752812881369D+00    -.50 -2.709803623646D+00
3           2,754029020105D+00   8.652676309669D-01    -.25 -9,242832165896D-01
4          -5,306520543709D-01  -2.571831172856D-01     .25  3,461105624845D-01
                                                        .50  3,473647216345D-01
0   -8,5    1,777760985039D-04   8.888804925193D-05     .75  1,873420615793D-01
1   -7,5   -1.280199940397D-03  -6.400999701986D-04    1.25 -1,297605482691D-01
2   -6,5    4.285418536306D-03   2.142709268153D-03    1.50 -1,642399987451D-01
3   -5,5   -9,023011996795D-03  -4.511505998397D-03    1.75 -1,082950524919D-01
4   -4,5    1.366975252141D-02   6.834876260707D-03    2.25  1,046714699351D-01
5   -3.5   -1.610602023946D-02  -8.053010119731D-03    2.50  1,523678228793D-01
6   -2,5    1.548937736936D-02   7.744688684679D-03    2.75  1,139699066666D-01
```

7	-1.5	-1.260141343139D-02	-6.300706715696D-03	3.25	-1.379186680948D-01
8	-.5	9.000258489066D-03	4.500129244533D-03	3.50	-2.236169671415D-01
9	.5	-5.888592660449D-03	-2.944296330225D-03	3.75	-1.874534078347D-01
10	1.5	3.670472407930D-03	1.835236203965D-03	4.25	3.074488458749D-01
11	2.5	-2.231143710500D-03	-1.115571855250D-03	4.50	6.476788859610D-01
12	3.5	1.325104030926D-03	6.625520154628D-04	4.75	9.088709197037D-01
13	4.5	-7.512439206073D-04	-3.756219603037D-04	5.25	8.897116236350D-01
14	5.5	3.824810434836D-04	1.912405217418D-04	5.50	6.199078143541D-01
15	6.5	-1.564188593690D-04	-7.820942968450D-05	5.75	2.870001586051D-01
16	7.5	4.315903405364D-05	2.157951702682D-05	6.25	-1.642779956244D-01
17	8.5	-5.754772068934D-06	-2.877386034467D-06	6.50	-1.880152261145D-01
				6.75	-1.102332466505D-01
	0	-1.390372770063D+03	-6.737284932627D+02	7.25	7.928721239931D-02
	1	-7.823099839539D+02	-4.032088370430D+02	7.50	9.649811178550D-02
	2	-1.644021410420D+02	-7.967174387230D+01	7.75	5.910077150463D-02
	3	-1.528814082850D+01	-7.878958356745D+00	8.25	-4.491965000278D-02
	4	-5.306520543709D-01	-2.571831172856D-01	8.50	-5.563847297853D-02

I= 6 H= -2.5

0		0.	-8.334316217209D+02	-1.00	6.359392708110D+00
1		-1.465872687483D+00	2.120392176799D+02	-.75	3.529369588570D+00
2		2.848004023932D+00	-1.892789696227D+01	-.50	1.679892477987D+00
3		-1.713766055502D+00	7.511919299381D-03	-.25	5.725419161916D-01
4		3.317499411923D-01	1.521207223474D-01	.25	-2.139496172241D-01
				.50	-2.144219484153D-01
0	-8.5	-1.152221382854D-04	-5.761106914268D-05	.75	-1.154382034038D-01
1	-7.5	8.523989834096D-04	4.261994917048D-04	1.25	7.954401217392D-02
2	-6.5	-2.990822586083D-03	-1.495411293041D-03	1.50	1.002958013587D-01
3	-5.5	6.793166239488D-03	3.396583119744D-03	1.75	6.579883888539D-02
4	-4.5	-1.147388937220D-02	-5.736944686098D-03	2.25	-6.260651703303D-02
5	-3.5	1.548937736936D-02	7.744688684679D-03	2.50	-9.004344539066D-02
6	-2.5	-1.728351112805D-02	-8.641755564026D-03	2.75	-6.626784509658D-02
7	-1.5	1.621701234644D-02	8.108506173219D-03	3.25	7.612385662901D-02
8	-.5	-1.304140087462D-02	-6.520700437308D-03	3.50	1.183744610168D-01
9	.5	9.259413895847D-03	4.629706947924D-03	3.75	9.351700817616D-02
10	1.5	-6.031573132617D-03	-3.015786566309D-03	4.25	-1.227640544008D-01
11	2.5	3.733462186974D-03	1.866731093487D-03	4.50	-2.050115328366D-01
12	3.5	-2.231143710500D-03	-1.115571855250D-03	4.75	-1.759860544570D-01
13	4.5	1.267529264106D-03	6.337646320528D-04	5.25	2.986615763203D-01
14	5.5	-6.457861850470D-04	-3.228930925235D-04	5.50	6.367659280234D-01
15	6.5	2.641631080228D-04	1.320815540114D-04	5.75	9.020840802156D-01
16	7.5	-7.289424844535D-05	-3.644712422267D-05	6.25	8.949783947158D-01
17	8.5	9.719982196883D-06	4.859991098441D-06	6.50	6.264782296228D-01
				6.75	2.911017217052D-01
	0	8.725972214454D+02	4.000541086722D+02	7.25	-1.674808733131D-01
	1	4.904351337282D+02	2.655777322758D+02	7.50	-1.920225855287D-01
	2	1.029605691155D+02	4.720799011746D+01	7.75	-1.127424761918D-01
	3	9.565731945037D+00	5.179616479111D+00	8.25	8.126098029235D-02
	4	3.317499411923D-01	1.521207223474D-01	8.50	9.897921009791D-02

I= 7 H= -1.5

0		0.	1.384226686257D+03	-1.00	-3.892219310912D+00
1		8.963132581499D-01	-3.529406257009D+02	-.75	-2.159725140823D+00
2		-1.742627982022D+00	3.293077992646D+01	-.50	-1.027753544152D+00
3		1.049760642401D+00	-9.107795231100D-01	-.25	-3.501900634058D-01

4		-2.035174283390D-01	-7.891446797635D-02	.25	1.307715860170D-01
				.50	1.310000142660D-01
0	-8.5	7.150981004391D-05	3.575490502195D-05	.75	7.048565771930D-02
1	-7.5	-5.338967375212D-04	-2.669483687606D-04	1.25	-4.848682849676D-02
2	-6.5	1.913952901564D-03	9.569764507819D-04	1.50	-6.105985062830D-02
3	-5.5	-4.546824234073D-03	-2.273412117037D-03	1.75	-3.999202749637D-02
4	-4.5	8.313336696587D-03	4.156668348293D-03	2.25	3.785733229171D-02
5	-3.5	-1.260141343139D-02	-6.300706715696D-03	2.50	5.423598163800D-02
6	-2.5	1.621701234644D-02	8.108506173219D-03	2.75	3.970812651838D-02
7	-1.5	-1.772978629901D-02	-8.864893149504D-03	3.25	-4.488200717184D-02
8	-.5	1.648125553323D-02	8.240627766613D-03	3.50	-6.893609955946D-02
9	.5	-1.318768344744D-02	-6.593841723721D-03	3.75	-5.356680766808D-02
10	1.5	9.324232130534D-03	4.662116065267D-03	4.25	6.671016734998D-02
11	2.5	-6.031573132617D-03	-3.015786566309D-03	4.50	1.068147074075D-01
12	3.5	3.670472407930D-03	1.835236203965D-03	4.75	8.639025076191D-02
13	4.5	-2.098224481342D-03	-1.049112240671D-03	5.25	-1.172985128900D-01
14	5.5	1.071251769665D-03	5.356258848327D-04	5.50	-1.982199112516D-01
15	6.5	-4.385194075761D-04	-2.192597037880D-04	5.75	-1.717591661365D-01
16	7.5	1.210391378350D-04	6.051956891750D-05	6.25	2.953742887680D-01
17	8.5	-1.614156285146D-05	-8.070781425731D-06	6.50	6.326585933375D-01
				6.75	8.995150010331D-01
	0	-5.359756502614D+02	-2.077907121344D+02	7.25	8.969961456300D-01
	1	-3.011333059107D+02	-1.843822191690D+02	7.50	6.290131570942D-01
	2	-6.319853678577D+01	-2.450351978059D+01	7.75	2.926972574745D-01
	3	-5.869831921126D+00	-3.593871434306D+00	8.25	-1.687547556015D-01
	4	-2.035174283390D-01	-7.891446797635D-02	8.50	-1.936406752119D-01

I= 8 H= -.5

0		0.	-2.284225827034D+03	-1.00	2.370615109117D+00
1		-5.457400874408D-01	5.828858402426D+02	-.75	1.315333212095D+00
2		1.061279330306D+00	-5.524951454282D+01	-.50	6.258862064117D-01
3		-6.395455904671D-01	2.049361459504D+00	-.25	2.132424505620D-01
4		1.240501009036D-01	2.432717476482D-02	.25	-7.961339302737D-02
				.50	-7.974036410149D-02
0	-8.5	-4.375330166355D-05	-2.187665083177D-05	.75	-4.289679622571D-02
1	-7.5	3.276533884067D-04	1.638266942034D-04	1.25	2.949206832968D-02
2	-6.5	-1.183239233427D-03	-5.916196167133D-04	1.50	3.712424607362D-02
3	-5.5	2.865378229188D-03	1.432689114594D-03	1.75	2.430177256363D-02
4	-4.5	-5.475437506139D-03	-2.737718753070D-03	2.25	-2.296561790911D-02
5	-3.5	9.000258489066D-03	4.500129244533D-03	2.50	-3.285900296171D-02
6	-2.5	-1.304140087462D-02	-6.520700437308D-03	2.75	-2.401589618541D-02
7	-1.5	1.648125553323D-02	8.240627766613D-03	3.25	2.700076317437D-02
8	-.5	-1.787683247699D-02	-8.938416238494D-03	3.50	4.130454135468D-02
9	.5	1.654662415998D-02	8.273312079990D-03	3.75	3.192464860160D-02
10	1.5	-1.318768344744D-02	-6.593841723721D-03	4.25	-3.910839696916D-02
11	2.5	9.259413895847D-03	4.629706947924D-03	4.50	-6.184212925553D-02
12	3.5	-5.888592660449D-03	-2.944296330225D-03	4.75	-4.919000968861D-02
13	4.5	3.428005143731D-03	1.714002571866D-03	5.25	6.334628183033D-02
14	5.5	-1.761213224913D-03	-8.806066124564D-04	5.50	1.026281457734D-01
15	6.5	7.225294478462D-04	3.612647239231D-04	5.75	8.377948540219D-02
16	7.5	-1.995915175056D-04	-9.979575875278D-05	6.25	-1.152563542355D-01
17	8.5	2.662595585055D-05	1.331297792527D-05	6.50	-1.956578140937D-01
				6.75	-1.701482369271D-01
	0	3.268269849740D+02	6.407986834787D+01	7.25	2.940900470708D-01
	1	1.836035746638D+02	1.476028941718D+02	7.50	6.310282468695D-01
	2	3.852858551511D+01	7.555032935073D+00	7.75	8.984753620733D-01

```
        3   3.578157840256D+00   2.876485401508D+00    8.25  8.978566684228D-01
        4   1.240501009036D-01   2.432717476482D-02    8.50  6.301329389904D-01


I= 9      H=    .5
0            0.                  3.758066568289D+03   -1.00 -1.440459507261D+00
1            3.315738566962D-01 -9.592763696546D+02    -.75 -7.992212844064D-01
2           -6.448473152345D-01  9.145124093406D+01    -.50 -3.802913146122D-01
3            3.886425839564D-01 -3.703609343511D+00    -.25 -1.295634764043D-01
4           -7.539575137399D-02  2.432717476482D-02     .25  4.836853279398D-02
                                                        .50  4.844324007698D-02
0   -8.5     2.662595585055D-05  1.331297792527D-05     .75  2.605868005274D-02
1   -7.5    -1.995915175056D-04 -9.979575875278D-05    1.25 -1.791234295300D-02
2   -6.5     7.225294478462D-04  3.612647239231D-04    1.50 -2.254474334890D-02
3   -5.5    -1.761213224913D-03 -8.806066124564D-04    1.75 -1.475526105893D-02
4   -4.5     3.428005143731D-03  1.714002571866D-03    2.25  1.393614747654D-02
5   -3.5    -5.888592660449D-03 -2.944296330225D-03    2.50  1.993116153007D-02
6   -2.5     9.259413895847D-03  4.629706947924D-03    2.75  1.455890380761D-02
7   -1.5    -1.318768344744D-02 -6.593841723721D-03    3.25 -1.633939621181D-02
8    -.5     1.654662415998D-02  8.273312079990D-03    3.50 -2.496174772037D-02
9     .5    -1.787683247699D-02 -8.938416238494D-03    3.75 -1.925893540640D-02
10   1.5     1.648125553323D-02  8.240627766613D-03    4.25  2.346443400394D-02
11   2.5    -1.304140087462D-02 -6.520700437308D-03    4.50  3.695269484215D-02
12   3.5     9.000258489066D-03  4.500129244533D-03    4.75  2.923431115793D-02
13   4.5    -5.475437506139D-03 -2.737718753070D-03    5.25 -3.702869075468D-02
14   5.5     2.865378229188D-03  1.432689114594D-03    5.50 -5.924313791451D-02
15   6.5    -1.183239233427D-03 -5.916196167133D-04    5.75 -4.756078313797D-02
16   7.5     3.276533884067D-04  1.638266942034D-04    6.25  6.205271452665D-02
17   8.5    -4.375330166355D-05 -2.187665083177D-05    6.50  1.009881854047D-01
                                                       6.75  8.273481500138D-02
        0  -1.986672482782D+02   6.407986834787D+01    7.25 -1.143929463771D-01
        1  -1.116022136799D+02  -1.476028941718D+02    7.50 -1.945348808856D-01
        2  -2.341851964497D+01   7.555032935073D+00    7.75 -1.694110511659D-01
        3  -2.174812962759D+00  -2.876485401508D+00    8.25  2.934327080910D-01
        4  -7.539575137399D-02   2.432717476482D-02    8.50  6.301329389904D-01


I=10      H=   1.5
0            0.                 -6.164443628803D+03   -1.00  8.728198696223D-01
1           -2.009040703789D-01  1.573767079409D+03    -.75  4.842701602033D-01
2            3.907295132062D-01 -1.503566632231D+02    -.50  2.304271684715D-01
3           -2.354977936508D-01  6.276963345501D+00    -.25  7.850473586930D-02
4            4.568849238635D-02 -7.891446797635D-02     .25 -2.930660543332D-02
                                                        .50 -2.935138184659D-02
0   -8.5    -1.614156285146D-05 -8.070781425731D-06     .75 -1.578842074218D-02
1   -7.5     1.210391378350D-04  6.051956891750D-05    1.25  1.085205549869D-02
2   -6.5    -4.385194075761D-04 -2.192597037880D-04    1.50  1.365793878302D-02
3   -5.5     1.071251769665D-03  5.356258848327D-04    1.75  8.938417724679D-03
4   -4.5    -2.098224481342D-03 -1.049112240671D-03    2.25 -8.440639966578D-03
5   -3.5     3.670472407930D-03  1.835236203965D-03    2.50 -1.206989144205D-02
6   -2.5    -6.031573132617D-03 -3.015786566309D-03    2.75 -8.814890765038D-03
7   -1.5     9.324232130534D-03  4.662116065267D-03    3.25  9.887086424805D-03
8    -.5    -1.318768344744D-02 -6.593841723721D-03    3.50  1.509779234041D-02
9     .5     1.648125553323D-02  8.240627766613D-03    3.75  1.164163832342D-02
10   1.5    -1.772978629901D-02 -8.864893149504D-03    4.25 -1.415806358606D-02
11   2.5     1.621701234644D-02  8.108506173219D-03    4.50 -2.226627577024D-02
12   3.5    -1.260141343139D-02 -6.300706715696D-03    4.75 -1.758384185594D-02
```

13	4.5	8.313336696587D-03	4.156668348293D-03	5.25	2.215003782560D-02
14	5.5	-4.546824234073D-03	-2.273412117037D-03	5.50	3.529284723858D-02
15	6.5	1.913952901564D-03	9.569764507819D-04	5.75	2.818014504768D-02
16	7.5	-5.338967375212D-04	-2.669483687606D-04	6.25	-3.616098060191D-02
17	8.5	7.150981004391D-05	3.575490502195D-05	6.50	-5.811615796871D-02
				6.75	-4.682173503685D-02
	0	1.203942259926D+02	-2.077907121344D+02	7.25	6.139472748326D-02
	1	6.763113242738D+01	1.843822191690D+02	7.50	1.000925030333D-01
	2	1.419149722459D+01	-2.450351978059D+01	7.75	8.211616077025D-02
	3	1.317910947485D+00	3.593871434306D+00	8.25	-1.137747906444D-01
	4	4.568849238635D-02	-7.891446797635D-02	8.50	-1.936406752119D-01

I=11 H= 2.5

0		0.	1.004325376744D+04	-1.00	-5.254998251079D-01
1		1.209571848314D-01	-2.564479990566D+03	-.75	-2.915645358194D-01
2		-2.352462760475D-01	2.452325434724D+02	-.50	-1.387329259251D-01
3		1.417878677314D-01	-1.035172103892D+01	-.25	-4.726507895857D-02
4		-2.750849649754D-02	1.521207223474D-01	.25	1.764438436082D-02
				.50	1.767124482351D-02
0	-8.5	9.719982196883D-06	4.859991098441D-06	.75	9.505484647928D-03
1	-7.5	-7.289424844535D-05	-3.644712422267D-05	1.25	-6.533392166511D-03
2	-6.5	2.641631080228D-04	1.320815540114D-04	1.50	-8.222527387839D-03
3	-5.5	-6.457861850470D-04	-3.228930925235D-04	1.75	-5.381133881158D-03
4	-4.5	1.267529264106D-03	6.337646320528D-04	2.25	5.081125742125D-03
5	-3.5	-2.231143710500D-03	-1.115571855250D-03	2.50	7.265531312738D-03
6	-2.5	3.733462186974D-03	1.866731093487D-03	2.75	5.305832670064D-03
7	-1.5	-6.031573132617D-03	-3.015786566309D-03	3.25	-5.950040374619D-03
8	-.5	9.259413895847D-03	4.629706947924D-03	3.50	-9.084490852153D-03
9	.5	-1.304140087462D-02	-6.520700437308D-03	3.75	-7.003512641361D-03
10	1.5	1.621701234644D-02	8.108506173219D-03	4.25	8.512232170551D-03
11	2.5	-1.728351112805D-02	-8.641755564026D-03	4.50	1.338104274889D-02
12	3.5	1.548937736936D-02	7.744688684679D-03	4.75	1.056078235344D-02
13	4.5	-1.147388937220D-02	-5.736944686098D-03	5.25	-1.327889001234D-02
14	5.5	6.793166239488D-03	3.396583119744D-03	5.50	-2.112896034917D-02
15	6.5	-2.990822586083D-03	-1.495411293041D-03	5.75	-1.684032704937D-02
16	7.5	8.523989834096D-04	4.261994917048D-04	6.25	2.149094298634D-02
17	8.5	-1.152221382854D-04	-5.761106914268D-05	6.50	3.439703233696D-02
				6.75	2.756217021065D-02
	0	-7.248900410098D+01	4.000541086722D+02	7.25	-3.554466867366D-02
	1	-4.072033082339D+01	-2.655777322758D+02	7.50	-5.722526922886D-02
	2	-8.544588880578D+00	4.720799011746D+01	7.75	-4.616828314311D-02
	3	-7.935010131848D-01	-5.179616479111D+00	8.25	6.066321664843D-02
	4	-2.750849649754D-02	1.521207223474D-01	8.50	9.897921009791D-02

I=12 H= 3.5

0		0.	-1.603846434991D+04	-1.00	3.111081115623D-01
1		-7.160919019369D-02	4.097163034309D+03	-.75	1.726128553580D-01
2		1.392711133676D-01	-3.920740633126D+02	-.50	8.213298570139D-02
3		-8.394198820141D-02	1.662318434446D+01	-.25	2.798195218314D-02
4		1.628581979960D-02	-2.571831172856D-01	.25	-1.044583006696D-02
				.50	-1.046171278244D-02
0	-8.5	-5.754772068934D-06	-2.877386034467D-06	.75	-5.627414572769D-03
1	-7.5	4.315903405364D-05	2.157951702682D-05	1.25	3.867857205476D-03
2	-6.5	-1.564188593690D-04	-7.820942968450D-05	1.50	4.867823866438D-03
3	-5.5	3.824810434836D-04	1.912405217418D-04	1.75	3.185655355568D-03

4	-4.5	-7.512439206073D-04	-3.756219603037D-04	2.25	-3.007998667907D-03
5	-3.5	1.325104030926D-03	6.625520154628D-04	2.50	-4.301087186848D-03
6	-2.5	-2.231143710500D-03	-1.115571855250D-03	2.75	-3.140908687309D-03
7	-1.5	3.670472407930D-03	1.835236203965D-03	3.25	3.522032916740D-03
8	-.5	-5.888592660449D-03	-2.944296330225D-03	3.50	5.377156394007D-03
9	.5	9.000258489066D-03	4.500129244533D-03	3.75	4.145143991956D-03
10	1.5	-1.260141343139D-02	-6.300706715696D-03	4.25	-5.037092797384D-03
11	2.5	1.548937736936D-02	7.744688684679D-03	4.50	-7.917003261707D-03
12	3.5	-1.610602023946D-02	-8.053010119731D-03	4.75	-6.247127890239D-03
13	4.5	1.366975252141D-02	6.834876260707D-03	5.25	7.850209700404D-03
14	5.5	-9.023011996795D-03	-4.511505998397D-03	5.50	1.248530002473D-02
15	6.5	4.285418536306D-03	2.142709268153D-03	5.75	9.945115975907D-03
16	7.5	-1.280199940397D-03	-6.400999701986D-04	6.25	-1.266824671551D-02
17	8.5	1.777760985039D-04	8.888804925193D-05	6.50	-2.024809989206D-02
				6.75	-1.619535642349D-02
	0	4.291578353763D+01	-6.737284932627D+02	7.25	2.077086610901D-02
	1	2.410769013212D+01	4.032088370430D+02	7.50	3.330226543467D-02
	2	5.058653297359D+00	-7.967174387230D+01	7.75	2.672194419638D-02
	3	4.697758849851D-01	7.878958356745D+00	8.25	-3.453177049989D-02
	4	1.628581979960D-02	-2.571831172856D-01	8.50	-5.563847297853D-02

I=13 H= 4.5

0		0.	2.404735893614D+04	-1.00	-1.761902036623D-01
1		4.055446068472D-02	-6.151921142785D+03	-.75	-9.775600682835D-02
2		-7.887353655871D-02	5.895705448469D+02	-.50	-4.651444060593D-02
3		4.753901156333D-02	-2.506606367879D+01	-.25	-1.584703636668D-02
4		-9.223194855536D-03	3.953442334787D-01	.25	5.915788099466D-03
				.50	5.924779335186D-03
0	-8.5	3.259166201446D-06	1.629583100723D-06	.75	3.186969918755D-03
1	-7.5	-2.444304776906D-05	-1.222152388453D-05	1.25	-2.190476035519D-03
2	-6.5	8.859017753639D-05	4.429508876820D-05	1.50	-2.756780602728D-03
3	-5.5	-2.166412073760D-04	-1.083206036880D-04	1.75	-1.804118992640D-03
4	-4.5	4.256107382714D-04	2.128053691357D-04	2.25	1.703495727953D-03
5	-3.5	-7.512439206073D-04	-3.756219603037D-04	2.50	2.435787407412D-03
6	-2.5	1.267529264106D-03	6.337646320528D-04	2.75	1.778743855601D-03
7	-1.5	-2.098224481342D-03	-1.049112240671D-03	3.25	-1.994537011839D-03
8	-.5	3.428005143731D-03	1.714002571866D-03	3.50	-3.045048557565D-03
9	.5	-5.475437506139D-03	-2.737718753070D-03	3.75	-2.347316900667D-03
10	1.5	8.313336696587D-03	4.156668348293D-03	4.25	2.852219880930D-03
11	2.5	-1.147388937220D-02	-5.736944686098D-03	4.50	4.482724090744D-03
12	3.5	1.366975252141D-02	6.834876260707D-03	4.75	3.536981077130D-03
13	4.5	-1.329689469502D-02	-6.648447347511D-03	5.25	-4.443706005391D-03
14	5.5	9.887467910835D-03	4.943733955418D-03	5.50	-7.066376968655D-03
15	6.5	-5.153472156026D-03	-2.576736078013D-03	5.75	-5.627563743543D-03
16	7.5	1.646007402288D-03	8.230037011441D-04	6.25	7.164082438527D-03
17	8.5	-2.393126344913D-04	-1.196563172456D-04	6.50	1.144533830597D-02
				6.75	9.148967107965D-03
	0	-2.430465821980D+01	1.022526338688D+03	7.25	-1.171209083078D-02
	1	-1.365299306709D+01	-5.911280732252D+02	7.50	-1.875225623727D-02
	2	-2.864883711569D+00	1.217676462508D+02	7.75	-1.501961244015D-02
	3	-2.660496135249D-01	-1.162435974051D+01	8.25	1.930210276147D-02
	4	-9.223194855536D-03	3.953442334787D-01	8.50	3.097144041043D-02

I=14 H= 5.5

0		0.	-2.921256559669D+04	-1.00	8.967185461688D-02

CARDINAL SPLINES N=5

1		-2.064014949090D-02	7.513163992579D+03	-.75	4.975283260964D-02
2		4.014261464759D-02	-7.237163239823D+02	-.50	2.367348083959D-02
3		-2.419494843750D-02	3.093470094216D+01	-.25	8.065333349883D-03
4		4.694142040889D-03	-4.932552337820D-01	.25	-3.010833540567D-03
				.50	-3.015409000451D-03
0	-8.5	-1.658760074722D-06	-8.293800373609D-07	.75	-1.622003918333D-03
1	-7.5	1.244039550520D-05	6.220197752601D-06	1.25	1.114838895077D-03
2	-6.5	-4.508880292967D-05	-2.254440146483D-05	1.50	1.403057814130D-03
3	-5.5	1.102645337781D-04	5.513226688905D-05	1.75	9.182019674127D-04
4	-4.5	-2.166412073760D-04	-1.083206036880D-04	2.25	-8.669880049052D-04
5	-3.5	3.824810434836D-04	1.912405217418D-04	2.50	-1.239683033848D-03
6	-2.5	-6.457861850470D-04	-3.228930925235D-04	2.75	-9.052815544942D-04
7	-1.5	1.071251769665D-03	5.356258848327D-04	3.25	1.015100866620D-03
8	-.5	-1.761213224913D-03	-8.806066124564D-04	3.50	1.549740297745D-03
9	.5	2.865378229188D-03	1.432689114594D-03	3.75	1.194629560303D-03
10	1.5	-4.546824234073D-03	-2.273412117037D-03	4.25	-1.451559291413D-03
11	2.5	6.793166239488D-03	3.396583119744D-03	4.50	-2.281321246136D-03
12	3.5	-9.023011996795D-03	-4.511505998397D-03	4.75	-1.799979201674D-03
13	4.5	9.887467910835D-03	4.943733955418D-03	5.25	2.261259601035D-03
14	5.5	-8.204983892687D-03	-4.102491946343D-03	5.50	3.595668611533D-03
15	6.5	4.678336675231D-03	2.339168337615D-03	5.75	2.863346929172D-03
16	7.5	-1.600393262279D-03	-8.001966311396D-04	6.25	-3.644387643042D-03
17	8.5	2.448147690008D-04	1.224073845004D-04	6.50	-5.821373832324D-03
				6.75	-4.652437422731D-03
	0	1.236985476575D+01	-1.216226000177D+03	7.25	5.952139388523D-03
	1	6.948689149133D+00	7.034014323084D+02	7.50	9.525550726587D-03
	2	1.458082004217D+00	-1.487075938017D+02	7.75	7.624816661843D-03
	3	1.354058809527D-01	1.416402299357D+01	8.25	-9.780576840150D-03
	4	4.694142040889D-03	-4.932552337820D-01	8.50	-1.567164469731D-02

I=15 H= 6.5

0		0.	2.392595773430D+04	-1.00	-3.666755655985D-02
1		8.439925101709D-03	-6.208740941339D+03	-.75	-2.034434059178D-02
2		-1.641464356912D-02	6.034336760848D+02	-.50	-9.680279813524D-03
3		9.893514037199D-03	-2.602724283989D+01	-.25	-3.297980600062D-03
4		-1.919473851819D-03	4.195830390578D-01	.25	1.231154267042D-03
				.50	1.233025122254D-03
0	-8.5	6.782820356788D-07	3.391410178394D-07	.75	6.632504582593D-04
1	-7.5	-5.086997671168D-06	-2.543498835584D-06	1.25	-4.558664786907D-04
2	-6.5	1.843730935095D-05	9.218654675473D-06	1.50	-5.737213727422D-04
3	-5.5	-4.508880292967D-05	-2.254440146483D-05	1.75	-3.754599083125D-04
4	-4.5	8.859017753639D-05	4.429508876820D-05	2.25	3.545178396109D-04
5	-3.5	-1.564188593690D-04	-7.820942968450D-05	2.50	5.069152993729D-04
6	-2.5	2.641631080228D-04	1.320815540114D-04	2.75	3.701758321208D-04
7	-1.5	-4.385194075761D-04	-2.192597037880D-04	3.25	-4.150806582304D-04
8	-.5	7.225294478462D-04	3.612647239231D-04	3.50	-6.336966290515D-04
9	.5	-1.183239233427D-03	-5.916196167133D-04	3.75	-4.884888236030D-04
10	1.5	1.913952901564D-03	9.569764507819D-04	4.25	5.935438169952D-04
11	2.5	-2.990822586083D-03	-1.495411293041D-03	4.50	9.328287316621D-04
12	3.5	4.285418536306D-03	2.142709268153D-03	4.75	7.360030429022D-04
13	4.5	-5.153472156026D-03	-2.576736078013D-03	5.25	-9.245966261978D-04
14	5.5	4.678336675231D-03	2.339168337615D-03	5.50	-1.470191312607D-03
15	6.5	-2.876025962084D-03	-1.438012981042D-03	5.75	-1.170733476864D-03
16	7.5	1.043550544248D-03	5.217752721241D-04	6.25	1.489970537935D-03
17	8.5	-1.669829769747D-04	-8.349148848733D-05	6.50	2.379881863364D-03
				6.75	1.901865920288D-03

	0	-5.058137800022D+00	9.560120088187D+02	7.25	-2.432647260672D-03
	1	-2.841377365004D+00	-5.610675979980D+02	7.50	-3.892478488677D-03
	2	-5.962219503842D-01	1.216282310992D+02	7.75	-3.115110878608D-03
	3	-5.536859692465D-02	-1.176141951193D+01	8.25	3.993261448155D-03
	4	-1.919473851819D-03	4.195830390578D-01	8.50	6.395398079205D-03

I=16	H= 7.5				
0		0.	-1.111968196444D+04	-1.00	1.011683929916D-02
1		-2.328635155612D-03	2.915976103161D+03	-.75	5.613148013148D-03
2		4.528916681815D-03	-2.865276952166D+02	-.50	2.670857897273D-03
3		-2.729690922478D-03	1.250054232862D+01	-.25	9.099362386617D-04
4		5.295965392531D-04	-2.040747472685D-01	.25	-3.396841811858D-04
				.50	-3.402003544724D-04
0	-8.5	-1.871429782903D-07	-9.357148914516D-08	.75	-1.829954876822D-04
1	-7.5	1.403540619633D-06	7.017703098167D-07	1.25	1.257767702787D-04
2	-6.5	-5.086997671168D-06	-2.543498835584D-06	1.50	1.582937537939D-04
3	-5.5	1.244039550520D-05	6.220197752601D-06	1.75	1.035920143202D-04
4	-4.5	-2.444304776906D-05	-1.222152388453D-05	2.25	-9.781392236093D-05
5	-3.5	4.315903405364D-05	2.157951702682D-05	2.50	-1.398614034929D-04
6	-2.5	-7.289424844535D-05	-3.644712422267D-05	2.75	-1.021340172505D-04
7	-1.5	1.210391378350D-04	6.051956891750D-05	3.25	1.145234563196D-04
8	-.5	-1.995915175056D-04	-9.979575875278D-05	3.50	1.748409030731D-04
9	.5	3.276533884067D-04	1.638266942034D-04	3.75	1.347770266000D-04
10	1.5	-5.338967375212D-04	-2.669483687606D-04	4.25	-1.637618664521D-04
11	2.5	8.523989834096D-04	4.261994917048D-04	4.50	-2.573718067654D-04
12	3.5	-1.280199940397D-03	-6.400999701986D-04	4.75	-2.030661012833D-04
13	4.5	1.646007402288D-03	8.230037011441D-04	5.25	2.550975813501D-04
14	5.5	-1.600393262279D-03	-8.001966311396D-04	5.50	4.056253121731D-04
15	6.5	1.043550544248D-03	5.217752721241D-04	5.75	3.230022081086D-04
16	7.5	-3.968106207247D-04	-1.984053103623D-04	6.25	-4.110680269465D-04
17	8.5	6.585108892546D-05	3.292554446273D-05	6.50	-6.565727309385D-04
				6.75	-5.246818566816D-04
	0	1.395576428357D+00	-4.238984448219D+02	7.25	6.710591480017D-04
	1	7.839563396635D-01	2.531882175414D+02	7.50	1.073697727421D-03
	2	1.645018979249D-01	-5.623026877771D+01	7.75	8.592020336880D-04
	3	1.527659141213D-02	5.562000921491D+00	8.25	-1.101147905819D-03
	4	5.295965392531D-04	-2.040747472685D-01	8.50	-1.763224790840D-03

I=17	H= 8.5				
0		0.	2.182499991285D+03	-1.00	-1.348941667839D-03
1		3.104915307175D-04	-5.784511496627D+02	-.75	-7.484362443763D-04
2		-6.038688767335D-04	5.749843543619D+01	-.50	-3.561222399866D-04
3		3.639668267241D-04	-2.540153865104D+00	-.25	-1.213275067743D-04
4		-7.061443366376D-05	4.205239666378D-02	.25	4.529222201477D-05
				.50	4.536104614813D-05
0	-8.5	2.495295567299D-08	1.247647783650D-08	.75	2.439993533317D-05
1	-7.5	-1.871429782903D-07	-9.357148914516D-08	1.25	-1.677060444715D-05
2	-6.5	6.782820356788D-07	3.391410178394D-07	1.50	-2.110629687428D-05
3	-5.5	-1.658760074722D-06	-8.293800373609D-07	1.75	-1.381257106313D-05
4	-4.5	3.259166201446D-06	1.629583100723D-06	2.25	1.304214035654D-05
5	-3.5	-5.754772068934D-06	-2.877386034467D-06	2.50	1.864859155869D-05
6	-2.5	9.719982196883D-06	4.859991098441D-06	2.75	1.361816262781D-05
7	-1.5	-1.614156285146D-05	-8.070781425731D-06	3.25	-1.527011764069D-05
8	-.5	2.662595585055D-05	1.331297792527D-05	3.50	-2.331260750350D-05
9	.5	-4.375330166355D-05	-2.187665083177D-05	3.75	-1.797063881636D-05

CARDINAL SPLINES N=5

10	1.5	7.150981004391D-05	3.575490502195D-05	4.25	2.183533771423D-05
11	2.5	-1.152221382854D-04	-5.761106914268D-05	4.50	3.431687403562D-05
12	3.5	1.777760985039D-04	8.888804925193D-05	4.75	2.707594808264D-05
13	4.5	-2.393126344913D-04	-1.196563172456D-04	5.25	-3.401347702086D-05
14	5.5	2.448147690008D-04	1.224073845004D-04	5.50	-5.408397104993D-05
15	6.5	-1.669829769747D-04	-8.349148848733D-05	5.75	-4.306728346651D-05
16	7.5	6.585108892546D-05	3.292554446273D-05	6.25	5.480888820284D-05
17	8.5	-1.124681632598D-05	-5.623408162991D-06	6.50	8.754203058144D-05
				6.75	6.995603509589D-05
	0	-1.860809779979D-01	7.947132586911D+01	7.25	-8.946965970305D-05
	1	-1.045298259763D-01	-4.825438510417D+01	7.50	-1.431483702772D-04
	2	-2.193407178851D-02	1.095422582979D+01	7.75	-1.145475230396D-04
	3	-2.036923917844D-03	-1.110372378535D+00	8.25	1.467889670965D-04
	4	-7.061443366376D-05	4.205239666378D-02	8.50	2.350301006939D-04

N= 5 M=18 P= 9.0 Q= 9

I= 0 H= -9.0

0		1.000000000000D+00	2.165981837378D+00	-1.00	5.855091175997D+00
1		-2.284283635343D+00	-1.541883557565D+00	-.75	4.048118990820D+00
2		1.843375745741D+00	9.576278243864D-01	-.50	2.688653761866D+00
3		-6.432563292371D-01	-3.230631202047D-01	-.25	1.696661583502D+00
4		8.417546567581D-02	4.210919951760D-02	.25	5.344180054986D-01
				.50	2.435560222474D-01
0	-9.0	-1.124683679122D-05	-5.623418395609D-06	.75	7.894516554646D-02
1	-8.0	6.585124241134D-05	3.292562120567D-05	1.25	-2.601724671467D-02
2	-7.0	-1.669835332701D-04	-8.349176663507D-05	1.50	-2.411407886130D-02
3	-6.0	2.448161294418D-04	1.224080647209D-04	1.75	-1.186258648118D-02
4	-5.0	-2.393153075304D-04	-1.196576537652D-04	2.25	6.554047014018D-03
5	-4.0	1.777808184267D-04	8.889040921335D-05	2.50	7.241245439470D-03
6	-3.0	-1.152301108149D-04	-5.761505540746D-05	2.75	4.101568623811D-03
7	-2.0	7.152305187937D-05	3.576152593968D-05	3.25	-2.785864510162D-03
8	-1.0	-4.377515534545D-05	-2.188757767272D-05	3.50	-3.317136632187D-03
9	0.0	2.666192122704D-05	1.333096061352D-05	3.75	-1.995709578385D-03
10	1.0	-1.620061298252D-05	-8.100306491261D-06	4.25	1.478459085830D-03
11	2.0	9.816462621774D-06	4.908231310887D-06	4.50	1.814881269736D-03
12	3.0	-5.910210766802D-06	-2.955105383401D-06	4.75	1.118406961721D-03
13	4.0	3.498982155165D-06	1.749491077582D-06	5.25	-8.566056535899D-04
14	5.0	-1.981582464801D-06	-9.907912324006D-07	5.50	-1.063453939487D-03
15	6.0	1.008524713642D-06	5.042623568209D-07	5.75	-6.610747953918D-04
16	7.0	-4.123940241614D-07	-2.061970120807D-07	6.25	5.123450130565D-04
17	8.0	1.137824404406D-07	5.689122022030D-08	6.50	6.385642690176D-04
18	9.0	-1.517132677906D-08	-7.585663389528D-09	6.75	3.981446164328D-04
				7.25	-3.098216286997D-04
	0	2.130962489721D+02	1.066223270004D+02	7.50	-3.866681013509D-04
	1	1.200408496940D+02	5.998150486498D+01	7.75	-2.413376511261D-04
	2	2.538473117478D+01	1.269999454442D+01	8.25	1.880702886891D-04
	3	2.387060435092D+00	1.192868062429D+00	8.50	2.348374292130D-04
	4	8.417546567581D-02	4.210919951760D-02	8.75	1.466347159926D-04

CARDINAL SPLINES N=5

I= 1	H= -8.0				
0		0.	-1.249458917394D+01	-1.00	-1.484716366823D+01
1		5.137182824706D+00	5.566588574515D+00	-.75	-8.822345135519D+00
2		-6.514935234310D+00	-3.527011330017D+00	-.50	-4.571163451743D+00
3		2.786366083788D+00	1.403944972296D+00	-.25	-1.736612532776D+00
4		-4.086795254262D-01	-2.045007592731D-01	.25	9.190528199463D-01
				.50	1.262611022525D+00
0	-9.0	6.585124241134D-05	3.292562120567D-05	.75	1.234430429156D+00
1	-8.0	-3.968117718434D-04	-1.984058859217D-04	1.25	6.867513655241D-01
2	-7.0	1.043554716372D-03	5.217773581861D-04	1.50	3.847461631301D-01
3	-6.0	-1.600403465362D-03	-8.002017326808D-04	1.75	1.484016978646D-01
4	-5.0	1.646027449640D-03	8.230137248198D-04	2.25	-6.439179605698D-02
5	-4.0	-1.280235339037D-03	-6.401176695187D-04	2.50	-6.651224043628D-02
6	-3.0	8.524587760621D-04	4.262293880311D-04	2.75	-3.587765152889D-02
7	-2.0	-5.339960490960D-04	-2.669980245480D-04	3.25	2.286554533942D-02
8	-1.0	3.278172874048D-04	1.639086437024D-04	3.50	2.666339448897D-02
9	0.0	-1.998612518780D-04	-9.993062593901D-05	3.75	1.578663912709D-02
10	1.0	1.214820040469D-04	6.074100202346D-05	4.25	-1.144238814581D-02
11	2.0	-7.361783566739D-05	-3.680891783370D-05	4.50	-1.394344905209D-02
12	3.0	4.432479856718D-05	2.216239928359D-05	4.75	-8.544097812421D-03
13	4.0	-2.624162773955D-05	-1.312081386978D-05	5.25	6.493980250008D-03
14	5.0	1.486151001332D-05	7.430755006658D-06	5.50	8.041528015278D-03
15	6.0	-7.563763110572D-06	-3.781881555286D-06	5.75	4.989081637763D-03
16	7.0	3.092886191333D-06	1.546443095667D-06	6.25	-3.856459747656D-03
17	8.0	-8.533494154157D-07	-4.266747077079D-07	6.50	-4.802349606206D-03
18	9.0	1.137824404406D-07	5.689122022030D-08	6.75	-2.992282706310D-03
				7.25	2.326423504533D-03
	0	-1.131560599797D+03	-5.663368065219D+02	7.50	2.902612623654D-03
	1	-6.267541891752D+02	-3.130852011383D+02	7.75	1.811255671748D-03
	2	-1.299013003292D+02	-6.500786608475D+01	8.25	-1.411063478629D-03
	3	-1.192609683156D+01	-5.958082361535D+00	8.50	-1.761779786615D-03
	4	-4.086795254262D-01	-2.045007592731D-01	8.75	-1.099991048064D-03

I= 2	H= -7.0				
0		0.	4.528550239305D+01	-1.00	2.176758427391D+01
1		-5.754262191031D+00	-1.374309924416D+01	-.75	1.239487562230D+01
2		1.004278850249D+01	5.998330935010D+00	-.50	6.081576973297D+00
3		-5.129446454156D+00	-2.603728865567D+00	-.25	2.149672926596D+00
4		8.410871262288D-01	4.211270805128D-01	.25	-8.877533712484D-01
				.50	-9.550471574125D-01
0	-9.0	-1.669835332701D-04	-8.349176663507D-05	.75	-5.645006478418D-01
1	-8.0	1.043554716372D-03	5.217773581861D-04	1.25	5.327704370582D-01
2	-7.0	-2.876041083565D-03	-1.438020541783D-03	1.50	9.045852680300D-01
3	-6.0	4.678373655368D-03	2.339186827684D-03	1.75	1.058304614485D+00
4	-5.0	-5.153544815813D-03	-2.576772407907D-03	2.25	7.840604230054D-01
5	-4.0	4.285546835432D-03	2.142773417716D-03	2.50	4.922581890735D-01
6	-3.0	-2.991039299070D-03	-1.495519649535D-03	2.75	2.092984094478D-01
7	-2.0	1.914312847262D-03	9.571564236311D-04	3.25	-1.057540395038D-01
8	-1.0	-1.183833270319D-03	-5.919166351597D-04	3.50	-1.157623981083D-01
9	0.0	7.235070753508D-04	3.617535376754D-04	3.75	-6.550834547967D-02
10	1.0	-4.401245355606D-04	-2.200622677803D-04	4.25	4.481655102271D-02
11	2.0	2.667856835729D-04	1.333928417865D-04	4.50	5.360936207620D-02
12	3.0	-1.606440659747D-04	-8.032203298735D-05	4.75	3.239191159290D-02
13	4.0	9.510896575131D-05	4.755448287566D-05	5.25	-2.416067968499D-02

```
14   5.0  -5.386391050418D-05  -2.693195525209D-05    5.50 -2.973292600360D-02
15   6.0   2.741411862402D-05   1.370705931201D-05    5.75 -1.835934455451D-02
16   7.0  -1.120987582192D-05  -5.604937910958D-06    6.25  1.410105547684D-02
17   8.0   3.092886191333D-06   1.546443095667D-06    6.50  1.752272359699D-02
18   9.0  -4.123940241614D-07  -2.061970120807D-07    6.75  1.090066151933D-02
                                                      7.25 -8.456766723116D-03
       0   2.540683679090D+03   1.272358847177D+03    7.50 -1.054380942827D-02
       1   1.381170502577D+03   6.895273100285D+02    7.75 -6.575890580718D-03
       2   2.803160775875D+02   1.403654126939D+02    8.25  5.119282565917D-03
       3   2.514969009008D+01   1.255684603289D+01    8.50  6.390160793863D-03
       4   8.410871262288D-01   4.211270805128D-01    8.75  3.989068523280D-03


I= 3     H= -6.0
0          0.                  -1.107474194770D+02   -1.00 -2.127852827572D+01
1          5.130805265874D+00   2.913854142968D+01    -.75 -1.191265812037D+01
2         -9.648178086405D+00  -7.213224912550D+00    -.50 -5.727939741248D+00
3          5.508336463921D+00   2.849558007105D+00    -.25 -1.975652112163D+00
4         -9.912084595189D-01  -4.970312437992D-01     .25  7.618860362058D-01
                                                       .50  7.799501187623D-01
0   -9.0   2.448161294418D-04   1.224080647209D-04     .75  4.312270518167D-01
1   -8.0  -1.600403465362D-03  -8.002017326808D-04    1.25 -3.214205334940D-01
2   -7.0   4.678373655368D-03   2.339186827684D-03    1.50 -4.301416275257D-01
3   -6.0  -8.205074328968D-03  -4.102537164484D-03    1.75 -3.062806096833D-01
4   -5.0   9.887645603010D-03   4.943822801505D-03    2.25  3.901044895654D-01
5   -4.0  -9.023325757014D-03  -4.511662878507D-03    2.50  7.469605949055D-01
6   -3.0   6.793696219048D-03   3.396848109524D-03    2.75  9.690231892236D-01
7   -2.0  -4.547704494482D-03  -2.273852247241D-03    3.25  8.447021704850D-01
8   -1.0   2.866830968200D-03   1.433415484100D-03    3.50  5.644645562535D-01
9    0.0  -1.763604048837D-03  -8.818020244185D-04    3.75  2.527404294644D-01
10   1.0   1.075177169031D-03   5.375885845157D-04    4.25 -1.379368182164D-01
11   2.0  -6.521997897569D-04  -3.260998948784D-04    4.50 -1.552684442403D-01
12   3.0   3.928139412280D-04   1.964069706140D-04    4.75 -8.985373026979D-02
13   4.0  -2.325831430796D-04  -1.162915715398D-04    5.25  6.346324780582D-02
14   5.0   1.317243812426D-04   6.586219062128D-05    5.50  7.675891839215D-02
15   6.0  -6.704191929684D-05  -3.352095964842D-05    5.75  4.678249665279D-02
16   7.0   2.741411862402D-05   1.370705931201D-05    6.25 -3.531393580499D-02
17   8.0  -7.563763110572D-06  -3.781881555286D-06    6.50 -4.363411405515D-02
18   9.0   1.008524713642D-06   5.042623568209D-07    6.75 -2.702693918304D-02
                                                      7.25  2.084633642146D-02
       0  -3.223066598311D+03  -1.616465967914D+03    7.50  2.594149399307D-02
       1  -1.720374507514D+03  -8.576000181882D+02    7.75  1.615555139981D-02
       2  -3.426504048867D+02  -1.718323432071D+02    8.25 -1.255259130923D-02
       3  -3.017516807876D+01  -1.504356676967D+01    8.50 -1.565883672389D-02
       4  -9.912084595189D-01  -4.970312437992D-01    8.75 -9.770322888561D-03


I= 4     H= -5.0
0          0.                   2.176005191355D+02   -1.00  1.580732848947D+01
1         -3.689178963681D+00  -5.405645081153D+01    -.75  8.793843043139D+00
2          7.103864676136D+00   8.246192732692D+00    -.50  4.197346305440D+00
3         -4.214365623398D+00  -2.294608088433D+00    -.25  1.435260430522D+00
4          7.999192262504D-01   4.027634619178D-01     .25 -5.410279779628D-01
                                                       .50 -5.454245315034D-01
0   -9.0  -2.393153075304D-04  -1.196576537652D-04     .75 -2.958143660393D-01
1   -8.0   1.646027449640D-03   8.230137248198D-04    1.25  2.082767604175D-01
2   -7.0  -5.153544815813D-03  -2.576772407907D-03    1.50  2.668373102356D-01
```

3	-6.0	9.887645603010D-03	4.943822801505D-03	1.75	1.788082120933D-01
4	-5.0	-1.329724383044D-02	-6.648621915220D-03	2.25	-1.819596264638D-01
5	-4.0	1.367036900779D-02	6.835184503896D-03	2.50	-2.760579635254D-01
6	-3.0	-1.147493069343D-02	-5.737465346717D-03	2.75	-2.190048143308D-01
7	-2.0	8.315066261135D-03	4.157533130568D-03	3.25	3.308248866389D-01
8	-1.0	-5.478291895440D-03	-2.739145947720D-03	3.50	6.763760365160D-01
9	0.0	3.432702713062D-03	1.716351356531D-03	3.75	9.265568308389D-01
10	1.0	-2.105937234878D-03	-1.052968617439D-03	4.25	8.761623277435D-01
11	2.0	1.280130925183D-03	6.400654625914D-04	4.50	6.030836962000D-01
12	3.0	-7.715463372998D-04	-3.857731686499D-04	4.75	2.765393811311D-01
13	4.0	4.569339768773D-04	2.284669884386D-04	5.25	-1.561652533010D-01
14	5.0	-2.588062201493D-04	-1.294031100746D-04	5.50	-1.778989717033D-01
15	6.0	1.317243812426D-04	6.586219062128D-05	5.75	-1.039218565531D-01
16	7.0	-5.386391050418D-05	-2.693195525209D-05	6.25	7.436701237128D-02
17	8.0	1.486151001332D-05	7.430755006658D-06	6.50	9.034952790348D-02
18	9.0	-1.981582464801D-06	-9.907912324006D-07	6.75	5.525674683548D-02
				7.25	-4.190944938324D-02
	0	2.718207932066D+03	1.368795850355D+03	7.50	-5.186658631364D-02
	1	1.432654002467D+03	7.112435078400D+02	7.75	-3.216604523021D-02
	2	2.820767368021D+02	1.420348168371D+02	8.25	2.485305888358D-02
	3	2.458272652162D+01	1.220487654061D+01	8.50	3.094627488380D-02
	4	7.999192262504D-01	4.027634619178D-01	8.75	1.928195546077D-02

I= 5 H= -4.0

0		0.	-3.842302980569D+02	-1.00	-1.024581245842D+01
1		2.368719473364D+00	9.337798256924D+01	-.75	-5.689515955733D+00
2		-4.592329708334D+00	-1.058510313590D+01	-.50	-2.709870985151D+00
3		2.754097865438D+00	1.707997048700D+00	-.25	-9.243061660223D-01
4		-5.306654112872D-01	-2.702836308270D-01	.25	3.461191296330D-01
				.50	3.473733018012D-01
0	-9.0	1.777808184267D-04	8.889040921335D-05	.75	1.873466768941D-01
1	-8.0	-1.280235339037D-03	-6.401176695187D-04	1.25	-1.297637204749D-01
2	-7.0	4.285546835432D-03	2.142773417716D-03	1.50	-1.642439910587D-01
3	-6.0	-9.023325757014D-03	-4.511662878507D-03	1.75	-1.082976651772D-01
4	-5.0	1.367036900779D-02	6.835184503896D-03	2.25	1.046739368909D-01
5	-4.0	-1.610710880119D-02	-8.053554400596D-03	2.50	1.523713503099D-01
6	-3.0	1.549121608387D-02	7.745608041934D-03	2.75	1.139724825778D-01
7	-2.0	-1.260446741254D-02	-6.302233706270D-03	3.25	-1.379215564759D-01
8	-1.0	9.005298631177D-03	4.502649315588D-03	3.50	-2.236213767778D-01
9	0.0	-5.896887400413D-03	-2.948443700206D-03	3.75	-1.874568070227D-01
10	1.0	3.684091213843D-03	1.842045606921D-03	4.25	3.074529760724D-01
11	2.0	-2.253395112103D-03	-1.126697556052D-03	4.50	6.476853770571D-01
12	3.0	1.360953053096D-03	6.804765265482D-04	4.75	9.088760411571D-01
13	4.0	-8.065529753001D-04	-4.032764876500D-04	5.25	8.897051899732D-01
14	5.0	4.569339768773D-04	2.284669884386D-04	5.50	6.198975843925D-01
15	6.0	-2.325831430796D-04	-1.162915715398D-04	5.75	2.869920124891D-01
16	7.0	9.510896575131D-05	4.755448287566D-05	6.25	-1.642676287699D-01
17	8.0	-2.624162773955D-05	-1.312081386978D-05	6.50	-1.879986681365D-01
18	9.0	3.498982155165D-06	1.749491077582D-06	6.75	-1.102200151573D-01
				7.25	7.927029093142D-02
	0	-1.824618650666D+03	-9.294228622952D+02	7.50	9.647103902141D-02
	1	-9.584677732887D+02	-4.702576585343D+02	7.75	5.907910889603D-02
	2	-1.881350772271D+02	-9.582702740291D+01	8.25	-4.489189419274D-02
	3	-1.634985694090D+01	-8.022213661071D+00	8.50	-5.559403691283D-02
	4	-5.306654112872D-01	-2.702836308270D-01	8.75	-3.449619537402D-02

CARDINAL SPLINES N=5

I= 6 H= -3.0

0		0.	6.490216220796D+02	-1.00	6.359823698019D+00
1		-1.465971890225D+00	-1.564616341695D+02	-.75	3.529608715638D+00
2		2.848196961400D+00	1.542534203386D+01	-.50	1.680006259845D+00
3		-1.713882343729D+00	-1.415961610234D+00	-.25	5.725806806031D-01
4		3.317725026652D-01	1.742490993293D-01	.25	-2.139640881906D-01
				.50	-2.144364413711D-01
0	-9.0	-1.152301108149D-04	-5.761505540746D-05	.75	-1.154459992375D-01
1	-8.0	8.524587760621D-04	4.262293880311D-04	1.25	7.954937041914D-02
2	-7.0	-2.991039299070D-03	-1.495519649535D-03	1.50	1.003025448670D-01
3	-6.0	6.793696219048D-03	3.396848109524D-03	1.75	6.580325203194D-02
4	-5.0	-1.147493069343D-02	-5.737465346717D-03	2.25	-6.261068402453D-02
5	-4.0	1.549121608387D-02	7.745608041934D-03	2.50	-9.004940365439D-02
6	-3.0	-1.728661694278D-02	-8.643308471388D-03	2.75	-6.627219612701D-02
7	-2.0	1.622217089583D-02	8.111085447914D-03	3.25	7.612873545958D-02
8	-1.0	-1.304991429390D-02	-6.524957146952D-03	3.50	1.183819094344D-01
9	0.0	9.273424730789D-03	4.636712365394D-03	3.75	9.352274982237D-02
10	1.0	-6.054576969095D-03	-3.027288484548D-03	4.25	-1.227710308117D-01
11	2.0	3.771047538228D-03	1.885523769114D-03	4.50	-2.050224970954D-01
12	3.0	-2.291697113580D-03	-1.145848556790D-03	4.75	-1.759947052210D-01
13	4.0	1.360953053096D-03	6.804765265482D-04	5.25	2.986724435655D-01
14	5.0	-7.715463372998D-04	-3.857731686499D-04	5.50	6.367832076856D-01
15	6.0	3.928139412280D-04	1.964069706140D-04	5.75	9.020978400064D-01
16	7.0	-1.606440659747D-04	-8.032203298735D-05	6.25	8.949608838248D-01
17	8.0	4.432479856718D-05	2.216239928359D-05	6.50	6.264502611636D-01
18	9.0	-5.910210766802D-06	-2.955105383401D-06	6.75	2.910793720874D-01
				7.25	-1.674522908747D-01
	0	1.144849368269D+03	6.013319461358D+02	7.50	-1.919768563039D-01
	1	6.007767816606D+02	2.852262247975D+02	7.75	-1.127058853843D-01
	2	1.178148099760D+02	6.187944083161D+01	8.25	8.121409731895D-02
	3	1.022992775222D+01	4.857005965622D+00	8.50	9.890415212325D-02
	4	3.317725026652D-01	1.742490993293D-01	8.75	6.060312588985D-02

I= 7 H= -2.0

0		0.	-1.078028456663D+03	-1.00	-3.892935156166D+00
1		8.964780272326D-01	2.591145075257D+02	-.75	-2.160122314802D+00
2		-1.742948438181D+00	-2.412763978957D+01	-.50	-1.027942528166D+00
3		1.049953789324D+00	1.453514280799D+00	-.25	-3.502544484864D-01
4		-2.035549014282D-01	-1.156682140092D-01	.25	1.307956213192D-01
				.50	1.310240860909D-01
0	-9.0	7.152305187937D-05	3.576152593968D-05	.75	7.049860607494D-02
1	-8.0	-5.339960490960D-04	-2.669980245480D-04	1.25	-4.849572818172D-02
2	-7.0	1.914312847262D-03	9.571564236311D-04	1.50	-6.107105114199D-02
3	-6.0	-4.547704494482D-03	-2.273852247241D-03	1.75	-3.999935743602D-02
4	-5.0	8.315066261135D-03	4.157533130568D-03	2.25	3.786425338437D-02
5	-4.0	-1.260446741254D-02	-6.302233706270D-03	2.50	5.424587791335D-02
6	-3.0	1.622217089583D-02	8.111085447914D-03	2.75	3.971535328732D-02
7	-2.0	-1.773835430263D-02	-8.869177151317D-03	3.25	-4.989011058132D-02
8	-1.0	1.649539575014D-02	8.247697875069D-03	3.50	-6.894847088013D-02
9	0.0	-1.321095450205D-02	-6.605477251025D-03	3.75	-5.357634415630D-02
10	1.0	9.362439955915D-03	4.681219977958D-03	4.25	6.672175469934D-02
11	2.0	-6.093999873433D-03	-3.046999936717D-03	4.50	1.068329183040D-01
12	3.0	3.771047538228D-03	1.885523769114D-03	4.75	8.640461909910D-02
13	4.0	-2.253395112103D-03	-1.126697556052D-03	5.25	-1.173165626536D-01
14	5.0	1.280130925183D-03	6.400654625914D-04	5.50	-1.982486116084D-01

CARDINAL SPLINES N=5

15	6.0	-6.521997897569D-04	-3.260998948784D-04	5.75	-1.717820202237D-01
16	7.0	2.667856835729D-04	1.333928417865D-04	6.25	2.954033731802D-01
17	8.0	-7.361783566739D-05	-3.680891783370D-05	6.50	6.327050470673D-01
18	9.0	9.816462621774D-06	4.908231310887D-06	6.75	8.995521222451D-01
				7.25	8.969486721221D-01
	0	-7.032179171004D+02	-3.996239532989D+02	7.50	6.289372039270D-01
	1	-3.689039156188D+02	-1.592675505033D+02	7.75	2.926364826039D-01
	2	-7.232187822052D+01	-4.109750621647D+01	8.25	-1.686768861356D-01
	3	-6.278022662090D+00	-2.710541423532D+00	8.50	-1.935160089637D-01
	4	-2.035549014282D-01	-1.156682140092D-01	8.75	-1.136749839355D-01

I= 8	H= -1.0				
0		0.	1.779360433274D+03	-1.00	2.371796505346D+00
1		-5.460120143297D-01	-4.272202152091D+02	-.75	1.315988688734D+00
2		1.061808195582D+00	3.891692919868D+01	-.50	6.261980964478D-01
3		-6.398643507659D-01	-1.852554166309D+00	-.25	2.133487085709D-01
4		1.241119446693D-01	8.498377155311D-02	.25	-7.965305972242D-02
				.50	-7.978009107171D-02
0	-9.0	-4.377515534545D-05	-2.188757767272D-05	.75	-4.291816556277D-02
1	-8.0	3.278172874048D-04	1.639086437024D-04	1.25	2.950675593737D-02
2	-7.0	-1.183833270319D-03	-5.919166351597D-04	1.50	3.714273085724D-02
3	-6.0	2.866830968200D-03	1.433415484100D-03	1.75	2.431386953989D-02
4	-5.0	-5.478291895440D-03	-2.739145947720D-03	2.25	-2.297704014417D-02
5	-4.0	9.005298631177D-03	4.502649315588D-03	2.50	-3.287533529375D-02
6	-3.0	-1.304991429390D-02	-6.524957146952D-03	2.75	-2.402782289375D-02
7	-2.0	1.649539575014D-02	8.247697875069D-03	3.25	2.701413664768D-02
8	-1.0	-1.790016880398D-02	-8.950084401991D-03	3.50	4.132495838127D-02
9	0.0	1.658502957789D-02	8.292514788946D-03	3.75	3.194038715842D-02
10	1.0	-1.325073978614D-02	-6.625369893070D-03	4.25	-3.912752016758D-02
11	2.0	9.362439955915D-03	4.681219977958D-03	4.50	-6.187218363438D-02
12	3.0	-6.054576969095D-03	-3.027288484548D-03	4.75	-4.921372249422D-02
13	4.0	3.684091213843D-03	1.842045606921D-03	5.25	6.337607028324D-02
14	5.0	-2.105937234878D-03	-1.052968617439D-03	5.50	1.026755114479D-01
15	6.0	1.075177169031D-03	5.375885845157D-04	5.75	8.381720267733D-02
16	7.0	-4.401245355606D-04	-2.200622677803D-04	6.25	-1.153043537363D-01
17	8.0	1.214820040469D-04	6.074100202346D-05	6.50	-1.957344790724D-01
18	9.0	-1.620061298252D-05	-8.100306491261D-06	6.75	-1.702094999719D-01
				7.25	2.941683950418D-01
	0	4.289297129804D+02	2.937162994052D+02	7.50	6.311535962866D-01
	1	2.249899289258D+02	7.092652580284D+01	7.75	8.985756619669D-01
	2	4.410387583420D+01	3.020007968315D+01	8.25	8.977281564395D-01
	3	3.828165657330D+00	1.206861609603D+00	8.50	6.299271958711D-01
	4	1.241119446693D-01	8.498377155311D-02	8.75	2.932679919477D-01

I= 9	H= 0.0				
0		0.	-2.929520925374D+03	-1.00	-1.442403772779D+00
1		3.320213763857D-01	7.030908574658D+02	-.75	-8.003000254727D-01
2		-6.457176875059D-01	-6.352180286215D+01	-.50	-3.808046030651D-01
3		3.891671790434D-01	2.717911074400D+00	-.25	-1.297383489641D-01
4		-7.549752984446D-02	-7.549752984446D-02	.25	4.843381367562D-02
				.50	4.850862015561D-02
0	-9.0	2.666192122704D-05	1.333096061352D-05	.75	2.609384832581D-02
1	-8.0	-1.998612518780D-04	-9.993062593901D-05	1.25	-1.793651486794D-02
2	-7.0	7.235070753508D-04	3.617535376754D-04	1.50	-2.257516441045D-02
3	-6.0	-1.763604048837D-03	-8.818020244185D-04	1.75	-1.477516948021D-02

CARDINAL SPLINES N=5

4	-5.0	3.432702713062D-03	1.716351356531D-03	2.25	1.395494545239D-02
5	-4.0	-5.896887400413D-03	-2.948443700206D-03	2.50	1.995804022599D-02
6	-3.0	9.273424730789D-03	4.636712365394D-03	2.75	1.457853201294D-02
7	-2.0	-1.321095450205D-02	-6.605477251025D-03	3.25	-1.636140540937D-02
8	-1.0	1.658502957789D-02	8.292514788946D-03	3.50	-2.499534874244D-02
9	0.0	-1.794003763029D-02	-8.970018815144D-03	3.75	-1.928483690549D-02
10	1.0	1.658502957789D-02	8.292514788946D-03	4.25	2.349590572729D-02
11	2.0	-1.321095450205D-02	-6.605477251025D-03	4.50	3.700215639539D-02
12	3.0	9.273424730789D-03	4.636712365394D-03	4.75	2.927333616000D-02
13	4.0	-5.896887400413D-03	-2.948443700206D-03	5.25	-3.707771466421D-02
14	5.0	3.432702713062D-03	1.716351356531D-03	5.50	-5.932108927873D-02
15	6.0	-1.763604048837D-03	-8.818020244185D-04	5.75	-4.762285579039D-02
16	7.0	7.235070753508D-04	3.617535376754D-04	6.25	6.213170900130D-02
17	8.0	-1.998612518780D-04	-9.993062593901D-05	6.50	1.011143556691D-01
18	9.0	2.666192122704D-05	1.333096061352D-05	6.75	8.283563775886D-02
				7.25	-1.145218864013D-01
	0	-2.609513600874D+02	-2.609513600874D+02	7.50	-1.947411728504D-01
	1	-1.368740695176D+02	-2.772778922748D-19	7.75	-1.695761182450D-01
	2	-2.683000335774D+01	-2.683000335774D+01	8.25	2.936442048029D-01
	3	-2.328743895357D+00	-5.211442535396D-21	8.50	6.304715377116D-01
	4	-7.549752984446D-02	-7.549752984446D-02	8.75	8.981277476000D-01

I=10 H= 1.0

0		0.	4.815642124526D+03	-1.00	8.760120823805D-01
1		-2.016388352806D-01	-1.155608009082D+03	-.75	4.860413025509D-01
2		3.921585430599D-01	1.040874561172D+02	-.50	2.312699165080D-01
3		-2.363591056032D-01	-4.266277385515D+00	-.25	7.879185221784D-02
4		4.585559843687D-02	8.498377155311D-02	.25	-2.941378753437D-02
				.50	-2.945872681525D-02
0	-9.0	-1.620061298252D-05	-8.100306491261D-06	.75	-1.584616214064D-02
1	-8.0	1.214820040469D-04	6.074100202346D-05	1.25	1.089174241107D-02
2	-7.0	-4.401245355606D-04	-2.200622677803D-04	1.50	1.370788592238D-02
3	-6.0	1.075177169031D-03	5.375885845157D-04	1.75	8.971104575214D-03
4	-5.0	-2.105937234878D-03	-1.052968617439D-03	2.25	-8.471503620616D-03
5	-4.0	3.684091213843D-03	1.842045606921D-03	2.50	-1.211402251121D-02
6	-3.0	-6.054576969095D-03	-3.027288484548D-03	2.75	-8.847117540081D-03
7	-2.0	9.362439955915D-03	4.681219977958D-03	3.25	9.923222456940D-03
8	-1.0	-1.325073978614D-02	-6.625369893070D-03	3.50	1.515296053135D-02
9	0.0	1.658502957789D-02	8.292514788946D-03	3.75	1.168416497169D-02
10	1.0	-1.790016880398D-02	-8.950084401991D-03	4.25	-1.420973576553D-02
11	2.0	1.649539575014D-02	8.247697875069D-03	4.50	-2.234748474083D-02
12	3.0	-1.304991429390D-02	-6.524957146952D-03	4.75	-1.764791546540D-02
13	4.0	9.005298631177D-03	4.502649315588D-03	5.25	2.223052824626D-02
14	5.0	-5.478291895440D-03	-2.739145947720D-03	5.50	3.542083250449D-02
15	6.0	2.866830968200D-03	1.433415484100D-03	5.75	2.828205968389D-02
16	7.0	-1.183833270319D-03	-5.919166351597D-04	6.25	-3.629067850963D-02
17	8.0	3.278172874048D-04	1.639086437024D-04	6.50	-5.832331194221D-02
18	9.0	-4.377515534545D-05	-2.188757767272D-05	6.75	-4.698727193996D-02
				7.25	6.160642901604D-02

0	1.585028858299D+02	2.937162994052D+02	7.50	1.004312056620D-01
1	8.313687732015D+01	-7.092652580284D+01	7.75	8.238717788736D-02
2	1.629628353209D+01	3.020007968315D+01	8.25	-1.141220387417D-01
3	1.414442438124D+00	-1.206861609603D+00	8.50	-1.941966070781D-01
4	4.585559843687D-02	8.498377155311D-02	8.75	-1.691998297952D-01

CARDINAL SPLINES N=5

I=11	H= 2.0				
0		0.	-7.896813761233D+03	-1.00	-5.307154958899D-01
1		1.221576974914D-01	1.894972740369D+03	-.75	-2.944583577834D-01
2		-2.375811295860D-01	-1.704968766603D+02	-.50	-1.401098693319D-01
3		1.431951422223D-01	6.874597127864D+00	-.25	-4.773419065744D-02
4		-2.778152659021D-02	-1.156682140092D-01	.25	1.781950632014D-02
				.50	1.784663288784D-02
0	-9.0	9.816462621774D-06	4.908231310887D-06	.75	9.599826769725D-03
1	-8.0	-7.361783566739D-05	-3.680891783370D-05	1.25	-6.598235547891D-03
2	-7.0	2.667856835729D-04	1.333928417865D-04	1.50	-8.304134678814D-03
3	-6.0	-6.521997897569D-04	-3.260998948784D-04	1.75	-5.434519556294D-03
4	-5.0	1.280130925183D-03	6.400654625914D-04	2.25	5.131553038420D-03
5	-4.0	-2.253395112103D-03	-1.126697556052D-03	2.50	7.337635882640D-03
6	-3.0	3.771047538228D-03	1.885523769114D-03	2.75	5.358487133197D-03
7	-2.0	-6.093999873433D-03	-3.046999936717D-03	3.25	-6.009082068033D-03
8	-1.0	9.362439955915D-03	4.681219977958D-03	3.50	-9.174628679229D-03
9	0.0	-1.321095450205D-02	-6.605477251025D-03	3.75	-7.072995791028D-03
10	1.0	1.649539575014D-02	8.247697875069D-03	4.25	8.596657958269D-03
11	2.0	-1.773835430263D-02	-8.869177151317D-03	4.50	1.351372790718D-02
12	3.0	1.622217089583D-02	8.111085447914D-03	4.75	1.066547050491D-02
13	4.0	-1.260446741254D-02	-6.302233706270D-03	5.25	-1.341040115118D-02
14	5.0	8.315066261135D-03	4.157533130568D-03	5.50	-2.133807204143D-02
15	6.0	-4.547704494482D-03	-2.273852247241D-03	5.75	-1.700684263922D-02
16	7.0	1.914312847262D-03	9.571564236311D-04	6.25	2.170285291811D-02
17	8.0	-5.339960490960D-04	-2.669980245480D-04	6.50	3.473549565655D-02
18	9.0	7.152305187937D-05	3.576152593968D-05	6.75	2.783263651547D-02
				7.25	-3.589056212834D-02
	0	-9.602998949741D+01	-3.996239532989D+02	7.50	-5.777866636690D-02
	1	-5.036881461211D+01	1.592675505033D+02	7.75	-4.661109073966D-02
	2	-9.873134212429D+00	-4.109750621647D+01	8.25	6.123057599841D-02
	3	-8.569398150255D-01	2.710541423532D+00	8.50	9.988753225893D-02
	4	-2.778152659021D-02	-1.156682140092D-01	8.75	8.198817507655D-02
I=12	H= 3.0				
0		0.	1.286460836631D+04	-1.00	3.195110293125D-01
1		-7.354332475878D-02	-3.087418983057D+03	-.75	1.772750644895D-01
2		1.430327737681D-01	2.777036641775D+02	-.50	8.435136617003D-02
3		-8.620923479208D-02	-1.112997354148D+01	-.25	2.873773359380D-02
4		1.672569599348D-02	1.742490993293D-01	.25	-1.072796739538D-02
				.50	-1.074427883015D-02
0	-9.0	-5.910210766802D-06	-2.955105383401D-06	.75	-5.779408270794D-03
1	-8.0	4.432479856718D-05	2.216239928359D-05	1.25	3.972325762475D-03
2	-7.0	-1.606440659747D-04	-8.032203298735D-05	1.50	4.999300599256D-03
3	-6.0	3.928139412280D-04	1.964069706140D-04	1.75	3.271697526741D-03
4	-5.0	-7.715463372998D-04	-3.857731686499D-04	2.25	-3.089241606896D-03
5	-4.0	1.360953053096D-03	6.804765265482D-04	2.50	-4.417254175835D-03
6	-3.0	-2.291697113580D-03	-1.145848556790D-03	2.75	-3.225739793677D-03
7	-2.0	3.771047538228D-03	1.885523769114D-03	3.25	3.617154429261D-03
8	-1.0	-6.054576969095D-03	-3.027288484548D-03	3.50	5.522376592665D-03
9	0.0	9.273424730789D-03	4.636712365394D-03	3.75	4.257087635657D-03
10	1.0	-1.304991429390D-02	-6.524957146952D-03	4.25	-5.173110383586D-03
11	2.0	1.622217089583D-02	8.111085447914D-03	4.50	-8.130771062315D-03
12	3.0	-1.728661694278D-02	-8.643308471388D-03	4.75	-6.415789978577D-03
13	4.0	1.549121608387D-02	7.745608041934D-03	5.25	8.062086049403D-03
14	5.0	-1.147493069343D-02	-5.737465346717D-03	5.50	1.282219788968D-02
15	6.0	6.793696219048D-03	3.396848109524D-03	5.75	1.021338766408D-02

16	7.0	-2.991039299070D-03	-1.495519649535D-03	6.25	-1.300965279759D-02
17	8.0	8.524587760621D-04	4.262293880311D-04	6.50	-2.079339493748D-02
18	9.0	-1.152301108149D-04	-5.761505540746D-05	6.75	-1.663110212306D-02
				7.25	2.132813177464D-02
	0	5.781452400217D+01	6.013319461358D+02	7.50	3.419383831732D-02
	1	3.032433206557D+01	-2.852262247975D+02	7.75	2.743534731719D-02
	2	5.944071687212D+00	6.187944083161D+01	8.25	-3.544583776984D-02
	3	5.159158209731D-01	-4.857005965622D+00	8.50	-5.710186220088D-02
	4	1.672569599348D-02	1.742490993293D-01	8.75	-4.609122662550D-02

I=13 H= 4.0

0		0.	-2.054525566952D+04	-1.00	-1.891545025269D-01
1		4.353850733274D-02	4.932689138918D+03	-.75	-1.049490169645D-01
2		-8.467715038785D-02	-4.437846408337D+02	-.50	-4.993703121620D-02
3		5.103699443963D-02	1.775242437084D+01	-.25	-1.701308087354D-02
4		-9.901850366675D-03	-2.702836308270D-01	.25	6.351078882417D-03
				.50	6.360731560395D-03
0	-9.0	3.498982155165D-06	1.749491077582D-06	.75	3.421470813006D-03
1	-8.0	-2.624162773955D-05	-1.312081386978D-05	1.25	-2.351653573506D-03
2	-7.0	9.510896575131D-05	4.755448287566D-05	1.50	-2.959627245842D-03
3	-6.0	-2.325831430796D-04	-1.162915715398D-04	1.75	-1.936867694365D-03
4	-5.0	4.569339768773D-04	2.284669884386D-04	2.25	1.828840015855D-03
5	-4.0	-8.065529753001D-04	-4.032764876500D-04	2.50	2.615013672406D-03
6	-3.0	1.360953053096D-03	6.804765265482D-04	2.75	1.909624086572D-03
7	-2.0	-2.253395112103D-03	-1.126697556052D-03	3.25	-2.141293621082D-03
8	-1.0	3.684091213843D-03	1.842045606921D-03	3.50	-3.269099073414D-03
9	0.0	-5.896887400413D-03	-2.948443700206D-03	3.75	-2.520027249151D-03
10	1.0	9.005298631177D-03	4.502649315588D-03	4.25	3.062072302526D-03
11	2.0	-1.260446741254D-02	-6.302233706270D-03	4.50	4.812532109470D-03
12	3.0	1.549121608387D-02	7.745608041934D-03	4.75	3.797198514137D-03
13	4.0	-1.610710880119D-02	-8.053554400596D-03	5.25	-4.770595830142D-03
14	5.0	1.367036900779D-02	6.835184503896D-03	5.50	-7.586154114859D-03
15	6.0	-9.023325757014D-03	-4.511662878507D-03	5.75	-6.041462159043D-03
16	7.0	4.285546835432D-03	2.142773417716D-03	6.25	7.690815010924D-03
17	8.0	-1.280235339037D-03	-6.401176695187D-04	6.50	1.228663750089D-02
18	9.0	1.777808184267D-04	8.889040921335D-05	6.75	9.821249956160D-03
				7.25	-1.257185869368D-02
	0	-3.422707392469D+01	-9.294228622952D+02	7.50	-2.012780430924D-02
	1	-1.795245622004D+01	4.702576585343D+02	7.75	-1.612027432260D-02
	2	-3.518977578722D+00	-9.582702740291D+01	8.25	2.071235591666D-02
	3	-3.054296187607D-01	8.022213661071D+00	8.50	3.322920560554D-02
	4	-9.901850366675D-03	-2.702836308270D-01	8.75	2.667632494680D-02

I=14 H= 5.0

0		0.	3.081469365646D+04	-1.00	1.071234289608D-01
1		-2.465705101929D-02	-7.408113465227D+03	-.75	5.943552755975D-02
2		4.795500768638D-02	6.673095259255D+02	-.50	2.828071761405D-02
3		-2.890367266988D-02	-2.670436116965D+01	-.25	9.634975689380D-03
4		5.607697585254D-03	4.027634619178D-01	.25	-3.596789598757D-03
				.50	-3.602255442984D-03
0	-9.0	-1.981582464801D-06	-9.907912324006D-07	.75	-1.937671570897D-03
1	-8.0	1.486151001332D-05	7.430755006658D-06	1.25	1.331804102075D-03
2	-7.0	-5.386391050418D-05	-2.693195525209D-05	1.50	1.676114869485D-03
3	-6.0	1.317243812426D-04	6.586219062128D-05	1.75	1.096898390675D-03
4	-5.0	-2.588062201493D-04	-1.294031100746D-04	2.25	-1.035717157307D-03

5	-4.0	4.569339768773D-04	2.284669884386D-04	2.50	-1.480944093980D-03
6	-3.0	-7.715463372998D-04	-3.857731686499D-04	2.75	-1.081462781476D-03
7	-2.0	1.280130925183D-03	6.400654625914D-04	3.25	1.212653692532D-03
8	-1.0	-2.105937234878D-03	-1.052968617439D-03	3.50	1.851340428611D-03
9	0.0	3.432702713062D-03	1.716351356531D-03	3.75	1.427119378535D-03
10	1.0	-5.478291895440D-03	-2.739145947720D-03	4.25	-1.734047004400D-03
11	2.0	8.315066261135D-03	4.157533130568D-03	4.50	-2.725284257027D-03
12	3.0	-1.147493069343D-02	-5.737465346717D-03	4.75	-2.150264550304D-03
13	4.0	1.367036900779D-02	6.835184503896D-03	5.25	2.701294356365D-03
14	5.0	-1.329724383044D-02	-6.648621915220D-03	5.50	4.295353924561D-03
15	6.0	9.887645603010D-03	4.943822801505D-03	5.75	3.420506174294D-03
16	7.0	-5.153544815813D-03	-2.576772407907D-03	6.25	-4.353435833196D-03
17	8.0	1.646027449640D-03	8.230137248198D-04	6.50	-6.953868200046D-03
18	9.0	-2.393153075304D-04	-1.196576537652D-04	6.75	-5.557414562487D-03
				7.25	7.109494902667D-03
	0	1.938376864393D+01	1.368795850355D+03	7.50	1.137721117444D-02
	1	1.016698678716D+01	-7.112435078400D+02	7.75	9.106445778061D-03
	2	1.992896872033D+00	1.420348168371D+02	8.25	-1.167895468832D-02
	3	1.729734403993D-01	-1.220487654061D+01	8.50	-1.871088018278D-02
	4	5.607697585254D-03	4.027634619178D-01	8.75	-1.499377683967D-02

I=15 H= 6.0

0		0.	-3.748106809704D+04	-1.00	-5.452026697622D-02
1		1.254915779007D-02	9.055512027864D+03	-.75	-3.024959874143D-02
2		-2.440660967096D-02	-8.195658304745D+02	-.50	-1.439341699720D-02
3		1.471047143568D-02	3.293669154644D+01	-.25	-4.903702215320D-03
4		-2.854028079501D-03	-4.970312437992D-01	.25	1.830578915926D-03
				.50	1.833360621560D-03
0	-9.0	1.008524713642D-06	5.042623568209D-07	.75	9.861739422950D-04
1	-8.0	-7.563763110572D-06	-3.781881555286D-06	1.25	-6.778187756375D-04
2	-7.0	2.741411862402D-05	1.370705931201D-05	1.50	-8.530548265807D-04
3	-6.0	-6.704191929684D-05	-3.352095964842D-05	1.75	-5.582637880121D-04
4	-5.0	1.317243812426D-04	6.586219062128D-05	2.25	5.271253440837D-04
5	-4.0	-2.325831430796D-04	-1.162915715398D-04	2.50	7.537219058366D-04
6	-3.0	3.928139412280D-04	1.964069706140D-04	2.75	5.504067020772D-04
7	-2.0	-6.521997897569D-04	-3.260998948784D-04	3.25	-6.171743675676D-04
8	-1.0	1.075177169031D-03	5.375885845157D-04	3.50	-9.422292400223D-04
9	0.0	-1.763604048837D-03	-8.818020244185D-04	3.75	-7.263225753237D-04
10	1.0	2.866830968200D-03	1.433415484100D-03	4.25	8.825246984186D-04
11	2.0	-4.547704494482D-03	-2.273852247241D-03	4.50	1.386996528379D-03
12	3.0	6.793696219048D-03	3.396848109524D-03	4.75	1.094339933829D-03
13	4.0	-9.023325757014D-03	-4.511662878507D-03	5.25	-1.374745873781D-03
14	5.0	9.887645603010D-03	4.943822801505D-03	5.50	-2.185959359082D-03
15	6.0	-8.205074328968D-03	-4.102537164484D-03	5.75	-1.740699398731D-03
16	7.0	4.678373655368D-03	2.339186827684D-03	6.25	2.215316673510D-03
17	8.0	-1.600403465362D-03	-8.002017326808D-04	6.50	3.538407369238D-03
18	9.0	2.448161294418D-04	1.224080647209D-04	6.75	2.827644563010D-03
				7.25	-3.616605362996D-03
	0	-9.865337516231D+00	-1.616465967914D+03	7.50	-5.786700586555D-03
	1	-5.174471137241D+00	8.576000181882D+02	7.75	-4.630796227966D-03
	2	-1.014281527545D+00	-1.718323432071D+02	8.25	5.935274776938D-03
	3	-8.803453942635D-02	1.504356676967D+01	8.50	9.504492418776D-03
	4	-2.854028079501D-03	-4.970312437992D-01	8.75	7.611667654908D-03

I=16 H= 7.0

CARDINAL SPLINES N=5

0		0.	3.076465859886D+04	-1.00	2.229376845236D-02
1		-5.131449638858D-03	-7.495424891287D+03	-.75	1.236930009983D-02
2		9.980055626391D-03	6.840680167112D+02	-.50	5.885581949615D-03
3		-6.015228390310D-03	-2.771742093135D+01	-.25	2.005162559637D-03
4		1.167034796801D-03	4.211270805128D-01	.25	-7.485381485617D-04
				.50	-7.496755922768D-04
0	-9.0	-4.123940241614D-07	-2.061970120807D-07	.75	-4.032542772609D-04
1	-8.0	3.092886191333D-06	1.546443095667D-06	1.25	2.771654016002D-04
2	-7.0	-1.120987582192D-05	-5.604937910958D-06	1.50	3.488207717190D-04
3	-6.0	2.741411862402D-05	1.370705931201D-05	1.75	2.282783975146D-04
4	-5.0	-5.386391050418D-05	-2.693195525209D-05	2.25	-2.155455891083D-04
5	-4.0	9.510896575131D-05	4.755448287566D-05	2.50	-3.082025960267D-04
6	-3.0	-1.606440659747D-04	-8.032203298735D-05	2.75	-2.250654088148D-04
7	-2.0	2.667856835729D-04	1.333928417865D-04	3.25	2.523669951165D-04
8	-1.0	-4.401245355606D-04	-2.200622677803D-04	3.50	3.852840042954D-04
9	0.0	7.235070753508D-04	3.617535376754D-04	3.75	2.969980382784D-04
10	1.0	-1.183833270319D-03	-5.919166351597D-04	4.25	-3.608691760793D-04
11	2.0	1.914312847262D-03	9.571564236311D-04	4.50	-5.671493674440D-04
12	3.0	-2.991039299070D-03	-1.495519649535D-03	4.75	-4.474795730772D-04
13	4.0	4.285546835432D-03	2.142773417716D-03	5.25	5.621341807428D-04
14	5.0	-5.153544815813D-03	-2.576772407907D-03	5.50	8.938344741872D-04
15	6.0	4.678373655368D-03	2.339186827684D-03	5.75	7.117630454519D-04
16	7.0	-2.876041083565D-03	-1.438020541783D-03	6.25	-9.058101864534D-04
17	8.0	1.043554716372D-03	5.217773581861D-04	6.50	-1.446776721156D-03
18	9.0	-1.669835332701D-04	-8.349176663507D-05	6.75	-1.156134486622D-03
				7.25	1.478610135842D-03
	0	4.034015264261D+00	1.272358847177D+03	7.50	2.365703688887D-03
	1	2.115882520261D+00	-6.895273100285D+02	7.75	1.893016658315D-03
	2	4.147478003331D-01	1.403654126939D+02	8.25	-2.425751182725D-03
	3	3.599802429451D-02	-1.255684603289D+01	8.50	-3.883867573959D-03
	4	1.167034796801D-03	4.211270805128D-01	8.75	-3.109734141866D-03

I=17 H= 8.0

0		0.	-1.433491229278D+04	-1.00	-6.151007463128D-03
1		1.415803008776D-03	3.527365018557D+03	-.75	-3.412776853985D-03
2		-2.753567502863D-03	-3.252634588529D+02	-.50	-1.623873429045D-03
3		1.659643831568D-03	1.332010969537D+01	-.25	-5.532384416158D-04
4		-3.219931199209D-04	-2.045007592731D-01	.25	2.065269329425D-04
				.50	2.068407598542D-04
0	-9.0	1.137824404406D-07	5.689122022030D-08	.75	1.112606853828D-04
1	-8.0	-8.533494154157D-07	-4.266747077079D-07	1.25	-7.647187716689D-05
2	-7.0	3.092886191333D-06	1.546443095667D-06	1.50	-9.624209378544D-05
3	-6.0	-7.563763110572D-06	-3.781881555286D-06	1.75	-6.298360657979D-05
4	-5.0	1.486151001332D-05	7.430755006658D-06	2.25	5.947052987880D-05
5	-4.0	-2.624162773955D-05	-1.312081386978D-05	2.50	8.503523735701D-05
6	-3.0	4.432479856718D-05	2.216239928359D-05	2.75	6.209710367865D-05
7	-2.0	-7.361783566739D-05	-3.680891783370D-05	3.25	-6.962977880239D-05
8	-1.0	1.214820040469D-04	6.074100202346D-05	3.50	-1.063024652517D-04
9	0.0	-1.998612518780D-04	-9.993062593901D-05	3.75	-8.194374415820D-05
10	1.0	3.278172874048D-04	1.639086437024D-04	4.25	9.956612333790D-05
11	2.0	-5.339960490960D-04	-2.669980245480D-04	4.50	1.564800409201D-04
12	3.0	8.524587760621D-04	4.262293880311D-04	4.75	1.234622830972D-04
13	4.0	-1.280235339037D-03	-6.401176695187D-04	5.25	-1.550957247120D-04
14	5.0	1.646027449640D-03	8.230137248198D-04	5.50	-2.466130141575D-04
15	6.0	-1.600403465362D-03	-8.002017326808D-04	5.75	-1.963781138136D-04
16	7.0	1.043554716372D-03	5.217773581861D-04	6.25	2.499142691873D-04

17	8,0	-3,968117718434D-04	-1,984058859217D-04	6.50	3,991650600813D-04
18	9,0	6,585124241134D-05	3,292562120567D-05	6,75	3,189742430063D-04
				7.25	-4,079336113175D-04
	0	-1,113013247241D+00	-5,663368065219D+02	7,50	-6,526608718950D-04
	1	-5,837868986609D-01	3,130852011383D+02	7,75	-5,222402824483D-04
	2	-1,144318403321D-01	-6,500786608475D+01	8,25	6,691564711141D-04
	3	-9,932108485583D-03	5,958082361535D+00	8,50	1,071321914888D-03
	4	-3,219931199209D-04	-2,045007592731D-01	8,75	8,577185538306D-04

I=18 H= 9.0

0		0.	2.821034704429D+03	-1,00	8,201523355162D-04
1		-1,887778778649D-04	-7,012387716281D+02	-.75	4,550469041314D-04
2		3,671503940254D-04	6,537250319556D+01	-.50	2,165212104298D-04
3		-2,212907042298D-04	-2,708799245063D+00	-,25	7,3766744478154D-05
4		4,293335939608D-05	4,210919951760D-02	.25	-2,753752871596D-05
				.50	-2,757937312406D-05
0	-9,0	-1,517132677906D-08	-7,585663389528D-09	.75	-1,483508349422D-05
1	-8,0	1,137824404406D-07	5,689122022030D-08	1,25	1,019647383891D-05
2	-7,0	-4,123940241614D-07	-2,061970120807D-07	1,50	1,283256039057D-05
3	-6,0	1,008524713642D-06	5,042623568209D-07	1,75	8,397998140363D-06
4	-5,0	-1,981582464801D-06	-9,907912324006D-07	2,25	-7,929577336964D-06
5	-4,0	3,498982155165D-06	1,749491077582D-06	2,50	-1,133827917393D-05
6	-3,0	-5,910210766802D-06	-2,955105383401D-06	2,75	-8,279794242970D-06
7	-2,0	9,816462621774D-06	4,908231310887D-06	3,25	9,284171637834D-06
8	-1,0	-1,620061298252D-05	-8,100306491261D-06	3,50	1,417396773762D-05
9	0,0	2,666192122704D-05	1,333096061352D-05	3,75	1,092606682062D-05
10	1,0	-4,377515534545D-05	-2,188757767272D-05	4,25	-1,327576257562D-05
11	2,0	7,152305187937D-05	3,576152593968D-05	4,50	-2,086443853657D-05
12	3,0	-1,152301108149D-04	-5,761505540746D-05	4,75	-1,646197306410D-05
13	4,0	1,777808184267D-04	8,889040921335D-05	5,25	2,067982665520D-05
14	5,0	-2,393153075304D-04	-1,196576537652D-04	5,50	3,288233670482D-05
15	6,0	2,448161294418D-04	1,224080647209D-04	5,75	2,618419674558D-05
16	7,0	-1,669835332701D-04	-8,349176663507D-05	6,25	-3,332235348444D-05
17	8,0	6,585124241134D-05	3,292562120567D-05	6,50	-5,322258509986D-05
18	9,0	-1,124683679122D-05	-5,623418395609D-06	6,75	-4,253020960357D-05
				7,25	5,439095689829D-05
	0	1,484050286294D-01	1,066223270004D+02	7,50	8,702043786790D-05
	1	7,783996408571D-02	-5,998150486498D+01	7,75	6,963048468227D-05
	2	1,525791404631D-02	1,269999454442D+01	8,25	-8,921596385416D-05
	3	1,324310234029D-03	-1,192868062429D+00	8,50	-1,428315882837D-04
	4	4,293335939608D-05	4,210919951760D-02	8,75	-1,143497213612D-04

N= 5 M=19 P= 9.5 Q= 9

I= 0 H= -9.5

0		1.000000000000D+00	-7.786910792828D-01	-1.00	5,855091584967D+00
1		-2.284283729477D+00	-8.527276645832D-01	-.75	4,048119217730D+00
2		1.843375928821D+00	8.971409168419D-01	-.50	2,688653869835D+00

CARDINAL SPLINES N=5

3		-6,432564395838D-01	-3.207035640653D-01	-.25	1,696661620286D+00
4		8,417548708455D-02	4.207469186081D-02	.25	5,344179917670D-01
				.50	2,435560084949D-01
0	-9,5	-1,124684435641D-05	-5.623422178203D-06	.75	7,894515814894D-02
1	-8,5	6,585129914900D-05	3.292564957450D-05	1.25	-2,601724163020D-02
2	-7,5	-1,669837389107D-04	-8,349186945533D-05	1.50	-2,411407246234D-02
3	-6,5	2,448166323436D-04	1.224083161718D-04	1.75	-1,186258229352D-02
4	-5,5	-2,393162956504D-04	-1.196581478252D-04	2.25	6,554043059932D-03
5	-4,5	1,777825632116D-04	8.889128160580D-05	2.50	7,241239785634D-03
6	-3,5	-1,152330580248D-04	-5,761652901241D-05	2.75	4,101564495090D-03
7	-2,5	7,152794726579D-05	3,576397363290D-05	3.25	-2,785859880609D-03
8	-1,5	-4,378323579098D-05	-2.189161789549D-05	3.50	-3,317129564337D-03
9	-,5	2,667522614070D-05	1.333761307035D-05	3.75	-1,995704130102D-03
10	,5	-1,622249073154D-05	-8.111245365771D-06	4,25	1,478452465876D-03
11	1,5	9,852371544864D-06	4.926185772432D-06	4.50	1,814870865696D-03
12	2,5	-5,968874579806D-06	-2,984437289903D-06	4.75	1,118398752972D-03
13	3,5	3.593490816832D-06	1.796745408416D-06	5.25	-8,565953416310D-04
14	4.5	-2,127391249233D-06	-1.063695624617D-06	5.50	-1,063437542790D-03
15	5,5	1,204800532447D-06	6.024002662233D-07	5.75	-6,610617387293D-04
16	6,5	-6,131809569512D-07	-3.065904784756D-07	6.25	5,123283970660D-04
17	7,5	2,507345439014D-07	1.253672719507D-07	6.50	6,385377300046D-04
18	8,5	-6,917942357289D-08	-3.458971178645D-08	6.75	3,981234092004D-04
19	9,5	9,224125666094D-09	4.612062833047D-09	7.25	-3,097945076769D-04
				7,50	-3,866247107865D-04
	0	2,797665955295D+02	1.398251404862D+02	7.75	-2,413029321877D-04
	1	1.472580108573D+02	7.365761552138D+01	8.25	1,880258062943D-04
	2	2,909159365696D+01	1.454053498361D+01	8.50	2,347662171131D-04
	3	2,555412069629D+00	1.278134726645D+00	8.75	1,465777069835D-04
	4	8,417548708455D-02	4,207469186081D-02	9.25	-1,142766275483D-04
				9.50	-1,427145406199D-04

I= 1 H= -8.5

0		0.	9.589972806189D+00	-1.00	-1,484716673543D+01
1		5,137183530697D+00	3.980328594071D-01	-.75	-8,822346837301D+00
2		-6,514936607378D+00	-3,073369483721D+00	-.50	-4,571164261488D+00
3		2,786366911370D+00	1.386248689795D+00	-.25	-1,736612808649D+00
4		-4,086796859882D-01	-2.042419575294D-01	.25	9,190529229310D-01
				.50	1,262611125667D+00
0	-9,5	6,585129914900D-05	3,292564957450D-05	.75	1,234430484636D+00
1	-8,5	-3,968121973665D-04	-1,984060986833D-04	1.25	6,867513273914D-01
2	-7,5	1,043556258642D-03	5.217781293211D-04	1.50	3,847461151389D-01
3	-6,5	-1,600407237042D-03	-8,002036185211D-04	1.75	1,484016664578D-01
4	-5,5	1,646034860377D-03	8,230174301883D-04	2.25	-6,439176640199D-02
5	-4,5	-1,280248424637D-03	-6,401242123184D-04	2,50	-6,651219803345D-02
6	-3,5	8,524808796510D-04	4.262404398255D-04	2.75	-3,587762056416D-02
7	-2,5	-5,340327636881D-04	-2.670163818440D-04	3,25	2,286551061854D-02
8	-1,5	3,278778894157D-04	1.639389447079D-04	3,50	2,666334148126D-02
9	-,5	-1,999610365396D-04	-9,998051826979D-05	3,75	1,578659826587D-02
10	,5	1.216460835620D-04	6,082304178100D-05	4.25	-1,144233849724D-02
11	1,5	-7,388714667751D-05	-3,694357333875D-05	4,50	-1,394337102350D-02
12	2,5	4,476476750463D-05	2.238238375232D-05	4,75	-8,544036248159D-03
13	3,5	-2,695042713946D-05	-1,347521356973D-05	5,25	6,493902912014D-03
14	4,5	1,595505188644D-05	7,977525943222D-06	5,50	8,041405042751D-03
15	5,5	-9,035799431166D-06	-4,517899715583D-06	5,75	4,988983714944D-03
16	6,5	4,598755123976D-06	2,299377561988D-06	6,25	-3,856335130464D-03
17	7,5	-1,880467639676D-06	-9,402338198382D-07	6,50	-4,802150567979D-03

18	8.5	5.188342729628D-07	2.594171364814D-07	6.75	-2.992123655559D-03
19	9.5	-6.917942357289D-08	-3.458971178645D-08	7.25	2.326220101328D-03
				7.50	2.902287201566D-03
	0	-1.478930039492D+03	-7.390288497222D+02	7.75	1.810995285427D-03
	1	-7.658047538525D+02	-3.831169479166D+02	8.25	-1.410729867993D-03
	2	-1.484035295959D+02	-7.416230182676D+01	8.50	-1.761245707592D-03
	3	-1.274346115618D+01	-6.374945696324D+00	8.75	-1.099563489883D-03
	4	-4.086796859882D-01	-2.042419575294D-01	9.25	8.571703622696D-04
				9.50	1.070444076683D-03

I= 2 H= -7.5

0		0.	-3.475798871114D+01	-1.00	2.176759539070D+01
1		-5.754264749830D+00	4.989861988034D+00	-.75	1.239488179026D+01
2		1.004279347905D+01	4.354147223276D+00	-.50	6.081579908144D+00
3		-5.129449453651D+00	-2.539590295275D+00	-.25	2.149673926471D+00
4		8.410877081711D-01	4.201890770747D-01	.25	-8.877537445071D-01
				.50	-9.550475312385D-01
0	-9.5	-1.669837389107D-04	-8.349186945533D-05	.75	-5.645008489246D-01
1	-8.5	1.043556258642D-03	5.217781293211D-04	1.25	5.327705752668D-01
2	-7.5	-2.876046673383D-03	-1.438023336692D-03	1.50	9.045854419696D-01
3	-6.5	4.678387325482D-03	2.339193662741D-03	1.75	1.058304728316D+00
4	-5.5	-5.153571675358D-03	-2.576785837679D-03	2.25	7.840603155235D-01
5	-4.5	4.285594263000D-03	2.142797131500D-03	2.50	4.922580353883D-01
6	-3.5	-2.991119411523D-03	-1.495559705761D-03	2.75	2.092982972190D-01
7	-2.5	1.914445915970D-03	9.572229579850D-04	3.25	-1.057539136611D-01
8	-1.5	-1.184052916809D-03	-5.920264584047D-04	3.50	-1.157622059867D-01
9	-.5	7.238687358028D-04	3.619343679014D-04	3.75	-6.550819738188D-02
10	.5	-4.407192268756D-04	-2.203596134378D-04	4.25	4.481637107598D-02
11	1.5	2.677617768937D-04	1.338808884469D-04	4.50	5.360907926866D-02
12	2.5	-1.622386934689D-04	-8.111934673446D-05	4.75	3.239168845882D-02
13	3.5	9.767794486497D-05	4.883897243249D-05	5.25	-2.416039938045D-02
14	4.5	-5.782735380328D-05	-2.891367690164D-05	5.50	-2.973248030083D-02
15	5.5	3.274938070612D-05	1.637469035306D-05	5.75	-1.835898964214D-02
16	6.5	-1.666776113667D-05	-8.333880568334D-06	6.25	1.410060381313D-02
17	7.5	6.815583001917D-06	3.407791500958D-06	6.50	1.752200220100D-02
18	8.5	-1.880467639676D-06	-9.402338198382D-07	6.75	1.090008505432D-02
19	9.5	2.507345439014D-07	1.253672719507D-07	7.25	-8.456029506655D-03
				7.50	-1.054262996544D-02
	0	3.304546822072D+03	1.650692552256D+03	7.75	-6.574946834125D-03
	1	1.680770666949D+03	8.411630266125D+02	8.25	5.118073424436D-03
	2	3.193024780247D+02	1.595082090439D+02	8.50	6.388225072904D-03
	3	2.683188345685D+01	1.342759463357D+01	8.75	3.987518877446D-03
	4	8.410877081711D-01	4.201890770747D-01	9.25	-3.107747271292D-03
				9.50	-3.880685929036D-03

I= 3 H= -6.5

0		0.	8.500201220993D+01	-1.00	-2.127855546227D+01
1		5.130811523515D+00	-1.667363473143D+01	-.75	-1.191267320434D+01
2		-9.648190256769D+00	-3.192310498136D+00	-.50	-5.727946918530D+00
3		5.508343799303D+00	2.692704670115D+00	-.25	-1.975654557395D+00
4		-9.912098826812D-01	-4.947373203660D-01	.25	7.618869490245D-01
				.50	7.799510329680D-01
0	-9.5	2.448166323436D-04	1.224083161718D-04	.75	4.312275435725D-01
1	-8.5	-1.600407237042D-03	-8.002036185211D-04	1.25	-3.214208714885D-01
2	-7.5	4.678387325482D-03	2.339193662741D-03	1.50	-4.301420529016D-01

3	-6.5	-8.205107759758D-03	-4.102553879879D-03	1.75	-3.062808880615D-01
4	-5.5	9.887711289056D-03	4.943855644528D-03	2.25	3.901047524163D-01
5	-4.5	-9.023441742953D-03	-4.511720871476D-03	2.50	7.469609707487D-01
6	-3.5	6.793892137127D-03	3.396946068563D-03	2.75	9.690234636835D-01
7	-2.5	-4.548029919118D-03	-2.274014959559D-03	3.25	8.447018627318D-01
8	-1.5	2.867368122127D-03	1.433684061063D-03	3.50	5.644640864127D-01
9	-.5	-1.764488503363D-03	-8.822442516813D-04	3.75	2.527400672856D-01
10	.5	1.076631509482D-03	5.383157547411D-04	4.25	-1.379363781498D-01
11	1.5	-6.545868634613D-04	-3.272934317306D-04	4.50	-1.552677526236D-01
12	2.5	3.967136640092D-04	1.983568320046D-04	4.75	-8.985318458683D-02
13	3.5	-2.388656801676D-04	-1.194328400838D-04	5.25	6.346256231030D-02
14	4.5	1.414171340526D-04	7.070856702628D-05	5.50	7.675782840892D-02
15	5.5	-8.008950762745D-05	-4.004475381373D-05	5.75	4.678162870095D-02
16	6.5	4.076158677960D-05	2.038079338980D-05	6.25	-3.531283124404D-02
17	7.5	-1.666776113667D-05	-8.333880568334D-06	6.50	-4.363234985354D-02
18	8.5	4.598755123976D-06	2.299377561988D-06	6.75	-2.702552941570D-02
19	9.5	-6.131809569512D-07	-3.065904784756D-07	7.25	2.084453353029D-02
				7.50	2.593860957141D-02
	0	-4.172756641329D+03	-2.082512269119D+03	7.75	1.615324343127D-02
	1	-2.086155012349D+03	-1.044979384958D+03	8.25	-1.254963430730D-02
	2	-3.894005434485D+02	-1.943504863780D+02	8.50	-1.565410284394D-02
	3	-3.215763174258D+01	-1.610731350379D+01	8.75	-9.766533170157D-03
	4	-9.912098826812D-01	-4.947373203660D-01	9.25	7.606808686611D-03
				9.50	9.496711583920D-03

I= 4 H= -5.5

0		0.	-1.670151752187D+02	-1.00	1.580738190660D+01
1		-3.689191258925D+00	3.595700052897D+01	-.75	8.793872680683D+00
2		7.103888588915D+00	3.457519696666D-01	-.50	4.197360407628D+00
3		-4.214380036226D+00	-1.986416873786D+00	-.25	1.435265235005D+00
4		7.999220225321D-01	3.982562766532D-01	.25	-5.410297715025D-01
				.50	-5.454263277685D-01
0	-9.5	-2.393162956504D-04	-1.196581478252D-04	.75	-2.958153322593D-01
1	-8.5	1.646034860377D-03	8.230174301883D-04	1.25	2.082774245214D-01
2	-7.5	-5.153571675358D-03	-2.576785837679D-03	1.50	2.668381460298D-01
3	-6.5	9.887711289056D-03	4.943855644528D-03	1.75	1.788087590611D-01
4	-5.5	-1.329737289280D-02	-6.648686446402D-03	2.25	-1.819601429230D-01
5	-4.5	1.367059690124D-02	6.835298450619D-03	2.50	-2.760587019958D-01
6	-3.5	-1.147531564049D-02	-5.737657820246D-03	2.75	-2.190053535998D-01
7	-2.5	8.315705667447D-03	4.157852833724D-03	3.25	3.308254913237D-01
8	-1.5	-5.479347315263D-03	-2.739673657632D-03	3.50	6.763769596766D-01
9	-.5	3.434440521896D-03	1.717220260948D-03	3.75	9.265575424612D-01
10	.5	-2.108794776571D-03	-1.054397388285D-03	4.25	8.761614630844D-01
11	1.5	1.284821135532D-03	6.424105677659D-04	4.50	6.030823372860D-01
12	2.5	-7.792086562775D-04	-3.896043281387D-04	4.75	2.765383089531D-01
13	3.5	4.692781371825D-04	2.346390685913D-04	5.25	-1.561639064141D-01
14	4.5	-2.778508979299D-04	-1.389254489650D-04	5.50	-1.778968300641D-01
15	5.5	1.573607634388D-04	7.868038171941D-05	5.75	-1.039201511693D-01
16	6.5	-8.008950762745D-05	-4.004475381373D-05	6.25	7.436484208924D-02
17	7.5	3.274938070612D-05	1.637469035306D-05	6.50	9.034606153516D-02
18	8.5	-9.035799431166D-06	-4.517899715583D-06	6.75	5.525397687268D-02
19	9.5	1.204800532447D-06	6.024002662233D-07	7.25	-4.190590699625D-02
				7.50	-5.186091889571D-02
	0	3.508189413281D+03	1.746498542264D+03	7.75	-3.216151044861D-02
	1	1.733573873396D+03	8.705238201572D+02	8.25	2.484724885726D-02
	2	3.201518327576D+02	1.593886448745D+02	8.50	3.093697358209D-02

	3	2.618265681999D+01	1.314732163904D+01	8.75	1.927450928245D-02
	4	7.999220225321D-01	3.982562766532D-01	9.25	-1.498422975977D-02
				9.50	-1.869559211183D-02

I= 5 H= -4.5

0		0.	2.949095599121D+02	-1.00	-1.024590678038D+01
1		2.368741183841D+00	-6.556436808376D+01	-.75	-5.689568288588D+00
2		-4.592371932615D+00	3.365196595637D+00	-.50	-2.709895886261D+00
3		2.754123315066D+00	1.163804648818D+00	-.25	-9.243146495968D-01
4		-5.306703488559D-01	-2.623250134191D-01	.25	3.461222965977D-01
				.50	3.473764735782D-01
0	-9.5	1.777825632116D-04	8.889128160580D-05	.75	1.873483830089D-01
1	-8.5	-1.280248424637D-03	-6.401242123184D-04	1.25	-1.297648931244D-01
2	-7.5	4.285594263000D-03	2.142797131500D-03	1.50	-1.642454668724D-01
3	-6.5	-9.023441742953D-03	-4.511720871476D-03	1.75	-1.082986309924D-01
4	-5.5	1.367059690124D-02	6.835298450619D-03	2.25	1.046748488351D-01
5	-4.5	-1.610751120683D-02	-8.053755603413D-03	2.50	1.523726542731D-01
6	-3.5	1.549189580884D-02	7.745947904418D-03	2.75	1.139734347986D-01
7	-2.5	-1.260559645204D-02	-6.302798226021D-03	3.25	-1.379226242054D-01
8	-1.5	9.007162251615D-03	4.503581125808D-03	3.50	-2.236230068598D-01
9	-.5	-5.899559957577D-03	-2.949779978788D-03	3.75	-1.874580635785D-01
10	.5	3.689136952973D-03	1.844568476487D-03	4.25	3.074545028548D-01
11	1.5	-2.261676908288D-03	-1.130838454144D-03	4.50	6.476877765760D-01
12	2.5	1.374482886378D-03	6.872414431891D-04	4.75	9.088779343685D-01
13	3.5	-8.283498270654D-04	-4.141749135327D-04	5.25	8.897028116912D-01
14	4.5	4.905623491795D-04	2.452811745898D-04	5.50	6.198938027669D-01
15	5.5	-2.778508979299D-04	-1.389254489650D-04	5.75	2.869890011870D-01
16	6.5	1.414171340526D-04	7.070856702628D-05	6.25	-1.642637965679D-01
17	7.5	-5.782735380328D-05	-2.891367690164D-05	6.50	-1.879925473544D-01
18	8.5	1.595505188644D-05	7.977525943222D-06	6.75	-1.102151240617D-01
19	9.5	-2.127391249233D-06	-1.063695624617D-06	7.25	7.926403591836D-02
				7.50	9.646103170902D-02
	0	-2.352985206745D+03	-1.163079562958D+03	7.75	5.907110155004D-02
	1	-1.159141399383D+03	-5.861691576197D+02	8.25	-4.488163506783D-02
	2	-2.134578513587D+02	-1.055153656795D+02	8.50	-5.557761302577D-02
	3	-1.741134994146D+01	-8.804545861106D+00	8.75	-3.448304719373D-02
	4	-5.306703488559D-01	-2.623250134191D-01	9.25	2.665946707390D-02
				9.50	3.320221050761D-02

I= 6 H= -3.5

0		0.	-4.981499872030D+02	-1.00	6.359983022296D+00
1		-1.466008562557D+00	1.120164267810D+02	-.75	3.529697113873D+00
2		2.848268284700D+00	-8.138858484219D+00	-.50	1.680048321647D+00
3		-1.713925332062D+00	-4.967370001538D-01	-.25	5.725950106644D-01
4		3.317808429775D-01	1.608057712249D-01	.25	-2.139694376807D-01
				.50	-2.144417989899D-01
0	-9.5	-1.152330580248D-04	-5.761652901241D-05	.75	-1.154488811276D-01
1	-8.5	8.524808796510D-04	4.262404398255D-04	1.25	7.955135120452D-02
2	-7.5	-2.991119411523D-03	-1.495559705761D-03	1.50	1.003050377432D-01
3	-6.5	6.793892137127D-03	3.396946068563D-03	1.75	6.580488344213D-02
4	-5.5	-1.147531564049D-02	-5.737657820246D-03	2.25	-6.261222443836D-02
5	-4.5	1.549189580884D-02	7.745947904418D-03	2.50	-9.005160624861D-02
6	-3.5	-1.728776510273D-02	-8.643882551363D-03	2.75	-6.627380457443D-02
7	-2.5	1.622407801709D-02	8.112039008545D-03	3.25	7.613053901878D-02
8	-1.5	-1.305306223528D-02	-6.526531117639D-03	3.50	1.183846628934D-01

9	-.5	9.278607995501D-03	4.639303997750D-03	3.75	9.352487233824D-02
10	.5	-6.063099998057D-03	-3.031549999028D-03	4.25	-1.227736097818D-01
11	1.5	3.785036764962D-03	1.892518382481D-03	4.50	-2.050265502516D-01
12	2.5	-2.314551081636D-03	-1.157275540818D-03	4.75	-1.759979031460D-01
13	3.5	1.397771286077D-03	6.988856430384D-04	5.25	2.986764608493D-01
14	4.5	-8.283498270654D-04	-4.141749135327D-04	5.50	6.367895954324D-01
15	5.5	4.692781371825D-04	2.346390685913D-04	5.75	9.021029265586D-01
16	6.5	-2.388656801676D-04	-1.194328400838D-04	6.25	8.949544106466D-01
17	7.5	9.767794486497D-05	4.883897243249D-05	6.50	6.264399222217D-01
18	8.5	-2.695042713946D-05	-1.347521356973D-05	6.75	2.910711102751D-01
19	9.5	3.593490816832D-06	1.796745408416D-06	7.25	-1.674417251962D-01
				7.50	-1.919599524150D-01
	0	1.476028102128D+03	7.153572604963D+02	7.75	-1.126923597462D-01
	1	7.264482061824D+02	3.743699652049D+02	8.25	8.119676808006D-02
	2	1.336607227933D+02	6.478046212967D+01	8.50	9.887640965353D-02
	3	1.089374670108D+01	5.613882306392D+00	8.75	6.058091659237D-02
	4	3.317808429775D-01	1.608057712249D-01	9.25	-4.606275108698D-02
				9.50	-5.705626333111D-02

I= 7 H= -2.5

0		0.	8.274511469720D+02	-1.00	-3.893199797619D+00
1		8.965389408571D-01	-1.868340018561D+02	-.75	-2.160269146395D+00
2		-1.743066907896D+00	1.501306339714D+01	-.50	-1.028012393830D+00
3		1.050025193979D+00	-7.333988076923D-02	-.25	-3.502782510624D-01
4		-2.035687548874D-01	-9.333852305388D-02	.25	1.308045069508D-01
				.50	1.310329852245D-01
0	-9.5	7.152794726579D-05	3.576397363290D-05	.75	7.050339296363D-02
1	-8.5	-5.340327636881D-04	-2.670163818440D-04	1.25	-4.849901831382D-02
2	-7.5	1.914445915970D-03	9.572229579850D-04	1.50	-6.107519186935D-02
3	-6.5	-4.548029919118D-03	-2.274014959559D-03	1.75	-4.000206724754D-02
4	-5.5	8.315705667447D-03	4.157852833724D-03	2.25	3.786681204875D-02
5	-4.5	-1.260559645204D-02	-6.302798226021D-03	2.50	5.424953647527D-02
6	-3.5	1.622407801709D-02	8.112039008545D-03	2.75	3.971802495712D-02
7	-2.5	-1.774152207687D-02	-8.870761038434D-03	3.25	-4.489310633647D-02
8	-1.5	1.650062455639D-02	8.250312278194D-03	3.50	-6.895304444172D-02
9	-.5	-1.321956402926D-02	-6.609782014632D-03	3.75	-5.357986970615D-02
10	.5	9.376596911784D-03	4.688298455892D-03	4.25	6.672603843071D-02
11	1.5	-6.117236315128D-03	-3.058618157564D-03	4.50	1.068396506939D-01
12	2.5	3.809008528231D-03	1.904504264115D-03	4.75	8.640993092938D-02
13	3.5	-2.314551081636D-03	-1.157275540818D-03	5.25	-1.173232354584D-01
14	4.5	1.374482886378D-03	6.872414431891D-04	5.50	-1.982592218094D-01
15	5.5	-7.792086562775D-04	-3.896043281387D-04	5.75	-1.717904691092D-01
16	6.5	3.967136640092D-04	1.983568320046D-04	6.25	2.954141252847D-01
17	7.5	-1.622386934689D-04	-8.111934673446D-05	6.50	6.327222202982D-01
18	8.5	4.476476750463D-05	2.238238375232D-05	6.75	8.995658453138D-01
19	9.5	-5.968874579806D-06	-2.984437289903D-06	7.25	8.969311222763D-01
				7.50	6.289091261615D-01
	0	-9.066095494168D+02	-4.156707832750D+02	7.75	2.926140161940D-01
	1	-4.460664559257D+02	-2.415470348420D+02	8.25	-1.686481018534D-01
	2	-8.204982965102D+01	-3.761993343846D+01	8.50	-1.934699280550D-01
	3	-6.685587491742D+00	-3.620203756817D+00	8.75	-1.136380937588D-01
	4	-2.035687548874D-01	-9.333852305388D-02	9.25	8.194087652296D-02
				9.50	9.981179144074D-02

I= 8 H= -1.5

CARDINAL SPLINES N=5

0		0.	-1.365871180351D+03	-1.00	2.372233329053D+00
1		-5.461125598600D-01	3.088733892991D+02	-.75	1.316231052581D+00
2		1.062003744614D+00	-2.568968050081D+01	-.50	6.263134184353D-01
3		-6.399822130489D-01	6.677089575780D-01	-.25	2.133879976898D-01
4		1.241348115307D-01	4.812583312060D-02	.25	-7.966772656499D-02
				.50	-7.979478020107D-02
0	-9.5	-4.378323579098D-05	-2.189161789549D-05	.75	-4.292606691926D-02
1	-8.5	3.278778894157D-04	1.639389447079D-04	1.25	2.951218671052D-02
2	-7.5	-1.184052916809D-03	-5.920264584047D-04	1.50	3.714956564391D-02
3	-6.5	2.867368122127D-03	1.433684061063D-03	1.75	2.431834242160D-02
4	-5.5	-5.479347315263D-03	-2.739673657632D-03	2.25	-2.298126353869D-02
5	-4.5	9.007162251615D-03	4.503581125808D-03	2.50	-3.288137420626D-02
6	-3.5	-1.305306223528D-02	-6.526531117639D-03	2.75	-2.403223281780D-02
7	-2.5	1.650062455639D-02	8.250312278194D-03	3.25	2.701908151499D-02
8	-1.5	-1.790879960050D-02	-8.954399800248D-03	3.50	4.133250761478D-02
9	-.5	1.659924067578D-02	8.299620337890D-03	3.75	3.194620651794D-02
10	.5	-1.327410760654D-02	-6.637053803270D-03	4.25	-3.913459100019D-02
11	1.5	9.400794599474D-03	4.700397299737D-03	4.50	-6.188329628316D-02
12	2.5	-6.117236315128D-03	-3.058618157564D-03	4.75	-4.922249033224D-02
13	3.5	3.785036764962D-03	1.892518382481D-03	5.25	6.338708457930D-02
14	4.5	-2.261676908288D-03	-1.130838454144D-03	5.50	1.026930249073D-01
15	5.5	1.284821135532D-03	6.424105677659D-04	5.75	8.383114861600D-02
16	6.5	-6.545868634613D-04	-3.272934317306D-04	6.25	-1.153221014252D-01
17	7.5	2.677617768937D-04	1.338808884469D-04	6.50	-1.957628256307D-01
18	8.5	-7.388714667751D-05	-3.694357333875D-05	6.75	-1.702321516075D-01
19	9.5	9.852371544864D-06	4.926185772432D-06	7.25	2.941973632501D-01
				7.50	6.311999421385D-01
	0	5.530388170629D+02	2.143972389278D+02	7.75	8.986127455771D-01
	1	2.720771105492D+02	1.665992047350D+02	8.25	8.976806443937D-01
	2	5.004151111657D+01	1.940016342496D+01	8.50	6.298511335870D-01
	3	4.077140625116D+00	2.496490616161D+00	8.75	2.932071001125D-01
	4	1.241348115307D-01	4.812583312060D-02	9.25	-1.691217576365D-01
				9.50	-1.940715874161D-01

I= 9 H= -.5

0		0.	2.249282012400D+03	-1.00	-1.443123027885D+00
1		3.321869303263D-01	-5.089291746968D+02	-.75	-8.006990913465D-01
2		-6.460396701221D-01	4.285665863555D+01	-.50	-3.809944872837D-01
3		3.893612460033D-01	-1.431845660019D+00	-.25	-1.298030407355D-01
4		-7.553518143363D-02	-1.480883556589D-02	.25	4.845796346703D-02
				.50	4.853280664348D-02
0	-9.5	2.667522614070D-05	1.333761307035D-05	.75	2.610685835918D-02
1	-8.5	-1.999610365396D-04	-9.998051826979D-05	1.25	-1.794545694516D-02
2	-7.5	7.238687358028D-04	3.619343679014D-04	1.50	-2.258641827587D-02
3	-6.5	-1.764488503363D-03	-8.822442516813D-04	1.75	-1.478253433462D-02
4	-5.5	3.434440521896D-03	1.717220260948D-03	2.25	1.396189951193D-02
5	-4.5	-5.899559557577D-03	-2.949977978788D-03	2.50	1.996798363930D-02
6	-3.5	9.278607995501D-03	4.639303997750D-03	2.75	1.458579320399D-02
7	-2.5	-1.321956402926D-02	-6.609782014632D-03	3.25	-1.636954741496D-02
8	-1.5	1.659924067578D-02	8.299620337890D-03	3.50	-2.500777898527D-02
9	-.5	-1.796343701212D-02	-8.981718506061D-03	3.75	-1.929441881227D-02
10	.5	1.662350602472D-02	8.311753012359D-03	4.25	2.350754825579D-02
11	1.5	-1.327410760654D-02	-6.637053803270D-03	4.50	3.702045400435D-02
12	2.5	9.376596911784D-03	4.688298455892D-03	4.75	2.928777290454D-02
13	3.5	-6.063099998057D-03	-3.031549999028D-03	5.25	-3.709585032980D-02
14	4.5	3.689136952973D-03	1.844568476487D-03	5.50	-5.934992618674D-02

15	5.5	-2.108794776571D-03	-1.054397388285D-03	5.75	-4.764581857248D-02
16	6.5	1.076631509482D-03	5.383157547411D-04	6.25	6.216093158149D-02
17	7.5	-4.407192268756D-04	-2.203596134378D-04	6.50	1.011610298914D-01
18	8.5	1.216460835620D-04	6.082304178100D-05	6.75	8.287293496669D-02
19	9.5	-1.622249073154D-05	-8.111245365771D-06	7.25	-1.145695842057D-01
				7.50	-1.948174839345D-01
	0	-3.365594798241D+02	-6.597926935570D+01	7.75	-1.696371785206D-01
	1	-1.655709141732D+02	-1.331117746446D+02	8.25	2.937224360915D-01
	2	-3.045154490534D+01	-5.969927133918D+00	8.50	6.305967785932D-01
	3	-2.480975648475D+00	-1.994581411523D+00	8.75	8.982280094737D-01
	4	-7.553518143363D-02	-1.480883556589D-02	9.25	8.979991973453D-01
				9.50	6.302656857198D-01

I=10 H= .5

0		0.	-3.700050181257D+03	-1.00	8.771947794718D-01
1		-2.019110615913D-01	8.373602089321D+02	-.75	4.866975009490D-01
2		3.926879906794D-01	-7.083448182125D+01	-.50	2.315821499718D-01
3		-2.366782168992D-01	2.557317163027D+00	-.25	7.889822722896D-02
4		4.591751030185D-02	-1.480883556589D-02	.25	-2.945349790669D-02
				.50	-2.949849752890D-02
0	-9.5	-1.622249073154D-05	-8.111245365771D-06	.75	-1.586755500736D-02
1	-8.5	1.216460835620D-04	6.082304178100D-05	1.25	1.090644619091D-02
2	-7.5	-4.407192268756D-04	-2.203596134378D-04	1.50	1.372639105886D-02
3	-6.5	1.076631509482D-03	5.383157547411D-04	1.75	8.983214870739D-03
4	-5.5	-2.108794776571D-03	-1.054397388285D-03	2.25	-8.482938431332D-03
5	-4.5	3.689136952973D-03	1.844568476487D-03	2.50	-1.213037282405D-02
6	-3.5	-6.063099998057D-03	-3.031549999028D-03	2.75	-8.859057378195D-03
7	-2.5	9.376596911784D-03	4.688298455892D-03	3.25	9.936610650163D-03
8	-1.5	-1.327410760654D-02	-6.637053803270D-03	3.50	1.517340002759D-02
9	-.5	1.662350602472D-02	8.311753012359D-03	3.75	1.169992084633D-02
10	.5	-1.796343701212D-02	-8.981718506061D-03	4.25	-1.422887999471D-02
11	1.5	1.659924067578D-02	8.299620337890D-03	4.50	-2.237757215866D-02
12	2.5	-1.321956402926D-02	-6.609782014632D-03	4.75	-1.767165432473D-02
13	3.5	9.278607995501D-03	4.639303997750D-03	5.25	2.226034937478D-02
14	4.5	-5.899955957577D-03	-2.949977978788D-03	5.50	3.546825007156D-02
15	5.5	3.434440521896D-03	1.717220260948D-03	5.75	2.831981821402D-02
16	6.5	-1.764488503363D-03	-8.822442516813D-04	6.25	-3.633873025138D-02
17	7.5	7.238687358028D-04	3.619343679014D-04	6.50	-5.840006004791D-02
18	8.5	-1.999610365396D-04	-9.998051826979D-05	6.75	-4.704860108293D-02
19	9.5	2.667522614070D-05	1.333761307035D-05	7.25	6.168486023464D-02
				7.50	1.005566867276D-01
	0	2.046009411127D+02	-6.597926935570D+01	7.75	8.248758149909D-02
	1	1.006526351160D+02	1.331117746446D+02	8.25	-1.142506772690D-01
	2	1.851169063750D+01	-5.969927133918D+00	8.50	-1.944025451732D-01
	3	1.508187174571D+00	1.994581411523D+00	8.75	-1.693646940074D-01
	4	4.591751030185D-02	-1.480883556589D-02	9.25	2.934793717642D-01
				9.50	6.302656857198D-01

I=11 H= 1.5

0		0.	6.080370993676D+03	-1.00	-5.326567092593D-01
1		1.226045146500D-01	-1.376174688822D+03	-.75	-2.955354054169D-01
2		-2.384501355260D-01	1.166102846204D+02	-.50	-1.406223520113D-01
3		1.437189137939D-01	-4.325272274744D+00	-.25	-4.790878869719D-02
4		-2.788314528946D-02	4.812583312060D-02	.25	1.788468472146D-02
				.50	1.791191033007D-02

CARDINAL SPLINES N=5

0	-9.5	9.852371544864D-06	4.926185772432D-06	.75	9.634939833929D-03
1	-8.5	-7.388714667751D-05	-3.694357333875D-05	1.25	-6.622369516100D-03
2	-7.5	2.677617768937D-04	1.338808884469D-04	1.50	-8.334507982843D-03
3	-6.5	-6.545868634613D-04	-3.272934317306D-04	1.75	-5.454396723283D-03
4	-5.5	1.284821135532D-03	6.424105677659D-04	2.25	5.150321502130D-03
5	-4.5	-2.261676908288D-03	-1.130838454144D-03	2.50	7.364472378751D-03
6	-3.5	3.785036764962D-03	1.892518382481D-03	2.75	5.378084520852D-03
7	-2.5	-6.117236315128D-03	-3.058618157564D-03	3.25	-6.031056705384D-03
8	-1.5	9.400794599474D-03	4.700397299737D-03	3.50	-9.208176934026D-03
9	-.5	-1.327410760654D-02	-6.637053803270D-03	3.75	-7.098856609255D-03
10	.5	1.659924067578D-02	8.299620337890D-03	4.25	8.628080233607D-03
11	1.5	-1.790879960050D-02	-8.954399800248D-03	4.50	1.356311172504D-02
12	2.5	1.650062455639D-02	8.250312278194D-03	4.75	1.070443415117D-02
13	3.5	-1.305306223528D-02	-6.526531117639D-03	5.25	-1.345934789655D-02
14	4.5	9.007162251615D-03	4.503581125808D-03	5.50	-2.141590060449D-02
15	5.5	-5.479347315263D-03	-2.739673657632D-03	5.75	-1.706881739507D-02
16	6.5	2.867368122127D-03	1.433684061063D-03	6.25	2.178172238028D-02
17	7.5	-1.184052916809D-03	-5.920264584047D-04	6.50	3.486146573795D-02
18	8.5	3.278778894157D-04	1.639389447079D-04	6.75	2.793329876716D-02
19	9.5	-4.378323579098D-05	-2.189161789549D-05	7.25	-3.601929477785D-02
				7.50	-5.798462402421D-02
	0	-1.242443392072D+02	2.143972389278D+02	7.75	-4.677587765585D-02
	1	-6.112129892085D+01	-1.665992047350D+02	8.25	6.144171613793D-02
	2	-1.124118426664D+01	1.940016342496D+01	8.50	1.002255476193D-01
	3	-9.158406072055D-01	-2.496490616161D+00	8.75	8.225877404503D-02
	4	-2.788314528946D-02	4.812583312060D-02	9.25	-1.140219310353D-01
				9.50	-1.940715874161D-01

I=12 H= 2.5

0		0.	-9.969686907026D+03	-1.00	3.226823574478D-01
1		-7.427328255163D-02	2.256600402144D+03	-.75	1.790346193726D-01
2		1.444524543815D-01	-1.913385507414D+02	-.50	8.518860063680D-02
3		-8.706491173496D-02	7.167067632864D+00	-.25	2.902297152003D-02
4		1.689170877964D-02	-9.333852305388D-02	.25	-1.083444827027D-02
				.50	-1.085092150653D-02
0	-9.5	-5.968874579806D-06	-2.984437289903D-06	.75	-5.836771902314D-03
1	-8.5	4.476476750463D-05	2.238238375232D-05	1.25	4.011753026682D-03
2	-7.5	-1.622386934689D-04	-8.111934673446D-05	1.50	5.048920963071D-03
3	-6.5	3.967136640092D-04	1.983568320046D-04	1.75	3.304170525287D-03
4	-5.5	-7.792086562775D-04	-3.896043281387D-04	2.25	-3.119903335255D-03
5	-4.5	1.374482886378D-03	6.872414431891D-04	2.50	-4.461096515123D-03
6	-3.5	-2.314551081636D-03	-1.157275540818D-03	2.75	-3.257755721358D-03
7	-2.5	3.809008528231D-03	1.904504264115D-03	3.25	3.653054030403D-03
8	-1.5	-6.117236315128D-03	-3.058618157564D-03	3.50	5.577183820971D-03
9	-.5	9.376596911784D-03	4.688298455892D-03	3.75	4.299336026141D-03
10	.5	-1.321956402926D-02	-6.609782014632D-03	4.25	-5.224444434456D-03
11	1.5	1.650062455639D-02	8.250312278194D-03	4.50	-8.211448587765D-03
12	2.5	-1.774152207687D-02	-8.870761038434D-03	4.75	-6.479444242242D-03
13	3.5	1.622407801709D-02	8.112039008545D-03	5.25	8.142049536758D-03
14	4.5	-1.260559645204D-02	-6.302798226021D-03	5.50	1.294934512431D-02
15	5.5	8.315705667447D-03	4.157852833724D-03	5.75	1.031463479642D-02
16	6.5	-4.548029919118D-03	-2.274014959559D-03	6.25	-1.313850053187D-02
17	7.5	1.914445915970D-03	9.572229579850D-04	6.50	-2.099919017341D-02
18	8.5	-5.340327636881D-04	-2.670163818440D-04	6.75	-1.679552237668D-02
19	9.5	7.152794726579D-05	3.576397363290D-05	7.25	2.153844017234D-02
				7.50	3.453030793144D-02

	0	7.526798286686D+01	-4.156707832750D+02	7.75	2.770457331016D-02
	1	3.702761375822D+01	2.415470348420D+02	8.25	-3.579077391932D-02
	2	6.809962774108D+00	-3.761993343846D+01	8.50	-5.765407229756D-02
	3	5.548200218912D-01	3.620203756817D+00	8.75	-4.653329967921D-02
	4	1.689170877964D-02	-9.333852305388D-02	9.25	6.116992762135D-02
				9.50	9.981179144074D-02

I=13 H= 3.5

0		0.	1.624128403658D+04	-1.00	-1.942635802091D-01
1		4.471448500182D-02	-3.676640772540D+03	-.75	-1.077836976409D-01
2		-8.696428632217D-02	3.118524329801D+02	-.50	-5.128583390910D-02
3		5.241550835731D-02	-1.172450161294D+01	-.25	-1.747260529386D-02
4		-1.016930052779D-02	1.608057712249D-01	.25	6.522621856925D-03
				.50	6.532535200584D-03
0	-9.5	3.593490816832D-06	1.796745408416D-06	.75	3.513884854400D-03
1	-8.5	-2.695042713946D-05	-1.347521356973D-05	1.25	-2.415171742246D-03
2	-7.5	9.767794486497D-05	4.883897243249D-05	1.50	-3.039566716793D-03
3	-6.5	-2.388656801676D-04	-1.194328400838D-04	1.75	-1.989182391889D-03
4	-5.5	4.692781371825D-04	2.346390685913D-04	2.25	1.878236718238D-03
5	-4.5	-8.283498270654D-04	-4.141749135327D-04	2.50	2.685644621850D-03
6	-3.5	1.397771286077D-03	6.988856430384D-04	2.75	1.961202433123D-03
7	-2.5	-2.314551081636D-03	-1.157275540818D-03	3.25	-2.199128648922D-03
8	-1.5	3.785036764962D-03	1.892518382481D-03	3.50	-3.357394694370D-03
9	-.5	-6.063099998057D-03	-3.031549999028D-03	3.75	-2.588090312865D-03
10	.5	9.278607995501D-03	4.639303997750D-03	4.25	3.144772560099D-03
11	1.5	-1.305306223528D-02	-6.526531117639D-03	4.50	4.942505334298D-03
12	2.5	1.622407801709D-02	8.112039008545D-03	4.75	3.899746898913D-03
13	3.5	-1.728776510273D-02	-8.643882551363D-03	5.25	-4.899418724277D-03
14	4.5	1.549189580884D-02	7.745947904418D-03	5.50	-7.790991038409D-03
15	5.5	-1.147531564049D-02	-5.737657820246D-03	5.75	-6.204573461997D-03
16	6.5	6.793892137127D-03	3.396946068563D-03	6.25	7.898391475911D-03
17	7.5	-2.991119411523D-03	-1.495559705761D-03	6.50	1.261817804231D-02
18	8.5	8.524808796510D-04	4.262404398255D-04	6.75	1.008618284381D-02
19	9.5	-1.152330580248D-04	-5.761652901241D-05	7.25	-1.291067003556D-02
				7.50	-2.066986407818D-02
	0	-4.531358113530D+01	7.153572604963D+02	7.75	-1.655400317483D-02
	1	-2.229172422742D+01	-3.743699652049D+02	8.25	2.126805545574D-02
	2	-4.099798533934D+00	6.478046212967D+01	8.50	3.411882792164D-02
	3	-3.340179116985D-01	-5.613882306392D+00	8.75	2.738851412316D-02
	4	-1.016930052779D-02	1.608057712249D-01	9.25	-3.540932512695D-02
				9.50	-5.705626333111D-02

I=14 H= 4.5

0		0.	-2.593989945019D+04	-1.00	1.150057584031D-01
1		-2.647135958553D-02	5.874435530944D+03	-.75	6.380889753864D-02
2		5.148362087939D-02	-4.984939174874D+02	-.50	3.036166212877D-02
3		-3.103045592037D-02	1.877289637103D+01	-.25	1.034393395798D-02
4		6.020322017761D-03	-2.623250134191D-01	.25	-3.861447590411D-03
				.50	-3.867315591916D-03
0	-9.5	-2.127391249233D-06	-1.063695624617D-06	.75	-2.080248756971D-03
1	-8.5	1.595505188644D-05	7.977525943222D-06	1.25	1.429800484735D-03
2	-7.5	-5.782735380328D-05	-2.891367690164D-05	1.50	1.799446179884D-03
3	-6.5	1.414171340526D-04	7.070856702628D-05	1.75	1.177609960592D-03
4	-5.5	-2.778508979299D-04	-1.389254489650D-04	2.25	-1.111926818964D-03
5	-4.5	4.905623491795D-04	2.452811745898D-04	2.50	-1.589914136513D-03

CARDINAL SPLINES N=5

6	-3,5	-8,283498270654D-04	-4,141749135327D-04	2.75	-1,161038302714D-03
7	-2,5	1,374482886378D-03	6,872414431891D-04	3.25	1,301882076029D-03
8	-1,5	-2,261676908288D-03	-1,130838454144D-03	3.50	1,987563679847D-03
9	-,5	3,689136952973D-03	1,844568476487D-03	3.75	1,532127664651D-03
10	,5	-5,899955957577D-03	-2,949977978788D-03	4.25	-1,861637680389D-03
11	1,5	9,007162251615D-03	4,503581125808D-03	4.50	-2,925808077663D-03
12	2,5	-1,260559645204D-02	-6,302798226021D-03	4.75	-2,308477089406D-03
13	3,5	1,549189580884D-02	7,745947904418D-03	5.25	2,900043436812D-03
14	4,5	-1,610751120683D-02	-8,053755603413D-03	5.50	4,611378110095D-03
15	5,5	1,367059690124D-02	6,835298450619D-03	5.75	3,672155709809D-03
16	6,5	-9,023441742953D-03	-4,511720871476D-03	6.25	-4,673686607065D-03
17	7,5	4,285594263000D-03	2,142797131500D-03	6.50	-7,465371828196D-03
18	8,5	-1,280248424637D-03	-6,401242123184D-04	6.75	-5,966155323632D-03
19	9,5	1,777825632116D-04	8,889128160580D-05	7.25	7,632215984938D-03
				7.50	1,221350569606D-02
	0	2,682608082837D+01	-1,163079562958D+03	7.75	9,775606420063D-03
	1	1,319691585659D+01	5,861691576197D+02	8.25	-1,253629276583D-02
	2	2,427119999766D+00	-1,055153656795D+02	8.50	-2,008339722223D-02
	3	1,977417807545D-01	8,804545861106D+00	8.75	-1,609254848935D-02
	4	6,020322017761D-03	-2,623250134191D-01	9.25	2,069073997802D-02
				9.50	3,320221050761D-02

I=15 H= 5.5

0		0.	3,891730718830D+04	-1.00	-6,513081195396D-02
1		1,499143107827D-02	-8,824365307323D+03	-.75	-3,613667047642D-02
2		-2,915653915156D-02	7,497430853947D+02	-.50	-1,719461371594D-02
3		1,757337249844D-02	-2,828106015186D+01	-.25	-5,858043650992D-03
4		-3,409469225688D-03	3,982562766532D-01	.25	2,186839783317D-03
				.50	2,190162840073D-03
0	-9,5	1,204800532447D-06	6,024002662233D-07	.75	1,178099648613D-03
1	-8,5	-9,035799431166D-06	-4,517899715583D-06	1.25	-8,097334632285D-04
2	-7,5	3,274938070612D-05	1,637469035306D-05	1.50	-1,019073314610D-03
3	-6,5	-8,008950762745D-05	-4,004475381373D-05	1.75	-6,669110792860D-04
4	-5,5	1,573607634388D-04	7,868038171941D-05	2.25	6,297125359387D-04
5	-4,5	-2,778508979299D-04	-1,389254489650D-04	2.50	9,004084337208D-04
6	-3,5	4,692781371825D-04	2,346390685913D-04	2.75	6,575247378567D-04
7	-2,5	-7,792086562775D-04	-3,896043281387D-04	3.25	-7,372862942917D-04
8	-1,5	1,284821135532D-03	6,424105677659D-04	3.50	-1,125601800741D-03
9	-,5	-2,108794776571D-03	-1,054397388285D-03	3.75	-8,676761118686D-04
10	,5	3,434440521896D-03	1,717220260948D-03	4.25	1,054276794832D-03
11	1,5	-5,479347315263D-03	-2,739673657632D-03	4.50	1,656925238357D-03
12	2,5	8,315705667447D-03	4,157852833724D-03	4.75	1,307312669171D-03
13	3,5	-1,147531564049D-02	-5,737657820246D-03	5.25	-1,642285574173D-03
14	4,5	1,367059690124D-02	6,835298450619D-03	5.50	-2,611365180958D-03
15	5,5	-1,329737289280D-02	-6,648686446402D-03	5.75	-2,079449350096D-03
16	6,5	9,887711289056D-03	4,943855644528D-03	6.25	2,646411981718D-03
17	7,5	-5,153571675358D-03	-2,576785837679D-03	6.50	4,226951573402D-03
18	8,5	1,646034860377D-03	8,230174301883D-04	6.75	3,377857829551D-03
19	9,5	-2,393162956504D-04	-1,196581478252D-04	7.25	-4,320249583260D-03
				7.50	-6,912451635305D-03
	0	-1,519232875239D+01	1,746498542264D+03	7.75	-5,531565364972D-03
	1	-7,473766918347D+00	-8,705238201572D+02	8.25	7,089352933564D-03
	2	-1,374543008656D+00	1,593886448745D+02	8.50	1,135206221226D-02
	3	-1,119864580777D-01	-1,314732163904D+01	8.75	9,090743871480D-03
	4	-3,409469225688D-03	3,982562766532D-01	9.25	-1,166671298412D-02
				9.50	-1,869559211183D-02

I=16 H= 6.5

0		0.	-4.738962213261D+04
1		-7.629855023481D-03	1.079539539749D+04
2		1.483915532183D-02	-9.213091802143D+02
3		-8.943929066593D-03	3.490733167770D+01
4		1.735241949202D-03	-4.947373203660D-01
0	-9.5	-6.131809569512D-07	-3.065904784756D-07
1	-8.5	4.598755123976D-06	2.299377561988D-06
2	-7.5	-1.666776113667D-05	-8.333880568334D-06
3	-6.5	4.076158677960D-05	2.038079338980D-05
4	-5.5	-8.008950762745D-05	-4.004475381373D-05
5	-4.5	1.414171340526D-04	7.070856702628D-05
6	-3.5	-2.388656801676D-04	-1.194328400838D-04
7	-2.5	3.967136640092D-04	1.983568320046D-04
8	-1.5	-6.545868634613D-04	-3.272934317306D-04
9	-.5	1.076631509482D-03	5.383157547411D-04
10	.5	-1.764488503363D-03	-8.822442516813D-04
11	1.5	2.867368122127D-03	1.433684061063D-03
12	2.5	-4.548029919118D-03	-2.274014959559D-03
13	3.5	6.793892137127D-03	3.396946068563D-03
14	4.5	-9.023441742953D-03	-4.511720871476D-03
15	5.5	9.887711289056D-03	4.943855644528D-03
16	6.5	-8.205107759758D-03	-4.102553879879D-03
17	7.5	4.678387325482D-03	2.339193662741D-03
18	8.5	-1.600407237042D-03	-8.002036185211D-04
19	9.5	2.448166323436D-04	1.224083161718D-04
	0	7.732103090470D+00	-2.082512269119D+03
	1	3.803757566098D+00	1.044979384958D+03
	2	6.995706924165D-01	-1.943504863780D+02
	3	5.699526500306D-02	1.610731350379D+01
	4	1.735241949202D-03	-4.947373203660D-01

-1.00	3.314818136110D-02
-.75	1.839167760910D-02
-.50	8.751160097347D-03
-.25	2.981438144014D-03
.25	-1.112987153396D-03
.50	-1.114678390401D-03
.75	-5.995911165608D-04
1.25	4.121119562630D-04
1.50	5.186549530293D-04
1.75	3.394227886252D-04
2.25	-3.204905982363D-04
2.50	-4.582605002291D-04
2.75	-3.346453967197D-04
3.25	3.752395196346D-04
3.50	5.728711163943D-04
3.75	4.416003800985D-04
4.25	-5.365687444164D-04
4.50	-8.432819958972D-04
4.75	-6.653471767258D-04
5.25	8.358228917023D-04
5.50	1.329017629633D-03
5.75	1.058298671346D-03
6.25	-1.346813592793D-03
6.50	-2.151146111568D-03
6.75	-1.718993604172D-03
7.25	2.198426594244D-03
7.50	3.517328486786D-03
7.75	2.814488662725D-03
8.25	-3.606354145839D-03
8.50	-5.773901069721D-03
8.75	-4.622804772248D-03
9.25	5.929044384695D-03
9.50	9.496711583920D-03

I=17 H= 7.5

0		0.	3.897230741483D+04
1		3.119907294686D-03	-8.948378685313D+03
2		-6.067846549662D-03	7.697270413365D+02
3		3.657242542058D-03	-2.939477956241D+01
4		-7.095540216268D-04	4.201890770747D-01
0	-9.5	2.507345439014D-07	1.253672719507D-07
1	-8.5	-1.880467639676D-06	-9.402338198382D-07
2	-7.5	6.815583001917D-06	3.407791500958D-06
3	-6.5	-1.666776113667D-05	-8.333880568334D-06
4	-5.5	3.274938070612D-05	1.637469035306D-05
5	-4.5	-5.782735380328D-05	-2.891367690164D-05
6	-3.5	9.767794486497D-05	4.883897243249D-05
7	-2.5	-1.622386934689D-04	-8.111934673446D-05
8	-1.5	2.677617768937D-04	1.338808884469D-04
9	-.5	-4.407192268756D-04	-2.203596134378D-04
10	.5	7.238687358028D-04	3.619343679014D-04
11	1.5	-1.184052916809D-03	-5.920264584047D-04

-1.00	-1.355455040803D-02
-.75	-7.520500679786D-03
-.50	-3.578417728868D-03
-.25	-1.219133343142D-03
.25	4.551091345969D-04
.50	4.558006910493D-04
.75	2.451774834301D-04
1.25	-1.685157878890D-04
1.50	-2.120820443581D-04
1.75	-1.387926143623D-04
2.25	1.310510831953D-04
2.50	1.873862486300D-04
2.75	1.368390683686D-04
3.25	-1.534382739892D-04
3.50	-2.342512775929D-04
3.75	-1.805736430389D-04
4.25	2.194067181744D-04
4.50	3.448236934447D-04

12	2,5	1.914445915970D-03	9.572229579850D-04	4.75	2,720647154422D-04
13	3,5	-2,991119411523D-03	-1.495559705761D-03	5.25	-3,417724378756D-04
14	4,5	4,285594263000D-03	2,142797131500D-03	5,50	-5,434413371015D-04
15	5,5	-5,153571675358D-03	-2,576785837679D-03	5,75	-4,327420157412D-04
16	6,5	4.678387325482D-03	2,339193662741D-03	6.25	5,507124500870D-04
17	7,5	-2,876046673383D-03	-1,438023336692D-03	6,50	8,795990618863D-04
18	8,5	1,043556258642D-03	5,217781293211D-04	6.75	7,028873743517D-04
19	9,5	-1,669837389107D-04	-8,349186945533D-05	7.25	-8,989036994315D-04
				7,50	-1,438157399376D-03
	0	-3,161717560587D+00	1,650692552256D+03	7.75	-1,150754934925D-03
	1	-1,555386276056D+00	-8.411630266125D+02	8.25	1,474418332053D-03
	2	-2,860599368119D-01	1,595082090439D+02	8,50	2,360469865203D-03
	3	-2,330581027976D-02	-1,342759463357D+01	8,75	1,889748888856D-03
	4	-7,095540216268D-04	4.201890770747D-01	9,25	-2,423203526051D-03
				9,50	-3,880685929036D-03

I=18 H= 8.5

0		0.	-1.820107017038D+04	-1.00	3,739795436143D-03
1		-8,608042790263D-04	4.218665023252D+03	-,75	2,074958832316D-03
2		1.674161378435D-03	-3,664452741742D+02	-,50	9,873105233728D-04
3		-1,009058849334D-03	1.413614008244D+01	-,25	3,363674306224D-04
4		1.957709293486D-04	-2,042419575294D-01	.25	-1,255677981964D-04
				.50	-1,257586031029D-04
0	-9,5	-6,917942357289D-08	-3,458971178645D-08	.75	-6,764618466385D-05
1	-8,5	5,188342729628D-07	2,594171364814D-07	1.25	4,649468550518D-05
2	-7,5	-1,880467639676D-06	-9,402338198382D-07	1.50	5,851492017199D-05
3	-6,5	4,598755123976D-06	2,299377561988D-06	1.75	3,829385316337D-05
4	-5,5	-9,035799431166D-06	-4,517899715583D-06	2.25	-3,615790939159D-05
5	-4,5	1,595505188644D-05	7.977525943222D-06	2.50	-5,170117382170D-05
6	-3,5	-2,695042713946D-05	-1,347521356973D-05	2.75	-3,775485275999D-05
7	-2,5	4,476476750463D-05	2,238238375232D-05	3.25	4,233468496195D-05
8	-1,5	-7,388714667751D-05	-3,694357333875D-05	3.50	6,463154865060D-05
9	-,5	1,216460835620D-04	6,082304178100D-05	3,75	4,982151268019D-05
10	,5	-1,999610365396D-04	-9,998051826979D-05	4.25	-6,053580665091D-05
11	1,5	3,278778894157D-04	1.639389447079D-04	4.50	-9,513918101221D-05
12	2,5	-5,340327636881D-04	-2,670163818440D-04	4.75	-7,506446508192D-05
13	3,5	8,524808796510D-04	4.262404398255D-04	5.25	9,429719899865D-05
14	4,5	-1,280248424637D-03	-6.401242123184D-04	5.50	1,499388286578D-04
15	5,5	1,646034860377D-03	8.230174301883D-04	5.75	1.193960874087D-04
16	6,5	-1,600407237042D-03	-8.002036185211D-04	6.25	-1,519443889181D-04
17	7,5	1.043556258642D-03	5.217781293211D-04	6,50	-2,426853638148D-04
18	8,5	-3,968121973665D-04	-1,984060986833D-04	6,75	-1,939292537523D-04
19	9,5	6,585129914900D-05	3,292564957450D-05	7,25	2,480087208585D-04
				7,50	3,967869285998D-04
	0	8,723400480333D-01	-7,390288497222D+02	7,75	3,174899868696D-04
	1	4,291419806553D-01	3,831169479166D+02	8,25	-4,067770631798D-04
	2	7,892594241471D-02	-7,416230182676D+01	8,50	-6,512168231414D-04
	3	6,430236465914D-03	6,374945696324D+00	8,75	-5,213386818787D-04
	4	1,957709293486D-04	-2,042419575294D-01	9,25	6,684535547252D-04
				9,50	1,070444076683D-03

I=19 H= 9.5

0		0,	3,590397526342D+03	-1,00	-4,986503220643D-04
1		1,147764195230D-04	-8,402779131858D+02	-,75	-2,766672421474D-04
2		-2,232264101655D-04	7,375082033563D+01	-,50	-1,316442886667D-04

CARDINAL SPLINES N=5

3		1.345441294464D-04	-2.876973017356D+00	-.25	-4.484997380012D-05
4		-2.610336292953D-05	4.207469186081D-02	.25	1.674274004174D-05
				.50	1.676818123368D-05
0	-9.5	9.224125666094D-09	4.612062833047D-09	.75	9.019688947992D-06
1	-8.5	-6.917942357289D-08	-3.458971178645D-08	1.25	-6.199427229259D-06
2	-7.5	2.507345439014D-07	1.253672719507D-07	1.50	-7.802160275264D-06
3	-6.5	-6.131809569512D-07	-3.065904784756D-07	1.75	-5.105958920940D-06
4	-5.5	1.204800532447D-06	6.024002662233D-07	2.25	4.821160116232D-06
5	-4.5	-2.127391249233D-06	-1.063695624617D-06	2.50	6.893640668353D-06
6	-3.5	3.593490816832D-06	1.796745408416D-06	2.75	5.034090440445D-06
7	-2.5	-5.968874579806D-06	-2.984437289903D-06	3.25	-5.644747849462D-06
8	-1.5	9.852371544864D-06	4.926185772432D-06	3.50	-8.617727614530D-06
9	-.5	-1.622249073154D-05	-8.111245365771D-06	3.75	-6.643012839668D-06
10	.5	2.667522614070D-05	1.333761307035D-05	4.25	8.071615394996D-06
11	1.5	-4.378323579098D-05	-2.189161789549D-05	4.50	1.268549715846D-05
12	2.5	7.152794726579D-05	3.576397363290D-05	4.75	1.000880947133D-05
13	3.5	-1.152330580248D-04	-5.761652901241D-05	5.25	-1.257322390559D-05
14	4.5	1.777825632116D-04	8.889128160580D-05	5.50	-1.999225775218D-05
15	5.5	-2.393162956504D-04	-1.196581478252D-04	5.75	-1.591980173516D-05
16	6.5	2.448166323436D-04	1.224083161718D-04	6.25	2.025963932380D-05
17	7.5	-1.669837389107D-04	-8.349186945533D-05	6.50	3.235863849424D-05
18	8.5	6.585129914900D-05	3.292564957450D-05	6.75	2.585767490170D-05
19	9.5	-1.124684435641D-05	-5.623422178203D-06	7.25	-3.306827479544D-05
				7.50	-5.290549394034D-05
	0	-1.163145570691D-01	1.398251404862D+02	7.75	-4.233230444017D-05
	1	-5.722018549284D-02	-7.365761552138D+01	8.25	5.423674709412D-05
	2	-1.052368974728D-02	1.454053498361D+01	8.50	8.682789380664D-05
	3	-8.573836618758D-04	-1.278134726645D+00	8.75	6.951026863355D-05
	4	-2.610336292953D-05	4.207469186081D-02	9.25	-8.912223961118D-05
				9.50	-1.427145406199D-04

N= 5 M=20 P= 10.0 Q= 9

I= 0		H=-10.0			
0		1.000000000000D+00	1.468898475238D+00	-1.00	5.855091736146D+00
1		-2.284283764275D+00	-1.349672325641D+00	-.75	4.048119301609D+00
2		1.843375996498D+00	9.383467143936D-01	-.50	2.688653909746D+00
3		-6.432564803746D-01	-3.222221699428D-01	-.25	1.696661633883D+00
4		8.417549499851D-02	4.209568288664D-02	.25	5.344179866910D-01
				.50	2.435560034112D-01
0	-10.0	-1.124684715296D-05	-5.623423576479D-06	.75	7.894515541437D-02
1	-9.0	6.585132012269D-05	3.292566006134D-05	1.25	-2.601723975067D-02
2	-8.0	-1.669838149279D-04	-8.349190746395D-05	1.50	-2.411407009689D-02
3	-7.0	2.448168182466D-04	1.224084091233D-04	1.75	-1.186258074550D-02
4	-6.0	-2.393166609199D-04	-1.196583304599D-04	2.25	6.554041598262D-03
5	-5.0	1.777832081919D-04	8.889160409597D-05	2.50	7.241237695634D-03
6	-4.0	-1.152341475024D-04	-5.761707375118D-05	2.75	4.101562968864D-03
7	-3.0	7.152975694717D-05	3.576487847359D-05	3.25	-2.785858169245D-03
8	-2.0	-4.378622305975D-05	-2.189311152987D-05	3.50	-3.317126951632D-03

CARDINAL SPLINES N=5

9	-1.0	2.668014566302D-05	1.334007283151D-05	3.75	-1.995702116087D-03
10	0.0	-1.623058411754D-05	-8.115292058770D-06	4.25	1.478450018741D-03
11	1.0	9.865675608092D-06	4.932837804046D-06	4.50	1.814867019734D-03
12	2.0	-5.990708594037D-06	-2.995354297018D-06	4.75	1.118395718523D-03
13	3.0	3.629159153171D-06	1.814579576586D-06	5.25	-8.565915297107D-04
14	4.0	-2.184852793998D-06	-1.092426396999D-06	5.50	-1.063431481588D-03
15	5.0	1.293452186469D-06	6.467260932346D-07	5.75	-6.610569122064D-04
16	6.0	-7.325162869942D-07	-3.662581434971D-07	6.25	5.123222548138D-04
17	7.0	3.728125291669D-07	1.864062645834D-07	6.50	6.385279196302D-04
18	8.0	-1.524459566662D-07	-7.622297833311D-08	6.75	3.981155697784D-04
19	9.0	4.206090765363D-08	2.103045382682D-08	7.25	-3.097844822298D-04
20	10.0	-5.608243955031D-09	-2.804121977515D-09	7.50	-3.866086712721D-04
				7.75	-2.412900982471D-04
	0	3.609932316176D+02	1.805415055818D+02	8.25	1.880093635374D-04
	1	1.783082720474D+02	8.913334252594D+01	8.50	2.347398941401D-04
	2	3.305097858437D+01	1.652909134809D+01	8.75	1.465566344079D-04
	3	2.723763319566D+00	1.361605145523D+00	9.25	-1.142496107853D-04
	4	8.417549499851D-02	4.209568288664D-02	9.50	-1.426712793382D-04
				9.75	-8.908760044516D-05

I= 1 H= -9.0

0		0.	-7.266579343469D+00	-1.00	-1.484716786926D+01
1		5.137183791674D+00	4.125036118417D+00	-.75	-8.822347466383D+00
2		-6.514937114947D+00	-3.382406191024D+00	-.50	-4.571164560819D+00
3		2.786367217295D+00	1.397637984213D+00	-.25	-1.736612910628D+00
4		-4.086797453417D-01	-2.043993867722D-01	.25	9.190529610004D-01
				.50	1.262611163794D+00
0	-10.0	6.585132012269D-05	3.292566006134D-05	.75	1.234430505145D+00
1	-9.0	-3.968123546657D-04	-1.984061773329D-04	1.25	6.867513132952D-01
2	-8.0	1.043556828759D-03	5.217784143795D-04	1.50	3.847460973985D-01
3	-7.0	-1.600408631285D-03	-8.002043156424D-04	1.75	1.484016548479D-01
4	-6.0	1.646037599838D-03	8.230187999189D-04	2.25	-6.439175543971D-02
5	-5.0	-1.280253261883D-03	-6.401266309417D-04	2.50	-6.651218235879D-02
6	-4.0	8.524890505535D-04	4.262445252767D-04	2.75	-3.587760911772D-02
7	-3.0	-5.340463360009D-04	-2.670231680005D-04	3.25	2.286549778359D-02
8	-2.0	3.279002934404D-04	1.639501467202D-04	3.50	2.666332188640D-02
9	-1.0	-1.999979321482D-04	-9.999896607410D-05	3.75	1.578658316109D-02
10	0.0	1.217067826264D-04	6.085339131320D-05	4.25	-1.144232014413D-02
11	1.0	-7.398692496449D-05	-3.699346248225D-05	4.50	-1.394334217941D-02
12	2.0	4.492851902180D-05	2.246425951090D-05	4.75	-8.544013490291D-03
13	3.0	-2.721793379804D-05	-1.360896689902D-05	5.25	6.493874323239D-03
14	4.0	1.638600402536D-05	8.193002012681D-06	5.50	8.041359584727D-03
15	5.0	-9.700672261785D-06	-4.850336130893D-06	5.75	4.988947516816D-03
16	6.0	5.493750480277D-06	2.746875240139D-06	6.25	-3.856289064583D-03
17	7.0	-2.796032459246D-06	-1.398016229623D-06	6.50	-4.802076991783D-03
18	8.0	1.143319581940D-06	5.716597909700D-07	6.75	-2.992064861183D-03
19	9.0	-3.154498900317D-07	-1.577249450159D-07	7.25	2.326144912123D-03
20	10.0	4.206090765363D-08	2.103045382682D-08	7.50	2.902166907845D-03
				7.75	1.810899032983D-03
	0	-1.900552109700D+03	-9.506127207702D+02	8.25	-1.410606550019D-03
	1	-9.239703746854D+02	-4.618292395268D+02	8.50	-1.761048289622D-03
	2	-1.681317678011D+02	-8.409289872793D+01	8.75	-1.099405449030D-03
	3	-1.356082259637D+01	-6.778337486673D+00	9.25	8.569677409888D-04
	4	-4.086797453417D-01	-2.043993867722D-01	9.50	1.070119624183D-03
				9.75	6.681937666748D-04

CARDINAL SPLINES N=5

I= 2	H= -8.0				
0		0.	2.633706039502D+01	-1.00	2.176759950014D+01
1		-5.754265695717D+00	-8.518324330642D+00	-.75	1.239488407031D+01
2		1.004279531869D+01	5.474222825966D+00	-.50	6.081580993042D+00
3		-5.129450562447D+00	-2.580869763082D+00	-.25	2.149674296086D+00
4		8.410879232923D-01	4.207596651304D-01	.25	-8.877538824862D-01
				.50	-9.550476694273D-01
0	-10.0	-1.669838149279D-04	-8.349190746395D-05	.75	-5.645009232570D-01
1	-9.0	1.043556828759D-03	5.217784143795D-04	1.25	5.327706263571D-01
2	-8.0	-2.876048739720D-03	-1.438024369860D-03	1.50	9.045855062682D-01
3	-7.0	4.678392378788D-03	2.339196189394D-03	1.75	1.058304770395D+00
4	-6.0	-5.153581604287D-03	-2.576790802143D-03	2.25	7.840602757917D-01
5	-5.0	4.285611795163D-03	2.142805897581D-03	2.50	4.922579785769D-01
6	-4.0	-2.991149026220D-03	-1.495574513110D-03	2.75	2.092982557324D-01
7	-3.0	1.914495107591D-03	9.572475537953D-04	3.25	-1.057538671420D-01
8	-2.0	-1.184134118171D-03	-5.920670590855D-04	3.50	-1.157621349669D-01
9	-1.0	7.240024606002D-04	3.620012303001D-04	3.75	-6.550814263596D-02
10	0.0	-4.409392251462D-04	-2.204696125731D-04	4.25	4.481630455678D-02
11	1.0	2.681234142730D-04	1.340617071365D-04	4.50	5.360897472589D-02
12	2.0	-1.628321960366D-04	-8.141609801828D-05	4.75	3.239160597498D-02
13	3.0	9.864749856119D-05	4.932374928059D-05	5.25	-2.416029576302D-02
14	4.0	-5.938930086433D-05	-2.969465043216D-05	5.50	-2.973231554234D-02
15	5.0	3.515915216326D-05	1.757957608163D-05	5.75	-1.835885844530D-02
16	6.0	-1.991159088059D-05	-9.955795440295D-06	6.25	1.410043685151D-02
17	7.0	1.013396490558D-05	5.066982452788D-06	6.50	1.752173553073D-02
18	8.0	-4.143858176442D-06	-2.071929088221D-06	6.75	1.089987195942D-02
19	9.0	1.143319581940D-06	5.716597909700D-07	7.25	-8.455756990180D-03
20	10.0	-1.524459566662D-07	-7.622297833311D-08	7.50	-1.054219397172D-02
				7.75	-6.574597975843D-03
	0	4.228165545388D+03	2.115302987908D+03	8.25	5.117626469590D-03
	1	2.020618165113D+03	1.009743863786D+03	8.50	6.387509549322D-03
	2	3.608120324207D+02	1.805039290118D+02	8.75	3.986946072665D-03
	3	2.851406636925D+01	1.424951684213D+01	9.25	-3.107012888780D-03
	4	8.410879232923D-01	4.207596651304D-01	9.50	-3.879509980285D-03
				9.75	-2.422261947752D-03

I= 3	H= -7.0				
0		0.	-6.440831178255D+01	-1.00	-2.127856551207D+01
1		5.130813836720D+00	1.636116144194D+01	-.75	-1.191267878029D+01
2		-9.648194755675D+00	-5.931499043564D+00	-.50	-5.727949571690D+00
3		5.508346510906D+00	2.793655217655D+00	-.25	-1.975655461302D+00
4		-9.912104087685D-01	-4.961327156968D-01	.25	7.618872864577D-01
				.50	7.799513709140D-01
0	-10.0	2.448168182466D-04	1.224084091233D-04	.75	4.312277253554D-01
1	-9.0	-1.600408631285D-03	-8.002043156424D-04	1.25	-3.214209964317D-01
2	-8.0	4.678392378788D-03	2.339196189394D-03	1.50	-4.301422101463D-01
3	-7.0	-8.205120117815D-03	-4.102560058907D-03	1.75	-3.062809909670D-01
4	-6.0	9.887735570639D-03	4.943867785320D-03	2.25	3.901048495820D-01
5	-5.0	-9.023484618540D-03	-4.511742309270D-03	2.50	7.469611096831D-01
6	-4.0	6.793964561024D-03	3.396982280512D-03	2.75	9.690235651406D-01
7	-3.0	-4.548150219144D-03	-2.274075109572D-03	3.25	8.447017489675D-01
8	-2.0	2.867566703222D-03	1.433783351611D-03	3.50	5.644639127310D-01
9	-1.0	-1.764815532567D-03	-8.824077662835D-04	3.75	2.527399334023D-01
10	0.0	1.077169523828D-03	5.385847619141D-04	4.25	-1.379362154745D-01
11	1.0	-6.554712617684D-04	-3.277356308842D-04	4.50	-1.552674969602D-01

12	2.0	3.981650976723D-04	1.990825488362D-04	4.75	-8.985298286939D-02
13	3.0	-2.412367615236D-04	-1.206183807618D-04	5.25	6.346230890983D-02
14	4.0	1.452369364730D-04	7.261846823648D-05	5.50	7.675742548559D-02
15	5.0	-8.598269765556D-05	-4.299134882778D-05	5.75	4.678130785396D-02
16	6.0	4.869449885851D-05	2.434724942925D-05	6.25	-3.531242293289D-02
17	7.0	-2.478299342816D-05	-1.239149671408D-05	6.50	-4.363169770101D-02
18	8.0	1.013396490558D-05	5.066982452788D-06	6.75	-2.702500828381D-02
19	9.0	-2.796032459246D-06	-1.398016229623D-06	7.25	2.084386708063D-02
20	10.0	3.728125291669D-07	1.864062645834D-07	7.50	2.593754333178D-02
				7.75	1.615239028474D-02
	0	-5.317268913980D+03	-2.661618541033D+03	8.25	-1.254854126180D-02
	1	-2.500170763079D+03	-1.248703116920D+03	8.50	-1.565235300313D-02
	2	-4.391240446896D+02	-2.198014719320D+02	8.75	-9.765132353738D-03
	3	-3.414006983983D+01	-1.705165341022D+01	9.25	7.605012725554D-03
	4	-9.912104087685D-01	-4.961327156968D-01	9.50	9.493835755412D-03
				9.75	5.926741718264D-03

I= 4 H= -6.0

0		0.	1.265519321265D+02	-1.00	1.580740165283D+01
1		-3.689195803998D+00	-2.895102788039D+01	-.75	8.793883636528D+00
2		7.103897428538D+00	5.727814174384D+00	-.50	4.197365620657D+00
3		-4.214385364088D+00	-2.184768362721D+00	-.25	1.435267011035D+00
4		7.999230562086D-01	4.009980026594D-01	.25	-5.410304345043D-01
				.50	-5.454269917777D-01
0	-10.0	-2.393166609199D-04	-1.196583304599D-04	.75	-2.958156894332D-01
1	-9.0	1.646037599838D-03	8.230187999189D-04	1.25	2.082776700148D-01
2	-8.0	-5.153581604287D-03	-2.576790802143D-03	1.50	2.668384549903D-01
3	-7.0	9.887735570639D-03	4.943867785320D-03	1.75	1.788089612538D-01
4	-6.0	-1.329742060219D-02	-6.648710301093D-03	2.25	-1.819603338378D-01
5	-5.0	1.367068114483D-02	6.835340572414D-03	2.50	-2.760589749796D-01
6	-4.0	-1.147545794173D-02	-5.737728970865D-03	2.75	-2.190055529466D-01
7	-3.0	8.315942037530D-03	4.157971018765D-03	3.25	3.308257148520D-01
8	-2.0	-5.479737494978D-03	-2.739868747489D-03	3.50	6.763773009330D-01
9	-1.0	3.435083081358D-03	1.717541540679D-03	3.75	9.265578055202D-01
10	0.0	-2.109851887693D-03	-1.054925943846D-03	4.25	8.761611434538D-01
11	1.0	1.286558835082D-03	6.432794175409D-04	4.50	6.030818349488D-01
12	2.0	-7.820604885378D-04	-3.910302442689D-04	4.75	2.765379126109D-01
13	3.0	4.739369283314D-04	2.369684641657D-04	5.25	-1.561634085232D-01
14	4.0	-2.853561915216D-04	-1.426780957608D-04	5.50	-1.778960383849D-01
15	5.0	1.689399281401D-04	8.446996407007D-05	5.75	-1.039195207569D-01
16	6.0	-9.567639606727D-05	-4.783819803364D-05	6.25	7.436403982391D-02
17	7.0	4.869449885851D-05	2.434724942925D-05	6.50	9.034478016098D-02
18	8.0	-1.991159088059D-05	-9.955795440295D-06	6.75	5.525295293287D-02
19	9.0	5.493750480277D-06	2.746875240139D-06	7.25	-4.190459753052D-02
20	10.0	-7.325162869942D-07	-3.662581434971D-07	7.50	-5.185882390740D-02
				7.75	-3.215983415374D-02
	0	4.458342982812D+03	2.235034734633D+03	8.25	2.484510119974D-02
	1	2.073765368375D+03	1.034166757428D+03	8.50	3.093353542808D-02
	2	3.606261702310D+02	1.807835648884D+02	8.75	1.927175690488D-02
	3	2.778253688425D+01	1.385515174365D+01	9.25	-1.498070098693D-02
	4	7.999230562086D-01	4.009980026594D-01	9.50	-1.868994157421D-02
				9.75	-1.166218861723D-02

I= 5 H= -5.0

0		0.	-2.234611898077D+02	-1.00	-1.024594164760D+01

CARDINAL SPLINES N=5

1		2,368749209375D+00	4,904800676184D+01	-,75	-5,689587634045D+00
2		-4,592387541318D+00	-6,138264624277D+00	-,50	-2,709905091250D+00
3		2,754132722821D+00	1,514046918660D+00	-,25	-9,243177856501D-01
4		-5,306721740864D-01	-2,671662593549D-01	,25	3,461234673035D-01
				,50	3,473776460629D-01
0	-10,0	1,777832081919D-04	8,889160409597D-05	,75	1,873490136943D-01
1	-9,0	-1,280253261883D-03	-6,401266309417D-04	1,25	-1,297653266082D-01
2	-8,0	4,285611795163D-03	2,142805897581D-03	1,50	-1,642460124244D-01
3	-7,0	-9,023484618540D-03	-4,511742309270D-03	1,75	-1,082989880173D-01
4	-6,0	1,367068114483D-02	6,835340572414D-03	2,25	1,046751859460D-01
5	-5,0	-1,610765996128D-02	-8,053829980638D-03	2,50	1,523731362984D-01
6	-4,0	1,549214707949D-02	7,746073539744D-03	2,75	1,139737867983D-01
7	-3,0	-1,260601382624D-02	-6,303006913121D-03	3,25	-1,379230189041D-01
8	-2,0	9,007851217600D-03	4,503925608800D-03	3,50	-2,236236094387D-01
9	-1,0	-5,901090567082D-03	-2,950545283541D-03	3,75	-1,874585280790D-01
10	0,0	3,691003563609D-03	1,845501781804D-03	4,25	3,074530672476D-01
11	1,0	-2,264745278674D-03	-1,132372639337D-03	4,50	6,476886635859D-01
12	2,0	1,379518554169D-03	6,897592770844D-04	4,75	9,088786342158D-01
13	3,0	-8,365761610115D-04	-4,182880805057D-04	5,25	8,897019325325D-01
14	4,0	5,038149397070D-04	2,519074698535D-04	5,50	6,198924048470D-01
15	5,0	-2,982969906194D-04	-1,491484953097D-04	5,75	2,869878880263D-01
16	6,0	1,689399281401D-04	8,446996407007D-05	6,25	-1,642623799552D-01
17	7,0	-8,598269765556D-05	-4,299134882778D-05	6,50	-1,879902847477D-01
18	8,0	3,515915216326D-05	1,757957608163D-05	6,75	-1,102133160238D-01
19	9,0	-9,700672261785D-06	-4,850336130893D-06	7,25	7,926172370859D-02
20	10,0	1,293452186469D-06	6,467260932346D-07	7,50	9,645733245036D-02
				7,75	5,906814160595D-02
	0	-2,988140280080D+03	-1,504423259507D+03	8,25	-4,487784280770D-02
	1	-1,385927881116D+03	-6,881682475455D+02	8,50	-5,557154205110D-02
	2	-2,403717103085D+02	-1,210166126774D+02	8,75	-3,447818713964D-02
	3	-1,847275424063D+01	-9,172603455537D+00	9,25	2,665323608762D-02
	4	-5,306721740864D-01	-2,671662593549D-01	9,50	3,319223298169D-02
				9,75	2,068275100583D-02

I= 6 H= -4.0

0		0.	3,774631917764D+02	-1,00	6,360041918751D+00
1		-1,466022118999D+00	-8,158266410748D+01	-,75	3,529729791522D+00
2		2,848294650358D+00	7,914045532018D+00	-,50	1,680063870383D+00
3		-1,713941223303D+00	-1,088353613310D+00	-,25	5,726003079726D-01
4		3,317839260911D-01	1,689834292938D-01	,25	-2,139714151948D-01
				,50	-2,144437795089D-01
0	-10,0	-1,152341475024D-04	-5,761707375118D-05	,75	-1,154499464587D-01
1	-9,0	8,524890505535D-04	4,262445252767D-04	1,25	7,955208342961D-02
2	-8,0	-2,991149026220D-03	-1,495574513110D-03	1,50	1,003059592699D-01
3	-7,0	6,793964561024D-03	3,396982280512D-03	1,75	6,580548651579D-02
4	-6,0	-1,147545794173D-02	-5,737728970865D-03	2,25	-6,261279387392D-02
5	-5,0	1,549214707949D-02	7,746073539744D-03	2,50	-9,005242046842D-02
6	-4,0	-1,728818954005D-02	-8,644094770026D-03	2,75	-6,627439915954D-02
7	-3,0	1,622478303054D-02	8,112391515270D-03	3,25	7,613120572964D-02
8	-2,0	-1,305422601178D-02	-6,527113005888D-03	3,50	1,183856807480D-01
9	-1,0	9,280524536965D-03	4,640262268483D-03	3,75	9,352565695581D-02
10	0,0	-6,066253009436D-03	-3,033126504718D-03	4,25	-1,227745631339D-01
11	1,0	3,790219745579D-03	1,895109872789D-03	4,50	-2,050280485568D-01
12	2,0	-2,323057150091D-03	-1,161528575045D-03	4,75	-1,759990853029D-01
13	3,0	1,411666912753D-03	7,058334563764D-04	5,25	2,986779458925D-01
14	4,0	-8,507356251215D-04	-4,253678125608D-04	5,50	6,367919567483D-01

15	5,0	5.038149397070D-04	2.519074698535D-04	5.75	9,021048068695D-01
16	6,0	-2,853561915216D-04	-1.426780957608D-04	6.25	8,949520177556D-01
17	7,0	1.452369364730D-04	7.261846823648D-05	6.50	6,264361003081D-01
18	8,0	-5,938930086433D-05	-2,969465043216D-05	6.75	2,910680562027D-01
19	9.0	1.638600402536D-05	8.193002012681D-06	7.25	-1,674378194948D-01
20	10.0	-2,184852793998D-06	-1.092426396999D-06	7.50	-1,919537037607D-01
				7,75	-1,126873599153D-01
	0	1.874067281454D+03	9.545217835309D+02	8.25	8,119036233100D-02
	1	8,684532082616D+02	4,261258797149D+02	8.50	9,886615478202D-02
	2	1.505004136059D+02	7.665349470898D+01	8.75	6,057270716423D-02
	3	1,155741582034D+01	5.670983558441D+00	9.25	-4,605222592963D-02
	4	3.317839260911D-01	1.689834292938D-01	9.50	-5,703940965411D-02
				9.75	-3,539583044318D-02

I= 7 H= -3.0

0		0.	-6.269898990608D+02	-1.00	-3,893297627830D+00
1		8,965614588436D-01	1.347446027408D+02	-.75	-2,160323425743D+00
2		-1,743110702687D+00	-1.165168202898D+01	-.50	-1,028038221124D+00
3		1,050051590193D+00	9.093674451904D-01	-.25	-3,502870501791D-01
4		-2,035738761064D-01	-1,069220572461D-01	.25	1,308077917091D-01
				.50	1,310362749741D-01
0	-10,0	7,152975694717D-05	3.576487847359D-05	.75	7,050516253646D-02
1	-9,0	-5,340463360009D-04	-2,670231680005D-04	1.25	-4,850023457948D-02
2	-8,0	1,914495107591D-03	9.572475537953D-04	1.50	-6,107672257520D-02
3	-7,0	-4,548150219144D-03	-2.274075109572D-03	1.75	-4,000306898563D-02
4	-6,0	8,315942037530D-03	4,157971018765D-03	2,25	3,786775791214D-02
5	-5,0	-1,260601382624D-02	-6,303006913121D-03	2,50	5,425088893860D-02
6	-4,0	1,622478303054D-02	8,112391515270D-03	2,75	3,971901259528D-02
7	-3,0	-1,774269314250D-02	-8,871346571249D-03	3.25	-4,489421377944D-02
8	-2,0	1,650255765236D-02	8,251278826180D-03	3,50	-6,895473515343D-02
9	-1,0	-1,322274750873D-02	-6,611373754363D-03	3,75	-5,358117299856D-02
10	0,0	9,381834234930D-03	4,690917117465D-03	4,25	6,672762200027D-02
11	1,0	-6,125845527427D-03	-3,062922763714D-03	4,50	1,068421394602D-01
12	2,0	3,823137569912D-03	1,911568784956D-03	4,75	8,641189455622D-02
13	3,0	-2,337632472235D-03	-1,168816236117D-03	5,25	-1,173257021958D-01
14	4,0	1,411666912753D-03	7,058334563764D-04	5,50	-1,982631440834D-01
15	5,0	-8,365761610115D-04	-4,182880805057D-04	5,75	-1,717935924076D-01
16	6,0	4,739369283314D-04	2,369684641657D-04	6,25	2,954181000067D-01
17	7,0	-2,412367615236D-04	-1,206183807618D-04	6,50	6,327285687043D-01
18	8,0	9,864749856119D-05	4,932374928059D-05	6,75	8,995709182939D-01
19	9,0	-2,721793379804D-05	-1,360896689902D-05	7,25	8,969246346942D-01
20	10,0	3,629159153171D-06	1,814579576586D-06	7,50	6,288987468073D-01
				7,75	2,926057112034D-01
	0	-1,151032626552D+03	-6,045652018212D+02	8,25	-1,686374615563D-01
	1	-5,332456799627D+02	-2,531670332661D+02	8,50	-1,934528941564D-01
	2	-9,238588866075D+01	-4,852389302093D+01	8,75	-1,136244574528D-01
	3	-7,092903454064D+00	-3,367514844654D+00	9,25	8,192339366496D-02
	4	-2,035738761064D-01	-1,069220572461D-01	9,50	9,978379656789D-02
				9,75	6,114751221870D-02

I= 8 H= -2.0

0		0.	1.034996928829D+03	-1.00	2,372394818880D+00
1		-5,461497306452D-01	-2.219613382255D+02	-.75	1,316320652333D+00
2		1,062076037346D+00	1.832622313970D+01	-.50	6,263560519449D-01
3		-6,400257856833D-01	-9,544610753634D-01	-.25	2,134025225264D-01

4		1.241432652053D-01	7.054838104102D-02	.25	-7.967314876578D-02
				.50	-7.980021064108D-02
0	-10.0	-4.378622305975D-05	-2.189311152987D-05	.75	-4.292898798023D-02
1	-9.0	3.279002934404D-04	1.639501467202D-04	1.25	2.951419441887D-02
2	-8.0	-1.184134118171D-03	-5.920670590855D-04	1.50	3.715209240350D-02
3	-7.0	2.867566703222D-03	1.433783351611D-03	1.75	2.431999600601D-02
4	-6.0	-5.479737494978D-03	-2.739868747489D-03	2.25	-2.298282488988D-02
5	-5.0	9.007851217600D-03	4.503925608800D-03	2.50	-3.288360673819D-02
6	-4.0	-1.305422601178D-02	-6.527113005888D-03	2.75	-2.403386312723D-02
7	-3.0	1.650255765236D-02	8.251278826180D-03	3.25	2.702090958805D-02
8	-2.0	-1.791199059162D-02	-8.955995295811D-03	3.50	4.133529849850D-02
9	-1.0	1.660449569408D-02	8.302247847042D-03	3.75	3.194835788266D-02
10	0.0	-1.328275293607D-02	-6.641376468033D-03	4.25	-3.913720502272D-02
11	1.0	9.415005958065D-03	4.707502979033D-03	4.50	-6.188740452778D-02
12	2.0	-6.140559340614D-03	-3.070279670307D-03	4.75	-4.922573172114D-02
13	3.0	3.823137569912D-03	1.911568784956D-03	5.25	6.339115646060D-02
14	4.0	-2.323057150091D-03	-1.161528575045D-03	5.50	1.026994994650D-01
15	5.0	1.379518554169D-03	6.897592770844D-04	5.75	8.383630429256D-02
16	6.0	-7.820604885378D-04	-3.910302442689D-04	6.25	-1.153286625596D-01
17	7.0	3.981650976723D-04	1.990825488362D-04	6.50	-1.957733050419D-01
18	8.0	-1.628321960366D-04	-8.141609801828D-05	6.75	-1.702405256535D-01
19	9.0	4.492851902180D-05	2.246425951090D-05	7.25	2.942080724012D-01
20	10.0	-5.990708594037D-06	-2.995354297018D-06	7.50	6.312170754975D-01
				7.75	8.986264547522D-01
	0	7.021529727980D+02	3.990285955911D+02	8.25	8.976630802923D-01
	1	3.252606961325D+02	1.404183261235D+02	8.50	6.298230154699D-01
	2	5.634726159004D+01	3.202141950341D+01	8.75	2.931845904534D-01
	3	4.325704822529D+00	1.867474166277D+00	9.25	-1.690928984150D-01
	4	1.241432652053D-01	7.054838104102D-02	9.50	-1.940253758508D-01
				9.75	-1.139849295868D-01

I= 9 H= -1.0

0		0.	-1.704538403195D+03	-1.00	-1.443388974095D+00
1		3.322481442717D-01	3.652651115152D+02	-.75	-8.008466468600D-01
2		-6.461587239259D-01	-2.963002972492D+01	-.50	-3.810646974048D-01
3		3.894330027029D-01	1.239591795994D+00	-.25	-1.298269606649D-01
4		-7.554910319439D-02	-5.173494883235D-02	.25	4.846689290722D-02
				.50	4.854174965226D-02
0	-10.0	2.668014566302D-05	1.334007283151D-05	.75	2.611166884853D-02
1	-9.0	-1.999979321482D-04	-9.999896607410D-05	1.25	-1.794876329848D-02
2	-8.0	7.240024606002D-04	3.620012303001D-04	1.50	-2.259057941809D-02
3	-7.0	-1.764815532567D-03	-8.824077662835D-04	1.75	-1.478525750620D-02
4	-6.0	3.435083081358D-03	1.717541540679D-03	2.25	1.396447079112D-02
5	-5.0	-5.901090567082D-03	-2.950545283541D-03	2.50	1.997166023873D-02
6	-4.0	9.280524536965D-03	4.640262268483D-03	2.75	1.458847804565D-02
7	-3.0	-1.322274750873D-02	-6.611373754363D-03	3.25	-1.637255793959D-02
8	-2.0	1.660449569408D-02	8.302247847042D-03	3.50	-2.501237509490D-02
9	-1.0	-1.797209113126D-02	-8.986045565631D-03	3.75	-1.929796174316D-02
10	0.0	1.663774340837D-02	8.318871704187D-03	4.25	2.351185310520D-02
11	1.0	-1.329751129103D-02	-6.648755645515D-03	4.50	3.702721958277D-02
12	2.0	9.415005958065D-03	4.707502979033D-03	4.75	2.929311091938D-02
13	3.0	-6.125845527427D-03	-3.062922763714D-03	5.25	-3.710255602403D-02
14	4.0	3.790219745579D-03	1.895109872789D-03	5.50	-5.936058867938D-02
15	5.0	-2.264745278674D-03	-1.132372639337D-03	5.75	-4.765430909270D-02
16	6.0	1.286558835082D-03	6.432794175409D-04	6.25	6.217173665121D-02
17	7.0	-6.554712617684D-04	-3.277356308842D-04	6.50	1.011782876948D-01

```
18   8.0   2.681234142730D-04   1.340617071365D-04     6.75  8.288672559256D-02
19   9.0  -7.398692496449D-05  -3.699346248225D-05     7.25 -1.145872203516D-01
20  10.0   9.865675608092D-06   4.932837804046D-06     7.50 -1.948456996554D-01
                                                       7.75 -1.696597551945D-01
       0  -4.273514201909D+02  -2.926479528645D+02     8.25  2.937513611717D-01
       1  -1.979574383010D+02  -6.239773951435D+01     8.50  6.306430843379D-01
       2  -3.429263055948D+01  -2.348324514451D+01     8.75  8.982650790443D-01
       3  -2.632531125073D+00  -8.298061573001D-01     9.25  8.979516711283D-01
       4  -7.554910319439D-02  -5.173494883235D-02     9.50  6.301895831510D-01
                                                       9.75  2.934184366875D-01


I=10      H=  0.0
0          0.                   2.804604683772D+03    -1.00  8.776323027085D-01
1         -2.020117681349D-01  -6.008265368189D+02     -.75  4.869402529173D-01
2          3.928838528640D-01   4.841749235389D+01     -.50  2.316976566358D-01
3         -2.367962679273D-01  -1.837616551295D+00     -.25  7.893757926542D-02
4          4.594041378237D-02   4.594041378237D-02      .25 -2.946818823667D-02
                                                        .50 -2.951321018134D-02
0  -10.0  -1.623058411754D-05  -8.115292058770D-06      .75 -1.587546901702D-02
1   -9.0   1.217067826264D-04   6.085339131320D-05     1.25  1.091188566082D-02
2   -8.0  -4.409392251462D-04  -2.204696125731D-04     1.50  1.373323679062D-02
3   -7.0   1.077169523828D-03   5.385847619141D-04     1.75  8.987694915199D-03
4   -6.0  -2.109851887693D-03  -1.054925943846D-03     2.25 -8.487168588991D-03
5   -5.0   3.691003563609D-03   1.845501781804D-03     2.50 -1.213642140691D-02
6   -4.0  -6.066253009436D-03  -3.033126504718D-03     2.75 -8.863474363933D-03
7   -3.0   9.381834234930D-03   4.690917117465D-03     3.25  9.941563435468D-03
8   -2.0  -1.328275293607D-02  -6.641376468033D-03     3.50  1.518096134898D-02
9   -1.0   1.663774340837D-02   8.318871704187D-03     3.75  1.170574952346D-02
10   0.0  -1.798685974182D-02  -8.993429870909D-03     4.25 -1.423596214734D-02
11   1.0   1.663774340837D-02   8.318871704187D-03     4.50 -2.238870259654D-02
12   2.0  -1.328275293607D-02  -6.641376468033D-03     4.75 -1.768043619658D-02
13   3.0   9.381834234930D-03   4.690917117465D-03     5.25  2.227138129377D-02
14   4.0  -6.066253009436D-03  -3.033126504718D-03     5.50  3.548579154470D-02
15   5.0   3.691003563609D-03   1.845501781804D-03     5.75  2.833378645179D-02
16   6.0  -2.109851887693D-03  -1.054925943846D-03     6.25 -3.635650628618D-02
17   7.0   1.077169523828D-03   5.385847619141D-04     6.50 -5.842845184197D-02
18   8.0  -4.409392251462D-04  -2.204696125731D-04     6.75 -4.707128882597D-02
19   9.0   1.217067826264D-04   6.085339131320D-05     7.25  6.171387446150D-02
20  10.0  -1.623058411754D-05  -8.115292058770D-06     7.50  1.006031060151D-01
                                                       7.75  8.252472366951D-02
       0   2.598761375015D+02   2.598761375015D+02     8.25 -1.142982635637D-01
       1   1.203784400405D+02  -2.464978207414D-18     8.50 -1.944787253877D-01
       2   2.085324408447D+01   2.085324408447D+01     8.75 -1.694256792666D-01
       3   1.600820283368D+00  -3.575973227177D-20     9.25  2.935575598473D-01
       4   4.594041378237D-02   4.594041378237D-02     9.50  6.303908863846D-01
                                                       9.75  8.980994451060D-01


I=11      H=  1.0
0          0.                  -4.612105508083D+03    -1.00 -5.333759183526D-01
1          1.227700579991D-01   9.879442849240D+02     -.75 -2.959344457613D-01
2         -2.387720975438D-01  -7.941839916292D+01     -.50 -1.408122240823D-01
3          1.439129683395D-01   2.899204110594D+00     -.25 -4.797347632996D-02
4         -2.792079447030D-02  -5.173494883235D-02      .25  1.790883296782D-02
                                                        .50  1.793609527051D-02
0  -10.0   9.865675608092D-06   4.932837804046D-06      .75  9.647949034893D-03
```

CARDINAL SPLINES N=5

1	-9.0	-7.398692496449D-05	-3.699346248225D-05	1.25	-6.631311021136D-03
2	-8.0	2.681234142730D-04	1.340617071365D-04	1.50	-8.345761128090D-03
3	-7.0	-6.554712617684D-04	-3.277356308842D-04	1.75	-5.461761106347D-03
4	-6.0	1.286558835082D-03	6.432794175409D-04	2.25	5.157275116474D-03
5	-5.0	-2.264745278674D-03	-1.132372639337D-03	2.50	7.374415155332D-03
6	-4.0	3.790219745579D-03	1.895109872789D-03	2.75	5.385345246777D-03
7	-3.0	-6.125845527427D-03	-3.062922763714D-03	3.25	-6.039198188921D-03
8	-2.0	9.415005958065D-03	4.707502979033D-03	3.50	-9.220606379236D-03
9	-1.0	-1.329751129103D-02	-6.648755645515D-03	3.75	-7.108437900583D-03
10	0.0	1.663774340837D-02	8.318871704187D-03	4.25	8.639722012014D-03
11	1.0	-1.797209113126D-02	-8.986045565631D-03	4.50	1.358140815244D-02
12	2.0	1.660449569408D-02	8.302247847042D-03	4.75	1.071886996066D-02
13	3.0	-1.322274750873D-02	-6.611373754363D-03	5.25	-1.347748237671D-02
14	4.0	9.280524536965D-03	4.640262268483D-03	5.50	-2.144473561468D-02
15	5.0	-5.901090567082D-03	-2.950545283541D-03	5.75	-1.709177865241D-02
16	6.0	3.435083081358D-03	1.717541540679D-03	6.25	2.181094296745D-02
17	7.0	-1.764815532567D-03	-8.824077662835D-04	6.50	3.490813671391D-02
18	8.0	7.240024606002D-04	3.620012303001D-04	6.75	2.797059331458D-02
19	9.0	-1.999979321482D-04	-9.999896607410D-05	7.25	-3.606698892080D-02
20	10.0	2.668014566302D-05	1.334007283151D-05	7.50	-5.806092893977D-02
				7.75	-4.683694266841D-02
	0	-1.579444855380D+02	-2.926479528645D+02	8.25	6.151993940245D-02
	1	-7.316195927226D+01	6.239773951435D+01	8.50	1.003507741180D-01
	2	-1.267385972954D+01	-2.348324514451D+01	8.75	8.235902278449D-02
	3	-9.729188104727D-01	8.298061573001D-01	9.25	-1.141504581068D-01
	4	-2.792079447030D-02	-5.173494883235D-02	9.50	-1.942773946717D-01
				9.75	-1.692865468294D-01

I=12 H= 2.0

0		0.	7.578311783855D+03	-1.00	3.238626902010D-01
1		-7.454496465182D-02	-1.623282490239D+03	-.75	1.796895059594D-01
2		1.449808435781D-01	1.303746731163D+02	-.50	8.550020991202D-02
3		-8.738338509440D-02	-4.689409407918D+00	-.25	2.912913387586D-02
4		1.695349687672D-02	7.054838104102D-02	.25	-1.087407925710D-02
				.50	-1.089061271400D-02
0	-10.0	-5.990708594037D-06	-2.995354297018D-06	.75	-5.858122002214D-03
1	-9.0	4.492851902180D-05	2.246425951090D-05	1.25	4.026427411858D-03
2	-8.0	-1.628321960366D-04	-8.141609801828D-05	1.50	5.067389105415D-03
3	-7.0	3.981650976723D-04	1.990825488362D-04	1.75	3.316256610707D-03
4	-6.0	-7.820604885378D-04	-3.910302442689D-04	2.25	-3.131315286021D-03
5	-5.0	1.379518554169D-03	6.897592770844D-04	2.50	-4.477414140922D-03
6	-4.0	-2.323057150091D-03	-1.161528575045D-03	2.75	-3.269671689469D-03
7	-3.0	3.823137569912D-03	1.911568784956D-03	3.25	3.666415457237D-03
8	-2.0	-6.140559340614D-03	-3.070279670307D-03	3.50	5.597582452500D-03
9	-1.0	9.415005958065D-03	4.707502979033D-03	3.75	4.315060399036D-03
10	0.0	-1.328275293607D-02	-6.641376468033D-03	4.25	-5.243550383597D-03
11	1.0	1.660449569408D-02	8.302247847042D-03	4.50	-8.241475839551D-03
12	2.0	-1.791199059162D-02	-8.955995295811D-03	4.75	-6.503135626176D-03
13	3.0	1.650255765236D-02	8.251278826180D-03	5.25	8.171811008212D-03
14	4.0	-1.305422601178D-02	-6.527113005888D-03	5.50	1.299666781153D-02
15	5.0	9.007851217600D-03	4.503925608800D-03	5.75	1.035231775168D-02
16	6.0	-5.479737494978D-03	-2.739868747489D-03	6.25	-1.318645601022D-02
17	7.0	2.867566703222D-03	1.433783351611D-03	6.50	-2.107578442405D-02
18	8.0	-1.184134118171D-03	-5.920670590855D-04	6.75	-1.685675846577D-02
19	9.0	3.279002934404D-04	1.639501467202D-04	7.25	2.161671359740D-02
20	10.0	-4.378622305975D-05	-2.189311152987D-05	7.50	3.465553603534D-02

				7.75	2,780477397745D-02
	0	9.590421838414D+01	3.990285955911D+02	8.25	-3,591915032896D-02
	1	4.442404388549D+01	-1.404183261235D+02	8.50	-5,785958824467D-02
	2	7.695577416781D+00	3.202141950341D+01	8.75	-4,669782328125D-02
	3	5.907564899746D-01	-1.867474166277D+00	9,25	6,138086031621D-02
	4	1.695349687672D-02	7.054838104102D-02	9.50	1,001495528051D-01
				9,75	8,221132093694D-02
I=13	H= 3.0				
0		0.	-1.242536025369D+04	-1.00	-1,961917875094D-01
1		4.515830849778D-02	2.661587576065D+03	-.75	-1,088535290881D-01
2		-8.782746994846D-02	-2.137025727082D+02	-.50	-5,179488296979D-02
3		5.293577067734D-02	7.644397134499D+00	-.25	-1,764603353175D-02
4		-1.027023838582D-02	-1.069220572461D-01	.25	6,587363564649D-03
				.50	6,597375285531D-03
0	-10.0	3.629159153171D-06	1.814579576586D-06	.75	3,548762661845D-03
1	-9.0	-2.721793379804D-05	-1.360896689902D-05	1.25	-2,439144013240D-03
2	-8.0	9.864749856119D-05	4.932374928059D-05	1.50	-3,069736520007D-03
3	-7.0	-2.412367615236D-04	-1.206183807618D-04	1.75	-2,008926381942D-03
4	-6.0	4.739369283314D-04	2.369684641657D-04	2.25	1,896879432944D-03
5	-5.0	-8.365761610115D-04	-4.182880805057D-04	2.50	2,712301312604D-03
6	-4.0	1.411666912753D-03	7.058334563764D-04	2.75	1,980668517274D-03
7	-3.0	-2.337632472235D-03	-1.168816236117D-03	3.25	-2,220956053688D-03
8	-2.0	3.823137569912D-03	1.911568784956D-03	3.50	-3,390718170417D-03
9	-1.0	-6.125845527427D-03	-3.062922763714D-03	3.75	-2,613777857776D-03
10	0.0	9.381834234930D-03	4.690917117465D-03	4.25	3.175984292987D-03
11	1.0	-1.322274750873D-02	-6.611373754363D-03	4.50	4.991558252589D-03
12	2.0	1.650255765236D-02	8.251278826180D-03	4.75	3.938449459128D-03
13	3.0	-1.774269314250D-02	-8.871346571249D-03	5.25	-4.948037460279D-03
14	4.0	1.622478303054D-02	8.112391515270D-03	5.50	-7.868298010302D-03
15	5.0	-1.260601382624D-02	-6.303006913121D-03	5.75	-6.266132839404D-03
16	6.0	8.315942037530D-03	4.157971018765D-03	6.25	7.976732183802D-03
17	7.0	-4.548150219144D-03	-2.274075109572D-03	6.50	1,274330342991D-02
18	8.0	1.914495107591D-03	9.572475537953D-04	6.75	1,018616992583D-02
19	9.0	-5.340463360009D-04	-2.670231680005D-04	7.25	-1,303853854484D-02
20	10.0	7.152975694717D-05	3.576487847359D-05	7.50	-2,087443837595D-02
				7.75	-1,671769231895D-02
	0	-5.809777709071D+01	-6.045652018212D+02	8.25	2,147777286769D-02
	1	-2.691161343054D+01	2.531670332661D+02	8.50	3,445456150901D-02
	2	-4.661897381119D+00	-4.852389302093D+01	8,75	2,765728206987D-02
	3	-3.578737647554D-01	3.367514844654D+00	9,25	-3,575390758557D-02
	4	-1.027023838582D-02	-1.069220572461D-01	9,50	-5,760803479281D-02
				9,75	-4,650455268228D-02
I=14	H= 4.0				
0		0.	2.024194790296D+04	-1.00	1,181120925211D-01
1		-2.718635743231D-02	-4.336424558602D+03	-.75	6.553239164859D-02
2		5.287420619043D-02	3.481730590385D+02	-.50	3,118173809502D-02
3		-3.186859640182D-02	-1.243032073019D+01	-.25	1,062332614382D-02
4		6.182932496499D-03	1.689834292938D-01	.25	-3,965746218225D-03
				.50	-3,971772705037D-03
0	-10.0	-2.184852793998D-06	-1.092426396999D-06	.75	-2,136436762903D-03
1	-9.0	1.638600402536D-05	8.193002012681D-06	1.25	1,468419715802D-03
2	-8.0	-5.938930086433D-05	-2.969465043216D-05	1.50	1,848049610148D-03
3	-7.0	1.452369364730D-04	7.261846823648D-05	1.75	1,209417448260D-03

CARDINAL SPLINES N=5

4	-6.0	-2.853561915216D-04	-1.426780957608D-04	2.25	-1.141960156529D-03
5	-5.0	5.038149397070D-04	2.519074698535D-04	2.50	-1.632857956894D-03
6	-4.0	-8.507356251215D-04	-4.253678125608D-04	2.75	-1.192398085105D-03
7	-3.0	1.411666912753D-03	7.058334563764D-04	3.25	1.337045936325D-03
8	-2.0	-2.323057150091D-03	-1.161528575045D-03	3.50	2.041247664218D-03
9	-1.0	3.790219745579D-03	1.895109872789D-03	3.75	1.573510195021D-03
10	0.0	-6.066253009436D-03	-3.033126504718D-03	4.25	-1.911919655122D-03
11	1.0	9.280524536965D-03	4.640262268483D-03	4.50	-3.004832129514D-03
12	2.0	-1.305422601178D-02	-6.527113005888D-03	4.75	-2.370826756494D-03
13	3.0	1.622478303054D-02	8.112391515270D-03	5.25	2.978368022761D-03
14	4.0	-1.728818954005D-02	-8.644094770026D-03	5.50	4.735919327377D-03
15	5.0	1.549214707949D-02	7.746073539744D-03	5.75	3.771327616426D-03
16	6.0	-1.147545794173D-02	-5.737728970865D-03	6.25	-4.799893168836D-03
17	7.0	6.793964561024D-03	3.396982280512D-03	6.50	-7.666948318678D-03
18	8.0	-2.991149026220D-03	-1.495574513110D-03	6.75	-6.127234106075D-03
19	9.0	8.524890505535D-04	4.262445252767D-04	7.25	7.838211633299D-03
20	10.0	-1.152341475024D-04	-5.761707375118D-05	7.50	1.254307405769D-02
				7.75	1.003930896546D-02
	0	3.497628560788D+01	9.545217835309D+02	8.25	-1.287414666349D-02
	1	1.620144883182D+01	-4.261258797149D+02	8.50	-2.062426266592D-02
	2	2.806575812035D+00	7.665349470898D+01	8.75	-1.652553255832D-02
	3	2.154487034581D-01	-5.670983558441D+00	9.25	2.124586091731D-02
	4	6.182932496499D-03	1.689834292938D-01	9.50	3.409111208813D-02
				9.75	2.737120764694D-02

I=15 H= 5.0

0		0.	-3.233203305179D+04	-1.00	-6.992326329921D-02
1		1.609452954176D-02	6.928946575175D+03	-.75	-3.879567667214D-02
2		-3.130193375220D-02	-5.564944719565D+02	-.50	-1.845982667062D-02
3		1.886645538175D-02	1.985925382973D+01	-.25	-6.289089831479D-03
4		-3.660344623495D-03	-2.671662593549D-01	.25	2.347751675016D-03
				.50	2.351319242854D-03
0	-10.0	1.293452186469D-06	6.467260932346D-07	.75	1.264786487307D-03
1	-9.0	-9.700672261785D-06	-4.850336130893D-06	1.25	-8.693152013310D-04
2	-8.0	3.515915216326D-05	1.757957608163D-05	1.50	-1.094058672239D-03
3	-7.0	-8.598269765556D-05	-4.299134882778D-05	1.75	-7.159836617171D-04
4	-6.0	1.689399281401D-04	8.446996407007D-05	2.25	6.760479601215D-04
5	-5.0	-2.982969906194D-04	-1.491484953097D-04	2.50	9.666621469208D-04
6	-4.0	5.038149397070D-04	2.519074698535D-04	2.75	7.059066007196D-04
7	-3.0	-8.365761610115D-04	-4.182880805057D-04	3.25	-7.915370874071D-04
8	-2.0	1.379518554169D-03	6.897592770844D-04	3.50	-1.208425435721D-03
9	-1.0	-2.264745278674D-03	-1.132372639337D-03	3.75	-9.315210673070D-04
10	0.0	3.691003563609D-03	1.845501781804D-03	4.25	1.131851810362D-03
11	1.0	-5.901090567082D-03	-2.950545283541D-03	4.50	1.778843521841D-03
12	2.0	9.007851217600D-03	4.503925608800D-03	4.75	1.403505716848D-03
13	3.0	-1.260601382624D-02	-6.303006913121D-03	5.25	-1.763124721873D-03
14	4.0	1.549214707949D-02	7.746073539744D-03	5.50	-2.803507333613D-03
15	5.0	-1.610765996128D-02	-8.053829980638D-03	5.75	-2.232451737870D-03
16	6.0	1.367068114483D-02	6.835340572414D-03	6.25	2.841123427351D-03
17	7.0	-9.023484618540D-03	-4.511742309270D-03	6.50	4.537943722437D-03
18	8.0	4.285611795163D-03	2.142805897581D-03	6.75	3.626370126774D-03
19	9.0	-1.280253261883D-03	-6.401266309417D-04	7.25	-4.638059607745D-03
20	10.0	1.777832081919D-04	8.889160409597D-05	7.50	-7.420909604751D-03
				7.75	-5.938405572887D-03
	0	-2.070623893301D+01	-1.504423259507D+03	8.25	7.610593825498D-03
	1	-9.591386024959D+00	6.881682475455D+02	8.50	1.218650925650D-02

	2	-1,661515046397D+00	-1,210166126774D+02	8.75	9,758751563174D-03
	3	-1.275473295581D-01	9.172603455537D+00	9.25	-1,252315340270D-02
	4	-3,660344623495D-03	-2,671662593549D-01	9.50	-2,006698920355D-02
				9.75	-1,608230290230D-02

I=16 H= 6.0

0		0.	4.852019056800D+04	-1.00	3,959937247882D-02
1		-9.114752521015D-03	-1,041037558893D+04	-.75	2,197100599784D-02
2		1.772710338736D-02	8.370369187936D+02	-.50	1,045428235927D-02
3		-1,068456746025D-02	-2,989507185003D+01	-.25	3,561675915992D-03
4		2.072949110191D-03	4.009980026594D-01	.25	-1,329593080443D-03
				.50	-1,331613457507D-03
0	-10.0	-7,325162869942D-07	-3,662581434971D-07	.75	-7,162815790226D-04
1	-9.0	5,493750480277D-06	2.746875240139D-06	1.25	4,923158335555D-04
2	-8.0	-1,991159088059D-05	-9,955795440295D-06	1.50	6,195938751477D-04
3	-7.0	4,869449885851D-05	2.434724942925D-05	1.75	4,054801327892D-04
4	-6.0	-9,567639606727D-05	-4.783819803364D-05	2.25	-3,828634123537D-04
5	-5.0	1,689399281401D-04	8.446996407007D-05	2.50	-5,474456239413D-04
6	-4.0	-2,853561915216D-04	-1.426780957608D-04	2.75	-3,997729528436D-04
7	-3.0	4,739369283314D-04	2.369684641657D-04	3.25	4,482673313872D-04
8	-2.0	-7,820604885378D-04	-3,910302442689D-04	3,50	6,843612620067D-04
9	-1.0	1.286558835082D-03	6.432794175409D-04	3,75	5,275430386178D-04
10	0.0	-2,109851887693D-03	-1,054925943846D-03	4.25	-6,409936441378D-04
11	1.0	3.435083081358D-03	1.717541540679D-03	4.50	-1,007398038157D-03
12	2.0	-5.479737494978D-03	-2.739868747489D-03	4,75	-7,948340906353D-04
13	3.0	8.315942037530D-03	4.157971018765D-03	5.25	9,984862929889D-04
14	4.0	-1.147545794173D-02	-5.737728970865D-03	5.50	1,587663082503D-03
15	5.0	1.367068114483D-02	6.835340572414D-03	5.75	1,264257495731D-03
16	6.0	-1.329742060219D-02	-6.648710301093D-03	6.25	-1,608917609735D-03
17	7.0	9.887735570639D-03	4.943867785320D-03	6.50	-2,569777339097D-03
18	8.0	-5,153581604287D-03	-2,576790802143D-03	6.75	-2,053519759785D-03
19	9.0	1.646037599838D-03	8.230187999189D-04	7.25	2,626235466859D-03
20	10.0	-2.393166609199D-04	-1.196583304599D-04	7.50	4,201771436733D-03
				7.75	3,362142409535D-03
	0	1.172648645518D+01	2.235034734633D+03	8.25	-4,308004380758D-03
	1	5.431853517915D+00	-1.034166757428D+03	8.50	-6,897162835402D-03
	2	9.409595456945D-01	1.807835648884D+02	8.75	-5,522020012376D-03
	3	7.223339694739D-02	-1.385515174365D+01	9.25	7,081911763648D-03
	4	2.072949110191D-03	4.009980026594D-01	9.50	1,134276991600D-02
				9.75	9.084941523728D-03

I=17 H= 7.0

0		0.	-5.914177747062D+04	-1.00	-2,015400770144D-02
1		4.638931834689D-03	1.274475944141D+04	-.75	-1,118209191303D-02
2		-9.022167631814D-03	-1.029030703657D+03	-.50	-5,320682440592D-03
3		5.437885609765D-03	3.689696203809D+01	-.25	-1,812706580443D-03
4		-1.055022625170D-03	-4.961327156968D-01	.25	6,766932636291D-04
				.50	6,777215246882D-04
0	-10.0	3.728125291669D-07	1.864062645834D-07	.75	3,645498147581D-04
1	-9.0	-2.796032459246D-06	-1.398016229623D-06	1.25	-2,505629716259D-04
2	-8.0	1.013396490558D-05	5.066982452788D-06	1.50	-3,153408214749D-04
3	-7.0	-2.478299342816D-05	-1.239149671408D-05	1.75	-2,063681385982D-04
4	-6.0	4.869449885851D-05	2.434724942925D-05	2.25	1,948573967156D-04
5	-5.0	-8.598269765556D-05	-4.299134882778D-05	2.50	2,786210926156D-04
6	-4.0	1.452369364730D-04	7.261846823648D-05	2.75	2,034634354693D-04

7	-3.0	-2.412367615236D-04	-1.206183807618D-04	3.25	-2.281444663959D-04
8	-2.0	3.981650976723D-04	1.990825488362D-04	3.50	-3.483037744386D-04
9	-1.0	-6.554712617684D-04	-3.277356308842D-04	3.75	-2.684915010329D-04
10	0.0	1.077169523828D-03	5.385847619141D-04	4.25	3.262315902801D-04
11	1.0	-1.764815532567D-03	-8.824077662835D-04	4.50	5.127115752778D-04
12	2.0	2.867566703222D-03	1.433783351611D-03	4.75	4.045275959874D-04
13	3.0	-4.548150219144D-03	-2.274075109572D-03	5.25	-5.081742930372D-04
14	4.0	6.793964561024D-03	3.396982280512D-03	5.50	-8.080311762804D-04
15	5.0	-9.023484618540D-03	-4.511742309270D-03	5.75	-6.434343423100D-04
16	6.0	9.887735570639D-03	4.943867785320D-03	6.25	8.188403406844D-04
17	7.0	-8.205120117815D-03	-4.102560058907D-03	6.50	1.307851590349D-03
18	8.0	4.678392378788D-03	2.339196189394D-03	6.75	1.045101864236D-03
19	9.0	-1.600408631285D-03	-8.002043156424D-04	7.25	-1.336544800593D-03
20	10.0	2.448168182466D-04	1.224084091233D-04	7.50	-2.138330737217D-03
				7.75	-1.710995276235D-03
	0	-5.968168086769D+00	-2.661618541033D+03	8.25	2.192194425764D-03
	1	-2.764529238552D+00	1.248703116920D+03	8.50	3.509547286203D-03
	2	-4.788991744408D-01	-2.198014719320D+02	8.75	2.809630576828D-03
	3	-3.676301939703D-02	1.705165341022D+01	9.25	-3.602566979020D-03
	4	-1.055022625170D-03	-4.961327156968D-01	9.50	-5.769171776278D-03
				9.75	-4.619851680312D-03

I=18 H= 8.0

0		0.	4.872024802038D+04	-1.00	8.241130610740D-03
1		-1.896895322425D-03	-1.057771615718D+04	-.75	4.572444410358D-03
2		3.689234559897D-03	8.604452333540D+02	-.50	2.175668456711D-03
3		-2.223593759966D-03	-3.107990344735D+01	-.25	7.412298265699D-04
4		4.314069684517D-04	4.207596651304D-01	.25	-2.767051402233D-04
				.50	-2.771256034520D-04
0	-10.0	-1.524459566662D-07	-7.622297833311D-08	.75	-1.490672546081D-04
1	-9.0	1.143319581940D-06	5.716597909700D-07	1.25	1.024571469186D-04
2	-8.0	-4.143858176442D-06	-2.071929088221D-06	1.50	1.289453123781D-04
3	-7.0	1.013396490558D-05	5.066982452788D-06	1.75	8.438553474587D-05
4	-6.0	-1.991159088059D-05	-9.955795440295D-06	2.25	-7.967870066899D-05
5	-5.0	3.515915216326D-05	1.757957608163D-05	2.50	-1.139303229960D-04
6	-4.0	-5.938930086433D-05	-2.969465043216D-05	2.75	-8.319777234907D-05
7	-3.0	9.864749856119D-05	4.932374928059D-05	3.25	9.329002821058D-05
8	-2.0	-1.628321960366D-04	-8.141609801828D-05	3.50	1.424240841543D-04
9	-1.0	2.681234142730D-04	1.340617071365D-04	3.75	1.097882234367D-04
10	0.0	-4.409392251462D-04	-2.204696125731D-04	4.25	-1.333985486845D-04
11	1.0	7.240024606002D-04	3.620012303001D-04	4.50	-2.096515710900D-04
12	2.0	-1.184134118171D-03	-5.920670590855D-04	4.75	-1.654142918961D-04
13	3.0	1.914495107591D-03	9.572475537953D-04	5.25	2.077960051059D-04
14	4.0	-2.991149026220D-03	-1.495574513110D-03	5.50	3.304093578220D-04
15	5.0	4.285611795163D-03	2.142805897581D-03	5.75	2.631043862031D-04
16	6.0	-5.153581604287D-03	-2.576790802143D-03	6.25	-3.348281400476D-04
17	7.0	4.678392378788D-03	2.339196189394D-03	6.50	-5.347863782280D-04
18	8.0	-2.876048739720D-03	-1.438024369860D-03	6.75	-4.273457381825D-04
19	9.0	1.043556828759D-03	5.217784143795D-04	7.25	5.465134607070D-04
20	10.0	-1.669838149279D-04	-8.349190746395D-05	7.50	8.743587551634D-04
				7.75	6.996167955521D-04
	0	2.440430427315D+00	2.115302987908D+03	8.25	-8.963553170362D-04
	1	1.130437541692D+00	-1.009743863786D+03	8.50	-1.434975605648D-03
	2	1.958256028319D-01	1.805039290118D+02	8.75	-1.148768425636D-03
	3	1.503268497810D-02	-1.424951684213D+01	9.25	1.472869729945D-03
	4	4.314069684517D-04	4.207596651304D-01	9.50	2.358536020155D-03

CARDINAL SPLINES N=5

```
                                                          9.75  1.888541346485D-03

I=19     H=  9.0
0            0.                  -2.280052634323D+04    -1.00 -2.273785617002D-03
1            5.233666952007D-04   4.994786007176D+03     -.75 -1.261569415885D-03
2           -1.017885636273D-03  -4.100826553914D+02     -.50 -6.002821546921D-04
3            6.135050828466D-04   1.495431295756D+01     -.25 -2.045104969035D-04
4           -1.190282026819D-04  -2.043993867722D-01      .25  7.634488469632D-05
                                                          .50  7.646089337051D-05
0  -10.0   4.206090765363D-08   2.103045382682D-08      .75  4.112869869809D-05
1   -9.0  -3.154498900317D-07  -1.577249450159D-07     1.25 -2.826864371578D-05
2   -8.0   1.143319581940D-06   5.716597909700D-07     1.50 -3.557691380772D-05
3   -7.0  -2.796032459246D-06  -1.398016229623D-06     1.75 -2.328255933095D-05
4   -6.0   5.493750480277D-06   2.746875240139D-06     2.25  2.198391064771D-05
5   -5.0  -9.700672261785D-06  -4.850336130893D-06     2.50  3.143417248583D-05
6   -4.0   1.638600402536D-05   8.193002012681D-06     2.75  2.295484669135D-05
7   -3.0  -2.721793379804D-05  -1.360896689902D-05     3.25 -2.573936972450D-05
8   -2.0   4.492851902180D-05   2.246425951090D-05     3.50 -3.929579758688D-05
9   -1.0  -7.398692496449D-05  -3.699346248225D-05     3.75 -3.029133509435D-05
10   0.0   1.217067826264D-04   6.085339131320D-05     4.25  3.680558432930D-05
11   1.0  -1.999979321482D-04  -9.999896607410D-05     4.50  5.784431725300D-05
12   2.0   3.279002934404D-04   1.639501467202D-04     4.75  4.563894147626D-05
13   3.0  -5.340463360009D-04  -2.670231680005D-04     5.25 -5.733232955746D-05
14   4.0   8.524890505535D-04   4.262445252767D-04     5.50 -9.116216481072D-05
15   5.0  -1.280253261883D-03  -6.401266309417D-04     5.75 -7.259224745862D-05
16   6.0   1.646037599838D-03   8.230187999189D-04     6.25  9.238121856826D-05
17   7.0  -1.600408631285D-03  -8.002043156424D-04     6.50  1.475508648582D-04
18   8.0   1.043556828759D-03   5.217784143795D-04     6.75  1.179072166178D-04
19   9.0  -3.968123546657D-04  -1.984061773329D-04     7.25 -1.507858583253D-04
20  10.0   6.585132012269D-05   3.292566006134D-05     7.50 -2.412395264704D-04
                                                         7.75 -1.930268781441D-04
         0  -6.733318406478D-01  -9.506127207702D+02     8.25  2.473056042988D-04
         1  -3.118956319039D-01   4.618292395268D+02     8.50  3.959090494150D-04
         2  -5.402965476002D-02  -8.409289872793D+01     8.75  3.169418950371D-04
         3  -4.147623024430D-03   6.778337486673D+00     9.25 -4.063497930068D-04
         4  -1.190282026819D-04  -2.043993867722D-01     9.50 -6.506832617421D-04
                                                         9.75 -5.210055124772D-04

I=20     H= 10.0
0            0.                   4.507346040040D+03    -1.00  3.031780577943D-04
1           -6.978375483090D-05  -9.965794446912D+02     -.75  1.682129407277D-04
2            1.357210582577D-04   8.263465544576D+01     -.50  8.003937414551D-05
3           -8.180246994428D-05  -3.045432460988D+00     -.25  2.726866365562D-05
4            1.587077476140D-05   4.209568288664D-02      .25 -1.017954096698D-05
                                                          .50 -1.019500912507D-05
0  -10.0  -5.608243955031D-09  -2.804121977515D-09      .75 -5.483946627332D-06
1   -9.0   4.206090765363D-08   2.103045382682D-08     1.25  3.769235060846D-06
2   -8.0  -1.524459566662D-07  -7.622297833311D-08     1.50  4.743692413586D-06
3   -7.0   3.728125291669D-07   1.864062645834D-07     1.75  3.104409238734D-06
4   -6.0  -7.325162869942D-07  -3.662581434971D-07     2.25 -2.931252275462D-06
5   -5.0   1.293452186469D-06   6.467260932346D-07     2.50 -4.191314766192D-06
6   -4.0  -2.184852793998D-06  -1.092426396999D-06     2.75 -3.060713225549D-06
7   -3.0   3.629159153171D-06   1.814579576586D-06     3.25  3.431991042913D-06
8   -2.0  -5.990708594037D-06  -2.995354297018D-06     3.50  5.239554245593D-06
9   -1.0   9.865675608092D-06   4.932837804046D-06     3.75  4.038932937514D-06
```

CARDINAL SPLINES N=5

10	0,0	-1.623058411754D-05	-8,115292058770D-06
11	1.0	2.668014566302D-05	1,334007283151D-05
12	2,0	-4.378622305975D-05	-2.189311152987D-05
13	3,0	7.152975694717D-05	3.576487847359D-05
14	4,0	-1.152341475024D-04	-5.761707375118D-05
15	5,0	1.777832081919D-04	8,889160409597D-05
16	6,0	-2.393166609199D-04	-1,196583304599D-04
17	7,0	2.448168182466D-04	1.224084091233D-04
18	8,0	-1.669838149279D-04	-8,349190746395D-05
19	9,0	6.585132012269D-05	3.292566006134D-05
20	10,0	-1.124684715296D-05	-5,623423576479D-06

0	8.977954594720D-02	1.805415055818D+02
1	4,158699547265D-02	-8,913334252594D+01
2	7.204111816770D-03	1.652909134809D+01
3	5,530285205118D-04	-1.361605145523D+00
4	1,587077476140D-05	4.209568288664D-02

4.25	-4,907518271461D-06
4.50	-7,712743590636D-06
4.75	-6,085324454431D-06
5.25	7,644475752828D-06
5.50	1,215521670767D-05
5.75	9,679173229155D-06
6,25	-1,231775424822D-05
6.50	-1,967385559940D-05
6,75	-1,572128127807D-05
7.25	2,010516519196D-05
7.50	3,216585595580D-05
7.75	2,573735551650D-05
8.25	-3,297452386576D-05
8.50	-5,278844081653D-05
8,75	-4,225922392723D-05
9,25	5,417977648891D-05
9.50	8,675675072725D-05
9.75	6,946584507289D-05

CARDINAL SPLINES N=8

N= 8 M= 8 P= 4.0 Q=15

I= 0 H= -4.0

0		1.000000000000D+00	-8.758298238953D+00	-1.00	8.695777854913D+00
1		-2.687092977779D+00	1.144326917432D+01	-.75	5.456495605164D+00
2		2.842247287290D+00	-6.142668712430D+00	-.50	3.284279017996D+00
3		-1.569127994575D+00	1.690305173345D+00	-.25	1.875962182068D+00
4		4.965024596095D-01	-2.339785159775D-01	.25	4.832021167995D-01
5		-9.127887535168D-02	1.017681570207D-02	.50	1.991930324927D-01
6		9.139180555973D-03	1.022725271673D-03	.75	5.847871133715D-02
7		-3.890797522171D-04	-9.920634920635D-05	1.25	-1.594683287001D-02
				1.50	-1.354212271498D-02
0	-4.0	2.234443049317D-12	1.117221524659D-12	1.75	-6.147084117553D-03
1	-3.0	-1.787554439454D-11	-8.937772197269D-12	2.25	2.976078351849D-03
2	-2.0	6.256440538088D-11	3.128220269044D-11	2.50	3.134734695895D-03
3	-1.0	-1.251288107618D-10	-6.256440538088D-11	2.75	1.717859284470D-03
4	0.0	1.564110134522D-10	7.820550672611D-11	3.25	-1.149698419853D-03
5	1.0	-1.251288107618D-10	-6.256440538088D-11	3.50	-1.398585945673D-03
6	2.0	6.256440538088D-11	3.128220269044D-11	3.75	-8.780136454124D-04
7	3.0	-1.787554439454D-11	-8.937772197269D-12	4.25	7.584914815407D-04
8	4.0	2.234443049317D-12	1.117221524659D-12	4.50	1.042820304327D-03
				4.75	7.385767266316D-04
	0	-2.144770416613D-03	-2.144770416613D-03	5.25	-8.112850026389D-04
	1	-4.158528842191D-03	3.571428571429D-03	5.50	-1.259796554105D-03
	2	-1.385040691162D-02	-1.385040691162D-02	5.75	-1.010280290729D-03
	3	-1.771216531125D-02	-4.861111111111D-03	6.25	1.440348743775D-03
	4	-7.210358956819D-03	-7.210358956819D-03	6.50	2.571158535020D-03
	5	-2.669338753261D-03	1.388888888889D-03	6.75	2.390416885846D-03
	6	-1.755052506105D-03	-1.755052506105D-03	7.25	-4.749895840589D-03
	7	-3.890797522171D-04	-9.920634920635D-05	7.50	-1.027962375732D-02

I= 1 H= -3.0

0		0.	7.806638591162D+01	-1.00	-3.356622283930D+01
1		7.753886679374D+00	-1.052890105374D+02	-.75	-1.807150143189D+01
2		-1.304353385387D+01	5.883579414389D+01	-.50	-8.469056362717D+00
3		8.897468401044D+00	-1.717799694232D+01	-.25	-2.905742042725D+00
4		-3.173408565765D+00	2.670439238931D+00	.25	1.250475747489D+00
5		6.288421139246D-01	-1.828034145054D-01	.50	1.548545583809D+00
6		-6.616900000334D-02	-1.237357728941D-03	.75	1.365799917457D+00
7		2.914225319324D-03	5.952380952381D-04	1.25	6.226804063683D-01
				1.50	3.178096379698D-01
0	-4.0	-1.787554439454D-11	-8.937772197269D-12	1.75	1.124052801182D-01
1	-3.0	1.430043551563D-10	7.150217757815D-11	2.25	-4.214587657065D-02
2	-2.0	-5.005152430471D-10	-2.502576215235D-10	2.50	-4.119115881716D-02
3	-1.0	1.001030486094D-09	5.005152430471D-10	2.75	-2.133030713709D-02
4	0.0	-1.251288107618D-09	-6.256440538088D-10	3.25	1.318394600140D-02
5	1.0	1.001030486094D-09	5.005152430471D-10	3.50	1.558321881538D-02
6	2.0	-5.005152430471D-10	-2.502576215235D-10	3.75	9.553253450409D-03

CARDINAL SPLINES N=8

7	3.0	1.430043551563D-10	7.150217757815D-11	4.25	-7.956206998810D-03
8	4.0	-1.787554439454D-11	-8.937772197269D-12	4.50	-1.078396868462D-02
				4.75	-7.545118940006D-03
	0	1.715816333291D-02	1.715816333291D-02	5.25	8.126785148065D-03
	1	2.374442121371D-02	-3.809523809524D-02	5.50	1.251977868284D-02
	2	1.163588108485D-01	1.163588108485D-01	5.75	9.970517472320D-03
	3	1.528084336011D-01	5.000000000000D-02	6.25	-1.405193423731D-02
	4	5.073842721011D-02	5.073842721011D-02	6.50	-2.496379953016D-02
	5	1.996582113720D-02	-1.250000000000D-02	6.75	-2.310969372935D-02
	6	1.542930893773D-02	1.542930893773D-02	7.25	4.558659958604D-02
	7	2.914225319324D-03	5.952380952381D-04	7.50	9.835027130860D-02

I= 2 H= -2.0

0		0.	-2.732323506907D+02	-1.00	7.548177993757D+01
1		-1.313860337781D+01	3.825115368809D+02	-.75	3.883253993899D+01
2		2.860236848856D+01	-2.229752795036D+02	-.50	1.717807422263D+01
3		-2.270086162588D+01	6.856326707589D+01	-.25	5.463781155275D+00
4		8.929846646844D+00	-1.152362066959D+01	.25	-1.818618775268D+00
5		-1.891919620958D+00	9.488397285468D-01	.50	-1.754138058955D+00
6		2.086748333450D-01	-1.858591461537D-02	.75	-9.296775339759D-01
7		-9.505344173189D-03	-1.388888888889D-03	1.25	7.087354142588D-01
				1.50	1.087129157731D+00
0	-4.0	6.256440538088D-11	3.128220269044D-11	1.75	1.155682395441D+00
1	-3.0	-5.005152430471D-10	-2.502576215235D-10	2.25	7.251339353068D-01
2	-2.0	1.751803350665D-09	8.759016753324D-10	2.50	4.261514777351D-01
3	-1.0	-3.503606701330D-09	-1.751803350665D-09	2.75	1.720281300335D-01
4	0.0	4.379508376662D-09	2.189754188331D-09	3.25	-8.241967465236D-02
5	1.0	-3.503606701330D-09	-1.751803350665D-09	3.50	-9.042993772884D-02
6	2.0	1.751803350665D-09	8.759016753324D-10	3.75	-5.240496922975D-02
7	3.0	-5.005152430471D-10	-2.502576215235D-10	4.25	4.033032129759D-02
8	4.0	6.256440538088D-11	3.128220269044D-11	4.50	5.312474977116D-02
				4.75	3.630133364842D-02
	0	-6.005357166517D-02	-6.005357166517D-02	5.25	-3.769629623600D-02
	1	-1.643880758133D-02	2.000000000000D-01	5.50	-5.724695976495D-02
	2	-4.628113935254D-01	-4.628113935254D-01	5.75	-4.503055443925D-02
	3	-5.945517398262D-01	-2.347222222222D-01	6.25	6.219942424953D-02
	4	-1.185567174576D-01	-1.185567174576D-01	6.50	1.095900952306D-01
	5	-7.751926286909D-02	3.611111111111D-02	6.75	1.007108170908D-01
	6	-5.747480350426D-02	-5.747480350426D-02	7.25	-1.962256907142D-01
	7	-9.505344173189D-03	-1.388888888889D-03	7.50	-4.211302464551D-01

I= 3 H= -1.0

0		0.	5.464647013814D+02	-1.00	-1.089635598751D+02
1		1.694387342228D+01	-7.743564070951D+02	-.75	-5.487521247686D+01
2		-3.961584808824D+01	4.635394478961D+02	-.50	-2.367304297652D+01
3		3.427116769620D+01	-1.482570897073D+02	-.25	-7.307319238040D+00
4		-1.449163773813D+01	2.641529689474D+01	.25	2.241939546385D+00
5		3.249117019694D+00	-2.432401679316D+00	.50	2.041967524160D+00
6		-3.742941111345D-01	8.022738478630D-02	.75	1.003661250813D+00
7		1.762179945749D-02	1.388888888889D-03	1.25	-5.922945895039D-01
				1.50	-7.079497217111D-01
0	-4.0	-1.251288107618D-10	-6.256440538088D-11	1.75	-4.524437398565D-01
1	-3.0	1.001030486094D-09	5.005152430471D-10	2.25	4.751433537516D-01
2	-2.0	-3.503606701330D-09	-1.751803350665D-09	2.50	8.395915757799D-01
3	-1.0	7.007213402659D-09	3.503606701330D-09	2.75	1.019152510685D+00

CARDINAL SPLINES N=8

```
4   0.0  -8.759016753324D-09  -4.379508376662D-09    3.25  8.178048949590D-01
5   1.0   7.007213402659D-09   3.503606701330D-09    3.50  5.397465942077D-01
6   2.0  -3.503606701330D-09  -1.751803350665D-09    3.75  2.439128742505D-01
7   3.0   1.001030486094D-09   5.005152430471D-10    4.25 -1.455753459643D-01
8   4.0  -1.251288107618D-10  -6.256440538088D-11    4.50 -1.780268432923D-01
                                                     4.75 -1.150030274043D-01
      0   1.201071433303D-01   1.201071433303D-01    5.25  1.103466635169D-01
      1  -3.671223848373D-01  -8.000000000000D-01    5.50  1.628337632799D-01
      2   1.525622787051D+00   1.525622787051D+00    5.75  1.250685856851D-01
      3   1.058547924097D+00   3.388888888889D-01    6.25 -1.664558858526D-01
      4   1.329467682485D-01   1.329467682485D-01    6.50 -2.890044092111D-01
      5   1.869829701826D-01  -4.027777777778D-02    6.75 -2.622240938984D-01
      6   1.191162736752D-01   1.191162736752D-01    7.25  5.003119466140D-01
      7   1.762179945749D-02   1.388888888889D-03    7.50  1.064428461660D+00


I= 4     H=  0.0
0         0.                  -6.830808767267D+02   -1.00  1.047044498439D+02
1        -1.534650844452D+01   9.737788422022D+02    -.75  5.210134839789D+01
2         3.706842122141D+01  -5.918756987590D+02    -.50  2.217248047846D+01
3        -3.349868184247D+01   1.946616399119D+02    -.25  6.738399955448D+00
4         1.486628328378D+01  -3.626738500731D+01     .25 -1.988500597227D+00
5        -3.486743496840D+00   3.615154876922D+00     .50 -1.766936944263D+00
6         4.175204166959D-01  -1.506314532051D-01     .75 -8.431853052768D-01
7        -2.029113821075D-02  -3.975287038158D-26    1.25  4.590581553979D-01
                                                     1.50  5.183600037014D-01
0  -4.0   1.564110134522D-10   7.820550672611D-11    1.75  3.073711955115D-01
1  -3.0  -1.251288107618D-09  -6.256440538088D-10    2.25 -2.501057592671D-01
2  -2.0   4.379508376662D-09   2.189754188331D-09    2.50 -3.445334150374D-01
3  -1.0  -8.759016753324D-09  -4.379508376662D-09    2.75 -2.515340602921D-01
4   0.0   1.094877094165D-08   5.474385470827D-09    3.25  3.380887680810D-01
5   1.0  -8.759016753324D-09  -4.379508376662D-09    3.50  6.711419525529D-01
6   2.0   4.379508376662D-09   2.189754188331D-09    3.75  9.122595953582D-01
7   3.0  -1.251288107618D-09  -6.256440538088D-10    4.25  9.122595953582D-01
8   4.0   1.564110134522D-10   7.820550672611D-11    4.50  6.711419525529D-01
                                                     4.75  3.380887680810D-01
      0   8.498660708371D-01   8.498660708371D-01    5.25 -2.515340602921D-01
      1  -5.410970189533D-01   3.377759835210D-23    5.50 -3.445334150374D-01
      2  -2.330639594925D+00  -2.330639594925D+00    5.75 -2.501057592671D-01
      3  -8.995737940099D-01   4.206814960480D-24    6.25  3.073711955115D-01
      4  -1.158362380884D-01  -1.158362380884D-01    6.50  5.183600037014D-01
      5  -2.840759349505D-01   4.950604355113D-25    6.75  4.590581553979D-01
      6  -1.506314532051D-01  -1.506314532051D-01    7.25 -8.431853052768D-01
      7  -2.029113821075D-02  -3.975287038158D-26    7.50 -1.766936944263D+00


I= 5     H=  1.0
0         0.                   5.464647013814D+02   -1.00 -6.696355987513D+01
1         9.477206755615D+00  -7.818230737617D+02    -.75 -3.307620887578D+01
2        -2.330473697712D+01   4.798505590072D+02    -.50 -1.396112891402D+01
3         2.159338991842D+01  -1.609348674851D+02    -.25 -4.204253747317D+00
4        -9.880526627022D+00   3.102640800585D+01     .25  1.213825227537D+00
5         2.396339241917D+00  -3.285179457094D+00     .50  1.064428461660D+00
6        -2.965163333567D-01   1.580051625641D-01     .75  5.003119466140D-01
7         1.484402167971D-02  -1.388888888889D-03    1.25 -2.622240938984D-01
                                                     1.50 -2.890044092111D-01
0  -4.0  -1.251288107618D-10  -6.256440538088D-11    1.75 -1.664558858526D-01
```

CARDINAL SPLINES N=8

1	-3.0	1.001030486094D-09	5.005152430471D-10	2.25	1.250685856851D-01
2	-2.0	-3.503606701330D-09	-1.751803350665D-09	2.50	1.628337632799D-01
3	-1.0	7.007213402659D-09	3.503606701330D-09	2.75	1.103466635169D-01
4	0.0	-8.759016753324D-09	-4.379508376662D-09	3.25	-1.150030274043D-01
5	1.0	7.007213402659D-09	3.503606701330D-09	3.50	-1.780268432923D-01
6	2.0	-3.503606701330D-09	-1.751803350665D-09	3.75	-1.455753459643D-01
7	3.0	1.001030486094D-09	5.005152430471D-10	4.25	2.439128742505D-01
8	4.0	-1.251288107618D-10	-6.256440538088D-11	4.50	5.397465942077D-01
				4.75	8.178048949590D-01
	0	1.201071433303D-01	1.201071433303D-01	5.25	1.019152510685D+00
	1	1.232877615163D+00	8.000000000000D-01	5.50	8.395915757799D-01
	2	1.525622787051D+00	1.525622787051D+00	5.75	4.751433537516D-01
	3	3.807701463190D-01	-3.388888888889D-01	6.25	-4.524437398565D-01
	4	1.329467682485D-01	1.329467682485D-01	6.50	-7.079497217111D-01
	5	2.675385257382D-01	4.027777777778D-02	6.75	-5.922945895039D-01
	6	1.191162736752D-01	1.191162736752D-01	7.25	1.003661250813D+00
	7	1.484402167971D-02	-1.388888888889D-03	7.50	2.041967524160D+00

I= 6 H= 2.0

0		0.	-2.732323506907D+02	-1.00	2.748177993757D+01
1		-3.805270044474D+00	3.918448702142D+02	-.75	1.351046504885D+01
2		9.457924044118D+00	-2.421197239480D+02	-.50	5.673191410134D+00
3		-8.898083848098D+00	8.236604485367D+01	-.25	1.698728359865D+00
4		4.152068869067D+00	-1.630139844737D+01	.25	-4.839859017324D-01
5		-1.030808509847D+00	1.809950839658D+00	.50	-4.211302464551D-01
6		1.308970555672D-01	-9.636369239315D-02	.75	-1.962256907142D-01
7		-6.727566395411D-03	1.388888888889D-03	1.25	1.007108170908D-01
				1.50	1.095900952306D-01
0	-4.0	6.256440538088D-11	3.128220269044D-11	1.75	6.219942424953D-02
1	-3.0	-5.005152430471D-10	-2.502576215235D-10	2.25	-4.503055443925D-02
2	-2.0	1.751803350665D-09	8.759016753324D-10	2.50	-5.724695976495D-02
3	-1.0	-3.503606701330D-09	-1.751803350665D-09	2.75	-3.769629623600D-02
4	0.0	4.379508376662D-09	2.189754188331D-09	3.25	3.630133364842D-02
5	1.0	-3.503606701330D-09	-1.751803350665D-09	3.50	5.312474977116D-02
6	2.0	1.751803350665D-09	8.759016753324D-10	3.75	4.033032129759D-02
7	3.0	-5.005152430471D-10	-2.502576215235D-10	4.25	-5.240496922975D-02
8	4.0	6.256440538088D-11	3.128220269044D-11	4.50	-9.042993772884D-02
				4.75	-8.241967465236D-02
	0	-6.005357166517D-02	-6.005357166517D-02	5.25	1.720281300335D-01
	1	-4.164388075813D-01	-2.000000000000D-01	5.50	4.261514777351D-01
	2	-4.628113935254D-01	-4.628113935254D-01	5.75	7.251339353068D-01
	3	-1.251072953817D-01	2.347222222222D-01	6.25	1.155682395441D+00
	4	-1.185567174576D-01	-1.185567174576D-01	6.50	1.087129157731D+00
	5	-1.497414850913D-01	-3.611111111111D-02	6.75	7.087354142588D-01
	6	-5.747480350426D-02	-5.747480350426D-02	7.25	-9.296775339759D-01
	7	-6.727566395411D-03	1.388888888889D-03	7.50	-1.754138058955D+00

I= 7 H= 3.0

0		0.	7.806638591162D+01	-1.00	-6.566222839305D+00
1		8.967438222307D-01	-1.121461533945D+02	-.75	-3.218329129646D+00
2		-2.243533853875D+00	6.963579414389D+01	-.50	-1.346986050217D+00
3		2.130801734377D+00	-2.394466360898D+01	-.25	-4.018891984867D-01
4		-1.006741899098D+00	4.837105905598D+00	.25	1.135662626253D-01
5		2.538421139246D-01	-5.578034145054D-01	.50	9.835027130860D-02
6		-3.283566667001D-02	3.209597560439D-02	.75	4.558659958604D-02

7		1.723749128848D-03	-5.952380952381D-04	1.25	-2.310969372935D-02
				1.50	-2.496379953016D-02
0	-4.0	-1.787554439454D-11	-8.937772197269D-12	1.75	-1.405193423731D-02
1	-3.0	1.430043551563D-10	7.150217757815D-11	2.25	9.970517472320D-03
2	-2.0	-5.005152430471D-10	-2.502576215235D-10	2.50	1.251977868284D-02
3	-1.0	1.001030486094D-09	5.005152430471D-10	2.75	8.126785148065D-03
4	0.0	-1.251288107618D-09	-6.256440538088D-10	3.25	-7.545118940006D-03
5	1.0	1.001030486094D-09	5.005152430471D-10	3.50	-1.078396868462D-02
6	2.0	-5.005152430471D-10	-2.502576215235D-10	3.75	-7.956206998810D-03
7	3.0	1.430043551563D-10	7.150217757815D-11	4.25	9.553253450409D-03
8	4.0	-1.787554439454D-11	-8.937772197269D-12	4.50	1.558321881538D-02
				4.75	1.318394600140D-02
	0	1.715816333291D-02	1.715816333291D-02	5.25	-2.133030713709D-02
	1	9.993489740419D-02	3.809523809524D-02	5.50	-4.119115881716D-02
	2	1.163588108485D-01	1.163588108485D-01	5.75	-4.214587657065D-02
	3	5.280843360113D-02	-5.000000000000D-02	6.25	1.124052801182D-01
	4	5.073842721011D-02	5.073842721011D-02	6.50	3.178096379698D-01
	5	4.496582113720D-02	1.250000000000D-02	6.75	6.226804063683D-01
	6	1.542930893773D-02	1.542930893773D-02	7.25	1.365799917457D+00
	7	1.723749128848D-03	-5.952380952381D-04	7.50	1.548545583809D+00

I= 8 H= 4.0

0		0.	-9.758298238953D+00	-1.00	6.957778549131D-01
1		-9.423583492170D-02	1.403612631717D+01	-.75	3.404029232797D-01
2		2.366917317344D-01	-8.748224267986D+00	-.50	1.421891742459D-01
3		-2.260724390194D-01	3.033360728901D+00	-.25	4.233257391362D-02
4		1.076135707206D-01	-6.228674048664D-01	.25	-1.190362660868D-02
5		-2.738998646279D-02	7.406570459096D-02	.50	-1.027962375732D-02
6		3.583625000418D-03	-4.532830283882D-03	.75	-4.749895840589D-03
7		-1.906670538044D-04	9.920634920635D-05	1.25	2.390416885846D-03
				1.50	2.571158535020D-03
0	-4.0	2.234443049317D-12	1.117221524659D-12	1.75	1.440348743775D-03
1	-3.0	-1.787554439454D-11	-8.937772197269D-12	2.25	-1.010280290729D-03
2	-2.0	6.256440538088D-11	3.128220269044D-11	2.50	-1.259796554105D-03
3	-1.0	-1.251288107618D-10	-6.256440538088D-11	2.75	-8.112850026389D-04
4	0.0	1.564110134522D-10	7.820550672611D-11	3.25	7.385767266316D-04
5	1.0	-1.251288107618D-10	-6.256440538088D-11	3.50	1.042820304327D-03
6	2.0	6.256440538088D-11	3.128220269044D-11	3.75	7.584914815407D-04
7	3.0	-1.787554439454D-11	-8.937772197269D-12	4.25	-8.780136454124D-04
8	4.0	2.234443049317D-12	1.117221524659D-12	4.50	-1.398585945673D-03
				4.75	-1.149698419853D-03
	0	-2.144770416613D-03	-2.144770416613D-03	5.25	1.717859284470D-03
	1	-1.130138598505D-02	-3.571428571429D-03	5.50	3.134734695895D-03
	2	-1.385040691162D-02	-1.385040691162D-02	5.75	2.976078351849D-03
	3	-7.989943089031D-03	4.861111111111D-03	6.25	-6.147084117553D-03
	4	-7.210358956819D-03	-7.210358956819D-03	6.50	-1.354212271498D-02
	5	-5.447116531039D-03	-1.388888888889D-03	6.75	-1.594683287001D-02
	6	-1.755052506105D-03	-1.755052506105D-03	7.25	5.847871133715D-02
	7	-1.906670538044D-04	9.920634920635D-05	7.50	1.991930324927D-01

CARDINAL SPLINES N=8

```
                N= 8        M= 9                       P=  4.5       Q=15

I= 0      H= -4.5
0            1.000000000000D+00    3.195009279802D+01    -1.00  9.038961061071D+00
1           -2.732331325950D+00   -3.651650505123D+01     -.75  5.623399396367D+00
2            2.957322563038D+00    1.829863647616D+01     -.50  3.353547680067D+00
3           -1.681058238545D+00   -5.300661229416D+00     -.25  1.896440522042D+00
4            5.510759441156D-01    9.823521195252D-01      .25  4.775354902956D-01
5           -1.055939293497D-01   -1.196547249443D-01      .50  1.943444076977D-01
6            1.108202337937D-02    9.032387867167D-03      .75  5.626095707867D-02
7           -4.970366931428D-04   -3.260983817532D-04     1.25 -1.485686074173D-02
                                                          1.50 -1.238587451331D-02
0   -4.5     4.331212643650D-12    2.165606321825D-12     1.75 -5.509302908772D-03
1   -3.5    -3.766326299157D-11   -1.883163149579D-11     2.25  2.544866414569D-03
2   -2.5     1.453824487612D-10    7.269122438059D-11     2.50  2.608639902147D-03
3   -1.5    -3.269276396308D-10   -1.634638198154D-10     2.75  1.387295970649D-03
4    -.5     4.719443482283D-10    2.359721741142D-10     3.25 -8.660082873000D-04
5     .5    -4.534972370104D-10   -2.267486185052D-10     3.50 -1.011929262291D-03
6    1.5     2.900334171950D-10    1.450167085975D-10     3.75 -6.077605224630D-04
7    2.5    -1.190294327356D-10   -5.951471636780D-11     4.25  4.741410859558D-04
8    3.5     2.843970738262D-11    1.421985369131D-11     4.50  6.149510098305D-04
9    4.5    -3.013561842371D-12   -1.506780921186D-12     4.75  4.086798589192D-04
                                                          5.25 -3.884186900188D-04
         0  -2.127779941773D-02   -2.868736099857D-03     5.50 -5.557185826644D-04
         1  -6.832161390049D-02   -5.731300858626D-02     5.75 -4.077309022549D-04
         2  -9.949368589317D-02   -1.883027703247D-02     6.25  4.750989291877D-04
         3  -8.103814323305D-02   -6.710310963998D-02     6.50  7.561246596637D-04
         4  -4.385926796012D-02   -6.341403373744D-03     6.75  6.197551908777D-04
         5  -1.774415186560D-02   -1.445358937135D-02     7.25 -9.166448806568D-04
         6  -4.574632454626D-03   -1.239711158060D-03     7.50 -1.661806557233D-03
         7  -4.970366931428D-04   -3.260983817532D-04     7.75 -1.566551911054D-03


I= 1      H= -3.5
0            0.                   -3.061084841146D+02    -1.00 -3.680492523609D+01
1            8.180811843600D+00    3.473189054289D+02     -.75 -1.964661262295D+01
2           -1.412952688818D+01   -1.718226771174D+02     -.50 -9.122761374632D+00
3            9.953781014580D+00    4.879743320177D+01     -.25 -3.099001058800D+00
4           -3.688431655181D+00   -8.808365299885D+00      .25  1.303953063717D+00
5            7.639367189682D-01    1.042448038145D+00      .50  1.594303221312D+00
6           -8.450407466849D-02   -7.682646867618D-02      .75  1.386729397407D+00
7            3.933040917094D-03    2.736472737367D-03     1.25  6.123940774310D-01
                                                          1.50  3.068978457992D-01
0   -4.5    -3.766326299157D-11   -1.883163149579D-11     1.75  1.063863855673D-01
1   -3.5     3.297458113152D-10    1.648729056576D-10     2.25 -3.807642588153D-02
2   -2.5    -1.282089022825D-09   -6.410445114125D-10     2.50 -3.622627580751D-02
3   -1.5     2.905454534242D-09    1.452727267121D-09     2.75 -1.821070169733D-02
4    -.5    -4.229052022837D-09   -2.114526011419D-09     3.25  1.050669403872D-02
5     .5     4.099922244312D-09    2.049961122156D-09     3.50  1.193424676567D-02
6    1.5    -2.647194977191D-09   -1.323597488595D-09     3.75  7.002809776794D-03
```

7	2,5	1.097617910646D-09	5.488089553230D-10	4.25	-5,272723975212D-03
8	3,5	-2.651809220525D-10	-1.325904610263D-10	4.50	-6,746063318814D-03
9	4,5	2,843970738262D-11	1.421985369131D-11	4.75	-4,431802912351D-03
				5.25	4,136094124640D-03
	0	1.833213163099D-01	1.763974644900D-02	5.50	5,875225649023D-03
	1	5,935748591188D-01	5.165146219192D-01	5.75	4.284114077224D-03
	2	8,690239384180D-01	1.430532586718D-01	6.25	-4,942639749962D-03
	3	6,982614163557D-01	6.007161812042D-01	6.50	-7,834888625162D-03
	4	3,759742160047D-01	3,831343472732D-02	6.75	-6,399533755031D-03
	5	1,548523529130D-01	1.318184154533D-01	7.25	9,411288997810D-03
	6	3,938671421996D-02	9.372422550865D-03	7.50	1,702185919455D-02
	7	3,933040917094D-03	2.736472737367D-03	7.75	1,601280430257D-02
I= 2	H= -2.5				
0		0.	1.334664532188D+03	-1.00	8,903680369019D+01
1		-1.492542415746D+01	-1.511799918730D+03	-.75	4,542489298414D+01
2		3,314760283605D+01	7.424055037977D+02	-.50	1,991404248341D+01
3		-2,712187375125D+01	-2.075653835774D+02	-.25	6,272633044818D+00
4		1,108538591995D+01	3.651891276284D+01	.25	-2,042438750953D+00
5		-2,457334517379D+00	-4,179236429388D+00	.50	-1,945648666364D+00
6		2,854130890238D-01	2.977791195248D-01	.75	-1,017274220270D+00
7		-1,376941908202D-02	-1,035065285423D-02	1.25	7,517870491392D-01
				1.50	1,132798559835D+00
0	-4,5	1,453824487612D-10	7.269122438059D-11	1.75	1,180873428332D+00
1	-3,5	-1.282089022825D-09	-6.410445114125D-10	2.25	7,081019595360D-01
2	-2,5	5,022944027198D-09	2.511472013599D-09	2.50	4,053718262328D-01
3	-1,5	-1.147424124722D-08	-5.737120623612D-09	2.75	1,589715658130D-01
4	-,5	1.684241964648D-08	8.421209823238D-09	3.25	-7,121450372285D-02
5	,5	-1.647347742212D-08	-8,236738711059D-09	3.50	-7,515780193284D-02
6	1,5	1.073635679851D-08	5.368178399253D-09	3.75	-4,173053226702D-02
7	2,5	-4,495883706686D-09	-2,247941853343D-09	4.25	2,909907134986D-02
8	3,5	1.097617910646D-09	5.488089553230D-10	4.50	3,622480144661D-02
9	4,5	-1,190294327356D-10	-5,951471636780D-11	4.75	2,327109266482D-02
				5.25	-2,099395530293D-02
	0	-6,961752201907D-01	-3,344894074715D-02	5.50	-2,943734246627D-02
	1	-2,293010727890D+00	-2,072838621605D+00	5.75	-2,123110526357D-02
	2	-3,388778190190D+00	-4,848954712050D-01	6.25	2,407406167589D-02
	3	-2,651096646678D+00	-2,372395974817D+00	6.50	3,790002750341D-02
	4	-1,426255914806D+00	-7,561278969711D-02	6.75	3,077335831514D-02
	5	-6,066265783648D-01	-5,408153284798D-01	7.25	-4,482024056595D-02
	6	-1,483236120598D-01	-2,826644538342D-02	7.50	-8,074436874605D-02
	7	-1.376941908202D-02	-1.035065285423D-02	7.75	-7,569292426514D-02
I= 3	H= -1.5				
0		0.	-3,371421965140D+03	-1.00	-1,419924483507D+02
1		2.129773552786D+01	3.841423170069D+03	-.75	-7,093848890436D+01
2		-5,069100672300D+01	-1.888758475738D+03	-.50	-3,033964808228D+01
3		4,504363988827D+01	5.245725986633D+02	-,25	-9,278210572539D+00
4		-1,974393884475D+01	-9,064768342765D+01	.25	2,787311148119D+00
5		4,626836889250D+00	1.006294016631D+01	.50	2,508612408702D+00
6		-5,612786074275D-01	-6,906444056934D-01	.75	1,217103973986D+00
7		2,801187011951D-02	2,322559740061D-02	1.25	-6,971964821973D-01
				1.50	-8,192302046073D-01
0	-4,5	-3,269276396308D-10	-1.634638198154D-10	1.75	-5,138255442066D-01
1	-3,5	2,905454534242D-09	1.452727267121D-09	2.25	5,166443675838D-01

2	-2.5	-1.147424124722D-08	-5.737120623612D-09	2.50	8.902243747003D-01
3	-1.5	2.642888350079D-08	1.321444175039D-08	2.75	1.050966826276D+00
4	-.5	-3.912680613708D-08	-1.956340306854D-08	3.25	7.905017822729D-01
5	.5	3.861028702298D-08	1.930514351149D-08	3.50	5.025336996237D-01
6	1.5	-2.539584527258D-08	-1.269792263629D-08	3.75	2.179029760448D-01
7	2.5	1.073635679851D-08	5.368178399253D-09	4.25	-1.182086877618D-01
8	3.5	-2.647194977191D-09	-1.323597488595D-09	4.50	-1.368475345253D-01
9	4.5	2.900334171950D-10	1.450167085975D-10	4.75	-8.325285240560D-02
				5.25	6.964885538123D-02
	0	1.507057023248D+00	-3.930429545399D-02	5.50	9.507137753504D-02
	1	5.202286274211D+00	4.894045325412D+00	5.75	6.707758788617D-02
	2	7.810395976208D+00	1.034669631909D+00	6.25	-7.355760905752D-02
	3	5.688560900364D+00	5.298379959758D+00	6.50	-1.143205949310D-01
	4	3.211872419720D+00	6.037179446484D-02	6.75	-9.181077937225D-02
	5	1.384362257033D+00	1.292226507194D+00	7.25	1.313894114122D-01
	6	3.210953013373D-01	4.096191242568D-02	7.50	2.350262120824D-01
	7	2.801187011951D-02	2.322559740061D-02	7.75	2.189569437018D-01

I= 4 H= -.5

0		0.	5.442940032763D+03	-1.00	1.563485321364D+02
1		-2.215422203250D+01	-6.243469794652D+03	-.75	7.721794202238D+01
2		5.438557690951D+01	3.086185330588D+03	-.50	3.259640688301D+01
3		-5.034255597988D+01	-8.573771479133D+02	-.25	9.820092345543D+00
4		2.307879914775D+01	1.467726889463D+02	.25	-2.841245487530D+00
5		-5.640950968036D+00	-1.592260410718D+01	.50	-2.496584471956D+00
6		7.098900106990D-01	1.054706298744D+00	.75	-1.176925072767D+00
7		-3.653708801592D-02	-3.414395165646D-02	1.25	6.230831113968D-01
				1.50	6.923585450235D-01
0	-4.5	4.719443482283D-10	2.359721741142D-10	1.75	4.033479934429D-01
1	-3.5	-4.229052022837D-09	-2.114526011419D-09	2.25	-3.149968855658D-01
2	-2.5	1.684241964648D-08	8.421209823238D-09	2.50	-4.237030315409D-01
3	-1.5	-3.912680613708D-08	-1.956340306854D-08	2.75	-3.012790322942D-01
4	-.5	5.843194964857D-08	2.921597482428D-08	3.25	3.807800078860D-01
5	.5	-5.817369009152D-08	-2.908684504576D-08	3.50	7.293281610188D-01
6	1.5	3.861028702298D-08	1.930514351149D-08	3.75	9.529287606058D-01
7	2.5	-1.647347742212D-08	-8.236738711059D-09	4.25	8.694689955678D-01
8	3.5	4.099922244312D-09	2.049961122156D-09	4.50	6.067538453876D-01
9	4.5	-4.534972370104D-10	-2.267486185052D-10	4.75	2.884440865276D-01
				5.25	-1.878988291508D-01
	0	-1.761559752200D+00	5.579822258520D-01	5.50	-2.385799183482D-01
	1	-8.571848618768D+00	-8.417728144369D+00	5.75	-1.594308379851D-01
	2	-1.083758665879D+01	-6.739971423435D-01	6.25	1.621151341341D-01
	3	-7.829335757162D+00	-7.634245286859D+00	6.50	2.452237979062D-01
	4	-4.743981974004D+00	-1.673103612131D-02	6.75	1.925993175343D-01
	5	-2.011317357932D+00	-1.965249483013D+00	7.25	-2.663367751903D-01
	6	-4.410282618024D-01	-2.082817843506D-02	7.50	-4.700807675111D-01
	7	-3.653708801592D-02	-3.414395165646D-02	7.75	-4.329211668561D-01

I= 5 H= .5

0		0.	-5.828747117816D+03	-1.00	-1.207083917468D+02
1		1.656184077320D+01	6.729004834201D+03	-.75	-5.921448148350D+01
2		-4.132631040118D+01	-3.347824614236D+03	-.50	-2.480907406448D+01
3		3.912242990514D+01	9.338981756095D+02	-.25	-7.411301525758D+00
4		-1.842710669505D+01	-1.594592694179D+02	.25	2.101257605543D+00
5		4.638174379975D+00	1.704732574267D+01	.50	1.823756190234D+00

6		-6.007787769235D-01	-1.096362655614D+00	.75	8.476273968340D-01
7		3.175081529701D-02	3.414395165646D-02	1.25	-4.329211668561D-01
				1.50	-4.700807675111D-01
0	-4.5	-4.534972370104D-10	-2.267486185052D-10	1.75	-2.663367751903D-01
1	-3.5	4.099922244312D-09	2.049961122156D-09	2.25	1.925993175343D-01
2	-2.5	-1.647347742212D-08	-8.236738711059D-09	2.50	2.452237979062D-01
3	-1.5	3.861028702298D-08	1.930514351149D-08	2.75	1.621151341341D-01
4	-.5	-5.817369009152D-08	-2.908684504576D-08	3.25	-1.594308379851D-01
5	.5	5.843194964857D-08	2.921597482428D-08	3.50	-2.385799183482D-01
6	1.5	-3.912680613708D-08	-1.956340306854D-08	3.75	-1.878988291508D-01
7	2.5	1.684241964648D-08	8.421209823238D-09	4.25	2.884440865276D-01
8	3.5	-4.229052022837D-09	-2.114526011419D-09	4.50	6.067538453876D-01
9	4.5	4.719443482283D-10	2.359721741142D-10	4.75	8.694689955678D-01
				5.25	9.529287606058D-01
	0	2.877524203904D+00	5.579822258520D-01	5.50	7.293281610188D-01
	1	8.263607669969D+00	8.417728144369D+00	5.75	3.807800078860D-01
	2	9.489592374105D+00	-6.739971423435D-01	6.25	-3.012790322942D-01
	3	7.439154816556D+00	7.634245286859D+00	6.50	-4.237030315409D-01
	4	4.710519901761D+00	-1.673103612131D-02	6.75	-3.149968855658D-01
	5	1.919181608093D+00	1.965249483013D+00	7.25	4.033479934429D-01
	6	3.993719049323D-01	-2.082817843506D-02	7.50	6.923585450235D-01
	7	3.175081529701D-02	3.414395165646D-02	7.75	6.230831113968D-01

I= 6 H= 1.5

0		0.	4.143036135246D+03	-1.00	6.471216757142D+01
1		-8.712973009265D+00	-4.811093249167D+03	-.75	3.161709944282D+01
2		2.194191815079D+01	2.409406487478D+03	-.50	1.318783400772D+01
3		-2.104088773878D+01	-6.760521540556D+02	-.25	3.920330471056D+00
4		1.007249838379D+01	1.156527888152D+02	.25	-1.098732478872D+00
5		-2.583783713127D+00	-1.227488343729D+01	.50	-9.471355327581D-01
6		3.416672509876D-01	7.725682305448D-01	.75	-4.368197793478D-01
7		-1.843932468170D-02	-2.322559740061D-02	1.25	2.189569437018D-01
				1.50	2.350262120824D-01
0	-4.5	2.900334171950D-10	1.450167085975D-10	1.75	1.313894114122D-01
1	-3.5	-2.647194977191D-09	-1.323597488595D-09	2.25	-9.181077937225D-02
2	-2.5	1.073635679851D-08	5.368178399253D-09	2.50	-1.143205949310D-01
3	-1.5	-2.539584527258D-08	-1.269792263629D-08	2.75	-7.355760905752D-02
4	-.5	3.861028702298D-08	1.930514351149D-08	3.25	6.707758788617D-02
5	.5	-3.912680613708D-08	-1.956340306854D-08	3.50	9.507137753504D-02
6	1.5	2.642888350079D-08	1.321444175039D-08	3.75	6.964885538123D-02
7	2.5	-1.147424124722D-08	-5.737120623612D-09	4.25	-8.325285240560D-02
8	3.5	2.905454534242D-09	1.452727267121D-09	4.50	-1.368475345253D-01
9	4.5	-3.269276396308D-10	-1.634638198154D-10	4.75	-1.182086877618D-01
				5.25	2.179029760448D-01
	0	-1.585665614156D+00	-3.930429545399D-02	5.50	5.025336996237D-01
	1	-4.585804376614D+00	-4.894045325412D+00	5.75	7.905017822729D-01
	2	-5.741056712389D+00	1.034669631909D+00	6.25	1.050966826276D+00
	3	-4.908199019152D+00	-5.298379959758D+00	6.50	8.902243747003D-01
	4	-3.091128830790D+00	6.037179446484D-02	6.75	5.166443675838D-01
	5	-1.200090757355D+00	-1.292226507194D+00	7.25	-5.138255442066D-01
	6	-2.391714764859D-01	4.096191242568D-02	7.50	-8.192302046073D-01
	7	-1.843932468170D-02	-2.322559740061D-02	7.75	-6.971964821973D-01

I= 7 H= 2.5

0		0.	-1.885817510835D+03	-1.00	-2.312231741929D+01

CARDINAL SPLINES N=8

1		3.079165215599D+00	2.201564260943D+03	-.75	-1,127022357941D+01
2		-7,795079252722D+00	-1.109479480438D+03	-.50	-4,688695376586D+00
3		7,530622215900D+00	3.133186374291D+02	-.25	-1,389820214239D+00
4		-3,639515749421D+00	-5.384200391266D+01	.25	3,869397921714D-01
5		9,444393915766D-01	5.705624480093D+00	.50	3,322611656900D-01
6		-1.265637074398D-01	-3.543120102916D-01	.75	1,525771183519D-01
7		6.931886626436D-03	1.035065285423D-02	1.25	-7,569292426514D-02
				1.50	-8,074436874605D-02
0	-4,5	-1.190294327356D-10	-5.951471636780D-11	1.75	-4,482024056595D-02
1	-3,5	1,097617910646D-09	5.488089553230D-10	2.25	3,077335831514D-02
2	-2,5	-4,495883706686D-09	-2,247941853343D-09	2.50	3,790002750341D-02
3	-1.5	1,073635679851D-08	5,368178399253D-09	2.75	2,407406167589D-02
4	-,5	-1,647347742212D-08	-8,236738711059D-09	3,25	-2,123110526357D-02
5	,5	1,684241964648D-08	8.421209823238D-09	3,50	-2,943734246627D-02
6	1.5	-1,147424124722D-08	-5.737120623612D-09	3,75	-2,099395530293D-02
7	2,5	5,022944027198D-09	2.511472013599D-09	4,25	2,327109266482D-02
8	3,5	-1,282089022825D-09	-6.410445114125D-10	4,50	3,622480144661D-02
9	4,5	1,453824487612D-10	7.269122438059D-11	4.75	2,909907134986D-02
				5,25	-4,173053226702D-02
	0	6,292773386964D-01	-3,344894074715D-02	5.50	-7,515780193284D-02
	1	1,852666515321D+00	2.072838621605D+00	5.75	-7,121450372285D-02
	2	2,418987247780D+00	-4,848954712050D-01	6.25	1,589715658130D-01
	3	2,093695302955D+00	2.372395974817D+00	6.50	4,053718262328D-01
	4	1,275030335412D+00	-7,561278969711D-02	6.75	7,081019595360D-01
	5	4,750040785948D-01	5.408153284798D-01	7,25	1,180873428332D+00
	6	9,179072129297D-02	-2,826644538342D-02	7.50	1,132798559835D+00
	7	6,931886626436D-03	1.035065285423D-02	7,75	7,517870491392D-01

I= 8 H= 3.5

0		0.	4.990120266411D+02	-1,00	4,984855041276D+00
1		-6,596212139498D-01	-5,853614252035D+02	-.75	2,426353605952D+00
2		1.674893634014D+00	2,967173189414D+02	-.50	1,007903559118D+00
3		-1.624967977207D+00	-8.433919704986D+01	-.25	2,982692841622D-01
4		7.896687621622D-01	1.457873886899D+01	.25	-8,272468668825D-02
5		-2,062984249374D-01	-1,548558855892D+00	.50	-7,087736170129D-02
6		2.786512444741D-02	9,557131377791D-02	.75	-3,246721715566D-02
7		-1.539904557639D-03	-2.736472737367D-03	1.25	1,601280430257D-02
				1.50	1,702185919455D-02
0	-4.5	2.843970738262D-11	1.421985369131D-11	1.75	9,411288997810D-03
1	-3,5	-2,651809220525D-10	-1.325904610263D-10	2.25	-6,399533755031D-03
2	-2,5	1.097617910646D-09	5.488089553230D-10	2.50	-7,834888625162D-03
3	-1.5	-2,647194977191D-09	-1.323597488595D-09	2,75	-4,942639749962D-03
4	-,5	4.099922244312D-09	2.049961122156D-09	3,25	4,284114077224D-03
5	,5	-4,229052022837D-09	-2.114526011419D-09	3.50	5,875225649023D-03
6	1.5	2.905454534242D-09	1.452727267121D-09	3.75	4,136094124640D-03
7	2.5	-1.282089022825D-09	-6.410445114125D-10	4,25	-4,431802912351D-03
8	3.5	3.297458113152D-10	1.648729056576D-10	4.50	-6,746063318814D-03
9	4,5	-3,766326299157D-11	-1.883163149579D-11	4.75	-5,272723975212D-03
				5,25	7,002809776794D-03
	0	-1.480418234119D-01	1.763974644900D-02	5.50	1,193424676567D-02
	1	-4,394543847196D-01	-5,165146219192D-01	5.75	1,050669403872D-02
	2	-5,829174210745D-01	1.430532586718D-01	6,25	-1,821070169733D-02
	3	-5,031709460527D-01	-6,007161812042D-01	6.50	-3,622627580751D-02
	4	-2,993473465500D-01	3.831343472732D-02	6,75	-3,807642588153D-02
	5	-1,087844779935D-01	-1,318184154533D-01	7.25	1,063863855673D-01
	6	-2,064186911823D-02	9.372422550865D-03	7.50	3,068978457992D-01

7		-1.539904557639D-03	-2.736472737367D-03	7.75	6,123940774310D-01

I= 9	H= 4.5				
0		0.	-5.850774173039D+01	-1.00	-4,932367475257D-01
1		6.501837885712D-02	6.892972216189D+01	-.75	-2,398808614336D-01
2		-1.653908283164D-01	-3.512802975260D+01	-,50	-9,955571535015D-02
3		1.608706617773D-01	1.004769892200D+01	-,25	-2,943229628434D-02
4		-7.843521336693D-02	-1.748159454794D+00	.25	8,144304196957D-03
5		2.057417305957D-02	1.865991274795D-01	.50	6,968639143619D-03
6		-2.792332077949D-03	-1.151181018329D-02	.75	3,187445882073D-03
7		1.551600703636D-04	3.260983817532D-04	1.25	-1,566551911054D-03
				1.50	-1,661806557233D-03
0	-4.5	-3.013561842371D-12	-1.506780921186D-12	1.75	-9,166448806568D-04
1	-3,5	2.843970738262D-11	1.421985369131D-11	2.25	6,197551908777D-04
2	-2,5	-1.190294327356D-10	-5.951471636780D-11	2.50	7,561246596637D-04
3	-1,5	2.900334171950D-10	1.450167085975D-10	2.75	4,750989291877D-04
4	-,5	-4.534972370104D-10	-2.267486185052D-10	3.25	-4,077309022549D-04
5	,5	4.719443482283D-10	2.359721741142D-10	3,50	-5,557185826644D-04
6	1,5	-3.269276396308D-10	-1.634638198154D-10	3,75	-3,884186900188D-04
7	2,5	1.453824487612D-10	7.269122438059D-11	4.25	4,086798589192D-04
8	3.5	-3.766326299157D-11	-1.883163149579D-11	4.50	6,149510098305D-04
9	4,5	4.331212643650D-12	2.165606321825D-12	4.75	4,741410859558D-04
				5,25	-6,077605224630D-04
	0	1.554032721802D-02	-2,868736099857D-03	5.50	-1,011929262291D-03
	1	4.630440327203D-02	5.731300858626D-02	5.75	-8,660082873000D-04
	2	6.183313182823D-02	-1.883027703247D-02	6.25	1,387295970649D-03
	3	5,316807604690D-02	6.710310963998D-02	6.50	2,608639902147D-03
	4	3,117646121263D-02	-6.341403373744D-03	6,75	2,544866414569D-03
	5	1.116302687710D-02	1.445358937135D-02	7.25	-5,509302908772D-03
	6	2,095210138506D-03	-1.239711158060D-03	7.50	-1,238587451331D-02
	7	1.551600703636D-04	3.260983817532D-04	7.75	-1,485686074173D-02

N= 8 M=10 P= 5.0 Q=15

I= 0	H= -5.0				
0		1.000000000000D+00	-7.155802400008D+01	-1.00	9,213565368540D+00
1		-2,755028923439D+00	6,732648225670D+01	-.75	5,708059376157D+00
2		3,015438674649D+00	-2.653294550528D+01	-.50	3,388567316044D+00
3		-1.738110733438D+00	5.476052734528D+00	-,25	1,906756163327D+00
4		5,792271655968D-01	-5.694229860336D-01	.25	4,747047996627D-01
5		-1,130873298752D-01	1.298876572828D-02	.50	1,919340683552D-01
6		1.211684402186D-02	2.944869235658D-03	.75	5,516439937812D-02
7		-5,556975207249D-04	-2,183835766790D-04	1.25	-1,432475193551D-02
				1.50	-1,182566212443D-02
0	-5,0	5,723684970042D-12	2.861842485021D-12	1.75	-5,202915767540D-03
1	-4,0	-5,173616557005D-11	-2,586808278502D-11	2,25	2,341937935393D-03
2	-3,0	2,106597973973D-10	1,053298986987D-10	2.50	2,364126394718D-03
3	-2.0	-5,096186150621D-10	-2,548093075311D-10	2,75	1,235828453312D-03

CARDINAL SPLINES N=8

4	-1,0	8,127200424843D-10	4.063600212421D-10	3,25	-7,404642348373D-04
5	0,0	-8,948897434772D-10	-4,474448717386D-10	3,50	-8,444112700434D-04
6	1.0	6,908968075989D-10	3.454484037994D-10	3.75	-4,934790942503D-04
7	2,0	-3,703920609074D-10	-1,851960304537D-10	4,25	3,608126122794D-04
8	3.0	1.323448606853D-10	6.617243034265D-11	4.50	4,504325308616D-04
9	4,0	-2,853173987759D-11	-1.426586993880D-11	4.75	2,868937013506D-04
10	5,0	2,823131758485D-12	1.411565879243D-12	5,25	-2,469121738790D-04
				5,50	-3,335487061624D-04
	0	-1,221277579216D-01	-8,976702364103D-02	5,75	-2,297648629080D-04
	1	-3,318611659107D-01	-9,002335315517D-02	6,25	2,316969057677D-04
	2	-4,035469240633D-01	-2,929183673366D-01	6,50	3,396299912535D-04
	3	-2,891731015119D-01	-8,018320478617D-02	6,75	2,545310200274D-04
	4	-1,353162262569D-01	-9,580602742595D-02	7,25	-3,070888738456D-04
	5	-4,132320759991D-02	-1,331653495850D-02	7,50	-4,957325435627D-04
	6	-7,332569203507D-03	-4,698555948109D-03	7,75	-4,114220608944D-04
	7	-5,556975207249D-04	-2,183835766790D-04	8,25	6,211766312419D-04

I= 1 H= -4.0

0		0,	7,399874534295D+02	-1.00	-3,856954874941D+01
1		8,410203173086D+00	-7,021613723077D+02	-,75	-2,050222147441D+01
2		-1,471687240409D+01	2,812638752727D+02	-,50	-9,476684333179D+00
3		1,053037717903D+01	-6,011650559343D+01	-,25	-3,203255205970D+00
4		-3,972939569529D+00	6,874517203652D+00	.25	1,332561196713D+00
5		8,396681310609D-01	-2,981020658134D-01	.50	1,618663109911D+00
6		-9,496240106531D-02	-1,530348181161D-02	.75	1,397811664106D+00
7		4,525891558423D-03	1,647862264226D-03	1.25	6,070163652502D-01
				1.50	3,012361071618D-01
0	-5.0	-5,173616557005D-11	-2,586808278502D-11	1.75	1,032899101055D-01
1	-4,0	4,719723986804D-10	2,359861993402D-10	2.25	-3,602554654719D-02
2	-3.0	-1,941808968799D-09	-9,709044843996D-10	2,50	-3,375512093861D-02
3	-2,0	4,751805266304D-09	2,375902633152D-09	2,75	-1,667990817642D-02
4	-1.0	-7,673072457198D-09	-3,836536228599D-09	3,25	9,237893800379D-03
5	0,0	8,560817893470D-09	4,280408946735D-09	3,50	1,024124049160D-02
6	1,0	-6,698486578115D-09	-3,349243289057D-09	3,75	5,847834288030D-03
7	2,0	3,637992833067D-09	1,818996416533D-09	4,25	-4,127379440036D-03
8	3,0	-1,315289475103D-09	-6,576447375514D-10	4,50	-5,083371369389D-03
9	4,0	2,863369931408D-10	1,431684965704D-10	4,75	-3,200981517637D-03
10	5,0	-2,853173987759D-11	-1,426586993880D-11	5,25	2,705970611153D-03
				5,50	3,629884776803D-03
	0	1,099793107258D+00	8,409072330134D-01	5,75	2,485515507947D-03
	1	2,987261587636D+00	7,237582538128D-01	6,25	-2,482717999224D-03
	2	3,622137212175D+00	2,737108758361D+00	6,50	-3,625620848776D-03
	3	2,586493730869D+00	6,366045271782D-01	6,75	-2,708426881117D-03
	4	1.208638875599D+00	8,925572849511D-01	7,25	3,250863302166D-03
	5	3,668891672734D-01	1,079211685569D-01	7,50	5,237031817632D-03
	6	6,344380347948D-02	4,237169743630D-02	7,75	4,338583178968D-03
	7	4.525891558423D-03	1.647862264226D-03	8,25	-6,531436262500D-03

I= 2 H= -3.0

0		0,	-3,517665599508D+03	-1,00	9,722203368438D+01
1		-1,598945893674D+01	3,356228520647D+03	-,75	4,939364619490D+01
2		3,587201294249D+01	-1,359242587476D+03	-,50	2,155571890564D+01
3		-2,979642284579D+01	2,976333921853D+02	-,25	6,756217314207D+00
4		1,240507972236D+01	-3,622634949524D+01	.25	-2,175137962149D+00
5		-2,808615694970D+00	2,038923329225D+00	.50	-2,058642338561D+00

6		3.339241772339D-01	1.240391421070D-02	.75	-1.068679463995D+00
7		-1.651936478707D-02	-5.301118475096D-03	1.25	7.767316397047D-01
				1.50	1.159060610955D+00
0	-5.0	2.106597973973D-10	1.053298986987D-10	1.75	1.195236471573D+00
1	-4.0	-1.941808968799D-09	-9.709044843996D-10	2.25	6.985889280507D-01
2	-3.0	8.083063904775D-09	4.041531952387D-09	2.50	3.939093410055D-01
3	-2.0	-2.003856404911D-08	-1.001928202455D-08	2.75	1.518709593960D-01
4	-1.0	3.281755493575D-08	1.640877746787D-08	3.25	-6.532915668472D-02
5	0.0	-3.716540212606D-08	-1.858270106303D-08	3.50	-6.730476946997D-02
6	1.0	2.952832759384D-08	1.476416379692D-08	3.75	-3.637316286679D-02
7	2.0	-1.627944708693D-08	-8.139723543464D-09	4.25	2.378637525790D-02
8	3.0	5.968560613550D-09	2.984280306775D-09	4.50	2.851238238793D-02
9	4.0	-1.315289475103D-09	-6.576447375514D-10	4.75	1.756191100571D-02
10	5.0	1.323448606853D-10	6.617243034265D-11	5.25	-1.436030807471D-02
				5.50	-1.902229851172D-02
	0	-4.459151862247D+00	-3.585412036671D+00	5.75	-1.288828151356D-02
	1	-1.211242899742D+01	-2.468725773285D+00	6.25	1.266368141886D-02
	2	-1.462614268617D+01	-1.163917165455D+01	6.50	1.837528208504D-02
	3	-1.039941377335D+01	-2.114946529226D+00	6.75	1.365210923960D-02
	4	-4.860967132581D+00	-3.794191764145D+00	7.25	-1.624502279269D-02
	5	-1.463556891162D+00	-3.720464438794D-01	7.50	-2.608028782808D-02
	6	-2.442535903134D-01	-1.731352324177D-01	7.75	-2.154189194673D-02
	7	-1.651936478707D-02	-5.301118475096D-03	8.25	3.227364846908D-02

I= 3 H= -2.0

0		0.	1.020873935261D+04	-1.00	-1.649003587210D+02
1		2.427563759929D+01	-9.782672823383D+03	-.75	-8.204579351258D+01
2		-5.831578239512D+01	3.993100206180D+03	-.50	-3.493418926977D+01
3		5.252887011177D+01	-8.893214864843D+02	-.25	-1.063161238379D+01
4		-2.343735092519D+01	1.129436571767D+02	.25	3.158694925784D+00
5		5.609963537705D+00	-7.339753040802D+00	.50	2.824846517816D+00
6		-6.970460399213D-01	1.080319454758D-01	.75	1.360971246702D+00
7		3.570811198188D-02	9.093522791705D-03	1.25	-7.670086249600D-01
				1.50	-8.927295092849D-01
0	-5.0	-5.096186150621D-10	-2.548093075311D-10	1.75	-5.540232304590D-01
1	-4.0	4.751805266304D-09	2.375902633152D-09	2.25	5.432683809310D-01
2	-3.0	-2.003856404911D-08	-1.001928202455D-08	2.50	9.223043021300D-01
3	-2.0	5.039775661335D-08	2.519887830667D-08	2.75	1.070839212940D+00
4	-1.0	-8.383623612498D-08	-4.191811806249D-08	3.25	7.740305282528D-01
5	0.0	9.652054206082D-08	4.826027103041D-08	3.50	4.805555067174D-01
6	1.0	-7.798872085048D-08	-3.899436042524D-08	3.75	2.029093670530D-01
7	2.0	4.371488201392D-08	2.185744100696D-08	4.25	-1.033401054513D-01
8	3.0	-1.627944708693D-08	-8.139723543464D-09	4.50	-1.152628747588D-01
9	4.0	3.637992833067D-09	1.818996416533D-09	4.75	-6.727463043412D-02
10	5.0	-3.703920609074D-10	-1.851960304537D-10	5.25	5.108334210425D-02
				5.50	6.592291211831D-02
	0	1.073599398728D+01	9.182678741810D+00	5.75	4.372862165847D-02
	1	2.920516425760D+01	4.521120951448D+00	6.25	-4.162350721740D-02
	2	3.500727718314D+01	2.969710646026D+01	6.50	-5.967691132480D-02
	3	2.477258583473D+01	3.614071607514D+00	6.75	-4.389373372681D-02
	4	1.164246246770D+01	9.745972923813D+00	7.25	5.141627345193D-02
	5	3.445341130556D+00	6.753047891189D-01	7.50	8.203846866743D-02
	6	5.527378794446D-01	4.263052431855D-01	7.75	6.740506358480D-02
	7	3.570811198188D-02	9.093522791705D-03	8.25	-1.001278223388D-01

CARDINAL SPLINES N=8

I= 4	H= -1.0				
0		0.	-1,988829973012D+04	-1.00	1,990789354415D+02
1		-2,770893891668D+01	1.916972185216D+04	-.75	9,793653214563D+01
2		6,860816353196D+01	-7,885317071837D+03	-.50	4,116666033565D+01
3		-6,430484678300D+01	1.779976653133D+03	-.25	1,234460917374D+01
4		2,996816553772D+01	-2,329887101877D+02	.25	-3,533992156437D+00
5		-7,474789040102D+00	1,653885079269D+01	.50	-3,086459794622D+00
6		9,631386506836D-01	-4,350743946540D-01	.75	-1,445282443045D+00
7		-5,089298139061D-02	-7,783221331631D-03	1.25	7,533045487792D-01
				1.50	8,294576902863D-01
0	-5,0	8,127200424843D-10	4.063600212421D-10	1.75	4,783292255908D-01
1	-4,0	-7,673072457198D-09	-3,836536228599D-09	2.25	-3,646589808655D-01
2	-3,0	3,281755493575D-08	1.640877746787D-08	2.50	-4,835421096108D-01
3	-2,0	-8,383623612498D-08	-4.191811806249D-08	2,75	-3,383472363794D-01
4	-1.0	1.418289921568D-07	7.091449607840D-08	3,25	4,115040379631D-01
5	0,0	-1,661943918173D-07	-8,309719590866D-08	3,50	7,703243510520D-01
6	1.0	1.367124162916D-07	6.835620814581D-08	3.75	9,808965214737D-01
7	2,0	-7,798872085048D-08	-3,899436042524D-08	4,25	8,417344484863D-01
8	3,0	2,952832759384D-08	1.476416379692D-08	4.50	5,664917174975D-01
9	4,0	-6,698486578115D-09	-3,349243289057D-09	4.75	2,586397151160D-01
10	5,0	6,908968075989D-10	3.454484037994D-10	5.25	-1,532683518585D-01
				5,50	-1,842089319575D-01
	0	-1.703149761418D+01	-1.567234677439D+01	5,75	-1,158777278349D-01
	1	-4,669393306214D+01	-4.580980832517D+00	6,25	1,025480663065D-01
	2	-5,465306128294D+01	-5,000666190042D+01	6,50	1,432962715264D-01
	3	-3,907613726479D+01	-3,028805712130D+00	6.75	1,032190706011D-01
	4	-1,838136004241D+01	-1.672193169151D+01	7.25	-1,171619101161D-01
	5	-5,299444749666D+00	-5,995722460346D-01	7.50	-1,847108725186D-01
	6	-8,181156979878D-01	-7,074871412611D-01	7,75	-1,502296051041D-01
	7	-5,089298139061D-02	-7,783221331631D-03	8,25	2,195918021404D-01

I= 5	H= 0.0				
0		0.	2,698174846421D+04	-1.00	-1,760552962799D+02
1		2,375663327029D+01	-2,618763959609D+04	-.75	-8,605040487565D+01
2		-5,974823210376D+01	1.086310364955D+04	-.50	-3,590976497724D+01
3		5,720720131779D+01	-2,482155915998D+03	-.25	-1,068120260497D+01
4		-2,735061488561D+01	3,324297903024D+02	.25	2,998543397016D+00
5		7,013468148654D+00	-2,499863934030D+01	.50	2,587797028718D+00
6		-9,288011501334D-01	8,332879780101D-01	.75	1,195219476589D+00
7		5,034540366125D-02	-9,060120323966D-26	1.25	-6,015915491949D-01
				1.50	-6,476595572605D-01
0	-5,0	-8,948897434772D-10	-4,474448717386D-10	1.75	-3,634568296824D-01
1	-4,0	8,560817893470D-09	4,280408946735D-09	2.25	2,569245528436D-01
2	-3,0	-3,716540212606D-08	-1,858270106303D-08	2.50	3,227308534245D-01
3	-2,0	9,652054206082D-08	4,826027103041D-08	2.75	2,101280287953D-01
4	-1,0	-1,661943918173D-07	-8,309719590866D-08	3,25	-1,992263893495D-01
5	0,0	1,983466474651D-07	9,917332373257D-08	3,50	-2,916805689502D-01
6	1,0	-1,661943918173D-07	-8,309719590866D-08	3,75	-2,241243002681D-01
7	2,0	9,652054206082D-08	4,826027103041D-08	4.25	3,243674855798D-01
8	3,0	-3,716540212606D-08	-1,858270106303D-08	4,50	6,589036959921D-01
9	4,0	8,560817893470D-09	4,280408946735D-09	4.75	9,080733515204D-01
10	5,0	-8,948897434772D-10	-4,474448717386D-10	5,25	9,080733515204D-01
				5,50	6,589036959921D-01
	0	1,964787971976D+01	1,964787971976D+01	5.75	3,243674855798D-01
	1	5,010189325310D+01	-6,848274190308D-23	6,25	-2,241243002681D-01
	2	5,900907340738D+01	5,900907340738D+01	6.50	-2,916805689502D-01

	3	4.286477052521D+01	-5.068477164405D-23	6.75	-1.992263893495D-01
	4	1.994679854864D+01	1.994679854864D+01	7.25	2.101280287953D-01
	5	5.580770566804D+00	-6.225637318718D-24	7.50	3.227308534245D-01
	6	8.332879780101D-01	8.332879780101D-01	7.75	2.569245528436D-01
	7	5.034540366125D-02	-9.060120323966D-26	8.25	-3.634568296824D-01

I= 6 H= 1.0

0		0.	-2.565476583726D+04	-1.00	1.149770621368D+02
1		-1.524713222443D+01	2.508311087465D+04	-.75	5.598892061783D+01
2		3.867231996609D+01	-1.049665965806D+04	-.50	2.326924800981D+01
3		-3.746509675217D+01	2.426336320833D+03	-.25	6.889985854173D+00
4		1.817664083105D+01	-3.310705091412D+02	.25	-1.913628561748D+00
5		-4.740975553423D+00	2.591037768297D+01	.50	-1.641021351920D+00
6		6.395702709228D-01	-9.798998878682D-01	.75	-7.524955951501D-01
7		-3.532653872735D-02	7.783221331631D-03	1.25	3.721398372363D-01
				1.50	3.962995338535D-01
0	-5.0	6.908968075989D-10	3.454484037994D-10	1.75	2.195918021404D-01
1	-4.0	-6.698486578115D-09	-3.349243289057D-09	2.25	-1.502296051041D-01
2	-3.0	2.952832759384D-08	1.476416379692D-08	2.50	-1.847108725186D-01
3	-2.0	-7.798872085048D-08	-3.899436042524D-08	2.75	-1.171619101161D-01
4	-1.0	1.367124162916D-07	6.835620814581D-08	3.25	1.032190706011D-01
5	0.0	-1.661943918173D-07	-8.309719590866D-08	3.50	1.432962715264D-01
6	1.0	1.418289921568D-07	7.091449607840D-08	3.75	1.025480663065D-01
7	2.0	-8.383623612498D-08	-4.191811806249D-08	4.25	-1.158777278349D-01
8	3.0	3.281755493575D-08	1.640877746787D-08	4.50	-1.842089319575D-01
9	4.0	-7.673072457198D-09	-3.836536228599D-09	4.75	-1.532683518585D-01
10	5.0	8.127200424843D-10	4.063600212421D-10	5.25	2.586397151160D-01
				5.50	5.664917174975D-01
	0	-1.431319593460D+01	-1.567234677439D+01	5.75	8.417344484863D-01
	1	-3.753197139711D+01	4.580980832517D+00	6.25	9.808965214737D-01
	2	-4.536026251790D+01	-5.000666190042D+01	6.50	7.703243510520D-01
	3	-3.301852584053D+01	3.028805712130D+00	6.75	4.115040379631D-01
	4	-1.506250334061D+01	-1.672193169151D+01	7.25	-3.383472363794D-01
	5	-4.100300257597D+00	5.995722460346D-01	7.50	-4.835421096108D-01
	6	-5.968585845344D-01	-7.074871412611D-01	7.75	-3.646589808655D-01
	7	-3.532653872735D-02	7.783221331631D-03	8.25	4.783292255908D-01

I= 7 H= 2.0

0		0.	1.679898633220D+04	-1.00	-5.464107494416D+01
1		7.176429951007D+00	-1.654368884909D+04	-.75	-2.655264938964D+01
2		-1.828592943095D+01	6.983309193614D+03	-.50	-1.101027719203D+01
3		1.782947325114D+01	-1.632043646396D+03	-.25	-3.251951802541D+00
4		-8.721243323924D+00	2.262772210601D+02	.25	8.979229000884D-01
5		2.297113203723D+00	-1.823856155033D+01	.50	7.673644136569D-01
6		-3.133647170201D-01	7.445785408952D-01	.75	3.505226145231D-01
7		1.752106639847D-02	-9.093522791705D-03	1.25	-1.717467312017D-01
				1.50	-1.818713037189D-01
0	-5.0	-3.703920609074D-10	-1.851960304537D-10	1.75	-1.001278223388D-01
1	-4.0	3.637992833067D-09	1.818996416533D-09	2.25	6.740506358480D-02
2	-3.0	-1.627944708693D-08	-8.139723543464D-09	2.50	8.203846866743D-02
3	-2.0	4.371488201392D-08	2.185744100696D-08	2.75	5.141627345193D-02
4	-1.0	-7.798872085048D-08	-3.899436042524D-08	3.25	-4.389373372681D-02
5	0.0	9.652054206082D-08	4.826027103041D-08	3.50	-5.967691132480D-02
6	1.0	-8.383623612498D-08	-4.191811806249D-08	3.75	-4.162350721740D-02
7	2.0	5.039775661335D-08	2.519887830667D-08	4.25	4.372862165847D-02

CARDINAL SPLINES N=8

```
 8   3,0  -2,003856404911D-08  -1,001928202455D-08    4.50  6,592291211831D-02
 9   4,0   4,751805266304D-09   2,375902633152D-09    4.75  5,108334210425D-02
10   5,0  -5,096186150621D-10  -2,548093075311D-10    5.25 -6,727463043412D-02
                                                      5.50 -1,152628747588D-01
       0   7,629363496341D+00   9,182678741810D+00    5.75 -1,033401054513D-01
       1   2,016292235470D+01  -4,521120951448D+00    6,25  2,029093670530D-01
       2   2,438693573738D+01   2,969710646026D+01    6,50  4,805555067174D-01
       3   1,754444261970D+01  -3,614071607514D+00    6.75  7,740305282528D-01
       4   7,849483379926D+00   9,745972923813D+00    7.25  1,070839212940D+00
       5   2,094731552318D+00  -6,753047891189D-01    7.50  9,223043021300D-01
       6   2,998726069264D-01   4,263052431855D-01    7.75  5,432683809310D-01
       7   1,752106639847D-02  -9,093522791705D-03    8.25 -5,540232304590D-01


I= 8     H= 3.0
0              0.              -7,224679525527D+03   -1.00  1,801368655991D+01
1             -2,353297491727D+00   7,163317892251D+03    -.75  8,743612492834D+00
2              6,011470650146D+00  -3,048578892908D+03    -.50  3,621040334566D+00
3             -5,882178778355D+00   7,198104404689D+02    -.25  1,068014053142D+00
4              2,890290029976D+00  -1,012134583463D+02     .25 -2,939485211515D-01
5             -7,654498821044D-01   8,349190615835D+00     .50 -2,507349245033D-01
6              1,050825997687D-01  -3,586743790460D-01     .75 -1,142914615743D-01
7             -5,917127836874D-03   5,301118475096D-03    1.25  5,571833096595D-02
                                                      1.50  5,882445235572D-02
0  -5,0   1,323448606853D-10   6,617243034265D-11    1.75  3,227364846908D-02
1  -4,0  -1,315289475103D-09  -6,576447375514D-10    2.25 -2,154189194673D-02
2  -3,0   5,968560613550D-09   2,984280306775D-09    2.50 -2,608028782808D-02
3  -2,0  -1,627944708693D-08  -8,139723543464D-09    2.75 -1,624502279269D-02
4  -1,0   2,952832759384D-08   1,476416379692D-08    3.25  1,365210923960D-02
5   0,0  -3,716540212606D-08  -1,858270106303D-08    3.50  1,837528208504D-02
6   1,0   3,281755493575D-08   1,640877746787D-08    3.75  1,266368141886D-02
7   2,0  -2,003856404911D-08  -1,001928202455D-08    4.25 -1,288828151356D-02
8   3,0   8,083063904775D-09   4,041531952387D-09    4,50 -1,902229851172D-02
9   4,0  -1,941808968799D-09  -9,709044843996D-10    4,75 -1,436030807471D-02
10  5,0   2,106597973973D-10   1,053298986987D-10    5,25  1,756191100571D-02
                                                      5,50  2,851238238793D-02
       0  -2,711672211095D+00  -3,585412036671D+00    5,75  2,378637525790D-02
       1  -7,174977450853D+00   2,468725773285D+00    6.25 -3,637316286679D-02
       2  -8,652200622932D+00  -1,163917165455D+01    6,50 -6,730476946997D-02
       3  -6,169520714897D+00   2,114946529226D+00    6,75 -6,532915668472D-02
       4  -2,727416395709D+00  -3,794191764145D+00    7.25  1,518709593960D-01
       5  -7,194640034033D-01   3,720464438794D-01    7.50  3,939093410055D-01
       6  -1,020168745219D-01  -1,731352324177D-01    7,75  6,985889280507D-01
       7  -5,917127836874D-03   5,301118475096D-03    8.25  1,195236471573D+00


I= 9     H= 4.0
0              0.               1,838361950028D+03   -1.00 -3,693001453276D+00
1              4,809701523408D-01  -1,834076741672D+03    -.75 -1,791343246561D+00
2             -1,230415428576D+00   7,864468546599D+02    -.50 -7,413187785547D-01
3              1,206403628181D+00  -1,874442729861D+02    -.25 -2,184747396296D-01
4             -5,943290433917D-01   2,668937044347D+01     .25  6,001899223493D-02
5              1,578967791008D-01  -2,244199780364D+00     .50  5,114005088480D-02
6             -2,175625465586D-02   1,000468766842D-01     .75  2,328274982325D-02
7              1,230167029971D-03  -1,647862264226D-03    1.25 -1,131787478747D-02
                                                      1.50 -1,192815024389D-02
0  -5,0  -2,853173987759D-11  -1,426586993880D-11    1.75 -6,531436262500D-03
```

CARDINAL SPLINES N=8

1	-4.0	2.863369931408D-10	1.431684965704D-10	2.25	4.338583178968D-03
2	-3.0	-1.315289475103D-09	-6.576447375514D-10	2.50	5.237031817632D-03
3	-2.0	3.637992833067D-09	1.818996416533D-09	2.75	3.250863302166D-03
4	-1.0	-6.698486578115D-09	-3.349243289057D-09	3.25	-2.708426881117D-03
5	0.0	8.560817893470D-09	4.280408946735D-09	3.50	-3.625620848776D-03
6	1.0	-7.673072457198D-09	-3.836536228599D-09	3.75	-2.482717999224D-03
7	2.0	4.751805266304D-09	2.375902633152D-09	4.25	2.485515507947D-03
8	3.0	-1.941808968799D-09	-9.709044843996D-10	4.50	3.629884776803D-03
9	4.0	4.719723986804D-10	2.359861993402D-10	4.75	2.705970611153D-03
10	5.0	-5.173616557005D-11	-2.586808278502D-11	5.25	-3.200981517637D-03
				5.50	-5.083371369389D-03
	0	5.820213587686D-01	8.409072330134D-01	5.75	-4.127379440036D-03
	1	1.539745080010D+00	-7.237582538128D-01	6.25	5.847834288030D-03
	2	1.852080304547D+00	2.737108758361D+00	6.50	1.024124049160D-02
	3	1.313284676512D+00	-6.366045271782D-01	6.75	9.237893800379D-03
	4	5.764756943034D-01	8.925572849511D-01	7.25	-1.667990817642D-02
	5	1.510468301596D-01	-1.079211685569D-01	7.50	-3.375512093861D-02
	6	2.129959139311D-02	4.237169743630D-02	7.75	-3.602554654719D-02
	7	1.230167029971D-03	-1.647862264226D-03	8.25	1.032899101055D-01

I=10 H= 5.0

0		0.	-2.098548360749D+02	-1.00	3.539969565231D-01
1		-4.601765298836D-02	2.105337605701D+02	-.75	1.716416714980D-01
2		1.178259971543D-01	-9.089262348423D+01	-.50	7.099964906017D-02
3		-1.156695951381D-01	2.184896810306D+01	-.25	2.091417830745D-02
4		5.707446094072D-02	-3.146106029900D+00	.25	-5.739010013152D-03
5		-1.519229976913D-02	2.689245911583D-01	.50	-4.886779735275D-03
6		2.098020165125D-03	-1.234198113188D-02	.75	-2.223187355870D-03
7		-1.189303673668D-04	2.183835766790D-04	1.25	1.078810143172D-03
				1.50	1.135788020028D-03
0	-5.0	2.823131758485D-12	1.411565879243D-12	1.75	6.211766312419D-04
1	-4.0	-2.853173987759D-11	-1.426586993880D-11	2.25	-4.114220608944D-04
2	-3.0	1.323448606853D-10	6.617243034265D-11	2.50	-4.957325435627D-04
3	-2.0	-3.703920609074D-10	-1.851960304537D-10	2.75	-3.070888738456D-04
4	-1.0	6.908968075989D-10	3.454484037994D-10	3.25	2.545310200274D-04
5	0.0	-8.948897434772D-10	-4.474448717386D-10	3.50	3.396299912535D-04
6	1.0	8.127200424843D-10	4.063600212421D-10	3.75	2.316969057677D-04
7	2.0	-5.096186150621D-10	-2.548093075311D-10	4.25	-2.297648629080D-04
8	3.0	2.106597973973D-10	1.053298986987D-10	4.50	-3.335487061624D-04
9	4.0	-5.173616557005D-11	-2.586808278502D-11	4.75	-2.469121738790D-04
10	5.0	5.723684970042D-12	2.861842485021D-12	5.25	2.868937013506D-04
				5.50	4.504325308616D-04
	0	-5.740628936043D-02	-8.976702364103D-02	5.75	3.608126122794D-04
	1	-1.518144596004D-01	9.002335315517D-02	6.25	-4.934790942503D-04
	2	-1.822898106099D-01	-2.929183673366D-01	6.50	-8.444112700434D-04
	3	-1.288066919396D-01	8.018320478617D-02	6.75	-7.404642348373D-04
	4	-5.629582859498D-02	-9.580602742595D-02	7.25	1.235828453312D-03
	5	-1.469013768291D-02	1.331653495850D-02	7.50	2.364126394718D-03
	6	-2.064542692711D-03	-4.698555948109D-03	7.75	2.341937935393D-03
	7	-1.189303673668D-04	2.183835766790D-04	8.25	-5.202915767540D-03

CARDINAL SPLINES N=8

N= 8 M=11 P= 5.5 Q=15

I= 0 H= -5.5

0		1,000000000000D+00	1.353420266183D+02	-1.00	9,304235958091D+00
1		-2,766730573830D+00	-1.140000353880D+02	-.75	5,751953788631D+00
2		3.045502273168D+00	4.178365742382D+01	-.50	3,406693152668D+00
3		-1,767765077831D+00	-8,866419532541D+00	-.25	1,912085398369D+00
4		5,939489404389D-01	1.244472096828D+00	.25	4,732487968185D-01
5		-1,170351493127D-01	-1.258134847293D-01	.50	1,906974378908D-01
6		1,266676543539D-02	8,980412918899D-03	.75	5,460340677454D-02
7		-5,871780750577D-04	-3.380534795992D-04	1.25	-1,405436657518D-02
				1.50	-1,154214790741D-02
0	-5.5	6,543701347978D-12	3.271850673989D-12	1.75	-5,048569919265D-03
1	-4,5	-6,029082012180D-11	-3,014541006090D-11	2.25	2,240852238657D-03
2	-3,5	2.523876081424D-10	1.261938040712D-10	2.50	2,243159882739D-03
3	-2,5	-6,361919528736D-10	-3,180959764368D-10	2.75	1,161483820049D-03
4	-1,5	1.079058699645D-09	5,395293498223D-10	3.25	-6,800405770559D-04
5	-,5	-1,302713098203D-09	-6,513565491015D-10	3.50	-7,647582561162D-04
6	,5	1,152053279586D-09	5,760266397928D-10	3.75	-4,398955733179D-04
7	1,5	-7,523755564913D-10	-3,761877782457D-10	4,25	3,094931872100D-04
8	2,5	3,575548757671D-10	1.787774378835D-10	4.50	3,775120319583D-04
9	3,5	-1,178761850242D-10	-5,893809251208D-11	4,75	2,342253855913D-04
10	4,5	2,416589858789D-11	1.208294929394D-11	5.25	-1,892977392980D-04
11	5,5	-2,316450361580D-12	-1.158225180790D-12	5,50	-2,463437635923D-04
				5,75	-1,627540393198D-04
	0	-4,915837890438D-01	-1,658130256936D-01	6.25	1,487818321280D-04
	1	-1.165397534344D+00	-7.678731933568D-01	6.50	2,059918808894D-04
	2	-1.233573406072D+00	-4,269552394423D-01	6.75	1,449713736904D-04
	3	-7.610213747688D-01	-4.911907214411D-01	7.25	-1,513671797031D-04
	4	-2,961841776729D-01	-1.090640396672D-01	7.50	-2,250104484216D-04
	5	-7.203676212529D-02	-4,420833132098D-02	7.75	-1,706897663044D-04
	6	-9,939590454334D-03	-4.034646045670D-03	8.25	2,100148938938D-04
	7	-5,871780750577D-04	-3,380534795992D-04	8.50	3,417162485435D-04

I= 1 H= -4.5

0		0.	-1.418455362354D+03	-1.00	-3,951545128561D+01
1		8,532278267530D+00	1.189490747406D+03	-.75	-2,096014103013D+01
2		-1,503050479831D+01	-4.314352291997D+02	-.50	-9,665778447574D+00
3		1.083974011258D+01	8.950842747402D+01	-.25	-3,258851369694D+00
4		-4,126521499782D+00	-1,204857536440D+01	.25	1,347750651086D+00
5		8,808529566675D-01	1.149924261001D+00	.50	1.631564006424D+00
6		-1,006993446800D-01	-7,826806719235D-02	.75	1.403664105665D+00
7		4,854306056793D-03	2.896294255671D-03	1.25	6,041956248165D-01
				1.50	2,982784025934D-01
0	-5,5	-6,029082012180D-11	-3,014541006090D-11	1.75	1,016797283469D-01
1	-4,5	5,612170972852D-10	2,806085486426D-10	2.25	-3,497099051762D-02
2	-3,5	-2,377125891400D-09	-1,188562945700D-09	2.50	-3,249316233613D-02
3	-2,5	6,072255942138D-09	3,036127971069D-09	2.75	-1,590432286806D-02
4	-1,5	-1,045159648073D-08	-5,225798240365D-09	3,25	8,607536245357D-03

5	-.5	1.281535232923D-08	6.407676164615D-09	3.50	9.410276578575D-03
6	.5	-1.150940794298D-08	-5.754703971490D-09	3.75	5.288835073408D-03
7	1.5	7.622958179024D-09	3.811479089512D-09	4.25	-3.591999946314D-03
8	2.5	-3.664747281628D-09	-1.832373640814D-09	4.50	-4.322643053901D-03
9	3.5	1.218404796800D-09	6.092023983998D-10	4.75	-2.651529995368D-03
10	4.5	-2.511858262066D-10	-1.255929131033D-10	5.25	2.104919707418D-03
11	5.5	2.416589858789D-11	1.208294929394D-11	5.50	2.720136899852D-03
				5.75	1.786438681618D-03
	0	4.501025962267D+00	1.414123669072D+00	6.25	-1.617723309841D-03
	1	1.064833098310D+01	7.262206162410D+00	6.50	-2.231468366125D-03
	2	1.123799200492D+01	3.629811493465D+00	6.75	-1.565468090972D-03
	3	6.906987349973D+00	4.615024008884D+00	7.25	1.626328310614D-03
	4	2.671838116999D+00	9.256898121441D-01	7.50	2.412778466390D-03
	5	6.414725048061D-01	4.069489695682D-01	7.75	1.827192339614D-03
	6	8.619143850655D-02	3.323926165097D-02	8.25	-2.242074827103D-03
	7	4.854306056793D-03	2.896294255671D-03	8.50	-3.644137537010D-03

I= 2 H= -3.5

0		0.	7.010765768029D+03	-1.00	1.018359478728D+02
1		-1.658491564476D+01	-5.870853785182D+03	-.75	5.162728174873D+01
2		3.740184598375D+01	2.117154170535D+03	-.50	2.247808033716D+01
3		-3.130543035332D+01	-4.322056506438D+02	-.25	7.027403740729D+00
4		1.315422016756D+01	5.607652733219D+01	.25	-2.249228934916D+00
5		-3.009506636630D+00	-5.024245399994D+00	.50	-2.121570206061D+00
6		3.619077849833D-01	3.195319562041D-01	.75	-1.097226446496D+00
7		-1.812130183210D-02	-1.139070789708D-02	1.25	7.904906199495D-01
				1.50	1.173487673524D+00
0	-5.5	2.523876081424D-10	1.261938040712D-10	1.75	1.203090600490D+00
1	-4.5	-2.377125891400D-09	-1.188562945700D-09	2.25	6.934450249093D-01
2	-3.5	1.020644855468D-08	5.103224277339D-09	2.50	3.877537716118D-01
3	-2.5	-2.647944550914D-08	-1.323972275457D-08	2.75	1.480878169245D-01
4	-1.5	4.637061238572D-08	2.318530619286D-08	3.25	-6.225440473553D-02
5	-.5	-5.791812880483D-08	-2.895906440242D-08	3.50	-6.325150177940D-02
6	.5	5.299499302236D-08	2.649749651118D-08	3.75	-3.364648193616D-02
7	1.5	-3.571727228207D-08	-1.785863614104D-08	4.25	2.117490637208D-02
8	2.5	1.742872304455D-08	8.714361522277D-09	4.50	2.480170921170D-02
9	3.5	-5.861720739787D-09	-2.930860369894D-09	4.75	1.488180171594D-02
10	4.5	1.218404796800D-09	6.092023983998D-10	5.25	-1.142850782404D-02
11	5.5	-1.178761850242D-10	-5.893809251208D-11	5.50	-1.458473916838D-02
				5.75	-9.478331371347D-03
	0	-1.870674942167D+01	-5.258542370335D+00	6.25	8.444418738439D-03
	1	-4.414492960651D+01	-3.157273740934D+01	6.50	1.157489866875D-02
	2	-4.642511770185D+01	-1.344269243043D+01	6.75	8.076995972770D-03
	3	-2.841452123655D+01	-1.992603614210D+01	7.25	-8.320881815843D-03
	4	-1.091416063466D+01	-3.432111963215D+00	7.50	-1.230417230830D-02
	5	-2.578106721024D+00	-1.715638036880D+00	7.75	-9.291854143620D-03
	6	-3.357623355526D-01	-1.190102978336D-01	8.25	1.135104299031D-02
	7	-1.812130183210D-02	-1.139070789708D-02	8.50	1.841687332430D-02

I= 3 H= -2.5

0		0.	-2.172724473841D+04	-1.00	-1.788937854827D+02
1		2.608184170418D+01	1.820591763598D+04	-.75	-8.882110001572D+01
2		-6.295623850504D+01	-6.551884861609D+03	-.50	-3.773199636704D+01
3		5.710615594584D+01	1.324505744480D+03	-.25	-1.145420457257D+01
4		-2.570972527089D+01	-1.670394763986D+02	.25	3.383435729513D+00

5		6.219327809831D+00	1.408501920628D+01	.50	3.015726175497D+00
6		-7.819289650879D-01	-8.235822637645D-01	.75	1.447563061490D+00
7		4.056728180492D-02	2.756512830562D-02	1.25	-8.087438615651D-01
				1.50	-9.364912459174D-01
0	-5.5	-6.361919528736D-10	-3.180959764368D-10	1.75	-5.778472292463D-01
1	-4.5	6.072255942138D-09	3.036127971069D-09	2.25	5.588714278763D-01
2	-3.5	-2.647944550914D-08	-1.323972275457D-08	2.50	9.409760452763D-01
3	-2.5	6.993493937678D-08	3.496746968839D-08	2.75	1.082314652496D+00
4	-1.5	-1.249468452934D-07	-6.247342264669D-08	3.25	7.647038559400D-01
5	-.5	1.594699716503D-07	7.973498582516D-08	3.50	4.682606939420D-01
6	.5	-1.491703648801D-07	-7.458518244006D-08	3.75	1.946385016428D-01
7	1.5	1.026758092903D-07	5.133790464517D-08	4.25	-9.541871375673D-02
8	2.5	-5.104165926277D-08	-2.552082963138D-08	4.50	-1.040072569573D-01
9	3.5	1.742872304455D-08	8.714361522277D-09	4.75	-5.914503119448D-02
10	4.5	-3.664747281628D-09	-1.832373640814D-09	5.25	4.219028644431D-02
11	5.5	3.575548757671D-10	1.787774378835D-10	5.50	5.246242406984D-02
				5.75	3.338518969873D-02
	0	4.677146768595D+01	1.089603146487D+01	6.25	-2.882518036948D-02
	1	1.100335844789D+02	8.405114386882D+01	6.50	-3.904924809425D-02
	2	1.152233417184D+02	2.765763979730D+01	6.75	-2.698269328976D-02
	3	7.022617186152D+01	5.271912363915D+01	7.25	2.737990646266D-02
	4	2.674987445108D+01	7.113287355599D+00	7.50	4.025125548082D-02
	5	6.186037728504D+00	4.417552258198D+00	7.75	3.024691544882D-02
	6	7.799113844014D-01	2.376751760019D-01	8.25	-3.666309788105D-02
	7	4.056728180492D-02	2.756512830562D-02	8.50	-5.930263645447D-02

I= 4   H= -1.5

0		0.	4.731216495668D+04	-1.00	2.285284485424D+02
1		-3.150959697131D+01	-3.972454235580D+04	-.75	1.121932948027D+02
2		7.837272272189D+01	1.430369375217D+04	-.50	4.705387322079D+01
3		-7.393648156141D+01	-2.878411772698D+03	-.25	1.407552741244D+01
4		3.474975020841D+01	3.561585234503D+02	.25	-4.006897152933D+00
5		-8.757027975770D+00	-2.854366949882D+01	.50	-3.488113354407D+00
6		1.141751340335D+00	1.525250597723D+00	.75	-1.627491019279D+00
7		-6.111776322822D-02	-4.665161726108D-02	1.25	8.411248359414D-01
				1.50	9.215421874737D-01
0	-5.5	1.079058699645D-09	5.395293498223D-10	1.75	5.284602559872D-01
1	-4.5	-1.045159648073D-08	-5.225798240365D-09	2.25	-3.974912869648D-01
2	-3.5	4.637061238572D-08	2.318530619286D-08	2.50	-5.228316399528D-01
3	-2.5	-1.249468452934D-07	-6.247342264669D-08	2.75	-3.624941318843D-01
4	-1.5	2.283349232657D-07	1.141674616329D-07	3.25	4.311294458296D-01
5	-.5	-2.986540933292D-07	-1.493270466646D-07	3.50	7.961953913616D-01
6	.5	2.864945411301D-07	1.432472705651D-07	3.75	9.983002750999D-01
7	1.5	-2.020557200088D-07	-1.010278600044D-07	4.25	8.250660658454D-01
8	2.5	1.026758092903D-07	5.133790464517D-08	4.50	5.428073762088D-01
9	3.5	-3.571727228207D-08	-1.785863614104D-08	4.75	2.415332171586D-01
10	4.5	7.622958179024D-09	3.811479089512D-09	5.25	-1.345553705666D-01
11	5.5	-7.523755564913D-10	-3.761877782457D-10	5.50	-1.558850497439D-01
				5.75	-9.411282990125D-02
	0	-7.864534067618D+01	-1.285338147791D+01	6.25	7.561751959331D-02
	1	-1.842414399236D+02	-1.538288881843D+02	6.50	9.989104717058D-02
	2	-1.919165231370D+02	-3.196055115237D+01	6.75	6.763445318430D-02
	3	-1.166954171716D+02	-9.623300946023D+01	7.25	-6.658400944902D-02
	4	-4.389522344642D+01	-8.368127813941D+00	7.50	-9.678121498549D-02
	5	-9.904292835431D+00	-7.845839639067D+00	7.75	-7.204053742040D-02
	6	-1.211282543951D+00	-2.708366668287D-01	8.25	8.604780479176D-02

	7	-6,111776322822D-02	-4,665161726108D-02	8,50	1,383876006330D-01

I= 5 H= -.5

0		0.	-7.591701673908D+04	-1.00	-2,211490102599D+02
1		2,957628083677D+01	6,399249669910D+04	-,75	-1,078806597712D+02
2		-7,469993054932D+01	-2,311317650735D+04	-,50	-4,492438919440D+01
3		7,195536310814D+01	4,650866438673D+03	-,25	-1,333162102485D+01
4		-3,467227792205D+01	-5,696848475476D+02	.25	3,722665476004D+00
5		8.976859381001D+00	4.403266411674D+01	.50	3,202817408920D+00
6		-1,202296657947D+00	-2,168402889053D+00	.75	1,474221081582D+00
7		6,600180470649D-02	5,951610550929D-02	1,25	-7,360638203452D-01
				1,50	-7,886612765062D-01
0	-5,5	-1,302713098203D-09	-6,513565491015D-10	1,75	-4,402185180159D-01
1	-4,5	1,281535232923D-08	6,407676164615D-09	2,25	3,071980706256D-01
2	-3,5	-5,791812880483D-08	-2,895906440242D-08	2,50	3,828918088055D-01
3	-2,5	1,594699716503D-07	7,973498582516D-08	2,75	2,471022631629D-01
4	-1,5	-2,986540933292D-07	-1,493270466646D-07	3,25	-2,292772267282D-01
5	-,5	4,011717289328D-07	2,005858644664D-07	3,50	-3,312948501131D-01
6	,5	-3,955439322891D-07	-1,977719661446D-07	3,75	-2,507732939378D-01
7	1,5	2,864945411301D-07	1,432472705651D-07	4,25	3,498904637270D-01
8	2,5	-1,491703648801D-07	-7,458518244006D-08	4,50	6,951696576133D-01
9	3,5	5,299499302236D-08	2,649749651118D-08	4,75	9,342671810568D-01
10	4,5	-1,150940794298D-08	-5,754703971490D-09	5,25	8,794196409072D-01
11	5,5	1,152053279586D-09	5,760266397928D-10	5,50	6,155335717062D-01
				5,75	2,910406160026D-01
	0	9,440508534684D+01	6,467581740000D+00	6,25	-1,828876806116D-01
	1	2,183405595861D+02	2,043268567394D+02	6,50	-2,252175763145D-01
	2	2,278583060096D+02	1,454274753147D+01	6,75	-1,447384745787D-01
	3	1,378752569313D+02	1,284535450356D+02	7,25	1,326820833025D-01
	4	5,098600554362D+01	3,870326649080D+00	7,50	1,880911120614D-01
	5	1,122871610855D+01	1,028297480279D+01	7,75	1,371998075677D-01
	6	1,338772823253D+00	1,229671730551D-01	8,25	-1,589714516194D-01
	7	6,600180470649D-02	5,951610550929D-02	8,50	-2,527393795017D-01

I= 6 H= .5

0		0.	9,070058993888D+04	-1,00	1,659679085580D+02
1		-2,182784445895D+01	-7,689033673510D+04	-,75	8,067402806332D+01
2		5,557932743750D+01	2,792287003984D+04	-,50	3,346276010787D+01
3		-5,414195005328D+01	-5,639507347915D+03	-,25	9,887012860202D+00
4		2,645579467508D+01	6,890182103932D+02	.25	-2,732447737659D+00
5		-6,961129361063D+00	-5,214849753838D+01	.50	-2,336471049558D+00
6		9,488321658729D-01	2,414337235163D+00	.75	-1,067983647382D+00
7		-5,303040631210D-02	-5,951610550929D-02	1,25	5,241977255969D-01
				1,50	5,557407591731D-01
0	-5,5	1,152053279586D-09	5,760266397928D-10	1,75	3,063920036626D-01
1	-4,5	-1,150940794298D-08	-5,754703971490D-09	2,25	-2,070776443987D-01
2	-3,5	5,299499302236D-08	2,649749651118D-08	2,50	-2,527393795017D-01
3	-2,5	-1,491703648801D-07	-7,458518244006D-08	2,75	-1,589714516194D-01
4	-1,5	2,864945411301D-07	1.432472705651D-07	3,25	1,371998075677D-01
5	-,5	-3,955439322891D-07	-1.977719661446D-07	3,50	1,880911120614D-01
6	,5	4,011717289328D-07	2,005858644664D-07	3,75	1,326820833025D-01
7	1,5	-2,986540933292D-07	-1,493270466646D-07	4,25	-1,447384745787D-01
8	2,5	1,594699716503D-07	7,973498582516D-08	4,50	-2,252175763145D-01
9	3,5	-5,791812880483D-08	-2,895906440242D-08	4,75	-1,828876806116D-01
10	4,5	1,281535232923D-08	6,407676164615D-09	5,25	2,910406160026D-01

11	5.5	-1.302713098203D-09	-6.513565491015D-10	5.50	6.155335717062D-01
				5.75	8.794196409072D-01
	0	-8.146992186684D+01	6.467581740000D+00	6.25	9.342671810568D-01
	1	-1.903131538927D+02	-2.043268567394D+02	6.50	6.951696576133D-01
	2	-1.987728109467D+02	1.454274753147D+01	6.75	3.498904637270D-01
	3	-1.190318331398D+02	-1.284535450356D+02	7.25	-2.507732939378D-01
	4	-4.324535224546D+01	3.870326649080D+00	7.50	-3.312948501131D-01
	5	-9.337233497017D+00	-1.028297480279D+01	7.75	-2.292772267282D-01
	6	-1.092838477143D+00	1.229671730551D-01	8.25	2.471022631629D-01
	7	-5.303040631210D-02	-5.951610550929D-02	8.50	3.828918088055D-01

I= 7 H= 1.5

0		0.	-7.958006986496D+04	-1.00	-9.687763225594D+01
1		1.262734220199D+01	6.792259611254D+04	-.75	-4.699973016089D+01
2		-3.229028197960D+01	-2.484021957413D+04	-.50	-1.945373103042D+01
3		3.164318531258D+01	5.049027414256D+03	-.25	-5.734438597512D+00
4		-1.557900271347D+01	-6.186790542252D+02	.25	1.576164299842D+00
5		4.136103142481D+00	4.641888950951D+01	.50	1.343416844109D+00
6		-5.695314345228D-01	-2.066923931380D+00	.75	6.118465605744D-01
7		3.218547129394D-02	4.665161726108D-02	1.25	-2.976987827405D-01
				1.50	-3.139390947777D-01
0	-5.5	-7.523755564913D-10	-3.761877782457D-10	1.75	-1.720258578407D-01
1	-4.5	7.622958179024D-09	3.811479089512D-09	2.25	1.144932301529D-01
2	-3.5	-3.571727228207D-08	-1.785863614104D-08	2.50	1.383876006330D-01
3	-2.5	1.026758092903D-07	5.133790464517D-08	2.75	8.604780479176D-02
4	-1.5	-2.020557200088D-07	-1.010278600044D-07	3.25	-7.204053742040D-02
5	-.5	2.864945411301D-07	1.432472705651D-07	3.50	-9.678121498549D-02
6	.5	-2.986540933292D-07	-1.493270466646D-07	3.75	-6.658400944902D-02
7	1.5	2.283349232657D-07	1.141674616329D-07	4.25	6.763445318430D-02
8	2.5	-1.249468452934D-07	-6.247342264669D-08	4.50	9.989104717058D-02
9	3.5	4.637061238572D-08	2.318530619286D-08	4.75	7.561751959331D-02
10	4.5	-1.045159648073D-08	-5.225798240365D-09	5.25	-9.411282990125D-02
11	5.5	1.079058699645D-09	5.395293498223D-10	5.50	-1.558850497439D-01
				5.75	-1.345553705666D-01
	0	5.293857772035D+01	-1.285338147791D+01	6.25	2.415332171586D-01
	1	1.234163364450D+02	1.538288881843D+02	6.50	5.428073762088D-01
	2	1.279954208322D+02	-3.196055115237D+01	6.75	8.250660658454D-01
	3	7.577060174882D+01	9.623300946023D+01	7.25	9.983002750999D-01
	4	2.715896781854D+01	-8.368127813941D+00	7.50	7.961953913616D-01
	5	5.787386442703D+00	7.845839639067D+00	7.75	4.311294458296D-01
	6	6.696092102938D-01	-2.708366668287D-01	8.25	-3.624941318843D-01
	7	3.218547129394D-02	4.665161726108D-02	8.50	-5.228316399528D-01

I= 8 H= 2.5

0		0.	4.959853129552D+04	-1.00	4.291553655908D+01
1		-5.567048987395D+00	-4.263635547067D+04	-.75	2.079881231719D+01
2		1.426816363598D+01	1.571395267349D+04	-.50	8.599135954586D+00
3		-1.402647385303D+01	-3.219218684419D+03	-.25	2.531639824933D+00
4		6.933491145737D+00	3.969562733315D+02	.25	-6.938264154414D-01
5		-1.849682464483D+00	-2.977158082240D+01	.50	-5.903641965786D-01
6		2.561134976456D-01	1.298932615768D+00	.75	-2.683629607669D-01
7		-1.456297480632D-02	-2.756512830562D-02	1.25	1.299772027616D-01
				1.50	1.366890448316D-01
0	-5.5	3.575548757671D-10	1.787774378835D-10	1.75	7.466332720371D-02
1	-4.5	-3.664747281628D-09	-1.832373640814D-09	2.25	-4.930415609051D-02

2	-3.5	1.742872304455D-08	8.714361522277D-09	2.50	-5.930263645447D-02
3	-2.5	-5.104165926277D-08	-2.552082963138D-08	2.75	-3.666309788105D-02
4	-1.5	1.026758092903D-07	5.133790464517D-08	3.25	3.024691544882D-02
5	-.5	-1.491703648801D-07	-7.458518244006D-08	3.50	4.025125548082D-02
6	.5	1.594699716503D-07	7.973498582516D-08	3.75	2.737990646266D-02
7	1.5	-1.249468452934D-07	-6.247342264669D-08	4.25	-2.698269328976D-02
8	2.5	6.993493937678D-08	3.496746968839D-08	4.50	-3.904924809425D-02
9	3.5	-2.647944550914D-08	-1.323972275457D-08	4.75	-2.882518036948D-02
10	4.5	6.072255942138D-09	3.036127971069D-09	5.25	3.338518969873D-02
11	5.5	-6.361919528736D-10	-3.180959764368D-10	5.50	5.246242406984D-02
				5.75	4.219028644431D-02
	0	-2.497940475622D+01	1.089603146487D+01	6.25	-5.914503119448D-02
	1	-5.806870325875D+01	-8.405114386882D+01	6.50	-1.040072569573D-01
	2	-5.990806212379D+01	2.765763979730D+01	6.75	-9.541871375673D-02
	3	-3.521207541677D+01	-5.271912363915D+01	7.25	1.946385016428D-01
	4	-1.252329973988D+01	7.113287355599D+00	7.50	4.682606939420D-01
	5	-2.649066787892D+00	-4.417552258198D+00	7.75	7.647038559400D-01
	6	-3.045610323977D-01	2.376751760019D-01	8.25	1.082314652496D+00
	7	-1.456297480632D-02	-2.756512830562D-02	8.50	9.409760452763D-01

I= 9 H= 3.5

0		0.	-2.070432222150D+04	-1.00	-1.357196696238D+01
1		1.755917201721D+00	1.792225684534D+04	-.75	-6.573835441890D+00
2		-4.505978699139D+00	-6.656952017884D+03	-.50	-2.716209597224D+00
3		4.437376834648D+00	1.375232109108D+03	-.25	-7.991186885031D-01
4		-2.198332134340D+00	-1.709425965426D+02	.25	2.186570020887D-01
5		5.880293309203D-01	1.287892505701D+01	.50	1.858764606226D-01
6		-8.167264765423D-02	-5.575525518714D-01	.75	8.440539840201D-02
7		4.660113962068D-03	1.139070789708D-02	1.25	-4.077566868814D-02
				1.50	-4.281828987111D-02
0	-5.5	-1.178761850242D-10	-5.893809251208D-11	1.75	-2.334810547285D-02
1	-4.5	1.218404796800D-09	6.092023983998D-10	2.25	1.535232111171D-02
2	-3.5	-5.861720739787D-09	-2.930860369894D-09	2.50	1.841687332430D-02
3	-2.5	1.742872304455D-08	8.714361522277D-09	2.75	1.135104299031D-02
4	-1.5	-3.571727228207D-08	-1.785863614104D-08	3.25	-9.291854143620D-03
5	-.5	5.299499302236D-08	2.649749651118D-08	3.50	-1.230417230830D-02
6	.5	-5.791812880483D-08	-2.895906440242D-08	3.75	-8.320881815843D-03
7	1.5	4.637061238572D-08	2.318530619286D-08	4.25	8.076995972770D-03
8	2.5	-2.647944550914D-08	-1.323972275457D-08	4.50	1.157489866875D-02
9	3.5	1.020644855468D-08	5.103224277339D-09	4.75	8.444418738439D-03
10	4.5	-2.377125891400D-09	-1.188562945700D-09	5.25	-9.478331371347D-03
11	5.5	2.523876081424D-10	1.261938040712D-10	5.50	-1.458473916838D-02
				5.75	-1.142850782404D-02
	0	8.189664681002D+00	-5.258542370335D+00	6.25	1.488180171594D-02
	1	1.900054521218D+01	3.157273740934D+01	6.50	2.480170921170D-02
	2	1.953973284100D+01	-1.344269243043D+01	6.75	2.117490637208D-02
	3	1.143755104765D+01	1.992603614210D+01	7.25	-3.364648193616D-02
	4	4.049936708233D+00	-3.432111963215D+00	7.50	-6.325150177940D-02
	5	8.531693527348D-01	1.715638036880D+00	7.75	-6.225440473553D-02
	6	9.774173988541D-02	-1.190102978336D-01	8.25	1.480878169245D-01
	7	4.660113962068D-03	1.139070789708D-02	8.50	3.877537716118D-01

I=10 H= 4.5

0		0.	5.175183385299D+03	-1.00	2.713902567115D+00
1		-3.505793697254D-01	-4.508895863283D+03	-.75	1.314092280720D+00

CARDINAL SPLINES N=8

2		9.002986288202D-01	1.687200309930D+03	-.50	5.427652321048D-01
3		-8.874904695589D-01	-3.514460654956D+02	-.25	1.596194821613D-01
4		4.402416954296D-01	4.406458493695D+01	.25	-4.363475214950D-02
5		-1.179431618306D-01	-3.343715529965D+00	.50	-3.707286328805D-02
6		1.641095929553D-02	1.447465904943D-01	.75	-1.682427905745D-02
7		-9.382824545489D-04	-2.896294255671D-03	1.25	8.116196047938D-03
				1.50	8.514881888816D-03
0	-5.5	2.416589858789D-11	1.208294929394D-11	1.75	4.638373774442D-03
1	-4.5	-2.511858262066D-10	-1.255929131033D-10	2.25	-3.042404213539D-03
2	-3.5	1.218404796800D-09	6.092023983998D-10	2.50	-3.644157537010D-03
3	-2.5	-3.664747281628D-09	-1.832373640814D-09	2.75	-2.242074827103D-03
4	-1.5	7.622958179024D-09	3.811479089512D-09	3.25	1.827192339614D-03
5	-.5	-1.150940794298D-08	-5.754703971490D-09	3.50	2.412778466390D-03
6	.5	1.281535232923D-08	6.407676164615D-09	3.75	1.626328310614D-03
7	1.5	-1.045159648073D-08	-5.225798240365D-09	4.25	-1.565468090972D-03
8	2.5	6.072255942138D-09	3.036127971069D-09	4.50	-2.231468366125D-03
9	3.5	-2.377125891400D-09	-1.188562945700D-09	4.75	-1.617723309841D-03
10	4.5	5.612170972852D-10	2.806085486426D-10	5.25	1.786438681618D-03
11	5.5	-6.029082012180D-11	-3.014541006090D-11	5.50	2.720136899852D-03
				5.75	2.104919707418D-03
	0	-1.672778624122D+00	1.414123669072D+00	6.25	-2.651529995368D-03
	1	-3.876081341724D+00	-7.262206162410D+00	6.50	-4.322643053901D-03
	2	-3.978369017993D+00	3.629811493465D+00	6.75	-3.591999946314D-03
	3	-2.323060667795D+00	-4.615024008884D+00	7.25	5.288835073408D-03
	4	-8.204584927108D-01	9.256898121441D-01	7.50	9.410276578575D-03
	5	-1.724254343303D-01	-4.069489695682D-01	7.75	8.607536245357D-03
	6	-1.971291520460D-02	3.323926165097D-02	8.25	-1.590432286806D-02
	7	-9.382824545489D-04	-2.896294255671D-03	8.50	-3.249316233613D-02

I=11 H= 5.5

0		0.	-5.844684447306D+02	-1.00	-2.561338109837D-01
1		3.305579377306D-02	5.122262050658D+02	-.75	-1.239965815105D-01
2		-8.492614968986D-02	-1.929864132169D+02	-.50	-5.120336853044D-02
3		8.377005464748D-02	4.051580671184D+01	-.25	-1.505446570417D-02
4		-4.158729212441D-02	-5.124041462608D+00	.25	4.113037746033D-03
5		1.115212818833D-02	3.921001237435D-01	.50	3.493336430072D-03
6		-1.553463676426D-03	-1.704970501024D-02	.75	1.584738493465D-03
7		8.892888414075D-05	3.380534795992D-04	1.25	-7.638070196794D-04
				1.50	-8.008945043126D-04
0	-5.5	-2.316450361580D-12	-1.158225180790D-12	1.75	-4.360089696565D-04
1	-4.5	2.416589858789D-11	1.208294929394D-11	2.25	2.855552706695D-04
2	-3.5	-1.178761850242D-10	-5.893809251208D-11	2.50	3.417162485435D-04
3	-2.5	3.575548757671D-10	1.787774378835D-10	2.75	2.100148938938D-04
4	-1.5	-7.523755564913D-10	-3.761877782457D-10	3.25	-1.706897663044D-04
5	-.5	1.152053279586D-09	5.760266397928D-10	3.50	-2.250104484216D-04
6	.5	-1.302713098203D-09	-6.513565491015D-10	3.75	-1.513671797031D-04
7	1.5	1.079058699645D-09	5.395293498223D-10	4.25	1.449713736904D-04
8	2.5	-6.361919528736D-10	-3.180959764368D-10	4.50	2.059918808894D-04
9	3.5	2.523876081424D-10	1.261938040712D-10	4.75	1.487818321280D-04
10	4.5	-6.029082012180D-11	-3.014541006090D-11	5.25	-1.627540393198D-04
11	5.5	6.543701347978D-12	3.271850673989D-12	5.50	-2.463437635923D-04
				5.75	-1.892977392980D-04
	0	1.599577376567D-01	-1.658130256936D-01	6.25	2.342253855913D-04
	1	3.703488523696D-01	7.678731933568D-01	6.50	3.775120319583D-04
	2	3.796629271871D-01	-4.269552394423D-01	6.75	3.094931872100D-04
	3	2.213600681134D-01	4.911907214411D-01	7.25	-4.398955733179D-04

4	7.805609833835D-02	-1.090640396672D-01	7.50	-7.647582561162D-04
5	1.637990051668D-02	4.420833132098D-02	7.75	-6.800405770559D-04
6	1.870298362993D-03	-4.034646045670D-03	8.25	1.161483820049D-03
7	8.892888414075D-05	3.380534795992D-04	8.50	2.243159882739D-03

N= 8 M=12 P= 6.0 Q=15

I= 0	H= -6.0				
0		1.000000000000D+00	-2.213021117809D+02	-1.00	9.351955212827D+00
1		-2.772865834816D+00	1.637103710195D+02	-.75	5.775036256169D+00
2		3.061292871309D+00	-5.107618721693D+01	-.50	3.416216350078D+00
3		-1.783379370189D+00	8.407019737350D+00	-.25	1.914882591625D+00
4		6.017250971498D-01	-6.834946408282D-01	.25	4.724863225587D-01
5		-1.191283741042D-01	2.822119010684D-03	.50	1.900507114064D-01
6		1.295963795118D-02	4.288653617318D-03	.75	5.431046134471D-02
7		-6.040273081911D-04	-2.691221879733D-04	1.25	-1.391367889492D-02
				1.50	-1.139494642948D-02
0	-6.0	7.003138983410D-12	3.501569491705D-12	1.75	-4.968629401846D-03
1	-5.0	-6.515885754895D-11	-3.257942877448D-11	2.25	2.188811375491D-03
2	-4.0	2.767321973453D-10	1.383660986726D-10	2.50	2.181114725710D-03
3	-3.0	-7.131712964608D-10	-3.565856482304D-10	2.75	1.123514843291D-03
4	-2.0	1.252832049281D-09	6.264160246407D-10	3.25	-6.495110863146D-04
5	-1.0	-1.600416486805D-09	-8.002082434026D-10	3.50	-7.247821636705D-04
6	0.0	1.548737254894D-09	7.743686274470D-10	3.75	-4.132122535030D-04
7	1.0	-1.162296330418D-09	-5.811481652090D-10	4.25	2.844339202787D-04
8	2.0	6.784475230721D-10	3.392237615360D-10	4.50	3.423381836330D-04
9	3.0	-3.005434670708D-10	-1.502717335354D-10	4.75	2.091781682267D-04
10	4.0	9.504637091583D-11	4.752318545791D-11	5.25	-1.628493219571D-04
11	5.0	-1.900583686494D-11	-9.502918432469D-12	5.50	-2.071738189484D-04
12	6.0	1.793740676669D-12	8.968703383347D-13	5.75	-1.333974591460D-04
				6.25	1.146142325413D-04
	0	-1.591227026042D+00	-9.761849353487D-01	6.50	1.529276611061D-04
	1	-3.329322003694D+00	-1.299512881280D+00	6.75	1.032601562415D-04
	2	-3.084084420227D+00	-1.864047787080D+00	7.25	-9.778957830058D-05
	3	-1.641234466558D+00	-6.612876183377D-01	7.50	-1.373260648024D-04
	4	-5.403680822643D-01	-3.175218582344D-01	7.75	-9.783662233035D-05
	5	-1.092260528543D-01	-4.624272487371D-02	8.25	1.041772646589D-04
	6	-1.240950899285D-02	-7.014478277562D-03	8.50	1.560750308295D-04
	7	-6.040273081911D-04	-2.691221879733D-04	8.75	1.191935208556D-04

I= 1	H= -5.0				
0		0.	2.360418809044D+03	-1.00	-4.002106750016D+01
1		8.597285306143D+00	-1.753029734248D+03	-.75	-2.120471462886D+01
2		-1.519781634795D+01	5.524745864725D+02	-.50	-9.766682852966D+00
3		1.100518358898D+01	-9.351477823794D+01	-.25	-3.288489432427D+00
4		-4.208914884627D+00	8.379473000400D+00	.25	1.355829556574D+00
5		9.030320204679D-01	-2.130527464915D-01	.50	1.638416489740D+00
6		-1.038025174684D-01	-2.855585792195D-02	.75	1.406768051025D+00

7		5.032834525483D-03	2.165922858862D-03	1.25	6.027049483020D-01
				1.50	2.967187082005D-01
0	-6.0	-6.515885754895D-11	-3.257942877448D-11	1.75	1.008327071176D-01
1	-5.0	6.127970867431D-10	3.063985433716D-10	2.25	-3.441958408050D-02
2	-4.0	-2.635072481423D-09	-1.317536240712D-09	2.50	-3.183575396708D-02
3	-3.0	6.887901628951D-09	3.443950814476D-09	2.75	-1.550201712366D-02
4	-2.0	-1.229283695525D-08	-6.146418477624D-09	3.25	8.284056643124D-03
5	-1.0	1.596971132410D-08	7.984855662050D-09	3.50	8.986704151745D-03
6	0.0	-1.571252995482D-08	-7.856264977411D-09	3.75	5.006108127415D-03
7	1.0	1.196633259102D-08	5.983166295509D-09	4.25	-3.326480885814D-03
8	2.0	-7.064811396336D-09	-3.532405698168D-09	4.50	-3.949953494102D-03
9	3.0	3.153882209057D-09	1.576941104529D-09	4.75	-2.386138607781D-03
10	4.0	-1.002210046985D-09	-5.011050234926D-10	5.25	1.824681704314D-03
11	5.0	2.010006893603D-10	1.005003446801D-10	5.50	2.305106127793D-03
12	6.0	-1.900583686494D-11	-9.502918432469D-12	5.75	1.475386821696D-03
				6.25	-1.255695603018D-03
	0	1.466658993058D+01	9.249943186095D+00	6.50	-1.669218807042D-03
	1	3.060158739476D+01	1.142309992084D+01	6.75	-1.123510898060D-03
	2	2.823984447225D+01	1.754147501245D+01	7.25	1.058639142879D-03
	3	1.494525233876D+01	5.778539689896D+00	7.50	1.483705991007D-03
	4	4.876915309123D+00	2.942104140802D+00	7.75	1.055266393195D-03
	5	9.709642928706D-01	3.963740496183D-01	8.25	-1.120657031558D-03
	6	1.075765326019D-01	6.241290215027D-02	8.50	-1.677169362426D-03
	7	5.032834525483D-03	2.165922858862D-03	8.75	-1.279694409020D-03

I= 2 H= -4.0

0		0.	-1.188702224790D+04	-1.00	1.043644861640D+02
1		-1.691000962237D+01	8.844409484421D+03	-.75	5.285037090528D+01
2		3.823855502838D+01	-2.803284637431D+03	-.50	2.298269360992D+01
3		-3.213279734247D+01	4.830758820728D+02	-.25	7.175620855491D+00
4		1.356626159845D+01	-4.608218715986D+01	.25	-2.289630767945D+00
5		-3.120422011708D+00	1.791872149813D+00	.50	-2.155838819201D+00
6		3.774264550614D-01	7.092595611393D-02	.75	-1.112748980128D+00
7		-1.901410561521D-02	-7.738190453066D-03	1.25	7.979453505106D-01
				1.50	1.181287555888D+00
0	-6.0	2.767321973453D-10	1.383660986726D-10	1.75	1.207326472580D+00
1	-5.0	-2.635072481423D-09	-1.317536240712D-09	2.25	6.906874940981D-01
2	-4.0	1.149641476064D-08	5.748207380321D-09	2.50	3.844661352855D-01
3	-3.0	-3.055841151499D-08	-1.527920575749D-08	2.75	1.460759244056D-01
4	-2.0	5.557847975457D-08	2.778923987728D-08	3.25	-6.063671420834D-02
5	-1.0	-7.369277620177D-08	-3.684638810088D-08	3.50	-6.113325661733D-02
6	0.0	7.401440387920D-08	3.700720193960D-08	3.75	-3.223259154198D-02
7	1.0	-5.743807196705D-08	-2.871903598353D-08	4.25	1.984707096611D-02
8	2.0	3.443211822682D-08	1.721605911341D-08	4.50	2.293792439713D-02
9	3.0	-1.554085801387D-08	-7.770429006937D-09	4.75	1.355460478998D-02
10	4.0	4.974205036598D-09	2.487102518299D-09	5.25	-1.002706439500D-02
11	5.0	-1.002210046985D-09	-5.011050234926D-10	5.50	-1.250921000421D-02
12	6.0	9.504637091583D-11	4.752318545791D-11	5.75	-7.922790793877D-03
				6.25	6.633952830043D-03
	0	-6.160681629519D+01	-4.041515760427D+01	6.50	8.763142442511D-03
	1	-1.281201708162D+02	-4.459876421208D+01	6.75	5.866810355322D-03
	2	-1.177079413085D+02	-7.604230442334D+01	7.25	-5.481922627418D-03
	3	-6.189198803530D+01	-2.242682437013D+01	7.50	-7.657969790070D-03
	4	-1.998275147063D+01	-1.252672618913D+01	7.75	-5.431526374615D-03
	5	-3.907733474597D+00	-1.504865412603D+00	8.25	5.742939937311D-03
	6	-4.211659807775D-01	-2.540780429148D-01	8.50	8.580153744034D-03

	7	-1.901410561521D-02	-7.738190453066D-03	8.75	6,537178106541D-03

I= 3 H= -3.0

0		0.	3,802892138925D+04	-1.00	-1,868912055849D+02
1		2,710981226437D+01	-2.832480415058D+04	-.75	-9,268859597697D+01
2		-6,560197273751D+01	9.006896863635D+03	-.50	-3,932761986179D+01
3		5,972234992635D+01	-1.569680351612D+03	-.25	-1,192287775937D+01
4		-2,701262993754D+01	1.559937330664D+02	.25	3,511189234014D+00
5		6,570050210445D+00	-7,468035167520D+00	.50	3,124085997188D+00
6		-8,310001170390D-01	-3.747219473408D-02	.75	1,496646430279D+00
7		4,339039163862D-02	1.601560520879D-02	1.25	-8,323162562534D-01
				1.50	-9,611550354547D-01
0	-6.0	-7,131712964608D-10	-3,565856482304D-10	1.75	-5,912413617299D-01
1	-5.0	6.887901628951D-09	3.443950814476D-09	2.25	5,675909388115D-01
2	-4.0	-3,055841151499D-08	-1.527920575749D-08	2.50	9,513717880115D-01
3	-3.0	8,283292352546D-08	4.141646176273D-08	2.75	1,088676401474D+00
4	-2.0	-1,540627847732D-07	-7,703139238659D-08	3.25	7,595886020381D-01
5	-1.0	2,093505423108D-07	1.046752711554D-07	3.50	4,615626502505D-01
6	0.0	-2,156352551629D-07	-1,078176275815D-07	3.75	1,901676784327D-01
7	1.0	1.713585415396D-07	8.567927076979D-08	4.25	-9,122000261429D-02
8	2.0	-1,048076188761D-07	-5,240380943805D-08	4.50	-9,811383520273D-02
9	3.0	4,803485189070D-08	2.401742594535D-08	4.75	-5,494833897183D-02
10	4.0	-1,554085801387D-08	-7,770429006937D-09	5.25	3,775882143403D-02
11	5.0	3.153882209057D-09	1.576941104529D-09	5.50	4,589945143875D-02
12	6.0	-3,005434670708D-10	-1,502717335354D-10	5.75	2,846645841191D-02
				6.25	-2,310035683461D-02
	0	1.567486896700D+02	1.089059620799D+02	6.50	-3,015827249309D-02
	1	3,247263545755D+02	1.004635416198D+02	6.75	-1,999392725969D-02
	2	2,968158317706D+02	2.031657176594D+02	7.25	1,840291313676D-02
	3	1,549049663048D+02	5,026455269317D+01	7.50	2,555962864555D-02
	4	4,938017396270D+01	3,279566826280D+01	7.75	1,804028114446D-02
	5	9,457182075836D+00	3.290763359899D+00	8.25	-1,892987230918D-02
	6	9,913963317830D-01	6,351832240351D-01	8.50	-2,819822101211D-02
	7	4,339039163862D-02	1.601560520879D-02	8.75	-2,143003932025D-02

I= 4 H= -2.0

0		0.	-8,758154373949D+04	-1.00	2,465773285168D+02
1		-3,383014010987D+01	6,531401745893D+04	-.75	1,209237892000D+02
2		8.434520934494D+01	-2.081873637879D+04	-.50	5,065583741693D+01
3		-7,984228395084D+01	3,654930666385D+03	-.25	1,513351135669D+01
4		3,769093023973D+01	-3,730573965214D+02	.25	-4,295288211761D+00
5		-9,548749540380D+00	2,011024568043D+01	.50	-3,732725068457D+00
6		1,252524673125D+00	-2,493161043706D-01	.75	-1,738291930386D+00
7		-6,749065794692D-02	-2,057969690269D-02	1.25	8,943372146531D-01
				1.50	9,772182834614D-01
0	-6.0	1.252832049281D-09	6.264160246407D-10	1.75	5,586962018747D-01
1	-5.0	-1.229283695525D-08	-6,146418477624D-09	2.25	-4,171747312732D-01
2	-4.0	5.557847975457D-08	2.778923987728D-08	2.50	-5,462990138337D-01
3	-3.0	-1.540627847732D-07	-7,703139238659D-08	2.75	-3,768551588543D-01
4	-2.0	2.940613129109D-07	1.470306564554D-07	3.25	4,426766319104D-01
5	-1.0	-4.112546097998D-07	-2,056273048999D-07	3.50	8,113155708026D-01
6	0.0	4.365325393708D-07	2.182662696854D-07	3.75	1,008392721835D+00
7	1.0	-3,571002798739D-07	-1.785501399369D-07	4.25	8,155878855261D-01
8	2.0	2,240472118391D-07	1.120236059195D-07	4.50	5,295035522038D-01
9	3.0	-1.048076188761D-07	-5,240380943805D-08	4.75	2,320595943535D-01

CARDINAL SPLINES N=8

10	4,0	3,443211822682D-08	1,721605911341D-08	5,25	-1,245517711236D-01
11	5,0	-7,064811396336D-09	-3,532405698168D-09	5,50	-1,410697800295D-01
12	6,0	6,784475230721D-10	3,392237615360D-10	5,75	-8,300927442981D-02
				6,25	6,269428960229D-02
	0	-2,713911466539D+02	-2,051303614869D+02	6,50	7,982053812544D-02
	1	-5,597054000924D+02	-1,387622385972D+02	6,75	5,185799645861D-02
	2	-5,087488117650D+02	-3,795386171357D+02	7,25	-4,631932379158D-02
	3	-2,632794493075D+02	-6,929902756088D+01	7,50	-6,361630232433D-02
	4	-8,263760656301D+01	-5,996323105299D+01	7,75	-4,448525269052D-02
	5	-1,548079871576D+01	-4,423384935345D+00	8,25	4,601678005325D-02
	6	-1,582082960646D+00	-1,113663374283D+00	8,50	6,817242071530D-02
	7	-6,749065794692D-02	-2,057969690269D-02	8,75	5,157013398672D-02

I= 5 H= -1.0

0		0.	1,551789151969D+05	-1,00	-2,520698202697D+02
1		3,355176705022D+01	-1,159564727838D+05	-,75	-1,228374858419D+02
2		-8,493181886693D+01	3,705753717102D+04	-,50	-5,109516825260D+01
3		8,207300995919D+01	-6,541863557487D+03	-,25	-1,514412798260D+01
4		-3,971102099538D+01	6,795864063699D+02	,25	4,216728550895D+00
5		1,033321346583D+01	-3,931979152676D+01	,50	3,621878998725D+00
6		-1,392070271721D+00	8,717325610493D-01	,75	1,664041941718D+00
7		7,691966038992D-02	1,485045882479D-02	1,25	-8,272257015610D-01
				1,50	-8,840439256542D-01
0	-6,0	-1,600416486805D-09	-8,002082434026D-10	1,75	-4,920178506933D-01
1	-5,0	1,596971132410D-08	7,984855662050D-09	2,25	3,409191681485D-01
2	-4,0	-7,369277620177D-08	-3,684638810088D-08	2,50	4,230954231143D-01
3	-3,0	2,093505423108D-07	1,046752711554D-07	2,75	2,717051519356D-01
4	-2,0	-4,112546097998D-07	-2,056273048999D-07	3,25	-2,490595260983D-01
5	-1,0	5,940756200170D-07	2,970378100085D-07	3,50	-3,571982965580D-01
6	0,0	-6,525846153898D-07	-3,262923076949D-07	3,75	-2,680633766726D-01
7	1,0	5,521123182670D-07	2,760561591335D-07	4,25	3,661282033155D-01
8	2,0	-3,571002798739D-07	-1,785501399369D-07	4,50	7,179613773791D-01
9	3,0	1,713585415396D-07	8,567927076979D-08	4,75	9,504971128461D-01
10	4,0	-5,743807196705D-08	-2,871903598353D-08	5,25	8,622817687384D-01
11	5,0	1,196633259102D-08	5,983166295509D-09	5,50	5,901524876317D-01
12	6,0	-1,162296330418D-09	-5,811481652090D-10	5,75	2,720183315367D-01
				6,25	-1,607479833102D-01
	0	3,412714280663D+02	2,909043869175D+02	6,50	-1,908333708757D-01
	1	7,000712634705D+02	1,047104868667D+02	6,75	-1,177107132033D-01
	2	6,333696923703D+02	5,354087254049D+02	7,25	9,796522020596D-02
	3	3,242975752221D+02	5,258672178329D+01	7,50	1,312739597043D-01
	4	1,000800687981D+02	8,299771224924D+01	7,75	8,999290469608D-02
	5	1,836994693866D+01	3,289527542559D+00	8,25	-9,039147811491D-02
	6	1,838555464656D+00	1,495451831690D+00	8,50	-1,324487996372D-01
	7	7,691966038992D-02	1,485045882479D-02	8,75	-9,927964781782D-02

I= 6 H= 0.0

0		0.	-2,172302448036D+05	-1,00	2,071692865202D+02
1		-2,712510251953D+01	1,628881652909D+05	-,75	1,006037079087D+02
2		6,921311962171D+01	-5,225343568225D+04	-,50	4,168520318805D+01
3		-6,762351770517D+01	9,274587685656D+03	-,25	1,230214313013D+01
4		3,316982182227D+01	-9,756114384970D+02	,25	-3,390777175966D+00
5		-8,768444799498D+00	5,891702849280D+01	,50	-2,894862467686D+00
6		1,201701815338D+00	-1,636584124800D+00	,75	-1,320916251707D+00
7		-6,757823666996D-02	1,067524825311D-23	1,25	6,456691593183D-01

CARDINAL SPLINES N=8

				1.50	6,828362840749D-01
0	-6.0	1.548737254894D-09	7.743686274470D-10	1.75	3,754136070151D-01
1	-5,0	-1,571252995482D-08	-7,856264977411D-09	2.25	-2,520103510096D-01
2	-4,0	7,401440387920D-08	3,700720193960D-08	2.50	-3,063099136521D-01
3	-3.0	-2.156352551629D-07	-1.078176275815D-07	2.75	-1,917543224097D-01
4	-2,0	4,365325393708D-07	2.182662696854D-07	3.25	1,635593367343D-01
5	-1,0	-6.525846153898D-07	-3.262923076949D-07	3.50	2,226069506557D-01
6	0,0	7.436734400053D-07	3.718367200026D-07	3,75	1,557207822735D-01
7	1,0	-6,525846153898D-07	-3,262923076949D-07	4,25	-1,663749467742D-01
8	2,0	4,365325393708D-07	2.182662696854D-07	4,50	-2,555870996285D-01
9	3,0	-2.156352551629D-07	-1.078176275815D-07	4.75	-2,045137490665D-01
10	4,0	7,401440387920D-08	3,700720193960D-08	5,25	3,138764973547D-01
11	5,0	-1,571252995482D-08	-7,856264977411D-09	5,50	6,493533726011D-01
12	6,0	1.548737254894D-09	7.743686274470D-10	5,75	9,047664649954D-01
				6,25	9,047664649954D-01
	0	-3.240771763138D+02	-3.240771763138D+02	6.50	6,493533726011D-01
	1	-6,646153996234D+02	6,615125109169D-20	6,75	3,138764973547D-01
	2	-5,973418974613D+02	-5,973418974613D+02	7,25	-2,045137490665D-01
	3	-3,021848948786D+02	3,323761520728D-20	7,50	-2,555870996285D-01
	4	-9,185601110497D+01	-9,185601110497D+01	7,75	-1,663749467742D-01
	5	-1,659632636981D+01	2,199595174010D-21	8,25	1,557207822735D-01
	6	-1,636584124800D+00	-1,636584124800D+00	8,50	2,226069506557D-01
	7	-6,757823666996D-02	1,067524825311D-23	8,75	1,635593367343D-01

I= 7 H= 1.0

0		0.	2.386259932748D+05	-1.00	-1,394538435283D+02
1		1,810136247769D+01	-1,798569847113D+05	-,75	-6,759443600947D+01
2		-4,637901505575D+01	5,801145918758D+04	-.50	-2,795054572623D+01
3		4,557461433315D+01	-1,036273045628D+04	-.25	-8,230158442543D+00
4		-2,251706771144D+01	1,101496996354D+03	.25	2,256461285975D+00
5		6,003726211138D+00	-6,835274035496D+01	.50	1,920441015793D+00
6		-8,308389963690D-01	2,119171102332D+00	.75	8,732191779971D-01
7		4,721874274034D-02	-1,485045882479D-02	1,25	-4,232235511215D-01
				1.50	-4,452756225259D-01
0	-6.0	-1.162296330418D-09	-5.811481652090D-10	1.75	-2,433506171580D-01
1	-5,0	1,196633259102D-08	5,983166295509D-09	2,25	1,609252793785D-01
2	-4,0	-5,743807196705D-08	-2.871903598353D-08	2,50	1,937457098218D-01
3	-3,0	1.713585415396D-07	8.567927076979D-08	2.75	1,199245948876D-01
4	-2.0	-3,571002798739D-07	-1.785501399369D-07	3.25	-9,927964781782D-02
5	-1.0	5,521123182670D-07	2.760561591335D-07	3.50	-1,324487996372D-01
6	0,0	-6,525846153898D-07	-3.262923076949D-07	3.75	-9,039147811491D-02
7	1,0	5,940756200170D-07	2.970378100085D-07	4.25	8,999290469608D-02
8	2,0	-4,112546097998D-07	-2,056273048999D-07	4.50	1,312739597043D-01
9	3,0	2.093505423108D-07	1.046752711554D-07	4.75	9,796522020596D-02
10	4,0	-7,369277620177D-08	-3,684638810088D-08	5,25	-1,177107132033D-01
11	5,0	1,596971132410D-08	7,984855662050D-09	5,50	-1,908333708757D-01
12	6,0	-1,600416486805D-09	-8,002082434026D-10	5,75	-1,607479833102D-01
				6,25	2,720183315367D-01
	0	2,405373457688D+02	2,909043869175D+02	6,50	5,901524876317D-01
	1	4,906502897371D+02	-1,047104868667D+02	6,75	8,622817687384D-01
	2	4,374477584395D+02	5,354087254049D+02	7,25	9,504971128461D-01
	3	2,191241316555D+02	-5,258672178329D+01	7,50	7,179613773791D-01
	4	6,591535570034D+01	8,299771224924D+01	7,75	3,661282033155D-01
	5	1,179089185355D+01	-3,289527542559D+00	8,25	-2,680633766726D-01
	6	1,152348198725D+00	1,495451831690D+00	8,50	-3,571982965580D-01
	7	4,721874274034D-02	-1,485045882479D-02	8,75	-2,490595260983D-01

CARDINAL SPLINES N=8

I= 8	H= 2.0				
0		0.	-1.994983490877D+05	-1.00	7,624488704923D+01
1		-9,852201068963D+00	1.513295293930D+05	-.75	3,692068293359D+01
2		2,529705297641D+01	-4,914368866251D+04	-.50	1,525058066006D+01
3		-2,493222302093D+01	8.845355977967D+03	-.25	4,485329901024D+00
4		1,236472124185D+01	-9.496255098107D+02	.25	-1,226373958086D+00
5		-3,311688170923D+00	6.007351726798D+01	.50	-1,042068100816D+00
6		4,606693060240D-01	-1,978010644196D+00	.75	-4,729696957084D-01
7		-2,633126414155D-02	2.057969690269D-02	1,25	2,282400325331D-01
				1.50	2,395014143373D-01
0	-6,0	6,784475230721D-10	3.392237615360D-10	1.75	1,304975085663D-01
1	-5,0	-7,064811396336D-09	-3,532405698168D-09	2,25	-8,565191922798D-02
2	-4,0	3,443211822682D-08	1.721605911341D-08	2,50	-1,026378644724D-01
3	-3,0	-1,048076188761D-07	-5.240380943805D-08	2,75	-6,318240001822D-02
4	-2,0	2,240472118391D-07	1,120236059195D-07	3,25	5,157013398672D-02
5	-1,0	-3,571002798739D-07	-1,785501399369D-07	3,50	6,817242071530D-02
6	0,0	4,365325393708D-07	2,182662696854D-07	3,75	4,601678005325D-02
7	1,0	-4,112546097998D-07	-2,056273048999D-07	4,25	-4,448525269052D-02
8	2,0	2,940613129109D-07	1,470306564554D-07	4,50	-6,361630232433D-02
9	3,0	-1,540627847732D-07	-7,703139238659D-08	4,75	-4,631932379158D-02
10	4,0	5,557847975457D-08	2,778923987728D-08	5,25	5,185799645861D-02
11	5.0	-1,229283695525D-08	-6,146418477624D-09	5,50	7,982053812544D-02
12	6,0	1,252832049281D-09	6,264160246407D-10	5,75	6,269428960229D-02
				6,25	-8,300927442981D-02
	0	-1,388695763200D+02	-2,051303614869D+02	6,50	-1,410697800295D-01
	1	-2,821809228980D+02	1,387622385972D+02	6,75	-1,245517711236D-01
	2	-2,503284225064D+02	-3,795386171357D+02	7,25	2,320595943535D-01
	3	-1,246813941857D+02	6,929902756088D+01	7,50	5,295035522038D-01
	4	-3,728885554297D+01	-5,996323105299D+01	7,75	8,155878855261D-01
	5	-6,634028845068D+00	4,423384935345D+00	8,25	1,008392721835D+00
	6	-6,452437879210D-01	-1,113663374283D+00	8,50	8,113155708026D-01
	7	-2,633126414155D-02	2,057969690269D-02	8,75	4,426766319104D-01

I= 9	H= 3.0				
0		0.	1.210934113447D+05	-1.00	-3,254461067220D+01
1		4,195228574141D+00	-9,249230483253D+04	-.75	-1,575116721231D+01
2		-1,078414448676D+01	3,026308701170D+04	-.50	-6,502526723022D+00
3		1,064544540809D+01	-5,492494780666D+03	-.25	-1,911251544018D+00
4		-5,290045354193D+00	5,955954854172D+02	.25	5,218082677615D-01
5		1,420272172803D+00	-3,826515696301D+01	,50	4,430077011478D-01
6		-1,981154949963D-01	1.307838642804D+00	.75	2,008772355541D-01
7		1,135918122104D-02	-1,601560520879D-02	1.25	-9,671342015528D-02
				1.50	-1,013439564954D-01
0	-6,0	-3,005434670708D-10	-1,502717335354D-10	1.75	-5,513156402344D-02
1	-5,0	3,153882209057D-09	1,576941104529D-09	2,25	3,604318815575D-02
2	-4,0	-1,554085801387D-08	-7,770429006937D-09	2,50	4,308533608089D-02
3	-3,0	4,803485189070D-08	2,401742594535D-08	2,75	2,644708487178D-02
4	-2,0	-1,048076188761D-07	-5,240380943805D-08	3,25	-2,143003932025D-02
5	-1,0	1,713585415396D-07	8,567927076979D-08	3,50	-2,819822101211D-02
6	0,0	-2,156352551629D-07	-1,078176275815D-07	3.75	-1,892987230918D-02
7	1,0	2,093505423108D-07	1,046752711554D-07	4,25	1,804028114446D-02
8	2,0	-1,540627847732D-07	-7,703139238659D-08	4,50	2,555962864555D-02
9	3,0	8,283292352546D-08	4,141646176273D-08	4,75	1,840291313676D-02
10	4,0	-3,055841151499D-08	-1,527920575749D-08	5,25	-1,999392725969D-02

CARDINAL SPLINES N=8

11	5.0	6.887901628951D-09	3.443950814476D-09	5.50	-3.015827249309D-02
12	6.0	-7.131712964608D-10	-3.565856482304D-10	5.75	-2.310035683461D-02
				6.25	2.846645841191D-02
	0	6.106323448968D+01	1.089059620799D+02	6.50	4.589945143875D-02
	1	1.237992713359D+02	-1.004635416198D+02	6.75	3.775882143403D-02
	2	1.095156035483D+02	2.031657176594D+02	7.25	-5.494833897183D-02
	3	5.437586091844D+01	-5.026455269317D+01	7.50	-9.811383520273D-02
	4	1.621116256291D+01	3.279566826280D+01	7.75	-9.122000261429D-02
	5	2.875655356039D+00	-3.290763359899D+00	8.25	1.901676784327D-01
	6	2.789701162873D-01	6.351832240351D-01	8.50	4.615626502505D-01
	7	1.135918122104D-02	-1.601560520879D-02	8.75	7.595886020381D-01
I=10	H= 4.0				
0		0.	-4.984665860849D+04	-1.00	1.007586652910D+01
1		-1.297106521686D+00	3.833531901703D+04	-.75	4.875176692275D+00
2		3.336418285246D+00	-1.263889999365D+04	-.50	2.011971667176D+00
3		-3.296410270628D+00	2.313441265790D+03	-.25	5.911609230098D-01
4		1.639920646528D+00	-2.533755515664D+02	.25	-1.612666819453D-01
5		-4.408787475755D-01	1.650174694006D+01	.50	-1.368476208390D-01
6		6.159433272916D-02	-5.790820419436D-01	.75	-6.201890142083D-02
7		-3.537724709080D-03	7.738190453066D-03	1.25	2.982101180168D-02
				1.50	3.122462414348D-02
0	-6.0	9.504637091583D-11	4.752318545791D-11	1.75	1.697132431183D-02
1	-5.0	-1.002210046985D-09	-5.011050234926D-10	2.25	-1.107109119308D-02
2	-4.0	4.974205036598D-09	2.487102518299D-09	2.50	-1.321627287261D-02
3	-3.0	-1.554085801387D-08	-7.770429006937D-09	2.75	-8.099799137949D-03
4	-2.0	3.443211822682D-08	1.721605911341D-08	3.25	6.537178106541D-03
5	-1.0	-5.743807196705D-08	-2.871903598353D-08	3.50	8.580153744034D-03
6	0.0	7.401440387920D-08	3.700720193960D-08	3.75	5.742939937311D-03
7	1.0	-7.369277620177D-08	-3.684638810088D-08	4.25	-5.431526374615D-03
8	2.0	5.557847975457D-08	2.778923987728D-08	4.50	-7.657969790070D-03
9	3.0	-3.055841151499D-08	-1.527920575749D-08	4.75	-5.481922627418D-03
10	4.0	1.149641476064D-08	5.748207380321D-09	5.25	5.866810355322D-03
11	5.0	-2.635072481423D-09	-1.317536240712D-09	5.50	8.763142442511D-03
12	6.0	2.767321973453D-10	1.383660986726D-10	5.75	6.633952830043D-03
				6.25	-7.922790793877D-03
	0	-1.922349891335D+01	-4.041515760427D+01	6.50	-1.250921000421D-02
	1	-3.892264239200D+01	4.459876421208D+01	6.75	-1.002706439500D-02
	2	-3.437666753821D+01	-7.604230442334D+01	7.25	1.355460478998D-02
	3	-1.703833929504D+01	2.242682437013D+01	7.50	2.293792439713D-02
	4	-5.070700907636D+00	-1.252672618913D+01	7.75	1.984707096611D-02
	5	-8.980026493902D-01	1.504865412603D+00	8.25	-3.223259154198D-02
	6	-8.699010505220D-02	-2.540780429148D-01	8.50	-6.113325661733D-02
	7	-3.537724709080D-03	7.738190453066D-03	8.75	-6.063671420834D-02
I=11	H= 5.0				
0		0.	1.237087393663D+04	-1.00	-1.989568393882D+00
1		2.559233475036D-01	-9.575794346803D+03	-.75	-9.624830103332D-01
2		-6.585304327767D-01	3.180210987981D+03	-.50	-3.971400312360D-01
3		6.509699094907D-01	-5.869536950989D+02	-.25	-1.166644248649D-01
4		-3.240615059274D-01	6.491066960350D+01	.25	3.181043700170D-02
5		8.718995114819D-02	-4.280676208328D+00	.50	2.698611853714D-02
6		-1.219225822721D-02	1.533816622225D-01	.75	1.222618169669D-02
7		7.009888077584D-04	-2.165922858862D-03	1.25	-5.874383278572D-03
				1.50	-6.148088920938D-03

CARDINAL SPLINES N=8

0	-6.0	-1.900583686494D-11	-9.502918432469D-12	1.75	-3.339902968740D-03
1	-5.0	2.010006893603D-10	1.005003446801D-10	2.25	2.175975237062D-03
2	-4.0	-1.002210046985D-09	-5.011050234926D-10	2.50	2.595549035541D-03
3	-3.0	3.153882209057D-09	1.576941104529D-09	2.75	1.589263957042D-03
4	-2.0	-7.064811396336D-09	-3.532405698168D-09	3.25	-1.279694409020D-03
5	-1.0	1.196633259102D-08	5.983166295509D-09	3.50	-1.677169362426D-03
6	0.0	-1.571252995482D-08	-7.856264977411D-09	3.75	-1.120657031558D-03
7	1.0	1.596971132410D-08	7.984855662050D-09	4.25	1.055266393195D-03
8	2.0	-1.229283695525D-08	-6.146418477624D-09	4.50	1.483705991007D-03
9	3.0	6.887901628951D-09	3.443950814476D-09	4.75	1.058639142879D-03
10	4.0	-2.635072481423D-09	-1.317536240712D-09	5.25	-1.123510898060D-03
11	5.0	6.127970867431D-10	3.063985433716D-10	5.50	-1.669218807042D-03
12	6.0	-6.515885754895D-11	-3.257942877448D-11	5.75	-1.255695603018D-03
				6.25	1.475386821696D-03
	0	3.833296441612D+00	9.249943186095D+00	6.50	2.305106127793D-03
	1	7.755387553084D+00	-1.142309992084D+01	6.75	1.824681704314D-03
	2	6.843105552655D+00	1.754147501245D+01	7.25	-2.386138607781D-03
	3	3.388172958971D+00	-5.778539689896D+00	7.50	-3.949953494102D-03
	4	1.007292972480D+00	2.942104140802D+00	7.75	-3.326480885814D-03
	5	1.782161936341D-01	-3.963740496183D-01	8.25	5.006108127415D-03
	6	1.724927169865D-02	6.241290215027D-02	8.50	8.986704151745D-03
	7	7.009888077584D-04	-2.165922858862D-03	8.75	8.284056643124D-03

I=12 H= 6.0

0		0.	-1.392413352339D+03	-1.00	1.863059568465D-01
1		-2.395334283593D-02	1.084239543935D+03	-.75	9.011878380981D-02
2		6.164979969021D-02	-3.625442665461D+02	-.50	3.718055563764D-02
3		-6.096146501422D-02	6.743912177604D+01	-.25	1.092082784857D-02
4		3.035974314015D-02	-7.527185615408D+00	.25	-2.976859075847D-03
5		-8.172387641754D-03	5.022203169738D-01	.50	-2.524955537767D-03
6		1.143435591993D-03	-1.831761017244D-02	.75	-1.143720263669D-03
7		-6.578293224447D-05	2.691221879733D-04	1.25	5.492741458742D-04
				1.50	5.747053751335D-04
0	-6.0	1.793740676669D-12	8.968703383347D-13	1.75	3.121045094045D-04
1	-5.0	-1.900583686494D-11	-9.502918432469D-12	2.25	-2.031784205503D-04
2	-4.0	9.504637091583D-11	4.752318545791D-11	2.50	-2.422372774238D-04
3	-3.0	-3.005434670708D-10	-1.502717335354D-10	2.75	-1.482388311475D-04
4	-2.0	6.784475230721D-10	3.392237615360D-10	3.25	1.191935208556D-04
5	-1.0	-1.162296330418D-09	-5.811481652090D-10	3.50	1.560750308295D-04
6	0.0	1.548737254894D-09	7.743686274470D-10	3.75	1.041772646589D-04
7	1.0	-1.600416486805D-09	-8.002082434026D-10	4.25	-9.783662233035D-05
8	2.0	1.252832049281D-09	6.264160246407D-10	4.50	-1.373260648024D-04
9	3.0	-7.131712964608D-10	-3.565856482304D-10	4.75	-9.778957830058D-05
10	4.0	2.767321973453D-10	1.383660986726D-10	5.25	1.032601562415D-04
11	5.0	-6.515885754895D-11	-3.257942877448D-11	5.50	1.529276611061D-04
12	6.0	7.003138983410D-12	3.501569491705D-12	5.75	1.146142325413D-04
				6.25	-1.333974591460D-04
	0	-3.611428446558D-01	-9.761849353487D-01	6.50	-2.071738189484D-04
	1	-7.302962411344D-01	1.299512881280D+00	6.75	-1.628493219571D-04
	2	-6.440111539344D-01	-1.864047787080D+00	7.25	2.091781682267D-04
	3	-3.186592298822D-01	6.612876183377D-01	7.50	3.423381836330D-04
	4	-9.467563420454D-02	-3.175218582344D-01	7.75	2.844339202787D-04
	5	-1.674060310683D-02	4.624272487371D-02	8.25	-4.132122535030D-04
	6	-1.619447562275D-03	-7.014478277562D-03	8.50	-7.247821636705D-04
	7	-6.578293224447D-05	2.691221879733D-04	8.75	-6.495110863146D-04

CARDINAL SPLINES N=8

N= 8 M=13 P= 6.5 Q=15

I= 0 H= -6.5

0		1,000000000000D+00	3.332893754788D+02	-1,00	9,377288098090D+00
1		-2,776116443049D+00	-2,249782302902D+02	-,75	5,787284900658D+00
2		3,069666899953D+00	6.584856388869D+01	-,50	3,421267443515D+00
3		-1,791670634248D+00	-1,115975745176D+01	-,25	1,916365455773D+00
4		6,058610724301D-01	1,284218225260D+00	,25	4,720826006841D-01
5		-1,202439212696D-01	-1,163106495610D-01	,50	1,897085179110D-01
6		1.311607665873D-02	8,341851899784D-03	,75	5,415558139503D-02
7		-6,130504828691D-04	-3,307412537154D-04	1,25	-1,383943795617D-02
				1,50	-1,131735659589D-02
0	-6,5	7,254957916529D-12	3.627478958265D-12	1.75	-4,926547522697D-03
1	-5,5	-6,784804150633D-11	-3,392402075316D-11	2,25	2,161503936308D-03
2	-4,5	2,903511604883D-10	1.451755802441D-10	2.50	2,148622172157D-03
3	-3,5	-7,571599551003D-10	-3,785799775502D-10	2,75	1,103676422847D-03
4	-2,5	1,355887831488D-09	6,779439157439D-10	3,25	-6,336517305201D-04
5	-1,5	-1,788908141560D-09	-8,944540707799D-10	3,50	-7,040908454824D-04
6	-,5	1,828755938083D-09	9,143779690417D-10	3,75	-3,994596857895D-04
7	,5	-1,503004455301D-09	-7,515022276506D-10	4,25	2,716566345412D-04
8	1,5	1,011731766096D-09	5,058658830479D-10	4,50	3,245246986122D-04
9	2,5	-5,531534468927D-10	-2,765767234464D-10	4,75	1,965930502463D-04
10	3,5	2,361097808367D-10	1,180548904183D-10	5,25	-1,498220875268D-04
11	4,5	-7,309998764120D-11	-3,654999382060D-11	5,50	-1,881187394786D-04
12	5,5	1,443423496967D-11	7,217117484833D-12	5,75	-1,193202741139D-04
13	6,5	-1,351641876520D-12	-6,758209382600D-13	6,25	9,880579037628D-05
				6,50	1,289108094109D-04
	0	-4,399309900794D+00	-1,851138121164D+00	6,75	8,485204892344D-05
	1	-8,182349298567D+00	-4,711946517159D+00	7,25	-7,555950046061D-05
	2	-6,683261398480D+00	-2,863357528678D+00	7,50	-1,022784230431D-04
	3	-3,103958404603D+00	-1,757494797343D+00	7,75	-6,992387495693D-05
	4	-8,823173963917D-01	-3,882728320410D-01	8,25	6,755080832341D-05
	5	-1,526459725049D-01	-8,442860282842D-02	8,50	9,561414633425D-05
	6	-1,477772031182D-02	-6,706875144266D-03	8,75	6,858227162366D-05
	7	-6,130504828691D-04	-3,307412537154D-04	9,25	-7,381711216493D-05

I= 1 H= -5.5

0		0,	-3,562084761127D+03	-1,00	-4,029159834469D+01
1		8.631998675000D+00	2,397790611582D+03	-,75	-2,133551836887D+01
2		-1,528724291810D+01	-6,961692744098D+02	-,50	-9,820623671749D+00
3		1,109372631298D+01	1,154395790167D+02	-,25	-3,304324995022D+00
4		-4,253083121837D+00	-1,263380748117D+01	,25	1,360140917817D+00
5		9,149449910452D-01	1.059170616829D+00	,50	1,642070787083D+00
6		-1,054731323719D-01	-7,184011642569D-02	,75	1,408422019894D+00
7		5,129193350803D-03	2,823955203807D-03	1,25	6,019121264955D-01
				1.50	2,958901234187D-01
0	-6,5	-6,784804150633D-11	-3,392402075316D-11	1.75	1,003833131308D-01
1	-5,5	6,415149844286D-10	3,207574922143D-10	2,25	-3,412796689962D-02

CARDINAL SPLINES N=8

2	-4.5	-2.780509906427D-09	-1.390254953214D-09	2.50	-3.148876475279D-02
3	-3.5	7.357658186390D-09	3.678829093195D-09	2.75	-1.529016187777D-02
4	-2.5	-1.339337357015D-08	-6.696686785076D-09	3.25	8.114693979846D-03
5	-1.5	1.798262086901D-08	8.991310434505D-09	3.50	8.765740778423D-03
6	-.5	-1.870286011495D-08	-9.351430057476D-09	3.75	4.859243934124D-03
7	.5	1.560476760555D-08	7.802383802776D-09	4.25	-3.190031765811D-03
8	1.5	-1.062396649434D-08	-5.311983247170D-09	4.50	-3.759722610548D-03
9	2.5	5.851513784234D-09	2.925756892117D-09	4.75	-2.251741654559D-03
10	3.5	-2.508631562766D-09	-1.254315781383D-09	5.25	1.685563372722D-03
11	4.5	7.786741813945D-10	3.893370906973D-10	5.50	2.101616209930D-03
12	5.5	-1.539941558327D-10	-7.699707791634D-11	5.75	1.325056027722D-03
13	6.5	1.443423496967D-11	7.217117484833D-12	6.25	-1.086876648324D-03
				6.50	-1.412741935009D-03
	0	4.061916746804D+01	1.658347378301D+01	6.75	-9.269300203947D-04
	1	7.528311235958D+01	4.428007784279D+01	7.25	8.212432967525D-04
	2	6.119407147473D+01	2.548099616372D+01	7.50	1.109430892644D-03
	3	2.822421901090D+01	1.631110940838D+01	7.75	7.571851031460D-04
	4	7.940196785175D+00	3.404068205587D+00	8.25	-7.295217123314D-04
	5	1.352369629040D+00	7.629603308051D-01	8.50	-1.031505279033D-03
	6	1.279051650896D-01	5.664984534755D-02	8.75	-7.392149549488D-04
	7	5.129193350803D-03	2.823955203807D-03	9.25	7.945115915656D-04
I= 2		H= -4.5			
0		0.	1.810659630509D+04	-1.00	1.057345484800D+02
1		-1.708581019924D+01	-1.217678884154D+04	-.75	5.351280655210D+01
2		3.869144229163D+01	3.520282284583D+03	-.50	2.325586868256D+01
3		-3.258120850136D+01	-5.751416795084D+02	-.25	7.255817653752D+00
4		1.378994491806D+01	6.033637584202D+01	.25	-2.311465001225D+00
5		-3.180753439303D+00	-4.651109575993D+00	.50	-2.174345452295D+00
6		3.858870301495D-01	2.901325000236D-01	.75	-1.121125253906D+00
7		-1.950210023043D-02	-1.107069518117D-02	1.25	8.019604759661D-01
				1.50	1.185483797493D+00
0	-6.5	2.903511604883D-10	1.451755802441D-10	1.75	1.209602360076D+00
1	-5.5	-2.780509906427D-09	-1.390254953214D-09	2.25	6.892106432388D-01
2	-4.5	1.223296047939D-08	6.116480239697D-09	2.50	3.827088611848D-01
3	-3.5	-3.293742215291D-08	-1.646871107646D-08	2.75	1.450030157347D-01
4	-2.5	6.115198005557D-08	3.057599002779D-08	3.25	-5.977900275924D-02
5	-1.5	-8.388685044020D-08	-4.194342522010D-08	3.50	-6.001422122038D-02
6	-.5	8.915847616346D-08	4.457923808173D-08	3.75	-3.148882017224D-02
7	.5	-7.586437271348D-08	-3.793218635674D-08	4.25	1.915604515096D-02
8	1.5	5.245691789400D-08	2.622845894700D-08	4.50	2.197452902037D-02
9	2.5	-2.920260283238D-08	-1.460130141619D-08	4.75	1.287397186138D-02
10	3.5	1.260324768537D-08	6.301623842685D-09	5.25	-9.322520763740D-03
11	4.5	-3.927749586635D-09	-1.963874793317D-09	5.50	-1.147866626482D-02
12	5.5	7.786741813945D-10	3.893370906973D-10	5.75	-7.161463352230D-03
13	6.5	-7.309998764120D-11	-3.654999382060D-11	6.25	5.778994910370D-03
				6.50	7.464254332728D-03
	0	-1.712313715479D+02	-6.671062608162D+01	6.75	4.871256399188D-03
	1	-3.160503228579D+02	-1.917202506458D+02	7.25	-4.279667471231D-03
	2	-2.554500717865D+02	-1.017461800966D+02	7.50	-5.762510479460D-03
	3	-1.168628283816D+02	-6.960284568242D+01	7.75	-3.921939017584D-03
	4	-3.248038615441D+01	-1.336335173228D+01	8.25	3.762094602480D-03
	5	-5.434397692921D+00	-3.158416374564D+00	8.50	5.310286195421D-03
	6	-5.014585303350D-01	-2.135841307196D-01	8.75	3.800002115160D-03
	7	-1.950210023043D-02	-1.107069518117D-02	9.25	-4.074896660178D-03

CARDINAL SPLINES N=8

I= 3	H= -3.5				
0		0.	-5.884916206277D+04	-1.00	-1.913164472790D+02
1		2.767764048180D+01	3.957275219415D+04	-.75	-9.482823430573D+01
2		-6.706477890096D+01	-1.141794927165D+04	-.50	-4.020996346557D+01
3		6.117069846737D+01	1.848316348591D+03	-.25	-1.218190992985D+01
4		-2.773511733153D+01	-1.877335971031D+02	.25	3.581712857900D+00
5		6.764918097662D+00	1.334251559046D+01	.50	3.183861617194D+00
6		-8.583274066086D-01	-7.454997978922D-01	.75	1.523701430284D+00
7		4.496659302205D-02	2.677945054419D-02	1.25	-8.452849368501D-01
				1.50	-9.747087133363D-01
0	-6.5	-7.571599551003D-10	-3.785799775502D-10	1.75	-5.985923793594D-01
1	-5.5	7.357658186390D-09	3.678829093195D-09	2.25	5.723611028562D-01
2	-4.5	-3.293742215291D-08	-1.646871107646D-08	2.50	9.570477069301D-01
3	-3.5	9.051702441868D-08	4.525851220934D-08	2.75	1.092141849818D+00
4	-2.5	-1.720649483443D-07	-8.603247417215D-08	3.25	7.568182313608D-01
5	-1.5	2.422769587431D-07	1.211384793716D-07	3.50	4.579482145871D-01
6	-.5	-2.645499499993D-07	-1.322749749996D-07	3.75	1.877653292563D-01
7	.5	2.308746915590D-07	1.154373457795D-07	4.25	-8.898801928529D-02
8	1.5	-1.630269378715D-07	-8.151346893574D-08	4.50	-9.500211003553D-02
9	2.5	9.216169348165D-08	4.608084674082D-08	4.75	-5.274992421435D-02
10	3.5	-4.018233396920D-08	-2.009116698460D-08	5.25	3.548317614687D-02
11	4.5	1.260324768537D-08	6.301623842685D-09	5.50	4.257083998064D-02
12	5.5	-2.508631562766D-09	-1.254315781383D-09	5.75	2.600740388683D-02
13	6.5	2.361097808367D-10	1.180548904183D-10	6.25	-2.033887993766D-02
				6.50	-2.596292038937D-02
	0	4.387612620964D+02	1.581302774187D+02	6.75	-1.677833127974D-02
	1	8.059625576036D+02	5.122231849510D+02	7.25	1.451968127043D-02
	2	6.471533570340D+02	2.392886418206D+02	7.50	1.943737750907D-02
	3	2.932585267863D+02	1.829027737529D+02	7.75	1.316437963580D-02
	4	8.037299820804D+01	3.083839387210D+01	8.25	-1.253182802806D-02
	5	1.318675889874D+01	8.028090967999D+00	8.50	-1.763669104226D-02
	6	1.187652575894D+00	4.729652018687D-01	8.75	-1.258907991258D-02
	7	4.496659302205D-02	2.677945054419D-02	9.25	1.344682491852D-02

I= 4	H= -2.5				
0		0.	1.393825642274D+05	-1.00	2.569446995242D+02
1		-3.516043712087D+01	-9.375506978767D+04	-.75	1.259364926656D+02
2		8.777224344993D+01	2.703219974910D+04	-.50	5.272297501726D+01
3		-8.323544701647D+01	-4.352686169670D+03	-.25	1.574036693655D+01
4		3.938355979072D+01	4.322203721535D+02	.25	-4.460509599464D+00
5		-1.000528227325D+01	-2.864431180855D+01	.50	-3.872766239685D+00
6		1.316546520784D+00	1.409437327166D+00	.75	-1.801675863298D+00
7		-7.118335218919D-02	-4.579702643474D-02	1.25	9.247199897686D-01
				1.50	1.008971578948D+00
0	-6.5	1.355887831488D-09	6.779439157439D-10	1.75	5.759180243910D-01
1	-5.5	-1.339337357015D-08	-6.696686785076D-09	2.25	-4.283501797202D-01
2	-4.5	6.115198005557D-08	3.057599002779D-08	2.50	-5.595964473537D-01
3	-3.5	-1.720649483443D-07	-8.603247417215D-08	2.75	-3.849739444883D-01
4	-2.5	3.362364349901D-07	1.681182174950D-07	3.25	4.491670029736D-01
5	-1.5	-4.883939840299D-07	-2.441969920150D-07	3.50	8.197834010580D-01
6	-.5	5.511289442998D-07	2.755644721499D-07	3.75	1.014020899214D+00
7	.5	-4.965335698136D-07	-2.482667849068D-07	4.25	8.103588379726D-01
8	1.5	3.604423110338D-07	1.802211555169D-07	4.50	5.222134624162D-01
9	2.5	-2.081871334393D-07	-1.040935667197D-07	4.75	2.269091906028D-01
10	3.5	9.216169348165D-08	4.608084674082D-08	5.25	-1.192204331754D-01

11	4,5	-2,920260283238D-08	-1,460130141619D-08	5,50	-1,332715732170D-01
12	5.5	5.851513784234D-09	2.925756892117D-09	5.75	-7,724824885169D-02
13	6.5	-5,531534468927D-10	-2,765767234464D-10	6.25	5,622475471622D-02
				6,50	6,999174802248D-02
	0	-7,701601060510D+02	-2.408221910430D+02	6.75	4,432456013156D-02
	1	-1.407035735963D+03	-9,611119424922D+02	7,25	-3,722176296908D-02
	2	-1,121434655222D+03	-3,617024297096D+02	7,50	-4,927321028094D-02
	3	-5,026985291003D+02	-3,371131208409D+02	7.75	-3,306208454956D-02
	4	-1,356322398663D+02	-4,568414899561D+01	8,25	3,102756514256D-02
	5	-2,181739719252D+01	-1.430966775331D+01	8,50	4,342907172493D-02
	6	-1,922296003824D+00	-6,743273756149D-01	8,75	3,085770456995D-02
	7	-7,118335218919D-02	-4,579702643474D-02	9,25	-3,272564139448D-02

I= 5 H= -1.5

0		0.	-2,599442284508D+05	-1,00	-2,710320063285D+02
1		3.598491422111D+01	1.749849458683D+05	-,75	-1,320058483616D+02
2		-9,119995167346D+01	-5,046304575215D+04	-,50	-5,487601565923D+01
3		8,827919177062D+01	8,104271921322D+03	-,25	-1,625408231991D+01
4		-4,280688362552D+01	-7,932871763146D+02	,25	4,518922684036D+00
5		1,116822348104D+01	4,985353791851D+01	.50	3,878017873549D+00
6		-1,509167864378D+00	-2,162169782384D+00	.75	1,779972772027D+00
7		8,367369236490D-02	6,097359786876D-02	1.25	-8,827965723550D-01
				1.50	-9,421215131716D-01
0	-6,5	-1,788908141560D-09	-8,944540707799D-10	1.75	-5,235170034288D-01
1	-5,5	1.798262086901D-08	8.991310434505D-09	2.25	3,613593484596D-01
2	-4,5	-8,388685044020D-08	-4,194342522010D-08	2,50	4,474167682587D-01
3	-3,5	2,422769587431D-07	1.211384793716D-07	2,75	2,865546179821D-01
4	-2,5	-4,883939840299D-07	-2,441969920150D-07	3,25	-2,609305803515D-01
5	-1,5	7,351655034187D-07	3,675827517093D-07	3,50	-3,726861741215D-01
6	-,5	-8,621843635217D-07	-4,310921817609D-07	3,75	-2,783574566821D-01
7	,5	8,071393627079D-07	4,035696813540D-07	4,25	3,756922646913D-01
8	1,5	-6,065703959334D-07	-3,032851979667D-07	4,50	7,312951375811D-01
9	2,5	3,604423110338D-07	1,802211555169D-07	4,75	9,599173327002D-01
10	3,5	-1.630269378715D-07	-8,151346893574D-08	5.25	8,525306156213D-01
11	4,5	5,245691789400D-08	2,622845894700D-08	5,50	5,758893684283D-01
12	5,5	-1,062396649434D-08	-5,311983247170D-09	5,75	2,614812687824D-01
13	6,5	1,011731766096D-09	5,058658830479D-10	6,25	-1,489150389565D-01
				6,50	-1,728562630310D-01
	0	9,942490960441D+02	2,308504632323D+02	6,75	-1,039318659725D-01
	1	1.805935968171D+03	1.379467160896D+03	7.25	8,132554936993D-02
	2	1,428066780132D+03	3,450028048179D+02	7.50	1,050400784443D-01
	3	6,324671098508D+02	4,756664845664D+02	7.75	6,909963874503D-02
	4	1,679863172587D+02	4,274830746610D+01	8,25	-6,297582164712D-02
	5	2,655016032106D+01	1,962774111461D+01	8,50	-8,719258309584D-02
	6	2,297985138225D+00	6,121289206451D-01	8,75	-6,139608601110D-02
	7	8,367369236490D-02	6,097359786876D-02	9,25	6,421840636971D-02

I= 6 H= -.5

0		0.	3,994667447640D+05	-1.00	2,353390555081D+02
1		-3,073972762156D+01	-2,693274308974D+05	-,75	1,142240074552D+02
2		7,852490743973D+01	7,776504399019D+04	-,50	4,730193944899D+01
3		-7,684327261016D+01	-1,248336003852D+04	-,25	1,395106468233D+01
4		3,776896112186D+01	1,212454243589D+03	,25	-3,839709535014D+00
5		-1,000891565862D+01	-7,355673178802D+01	,50	-3,275376239158D+00
6		1,375659191559D+00	2,870508458370D+00	.75	-1,493140313756D+00

CARDINAL SPLINES N=8

7		-7.761186464313D-02	-6.851942955501D-02	1.25	7,282239126505D-01
				1.50	7,691149556726D-01
0	-6,5	1,828755938083D-09	9.143779690417D-10	1,75	4,222079931222D-01
1	-5.5	-1.870286011495D-08	-9,351430057476D-09	2,25	-2,823757928235D-01
2	-4,5	8,915847616346D-08	4.457923808173D-08	2,50	-3,424411208910D-01
3	-3,5	-2,645499499993D-07	-1.322749749996D-07	2,75	-2,138143331504D-01
4	-2,5	5,511289442998D-07	2.755644721499D-07	3,25	1,811946903227D-01
5	-1,5	-8,621843635217D-07	-4,310921817609D-07	3,50	2,456153702400D-01
6	-,5	1,055049800647D-06	5.275249003233D-07	3,75	1,710134207336D-01
7	,5	-1.031446678302D-06	-5,157233391511D-07	4,25	-1,805830870544D-01
8	1,5	8,071393627079D-07	4,035696813540D-07	4,50	-2,753954141058D-01
9	2,5	-4,965335698136D-07	-2,482667849068D-07	4,75	-2,185082018307D-01
10	3,5	2,308746915590D-07	1.154373457795D-07	5,25	3,283625766402D-01
11	4,5	-7,586437271348D-08	-3,793218635674D-08	5,50	6,705423205294D-01
12	5,5	1,560476760555D-08	7,802383802776D-09	5,75	9,204200727957D-01
13	6,5	-1,503004455301D-09	-7,515022276506D-10	6,25	8,871877265602D-01
				6,50	6,226470121907D-01
	0	-9,941831635305D+02	-9,568025918821D+01	6,75	2,934069722584D-01
	1	-1,797234306599D+03	-1,619747360618D+03	7,25	-1,797942530109D-01
	2	-1,408894333248D+03	-1,434604754674D+02	7,50	-2,166146745896D-01
	3	-6,167800591047D+02	-5,519024257503D+02	7,75	-1,353364117963D-01
	4	-1,616923265995D+02	-1,755499598386D+01	8,25	1,149927403191D-01
	5	-2,521933409244D+01	-2,240076578426D+01	8,50	1,553754003860D-01
	6	-2,155680649704D+00	-2,471255863825D-01	8,75	1,072804255779D-01
	7	-7,761186464313D-02	-6,851942955501D-02	9,25	-1,089985443111D-01

I= 7	H=	.5			
0		0.	-5.117299206363D+05	-1.00	-1,737289463927D+02
1		2,249939808960D+01	3,460342295654D+05	-.75	-8,416671135195D+01
2		-5,770897965187D+01	-1,001863864162D+05	-.50	-3,478461714472D+01
3		5,679259939681D+01	1,611089439609D+04	-.25	-1,023645660712D+01
4		-2,811299477243D+01	-1,560795916296D+03	.25	2,802692346376D+00
5		7,513048716227D+00	9,283252752586D+01	.50	2,383424885634D+00
6		-1,042498771305D+00	-3,364759631135D+00	.75	1,082769971437D+00
7		5,942699446689D-02	6.851942955501D-02	1.25	-5,236706887469D-01
				1.50	-5,502537764214D-01
0	-6,5	-1,503004455301D-09	-7,515022276506D-10	1,75	-3,002869164221D-01
1	-5,5	1,560476760555D-08	7,802383802776D-09	2,25	1,978719310227D-01
2	-4,5	-7,586437271348D-08	-3,793218635674D-08	2,50	2,377077616676D-01
3	-3,5	2,308746915590D-07	1.154373457795D-07	2,75	1,467657499447D-01
4	-2,5	-4,965335698136D-07	-2,482667849068D-07	3,25	-1,207371740955D-01
5	-1,5	8,071393627079D-07	4,035696813540D-07	3,50	-1,604439155481D-01
6	-,5	-1.031446678302D-06	-5,157233391511D-07	3,75	-1,089985443111D-01
7	,5	1,055049800647D-06	5,275249003233D-07	4,25	1,072804255779D-01
8	1,5	-8,621843635217D-07	-4,310921817609D-07	4,50	1,553754003860D-01
9	2,5	5,511289442998D-07	2,755644721499D-07	4,75	1,149927403191D-01
10	3,5	-2,645499499993D-07	-1,322749749996D-07	5,25	-1,353364117963D-01
11	4,5	8,915847616346D-08	4,457923808173D-08	5,50	-2,166146745896D-01
12	5,5	-1,870286011495D-08	-9,351430057476D-09	5,75	-1,797942530109D-01
13	6,5	1,828755938083D-09	9.143779690417D-10	6,25	2,934069722584D-01
				6,50	6,226470121907D-01
	0	8,028226451541D+02	-9,568025918821D+01	6,75	8,871877265602D-01
	1	1,442260414638D+03	1,619747360618D+03	7,25	9,204200727957D-01
	2	1,121973382313D+03	-1,434604754674D+02	7,50	6,705423205294D-01
	3	4,870247923959D+02	5,519024257503D+02	7,75	3,283625766402D-01
	4	1,265823346318D+02	-1,755499598386D+01	8,25	-2,185082018307D-01

	5	1.958219747608D+01	2.240076578426D+01	8.50	-2.753954141058D-01
	6	1.661429476939D+00	-2.471255863825D-01	8.75	-1.805830870544D-01
	7	5.942699446689D-02	6.851942955501D-02	9.25	1.710134207336D-01

I= 8 H= 1.5

0		0.	5.345076358059D+05	-1.00	1.097731503293D+02
1		-1.415440539770D+01	-3.631027440465D+05	-.75	5.313185560402D+01
2		3.638014257163D+01	1.056070946894D+05	-.50	2.193574063464D+01
3		-3.590577306900D+01	-1.705142010285D+04	-.25	6.447911701181D+00
4		1.783871548597D+01	1.654657198164D+03	.25	-1.760702881249D+00
5		-4.788123195065D+00	-9.759959372882D+01	.50	-1.494963753495D+00
6		6.677171065198D-01	3.386427623674D+00	.75	-6.779544684980D-01
7		-3.827350337263D-02	-6.097359786876D-02	1.25	3.264984707163D-01
				1.50	3.421921399167D-01
0	-6.5	1.011731766096D-09	5.058658830479D-10	1.75	1.861931906324D-01
1	-5.5	-1.062396649434D-08	-5.311983247170D-09	2.25	-1.217935194018D-01
2	-4.5	5.245691789400D-08	2.622845894700D-08	2.50	-1.456420023285D-01
3	-3.5	-1.630269378715D-07	-8.151346893574D-08	2.75	-8.943869798038D-02
4	-2.5	3.604423110338D-07	1.802211555169D-07	3.25	7.256011009914D-02
5	-1.5	-6.065703959334D-07	-3.032851979667D-07	3.50	9.555753522624D-02
6	-.5	8.071393627079D-07	4.035696813540D-07	3.75	6.421840636971D-02
7	.5	-8.621843635217D-07	-4.310921817609D-07	4.25	-6.139608601110D-02
8	1.5	7.351655034187D-07	3.675827517093D-07	4.50	-8.719258309584D-02
9	2.5	-4.883939840299D-07	-2.441969920150D-07	4.75	-6.297582164712D-02
10	3.5	2.422769587431D-07	1.211384793716D-07	5.25	6.909963874503D-02
11	4.5	-8.388685044020D-08	-4.194342522010D-08	5.50	1.050400784443D-01
12	5.5	1.798262086901D-08	8.991310434505D-09	5.75	8.132554936993D-02
13	6.5	-1.788908141560D-09	-8.944540707799D-10	6.25	-1.039318659725D-01
				6.50	-1.728562630310D-01
	0	-5.325481695795D+02	2.308504632323D+02	6.75	-1.489150389565D-01
	1	-9.529983536199D+02	-1.379467160896D+03	7.25	2.614812687824D-01
	2	-7.380611704961D+02	3.450028048179D+02	7.50	5.758893684283D-01
	3	-3.188658592821D+02	-4.756664845664D+02	7.75	8.525306156213D-01
	4	-8.248970232648D+01	4.274830746610D+01	8.25	9.599173327002D-01
	5	-1.270532190816D+01	-1.962774111461D+01	8.50	7.312951375811D-01
	6	-1.073727296935D+00	6.121289206451D-01	8.75	3.756922646913D-01
	7	-3.827350337263D-02	-6.097359786876D-02	9.25	-2.783574566821D-01

I= 9 H= 2.5

0		0.	-4.352402315268D+05	-1.00	-5.795707492130D+01
1		7.456048044068D+00	2.974172961563D+05	-.75	-2.803828875160D+01
2		-1.918447872883D+01	-8.702896347730D+04	-.50	-1.156948731786D+01
3		1.896275507466D+01	1.413574813857D+04	-.25	-3.398773859620D+00
4		-9.439013103187D+00	-1.378298618737D+03	.25	9.267983665248D-01
5		2.539323641522D+00	8.124184710652D+01	.50	7.862761396635D-01
6		-3.550456283590D-01	-2.758092078396D+00	.75	3.562437144743D-01
7		2.041070068029D-02	4.579702643474D-02	1.25	-1.711875742935D-01
				1.50	-1.791775254670D-01
0	-6.5	-5.531534468927D-10	-2.765767234464D-10	1.75	-9.734563629771D-02
1	-5.5	5.851513784234D-09	2.925756892117D-09	2.25	6.343640905276D-02
2	-4.5	-2.920260283238D-08	-1.460130141619D-08	2.50	7.567995950963D-02
3	-3.5	9.216169348165D-08	4.608084674082D-08	2.75	4.634782436741D-02
4	-2.5	-2.081871334393D-07	-1.040935667197D-07	3.25	-3.733921460592D-02
5	-1.5	3.604423110338D-07	1.802211555169D-07	3.50	-4.895453748619D-02
6	-.5	-4.965335698136D-07	-2.482667849068D-07	3.75	-3.272564139448D-02

```
 7     ,5   5,511289442998D-07   2,755644721499D-07    4.25  3,085770456995D-02
 8    1,5  -4,883939840299D-07  -2,441969920150D-07    4,50  4,342907172493D-02
 9    2,5   3,362364349901D-07   1,681182174950D-07    4,75  3,102756514256D-02
10    3,5  -1,720649483443D-07  -8,603247417215D-08    5,25 -3,306208454956D-02
11    4,5   6,115198005557D-08   3,057599002779D-08    5,50 -4,927321028094D-02
12    5,5  -1,339337357015D-08  -6,696686785076D-09    5,75 -3,722176296908D-02
13    6,5   1,355887831488D-09   6.779439157439D-10    6,25  4,432456013156D-02
                                                       6.50  6,999174802248D-02
        0   2,885157239650D+02  -2,408221910430D+02    6,75  5,622475471622D-02
        1   5,151881490213D+02   9.611119424922D+02    7,25 -7,724824885169D-02
        2   3.980297958026D+02  -3,617024297096D+02    7,50 -1,332715732170D-01
        3   1,715277125815D+02   3,371131208409D+02    7,75 -1,192204331754D-01
        4   4,426394187508D+01  -4,568414899561D+01    8,25  2,269091906028D-01
        5   6,801938314105D+00   1,430966775331D+01    8,50  5,222134624162D-01
        6   5,736412525940D-01  -6,743273756149D-01    8,75  8,103588379726D-01
        7   2,041070068029D-02   4,579702643474D-02    9,25  1,014020899214D+00


I=10      H=  3,5
0           0.                   2.608232554324D+05   -1,00  2,426678988057D+01
1          -3,118025528483D+00  -1,793994589800D+05    -,75  1,173659706714D+01
2           8,027364357513D+00   5,285976561827D+04    -,50  4,841482757541D+00
3          -7,940993448859D+00  -8,647435572468D+03    -,25  1,421828699161D+00
4           3,956802737795D+00   8,488937782158D+02     .25 -3,874227679242D-01
5          -1,065783654997D+00  -5,023380133621D+01     .50 -3,285368664079D-01
6           1,492278448615D-01   1,691430201630D+00     .75 -1,487792317899D-01
7          -8,592308066343D-03  -2,677945054419D-02    1.25  7,140914755592D-02
                                                       1,50  7,468873689619D-02
0    -6,5   2,361097808367D-10   1.180548904183D-10    1,75  4,054466446986D-02
1    -5,5  -2,508631562766D-09  -1.254315781383D-09    2,25 -2,636811608169D-02
2    -4,5   1,260324768537D-08   6,301623842685D-09    2,50 -3,141788442092D-02
3    -3,5  -4,018233396920D-08  -2,009116698460D-08    2,75 -1,921284453378D-02
4    -2,5   9,216169348165D-08   4,608084674082D-08    3,25  1,542123928421D-02
5    -1,5  -1,630269378715D-07  -8,151346893574D-08    3,50  2,017097363680D-02
6     -,5   2,308746915590D-07   1,154373457795D-07    3,75  1,344682491852D-02
7      ,5  -2,645499499993D-07  -1,322749749996D-07    4,25 -1,258907991258D-02
8     1,5   2,422769587431D-07   1,211384793716D-07    4,50 -1,763669104226D-02
9     2,5  -1,720649483443D-07  -8,603247417215D-08    4,75 -1,253182802806D-02
10    3,5   9,051702441868D-08   4,525851220934D-08    5,25  1,316437963580D-02
11    4,5  -3,293742215291D-08  -1,646871107646D-08    5,50  1,943737750907D-02
12    5,5   7,357658186390D-09   3,678829093195D-09    5,75  1,451968127043D-02
13    6,5  -7,571599551003D-10  -3,785799775502D-10    6,25 -1,677833127974D-02
                                                       6,50 -2,596292038937D-02
        0  -1,225007072589D+02   1,581302774187D+02    6,75 -2,033887993766D-02
        1  -2,184838122985D+02  -5,122231849510D+02    7,25  2,600740388683D-02
        2  -1,685760733929D+02   2,392886418206D+02    7,50  4,257083998064D-02
        3  -7,254702071957D+01  -1,829027737529D+02    7,75  3,548317614687D-02
        4  -1,869621046384D+01   3,083839387210D+01    8,25 -5,274992421435D-02
        5  -2,869423037263D+00  -8,028090967999D+00    8,50 -9,500211003553D-02
        6  -2,417221721571D-01   4,729652018687D-01    8,75 -8,898801928529D-02
        7  -8,592308066343D-03  -2,677945054419D-02    9,25  1,877653292563D-01


I=11      H=  4,5
0           0.                  -1,067629622666D+05   -1,00 -7,431418607624D+00
1           9,541984520137D-01   7,391983302261D+04    -,75 -3,593659362200D+00
2          -2,457386310845D+00  -2,193682464332D+04    -,50 -1,482183981550D+00
```

CARDINAL SPLINES N=8

3		2.432046822895D+00	3.616257645542D+03	-.25	-4.352039238981D-01
4		-1.212525560824D+00	-3.577809649937D+02	.25	1.185354170712D-01
5		3.268247388077D-01	2.131067177212D+01	.50	1.004939612712D-01
6		-4.579743210610D-02	-7.173007614628D-01	.75	4.549651306764D-02
7		2.639290131911D-03	1.107069518117D-02	1.25	-2.182235226115D-02
				1.50	-2.281544631660D-02
0	-6.5	-7.309998764120D-11	-3.654999382060D-11	1.75	-1.237966613287D-02
1	-5.5	7.786741813945D-10	3.893370906973D-10	2.25	8.041986637304D-03
2	-4.5	-3.927749586635D-09	-1.963874793317D-09	2.50	9.575393915355D-03
3	-3.5	1.260324768537D-08	6.301623842685D-09	2.75	5.850827983907D-03
4	-2.5	-2.920260283238D-08	-1.460130141619D-08	3.25	-4.686500930426D-03
5	-1.5	5.245691789400D-08	2.622845894700D-08	3.50	-6.121947489168D-03
6	-.5	-7.586437271348D-08	-3.793218635674D-08	3.75	-4.074896660178D-03
7	.5	8.915847616346D-08	4.457923808173D-08	4.25	3.800002115160D-03
8	1.5	-8.388685044020D-08	-4.194342522010D-08	4.50	5.310286195421D-03
9	2.5	6.115198005557D-08	3.057599002779D-08	4.75	3.762094602480D-03
10	3.5	-3.293742215291D-08	-1.646871107646D-08	5.25	-3.921939017584D-03
11	4.5	1.223296047939D-08	6.116480239697D-09	5.50	-5.762510479460D-03
12	5.5	-2.780509906427D-09	-1.390254953214D-09	5.75	-4.279667471231D-03
13	6.5	2.903511604883D-10	1.451755802441D-10	6.25	4.871256399188D-03
				6.50	7.464254332728D-03
	0	3.781011938468D+01	-6.671062608162D+01	6.75	5.778994910370D-03
	1	6.739017843371D+01	1.917202506458D+02	7.25	-7.161463352230D-03
	2	5.195771159317D+01	-1.017461800966D+02	7.50	-1.147866626482D-02
	3	2.234286298324D+01	6.960284568242D+01	7.75	-9.322520763740D-03
	4	5.753682689843D+00	-1.336335173228D+01	8.25	1.287397186138D-02
	5	8.824350562075D-01	3.158416374564D+00	8.50	2.197452902037D-02
	6	7.429026889583D-02	-2.135841307196D-01	8.75	1.915604515096D-02
	7	2.639290131911D-03	1.107069518117D-02	9.25	-3.148882017224D-02

I=12 H= 5.5

0		0.	2.644628186738D+04	-1.00	1.457934692727D+00
1		-1.871233408272D-01	-1.842666849525D+04	-.75	7.049610390697D-01
2		4.819988972239D-01	5.506699283791D+03	-.50	2.907290754827D-01
3		-4.771560486467D-01	-9.147496768501D+02	-.25	8.535580208964D-02
4		2.379722974422D-01	9.124562287036D+01	.25	-2.324238871070D-02
5		-6.416923984011D-02	-5.477858553937D+00	.50	-1.970195995020D-02
6		8.996151690418D-03	1.851398071208D-01	.75	-8.918191769103D-03
7		-5.187170568118D-04	-2.823955203807D-03	1.25	4.275928655027D-03
				1.50	4.469463551681D-03
0	-6.5	1.443423496967D-11	7.217117484833D-12	1.75	2.424478454580D-03
1	-5.5	-1.539941558327D-10	-7.699707791634D-11	2.25	-1.573923363642D-03
2	-4.5	7.786741813945D-10	3.893370906973D-10	2.50	-1.873258156820D-03
3	-3.5	-2.508631562766D-09	-1.254315781383D-09	2.75	-1.144063240771D-03
4	-2.5	5.851513784234D-09	2.925756892117D-09	3.25	9.152817812905D-04
5	-1.5	-1.062396649434D-08	-5.311983247170D-09	3.50	1.194712142947D-03
6	-.5	1.560476760555D-08	7.802383802776D-09	3.75	7.945115915656D-04
7	.5	-1.870286011495D-08	-9.351430057476D-09	4.25	-7.392149549488D-04
8	1.5	1.798262086901D-08	8.991310434505D-09	4.50	-1.031505279033D-03
9	2.5	-1.339337357015D-08	-6.696686785076D-09	4.75	-7.295217123314D-04
10	3.5	7.357658186390D-09	3.678829093195D-09	5.25	7.571851031460D-04
11	4.5	-2.780509906427D-09	-1.390254953214D-09	5.50	1.109430892644D-03
12	5.5	6.415149844286D-10	3.207574922143D-10	5.75	8.212432967525D-04
13	6.5	-6.784804150633D-11	-3.392402075316D-11	6.25	-9.269300203947D-04
				6.50	-1.412741935009D-03
	0	-7.452219902023D+00	1.658347378301D+01	6.75	-1.086876648324D-03

CARDINAL SPLINES N=8

```
         1  -1,327704332599D+01  -4.428007784279D+01     7.25  1,325056027722D-03
         2  -1.023207914728D+01   2.548099616372D+01     7.50  2,101616209930D-03
         3  -4,397999805854D+00  -1.631110940838D+01     7.75  1,685563372722D-03
         4  -1,132060374002D+00   3.404068205587D+00     8,25 -2,251741654559D-03
         5  -1,735510325701D-01  -7.629603308051D-01     8.50 -3,759722610548D-03
         6  -1,460547439452D-02   5.664984534755D-02     8.75 -3,190031765811D-03
         7  -5,187170568118D-04  -2,823955203807D-03     9,25  4,859243934124D-03


I=13     H=  6.5
0            0.                  -2,976778073267D+03    -1,00 -1,359746392002D-01
1            1.744768813344D-02   2.086291860382D+03     -.75 -6,574478184161D-02
2           -4,494772354738D-02  -6.275953441572D+02     -,50 -2,711181930141D-02
3            4,450348340651D-02   1,050249681777D+02     -,25 -7,959295414394D-03
4           -2,219990895778D-02  -1,056172813470D+01      .25  2,166983178775D-03
5            5,987716036273D-03   6.394469108138D-01      .50  1,836728686304D-03
6           -8,396870942724D-04  -2.175560218832D-02      .75  8,313204381871D-04
7            4,843202456167D-05   3,307412537154D-04     1.25 -3,984893452057D-04
                                                         1,50 -4,164645881970D-04
0   -6,5    -1,351641876520D-12  -6.758209382600D-13     1.75 -2,258751135034D-04
1   -5,5     1.443423496967D-11   7.217117484833D-12     2,25  1,465730867928D-04
2   -4,5    -7,309998764120D-11  -3,654999382060D-11     2,50  1,744042654481D-04
3   -3,5     2,361097808367D-10   1.180548904183D-10     2,75  1,064830173970D-04
4   -2,5    -5,531534468927D-10  -2,765767234464D-10     3,25 -8,512532858830D-05
5   -1,5     1,011731766096D-09   5,058658830479D-10     3,50 -1,110609587494D-04
6    -,5    -1,503004455301D-09  -7,515022276506D-10     3,75 -7,381711216493D-05
7     ,5     1,828755938083D-09   9.143779690417D-10     4,25  6,858227162366D-05
8    1,5    -1,788908141560D-09  -8,944540707799D-10     4,50  9,561414633425D-05
9    2,5     1,355887831488D-09   6,779439157439D-10     4,75  6,755080832341D-05
10   3,5    -7,571599551003D-10  -3,785799775502D-10     5,25 -6,992387495693D-05
11   4,5     2,903511604883D-10   1,451755802441D-10     5,50 -1,022784230431D-04
12   5,5    -6,784804150633D-11  -3,392402075316D-11     5,75 -7,555950046061D-05
13   6,5     7,254957916529D-12   3,627478958265D-12     6,25  8,485204892344D-05
                                                         6,50  1,289108094109D-04
         0   6,970336584651D-01  -1,851138121164D+00     6,75  9,880579037628D-05
         1   1,241543735751D+00   4,711946517159D+00     7,25 -1,193202741139D-04
         2   9,565463411245D-01  -2,863357528678D+00     7,50 -1,881187394786D-04
         3   4,110311900824D-01   1,757494797343D+00     7,75 -1,498220875268D-04
         4   1,057717323097D-01  -3,882728320410D-01     8,25  1,965930502463D-04
         5   1,621123315199D-02   8,442860282842D-02     8,50  3,245246986122D-04
         6   1,363970023284D-03  -6,706875144266D-03     8,75  2,716566345412D-04
         7   4,843202456167D-05   3,307412537154D-04     9,25 -3,994596857895D-04




                    N= 8       M=14                        P=  7,0      Q=15

I= 0     H= -7,0
0            1,000000000000D+00  -4,648991618107D+02    -1,00  9,390811224772D+00
1           -2,777849873046D+00   2,836628369700D+02     -.75  5,793821974956D+00
2            3,074134643446D+00  -7,321594643827D+01     -.50  3,423962539617D+00
```

CARDINAL SPLINES N=8

3		-1.796097224685D+00	9.980232808228D+00	-.25	1.917156449157D+00
4		6.080711136392D-01	-6.445154617177D-01	.25	4.718673820904D-01
5		-1.208406254870D-01	-1.086936964054D-02	.50	1.895261667887D-01
6		1.319985529724D-02	5.164318395329D-03	.75	5.407308182543D-02
7		-6.178891720426D-04	-2.911519696850D-04	1.25	-1.379993154927D-02
				1.50	-1.127609291319D-02
0	-7.0	7.391622380403D-12	3.695811190202D-12	1.75	-4.904182967650D-03
1	-6.0	-6.931337257333D-11	-3.465668628667D-11	2.25	2.147015888207D-03
2	-5.0	2.978202800668D-10	1.489101400334D-10	2.50	2.131401246159D-03
3	-4.0	-7.815498549264D-10	-3.907749274632D-10	2.75	1.093174959670D-03
4	-3.0	1.414138933446D-09	7.070694667232D-10	3.25	-6.252824186606D-04
5	-2.0	-1.899238191529D-09	-9.496190957646D-10	3.50	-6.931928011701D-04
6	-1.0	2.003366243783D-09	1.001683121892D-09	3.75	-3.922327270138D-04
7	0.0	-1.739537429677D-09	-8.697687148384D-10	4.25	2.649810097509D-04
8	1.0	1.284138768280D-09	6.420693841401D-10	4.50	3.152519504438D-04
9	2.0	-8.114016747412D-10	-4.057008373706D-10	4.75	1.900700191196D-04
10	3.0	4.283968120910D-10	2.141984060455D-10	5.25	-1.431432276971D-04
11	4.0	-1.793709471818D-10	-8.968547359090D-11	5.50	-1.784163995279D-04
12	5.0	5.493024009826D-11	2.746512004913D-11	5.75	-1.122097346098D-04
13	6.0	-1.077650270393D-11	-5.388251351965D-12	6.25	9.098005629432D-05
14	7.0	1.005073186280D-12	5.025365931398D-13	6.50	1.171699041073D-04
				6.75	7.598272551078D-05
	0	-1.077187490811D+01	-5.983216084626D+00	7.25	-6.522777677383D-05
	1	-1.794289247921D+01	-8.024157316205D+00	7.50	-8.634628760699D-05
	2	-1.304368616031D+01	-7.168828314195D+00	7.75	-5.755289462789D-05
	3	-5.355321764970D+00	-2.431922824258D+00	8.25	5.229313014509D-05
	4	-1.337216645308D+00	-7.244487746386D-01	8.50	7.135404078104D-05
	5	-2.022546610349D-01	-9.356337384262D-02	8.75	4.911842288359D-05
	6	-1.707671413285D-02	-9.102128119238D-03	9.25	-4.797240989236D-05
	7	-6.178891720426D-04	-2.911519696850D-04	9.50	-6.819468853062D-05
I= 1	H= -6.0				
0		0.	4.996178505437D+03	-1.00	-4.043659476210D+01
1		8.650584697700D+00	-3.055913567861D+03	-.75	-2.140560958193D+01
2		-1.533514654392D+01	7.948953485620D+02	-.50	-9.849520781763D+00
3		1.114118869131D+01	-1.112256681591D+02	-.25	-3.312806111090D+00
4		-4.276779421162D+00	8.046282363642D+00	.25	1.362448514705D+00
5		9.213429177825D-01	-7.138211672376D-02	.50	1.644025975411D+00
6		-1.063714159365D-01	-3.777026074748D-02	.75	1.409306589144D+00
7		5.181074296065D-03	2.399474646348D-03	1.25	6.014885345555D-01
				1.50	2.954476897788D-01
0	-7.0	-6.931337257333D-11	-3.465668628667D-11	1.75	1.001435179684D-01
1	-6.0	6.572264215869D-10	3.286132107935D-10	2.25	-3.397262449106D-02
2	-5.0	-2.860594609147D-09	-1.430297304573D-09	2.50	-3.130412013374D-02
3	-4.0	7.619169313154D-09	3.809584656577D-09	2.75	-1.517756406114D-02
4	-3.0	-1.401794814475D-08	-7.008974072377D-09	3.25	8.024957318760D-03
5	-2.0	1.916559152035D-08	9.582795760176D-09	3.50	8.648890776252D-03
6	-1.0	-2.057505057550D-08	-1.028752528775D-08	3.75	4.781755705555D-03
7	0.0	1.814089957357D-08	9.070449786787D-09	4.25	-3.118454999477D-03
8	1.0	-1.354474364339D-08	-6.772371821696D-09	4.50	-3.660299208279D-03
9	2.0	8.620479033417D-09	4.310239516708D-09	4.75	-2.181801013883D-03
10	3.0	-4.570353774174D-09	-2.285176887087D-09	5.25	1.613951919948D-03
11	4.0	1.918122828988D-09	9.590614144938D-10	5.50	1.997586678409D-03
12	5.0	-5.881966743398D-10	-2.940983371699D-10	5.75	1.248816059092D-03
13	6.0	1.154886055143D-10	5.774430275713D-11	6.25	-1.002968286579D-03
14	7.0	-1.077650270393D-11	-5.388251351965D-12	6.50	-1.286854686718D-03

CARDINAL SPLINES N=8

				6,75	-8,318321817066D-04
0		9,936941780635D+01	5.609329271441D+01	7.25	7,104654451824D-04
1		1.648247759891D+02	7.224709865387D+01	7,50	9,386048245022D-04
2		1,191409587673D+02	6.651410895427D+01	7,75	6,245421227282D-04
3		4.853305975786D+01	2.162886400642D+01	8,25	-5,659272473821D-04
4		1,198602891218D+01	6,592459758319D+00	8,50	-7,713858204098D-04
5		1.785068899102D+00	8.113263429740D-01	8,75	-5,305214772602D-04
6		1.475012245707D-01	7.980399692356D-02	9,25	5,174019166438D-04
7		5.181074296065D-03	2.399474646348D-03	9,50	7,350953874454D-04
I= 2	H= -5.0				
0		0.	-2,551678216676D+04	-1,00	1,064736275895D+02
1		-1,718054730621D+01	1,562195888136D+04	-,75	5,387007711271D+01
2		3,893561776674D+01	-4,080006973638D+03	-.50	2,340316370348D+01
3		-3,282313483950D+01	5,802216875659D+02	-,25	7,299047795734D+00
4		1,391073024584D+01	-4,507465885430D+01	.25	-2,323227337725D+00
5		-3,213365100198D+00	1.111570098846D+00	.50	-2,184311483692D+00
6		3,904657820166D-01	1.164708396563D-01	.75	-1,125634100852D+00
7		-1,976654899067D-02	-8,907023016006D-03	1,25	8,041196186496D-01
				1.50	1,187738980527D+00
0	-7,0	2,978202800668D-10	1.489101400334D-10	1,75	1,210824649552D+00
1	-6,0	-2,860594609147D-09	-1,430297304573D-09	2,25	6,884188283028D-01
2	-5,0	1,264117008778D-08	6.320585043891D-09	2,50	3,817676863403D-01
3	-4,0	-3,427040274747D-08	-1,713520137373D-08	2,75	1,444290795263D-01
4	-3,0	6,433557609639D-08	3,216778804820D-08	3,25	-5,932159496429D-02
5	-2,0	-8,991671594413D-08	-4,495835797206D-08	3.50	-5,941861067296D-02
6	-1,0	9,870144890083D-08	4,935072445041D-08	3,75	-3,109384537225D-02
7	0,0	-8,879160353704D-08	-4,439580176852D-08	4.25	1,879120239370D-02
8	1.0	6,734477107594D-08	3,367238553797D-08	4,50	2,146774574308D-02
9	2,0	-4,331663686922D-08	-2,165831843461D-08	4,75	1,251746880186D-02
10	3,0	2,311230606915D-08	1,155615303457D-08	5,25	-8,957501201945D-03
11	4,0	-9,735773721320D-09	-4,867886860660D-09	5,50	-1,094840451843D-02
12	5,0	2,991901352411D-09	1,495950676205D-09	5,75	-6,772851211819D-03
13	6,0	-5,881966743398D-10	-2,940983371699D-10	6,25	5,351295258309D-03
14	7,0	5,493024009826D-11	2,746512004913D-11	6.50	6,822578924306D-03
				6,75	4,386521485137D-03
0		-4,188119962417D+02	-2.421087692205D+02	7,25	-3,715008031012D-03
1		-6,912615158677D+02	-2,937476318791D+02	7,50	-4,891771873824D-03
2		-4,963568425463D+02	-2,836255610821D+02	7,75	-3,245828134316D-03
3		-2,003582668498D+02	-8,671113102740D+01	8,25	2,928217092745D-03
4		-4,886211911184D+01	-2,749244955443D+01	8,50	3,984399244064D-03
5		-7,153581166897D+00	-3.161981319058D+00	8,75	2,736244869296D-03
6		-5,780951185261D-01	-3.199732881280D-01	9,25	-2,662406783341D-03
7		-1,976654899067D-02	-8,907023016006D-03	9,50	-3,779178278550D-03
I= 3	H= -4.0				
0		0.	8.360000076211D+04	-1,00	-1,937298596303D+02
1		2,798699803384D+01	-5,120215977792D+04	-,75	-9,599487691900D+01
2		-6,786211723721D+01	1.340027618420D+04	-,50	-4,069094529581D+01
3		6,196069240046D+01	-1.924443812671D+03	-,25	-1,232307500578D+01
4		-2,812953356947D+01	1.564789368868D+02	.25	3,620121964348D+00
5		6.871409248863D+00	-5,475121283735D+00	.50	3,216405012019D+00
6		-8,732790080917D-01	-1.784195074451D-01	.75	1,538424762019D+00
7		4.583013238616D-02	1.971412687253D-02	1,25	-8,523354698088D-01
				1,50	-9,820728595000D-01

CARDINAL SPLINES N=8

0	-7.0	-7.815498549264D-10	-3.907749274632D-10	1.75	-6.025836821589D-01
1	-6.0	7.619169313154D-09	3.809584656577D-09	2.25	5.749467204472D-01
2	-5.0	-3.427040274747D-08	-1.713520137373D-08	2.50	9.601210491006D-01
3	-4.0	9.486978150462D-08	4.743489075231D-08	2.75	1.094015999306D+00
4	-3.0	-1.824607637688D-07	-9.123038188441D-08	3.25	7.553245974484D-01
5	-2.0	2.619670726928D-07	1.309835363464D-07	3.50	4.560032890261D-01
6	-1.0	-2.957118756666D-07	-1.478559378333D-07	3.75	1.864755660246D-01
7	0.0	2.730876808544D-07	1.365438404272D-07	4.25	-8.779665018102D-02
8	1.0	-2.116422055632D-07	-1.058211027816D-07	4.50	-9.334724385799D-02
9	2.0	1.382501076086D-07	6.912505380432D-08	4.75	-5.158578782343D-02
10	3.0	-7.449894632908D-08	-3.724947316454D-08	5.25	3.429122969948D-02
11	4.0	3.156895379606D-08	1.578447689803D-08	5.50	4.083930646588D-02
12	5.0	-9.735773721320D-09	-4.867886860660D-09	5.75	2.473841748540D-02
13	6.0	1.918122828988D-09	9.590614144938D-10	6.25	-1.894225593670D-02
14	7.0	-1.793709471818D-10	-8.968547359090D-11	6.50	-2.386757317105D-02
				6.75	-1.519546252639D-02
	0	1.074630573038D+03	6.443062013597D+02	7.25	1.267582443590D-02
	1	1.763570096587D+03	7.119879528666D+02	7.50	1.659404006577D-02
	2	1.256627536226D+03	7.443609362498D+02	7.75	1.095658572638D-02
	3	5.019654639605D+02	2.068758217879D+02	8.25	-9.808857975727D-03
	4	1.207004584892D+02	7.037944708863D+01	8.50	-1.330709776367D-02
	5	1.735289713437D+01	7.317095955406D+00	8.75	-9.115453283229D-03
	6	1.372397478830D+00	7.875727093090D-01	9.25	8.834435701946D-03
	7	4.583013238616D-02	1.971412687253D-02	9.50	1.252082079594D-02

I= 4 H= -3.0

0		0.	-2.008328842497D+05	-1.00	2.627087219249D+02
1		-3.589928469442D+01	1.230452733987D+05	-.75	1.287228189031D+02
2		8.967654961977D+01	-3.224188467811D+04	-.50	5.387171780971D+01
3		-8.512221233619D+01	4.657905853522D+03	-.25	1.607751556321D+01
4		4.032555547967D+01	-3.898723611450D+02	.25	-4.552243177369D+00
5		-1.025961816436D+01	1.629839365545D+01	.50	-3.950490569164D+00
6		1.352255862300D+00	5.506309631019D-02	.75	-1.836840021586D+00
7		-7.324576818529D-02	-2.892270991422D-02	1.25	9.415589812022D-01
				1.50	1.026559581855D+00
0	-7.0	1.414138933446D-09	7.070694667232D-10	1.75	5.854505681070D-01
1	-6.0	-1.401794814475D-08	-7.008974072377D-09	2.25	-4.345254848668D-01
2	-5.0	6.433557609639D-08	3.216778804820D-08	2.50	-5.669365991953D-01
3	-4.0	-1.824607637688D-07	-9.123038188441D-08	2.75	-3.894500298210D-01
4	-3.0	3.610650611245D-07	1.805325305622D-07	3.25	4.527342919720D-01
5	-2.0	-5.354204516842D-07	-2.677102258421D-07	3.50	8.244285229309D-01
6	-1.0	6.255538711023D-07	3.127769355512D-07	3.75	1.017101277980D+00
7	0.0	-5.973520709538D-07	-2.986760354769D-07	4.25	8.075134567428D-01
8	1.0	4.765515585195D-07	2.382757792598D-07	4.50	5.182610977438D-01
9	2.0	-3.182614235992D-07	-1.591307117996D-07	4.75	2.241288500485D-01
10	3.0	1.741210495965D-07	8.706052479827D-08	5.25	-1.163736730603D-01
11	4.0	-7.449894632908D-08	-3.724947316454D-08	5.50	-1.291361017295D-01
12	5.0	2.311230606915D-08	1.155615303457D-08	5.75	-7.421749199000D-02
13	6.0	-4.570353774174D-09	-2.285176887087D-09	6.25	5.288915729033D-02
14	7.0	4.283968120910D-10	2.141984060455D-10	6.50	6.498736976246D-02
				6.75	4.054414898613D-02
	0	-1.895296399306D+03	-1.204025943293D+03	7.25	-3.281802645393D-02
	1	-3.090300750957D+03	-1.138037817180D+03	7.50	-4.248238537007D-02
	2	-2.183046165105D+03	-1.369931975073D+03	7.75	-2.778914661057D-02
	3	-8.619524735101D+02	-3.250944543214D+02	8.25	2.452421714688D-02
	4	-2.041684685464D+02	-1.261743399365D+02	8.50	3.308857909030D-02

	5	-2,883476741040D+01	-1.115042480125D+01	8,75	2,256154179783D-02
	6	-2,236786778778D+00	-1.362149689486D+00	9.25	-2,170973905607D-02
	7	-7,324576818529D-02	-2,892270991422D-02	9.50	-3,068345657523D-02

I= 5 H= -2.0

0		0.	3,844382152926D+05	-1.00	-2,819493084586D+02
1		3,738432276393D+01	-2,356440704571D+05	-.75	-1,372832680614D+02
2		-9,480678807331D+01	6.180457626455D+04	-.50	-5,705178307703D+01
3		9,185280505871D+01	-8.962170338930D+03	-.25	-1,689265607347D+01
4		-4,459106328803D+01	7,637912186787D+02	.25	4,692669959089D+00
5		1,164994641214D+01	-3.526984641037D+01	.50	4,025231039772D+00
6		-1,576802867003D+00	4.030722738737D-01	.75	1,846575174567D+00
7		8,757999546056D-02	2.901292893187D-02	1.25	-9,146903340837D-01
				1.50	-9,754339342976D-01
0	-7.0	-1,899238191529D-09	-9.496190957646D-10	1.75	-5,415720442633D-01
1	-6,0	1,916559152035D-08	9.582795760176D-09	2.25	3,730556374995D-01
2	-5.0	-8,991671594413D-08	-4.495835797206D-08	2,50	4,613193261887D-01
3	-4,0	2,619670726928D-07	1.309835363464D-07	2,75	2,950325128512D-01
4	-3.0	-5,354204516842D-07	-2.677102258421D-07	3,25	-2,676871765397D-01
5	-2,0	8,242356214374D-07	4.121178107187D-07	3,50	-3,814842313447D-01
6	-1.0	-1,003148321576D-06	-5.015741607879D-07	3,75	-2,841918243022D-01
7	0.0	9,980938664792D-07	4.990469332396D-07	4,25	3,810815370436D-01
8	1.0	-8,264862197768D-07	-4.132431098884D-07	4,50	7,387810833768D-01
9	2,0	5,689276705436D-07	2,844638352718D-07	4,75	9,651834152630D-01
10	3.0	-3,182614235992D-07	-1.591307117996D-07	5.25	8,471387316022D-01
11	4,0	1,382501076086D-07	6.912505380432D-08	5,50	5,680566105831D-01
12	5,0	-4,331663686922D-08	-2,165831843461D-08	5,75	2,557408871458D-01
13	6.0	8,620479033417D-09	4.310239516708D-09	6.25	-1,425972764027D-01
14	7,0	-8,114016747412D-10	-4,057008373706D-10	6.50	-1,633777590211D-01
				6,75	-9,677160742709D-02
	0	2,475787874755D+03	1.726384711163D+03	7.25	7,298468622001D-02
	1	4,009056967232D+03	1.228012067792D+03	7.50	9,217796894669D-02
	2	2,806187107840D+03	1.934718422266D+03	7.75	5,911247134847D-02
	3	1,094694025833D+03	3.449363245557D+02	8.25	-5,065820558139D-02
	4	2,556047993940D+02	1.739049274400D+02	8.50	-6,760725284926D-02
	5	3,554404132695D+01	1.151349296321D+01	8.75	-4,568280295849D-02
	6	2,714616910565D+00	1.824705791535D+00	9.25	4,335382158780D-02
	7	8,757999546056D-02	2.901292893187D-02	9.50	6,095133547858D-02

I= 6 H= -1.0

0		0.	-6,203444084162D+05	-1,00	2,526169726566D+02
1		-3,295445655106D+01	3,805413771901D+05	-.75	1,225761460371D+02
2		8,423315108290D+01	-9,991169780388D+04	-.50	5,074534778660D+01
3		-8,249893676385D+01	1.452629599898D+04	-.25	1,496168289877D+01
4		4,059263569086D+01	-1,251805807267D+03	.25	-4,114685072685D+00
5		-1,077129896475D+01	6.116105510685D+01	,50	-3,508358422747D+00
6		1,482699553523D+00	-1.189289391460D+00	.75	-1,598546464401D+00
7		-8,379404962247D-02	-1.793790462147D-02	1.25	7,786995490425D-01
				1.50	8,218357877399D-01
0	-7,0	2,003366243783D-09	1.001683121892D-09	1.75	4,507822229030D-01
1	-6,0	-2,057505057550D-08	-1.028752528775D-08	2.25	-3,008865485550D-01
2	-5,0	9,870144890083D-08	4.935072445041D-08	2,50	-3,644435571342D-01
3	-4,0	-2,957118756666D-07	-1.478559378333D-07	2,75	-2,272316008413D-01
4	-3,0	6,255538711023D-07	3.127769355512D-07	3,25	1,918878000080D-01
5	-2.0	-1,003148321576D-06	-5.015741607879D-07	3,50	2,595393325851D-01

6	-1.0	1.278141875882D-06	6.390709379410D-07	3.75	1,802469950318D-01
7	0.0	-1.333654683171D-06	-6.668273415853D-07	4.25	-1,891122456398D-01
8	1.0	1,155182069479D-06	5,775910347395D-07	4,50	-2,872428054938D-01
9	2.0	-8,264862197768D-07	-4.132431098884D-07	4.75	-2,268423980572D-01
10	3.0	4,765515585195D-07	2.382757792598D-07	5.25	3,368958684963D-01
11	4.0	-2,116422055632D-07	-1,058211027816D-07	5,50	6,829385828551D-01
12	5,0	6.734477107594D-08	3.367238553797D-08	5.75	9,295049031994D-01
13	6.0	-1,354474364339D-08	-6.772371821696D-09	6.25	8,771891228146D-01
14	7.0	1.284138768280D-09	6.420693841401D-10	6.50	6,076461626199D-01
				6.75	2,820750193239D-01
	0	-2,540578745369D+03	-2.051666295830D+03	7.25	-1,665938541639D-01
	1	-4,086996581483D+03	-7.989378268243D+02	7.50	-1,962588706001D-01
	2	-2,834999088899D+03	-2.271357201307D+03	7.75	-1,195305424735D-01
	3	-1,094155653002D+03	-2,212866424302D+02	8,25	9,549866135035D-02
	4	-2,525662219535D+02	-2.006411262312D+02	8.50	1,243793072769D-01
	5	-3.472199477829D+01	-7.247203189947D+00	8.75	8,241230344686D-02
	6	-2,623208877978D+00	-2,068246717912D+00	9.25	-7,597788014866D-02
	7	-8,379404962247D-02	-1,793790462147D-02	9,50	-1,056755263492D-01

I= 7 H= 0.0

0		0.	8.697404593404D+05	-1.00	-1,971341922501D+02
1		2,549954415218D+01	-5,342998464487D+05	-.75	-9,548079830260D+01
2		-6,544155787494D+01	1.405004851030D+05	-.50	-3,944917352827D+01
3		6,445395168057D+01	-2.047729124279D+04	-.25	-1.160547392677D+01
4		-3,193803901479D+01	1,777373369460D+03	.25	3,175183428990D+00
5		8,545798703640D+00	-8,966070337373D+01	.50	2,699030362870D+00
6		-1,187499234435D+00	2.134778651755D+00	.75	1,225556677123D+00
7		6,780158951409D-02	6.019945952970D-24	1.25	-5,920466781334D-01
				1.50	-6,216711831458D-01
0	-7.0	-1,739537429677D-09	-8.697687148384D-10	1.75	-3,389945261101D-01
1	-6.0	1,814089957357D-08	9.070449786787D-09	2.25	2,229472213589D-01
2	-5.0	-8,879160353704D-08	-4.439580176852D-08	2.50	2,675129994928D-01
3	-4.0	2,730876808544D-07	1.365438404272D-07	2,75	1,649412306083D-01
4	-3.0	-5,973520709538D-07	-2,986760354769D-07	3,25	-1,352224189901D-01
5	-2.0	9,980938664792D-07	4.990469332396D-07	3.50	-1,793057817786D-01
6	-1.0	-1.333654683171D-06	-6.668273415853D-07	3.75	-1,215066536356D-01
7	0.0	1.464430896368D-06	7.322154481839D-07	4.25	1,188343097604D-01
8	1.0	-1.333654683171D-06	-6.668273415853D-07	4.50	1,714242739755D-01
9	2.0	9,980938664792D-07	4.990469332396D-07	4,75	1,262825219536D-01
10	3.0	-5,973520709538D-07	-2,986760354769D-07	5.25	-1,468958950458D-01
11	4.0	2,730876808544D-07	1,365438404272D-07	5,50	-2,334070671537D-01
12	5.0	-8,879160353704D-08	-4,439580176852D-08	5,75	-1,921008688688D-01
13	6.0	1.814089957357D-08	9,070449786787D-09	6.25	3,069514161218D-01
14	7.0	-1,739537429677D-09	-8,697687148384D-10	6.50	6,429676659628D-01
				6,75	9,025383699390D-01
	0	2,155000038382D+03	2.155000038382D+03	7,25	9,025383699390D-01
	1	3,445099174183D+03	1.002868730155D-20	7.50	6,429676659628D-01
	2	2,372980196614D+03	2,372980196614D+03	7,75	3,069514161218D-01
	3	9.090920506437D+02	1.291446840092D-20	8,25	-1,921008688688D-01
	4	2,083110604196D+02	2,083110604196D+02	8.50	-2,334070671537D-01
	5	2,843866646737D+01	1.152285944258D-21	8.75	-1,468958950458D-01
	6	2,134778651755D+00	2.134778651755D+00	9.25	1,262825219536D-01
	7	6,780158951409D-02	6,019945952970D-24	9,50	1,714242739755D-01

I= 8 H= 1.0

CARDINAL SPLINES N=8

0		0.	-1,056484934297D+06	-1.00	1,367281779580D+02
1		-1,760957183772D+01	6.507481458802D+05	-,75	6,616190571359D+01
2		4.528549058269D+01	-1.715838099138D+05	-,50	2,730775197604D+01
3		-4.472909258210D+01	2.508595203972D+04	-,25	8,024562504177D+00
4		2,224388917490D+01	-2,189799120525D+03	.25	-2,189688224300D+00
5		-5,977506319425D+00	1.125716691977D+02	.50	-1,858435874682D+00
6		8,347092207516D-01	-2.947204044363D+00	.75	-8,423970733832D-01
7		-4,791824037953D-02	1,793790462147D-02	1.25	4,052447780307D-01
				1,50	4,244411443988D-01
0	-7,0	1,284138768280D-09	6,420693841401D-10	1.75	2,307714315506D-01
1	-6,0	-1,354474364339D-08	-6,772371821696D-09	2,25	-1,506718806003D-01
2	-5,0	6,734477107594D-08	3,367238553797D-08	2,50	-1,799676835862D-01
3	-4,0	-2,116422055632D-07	-1,058211027816D-07	2,75	-1,103707824557D-01
4	-3,0	4,765515585195D-07	2,382757792598D-07	3,25	8,924227523096D-02
5	-2,0	-8,264862197768D-07	-4,132431098884D-07	3,50	1,172801066733D-01
6	-1,0	1,155182069479D-06	5.775910347395D-07	3,75	7,862357153983D-02
7	0,0	-1,333654683171D-06	-6,668273415853D-07	4,25	-7.470230245737D-02
8	1,0	1,278141875882D-06	6,390709379410D-07	4.50	-1,056755263492D-01
9	2,0	-1,003148321576D-06	-5,015741607879D-07	4,75	-7,597788014866D-02
10	3,0	6,255538711023D-07	3,127769355512D-07	5,25	8,241230344686D-02
11	4,0	-2,957118756666D-07	-1,478559378333D-07	5,50	1,243793072769D-01
12	5,0	9,870144890083D-08	4,935072445041D-08	5,75	9,549866135035D-02
13	6,0	-2,057505057550D-08	-1,028752528775D-08	6,25	-1,195305424735D-01
14	7,0	2,003366243783D-09	1,001683121892D-09	6,50	-1,962588706001D-01
				6,75	-1,665938541639D-01
	0	-1,562753846291D+03	-2,051666295830D+03	7.25	2,820750193239D-01
	1	-2.489120927834D+03	7.989378268243D+02	7.50	6.076461626199D-01
	2	-1,707715313716D+03	-2.271357201307D+03	7,75	8,771891228146D-01
	3	-6,515823681415D+02	2,212866424302D+02	8,25	9,295049031994D-01
	4	-1,487160305089D+02	-2,006411262312D+02	8,50	6,829385828551D-01
	5	-2,022758839840D+01	7,247203189947D+00	8,75	3,368958684963D-01
	6	-1,513284557846D+00	-2,068246717912D+00	9,25	-2,268423980572D-01
	7	-4,791824037953D-02	1.793790462147D-02	9,50	-2,872428054938D-01

I= 9 H= 2.0

0		0.	1,073058044415D+06	-1.00	-8,351107339014D+01
1		1.073162687612D+01	-6.637371291215D+05	-,75	-4,039108189786D+01
2		-2,762695753366D+01	1,757545160790D+05	-,50	-1,666228007933D+01
3		2,732746893930D+01	-2,581146905757D+04	-,25	-4,893475803933D+00
4		-1,361522109127D+01	2,266336149758D+03	.25	1,333486534331D+00
5		3,666886742455D+00	-1,180054400786D+02	.50	1,130856236416D+00
6		-5,133580697398D-01	3,246339309196D+00	.75	5,121391611913D-01
7		2.955413759683D-02	-2.901292893187D-02	1.25	-2,458409207584D-01
				1.50	-2,571515110087D-01
0	-7,0	-8,114016747412D-10	-4,057008373706D-10	1,75	-1,396068543480D-01
1	-6,0	8.620479033417D-09	4.310239516708D-09	2,25	9,081377286083D-02
2	-5,0	-4,331663686922D-08	-2,165831843461D-08	2.50	1,082215105102D-01
3	-4,0	1,382501076086D-07	6,912505380432D-08	2,75	6,619193144833D-02
4	-3,0	-3,182614235992D-07	-1,591307117996D-07	3,25	-5,315429845994D-02
5	-2,0	5,689276705436D-07	2,844638352718D-07	3,50	-6,954804475852D-02
6	-1,0	-8,264862197768D-07	-4,132431098884D-07	3,75	-4,638207587384D-02
7	0,0	9,980938664792D-07	4,990469332396D-07	4,25	4,347230988135D-02
8	1.0	-1,003148321576D-06	-5,015741607879D-07	4,50	6,095133547858D-02
9	2,0	8,242356214374D-07	4,121178107187D-07	4,75	4,335382158780D-02
10	3,0	-5,354204516842D-07	-2,677102258421D-07	5,25	-4,568280295849D-02
11	4,0	2,619670726928D-07	1,309835363464D-07	5,50	-6,760725284926D-02

12	5.0	-8.991671594413D-08	-4.495835797206D-08	5.75	-5.065820558139D-02
13	6.0	1.916559152035D-08	9.582795760176D-09	6.25	5.911247134847D-02
14	7.0	-1.899238191529D-09	-9.496190957646D-10	6.50	9.217796894669D-02
				6.75	7.298468622001D-02
	0	9.769815475699D+02	1.726384711163D+03	7.25	-9.677160742709D-02
	1	1.553032831648D+03	-1.228012067792D+03	7.50	-1.633777590211D-01
	2	1.063249736691D+03	1.934718422266D+03	7.75	-1.425972764027D-01
	3	4.048213767218D+02	-3.449363245557D+02	8.25	2.557408871458D-01
	4	9.220505548590D+01	1.739049274400D+02	8.50	5.680566105831D-01
	5	1.251705540053D+01	-1.151349296321D+01	8.75	8.471387316022D-01
	6	9.347946725050D-01	1.824705791535D+00	9.25	9.651834152630D-01
	7	2.955413759683D-02	-2.901292893187D-02	9.50	7.387810833768D-01
I=10	H= 3.0				
0		0.	-8.622287792058D+05	-1.00	4.329384204348D+01
1		-5.556963191297D+00	5.362590086634D+05	-.75	2.093426685013D+01
2		1.431348372156D+01	-1.428041332884D+05	-.50	8.633485549023D+00
3		-1.416921042338D+01	2.109655092268D+04	-.25	2.534757147409D+00
4		7.066332830678D+00	-1.864836362273D+03	.25	-6.902355348702D-01
5		-1.905347058924D+00	9.812218026141D+01	.50	-5.851050697442D-01
6		2.671044692912D-01	-2.779362475283D+00	.75	-2.648562050379D-01
7		-1.540034835685D-02	2.892270991422D-02	1.25	1.269946998152D-01
				1.50	1.327467778741D-01
0	-7.0	4.283968120910D-10	2.141984060455D-10	1.75	7.201161479043D-02
1	-6.0	-4.570353774174D-09	-2.285176887087D-09	2.25	-4.675281360346D-02
2	-5.0	2.311230606915D-08	1.155615303457D-08	2.50	-5.564774375604D-02
3	-4.0	-7.449894632908D-08	-3.724947316454D-08	2.75	-3.398841325801D-02
4	-3.0	1.741210495965D-07	8.706052479827D-08	3.25	2.719686906747D-02
5	-2.0	-3.182614235992D-07	-1.591307117996D-07	3.50	3.550453239730D-02
6	-1.0	4.765515585195D-07	2.382757792598D-07	3.75	2.361516282129D-02
7	0.0	-5.973520709538D-07	-2.986760354769D-07	4.25	-2.198169037107D-02
8	1.0	6.255538711023D-07	3.127769355512D-07	4.50	-3.068345657523D-02
9	2.0	-5.354204516842D-07	-2.677102258421D-07	4.75	-2.170973905607D-02
10	3.0	3.610650611245D-07	1.805325305622D-07	5.25	2.256154179783D-02
11	4.0	-1.824607637688D-07	-9.123038188441D-08	5.50	3.308857909030D-02
12	5.0	6.433557609639D-08	3.216778804820D-08	5.75	2.452421714688D-02
13	6.0	-1.401794814475D-08	-7.008974072377D-09	6.25	-2.778914661057D-02
14	7.0	1.414138933446D-09	7.070694667232D-10	6.50	-4.248238537007D-02
				6.75	-3.281802645393D-02
	0	-5.127554872792D+02	-1.204025943293D+03	7.25	4.054414898613D-02
	1	-8.142251165967D+02	1.138037817180D+03	7.50	6.498736976246D-02
	2	-5.568177850406D+02	-1.369931975073D+03	7.75	5.288915729033D-02
	3	-2.117635648672D+02	3.250944543214D+02	8.25	-7.421749199000D-02
	4	-4.818021132659D+01	-1.261743399365D+02	8.50	-1.291361017295D-01
	5	-6.533917807891D+00	1.115042480125D+01	8.75	-1.163736730603D-01
	6	-4.875126001944D-01	-1.362149689486D+00	9.25	2.241288500485D-01
	7	-1.540034835685D-02	2.892270991422D-02	9.50	5.182610977438D-01
I=11	H= 4.0				
0		0.	5.139123716677D+05	-1.00	-1.794706845354D+01
1		2.302122147107D+00	-3.216019635525D+05	-.75	-8.676920500507D+00
2		-5.931525633949D+00	8.620042282794D+04	-.50	-3.577903971623D+00
3		5.874185326940D+00	-1.282230279601D+04	-.25	-1.050284271875D+00
4		-2.931064479104D+00	1.142011839975D+03	.25	2.858904544576D-01
5		7.908248741701D-01	-6.068098629822D+01	.50	2.422910912092D-01

CARDINAL SPLINES N=8

6		-1.109441136261D-01	1.753564926063D+00	.75	1.096485843080D-01
7		6.401878641097D-03	-1.971412687253D-02	1.25	-5.254272865188D-02
				1.50	-5.490228900962D-02
0	-7.0	-1.793709471818D-10	-8.968547359090D-11	1.75	-2.977045417806D-02
1	-6.0	1.918122828988D-09	9.590614144938D-10	2.25	1.930796412654D-02
2	-5.0	-9.735773721320D-09	-4.867886860660D-09	2.50	2.296647070446D-02
3	-4.0	3.156895379606D-08	1.578447689803D-08	2.75	1.401681742084D-02
4	-3.0	-7.449894632908D-08	-3.724947316454D-08	3.25	-1.119451901204D-02
5	-2.0	1.382501076086D-07	6.912505380432D-08	3.50	-1.459632036820D-02
6	-1.0	-2.116422055632D-07	-1.058211027816D-07	3.75	-9.694615363054D-03
7	0.0	2.730876808544D-07	1.365438404272D-07	4.25	8.991000798238D-03
8	1.0	-2.957118756666D-07	-1.478559378333D-07	4.50	1.252082079594D-02
9	2.0	2.619670726928D-07	1.309835363464D-07	4.75	8.834435701946D-03
10	3.0	-1.824607637688D-07	-9.123038188441D-08	5.25	-9.115453283229D-03
11	4.0	9.486978150462D-08	4.743489075231D-08	5.50	-1.330709776367D-02
12	5.0	-3.427040274747D-08	-1.713520137373D-08	5.75	-9.808857975727D-03
13	6.0	7.619169313154D-09	3.809584656577D-09	6.25	1.095658572638D-02
14	7.0	-7.815498549264D-10	-3.907749274632D-10	6.50	1.659404006577D-02
				6.75	1.267582443590D-02
	0	2.139818296814D+02	6.443062013597D+02	7.25	-1.519546252639D-02
	1	3.395941908539D+02	-7.119879528666D+02	7.50	-2.386757317105D-02
	2	2.320943362735D+02	7.443609362498D+02	7.75	-1.894225593670D-02
	3	8.821382038483D+01	-2.068758217879D+02	8.25	2.473841748540D-02
	4	2.005843568803D+01	7.037944708863D+01	8.50	4.083930646588D-02
	5	2.718705223562D+00	-7.317095955406D+00	8.75	3.429122969948D-02
	6	2.027479397877D-01	7.875727093090D-01	9.25	-5.158578782343D-02
	7	6.401878641097D-03	-1.971412687253D-02	9.50	-9.334724385799D-02
I=12	H= 5.0				
0		0.	-2.100705578680D+05	-1.00	5.465067139178D+00
1		-7.007681324564D-01	1.322923310491D+05	-.75	2.642007340493D+00
2		1.805867284731D+00	-3.570048043039D+04	-.50	1.089331890710D+00
3		-1.788830085615D+00	5.349389000598D+03	-.25	3.197405957702D-01
4		8.928450975860D-01	-4.802709738027D+02	.25	-8.701531860229D-02
5		-2.409828485864D-01	2.576618610390D+01	.50	-7.373569904515D-02
6		3.382118724434D-02	-7.564174159123D-01	.75	-3.336421744903D-02
7		-1.952502958656D-03	8.907023016006D-03	1.25	1.598234897201D-02
				1.50	1.669659457481D-02
0	-7.0	5.493024009826D-11	2.746512004913D-11	1.75	9.051476506661D-03
1	-6.0	-5.881966743398D-10	-2.940983371699D-10	2.25	-5.866978438659D-03
2	-5.0	2.991901352411D-09	1.495950676205D-09	2.50	-6.976111435584D-03
3	-4.0	-9.735773721320D-09	-4.867886860660D-09	2.75	-4.255825396184D-03
4	-3.0	2.311230606915D-08	1.155615303457D-08	3.25	3.395251236249D-03
5	-2.0	-4.331663686922D-08	-2.165831843461D-08	3.50	4.423988034068D-03
6	-1.0	6.734477107594D-08	3.367238553797D-08	3.75	2.935982397858D-03
7	0.0	-8.879160353704D-08	-4.439580176852D-08	4.25	-2.717316123896D-03
8	1.0	9.870144890083D-08	4.935072445041D-08	4.50	-3.779178278550D-03
9	2.0	-8.991671594413D-08	-4.495835797206D-08	4.75	-2.662406783341D-03
10	3.0	6.433557609639D-08	3.216778804820D-08	5.25	2.736244869296D-03
11	4.0	-3.427040274747D-08	-1.713520137373D-08	5.50	3.984399244064D-03
12	5.0	1.264117008778D-08	6.320585043891D-09	5.75	2.928217092745D-03
13	6.0	-2.860594609147D-09	-1.430297304573D-09	6.25	-3.245828134316D-03
14	7.0	2.978202800668D-10	1.489101400334D-10	6.50	-4.891771873824D-03
				6.75	-3.715008031012D-03
	0	-6.540554219932D+01	-2.421087692205D+02	7.25	4.386521485137D-03
	1	-1.037662521096D+02	2.937476318791D+02	7.50	6.822578924306D-03

	2	-7.089427961796D+01	-2.836255610821D+02	7.75	5.351295258309D-0
	3	-2.693600479501D+01	8.671113102740D+01	8.25	-6.772851211819D-03
	4	-6.122779997013D+00	-2.749244955443D+01	8.50	-1.094840451843D-02
	5	-8.296185287812D-01	3.161981319058D+00	8.75	-8.957501201945D-03
	6	-6.185145772981D-02	-3.199732881280D-01	9.25	1.251746880186D-02
	7	-1.952502958656D-03	8.907023016006D-03	9.50	2.146774574308D-02

I=13	H= 6.0				
0		0.	5.206910338508D+04	-1.00	-1.068576892308D+00
1		1.369911013777D-01	-3.299138037522D+04	-.75	-5.165643564933D-01
2		-3.530589870274D-01	8.962781105642D+03	-.50	-2.129751005312D-01
3		3.497766877657D-01	-1.352862916098D+03	-.25	-6.250897391936D-02
4		-1.746121032326D-01	1.224505126306D+02	.25	1.700921235362D-02
5		4.713853201266D-02	-6.632153624855D+00	.50	1.441230174190D-02
6		-6.617355888720D-03	1.973782545946D-01	.75	6.520780153657D-03
7		3.821250033697D-04	-2.399474646348D-03	1.25	-3.122989747761D-03
				1.50	-3.262152893638D-03
0	-7.0	-1.077650270393D-11	-5.388251351965D-12	1.75	-1.768213281084D-03
1	-6.0	1.154886055143D-10	5.774430275713D-11	2.25	1.145719705773D-03
2	-5.0	-5.881966743398D-10	-2.940983371699D-10	2.50	1.362019730784D-03
3	-4.0	1.918122828988D-09	9.590614144938D-10	2.75	8.307007556563D-04
4	-3.0	-4.570353774174D-09	-2.285176887087D-09	3.25	-6.623024327247D-04
5	-2.0	8.620479033417D-09	4.310239516708D-09	3.50	-8.626283232457D-04
6	-1.0	-1.354474364339D-08	-6.772371821696D-09	3.75	-5.722135404343D-04
7	0.0	1.814089957357D-08	9.070449786787D-09	4.25	5.289567742608D-04
8	1.0	-2.057505057550D-08	-1.028752528775D-08	4.50	7.350953874454D-04
9	2.0	1.916559152035D-08	9.582795760176D-09	4.75	5.174019166438D-04
10	3.0	-1.401794814475D-08	-7.008974072377D-09	5.25	-5.305214772602D-04
11	4.0	7.619169313154D-09	3.809584656577D-09	5.50	-7.713858204098D-04
12	5.0	-2.860594609147D-09	-1.430297304573D-09	5.75	-5.659272473821D-04
13	6.0	6.572264215869D-10	3.286132107935D-10	6.25	6.245421227282D-04
14	7.0	-6.931337257333D-11	-3.465668628667D-11	6.50	9.386048245022D-04
				6.75	7.104654451824D-04
	0	1.281716762247D+01	5.609329271441D+01	7.25	-8.318321817066D-04
	1	2.033057868131D+01	-7.224709865387D+01	7.50	-1.286854686718D-03
	2	1.388725914120D+01	6.651410895427D+01	7.75	-1.002968286579D-03
	3	5.275331745015D+00	-2.162886400642D+01	8.25	1.248816059092D-03
	4	1.198890604455D+00	6.592459758319D+00	8.50	1.997586678409D-03
	5	1.624162131539D-01	-8.113263429740D-01	8.75	1.613951919948D-03
	6	1.210676927640D-02	7.980399692356D-02	9.25	-2.181801013883D-03
	7	3.821250033697D-04	-2.399474646348D-03	9.50	-3.660299208279D-03

I=14	H= 7.0				
0		0.	-5.870128003165D+03	-1.00	9.945330071472D-02
1		-1.274818604204D-02	3.740705401036D+03	-.75	4.807568776132D-02
2		3.285718218925D-02	-1.022723878106D+03	-.50	1.982057917871D-02
3		-3.255452974953D-02	1.554700963675D+02	-.25	5.817212602871D-03
4		1.625333387228D-02	-1.418451042284D+01	.25	-1.582784811514D-03
5		-4.388349333617D-03	7.754481316565D-01	.50	-1.341067153323D-03
6		6.161342953326D-04	-2.336857463380D-02	.75	-6.067276227365D-04
7		-3.558523267255D-05	2.911519696850D-04	1.25	2.905424653755D-04
				1.50	3.034660208446D-04
0	-7.0	1.005073186280D-12	5.025365931398D-13	1.75	1.644759285891D-04
1	-6.0	-1.077650270393D-11	-5.388251351965D-12	2.25	-1.065496344496D-04
2	-5.0	5.493024009826D-11	2.746512004913D-11	2.50	-1.266480727597D-04

CARDINAL SPLINES N=8

3	-4.0	-1.793709471818D-10	-8.968547359090D-11	2.75	-7.723104277950D-05
4	-3.0	4.283968120910D-10	2.141984060455D-10	3.25	6.155053551719D-05
5	-2.0	-8.114016747412D-10	-4.057008373706D-10	3.50	8.014762441279D-05
6	-1.0	1.284138768280D-09	6.420693841401D-10	3.75	5.314931385876D-05
7	0.0	-1.739537429677D-09	-8.697687148384D-10	4.25	-4.909463139251D-05
8	1.0	2.003366243783D-09	1.001683121892D-09	4.50	-6.819468853062D-05
9	2.0	-1.899238191529D-09	-9.496190957646D-10	4.75	-4.797240989236D-05
10	3.0	1.414138933446D-09	7.070694667232D-10	5.25	4.911842288359D-05
11	4.0	-7.815498549264D-10	-3.907749274632D-10	5.50	7.135404078104D-05
12	5.0	2.978202800668D-10	1.489101400334D-10	5.75	5.229313014509D-05
13	6.0	-6.931337257333D-11	-3.465668628667D-11	6.25	-5.755289462789D-05
14	7.0	7.391622380403D-12	3.695811190202D-12	6.50	-8.634628760699D-05
				6.75	-6.522777677383D-05
	0	-1.194557261140D+00	-5.983216084626D+00	7.25	7.598272551078D-05
	1	-1.894577846802D+00	8.024157316205D+00	7.50	1.171699041073D-04
	2	-1.293970468083D+00	-7.168828314195D+00	7.75	9.098005629432D-05
	3	-4.914761164542D-01	2.431922824258D+00	8.25	-1.122097346098D-04
	4	-1.116809039688D-01	-7.244487746386D-01	8.50	-1.784163995279D-04
	5	-1.512791334970D-02	9.356337384262D-02	8.75	-1.431432276971D-04
	6	-1.127542105622D-03	-9.102128119238D-03	9.25	1.900700191196D-04
	7	-3.558523267255D-05	2.911519696850D-04	9.50	3.152519504438D-04

N= 8 M=15 P= 7.5 Q=15

I= 0	H= -7.5				
0		1.000000000000D+00	6.156851089787D+02	-1.00	9.398055229021D+00
1		-2.778777922729D+00	-3.476564282708D+02	-.75	5.797323313153D+00
2		3.076527212757D+00	8.499489911241D+01	-.50	3.425405880704D+00
3		-1.798468600216D+00	-1.206258131850D+01	-.25	1.917580000387D+00
4		6.092555926601D-01	1.199524436091D+00	.25	4.717521777173D-01
5		-1.211606042510D-01	-1.035671925184D-01	.50	1.894285750408D-01
6		1.324480908994D-02	7.767924671038D-03	.75	5.402893885136D-02
7		-6.204873178316D-04	-3.233057984845D-04	1.25	-1.377880396080D-02
				1.50	-1.125403253294D-02
0	-7.5	7.465462968715D-12	3.732731484357D-12	1.75	-4.892230756881D-03
1	-6.5	-7.010675344473D-11	-3.505337672236D-11	2.25	2.139280011965D-03
2	-5.5	3.018779007774D-10	1.509389503887D-10	2.50	2.122211253275D-03
3	-4.5	-7.948749239141D-10	-3.974374619571D-10	2.75	1.087574439904D-03
4	-3.5	1.446283093005D-09	7.231415465023D-10	3.25	-6.208262730518D-04
5	-2.5	-1.961242235089D-09	-9.806211175444D-10	3.50	-6.873962403509D-04
6	-1.5	2.104882121352D-09	1.052441060676D-09	3.75	-3.883934283766D-04
7	-.5	-1.885902654637D-09	-9.429513273183D-10	4.25	2.614455530853D-04
8	.5	1.471698012110D-09	7.358490060552D-10	4.50	3.103506704154D-04
9	1.5	-1.020739149325D-09	-5.103695746625D-10	4.75	1.866300950482D-04
10	2.5	6.234488258717D-10	3.117244129359D-10	5.25	-1.396417958383D-04
11	3.5	-3.232085745490D-10	-1.616042872745D-10	5.50	-1.733487985252D-04
12	4.5	1.339831749369D-10	6.699158746844D-11	5.75	-1.085120079266D-04
13	5.5	-4.079925634907D-11	-2.039962817454D-11	6.25	8.695513670666D-05
14	6.5	7.977421218751D-12	3.988710609375D-12	6.50	1.111731179867D-04

CARDINAL SPLINES N=8

15	7.5	-7.424649336022D-13	-3,712324668011D-13	6.75	7,148909684060D-05
				7,25	-6,009838207182D-05
	0	-2,394940555327D+01	-1.104170460505D+01	7,50	-7,853565726790D-05
	1	-3,600296863597D+01	-1.933284436868D+01	7.75	-5,157591216324D-05
	2	-2,351404955351D+01	-1,094095059391D+01	8.25	4,518330349918D-05
	3	-8,634686907961D+00	-4,595120971136D+00	8,50	6,029750724651D-05
	4	-1,920842449477D+00	-9,038710229108D-01	8,75	4,047044752401D-05
	5	-2,580948393925D-01	-1.359155567816D-01	9,25	-3,718053253688D-05
	6	-1,933077509622D-02	-9,205629749400D-03	9,50	-5,095457834236D-05
	7	-6,204873178316D-04	-3,233057984845D-04	9,75	-3,520679387866D-05
I= 1	H= -6.5				
0		0.	-6,614169000971D+03	-1,00	-4,051442803151D+01
1		8,660556135944D+00	3,727302008265D+03	-.75	-2,144322974198D+01
2		-1,536085352660D+01	-9,050025176028D+02	-,50	-9,865028773173D+00
3		1,116666795723D+01	1.256135483174D+02	-,25	-3,317356961120D+00
4		-4,289506066753D+00	-1,176702105748D+01	.25	1,363686329138D+00
5		9,247809325174D-01	9,246104826445D-01	.50	1,645074550832D+00
6		-1,068544223702D-01	-6,574473201084D-02	.75	1,409780883730D+00
7		5,208990097079D-03	2,744951751757D-03	1,25	6,012615289894D-01
				1,50	2,952106618230D-01
0	-7,5	-7,010675344473D-11	-3,505337672236D-11	1,75	1,000150973313D-01
1	-6,5	6,657509095615D-10	3,328754547808D-10	2,25	-3,388950629856D-02
2	-5,5	-2,904191751705D-09	-1,452095875852D-09	2,50	-3,120537817179D-02
3	-4,5	7,762340637849D-09	3,881170318925D-09	2,75	-1,511738922567D-02
4	-3,5	-1,436332135891D-08	-7.181660679456D-09	3,25	7,977078221816D-03
5	-2,5	1,983179452922D-08	9,915897264608D-09	3,50	8,586609575912D-03
6	-1,5	-2,166578884355D-08	-1,083289442177D-08	3,75	4,740504326497D-03
7	-,5	1,971352206177D-08	9,856761030886D-09	4.25	-3,080468253241D-03
8	,5	-1,555997564128D-08	-7,779987820639D-09	4,50	-3,607637360523D-03
9	1,5	1,086970742933D-08	5,434853714663D-09	4,75	-2,144840718473D-03
10	2,5	-6,666091906808D-09	-3,333045953404D-09	5,25	1,576330753557D-03
11	3,5	3,463587521908D-09	1,731793760954D-09	5,50	1,943137793435D-03
12	4,5	-1,437581650310D-09	-7,187908251548D-10	5,75	1,209085801180D-03
13	5,5	4,380683545835D-10	2,190341772917D-10	6,25	-9,597225015336D-04
14	6,5	-8,569095942971D-11	-4,284547971486D-11	6,50	-1,222422163811D-03
15	7,5	7,977421218751D-12	3,988710609375D-12	6,75	-7,835503476693D-04
				7,25	6,553526172450D-04
	0	2,204348939298D+02	1,001887152545D+02	7,50	8,546834359235D-04
	1	3,297899223925D+02	1,792064814212D+02	7,75	5,603223803812D-04
	2	2.140366864027D+02	9,824539644791D+01	8,25	-4,895356504462D-04
	3	7,794152976350D+01	4,195709831071D+01	8,50	-6,525887954833D-04
	4	1,714535442998D+01	7,964682617085D+00	8,75	-4,376032275079D-04
	5	2,269451478033D+00	1,208571798920D+00	9,25	4,014484924354D-04
	6	1,666175577265D-01	7,836523495642D-02	9,50	5,498588670320D-04
	7	5,208990097079D-03	2,744951751757D-03	9,75	3,797387780276D-04
I= 2	H= -5.5				
0		0.	3,386249983760D+04	-1,00	1,068716935088D+02
1		-1,723154464657D+01	-1,906972199804D+04	-,75	5,406247945823D+01
2		3,906709205492D+01	4,613851217764D+03	-,50	2,348247686743D+01
3		-3,295344450643D+01	-6,310549445146D+02	-,25	7,322322396870D+00
4		1,397581865746D+01	5,625734095287D+01	,25	-2,329557943441D+00
5		-3,230948281611D+00	-3,982276179183D+00	,50	-2,189674256476D+00
6		3,929360418622D-01	2,595418387934D-01	,75	-1,128059805331D+00

CARDINAL SPLINES N=8

7		-1.990931993012D-02	-1.067391091121D-02	1.25	8.052806026352D-01
				1.50	1.188951222437D+00
0	-7.5	3.018779007774D-10	1.509389503887D-10	1.75	1.211481436545D+00
1	-6.5	-2.904191751705D-09	-1.452095875852D-09	2.25	6.879937334820D-01
2	-5.5	1.286414076296D-08	6.432070381479D-09	2.50	3.812626862271D-01
3	-4.5	-3.500262979402D-08	-1.750131489701D-08	2.75	1.441213248684D-01
4	-3.5	6.610193265626D-08	3.305096632813D-08	3.25	-5.907672491213D-02
5	-2.5	-9.332390563536D-08	-4.666195281768D-08	3.50	-5.910008334581D-02
6	-1.5	1.042798568835D-07	5.213992844174D-08	3.75	-3.088287173267D-02
7	-.5	-9.683453198182D-08	-4.841726599091D-08	4.25	1.859692520122D-02
8	.5	7.765135570148D-08	3.882567785074D-08	4.50	2.119841506999D-02
9	1.5	-5.481995898246D-08	-2.740997949123D-08	4.75	1.232844122943D-02
10	2.5	3.383062656010D-08	1.691531328005D-08	5.25	-8.765093709643D-03
11	3.5	-1.763980792685D-08	-8.819903963424D-09	5.50	-1.066993432674D-02
12	4.5	7.335946178850D-09	3.667973089425D-09	5.75	-6.569657105235D-03
13	5.5	-2.237979669922D-09	-1.118989834961D-09	6.25	5.130121545438D-03
14	6.5	4.380683545835D-10	2.190341772917D-10	6.50	6.493048998879D-03
15	7.5	-4.079925634907D-11	-2.039962817454D-11	6.75	4.139591697944D-03
				7.25	-3.433142208275D-03
	0	-9.270611624837D+02	-4.122063689936D+02	7.50	-4.462569234036D-03
	1	-1.379216111150D+03	-7.628571468345D+02	7.75	-2.917386441337D-03
	2	-8.886041343184D+02	-3.993961992330D+02	8.25	2.537524378792D-03
	3	-3.204910806750D+02	-1.755330453870D+02	8.50	3.376830690195D-03
	4	-6.961850867484D+01	-3.169655508288D+01	8.75	2.261029209098D-03
	5	-9.066710565262D+00	-4.911450697346D+00	9.25	-2.069381374615D-03
	6	-6.523032544689D-01	-3.008384840451D-01	9.50	-2.831815456114D-03
	7	-1.990931993012D-02	-1.067391091121D-02	9.75	-1.954164334204D-03

I= 3 H= -4.5

0		0.	-1.113992575662D+05	-1.00	-1.950370926653D+02
1		2.815447132102D+01	6.272397427584D+04	-.75	-9.662671875694D+01
2		-6.829387369871D+01	-1.515001681547D+04	-.50	-4.095140664905D+01
3		6.238862429097D+01	2.053341652447D+03	-.25	-1.239950789291D+01
4		-2.834328138795D+01	-1.762914190409D+02	.25	3.640911427908D+00
5		6.929151734439D+00	1.125287226427D+01	.50	3.234016149780D+00
6		-8.813912455845D-01	-6.482591177990D-01	.75	1.546390681419D+00
7		4.629898661214D-02	2.551651816936D-02	1.25	-8.561480961739D-01
				1.50	-9.860538148994D-01
0	-7.5	-7.948749239141D-10	-3.974374619571D-10	1.75	-6.047405451618D-01
1	-6.5	7.762340637849D-09	3.881170318925D-09	2.25	5.763427153460D-01
2	-5.5	-3.500262979402D-08	-1.750131489701D-08	2.50	9.617794498794D-01
3	-4.5	9.727438671684D-08	4.863719335842D-08	2.75	1.095026653662D+00
4	-3.5	-1.882614101810D-07	-9.413070509051D-08	3.25	7.545204536975D-01
5	-2.5	2.731561514480D-07	1.365780757240D-07	3.50	4.549572576418D-01
6	-1.5	-3.140311510528D-07	-1.570155755264D-07	3.75	1.857827367775D-01
7	-.5	2.995003458206D-07	1.497501729103D-07	4.25	-8.715865141842D-02
8	.5	-2.454886296027D-07	-1.227443148013D-07	4.50	-9.246277237692D-02
9	1.5	1.760265711272D-07	8.801328556359D-08	4.75	-5.096502860940D-02
10	2.5	-1.096974949762D-07	-5.484874748809D-08	5.25	3.365937095971D-02
11	3.5	5.752549546941D-08	2.876274773470D-08	5.50	3.992482116035D-02
12	4.5	-2.400144839281D-08	-1.200072419640D-08	5.75	2.407113592283D-02
13	5.5	7.335946178850D-09	3.667973089425D-09	6.25	-1.821593004464D-02
14	6.5	-1.437581650310D-09	-7.187908251548D-10	6.50	-2.278540968091D-02
15	7.5	1.339831749369D-10	6.699158746844D-11	6.75	-1.438455468553D-02
				7.25	1.175018800976D-02
	0	2.374915210938D+03	1.018723495720D+03	7.50	1.518455524874D-02

CARDINAL SPLINES N=8

```
        1   3.510495097513D+03   1.996099778890D+03    7.75  9.877995949394D-03
        2   2.242926518425D+03   9.737835070638D+02    8.25 -8.525838261047D-03
        3   8.002504551171D+02   4.504084629007D+02    8.50 -1.131186620496D-02
        4   1.714595193865D+02   7.548999882077D+01    8.75 -7.554863492368D-03
        5   2.195722361873D+01   1.222259905087D+01    9.25  6.886963267655D-03
        6   1.549305551553D+00   6.913580860924D-01    9.50  9.409718042467D-03
        7   4.629898661214D-02   2.551651816936D-02    9.75  6.484839929353D-03


I= 4     H= -3.5
0            0.                  2.695652262004D+05   -1.00  2.658621695577D+02
1          -3.630328170711D+01  -1.517795574709D+05    -.75  1.302470155037D+02
2           9.071807883072D+01   3.663019204144D+04    -.50  5.450003060156D+01
3          -8.615451550709D+01  -4.937734574079D+03    -.25  1.626189516543D+01
4           4.084118087635D+01   4.128719568802D+02     .25 -4.602393749531D+00
5          -1.039891077675D+01  -2.405466524365D+01     .50 -3.992974043186D+00
6           1.371825071029D+00   1.188460560586D+00     .75 -1.856056266127D+00
7          -7.437678859976D-02  -4.291986008555D-02    1.25  9.507562071476D-01
                                                       1.50  1.036162869137D+00
0   -7.5    1.446283093005D-09   7.231415465023D-10    1.75  5.906535842096D-01
1   -6.5   -1.436332135891D-08  -7.181660679456D-09    2.25 -4.378930533808D-01
2   -5.5    6.610193265626D-08   3.305096632813D-08    2.50 -5.709371712971D-01
3   -4.5   -1.882614101810D-07  -9.413070509051D-08    2.75 -3.918880385788D-01
4   -3.5    3.750580020981D-07   1.875290010491D-07    3.25  4.546741337141D-01
5   -2.5   -5.624119466736D-07  -2.812059733368D-07    3.50  8.269518719470D-01
6   -1.5    6.697455906008D-07   3.348727953004D-07    3.75  1.018772594945D+00
7    -.5   -6.610675322259D-07  -3.305337661129D-07    4.25  8.059744077486D-01
8     .5    5.581995301587D-07   2.790997650794D-07    4.50  5.161274808418D-01
9    1.5   -4.093898576811D-07  -2.046949288406D-07    4.75  2.226313881411D-01
10   2.5    2.590307611315D-07   1.295153805657D-07    5.25 -1.148494356873D-01
11   3.5   -1.371140981078D-07  -6.855704905389D-08    5.50 -1.269300822634D-01
12   4.5    5.752549546941D-08   2.876274773470D-08    5.75 -7.260780388726D-02
13   5.5   -1.763980792685D-08  -8.819903963424D-09    6.25  5.113703612780D-02
14   6.5    3.463587521908D-09   1.731793760954D-09    6.50  6.237685900812D-02
15   7.5   -3.232085745490D-10  -1.616042872745D-10    6.75  3.858799018143D-02
                                                       7.25 -3.058510705949D-02
        0  -4.188435663977D+03  -1.686382693948D+03    7.50 -3.908227502953D-02
        1  -6.146613162536D+03  -3.655613987550D+03    7.75 -2.518725675008D-02
        2  -3.891078066129D+03  -1.588393406505D+03    8.25  2.142917948043D-02
        3  -1.372180633096D+03  -8.077282803442D+02    8.50  2.827546768070D-02
        4  -2.898603387395D+02  -1.201529508375D+02    8.75  1.879691984439D-02
        5  -3.652436411392D+01  -2.127302474332D+01    9.25 -1.701183732508D-02
        6  -2.532956330458D+00  -1.064832093905D+00    9.50 -2.317852107345D-02
        7  -7.437678859976D-02  -4.291986008555D-02    9.75 -1.593606051938D-02


I= 5     H= -2.5
0            0.                 -5.229298830370D+05   -1.00 -2.880321087891D+02
1           3.816360735923D+01   2.944755848308D+05    -.75 -1.402233465615D+02
2          -9.681583181414D+01  -7.104528833471D+04    -.50 -5.826375857275D+01
3           9.384405235386D+01   9.547210999183D+03    -.25 -1.724831262335D+01
4          -4.558567195087D+01  -7.846516196537D+02     .25  4.789407229671D+00
5           1.191863302038D+01   4.256864294926D+01     .50  4.107178964684D+00
6          -1.614550628559D+00  -1.783179496628D+00     .75  1.883642090610D+00
7           8.976166205287D-02   5.601254319656D-02    1.25 -9.324311991734D-01
                                                       1.50 -9.939580659342D-01
0   -7.5   -1.961242235089D-09  -9.806211175444D-10    1.75 -5.516083320836D-01
```

CARDINAL SPLINES N=8

1	-6.5	1.983179452922D-08	9.915897264608D-09	2.25	3.795514634194D-01
2	-5.5	-9.332390563536D-08	-4.666195281768D-08	2.50	4.690361758520D-01
3	-4.5	2.731561514480D-07	1.365780757240D-07	2.75	2.997352770018D-01
4	-3.5	-5.624119466736D-07	-2.812059733368D-07	3.25	-2.714290081360D-01
5	-2.5	8.763005020898D-07	4.381502510449D-07	3.50	-3.863516114367D-01
6	-1.5	-1.088391343587D-06	-5.441956717936D-07	3.75	-2.874156886481D-01
7	-.5	1.120996947177D-06	5.604984735883D-07	4.25	3.840502648681D-01
8	.5	-9.839799747748D-07	-4.919899873874D-07	4.50	7.428966949575D-01
9	1.5	7.447086358574D-07	3.723543179287D-07	4.75	9.680719242367D-01
10	2.5	-4.820468678295D-07	-2.410234339147D-07	5.25	8.441985744547D-01
11	3.5	2.590307611315D-07	1.295153805657D-07	5.50	5.638013390516D-01
12	4.5	-1.096974949762D-07	-5.484874748809D-08	5.75	2.526359009644D-01
13	5.5	3.383062656010D-08	1.691531328005D-08	6.25	-1.392175458894D-01
14	6.5	-6.666091906808D-09	-3.333045953404D-09	6.50	-1.583422494682D-01
15	7.5	6.234488258717D-10	3.117244129359D-10	6.75	-9.299830125256D-02
				7.25	6.867752643418D-02
	0	5.491660397483D+03	1.956340379882D+03	7.50	8.561937190884D-02
	1	7.998059690595D+03	5.126835054283D+03	7.75	5.409359095264D-02
	2	5.014970287015D+03	1.815411339871D+03	8.25	-4.468807436855D-02
	3	1.748136848698D+03	1.109898626762D+03	8.50	-5.832306643406D-02
	4	3.644730147163D+02	1.341749988008D+02	8.75	-3.842108615533D-02
	5	4.529481803519D+01	2.849038225196D+01	9.25	3.429186733013D-02
	6	3.097936629217D+00	1.157479021192D+00	9.50	4.647479124284D-02
	7	8.976166205287D-02	5.601254319656D-02	9.75	3.181067362077D-02

I= 6   H= -1.5

0		0.	8.652404581825D+05	-1.00	2.625760136832D+02
1		-3.423033720378D+01	-4.873949529851D+05	-.75	1.273897780508D+02
2		8.752245008206D+01	1.175962228974D+05	-.50	5.272964992204D+01
3		-8.575909860200D+01	-1.577811613169D+04	-.25	1.554398019266D+01
4		4.222105483015D+01	1.283376167561D+03	.25	-4.273067790214D+00
5		-1.121120506217D+01	-6.627970685562D+01	.50	-3.642527341978D+00
6		1.544501928347D+00	2.390142760508D+00	.75	-1.659234127574D+00
7		-8.736597470939D-02	-6.214291737150D-02	1.25	8.077457109543D-01
				1.50	8.521643497390D-01
0	-7.5	2.104882121352D-09	1.052441060676D-09	1.75	4.672140958441D-01
1	-6.5	-2.166578884355D-08	-1.083289442177D-08	2.25	-3.115218141108D-01
2	-5.5	1.042798568835D-07	5.213992844174D-08	2.50	-3.770779390337D-01
3	-4.5	-3.140311510528D-07	-1.570155755264D-07	2.75	-2.349311830353D-01
4	-3.5	6.697455906008D-07	3.348727953004D-07	3.25	1.980140991396D-01
5	-2.5	-1.088391343587D-06	-5.441956717936D-07	3.50	2.675084315847D-01
6	-1.5	1.417705679940D-06	7.088528399698D-07	3.75	1.855252543105D-01
7	-.5	-1.534877270871D-06	-7.674386354354D-07	4.25	-1.939727836477D-01
8	.5	1.413038103254D-06	7.065190516268D-07	4.50	-2.939810744415D-01
9	1.5	-1.114282917046D-06	-5.571414585228D-07	4.75	-2.315715980658D-01
10	2.5	7.447086358574D-07	3.723543179287D-07	5.25	3.417096292750D-01
11	3.5	-4.093898576811D-07	-2.046949288406D-07	5.50	6.899055095105D-01
12	4.5	1.760265711272D-07	8.801328556359D-08	5.75	9.345885296686D-01
13	5.5	-5.481995898246D-08	-2.740997949123D-08	6.25	8.716556722635D-01
14	6.5	1.086970742933D-08	5.434853714663D-09	6.50	5.994017943186D-01
15	7.5	-1.020739149325D-09	-5.103695746625D-10	6.75	2.758971885623D-01
				7.25	-1.595419736673D-01
	0	-5.701642638529D+03	-1.549722189659D+03	7.50	-1.855208332771D-01
	1	-8.244316028991D+03	-5.982579667559D+03	7.75	-1.113134001946D-01
	2	-5.124159315170D+03	-1.419322371501D+03	8.25	8.572408739547D-02
	3	-1.768794433130D+03	-1.274179276708D+03	8.50	1.091788094166D-01

CARDINAL SPLINES N=8

	4	-3,650388532769D+02	-1.030088996594D+02	8.75	7,052308604088D-02
	5	-4,490967591203D+01	-3,212960377783D+01	9,25	-6,114123101829D-02
	6	-3,042211743896D+00	-8,723604014957D-01	9,50	-8,197386097569D-02
	7	-8,736597470939D-02	-6,214291737150D-02	9,75	-5,560933414562D-02
I= 7	H= -.5				
0		0.	-1,272170425963D+06	-1.00	-2,114931015670D+02
1		2,733910426068D+01	7.170876139048D+05	-.75	-1,024210755225D+02
2		-7,018405727012D+01	-1,731016479454D+05	-.50	-4,231013316725D+01
3		6,915444122242D+01	2,321550051539D+04	-,25	-1,244502805815D+01
4		-3,428588783762D+01	-1,877842828685D+03	.25	3,403539058041D+00
5		9,180053723172D+00	9,408292609911D+01	,50	2,892474625148D+00
6		-1,276605674833D+00	-3,026033673331D+00	.75	1,313055930297D+00
7		7,295157817996D-02	6,373462743401D-02	1.25	-6,339253292788D-01
				1.50	-6,653987940969D-01
0	-7,5	-1,885902654637D-09	-9,429513273183D-10	1.75	-3,626859410605D-01
1	-6,5	1,971352206177D-08	9,856761030886D-09	2,25	2,382811087552D-01
2	-5,5	-9,683453198182D-08	-4,841726599091D-08	2,50	2,857292055998D-01
3	-4,5	2,995003458206D-07	1,497501729103D-07	2,75	1,760424603559D-01
4	-3,5	-6,610675322259D-07	-3,305337661129D-07	3.25	-1,440552949110D-01
5	-2,5	1,120996947177D-06	5,604984735883D-07	3,50	-1,907955998813D-01
6	-1,5	-1,534877270871D-06	-7,674386354354D-07	3.75	-1,291168287651D-01
7	-,5	1,754552895221D-06	8,772764476107D-07	4,25	1,258422158731D-01
8	,5	-1,705430579655D-06	-8,527152898273D-07	4,50	1,811394857629D-01
9	1,5	1,413038103254D-06	7,065190516268D-07	4,75	1,331010654141D-01
10	2,5	-9,839799747748D-07	-4,919899873874D-07	5.25	-1,538363579188D-01
11	3,5	5,581995301587D-07	2,790997650794D-07	5,50	-2,434519567967D-01
12	4,5	-2,454886296027D-07	-1,227443148013D-07	5,75	-1,994304231179D-01
13	5,5	7,765135570148D-08	3,882567785074D-08	6,25	3,149295251045D-01
14	6,5	-1,555997564128D-08	-7,779987820639D-09	6,50	6,548543663485D-01
15	7,5	1,471698012110D-09	7,358490060552D-10	6,75	9,114455439775D-01
				7,25	8,923709941552D-01
	0	4,966224898032D+03	5,846003663490D+02	7,50	6,274856026045D-01
	1	7,138261885431D+03	6,288700656336D+03	7,75	2,951039701166D-01
	2	4,408431295773D+03	5,306126844502D+02	8,25	-1,780079235085D-01
	3	1,511814993197D+03	1,327814528405D+03	8,50	-2,114910443395D-01
	4	3,100057352041D+02	3,813259636395D+01	8,75	-1,297540644091D-01
	5	3,790685008075D+01	3,319793945562D+01	9,25	1,048910949216D-01
	6	2,553352179614D+00	3,200342669542D-01	9,50	1,372512986007D-01
	7	7,295157817996D-02	6,373462743401D-02	9,75	9,130613713344D-02
I= 8	H= .5				
0		0.	1.688259837625D+06	-1.00	1,551283558312D+02
1		-1,996687018524D+01	-9,528382652173D+05	-.75	7,505550149566D+01
2		5,136275108407D+01	2,302806285435D+05	-.50	3,097391944624D+01
3		-5,075251969290D+01	-3,090403455208D+04	-,25	9,100406428712D+00
4		2,525253245038D+01	2,494165846898D+03	.25	-2,482313806101D+00
5		-6,790270361511D+00	-1,228860101250D+02	.50	-2,106324398518D+00
6		9,488943804175D-01	3,666102207240D+00	.75	-9,545227081812D-01
7		-5,451767668807D-02	-6,373462743401D-02	1.25	4,589100400055D-01
				1,50	4,804757499326D-01
0	-7,5	1,471698012110D-09	7,358490060552D-10	1,75	2,611307185642D-01
1	-6,5	-1.555997564128D-08	-7,779987820639D-09	2,25	-1,703214409779D-01
2	-5,5	7,765135570148D-08	3,882567785074D-08	2,50	-2,033107819474D-01
3	-4,5	-2,454886296027D-07	-1,227443148013D-07	2,75	-1,245964173879D-01

4	-3,5	5.581995301587D-07	2.790997650794D-07	3.25	1,005611354932D-01
5	-2,5	-9,839799747748D-07	-4,919898873874D-07	3,50	1,320036968955D-01
6	-1,5	1,413038103254D-06	7,065190516268D-07	3,75	8,837560583004D-02
7	-,5	-1,705430579655D-06	-8,527152898273D-07	4,25	-8,368256106526D-02
8	,5	1,754552895221D-06	8,772764476107D-07	4,50	-1,181250530520D-01
9	1,5	-1,534877270871D-06	-7,674386354354D-07	4,75	-8,471548059869D-02
10	2,5	1,120996947177D-06	5,604984735883D-07	5,25	9,130613713344D-02
11	3,5	-6,610675322259D-07	-3,305337661129D-07	5,50	1,372512986007D-01
12	4,5	2,995003458206D-07	1.497501729103D-07	5,75	1,048910949216D-01
13	5,5	-9,683453198182D-08	-4,841726599091D-08	6,25	-1,297540644091D-01
14	6,5	1,971352206177D-08	9,856761030886D-09	6,50	-2,114910443395D-01
15	7,5	-1,885902654637D-09	-9,429513273183D-10	6,75	-1,780079235085D-01
				7,25	2,951039701166D-01
	0	-3,797024165334D+03	5.846003663490D+02	7,50	6,274856026045D-01
	1	-5,439139427242D+03	-6,288700656336D+03	7,75	8,923709941552D-01
	2	-3,347205926873D+03	5,306126844502D+02	8,25	9,114455439775D-01
	3	-1,143814063613D+03	-1,327814528405D+03	8,50	6,548543663485D-01
	4	-2,337405424762D+02	3,813259636395D+01	8,75	3,149295251045D-01
	5	-2,848902883050D+01	-3,319793945562D+01	9,25	-1,994304231179D-01
	6	-1,913283645706D+00	3.200342669542D-01	9,50	-2,434519567967D-01
	7	-5,451767668807D-02	-6,373462743401D-02	9,75	-1,538363579188D-01

I= 9   H= 1.5

0		0.	-1,990389681209D+06	-1.00	-1,040477672327D+02
1		1,336264027131D+01	1.126047915438D+06	-.75	-5,031734755647D+01
2		-3,440987230644D+01	-2,727719598961D+05	-.50	-2,075414046404D+01
3		3,405029951788D+01	3,667973188649D+04	-,25	-6,094240136362D+00
4		-1,697320959852D+01	-2,961502144404D+03	,25	1,660090006608D+00
5		4,574023967623D+00	1.447921429902D+02	.50	1,407528057535D+00
6		-6,408017108597D-01	-4,134863563500D+00	.75	6,372841380285D-01
7		3,691986003361D-02	6,214291737150D-02	1.25	-3,057374643000D-01
				1,50	-3,196925117731D-01
0	-7,5	-1,020739149325D-09	-5,103695746625D-10	1,75	-1,734912758449D-01
1	-6,5	1,086970742933D-08	5,434853714663D-09	2,25	1,127449198528D-01
2	-5,5	-5,481995898246D-08	-2,740997949123D-08	2,50	1,342750664966D-01
3	-4,5	1,760265711272D-07	8,801328556359D-08	2,75	8,206935998898D-02
4	-3,5	-4,093898576811D-07	-2,046949288406D-07	3,25	-6,578743545745D-02
5	-2,5	7,447086358574D-07	3,723543179287D-07	3,50	-8,598124810790D-02
6	-1,5	-1,114282917046D-06	-5,571414585228D-07	3,75	-5,726645666297D-02
7	-,5	1,413038103254D-06	7,065190516268D-07	4,25	5,349530112239D-02
8	,5	-1,534877270871D-06	-7,674386354354D-07	4,50	7,484642497937D-02
9	1,5	1,417705679940D-06	7,088528399698D-07	4,75	5,310597869176D-02
10	2,5	-1,088391343587D-06	-5,441956717936D-07	5,25	-5,560933414562D-02
11	3,5	6,697455906008D-07	3,348727953004D-07	5,50	-8,197386097569D-02
12	4,5	-3,140311510528D-07	-1,570155755264D-07	5,75	-6,114123101829D-02
13	5,5	1,042798568835D-07	5,213992844174D-08	6,25	7,052308604088D-02
14	6,5	-2,166578884355D-08	-1,083289442177D-08	6,50	1,091788094166D-01
15	7,5	2,104882121352D-09	1.052441060676D-09	6,75	8,572408739547D-02
				7,25	-1,113134001946D-01
	0	2,602198259211D+03	-1,549722189659D+03	7,50	-1,855208332771D-01
	1	3,720843306126D+03	5.982579667559D+03	7,75	-1,595419736673D-01
	2	2,285514572168D+03	-1.419322371501D+03	8,25	2,758971885623D-01
	3	7,795641202870D+02	1.274179276708D+03	8,50	5,994017943186D-01
	4	1,590210539582D+02	-1,030088996594D+02	8,75	8,716556722635D-01
	5	1,934953164364D+01	3,212960377783D+01	9,25	9,345885296686D-01
	6	1,297490940905D+00	-8,723604014957D-01	9,50	6,899055095105D-01

CARDINAL SPLINES N=8

	7	3.691986003361D-02	6.214291737150D-02	9.75	3.417096292750D-01
I=10	H= 2.5				
0		0.	1.992165299936D+06	-1.00	6.242908525947D+01
1		-8.008432819640D+00	-1.131388782874D+06	-.75	3.018315126456D+01
2		2.063352359194D+01	2.751143141697D+05	-.50	1.244611207341D+01
3		-2.043326632289D+01	-3.713016940985D+04	-.25	3.653579764247D+00
4		1.019516784646D+01	3.006247465517D+03	.25	-9.945511625527D-01
5		-2.750580063173D+00	-1.467417548565D+02	.50	-8.428964469113D-01
6		3.858511910304D-01	4.098137539012D+00	.75	-3.814611263265D-01
7		-2.226342434025D-02	-5.601254319656D-02	1.25	1.828038255320D-01
				1.50	1.910198996225D-01
0	-7.5	6.234488258717D-10	3.117244129359D-10	1.75	1.035837194119D-01
1	-6.5	-6.666091906808D-09	-3.333045953404D-09	2.25	-6.718735067717D-02
2	-5.5	3.383062656010D-08	1.691531328005D-08	2.50	-7.992337128709D-02
3	-4.5	-1.096974949762D-07	-5.484874748809D-08	2.75	-4.878234547997D-02
4	-3.5	2.590307611315D-07	1.295153805657D-07	3.25	3.896790440947D-02
5	-2.5	-4.820468678295D-07	-2.410234339147D-07	3.50	5.081631263415D-02
6	-1.5	7.447086358574D-07	3.723543179287D-07	3.75	3.375677934693D-02
7	-.5	-9.839799747748D-07	-4.919899873874D-07	4.25	-3.132069968871D-02
8	.5	1.120996947177D-06	5.604984735883D-07	4.50	-4.363032714478D-02
9	1.5	-1.088391343587D-06	-5.441956717936D-07	4.75	-3.079639630946D-02
10	2.5	8.763005020898D-07	4.381502510449D-07	5.25	3.181067362077D-02
11	3.5	-5.624119466736D-07	-2.812059733368D-07	5.50	4.647479124284D-02
12	4.5	2.731561514480D-07	1.365780757240D-07	5.75	3.429186733013D-02
13	5.5	-9.332390563536D-08	-4.666195281768D-08	6.25	-3.842108615533D-02
14	6.5	1.983179452922D-08	9.915897264608D-09	6.50	-5.832306643406D-02
15	7.5	-1.961242235089D-09	-9.806211175444D-10	6.75	-4.468807436855D-02
				7.25	5.409359095264D-02
	0	-1.578979637719D+03	1.956340379882D+03	7.50	8.561937190884D-02
	1	-2.255610417971D+03	-5.126835054283D+03	7.75	6.867752643418D-02
	2	-1.384147607273D+03	1.815411339871D+03	8.25	-9.299830125256D-02
	3	-4.716604048255D+02	-1.109898626762D+03	8.50	-1.583422494682D-01
	4	-9.612301711471D+01	1.341749988008D+02	8.75	-1.392175458894D-01
	5	-1.168594646873D+01	-2.849038225196D+01	9.25	2.526359009644D-01
	6	-7.829785868330D-01	1.157479021192D+00	9.50	5.638013390516D-01
	7	-2.226342434025D-02	-5.601254319656D-02	9.75	8.441985744547D-01
I=11	H= 3.5				
0		0.	-1.591009612765D+06	-1.00	-3.205801214457D+01
1		4.109914684436D+00	9.081751454834D+05	-.75	-1.549734528967D+01
2		-1.059212627170D+01	-2.219860795834D+05	-.50	-6.389457375635D+00
3		1.049350158427D+01	3.011595320897D+04	-.25	-1.875340048185D+00
4		-5.238368035575D+00	-2.450082017020D+03	.25	5.103025843719D-01
5		1.414126872171D+00	1.198895536951D+02	.50	4.323947426851D-01
6		-1.985117648465D-01	-3.318124748396D+00	.75	1.956368030011D-01
7		1.146293157134D-02	4.291986008555D-02	1.25	-9.369817264315D-02
				1.50	-9.787476291088D-02
0	-7.5	-3.232085745490D-10	-1.616042872745D-10	1.75	-5.305273940918D-02
1	-6.5	3.463587521908D-09	1.731793760954D-09	2.25	3.437704889044D-02
2	-5.5	-1.763980792685D-08	-8.819903963424D-09	2.50	4.086809909186D-02
3	-4.5	5.752549546941D-08	2.876274773470D-08	2.75	2.492633877872D-02
4	-3.5	-1.371140981078D-07	-6.855704905389D-08	3.25	-1.987485900151D-02
5	-2.5	2.590307611315D-07	1.295153805657D-07	3.50	-2.588771950500D-02
6	-1.5	-4.093898576811D-07	-2.046949288406D-07	3.75	-1.717336951914D-02

7	-.5	5.581995301587D-07	2.790997650794D-07	4.25	1.587788658923D-02
8	.5	-6.610675322259D-07	-3.305337661129D-07	4.50	2.206825949425D-02
9	1.5	6.697455906008D-07	3.348727953004D-07	4.75	1.553522896571D-02
10	2.5	-5.624119466736D-07	-2.812059733368D-07	5.25	-1.593606051938D-02
11	3.5	3.750580020981D-07	1.875290010491D-07	5.50	-2.317852107345D-02
12	4.5	-1.882614101810D-07	-9.413070509051D-08	5.75	-1.701183732508D-02
13	5.5	6.610193265626D-08	3.305096632813D-08	6.25	1.879691984439D-02
14	6.5	-1.436332135891D-08	-7.181660679456D-09	6.50	2.827546768070D-02
15	7.5	1.446283093005D-09	7.231415465023D-10	6.75	2.142917948043D-02
				7.25	-2.518725675008D-02
	0	8.156702760805D+02	-1.686382693948D+03	7.50	-3.908227502953D-02
	1	1.164614812563D+03	3.655613987550D+03	7.75	-3.058510705949D-02
	2	7.142912531183D+02	-1.588393406505D+03	8.25	3.858799018143D-02
	3	2.432759275929D+02	8.077282803442D+02	8.50	6.237685900812D-02
	4	4.955443706456D+01	-1.201529508375D+02	8.75	5.113703612780D-02
	5	6.021685372716D+00	2.127302474332D+01	9.25	-7.260780388726D-02
	6	4.032921426485D-01	-1.064832093905D+00	9.50	-1.269300822634D-01
	7	1.146293157134D-02	4.291986008555D-02	9.75	-1.148494356873D-01

I=12 H= 4.5

0		0.	9.467912697462D+05	-1.00	1.322041970260D+01
1		-1.694328081067D+00	-5.435912429928D+05	-.75	6.390502127971D+00
2		4.367326099569D+00	1.336783388619D+05	-.50	2.634557240370D+00
3		-4.327599069598D+00	-1.824940928450D+04	-.25	7.731899588941D-01
4		2.160935520386D+00	1.493938186963D+03	.25	-2.103518739101D-01
5		-5.835486509751D-01	-7.347510001259D+01	.50	-1.782163683094D-01
6		8.194823127731D-02	2.030975289984D+00	.75	-8.062320783739D-02
7		-4.734049726579D-03	-2.551651816936D-02	1.25	3.860131656518D-02
				1.50	4.031419896214D-02
0	-7.5	1.339831749369D-10	6.699158746844D-11	1.75	2.184738402377D-02
1	-6.5	-1.437581650310D-09	-7.187908251548D-10	2.25	-1.414892383586D-02
2	-5.5	7.335946178850D-09	3.667973089425D-09	2.50	-1.681481837422D-02
3	-4.5	-2.400144839281D-08	-1.200072419640D-08	2.75	-1.025168125679D-02
4	-3.5	5.752549546941D-08	2.876274773470D-08	3.25	8.165952506324D-03
5	-2.5	-1.096974949762D-07	-5.484874748809D-08	3.50	1.062972332418D-02
6	-1.5	1.760265711272D-07	8.801328556359D-08	3.75	7.046293974856D-03
7	-.5	-2.454886296027D-07	-1.227443148013D-07	4.25	-6.502337773942D-03
8	.5	2.995003458206D-07	1.497501729103D-07	4.50	-9.026435641650D-03
9	1.5	-3.140311510528D-07	-1.570155755264D-07	4.75	-6.345152271733D-03
10	2.5	2.731561514480D-07	1.365780757240D-07	5.25	6.484839929353D-03
11	3.5	-1.882614101810D-07	-9.413070509051D-08	5.50	9.409718042467D-03
12	4.5	9.727438671684D-08	4.863719335842D-08	5.75	6.886963267655D-03
13	5.5	-3.500262979402D-08	-1.750131489701D-08	6.25	-7.554863492368D-03
14	6.5	7.762340637849D-09	3.881170318925D-09	6.50	-1.131186620496D-02
15	7.5	-7.948749239141D-10	-3.974374619571D-10	6.75	-8.525838261047D-03
				7.25	9.877995949394D-03
	0	-3.374682194973D+02	1.018723495720D+03	7.50	1.518455524874D-02
	1	-4.817044602661D+02	-1.996099778890D+03	7.75	1.175018800976D-02
	2	-2.953595042972D+02	9.737835070638D+02	8.25	-1.438455468553D-02
	3	-1.005664706843D+02	-4.504084629007D+02	8.50	-2.278540968091D-02
	4	-2.047952174496D+01	7.548999882077D+01	8.75	-1.821593004464D-02
	5	-2.487974483017D+00	-1.222259905087D+01	9.25	2.407113592283D-02
	6	-1.665893793681D-01	6.913580860924D-01	9.50	3.992482116035D-02
	7	-4.734049726579D-03	-2.551651816936D-02	9.75	3.365937095971D-02

CARDINAL SPLINES N=8

```
I=13      H=  5.5
0              0.                  -3.872843212896D+05    -1.00 -4.013907637057D+00
1              5.143256763647D-01   2.236959403129D+05     -.75 -1.940169131879D+00
2             -1.325850789470D+00  -5.536397127490D+04     -.50 -7.998213712454D-01
3              1.313951337621D+00   7.609497667749D+03     -.25 -2.347201454263D-01
4             -6.562079660707D-01  -6.273153929448D+02      .25  6.385001621506D-02
5              1.772383026709D-01   3.105773974324D+01      .50  5.409201003362D-02
6             -2.489506296716D-02  -8.612188068836D-01      .75  2.446881748895D-02
7              1.438501892302D-03   1.067391091121D-02     1.25 -1.171322985894D-02
                                                           1.50 -1.223165566614D-02
0   -7.5      -4.079925634907D-11  -2.039962817454D-11     1.75 -6.627847902320D-03
1   -6.5       4.380683545835D-10   2.190341772917D-10     2.25  4.291040077120D-03
2   -5.5      -2.237979669922D-09  -1.118989834961D-09     2.50  5.098567592143D-03
3   -4.5       7.335946178850D-09   3.667973089425D-09     2.75  3.107809173734D-03
4   -3.5      -1.763980792685D-08  -8.819903963424D-09     3.25 -2.474121174665D-03
5   -2.5       3.383062656010D-08   1.691531328005D-08     3.50 -3.219444830848D-03
6   -1.5      -5.481995898246D-08  -2.740997949123D-08     3.75 -2.133229233690D-03
7    -.5       7.765135570148D-08   3.882567785074D-08     4.25  1.966433735399D-03
8     .5      -9.683453198182D-08  -4.841726599091D-08     4.50  2.727900783559D-03
9    1.5       1.042798568835D-07   5.213992844174D-08     4.75  1.916036398229D-03
10   2.5      -9.332390563536D-08  -4.666195281768D-08     5.25 -1.954164334204D-03
11   3.5       6.610193265626D-08   3.305096632813D-08     5.50 -2.831815456114D-03
12   4.5      -3.500262979402D-08  -1.750131489701D-08     5.75 -2.069381374615D-03
13   5.5       1.286414076296D-08   6.432070381479D-09     6.25  2.261029209098D-03
14   6.5      -2.904191751705D-09  -1.452095875852D-09     6.50  3.376830690195D-03
15   7.5       3.018779007774D-10   1.509389503887D-10     6.75  2.537524378792D-03
                                                           7.25 -2.917386441337D-03
          0    1.026484244965D+02  -4.122063689936D+02     7.50 -4.462569234036D-03
          1    1.464981825190D+02   7.628571468345D+02     7.75 -3.433142208275D-03
          2    8.981173585234D+01  -3.993961992330D+02     8.25  4.139591697944D-03
          3    3.057501009895D+01   1.755330453870D+02     8.50  6.493048998879D-03
          4    6.225398509075D+00  -3.169655508288D+01     8.75  5.130121545438D-03
          5    7.561908294307D-01   4.911450697346D+00     9.25 -6.569657105235D-03
          6    5.062628637870D-02  -3.008384840451D-01     9.50 -1.066993432674D-02
          7    1.438501892302D-03   1.067391091121D-02     9.75 -8.765093709643D-03


I=14      H=  6.5
0              0.                   9.616331710641D+04    -1.00  7.834635439407D-01
1             -1.003786550054D-01  -5.587119284500D+04     -.75  3.786871990171D-01
2              2.587739348608D-01   1.391622911391D+04     -.50  1.561070933786D-01
3             -2.564700740176D-01  -1.925907845232D+03     -.25  4.581075340255D-02
4              1.280969710814D-01   1.599377202806D+02      .25 -1.246088048864D-02
5             -3.460212954498D-02  -7.977481628722D+00      .50 -1.055610249974D-02
6              4.860866024230D-03   2.224752019237D-01      .75 -4.774898346867D-03
7             -2.809134064354D-04  -2.744951751757D-03     1.25  2.285500837467D-03
                                                           1.50  2.386502631919D-03
0   -7.5       7.977421218751D-12   3.988710609375D-12     1.75  1.293055415014D-03
1   -6.5      -8.569095942971D-11  -4.284547971486D-11     2.25 -8.370045607984D-04
2   -5.5       4.380683545835D-10   2.190341772917D-10     2.50 -9.944069877917D-04
3   -4.5      -1.437581650310D-09  -7.187908251548D-10     2.75 -6.060563659382D-04
4   -3.5       3.463587521908D-09   1.731793760954D-09     3.25  4.823190962521D-04
5   -2.5      -6.666091906808D-09  -3.333045953404D-09     3.50  6.274839924855D-04
6   -1.5       1.086970742933D-08   5.434853714663D-09     3.75  4.156725129687D-04
7    -.5      -1.555997564128D-08  -7.779987820639D-09     4.25 -3.829277561011D-04
8     .5       1.971352206177D-08   9.856761030886D-09     4.50 -5.309947683274D-04
9    1.5      -2.166578884355D-08  -1.083289442177D-08     4.75 -3.727849334565D-04
```

10	2.5	1.983179452922D-08	9.915897264608D-09	5.25	3.797387780276D-04
11	3.5	-1.436332135891D-08	-7.181660679456D-09	5.50	5.498588670320D-04
12	4.5	7.762340637849D-09	3.881170318925D-09	5.75	4.014484924354D-04
13	5.5	-2.904191751705D-09	-1.452095875852D-09	6.25	-4.376032275079D-04
14	6.5	6.657509095615D-10	3.328754547808D-10	6.50	-6.525887954833D-04
15	7.5	-7.010675344473D-11	-3.505337672236D-11	6.75	-4.895356504462D-04
				7.25	5.603223803812D-04
	0	-2.005746342072D+01	1.001887152545D+02	7.50	8.546834359235D-04
	1	-2.862304044995D+01	-1.792064814212D+02	7.75	6.553526172450D-04
	2	-1.754589350689D+01	9.824539644791D+01	8.25	-7.835503476693D-04
	3	-5.972666857919D+00	-4.195709831071D+01	8.50	-1.222422163811D-03
	4	-1.215989195809D+00	7.964682617085D+00	8.75	-9.597225015336D-04
	5	-1.476921198064D-01	-1.208571798920D+00	9.25	1.209085801180D-03
	6	-9.887087813629D-03	7.836523495642D-02	9.50	1.943137793435D-03
	7	-2.809134064354D-04	-2.744951751757D-03	9.75	1.576330753557D-03

I=15 H= 7.5

0		0.	-1.086524291324D+04	-1.00	-7.283824867245D-02
1		9.331512141588D-03	6.347896557519D+03	-.75	-3.520585211889D-02
2		-2.405721372351D-02	-1.590805377134D+03	-.50	-1.451275198040D-02
3		2.384411089826D-02	2.216398447143D+02	-.25	-4.258795096868D-03
4		-1.190990155505D-02	-1.854176668402D+01	.25	1.158376567983D-03
5		3.217377017617D-03	9.320738699644D-01	.50	9.812821412179D-04
6		-4.520090572900D-04	-2.617918416984D-02	.75	4.438562999898D-04
7		2.612427913743D-05	3.233057984845D-04	1.25	-2.124372777114D-04
				1.50	-2.218164715716D-04
0	-7.5	-7.424649336022D-13	-3.712324668011D-13	1.75	-1.201791261066D-04
1	-6.5	7.977421218751D-12	3.988710609375D-12	2.25	7.778400703253D-05
2	-5.5	-4.079925634907D-11	-2.039962817454D-11	2.50	9.240510689246D-05
3	-4.5	1.339831749369D-10	6.699158746844D-11	2.75	5.631306076772D-05
4	-3.5	-3.232085745490D-10	-1.616042872745D-10	3.25	-4.480641242419D-05
5	-2.5	6.234488258717D-10	3.117244129359D-10	3.50	-5.828424775794D-05
6	-1.5	-1.020739149325D-09	-5.103695746625D-10	3.75	-3.860403435154D-05
7	-.5	1.471698012110D-09	7.358490060552D-10	4.25	3.554891230567D-05
8	.5	-1.885902654637D-09	-9.429513273183D-10	4.50	4.928222586096D-05
9	1.5	2.104882121352D-09	1.052441060676D-09	4.75	3.458833489351D-05
10	2.5	-1.961242235089D-09	-9.806211175444D-10	5.25	-3.520679387866D-05
11	3.5	1.446283093005D-09	7.231415465023D-10	5.50	-5.095457834236D-05
12	4.5	-7.948749239141D-10	-3.974374619571D-10	5.75	-3.718053253688D-05
13	5.5	3.018779007774D-10	1.509389503887D-10	6.25	4.047044752401D-05
14	6.5	-7.010675344473D-11	-3.505337672236D-11	6.50	6.029750724651D-05
15	7.5	7.465462968715D-12	3.732731484357D-12	6.75	4.518330349918D-05
				7.25	-5.157591216324D-05
	0	1.865996343184D+00	-1.104170460505D+01	7.50	-7.853565726790D-05
	1	2.662720101388D+00	1.933284436868D+01	7.75	-6.009838207182D-05
	2	1.632148365683D+00	-1.094095059391D+01	8.25	7.148909684060D-05
	3	5.555550343116D-01	4.595120971136D+00	8.50	1.111731179867D-04
	4	1.131004036558D-01	-9.038710229108D-01	8.75	8.695513670666D-05
	5	1.373627417066D-02	1.359155567816D-01	9.25	-1.085120079266D-04
	6	9.195155974251D-04	-9.205629749400D-03	9.50	-1.733487985252D-04
	7	2.612427913743D-05	3.233057984845D-04	9.75	-1.396417958383D-04

N= 8 M=16 P= 8.0 Q=15

I= 0 H= -8.0

0		1.000000000000D+00	-7.756406227169D+02	-1.00	9.401944018923D+00
1		-2.779275983926D+00	4.027779299633D+02	-.75	5.799202816472D+00
2		3.077811418223D+00	-8.858945873071D+01	-.50	3.426180607612D+00
3		-1.799741668070D+00	1.025575446158D+01	-.25	1.917807328551D+00
4		6.098916283251D-01	-5.226784466621D-01	.25	4.716903560594D-01
5		-1.213324736495D-01	-2.386758638887D-02	.50	1.893762101311D-01
6		1.326896290938D-02	5.726617604397D-03	.75	5.400525572418D-02
7		-6.218838200211D-04	-3.013555854152D-04	1.25	-1.376747193001D-02
				1.50	-1.124220215735D-02
0	-8.0	7.505281568124D-12	3.752640784062D-12	1.75	-4.885822334293D-03
1	-7.0	-7.053505068658D-11	-3.526752534329D-11	2.25	2.135134216128D-03
2	-6.0	3.040721899480D-10	1.520360949740D-10	2.50	2.117287609812D-03
3	-5.0	-8.021021514457D-10	-4.010510757229D-10	2.75	1.084574917956D-03
4	-4.0	1.463808296681D-09	7.319041483407D-10	3.25	-6.184417093244D-04
5	-3.0	-1.995370440843D-09	-9.976852204213D-10	3.50	-6.842960858145D-04
6	-2.0	2.161762077123D-09	1.080881038561D-09	3.75	-3.863413832726D-04
7	-1.0	-1.970706915728D-09	-9.853534578641D-10	4.25	2.595589995085D-04
8	0.0	1.587315276976D-09	7.936576384882D-10	4.50	3.077380244202D-04
9	1.0	-1.164281989713D-09	-5.821409948567D-10	4.75	1.847986776334D-04
10	2.0	7.809077896158D-10	3.904538948079D-10	5.25	-1.377834732050D-04
11	3.0	-4.685478629127D-10	-2.342739314564D-10	5.50	-1.706646133292D-04
12	4.0	2.406044076594D-10	1.203022038297D-10	5.75	-1.065579880539D-04
13	5.0	-9.922297231679D-11	-4.961148615839D-11	6.25	8.484085103852D-05
14	6.0	3.012550700456D-11	1.506275350228D-11	6.50	1.080348449359D-04
15	7.0	-5.880114495636D-12	-2.940057247818D-12	6.75	6.914778937159D-05
16	8.0	5.466715652158D-13	2.733357826079D-13	7.25	-5.745541008539D-05
				7.50	-7.453915324477D-05
	0	-4.923428209277D+01	-2.597026914522D+01	7.75	-4.854237670069D-05
	1	-6.729322421585D+01	-3.189325722995D+01	8.25	4.164756179391D-05
	2	-3.983045657530D+01	-2.088974814864D+01	8.50	5.486820586664D-05
	3	-1.321507694353D+01	-6.306983576594D+00	8.75	3.628549607664D-05
	4	-2.649360979429D+00	-1.380121092637D+00	9.25	-3.214392865948D-05
	5	-3.202341081077D-01	-1.540118481759D-01	9.50	-4.308550944163D-05
	6	-2.155653101180D-02	-1.114929517886D-02	9.75	-2.902713487314D-05
	7	-6.218838200211D-04	-3.013555854152D-04	10.25	2.682683766160D-05

I= 1 H= -7.0

0	0.	8.351223610705D+03	-1.00	-4.055625667476D+01
1	8.665913387076D+00	-4.344528005673D+03	-.75	-2.146344607553D+01
2	-1.537466671107D+01	9.621073957618D+02	-.50	-9.873361898969D+00
3	1.118036134320D+01	-1.144471720296D+02	-.25	-3.319802150728D+00
4	-4.296347400329D+00	6.757355880445D+00	.25	1.364351295911D+00
5	9.266295959666D-01	6.734473275320D-02	.50	1.645637798830D+00
6	-1.071142259394D-01	-4.378800326117D-02	.75	1.410035624434D+00
7	5.224011168761D-03	2.508850636822D-03	1.25	6.011396392794D-01
			1.50	2.950834118114D-01

CARDINAL SPLINES N=8

```
 0  -8.0  -7.053505068658D-11  -3.526752534329D-11    1.75   9.994616698823D-02
 1  -7.0   6.703577649030D-10   3.351788824515D-10    2.25  -3.384491324535D-02
 2  -6.0  -2.927793988185D-09  -1.463896994093D-09    2.50  -3.115241842567D-02
 3  -5.0   7.840078219381D-09   3.920039109690D-09    2.75  -1.508512573584D-02
 4  -4.0  -1.455182614082D-08  -7.275913070412D-09    3.25   7.951429352126D-03
 5  -3.0   2.019888469840D-08   1.009944234920D-08    3.50   8.553263660763D-03
 6  -2.0  -2.227760162789D-08  -1.113880081395D-08    3.75   4.718432097156D-03
 7  -1.0   2.062569455558D-08   1.031284727779D-08    4.25  -3.060176085622D-03
 8   0.0  -1.680357929353D-08  -8.401789646765D-09    4.50  -3.579535189922D-03
 9   1.0   1.241368445359D-08   6.206842226796D-09    4.75  -2.125141607026D-03
10   2.0  -8.359753684907D-09  -4.179876842454D-09    5.25   1.556342243913D-03
11   3.0   5.026887518965D-09   2.513443759483D-09    5.50   1.914266132123D-03
12   4.0  -2.584422086077D-09  -1.292211043039D-09    5.75   1.188067951873D-03
13   5.0   1.066486148758D-09   5.332430743788D-10    6.25  -9.369807995754D-04
14   6.0  -3.239204349802D-10  -1.619602174901D-10    6.50  -1.188666237891D-03
15   7.0   6.323906200170D-11   3.161953100085D-11    6.75  -7.583667514055D-04
16   8.0  -5.880114495636D-12  -2.940057247818D-12    7.25   6.269242538952D-04
                                                      7.50   8.116961969056D-04
       0   4.518471201237D+02   2.404815079016D+02    7.75   5.276930340216D-04
       1   6.143438266980D+02   2.883075109713D+02    8.25  -4.515044679650D-04
       2   3.611904509684D+02   1.910804074872D+02    8.50  -5.941900863198D-04
       3   1.188047534853D+02   5.616851900703D+01    8.75  -3.925890083039D-04
       4   2.355345968070D+01   1.237326547169D+01    9.25   3.472737203187D-04
       5   2.806217761690D+00   1.337415832105D+00    9.50   4.652175045494D-04
       6   1.854303995112D-01   9.670763240084D-02    9.75   3.132690642020D-04
       7   5.224011168761D-03   2.508850636822D-03   10.25  -2.893006345006D-04


I= 2      H= -6.0
 0            0.                 -4.280948338376D+04   -1.00   1.070859936021D+02
 1           -1.725899137513D+01  2.228456996430D+04    -.75   5.416605351252D+01
 2            3.913786094659D+01 -4.951886458789D+03    -.50   2.352516985235D+01
 3           -3.302359963611D+01  5.988447313676D+02    -.25   7.334849802031D+00
 4            1.401086876451D+01 -3.864833607309D+01     .25  -2.332964758257D+00
 5           -3.240419512781D+00  4.097412504586D-01     .50  -2.192559936938D+00
 6            3.942670897950D-01  1.470512425242D-01     .75  -1.129364914759D+00
 7           -1.998627717313D-02 -9.464297425442D-03    1.25   8.059050784537D-01
                                                        1.50   1.189603160612D+00
 0  -8.0   3.040721899480D-10   1.520360949740D-10    1.75   1.211834586392D+00
 1  -7.0  -2.927793988185D-09  -1.463896994093D-09    2.25   6.877652705264D-01
 2  -6.0   1.298506176559D-08   6.492530882797D-09    2.50   3.809913583127D-01
 3  -5.0  -3.540090164037D-08  -1.770045082019D-08    2.75   1.439560297896D-01
 4  -4.0   6.706769652567D-08   3.353384826284D-08    3.25  -5.894531842266D-02
 5  -3.0  -9.520461348063D-08  -4.760230674032D-08    3.50  -5.892924269276D-02
 6  -2.0   1.074143486241D-07   5.370717431203D-08    3.75  -3.076978939386D-02
 7  -1.0  -1.015078523520D-07  -5.075392617601D-08    4.25   1.849296262733D-02
 8   0.0   8.402269261415D-08   4.201134630708D-08    4.50   2.105443961884D-02
 9   1.0  -6.273019452191D-08  -3.136509726095D-08    4.75   1.222751705180D-02
10   2.0   4.250773987096D-08   2.125386993548D-08    5.25  -8.662686862338D-03
11   3.0  -2.564904057810D-08  -1.282452028905D-08    5.50  -1.052201655477D-02
12   4.0   1.321153748009D-08   6.605768740044D-09    5.75  -6.461976656667D-03
13   5.0  -5.457543621311D-09  -2.728771810655D-09    6.25   5.013609307089D-03
14   6.0   1.658586040033D-09   8.292930200163D-10    6.50   6.320107743485D-03
15   7.0  -3.239204349802D-10  -1.619602174901D-10    6.75   4.010568937213D-03
16   8.0   3.012550700456D-11   1.506275350228D-11    7.25  -3.287495578459D-03
                                                      7.50  -4.242333323476D-03
       0  -1.894389138420D+03  -1.021807186069D+03    7.75  -2.750216974988D-03
```

CARDINAL SPLINES N=8

	1	-2.560031250876D+03	-1.183327221901D+03	8.25	2.342679762232D-03
	2	-1.493549771767D+03	-8.005184737723D+02	8.50	3.077637413726D-03
	3	-4.864819833895D+02	-2.266645781415D+02	8.75	2.030408499766D-03
	4	-9.526359248589D+01	-5.068970309545D+01	9.25	-1.791828534771D-03
	5	-1.117715572330D+01	-5.251814848175D+00	9.50	-2.398173566957D-03
	6	-7.249644319000D-01	-3.829494133006D-01	9.75	-1.613620994287D-03
	7	-1.998627717313D-02	-9.464297425442D-03	10.25	1.488322476189D-03
I= 3	H= -5.0				
0		0.	1.411316647348D+05	-1.00	-1.957429228517D+02
1		2.824487132617D+01	-7.348272146337D+04	-.75	-9.696785577151D+01
2		-6.852696188467D+01	1.635620562011D+04	-.50	-4.109202253514D+01
3		6.261969095780D+01	-1.997521341295D+03	-.25	-1.244076882091D+01
4		-2.845872428494D+01	1.362949670679D+02	.25	3.652132294401D+00
5		6.960346685215D+00	-3.212909963732D+00	.50	3.243520580446D+00
6		-8.857752558575D-01	-2.777541062873D-01	.75	1.550689259252D+00
7		4.655245708125D-02	2.153247085045D-02	1.25	-8.582049029471D-01
				1.50	-9.882010731521D-01
0	-8.0	-8.021021514457D-10	-4.010510757229D-10	1.75	-6.059036982825D-01
1	-7.0	7.840078219381D-09	3.920039109690D-09	2.25	5.770951930890D-01
2	-6.0	-3.540090164037D-08	-1.770045082019D-08	2.50	9.626731099308D-01
3	-5.0	9.858615605501D-08	4.929307802750D-08	2.75	1.095571078326D+00
4	-4.0	-1.914423014326D-07	-9.572115071628D-08	3.25	7.540876462939D-01
5	-3.0	2.793505508465D-07	1.396752754232D-07	3.50	4.543945677795D-01
6	-2.0	-3.243550797805D-07	-1.621775398903D-07	3.75	1.854102827713D-01
7	-1.0	3.148926423768D-07	1.574463211884D-07	4.25	-8.681623475505D-02
8	0.0	-2.664736037647D-07	-1.332368018823D-07	4.50	-9.198856717264D-02
9	1.0	2.020801436639D-07	1.010400718320D-07	4.75	-5.063261936954D-02
10	2.0	-1.382768969329D-07	-6.913844846646D-08	5.25	3.332207832017D-02
11	3.0	8.390513015951D-08	4.195256507975D-08	5.50	3.943763131850D-02
12	4.0	-4.335360843908D-08	-2.167680421954D-08	5.75	2.371647386965D-02
13	5.0	1.794007326532D-08	8.970036632660D-09	6.25	-1.783217913886D-02
14	6.0	-5.457543621311D-09	-2.728771810655D-09	6.50	-2.221580116300D-02
15	7.0	1.066486148758D-09	5.332430743788D-10	6.75	-1.395959845777D-02
16	8.0	-9.922297231679D-11	-4.961148615839D-11	7.25	1.127047852251D-02
				7.50	1.445917453886D-02
	0	4.838130491756D+03	2.664969260647D+03	7.75	9.327397702888D-03
	1	6.492856582841D+03	2.927744368990D+03	8.25	-7.884087667702D-03
	2	3.755187990510D+03	2.054745579887D+03	8.50	-1.032642731479D-02
	3	1.209984019565D+03	5.503482008282D+02	8.75	-6.795278856854D-03
	4	2.338309283965D+02	1.269965041229D+02	9.25	5.972800470656D-03
	5	2.700963672126D+01	1.239453375748D+01	9.50	7.981452049910D-03
	6	1.721162340693D+00	9.280642613378D-01	9.75	5.363208270698D-03
	7	4.655245708125D-02	2.153247085045D-02	10.25	-4.936410946996D-03
I= 4	H= -4.0				
0		0.	-3.427935003847D+05	-1.00	2.675737273749D+02
1		-3.652249142421D+01	1.785061675217D+05	-.75	1.310742339130D+02
2		9.128329113595D+01	-3.976880921498D+04	-.50	5.484100811101D+01
3		-8.671482585866D+01	4.885146845021D+03	-.25	1.636194821674D+01
4		4.112111675409D+01	-3.451144299010D+02	.25	-4.629603072864D+00
5		-1.047455497969D+01	1.102320919440D+01	.50	-4.016021205121D+00
6		1.382455796659D+00	2.900281033829D-01	.75	-1.866479827819D+00
7		-7.499142562036D-02	-3.325899906847D-02	1.25	9.557437293290D-01
				1.50	1.041369725855D+00

CARDINAL SPLINES N=8

0	-8.0	1.463808296681D-09	7.319041483407D-10	1.75	5.934740980592D-01
1	-7.0	-1.455182614082D-08	-7.275913070412D-09	2.25	-4.397177262101D-01
2	-6.0	6.706769652567D-08	3.353384826284D-08	2.50	-5.731041951456D-01
3	-5.0	-1.914423014326D-07	-9.572115071628D-08	2.75	-3.932082063860D-01
4	-4.0	3.827713011209D-07	1.913856505604D-07	3.25	4.557236423764D-01
5	-3.0	-5.774326597532D-07	-2.887163298766D-07	3.50	8.283163307799D-01
6	-2.0	6.947799420041D-07	3.473899710021D-07	3.75	1.019675753492D+00
7	-1.0	-6.983920983918D-07	-3.491960491959D-07	4.25	8.051440863367D-01
8	0.0	6.090857025853D-07	3.045428512927D-07	4.50	5.149775872273D-01
9	1.0	-4.725668039189D-07	-2.362834019595D-07	4.75	2.218253335915D-01
10	2.0	3.283325560535D-07	1.641662780268D-07	5.25	-1.140315394497D-01
11	3.0	-2.010817079332D-07	-1.005408539666D-07	5.50	-1.257487023815D-01
12	4.0	1.044522796217D-07	5.222613981087D-08	5.75	-7.174778879970D-02
13	5.0	-4.335360843908D-08	-2.167680421954D-08	6.25	5.020648388552D-02
14	6.0	1.321153748009D-08	6.605768740044D-09	6.50	6.099562322318D-02
15	7.0	-2.584422086077D-09	-1.292211043039D-09	6.75	3.755751972609D-02
16	8.0	2.406044076594D-10	1.203022038297D-10	7.25	-2.942186618559D-02
				7.50	-3.732330941951D-02
	0	-8.512089351027D+03	-4.853000592811D+03	7.75	-2.385211870830D-02
	1	-1.133607194486D+04	-4.894297822734D+03	8.25	1.987300737864D-02
	2	-6.494700774428D+03	-3.677516780299D+03	8.50	2.588589060073D-02
	3	-2.068977695870D+03	-9.017833552121D+02	8.75	1.695501361010D-02
	4	-3.945498647572D+02	-2.217603461845D+02	9.25	-1.479509665363D-02
	5	-4.490515277380D+01	-1.975553659125D+01	9.50	-1.971513870246D-02
	6	-2.817064038081D+00	-1.572475844452D+00	9.75	-1.321623148643D-02
	7	-7.499142562036D-02	-3.325899906847D-02	10.25	1.211919169058D-02

I= 5 H= -3.0

0		0.	6.695643670418D+05	-1.00	-2.913651597723D+02
1		3.859049166063D+01	-3.487157519421D+05	-.75	-1.418342539492D+02
2		-9.791651413449D+01	7.773248808095D+04	-.50	-5.892777086125D+01
3		9.493518871478D+01	-9.581657591957D+03	-.25	-1.744315378952D+01
4		-4.613081307502D+01	6.914348124209D+02	.25	4.842394082336D+00
5		1.206594091873D+01	-2.574125736951D+01	.50	4.152060513513D+00
6		-1.635252675640D+00	-3.359139934679D-02	.75	1.903940711081D+00
7		9.095859300385D-02	3.719918966126D-02	1.25	-9.421437928953D-01
				1.50	-1.004097787028D+00
0	-8.0	-1.995370440843D-09	-9.976852204213D-10	1.75	-5.571009402589D-01
1	-7.0	2.019888469840D-08	1.009944234920D-08	2.25	3.831047921510D-01
2	-6.0	-9.520461348063D-08	-4.760230674032D-08	2.50	4.732561916154D-01
3	-5.0	2.793505508465D-07	1.396752754232D-07	2.75	3.023061434741D-01
4	-4.0	-5.774326597532D-07	-2.887163298766D-07	3.25	-2.734727987949D-01
5	-3.0	9.055151165717D-07	4.527757582859D-07	3.50	-3.890087293027D-01
6	-2.0	-1.137142702565D-06	-5.685713512825D-07	3.75	-2.891744802321D-01
7	-1.0	1.193682006779D-06	5.968410033893D-07	4.25	3.856672149765D-01
8	0.0	-1.083074615535D-06	-5.415373077674D-07	4.50	7.451359731192D-01
9	1.0	8.677380659208D-07	4.338690329604D-07	4.75	9.696416175780D-01
10	2.0	-6.170036968529D-07	-3.085018484264D-07	5.25	8.426058208641D-01
11	3.0	3.835999125071D-07	1.917999562535D-07	5.50	5.615007452082D-01
12	4.0	-2.010817079332D-07	-1.005408539666D-07	5.75	2.509611260288D-01
13	5.0	8.390513015951D-08	4.195256507975D-08	6.25	-1.374054084115D-01
14	6.0	-2.564904057810D-08	-1.282452028905D-08	6.50	-1.556524605258D-01
15	7.0	5.026887518965D-09	2.513443759483D-09	6.75	-9.099158518971D-02
16	8.0	-4.685478629127D-10	-2.342739314564D-10	7.25	6.641225610593D-02
				7.50	8.219400000676D-02
	0	1.115614315253D+04	6.742943002093D+03	7.75	5.149357176656D-02

CARDINAL SPLINES N=8

```
        1  1,474003511526D+04   5,859817386957D+03    8,25 -4,165761819480D-02
        2  8,365143593338D+03   5.022384519064D+03    8.50 -5,366965528097D-02
        3  2,635787852779D+03   1.058751589551D+03    8,75 -3,483419746049D-02
        4  4,966422416887D+02   2.961462529973D+02    9,25  2,997503409195D-02
        5  5,582216148518D+01   2.264206636658D+01    9.50  3,973027469735D-02
        6  3,458428532575D+00   2.049563221684D+00    9,75  2,651413689780D-02
        7  9,095859300385D-02   3.719918966126D-02   10.25 -2,414407858816D-02


I= 6     H= -2,0
0          0.                  -1.122236430880D+06   -1.00  2,681310608449D+02
1         -3,494180619684D+01   5,845833118969D+05    -.75  1,300746054227D+02
2          8,935690812976D+01  -1,303650500502D+05    -.50  5,383632953693D+01
3         -8,757764682561D+01   1.610311452103D+04    -.25  1,586871325931D+01
4          4,312961718629D+01  -1.176751141072D+03     .25 -4,361378610236D+00
5         -1.145671655502D+01   4.756935209686D+01     .50 -3,717329414220D+00
6          1.579005105303D+00  -5,258175541804D-01     .75 -1,693064931424D+00
7         -8,936084604962D-02  -3.078754156595D-02    1.25  8,239332569772D-01
                                                      1.50  8,690637698694D-01
0   -8,0   2,161762077123D-09   1.080881038561D-09    1.75  4,763683804945D-01
1   -7,0  -2,227760162789D-08  -1.113880081395D-08    2,25 -3,174439883543D-01
2   -6.0   1,074143486241D-07   5,370717431203D-08    2.50 -3,841112507672D-01
3   -5,0  -3,243550797805D-07  -1.621775398903D-07    2,75 -2,392159313250D-01
4   -4,0   6,947799420041D-07   3,473899710021D-07    3,25  2,014203937196D-01
5   -3,0  -1,137142702565D-06  -5,685713512825D-07    3,50  2,719369312188D-01
6   -2,0   1,498957391864D-06   7.494786959319D-07    3,75  1,884565536653D-01
7   -1,0  -1,656018212330D-06  -8,280091061648D-07    4,25 -1,966676821524D-01
8    0,0   1,578194713717D-06   7.890973568587D-07    4,50 -2,977131793084D-01
9    1,0  -1,319330571497D-06  -6,596652857485D-07    4,75 -2,341877358279D-01
10   2,0   9,696351532703D-07   4,848175766352D-07    5,25  3,443642005245D-01
11   3,0  -6,170036968529D-07  -3,085018484264D-07    5,50  6,937398064846D-01
12   4,0   3,283325560535D-07   1,641662780268D-07    5,75  9,373798022292D-01
13   5,0  -1,382768969329D-07  -6,913844846646D-08    6,25  8,686354636908D-01
14   6,0   4,250773987096D-08   2,125386993548D-08    6,50  5,949188432612D-01
15   7,0  -8,359753684907D-09  -4,179876842454D-09    6,75  2,725526845552D-01
16   8,0   7,809077896158D-10   3,904538948079D-10    7,25 -1,557665488176D-01
                                                      7,50 -1,798119189647D-01
        0 -1,163178627369D+04  -7,764719767363D+03    7,75 -1,069800643793D-01
        1 -1,525707016016D+04  -5,100829607504D+03    8,25  8,067336148360D-02
        2 -8,586748978419D+03  -5,699656912859D+03    8,50  1,014231769501D-01
        3 -2,681487103440D+03  -9,066103649895D+02    8,75  6,454497890605D-02
        4 -5,006405051322D+02  -3,304746540726D+02    9,25 -5,394655792510D-02
        5 -5,576544859113D+01  -1,904834636844D+01    9,50 -7,073307657678D-02
        6 -3,425202273475D+00  -2,249919881874D+00    9,75 -4,678183302494D-02
        7 -8,936084604962D-02  -3,078754156595D-02   10,25  4,201885795395D-02


I= 7     H= -1.0
0          0.                   1,691026481120D+06   -1.00 -2,197753103608D+02
1          2,839985758633D+01  -8,811612494857D+05    -.75 -1,064239759881D+02
2         -7,291911318097D+01   1,965922494369D+05    -.50 -4,396011945389D+01
3          7,186577663530D+01  -2,431731085164D+04    -.25 -1,292918363450D+01
4         -3,564049436333D+01   1,790044668317D+03     .25  3,535204663910D+00
5          9,546095160056D+00  -7,565850899324D+01     .50  3,003999579391D+00
6         -1,328047635142D+00   1,321470759658D+00     .75  1,363495426667D+00
7          7,592579972583D-02   1,698583094800D-02    1.25 -6,580598923383D-01
                                                      1.50 -6,905947144810D-01
```

CARDINAL SPLINES N=8

0	-8,0	-1.970706915728D-09	-9.853534578641D-10	1.75	-3.763343754709D-01
1	-7,0	2,062569455558D-08	1.031284727779D-08	2.25	2,471106798048D-01
2	-6,0	-1,015078523520D-07	-5,075392617601D-08	2.50	2,962154093338D-01
3	-5.0	3,148926423768D-07	1.574463211884D-07	2.75	1,824307373465D-01
4	-4,0	-6,983920983918D-07	-3,491960491959D-07	3.25	-1,491338553810D-01
5	-3,0	1,193682006779D-06	5,968410033893D-07	3.50	-1,973982006415D-01
6	-2,0	-1,656018212330D-06	-8,280091061648D-07	3,75	-1,334872026947D-01
7	-1,0	1,935166046009D-06	9,675830230047D-07	4.25	1,298601317331D-01
8	0,0	-1,951668188758D-06	-9,758340943788D-07	4,50	1,867038078737D-01
9	1,0	1,718750626916D-06	8,593753134579D-07	4,75	1,370015542036D-01
10	2,0	-1,319330571497D-06	-6,596652857485D-07	5,25	-1,577941485021D-01
11	3,0	8,677380659208D-07	4,338690329604D-07	5,50	-2,491686405967D-01
12	4,0	-4,725668039189D-07	-2,362834019595D-07	5.75	-2,035920263365D-01
13	5,0	2,020801436639D-07	1.010400718320D-07	6,25	3,194324568118D-01
14	6,0	-6,273019452191D-08	-3,136509726095D-08	6.50	6,615381505884D-01
15	7,0	1,241368445359D-08	6,206842226796D-09	6,75	9,164319787804D-01
16	8,0	-1,164281989713D-09	-5,821409948567D-10	7,25	8,867420848083D-01
				7,50	6,189739880410D-01
	0	1,026685802682D+04	8,059732056015D+03	7,75	2,886432523311D-01
	1	1,339188003578D+04	2,897133312568D+03	8,25	-1,704776244228D-01
	2	7,493338920637D+03	5,861964937924D+03	8,50	-1,999279080516D-01
	3	2,326385724284D+03	5.096220824453D+02	8,75	-1,208411011090D-01
	4	4,318679133893D+02	3,367023284479D+02	9,25	9,416431204700D-02
	5	4,784408350475D+01	1,060104426446D+01	9,50	1,204920307333D-01
	6	2,923797149504D+00	2,272677292746D+00	9,75	7,814491540451D-02
	7	7,592579972583D-02	1,698583094800D-02	10,25	-6,818531511120D-02

I= 8 H= 0.0

0		0.	-2,351592844851D+06	-1.00	1,664198442006D+02
1		-2,141304042044D+01	1.226122553282D+06	-.75	8,051282636833D+01
2		5,509156941935D+01	-2.737388240500D+05	-.50	3,322341609974D+01
3		-5,444899886194D+01	3.389947379653D+04	-.25	9,760476368441D+00
4		2,709932528197D+01	-2,506421562047D+03	.25	-2,661819140285D+00
5		-7,289310258981D+00	1.085297254429D+02	.50	-2,258371123622D+00
6		1,019027398189D+00	-2.261035946727D+00	.75	-1,023289024124D+00
7		-5,857255973065D-02	-3,650514413980D-23	1.25	4,918137188272D-01
				1.50	5,148264224471D-01
0	-8,0	1,587315276976D-09	7,936576384882D-10	1,75	2,797382111115D-01
1	-7,0	-1,680357929353D-08	-8,401789646765D-09	2,25	-1,823591716480D-01
2	-6,0	8,402269261415D-08	4,201134630708D-08	2,50	-2.176070705884D-01
3	-5,0	-2,664736037647D-07	-1.332368018823D-07	2,75	-1,333058277802D-01
4	-4,0	6,090857025853D-07	3.045428512927D-07	3,25	1,074849535118D-01
5	-3,0	-1,083074615535D-06	-5,415373077674D-07	3,50	1,410053040207D-01
6	-2,0	1,578194713717D-06	7,890973568587D-07	3,75	9,433392294386D-02
7	-1,0	-1,951668188758D-06	-9,758340943788D-07	4.25	-8,916035706529D-02
8	0,0	2,090259126314D-06	1,045129563157D-06	4,50	-1,257111306266D-01
9	1,0	-1,951668188758D-06	-9,758340943788D-07	4,75	-9,003318326687D-02
10	2,0	1,578194713717D-06	7,890973568587D-07	5,25	9,670196178031D-02
11	3,0	-1,083074615535D-06	-5,415373077674D-07	5,50	1,450450973621D-01
12	4,0	6,090857025853D-07	3,045428512927D-07	5.75	1,105647860239D-01
13	5,0	-2,664736037647D-07	-1,332368018823D-07	6,25	-1,358931031930D-01
14	6,0	8,402269261415D-08	4,201134630708D-08	6,50	-2,206033323639D-01
15	7,0	-1,680357929353D-08	-8,401789646765D-09	6,75	-1,848061426772D-01
16	8,0	1,587315276976D-09	7,936576384882D-10	7,25	3,027781022237D-01
				7,50	6,390898496406D-01
	0	-8,084256022536D+03	-8,084256022536D+03	7,75	9,011791661717D-01

CARDINAL SPLINES N=8

1	-1.051198862071D+04	-6.578959448792D-18	8.25	9.011791661717D-01
2	-5.863187058564D+03	-5.863187058564D+03	8.50	6.390898496406D-01
3	-1.814550761116D+03	-1.197997592760D-18	8.75	3.027781022237D-01
4	-3.358270531889D+02	-3.358270531889D+02	9.25	-1.848061426772D-01
5	-3.709751542390D+01	-2.510924166211D-20	9.50	-2.206033323639D-01
6	-2.261035946727D+00	-2.261035946727D+00	9.75	-1.358931031930D-01
7	-5.857255973065D-02	-3.650514413980D-23	10.25	1.105647860239D-01

I= 9 H= 1.0

0		0.	3.025227402150D+06	-1.00	-1.180665412854D+02
1		1.515811072710D+01	-1.579207440195D+06	-.75	-5.709280547838D+01
2		-3.903932947063D+01	3.529856544195D+05	-.50	-2.354696850283D+01
3		3.863960648666D+01	-4.377606912446D+04	-.25	-6.913739837496D+00
4		-1.926606688902D+01	3.246900390651D+03	.25	1.882952098732D+00
5		5.193599291715D+00	-1.425185111104D+02	.50	1.596299335064D+00
6		-7.278742824832D-01	3.223883825834D+00	.75	7.226599019760D-01
7		4.195413782984D-02	-1.698583094800D-02	1.25	-3.465885216465D-01
				1.50	-3.623400619266D-01
0	-8.0	-1.164281989713D-09	-5.821409948567D-10	1.75	-1.965931225789D-01
1	-7.0	1.241368445359D-08	6.206842226796D-09	2.25	1.276901795786D-01
2	-6.0	-6.273019452191D-08	-3.136509726095D-08	2.50	1.520244042007D-01
3	-5.0	2.020801436639D-07	1.010400718320D-07	2.75	9.288239476595D-02
4	-4.0	-4.725668039189D-07	-2.362834019595D-07	3.25	-7.438359543298D-02
5	-3.0	8.677380659208D-07	4.338690329604D-07	3.50	-9.715705525328D-02
6	-2.0	-1.319330571497D-06	-6.596652857485D-07	3.75	-6.466391381128D-02
7	-1.0	1.718750626916D-06	8.593753134579D-07	4.25	6.029617465821D-02
8	0.0	-1.951668188758D-06	-9.758340943788D-07	4.50	8.426480314043D-02
9	1.0	1.935166046009D-06	9.675830230047D-07	4.75	5.970809080899D-02
10	2.0	-1.656018212330D-06	-8.280091061648D-07	5.25	-6.230843740625D-02
11	3.0	1.193682006779D-06	5.968410033893D-07	5.50	-9.165013218769D-02
12	4.0	-6.983920983918D-07	-3.491960491959D-07	5.75	-6.818531511120D-02
13	5.0	3.148926423768D-07	1.574463211884D-07	6.25	7.814491540451D-02
14	6.0	-1.015078523520D-07	-5.075392617601D-08	6.50	1.204920307333D-01
15	7.0	2.062569455558D-08	1.031284727779D-08	6.75	9.416431204700D-02
16	8.0	-1.970706915728D-09	-9.853534578641D-10	7.25	-1.208411011090D-01
				7.50	-1.999279080516D-01
	0	5.852606085209D+03	8.059732056015D+03	7.75	-1.704776244228D-01
	1	7.597613410649D+03	-2.897133312568D+03	8.25	2.886432523311D-01
	2	4.230590955211D+03	5.861964937924D+03	8.50	6.189739880410D-01
	3	1.307141559393D+03	-5.096220824453D+02	8.75	8.867420848083D-01
	4	2.415367435064D+02	3.367023284479D+02	9.25	9.164319787804D-01
	5	2.664199497582D+01	-1.060104426446D+01	9.50	6.615381505884D-01
	6	1.621557435988D+00	2.272677292746D+00	9.75	3.194324568118D-01
	7	4.195413782984D-02	-1.698583094800D-02	10.25	-2.035920263365D-01

I=10 H= 2.0

0	0.	-3.509703454692D+06	-1.00	7.780694478447D+01
1	-9.977969695436D+00	1.836134388667D+06	-.75	3.761547307447D+01
2	2.571179517831D+01	-4.113089531879D+05	-.50	1.550969788316D+01
3	-2.546749525171D+01	5.112562252040D+04	-.25	4.552527934564D+00
4	1.271031198042D+01	-3.804044340271D+03	.25	-1.239019183311D+00
5	-3.430221682560D+00	1.684229565630D+02	.50	-1.049968632914D+00
6	4.813652331190D-01	-3.974022209567D+00	.75	-4.751138594317D-01
7	-2.778576291771D-02	3.078754156595D-02	1.25	2.276152918527D-01
			1.50	2.378020242847D-01

CARDINAL SPLINES N=8

0	-8,0	7,809077896158D-10	3,904538948079D-10	1.75	1,289252330498D-01
1	-7,0	-8,359753684907D-09	-4,179876842454D-09	2,25	-8,358151636977D-02
2	-6,0	4,250773987096D-08	2,125386993548D-08	2,50	-9,939346338277D-02
3	-5,0	-1,382768969329D-07	-6,913844846646D-08	2,75	-6,064367723505D-02
4	-4,0	3,283325560535D-07	1,641662780268D-07	3,25	4,839744083403D-02
5	-3,0	-6,170036968529D-07	-3,085018484264D-07	3,50	6,307558664151D-02
6	-2,0	9,696351532703D-07	4,848175766352D-07	3,75	4,187140166988D-02
7	-1,0	-1,319330571497D-06	-6,596652857485D-07	4,25	-3,878090109345D-02
8	0,0	1,578194713717D-06	7,890973568587D-07	4,50	-5,396179377438D-02
9	1,0	-1,656018212330D-06	-8,280091061648D-07	4,75	-3,803856686766D-02
10	2,0	1,498957391864D-06	7,494786959319D-07	5,25	3,915923837385D-02
11	3,0	-1,137142702565D-06	-5,685713512825D-07	5,50	5,708915301684D-02
12	4,0	6,947799420041D-07	3,473899710021D-07	5,75	4,201885795395D-02
13	5,0	-3,243550797805D-07	-1,621775398903D-07	6,25	-4,678183302494D-02
14	6,0	1,074143486241D-07	5,370717431203D-08	6,50	-7,073307657678D-02
15	7,0	-2,227760162789D-08	-1,113880081395D-08	6,75	-5,394655792510D-02
16	8,0	2,161762077123D-09	1,080881038561D-09	7,25	6,454497890605D-02
				7,50	1,014231769501D-01
	0	-3,897653261034D+03	-7,764719767363D+03	7,75	8,067336148360D-02
	1	-5,055410945156D+03	5,100829607504D+03	8,25	-1,069800643793D-01
	2	-2,812564847299D+03	-5,699656912859D+03	8,50	-1,798119189647D-01
	3	-8,682663734614D+02	9,066103649895D+02	8,75	-1,557665488176D-01
	4	-1,603088030131D+02	-3,304746540726D+02	9,25	2,725526845552D-01
	5	-1,766875585425D+01	1,904834636844D+01	9,50	5,949188432612D-01
	6	-1,074637490273D+00	-2,249919881874D+00	9,75	8,686354636908D-01
	7	-2,778576291771D-02	3,078754156595D-02	10,25	9,373798022292D-01
I=11	H= 3.0				
0		0.	3,487378244326D+06	-1,00	-4,625223206560D+01
1		5,927855569653D+00	-1,830936656722D+06	-.75	-2,235759850776D+01
2		-1,527952139314D+01	4,116029319396D+05	-.50	-9,217237838827D+00
3		1,514024402799D+01	-5,134675737995D+04	-.25	-2,705095841333D+00
4		-7,559920584061D+00	3,836019079207D+03	.25	7,359538100614D-01
5		2,041456221707D+00	-1,710168119121D+02	.50	6,235285050189D-01
6		-2,866740553622D-01	4,132717842714D+00	.75	2,820810501825D-01
7		1,656021368133D-02	-3,719918966126D-02	1,25	-1,350604836270D-01
				1,50	-1,410560499260D-01
0	-8,0	-4,685478629127D-10	-2,342739314564D-10	1,75	-7,644370725630D-02
1	-7,0	5,026887518965D-09	2,513443759483D-09	2,25	4,950934951291D-02
2	-6,0	-2,564904057810D-08	-1,282452028905D-08	2,50	5,883957091078D-02
3	-5,0	8,390513015951D-08	4,195256507975D-08	2,75	3,587469938860D-02
4	-4,0	-2,010817079332D-07	-1,005408539666D-07	3,25	-2,857860047775D-02
5	-3,0	3,835999125071D-07	1,917999562535D-07	3,50	-3,720339260305D-02
6	-2,0	-6,170036968529D-07	-3,085018484264D-07	3,75	-2,466340632439D-02
7	-1,0	8,677380659208D-07	4,338690329604D-07	4,25	2,276387349929D-02
8	0,0	-1,083074615535D-06	-5,415373077674D-07	4,50	3,160450927004D-02
9	1,0	1,193682006779D-06	5,968410033893D-07	4,75	2,221996694511D-02
10	2,0	-1,137142702565D-06	-5,685713512825D-07	5,25	-2,271900349265D-02
11	3,0	9,055515165717D-07	4,527757582859D-07	5,50	-3,297589144847D-02
12	4,0	-5,774326597532D-07	-2,887163298766D-07	5,75	-2,414407858816D-02
13	5,0	2,793505508465D-07	1,396752754232D-07	6,25	2,651413689780D-02
14	6,0	-9,520461348063D-08	-4,760230674032D-08	6,50	3,973027469735D-02
15	7,0	2,019888469840D-08	1,009944234920D-08	6,75	2,997503409195D-02
16	8,0	-1,995370440843D-09	-9,976852204213D-10	7,25	-3,483419746049D-02
				7,50	-5,366965528097D-02
	0	2,329742851659D+03	6,742943002093D+03	7,75	-4,165761819480D-02

CARDINAL SPLINES N=8

	1	3.020400341348D+03	-5.859817386957D+03	8.25	5.149357176656D-02
	2	1.679625444790D+03	5.022384519064D+03	8.50	8.219400000676D-02
	3	5.182846736767D+02	-1.058751589551D+03	8.75	6.641225610593D-02
	4	9.565026430590D+01	2.961462529973D+02	9.25	-9.099158518971D-02
	5	1.053802875203D+01	-2.264206636658D+01	9.50	-1.556524605258D-01
	6	6.406979107921D-01	2.049563221684D+00	9.75	-1.374054084115D-01
	7	1.656021368133D-02	-3.719918966126D-02	10.25	2.509611260288D-01

I=12 H= 4.0

0		0.	-2.778725620159D+06	-1.00	2.363333174140D+01
1		-3.027973651928D+00	1.465827409071D+06	-.75	1.142319950207D+01
2		7.806009828737D+00	-3.311240019815D+05	-.50	4.709023423936D+00
3		-7.736459901871D+00	4.151182060516D+04	-.25	1.381900697195D+00
4		3.864031753595D+00	-3.117559883815D+03	.25	-3.758901266373D-01
5		-1.043758889538D+00	1.399344718729D+02	.50	-3.184325313603D-01
6		1.466242882488D-01	-3.434979792286D+00	.75	-1.440389043088D-01
7		-8.473427483413D-03	3.325899906847D-02	1.25	6.894480203977D-02
				1.50	7.199208939306D-02
0	-8.0	2.406044076594D-10	1.203022038297D-10	1.75	3.900705127210D-02
1	-7.0	-2.584422086077D-09	-1.292211043039D-09	2.25	-2.525001401222D-02
2	-6.0	1.321153748009D-08	6.605768740044D-09	2.50	-2.999873100199D-02
3	-5.0	-4.335360843908D-08	-2.167680421954D-08	2.75	-1.828342351766D-02
4	-4.0	1.044522796217D-07	5.222613981087D-08	3.25	1.455103706221D-02
5	-3.0	-2.010817079332D-07	-1.005408539666D-07	3.50	1.893092686234D-02
6	-2.0	3.283325560535D-07	1.641662780268D-07	3.75	1.254100202182D-02
7	-1.0	-4.725668039189D-07	-2.362834019595D-07	4.25	-1.155391344304D-02
8	0.0	6.090857025853D-07	3.045428512927D-07	4.50	-1.602225018079D-02
9	1.0	-6.983920983918D-07	-3.491960491959D-07	4.75	-1.124909126727D-02
10	2.0	6.947799420041D-07	3.473899710021D-07	5.25	1.146082233190D-02
11	3.0	-5.774326597532D-07	-2.887163298766D-07	5.50	1.659709124750D-02
12	4.0	3.827713011209D-07	1.913856505604D-07	5.75	1.211919169058D-02
13	5.0	-1.914423014326D-07	-9.572115071628D-08	6.25	-1.321623148643D-02
14	6.0	6.706769652567D-08	3.353384826284D-08	6.50	-1.971513870246D-02
15	7.0	-1.455182614082D-08	-7.275913070412D-09	6.75	-1.479509665363D-02
16	8.0	1.463808296681D-09	7.319041483407D-10	7.25	1.695501361010D-02
				7.50	2.588589060073D-02
	0	-1.193911834596D+03	-4.853000592811D+03	7.75	1.987300737864D-02
	1	-1.547476299391D+03	4.894297822734D+03	8.25	-2.385211870830D-02
	2	-8.603327861701D+02	-3.677516780299D+03	8.50	-3.732330941951D-02
	3	-2.654109854455D+02	9.017833552121D+02	8.75	-2.942186618559D-02
	4	-4.897082761183D+01	-2.217603461845D+02	9.25	3.755751972609D-02
	5	-5.394079591302D+00	1.975553659125D+01	9.50	6.099562322318D-02
	6	-3.278876508223D-01	-1.572475844452D+00	9.75	5.020648388552D-02
	7	-8.473427483413D-03	3.325899906847D-02	10.25	-7.174778879970D-02

I=13 H= 5.0

0		0.	1.654134025235D+06	-1.00	-9.719722509178D+00
1		1.245104415684D+00	-8.773765345665D+05	-.75	-4.697864521699D+00
2		-3.210097260514D+00	1.993271371392D+05	-.50	-1.936537014074D+00
3		3.181856181271D+00	-2.513701099477D+04	-.25	-5.682666855752D-01
4		-1.589429314708D+00	1.899581422946D+03	.25	1.545576511971D-01
5		4.294131423038D-01	-8.588125912470D+01	.50	1.309242607495D-01
6		-6.033467931684D-02	2.133882628963D+00	.75	5.921781207416D-02
7		3.487515380356D-03	-2.153247085045D-02	1.25	-2.834011675165D-02
				1.50	-2.958973740779D-02

CARDINAL SPLINES N=8

0	-8.0	-9.922297231679D-11	-4.961148615839D-11	1.75	-1.603058598571D-02
1	-7.0	1.066486148758D-09	5.332430743788D-10	2.25	1.037394610658D-02
2	-6.0	-5.457543621311D-09	-2.728771810655D-09	2.50	1.232276812932D-02
3	-5.0	1.794007326532D-08	8.970036632660D-09	2.75	7.508848409477D-03
4	-4.0	-4.335360843908D-08	-2.167680421954D-08	3.25	-5.972864876998D-03
5	-3.0	8.390513015951D-08	4.195256507975D-08	3.50	-7.768136918905D-03
6	-2.0	-1.382768969329D-07	-6.913844846646D-08	3.75	-5.144086021443D-03
7	-1.0	2.020801436639D-07	1.010400718320D-07	4.25	4.734473595235D-03
8	0.0	-2.664736037647D-07	-1.332368018823D-07	4.50	6.561297482897D-03
9	1.0	3.148926423768D-07	1.574463211884D-07	4.75	4.603177896057D-03
10	2.0	-3.243550797805D-07	-1.621775398903D-07	5.25	-4.680782430089D-03
11	3.0	2.793505508465D-07	1.396752754232D-07	5.50	-6.770177825057D-03
12	4.0	-1.914423014326D-07	-9.572115071628D-08	5.75	-4.936410946996D-03
13	5.0	9.858615605501D-08	4.929307802750D-08	6.25	5.363208270698D-03
14	6.0	-3.540090164037D-08	-1.770045082019D-08	6.50	7.981452049910D-03
15	7.0	7.840078219381D-09	3.920039109690D-09	6.75	5.972800470656D-03
16	8.0	-8.021021514457D-10	-4.010510757229D-10	7.25	-6.795278856854D-03
				7.50	-1.032642731479D-02
	0	4.918080295388D+02	2.664969260647D+03	7.75	-7.884087667702D-03
	1	6.373678448611D+02	-2.927744368990D+03	8.25	9.327397702888D-03
	2	3.543031692642D+02	2.054745579887D+03	8.50	1.445917453886D-02
	3	1.092876179084D+02	-5.503482008282D+02	8.75	1.127047852251D-02
	4	2.016207984925D+01	1.269965041229D+02	9.25	-1.395959845777D-02
	5	2.220569206294D+00	-1.239453375748D+01	9.50	-2.221580116300D-02
	6	1.349661819831D-01	9.280642613378D-01	9.75	-1.783217913886D-02
	7	3.487515380356D-03	-2.153247085045D-02	10.25	2.371647386965D-02

I=14 H= 6.0

0		0.	-6.777263253897D+05	-1.00	2.946504274447D+00
1		-3.774125605144D-01	3.615388815758D+05	-.75	1.424113239678D+00
2		9.730806473155D-01	-8.263566637244D+04	-.50	5.870292823291D-01
3		-9.645817720092D-01	1.048810025114D+04	-.25	1.722562774918D-01
4		4.818756745483D-01	-7.979939436549D+02	.25	-4.684760936635D-02
5		-1.302001223926D-01	3.635340242639D+01	.50	-3.968275512335D-02
6		1.829581534452D-02	-9.129500691253D-01	.75	-1.794803703372D-02
7		-1.057682322243D-03	9.464297425442D-03	1.25	8.588655492236D-03
				1.50	8.966848936895D-03
0	-8.0	3.012550700456D-11	1.506275350228D-11	1.75	4.857577908027D-03
1	-7.0	-3.239204349802D-10	-1.619602174901D-10	2.25	-3.142998322680D-03
2	-6.0	1.658586040033D-09	8.292930200163D-10	2.50	-3.733058759269D-03
3	-5.0	-5.457543621311D-09	-2.728771810655D-09	2.75	-2.274464356177D-03
4	-4.0	1.321153748009D-08	6.605768740044D-09	3.25	1.808672176988D-03
5	-3.0	-2.564904057810D-08	-1.282452028905D-08	3.50	2.351866307284D-03
6	-2.0	4.250773987096D-08	2.125386993548D-08	3.75	1.557070544905D-03
7	-1.0	-6.273019452191D-08	-3.136509726095D-08	4.25	-1.432275324271D-03
8	0.0	8.402269261415D-08	4.201134630708D-08	4.50	-1.984212824414D-03
9	1.0	-1.015078523520D-07	-5.075392617601D-08	4.75	-1.391464410219D-03
10	2.0	1.074143486241D-07	5.370717431203D-08	5.25	1.413383600041D-03
11	3.0	-9.520461348063D-08	-4.760230674032D-08	5.50	2.042868788742D-03
12	4.0	6.706769652567D-08	3.353384826284D-08	5.75	1.488322476189D-03
13	5.0	-3.540090164037D-08	-1.770045082019D-08	6.25	-1.613620994287D-03
14	6.0	1.298506176559D-08	6.492530882797D-09	6.50	-2.398173566957D-03
15	7.0	-2.927793988185D-09	-1.463896994093D-09	6.75	-1.791828534771D-03
16	8.0	3.040721899480D-10	1.520360949740D-10	7.25	2.030408499766D-03
				7.50	3.077637413726D-03
	0	-1.492252337187D+02	-1.021807186069D+03	7.75	2.342679762232D-03

CARDINAL SPLINES N=8

	1	-1,933768070743D+02	1.183327221901D+03	8.25	-2,750216974988D-03
	2	-1.074871757777D+02	-8,005184737723D+02	8.50	-4,242333323476D-03
	3	-3,315282710663D+01	2,266645781415D+02	8.75	-3,287495578459D-03
	4	-6,115813705015D+00	-5,068970309545D+01	9,25	4,010568937213D-03
	5	-6,735260269505D-01	5,251814848175D+00	9.50	6,320107743485D-03
	6	-4,093439470110D-02	-3,829494133006D-01	9.75	5,013609307089D-03
	7	-1,057682322243D-03	9,464297425442D-03	10.25	-6,461976656667D-03

I=15 H= 7.0

0		0.	1,686524734837D+05	-1.00	-5,745947150875D-01
1		7,359454313539D-02	-9,047792878809D+04	-.75	-2,777113531308D-01
2		-1,897535905522D-01	2,080615517089D+04	-.50	-1,144729569725D-01
3		1,881034419411D-01	-2,658014129728D+03	-.25	-3,359012412190D-02
4		-9,397527815242D-02	2.036678292725D+02	.25	9.135001016112D-03
5		2.539306257503D-02	-9,351277443234D+00	.50	7,737736651209D-03
6		-3,568488836122D-03	2,372032680629D-01	.75	3,499604361452D-03
7		2,063098951177D-04	-2,508850636822D-03	1.25	-1,674568139253D-03
				1.50	-1,748246955846D-03
0	-8,0	-5,880114495636D-12	-2,940057247818D-12	1.75	-9,470346657767D-04
1	-7.0	6,323906200170D-11	3.161953100085D-11	2.25	6,127010177104D-04
2	-6.0	-3,239204349802D-10	-1,619602174901D-10	2,50	7,276849775938D-04
3	-5,0	1.066486148758D-09	5,332430743788D-10	2.75	4,433305061233D-04
4	-4,0	-2,584422086077D-09	-1,292211043039D-09	3.25	-3,524780274904D-04
5	-3,0	5,026887518965D-09	2.513443759483D-09	3.50	-4,582859512947D-04
6	-2.0	-8,359753684907D-09	-4.179876842454D-09	3.75	-3,033719766285D-04
7	-1.0	1.241368445359D-08	6.206842226796D-09	4.25	2,789640826354D-04
8	0,0	-1,680357929353D-08	-8,401789646765D-09	4.50	3,863824692679D-04
9	1,0	2.062569455558D-08	1.031284727779D-08	4.75	2,708894860745D-04
10	2.0	-2,227760162789D-08	-1,113880081395D-08	5,25	-2,749794278874D-04
11	3,0	2.019888469840D-08	1.009944234920D-08	5.50	-3,972852655433D-04
12	4.0	-1.455182614082D-08	-7.275913070412D-09	5.75	-2,893006345006D-04
13	5,0	7.840078219381D-09	3.920039109690D-09	6.25	3,132690642020D-04
14	6,0	-2,927793988185D-09	-1.463896994093D-09	6.50	4,652175045494D-04
15	7.0	6,703577649030D-10	3,351788824515D-10	6.75	3,472737203187D-04
16	8,0	-7,053505068658D-11	-3,526752534329D-11	7,25	-3,925890083039D-04
				7,50	-5,941900863198D-04
	0	2.911589567945D+01	2.404815079016D+02	7.75	-4,515044679650D-04
	1	3.772880475538D+01	-2.883075109713D+02	8,25	5,276930340216D-04
	2	2,097036400590D+01	1,910804074872D+02	8,50	8,116961969056D-04
	3	6,467715471271D+00	-5,616851900703D+01	8.75	6,269242538952D-04
	4	1.193071262681D+00	1,237326547169D+01	9.25	-7,583667514055D-04
	5	1.313860974794D-01	-1,337415832105D+00	9.50	-1,188666237891D-03
	6	7,984865290471D-03	9,670763240084D-02	9.75	-9,369807995754D-04
	7	2,063098951177D-04	-2.508850636822D-03	10.25	1,188067951873D-03

I=16 H= 8.0

0		0.	-1.910158134021D+04	-1.00	5,338939325081D-02
1		-6,837907357190D-03	1.030275125928D+04	-.75	2,580379611550D-02
2		1,763092180147D-02	-2,383148428290D+03	-.50	1,063626490334D-02
3		-1,747801295481D-02	3,064095607302D+02	-,25	3,120999856912D-03
4		8,732165814897D-03	-2,364421048202D+01	.25	-8,487526678389D-04
5		-2,359603664483D-03	1.094199923559D+00	.50	-7,189204942239D-04
6		3,316090087669D-04	-2,802520796211D-02	.75	-3,251468508108D-04
7		-1,917264919064D-05	3,013555854152D-04	1,25	1,555780239540D-04
				1.50	1.624198248550D-04

CARDINAL SPLINES N=8

0	-8.0	5.466715652158D-13	2.733357826079D-13	1.75	8.798155783516D-05
1	-7.0	-5.880114495636D-12	-2.940057247818D-12	2.25	-5.691784072795D-05
2	-6.0	3.012550700456D-11	1.506275350228D-11	2.50	-6.759695013705D-05
3	-5.0	-9.922297231679D-11	-4.961148615839D-11	2.75	-4.118058852863D-05
4	-4.0	2.406044076594D-10	1.203022038297D-10	3.25	3.273779601934D-05
5	-3.0	-4.685478629127D-10	-2.342739314564D-10	3.50	4.256217842831D-05
6	-2.0	7.809077896158D-10	3.904538948079D-10	3.75	2.817263102045D-05
7	-1.0	-1.164281989713D-09	-5.821409948567D-10	4.25	-2.590058947282D-05
8	0.0	1.587315276976D-09	7.936576384882D-10	4.50	-3.586914900925D-05
9	1.0	-1.970706915728D-09	-9.853534578641D-10	4.75	-2.514362231574D-05
10	2.0	2.161762077123D-09	1.080881038561D-09	5.25	2.551300542141D-05
11	3.0	-1.995370440843D-09	-9.976852204213D-10	5.50	3.685131431570D-05
12	4.0	1.463808296681D-09	7.319041483407D-10	5.75	2.682683766160D-05
13	5.0	-8.021021514457D-10	-4.010510757229D-10	6.25	-2.902713487314D-05
14	6.0	3.040721899480D-10	1.520360949740D-10	6.50	-4.308550944163D-05
15	7.0	-7.053505068658D-11	-3.526752534329D-11	6.75	-3.214392865948D-05
16	8.0	7.505281568124D-12	3.752640784062D-12	7.25	3.628549607664D-05
				7.50	5.486820586664D-05
	0	-2.706256197665D+00	-2.597026914522D+01	7.75	4.164756179391D-05
	1	-3.506709755948D+00	3.189325722995D+01	8.25	-4.854237670069D-05
	2	-1.949039721969D+00	-2.088974814864D+01	8.50	-7.453915324477D-05
	3	-6.011097903450D-01	6.306983576594D+00	8.75	-5.745541008539D-05
	4	-1.108812058445D-01	-1.380121092637D+00	9.25	6.914778937159D-05
	5	-1.221041175590D-02	1.540118481759D-01	9.50	1.080348449359D-04
	6	-7.420593459091D-04	-1.114929517886D-02	9.75	8.484085103852D-05
	7	-1.917264919064D-05	3.013555854152D-04	10.25	-1.065579880539D-04

N= 8 M=17 P= 8.5 Q=15

I= 0 H= -8.5

0		1.000000000000D+00	9.418396493813D+02	-1.00	9.404034375861D+00
1		-2.779543668518D+00	-4.574810676140D+02	-.75	5.800213080794D+00
2		3.078501667435D+00	9.617766754861D+01	-.50	3.426597021429D+00
3		-1.800425998144D+00	-1.180097890959D+01	-.25	1.917929512098D+00
4		6.102335680592D-01	1.057786954208D+00	.25	4.716571314103D-01
5		-1.214248862928D-01	-9.186347267620D-02	.50	1.893480693096D-01
6		1.328195243282D-02	7.356459850302D-03	.75	5.399252921539D-02
7		-6.226349796650D-04	-3.183522171277D-04	1.25	-1.376138336303D-02
				1.50	-1.123584639179D-02
0	-8.5	7.526735981935D-12	3.763367990968D-12	1.75	-4.882379809967D-03
1	-7.5	-7.076595034737D-11	-3.538297517368D-11	2.25	2.132907698182D-03
2	-6.5	3.052562364065D-10	1.526281182032D-10	2.50	2.114643752217D-03
3	-5.5	-8.060080012062D-10	-4.030040006031D-10	2.75	1.082964547690D-03
4	-4.5	1.473305300637D-09	7.366526503186D-10	3.25	-6.171620759799D-04
5	-3.5	-2.013956801653D-09	-1.006978400826D-09	3.50	-6.826329211835D-04
6	-2.5	2.193028631940D-09	1.096514315970D-09	3.75	-3.852408777569D-04
7	-1.5	-2.018151703535D-09	-1.009075851768D-09	4.25	2.585481210683D-04
8	-.5	1.654190736434D-09	8.270953682170D-10	4.50	3.063388521693D-04
9	.5	-1.252598229453D-09	-6.262991147264D-10	4.75	1.838185192847D-04

CARDINAL SPLINES N=8

10	1.5	8.886556845557D-10	4.443278422778D-10	5.25	-1.367905680193D-04
11	2.5	-5.856323309537D-10	-2.928161654768D-10	5.50	-1.692319646412D-04
12	3.5	3.481283393329D-10	1.740641696665D-10	5.75	-1.055163520060D-04
13	4.5	-1.778818483910D-10	-8.894092419551D-11	6.25	8.371735740106D-05
14	5.5	7.315778514425D-11	3.657889257213D-11	6.50	1.063705772225D-04
15	6.5	-2.217755974340D-11	-1.108877987170D-11	6.75	6.790909071288D-05
16	7.5	4.324822717079D-12	2.162411358539D-12	7.25	-5.606547954083D-05
17	8.5	-4.018478672345D-13	-2.009239336173D-13	7.50	-7.244534422557D-05
				7.75	-4.696008817074D-05
	0	-9.486350057581D+01	-4.558518915485D+01	8.25	3.982381606754D-05
	1	-1.186977220568D+02	-6.153909046464D+01	8.50	5.208729669976D-05
	2	-6.416690643128D+01	-3.097009557962D+01	8.75	3.415932669986D-05
	3	-1.940315707342D+01	-1.001550982844D+01	9.25	-2.963675804504D-05
	4	-3.539157891356D+00	-1.716629208790D+00	9.50	-3.921769646958D-05
	5	-3.887382351158D-01	-1.997039217478D-01	9.75	-2.603383293813D-05
	6	-2.376482885725D-02	-1.158549706880D-02	10.25	2.320121251223D-05
	7	-6.226349796650D-04	-3.183522171277D-04	10.50	3.117225373805D-05

I= 1 H= -7.5

0		0.	-1.013288020704D+04	-1.00	-4.057875380342D+01
1		8.668794299236D+00	4.913870404894D+03	-.75	-2.147431888200D+01
2		-1.538209540664D+01	-1.026418935363D+03	-.50	-9.877843485362D+00
3		1.118772633501D+01	1.229348556632D+02	-.25	-3.321117131406D+00
4		-4.300027471447D+00	-1.025214753257D+01	.25	1.364708870826D+00
5		9.276241721032D-01	7.991394681765D-01	.50	1.645940659872D+00
6		-1.072540235858D-01	-6.132891698770D-02	.75	1.410172591427D+00
7		5.232095403013D-03	2.691774138320D-03	1.25	6.010741120616D-01
				1.50	2.950150089119D-01
0	-8.5	-7.076595034737D-11	-3.538297517368D-11	1.75	9.990911737661D-02
1	-7.5	6.728427851800D-10	3.364213925900D-10	2.25	-3.382095070585D-02
2	-6.5	-2.940537096807D-09	-1.470268548404D-09	2.50	-3.112396433565D-02
3	-5.5	7.882114296092D-09	3.941057148046D-09	2.75	-1.506779438615D-02
4	-4.5	-1.465403611082D-08	-7.327018055410D-09	3.25	7.937657505093D-03
5	-3.5	2.039891740235D-08	1.019945870118D-08	3.50	8.535364120252D-03
6	-2.5	-2.261410287024D-08	-1.130705143512D-08	3.75	4.706588084427D-03
7	-1.5	2.113631141546D-08	1.056815570773D-08	4.25	-3.049296669807D-03
8	-.5	-1.752331561630D-08	-8.761657808151D-09	4.50	-3.564476824732D-03
9	.5	1.336417370849D-08	6.682086854246D-09	4.75	-2.114592811229D-03
10	1.5	-9.519372988004D-09	-4.759686494002D-09	5.25	1.545656262343D-03
11	2.5	6.286990180914D-09	3.143495090457D-09	5.50	1.898847482189D-03
12	3.5	-3.741631020722D-09	-1.870815510361D-09	5.75	1.176857512275D-03
13	4.5	1.913039588214D-09	9.565197941071D-10	6.25	-9.248893809020D-04
14	5.5	-7.870483727915D-10	-3.935241863957D-10	6.50	-1.170754825636D-03
15	6.5	2.386376821884D-10	1.193188410942D-10	6.75	-7.450354573124D-04
16	7.5	-4.654185557337D-11	-2.327092778669D-11	7.25	6.119653510755D-04
17	8.5	4.324822717079D-12	2.162411358539D-12	7.50	7.891619157763D-04
				7.75	5.106639094766D-04
	0	8.678520126344D+02	4.140500753542D+02	8.25	-4.318766993723D-04
	1	1.079843843765D+03	5.634973261360D+02	8.50	-5.642610009130D-04
	2	5.796922710038D+02	2.778619105882D+02	8.75	-3.697064540569D-04
	3	1.737629872964D+02	9.026070573877D+01	9.25	3.202907018357D-04
	4	3.134807240992D+01	1.510406882386D+01	9.50	4.235907926247D-04
	5	3.396065719446D+00	1.755459013170D+00	9.75	2.810541359590D-04
	6	2.040556528934D-01	9.883164424232D-02	10.25	-2.502804298706D-04
	7	5.232095403013D-03	2.691774138320D-03	10.50	-3.361652940304D-04

CARDINAL SPLINES N=8

I= 2 H= -6.5

0		0.	5.197644046523D+04
1		-1.727376460533D+01	-2.519221632517D+04
2		3.917595507220D+01	5.245216265374D+03
3		-3.306136709073D+01	-6.184431080570D+02
4		1.402974005760D+01	4.857588161755D+01
5		-3.245519669049D+00	-3.342879961197D+00
6		3.949839678866D-01	2.370005161875D-01
7		-2.002773288122D-02	-1.040232359399D-02
0	-8.5	3.052562364065D-10	1.526281182032D-10
1	-7.5	-2.940537096807D-09	-1.470268548404D-09
2	-6.5	1.305040804017D-08	6.525204020087D-09
3	-5.5	-3.561646136701D-08	-1.780823068350D-08
4	-4.5	6.759182615230D-08	3.379591307615D-08
5	-3.5	-9.623037512053D-08	-4.811518756027D-08
6	-2.5	1.091399167905D-07	5.456995839523D-08
7	-1.5	-1.041262801251D-07	-5.206314006254D-08
8	-.5	8.771347865140D-08	4.385673932570D-08
9	.5	-6.760427459756D-08	-3.380213729878D-08
10	1.5	4.845423249191D-08	2.422711624596D-08
11	2.5	-3.211080881610D-08	-1.605540440805D-08
12	3.5	1.914566980471D-08	9.572834902354D-09
13	4.5	-9.798643987203D-09	-4.899321993602D-09
14	5.5	4.033492060825D-09	2.016746030412D-09
15	6.5	-1.223359240353D-09	-6.116796201767D-10
16	7.5	2.386376821884D-10	1.193188410942D-10
17	8.5	-2.217755974340D-11	-1.108877987170D-11
	0	-3.625798999625D+03	-1.710954816216D+03
	1	-4.482612325213D+03	-2.362228204239D+03
	2	-2.387393376451D+03	-1.132226371203D+03
	3	-7.086581885165D+02	-3.716623406569D+02
	4	-1.263245716285D+02	-6.023865151567D+01
	5	-1.348841502086D+01	-7.038779108614D+00
	6	-7.966661385457D-01	-3.819377376548D-01
	7	-2.002773288122D-02	-1.040232359399D-02

-1.00	1.072013581957D+02
-.75	5.422180893438D+01
-.50	2.354815129148D+01
-.25	7.341592983068D+00
.25	-2.334798391578D+00
.50	-2.194112999180D+00
.75	-1.130067277350D+00
1.25	8.062411000399D-01
1.50	1.189953928606D+00
1.75	1.212024575677D+00
2.25	6.876423913505D-01
2.50	3.808454466019D-01
2.75	1.438671551540D-01
3.25	-5.887469680871D-02
3.50	-5.883745439183D-02
3.75	-3.070905365576D-02
4.25	1.843717331295D-02
4.50	2.097722077879D-02
4.75	1.217342314684D-02
5.25	-8.607889472899D-03
5.50	-1.044295018548D-02
5.75	-6.404489862364D-03
6.25	4.951604878802D-03
6.50	6.228258564601D-03
6.75	3.942206465373D-03
7.25	-3.210786778427D-03
7.50	-4.126778213194D-03
7.75	-2.662892140727D-03
8.25	2.242029160099D-03
8.50	2.924161971355D-03
8.75	1.913067455498D-03
9.25	-1.653460434247D-03
9.50	-2.184713050533D-03
9.75	-1.448423819171D-03
10.25	1.288228050130D-03
10.50	1.729447428280D-03

I= 3 H= -5.5

0		0.	-1.715415207678D+05
1		2.829360422614D+01	8.313038524859D+04
2		-6.865262412563D+01	-1.728128465473D+04
3		6.274427559933D+01	2.017982764915D+03
4		-2.852097558858D+01	-1.514341902883D+02
5		6.977170724505D+00	9.165974796649D+00
6		-8.881400432374D-01	-5.744724879613D-01
7		4.668920827573D-02	2.462676606111D-02
0	-8.5	-8.060080012062D-10	-4.030040006031D-10
1	-7.5	7.882114296092D-09	3.941057148046D-09
2	-6.5	-3.561646136701D-08	-1.780823068350D-08
3	-5.5	9.929722942599D-08	4.964861471300D-08
4	-4.5	-1.931712636383D-07	-9.658563181917D-08
5	-3.5	2.827342617024D-07	1.413671308512D-07
6	-2.5	-3.300472635298D-07	-1.650236317649D-07

-1.00	-1.961234795157D+02
-.75	-9.715177786303D+01
-.50	-4.116783210144D+01
-.25	-1.246301275555D+01
.25	3.658180955934D+00
.50	3.248643713698D+00
.75	1.553006163911D+00
1.25	-8.593133474870D-01
1.50	-9.893581621127D-01
1.75	-6.065304216659D-01
2.25	5.775005383323D-01
2.50	9.631544332914D-01
2.75	1.095864251767D+00
3.25	7.538546846462D-01
3.50	4.540917829431D-01
3.75	1.852099319604D-01

CARDINAL SPLINES N=8

7	-1.5	3.235301290369D-07	1.617650645185D-07	4.25	-8.663220086122D-02
8	-.5	-2.786485109704D-07	-1.393242554852D-07	4.50	-9.173384305624D-02
9	.5	2.181584185993D-07	1.090792092997D-07	4.75	-5.045417817394D-02
10	1.5	-1.578927714727D-07	-7.894638573635D-08	5.25	3.314131651908D-02
11	2.5	1.052207599167D-07	5.261037995833D-08	5.50	3.917681270168D-02
12	3.5	-6.292870969662D-08	-3.146435484831D-08	5.75	2.352684044762D-02
13	4.5	3.226019195882D-08	1.613009597941D-08	6.25	-1.762764326137D-02
14	5.5	-1.329171773321D-08	-6.645858866606D-09	6.50	-2.191281550668D-02
15	6.5	4.033492060825D-09	2.016746030412D-09	6.75	-1.373408911027D-02
16	7.5	-7.870483727915D-10	-3.935241863957D-10	7.25	1.101743689050D-02
17	8.5	7.315778514425D-11	3.657889257213D-11	7.50	1.407798941173D-02
				7.75	9.039336638619D-03
	0	9.225985647894D+03	4.276585256018D+03	8.25	-7.552068491079D-03
	1	1.132379321802D+04	6.060941893305D+03	8.50	-9.820153244617D-03
	2	5.977959652649D+03	2.786119260323D+03	8.75	-6.408202415421D-03
	3	1.755677357597D+03	9.350427827213D+02	9.25	5.516361444630D-03
	4	3.090423594759D+02	1.448721244960D+02	9.50	7.277303406956D-03
	5	3.252122977574D+01	1.723283871684D+01	9.75	4.818267367217D-03
	6	1.889867849168D+00	8.908200926748D-01	10.25	-4.276353429416D-03
	7	4.668920827573D-02	2.462676606111D-02	10.50	-5.736285331313D-03

I= 4 H= -4.5

0		0.	4.174657675775D+05	-1.00	2.684990441087D+02
1		-3.664098460038D+01	-2.022958052177D+05	-.75	1.315214371947D+02
2		9.158883663122D+01	4.202014781158D+04	-.50	5.502533771791D+01
3		-8.701775119016D+01	-4.878479694873D+03	-.25	1.641603394668D+01
4		4.127247968744D+01	3.544938789836D+02	.25	-4.644310285386D+00
5		-1.051546233083D+01	-1.907582808310D+01	.50	-4.028478012153D+00
6		1.388205734952D+00	1.011493550699D+00	.75	-1.872113340132D+00
7		-7.532393370811D-02	-4.078272262646D-02	1.25	9.584388924372D-01
				1.50	1.044183167004D+00
0	-8.5	1.473305300637D-09	7.366526503186D-10	1.75	5.949979647763D-01
1	-7.5	-1.465403611082D-08	-7.327018055410D-09	2.25	-4.407033159319D-01
2	-6.5	6.759182615230D-08	3.379591307615D-08	2.50	-5.742745243030D-01
3	-5.5	-1.931712636383D-07	-9.658563181917D-08	2.75	-3.939210523728D-01
4	-4.5	3.869752418547D-07	1.934876209273D-07	3.25	4.562900844635D-01
5	-3.5	-5.856600922852D-07	-2.928300461426D-07	3.50	8.290525467036D-01
6	-2.5	7.086203851600D-07	3.543101925800D-07	3.75	1.020162902912D+00
7	-1.5	-7.193939929888D-07	-3.596969964944D-07	4.25	8.046966112097D-01
8	-.5	6.386887720807D-07	3.193443860403D-07	4.50	5.143582300862D-01
9	.5	-5.116608423566D-07	-2.558304211783D-07	4.75	2.213914570104D-01
10	1.5	3.760282047677D-07	1.880141023838D-07	5.25	-1.135920203577D-01
11	2.5	-2.529102811103D-07	-1.264551405551D-07	5.50	-1.251145265714D-01
12	3.5	1.520487888281D-07	7.602439441403D-08	5.75	-7.128669851984D-02
13	4.5	-7.817272136541D-08	-3.908636068270D-08	6.25	4.970915855128D-02
14	5.5	3.226019195882D-08	1.613009597941D-08	6.50	6.025891900946D-02
15	6.5	-9.798643987203D-09	-4.899321993602D-09	6.75	3.700919777399D-02
16	7.5	1.913039588214D-09	9.565197941071D-10	7.25	-2.880659997526D-02
17	8.5	-1.778818483910D-10	-8.894092419551D-11	7.50	-3.639646458871D-02
				7.75	-2.315170337368D-02
	0	-1.617581076779D+04	-7.269423932789D+03	8.25	1.906570867868D-02
	1	-1.969787502322D+04	-1.081987811466D+04	8.50	2.465489423832D-02
	2	-1.030143846771D+04	-4.657096232559D+03	8.75	1.601384415377D-02
	3	-2.992385495880D+03	-1.635396699066D+03	9.25	-1.368527331129D-02
	4	-5.202075817166D+02	-2.366218126320D+02	9.50	-1.800301384926D-02
	5	-5.400220826694D+01	-2.936724290244D+01	9.75	-1.189121740920D-02

```
        6  -3.093568320681D+00  -1.415078445575D+00   10.25  1.051427362150D-02
        7  -7.532393370811D-02  -4.078272262646D-02   10.50  1.408320010847D-02


I= 5     H= -3.5
0           0.                  -8.183209838928D+05   -1.00 -2.931760751682D+02
1           3.882239184054D+01   3.965427417356D+05    -.75 -1.427094648671D+02
2          -9.851448996407D+01  -8.233473655706D+04    -.50 -5.928851795558D+01
3           9.552803668411D+01   9.526503943045D+03    -.25 -1.754900368659D+01
4          -4.642704187576D+01  -6.777519874608D+02     .25  4.871177218546D+00
5           1.214599972378D+01   3.316485338723D+01     .50  4.176439434602D+00
6          -1.646505743264D+00  -1.445554285567D+00     .75  1.914965924185D+00
7           9.160933668260D-02   5.192369028645D-02    1.25 -9.474184325823D-01
                                                       1.50 -1.009603905858D+00
0   -8.5   -2.013956801653D-09  -1.006978400826D-09    1.75 -5.600832636066D-01
1   -7.5    2.039891740235D-08   1.019945870118D-08    2.25  3.850336663797D-01
2   -6.5   -9.623037512053D-08  -4.811518756027D-08    2.50  4.755466150046D-01
3   -5.5    2.827342617024D-07   1.413671308512D-07    2.75  3.037012374189D-01
4   -4.5   -5.856600922852D-07  -2.928300461426D-07    3.25 -2.745813691457D-01
5   -3.5    9.216532293112D-07   4.608266146556D-07    3.50 -3.904495599957D-01
6   -2.5   -1.164229505041D-06  -5.821147525207D-07    3.75 -2.901278687874D-01
7   -1.5    1.234784315727D-06   6.173921578635D-07    4.25  3.865429579164D-01
8    -.5   -1.141010079528D-06  -5.705050397642D-07    4.50  7.463481022652D-01
9     .5    9.442480797431D-07   4.721240398716D-07    4.75  9.704907471269D-01
10   1.5   -7.103477177884D-07  -3.551738588942D-07    5.25  8.417456484914D-01
11   2.5    4.850323817181D-07   2.425161908590D-07    5.50  5.602596147973D-01
12   3.5   -2.942317052934D-07  -1.471158526467D-07    5.75  2.500587371968D-01
13   4.5    1.520487888281D-07   7.602439441403D-08    6.25 -1.364321048154D-01
14   5.5   -6.292870969662D-08  -3.146435484831D-08    6.50 -1.542106742119D-01
15   6.5    1.914566980471D-08   9.572834902354D-09    6.75 -8.991847732439D-02
16   7.5   -3.741631020722D-09  -1.870815510361D-09    7.25  6.520813322332D-02
17   8.5    3.481283393329D-10   1.740641696665D-10    7.50  8.038009400857D-02
                                                       7.75  5.012280555579D-02
        0   2.115057456940D+04   8.971624046917D+03    8.25 -4.007767308907D-02
        1   2.555104954876D+04   1.467812302546D+04    8.50 -5.126050154634D-02
        2   1.323985895673D+04   5.652543695491D+03    8.75 -3.299225707800D-02
        3   3.806512829038D+03   2.176094227584D+03    9.25  2.780302519606D-02
        4   6.544627832799D+02   2.812020949143D+02    9.50  3.637951582416D-02
        5   6.716847289901D+01   3.822280391044D+01    9.75  2.392098332721D-02
        6   3.804249789351D+00   1.643905286477D+00   10.25 -2.100313156515D-02
        7   9.160933668260D-02   5.192369028645D-02   10.50 -2.805601482583D-02


I= 6     H= -2.5
0           0.                   1.380730629858D+06   -1.00  2.711774391044D+02
1          -3.533191589939D+01  -6.691137288318D+05    -.75  1.315469124774D+02
2           9.036284168749D+01   1.389050187229D+05    -.50  5.444318953041D+01
3          -8.857495413716D+01  -1.604122972117D+04    -.25  1.604677725167D+01
4           4.362794249759D+01   1.126537521558D+03     .25 -4.409798498095D+00
5          -1.159139396940D+01  -5.152434138408D+01     .50 -3.758340394284D+00
6           1.597935366010D+00   1.849430372666D+00     .75 -1.711611888164D+00
7          -9.045554732608D-02  -5.555755494602D-02    1.25  8.328064200778D-01
                                                       1.50  8.783263343912D-01
0   -8.5    2.193028631940D-09   1.096514315970D-09    1.75  4.813853376808D-01
1   -7.5   -2.261410287024D-08  -1.130705143512D-08    2.25 -3.206888006349D-01
2   -6.5    1.091399167905D-07   5.456995839523D-08    2.50 -3.879642723019D-01
3   -5.5   -3.300472635298D-07  -1.650236317649D-07    2.75 -2.415628017935D-01
```

4	-4.5	7.086203851600D-07	3.543101925800D-07	3.25	2,032852652777D-01
5	-3.5	-1.164229505041D-06	-5.821147525207D-07	3.50	2,743607414690D-01
6	-2.5	1.544523654302D-06	7.722618271511D-07	3.75	1,900603735756D-01
7	-1.5	-1.725161796110D-06	-8.625808980551D-07	4.25	-1,981408841910D-01
8	-.5	1.675655554993D-06	8.378277774967D-07	4.50	-2,997522606959D-01
9	.5	-1.448038098790D-06	-7.240190493949D-07	4.75	-2,356161679916D-01
10	1.5	1.126661371434D-06	5.633306857170D-07	5.25	3,458112092702D-01
11	2,5	-7,876365545290D-07	-3,938182772645D-07	5.50	6,958276747030D-01
12	3.5	4.850323817181D-07	2.425161908590D-07	5.75	9,388978288172D-01
13	4,5	-2,529102811103D-07	-1,264551405551D-07	6.25	8,669981420789D-01
14	5.5	1.052207599167D-07	5.261037995833D-08	6.50	5,924934254357D-01
15	6,5	-3,211080881610D-08	-1.605540440805D-08	6,75	2,707474690944D-01
16	7.5	6,286990180914D-09	3.143495090457D-09	7.25	-1,537409358156D-01
17	8,5	-5,856323309537D-10	-2,928161654768D-10	7.50	-1,767605098209D-01
				7.75	-1,046741187825D-01
	0	-2,206263196531D+04	-8,327816121622D+03	8.25	7,801552863719D-02
	1	-2,646416354231D+04	-1,644022807062D+04	8.50	9,737042353179D-02
	2	-1.360617088443D+04	-5,171535861893D+03	8.75	6,144640944818D-02
	3	-3,879792287080D+03	-2,400148906717D+03	9,25	-5,029273693773D-02
	4	-6,615293033458D+02	-2,531017409814D+02	9.50	-6,509632567873D-02
	5	-6,734036948340D+01	-4,149809261996D+01	9.75	-4,241954935400D+02
	6	-3,784169699892D+00	-1.456244146622D+00	10.25	3,673505908249D-02
	7	-9,045554732608D-02	-5,555755494602D-02	10.50	4,881278338820D-02

I= 7 H= -1.5

0		0.	-2,107049187206D+06	-1,00	-2,243979741288D+02
1		2,899182149688D+01	1.021235430254D+06	-.75	-1,086580979182D+02
2		-7,444554628091D+01	-2,120060547350D+05	-,50	-4,488098661651D+01
3		7,337912001615D+01	2,445946038745D+04	-,25	-1,319938316186D+01
4		-3,639666781421D+01	-1,705033139094D+03	.25	3,608678422644D+00
5		9.750458620789D+00	7.470916943062D+01	.50	3,066230844019D+00
6		-1,356772968252D+00	-2,282800146943D+00	.75	1,391639123090D+00
7		7,758693156709D-02	5.457257613482D-02	1,25	-6,715242904128D-01
				1,50	-7,046500016992D-01
0	-8,5	-2,018151703535D-09	-1.009075851768D-09	1,75	-3,839472535295D-01
1	-7,5	2.113631141546D-08	1.056815570773D-08	2,25	2.520344531794D-01
2	-6,5	-1,041262801251D-07	-5,206314006254D-08	2,50	3,020620972787D-01
3	-5,5	3,235301290369D-07	1.617650645185D-07	2.75	1,859919474650D-01
4	-4,5	-7,193939929888D-07	-3,596969964944D-07	3,25	-1,519636662086D-01
5	-3,5	1.234784315727D-06	6,173921578635D-07	3,50	-2,010761614426D-01
6	-2,5	-1,725161796110D-06	-8,625808980551D-07	3,75	-1,359208860912D-01
7	-1,5	2,040086549079D-06	1,020043274539D-06	4,25	1,320956117381D-01
8	-,5	-2,099558129329D-06	-1,049779064665D-06	4,50	1,897979698086D-01
9	,5	1,914055205793D-06	9,570276028964D-07	4,75	1,391690990865D-01
10	1,5	-1,557606762923D-06	-7,788033814614D-07	5,25	-1,599898820358D-01
11	2,5	1,126661371434D-06	5,633306857170D-07	5,50	-2,523368331003D-01
12	3,5	-7,103477177884D-07	-3,551738588942D-07	5,75	-2,058955244281D-01
13	4,5	3,760282047677D-07	1,880141023838D-07	6,25	3,219169766734D-01
14	5,5	-1,578927714727D-07	-7,894638573635D-08	6,50	6,652185507713D-01
15	6,5	4,845423249191D-08	2,422711624596D-08	6,75	9,191712656954D-01
16	7,5	-9,519372988004D-09	-4,759686494002D-09	7,25	8,836683601969D-01
17	8,5	8,886556845557D-10	4,443278422778D-10	7,50	6,143436902616D-01
				7,75	2,851441428138D-01
	0	1,959149831261D+04	5,676817911176D+03	8,25	-1,664445508667D-01
	1	2,337863562709D+04	1,657911435271D+04	8,50	-1,937781410853D-01
	2	1,195621593826D+04	3,490312573142D+03	8,75	-1,161392408851D-01

CARDINAL SPLINES N=8

```
     3   3.391336904016D+03    2.397744804393D+03      9.25   8.861989685822D-02
     4   5.752777214284D+02    1.691253186159D+02      9.50   1.119386595084D-01
     5   5.827380916010D+01    4.108660307708D+01      9.75   7.152545814961D-02
     6   3.259649459990D+00    9.642681330786D-01     10.25  -6.016752364889D-02
     7   7.758693156709D-02    5.457257613482D-02     10.50  -7.909132604049D-02


I= 8      H=  -.5
0            0.                3.001957399810D+06     -1.00   1.729356866711D+02
1          -2.224743889349D+01  -1.455387005676D+06     -.75   8.366191686826D+01
2           5.724314221812D+01   3.021979771859D+05     -.50   3.452141778644D+01
3          -5.658212114130D+01  -3.485347332837D+04     -.25   1.014133416910D+01
4           2.816518413168D+01   2.420040226752D+03      .25  -2.765383548527D+00
5          -7.577369305390D+00  -1.034199543195D+02      .50  -2.346088755146D+00
6           1.059516984898D+00   2.819338189780D+00      .75  -1.062958769025D+00
7          -6.091399617157D-02  -5.298012637531D-02     1.25   5.107923643048D-01
                                                        1.50   5.346379514678D-01
0   -8.5   1.654190736434D-09    8.270953682170D-10     1.75   2.904688886314D-01
1   -7.5  -1.752331561630D-08   -8.761657808151D-09     2.25  -1.892994409992D-01
2   -6.5   8.771347865140D-08    4.385673932570D-08     2.50  -2.258482275350D-01
3   -5.5  -2.786485109704D-07   -1.393242554852D-07     2.75  -1.383255058925D-01
4   -4.5   6.386887720807D-07    3.193443860403D-07     3.25   1.114736930133D-01
5   -3.5  -1.141010079528D-06   -5.705050397642D-07     3.50   1.461895472182D-01
6   -2.5   1.675655554993D-06    8.378277774967D-07     3.75   9.776430387378D-02
7   -1.5  -2.099558129329D-06   -1.049779064665D-06     4.25  -9.231136182998D-02
8    -.5   2.298716328505D-06    1.149358164253D-06     4.50  -1.300724843304D-01
9     .5  -2.226958353393D-06   -1.113479176697D-06     4.75  -9.308843056882D-02
10   1.5   1.914055205793D-06    9.570276028964D-07     5.25   9.979694219221D-02
11   2.5  -1.448038098790D-06   -7.240190493949D-07     5.50   1.495108003217D-01
12   3.5   9.442480797431D-07    4.721240398716D-07     5.75   1.138116651934D-01
13   4.5  -5.116608423566D-07   -2.558304211783D-07     6.25  -1.393951402967D-01
14   5.5   2.181584185993D-07    1.090792092997D-07     6.50  -2.257910139743D-01
15   6.5  -6.760427459756D-08   -3.380213729878D-08     6.75  -1.886672848488D-01
16   7.5   1.336417370849D-08    6.682086854246D-09     7.25   3.071106486452D-01
17   8.5  -1.252598229453D-09   -6.262991147264D-10     7.50   6.456164525770D-01
                                                        7.75   9.061113108281D-01
     0  -1.566965904928D+04   -1.984797229683D+03     8.25   8.954943763443D-01
     1  -1.864502088397D+04   -1.627285037677D+04     8.50   6.304214898390D-01
     2  -9.507853442395D+03   -1.215008878309D+03     8.75   2.961506289583D-01
     3  -2.689218166712D+03   -2.344086281483D+03     9.25  -1.769910521202D-01
     4  -4.549297554245D+02   -5.862477251216D+01     9.50  -2.085469894648D-01
     5  -4.596376376691D+01   -4.001780338370D+01     9.75  -1.265626948228D-01
     6  -2.564865787311D+00   -3.329793295516D-01    10.25   9.926336572762D-02
     7  -6.091399617157D-02   -5.298012637531D-02    10.50   1.273429224732D-01


I= 9      H=   .5
0            0.               -4.044711716015D+06     -1.00  -1.266714123456D+02
1           1.626002368862D+01   1.962014451860D+06     -.75  -6.125151831278D+01
2          -4.188071285183D+01  -4.076008600546D+05     -.50  -2.526111938980D+01
3           4.145662397178D+01   4.701959508963D+04     -.25  -7.416703542004D+00
4          -2.067364818211D+01  -3.259022468579D+03      .25   2.019720042922D+00
5           5.574012256472D+00   1.373838459338D+02      .50   1.712139905550D+00
6          -7.813451382459D-01  -3.485296848883D+00      .75   7.750480684195D-01
7           4.504625657905D-02   5.298012637531D-02     1.25  -3.716518650242D-01
                                                        1.50  -3.885033176378D-01
0   -8.5  -1.252598229453D-09   -6.262991147264D-10     1.75  -2.107641368210D-01
```

CARDINAL SPLINES N=8

1	-7.5	1.336417370849D-08	6.682086854246D-09	2.25	1.368555200014D-01
2	-6.5	-6.760427459756D-08	-3.380213729878D-08	2.50	1.629077386635D-01
3	-5.5	2.181584185993D-07	1.090792092997D-07	2.75	9.951141981011D-02
4	-4.5	-5.116608423566D-07	-2.558304211783D-07	3.25	-7.965115511688D-02
5	-3.5	9.442480797431D-07	4.721240398716D-07	3.50	-1.040034061988D-01
6	-2.5	-1.448038098790D-06	-7.240190493949D-07	3.75	-6.919410092419D-02
7	-1.5	1.914055205793D-06	9.570276028964D-07	4.25	6.445741548521D-02
8	-.5	-2.226958353393D-06	-1.113479176697D-06	4.50	9.002443997012D-02
9	.5	2.298716328505D-06	1.149358164253D-06	4.75	6.374287360331D-02
10	1.5	-2.099558129329D-06	-1.049779064665D-06	5.25	-6.639569204740D-02
11	2.5	1.675655554993D-06	8.378277774967D-07	5.50	-9.754757343720D-02
12	3.5	-1.141010079528D-06	-5.705050397642D-07	5.75	-7.247316840543D-02
13	4.5	6.386887720807D-07	3.193443860403D-07	6.25	8.276973220473D-02
14	5.5	-2.786485109704D-07	-1.393242554852D-07	6.50	1.273429224732D-01
15	6.5	8.771347865140D-08	4.385673932570D-08	6.75	9.926336572762D-02
16	7.5	-1.752331561630D-08	-8.761657808151D-09	7.25	-1.265626948228D-01
17	8.5	1.654190736434D-09	8.270953682170D-10	7.50	-2.085469894648D-01
				7.75	-1.769910521202D-01
	0	1.170006458991D+04	-1.984797229683D+03	8.25	2.961506289583D-01
	1	1.390067986957D+04	1.627285037677D+04	8.50	6.304214898390D-01
	2	7.077835685777D+03	-1.215008878309D+03	8.75	8.954943763443D-01
	3	1.998954396254D+03	2.344086281483D+03	9.25	9.061113108281D-01
	4	3.376802104002D+02	-5.862477251216D+01	9.50	6.456164525770D-01
	5	3.407184300049D+01	4.001780338370D+01	9.75	3.071106486452D-01
	6	1.898907128208D+00	-3.329793295516D-01	10.25	-1.886672848488D-01
	7	4.504625657905D-02	5.298012637531D-02	10.50	-2.257910139743D-01

I=10 H= 1.5

0		0.	5.115788975508D+06	-1.00	8.830509037697D+01
1		-1.132232949111D+01	-2.484239820086D+06	-.75	4.268920079306D+01
2		2.917834990194D+01	5.166245287641D+05	-.50	1.760100238417D+01
3		-2.890432300758D+01	-5.964722874240D+04	-.25	5.166155513725D+00
4		1.442759434814D+01	4.133334954773D+03	.25	-1.405879300149D+00
5		-3.894334469533D+00	-1.730645190046D+02	.50	-1.191296860377D+00
6		5.466009379687D-01	4.211336413100D+00	.75	-5.390286577826D-01
7		-3.155822070254D-02	-5.457257613482D-02	1.25	2.581931617560D-01
				1.50	2.697218131703D-01
0	-8.5	8.886556845557D-10	4.443278422778D-10	1.75	1.462142045714D-01
1	-7.5	-9.519372988004D-09	-4.759686494002D-09	2.25	-9.476348684107D-02
2	-6.5	4.845423249191D-08	2.422711624596D-08	2.50	-1.126713880516D-01
3	-5.5	-1.578927714727D-07	-7.894638573635D-08	2.75	-6.873124409839D-02
4	-4.5	3.760282047677D-07	1.880141023838D-07	3.25	5.482398785510D-02
5	-3.5	-7.103477177884D-07	-3.551738588942D-07	3.50	7.142829622899D-02
6	-2.5	1.126661371434D-06	5.633306857170D-07	3.75	4.739833676744D-02
7	-1.5	-1.557606762923D-06	-7.788033814614D-07	4.25	-4.385771302260D-02
8	-.5	1.914055205793D-06	9.570276028964D-07	4.50	-6.098868651629D-02
9	.5	-2.099558129329D-06	-1.049779064665D-06	4.75	-4.296109701512D-02
10	1.5	2.040086549079D-06	1.020043274539D-06	5.25	4.414578541170D-02
11	2.5	-1.725161796110D-06	-8.625808980551D-07	5.50	6.428417040031D-02
12	3.5	1.234784315727D-06	6.173921578635D-07	5.75	4.725014007859D-02
13	4.5	-7.193939929888D-07	-3.596969964944D-07	6.25	-5.242421857235D-02
14	5.5	3.235301290369D-07	1.617650645185D-07	6.50	-7.909132604049D-02
15	6.5	-1.041262801251D-07	-5.206314006254D-08	6.75	-6.016752364889D-02
16	7.5	2.113631141546D-08	1.056815570773D-08	7.25	7.152545814961D-02
17	8.5	-2.018151703535D-09	-1.009075851768D-09	7.50	1.119386595084D-01
				7.75	8.861989685822D-02

CARDINAL SPLINES N=8

	0	-8,237862490257D+03	5,676817911176D+03	8,25	-1,161392408851D-01
	1	-9,779593078328D+03	-1,657911435271D+04	8,50	-1,937781410853D-01
	2	-4,975590791974D+03	3,490312573142D+03	8,75	-1,664445508667D-01
	3	-1,404152704770D+03	-2,397744804393D+03	9,25	2,851441428138D-01
	4	-2,370270841967D+02	1,691253186159D+02	9.50	6,143436902616D-01
	5	-2,389939699406D+01	-4,108660307708D+01	9.75	8,836683601969D-01
	6	-1,331113193833D+00	9,642681330786D-01	10,25	9,191712656954D-01
	7	-3,155822070254D-02	-5,457257613482D-02	10,50	6,652185507713D-01
I=11	H= 2.5				
0		0,	-5,885530556769D+06	-1,00	-5,766006321442D+01
1		7,388706861643D+00	2,863806718562D+06	-,75	-2,787097499875D+01
2		-1,904646009740D+01	-5,967378636964D+05	-,50	-1,148975810616D+01
3		1,887487986259D+01	6,902478556968D+04	-,25	-3,371895491772D+00
4		-9,426008943491D+00	-4,789150191324D+03	,25	9,172726936294D-01
5		2,545785320702D+00	2,000612443395D+02	,50	7,771031093381D-01
6		-3,575625660274D-01	-4,761918665911D+00	,75	3,515341951283D-01
7		2,065956256597D-02	5,555755494602D-02	1,25	-1,682879878252D-01
				1,50	-1,757417531014D-01
0	-8,5	-5,856323309537D-10	-2,928161654768D-10	1,75	-9,523080314075D-02
1	-7,5	6,286990180914D-09	3,143495090457D-09	2,25	6,166026034984D-02
2	-6,5	-3,211080881610D-08	-1,605540440805D-08	2,50	7,326805461712D-02
3	-5,5	1,052207599167D-07	5,261037995833D-08	2,75	4,466307028410D-02
4	-4,5	-2,529102811103D-07	-1,264551405551D-07	3,25	-3,556202080093D-02
5	-3,5	4,850323817181D-07	2,425161908590D-07	3.50	-4,627988137619D-02
6	-2,5	-7,876365545290D-07	-3,938182772645D-07	3,75	-3,066926153070D-02
7	-1,5	1,126661371434D-06	5,633306857170D-07	4,25	2,828060145394D-02
8	-,5	-1,448038098790D-06	-7,240190493949D-07	4,50	3,924029648544D-02
9	,5	1,675655554993D-06	8,378277774967D-07	4,75	2,756904428953D-02
10	1,5	-1,725161796110D-06	-8,625808980551D-07	5,25	-2,813764492053D-02
11	2,5	1,544523654302D-06	7,722618271511D-07	5,50	-4,079437164578D-02
12	3,5	-1,164229505041D-06	-5,821147525207D-07	5,75	-2,982866189327D-02
13	4,5	7,086203851600D-07	3,543101925800D-07	6,25	3,264544656904D-02
14	5,5	-3,300472635298D-07	-1,650236317649D-07	6.50	4,881278338820D-02
15	6,5	1,091399167905D-07	5,456995839523D-08	6,75	3,673505908249D-02
16	7,5	-2,261410287024D-08	-1,130705143512D-08	7,25	-4,241954935400D-02
17	8,5	2,193028631940D-09	1,096514315970D-09	7,50	-6,509632567873D-02
				7,75	-5,029273693773D-02
	0	5,406999722063D+03	-8,327816121622D+03	8,25	6,144640944818D-02
	1	6,416292598924D+03	1,644022807062D+04	8,50	9,737042353179D-02
	2	3,263099160640D+03	-5,171535861893D+03	8,75	7,801552863719D-02
	3	9,205055263547D+02	2,400148906717D+03	9,25	-1,046741187825D-01
	4	1,553258213830D+02	-2,531017409814D+02	9,50	-1,767605098209D-01
	5	1,565581575652D+01	4,149809261996D+01	9,75	-1,537409358156D-01
	6	8,716814066477D-01	-1,456244146622D+00	10,25	2,707474690944D-01
	7	2,065956256597D-02	5,555755494602D-02	10,50	5,924934254357D-01
I=12	H= 3.5				
0		0,	5,828837982208D+06	-1,00	3,410965603542D+01
1		-4,369539080147D+00	-2,845566530324D+06	-,75	1,648638104123D+01
2		1,126535902024D+01	5,948806904234D+05	-,50	6,795980968690D+00
3		-1,116614391568D+01	-6,903077975952D+04	-,25	1,994252798660D+00
4		5,577744602221D+00	4,803320885728D+03	,25	-5,424034103726D-01
5		-1,506906978168D+00	-2,008431926079D+02	,50	-4,594669959474D-01
6		2,117243950756D-01	4,733364858521D+00	,75	-2,078208502545D-01

7		-1.223804389030D-02	-5.192369028645D-02	1.25	9.945911321073D-02
				1.50	1.038455302523D-01
0	-8.5	3.481283393329D-10	1.740641696665D-10	1.75	5.626008617950D-02
1	-7.5	-3.741631020722D-09	-1.870815510361D-09	2.25	-3.640874179622D-02
2	-6.5	1.914566980471D-08	9.572834902354D-09	2.50	-4.324905636241D-02
3	-5.5	-6.292870969662D-08	-3.146435484831D-08	2.75	-2.635417967773D-02
4	-4.5	1.520487888281D-07	7.602439441403D-08	3.25	2.096422595260D-02
5	-3.5	-2.942317052934D-07	-1.471158526467D-07	3.50	2.726627462459D-02
6	-2.5	4.850323817181D-07	2.425161908590D-07	3.75	1.805644890944D-02
7	-1.5	-7.103477177884D-07	-3.551738588942D-07	4.25	-1.662017278197D-02
8	-.5	9.442480797431D-07	4.721240398716D-07	4.50	-2.303453692181D-02
9	.5	-1.141010079528D-06	-5.705050397642D-07	4.75	-1.616138951247D-02
10	1.5	1.234784315727D-06	6.173921578635D-07	5.25	1.643700440288D-02
11	2.5	-1.164229505041D-06	-5.821147525207D-07	5.50	2.377715316854D-02
12	3.5	9.216532293112D-07	4.608266146556D-07	5.75	1.733960014542D-02
13	4.5	-5.856600922852D-07	-2.928300461426D-07	6.25	-1.884688885021D-02
14	5.5	2.827342617024D-07	1.413671308512D-07	6.50	-2.805601482583D-02
15	6.5	-9.623037512053D-08	-4.811518756027D-08	6.75	-2.100313156515D-02
16	7.5	2.039891740235D-08	1.019945870118D-08	7.25	2.392098332721D-02
17	8.5	-2.013956801653D-09	-1.006978400826D-09	7.50	3.637951582416D-02
				7.75	2.780302519606D-02
	0	-3.207326475565D+03	8.971624046917D+03	8.25	-3.299225707800D-02
	1	-3.805196502165D+03	-1.467812302546D+04	8.50	-5.126050154634D-02
	2	-1.934771565747D+03	5.652543695491D+03	8.75	-4.007767308907D-02
	3	-5.456756261298D+02	-2.176094227584D+03	9.25	5.012280555579D-02
	4	-9.205859345134D+01	2.812020949143D+02	9.50	8.038009400857D-02
	5	-9.277134921871D+00	-3.822280391044D+01	9.75	6.520813322332D-02
	6	-5.164392163972D-01	1.643905286477D+00	10.25	-8.991847732439D-02
	7	-1.223804389030D-02	-5.192369028645D-02	10.50	-1.542106742119D-01

I=13 H= 4.5

0		0.	-4.642708623292D+06	-1.00	-1.738365258995D+01
1		2.226523408845D+00	2.276613431273D+06	-.75	-8.401822915257D+00
2		-5.740775920758D+00	-4.780893896778D+05	-.50	-3.463245834426D+00
3		5.690833351438D+00	5.573016496940D+04	-.25	-1.016231406667D+00
4		-2.843091866745D+00	-3.894920035032D+03	.25	2.763700472487D-01
5		7.682280232378D-01	1.634138295318D+02	.50	2.340976873484D-01
6		-1.079585073856D-01	-3.841650441850D+00	.75	1.058773426905D-01
7		6.241511544811D-03	4.078272262646D-02	1.25	-5.066278839812D-02
				1.50	-5.289204618769D-02
0	-8.5	-1.778818483910D-10	-8.894092419551D-11	1.75	-2.865200257234D-02
1	-7.5	1.913039588214D-09	9.565197941071D-10	2.25	1.853708683484D-02
2	-6.5	-9.798643987203D-09	-4.899321993602D-09	2.50	2.201601208204D-02
3	-5.5	3.226019195882D-08	1.613009597941D-08	2.75	1.341299085236D-02
4	-4.5	-7.817272136541D-08	-3.908636068270D-08	3.25	-1.066441793238D-02
5	-3.5	1.520487888281D-07	7.602439441403D-08	3.50	-1.386584071806D-02
6	-2.5	-2.529102811103D-07	-1.264551405551D-07	3.75	-9.178898053282D-03
7	-1.5	3.760282047677D-07	1.880141023838D-07	4.25	8.440683545882D-03
8	-.5	-5.116608423566D-07	-2.558304211783D-07	4.50	1.169111914374D-02
9	.5	6.386887720807D-07	3.193443860403D-07	4.75	8.196757975347D-03
10	1.5	-7.193939929888D-07	-3.596969964944D-07	5.25	-8.321096569942D-03
11	2.5	7.086203851600D-07	3.543101925800D-07	5.50	-1.202273501918D-02
12	3.5	-5.856600922852D-07	-2.928300461426D-07	5.75	-8.755388316169D-03
13	4.5	3.869752418547D-07	1.934876209273D-07	6.25	9.482302251925D-03
14	5.5	-1.931712636383D-07	-9.658563181917D-08	6.50	1.408320010847D-02
15	6.5	6.759182615230D-08	3.379591307615D-08	6.75	1.051427362150D-02

16	7.5	-1.465403611082D-08	-7.327018055410D-09	7.25	-1.189121740920D-02
17	8.5	1.473305300637D-09	7.366526503186D-10	7.50	-1.800301384926D-02
				7.75	-1.368527331129D-02
	0	1.636962902213D+03	-7.269423932789D+03	8.25	1.601384415377D-02
	1	1.941881206098D+03	1.081987811466D+04	8.50	2.465489423832D-02
	2	9.872460025935D+02	-4.657096232559D+03	8.75	1.906570867868D-02
	3	2.784079022511D+02	1.635396699066D+03	9.25	-2.315170337368D-02
	4	4.696395645270D+01	-2.366218126320D+02	9.50	-3.639646458871D-02
	5	4.732277537936D+00	2.936724290244D+01	9.75	-2.880659997526D-02
	6	2.634114295306D-01	-1.415078445575D+00	10.25	3.700919777399D-02
	7	6.241511544811D-03	4.078272262646D-02	10.50	6.025891900946D-02

I=14 H= 5.5

0		0.	2.767116746073D+06	-1.00	7.139246242535D+00
1		-9.143220257020D-01	-1.363929116870D+06	-.75	3.450455030515D+00
2		2.357550823155D+00	2.879617519139D+05	-.50	1.422252996100D+00
3		-2.337179612473D+00	-3.375228280720D+04	-.25	4.173264264298D-01
4		1.167722693029D+00	2.372030990153D+03	.25	-1.134880849156D-01
5		-3.155571592826D-01	-1.000296242495D+02	.50	-9.612632273599D-02
6		4.434960504778D-02	2.356112673311D+00	.75	-4.347428403291D-02
7		-2.564323846496D-03	-2.462676606111D-02	1.25	2.080082356600D-02
				1.50	2.171495166554D-02
0	-8.5	7.315778514425D-11	3.657889257213D-11	1.75	1.176243513686D-02
1	-7.5	-7.870483727915D-10	-3.935241863957D-10	2.25	-7.608845787630D-03
2	-6.5	4.033492060825D-09	2.016746030412D-09	2.50	-9.035986930477D-03
3	-5.5	-1.329171773321D-08	-6.645858866606D-09	2.75	-5.504471094881D-03
4	-4.5	3.226019195882D-08	1.613009597941D-08	3.25	4.375301984867D-03
5	-3.5	-6.292870969662D-08	-3.146435484831D-08	3.50	5.687765559786D-03
6	-2.5	1.052207599167D-07	5.261037995833D-08	3.75	3.764413978488D-03
7	-1.5	-1.578927714727D-07	-7.894638573635D-08	4.25	-3.459848884944D-03
8	-.5	2.181584185993D-07	1.090792092997D-07	4.50	-4.790608282039D-03
9	.5	-2.786485109704D-07	-1.393242554852D-07	4.75	-3.357421021948D-03
10	1.5	3.235301290369D-07	1.617650645185D-07	5.25	3.404907232214D-03
11	2.5	-3.300472635298D-07	-1.650236317649D-07	5.50	4.916409746902D-03
12	3.5	2.827342617024D-07	1.413671308512D-07	5.75	3.577588234378D-03
13	4.5	-1.931712636383D-07	-9.658563181917D-08	6.25	-3.867072946949D-03
14	5.5	9.929722942599D-08	4.964861471300D-08	6.50	-5.736285331313D-03
15	6.5	-3.561646136701D-08	-1.780823068350D-08	6.75	-4.276353429416D-03
16	7.5	7.882114296092D-09	3.941057148046D-09	7.25	4.818267367217D-03
17	8.5	-8.060080012062D-10	-4.030040006031D-10	7.50	7.277303406956D-03
				7.75	5.516361444630D-03
	0	-6.728151358571D+02	4.276585256018D+03	8.25	-6.408202415421D-03
	1	-7.980905685950D+02	-6.060941893305D+03	8.50	-9.820153244617D-03
	2	-4.057211320032D+02	2.786119260323D+03	8.75	-7.552068491079D-03
	3	-1.144082078451D+02	-9.350427827213D+02	9.25	9.039336638619D-03
	4	-1.929811048398D+01	1.448721244960D+02	9.50	1.407798941173D-02
	5	-1.944447657942D+00	-1.723283871684D+01	9.75	1.101743689050D-02
	6	-1.082276638187D-01	8.908200926748D-01	10.25	-1.373408911027D-02
	7	-2.564323846496D-03	-2.462676606111D-02	10.50	-2.191281550668D-02

I=15 H= 6.5

0		0.	-1.135999377902D+06	-1.00	-2.162495370524D+00
1		2.769361506491D-01	5.630017717325D+05	-.75	-1.045139800266D+00
2		-7.140885284948D-01	-1.195487457219D+05	-.50	-4.307939451531D-01
3		7.079420170579D-01	1.409697193661D+04	-.25	-1.264045765944D-01

CARDINAL SPLINES N=8

4		-3,537234187400D-01	-9,969032310157D+02	.25	3,437348590466D-02
5		9,559260325929D-02	4.230052920199D+01	.50	2,911438675022D-02
6		-1,343573801608D-02	-1.000875991497D+00	.75	1,316705937045D-02
7		7,769143067600D-04	1.040232359399D-02	1.25	-6,299634464875D-03
				1.50	-6,576285702912D-03
0	-8,5	-2,217755974340D-11	-1,108877987170D-11	1.75	-3,562084054175D-03
1	-7,5	2,386376821884D-10	1.193188410942D-10	2.25	2,304033842541D-03
2	-6,5	-1.223359240353D-09	-6.116796201767D-10	2.50	2,736042082057D-03
3	-5,5	4,033492060825D-09	2,016746030412D-09	2,75	1,666618239184D-03
4	-4,5	-9,798643987203D-09	-4,899321993602D-09	3,25	-1,324527627779D-03
5	-3,5	1,914566980471D-08	9,572834902354D-09	3,50	-1,721677939124D-03
6	-2,5	-3,211080881610D-08	-1.605540440805D-08	3.75	-1,139350463174D-03
7	-1,5	4,845423249191D-08	2,422711624596D-08	4,25	1,046859033110D-03
8	-.5	-6,760427459756D-08	-3,380213729878D-08	4.50	1,449237569938D-03
9	,5	8,771347865140D-08	4,385673932570D-08	4.75	1,015448492110D-03
10	1,5	-1.041262801251D-07	-5,206314006254D-08	5,25	-1,029221330898D-03
11	2,5	1.091399167905D-07	5,456995839523D-08	5,50	-1,485570117226D-03
12	3,5	-9,623037512053D-08	-4,811518756027D-08	5,75	-1,080560028009D-03
13	4,5	6,759182615230D-08	3,379591307615D-08	6,25	1,166710048613D-03
14	5,5	-3,561646136701D-08	-1,780823068350D-08	6,50	1,729447428280D-03
15	6,5	1,305040804017D-08	6,525204020087D-09	6,75	1,288228050130D-03
16	7,5	-2,940537096807D-09	-1,470268548404D-09	7,25	-1,448423819171D-03
17	8,5	3,052562364065D-10	1,526281182032D-10	7,50	-2,184713050533D-03
				7,75	-1,653460434247D-03
	0	2,038893671930D+02	-1,710954816216D+03	8,25	1,913067455498D-03
	1	2,418440832650D+02	2,362228204239D+03	8,50	2,924161971355D-03
	2	1.229406340451D+02	-1.132226371203D+03	8,75	2,242029160099D-03
	3	3,466649279730D+01	3,716623406569D+02	9,25	-2,662892140727D-03
	4	5.847268597214D+00	-6,023865151567D+01	9,50	-4,126778213194D-03
	5	5,891431963706D-01	7,038779108614D+00	9.75	-3,210786778427D-03
	6	3,279066323614D-02	-3,819377376548D-01	10,25	3,942206465373D-03
	7	7,769143067600D-04	1.040232359399D-02	10,50	6,228258564601D-03

I=16 H= 7.5

0		0.	2,833490064763D+05	-1.00	4,215040639353D-01
1		-5,397752445484D-02	-1,411900162086D+05	-.75	2,037127031518D-01
2		1,391847371162D-01	3.015459122450D+04	-.50	8,396729855211D-02
3		-1,379894600866D-01	-3,577810876498D+03	-,25	2,463768699329D-02
4		6,894821327728D-02	2,546778740755D+02	.25	-6,699658089105D-03
5		-1,863359455473D-02	-1,087996718089D+01	.50	-5,674557035209D-03
6		2.619081571995D-03	2,589922054723D-01	.75	-2,566302180530D-03
7		-1,514528736264D-04	-2,691774138320D-03	1,25	1,227782863420D-03
				1,50	1,281678694951D-03
0	-8,5	4,324822717079D-12	2,162411358539D-12	1,75	6,942147255955D-04
1	-7,5	-4,654185557337D-11	-2,327092778669D-11	2,25	-4,490106481079D-04
2	-6,5	2,386376821884D-10	1,193188410942D-10	2,50	-5,331838315300D-04
3	-5,5	-7,870483727915D-10	-3,935241863957D-10	2,75	-3,247689600469D-04
4	-4,5	1,913039588214D-09	9,565197941071D-10	3,25	2,580829505000D-04
5	-3,5	-3,741631020722D-09	-1,870815510361D-09	3,50	3,354477087371D-04
6	-2,5	6,286990180914D-09	3,143495090457D-09	3,75	2,219731337286D-04
7	-1,5	-9,519372988004D-09	-4,759686494002D-09	4,25	-2,039176435287D-04
8	-,5	1,336417370849D-08	6,682086854246D-09	4,50	-2,822652658122D-04
9	,5	-1,752331561630D-08	-8,761657808151D-09	4,75	-1,977508269123D-04
10	1,5	2,113631141546D-08	1,056815570773D-08	5,25	2,003649451777D-04
11	2,5	-2,261410287024D-08	-1,130705143512D-08	5,50	2,891426767612D-04
12	3,5	2,039891740235D-08	1,019945870118D-08	5,75	2,102603478565D-04

13	4,5	-1.465403611082D-08	-7.327018055410D-09	6.25	-2,268758767636D-04
14	5.5	7.882114296092D-09	3.941057148046D-09	6.50	-3,361652940304D-04
15	6,5	-2.940537096807D-09	-1.470268548404D-09	6.75	-2.502804298706D-04
16	7.5	6,728427851800D-10	3.364213925900D-10	7.25	2,810541359590D-04
17	8,5	-7.076595034737D-11	-3.538297517368D-11	7,50	4,235907926247D-04
				7,75	3,202907018357D-04
	0	-3,975186192614D+01	4,140500753542D+02	8.25	-3,697064540569D-04
	1	-4.715080850679D+01	-5,634973261360D+02	8.50	-5,642610009130D-04
	2	-2.396844982742D+01	2.778619105882D+02	8.75	-4,318766993723D-04
	3	-6,758424181125D+00	-9.026070573877D+01	9.25	5,106639094766D-04
	4	-1.139934762202D+00	1.510406882386D+01	9.50	7,891619157763D-04
	5	-1,148523068926D-01	-1.755459013170D+00	9.75	6,119653510755D-04
	6	-6,392364408775D-03	9.883164424232D-02	10.25	-7,450354573124D-04
	7	-1.514528736264D-04	-2,691774138320D-03	10.50	-1,170754825636D-03

I=17 H= 8.5

0		0.	-3,216894157291D+04	-1.00	-3,915303791893D-02
1		5,013815972822D-03	1.611291953696D+04	-.75	-1,892256608615D-02
2		-1,292858318132D-02	-3,460745946509D+03	-.50	-7,799560761176D-03
3		1,281771587762D-02	4,131295006023D+02	-.25	-2,288535976503D-03
4		-6,404637949469D-03	-2,960261026841D+01	.25	6,223080470901D-04
5		1,730917654635D-03	1,273584173693D+00	.50	5,270863696924D-04
6		-2,432978284730D-04	-3,052745398790D-02	.75	2,383714820263D-04
7		1,406945459040D-05	3.183522171277D-04	1.25	-1,140407597948D-04
				1,50	-1,190454727068D-04
0	-8,5	-4,018478672345D-13	-2,009239336173D-13	1,75	-6,447955513670D-05
1	-7,5	4,324822717079D-12	2.162411358539D-12	2,25	4,170337608805D-05
2	-6,5	-2.217755974340D-11	-1,108877987170D-11	2.50	4,952027798250D-05
3	-5,5	7,315778514425D-11	3,657889257213D-11	2,75	3,016273772185D-05
4	-4,5	-1,778818483910D-10	-8,894092419551D-11	3,25	-2,396793195398D-05
5	-3,5	3,481283393329D-10	1,740641696665D-10	3,50	-3,115159266204D-05
6	-2,5	-5,856323309537D-10	-2,928161654768D-10	3,75	-2,061281180978D-05
7	-1,5	8,886556845557D-10	4,443278422778D-10	4,25	1,893406870937D-05
8	-,5	-1,252598229453D-09	-6,262991147264D-10	4,50	2,620693298184D-05
9	,5	1,654190736434D-09	8,270953682170D-10	4,75	1,835867180757D-05
10	1,5	-2,018151703535D-09	-1,009075851768D-09	5,25	-1,859742404383D-05
11	2,5	2,193028631940D-09	1,096514315970D-09	5,50	-2,683395709842D-05
12	3,5	-2,013956801653D-09	-1,006978400826D-09	5,75	-1,951016830410D-05
13	4,5	1,473305300637D-09	7,366526503186D-10	6,25	2,104338650520D-05
14	5,5	-8,060080012062D-10	-4,030040006031D-10	6,50	3,117225373805D-05
15	6,5	3,052562364065D-10	1,526281182032D-10	6,75	2,320121251223D-05
16	7,5	-7,076595034737D-11	-3,538297517368D-11	7,25	-2,603383293813D-05
17	8,5	7,526735981935D-12	3,763367990968D-12	7,50	-3,921769646958D-05
				7,75	-2,963675804504D-05
	0	3,693122266111D+00	-4,558518915485D+01	8,25	3,415932669986D-05
	1	4,380458872493D+00	6,153909046464D+01	8,50	5,208729669976D-05
	2	2,226715272038D+00	-3,097009557962D+01	8,75	3,982381606754D-05
	3	6,278625834492D-01	1,001550982844D+01	9,25	-4,696008817074D-05
	4	1,058994737766D-01	-1,716629208790D+00	9,50	-7,244534422557D-05
	5	1,066960837980D-02	1,997039217478D-01	9,75	-5,606547954083D-05
	6	5,938347196560D-04	-1,158549706880D-02	10,25	6,790909071288D-05
	7	1,406945459040D-05	3,183522171277D-04	10,50	1,063705772225D-04

CARDINAL SPLINES N=8

N= 8 M=18 P= 9.0 Q=15

I= 0		H= -9.0			
0		1.000000000000D+00	-1.103117020928D+03	-1.00	9.405158903111D+00
1		-2.779687660596D+00	4.985689386235D+02	-.75	5.800756552786D+00
2		3.078872977988D+00	-9.546424766010D+01	-.50	3.426821027184D+00
3		-1.800794143575D+00	9.548049365341D+00	-.25	1.917995238194D+00
4		6.104175315914D-01	-3.694999385517D-01	.25	4.716392597609D-01
5		-1.214746081753D-01	-3.461864327278D-02	.50	1.893329326955D-01
6		1.328894197200D-02	6.083291536640D-03	.75	5.398568399408D-02
7		-6.230392122024D-04	-3.063573944462D-04	1.25	-1.375810874806D-02
				1.50	-1.123242822660D-02
0	-9.0	7.538291884714D-12	3.769145942357D-12	1.75	-4.880528499107D-03
1	-8.0	-7.089035601913D-11	-3.544517800956D-11	2.25	2.131710483699D-03
2	-7.0	3.058944904882D-10	1.529472452441D-10	2.50	2.113222246284D-03
3	-6.0	-8.081151285168D-10	-4.040575642584D-10	2.75	1.082098791796D-03
4	-5.0	1.478436034035D-09	7.392180170175D-10	3.25	-6.164742927796D-04
5	-4.0	-2.024024165258D-09	-1.012012082629D-09	3.50	-6.817391309379D-04
6	-3.0	2.210046828233D-09	1.105023414116D-09	3.75	-3.846495674896D-04
7	-2.0	-2.044214688518D-09	-1.022107344259D-09	4.25	2.580052154924D-04
8	-1.0	1.691577427321D-09	8.457887136606D-10	4.50	3.055876264040D-04
9	0.0	-1.303639683410D-09	-6.518198417049D-10	4.75	1.832924451384D-04
10	1.0	9.548860797104D-10	4.774430398552D-10	5.25	-1.362581200084D-04
11	2.0	-6.656695461677D-10	-3.328347730838D-10	5.50	-1.684641324406D-04
12	3.0	4.346583866013D-10	2.173291933007D-10	5.75	-1.049584513799D-04
13	4.0	-2.571285432822D-10	-1.285642716411D-10	6.25	8.311662624709D-05
14	5.0	1.310430472455D-10	6.552152362277D-11	6.50	1.054816454659D-04
15	6.0	-5.381772201308D-11	-2.690886100654D-11	6.75	6.724829732877D-05
16	7.0	1.630152130235D-11	8.150760651176D-12	7.25	-5.532638017974D-05
17	8.0	-3.177420664538D-12	-1.588710332269D-12	7.50	-7.133420830790D-05
18	9.0	2.951470280062D-13	1.475735140031D-13	7.75	-4.612239402007D-05
				8.25	3.886408952873D-05
	0	-1.731007489705D+02	-8.894539293661D+01	8.50	5.062941763400D-05
	1	-1.995176167903D+02	-9.714040561163D+01	8.75	3.304962134837D-05
	2	-9.918500923624D+01	-5.081871128991D+01	9.25	-2.834276496718D-05
	3	-2.753903117053D+01	-1.345103914292D+01	9.50	-3.723540096114D-05
	4	-4.606720839655D+00	-2.352848588105D+00	9.75	-2.451218576713D-05
	5	-4.636614416433D-01	-2.272348282473D-01	10.25	2.139538067025D-05
	6	-2.596252839675D-02	-1.321722431347D-02	10.50	2.837930277525D-05
	7	-6.230392122024D-04	-3.063573944462D-04	10.75	1.887719064411D-05

I= 1	H= -8.0			
0	0.	1.188220707410D+04	-1.00	-4.059085995995D+01
1	8.670344453430D+00	-5.378535609397D+03	-.75	-2.148016965797D+01
2	-1.538609276970D+01	1.036712059685D+03	-.50	-9.880255030960D+00
3	1.119168962378D+01	-1.068992174319D+02	-.25	-3.321824709113D+00
4	-4.302007940410D+00	5.113383666962D+00	.25	1.364901268996D+00
5	9.281594556120D-01	1.828672830452D-01	.50	1.646103613878D+00
6	-1.073292698331D-01	-4.762255735875D-02	.75	1.410246284013D+00

7		5.236447189396D-03	2.562643251544D-03	1.25	6.010388590239D-01
				1.50	2.949782104769D-01
0	-9.0	-7.089035601913D-11	-3.544517800956D-11	1.75	9.988918699336D-02
1	-8.0	6.741820809029D-10	3.370910404514D-10	2.25	-3.380806203113D-02
2	-7.0	-2.947408254285D-09	-1.473704127142D-09	2.50	-3.110866103968D-02
3	-6.0	7.904798684633D-09	3.952399342316D-09	2.75	-1.505847404609D-02
4	-5.0	-1.470927128778D-08	-7.354635643892D-09	3.25	7.930253139317D-03
5	-4.0	2.050729812830D-08	1.025364906415D-08	3.50	8.525741974927D-03
6	-3.0	-2.279731314662D-08	-1.139865657331D-08	3.75	4.700222303023D-03
7	-2.0	2.141689383440D-08	1.070844691720D-08	4.25	-3.043451991622D-03
8	-1.0	-1.792580397949D-08	-8.962901989747D-09	4.50	-3.556389464724D-03
9	0.0	1.391366313296D-08	6.956831566479D-09	4.75	-2.108929332674D-03
10	1.0	-1.023237974808D-08	-5.116189874040D-09	5.25	1.539924165617D-03
11	2.0	7.148634979681D-09	3.574317489840D-09	5.50	1.890581344741D-03
12	3.0	-4.673174740309D-09	-2.336587370155D-09	5.75	1.170851404098D-03
13	4.0	2.766173999374D-09	1.383086999687D-09	6.25	-9.184221784526D-04
14	5.0	-1.410215169406D-09	-7.051075847028D-10	6.50	-1.161184984625D-03
15	6.0	5.792614929858D-10	2.896307464929D-10	6.75	-7.379216518039D-04
16	7.0	-1.754776264496D-10	-8.773881322478D-11	7.25	6.040085385033D-04
17	8.0	3.420539587296D-11	1.710269793648D-11	7.50	7.771999243248D-04
18	9.0	-3.177420664538D-12	-1.588710332269D-12	7.75	5.016456695747D-04
				8.25	-4.215447134555D-04
	0	1.578505006526D+03	8.150380765472D+02	8.50	-5.485661280856D-04
	1	1.808839829508D+03	8.762291309903D+02	8.75	-3.577598634400D-04
	2	8.928323806406D+02	4.596181106745D+02	9.25	3.063601521259D-04
	3	2.457393589852D+02	1.194407906231D+02	9.50	4.022502873373D-04
	4	4.066805475236D+01	2.086684677626D+01	9.75	2.646727642815D-04
	5	4.039575553788D+00	1.970305356549D+00	10.25	-2.308396533527D-04
	6	2.225669030989D-01	1.138239674885D-01	10.50	-3.060976353310D-04
	7	5.236447189396D-03	2.562643251544D-03	10.75	-2.035558267750D-04

I= 2 H= -7.0

0		0.	-6.097033371926D+04	-1.00	1.072634679348D+02
1		-1.728171755657D+01	2.761221016044D+04	-.75	5.425182590699D+01
2		3.919646324717D+01	-5.339524446069D+03	-.50	2.356052354749D+01
3		-3.308170045015D+01	5.607035754090D+02	-.25	7.345223158070D+00
4		1.403990070688D+01	-3.025583775998D+01	.25	-2.335785476137D+00
5		-3.248265901431D+00	-1.811411753672D-01	.50	-2.194949022632D+00
6		3.953700131830D-01	1.666810536238D-01	.75	-1.130445351701D+00
7		-2.005005939883D-02	-9.739827146772D-03	1.25	8.064219631376D-01
				1.50	1.190142720251D+00
0	-9.0	3.058944904882D-10	1.529472452441D-10	1.75	1.212126827032D+00
1	-8.0	-2.947408254285D-09	-1.473704127142D-09	2.25	6.875762669598D-01
2	-7.0	1.308566000452D-08	6.542830002262D-09	2.50	3.807669341758D-01
3	-6.0	-3.573284194171D-08	-1.786642097086D-08	2.75	1.438193378399D-01
4	-5.0	6.787520613531D-08	3.393760306766D-08	3.25	-5.883670925869D-02
5	-4.0	-9.678641440501D-08	-4.839320720251D-08	3.50	-5.878808868814D-02
6	-3.0	1.100798635392D-07	5.503993176962D-08	3.75	-3.067639448591D-02
7	-2.0	-1.055657874370D-07	-5.278289371848D-08	4.25	1.840718762456D-02
8	-1.0	8.977841537815D-08	4.488920768907D-08	4.50	2.093572917747D-02
9	0.0	-7.042338937776D-08	-3.521169468888D-08	4.75	1.214436708972D-02
10	1.0	5.211226084738D-08	2.605613042369D-08	5.25	-8.578481375344D-03
11	2.0	-3.653141361638D-08	-1.826570680819D-08	5.50	-1.040054137975D-02
12	3.0	2.392488583940D-08	1.196244291970D-08	5.75	-6.373675969362D-03
13	4.0	-1.417558687440D-08	-7.087793437201D-09	6.25	4.918425375795D-03
14	5.0	7.230603140041D-09	3.615301570021D-09	6.50	6.179161204329D-03

CARDINAL SPLINES N=8

15	6.0	-2.970904464621D-09	-1.485452232311D-09	6.75	3.905709613278D-03
16	7.0	9.001331017378D-10	4.500665508689D-10	7.25	-3.169964941050D-03
17	8.0	-1.754776264496D-10	-8.773881322478D-11	7.50	-4.065408102324D-03
18	9.0	1.630152130235D-11	8.150760651176D-12	7.75	-2.616624729077D-03
				8.25	2.189021671856D-03
	0	-6.571223615607D+03	-3.417918230785D+03	8.50	2.843640589157D-03
	1	-7.480352891789D+03	-3.595527348818D+03	8.75	1.851776357469D-03
	2	-3.662791385651D+03	-1.899136454402D+03	9.25	-1.581990781084D-03
	3	-9.984412530022D+02	-4.816263810115D+02	9.50	-2.075227169287D-03
	4	-1.633347644012D+02	-8.440140014845D+01	9.75	-1.364380405485D-03
	5	-1.600343622695D+01	-7.747810256340D+00	10.25	1.188488586584D-03
	6	-8.677837289431D-01	-4.469280566228D-01	10.50	1.575187533054D-03
	7	-2.005005939883D-02	-9.739827146772D-03	10.75	1.047067007607D-03

I= 3 H= -6.0

0		0.	2.013401252205D+05	-1.00	-1.963285281496D+02
1		2.831986004061D+01	-9.119775830034D+04	-.75	-9.725087567467D+01
2		-6.872032966300D+01	1.766310014369D+04	-.50	-4.120867777606D+01
3		6.281140400191D+01	-1.874843385873D+03	-.25	-1.247499738849D+01
4		-2.855451988106D+01	1.088202657581D+02	.25	3.661439709606D+00
5		6.986237115805D+00	-1.272166276062D+00	.50	3.251403755348D+00
6		-8.894145303274D-01	-3.423203336827D-01	.75	1.554254335780D+00
7		4.676291687605D-02	2.243960518394D-02	1.25	-8.599104475818D-01
				1.50	-9.899814374530D-01
0	-9.0	-8.081151285168D-10	-4.040575642584D-10	1.75	-6.068679935328D-01
1	-8.0	7.904798684633D-09	3.952399342316D-09	2.25	5.777188409074D-01
2	-7.0	-3.573284194171D-08	-1.786642097086D-08	2.50	9.634136336352D-01
3	-6.0	9.968144739566D-08	4.984072369783D-08	2.75	1.096022115495D+00
4	-5.0	-1.941068122210D-07	-9.705340611050D-08	3.25	7.537292728292D-01
5	-4.0	2.845699656889D-07	1.422849828444D-07	3.50	4.539288073736D-01
6	-3.0	-3.331503967831D-07	-1.665751983916D-07	3.75	1.851021112183D-01
7	-2.0	3.282825078432D-07	1.641412539216D-07	4.25	-8.653320633107D-02
8	-1.0	-2.854656778270D-07	-1.427328389135D-07	4.50	-9.159686299013D-02
9	0.0	2.274654233107D-07	1.137327116554D-07	4.75	-5.035825272160D-02
10	1.0	-1.699693592647D-07	-8.498467963233D-08	5.25	3.304422884304D-02
11	2.0	1.198149119506D-07	5.990745597529D-08	5.50	3.903680458368D-02
12	3.0	-7.870677821091D-08	-3.935338910545D-08	5.75	2.342511168888D-02
13	4.0	4.671019887544D-08	2.335509943772D-08	6.25	-1.751810469523D-02
14	5.0	-2.384663663403D-08	-1.192331831701D-08	6.50	-2.175072584414D-02
15	6.0	9.802824955625D-09	4.901412477812D-09	6.75	-1.361359867236D-02
16	7.0	-2.970904464621D-09	-1.485452232311D-09	7.25	1.088266797812D-02
17	8.0	5.792614929858D-10	2.896307464929D-10	7.50	1.387538258102D-02
18	9.0	-5.381772201308D-11	-2.690886100654D-11	7.75	8.886589747561D-03
				8.25	-7.377069960945D-03
	0	1.665639042256D+04	8.765076573357D+03	8.50	-9.554320548216D-03
	1	1.882094125423D+04	8.932825011575D+03	8.75	-6.205856437126D-03
	2	9.135001539128D+03	4.791191268324D+03	9.25	5.280412059601D-03
	3	2.464439316743D+03	1.174119769128D+03	9.50	6.915847527350D-03
	4	3.983433200747D+02	2.082001041790D+02	9.75	4.540807062079D-03
	5	3.850157408428D+01	1.841230412295D+01	10.25	-3.947074301211D-03
	6	2.056649232864D+00	1.071374792905D+00	10.50	-5.227012852686D-03
	7	4.676291687605D-02	2.243960518394D-02	10.75	-3.472084295138D-03

I= 4 H= -5.0

0		0.	-4.904795211578D+05	-1.00	2.689983256988D+02

CARDINAL SPLINES N=8

1		-3.670491599338D+01	2.221831637315D+05	-.75	1.317627346470D+02
2		9.175369570740D+01	-4.306741250902D+04	-.50	5.512479457838D+01
3		-8.718120497641D+01	4.600327324724D+03	-.25	1.644521583624D+01
4		4.135415810377D+01	-2.792107020327D+02	.25	-4.652245162514D+00
5		-1.053753847042D+01	6.340443036931D+00	.50	-4.035198554287D+00
6		1.391309037464D+00	4.462164526851D-01	.75	-1.875152566600D+00
7		-7.550340989613D-02	-3.545711211305D-02	1.25	9.598927967088D-01
				1.50	1.045700806488D+00
0	-9.0	1.478436034035D-09	7.392180170175D-10	1.75	5.958199327906D-01
1	-8.0	-1.470927128778D-08	-7.354635643892D-09	2.25	-4.412348701039D-01
2	-7.0	6.787520613531D-08	3.393760306766D-08	2.50	-5.749056621816D-01
3	-6.0	-1.941068122210D-07	-9.705340611050D-08	2.75	-3.943054414392D-01
4	-5.0	3.892532486623D-07	1.946266243312D-07	3.25	4.565954550004D-01
5	-4.0	-5.901299255526D-07	-2.950649627763D-07	3.50	8.294493828099D-01
6	-3.0	7.161763355480D-07	3.580881677740D-07	3.75	1.020425440197D+00
7	-2.0	-7.309657611191D-07	-3.654828805595D-07	4.25	8.044555652418D-01
8	-1.0	6.552881799533D-07	3.276440899767D-07	4.50	5.140246915304D-01
9	0.0	-5.343228617155D-07	-2.671614308578D-07	4.75	2.211578840699D-01
10	1.0	4.054339990932D-07	2.027169995466D-07	5.25	-1.133556174695D-01
11	2.0	-2.884461990747D-07	-1.442230995374D-07	5.50	-1.247736148841D-01
12	3.0	1.904674750976D-07	9.523373754879D-08	5.75	-7.103899486314D-02
13	4.0	-1.133576542879D-07	-5.667882714396D-08	6.25	4.944243846426D-02
14	5.0	5.796081035095D-08	2.898040517548D-08	6.50	5.986424003549D-02
15	6.0	-2.384663663403D-08	-1.192331831701D-08	6.75	3.671581051125D-02
16	7.0	7.230603140041D-09	3.615301570021D-09	7.25	-2.847844545123D-02
17	8.0	-1.410215169406D-09	-7.051075847028D-10	7.50	-3.590312864852D-02
18	9.0	1.310430472455D-10	6.552152362277D-11	7.75	-2.277977350908D-02
				8.25	1.863959735708D-02
	0	-2.909167388906D+04	-1.560998646477D+04	8.50	2.400760696396D-02
	1	-3.261220518506D+04	-1.514262476546D+04	8.75	1.552114337414D-02
	2	-1.568341239765D+04	-8.386601275571D+03	9.25	-1.311075017554D-02
	3	-4.186777439551D+03	-1.951857148471D+03	9.50	-1.712288964228D-02
	4	-6.688640960458D+02	-3.564259909228D+02	9.75	-1.121561757861D-02
	5	-6.383815068066D+01	-2.987641622236D+01	10.25	9.712497947242D-03
	6	-3.365405785992D+00	-1.787581610437D+00	10.50	1.284315101347D-02
	7	-7.550340989613D-02	-3.545711211305D-02	10.75	8.520527679821D-03
I= 5		H= -4.0			
0		0.	9.632206879443D+05	-1.00	-2.941557497914D+02
1		3.894783600791D+01	-4.363565324649D+05	-.75	-1.431829311343D+02
2		-9.883797125511D+01	8.462139653466D+04	-.50	-5.948366907721D+01
3		9.584876055948D+01	-9.072512885641D+03	-.25	-1.760626347191D+01
4		-4.658730869381D+01	5.656831982509D+02	.25	4.886746784730D+00
5		1.218931683013D+01	-1.670615393537D+01	.50	4.189626270878D+00
6		-1.652594945791D+00	-3.363853535378D-01	.75	1.920929398714D+00
7		9.196149921046D-02	4.147394493798D-02	1.25	-9.502712377885D-01
				1.50	-1.012581770292D+00
0	-9.0	-2.024024165258D-09	-1.012012082629D-09	1.75	-5.616961033763D-01
1	-8.0	2.050729812830D-08	1.025364906415D-08	2.25	3.860766652475D-01
2	-7.0	-9.678641440501D-08	-4.839320720251D-08	2.50	4.767850138876D-01
3	-6.0	2.845699656889D-07	1.422849828444D-07	2.75	3.044554735479D-01
4	-5.0	-5.901299255526D-07	-2.950649627763D-07	3.25	-2.751805576030D-01
5	-4.0	9.304237954794D-07	4.652118977397D-07	3.50	-3.912282193147D-01
6	-3.0	-1.179055553101D-06	-5.895277765506D-07	3.75	-2.906430111850D-01
7	-2.0	1.257490074980D-06	6.287450374900D-07	4.25	3.870159307287D-01
8	-1.0	-1.173580915260D-06	-5.867904576302D-07	4.50	7.470025611227D-01

9	0.0	9.887147809384D-07	4.943573904692D-07	4.75	9.709490566001D-01
10	1.0	-7.680468419916D-07	-3.840234209958D-07	5.25	8.412817861840D-01
11	2.0	5.547598416060D-07	2.773799208030D-07	5.50	5.595906886131D-01
12	3.0	-3.696156424266D-07	-1.848078212133D-07	5.75	2.495727008770D-01
13	4.0	2.210875569124D-07	1.105437784562D-07	6.25	-1.359087550546D-01
14	5.0	-1.133576542879D-07	-5.667882714396D-08	6.50	-1.534362475500D-01
15	6.0	4.671019887544D-08	2.335509943772D-08	6.75	-8.934280207056D-02
16	7.0	-1.417558687440D-08	-7.087793437201D-09	7.25	6.456423874353D-02
17	8.0	2.766173999374D-09	1.383086999687D-09	7.50	7.941208575238D-02
18	9.0	-2.571285432822D-10	-1.285642716411D-10	7.75	4.939301648038D-02
				8.25	-3.924157086388D-02
	0	3.791832779357D+04	2.104852570620D+04	8.50	-4.999041482722D-02
	1	4.217336361991D+04	1.880217115416D+04	8.75	-3.202549511387D-02
	2	2.010256054474D+04	1.112125068252D+04	9.25	2.667571397976D-02
	3	5.314796841550D+03	2.379468454993D+03	9.50	3.465256380339D-02
	4	8.404267418810D+02	4.634057717034D+02	9.75	2.259534260126D-02
	5	7.937569991442D+01	3.567621731309D+01	10.25	-1.942991256959D-02
	6	4.140979504468D+00	2.276473177555D+00	10.50	-2.562282951689D-02
	7	9.196149921046D-02	4.147394493798D-02	10.75	-1.695972505260D-02

I= 6 H= -3.0

0		0.	-1.630844899094D+06	-1.00	2.728335126997D+02
1		-3.554397076776D+01	7.388460803136D+05	-.75	1.323472751333D+02
2		9.090966489997D+01	-1.433230160045D+05	-.50	5.477307928356D+01
3		-8.911711612523D+01	1.539914848007D+04	-.25	1.614357104044D+01
4		4.389886269832D+01	-9.754054396000D+02	.25	-4.436117795171D+00
5		-1.166461860373D+01	3.277921863429D+01	.50	-3.780631847838D+00
6		1.608228750448D+00	-2.554458075602D-02	.75	-1.721692737937D+00
7		-9.105085422678D-02	-3.789296821605D-02	1.25	8.376288940529D-01
				1.50	8.833602125204D-01
0	-9.0	2.210046828233D-09	1.105023414116D-09	1.75	4.841117340707D-01
1	-8.0	-2.279731314662D-08	-1.139865657331D-08	2.25	-3.224519195765D-01
2	-7.0	1.100798635392D-07	5.503993176962D-08	2.50	-3.900577017344D-01
3	-6.0	-3.331503967831D-07	-1.665751983916D-07	2.75	-2.428377868836D-01
4	-5.0	7.161763355480D-07	3.580881677740D-07	3.25	2.042981527805D-01
5	-4.0	-1.179055553101D-06	-5.895277765506D-07	3.50	2.756770123077D-01
6	-3.0	1.569586084350D-06	7.847930421752D-07	3.75	1.909311869072D-01
7	-2.0	-1.763544344222D-06	-8.817721721111D-07	4.25	-1.989404126947D-01
8	-1.0	1.730714346712D-06	8.653571733561D-07	4.50	-3.008585790666D-01
9	0.0	-1.523206045344D-06	-7.616030226720D-07	4.75	-2.363909091138D-01
10	1.0	1.224197832825D-06	6.120989164127D-07	5.25	3.465953370826D-01
11	2.0	-9.055061026928D-07	-4.527530513464D-07	5.50	6.969584491097D-01
12	3.0	6.124638215408D-07	3.062319107704D-07	5.75	9.397194402956D-01
13	4.0	-3.696156424266D-07	-1.848078212133D-07	6.25	8.661134547557D-01
14	5.0	1.904674750976D-07	9.523373754879D-08	6.50	5.911843096173D-01
15	6.0	-7.870677821091D-08	-3.935338910545D-08	6.75	2.697743290694D-01
16	7.0	2.392488583940D-08	1.196244291970D-08	7.25	-1.526524758098D-01
17	8.0	-4.673174740309D-09	-2.336587370155D-09	7.50	-1.751241574289D-01
18	9.0	4.346583866013D-10	2.173291933007D-10	7.75	-1.034404597806D-01
				8.25	7.660215445633D-02
	0	-3.950093412640D+04	-2.328656857716D+04	8.50	9.522342795856D-02
	1	-4.363466112461D+04	-1.795055183112D+04	8.75	5.981216382850D-02
	2	-2.064734206088D+04	-1.213489624320D+04	9.25	-4.838709368157D-02
	3	-5.417586856835D+03	-2.238271995467D+03	9.50	-6.217703025745D-02
	4	-8.501735882710D+02	-4.982163507081D+02	9.75	-4.017864349500D-02
	5	-7.969776911927D+01	-3.305612766204D+01	10.25	3.407563892769D-02

	6	-4,127975065839D+00	-2.412801578367D+00	10.50	4,469964842781D-02
	7	-9,105085422678D-02	-3.789296821605D-02	10.75	2,945532399045D-02

I= 7 H= -2.0

0		0.	2.505111022752D+06	-1.00	-2,269342136046D+02
1		2,931657875993D+01	-1,135023360712D+06	-.75	-1,098838353302D+02
2		-7,528299373598D+01	2.202198348003D+05	-.50	-4,538620535617D+01
3		7,420942891083D+01	-2,369077191846D+04	-.25	-1,334762067304D+01
4		-3,681157601023D+01	1.514045245577D+03	.25	3,648985834350D+00
5		9,862600501878D+00	-5,439983724597D+01	.50	3,100369703934D+00
6		-1,372537055083D+00	5,886818370572D-01	.75	1,407077717859D+00
7		7,849863070057D-02	2,751965873059D-02	1.25	-6,789098008766D-01
				1.50	-7,123592728717D-01
0	-9,0	-2,044214688518D-09	-1.022107344259D-09	1.75	-3,881226683001D-01
1	-8,0	2,141689383440D-08	1.070844691720D-08	2.25	2,547346301963D-01
2	-7,0	-1,055657874370D-07	-5,278289371848D-08	2.50	3,052681373856D-01
3	-6,0	3,282825078432D-07	1.641412539216D-07	2.75	1,879445584636D-01
4	-5,0	-7,309657611191D-07	-3,654828805595D-07	3.25	-1,535148806452D-01
5	-4,0	1,257490074980D-06	6,287450374900D-07	3,50	-2,030920006369D-01
6	-3,0	-1,763544344222D-06	-8,817721721111D-07	3.75	-1,372545171290D-01
7	-2,0	2.098868558514D-06	1.049434279257D-06	4.25	1,333200716649D-01
8	-1,0	-2.183879430898D-06	-1.091939715449D-06	4.50	1,914922715188D-01
9	0,0	2.029173225855D-06	1.014586612928D-06	4.75	1,403555976958D-01
10	1,0	-1,706981660522D-06	-8,534908302610D-07	5.25	-1,611907561499D-01
11	2,0	1.307175934253D-06	6,535879671267D-07	5,50	-2,540685886797D-01
12	3,0	-9,055061026928D-07	-4,527530513464D-07	5,75	-2,071538039340D-01
13	4,0	5,547598416060D-07	2,773799208030D-07	6,25	3,232718554176D-01
14	5,0	-2,884461990747D-07	-1,442230995374D-07	6,50	6,672234322130D-01
15	6,0	1,198149119506D-07	5,990745597529D-08	6,75	9,206616077635D-01
16	7,0	-3,653141361638D-08	-1,826570680819D-08	7,25	8,820014081710D-01
17	8,0	7,148634979681D-09	3,574317489840D-09	7,50	6,118376533652D-01
18	9,0	-6,656695461677D-10	-3,328347730838D-10	7,75	2,832548217901D-01
				8,25	-1,642800000872D-01
	0	3,515459931762D+04	2,300487207505D+04	8,50	-1,904900656382D-01
	1	3,864725992867D+04	1,338733222428D+04	8,75	-1,136364304631D-01
	2	1,819864755761D+04	1,188046671943D+04	9,25	8,570144705813D-02
	3	4,752141896884D+03	1,653446769981D+03	9,50	1,074678242085D-01
	4	7,422654869739D+02	4,834650940441D+02	9,75	6,809356121136D-02
	5	6,927177034908D+01	2,419992145585D+01	10,25	-5,609468147655D-02
	6	3,572876679053D+00	2,322420337084D+00	10,50	-7,279215232889D-02
	7	7,849863070057D-02	2,751965873059D-02	10,75	-4,753861757035D-02

I= 8 H= -1.0

0		0.	-3,614069436986D+06	-1.00	1,765738578691D+02
1		-2,271329495007D+01	1,637711537889D+06	-,75	8,542020609586D+01
2		5,844443935112D+01	-3,178190982292D+05	-,50	3,524614125396D+01
3		-5,777317819049D+01	3,421681361570D+04	-,25	1,035397711894D+01
4		2,876035941499D+01	-2,197646066268D+03	,25	-2,823203508247D+00
5		-7,738233989407D+00	8,178364099361D+01	,50	-2,395060083784D+00
6		1,082130167611D+00	-1,299729883030D+00	,75	-1,085105041630D+00
7		-6,222180545390D-02	-1,417340253828D-02	1,25	5,213866917365D-01
				1,50	5,456967054241D-01
0	-9,0	1,691577427321D-09	8,457887136606D-10	1,75	2,964584152132D-01
1	-8,0	-1,792580397949D-08	-8,962901989747D-09	2,25	-1,931727764416D-01
2	-7,0	8,977841537815D-08	4,488920768907D-08	2,50	-2,304472107493D-01

3	-6.0	-2.854656778270D-07	-1.427328389135D-07	2.75	-1.411264768433D-01
4	-5.0	6.552881799533D-07	3.276440899767D-07	3.25	1.136988707772D-01
5	-4.0	-1.173580915260D-06	-5.867904576302D-07	3.50	1.490812174119D-01
6	-3.0	1.730714346712D-06	8.653571733561D-07	3.75	9.967736377084D-02
7	-2.0	-2.183879430898D-06	-1.091939715449D-06	4.25	-9.406781853740D-02
8	-1.0	2.419673096851D-06	1.209836548426D-06	4.50	-1.325029171587D-01
9	0.0	-2.392092233280D-06	-1.196046116640D-06	4.75	-9.479043272909D-02
10	1.0	2.128329707604D-06	1.064164853802D-06	5.25	1.015195656496D-01
11	2.0	-1.706981660522D-06	-8.534908302610D-07	5.50	1.519949597756D-01
12	3.0	1.224197832825D-06	6.120989164127D-07	5.75	1.156166352292D-01
13	4.0	-7.680468419916D-07	-3.840234209958D-07	6.25	-1.413386794895D-01
14	5.0	4.054339990932D-07	2.027169995466D-07	6.50	-2.286669655500D-01
15	6.0	-1.699693592647D-07	-8.498467963233D-08	6.75	-1.908051427439D-01
16	7.0	5.211226084738D-08	2.605613042369D-08	7.25	3.095018490452D-01
17	8.0	-1.023237974808D-08	-5.116189874040D-09	7.50	6.492112989409D-01
18	9.0	9.548860797104D-10	4.774430398552D-10	7.75	9.088214939153D-01
				8.25	8.923893831535D-01
	0	-2.834195589160D+04	-2.198974418910D+04	8.50	6.257048290100D-01
	1	-3.107545763746D+04	-6.982897769678D+03	8.75	2.925604108901D-01
	2	-1.459462522868D+04	-1.130672258254D+04	9.25	-1.728046099065D-01
	3	-3.801216222314D+03	-8.584665515664D+02	9.50	-2.021336896269D-01
	4	-5.922613826177D+02	-4.581883952008D+02	9.75	-1.216397257127D-01
	5	-5.514249601552D+01	-1.251073040763D+01	10.25	9.342097693634D-02
	6	-2.837843575985D+00	-2.192654242942D+00	10.50	1.183069175583D-01
	7	-6.222180545390D-02	-1.417340253828D-02	10.75	7.576146148740D-02

I= 9 H= 0.0

0		0.	4.987690487992D+06	-1.00	-1.316383549471D+02
1		1.689602462635D+01	-2.260778289355D+06	-.75	-6.365198854762D+01
2		-4.352076044225D+01	4.388654159518D+05	-.50	-2.625053403499D+01
3		4.308269148654D+01	-4.727727332016D+04	-.25	-7.707010201300D+00
4		-2.148619968370D+01	3.045184101218D+03	.25	2.098657620489D+00
5		5.793629643510D+00	-1.154617672661D+02	.50	1.778997061308D+00
6		-8.122173469614D-01	2.138180875298D+00	.75	8.052828371757D-01
7		4.683171781364D-02	-5.375855602349D-21	1.25	-3.861155648859D-01
				1.50	-4.036010667965D-01
0	-9.0	-1.303639683410D-09	-6.518198417049D-10	1.75	-2.189412217139D-01
1	-8.0	1.391366313296D-08	6.956831566479D-09	2.25	1.421435480224D-01
2	-7.0	-7.042338937776D-08	-3.521169468888D-08	2.50	1.691864112182D-01
3	-6.0	2.274654233107D-07	1.137327116554D-07	2.75	1.033353910263D-01
4	-5.0	-5.343228617155D-07	-2.671614308578D-07	3.25	-8.268903586229D-02
5	-4.0	9.887147809384D-07	4.943573904692D-07	3.50	-1.079512027957D-01
6	-3.0	-1.523206045344D-06	-7.616030226720D-07	3.75	-7.180586881099D-02
7	-2.0	2.029173225855D-06	1.014586612928D-06	4.25	6.685538390745D-02
8	-1.0	-2.392092233280D-06	-1.196046116640D-06	4.50	9.334254120780D-02
9	0.0	2.524162152327D-06	1.262081076164D-06	4.75	6.606649901216D-02
10	1.0	-2.392092233280D-06	-1.196046116640D-06	5.25	-6.874747028183D-02
11	2.0	2.029173225855D-06	1.014586612928D-06	5.50	-1.009390239073D-01
12	3.0	-1.523206045344D-06	-7.616030226720D-07	5.75	-7.493736870751D-02
13	4.0	9.887147809384D-07	4.943573904692D-07	6.25	8.542311135839D-02
14	5.0	-5.343228617155D-07	-2.671614308578D-07	6.50	1.312692595319D-01
15	6.0	2.274654233107D-07	1.137327116554D-07	6.75	1.021820347173D-01
16	7.0	-7.042338937776D-08	-3.521169468888D-08	7.25	-1.298272347493D-01
17	8.0	1.391366313296D-08	6.956831566479D-09	7.50	-2.134547836775D-01
18	9.0	-1.303639683410D-09	-6.518198417049D-10	7.75	-1.806910769680D-01
				8.25	3.003896606357D-01

CARDINAL SPLINES N=8

0		2.152030084721D+04	2.152030084721D+04	8.50	6.368608194727D-01
1		2.356321044740D+04	-2.057763811175D-15	8.75	9.003958518660D-01
2		1.105129697212D+04	1.105129697212D+04	9.25	9.003958518660D-01
3		2.874492115553D+03	-2.728038907323D-16	9.50	6.368608194727D-01
4		4.472943377311D+02	4.472943377311D+02	9.75	3.003896606357D-01
5		4.159464490859D+01	-4.332816100458D-18	10.25	-1.806910769680D-01
6		2.138180875298D+00	2.138180875298D+00	10.50	-2.134547836775D-01
7		4.683171781364D-02	-5.375855602349D-21	10.75	-1.298272347493D-01

I=10 H= 1.0

0		0.	-6.604479958635D+06	-1.00	9.475009822262D+01
1		-1.214759190350D+01	2.995173680591D+06	-.75	4.580400415159D+01
2		3.130644366817D+01	-5.817336017093D+05	-.50	1.888484753046D+01
3		-3.101427650032D+01	6.271054856294D+04	-.25	5.542851772757D+00
4		1.548194532879D+01	-4.046880534483D+03	.25	-1.508307161221D+00
5		-4.179305707589D+00	1.550230172441D+02	.50	-1.278049401722D+00
6		5.866601138738D-01	-3.085578602853D+00	.75	-5.782607040290D-01
7		-3.387500037734D-02	1.417340253828D-02	1.25	2.769609766000D-01
				1.50	2.893123580158D-01
0	-9.0	9.548860797104D-10	4.774430398552D-10	1.75	1.568246304445D-01
1	-8.0	-1.023237974808D-08	-5.116189874040D-09	2.25	-1.016250873420D-01
2	-7.0	5.211226084738D-08	2.605613042369D-08	2.50	-1.208184711008D-01
3	-6.0	-1.699693592647D-07	-8.498467963233D-08	2.75	-7.369315456669D-02
4	-5.0	4.054339990932D-07	2.027169995466D-07	3.25	5.876588266026D-02
5	-4.0	-7.680468419916D-07	-3.840234209958D-07	3.50	7.655088009626D-02
6	-3.0	1.224197832825D-06	6.120989164127D-07	3.75	5.078731583806D-02
7	-2.0	-1.706981660522D-06	-8.534908302610D-07	4.25	-4.696927007791D-02
8	-1.0	2.128329707604D-06	1.064164853802D-06	4.50	-6.529418996687D-02
9	0.0	-2.392092233280D-06	-1.196046116640D-06	4.75	-4.597618802622D-02
10	1.0	2.419673096851D-06	1.209836548426D-06	5.25	4.719740698018D-02
11	2.0	-2.183879430898D-06	-1.091939715449D-06	5.50	6.868485038429D-02
12	3.0	1.730714346712D-06	8.653571733561D-07	5.75	5.044763832941D-02
13	4.0	-1.173580915260D-06	-5.867904576302D-07	6.25	-5.586719161127D-02
14	5.0	6.552881799533D-07	3.276440899767D-07	6.50	-8.418606442950D-02
15	6.0	-2.854656778270D-07	-1.427328389135D-07	6.75	-6.395473160495D-02
16	7.0	8.977841537815D-08	4.488920768907D-08	7.25	7.576146148740D-02
17	8.0	-1.792580397949D-08	-8.962901989747D-09	7.50	1.183069175583D-01
18	9.0	1.691577427321D-09	8.457887136606D-10	7.75	9.342097693634D-02
				8.25	-1.216397257127D-01
0		-1.563753248659D+04	-2.198974418910D+04	8.50	-2.021336896269D-01
1		-1.710966209810D+04	6.982897769678D+03	8.75	-1.728046099065D-01
2		-8.018819936397D+03	-1.130672258254D+04	9.25	2.925604108901D-01
3		-2.084283119182D+03	8.584665515664D+02	9.50	6.257048290100D-01
4		-3.241154077839D+02	-4.581883952008D+02	9.75	8.923893831535D-01
5		-3.012103520026D+01	1.251073040763D+01	10.25	9.088214939153D-01
6		-1.547464909899D+00	-2.192654242942D+00	10.50	6.492112989409D-01
7		-3.387500037734D-02	1.417340253828D-02	10.75	3.095018490452D-01

I=11 H= 2.0

0	0.	8.278022066712D+06	-1.00	-6.544863942316D+01
1	8.386008858019D+00	-3.757881051466D+06	-.75	-3.163511057689D+01
2	-2.161819016734D+01	7.305911649977D+05	-.50	-1.304124198354D+01
3	2.142468801933D+01	-7.884049188210D+04	-.25	-3.827120309380D+00
4	-1.070015680494D+01	5.096366361626D+03	.25	1.041053334547D+00
5	2.890163521328D+00	-1.964215591591D+02	.50	8.819406511620D-01

6		-4.059727389658D-01	4.056158837112D+00	.75	3,989448095656D-01
7		2.345931323939D-02	-2.751965873059D-02	1.25	-1,909682637357D-01
				1.50	-1,994162706744D-01
0	-9.0	-6.656695461677D-10	-3.328347730838D-10	1.75	-1,080531475816D-01
1	-8.0	7.148634979681D-09	3.574317489840D-09	2.25	6,995227483542D-02
2	-7.0	-3.653141361638D-08	-1.826570680819D-08	2.50	8,311353172562D-02
3	-6.0	1.198149119506D-07	5.990745597529D-08	2.75	5,065937296501D-02
4	-5.0	-2.884461990747D-07	-1.442230995374D-07	3.25	-4,032566870209D-02
5	-4.0	5.547598416060D-07	2.773799208030D-07	3.50	-5,247035248056D-02
6	-3.0	-9.055061026928D-07	-4.527530513464D-07	3.75	-3,476472927776D-02
7	-2.0	1.307175934253D-06	6.535879671267D-07	4.25	3,204081398332D-02
8	-1.0	-1.706981660522D-06	-8.534908302610D-07	4.50	4,444335326131D-02
9	0.0	2.029173225855D-06	1.014586612928D-06	4.75	3,121268083407D-02
10	1.0	-2.183879430898D-06	-1.091939715449D-06	5.25	-3,182542742066D-02
11	2.0	2.098868558514D-06	1.049434279257D-06	5.50	-4,611244607585D-02
12	3.0	-1.763544344222D-06	-8.817721721111D-07	5.75	-3,369273148020D-02
13	4.0	1.257490074980D-06	6.287450374900D-07	6.25	3,680616421912D-02
14	5.0	-7.309657611191D-07	-3.654828805595D-07	6.50	5,496960416433D-02
15	6.0	3.282825078432D-07	1.641412539216D-07	6.75	4,131177310527D-02
16	7.0	-1.055657874370D-07	-5.278289371848D-08	7.25	-4,753861757035D-02
17	8.0	2.141689383440D-08	1.070844691720D-08	7.50	-7,279215232889D-02
18	9.0	-2.044214688518D-09	-1.022107344259D-09	7.75	-5,609468147655D-02
				8.25	6,809356121136D-02
	0	1.085514483247D+04	2.300487207505D+04	8.50	1,074678242085D-01
	1	1.187259548012D+04	-1.338733222428D+04	8.75	8,570144705813D-02
	2	5.562285881239D+03	1.188046671943D+04	9.25	-1,136364304631D-01
	3	1.445248356923D+03	-1.653446769981D+03	9.50	-1,904900656382D-01
	4	2.246647011144D+02	4.834650940441D+02	9.75	-1,642800000872D-01
	5	2.087192743737D+01	-2.419992145585D+01	10.25	2,832548217901D-01
	6	1.071963995116D+00	2.322420337084D+00	10.50	6,118376533652D-01
	7	2.345931323939D-02	-2.751965873059D-02	10.75	8,820014081710D-01

I=12 H= 3.0

0		0.	-9.483699750668D+06	-1.00	4,253006228887D+01
1		-5.447744870499D+00	4.313290186583D+06	-.75	2,055587332217D+01
2		1.404571468252D+01	-8.401248040402D+05	-.50	8,473325350138D+00
3		-1.392279930078D+01	9.082972279610D+04	-.25	2,486406665833D+00
4		6.955254731524D+00	-5.884135097248D+03	.25	-6,762254665362D-01
5		-1.879222055581D+00	2.278033518294D+02	.50	-5,728092383689D-01
6		2.640617301693D-01	-4.800058575978D+00	.75	-2,590775399996D-01
7		-1.526491779468D-02	3.789296821605D-02	1.25	1,239792735055D-01
				1.50	1,294405877412D-01
0	-9.0	4.346583866013D-10	2.173291933007D-10	1.75	7,012261334999D-02
1	-8.0	-4.673174740309D-09	-2.336587370155D-09	2.25	-4,537342658028D-02
2	-7.0	2.392488583940D-08	1.196244291970D-08	2.50	-5,389322479456D-02
3	-6.0	-7.870677821091D-08	-3.935338910545D-08	2.75	-3,283691840155D-02
4	-5.0	1.904674750976D-07	9.523373754879D-08	3.25	2,611431365811D-02
5	-4.0	-3.696156424266D-07	-1.848078212133D-07	3.50	3,395893323150D-02
6	-3.0	6.124638215408D-07	3.062319107704D-07	3.75	2,248415190725D-02
7	-2.0	-9.055061026928D-07	-4.527530513464D-07	4.25	-2,068542376599D-02
8	-1.0	1.224197832825D-06	6.120989164127D-07	4.50	-2,865967952019D-02
9	0.0	-1.523206045344D-06	-7.616030226720D-07	4.75	-2,010060755729D-02
10	1.0	1.730714346712D-06	8.653571733561D-07	5.25	2,042394964078D-02
11	2.0	-1.763544344222D-06	-8.817721721111D-07	5.50	2,952664395955D-02
12	3.0	1.569586084350D-06	7.847930421752D-07	5.75	2,151713334808D-02
13	4.0	-1.179055553101D-06	-5.895277765506D-07	6.25	-2,334513499696D-02

14	5.0	7.161763355480D-07	3.580881677740D-07	6.50	-3.471229330026D-02
15	6.0	-3.331503967831D-07	-1.665751983916D-07	6.75	-2.595112081823D-02
16	7.0	1.100798635392D-07	5.503993176962D-08	7.25	2.945532399045D-02
17	8.0	-2.279731314662D-08	-1.139865657331D-08	7.50	4.469964842781D-02
18	9.0	2.210046828233D-09	1.105023414116D-09	7.75	3.407563892769D-02
				8.25	-4.017864349500D-02
	0	-7.072203027923D+03	-2.328656857716D+04	8.50	-6.217703025745D-02
	1	-7.733557462361D+03	1.795055183112D+04	8.75	-4.838709368157D-02
	2	-3.622450425516D+03	-1.213489624320D+04	9.25	5.981216382850D-02
	3	-9.410428659002D+02	2.238271995467D+03	9.50	9.522342795856D-02
	4	-1.462591131452D+02	-4.982163507081D+02	9.75	7.660215445633D-02
	5	-1.358551379519D+01	3.305612766204D+01	10.25	-1.034404597806D-01
	6	-6.976280908956D-01	-2.412801578367D+00	10.50	-1.751241574289D-01
	7	-1.526491779468D-02	3.789296821605D-02	10.75	-1.526524758098D-01

I=13 H= 4.0

0		0.	9.380951874233D+06	-1.00	-2.509530173718D+01
1		3.213975143765D+00	-4.279672528617D+06	-.75	-1.212878046735D+01
2		-8.287105245813D+00	8.361297304642D+05	-.50	-4.999405846124D+00
3		8.215457284040D+00	-9.067466053450D+04	-.25	-1.466960009998D+00
4		-4.104655085216D+00	5.892958166614D+03	.25	3.989281179916D-01
5		1.109204831579D+00	-2.291529492406D+02	.50	3.378997582706D-01
6		-1.558905374321D-01	4.889331708648D+00	.75	1.528196867847D-01
7		9.013609334502D-03	-4.147394493798D-02	1.25	-7.311905327838D-02
				1.50	-7.633273269571D-02
0	-9.0	-2.571285432822D-10	-1.285642716411D-10	1.75	-4.134770187889D-02
1	-8.0	2.766173999374D-09	1.383086999687D-09	2.25	2.674720184465D-02
2	-7.0	-1.417558687440D-08	-7.087793437201D-09	2.50	3.176424630520D-02
3	-6.0	4.671019887544D-08	2.335509943772D-08	2.75	1.935006859500D-02
4	-5.0	-1.133576542879D-07	-5.667882714396D-08	3.25	-1.538101571483D-02
5	-4.0	2.210875569124D-07	1.105437784562D-07	3.50	-1.999516910023D-02
6	-3.0	-3.696156424266D-07	-1.848078212133D-07	3.75	-1.323391523648D-02
7	-2.0	5.547598416060D-07	2.773799208030D-07	4.25	1.216375679681D-02
8	-1.0	-7.680468419916D-07	-3.840234209958D-07	4.50	1.684278579927D-02
9	0.0	9.887147809384D-07	4.943573904692D-07	4.75	1.180440665187D-02
10	1.0	-1.173580915260D-06	-5.867904576302D-07	5.25	-1.197245517646D-02
11	2.0	1.257490074980D-06	6.287450374900D-07	5.50	-1.728828330995D-02
12	3.0	-1.179055553101D-06	-5.895277765506D-07	5.75	-1.258129283790D-02
13	4.0	9.304237954794D-07	4.652118977397D-07	6.25	1.360192487076D-02
14	5.0	-5.901299255526D-07	-2.950649627763D-07	6.50	2.017921052365D-02
15	6.0	2.845699656889D-07	1.422849828444D-07	6.75	1.504578385457D-02
16	7.0	-9.678641440501D-08	-4.839320720251D-08	7.25	-1.695972505260D-02
17	8.0	2.050729812830D-08	1.025364906415D-08	7.50	-2.562282951689D-02
18	9.0	-2.024024165258D-09	-1.012012082629D-09	7.75	-1.942991256959D-02
				8.25	2.259534260126D-02
	0	4.178723618827D+03	2.104852570620D+04	8.50	3.465256380339D-02
	1	4.569021311588D+03	-1.880217115416D+04	8.75	2.667571397976D-02
	2	2.139940820292D+03	1.112125068252D+04	9.25	-3.202549511387D-02
	3	5.558599315637D+02	-2.379468454993D+03	9.50	-4.999041482722D-02
	4	8.638480152568D+01	4.634057717034D+02	9.75	-3.924157086388D-02
	5	8.023265288235D+00	-3.567621731309D+01	10.25	4.939301648038D-02
	6	4.119668506416D-01	2.276473177555D+00	10.50	7.941208575238D-02
	7	9.013609334502D-03	-4.147394493798D-02	10.75	6.456423874353D-02

I=14 H= 5.0

0		0.	-7.476380028568D+06	-1.00	1.277217305082D+01
1		-1.635600087414D+00	3.425069739427D+06	-.75	6.172788309389D+00
2		4.217501456798D+00	-6.720015416014D+05	-.50	2.544331652235D+00
3		-4.181275675363D+00	7.318822378206D+04	-.25	7.465583694927D-01
4		2.089223806215D+00	-4.777464593174D+03	.25	-2.030098757905D-01
5		-5.646215770274D-01	1.867183708902D+02	.50	-1.719479086108D-01
6		7.936126233009D-02	-4.021379673559D+00	.75	-7.776303128317D-02
7		-4.589185670033D-03	3.545711211305D-02	1.25	3.720386447761D-02
				1.50	3.883705731567D-02
0	-9.0	1.310430472455D-10	6.552152362277D-11	1.75	2.103593080665D-02
1	-8.0	-1.410215169406D-09	-7.051075847028D-10	2.25	-1.360587392713D-02
2	-7.0	7.230603140041D-09	3.615301570021D-09	2.50	-1.615652481293D-02
3	-6.0	-2.384663663403D-08	-1.192331831701D-08	2.75	-9.841173107752D-03
4	-5.0	5.796081035095D-08	2.898040517548D-08	3.25	7.820511898521D-03
5	-4.0	-1.133576542879D-07	-5.667882714396D-08	3.50	1.016489587284D-02
6	-3.0	1.904674750976D-07	9.523373754879D-08	3.75	6.726376402248D-03
7	-2.0	-2.884461990747D-07	-1.442230995374D-07	4.25	-6.179344899863D-03
8	-1.0	4.054339990932D-07	2.027169995466D-07	4.50	-8.553611584864D-03
9	0.0	-5.343228617155D-07	-2.671614308578D-07	4.75	-5.992605864633D-03
10	1.0	6.552881799533D-07	3.276440899767D-07	5.25	6.072019725096D-03
11	2.0	-7.309657611191D-07	-3.654828805595D-07	5.50	8.762597191527D-03
12	3.0	7.161763355480D-07	3.580881677740D-07	5.75	6.372196720179D-03
13	4.0	-5.901299255526D-07	-2.950649627763D-07	6.25	-6.876226014814D-03
14	5.0	3.892532486623D-07	1.946266243312D-07	6.50	-1.018907873978D-02
15	6.0	-1.941068122210D-07	-9.705340611050D-08	6.75	-7.586367278140D-03
16	7.0	6.787520613531D-08	3.393760306766D-08	7.25	8.520527679821D-03
17	8.0	-1.470927128778D-08	-7.354635643892D-09	7.50	1.284315101347D-02
18	9.0	1.478436034035D-09	7.392180170175D-10	7.75	9.712497947242D-03
				8.25	-1.121561757861D-02
	0	-2.128299040487D+03	-1.560998646477D+04	8.50	-1.712288964228D-02
	1	-2.326955654139D+03	1.514262476546D+04	8.75	-1.311075017554D-02
	2	-1.089790153497D+03	-8.386601275571D+03	9.25	1.552114337414D-02
	3	-2.830631426092D+02	1.951857148471D+03	9.50	2.400760696396D-02
	4	-4.398788579986D+01	-3.564259909228D+02	9.75	1.863959735708D-02
	5	-4.085318235930D+00	2.987641622236D+01	10.25	-2.277977350908D-02
	6	-2.097574348820D-01	-1.787581610437D+00	10.50	-3.590312864852D-02
	7	-4.589185670033D-03	3.545711211305D-02	10.75	-2.847844545123D-02

I=15 H= 6.0

0		0.	4.463109763064D+06	-1.00	-5.241460754760D+00
1		6.711877111622D-01	-2.054671456687D+06	-.75	-2.533170839064D+00
2		-1.730740051280D+00	4.051684841425D+05	-.50	-1.044123650689D+00
3		1.715927410103D+00	-4.435685307614D+04	-.25	-3.063632003395D-01
4		-8.574171690056D-01	2.911020689360D+03	.25	8.330621743484D-02
5		2.317315496332D-01	-1.144363113577D+02	.50	7.055856776517D-02
6		-3.257315706766D-02	2.485069919493D+00	.75	3.190933482433D-02
7		1.883706508172D-03	-2.243960518394D-02	1.25	-1.526558873160D-02
				1.50	-1.593525172471D-02
0	-9.0	-5.381772201308D-11	-2.690886100654D-11	1.75	-8.630989284074D-03
1	-8.0	5.792614929858D-10	2.896307464929D-10	2.25	5.582017505077D-03
2	-7.0	-2.970904464621D-09	-1.485452232311D-09	2.50	6.628137684114D-03
3	-6.0	9.802824955625D-09	4.901412477812D-09	2.75	4.037065403480D-03
4	-5.0	-2.384663663403D-08	-1.192331831701D-08	3.25	-3.207684009284D-03
5	-4.0	4.671019887544D-08	2.335509943772D-08	3.50	-4.168883401539D-03
6	-3.0	-7.870677821091D-08	-3.935338910545D-08	3.75	-2.758363115487D-03
7	-2.0	1.198149119506D-07	5.990745597529D-08	4.25	2.533339219608D-03

CARDINAL SPLINES N=8

8	-1.0	-1.699693592647D-07	-8.498467963233D-08	4.50	3.506100246216D-03
9	0.0	2.274654233107D-07	1.137327116554D-07	4.75	2.455844078249D-03
10	1.0	-2.854656778270D-07	-1.427328389135D-07	5.25	-2.487068613659D-03
11	2.0	3.282825078432D-07	1.641412539216D-07	5.50	-3.587901357510D-03
12	3.0	-3.331503967831D-07	-1.665751983916D-07	5.75	-2.608096792311D-03
13	4.0	2.845699656889D-07	1.422849828444D-07	6.25	2.811517170587D-03
14	5.0	-1.941068122210D-07	-9.705340611050D-08	6.50	4.163350305853D-03
15	6.0	9.968144739566D-08	4.984072369783D-08	6.75	3.097486080357D-03
16	7.0	-3.573284194171D-08	-1.786642097086D-08	7.25	-3.472084295138D-03
17	8.0	7.904798684633D-09	3.952399342316D-09	7.50	-5.227012852686D-03
18	9.0	-8.081151285168D-10	-4.040575642584D-10	7.75	-3.947074301211D-03
				8.25	4.540807062079D-03
	0	8.737627241488D+02	8.765076573357D+03	8.50	6.915847527350D-03
	1	9.552912310773D+02	-8.932825011575D+03	8.75	5.280412059601D-03
	2	4.473809975193D+02	4.791191268324D+03	9.25	-6.205856437126D-03
	3	1.161997784864D+02	-1.174119769128D+03	9.50	-9.554320548216D-03
	4	1.805688828329D+01	2.082001041790D+02	9.75	-7.377069960945D-03
	5	1.676965838380D+00	-1.841230412295D+01	10.25	8.886589747561D-03
	6	8.610035294719D-02	1.071374792905D+00	10.50	1.387538258102D-02
	7	1.883706508172D-03	-2.243960518394D-02	10.75	1.088266797812D-02

I=16 H= 7.0

0		0.	-1.836072567783D+06	-1.00	1.586979766847D+00
1		-2.032129192292D-01	8.496735256500D+05	-.75	7.669746448993D-01
2		5.240161659463D-01	-1.684657587921D+05	-.50	3.161299832435D-01
3		-5.195405130796D-01	1.854861936640D+04	-.25	9.275712886106D-02
4		2.596105801388D-01	-1.224582140130D+03	.25	-2.522208447541D-02
5		-7.016604601753D-02	4.844937129063D+01	.50	-2.136235471143D-02
6		9.863137330220D-03	-1.060537166869D+00	.75	-9.660784882611D-03
7		-5.704051052826D-04	9.739827146772D-03	1.25	4.621639429820D-03
				1.50	4.824312710464D-03
0	-9.0	1.630152130235D-11	8.150760651176D-12	1.75	2.612939085426D-03
1	-8.0	-1.754776264496D-10	-8.773881322478D-11	2.25	-1.689820412578D-03
2	-7.0	9.001331017378D-10	4.500665508689D-10	2.50	-2.006452378436D-03
3	-6.0	-2.970904464621D-09	-1.485452232311D-09	2.75	-1.222050426466D-03
4	-5.0	7.230603140041D-09	3.615301570021D-09	3.25	9.709110383582D-04
5	-4.0	-1.417558687440D-08	-7.087793437201D-09	3.50	1.261784367499D-03
6	-3.0	2.392488583940D-08	1.196244291970D-08	3.75	8.348153302138D-04
7	-2.0	-3.653141361638D-08	-1.826570680819D-08	4.25	-7.665925437055D-04
8	-1.0	5.211226084738D-08	2.605613042369D-08	4.50	-1.060846111736D-03
9	0.0	-7.042338937776D-08	-3.521169468888D-08	4.75	-7.429813991106D-04
10	1.0	8.977841537815D-08	4.488920768907D-08	5.25	7.522014793955D-04
11	2.0	-1.055657874370D-07	-5.278289371848D-08	5.50	1.084934663436D-03
12	3.0	1.100798635392D-07	5.503993176962D-08	5.75	7.884763269036D-04
13	4.0	-9.678641440501D-08	-4.839320720251D-08	6.25	-8.494820114254D-04
14	5.0	6.787520613531D-08	3.393760306766D-08	6.50	-1.257466547231D-03
15	6.0	-3.573284194171D-08	-1.786642097086D-08	6.75	-9.351358963603D-04
16	7.0	1.308566000452D-08	6.542830002262D-09	7.25	1.047067007607D-03
17	8.0	-2.947408254285D-09	-1.473704127142D-09	7.50	1.575187533054D-03
18	9.0	3.058944904882D-10	1.529472452441D-10	7.75	1.188488586584D-03
				8.25	-1.364380405485D-03
	0	-2.646128459640D+02	-3.417918230785D+03	8.50	-2.075227169287D-03
	1	-2.892981941534D+02	3.595527348818D+03	8.75	-1.581990781084D-03
	2	-1.354815231524D+02	-1.899136454402D+03	9.25	1.851776357469D-03
	3	-3.518849097924D+01	4.816263810115D+02	9.50	2.843640589157D-03
	4	-5.468035895718D+00	-8.440140014845D+01	9.75	2.189021671856D-03

	5	-5.078157142713D-01	7.747810256340D+00	10.25	-2.616624729077D-03
	6	-2.607238430258D-02	-4.469280566228D-01	10.50	-4.065408102324D-03
	7	-5.704051052826D-04	9.739827146772D-03	10.75	-3.169964941050D-03
I=17	H= 8.0				
0		0.	4.590022126184D+05	-1.00	-3.092493949314D-01
1		3.959878103487D-02	-2.135174642488D+05	-.75	-1.494572465859D-01
2		-1.021123960495D-01	4.256907074895D+04	-.50	-6.160273762083D-02
3		1.012413090681D-01	-4.714620642424D+03	-.25	-1.807506249460D-02
4		-5.059021005536D-02	3.132125508826D+02	.25	4.914838451179D-03
5		1.367344667798D-02	-1.247585577180D+01	.50	4.162697984170D-03
6		-1.922091359221D-03	2.752704923358D-01	.75	1.882501795454D-03
7		1.111606863080D-04	-2.562643251544D-03	1.25	-9.005593397338D-04
				1.50	-9.400428898373D-04
0	-9.0	-3.177420664538D-12	-1.588710332269D-12	1.75	-5.091395840045D-04
1	-8.0	3.420539587296D-11	1.710269793648D-11	2.25	3.292582314074D-04
2	-7.0	-1.754776264496D-10	-8.773881322478D-11	2.50	3.909469301437D-04
3	-6.0	5.792614929858D-10	2.896307464929D-10	2.75	2.381056872740D-04
4	-5.0	-1.410215169406D-09	-7.051075847028D-10	3.25	-1.891642283896D-04
5	-4.0	2.766173999374D-09	1.383086999687D-09	3.50	-2.458280172960D-04
6	-3.0	-4.673174740309D-09	-2.336587370155D-09	3.75	-1.626376041591D-04
7	-2.0	7.148634979681D-09	3.574317489840D-09	4.25	1.493327036534D-04
8	-1.0	-1.023237974808D-08	-5.116189874040D-09	4.50	2.066412823195D-04
9	0.0	1.391366313296D-08	6.956831566479D-09	4.75	1.447146232141D-04
10	1.0	-1.792580397949D-08	-8.962901989747D-09	5.25	-1.464842922010D-04
11	2.0	2.141689383440D-08	1.070844691720D-08	5.50	-2.112569493612D-04
12	3.0	-2.279731314662D-08	-1.139865657331D-08	5.75	-1.535104131113D-04
13	4.0	2.050729812830D-08	1.025364906415D-08	6.25	1.653309494340D-04
14	5.0	-1.470927128778D-08	-7.354635643892D-09	6.50	2.446817345267D-04
15	6.0	7.904798684633D-09	3.952399342316D-09	6.75	1.819149172397D-04
16	7.0	-2.947408254285D-09	-1.473704127142D-09	7.25	-2.035558267500D-04
17	8.0	6.741820809029D-10	3.370910404514D-10	7.50	-3.060976353310D-04
18	9.0	-7.089035601913D-11	-3.544517800956D-11	7.75	-2.308396533527D-04
				8.25	2.646727642815D-04
	0	5.157114656885D+01	8.150380765472D+02	8.50	4.022502873373D-04
	1	5.638156752698D+01	-8.762291309903D+02	8.75	3.063601521259D-04
	2	2.640384070843D+01	4.596181106745D+02	9.25	-3.577598634400D-04
	3	6.857777739131D+00	-1.194407906231D+02	9.50	-5.485661280856D-04
	4	1.065638800148D+00	2.086684677626D+01	9.75	-4.215447134555D-04
	5	9.896484068992D-02	-1.970305356549D+00	10.25	5.016456695747D-04
	6	5.081031878182D-03	1.138239674885D-01	10.50	7.771999243248D-04
	7	1.111606863080D-04	-2.562643251544D-03	10.75	6.040085385033D-04
I=18	H= 9.0				
0		0.	-5.222983397977D+04	-1.00	2.872132812549D-02
1		-3.677673185758D-03	2.441828417548D+04	-.75	1.388071069102D-02
2		9.483569423898D-03	-4.894688264178D+03	-.50	5.721286706241D-03
3		-9.402729685621D-03	5.452713099591D+02	-.25	1.678697236435D-03
4		4.698576200371D-03	-3.645405231940D+01	.25	-4.564562646758D-04
5		-1.269936768202D-03	1.462078869128D+00	.50	-3.866012680483D-04
6		1.785184383265D-04	-3.251774016358D-02	.75	-1.748324440038D-04
7		-1.032442330994D-05	3.063573944462D-04	1.25	8.363629358458D-05
				1.50	8.730268141489D-05
0	-9.0	2.951470280062D-13	1.475735140031D-13	1.75	4.728396463640D-05
1	-8.0	-3.177420664538D-12	-1.588710332269D-12	2.25	-3.057781839436D-05

CARDINAL SPLINES N=8

2	-7.0	1.630152130235D-11	8.150760651176D-12	2.50	-3.630640196696D-05
3	-6.0	-5.381772201308D-11	-2.690886100654D-11	2.75	-2.211210010949D-05
4	-5.0	1.310430472455D-10	6.552152362277D-11	3.25	1.756653472680D-05
5	-4.0	-2.571285432822D-10	-1.285642716411D-10	3.50	2.282811993807D-05
6	-3.0	4.346583866013D-10	2.173291933007D-10	3.75	1.510253862043D-05
7	-2.0	-6.656695461677D-10	-3.328347730838D-10	4.25	-1.386624396852D-05
8	-1.0	9.548860797104D-10	4.774430398552D-10	4.50	-1.918690873789D-05
9	0.0	-1.303639683410D-09	-6.518198417049D-10	4.75	-1.343635575502D-05
10	1.0	1.691577427321D-09	8.457887136606D-10	5.25	1.359914936992D-05
11	2.0	-2.044214688518D-09	-1.022107344259D-09	5.50	1.961105041642D-05
12	3.0	2.210046828233D-09	1.105023414116D-09	5.75	1.424922958078D-05
13	4.0	-2.024024165258D-09	-1.012012082629D-09	6.25	-1.534315562469D-05
14	5.0	1.478436034035D-09	7.392180170175D-10	6.50	-2.270403023374D-05
15	6.0	-8.081151285168D-10	-4.040575642584D-10	6.75	-1.687719316880D-05
16	7.0	3.058944904882D-10	1.529472452441D-10	7.25	1.887719064411D-05
17	8.0	-7.089035601913D-11	-3.544517800956D-11	7.50	2.837930277525D-05
18	9.0	7.538291884714D-12	3.769145942357D-12	7.75	2.139538067025D-05
				8.25	-2.451218576713D-05
	0	-4.790036902696D+00	-8.894539293661D+01	8.50	-3.723540096114D-05
	1	-5.236805567077D+00	9.714040561163D+01	8.75	-2.834276496718D-05
	2	-2.452413343591D+00	-5.081871128991D+01	9.25	3.304962134837D-05
	3	-6.369528846920D-01	1.345103914292D+01	9.50	5.062941763400D-05
	4	-9.897633655497D-02	-2.352848588105D+00	9.75	3.886408952873D-05
	5	-9.191785148768D-03	2.272348282473D-01	10.25	-4.612239402007D-05
	6	-4.719202301994D-04	-1.321722431347D-02	10.50	-7.133420830790D-05
	7	-1.032442330994D-05	3.063573944462D-04	10.75	-5.532638017974D-05

N= 8 M=19 P= 9.5 Q=15

I= 0 H= -9.5

0		1.000000000000D+00	1.256688199681D+03	-1.00	9.405764138310D+00
1		-2.779765155789D+00	-5.357488693374D+02	-.75	5.801049053872D+00
2		3.079072817769D+00	9.889820659050D+01	-.50	3.426941587738D+00
3		-1.800992285258D+00	-1.074868427348D+01	-.25	1.918030611790D+00
4		6.105165470674D-01	9.025348635825D-01	.25	4.716296415337D-01
5		-1.215013712966D-01	-8.246801422173D-02	.50	1.893247865402D-01
6		1.329270431436D-02	7.084681418295D-03	.75	5.398200012479D-02
7		-6.232568147545D-04	-3.154165774638D-04	1.25	-1.375634653023D-02
				1.50	-1.123058880240D-02
0	-9.5	7.544515466346D-12	3.772257733173D-12	1.75	-4.879532276968D-03
1	-8.5	-7.095736680796D-11	-3.547868340398D-11	2.25	2.131066286151D-03
2	-7.5	3.062383707601D-10	1.531191853801D-10	2.50	2.112457394460D-03
3	-6.5	-8.092508928079D-10	-4.046254464040D-10	2.75	1.081632988473D-03
4	-5.5	1.481203617110D-09	7.406018085548D-10	3.25	-6.161042907037D-04
5	-4.5	-2.029462028655D-09	-1.014731014328D-09	3.50	-6.812583428682D-04
6	-3.5	2.219262603816D-09	1.109631301908D-09	3.75	-3.843315193227D-04
7	-2.5	-2.058396656300D-09	-1.029198328150D-09	4.25	2.577132727488D-04
8	-1.5	1.712108684383D-09	8.560543421916D-10	4.50	3.051837230168D-04
9	-.5	-1.332164201646D-09	-6.660821008231D-10	4.75	1.830096475177D-04

10	.5	9.931476048629D-10	4.965738024314D-10	5.25	-1.359720283890D-04
11	1.5	-7.148435814783D-10	-3.574217907392D-10	5.50	-1.680516872444D-04
12	2.5	4.937782723793D-10	2.468891361897D-10	5.75	-1.046588760416D-04
13	3.5	-3.208687127230D-10	-1.604343563615D-10	6.25	8.279433888215D-05
14	4.5	1.893318920888D-10	9.466594604438D-11	6.50	1.050050102274D-04
15	5.5	-9.635985712074D-11	-4.817992856037D-11	6.75	6.689422255726D-05
16	6.5	3.954428173053D-11	1.977214086527D-11	7.25	-5.493101777602D-05
17	7.5	-1.197299339575D-11	-5.986496697877D-12	7.50	-7.074047326130D-05
18	8.5	2.333136843018D-12	1.166568421509D-12	7.75	-4.567533687356D-05
19	9.5	-2.166885042567D-13	-1.083442521284D-13	8.25	3.835354994889D-05
				8.50	4.985545286243D-05
	0	-3.015719154567D+02	-1.478106969674D+02	8.75	3.246189589454D-05
	1	-3.219833482700D+02	-1.640078669354D+02	9.25	-2.766156086457D-05
	2	-1.480831768022D+02	-7.273221286559D+01	9.50	-3.619580924699D-05
	3	-3.799600370177D+01	-1.931323846649D+01	9.75	-2.371770900108D-05
	4	-5.868568528207D+00	-2.888868425538D+00	10.25	2.046303190668D-05
	5	-5.450447035413D-01	-2.764344418171D-01	10.50	2.694738795475D-05
	6	-2.815387386681D-02	-1.389052098305D-02	10.75	1.777562575125D-05
	7	-6.232568147545D-04	-3.154165774638D-04	11.25	-1.556533683122D-05

I= 1		H= -8.5			
0		0.	-1.352637895788D+04	-1.00	-4.059737667179D+01
1		8.671178862671D+00	5.758211092606D+03	-.75	-2.148331908698D+01
2		-1.538824449232D+01	-1.056034874293D+03	-.50	-9.881553135215D+00
3		1.119382306257D+01	1.116405579388D+02	-.25	-3.322205585056D+00
4		-4.303074063672D+00	-8.582918647923D+00	.25	1.365004830744D+00
5		9.284476205270D-01	6.980728804034D-01	.50	1.646191325476D+00
6		-1.073697798712D-01	-5.840476121922D-02	.75	1.410285949113D+00
7		5.238790168010D-03	2.660185637276D-03	1.25	6.010198848040D-01
				1.50	2.949584049575D-01
0	-9.5	-7.095736680796D-11	-3.547868340398D-11	1.75	9.987846043182D-02
1	-8.5	6.749036020599D-10	3.374518010299D-10	2.25	-3.380112580237D-02
2	-7.5	-2.951110895248D-09	-1.475555447624D-09	2.50	-3.110042569754D-02
3	-6.5	7.917027729831D-09	3.958513864915D-09	2.75	-1.505345863054D-02
4	-5.5	-1.473907051529D-08	-7.369535257647D-09	3.25	7.926269238656D-03
5	-4.5	2.056584890137D-08	1.028292445068D-08	3.50	8.520565215029D-03
6	-3.5	-2.289654160188D-08	-1.144827080094D-08	3.75	4.696797802500D-03
7	-2.5	2.156959446637D-08	1.078479723318D-08	4.25	-3.040308574423D-03
8	-1.5	-1.814686892454D-08	-9.073434462269D-09	4.50	-3.552040540554D-03
9	-.5	1.422079342884D-08	7.110396714419D-09	4.75	-2.105884383195D-03
10	.5	-1.064435072139D-08	-5.322175360695D-09	5.25	1.536843748870D-03
11	1.5	7.678103554383D-09	3.839051777191D-09	5.50	1.886140448873D-03
12	2.5	-5.309732660799D-09	-2.654866330399D-09	5.75	1.167625804931D-03
13	3.5	3.452479616798D-09	1.726239808399D-09	6.25	-9.149520334729D-04
14	4.5	-2.037825073804D-09	-1.018912536902D-09	6.50	-1.156052939252D-03
15	5.5	1.037322815347D-09	5.186614076736D-10	6.75	-7.341092442310D-04
16	6.5	-4.257379753981D-10	-2.128689876991D-10	7.25	5.997515771402D-04
17	7.5	1.289094267353D-10	6.445471336766D-11	7.50	7.708070373697D-04
18	8.5	-2.512094341582D-11	-1.256047170791D-11	7.75	4.968320985777D-04
19	9.5	2.333136843018D-12	1.166568421509D-12	8.25	-4.160476119474D-04
				8.50	-5.402326646656D-04
	0	2.741362140169D+03	1.338644448423D+03	8.75	-3.514316832044D-04
	1	2.909480939752D+03	1.487217934341D+03	9.25	2.990254649681D-04
	2	1.328514910764D+03	6.501584473526D+02	9.50	3.910567552286D-04
	3	3.379372528257D+02	1.723630205777D+02	9.75	2.561184433212D-04
	4	5.165261862110D+01	2.533728079730D+01	10.25	-2.208008316890D-04

5		4.737187233788D+00	2.410718319955D+00	10.50	-2.906798666993D-04
6		2.410097663014D-01	1.184975836596D-01	10.75	-1.916947705599D-04
7		5.238790168010D-03	2.660185637276D-03	11.25	1.677898469483D-04

I= 2 H= -7.5

0		0.	6.941928901835D+04	-1.00	1.072969098429D+02
1		-1.728599950701D+01	-2.953840023444D+04	-.75	5.426798787915D+01
2		3.920750527448D+01	5.399856441289D+03	-.50	2.356718504860D+01
3		-3.309264865002D+01	-5.607802737070D+02	-.25	7.347177704872D+00
4		1.404537174765D+01	4.002968263420D+01	.25	-2.336316925513D+00
5		-3.249744681655D+00	-2.825029480969D+00	.50	-2.195399133561D+00
6		3.955778991545D-01	2.220122511913D-01	.75	-1.130648901690D+00
7		-2.006208289739D-02	-1.024038687334D-02	1.25	8.065193334273D-01
				1.50	1.190244356532D+00
0	-9.5	3.062383707601D-10	1.531191853801D-10	1.75	1.212181872689D+00
1	-8.5	-2.951110895248D-09	-1.475555447624D-09	2.25	6.875406722105D-01
2	-7.5	1.310466090279D-08	6.552330451395D-09	2.50	3.807246727468D-01
3	-6.5	-3.579559791672D-08	-1.789779895836D-08	2.75	1.437936001565D-01
4	-5.5	6.802812727890D-08	3.401406363945D-08	3.25	-5.881626501580D-02
5	-4.5	-9.708688028697D-08	-4.854344014349D-08	3.50	-5.876152303159D-02
6	-3.5	1.105890756942D-07	5.529453784708D-08	3.75	-3.065882092512D-02
7	-2.5	-1.063494035645D-07	-5.317470178227D-08	4.25	1.839105650329D-02
8	-1.5	9.091285767842D-08	4.545642883921D-08	4.50	2.091341173796D-02
9	-.5	-7.199949453217D-08	-3.599974726609D-08	4.75	1.212874127673D-02
10	.5	5.422637848151D-08	2.711318924075D-08	5.25	-8.562673554432D-03
11	1.5	-3.924849545714D-08	-1.962424772857D-08	5.50	-1.037775196770D-02
12	2.5	2.719151968510D-08	1.359575984255D-08	5.75	-6.357123113646D-03
13	3.5	-1.769751172398D-08	-8.848755861989D-09	6.25	4.900617580659D-03
14	4.5	1.045131831710D-08	5.225659158551D-09	6.50	6.152825009972D-03
15	5.5	-5.321544635555D-09	-2.660772317777D-09	6.75	3.886145424050D-03
16	6.5	2.184397894927D-09	1.092198947463D-09	7.25	-3.148119428457D-03
17	7.5	-6.614717246780D-10	-3.307358623390D-10	7.50	-4.032601628275D-03
18	8.5	1.289094267353D-10	6.445471336766D-11	7.75	-2.591922854462D-03
19	9.5	-1.197299339575D-11	-5.986496697877D-12	8.25	2.160812113446D-03
				8.50	2.800875629704D-03
	0	-1.137206866772D+04	-5.521620996933D+03	8.75	1.819301939864D-03
	1	-1.198838184953D+04	-6.160867510492D+03	9.25	-1.544351256783D-03
	2	-5.430442317101D+03	-2.642863725561D+03	9.50	-2.017785152367D-03
	3	-1.368333239461D+03	-7.015870516413D+02	9.75	-1.320482068074D-03
	4	-2.068294109958D+02	-1.009049420050D+02	10.25	1.136972214606D-03
	5	-1.872446704112D+01	-9.578424384752D+00	10.50	1.496067938095D-03
	6	-9.385506135218D-01	-4.589734758855D-01	10.75	9.862006998510D-04
	7	-2.006208289739D-02	-1.024038687334D-02	11.25	-8.626500608980D-04

I= 3 H= -6.5

0		0.	-2.293094045874D+05	-1.00	-1.964389797460D+02
1		2.833400242331D+01	9.755870082612D+04	-.75	-9.730425527319D+01
2		-6.875679915933D+01	-1.780681864636D+04	-.50	-4.123067931469D+01
3		6.284756360595D+01	1.829182150583D+03	-.25	-1.248145284594D+01
4		-2.857258957927D+01	-1.233180415286D+02	.25	3.663194975240D+00
5		6.991121216203D+00	7.460041331926D+00	.50	3.252890377181D+00
6		-8.901011340155D-01	-5.250676544883D-01	.75	1.554926618629D+00
7		4.680262796573D-02	2.409284879611D-02	1.25	-8.602320411916D-01
				1.50	-9.903171207375D-01
0	-9.5	-8.092508928079D-10	-4.046254464040D-10	1.75	-6.070497977732D-01

CARDINAL SPLINES N=8

1	-8.5	7.917027729831D-09	3.958513864915D-09	2.25	5.778364028861D-01
2	-7.5	-3.579559791672D-08	-1.789779895836D-08	2.50	9.635532142563D-01
3	-6.5	9.988871719642D-08	4.994435859821D-08	2.75	1.096107121655D+00
4	-5.5	-1.946118786304D-07	-9.730593931522D-08	3.25	7.536617497902D-01
5	-4.5	2.855623413709D-07	1.427811706854D-07	3.50	4.538410665913D-01
6	-3.5	-3.348322175483D-07	-1.674161087742D-07	3.75	1.850440694391D-01
7	-2.5	3.308706272794D-07	1.654353136397D-07	4.25	-8.647992862655D-02
8	-1.5	-2.892125024052D-07	-1.446062512026D-07	4.50	-9.152315317617D-02
9	-.5	2.326709675053D-07	1.163354837526D-07	4.75	-5.030664394417D-02
10	.5	-1.769518456465D-07	-8.847592282324D-08	5.25	3.299201893170D-02
11	1.5	1.287888624277D-07	6.439443121383D-08	5.50	3.896153594360D-02
12	2.5	-8.949578350264D-08	-4.474789175132D-08	5.75	2.337044108411D-02
13	3.5	5.834237671784D-08	2.917118835892D-08	6.25	-1.745928928921D-02
14	4.5	-3.448398221050D-08	-1.724199110525D-08	6.50	-2.166374292726D-02
15	5.5	1.756649558451D-08	8.783247792256D-09	6.75	-1.354898226566D-02
16	6.5	-7.212561260029D-09	-3.606280630014D-09	7.25	1.081051683964D-02
17	7.5	2.184397894927D-09	1.092198947463D-09	7.50	1.376702968968D-02
18	8.5	-4.257379753981D-10	-2.128689876991D-10	7.75	8.805004645426D-03
19	9.5	3.954428173053D-11	1.977214086527D-11	8.25	-7.283899716178D-03
				8.50	-9.413076871808D-03
	0	2.871598790625D+04	1.381491270663D+04	8.75	-6.098600258463D-03
	1	3.004748493166D+04	1.557430413899D+04	9.25	5.156096618679D-03
	2	1.349298733975D+04	6.507482558956D+03	9.50	6.726128613696D-03
	3	3.365928409765D+03	1.740507842882D+03	9.75	4.395820076392D-03
	4	5.029903683410D+02	2.432048027543D+02	10.25	-3.776926547855D-03
	5	4.495803722937D+01	2.319315670693D+01	10.50	-4.965697453097D-03
	6	2.222273625705D+00	1.077106790453D+00	10.75	-3.271055668275D-03
	7	4.680262796573D-02	2.409284879611D-02	11.25	2.858080594673D-03

I= 4 H= -5.5

0		0.	5.589092993088D+05	-1.00	2.692674695442D+02
1		-3.673937756194D+01	-2.377707066153D+05	-.75	1.318928078273D+02
2		9.184256305563D+01	4.336420685845D+04	-.50	5.517840701048D+01
3		-8.726931719204D+01	-4.425487681768D+03	-.25	1.646094622582D+01
4		4.139818959105D+01	2.864542473226D+02	.25	-4.656522320633D+00
5		-1.054943984266D+01	-1.493783641668D+01	.50	-4.038821092391D+00
6		1.392982124707D+00	8.915275052925D-01	.75	-1.876790757420D+00
7		-7.560017622391D-02	-3.948566722351D-02	1.25	9.606764426101D-01
				1.50	1.046518785521D+00
0	-9.5	1.481203617110D-09	7.406018085548D-10	1.75	5.962629457843D-01
1	-8.5	-1.473907051529D-08	-7.369535257647D-09	2.25	-4.415213402324D-01
2	-7.5	6.802812727890D-08	3.401406363945D-08	2.50	-5.752457864200D-01
3	-6.5	-1.946118786304D-07	-9.730593931522D-08	2.75	-3.945125809078D-01
4	-5.5	3.904839734326D-07	1.952419867163D-07	3.25	4.567599923269D-01
5	-4.5	-5.925481052424D-07	-2.962740526212D-07	3.50	8.296631858903D-01
6	-3.5	7.202745262735D-07	3.601372631368D-07	3.75	1.020566873984D+00
7	-2.5	-7.372723826626D-07	-3.686361913313D-07	4.25	8.043257403526D-01
8	-1.5	6.644182858802D-07	3.322091429401D-07	4.50	5.138450785287D-01
9	-.5	-5.470075147898D-07	-2.735037573949D-07	4.75	2.210321259526D-01
10	.5	4.224486308088D-07	2.112243154044D-07	5.25	-1.132283945342D-01
11	1.5	-3.103135474735D-07	-1.551567737367D-07	5.50	-1.245902033999D-01
12	2.5	2.167576733991D-07	1.083788366995D-07	5.75	-7.090577581264D-02
13	3.5	-1.417024603091D-07	-7.085123015454D-08	6.25	4.929911953489D-02
14	4.5	8.388145058100D-08	4.194072529050D-08	6.50	5.965228368995D-02
15	5.5	-4.276482555525D-08	-2.138241277762D-08	6.75	3.655835594526D-02
16	6.5	1.756649558451D-08	8.783247792256D-09	7.25	-2.830263056496D-02

17	7.5	-5.321544635555D-09	-2.660772317777D-09	7.50	-3.563909884014D-02
18	8.5	1.037322815347D-09	5.186614076736D-10	7.75	-2.258097033348D-02
19	9.5	-9.635985712074D-11	-4.817992856037D-11	8.25	1.841256398890D-02
				8.50	2.366343026173D-02
	0	-4.996182997411D+04	-2.365730597994D+04	8.75	1.525978598948D-02
	1	-5.186776807705D+04	-2.727574418578D+04	9.25	-1.280782349181D-02
	2	-2.308332414828D+04	-1.095912222234D+04	9.50	-1.666059050286D-02
	3	-5.700732199731D+03	-2.990618711197D+03	9.75	-1.086231933639D-02
	4	-8.425701897620D+02	-4.010784599314D+02	10.25	9.297888995319D-03
	5	-7.443069272272D+01	-3.895597942036D+01	10.50	1.220638854198D-02
	6	-3.634429594183D+00	-1.734269365071D+00	10.75	8.030669502760D-03
	7	-7.560017622391D-02	-3.948566722351D-02	11.25	-7.002993971437D-03
I= 5	H= -4.5				
0		0.	-1.098662515834D+06	-1.00	-2.946845749167D+02
1		3.901554754439D+01	4.673801390953D+05	-.75	-1.434385043789D+02
2		-9.901258155121D+01	-8.520307726707D+04	-.50	-5.958900903425D+01
3		9.602188713935D+01	8.661787308279D+03	-.25	-1.763717120166D+01
4		-4.667382360828D+01	-5.457590386916D+02	.25	4.895150724541D+00
5		1.221270114586D+01	2.510230271579D+01	.50	4.196743985804D+00
6		-1.655882297835D+00	-1.211351202154D+00	.75	1.924148184803D+00
7		9.215162975317D-02	4.938941912234D-02	1.25	-9.518109781498D-01
				1.50	-1.014188969832D+00
0	-9.5	-2.029462028655D-09	-1.014731014328D-09	1.75	-5.625665539024D-01
1	-8.5	2.056584890137D-08	1.028292445068D-08	2.25	3.866395337810D-01
2	-7.5	-9.708688028697D-08	-4.854344014349D-08	2.50	4.774533042683D-01
3	-6.5	2.855623413709D-07	1.427811706854D-07	2.75	3.048624698708D-01
4	-5.5	-5.925481052424D-07	-2.962740526212D-07	3.25	-2.755038474487D-01
5	-4.5	9.351751364399D-07	4.675875682199D-07	3.50	-3.916483085818D-01
6	-3.5	-1.187107850265D-06	-5.935539251325D-07	3.75	-2.909209062281D-01
7	-2.5	1.269881589717D-06	6.349407948584D-07	4.25	3.872710161206D-01
8	-1.5	-1.191520130632D-06	-5.957600653162D-07	4.50	7.473554722878D-01
9	-.5	1.013638119752D-06	5.068190598760D-07	4.75	9.711961514277D-01
10	.5	-8.014779045954D-07	-4.007389522977D-07	5.25	8.410318132203D-01
11	1.5	5.977257256474D-07	2.988628628237D-07	5.50	5.592303140300D-01
12	2.5	-4.212717276112D-07	-2.106358638056D-07	5.75	2.493109464946D-01
13	3.5	2.767806215892D-07	1.383903107946D-07	6.25	-1.356271560062D-01
14	4.5	-1.642876162612D-07	-8.214380813059D-08	6.50	-1.530197868252D-01
15	5.5	8.388145058100D-08	4.194072529050D-08	6.75	-8.903342872996D-02
16	6.5	-3.448398221050D-08	-1.724199110525D-08	7.25	6.421879027494D-02
17	7.5	1.045131831710D-08	5.225659158551D-09	7.50	7.889330890192D-02
18	8.5	-2.037825073804D-09	-1.018912536902D-09	7.75	4.900239965596D-02
19	9.5	1.893318920888D-10	9.466594604438D-11	8.25	-3.879548617209D-02
				8.50	-4.931416199547D-02
	0	6.489940206124D+04	2.984811710684D+04	8.75	-3.151196915372D-02
	1	6.685921110177D+04	3.606713316463D+04	9.25	2.608051091586D-02
	2	2.950427641110D+04	1.360708187721D+04	9.50	3.374421904871D-02
	3	7.220465105614D+03	3.885923995720D+03	9.75	2.190116738484D-02
	4	1.057076444713D+03	4.888175130931D+02	10.25	-1.861527149716D-02
	5	9.247778645896D+01	4.966058078463D+01	10.50	-2.437169188406D-02
	6	4.472201080751D+00	2.073045169482D+00	10.75	-1.599723114394D-02
	7	9.215162975317D-02	4.938941912234D-02	11.25	1.389954749378D-02
I= 6	H= -3.5				
0		0.	1.863515200953D+06	-1.00	2.737297348592D+02

CARDINAL SPLINES N=8

1		-3.565872435368D+01	-7.927544294839D+05	-.75	1.327804058494D+02
2		9.120558430541D+01	1.444856552999D+05	-.50	5.495160333485D+01
3		-8.941052101444D+01	-1.465591520459D+04	-.25	1.619595166870D+01
4		4.404548315899D+01	9.082024190619D+02	.25	-4.450360305080D+00
5		-1.170424898615D+01	-3.807532952765D+01	.50	-3.792694538390D+00
6		1.613799964083D+00	1.457296806528D+00	.75	-1.727147749965D+00
7		-9.137307644626D-02	-5.130765490253D-02	1.25	8.402383567387D-01
				1.50	8.860840011104D-01
0	-9.5	2.219262603816D-09	1.109631301908D-09	1.75	4.855869231512D-01
1	-8.5	-2.289654160188D-08	-1.144827080094D-08	2.25	-3.234058365304D-01
2	-7.5	1.105890756942D-07	5.529453784708D-08	2.50	-3.911902815222D-01
3	-6.5	-3.348322175483D-07	-1.674161087742D-07	2.75	-2.435275406528D-01
4	-5.5	7.202745262735D-07	3.601372631368D-07	3.25	2.048460456658D-01
5	-4.5	-1.187107850265D-06	-5.935539251325D-07	3.50	2.763889552395D-01
6	-3.5	1.583232650689D-06	7.916163253445D-07	3.75	1.914021473130D-01
7	-2.5	-1.784544764903D-06	-8.922723824517D-07	4.25	-1.993727166264D-01
8	-1.5	1.761116688759D-06	8.805583443796D-07	4.50	-3.014566724466D-01
9	-.5	-1.565444674841D-06	-7.827223374204D-07	4.75	-2.368096711004D-01
10	.5	1.280854859300D-06	6.404274296502D-07	5.25	3.470189767672D-01
11	1.5	-9.783221916849D-07	-4.891610958425D-07	5.50	6.975691910576D-01
12	2.5	7.000075595484D-07	3.500037797742D-07	5.75	9.401630464457D-01
13	3.5	-4.640010192022D-07	-2.320005096011D-07	6.25	8.656362170163D-01
14	4.5	2.767806215892D-07	1.383903107946D-07	6.50	5.904785161284D-01
15	5.5	-1.417024603091D-07	-7.085123015454D-08	6.75	2.692500210700D-01
16	6.5	5.834237671784D-08	2.917118835892D-08	7.25	-1.520670297753D-01
17	7.5	-1.769751172398D-08	-8.848755861989D-09	7.50	-1.742449645028D-01
18	8.5	3.452479616798D-09	1.726239808399D-09	7.75	-1.027784650356D-01
19	9.5	-3.208687127230D-10	-1.604343563615D-10	8.25	7.584615598591D-02
				8.50	9.407735387032D-02
	0	-6.746606622751D+04	-2.932116392595D+04	8.75	5.894186980689D-02
	1	-6.905252736447D+04	-3.899931162764D+04	9.25	-4.737837804040D-02
	2	-3.026391031486D+04	-1.318993024720D+04	9.50	-6.063762023558D-02
	3	-7.354511633379D+03	-4.144853145832D+03	9.75	-3.900219558842D-02
	4	-1.069159341939D+03	-4.672067034356D+02	10.25	3.269503247328D-02
	5	-9.289247416819D+01	-5.225024450954D+01	10.50	4.257929291214D-02
	6	-4.462509619593D+00	-1.954662244490D+00	10.75	2.782414512213D-02
	7	-9.137307644626D-02	-5.130765490253D-02	11.25	-2.400854589593D-02

I= 7 H= -2.5

0		0.	-2.872287972089D+06	-1.00	-2.283133914803D+02
1		2.949317069661D+01	1.221925209577D+06	-.75	-1.105503712941D+02
2		-7.573837803177D+01	-2.226830187888D+05	-.50	-4.566093235238D+01
3		7.466094366922D+01	2.256034834978D+04	-.25	-1.342822814613D+01
4		-3.703720722603D+01	-1.384600204760D+03	.25	3.670903340353D+00
5		9.923586883501D+00	5.463677179177D+01	.50	3.118932730695D+00
6		-1.381110481330D+00	-1.693232320955D+00	.75	1.415472323841D+00
7		7.899449185670D-02	4.816324457710D-02	1.25	-6.829254494136D-01
				1.50	-7.165508551948D-01
0	-9.5	-2.058396656300D-09	-1.029198328150D-09	1.75	-3.903928064804D-01
1	-8.5	2.156959446637D-08	1.078479723318D-08	2.25	2.562025934063D-01
2	-7.5	-1.063494035645D-07	-5.317470178227D-08	2.50	3.070110411615D-01
3	-6.5	3.308706272794D-07	1.654353136397D-07	2.75	1.890060063737D-01
4	-5.5	-7.372723826626D-07	-3.686361913313D-07	3.25	-1.543580217544D-01
5	-4.5	1.269881589717D-06	6.349407948584D-07	3.50	-2.041875949937D-01
6	-3.5	-1.784544764903D-06	-8.922723824517D-07	3.75	-1.379792684256D-01
7	-2.5	2.131185674618D-06	1.065592837309D-06	4.25	1.339853353071D-01

CARDINAL SPLINES N=8

8	-1.5	-2.230664970706D-06	-1.115332485353D-06	4.50	1.924126651382D-01
9	-.5	2.094173385340D-06	1.047086692670D-06	4.75	1.410000219170D-01
10	.5	-1.794169995008D-06	-8.970849975040D-07	5.25	-1.618426865605D-01
11	1.5	1.419231118191D-06	7.096155590955D-07	5.50	-2.550084469206D-01
12	2.5	-1.040225363241D-06	-5.201126816204D-07	5.75	-2.078364603281D-01
13	3.5	7.000075595484D-07	3.500037797742D-07	6.25	3.240062667737D-01
14	4.5	-4.212717276112D-07	-2.106358638056D-07	6.50	6.683095633183D-01
15	5.5	2.167576733991D-07	1.083788366995D-07	6.75	9.214684545760D-01
16	6.5	-8.949578350264D-08	-4.474789175132D-08	7.25	8.811004772917D-01
17	7.5	2.719151968510D-08	1.359575984255D-08	7.50	6.104846814272D-01
18	8.5	-5.309732660799D-09	-2.654866330399D-09	7.75	2.822360916723D-01
19	9.5	4.937782723793D-10	2.468891361897D-10	8.25	-1.631166095669D-01
				8.50	-1.887263957606D-01
	0	6.005997522117D+04	2.323233602332D+04	8.75	-1.122971528664D-01
	1	6.119828401088D+04	3.749170925306D+04	9.25	8.414915527377D-02
	2	2.670164312754D+04	1.035805537184D+04	9.50	1.050988577404D-01
	3	6.460219819333D+03	3.950768192292D+03	9.75	6.628314968845D-02
	4	9.351314415871D+02	3.637218645308D+02	10.25	-5.397009455715D-02
	5	8.091460013910D+01	4.940391878207D+01	10.50	-6.952918077122D-02
	6	3.872023227141D+00	1.509623443422D+00	10.75	-4.502842990915D-02
	7	7.899449185670D-02	4.816324457710D-02	11.25	3.832238445170D-02

I= 8		H= -1.5			
0		0.	4.170799544769D+06	-1.00	1.785704958488D+02
1		-2.296894737453D+01	-1.774446614043D+06	-.75	8.638515125898D+01
2		5.910369991231D+01	3.233720594559D+05	-.50	3.564386395842D+01
3		-5.842683681631D+01	-3.274100593480D+04	-.25	1.047067268451D+01
4		2.908700608096D+01	1.998727510825D+03	.25	-2.854933516158D+00
5		-7.826524069995D+00	-7.606882938991D+01	.50	-2.421933806820D+00
6		1.094541930272D+00	2.003800832243D+00	.75	-1.097257925912D+00
7		-6.293966442946D-02	-4.405915423401D-02	1.25	5.272001524389D-01
				1.50	5.517648657454D-01
0	-9.5	1.712108684383D-09	8.560543421916D-10	1.75	2.997448978270D-01
1	-8.5	-1.814686892454D-08	-9.073434462269D-09	2.25	-1.952979490895D-01
2	-7.5	9.091285767842D-08	4.545642883921D-08	2.50	-2.329704152867D-01
3	-6.5	-2.892125024052D-07	-1.446062512026D-07	2.75	-1.426631366521D-01
4	-5.5	6.644182858802D-07	3.322091429401D-07	3.25	1.149194874870D-01
5	-4.5	-1.191520130632D-06	-5.957600653162D-07	3.50	1.506673110852D-01
6	-3.5	1.761116688759D-06	8.805583443796D-07	3.75	1.007265873626D-01
7	-2.5	-2.230664970706D-06	-1.115332485353D-06	4.25	-9.503092175849D-02
8	-1.5	2.487404596410D-06	1.243702298205D-06	4.50	-1.338353724491D-01
9	-.5	-2.486193066391D-06	-1.243096533195D-06	4.75	-9.572336668757D-02
10	.5	2.254552396558D-06	1.127276198279D-06	5.25	1.024633663302D-01
11	1.5	-1.869204126535D-06	-9.346020632676D-07	5.50	1.533555940290D-01
12	2.5	1.419231118191D-06	7.096155590955D-07	5.75	1.166049179646D-01
13	3.5	-9.783221916849D-07	-4.891610958425D-07	6.25	-1.424018879648D-01
14	4.5	5.977257256474D-07	2.988628628237D-07	6.50	-2.302393592592D-01
15	5.5	-3.103135474735D-07	-1.551567737367D-07	6.75	-1.919732161433D-01
16	6.5	1.287888624277D-07	6.439443121383D-08	7.25	3.108061280960D-01
17	7.5	-3.924849545714D-08	-1.962424772857D-08	7.50	6.511699985434D-01
18	8.5	7.678103554383D-09	3.839051777191D-09	7.75	9.102963111089D-01
19	9.5	-7.148435814783D-10	-3.574217907392D-10	8.25	8.907051408676D-01
				8.50	6.231515613545D-01
	0	-4.860608991080D+04	-1.443916608848D+04	8.75	2.906215366070D-01
	1	-4.940646578356D+04	-3.470664824104D+04	9.25	-1.705573546490D-01
	2	-2.150459798871D+04	-6.408024017384D+03	9.50	-1.987041331455D-01

3		-5.190563347313D+03	-3.641564049610D+03	9.75	-1.190187901038D-01
4		-7.496380667952D+02	-2.240291161951D+02	10.25	9.034520911184D-02
5		-6.472403305442D+01	-4.535529401407D+01	10.50	1.135831085027D-01
6		-3.090945754287D+00	-9.261329243188D-01	10.75	7.212745886255D-02
7		-6.293966442946D-02	-4.405915423401D-02	11.25	-5.962698894983D-02
I= 9	H= -.5				
0		0.	-5.827995885540D+06	-1.00	-1.344123271540D+02
1		1.725120805333D+01	2.479806621538D+06	-.75	-6.499260767382D+01
2		-4.443668535826D+01	-4.519552880646D+05	-.50	-2.680309876602D+01
3		4.399083351292D+01	4.574866946846D+04	-.25	-7.869137867330D+00
4		-2.194001694156D+01	-2.784928213116D+03	.25	2.142740804804D+00
5		5.916292956633D+00	1.038460742604D+02	.50	1.816333304310D+00
6		-8.294612765132D-01	-2.451485595064D+00	.75	8.221671014113D-01
7		4.782905477054D-02	4.152091938038D-02	1.25	-3.941923312194D-01
				1.50	-4.120316928035D-01
0	-9.5	-1.332164201646D-09	-6.660821008231D-10	1.75	-2.235072028892D-01
1	-8.5	1.422079342884D-08	7.110396714419D-09	2.25	1.450960962159D-01
2	-7.5	-7.199949453217D-08	-3.599974726609D-08	2.50	1.726919537002D-01
3	-6.5	2.326709675053D-07	1.163354837526D-07	2.75	1.054703056400D-01
4	-5.5	-5.470075147898D-07	-2.735037573949D-07	3.25	-8.438486498236D-02
5	-4.5	1.013638119752D-06	5.068190598760D-07	3.50	-1.101547969435D-01
6	-3.5	-1.565444674841D-06	-7.827223374204D-07	3.75	-7.326357777581D-02
7	-2.5	2.094173385340D-06	1.047086692670D-06	4.25	6.819344398386D-02
8	-1.5	-2.486193066391D-06	-1.243096533195D-06	4.50	9.519375007996D-02
9	-.5	2.654898469251D-06	1.327449234626D-06	4.75	6.736264428137D-02
10	.5	-2.567456137305D-06	-1.283728068652D-06	5.25	-7.005871292246D-02
11	1.5	2.254552396558D-06	1.127276198279D-06	5.50	-1.028293824206D-01
12	2.5	-1.794169995008D-06	-8.970849975040D-07	5.75	-7.631041122617D-02
13	3.5	1.280854859300D-06	6.404274296502D-07	6.25	8.690024982345D-02
14	4.5	-8.014779045954D-07	-4.007389522977D-07	6.50	1.334538200241D-01
15	5.5	4.224486308088D-07	2.112243154044D-07	6.75	1.038048642831D-01
16	6.5	-1.769518456465D-07	-8.847592282324D-08	7.25	-1.316392977641D-01
17	7.5	5.422637848151D-08	2.711318924075D-08	7.50	-2.161760472788D-01
18	8.5	-1.064435072139D-08	-5.322175360695D-09	7.75	-1.827400723027D-01
19	9.5	9.931476048629D-10	4.965738024314D-10	8.25	3.027296147678D-01
				8.50	6.404081292932D-01
0		3.723664819632D+04	4.853557403047D+03	8.75	9.030895717235D-01
1		3.780152134962D+04	3.286507085167D+04	9.25	8.972736916700D-01
2		1.643273331908D+04	2.149894169999D+03	9.50	6.320960626875D-01
3		3.961509758854D+03	3.442297068081D+03	9.75	2.967483382683D-01
4		5.714609496050D+02	7.502662881696D+01	10.25	-1.764178463940D-01
5		4.928501624925D+01	4.280391779743D+01	10.50	-2.068918938775D-01
6		2.351170865728D+00	3.096555437310D-01	10.75	-1.247784365276D-01
7		4.782905477054D-02	4.152091938038D-02	11.25	9.616940850966D-02
I=10	H= .5				
0		0.	7.903202617955D+06	-1.00	9.847098165652D+01
1		-1.262401921816D+01	-3.363636915370D+06	-.75	4.760225125930D+01
2		3.253502477588D+01	6.131738281282D+05	-.50	1.962603338257D+01
3		-3.223241798466D+01	-6.207033013080D+04	-.25	5.760322641865D+00
4		1.609067569480D+01	3.773373855402D+03	.25	-1.567438391817D+00
5		-4.343840841942D+00	-1.391468062458D+02	.50	-1.328130584125D+00
6		6.097903570845D-01	3.070796682527D+00	.75	-6.009085081173D-01
7		-3.521278399023D-02	-4.152091938038D-02	1.25	2.877947931665D-01

CARDINAL SPLINES N=8

0	-9.5	9.931476048629D-10	4.965738024314D-10
1	-8.5	-1.064435072139D-08	-5.322175360695D-09
2	-7.5	5.422637848151D-08	2.711318924075D-08
3	-6.5	-1.769518456465D-07	-8.847592282324D-08
4	-5.5	4.224486308088D-07	2.112243154044D-07
5	-4.5	-8.014779045954D-07	-4.007389522977D-07
6	-3.5	1.280854859300D-06	6.404274296502D-07
7	-2.5	-1.794169995008D-06	-8.970849975040D-07
8	-1.5	2.254552396558D-06	1.127276198279D-06
9	-.5	-2.567456137305D-06	-1.283728068652D-06
10	.5	2.654898469251D-06	1.327449234626D-06
11	1.5	-2.486193066391D-06	-1.243096533195D-06
12	2.5	2.094173385340D-06	1.047086692670D-06
13	3.5	-1.565444674841D-06	-7.827223374204D-07
14	4.5	1.013638119752D-06	5.068190598760D-07
15	5.5	-5.470075147898D-07	-2.735037573949D-07
16	6.5	2.326709675053D-07	1.163354837526D-07
17	7.5	-7.199949453217D-08	-3.599974726609D-08
18	8.5	1.422079342884D-08	7.110396714419D-09
19	9.5	-1.332164201646D-09	-6.660821008231D-10

0	-2.752953339022D+04	4.853557403047D+03
1	-2.792862035372D+04	-3.286507085167D+04
2	-1.213294497908D+04	2.149894169999D+03
3	-2.923084377308D+03	-3.442297068081D+03
4	-4.214076919711D+02	7.502662881696D+01
5	-3.632281934561D+01	-4.280391779743D+01
6	-1.731859778266D+00	3.096555437310D-01
7	-3.521278399023D-02	-4.152091938038D-02

1.50	3.006208262727D-01
1.75	1.629492353243D-01
2.25	-1.055855046937D-01
2.50	-1.255206504940D-01
2.75	-7.655683445477D-02
3.25	6.104059271827D-02
3.50	7.950668366900D-02
3.75	5.274262206832D-02
4.25	-4.876408458959D-02
4.50	-6.777731954114D-02
4.75	-4.771477987231D-02
5.25	4.895624977149D-02
5.50	7.122049355286D-02
5.75	5.228937673942D-02
6.25	-5.784855971245D-02
6.50	-8.711633709977D-02
6.75	-6.613152329461D-02
7.25	7.819208254278D-02
7.50	1.219571000036D-01
7.75	9.616940850966D-02
8.25	-1.247784365276D-01
8.50	-2.068918938775D-01
8.75	-1.764178463940D-01
9.25	2.967483382683D-01
9.50	6.320960626875D-01
9.75	8.972736916700D-01
10.25	9.030895717235D-01
10.50	6.404081292932D-01
10.75	3.027296147678D-01
11.25	-1.827400723027D-01

I=11 H= 1.5

0	0.	-1.036737412316D+07
1	8.998317267074D+00	4.414516037600D+06
2	-2.319717298446D+01	-8.051139568864D+05
3	2.299025374431D+01	8.152894740514D+04
4	-1.148250221312D+01	-4.954290635808D+03
5	3.101625465781D+00	1.816479827623D+02
6	-4.356999229150D-01	-3.856066680881D+00
7	2.517864403856D-02	4.405915423401D-02

0	-9.5	-7.148435814783D-10	-3.574217907392D-10
1	-8.5	7.678103554383D-09	3.839051777191D-09
2	-7.5	-3.924849545714D-08	-1.962424772857D-08
3	-6.5	1.287888624277D-07	6.439443121383D-08
4	-5.5	-3.103135474735D-07	-1.551567737367D-07
5	-4.5	5.977257256474D-07	2.988628628237D-07
6	-3.5	-9.783221916849D-07	-4.891610958425D-07
7	-2.5	1.419231118191D-06	7.096155590955D-07
8	-1.5	-1.869204126535D-06	-9.346020632676D-07
9	-.5	2.254552396558D-06	1.127276198279D-06
10	.5	-2.486193066391D-06	-1.243096533195D-06
11	1.5	2.487404596410D-06	1.243702298205D-06
12	2.5	-2.230664970706D-06	-1.115332485353D-06
13	3.5	1.761116688759D-06	8.805583443796D-07
14	4.5	-1.191520130632D-06	-5.957600653162D-07

-1.00	-7.023075024170D+01
-.75	-3.394623294902D+01
-.50	-1.399382030127D+01
-.25	-4.106615707275D+00
.25	1.117049291575D+00
.50	9.463054098080D-01
.75	4.280519586797D-01
1.25	-2.048919762972D-01
1.50	-2.139500095977D-01
1.75	-1.159245415059D-01
2.25	7.504223666829D-02
2.50	8.915681239968D-02
2.75	5.433979854589D-02
3.25	-4.324914528249D-02
3.50	-5.626917619911D-02
3.75	-3.727770536082D-02
4.25	3.434752475404D-02
4.50	4.763469236448D-02
4.75	3.344713376008D-02
5.25	-3.408590703252D-02
5.50	-4.937127609399D-02
5.75	-3.605974924141D-02
6.25	3.935263523883D-02
6.50	5.873561535138D-02

15	5.5	6.644182858802D-07	3.322091429401D-07	6.75	4.410940417158D-02
16	6.5	-2.892125024052D-07	-1.446062512026D-07	7.25	-5.066247228153D-02
17	7.5	9.091285767842D-08	4.545642883921D-08	7.50	-7.748339763968D-02
18	8.5	-1.814686892454D-08	-9.073434462269D-09	7.75	-5.962698894983D-02
19	9.5	1.712108684383D-09	8.560543421916D-10	8.25	7.212745886255D-02
				8.50	1.135831085027D-01
	0	1.972775773384D+04	-1.443916608848D+04	8.75	9.034520911184D-02
	1	2.000683069852D+04	3.470664824104D+04	9.25	-1.190187901038D-01
	2	8.688549953944D+03	-6.408024017384D+03	9.50	-1.987041331455D-01
	3	2.092564751906D+03	3.641564049610D+03	9.75	-1.705573546490D-01
	4	3.015798344051D+02	-2.240291161951D+02	10.25	2.906215366070D-01
	5	2.598655497372D+01	4.535529401407D+01	10.50	6.231515613545D-01
	6	1.238679905650D+00	-9.261329243188D-01	10.75	8.907051408676D-01
	7	2.517864403856D-02	4.405915423401D-02	11.25	9.102963111089D-01

I=12 H= 2.5

0		0.	1.293288017951D+07	-1.00	4.827939409748D+01
1		-6.183897658764D+00	-5.512040994511D+06	-.75	2.333443907941D+01
2		1.594405967707D+01	1.006189227763D+06	-.50	9.618570416621D+00
3		-1.580501348666D+01	-1.019757460463D+05	-.25	2.822432291850D+00
4		7.895835891437D+00	6.199349406887D+03	.25	-7.675922268732D-01
5		-2.133453906263D+00	-2.267338443419D+02	.50	-6.501922954078D-01
6		2.998014799880D-01	4.712479207800D+00	.75	-2.940718478712D-01
7		-1.733199729751D-02	-4.816324457710D-02	1.25	1.407191707102D-01
				1.50	1.469138941242D-01
0	-9.5	4.937782723793D-10	2.468891361897D-10	1.75	7.958606101662D-02
1	-8.5	-5.309732660799D-09	-2.654866330399D-09	2.25	-5.149287478649D-02
2	-7.5	2.719151968510D-08	1.359575984255D-08	2.50	-6.115880836764D-02
3	-6.5	-8.949578350264D-08	-4.474789175132D-08	2.75	-3.726174013077D-02
4	-5.5	2.167576733991D-07	1.083788366995D-07	3.25	2.962908726609D-02
5	-4.5	-4.212717276112D-07	-2.106358638056D-07	3.50	3.852610008921D-02
6	-3.5	7.000075595484D-07	3.500037797742D-07	3.75	2.550539793790D-02
7	-2.5	-1.040225363241D-06	-5.201126816204D-07	4.25	-2.345868563938D-02
8	-1.5	1.419231118191D-06	7.096155590955D-07	4.50	-3.249649303502D-02
9	-.5	-1.794169995008D-06	-8.970849975040D-07	4.75	-2.278699684642D-02
10	.5	2.094173385340D-06	1.047086692670D-06	5.25	2.314162972565D-02
11	1.5	-2.230664970706D-06	-1.115332485353D-06	5.50	3.344459897839D-02
12	2.5	2.131185674618D-06	1.065592837309D-06	5.75	2.436289979359D-02
13	3.5	-1.784544764903D-06	-8.922723824517D-07	6.25	-2.640665058889D-02
14	4.5	1.269881589717D-06	6.349407948584D-07	6.50	-3.924001102385D-02
15	5.5	-7.372723826626D-07	-3.686361913313D-07	6.75	-2.931459562855D-02
16	6.5	3.308706272794D-07	1.654353136397D-07	7.25	3.321100384355D-02
17	7.5	-1.063494035645D-07	-5.317470178227D-08	7.50	5.033973643723D-02
18	8.5	2.156959446637D-08	1.078479723318D-08	7.75	3.832238445170D-02
19	9.5	-2.058396656300D-09	-1.029198328150D-09	8.25	-4.502842990915D-02
				8.50	-6.952918077122D-02
	0	-1.359530317452D+04	2.323233602332D+04	8.75	-5.397009455715D-02
	1	-1.378513449524D+04	-3.749170925306D+04	9.25	6.628314968845D-02
	2	-5.985532383864D+03	1.035805537184D+04	9.50	1.050988577404D-01
	3	-1.441316565251D+03	-3.950768192292D+03	9.75	8.414915527377D-02
	4	-2.076877125255D+02	3.637218645308D+02	10.25	-1.122971528664D-01
	5	-1.789323742505D+01	-4.940391878207D+01	10.50	-1.887263957606D-01
	6	-8.527763402963D-01	1.509623443422D+00	10.75	-1.631166095669D-01
	7	-1.733199729751D-02	-4.816324457710D-02	11.25	2.822360916723D-01

CARDINAL SPLINES N=8

I=13	H= 3.5				
0		0.	-1.478750826790D+07	-1.00	-3.129395010786D+01
1		4.007659076690D+00	6.313519018586D+06	-.75	-1.512449417413D+01
2		-1.033380797565D+01	-1.154475760048D+06	-.50	-6.234153065748D+00
3		1.024476856490D+01	1.171987663405D+05	-.25	-1.829246405800D+00
4		-5.118743690374D+00	-7.134863852889D+03	.25	4.974352904684D-01
5		1.383305179142D+00	2.609068253995D+02	.50	4.213303857592D-01
6		-1.944233877453D-01	-5.366621295508D+00	.75	1.905488380483D-01
7		1.124223335879D-02	5.130765490253D-02	1.25	-9.116719173821D-02
				1.50	-9.517159706236D-02
0	-9.5	-3.208687127230D-10	-1.604343563615D-10	1.75	-5.155072831960D-02
1	-8.5	3.452479616798D-09	1.726239808399D-09	2.25	3.334489161375D-02
2	-7.5	-1.769751172398D-08	-8.848755861989D-09	2.50	3.959764317226D-02
3	-6.5	5.834237671784D-08	2.917118835892D-08	2.75	2.412069497653D-02
4	-5.5	-1.417024603091D-07	-7.085123015454D-08	3.25	-1.917047272022D-02
5	-4.5	2.767806215892D-07	1.383903107946D-07	3.50	-2.491926503757D-02
6	-3.5	-4.640010192022D-07	-2.320005096011D-07	3.75	-1.649127494518D-02
7	-2.5	7.000075595484D-07	3.500037797742D-07	4.25	1.515375211543D-02
8	-1.5	-9.783221916849D-07	-4.891610958425D-07	4.50	2.097945048205D-02
9	-.5	1.280854859300D-06	6.404274296502D-07	4.75	1.470074019886D-02
10	.5	-1.565444674841D-06	-7.827223374204D-07	5.25	-1.490252492906D-02
11	1.5	1.761116688759D-06	8.805583443796D-07	5.50	-2.151243079394D-02
12	2.5	-1.784544764903D-06	-8.922723824517D-07	5.75	-1.564945904016D-02
13	3.5	1.583232650689D-06	7.916163253445D-07	6.25	1.690270125095D-02
14	4.5	-1.187107850265D-06	-5.935539251325D-07	6.50	2.506077433393D-02
15	5.5	7.202745262735D-07	3.601372631368D-07	6.75	1.867211789596D-02
16	6.5	-3.348322175483D-07	-1.674161087742D-07	7.25	-2.100891538935D-02
17	7.5	1.105890756942D-07	5.529453784708D-08	7.50	-3.170369656611D-02
18	8.5	-2.289654160188D-08	-1.144827080094D-08	7.75	-2.400854589593D-02
19	9.5	2.219262603816D-09	1.109631301908D-09	8.25	2.782414512213D-02
				8.50	4.257929291214D-02
	0	8.823738375617D+03	-2.932116392595D+04	8.75	3.269503247328D-02
	1	8.946095890810D+03	3.899931162764D+04	9.25	-3.900219558842D-02
	2	3.884049820459D+03	-1.318993024720D+04	9.50	-6.063762023558D-02
	3	9.351946582860D+02	4.144853145832D+03	9.75	-4.737837804040D-02
	4	1.347459350684D+02	-4.672067034356D+02	10.25	5.894186980689D-02
	5	1.160801485090D+01	5.225024450954D+01	10.50	9.407735387032D-02
	6	5.531851306141D-01	-1.954662244490D+00	10.75	7.584615598591D-02
	7	1.124223335879D-02	5.130765490253D-02	11.25	-1.027784650356D-01

I=14	H= 4.5				
0		0.	1.462509280533D+07	-1.00	1.844068721208D+01
1		-2.361404866829D+00	-6.262147974401D+06	-.75	8.912296110849D+00
2		6.089161650147D+00	1.148359081824D+06	-.50	3.673478150223D+00
3		-6.037031808838D+00	-1.169070023907D+05	-.25	1.077860524774D+00
4		3.016583332898D+00	7.136163861250D+03	.25	-2.930923048587D-01
5		-8.152797219740D-01	-2.614294520367D+02	.50	-2.482432015715D-01
6		1.145986228985D-01	5.357441541118D+00	.75	-1.122654284899D-01
7		-6.627208491515D-03	-4.938941912234D-02	1.25	5.370845101422D-02
				1.50	5.606474358830D-02
0	-9.5	1.893318920888D-10	9.466594604438D-11	1.75	3.036635195789D-02
1	-8.5	-2.037825073804D-09	-1.018912536902D-09	2.25	-1.963930180548D-02
2	-7.5	1.045131831710D-08	5.225659158551D-09	2.50	-2.331997697481D-02
3	-6.5	-3.448398221050D-08	-1.724199110525D-08	2.75	-1.420379566426D-02
4	-5.5	8.388145058100D-08	4.194072529050D-08	3.25	1.128587876787D-02
5	-4.5	-1.642876162612D-07	-8.214380813059D-08	3.50	1.466786262995D-02

CARDINAL SPLINES N=8

6	-3.5	2.767806215892D-07	1.383903107946D-07	3.75	9.705153150485D-03
7	-2.5	-4.212717276112D-07	-2.106358638056D-07	4.25	-8.913623373521D-03
8	-1.5	5.977257256474D-07	2.988628628237D-07	4.50	-1.233649148585D-02
9	-.5	-8.014779045954D-07	-4.007389522977D-07	4.75	-8.641232911213D-03
10	.5	1.013638119752D-06	5.068190598760D-07	5.25	8.751497716568D-03
11	1.5	-1.191520130632D-06	-5.957600653162D-07	5.50	1.262547800043D-02
12	2.5	1.269881589717D-06	6.349407948584D-07	5.75	9.177960578136D-03
13	3.5	-1.187107850265D-06	-5.935539251325D-07	6.25	-9.894706260079D-03
14	4.5	9.351751364399D-07	4.675875682199D-07	6.50	-1.465315089438D-02
15	5.5	-5.925481052424D-07	-2.962740526212D-07	6.75	-1.090256213835D-02
16	6.5	2.855623413709D-07	1.427811706854D-07	7.25	1.222341440495D-02
17	7.5	-9.708688028697D-08	-4.854344014349D-08	7.50	1.840395700212D-02
18	8.5	2.056584890137D-08	1.028292445068D-08	7.75	1.389954749378D-02
19	9.5	-2.029462028655D-09	-1.014731014328D-09	8.25	-1.599723114394D-02
				8.50	-2.437169188406D-02
	0	-5.203167847567D+03	2.984811710684D+04	8.75	-1.861527149716D-02
	1	-5.275055227496D+03	-3.606713316463D+04	9.25	2.190116738484D-02
	2	-2.290112656685D+03	1.360708187721D+04	9.50	3.374421904871D-02
	3	-5.513828858260D+02	-3.885923995720D+03	9.75	2.608051091586D-02
	4	-7.944141852651D+01	4.888175130931D+02	10.25	-3.151196915372D-02
	5	-6.843375110305D+00	-4.966058078463D+01	10.50	-4.931416199547D-02
	6	-3.261107417873D-01	2.073045169482D+00	10.75	-3.879548617209D-02
	7	-6.627208491515D-03	-4.938941912234D-02	11.25	4.900239965596D-02

I=15 H= 5.5

0		0.	-1.166765841234D+07	-1.00	-9.378627925537D+00
1		1.200916599609D+00	5.015547997952D+06	-.75	-4.532601627780D+00
2		-3.096771940147D+00	-9.234223945189D+05	-.50	-1.868231642586D+00
3		3.070351714952D+00	9.438421851166D+04	-.25	-5.481642017035D-01
4		-1.534250864778D+00	-5.784145473115D+03	.25	1.490529118247D-01
5		4.146744900335D-01	2.126445440348D+02	.50	1.262427157345D-01
6		-5.829115779608D-02	-4.360066235434D+00	.75	5.709092226965D-02
7		3.371158223112D-03	3.948566722351D-02	1.25	-2.731143735940D-02
				1.50	-2.850888496403D-02
0	-9.5	-9.635985712074D-11	-4.817992856037D-11	1.75	-1.544080063468D-02
1	-8.5	1.037322815347D-09	5.186614076736D-10	2.25	9.985517088116D-03
2	-7.5	-5.321544635555D-09	-2.660772317777D-09	2.50	1.185638591116D-02
3	-6.5	1.756649558451D-08	8.783247792256D-09	2.75	7.221127100397D-03
4	-5.5	-4.276482555525D-08	-2.138241277762D-08	3.25	-5.736883308912D-03
5	-4.5	8.388145058100D-08	4.194072529050D-08	3.50	-7.455375365507D-03
6	-3.5	-1.417024603091D-07	-7.085123015454D-08	3.75	-4.932424434080D-03
7	-2.5	2.167576733991D-07	1.083788366995D-07	4.25	4.528953381480D-03
8	-1.5	-3.103135474735D-07	-1.551567737367D-07	4.50	6.267036550179D-03
9	-.5	4.224486308088D-07	2.112243154044D-07	4.75	4.388945514183D-03
10	.5	-5.470075147898D-07	-2.735037573949D-07	5.25	-4.442686623208D-03
11	1.5	6.644182858802D-07	3.322091429401D-07	5.50	-6.407226353896D-03
12	2.5	-7.372723826626D-07	-3.686361913313D-07	5.75	-4.655884588525D-03
13	3.5	7.202745262735D-07	3.601372631368D-07	6.25	5.014556101265D-03
14	4.5	-5.925481052424D-07	-2.962740526212D-07	6.50	7.421455029701D-03
15	5.5	3.904839734326D-07	1.952419867163D-07	6.75	5.517812133578D-03
16	6.5	-1.946118786304D-07	-9.730593931522D-08	7.25	-6.174637549921D-03
17	7.5	6.802812727890D-08	3.401406363945D-08	7.50	-9.285569173593D-03
18	8.5	-1.473907051529D-08	-7.369535257647D-09	7.75	-7.002993971437D-03
19	9.5	1.481203617110D-09	7.406018085548D-10	8.25	8.030669502760D-03
				8.50	1.220638854198D-02
	0	2.647218014240D+03	-2.365730597994D+04	8.75	9.297888995319D-03

CARDINAL SPLINES N=8

1		2.683720294516D+03	2.727574418578D+04	9.25	-1.086231933639D-02
2		1.165079703602D+03	-1.095912222234D+04	9.50	-1.666059050286D-02
3		2.805052226627D+02	2.990618711197D+03	9.75	-1.280782349181D-02
4		4.041326989930D+01	-4.010784599314D+02	10.25	1.525978598948D-02
5		3.481266118011D+00	3.895597942036D+01	10.50	2.366343026173D-02
6		1.658908640409D-01	-1.734269365071D+00	10.75	1.841256398890D-02
7		3.371158223112D-03	3.948566722351D-02	11.25	-2.258097033348D-02

I=16 H= 6.5

0		0.	6.976921403359D+06	-1.00	3.847307913479D+00
1		-4.926286153357D-01	-3.013118422684D+06	-.75	1.859357307341D+00
2		1.270343350744D+00	5.574056489697D+05	-.50	7.663787677518D-01
3		-1.259525931207D+00	-5.725212453851D+04	-.25	2.248643381437D-01
4		6.293965206125D-01	3.525994282190D+03	.25	-6.114258218389D-02
5		-1.701163405899D-01	-1.302502154436D+02	.50	-5.178521213559D-02
6		2.391408536350D-02	2.679281235395D+00	.75	-2.341866522198D-02
7		-1.383069626500D-03	-2.409284879611D-02	1.25	1.120286696775D-02
				1.50	1.169387730457D-02
0	-9.5	3.954428173053D-11	1.977214086527D-11	1.75	6.333457893021D-03
1	-8.5	-4.257379753981D-10	-2.128689876991D-10	2.25	-4.095658426941D-03
2	-7.5	2.184397894927D-09	1.092198947463D-09	2.50	-4.862889195550D-03
3	-6.5	-7.212561260029D-09	-3.606280630014D-09	2.75	-2.961652423938D-03
4	-5.5	1.756649558451D-08	8.783247792256D-09	3.25	2.352731044514D-03
5	-4.5	-3.448398221050D-08	-1.724199110525D-08	3.50	3.057348816118D-03
6	-3.5	5.834237671784D-08	2.917118835892D-08	3.75	2.022606828133D-03
7	-2.5	-8.949578350264D-08	-4.474789175132D-08	4.25	-1.856890004190D-03
8	-1.5	1.287888624277D-07	6.439443121383D-08	4.50	-2.569274894355D-03
9	-.5	-1.769518456465D-07	-8.847592282324D-08	4.75	-1.799125230281D-03
10	.5	2.326709675053D-07	1.163354837526D-07	5.25	1.820647169062D-03
11	1.5	-2.892125024052D-07	-1.446062512026D-07	5.50	2.625263922753D-03
12	2.5	3.308706272794D-07	1.654353136397D-07	5.75	1.907278686839D-03
13	3.5	-3.348322175483D-07	-1.674161087742D-07	6.25	-2.053105335692D-03
14	4.5	2.855623413709D-07	1.427811706854D-07	6.50	-3.037521717350D-03
15	5.5	-1.946118786304D-07	-9.730593931522D-08	6.75	-2.257473350057D-03
16	6.5	9.988871719642D-08	4.994435859821D-08	7.25	2.523598396695D-03
17	7.5	-3.579559791672D-08	-1.789779895836D-08	7.50	3.792566891481D-03
18	8.5	7.917027729831D-09	3.958513864915D-09	7.75	2.858080594673D-03
19	9.5	-8.092508928079D-10	-4.046254464040D-10	8.25	-3.271055668275D-03
				8.50	-4.965697453097D-03
	0	-1.086162492988D+03	1.381491270663D+04	8.75	-3.776926547855D-03
	1	-1.101123346324D+03	-1.557430413899D+04	9.25	4.395820076392D-03
	2	-4.780222218339D+02	6.507482558956D+03	9.50	6.726128613696D-03
	3	-1.150872759991D+02	-1.740507842882D+03	9.75	5.156096618679D-03
	4	-1.658076283229D+01	2.432048027543D+02	10.25	-6.098600258463D-03
	5	-1.428276184495D+00	-2.319315670693D+01	10.50	-9.413076871808D-03
	6	-6.806004479876D-02	1.077106790453D+00	10.75	-7.283899716178D-03
	7	-1.383069626500D-03	-2.409284879611D-02	11.25	8.805004645426D-03

I=17 H= 7.5

0		0.	-2.876028912126D+06	-1.00	-1.164607374959D+00
1		1.491200581021D-01	1.248248130768D+06	-.75	-5.628389181611D-01
2		-3.845390490909D-01	-2.321166654917D+05	-.50	-2.319867995106D-01
3		3.812680978523D-01	2.397005122158D+04	-.25	-6.806734164551D-02
4		-1.905253586565D-01	-1.484510252604D+03	.25	1.850794634009D-02
5		5.149684068209D-02	5.514800573191D+01	.50	1.567537764167D-02

6		-7,239279726374D-03	-1.139959202962D+00	.75	7,088786889096D-03
7		4,186908492822D-04	1.024038687334D-02	1.25	-3,391040866205D-03
				1.50	-3,539637512256D-03
0	-9,5	-1,197299339575D-11	-5.986496697877D-12	1.75	-1,917065887387D-03
1	-8,5	1,289094267353D-10	6.445471336766D-11	2.25	1,239680351888D-03
2	-7,5	-6,614717246780D-10	-3.307358623390D-10	2.50	1,471885567131D-03
3	-6,5	2,184397894927D-09	1.092198947463D-09	2.75	8,964094175649D-04
4	-5,5	-5,321544635555D-09	-2,660772317777D-09	3,25	-7,120753689640D-04
5	-4,5	1,045131831710D-08	5,225659158551D-09	3,50	-9,253091723081D-04
6	-3,5	-1,769751172398D-08	-8,848755861989D-09	3,75	-6,121240815399D-04
7	-2,5	2,719151968510D-08	1,359575984255D-08	4.25	5,619252888579D-04
8	-1,5	-3,924849545714D-08	-1,962424772857D-08	4,50	7.774639739005D-04
9	-,5	5,422637848151D-08	2,711318924075D-08	4,75	5,443827224655D-04
10	,5	-7,199949453217D-08	-3,599974726609D-08	5,25	-5,508076857949D-04
11	1,5	9,091285767842D-08	4,545642883921D-08	5,50	-7,941514842945D-04
12	2,5	-1,063494035645D-07	-5,317470178227D-08	5,75	-5,768898804116D-04
13	3,5	1,105890756942D-07	5,529453784708D-08	6.25	6,208085387801D-04
14	4,5	-9,708688028697D-08	-4,854344014349D-08	6,50	9,182938367872D-04
15	5,5	6,802812727890D-08	3,401406363945D-08	6,75	6,823165491863D-04
16	6,5	-3,579559791672D-08	-1,789779895836D-08	7,25	-7,623075976426D-04
17	7,5	1,310466090279D-08	6,552330451395D-09	7,50	-1,145202847468D-03
18	8,5	-2,951110895248D-09	-1,475555447624D-09	7,75	-8,626500608980D-04
19	9,5	3,062383707601D-10	1,531191853801D-10	8,25	9,862006998510D-04
				8,50	1,496067938095D-03
	0	3,288266738490D+02	-5,521620996933D+03	8.75	1,136972214606D-03
	1	3,333531714519D+02	6,160867510492D+03	9,25	-1,320482068074D-03
	2	1,447148659793D+02	-2,642863725561D+03	9,50	-2,017785152367D-03
	3	3,484086382184D+01	7,015870516413D+02	9,75	-1,544351256783D-03
	4	5,019526985780D+00	-1,009049420050D+02	10,25	1,819301939864D-03
	5	4,323817283808D-01	9,578424384752D+00	10,50	2,800875629704D-03
	6	2,060366175089D-02	-4,589734758855D-01	10,75	2,160812113446D-03
	7	4,186908492822D-04	1,024038687334D-02	11,25	-2,591922854462D-03

I=18 H= 8.5

0		0.	7,205179645637D+05	-1,00	2,269131804081D-01
1		-2,905444557010D-02	-3,142820467315D+05	-.75	1,096638577501D-01
2		7,492361067822D-02	5,875173362268D+04	-,50	4,520035116095D-02
3		-7,428670721754D-02	-6,101148530141D+03	-,25	1,326222440096D-02
4		3,712243492183D-02	3,800896880010D+02	.25	-3,606065309876D-03
5		-1,003385631104D-02	-1,420679741760D+01	.50	-3,054161937771D-03
6		1,410544602842D-03	2,953999285385D-01	.75	-1,381161618821D-03
7		-8,158110654172D-05	-2,660185637276D-03	1,25	6,606966102380D-04
				1.50	6,896451923062D-04
0	-9,5	2,331368430180D-12	1.166568421509D-12	1.75	3,735093912534D-04
1	-8,5	-2,512094341582D-11	-1,256047170791D-11	2,25	-2,415283801828D-04
2	-7,5	1,289094267353D-10	6,445471336766D-11	2.50	-2,867667187823D-04
3	-6,5	-4,257379753981D-10	-2,128689876991D-10	2,75	-1,746452367813D-04
4	-5,5	1,037322815347D-09	5,186614076736D-10	3,25	1,387283522927D-04
5	-4,5	-2,037825073804D-09	-1,018912536902D-09	3,50	1,802681998197D-04
6	-3,5	3,452479616798D-09	1,726239808399D-09	3,75	1,192513974704D-04
7	-2,5	-5,309732660799D-09	-2,654866330399D-09	4,25	-1,094665547262D-04
8	-1,5	7,678103554383D-09	3,839051777191D-09	4,50	-1,514501356101D-04
9	-,5	-1,064435072139D-08	-5,322175360695D-09	4,75	-1,060419813977D-04
10	,5	1,422079342884D-08	7,110396714419D-09	5,25	1,072834361465D-04
11	1,5	-1,814686892454D-08	-9,073434462269D-09	5,50	1,546713997219D-04
12	2,5	2,156959446637D-08	1,078479723318D-08	5,75	1,123489291240D-04

13	3.5	-2.289654160188D-08	-1.144827080094D-08	6.25	-1.208801929980D-04
14	4.5	2.056584890137D-08	1.028292445068D-08	6.50	-1.787842134836D-04
15	5.5	-1.473907051529D-08	-7.369535257647D-09	6.75	-1.328234152611D-04
16	6.5	7.917027729831D-09	3.958513864915D-09	7.25	1.483435654238D-04
17	7.5	-2.951110895248D-09	-1.475555447624D-09	7.50	2.228052859735D-04
18	8.5	6.749036020599D-10	3.374518010299D-10	7.75	1.677898469483D-04
19	9.5	-7.095736680796D-11	-3.547868340398D-11	8.25	-1.916947705599D-04
				8.50	-2.906798666993D-04
	0	-6.407324332180D+01	1.338644448423D+03	8.75	-2.208008316890D-04
	1	-6.495492893023D+01	-1.487217934341D+03	9.25	2.561184433212D-04
	2	-2.819801605833D+01	6.501584473526D+02	9.50	3.910567552286D-04
	3	-6.788788329718D+00	-1.723630205777D+02	9.75	2.990254649681D-04
	4	-9.780570264973D-01	2.533728079730D+01	10.25	-3.514316832044D-04
	5	-8.424940612223D-02	-2.410718319955D+00	10.50	-5.402326646656D-04
	6	-4.014598982182D-03	1.184975836596D-01	10.75	-4.160476119474D-04
	7	-8.158110654172D-05	-2.660185637276D-03	11.25	4.968320985777D-04

I=19 H= 9.5

0		0.	-8.216212043881D+04	-1.00	-2.107267451027D-02
1		2.698175825180D-03	3.601218590734D+04	-.75	-1.018410725920D-02
2		-6.957887892003D-03	-6.767181984113D+03	-.50	-4.197596751039D-03
3		6.898764624855D-03	7.066781017279D+02	-.25	-1.231614190984D-03
4		-3.447454642232D-03	-4.428885727626D+01	.25	3.348810028248D-04
5		9.318204650365D-04	1.665987406289D+00	.50	2.836273897339D-04
6		-1.309947207918D-04	-3.486572338439D-02	.75	1.282624979806D-04
7		7.576340173074D-06	3.154165774638D-04	1.25	-6.135572217314D-05
				1.50	-6.404384201230D-05
0	-9.5	-2.166885042567D-13	-1.083442521284D-13	1.75	-3.468579637810D-05
1	-8.5	2.333136843018D-12	1.166568421509D-12	2.25	2.242923952287D-05
2	-7.5	-1.197299339575D-11	-5.986496697877D-12	2.50	2.663009943750D-05
3	-6.5	3.954428173053D-11	1.977214086527D-11	2.75	1.621802866311D-05
4	-5.5	-9.635985712074D-11	-4.817992856037D-11	3.25	-1.288248490153D-05
5	-4.5	1.893318920888D-10	9.466594604438D-11	3.50	-1.673975756322D-05
6	-3.5	-3.208687127230D-10	-1.604343563615D-10	3.75	-1.107358843405D-05
7	-2.5	4.937782723793D-10	2.468891361897D-10	4.25	1.016466725789D-05
8	-1.5	-7.148435814783D-10	-3.574217907392D-10	4.50	1.406283809075D-05
9	-.5	9.931476048629D-10	4.965738024314D-10	4.75	9.846258483949D-06
10	.5	-1.332164201646D-09	-6.660821008231D-10	5.25	-9.960946725269D-06
11	1.5	1.712108684383D-09	8.560543421916D-10	5.50	-1.436024108650D-05
12	2.5	-2.058396656300D-09	-1.029198328150D-09	5.75	-1.043041383499D-05
13	3.5	2.219262603816D-09	1.109631301908D-09	6.25	1.122118599595D-05
14	4.5	-2.029462028655D-09	-1.014731014328D-09	6.50	1.659516706286D-05
15	5.5	1.481203617110D-09	7.406018085548D-10	6.75	1.232793866502D-05
16	6.5	-8.092508928079D-10	-4.046254464040D-10	7.25	-1.376546383888D-05
17	7.5	3.062383707601D-10	1.531191853801D-10	7.50	-2.067226988945D-05
18	8.5	-7.095736680796D-11	-3.547868340398D-11	7.75	-1.556533683122D-05
19	9.5	7.544515466346D-12	3.772257733173D-12	8.25	1.777562575125D-05
				8.50	2.694738795475D-05
	0	5.950521521954D+00	-1.478106969674D+02	8.75	2.046303190668D-05
	1	6.032385600829D+00	1.640078669354D+02	9.25	-2.371770900108D-05
	2	2.618751071039D+00	-7.273221286559D+01	9.50	-3.619580924699D-05
	3	6.304732312211D-01	1.931323846649D+01	9.75	-2.766156086457D-05
	4	9.083167713122D-02	-2.888868425538D+00	10.25	3.246189589454D-05
	5	7.824180092922D-03	2.764344418171D-01	10.50	4.985545286243D-05
	6	3.728319007176D-04	-1.389052098305D-02	10.75	3.835354994889D-05
	7	7.576340173074D-06	3.154165774638D-04	11.25	-4.567533687356D-05

CARDINAL SPLINES N=8

```
                N= 8      M=20                            P= 10.0      Q=15

I= 0     H=-10.0
0            1.000000000000D+00  -1.392642260593D+03    -1.00  9.406089974316D+00
1           -2.779806875395D+00   5.571647829816D+02     -.75  5.801206524784D+00
2            3.079180402795D+00  -9.438305752837D+01     -.50  3.427006492394D+00
3           -1.801098957621D+00   8.245708630498D+00     -.25  1.918049655315D+00
4            6.105698544039D-01  -2.176338480404D-01      .25  4.716244635889D-01
5           -1.215157801709D-01  -4.283235957508D-02      .50  1.893204011151D-01
6            1.329472995502D-02   6.306241016067D-03      .75  5.398001695719D-02
7           -6.233739749621D-04  -3.089070791313D-04     1.25 -1.375539788098D-02
                                                         1.50 -1.122959860346D-02
0  -10.0     7.547867154203D-12   3.773933577101D-12     1.75 -4.878995998403D-03
1   -9.0    -7.099345819973D-11  -3.549672909986D-11     2.25  2.130719519208D-03
2   -8.0     3.064236059278D-10   1.532118029639D-10     2.50  2.112045689386D-03
3   -7.0    -8.098628215504D-10  -4.049314107752D-10     2.75  1.081382262000D-03
4   -6.0     1.482695323030D-09   7.413476615151D-10     3.25 -6.159051440066D-04
5   -5.0    -2.032395089068D-09  -1.016197544534D-09     3.50 -6.809995784965D-04
6   -4.0     2.224240023864D-09   1.112120011932D-09     3.75 -3.841603514113D-04
7   -3.0    -2.066075674872D-09  -1.033037837436D-09     4.25  2.575561741170D-04
8   -2.0     1.723279148419D-09   8.616395742095D-10     4.50  3.049663942027D-04
9   -1.0    -1.347826380327D-09  -6.739131901635D-10     4.75  1.828574966560D-04
10   0.0     1.014526389246D-09   5.072631946228D-10     5.25 -1.358181427444D-04
11   1.0    -7.432458880873D-10  -3.716229440437D-10     5.50 -1.678298717665D-04
12   2.0     5.300923729643D-10   2.650461864822D-10     5.75 -1.044977920814D-04
13   3.0    -3.644064529332D-10  -1.822032264666D-10     6.25  8.262112363277D-05
14   4.0     2.362025781444D-10   1.181012890722D-10     6.50  1.047489160945D-04
15   5.0    -1.391878573090D-10  -6.959392865448D-11     6.75  6.670404628095D-05
16   6.0     7.078863767065D-11   3.539431883532D-11     7.25 -5.471885570025D-05
17   7.0    -2.903896903226D-11  -1.451948451613D-11     7.50 -7.042204015025D-05
18   8.0     8.790301752999D-12   4.395150876499D-12     7.75 -4.543573031655D-05
19   9.0    -1.712710541258D-12  -8.563552706290D-13     8.25  3.808038469458D-05
20  10.0     1.590537477873D-13   7.952687389364D-14     8.50  4.944178784986D-05
                                                         8.75  3.214816825664D-05
           0 -5.048682537442D+02  -2.559957662750D+02     9.25 -2.729910326348D-05
           1 -5.018153287564D+02  -2.475420936314D+02     9.50 -3.564378360511D-05
           2 -2.146455315596D+02  -1.086844169543D+02     9.75 -2.329684333081D-05
           3 -5.118037708868D+01  -2.528466224083D+01    10.25  1.997210511006D-05
           4 -7.341213345284D+00  -3.711638072291D+00    10.50  2.619627810452D-05
           5 -6.329173302902D-01  -3.131627647869D-01    10.75  1.720036063376D-05
           6 -3.034144829233D-02  -1.531725452313D-02    11.25 -1.488788894544D-05
           7 -6.233739749621D-04  -3.089070791313D-04    11.50 -1.963039918688D-05

I= 1     H= -9.0
0            0.                   1.500194099504D+04    -1.00 -4.060088531444D+01
1            8.671628104569D+00  -6.010419065767D+03     -.75 -2.148501475334D+01
2           -1.538940298126D+01   1.025241820468D+03     -.50 -9.882252036628D+00
```

CARDINAL SPLINES N=8

3		1.119497172384D+01	-9.289342761030D+01	-.25	-3.322410648110D+00
4		-4.303648083703D+00	3.479197379358D+00	.25	1.365060587498D+00
5		9.286027770771D-01	2.712712207922D-01	.50	1.646238548279D+00
6		-1.073915922218D-01	-5.002241813592D-02	.75	1.410307304108D+00
7		5.240051763700D-03	2.590090549500D-03	1.25	6.010096696308D-01
				1.50	2.949477423722D-01
0	-10.0	-7.099345819973D-11	-3.549672909986D-11	1.75	9.987268571753D-02
1	-9.0	6.752922386559D-10	3.376461193280D-10	2.25	-3.379739177279D-02
2	-8.0	-2.953105530575D-09	-1.476552765288D-09	2.50	-3.109599240609D-02
3	-7.0	7.923617054650D-09	3.961808527325D-09	2.75	-1.505075877669D-02
4	-6.0	-1.475513339021D-08	-7.377566695103D-09	3.25	7.924124802249D-03
5	-5.0	2.059743246074D-08	1.029871623037D-08	3.50	8.517778808086D-03
6	-4.0	-2.295013908011D-08	-1.147506954005D-08	3.75	4.694954645138D-03
7	-3.0	2.165228309356D-08	1.082614154678D-08	4.25	-3.038616916814D-03
8	-2.0	-1.826715387080D-08	-9.133576935398D-09	4.50	-3.549700316850D-03
9	-1.0	1.438944571713D-08	7.194722858565D-09	4.75	-2.104246003791D-03
10	0.0	-1.087456013098D-08	-5.437280065490D-09	5.25	1.535186689106D-03
11	1.0	7.983943124981D-09	3.991971562490D-09	5.50	1.883751912216D-03
12	2.0	-5.700767413098D-09	-2.850383706549D-09	5.75	1.165891232809D-03
13	3.0	3.921299419147D-09	1.960649709573D-09	6.25	-9.130868301210D-04
14	4.0	-2.542534451610D-09	-1.271267225805D-09	6.50	-1.153295285773D-03
15	5.0	1.498500048853D-09	7.492500244267D-10	6.75	-7.320614024311D-04
16	6.0	-7.621810867231D-10	-3.810905433616D-10	7.25	5.974669895169D-04
17	7.0	3.126779752676D-10	1.563389876338D-10	7.50	7.673781100192D-04
18	8.0	-9.465249838903D-11	-4.732624919451D-11	7.75	4.942519853666D-04
19	9.0	1.844248823129D-11	9.221244115644D-12	8.25	-4.131061345094D-04
20	10.0	-1.712710541258D-12	-8.563552706290D-13	8.50	-5.357782683534D-04
				8.75	-3.480534249742D-04
	0	4.575469992658D+03	2.326088005805D+03	9.25	2.951224764325D-04
	1	4.520328617503D+03	2.223922582504D+03	9.50	3.851124744491D-04
	2	1.919668539785D+03	9.744970612584D+02	9.75	2.515865094313D-04
	3	4.538380983121D+02	2.236290179628D+02	10.25	-2.155144708753D-04
	4	6.444091416697D+01	3.266230044758D+01	10.50	-2.825918223997D-04
	5	5.489215947540D+00	2.709116286587D+00	10.75	-1.855002442161D-04
	6	2.594120312372D-01	1.312839203291D-01	11.25	1.604950038449D-04
	7	5.240051763700D-03	2.590090549500D-03	11.50	2.115862926474D-04
I= 2		H= -8.0			
0		0.	-7.699922996156D+04	-1.00	1.073149175721D+02
1		-1.728830519250D+01	3.086282216683D+04	-.75	5.427669070351D+01
2		3.921345109377D+01	-5.282071262032D+03	-.50	2.357077208457D+01
3		-3.309854402990D+01	4.889683022543D+02	-.25	7.348230168835D+00
4		1.404831784342D+01	-2.187782427096D+01	.25	-2.336603091028D+00
5		-3.250541005880D+00	-6.345160893097D-01	.50	-2.195641499489D+00
6		3.956898486864D-01	1.789907812322D-01	.75	-1.130758503895D+00
7		-2.006855790092D-02	-9.880631410802D-03	1.25	8.065717616994D-01
				1.50	1.190299081099D+00
0	-10.0	3.064236059278D-10	1.532118029639D-10	1.75	1.212211510786D+00
1	-9.0	-2.953105530575D-09	-1.476552765288D-09	2.25	6.875215077072D-01
2	-8.0	1.311489815304D-08	6.557449076520D-09	2.50	3.807019193575D-01
3	-7.0	-3.582941691421D-08	-1.791470845711D-08	2.75	1.437797434484D-01
4	-6.0	6.811056824839D-08	3.405528412420D-08	3.25	-5.880525892770D-02
5	-5.0	-9.724897949210D-08	-4.862448974605D-08	3.50	-5.874722209909D-02
6	-4.0	1.108641589586D-07	5.543207947932D-08	3.75	-3.064936111916D-02
7	-3.0	-1.067737940071D-07	-5.338689700355D-08	4.25	1.838237425343D-02
8	-2.0	9.153020716538D-08	4.576510358269D-08	4.50	2.090140079272D-02

9	-1.0	-7.286508417485D-08	-3.643254208742D-08	4.75	1.212033247151D-02
10	0.0	5.540790339651D-08	2.770395169826D-08	5.25	-8.554168874281D-03
11	1.0	-4.081818399983D-08	-2.040909199992D-08	5.50	-1.036549306144D-02
12	2.0	2.919846329046D-08	1.459923164523D-08	5.75	-6.348220609693D-03
13	3.0	-2.010367869216D-08	-1.005183934608D-08	6.25	4.891044626036D-03
14	4.0	1.304168465257D-08	6.520842326287D-09	6.50	6.138671651523D-03
15	5.0	-7.688486945115D-09	-3.844243472557D-09	6.75	3.875635097331D-03
16	6.0	3.911155802946D-09	1.955577901473D-09	7.25	-3.136394029385D-03
17	7.0	-1.604643939361D-09	-8.023219696806D-10	7.50	-4.015003029228D-03
18	8.0	4.857726180605D-10	2.428863090303D-10	7.75	-2.578680702250D-03
19	9.0	-9.465249838903D-11	-4.732624919451D-11	8.25	2.145715298393D-03
20	10.0	8.790301752999D-12	4.395150876499D-12	8.50	2.778013922037D-03
				8.75	1.801963394601D-03
	0	-1.891673444909D+04	-9.655216758488D+03	9.25	-1.524319589937D-03
	1	-1.856235750723D+04	-9.095165083120D+03	9.50	-1.987276773690D-03
	2	-7.820656511529D+03	-3.985586427927D+03	9.75	-1.297222407169D-03
	3	-1.831905127768D+03	-8.990661270301D+02	10.25	1.109840539050D-03
	4	-2.573434859532D+02	-1.309395562662D+02	10.50	1.454556924837D-03
	5	-2.165312167663D+01	-1.064439517806D+01	10.75	9.544079628605D-04
	6	-1.009109204378D+00	-5.126534175239D-01	11.25	-8.252100670549D-04
	7	-2.006855790092D-02	-9.880631410802D-03	11.50	-1.087622576707D-03

I= 3 H= -7.0

0		0.	2.543876240108D+05	-1.00	-1.964984687039D+02
1		2.834161930968D+01	-1.019781610595D+05	-.75	-9.733300525859D+01
2		-6.877644131349D+01	1.748118129923D+04	-.50	-4.124252917256D+01
3		6.286703913231D+01	-1.638687013135D+03	-.25	-1.248492968545D+01
4		-2.858232207610D+01	8.119486671943D+01	.25	3.664140329765D+00
5		6.993751892051D+00	2.236287112274D-01	.50	3.253691038738D+00
6		-8.904709619328D-01	-3.829452144436D-01	.75	1.555288692100D+00
7		4.682401829251D-02	2.290438814604D-02	1.25	-8.604052392243D-01
				1.50	-9.904979046340D-01
0	-10.0	-8.098628215504D-10	-4.049314107752D-10	1.75	-6.071477079253D-01
1	-9.0	7.923617054650D-09	3.961808527325D-09	2.25	5.778997132743D-01
2	-8.0	-3.582941691421D-08	-1.791470845711D-08	2.50	9.636283806140D-01
3	-7.0	1.000004390518D-07	5.000021952589D-08	2.75	1.096152897617D+00
4	-6.0	-1.948842243163D-07	-9.744211215817D-08	3.25	7.536253909187D-01
5	-5.0	2.860978399157D-07	1.430489199579D-07	3.50	4.537938231237D-01
6	-4.0	-3.357409615964D-07	-1.678704807982D-07	3.75	1.850128187471D-01
7	-3.0	3.322726111095D-07	1.661363055547D-07	4.25	-8.645124660951D-02
8	-2.0	-2.912519309865D-07	-1.456259654933D-07	4.50	-9.148347473546D-02
9	-1.0	2.355304632922D-07	1.177652316461D-07	4.75	-5.027886525791D-02
10	0.0	-1.808550410523D-07	-9.042752052613D-08	5.25	3.296392352036D-02
11	1.0	1.339743655267D-07	6.698718276335D-08	5.50	3.892103835783D-02
12	2.0	-9.612576900748D-08	-4.806288450374D-08	5.75	2.334103145246D-02
13	3.0	6.629120601163D-08	3.314560300581D-08	6.25	-1.742766480797D-02
14	4.0	-4.304131845579D-08	-2.152065922789D-08	6.50	-2.161698697402D-02
15	5.0	2.538574529400D-08	1.269287264700D-08	6.75	-1.351426115361D-02
16	6.0	-1.291694670960D-08	-6.458473354801D-09	7.25	1.077178170743D-02
17	7.0	5.300189537111D-09	2.650094768555D-09	7.50	1.370889230293D-02
18	8.0	-1.604643939361D-09	-8.023219696806D-10	7.75	8.761258883469D-03
19	9.0	3.126779752676D-10	1.563389876338D-10	8.25	-7.234027031813D-03
20	10.0	-2.903896903226D-11	-1.451948451613D-11	8.50	-9.337552681671D-03
				8.75	-6.041321965509D-03
	0	4.759400063052D+04	2.444733553797D+04	9.25	5.089921535098D-03
	1	4.635678171881D+04	2.256385177989D+04	9.50	6.625343465640D-03

CARDINAL SPLINES N=8

2		1.936515453187D+04	9.931210989262D+03	9.75	4.318981238353D-03
3		4.492313211864D+03	2.190367929112D+03	10.25	-3.687296418149D-03
4		6.242394698653D+02	3.196120657268D+02	10.50	-4.828564842367D-03
5		5.189593259036D+01	2.534613095130D+01	10.75	-3.166027612259D-03
6		2.387210318543D+00	1.220361955779D+00	11.25	2.734396692536D-03
7		4.682401829251D-02	2.290438814604D-02	11.50	3.602349109829D-03

I= 4 H= -6.0

0		0.	-6.202046582759D+05	-1.00	2.694124864791D+02
1		-3.675794533577D+01	2.486426575186D+05	-.75	1.319628920062D+02
2		9.189044496675D+01	-4.265776725142D+04	-.50	5.520729354891D+01
3		-8.731679291261D+01	4.028177923472D+03	-.25	1.646942175854D+01
4		4.142191461347D+01	-2.120892837806D+02	.25	-4.658826822521D+00
5		-1.055585267223D+01	2.702452068530D+00	.50	-4.040772874523D+00
6		1.393883658588D+00	5.450739551677D-01	.75	-1.877673388193D+00
7		-7.565231967607D-02	-3.658854266477D-02	1.25	9.610986494993D-01
				1.50	1.046959484560D+00
0	-10.0	1.482695323030D-09	7.413476615151D-10	1.75	5.965016225180D-01
1	-9.0	-1.475513339021D-08	-7.377566695103D-09	2.25	-4.416756727134D-01
2	-8.0	6.811056824839D-08	3.405528412420D-08	2.50	-5.754290203375D-01
3	-7.0	-1.948842243163D-07	-9.744211215817D-08	2.75	-3.946241695105D-01
4	-6.0	3.911478737267D-07	1.955739368634D-07	3.25	4.568486247771D-01
5	-5.0	-5.938534963679D-07	-2.969267481839D-07	3.50	8.297783518490D-01
6	-4.0	7.224897823690D-07	3.612448911845D-07	3.75	1.020643054165D+00
7	-3.0	-7.406900151984D-07	-3.703450075992D-07	4.25	8.042558218610D-01
8	-2.0	6.693898250124D-07	3.346949125062D-07	4.50	5.137483539253D-01
9	-1.0	-5.539781414314D-07	-2.769890707157D-07	4.75	2.209644095211D-01
10	0.0	4.319634963341D-07	2.159817481671D-07	5.25	-1.131599060183D-01
11	1.0	-3.229543096028D-07	-1.614771548014D-07	5.50	-1.244914819569D-01
12	2.0	2.329196673439D-07	1.164598336720D-07	5.75	-7.083408360635D-02
13	3.0	-1.610794153297D-07	-8.053970766485D-08	6.25	4.922202816359D-02
14	4.0	1.047417701819D-07	5.237088509095D-08	6.50	5.953830615070D-02
15	5.0	-6.182590289331D-08	-3.091295144665D-08	6.75	3.647371588047D-02
16	6.0	3.147214334615D-08	1.573607167307D-08	7.25	-2.820820547584D-02
17	7.0	-1.291694670960D-08	-6.458473354801D-09	7.50	-3.549737664601D-02
18	8.0	3.911155802946D-09	1.955577901473D-09	7.75	-2.247433077305D-02
19	9.0	-7.621810867231D-10	-3.810905433616D-10	8.25	1.829098875738D-02
20	10.0	7.078863767065D-11	3.539431883532D-11	8.50	2.347932405173D-02
				8.75	1.512015801847D-02
0		-8.250098713080D+04	-4.283598719426D+04	9.25	-1.264650770924D-02
1		-7.973500465739D+04	-3.836926771723D+04	9.50	-1.641490535749D-02
2		-3.302031382825D+04	-1.711632545123D+04	9.75	-1.067500839367D-02
3		-7.586931595477D+03	-3.657452188541D+03	10.25	9.079396569880D-03
4		-1.043376419779D+03	-5.399547408697D+02	10.50	1.187209875769D-02
5		-8.579270447672D+01	-4.142905021743D+01	10.75	7.774641370551D-03
6		-3.901778718737D+00	-2.016124031366D+00	11.25	-6.701488262870D-03
7		-7.565231967607D-02	-3.658854266477D-02	11.50	-8.821873041376D-03

I= 5 H= -5.0

0	0.	1.219765281052D+06	-1.00	-2.949697138472D+02
1	3.905205635059D+01	-4.890280663462D+05	-.75	-1.435763070971D+02
2	-9.910672915500D+01	8.393726293196D+04	-.50	-5.964580706740D+01
3	9.611523607270D+01	-7.960196681217D+03	-.25	-1.765383618177D+01
4	-4.672047283257D+01	4.345000558872D+02	.25	4.899681941563D+00
5	1.222531034470D+01	-9.582839382733D+00	.50	4.200581669110D+00

CARDINAL SPLINES N=8

6		-1.657654934988D+00	-5.301383848754D-01	.75	1.925883653801D+00
7		9.225415659388D-02	4.369296029849D-02	1.25	-9.526411406475D-01
				1.50	-1.015055492440D+00
0	-10.0	-2.032395089068D-09	-1.016197544534D-09	1.75	-5.630358510134D-01
1	-9.0	2.059743246074D-08	1.029871623037D-08	2.25	3.869429893644D-01
2	-8.0	-9.724897949210D-08	-4.862448974605D-08	2.50	4.778135871782D-01
3	-7.0	2.860978399157D-07	1.430489199579D-07	2.75	3.050818804856D-01
4	-6.0	-5.938534963679D-07	-2.969267481839D-07	3.25	-2.756781206265D-01
5	-5.0	9.377418561802D-07	4.688709280901D-07	3.50	-3.918747531575D-01
6	-4.0	-1.191463588127D-06	-5.957317940636D-07	3.75	-2.910706951840D-01
7	-3.0	1.276601495133D-06	6.383007475667D-07	4.25	3.874084930569D-01
8	-2.0	-1.201295398276D-06	-6.006476991379D-07	4.50	7.475456566273D-01
9	-1.0	1.027344084696D-06	5.136720423479D-07	4.75	9.713292985727D-01
10	0.0	-8.201864684078D-07	-4.100932342039D-07	5.25	8.408971479676D-01
11	1.0	6.225805703432D-07	3.112902851716D-07	5.50	5.590362034132D-01
12	2.0	-4.530501793663D-07	-2.265250896831D-07	5.75	2.491699819993D-01
13	3.0	3.148804768660D-07	1.574402384330D-07	6.25	-1.354755754257D-01
14	4.0	-2.053041308258D-07	-1.026520654129D-07	6.50	-1.527956789741D-01
15	5.0	1.213602127099D-07	6.068010635493D-08	6.75	-8.886700556308D-02
16	6.0	-6.182590289331D-08	-3.091295144665D-08	7.25	6.403312734565D-02
17	7.0	2.538574529400D-08	1.269287264700D-08	7.50	7.861464824136D-02
18	8.0	-7.688486945115D-09	-3.844243472557D-09	7.75	4.879272007449D-02
19	9.0	1.498500048853D-09	7.492500244267D-10	8.25	-3.855643939094D-02
20	10.0	-1.391878573090D-10	-6.959392865448D-11	8.50	-4.895216394421D-02
				8.75	-3.123742625043D-02
	0	1.068080208162D+05	5.652206827632D+04	9.25	2.576332444548D-02
	1	1.024612493644D+05	4.828413350413D+04	9.50	3.326114167807D-02
	2	4.209065869949D+04	2.223746109594D+04	9.75	2.153286804154D-02
	3	9.588462775575D+03	4.526734578500D+03	10.25	-1.818566169626D-02
	4	1.306958122707D+03	6.894041198845D+02	10.50	-2.371439601954D-02
	5	1.064997430926D+02	5.036407415156D+01	10.75	-1.549381691949D-02
	6	4.800136026584D+00	2.528368836019D+00	11.25	1.330671317851D-02
	7	9.225415659388D-02	4.369296029849D-02	11.50	1.749221645718D-02

I= 6 H= -4.0

0		0.	-2.070869834161D+06	-1.00	2.742136172309D+02
1		-3.572068000353D+01	8.302756842872D+05	-.75	1.330142578386D+02
2		9.136535331955D+01	-1.425464550841D+05	-.50	5.504798991715D+01
3		-8.956893468034D+01	1.355168528658D+04	-.25	1.622423223384D+01
4		4.412464715547D+01	-7.553027879694D+02	.25	-4.458049805898D+00
5		-1.172564686763D+01	2.078555205623D+01	.50	-3.799207108584D+00
6		1.616808139320D+00	3.012748171313D-01	.75	-1.730092850635D+00
7		-9.154706507267D-02	-4.164073218023D-02	1.25	8.416471471750D-01
				1.50	8.875544948597D-01
0	-10.0	2.224240023864D-09	1.112120011932D-09	1.75	4.863833229930D-01
1	-9.0	-2.295013908011D-08	-1.147506954005D-08	2.25	-3.239208023653D-01
2	-8.0	1.108641589586D-07	5.543207947932D-08	2.50	-3.918016836504D-01
3	-7.0	-3.357409615964D-07	-1.678704807982D-07	2.75	-2.438998817026D-01
4	-6.0	7.224897823690D-07	3.612448911845D-07	3.25	2.051417882305D-01
5	-5.0	-1.191463588127D-06	-5.957317940636D-07	3.50	2.767732329606D-01
6	-4.0	1.590624362385D-06	7.953121811924D-07	3.75	1.916563400153D-01
7	-3.0	-1.795948482260D-06	-8.979742411302D-07	4.25	-1.996060157619D-01
8	-2.0	1.777705373031D-06	8.888526865157D-07	4.50	-3.017794163423D-01
9	-1.0	-1.588703774581D-06	-7.943518872903D-07	4.75	-2.370356225557D-01
10	0.0	1.312603397509D-06	6.563016987543D-07	5.25	3.472475044600D-01
11	1.0	-1.020501003185D-06	-5.102505015925D-07	5.50	6.978985978680D-01

12	2.0	7.539357712410D-07	3.769678856205D-07	5.75	9.404022639885D-01
13	3.0	-5.286566882907D-07	-2.643283441454D-07	6.25	8.653789839172D-01
14	4.0	3.463858779239D-07	1.731929389619D-07	6.50	5.900982038423D-01
15	5.0	-2.053041308258D-07	-1.026520654129D-07	6.75	2.689676000181D-01
16	6.0	1.047417701819D-07	5.237088509095D-08	7.25	-1.517519587381D-01
17	7.0	-4.304131845579D-08	-2.152065922789D-08	7.50	-1.737720757917D-01
18	8.0	1.304168465257D-08	6.520842326287D-09	7.75	-1.024226376043D-01
19	9.0	-2.542534451610D-09	-1.271267225805D-09	8.25	7.544049225133D-02
20	10.0	2.362025781444D-10	1.181012890722D-10	8.50	9.346304113757D-02
				8.75	5.847596895302D-02
	0	-1.107703327634D+05	-6.067839186397D+04	9.25	-4.684011081289D-02
	1	-1.056074206941D+05	-4.780161547059D+04	9.50	-5.981783521556D-02
	2	-4.310499882473D+04	-2.357636371491D+04	9.75	-3.837718952756D-02
	3	-9.755539905129D+03	-4.423634096425D+03	10.25	3.196598222041D-02
	4	-1.321092764790D+03	-7.215385857691D+02	10.50	4.146385812634D-02
	5	-1.069659951610D+02	-4.858349649437D+01	10.75	2.696984834661D-02
	6	-4.791486415767D+00	-2.613576435485D+00	11.25	-2.300250272245D-02
	7	-9.154706507267D-02	-4.164073218023D-02	11.50	-3.015646772389D-02

I= 7 H= -3.0

0		0.	3.197566584041D+06	-1.00	-2.290599110996D+02
1		2.958875406737D+01	-1.282038343274D+06	-.75	-1.109115132700D+02
2		-7.598486500998D+01	2.201417549487D+05	-.50	-4.580963476215D+01
3		7.490533965567D+01	-2.095751551732D+04	-.25	-1.347185857774D+01
4		-3.715933913239D+01	1.181807142240D+03	.25	3.682766478088D+00
5		9.956598911356D+00	-3.617208068989D+01	.50	3.128980134167D+00
6		-1.385751406405D+00	9.024470955991D-02	.75	1.420015939313D+00
7		7.926291643825D-02	3.324939796422D-02	1.25	-6.850988902366D-01
				1.50	-7.188194900906D-01
0	-10.0	-2.066075674872D-09	-1.033037837436D-09	1.75	-3.916214689434D-01
1	-9.0	2.165228309356D-08	1.082614154678D-08	2.25	2.569970676865D-01
2	-8.0	-1.067737940071D-07	-5.338689700355D-08	2.50	3.079542945443D-01
3	-7.0	3.322726111095D-07	1.661363055547D-07	2.75	1.895804432926D-01
4	-6.0	-7.406900151984D-07	-3.703450075992D-07	3.25	-1.548142847631D-01
5	-5.0	1.276601495133D-06	6.383007475667D-07	3.50	-2.047804474611D-01
6	-4.0	-1.795948482260D-06	-8.979742411302D-07	3.75	-1.383714295176D-01
7	-3.0	2.148778997371D-06	1.074389498685D-06	4.25	1.343452624134D-01
8	-2.0	-2.256257509289D-06	-1.128128754644D-06	4.50	1.929105850134D-01
9	-1.0	2.130056846482D-06	1.065028423241D-06	4.75	1.413486132364D-01
10	0.0	-1.843150714377D-06	-9.215753571884D-07	5.25	-1.621952524239D-01
11	1.0	1.484303359182D-06	7.421516795909D-07	5.50	-2.555166461465D-01
12	2.0	-1.123424236387D-06	-5.617121181937D-07	5.75	-2.082055181806D-01
13	3.0	7.997564412908D-07	3.998782206454D-07	6.25	3.244031185015D-01
14	4.0	-5.286566882907D-07	-2.643283441454D-07	6.50	6.688962980282D-01
15	5.0	3.148804768660D-07	1.574402384330D-07	6.75	9.219041655432D-01
16	6.0	-1.610794153297D-07	-8.053970766485D-08	7.25	8.806143948790D-01
17	7.0	6.629120601163D-08	3.314560300581D-08	7.50	6.097551225069D-01
18	8.0	-2.010367869216D-08	-1.005183934608D-08	7.75	2.816871314827D-01
19	9.0	3.921299419147D-09	1.960649709573D-09	8.25	-1.624907633811D-01
20	10.0	-3.644064529332D-10	-1.822032264666D-10	8.50	-1.877786519819D-01
				8.75	-1.115783746087D-01
	0	9.854699848450D+04	5.744517145726D+04	9.25	8.331873228412D-02
	1	9.356365361353D+04	3.905924645411D+04	9.50	1.038341173059D-01
	2	3.803097451835D+04	2.214021003305D+04	9.75	6.531890855489D-02
	3	8.572123311002D+03	3.584872961068D+03	10.25	-5.284533707676D-02
	4	1.156245572167D+03	6.722991008332D+02	10.50	-6.780832049542D-02

	5	9,326363904737D+01	3,906633760857D+01	10,75	-4,371044574418D-02
	6	4,162652744272D+00	2,417702567055D+00	11,25	3,677029036693D-02
	7	7,926291643825D-02	3,324939796422D-02	11,50	4,795271687674D-02

I= 8 H= -2.0

0		0.	-4,658856404812D+06	-1,00	1,796564380797D+02
1		-2,310798996122D+01	1,868002554635D+06	-,75	8,690996837513D+01
2		5,946225796234D+01	-3,207933586602D+05	-,50	3,586017739807D+01
3		-5,878235315625D+01	3,056327294931D+04	-,25	1,053414071269D+01
4		2,926466811613D+01	-1,734556984041D+03	,25	-2,872190507032D+00
5		-7,874545788714D+00	5,602839257393D+01	,50	-2,436549497365D+00
6		1,101292960519D+00	-5,905757591245D-01	,75	-1,103867402468D+00
7		-6,333013452748D-02	-2,236437831165D-02	1,25	5,303617989852D-01
				1,50	5,550649885769D-01
0	-10,0	1,723279148419D-09	8,616395742095D-10	1,75	3,015322004265D-01
1	-9,0	-1,826715387080D-08	-9,133576935398D-09	2,25	-1,964536496899D-01
2	-8,0	9,153020716538D-08	4,576510358269D-08	2,50	-2,343425408898D-01
3	-7,0	-2,912519309865D-07	-1,456259654933D-07	2,75	-1,434987547565D-01
4	-6,0	6,693898250124D-07	3,346949125062D-07	3,25	1,155832011480D-01
5	-5,0	-1,201295398276D-06	-6,006476991379D-07	3,50	1,515297177983D-01
6	-4,0	1,777705373031D-06	8,888526865157D-07	3,75	1,012970536707D-01
7	-3,0	-2,256257509289D-06	-1,128128754644D-06	4,25	-9,555449814506D-02
8	-2,0	2,524633381461D-06	1,262316690730D-06	4,50	-1,345596832457D-01
9	-1,0	-2,538391782725D-06	-1,269195891363D-06	4,75	-9,623045320644D-02
10	0,0	2,325803346015D-06	1,162901673008D-06	5,25	1,029762345126D-01
11	1,0	-1,963862984421D-06	-9,819314922105D-07	5,50	1,540948579235D-01
12	2,0	1,540258305660D-06	7,701291528301D-07	5,75	1,171417766047D-01
13	3,0	-1,123424236387D-06	-5,617121181937D-07	6,25	-1,429791776160D-01
14	4,0	7,539357712410D-07	3,769678856205D-07	6,50	-2,310928666332D-01
15	5,0	-4,530501793663D-07	-2,265250896831D-07	6,75	-1,926070332963D-01
16	6,0	2,329196673439D-07	1,164598336720D-07	7,25	3,115132192509D-01
17	7,0	-9,612576900748D-08	-4,806288450374D-08	7,50	6,522312684966D-01
18	8,0	2,919846329046D-08	1,459923164523D-08	7,75	9,110948688930D-01
19	9,0	-5,700767413098D-09	-2,850383706549D-09	8,25	8,897947390702D-01
20	10,0	5,300923729643D-10	2,650461864822D-10	8,50	6,217729036920D-01
				8,75	2,895759489969D-01
	0	-7,988348972584D+04	-5,184390043148D+04	9,25	-1,693493604173D-01
	1	-7,567235112992D+04	-2,658714489987D+04	9,50	-1,968643488806D-01
	2	-3,069000378409D+04	-1,989699318933D+04	9,75	-1,176161341780D-01
	3	-6,902429291469D+03	-2,429661429964D+03	10,25	8,870905432645D-02
	4	-9,290778890033D+02	-6,017542349388D+02	10,50	1,110798188386D-01
	5	-7,479025066530D+01	-2,637134742801D+01	10,75	7,021022235943D-02
	6	-3,331816456405D+00	-2,156082240940D+00	11,25	-5,736919897427D-02
	7	-6,333013452748D-02	-2,236437831165D-02	11,50	-7,401106378569D-02

I= 9 H= -1.0

0		0.	6,552120887417D+06	-1,00	-1,359349336637D+02
1		1,744616054859D+01	-2,627294527735D+06	-,75	-6,572845709078D+01
2		-4,493942188101D+01	4,512331422564D+05	-,50	-2,710639321574D+01
3		4,448930523462D+01	-4,301066263781D+04	-,25	-7,958126793365D+00
4		-2,218911801119D+01	2,449533724815D+03	,25	2,166936941482D+00
5		5,983624513917D+00	-8,136828093459D+01	,50	1,836826056994D+00
6		-8,389269389654D-01	1,186106326700D+00	,75	8,314342905331D-01
7		4,837653537877D-02	1,110253605767D-02	1,25	-3,986252959533D-01
				1,50	-4,166588160179D-01

CARDINAL SPLINES N=8

0	-10.0	-1.347826380327D-09	-6.739131901635D-10	1.75	-2.260131912331D-01
1	-9.0	1.438944571713D-08	7.194722858565D-09	2.25	1.467165113792D-01
2	-8.0	-7.286508417485D-08	-3.643254208742D-08	2.50	1.746158197364D-01
3	-7.0	2.355304632922D-07	1.177652316461D-07	2.75	1.066419310996D-01
4	-6.0	-5.539781414314D-07	-2.769890707157D-07	3.25	-8.531546212841D-02
5	-5.0	1.027344084696D-06	5.136720423479D-07	3.50	-1.113639828818D-01
6	-4.0	-1.588703774581D-06	-7.943518872903D-07	3.75	-7.406343221515D-02
7	-3.0	2.130056846482D-06	1.065028423241D-06	4.25	6.892755376599D-02
8	-2.0	-2.538391782725D-06	-1.269195891363D-06	4.50	9.620931085523D-02
9	-1.0	2.728086621463D-06	1.364043310732D-06	4.75	6.807363351510D-02
10	0.0	-2.667357546770D-06	-1.333678773385D-06	5.25	-7.077780866336D-02
11	1.0	2.387274183332D-06	1.193637091666D-06	5.50	-1.038659090295D-01
12	2.0	-1.963862984421D-06	-9.819314922105D-07	5.75	-7.706314414049D-02
13	3.0	1.484303359182D-06	7.421516795909D-07	6.25	8.770967131552D-02
14	4.0	-1.020501003185D-06	-5.102505015925D-07	6.50	1.346505281424D-01
15	5.0	6.225805703432D-07	3.112902851716D-07	6.75	1.046935433450D-01
16	6.0	-3.229543096028D-07	-1.614771548014D-07	7.25	-1.326307147701D-01
17	7.0	1.339743655267D-07	6.698718276335D-08	7.50	-2.176640606309D-01
18	8.0	-4.081818399983D-08	-2.040909199992D-08	7.75	-1.838597352514D-01
19	9.0	7.983943124981D-09	3.991971562490D-09	8.25	3.040060949190D-01
20	10.0	-7.432458880873D-10	-3.716229440437D-10	8.50	6.423411539714D-01
				8.75	9.045555967627D-01
	0	6.147951075419D+04	4.746803985939D+04	9.25	8.955799552736D-01
	1	5.816978721659D+04	1.327192700202D+04	9.50	6.295164894668D-01
	2	2.356419751823D+04	1.818196339102D+04	9.75	2.947816654723D-01
	3	5.293797701966D+03	1.210419574371D+03	10.25	-1.741237833617D-01
	4	7.117804374935D+02	5.488679301540D+02	10.50	-2.033820155160D-01
	5	5.723873247141D+01	1.311342438854D+01	10.75	-1.220902670309D-01
	6	2.547430537548D+00	1.963283850737D+00	11.25	9.300374682297D-02
	7	4.837653537877D-02	1.110253605767D-02	11.50	1.170885186192D-01

I=10 H= 0.0

0		0.	-8.995586019044D+06	-1.00	1.005493308398D+02
1		-1.289012825995D+01	3.607526963945D+06	-.75	4.860668150773D+01
2		3.322125725607D+01	-6.196732171208D+05	-.50	2.004002857297D+01
3		-3.291282903578D+01	5.908565089367D+04	-.25	5.881792017290D+00
4		1.643069723111D+01	-3.371636772505D+03	.25	-1.600465980090D+00
5		-4.435748032804D+00	1.136697300098D+02	.50	-1.356103074413D+00
6		6.227109322311D-01	-1.894495500164D+00	.75	-6.135581683024D-01
7		-3.596009189135D-02	2.528054023985D-20	1.25	2.938457646919D-01
				1.50	3.069368230141D-01
0	-10.0	1.014526389246D-09	5.072631946228D-10	1.75	1.663698950909D-01
1	-9.0	-1.087456013098D-08	-5.437280065490D-09	2.25	-1.077973621302D-01
2	-8.0	5.540790339651D-08	2.770395169826D-08	2.50	-1.281467166376D-01
3	-7.0	-1.808550410523D-07	-9.042752052613D-08	2.75	-7.815609651072D-02
4	-6.0	4.319634963341D-07	2.159817481671D-07	3.25	6.231085250383D-02
5	-5.0	-8.201864684078D-07	-4.100932342039D-07	3.50	8.115721556360D-02
6	-4.0	1.312603397509D-06	6.563016987543D-07	3.75	5.383441880649D-02
7	-3.0	-1.843150714377D-06	-9.215753571884D-07	4.25	-4.976614024641D-02
8	-2.0	2.325803346015D-06	1.162901673008D-06	4.50	-6.916355419500D-02
9	-1.0	-2.667357546770D-06	-1.333678773385D-06	4.75	-4.868527611272D-02
10	0.0	2.791263322189D-06	1.395631661094D-06	5.25	4.993781134786D-02
11	1.0	-2.667357546770D-06	-1.333678773385D-06	5.50	7.263534644986D-02
12	2.0	2.325803346015D-06	1.162901673008D-06	5.75	5.331685286526D-02
13	3.0	-1.843150714377D-06	-9.215753571884D-07	6.25	-5.895341542290D-02
14	4.0	1.312603397509D-06	6.563016987543D-07	6.50	-8.874983684146D-02

15	5.0	-8.201864684078D-07	-4.100932342039D-07	6.75	-6.734456513496D-02
16	6.0	4.319634963341D-07	2.159817481671D-07	7.25	7.954536108958D-02
17	7.0	-1.808550410523D-07	-9.042752052613D-08	7.50	1.239882297246D-01
18	8.0	5.540790339651D-08	2.770395169826D-08	7.75	9.769774203531D-02
19	9.0	-1.087456013098D-08	-5.437280065490D-09	8.25	-1.265208246384D-01
20	10.0	1.014526389246D-09	5.072631946228D-10	8.50	-2.095304615152D-01
				8.75	-1.784189621935D-01
	0	-4.587742224454D+04	-4.587742224454D+04	9.25	2.990602787689D-01
	1	-4.338101031047D+04	2.307962513861D-14	9.50	6.356171653903D-01
	2	-1.756277874037D+04	-1.756277874037D+04	9.75	8.999581887960D-01
	3	-3.943246489948D+03	2.279679047069D-15	10.25	8.999581887960D-01
	4	-5.298935222599D+02	-5.298935222599D+02	10.50	6.356171653903D-01
	5	-4.258928507078D+01	2.703865006364D-17	10.75	2.990602787689D-01
	6	-1.894495500164D+00	-1.894495500164D+00	11.25	-1.784189621935D-01
	7	-3.596009189135D-02	2.528054023985D-20	11.50	-2.095304615152D-01

I=11 H= 1.0

0		0.	1.208313417506D+07	-1.00	-7.299189464023D+01
1		9.351850492398D+00	-4.846868070023D+06	-.75	-3.528064632690D+01
2		-2.410885182263D+01	8.327574559316D+05	-.50	-1.454382436707D+01
3		2.389419865289D+01	-7.943012580400D+04	-.25	-4.267991133616D+00
4		-1.193423023260D+01	4.538053687705D+03	.25	1.160927356478D+00
5		3.223726711938D+00	-1.542257811539D+02	.50	9.834676371613D-01
6		-4.528652645130D-01	2.740461374775D+00	.75	4.448573822550D-01
7		2.617146326342D-02	-1.110253605767D-02	1.25	-2.129308594007D-01
				1.50	-2.223409860340D-01
0	-10.0	-7.432458880873D-10	-3.716229440437D-10	1.75	-1.204689826564D-01
1	-9.0	7.983943124981D-09	3.991971562490D-09	2.25	7.798075047538D-02
2	-8.0	-4.081818399983D-08	-2.040909199992D-08	2.50	9.264561393547D-02
3	-7.0	1.339743655267D-07	6.698718276335D-08	2.75	5.646446243922D-02
4	-6.0	-3.229543096028D-07	-1.614771548014D-07	3.25	-4.493672055604D-02
5	-5.0	6.225805703432D-07	3.112902851716D-07	3.50	-5.846195349174D-02
6	-4.0	-1.020501003185D-06	-5.102505015925D-07	3.75	-3.872818753779D-02
7	-3.0	1.484303359182D-06	7.421516795909D-07	4.25	3.567878342127D-02
8	-2.0	-1.963862984421D-06	-9.819314922105D-07	4.50	4.947634345960D-02
9	-1.0	2.387274183332D-06	1.193637091666D-06	4.75	3.473646486934D-02
10	0.0	-2.667357546770D-06	-1.333678773385D-06	5.25	-3.538993874679D-02
11	1.0	2.728086621463D-06	1.364043310732D-06	5.50	-5.125094731700D-02
12	2.0	-2.538391782725D-06	-1.269195891363D-06	5.75	-3.742477970512D-02
13	3.0	2.130056846482D-06	1.065028423241D-06	6.25	4.082046662181D-02
14	4.0	-1.588703774581D-06	-7.943518872903D-07	6.50	6.090576495758D-02
15	5.0	1.027344084696D-06	5.136720423479D-07	6.75	4.572096382168D-02
16	6.0	-5.539781414314D-07	-2.769890707157D-07	7.25	-5.246034027652D-02
17	7.0	2.355304632922D-07	1.177652316461D-07	7.50	-8.018180967592D-02
18	8.0	-7.286508417485D-08	-3.643254208742D-08	7.75	-6.165742232806D-02
19	9.0	1.438944571713D-08	7.194722858565D-09	8.25	7.444226968005D-02
20	10.0	-1.347826380327D-09	-6.739131901635D-10	8.50	1.170885186192D-01
				8.75	9.300374682297D-02
	0	3.345656896458D+04	4.746803985939D+04	9.25	-1.220902670309D-01
	1	3.162593321256D+04	-1.327192700202D+04	9.50	-2.033820155160D-01
	2	1.279972926382D+04	1.818196339102D+04	9.75	-1.741237833617D-01
	3	2.872958553224D+03	-1.210419574371D+03	10.25	2.947816654723D-01
	4	3.859554228145D+02	5.488679301540D+02	10.50	6.295164894668D-01
	5	3.101188369434D+01	-1.311342438854D+01	10.75	8.955799552736D-01
	6	1.379137163926D+00	1.963283850737D+00	11.25	9.045559967627D-01
	7	2.617146326342D-02	-1.110253605767D-02	11.50	6.423411539714D-01

I=12 H= 2.0

0		0.	-1.577147921457D+07	-1.00	5.180968749225D+01
1		-6.635911689923D+00	6.329209741577D+06	-.75	2.504056887702D+01
2		1.710969729539D+01	-1.087930381927D+06	-.50	1.032178453379D+01
3		-1.696076279904D+01	1.038203554834D+05	-.25	3.028760726536D+00
4		8.473398022651D+00	-5.937198208657D+03	.25	-8.236930388360D-01
5		-2.289567879523D+00	2.027014763389D+02	.50	-6.977064948934D-01
6		3.217484278168D-01	-3.721588722756D+00	.75	-3.155586183117D-01
7		-1.860137790417D-02	2.236437831165D-02	1.25	1.509973785202D-01
				1.50	1.576422749740D-01
0	-10.0	5.300923729643D-10	2.650461864822D-10	1.75	8.539640928565D-02
1	-9.0	-5.700767413098D-09	-2.850383706549D-09	2.25	-5.524994590856D-02
2	-8.0	2.919846329046D-08	1.459923164523D-08	2.50	-6.561945628740D-02
3	-7.0	-9.612576900748D-08	-4.806288450374D-08	2.75	-3.997825394755D-02
4	-6.0	2.329196673439D-07	1.164598336720D-07	3.25	3.178675632699D-02
5	-5.0	-4.530501793663D-07	-2.265250896831D-07	3.50	4.132970108759D-02
6	-4.0	7.539357712410D-07	3.769678856205D-07	3.75	2.735992885316D-02
7	-3.0	-1.123424236387D-06	-5.617121181937D-07	4.25	-2.516078194487D-02
8	-2.0	1.540258305660D-06	7.701291528301D-07	4.50	-3.485115753574D-02
9	-1.0	-1.963862984421D-06	-9.819314922105D-07	4.75	-2.443548618335D-02
10	0.0	2.325803346015D-06	1.162901673008D-06	5.25	2.480891469266D-02
11	1.0	-2.538391782725D-06	-1.269195891363D-06	5.50	3.584787455735D-02
12	2.0	2.524633381461D-06	1.262316690730D-06	5.75	2.610817542360D-02
13	3.0	-2.256257509289D-06	-1.128128754644D-06	6.25	-2.828336353398D-02
14	4.0	1.777705373031D-06	8.888526865157D-07	6.50	-4.201468112967D-02
15	5.0	-1.201295398276D-06	-6.006476991379D-07	6.75	-3.137507402694D-02
16	6.0	6.693898250124D-07	3.346949125062D-07	7.25	3.550968895063D-02
17	7.0	-2.912519309865D-07	-1.456259654933D-07	7.50	5.378982263808D-02
18	8.0	9.153020716538D-08	4.576510358269D-08	7.75	4.091841874367D-02
19	9.0	-1.826715387080D-08	-9.133576935398D-09	8.25	-4.798805829494D-02
20	10.0	1.723279148419D-09	8.616395742095D-10	8.50	-7.401106378569D-02
				8.75	-5.736919897427D-02
	0	-2.380431113712D+04	-5.184390043148D+04	9.25	7.021022235943D-02
	1	-2.249806133017D+04	2.658714489987D+04	9.50	1.110798188386D-01
	2	-9.103982594560D+03	-1.989699318933D+04	9.75	8.870905432645D-02
	3	-2.043106431541D+03	2.429661429964D+03	10.25	-1.176161341780D-01
	4	-2.744305808744D+02	-6.017542349388D+02	10.50	-1.968643488806D-01
	5	-2.204755580928D+01	2.637134742801D+01	10.75	-1.693493604173D-01
	6	-9.803480254754D-01	-2.156082240940D+00	11.25	2.895759489969D-01
	7	-1.860137790417D-02	2.236437831165D-02	11.50	6.217729036920D-01

I=13 H= 3.0

0		0.	1.962675291626D+07	-1.00	-3.552649323053D+01
1		4.549588221098D+00	-7.883205945179D+06	-.75	-1.717000913103D+01
2		-1.173131547241D+01	1.356208356234D+06	-.50	-7.077251285233D+00
3		1.163042076335D+01	-1.295345152315D+05	-.25	-2.076617912808D+00
4		-5.811195096824D+00	7.415898760593D+03	.25	5.646957357904D-01
5		1.570473510493D+00	-2.539522273568D+02	.50	4.782961604906D-01
6		-2.207360458474D-01	4.745160424551D+00	.75	2.163097797850D-01
7		1.276412050980D-02	-3.324939796422D-02	1.25	-1.034899516098D-01
				1.50	-1.080340788750D-01
0	-10.0	-3.644064529332D-10	-1.822032264666D-10	1.75	-5.851687726431D-02
1	-9.0	3.921299419147D-09	1.960649709573D-09	2.25	3.784932333115D-02
2	-8.0	-2.010367869216D-08	-1.005183934608D-08	2.50	4.494560793446D-02

CARDINAL SPLINES N=8

3	-7.0	6.629120601163D-08	3.314560300581D-08	2.75	2.737758061950D-02
4	-6.0	-1.610794153297D-07	-8.053970766485D-08	3.25	-2.175734761233D-02
5	-5.0	3.148804768660D-07	1.574402384330D-07	3.50	-2.828056133718D-02
6	-4.0	-5.286566882907D-07	-2.643283441454D-07	3.75	-1.871471112836D-02
7	-3.0	7.997564412908D-07	3.998782206454D-07	4.25	1.719443123366D-02
8	-2.0	-1.123424236387D-06	-5.617121181937D-07	4.50	2.380250735701D-02
9	-1.0	1.484303359182D-06	7.421516795909D-07	4.75	1.667714879317D-02
10	0.0	-1.843150714377D-06	-9.215753571884D-07	5.25	-1.690146800024D-02
11	1.0	2.130056846482D-06	1.065028423241D-06	5.50	-2.439376851467D-02
12	2.0	-2.256257509289D-06	-1.128128754644D-06	5.75	-1.774190676021D-02
13	3.0	2.148778997371D-06	1.074389498685D-06	6.25	1.915273193700D-02
14	4.0	-1.795948482260D-06	-8.979742411302D-07	6.50	2.838738477698D-02
15	5.0	1.276601495133D-06	6.383007475667D-07	6.75	2.114246885530D-02
16	6.0	-7.406900151984D-07	-3.703450075992D-07	7.25	-2.376485738997D-02
17	7.0	3.322726111095D-07	1.661363055547D-07	7.50	-3.584007759429D-02
18	8.0	-1.067737940071D-07	-5.338689700355D-08	7.75	-2.712098609849D-02
19	9.0	2.165228309356D-08	1.082614154678D-08	8.25	3.137250594350D-02
20	10.0	-2.066075674872D-09	-1.033037837436D-09	8.50	4.795271687674D-02
				8.75	3.677029036693D-02
	0	1.634334443002D+04	5.744517145726D+04	9.25	-4.371044574418D-02
	1	1.544516070532D+04	-3.905924645411D+04	9.50	-6.780832049542D-02
	2	6.249445547741D+03	2.214021003305D+04	9.75	-5.284533707676D-02
	3	1.402377388866D+03	-3.584872961068D+03	10.25	6.531890855489D-02
	4	1.883526294998D+02	6.722991008332D+02	10.50	1.038341173059D-01
	5	1.513096383023D+01	-3.906633760857D+01	10.75	8.331873228412D-02
	6	6.727523898387D-01	2.417702567055D+00	11.25	-1.115783746087D-01
	7	1.276412050980D-02	-3.324939796422D-02	11.50	-1.877786519819D-01

I=14 H= 4.0

0		0.	-2.242368427890D+07	-1.00	2.299724432694D+01
1		-2.944820328203D+00	9.021379273044D+06	-.75	1.111440143153D+01
2		7.593652577442D+00	-1.554525505914D+06	-.50	4.581118091312D+00
3		-7.528759878836D+00	1.487144589943D+05	-.25	1.344169069685D+00
4		3.762043991255D+00	-8.528503690022D+03	.25	-3.655017409349D-01
5		-1.016776355835D+00	2.928436202019D+02	.50	-3.095698788511D-01
6		1.429255946530D-01	-5.528427688100D+00	.75	-1.399984483207D-01
7		-8.265600712215D-03	4.164073218023D-02	1.25	6.697455546726D-02
				1.50	6.991188733128D-02
0	-10.0	2.362025781444D-10	1.181012890722D-10	1.75	3.786578069782D-02
1	-9.0	-2.542534451610D-09	-1.271267225805D-09	2.25	-2.448856141027D-02
2	-8.0	1.304168465257D-08	6.520842326287D-09	2.50	-2.907734463321D-02
3	-7.0	-4.304131845579D-08	-2.152065922789D-08	2.75	-1.771005576505D-02
4	-6.0	1.047417701819D-07	5.237088509095D-08	3.25	1.407078676604D-02
5	-5.0	-2.053041308258D-07	-1.026520654129D-07	3.50	1.828647633932D-02
6	-4.0	3.463858779239D-07	1.731929389619D-07	3.75	1.209880010605D-02
7	-3.0	-5.286566882907D-07	-2.643283441454D-07	4.25	-1.111052265743D-02
8	-2.0	7.539357712410D-07	3.769678856205D-07	4.50	-1.537566191281D-02
9	-1.0	-1.020501003185D-06	-5.102505015925D-07	4.75	-1.076894156798D-02
10	0.0	1.312603397509D-06	6.563016987543D-07	5.25	1.090346593265D-02
11	1.0	-1.588703774581D-06	-7.943518872903D-07	5.50	1.572739084838D-02
12	2.0	1.777705373031D-06	8.888526865157D-07	5.75	1.143059150709D-02
13	3.0	-1.795948482260D-06	-8.979742411302D-07	6.25	-1.231698360824D-02
14	4.0	1.590624362385D-06	7.953121811924D-07	6.50	-1.823442343983D-02
15	5.0	-1.191463588127D-06	-5.957317940636D-07	6.75	-1.356202594385D-02
16	6.0	7.224897823690D-07	3.612448911845D-07	7.25	1.519033211071D-02
17	7.0	-3.357409615964D-07	-1.678704807982D-07	7.50	2.285699052298D-02

18	8,0	1,108641589586D-07	5,543207947932D-08	7.75	1,725025441832D-02
19	9,0	-2,295013908011D-08	-1.147506954005D-08	8.25	-1,981722973063D-02
20	10,0	2,224240023864D-09	1.112120011932D-09	8.50	-3,015646772389D-02
				8.75	-2,300250272245D-02
	0	-1,058645096451D+04	-6,067839186397D+04	9.25	2,696984834661D-02
	1	-1,000418975288D+04	4,780161547059D+04	9.50	4,146385812634D-02
	2	-4,047728605089D+03	-2,357636371491D+04	9.75	3,196598222041D-02
	3	-9,082717122791D+02	4,423634096425D+03	10.25	-3,837718952756D-02
	4	-1,219844067485D+02	-7,215385857691D+02	10.50	-5,981783521556D-02
	5	-9,799002172306D+00	4,858349649437D+01	10.75	-4,684011081289D-02
	6	-4,356664552020D-01	-2,613576435485D+00	11.25	5,847596895302D-02
	7	-8,265600712215D-03	4,164073218023D-02	11.50	9,346304113757D-02

I=15 H= 5.0

0		0.	2.218559114442D+07	-1.00	-1,354217331810D+01
1		1,734011367198D+00	-8,949745939790D+06	-.75	-6,544771173933D+00
2		-4,471497661911D+00	1.546333253927D+06	-.50	-2,697585911514D+00
3		4.433415400147D+00	-1.483268863503D+05	-.25	-7,915031278295D-01
4		-2.215414122737D+00	8.529414691938D+03	.25	2,152168962439D-01
5		5,987916557756D-01	-2.938214209395D+02	.50	1,822798503939D-01
6		-8,417487433047D-02	5,586876056913D+00	.75	8,243191636404D-02
7		4,868235996910D-03	-4,369296029849D-02	1.25	-3,943331505792D-02
				1.50	-4,116168620802D-02
0	-10,0	-1,391878573090D-10	-6,959392865448D-11	1.75	-2,229338937436D-02
1	-9,0	1,498500048853D-09	7.492500244267D-10	2.25	1,441651882384D-02
2	-8,0	-7,688486945115D-09	-3,844243472557D-09	2.50	1,711716965272D-02
3	-7,0	2,538574529400D-08	1,269287264700D-08	2.75	1,042491990766D-02
4	-6,0	-6,182590289331D-08	-3,091295144665D-08	3.25	-8,281587692544D-03
5	-5,0	1,213602127099D-07	6.068010635493D-08	3.50	-1,076187675620D-02
6	-4,0	-2,053041308258D-07	-1,026520654129D-07	3.75	-7,119614784607D-03
7	-3,0	3,148804768660D-07	1,574402384330D-07	4.25	6,536365921442D-03
8	-2,0	-4,530501793663D-07	-2,265250896831D-07	4.50	9,044072744264D-03
9	-1,0	6,225805703432D-07	3,112902851716D-07	4.75	6,333135250268D-03
10	0,0	-8,201864684078D-07	-4,100932342039D-07	5.25	-6,409043485446D-03
11	1,0	1,027344084696D-06	5,136720423479D-07	5.50	-9,241593315758D-03
12	2,0	-1,201295398276D-06	-6,006476991379D-07	5.75	-6,714221813088D-03
13	3,0	1,276601495133D-06	6,383007475667D-07	6.25	7,227907418546D-03
14	4,0	-1,191463588127D-06	-5,957317940636D-07	6.50	1,069383600292D-02
15	5,0	9,377418561802D-07	4,688709280901D-07	6.75	7,947892124889D-03
16	6,0	-5,938534963679D-07	-2,969267481839D-07	7.25	-8,885652957508D-03
17	7,0	2,860978399157D-07	1,430489199579D-07	7.50	-1,335452007914D-02
18	8,0	-9,724897949210D-08	-4,862448974605D-08	7.75	-1,006469604733D-02
19	9,0	2,059743246074D-08	1,029871623037D-08	8.25	1,152118594188D-02
20	10,0	-2,032395089068D-09	-1,016197544534D-09	8.50	1,749221645718D-02
				8.75	1,330671317851D-02
	0	6,236115736451D+03	5,652206827632D+04	9.25	-1,549381691949D-02
	1	5,892982356096D+03	-4,828413350413D+04	9.50	-2,371439601954D-02
	2	2,384263492398D+03	2.223746109594D+04	9.75	-1,818566169626D-02
	3	5,349936185755D+02	-4,526734578500D+03	10.25	2,153286804154D-02
	4	7,185011706221D+01	6,894041198845D+02	10.50	3,326114167807D-02
	5	5,771594789459D+00	-5,036407415156D+01	10.75	2,576332444548D-02
	6	2,566016454533D-01	2,528368836019D+00	11.25	-3,123742625043D-02
	7	4,868235996910D-03	-4,369296029849D-02	11.50	-4,895216394421D-02

I=16 H= 6.0

CARDINAL SPLINES N=8

0		0.	-1.772007528648D+07	-1.00	6.884743554796D+00
1		-8.815378008027D-01	7.174997125127D+06	-.75	3.327297468814D+00
2		2.273248449793D+00	-1.244357782105D+06	-.50	1.371418458257D+00
3		-2.253923212476D+00	1.198131626008D+05	-.25	4.023876403888D-01
4		1.126326267308D+00	-6.916192292058D+03	.25	-1.094112631992D-01
5		-3.044355298085D-01	2.392324316954D+02	.50	-9.266604181397D-02
6		4.279706026075D-02	-4.577322017900D+00	.75	-4.190570724781D-02
7		-2.475234346520D-03	3.658854266477D-02	1.25	2.004615313126D-02
				1.50	2.092448867959D-02
0	-10.0	7.078863767065D-11	3.539431883532D-11	1.75	1.133263415044D-02
1	-9.0	-7.621810867231D-10	-3.810905433616D-10	2.25	-7.328211775071D-03
2	-8.0	3.911155802946D-09	1.955577901473D-09	2.50	-8.700794281550D-03
3	-7.0	-1.291694670960D-08	-6.458473354801D-09	2.75	-5.298918754111D-03
4	-6.0	3.147214334615D-08	1.573607167307D-08	3.25	4.209171905054D-03
5	-5.0	-6.182590289331D-08	-3.091295144665D-08	3.50	5.469544246990D-03
6	-4.0	1.047417701819D-07	5.237088509095D-08	3.75	3.618230100767D-03
7	-3.0	-1.610794153297D-07	-8.053970766485D-08	4.25	-3.321359781943D-03
8	-2.0	2.329196673439D-07	1.164598336720D-07	4.50	-4.595209024487D-03
9	-1.0	-3.229543096028D-07	-1.614771548014D-07	4.75	-3.217472009547D-03
10	0.0	4.319634963341D-07	2.159817481671D-07	5.25	3.255165555143D-03
11	1.0	-5.539781414314D-07	-2.769890707157D-07	5.50	4.693022628090D-03
12	2.0	6.693898250124D-07	3.346949125062D-07	5.75	3.408899603212D-03
13	3.0	-7.406900151984D-07	-3.703450075992D-07	6.25	-3.667813845866D-03
14	4.0	7.224897823690D-07	3.612448911845D-07	6.50	-5.424825244243D-03
15	5.0	-5.938534963679D-07	-2.969267481839D-07	6.75	-4.030292157329D-03
16	6.0	3.911478737267D-07	1.955739368634D-07	7.25	4.501368313141D-03
17	7.0	-1.948842243163D-07	-9.744211215817D-08	7.50	6.760992917374D-03
18	8.0	6.811056824839D-08	3.405528412420D-08	7.75	5.091687324414D-03
19	9.0	-1.475513339021D-08	-7.377566695103D-09	8.25	-5.817495790609D-03
20	10.0	1.482695323030D-09	7.413476615151D-10	8.50	-8.821873041376D-03
				8.75	-6.701488262870D-03
	0	-3.170987257722D+03	-4.283598719426D+04	9.25	7.774641370551D-03
	1	-2.996469222931D+03	3.836926771723D+04	9.50	1.187209875769D-02
	2	-1.212337074205D+03	-1.711632545123D+04	9.75	9.079396569880D-03
	3	-2.720272183958D+02	3.657452188541D+03	10.25	-1.067500839367D-02
	4	-3.653306196021D+01	-5.399547408697D+02	10.50	-1.641490535749D-02
	5	-2.934604041856D+00	4.142905021743D+01	10.75	-1.264650770924D-02
	6	-1.304693439957D-01	-2.016124031366D+00	11.25	1.512015801847D-02
	7	-2.475234346520D-03	3.658854266477D-02	11.50	2.347932405173D-02

I=17 H= 7.0

0		0.	1.061371447101D+07	-1.00	-2.823684705646D+00
1		3.615460942384D-01	-4.316601151261D+06	-.75	-1.364642324817D+00
2		-9.323357141558D-01	7.520243062853D+05	-.50	-5.624654509853D-01
3		9.244177088621D-01	-7.274475647619D+04	-.25	-1.650323217153D-01
4		-4.619532916380D-01	4.219115132072D+03	.25	4.487277638467D-02
5		1.248633077365D-01	-1.466670634048D+02	.50	3.800489035841D-02
6		-1.755334701452D-02	2.823669126002D+00	.75	1.718659173450D-02
7		1.015242000429D-03	-2.290438814604D-02	1.25	-8.221331063332D-03
				1.50	-8.581488563307D-03
0	-10.0	-2.903896903226D-11	-1.451948451613D-11	1.75	-4.647665307494D-03
1	-9.0	3.126779752676D-10	1.563389876338D-10	2.25	3.005332900099D-03
2	-8.0	-1.604643939361D-09	-8.023219696806D-10	2.50	3.568187210167D-03
3	-7.0	5.300189537111D-09	2.650094768555D-09	2.75	2.173047358933D-03
4	-6.0	-1.291694670960D-08	-6.458473354801D-09	3.25	-1.726081692452D-03
5	-5.0	2.538574529400D-08	1.269287264700D-08	3.50	-2.242874128124D-03

6	-4.0	-4.304131845579D-08	-2.152065922789D-08	3.75	-1.483669263908D-03
7	-3.0	6.629120601163D-08	3.314560300581D-08	4.25	1.361833137669D-03
8	-2.0	-9.612576900748D-08	-4.806288450374D-08	4.50	1.884049194356D-03
9	-1.0	1.339743655267D-07	6.698718276335D-08	4.75	1.319097734209D-03
10	0.0	-1.808550410523D-07	-9.042752052613D-08	5.25	-1.334355788815D-03
11	1.0	2.355304632922D-07	1.177652316461D-07	5.50	-1.923581700589D-03
12	2.0	-2.912519309865D-07	-1.456259654933D-07	5.75	-1.397090047865D-03
13	3.0	3.322726111095D-07	1.661363055547D-07	6.25	1.502778266289D-03
14	4.0	-3.357409615964D-07	-1.678704807982D-07	6.50	2.222262588935D-03
15	5.0	2.860978399157D-07	1.430489199579D-07	6.75	1.650647681721D-03
16	6.0	-1.948842243163D-07	-9.744211215817D-08	7.25	-1.842585049140D-03
17	7.0	1.000004390518D-07	5.000021952589D-08	7.50	-2.766586444997D-03
18	8.0	-3.582941691421D-08	-1.791470845711D-08	7.75	-2.082668105256D-03
19	9.0	7.923617054650D-09	3.961808527325D-09	8.25	2.377091431312D-03
20	10.0	-8.098628215504D-10	-4.049314107752D-10	8.50	3.602349109829D-03
				8.75	2.734396692536D-03
	0	1.300670445430D+03	2.444733553797D+04	9.25	-3.166027612259D-03
	1	1.229078159034D+03	-2.256385177989D+04	9.50	-4.828564842367D-03
	2	4.972674466568D+02	9.931210989262D+03	9.75	-3.687296418149D-03
	3	1.115773536396D+02	-2.190367929112D+03	10.25	4.318981238353D-03
	4	1.498466158842D+01	3.196120657268D+02	10.50	6.625343465640D-03
	5	1.203670687766D+00	-2.534613095130D+01	10.75	5.089921535098D-03
	6	5.351359301550D-02	1.220361955779D+00	11.25	-6.041321965509D-03
	7	1.015242000429D-03	-2.290438814604D-02	11.50	-9.337552681671D-03

I=18 H= 8.0

0		0.	-4.383526449512D+06	-1.00	8.546496867077D-01
1		-1.094289858672D-01	1.791261186108D+06	-.75	4.130380645599D-01
2		2.821906007878D-01	-3.136125943704D+05	-.50	1.702418487731D-01
3		-2.797954155261D-01	3.049233290000D+04	-.25	4.995036226622D-02
4		1.398212176793D-01	-1.777961540833D+03	.25	-1.358158965419D-02
5		-3.779314688567D-02	6.215292619218D+01	.50	-1.150286414165D-02
6		5.313024882361D-03	-1.204297616280D+00	.75	-5.201815796724D-03
7		-3.072950793165D-04	9.880631410802D-03	1.25	2.488308522398D-03
				1.50	2.597304290246D-03
0	-10.0	8.790301752999D-12	4.395150876499D-12	1.75	1.406672166958D-03
1	-9.0	-9.465249838903D-11	-4.732624919451D-11	2.25	-9.095892001655D-04
2	-8.0	4.857726180605D-10	2.428863090303D-10	2.50	-1.079933526146D-03
3	-7.0	-1.604643939361D-09	-8.023219696806D-10	2.75	-6.576801299120D-04
4	-6.0	3.911155802946D-09	1.955577901473D-09	3.25	5.223926789028D-04
5	-5.0	-7.688486945115D-09	-3.844243472557D-09	3.50	6.787884356695D-04
6	-4.0	1.304168465257D-08	6.520842326287D-09	3.75	4.490134391442D-04
7	-3.0	-2.010367869216D-08	-1.005183934608D-08	4.25	-4.121235491210D-04
8	-2.0	2.919846329046D-08	1.459923164523D-08	4.50	-5.701430605048D-04
9	-1.0	-4.081818399983D-08	-2.040909199992D-08	4.75	-3.991668923676D-04
10	0.0	5.540790339651D-08	2.770395169826D-08	5.25	4.037504781645D-04
11	1.0	-7.286508417485D-08	-3.643254208742D-08	5.50	5.820080675027D-04
12	2.0	9.153020716538D-08	4.576510358269D-08	5.75	4.226838231406D-04
13	3.0	-1.067737940071D-07	-5.338689700355D-08	6.25	-4.545865158848D-04
14	4.0	1.108641589586D-07	5.543207947932D-08	6.50	-6.721601240918D-04
15	5.0	-9.724897949210D-08	-4.862448974605D-08	6.75	-4.992058561731D-04
16	6.0	6.811056824839D-08	3.405528412420D-08	7.25	5.570825558822D-04
17	7.0	-3.582941691421D-08	-1.791470845711D-08	7.50	8.362798094685D-04
18	8.0	1.311489815304D-08	6.557449076520D-09	7.75	6.294017625730D-04
19	9.0	-2.953105530575D-09	-1.476552765288D-09	8.25	-7.179589007860D-04
20	10.0	3.064236059278D-10	1.532118029639D-10	8.50	-1.087622576707D-03

				8.75	-8,252100670549D-04
	0	-3,936990678845D+02	-9,655216758488D+03	9.25	9,544079628605D-04
	1	-3,720273409934D+02	9,095165083120D+03	9,50	1,454556924837D-03
	2	-1,505163443247D+02	-3,985586427927D+03	9,75	1,109840539050D-03
	3	-3,377287370759D+01	8,990661270301D+02	10,25	-1,297222407169D-03
	4	-4,535626579141D+00	-1,309395562662D+02	10,50	-1,987276773690D-03
	5	-3,643313205087D-01	1,064439517806D+01	10,75	-1,524319589937D-03
	6	-1,619763066979D-02	-5,126534175239D-01	11.25	1,801963394601D-03
	7	-3,072950793165D-04	9,880631410802D-03	11,50	2,778013922037D-03
I=19	H= 9.0				
0		0.	1,100363496878D+06	-1,00	-1,665092027617D-01
1		2,131967383160D-02	-4,518085713601D+05	-,75	-8,047106835938D-02
2		-5,497834099638D-02	7,950368893787D+04	-,50	-3,316771577003D-02
3		5,451185190253D-02	-7,771447421359D+03	-,25	-9,731670097048D-03
4		-2,724112327549D-02	4,556971645030D+02	,25	2,646050758870D-03
5		7,363204982873D-03	-1,602534166028D+01	,50	2,241056827773D-03
6		-1,035137108126D-03	3,125902587940D-01	,75	1,013447083722D-03
7		5,987066470059D-05	-2,590090549500D-03	1,25	-4,847842289241D-04
				1,50	-5,060180039535D-04
0	-10,0	-1,712710541258D-12	-8,563552706290D-13	1,75	-2,740531196247D-04
1	-9,0	1,844248823129D-11	9,221244115644D-12	2,25	1,772082675769D-04
2	-8,0	-9,465249838903D-11	-4,732624919451D-11	2,50	2,103941877605D-04
3	-7,0	3,126779752676D-10	1,563389876338D-10	2,75	1,281295029461D-04
4	-6,0	-7,621810867231D-10	-3,810905433616D-10	3,25	-1,017713851266D-04
5	-5,0	1,498500048853D-09	7,492500244267D-10	3,50	-1,322389385691D-04
6	-4,0	-2,542534451610D-09	-1,271267225805D-09	3,75	-8,747418965583D-05
7	-3,0	3,921299419147D-09	1,960649709573D-09	4,25	8,028546169002D-05
8	-2,0	-5,700767413098D-09	-2,850383706549D-09	4,50	1,110673053936D-04
9	-1,0	7,983943124981D-09	3,991971562490D-09	4,75	7,775862134266D-05
10	0,0	-1,087456013098D-08	-5,437280065490D-09	5,25	-7,864762799897D-05
11	1,0	1,438944571713D-08	7,194722858565D-09	5,50	-1,133673256401D-04
12	2,0	-1,826715387080D-08	-9,133576935398D-09	5,75	-8,233005505085D-05
13	3,0	2,165228309356D-08	1,082614154678D-08	6,25	8,853561421409D-05
14	4,0	-2,295013908011D-08	-1,147506954005D-08	6,50	1,309024900266D-04
15	5,0	2,059743246074D-08	1,029871623037D-08	6,75	9,721290183633D-05
16	6,0	-1,475513339021D-08	-7,377566695103D-09	7,25	-1,084637634227D-04
17	7,0	7,923617054650D-09	3,961808527325D-09	7,50	-1,628045243823D-04
18	8,0	-2,953105530575D-09	-1,476552765288D-09	7,75	-1,225134491908D-04
19	9,0	6,752922386559D-10	3,376461193280D-10	8,25	1,397026242215D-04
20	10,0	-7,099345819973D-11	-3,549672909986D-11	8,50	2,115862926474D-04
				8,75	1,604950038449D-04
	0	7,670601895316D+01	2,326088005805D+03	9,25	-1,855002442161D-04
	1	7,248345249467D+01	-2,223922582504D+03	9,50	-2,825918223997D-04
	2	2,932558273180D+01	9,744970612584D+02	9,75	-2,155144708753D-04
	3	6,580062386435D+00	-2,236290179628D+02	10,25	2,515865094313D-04
	4	8,836867281992D-01	3,266230044758D+01	10,50	3,851124744491D-04
	5	7,098337436653D-02	-2,709116286587D+00	10,75	2,951224764325D-04
	6	3,155809420915D-03	1,312839203291D-01	11,25	-3,480534249742D-04
	7	5,987066470059D-05	-2,590090549500D-03	11,50	-5,357782683534D-04
I=20	H= 10.0				
0		0.	-1,257235031549D+05	-1,00	1,546248939616D-02
1		-1,979796402780D-03	5,186402190127D+04	-,75	7,472753948178D-03
2		5,105428160673D-03	-9,172127820067D+03	-,50	3,080038856279D-03

CARDINAL SPLINES N=8

3		-5.062117928701D-03	9.013755180778D+02	-.25	9.037071077983D-04
4		2.529690122288D-03	-5.315740586592D+01	.25	-2.457184469077D-04
5		-6.837705499802D-04	1.880902902350D+00	.50	-2.081095601862D-04
6		9.612641504075D-05	-3.694075006232D-02	.75	-9.411086358587D-05
7		-5.559816699477D-06	3.089070791313D-04	1.25	4.501798051966D-05
				1.50	4.698971348229D-05
0	-10.0	1.590537477873D-13	7.952687389364D-14	1.75	2.544900336930D-05
1	-9.0	-1.712710541258D-12	-8.563552706290D-13	2.25	-1.645576327218D-05
2	-8.0	8.790301752999D-12	4.395150876499D-12	2.50	-1.953739064905D-05
3	-7.0	-2.903896903226D-11	-1.451948451613D-11	2.75	-1.189817994980D-05
4	-6.0	7.078863767065D-11	3.539431883532D-11	3.25	9.450470894014D-06
5	-5.0	-1.391878573090D-10	-6.959392865448D-11	3.50	1.227961697821D-05
6	-4.0	2.362025781444D-10	1.181012890722D-10	3.75	8.122742618072D-06
7	-3.0	-3.644064529332D-10	-1.822032264666D-10	4.25	-7.455087476256D-06
8	-2.0	5.300923729643D-10	2.650461864822D-10	4.50	-1.031330000161D-05
9	-1.0	-7.432458880873D-10	-3.716229440437D-10	4.75	-7.220291925258D-06
10	0.0	1.014526389246D-09	5.072631946228D-10	5.25	7.302615743248D-06
11	1.0	-1.347826380327D-09	-6.739131901635D-10	5.50	1.052621382991D-05
12	2.0	1.723279148419D-09	8.616395742095D-10	5.75	7.644210524376D-06
13	3.0	-2.066075674872D-09	-1.033037837436D-09	6.25	-8.219898679283D-06
14	4.0	2.224240023864D-09	1.112120011932D-09	6.50	-1.215290127234D-05
15	5.0	-2.032395089068D-09	-1.016197544534D-09	6.75	-9.024781179189D-06
16	6.0	1.482695323030D-09	7.413476615151D-10	7.25	1.006811336020D-05
17	7.0	-8.098628215504D-10	-4.049314107752D-10	7.50	1.511118633242D-05
18	8.0	3.064236059278D-10	1.532118029639D-10	7.75	1.137048630892D-05
19	9.0	-7.099345819973D-11	-3.549672909986D-11	8.25	-1.296300828698D-05
20	10.0	7.547867154203D-12	3.773933577101D-12	8.50	-1.963039918688D-05
				8.75	-1.488788894544D-05
	0	-7.123278805824D+00	-2.559957662750D+02	9.25	1.720036063376D-05
	1	-6.731141493548D+00	2.475420936314D+02	9.50	2.619627810452D-05
	2	-2.723302348919D+00	-1.086844169543D+02	9.75	1.997210511006D-05
	3	-6.110526070193D-01	2.528466224083D+01	10.25	-2.329684333081D-05
	4	-8.206279929729D-02	-3.711638072291D+00	10.50	-3.564378360511D-05
	5	-6.591800716437D-03	3.131627647869D-01	10.75	-2.729910326348D-05
	6	-2.930607539226D-04	-1.531725452313D-02	11.25	3.214816825664D-05
	7	-5.559816699477D-06	3.089070791313D-04	11.50	4.944178784986D-05

Table 2. The matrix $B = \begin{pmatrix} B_{1,1} & B_{1,2} \\ B_{2,1} & B_{2,2} \end{pmatrix} = B^*$*; centered observations*

$n = 1, 2, 3, 4, 5$, and 8.
$m = n - 1, n, n + 1, \ldots, 20$.

In the case $n = 1$ the $(2 + m) \times (2 + m)$ matrix B is as follows: if $m = 0$, then

$$B = \begin{pmatrix} 0 & 1 \\ 1 & 0 \end{pmatrix}.$$

If $m = 1$, then

$$B = \begin{pmatrix} -0.5 & & \\ 0.5 & -0.5 & \\ 0.5 & 0.5 & -0.5 \end{pmatrix} = B^*.$$

If $m > 1$, the main diagonal of B is

$$-0.5m, \ -0.5, \ -1, \ -1, \ -1, \ldots, \ -1, \ -0.5;$$

the sub and super diagonals of B are

$$0.5, 0.5, \ldots, 0.5$$

both elements in the remaining corners of B are 0.5, and all other elements are 0. For example, if $n = 1$, $m = 3$, then

$$B = \begin{pmatrix} -1.5 & & & & \\ 0.5 & -0.5 & & & \\ 0 & 0.5 & -1 & & \\ 0 & 0 & 0.5 & -1 & \\ 0.5 & 0 & 0 & 0.5 & -0.5 \end{pmatrix} = B^*.$$

The other cases are in the following pages.

THE MATRIX B   N=2

```
                N=  2            M=  1

-5,00000D-01

 0,            2,00000D+00

 5,00000D-01 -1,00000D+00  0,

 5,00000D-01  1,00000D+00  0,           0,

                N=  2            M=  2

 5,00000D-01

 0,            4,00000D+00

-2,50000D-01 -5,00000D-01  1,25000D-01

 1,50000D+00  0,           -2,50000D-01  5,00000D-01

-2,50000D-01  5,00000D-01  1,25000D-01 -2,50000D-01  1,25000D-01

                N=  2            M=  3

 2,70000D+00

 0,            7,33333D+00

-4,00000D-01 -6,66667D-01  1,33333D-01

 9,00000D-01  1,00000D+00 -3,00000D-01  8,00000D-01

 9,00000D-01 -1,00000D+00  2,00000D-01 -7,00000D-01  8,00000D-01

-4,00000D-01  6,66667D-01 -3,33333D-02  2,00000D-01 -3,00000D-01  1,33333D-01

                N=  2            M=  4

 9,14286D+00

 0,            1,02500D+01

-7,85714D-01 -6,25000D-01  1,33929D-01

 1,71429D+00  7,50000D-01 -3,03571D-01  8,21429D-01

-8,57143D-01  3,56462D-28  2,14286D-01 -7,85714D-01  1,14286D+00

 1,71429D+00 -7,50000D-01 -5,35714D-02  3,21429D-01 -7,85714D-01  8,21429D-01
```

THE MATRIX B N=2

-7,85714D-01 6,25000D-01 8,92857D-03 -5,35714D-02 2,14286D-01 -3,03571D-01
1,33929D-01

N= 2 M= 5

2,03947D+01

0, 1,32727D+01

-1,07895D+00 -6,36364D-01 1,33971D-01

1,97368D+00 8,18182D-01 -3,03828D-01 8,22967D-01

-3,94737D-01 -2,72727D-01 2,15311D-01 -7,91866D-01 1,16746D+00

-3,94737D-01 2,72727D-01 -5,74163D-02 3,44498D-01 -8,77990D-01 1,16746D+00

1,97368D+00 -8,18182D-01 1,43541D-02 -8,61244D-02 3,44498D-01 -7,91866D-01
8,22967D-01

-1,07895D+00 6,36364D-01 -2,39234D-03 1,43541D-02 -5,74163D-02 2,15311D-01
-3,03828D-01 1,33971D-01

N= 2 M= 6

3,84231D+01

0, 1,62667D+01

-1,40385D+00 -6,33333D-01 1,33974D-01

2,42308D+00 8,00000D-01 -3,03846D-01 8,23077D-01

-6,92308D-01 -2,00000D-01 2,15385D-01 -7,92308D-01 1,16923D+00

3,46154D-01 -3,41721D-28 -5,76923D-02 3,46154D-01 -8,84615D-01 1,19231D+00

-6,92308D-01 2,00000D-01 1,53846D-02 -9,23077D-02 3,69231D-01 -8,84615D-01
1,16923D+00

2,42308D+00 -8,00000D-01 -3,84615D-03 2,30769D-02 -9,23077D-02 3,46154D-01
-7,92308D-01 8,23077D-01

-1,40385D+00 6,33333D-01 6,41026D-04 -3,84615D-03 1,53846D-02 -5,76923D-02
2,15385D-01 -3,03846D-01 1,33974D-01

N= 2 M= 7

6,45282D+01

0, 1,92683D+01

-1,71831D+00 -6,34146D-01 1,33975D-01

THE MATRIX B N=2

```
 2,80986D+00  8,04878D-01 -3,03847D-01  8,23085D-01

-7,39437D-01 -2,19512D-01  2,15390D-01 -7,92339D-01  1,16936D+00

 1,47887D-01  7,31707D-02 -5,77121D-02  3,46273D-01 -8,85091D-01  1,19409D+00

 1,47887D-01 -7,31707D-02  1,54586D-02 -9,27516D-02  3,71007D-01 -8,91274D-01
 1,19409D+00

-7,39437D-01  2,19512D-01 -4,12229D-03  2,47338D-02 -9,89351D-02  3,71007D-01
-8,85091D-01  1,16936D+00

 2,80986D+00 -8,04878D-01  1,03057D-03 -6,18344D-03  2,47338D-02 -9,27516D-02
 3,46273D-01 -7,92339D-01  8,23085D-01

-1,71831D+00  6,34146D-01 -1,71762D-04  1,03057D-03 -4,12229D-03  1,54586D-02
-5,77121D-02  2,15390D-01 -3,03847D-01  1,33975D-01
```

N= 2 M= 8

```
 1,00289D+02

 0,           2,22679D+01

-2,03608D+00 -6,33929D-01  1,33975D-01

 3,21649D+00  8,03571D-01 -3,03848D-01  8,23085D-01

-8,65979D-01 -2,14286D-01  2,15390D-01 -7,92342D-01  1,16937D+00

 2,47423D-01  5,35714D-02 -5,77135D-02  3,46281D-01 -8,85125D-01  1,19422D+00

-1,23711D-01  1,50586D-27  1,54639D-02 -9,27835D-02  3,71134D-01 -8,91753D-01
 1,19588D+00

 2,47423D-01 -5,35714D-02 -4,14212D-03  2,48527D-02 -9,94109D-02  3,72791D-01
-8,91753D-01  1,19422D+00

-8,65979D-01  2,14286D-01  1,10457D-03 -6,62739D-03  2,65096D-02 -9,94109D-02
 3,71134D-01 -8,85125D-01  1,16937D+00

 3,21649D+00 -8,03571D-01 -2,76141D-04  1,65685D-03 -6,62739D-03  2,48527D-02
-9,27835D-02  3,46281D-01 -7,92342D-01  8,23085D-01

-2,03608D+00  6,33929D-01  4,60236D-05 -2,76141D-04  1,10457D-03 -4,14212D-03
 1,54639D-02 -5,77135D-02  2,15390D-01 -3,03848D-01  1,33975D-01
```

N= 2 M= 9

```
 1,47175D+02

 0,           2,52680D+01

-2,35283D+00 -6,33987D-01  1,33975D-01
```

THE MATRIX B N=2

3,61698D+00 8,03922D-01 -3,03848D-01 8,23085D-01

-9,67925D-01 -2,15686D-01 2,15390D-01 -7,92342D-01 1,16937D+00

2,54717D-01 5,88235D-02 -5,77137D-02 3,46282D-01 -8,85128D-01 1,19423D+00

-5,09434D-02 -1,96078D-02 1,54643D-02 -9,27858D-02 3,71143D-01 -8,91787D-01
1,19600D+00

-5,09434D-02 1,96078D-02 -4,14354D-03 2,48613D-02 -9,94451D-02 3,72919D-01
-8,92231D-01 1,19600D+00

2,54717D-01 -5,88235D-02 1,10988D-03 -6,65927D-03 2,66371D-02 -9,98890D-02
3,72919D-01 -8,91787D-01 1,19423D+00

-9,67925D-01 2,15686D-01 -2,95967D-04 1,77580D-03 -7,10322D-03 2,66371D-02
-9,94451D-02 3,71143D-01 -8,85128D-01 1,16937D+00

3,61698D+00 -8,03922D-01 7,39919D-05 -4,43951D-04 1,77580D-03 -6,65927D-03
2,48613D-02 -9,27858D-02 3,46282D-01 -7,92342D-01 8,23085D-01

-2,35283D+00 6,33987D-01 -1,23320D-05 7,39919D-05 -2,95967D-04 1,10988D-03
-4,14354D-03 1,54643D-02 -5,77137D-02 2,15390D-01 -3,03848D-01 1,33975D-01

N= 2 M= 10

2,06699D+02

0, 2,82679D+01

-2,66989D+00 -6,33971D-01 1,33975D-01

4,01934D+00 8,03828D-01 -3,03848D-01 8,23085D-01

-1,07735D+00 -2,15311D-01 2,15390D-01 -7,92342D-01 1,16937D+00

2,90055D-01 5,74163D-02 -5,77137D-02 3,46282D-01 -8,85128D-01 1,19423D+00

-8,28729D-02 -1,43541D-02 1,54643D-02 -9,27860D-02 3,71144D-01 -8,91789D-01
1,19601D+00

4,14365D-02 -6,97830D-28 -4,14365D-03 2,48619D-02 -9,94475D-02 3,72928D-01
-8,92265D-01 1,19613D+00

-8,28729D-02 1,43541D-02 1,11026D-03 -6,66156D-03 2,66462D-02 -9,99233D-02
3,73047D-01 -8,92265D-01 1,19601D+00

2,90055D-01 -5,74163D-02 -2,97391D-04 1,78435D-03 -7,13738D-03 2,67652D-02
-9,99233D-02 3,72928D-01 -8,91789D-01 1,19423D+00

-1,07735D+00 2,15311D-01 7,93042D-05 -4,75825D-04 1,90330D-03 -7,13738D-03
2,66462D-02 -9,94475D-02 3,71144D-01 -8,85128D-01 1,16937D+00

4,01934D+00 -8,03828D-01 -1,98261D-05 1,18956D-04 -4,75825D-04 1,78435D-03
-6,66156D-03 2,48619D-02 -9,27860D-02 3,46282D-01 -7,92342D-01 8,23085D-01

THE MATRIX B   N=2

-2,66989D+00 6,33971D-01 3,30434D-06 -1,98261D-05 7,93042D-05 -2,97391D-04
1,11026D-03 -4,14365D-03 1,54643D-02 -5,77137D-02 2,15390D-01 -3,03848D-01
1,33975D-01

N= 2   M= 11

2,80355D+02

0,   3,12680D+01

-2,98686D+00 -6,33975D-01 1,33975D-01

4,42113D+00 8,03853D-01 -3,03848D-01 8,23085D-01

-1,18453D+00 -2,15412D-01 2,15390D-01 -7,92342D-01 1,16937D+00

3,16987D-01 5,77933D-02 -5,77137D-02 3,46282D-01 -8,85128D-01 1,19423D+00

-8,34176D-02 -1,57618D-02 1,54643D-02 -9,27860D-02 3,71144D-01 -8,91790D-01
1,19601D+00

1,66835D-02 5,25394D-03 -4,14365D-03 2,48619D-02 -9,94477D-02 3,72929D-01
-8,92268D-01 1,19614D+00

1,66835D-02 -5,25394D-03 1,11029D-03 -6,66172D-03 2,66469D-02 -9,99258D-02
3,73056D-01 -8,92300D-01 1,19614D+00

-8,34176D-02 1,57618D-02 -2,97493D-04 1,78496D-03 -7,13983D-03 2,67744D-02
-9,99577D-02 3,73056D-01 -8,92268D-01 1,19601D+00

3,16987D-01 -5,77933D-02 7,96856D-05 -4,78114D-04 1,91246D-03 -7,17171D-03
2,67744D-02 -9,99258D-02 3,72929D-01 -8,91790D-01 1,19423D+00

-1,18453D+00 2,15412D-01 -2,12495D-05 1,27497D-04 -5,09988D-04 1,91246D-03
-7,13983D-03 2,66469D-02 -9,94477D-02 3,71144D-01 -8,85128D-01 1,16937D+00

4,42113D+00 -8,03853D-01 5,31238D-06 -3,18743D-05 1,27497D-04 -4,78114D-04
1,78496D-03 -6,66172D-03 2,48619D-02 -9,27860D-02 3,46282D-01 -7,92342D-01
8,23085D-01

-2,98686D+00 6,33975D-01 -8,85396D-07 5,31238D-06 -2,12495D-05 7,96856D-05
-2,97493D-04 1,11029D-03 -4,14365D-03 1,54643D-02 -5,77137D-02 2,15390D-01
-3,03848D-01 1,33975D-01

N= 2   M= 12

3,69646D+02

0,   3,42679D+01

-3,30385D+00 -6,33974D-01 1,33975D-01

4,82309D+00 8,03846D-01 -3,03848D-01 8,23085D-01

-1,29238D+00 -2,15385D-01 2,15390D-01 -7,92342D-01 1,16937D+00

THE MATRIX B   N=2

```
 3,46410D-01  5,76923D-02 -5,77137D-02  3,46282D-01 -8,85128D-01  1,19423D+00

-9,32642D-02 -1,53846D-02  1,54643D-02 -9,27860D-02  3,71144D-01 -8,91790D-01
 1,19601D+00

 2,66469D-02  3,84615D-03 -4,14365D-03  2,48619D-02 -9,94477D-02  3,72929D-01
-8,92268D-01  1,19614D+00

-1,33235D-02  6,14672D-27  1,11029D-03 -6,66173D-03  2,66469D-02 -9,99260D-02
 3,73057D-01 -8,92302D-01  1,19615D+00

 2,66469D-02 -3,84615D-03 -2,97500D-04  1,78500D-03 -7,14001D-03  2,67750D-02
-9,99601D-02  3,73066D-01 -8,92302D-01  1,19614D+00

-9,32642D-02  1,53846D-02  7,97130D-05 -4,78278D-04  1,91311D-03 -7,17417D-03
 2,67836D-02 -9,99601D-02  3,73057D-01 -8,92268D-01  1,19601D+00

 3,46410D-01 -5,76923D-02 -2,13517D-05  1,28110D-04 -5,12441D-04  1,92165D-03
-7,17417D-03  2,67750D-02 -9,99260D-02  3,72929D-01 -8,91790D-01  1,19423D+00

-1,29238D+00  2,15385D-01  5,69379D-06 -3,41627D-05  1,36651D-04 -5,12441D-04
 1,91311D-03 -7,14001D-03  2,66469D-02 -9,94477D-02  3,71144D-01 -8,85128D-01
 1,16937D+00

 4,82309D+00 -8,03846D-01 -1,42345D-06  8,54068D-06 -3,41627D-05  1,28110D-04
-4,78278D-04  1,78500D-03 -6,66173D-03  2,48619D-02 -9,27860D-02  3,46282D-01
-7,92342D-01  8,23085D-01

-3,30385D+00  6,33974D-01  2,37241D-07 -1,42345D-06  5,69379D-06 -2,13517D-05
 7,97130D-05 -2,97500D-04  1,11029D-03 -4,14365D-03  1,54643D-02 -5,77137D-02
 2,15390D-01 -3,03848D-01  1,33975D-01
```

N= 2   M= 13

```
 4,76071D+02

 0,           3,72679D+01

-3,62083D+00 -6,33975D-01  1,33975D-01

 5,22501D+00  8,03848D-01 -3,03848D-01  8,23085D-01

-1,40003D+00 -2,15392D-01  2,15390D-01 -7,92342D-01  1,16937D+00

 3,75102D-01  5,77194D-02 -5,77137D-02  3,46282D-01 -8,85128D-01  1,19423D+00

-1,00379D-01 -1,54857D-02  1,54643D-02 -9,27860D-02  3,71144D-01 -8,91790D-01
 1,19601D+00

 2,64156D-02  4,22337D-03 -4,14365D-03  2,48619D-02 -9,94477D-02  3,72929D-01
-8,92268D-01  1,19614D+00

-5,28312D-03 -1,40779D-03  1,11029D-03 -6,66173D-03  2,66469D-02 -9,99260D-02
 3,73057D-01 -8,92302D-01  1,19615D+00
```

THE MATRIX B N=2

```
-5,28312D-03  1,40779D-03 -2,97501D-04  1,78501D-03 -7,14002D-03  2,67751D-02
-9,99603D-02  3,73066D-01 -8,92304D-01  1,19615D+00

 2,64156D-02 -4,22337D-03  7,97150D-05 -4,78290D-04  1,91316D-03 -7,17435D-03
 2,67842D-02 -9,99626D-02  3,73066D-01 -8,92302D-01  1,19614D+00

-1,00379D-01  1,54857D-02 -2,13590D-05  1,28154D-04 -5,12617D-04  1,92231D-03
-7,17664D-03  2,67842D-02 -9,99603D-02  3,73057D-01 -8,92268D-01  1,19601D+00

 3,75102D-01 -5,77194D-02  5,72117D-06 -3,43270D-05  1,37308D-04 -5,14905D-04
 1,92231D-03 -7,17435D-03  2,67751D-02 -9,99260D-02  3,72929D-01 -8,91790D-01
 1,19423D+00

-1,40003D+00  2,15392D-01 -1,52565D-06  9,15388D-06 -3,66155D-05  1,37308D-04
-5,12617D-04  1,91316D-03 -7,14002D-03  2,66469D-02 -9,94477D-02  3,71144D-01
-8,85128D-01  1,16937D+00

 5,22501D+00 -8,03848D-01  3,81411D-07 -2,28847D-06  9,15388D-06 -3,43270D-05
 1,28154D-04 -4,78290D-04  1,78501D-03 -6,66173D-03  2,48619D-02 -9,27860D-02
 3,46282D-01 -7,92342D-01  8,23085D-01

-3,62083D+00  6,33975D-01 -6,35686D-08  3,81411D-07 -1,52565D-06  5,72117D-06
-2,13590D-05  7,97150D-05 -2,97501D-04  1,11029D-03 -4,14365D-03  1,54643D-02
-5,77137D-02  2,15390D-01 -3,03848D-01  1,33975D-01
```

N= 2 M= 14

```
 6,01130D+02

 0,           4,02679D+01

-3,93782D+00 -6,33975D-01  1,33975D-01

 5,62693D+00  8,03847D-01 -3,03848D-01  8,23085D-01

-1,50774D+00 -2,15390D-01  2,15390D-01 -7,92342D-01  1,16937D+00

 4,04006D-01  5,77121D-02 -5,77137D-02  3,46282D-01 -8,85128D-01  1,19423D+00

-1,08290D-01 -1,54586D-02  1,54643D-02 -9,27860D-02  3,71144D-01 -8,91790D-01
 1,19601D+00

 2,91551D-02  4,12229D-03 -4,14365D-03  2,48619D-02 -9,94477D-02  3,72929D-01
-8,92268D-01  1,19614D+00

-8,33003D-03 -1,03057D-03  1,11029D-03 -6,66173D-03  2,66469D-02 -9,99260D-02
 3,73057D-01 -8,92302D-01  1,19615D+00

 4,16501D-03  3,76773D-27 -2,97501D-04  1,78501D-03 -7,14002D-03  2,67751D-02
-9,99603D-02  3,73066D-01 -8,92305D-01  1,19615D+00

-8,33003D-03  1,03057D-03  7,97151D-05 -4,78291D-04  1,91316D-03 -7,17436D-03
 2,67843D-02 -9,99628D-02  3,73067D-01 -8,92305D-01  1,19615D+00

 2,91551D-02 -4,12229D-03 -2,13596D-05  1,28157D-04 -5,12630D-04  1,92236D-03
-7,17682D-03  2,67849D-02 -9,99628D-02  3,73066D-01 -8,92302D-01  1,19614D+00
```

THE MATRIX B N=2

```
-1,08290D-01  1,54586D-02  5,72314D-06 -3,43388D-05  1,37355D-04 -5,15082D-04
 1,92297D-03 -7,17682D-03  2,67843D-02 -9,99603D-02  3,73057D-01 -8,92268D-01
 1,19601D+00

 4,04006D-01 -5,77121D-02 -1,53298D-06  9,19790D-06 -3,67916D-05  1,37969D-04
-5,15082D-04  1,92236D-03 -7,17436D-03  2,67751D-02 -9,99260D-02  3,72929D-01
-8,91790D-01  1,19423D+00

-1,50774D+00  2,15390D-01  4,08796D-07 -2,45277D-06  9,81109D-06 -3,67916D-05
 1,37355D-04 -5,12630D-04  1,91316D-03 -7,14002D-03  2,66469D-02 -9,94477D-02
 3,71144D-01 -8,85128D-01  1,16937D+00

 5,62693D+00 -8,03847D-01 -1,02199D-07  6,13193D-07 -2,45277D-06  9,19790D-06
-3,43388D-05  1,28157D-04 -4,78291D-04  1,78501D-03 -6,66173D-03  2,48619D-02
-9,27860D-02  3,46282D-01 -7,92342D-01  8,23085D-01

-3,93782D+00  6,33975D-01  1,70331D-08 -1,02199D-07  4,08796D-07 -1,53298D-06
 5,72314D-06 -2,13596D-05  7,97151D-05 -2,97501D-04  1,11029D-03 -4,14365D-03
 1,54643D-02 -5,77137D-02  2,15390D-01 -3,03848D-01  1,33975D-01
```

N= 2 M= 15

```
 7,46322D+02

 0,           4,32679D+01

-4,25481D+00 -6,33975D-01  1,33975D-01

 6,02886D+00  8,03848D-01 -3,03848D-01  8,23085D-01

-1,61543D+00 -2,15390D-01  2,15390D-01 -7,92342D-01  1,16937D+00

 4,32849D-01  5,77141D-02 -5,77137D-02  3,46282D-01 -8,85128D-01  1,19423D+00

-1,15971D-01 -1,54659D-02  1,54643D-02 -9,27860D-02  3,71144D-01 -8,91790D-01
 1,19601D+00

 3,10345D-02  4,14938D-03 -4,14365D-03  2,48619D-02 -9,94477D-02  3,72929D-01
-8,92268D-01  1,19614D+00

-8,16697D-03 -1,13165D-03  1,11029D-03 -6,66173D-03  2,66469D-02 -9,99260D-02
 3,73057D-01 -8,92302D-01  1,19615D+00

 1,63339D-03  3,77216D-04 -2,97501D-04  1,78501D-03 -7,14002D-03  2,67751D-02
-9,99603D-02  3,73066D-01 -8,92305D-01  1,19615D+00

 1,63339D-03 -3,77216D-04  7,97152D-05 -4,78291D-04  1,91316D-03 -7,17436D-03
 2,67843D-02 -9,99628D-02  3,73067D-01 -8,92305D-01  1,19615D+00

-8,16697D-03  1,13165D-03 -2,13596D-05  1,28158D-04 -5,12631D-04  1,92236D-03
-7,17683D-03  2,67849D-02 -9,99630D-02  3,73067D-01 -8,92305D-01  1,19615D+00

 3,10345D-02 -4,14938D-03  5,72328D-06 -3,43397D-05  1,37359D-04 -5,15095D-04
 1,92302D-03 -7,17699D-03  2,67849D-02 -9,99628D-02  3,73066D-01 -8,92302D-01
 1,19614D+00
```

THE MATRIX B   N=2

```
-1,15971D-01  1,54659D-02 -1,53351D-06  9,20106D-06 -3,68042D-05  1,38016D-04
-5,15259D-04  1,92302D-03 -7,17683D-03  2,67843D-02 -9,99603D-02  3,73057D-01
-8,92268D-01  1,19601D+00

 4,32849D-01 -5,77141D-02  4,10762D-07 -2,46457D-06  9,85828D-06 -3,69686D-05
 1,38016D-04 -5,15095D-04  1,92236D-03 -7,17436D-03  2,67751D-02 -9,99260D-02
 3,72929D-01 -8,91790D-01  1,19423D+00

-1,61543D+00  2,15390D-01 -1,09536D-07  6,57219D-07 -2,62887D-06  9,85828D-06
-3,68042D-05  1,37359D-04 -5,12631D-04  1,91316D-03 -7,14002D-03  2,66469D-02
-9,94477D-02  3,71144D-01 -8,85128D-01  1,16937D+00

 6,02886D+00 -8,03848D-01  2,73841D-08 -1,64305D-07  6,57219D-07 -2,46457D-06
 9,20106D-06 -3,43397D-05  1,28158D-04 -4,78291D-04  1,78501D-03 -6,66173D-03
 2,48619D-02 -9,27860D-02  3,46282D-01 -7,92342D-01  8,23085D-01

-4,25481D+00  6,33975D-01 -4,56402D-09  2,73841D-08 -1,09536D-07  4,10762D-07
-1,53351D-06  5,72328D-06 -2,13596D-05  7,97152D-05 -2,97501D-04  1,11029D-03
-4,14365D-03  1,54643D-02 -5,77137D-02  2,15390D-01 -3,03848D-01  1,33975D-01
```

N= 2   M= 16

```
 9,13149D+02

 0.           4,62679D+01

-4,57180D+00 -6,33975D-01  1,33975D-01

 6,43078D+00  8,03848D-01 -3,03848D-01  8,23085D-01

-1,72312D+00 -2,15390D-01  2,15390D-01 -7,92342D-01  1,16937D+00

 4,61710D-01  5,77135D-02 -5,77137D-02  3,46282D-01 -8,85128D-01  1,19423D+00

-1,23718D-01 -1,54639D-02  1,54643D-02 -9,27860D-02  3,71144D-01 -8,91790D-01
 1,19601D+00

 3,31615D-02  4,14212D-03 -4,14365D-03  2,48619D-02 -9,94477D-02  3,72929D-01
-8,92268D-01  1,19614D+00

-8,92810D-03 -1,10457D-03  1,11029D-03 -6,66173D-03  2,66469D-02 -9,99260D-02
 3,73057D-01 -8,92302D-01  1,19615D+00

 2,55088D-03  2,76141D-04 -2,97501D-04  1,78501D-03 -7,14002D-03  2,67751D-02
-9,99603D-02  3,73066D-01 -8,92305D-01  1,19615D+00

-1,27544D-03 -5,40394D-27  7,97152D-05 -4,78291D-04  1,91316D-03 -7,17436D-03
 2,67843D-02 -9,99628D-02  3,73067D-01 -8,92305D-01  1,19615D+00

 2,55088D-03 -2,76141D-04 -2,13596D-05  1,28158D-04 -5,12631D-04  1,92236D-03
-7,17683D-03  2,67850D-02 -9,99630D-02  3,73067D-01 -8,92305D-01  1,19615D+00

-8,92810D-03  1,10457D-03  5,72329D-06 -3,43397D-05  1,37359D-04 -5,15096D-04
 1,92303D-03 -7,17701D-03  2,67850D-02 -9,99630D-02  3,73067D-01 -8,92305D-01
 1,19615D+00
```

THE MATRIX B   N=2

```
 3,31615D-02 -4,14212D-03 -1,53355D-06  9,20129D-06 -3,68052D-05  1,38019D-04
-5,15272D-04  1,92307D-03 -7,17701D-03  2,67850D-02 -9,99628D-02  3,73066D-01
-8,92302D-01  1,19614D+00

-1,23718D-01  1,54639D-02  4,10903D-07 -2,46542D-06  9,86167D-06 -3,69813D-05
 1,38063D-04 -5,15272D-04  1,92303D-03 -7,17683D-03  2,67843D-02 -9,99603D-02
 3,73057D-01 -8,92268D-01  1,19601D+00

 4,61710D-01 -5,77135D-02 -1,10063D-07  6,60380D-07 -2,64152D-06  9,90569D-06
-3,69813D-05  1,38019D-04 -5,15096D-04  1,92236D-03 -7,17436D-03  2,67751D-02
-9,99260D-02  3,72929D-01 -8,91790D-01  1,19423D+00

-1,72312D+00  2,15390D-01  2,93502D-08 -1,76101D-07  7,04405D-07 -2,64152D-06
 9,86167D-06 -3,68052D-05  1,37359D-04 -5,12631D-04  1,91316D-03 -7,14002D-03
 2,66469D-02 -9,94477D-02  3,71144D-01 -8,85128D-01  1,16937D+00

 6,43078D+00 -8,03848D-01 -7,33755D-09  4,40253D-08 -1,76101D-07  6,60380D-07
-2,46542D-06  9,20129D-06 -3,43397D-05  1,28158D-04 -4,78291D-04  1,78501D-03
-6,66173D-03  2,48619D-02 -9,27860D-02  3,46282D-01 -7,92342D-01  8,23085D-01

-4,57180D+00  6,33975D-01  1,22293D-09 -7,33755D-09  2,93502D-08 -1,10063D-07
 4,10903D-07 -1,53355D-06  5,72329D-06 -2,13596D-05  7,97152D-05 -2,97501D-04
 1,11029D-03 -4,14365D-03  1,54643D-02 -5,77137D-02  2,15390D-01 -3,03848D-01
 1,33975D-01
```

N= 2   M= 17

```
 1,10311D+03

 0,           4,92679D+01

-4,88878D+00 -6,33975D-01  1,33975D-01

 6,83270D+00  8,03848D-01 -3,03848D-01  8,23085D-01

-1,83082D+00 -2,15390D-01  2,15390D-01 -7,92342D-01  1,16937D+00

 4,90566D-01  5,77137D-02 -5,77137D-02  3,46282D-01 -8,85128D-01  1,19423D+00

-1,31446D-01 -1,54644D-02  1,54643D-02 -9,27860D-02  3,71144D-01 -8,91790D-01
 1,19601D+00

 3,52176D-02  4,14407D-03 -4,14365D-03  2,48619D-02 -9,94477D-02  3,72929D-01
-8,92268D-01  1,19614D+00

-9,42442D-03 -1,11182D-03  1,11029D-03 -6,66173D-03  2,66469D-02 -9,99260D-02
 3,73057D-01 -8,92302D-01  1,19615D+00

 2,48011D-03  3,03224D-04 -2,97501D-04  1,78501D-03 -7,14002D-03  2,67751D-02
-9,99603D-02  3,73066D-01 -8,92305D-01  1,19615D+00

-4,96022D-04 -1,01075D-04  7,97152D-05 -4,78291D-04  1,91316D-03 -7,17436D-03
 2,67843D-02 -9,99628D-02  3,73067D-01 -8,92305D-01  1,19615D+00

-4,96022D-04  1,01075D-04 -2,13596D-05  1,28158D-04 -5,12631D-04  1,92236D-03
-7,17683D-03  2,67850D-02 -9,99630D-02  3,73067D-01 -8,92305D-01  1,19615D+00
```

THE MATRIX B   N=2

```
 2,48011D-03 -3,03224D-04  5,72329D-06 -3,43397D-05  1,37359D-04 -5,15096D-04
 1,92303D-03 -7,17701D-03  2,67850D-02 -9,99630D-02  3,73067D-01 -8,92305D-01
 1,19615D+00

-9,42442D-03  1,11182D-03 -1,53355D-06  9,20130D-06 -3,68052D-05  1,38020D-04
-5,15273D-04  1,92307D-03 -7,17702D-03  2,67850D-02 -9,99630D-02  3,73067D-01
-8,92305D-01  1,19615D+00

 3,52176D-02 -4,14407D-03  4,10913D-07 -2,46548D-06  9,86191D-06 -3,69822D-05
 1,38067D-04 -5,15285D-04  1,92307D-03 -7,17701D-03  2,67850D-02 -9,99628D-02
 3,73066D-01 -8,92302D-01  1,19614D+00

-1,31446D-01  1,54644D-02 -1,10101D-07  6,60607D-07 -2,64243D-06  9,90910D-06
-3,69940D-05  1,38067D-04 -5,15273D-04  1,92303D-03 -7,17683D-03  2,67843D-02
-9,99603D-02  3,73057D-01 -8,92268D-01  1,19601D+00

 4,90566D-01 -5,77137D-02  2,94914D-08 -1,76948D-07  7,07793D-07 -2,65422D-06
 9,90910D-06 -3,69822D-05  1,38020D-04 -5,15096D-04  1,92236D-03 -7,17436D-03
 2,67751D-02 -9,99260D-02  3,72929D-01 -8,91790D-01  1,19423D+00

-1,83082D+00  2,15390D-01 -7,86436D-09  4,71862D-08 -1,88745D-07  7,07793D-07
-2,64243D-06  9,86191D-06 -3,68052D-05  1,37359D-04 -5,12631D-04  1,91316D-03
-7,14002D-03  2,66469D-02 -9,94477D-02  3,71144D-01 -8,85128D-01  1,16937D+00

 6,83270D+00 -8,03848D-01  1,96609D-09 -1,17965D-08  4,71862D-08 -1,76948D-07
 6,60607D-07 -2,46548D-06  9,20130D-06 -3,43397D-05  1,28158D-04 -4,78291D-04
 1,78501D-03 -6,66173D-03  2,48619D-02 -9,27860D-02  3,46282D-01 -7,92342D-01
 8,23085D-01

-4,88878D+00  6,33975D-01 -3,27682D-10  1,96609D-09 -7,86436D-09  2,94914D-08
-1,10101D-07  4,10913D-07 -1,53355D-06  5,72329D-06 -2,13596D-05  7,97152D-05
-2,97501D-04  1,11029D-03 -4,14365D-03  1,54643D-02 -5,77137D-02  2,15390D-01
-3,03848D-01  1,33975D-01
```

N= 2   M= 18

```
 1,31770D+03

 0,           5,22679D+01

-5,20577D+00 -6,33975D-01  1,33975D-01

 7,23463D+00  8,03848D-01 -3,03848D-01  8,23085D-01

-1,93851D+00 -2,15390D-01  2,15390D-01 -7,92342D-01  1,16937D+00

 5,19423D-01  5,77137D-02 -5,77137D-02  3,46282D-01 -8,85128D-01  1,19423D+00

-1,39179D-01 -1,54643D-02  1,54643D-02 -9,27860D-02  3,71144D-01 -8,91790D-01
 1,19601D+00

 3,72939D-02  4,14354D-03 -4,14365D-03  2,48619D-02 -9,94477D-02  3,72929D-01
-8,92268D-01  1,19614D+00

-9,99630D-03 -1,10988D-03  1,11029D-03 -6,66173D-03  2,66469D-02 -9,99260D-02
 3,73057D-01 -8,92302D-01  1,19615D+00
```

THE MATRIX B   N=2

```
 2,69131D-03  2,95967D-04 -2,97501D-04  1,78501D-03 -7,14002D-03  2,67751D-02
-9,99603D-02  3,73066D-01 -8,92305D-01  1,19615D+00

-7,68946D-04 -7,39919D-05  7,97152D-05 -4,78291D-04  1,91316D-03 -7,17436D-03
 2,67843D-02 -9,99628D-02  3,73067D-01 -8,92305D-01  1,19615D+00

 3,84473D-04 -4,02890D-28 -2,13596D-05  1,28158D-04 -5,12631D-04  1,92236D-03
-7,17683D-03  2,67850D-02 -9,99630D-02  3,73067D-01 -8,92305D-01  1,19615D+00

-7,68946D-04  7,39919D-05  5,72329D-06 -3,43397D-05  1,37359D-04 -5,15096D-04
 1,92303D-03 -7,17701D-03  2,67850D-02 -9,99630D-02  3,73067D-01 -8,92305D-01
 1,19615D+00

 2,69131D-03 -2,95967D-04 -1,53355D-06  9,20131D-06 -3,68052D-05  1,38020D-04
-5,15273D-04  1,92307D-03 -7,17702D-03  2,67850D-02 -9,99630D-02  3,73067D-01
-8,92305D-01  1,19615D+00

-9,99630D-03  1,10988D-03  4,10914D-07 -2,46548D-06  9,86193D-06 -3,69822D-05
 1,38067D-04 -5,15286D-04  1,92308D-03 -7,17702D-03  2,67850D-02 -9,99630D-02
 3,73067D-01 -8,92305D-01  1,19615D+00

 3,72939D-02 -4,14354D-03 -1,10104D-07  6,60623D-07 -2,64249D-06  9,90934D-06
-3,69949D-05  1,38070D-04 -5,15286D-04  1,92307D-03 -7,17701D-03  2,67850D-02
-9,99628D-02  3,73066D-01 -8,92302D-01  1,19614D+00

-1,39179D-01  1,54643D-02  2,95015D-08 -1,77009D-07  7,08036D-07 -2,65513D-06
 9,91250D-06 -3,69949D-05  1,38067D-04 -5,15273D-04  1,92303D-03 -7,17683D-03
 2,67843D-02 -9,99603D-02  3,73057D-01 -8,92268D-01  1,19601D+00

 5,19423D-01 -5,77137D-02 -7,90219D-09  4,74131D-08 -1,89652D-07  7,11197D-07
-2,65513D-06  9,90934D-06 -3,69822D-05  1,38020D-04 -5,15096D-04  1,92236D-03
-7,17436D-03  2,67751D-02 -9,99260D-02  3,72929D-01 -8,91790D-01  1,19423D+00

-1,93851D+00  2,15390D-01  2,10725D-09 -1,26435D-08  5,05740D-08 -1,89652D-07
 7,08036D-07 -2,64249D-06  9,86193D-06 -3,68052D-05  1,37359D-04 -5,12631D-04
 1,91316D-03 -7,14002D-03  2,66469D-02 -9,94477D-02  3,71144D-01 -8,85128D-01
 1,16937D+00

 7,23463D+00 -8,03848D-01 -5,26812D-10  3,16087D-09 -1,26435D-08  4,74131D-08
-1,77009D-07  6,60623D-07 -2,46548D-06  9,20131D-06 -3,43397D-05  1,28158D-04
-4,78291D-04  1,78501D-03 -6,66173D-03  2,48619D-02 -9,27860D-02  3,46282D-01
-7,92342D-01  8,23085D-01

-5,20577D+00  6,33975D-01  8,78021D-11 -5,26812D-10  2,10725D-09 -7,90219D-09
 2,95015D-08 -1,10104D-07  4,10914D-07 -1,53355D-06  5,72329D-06 -2,13596D-05
 7,97152D-05 -2,97501D-04  1,11029D-03 -4,14365D-03  1,54643D-02 -5,77137D-02
 2,15390D-01 -3,03848D-01  1,33975D-01


                 N=  2            M= 19

 1,55843D+03

 0,           5,52679D+01

-5,52276D+00 -6,33975D-01  1,33975D-01
```

THE MATRIX B   N=2

```
 7,63655D+00  8,03848D-01 -3,03848D-01  8,23085D-01

-2,04621D+00 -2,15390D-01  2,15390D-01 -7,92342D-01  1,16937D+00

 5,48280D-01  5,77137D-02 -5,77137D-02  3,46282D-01 -8,85128D-01  1,19423D+00

-1,46911D-01 -1,54643D-02  1,54643D-02 -9,27860D-02  3,71144D-01 -8,91790D-01
 1,19601D+00

 3,93644D-02  4,14368D-03 -4,14365D-03  2,48619D-02 -9,94477D-02  3,72929D-01
-8,92268D-01  1,19614D+00

-1,05467D-02 -1,11040D-03  1,11029D-03 -6,66173D-03  2,66469D-02 -9,99260D-02
 3,73057D-01 -8,92302D-01  1,19615D+00

 2,82236D-03  2,97912D-04 -2,97501D-04  1,78501D-03 -7,14002D-03  2,67751D-02
-9,99603D-02  3,73066D-01 -8,92305D-01  1,19615D+00

-7,42725D-04 -8,12487D-05  7,97152D-05 -4,78291D-04  1,91316D-03 -7,17436D-03
 2,67843D-02 -9,99628D-02  3,73067D-01 -8,92305D-01  1,19615D+00

 1,48545D-04  2,70829D-05 -2,13596D-05  1,28158D-04 -5,12631D-04  1,92236D-03
-7,17683D-03  2,67850D-02 -9,99630D-02  3,73067D-01 -8,92305D-01  1,19615D+00

 1,48545D-04 -2,70829D-05  5,72329D-06 -3,43397D-05  1,37359D-04 -5,15096D-04
 1,92303D-03 -7,17701D-03  2,67850D-02 -9,99630D-02  3,73067D-01 -8,92305D-01
 1,19615D+00

-7,42725D-04  8,12487D-05 -1,53355D-06  9,20131D-06 -3,68052D-05  1,38020D-04
-5,15273D-04  1,92307D-03 -7,17702D-03  2,67850D-02 -9,99630D-02  3,73067D-01
-8,92305D-01  1,19615D+00

 2,82236D-03 -2,97912D-04  4,10914D-07 -2,46548D-06  9,86193D-06 -3,69822D-05
 1,38067D-04 -5,15286D-04  1,92308D-03 -7,17702D-03  2,67850D-02 -9,99630D-02
 3,73067D-01 -8,92305D-01  1,19615D+00

-1,05467D-02  1,11040D-03 -1,10104D-07  6,60624D-07 -2,64250D-06  9,90936D-06
-3,69949D-05  1,38070D-04 -5,15287D-04  1,92308D-03 -7,17702D-03  2,67850D-02
-9,99630D-02  3,73067D-01 -8,92305D-01  1,19615D+00

 3,93644D-02 -4,14368D-03  2,95022D-08 -1,77013D-07  7,08053D-07 -2,65520D-06
 9,91275D-06 -3,69958D-05  1,38070D-04 -5,15286D-04  1,92307D-03 -7,17701D-03
 2,67850D-02 -9,99628D-02  3,73066D-01 -8,92302D-01  1,19614D+00

-1,46911D-01  1,54643D-02 -7,90490D-09  4,74294D-08 -1,89718D-07  7,11441D-07
-2,65605D-06  9,91275D-06 -3,69949D-05  1,38067D-04 -5,15273D-04  1,92303D-03
-7,17683D-03  2,67843D-02 -9,99603D-02  3,73057D-01 -8,92268D-01  1,19601D+00

 5,48280D-01 -5,77137D-02  2,11738D-09 -1,27043D-08  5,08172D-08 -1,90565D-07
 7,11441D-07 -2,65520D-06  9,90936D-06 -3,69822D-05  1,38020D-04 -5,15096D-04
 1,92236D-03 -7,17436D-03  2,67751D-02 -9,99260D-02  3,72929D-01 -8,91790D-01
 1,19423D+00

-2,04621D+00  2,15390D-01 -5,64636D-10  3,38782D-09 -1,35513D-08  5,08172D-08
-1,89718D-07  7,08053D-07 -2,64250D-06  9,86193D-06 -3,68052D-05  1,37359D-04
-5,12631D-04  1,91316D-03 -7,14002D-03  2,66469D-02 -9,94477D-02  3,71144D-01
-8,85128D-01  1,16937D+00
```

THE MATRIX B N=2

```
 7,63655D+00 -8,03848D-01  1,41159D-10 -8,46954D-10  3,38782D-09 -1,27043D-08
 4,74294D-08 -1,77013D-07  6,60624D-07 -2,46548D-06  9,20131D-06 -3,43397D-05
 1,28158D-04 -4,78291D-04  1,78501D-03 -6,66173D-03  2,48619D-02 -9,27860D-02
 3,46282D-01 -7,92342D-01  8,23085D-01

-5,52276D+00  6,33975D-01 -2,35265D-11  1,41159D-10 -5,64636D-10  2,11738D-09
-7,90490D-09  2,95022D-08 -1,10104D-07  4,10914D-07 -1,53355D-06  5,72329D-06
-2,13596D-05  7,97152D-05 -2,97501D-04  1,11029D-03 -4,14365D-03  1,54643D-02
-5,77137D-02  2,15390D-01 -3,03848D-01  1,33975D-01

             N=  2              M= 20

 1,82679D+03

 0,           5,82679D+01

-5,83975D+00 -6,33975D-01  1,33975D-01

 8,03848D+00  8,03848D-01 -3,03848D-01  8,23085D-01

-2,15390D+00 -2,15390D-01  2,15390D-01 -7,92342D-01  1,16937D+00

 5,77137D-01  5,77137D-02 -5,77137D-02  3,46282D-01 -8,85128D-01  1,19423D+00

-1,54643D-01 -1,54643D-02  1,54643D-02 -9,27860D-02  3,71144D-01 -8,91790D-01
 1,19601D+00

 4,14366D-02  4,14365D-03 -4,14365D-03  2,48619D-02 -9,94477D-02  3,72929D-01
-8,92268D-01  1,19614D+00

-1,11032D-02 -1,11026D-03  1,11029D-03 -6,66173D-03  2,66469D-02 -9,99260D-02
 3,73057D-01 -8,92302D-01  1,19615D+00

 2,97611D-03  2,97391D-04 -2,97501D-04  1,78501D-03 -7,14002D-03  2,67751D-02
-9,99603D-02  3,73066D-01 -8,92305D-01  1,19615D+00

-8,01261D-04 -7,93042D-05  7,97152D-05 -4,78291D-04  1,91316D-03 -7,17436D-03
 2,67843D-02 -9,99628D-02  3,73067D-01 -8,92305D-01  1,19615D+00

 2,28932D-04  1,98261D-05 -2,13596D-05  1,28158D-04 -5,12631D-04  1,92236D-03
-7,17683D-03  2,67850D-02 -9,99630D-02  3,73067D-01 -8,92305D-01  1,19615D+00

-1,14466D-04  2,77379D-26  5,72329D-06 -3,43397D-05  1,37359D-04 -5,15096D-04
 1,92303D-03 -7,17701D-03  2,67850D-02 -9,99630D-02  3,73067D-01 -8,92305D-01
 1,19615D+00

 2,28932D-04 -1,98261D-05 -1,53355D-06  9,20131D-06 -3,68052D-05  1,38020D-04
-5,15273D-04  1,92307D-03 -7,17702D-03  2,67850D-02 -9,99630D-02  3,73067D-01
-8,92305D-01  1,19615D+00

-8,01261D-04  7,93042D-05  4,10914D-07 -2,46548D-06  9,86193D-06 -3,69822D-05
 1,38067D-04 -5,15286D-04  1,92308D-03 -7,17702D-03  2,67850D-02 -9,99630D-02
 3,73067D-01 -8,92305D-01  1,19615D+00
```

THE MATRIX B N=2

```
 2,97611D-03 -2,97391D-04 -1,10104D-07  6,60624D-07 -2,64250D-06  9,90936D-06
-3,69949D-05  1,38070D-04 -5,15287D-04  1,92308D-03 -7,17702D-03  2,67850D-02
-9,99630D-02  3,73067D-01 -8,92305D-01  1,19615D+00

-1,11032D-02  1,11026D-03  2,95023D-08 -1,77014D-07  7,08055D-07 -2,65520D-06
 9,91276D-06 -3,69959D-05  1,38071D-04 -5,15287D-04  1,92308D-03 -7,17702D-03
 2,67850D-02 -9,99630D-02  3,73067D-01 -8,92305D-01  1,19615D+00

 4,14366D-02 -4,14365D-03 -7,90510D-09  4,74306D-08 -1,89722D-07  7,11459D-07
-2,65611D-06  9,91299D-06 -3,69959D-05  1,38070D-04 -5,15286D-04  1,92307D-03
-7,17701D-03  2,67850D-02 -9,99628D-02  3,73066D-01 -8,92302D-01  1,19614D+00

-1,54643D-01  1,54643D-02  2,11811D-09 -1,27087D-08  5,08347D-08 -1,90630D-07
 7,11686D-07 -2,65611D-06  9,91276D-06 -3,69949D-05  1,38067D-04 -5,15273D-04
 1,92303D-03 -7,17683D-03  2,67843D-02 -9,99603D-02  3,73057D-01 -8,92268D-01
 1,19601D+00

 5,77137D-01 -5,77137D-02 -5,67351D-10  3,40411D-09 -1,36164D-08  5,10616D-08
-1,90630D-07  7,11459D-07 -2,65520D-06  9,90936D-06 -3,69822D-05  1,38020D-04
-5,15096D-04  1,92236D-03 -7,17436D-03  2,67751D-02 -9,99260D-02  3,72929D-01
-8,91790D-01  1,19423D+00

-2,15390D+00  2,15390D-01  1,51294D-10 -9,07762D-10  3,63105D-09 -1,36164D-08
 5,08347D-08 -1,89722D-07  7,08055D-07 -2,64250D-06  9,86193D-06 -3,68052D-05
 1,37359D-04 -5,12631D-04  1,91316D-03 -7,14002D-03  2,66469D-02 -9,94477D-02
 3,71144D-01 -8,85128D-01  1,16937D+00

 8,03848D+00 -8,03848D-01 -3,78234D-11  2,26941D-10 -9,07762D-10  3,40411D-09
-1,27087D-08  4,74306D-08 -1,77014D-07  6,60624D-07 -2,46548D-06  9,20131D-06
-3,43397D-05  1,28158D-04 -4,78291D-04  1,78501D-03 -6,66173D-03  2,48619D-02
-9,27860D-02  3,46282D-01 -7,92342D-01  8,23085D-01

-5,83975D+00  6,33975D-01  6,30391D-12 -3,78234D-11  1,51294D-10 -5,67351D-10
 2,11811D-09 -7,90510D-09  2,95023D-08 -1,10104D-07  4,10914D-07 -1,53355D-06
 5,72329D-06 -2,13596D-05  7,97152D-05 -2,97501D-04  1,11029D-03 -4,14365D-03
 1,54643D-02 -5,77137D-02  2,15390D-01 -3,03848D-01  1,33975D-01
```

THE MATRIX B N=3

```
                    N=  3                 M=  2

 0.

 0.            1.60000D+01

-1.00000D+00  0.            -1.40000D+01

 0.           -5.00000D-01  5.00000D-01  0.

 1.00000D+00  0.            -1.00000D+00  0.             0.

 0.            5.00000D-01  5.00000D-01  0.             0.             0.

                    N=  3                 M=  3

 2.10938D+00

 0.           -2.17576D+01

-2.43750D+00  0.            -2.22500D+01

-6.25000D-02  4.24242D-01  2.50000D-01 -7.57576D-03

 5.62500D-01 -2.27273D+00 -2.50000D-01  2.27273D-02 -6.81818D-02

 5.62500D-01  2.27273D+00 -2.50000D-01 -2.27273D-02  6.81818D-02 -6.81818D-02

-6.25000D-02 -4.24242D-01  2.50000D-01  7.57576D-03 -2.27273D-02  2.27273D-02
-7.57576D-03

                    N=  3                 M=  4

-1.96000D+01

 0.           -7.00978D+01

-2.41000D+01  0.            -4.12250D+01

 3.50000D-01  4.83696D-01  4.12500D-01 -8.96739D-03

-1.40000D+00 -1.46739D+00 -1.15000D+00  3.04348D-02 -1.10870D-01

 3.10000D+00  4.41461D-28  1.47500D+00 -3.75000D-02  1.50000D-01 -2.25000D-01

-1.40000D+00  1.46739D+00 -1.15000D+00  1.95652D-02 -8.91304D-02  1.50000D-01
-1.10870D-01

 3.50000D-01 -4.83696D-01  4.12500D-01 -3.53261D-03  1.95652D-02 -3.75000D-02
 3.04348D-02 -8.96739D-03
```

THE MATRIX B   N=3

```
                 N=  3           M=  5

-9,02945D+01

 0,           -2,36481D+02

-5,72837D+01   0,             -5,48038D+01

 5,96154D-01   1,07427D+00   3,53846D-01  -9,22075D-03

-1,85096D+00  -3,64495D+00  -8,11538D-01   3,18968D-02  -1,19306D-01

 1,75481D+00   4,56352D+00   4,57692D-01  -4,18943D-02   1,75357D-01  -3,01215D-01

 1,75481D+00  -4,56352D+00   4,57692D-01   2,65096D-02  -1,29203D-01   2,70446D-01
-3,01215D-01

-1,85096D+00   3,64495D+00  -8,11538D-01  -8,81984D-03   5,00752D-02  -1,29203D-01
 1,75357D-01  -1,19306D-01

 5,96154D-01  -1,07427D+00   3,53846D-01   1,52844D-03  -8,81984D-03   2,65096D-02
-4,18943D-02   3,18968D-02  -9,22075D-03

                 N=  3           M=  6

-3,68111D+02

 0,           -4,83141D+02

-1,25037D+02   0,             -7,04586D+01

 1.31514D+00   1,33343D+00   3,80893D-01  -9,26748D-03

-4,41737D+00  -3,90792D+00  -9,68238D-01   3,21675D-02  -1,20874D-01

 5,83325D+00   3,31555D+00   9,44913D-01  -4,27360D-02   1,80234D-01  -3,16379D-01

-4,46203D+00  -6,87918D-27  -7,15136D-01   2,85360D-02  -1,40943D-01   3,06948D-01
-3,89082D-01

 5,83325D+00  -3,31555D+00   9,44913D-01  -1,18545D-02   6,76568D-02  -1,83869D-01
 3,06948D-01  -3,16379D-01

-4,41737D+00   3,90792D+00  -9,68238D-01   3,81265D-03  -2,20536D-02   6,76568D-02
-1,40943D-01   1,80234D-01  -1,20874D-01

 1,31514D+00  -1,33343D+00   3,80893D-01  -6,58081D-04   3,81265D-03  -1,18545D-02
 2,85360D-02  -4,27360D-02   3,21675D-02  -9,26748D-03

                 N=  3           M=  7

-9,93748D+02

 0,           -8,90701D+02
```

THE MATRIX B N=3

```
-2,22959D+02  0,           -8,51853D+01

 2,01551D+00  1,76833D+00  3,69604D-01 -9,27613D-03

-6,25659D+00 -5,18920D+00 -9,02809D-01  3,22177D-02 -1,21165D-01

 6,61419D+00  5,23702D+00  7,40800D-01 -4,28925D-02  1,81141D-01 -3,19208D-01

-1,87311D+00 -3,14341D+00 -2,07595D-01  2,89250D-02 -1,43198D-01  3,13982D-01
-4,06574D-01

-1,87311D+00  3,14341D+00 -2,07595D-01 -1,27380D-02  7,27773D-02 -1,99843D-01
 3,46668D-01 -4,06574D-01

 6,61419D+00 -5,23702D+00  7,40800D-01  5,12273D-03 -2,96467D-02  9,13441D-02
-1,99843D-01  3,13982D-01 -3,19208D-01

-6,25659D+00  5,18920D+00 -9,02809D-01 -1,64212D-03  9,51602D-03 -2,96467D-02
 7,27773D-02 -1,43198D-01  1,81141D-01 -1,21165D-01

 2,01551D+00 -1,76833D+00  3,69604D-01  2,83325D-04 -1,64212D-03  5,12273D-03
-1,27380D-02  2,89250D-02 -4,28925D-02  3,22177D-02 -9,27613D-03
```

N= 3 M= 8

```
-2,32406D+03

 0,          -1,45584D+03

-3,66650D+02  0,           -1,00305D+02

 3,03195D+00  2,11019D+00  3,74529D-01 -9,27773D-03

-9,30796D+00 -5,93119D+00 -9,31352D-01  3,22270D-02 -1,21219D-01

 1,01390D+01  5,47597D+00  8,29872D-01 -4,29215D-02  1,81309D-01 -3,19733D-01

-5,29542D+00 -2,09914D+00 -4,29784D-01  2,89974D-02 -1,43617D-01  3,15291D-01
-4,09839D-01

 3,86491D+00  1,17351D-25  3,13470D-01 -1,29077D-02  7,37610D-02 -2,02913D-01
 3,54325D-01 -4,24532D-01

-5,29542D+00  2,09914D+00 -4,29784D-01  5,50401D-03 -3,18566D-02  9,82405D-02
-2,17046D-01  3,54325D-01 -4,09839D-01

 1,01390D+01 -5,47597D+00  8,29872D-01 -2,20648D-03  1,27871D-02 -3,98545D-02
 9,82405D-02 -2,02913D-01  3,15291D-01 -3,19733D-01

-9,30796D+00  5,93119D+00 -9,31352D-01  7,07064D-04 -4,09816D-03  1,27871D-02
-3,18566D-02  7,37610D-02 -1,43617D-01  1,81309D-01 -1,21219D-01

 3,03195D+00 -2,11019D+00  3,74529D-01 -1,21989D-04  7,07064D-04 -2,20648D-03
 5,50401D-03 -1,29077D-02  2,89974D-02 -4,29215D-02  3,22270D-02 -9,27773D-03
```

THE MATRIX B   N=3

```
              N=  3            M=  9

-4,74027D+03

 0,           -2,22917D+03

-5,58594D+02  0,           -1,15253D+02

 4,15714D+00  2,49886D+00  3,72420D-01 -9,27803D-03

-1,23700D+01 -6,94453D+00 -9,19131D-01  3,22287D-02 -1,21229D-01

 1,26587D+01  6,56173D+00  7,91736D-01 -4,29269D-02  1,81340D-01 -3,19830D-01

-5,39020D+00 -3,16743D+00 -3,34624D-01  2,90108D-02 -1,43695D-01  3,15534D-01
-4,10445D-01

 1,44431D+00  1,81560D+00  8,95990D-02 -1,29393D-02  7,39440D-02 -2,03484D-01
 3,55750D-01 -4,27884D-01

 1,44431D+00 -1,81560D+00  8,95990D-02  5,57726D-03 -3,22812D-02  9,95655D-02
-2,20352D-01  3,62103D-01 -4,27884D-01

-5,39020D+00  3,16743D+00 -3,34624D-01 -2,37071D-03  1,37391D-02 -4,28252D-02
 1,05653D-01 -2,20352D-01  3,55750D-01 -4,10445D-01

 1,26587D+01 -6,56173D+00  7,91736D-01  9,50084D-04 -5,50674D-03  1,71829D-02
-4,28252D-02  9,95655D-02 -2,03484D-01  3,15534D-01 -3,19830D-01

-1,23700D+01  6,94453D+00 -9,19131D-01 -3,04444D-04  1,76460D-03 -5,50674D-03
 1,37391D-02 -3,22812D-02  7,39440D-02 -1,43695D-01  1,81340D-01 -1,21229D-01

 4,15714D+00 -2,49886D+00  3,72420D-01  5,25252D-05 -3,04444D-04  9,50084D-04
-2,37071D-03  5,57726D-03 -1,29393D-02  2,90108D-02 -4,29269D-02  3,22287D-02
-9,27803D-03


              N=  3            M= 10

-8,86000D+03

 0,           -3,22868D+03

-8,09243D+02  0,           -1,30276D+02

 5,51311D+00  2,86430D+00  3,73331D-01 -9,27809D-03

-1,61499D+01 -7,82320D+00 -9,24406D-01  3,22290D-02 -1,21231D-01

 1,63806D+01  7,22717D+00  8,08196D-01 -4,29279D-02  1,81346D-01 -3,19848D-01

-7,66415D+00 -3,18691D+00 -3,75699D-01  2,90133D-02 -1,43710D-01  3,15579D-01
-4,10557D-01

 3,80196D+00  1,16358D+00  1,86261D-01 -1,29451D-02  7,39780D-02 -2,03590D-01
 3,56015D-01 -4,28506D-01
```

THE MATRIX B   N=3

```
-2,76324D+00  3,49988D-26 -1,35366D-01  5,59089D-03 -3,23602D-02  9,98120D-02
-2,20967D-01  3,63551D-01 -4,31253D-01

 3,80196D+00 -1,16358D+00  1,86261D-01 -2,40227D-03  1,39219D-02 -4,33960D-02
 1,07078D-01 -2,23704D-01  3,63551D-01 -4,28506D-01

-7,66415D+00  3,18691D+00 -3,75699D-01  1,02081D-03 -5,91666D-03  1,84621D-02
-4,60174D-02  1,07078D-01 -2,20967D-01  3,56015D-01 -4,10557D-01

 1,63806D+01 -7,22717D+00  8,08196D-01 -4,09084D-04  2,37111D-03 -7,39949D-03
 1,84621D-02 -4,33960D-02  9,98120D-02 -2,03590D-01  3,15579D-01 -3,19848D-01

-1,61499D+01  7,82320D+00 -9,24406D-01  1,31086D-04 -7,59793D-04  2,37111D-03
-5,91666D-03  1,39219D-02 -3,23602D-02  7,39780D-02 -1,43710D-01  1,81346D-01
-1,21231D-01

 5,51311D+00 -2,86430D+00  3,73331D-01 -2,26160D-05  1,31086D-04 -4,09084D-04
 1,02081D-03 -2,40227D-03  5,59089D-03 -1,29451D-02  2,90133D-02 -4,29279D-02
 3,22290D-02 -9,27809D-03
```

N= 3   M= 11

```
-1,54084D+04

 0,          -4,49112D+03

-1,12444D+03  0,          -1,45266D+02

 7,03122D+00  3,24104D+00  3,72939D-01 -9,27810D-03

-2,02498D+01 -8,76732D+00 -9,22137D-01  3,22291D-02 -1,21231D-01

 2,00628D+01  8,09690D+00  8,01116D-01 -4,29281D-02  1,81347D-01 -3,19851D-01

-9,01844D+00 -3,71615D+00 -3,58031D-01  2,90138D-02 -1,43712D-01  3,15588D-01
-4,10578D-01

 3,64642D+00  1,71118D+00  1,44683D-01 -1,29462D-02  7,39843D-02 -2,03609D-01
 3,56064D-01 -4,28622D-01

-9,72119D-01 -9,76581D-01 -3,85697D-02  5,59341D-03 -3,23748D-02  9,98577D-02
-2,21081D-01  3,63819D-01 -4,31878D-01

-9,72119D-01  9,76581D-01 -3,85697D-02 -2,40814D-03  1,39560D-02 -4,35022D-02
 1,07343D-01 -2,24327D-01  3,65003D-01 -4,31878D-01

 3,64642D+00 -1,71118D+00  1,44683D-01  1,03439D-03 -5,99541D-03  1,87079D-02
-4,66307D-02  1,08521D-01 -2,24327D-01  3,63819D-01 -4,28622D-01

-9,01844D+00  3,71615D+00 -3,58031D-01 -4,39535D-04  2,54761D-03 -7,95031D-03
 1,98366D-02 -4,66307D-02  1,07343D-01 -2,21081D-01  3,56064D-01 -4,10578D-01

 2,00628D+01 -8,09690D+00  8,01116D-01  1,76141D-04 -1,02094D-03  3,18608D-03
-7,95031D-03  1,87079D-02 -4,35022D-02  9,98577D-02 -2,03609D-01  3,15588D-01
-3,19851D-01
```

THE MATRIX B   N=3

```
-2,02498D+01  8,76732D+00 -9,22137D-01 -5,64423D-05  3,27148D-04 -1,02094D-03
 2,54761D-03 -5,99541D-03  1,39560D-02 -3,23748D-02  7,39843D-02 -1,43712D-01
 1,81347D-01 -1,21231D-01

 7,03122D+00 -3,24104D+00  3,72939D-01  9,73791D-06 -5,64423D-05  1,76141D-04
-4,39535D-04  1,03439D-03 -2,40814D-03  5,59341D-03 -1,29462D-02  2,90138D-02
-4,29281D-02  3,22291D-02 -9,27810D-03
```

N= 3   M= 12

```
-2,53524D+04

 0,          -6,04282D+03

-1,51258D+03  0,          -1,60270D+02

 8,74892D+00  3,61234D+00  3,73108D-01 -9,27810D-03

-2,48870D+01 -9,67999D+00 -9,23114D-01  3,22291D-02 -1,21231D-01

 2,43830D+01  8,86849D+00  8,04166D-01 -4,29281D-02  1,81347D-01 -3,19852D-01

-1,11445D+01 -4,00050D+00 -3,65642D-01  2,90139D-02 -1,43713D-01  3,15589D-01
-4,10582D-01

 4,95816D+00  1,68245D+00  1,62594D-01 -1,29464D-02  7,39854D-02 -2,03613D-01
 3,56073D-01 -4,28643D-01

-2,44783D+00 -6,11434D-01 -8,02683D-02  5,59388D-03 -3,23775D-02  9,98662D-02
-2,21102D-01  3,63869D-01 -4,31994D-01

 1,77835D+00  1,72116D-24  5,83146D-02 -2,40923D-03  1,39623D-02 -4,35219D-02
 1,07392D-01 -2,24443D-01  3,65272D-01 -4,32504D-01

-2,44783D+00  6,11434D-01 -8,02683D-02  1,03692D-03 -6,01007D-03  1,87536D-02
-4,67448D-02  1,08789D-01 -2,24952D-01  3,65272D-01 -4,31994D-01

 4,95816D+00 -1,68245D+00  1,62594D-01 -4,45386D-04  2,58152D-03 -8,05614D-03
 2,01007D-02 -4,72522D-02  1,08789D-01 -2,24443D-01  3,63869D-01 -4,28643D-01

-1,11445D+01  4,00050D+00 -3,65642D-01  1,89253D-04 -1,09694D-03  3,42325D-03
-8,54213D-03  2,01007D-02 -4,67448D-02  1,07392D-01 -2,21102D-01  3,56073D-01
-4,10582D-01

 2,43830D+01 -8,86849D+00  8,04166D-01 -7,58422D-05  4,39592D-04 -1,37185D-03
 3,42325D-03 -8,05614D-03  1,87536D-02 -4,35219D-02  9,98662D-02 -2,03613D-01
 3,15589D-01 -3,19852D-01

-2,48870D+01  9,67999D+00 -9,23114D-01  2,43027D-05 -1,40862D-04  4,39592D-04
-1,09694D-03  2,58152D-03 -6,01007D-03  1,39623D-02 -3,23775D-02  7,39854D-02
-1,43713D-01  1,81347D-01 -1,21231D-01

 8,74892D+00 -3,61234D+00  3,73108D-01 -4,19290D-06  2,43027D-05 -7,58422D-05
 1,89253D-04 -4,45386D-04  1,03692D-03 -2,40923D-03  5,59388D-03 -1,29464D-02
 2,90139D-02 -4,29281D-02  3,22291D-02 -9,27810D-03
```

THE MATRIX B   N=3

```
               N= 3           M= 13

-3,98457D+04

 0,           -7,91572D+03

-1,98069D+03   0,           -1,75268D+02

 1,06463D+01   3,98623D+00   3,73035D-01  -9,27810D-03

-2,99457D+01  -1,06076D+01  -9,22694D-01   3,22291D-02  -1,21231D-01

 2,89808D+01   9,68668D+00   8,02853D-01  -4,29281D-02   1,81347D-01  -3,19852D-01

-1,31426D+01  -4,40115D+00  -3,62366D-01   2,90139D-02  -1,43713D-01   3,15589D-01
-4,10582D-01

 5,61998D+00   1,92744D+00   1,54883D-01  -1,29465D-02   7,39856D-02  -2,03614D-01
 3,56074D-01  -4,28647D-01

-2,26130D+00  -8,83525D-01  -6,23173D-02   5,59397D-03  -3,23780D-02   9,98677D-02
-2,21106D-01   3,63878D-01  -4,32016D-01

 6,02556D-01   5,04017D-01   1,66053D-02  -2,40943D-03   1,39635D-02  -4,35255D-02
 1,07401D-01  -2,24464D-01   3,65322D-01  -4,32620D-01

 6,02556D-01  -5,04017D-01   1,66053D-02   1,03739D-03  -6,01278D-03   1,87621D-02
-4,67660D-02   1,08839D-01  -2,25068D-01   3,65541D-01  -4,32620D-01

-2,26130D+00   8,83525D-01  -6,23173D-02  -4,46475D-04   2,58783D-03  -8,07583D-03
 2,01498D-02  -4,73678D-02   1,09059D-01  -2,25068D-01   3,65322D-01  -4,32016D-01

 5,61998D+00  -1,92744D+00   1,54883D-01   1,91772D-04  -1,11154D-03   3,46882D-03
-8,65584D-03   2,03683D-02  -4,73678D-02   1,08839D-01  -2,24464D-01   3,63878D-01
-4,28647D-01

-1,31426D+01   4,40115D+00  -3,62366D-01  -8,14878D-05   4,72315D-04  -1,47397D-03
 3,67807D-03  -8,65584D-03   2,01498D-02  -4,67660D-02   1,07401D-01  -2,21106D-01
 3,56074D-01  -4,10582D-01

 2,89808D+01  -9,68668D+00   8,02853D-01   3,26558D-05  -1,89278D-04   5,90685D-04
-1,47397D-03   3,46882D-03  -8,07583D-03   1,87621D-02  -4,35255D-02   9,98677D-02
-2,03614D-01   3,15589D-01  -3,19852D-01

-2,99457D+01   1,06076D+01  -9,22694D-01  -1,04641D-05   6,06516D-05  -1,89278D-04
 4,72315D-04  -1,11154D-03   2,58783D-03  -6,01278D-03   1,39635D-02  -3,23780D-02
 7,39856D-02  -1,43713D-01   1,81347D-01  -1,21231D-01

 1,06463D+01  -3,98623D+00   3,73035D-01   1,80536D-06  -1,04641D-05   3,26558D-05
-8,14878D-05   1,91772D-04  -4,46475D-04   1,03739D-03  -2,40943D-03   5,59397D-03
-1,29465D-02   2,90139D-02  -4,29281D-02   3,22291D-02  -9,27810D-03


               N= 3           M= 14

-6,03024D+04
```

THE MATRIX B N=3

```
 0,           -1,01388D+04

-2,53653D+03  0,           -1,90269D+02

 1,27337D+01  4,35890D+00  3,73066D-01 -9,27810D-03

-3,54864D+01 -1,15282D+01 -9,22875D-01  3,22291D-02 -1,21231D-01

 3,40442D+01  1,04829D+01  8,03418D-01 -4,29281D-02  1,81347D-01 -3,19852D-01

-1,54820D+01 -4,74703D+00 -3,63777D-01  2,90139D-02 -1,43713D-01  3,15589D-01
-4,10583D-01

 6,73575D+00  2,04353D+00  1,58204D-01 -1,29465D-02  7,39857D-02 -2,03614D-01
 3,56075D-01 -4,28648D-01

-2,98249D+00 -8,55527D-01 -7,00472D-02  5,59399D-03 -3,23781D-02  9,98680D-02
-2,21107D-01  3,63880D-01 -4,32020D-01

 1,47176D+00  3,10772D-01  3,45658D-02 -2,40947D-03  1,39637D-02 -4,35262D-02
 1,07402D-01 -2,24468D-01  3,65331D-01 -4,32641D-01

-1,06919D+00  7,60351D-25 -2,51110D-02  1,03748D-03 -6,01329D-03  1,87637D-02
-4,67699D-02  1,08848D-01 -2,25090D-01  3,65591D-01 -4,32736D-01

 1,47176D+00 -3,10772D-01  3,45658D-02 -4,46677D-04  2,58900D-03 -8,07948D-03
 2,01589D-02 -4,73892D-02  1,09109D-01 -2,25184D-01  3,65591D-01 -4,32641D-01

-2,98249D+00  8,55527D-01 -7,00472D-02  1,92241D-04 -1,11426D-03  3,47729D-03
-8,67699D-03  2,04181D-02 -4,74837D-02  1,09109D-01 -2,25090D-01  3,65331D-01
-4,32020D-01

 6,73575D+00 -2,04353D+00  1,58204D-01 -8,25725D-05  4,78602D-04 -1,49359D-03
 3,72703D-03 -8,77106D-03  2,04181D-02 -4,73892D-02  1,08848D-01 -2,24468D-01
 3,63880D-01 -4,28648D-01

-1,54820D+01  4,74703D+00 -3,63777D-01  3,50866D-05 -2,03367D-04  6,34655D-04
-1,58369D-03  3,72703D-03 -8,67699D-03  2,01589D-02 -4,67699D-02  1,07402D-01
-2,21107D-01  3,56075D-01 -4,10583D-01

 3,40442D+01 -1,04829D+01  8,03418D-01 -1,40608D-05  8,14982D-05 -2,54334D-04
 6,34655D-04 -1,49359D-03  3,47729D-03 -8,07948D-03  1,87637D-02 -4,35262D-02
 9,98680D-02 -2,03614D-01  3,15589D-01 -3,19852D-01

-3,54864D+01  1,15282D+01 -9,22875D-01  4,50560D-06 -2,61151D-05  8,14982D-05
-2,03367D-04  4,78602D-04 -1,11426D-03  2,58900D-03 -6,01329D-03  1,39637D-02
-3,23781D-02  7,39857D-02 -1,43713D-01  1,81347D-01 -1,21231D-01

 1,27337D+01 -4,35890D+00  3,73066D-01 -7,77344D-07  4,50560D-06 -1,40608D-05
 3,50866D-05 -8,25725D-05  1,92241D-04 -4,46677D-04  1,03748D-03 -2,40947D-03
 5,59399D-03 -1,29465D-02  2,90139D-02 -4,29281D-02  3,22291D-02 -9,27810D-03
```

N= 3 M= 15

-8,83870D+04

THE MATRIX B   N=3

```
 0,           -1,27426D+04

-3,18745D+03  0,           -2,05269D+02

 1,50059D+01  4,73213D+00  3,73053D-01 -9,27810D-03

-4,14781D+01 -1,24520D+01 -9,22797D-01  3,22291D-02 -1,21231D-01

 3,94767D+01  1,12894D+01  8,03175D-01 -4,29281D-02  1,81347D-01 -3,19852D-01

-1,79220D+01 -5,11852D+00 -3,63169D-01  2,90139D-02 -1,43713D-01  3,15589D-01
-4,10583D-01

 7,73956D+00  2,21987D+00  1,56774D-01 -1,29465D-02  7,39857D-02 -2,03614D-01
 3,56075D-01 -4,28648D-01

-3,29389D+00 -9,67792D-01 -6,67190D-02  5,59399D-03 -3,23782D-02  9,98681D-02
-2,21107D-01  3,63880D-01 -4,32020D-01

 1,32472D+00  4,43431D-01  2,68327D-02 -2,40948D-03  1,39637D-02 -4,35264D-02
 1,07403D-01 -2,24469D-01  3,65333D-01 -4,32645D-01

-3,52975D-01 -2,52950D-01 -7,14962D-03  1,03749D-03 -6,01338D-03  1,87640D-02
-4,67706D-02  1,08850D-01 -2,25094D-01  3,65601D-01 -4,32757D-01

-3,52975D-01  2,52950D-01 -7,14962D-03 -4,46714D-04  2,58922D-03 -8,08015D-03
 2,01606D-02 -4,73932D-02  1,09118D-01 -2,25206D-01  3,65641D-01 -4,32757D-01

 1,32472D+00 -4,43431D-01  2,68327D-02  1,92328D-04 -1,11476D-03  3,47887D-03
-8,68091D-03  2,04273D-02 -4,75052D-02  1,09158D-01 -2,25206D-01  3,65601D-01
-4,32645D-01

-3,29389D+00  9,67792D-01 -6,67190D-02 -8,27743D-05  4,79772D-04 -1,49724D-03
 3,73614D-03 -8,79249D-03  2,04680D-02 -4,75052D-02  1,09118D-01 -2,25094D-01
 3,65333D-01 -4,32020D-01

 7,73956D+00 -2,21987D+00  1,56774D-01  3,55537D-05 -2,06074D-04  6,43102D-04
-1,60477D-03  3,77664D-03 -8,79249D-03  2,04273D-02 -4,73932D-02  1,08850D-01
-2,24469D-01  3,63880D-01 -4,28648D-01

-1,79220D+01  5,11852D+00 -3,63169D-01 -1,51074D-05  8,75649D-05 -2,73267D-04
 6,81898D-04 -1,60477D-03  3,73614D-03 -8,68091D-03  2,01606D-02 -4,67706D-02
 1,07403D-01 -2,21107D-01  3,56075D-01 -4,10583D-01

 3,94767D+01 -1,12894D+01  8,03175D-01  6,05422D-06 -3,50911D-05  1,09510D-04
-2,73267D-04  6,43102D-04 -1,49724D-03  3,47887D-03 -8,08015D-03  1,87640D-02
-4,35264D-02  9,98681D-02 -2,03614D-01  3,15589D-01 -3,19852D-01

-4,14781D+01  1,24520D+01 -9,22797D-01 -1,94000D-06  1,12445D-05 -3,50911D-05
 8,75649D-05 -2,06074D-04  4,79772D-04 -1,11476D-03  2,58922D-03 -6,01338D-03
 1,39637D-02 -3,23782D-02  7,39857D-02 -1,43713D-01  1,81347D-01 -1,21231D-01

 1,50059D+01 -4,73213D+00  3,73053D-01  3,34705D-07 -1,94000D-06  6,05422D-06
-1,51074D-05  3,55537D-05 -8,27743D-05  1,92328D-04 -4,46714D-04  1,03749D-03
-2,40948D-03  5,59399D-03 -1,29465D-02  2,90139D-02 -4,29281D-02  3,22291D-02
-9,27810D-03
```

THE MATRIX B   N=3

```
              N=  3            M= 16

-1,26057D+05

 0,           -1,57569D+04

-3,94103D+03   0,           -2,20269D+02

 1,74655D+01   5,10511D+00   3,73059D-01  -9,27810D-03

-4,79363D+01  -1,33744D+01  -9,22831D-01   3,22291D-02  -1,21231D-01

 4,53271D+01   1,20912D+01   8,03280D-01  -4,29281D-02   1,81347D-01  -3,19852D-01

-2,05843D+01  -5,47810D+00  -3,63431D-01   2,90139D-02  -1,43713D-01   3,15589D-01
-4,10583D-01

 8,91751D+00   2,36822D+00   1,57390D-01  -1,29465D-02   7,39857D-02  -2,03614D-01
 3,56075D-01  -4,28648D-01

-3,86155D+00  -1,01490D+00  -6,81521D-02   5,59399D-03  -3,23782D-02   9,98681D-02
-2,21107D-01   3,63881D-01  -4,32021D-01

 1,70904D+00   4,24698D-01   3,01625D-02  -2,40948D-03   1,39637D-02  -4,35264D-02
 1,07403D-01  -2,24469D-01   3,65333D-01  -4,32646D-01

-8,43314D-01  -1,54265D-01  -1,48835D-02   1,03750D-03  -6,01340D-03   1,87640D-02
-4,67707D-02   1,08850D-01  -2,25095D-01   3,65602D-01  -4,32761D-01

 6,12640D-01   1,31884D-23   1,08124D-02  -4,46721D-04   2,58926D-03  -8,08028D-03
 2,01609D-02  -4,73940D-02   1,09120D-01  -2,25210D-01   3,65651D-01  -4,32779D-01

-8,43314D-01   1,54265D-01  -1,48835D-02   1,92344D-04  -1,11485D-03   3,47916D-03
-8,68164D-03   2,04290D-02  -4,75092D-02   1,09168D-01  -2,25227D-01   3,65651D-01
-4,32761D-01

 1,70904D+00  -4,24698D-01   3,01625D-02  -8,28117D-05   4,79989D-04  -1,49792D-03
 3,73783D-03  -8,79647D-03   2,04772D-02  -4,75267D-02   1,09168D-01  -2,25210D-01
 3,65602D-01  -4,32646D-01

-3,86155D+00   1,01490D+00  -6,81521D-02   3,56406D-05  -2,06578D-04   6,44674D-04
-1,60869D-03   3,78587D-03  -8,81398D-03   2,04772D-02  -4,75092D-02   1,09120D-01
-2,25095D-01   3,65333D-01  -4,32021D-01

 8,91751D+00  -2,36822D+00   1,57390D-01  -1,53085D-05   8,87305D-05  -2,76904D-04
 6,90974D-04  -1,62613D-03   3,78587D-03  -8,79647D-03   2,04290D-02  -4,73940D-02
 1,08850D-01  -2,24469D-01   3,63881D-01  -4,28648D-01

-2,05843D+01   5,47810D+00  -3,63431D-01   6,50489D-06  -3,77033D-05   1,17662D-04
-2,93608D-04   6,90974D-04  -1,60869D-03   3,73783D-03  -8,68164D-03   2,01609D-02
-4,67707D-02   1,07403D-01  -2,21107D-01   3,56075D-01  -4,10583D-01

 4,53271D+01  -1,20912D+01   8,03280D-01  -2,60680D-06   1,51094D-05  -4,71523D-05
 1,17662D-04  -2,76904D-04   6,44674D-04  -1,49792D-03   3,47916D-03  -8,08028D-03
 1,87640D-02  -4,35264D-02   9,98681D-02  -2,03614D-01   3,15589D-01  -3,19852D-01
```

THE MATRIX B   N=3

```
-4,79363D+01  1,33744D+01 -9,22831D-01  8,35316D-07 -4,84161D-06  1,51094D-05
-3,77033D-05  8,87305D-05 -2,06578D-04  4,79989D-04 -1,11485D-03  2,58926D-03
-6,01340D-03  1,39637D-02 -3,23782D-02  7,39857D-02 -1,43713D-01  1,81347D-01
-1,21231D-01

 1,74655D+01 -5,10511D+00  3,73059D-01 -1,44116D-07  8,35316D-07 -2,60680D-06
 6,50489D-06 -1,53085D-05  3,56406D-05 -8,28117D-05  1,92344D-04 -4,46721D-04
 1,03750D-03 -2,40948D-03  5,59399D-03 -1,29465D-02  2,90139D-02 -4,29281D-02
 3,22291D-02 -9,27810D-03
```

N= 3   M= 17

```
-1,75574D+05

 0.          -1,92117D+04

-4,80473D+03  0.          -2,35269D+02

 2,01112D+01  5,47820D+00  3,73056D-01 -9,27810D-03

-5,48535D+01 -1,42974D+01 -9,22816D-01  3,22291D-02 -1,21231D-01

 5,15711D+01  1,28951D+01  8,03235D-01 -4,29281D-02  1,81347D-01 -3,19852D-01

-2,34081D+01 -5,84319D+00 -3,63318D-01  2,90139D-02 -1,43713D-01  3,15589D-01
-4,10583D-01

 1,01267D+01  2,52950D+00  1,57124D-01 -1,29465D-02  7,39857D-02 -2,03614D-01
 3,56075D-01 -4,28648D-01

-4,35278D+00 -1,09211D+00 -6,75350D-02  5,59399D-03 -3,23782D-02  9,98681D-02
-2,21107D-01  3,63881D-01 -4,32021D-01

 1,85164D+00  4,75914D-01  2,87288D-02 -2,40948D-03  1,39637D-02 -4,35264D-02
 1,07403D-01 -2,24469D-01  3,65333D-01 -4,32646D-01

-7,44650D-01 -2,18049D-01 -1,15535D-02  1,03750D-03 -6,01340D-03  1,87640D-02
-4,67708D-02  1,08851D-01 -2,25095D-01  3,65603D-01 -4,32762D-01

 1,98412D-01  1,24383D-01  3,07843D-03 -4,46722D-04  2,58926D-03 -8,08030D-03
 2,01610D-02 -4,73941D-02  1,09120D-01 -2,25211D-01  3,65652D-01 -4,32783D-01

 1,98412D-01 -1,24383D-01  3,07843D-03  1,92347D-04 -1,11487D-03  3,47921D-03
-8,68178D-03  2,04293D-02 -4,75099D-02  1,09169D-01 -2,25231D-01  3,65660D-01
-4,32783D-01

-7,44650D-01  2,18049D-01 -1,15535D-02 -8,28186D-05  4,80029D-04 -1,49804D-03
 3,73814D-03 -8,79720D-03  2,04789D-02 -4,75307D-02  1,09177D-01 -2,25231D-01
 3,65652D-01 -4,32762D-01

 1,85164D+00 -4,75914D-01  2,87288D-02  3,56567D-05 -2,06671D-04  6,44966D-04
-1,60942D-03  3,78758D-03 -8,81797D-03  2,04865D-02 -4,75307D-02  1,09169D-01
-2,25211D-01  3,65603D-01 -4,32646D-01
```

THE MATRIX B   N=3

```
-4,35278D+00  1,09211D+00 -6,75350D-02 -1,53459D-05  8,89473D-05 -2,77581D-04
 6,92663D-04 -1,63011D-03  3,79512D-03 -8,81797D-03  2,04789D-02 -4,75099D-02
 1,09120D-01 -2,25095D-01  3,65333D-01 -4,32021D-01

 1,01267D+01 -2,52950D+00  1,57124D-01  6,59148D-06 -3,82052D-05  1,19228D-04
-2,97517D-04  7,00172D-04 -1,63011D-03  3,78758D-03 -8,79720D-03  2,04293D-02
-4,73941D-02  1,08851D-01 -2,24469D-01  3,63881D-01 -4,28648D-01

-2,34081D+01  5,84319D+00 -3,63318D-01 -2,80084D-06  1,62341D-05 -5,06623D-05
 1,26420D-04 -2,97517D-04  6,92663D-04 -1,60942D-03  3,73814D-03 -8,68178D-03
 2,01610D-02 -4,67708D-02  1,07403D-01 -2,21107D-01  3,56075D-01 -4,10583D-01

 5,15711D+01 -1,28951D+01  8,03235D-01  1,12242D-06 -6,50573D-06  2,03026D-05
-5,06623D-05  1,19228D-04 -2,77581D-04  6,44966D-04 -1,49804D-03  3,47921D-03
-8,08030D-03  1,87640D-02 -4,35264D-02  9,98681D-02 -2,03614D-01  3,15589D-01
-3,19852D-01

-5,48535D+01  1,42974D+01 -9,22816D-01 -3,59666D-07  2,08468D-06 -6,50573D-06
 1,62341D-05 -3,82052D-05  8,89473D-05 -2,06671D-04  4,80029D-04 -1,11487D-03
 2,58926D-03 -6,01340D-03  1,39637D-02 -3,23782D-02  7,39857D-02 -1,43713D-01
 1,81347D-01 -1,21231D-01

 2,01112D+01 -5,47820D+00  3,73056D-01  6,20527D-08 -3,59666D-07  1,12242D-06
-2,80084D-06  6,59148D-06 -1,53459D-05  3,56567D-05 -8,28186D-05  1,92347D-04
-4,46722D-04  1,03750D-03 -2,40948D-03  5,59399D-03 -1,29465D-02  2,90139D-02
-4,29281D-02  3,22291D-02 -9,27810D-03
```

N= 3   M= 18

```
-2,39530D+05

 0,          -2,31371D+04

-5,78608D+03  0,          -2,50269D+02

 2,29436D+01  5,85124D+00  3,73057D-01 -9,27810D-03

-6,22333D+01 -1,52201D+01 -9,22822D-01  3,22291D-02 -1,21231D-01

 5,82207D+01  1,36980D+01  8,03254D-01 -4,29281D-02  1,81347D-01 -3,19852D-01

-2,64235D+01 -6,20575D+00 -3,63366D-01  2,90139D-02 -1,43713D-01  3,15589D-01
-4,10583D-01

 1,14377D+01  2,68483D+00  1,57239D-01 -1,29465D-02  7,39857D-02 -2,03614D-01
 3,56075D-01 -4,28648D-01

-4,93206D+00 -1,15548D+00 -6,78007D-02  5,59399D-03 -3,23782D-02  9,98681D-02
-2,21107D-01  3,63881D-01 -4,32021D-01

 2,13474D+00  4,94960D-01  2,93461D-02 -2,40948D-03  1,39637D-02 -4,35264D-02
 1,07403D-01 -2,24469D-01  3,65333D-01 -4,32646D-01

-9,44745D-01 -2,07113D-01 -1,29873D-02  1,03750D-03 -6,01340D-03  1,87640D-02
-4,67708D-02  1,08851D-01 -2,25095D-01  3,65603D-01 -4,32762D-01
```

THE MATRIX B N=3

```
 4,66177D-01  7,52306D-02  6,40848D-03 -4,46722D-04  2,58926D-03 -8,08031D-03
 2,01610D-02 -4,73941D-02  1,09120D-01 -2,25211D-01  3,65653D-01 -4,32784D-01

-3,38662D-01 -5,02655D-24 -4,65555D-03  1,92348D-04 -1,11487D-03  3,47922D-03
-8,68180D-03  2,04294D-02 -4,75101D-02  1,09170D-01 -2,25232D-01  3,65662D-01
-4,32787D-01

 4,66177D-01 -7,52306D-02  6,40848D-03 -8,28199D-05  4,80036D-04 -1,49806D-03
 3,73820D-03 -8,79734D-03  2,04793D-02 -4,75314D-02  1,09179D-01 -2,25235D-01
 3,65662D-01 -4,32784D-01

-9,44745D-01  2,07113D-01 -1,29873D-02  3,56597D-05 -2,06689D-04  6,45020D-04
-1,60955D-03  3,78790D-03 -8,81870D-03  2,04882D-02 -4,75347D-02  1,09179D-01
-2,25232D-01  3,65653D-01 -4,32762D-01

 2,13474D+00 -4,94960D-01  2,93461D-02 -1,53529D-05  8,89875D-05 -2,77706D-04
 6,92976D-04 -1,63084D-03  3,79684D-03 -8,82195D-03  2,04882D-02 -4,75314D-02
 1,09170D-01 -2,25211D-01  3,65603D-01 -4,32646D-01

-4,93206D+00  1,15548D+00 -6,78007D-02  6,60759D-06 -3,82985D-05  1,19519D-04
-2,98244D-04  7,01883D-04 -1,63409D-03  3,79684D-03 -8,81870D-03  2,04793D-02
-4,75101D-02  1,09120D-01 -2,25095D-01  3,65333D-01 -4,32021D-01

 1,14377D+01 -2,68483D+00  1,57239D-01 -2,83813D-06  1,64502D-05 -5,13367D-05
 1,28103D-04 -3,01477D-04  7,01883D-04 -1,63084D-03  3,78790D-03 -8,79734D-03
 2,04294D-02 -4,73941D-02  1,08851D-01 -2,24469D-01  3,63881D-01 -4,28648D-01

-2,64235D+01  6,20575D+00 -3,63366D-01  1,20597D-06 -6,99001D-06  2,18139D-05
-5,44335D-05  1,28103D-04 -2,98244D-04  6,92976D-04 -1,60955D-03  3,73820D-03
-8,68180D-03  2,01610D-02 -4,67708D-02  1,07403D-01 -2,21107D-01  3,56075D-01
-4,10583D-01

 5,82207D+01 -1,36980D+01  8,03254D-01 -4,83288D-07  2,80121D-06 -8,74181D-06
 2,18139D-05 -5,13367D-05  1,19519D-04 -2,77706D-04  6,45020D-04 -1,49806D-03
 3,47922D-03 -8,08031D-03  1,87640D-02 -4,35264D-02  9,98681D-02 -2,03614D-01
 3,15589D-01 -3,19852D-01

-6,22333D+01  1,52201D+01 -9,22822D-01  1,54864D-07 -8,97611D-07  2,80121D-06
-6,99001D-06  1,64502D-05 -3,82985D-05  8,89875D-05 -2,06689D-04  4,80036D-04
-1,11487D-03  2,58926D-03 -6,01340D-03  1,39637D-02 -3,23782D-02  7,39857D-02
-1,43713D-01  1,81347D-01 -1,21231D-01

 2,29436D+01 -5,85124D+00  3,73057D-01 -2,67184D-08  1,54864D-07 -4,83288D-07
 1,20597D-06 -2,83813D-06  6,60759D-06 -1,53529D-05  3,56597D-05 -8,28199D-05
 1,92348D-04 -4,46722D-04  1,03750D-03 -2,40948D-03  5,59399D-03 -1,29465D-02
 2,90139D-02 -4,29281D-02  3,22291D-02 -9,27810D-03


                N= 3            M= 19

-3,20873D+05

 0,          -2,75630D+04

-6,89255D+03  0,          -2,65269D+02

 2,59625D+01  6,22431D+00  3,73057D-01 -9,27810D-03
```

THE MATRIX B   N=3

```
-7,00739D+01 -1,61430D+01 -9,22820D-01  3,22291D-02 -1,21231D-01

 6,52699D+01  1,45014D+01  8,03246D-01 -4,29281D-02  1,81347D-01 -3,19852D-01

-2,96158D+01 -6,56946D+00 -3,63346D-01  2,90139D-02 -1,43713D-01  3,15589D-01
-4,10583D-01

 1,28160D+01  2,84289D+00  1,57189D-01 -1,29465D-02  7,39857D-02 -2,03614D-01
 3,56075D-01 -4,28648D-01

-5,51879D+00 -1,22519D+00 -6,76863D-02  5,59399D-03 -3,23782D-02  9,98681D-02
-2,21107D-01  3,63881D-01 -4,32021D-01

 2,37106D+00  5,28741D-01  2,90803D-02 -2,40948D-03  1,39637D-02 -4,35264D-02
 1,07403D-01 -2,24469D-01  3,65333D-01 -4,32646D-01

-1,00858D+00 -2,30401D-01 -1,23699D-02  1,03750D-03 -6,01340D-03  1,87640D-02
-4,67708D-02  1,08851D-01 -2,25095D-01  3,65603D-01 -4,32762D-01

 4,05608D-01  1,05562D-01  4,97465D-03 -4,46722D-04  2,58926D-03 -8,08031D-03
 2,01610D-02 -4,73941D-02  1,09120D-01 -2,25211D-01  3,65653D-01 -4,32784D-01

-1,08074D-01 -6,02163D-02 -1,32549D-03  1,92348D-04 -1,11488D-03  3,47922D-03
-8,68181D-03  2,04294D-02 -4,75101D-02  1,09170D-01 -2,25232D-01  3,65662D-01
-4,32788D-01

-1,08074D-01  6,02163D-02 -1,32549D-03 -8,28201D-05  4,80038D-04 -1,49807D-03
 3,73821D-03 -8,79737D-03  2,04793D-02 -4,75316D-02  1,09179D-01 -2,25236D-01
 3,65663D-01 -4,32788D-01

 4,05608D-01 -1,05562D-01  4,97465D-03  3,56602D-05 -2,06692D-04  6,45030D-04
-1,60958D-03  3,78796D-03 -8,81884D-03  2,04885D-02 -4,75354D-02  1,09180D-01
-2,25236D-01  3,65662D-01 -4,32784D-01

-1,00858D+00  2,30401D-01 -1,23699D-02 -1,53542D-05  8,89950D-05 -2,77730D-04
 6,93034D-04 -1,63098D-03  3,79716D-03 -8,82269D-03  2,04899D-02 -4,75354D-02
 1,09179D-01 -2,25232D-01  3,65653D-01 -4,32762D-01

 2,37106D+00 -5,28741D-01  2,90803D-02  6,61057D-06 -3,83158D-05  1,19573D-04
-2,98378D-04  7,02201D-04 -1,63483D-03  3,79856D-03 -8,82269D-03  2,04885D-02
-4,75316D-02  1,09170D-01 -2,25211D-01  3,65603D-01 -4,32646D-01

-5,51879D+00  1,22519D+00 -6,76863D-02 -2,84506D-06  1,64904D-05 -5,14621D-05
 1,28416D-04 -3,02214D-04  7,03599D-04 -1,63483D-03  3,79716D-03 -8,81884D-03
 2,04793D-02 -4,75101D-02  1,09120D-01 -2,25095D-01  3,65333D-01 -4,32021D-01

 1,28160D+01 -2,84289D+00  1,57189D-01  1,22203D-06 -7,08305D-06  2,21043D-05
-5,51581D-05  1,29808D-04 -3,02214D-04  7,02201D-04 -1,63098D-03  3,78796D-03
-8,79737D-03  2,04294D-02 -4,73941D-02  1,08851D-01 -2,24469D-01  3,63881D-01
-4,28648D-01

-2,96158D+01  6,56946D+00 -3,63346D-01 -5,19263D-07  3,00972D-06 -9,39254D-06
 2,34377D-05 -5,51581D-05  1,28416D-04 -2,98378D-04  6,93034D-04 -1,60958D-03
 3,73821D-03 -8,68181D-03  2,01610D-02 -4,67708D-02  1,07403D-01 -2,21107D-01
 3,56075D-01 -4,10583D-01
```

THE MATRIX B N=3

```
 6,52699D+01 -1,45014D+01  8,03246D-01  2,08092D-07 -1,20613D-06  3,76401D-06
-9,39254D-06  2,21043D-05 -5,14621D-05  1.19573D-04 -2,77730D-04  6,45030D-04
-1,49807D-03  3,47922D-03 -8,08031D-03  1,87640D-02 -4,35264D-02  9,98681D-02
-2,03614D-01  3,15589D-01 -3,19852D-01

-7,00739D+01  1,61430D+01 -9,22820D-01 -6,66804D-08  3,86489D-07 -1,20613D-06
 3,00972D-06 -7,08305D-06  1,64904D-05 -3,83158D-05  8,89950D-05 -2,06692D-04
 4,80038D-04 -1,11488D-03  2,58926D-03 -6,01340D-03  1,39637D-02 -3,23782D-02
 7,39857D-02 -1,43713D-01  1,81347D-01 -1,21231D-01

 2,59625D+01 -6,22431D+00  3,73057D-01  1,15043D-08 -6,66804D-08  2,08092D-07
-5,19263D-07  1,22203D-06 -2,84506D-06  6,61057D-06 -1,53542D-05  3,56602D-05
-8,28201D-05  1,92348D-04 -4,46722D-04  1,03750D-03 -2,40948D-03  5,59399D-03
-1,29465D-02  2,90139D-02 -4,29281D-02  3,22291D-02 -9,27810D-03


              N=  3             M= 20

-4,22923D+05

 0,          -3,25194D+04

-8,13166D+03  0,          -2,80269D+02

 2,91679D+01  6,59736D+00  3,73057D-01 -9,27810D-03

-7,83762D+01 -1,70658D+01 -9,22821D-01  3,22291D-02 -1,21231D-01

 7,27217D+01  1,53046D+01  8,03249D-01 -4,29281D-02  1,81347D-01 -3,19852D-01

-3,29921D+01 -6,93265D+00 -3,63355D-01  2,90139D-02 -1,43713D-01  3,15589D-01
-4,10583D-01

 1,42784D+01  2,99971D+00  1,57211D-01 -1,29465D-02  7,39857D-02 -2,03614D-01
 3,56075D-01 -4,28648D-01

-6,15216D+00 -1,29201D+00 -6,77356D-02  5,59399D-03 -3,23782D-02  9,98681D-02
-2,21107D-01  3,63881D-01 -4,32021D-01

 2,65165D+00  5,55798D-01  2,91947D-02 -2,40948D-03  1,39637D-02 -4,35264D-02
 1,07403D-01 -2,24469D-01  3,65333D-01 -4,32646D-01

-1,14766D+00 -2,38070D-01 -1,26358D-02  1,03750D-03 -6,01340D-03  1,87640D-02
-4,67708D-02  1,08851D-01 -2,25095D-01  3,65603D-01 -4,32762D-01

 5,07903D-01  9,96187D-02  5,59202D-03 -4,46722D-04  2,58926D-03 -8,08031D-03
 2,01610D-02 -4,73941D-02  1,09120D-01 -2,25211D-01  3,65653D-01 -4,32784D-01

-2,50620D-01 -3,61849D-02 -2,75934D-03  1,92348D-04 -1,11488D-03  3,47922D-03
-8,68181D-03  2,04294D-02 -4,75101D-02  1,09170D-01 -2,25232D-01  3,65662D-01
-4,32788D-01

 1,82068D-01  2,23607D-23  2,00457D-03 -8,28202D-05  4,80038D-04 -1,49807D-03
 3,73821D-03 -8,79737D-03  2,04793D-02 -4,75316D-02  1,09179D-01 -2,25236D-01
 3,65664D-01 -4,32788D-01
```

THE MATRIX B   N=3

```
-2,50620D-01  3,61849D-02 -2,75934D-03  3,56603D-05 -2,06692D-04  6,45032D-04
-1,60958D-03  3,78797D-03 -8,81887D-03  2,04886D-02 -4,75356D-02  1,09181D-01
-2,25237D-01  3,65664D-01 -4,32788D-01

 5,07903D-01 -9,96187D-02  5,59202D-03 -1,53544D-05  8,89964D-05 -2,77734D-04
 6,93045D-04 -1,63100D-03  3,79722D-03 -8,82283D-03  2,04902D-02 -4,75362D-02
 1,09181D-01 -2,25236D-01  3,65662D-01 -4,32784D-01

-1,14766D+00  2,38070D-01 -1,26358D-02  6,61113D-06 -3,83191D-05  1,19584D-04
-2,98403D-04  7,02259D-04 -1,63496D-03  3,79888D-03 -8,82343D-03  2,04902D-02
-4,75356D-02  1,09179D-01 -2,25232D-01  3,65653D-01 -4,32762D-01

 2,65165D+00 -5,55798D-01  2,91947D-02 -2,84635D-06  1,64979D-05 -5,14854D-05
 1,28474D-04 -3,02350D-04  7,03917D-04 -1,63557D-03  3,79888D-03 -8,82283D-03
 2,04886D-02 -4,75316D-02  1,09170D-01 -2,25211D-01  3,65603D-01 -4,32646D-01

-6,15216D+00  1,29201D+00 -6,77356D-02  1,22501D-06 -7,10036D-06  2,21583D-05
-5,52929D-05  1,30126D-04 -3,02952D-04  7,03917D-04 -1,63496D-03  3,79722D-03
-8,81887D-03  2,04793D-02 -4,75101D-02  1,09120D-01 -2,25095D-01  3,65333D-01
-4,32021D-01

 1,42784D+01 -2,99971D+00  1,57211D-01 -5,26175D-07  3,04979D-06 -9,51757D-06
 2,37497D-05 -5,58923D-05  1,30126D-04 -3,02350D-04  7,02259D-04 -1,63100D-03
 3,78797D-03 -8,79737D-03  2,04294D-02 -4,73941D-02  1,08851D-01 -2,24469D-01
 3,63881D-01 -4,28648D-01

-3,29921D+01  6,93265D+00 -3,63355D-01  2,23582D-07 -1,29591D-06  4,04420D-06
-1,00917D-05  2,37497D-05 -5,52929D-05  1,28474D-04 -2,98403D-04  6,93045D-04
-1,60958D-03  3,73821D-03 -8,68181D-03  2,01610D-02 -4,67708D-02  1,07403D-01
-2,21107D-01  3,56075D-01 -4,10583D-01

 7,27217D+01 -1,53046D+01  8,03249D-01 -8,95992D-08  5,19330D-07 -1,62069D-06
 4,04420D-06 -9,51757D-06  2,21583D-05 -5,14854D-05  1,19584D-04 -2,77734D-04
 6,45032D-04 -1,49807D-03  3,47922D-03 -8,08031D-03  1,87640D-02 -4,35264D-02
 9,98681D-02 -2,03614D-01  3,15589D-01 -3,19852D-01

-7,83762D+01  1,70658D+01 -9,22821D-01  2,87109D-08 -1,66413D-07  5,19330D-07
-1,29591D-06  3,04979D-06 -7,10036D-06  1,64979D-05 -3,83191D-05  8,89964D-05
-2,06692D-04  4,80038D-04 -1,11488D-03  2,58926D-03 -6,01340D-03  1,39637D-02
-3,23782D-02  7,39857D-02 -1,43713D-01  1,81347D-01 -1,21231D-01

 2,91679D+01 -6,59736D+00  3,73057D-01 -4,95345D-09  2,87109D-08 -8,95992D-08
 2,23582D-07 -5,26175D-07  1,22501D-06 -2,84635D-06  6,61113D-06 -1,53544D-05
 3,56603D-05 -8,28202D-05  1,92348D-04 -4,46722D-04  1,03750D-03 -2,40948D-03
 5,59399D-03 -1,29465D-02  2,90139D-02 -4,29281D-02  3,22291D-02 -9,27810D-03
```

THE MATRIX B   N=4

```
           N=  4              M=  3

4,21875D-01

0,          -1,36875D+01

2,83125D+01  0,           -2,41250D+02

0,           2,14167D+01  0,           7,96667D+01

-6,25000D-02  4,16667D-02  2,50000D-01 -1,66667D-01  0,

5,62500D-01 -1,12500D+00 -2,50000D-01  5,00000D-01  0,           0,

5,62500D-01  1,12500D+00 -2,50000D-01 -5,00000D-01  0,           0,
0,

-6,25000D-02 -4,16667D-02  2,50000D-01  1,66667D-01  0,           0,
0,           0,

           N=  4              M=  4

1,27285D+01

0,          -1,44444D+02

7,25993D+01  0,            3,63570D+02

0,           4,76111D+01  0,            1,13222D+02

-5,13245D-02  8,33333D-02 -2,96772D-01 -8,33333D-02  2,06954D-04

2,05298D-01 -6,66667D-01  1,68709D+00  1,66667D-01 -8,27815D-04  3,31126D-03

6,92053D-01  6,30716D-29 -2,78063D+00  1,08587D-29  1,24172D-03 -4,96689D-03
7,45033D-03

2,05298D-01  6,66667D-01  1,68709D+00 -1,66667D-01 -8,27815D-04  3,31126D-03
-4,96689D-03  3,31126D-03

-5,13245D-02 -8,33333D-02 -2,96772D-01  8,33333D-02  2,06954D-04 -8,27815D-04
1,24172D-03 -8,27815D-04  2,06954D-04

           N=  4              M=  5

5,79434D+01

0,           1,43756D+03

2,85676D+02  0,            9,10245D+02

0,           4,59693D+02  0,            2,12026D+02
```

THE MATRIX B N=4

```
-6,31411D-02 -5,44218D-01 -2,82506D-01 -1,64354D-01  2,73391D-04

 1,26923D-01  2,76276D+00  1,09752D+00  6,55102D-01 -1,22834D-03  5,72583D-03

 4,36218D-01 -6,56718D+00 -8,15012D-01 -1,14354D+00  2,17944D-03 -1,06199D-02
 2,06854D-02

 4,36218D-01  6,56718D+00 -8,15012D-01  1,14354D+00 -1,90220D-03  9,78822D-03
-2,01309D-02  2,06854D-02

 1,26923D-01 -2,76276D+00  1,09752D+00 -6,55102D-01  8,12479D-04 -4,47825D-03
 9,78822D-03 -1,06199D-02  5,72583D-03

-6,31411D-02  5,44218D-01 -2,82506D-01  1,64354D-01 -1,34772D-04  8,12479D-04
-1,90220D-03  2,17944D-03 -1,22834D-03  2,73391D-04
```

N= 4 M= 6

```
 1,79924D+03

 0,           5,11447D+03

 2,28903D+03  0,           3,11337D+03

 0,           9,57434D+02  0,           2,70627D+02

-5,72694D-01 -7,56715D-01 -7,10510D-01 -1,30916D-01  2,92471D-04

 2,80409D+00  3,11019D+00  3,24578D+00  4,40331D-01 -1,35089D-03  6,51296D-03

-6,06210D+00 -4,45024D+00 -6,08853D+00 -4,87914D-01  2,55354D-03 -1,30228D-02
 2,80205D-02

 8,66142D+00 -4,02091D-26  7,10652D+00 -6,05235D-27 -2,55470D-03  1,39793D-02
-3,29248D-02  4,30004D-02

-6,06210D+00  4,45024D+00 -6,08853D+00  4,87914D-01  1,46468D-03 -8,66739D-03
 2,25762D-02 -3,29248D-02  2,80205D-02

 2,80409D+00 -3,11019D+00  3,24578D+00 -4,40331D-01 -4,79806D-04  3,02864D-03
-8,66739D-03  1,39793D-02 -1,30228D-02  6,51296D-03

-5,72694D-01  7,56715D-01 -7,10510D-01  1,30916D-01  7,47008D-05 -4,79806D-04
 1,46468D-03 -2,55470D-03  2,55354D-03 -1,35089D-03  2,92471D-04
```

N= 4 M= 7

```
 7,95680D+03

 0,           2,08120D+04

 6,54164D+03  0,           5,86502D+03

 0,           2,07999D+03  0,           3,47532D+02
```

THE MATRIX B   N=4

```
-1,00878D+00 -1,80990D+00 -7,98960D-01 -1,51362D-01  2,97907D-04

 4,33861D+00  8,27773D+00  3,20372D+00  5,73430D-01 -1,38627D-03  6,74331D-03

-7,02568D+00 -1,60085D+01 -4,56741D+00 -9,14742D-01  2,66701D-03 -1,37615D-02
 3,03894D-02

 4,19585D+00  1,83060D+01  2,16265D+00  9,36614D-01 -2,80371D-03  1,56003D-02
-3,81230D-02  5,44073D-02

 4,19585D+00 -1,83060D+01  2,16265D+00 -9,36614D-01  1,84342D-03 -1,11328D-02
 3,04825D-02 -5,02739D-02  5,44073D-02

-7,02568D+00  1,60085D+01 -4,56741D+00  9,14742D-01 -8,40070D-04  5,37385D-03
-1,61881D-02  3,04825D-02 -3,81230D-02  3,03894D-02

 4,33861D+00 -8,27773D+00  3,20372D+00 -5,73430D-01  2,61961D-04 -1,69881D-03
 5,37385D-03 -1,11328D-02  1,56003D-02 -1,37615D-02  6,74331D-03

-1,00878D+00  1,80990D+00 -7,98960D-01  1,51362D-01 -4,02418D-05  2,61961D-04
-8,40070D-04  1,84342D-03 -2,80371D-03  2,66701D-03 -1,38627D-03  2,97907D-04

                N=  4           M=  8

 3,46947D+04

 0,           5,15284D+04

 1,81168D+04  0,           1,07289D+04

 0,           3,53802D+03  0,           4,14106D+02

-2,29784D+00 -2,53952D+00 -1,10561D+00 -1,41195D-01  2,99459D-04

 1,04304D+01  1,07176D+01  4,58368D+00  5,07046D-01 -1,39641D-03  6,80951D-03

-1,97880D+01 -1,73917D+01 -7,62446D+00 -6,99563D-01  2,69987D-03 -1,39761D-02
 3,10849D-02

 2,20435D+01  1,22889D+01  7,43456D+00  4,42770D-01 -2,87913D-03  1,60927D-02
-3,97192D-02  5,80706D-02

-1,97762D+01  1,98919D-24 -6,57633D+00  9,27740D-26  1,98645D-03 -1,20667D-02
 3,35098D-02 -5,72217D-02  6,75843D-02

 2,20435D+01 -1,22889D+01  7,43456D+00 -4,42770D-01 -1,04739D-03  6,72749D-03
-2,05759D-02  4,05525D-02 -5,72217D-02  5,80706D-02

-1,97880D+01  1,73917D+01 -7,62446D+00  6,99563D-01  4,56370D-04 -2,96817D-03
 9,48840D-03 -2,05759D-02  3,35098D-02 -3,97192D-02  3,10849D-02

 1,04304D+01 -1,07176D+01  4,58368D+00 -5,07046D-01 -1,40793D-04  9,18491D-04
-2,96817D-03  6,72749D-03 -1,20667D-02  1,60927D-02 -1,39761D-02  6,80951D-03
```

THE MATRIX B   N=4

```
-2,29784D+00  2,53952D+00 -1,10561D+00  1,41195D-01  2,15631D-05 -1,40793D-04
 4,56370D-04 -1,04739D-03  1,98645D-03 -2,87913D-03  2,69987D-03 -1,39641D-03
 2,99459D-04


                N=  4            M=  9

 1,03861D+05

 0,           1,18545D+05

 3,92593D+04  0,           1,70267D+04

 0,           5,75881D+03  0,           4,86010D+02

-3,71852D+00 -3,90119D+00 -1,26858D+00 -1,46848D-01  2,99904D-04

 1,60633D+01  1,66627D+01  5,02521D+00  5,43977D-01 -1,39931D-03  6,82848D-03

-2,74387D+01 -2,87289D+01 -7,63713D+00 -8,19565D-01  2,70931D-03 -1,40377D-02
 3,12852D-02

 2,30588D+01  2,74446D+01  5,69596D+00  7,20907D-01 -2,90100D-03  1,62356D-02
-4,01834D-02  5,91465D-02

-7,46496D+00 -2,12172D+01 -1,81545D+00 -5,51108D-01  2,02978D-03 -1,23498D-02
 3,44295D-02 -5,93534D-02  7,18083D-02

-7,46496D+00  2,12172D+01 -1,81545D+00  5,51108D-01 -1,12552D-03  7,23797D-03
-2,22346D-02  4,43970D-02 -6,48392D-02  7,18083D-02

 2,30588D+01 -2,74446D+01  5,69596D+00 -7,20907D-01  5,68033D-04 -3,69776D-03
 1,18591D-02 -2,60705D-02  4,43970D-02 -5,93534D-02  5,91465D-02

-2,74387D+01  2,87289D+01 -7,63713D+00  8,19565D-01 -2,45077D-04  1,59987D-03
-5,18218D-03  1,18591D-02 -2,22346D-02  3,44295D-02 -4,01834D-02  3,12852D-02

 1,60633D+01 -1,66627D+01  5,02521D+00 -5,43977D-01  7,54244D-05 -4,92713D-04
 1,59987D-03 -3,69776D-03  7,23797D-03 -1,23498D-02  1,62356D-02 -1,40377D-02
 6,82848D-03

-3,71852D+00  3,90119D+00 -1,26858D+00  1,46848D-01 -1,15437D-05  7,54244D-05
-2,45077D-04  5,68033D-04 -1,12552D-03  2,02978D-03 -2,90100D-03  2,70931D-03
-1,39931D-03  2,99904D-04


                N=  4            M= 10

 2,80198D+05

 0,           2,33686D+05

 7,89894D+04  0,           2,58224D+04

 0,           8,60428D+03  0,           5,55012D+02

-6,13088D+00 -5,17336D+00 -1,52016D+00 -1,43884D-01  3,00031D-04
```

THE MATRIX B   N=4

```
 2,65441D+01  2,13996D+01  6,04576D+00  5,24613D-01 -1,40014D-03  6,83391D-03

-4,64974D+01 -3,47171D+01 -9,53162D+00 -7,56609D-01  2,71201D-03 -1,40554D-02
 3,13426D-02

 4,45267D+01  2,84230D+01  8,32241D+00  5,74663D-01 -2,90728D-03  1,62766D-02
-4,03169D-02  5,94565D-02

-3,10631D+01 -1,29263D+01 -5,73326D+00 -2,58528D-01  2,04234D-03 -1,24319D-02
 3,46965D-02 -5,99735D-02  7,30489D-02

 2,62412D+01 -1,06563D-23  4,83374D+00 -2,67163D-25 -1,14918D-03  7,39263D-03
-2,27374D-02  4,55650D-02 -6,71760D-02  7,62099D-02

-3,10631D+01  1,29263D+01 -5,73326D+00  2,58528D-01  6,10094D-04 -3,97262D-03
 1,27527D-02 -2,81463D-02  4,85500D-02 -6,71760D-02  7,30489D-02

 4,45267D+01 -2,84230D+01  8,32241D+00 -5,74663D-01 -3,04952D-04  1,99114D-03
-6,45425D-03  1,48140D-02 -2,81463D-02  4,55650D-02 -5,99735D-02  5,94565D-02

-4,64974D+01  3,47171D+01 -9,53162D+00  7,56609D-01  1,31278D-04 -8,57706D-04
 2,78650D-03 -6,45425D-03  1,27527D-02 -2,27374D-02  3,46965D-02 -4,03169D-02
 3,13426D-02

 2,65441D+01 -2,13996D+01  6,04576D+00 -5,24613D-01 -4,03793D-05  2,63861D-04
-8,57706D-04  1,99114D-03 -3,97262D-03  7,39263D-03 -1,24319D-02  1,62766D-02
-1,40554D-02  6,83391D-03

-6,13088D+00  5,17336D+00 -1,52016D+00  1,43884D-01  6,17906D-06 -4,03793D-05
 1,31278D-04 -3,04952D-04  6,10094D-04 -1,14918D-03  2,04234D-03 -2,90728D-03
 2,71201D-03 -1,40014D-03  3,00031D-04


              N=  4              M= 11

 6,51724D+05

 0,           4,30346D+05

 1,43595D+05  0,           3,68966D+04

 0,           1,23550D+04  0,           6,25552D+02

-9,01621D+00 -6,87723D+00 -1,71706D+00 -1,45488D-01  3,00068D-04

 3,81714D+01  2,83342D+01  6,70899D+00  5,35093D-01 -1,40038D-03  6,83547D-03

-6,42217D+01 -4,64269D+01 -1,02644D+01 -7,90685D-01  2,71279D-03 -1,40604D-02
 3,13591D-02

 5,59826D+01  4,09201D+01  8,24972D+00  6,53860D-01 -2,90908D-03  1,62884D-02
-4,03551D-02  5,95454D-02

-2,91513D+01 -2,63595D+01 -4,24700D+00 -4,17319D-01  2,04595D-03 -1,24555D-02
 3,47732D-02 -6,01518D-02  7,34064D-02
```

THE MATRIX B   N=4

```
 8,73514D+00  1,91085D+01  1,26972D+00  3,01983D-01 -1,15605D-03  7,437490-03
-2,28833D-02  4,59041D-02 -6,78558D-02  7,75027D-02

 8,73514D+00 -1,91085D+01  1,26972D+00 -3,01983D-01  6,22837D-04 -4,05590D-03
 1,30234D-02 -2,87756D-02  4,98117D-02 -6,95756D-02  7,75027D-02

-2,91513D+01  2,63595D+01 -4,24700D+00  4,17319D-01 -3,27503D-04  2,13852D-03
-6,93345D-03  1,59277D-02 -3,03793D-02  4,98117D-02 -6,78558D-02  7,34064D-02

 5,59826D+01 -4,09201D+01  8,24972D+00 -6,53860D-01  1,63343D-04 -1,06726D-03
 3,46786D-03 -8,03782D-03  1,59277D-02 -2,87756D-02  4,59041D-02 -6,01518D-02
 5,95454D-02

-6,42217D+01  4,64269D+01 -1,02644D+01  7,90685D-01 -7,02810D-05  4,59271D-04
-1,49309D-03  3,46786D-03 -6,93345D-03  1,30234D-02 -2,28833D-02  3,47732D-02
-4,03551D-02  3,13591D-02

 3,81714D+01 -2,83342D+01  6,70899D+00 -5,35093D-01  2,16148D-05 -1,41253D-04
 4,59271D-04 -1,06726D-03  2,13852D-03 -4,05590D-03  7,43749D-03 -1,24555D-02
 1,62684D-02 -1,40604D-02  6,83547D-03

-9,01621D+00  6,87723D+00 -1,71706D+00  1,45488D-01 -3,30749D-06  2,16148D-05
-7,02810D-05  1,63343D-04 -3,27503D-04  6,22837D-04 -1,15605D-03  2,04595D-03
-2,90908D-03  2,71279D-03 -1,40038D-03  3,00068D-04
```

N= 4   M= 12

```
 1,39632D+06

 0,           7,35606D+05

 2,46404D+05  0,           5,09339D+04

 0,           1,69945D+04  0,           6,95264D+02

-1,29788D+01 -8,65236D+00 -1,94678D+00 -1,44635D-01  3,00078D-04

 5,45831D+01  3,51109D+01  7,58669D+00  5,29516D-01 -1,40045D-03  6,83591D-03

-9,17272D+01 -5,61996D+01 -1,16945D+01 -7,72552D-01  2,71301D-03 -1,40619D-02
 3,13638D-02

 8,23690D+01  4,71387D+01  9,79783D+00  6,11711D-01 -2,90959D-03  1,62917D-02
-4,03661D-02  5,95709D-02

-5,10321D+01 -2,58614D+01 -6,01103D+00 -3,32768D-01  2,04699D-03 -1,24623D-02
 3,47952D-02 -6,02029D-02  7,35089D-02

 3,31051D+01  1,09646D+01  3,89172D+00  1,40837D-01 -1,15802D-03  7,45038D-03
-2,29252D-02  4,60015D-02 -6,80513D-02  7,78752D-02

-2,76380D+01  1,23931D-22 -3,24794D+00  1,76414D-24  6,26533D-04 -4,08005D-03
 1,31020D-02 -2,89582D-02  5,01780D-02 -7,02737D-02  7,88109D-02

 3,31051D+01 -1,09646D+01  3,89172D+00 -1,40837D-01 -3,34336D-04  2,18317D-03
-7,07863D-03  1,62652D-02 -3,10563D-02  5,11019D-02 -7,02737D-02  7,78752D-02
```

THE MATRIX B   N=4

```
-5,10321D+01  2,58614D+01 -6,01103D+00  3,32768D-01  1,75420D-04 -1,14618D-03
 3,72449D-03 -8,63434D-03  1,71244D-02 -3,10563D-02  5,01780D-02 -6,80513D-02
 7,35089D-02

 8,23690D+01 -4,71387D+01  9,79783D+00 -6,11711D-01 -8,74473D-05  5,71456D-04
-1,85787D-03  4,31576D-03 -8,63434D-03  1,62652D-02 -2,89582D-02  4,60015D-02
-6,02029D-02  5,95709D-02

-9,17272D+01  5,61996D+01 -1,16945D+01  7,72552D-01  3,76212D-05 -2,45858D-04
 7,99405D-04 -1,85787D-03  3,72449D-03 -7,07863D-03  1,31020D-02 -2,29252D-02
 3,47952D-02 -4,03661D-02  3,13638D-02

 5,45831D+01 -3,51109D+01  7,58669D+00 -5,29516D-01 -1,15700D-05  7,56115D-05
-2,45858D-04  5,71456D-04 -1,14618D-03  2,18317D-03 -4,08005D-03  7,45038D-03
-1,24623D-02  1,62917D-02 -1,40619D-02  6,83591D-03

-1,29788D+01  8,65236D+00 -1,94678D+00  1,44635D-01  1,77043D-06 -1,15700D-05
 3,76212D-05 -8,74473D-05  1,75420D-04 -3,34336D-04  6,26533D-04 -1,15802D-03
 2,04699D-03 -2,90959D-03  2,71301D-03 -1,40045D-03  3,00078D-04
```

N= 4   M= 13

```
 2,75911D+06

 0,           1,19692D+06

 3,99624D+05  0,           6,80017D+04

 0,           2,27096D+04  0,           7,65419D+02

-1,77537D+01 -1,07415D+01 -2,15692D+00 -1,45093D-01  3,00081D-04

 7,37350D+01  4,33164D+01  8,33645D+00  5,32511D-01 -1,40047D-03  6,83604D-03

-1,21656D+02 -6,91943D+01 -1,27085D+01 -7,82289D-01  2,71307D-03 -1,40623D-02
 3,13652D-02

 1,05695D+02  5,90710D+01  1,03789D+01  6,34345D-01 -2,90974D-03  1,62927D-02
-4,03692D-02  5,95782D-02

-5,99510D+01 -3,54837D+01 -5,83515D+00 -3,78176D-01  2,04728D-03 -1,24642D-02
 3,48015D-02 -6,02176D-02  7,35383D-02

 2,89831D+01  2,13721D+01  2,81579D+00  2,27423D-01 -1,15859D-03  7,45408D-03
-2,29372D-02  4,60294D-02 -6,81073D-02  7,79821D-02

-8,55238D+00 -1,52866D+01 -8,30559D-01 -1,62614D-01  6,27596D-04 -4,08699D-03
 1,31246D-02 -2,90107D-02  5,02832D-02 -7,04744D-02  7,91878D-02

-8,55238D+00  1,52866D+01 -8,30559D-01  1,62614D-01 -3,36318D-04  2,19612D-03
-7,12075D-03  1,63631D-02 -3,12527D-02  5,14765D-02 -7,09770D-02  7,91878D-02

 2,89831D+01 -2,13721D+01  2,81579D+00 -2,27423D-01  1,79079D-04 -1,17009D-03
 3,80224D-03 -8,81507D-03  1,74870D-02 -3,17477D-02  5,14765D-02 -7,04744D-02
 7,79821D-02
```

THE MATRIX B   N=4

```
-5,99510D+01  3,54837D+01 -5,83515D+00  3,78176D-01 -9,39127D-05  6,13708D-04
-1,99526D-03  4,63512D-03 -9,27505D-03  1,74870D-02 -3,12527D-02  5,02832D-02
-6,81073D-02  7,35383D-02

 1,05695D+02 -5,90710D+01  1,03789D+01 -6,34345D-01  4,68104D-05 -3,05910D-04
 9,94674D-04 -2,31176D-03  4,63512D-03 -8,81507D-03  1,63631D-02 -2,90107D-02
 4,60294D-02 -6,02176D-02  5,95782D-02

-1,21656D+02  6,91943D+01 -1,27085D+01  7,82289D-01 -2,01380D-05  1,31605D-04
-4,27928D-04  9,94674D-04 -1,99526D-03  3,80224D-03 -7,12075D-03  1,31246D-02
-2,29372D-02  3,48015D-02 -4,03692D-02  3,13652D-02

 7,37350D+01 -4,33164D+01  8,33645D+00 -5,32511D-01  6,19319D-06 -4,04735D-05
 1,31605D-04 -3,05910D-04  6,13708D-04 -1,17009D-03  2,19612D-03 -4,08699D-03
 7,45408D-03 -1,24642D-02  1,62927D-02 -1,40623D-02  6,83604D-03

-1,77537D+01  1,07415D+01 -2,15692D+00  1,45093D-01 -9,47673D-07  6,19319D-06
-2,01380D-05  4,68104D-05 -9,39127D-05  1,79079D-04 -3,36318D-04  6,27596D-04
-1,15859D-03  2,04728D-03 -2,90974D-03  2,71307D-03 -1,40047D-03  3,00081D-04
```

N= 4   M= 14

```
 5,14052D+06

 0,           1,86133D+06

 6,21830D+05  0,           8,85775D+04

 0,           2,95516D+04  0,           8,35336D+02

-2,37220D+01 -1,29853D+01 -2,37858D+00 -1,44848D-01  3,00082D-04

 9,78059D+01  5,19088D+01  9,16150D+00  5,30911D-01 -1,40048D-03  6,83608D-03

-1,60380D+02 -8,20231D+01 -1,39674D+01 -7,77086D-01  2,71309D-03 -1,40624D-02
 3,13655D-02

 1,39885D+02  6,88414D+01  1,15291D+01  6,22250D-01 -2,90978D-03  1,62930D-02
-4,03701D-02  5,95802D-02

-8,31729D+01 -3,94271D+01 -6,80089D+00 -3,53910D-01  2,04737D-03 -1,24648D-02
 3,48033D-02 -6,02218D-02  7,35467D-02

 4,79803D+01  2,02099D+01  3,91684D+00  1,81147D-01 -1,15875D-03  7,45514D-03
-2,29406D-02  4,60374D-02 -6,81234D-02  7,80127D-02

-3,06640D+01 -8,44396D+00 -2,50235D+00 -7,56610D-02  6,27900D-04 -4,08898D-03
 1,31310D-02 -2,90257D-02  5,03134D-02 -7,05319D-02  7,92959D-02

 2,55351D+01 -9,06862D-22  2,08367D+00 -8,78347D-24 -3,36887D-04  2,19984D-03
-7,13285D-03  1,63912D-02 -3,13092D-02  5,15841D-02 -7,11793D-02  7,95660D-02

-3,06640D+01  8,44396D+00 -2,50235D+00  7,56610D-02  1,80140D-04 -1,17703D-03
 3,82480D-03 -8,86750D-03  1,75921D-02 -3,19483D-02  5,18534D-02 -7,11793D-02
 7,92959D-02
```

THE MATRIX B N=4

```
 4,79803D+01 -2,02099D+01  3,91684D+00 -1,81147D-01 -9,58715D-05  6,26509D-04
-2,03688D-03  4,73187D-03 -9,46917D-03  1,78572D-02 -3,19483D-02  5,15841D-02
-7,05319D-02  7,80127D-02

-8,31729D+01  3,94271D+01 -6,80089D+00  3,53910D-01  5,02713D-05 -3,28528D-04
 1,06822D-03 -2,48272D-03  4,97810D-03 -9,46917D-03  1,75921D-02 -3,13092D-02
 5,03134D-02 -6,81234D-02  7,35467D-02

 1,39885D+02 -6,88414D+01  1,15291D+01 -6,22250D-01 -2,50569D-05  1,63750D-04
-5,32453D-04  1,23764D-03 -2,48272D-03  4,73187D-03 -8,86750D-03  1,63912D-02
-2,90257D-02  4,60374D-02 -6,02218D-02  5,95802D-02

-1,60380D+02  8,20231D+01 -1,39674D+01  7,77086D-01  1,07795D-05 -7,04458D-05
 2,29064D-04 -5,32453D-04  1,06822D-03 -2,03688D-03  3,82480D-03 -7,13285D-03
 1,31310D-02 -2,29406D-02  3,48033D-02 -4,03701D-02  3,13655D-02

 9,78059D+01 -5,19088D+01  9,16150D+00 -5,30911D-01 -3,31509D-06  2,16647D-05
-7,04458D-05  1,63750D-04 -3,28528D-04  6,26509D-04 -1,17703D-03  2,19984D-03
-4,08898D-03  7,45514D-03 -1,24648D-02  1,62930D-02 -1,40624D-02  6,83608D-03

-2,37220D+01  1,29853D+01 -2,37858D+00  1,44848D-01  5,07270D-07 -3,31509D-06
 1,07795D-05 -2,50569D-05  5,02713D-05 -9,58715D-05  1,80140D-04 -3,36887D-04
 6,27900D-04 -1,15875D-03  2,04737D-03 -2,90978D-03  2,71309D-03 -1,40048D-03
 3,00082D-04
```

N= 4 M= 15

```
 9,08717D+06

 0,           2,79393D+06

 9,32484D+05  0,           1,12872D+05

 0,           3,76602D+04  0,           9,05381D+02

-3,07995D+01 -1,54865D+01 -2,59350D+00 -1,44979D-01  3,00082D-04

 1,25935D+02  6,15601D+01  9,94253D+00  5,31768D-01 -1,40048D-03  6,83609D-03

-2,04388D+02 -9,68711D+01 -1,50832D+01 -7,79873D-01  2,71310D-03 -1,40624D-02
 3,13657D-02

 1,75889D+02  8,15293D+01  1,23465D+01  6,28730D-01 -2,90979D-03  1,62931D-02
-4,03704D-02  5,95808D-02

-1,01869D+02 -4,78822D+01 -7,09918D+00 -3,66911D-01  2,04739D-03 -1,24649D-02
 3,48038D-02 -6,02230D-02  7,35491D-02

 5,38198D+01  2,69111D+01  3,74503D+00  2,05941D-01 -1,15879D-03  7,45544D-03
-2,29416D-02  4,60397D-02 -6,81280D-02  7,80215D-02

-2,56228D+01 -1,59801D+01 -1,78235D+00 -1,22254D-01  6,27987D-04 -4,08955D-03
 1,31329D-02 -2,90300D-02  5,03220D-02 -7,05484D-02  7,93269D-02

 7,53566D+00  1,13960D+01  5,24149D-01  8,71789D-02 -3,37050D-04  2,20091D-03
-7,13632D-03  1,63993D-02 -3,13253D-02  5,16150D-02 -7,12373D-02  7,96745D-02
```

THE MATRIX B   N=4

```
 7,53566D+00 -1,13960D+01  5,24149D-01 -8,71789D-02  1,80445D-04 -1,17902D-03
 3,83128D-03 -8,88256D-03  1,76224D-02 -3,20060D-02  5,19617D-02 -7,13820D-02
 7,96745D-02

-2,56228D+01  1,59801D+01 -1,78235D+00  1,22254D-01 -9,64397D-05  6,30223D-04
-2,04896D-03  4,75994D-03 -9,52548D-03  1,79646D-02 -3,21501D-02  5,19617D-02
-7,12373D-02  7,93269D-02

 5,38198D+01 -2,69111D+01  3,74503D+00 -2,05941D-01  5,13199D-05 -3,35380D-04
 1,09050D-03 -2,53451D-03  5,08202D-03 -9,66735D-03  1,79646D-02 -3,20060D-02
 5,16150D-02 -7,05484D-02  7,80215D-02

-1,01869D+02  4,78822D+01 -7,09918D+00  3,66911D-01 -2,69095D-05  1,75857D-04
-5,71821D-04  1,32915D-03 -2,66632D-03  5,08202D-03 -9,52548D-03  1,76224D-02
-3,13253D-02  5,03220D-02 -6,81280D-02  7,35491D-02

 1,75889D+02 -8,15293D+01  1,23465D+01 -6,28730D-01  1,34125D-05 -8,76526D-05
 2,85015D-04 -6,62509D-04  1,32915D-03 -2,53451D-03  4,75994D-03 -8,88256D-03
 1,63993D-02 -2,90300D-02  4,60397D-02 -6,02230D-02  5,95808D-02

-2,04388D+02  9,68711D+01 -1,50832D+01  7,79873D-01 -5,77006D-06  3,77083D-05
-1,22614D-04  2,85015D-04 -5,71821D-04  1,09050D-03 -2,04896D-03  3,83128D-03
-7,13632D-03  1,31329D-02 -2,29416D-02  3,48038D-02 -4,03704D-02  3,13657D-02

 1,25935D+02 -6,15601D+01  9,94253D+00 -5,31768D-01  1,77451D-06 -1,15967D-05
 3,77083D-05 -8,76526D-05  1,75857D-04 -3,35380D-04  6,30223D-04 -1,17902D-03
 2,20091D-03 -4,08955D-03  7,45544D-03 -1,24649D-02  1,62931D-02 -1,40624D-02
 6,83609D-03

-3,07995D+01  1,54865D+01 -2,59350D+00  1,44979D-01 -2,71532D-07  1,77451D-06
-5,77006D-06  1,34125D-05 -2,69095D-05  5,13199D-05 -9,64397D-05  1,80445D-04
-3,37050D-04  6,27987D-04 -1,15879D-03  2,04739D-03 -2,90979D-03  2,71310D-03
-1,40048D-03  3,00082D-04
```

N= 4   M= 16

```
 1,53918D+07

 0,           4,06520D+06

 1,35676D+06  0,           1,41266D+05

 0,           4,71183D+04  0,           9,75357D+02

-3,92283D+01 -1,81800D+01 -2,81233D+00 -1,44909D-01  3,00082D-04

 1,59391D+02  7,18439D+01  1,07491D+01  5,31309D-01 -1,40048D-03  6,83609D-03

-2,57096D+02 -1,12352D+02 -1,62819D+01 -7,78382D-01  2,71310D-03 -1,40625D-02
 3,13657D-02

 2,20758D+02  9,39124D+01  1,33569D+01  6,25263D-01 -2,90980D-03  1,62931D-02
-4,03705D-02  5,95810D-02

-1,29514D+02 -5,43839D+01 -7,78454D+00 -3,59955D-01  2,04740D-03 -1,24650D-02
 3,48040D-02 -6,02233D-02  7,35498D-02
```

THE MATRIX B   N=4

```
 7,18284D+01  2,91463D+01  4,31139D+00  1,92675D-01 -1,15881D-03  7,45553D-03
-2,29419D-02  4,60404D-02 -6,81293D-02  7,80240D-02

-4,08222D+01 -1,47266D+01 -2,44954D+00 -9,73250D-02  6,28012D-04 -4,08972D-03
 1,31334D-02 -2,90313D-02  5,03245D-02 -7,05531D-02  7,93358D-02

 2,60040D+01  6,13303D+00  1,56027D+00  4,05295D-02 -3,37097D-04  2,20121D-03
-7,13731D-03  1,64016D-02 -3,13300D-02  5,16238D-02 -7,12539D-02  7,97056D-02

-2,16424D+01  2,34018D-21 -1,29855D+00  1,64038D-23  1,80533D-04 -1,17959D-03
 3,83314D-03 -8,88688D-03  1,76310D-02 -3,20225D-02  5,19928D-02 -7,14401D-02
 7,97831D-02

 2,60040D+01 -6,13303D+00  1,56027D+00 -4,05295D-02 -9,66030D-05  6,31290D-04
-2,05243D-03  4,76801D-03 -9,54167D-03  1,79954D-02 -3,22081D-02  5,20702D-02
-7,14401D-02  7,97056D-02

-4,08222D+01  1,47266D+01 -2,44954D+00  9,73250D-02  5,16240D-05 -3,37368D-04
 1,09697D-03 -2,54954D-03  5,11216D-03 -9,72484D-03  1,80726D-02 -3,22081D-02
 5,19928D-02 -7,12539D-02  7,93358D-02

 7,18284D+01 -2,91463D+01  4,31139D+00 -1,92675D-01 -2,74707D-05  1,79525D-04
-5,83748D-04  1,35688D-03 -2,72194D-03  5,18810D-03 -9,72484D-03  1,79954D-02
-3,20225D-02  5,16238D-02 -7,05531D-02  7,80240D-02

-1,29514D+02  5,43839D+01 -7,78454D+00  3,59955D-01  1,44041D-05 -9,41333D-05
 3,06088D-04 -7,11493D-04  1,42743D-03 -2,72194D-03  5,11216D-03 -9,54167D-03
 1,76310D-02 -3,13300D-02  5,03245D-02 -6,81293D-02  7,35498D-02

 2,20758D+02 -9,39124D+01  1,33569D+01 -6,25263D-01 -7,17943D-06  4,69188D-05
-1,52563D-04  3,54631D-04 -7,11493D-04  1,35688D-03 -2,54954D-03  4,76801D-03
-8,88688D-03  1,64016D-02 -2,90313D-02  4,60404D-02 -6,02233D-02  5,95810D-02

-2,57096D+02  1,12352D+02 -1,62819D+01  7,78382D-01  3,08860D-06 -2,01845D-05
 6,56329D-05 -1,52563D-04  3,06088D-04 -5,83748D-04  1,09697D-03 -2,05243D-03
 3,83314D-03 -7,13731D-03  1,31334D-02 -2,29419D-02  3,48040D-02 -4,03705D-02
 3,13657D-02

 1,59391D+02 -7,18439D+01  1,07491D+01 -5,31309D-01 -9,49858D-07  6,20748D-06
-2,01845D-05  4,69188D-05 -9,41333D-05  1,79525D-04 -3,37368D-04  6,31290D-04
-1,17959D-03  2,20121D-03 -4,08972D-03  7,45553D-03 -1,24650D-02  1,62931D-02
-1,40625D-02  6,83609D-03

-3,92283D+01  1,81800D+01 -2,81233D+00  1,44909D-01  1,45346D-07 -9,49858D-07
 3,08860D-06 -7,17943D-06  1,44041D-05 -2,74707D-05  5,16240D-05 -9,66030D-05
 1,80533D-04 -3,37097D-04  6,28012D-04 -1,15881D-03  2,04740D-03 -2,90980D-03
 2,71310D-03 -1,40048D-03  3,00082D-04
```

N= 4   M= 17

```
 2,51111D+07

 0,           5,76223D+06

 1,92245D+06  0,           1,74033D+05
```

THE MATRIX B   N=4

```
 0.            5,80447D+04  0.            1,04537D+03

-4,90268D+01 -2,11065D+01 -3,02891D+00 -1,44947D-01  3,00082D-04

 1,97984D+02  8,30272D+01  1,15409D+01  5,31555D-01 -1,40048D-03  6,83609D-03

-3,17170D+02 -1,29334D+02 -1,74328D+01 -7,79180D-01  2,71310D-03 -1,40625D-02
 3,13657D-02

 2,70503D+02  1,08009D+02  1,42560D+01  6,27119D-01 -2,90980D-03  1,62931D-02
-4,03705D-02  5,95811D-02

-1,57437D+02 -6,29807D+01 -8,24665D+00 -3,63679D-01  2,04740D-03 -1,24650D-02
 3,48040D-02 -6,02234D-02  7,35500D-02

 8,51011D+01  3,46364D+01  4,45198D+00  1,99778D-01 -1,15881D-03  7,45555D-03
-2,29420D-02  4,60406D-02 -6,81297D-02  7,80247D-02

-4,42954D+01 -1,91927D+01 -2,31661D+00 -1,10673D-01  6,28019D-04 -4,08976D-03
 1,31336D-02 -2,90316D-02  5,03252D-02 -7,05545D-02  7,93384D-02

 2,10172D+01  1,13609D+01  1,09911D+00  6,55073D-02 -3,37111D-04  2,20130D-03
-7,13760D-03  1,64023D-02 -3,13313D-02  5,16263D-02 -7,12586D-02  7,97145D-02

-6,17658D+00 -8,09647D+00 -3,23003D-01 -4,66841D-02  1,80558D-04 -1,17976D-03
 3,83367D-03 -8,88812D-03  1,76335D-02 -3,20272D-02  5,20017D-02 -7,14567D-02
 7,98143D-02

-6,17658D+00  8,09647D+00 -3,23003D-01  4,66841D-02 -9,66498D-05  6,31596D-04
-2,05342D-03  4,77032D-03 -9,54630D-03  1,80043D-02 -3,22248D-02  5,21013D-02
-7,14982D-02  7,98143D-02

 2,10172D+01 -1,13609D+01  1,09911D+00 -6,55073D-02  5,17114D-05 -3,37939D-04
 1,09882D-03 -2,55386D-03  5,12082D-03 -9,74136D-03  1,81036D-02 -3,22662D-02
 5,21013D-02 -7,14567D-02  7,97145D-02

-4,42954D+01  1,91927D+01 -2,31661D+00  1,10673D-01 -2,76335D-05  1,80589D-04
-5,87208D-04  1,36492D-03 -2,73808D-03  5,21887D-03 -9,78267D-03  1,81036D-02
-3,22248D-02  5,20017D-02 -7,12586D-02  7,93384D-02

 8,51011D+01 -3,46364D+01  4,45198D+00 -1,99778D-01  1,47046D-05 -9,60967D-05
 3,12472D-04 -7,26334D-04  1,45720D-03 -2,77872D-03  5,21887D-03 -9,74136D-03
 1,80043D-02 -3,20272D-02  5,16263D-02 -7,05545D-02  7,80247D-02

-1,57437D+02  6,29807D+01 -8,24665D+00  3,63679D-01 -7,71025D-06  5,03877D-05
-1,63843D-04  3,80851D-04 -7,64099D-04  1,45720D-03 -2,73808D-03  5,12082D-03
-9,54630D-03  1,76335D-02 -3,13313D-02  5,03252D-02 -6,81297D-02  7,35500D-02

 2,70503D+02 -1,08009D+02  1,42560D+01 -6,27119D-01  3,84301D-06 -2,51147D-05
 8,16641D-05 -1,89828D-04  3,80851D-04 -7,26334D-04  1,36492D-03 -2,55386D-03
 4,77032D-03 -8,88812D-03  1,64023D-02 -2,90316D-02  4,60406D-02 -6,02234D-02
 5,95811D-02

-3,17170D+02  1,29334D+02 -1,74328D+01  7,79180D-01 -1,65327D-06  1,08044D-05
-3,51320D-05  8,16641D-05 -1,63843D-04  3,12472D-04 -5,87208D-04  1,09882D-03
-2,05342D-03  3,83367D-03 -7,13760D-03  1,31336D-02 -2,29420D-02  3,48040D-02
-4,03705D-02  3,13657D-02
```

THE MATRIX B N=4

```
 1,97984D+02 -8,30272D+01  1,15409D+01 -5,31555D-01  5,08440D-07 -3,32274D-06
 1,08044D-05 -2,51147D-05  5,03877D-05 -9,60967D-05  1,80589D-04 -3,37939D-04
 6,31596D-04 -1,17976D-03  2,20130D-03 -4,08976D-03  7,45555D-03 -1,24650D-02
 1,62931D-02 -1,40625D-02  6,83609D-03

-4,90268D+01  2,11065D+01 -3,02891D+00  1,44947D-01 -7,78007D-08  5,08440D-07
-1,65327D-06  3,84301D-06 -7,71025D-06  1,47046D-05 -2,76335D-05  5,17114D-05
-9,66498D-05  1,80558D-04 -3,37111D-04  6,28019D-04 -1,15881D-03  2,04740D-03
-2,90980D-03  2,71310D-03 -1,40048D-03  3,00082D-04
```

N= 4 M= 18

```
 3,96736D+07

 0,           7,98209D+06

 2,66281D+06  0,           2,11514D+05

 0,           7,05360D+04  0,           1,11536D+03

-6,03637D+01 -2,42407D+01 -3,24678D+00 -1,44927D-01  3,00082D-04

 2,42504D+02  9,49448D+01  1,23412D+01  5,31423D-01 -1,40048D-03  6,83609D-03

-3,86470D+02 -1,47280D+02 -1,86112D+01 -7,78753D-01  2,71310D-03 -1,40625D-02
 3,13657D-02

 3,28564D+02  1,22569D+02  1,52189D+01  6,26126D-01 -2,90980D-03  1,62931D-02
-4,03705D-02  5,95811D-02

-1,91865D+02 -7,11667D+01 -8,83683D+00 -3,61686D-01  2,04740D-03 -1,24650D-02
 3,48040D-02 -6,02235D-02  7,35501D-02

 1,05141D+02  3,86024D+01  4,83681D+00  1,95977D-01 -1,15881D-03  7,45556D-03
-2,29420D-02  4,60406D-02 -6,81298D-02  7,80249D-02

-5,74602D+01 -2,03975D+01 -2,64266D+00 -1,03529D-01  6,28022D-04 -4,08978D-03
 1,31336D-02 -2,90317D-02  5,03254D-02 -7,05549D-02  7,93391D-02

 3,25487D+01  1,02731D+01  1,49686D+00  5,21389D-02 -3,37114D-04  2,20133D-03
-7,13768D-03  1,64025D-02 -3,13317D-02  5,16271D-02 -7,12600D-02  7,97171D-02

-2,07186D+01 -4,27520D+00 -9,52800D-01 -2,16977D-02  1,80565D-04 -1,17980D-03
 3,83382D-03 -8,88847D-03  1,76342D-02 -3,20286D-02  5,20042D-02 -7,14615D-02
 7,98232D-02

 1,72413D+01  1,06582D-21  7,92884D-01  5,12210D-24 -9,66632D-05  6,31683D-04
-2,05371D-03  4,77098D-03 -9,54763D-03  1,80068D-02 -3,22295D-02  5,21103D-02
-7,15149D-02  7,98454D-02

-2,07186D+01  4,27520D+00 -9,52800D-01  2,16977D-02  5,17365D-05 -3,38103D-04
 1,09935D-03 -2,55509D-03  5,12331D-03 -9,74609D-03  1,81125D-02 -3,22829D-02
 5,21325D-02 -7,15149D-02  7,98232D-02
```

THE MATRIX B N=4

```
 3.25487D+01 -1.02731D+01  1.49686D+00 -5.21389D-02 -2.76803D-05  1.80895D-04
-5.88202D-04  1.36723D-03 -2.74271D-03  5.22771D-03 -9.79929D-03  1.81347D-02
-3.22829D-02  5.21103D-02 -7.14615D-02  7.97171D-02

-5.74602D+01  2.03975D+01 -2.64266D+00  1.03529D-01  1.47917D-05 -9.66663D-05
 3.14324D-04 -7.30639D-04  1.46584D-03 -2.79520D-03  5.24983D-03 -9.79929D-03
 1.81125D-02 -3.22295D-02  5.20042D-02 -7.12600D-02  7.93391D-02

 1.05141D+02 -3.86024D+01  4.83681D+00 -1.95977D-01 -7.87107D-06  5.14387D-05
-1.67260D-04  3.88795D-04 -7.80036D-04  1.48760D-03 -2.79520D-03  5.22771D-03
-9.74609D-03  1.80068D-02 -3.20286D-02  5.16271D-02 -7.05549D-02  7.80249D-02

-1.91865D+02  7.11667D+01 -8.83683D+00  3.61686D-01  4.12715D-06 -2.69716D-05
 8.77020D-05 -2.03863D-04  4.09010D-04 -7.80036D-04  1.46584D-03 -2.74271D-03
 5.12331D-03 -9.54763D-03  1.76342D-02 -3.13317D-02  5.03254D-02 -6.81298D-02
 7.35501D-02

 3.28564D+02 -1.22569D+02  1.52189D+01 -6.26126D-01 -2.05709D-06  1.34434D-05
-4.37132D-05  1.01611D-04 -2.03863D-04  3.88795D-04 -7.30639D-04  1.36723D-03
-2.55509D-03  4.77098D-03 -8.88847D-03  1.64025D-02 -2.90317D-02  4.60406D-02
-6.02235D-02  5.95811D-02

-3.86470D+02  1.47280D+02 -1.86112D+01  7.78753D-01  8.84962D-07 -5.78337D-06
 1.88055D-05 -4.37132D-05  8.77020D-05 -1.67260D-04  3.14324D-04 -5.88202D-04
 1.09935D-03 -2.05371D-03  3.83382D-03 -7.13768D-03  1.31336D-02 -2.29420D-02
 3.48040D-02 -4.03705D-02  3.13657D-02

 2.42504D+02 -9.49448D+01  1.23412D+01 -5.31423D-01 -2.72158D-07  1.77860D-06
-5.78337D-06  1.34434D-05 -2.69716D-05  5.14387D-05 -9.66663D-05  1.80895D-04
-3.38103D-04  6.31683D-04 -1.17980D-03  2.20133D-03 -4.08978D-03  7.45556D-03
-1.24650D-02  1.62931D-02 -1.40625D-02  6.83609D-03

-6.03637D+01  2.42407D+01 -3.24678D+00  1.44927D-01  4.16452D-08 -2.72158D-07
 8.84962D-07 -2.05709D-06  4.12715D-06 -7.87107D-06  1.47917D-05 -2.76803D-05
 5.17365D-05 -9.66632D-05  1.80565D-04 -3.37114D-04  6.28022D-04 -1.15881D-03
 2.04740D-03 -2.90980D-03  2.71310D-03 -1.40048D-03  3.00082D-04
```

N= 4 M= 19

```
 6.09315D+07

 0.           1.08382D+07

 3.61501D+06  0.           2.54008D+05

 0.           8.47022D+04  0.           1.18537D+03

-7.33084D+01 -2.75982D+01 -3.46391D+00 -1.44937D-01  3.00082D-04

 2.93092D+02  1.07698D+02  1.31366D+01  5.31494D-01 -1.40048D-03  6.83609D-03

-4.64745D+02 -1.66519D+02 -1.97738D+01 -7.78982D-01  2.71310D-03 -1.40625D-02
 3.13657D-02

 3.93462D+02  1.38359D+02  1.61454D+01  6.26658D-01 -2.90980D-03  1.62931D-02
-4.03705D-02  5.95811D-02
```

THE MATRIX B   N=4

```
-2,29163D+02 -8,04780D+01 -9,35385D+00 -3,62753D-01  2,04740D-03 -1,24650D-02
 3,48040D-02 -6,02235D-02  7,35501D-02

 1,24646D+02  4,39737D+01  5,08214D+00  1,98012D-01 -1,15881D-03  7,45556D-03
-2,29420D-02  4,60406D-02 -6,81298D-02  7,80250D-02

-6,63983D+01 -2,38459D+01 -2,70655D+00 -1,07353D-01  6,28022D-04 -4,08978D-03
 1,31336D-02 -2,90317D-02  5,03255D-02 -7,05550D-02  7,93393D-02

 3,44463D+01  1,31717D+01  1,40404D+00  5,92953D-02 -3,37115D-04  2,20133D-03
-7,13770D-03  1,64025D-02 -3,13318D-02  5,16273D-02 -7,12604D-02  7,97178D-02

-1,63317D+01 -7,79124D+00 -6,65672D-01 -3,50735D-02  1,80567D-04 -1,17982D-03
 3,83387D-03 -8,88858D-03  1,76344D-02 -3,20290D-02  5,20050D-02 -7,14629D-02
 7,98257D-02

 4,79878D+00  5,55170D+00  1,95596D-01  2,49918D-02 -9,66671D-05  6,31708D-04
-2,05379D-03  4,77117D-03 -9,54801D-03  1,80075D-02 -3,22309D-02  5,21128D-02
-7,15197D-02  7,98543D-02

 4,79878D+00 -5,55170D+00  1,95596D-01 -2,49918D-02  5,17437D-05 -3,38150D-04
 1,09951D-03 -2,55545D-03  5,12402D-03 -9,74745D-03  1,81151D-02 -3,22877D-02
 5,21414D-02 -7,15316D-02  7,98543D-02

-1,63317D+01  7,79124D+00 -6,65672D-01  3,50735D-02 -2,76937D-05  1,80983D-04
-5,88487D-04  1,36789D-03 -2,74404D-03  5,23025D-03 -9,80405D-03  1,81436D-02
-3,22995D-02  5,21414D-02 -7,15197D-02  7,98257D-02

 3,44463D+01 -1,31717D+01  1,40404D+00 -5,92953D-02  1,48168D-05 -9,68299D-05
 3,14856D-04 -7,31876D-04  1,46832D-03 -2,79993D-03  5,25872D-03 -9,81593D-03
 1,81436D-02 -3,22877D-02  5,21128D-02 -7,14629D-02  7,97178D-02

-6,63983D+01  2,38459D+01 -2,70655D+00  1,07353D-01 -7,91772D-06  5,17436D-05
-1,68252D-04  3,91100D-04 -7,84660D-04  1,49642D-03 -2,81177D-03  5,25872D-03
-9,80405D-03  1,81151D-02 -3,22309D-02  5,20050D-02 -7,12604D-02  7,93393D-02

 1,24646D+02 -4,39737D+01  5,08214D+00 -1,98012D-01  4,21323D-06 -2,75341D-05
 8,95313D-05 -2,08115D-04  4,17541D-04 -7,96306D-04  1,49642D-03 -2,79993D-03
 5,23025D-03 -9,74745D-03  1,80075D-02 -3,20290D-02  5,16273D-02 -7,05550D-02
 7,80250D-02

-2,29163D+02  8,04780D+01 -9,35385D+00  3,62753D-01 -2,20918D-06  1,44374D-05
-4,69452D-05  1,09124D-04 -2,18935D-04  4,17541D-04 -7,84660D-04  1,46832D-03
-2,74404D-03  5,12402D-03 -9,54801D-03  1,76344D-02 -3,13318D-02  5,03255D-02
-6,81298D-02  7,35501D-02

 3,93462D+02 -1,38359D+02  1,61454D+01 -6,26658D-01  1,10112D-06 -7,19599D-06
 2,33988D-05 -5,43904D-05  1,09124D-04 -2,08115D-04  3,91100D-04 -7,31876D-04
 1,36789D-03 -2,55545D-03  4,77117D-03 -8,88858D-03  1,64025D-02 -2,90317D-02
 4,60406D-02 -6,02235D-02  5,95811D-02

-4,64745D+02  1,66519D+02 -1,97738D+01  7,78982D-01 -4,73703D-07  3,09573D-06
-1,00662D-05  2,33988D-05 -4,69452D-05  8,95313D-05 -1,68252D-04  3,14856D-04
-5,88487D-04  1,09951D-03 -2,05379D-03  3,83387D-03 -7,13770D-03  1,31336D-02
-2,29420D-02  3,48040D-02 -4,03705D-02  3,13657D-02
```

THE MATRIX B   N=4

```
 2,93092D+02 -1,07698D+02  1,31366D+01 -5,31494D-01  1,45681D-07 -9,52049D-07
 3,09573D-06 -7,19599D-06  1,44374D-05 -2,75341D-05  5,17436D-05 -9,68299D-05
 1,80983D-04 -3,38150D-04  6,31708D-04 -1,17982D-03  2,20133D-03 -4,08978D-03
 7,45556D-03 -1,24650D-02  1,62931D-02 -1,40625D-02  6,83609D-03

-7,33084D+01  2,75982D+01 -3,46391D+00  1,44937D-01 -2,22919D-08  1,45681D-07
-4,73703D-07  1,10112D-06 -2,20918D-06  4,21323D-06 -7,91772D-06  1,48168D-05
-2,76937D-05  5,17437D-05 -9,66671D-05  1,80567D-04 -3,37115D-04  6,28022D-04
-1,15881D-03  2,04740D-03 -2,90980D-03  2,71310D-03 -1,40048D-03  3,00082D-04

                N=  4            M= 20

 9,12784D+07

 0,           1,44575D+07

 4,82180D+06  0,           3,01840D+05

 0,           1,00645D+05  0,           1,25536D+03

-8,79952D+01 -3,11696D+01 -3,68147D+00 -1,44932D-01  3,00082D-04

 3,50315D+02  1,21225D+02  1,39348D+01  5,31456D-01 -1,40048D-03  6,83609D-03

-5,53119D+02 -1,86850D+02 -2,09454D+01 -7,78859D-01  2,71310D-03 -1,40625D-02
 3,13657D-02

 4,66929D+02  1,54913D+02  1,70926D+01  6,26373D-01 -2,90980D-03  1,62931D-02
-4,03705D-02  5,95811D-02

-2,72132D+02 -8,99811D+01 -9,91249D+00 -3,62182D-01  2,04740D-03 -1,24650D-02
 3,48040D-02 -6,02235D-02  7,35501D-02

 1,48590D+02  4,89701D+01  5,40683D+00  1,96922D-01 -1,15881D-03  7,45556D-03
-2,29420D-02  4,60406D-02 -6,81298D-02  7,80250D-02

-8,02541D+01 -2,61928D+01 -2,91958D+00 -1,05306D-01  6,28022D-04 -4,08978D-03
 1,31336D-02 -2,90318D-02  5,03255D-02 -7,05550D-02  7,93394D-02

 4,37163D+01  1,37964D+01  1,59028D+00  5,54648D-02 -3,37116D-04  2,20134D-03
-7,13771D-03  1,64025D-02 -3,13318D-02  5,16273D-02 -7,12605D-02  7,97180D-02

-2,47450D+01 -6,94345D+00 -9,00145D-01 -2,79139D-02  1,80567D-04 -1,17982D-03
 3,83388D-03 -8,88861D-03  1,76345D-02 -3,20291D-02  5,20052D-02 -7,14633D-02
 7,98265D-02

 1,57486D+01  2,88908D+00  5,72883D-01  1,16146D-02 -9,66682D-05  6,31716D-04
-2,05381D-03  4,77123D-03 -9,54812D-03  1,80077D-02 -3,22313D-02  5,21135D-02
-7,15210D-02  7,98569D-02

-1,31051D+01  2,43631D-22 -4,76719D-01 -6,68125D-25  5,17457D-05 -3,38164D-04
 1,09955D-03 -2,55555D-03  5,12422D-03 -9,74784D-03  1,81158D-02 -3,22890D-02
 5,21440D-02 -7,15363D-02  7,98632D-02
```

THE MATRIX B N=4

```
 1,57486D+01 -2,88908D+00  5,72883D-01 -1,16146D-02 -2,76976D-05  1,81008D-04
-5,88569D-04  1,36808D-03 -2,74442D-03  5,23098D-03 -9,80542D-03  1,81462D-02
-3,23043D-02  5,21503D-02 -7,15363D-02  7,98569D-02

-2,47450D+01  6,94345D+00 -9,00145D-01  2,79139D-02  1,48239D-05 -9,68769D-05
 3,15009D-04 -7,32230D-04  1,46903D-03 -2,80129D-03  5,26127D-03 -9,82070D-03
 1,81526D-02 -3,23043D-02  5,21440D-02 -7,15210D-02  7,98265D-02

 4,37163D+01 -1,37964D+01  1,59028D+00 -5,54648D-02 -7,93113D-06  5,18312D-05
-1,68537D-04  3,91762D-04 -7,85988D-04  1,49895D-03 -2,81653D-03  5,26763D-03
-9,82070D-03  1,81462D-02 -3,22890D-02  5,21135D-02 -7,14633D-02  7,97180D-02

-8,02541D+01  2,61928D+01 -2,91958D+00  1,05306D-01  4,23820D-06 -2,76973D-05
 9,00620D-05 -2,09348D-04  4,20016D-04 -8,01026D-04  1,50528D-03 -2,81653D-03
 5,26127D-03 -9,80542D-03  1,81158D-02 -3,22313D-02  5,20052D-02 -7,12605D-02
 7,93394D-02

 1,48590D+02 -4,89701D+01  5,40683D+00 -1,96922D-01 -2,25526D-06  1,47385D-05
-4,79244D-05  1,11400D-04 -2,23502D-04  4,26250D-04 -8,01026D-04  1,49895D-03
-2,80129D-03  5,23098D-03 -9,74784D-03  1,80077D-02 -3,20291D-02  5,16273D-02
-7,05550D-02  7,80250D-02

-2,72132D+02  8,99811D+01 -9,91249D+00  3,62182D-01  1,18253D-06 -7,72804D-06
 2,51288D-05 -5,84118D-05  1.17192D-04 -2,23502D-04  4,20016D-04 -7,85988D-04
 1,46903D-03 -2,74442D-03  5,12422D-03 -9,54812D-03  1,76345D-02 -3,13318D-02
 5,03255D-02 -6,81298D-02  7,35501D-02

 4,66929D+02 -1,54913D+02  1,70926D+01 -6,26373D-01 -5,89407D-07  3,85187D-06
-1,25249D-05  2,91141D-05 -5,84118D-05  1,11400D-04 -2,09348D-04  3,91762D-04
-7,32230D-04  1,36808D-03 -2,55555D-03  4,77123D-03 -8,88861D-03  1,64025D-02
-2,90318D-02  4,60406D-02 -6,02235D-02  5,95811D-02

-5,53119D+02  1,86850D+02 -2,09454D+01  7,78859D-01  2,53564D-07 -1,65708D-06
 5,38824D-06 -1,25249D-05  2,51288D-05 -4,79244D-05  9,00620D-05 -1,68537D-04
 3,15009D-04 -5,88569D-04  1,09955D-03 -2,05381D-03  3,83388D-03 -7,13771D-03
 1,31336D-02 -2,29420D-02  3,48040D-02 -4,03705D-02  3,13657D-02

 3,50315D+02 -1,21225D+02  1,39348D+01 -5,31456D-01 -7,79802D-08  5,09613D-07
-1,65708D-06  3,85187D-06 -7,72804D-06  1,47385D-05 -2,76973D-05  5,18312D-05
-9,68769D-05  1,81008D-04 -3,38164D-04  6,31716D-04 -1,17982D-03  2,20134D-03
-4,08978D-03  7,45556D-03 -1,24650D-02  1,62931D-02 -1,40625D-02  6,83609D-03

-8,79952D+01  3,11696D+01 -3,68147D+00  1,44932D-01  1,19324D-08 -7,79802D-08
 2,53564D-07 -5,89407D-07  1,18253D-06 -2,25526D-06  4,23820D-06 -7,93113D-06
 1.48239D-05 -2,76976D-05  5,17457D-05 -9,66682D-05  1,80567D-04 -3,37116D-04
 6,28022D-04 -1,15881D-03  2,04740D-03 -2,90980D-03  2,71310D-03 -1.40048D-03
 3,00082D-04
```

THE MATRIX B   N=5

```
                N=  5             M=  4

 0.

 0.           -2.77778D+02

 4.13333D+01   0.            7.18444D+02

 0.           -1.02106D+03   0.            2.57589D+03

-4.23333D+01   0.           -2.79694D+02   0.           -4.13056D+02

 0.            8.33333D-02  -4.16667D-02  -8.33333D-02   4.16667D-02   0.

 0.           -6.66667D-01   6.66667D-01   1.66667D-01  -1.66667D-01   0.
 0.

 1.00000D+00   0.           -1.25000D+00   0.            2.50000D-01   0.
 0.            0.

 0.            6.66667D-01   6.66667D-01  -1.66667D-01  -1.66667D-01   0.
 0.            0.            0.

 0.           -8.33333D-02  -4.16667D-02   8.33333D-02   4.16667D-02   0.
 0.            0.            0.            0.


                N=  5             M=  5

-1.49803D+02

 0.           -1.76863D+03

-9.57967D+01   0.            3.89835D+03

 0.           -3.02418D+03   0.           -4.10990D+03

-2.14338D+02   0.           -5.60324D+02   0.           -5.48637D+02

 1.17187D-02   7.47124D-02  -5.20833D-02   1.26939D-01   2.08333D-02  -3.20123D-06

-9.76562D-02  -3.31895D-01   4.06250D-01  -8.01364D-01  -6.25000D-02   1.60061D-05
-8.00307D-05

 5.85937D-01  -3.77876D-01  -3.54167D-01   1.76939D+00   4.16667D-02  -3.20123D-05
 1.60061D-04  -3.20123D-04

 5.85938D-01   3.77876D-01  -3.54167D-01  -1.76939D+00   4.16667D-02   3.20123D-05
-1.60061D-04   3.20123D-04  -3.20123D-04

-9.76563D-02   3.31895D-01   4.06250D-01   8.01364D-01  -6.25000D-02  -1.60061D-05
 8.00307D-05  -1.60061D-04   1.60061D-04  -8.00307D-05
```

THE MATRIX B   N=5

```
 1,17188D-02 -7,47124D-02 -5,20833D-02 -1,26939D-01  2,08333D-02  3,20123D-06
-1,60061D-05  3,20123D-05 -3,20123D-05  1,60061D-05 -3,20123D-06
```

N= 5   M= 6

```
-4,06769D+03

 0,          -6,84436D+03

-1,29187D+04  0,          -4,05736D+04

 0,          -9,65391D+03  0,          -8,77341D+03

-1,63455D+03  0,          -5,35265D+03  0,          -1,03630D+03

 1,22329D-01  8,19536D-02  3,89753D-01  1,07741D-01  4,78833D-02 -4,70166D-06

-7,33975D-01 -2,44481D-01 -2,38018D+00 -5,14299D-01 -2,45633D-01  2,61644D-05
-1,48804D-04

 1,83494D+00 -2,56899D-01  6,51296D+00  7,05374D-01  5,51583D-01 -6,02968D-05
 3,51553D-04 -8,53312D-04

-1,44658D+00 -1,17399D-27 -9,04506D+00 -1,50123D-27 -7,07666D-01  7,35770D-05
-4,41462D-04  1,10366D-03 -1,47154D-03

 1,83494D+00  2,56899D-01  6,51296D+00 -7,05374D-01  5,51583D-01 -5,00687D-05
 3,10640D-04 -8,02171D-04  1,10366D-03 -8,53312D-04

-7,33975D-01  2,44481D-01 -2,38018D+00  5,14299D-01 -2,45633D-01  1,79819D-05
-1,16074D-04  3,10640D-04 -4,41462D-04  3,51553D-04 -1,48804D-04

 1,22329D-01 -8,19536D-02  3,89753D-01 -1,07741D-01  4,78833D-02 -2,65604D-06
 1,79819D-05 -5,00687D-05  7,35770D-05 -6,02968D-05  2,61644D-05 -4,70166D-06
```

N= 5   M= 7

```
-2,47019D+04

 0,          -2,13168D+05

-5,52920D+04  0,          -1,20190D+05

 0,          -7,76554D+04  0,          -3,04573D+04

-4,45539D+03  0,          -1,00947D+04  0,          -1,28122D+03

 2,08214D-01  8,62106D-01  4,68092D-01  3,04353D-01  3,59662D-02 -5,28151D-06

-1,02935D+00 -5,20985D+00 -2,39254D+00 -1,73869D+00 -1,58998D-01  3,03797D-05
-1,79449D-04

 1,77627D+00  1,40214D+01  4,61907D+00  4,26585D+00  2,61196D-01 -7,44260D-05
 4,54270D-04 -1,19760D-03
```

THE MATRIX B N=5

```
-4,55134D-01 -2,30498D+01 -2,69462D+00 -6,23457D+00 -1,38164D-01  1,01287D-04
-6,42909D-04  1,77887D-03 -2,79576D-03

-4,55134D-01  2,30498D+01 -2,69462D+00  6,23457D+00 -1,38164D-01 -8,36293D-05
 5,54621D-04 -1,61995D-03  2,70747D-03 -2,79576D-03

 1,77627D+00 -1,40214D+01  4,61907D+00 -4,26585D+00  2,61196D-01  4,26423D-05
-2,95351D-04  9,11548D-04 -1,61995D-03  1,77887D-03 -1,19760D-03

-1,02935D+00  5,20985D+00 -2,39254D+00  1,73869D+00 -1,58998D-01 -1,27221D-05
 9,11607D-05 -2,95351D-04  5,54621D-04 -6,42909D-04  4,54270D-04 -1,79449D-04

 2,08214D-01 -8,62106D-01  4,68092D-01 -3,04353D-01  3,59662D-02  1,74998D-06
-1,27221D-05  4,26423D-05 -8,36293D-05  1,01287D-04 -7,44260D-05  3,03797D-05
-5,28151D-06
```

N= 5 M= 8

```
-3,03705D+05

 0,          -7,92349D+05

-3,81011D+05  0,          -4,92761D+05

 0,          -1,97148D+05  0,          -5,35267D+04

-1,51634D+04  0,          -2,19723D+04  0,          -1,64713D+03

 9,93328D-01  1,31582D+00  1,22026D+00  3,14077D-01  4,48494D-02 -5,49716D-06

-5,92052D+00 -7,05080D+00 -7,03365D+00 -1,59479D+00 -2,24955D-01  3,19810D-05
-1,91338D-04

 1,56566D+01  1,51283D+01  1,77549D+01  3,15506D+00  4,94410D-01 -8,00878D-05
 4,96308D-04 -1,34624D-03

-2,52348D+01 -1,48675D+01 -2,67411D+01 -2,78206D+00 -6,70633D-01  1,14214D-04
-7,38889D-04  2,11824D-03 -3,57060D-03

 3,00109D+01 -2,85532D-25  2,95992D+01 -7,29003D-26  7,12657D-01 -1,04285D-04
 7,07986D-04 -2,16222D-03  3,94557D-03 -4,77410D-03

-2,52348D+01  1,48675D+01 -2,67411D+01  2,78206D+00 -6,70633D-01  6,52643D-05
-4,63319D-04  1,50545D-03 -2,97594D-03  3,94557D-03 -3,57060D-03

 1,56566D+01 -1,51283D+01  1,77549D+01 -3,15506D+00  4,94410D-01 -2,85849D-05
 2,08941D-04 -7,11802D-04  1,50545D-03 -2,16222D-03  2,11824D-03 -1,34624D-03

-5,92052D+00  7,05080D+00 -7,03365D+00  1,59479D+00 -2,24955D-01  8,08437D-06
-5,97548D-05  2,08941D-04 -4,63319D-04  7,07986D-04 -7,38889D-04  4,96308D-04
-1,91338D-04

 9,93328D-01 -1,31582D+00  1,22026D+00 -3,14077D-01  4,48494D-02 -1,08881D-06
 8,08437D-06 -2,85849D-05  6,52643D-05 -1,04285D-04  1,14214D-04 -8,00878D-05
 3,19810D-05 -5,49716D-06
```

THE MATRIX B N=5

```
                 N=  5           M=  9

-1,32253D+06

 0,           -3,43897D+06

-1,19744D+06   0,           -1,12533D+06

 0,           -5,43821D+05   0,           -9,78237D+04

-3,35826D+04   0,           -3,57280D+04   0,           -1,93492D+03

 1,78505D+00   3,17329D+00   1,57093D+00   4,58577D-01   4,00619D-02  -5,57681D-06

-9,82316D+00  -1,82277D+01  -8,35997D+00  -2,45435D+00  -1,89168D-01   3,25763D-05
-1,95788D-04

 2,23517D+01   4,56883D+01   1,81839D+01   5,61668D+00   3,65744D-01  -8,22281D-05
 5,12308D-04  -1,40376D-03

-2,60293D+01  -6,84103D+01  -1,98511D+01  -7,64648D+00  -3,62844D-01   1,19334D-04
-7,77163D-04   2,25584D-03  -3,89977D-03

 1,22156D+01   7,48238D+01   8,45624D+00   7,90931D+00   1,46206D-01  -1,13708D-04
 7,78425D-04  -2,41547D-03   4,55138D-03  -5,88903D-03

 1,22156D+01  -7,48238D+01   8,45624D+00  -7,90931D+00   1,46206D-01   7,88527D-05
-5,64894D-04   1,87064D-03  -3,84953D-03   5,55333D-03  -5,88903D-03

-2,60293D+01   6,84103D+01  -1,98511D+01   7,64648D+00  -3,62844D-01  -4,28455D-05
 3,15542D-04  -1,09506D-03   2,42227D-03  -3,84953D-03   4,55138D-03  -3,89977D-03

 2,23517D+01  -4,56883D+01   1,81839D+01  -5,61668D+00   3,65744D-01   1,79108D-05
-1,33209D-04   4,73030D-04  -1,09506D-03   1,87064D-03  -2,41547D-03   2,25584D-03
-1,40376D-03

-9,82316D+00   1,82277D+01  -8,35997D+00   2,45435D+00  -1,89168D-01  -4,98175D-06
 3,71847D-05  -1,33209D-04   3,15542D-04  -5,64894D-04   7,78425D-04  -7,77163D-04
 5,12308D-04  -1,95788D-04

 1,78505D+00  -3,17329D+00   1,57093D+00  -4,58577D-01   4,00619D-02   6,66441D-07
-4,98175D-06   1,79108D-05  -4,28455D-05   7,88527D-05  -1,13708D-04   1,19334D-04
-8,22281D-05   3,25763D-05  -5,57681D-06

                 N=  5           M= 10

-5,95806D+06

 0,           -9,51832D+06

-3,80960D+06   0,           -2,57173D+06

 0,           -1,11738D+06   0,           -1,50573D+05

-7,43970D+04   0,           -5,78893D+04   0,           -2,26765D+03
```

THE MATRIX B   N=5

```
 4,12280D+00  4,76325D+00  2,49991D+00  5,02325D-01  4,31894D-02 -5,60620D-06

-2,35822D+01 -2,59390D+01 -1,36814D+01 -2,56017D+00 -2,12598D-01  3,27965D-05
-1,97438D-04

 5,85516D+01  5,89267D+01  3,20599D+01  5,36310D+00  4,50439D-01 -8,30242D-05
 5,18272D-04 -1,42532D-03

-8,58051D+01 -7,14989D+01 -4,37676D+01 -5,98051D+00 -5,68477D-01  1,21267D-04
-7,91644D-04  2,30819D-03 -4,02686D-03

 9,05013D+01  4,56572D+01  4,34369D+01  3,60078D+00  5,41790D-01 -1,17426D-04
 8,06282D-04 -2,51617D-03  4,79587D-03 -6,35935D-03

-8,65769D+01  6,33161D-23 -4,10953D+01  5,97518D-24 -5,08687D-01  8,50084D-05
-6,11011D-04  2,03734D-03 -4,25427D-03  6,33195D-03 -7,17804D-03

 9,05013D+01 -4,56572D+01  4,34369D+01 -3,60078D+00  5,41790D-01 -5,13488D-05
 3,79245D-04 -1,32534D-03  2,98136D-03 -4,92507D-03  6,33195D-03 -6,35935D-03

-8,58051D+01  7,14989D+01 -4,37676D+01  5,98051D+00 -5,68477D-01  2,66922D-05
-1,98996D-04  7,10836D-04 -1,67244D-03  2,98136D-03 -4,25427D-03  4,79587D-03
-4,02686D-03

 5,85516D+01 -5,89267D+01  3,20599D+01 -5,36310D+00  4,50439D-01 -1,09938D-05
 8,22253D-05 -2,96021D-04  7,10836D-04 -1,32534D-03  2,03734D-03 -2,51617D-03
 2,30819D-03 -1,42532D-03

-2,35822D+01  2,59390D+01 -1,36814D+01  2,56017D+00 -2,12598D-01  3,04136D-06
-2,27738D-05  8,22253D-05 -1,98996D-04  3,79245D-04 -6,11011D-04  8,06282D-04
-7,91644D-04  5,18272D-04 -1,97438D-04

 4,12280D+00 -4,76325D+00  2,49991D+00 -5,02325D-01  4,31894D-02 -4,05966D-07
 3,04136D-06 -1,09938D-05  2,66922D-05 -5,13488D-05  8,50084D-05 -1,17426D-04
 1,21267D-04 -8,30242D-05  3,27965D-05 -5,60620D-06
```

N= 5   M= 11

```
-1,90124D+07

 0,          -2,53425D+07

-9,35574D+06  0,          -4,88783D+06

 0,          -2,22688D+06  0,          -2,27148D+05

-1,39771D+05  0,          -8,47005D+04  0,          -2,57235D+03

 6,89189D+00  8,01383D+00  3,17844D+00  6,15777D-01  4,13706D-02 -5,61706D-06

-3,80338D+01 -4,42883D+01 -1,67929D+01 -3,18834D+00 -1,98962D-01  3,28779D-05
-1,98048D-04

 8,85024D+01  1,04799D+02  3,69864D+01  6,99896D+00  4,01060D-01 -8,33190D-05
 5,20482D-04 -1,43333D-03
```

THE MATRIX B N=5

```
-1,13210D+02 -1,43372D+02 -4,43354D+01 -8,91061D+00 -4,48006D-01  1,21986D-04
-7,97035D-04  2,32771D-03 -4,07449D-03

 8,55129D+01  1,37601D+02  3,15906D+01  8,18325D+00  3,06720D-01 -1,18829D-04
 8,16802D-04 -2,55426D-03  4,88881D-03 -6,54070D-03

-2,91632D+01 -1,20093D+02 -1,06271D+01 -7,06795D+00 -1,02183D-01  8,74350D-05
-6,29203D-04  2,10322D-03 -4,41499D-03  6,64556D-03 -7,72036D-03

-2,91632D+01  1,20093D+02 -1,06271D+01  7,06795D+00 -1,02183D-01 -5,51929D-05
 4,08065D-04 -1,42970D-03  3,23597D-03 -5,42188D-03  7,19108D-03 -7,72036D-03

 8,55129D+01 -1,37601D+02  3,15906D+01 -8,18325D+00  3,06720D-01  3,19165D-05
-2,38163D-04  8,52669D-04 -2,01847D-03  3,65655D-03 -5,42188D-03  6,64556D-03
-6,54070D-03

-1,13210D+02  1,43372D+02 -4,43354D+01  8,91061D+00 -4,48006D-01 -1,63569D-05
 1,22433D-04 -4,41622D-04  1,06606D-03 -2,01847D-03  3,23597D-03 -4,41499D-03
 4,88881D-03 -4,07449D-03

 8,85024D+01 -1,04799D+02  3,69864D+01 -6,99896D+00  4,01060D-01  6,70446D-06
-5,02366D-05  1,81674D-04 -4,41622D-04  8,52669D-04 -1,42970D-03  2,10322D-03
-2,55426D-03  2,32771D-03 -1,43333D-03

-3,80338D+01  4,42883D+01 -1,67929D+01  3,18834D+00 -1,98962D-01 -1,85144D-06
 1,38783D-05 -5,02366D-05  1,22433D-04 -2,38163D-04  4,08065D-04 -6,29203D-04
 8,16802D-04 -7,97035D-04  5,20482D-04 -1,98048D-04

 6,89189D+00 -8,01383D+00  3,17844D+00 -6,15777D-01  4,13706D-02  2,46953D-07
-1,85144D-06  6,70446D-06 -1,63569D-05  3,19165D-05 -5,51929D-05  8,74350D-05
-1,18829D-04  1,21986D-04 -8,33190D-05  3,28779D-05 -5,61706D-06
```

N= 5 M= 12

```
-5,66633D+07

 0,          -5,67865D+07

-2,17443D+07  0,          -8,91534D+06

 0,          -3,94549D+06  0,          -3,19783D+05

-2,52577D+05  0,          -1,20912D+05  0,          -2,89378D+03

 1,20961D+01  1,13710D+01  4,27846D+00  6,80343D-01  4,25061D-02 -5,62107D-06

-6,72104D+01 -6,12940D+01 -2,27310D+01 -3,44996D+00 -2,07477D-01  3,29080D-05
-1,98274D-04

 1,59829D+02  1,38982D+02  5,11921D+01  7,30724D+00  4,31914D-01 -8,34280D-05
 5,21299D-04 -1,43629D-03

-2,18578D+02 -1,74445D+02 -6,59623D+01 -8,59986D+00 -5,23401D-01  1,22252D-04
-7,99032D-04  2,33495D-03 -4,09218D-03
```

THE MATRIX B N=5

```
 2,05313D+02  1,35943D+02  5,92025D+01  6,43102D+00  4,54521D-01 -1,19351D-04
 8,20718D-04 -2,56845D-03  4,92348D-03 -6,60867D-03

-1,66517D+02 -6,67327D+01 -4,74720D+01 -3,12365D+00 -3,61381D-01  8,83507D-05
-6,36070D-04  2,12810D-03 -4,47579D-03  6,76474D-03 -7,92938D-03

 1,51134D+02 -9,15659D-22  4,29845D+01 -4,85051D-23  3,26637D-01 -5,67078D-05
 4,19425D-04 -1,47086D-03  3,33656D-03 -5,61906D-03  7,53688D-03 -8,29245D-03

-1,66517D+02  6,67327D+01 -4,74720D+01  3,12365D+00 -3,61381D-01  3,42768D-05
-2,55863D-04  9,16801D-04 -2,17518D-03  3,96376D-03 -5,96064D-03  7,53688D-03
-7,92938D-03

 2,05313D+02 -1,35943D+02  5,92025D+01 -6,43102D+00  4,54521D-01 -1,95454D-05
 1,46343D-04 -5,28256D-04  1,27776D-03 -2,43347D-03  3,96376D-03 -5,61906D-03
 6,76474D-03 -6,60867D-03

-2,18578D+02  1,74445D+02 -6,59623D+01  8,59986D+00 -5,23401D-01  9,97041D-06
-7,47277D-05  2,70414D-04 -6,58467D-04  1,27776D-03 -2,17518D-03  3,33656D-03
-4,47579D-03  4,92348D-03 -4,09218D-03

 1,59829D+02 -1,38982D+02  5,11921D+01 -7,30724D+00  4,31914D-01 -4,08020D-06
 3,05916D-05 -1,10795D-04  2,70414D-04 -5,28256D-04  9,16801D-04 -1,47086D-03
 2,12810D-03 -2,56845D-03  2,33495D-03 -1,43629D-03

-6,72104D+01  6,12940D+01 -2,27310D+01  3,44996D+00 -2,07477D-01  1,12609D-06
-8,44400D-06  3,05916D-05 -7,47277D-05  1,46343D-04 -2,55863D-04  4,19425D-04
-6,36070D-04  8,20718D-04 -7,99032D-04  5,21299D-04 -1,98274D-04

 1,20961D+01 -1,13710D+01  4,27846D+00 -6,80343D-01  4,25061D-02 -1,50166D-07
 1,12609D-06 -4,08020D-06  9,97041D-06 -1,95454D-05  3,42768D-05 -5,67078D-05
 8,83507D-05 -1,19351D-04  1,22252D-04 -8,34280D-05  3,29080D-05 -5,62107D-06
```

N= 5 M= 13

```
-1,44594D+08

 0,          -1,20038D+08

-4,49154D+07  0,          -1,49569D+07

 0,          -6,70596D+06  0,          -4,39007D+05

-4,21624D+05  0,          -1,64510D+05  0,          -3,20493D+03

 1,86363D+01  1,64481D+01  5,27704D+00  7,77988D-01  4,18269D-02 -5,62255D-06

-1,01715D+02 -8,87193D+01 -2,75745D+01 -3,95966D+00 -2,02384D-01  3,29191D-05
-1,98357D-04

 2,34833D+02  2,03129D+02  6,04718D+01  8,51443D+00  4,13454D-01 -8,34683D-05
 5,21601D-04 -1,43738D-03

-3,03289D+02 -2,64868D+02 -7,39530D+01 -1,04862D+01 -4,78268D-01  1,22351D-04
-7,99771D-04  2,33763D-03 -4,09872D-03
```

THE MATRIX B   N=5

```
 2,49436D+02  2,35042D+02  5,83511D+01  8,98827D+00  3,65908D-01 -1,19545D-04
 8,22168D-04 -2,57371D-03  4,93633D-03 -6,63390D-03

-1,42607D+02 -1,78119D+02 -3,30035D+01 -6,75018D+00 -2,05266D-01  8,86914D-05
-6,38625D-04  2,13737D-03 -4,49844D-03  6,80920D-03 -8,00770D-03

 4,52054D+01  1,46141D+02  1,04311D+01  5,52483D+00  6,47288D-02 -5,72795D-05
 4,23713D-04 -1,48640D-03  3,37455D-03 -5,69365D-03  7,66828D-03 -8,51291D-03

 4,52054D+01 -1,46141D+02  1,04311D+01 -5,52483D+00  6,47288D-02  3,52069D-05
-2,62839D-04  9,42081D-04 -2,23699D-03  4,08511D-03 -6,17443D-03  7,89555D-03
-8,51291D-03

-1,42607D+02  1,78119D+02 -3,30035D+01  6,75018D+00 -2,05266D-01 -2,09856D-05
 1,57144D-04 -5,67400D-04  1,37346D-03 -2,62137D-03  4,29480D-03 -6,17443D-03
 7,66828D-03 -8,00770D-03

 2,49436D+02 -2,35042D+02  5,83511D+01 -8,98827D+00  3,65908D-01  1,19117D-05
-8,92862D-05  3,23175D-04 -7,87462D-04  1,53102D-03 -2,62137D-03  4,08511D-03
-5,69365D-03  6,80920D-03 -6,63390D-03

-3,03289D+02  2,64868D+02 -7,39530D+01  1,04862D+01 -4,78268D-01 -6,06702D-06
 4,54918D-05 -1,64794D-04  4,02436D-04 -7,87462D-04  1,37346D-03 -2,23699D-03
 3,37455D-03 -4,49844D-03  4,93633D-03 -4,09872D-03

 2,34833D+02 -2,03129D+02  6,04718D+01 -8,51443D+00  4,13454D-01  2,48150D-06
-1,86090D-05  6,74302D-05 -1,64794D-04  3,23175D-04 -5,67400D-04  9,42081D-04
-1,48640D-03  2,13737D-03 -2,57371D-03  2,33763D-03 -1,43738D-03

-1,01715D+02  8,87193D+01 -2,75745D+01  3,95966D+00 -2,02384D-01 -6,84730D-07
 5,13507D-06 -1,86090D-05  4,54918D-05 -8,92862D-05  1,57144D-04 -2,62839D-04
 4,23713D-04 -6,38625D-04  8,22168D-04 -7,99771D-04  5,21601D-04 -1,98357D-04

 1,86363D+01 -1,64481D+01  5,27704D+00 -7,77988D-01  4,18269D-02  9,13030D-08
-6,84730D-07  2,48150D-06 -6,06702D-06  1,19117D-05 -2,09856D-05  3,52069D-05
-5,72795D-05  8,86914D-05 -1,19545D-04  1,22351D-04 -8,34683D-05  3,29191D-05
-5,62255D-06
```

N= 5   M= 14

```
-3,43971D+08

 0,          -2,32413D+08

-8,78504D+07  0,          -2,41256D+07

 0,          -1,07225D+07  0,          -5,81290D+05

-6,77521D+05  0,          -2,18687D+05  0,          -3,52229D+03

 2,85614D+01  2,21112D+01  6,57164D+00  8,53219D-01  4,22439D-02 -5,62310D-06

-1,55590D+02 -1,17728D+02 -3,43041D+01 -4,30126D+00 -2,05511D-01  3,29233D-05
-1,98388D-04
```

THE MATRIX B   N=5

```
 3,60270D+02  2,64260D+02  7,56270D+01  9,11239D+00  4,24789D-01 -8,34832D-05
 5,21713D-04 -1,43779D-03

-4,74293D+02 -3,32416D+02 -9,47118D+01 -1,08828D+01 -5,05987D-01  1,22387D-04
-8,00045D-04  2,33862D-03 -4,10114D-03

 4,19505D+02  2,71817D+02  8,08529D+01  8,62033D+00  4,20359D-01 -1,19616D-04
 8,22705D-04 -2,57565D-03  4,94108D-03 -6,64324D-03

-3,05730D+02 -1,66001D+02 -5,83824D+01 -5,22026D+00 -3,01341D-01  8,88177D-05
-6,39572D-04  2,14080D-03 -4,50683D-03  6,82569D-03 -8,03679D-03

 2,31054D+02  7,59965D+01  4,40043D+01  2,38392D+00  2,26649D-01 -5,74923D-05
 4,25309D-04 -1,49218D-03  3,38869D-03 -5,72143D-03  7,71730D-03 -8,59552D-03

-2,06553D+02  1,15248D-20 -3,93152D+01  4,04210D-22 -2,02403D-01  3,55579D-05
-2,65471D-04  9,51623D-04 -2,26032D-03  4,13095D-03 -6,25529D-03  8,03184D-03
-8,73776D-03

 2,31054D+02 -7,59965D+01  4,40043D+01 -2,38392D+00  2,26649D-01 -2,15532D-05
 1,61401D-04 -5,82827D-04  1,41119D-03 -2,69547D-03  4,42555D-03 -6,39479D-03
 8,03184D-03 -8,59552D-03

-3,05730D+02  1,66001D+02 -5,83824D+01  5,22026D+00 -3,01341D-01  1,27885D-05
-9,58622D-05  3,47008D-04 -8,45742D-04  1,64550D-03 -2,82337D-03  4,42555D-03
-6,25529D-03  7,71730D-03 -8,03679D-03

 4,19505D+02 -2,71817D+02  8,08529D+01 -8,62033D+00  4,20359D-01 -7,24789D-06
 5,43481D-05 -1,96892D-04  4,80926D-04 -9,41642D-04  1,64550D-03 -2,69547D-03
 4,13095D-03 -5,72143D-03  6,82569D-03 -6,64324D-03

-4,74293D+02  3,32416D+02 -9,47118D+01  1,08828D+01 -5,05987D-01  3,68973D-06
-2,76704D-05  1,00271D-04 -2,45101D-04  4,80926D-04 -8,45742D-04  1,41119D-03
-2,26032D-03  3,38869D-03 -4,50683D-03  4,94108D-03 -4,10114D-03

 3,60270D+02 -2,64260D+02  7,56270D+01 -9,11239D+00  4,24789D-01 -1,50889D-06
 1,13160D-05 -4,10107D-05  1,00271D-04 -1,96892D-04  3,47008D-04 -5,82827D-04
 9,51623D-04 -1,49218D-03  2,14080D-03 -2,57565D-03  2,33862D-03 -1,43779D-03

-1,55590D+02  1,17728D+02 -3,43041D+01  4,30126D+00 -2,05511D-01  4,16327D-07
-3,12232D-06  1,13160D-05 -2,76704D-05  5,43481D-05 -9,58622D-05  1,61401D-04
-2,65471D-04  4,25309D-04 -6,39572D-04  8,22705D-04 -8,00045D-04  5,21713D-04
-1,98388D-04

 2,85614D+01 -2,21112D+01  6,57164D+00 -8,53219D-01  4,22439D-02 -5,55122D-08
 4,16327D-07 -1,50889D-06  3,68973D-06 -7,24789D-06  1,27885D-05 -2,15532D-05
 3,55579D-05 -5,74923D-05  8,88177D-05 -1,19616D-04  1,22387D-04 -8,34832D-05
 3,29233D-05 -5,62310D-06
```

N= 5   M= 15

```
-7,49671D+08

 0,          -4,28148D+08

-1,60680D+08  0,          -3,71051D+07
```

THE MATRIX B   N=5

```
 0,           -1,65518D+07  0,           -7,53639D+05

-1,03949D+06  0,           -2,82722D+05  0,           -3,83587D+03

 4,11224D+01  2,95211D+01  7,86761D+00  9,43369D-01  4,19919D-02 -5,62331D-06

-2,21730D+02 -1,56689D+02 -4,07102D+01 -4,75475D+00 -2,03621D-01  3,29248D-05
-1,98400D-04

 5,05763D+02  3,51750D+02  8,86492D+01  1,01159D+01  4,17939D-01 -8,34887D-05
 5,21754D-04 -1,43794D-03

-6,49164D+02 -4,47321D+02 -1,08656D+02 -1,22712D+01 -4,89234D-01  1,22401D-04
-8,00145D-04  2,33898D-03 -4,10204D-03

 5,44497D+02  3,82486D+02  8,82469D+01  1,02005D+01  3,87443D-01 -1,19643D-04
 8,22903D-04 -2,57637D-03  4,94284D-03 -6,64670D-03

-3,47094D+02 -2,69299D+02 -5,57780D+01 -7,12833D+00 -2,43234D-01  8,88644D-05
-6,39923D-04  2,14207D-03 -4,50993D-03  6,83179D-03 -8,04756D-03

 1,84318D+02  1,90853D+02  2,95442D+01  5,04059D+00  1,28570D-01 -5,75711D-05
 4,25900D-04 -1,49433D-03  3,39393D-03 -5,73173D-03  7,73548D-03 -8,62620D-03

-5,72131D+01 -1,53690D+02 -9,16383D+00 -4,05652D+00 -3,98552D-02  3,56886D-05
-2,66451D-04  9,55174D-04 -2,26900D-03  4,14801D-03 -6,28541D-03  8,08268D-03
-8,82202D-03

-5,72131D+01  1,53690D+02 -9,16383D+00  4,05652D+00 -3,98552D-02 -2,17674D-05
 1,63007D-04 -5,88650D-04  1,42543D-03 -2,72345D-03  4,47493D-03 -6,47815D-03
 8,16999D-03 -8,82202D-03

 1,84318D+02 -1,90853D+02  2,95442D+01 -5,04059D+00  1,28570D-01  1,31340D-05
-9,84537D-05  3,56401D-04 -8,68711D-04  1,69063D-03 -2,90303D-03  4,56001D-03
-6,47815D-03  8,08268D-03 -8,62620D-03

-3,47094D+02  2,69299D+02 -5,57780D+01  7,12833D+00 -2,43234D-01 -7,78126D-06
 5,83482D-05 -2,11390D-04  5,16380D-04 -1,01130D-03  1,76847D-03 -2,90303D-03
 4,47493D-03 -6,28541D-03  7,73548D-03 -8,04756D-03

 5,44497D+02 -3,82486D+02  8,82469D+01 -1,02005D+01  3,87443D-01  4,40783D-06
-3,30560D-05  1,19791D-04 -2,92836D-04  5,74708D-04 -1,01130D-03  1,69063D-03
-2,72345D-03  4,14801D-03 -5,73173D-03  6,83179D-03 -6,64670D-03

-6,49164D+02  4,47321D+02 -1,08656D+02  1,22712D+01 -4,89234D-01 -2,24354D-06
 1,68258D-05 -6,09800D-05  1,49106D-04 -2,92836D-04  5,16380D-04 -8,68711D-04
 1,42543D-03 -2,26900D-03  3,39393D-03 -4,50993D-03  4,94284D-03 -4,10204D-03

 5,05763D+02 -3,51750D+02  8,86492D+01 -1,01159D+01  4,17939D-01  9,17428D-07
-6,88048D-06  2,49370D-05 -6,09800D-05  1,19791D-04 -2,11390D-04  3,56401D-04
-5,88650D-04  9,55174D-04 -1,49433D-03  2,14207D-03 -2,57637D-03  2,33898D-03
-1,43794D-03

-2,21730D+02  1,56689D+02 -4,07102D+01  4,75475D+00 -2,03621D-01 -2,53127D-07
 1,89840D-06 -6,88048D-06  1,68258D-05 -3,30560D-05  5,83482D-05 -9,84537D-05
 1,63007D-04 -2,66451D-04  4,25900D-04 -6,39923D-04  8,22903D-04 -8,00145D-04
 5,21754D-04 -1,98400D-04
```

THE MATRIX B   N=5

```
 4,11224D+01 -2,95211D+01  7,86761D+00 -9,43369D-01  4,19919D-02  3,37512D-08
-2,53127D-07  9,17428D-07 -2,24354D-06  4,40783D-06 -7,78126D-06  1,31340D-05
-2,17674D-05  3,56886D-05 -5,75711D-05  8,88644D-05 -1,19643D-04  1,22401D-04
-8,34887D-05  3,29248D-05 -5,62331D-06
```

N= 5   M= 16

```
-1,54358D+09

 0,          -7,47285D+08

-2,81244D+08  0,          -5,53034D+07

 0,          -2,45995D+07  0,          -9,55310D+05

-1,54648D+06  0,          -3,58775D+05  0,          -4,15174D+03

 5,81828D+01  3,79787D+01  9,37988D+00  1,02362D+00  4,21457D-02 -5,62338D-06

-3,12451D+02 -2,00028D+02 -4,84044D+01 -5,13398D+00 -2,04774D-01  3,29253D-05
-1,98404D-04

 7,10863D+02  4,44428D+02  1,05380D+02  1,08502D+01  4,22119D-01 -8,34907D-05
 5,21770D-04 -1,43799D-03

-9,15667D+02 -5,56212D+02 -1,30072D+02 -1,30013D+01 -4,99457D-01  1,22406D-04
-8,00183D-04  2,33912D-03 -4,10237D-03

 7,86832D+02  4,60476D+02  1,08599D+02  1,04875D+01  4,07530D-01 -1,19653D-04
 8,22977D-04 -2,57664D-03  4,94349D-03 -6,64798D-03

-5,45556D+02 -2,98585D+02 -7,47286D+01 -6,75340D+00 -2,78700D-01  8,88817D-05
-6,40052D-04  2,14254D-03 -4,51108D-03  6,83404D-03 -8,05154D-03

 3,70453D+02  1,70270D+02  5,06278D+01  3,84306D+00  1,88464D-01 -5,76003D-05
 4,26119D-04 -1,49512D-03  3,39587D-03 -5,73554D-03  7,74220D-03 -8,63756D-03

-2,74100D+02 -7,63324D+01 -3,74337D+01 -1,72171D+00 -1,39269D-01  3,57370D-05
-2,66814D-04  9,56490D-04 -2,27222D-03  4,15433D-03 -6,29658D-03  8,10153D-03
-8,85331D-03

 2,43889D+02  2,78780D-20  3,33025D+01  7,40520D-22  1,23883D-01 -2,18471D-05
 1,63605D-04 -5,90817D-04  1,43073D-03 -2,73386D-03  4,49332D-03 -6,50919D-03
 8,22152D-03 -8,90690D-03

-2,74100D+02  7,63324D+01 -3,74337D+01  1,72171D+00 -1,39269D-01  1,32644D-05
-9,94317D-05  3,59945D-04 -8,77380D-04  1,70766D-03 -2,93310D-03  4,61080D-03
-6,56244D-03  8,22152D-03 -8,85331D-03

 3,70453D+02 -1,70270D+02  5,06278D+01 -3,84306D+00  1,88464D-01 -7,99145D-06
 5,99246D-05 -2,17103D-04  5,30352D-04 -1,03875D-03  1,81694D-03 -2,98489D-03
 4,61080D-03 -6,50919D-03  8,10153D-03 -8,63756D-03

-5,45556D+02  2,98585D+02 -7,47286D+01  6,75340D+00 -2,78700D-01  4,73217D-06
-3,54885D-05  1,28607D-04 -3,14396D-04  6,17070D-04 -1,08609D-03  1,81694D-03
-2,93310D-03  4,49332D-03 -6,29658D-03  7,74220D-03 -8,05154D-03
```

THE MATRIX B   N=5

```
 7,86832D+02 -4,60476D+02  1,08599D+02 -1,04875D+01  4,07530D-01 -2,68017D-06
 2,01005D-05 -7,28487D-05  1,78131D-04 -3,49863D-04  6,17070D-04 -1,03875D-03
 1,70766D-03 -2,73386D-03  4,15433D-03 -5,73554D-03  6,83404D-03 -6,64798D-03

-9,15667D+02  5,56212D+02 -1,30072D+02  1,30013D+01 -4,99457D-01  1,36411D-06
-1,02305D-05  3,70788D-05 -9,06728D-05  1,78131D-04 -3,14396D-04  5,30352D-04
-8,77380D-04  1,43073D-03 -2,27222D-03  3,39587D-03 -4,51108D-03  4,94349D-03
-4,10237D-03

 7,10863D+02 -4,44428D+02  1,05380D+02 -1,08502D+01  4,22119D-01 -5,57799D-07
 4,18338D-06 -1,51622D-05  3,70788D-05 -7,28487D-05  1,28607D-04 -2,17103D-04
 3,59945D-04 -5,90817D-04  9,56490D-04 -1,49512D-03  2,14254D-03 -2,57664D-03
 2,33912D-03 -1,43799D-03

-3,12451D+02  2,00028D+02 -4,84044D+01  5,13398D+00 -2,04774D-01  1,53901D-07
-1,15423D-06  4,18338D-06 -1,02305D-05  2,01005D-05 -3,54885D-05  5,99246D-05
-9,94317D-05  1,63605D-04 -2,66814D-04  4,26119D-04 -6,40052D-04  8,22977D-04
-8,00183D-04  5,21770D-04 -1,98404D-04

 5,81828D+01 -3,79787D+01  9,37988D+00 -1,02362D+00  4,21457D-02 -2,05206D-08
 1,53901D-07 -5,57799D-07  1,36411D-06 -2,68017D-06  4,73217D-06 -7,99145D-06
 1,32644D-05 -2,18471D-05  3,57370D-05 -5,76003D-05  8,88817D-05 -1,19653D-04
 1,22406D-04 -8,34907D-05  3,29253D-05 -5,62338D-06
```

N= 5   M= 17

```
-3,00088D+09

 0.          -1,25445D+09

-4,70984D+08  0.          -7,98770D+07

 0.          -3,55658D+07  0.          -1,19117D+06

-2,23073D+06  0.          -4,46920D+05  0.          -4,46621D+03

 7,94713D+01  4,82544D+01  1,09542D+01  1,11037D+00  4,20524D-02 -5,62341D-06

-4,23898D+02 -2,53188D+02 -5,62303D+01 -5,56200D+00 -2,04075D-01  3,29255D-05
-1,98405D-04

 9,56012D+02  5,61068D+02  1,21628D+02  1,17614D+01  4,19583D-01 -8,34915D-05
 5,21775D-04 -1,43801D-03

-1,21623D+03 -7,03401D+02 -1,48708D+02 -1,41640D+01 -4,93255D-01  1,22407D-04
-8,00197D-04  2,33917D-03 -4,10249D-03

 1,02253D+03  5,91128D+02  1,21768D+02  1,16244D+01  3,95344D-01 -1,19656D-04
 8,23004D-04 -2,57674D-03  4,94373D-03 -6,64845D-03

-6,73728D+02 -4,03209D+02 -7,96717D+01 -7,87896D+00 -2,57183D-01  8,88880D-05
-6,40100D-04  2,14271D-03 -4,51151D-03  6,83488D-03 -8,05301D-03

 4,00054D+02  2,65578D+02  4,72080D+01  5,17962D+00  1,52121D-01 -5,76111D-05
 4,26199D-04 -1,49541D-03  3,39658D-03 -5,73694D-03  7,74469D-03 -8,64176D-03
```

THE MATRIX B   N=5

```
-2,07791D+02 -1,84382D+02 -2,45035D+01 -3,59387D+00 -7,89145D-02  3,57549D-05
-2,66948D-04  9,56976D-04 -2,27341D-03  4,15667D-03 -6,30071D-03  8,10851D-03
-8,86489D-03

 6,40799D+01  1,47603D+02  7,55503D+00  2,87649D+00  2,43272D-02 -2,18767D-05
 1,63827D-04 -5,91620D-04  1,43269D-03 -2,73772D-03  4,50013D-03 -6,52070D-03
 8,24063D-03 -8,93842D-03

 6,40799D+01 -1,47603D+02  7,55503D+00 -2,87649D+00  2,43272D-02  1,33130D-05
-9,97958D-05  3,61265D-04 -8,80607D-04  1,71400D-03 -2,94430D-03  4,62971D-03
-6,59384D-03  8,27331D-03 -8,93842D-03

-2,07791D+02  1,84382D+02 -2,45035D+01  3,59387D+00 -7,89145D-02 -8,07078D-06
 6,05196D-05 -2,19260D-04  5,35626D-04 -1,04911D-03  1,83524D-03 -3,01579D-03
 4,66212D-03 -6,59384D-03  8,24063D-03 -8,86489D-03

 4,00054D+02 -2,65578D+02  4,72080D+01 -5,17962D+00  1,52121D-01  4,85999D-06
-3,64471D-05  1,32082D-04 -3,22893D-04  6,33765D-04 -1,11557D-03  1,86673D-03
-3,01579D-03  4,62971D-03 -6,52070D-03  8,10851D-03 -8,64176D-03

-6,73728D+02  4,03209D+02 -7,96717D+01  7,87896D+00 -2,57183D-01 -2,87739D-06
 2,15795D-05 -7,82094D-05  1,91241D-04 -3,75622D-04  6,62552D-04 -1,11557D-03
 1,83524D-03 -2,94430D-03  4,50013D-03 -6,30071D-03  7,74469D-03 -8,05301D-03

 1,02253D+03 -5,91128D+02  1,21768D+02 -1,16244D+01  3,95344D-01  1,62958D-06
-1,22215D-05  4,42951D-05 -1,08321D-04  2,12805D-04 -3,75622D-04  6,33765D-04
-1,04911D-03  1,71400D-03 -2,73772D-03  4,15667D-03 -5,73694D-03  6,83488D-03
-6,64845D-03

-1,21623D+03  7,03401D+02 -1,48708D+02  1,41640D+01 -4,93255D-01 -8,29380D-07
 6,22020D-06 -2,25444D-05  5,51323D-05 -1,08321D-04  1,91241D-04 -3,22893D-04
 5,35626D-04 -8,80607D-04  1,43269D-03 -2,27341D-03  3,39658D-03 -4,51151D-03
 4,94373D-03 -4,10249D-03

 9,56012D+02 -5,61068D+02  1,21628D+02 -1,17614D+01  4,19583D-01  3,39141D-07
-2,54350D-06  9,21865D-06 -2,25444D-05  4,42951D-05 -7,82094D-05  1,32082D-04
-2,19260D-04  3,61265D-04 -5,91620D-04  9,56976D-04 -1,49541D-03  2,14271D-03
-2,57674D-03  2,33917D-03 -1,43801D-03

-4,23898D+02  2,53188D+02 -5,62303D+01  5,56200D+00 -2,04075D-01 -9,35715D-08
 7,01770D-07 -2,54350D-06  6,22020D-06 -1,22215D-05  2,15795D-05 -3,64471D-05
 6,05196D-05 -9,97958D-05  1,63827D-04 -2,66948D-04  4,26199D-04 -6,40100D-04
 8,23004D-04 -8,00197D-04  5,21775D-04 -1,98405D-04

 7,94713D+01 -4,82544D+01  1,09542D+01 -1,11037D+00  4,20524D-02  1,24765D-08
-9,35715D-08  3,39141D-07 -8,29380D-07  1,62958D-06 -2,87739D-06  4,85999D-06
-8,07078D-06  1,33130D-05 -2,18767D-05  3,57549D-05 -5,76111D-05  8,88880D-05
-1,19656D-04  1,22407D-04 -8,34915D-05  3,29255D-05 -5,62341D-06
```

N= 5   M= 18

```
-5,58134D+09

 0,           -2,02858D+09

-7,62190D+08  0,           -1,12590D+08
```

THE MATRIX B N=5

```
 0.           -5,00811D+07  0.           -1,46207D+06

-3,14114D+06  0.           -5,48719D+05  0.           -4,78153D+03

 1,06622D+02  5,99815D+01  1,27000D+01  1,19287D+00  4,21092D-02 -5,62342D-06

-5,66337D+02 -3,13085D+02 -6,50079D+01 -5,95808D+00 -2,04501D-01  3,29256D-05
-1,98406D-04

 1,27236D+03  6,89527D+02  1,40365D+02  1,25568D+01  4,21127D-01 -8,34918D-05
 5,21777D-04 -1,43802D-03

-1,61647D+03 -8,57600D+02 -1,71832D+02 -1,50436D+01 -4,97031D-01  1,22408D-04
-8,00202D-04  2,33919D-03 -4,10254D-03

 1,36880D+03  7,11244D+02  1,42035D+02  1,22049D+01  4,02763D-01 -1,19658D-04
 8,23014D-04 -2,57677D-03  4,94382D-03 -6,64862D-03

-9,29423D+02 -4,70258D+02 -9,58270D+01 -8,02221D+00 -2,70284D-01  8,88904D-05
-6,40118D-04  2,14277D-03 -4,51166D-03  6,83518D-03 -8,05355D-03

 6,01332D+02  2,85226D+02  6,18794D+01  4,85701D+00  1,74249D-01 -5,76151D-05
 4,26229D-04 -1,49552D-03  3,39685D-03 -5,73747D-03  7,74561D-03 -8,64331D-03

-3,99624D+02 -1,59268D+02 -4,10975D+01 -2,71054D+00 -1,15668D-01  3,57615D-05
-2,66998D-04  9,57156D-04 -2,27385D-03  4,15753D-03 -6,30223D-03  8,11109D-03
-8,86918D-03

 2,93716D+02  7,09265D+01  3,02001D+01  1,20686D+00  8,49838D-02 -2,18876D-05
 1,63909D-04 -5,91917D-04  1,43342D-03 -2,73915D-03  4,50265D-03 -6,52496D-03
 8,24770D-03 -8,95008D-03

-2,60951D+02  1,15690D-19 -2,68300D+01  2,06945D-21 -7,54975D-02  1,33310D-05
-9,99306D-05  3,61754D-04 -8,81802D-04  1,71635D-03 -2,94844D-03  4,63671D-03
-6,60548D-03  8,29251D-03 -8,97002D-03

 2,93716D+02 -7,09265D+01  3,02001D+01 -1,20686D+00  8,49838D-02 -8,10031D-06
 6,07410D-05 -2,20062D-04  5,37589D-04 -1,05297D-03  1,84205D-03 -3,02729D-03
 4,68122D-03 -6,62537D-03  8,29251D-03 -8,95008D-03

-3,99624D+02  1,59268D+02 -4,10975D+01  2,71054D+00 -1,15668D-01  4,90823D-06
-3,68089D-05  1,33393D-04 -3,26100D-04  6,40065D-04 -1,12670D-03  1,88552D-03
-3,04700D-03  4,68122D-03 -6,60548D-03  8,24770D-03 -8,86918D-03

 6,01332D+02 -2,85226D+02  6,18794D+01 -4,85701D+00  1,74249D-01 -2,95511D-06
 2,21624D-05 -8,03220D-05  1,96407D-04 -3,85773D-04  6,80477D-04 -1,14585D-03
 1,88552D-03 -3,02729D-03  4,63671D-03 -6,52496D-03  8,11109D-03 -8,64331D-03

-9,29423D+02  4,70258D+02 -9,58270D+01  8,02221D+00 -2,70284D-01  1,74949D-06
-1,31208D-05  4,75545D-05 -1,16292D-04  2,28467D-04 -4,03276D-04  6,80477D-04
-1,12670D-03  1,84205D-03 -2,94844D-03  4,50265D-03 -6,30223D-03  7,74561D-03
-8,05355D-03

 1,36880D+03 -7,11244D+02  1,42035D+02 -1,22049D+01  4,02763D-01 -9,90791D-07
 7,43076D-06 -2,69320D-05  6,58622D-05 -1,29403D-04  2,28467D-04 -3,85773D-04
 6,40065D-04 -1,05297D-03  1,71635D-03 -2,73915D-03  4,15753D-03 -5,73747D-03
 6,83518D-03 -6,64862D-03
```

THE MATRIX B   N=5

```
-1,61647D+03  8,57600D+02 -1,71832D+02  1,50436D+01 -4,97031D-01  5,04262D-07
-3,78188D-06  1,37071D-05 -3,35210D-05  6,58622D-05 -1,16292D-04  1,96407D-04
-3,26100D-04  5,37589D-04 -8,81802D-04  1,43342D-03 -2,27385D-03  3,39685D-03
-4,51166D-03  4,94382D-03 -4,10254D-03

 1,27236D+03 -6,89527D+02  1,40365D+02 -1,25568D+01  4,21127D-01 -2,06197D-07
 1,54644D-06 -5,60494D-06  1,37071D-05 -2,69320D-05  4,75545D-05 -8,03220D-05
 1,33393D-04 -2,20062D-04  3,61754D-04 -5,91917D-04  9,57156D-04 -1,49552D-03
 2,14277D-03 -2,57677D-03  2,33919D-03 -1,43802D-03

-5,66337D+02  3,13085D+02 -6,50079D+01  5,95808D+00 -2,04501D-01  5,68912D-08
-4,26675D-07  1,54644D-06 -3,78188D-06  7,43076D-06 -1,31208D-05  2,21624D-05
-3,68089D-05  6,07410D-05 -9,99306D-05  1,63909D-04 -2,66998D-04  4,26229D-04
-6,40118D-04  8,23014D-04 -8,00202D-04  5,21777D-04 -1,98406D-04

 1,06622D+02 -5,99815D+01  1,27000D+01 -1,19287D+00  4,21092D-02 -7,58566D-09
 5,68912D-08 -2,06197D-07  5,04262D-07 -9,90791D-07  1,74949D-06 -2,95511D-06
 4,90823D-06 -8,10031D-06  1,33310D-05 -2,18876D-05  3,57615D-05 -5,76151D-05
 8,88904D-05 -1,19658D-04  1,22408D-04 -8,34918D-05  3,29256D-05 -5,62342D-06
```

N= 5   M= 19

```
-9,96003D+09

 0,          -3,18246D+09

-1,19467D+09  0,          -1,55134D+08

 0,          -6,90213D+07  0,          -1,77170D+06

-4,32480D+06  0,          -6,64684D+05  0,          -3,09633D+03

 1,39825D+02  7,36576D+01  1,45405D+01  1,27813D+00  4,20747D-02 -5,62342D-06

-7,39029D+02 -3,83117D+02 -7,41623D+01 -6,37495D+00 -2,04242D-01  3,29256D-05
-1,98406D-04

 1,65069D+03  8,41163D+02  1,59508D+02  1,34276D+01  4,20189D-01 -8,34919D-05
 5,21778D-04 -1,43802D-03

-2,08251D+03 -1,04498D+03 -1,94350D+02 -1,61073D+01 -4,94737D-01  1,22408D-04
-8,00204D-04  2,33919D-03 -4,10255D-03

 1,74650D+03  8,70524D+02  1,59389D+02  1,31473D+01  3,98256D-01 -1,19658D-04
 8,23017D-04 -2,57679D-03  4,94386D-03 -6,64869D-03

-1,16308D+03 -5,86169D+02 -1,05515D+02 -8,80455D+00 -2,62325D-01  8,88913D-05
-6,40124D-04  2,14280D-03 -4,51172D-03  6,83530D-03 -8,05376D-03

 7,15357D+02  3,74370D+02  6,47805D+01  5,61388D+00  1,60806D-01 -5,76165D-05
 4,26240D-04 -1,49556D-03  3,39695D-03 -5,73766D-03  7,74595D-03 -8,64388D-03

-4,15671D+02 -2,41547D+02 -3,76199D+01 -3,62020D+00 -9,33385D-02  3,57640D-05
-2,67016D-04  9,57223D-04 -2,27401D-03  4,15785D-03 -6,30280D-03  8,11204D-03
-8,87076D-03
```

THE MATRIX B   N=5

```
 2,14397D+02  1,66599D+02  1,94002D+01  2,49649D+00  4,81258D-02 -2,18916D-05
 1,63939D-04 -5,92026D-04  1,43368D-03 -2,73967D-03  4,50358D-03 -6,52653D-03
 8,25031D-03 -8,95440D-03

-6,59793D+01 -1,33112D+02 -5,96993D+00 -1,99458D+00 -1,48088D-02  1,33376D-05
-9,99805D-05  3,61934D-04 -8,82244D-04  1,71722D-03 -2,94998D-03  4,63930D-03
-6,60978D-03  8,29962D-03 -8,98172D-03

-6,59793D+01  1,33112D+02 -5,96993D+00  1,99458D+00 -1,48088D-02 -8,11125D-06
 6,08230D-05 -2,20360D-04  5,38316D-04 -1,05440D-03  1,84457D-03 -3,03155D-03
 4,68830D-03 -6,63705D-03  8,31175D-03 -8,98172D-03

 2,14397D+02 -1,66599D+02  1,94002D+01 -2,49649D+00  4,81258D-02  4,92619D-06
-3,69436D-05  1,33881D-04 -3,27293D-04  6,42411D-04 -1,13084D-03  1,89252D-03
-3,05862D-03  4,70040D-03 -6,63705D-03  8,29962D-03 -8,95440D-03

-4,15671D+02  2,41547D+02 -3,76199D+01  3,62020D+00 -9,33385D-02 -2,98444D-06
 2,23824D-05 -8,11193D-05  1,98357D-04 -3,89604D-04  6,87241D-04 -1,15728D-03
 1,90450D-03 -3,05862D-03  4,68830D-03 -6,60978D-03  8,25031D-03 -8,87076D-03

 7,15357D+02 -3,74370D+02  6,47805D+01 -5,61388D+00  1,60806D-01  1,79675D-06
-1,34752D-05  4,88390D-05 -1,19433D-04  2,34639D-04 -4,14175D-04  6,98886D-04
-1,15728D-03  1,89252D-03 -3,03155D-03  4,63930D-03 -6,52653D-03  8,11204D-03
-8,64388D-03

-1,16308D+03  5,86169D+02 -1,05515D+02  8,80455D+00 -2,62325D-01 -1,06370D-06
 7,97753D-06 -2,89137D-05  7,07086D-05 -1,38925D-04  2,45281D-04 -4,14175D-04
 6,87241D-04 -1,13084D-03  1,84457D-03 -2,94998D-03  4,50358D-03 -6,30280D-03
 7,74595D-03 -8,05376D-03

 1,74650D+03 -8,70524D+02  1,59389D+02 -1,31473D+01  3,98256D-01  6,02400D-07
-4,51790D-06  1,63747D-05 -4,00448D-05  7,86804D-05 -1,38925D-04  2,34639D-04
-3,89604D-04  6,42411D-04 -1,05440D-03  1,71722D-03 -2,73967D-03  4,15785D-03
-5,73766D-03  6,83530D-03 -6,64869D-03

-2,08251D+03  1,04498D+03 -1,94350D+02  1,61073D+01 -4,94737D-01 -3,06590D-07
 2,29938D-06 -8,33388D-06  2,03808D-05 -4,00448D-05  7,07086D-05 -1,19433D-04
 1,98357D-04 -3,27293D-04  5,38316D-04 -8,82244D-04  1,43368D-03 -2,27401D-03
 3,39695D-03 -4,51172D-03  4,94386D-03 -4,10255D-03

 1,65069D+03 -8,41163D+02  1,59508D+02 -1,34276D+01  4,20189D-01  1,25367D-07
-9,40234D-07  3,40779D-06 -8,33388D-06  1,63747D-05 -2,89137D-05  4,88390D-05
-8,11193D-05  1,33881D-04 -2,20360D-04  3,61934D-04 -5,92026D-04  9,57223D-04
-1,49556D-03  2,14280D-03 -2,57679D-03  2,33919D-03 -1,43802D-03

-7,39029D+02  3,83117D+02 -7,41623D+01  6,37495D+00 -2,04242D-01 -3,45897D-08
 2,59417D-07 -9,40234D-07  2,29938D-06 -4,51790D-06  7,97753D-06 -1,34752D-05
 2,23824D-05 -3,69436D-05  6,08230D-05 -9,99805D-05  1,63939D-04 -2,67016D-04
 4,26240D-04 -6,40124D-04  8,23017D-04 -8,00204D-04  5,21778D-04 -1,98406D-04

 1,39825D+02 -7,36576D+01  1,45405D+01 -1,27813D+00  4,20747D-02  4,61206D-09
-3,45897D-08  1,25367D-07 -3,06590D-07  6,02400D-07 -1,06370D-06  1,79675D-06
-2,98444D-06  4,92619D-06 -8,11125D-06  1,33376D-05 -2,18916D-05  3,57640D-05
-5,76165D-05  8,88913D-05 -1,19658D-04  1,22408D-04 -8,34919D-05  3,29256D-05
-5,62342D-06
```

THE MATRIX B   N=5

N= 5   M= 20

```
-1,71682D+10

 0,          -4,85559D+09

-1,82300D+09  0,          -2,09710D+08

 0,          -9,32665D+07  0,          -2,12177D+06

-5,84278D+06  0,          -7,96062D+05  0,          -5,41145D+03

 1,80542D+02  8,91333D+01  1,65291D+01  1,36161D+00  4,20957D-02 -5,62342D-06

-9,50613D+02 -4,61829D+02 -8,40929D+01 -6,77834D+00 -2,04399D-01  3,29257D-05
-1,98406D-04

 2,11530D+03  1,00974D+03  1,80504D+02  1,42495D+01  4,20760D-01 -8,34919D-05
 5,21778D-04 -1,43802D-03

-2,66162D+03 -1,24870D+03 -2,19801D+02 -1,70517D+01 -4,96133D-01  1,22408D-04
-8,00204D-04  2,33920D-03 -4,10256D-03

 2,23503D+03  1,03417D+03  1,80784D+02  1,38552D+01  4,00998D-01 -1,19658D-04
 8,23019D-04 -2,57679D-03  4,94387D-03 -6,64871D-03

-1,50442D+03 -6,88168D+02 -1,21017D+02 -9,17260D+00 -2,67166D-01  8,88916D-05
-6,40127D-04  2,14281D-03 -4,51174D-03  6,83534D-03 -8,05383D-03

 9,54522D+02  4,26126D+02  7,66535D+01  5,67098D+00  1,68983D-01 -5,76171D-05
 4,26245D-04 -1,49557D-03  3,39698D-03 -5,73773D-03  7,74607D-03 -8,64409D-03

-6,04565D+02 -2,53167D+02 -4,85239D+01 -3,36751D+00 -1,06922D-01  3,57649D-05
-2,67023D-04  9,57248D-04 -2,27408D-03  4,15797D-03 -6,30301D-03  8,11239D-03
-8,87135D-03

 3,99029D+02  1,40418D+02  3,20214D+01  1,86747D+00  7,05484D-02 -2,18931D-05
 1,63950D-04 -5,92067D-04  1,43378D-03 -2,73987D-03  4,50393D-03 -6,52711D-03
 8,25128D-03 -8,95600D-03

-2,92648D+02 -6,23977D+01 -2,34832D+01 -8,29806D-01 -5,17349D-02  1,33401D-05
-9,99990D-05  3,62001D-04 -8,82408D-04  1,71754D-03 -2,95055D-03  4,64026D-03
-6,61137D-03  8,30225D-03 -8,98605D-03

 2,59876D+02 -4,54785D-19  2,08532D+01 -5,76177D-21  4,59404D-02 -8,11529D-06
 6,08534D-05 -2,20470D-04  5,38585D-04 -1,05493D-03  1,84550D-03 -3,03313D-03
 4,69092D-03 -6,64138D-03  8,31887D-03 -8,99343D-03

-2,92648D+02  6,23977D+01 -2,34832D+01  8,29806D-01 -5,17349D-02  4,93284D-06
-3,69935D-05  1,34062D-04 -3,27736D-04  6,43279D-04 -1,13237D-03  1,89511D-03
-3,06292D-03  4,70750D-03 -6,64876D-03  8,31887D-03 -8,98605D-03

 3,99029D+02 -1,40418D+02  3,20214D+01 -1,86747D+00  7,05484D-02 -2,99535D-06
 2,24643D-05 -8,14161D-05  1,99083D-04 -3,91030D-04  6,89759D-04 -1,16153D-03
 1,91157D-03 -3,07028D-03  4,70750D-03 -6,64138D-03  8,30225D-03 -8,95600D-03
```

THE MATRIX B   N=5

```
-6,04565D+02  2,53167D+02 -4,85239D+01  3,36751D+00 -1,06922D-01  1,81458D-06
-1,36090D-05  4,93237D-05 -1,20618D-04  2,36968D-04 -4,18288D-04  7,05833D-04
-1,16882D-03  1,91157D-03 -3,06292D-03  4,69092D-03 -6,61137D-03  8,25128D-03
-8,87135D-03

 9,54522D+02 -4,26126D+02  7,66535D+01 -5,67098D+00  1,68983D-01 -1,09243D-06
 8,19300D-06 -2,96947D-05  7,26185D-05 -1,42678D-04  2,51907D-04 -4,25368D-04
 7,05833D-04 -1,16153D-03  1,89511D-03 -3,03313D-03  4,64026D-03 -6,52711D-03
 8,11239D-03 -8,64409D-03

-1,50442D+03  6,88168D+02 -1,21017D+02  9,17260D+00 -2,67166D-01  6,46726D-07
-4,85034D-06  1,75796D-05 -4,29913D-05  8,44700D-05 -1,49148D-04  2,51907D-04
-4,18288D-04  6,89759D-04 -1,13237D-03  1,84550D-03 -2,95055D-03  4,50393D-03
-6,30301D-03  7,74607D-03 -8,05383D-03

 2,23503D+03 -1,03417D+03  1,80784D+02 -1,38552D+01  4,00998D-01 -3,66258D-07
 2,74688D-06 -9,95580D-06  2,43472D-05 -4,78382D-05  8,44700D-05 -1,42678D-04
 2,36968D-04 -3,91030D-04  6,43279D-04 -1,05493D-03  1,71754D-03 -2,73987D-03
 4,15797D-03 -5,73773D-03  6,83534D-03 -6,64871D-03

-2,66162D+03  1,24870D+03 -2,19801D+02  1,70517D+01 -4,96133D-01  1,86406D-07
-1,39802D-06  5,06698D-06 -1,23915D-05  2,43472D-05 -4,29913D-05  7,26185D-05
-1,20618D-04  1,99083D-04 -3,27736D-04  5,38585D-04 -8,82408D-04  1,43378D-03
-2,27408D-03  3,39698D-03 -4,51174D-03  4,94387D-03 -4,10256D-03

 2,11530D+03 -1,00974D+03  1,80504D+02 -1,42495D+01  4,20760D-01 -7,62230D-08
 5,71660D-07 -2,07193D-06  5,06698D-06 -9,95580D-06  1,75796D-05 -2,96947D-05
 4,93237D-05 -8,14161D-05  1,34062D-04 -2,20470D-04  3,62001D-04 -5,92067D-04
 9,57248D-04 -1,49557D-03  2,14281D-03 -2,57679D-03  2,33920D-03 -1,43802D-03

-9,50613D+02  4,61829D+02 -8,40929D+01  6,77834D+00 -2,04399D-01  2,10305D-08
-1,57725D-07  5,71660D-07 -1,39802D-06  2,74688D-06 -4,85034D-06  8,19300D-06
-1,36090D-05  2,24643D-05 -3,69935D-05  6,08534D-05 -9,99990D-05  1,63950D-04
-2,67023D-04  4,26245D-04 -6,40127D-04  8,23019D-04 -8,00204D-04  5,21778D-04
-1,98406D-04

 1,80542D+02 -8,91333D+01  1,65291D+01 -1,36161D+00  4,20957D-02 -2,80412D-09
 2,10305D-08 -7,62230D-08  1,86406D-07 -3,66258D-07  6,46726D-07 -1,09243D-06
 1,81458D-06 -2,99535D-06  4,93284D-06 -8,11529D-06  1,33401D-05 -2,18931D-05
 3,57649D-05 -5,76171D-05  8,88916D-05 -1,19658D-04  1,22408D-04 -8,34919D-05
 3,29257D-05 -5,62342D-06
```

THE MATRIX B   N=8

```
              N=  8              M=  7

 9,64453D+04

 0,          -1,54274D+06

 8,04480D+06  0,          -7,62870D+07

 0,           1,98013D+06  0,           1.46215D+07

 2,95784D+06  0,           7,78995D+06  0,           1,47686D+07

 0,          -2,63458D+06  0,          -9,38104D+05  0,          -3,69090D+06

 1,52243D+06  0,          -9,64705D+06  0,           3,23711D+06  0,
-1,20788D+06

 0,           2,06079D+05  0,           6,51954D+05  0,           1,78182D+05
 0,           4,45269D+04

-2,44141D-03  6,97545D-04  1,12413D-02 -3,21181D-03 -6,07639D-03  1,73611D-03
 6,94444D-04 -1,98413D-04  0,

 2,39258D-02 -9,57031D-03 -1,08290D-01  4,33160D-02  5,12153D-02 -2,04861D-02
-3,47222D-03  1,38889D-03  0,           0,

-1,19629D-01  7,97526D-02  5,07422D-01 -3,38281D-01 -1,17187D-01  7,81250D-02
 6,25000D-03 -4,16667D-03  0,           0,           0,

 5,98145D-01 -1,19629D+00 -4,10373D-01  8,20747D-01  7,20486D-02 -1,44097D-01
-3,47222D-03  6,94444D-03  0,           0,           0,           0,

 5,98145D-01  1,19629D+00 -4,10373D-01 -8,20747D-01  7,20486D-02  1,44097D-01
-3,47222D-03 -6,94444D-03  0,           0,           0,           0,
 0,

-1,19629D-01 -7,97526D-02  5,07422D-01  3,38281D-01 -1,17188D-01 -7,81250D-02
 6,25000D-03  4,16667D-03  0,           0,           0,           0,
 0,           0,

 2,39258D-02  9,57031D-03 -1,08290D-01 -4,33160D-02  5,12153D-02  2,04861D-02
-3,47222D-03 -1,38889D-03  0,           0,           0,           0,
 0,           0,           0,

-2,44141D-03 -6,97545D-04  1,12413D-02  3,21181D-03 -6,07639D-03 -1,73611D-03
 6,94444D-04  1,98413D-04  0,           0,           0,           0,
 0,           0,           0,           0,

              N=  8              M=  8

 4,11739D+06

 0,          -4,90712D+07
```

THE MATRIX B N=8

```
 2,64346D+07  0,           1,67181D+08

 0,          -1,41746D+06  0,           9,69245D+07

 1,40358D+07  0,           9,07728D+07  0,           5,02774D+07

 0,          -2,08303D+07  0,           8,17573D+05  0,          -9,73701D+06

 3,32993D+06  0,           2,01385D+07  0,           1,22753D+07  0,
 2,09460D+06

 0,           4,38464D+05  0,           1,62647D+06  0,           2,88619D+05
 0,           5,33361D+04

-2,14477D-03  3,57143D-03 -1,38504D-02 -4,86111D-03 -7,21036D-03  1,38889D-03
-1,75505D-03 -9,92063D-05  1,11722D-12

 1,71582D-02 -3,80952D-02  1,16359D-01  5,00000D-02  5,07384D-02 -1,25000D-02
 1,54293D-02  5,95238D-04 -8,93777D-12  7,15022D-11

-6,00536D-02  2,00000D-01 -4,62811D-01 -2,34722D-01 -1,18557D-01  3,61111D-02
-5,74748D-02 -1,38889D-03  3,12822D-11 -2,50258D-10  8,75902D-10

 1,20107D-01 -8,00000D-01  1,52562D+00  3,38889D-01  1,32947D-01 -4,02778D-02
 1,19116D-01  1,38889D-03 -6,25644D-11  5,00515D-10 -1,75180D-09  3,50361D-09

 8,49866D-01 -2,81886D-28 -2,33064D+00  1,32227D-28 -1,15836D-01 -8,69220D-29
-1,50631D-01  2,46750D-30  7,82055D-11 -6,25644D-10  2,18975D-09 -4,37951D-09
 5,47439D-09

 1,20107D-01  8,00000D-01  1,52562D+00 -3,38889D-01  1,32947D-01  4,02778D-02
 1,19116D-01 -1,38889D-03 -6,25644D-11  5,00515D-10 -1,75180D-09  3,50361D-09
-4,37951D-09  3,50361D-09

-6,00536D-02 -2,00000D-01 -4,62811D-01  2,34722D-01 -1,18557D-01 -3,61111D-02
-5,74748D-02  1,38889D-03  3,12822D-11 -2,50258D-10  8,75902D-10 -1,75180D-09
 2,18975D-09 -1,75180D-09  8,75902D-10

 1,71582D-02  3,80952D-02  1,16359D-01 -5,00000D-02  5,07384D-02  1,25000D-02
 1,54293D-02 -5,95238D-04 -8,93777D-12  7,15022D-11 -2,50258D-10  5,00515D-10
-6,25644D-10  5,00515D-10 -2,50258D-10  7,15022D-11

-2,14477D-03 -3,57143D-03 -1,38504D-02  4,86111D-03 -7,21036D-03 -1,38889D-03
-1,75505D-03  9,92063D-05  1,11722D-12 -8,93777D-12  3,12822D-11 -6,25644D-11
 7,82055D-11 -6,25644D-11  3,12822D-11 -8,93777D-12  1,11722D-12
```

N= 8 M= 9

```
 2,50792D+07

 0,           1,78736D+09

 1,72031D+08  0,           1,00011D+09

 0,           2,09647D+09  0,           2,46688D+09
```

THE MATRIX B   N=8

```
 5,81828D+07  0,           3,70285D+08  0,           1,36845D+08

 0,           4,48504D+08  0,           5,27264D+08  0,           1,10080D+08

 1,23186D+07  0,           6,12188D+07  0,           2,71023D+07  0,
 3,45765D+06

 0,           1,03846D+07  0,           1,25691D+07  0,           2,74506D+06
 0,           1,02440D+05

-2,86874D-03 -5,73130D-02 -1,88303D-02 -6,71031D-02 -6,34140D-03 -1,44536D-02
-1,23971D-03 -3,26098D-04  2,16561D-12

 1,76397D-02  5,16515D-01  1,43053D-01  6,00716D-01  3,83134D-02  1,31818D-01
 9,37242D-03  2,73647D-03 -1,88316D-11  1,64873D-10

-3,34489D-02 -2,07284D+00 -4,84895D-01 -2,37240D+00 -7,56128D-02 -5,40815D-01
-2,82664D-02 -1,03507D-02  7,26912D-11 -6,41045D-10  2,51147D-09

-3,93043D-02  4,89405D+00  1,03467D+00  5,29838D+00  6,03718D-02  1,29223D+00
 4,09619D-02  2,32256D-02 -1,63464D-10  1,45273D-09 -5,73712D-09  1,32144D-08

 5,57982D-01 -8,41773D+00 -6,73997D-01 -7,63425D+00 -1,67310D-02 -1,96525D+00
-2,08282D-02 -3,41440D-02  2,35972D-10 -2,11453D-09  8,42121D-09 -1,95634D-08
 2,92160D-08

 5,57982D-01  8,41773D+00 -6,73997D-01  7,63425D+00 -1,67310D-02  1,96525D+00
-2,08282D-02  3,41440D-02 -2,26749D-10  2,04996D-09 -8,23674D-09  1,93051D-08
-2,90868D-08  2,92160D-08

-3,93043D-02 -4,89405D+00  1,03467D+00 -5,29838D+00  6,03718D-02 -1,29223D+00
 4,09619D-02 -2,32256D-02  1,45017D-10 -1,32360D-09  5,36818D-09 -1,26979D-08
 1,93051D-08 -1,95634D-08  1,32144D-08

-3,34489D-02  2,07284D+00 -4,84895D-01  2,37240D+00 -7,56128D-02  5,40815D-01
-2,82664D-02  1,03507D-02 -5,95147D-11  5,48809D-10 -2,24794D-09  5,36818D-09
-8,23674D-09  8,42121D-09 -5,73712D-09  2,51147D-09

 1,76397D-02 -5,16515D-01  1,43053D-01 -6,00716D-01  3,83134D-02 -1,31818D-01
 9,37242D-03 -2,73647D-03  1,42199D-11 -1,32590D-10  5,48809D-10 -1,32360D-09
 2,04996D-09 -2,11453D-09  1,45273D-09 -6,41045D-10  1,64873D-10

-2,86874D-03  5,73130D-02 -1,88303D-02  6,71031D-02 -6,34140D-03  1,44536D-02
-1,23971D-03  3,26098D-04 -1,50678D-12  1,42199D-11 -5,95147D-11  1,45017D-10
-2,26749D-10  2,35972D-10 -1,63464D-10  7,26912D-11 -1,88316D-11  2,16561D-12
```

N= 8   M= 10

```
 3,77540D+09

 0,           1,11270D+10

 1,23325D+10  0,           4,03230D+10

 0,           9,95304D+09  0,           8,96330D+09
```

THE MATRIX B N=8

```
 4,03903D+09  0,           1,32247D+10  0,           4,34605D+09

 0,           1,63237D+09  0,           1,47331D+09  0,           2,34810D+08

 2,00725D+08  0,           6,64258D+08  0,           2,22950D+08  0,
 1,24266D+07

 0,           2,75500D+07  0,           2,57745D+07  0,           4,29904D+06
 0,           1,19105D+05

-8,97670D-02 -9,00234D-02 -2,92918D-01 -8,01832D-02 -9,58060D-02 -1,33165D-02
-4,69856D-03 -2,18384D-04  2,86184D-12

 8,40907D-01  7,23758D-01  2,73711D+00  6,36605D-01  8,92557D-01  1,07921D-01
 4,23717D-02  1,64786D-03 -2,58681D-11  2,35986D-10

-3,58541D+00 -2,46873D+00 -1,16392D+01 -2,11495D+00 -3,79419D+00 -3,72046D-01
-1,73135D-01 -5,30112D-03  1,05330D-10 -9,70904D-10  4,04153D-09

 9,18268D+00  4,52112D+00  2,96971D+01  3,61407D+00  9,74597D+00  6,75305D-01
 4,26305D-01  9,09352D-03 -2,54809D-10  2,37590D-09 -1,00193D-08  2,51989D-08

-1,56723D+01 -4,58098D+00 -5,00067D+01 -3,02881D+00 -1,67219D+01 -5,99572D-01
-7,07487D-01 -7,78322D-03  4,06360D-10 -3,83654D-09  1,64088D-08 -4,19181D-08
 7,09145D-08

 1,96479D+01 -4,38780D-24  5,90091D+01 -3,82914D-24  1,99468D+01 -6,54008D-25
 8,33288D-01 -9,89364D-27 -4,47445D-10  4,28041D-09 -1,85827D-08  4,82603D-08
-8,30972D-08  9,91733D-08

-1,56723D+01  4,58098D+00 -5,00067D+01  3,02881D+00 -1,67219D+01  5,99572D-01
-7,07487D-01  7,78322D-03  3,45448D-10 -3,34924D-09  1,47642D-08 -3,89944D-08
 6,83562D-08 -8,30972D-08  7,09145D-08

 9,18268D+00 -4,52112D+00  2,96971D+01 -3,61407D+00  9,74597D+00 -6,75305D-01
 4,26305D-01 -9,09352D-03 -1,85196D-10  1,81900D-09 -8,13972D-09  2,18574D-08
-3,89944D-08  4,82603D-08 -4,19181D-08  2,51989D-08

-3,58541D+00  2,46873D+00 -1,16392D+01  2,11495D+00 -3,79419D+00  3,72046D-01
-1,73135D-01  5,30112D-03  6,61724D-11 -6,57645D-10  2,98428D-09 -8,13972D-09
 1,47642D-08 -1,85827D-08  1,64088D-08 -1,00193D-08  4,04153D-09

 8,40907D-01 -7,23758D-01  2,73711D+00 -6,36605D-01  8,92557D-01 -1,07921D-01
 4,23717D-02 -1,64786D-03 -1,42659D-11  1,43168D-10 -6,57645D-10  1,81900D-09
-3,34924D-09  4,28041D-09 -3,83654D-09  2,37590D-09 -9,70904D-10  2,35986D-10

-8,97670D-02  9,00234D-02 -2,92918D-01  8,01832D-02 -9,58060D-02  1,33165D-02
-4,69856D-03  2,18384D-04  1,41157D-12 -1,42659D-11  6,61724D-11 -1,85196D-10
 3,45448D-10 -4,47445D-10  4,06360D-10 -2,54809D-10  1,05330D-10 -2,58681D-11
 2,86184D-12
```

N= 8 M= 11

```
 2,60410D+10

 0,           2,67981D+11
```

THE MATRIX B   N=8

```
 6,71609D+10  0,           1,73491D+11

 0,           1,72374D+11  0,           1,11390D+11

 1,71702D+10  0,           4,44325D+10  0,           1,13924D+10

 0,           1,57739D+10  0,           1,03306D+10  0,           9,92402D+08

 6,45351D+08  0,           1,69123D+09  0,           4,43481D+08  0,
 1,88609D+07

 0,           1,27575D+08  0,           8,75334D+07  0,           9,49203D+06
 0,           1,54033D+05

-1,65813D-01 -7,67873D-01 -4,26955D-01 -4,91191D-01 -1,09064D-01 -4,42083D-02
-4,03465D-03 -3,38053D-04  3,27185D-12

 1,41412D+00  7,26221D+00  3,62981D+00  4,61502D+00  9,25690D-01  4,06949D-01
 3,32393D-02  2,89629D-03 -3,01454D-11  2,80609D-10

-5,25854D+00 -3,15727D+01 -1,34427D+01 -1,99260D+01 -3,43211D+00 -1,71564D+00
-1,19010D-01 -1,13907D-02  1,26194D-10 -1,18856D-09  5,10322D-09

 1,08960D+01  8,40511D+01  2,76576D+01  5,27191D+01  7,11329D+00  4,41755D+00
 2,37675D-01  2,75651D-02 -3,18096D-10  3,03613D-09 -1,32397D-08  3,49675D-08

-1,28534D+01 -1,53829D+02 -3,19606D+01 -9,62330D+01 -8,36813D+00 -7,84584D+00
-2,70837D-01 -4,66516D-02  5,39529D-10 -5,22580D-09  2,31853D-08 -6,24734D-08
 1,14167D-07

 6,46758D+00  2,04327D+02  1,45427D+01  1,28454D+02  3,87033D+00  1,02830D+01
 1,22967D-01  5,95161D-02 -6,51357D-10  6,40768D-09 -2,89591D-08  7,97350D-08
-1,49327D-07  2,00586D-07

 6,46758D+00 -2,04327D+02  1,45427D+01 -1,28454D+02  3,87033D+00 -1,02830D+01
 1,22967D-01 -5,95161D-02  5,76027D-10 -5,75470D-09  2,64975D-08 -7,45852D-08
 1,43247D-07 -1,97772D-07  2,00586D-07

-1,28534D+01  1,53829D+02 -3,19606D+01  9,62330D+01 -8,36813D+00  7,84584D+00
-2,70837D-01  4,66516D-02 -3,76188D-10  3,81148D-09 -1,78586D-08  5,13379D-08
-1,01028D-07  1,43247D-07 -1,49327D-07  1,14167D-07

 1,08960D+01 -8,40511D+01  2,76576D+01 -5,27191D+01  7,11329D+00 -4,41755D+00
 2,37675D-01 -2,75651D-02  1,78777D-10 -1,83237D-09  8,71436D-09 -2,55208D-08
 5,13379D-08 -7,45852D-08  7,97350D-08 -6,24734D-08  3,49675D-08

-5,25854D+00  3,15727D+01 -1,34427D+01  1,99260D+01 -3,43211D+00  1,71564D+00
-1,19010D-01  1,13907D-02 -5,89381D-11  6,09202D-10 -2,93086D-09  8,71436D-09
-1,78586D-08  2,64975D-08 -2,89591D-08  2,31853D-08 -1,32397D-08  5,10322D-09

 1,41412D+00 -7,26221D+00  3,62981D+00 -4,61502D+00  9,25690D-01 -4,06949D-01
 3,32393D-02 -2,89629D-03  1,20829D-11 -1,25593D-10  6,09202D-10 -1,83237D-09
 3,81148D-09 -5,75470D-09  6,40768D-09 -5,22580D-09  3,03613D-09 -1,18856D-09
 2,80609D-10
```

THE MATRIX B   N=8

```
-1,65813D-01  7,67873D-01 -4,26955D-01  4,91191D-01 -1,09064D-01  4,42083D-02
-4,03465D-03  3,38053D-04 -1,15823D-12  1,20829D-11 -5,89381D-11  1,78777D-10
-3,76188D-10  5,76027D-10 -6,51357D-10  5,39529D-10 -3,18096D-10  1,26194D-10
-3,01454D-11  3,27185D-12

              N=  8             M= 12

 4,37082D+11

 0,           1,30781D+12

 8,39742D+11  0,            1,62028D+12

 0,           6,69800D+11  0,            3,44742D+11

 1,44980D+11  0,            2,82355D+11  0,            5,01896D+10

 0,           4,77744D+10  0,            2,49538D+10  0,            1,87676D+09

 3,31430D+09  0,            6,60969D+09  0,            1,23570D+09  0,
 3,47995D+07

 0,           2,96086D+08  0,            1,62425D+08  0,            1,38548D+07
 0,           1,74717D+05

-9,76185D-01 -1,29951D+00 -1,86405D+00 -6,61288D-01 -3,17522D-01 -4,62427D-02
-7,01448D-03 -2,69122D-04  3,50157D-12

 9,24994D+00  1,14231D+01  1,75415D+01  5,77854D+00  2,94210D+00  3,96374D-01
 6,24129D-02  2,16592D-03 -3,25794D-11  3,06399D-10

-4,04152D+01 -4,45988D+01 -7,60423D+01 -2,24268D+01 -1,25267D+01 -1,50487D+00
-2,54078D-01 -7,73819D-03  1,38366D-10 -1,31754D-09  5,74821D-09

 1,08906D+02  1,00464D+02  2,03166D+02  5,02646D+01  3,27957D+01  3,29076D+00
 6,35183D-01  1,60156D-02 -3,56586D-10  3,44395D-09 -1,52792D-08  4,14165D-08

-2,05130D+02 -1,38762D+02 -3,79539D+02 -6,92990D+01 -5,99632D+01 -4,42338D+00
-1,11366D+00 -2,05797D-02  6,26416D-10 -6,14642D-09  2,77892D-08 -7,70314D-08
 1,47031D-07

 2,90904D+02  1,04710D+02  5,35409D+02  5,25867D+01  8,29977D+01  3,28953D+00
 1,49545D+00  1,48505D-02 -8,00208D-10  7,98486D-09 -3,68464D-08  1,04675D-07
-2,05627D-07  2,97038D-07

-3,24077D+02 -3,68792D-21 -5,97342D+02 -1,86003D-21 -9,18560D+01 -1,25887D-22
-1,63658D+00 -6,65310D-25  7,74369D-10 -7,85626D-09  3,70072D-08 -1,07818D-07
 2,18266D-07 -3,26292D-07  3,71837D-07

 2,90904D+02 -1,04710D+02  5,35409D+02 -5,25867D+01  8,29977D+01 -3,28953D+00
 1,49545D+00 -1,48505D-02 -5,81148D-10  5,98317D-09 -2,87190D-08  8,56793D-08
-1,78550D-07  2,76056D-07 -3,26292D-07  2,97038D-07

-2,05130D+02  1,38762D+02 -3,79539D+02  6,92990D+01 -5,99632D+01  4,42338D+00
-1,11366D+00  2,05797D-02  3,39224D-10 -3,53241D-09  1,72161D-08 -5,24038D-08
 1,12024D-07 -1,78550D-07  2,18266D-07 -2,05627D-07  1,47031D-07
```

THE MATRIX B N=8

```
 1,08906D+02 -1,00464D+02  2,03166D+02 -5,02646D+01  3,27957D+01 -3,29076D+00
 6,35183D-01 -1,60156D-02 -1,50272D-10  1,57694D-09 -7,77043D-09  2,40174D-08
-5,24038D-08  8,56793D-08 -1,07818D-07  1,04675D-07 -7,70314D-08  4,14165D-08

-4,04152D+01  4,45988D+01 -7,60423D+01  2,24268D+01 -1,25267D+01  1,50487D+00
-2,54078D-01  7,73819D-03  4,75232D-11 -5,01105D-10  2,48710D-09 -7,77043D-09
 1,72161D-08 -2,87190D-08  3,70072D-08 -3,68464D-08  2,77892D-08 -1,52792D-08
 5,74821D-09

 9,24994D+00 -1,14231D+01  1,75415D+01 -5,77854D+00  2,94210D+00 -3,96374D-01
 6,24129D-02 -2,16592D-03 -9,50292D-12  1,00500D-10 -5,01105D-10  1,57694D-09
-3,53241D-09  5,98317D-09 -7,85626D-09  7,98486D-09 -6,14642D-09  3,44395D-09
-1,31754D-09  3,06399D-10

-9,76185D-01  1,29951D+00 -1,86405D+00  6,61288D-01 -3,17522D-01  4,62427D-02
-7,01448D-03  2,69122D-04  8,96870D-13 -9,50292D-12  4,75232D-11 -1,50272D-10
 3,39224D-10 -5,81148D-10  7,74369D-10 -8,00208D-10  6,26416D-10 -3,56586D-10
 1,38366D-10 -3,25794D-11  3,50157D-12
```

N= 8 M= 13

```
 2,34793D+12

 0,           1,05868D+13

 3,65868D+12  0,           5,72889D+12

 0,           4,02115D+12  0,           1,54693D+12

 5,04182D+11  0,           7,97747D+11  0,           1,13542D+11

 0,           2,00666D+11  0,           7,92929D+10  0,           4,30530D+09

 9,05781D+09  0,           1,47037D+10  0,           2,20798D+09  0,
 4,93402D+07

 0,           8,51873D+08  0,           3,57687D+08  0,           2,24700D+07
 0,           2,04873D+05

-1,85114D+00 -4,71195D+00 -2,86336D+00 -1,75749D+00 -3,88273D-01 -8,44286D-02
-6,70688D-03 -3,30741D-04  3,62748D-12

 1,65835D+01  4,42801D+01  2,54810D+01  1,63111D+01  3,40407D+00  7,62960D-01
 5,66498D-02  2,82396D-03 -3,39240D-11  3,20757D-10

-6,67106D+01 -1,91720D+02 -1,01746D+02 -6,96028D+01 -1,33634D+01 -3,15842D+00
-2,13584D-01 -1,10707D-02  1,45176D-10 -1,39025D-09  6,11648D-09

 1,58130D+02  5,12223D+02  2,39289D+02  1,82903D+02  3,08384D+01  8,02809D+00
 4,72965D-01  2,67795D-02 -3,78580D-10  3,67883D-09 -1,64687D-08  4,52585D-08

-2,40822D+02 -9,61112D+02 -3,61702D+02 -3,37113D+02 -4,56841D+01 -1,43097D+01
-6,74327D-01 -4,57970D-02  6,77944D-10 -6,69669D-09  3,05760D-08 -8,60325D-08
 1,68118D-07
```

THE MATRIX B N=8

```
 2,30850D+02  1,37947D+03  3,45003D+02  4,75666D+02  4,27483D+01  1,96277D+01
 6,12129D-01  6,09736D-02 -8,94454D-10  8,99131D-09 -4,19434D-08  1,21138D-07
-2,44197D-07  3,67583D-07

-9,56803D+01 -1,61975D+03 -1,43460D+02 -5,51902D+02 -1,75550D+01 -2,24008D+01
-2,47126D-01 -6,85194D-02  9,14378D-10 -9,35143D-09  4,45792D-08 -1,32275D-07
 2,75564D-07 -4,31092D-07  5,27525D-07

-9,56803D+01  1,61975D+03 -1,43460D+02  5,51902D+02 -1,75550D+01  2,24008D+01
-2,47126D-01  6,85194D-02 -7,51502D-10  7,80238D-09 -3,79322D-08  1,15437D-07
-2,48267D-07  4,03570D-07 -5,15723D-07  5,27525D-07

 2,30850D+02 -1,37947D+03  3,45003D+02 -4,75666D+02  4,27483D+01 -1,96277D+01
 6,12129D-01 -6,09736D-02  5,05866D-10 -5,31198D-09  2,62285D-08 -8,15135D-08
 1,80221D-07 -3,03285D-07  4,03570D-07 -4,31092D-07  3,67583D-07

-2,40822D+02  9,61112D+02 -3,61702D+02  3,37113D+02 -4,56841D+01  1,43097D+01
-6,74327D-01  4,57970D-02 -2,76577D-10  2,92576D-09 -1,46013D-08  4,60808D-08
-1,04094D-07  1,80221D-07 -2,48267D-07  2,75564D-07 -2,44197D-07  1,68118D-07

 1,58130D+02 -5,12223D+02  2,39289D+02 -1,82903D+02  3,08384D+01 -8,02809D+00
 4,72965D-01 -2,67795D-02  1,18055D-10 -1,25432D-09  6,30162D-09 -2,00912D-08
 4,60808D-08 -8,15135D-08  1,15437D-07 -1,32275D-07  1,21138D-07 -8,60325D-08
 4,52585D-08

-6,67106D+01  1,91720D+02 -1,01746D+02  6,96028D+01 -1,33634D+01  3,15842D+00
-2,13584D-01  1,10707D-02 -3,65500D-11  3,89337D-10 -1,96387D-09  6,30162D-09
-1,46013D-08  2,62285D-08 -3,79322D-08  4,45792D-08 -4,19434D-08  3,05760D-08
-1,64687D-08  6,11648D-09

 1,65835D+01 -4,42801D+01  2,54810D+01 -1,63111D+01  3,40407D+00 -7,62960D-01
 5,66498D-02 -2,82396D-03  7,21712D-12 -7,69971D-11  3,89337D-10 -1,25432D-09
 2,92576D-09 -5,31198D-09  7,80238D-09 -9,35143D-09  8,99131D-09 -6,69669D-09
 3,67883D-09 -1,39025D-09  3,20757D-10

-1,85114D+00  4,71195D+00 -2,86336D+00  1,75749D+00 -3,88273D-01  8,44286D-02
-6,70688D-03  3,30741D-04 -6,75821D-13  7,21712D-12 -3,65500D-11  1,18055D-10
-2,76577D-10  5,05866D-10 -7,51502D-10  9,14378D-10 -8,94454D-10  6,77944D-10
-3,78580D-10  1,45176D-10 -3,39240D-11  3,62748D-12


                N=  8           M= 14

 1,74909D+13

 0,           4,16345D+13

 2,12729D+13  0,           2,61023D+13

 0,           1,28918D+13  0,           4,04893D+12

 2,20986D+12  0,           2,75640D+12  0,           3,00205D+11

 0,           5,17992D+11  0,           1,67465D+11  0,           7,35960D+09

 2,92578D+10  0,           3,76941D+10  0,           4,37694D+09  0,
 7,42668D+07
```

THE MATRIX B   N=8

```
 0.            1,75932D+09  0.            6,05894D+08  0.            3,09164D+07
 0.            2,27810D+05

-5,98322D+00  -8,02416D+00 -7,16883D+00  -2,43192D+00 -7,24449D-01  -9,35634D-02
-9,10213D-03  -2,91152D-04  3,69581D-12

 5,60933D+01   7,22471D+01  6,65141D+01   2,16289D+01  6,59246D+00   8,11326D-01
 7,98040D-02   2,39947D-03 -3,46567D-11   3,28613D-10

-2,42109D+02  -2,93748D+02 -2,83626D+02  -8,67111D+01 -2,74924D+01  -3,16198D+00
-3,19973D-01  -8,90702D-03  1,48910D-10  -1,43030D-09  6,32059D-09

 6,44306D+02   7,11988D+02  7,44361D+02   2,06876D+02  7,03794D+01   7,31710D+00
 7,87573D-01   1,97141D-02 -3,90775D-10   3,80958D-09 -1,71352D-08   4,74349D-08

-1,20403D+03  -1,13804D+03 -1,36993D+03  -3,25094D+02 -1,26174D+02  -1,11504D+01
-1,36215D+00  -2,89227D-02  7,07069D-10  -7,00897D-09  3,21678D-08  -9,12304D-08
 1,80533D-07

 1,72638D+03   1,22801D+03  1,93472D+03   3,44936D+02  1,73905D+02   1,15135D+01
 1,82471D+00   2,90129D-02 -9,49619D-10   9,58280D-09 -4,49584D-08   1,30984D-07
-2,67710D-07   4,12118D-07

-2,05167D+03  -7,98938D+02 -2,27136D+03  -2,21287D+02 -2,00641D+02  -7,24720D+00
-2,06825D+00  -1,79379D-02  1,00168D-09  -1,02875D-08  4,93507D-08  -1,47856D-07
 3,12777D-07  -5,01574D-07  6,39071D-07

 2,15500D+03   8,06645D-20  2,37298D+03   2,23901D-20  2,08311D+02   7,27564D-22
 2,13478D+00   1,75967D-24 -8,69769D-10   9,07045D-09 -4,43958D-08   1,36544D-07
-2,98676D-07   4,99047D-07 -6,66827D-07   7,32215D-07

-2,05167D+03   7,98938D+02 -2,27136D+03   2,21287D+02 -2,00641D+02   7,24720D+00
-2,06825D+00   1,79379D-02  6,42069D-10  -6,77237D-09  3,36724D-08  -1,05821D-07
 2,38276D-07  -4,13243D-07  5,77591D-07  -6,66827D-07  6,39071D-07

 1,72638D+03  -1,22801D+03  1,93472D+03  -3,44936D+02  1,73905D+02  -1,15135D+01
 1,82471D+00  -2,90129D-02 -4,05701D-10   4,31024D-09 -2,16583D-08   6,91251D-08
-1,59131D-07   2,84464D-07 -4,13243D-07   4,99047D-07 -5,01574D-07   4,12118D-07

-1,20403D+03   1,13804D+03 -1,36993D+03   3,25094D+02 -1,26174D+02   1,11504D+01
-1,36215D+00   2,89227D-02  2,14198D-10  -2,28518D-09  1,15562D-08  -3,72495D-08
 8,70605D-08  -1,59131D-07  2,38276D-07  -2,98676D-07  3,12777D-07  -2,67710D-07
 1,80533D-07

 6,44306D+02  -7,11988D+02  7,44361D+02  -2,06876D+02  7,03794D+01  -7,31710D+00
 7,87573D-01  -1,97141D-02 -8,96855D-11   9,59061D-10 -4,86789D-09   1,57845D-08
-3,72495D-08   6,91251D-08 -1,05821D-07   1,36544D-07 -1,47856D-07   1,30984D-07
-9,12304D-08   4,74349D-08

-2,42109D+02   2,93748D+02 -2,83626D+02   8,67111D+01 -2,74924D+01   3,16198D+00
-3,19973D-01   8,90702D-03  2,74651D-11  -2,94098D-10  1,49595D-09  -4,86789D-09
 1,15562D-08  -2,16583D-08  3,36724D-08  -4,43958D-08  4,93507D-08  -4,49584D-08
 3,21678D-08  -1,71352D-08  6,32059D-09
```

THE MATRIX B   N=8

```
 5,60933D+01 -7,22471D+01  6,65141D+01 -2,16289D+01  6,59246D+00 -8,11326D-01
 7,98040D-02 -2,39947D-03 -5,38825D-12  5,77443D-11 -2,94098D-10  9,59061D-10
-2,28518D-09  4,31024D-09 -6,77237D-09  9,07045D-09 -1,02875D-08  9,58280D-09
-7,00897D-09  3,80958D-09 -1,43030D-09  3,28613D-10

-5,98322D+00  8,02416D+00 -7,16883D+00  2,43192D+00 -7,24449D-01  9,35634D-02
-9,10213D-03  2,91152D-04  5,02537D-13 -5,38825D-12  2,74651D-11 -8,96855D-11
 2,14198D-10 -4,05701D-10  6,42069D-10 -8,69769D-10  1,00168D-09 -9,49619D-10
 7,07069D-10 -3,90775D-10  1,48910D-10 -3,46567D-11  3,69581D-12
```

N= 8   M= 15

```
 7,48865D+13

 0,           1,92732D+14

 7,55354D+13  0,           7,69497D+13

 0,           4,73143D+13  0,           1,18371D+13

 6,44217D+12  0,           6,68156D+12  0,           5,99625D+11

 0,           1,48120D+12  0,           3,83656D+11  0,           1,33078D+10

 6,95343D+10  0,           7,46284D+10  0,           7,15857D+09  0,
 9,97603D+07

 0,           3,89791D+09  0,           1,08168D+09  0,           4,38845D+07
 0,           2,55812D+05

-1,10417D+01 -1,93328D+01 -1,09410D+01 -4,59512D+00 -9,03871D-01 -1,35916D-01
-9,20563D-03 -3,23306D-04  3,73273D-12

 1,00189D+02  1,79206D+02  9,82454D+01  4,19571D+01  7,96468D+00  1,20857D+00
 7,83652D-02  2,74495D-03 -3,50534D-11  3,32875D-10

-4,12206D+02 -7,62857D+02 -3,99396D+02 -1,75533D+02 -3,16966D+01 -4,91145D+00
-3,00838D-01 -1,06739D-02  1,50939D-10 -1,45210D-09  6,43207D-09

 1,01872D+03  1,99610D+03  9,73784D+02  4,50408D+02  7,54900D+01  1,22226D+01
 6,91358D-01  2,55165D-02 -3,97437D-10  3,88117D-09 -1,75013D-08  4,86372D-08

-1,68638D+03 -3,65561D+03 -1,58839D+03 -8,07728D+02 -1,20153D+02 -2,12730D+01
-1,06483D+00 -4,29199D-02  7,23142D-10 -7,18166D-09  3,30510D-08 -9,41307D-08
 1,87529D-07

 1,95634D+03  5,12684D+03  1,81541D+03  1,10990D+03  1,34175D+02  2,84904D+01
 1,15748D+00  5,60125D-02 -9,80621D-10  9,91590D-09 -4,66620D-08  1,36578D-07
-2,81206D-07  4,38150D-07

-1,54972D+03 -5,98258D+03 -1,41932D+03 -1,27418D+03 -1,03009D+02 -3,21296D+01
-8,72360D-01 -6,21429D-02  1,05244D-09 -1,08329D-08  5,21399D-08 -1,57016D-07
 3,34873D-07 -5,44196D-07  7,08853D-07
```

THE MATRIX B   N=8

```
 5,84600D+02  6,28870D+03  5,30613D+02  1,32781D+03  3,81326D+01  3,31979D+01
 3,20034D-01  6,37346D-02 -9,42951D-10  9,85676D-09 -4,84173D-08  1,49750D-07
-3,30534D-07  5,60498D-07 -7,67439D-07  8,77276D-07

 5,84600D+02 -6,28870D+03  5,30613D+02 -1,32781D+03  3,81326D+01 -3,31979D+01
 3,20034D-01 -6,37346D-02  7,35849D-10 -7,77999D-09  3,88257D-08 -1,22744D-07
 2,79100D-07 -4,91990D-07  7,06519D-07 -8,52715D-07  8,77276D-07

-1,54972D+03  5,98258D+03 -1,41932D+03  1,27418D+03 -1,03009D+02  3,21296D+01
-8,72360D-01  6,21429D-02 -5,10370D-10  5,43485D-09 -2,74100D-08  8,80133D-08
-2,04695D-07  3,72354D-07 -5,57141D-07  7,06519D-07 -7,67439D-07  7,08853D-07

 1,95634D+03 -5,12684D+03  1,81541D+03 -1,10990D+03  1,34175D+02 -2,84904D+01
 1,15748D+00 -5,60125D-02  3,11724D-10 -3,33305D-09  1,69153D-08 -5,48487D-08
 1,29515D-07 -2,41023D-07  3,72354D-07 -4,91990D-07  5,60498D-07 -5,44196D-07
 4,38150D-07

-1,68638D+03  3,65561D+03 -1,58839D+03  8,07728D+02 -1,20153D+02  2,12730D+01
-1,06483D+00  4,29199D-02 -1,61604D-10  1,73179D-09 -8,81990D-09  2,87627D-08
-6,85570D-08  1,29515D-07 -2,04695D-07  2,79100D-07 -3,30534D-07  3,34873D-07
-2,81206D-07  1,87529D-07

 1,01872D+03 -1,99610D+03  9,73784D+02 -4,50408D+02  7,54900D+01 -1,22226D+01
 6,91358D-01 -2,55165D-02  6,69916D-11 -7,18791D-10  3,66797D-09 -1,20007D-08
 2,87627D-08 -5,48487D-08  8,80133D-08 -1,22744D-07  1,49750D-07 -1,57016D-07
 1,36578D-07 -9,41307D-08  4,86372D-08

-4,12206D+02  7,62857D+02 -3,99396D+02  1,75533D+02 -3,16966D+01  4,91145D+00
-3,00838D-01  1,06739D-02 -2,03996D-11  2,19034D-10 -1,11899D-09  3,66797D-09
-8,81990D-09  1,69153D-08 -2,74100D-08  3,88257D-08 -4,84173D-08  5,21399D-08
-4,66620D-08  3,30510D-08 -1,75013D-08  6,43207D-09

 1,00189D+02 -1,79206D+02  9,82454D+01 -4,19571D+01  7,96468D+00 -1,20857D+00
 7,83652D-02 -2,74495D-03  3,98871D-12 -4,28455D-11  2,19034D-10 -7,18791D-10
 1,73179D-09 -3,33305D-09  5,43485D-09 -7,77999D-09  9,85676D-09 -1,08329D-08
 9,91590D-09 -7,18166D-09  3,88117D-09 -1,45210D-09  3,32875D-10

-1,10417D+01  1,93328D+01 -1,09410D+01  4,59512D+00 -9,03871D-01  1,35916D-01
-9,20563D-03  3,23306D-04 -3,71232D-13  3,98871D-12 -2,03996D-11  6,69916D-11
-1,61604D-10  3,11724D-10 -5,10370D-10  7,35849D-10 -9,42951D-10  1,05244D-09
-9,80621D-10  7,23142D-10 -3,97437D-10  1,50939D-10 -3,50534D-11  3,73273D-12
```

N= 8   M= 16

```
 3,52542D+14

 0,           6,26337D+14

 2,90779D+14  0,           2,42885D+14

 0,           1,28519D+14  0,           2,69144D+13

 2,00153D+13  0,           1,70786D+13  0,           1,24643D+12

 0,           3,34151D+12  0,           7,25824D+11  0,           2,09943D+10
```

THE MATRIX B   N=8

```
 1,73099D+11  0,           1,53396D+11  0,           1,20210D+10  0,
 1,36019D+08

 0,           7,28410D+09  0,           1,69844D+09  0,           5,75955D+07
 0,           2,80013D+05

-2,59703D+01 -3,18933D+01 -2,08897D+01 -6,30698D+00 -1,38012D+00 -1,54012D-01
-1,11493D-02 -3,01356D-04  3,75264D-12

 2,40482D+02  2,88308D+02  1,91080D+02  5,61685D+01  1,23733D+01  1,33742D+00
 9,67076D-02  2,50885D-03 -3,52675D-11  3,35179D-10

-1,02181D+03 -1,18333D+03 -8,00518D+02 -2,26665D+02 -5,06897D+01 -5,25181D+00
-3,82949D-01 -9,46430D-03  1,52036D-10 -1,46390D-09  6,49253D-09

 2,66497D+03  2,92774D+03  2,05475D+03  5,50348D+02  1,26997D+02  1,23945D+01
 9,28064D-01  2,15325D-02 -4,01051D-10  3,92004D-09 -1,77005D-08  4,92931D-08

-4,85300D+03 -4,89430D+03 -3,67752D+03 -9,01783D+02 -2,21760D+02 -1,97555D+01
-1,57248D+00 -3,32590D-02  7,31904D-10 -7,27591D-09  3,35338D-08 -9,57212D-08
 1,91386D-07

 6,74294D+03  5,85982D+03  5,02238D+03  1,05875D+03  2,96146D+02  2,26421D+01
 2,04956D+00  3,71992D-02 -9,97685D-10  1,00994D-08 -4,76023D-08  1,39675D-07
-2,88716D-07  4,52776D-07

-7,76472D+03 -5,10083D+03 -5,69966D+03 -9,06610D+02 -3,30475D+02 -1,90483D+01
-2,24992D+00 -3,07875D-02  1,08088D-09 -1,11388D-08  5,37072D-08 -1,62178D-07
 3,47390D-07 -5,68571D-07  7,49479D-07

 8,05973D+03  2,89713D+03  5,86196D+03  5,09622D+02  3,36702D+02  1,06010D+01
 2,27268D+00  1,69858D-02 -9,85353D-10  1,03128D-08 -5,07539D-08  1,57446D-07
-3,49196D-07  5,96841D-07 -8,28009D-07  9,67583D-07

-8,08426D+03  8,78772D-17 -5,86319D+03  1,66887D-17 -3,35827D+02  3,81423D-19
-2,26104D+00  6,75215D-22  7,93658D-10 -8,40179D-09  4,20113D-08 -1,33237D-07
 3,04543D-07 -5,41537D-07  7,89097D-07 -9,75834D-07  1,04513D-06

 8,05973D+03 -2,89713D+03  5,86196D+03 -5,09622D+02  3,36702D+02 -1,06010D+01
 2,27268D+00 -1,69858D-02 -5,82141D-10  6,20684D-09 -3,13651D-08  1,01040D-07
-2,36283D-07  4,33869D-07 -6,59665D-07  8,59375D-07 -9,75834D-07  9,67583D-07

-7,76472D+03  5,10083D+03 -5,69966D+03  9,06610D+02 -3,30475D+02  1,90483D+01
-2,24992D+00  3,07875D-02  3,90454D-10 -4,17988D-09  2,12539D-08 -6,91384D-08
 1,64166D-07 -3,08502D-07  4,84818D-07 -6,59665D-07  7,89097D-07 -8,28009D-07
 7,49479D-07

 6,74294D+03 -5,85982D+03  5,02238D+03 -1,05875D+03  2,96146D+02 -2,26421D+01
 2,04956D+00 -3,71992D-02 -2,34274D-10  2,51344D-09 -1,28245D-08  4,19526D-08
-1,00541D-07  1,91800D-07 -3,08502D-07  4,33869D-07 -5,41537D-07  5,96841D-07
-5,68571D-07  4,52776D-07

-4,85300D+03  4,89430D+03 -3,67752D+03  9,01783D+02 -2,21760D+02  1,97555D+01
-1,57248D+00  3,32590D-02  1,20302D-10 -1,29221D-09  6,60577D-09 -2,16768D-08
 5,22261D-08 -1,00541D-07  1,64166D-07 -2,36283D-07  3,04543D-07 -3,49196D-07
 3,47390D-07 -2,88716D-07  1,91386D-07
```

THE MATRIX B N=8

```
 2,66497D+03 -2,92774D+03  2,05475D+03 -5,50348D+02  1,26997D+02 -1,23945D+01
 9,28064D-01 -2,15325D-02 -4,96115D-11  5,33243D-10 -2,72877D-09  8,97004D-09
-2,16768D-08  4,19526D-08 -6,91384D-08  1,01040D-07 -1,33237D-07  1,57446D-07
-1,62178D-07  1,39675D-07 -9,57212D-08  4,92931D-08

-1,02181D+03  1,18333D+03 -8,00518D+02  2,26665D+02 -5,06897D+01  5,25181D+00
-3,82949D-01  9,46430D-03  1,50628D-11 -1,61960D-10  8,29293D-10 -2,72877D-09
 6,60577D-09 -1,28245D-08  2,12539D-08 -3,13651D-08  4,20113D-08 -5,07539D-08
 5,37072D-08 -4,76023D-08  3,35338D-08 -1,77005D-08  6,49253D-09

 2,40482D+02 -2,88308D+02  1,91080D+02 -5,61685D+01  1,23733D+01 -1,33742D+00
 9,67076D-02 -2,50885D-03 -2,94006D-12  3,16195D-11 -1,61960D-10  5,33243D-10
-1,29221D-09  2,51344D-09 -4,17988D-09  6,20684D-09 -8,40179D-09  1,03128D-08
-1,11388D-08  1,00994D-08 -7,27591D-09  3,92004D-09 -1,46390D-09  3,35179D-10

-2,59703D+01  3,18933D+01 -2,08897D+01  6,30698D+00 -1,38012D+00  1,54012D-01
-1,11493D-02  3,01356D-04  2,73336D-13 -2,94006D-12  1,50628D-11 -4,96115D-11
 1,20302D-10 -2,34274D-10  3,90454D-10 -5,82141D-10  7,93658D-10 -9,85353D-10
 1,08088D-09 -9,97685D-10  7,31904D-10 -4,01051D-10  1,52036D-10 -3,52675D-11
 3,75264D-12
```

N= 8 M= 17

```
 1,24151D+15

 0,           2,10332D+15

 8,68125D+14  0,           6,15427D+14

 0,           3,58263D+14  0,           6,24440D+13

 5,03743D+13  0,           3,65286D+13  0,           2,25400D+12

 0,           7,68135D+12  0,           1,39273D+12  0,           3,34283D+10

 3,66031D+11  0,           2,76065D+11  0,           1,83239D+10  0,
 1,75115D+08

 0,           1,37817D+10  0,           2,69027D+09  0,           7,59590D+07
 0,           3,06943D+05

-4,55852D+01 -6,15391D+01 -3,09701D+01 -1,00155D+01 -1,71663D+00 -1,99704D-01
-1,15855D-02 -3,18352D-04  3,76337D-12

 4,14050D+02  5,63497D+02  2,77862D+02  9,02607D+01  1,51041D+01  1,75546D+00
 9,88316D-02  2,69177D-03 -3,53830D-11  3,36421D-10

-1,71095D+03 -2,36223D+03 -1,13223D+03 -3,71662D+02 -6,02387D+01 -7,03878D+00
-3,81938D-01 -1,04023D-02  1,52628D-10 -1,47027D-09  6,52520D-09

 4,27659D+03  6,06094D+03  2,78612D+03  9,35043D+02  1,44872D+02  1,72328D+01
 8,90820D-01  2,46268D-02 -4,03004D-10  3,94106D-09 -1,78082D-08  4,96486D-08

-7,26942D+03 -1,08199D+04 -4,65710D+03 -1,63540D+03 -2,36622D+02 -2,93672D+01
-1,41508D+00 -4,07827D-02  7,36653D-10 -7,32702D-09  3,37959D-08 -9,65856D-08
 1,93488D-07
```

THE MATRIX B N=8

```
 8,97162D+03  1,46781D+04  5,65254D+03  2,17609D+03  2,81202D+02  3,82228D+01
 1,64391D+00  5,19237D-02 -1,00698D-09  1,01995D-08 -4,81152D-08  1,41367D-07
-2,92830D-07  4,60827D-07

-8,32782D+03 -1,64402D+04 -5,17154D+03 -2,40015D+03 -2,53102D+02 -4,14981D+01
-1,45624D+00 -5,55576D-02  1,09651D-09 -1,13071D-08  5,45700D-08 -1,65024D-07
 3,54310D-07 -5,82115D-07  7,72262D-07

 5,67682D+03  1,65791D+04  3,49031D+03  2,39774D+03  1,69125D+02  4,10866D+01
 9,64268D-01  5,45726D-02 -1,00908D-09  1,05682D-08 -5,20631D-08  1,61765D-07
-3,59697D-07  6,17392D-07 -8,62581D-07  1,02004D-06

-1,98480D+03 -1,62729D+04 -1,21501D+03 -2,34409D+03 -5,86248D+01 -4,00178D+01
-3,32979D-01 -5,29801D-02  8,27095D-10 -8,76166D-09  4,38567D-08 -1,39324D-07
 3,19344D-07 -5,70505D-07  8,37828D-07 -1,04978D-06  1,14936D-06

-1,98480D+03  1,62729D+04 -1,21501D+03  2,34409D+03 -5,86248D+01  4,00178D+01
-3,32979D-01  5,29801D-02 -6,26299D-10  6,68209D-09 -3,38021D-08  1,09079D-07
-2,55830D-07  4,72124D-07 -7,24019D-07  9.57028D-07 -1,11348D-06  1,14936D-06

 5,67682D+03 -1,65791D+04  3,49031D+03 -2,39774D+03  1,69125D+02 -4,10866D+01
 9,64268D-01 -5,45726D-02  4,44328D-10 -4,75969D-09  2,42271D-08 -7,89464D-08
 1,88014D-07 -3,55174D-07  5,63331D-07 -7,78803D-07  9,57028D-07 -1,04978D-06
 1,02004D-06

-8,32782D+03  1,64402D+04 -5,17154D+03  2,40015D+03 -2,53102D+02  4,14981D+01
-1,45624D+00  5,55576D-02 -2,92816D-10  3,14350D-09 -1,60554D-08  5,26104D-08
-1,26455D-07  2,42516D-07 -3,93818D-07  5,63331D-07 -7,24019D-07  8,37828D-07
-8,62581D-07  7,72262D-07

 8,97162D+03 -1,46781D+04  5,65254D+03 -2,17609D+03  2,81202D+02 -3,82228D+01
 1,64391D+00 -5,19237D-02  1,74064D-10 -1,87082D-09  9,57283D-09 -3,14644D-08
 7,60244D-08 -1,47116D-07  2,42516D-07 -3,55174D-07  4,72124D-07 -5,70505D-07
 6,17392D-07 -5,82115D-07  4,60827D-07

-7,26942D+03  1,08199D+04 -4,65710D+03  1,63540D+03 -2,36622D+02  2,93672D+01
-1,41508D+00  4,07827D-02 -8,89409D-11  9,56520D-10 -4,89932D-09  1,61301D-08
-3,90864D-08  7,60244D-08 -1,26455D-07  1,88014D-07 -2,55830D-07  3,19344D-07
-3,59697D-07  3,54310D-07 -2,92830D-07  1,93488D-07

 4,27659D+03 -6,06094D+03  2,78612D+03 -9,35043D+02  1,44872D+02 -1,72328D+01
 8,90820D-01 -2,46268D-02  3,65789D-11 -3,93524D-10  2,01675D-09 -6,64586D-09
 1,61301D-08 -3,14644D-08  5,26104D-08 -7,89464D-08  1,09079D-07 -1,39324D-07
 1,61765D-07 -1,65024D-07  1,41367D-07 -9,65856D-08  4,96486D-08

-1,71095D+03  2,36223D+03 -1,13223D+03  3,71662D+02 -6,02387D+01  7,03878D+00
-3,81938D-01  1,04023D-02 -1,10888D-11  1,19319D-10 -6,11680D-10  2,01675D-09
-4,89932D-09  9,57283D-09 -1,60554D-08  2,42271D-08 -3,38021D-08  4,38567D-08
-5,20631D-08  5,45700D-08 -4,81152D-08  3,37959D-08 -1,78082D-08  6,52520D-09

 4,14050D+02 -5,63497D+02  2,77862D+02 -9,02607D+01  1,51041D+01 -1,75546D+00
 9,88316D-02 -2,69177D-03  2,16241D-12 -2,32709D-11  1,19319D-10 -3,93524D-10
 9,56520D-10 -1,87082D-09  3,14350D-09 -4,75969D-09  6,68209D-09 -8,76166D-09
 1,05682D-08 -1,13071D-08  1.01995D-08 -7,32702D-09  3,94106D-09 -1,47027D-09
 3,36421D-10
```

THE MATRIX B   N=8

```
-4,55852D+01  6,15391D+01 -3,09701D+01  1,00155D+01 -1,71663D+00  1,99704D-01
-1,15855D-02  3,18352D-04 -2,00924D-13  2,16241D-12 -1,10888D-11  3,65789D-11
-8,89409D-11  1,74064D-10 -2,92816D-10  4,44328D-10 -6,26299D-10  8,27095D-10
-1,00908D-09  1,09651D-09 -1,00698D-09  7,36653D-10 -4,03004D-10  1,52628D-10
-3,53830D-11  3,76337D-12
```

N= 8   M= 18

```
 4,39229D+15

 0,           5,83748D+15

 2,59888D+15  0,           1,56161D+15

 0,           8,50105D+14  0,           1,26833D+14

 1,26878D+14  0,           7,81341D+13  0,           4,07323D+12

 0,           1,55316D+13  0,           2,41379D+12  0,           4,95099D+10

 7,73403D+11  0,           4,96378D+11  0,           2,79007D+10  0,
 2,25224D+08

 0,           2,37188D+10  0,           3,97415D+09  0,           9,60401D+07
 0,           3,31844D+05

-8,89454D+01 -9,71404D+01 -5,08187D+01 -1,34510D+01 -2,35285D+00 -2,27235D-01
-1,32172D-02 -3,06357D-04  3,76915D-12

 8,15038D+02  8,76229D+02  4,59618D+02  1,19441D+02  2,08668D+01  1,97031D+00
 1,13824D-01  2,56264D-03 -3,54452D-11  3,37091D-10

-3,41792D+03 -3,59553D+03 -1,89914D+03 -4,81626D+02 -8,44014D+01 -7,74781D+00
-4,46928D-01 -9,73983D-03  1,52947D-10 -1,47370D-09  6,54283D-09

 8,76508D+03  8,93283D+03  4,79119D+03  1,17412D+03  2,08200D+02  1,84123D+01
 1,07137D+00  2,24396D-02 -4,04058D-10  3,95240D-09 -1,78664D-08  4,98407D-08

-1,56100D+04 -1,51426D+04 -8,38660D+03 -1,95186D+03 -3,56426D+02 -2,98764D+01
-1,78758D+00 -3,54571D-02  7,39218D-10 -7,35464D-09  3,39376D-08 -9,70534D-08
 1,94627D-07

 2,10485D+04  1,88022D+04  1,11213D+04  2,37947D+03  4,63406D+02  3,56762D+01
 2,27647D+00  4,14739D-02 -1,01201D-09  1,02536D-08 -4,83932D-08  1,42285D-07
-2,95065D-07  4,65212D-07

-2,32866D+04 -1,79506D+04 -1,21349D+04 -2,23827D+03 -4,98216D+02 -3,30561D+01
-2,41280D+00 -3,78930D-02  1,10502D-09 -1,13987D-08  5,50399D-08 -1,66575D-07
 3,58088D-07 -5,89528D-07  7,84793D-07

 2,30049D+04  1,33873D+04  1,18805D+04  1,65345D+03  4,83465D+02  2,41999D+01
 2,32242D+00  2,75197D-02 -1,02211D-09  1,07084D-08 -5,27829D-08  1,64141D-07
-3,65483D-07  6,28745D-07 -8,81772D-07  1,04943D-06
```

THE MATRIX B N=8

```
-2,19897D+04 -6,98290D+03 -1,13067D+04 -8,58467D+02 -4,58188D+02 -1,25107D+01
-2,19265D+00 -1,41734D-02  8,45789D-10 -8,96290D-09  4,48892D-08 -1,42733D-07
 3,27644D-07 -5,86790D-07  8,65357D-07 -1,09194D-06  1,20984D-06

 2,15203D+04 -2,15703D-15  1,10513D+04 -2,84388D-16  4,47294D+02 -4,46886D-18
 2,13818D+00 -5,44691D-21 -6,51820D-10  6,95683D-09 -3,52117D-08  1,13733D-07
-2,67161D-07  4,94357D-07 -7,61603D-07  1,01459D-06 -1,19605D-06  1,26208D-06

-2,19897D+04  6,98290D+03 -1,13067D+04  8,58467D+02 -4,58188D+02  1,25107D+01
-2,19265D+00  1,41734D-02  4,77443D-10 -5,11619D-09  2,60561D-08 -8,49847D-08
 2,02717D-07 -3,84023D-07  6,12099D-07 -8,53491D-07  1,06416D-06 -1,19605D-06
 1,20984D-06

 2,30049D+04 -1,33873D+04  1,18805D+04 -1,65345D+03  4,83465D+02 -2,41999D+01
 2,32242D+00 -2,75197D-02 -3,32835D-10  3,57432D-09 -1,82657D-08  5,99075D-08
-1,44223D-07  2,77380D-07 -4,52753D-07  6,53588D-07 -8,53491D-07  1,01459D-06
-1,09194D-06  1,04943D-06

-2,32866D+04  1,79506D+04 -1,21349D+04  2,23827D+03 -4,98216D+02  3,30561D+01
-2,41280D+00  3,78930D-02  2,17329D-10 -2,33659D-09  1,19624D-08 -3,93534D-08
 9,52337D-08 -1,84808D-07  3,06232D-07 -4,52753D-07  6,12099D-07 -7,61603D-07
 8,65357D-07 -8,81772D-07  7,84793D-07

 2,10485D+04 -1,88022D+04  1,11213D+04 -2,37947D+03  4,63406D+02 -3,56762D+01
 2,27647D+00 -4,14739D-02 -1,28564D-10  1,38309D-09 -7,08779D-09  2,33551D-08
-5,66788D-08  1,10544D-07 -1,84808D-07  2,77380D-07 -3,84023D-07  4,94357D-07
-5,86790D-07  6,28745D-07 -5,89528D-07  4,65212D-07

-1,56100D+04  1,51426D+04 -8,38660D+03  1,95186D+03 -3,56426D+02  2,98764D+01
-1,78758D+00  3,54571D-02  6,55215D-11 -7,05108D-10  3,61530D-09 -1,19233D-08
 2,89804D-08 -5,66788D-08  9,52337D-08 -1,44223D-07  2,02717D-07 -2,67161D-07
 3,27644D-07 -3,65483D-07  3,58088D-07 -2,95065D-07  1,94627D-07

 8,76508D+03 -8,93283D+03  4,79119D+03 -1,17412D+03  2,08200D+02 -1,84123D+01
 1,07137D+00 -2,24396D-02 -2,69089D-11  2,89631D-10 -1,48545D-09  4,90141D-09
-1,19233D-08  2,33551D-08 -3,93534D-08  5,99075D-08 -8,49847D-08  1,13733D-07
-1,42733D-07  1,64141D-07 -1,66575D-07  1,42285D-07 -9,70534D-08  4,98407D-08

-3,41792D+03  3,59553D+03 -1,89914D+03  4,81626D+02 -8,44014D+01  7,74781D+00
-4,46928D-01  9,73983D-03  8,15076D-12 -8,77388D-11  4,50067D-10 -1,48545D-09
 3,61530D-09 -7,08779D-09  1,19624D-08 -1,82657D-08  2,60561D-08 -3,52117D-08
 4,48892D-08 -5,27829D-08  5,50399D-08 -4,83932D-08  3,39376D-08 -1,78664D-08
 6,54283D-09

 8,15038D+02 -8,76229D+02  4,59618D+02 -1,19441D+02  2,08668D+01 -1,97031D+00
 1,13824D-01 -2,56264D-03 -1,58871D-12  1,71027D-11 -8,77388D-11  2,89631D-10
-7,05108D-10  1,38309D-09 -2,33659D-09  3,57432D-09 -5,11619D-09  6,95683D-09
-8,96290D-09  1,07084D-08 -1,13987D-08  1,02536D-08 -7,35464D-09  3,95240D-09
-1,47370D-09  3,37091D-10

-8,89454D+01  9,71404D+01 -5,08187D+01  1,34510D+01 -2,35285D+00  2,27235D-01
-1,32172D-02  3,06357D-04  1,47574D-13 -1,58871D-12  8,15076D-12 -2,69089D-11
 6,55215D-11 -1,28564D-10  2,17329D-10 -3,32835D-10  4,77443D-10 -6,51820D-10
 8,45789D-10 -1,02211D-09  1,10502D-09 -1,01201D-09  7,39218D-10 -4,04058D-10
 1,52947D-10 -3,54452D-11  3,76915D-12
```

THE MATRIX B   N=8

N= 8   M= 19

```
1,31676D+16

0,           1,59664D+16

6,73649D+15  0,           3,50304D+15

0,           1,99075D+15  0,           2,54683D+14

2,83461D+14  0,           1,51222D+14  0,           6,80955D+12

0,           3,10475D+13  0,           4,14429D+12  0,           7,28119D+10

1,48678D+12  0,           8,27546D+11  0,           4,02283D+10  0,
2,80438D+08

0,           4,04429D+10  0,           5,82985D+09  0,           1,20896D+08
0,           3,58217D+05

-1,47811D+02 -1,64008D+02 -7,27322D+01 -1,93132D+01 -2,88887D+00 -2,76434D-01
-1,38905D-02 -3,15417D-04  3,77226D-12

 1,33864D+03  1,48722D+03  6,50158D+02  1,72363D+02  2,53373D+01  2,41072D+00
 1,18498D-01  2,66019D-03 -3,54787D-11  3,37452D-10

-5,52162D+03 -6,16087D+03 -2,64286D+03 -7,01587D+02 -1,00905D+02 -9,57842D+00
-4,58973D-01 -1,02404D-02  1,53119D-10 -1,47556D-09  6,55233D-09

 1,38149D+04  1,55743D+04  6,50748D+03  1,74051D+03  2,43205D+02  2,31932D+01
 1,07711D+00  2,40928D-02 -4,04625D-10  3,95851D-09 -1,78978D-08  4,99444D-08

-2,36573D+04 -2,72757D+04 -1,09591D+04 -2,99062D+03 -4,01078D+02 -3,89560D+01
-1,73427D+00 -3,94857D-02  7,40602D-10 -7,36954D-09  3,40141D-08 -9,73059D-08
 1,95242D-07

 2,98481D+04  3,60671D+04  1,36071D+04  3,88592D+03  4,88818D+02  4,96606D+01
 2,07305D+00  4,93894D-02 -1,01473D-09  1,02829D-08 -4,85434D-08  1,42781D-07
-2,96274D-07  4,67588D-07

-2,93212D+04 -3,89993D+04 -1,31899D+04 -4,14485D+03 -4,67207D+02 -5,22502D+01
-1,95466D+00 -5,13077D-02  1,10963D-09 -1,14483D-08  5,52945D-08 -1,67416D-07
 3,60137D-07 -5,93554D-07  7,91616D-07

 2,32323D+04  3,74917D+04  1,03581D+04  3,95077D+03  3,63722D+02  4,94039D+01
 1,50962D+00  4,81632D-02 -1,02920D-09  1,07848D-08 -5,31747D-08  1,65435D-07
-3,68636D-07  6,34941D-07 -8,92272D-07  1,06559D-06

-1,44392D+04 -3,47066D+04 -6,40802D+03 -3,64156D+03 -2,24029D+02 -4,53553D+01
-9,26133D-01 -4,40592D-02  8,56054D-10 -9,07343D-09  4,54564D-08 -1,44606D-07
 3,32209D-07 -5,95760D-07  8,80558D-07 -1,11533D-06  1,24370D-06

 4,85356D+03  3,28651D+04  2,14989D+03  3,44230D+03  7,50266D+01  4,28039D+01
 3,09656D-01  4,15209D-02 -6,66082D-10  7,11040D-09 -3,59997D-08  1,16335D-07
-2,73504D-07  5,06819D-07 -7,82722D-07  1,04709D-06 -1,24310D-06  1,32745D-06
```

THE MATRIX B   N=8

```
 4,85356D+03 -3,28651D+04  2,14989D+03 -3,44230D+03  7,50266D+01 -4,28039D+01
 3,09656D-01 -4,15209D-02  4,96574D-10 -5,32218D-09  2,71132D-08 -8,84759D-08
 2,11224D-07 -4,00739D-07  6,40427D-07 -8,97085D-07  1,12728D-06 -1,28373D-06
 1,32745D-06

-1,44392D+04  3,47066D+04 -6,40802D+03  3,64156D+03 -2,24029D+02  4,53553D+01
-9,26133D-01  4,40592D-02 -3,57422D-10  3,83905D-09 -1,96242D-08  6,43944D-08
-1,55157D-07  2,98863D-07 -4,89161D-07  7,09616D-07 -9,34602D-07  1,12728D-06
-1,24310D-06  1,24370D-06

 2,32323D+04 -3,74917D+04  1,03581D+04 -3,95077D+03  3,63722D+02 -4,94039D+01
 1,50962D+00 -4,81632D-02  2,46889D-10 -2,65487D-09  1,35958D-08 -4,47479D-08
 1,08379D-07 -2,10636D-07  3,50004D-07 -5,20113D-07  7,09616D-07 -8,97085D-07
 1,04709D-06 -1,11533D-06  1,06559D-06

-2,93212D+04  3,89993D+04 -1,31899D+04  4,14485D+03 -4,67207D+02  5,22502D+01
-1,95466D+00  5,13077D-02 -1,60434D-10  1,72624D-09 -8,84876D-09  2,91712D-08
-7,08512D-08  1,38390D-07 -2,32001D-07  3,50004D-07 -4,89161D-07  6,40427D-07
-7,82722D-07  8,80558D-07 -8,92272D-07  7,91616D-07

 2,98481D+04 -3,60671D+04  1,36071D+04 -3,88592D+03  4,88818D+02 -4,96606D+01
 2,07305D+00 -4,93894D-02  9,46659D-11 -1,01891D-09  5,22566D-09 -1,72420D-08
 4,19407D-08 -8,21438D-08  1,38390D-07 -2,10636D-07  2,98863D-07 -4,00739D-07
 5,06819D-07 -5,95760D-07  6,34941D-07 -5,93554D-07  4,67588D-07

-2,36573D+04  2,72757D+04 -1,09591D+04  2,99062D+03 -4,01078D+02  3,89560D+01
-1,73427D+00  3,94857D-02 -4,81799D-11  5,18661D-10 -2,66077D-09  8,78325D-09
-2,13824D-08  4,19407D-08 -7,08512D-08  1,08379D-07 -1,55157D-07  2,11224D-07
-2,73504D-07  3,32209D-07 -3,68636D-07  3,60137D-07 -2,96274D-07  1,95242D-07

 1,38149D+04 -1,55743D+04  6,50748D+03 -1,74051D+03  2,43205D+02 -2,31932D+01
 1,07711D+00 -2,40928D-02  1,97721D-11 -2,12869D-10  1,09220D-09 -3,60628D-09
 8,78325D-09 -1,72420D-08  2,91712D-08 -4,47479D-08  6,43944D-08 -8,84759D-08
 1,16335D-07 -1,44606D-07  1,65435D-07 -1,67416D-07  1,42781D-07 -9,73059D-08
 4,99444D-08

-5,52162D+03  6,16087D+03 -2,64286D+03  7,01587D+02 -1,00905D+02  9,57842D+00
-4,58973D-01  1,02404D-02 -5,98650D-12  6,44547D-11 -3,30736D-10  1,09220D-09
-2,66077D-09  5,22566D-09 -8,84876D-09  1,35958D-08 -1,96242D-08  2,71132D-08
-3,59997D-08  4,54564D-08 -5,31747D-08  5,52945D-08 -4,85434D-08  3,40141D-08
-1,78978D-08  6,55233D-09

 1,33864D+03 -1,48722D+03  6,50158D+02 -1,72363D+02  2,53373D+01 -2,41072D+00
 1,18498D-01 -2,66019D-03  1,16657D-12 -1,25605D-11  6,44547D-11 -2,12869D-10
 5,18661D-10 -1,01891D-09  1,72624D-09 -2,65487D-09  3,83905D-09 -5,32218D-09
 7,11040D-09 -9,07343D-09  1,07848D-08 -1,14483D-08  1,02829D-08 -7,36954D-09
 3,95851D-09 -1,47556D-09  3,37452D-10

-1,47811D+02  1,64008D+02 -7,27322D+01  1,93132D+01 -2,88887D+00  2,76434D-01
-1,38905D-02  3,15417D-04 -1,08344D-13  1,16657D-12 -5,98650D-12  1,97721D-11
-4,81799D-11  9,46659D-11 -1,60434D-10  2,46889D-10 -3,57422D-10  4,96574D-10
-6,66082D-10  8,56054D-10 -1,02920D-09  1,10963D-09 -1,01473D-09  7,40602D-10
-4,04625D-10  1,53119D-10 -3,54787D-11  3,77226D-12
```

N= 8   M= 20

THE MATRIX B   N=8

```
 3,84053D+16

 0,           3,90104D+16

 1,70393D+16  0,           7,69236D+15

 0,           4,23848D+15  0,           4,72931D+14

 6,20026D+14  0,           2,87503D+14  0,           1,12233D+13

 0,           5,74983D+13  0,           6,70039D+12  0,           1,02601D+11

 2,80854D+12  0,           1,36040D+12  0,           5,74077D+10  0,
 3,46994D+08

 0,           6,51151D+10  0,           8,20249D+09  0,           1,48410D+08
 0,           3,83502D+05

-2,55996D+02 -2,47542D+02 -1,08684D+02 -2,52847D+01 -3,71164D+00 -3,13163D-01
-1,53173D-02 -3,08907D-04  3,77393D-12

 2,32609D+03  2,22392D+03  9,74497D+02  2,23629D+02  3,26623D+01  2,70912D+00
 1,31284D-01  2,59009D-03 -3,54967D-11  3,37646D-10

-9,65522D+03 -9,09517D+03 -3,98559D+03 -8,99066D+02 -1,30940D+02 -1,06444D+01
-5,12653D-01 -9,88063D-03  1,53212D-10 -1,47655D-09  6,55745D-09

 2,44473D+04  2,25639D+04  9,93121D+03  2,19037D+03  3,19612D+02  2,53461D+01
 1,22036D+00  2,29044D-02 -4,04931D-10  3,96181D-09 -1,79147D-08  5,00002D-08

-4,28360D+04 -3,83693D+04 -1,71163D+04 -3,65745D+03 -5,39955D+02 -4,14291D+01
-2,01612D+00 -3,65885D-02  7,41348D-10 -7,37757D-09  3,40553D-08 -9,74421D-08
 1,95574D-07

 5,65221D+04  4,82841D+04  2,22375D+04  4,52673D+03  6,89404D+02  5,03641D+01
 2,52837D+00  4,36930D-02 -1,01620D-09  1,02987D-08 -4,86245D-08  1,43049D-07
-2,96927D-07  4,68871D-07

-6,06784D+04 -4,78016D+04 -2,35764D+04 -4,42363D+03 -7,21539D+02 -4,85835D+01
-2,61358D+00 -4,16407D-02  1,11212D-09 -1,14751D-08  5,54321D-08 -1,67870D-07
 3,61245D-07 -5,95732D-07  7,95312D-07

 5,74452D+04  3,90592D+04  2,21402D+04  3,58487D+03  6,72299D+02  3,90663D+01
 2,41770D+00  3,32494D-02 -1,03304D-09  1,08261D-08 -5,33869D-08  1,66136D-07
-3,70345D-07  6,38301D-07 -8,97974D-07  1,07439D-06

-5,18439D+04 -2,65871D+04 -1,98970D+04 -2,42966D+03 -6,01754D+02 -2,63713D+01
-2,15608D+00 -2,23644D-02  8,61640D-10 -9,13358D-09  4,57651D-08 -1,45626D-07
 3,34695D-07 -6,00648D-07  8,88853D-07 -1,12813D-06  1,26232D-06

 4,74680D+04  1,32719D+04  1,81820D+04  1,21042D+03  5,48868D+02  1,31134D+01
 1,96328D+00  1,11025D-02 -6,73913D-10  7,19472D-09 -3,64325D-08  1,17765D-07
-2,76989D-07  5,13672D-07 -7,94352D-07  1,06503D-06 -1,26920D-06  1,36404D-06
```

THE MATRIX B N=8

```
-4,58774D+04  3,96789D-15 -1,75628D+04  3,41620D-16 -5,29894D+02  3,43038D-18
-1,89450D+00  2,66029D-21  5,07263D-10 -5,43728D-09  2,77040D-08 -9,04275D-08
 2,15982D-07 -4,10093D-07  6,56302D-07 -9,21575D-07  1,16290D-06 -1,33368D-06
 1,39563D-06

 4,74680D+04 -1,32719D+04  1,81820D+04 -1,21042D+03  5,48868D+02 -1,31134D+01
 1,96328D+00 -1,11025D-02 -3,71623D-10  3,99197D-09 -2,04091D-08  6,69872D-08
-1,61477D-07  3,11290D-07 -5,10251D-07  7,42152D-07 -9,81931D-07  1,19364D-06
-1,33368D-06  1,36404D-06

-5,18439D+04  2,65871D+04 -1,98970D+04  2,42966D+03 -6,01754D+02  2,63713D+01
-2,15608D+00  2,23644D-02  2,65046D-10 -2,85038D-09  1,45992D-08 -4,80629D-08
 1,16460D-07 -2,26525D-07  3,76968D-07 -5,61712D-07  7,70129D-07 -9,81931D-07
 1,16290D-06 -1,26920D-06  1,26232D-06

 5,74452D+04 -3,90592D+04  2,21402D+04 -3,58487D+03  6,72299D+02 -3,90663D+01
 2,41770D+00 -3,32494D-02 -1,82203D-10  1,96065D-09 -1,00518D-08  3,31456D-08
-8,05397D-08  1,57440D-07 -2,64328D-07  3,99878D-07 -5,61712D-07  7,42152D-07
-9,21575D-07  1,06503D-06 -1,12813D-06  1,07439D-06

-6,06784D+04  4,78016D+04 -2,35764D+04  4,42363D+03 -7,21539D+02  4,85835D+01
-2,61358D+00  4,16407D-02  1,18101D-10 -1,27127D-09  6,52084D-09 -2,15207D-08
 5,23709D-08 -1,02652D-07  1,73193D-07 -2,64328D-07  3,76968D-07 -5,10251D-07
 6,56302D-07 -7,94352D-07  8,88853D-07 -8,97974D-07  7,95312D-07

 5,65221D+04 -4,82841D+04  2,22375D+04 -4,52673D+03  6,89404D+02 -5,03641D+01
 2,52837D+00 -4,36930D-02 -6,95939D-11  7,49250D-10 -3,84424D-09  1,26929D-08
-3,09130D-08  6,06801D-08 -1,02652D-07  1,57440D-07 -2,26525D-07  3,11290D-07
-4,10093D-07  5,13672D-07 -6,00648D-07  6,38301D-07 -5,95732D-07  4,68871D-07

-4,28360D+04  3,83693D+04 -1,71163D+04  3,65745D+03 -5,39955D+02  4,14291D+01
-2,01612D+00  3,65885D-02  3,53943D-11 -3,81091D-10  1,95558D-09 -6,45847D-09
 1,57361D-08 -3,09130D-08  5,23709D-08 -8,05397D-08  1,16460D-07 -1,61477D-07
 2,15982D-07 -2,76989D-07  3,34695D-07 -3,70345D-07  3,61245D-07 -2,96927D-07
 1,95574D-07

 2,44473D+04 -2,25639D+04  9,93121D+03 -2,19037D+03  3,19612D+02 -2,53461D+01
 1,22036D+00 -2,29044D-02 -1,45195D-11  1,56339D-10 -8,02322D-10  2,65009D-09
-6,45847D-09  1,26929D-08 -2,15207D-08  3,31456D-08 -4,80629D-08  6,69872D-08
-9,04275D-08  1,17765D-07 -1,45626D-07  1,66136D-07 -1,67870D-07  1,43049D-07
-9,74421D-08  5,00002D-08

-9,65522D+03  9,09517D+03 -3,98559D+03  8,99066D+02 -1,30940D+02  1,06444D+01
-5,12653D-01  9,88063D-03  4,39515D-12 -4,73262D-11  2,42886D-10 -8,02322D-10
 1,95558D-09 -3,84424D-09  6,52084D-09 -1,00518D-08  1,45992D-08 -2,04091D-08
 2,77040D-08 -3,64325D-08  4,57651D-08 -5,33869D-08  5,54321D-08 -4,86245D-08
 3,40553D-08 -1,79147D-08  6,55745D-09

 2,32609D+03 -2,22392D+03  9,74497D+02 -2,23629D+02  3,26623D+01 -2,70912D+00
 1,31284D-01 -2,59009D-03 -8,56355D-13  9,22124D-12 -4,73262D-11  1,56339D-10
-3,81091D-10  7,49250D-10 -1,27127D-09  1,96065D-09 -2,85038D-09  3,99197D-09
-5,43728D-09  7,19472D-09 -9,13358D-09  1,08261D-08 -1,14751D-08  1,02987D-08
-7,37757D-09  3,96181D-09 -1,47655D-09  3,37646D-10
```

THE MATRIX B   N=8

```
-2.55996D+02  2.47542D+02 -1.08684D+02  2.52847D+01 -3.71164D+00  3.13163D-01
-1.53173D-02  3.08907D-04  7.95269D-14 -8.56355D-13  4.39515D-12 -1.45195D-11
 3.53943D-11 -6.95939D-11  1.18101D-10 -1.82203D-10  2.65046D-10 -3.71623D-10
 5.07263D-10 -6.73913D-10  8.61640D-10 -1.03304D-09  1.11212D-09 -1.01620D-09
 7.41348D-10 -4.04931D-10  1.53212D-10 -3.54967D-11  3.77393D-12
```

Table 3. Lagrangians

$n = 2, 3, \ldots, 10.$
$m = n - 1.$

```
            N=  2           M=  1           P=    .5


      I=  0                                H=   -.5

0      1,0000000000000000000D+00            5,0000000000000000000D-01       0
1     -1,0000000000000000000D+00           -1,0000000000000000000D+00       1


      I=  1                                H=    .5

0      0,                                   5,0000000000000000000D-01       0
1      1,0000000000000000000D+00            1,0000000000000000000D+00       1



            N=  3           M=  2           P=   1,0


      I=  0                                H=  -1,0

0      1,0000000000000000000D+00            0,                              0
1     -1,5000000000000000000D+00           -5,0000000000000000000D-01       1
2      5,0000000000000000000D-01            5,0000000000000000000D-01       2


      I=  1                                H=   0,0

0      0,                                   1,0000000000000000000D+00       0
1      2,0000000000000000000D+00            0,                              1
2     -1,0000000000000000000D+00           -1,0000000000000000000D+00       2


      I=  2                                H=   1,0

0      0,                                   0,                              0
1     -5,0000000000000000000D-01            5,0000000000000000000D-01       1
2      5,0000000000000000000D-01            5,0000000000000000000D-01       2
```

N= 4 M= 3 P= 1,5

	I= 0	H= -1,5	
0	1,000000000000000000D+00	-6,250000000000000000D-02	0
1	-1,833333333333333333D+00	4,166666666666666667D-02	1
2	1,000000000000000000D+00	2,500000000000000000D-01	2
3	-1,666666666666666667D-01	-1,666666666666666667D-01	3

	I= 1	H= -,5	
0	0,	5,625000000000000000D-01	0
1	3,000000000000000000D+00	-1,125000000000000000D+00	1
2	-2,500000000000000000D+00	-2,500000000000000000D-01	2
3	5,000000000000000000D-01	5,000000000000000000D-01	3

	I= 2	H= ,5	
0	0,	5,625000000000000000D-01	0
1	-1,500000000000000000D+00	1,125000000000000000D+00	1
2	2,000000000000000000D+00	-2,500000000000000000D-01	2
3	-5,000000000000000000D-01	-5,000000000000000000D-01	3

	I= 3	H= 1,5	
0	0,	-6,250000000000000000D-02	0
1	3,333333333333333333D-01	-4,166666666666666667D-02	1
2	-5,000000000000000000D-01	2,500000000000000000D-01	2
3	1,666666666666666667D-01	1,666666666666666667D-01	3

N= 5 M= 4 P= 2,0

	I= 0	H= -2,0	
0	1,000000000000000000D+00	0,	0
1	-2,083333333333333333D+00	8,333333333333333333D-02	1
2	1,458333333333333333D+00	-4,166666666666666667D-02	2
3	-4,166666666666666667D-01	-8,333333333333333333D-02	3

4	4,16666666666666667D-02	4,16666666666666667D-02	4

	I= 1	H= -1,0	
0	0,	0,	0
1	4,00000000000000000D+00	-6,66666666666666667D-01	1
2	-4,33333333333333333D+00	6,66666666666666667D-01	2
3	1,50000000000000000D+00	1,66666666666666667D-01	3
4	-1,66666666666666667D-01	-1,66666666666666667D-01	4

	I= 2	H= 0,0	
0	0,	1,00000000000000000D+00	0
1	-3,00000000000000000D+00	0,	1
2	4,75000000000000000D+00	-1,25000000000000000D+00	2
3	-2,00000000000000000D+00	0,	3
4	2,50000000000000000D-01	2,50000000000000000D-01	4

	I= 3	H= 1,0	
0	0,	0,	0
1	1,33333333333333333D+00	6,66666666666666667D-01	1
2	-2,33333333333333333D+00	6,66666666666666667D-01	2
3	1,16666666666666667D+00	-1,66666666666666667D-01	3
4	-1,66666666666666667D-01	-1,66666666666666667D-01	4

	I= 4	H= 2,0	
0	0,	0,	0
1	-2,50000000000000000D-01	-8,33333333333333333D-02	1
2	4,58333333333333333D-01	-4,16666666666666667D-02	2
3	-2,50000000000000000D-01	8,33333333333333333D-02	3
4	4,16666666666666667D-02	4,16666666666666667D-02	4

N= 6 M= 5 P= 2,5

	I= 0	H= -2,5	
0	1,00000000000000000D+00	1,17187500000000000D-02	0
1	-2,28333333333333333D+00	-4,68750000000000000D-03	1
2	1,87500000000000000D+00	-5,20833333333333333D-02	2

3	-7,08333333333333333D-01	2,08333333333333333D-02	3
4	1,25000000000000000D-01	2,08333333333333333D-02	4
5	-8,33333333333333333D-03	-8,33333333333333333D-03	5

I= 1 H= -1,5

0	0.	-9,76562500000000000D-02	0
1	5,00000000000000000D+00	6,51041666666666667D-02	1
2	-6,41666666666666667D+00	4,06250000000000000D-01	2
3	2,95833333333333333D+00	-2,70833333333333333D-01	3
4	-5,83333333333333333D-01	-6,25000000000000000D-02	4
5	4,16666666666666667D-02	4,16666666666666667D-02	5

I= 2 H= -,5

0	0.	5,85937500000000000D-01	0
1	-5,00000000000000000D+00	-1,17187500000000000D+00	1
2	8,91666666666666667D+00	-3,54166666666666667D-01	2
3	-4,91666666666666667D+00	7,08333333333333333D-01	3
4	1,08333333333333333D+00	4,16666666666666667D-02	4
5	-8,33333333333333333D-02	-8,33333333333333333D-02	5

I= 3 H= ,5

0	0.	5,85937500000000000D-01	0
1	3,33333333333333333D+00	1,17187500000000000D+00	1
2	-6,50000000000000000D+00	-3,54166666666666667D-01	2
3	4,08333333333333333D+00	-7,08333333333333333D-01	3
4	-1,00000000000000000D+00	4,16666666666666667D-02	4
5	8,33333333333333333D-02	8,33333333333333333D-02	5

I= 4 H= 1,5

0	0.	-9,76562500000000000D-02	0
1	-1,25000000000000000D+00	-6,51041666666666667D-02	1
2	2,54166666666666667D+00	4,06250000000000000D-01	2
3	-1,70833333333333333D+00	2,70833333333333333D-01	3
4	4,58333333333333333D-01	-6,25000000000000000D-02	4
5	-4,16666666666666667D-02	-4,16666666666666667D-02	5

I= 5 H= 2,5

0	0.	1,17187500000000000D-02	0
1	2,00000000000000000D-01	4,68750000000000000D-03	1
2	-4,16666666666666667D-01	-5,20833333333333333D-02	2
3	2,91666666666666667D-01	-2,08333333333333333D-02	3
4	-8,33333333333333333D-02	2,08333333333333333D-02	4
5	8,33333333333333333D-03	8,33333333333333333D-03	5

N= 7 M= 6 P= 3,0

	I= 0	H= -3,0	
0	1,000000000000000000D+00	0,	0
1	-2,450000000000000000D+00	-1,666666666666666667D-02	1
2	2,255555555555555556D+00	5,555555555555555556D-03	2
3	-1,020833333333333333D+00	2,083333333333333333D-02	3
4	2,430555555555555556D-01	-6,944444444444444444D-03	4
5	-2,916666666666666667D-02	-4,166666666666666667D-03	5
6	1,388888888888888889D-03	1,388888888888888889D-03	6

	I= 1	H= -2,0	
0	0,	0,	0
1	6,000000000000000000D+00	1,500000000000000000D-01	1
2	-8,700000000000000000D+00	-7,500000000000000000D-02	2
3	4,833333333333333333D+00	-1,666666666666666667D-01	3
4	-1,291666666666666667D+00	8,333333333333333333D-02	4
5	1,666666666666666667D-01	1,666666666666666667D-02	5
6	-8,333333333333333333D-03	-8,333333333333333333D-03	6

	I= 2	H= -1,0	
0	0,	0,	0
1	-7,500000000000000000D+00	-7,500000000000000000D-01	1
2	1,462500000000000000D+01	7,500000000000000000D-01	2
3	-9,604166666666666667D+00	2,708333333333333333D-01	3
4	2,854166666666666667D+00	-2,708333333333333333D-01	4
5	-3,958333333333333333D-01	-2,083333333333333333D-02	5
6	2,083333333333333333D-02	2,083333333333333333D-02	6

	I= 3	H= 0,0	
0	0,	1,000000000000000000D+00	0
1	6,666666666666666667D+00	0,	1
2	-1,411111111111111111D+01	-1,361111111111111111D+00	2
3	1,033333333333333333D+01	0,	3
4	-3,361111111111111111D+00	3,888888888888888889D-01	4
5	5,000000000000000000D-01	0,	5
6	-2,777777777777777778D-02	-2,777777777777777778D-02	6

I= 4 H= 1.0

0	0.	0.	0
1	-3.750000000000000000D+00	7.500000000000000000D-01	1
2	8.250000000000000000D+00	7.500000000000000000D-01	2
3	-6.395833333333333333D+00	-2.708333333333333333D-01	3
4	2.229166666666666667D+00	-2.708333333333333333D-01	4
5	-3.541666666666666667D-01	2.083333333333333333D-02	5
6	2.083333333333333333D-02	2.083333333333333333D-02	6

I= 5 H= 2.0

0	0.	0.	0
1	1.200000000000000000D+00	-1.500000000000000000D-01	1
2	-2.700000000000000000D+00	-7.500000000000000000D-02	2
3	2.166666666666666667D+00	1.666666666666666667D-01	3
4	-7.916666666666666667D-01	8.333333333333333333D-02	4
5	1.333333333333333333D-01	-1.666666666666666667D-02	5
6	-8.333333333333333333D-03	-8.333333333333333333D-03	6

I= 6 H= 3.0

0	0.	0.	0
1	-1.666666666666666667D-01	1.666666666666666667D-02	1
2	3.805555555555555556D-01	5.555555555555555556D-03	2
3	-3.125000000000000000D-01	-2.083333333333333333D-02	3
4	1.180555555555555556D-01	-6.944444444444444444D-03	4
5	-2.083333333333333333D-02	4.166666666666666667D-03	5
6	1.388888888888888889D-03	1.388888888888888869D-03	6

N= 8 M= 7 P= 3.5

I= 0 H= -3.5

0	1.000000000000000000D+00	-2.441406250000000000D-03	0
1	-2.592857142857142857D+00	6.975446428571428571D-04	1
2	2.605555555555555556D+00	1.124131944444444444D-02	2
3	-1.343055555555555556D+00	-3.211805555555555556D-03	3
4	3.888888888888888889D-01	-6.076388888888888889D-03	4
5	-6.388888888888888889D-02	1.736111111111111111D-03	5
6	5.555555555555555556D-03	6.944444444444444444D-04	6
7	-1.984126984126984127D-04	-1.984126984126984127D-04	7

	I= 1	H= -2,5	
0	0,	2,39257812500000000D-02	0
1	7,00000000000000000D+00	-9,57031250000000000D-03	1
2	-1,11500000000000000D+01	-1,08289930555555556D-01	2
3	7,08888888888888889D+00	4,33159722222222222D-02	3
4	-2,31250000000000000D+00	5,12152777777777778D-02	4
5	4,09722222222222222D-01	-2,04861111111111111D-02	5
6	-3,75000000000000000D-02	-3,47222222222222222D-03	6
7	1,38888888888888889D-03	1,38888888888888889D-03	7

	I= 2	H= -1,5	
0	0,	-1,19628906250000000D-01	0
1	-1,05000000000000000D+01	7,97526041666666667D-02	1
2	2,19750000000000000D+01	5,07421875000000000D-01	2
3	-1,63708333333333333D+01	-3,38281250000000000D-01	3
4	5,91666666666666667D+00	-1,17187500000000000D-01	4
5	-1,12500000000000000D+00	7,81250000000000000D-02	5
6	1,08333333333333333D-01	6,25000000000000000D-03	6
7	-4,16666666666666667D-03	-4,16666666666666667D-03	7

	I= 3	H= -,5	
0	0,	5,98144531250000000D-01	0
1	1,16666666666666667D+01	-1,19628906250000000D+00	1
2	-2,63611111111111111D+01	-4,10373263888888889D-01	2
3	2,16111111111111111D+01	8,20746527777777778D-01	3
4	-8,46527777777777778D+00	7,20486111111111111D-02	4
5	1,71527777777777778D+00	-1,44097222222222222D-01	5
6	-1,73611111111111111D-01	-3,47222222222222222D-03	6
7	6,94444444444444444D-03	6,94444444444444444D-03	7

	I= 4	H= ,5	
0	0,	5,98144531250000000D-01	0
1	-8,75000000000000000D+00	1,19628906250000000D+00	1
2	2,05000000000000000D+01	-4,10373263888888889D-01	2
3	-1,76736111111111111D+01	-8,20746527777777778D-01	3
4	7,33333333333333333D+00	7,20486111111111111D-02	4
5	-1,56944444444444444D+00	1,44097222222222222D-01	5
6	1,66666666666666667D-01	-3,47222222222222222D-03	6
7	-6,94444444444444444D-03	-6,94444444444444444D-03	7

	I= 5	H= 1,5	
0	0,	-1,19628906250000000D-01	0
1	4,20000000000000000D+00	-7,97526041666666667D-02	1
2	-1,00500000000000000D+01	5,07421875000000000D-01	2
3	8,93333333333333333D+00	3,38281250000000000D-01	3
4	-3,85416666666666667D+00	-1,17187500000000000D-01	4
5	8,62500000000000000D-01	-7,81250000000000000D-02	5

6	-9,583333333333333333D-02	6,250000000000000000D-03	6
7	4,166666666666666667D-03	4,166666666666666667D-03	7

I= 6 H= 2,5

0	0,	2,392578125000000000D-02	0
1	-1,166666666666666667D+00	9,570312500000000000D-03	1
2	2,830555555555555556D+00	-1,082899305555555556D-01	2
3	-2,568055555555555556D+00	-4,331597222222222222D-02	3
4	1,138888888888888889D+00	5,121527777777777778D-02	4
5	-2,638888888888888889D-01	2,048611111111111111D-02	5
6	3,055555555555555556D-02	-3,472222222222222222D-03	6
7	-1,388888888888888889D-03	-1,388888888888888889D-03	7

I= 7 H= 3,5

0	0,	-2,441406250000000000D-03	0
1	1,428571428571428571D-01	-6,975446428571428571D-04	1
2	-3,500000000000000000D-01	1,124131944444444444D-02	2
3	3,222222222222222222D-01	3,211805555555555556D-03	3
4	-1,458333333333333333D-01	-6,076388888888888889D-03	4
5	3,472222222222222222D-02	-1,736111111111111111D-03	5
6	-4,166666666666666667D-03	6,944444444444444444D-04	6
7	1,984126984126984127D-04	1,984126984126984127D-04	7

N= 9 M= 8 P= 4,0

I= 0 H= -4,0

0	1,000000000000000000D+00	0,	0
1	-2,717857142857142857D+00	3,571428571428571429D-03	1
2	2,929662698412698413D+00	-8,928571428571428571D-04	2
3	-1,668750000000000000D+00	-4,861111111111111111D-03	3
4	5,567708333333333333D-01	1,215277777777777778D-03	4
5	-1,125000000000000000D-01	1,388888888888888889D-03	5
6	1,354166666666666667D-02	-3,472222222222222222D-04	6
7	-8,928571428571428571D-04	-9,920634920634920635D-05	7
8	2,480158730158730159D-05	2,480158730158730159D-05	8

I= 1 H= -3,0

0	0,	0,	0
1	8,000000000000000000D+00	-3,809523809523809524D-02	1

2	-1.374285714285714286D+01	1,269841269841269841D-02	2
3	9,694444444444444444D+00	5,000000000000000000D-02	3
4	-3,655555555555555556D+00	-1,666666666666666667D-02	4
5	7,986111111111111111D-01	-1,250000000000000000D-02	5
6	-1,013888888888888889D-01	4,166666666666666667D-03	6
7	6,944444444444444444D-03	5,952380952380952381D-04	7
8	-1,984126984126984127D-04	-1,984126984126984127D-04	8

I= 2 H= -2,0

0	0.	0.	0
1	-1.400000000000000000D+01	2,000000000000000000D-01	1
2	3,105000000000000000D+01	-1,000000000000000000D-01	2
3	-2,549027777777777778D+01	-2,347222222222222222D-01	3
4	1,061736111111111111D+01	1,173611111111111111D-01	4
5	-2,486111111111111111D+00	3,611111111111111111D-02	5
6	3,319444444444444444D-01	-1,805555555555555556D-02	6
7	-2,361111111111111111D-02	-1,388888888888888889D-03	7
8	6,944444444444444444D-04	6,944444444444444444D-04	8

I= 3 H= -1,0

0	0.	0.	0
1	1,866666666666666667D+01	-8,000000000000000000D-01	1
2	-4,451111111111111111D+01	8,000000000000000000D-01	2
3	3,985000000000000000D+01	3,388888888888888889D-01	3
4	-1,786666666666666667D+01	-3,388888888888888889D-01	4
5	4,437500000000000000D+00	-4,027777777777777778D-02	5
6	-6,208333333333333333D-01	4,027777777777777778D-02	6
7	4,583333333333333333D-02	1,388888888888888889D-03	7
8	-1,388888888888888889D-03	-1,388888888888888889D-03	8

I= 4 H= 0,0

0	0.	1,000000000000000000D+00	0
1	-1,750000000000000000D+01	0.	1
2	4.318750000000000000D+01	-1,423611111111111111D+00	2
3	-4,047222222222222222D+01	0.	3
4	1,908506944444444444D+01	4,739583333333333333D-01	4
5	-4.972222222222222222D+00	0.	5
6	7,256944444444444444D-01	-5,208333333333333333D-02	6
7	-5,555555555555555556D-02	0.	7
8	1,736111111111111111D-03	1,736111111111111111D-03	8

I= 5 H= 1,0

0	0.	0.	0
1	1,120000000000000000D+01	8,000000000000000000D-01	1
2	-2.820000000000000000D+01	8,000000000000000000D-01	2
3	2,717222222222222222D+01	-3,388888888888888889D-01	3
4	-1,325555555555555556D+01	-3,388888888888888889D-01	4
5	3,584722222222222222D+00	4,027777777777777778D-02	5

6	-5,430555555555555556D-01	4,027777777777777778D-02	6
7	4,305555555555555556D-02	-1,388888888888888889D-03	7
8	-1,388888888888888889D-03	-1,388888888888888889D-03	8

I= 6 H= 2,0

0	0,	0,	0
1	-4,666666666666666667D+00	-2,000000000000000000D-01	1
2	1,190555555555555556D+01	-1,000000000000000000D-01	2
3	-1,168750000000000000D+01	2,347222222222222222D-01	3
4	5,839583333333333333D+00	1,173611111111111111D-01	4
5	-1,625000000000000000D+00	-3,611111111111111111D-02	5
6	2,541666666666666667D-01	-1,805555555555555556D-02	6
7	-2,083333333333333333D-02	1,388888888888888889D-03	7
8	6,944444444444444444D-04	6,944444444444444444D-04	8

I= 7 H= 3,0

0	0,	0,	0
1	1,142857142857142857D+00	3,809523809523809524D-02	1
2	-2,942857142857142857D+00	1,269841269841269841D-02	2
3	2,927777777777777778D+00	-5,000000000000000000D-02	3
4	-1,488888888888888889D+00	-1,666666666666666667D-02	4
5	4,236111111111111111D-01	1,250000000000000000D-02	5
6	-6,805555555555555556D-02	4,166666666666666667D-03	6
7	5,753968253968253968D-03	-5,952380952380952381D-04	7
8	-1,984126984126984127D-04	-1,984126984126984127D-04	8

I= 8 H= 4,0

0	0,	0,	0
1	-1,250000000000000000D-01	-3,571428571428571429D-03	1
2	3,241071428571428571D-01	-8,928571428571428571D-04	2
3	-3,256944444444444444D-01	4,861111111111111111D-03	3
4	1,678819444444444444D-01	1,215277777777777778D-03	4
5	-4,861111111111111111D-02	-1,388888888888888889D-03	5
6	7,986111111111111111D-03	-3,472222222222222222D-04	6
7	-6,944444444444444444D-04	9,920634920634920635D-05	7
8	2,480158730158730159D-05	2,480158730158730159D-05	8

N= 10 M= 9 P= 4,5

I= 0 H= -4,5

Lagrangians

0	1,000000000000000000D+00	5,340576171875000000D-04	0
1	-2,828968253968253968D+00	-1,186794704861111111D-04	1
2	3,231646825396825397D+00	-2,502635168650793651D-03	2
3	-1.994268077601410935D+00	5,561411485890652557D-04	3
4	7,421875000000000000D-01	1,529947916666666667D-03	4
5	-1,743634259259259259D-01	-3,399884259259259259D-04	5
6	2,604166666666666667D-02	-2,604166666666666667D-04	6
7	-2,397486772486772487D-03	5,787037037037037037D-05	7
8	1,240079365079365079D-04	1,240079365079365079D-05	8
9	-2,755731922398589065D-06	-2,755731922398589065D-06	9

I= 1 H= -3,5

0	0,	-6,179809570312500000D-03	0
1	9,000000000000000000D+00	1,765659877232142857D-03	1
2	-1.646071428571428571D+01	2,875976562500000000D-02	2
3	1.262410714285714286D+01	-8,217075892857142857D-03	3
4	-5,324305555555555556D+00	-1,678602430555555556D-02	4
5	1,355381944444444444D+00	4,796006944444444444D-03	5
6	-2,138888888888888889D-01	2,517361111111111111D-03	6
7	2,048611111111111111D-02	-7,192460317460317460D-04	7
8	-1,091269841269841270D-03	-6,805555555555555556D-05	8
9	2,480158730158730159D-05	2,480158730158730159D-05	9

I= 2 H= -2,5

0	0,	3,460693359375000000D-02	0
1	-1,800000000000000000D+01	-1,384277343750000000D-02	1
2	4,192142857142857143D+01	-1,583426339285714286D-01	2
3	-3,720892857142857143D+01	6,333705357142857143D-02	3
4	1,729236111111111111D+01	8,181423611111111111D-02	4
5	-4,713194444444444444D+00	-3,272569444444444444D-02	5
6	7,819444444444444444D-01	-8,680555555555555556D-03	6
7	-7,777777777777777778D-02	3,472222222222222222D-03	7
8	4,265873015873015873D-03	2,480158730158730159D-04	8
9	-9,920634920634920635D-05	-9,920634920634920635D-05	9

I= 3 H= -1,5

0	0,	-1,345825195312500000D-01	0
1	2,800000000000000000D+01	8,972167968750000000D-02	1
2	-6,987777777777777778D+01	5,774956597222222222D-01	2
3	6,719351851851851852D+01	-3,849971064814814815D-01	3
4	-3,344166666666666667D+01	-1,600260416666666667D-01	4
5	9,634027777777777778D+00	1,066840277777777778D-01	5
6	-1,670833333333333333D+00	1,354166666666666667D-02	6
7	1,722222222222222222D-01	-9,027777777777777778D-03	7
8	-9.722222222222222222D-03	-3,472222222222222222D-04	8
9	2,314814814814814815D-04	2,314814814814814815D-04	9

I= 4 H= -,5

0	0.	6.056213378906250000D-01	0
1	-3.150000000000000000D+01	-1.211242675781250000D+00	1
2	8.123750000000000000D+01	-4.454101562500000000D-01	2
3	-8.148750000000000000D+01	8.908203125000000000D-01	3
4	4.244756944444444444D+01	9.346788194444444444D-02	4
5	-1.276701388888888889D+01	-1.869357638888888889D-01	5
6	2.300694444444444444D+00	-7.118055555555555556D-03	6
7	-2.451388888888888889D-01	1.423611111111111111D-02	7
8	1.423611111111111111D-02	1.736111111111111111D-04	8
9	-3.472222222222222222D-04	-3.472222222222222222D-04	9

I= 5 H= .5

0	0.	6.056213378906250000D-01	0
1	2.520000000000000000D+01	1.211242675781250000D+00	1
2	-6.625000000000000000D+01	-4.454101562500000000D-01	2
3	6.818750000000000000D+01	-8.908203125000000000D-01	3
4	-3.661805555555555556D+01	9.346788194444444444D-02	4
5	1.137951388888888889D+01	1.869357638888888889D-01	5
6	-2.118055555555555556D+00	-7.118055555555555556D-03	6
7	2.326388888888888889D-01	-1.423611111111111111D-02	7
8	-1.388888888888888889D-02	1.736111111111111111D-04	8
9	3.472222222222222222D-04	3.472222222222222222D-04	9

I= 6 H= 1.5

0	0.	-1.345825195312500000D-01	0
1	-1.400000000000000000D+01	-8.972167968750000000D-02	1
2	3.727222222222222222D+01	5.774956597222222222D-01	2
3	-3.903101851851851852D+01	3.849971064814814815D-01	3
4	2.141458333333333333D+01	-1.600260416666666667D-01	4
5	-6.821527777777777778D+00	-1.066840277777777778D-01	5
6	1.304166666666666667D+00	1.354166666666666667D-02	6
7	-1.472222222222222222D-01	9.027777777777777778D-03	7
8	9.027777777777777778D-03	-3.472222222222222222D-04	8
9	-2.314814814814814815D-04	-2.314814814814814815D-04	9

I= 7 H= 2.5

0	0.	3.460693359375000000D-02	0
1	5.142857142857142857D+00	1.384277343750000000D-02	1
2	-1.381428571428571429D+01	-1.583426339285714286D-01	2
3	1.464642857142857143D+01	-6.333705357142857143D-02	3
4	-8.163888888888888889D+00	8.181423611111111111D-02	4
5	2.650694444444444444D+00	3.272569444444444444D-02	5
6	-5.180555555555555556D-01	-6.680555555555555556D-03	6
7	5.992063492063492063D-02	-3.472222222222222222D-03	7
8	-3.769841269841269841D-03	2.480158730158730159D-04	8
9	9.920634920634920635D-05	9.920634920634920635D-05	9

I= 8

0	0,	-6,179809570312500000D-03	0
1	-1,125000000000000000D+00	-1,765659877232142857D-03	1
2	3,041964285714285714D+00	2,875976562500000000D-02	2
3	-3,255357142857142857D+00	8,217075892857142857D-03	3
4	1,836631944444444444D+00	-1,678602430555555556D-02	4
5	-6,053819444444444444D-01	-4,796006944444444444D-03	5
6	1.204861111111111111D-01	2,517361111111111111D-03	6
7	-1.423611111111111111D-02	7,192460317460317460D-04	7
8	9,176587301587301587D-04	-8,680555555555555556D-05	8
9	-2,480158730158730159D-05	-2,480158730158730159D-05	9

	I= 9	H= 4,5	
0	0,	5,340576171875000000D-04	0
1	1,111111111111111111D-01	1,186794704861111111D-04	1
2	-3,019841269841269841D-01	-2,502635168650793651D-03	2
3	3,255180776014109347D-01	-5,561411485890652557D-04	3
4	-1,854166666666666667D-01	1,529947916666666667D-03	4
5	6.186342592592592593D-02	3,399884259259259259D-04	5
6	-1,250000000000000000D-02	-2,604166666666666667D-04	6
7	1.504629629629629630D-03	-5,787037037037037037D-05	7
8	-9,920634920634920635D-05	1,240079365079365079D-05	8
9	2,755731922398589065D-06	2,755731922398589065D-06	9

Index